1 MONTH OF
FREE
READING

at

www.ForgottenBooks.com

By purchasing this book you are eligible for one month membership to ForgottenBooks.com, giving you unlimited access to our entire collection of over 1,000,000 titles via our web site and mobile apps.

To claim your free month visit:

www.forgottenbooks.com/free558754

ISBN 978-0-331-81984-7
PIBN 10558754

THE

EDINBURGH

JOURNAL OF NATURAL HISTORY,

AND OF

THE PHYSICAL SCIENCES.

CONDUCTED BY

WILLIAM MACGILLIVRAY, A. M.,

CONSERVATOR OF THE ROYAL COLLEGE OF SURGEONS, EDINBURGH; FELLOW OF THE ROYAL SOCIETY
OF EDINBURGH; MEMBER OF THE WERNERIAN NATURAL HISTORY SOCIETY; OF THE
ACADEMY OF NATURAL SCIENCES OF PHILADELPHIA, &c. &c.

ASSISTED BY SEVERAL SCIENTIFIC AND LITERARY MEN.

VOLUME I.

A.D. 1835—1839.

EDINBURGH:

PUBLISHED FOR THE PROPRIETOR, 13, HILL STREET.

GLASGOW AND THE WEST OF SCOTLAND: JOHN SMITH AND SON, 70, ST VINCENT STREET; AND
JOHN M'LEOD, 20, ARGYLE STREET. ABERDEEN: A. BROWN AND CO.

MANCHESTER: AINSWORTH AND SONS, 107, GREAT ANCOATS STREET. LEEDS: THOMAS AINSWORTH, JUNIOR,
50, QUEEN'S PLACE, COBURG STREET.

DUBLIN: GEORGE YOUNG, 9, SUFFOLK STREET.

PARIS: J. B. BALLIÈRE, RUE DE L'ÉCOLE-DE-MÉDECINE, 13, (BIS.)

THE EDINBURGH

JOURNAL OF NATURAL HISTORY,

AND OF

THE PHYSICAL SCIENCES.

N°. 1.   SATURDAY, OCTOBER 24, 1835.

ADDRESS TO THE PUBLIC.

THE taste for Natural History is now universally diffused throughout the empire. Within the last few years, various cheap publications on this subject, illustrated by engravings on wood, have led all classes to observe and to enjoy the ever-varied beauties of the creation. But no description, however correct, or no wood-cut, however well executed, can give that complete idea of a natural object, which is effected by an engraving on steel, when coloured with accuracy. The enormous price at which these illustrations of a higher order are usually sold, has alone prevented them from becoming extensively popular. Hitherto, coloured engravings, executed with beauty and correctness, have been accessible to the wealthier classes alone. It is proposed, in this work, to place elegant engravings of the choicest productions of Nature, within the reach of all classes of the community; and thus, in small towns and in the country, where museums are unknown, and libraries very scarce, the lover of Nature will be enabled to extend his knowledge of the works of the Creator beyond the limited sphere of his own observations.

The plan of this work is twofold.—An attempt is here made for the first time to combine the lighter character of a popular periodical, with the more solid utility of an eminent scientific work. One half-sheet of each Number will contain, " THE EDINBURGH JOURNAL OF NATURAL HISTORY, AND OF THE PHYSICAL SCIENCES;" and the other will be devoted to a new translation of " THE ANIMAL KINGDOM," by BARON CUVIER," with extensive Notes and Additions.

THE EDINBURGH JOURNAL OF NATURAL HISTORY, AND OF THE PHYSICAL SCIENCES.

IN this portion of the work, interesting articles will be given on every department of Natural History, taken in its most extensive sense. The physical peculiarities of the inhabitants of different countries will be recorded, in so far as they have been ascertained, the habits and instinct of animals will be explained, and such discoveries as may be made from time to time, both relating to Man and to Animals, will be regularly published. The wonders of the Vegetable Kingdom, the culture and management of Plants, and their economical uses, will form subjects of description, as well as the interesting and useful objects of the Mineral Kingdom. The novelties of Geology, and of Physical Geography, will also be embraced; and an account given of every remarkable occurrence in nature, whether proceeding from sudden and violent atmospheric changes, or from causes within the earth's surface. Elementary papers, on the different departments of Natural History, will aid the student in his inquiries. Popular articles will be written on the most interesting phenomena of Astronomy; while Mechanics, Optics, Chemistry, Electricity, and the higher branches of Physical Science, will not be altogether neglected. The novelties of Foreign Science and Literature will hold a prominent place in this Journal; and translations will appear of the more popular parts of important works, which would be otherwise inaccessible to most persons in this country. We shall explain the proceedings of Learned Societies —give Reviews of New Works—and record the demise of all who have been skilled in the Physical Sciences. This part of the work will be written in a style as simple as the subject will admit, avoiding, on all occasions, deep and abstruse reasoning; yet, at the same time, being intelligible and instructive to all classes of readers. A short and popular account of the objects illustrated in each plate will be given in the Number accompanying it, while a more detailed and scientific description will afterwards appear in its proper place.

THE ANIMAL KINGDOM, BY BARON CUVIER.

THE second portion of our publication contains an improved translation of the last edition of this justly celebrated work. It is well known, that the original was chiefly designed by its illustrious author as an outline, being more particularly intended to elucidate the new system of classification, of which he was the inventor. It is proposed, in the present translation, to fill up the masterly outline of Baron Cuvier with interesting and popular accounts of the habits and peculiarities of the whole Animal Kingdom. Care will, however, be taken to distinguish this additional matter from the original text of Baron Cuvier, by printing it in a smaller type; and, wherever it has become necessary, from more recent discoveries, or for other reasons, to supply a few words in the text, these are inclosed in brackets [ ]. The whole will run on in a continuous manner; and the plates are intended to be bound, at the completion of the work, with this Edition of the Animal Kingdom; the whole thus forming the most complete and economical publication on the subject of Natural History ever offered to the Public. It will be accompanied by a Portrait of Baron Cuvier, and by a Memoir of his Life. An historical article will be given on the rise and progress of Zoology; also, a treatise on the Geographical Distribution of Animals, and another

on Fossil remains. The work will conclude with an extensive Synopsis, or Index, of the entire Animal Kingdom.

Occasional wood-cuts illustrate matters of general interest. Throughout the entire work, the utmost caution is observed in excluding all passages which are calculated to offend delicacy; and those details, without which the work would be incomplete, appear in a foreign language. Supplementary plates will be published at intervals, for the convenience of those who may wish to possess figures of the whole Animal Kingdom.

In respect to the execution of the work, although this number is offered as a specimen of its general character and arrangement, yet various improvements will obviously suggest themselves in the course of this extensive undertaking; and unlike many similar works, the later numbers will exceed the first, in the quality and the beauty of their illustrations. The artists employed are all of the most respectable description; the designs are drawn from the objects themselves; or, when this is impracticable, they are copied from the best authorities in our own and in foreign countries; while the engravings are coloured in the most careful manner. The average number of objects upon each plate will be about ten, varying in number according to the dimensions of the animals. We have selected the present size, as being the smallest that can combine economy with accuracy of delineation. It will be seen in the plate of Trogons, accompanying this number, that the Golden Couroucoui, a bird of the size of a small pigeon, has been reduced to about one-third of its natural dimensions; and it will be readily seen, from the great length of its tail, that any smaller work could not possibly have given a correct delineation of an object of this description. Also, by placing many objects on one plate, great facilities are afforded for a comparison of the external forms of allied animals; while room is given for a tasteful approximation and contrast of forms and colours.

It is hoped that the typography and general execution of the work, entitle it to some consideration: the sheets are all carefully pressed, and are not issued in their rough state, as is usual with periodical publications.

In regard to economy this work is unrivalled. The quantity of matter in each number is equal to SIXTY-FIVE ordinary foolscap octavo pages; so that a quantity of letter-press, equal to a volume of the Waverley Novels, will be had for ONE SHILLING.

Thus, while the views of the lowest classes will be fully met by the cheapness of this work, it is hoped that its beauty and accuracy will render it worthy of the approbation even of the very highest. By spreading a taste for science, and for intellectual improvement among all classes of the community, we endeavour to promote the moral and intellectual progress of the species. We hope to make it appear, in the words of an illustrious living authority, " that the pleasures of science go hand in hand with the solid benefits derived from it; that they tend, unlike those other gratifications, not only to make our lives more agreeable, but better; and that a rational being is bound, by every motive of interest and of duty, to direct his mind towards pursuits, which are found to be the sure path of virtue, as well as of happiness."

No expense or trouble will be spared by the Proprietors of this work to obtain the latest information, and the most correct delineations of natural objects. Naturalists and men of science, in this and in foreign countries, are requested to favor the Editor with an early notice of their observations.

To ensure regularity in the publication of the work, a large portion of the illustrations are already completed. Arrangements have been made at New York, Hamburgh, Paris, Brussels, and various parts of the Continent, to render them available to foreigners, as well as to our own countrymen.

ZOOLOGY.

DESCRIPTION OF THE PLATE—THE TROGONS.

THE figures on this plate represent the genus Trogon of Linnæus, commonly called the Couroucoui. These birds are remarkable for the splendour of their plumage, in which the hue of the emerald is placed in agreeable contrast to those of the topaz and ruby; while their brilliant colours lead us to forget their rather inelegant forms. They are members of that class of birds called Scansores, or Climbers; their feet being provided with four toes, two of which are placed before and two behind.

These birds lead a lonely life, frequenting solitary woods. They feed on insects, they fly principally in the morning and evening, and they incubate in the holes of trees. In general, they are very silent birds; but sometimes they cry Couroucoui, from which their name is derived, and sometimes Pio, Pio. They are noted by naturalists for the plenitude of their plumage.

Fig. 1.—The Golden Couroucoui (Trogon Pavoninus), is one of the most beautiful of its tribe, and remarkable for the great development of its tail, the two centre feathers of which are frequently three feet in length. It inhabits South America and the tropical parts of North America. Our figure is taken from a splendid specimen in the Edinburgh University Royal Museum. The Mexicans celebrated this bird in their mythology; they considered it as sacred to the deity Vitzliputzli, and they adorned the brows of their priests with the elegant feathers of its tail. This idol Vitzliputzli was represented, as in the annexed wood-cut, crowned with a helmet in the form of a Golden Couroucoui, composed of feathers of various colours, excepting the beak and crest, which were of gold.

Most nations have some legends regarding the foundation of their capital cities; and the site of Mexico was said to have been fixed by this bird. The Mexican fable was as follows:—

The Navatlecas were savages residing in forests and mountains, without laws or government; they worshipped the sun, and sacrificed these birds to his honour. Mexi the great captain and legislator appeared, and conducted by Vitzliputzli, the god of their nation, he led the savages in search of distant lands. Aided by their deity, his arms were irresistible, and his empire soon became extensive. Being in doubt where to establish his capital, a priest announced that it should be built on the spot where a Couroucoui was seen perched on a tree, the roots of which were inserted on a solid rock. The spot pointed out was the present site of Mexico. Thus did a barbarous people account, by an idle prodigy, for the situation of a city on an island in the centre of a lake—a spot evidently pointed out by nature as the most secure situation for the capital of a mighty empire.

Fig. 2.—Reinwardt's Couroucoui (T. Reinwardtii), is a native of Java. The plumage of this species is more compact than that of its American congeners; and also differs from them in having its general form and bill more slender, while the toothing of the edges of the bill is less marked.

Fig. 3.—The Black-necked Couroucoui (T. Atricollis), is a native of Cayenne; and one of the most timid of its kind.

Fig. 4.—The Mexican Couroucoui (T. Mexicanus), is a recently discovered species, and a bird of great beauty. There are several fine specimens in the Edinburgh University Royal Museum, from one of which we figured our subject, by permission of Professor Jameson.

Fig. 5.—The Flower Couroucoui, male (T. Narina), and Fig. 6., the female. This species is more nearly allied to the Asiatic bird than to those of South America. It inhabits Southern Africa.

It will be seen that this genus is widely distributed, being found in Asia and Africa, as well as on the continent of America; but it is in the last situation that the species are most numerous.

POPULAR ERRORS REGARDING THE TORPIDITY OF SWALLOWS.

IT has long been, and continues to be, a popular opinion in this country, and in other parts of Europe, that swallows of a certain species pass the winter at the bottoms of deep lakes and wells. Buffon entertained this idea, and Goldsmith makes the following observations on this theory:—He says, " There is a circumstance attending the migration of swallows, which wraps this subject in great obscurity. It is agreed on all hands, that they are seen migrating into warmer climates, and in amazing numbers, at the approach of the European winter. Their return into Europe is also well attested about the beginning of summer; but, we have another account, which serves to prove, that numbers of them continue torpid here during the winter; and, like bats, make their retreat into old walls, the hollows of trees, or even sink into the deepest lakes, and find security for the winter season by remaining there in clusters at the bottom."

The analogy between birds of passage, and animals which remain in a state of torpidity during the winter, is most inaccurately drawn by Goldsmith, and we offer the following objections to the supposed constitutional connection.

These quadrupeds, birds, reptiles, and insects, which pass the winter in a state of insensibility, may be recalled to sensation and action at pleasure, by the application of a gentle degree of heat. Naturalists have been induced, from this constitutional singularity of these animals, to conclude, that the return of spring rouses them from their lethargic state to enjoy the pleasures of sensation and locomotion. The animals in question take up their abodes a little below the surface of the soil; some in the crevices of walls, or interstices of rocks; while others, such as frogs, toads, and water-newts, bury themselves in the mud of shallow ponds. In the first of these retreats, they are only covered by a thin layer of earth and moss, or leaves; and in the last, by the addition of a shallow sheet of water; consequently they are re-animated in due season, by the genial rays of the sun, after he has entered the northern half of the ecliptic.

The temperature of places, situate at great depths below the surface of the land and water, is a sufficient objection to the assertion that birds remain in a torpid state, during the winter, in deep and solitary caverns, or at the bottom of deep lakes. Dr Hale has proved, by experimental facts, that the bulb of a thermometer, buried sixteen inches below the earth's surface, stood at 25° of his scale in September, at 16° in October, and at 10° in November, during a severe frost; from which point it ascended again slowly, and reached 23° in the beginning of April. Now the end of September and beginning of October is the season when the hedgehog, shrew, bat, toad, and frog, disappear; and, about the middle of April, these animals re-appear: this agrees very well with the variations of temperature of the preceding theory.

It is a well-established fact, that all places situate eighty feet below the surface of the earth are constantly of the same temperature. Mr Boyle kept a thermometer for a year under a roof of earth, eighty feet in thickness, and found that the fluid in the instrument remained stationary all the time. Dr Withering made a similar experiment on a well eighty-four feet deep, and found that it remained at 49° for the entire year. Surely, then, this invariable temperature is inconsistent with the theory of birds remaining in a state of torpidity in deep lakes, or solitary caverns, where the sun has no influence; for what would call forth their dormant organs into action, the sun having no influence in places so situate? It is but reasonable to conclude, that the cold, which kept them benumbed by its soporific influence, would perpetuate their slumber.

The state of torpor to which hybernating animals are annually subjected, is obviously analogous to sleep, but it differs from sleep in being occasioned solely by temperature. Hybernating animals always assume this torpid condition, whenever the thermometer sinks to a certain point. Man, and almost all animals, seem to be susceptible of this state, at least to a certain extent; for the apparent death produced by cold is probably nothing else but a species of torpor, out of which the animal, in most cases, might be roused if the requisite precautions in applying heat were attended to; for death, in most cases, seems to be produced, not by the cold, but by the incautious and sudden application of heat, which bursts the blood-vessels in some particular part of the body, before the heat has had the power of stimulating the heart, and setting the blood in motion through the whole animal frame; and this bursting of the blood-vessels destroys the texture of the body. It is well known that if any part of the body be frost-bitten, an incautious application of heat infallibly produces mortification, and destroys the part.

In the 28th volume of the Philosophical Transactions, there is a remarkable example recorded of a woman, almost naked, lying buried for six days under the snow, and yet recovering. In this case it is scarcely possible to avoid supposing that the woman must have been in a state of torpor, otherwise she would have endeavoured to find her way home.

That a few stragglers of the swallow tribe do remain in this country long after their fellows have departed, there can be no doubt; and even some have been known to sojourn during the whole winter; but, it is equally true that the uniform habit of these birds is, to quit the north on the approach of winter, and to seek climates more congenial to their mode of existence, which is entirely maintained by insect food.

The Rev. Gilbert White, in his interesting Natural History of Selborne, remarks, " I cannot agree with those persons who assert that the swallow kind disappear some and some, gradually, as they come, for the bulk of them seem to withdraw at once; only some stragglers stay behind a long while, and do never, there is great reason to believe, leave this island. Swallows seem to lay themselves up, and to come forth in a warm day, after they have disappeared for weeks. For a very respectable gentleman assured me, that as he was walking with some friends under Merton-wall, on a remarkably hot noon, either in the last week of December or the first week in January, he espied three or four swallows huddled together on the moulding of one of the windows of that college. I have frequently remarked that swallows are seen later at Oxford than elsewhere. Is it owing to the vast massy buildings of that place, to the many waters round it, or to what else?" He also mentions that a friend of his saw a martem on the 26th November, in a sheltered bottom; the sun shone warm, and the bird was hawking briskly after flies. Mr Sweet mentions the circumstance of a house swallow having taken up its residence, late in the autumn, within St Mary's church at Warwick; it was regularly observed there by the congregation until Christmas-eve; after which, it disappeared and was seen no more.

HEIFER WHICH YIELDED MILK.—Mr Joseph Marshall of Edrington, by Berwick, in June 1830, had a heifer which yielded milk. At that time he possessed two heifers aged two years, one of which he observed for several months to suck the other. On this account he judged it necessary to separate them; and, on milking the nurse, she gave a full English quart of genuine milk, which, on being kept for thirty hours, threw up a good coat of cream. The cream was churned in a bottle, and produced as much and as good butter as any other cream would have done under similar management.

ADDERS.—Two adders, upwards of two feet long, were noticed on the moss of Ashvore, Monedie parish, Methven, on the 3d week of June 1835. This place has long been noted as the haunt of adders.

INSECTS IN INDIA.—During the rainy season in India, the houses are so infected with insects, that it is necessary to have little covers (usually of silver) for tumblers and tea cups. The air is so still and stagnant, that persons are compelled to keep their doors wide open; and, consequently, the tables are thickly covered with a variety of the most disgusting vermin. These, mingling with the blood-thirsty musquitoes, are tormenting in the extreme. At this season, also, the white ants are extremely numerous and destructive. In one night, they have been known to spread themselves over a large apartment, and to devour the whole matting. They frequently take possession of the beams that support the roofs of the houses, and destroy them in a few weeks. Nothing is secure against the depredations of these mischievous little creatures. Tents, carriages, beds, carpets, and clothes of all descriptions, are subject to their voracious appetites.

Worms in the Eyes of the Perch.—Dr Nordmann discovered parasitic worms in the eyes of several distinct species of the perch. These eye-worms were sometimes in such numbers as must have interfered with that distinct sight of passing objects, which appears necessary to enable predaceous animals to discover their prey, in time to dart upon and secure it; in a single eye the Doctor detected, in different parts, 360 of these minute animals! When numerous, they often produce cataracts in the eye of the fishes they infest. The little animals which he found appear something related to the *Planaria*, or pseudo-leech; and, to judge from Dr Nordmann's figures, seem able, like it, to change its form. Underneath the body, at the anterior extremity, is the mouth; and in the middle are what he denominates two sucking cups; these are prominent, and, viewed laterally, form a truncated cone; the anterior one is the smallest and least prominent, and more properly a sucker; the other, probably, has other functions, since be could never ascertain that it was used for prehension. A kind of metamorphosis seems to take place in these animals, for our author observed that they appeared under three different forms. These little pests, small as they are, have a parasite of their own to avenge the cause of the perch, for Dr Nordmann observed some very minute brown dots, or capsules, attached to the intestinal canal, which when extracted, by means of a scalpel formed of the thorns of the creeping cereus, and laid upon a piece of talc, the membrane that enclosed them burst, and forth issued living animalcules, belonging to the genus *Monas*, and smaller than *Monas atomus*, which immediately turned round upon their own axes with great velocity, and then jumped a certain distance in a straight line, when they again revolved, and again took a second leap. Looking over our author's list of eye-worms that infest fishes, we find that five out of seven are attached to different species of perch, and one cannot help feeling some commiseration for these poor animals; but, when we recollect that they form the most numerous body of predaceous fishes in our rivers, we may conjecture that thus their organs of vision are rendered less acute, and that thus thousands of roach, dace, carp, and tench, may escape destruction.—*Kirby's Bridgewater Treatise.*

BOTANY AND HORTICULTURE.

Fountain Tree.—There are few rivulets, and only three springs, in the Island of Ferro, one of the Canaries; and these are on a part of the beach which is nearly inaccessible. To supply the place of fountains, however, Nature bestowed upon this island a species of tree, supposed to be nearly allied to the *Laurus Indica*, possessing properties unknown to trees in all other parts of the world. These fountain trees were of moderate size, and their leaves were straight, long, and evergreen. Around the summit a small cloud perpetually rested, which so drenched the leaves with moisture, that they constantly distilled upon the ground a stream of fine clear water. To these trees, as to perennial springs, the inhabitants of Ferro resorted, and were thus supplied with a sufficient abundance of water for themselves and for their cattle. The last of these remarkable productions received the appellation of the Holy Tree, and it is said to have been destroyed by a dreadful hurricane in 1612. Its real existence has been completely established in the *Viagero Universal* di P. Estala, tome xi.; but by this account the water was merely condensed upon the leaves. Purchas, in his "Book of Pilgrimages, 1639," states that be had been told by Mr Lewis Jackson, of Holborn, London, who visited Ferro in 1618, that the fountain tree he had seen was as large as a middling-sized oak, six or seven yards high, with a white bark, like that of hardbeam; its leaves were like that of the bay, white underneath and green above. Parkinson, in his "Theatrum Botanicum," published at London in 1640, also mentions this tree. He says that the islanders called it *Garoe*, the Spaniards *Arbor sancti;* and that the ancient historians call it *Til;* and adds, "It is thought that Solinus, and Pliny in his Lib. 6, c. 32, meant this island, under the name of *Ombrion* and *Pluvialis:* for he there saith, that in the island *Ombrion* grow trees like unto *Ferula*, from whence water is wrung out; from the black ones cometh bitter water, and from the white that which is sweet and pleasant to drink."

Great Productiveness of the Orange Trees of St Michaels.—The St Michael oranges have long been celebrated for their delicious flavour, and the abundance and sweetness of their juice; when allowed to ripen before being pulled, they are inferior to none in the world. On the other hand, the lemons of that island are less esteemed than those of various other countries, on account of the small quantity of juice in them; and therefore they are not much in demand. The orange and lemon trees blossom in February and March. At this season, nothing can be more gratifying to the sight than the appearance presented by these trees; the glassy green of the old leaves, the light fresh tints of those just shooting forth, the brilliant yellow of the ripe fruit, and the delicate purple and white of the flower, form a delightful contrast. Both orange and lemon trees are from fifteen to twenty feet in height, and their common annual produce is from 6000 to 8000. Dr Webster mentions, that he has known, in a very abundant year, 26,000 oranges to be obtained from one tree, and 29,000 from another. These. however, are the greatest quantities which have ever been known to be gathered from a single tree in a year.

New Method of Blanching Celery.—In the March number of the *Irish Farmer's and Gardener's Magazine*, a Mr Coglan recommends the following method of cultivating celery, by which he states he has been successful for many years in preserving this favourite vegetable from what is called "rust," occasioned by the attack of grubs. In the month of October he plants the ground, designed for celery the ensuing year, with early York cabbage, which will be cleared away by the first week in June, the most proper season for planting. Previous to forming the drills, he collects the stalks and remaining leaves of the cabbages, and places them in small heaps on the bed. After lying a day or two, they will be found to have collected a great number of slugs and other vermin, which may be easily destroyed. The ground is then prepared and the plants put in; when ready for blanching, the loose leaves of each plant are tied up, and strong wheaten straw laid full length along the side of the drills, and staked down so much that it will completely exclude the light (excepting at the top, which is all that is requisite). By this treatment, he says, in the course of a month he has gathered celery perfectly free from either rust, grub, or insect.

MINERALOGY.

CHEMICAL COMPOSITION OF NATIVE GOLD.

We principally confine our remarks on the chemical composition of native gold, to that found in the Uralian mountains.

Gold being a simple substance, is of course unsusceptible of decomposition; but, like most other simple bodies, it is never found in the earth in a state of purity, being always more or less combined with silver.

Although the mines of Colombia have long been celebrated for the quantity of gold found in them, yet it has recently been ascertained that the native ore of that country contains a smaller proportion of gold than that of most others. According to the experiments of Boussingaull, who analyzed gold from different places in Colombia, he found it combined with silver in variable quantities, but invariably in definite proportions, namely, one atom of silver with 2, 3, 4, 5, 6, 8, and 12 atoms of gold. We are informed by Fordyce, that he examined gold from Konsberg in Norway, and found it to consist of 28 parts gold, and 72 of silver, in the 100 parts. Klaproth, in analyzing gold from Schlangenberg in the Altai, found that it contained 64 parts gold, and 36 of silver; and Lampadius found in gold ore, whose locality was unknown to him, 96.6 parts of gold, and the remainder consisted of silver and iron. Mr G. Rose, who accompanied Baron Humboldt into Siberia, made a collection of gold ores, for the express purpose of testing the position assumed by the French chemists.

Gold ore is found in the Uralian mountains in rocks, and also scattered amongst sand; in which last situation it is now almost entirely sought after, being much less laborious than extracting it from rock veins. Its discovery in this situation occurred in 1819, since which period rock mines have been abandoned.

The gold which is procured in rocks is universally found in quartz. At Beresow, it occurs in a crystallized form; at Newiansk, in plates; but at Czarewo Alexanderowsk, masses are met with which weigh from 18 to 96 lbs. troy.

The proportions of gold and silver from the different localities vary considerably. The following are the extreme limits of this variation:—

	SILVER.	GOLD.
Hiel.........................rock	87.40	12.60
Schaitansk.	95.10	4.90

The following important consequences were deduced by Rose from his analysis. He remarked that gold and platinum were never found associated.

1. Gold ore does not contain gold and silver in definite proportions.
2. From the above fact, he infers that gold and silver are isomorphous.
3. Native gold always contains silver, copper, or iron. The smallest quantity of silver in combination, was in a specimen from Schabrouski, which only contained 16 per cent. of silver, while 35 per cent. of copper was found.
4. The specific gravity is in the inverse ratio of the proportion of silver contained in the ore. Gold obtained by fusion has a greater density than in a native state; but this may arise from cavities contained in the latter.
5. He found a difference in composition in specimens from the same locality.
6. Gold found in veins varies in different parts of the same mine.
7. Gold from sand contains more silver than that found in veins; that from sand being 89.7 per cent. of silver, and from veins, 79.1.—*Poggendorff, Ann.* xxviii., 566.

Edible Rocks.—Near the Ural mountains in Siberia, a substance called "rock meal"—powdered gypsum—is found, which the natives mix with their bread, and eat. The Tartars likewise eat the lithomarge, or rock marrow; and use rock butter as a remedy for certain disorders.

GEOLOGY AND PHYSICAL GEOGRAPHY.

Threatened Eruption of Vesuvius.—According to late accounts from Naples, Vesuvius continued to throw out stones and cinders, and a grand eruption was expected. Some slight shocks of an earthquake were lately felt in the south of Italy.

Water of the Seine.—It was not thought that the basin of the Seine—I mean all that portion of ground in France watered by streams, great or small, which flow into that river—annually received in rain a quantity of water equal to the tribute carried by the Seine to the sea in the same space of time. Perraul and Mariotte were the first who studied the question, supported by experiments, and found, as is common in similar cases, that the vague ideas of their predecessors were the very contrary of the truth. According to Mariotte, the Seine discharges every year into the sea only a sixth of the quantity of water which falls in all the extent of its basin, in rain, snow, and dew. The other five-sixths must either be evaporated to form clouds, absorbed by the superficial earths in which plants find nourishment, or penetrate, by fissures in rocks, into the internal reservoirs from which fountains issue. Mariotte's calculation has been re-made on data more exact, especially as regards the gauging of the Seine. The following are the results, as they were stated in an excellent memoir, hitherto unpublished, by Mr Dausse, civil engineer:—"The basin of the Seine has an area. 4,327,000 hectors. Were the water falling into this basin not to evaporate, nor penetrate into the soil, and were the ground every where horizontal, it would form at the end of the year a liquid sheet of 53 centimetres (20 inches) deep. It is easy to see that such a sheat would contain a volume of 22,933 millions of cubic metres of water. Now, at the Bridge of the Revolution, the mean proportion of the water passing there is at the rate of 255 cubic metres in a second, or 22 millions of cubic metres in a day, or 8042 millions of ditto in a year. This last number is to 22,933 millions of cubic metres, which is the annual amount of the rain received by the basin of the river, as 100 is to 285, or almost as 1 to 3. Thus, the volume of water passing annually under the Paris bridges is scarcely the third of that which falls in rain into the basin of the Seine. Two-thirds of this rain either return to the atmosphere, by means of evaporation, or sustain vegetation and the life of animals, or run into the sea by subterranean communications."—*Literary Gazette, 20th June.*

METEOROLOGY.

EAINE.—On the 11th of April 1832, a remarkable substance fell from the atmosphere, thirteen versts from Wolokalomsk, and covered a considerable space of ground, to the depth of one or two inches. It was examined by Professor Hermann of Moscow. He found it to be transparent and of a wine-yellow colour, soft, and elastic (like gum), and its specific gravity to be 1.1; it smelt like rancid oil, and burnt with a blue flame, without smoke; it was insoluble in cold water, but was soluble in boiling water, upon which it swims; it was also soluble in boiling alcohol; it could be dissolved by the carbonate of soda, and acids separated from the solution a yellow viscid substance, soluble in cold alcohol, and which contained a peculiar acid. When analyzed by oxide of copper, it furnished, of carbon 61.5, hydrogen 7.0, and of oxygen 31.5; total 100. This extraordinary substance has been termed inflammable snow, but Hermann gives it the name of *Eaine*, signifying the *Oil of Heaven*.

INDIAN DEATH-BLAST.—At Bandah, in Bundalcund (one of the northern provinces of Hindostan), there are numerous rocky hills, which, during the hot winds, become so thoroughly heated as to retain their warmth from sunset to sunrise. The natives, at this sultry season, invariably wear large folds of cloth around their heads and faces, just leaving themselves sufficiently exposed to be able to see and breathe. This precaution is taken in consequence of the terrific blasts which occasionally rush in narrow streams from between the hills. Persons crossed by these scorching winds drop suddenly to the earth, as if shot by a musket ball. When medical assistance, or a supply of cold water, is instantaneously procured, a recovery may generally be expected; but if no immediate remedy be applied, an almost certain death is the result.

GENERAL SCIENCE.

COMETS.

THE extraordinary aspect of comets, their rapid and seemingly irregular motions, the unexpected manner in which they often burst upon us, and the imposing magnitudes which they occasionally assume, have in all ages rendered them objects of astonishment, not unmixed with superstitious dread, to the uninstructed, and an enigma to those most conversant with the wonders of creation and the operations of natural causes. Even now, that we have ceased to regard their movements as irregular, or as governed by other laws than those which retain the planets in their orbits, their intimate nature, and the offices they perform in the economy of our system, are as much unknown as ever. No rational or even plausible account has yet been rendered of those immensely voluminous appendages which they bear about with them, and which are known by the name of their tails (though improperly, since they often precede them in their motions), any more than of several other singularities which they present.

The number of comets which have been astronomically observed, or of which notices have been recorded in history, is very great, amounting to several hundreds; and when we consider that in the earlier ages of astronomy, and indeed in more recent times, before the invention of the telescope, only large and conspicuous ones were noticed; and that, since due attention has been paid to the subject, scarcely a year has passed without the observation of one or two of these bodies, and that sometimes two and even three have appeared at once—it will be easily supposed that their actual number must be at least many thousands. Multitudes, indeed, must escape all observation, by reason of their paths traversing only that part of the heavens which is above the horizon in the daytime. Comets so circumstanced can only become visible by the rare coincidence of a total eclipse of the sun—a coincidence which happened, as related by Seneca, 60 years before Christ, when a large comet was actually observed very near the sun. Several, however, stand on record as having been bright enough to be seen in the daytime, even at noon and in bright sunshine. Such were the comets of 1402 and 1532, and that which appeared a little before the assassination of Cæsar, and was *(afterwards)* supposed to have predicted his death.

That feelings of awe and astonishment should be excited by the sudden and unexpected appearance of a great comet, is no way surprising; being, in fact, according to the accounts we have of such events, one of the most brilliant and imposing of all natural phenomena. Comets consist for the most part of a large and splendid, but ill defined, nebulous mass of light, called the head, which is usually much brighter towards its centre, and offers the appearance of a vivid *nucleus*, like a star or planet. From the head, and in a direction *opposite to that in which the sun is situated* from the comet, appear to diverge two streams of light, which grow broader and more diffused at a distance from the head, and which sometimes close in and unite at a little distance behind it, sometimes continue distinct for a great part of their course; producing an effect like that of the trains left by some bright meteors, or like the diverging fire of a sky-rocket (only without sparks or perceptible motion). This is the tail. This magnificent appendage attains occasionally an immense apparent length. Aristotle relates of the tail of the comet of 371 A.C., that it occupied a third of the hemisphere, or 60°; that of A.D. 1618 is stated to have been attended by a train no less than 104° in length. The comet of 1680, the most celebrated of modern times, and on many accounts the most remarkable of all, with a head not exceeding in brightness a star of the second magnitude, covered with its tail an extent of more than 70° of the heavens, or, as some accounts state, 90°.—*Sir J. Herschel's Treatise on Astronomy.*

MISCELLANEOUS.

EXTRAORDINARY BALLOON EXCURSION.—On the 8th of April 1835, Mr Clayton made an ascent in a balloon from Cincinnati, and made the most extraordinary aeronautic excursion, in point of distance, on record. His balloon took a south-easterly direction, and the greatest altitude to which he ascended was two and a half miles, at which height the thermometer stood at 23 degrees. In his progress he descended to nearly the earth's surface; and, on throwing out his anchor, it caught firm hold of the top branch of a tree, in the midst of a dense forest, on a considerable elevation. The wind blowing powerfully, and finding it impossible to extricate himself, he cut away the cable, and soon ascended to an altitude as high as formerly; where he found the temperature intensely cold, but could not ascertain the height of the thermometer. He, however, conceived it to be as low as zero. He put on his gloves, wrapt himself up in two blankets, took some brandy, laid himself down in the bottom of the car, and fell fast asleep. He was awakened by the car striking on the tops of trees, where he landed in the midst of a forest, at half past two o'clock. The spot where he landed was the top of a mountain 3000 feet, as indicated by the barometer, above the level of the Monroe county, Virginia. The distance travelled was three hundred and fifty miles in a direct line, or more than four hundred by the ordinary route, and this in the short space of nine and a half hours, which is at the rate of thirty-seven miles an hour. But as it was calm for some time after his ascent, the speed during part of the excursion must have been much greater; and this is without taking into account the distance in ascending and descending twice.

REVIEWS.

Bridgewater Treatise.—On the Wisdom and Goodness of God, as manifested in the Creation of Animals, and in their History, Habits, and Instincts. By the Rev. W. Kirby, M.A., F.R.S. 2 vols. 8vo. London. 1835.

THOUGH interspersed with much interesting detail, especially regarding the lower classes of living beings, this work more resembles a lecture on divinity than a philosophical treatise. It has evidently the appearance of being *made to order;* and instead of being delighted with such accounts of the beautiful adaptations in the structure of animals, as are given in the philosophical pages of Cuvier, the reader is fatigued with discussions on the etymology of the Cherubim and Seraphim, and on the fall of Adam. The learned author gravely discusses whether Adam and Eve were troubled before the fall "with certain *personal pests;*" and argues, very plausibly, that parasitic worms were created after the fall, and that their eggs did not exist in the intestines of Adam and Eve before that event. His attack of Laplace and Lamarck seems out of place. We shall return to this work at a future opportunity.

OBITUARY.

BARON WILLIAM DE HUMBOLDT died at Berlin on the 8th April 1835. He was remarkable on account of his extraordinary colloquial knowledge of languages, as well as the philosophy of every tongue of which he could obtain any information. Besides all those of Europe and the East, he was skilled in the languages of North and South America; and he had also an intimate acquaintance with the customs of various countries. He was also eminent as a statesman. The following affectionate account of his last moments, communicated to M. Arago by his brother, the celebrated traveller, Baron Alexander De Humboldt, cannot fail to prove acceptable to our readers:—"While we labour under the burden of severe affliction, we are apt to think of those who are dearest to us, and I feel a solace in writing to you. For ten days we watched his death-bed. His debility had increased for some weeks previously, which was manifested by a ceaseless trembling in all his extremities; yet his mind was unimpaired, and his labours were unabated. He has left two works nearly completed—the one, on those languages of the Indian Archipelago which have sprung from the Sanscrit—and the other, on the origin and philosophy of languages in general. Both will be published. My brother has left the MS. of these works, and his valuable collection of books, to the Public Library. He died of an inflammation on the lungs; and, from its commencement, he traced its progress with an affecting certainty. His mind was of the highest order, and his soul noble and elevated. I continue sorrowfully isolated, but hope for the pleasure of embracing you this year."

LEARNED SOCIETIES.

ROYAL INSTITUTION.—In the course of a lecture on metals, on Saturday the 13th June 1835, Professor Faraday stated the following curious particulars respecting the gold coinage: A small ingot of gold (which he held in his hand, and which measured about ten inches in length by two in breath), which weighed about 20 pounds troy, was worth L.1000. Last year, 1834, the coinage was somewhat smaller than the average amount; only half sovereigns were coined, and those to the value of L.66,944, or in weight 1433 pounds troy. The quantity of gold that had passed through the Mint since the accession of Queen Elizabeth to the throne, in the year 1558, to the end of last year, was 3,353,568 pounds weight troy. Of this nearly one-half was coined in the reign of George III.—namely, 1,594,078 pounds troy. The value of the gold coined in the reign of that sovereign was L.74,501,586. The total value of the gold coin issued from the Mint since 1558, was L.154,702,385. This gold, if made into a cubic form, would measure on each side thirteen feet thirty-two hundreds. It was extremely difficult to account for the constant loss in the quantity of gold; it continued to be brought in great quantities every year into this country, and yet the value of it did not fall. It was true that population had greatly increased; but that was not alone sufficient to account for the increased consumption of gold. Between the years 1492 and 1823, the estimated value of the gold imported into Europe from the New World was not less than L.1,223,000,000. The average value of gold brought into England for the last few years was L.1,600,000. The far greater portion of this was used in manufactures and articles of jewellery. A considerable quantity was made into gold-leaves, the intrinsic value of each of which was about one halfpenny; and the wages of labour and the profit of the manufacturer were an additional farthing; making altogether a charge of three farthings for each leaf. On the average, nearly two millions of these leaves were manufactured every week in London. Not more than one-half of this was returned in another shape to the goldsmith; scarcely any portion, however, of the gold used in gilding frames was lost, as the Jews carefully looked after the old frames, and burnt the gold off. The ordinary wear and tear of the gold coins probably amounted annually to about one-fiftieth part of their value.

Edinburgh: Published for the PROPRIETORS, at their Office, No. 16, Hanover Street. London: CHARLES TILT, Fleet Street. Dublin: W. F. WAKEMAN, 9, D'Olier Street. Glasgow and the West of Scotland: JOHN SMITH and SON, 70, St Vincent Street.

THE EDINBURGH PRINTING COMPANY.

THE EDINBURGH

JOURNAL OF NATURAL HISTORY,

AND OF

THE PHYSICAL SCIENCES.

| Nº. 2. | SATURDAY, NOVEMBER 7, 1835. |

ON THE PLEASURES AND ADVANTAGES OF THE STUDY OF NATURAL HISTORY.*

" He," says the great Linnæus, " who does not make himself acquainted with God from the consideration of Nature, will scarcely acquire knowledge of him from any other source; for, if we have no faith in the things which are seen, how shall we believe those things which are not seen?'"

Natural History, or the study of Nature, may be considered as the parent of natural religion. The history of the world shows, that most nations have had some method of tracing the hand of the Great Author of the universe in his works, and have thence deduced some particular reasons for loving and reverencing him. From the study of Nature we are taught that sublime lesson, inculcated by the religion and morality of every civilised people, and not altogether unfelt even by the untutored savage, namely,—to know and acknowledge the divine Author of Nature.

Without considering, at present, that knowledge of the ALL-WISE, derived from revelation, we may remark, that those habits of study and reflection, which arise from seeking after the wonders, of what are termed *natural objects*, directly lead us to acquire a knowledge of the Great Being who formed them. As we advance in pursuit of these inquiries, we are led to admire, at every step, the astonishing skill and contrivance manifested in his works; and we are lost in wonder and admiration of the superhuman power and wisdom displayed in the general system of the world, and in its various details.

For these reasons, the knowledge of the Author of Nature, through his works, may be designated the universal religion; just as the love of rectitude—a regard for justice, for temperance, and for truth—may be termed the universal morality. Neither of these interfere with the religion or morality of any particular people.

This is, then, the first and chief use of the study of Nature. We are taught to look from Nature up to HIM who formed the universe, and who imparts the living principle even to the lowest degrees of animal and vegetable existence.

An extensive acquaintance with natural objects, either in individuals or in a nation, cannot exist without producing great and corresponding improvements in taste, in literature, and in the elegant arts. A correct knowledge of natural objects will elicit greater accuracy in the delineation of them, both in the artist and in the man of letters. It is well known that the public taste is gradually, nay rapidly, improving in respect to painting, sculpture, and architecture; and no inconsiderable portion of the improvement will be found to be attributable to the more correct representation of natural objects. This improvement is also extended itself to our manufactures, more especially to the figures printed on cotton, paper, and earthenware; the great superiority in these is acknowledged chiefly to consist in the more correct imitation of plants, animals, and general scenery.

> With what attractive charms this goodly frame
> Of Nature touches the consenting hearts
> Of mortal men! and what the pleasing stores
> Which beauteous imitation thence derives
> To deck the poet's or the painter's toil !
>
> AKENSIDE.

What rational pleasure, or instruction, can a reflecting people derive from the representations of beings, which never had an existence, except in the imaginations of heathen nations? Can there be any really solid taste in admiring a hippogriff, a pegasus, a phœnix, a griffon, a dragon, and fifty more such fictitious animals, which have so long held sway in the ornamental parts of architecture? To the mere student of antiquities, who knows nothing of the beauties of creation, these may call up certain associations, but they are looked at, thrown aside, and treated with contempt by the lover of Nature. As natural history consists in an accumulation of facts, and as it is the province and the delight of the disciples of Nature to trace the *true character* of every object in Nature; so every thing which is detected as departing from the truth, must create disgust rather than pleasure, in those who are accustomed to search after her beauties.

The study of mineral substances is of the greatest importance: for we are led by means of them to the improvement of all the useful arts. What would civilized man be without iron? An acquaintance with the different strata, which compose the crust of the earth, enables us to detect the localities of coal and other useful minerals; and hence the importance of this species of knowledge in working mines and quarries. A knowledge of geology adds greatly to the interest of the traveller in passing through a

* The principal part of this article was contributed, several years ago, by a writer in this Journal to Chambers's Edinburgh Journal; and it is inserted here by the kind permission of the proprietors of that deservedly popular periodical.

country; while we are enabled to learn the past history of the globe from the changes which have evidently taken place on the earth's surface.

Having thus shown the utility of the study of Nature, we now turn to the pleasure to be derived from a pursuit of it. We must in the first place premise, that we consider all knowledge to be pleasure, as well as power; and that in the pursuit of pleasure, the reward obtained will be commensurate with the labour bestowed. These are facts which the reason and experience of ages have incontrovertibly established, and they ought to be treasured up in the mind of every young person, as perpetual incitements to exertion.

From this, however, we would not wish the young student to imagine, that very great mental exertion is required in the study of natural history; for the very reverse is the fact. The principal thing required is a good memory and a correct eye, both of which can be wonderfully improved by practice. It is the want of attention alone which makes the discrimination of objects *appear* a difficult task; for no sooner do we become acquainted with the trivial distinctions, than we are surprised to find how easy it is to recollect them; and things which appeared wrapt in mystery, now become obvious and familiar to us. It is the mere want of knowledge of the plain and simple means pursued by the naturalist, that has all along prevented thousands from following this, one of the most delightful and instructive exercises of the reasoning faculties; and, such are the charms which it carries along with it, that almost all who once enter upon the study become enthusiastically devoted to the subject.

It is our intention to introduce a series of essays, containing elementary instruction in the different departments of the system of Nature, and rendered in language which can be understood by every body. A certain number of technical terms are, however, indispensable; but these can easily be acquired.

We have said, that want of attention alone makes the task of discriminating natural objects difficult, and we shall illustrate this position by a very simple and familiar fact.

There is scarcely a human being who is not acquainted with the general appearance of a sheep. We have looked upon hundreds of them hundreds of times, and yet, strange to tell, we have not acquired an intimate knowledge of their appearance; nor can we discriminate one from another, although they are as unlike each other as are individuals of the human race. Let one be picked out from a flock of five hundred, nay, even of one hundred, and let us examine it for half an hour most attentively, and then set it at liberty again amongst its fellows; the chances are five hundred to one against us, that we shall ever be able again to find out the identical sheep. But let the experiment be tried with a shepherd, and he will, in a few minutes, detect the sheep, although set at liberty amongst thousands. And the shepherd requires no uncommon sagacity to be able to do so; for, on the contrary, there is scarcely a man exercising the calling, who will not readily perform this easy task. So it is with the study of Nature; a little attention and experience will soon render any object familiar and comparatively simple.

The young student, who aspires to become a zoologist, a botanist, or a geologist, need not, therefore, be discouraged from attempting to obtain his share of the superior delight which scientific knowledge can afford, by the obstacles which, only in *appearance*, oppose the acquirement.

Every step in the pursuit produces a reward and gratification in exact proportion to the difficulty; and each advantage, thus gained, produces fresh excitement to proceed in the path of science. Let us draw our illustration from the vegetable kingdom. Every plant, for example, of which we acquire a knowledge by sight and name, so as to be able to recognise it in another locality, not only gives a distinct pleasure at first, but the pleasure is renewed and increased, when we meet it for the second and third time, probably under very different circumstances, either as relate to ourselves, or to the plant. Thus, even the simple knowledge of the name, which enables us to communicate our ideas, although in an indistinct manner, brings with it sensations of a pleasurable kind, and often proves a source of the most interesting associations. But the pleasure we derive from a knowledge of the trivial or popular names of plants, becomes greatly enhanced by more extended views regarding them, which are not strictly botanical. We are astonished when we study their geological relation in any particular district or country; their geographical distribution, relatively to the world itself, or their migration from one country to another; their connection with climate; their being domestic plants, which follow man in his improvement and change of soil, or wanderers seeking to inhabit distant regions, formerly uninhabited by their kinds, or by their being social, and living, like man, in large communities; their abundance or rarity; their mode of propagation; their natural enemies, or more kindly friends; and, lastly, their properties, functions, uses, and culture. It is in acquiring a knowledge of all these that real pleasure is experienced; and, as we acquire this knowledge, our desire to become still farther acquainted with them increases.

To know any natural object, however, does not merely consist in having seen it, or in recollecting its name. For we cannot be justly said to be acquainted with a plant until we know its rank in the vegetable kingdom, its structure and habit, with all the other circumstances already explained.

There is hardly a child who cannot at once name a ranunculus, or tulip; but how few, even who cultivate these deservedly-admired productions of the garden, are aware that these two plants, however nearly they may be allied as fine flowers, are very different, in point of rank, in the scale of vegetable creation! They belong to separate fundamental divisions of plants, and the organization of the one is much more perfect than that of the other. They display totally different characters of structure and physiological economy, from the seminal embryo, through every stage, to the perfect plant. The ranunculus belongs to a division of plants characterized by a reticulated, or net-like structure, in their parts. It will admit of portions of its leaves being broken, or cut off, without impeding the remainder of the leaf in the performance of its functions; or, in other words, the leaf will continue to grow, and arrive at a state of maturity, although deprived of a portion, or limb. Now, the tulip belongs to a division, the structure of whose fibres are parallel, and will not admit of part of the leaves, more particularly their extremities, being cut off, without impeding their functions, and, consequently, injuring the present health of the plant, and affecting its vigour for the following year. Here, then, we have another example of the utility of natural knowledge; for any one who has paid the slightest attention to the anatomy or physiology of plants, will at once be able to know the distinctive structures of these two divisions; and, if only a part of a leaf be presented to him, the division to which it belongs will immediately be detected by him, and, consequently, the fundamental arrangements, culture, and general management of the plant, as far as regards its most important organs; for leaves are analogous to the lungs of animals. Thus we have the increased pleasure of not only knowing the plant by its name, but also its rank in vegetable physiology, and the manner in which its various functions are exercised.

One of the most extraordinary phenomena in nature is the endless variety of forms in the distinct species of animals, plants, and minerals; and still more wonderful are the infinite modifications of form in the same species. For it is our conviction, that, since the creation of the world down to the present time, there never has been two individuals of the same kind, formed exactly alike in all their parts. This leads us naturally to an expression of our admiration of the works of Providence, in the words of the Psalmist, "O Lord, how manifold are thy works! in wisdom hast thou made them all!"

This idea is sublime; and, however erroneous it may appear to those who have not deeply studied Nature, we firmly believe that it is nevertheless true. Let us illustrate this by another example from the vegetable kingdom. Behold the stately oak of the forest, spreading his branches afar on every side, rearing and shedding his millions of leaves for a series of hundreds of years, but never producing two leaves exactly alike; nay, even consider the leaves of the countless oaks existing at the same moment on the face of the globe, ever varying in appearance; yet a general similarity of form has been, and will be, maintained to the end of time! Let any one who is sceptical on this point repair to the forest, and patiently examine every leaf which has clothed one of its largest oaks, and he will never be able to find two of them perfectly alike in size, shape, and particular structure; nay, he may extend his search to all the oaks of a forest, and he will discover that he has been seeking for that which, like the philosopher's stone, will never be found. So it is with all the works of creation, whether animate or inanimate. One uniform and fundamental plan has been established, alike in its grand leading principles, but exceedingly varied in its detail; and we are thence led to admire the profound wisdom of the Creator and Preserver of the universe. Let us for a moment suppose that all mankind were formed exactly similar. What would be the consequence? Endless monotony, confusion, and crime. The variety of form and intellect in the human species, creates in us those varied sensations of pleasure, which are derived from the admiration and love of one object beyond that of another, for some real or fancied quality. If all were alike, the love of one particular object could not exist, and a disgusting monotony would every where surround us. A man would not know his own wife, nor a child its parent; perpetual scenes of confusion would prevail, and crime could not be traced to its perpetrators. There would be a total want of those varied sentiments which hold their sway over the human heart, and from which emanate every thing that is pleasurable in existence. It has, however, pleased the Dispenser of good to order every thing otherwise; and we now behold the world one vast machine, infinitely varied in its parts, but all of them tending to the furtherance of one mighty design.

The study of Nature teaches us to discover that, in the animal kingdom, there seems to be one great chain of being, from man down to the lowest scale of animated existence; and it is not impossible but that this may prevail even through the vegetable and mineral kingdoms, although man has hitherto been unable to detect the connecting links.

> Each shell, each crawling insect holds a rank
> Important in the plan of him who form'd
> This scale of beings; holds a rank, which lost,
> Would break the chain, and leave behind a gap
> Which Nature's self would rue.
>
> STILLINGFLEET.

Natural history is a study calculated in an especial manner for elevating the character of the labouring classes of society. Indeed, it may be said to be a study in which most labourers and mechanics are already engaged. Their implements, and the materials which they manufacture, are all derived from the field of Nature, and are only modified by the experienced hand of man from his knowledge of the several qualities which appertain to each. Besides, it requires less preliminary information than almost any other branch of study; and even the humblest individual has within his reach the means of contemplating Nature in one form or another. It is a much more rational manner of spending time than in dissipation, which debases the mind and undermines the constitution. While other branches of study have the effect of improving the reasoning powers of the mind, natural history may be said to improve and humanize the whole man. The intimate connexion between moral conduct and the love of animals and plants, will be thought intimate or remote according to the ideas of different individuals; but the more we consider and trace the design and purpose of the works of creation, shall we not sympathize the more with the fitness of man to the ends of human conduct? The deeper we enter into the details of nature, shall we not increase our relish for facts? This is nothing less than laying the foundation for the love of order, of justice, and of honesty.

Even those who have no knowledge of scientific zoology, derive great pleasure from their observations on the manifest variety in the forms, habits, and instincts of animals; and mankind are accustomed from these observations to transfer to some of the higher quadrupeds many of the virtues of humanity. We speak of the courage of the horse, the generosity of the lion, the sagacity of the dog, and the innocence of the lamb; we are delighted with the melody of the songsters of the grove; the industry of the bee holds up to man a useful lesson; the gay attire of the butterfly pleases us; and the noxious and disgusting appearance of varions reptiles excite in us varied emotions. But all these are nothing when put in comparison with the pleasure derived by the scientific zoologist. He who can trace the varied degrees of power and intelligence, imparted by the Supreme Being to animals, from intellectual man down to the lowest animalcule,—who can trace the complicated organization of beings down to the minutest conferva or lichen, and who knows scientifically that man is the most perfect of all created beings, enjoys a degree of exalted pleasure which scientific knowledge can alone impart.

ZOOLOGY.

DESCRIPTION OF THE PLATE—THE DEER.

THE animals of this active and beautiful tribe belong to the genus *Cervus* of Linnæus; they principally inhabit wild and woody regions. The species of this genus vary much in size. In their contentions, both among themselves and with other animals, they not only use their horns, but also strike furiously with their feet. Some of them are used as beasts of draught. The flesh of the whole tribe is accounted particularly delicious, and well known by the name of venison.

The males only are provided with horns; these are solid—branched; and are shed and renewed annually.

Fig. 1.—The Fallow-Deer (*Cervus dama.*) It is this species which is kept in the deer parks of Britain. It is readily tamed, and feeds upon a variety of vegetables. This animal, like nearly the whole tribe, is gregarious; and in parks where numbers are kept, they frequently divide into two parties, and maintain obstinate battles for some favourite parts of the pasture. There are two varieties of Fallow-Deer in Britain, both of which are said to be of foreign origin: the beautiful spotted kind, which we have figured, is supposed to be a native of Bengal; and that variety without spots, which is now very common in our parks, was introduced from Norway by King James I.

Fig. 2.—The Virginian Fallow-Deer (*Cervus Virginianus*), associate in numerous herds, and supply a most palatable food to the inhabitants of the back settlements of North America, and to other wild tribes. This species can at once be distinguished from the former by its horns being arched forwards.

Fig. 3.—The Long-tailed Deer (*Cervus macrourus.*) This species is larger than the Red-deer, and is distinguished by the size of its tail, which is nearly eighteen inches in length. It inhabits the central and northern parts of North America.

Fig. 4.—The Axis (*Cervus Axis*), is a native of the warmer parts of eastern Asia, and is considered one of the most beautiful of the group of which it is a member. In its general form and markings, it somewhat resembles the Fallow-Deer, but is at once distinguished from it by the structure of its horns.

Fig. 5.—The Malayan Rusa Deer (*Cervus Equinus.*) This animal differs from most of its congeners by its neck and throat being furnished with a thick coating of long bristly hair. It is a native of Bengal.

Fig. 6.—The Guazopuco Deer (*Cervus paludosus*), inhabits South America, and is nearly equal to the European Stag in point of size. Its horns are, however, of comparatively small dimensions. A distinguishing mark of this animal, is the long and flowing hair which ornaments its abdomen, back of the thighs, and under side of the tail.

Fig. 7.—The Nepaul Stag (*Cervus Wallichii*), is fully larger than the stag of Europe, and differs from it in having two small antlers at the base of the horns, projecting forwards. Its colour is of a yellowish brown, mixed with gray; it is distinguished by having a very short tail, and a disk of white spreading above the croup.

HUMANITY OF A WREN.—In the end of June 1835, a person was shooting in the neighbourhood of Bandrakehead, in the parish of Colton, Westmoreland: he killed a brace of Blue Titmice (*Parus cæruleus*), which some time before had been observed to be constructing a nest, in the end of a house, belonging to a Mr Innes of the same place. In the course of the day, it was ascertained that the Titmice had completed the time of incubation, and that their death had consequently left their offspring in a state of utter destitution. This, however, was not long permitted to continue, for the chirping of the young birds attracted the attention and excited the compassion of a Wren; which, since that period, has adopted the nestlings; and was daily engaged in rearing and feeding them, with the affectionate kindness and unremitting assiduity of a parent bird.

FATAL BITE FROM AN ADDER.—In a moss in the neighbourhood of Bucklyvie, a farm servant, while engaged in cutting peats, a few months ago, was stung by an adder, and died in consequence of the wound in about ten days.

MIGRATION.—By wonderful instinct, birds will follow cultivation, and make themselves denizens of new regions. The crossbill has followed the introduction of the apple to England. Glenco, in the Highlands of Scotland, never knew the partridge till its farmers of late years introduced corn into their lands; nor did the sparrow appear in Siberia, until the Russians had made arable the vast wastes of those parts of their dominions.

BOTANY AND HORTICULTURE.

AGES OF TREES.—From examining the layers of wood, it has been satisfactorily ascertained, that olive trees will live in favourable situations for 300, and oaks for 600 years. Greuw, in the year 1400, cut his name on two *Boababs*, and Petiver did the same thing 149 years afterwards. In 1749, Adanson saw these trees, and at that period they had increased seven feet in circumference since the time of Petiver, being an interval of 200 years. These trees are, however, sometimes found to acquire a perimeter of 455 feet; from this it is inferred that they must live many thousand years. The long period required to ascertain the age of trees, renders our knowledge on this subject very imperfect; and it will probably long remain so until records are established by scientific institutions, to ascertain the ages of such trees as are public property.

MINERALOGY.

PLATINUM IN SIBERIA.—This valuable metal has been discovered in Siberia; it is found in fine sand. A piece was got at Nischde Tagil, in 1827, which weighed 8 pounds 13 ounces and 4 drachms, avoirdupois. In 1831-32, three pieces were obtained, the two first weighing 17 pounds 11 ounces, and the third 11 pounds 1 ounce and 1 drachm. It is accompanied with gold, osmium, iridium, magnetic iron, chromium, brown oxide of iron, oxide of titanium, epidote, garnet, rock crystal, and sometimes diamonds. The sand in which it occurs is composed of jasper, green stone, and quartz; and also small yellow crystals, of rhomboidal dodecahedrons, resembling chrysoberyl, the precise nature of which is not yet known. Among the class of rocks which accompany platinum in the Uralian chain of mountains, serpentine is the most remarkable. Gold appears generally to exist in the same rock with platinum.—*Records of General Science, April* 1835.

GEOLOGY AND PHYSICAL GEOGRAPHY.

LARGE FOSSIL OAK TREE.—In February 1632, in the course of excavating the basin of the harbour at Aberdeen, and within one hundred and fifty yards of the Trades' Hospital, there was found embedded within a few inches of the surface of what was formerly the Green Inches, an oak tree of very large dimensions. It was lying in a horizontal position, in a north-west and south-east direction. The trunk, from the root to where it separated into limbs, was 6 feet 6 inches long, and 20 feet 2 inches in circumference: one of the limbs measured 23 feet 6 inches in length, and 13 feet 10 inches in circumference; the other limb, 8 feet long and 9 feet 7 inches in circumference; and from these two limbs various large branches appear to have struck off, some part of which still remained. The cubical contents of the trunk and two limbs are upwards of 380 feet. As it does not appear that the tree could have grown in the place in which it was embedded, and as the timber had not the marks of decay which are frequently to be seen on very old trees, it could only have been a flood of the river, equal if not greater than that of 1829, which could have brought it from its native soil, while encumbered with its immense branches.

FLOOD OF THE GARONNE.—The city of Toulouse has lately been devastated in a manner which has not been surpassed during the last fifty years. In the night of the 29th May 1835, the waters of the Garonne increased so suddenly, that before any measures of security could be taken, several depots of wood in the Port Garau, and the suspension bridge of Mourst, were washed away. The handsome stone bridge at Puisaquel lost two of its arches, and a third received considerable damage. On the 1st June, the waters having partly subsided, the Garonne was confined to its natural bed; but above Toulouse it broke its bank, and the inundation extended far and wide. We learn also, that throughout the whole department, the sudden swelling of unnavigable rivers had occasioned great disasters, and had carried away several mills. In Toulouse, fifty houses were destroyed.

METEOROLOGY.

EFFECTS OF LIGHTNING.—At the conclusion of the evening service on Sunday, 21st June 1835, the church of Semur, in the Cote d'Or, in France, was struck by lightning. The electric fluid was divided by the conductor, but entered the windows of the vestry, tore up the floor, and gave a violent shock to a person who was in it. The Abbe Bernard, who was kneeling on a chair, had one of his shoes burnt. The people, who were in the square before the church, received the shock, being all affected in the legs. The river was so much swollen, that the water rose in the houses near the bridge up to the ceilings. A poor woman, while endeavouring to save some of her little property, was crushed by a beam falling upon her.

FIRE BALL.—On Wednesday the 24th June 1835, as some fishermen were pursuing their occupation, about six miles off the shore at Stonehaven, they were much alarmed at the sight of a large fire-ball, which spread a brilliant glare all around, and appeared to fall on the bosom of the deep. There had been a considerable thunder storm in that neighbourhood on the previous day.

REMARKABLE METEOR.—On the Chinese frontier, near Kiakha, while the weather was very calm and extremely cold, on the 11th of March 1835, there was observed, a few minutes after 9 P.M., at an extraordinary elevation in the north-west, a fiery meteor, in a serpentine form, of a most dazzling brightness. In an instant this meteor was converted into a brilliant cloud, resembling an immense blazing sheaf, which proceeded to precipitate itself obliquely, with extreme rapidity, towards the earth, assuming the form of an enormous sheet of flame, which filled all the visible space of the horizon with a light as clear as day, and, separating into three portions, disappeared. The roar of distant thunder was then heard, and this occasioned a tremulous motion to be felt in the houses. Two other sounds were perceived, but these were reckoned merely the echoes of the first reverberating from the neighbouring mountains. This phenomenon, which lasted but a few seconds, has not occasioned, as far as has yet been ascertained, the least damage.—*Russian paper.*

GENERAL SCIENCE.

MILITARY MORTALITY IN THE BRITISH COLONIES.

MAN possesses a greater power of adapting his frame to different temperatures and climates than any other animal. But this power has its limits; and we find that a person, who has long been inured to the chill blasts of an arctic region, or to the milder airs of Britain, is frequently incapable of resisting the burning heats of a tropical climate. A sudden transition from the one to the other is often attended with fatal consequences. Individuals may escape its immediate effects; but when large bodies of men are considered, a greater mortality is absolutely certain to occur. This fact has been long noticed, in public and private records, by the frequent premature decease of many a gallant officer or near relative; and the popular views on this subject are fully confirmed by the following

OFFICIAL RETURNS

Of the mortality among officers and soldiers in the several British colonies—chiefly for the seven years from 1820 to 1826—showing the annual deaths out of 10,000 men:—

		Excess per cent.
Great Britain (1824 and 1826)	144	
Malta	182	.38
Bermudas	209	.65
Mauritius	240	.96
Ionian Islands	236	1.12
MADRAS.		
Civil Service in 1816	290	1.46
Ditto in 1820	600	4.56
34th regiment, during eleven years, with one-half of the invalids	874	7.30
69th regiment, during fourteen years	834	6.90
Average of Civil Service	445	3.01
Ditto of Military Service	854	7.10
CEYLON.		
45th regiment, on the passage (1819)	440	2.96
Ditto, in the island	500	3.56
19th regiment, twenty years, with one-half of the invalids	840	6.96
83d regiment, three years, with one-half of the invalids	1170	10.26
73d regiment, three years, with one-half of the invalids	2800	26.86
Officers of 1st Ceylon regiment	1230	10.36
73d regiment, officers,	1130	9.66
Average on the passage to Ceylon	440	2.96
Ditto of the soldiers on the island	1338	11.84
Ditto of officers	1180	10.36
WEST INDIES.		
Grenada	391	2.47
Trinidad	425	2.81
Antigua	457	3.13
Barbadoes	563	4.19
St Vincent	574	4.30
St Kitts	617	4.73
Bahamas	640	4.96
St Lucia	686	5.42
Dominica	987	8.43
Tobago	1061	9.17
Jamaica	1306	11.62
Average of West Indies	701	5.57
Ditto of smaller islands, excluding Jamaica,	640	4.96
Demerara	1206	10.64

Other climates seem to exercise an opposite influence over the human frame; and longevity is often promoted by exchanging the moist and ungenial air of Britain for the purer climates of the South and West. The following returns, procured from the same source, and for a similar period with the preceding, point out the annual deaths in 10,000 men. For

Great Britain (1824 and 1826),	144
Lower Canada,	138
Cape of Good Hope,	133
Nova Scotia,	115
Upper Canada,	107
Gibraltar,	107
New South Wales,	68

These returns generally include the whole regiments stationed at the several colonies; and it has been thought probable, that if we could distinguish between officers and private soldiers, the mortality of the former would be found much less than that of the latter. Private soldiers are more exposed to severe fatigues, and to night guards; while their general habits are likely to lead them to more frequent excesses, and to neglect the first appearances of disease. For these reasons, it has been considered that the mortality among them should exceed that among officers alone, by about one-third.

How far the above-mentioned difficulties may necessarily attach to the military life in general, we shall not at present attempt to determine. But, we have no hesitation in asserting, that *the degree* in which the average duration of the life of the officer exceeds that of the private soldier in foreign climates, has been very greatly exaggerated.

It is well known that, unless a statement of the law of mortality, in any place, is founded upon a very wide induction of particulars, it is in an eminent degree calculated to mislead. Surprising and unaccountable differences will be found to prevail in the mortality of small classes of men, in succeeding years, and in comparison to the rest of the community. It is only when the observations are extended over a large community, that any fixed *law* is found to exist; and, whenever a small number of individuals are alone included in the inquiry, the results must be received with considerable caution and reserve.

As the number of officers always bears a small proportion to that of the privates of a regiment, it becomes almost impossible to distinguish the one from the other, in framing a table of mortality. Who would think of attempting to ascertain the duration of human life in London or Edinburgh, by recording the deaths among his acquaintances? Yet *few* persons of any importance can number less than 136, which is the total number of British officers who have resided in the island of Ceylon, for a period of nine years.

A writer in the *United Service Journal* for June 1835, in an article of some interest, gives the first two columns of the following table, as being the observed mortality at the several stations, among officers alone. We have added a third column, which is deduced from the others by calculating the mortality for a radix of 10,000 men, that we may be able to compare it with the tables already given, for officers and private soldiers collectively.

MORTALITY AMONG OFFICERS IN THE BRITISH COLONIES, DURING THE NINE YEARS FROM 1826 TO 1835.

Station.	*Officers.*	*Deaths.*	*Die annually out of* 10,000.
Corps at home	936	96	114
Bengal	388	138	395
Madras	398	158	441
Bombay	186	44	263
*Ceylon	136	15	123
*Mauritius	113	10	98
*Jamaica	178	45	281
Windward and Leeward Islands	283	45	177
*New South Wales	102	11	120
Cape of Good Hope	102	12	131
North America and Bermuda	351	28	89
Gibraltar	219	22	111
Malta	140	12	95
Ionian Isles	212	22	115

We have marked with an asterisk (*) those returns which appear particularly liable to suspicion: the first three on the list are probably not very far from the truth.

Upon comparing these results with the tables for officers and private soldiers collectively, we shall find the mortality to be represented exceedingly low—to a degree, in some cases, altogether extravagant. !

We particularly notice Ceylon, in which the mortality is stated to be very slightly greater than in Great Britain, being only as 123 to 114—a result completely at variance with all other documents. A good authority on this subject describes the climate as being exceedingly diversified according to situation. In some parts, it is hot and oppressive, and liable to frightful storms of thunder and lightning; in others, it is more temperate and salubrious. But the woody parts prove destructive to strangers, who frequently become victims to the putrid miasmata, which taint the atmosphere. Until lately, the kingdoms of the interior have been guarded from the attacks of Europeans by the insuperable barrier of an unhealthy climate; and those who escaped the hazards of war generally fell victims to the ravages of disease.

How far the healthiness attributed to this island in the returns, may arise from the peculiar facilities afforded by an insular situation to such officers, whose constitutions are broken down, to die in peace on the passage homeward, and thus to escape enumeration in the returns, we cannot at present determine. Ceylon was believed, and is yet considered by many of the Orientals, to have been the paradise where Adam was created. Hammalleel, or Adam's Peak, is still held in great veneration by the natives. But though the notion, that Ceylon *was* paradise, may be popular, we must protest against documents which would give rise to the belief, that it *is* so still, at least in respect to the duration of human life.

In conclusion, we earnestly entreat our military friends, who are anxiously looking out for a station congenial to health, to be certain that the data on which they found their calculations are derived from a sufficiently wide induction of particulars, before they trust themselves implicitly to their guidance. We would recommend the seniors of each rank rather to expend their money in purchasing the chance of promotion by death or vacancies in Ceylon, than in negotiating an exchange to this boasted land of promise.

MISCELLANEOUS.

THE EUPHRATES EXPEDITION.—*(From the Malta Gazette, June 3.)*—By his Majesty's brig Columbine, Commander Henderson, from the Orontes, we have received some account of the Euphrates expedition and its first proceedings. Colonel Chesney and the whole of the officers and men were quite well on the 3d of May; they were encamped on a spot near the mouth of that river, to which they have given the name of Amelia Island.

The George Canning was towed by the Columbine almost the whole way from Malta to the bay of the Orontes, where the expedition anchored on the 3d of April. On the 6th the landing of the packages and stores was commenced by means of a hawser, which was extended over the bar from the George Canning to the shore, a distance of 1,200 yards, by the officers and men of the brig of war. Captain Henderson likewise stationed Lieutenant Thomson and Mr Pritchard, with 25 men, at the camp established on shore; and everything being thus well disposed, nearly two-thirds of the whole of the equipments were landed by the boats of the two ships, eight in number, during the first week. The only accident that happened was the temporary loss of a cask, containing the valves and other parts of the steam-engines, which, by the breaking of the slings, sank to the bottom; but it was soon recovered by part of the apparatus of the diving-bell.

The attention of the officers of the expedition was then directed to other objects. To Captain Estcourt was allotted the repair of the road to the Euphrates; to Lieutenant Murphy and a party the survey of the bay of the Issus; to Lieutenant Cleaveland the landing of the stores and the preparation of the caravans; whilst Colonel Chesney and Lieutenant Lynch, of the Indian Navy (who had been waiting and preparing for the expedition some time in Syria), were employed in soliciting aid from the authorities of the country, and making arrangements with the Arabs near Bir, on the Euphrates, whither Lieutenant Lynch proceeded to receive the first section of light materials, which would have arrived there about the 17th, if it had been possible at once to procure camels.

During the second week the weather was so boisterous as to retard the landing a good deal, and the gig of the Columbine was upset on the bar with Captain Henderson and four men in her, who were all happily saved by a boat which immediately pushed off from the George Canning. But, notwithstanding the bad weather, by the 21st every thing was disembarked except a few coals.

The estuary of the Orontes appears to have been a happy selection for the disembarkation of the expedition; and the success with which it has been effected may be a favourable omen of its future progress. Amelia Island is described as presenting a scene of high interest. The people of the surrounding places constantly visited the camp, and viewed with wonder and amazement the operations of our sailors and mechanics. The landing of the boilers and large pieces of the iron steam-boats and engines, as well as the fishing up of the heavy cask from the bottom of the sea, caused the greatest possible surprise. In truth, the various costumes, the mixed nature of the stores, the general activity which connected the ships with the shore, and the beautiful scenery, with the crest of Mount Cassius towering above to the height of 5,618 feet in the back-ground, formed altogether a striking picture on the ancient coast of Syria. Since the above was published, a private letter has been received, which gives us later information. It is dated "*Orontes Camp, Amelia Island, May 23d*, 1835.—I wrote you on the 2d instant by his Majesty's brig Columbine. We have since put the Tigris together; we commenced laying her down on the 6th, and had her ready for launching on the 21st, being thirteen working days. Her draft of water, when launched, with the two bed-plates and air-pumps of her engines, was 6½ inches upon an even keel. The intention of putting her together at present is, that we hope to be able to carry the other boat, heavy weights, &c. up the Orontes, as far as Antioch, a distance of about twenty miles, which will enable us to avoid a range of mountains, that we should have to contend with in land carriage, after which we intend to take her to pieces in eight sections, for transportation to Bir. All the men are in excellent health and spirits."

OBITUARY.

In the month of May last, at Chamarande, near Paris, died T. R. Underwood, Esq., Fellow of the Geological Society of London. He was born in London, the 24th of February 1772. The disease under which he sank, after long and constant suffering, originated, many years ago, in the *antrum maxillare*. Mr Underwood enjoyed the acquaintance of several of the most distinguished men of the age. He was an excellent artist, and a perfect judge of the arts—of much patient industry and indefatigable research. Geology seems to have been his principal study, but his attention was by no means confined to it; for, as a naturalist, he had accumulated an immense variety of observations, which, if placed in some able hand, would contribute largely to the general stock of knowledge. Mr Underwood was a Protestant; and, on the occasion of his interment, it is gratifying to record an instance of liberality on the part of the Roman Catholic clergyman of the village of Chamarande. In the absence of a Protestant minister, none being within many miles, the rev. gentleman, in a spirit which reflects the highest honour on his feelings and his character, led the mournful procession which followed the remains of the deceased to their last resting-place. Mr Underwood was so great an enthusiast in science, that he wished only to live that he might witness the return of Halley's comet; and would have consented to an excruciatingly painful operation, if his surgical friends could have given him hopes that his life could have been prolonged by it only for a few weeks.

LEARNED SOCIETIES.

GEOGRAPHICAL SOCIETY.—The president, Sir John Barrow, in the chair. A letter, addressed to Dr Hancock from Mr Paterson, a gentleman residing on the river Demerara, British Guiana, was read, on the climate and productions of that country. The seasons, it appears, have, within the last twenty years, become far less regular than they formerly were; and it is now somewhat difficult to predict, with precision, when the wet or dry seasons will set in. The wet season used to commence about the middle of April, gradually increasing until June, when it rained incessantly, accompanied with tremendous thunder and lightning; by the middle of August, what was called the long rainy season, had quite terminated. It is now of shorter duration, and much less severe. August, September, October, and November are generally dry; December and January are wet; in February and March it is again dry. There is most thunder in June and July, and the wet season frequently terminates abruptly, after a violent thunder storm. A remarkable change has also taken place within the last fifteen years in the violence of the wind on the coast, which had an important influence on the tides, causing them to rise so high, that the camp-house, stores, &c. were frequently in imminent danger of being washed away; the tide of late years, from the decrease of the wind, does not approach within a considerable distance of the fort; and the water being now left to pursue its natural channel, has increased so much in depth, on the bar of the river, that vessels, drawing eighteen feet of water, pass without difficulty. The average rise and fall of the river at Christianburgh, sixty-five miles up, is about six and a half feet. The country affords an inexhaustible treasure to the mineralogist, ornithologist, and botanist; it abounds with valuable woods and drugs.—*Literary Gazette, 27th June.*

Edinburgh: Published for the PROPRIETORS, at their Office, 16, Hanover Street. London: CHARLES TILT, Fleet Street. Dublin: W. F. WAKEMAN, 9, D'Olier Street. Glasgow and the West of Scotland: JOHN SMITH and SON, 70, St Vincent Street.

THE EDINBURGH PRINTING COMPANY.

THE EDINBURGH

JOURNAL OF NATURAL HISTORY,

AND OF

THE PHYSICAL SCIENCES.

N°. 3.   SATURDAY, NOVEMBER 21, 1835.

ZOOLOGY.

DESCRIPTION OF THE PLATE.

THE RHINOCEROS.—The animals of this genus are nearly equal in size to the Elephant, although they appear less, owing to their legs being much shorter, in proportion to the dimensions of their bodies. They are generally of peaceable dispositions, if unmolested, living upon herbs and branches of trees, and frequenting marshy places. There have been five species of this genus ascertained by naturalists.

Fig. 1.—The Indian Rhinoceros. (*Rhinoceros Indicus, Cuvier.*) This species has been the longest known to mankind, by the name of the one-horned Rhinoceros. It inhabits India beyond the Ganges. In a wild state, it grows to twelve feet in length, its circumference being nearly equal to its length. Its skin is composed externally of numerous horny tubercles, which render the hide impervious to the claws of the Lion and Tiger; and this, together with the formidable horn on its nose, makes it more than a match for either of these desperate animals. It is but seldom the Elephant will dare to give the Rhinoceros battle; when he does, he generally meets death as a reward for his temerity. The hide of this animal is remarkable for the deep folds formed across and behind the shoulders, as well as on the front of the legs, thighs, and flanks. The sight of the Rhinoceros is but dull, but its sense of hearing is said to be extremely acute.

Many of the Indian princes drink out of cups made of the horn of this animal. They have a superstitious belief, that, when these hold any poisonous draught, the liquor will ferment till it runs quite over the top. Martial informs us, that the Roman ladies of fashion used these horns in the baths, to hold their essence-bottles and oils. The Javanese make shields of the Rhinoceros' hide; the flesh is used as an article of food; and the teeth, which are very white and solid, are used by dentists in making false teeth.

Fig. 2.—The lesser two-horned Rhinoceros. (*Rhinoceros Africanus, Cuvier.*) This animal inhabits Africa, and differs from the Indian species in having two horns, and also in the appearance of its skin, which is nearly smooth, having merely slight wrinkles across the shoulders and hinder parts, and a few fainter folds on the sides; he also differs in colour, being usually of a brownish black; while the other species is of a violet-tinged blackish gray.

Bruce says that these Rhinoceroses are very swift when pursued, and even to a degree which is astonishing, when we take into account their unwieldy forms. Le Vaillant informs us that whenever they are at rest, they place themselves in the direction of the wind, with their noses towards it, in order to discover by smell the approach of any enemies. Bruce mentions, that, "besides the trees capable of most resistance, there are in the vast forests, within the tropics, trees of a softer consistence, and of a very succulent quality, which seem to be destined for the principal food of this animal. For the purpose of gaining the highest branches of these, his upper lip is capable of being lengthened out, so as to increase the power of laying hold with it, in the same manner as an elephant does with his trunk." It is not true that the skin of the Rhinoceros is so hard and impenetrable as to resist a musket ball. In his wild state he is slain with javelins thrown from the hand, some of which enter his body to a great depth; and the Shangalla, an Abyssinian tribe, kill him with very clumsy arrows, and afterwards cut him to pieces with the worst of knives.

PACE OF A LOADED CAMEL.—I have made many long journeys on camels, and I certainly think that animal, when well taken care of, and not overloaded, fully capable of marching ten or eleven hours per day, at an average rate of two miles and a half an hour, in valleys or over rough roads, and three miles on plains, without being at all distressed. On the banks of rivers, and in districts where water and forage are plentiful, except urged on, the men are always inclined to move more slowly, and make a shorter day's journey, not so much to save their camels, as to lessen the fatigue to themselves; a few pairs more or less en *route* being generally a matter of indifference to them.—*Travels in Ethiopia, by G. A. Hoskins, Esq. London.* 1835.

ENRAGED ELEPHANT.—A very characteristic action of D'Jeck, the famous elephant of M. Huguet, was lately near costing the life of a young man, a native of Bruges. The elephant, it is well known, is very fond of sweetmeats, and this young man amused himself at Madame D'Jeck's expense, baulking her by offering her some, which, when ever she reached out her trunk to take, he immediately withdrew. This trick having been noticed by M. Huguet, he observed to the young man how foolish such conduct was towards an animal at once so tractable and vindictive. But not taking warning from this remark, the Belgian again invited the elephant to approach, and not only again deceived her, but gave the sweetmeats to Mademoiselle Betsey. Madame D'Jeck now lost her patience, and, regardless of the presence of her master and a numerous assemblage of spectators, lifted her trunk and knocked the young man down, tearing open his cheek, and rending his clothes to tatters. Happily M. Huguet

interposed his authority, and the elephant left her hold, but the imprudent sufferer was long confined to his bed from the effects of his absurdity.

SOCIAL HABITS OF BIRDS.—In June last (1835), a singular instance of the domestic and social habits of birds was to be seen at Auchmuty paper-mills, in the parish of Markinch, Fifeshire, where a monthly China rose-bush, trained to the front of the dwelling-house, was fixed upon by three distinct species of birds, even of different genera, to build their nests and rear up their little families. A blackbird, a yellow-hammer, and a sparrow, composed this small community. Delighted with the rich green foliage and expanding roses, amidst which they had placed their downy dwellings, they lived together in the greatest harmony and good fellowship, alternately singing their sweet notes, and watching their progeny.—*Caledonian Mercury, 13th June* 1835.

CULTURE OF BEES.—Mr Begbie, gardener to Sir John D. Erskine, Bart., Torry, purchased a hive (a second cast) last year, and which, towards the approach of winter, showed that the "store" was far from "complete," and quite inadequate to their wants during that season. He determined to try a novel expedient; and, in November last, buried the hive in the earth three feet below the surface, covering it carefully with straw, and placing a flag above, and then earth on the top. In April it was dug up and found to be in good condition, contrary to all expectation; and to crown the whole, this hive threw a capital swarm in June last, as a grateful testimony of the snug quarters enjoyed during winter.

DESCRIPTION OF A NEW BRITISH SHELL—THE CLOUDED SCALLOP.—*Pecten Nebulosus.*—Shell almost circular, ears nearly equal in size, with seven broad, unequal, and flattish ribs; external surface of both valves covered with very fine, parallel, longitudinal striæ, and also with minute, undulating, transverse striæ, which are hardly discernible to the naked eye, but feel rough to the touch. Both valves are somewhat inflated towards their base, with a longitudinal series of densely-set ribs, and the margin finely crenulated, two-thirds the length of the shell. The upper, or convex valve, is of a rich, reddish brown, irregularly clouded with white; under valve, cinereous, and immaculate; inside white, of a pearlaceous lustre, exhibiting slightly iridescent reflections. Length, one inch and seven-eighths; breadth, the same. The specimen from which the following figure was drawn was found at Largs, mouth of the Clyde, in July 1834.

We are indebted for the discovery of this beautiful species to that zealous naturalist, Mr John Blythe of Glasgow. He first noticed some fragments of this shell about seven years ago, while examining the shores at Millport, and afterwards found a perfect specimen. Mr Blythe informs us that the shell is not uncommon in Lochfine, and that his friend, Mr Drew, writer, Inverary, has procured several live specimens attached to the lines employed in cod fishing. They live in very deep water. We read an account of this shell before the British Association at Edinburgh in 1834, and were then doubtful whether it was an undescribed species. We thought that it might possibly be the *Pecten asperus* of Lamarck, as it is nearly allied to it in its form and markings. We thought that Lamarck's might be the young shell. Since that time, however, Mr Blythe has kindly presented us with two young specimens, in both of which the ribs, or rays, are seven in number; whereas Lamarck says his shell is five-ribbed; and we hardly think so acute an observer would have overlooked the transverse undulating striæ.

BOTANY AND HORTICULTURE.

Air Vessels of Plants.—M. de Mirbel, in an interesting paper, laid before the Paris Academy of Sciences, divided the vessels which deposit the bark of plants into two sorts; the one which forms cortical layers, and the other a cortical net-work. The former of these are renewed annually, while the net-work is only to be met with in the young branches, or stems.

Zostera.—M. M. Pasteur d'Estreillis and Adolphe Dommieu transmitted to the French Academy of Science a memoir on the plant called Zostera. This plant grows abundantly on the southern coast of France, and on the south of the Baltic, where it assists in binding the sands of these shores together, and has long been known as an excellent manure. Its leaves have commonly been employed in packing fragile objects, and it has more recently been ascertained that they make a most excellent material for stuffing beds.

Cowdie Tree.—The British government having satisfactory information of the fitness of the timber of the Cowdie tree of New Zealand for spars for the navy, sent the Buffalo ship to that country for specimens. That vessel returned in the end of April 1835, and brought a cargo far exceeding all expectations. Before the return of the Buffalo, an enterprising and experienced naval officer, who had formed an establishment of his own in New Zealand, had offered, and, we believe, contracted with the government, to furnish spars of this kind from that island, at a lower price and of better quality than those from the Baltic—a circumstance which, in the not impossible contingency of a war with Russia, may be of essential importance to this country.

New Species of Wheat.—It is said that a new species of wheat, which grows and ripens in seventy days, has been introduced into the department Du Nord.

Forest Trees.—North America has 140 species of forest trees, which reach 30 feet in height; but France has only 30 of this magnitude.

Management of Fruit Trees.—A valuable discovery in the management of fruit trees (it has been applied to thorn hedges before) has been made by M. Crozier, nurseryman, late of Alnwick, and now of Newcastle-upon-Tyne. The object is to obtain new wood where it may be wanted, and for this purpose he makes a nick above the eye where it is wished to produce new shoots; and after many trials, M. Crozier has found the experiment completely successful. In the garden of Mr Carr, at the Barras Bridge, near Newcastle, there is a pear tree which has sixteen shoots produced by the above means this season. It has been applied with equal success to apple, pear, and plum trees, and to the cherry to a certain extent.

MINERALOGY.

Discovery of Coal in Greece.—A Saxon engineer has just discovered in tne isle of Negropont a very rich mine of excellent coal, a discovery which is of great importance at a moment when it is in contemplation to establish steam communication throughout the Levant. The British government have given orders for the immediate construction of six fine steam-vessels, which are to be built on an entirely new principle. It is intended with these to open a direct communication between Great Britain and her Indian possessions, by way of the Mediterranean, the Isthmus of Suez and the Red Sea, through Alexandria in Egypt, by which means the voyage to India will not occupy more than eight or ten weeks, avoiding thereby the tedious route of the Cape of Good Hope, and the contrary winds which are so prevalent in the Indian sea. This grand project was one of the great ideas of the emperor Napoleon, who intended to have had a navigable canal cut from one port to the other, and thus facilitate his design of seizing our Indian possessions.

GEOLOGY AND PHYSICAL GEOGRAPHY.

Metalliferous Veins.—A new geological work has just appeared, entitled *Etudes sur les depôts Metallifères*, by M. Fournet. This author conceives veins to have been produced generally by more or less violent local dislocations, by first forming fissures, which have been afterwards filled with metallic, or other matter, either by means of sublimation or dissolution. He particularly refers to the successive modifications effected on mineral substances in veins, which have transformed the primitive matter even into a different species. M. Fournet demonstrates, in a clear manner, the importance of these decompositions, and the great influence possessed by them by their constant action and reaction, and their infinite division into veins, rocks, and strata; and thus throws considerable light on the more obscure parts of geology.

Vesuvius.—A letter from Naples, dated 3d April 1835, says, "Vesuvius, which had for the last fortnight given indications of an approaching eruption, burst forth last evening in all its fury. During the afternoon a storm of hail and rain had detained the crowd of visitors at Resina, who would otherwise have been inevitably sacrificed, as the very ground around the crater, where hundreds of human beings had been walking only the evening before, was carried up into the air at the first explosion."

Discovery of Bone Caves in New Holland.—Hitherto no quadrupeds of a life-large size have been found to inhabit New Holland; and yet Colonel Lindsay, of the 39th regiment, mentions the discovery of great quantities of fossil bones of animals, embedded in marle and other substances, in caves in New Holland; some of these, the bones of quadrupeds, are of large dimensions, and consequently must have belonged to large animals.

Tivoli.—Near this town there is a celebrated cascade, which has long been the admiration of travellers, who have for past ages flocked from all quarters to gaze on its beauties. This classic ground, together with the Grotto of Neptune, which have given rise to so many poems, and have been the subjects of so many descriptions, both oral and written, will soon disappear. The rock over which the river Arno precipitates itself to form this superb cascade, consists of a soft freestone; and in the lapse of ages, the waters have washed away a great part of the soft rocks, near the Grotto of Neptune, and threatened with destruction a portion of the town, and even the Temple of the Sybil. To prevent the impending danger, the Arno will be led into another channel some hundred paces farther up. The rock opposite to Tivoli is broken through—a work which will render illustrious the reign of Gregory XVI. The river precipitating itself into the valley, in a north-west direction from the present cascade, will form a new one, equal in elevation and volume to that of Terni. The channel cut in the rock is 400 feet in length, and has been completed for some months; and the preparations for conducting the river into its new bed will be shortly terminated. Perhaps some little delay may take place, as the Pope intends to be present, and will probably wait for the cool season.

Falls of Niagara.—A recent letter from New York announces the fall of the Table Rock, at the Falls of the Niagara. This immense mass of stone was on the Canada side of the river, projecting so as to afford the spectator a front view of the horse-shoe fall. It was considerably undermined, and several fissures on the surface had for some time past indicated the approaching disruption. A large mass was detached two or three years back. By the total fall of the Table Rock, the visitor is now deprived of the most favourable position for viewing the magnificent appearance presented by that stupendous fall of waters.

Earthquake.—Letters from Valparaisó, Chili, to the 1st March 1835, state that a very severe earthquake occurred at Conception on the 20th February, and its effects were felt throughout the whole province. It was unusually terrific; and the damage and loss of property must have been very great, particularly at Conception, from its being situated on a plain between two rivers, which on convulsions of this nature always rise to a considerable height above their banks. The shock was felt at Valparaiso for about two minutes. The old town of Conception, situated about nine miles from the present, was totally destroyed by an earthquake in 1751.

An express was received by the government, mentioning the total destruction, on the 20th February, of the cities of Talca and Carico, with the towns of Conquenes, Lenaies, and Chillaux In Conception, not a house is standing, and all the workmen who were repairing the cathedral of that city were buried in the ruins.

The Gulf Stream. — We copy the following account of the Gulf Stream, from that ably conducted periodical, *The United Service Journal*, for June 1835.—"Thrown overboard from the packet-ship, South America, in March 1833, in the Gulf Stream, off Cape Cod, in latitude 40° 30′, longitude 68–W. Any person finding this bottle is earnestly requested to publish the fact in the nearest newspaper, in order to confer a benefit upon science, by determining the currents of the ocean." Of all the experiments upon the currents of the Atlantic, none was ever more important and successful than this. The whole ocean, from America to Europe, a distance of 68 degrees of longitude, has been crossed by this bottle. Estimating the time occupied in traversing the Atlantic to be 500 days, and the distance about 3000 miles, it follows that a current, which averages about six miles per day, flows regularly over all the North Atlantic Ocean, from America to Europe. But, according to the best American charts, and even the Admiralty charts of this country, no current whatever is laid down as extending to the eastward beyond the 35th degree of west longitude, where the current of the Gulf Stream is supposed to end and to be lost. In consequence of this, navigators invariably cease to allow for any influence from currents after passing that longitude, which, from the perseverence of this bottle onwards to the land, is evidently a most serious mistake. For, allowing that a ship bound from the West Indies to Europe should be drifting at the rate of only six miles a day, for a period of twenty days, and this not allowed for in the reckoning, it follows, that the ship in that time would be nearer to the land, by a distance of 120 miles, than would be supposed by the navigator. Thus it is, that so many merchant vessels sail, in the night, dead upon the land upon the western coast of Ireland, because the commanders are wholly unprepared to suppose themselves within several degrees longitude from the shore. But six miles a-day, be it observed, is much too little to allow for the drifting of a ship, since a heavy body will float, by reason of its own impetus, very much faster than a light substance similar to a bottle; nor has it indeed been ever sufficiently dwelt upon, that the heavier the cargo, and the deeper in the water, the greater is the influence of the current on the ship. It is therefore probable, that a current of about ten miles per day should in general be allowed for, from the 35th degree of west longitude onwards to the European coasts. I have myself twice returned from America to England, and upon both occasions with very experienced and careful navigators; yet the commanders of both these ships were so extensively ahead as to be utterly astounded upon speaking vessels which had just left the land. Experience has now completely disproved the position, that the influence of the Gulf Stream is at an end in the midst of the Atlantic.

METEOROLOGY.

On the 13th November 1834, Mr W. H. White of London arose at half-past one in the morning, for the purpose of making meteorological observations, when a phenomenon presented itself, which, in all probability, was the falling of a meteoric stone. We shall give its description in his own words:—"In a few minutes another meteor, of a paler colour than any I had observed before, glided almost perpendicularly towards the earth; this was succeeded by another, of a more brilliant appearance, which took a westerly direction. This meteor cast a brilliant blue light, and had a short or truncated train, which was of a paler light than the meteor itself, and gradually shaded off into a yellowish red; it appeared, in fact, like a stream of light, which the meteor, in its velocity, left behind. Another remarkable circumstance attended this meteor, which I have never observed before, and that was, the meteor separating itself from its train. The latter immediately vanished, while the former continued its downward course with amazing velocity, gradually losing its bright blue light, and increasing in redness as it approached the earth. As this meteor continued its course till surrounding objects hid it from my view, I inferred that it was the falling of a meteoric stone.

"Had this beautiful meteor taken its course against the wind, which was blowing a strong breeze from the north east, I should have concluded that the train was under atmospheric influence; but, as its direction was nearly with the wind, the train must either have been outstripped in velocity by the meteor, or it must have been the result of Electricity."*

* Magazine of Natural History, viii. p. 98.

Meteoric Stones.—It has only recently been discovered that copper forms an ingredient of meteoric stones. Hofrath Stromeyer asserts that it exists in all meteoric masses. He has examined specimens from Agram. Lenard, Elbogen, Bilburg, Siberia, Gotha, Louisiana, Buenos Ayres, Brazil, and the Cape of Good Hope, and detected in them all an appreciable quantity of copper, varying from one-tenth to one-fifth per cent. He has come to the conclusion that the presence of copper must be considered as a constant character in these substances, as are the nickel and cobalt, which are found in greater proportions.*

H. Stromeyer examined a meteoric mass, found at Magdeburg in 1831, the specific gravity of which was 7.39, and which contained 4.32 per cent. of copper. Another mass was found near the Iron Works of Rothchutte, in the Hartz, containing 7.69 per cent. of copper.

The celebrated chemist Berzelius, who analyzed a mass of meteoric stone from Macedonia, did not detect copper as forming part of its substance. Its ingredients were, silica 39.56, protoxide of iron 13.83, and peroxide of iron 5.00, making a total of iron 18.83. Alumina 2.70, oxide of chromium 0.50, lime 1.86, magnesia 26.30, oxide of nickel 0.10, oxide of manganese 2.40, potash 2.08, soda 1.20; making a total of 95.53. Besides the above substances, Stromeyer has detected copper, cobalt, arsenic, phosphorus, sulphur, silicon, and carbon.

Lightning.—The *Medical Gazette* of St Petersburgh states, that the life of a soldier, struck by lightning, has been saved by copious bleeding.

GENERAL SCIENCE.

NO. I.—ON THE COMET OF HALLEY, SOMETIMES CALLED THE COMET OF 1759.

HISTORY OF THE COMET TO A.D. 1456.

Hast thou ne'er seen the comet's flaming flight?
Th' illustrious stranger passing, terror sheds
On gazing nations, from his fiery train
Of length enormous; takes his ample round
Through depths of ether; coasts unnumber'd worlds
Of more than solar glory; doubles wide
Heaven's mighty cape, and then revisits earth,
From the long travel of a thousand years.

Young.

[The following is the first of a series of papers, drawn up chiefly from the works of M. Arago, and of M. de Pontécoulant, upon this interesting heavenly body which has just visited us.]

The first appearances of this comet are unrecorded and unknown. Its origin, like that of the most powerful empires of the earth, or of the greater number of extraordinary men, who have astonished and enlightened the world, is enveloped in profound darkness. Historians, influenced by the universal terror which these bodies formerly inspired, have given an exaggerated tone to their descriptions; and that propensity to assign unusual events to supernatural causes, when the explanation of them surpasses the human understanding, has long induced mankind to consider these stars, and their long train of attendant light, as evident signs of celestial wrath. They have been viewed as the messengers of destruction—the forerunners of the three greatest plagues which can desolate the earth—war, pestilence, and famine. Some comets were even thought to have been appointed on the special mission of presiding at the birth or death of those distinguished mortals, whom we may, without impropriety, style meteors of a moral order. Such was the extraordinary comet that appeared during seven days at the death of Julius Cæsar; and also that equally remarkable one, seen at Constantinople in the year signalized by the birth of Mahomet. We may therefore infer, that historians of those times, being under the influence of popular superstitions, viewed with a partial eye the comets they describe; or, at least, they seem to have greatly exaggerated the circumstances attending their appearances. Indeed, ever since these bodies have been examined without prejudice, and for scientific purposes alone, we no longer see them extending from east to west, covering one entire portion of the celestial sphere. No longer does their fearful hue appear to reflect, by anticipation, the blood-stained field of battle, or the flames of approaching conflagrations. Since, then, we have no other guide than these faithless accounts of cotemporary writers, it becomes very difficult, at the present time, to recognize the more sober visitants of the last two centuries among these denizens of a fairy sphere. An account drawn up with fidelity, wherein the physical appearances of a comet are exactly described, may be sufficient to establish a general probability, that two of these bodies, observed at different periods, are identical; yet, as we shall hereafter show, it is wholly inadequate to *prove* their identity.

The period of a comet is that portion of time which elapses between its two consecutive returns. That of Halley, the subject of these papers, reappears at intervals of time, the mean, or average of which, is about seventy-five years and a half. By counting backwards, it has been thought to correspond sufficiently near with that which appeared at the birth of Mithridates (A.C. 130). The comet of Mithridates—the most extraordinary of any that have ever been seen or described—exhibited itself for eighty days: its splendour surpassed that of the sun, its size occupied a fourth part of the heavens, and it took four hours in rising, and as many in setting! An origin so illustrious would not have been unworthy of a comet, occupying a place so important in the history of astronomy. But, after we have made all the necessary deductions for the tone of exaggeration evidently impressed on this account, the identity of the comet of Mithridates with that of Halley is still very doubtful, as it has been assumed, in the calculation, that the comet of 1759 returned periodically at equal intervals of time.

In the year A.D. 323, a comet was observed in the constellation Virgo (the virgin), which appears to have had some resemblance to that of Halley. All the historians of the Constantinopolitan empire allude to a comet seen at an interval of 76 years after the former, in the year A.D. 399. The coincidence of the period would lead us to consider this body to be identical with the comet of 1759, and to infer that one of the periods of its return to the neighbourhood of the sun is here pointed out. This comet was said to have exhibited some extraordinary characters. Lubienietski (*Theatrum Cometarum*) describes it as being of prodigious size, of frightful aspect, and seeming to dart its hair even to the earth. "Cometa fuit prodigiosæ magnitudinis, horribilis aspectu, comam ad terram usque demittere visus." Again, we hear of a comet in A.D. 550;—and that nothing marvellous might be wanting to its history, the return coincided with the taking of Rome by the Goths, under Totila; and it was, of course, considered by a credulous people as one of the *causes* of that event. After an interval of 380 years (A.D. 930), historians speak of a star, which we may again suppose to be the same comet, after having performed five revolutions during that time. It reappeared at its following return (A.D. 1005); and after passing over three periods, we find, from the list of 415 comets which Lubienietski has described, that in the year 1230 a comet was observed, which may have been the same with that of Halley. But it is no longer possible to follow the comet through the obscurity and ignorance of the middle ages. In those times of barbarity, the physical appearances of the heavenly bodies were either not observed, or distorted by the delusive medium through which they were viewed. We have already explained, that our only means of establishing the identity of Halley's comet with those recorded in ancient writers, is the constant periodical return of the comet of 1759, at equal periods of time. But the periods of this star are continually, and must essentially vary, as we shall hereafter show, by the interval of about one year or eighteen months. We cannot, therefore, place implicit reliance upon the preceding calculations, deduced from a principle which is not in itself strictly exact. In addition to this, there are some years when a great number of comets are visible. At present, a year does not elapse, without some comet being remarked at the different observatories of Europe. It is therefore absolutely impossible to decide, whether the comet of 1759 is really identical with those, whose appearances are stated to have been so extraordinary in 550 and 399; or whether it did not pass, during these or neighbouring years, without attracting the attention of common observers, who would have been entirely engrossed with other more remarkable phenomena. It is, therefore, only with the utmost reserve that we have ventured to suggest these analogies between the present comet and similar bodies which have appeared before the thirteenth century. We have related these circumstances, more as matters of curious research than of positive truth, upon the faith of those learned men who have sought the annals of antiquity for traces of Halley's comet. Like some historians, who love to surround the cradle of their favourite hero with circles of light and lambent flames, we expose the legends of antiquity. It is, in fact, only about two centuries ago that comets were first observed with care; and, previous to that time, the tales related of them are exceedingly doubtful.

Let us return to the uncertain history of this comet.—In the year 1305, there appeared one of these bodies, which would correspond in character with the comet of Halley. The chroniclers of the times gave a tremendous description of it. "Cometa horrendæ magnitudinis visus est circa ferias Paschatis, quem secuta est pestilentia maxima." (A comet of dreadful magnitude was seen about the Easter holidays, which was followed by a pestilence of the worst description.) The concluding observation throws a doubt upon the exactness of the whole statement; and, as Lalande has observed, it is very possible that fear of the pestilence may have increased the impressions left by the comet. However, it is but right to state, that from the position which the comet of 1759 occupied in its orbit, it must have passed very near to the earth during the year 1305: this would have given it an appearance of unusual magnitude.

In the catalogues of Alstedius and of Lubienietski, there are two comets mentioned, one in 1379, and another in 1380, separated from the former by an interval of about 75 years; but as no details are given of the time, place, or form of their appearances, nothing satisfactory can be ascertained.

The same comet reappeared in 1456, and was accompanied by several remarkable circumstances. "Cometa inauditæ magnitudinis toto mense Junii apparuit cum prolongâ caudâ, ita ut duo ferè signa cœli comprehenderit." (A comet of unusual magnitude appeared during the entire month of June, with a tail of so great length that it covered nearly two entire signs of the heavens.) It is possible that the length of this tail may be a little exaggerated, especially as the visible part of the heavens is 180°, and a sign is 30°. Yet we must allow that there was something grand in its appearance; because it had the effect of spreading, at this epoch, as great a terror throughout all Christendom, as the rapid progress of Mahomet II., who had just made himself master of Constantinople. The Pope, Calixtus II., with an economy of labour remarkable for the age, exorcised the comet and the Turks in the same Bull, and ordered public prayers, with a formula, to be made, which might include them both. Notwithstanding these efforts of his Holiness, Mahomet and the comet continued their several courses; the former succeeded in converting the Cathedral of Sancta Sophia into his principle mosque; and the latter vanished peaceably into the infinity of space.

The great length of tail ascribed to the comet of 1456, and the consternation which it diffused throughout benighted Europe, have led scientific men to consider it in another point of view. They have ascertained the position which a comet must occupy, in order that it may appear of the greatest possible brilliancy; and the result is, that this comet, from its near approach to the sun, and from its situation relative to the earth, must have united all the conditions necessary to that appearance of splendour assigned to it by the chroniclers of the times.

Although no precise observations upon this phenomenon have been handed down to us, yet the time when the comet appeared, and the course which it held, are described with sufficient accuracy; so that Dr Halley did not hesitate to assert its identity with the comets of 1531, of 1607, and of 1682. This appearance of 1456 ought, therefore, to be clearly distinguished from all those preceding it, and to be considered as a remarkable epoch in the history of science.

* Ann. Der Physik., xxvii. p. 689.

MISCELLANEOUS.

Projected Voyages of Discovery.—At the instance of the Geographical Society, government have granted L.1000 towards the expenses of two voyages of discovery; the one into the interior of Africa from Delagoa bay, on the east coast, and the other to explore the high land which forms the boundary between British Guiana and the basin of the Amazon in America.

The French Discovery-Ship La Lilloise.—No tidings have been received regarding that vessel since the month of August 1834, which, under the command of Lieutenant de Blosseville, was sent on a voyage to the coasts of Iceland and Greenland.—The following is the proclamation issued by order of the King, dated June 17, 1835:—1. That a sum of 100,000f. (L.4,000) shall be given to any French or foreign mariners who will bring back to their country the whole or part of the officers and crew of La Lilloise.—2. That a pecuniary reward proportioned to the service shall be granted to those who would be the first to bring any positive intelligence respecting the said officers and crew, or procure to France the restitution of any papers and effects soever which had belonged to the above expedition.

Since writing the above, intelligence has been received by the French Government, from M. Gaymard, surgeon and naturalist to the discovery ship commanded by Lieut. de Blosseville, that from the information which has been obtained at Iceland, there is but little hopes of the object of the expedition being attained. M. Gaymard has been fortunate in obtaining numerous treasures in Natural History, which he purposes forwarding to the Jardin des Plantes at Paris.

Expedition of Discovery—Van Diemen's Land.—Hobart Town papers of the 25th February 1835, have been received, which announce that letters had been transmitted by Mr Frankland, giving a very satisfactory account of the progress of his expedition. He proposes to trace the source of the Heron, the Derwent, and other rivers, to penetrate the east coast between Port Davy and Macquarie Harbour, and to return by the hitherto unexplored country in the neighbourhood of Mount Wellington. The weather had been favourable for him, and there was every prospect of the undertaking proving successful.

REVIEWS.

A Selection from the most Remarkable and Interesting Fishes found on the coast of Ceylon, from Drawings made in the southern part of that Island. By John W. Bennett, Esq., F.L.S., &c. Royal quarto. Longman and Co. London. 1830.

Although this volume has been before the public for some time, we do not think it is so well known as it deserves to be. It contains thirty plates of fishes, all of which are remarkable for the splendour of their colours, or the singularity of their structure; and these are executed with a fidelity which reflects high credit on the author as an artist. We cannot too favourably speak of the beautiful manner in which the engravings have been executed by Mr J. Clark, and also of the careful and sparkling effect of the colouring. The author has introduced into his descriptions not only the scientific, but also the Cingalese names. This elegant book is well fitted to ornament the table of the drawing-room or saloon; and we especially recommend it to the notice of all those who have been resident in the East.

Excursions illustrative of the Geology and Natural History of the Environs of Edinburgh. By William Rhind, Member of the Royal College of Surgeons, and of the Medical and Physical Societies, Lecturer on Natural History, &c. Royal 18mo. Maclachlan and Stewart, Edinburgh; and Baldwin and Cradock, London. 1833.

To every person who has the slightest desire to become acquainted with the Natural History of the vicinity of Edinburgh, we can confidently recommend this excellent pocket treatise. Indeed, it ought to be the companion of all our walks; for in whatever direction we may bend our steps, Mr Rhind edifies us with the structure of the rocks, the botany of the meadow or wood, and the animals which we are likely to meet with in our rambles. Our author, in particular, gives a lucid description of the geological structure of Arthur's Seat, Salisbury Crags, Castle Hill, Calton Hill, Craigleith Quarry, Braid, Blackford, Pentland, and Corstorphine Hills; Cramond Island, Isle of May and Bass Rock; the sea beach at Newhaven; all that is interesting in the natural history of Duddingston Loch and Lochend; Roslin, Lasswade, Musselburgh, and Ratho; with a list of the mountain rocks, fishes, and land shells. The work is illustrated with coloured maps of the environs of Edinburgh, with a coloured geological section of the country, from the Calton Hill on the east to the Pentland Hills on the west, and with eleven wood-cuts.

To strangers visiting Edinburgh, this cannot fail to prove a useful companion. It is handsomely printed, and very cheap.

LEARNED SOCIETIES.

Royal Society.—The last meeting of the season was held on the 18th June, Sir John Rennie, V.P. in the chair: on which occasion, the following papers were read, viz.—1. Discussion of Tide Observations made at Liverpool, by J. W. Lubbock, Esq. 2. Experimental Researches in Electricity, by Michael Faraday, Esq., tenth series. 3. On the Distinction between certain Genera of Shells, by S. E. Gray, Esq. 4. On the Ova of Mammiferous Animals, by T. W. Jones, Esq. 5. On the supposed existence of Metamorphoses in the Crustacea, by J. O. Westwood, Esq. 6. On the Star-fish of the Comatula, by T. V. Thompson, Esq. 7. On the influence of Perspiration on the quantity of Blood in the Heart, by James Wardrop, Esq. 8. On Sound, by P. Cooper, Esq. 9. On the Tides, by P. Cooper, Esq. The Society then adjourned till the 19th November next.

Linnæan Society.—The Duke of Somerset in the chair. The president nominated Robert Brown, Esq., Edward Forster, Esq., Dr Horsfield, and A. B. Lambert, Esq., to be vice-presidents of the society for the present year. There was read an account of the galls found on a species of oak from the shores of the Dead Sea, which have been mistaken for the fruits of certain plants; and a note on the mustard plant of scripture, by Mr Lambert. Also, descriptions of five new species of the genus *pinus*, discovered by Dr Coulter in California, by Mr Don, Lib. L. S. At a meeting on Tuesday evening, a paper was read, being some observations on the screech-owl, by Mr Knight; and also a memoir on the metamorphosis in the macroura of the class *crustacea*, by J. V. Thompson, deputy-inspector of hospitals. Mr Christy exhibited a flowering specimen of the very rare British plant, *liparis læselii*, from Bottisham Fen, Cambridgeshire.

Royal Institution.—On Friday, 12th June 1835, the evening meetings of this institution closed with a lecture on the "History and Manufacture of Gunpowder," by Mr Henry Wilkinson. Mr Wilkinson quoted a variety of authors, ancient and modern, to prove that gunpowder had been known in China and India, beyond all periods of investigation; and observed, that in the Gentoo laws, supposed to be coeval with Moses, there was a prohibition of the use of gunpowder and fire-arms. He then minutely described all the progressive stages of the manufacture of gunpowder, and produced specimens of each ingredient in its various states; and concluded with a variety of interesting experiments, to show the quantity of permanently elastic fluid generated by the ignition of gunpowder. This he effected by firing gunpowder under water, and collecting the gases in the pneumatic trough. The enormous amount of gunpowder consumed in war, he illustrated by stating the quantities used at the sieges of San Sebastian, Badajos, and Ciudad Rodrigo; and he concluded with several curious experiments, showing the effect of fulminating powders on gunpowder. A train of fulminating powder was drawn across another of gunpowder, and the fulminating powder inflamed, which passed with such rapidity over the gunpowder, that it merely separated without igniting it. He then sent the flame of fulminating powder through a box of gunpowder without igniting it; and proved, by other experiments, that the ignition depended on the velocity with which the flame was transmitted. The whole lecture gave very great satisfaction to a numerous auditory. At its conclusion, Mr Faraday addressed the audience on topics of general interest to the institution, and of consequence to the scientific world.

Geological Society.—*13th May* 1835.—Charles Lyell, Esq. president, in the chair, by whom a paper was read "On cretaceous and tertiary strata of the Danish islands of Iceland and Moën." Afterwards a notice was read "On a peculiarity in the neck of Ichthyosauri, not hitherto noticed;" by Sir Philip De Malpas Grey Egerton.

May 27.—C. Lyell, Esq. in the chair. A paper was read "On certain lines of dislocation of the new red sandstone of North Salop and Staffordshire; with an account of trap-dykes, in that formation, at Acton Reynolds, near Shrewsbury;" by Roderick I. Murchison, Esq. Afterwards a paper was read "On the crag of part of Essex and Suffolk;" by Edward Charlesworth, Esq.

June 10th.—C. Lyell, Esq. in the chair. There were read—1. "Notes on the trappean rocks associated with the new red sandstone of Devonshire;" by Henry T. De la Beche, Esq. 2. "On the range of the carboniferous limestone flanking the primary Cumbrian mountains, and on the coal-fields of the north-west of Cumberland;" by Professor Sedgwick, and Williamson Peile, Esq. 3. "Notice of the occurrence, near Shrewsbury, of marine shells of existing species in transported gravel and sand resting upon peat which contains embedded trees;" by Joshua Trimmer, Esq. 4. "Description of some fossil crustacea and radiata found at Lyme Regis, in Dorsetshire;" by William John Broderip, Esq. 5. A letter from Sir Philip Grey Egerton, Bart., to the president, "On the discovery of fishes in the coal-field of North Staffordshire." 6. Two notices of Gideon Mantell, Esq., "On bones of birds from the strata of Tilgate Forest, and on the coffin-bone of a horse from the shingle bed of the newer pliocene of strata of the cliffs near Brighton." 7. Extract of a letter from Professor Daubeny, "On the saline contents of the mineral spring lately discovered near Oxford."

Eastern Literary and Scientific Institution.—On the 23d June, a lecture was given in the room of this institution, in the Hackney Road, by Mr Taylor. The subject of the lecture was the safety-lamp, and the object was to show, by actual experiment, that the lamp hitherto used in coal-mines, and invented by Sir H. Davy, is liable to some serious objections. Mr Roberts proposed an improvement on Sir H. Davy's lamp, to remedy the practical objections to which it is liable. Mr Taylor commenced his lecture by explaining the nature of combustion. He described carburetted hydrogen, or fire-damp. He then detailed the doctrine of flame and the progress of combustion. He pointed out by experiments the properties of nitrogen, carbonic acid, and oxygen, and showed the manner in which the safety of the invention of Mr Roberts is connected with them. Mr Roberts then displayed his lamp, and described the manner in which it differs from that of Sir H. Davy. He stated the objections to the lamp of Sir H. Davy, being the insecurity it affords to the currents of carburetted hydrogen or fire-damp, and the dangers arising from the ignition of the small particles of coal adhering to the wire gauze by which it is surrounded, from the oil clinging to the sides of the gauze when the lamp is upset or held in a horizontal position. His own lamp he stated to be free from the two last-mentioned defects. It is surrounded by a double tube of wire gauze, and also by a glass chimney, and is so contrived that a current of carbonic acid air, or nitrogen, passes continually between the external atmosphere and the flame of the lamp; the flame alone can burn, and any ignition from external explosive current of fire-damp is repelled by the carbonic acid of nitrogen, by which combustion is immediately destroyed. In the course of the lecture allusion was made to the late deplorable accident by which upwards of a hundred lives have been destroyed by the explosion of fire-damp; and many other instances were given of the loss of life occasioned, either by a too strong reliance on the safety of the lamp hitherto used in the mines, or by the neglect of proper means for ventilation.

Edinburgh: Published for the Proprietors, at their Office, No. 16, Hanover Street. London: Charles Tilt, Fleet Street. Dublin: W. F. Wakeman, 9, D'Olier Street. Glasgow and the West of Scotland: John Smith and Son, 70, St Vincent Street.

THE EDINBURGH PRINTING COMPANY.

THE EDINBURGH
JOURNAL OF NATURAL HISTORY,

AND OF

THE PHYSICAL SCIENCES.

Nᵒ. 4.   SATURDAY, DECEMBER 5, 1835.

ZOOLOGY.

DESCRIPTION OF THE PLATE—BUTTERFLIES.

THE Papilionaceous or Butterfly tribe of insects yields to no other in the Animal Kingdom in point of beauty. Many of the species wear a garb of the most gorgeous colours, exhibiting the tints of the rainbow in all the varied and dazzling brightness of iridescent splendour; and the wonderful changes through which they pass are no less calculated to excite our admiration and even astonishment.

The whole of this tribe undergo four changes in passing from the egg to the perfect insect or butterfly. These metamorphoses seem to have been known to the ancient Greeks, and most probably suggested to them their principles of metempsychosis. Nothing could appear to them more confirmatory of the doctrine than that an inert aurelia should be again transformed into a living body. The only method which they had for explaining this phenomenon was, that it had been tenanted by the soul of some wretch, whose misdeeds on earth had merited such a pilgrimage.

Butterflies are strictly oviparous animals, and the female, by an unerring foresight, uniformly deposits her eggs in the place where food is to be found for the future caterpillar after its exclusion from the egg.

The eggs are usually enveloped in an adhesive cement, by which they are attached to the spot where they are deposited. This wise provision is designed to prevent the eggs from being removed to a situation where the proper food of the species might not be found, and where the caterpillar would consequently die of hunger.

The eggs of butterflies are of many different shapes, and hardly two species produce them alike. The following cut, figure 1, represents one of the eggs of the small tortoise-shell butterfly (Vanessa Urtica); these are of a cylindrical form, with eight prominent ribs; while the eggs of the large tortoise-shell butterfly (Vanessa Polychloris) are shaped like a flask, and quite smooth, as in figure 2.

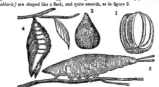

When the eggs have remained their proper time, the caterpillar, or larva, bursts from its confinement. At first it is exceedingly small, but increases daily, until it reaches its full size, as seen in figure 3, which represents the caterpillar of the Purple Emperor Butterfly (Apatura Iris).

The great proportional bulk at which many caterpillars arrive, in comparison to their original size when they emerge from the shell, is surprising. The larva of the Goat Moth (Cossus liniperda), on attaining its full magnitude, is seventy-two thousand times heavier than when it first bursts into life; and the maggot of the Blue Fly is, in twenty-four hours, one hundred and fifty-five times heavier than at its birth. Many caterpillars consume more than twice their own bulk of food every twenty-four hours. The cause assigned for this is, that their stomachs have not the power of dissolving vegetable matter, but merely the faculty of extracting their juices. When the larva has attained its full size, it soon afterwards ceases to eat, becomes excessively restless, and searches for a place fitted to its nature, to which it may retire, for the purpose of being transformed from one state of existence to another, and of assuming its pupa condition. Some spin for themselves a covering of silky filaments, while others simply attach themselves to the under side of a leaf or branch, as represented in figure 4, which is the pupa of the Purple Emperor. After remaining for some months in this state, the concealed animal bursts its casement, and emerges as the perfect Butterfly; in which condition it exists only for a very limited period; and, after having provided for the continuance of its race, speedily dies.

The transformation of insects, however, consists rather in a series of developments than in any absolute metamorphosis, being only a transition of changes in organs which lie concealed from human view. The caterpillar is compound in its nature, with the germs of the perfect insect hidden in a succession of cases. The first is the covering of the pupa, which is concealed within three or four mantles, the one over the other; these will, in succession, enrobe the larva; and as it enlarges the parts

become visible, and are alternately thrown off, until the perfect insect bursts from its confinement.

It is certainly wonderful that the simple caterpillar, when it first emerges from the egg, not thicker than a thread of silk, should contain its own covering threefold, and, in some instances, even eightfold, with the mask of a pupa and a butterfly, folded in the most astonishing manner over each other; and besides these, should possess different respiratory and digestive organs, a nervous system, and muscles of motion peculiar to every successive stage of its existence. And, what is truly wonderful, the stomach in its caterpillar state is fitted for the reception of vegetable food, while in its perfect condition of the butterfly, it is incapable of digesting ordinary vegetables, and is only fitted for containing honey, which the animal sips from flowers by means of a proboscis.

The whole of the figures given in the plate are of the natural size.

Fig. 1. Royal Butterfly (Endymion Regalis), and, 2. Do. Female, inhabit Brazil.—3, 4. Azure Blue (Polyommatus Argiolus), and, 5. Do. Female, inhabit Britain.—6. Mazarine Blue (Acis), inhabits Britain.—7. Silver Streak (Argynnis Paphia), inhabits Britain.—8. Nieppe (Pieris Nieippe), and, 9. Do., inhabits the United States.—10. Imperial Trojan (Papilio Priamus), one of the most beautiful of butterflies, inhabits Amboyna.—11. Merry (Aconthia Lubentina), and, 12. Do. Female, inhabit Java.—13. Painted Lady (Cynthia Cardui), inhabits Britain.—14. Amphinome (Amphinome), inhabits Surinam, and, 15. Do.—16. Elm (Vanessa Polychloris), inhabits Britain.—17. Oriental Emperor (Nymphalis Ripheus), inhabits China.

INSTINCT IN DOGS.—A singular instance of that power, commonly called instinct, possessed so remarkably by pigeons, and exhibited by dogs, lately occurred. The celebrated greyhound, Black-eyed Susan, was brought to Edinburgh from Glasgow in the boot of a coach, on the night of Wednesday the 13th May 1835. On the following Sunday evening she made her escape, and in forty-eight hours reached her kennel, eight miles beyond Glasgow, being fifty-two miles in all. The road between Glasgow and Edinburgh she had never travelled on foot, and from the time taken she cannot have come direct; but, by what route or process this animal made her point good, it is in vain to conjecture.

Another instance, of a similar nature, is recorded in Brown's Biographical Sketches and Anecdotes of Dogs. In the year 1816 a female greyhound was sent from the neighbourhood of Edinburgh by a carrier, viâ Dumfries, to the neighbourhood of Castle Douglas, in the stewartry of Kirkcudbright. She brought up a litter of pups there, and, in the following year, was returned by the same route to Edinburgh, from whence she went, by way of Douglas and Muirkirk, to the neighbourhood of Cumnock in Ayrshire. After remaining there five or six months, the found her way across the country to the house near Castle Douglas where she had brought up the pups. The fact of her crossing, and not pursuing her journey by the route she had been taken to Cumnock, was ascertained by shepherds, who saw her, accompanied by a pointer dog. The latter left her almost immediately, and found his way home again. This female greyhound was bred in East Lothian, and had never been previously either in Ayrshire or Dumfries-shire.

THE ELEPHANT'S LOVE OF SWEETMEATS.—The elephant has a natural partiality for sugar, which he finds abundant means to gratify in the plantations of sugar cane. A curious instance is recorded of his liking for sweetmeats, and of a method adopted in his savage state to gratify his propensity. It chanced that a Cooley, laden with jaggery, which is a coarse preparation of sugar, was surprised in a narrow pass in the kingdom of Candy by a wild elephant. The poor fellow, intent upon saving his life, threw down the burthen, which the elephant devoured, and being well pleased with the repast, determined not to allow any person egress or ingress who did not provide him with a similar banquet. The pass formed one of the principal thoroughfares to the capital, and the elephant, taking up a formidable position at the entrance, obliged every passenger to pay tribute. It soon became generally known that a donation of jaggery would ensure a safe conduct through the guarded portal, and no one presumed to attempt the passage without the expected offering.

SAGACITY OF TWO MULES.—About two miles from the town of Ballymahon, in the county of Longford, Ireland, resides a gentleman, who has in his possession two mules of the Spanish breed. They will regularly go to a pump placed in the yard, and while one applies his mouth to the spout, the other works the handle by alternately raising and depressing his shoulder. When one has satisfied his thirst, he exchanges with his companion, and returns the service he has received.

THE ELECTRICAL EEL.—(Gymnotus electricus.)—This rare fish was caught some time ago near Gravelines. The pilot of the vessel received a severe shock on taking it from the nets, and all the crew, on touching it, experienced a like sensation, which, however, weakened at every touch, and diminished gradually, till the animal expired.

BOTANY AND HORTICULTURE.

SUPPOSED FALL OF SULPHUR.—Captain Hafty communicated to the Academy of Sciences an account of a singular phenomenon which happened at Olerand, in the department of the Basses Pyrenées. On the 26th April 1835, there occurred a heavy fall of snow, which next day was covered with a fine yellow dust, having all the appearance of sulphur. The only probable explanation which can be given of this phenomenon is, that the dust must have been the pollen proceeding from the blossoms of numerous pine trees in the vicinity, and which flower at that season. This pollen is of a highly inflammable nature, and might easily be mistaken for sulphur.

TULIPA.—The sum of L.610 has lately been given for the bulb of a new variety of tulip, called the "Citadel of Antwerp." This enormous sum was paid by M. Vanderninck, of Amsterdam, a florist, but formerly a captain in the Dutch navy.

FOOD OF SILKWORMS.—There has lately been presented to the Institute of France a memoir on the leaves of the *Maclura Aurantiaca*, which, it is conceived, may replace those of the mulberry, for the food of silkworms, in climates wherein mulberry trees do not thrive. They stand the spring frosts, at Geneva, Paris, Turin, and Strasbourg, where they have been successfully cultivated for six years past. This tree is a native of North America, and is abundant on the banks of the Missouri, in the country of the Natchez.

THE KRUBUT, OR GREAT FLOWER OF SUMATRA.—This very wonderful vegetable production has been named, in scientific language, the *Rafflesia Arnoldi*. Its generic name is in honour of the late Sir Stamford Raffles, governor of Sumatra, and founder of the Zoological Society, and its specific name in memory of Dr Arnold, who discovered it in 1818.

In writing on this subject, Dr Arnold says, "At Pulo Lebbar, on the Manna River, I rejoice to tell you, I happened to meet with what I consider the greatest prodigy of the vegetable world. I had ventured some way beyond the party, when one of the Malay servants came running to me, with wonder in his eyes, and said, 'Come with me, Sir, come! a flower very large, beautiful, wonderful!' I went with the man about a hundred yards into the jungle, and he pointed to a flower growing close to the ground, under the bushes, which was truly astonishing. My first impulse was to cut it up, and carry it to the hut: I therefore seized the Malay's parang (a sort of instrument like a woodman's chopping-hook); and finding that it sprang from a small root, which run horizontally (about as large as two fingers or a little more), I soon detached it, and removed it to our hut." The following is a representation of the full blown flower.

The Krubut is a parasite, growing in the woods, on the roots and stems of those immense climbing plants, generally of the genus *vitis* (or vine), which are attached, like enormous cables, to the largest trees of the forest. The flower constitutes the whole of this plant, there being no leaves, and neither roots nor stems. Thus, the plant forms a complete anomaly in the history of vegetables. It grows out of another plant in the manner of the mistletoe, and not on the decayed surface of plants, as is the case with the common fern on the trunks of old oak pollards. In the latter case, the proper term is not *parasite*, but *epiphyte*.*

The flowers of this extraordinary plant are of one sex; and the male only has yet been sent to England. The breadth of a full flower exceeds three feet from the margin of the one petal *d* to that of the other *d*; the petals, or leaves of the flower, are roundish, and measure twelve inches from the base to the apex. It is about a foot from the insertion of one petal to the opposite one; and that part which is considered the nectarium, or central cup of the flower, would hold twelve pints of liquid. The pistils, which are abortive, and as large as cows' horns, are represented in fig 2. *b b*.

The weight of the whole flower is calculated at about fifteen pounds. It is of a very thick substance, the petals and nectary being in few places less than a quarter of an inch thick, and in some parts three-quarters of an inch; it is succulent in texture, but of a firm fleshy consistence. The flower, fully blown, was discovered in a jungle of Sumatra, growing close to the ground, under the bushes, with a swarm of flies hovering over the nectary, and apparently laying their eggs in its substance. The colour of the five petals, or flower leaves, of which it is composed, is a brick-red, covered with protuberances of a yellowish white. The inside of the cup is of an intense purple, and more or less densely yellow, with soft flexible spines of the same colour. Towards the mouth, it is marked with numerous depressed spots of the purest white, contrasting strongly with the purple of the surrounding substance, which is considerably elevated on the lower side. The smell is that of tainted beef.

* From (επι) *epi*, upon, and (Φυτον) *phyton*, a plant.

The structure of this plant is too imperfectly known to admit of determining its place in the natural system. That learned botanist, Mr Brown, however, thinks it will be found to approach near to *Asarina*, or to the *passiflores*, or passion flowers. Its first appearance is that of a round knob, proceeding from a crack or hollow in the stem or root, as represented in the following cut, Fig. 1.

This knob, when cut through, exhibits the infant flower enveloped in numerous bracteal sheaths. These successively open and wither away as the flower enlarges, until at the time of full expansion, when there are but a very few remaining, presenting somewhat the appearance of a broken calyx, as represented in fig. 2, *a a*. The female flower differs but little in appearance from the male, further than in being without the anthers, fig. 2, *c*. Fig. 3 represents one of the anthers a little larger than the natural size, and shewing a section of the cavity in which it is immersed. It takes three months from the first appearance of the bud to the full expansion of the flower. The blossoms decay not long after their expansion, and the seeds (sperm) are raised with the pulpy mass. The fruit has not yet been seen by botanists, but is said by the natives to be a many-seeded berry.

Mr Brown has made some interesting observations on the *Rafflesia Arnoldi*, wherein he remarks, that it is not common for parasitic plants to fix indiscriminately on the roots or branches of their stocks, as is supposed to be the case with the genus Rafflesia; and observes, that "plants parasitic on roots are chiefly distinguishable by the imperfect development of their leaves, and the entire absence of green colour; that their seeds are small, and their embryo not only minute, but apparently imperfectly developed." Mr Loudon says, that "the modes of union between a parasite and its supporter, or stock, vary in different genera and species of this class of vegetables. Some, as the mistletoe and Rafflesia, depend on the stock for nourishment during the whole of their existence; others, as the common broom-rape, are originated in the soil; and afterwards, when they have attached themselves to their stock, the original roots die. Other parasites, again, are originated on the stock, and in their more advanced state produce roots of their own. In some cases the nature of the connexion between the parasite and the stock is such, as can only be explained on the supposition that the germinating seed of the parasite excites a specific action in the stock, the result of which is the formation of a structure, either wholly or in part derived from the root, and adapted to the support and protection of the undeveloped parasite; analogous, therefore, to the production of galls by the puncture of insects. On this supposition may be explained the connexion between the flowers of the genus Rafflesia, and the root from whence it springs."

In Sumatra, all the vegetable productions seem to be on a gigantic scale. Sir Stamford Raffles, after describing this great flower, says, "There is nothing more striking in the Malayan forests than the grandeur of the vegetation. The magnitude of the flowers, creepers, and trees, contrasts strikingly with the stunted, and I had almost said, pigmy vegetation of England. Compared with our fruit trees, your largest oak is a mere dwarf. Here we have creepers and vines, entwining larger trees, and hanging suspended for more than 100 feet, in girth not less than a man's body, and many much thicker; the trees seldom under 100, and generally approaching 160 to 200 feet in height."

THE KNOWLE-PARK BEECH.—This most magnificent tree in Knowle-Park, Kent, is the largest undecayed and entire beech in the kingdom. It was measured in October 1835, and the following are its dimensions:—Circumference of the stem, at six inches from the ground, thirty-nine feet, five inches; at one foot, six inches; thirty feet, nine inches; four feet above, twenty-eight feet, one inch; seven feet above, twenty-five feet; one spiral limb, fourteen feet from the ground, fifteen feet. The mean height of the tree, eighty-nine feet; and the circumference of ground covered by branches, three-hundred and forty-seven feet.

THE TALLIPOT TREE is a native of the island of Ceylon, in the East Indies. This tree is remarkable on account of its leaves, which are of such a size as to cover ten men, and keep them from the rain; they are very light, and travellers carry them from place to place, and use them for huts.

MINERALOGY.

DISCOVERY OF MARBLE.—In the island of Tiree, on the west coast of Scotland, an engineer has lately discovered some beautiful blocks of white marble, and inexhaustible strata of variegated granite, in undulating streaks, of red, white, and black. At the Ross, in the island of Mull, comparatively pure red and white granite occurs in vast abundance. This is by far the most beautiful variety in this country, or perhaps in the world. One of the many blocks forming the *debris* of an adjoining mountain was found to measure 12 cubic feet to the ton—no less than 104 square tons of workable granite!

DEPTH OF MINES.—Kit's-puhi copper mine in the Tyrol mountains, 2764 feet; Samson's mine at Andreasburgh, in the Hartz, 2230 feet; Valenciana mine, Guanaxuato, Mexico, 2170 feet; Pearce's shaft, consolidated mines, Cornwall, 1650 feet; Monkwearmouth colliery, Durham, 1600 feet; Wheal Abraham's mine, 1452 feet; Dolwath mine, Cornwall, 1410; and Erton mine, Staffordshire, 1380 feet.

GEOLOGY AND PHYSICAL GEOGRAPHY.

GREAT ERUPTION OF THE VOLCANO OF COSIGÜEINA.

In America, between the 10th and 15th degrees of north latitude, there are at least twenty-one active volcanoes. All these are situate in the provinces of Guatimala and Nicaragua, which lie between Mexico and the Isthmus of Panama. The following account of the last great eruption of the volcano of Cosigüeina is translated from the official reports published by the government of Guatimala.

On the 20th of January 1835, at half-past 6 o'clock in the morning, the volcano of Cosigüeina broke out, and the vapour which arose was beautiful. At 11 o'clock it covered the whole of the territory around Nacaome, and at noon the obscurity was so intense as to exceed all description. We had then a night of 18 hours' duration, while tremulous movements of the earth, noises, tempests of thunder and lightning, caused by the combustible matter which filled the atmosphere, and an impetuous wind impelling a heavy shower of ashes, rendered that night a period of distress and horror. The morning of the 21st was melancholy, though the light penetrated through the dense vapours, and the sun sometimes showed a pale and saffron-coloured countenance. The 22d resembled the preceding day, and the night was passed rather quietly until 12 o'clock. There then commenced a hollow growling sound, vehement and alarming, which continued without interruption or diminution for at least 13 minutes. This noise was instantaneously followed by some terrific detonations, as loud as reports from artillery of the largest calibre. At a quarter past 12, a violent tremulous movement indicated a fresh eruption, which was soon confirmed by the ascent of a volume of smoke. At half-past 2 there was a sort of twilight, which served to interrupt a night of 36 hours, and the noises continued, being louder than on the 20th. A reflection of red light occasionally broke through the obscurity of the atmosphere; but so constant and terrible were the explosions, and the thunder and lightning, that it appeared to threaten the annihilation of the world itself. The 24th commenced much in the same manner as the 21st.

That a volcano should renew its eruptions, vomit forth lava and ashes, and occasion damage, might be expected to occur. But that the eruption of a hill, not one-eighth so high as Pacaya, should have darkened for several days the half of central America, and covered a space exceeding perhaps 15,000 square leagues with lava and ashes, to the height or thickness, in some places, of half a yard, in others a quarter, and no where less than two inches; that men should fly to the mountains, and wild beasts to the towns, as has happened at Nacaome, in Pesperi, Corpus, Ila del Tigre, Conchagua, el Puerto, &c.; that the fishes should have perished in the rivers, the birds be suffocated by dust, the reptiles and quadrupeds by slime; and that Man should remain unhurt amidst this convulsion of the elements—is a thing truly astonishing, and scarcely to be credited. Cosigüeina continued, like Isalco, to vomit exhalations until the 5th of February. The atmosphere cleared up slightly about 6 o'clock in the evening and 8 in the morning; but when the wind began to blow, clouds of dust enveloped Nacaome on all sides. Hitherto no bad effects had resulted except inflammation about the head, eyes, mouth and throat, which caused very severe coughing. It is extraordinary that the inhabitants of Nacaome were able to endure the showers of dust without being suffocated, especially as it has been found to be loaded with sulphur, iron, and antimony, and to be very inflammable. Cattle and flocks perished, and an unusual mortality was expected from the deficiency of pasture and from the deterioration of the water.

At San Marcos, from the 23d to the 29th of January, the atmosphere of the city was observed to be impregnated with smoke and ashes; and on the 24th particularly, so great a shower of ashes fell, that the roofs of all the houses were whitened with it. Until 9 o'clock in the morning, repeated explosions were heard, by which great alarm was created through the various towns, under the impression that they proceeded from the volcano of Quezaltenango; and the city was presently deserted by many of the merchants, who removed themselves to a distance, for the better security of their families.

On the 20th of January in the morning, the inhabitants of the town of Masaya heard towards the north-east some faint volcanic sounds, whilst those of the town of Viejo observed a sheet of fire rising perpendicularly to a considerable elevation, and afterwards declining towards the north. This was the same appearance which was observed in the department of Segovia, where at the same time some reports were heard, and some slight shocks were experienced.

In Leon, the capital, and in the department of Granada, the catastrophe had not been perceptibly felt until the dawn of the 25th, when the explosion developed itself to such a degree, that from 1 o'clock the sky was darkened with an opacity which continued to deepen till 11 in the morning, when the inhabitants of the capital were enveloped in a most frightful darkness, whilst terrific reports were heard, and showers of ashes were precipitated over all the face of the country.

This natural event produced an impression in the minds of the superstitious inhabitants that it proceeded from the Divine anger; and whilst the people ran in crowds to the temples to implore the mercy of Heaven, the garrison of the town diverted their consternation by discharges of cannon and musketry. This was done by order of the government, who, by the advice of some intelligent chemists, directed discharges of artillery to be fired, rockets to be let off, fires to be lighted, and the bells of all the churches to be rung, in order to dissipate the dense vapours with which the atmosphere was impregnated.

The quickness was astonishing with which, on the 23d, all the atmosphere was filled with volcanic matter, from Nicaragua, as far as the department of that name, towards the south-east. The murky clouds then gradually moved towards Nandayme, where, about 3 o'clock in the afternoon, the darkness reigned over the city, and extended to the town of Rivas. The same thing occurred in the department of Granada, the towns in which suffered nearly to the same extent as in Leon, whilst those of Matagalpa in Segovia experienced a night of 36 hours' duration.

Fortunately not a single life was lost, though in the immediate neighbourhood of the mountain where the eruption occurred some cattle were destroyed. It does not appear that the damage will be so great as was conceived at the time of the catastrophe, because the sand or ashes that have been scattered over the plains will wonderfully fertilize them—a fact which has been ascertained in some places watered a few days afterwards by the rain, where the plants showed a most luxuriant appearance, the pasture was rapidly rising, and every thing seemed to promise a forward spring.

The agitation of the air, when winds prevail, usually affects people with disagreeable sensations, and does great injury to cattle, on account of the dust which fills the atmosphere, to such a degree that it is impossible to see even for the distance of a league.

On the 9th of March, a commission went to observe the volcano, and they could not recognise the coast with perfect distinctness, or throughout its entire extent, in consequence of the cloud of smoke which covered the plains. A forest, which appeared to have survived many changes of the earth's surface, had disappeared. Two islands have been formed in the sea—one being 800 yards, and the other 200, in its greatest extent. Their composition consists of pumice-stone and scoriæ, with a number of pyrites of a golden colour, and having a strong metallic odour. Some shoals in the sea, from 500 to 600 yards long, were formed. In one of them a large tree was fixed with its branches downwards, and its roots raised up. The river Chiquito, which ran towards the north-west, was completely choked up, and another river, six yards broad, had sprung up in the opposite direction.

A party proceeded from the town of El Viejo to make another observation, by which it was ascertained that the farms of Sapasmapa and Cosigüeina, situate in the immediate neighbourhood of the volcano, had disappeared. From the first not a single head of cattle had escaped. In the latter 300 quadrupeds were found remaining, but in a weak and wretched condition, and they were not expected to survive. The remains of immense numbers of quadrupeds and birds were found lying in the immediate neighbourhood of the volcano. A vessel, which on the 20th of last month was near the coast, having a crew of seven men, was supposed to have been destroyed, since no information respecting it was received.

In the city of Leon the ravages done have been less, for the darkness there was not very great, and the same may be said of the showers of dust. The noise travelled to Costa Rica, where the cause was considered to be very near. The Colombian galley Boladora, which left Acapulco on the 20th ult. for the Realejo, experienced the darkness at 20 leagues from the shore, as well as such a copious shower of dust that the crew were apprehensive of being suffocated; and they were occupied for 48 hours in clearing the vessel with spades. Not being able to make for the Realejo on account of the darkness, they directed their course to Punta Arenas, with the full conviction that the whole state of Nicaragua had disappeared. The volcano continued vomiting fire and smoke, and causing at intervals a trembling of the earth.

Until further information arrives, it is impossible to calculate precisely the distance to which the showers of cinders extended and the noises were heard. The detonations were so loud as to be heard at Ciudad Real de Chiapas, which is 325 leagues from the mountain in one direction, and at El Peten, which is 322 leagues in another; and, as it is probable they could have been heard farther, we may estimate that the eruption affected the district, extending around the mountain 350 leagues in every direction. Even at Dolores, in the district of Peten, showers of ashes, volcanic reports, and earthquake shocks, were experienced.

In the time of the Roman Emperor Titus, in consequence of an eruption of Vesuvius, the ashes are stated to have been thrown into Africa, a story which has been considered incredible by some modern writers. This eruption of Cosigüeina shows the statement of the ancients to be by no means improbable.

Footmarks of an Extinct Animal in the Solid Rock.—Baron Alex. Von Humboldt has again arrived in Paris. At a meeting of the Academy of Sciences on the 17th August 1835, he directed the attention of the members to the prints of the footsteps of a quadruped in the variegated sandstone, or *bunte sandstein*, of Hildburghausen. It is an animal of the Plantigrada division, which had traversed the rock in various directions while soft. A stone, from ten to twelve feet long, and three to four wide, containing these impressions, has been sent to the Collection of Geology at Berlin, of which the Baron submitted to the Academy a beautiful drawing. There are four or five impressions of a smaller species, which cross those of the larger quadruped at right angles, and are remarkable for the unequal dimensions of the fore and hind feet; all of them have the impressions of five toes. The rock is covered with them as with a net-work, and here and there sinuous serpular concretions are visible.—perhaps the plants on which the animals walked, or probably some accidental defect in the process of drying. The great importance of this discovery consists in the position occupied by this sandstone in the chronological series of rocks.

Shock of an Earthquake at Chichester.—On Monday the 10th August 1835, between eleven and twelve o'clock, many of the inhabitants of Chichester were awakened from their sleep, and much alarmed by two shocks of earthquakes. It is nearly twelve months since the last of these awful phenomena, and people began to hope that they would have entirely ceased. These shocks, however, were considered generally to have been less violent than most of those which had occurred last year.

Floating Islands.—From the earliest times, authors have described those singular geological phenomena called floating islands. Pliny tells us of the floating islands of the Lago de Bassanello, near Rome. Near St Omer's, in the province of Artois in France, there is a large lake, in which there are several floating islands, some of which are inhabited; on one of them there was a church, and a religious convent of Bernardines. These islands are moved in different directions by the wind; and sometimes they are moored to the side with ropes. There are floating islands in Lochlomond, Scotland, and in the lake of Derwent-water, Cumberland: such islands appear and disappear. The latter, which had been under water for some time past, reappeared about the 7th September 1835, and attracted the attention of numerous visitors. Mr A. Pittingal, junior, in 1829, described a floating island about a mile southward of Newbury Port, in Massachusetts, Essex county, North America, 140 poles in length, and 120 in breadth. It is covered with trees; and in summer, when dry weather has long continued, it descends to the bottom of the lake.

METEOROLOGY.

TEMPERATURE.—Mr Warden made observations on the remarkable fall of the thermometer during the winter 1834-5, in the United States. It proved to be the most rigorous season known there for the last fifty years.—M. Gaymard, surgeon and naturalist to the discovery-ship sent by the French government to the coasts of Iceland and Greenland, has made daily meteorological observations, and has ascertained that a period of unusually cold weather occurred in Iceland at the time that the United States suffered from a remarkably low temperature.

On the night of midsummer-day 1835, several sheep perished of the cold, on Welland and Little Malvern Common.

RAIN.—M. Fleurau de Belle-Vue has addressed a letter to the Academy of Sciences, Paris, giving his opinion that the diminution of the springs of Poitou, for ten years past, has arisen from the decrease of the annual quantity of rain during that period. This fact was doubted by various members of the Academy, but M. Arago assured them that this has been the case, not only in Poitou, but in several other parts of France, where regularly-recorded observations have been made. This is very remarkable in the spring at Arcueil, which now yields very little water, and from the same cause.

MISCELLANEOUS.

SUBMARINE VESSEL.—According to the Paris papers, some curious experiments have lately been made at St Ouen, near Paris, with a submarine vessel, the invention of M. Villerni, the engineer. The vessel is of iron, and of the same shape as an animal of the cetaceous or whale tribe. Its movements and evolutions are performed by three or four men who are inside, and who have no communication with the surface of the water or the external air. With this machine navigation can be effected in spite of currents, any operations may be carried on under water, and it may be brought to the surface at will, and navigated like an ordinary vessel. It was with a machine similar to this that the project was formed, in 1821, for getting away Napoleon from St Helena. The Societé Generale des Naufrages (Protector, the King of France) appointed Admiral Sir Sidney Smith, Count Godde de Liancourt, the Baron de St Denis, and Dr Daniel St Antoine, to report on the experiments. These took place at St Ouen; when the vessel was repeatedly sunk to the depth of 10 or 12 feet, and reappeared on the surface at different points. M. Godde de Liancourt got into it, and remained there a quarter of an hour, without experiencing the slightest inconvenience, or any difficulty of respiration, during his voyage under water.

DIVERS, AND THEIR POWERS OF SUSPENDING RESPIRATION.—Surprising statements have been made by travellers—a privileged race—respecting the powers of pearl-fishers and others, in voluntarily suspending their respiration under water at considerable depths: some have mentioned half-hours, and some even longer periods, as within the bounds of possibility. Dr Lefevre of Rochefort, who was lately stationed at Navarino, had ample opportunities of putting the prowess of the best divers to the test. He witnessed the performance of those who were employed to fish up the relics of the Turkish fleet sunk in Navarino harbour. The depth to which they had to plunge was 100 feet; but though the Greek divers are, and always may have been, famous for their prowess, none of them could sustain submersion for two whole minutes together. Seventy-six seconds was the average period in fourteen instances, accurately noted; and frequently, after reaching the surface, blood issued from the mouth, eyes, and ears of the swimmer. But, in general, these people can repeat their task three or four times in an hour.—*Medical Gazette.*

LONGEVITY.—A woman, 110 years old, died lately at Fayence, in the department of the Var (France). She was born at Digne, in 1725, and had lived in service in one house since 1745.

ZOOLOGICAL GARDENS.—Such is the interest felt by all classes of the community in Britain, that the number of persons who visited the Zoological Gardens, London, amounted in July 1835, to 44,446, and the income derived therefrom L.1672, 9s., and in the month of August, 26,334, yielding an income of L.992, 16s.

STATUE OF CUVIER.—The inauguration of the statue of this eminent naturalist took place with great ceremony, at his native town of Montbelliard, on the 23d August 1835, being the day of his birth. Deputations from several learned bodies were present, and various orations were delivered in honour of the occasion. The house in which Cuvier first saw the light was very tastefully decorated, and the following inscription was placed on it:—*Ici naquit G. Cuvier, le* 23, *Août* 1769. The ceremony was succeeded by a banquet, a grand concert, and a ball.

EUPHRATES EXPEDITION.—Captain Chesney, with the expeditions under his command, had reached Bir, without encountering any obstacles, on the 11th August last.

SENSITIVE HAIR.—In the hospital of the royal guards at Paris was a private soldier who had received a violent kick on the back of the head from a horse. The excitement of the hair produced was extreme, and could only be kept under by almost innumerable bleedings, both local and general. Amongst a series of phenomena produced by this state of preternatural excitation, the sensibility acquired by the hairs of the head was not the least remarkable. The slightest touch was felt instantly, and cutting them gave exquisite pain, so that the patient would seldom allow any one to come near his head. Baron Larrey, on one occasion, to put him to the test, gave a hint to an assistant who was standing behind the patient, to clip one of his hairs without his perceiving it. This was done with great dexterity, but the soldier broke out into a sally of oaths, succeeded by complaints, and it was sometime before he could be appeased.—*Oracle of Health.*

CAPTURE OF THE LAST NATIVE INHABITANTS OF VAN DIEMEN'S LAND.—The following is a highly interesting extract of a letter from Launceston, Van Diemen's Land, dated the 31st of January 1835:—"I am just returned from seeing a very interesting but melancholy sight—the last of the unfortunate native inhabitants of this island, the remainder of those few unhappy savages, who so long kept us in terror! They were taken a few days since to the westward, and consist of three women, one man, and some little children, called piccaninies. One of the party, an old woman, spoke pretty good English, having probably learned it some years ago among the stock-keepers. They inform us that they are the last of their tribe, once 500 strong, which was long dreaded under the name of the Big-river tribe. They say that, by innumerable affrays with the white men, they were at last reduced to three men, exclusive of women and piccaninies, and that, a few months since, they were surprised, and two of the men were killed; that they wandered all over the island for the purpose of joining some other tribe, feeling themselves too weak to exist, and under constant dread that the remaining man would be killed, and the rest, who it appears could not get food themselves, starved. They wandered over the island in every direction, but found no traces of black men; they began to despond, and led a miserable existence, feeling themselves to be the last natives in the whole island, and that the white men had rooted them out. It makes my heart bleed to think of it, but they acknowledge having killed a great many white men, and said they were very glad when they were taken. This was effected by means of some Sydney natives. On a shot being fired they all fell on their faces, and did not attempt to escape. They are now merry and happy, and pointed to the vessel which was to take them off the island with great glee. They are there taught gardening, agriculture, and the arts of civilized life. We had long believed the natives were nearly, if not quite, extinct, and have not the slightest doubt of the truth of their simple story. To look on that fine, tall, and somewhat solemn-looking savage, the last of his tribe, filled me with emotions which it would be in vain to attempt to describe. *Sic vos non vobis.*"

SHORT'S POPULAR OBSERVATORY.—We congratulate our countrymen on the opening of this patriotic establishment, on the Calton Hill, Edinburgh, where every man who has a shilling in his pocket can behold the wonders of the solar system, through glasses of the first order. Among the many attractions are, "Short's large Gregorian Equatorial Reflecting Telescope; a Superb Achromatic Telescope, ten feet focal length, six inches aperture (the largest in the kingdom), by Tulley; an improved grand Solar Microscope of prodigious power, by Dollond; an improved grand Compound Microscope, with Achromatic Object Glasses (an exquisitely fine instrument), by Dollond; an elegant Orrery, with Planetarian, Tellurian, and Lunarian apparatus, by Dollond; a Camera Obscura, Camera Lucida, Phantasmagoria Lantern, Diagonal Mirror, &c. &c.

LEARNED SOCIETIES.

ROYAL ASIATIC SOCIETY.—A general meeting of this association was held on the 4th July 1835, the Right Honourable Sir Alexander Johnston in the chair; when the first part of a paper, by George Earl, Esq. was read, giving an account of a voyage he made in the year 1834, from Singapore to the western coast of the island of Borneo, accompanied by two interpreters, who were masters of the Tartar and Malay languages. The purpose of this voyage was to establish, if possible, a commercial intercourse with the Chinese colonies, who are in possession of rich gold and diamond mines on that island. Mr Earl sailed from Singapore on the 1st March; and on his arrival at Sinkawan, the principal seaport of the Chinese settlers, he proceeded to the court-house, the residence of the Chinese magistrates; but they declined giving Mr Earl permission to trade, lest it might give offence to the Dutch, who have two small settlements on the coast, and who are masters of the sea, and had prohibited all traffic but through the medium of their own ports. Under these circumstances, it became necessary for Mr Earl to sail for one of those ports, called Sambas, in latitude 1° 25′ north, situated on a small river, about fourteen miles inland. Here he found the habitations little better than huts, all built of wood, and for the most part erected on floats, and moored to large posts in the river. There is a fort here; and before the Dutch became masters of the place, the inhabitants lived entirely by piracy. This town is now inhabited by Chinese and Malays, the former being the most numerous. The latter are ruled by a kind of Rajah, who, however, is controlled by the Dutch resident. The Rajah derives his revenue principally from a monopoly in the sale of opium, which is here smoked to the greatest excess by all classes. The Dutch monopolize the sale of salt. The principal food of the inhabitants is rice, which they import from Java, in exchange for gold dust. The aboriginal inhabitants are called *Dyaks:* they are a ferocious people, divided into tribes, many of which retain their old-established customs. One of these is, that before a young man can be permitted to enter the marriage state, he must present the female of his choice with the head of a man which he has severed with his own hand! It was observed by Mr Earl that the natives of New Holland had had a similar mode of limiting population. On the death of a male of a tribe, they kill a male belonging to a neighbouring tribe, for the purpose of maintaining a balance of power.

NEW WORKS ON NATURAL HISTORY, AND THE PHYSICAL SCIENCES.

Reynolds' Voyages Round the World, 8vo, 22s.—The Sea-side Companion, or Marine Natural History, by Mary Roberts, 12mo, 6s. 6d.—Naturalist's Library, vol. 9, Pigeons, 6s.; vol. 10, British Diurnal Butterflies, 6s.—Cuvier's Work on Fishes, vol. 10.—Popular Illustrations of Natural History, &c., fcp., 6s. 6d.; the Language of Flowers. 3d edition.—Observations on certain curious Indentations in the old Red Sandstone, by Jabez Allies, Esq., 8vo, plates, 3s. 6d.—Barrow's Excursions in the North of Europe, 2d edition, 9 engravings, 8vo, 12s.—Lyell's Principles of Geology, 4th edition, 4 vols. 12mo, 24s.—Gleanings in Natural History, 3d Series, by Edward Jesse, post 8vo, 10s. 6d.—Zoological Journal, Part 20, 10s. 6d. coloured, 7s. 6d. plain; Part 5, the Supplementary Plates to ditto, 14s.; Parts 84 and 85, Conchological Illustrations; No. 42, Genera of Shells, by G. B. Sowerby.—The Shrubbery, 32mo, 3s. 6d.—The Earth, by R. Mudie, fcp. 5s.—An Account of New Zealand, by the Rev. W. Yates, royal 12mo, 10s. 6d.

Edinburgh: Published for the PROPRIETORS, at their Office, No. 16, Hanover Street. London: CHARLES TILT, Fleet Street. Dublin: W. F. WAKEMAN, 9, D'Olier Street. Glasgow and the West of Scotland: JOHN SMITH and SON, 70, St. Vincent Street.

THE EDINBURGH PRINTING COMPANY.

THE EDINBURGH

JOURNAL OF NATURAL HISTORY,

AND OF

THE PHYSICAL SCIENCES.

N°. 5. SATURDAY, DECEMBER 19, 1835.

ZOOLOGY.

DESCRIPTION OF THE PLATE—PIGEONS.

PIGEONS constitute a numerous family of the Gallinaceous order. Possessing a wider geographical range than almost any other tribe of birds, they are found in every quarter of the globe, from the southern boundary of ice, to the confines of the Arctic Circle. The general structure of the bill and feet being in all exceedingly characteristic, they form a well-marked family; and though modern naturalists have separated them into several sections and sub-genera, yet they all have such an affinity of form, as not easily to be mistaken. Their sizes are exceedingly various. The Goura or Crowned Pigeon (fig 9.), the largest of the tribe, measures about twenty-eight inches in length; while the Ground Dove (fig. 7.) is not larger than a sparrow, being only six inches and a half from the furthest extremity of the bill to the point of the tail.

The Domestic Pigeon, and Ring Turtle Dove, have been known to mankind from the remotest period of history, and are both frequently alluded to in the Sacred Writings. From the affectionate regard exhibited by the sexes, the ancients considered the dove as an emblem of love, and hence it was frequently depicted as an attendant in their representations of Venus and Cupid.

Many species of this tribe are remarkable for their powers of flight, and the short space of time in which they perform long journeys is almost incredible. To ascertain with some degree of exactness the speed of the Carrier Pigeon, a gentleman, some years ago, sent one from London by the coach to a friend at Bury St Edmunds, desiring that it might be set at liberty two days after its arrival, precisely as the town-clock struck nine in the morning. This request was strictly attended to, and the pigeon arrived at the Bull Inn, in Dishopsgate Street, at half-past eleven o'clock of the same morning, having thus flown seventy-two miles in two hours and a half.

During the breeding season, pigeons associate in pairs, and pay court to each other with their bills. Both the male and female assist in the labour of incubation. The female lays two eggs, and the young ones produced are generally a male and a female; these are attended to by both the parent birds. At first they are fed with a substance resembling curd, secreted within the crop, the coating of which becomes thickened and enlarged. The process is somewhat analogous to the secretion of milk within the mamma of quadrupeds. If the state of the crop be examined during incubation, it will be found to have a glandular and irregular appearance. Upon killing an old pigeon, when the young are just protruding from the egg, it will be observed to have within this cavity small pieces of white curd mixed with its ordinary food of pease, barley, and other grains. It is for a short time that the young are fed with this substance ; for, on the third day it is administered along with a mixture of common food, and in eight or nine days the secretion of curd completely fails in the old birds, from which time they are capable of ejecting common food alone. This singular disposition of Nature is very remarkable, and we cannot but admire the final cause by which the pigeon is assigned the power of casting up this curd alone, although other food be in the crop at the same time.

The plumage of nearly the whole species is of a close texture ; its tints are various, and its lustre remarkable.

Fig. 1.—The Blue-headed Ground Pigeon (*Columba Cyanocephala*) is a native of Jamaica, Cuba, and the Southern American Islands. It seldom resorts to trees, but is generally found upon the ground, where it forms its nest, and incubates. It runs with astonishing rapidity, and its habits are retired and solitary. Having but limited powers of flight, it seldom rises to a great height above the ground, but usually skims from one place to another, in nearly the same manner as the common Land Rail. Its size nearly approaches to the common Partridge.

Fig. 2.—The Zenaida Pigeon (*Columba Zenaida.*) This beautiful bird has been but recently discovered. It is about ten inches in length, and a native of Florida, in the United States of America. These pigeons also are generally found on the ground, in which situation they amuse themselves in dusting and seeking for gravel, which they swallow to assist in digesting their food.

Fig. 3.—The Purple-crowned Pigeon (*C. Purpurata*) is a native of the South Sea Islands, in many of which it greatly abounds. The first specimens which reached this country were brought from Tonga-Taboo. Their geographical range extends as far south as New Holland. This bird is from nine to ten inches in length, and frequents woods, feeding on different kinds of fruit, such as the Limonja bifoliata, and the Banana.

Fig. 4.—The Passenger Pigeon (*C. Migratoria.*) This elegant species is sixteen inches in length. In symmetry of form, and in the arrangement and contrast of its colours, it is exceeded by none of the genus. It inhabits North America, from the Stony Mountains to Hudson's Bay, and its range extends as far south as the Gulf of Mexico. Unlike most of the species in the genus Columba, these birds generally

associate together, both during their incubation and also during their migrations, in such vast numbers as to exceed all belief. They are sometimes seen in feeding parties, covering the country to the extent of two miles in length, and a quarter of a mile in breadth. Their migrations are occasioned by a scarcity of food, rather than by the beech-nut, and not by temperature, as is the case with some other birds. When they have fixed upon a resting-place, they do not remove from it even after they have exhausted all their food, but will extend their range for a distance of eighty miles to another forest, and return in the evening to their temporary home. When these roosting-places are discovered by the neighbouring inhabitants, they repair to them in the night. Vast numbers are first stupified with pots of burning sulphur, and then killed with poles, guns, and other instruments of destruction.

Fig. 5.—The Blue and Green Pigeon (*C. Cyanō-virens.*) Although displaying no great variety in its plumage, it is nevertheless a bird of great beauty. This species is a native of New Guinea, where it frequents the vast forests with which that country abounds. It is about eight inches in length.

Fig. 6.—The African Ground Pigeon (*C. Afra.*) This beautiful little species, as its name denotes, is a native of Africa.

Fig. 7.—The Ground Pigeon (*C. Passerina.*) The birds of this species are natives of North and South Carolina, and Georgia, in the United States. They also frequent several of the West India Islands, where the inhabitants frequently catch and keep them in cages. They congregate in small flocks of about fifteen or twenty, and are usually found on the ground. When disturbed, they rise to a short distance and then alight.

Fig. 8.—The Black-capped Pigeon (*C. Melano-cephala*) inhabits the island of Java, and builds its nest in trees. Its length is nine inches.

Fig. 9.—The Great-crowned Pigeon (*C. Coronata.*) This splendid bird is the largest of its tribe. Many attempts have been made to domesticate it, but they have invariably failed. This is to be regretted, as it would not fail to be an important acquisition for our poultry-yards. It has been found in the forests of Africa, New Guinea, and the Molucca Islands.

FLYING FISH.—The animals of the ocean seem to correspond in their general habits to those of the land—one portion depending upon an erratic mode of life for subsistence, like the wandering Arabs of the desert, and the other upon a sedentary life, like the domesticated ones of the plain. The erratic tribes of the ocean, however, have this advantage over those of the land, that, while the green oases scattered thinly over the deserts of the latter, and the caravans at chance intervals traversing them, afford but an uncertain supply to its roaming hordes, those of the ocean derive always an abundance in the variety of the finned fishes, and the gelatinous Mollusca and spawn which the latter contains ; the smaller finned tribes preying upon the Mollusca and spawn, and the larger again upon the smaller, until their eventual decease enables the Mollusca to prey in turn upon them. The animal species has, by an eminent naturalist, been compared to a circle, into which all are progressively united by successive connecting links ; and it may be only a high philosophic enthusiasm for practically demonstrating the truth of this circular theory, which induces them to eat each other in a circle also. Of all the smaller erratic fishes, the flying species is the most interesting, in consequence of its being one of the singular links connecting the fish with the bird tribe, its length seldom exceeding a foot, its shape roundish, and tapering from the head to the tail, with a long fin projecting out on each side of its centre of balance, to be applied either to swimming or flying, according as exigencies may require. It is not, however, a universal wanderer; like most of the other deep-sea fishes, its range of feeding-ground is confined to the latitudes of the trade winds, most probably in consequence of its slender filmy wings and delicate form rendering it unfit to encounter the rough buffetings of the stormy winds and waves of the seas beyond. Having so many enemies constantly in quest of them as a prey in their own element, no wonder that the Flying Fish should be by nature a timid race, always taking to the air for protection, when threatened by an enemy in the sea. The approach of the Porpoise, Dolphin, Albicore, and Bonetta, quickly scare them from their watery haunts; but the terror produced by the latter, is nothing in comparison to that excited by a huge ship suddenly plunging in among a shoal of them sunning themselves near the surface of the water, tumbling over and over in their hurried efforts to get up, or knocking each other down again into the sea in their haste to escape the fancied fangs of the nondescript monster that has thus unexpectedly invaded their domain. Strangers, on first seeing them, almost invariably take them for a flock of birds ; and, indeed, when viewed at a little distance on the wing, it frequently requires a practised eye to detect the deception, a fresh flock of them being often made to start up at every plunge of the ship, when sailing through a part of the sea where they are rife, Mounting suddenly upwards, with a squattering noise like a flock of ducks, they now flicker away in a covey together, with astonishing speed, their long, thin, tapering wings quivering in rapid vibrations as they dart through the air, resembling the wings of the sparrow tribe. They appear to have as perfect self-command in the air as the water, the body of the flock always following the motions of the leaders, just as seen in a flock of birds, soaring up and sinking down, or wheeling to either hand, according as the pilots of the band vary their own onward flight. They fly by night as well as by day, although their power of vision in the former must be very defective, as is shown by their frequently dropping on board ship during the night, an accident which never happens to them during the day. Like most other fishes, they are attracted by a glare of light, and it is by taking advantage of this that they are allured in such numbers into the nets constructed for them on the Barbadoes coast, as to constitute no inconsiderable item in the food of the inhabitants of that island. Ships have sometimes followed a similar plan with singular success. H.M.S. Prometheus, in running down the trades, by nailing hammock-cloths along her sides, supported out by handspikes, and illumined by a row of pursers' lanterns between, caught as many nightly as gave a daily meal to all on board. They are sweet, delicate, and juicy eating, contrary to that of most of the other deep-sea fishes, which are harsh, dry, and tasteless. Their manner of cooking them in Barbadoes is by frying with a little lard and flour, dusting until brown and dry, and in this state exposing them for sale, every boat that visits a ship having generally large platters-full of them piled up in cross layers over each other, which always find abundance of eager customers, particularly after a long salt-beef cruise. Nor is the peculiar oratory, playful motions, and merry smirking faces of the jetty belles who vend them, the least interesting part of the scene; dancing nimbly about on some convenient boat plank, wagging their heads laughingly to and fro, and snapping their fingers in cadence to the tune they are humming, until attaining the object they had in view, of attracting the attention of some one to their wares, they now simper out, in their best boarding-school English, some such speech as the following:—" Hye, buckra, do come buy him fine fish fo yam-chah! bady, what fo you no buy him all den off? I pop fo mee like a bottle o' pruce."

Eggs Preserved for Three Hundred Years.—Three eggs were found in the wall of a chapel, which was built upwards of 300 years ago, near the Lago Maggiore. These were embedded in the mortar of the wall, and, upon attentive examination, they were found to be quite fresh. It has been long known, that the eggs of birds, brought from India or America, when covered with a thin coating of wax, retain their vital principle, and have been hatched after the wax had been dissolved by alcohol.

Butterfly Feasts.—There is a certain mountain in New Holland, called the Bugong mountain, from multitudes of small moths called Bugong by the natives, which congregate at certain times, upon masses of granite, on this mountain. The months of November, December, and January, are quite a season of festivity amongst the people, who assemble from every quarter to collect these moths. They are stated also to form the principal summer food of those who inhabit to the south of the snow mountains. To collect these moths, or rather butterflies, the natives make smothered fires under the rocks on which they congregate; and suffocating them with smoke, collect them by bushels, and then bake them by placing them on heated ground. Thus they separate from them the down and the wings; they are then pounded and formed into cakes resembling lumps of fat, and often smoked, which preserves them for some time. When accustomed to this diet, they thrive and fatten exceedingly upon it. Millions of these animals were observed also, on the coast of New Holland, both by Captains Cook and King. Thus has a kind Providence provided an abundant supply of food for a race that, subsisting solely by hunting or fishing, must often be reduced to great streights.

BOTANY AND HORTICULTURE.

FAIRY RINGS.

The cause of these singular appearances was long unknown to Natural Philosophers, although many attempted to remove them, by conjecture, from the dominion of " fairy elves," into the soberer domains of Science. As is well known to shepherds and agricultural labourers, fairy rings are spots of grass, more luxuriant and green than any other part in a field. These spots are usually circular, either having within them a spot of grass, peculiarly luxuriant throughout its entire surface, or being only a circular zone of very luxuriant grass, inclosing within it a quantity of similarly coloured herbage, but not quite so luxuriant as the zone, although superior to the rest of the grass in the same field. Often the circle is incomplete, and consists only of an arch or segment of a circle, which is frequently bent in an irregular manner.

It is unnecessary at present to speak of the absurdity of the popular opinion on this subject. The belief in Fairies or Genii, as secondary causes of the various phenomena of nature, seems congenial to the human mind, and it has, accordingly, been found in almost every country. The Arabs call them Ginn, and the Persians Peri; and they were supposed to inhabit a fairy-land, which was called Ginnistân. In many districts of our own country, the belief in the existence of Fairies, Benshees, and Bogles, still retains its hold over the minds of the people.

Mr Jessop and Dr Priestley thought that fairy circles were caused by Electricity. " I have been often puzzled," says Mr Jessop in one of the earliest volumes of the Philosophical Transactions, " to give an account of those phenomena which are called Fairy Circles. I have seen many of them, and those of two sorts; one sort bare, of seven or eight yards in diameter, making a round path, something more than a foot broad, with green grass in the middle; the others like them, but of several bignesses, and encompassed with a circumference of grass, about the same breadth, much fresher and greener than that in the middle. But my worthy friend Mr Walker gave me full satisfaction from his own experience. It was his chance one day to walk out among some mowing grass (in which he had been but a little while before) after a great storm of thunder and lightning, which seemed by the noise and flashes to have been very near him; he presently observed a round circle, of about four or five yards diameter, the rim whereof was about a foot broad, newly burnt bare, as the colour and brittleness of the grass roots did plainly testify. He knew not what to ascribe it unto but to the lightning, which, besides the odd caprices remarkable in that fire in particular, might, without any wonder, like all other fires, move round and burn more in the extremities than the middle. After the grass was mowed, the next year it came up more fresh and green in the place burnt than in the middle, and at mowing time was much taller and ranker."

Dr Price suggested to Dr Priestley that fairy rings might be of an electric origin, and be produced in the same manner as those circular spots, which are procured by submitting metallic substances covered with water to the influence of an electric battery. " I have examined one of these rings," says Dr Priestley; " it was about a yard in diameter, the ring itself about a quarter of a yard broad, and equally so in the whole circumference; but there was no appearance of any thing to correspond with the central spot," observed in the electrical experiments.

Mr Cavallo, in his Treatise on Electricity, which appeared in 1777, was the first who called in question the electric origin of these circles. " This supposition," he observes, " is not very probable, for the spots in the fields, called *fairy circles*, have no central spot, no concentric circles, neither are they always of a circular figure; and, as I am informed, they seem to be rather beds of mushrooms than the effects of lightning."

Mr White, in the Natural History of Selborne, makes the following observations on the subject:—" The cause, occasion, call it what you will, of fairy rings, subsists in the turf, and is conveyable with it; for the turf of my garden-walks, brought from the Down above, abounds with those appearances, which vary their shape, and shift situation continually, discovering themselves now in circles, now in segments, and sometimes in irregular patches and spots. Wherever they obtain, puff-balls abound, the seeds of which are doubtless brought in the turf." Mr Johnson of Weatherly, in a paper in the fourth volume of the Philosophical Journal, attributes them " to the droppings of Starlings, which, when in large flights, frequently alight on the ground in circles, and sometimes are known to sit a considerable time in these annular congregations."

The suggestion made to Cavallo, that these circles are occasioned by mushrooms, has, since that time, been completely confirmed by the observations of Drs Withering and Wollaston. The former considered them to be caused by a species of Agaricus *(Ag. Orcades)*; but the latter showed that many other species of Agarici, and the Lycoperdon bovista, were capable of producing them.

In a valuable paper in the Philosophical Transactions for the year 1807, Dr Wollaston described the manner in which these circles were formed:—

" That which first attracted my notice," says he, " was the position of certain fungi, which are always to be found growing upon these circles, if examined in a proper season. In the case of mushrooms, I found them to be solely at the exterior margin of the dark ring of grass. The breadth of the ring in that instance, measured from them towards the centre, was about twelve or fourteen inches, while the mushrooms themselves covered an exterior ring about four or five inches broad.

" The position of these mushrooms leads me to conjecture that progressive increase, from a central point, was the probable mode of formation of the ring. I was the more inclined to this hypothesis, when I found that a second species of fungus presented a similar arrangement, with respect to the relative position of the ring and fungi; for I observed, that in all instances the present appearance of fungi was upon the exterior border of a dark ring of grass. I thought it not improbable that the soil, which had once contributed to the support of fungi, might be so exhausted of some peculiar *pabulum* necessary for their production, as to be rendered incapable of producing a second crop of that singular class of vegetables. The second year's crop would consequently appear in a small ring surrounding the original centre of vegetation, and at every succeeding year the defect of nutriment on one side would necessarily cause the new roots to extend themselves solely in the opposite direction, and would occasion the circle of fungi continually to proceed by annual enlargement from the centre outwards. An appearance of luxuriance of the grass would follow as a natural consequence, as the soil of an interior circle would always be enriched by the decayed roots of fungi of the preceding year's growth. During the growth of the fungi, they so entirely absorb all nutriment from the soil beneath, that the herbage is for a while destroyed, and a ring appears, bare of grass, surrounding the dark ring. If a transverse section be made of the soil beneath the ring at this time, the part beneath the fungi appears paler than the soil on either side of it; but that which is beneath the interior circle of dark grass is found, on the contrary, to be considerably darker than the general surrounding soil. But in the course of a few weeks after the fungi have ceased to appear, the soil where they stood grows darker, and the grass soon vegetates again with peculiar vigour; so that I have seen the surface covered with dark grass although the darkened soil has not exceeded half an inch in thickness, while that beneath has continued white with spawn of these mushrooms for about two inches in depth.

" For the purpose of observing the progress of various circles, I marked them three or four years in succession, by incisions of different kinds, by which I could distinguish clearly the successive annual increase, and I found it to vary in different circles from eight inches to as much as two feet. The broadest rings that I have seen were those of the common mushroom *(Agaricus campestris)*; the narrowest are the most frequent, and are those of the champignon *(Agaricus Orcades of Withering.)* The mushroom accordingly makes circles of largest diameter, but those of the champignon are most regular. There are, however, as many as three other fungi that exhibit the same mode of extension, and produce the same effect upon the herbage. These are the *Agaricus terreus*, *Agaricus procerus*, and the *Lycoperdon bovista*, the last of which is far more common than the two last mentioned agarici.

" There is one circumstance that may frequently be observed respecting these circles, which can satisfactorily be accounted for, according to the preceding hypothesis of the cause of their increase, and may be considered as a confirmation of its truth. Whenever two adjacent circles are found to interfere, they not only do not cross each other, but both circles are invariably obliterated between the points of contact, at least in more than twenty cases; I have seen no one instance to the contrary. The exhaustion occasioned by each obstructs the progress of the other, and both are starved."

MINERALOGY.

SINGULAR SECTIONS OF A KENTISH FLINT.

THE common flint is found in spherical masses embedded in the chalk formations, and the manner in which it has there been formed has given rise to considerable discussion among geologists. It is certainly a singular circumstance to find in extensive strata of chalk, which is a substance almost entirely composed of pure carbonate of lime, isolated masses of flint, formed of silica, or quartz. Most commonly the nucleus of these flint nodules consists of an animal or vegetable substance, as a shell-fish, coral, or piece of flustra, or sponge; but frequently no such matters are found. The nodules, too, assume various shapes, and seem to be moulded according to the cavities of the chalk in which they are surrounded. From the appearances of these nodules, it is evident that they have been in a fluid state, previous to their assuming such shapes, either from the agency of intense heat or of a liquid menstruum, or perhaps from the combination of both these means. The accurate experiments of modern chemists have shown us that silica may be liquified with the greatest ease in combination with either of the mineral alkalies. Thus, if silica and soda be subjected to the intense heat of a furnace, it liquifies into a glass; and if this glass be taken and again subjected to the agency of water, heated under a highly-condensed pressure, the silica will be dissolved, and will be deposited, on cooling, around the edge of the vessel. When combined with a large proportion of the alkali, and when under very minute mechanical division, silica may also be dissolved in water at the ordinary temperature of the atmosphere, thus forming the substance called liquor of flints. By one or other of such processes, then, taking place in nature, we may suppose that a quantity of liquified silica has been diffused among a bed of chalk. The singular circumstance however is, how it should have collected into the numerous separate nodules in which it is universally found; and this has been attempted to be explained, by supposing that a chemical attraction has taken place between the vegetable or animal remains strewed profusely among the chalk and the silicious matter, by which the latter has been accumulated around, and incorporated into, the minutest parts of the organized substances, thereby forming the petrifactions so generally found in the flint nodules. In those cases where no traces of organization are found, it may be supposed either that the organic structure has been entirely destroyed, or that the silicious matter has been from the first simply deposited in a cavity or fissure of the chalk. Flint nodules, although composed of the same materials, have not generally the compact and dense structure of the pebble or agate. Flint is of a more porous nature, and of an opaque and clouded appearance. If a piece of flint be broken, and immediately examined, minute drops of moisture will be seen to ooze out from its pores, affording a proof that water must have originally assisted at its fluidity. It must be remarked, however, that alkaline matter is not found to exist in any great quantity in its composition. From the opaque and porous nature of flints, it has been conjectured that they may have been formed partly from particles in a state of minute mechanical division, and partly from others in a state of chemical solution.

The fracture of flint nodules usually presents a dark opaque ground, clouded with whitish and dark gray spots and patches. Some of these often assume very fantastic imitations of figures of men and animals. In the British Museum is an agate on which is portrayed a very accurate likeness of the poet Chaucer; and during the French Revolution, immediately after the king was beheaded, a very remarkable portrait of this unfortunate monarch was discovered distinctly marked in a piece of Labrador spar. So accurate was the likeness, and so curious was this coincidence reckoned at the time, that a very large sum of money was obtained for it; and fac-similes were engraved from it, and worn as rings by the loyal inhabitants of Europe. In the annexed engravings we have given a fac-simile of three remarkable portraits found in a flint nodule, which may be seen in the Museum of Mr Robert Fraser, Jeweller, 17, South St Andrew Street, Edinburgh.

This mass of flint, weighing about sixteen ounces, was picked up by mere accident, on the Kent road, near London. On breaking off a small piece of it, the profile, No. 1, was discovered on the surface of the fracture, and immediately recognised as bearing a very striking resemblance to the

general contour of the features of the first warrior and general of the age. The portrait has somewhat the appearance of an enamel painting; the figure being of a whitish-gray substance, surrounded by a dark brown ground. As it was conjectured that, in all probability, the impression of the figure might penetrate deep into the stone, it was slit up nearly through the centre, when the figures No. 2 and 3 were displayed on each side of the exposed surfaces; and it will not require a very active fancy to discover in these the face and lineaments of a monarch endeared to the British nation. These two likenesses have actually been recognised and pointed out by different individuals, who had no previous knowledge that such a similarity had before been discovered; thus affording a test of the truth of the general resemblance. At the time that these likenesses were first discovered—about five years ago—it was looked on as a curious coincidence that the monarch and his prime minister should both be found depicted on one stone by the hand of nature, and by a process which, even with all the aids of modern chemistry, we fear we have but imperfectly conjectured, and endeavoured to explain to our readers.

Flint is not the only substance which is found to contain animal and vegetable matter in its nodules. Small portions of moss plants, and other cryptogamia, are frequently found, beautifully preserved, in the rock crystal, topaz, and agate, with all the minute lineaments of their original structure. This affords another proof of the fact that such crystals must have been in a fluid state, without any great increase of temperature, at the period when they assumed their solid form. Many of these stones as well as jaspers contain various figures assuming the forms of vegetation; being in reality, merely accidental admixtures of various metallic substances, which, in crystallizing, thus assume the appearance of leaves and stems of plants. Of this kind are the mocha stone, arborescent jasper, landscape marble, &c. Drops of water, however, and portions of air, are occasionally found in such crystals, under circumstances which led Dr Davy to suppose that these have been enveloped in the crystal while under great expansion by heat. On carefully opening the cavities, he found that the water and air diminished in volume, when exposed to the ordinary pressure of the atmosphere, in some instances to the extent of six to eight times their bulk—in the chalcedony to the extent of sixty times their volume.

MISCELLANEOUS.

NO. I.—ANIMAL MAGNETISM.

THE extraordinary influence of the imagination over the human body is strikingly illustrated in the History of Animal Magnetism, as well as the astonishing delusions by which an ignorant and credulous mind is liable to be influenced, when in the hands of a cunning and designing impostor. We shall, accordingly, give a few articles on this interesting and comparatively little-known subject.

In consequence of the extent to which the practice of Animal Magnetism, as it was called by its inventor, M. Mesmer, was carried in Paris, the French king appointed a committee, consisting of four physicians and five members of the Royal Academy of Sciences, to investigate the matter, in the year 1784. Among the latter were M. M. Bailly, Lavoisier, and Dr Franklin, who was at that time the American minister at Paris. This agent, which Mesmer pretended to have discovered, he affirmed, was "a fluid universally diffused, and filling all space, being the medium of a reciprocal influence between the celestial bodies, the earth, and living beings—it insinuated itself into the substance of the nerves; upon which, therefore, it had a direct operation. It was capable of being communicated from one body to other bodies, both animated and inanimate, and that at a considerable distance, without the assistance of any intermediate substance; and it exhibited in the human body some properties analogous to those of the loadstone, especially its two poles. This Animal Magnetism," he added, "was capable of curing directly all the disorders of the nervous system, and indirectly other maladies; it rendered perfect the operation of medicines, and excited and directed the salutary crises of diseases, so that it placed these crises in the power of the physician. Moreover, it enabled him to ascertain the state of health of each individual, and to form a correct judgment as to the origin, nature, and progress of the most complicated diseases," &c. In short, he said, "La nature offre dans le magnétisme un moyen universel de guerir et de preserver les hommes." (Nature supplies in Magnetism one universal cure for the maladies of the human race.)* Mons. Deslon, a pupil of Mesmer, also practised Animal Magnetism at Paris, and undertook to demonstrate its existence and properties to the commissioners. He commenced his instructions by reading a memoir, in which he maintained, that "there is but one nature, one disease, and one remedy, and that remedy is Animal Magnetism."

The first step of the commissioners was to examine the mode and instruments of operation, and the effects of the agent. It was observed that M. Deslon operated upon many individuals at the same time. In the middle of a large room was placed a circular chest of oak, raised about a foot from the floor, which was called the *Buquet;* the lid of this chest was pierced with a number of holes, through which there issued moveable and curved branches of iron. The patients were ranged in several circles round the chest, each at an iron branch, which, by means of its curvature, could be applied directly to the diseased part. A cord, which was passed round their bodies, connected them with one another, and sometimes a second chain of communication was formed by means of the hands, the thumb of each one's left hand being received and pressed between the forefingers and thumb of the right hand of his neighbour. Moreover, a piano-forte was placed in a corner of the room, on which different airs were played; sound being, according to the principles of Mesmer, a conductor of Magnetism. The patients, thus ranged in great numbers round the baquet, received the magnetic influence at once, by all these means of communication. By the branches of iron which transmitted to them the magnetism of the baquet—by the cord entwined round the body—by the union of thumbs, which conveyed to each the magnetism of his neighbour—and by the sound of the music, or of an agreeable voice, which diffused the principle through the air. The patients were, besides, directly magnetised by means of the finger of the magnetiser, and a rod of iron which he moved about before the face, above or behind the head, and over the diseased parts, always observing the distinction of the magnetic poles, and fixing his countenance upon the individual. But, above all, they were magnetised by the application of the hands, and by pressure with the fingers upon the hypochondria and the abdominal regions, which was often continued for a very long time, occasionally for several hours together.

The patients subjected to this treatment at length began to present very various appearances in their condition as the operation proceeded. Some of them were calm and tranquil, and felt nothing; others were affected with coughing and salivation; others again experienced slight pains, partial or universal heats, and considerable perspirations; and others were agitated and tortured with convulsions. These convulsions were extraordinary in their numbers, severity, and duration. The commissioners saw them, in some instances, continue for three hours, when they were accompanied with expectoration of a viscid phlegm, which was ejected by violent efforts, and sometimes streaked with blood; one young man often brought up blood copiously. The convulsions were characterized by violent involuntary motions of the limbs, and of the whole body, by spasms of the throat, by agitations of the epigastrium and hypochondres, and wandering motions of the eyes, accompanied by piercing shrieks, weeping, immoderate laughter, and hiccough. They were generally preceded or followed by a state of languor and rambling, or a degree of drowsiness, and even of coma (or profound sleep and torpitude). The least unexpected noise made the patients start; and it was remarked that even a change of measure in the air played upon the piano-forte affected them, so that a more lively movement increased their agitation and renewed the violence of their convulsions. Nothing can be more surprising or more inconceivable by those who have not witnessed it, than the spectacle of these convulsions, say the commissioners: all seem to be under the power of the magnetiser; a sign from him, his voice, his look, immediately rouses them from a state even of apparent stupor. In truth, they add, it was impossible not to recognise, in these constant effects, a great power or agency, which held the patients under its dominion, and of which the magnetiser appeared to be the sole depositary.†

* See Memoire sur la Decouverte du Magnetisme Animal. par M. Mesmer, Doct. en Med. de la Faculté de Vienne, 1779. Also his Precis Historique des Faits relatifs aux Mag. An jusques en Avril 1781.

† See Rapport des Comissaires chargés par le Roi, de l'Examen du Magnetisme Animal; à Paris, 1784.

Such, then, were the phenomena (of the reality of which they could not doubt) produced by the operation of this new agent, the nature and origin of which it was the duty of the commissioners to investigate. This convulsive and lethargic state, it may be noticed, was considered as a crisis, such as the constitution or the art of medicine is enabled to effect, for the purpose of curing diseases; and, for the sake of brevity, we shall adopt the term to express this occurrence, regardless of the hypothesis which led to its use.

A POLYPI-NESIAN WAR.—In those balmy regions of the South, where coral reefs arise, the germs of future continents, there dwelt two Polypi, celebrated in Polypinesian lore, by the names of Polycrates and Polydorus. Friendship had long united their minds, as much as nature had united their bodies; but chance, or a wayward fate, at length produced discord. A luckless worm, roaming for pleasure through the deep, became entangled in the tentacula of Polycrates, by the one extremity, while Polydorus at the same instant seized the other end. Each continued his repast; until at length, near the centre of the worm, a fierce conflict of hostile tentacula arose. Like some beings at the other end of the scale of creation, though *but one flesh*, they were of *two minds*. Polycrates being the larger and stronger, soon obtained the ascendancy; and, irritated by the contest, not only swallowed the worm, but Polydorus also! Polydorus, however, by no means disturbed by the novelty of his situation, continued, with philosophical resignation, within the stomach of Polycrates to complete the repast which had been thus strangely interrupted; until at length, finding nothing further worth his demolishing in that quarter, and being unwilling to imitate Jonas in the length of his captivity, he soon managed, by a retrograde movement of his tentacula, to release himself and worm from this apparently critical situation. Polycrates now obtains all the honour and glory of the contest; but Polydorus finds the truth of the popular adage, that "solid pudding is better than empty praise."

NEW HOLLAND.—This is New Holland—where it is summer with us when it is winter in Europe, and *vice versa*—where the barometer rises before bad weather, and falls before good—where the north is the hot wind, and the south the cold—where the humblest house is fitted up with cedar *(Cedrela toona)*—where the fields are fenced with mahogany *(Eucalyptus robusta)*, and myrtle trees *(Myrtaceæ)* are burnt for fire-wood—where the swans are black and the eagles white—where the kangaroo, an animal between the squirrel and the deer, has five claws on its fore-paws, and three talons on its hind legs, like a bird, and yet hops on its tail—where the mole *(Ornithorhynchus paradoxus)* lays eggs, and has a duck's bill—where there is a bird *(Meliphaga)*, with a broom in its mouth instead of a tongue—where there is a fish, one-half belonging to one genus *(Raja)*, and the other half to another *(Squalus)*—where the pears are of wood *(Xylomelum pyriforme)*, with the stalk at the broader end—and where the cherry *(Exocarpus cupressiformis)* grows with the stone on the outside.

ASTRONOMER ROYAL.—Mr Pond has retired from the situation of Astronomer Royal, and Professor Airy has been appointed, with a salary of L.800 a-year.

DREAMS.—By the kind attention of my friend Dr James Gregory, I have received a most interesting manuscript by his late eminent father, which contains a variety of curious matter on this subject. In this paper, Dr Gregory mentions of himself, that having on one occasion gone to bed with a vessel of hot water at his feet, he dreamt of walking up the crater of Mount Ætna, and of feeling the ground warm under him. He had, at an early period of his life, visited Mount Vesuvius, and actually felt a strong sensation of warmth in his feet, when walking up the side of the crater; but it was remarkable, that the dream was not of Vesuvius, but of Ætna, of which he had only read Brydon's description. This was probably from the latter impression having been the more recent. On another occasion, he dreamt of spending a winter at Hudson's Bay, and of suffering much distress from the intense frost. He found that he had thrown off the bed-clothes in his sleep, and, a few days before, he had been reading a very particular account of the colonies in that country during the winter. Again, when suffering from toothache, he dreamt of undergoing the operation of tooth-drawing, with the additional circumstance, that the operator drew a sound tooth, leaving the aching one in its place. But the most striking anecdote in this interesting document, is one in which similar dreams were produced in a gentleman and his wife, at the same time and by the same cause. It happened at the period when there was an alarm of French invasion, and almost every man in Edinburgh was a soldier. All things had been arranged in expectation of the landing of an enemy, the first notice of which was to be given by a gun from the Castle, and this was to be followed by a chain of signals, calculated to alarm the country in all directions. Farther, there had been recently in Edinburgh a splendid military spectacle, in which five thousand men had been drawn up in Princes' Street, fronting the Castle. The gentleman to whom the dream occurred, and who had been a most zealous volunteer, was in bed, between two and three o'clock in the morning, when he dreamt of hearing the signal-gun. He was immediately at the Castle, witnessed the proceedings for displaying the signals, and saw and heard a great bustle over the town, from troops and artillery assembling, especially in Princes' Street. At this time he was roused by his wife, who awoke in a fright, in consequence of a similar dream, connected with much noise and the landing of an enemy, and concluding with the death of a particular friend of her husband's, who had served with him as a volunteer during the late war. The origin of this remarkable concurrence, was ascertained in the morning to be the noise produced in the room above by the fall of a pair of tongs, which had been left in some very awkward position in support of a clothes screen.—Dr Reid relates of himself, that the dressing applied after a blister on his head having become ruffled, so as to produce considerable uneasiness, he dreamt of falling into the hands of savages, and being scalped by them.—*Abercrombie on the Intellectual Powers.*

Edinburgh: Published for the PROPRIETORS, at their Office, 16, Hanover Street. London: CHARLES TILT, Fleet Street. Dublin: W. F. WAKEMAN, 9, D'Olier Street. Glasgow and the West of Scotland: JOHN SMITH and SON, 70, St Vincent Street.

THE EDINBURGH PRINTING COMPANY.

THE EDINBURGH

JOURNAL OF NATURAL HISTORY,

AND OF

THE PHYSICAL SCIENCES.

N°. 6.   SATURDAY, JANUARY 2, 1836.

ZOOLOGY.

DESCRIPTION OF THE PLATE—THE CAT TRIBE.

THE animals of this tribe have long attracted the attention and admiration of mankind by their strength, magnanimity, and valour; while the extreme ferocity and cruelty of some have equally rendered them objects of fear. To conquer the Nemæan lion, was considered, in the earlier periods of human society, as a feat worthy of Hercules himself.

They are tolerably swift of foot, and hunt chiefly during the night. Lying in wait for their prey till it comes within reach, they spring forward upon it at one bound, and seize it by surprize. While eyeing their victim, they move the tail frequently from side to side. In a natural state, they never adopt vegetable food, except in cases of urgent necessity. Most of the species are very agile in climbing trees, for which purpose, the strength of their limbs and hooked claws admirably adapt them. They have the remarkable property of alighting on their feet, whenever they are thrown or fall from a height, by which means the injury attendant on such accidents is often averted.

Fig. 1. The African Lion (*Felis Leo*).—The Lion of Africa is usually about six feet in length from the muzzle to the insertion of the tail, and his height at the shoulders upwards of three feet; the tail is more than three feet long, and terminated by a tuft of blackish-brown hair. Attached to the last joint of the tail is a small dark brown horny prickle, which is concealed by the hair, and surrounded at its base by an annular fold of the skin. The Lion possesses a characteristic peculiarity which at once distinguishes him from all other species of his tribe, namely, a long and flowing mane; and, except in a very young state, his fur is totally divested of spots or stripes.

The courage of the Lion has been proverbial in all ages, yet we cannot ascribe this to any innate elevation of sentiment, but to the consciousness of his own physical powers. He well deserves the title of "the king of beasts," as no animal can singly overcome him; and he roams at large in the boundless desert, in the extensive plains, or skulks in the shades of the vast jungles of his native country. His head is peculiarly large, his jaws have immense strength, and his shoulders and chest have a depth far exceeding all other animals of his size. In his sense of smell and acuteness of hearing, he is far inferior to the dog. This deficiency seems to pervade the whole tribe, and appears to have been destined by Nature for limiting, in a certain degree, their means of destruction.

Fig. 2. The Lioness and Cubs.—The Lioness is at once distinguished from her mate by the want of the mane. Being inferior both in size and strength to the Lion, her make is at the same time more delicate, and she is more agile in her movements, while she greatly surpasses him in the liveliness of her disposition, and the constrained ardour of her passions. The sexes differ, in a remarkable degree, in the position and direction of their heads,—that of the Lion being uniformly elevated, impressing him with an air of dignity, while the Lioness carries her head always on a level with the line of her back, giving her a sullen and ferocious aspect. In tender and undivided attachment to her progeny, the Lioness is exceeded by no other animal. In a tame condition, and even when brought to a state of complete gentleness, her character becomes totally changed. When suckling the young, all her native ferocity seems to return with tenfold vigour. At this period she would tear to pieces the hand of her keeper, while at other times she would lick it with affectionate tenderness.

Fig. 3. The Puma (*F. Concolor*).—This animal is a native of the American continent. The colour of its fur is uniformly of a yellow-fawn, or brownish-red; the belly is white, or of a pale cream colour. It has been designated the American Lion, though destitute of a mane, or tuft at the point of its tail. It is about four feet in length, and its height somewhat more than two feet. The head is round, and the ears short. It inhabits the high and mountainous tracts of the United States, and is common in South America. Its power is greatly beyond what might be expected from its size. It leaps on the back of its victim, whom it seldom fails to vanquish, frequently overcoming even the Wild Ass. Sometimes the latter contrives to free itself from its assailant by lying down and rolling over the Puma, and thus crushing him to death.

Fig. 4. The Tiger (*F. Tigris*).—The Bengal Tiger measures six feet in length, besides the tail, which is generally about three feet. There are several varieties of this animal, but all of them have the black stripes on the body.

Such is the physical strength of the Tiger, that he can run at a considerable speed with the body of a man in his teeth; and his general movements are more nimble than those of the Lion. He is the most rapacious and destructive of all carnivorous animals. With strength nearly equal to the Lion, he is much more ferocious in disposition, and certainly more to be dreaded by the human species. He is a native of all the countries of Southern Asia which lie between the north of China, Chinese Tartary, and the Indies, abounding in Bengal, Tonquin, and Sumatra, and inhabiting most of the larger islands on that side of India. He frequently proves the scourge of many districts which are thickly covered with jungles and forests.

Fig. 5. The Clouded Tiger (*F. Nebulosus*).—A recently discovered species, for a knowledge of which we are indebted to the zeal of the late Sir T. Stamford Raffles. This animal was first observed in the extensive forests of Bencoolen, and was brought alive to England from Sumatra in August 1824. It was taken very young, and became completely domesticated The Clouded Tiger is equal in size to the Leopard. Its general aspect, even in a state of nature, indicates less ferocity than that of the Bengal Tiger or Leopard. The character of the eyes and general physiognomy bear a considerable resemblance to those of the domestic Cat. The prevailing colour of the fur is ash-coloured or whitish-gray; it is covered with spots and bands, defined posteriorly with a deep black margin. The back has most strikingly the appearance of velvet on the larger discolourations.

Fig. 6. The Leopard (*F. Leopardus*).—The general expression of the Leopard is ferocious and cruel; his eye is restless, his countenance forbidding, and all his motions short and precipitate. In his general habits he resembles the Panther, lying in ambush for prey, and then springing upon, and devouring almost every species of animal which he has strength to overcome. Occasionally Leopards have been known to congregate in large numbers, and, descending from their lurking places, to commit dreadful slaughter among the numerous herds of cattle which graze in the plains of Senegal and Guinea.

The ordinary size of the Leopard is four feet in length, exclusive of the tail, which is about two feet. He is distinguished from the Panther by the regularity of the spots which ornament his skin; these are always disposed in circles of from three to four, inclosing a central area of about an inch in extent, and of a somewhat deeper hue than the general ground-colour of the animal.

Fig. 7. The Ocelot (*F. Pardalis*).—This animal is one of the most beautiful of its tribe; it is about three feet in length, and in height about eighteen inches. There are several varieties of this species, remarkable for their great beauty and elegance of form, but all differing in the intensity of colouring in the fur, and also in the arrangement of the spots. They are natives of South America, and frequent the depths of the forests, where they prey on birds and the smaller quadrupeds. The disposition of the Ocelot is highly predatory, and less susceptible of the mild influence of domestication than most others of its congeners.

Fig. 8. The Sumatra Cat (*Felis Sumatrana*).—This is the Rimau Bulu of Sumatra, about the size of the Kowuk, or somewhat less than the Ocelot. The spots on its skin are very irregular, both in disposition and shape.

Fig. 9. The Neuwied Cat (*F. Macrourus*).—The ground-colour of this species is of an ochre-gray, with longitudinal patches, the upper part of the tail being only partly annulated. It is somewhat about the size of the Ocelot, but longer in the legs. It inhabits Brazil.

ANIMALCULES IN SNOW.—Dr J. E. Mure communicated the following information to Dr Silliman :—" When the winter had made considerable progress, without much frost, there happened a heavy fall of snow. Apprehending that I might not have an opportunity of filling my house with ice, I threw in snow, perhaps enough to fill it. There was afterwards severely cold weather, and I filled the remainder with ice. About August, the waste and consumption of the ice brought us down to the snow, when it was discovered that a glass of water which was cooled with it contained hundreds of animalcules. I then examined another glass of water out of the same pitcher, and, with the aid of a microscope, before the snow was put into it, found it perfectly clear and pure; the snow was then thrown into it, and, on solution, this water again exhibited the same phenomenon—hundreds of animalcules, visible to the naked eye with acute attention; and, when viewed through the microscope, resembling most diminutive shrimps, and, wholly unlike the eels discovered in the acetous acid, were seen in the full enjoyment of animated nature. I caused holes to be dug in several parts of the mass of snow in the ice-house, and to the centre of it, and, in the most unequivocal and repeated experiments, had similar results; so that my family did not again venture to introduce the snow into the water they drank. These little animals may class with the amphibia which have cold blood, and are generally capable, in a low temperature, of a torpid state of existence. Hence, their icy immersion did no violence to their constitution, and the possibility of their revival by heat is well sustained by analogy: but their generation, their parentage, and their extraordinary transmigration, are to me objects of profound astonishment."

ON THE AUTHENTICITY OF THE DODO.

Perhaps there is no Vertebrated animal, which has existed since the wreck of the former world, whose history is involved in such obscurity as that of the Dodo. Many have doubted that there ever was such a bird; but we hope to show that this is a mistake, although there is every reason to believe that it is now extinct.

This bird has been variously designated by Naturalists, as the Dodar, Didus, and Dodo; but the accounts of its earliest describers are so ambiguous, and their characters so ill defined, that there is much difficulty in tracing their specific distinctions.

There appear to be three distinct representations of this species, which have not been copied from each other; and although two of these are sufficiently rude, yet they bear evident traces of originality, and possess characters so peculiar, that their identity cannot be mistaken.

The above figure is taken from a plate in the "*Exotica*" of Clusius, published in 1605. He says it was copied from a rough sketch in the journal of a Dutch voyager, who had seen the bird in the Moluccas, in the year 1598. Clusius says, that he had himself seen only a leg of the Dodo, in the house of Peter Pauw, a professor of medicine at Leyden, which was brought from Mauritius. Clusius calls this bird "*Gallus gallinaceus peregrinus*," and mentions that the Dutch sailors called it Walgh-Vogel, "*Nauseam Movens avis*."

We copy the above from the travels of Herbert in Africa, Asia, and other places, published in 1634. It differs from the former in the shape of the bill, but possesses all the other characters sufficiently marked.

The third, and most perfect, representation of the Dodo, is taken from the "*Historia Naturalis et Medica India Orientalis*," by Jacob Bontius, which appeared in 1658. Ray, who published in 1676 and 1688 an edition of Willughby's Ornithology,

after quoting the accounts given of this kind by Clusius and Bontius, says, "We have seen this bird dried, or its skin stuffed, in Tradescant's cabinet," at Lambeth.

We now give a copy of the painting of the Dodo in the British Museum, which Edwards faithfully imitated in his "History of Uncommon Birds," plate 294, published in 1760. He says, "The original picture was drawn in Holland, from a living bird brought from St Maurice's Island, in the East Indies, in the early times of the discovery of the Indies, by the way of the Cape of Good Hope. It was the property of the late Sir Hans Sloane to the time of his death, and afterwards becoming my property, I deposited it in the British Museum, as a great curiosity. The above history of the picture I had from Sir Hans Sloane and the late Dr Mortimer, Secretary to the Royal Society."

There seems to be pretty clear evidence that an entire specimen of this bird was in the Museum of John Tradescant. It is mentioned in his printed catalogue of Stuffed Skins of Birds, in "Section 5, *Whole Birds*—Dodar, from the island Mauritius; it is not able to flie, being so big." This specimen was afterwards exhibited in the Ashmolean Museum, and is particularly alluded to by Hyde in his "*Religionis Veterum Persarum Historia*," printed in 1700, who states it to be then existing in the Museum at Oxford. It was destroyed at a later period than 1755, by order of the Visitors, in consequence of its state of decay. In a Catalogue of that Museum made subsequently to 1755, it is recorded that "the Nos. from 5 to 46, being decayed, were ordered to be removed at a meeting of the majority of the Visitors, January 8, 1755." The Dodo was one of these, as it stood No. 29 of the Catalogue, under the name of "*Gallus gallinaceus peregrinus Chusii*," &c. That there was such a bird in the Museum of Tradescant, there can be no doubt, as the head is still preserved in the Ashmolean Museum, and one of the feet in the British Museum, which are both well figured in Shaw's Naturalist's Miscellany,—the former in plate 166, and the latter in plate 143. We have carefully examined the latter, and from its construction, we would at once say that it is formed for walking, as the articulation of the hallux or hind toe is not constructed for grasping.

It will be observed, that our fourth figure exactly agrees with that of Bontius, figure 3d; and when we consider that the first and second figures were taken from rude sketches by travellers unacquainted with the art of drawing, we cannot doubt but that they are intended to represent the same bird. The general structure of the head (particularly its hooded appearance), the bill, and the curved and swelling neck, the rounded and clumsy shape of the body, terminating with a curious tuft of feathers on the rump, the short and clumsy legs, and feet with divided toes, all bear striking evidence of their being pictures of the same animal.

The testimonies of the existence of the Dodo are the following, which we arrange in chronological order:—

In the year 1497, Vasco de Gama, after doubling the Cape of Good Hope, discovered a bay at 60 leagues, Augra de San Blaz, near to an island, where he saw a number of birds which the Portuguese called Solitaries; they were of the form of geese, with wings like bats. On the return of the voyagers, in 1499, they landed and captured a number of these birds.

Castleton informs us, that during his voyage of 1614, he landed at the island of Bourbon, then "uninhabited, although occasionally visited by the early voyagers. Among the birds, he particularizes a kind of bird the size of a goose, very fat, with short wings, which do not permit them to fly. They have since been called the Giant, and the Isle of France also produces plenty of them. They are white, and naturally so tame as to allow themselves to be taken by the hand; or, at least, they were so little afraid at the sight of the sailors, it was easy for them to kill great numbers with sticks and stones."

In the year 1691, Leguat, with seven other individuals, was left upon the island of Rodrigos, with a view to colonize it; and so much struck was he with the appearance and habits of the Dodo, that he not only introduced figures of it into the frontispiece of his work, but also into his general chart of the island, and his plan of the small colony which was formed. No fewer than sixteen figures of the birds were introduced into the former, and twelve into the plan. He gives the following description of the Dodo:—

"Of all birds which inhabit the island, the most remarkable is that which has been designated Solitaire (the Solitary), as they are seldom seen in flocks, although there is abundance of them.

"The plumage of the male is generally grayish, or brown, with feet formed like those of a turkey-cock, as is also the bill, but a little more hooked. They have scarcely any tail, and their rump, covered with feathers, is as much rounded as the croup of a horse. They are higher than the turkey-cock, and have a straight neck, a little longer than that of the turkey-cock, when it raises its head. The eye is black and lively, and the head destitute of crest or tuft. They do not fly, their wings being too short to sustain the weight of their bodies; they are only used in beating their sides, and in whirling round. When they wish to call one another, they make, with rapidity, twenty or thirty rounds in the same direction, during the space of four or five minutes; in which action the noise made by their wings resembles that made by the Kestrel, and can be distinctly heard at 200 paces distant. The bone of the spurious wing enlarged at its point, forms, under the feathers, a little round mass like a musket bullet; and this and their bill are the principal weapons possessed by these birds. It is excessively difficult to catch them in the woods; but as a man runs swifter than they do in open places, it is not difficult to catch them; sometimes, indeed, they may be very easily approached. From the month of March until September, they are exceedingly fat, and their flesh of a very agreeable flavour, especially when young. The average weight of the males is about 45 lbs., and Herbert says he has known them of 50 lbs.

"The female is of admirable beauty. Some are of a blond, others of a brown colour. I mean by blond, the hue of flaxen hair. They have a band, not unlike the bandeau of a widow, above the beak, which is of a tan colour. One feather does not pass another over all the body, because they adjust and polish them with their bill; the feathers of the thighs are rounded and of a shell shape, and being very thickly clothed in these situations, produce an agreeable effect. Over the crop, they have two elevations, of whiter plumage than the rest of their bodies, which greatly resemble the female bosom. They walk with a graceful and stately air, which excites admiration, and even a love of the bird, which has frequently saved their lives."

Leguat says that they are incapable of being tamed, and, if taken, refuse all food, and die of hunger. This may account for their extirpation, for had they been susceptible of domestication, a bird so fitted for human food might have been widely spread over Europe at the present day. A remarkable peculiarity is stated by Leguat, that "there is always found in their gizzard a brown stone, the size of a hen's egg, slightly tuberculated, flat on one side, and rounded on the other, very heavy and hard. We supposed this stone to exist in the bird when hatched, because, however young they might be, they always had it, and never more than one; and besides this circumstance, the canal which passes from the crop to the gizzard is too small by one half to permit the passage of such a mass. We used them, in preference to all other stones, for sharping our knives."

It seems to be certain that this bird became extinct towards the end of the seventeenth or beginning of the eighteenth century. Mr John V. Thompson of Cork, a zealous naturalist, was unable to discover any traces of it during a recent residence in the East. He says, "Having resided some years amongst these islands, inclusive of Madagascar, and being curious to find whether any testimony could be obtained on the spot, as to the existence of the Dodo, in any of the islands of this or the neighbouring Archipelagos, I may venture to say, that no traces of any kind could be found, no more than of the truth of the beautiful tale of *Paul and Virginia*, although a very general belief prevailed as to both the one and the other."

At a late meeting of the Paris Academy of Sciences, the celebrated comparative anatomist, M. De Blainville, gave it as his opinion, that the extinct bird, the Dodo, was a large species of Vulture. He stated that he came to this conclusion, after a careful examination of a plaster cast, taken from the head of the Dodo, which is preserved in the Ashmolean Museum at Oxford. Cuvier seems not decided as to the place it should occupy in his system, but has placed it after the Cassowary of New Holland, and before the Bustards. The mutilated state of the bill prevented him from ascertaining with certainty its true character.

We cannot agree with De Blainville in the opinion which he entertains, as we think the extreme shortness of its wings, and consequent incapability of flight, must remove it from the Predatory order; and we are inclined to the opinion, that it must be nearly allied to the Gallinaceous order, from the construction of the feet, one of which we carefully examined in the British Museum.

MINERALOGY.

Gold Veins in North Carolina.—At a meeting of the Sheffield Literary and Philosophical Society, Dr Longstaff, who has been out, during the last twelve months, as the agent of a Company of British Mine Adventurers, to investigate the gold veins of North Carolina, stated that the gold region stretches from the shores of the Atlantic, in the direction of Carolina, through the country towards the Pacific Ocean; and that, judging from appearances, this immense tract promises to yield supplies such as have not been equalled by the most famous gold countries of antiquity. The precious metal is generally found in a matrix of quartz, and in veins, often running in the direction of N.E. and S.W., there being generally one leading vein, and on each side a parallel satellite. In some cases, rich branches pass off at right angles; or, in others, the ore is ramified in every way. It is sometimes enveloped in a rake of talcose slate, passing through the auriferous quartz; in other instances, disseminated in minute particles through oxide of iron; and contrary to what might have been supposed, judging from the effect of other metals, the sulphuret of iron, or martial pyrites, usually indicates a rich locality. The proportions of the precious metal to the quartzoze, or other matrix, are amazingly great, the minimum yield of the ore affording a large profit upon the capital invested, while some of the richer sorts (of which Dr Longstaff laid specimens on the table) gave almost incredible results.

Many of the inhabitants of Concord have pieces of pure gold of various weights, one of which weighs 28 lbs. The beds where the gold is discovered, in that locality, are of gravel, and very extensive, covered with water in the winter months, but dry in summer. The manner of searching for gold is, to take shovels and turn over the gravel, always advancing as it is turned back, and picking up what is discoverable to the eye, by which thousands of small grains are lost, as the fingers cannot separate them from the sand. By working this over again with quicksilver, large quantities may be obtained. No machinery is required, or smelting process. The first mine was found by a son of Mr Reed, who, in watering his horse at the creek, discovered a piece of gold quite pure. Two years after, Mr Reed, with two partners, pursued the search for gold, with six black boys, during the short period of only six weeks. In each of the two first years they obtained the value of 17,000 dollars, besides what was stolen from the streams, supposed to be half as much more. No attempt has been made to open the hills, as the persons there are totally unacquainted with the subject of mining. Messrs Morton and Bedford, of Baltimore, purchased a small tract of about 300 acres, joining the lower end of Reed's purchase and mine, for which they paid seven dollars an acre. Governor Mercer stated, that they had analyzed the sand and gravel, and found it worth a guinea a bushel, after the lump gold was picked out. The gold, as found, is worth 19 dollars an ounce, while the best East India and African gold dust is not worth more than from 12 to 16 dollars. Mr Thomas Moore got some hiccory nuts, and in looking for a stone to break the shell, he went to a tree that had been blown down, and picking up the first stone he met with in the fresh turned-up earth, perceiving it heavy, he washed it, and it turned out to be a piece of solid gold, which he sold for 450 dollars! He then set some men to work, and they made from two to five dollars a-day each. Some of it has been sent to the mint of the United States, where they exchange it for eagles ready coined, weight for weight; but the gold-beaters give a still better price, namely, four per cent., it is so pure and malleable.

Discovery of Coal on Mount Lebanon.—A bed of coal has been recently discovered at Carnayl, on Mount Lebanon, and the agents of Mahomet Ali, under the guidance of an English gentleman, are exploring it with all the energy which the nature of the country admits. It is about three miles north of the great road leading from Beirout to Damascus, and about 18 miles from the former city. It is the black bituminous coal, and burns readily, with a clear yellow flame.

GEOLOGY AND PHYSICAL GEOGRAPHY.

Temperature of the Earth.—For the purpose of ascertaining whether a constant stream of water could be obtained by means of an Artesian well, sunk on the south side of the Jura mountains, at the distance of about a league from Geneva, and at an elevation of about 297 feet above the level of the lake, M. Girond, at his country residence at Pregny, bored to the depth of 547 feet without any result. Despairing of success, he offered great facilities to any persons who might wish to prosecute the enterprise, for the purpose of scientific inquiry. On this occasion, MM. Aug. de la Rive and F. Marcet made a successful application to the friends of science, and also to the Government, and funds were obtained sufficient to enable them to continue the operation during eight months, and to extend the boring to the depth of 682 feet. The hole bored was about four and a half inches in diameter. Water began to appear in it at the depth of twenty feet; and, it is worthy of remark, that the height at which the water stood in the opening, as measured from the surface, was lower when the greatest depth was obtained than it was at half the depth. At 275 feet of depth, the water stood at 14 feet from the surface; at 500 feet, it sunk to 22 feet; at 550 feet, to 35 feet. It then rose—at 595 feet it stood at 24 feet 6 inches, but at 675 it again sunk to 35 feet 8 inches. Having attained the extraordinary depth above mentioned, the experimenters devised the means of ascertaining the temperature of this opening at different depths. As the common thermometer would not answer the purpose, they contrived a self-registering thermometer, constructed on a large scale, and whose accuracy was subjected to the most satisfactory tests. The following table exhibits the temperature of the bore hole at the depths specified:—

Depths below the surface in feet.	*Corresponding Temperature.* Reaumur.	Fahrenheit.
30	8.4	50.9
60	8.5	51.1
100	8.8	51.8
150	9.2	52.7
200	9.5	53.4
250	10.0	54.5
300	10.5	55.6
350	10.9	56.5
400	11.37	57.58
450	11.73	58.39
500	12.20	59.45
550	12.63	60.42
600	13.05	61.36
650	13.50	62.37
680	13.80	63.05

It thus appears that the increase of temperature below the depth of 100 feet from the surface, as far down as 680 feet, is precisely 0,875 of Reaumur (= 1.968 or 2 Fahrenheit very nearly) for every hundred feet. It will be observed that the increase, instead of moving *per saltum*, as in some other cases, moves with remarkable uniformity. This, the experimenters think, may be owing to the care which was taken in this case to remove and avoid every source of error.

Fossil Ferns are rarely found, but numerous impressions of them are met with in various countries. From the fact of the plants themselves not being met with in a mineralized state, there has been much difficulty in ascertaining their analogy with those now existing on the earth's surface. M. Gœpper has lately succeeded in ascertaining upwards of thirty species, which are analogous to those of the present day. His mode of investigation is, by taking impressions from recent ferns, which affords him a ready and easy mode of comparison.

Change of Currents.—The following singular fact was elicited during the examination of Captain Fitzroy, of the Beagle surveying ship, upon the recent naval court-martial, held at Portsmouth:—He stated that the late earthquakes in the western coast of South America have had the extraordinary effect of transforming what was once a current of two miles an hour to the northward, into a current of five miles an hour to the southward, and that the soundings along the whole coast have been materially changed. Since the middle of February last, not a day has intervened without a motion of the earth having been experienced in one quarter or another. It appears that the loss of the Challenger was occasioned by the ship being thus set, by a universal and unexpected current, 34 miles of latitude to the southward, between noon of the 17th of May last, to the time of the wreck, on the 19th of the same month. This latitude, by dead reckoning up to the time of taking the sights, was used to work the sights of the chronometer, and accordingly, on the morning of the 19th May, *the ship's situation was estimated* 60 *miles to the south-west of her actual position at that time.*

GENERAL SCIENCE.

SATELLITES OF JUPITER.—At a meeting of the Paris Academy of Sciences, in consequence of M. Paravey having asserted that the ancients had discovered some of the satellites of Jupiter, M. Arago endeavoured to ascertain if it were possible for him to observe any of these satellites with the aid of a magnifying lens, using only one that was darkened, in order to obscure the radiations. The experiment proved abortive, and it was in consequence to be repeated, as the moon at the time was above the horizon. It was suggested by M. Ampére, that it would require a peculiar organization in the visual organs to enable an observer to perceive the satellites without a telescope.

THE COMET OF HALLEY.—The following particulars respecting the orbit of this comet are contained in *Le Voleur*:—One of the most remarkable circumstances connected with this comet, is the size of its orbit. It is an elongated oval, the total length of which is about thirty-six times the distance of the sun from the earth, and the greatest breadth of the oval is about ten times that distance. The nearest extremity of its orbit is distant from the sun about half the distance of the earth from that planet; and its most distant extremity is thirty-five times and a half that of the earth from the sun. As the light and heat derived from the sun naturally decrease according to distance, hence it follows that the heat and light from the sun will be, at the most proximate point of the comet, four times more than on the earth, and at its most distant extremity 5000 times less than at the opposite point. At one of the extremities of the comet, the sun would appear four times larger than it does to our earth, while at the other its size would look like that of a star. The vicissitudes of temperature resulting from these positions are evident. Suppose the earth to be transported to the most distant extremity of the comet, liquids would congeal, and it is probable that atmospheric air, and all the permanent gases, would become liquid. On the contrary, were the earth at that part of the comet nearest to the sun, liquids would assume a gaseous form; metals would be liquified, would form a new ocean, and would occupy in the bed of the present one the place of the waters, which had become vapour."

MISCELLANEOUS.

AMBER.—This remarkable substance is found in Prussia, and has for more than three thousand years excited the curiosity of naturalists and the avidity of traders. It is uncertain even at present, whether it belongs to the animal or to the vegetable kingdom: almost all writers agree, that it forms no part of the mineral. It is a sort of solid bitumen, very light, of a vitreous fracture, and generally of a milky white, or yellow colour; although it is sometimes found brown or black, and sometimes quite opaque. It is combustible, evaporates, and diffuses an agreeable odour. *Succin* is that sort which is most crystallized and transparent; and what the Prussians term amber has a less vitreous fracture, and a more earthy appearance. From Ηλεκτρον (Electron), the Greek word for amber, is derived the term Electricity; so that an insignificant fossil has, from its power of attracting light bodies, given its name to the cause of the most imposing and terrible phenomena in nature,—

"The lightning's lurid glare, and thunder's awful crash."

Heinitz supposes that its formation must be attributed to forests submerged by the ocean, and afterwards covered with sand; the resinous particles, being distilled into amber, and the rest of the wood forming a residuum, or *caput mortuum*: and what strengthens this supposition is, that wood is generally found near it, which renders its vegetable origin probable. The supposition of its mineral origin is disproved by distillation, and by the foreign bodies found in its substance.

M. Schweigger, an eminent entomologist, has carefully examined the insects contained in the amber; and he has found that many of them would belong to genera of insects now existing, but that none of them were specifically the same. Professor Germar of Halle has been occupied in a similar investigation, and he also thinks that none of them are identical with analogous species now living.

M. Girtanner affirms, that amber is formed by a large ant (the *Formica rufa* of Linnæus); he conceives it to be a vegetable oil, rendered concrete by the acid of those animals, which inhabit old forests of fir trees where the fossil amber is found. The amber, when first dug, is ductile like wax, and becomes hard on exposure to air. Certain it is, that no insect is so commonly found in amber as the ant. Wallerius asserts, that the black and dark-coloured amber is often found in the bowels of cetaceous fishes. Others imagine that it is produced by a fish or an aquatic animal. It is certain, however, that amber must have passed from the fluid to the solid state; for foreign substances, such as leaves, insects, small fish, frogs, water, pieces of wood and straw, are often contained in it; and it is most esteemed when it contains any of these substances.

The Phœnicians were the first who navigated the North Seas in search of this substance. By the ancients it was considered as valuable as gold and precious stones. Its value, at present, is much diminished, though it is still required in some manufactories; for, at Stolpe in Pomerania, and Kœnigsberg in Prussia, workmen are employed in making from it small jewels, scented powder, spiritous acid, and a fine oil, that is used as a varnish. Amber is exported to Denmark and Italy, but Turkey is the chief market for the commodity; and a certain portion of it is carried every year to the Holy Kaaba at Mecca. This substance has long been regarded with superstitious veneration by several of the northern nations of Europe, as well as in Asia Minor; but what gave rise to this we have not been able to trace. Among the peasantry of Scotland amber beads have long been held as a complete antidote to the effects of witchcraft; and, in consequence, one or more beads of it were very commonly carried in the pocket: but, that it might have complete efficacy, it was considered necessary, that it should be accompanied by the following couplet, written on paper, wrapped round the bead, and secured by a *red* silk thread:—

"Lammar (amber) beads and red thread
Keep the witches at their speed."

A twig of the mountain ash, or rowan tree, was supposed to have precisely the same effect. Among the higher classes in Scotland, in former times, amber beads were much worn, and were always strung with red silk thread.

The quantity of amber annually found in Prussia amounts to more than two hundred tons, and the revenue derived from it by the crown is three or four thousand pounds.

Amber is obtained on the Prussian coast, between Pillau and Palmnicken, a tract of land about eighteen miles in length; and sometimes upon the surface of the water, where it is collected by means of nets. It is, however, only after violent north and north-west winds that any large quantity is drawn to the shore. Quarries, or pits, have been opened at Dirschkemen, on the hills near the coast, and their produce is less variable. In digging for it, the first stratum is found to be sand, then clay, then a layer of branches and trunks of trees, then a considerable quantity of pyrites, whence sulphuric acid is prepared, and lastly a bed of sand, through which the amber is dispersed in small pieces, or collected together in heaps. It assumes various shapes, as that of a pea, an almond, a pear, and letters, very well formed; and even Hebrew and Arabic characters. It is found in other places in the interior of Prussia; and the largest piece of amber which has yet been seen was found at Schleppacken, about twelve German miles from the Lithuanian frontier. It is fifteen inches in length, and seven or eight in breadth; it may be seen in the Museum at Berlin. Amber is also to be found in the high hills of Goldapp, seventy-five miles to the south-east of Kœnigsberg, and in the heights and valleys on the Vistula, in the neighbourhood of Thorn and Graudenz. A large piece of amber was cast ashore, about twenty-five years ago, at Peterhead, county of Aberdeen, in Scotland.

The most remarkable properties of amber are, that, being rubbed, it attracts light bodies. The friction which elicits the electric fluid also renders amber visible in the dark. Dr Wall remarked, that by rubbing amber upon a woollen substance in the dark, light was also produced in considerable quantities, accompanied with a crackling noise; and, what is still more extraordinary, he adds, "This light and crackling seems in some degree to represent thunder and lightning."

AMBERGRIS is a substance much of the same nature as amber, but differs from it by its particular consistence, which nearly approaches to that of bees' wax; sometimes it is granulated, and appears opaque, or of a dark gray. Experiments prove that it resembles amber in its nature. When analyzed, it is found to consist of phlegm, a volatile acid partly fluid, oil, and a little coaly matter. It dissolves more readily than amber in spirit of wine. It is most common in the Indian seas, on the eastern coast of Africa, Madagascar, &c. and it is found either floating on the sea, or cast on the sea shore. In this substance, animal and vegetable remains are sometimes found, as, for instance, the parts of birds. The origin of ambergris is probably the same with that of amber. According to M. Aublet (in his *Histoire de la Guiane*), it is nothing more than the juice of a tree, hardened by evaporation; and if this be true, it is a substance which belongs properly to the vegetable kingdom. The tree which is said to produce it grows in Guiana; it is called *cuma*, but has not been examined by other botanists. When a branch is broken by high winds, a large quantity of the juice exudes; and if it chance to have time to dry, various masses (some of which have been so large as to weigh one thousand two hundred pounds, and more) are carried into the rivers by heavy rains, and through them into the sea; afterwards they are either thrown on the shore, or eaten by fish, chiefly by the spermaceti whale (*Physeter macrocephalus.*) This fish swallows such large quantities of this gum resin, that it generally becomes sick, so that those employed in the catching of these whales always expect to find some ambergris in the bowels of the lean whales. Father Santes, who travelled to various places on the African coast, says, in his *Æthiopia Orientalis*, that some species of birds, of whales, and of fish, are fond of eating this substance; and the same assertion has been made by Bomare and various other authors. This accounts for the claws, beaks, bones, and feathers of birds, parts of vegetables, shells and bones of fish, and particularly for the beaks of the cuttle-fish (*Sepia octopodia*), which are sometimes found in the masses of this substance. M. Aublet brought specimens of this gum resin, which he collected on the spot, from the cuma tree at Guiana. It is of a whitish-brown colour, with a shade of yellow; while it melts and turns like wax in the fire. M. Pouelle examined very carefully this substance brought over by M. Aublet, and found that it produced exactly the same results as amber. These observations seem to place it beyond a doubt, that both amber and ambergris are vegetable products, and that naturalists were mistaken in supposing these substances to be of an animal nature, from having found them in the intestines of whales.

AURORA BOREALIS.—Sir John Ross states, that during his first Arctic expedition the Aurora Borealis sometimes appeared between the two ships, and also between the ships and the icebergs; and found, in his subsequent experience, both in Scotland and during his second voyage, proofs, *satisfactory to his own mind*, that the Aurora takes place within the cloudy regions of the earth's atmosphere. Under this belief, he founds the following extraordinary hypothesis on the subject:—"The Aurora is entirely occasioned by the action of the sun's rays upon the vast body of icy and snowy plains and mountains which surround the poles."

NEWTONIAN SYSTEM.—Mr Walsh, of Cork, addressed a pamphlet to the French Academy, entitled, "Appendix, containing some remarks, and a new theory of Physical Astronomy." In one of the author's marginal notes, he states, that, barring astrology, the greatest absurdity ever propagated is that of the Newtonian System!!"

Edinburgh: Published for the PROPRIETORS, at their Office, 16, Hanover Street. London: CHARLES TILT, Fleet Street. Dublin: W. F. WAKEMAN, 9, D'Olier Street. Glasgow and the West of Scotland: JOHN SMITH and SON, 70, St Vincent Street.

THE EDINBURGH PRINTING COMPANY.

THE EDINBURGH

JOURNAL OF NATURAL HISTORY,

AND OF

THE PHYSICAL SCIENCES.

N°. 7.   SATURDAY, JANUARY 16, 1836.

ZOOLOGY.

DESCRIPTION OF THE PLATE—CAMELS.

THE only two animals which constitute this genus are in general so mild and inoffensive in their disposition, that they prove extensively serviceable to mankind in these hot and sandy regions, where they are employed as beasts of burthen. Their pace is usually slow; but being able to sustain themselves, even on the longest journeys, with a very small portion of food, and but little water, they undergo fatigues which few, perhaps no other animals, could endure.

Their hair is a valuable article of commerce, and their flesh forms a palatable food. Like all, the other genera of the ruminating order, they are provided with four stomachs, in consequence of which they not only live solely on vegetable food, but ruminate or chew the cud. The food, after being swallowed without undergoing the process of mastication, is received into the first stomach, where it remains for some time to macerate; and afterwards, when the animal is at rest, by a peculiar action of the muscles, it is returned to the mouth in small quantities, chewed more fully, and then swallowed a second time for digestion.

Fig. 1. The Bactrian Camel *(Camelus Bactrianus, Linn.)*—This animal is at once distinguished from its congener the Dromedary, by having two haunches, one of which is situate on its shoulders, and the other at a little distance behind. It is a considerably-larger animal than the other; its legs are proportionally shorter, while the body is longer. An animal of this species, which was exhibited in London in 1829, measured eight feet from the part of the back between the humps to the ground.

Its original country is supposed to be the ancient Bactriana, now called Turkestan. The species has been spread over Persia, Thibet, and China. It is, however, rarely to be met with except in the great middle zone of Asia, to the north of Tsurus, and the great Himalaya range of mountains. It is capable of bearing a much colder and more moist climate than the Arabian species. The Camel has been known to exist in the neighbourhood of Lake Baikal, in Siberia, where its only food, during winter, was the bark and tender branches of the birch.

The Camel lives to a great age. One which was kept in the Menagerie of the *Jardin des Plantes* at Paris was supposed to be nearly fifty years old when it died. The food of this animal is hay or lucern, of which it consumes about thirty pounds a day; while a draught of six gallons of water will suffice for some days.

Fig. 2. The Dromedary *(Camelus Dromedarius, Linn.)*—The Arabian Camel or Dromedary has but one haunch, situate on the middle of the back. Its height at the shoulder varies from five to seven feet. Its hair, which is soft, woolly, and unequal, is longer on the nape of the neck, throat, and haunch, than on any other parts of the body; it is of a pale reddish-fawn colour.

The feet of both the Camel and Dromedary are very singular productions of Nature, being admirably fitted for treading on a smooth or soft surface. They are divided into two toes not separate, each covered with a broad nail; and thus the feet are intermediate between the hoofs of a horse and cloven feet. From the heel forwards, they are protected by a horny sole, uniting the middle part, and leaving the toes free. This sole is part of an elastic substance, which, being bedded in two cavities of the foot, yields to the pressure of the soil; whilst the toes spread open, touching the ground, in the same way as the foot of the Rein-Deer extends itself to present a large surface to the snow. By this formation, the Camel is prevented from sinking into the soft sand of the desert tracts, which it is so frequently obliged to cross.

ANTS.—Colonel Sykes relates an anecdote with regard to an Indian species of Ant, which he calls the *Large Black Ant,* instancing, in a wonderful manner, their perseverance in attaining a favourite object. This was witnessed by himself, his lady, and his whole household. When resident at Poonah, the dessert, consisting of fruits, cakes, and various preserves, always remained upon a small side table, in a verandah of the dining-room. To guard against inroads, the legs of the table were immersed in four basins filled with water, and, to keep off dust through open windows, was covered with a table-cloth. At first the ants did not attempt to cross the water, but as the strait was very narrow, from an inch to an inch and a half, amid the sweets were very tempting, they appear at length to have braved all risks, to have committed themselves to the deep, to have scrambled across the channel, and to have reached the object of their desires, for hundreds were found every morning revelling in enjoyment. Daily vengeance was executed upon them without lessening their numbers; at last, the legs of the table were painted, just above the water, with a circle of turpentine. This at first seemed to prove an effectual barrier, and for some days the sweets were unmolested, after which, they were again attacked by these resolute plunderers; but how they got at them seemed totally

unaccountable, till Colonel Sykes, who often passed the table, was surprised to see an ant drop from the wall, about a foot above the table, upon the cloth that covered it; another, and another succeeded. So that though the turpentine and the distance from the wall appeared effectual barriers, still the resources of the animal, when determined to carry its point, were not exhausted, and by ascending the wall to a certain height, with a slight effort against it, in falling managed to land in safety upon the table.

ON THE TAIL-GLAND IN BIRDS.—BY M. REAUMUR.

"WERE I tempted to explain why the hinder part of the hens without tails has not a secretion performed in it like that which is observed in other hens and in other kinds of birds—were I tempted, I say, to explain it, I should be aware of the danger of the possibility of committing mistakes, by the very obligation I think myself under of exposing, as an error, the notion which naturalists and philosophers have framed to themselves concerning the utility of the unctuous liquor that issues from the canals in the tail of birds. All the works of Nature being lavishly filled with wondrous characteristics, fit to raise in us a most just admiration of those who, from the best intentions, expose them to our eyes, in order to force us to acknowledge the AUTHOR of them, are, on account of the multitude of those wonders, liable to some reproach, when they happen to mention among them some that are not of the utmost certainty. They all have been of opinion, that the feathers of birds, in order to be sheltered against rain, wanted to be done over with a kind of oil or grease, that might cause the water to run off them without penetrating, and that this unction wanted to be repeated from time to time. I have elsewhere proved, in a memoir on feathers, that they have been wrong to entertain that notion. In consequence of it, they have pretended to make us admire a reservoir of unctuous matter placed in the hinder part of each bird, out of which he expresses, and takes it, with the end of his bill, to convey and spread it all over the feathers that want it.

"I shall not undertake to shew here how little the quantity of matter that may be daily supplied by that reservoir is in proportion to the extent of the surfaces resulting from the assemblage of the numberless feathers with which a hen or a duck is covered, nor how long a time would be necessary to enable the reservoir to supply a quantity of the said matter, sufficient to besmear the surface of only one of those feathers. In order to explode a notion that must needs have been pleasing, since it was universally espoused, I need only say that the feathers of our tail-less, or as I call them, rump-less hens, are as much proof against rain as those of other hens, and of many other birds that are provided with that part in which the secretion of an unctuous matter is made. It is, however, a fact that birds are sometimes seen pecking this part of their body; and this circumstance has been considered by the observers as conclusive evidence that they squeezed, from their tail-gland, the unctuous matter which was afterwards to be applied to their plumage; and in this hasty conclusion they forgot that the bill of the bird was insufficient to convey so much unctuous matter as was necessary to besmear their entire plumage, and render it greasy. A more natural idea would have been to suppose, that the bird pressed this reservoir or extraordinary canal to relieve an irritation of the skin, caused, in all probability, by the matter becoming too thick to flow, in its usual manner, through so small an orifice. Even school-boys are aware that an obstruction in this vessel occasionally takes place, and produces sickness in the birds; for their sparrows look poorly and droop when so afflicted, and almost the first thing they do, is to examine the state of the tail-gland; and when they think that it presents an unusually swelled appearance they press it, and even sometimes prick it with a needle, to allow the thickened matter to escape, so as to force it out. I do not know whether this operation is always attended with complete success, but I would conceive it better to endeavour to cure this obstruction in the excretory canal, produced by the inspissated unction, to moisten, or introduce into it some small solid body. So long as we remain in uncertainty why a secretion of a certain matter takes place in our ears, though in a very small quantity, we shall not think it incumbent on us to account for, how the secretion of a particular matter is effected in the tail-glands of birds."

In narrating the diseases of Birds, Dr Bechstein remarks, that "this gland, which contains the oil necessary for anointing the feathers, sometimes becomes hard and inflamed, and an abscess forms there. In this case the bird frequently pierces it itself. It may be softened by applying fresh butter without any salt; but it is better to use an ointment made of white lead, litharge, wax, and olive oil. The general method is to pierce or cut the hardened gland; but if this operation remove the obstruction, it also destroys the gland, and the bird will die in the next moulting for want of oil to soften the feathers."

When the gland is obstructed, the feathers which surround it are ruffled; and the bird is constantly pecking and adjusting them, and instead of being of a yellow, which is its natural colour, it becomes brown. This complaint is extremely rare among wild birds, for, being exposed to damp, and bathing often, they make more use of the oil in the gland, consequently it does not accumulate sufficiently to become inconvenient. This shows the necessity of providing birds, in a state of confinement, with water at all times for bathing, as nothing can be more conducive to their health. Dr Handel recommends that after the gland has been pierced, a little magnesia should be mixed with their drink.

MATERNAL SOLICITUDE.—A Cat, domesticated at a farm-house about a mile from Lanark, had numerous families, which were invariably taken away from her and drowned. At length the creature adopted an expedient for saving at least a portion of her offspring. She took away two or three of the kittens to a farm about half a mile distant, where she secluded them in the byre. As many, if not more, were deliberately left in her usual layer, and these she partially attended to, till they were taken away and drowned as usual. She then devoted the whole of her attention to the few which she had rescued, and thus succeeded in bringing them up. It was evident to the individuals under whose notice the circumstances occurred, that she left the few at home as a decoy, in order to lead away attention from the rest, as otherwise a search would have been made, which would probably have proved fatal.

WASPS.—Lieutenant Holman, in whom the loss of sight has been compensated by a wonderful acuteness of mental vision, relates the following anecdote in the second volume of his Travels:—"Eight miles from Grandie, the muleteers suddenly called out, 'Marambundas! Marambundas!' which indicated the approach of a host of *wasps*. In a moment all the animals, whether loaded or otherwise, lay down on their backs, kicking most violently, while the blacks, and all persons not already attacked, ran away in different directions, all being careful, by a wide sweep, to avoid the swarms of tormentors that come forward like a cloud. I never witnessed a panic so sudden and complete, and really believe that the bursting of a water-spout could hardly have produced more commotion. However, it must be confessed that the alarm was not without good reason, for so severe is the torture inflicted by these pigmy assailants, that the bravest travellers are not ashamed to fly the instant they perceive the terrific host approaching, which is of no uncommon occurrence on the Campos."

BOTANY AND HORTICULTURE.

ON THE CHANGE IN THE COLOURS OF THE FLOWERS OF THE HIBISCUS MUTABILIS.—The Changing Hibiscus has received this name on account of the remarkable and periodical variations which the colours of the flowers present. White in the morning, they become more or less red or carnation-coloured towards the middle of the day, and terminate in a rose colour when the sun is set. This fact has been long known, but we were totally ignorant of the cause. The following observation may assist to discover it, and give some useful ideas on the colouration of flowers:—M. Ramond de la Sagra remarked, on the 19th of October, in the Botanic Garden of Havannah, of which he is the director, that this flower remained white all day, and did not commence to redden till the next day towards noon. On consulting the meteorological tables, which he kept with care, he found, that on this very day, the 19th October, the temperature did not rise above 19° 5′ C., whilst ordinarily it was at least 30°, at the period of inflorescence of this plant.—It would appear, then, that the temperature holds a place of some importance in the colouration of certain flowers. The experiments of Mr Macaire have taught, that it seems to be connected with different degrees of oxygenation of the chromule, or colouring matter, contained in the parenchyma. Is this oxygenation altogether, or in part, determined by the temperature, and can the colour of certain petals be modified by variations of heat?

WARS OF THE PLANTS.—All the plants of a given country are at war, one with another. The first, which establish themselves by chance in a particular spot, tend, by the mere occupancy of space, to exclude other species—the greater choke the smaller, the longest livers replace those which last for a shorter period, the more prolific gradually make themselves masters of the ground, which species, multiplying more slowly, would otherwise fill.

GEOGRAPHICAL DISTRIBUTION OF PLANTS.—The most incurious observer in travelling through a country must be struck with the different vegetation that prevails in different parts of the country, and with the effect which this difference produces on the manners and on the health of the inhabitants. Thus, in some parts of England, the Apple and the Pear are seen growing spontaneously in every hedge-row, while in other parts, apple and pear trees will not flourish, even with the utmost care. Some situations are favourable to the Oak, others to the Beech, others to the Elm. Accordingly, these well-known and beautiful trees predominate in some districts, almost to the exclusion of every other, and thus constitute the leading feature in the landscape. These are familiar examples of partial changes among the larger vegetables of a country; while the general vegetation is supposed to remain nearly the same. Between such partial change, and the complete establishment of a peculiar vegetation, there exists among different localities every possible shade of diversity. Many of these differences in vegetation are obviously connected with differences in soil and in situation. Thus, some plants will thrive only on a calcareous soil; as a few of the Orchis tribe in our own country, and the *Teucrium montanum* in Switzerland. Others, like the *Salsolas* and the *Salicornias*, will only grow in salt marshes. Some plants flourish in sea water, some in fresh, while to others again, water, at least in excess, is so prejudicial that they can exist nowhere unless on bare rocks or in arid deserts. Mountainous situations are most favorable to the increase of some plants, while others abound in plains. The larger number of plants prefer sunshine, but some are most vigorous in the shade, and others are so impatient of light, that they are found only where there is absolute darkness. There are besides parasitic plants, like the Misletoe, whose nourishment is derived from the plants to which they are attached. In short, the varieties in the nature of plants are countless, nor is the enumeration of them requisite. What has been stated is more than enough to show the wonderful arrangements that have been made to ensure the clothing of every part of the earth's surface with vegetable organization. There is not a soil, however barren, nor a rock, however flinty, that has not its appropriate plant, which plant has no less wonderfully found its way to the spot adapted for it, nay, will perish if removed elsewhere. Saline plants, for instance, will grow only where saline matters are abundant; plants of the marsh, and of the bog, flourish only in marshy and boggy ground; those of the parched desert, and of the cloudy mountain, each in its fitting locality. Thus the soil and its occupant seem to have been made for each other; and hence one source of that astonishing variety exhibited in Nature.

There are still more remarkable deviations among the plants of different countries remote from one another, even where the circumstances of climate and of soil are in every respect alike. The plants of the Cape of Good Hope, for instance, differ exceedingly from those of the south of Europe, though the climate and much of the soil be not dissimilar. Often on the same continent, nay, on the same ridge of mountains, the plants on the opposite sides have no resemblance. Thus, in North America, on the east side of the Rocky Mountains, Azalias, Rhododendrons, Magnolias, Vacciniums, Actæas, and Oaks, form the principal features of the landscape; while, on the western side of the dividing ridge, these genera almost entirely disappear, and no longer constitute a striking characteristic of the vegetation.

In general, the plants of America are different from those of the Old World, except towards the north, where, as it might be expected, from the near approximation of the two continents, many individuals are common to both. The plants of islands, and those growing in isolated situations, are often quite peculiar. Thus the plants of New Holland, with comparatively few exceptions, differ from those of all the rest of the world; and, "of sixty-one native species in the little island of St Helena, only two or three are to be found in any other part of the globe." These facts are quite inexplicable upon any known principles, and are calculated to excite a more than ordinary degree of attention, as being solely referable to the will of the Great Creator, who has chosen to provide infinite diversity where all might have been uniform and monotonous, and has thus rendered more conspicuous his wisdom, his power, and his goodness.—*Prout's Bridgewater Treatise.*

GEOLOGY.

ERUPTION OF FISHES.—Baron Humboldt gives an account of a wonderful eruption of fishes that sometimes takes place from the volcanos of the kingdom of Quito. These fishes are ejected in the intervals of the igneous eruptions in such quantities as to occasion putrid fevers by the miasmata they produce. They sometimes issued from the crater of the volcano, and sometimes from lateral clefts, but constantly at the elevation of between two and three thousand toises above the level of the sea. In a few hours, millions are seen to descend from Cotopaxi with great masses of cold and fresh water. As they do not appear to be disfigured or mutilated, they cannot be exposed to the action of great heat. Humboldt thought they were identical with fishes that were found in the rivulets at the foot of the volcanos, and to which he assigns the name of Pimelodus Cyclopum.

GENERAL SCIENCE.

ON THE COLOUR OF THE SEA.—Persons who have spent their lives in the interior of great continents, and have only been accustomed to observe the flow of brooks and shallow rivers, the source of clear fountains, or the roll of muddy currents, must view with some emotion the first sight of the sea, with its waters of sparkling green—a colour which appears peculiar to itself. When we pour a portion of its water into a vessel, we are struck at its perfect limpidity, and its colourless appearance. So great is the transparency of the ocean, in situations where it is not subjected to the contamination of rivers or of impure substances, that the sand in the bottom of its bed, even at a considerable depth, can be distinctly seen, while stones and small shells are quite perceptible through the medium, and shine with resplendent brightness. Marine plants and corallines beam with dazzling splendour, exhibiting their varied tints, while immersed in the water; but no sooner are they removed from it than their beauty vanishes. The *Iridea* and *Alcyonia*, which there possess all the varied tints of the rainbow, or wear a garb of glittering purple or orange, soon change their vivid lustre when cast upon the strand, and become black, yellow, or of a dingy violet, from their exposure to atmospheric air. When we enjoy an aquatic excursion during a cloudless day, on the smooth, unruffled bosom of the sea, and behold the sunbeams penetrate the abyss of waters, the deepness of its green impresses us with the idea of a verdant and liquid meadow. In proportion as we recede from the shore, and the water deepens around, the tint changes into a blue; and when we have reached the open sea, where the depth exceeds fifty or sixty fathoms, the water assumes a tint of the finest azure blue. The green shade usually indicates danger, or an approach to shallow water; but along those coasts which are interspersed with peaks or mountains, and near which the sounding line descends to great depths, the azure blue invariably appears, and assumes a more lively hue, as the depth becomes more considerable. This blue, which is generally considered as a characteristic of the ocean, and which appearance is accounted for by the rays of the sun becoming more decomposed as they penetrate into the depths of the waters, is not, however, peculiar to it, as every deep bed of water presents the same appearance under similar circumstances. Deep fresh-water lakes, especially in mountainous regions, exhibit the azure blue tint, which even extends itself to the beds of torrents. At the bottom of the torrent where the water fills the hollow cavity of a rock, the serene sky produces, in a modified degree, this beautiful colouring effect upon the water.

Sir Isaac Newton has demonstrated that the colours of all bodies arise from their power of reflecting or transmitting to the eye certain rays of which white light is composed, and of stopping or absorbing the remaining rays. This followed from his observing that bodies, of whatever colour they may appear, exhibit this colour only in white light. Every thing appears red in red light; and the leaf of the rose then presents the same hue as its flower. Even the most transparent bodies in Nature possess this absorbing power when in large masses. Thus, the air which appears perfectly colourless when seen from one end of an apartment to the other, tinges with a soft bluish gray the distant hills; and at length, when we look into the deep and cloudless expanse of the heavens, it deepens into a bright and azure blue. If we descend in a diving bell to a considerable depth below the surface of the sea, the sun appears of a bright red. In great depths of water objects become nearly invisible;

but when we ascend to the top of a lofty mountain, the azure tint of the heavens deepens into a black, and a greater number of stars become visible to the eye. The cause of these different hues must be sought in that unknown constitution of bodies, by which the one set of coloured rays are absorbed in passing through the transparent medium, while the remainder of them are reflected to the eye.

MISCELLANEOUS.

Singular Structure of a Hen's Egg.—The Egg, of which a representation is given in the annexed wood-cut, was accidentally discovered, during breakfast, by a young lady, residing at Irvine, Ayrshire, who has kindly transmitted it to us for examination.

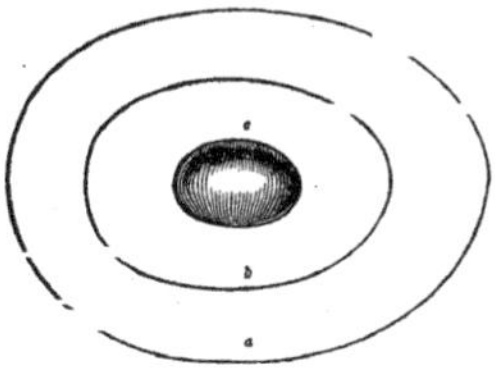

On breaking open the outward shell, *a*, a second egg, *b*, was found covered with shell, and, within this second body a third substance, c, consisting of a spherical membrane, one side of which was partially encrusted with shell, was also discovered. The first and second shells contained the usual albuminous matter or white of egg, and it is presumed that the substance in the third cavity was similar, although our informant does not state precisely whether a red yolk was present or not. On the inner side of both the first and second shells the usual membrane of the egg is distinctly visible; the matter of the shell differs in no respect from that of the common egg, except in being rather more porous and brittle. The inner membrane has all the appearance of being destined for the yolk, but, as we have remarked, there is reason to suspect that the contents within the membrane were of a white albuminous nature.

The most obvious explanation of this curious *Lusus Naturæ* is, that the three membraneous coats common to the egg have become ossified, or rather encrusted with carbonate of lime; that this has taken place in the second membrane from an unusual action of its secreting vessels; and that the same process had commenced, and was rapidly extending to the membrane of the yolk.

It would be a curious circumstance to ascertain whether the ossification began first on the yolk membrane, c, and being checked here by the production of the middle membrane, *b*, was subsequently covered by the membrane and shell, *a*,—or whether the ossification of the inner membranes proceeded subsequently to their development by the outer shell, *a*. If the latter were the case, it would show that the internal membranes of the egg have the power of secreting and depositing carbonate of lime, that is, of exercising a vital action when excluded from any of the vessels of the parent hen, and previous to, or independent of, the action which developes the chick. Such a circumstance would still farther tend to embarrass the curious question not yet solved, as to where the chick obtains the phosphate and carbonate of lime of its bones, for it has been ascertained that the albumen contains none, and it is probable that little or none is obtained by absorption from the shell.

Peculiarities of the Dayak.—The most numerous class of the inhabitants of Borneo, and probably the Aborigines, are the Dayak. Their manners are characterized by some strange peculiarities and uncommon features of barbarism; but the spirit of these traits has never been elucidated, nor the system of religious or superstitious opinion with which they are connected examined.

In appearance, the Dayak are fairer and handsomer than the Malays; they are of a more slender make, with higher foreheads and noses; their hair is long, straight, and coarse, generally cut short round their heads. The females are fair and handsome. Many of the Dayak have a rough, scaly scurf on their skin, like the Jakong of the Malay peninsula. This they consider as an ornament, and are said to acquire it by rubbing the juice of some plant on their skin. The female slaves of this race which are found among the Malays have no appearance of it.

With regard to their funeral ceremonies, the corpse is placed in a coffin, and remains in the house till the son, the father, or the nearest of blood, can procure or purchase a slave, who is beheaded at the time that the corpse is burnt, in order that he may become the slave of the deceased in the next world. The ashes of the deceased are then placed in an earthen urn, on which various figures are exhibited, and the head of the slave is dried, and prepared in a peculiar manner with camphor and drugs, and deposited near it. It is said that this practice often induces them to purchase a slave guilty of some capital crime, at five-fold his value, in order that they may be able to put him to death on such occasions.

With respect to marriage, the most brutal part of their customs is, that nobody can be permitted to marry till he can present a human head of some other tribe to his proposed bride, in which case she is not permitted to refuse him. It is not, however, necessary that this should be obtained entirely by his own personal prowess. When a person is determined to go a head hunting, as it is very often a very dangerous service, he consults with his friends and acquaintances, who frequently accompany him, or send their slaves along with him. The head hunter then proceeds with his party in the most cautious manner to the vicinity of the villages of another tribe, and lies in ambush till they surprise some heedless, unsuspecting wretch, who is instantly decapitated. Sometimes, too, they surprise a solitary fisherman in a river, or on the shore, who undergoes the same fate. When the hunter returns, the whole village is filled with joy, and old and young, men and women, hurry out to meet him, and conduct him, with the sound of brazen cymbals, dancing in long lines to the house of the female he admires, whose family likewise comes out to greet him with dances, and provide him a seat, and give him meat and drink. He still holds the bloody head in his hand, and puts part of the food into his mouth, after which, the females of the family receive the head from him, which they hang up to the ceiling over the door.

If a man's wife die, he is not permitted to make proposals of marriage to another, till he has provided another head of a different tribe, as if to revenge the death of his deceased wife. The heads procured in this manner they preserve with great care, and sometimes consult in divination. The religious opinions connected with this practice are by no means correctly understood: some assert, that they believe that every person whom a man kills in this world becomes his slave in the next. The Idaan, it is said, think that the entrance into Paradise is over a long tree, which serves for a bridge, over which it is impossible to pass without the assistance of a slave slain in this world.

The practice of stealing heads causes frequent wars among the different tribes of the Idaan. Many persons never can obtain a head, in which case they are generally despised by the warriors and the women. To such a height it is carried, however, that a person who had obtained eleven heads has been seen, and at the same time he pointed out his son, a young lad, who had procured three.

Walking on the Water.—A great inventor (in his own estimation) published to the world that he had solved the important problem of walking safely upon the water, and he invited a crowd to witness his first essay. He stepped boldly upon the wave, equipped in bulky cork boots, which he had previously tried in a butt of water at home; but it soon appeared that he had not pondered sufficiently on the centres of gravity and of flotation, for, in the next instant, all that was to be seen of him was a pair of legs sticking out of the water, the movements of which showed he was by no means at his ease. He was picked up by help at hand, and with his genius cooled, and schooled by the event, was conducted home.

The Tides.—The English Government has requested the co-operation of the Dutch Government in making simultaneous observations on the tides on their respective coasts. The King of Holland appointed Professor Moll, with the assistance of a certain number of naval officers, to make the necessary observations, from the 9th to the 30th July 1835.

Effects of Mercury.—There is a curious case mentioned in the *Lancet*, in which the secretary of a public institution was twice attacked with a very violent fit of salivation, so as to render medical aid indispensable, from his wafering 500 circulars with red wafers, which he had wetted in his mouth. Red wafers are coloured with vermilion, which is a preparation of mercury.

The Tatty.—The tatty is a trellise frame, very neatly made of split bamboo, thickly interwoven with a species of long grass, called kuss-kuss. During the hot winds in India, this is fixed at the door-way, and constantly wetted by a servant outside, who throws the water on it with a small jug from an earthen jar. The air passing through it is rendered delightfully cool and fragrant in the hottest weather. This is a luxury which is allowed even to the common European soldiers; their barracks being numerously attended by *beesties*, or water-carriers.

REVIEWS.

Wanderings in New South Wales, Batavia, Pedir Coast, Singapore, and China; being the Journal of a Naturalist, during 1832, 1833, *and* 1834. *By George Bennett, Esq., F.L.S, &c. &c. London.* 1834.

This is a work of unusual merit and interest. The excursions into the interior of New South Wales were made during the intervals of disengagement from professional duties, and at periods of the year best calculated for observations in Natural History. Though written from notes taken down at the instant of observation, and without any regard to studied composition, it is not deficient in many eloquent and vivid descriptions. Where the whole work is filled with interesting and popular matter, it becomes difficult to select a passage as a specimen of its style. We quote the following description of that splendid phenomenon, usually called

THE PHOSPHORESCENCE OF THE OCEAN:—

'Occasionally our attention was excited during the voyage by the remarkable luminosity assumed by the ocean in every direction, like rolling masses of liquid fire, as the waves broke and exhibited an appearance inconceivably grand and beautiful. The phosphoric light given out by the ocean, exists to a more extensive and brilliant degree in tropical regions, although in high latitudes it is occasionally visible, more especially during the warm months of the year. The cause of it has excited much speculation among naturalists; and although many of the marine Molluscous and Crustaceous animals, such as *Sulpa*, *Pyrosoma*, *Cancer*, and several *Medusæ*, have been found to occasion it; yet no doubt *debris*, from dead animal matter, with which sea water is usually loaded, is also one of the exciting causes.

"As the ship sails with a strong breeze through a luminous sea on a dark night, the effect produced is then seen to the greatest advantage. The wake of the vessel is one broad sheet of phosphoric matter, so brilliant as to cast a dull, pale light over the after-part of the ship; the foaming surges, as they gracefully curl on each side of the vessel's prow, are similar to rolling masses of liquid phosphorus; whilst in the distance, even to the horizon, it seems an ocean of fire, and the distant waves breaking, give out a light of an inconceivable beauty and brilliancy. In the combination the effect produces sensations of wonder and awe, and causes a reflection to arise on the reason of its appearance, as to which, as yet, no correct judgment has been formed, the whole being overwhelmed with mere hypothesis.

"Sometimes the luminosity is very visible without any disturbance of the water, its surface remaining smooth, unruffled even by a passing zephyr; whilst on other occasions no light is emitted unless the water is agitated by the winds, or by the passage of some heavy body through it. Perhaps the beauty of this luminous effect is seen to the greatest advantage when the ship, lying in a bay or harbour in tropical climates, the water around it has the resemblance of a sea of milk. An opportunity was afforded me when at Cavité, near Manilla, in 1830, of witnessing, for the first time, this beautiful scene. As far as the eye could reach over the extensive bay of Manilla, the surface of the tranquil water was one sheet of this dull, pale phosphorescence; and brilliant flashes were emitted instantly on any heavy body being cast into the water, or when fish sprang from it or swam about. The ship seemed, on looking over its side, to be anchored in a sea of liquid phosphorus, whilst in the distance the resemblance was that of an ocean of milk.

"The night to which I allude, when this magnificent appearance presented itself to my observation, was exceedingly dark, which, by the contrast, gave an increased sublimity to the scene. The canopy of the heavens was dark and gloomy, not even the glimmering of a star was to be seen; while the sea, of liquid fire, cast a deadly pale light over every part of the vessel, her masts, yards, and hull; the fish, meanwhile, sporting about in numbers, varying the scene by the brilliant flashes they occasioned. It would have formed, I thought at the time, a sublime and beautiful subject for an artist like Martin to execute, with his judgment and pencil; that is, if any artist could give the true effect of such a scene, on which I must express some doubts.

"It must not be for a moment conceived that the light described as brilliant, and like to a sea of 'liquid fire,' is of the same character as the flashes produced by the volcano, or by lightning, or by meteors. No; it is the light of phosphorus, as the matter truly is pale, dull, approaching to a white or very pale yellow, casting a melancholy light on objects around, only emitting flashes by collision. To read by it is possible, but not agreeable; and, on an attempt being made, it is almost always found that the eyes will not endure the peculiar light for any length of time, as headaches and sickness are often occasioned by it. I have frequently observed at Singapore, that, although the tranquil water exhibits no particular luminosity, yet, when disturbed by the passage of a boat, it gives out phosphoric matter, leaving a brilliant line in the boat's wake; and the blades of the oars, when raised from the water, seem to be dripping with liquid phosphorus.

"Even between the tropics, the phosphoric light is increased or diminished in its degree of brilliancy, in a very slight difference of latitude; on one day it would be seen to a most magnificent extent, on the next it would be perhaps merely a few luminous flashes. It might proceed from the shoals of marine animals, that caused the brilliancy, to be less extensively distributed over one part of the ocean than another. That I am correct in asserting that some of the animals which occasion the phosphoric light emitted by the ocean do travel in shoals, and are distributed in some latitudes only in a very limited range, I insert two facts which occurred during this voyage, and which will no doubt be regarded as interesting.

"On the 8th of June, being then in latitude 00° 30′ south, and longitude 27° 5′ west, having fine weather and a fresh south-easterly trade wind, and range of the thermometer being from 78° to 84°, late at night the mate of the watch came and called me to witness a very unusual appearance in the water, which he, on first seeing, considered to be breakers. On arriving upon the deck, this was found to be a very broad and extensive sheet of phosphorescence, extending in a direction from east to west as far as the eye could reach; the luminosity was confined to the range of animals on this shoal, for there was no similar light in any other direction. I immediately cast the towing net over the stern of the ship, as we approached nearer the luminous streak, to ascertain the cause of this extraordinary and so limited a phenomenon. The ship soon cleaved through the brilliant mass, from which, by the disturbance, strong masses of light were emitted; and the shoal (judging from the time the vessel took in passing through the mass) may have been a mile in breadth; the passage of the vessel through them increased the light around to a far stronger degree, illuminating the ship. On taking in the towing net, it was found half-filled with *Pyrosoma (Atlanticum)*, which shone with a beautiful pale greenish light; and there was also a few small fish in the net at the same time. After the mass had been passed through, the light was still seen astern, until it became invisible in the distance; and the whole of the ocean then became hidden in darkness as before this took place. The scene was as novel as it was beautiful and interesting, more so from having ascertained, by capturing the luminous animals, the cause of the phenomenon.

"The second was not exactly similar to the preceding; but, although also limited, was curious, as occurring in a high latitude during the winter season. It was on the 19th of August, the weather dark and gloomy, with light breezes from North-north-east, in latitude 40° 30′ South, and longitude 138° 3′ East, being then distant about 368 miles from King's Island (at the western entrance of Bass's Straits). My journal remarks the atmosphere to have been very chilly during the day, but much milder in the evening; the range of the thermometer during the day being from 49° to 56°. It was about eight o'clock P.M. when the ship's wake was perceived to be luminous, and scintillations of the same light were also abundant around. As this was unusual and had not been seen before, and it occasionally also appeared in larger or smaller detached masses, giving out a high degree of brilliancy, to ascertain the cause, so unusual in high latitudes during the winter season, I threw the towing net overboard, and in twenty minutes succeeded in capturing several *Pyrosoma*, giving out their usual pale green light; and it was no doubt detached groups of these animals that were the occasion of the light in question. The beautiful light given out by these Molluscous animals soon subsided (being seen emitted from every part of their bodies); but by moving them about, it could be reproduced for some length of time after. As long as the luminosity of the ocean was visible (which continued most part of the night) a number of *Pyrosoma Atlanticum*, two species of Phyllosoma, an animal apparently allied to Septocephalus, as well as several Crustaceous animals, all of which I had before considered as inter-tropical species, were caught and preserved. At half-past ten P.M. the temperature of the atmosphere on deck was 52°, and that of the water 51½°. The luminosity of the water gradually decreased during the night, and towards morning was no longer seen, nor on any subsequent night."

In a paper which Mr Bennett read before the Zoological Society, 25th June 1833, these luminous animals are more fully described.

Specimens taken from the sea, and placed in a glass vessel containing sea-water, ceased altogether to emit light, or emitted it but sparingly, while they remained at rest. On the water being agitated, or when one from the masses of animals was taken into the hand, the whole became instantly illuminated by myriads of bright dots, much resembling in hue the points on the elytra of a diamond beetle (*Curculio imperialis, Fab.*)

The Pyrosoma, thus enveloped throughout its whole extent in a flame of bright phosphorescent light, gleaming with its peculiar hue, presented a most splendid spectacle; the light shed by it was sufficient to render objects distinctly visible in every part of an otherwise dark room. If long retained in the hand, or returned to a quiescent state in the water, the luminous spots gradually faded, and no light was visible until the animal was again disturbed, when the illumination instantly returned with all its vivid splendour. After death it emitted no light.

The mass of Pyrosoma, of the usual cylindrical form and gelatinous substance, was about four inches in length, and one-and-a-half in circumference. The tube, passing along its middle, is described as being open at both ends; the orifice at the broader extremity being much better defined in its circular form, larger, and more distinct than that of the opposite end. The surface of the mass appeared to be studded with numerous prominent, rigid, and pearly tubercles, intermingled with small specks of a brown or red colour. In these latter the power of emitting light appear chiefly to be seated, these being frequently bright, while the remainder of the body exhibited only its natural white, or yellowish-white hue—a hue which changed after death into a red tinge. The brown specks, when removed from the body, did not emit light.

The extensive field of bright luminous matter, from which these specimens were taken, emitted so powerful a light as to illuminate the sails, and to permit a book of small print to be read with facility near the windows of the stern-cabins. Above this luminous field numerous sea-fowl were hovering in search of their prey. The light appeared to be entirely owing to the Pyrosomata.

The phosphorescence of the ocean often proceeds from other causes besides this one ascertained by Mr Bennett. Frequently it arises from putrescent particles of animal matter, dissolved by the sea water. Sometimes it is occasioned by minute animalcules, which possess the same remarkable property of shining in the dark with the fire-fly of tropical climates (*Elater noctilucus*), and the common glow-worm (*Lampyris noctiluca*). But many observers have examined sea-water, and were unable to perceive any animal matter in it. Professor Rennie states, that, being at Havre-de-Grace, he could not discover the slightest trace of animalcules, although the water which he examined was so strongly luminous, that it shone upon the skin of some night-bathers like scattered clouds of lambent flame, appearing more as a property of the water itself than any thing extraneous diffused through it.

In all these cases, the phosphorescent bodies merely emit, during the night, the solar rays which had been absorbed during the preceding day. The presence of salt seems necessary to phosphorescence, as the water of ponds and lakes is never luminous.' It may probably proceed from a similar cause with the luminous appearance observed by chemists during the crystallization of some salts, as the hydro-fluate of soda, and sulphate of potash.

A recent writer has described five varieties of these luminous appearances. The first shows itself in scattered sparkles in the spray of the sea, and in the foam created by the way of the ship when the water is agitated by the winds or currents; the second is a flash of pale light, of momentary duration, but often intense enough to illuminate the water to an extent of several feet; the third, of rare occurrence and peculiar to gulfs, bays, and shallows, in warm climates, is a diffused pale phosphorescence, resembling sometimes a sea of milk, or of some metal in a state of igneous liquefaction; the fourth presents itself to the astonished voyager, under the appearance of thick bars of metal, of about half a foot in length, ignited to whiteness, scattered over the surface of the ocean, some rising up and continuing luminous as long as they remain in view, while others decline and disappear; and the fifth variety is in distinct spots on the surface, of great beauty and brilliancy. The light of the first variety is more brilliant and condensed than that of any of the others, and very much resembles every way the red gold and silver rain of the pyrotechnist. The first and the third kind are produced by myriads of various minute Crustaceous animals, the smaller *Medusæ* and *Mollusca*, and perhaps some *Annelides*; the second appears to proceed from the gelatinous *Medusæ*, of a larger size; the *Pyrosomata* are the cause of the fourth kind, which may be often witnessed by vessels bound to India, or the eastward of the Cape of Good Hope, occurring in the calm latitudes near the line. The *Sapphirina indicator*, an insect somewhat resembling in appearance the wood-louse (*Oniscus*), and about one-third of an inch in length, emits the last variety enumerated, which appears to be limited to the seas situate on the north and west of a line drawn from the Cape of Good Hope to the southern extremity of the island of Ceylon. . .

Every page of Mr Bennett's excellent work is filled with interesting matter. There are some trifling inaccuracies, which we shall pass over in silence. We remark, however, that the author seems disposed to indulge in rather a satirical mode of expression towards the fair sex; and some remarks are made, which we think might have been omitted without prejudice to the work. Thus, in describing the town of Sydney, he observes—"Parrots are perhaps, of all the feathered tribe, the most numerous in the colony; and different species are lauded for speaking, whistling, and other noisy accomplishments. These birds are evidently gifted with the bump of *talkativeness*. It was once asserted that ladies kept the birds to converse with when alone, which served a double purpose—that of being to them both practice and amusement." We also find a chapter headed "Female Curiosity;" and, on examination, it turns out to be merely, that as the white men passed their dwellings, the ladies came forth to view the strangers, "with the usual feminine curiosity." Yet, immediately afterwards, we find that the *natives* (men as well as women) seemed to regard them as wonders.

We hope to return, on another occasion, to these otherwise interesting pages; and, in the meantime, we can recommend every student of Nature, to whom these books are accessible, to peruse them attentively.

Edinburgh: Published for the PROPRIETORS, at their Office, 16, Hanover Street. London: CHARLES TILT, Fleet Street. Dublin: W. F. WAKEMAN, 9, D'Olier Street. Glasgow and the West of Scotland: JOHN SMITH and SON, 70, St Vincent Street.

THE EDINBURGH PRINTING COMPANY.

THE EDINBURGH

JOURNAL OF NATURAL HISTORY,

AND OF

THE PHYSICAL SCIENCES.

N°. 8.   SATURDAY, JANUARY 30, 1836.

ZOOLOGY.

DESCRIPTION OF THE PLATE—THE HUMMING BIRDS.

THE genus Trochilus comprises some of the smallest, but, at the same time, some of the most beautiful of the feathered tribe. Observers of every description have been struck with admiration at the elegance and variety of the tints which adorn them; but the extreme delicacy of their constitution generally unfits them for enduring the variable climates of the temperate zone, or the restraints of confinement. They have almost always died on the passage homewards; and their admirers, in this country, are compelled to view only the preserved specimens in their cabinets, or such representations as we now offer.

All the objects on the accompanying plate are drawn the exact size of Nature; and we cannot fail to observe the striking contrast between the Gigantic Humming Bird, No. 7 (T. Gigas), which is about the size of a Sparrow, and the Least Humming Birds, Nos. 8 and 9 (T. Minimus). The latter scarcely surpasses the humble bee in magnitude, and is the smallest of Birds.

Those persons who have not seen them, numerous as Butterflies, sporting in the sunny prairies of America, would hesitate at first to believe that Birds of so minute a construction could exist. Yet we find the same perfection in the smallest as well as in the largest of Nature's works, and a structure prevails in these minute objects equally complicated with those of the Ostrich and Eagle.

It was long supposed, as they resemble the Butterfly in fluttering from flower to flower, that they also partook of the same food, and subsisted on honey. It seems now to be clearly ascertained that they do not feed on honey, but on the insects which prey upon it. This might have readily been discovered upon comparing the structure of their bills, which are long, pointed, and altogether incapable of sucking up a fluid, or saccharine matter, with the haustellum or sucker, used by certain insects for that purpose.

During their flight, they sometimes keep their bodies motionless in the air for hours together, emitting a loud humming noise, from which they derive their name. This sound is not emitted by the birds, but is occasioned solely by the exceedingly rapid vibration of their wings. They are generally confined to the tropical climates of America, although they have been found as far south as the Straits of Magellan, and as far north as the Elk River. They frequent the woods as well as prairies; and are often observed to enter the houses of the Americans in pursuit of Insects, sometimes venturing to insert their delicate bills into a bouquet of flowers, and rapidly retreating on being approached.

Figs. 1 and 2. The Tufted-necked Humming Bird (Trochilus Ornatus), Male and Female.—This species derives its name from the singular tuft of feathers which surrounds the neck of the male, but of which the female is altogether deprived.

Fig. 3. The Azure Blue Humming Bird (T. Lazulus) is distinguished by the brilliant hue of its breast.

Fig. 4. The Harlequin Humming Bird (T. Multicolor) is so singular and fantastic in its colours, that the specimen in the British Museum was long suspected to have been formed of feathers belonging to different species. This is now generally believed not to have been the case.

Fig. 5. The Ruby-crested Humming Bird (T. Moschitus) is very common in the West Indies and in tropical America.

Fig. 6. Gould's Humming Bird (T. Gouldii) possesses one of those singular tufts round the neck, which the French term Coquets, and have been not unaptly compared to the ruffs worn by ladies during the age of Queen Elizabeth.

Fig. 7. The Gigantic Humming Bird (T. Gigas).—This is the Patagonian of the Humming Bird genus. In strength and size it is equalled by none.

Fig. 8 and 9. The Least Humming Birds (T. Minimus), Male and Female, resemble the preceding one in the dulness of their colours, which are much inferior to their congeners. Yet we view these little creatures with singular interest, forming, as they do, one of the limits, in regard to size, of a numerous and interesting class of animated beings.

Fig. 10. The White Striped Humming Bird (T. Mesoleucus) differs but slightly from Fig. 11, the Evening Humming Bird (T. Vesper). Both these species have but a rudimentary tuft around the neck, which however is of a brilliant hue.

Fig. 12. The Tri-coloured Humming Bird (T. Tricolor) appears to be surpassed in beauty by few of its tribe.

REMARKABLE PECULIARITY OF THE FEMALE ASS.—It is universally known that many animals will continue to give milk not only after the young are removed, but even for years, when the impression of having had young must have been entirely forgotten. The Cow and Goat are instances of this kind; but in the Ass the secretion of milk is not continued after the mother has lost the impression of her foal's existence. This is a fact so well known to the keepers of Asses, that whenever an Ass's foal dies, they take every means in their power to keep up the impression, in the mother, of the foal being still alive, to keep her in milk. For this purpose they take off the skin of the foal and preserve it, so that it may be occasionally thrown over the back of another foal, and smelled by the mother, more particularly at the time they are milking her. The Ass, under the deception of having her own foal, gives down her milk, and the secretion is carried on as usual; but if this artifice be neglected she soon goes dry. To ascertain this fact more accurately, the celebrated Mr John Hunter put it to the test of experiment. He took an Ass, in milk, and kept her apart from her foal every night, but had the mother milked in the morning in presence of the foal. This was done for more than a month, without there being any diminution in the morning's milk. The foal was then taken away altogether, and the mother was milked instead of being sucked by the foal, particularly in the evening, at the same hour at which the foal had been taken from her, and again in the morning at the usual hour. The milk taken in the morning was always compared with that taken in the morning before, but in three mornings the quantity was lessened; and the fifth morning there was hardly any. The foal was then restored to her; but she would not allow it to suck. The experiment was repeated with similar results.

THE CANARY BIRD.—The Canary Bird is remarkable for its tractability and intelligence, as an instance of which the following anecdote may be given:—A bird-catcher in Prussia, who had rendered himself famous for educating and calling forth the talents of the feathered tribe, had a Canary Bird, which was introduced by the owner to a large party at Cleves, to amuse them with his wonderful feats. The Canary being produced, the owner harangued him in the following manner, placing him upon his fore-finger:—" Bijou (jewel), you are now in the presence of persons of great sagacity and honour; take heed, therefore, that you do not deceive the expectations they have conceived of you from the world's report. You have got laurels; beware of their withering: in a word, deport yourself like the bijou of Canary Birds, as you certainly are." All this time the bird seemed to listen, and indeed placed himself in the true attitude of attention. He sloped his head to the ear of the man, then distinctly nodded twice, when his master had left off speaking; and if ever nods were intelligible and promissory, these were of that nature. "That's good," said the master, pulling off his hat to the bird. "Now let us see if you are a Canary of honour? Give us a tune." The Canary sung. " Pshaw! that's too harsh: tis the note of a raven with a hoarseness upon him—something pathetic." The Canary whistled as if his little throat was changed to a lute. " Faster," says the man; " slower —very well. What the plague is this little foot about, and this little head? No wonder you are out, Mr Bijou, when you forget your time. That's a jewel; Bravo! bravo! my little man." All that he was ordered, or reminded of, did he to admiration. His head and foot beat time; humoured the variations both of tone and movement; and the sound was a just echo to the sense, according to the strictest laws of poetical, and (as it ought to be) of musical composition. " Bravo! bravo!" re-echoed from all parts of the room. The musicians declared the Canary was a greater master of music than any of their band. " And do you not show your sense of this civility, sir?" cried the bird-catcher with an angry air. The Canary bowed most respectfully, to the great delight of the company. His next achievement was going through the martial exercises with a straw gun; after which, " My poor Bijou," said the owner, " thou hast had hard work, and must be a little weary: a few performances more, and thou shalt repose. Show the ladies to make a curtsey." The bird here crossed his taper legs, and sunk and rose with an easy grace that would have put half our subscription assembly belles to the blush. " That's my fine bird! and now a bow, head and foot corresponding." Here the striplings for ten miles round London might have blushed also. " Let us finish with a hornpipe, my brave little fellow; that's it, keep it up, keep it up." The activity, glee, spirit, and accuracy with which this last order was obeyed, wound up the applause (in which all the musicians joined, as well with their instruments as their clappings) to the highest pitch of admiration. Bijou himself seemed to feel the sacred thirst of fame, and shook his little plumes, and carolled an Io pæan, that sounded like the conscious notes of victory. " Thou hast done all my biddings bravely," said the master, caressing his feathered servant; " now then take a nap, while I take thy place." Hereupon the Canary went into a counterfeit slumber, so like the effect of Morpheus, first shutting one eye, then the other, then nodding, then dropping so much on one side that the hands of several of the company were stretched out to save him from falling; and just at their hands approached his feathers, suddenly recovering, and dropping as much on the other. At length sleep seemed to fix him in a steady posture; whereupon the owner took him from his finger, and laid him flat on the table, where the man assured us he would remain in a good sound sleep, while he himself had the honour to do his best to fill up the interval. Accordingly, after drinking a glass of wine (in the progress of which he was interrupted by the Canary Bird springing suddenly up to assert his right to a share, really putting his little bill into the glass, and then laying himself down to sleep again), the owner called him a saucy fellow, and began to show off his own independent powers of entertainment, when a huge black cat, who had long been on the watch, sprang unobserved, from a corner, upon the table, seized the poor Canary in its mouth, and rushed out of the window in spite of opposition. And though the room was deserted in an instant, it was a vain pursuit; the life of the poor bird was gone; and its mangled body was brought in by the unfortunate owner, under such dismay, and accompanied by such looks and language, as would have awakened pity in a misanthrope.

ANECDOTE OF A RAVEN.—At the seat of the Earl of Aylesbury, in Wiltshire, a tame Raven, that had been taught to speak, used to ramble about in the park; there he was commonly attended and beset by crows, rooks, and others of his inquisitive tribe. When a considerable number of these were collected around him, he would lift up his head, and with a hoarse and hollow voice shout out the word Holla! This would instantly put to flight and disperse his sable brethren, while the Raven seemed to enjoy the fright he had occasioned.

CURIOUS MECHANISM IN THE FEET OF THE FLY AND LACERTA GECKO.—It is well known that the house-fly is capable of walking upon the ceiling of rooms, in which situation its body is not supported on the legs: The principle upon which it does so remained for a long time unexplained, because the animal is too small for the feet to be anatomically examined. Animals of a much larger size are endowed with the same power. The Lacerta Gecko, a native of the Island of Java, is in the habit of coming out of an evening from the roofs of the houses, and walking down the smooth, hard, polished chunam walls in search of flies that settle upon them, and then of running up again. Sir Joseph Banks, while at Batavia, used to catch this animal by standing close to the wall with a long flattened polo, which being made suddenly to scrape the surface, knocked it down. Sir Everard Home procured a specimen of a very large size, which enabled him to ascertain the peculiar mechanism by which the feet of this animal keep their hold of a smooth hard perpendicular wall, and carry up so large a weight as that of its own body. He found that the foot of this lizard was so constructed as to enable it to produce a number of small concavities which act like so many cupping glasses, and atmospheric pressure retains them in this position. It appears that the fly's foot possesses concave surfaces capable of acting in the same manner as those of the Lacerto Gecko, and therefore its progressive motion against gravity is effected by the same means.

BOTANY AND HORTICULTURE.

RAFFLESIA PATMA.

THIS is one of those anomalous vegetable productions which are so numerous in some of the Islands of the East Indian Archipelago, being without either root or leaves, as the flower constitutes the whole of the plant. Dr Blume, in his excellent work on the *Flora Javæ necnon Insularum adjacentium*, has given an engraving of this flower, from which the following has been reduced.

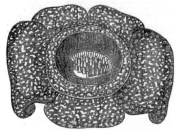

The Rafflesia Patma of Blume is found in the shady thickets of the little island of Nusa Kambagan, which adjoins Java on the south. It grows upon the roots of the *Cissus Scariosa* of Blume, and seems exceedingly partial to moist ground, where the diameter of the expanded flower is found to reach the size of two Dutch feet (about one foot seven inches English), but in dryer, and consequently less favorable situations, it does not exceed one English foot in diameter.

The plants of the genus Rafflesia have been found on the stems, as well as upon the roots, of the genera *Cissus* and *Vitis*; they form the only instances of parasites on roots, which likewise proceed from other parts of the plant. Iso_rr, in his *Reise nach Guinea*, p. 283, mentions a plant, which he had observed in equinoctial Africa, parasitic on the roots of trees, consisting almost entirely of a single flower of a red colour, which is probably allied to Rafflesia, the smaller species of which it seems to resemble in appearance.

Dr Blume was convinced, after a careful examination of the R. Patma, that it had no connexion whatever with the woody layers of the root of the Cissus Scariosa, but that it was only united with the substance of the bark of the root. It is a singular fact, that the growing bark, having its continuity interrupted by the collet of the Rafflesia entering into its substance, swells into a cup-shaped process round about the flower buds of the Rafflesia, and that this cup-like process varies in diameter according to the length of time which must elapse between the first rising of the flower bud, and the ultimate fall of the flower itself and of its remains.

The R. Patma differs from the R. Arnoldi, figured in a former number (4), in having the inside of the perianth red; and further, in having the columnar processes of its disc more numerous, stronger, and more unequal in length. It seems to depart still more widely from the general character of our ordinary diœcious plants.

The successive development of the flower, from its first appearance upon the original stock, until it is just ready to blow, is represented in the following cuts.

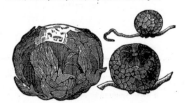

With respect to the place which the genus Rafflesia holds in the Vegetable Kingdom, the learned Dr Robert Brown has the following observations:—" As to which of the two primary divisions of phænogamous plants the genus belongs, it may, I think, without hesitation be referred to *Di-cotyledones*; yet, if the plant be parasitic, and consequently no argument on this subject to be derived from the structure of the root, which is exactly that of the Vine, the exclusion from *Mono-cotyledones* would rest on no other grounds, that I am able to state, than the quinary division of the perianthium, which, in other respects, also bears a considerable resemblance to that of certain Di-cotyledonous orders; the number of stamina and the ramification of vessels in the bractea.

" Assuming, however, that Rafflesia belongs to Di-cotyledones, and considering the foliaceous scales which cover the expanded flower, both from their indefinite number and imbricate insertion as bracteæ, and consequently the floral envelope as simple, its comparison with the families of this primary division would be limited to such as are apetalous, either absolutely, as Aserinæ; those of a nature intermediate between the apetalous and polypetalous, in which the segments of the perianthium are, generally, though not always, disposed in a double series, as Passifloreæ, Cucurbitaceæ, and Homalineæ; or those which have a simple-coloured floral envelope, but are decidedly related to polypetalous families, as Sterculieæ."

It is difficult to ascertain in what manner the impregnation of the female flower is effected, when the anthers are so completely concealed as those of Rafflesia seem to be in all states of the flower; for it does not appear either that they can ever become exposed by a change in the direction of the limb under which they are inserted, or even that this part of the column, in any stage, projects beyond the tube of the perianth.

It is probable, therefore, that the assistance of insects is absolutely necessary; and it is not unlikely, both as connected with that mode of impregnation, and from the structure of the anther itself, that in Rafflesia the same œconomy obtains as in the stamina of certain *Aroideæ*, in which it has been observed that a continued secretion and discharge of pollen takes place from the same cell; the whole quantity produced greatly exceeding the size of the secreting organ. The passage of the pollen to the bottom of the flower, where it is more easily accessible to insects, seems likewise to be provided for, not only by the direction of the anther, but also by the form of the corresponding cavities in the neck of the column, in the upper part of which they are immersed.

That insects are really necessary to the impregnation of the Rafflesia, is confirmed by Dr Arnold's statement in regard to the other species, R. Arnoldi. By the fact of the swarms actually seen hovering about and settling in the expanded flower, it is more than probable that they were attracted by its peculiar odour.

The modes of union between a parasite and the stock from which it springs, may be divided into such as are entirely dependent on the stock during the whole of their existence, and such as, in their more advanced state, produce roots of their own. Among those that are in all stages absolutely parasitic, to which division the genus Rafflesia probably belongs, very great differences exist in the mode of connexion. In some of these the nature of the connexion is such as can only be explained on the supposition that the germinating seed of the parasite excites a specific action on the stock, the result of which is the formation of a structure, either wholly or in part, derived from the root, and adapted to the support and protection of the undeveloped parasite; analogous therefore to the production of galls by the puncture of insects.

On this supposition the connexion between the flower of Rafflesia, and the root from which it springs, though considerably different from any other known plant, may also be explained. But until either precisely the same kind of union is observed in plants known to be parasite, or, which would be still more satisfactory, until the leaves and fructification belonging to the root to which Rafflesia is attached, shall have been found, its being a parasite, though highly probable, cannot be considered as absolutely ascertained.

MINERALOGY.

THE MATTAN DIAMOND.—The Rajah of Mattan, in the island of Borneo, possesses the finest and largest diamond which has hitherto been discovered. It weighs 367 carats, and is said to be of the finest water. The celebrated Pitt Diamond weighs only 127 carats. The Mattan Diamond is shaped like an egg, with an indented hollow near the smaller end. It was discovered at Landak about ninety years ago; and though the possession of it has occasioned numerous wars, it has been about eighty years in the possession of the Mattan family. Many years ago, the Governor of Batavia sent a Mr Stewart to ascertain the weight, quality, and value of this Diamond, and to endeavour to purchase it; and in this mission he was accompanied by the late Sultan of Pontiana. After examining it, Mr Stewart offered 150,000 dollars for the Diamond, the sum to which he was limited; and, in addition to this sum, two war-brigs, with their guns and ammunition, together with a certain number of great guns, and a quantity of powder and shot. The Rajah, however, refused to deprive his family of so valuable an hereditary possession, to which the Malays attach the miraculous power of curing all kinds of diseases, by means of the water in which it is dipped, and with which they imagine the fortune of the family is connected.

GEOLOGY AND PHYSICAL GEOGRAPHY.

INUNDATIONS OF MEXICO.—The Mexican Lakes are natural reservoirs, into which the torrents, rushing from the surrounding mountains, deposit their waters. The capital is situate in the centre of a valley, around which the encircling mountains rise in successive stages, until they are lost in the distance. The city of Mexico occupies the lowest part of the valley, and is scarcely more elevated than the level of the Lake Tezcuco. Several lakes rise above each other, such as those of Xaltocan, Xochimilco, Tzompango, and Chalco. In consequence of the different elevations of these natural reservoirs, the city of Mexico has for a long series of ages been exposed to the ravages of mighty inundations. The Lakes of Chalco and of Xochimilco must necessarily overflow their banks whenever a violent eruption of an adjoining volcano causes the snows which cover its summit to melt. "When I was at Guayaquil, on the borders of the province of Quito, in 1802," says the Baron Al. De Humboldt, "the cone of Cotopaxi was heated by subterranean fire to such a degree, that, in a single night, it lost the vast garment of snow with which it had long been covered." In the New World, eruptions and great earthquakes are often followed by heavy showers, which continue for several months. It may easily be imagined how dangerous the situation of Mexico must be in a climate where, in the driest years, the rain falls to the depth of 15 decimetres (about 59 inches).

The inhabitants of New Spain think that these violent inundations follow each other at nearly equal periods of time. Past experience seems to prove that a violent eruption of the waters occurs every 25 years. Since the arrival of the Spaniards, Mexico has experienced five very great inundations, in 1553, 1580, 1604, 1607, and 1629. In consequence of the opening of a canal, these evils were partially averted; and since that time their effects have been less violent, occurring at intervals of 27, 24, 3, 26, 19, 27, 32, 25, 16, 24, and 23 years. These numbers do not follow each other with so great a regularity as may be observed in the periods marking the returns of the earthquakes at Lima.

The situation of Mexico is rendered daily still more dangerous, by the circumstance that the difference of level between the surface of the lake Tezcuco and the soil on which the houses are constructed diminishes annually. This soil forms a fixed plane, especially since the streets of Mexico were paved. The bottom of the lake of Tezcuco, on the contrary, is continually elevated by the mud washed down by the surrounding torrents. It was to avoid a similar danger that the Venetians turned from their lagunes, the Brenta, Livenza, and other rivulets,* which form their alluvial deposits within them. If we could place much reliance upon the results of the levelling made in the sixteenth century, there would be no doubt that the *Plaza Mayor*, or Great Square of Mexico, had formerly been elevated eleven decimetres (43½ inches) above the level of the lake of Tezcuco, and that the mean level of the lake varies from year to year. On the one hand, the moisture of the atmosphere has diminished by the destruction of the forests, and consequently the sources of the streams flowing from the mountains which surround the valley have been lessened; but, on the other hand, the clearing of the ground has increased the quantity of alluvial deposit, and the rapidity of the inundations. General Andreossi, in his excellent work upon the Canal of Languedoc, has explained these causes, which are the same under all climates. The waters which glide over the declivities covered with a green sward, form less alluvial deposit than those which rush over the uncovered bank. This vegetable covering, which may be formed either by the grasses, as in Europe, or by the little Alpine plants, as we find them in Mexico, can only be preserved under the shadow of the forest. Again, the thickets and brushwood present obstacles to the waters which roll down the declivity of the mountains, upon the melting of the snows. When these declivities are destitute of vegetation, the rivulets of water are less retarded, they unite more rapidly in torrents, and their deposits swell the lakes adjoining the city of Mexico.

ASCENT OF THE ANDES.—Don Juan de Ulloa, who went to Peru in company with the French academicians, to measure a degree of the meridian, gives the following curious description of his ascent of the Andes:—

"After many days sailing up the river Guayaquil, I arrived at Caracol, a town situate at the foot of the Andes. Nothing can exceed the inconveniences we had experienced in this voyage from the flies and mosquitoes. We were the whole day in continual motion to keep them off, but at night our torments were excessive. Our gloves, indeed, were some defence to our hands, but our faces were entirely exposed; nor were our clothes a sufficient defence for the rest of our bodies, for the stings of these insects, penetrating through the cloth, caused a very painful itching. One night, on coming to anchor near a very handsome house that was uninhabited, we were no sooner seated in it than we were attacked on all sides by swarms of mosquitoes, so that it was impossible to have one moment's quiet. Those who had covered themselves with clothes made for this purpose found not the smallest defence; wherefore, hoping to find some relief in the open fields, they ventured out, although in danger of suffering in a more terrible manner from the serpents. But both places were equally obnoxious. On quitting this inhospitable retreat, we took up our quarters, the next night, in a house that was inhabited; the master of which being informed of the terrible manner we had passed the preceding night, told us gravely that the house we so greatly complained of had been forsaken on account of its being the purgatory of a soul; but we had more reason to believe that it was quitted on account of its being the purgatory of the body. After having journeyed upwards of three days, through boggy roads, in which the mules sank knee-deep at every step, we began at length to perceive an alteration in the climate; and after having been long accustomed to heat, we now felt it grown very sensibly colder.

"It is remarkable that at Taraguagua we often see instances of the effects of two opposite temperatures in two persons happening to meet; one of them leaving the plains below, and the other descending from the mountain. The former thinks the cold so severe that he wraps himself up in all the garments he can procure, while the latter finds the heat so great that he is scarcely able to bear any clothes whatever The one thinks the water so cold that he avoids being sprinkled by it, the other is so delighted with its warmth that he uses it as a bath.

"The ruggedness of the road from Taraguagua, leading up the mountain, is not easily described. The declivity is so great, in some parts, that the mules can scarcely keep their footing; and in others, the aclivity is equally difficult. The trouble of sending people before to mend the road, the pain arising from the many falls and bruises, and the being constantly wet to the skin, might be supported, were not these inconveniences augmented by the sight of such frightful precipices and deep abysses, as excite incessant terror. The road in some places is so steep, and yet so narrow, that the mules are obliged to slide down, without making any use whatever of their feet except as a support. On one side of the rider, in this situation, rises an eminence of several hundred yards, and on the other is an abyss of equal depth, so that, if he should give the least cheek to his mule, and thus destroy the equilibrium, they must both inevitably perish.

"Having travelled nine days in this manner, slowly winding along the side of a mountain, we began to find the whole country covered with a hear frost, and a hut, in which we reposed, had ice in it. At length, after a perilous journey of fifteen days, we arrived upon the plain, at the extremity of which stands the city of Quito, the capital of one of the most charming regions in the world. Here, in the centre of the torrid zone, the heat is not only very tolerable, but in some places the cold is even painful. Here the inhabitants enjoy all the temperature and advantages of perpetual spring; the fields being constantly covered with verdure, and enamelled with flowers of the most lively colours. However, although this beautiful region be more elevated than any other country in the world, and it took up so many days of painful journey in the ascent, it is overlooked, nevertheless, by tremendous mountains—their sides covered with snow, while their summits are flaming with volcanoes. These mountains seem piled one upon the other, and rise to an astonishing height, with great coldness. However, at a determined point above the surface of the sea, the congelation is found at the same height in all the mountains. Those parts which are not subject to a continual frost, have here and there growing upon them a rush, resembling the Genista, or broom, but much softer and more flexible. Toward the extremity of the part where the rush grows, and the cold begins to increase, is found a vegetable with a round bulbous head, which, when dried, has an amazing elasticity. Higher still, the earth is entirely bare of vegetation, and seems covered with eternal snow. The most remarkable of the Andes are the mountains of Cotopaxi, Chimborazo, and Pichincha. On the top of the latter was my station for measuring a degree of the meridian, where I suffered particular hardships, from the intenseness of the cold, and the violence of the storms. The sky around us, in general, was involved in thick fogs, which, when they cleared away, and the clouds, by their gravity, moved nearer to the surface of the earth, appeared surrounding the foot of the mountain, at a vast distance below, like a sea encompassing an island in the midst of it. When this happened, the horrid noises of tempests were heard from beneath, then discharging themselves on Quito and the neighbouring country. I saw the lightning issue from the clouds, and heard the thunders roll far beneath me. All this time, while the tempest is raging below, the mountain top where I was placed enjoyed a delightful serenity. The wind was abated, the sky clear, and the enlivening rays of the sun moderated the severity of the cold. However, this was of no very long duration; for the wind returned with all its violence, and with such velocity as to dazzle the sight, while my fears were increased by the dreadful concussions of the precipice, and the fall of enormous rocks, the only sounds that were heard in this frightful situation."

IRISH BOG.—The moving bog on Lord O'Neill's estate, near Randalstown, has changed its situation to a considerable extent. It has overspread the surrounding land, and precipitated its mass into the river Maine, so as to obstruct the course of the current, and lying in some places nearly twenty feet deep.

METEOROLOGY.

DEW-BUTTER.—The following singular fact is recorded in one of the first numbers of the Philosophical Transactions:—In the year 1695, during a great part of the winter and spring, a fatty substance, somewhat like butter, was deposited by the atmosphere, instead of the usual dew, in Ireland, and particularly in the provinces of Leinster and Munster. This substance is said to have been of a dark yellow colour, and felt clammy, whence the natives called it Dew-butter. It fell in the course of the night on the moorish low grounds; and it was found in the morning attached to the leaves of grass, to the thatches of houses, &c., in the form of pretty large lumps; and it is added, that it seldom fell twice in the same place. It had an offensive smell·

* Andreossi sur le Canal du Midi, p. 19.

yet it lay upon the ground a fortnight before it changed colour, after which it dried up and became black; but it never bred worms, nor did it prove noxious to cattle that fed in the fields where it fell. During the winter of the above-mentioned year some very disagreeable fogs were observed on the same places where the *dew-butter* fell.

During a great part of the year 1783, when the repeated earthquakes of Calabria and Sicily destroyed almost the whole of the former, and a great part of the latter place, a very large meteor was observed, in the month of September, by most European countries. At the same time the Hecla, in Iceland, made a vast eruption of ignited matter; and it is said that many persons observed, in Ireland, a peculiar kind of clamminess upon the leaves of trees, as if a dew of a glutinous nature had been deposited from the atmosphere; but we do not find that any particular experiments were made for the purpose of ascertaining its nature.

Snow in June.—On the morning of Tuesday the 23d June 1835, the Argyleshire mountains were covered with snow, particularly a hill on the Drimsynie estate, named Benevullo, the highest ridge of the mountains in Cowal, and Craigen Hill, or the Duke of Argyle's bowling green, on the western side of Loch Long.

MISCELLANEOUS.

NO. II.—ANIMAL MAGNETISM.

On witnessing the same experiments, frequently repeated, the commissioners remarked, that among the patients who fell into the *crises* there were always many women and very few men; that the *crises* were not effected in less than the space of an hour or two, and that as soon as one person was thus taken, the rest were similarly seized in a very short time. But they were unable to obtain any satisfactory results from experiments made upon so many persons at once. They resolved therefore to endeavour, by experiments on individuals, in a more private way, to ascertain the direct effects of the newly-discovered agent on the animal economy, in a state of health, which, if the agent existed, could of course be rendered manifest by its effects; and they determined to become themselves the subjects of the first experiments. No inquiry was ever conducted in a more philosophical manner, or terminated in a more complete and unequivocal development of the nature of the subject. Great and extraordinary as the powers of this new agent seemed to be, the phenomena were proved to be referable solely to the imagination of the parties magnetised.

The commissioners submitted to be magnetised together, excluding all strangers, by M. Deslon, once a-week, for the space of two hours and a half. They were ranged round the *baquet*, encircled by the cord of communication, with an iron branch from the *baquet* resting upon the left hypochondre of each, and forming from time to time the communication of thumbs: they were magnetised by the fingers, or the metallic rod, being moved about and presented to different parts of the body, as well as by the pressure of hands on the pit of the stomach and sides of the belly. The most irritable and delicate of the commissioners were magnetised the most frequently, and for the longest time; but none of them experienced any effects or sensations, or at least any that could be ascribed to magnetism. Three of them were valetudinarians, and some of their usual uneasy feelings were excited partly by the fatigue, and partly by the strong pressure made on the stomach. They submitted to the experiment on three days successively; still without any effect. The quiet and silence of the eight commissioners, thus magnetised, without any uneasiness or any new sensation, formed the most perfect contrast with the noise, agitation, and disorder of the public magnetism: here was the magnet without any influence, and the operator despoiled of his power. They were warranted, therefore, in concluding, "that magnetism has no agency in a state of health, or even in a state of slight indisposition."

They resolved, then, to make their next trials of its influence upon persons actually diseased, and seven persons, of the lower class, were magnetised by M. Deslon, in the presence of the commissioners, at Dr Franklin's house. Two women, the one asthmatic, the other with a swelling on the thigh, and two children, the one six, and the other nine years of age, felt nothing, and remained unaffected. One man, with diseased eyes, felt a pain in the ball of one of them, which also discharged tears when the finger of the magnetiser was brought near it, and moved quickly about for a considerable time; but when the other eye, which was most diseased, was magnetised, he felt nothing. A nervous, hysterical woman, to whom the pressure of the abdomen was painful, and who had a *hernia*, said she felt a pain in the head when the finger was pointed near the rupture, and that she lost her breath when it was brought opposite the face. When the finger of the magnetiser was repeatedly moved up and down, she experienced some catchings of the muscles of the head and shoulders, like one surprised and afraid. The seventh patient, a man, suffered some effects of the same sort, but much less marked.

Four persons, two ladies and two gentlemen, of good education, and in bad health, were afterwards magnetised. Three of these underwent the operation several times, and felt nothing; but the fourth, a nervous lady, being magnetised during an hour and twenty minutes, generally by the application of the hands, was several times on the point of falling asleep, and felt some degree of agitation and uneasiness. On a subsequent occasion, a large company assembled at Dr Franklin's (who was confined by an illness) were all magnetised, including some patients of M. Deslon, who had accompanied him thither. There were present several Americans, one of whom, an officer, had an intermittent fever; yet no person experienced any effects except M. Deslon's patients, who felt the same sensations to which they had been accustomed at his public magnetising.

These experiments, then, furnished some important facts. Of fourteen invalids, five experienced some effects from the operation, but nine felt none whatever. All the effects observed in the nervous lady, however, might be occasioned by the irksomeness of the same posture for so long a time, and by her attention being strongly fixed upon her feelings: for it is frequently sufficient to think of these nervous attacks, or to hear them mentioned, in order to reproduce them when they are habitual. The three other instances occurred among persons of the *lower class;* and this circumstance was remarked with surprise by the commissioners, that the only effects which could be ascribed to magnetism manifested themselves in the poor and ignorant, while those who were better able to observe and to describe their sensations felt nothing. At the same time, it was observed that *children*, although endowed with the peculiar sensibility of their age, likewise experienced no effect. The notion, that these effects might be explained by natural causes, therefore suggested itself to the commissioners: "If we figure to ourselves," they observe, "a poor, ignorant person, suffering from disease, and anxious to be relieved, brought before a large company, partly consisting of physicians, with some degree of preparation and ceremony, and subjected to a novel and mysterious treatment, the wonderful effects of which he is already persuaded that he is about to experience; and if, moreover, it is recollected that he is paid for his compliance, and supposes that the experimenters will be gratified in being told that he perceived certain operations, we shall have natural causes by which these effects may be explained, or at least very legitimate reasons for doubting that the real cause is magnetism."—*(Rapport des Commiss.* p. 30.)

Since the supposed effects of the *Animal Magnetism*, then, were not discoverable in those who were incredulous, there was great reason to suspect that the impressions which were produced were the result of a previous expectation of the mind, a mere effect of the *imagination*. The commissioners, therefore, now directed their experiments to a new point; namely, to determine how far the imagination could influence the sensations, and whether it could be the source of all the phenomena attributed to magnetism.

Greece.—M. M. Sauguy, Von Hammer, and other learned individuals, have resolved on making travels in Greece, for the purpose of geographical and historical discoveries. Their intention is fully to explore Eubœa, and other parts of Asia Minor, and more especially the shores of the Propontis.

Balls of Frozen Mercury and Almond Oil.—On one occasion, during Captain Ross's late detention in the northern regions, some of the officers fired a ball of frozen mercury through an inch plank; and at another time, they froze oil of almonds in a shot mould, at a temperature of 40 degrees below zero, and discharged this new species of projectile against a target, which it split, rebounding unbroken.

The Hedgehog proof against Poisons.—It has been said that the hedgehog is proof against poisons. M. Pallas states that it will eat a hundred cantharides, or Spanish flies, without receiving any injury. More recently, a German physician, who wished to dissect a hedgehog, gave it prussic acid, but the poison did not take effect; he then tried arsenic, opium, and corrosive sublimate, with the same results.

Health.—Without health what are we? And yet how little do we care about it, till, by some unexpected stroke, the vision of perpetual vigour vanishes, the blessing has fled. In vain we then wish that we had adopted some of those plain, easy, and agreeable precautions, which an acquaintance with medicine and good common sense would have ensured.

Persian Philosophy!—A meteoric stone fell near Bombay, 5th of November 1814, and an account of the phenomenon is given by the Persian philosopher, Syed Abdulla. After enumerating the number of stones found, he observes, "The causes of this may be, that in the course of working the ground, air being extricated, may have entered into combination and come near elemental fire, and from this fire had received a portion of heat. It may then have united with brimstone and terrene salt, as, for instance, saltpetre; when the mixture, from some cause, being ignited, the fire bestows its own property on the mass, and the stones which have been above it are blown into the air. God knows the truth."

Extinct Race of Men.—Mr J. B. Pentland, in a paper read before the British Association at Edinburgh, 1834, states the reasons which have led him to conclude that there existed, at a comparatively recent period, a race of men very different from any of those now inhabiting our globe, characterised principally by the anomalous forms of the cranium, in which two-thirds of the entire weight of the cerebral mass is placed behind the occipital foramen, and in which the bones of the face are very much elongated. Mr Pentland entered into details to prove that this extraordinary form cannot be attributed to pressure or any external force similar to that still employed by many American tribes, and adduced, in conformation of this view, the opinion of Cuvier, of Gall, and of many other celebrated naturalists and anatomists. The remains of this race are found in ancient tombs among the mountains of Peru and Bolivia, and principally in the great inter-alpine valley of Titicaca, and on the borders of the lake of the same name. These tombs present very remarkable architectural beauty, and appear not to date beyond seven or eight centuries before the present period.

The race of men to which these extraordinary remains belong, appears to Mr Pentland to have constituted the inhabitants of the elevated regions, situate between the 14th and 19th degrees of south latitude before the arrival of the present Indian population, which, in its physical characters, its customs, &c., offers many analogies with the Asiatic races of the Old World.

WORKS ON NATURAL HISTORY AND THE PHYSICAL SCIENCES.

Two Lectures on Comets, by John Drew, 8vo, 2s.—Animal Magnetism and Homœopathy, by Edwin Lee, 2s.—Naturalist's Library, vol. xi., Deer, 8vo, 6s.—A Familiar History of Birds, by the Reverend E. Stanley, 2 vols. 8vo, 7s.—Narrative of a Voyage Round the World, by T. B. Wilson, M.D., 8vo, 12s.—Minerals and Metals, 18mo, 2s. 6d.—The Little Library—Natural History of Birds, illustrated by Landseer, 4s.—Recherches sur les Poissons Fossiles par L. Agassiz, Livraisons, i. to iv., price, each liv., 30s.—Human Physiology, by John Elliotson, M.D., Part I., 10s. 6d.—Ornithological Biography, by J. J. Audubon, vol. iii, 25s.

Edinburgh: Published for the Proprietors, at their Office, 16, Hanover Street. London: Charles Tilt, Fleet Street. Dublin: W. F. Wakeman, 9, D'Olier Street. Glasgow and the West of Scotland: John Smith and Son, 70, St Vincent Street.

THE EDINBURGH PRINTING COMPANY.

THE EDINBURGH

JOURNAL OF NATURAL HISTORY,

AND OF

THE PHYSICAL SCIENCES.

SATURDAY, FEBRUARY 13, 1836.

ZOOLOGY.

DESCRIPTION OF THE PLATE—THE SQUIRRELS.

ALL the animals of this tribe are light, nimble, and elegant, climbing trees with the utmost agility, and springing from branch to branch with astonishing security. They reside almost exclusively upon trees, where many of the species form their nests of moss and other soft substances. They subsist upon fruits and nuts of various kinds, which they instinctively store up for their winter's food, in the fissures of the trees, or in some other place of security.

Most of the Squirrels may be rendered perfectly tame and docile, with the utmost facility, owing to their natural gentleness of disposition, and the cheerfulness and contentment which they manifest in a state of captivity. They are playful and frolicsome, and become extremely attached to those whom they are accustomed to see. The progressive motions of the Squirrels, while on the ground, are performed by a succession of leaps. While eating they sit erect, and hold the food in their fore-paws. They are widely diffused over every quarter of the globe, with the exception of New South Wales.

Few animals of the order Rodentia, of which they are members, can be compared to the Squirrels, for the elegance of their form, the beauty of their fur, and the rapidity of their movements.

Modern Naturalists divide the Squirrels into three sub-genera, and a new genus, under the name of Pteromys. The first sub-genus consists of the true Tree Squirrels, which form a pretty extensive group, distinguished by the absence of the lateral folds of the skin which characterise the Flying Squirrels, now grouped under the generic name of Pteromys, and being devoid of the cheek pouches which are found in the Tamias or Ground Squirrels of America. Further, they are distinguished from the Guerlinguets by their tail being distichous throughout, whereas those of the latter sub-genus are round and distichous only at the extremity.

Fig. 1. The Malabar Squirrel (Sciurus Maximus) is perhaps the largest of its tribe, being from 8 to 9 inches in height, and from 15 to 16 inches in length, exclusive of the tail, which is somewhat longer than the body. It inhabits the coast of Malabar, and resides chiefly on palm trees; its food consists of various nuts, but particularly of cocoa-nuts, and the milky juice contained within them; both of which it is said they are remarkably fond.

Fig. 2. The Grey Squirrel (S. Cinereus), is a native of the United States of America. Though considerably inferior in size to the last species, it is still larger than all our European Squirrels.

This species associates in numerous bodies; it is particularly abundant in North and South Carolina, and also in Pennsylvania, where it feeds upon the young shoots of trees, buds, acorns, various nuts, and even descends to the fields and destroys grain. In summer it builds its nest on the extreme branches of the trees, and in winter retreats to the hollow of some decayed tree, in which it had previously laid up its winter store.

Fig. 3. The American Black Squirrel (S. Niger).—This squirrel has a pretty wide geographical range, having been found in the woody regions of Mexico, Florida, North and South Carolina, and Pennsylvania. It feeds on nuts like the rest of its congeners, and is said to be very destructive to crops of grain. Its flesh is considered a delicate food, and the animal is hunted on this account as well as for its fur, which is of a beautiful glossy black.

Fig. 4. The Chickaree (S. Hudsonius) is a native of those extensive white spruce forests which are so numerous in the fur countries of North America. It burrows only at the root of the largest trees, and generally forms four openings to its retreat for egress and ingress. Its principal food consists in the cones of the pine, under which its retreat is constructed; and it appears seldom to quit the same tree. Its skin is of little value, and has never formed an article of commerce. Its flesh is eaten by the natives, but that of the male has a strong marine flavour.

Fig. 5. The Plantane Squirrel (S. Plantani) is usually seven inches in length, although many of them are found not to exceed six and a half inches; the tail is somewhat longer than the body and head together. It is called the Bajing by the natives of Java, in which island it is very abundant, both on the sea-coasts and in those districts of the interior which are but slightly elevated above the level of the ocean. It is often found on tamarind trees, and indeed on all fruit trees, but is notorious for the injury it occasions to the cocoa-nut tree. It is, consequently, hunted by the natives, as the preservation of the cocoa-nut is a great measure depends upon their being driven out of the district. The Javanese kill numbers, and suspend their skins about their dwellings as trophies. It is a very prolific animal.

Fig. 6. Javanese Squirrel (S. Insignis).—This is the Bokhol of the Javanese. Its height is three inches and a half; from one extremity to the other, exclusive of

the tail, it measures seven and a half inches; the tail is as long as the neck and body together. It is a very scarce animal in Java, and has only been met with in the extensive forests of Bambangan. It has also been discovered in Sumatra, but is rare on that island.

Fig. 7. The Two-coloured Squirrel (S. Bicolor).—This figure represents the animal in its common dress, in the eastern part of Java. On the continent of India, and in Cochin-China, it is found almost entirely black above, and a golden yellow below. It retires to the deepest forests, where its food consists of wild fruits of various kinds. It is not at all destructive to the cocoa-nut trees. The inhabitants occasionally feed on its flesh; and some of them keep it in a domesticated state.

Fig. 8. White-eared Squirrel (S. Leucotis).—This is another large-sized species, measuring twelve inches from the nose to the insertion of the tail, and the tail itself is thirteen inches. It is a native of Upper Canada, and is also found in the State of New York.

Fig. 9. Raffles' Squirrel (S. Rafflesii).—This beautiful and singularly-marked species was discovered by Sir Stamford Raffles, at Sumatra. Its length, exclusive of the tail, is eight inches: the tail itself, eight inches and a third; and its height at the shoulder, three inches and three quarters.

DESCRIPTION OF THE PLATE—THE ORIOLES.

ALL the birds which compose the genus Orioles, as it is now restricted by Naturalists, inhabit Asia, Africa, or Europe; and the Orioles of America form another and a separate genus.

The plumage of the Orioles consists of two prevailing colours, namely, black and yellow. In most of the male birds, these are the only tints, but the females are frequently of a greenish cast above. The birds of this genus are said to be but partially gregarious, and to live in pairs in thickly wooded districts; previous to their migration, they are sometimes seen in very small flocks before leaving the districts in which they have bred. Their common food is fruits, berries, insects, and larvæ. They are very shy birds, and permit no one to approach.

The Orioles build their nests with great art; and generally place them in the fork of a small branch, from which the nests are suspended by their rims.

Fig. 1. The Javanese Oriole (O. Leucogaster).—This species has nothing to distinguish it from the true character of the genus Orioles, in the shape of the body, or form of the bill and claws. In respect to size, it is somewhat smaller than the Asiatic variety of the Golden Oriole. This bird is of extremely limited distribution, being found only in a few circumscribed situations in the Island of Java, where it leads a solitary life. Dr Horsfield found it at Blitar, a district nearly covered with vast forests, the closest shades of which it seldom quits.

Fig. 2. The Golden Oriole, male (O. Galbula), and Fig. 3. Female.—This elegant bird is widely distributed, inhabiting Asia, Africa, and Europe. It breeds in the warmer parts of Europe, and solitary instances are recorded of pairs having been seen in Britain. Those which incubate in Europe invariably migrate, about September, to Asia or Africa, where they pass the winter. The song of the Golden Oriole is very sweet.

Fig. 4. The Kink Oriole (O. Sinensis) has been found in Cochin-China, and also in the woods of Senegal. These Orioles are said to be much devoted to their young. Like many other species, when in the act of incubation, their true character entirely forsakes them, and they will fly at any person who attempts to approach their nests.

Fig. 5. The Black-headed Oriole (O. Melanocephalus).—This species, by some authors, has been considered merely as a variety of the Golden Oriole; but in all those individuals which have come under our notice, we have observed that their bills were somewhat longer than those of the Golden Oriole. It inhabits Africa and Asia.

Fig. 6. The Two-coloured Oriole (O. Bicolor) inhabits South Africa, and is that described by Le Vaillant as the Coudougnan. The female bird is represented in the plate.

The specimens from which we took the last five birds are in the superb collection of the Edinburgh University Royal Museum.

FECUNDITY OF THE COD.—Three persons undertook to number the Roe in a very fine Cod. One of them took as much of the roe as weighed a drachm, and after having counted the eggs contained in it, passed it to the others, who did the same; and as the numbers all agreed, they wrote down the total of the whole drachm, after which they weighed all the mass of eggs, and repeated eight times the sum of one drachm for every ounce. The addition of all these sums produced a total of 9,344,000.

ON THE HYBERNATION OF ANIMALS.—NO. I.

It is a remarkable fact, and one which has attracted considerable attention, that while many animals migrate to a more genial climate when the cold of winter approaches, others betake themselves to some hiding place, where they fall into a deep lethargy, in which they remain until the revival of spring. This state is usually designated by the term Hybernation, and is best marked among the Insects, the Mollusca, the Reptiles, and a few of the Mammalia. In all probability it occurs also in a considerable number of Fishes; indeed it seems absolutely necessary for the preservation of many of them during the winter, in cold climates, when the surface of the lakes is frozen over for months together. This state is apparently intimately connected with the preservation of the individuals subject to its influence, under circumstances where they would otherwise find it difficult or impossible to exist, and it has therefore received from Mangili the appropriate term of the Conservative Lethargy. According to Humboldt, a lethargic state similar to hybernation is also induced in some animals under the influence of a tropical heat.

The hybernating animals belonging to the class Mammalia resemble the other animals of that class in every other respect. They possess no uniformity of structure or appearance, by which we might be enabled, *a priori*, to predict that they would become lethargic when placed under certain circumstances. Some belong to the Carnassiers, as the Bats, the Hedgehog, and Tanrec; a greater number to the Rodentia, as the Marmot, the Hamster, and the different kinds of Dormice. They differ also from each other in the nature of their food. The Bats live on Insects; the Hedgehogs on Worms and Snails; and the Marmots, Hamsters, and Dormice, on nuts, roots, and herbs. Most of them are crepuscular or nocturnal feeders.

Prunelle denies that the Bear of the Alps can be properly said to hybernate. "No doubt," he says, "he lies in a drowsy state for days together, but he is never so lethargic as to allow the hunters to approach him. He also adds that tame Bears never become lethargic. It appears, however, from numerous and authentic sources of information, that the Bears of more northern regions pass the winter in a state of actual hybernation.

When an animal is examined in this state, it is found to be under the influence of a lethargy more or less deep, and with all its vital actions nearly at a stand. This lethargy only occurs within a certain range of temperature, apparently varying to a small extent in the different hybernating animals of the same country, but probably to a considerable extent in different, or in the same hybernating animals, of dissimilar countries. The Dipus Sagitta is said to become equally lethargic, during the winter months, in Egypt and in Siberia. The Tanrec (*Centenes ecaudatus*) is also said to remain lethargic for six months of the year in the climates of India and Madagascar, of which places it is a native.

To produce this state in a Marmot, the temperature must not be raised much above 50° Fah., and should not be carried below the freezing point. When the temperature is lowered beyond this, the animal becomes lively, its respirations and the pulsations of the heart increase in frequency, its animal heat rises, and it endeavours to escape or protect itself from the cold. If it should remain exposed for some time to this diminished temperature, it proves fatal in the same manner as in the other Mammalia, viz. by inducing coldness, torpor, and complete cessation of the heart's action, and other vital functions. This torpor, which precedes death in the hybernating animals when exposed to extreme cold, is quite a different thing from the conservative lethargy peculiar to them; and these must never be confounded in any discussions or experiments on this subject. The lethargy resembles a deep sleep, from which the animal may be roused in perfect health,—the other is the failure of the vital functions, which precedes death in every animal placed under similar circumstances. Those who compare this lethargy to the irresistible torpor which creeps over the human species when exposed to extreme cold, must entertain very erroneous notions concerning its nature. Mangili, on placing the fat Dormouse (*Myoxus glis*) in a temperature equal to 17° Fah., found that it struggled to escape; but when placed in a temperature of 48°, it became lethargic. He killed a Muscardin (*Mus*, or *Myoxus avellanarius*) in twenty minutes, by exposing it to a temperature equal to about 8° Fah. He also found, that when Bats were placed in a temperature one degree above the freezing point, they were lively for some time, then became torpid, stiffened, and died. Prunelle killed a Hedgehog (*Erinaceus Europœus*) by confining it in a temperature of about 5° Fah. for twenty-two hours. He also found that a Hedgehog was lively and fed in a temperature of 41° Fah., and was dormant when it rose to 50°. Marshall Hall states, that when Dormice are supplied with cotton or wool, they become sooner lethargic than those to whom this is denied. In fact, all the hybernating animals can easily be killed by their exposure to severe cold in this manner, and nothing appears more effectual in rousing them from their lethargy than its application, as the experiments of John Hunter long ago proved. In this manner we can easily explain the circumstance of the Bat having been occasionally seen flying about houses in the middle of winter. The place to which it had retired for the winter had not been sufficiently sheltered to prevent the temperature falling below what was compatible with the lethargic state; it had become roused, and had betaken itself to the wing to look out for a warmer retreat.

All the hybernating animals instinctively take precautions against the fatal consequences of severe cold. They either retire into caverns, into holes, or make nests for themselves. The Bats retire into caverns, hang suspended by their claws to the roofs in clusters, and cover each other with their wings. The Marmots (*Arctomys marmotta*) retire to their holes at the end of September, and reappear at the end of April. As several of them retire to the same hole, and as they stuff the mouths of these with grass and other materials, the temperature can never fall very low. Prunelle states that the temperature at the bottom of their holes is from 46° to 48° Fah. The Hedgehog and Dormice make warm nests for themselves at the approach of winter.

This lethargy is not generally so profound as many have been led to believe. Mangili states that the slightest disturbance of the animal sometimes causes signs of irritability. Prunelle and Marshall Hall have dwelt upon this as one of the principal difficulties in making accurate observations upon the functions of the circulation and respiration in these animals, even while fairly under the influence of this lethargy. These signs of irritability are marked by muscular motion, increased respiration, pulsation, and animal heat. Prunelle also states that they can be roused by electricity, or by the fumes of ammonia applied to the nostrils. The stories told of dissecting these animals under a state of lethargy, without their exhibiting signs of sensation, must have been entirely a mistake; the authors of them must have confounded the state of torpor from extreme cold with that of lethargy.

This lethargy seems to vary in degree in different animals, and in different individuals of the same species. Mangili descended into the famous Grotto of Entrastico, in the month of December 1775, the temperature of the interior being about 53° Fah. He discharged a musket near 300 Bats, the *Vespertilio murinus* of Naturalists, without observing the slightest movement among them. He next discharged a loaded musket among them, and though many fell killed and wounded, those that remained untouched were perfectly quiescent. He returned into the same grotto in February 1804, the temperature of the interior being nearly the same as on the former occasion. Most of the Bats observed at this visit belonged to the *Vespertilio noctula*, which visits England during the summer, but retires to Italy to pass the winter in a lethargic state. The light of the torch, on this occasion, was sufficient to make some of them change their places. The same author preserved a Marmot for two years, which never became lethargic under all the different changes of temperature. M. Bossi kept two Marmots for two years, which were exposed for a time to a temperature from 18° to 20° Fah. without becoming lethargic.

Prunelle states, that having entered into the ancient aqueducts at Lyons and Vienne several times during the winter, he found, that while the greater number of Bats were in the lethargic state, some flew about as in the middle of summer. In the end of January 1807, he descended into a subterranean passage at Brunette, and found a great number of Bats collected upon the roof, in groups of from ten to twelve in number; others had placed themselves in holes, the greater part of them lethargic and cold as the stones upon which they rested; while others still flew about, but very feebly. The temperature of the place was 50° Fah.; the external air was nearly 39°; the temperature of the animals themselves was from 41° to 63°.

J. R.

Extraordinary!—The larvæ of a certain fly (*Eristalis tenax, Meig*) will admit of being pressed in a bookbinder's press, as broad and thin as a card, without being killed, when freed from its confinement and restored to its usual dwelling-place.—*Dr Hermann Bermeister's Manual of Entomology.*

BOTANY AND HORTICULTURE.

GENERAL REVIEW OF THE VEGETABLE KINGDOM.—NO. I.

BY C. F. BRISSEAU MIRBEL, MEMBER OF THE INSTITUTE OF FRANCE.

[The following subjects will be treated of briefly in a series of articles, of which this is the first. 1. The laws which regulate the distribution of the different tribes of plants over the globe. 2. The influence which climate, elevation, aspect, and soil, have upon these beings. 3. The effect which plants, in their turn, produce on the exterior bed of soil, on the temperature from latitude or position, as well as on the general constitution of the atmosphere.]

Multitudes of different species of plants are found spread over the whole surface of the globe. Like animals, these are endowed with the faculty of increasing their races to infinity; and differ from each other as much by their interior structure as external appearance; each has its peculiar wants, and, if we may be allowed the terms, its separate habits and instinct.

We see that some species belong to the mountains, others to the valleys, and others to the plains; some affect a clayey soil, some a chalky one, others one of a quartzose nature, while many will thrive in no place but where the soil is impregnated with soda and muriatic salt. There are some that confine themselves entirely to water; dividing themselves again into those of the marsh, the lake, the river, and the ocean. Some require the hottest climates, others delight in mild and temperate ones, others thrive nowhere but in the midst of ice and frost. A large proportion must have a constantly humid atmosphere, several do very well in a dry air, but the major part are equally averse to the extremes of both dryness and moisture. There are those which flourish when exposed to the action of a strong light, while others prefer the weaker action of that element. The result of this variety of wants is, that nearly the whole surface of the earth is occupied by vegetation.

Excess of heat, cold, or drought, a total privation of air or light, are the only bars to vegetation; and yet we find some agamous species (such as are presumed to propagate themselves without the intervention of the organs of fructification) which grow in caverns where the light has never gained admission.

Seeing that the forms of vegetables are infinitely various, and that certain species, genera, and even tribes, are attached exclusively to particular countries; and that this distribution of races, a consequence of the first order of creation, has maintained itself to our day by the effect of climate and situation, without perceptible deviation,—it must be admitted that the soil takes one of its distinctive features from the vegetation it bears.

Some species are confined to the narrowest limits. The *Origanum Tournefortii*, discovered by Tournefort in 1700, in the little island of Amorgos, upon one rock only, was found 80 years afterwards by Sibthorp, on the same island and upon the same rock; but no one has ever observed it any where else. Two of the *Orchideæ*, *Disa longicornis*, and *Cymbidium tabulare*, grow upon the Table Mountain, at the Cape of Good Hope; and Thunberg, who has described them, found them on no other spot.

Mountainous countries afford many of these local species, such as dwell secluded on the heights without ever migrating to the plains below. Thus we find that the Pyrenees, the Alps, the Appenines, &c., have their peculiar Floras, and that even some separate mountains of those great chains have species allotted to them alone, and which are not to be found on the adjoining summits.

Speculatively, we might presume that all the individuals of one species would establish themselves under the same, or nearly the same, degrees of latitude, as they would find a nearly similar climate. But, in reality, some species extend themselves in the direction of the longitude, and never swerve to the right or to the left. This is one of those anomalies of which it is not easy to trace the cause. The *Phalangium bicolor* begins to show itself in the country round Algiers; it crosses over to Spain, clears the Pyrenees, and terminates its career in Brittany. *Menziesia polifolia* belongs to Portugal, France, and Ireland. The heaths are confined exclusively to Europe and Africa; they extend themselves from the regions bordering on the Pole to the Cape of Good Hope, over a surface which is very narrow in proportion to its breadth. The *Ramonda pyrenaica*, as yet only found in the Pyrenees, follows, without deviating from its course, the valleys in those mountains which run from north to south, and so closely that not a single plant of it has been descried in those which skirt the chain in the other direction. But we will now quit insulated facts, and turn our attention to vegetation in general.

It may be observed that, with the exception of the Lichens, which bid defiance to all climates alike, a vastly greater proportion of species is calculated to endure a very high degree of warmth, than is calculated to bear severe cold. The progressive course of the proportion demonstrates itself most clearly if we direct our view from the polar towards the equinoctial regions. Botanists compute that at Spitzberg, which lies near the 80th degree of northern latitude, there are only about 30 species; in Lapland, which lies in the 70th degree, about 534; in Iceland, in the 65th degree, about 553; in Sweden, which reaches from the southern parts of Lapland to the 55th degree, 1300; in Brandenburgh, between the 52d and 54th degree, 2,000; in Piedmont, between the 43d and 46th degree, 2,800; nearly 4,000 in Jamaica, which is between the 17th and 19th degree; in Madagascar, situate between the 13th and 24th degree, under the tropic of Capricorn, more than 5,000. But such computations are very wide of the true proportion of species which belong to hot climates, as opposed in that respect to cold or temperate climates. To come at the real amount of the difference, we must first know the number of species spread over the whole globe; how many belong to the same space, under the same longitudes, at different latitudes; how many are common to several countries at the same time; how many belong exclusively to peculiar regions;—points that will require the lapse of ages for the Botanist to enable himself to resolve.

The general face of the vegetation of a country does not depend solely upon the number, it depends also upon the more or less remarkable characters of the species found there. The chief part of these characters are fixed, and are derived, as I have said before, from primitive creation, not from the effect of climate. As to the proposition, that certain vegetable forms are necessarily co-existent and dependent upon certain other animal forms in a given climate (an occult law of nature, of which some ingenious writers have endeavoured to find the proof, in those harmonies and contrasts which always result from the approximation of different beings), we do not presume to controvert it; but sound reasoning rejects its adoption as a doctrine, while the connection and reciprocal control of the phenomena of nature are unknown to us. Cautious and exact observers of those things which are the objects of our senses, let us leave to the fancy of the poet the bold task of unfolding the purposes of the Creator in his works, while we confine ourselves to the less presumptuous one of describing them as we find them.

Vegetation, within the tropics, fills the European traveller with amazement, by the majesty and vigour of its aspect. The proportion of the woody to the herbaceous species, is vastly more considerable towards the Equator than in Europe; and the difference is therefore in favor of the equinoctial regions, for trees give the character of grandeur to vegetation.

On the Duration of the Germinative Power of the Seeds of Plants.—The Society for the Improvement of Horticulture in Prussia, proposes from time to time certain questions, to which it directs the attention of Horticulturists. The following was proposed by the Society:—"Is it true that the seeds of the Melon and Cucumber, being preserved for some years, yield a greater abundance of fruit?" Most observers remark, that the plants obtained from the seeds of the preceding year produce many leaves, but few fruitful flowers, and almost entirely male ones; but that these same seeds, dried by the heat of the sun or of a stove, yield a greater number of fruitful plants, and that it is particularly at the end of some years that they acquire this property. These experiments vary from three to twenty years. The heat of the human body may be useful, but it must be used with discretion, or the germinating powers of the seeds will be destroyed.

The author of the preceding remarks made experiments of the same kind on balsams and gillyflowers. He sowed at the same time some seeds of the last, part of which were of the preceding year, others of some years previous. The first came up much sooner than the second, and gave only simple flowers; the others produced only sixteen out of several hundred plants.

M. Schmidt employs seed from five to twelve years old; those of twenty years do not grow. Professor Sprengel of Halle says, he obtained no fruit from seed a year old. M. d'Arenstorff, of Drebleau, obtained fruit, remarkable for flavour and size, from seed of twenty years old. The observations of Professor Treviranus, of Berlin, have afforded the same result. A vigorous vegetation induces, in numerous plants, male flowers in the greatest abundance, sometimes even exclusively. This has been proved as far as regards Cucumber seeds; but those which are too old produce an opposite result. He has seen seeds of five years old produce only female flowers; they were fecundated by male flowers of another bed, and yielded fruit.

M. Voss, head gardener at Sans Souci, sowed on the 17th February 1827, twenty-four seeds of a Spanish Melon of the year 1790, being consequently thirty years old, and he obtained eight plants which gave good fruit. This experiment, the most remarkable of all, will excuse our citing eleven others, which he made with seed of a less age, and of different species. Cucumber seeds of seventeen years afforded the same results. M. Voss adds, that some seeds of the *Althæa rosea*, of twenty-three years old, afforded very well-conditioned plants.

We admit, as incontestible, the above-mentioned observations. It is known that the seeds of different families retain for a greater or less time their germinative power. To cite only one example from among the Leguminous plants: About twenty-six years since, we believe, fruit was obtained in the Royal Garden, from a species of Phaseolus or Dolichos, taken from the herbarium of Tournefort.

The Rev. Mr White makes the following observations on this subject:—"The naked part of the Hanger at Selborne," says he, "is now covered with thistles of various kinds. The seeds of these thistles may have lain probably under the thick shade of the beeches for many years, but could not vegetate till the sun and air were admitted. When old beech trees are cleared away, the naked ground in a year or two becomes covered with strawberry plants, the seeds of which must have lain in the ground for an age at least. One of the *slidders*, or trenches, down the middle of the Hanger, close covered over with lofty beeches, near a century old, is still called *strawberry*-slidder, though no strawberries have grown there in the memory of man. That sort of fruit did once, no doubt, abound there, and will again, when the obstruction is removed."

Sir Thomas Dick Lauder, Bart., made some curious and interesting experiments in 1817, on the germination of seeds, which we shall describe in his own words:—"A friend of mine possesses an estate in Morayshire, a great part of which lying along the Moray Frith, was, at some period not very well ascertained, but certainly not less than sixty years ago, covered with sand, which had been blown from the westward, and overwhelmed the cultivated fields, so that the agriculturist was forced to abandon them altogether. My friend, soon after his purchase of the estate, began the arduous, but judicious, operation of trenching down the land, and bringing to the surface the original black mould. These operations of improvement were so productive, as to induce the very intelligent and enterprising proprietor to undertake lately a still more laborious task, viz., to trench down the superincumbent sand on a part of the property, where it was no less than eight feet deep.

"Conceiving this to be a favourable opportunity for trying some experiments relative to the length of time which seeds preserve their power of vegetation, even when immersed in the soil, I procured from my friend a quantity of the mould, taken fresh from under the sand, and carefully avoiding any mixture of the latter. This was instantly put into a jar, which was stopped up close, by means of a piece of bladder tied tightly over its mouth. Having prepared a couple of flower-pot-flats, by drilling small holes in the bottoms of them, so as to admit of the ascent of water, I filled the flats with some of the mould, and placing them in a very wide and shallow tub, made on purpose, I covered each of them with a large glass receiver. Each receiver, however, was provided with a brass rim, having little brass nobs on it, so as to raise its edge from the bottom of the tub, and leave a small opening for the admission of air. The whole apparatus was placed in my library, of which the door and windows were kept constantly shut.

"This was done on the 17th February last. It is now the 6th of May; and, on examining the flats, I find about forty-six plants in them, apparently of four different kinds; but as they are yet very young, I cannot determine their species with any degree of accuracy."

Sir Thomas has since informed us, that the seeds which germinated were all highly oleaginous; and the plants produced were the mouse-ear (*Myosotis scorpioides*), scorpion grass (*Lamium purpureum*), purple archangel, and (*Spergula arvensis*), corn-spurrey. The earth thus experimented upon was taken from the lands of Inveragie.

GEOLOGY.

The Mud Volcano of Grobogan.—Having received an extraordinary account of a natural phenomenon in the plains of Grobogan, 50 pals or miles N.E. of Solo, a party, of which I was one, set off from Solo on the 8th of September to examine it.

On approaching the village of Kuhoo, we saw, between two trees in a plain, an appearance like the surf breaking over rocks, with a strong spray falling to leeward. The spot was completely surrounded by huts for the manufacture of salt, and at a distance looked like a large village. Alighting, we went to the Bludugs, as the Javanese call them. They are situate in the village of Kuhoo, and by Europeans are called by that name. We found them to be on an elevated plain of mud, about two miles in circumference, in the centre of which immense bodies of salt mud were thrown up to the height of from ten to fifteen feet, in the forms of large globes, which bursting, emitted volumes of dense white smoke. These large globes or bubbles, of which there were two, continued throwing up and bursting seven or eight times in a minute, by the watch. At times they threw up two or three tons of mud. We got to leeward of the smoke, and found it to smell like the washing of a *gun* barrel. As the globes burst, they threw the mud out from the centre with a pretty loud noise, occasioned by the falling of the mud upon that which surrounded it, and of which the plain is composed. It was difficult and dangerous to approach the large globes or bubbles, as the ground was all a quagmire, except where the surface of the mud had become hardened by the sun. Upon this we approached cautiously to within fifty yards of the largest bubble or mud pudding, as it might very properly be called, for it was of the consistency of a custard pudding, and of very considerable diameter; here and there, where the foot accidentally rested on a spot not sufficiently hardened to bear, it sunk, to the no small distress of the walker.

We also got close to a small globe or bubble (the plain was full of them of different sizes), and observed it closely for some time. It appeared to heave and swell; and when the internal air had raised it to some height it burst, and the mud fell down in concentric circles, in which shape it remained quiet until a sufficient quantity of air was again formed internally to raise and burst another bubble. This continued at intervals, from about one half to two minutes. From various other parts of the quagmire, round the large globes or bubbles, there were occasionally small quantities of mud shot up like rockets, to the height of 20 or 30 feet, and accompanied by smoke. This was in parts where the mud was of too stiff a consistency to rise in globes or bubbles. The mud at all the places we came near was cold on the surface, but we were told it was warm beneath. The water which drains from the mud is collected by the Javanese, and by being exposed, in the hollows of split bamboos, to the rays of the sun, deposits crystals of salts. The salt thus made is reserved exclusively for the Emperor of Solo. In dry weather it yields 30 dudgins, of one hundred catties each, every month, but in wet or cloudy weather less.

In the afternoon we rode to a place in a forest, called Ramsam, to view a salt lake, a mud hillock, and various boiling, or rather bubbling pools. The lake was about half a mile in circumference, of a dirty-looking water, boiling up all over in gurgling bodies, but more particularly in the centre, which appeared like a strong spring; the water was quite cold, and tasted bitter, salt, and sour, and had an offensive smell. About 30 yards from the lake stood the mud hillock, which was about 15 feet high from the level of the earth. The diameter of its base was about 25 yards, and its top about eight feet, and is form an exact cone. The top is open, and the interior keeps constantly working and hearing up mud in globular forms, like the Bludugs. The hillock is entirely formed of mud which has flowed out of the top; every rise of the mud was accompanied by a rumbling noise from the bottom of the hillock, which was distinctly heard for some seconds before the bubbles burst. The outside of the hillock was quite firm. We stood on the edge of the opening and sounded it, and found it to be 11 fathoms deep. The mud was more liquid than the Bludugs, and no smoke was emitted from the lake, hillock, or pools.

Close to the foot of the hillock was a small pool of the same water as the lake; which appeared exactly like a pot of water boiling violently; it was shallow, except in the centre, into which we thrust a stick twelve feet long, but found no bottom. The hole not being perpendicular, we could not sound it with a line.

About 200 yards from the lake were several large pools or springs, two of which were eight and ten feet in diameter. They were like the small pool, but boiled more violently, and smelt excessively. The ground around them was hot to the feet, and the air which issued from them quite hot, so that it was most probably inflammable; but we did not ascertain this. We heard the boiling 30 yards before we came to the pools, resembling in noise a waterfall. The pools did not overflow; of course the bubbling was occasioned by the rising of air alone. The water of one of the pools appeared to contain a mixture of earth and lime, and, from the taste, to be combined with alkali. The water of the Bludugs and the lake is used medicinally by the Javanese, and cattle drinking of the water are poisoned.—*Extract of a letter from S. T. Good, Esq. East India Company's Service, Java.—May*, 1815.

GENERAL SCIENCE.
MOUNTAINS AND VOLCANOS OF THE MOON.

We are less ignorant regarding the physical constitution of the moon than of any other heavenly body, in consequence of the smallness of its distance from us. By the assistance of the telescope we are able to observe the nature of those spots and shadows with which its surface is always marked. Some parts cast a shadow on the side farthest from the sun, and these must be mountains; others again throw their shadows on the side next to the sun, and hence are cavities. The appearance of these spots on the moon varies according to the position of the sun, as the shadows may be thrown in different directions, or they may be entirely destroyed by the sun being exactly vertical to them. Further, the shadows appear exactly in proportion to the length they ought to have, if we consider the inclination of the sun's rays to that part of the moon on which they appear. Thus we must conclude that the spots on the moon are mountains and valleys analogous to those on our own globe.

"The convex outline of the limb turned towards the sun is always circular and very nearly smooth," observes Sir J. F. W. Herschell; "but the opposite border of the enlightened part, which (were the moon a perfect sphere) ought to be an exact and sharply-defined ellipse, is always observed to be extremely ragged, and indented with deep recesses and prominent points. The mountains near this edge cast long black shadows, as they should evidently do, when we consider that the sun is in the act of rising or setting, to the parts of the moon so circumstanced. But as the enlightened edge advances beyond them, i. e. as the sun to them gains altitude, their shadows shorten; and at the full moon, when all the light falls in our line of sight, no shadows are seen on any part of her surface. From micrometrical measures of the lengths of the shadows of many of the more conspicuous mountains, taken under the most favorable circumstances, the heights of many of them have been calculated, the highest being about 1¾ English miles in perpendicular altitude. The existence of such mountains is corroborated by their appearance, as small points or islands of light beyond the extreme edge of the enlightened part, which are their tops catching the sun-beams before the intermediate plain, and which, as the light advances, at length connect themselves with it, and appear as prominences from the general edge."

"The generality of the lunar mountains," continues the same accurate observer, "present a striking uniformity and singularity of aspect. They are wonderfully numerous, occupying by far the larger portion of the surface, and almost universally of an exactly circular or cup-shaped form, fore-shortened, however, into ellipses towards the limb; but the larger have, for the most part, flat bottoms within, from which rises centrally a small, steep, conical hill. They offer, in short, in the highest perfection, the true volcanic character, as it may be seen in the crater of Vesuvius, and in a map of the volcanic districts of the Campi Phlegræi or the Puy de Dôme. And in some of the principal ones decided marks of volcanic stratification, arising from successive deposits of ejected matter, may be clearly traced with powerful telescopes. This I state from my own observations. What is, moreover, extremely singular in the geology of the Moon is, that although nothing having the character of seas can be traced (for the dusky spots which are commonly called seas, when closely examined, present appearances incompatible with the supposition of deep water), yet there are large regions perfectly level, and apparently of a decided alluvial character."

Sir William Herschell, in 1787, observed three volcanos in the moon; one of which, he says, was an actual eruption of fire or luminous matter. Captain Kater remarked, on the 4th February 1821, a luminous appearance in the dark part of the moon, represented at A.

This spot is believed to be identical with the mountain called Aristarchus. Hevelius called it Mons Porphyrites, and considered it to be volcanic.

This volcano, as observed by Captain Kater, appeared like a small nebula, subtending an angle of about 3 or 4 seconds. Its brightness was very variable; a luminous point, like a small star of the sixth or seventh magnitude, would suddenly appear in its

centre, and as suddenly disappear, and these changes would sometimes take place in the course of a few seconds. On the evening of the 5th the same phenomena were observed as before, only in an inferior degree. It had become more faint on the 6th, and the star-like appearance less frequent; and on the 7th it was scarcely visible.

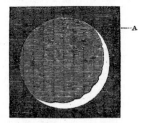
·····A

The existence of volcanos in the moon must therefore be considered as fairly established by observation. Independently of these observed facts, analogy would lead us to expect that the same laws which are found to regulate the internal constitution of our globe should also prevail in the other heavenly bodies.

MISCELLANEOUS.

Human Horns.—Excrescences of a horny nature have been occasionally observed on the human head. A portrait of a woman with excrescences of this description is deposited in the British Museum, and one of the horns is, we believe, still preserved in the Museum at Oxford.

A far more recent instance, however, was observed by Dr Wm. Roots, of Kingston-upon-Thames, who, in February 1811, amputated an excrescence of this sort, exactly resembling a *ram's horn*, from the head of a man, between fifty and sixty years of age, a drawing of which, in its growing state, as well as the horn itself, were presented by him to the collection of Sir Astley Cooper.

The account given by Dr Roots of this extraordinary case is, that John Kennedy, a gardener at Thames-Ditton, in Surrey, in the year 1796, had a tumour growing on the superior part of the occiput, which was taken off with the knife by the Doctor's father in about three years from its commencement. Soon after its removal, a horny substance began to make its appearance on the same place, which continued growing for four years, till it accidentally fell off in a most unexpected manner, being at that time not more than three inches in length; and it should be observed, that the surface of the part it grew from, on its dropping off, was perfectly smooth, without the slightest hemorrhage, and resembling the superficies of the stag's head when his horns have recently dropped. In a short time afterwards, a new horny sprout shot forth, which, as it grew, took on the exact form and crooked figure of the *ram's horn*. Having increased during seven years, without any disposition to fall off, to the great inconvenience of the poor man, he consented at length to its removal; in the performance of which, from the parts underneath being very vascular, a considerable hemorrhage ensued.

It appears probable from this, says Dr Roots, that had the horn been suffered to remain longer on the head it would have attained a much larger size, as Nature, in a playful mood, had most abundantly supplied it with vessels for that purpose. Its having likewise been observed, that the horn grew, *without bleeding*, induces Dr Roots to conclude, that as the sources of its nutriment continued open, it had not arrived at that state of perfection, when the gradual closing of the vessels would occasion spontaneous removal without any hemorrhage taking place, as is always observed to be the case with animals which drop their horns at regular and stated periods. It would appear that these horny excrescences, which are occasionally formed on the skin of the human subject, were originally encysted tumours, the cyst of which very curiously assumes the power of secreting horn instead of fat, a pappy substance, or a fluid like honey, as sometimes happens in particular cases. This case mentioned by Dr Roots tends to corroborate the above opinion, since, before any horn made its appearance, there was observed a tumour; and after the first horn dropped off, a surface, which was quite smooth, and did not bleed at all, presented itself to notice.

We may remark that the case of the human head assuming the power of secreting horn, is an example of that tendency, in the higher animals, to assume the characters of others lower than themselves in the scale of creation.

The Rogue Elephant is a curiosity in Natural History. He is an individual completely banished from the herd, which will not suffer him to approach or mingle with them. No cause can be assigned for the fact; but this animal is always more savage and ferocious than any of the rest. He attacks man, and faces every danger which the others would avoid.

To Correspondents.—We shall not have room for F. F. F.'s notices.

Edinburgh: Published for the Proprietors, at their Office, 16, Hanover Street. London: Smith, Elder, and Co., 65, Cornhill. Glasgow and the West of Scotland: John Smith and Son, 70, St Vincent Street. Dublin: W. F. Wakeman, 9, D'Olier Street.

THE EDINBURGH PRINTING COMPANY.

THE EDINBURGH
JOURNAL OF NATURAL HISTORY,

AND OF

THE PHYSICAL SCIENCES.

SATURDAY, MARCH 13, 1836.

ZOOLOGY.

DESCRIPTION OF THE PLATE—THE WHIDAH BUNTING.

FROM the remarkable characters of this section of Buntings, we have devoted a whole plate to the illustration of a single species (the *Emberiza Paradisea*), in its various states of plumage. This bird is an inhabitant of Africa, and is said to be plentiful in the neighbourhood of Fort Whidah, near Angola, whence it derives its name. It is very commonly designated the *Widow Bird*, being a corruption of its real appellation.

Like most of the feathered tribe, it undergoes two moults during the year; and when it has assumed its perfect garb, after each of these, presents appearances so very dissimilar, that it seems a totally distinct species. In its full summer dress it is adorned with four central tail feathers, which are both long and black, as exhibited at fig. 1. Its head and throat are also black; the sides of the neck are of a high orange brown. While in its winter attire it is denuded of these elegant appendages, and presents the more humble aspect of our common Corn Bunting. As the seasons in South Africa are opposite to those of Britain, the summer plumage of this bird appears during our winter, or about November and December; while its winter plumage appears during our summer, or in May and June. This bird appears to undergo an almost perpetual moult. After it has shed the long tail feathers, which usually takes place about the 6th December, it is something like fig. 3; but gradually changes till it assumes the Sparrow-like appearance of fig. 4; and finally acquires its winter plumage as seen at fig. 5. This is about the 20th of June in Britain. It remains but a very short time in this garb, when black feathers begin to shoot out in various places, and it soon acquires an appearance something like fig. 4. The moult continues; and the external appearance of the bird changes with all the varied tints of the rainbow. The order comprehends some of the largest species of Insects, while many of them are the most minute which have yet been discovered; the perfection of dress as represented by fig. 3; shortly after which the long central tail feathers begin to appear, and is seen, as in fig. 2, on or about the 1st of October. The central tail feathers increase in length until it has again acquired its full summer habit, as at fig. 1. Different as are the various appearances of the bird in the plate, it is subject to many more, during the transitions from one condition to another.

The bill undergoes considerable change, both in shape and colour, which are produced by exfoliation. It is deep bluish-black in summer, and pale lead-colour in winter.

The female Whidah Bunting, when young, has much the appearance of the male bird in its winter attire, but is considerably deeper in the tone of its plumage, which annually becomes darker till it arrives at mature age, which is said to be four years. It is represented at fig. 6.

Our figures of the male bird have been drawn from a living specimen which has been in the possession of Sir Patrick Walker of Drumsheugh since 1827. This specimen is supposed to have been two years old when he first procured it. Being a long-lived species, it has been known to survive in Europe to the age of sixteen years when in a captive state.

DESCRIPTION OF THE PLATE—THE BEETLES.

THE Insects which compose the order *Coleoptera* are the most perfect of this class of beings, and are, therefore, placed at its head. A striking characteristic which prevails generally throughout the order is, that their lower wings are covered by a scaly sheath. There are, however, a few species which possess all the other characters, but are destitute of this protection to their under wings.

Coleopterous Insects are remarkable for the perfection of their organic structure, the singularity of their forms, and the splendour of their colouring. Many of them exhibit hues of the most brilliant metallic lustre, sparkling with all the varied tints of the rainbow. The order comprehends some of the largest species of Insects, while many of them are the most minute which have yet been discovered; the perfection of whose structure is well calculated to excite our highest admiration of that infinite skill and adaptation manifested in all the works of the creation. Is it not truly wonderful that beings invisible to the naked eye, and measuring only the ninetieth part of an inch, should possess as complicated a form as their congeners which measure six inches in length!

Fig. 1. The Fiery Beetle (*Onthophagus Igneus*) is remarkable for the brilliancy of its head and thorax, which assume a fiery appearance, contrasted with the dull black of its lower wing covers, or *elytra*. It is a native of India.

Fig. 2. The Kangaroo Beetle (*Scarabæus Macropus*).—There is something very singular in the conformation of this Beetle. Only one specimen has yet been discovered, and is said to be in the cabinet of Mr Macleay. It is supposed to have been brought from South America.

Fig. 3. The Atlas Beetle (*S. Atlas*).—This large and curious Insect is a native of Java, and is said also to occur on the continent of Asia.

Fig. 4. The Gigantic Beetle (*S. Tityus*), Male, and fig. 5, the Female.—It will be remarked that there is a considerable difference between the male and female of this species; the male is provided with two strong horns, while the female is destitute of them. It is a very rare species, and inhabits Carolina, Virginia, and other states of North America.

Fig. 6. The Hercules Beetle (*S. Hercules*).—This gigantic species inhabits the Antilles, where it is very plentiful; it is sometimes found in several of the other American islands.

Fig. 7. The Elephant Beetle (*S. Elephas*) may be justly ranked as one of the most beautiful and interesting of its tribe. It is a native of the West Indies.

Fig. 8. The Golden Beetle (*Chrysophora Chrysoclora*).—This species *is* nearly allied, in many particulars, to the Kangaroo Beetle, and was first discovered in Peru, by Baron Humboldt. It lives in societies.

Fig. 9. The Prodigal Beetle (*Rutela Sumptuosa*).—This is a beautiful Insect, and forms a fine addition to a genus whose colours are almost all of a sombre hue. It is a native of Brazil.

Fig. 10. The Shining Beetle (*R. Nitescens*), so named from the glossy appearance of its body, is a native of Brazil.

Fig. 11. The Clubbed Beetle (*Macraspis Clavata*).—This is the appearance presented by the Insect when in a living state, but when dead it assumes a much darker colour. It is a native of Brazil.

Fig. 12. The Goliath Beetle (*Goliathus Magnus*).—This superb beetle is a native of the West Coast of Africa, and only one specimen has hitherto been found, which is preserved in the Hunterian Museum of Glasgow.

Fig. 13. The Hieroglyphic Beetle (*Gymnetis Hieroglyphica*) is a native of Brazil.

Fig. 14. Macleay's Beetle (*Euchlora Macleayana*) is a very beautiful Insect, and a native of Madras.

Fig. 15. The Peruvian Beetle (*Chryscina Peruviana*).—This very handsome beetle has one remarkable peculiarity which distinguishes it from its congeners. The green colour, at least on its upper surface, is merely superficial, and may be removed by friction, when it exposes the black colour beneath.

A CURIOUS FACT CONCERNING BEES.—As a small vessel was proceeding up the Channel from the coast of Cornwall, and running near the land, some of the sailors observed a swarm of Bees on an island; they steered for it, landed, and took the bees on board, succeeded in hiving them immediately, and proceeded on their voyage. As they sailed along the shore, the Bees constantly flew from the vessel to the land to collect honey, and returned again to their moving hive; and this was continued all the way up the Channel.

FIRE FLY.—At Baltimore I first saw the Fire Fly. They begin to appear about sunset, after which they are sparkling in all directions. In some places ladies wear them in their hair, and the effect is said to be very brilliant. Mischievous boys will sometimes catch a bull-frog and fasten them all over him. They show to great advantage; while the poor frog, who cannot understand the "new lights" that are breaking upon him, affords amusement to his tormentors by hopping about in a state of desperation.—*Vigne's Six Months in America.*

ATTACK OF SPARROWS ON A MOUSE.—In the summer of 1831, one of the residents of the Temple, London, turned a Mouse loose in the open grand space, and the little intruder had no sooner made its appearance than he was simultaneously attacked by the Sparrows. So furious was their onset, that he was compelled to run in all directions. Endeavouring to escape their fury, he leaped up in great agony from the severe pecks with which he was assailed, and in a very few minutes a period was put to his existence.

UTILITY OF CROWS.—In a field near a gentleman's house, about a mile from Caernarvon, there are some out-buildings much infested with Rats. Four or five traps are set on the premises every night, and it is the business of a servant-man to go to the spot between five and six in the morning. He is always punctually met by a company of Crows that station themselves at a little distance, and most narrowly watch all his proceedings. No sooner does he remove his captives from the traps, and throw them into the field, than the carnival begins. The Crows seize upon their booty, scientifically perforate the integuments, and scoop out and devour every particle of flesh, even to the head. In a very short time the skins are turned inside out, and a few clean-picked bones are the only memorials of the banquet. In hard winters Crows suffer severely; they have been observed to fall down in the fields and on the roads, exhausted with cold and hunger. In one of these winters, a few years ago, during a long-continued deep snow, more than six hundred were shot on the carcase of a dead horse, which was placed in the stable-yard, the discharges being made through a hole in the stable. The Crow is easily domesticated; and it is only when placed on terms of familiarity with man that the traits of his genius and native disposition fully develop themselves. In this state he soon learns to distinguish all the members of the family; flies towards the gate, screaming at the approach of a stranger; learns to open the door by alighting on the latch; and attends regularly at the hours of dinner and breakfast.

DEATH FEIGNED BY A CORNCRAKE.—Mr Ballard of Islington mentions a marked instance of a Corncrake feigning death, an instinct which that bird possesses in common with several other animals, and more particularly with certain insects. A gentleman had a Corncrake brought him by his dog, which was dead to all appearance. As it lay on the ground he turned it round with his foot; it showed no symptoms of life. Standing by, however, some time in silence, he suddenly saw it open one eye. He took it up, its head fell, its legs hung down, it appeared again to be quite dead. He attributed this to an involuntary muscular exertion, and put it into his pocket. It was not long before he felt it struggling to escape; he took it out, when it appeared as lifeless as before. He laid it on the ground, and retired to some distance to watch its motions; in about five minutes it warily raised its head, looked round, and decamped at full speed.

BOTANY AND HORTICULTURE.

AGE OF PLANTS.—Some of the minute Fungi, commonly termed mould, live only a few hours, and seldom exist above a few days. The greater number of mosses exist only one season, which is also the case with all plants called Annuals. These die of old age immediately after their seeds have ripened. Those termed Biennials survive but for two seasons; although the lives of several of them are occasionally prolonged to three years, if their flowering be prevented.

On the other hand, some trees live for centuries. The Olive may live three hundred years; the Oak six hundred; a Chesnut is said to have existed nine hundred and fifty; the Dragon's-blood tree of Teneriffe two thousand; and we are informed by Adanson that Banians are conjectured to be six thousand years old.

When the wood of the interior of trees becomes so close in its texture that the passage of sap or pulp is prevented, or the formation of new vessels cannot be admitted, then it dies; and as all its moisture passes off into the younger wood, the fibres shrink, and are ultimately reduced into dust. The centre of the tree loses its vitality, while the outer parts continue to exist, and may thus live for many years before a total dissolution takes place.

ON THE EFFECTS OF ELECTRICITY IN PLANTS.—Colonel Capper observes, that "many experiments in electrifying plants have been made by M. Nuneberg and the Abbé Nollet; according to the former, most of them increased in height, and flourished far beyond those not electrified." Some bulbous roots, he says, which had been frequently electrified, grew 82 lines and a half; whilst others of the same species, not electrified, grew only in the same period 52 lines and two-thirds. But the report of Abbé Nollet is not so favorable; he found that the plants electrified by him, at first, made vigorous shoots, but he thought their perspiration being, by these means, too much increased, their juices might have been too quickly dissipated, in consequence of which the plants became gradually weak, and at length prematurely perished. We do not hesitate to yield due credit to both of these reports, although they seem in some measure incompatible with each other. They may have been made on various plants, at different seasons. M. Nuneberg, therefore, might have succeeded, although those of the Abbé in some measure failed. Besides, when administered by art, either to animal or vegetable bodies, the electricity may be given when the plants may already have the proper natural quantity; and, therefore, in some instances, it may be too strong; in others not strong enough, not sufficiently diffused; or it may not be applied to the proper part. The various modifications of electricity cannot well be comprehended excepting by those who profess considerable knowledge of the theory, and likewise the practice of it, so as to judge of its effects, not only in the atmosphere, but likewise on all animal and vegetable bodies; some of them may benefit by the aura or mild state of the fluid, and yet be injured by a spark, or even killed by a severe shock.

HARMONIES OF COLOUR AND FORM IN PLANTS.—Dr Castles remarks that "Nature, in the creation of the universe, has very beautifully modulated the influence of colour. To the firmament she has given a beautiful azure tint; to the earth itself a variety of shades, all more or less harmonizing with the blue on high, and the agreeable green of plants. If she had given to plants a yellowish hue, they would have been confounded with the sky and waters. In the first case, all would have appeared earth; in the second, all would have been sea; but their verdure forms the most delightful contrast between them and the grounds of the grand picture, as well as consonances highly agreeable with the yellow colour of the earth, and with the azure of the heavens."

In giving to vegetable productions a green shade, though only one single colour is employed, there are certain tints which appear to be given according to the situation or circumstances under which a plant may grow. Those that are destined to grow immediately on the earth, on strands, or on dusky rocks, are entirely green, leaves and stem, as the greater part of reeds, grasses, mosses, taper-trees, and aloes. Such, on the contrary, as are intended to issue from amidst herbage, have stems of a brownish hue, like the trunks of most trees and shrubs. The elder, for example, which thrives in the midst of green turf, has the stem of an ash-gray; but the dwarf elder, which otherwise resembles it in every respect, and grows immediately on the ground, has the stem quite green.

Not only the green of the plant is grown to harmonize with other objects, but even the flower and fruit have their shades apparently proportioned accordingly.

It seems correct that the blue colour is not to be found in the flowers, or in the fruits of lofty trees, for, in that case, they would assimilate with the sky; but is very common on the ground in the flowers of herbs, as in the corn-bottle, the scabiosa, the violet, the liver-wort, and others. On the contrary, the colour of the earth is very common in the fruits of lofty trees, as in those of the walnut, the cocoa, the pine, and many others

In the form of flowers, the most perfect specimens of harmony might be selected which would faithfully show that, even in pleasing the sight, the greater object of utility is combined, if not increased.

This is very sweetly shown in the structure of compound flowers, particularly such as the sunflower and daisy. What would these flowers be in appearance without their radii? Yet are the radiated petals of the circumference, not only given to complete a pleasing harmony of light, to the tubular florets of the centre, but they answer an important purpose of moderating the influence of heat, &c. Thus is the double object of utility and beauty combined.

Another point, productive of some very pleasing deductions, is founded on the harmonies from contrast. Plants opposite in Nature are almost always associated.

Thus, round the faded trunks of trees, twines the creeping Ivy, or the great Convolvulus, compensating the apparent want of blossoms. The Fir rises in the forests of the north, like a lofty pyramid, of a dark green colour, and with motionless attitude. Near this tree you almost always find the Birch, which grows to the same height in the form of an inverted pyramid, of a lively green, and whose moveable foliage is incessantly playing with every breath of wind. The Reed, on the banks of rivers, raises erect into the air its radiated leaves and its embroidered stems, while the Nymphæa spreads at its feet its broad heart-shaped leaves, and its gold coloured flower; the dark blue violet is contrasted, in the spring, with the yellow tints of the cowslip and the primrose. On the herbiaged angles of the rock, the fungus, white and round, rises from amidst beds of moss of the most beautiful green.

MINERALOGY.

GOLD.—This is one of the few metals only found native, and in this state, is easily recognized by colour, malleability, &c.; it is found crystallized, filamentous, and disseminated in rounded lumps of various sizes in alluvial soils.

Geologists consider gold as one of the most ancient of the metals, for it is invariably found in primitive rocks. Its gangue is quartz, calcareous spar, felspar, carbonate of lime, and sulphate of barytes. Africa and America are the richest countries in Gold. In Africa, it always occurs in the beds of rivers and in the alluvial soils of the plains, either in small grains, or in masses of different sizes. The principal tracts, rich in this precious metal, are in the western parts of Africa, to the south of the great desert of Zara, and between Darfur and Abyssinia; and the sands of the Gambia, Niger, and Senegal, are all auriferous; it is supposed that Ophir, whence Solomon obtained Gold, was a country on the south-east coast; and Herodotus relates, that when the messengers of Cambyses waited upon the king of Æthiopia, they were shown the prisoners bound in chains of gold. As this metal is found in a ductile, tenacious, and workable state, it is almost the only one employed by savage nations, and various ornaments and utensils are frequently made of it. These untutored tribes have always regarded the eagerness of their European invaders to obtain it with the utmost astonishment; of this the history of America furnishes us with many curious instances. In one of the early incursions into the interior of that continent, the Spaniards contested with such eagerness about the division of some Gold, that they were on the point of proceeding to acts of mutual violence, when a young Casique, who was present, tumbled it out of the balance with indignation, and turning to the invaders, "Why," said he, "do you quarrel about such a trifle—if it be for Gold that you abandon the regions of your fathers, and disturb the peaceful tranquillity of these distant nations, I will conduct you to a region abundant in this mean object of your admiration and desire." The thirst for Gold was the principal incentive to the almost more than human enterprises performed by the followers of Columbus. Animated by the certain prospect of gain, they pursued discovery with greater eagerness than when excited only by curiosity and hope. The riches of Peru, Mexico, and Brazil, are well known, and the Gold is there principally found in the beds of rivers, although *veins* have been successfully worked.

Asia cannot at present be deemed rich in Gold, although it has been found in Ceylon, Borneo, Sumatra, and some of the Archipelago islands. Of the abundance of Gold which once enriched the Pactolus we now know nothing.

Nor can Europe boast of Golden treasures. According to Diodorus Siculus, and Pliny, the Phœnicians and Romans procured considerable quantities of Gold from Spain. The poets too found it in the sands of the Tagus. In France it has been found in the department of the Isere; in the Rhone, at its junction with the Araw; in the Rhine near Strasbourg and Germersheim, but neither above nor below it; and in the Garonne near Toulouse. In Piedmont, in the vallies at the foot of Mount Rosa, and of the Simplon; and also in the small streams that intersect the red alluvion, about Chivasso. The only important Gold mines of Europe are those of Hungary.

The metallurgic processes for obtaining Gold from its ores are sufficiently simple. They are broken in the stamping mill and washed, by which the lighter and earthy parts are separated; they are then submitted to the action of mercury, which dissolves the Gold, and this metal is afterwards obtained by distillation.—*Brande.*

GEOLOGY AND PHYSICAL GEOGRAPHY.

FILLING UP OF LAKE SUPERIOR.—This mighty Lake is the largest body of fresh water in the known world; its length is 480 miles, its breadth 161, its circumference about 1100 miles, and its depth 900 fathoms. Its waters are remarkable for their unrivalled transparency. About 1000 streams empty themselves into this lake, sweeping in sand, primitive boulder-stones, and drift timber, which sometimes accumulate so as to form islands in the estuaries. A lignite formation, indeed, is said to be now in progress similar to that of Bovey, in Devonshire. Within a mile from the shore the water is about 70 fathoms; within eight miles, 136 fathoms. From the above causes the lake is gradually filling up.

LAKE ERIE, from similar causes, is also filling up. This sheet of water is 270 miles in length, 60 in breadth, and 200 fathoms in depth. It is gradually becoming shallower. Long Point, for example, has, in three years, gained no less than three miles on the water. On its southern shore serious encroachments have been made in many places. For a considerable distance above the mouth of Black River the bank of the lake is low and without rock. Thirteen years ago the bank was generally sloping, with a wide beach; now the waves beat against a perpendicular bank. which, from continual abrasion, often falls off. From one to three roods in width are worn away annually.

HOT AND COLD SPRINGS.—In the Blue Mountains, about 37 miles from Batavia, in the Island of Java, there is a spring of water, so hot, that few persons can bear immersion in it; within little more than two feet of this almost boiling cauldron another spring arises, so cold, that it almost instantly benumbs those who attempt to use it. Those waters overflowing, join in a current, and supply a bath formed by the natives, of such a temperature as to be delightful at all seasons of the year.

GENERAL SCIENCE.

FIXED STARS.—Dr Brinkley, Bishop of Cloyne, has found, by computation, that the star Lyra has a parallax of 1."1; or, what is the same thing, that the radius of the earth's annual orbit would, if seen from that star, subtend an angle of 1."1; hence it follows that its distance is 20,150,665,000,000 miles, or 20 billions of miles. Sir William Herschel, from repeated measurements, considered the diameter of this star as three-tenths of a second: and, consequently, its diameter must be 3000 times greater than that of the sun, 2,659,000,000 miles, or three-fourths of the size of the whole Solar System, as circumscribed by the Georgium Sidus. It has, however, been thought probable, by many eminent Astronomers, that this apparent parallax is due only to the defects of the instruments employed; and we are inclined to attach much importance to this opinion, especially as we find that the amount of this observable parallax has always diminished with the improvements of the instruments.

Under any circumstances, the distance of the Fixed Stars must be enormous. Such is the amazing remoteness, even of the nearest Fixed Star, that its light would take three years in travelling to the earth; and light is computed to travel 195,072 English miles in a second. If the nearest of these Stars be at such an immense distance from us, what must be that of the smaller Stars? Astronomers conceive it possible that there may be Stars so remote, that the beams of their light may not have reached us since the creation; and others, that have been destroyed for many centuries, will continue to shine in the heavens till the last ray which they emitted shall have reached our earth.

MISCELLANEOUS.

PORCUPINE MEN.

IN the year 1731, Mr John Machin introduced to the notice of the Royal Society of London a boy, about fourteen years of age, the son of a country labourer residing in the neighbourhood of Euston Hall, in the county of Suffolk, exhibiting those singular characters peculiar to that rare variety of the human race commonly called Porcupine Men.

Instead of a skin, his body was enveloped in a dusky-coloured case, resembling a rugged bark or hide, with bristles in some places. This case fitted every part of his body excepting his face, the soles of the feet, and the palms of the hands. His body thus presented the appearance of being partially clothed. It would have been difficult to mention any other integument which resembled it exactly. Some persons considered it to be like one large wart, or number of warts uniting and spreading over the whole body. Others thought it like the hide of the Elephant, or the skin around the legs of the Rhinoceros; while others again compared it to Seal-skin, or to the bark of a tree. The bristly parts, which were chiefly near the abdomen, and on the sides of the body, made a rustling noise when he moved like the quills of a Hedgehog, and seemed as if shorn within an inch of the skin. The following is a representation of a portion of this extraordinary epidermis, which was probably nothing more than a prolongation of the nervous papillæ, grown to the size of common pack thread. These stood as close together as the bristles in a brush—and seemed, like them, to be all shorn off of the same length, being about half an inch above the skin, as in fig. 1. When magnified, these stumps or bristles appeared of various forms; some were concave, others were flat on the top, and others again were of a conical form, as in fig. 2.

This skin was callous and insensible to external injury in every part. But one very remarkable circumstance attending it was, that in every year about autumn it usually grew to the thickness of three quarters of an inch, and was then thrust off and shed by a new skin, which grew up beneath the former. This rugged covering gave the boy no pain or uneasiness, except after hard labour it was apt to cleave and start, so as to cause slight bleeding. His face was well featured and of a good complexion, if not rather too ruddy, while the palms of his hands were not harder or in a worse condition than is usual to workmen or labourers. His size was proportioned to his age—his body and limbs were straight, and otherwise well shaped—and there was nothing unusual either in his habits or disposition.

His father reported that, at birth, the skin of this boy resembled that of other children, and continued so for seven or eight weeks; when, without any apparent cause, and without his being even sick, it began to turn yellow, as if he had had the jaundice: that it afterwards changed gradually into black, then thickened, and finally appeared as we have already described.

When this boy grew up, he gained a subsistence by exhibiting himself publicly as "the Porcupine Man," along with a son of his, also in the same condition. His name was Edward Lambert, and at the age of forty years he was thus described by

Mr Henry Baker:—" He is a good-looking, well-shaped man, of a florid countenance —and, when his body and hands are covered, seems nothing different from other people; but, except his head and face, the palms of his hands, and the soles of his feet, his skin was covered in the same manner as in the year 1731. This covering seems to me most nearly to resemble an innumerable company of warts, of a dark brown colour, and a cylindrical figure rising to a like height, and growing as close as possible to one another, but so stiff and elastic, that when the hand is drawn over them they make a rustling noise.

" When I saw this man, in the month of September last, they were shedding off in several places, and young ones of a paler brown succeeding in their room, which he told me happens annually in some of the autumn or winter months; and then he commonly is let blood, to prevent some little sickness which he else is subject to whilst they are falling off. At other times he is incommoded by them no otherwise than by fretting out his linen, which, he says, they do very quickly; and when they come to their full growth, being then in many places near an inch in height, the pressure of his clothes is troublesome.

" He has had the small-pox, and been twice salivated, in hopes of getting rid of this disagreeable covering; during which disorders the warting came off, and his skin appeared white and smooth, like that of other people; but, on his recovery, soon became as it was before. His health at other times has been very good during his whole life.

" But the most extraordinary circumstance of this man's story, and indeed the only reason of my giving you this trouble, is, that he has had six children all with the same rugged covering as himself; the first appearance whereof in them, as well as in him, came on in about nine weeks after his birth. Only one of them is now living, a very pretty boy, and who is exactly in the same condition, which it is needless to repeat. He also has had the small-pox, and during that time was free from this disorder."

The annexed wood-cut exhibits the hand of this boy in such a manner as to show the palm free from these excrescences, and its other parts covered with them.

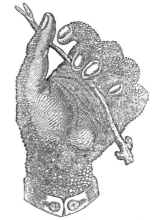

" It appears, therefore, past all doubt," continues Mr Baker, " that a race of people may be propagated by this man, having such rugged coats or coverings as himself: and, if this should ever happen, and the accidental original be forgotten, it is not improbable they might be deemed a different species of mankind; a consideration which would almost lead one to imagine, that if mankind were all produced from one nd the same stock, the black skin of the Negroes, and many other differences of the like kind, might possibly have been originally owing to some such accidental cause."

This young man afterwards married, and had two sons in all respects resembling himself, as well as their grandfather. They went over to Germany in 1801, where they exhibited themselves. Dr Blumenbach, who saw them, says that the palms of their hands and soles of their feet were of the usual appearance, but seemed to him rather red.

Dr Autenrieth (who endeavours to trace an analogy between these men and negroes, and even suspects them to be of African origin), rather thinks that the soles of the feet, of both brothers, are plain and flat, as we see them in children and adult negroes. The skin of the remaining parts of the body was covered with corneous excrescences, or pegs of greater or less size, differing in their horny consistence. The longest, strongest, and hardest, were on the fore-arm and thighs; the thinnest were on some parts of the abdomen. Those of the younger brother were in general smaller, and in several places the skin was soft, and comparable to black coarse morocco leather. The largest were from four to five lines long, and of an irregular prismatic form, with blunt edges; most of them seemed as if pressed flat. The thickest of them were about three lines in diameter at their extremities, either split or diverging like a fork. As to the cylindrical figure ascribed to them by Baker (who, besides, supposed them to be hollow), Dr Blumenbach scarcely observed one of that form.

Dr Telerius observed that these men look quite different in autumn from what they do at other seasons, because they then lose their outer skin, or oldest crust, and appeared spotted.

On examining the fragments, he found that those which he had broken off were softer to the touch than those which had fallen off spontaneously; probably on account of their being under the immediate influence of the exhaling vessels and the sebaceous glands. Where the excrescences were longest and thickest, they appeared to Dr Blumenbach to be like those of the Elephant, under the forehead and above the proboscis; their colour, in general, appeared of a chesnut or coffee-brown. This, however, was the case at their surface only; for the inferior parts, especially of the largest ones, were of a yellowish-gray. Some of the hair of the skin appeared as if grown into the horny substance of the excrescences. The skin on the top of the head before, especially in the eldest, formed a kind of broad callosity, somewhat like the top of the Camel. As for the perspiration of these men, it had nothing uncommon connected with it, nor any perceptible odour.

Of cases really analogous to that of the Porcupine Men, Dr Blumenbach mentions other two which came under his notice; the one was of the boy of Bifeglia; of whom Stalp Vanderwiel has given a description and figure in his Observations; the other, a female child at Vienna, described by Professor Brambilla, in his Memoirs of the Jos. Med. Chirurg. Academy. In both the face was free from any excrescences, but the palms of the hands and soles of the feet were most defaced by them.

CARGUERROES, OR MAN-CARRIERS OF QUINDIU.

The mountain of Quindiu is considered as the most difficult passage in the Cordilleras of the Andes. It is a thick uninhabited forest, which, in the finest season, cannot be traversed in less than ten or twelve days. Not even a hut is to be seen, nor can any means of subsistence be found.

Travellers, at all times of the year, furnish themselves with a month's provision, since it often happens, that by the melting of the snows, and the sudden swell of the torrents, they find themselves so circumstanced, that they can descend neither on the side of Carthago, nor on that of Ibague. The highest point of the road, the Garito del Paramo, is three thousand five hundred and five metres (11,500 feet) above the level of the sea. As the foot of the mountain, towards the banks of the Cauca, is only nine hundred and sixty metres (3,150 feet high), the climate there is in general mild and temperate. The pathway which forms the passage of the Cordilleras is only three or four decimetres in breadth (from a foot to a foot and a half), and has the appearance, in several places, of a gallery dug, and left open to the sky. In this part of the Andes, as almost in every other, the rock is covered with a thick stratum of clay. The streamlets, which flow down the mountains, have hollowed out gullies, six or seven metres deep (from 20 to 23 feet). Along these crevices, which are full of mud, the traveller is forced to grope his passage, the darkness of which is increased by the thick vegetation that covers the opening above. The oxen, which are the beasts of burden commonly made use of in this country, can scarcely force their way through these galleries, some of which are two thousand metres (2,200 yards) in length; and if, perchance, the traveller meets them in one of these passages, he finds no means of avoiding them but by turning back, and climbing the earthen wall which borders the crevice, and keeping himself suspended, by laying hold of the roots which penetrate to this depth from the surface of the ground.

We traversed the mountain of Quindiu in the month of October 1801, on foot, followed by twelve oxen, which carried our collections and instruments, amidst a deluge of rain, to which we were exposed during the last three or four days, in our descent on the western side of the Cordilleras. The road passes through a country full of bogs, and covered with bamboos. Our shoes were so torn by the prickles, which shoot out from the roots of these gigantic Gramina, that we were forced, like all other travellers who dislike being carried on men's backs, to go barefooted. This circumstance, the continual humidity, the length of the passage, the muscular force required to tread in a thick and muddy clay, the necessity of fording deep torrents of icy water, render this journey extremely fatiguing; but however painful, it is accompanied by none of those dangers with which the credulity of the people alarm travellers. The road is narrow, but the places where it skirts precipices are very rare. As the oxen are accustomed to put their feet in the same tracks, they form small furrows across the road, separated from each other by narrow ridges of earth. In very rainy seasons these ridges are covered with water, which renders the traveller's step doubly uncertain, since he knows not whether he places his foot on the ridge or in the furrow. As few persons in easy circumstances travel on foot in these climates, through roads so difficult, during fifteen or twenty days together, they are carried by men in a chair tied on their back; for, in the present state of the passage of Quindiu, it would be impossible to go on mules. They talk in this country of going on a man's back (*andar en carguero*), as we mention going on horseback, no humiliating idea is annexed to the trade of cargueroes; and the men who follow this occupation are not Indians but Mulattoes, and sometimes even whites. It is often curious to hear these men, with scarcely any covering, and following a profession which we should consider so disgraceful, quarrelling in the midst of the forest because one has refused the other, who pretends to have a whiter skin, the pompous title of *don*, or of *su merced*. The usual load of a carguero is six or seven arrobas (about 180 lbs.); those who are very strong carry as much as nine arrobas (about 250 lbs.) When we reflect on the enormous fatigue to which these miserable men are exposed, journeying eight or nine hours a day over a mountainous country; when we know that their backs are sometimes as raw as those of beasts of burden, and that travellers have often the cruelty to leave them in the forest when they are sick; that they earn by a journey from Ibague to Carthago only twelve or fourteen piastres (from L.2, 10s. to L.3), in a space of fifteen, and sometimes even twenty-five or thirty days, we are at a loss to conceive how this employment of a carguero, one of the most painful which can be undertaken by man, is eagerly embraced by all the robust young men who live at the foot of the mountains. The taste for a wandering and vagabond life, the idea of a certain independence amidst forests, leads them to prefer this employment to the sedentary and monotonous labour of cities.

The passage of the mountain of Quindiu is not the only part of South America which is traversed on the backs of men. The whole of the province of Antioquia is surrounded by mountains so difficult to pass, that they who dislike entrusting themselves to the skill of a carrier, and who are not strong enough to travel on foot from Santa Fé de Antioquia to Bocca de Nares or Rio Samana, must relinquish all thoughts of leaving the country. I was acquainted with an inhabitant of this province, so immensely bulky, that he had not met with more than two Mulattoes capable of carrying him; and it would have been impossible for him to return home if these two carriers had died, while he was on the banks of the Magdalena, at Mompox or Honda. The number of young men who undertake the employment of beasts of burden at Choco, Ibague, and Medellin, is so considerable, that we sometimes met a file of fifty or sixty. A few years ago, when a project was formed to make the passage from Nares to Antioquia passable for mules, the cargueroes presented formal remonstrances against mending the road, and the government was weak enough to yield to their clamours. We may here observe, that a class of men near the mines of Mexico, have no other employment than that of carrying other men on their backs. In these climates the indolence of the whites is so great, that every director of a mine has one or two Indians at his service, who are called his horses (cavallitoes), because they are saddled every morning, and supported by a small cane, and, bending forwards, they carry their master from one part of the mine to another. Among the cavallitoes or cargueroes, those who have a sure foot and easy step are known and recommended to travellers. It is distressing to hear the qualities of man spoken of in terms by which we are accustomed to denote the gait of mules and horses. The persons who are carried in a chair by a carguero must remain several hours motionless and leaning backwards; the least motion is sufficient to throw down the carrier; and his fall would be so much the more dangerous, as the carguero, too confident in his skill, chooses the most rapid declivities, or crosses a torrent on a narrow and slippery trunk of a tree. These accidents are, however, rare, and those which happen must be attributed to the imprudence of travellers, who, frightened at a false step of the carguero, leap down from their chairs.

When the cargueroes reach Ibague and prepare for their journey, they pluck in the neighbouring mountains several hundred leaves of the *Vijao*, a plant of the family of the Bananas, which forms a genus approaching the *Thalia*, and which must not be confounded with the *Heliconia bihai*. These leaves, which are membraneous and silky, like those of the Musa, are of an oval form, fifty-four centimetres (twenty inches) long, and thirty-seven centimetres (fourteen inches) in breadth. Their lower surface is a silvery white, and covered with a farinaceous substance which falls off in scales. This peculiar varnish enables them to resist the rain during a long time. In gathering these leaves, an incision is made in the middle rib, which is the continuation of the footstalk, and this serves as a hook to suspend them when the moveable roof is formed. On taking it down, they are spread out and carefully rolled up in a cylindrical bundle. It requires about a hundred weight of leaves (50 kilogrammes) to cover a hut large enough to hold six or eight persons. When the travellers reach a spot in the midst of the forest, where the ground is dry, and where they propose to pass the night, the cargueroes lop a few branches from the trees, with which they make a tent. In a few minutes the slight timber work is divided by the stalks of some climbing plant, or threads of the Agave, placed in parallel lines, three or four decimetres from each other. The Vijao leaves meanwhile have been unrolled, and are now spread over the above work, so as to cover each other like tiles of a house. These huts thus hastily built, are cool and commodious. If during the night the traveller feels the rain, he points out the spot where it enters, and a single leaf is sufficient to obviate the inconvenience. We passed several days in the valley of Boquia under one of these leafy tents, which was perfectly dry amidst violent and incessant rains. The mountain of Quindiu is one of the richest spots in useful and interesting plants. Here we found the Palm Tree (*Ceroxylon andicola*), the trunk of which is covered with a vegetable wax, the Passiflora in trees, and the majestic *Mutisia grandiflora*, with flowers of a scarlet colour, sixteen centimetres, or six inches long.—*Humboldt*.

The Sleeping Lassie of Dunninald.—Margaret Lyall, aged 21, daughter of John Lyall, labourer at Dunninald, near Montrose, was first seized with a sleeping fit on the 27th June 1815; next morning she was again found in a deep sleep. In this state she remained for seven days, without motion or food; but at the end of this time, by the moving of her left hand, and by plucking at the coverlet of the bed and pointing to her mouth, a wish for food being understood, it was given her. This she took, but still remained in her lethargic state till Tuesday the 8th of August, being six weeks from the time she was seized with the lethargy, without appearing to be awake, except on the afternoon of Friday the 30th June. For the first two weeks her pulse was generally about 50, the third week about 60, and previous to her recovery at 70 to 72. Though extremely feeble for some days after her recovery, she gained strength so rapidly, that before the end of August she began to work at the harvest on the lands of Mr Arkley, and continued, without inconvenience, to perform her labour. This account was drawn up by the Rev. James Brewster, Minister of Craig.

ERRATA.

Lettering of Plate LI., Deer, fig. 6, for *C. Macrourus* read *C. Paludosus*.
Animal Kingdom, page 31, col. 2, line 42, for *Aluntian* read *Aleutian*.
Journal, ... 29, 1, 11, for *Sparrow* read *Thrush*.

Edinburgh: Published for the Proprietors, at their Office, 16, Hanover Street. London: Smith, Elder, and Co., 65, Cornhill. Glasgow and the West of Scotland: John Smith and Son, 70, St Vincent Street. Dublin: W. F. Wakeman, 9, D'Olier Street.

THE EDINBURGH PRINTING COMPANY.

THE EDÍNBURGH

JOURNAL OF NATURAL HISTORY,

AND OF

THE PHYSICAL SCIENCES.

SATURDAY, APRIL 2, 1836.

ZOOLOGY.

DESCRIPTION OF THE PLATE—DOGS.

THE Dog has been the companion of Man from the earliest state of society; but the period when he was first domesticated, and the stock from which he was produced, are hidden by the cloud of antiquity. Innumerable attempts have been made to trace his origin, but these have resolved themselves into mere conjectures. The fidelity, caution, and perseverance of the Dog, have secured to him the friendship of Man, in all ages and in every country; and by his aid we are enabled to acquire the most necessary and certain means of conquest and dominion over all other creatures. His exquisite sense of smell enables him to pursue steadily, and with unerring certainty, all other beasts: their artifices and speed are unavailing; for by the strength and perseverance of combined numbers, he overtakes, overcomes, and destroys them.

It is almost universally believed that all Dogs are merely varieties of one species only; and many persons imagine the Shepherd's Dog to be the parent stock, whence all these varieties have sprung. Buffon, and more recently Dr Richardson, conceive that Dogs have been propagated from more than one common stock; that they have sprung from Wolves, Jackals, and other congenerous species. This opinion was also entertained by Desmoulins, who, in an ingenious paper in the "Memoirs of the Museum of Natural History of Paris," adduces a number of striking facts in illustration of his hypothesis. Dr Knox, during his residence in Africa, remarked that all the native Dogs in the neighbourhood of the Cape, bore the same general resemblance to the Hyæna that those of northern countries bear to the Wolf.

As we shall enter more deeply into this subject in the Animal Kingdom, it will be unnecessary to pursue it further at present, but proceed to a description of the Dogs figured in the accompanying plate.

Fig. 1. SPANISH POINTERS.—This breed, as its name implies, was introduced from Spain at an early period, and was long the chief companion of Sportsmen while in pursuit of game birds. It is one of the most steady Dogs used in field sports, and remarkable for the facility with which it can be trained to set all kinds of game. Indeed, it not unfrequently happens that puppies, when taken to the fields, instinctively set game when they have come upon the scent of them for the first time. In more recent times this Dog, however, has got into disuse, as its weight renders it too unwieldy for grouse shooting.

The beautiful Dogs from which our representations are taken, were a brace belonging to the late celebrated sportsman, Colonel Thornton; and as a proof of the steadiness and perseverance of both, they kept their point for upwards of an hour and a quarter, during the time that Mr Gilpin was engaged in sketching them.

Fig. 2. THE ENGLISH POINTER.—This Dog is sprung from the Spanish Pointer and Fox-hound, recrossed with the Harrier; he is much lighter in his form, and more rapid in his movements, than the Spanish Pointer. They have been produced of great variety in point of size, according to the tastes of sportsmen. It has, however, been found, that as they diverge from their Spanish progenitor, the difficulty of training them, and rendering them stanch for the field, increases in proportion to the remoteness of their lineage. The following is a beautiful instance of stanchness in a Pointer, and was communicated to us by James Webster, Esq. of Lively Bank, Forfarshire. In 1829, that gentleman was out on a shooting party, near Dundee, when a female Pointer, having traversed the field which the sportsmen were then in, proceeded to a wall, and, just as she made the leap, got the scent of some partridges on the opposite side of the wall. She hung by her fore-feet until the sportsmen came up; in which situation, while they were at some distance, it appeared to them that she had got her leg fastened among the stones of the wall, and was unable to extricate herself. But, on coming up to her, they found that this singular circumstance proceeded from her caution, lest she should flush the birds, and thus suspended herself in place of completing her leap. He adds, " It is impossible, adequately, to convey to you, in writing, a just idea of the beauty of this point."

Fig. 3. THE ENGLISH SETTER.—This beautiful and active Dog is the produce of the Spanish Pointer, the English Water Spaniel, and the Springer. He is remarkable for the elegance of his figure, the beauty of his fur, and the diversity of his colours. He possesses most of the excellent qualities of the Pointer, with a much greater degree of activity and speed, and a more buoyant vivacity of temper. But with all these recommendations, he is much more difficult to break in than the Pointer, and requires an annual training to preserve his education.

Fig. 4. THE OLD ENGLISH SETTER.—This breed was originally acquired by a mixture of the Spanish Pointer with the larger Water Spaniel, and was noted for its olfactory qualities and steadiness in the field; the hair over its whole frame was much more curled than the modern breed of Setters, and it was much less active than they were. The old English Setters were also famous for their sagacity.

Fig. 5. THE SPRINGER.—The chief difference between this dog and the Setter is in point of size, and in his head being larger, in proportion to the size of his body, than the latter variety; his ears are also longer, and he is more delicate in his general conformation. This dog is chiefly used in shooting Woodcocks, Pheasants, and Snipes, and, contrary to the practice of the Pointers and Setters, always gives tongue when in pursuit of game.

Fig. 6. THE COCKER differs from the Springer in his form being more compact, his head rounder, and muzzle shorter, and in being at least a third smaller than the Springer, while his habits and uses are exactly similar; the ears are also longer in proportion, and his tail more truncated. He is supposed to have had his origin in a cross between the Springer and the smallest Water Spaniel. The Cocker is a most affectionate dog, and in general of a mild and gentle disposition.

Fig. 7. KING CHARLES' SPANIEL is considerably smaller than the Cooker, differing from him in the greater proportional length of his ears, and in his tail being much more villous towards its point. His habits are nearly allied to those of the Springer and Cooker, but he is seldom used in field sports, owing to his diminutive size. He acquired his name from the circumstance of King Charles II. being much attached to him, so much so, that he was generally followed by eight to a dozen of them.

Fig. 8. THE COMFORTER.—This diminutive creature had its origin in the Maltese Dog and King Charles' Spaniel. He is principally used as a Lap-Dog, or as an attendant on the toilet and drawing-room. These Dogs were anciently denominated Spaniels-gentle. In the Life of Mary Queen of Scots, published at Glasgow some years ago, it is recorded, that, after she was beheaded, " her little favourite Lap-Dog, which had affectionately followed her, and, unobserved, had nestled among her clothes, now endeavoured by his caresses to restore her to life, and would not leave the body till he was forced away. He died two days afterwards, perhaps from loneliness and grief."

DESCRIPTION OF THE PLATE—THE WOODPECKERS.

THIS singular race of Birds live almost entirely on Insects and their larvæ, which they pick out of decayed trees, and also from the bark of such as are sound. These they transfix and draw from the crevices by means of their long extensile tongue, which is bony towards its point, and tipped with a barbed process. The tongue is provided with a curious muscular apparatus, which enables the birds to throw it forward with great force and rapidity. The bills are also strong, powerful, and generally wedge-shaped, by means of which they are enabled to perforate trees which are perfectly sound, and to make holes large enough for incubation. Most of the genus are provided with four claws, two placed before and two behind, by means of which they climb trees with great facility; and in this operation they are aided by their tails, the feathers of which are very strong, and generally sharp-pointed. Most of the species have a harsh, acute, and unpleasant voice.

Fig. 1. THE RED-HEADED WOODPECKER (*Picus Erythrocephalus*) is a native of North America, and notorious for its predatory habits. It is a bold and active bird, most abundant in the neighbourhood of all farms, and even frequents the vicinity of large cities. Wilson mentions instances of their nests being found within the boundaries of Philadelphia. These went to feed in the woods about a mile distant, and on returning to their nests they " preserved great silence and circumspection." Although Insects are the principal food of this Bird, yet it will occasionally feed on various kinds of fruit. Its total length is 9½ inches.

Fig. 2. THE YELLOW-BELLIED WOODPECKER (*P. Varius*) is a resident Bird of the United States of America, spending its winter in orchards, and retiring to the woods in summer for the purpose of incubation. It is 8½ inches long.

Fig. 3. THE DOWNY WOODPECKER (*P. Pubescens*).—This is one of the smallest of the North American Woodpeckers, being only 6½ inches in length. They generally build in apple, pear, or cherry trees; the direction of the hole bored downwards in an angle of thirty or forty degrees for a depth of six or eight inches, and then straight down for a distance of ten or twelve inches; it is made roomy and capacious, and as smoothly polished as if executed by the hands of an experienced carpenter. During this operation they carry the chips of wood to some distance to prevent detection.

Fig. 4. THE BENGAL WOODPECKER (*P. Bengalensis*).—This handsome Bird is a native of Bengal, where it builds in places remote from towns or villages. Its habits are similar to those of its congeners.

Fig. 5. THE RED-BELLIED WOODPECKER (*P. Carolinus*).—The habits of this Bird are solitary; it prefers the largest high-timbered woods of North America, and the tallest of the decayed trees of the forest, seldom appearing on the ground or near fences. It has a very hoarse voice, which resembles the bark of a small Lap-Dog. It is 10 inches in length.

Fig. 6. Lewis's Woodpecker (*P. Torquatus*).—This beautifully-coloured Bird was discovered in the remote regions of Louisiana, in North America.

Fig. 7. The Black-Breasted Woodpecker (*P. Multicolor*).—This Bird inhabits India. The bill is remarkably long in proportion to the size of its head.

Fig. 8. The Red Cockaded Woodpecker (*P. Querulus*) is a native of North and South Carolina, and also extends through Georgia. It is seven inches in length; and, although possessed of no variety of colours, is a beautiful Bird, from the strong contrast of the black and white markings.

Fig. 9. The Hairy Woodpecker (*P. Villosus*) is a resident Bird of North America; and haunts apple-trees in orchards, but retires to the woods in summer for incubation, although in some instances they have been known to remain and breed in their winter haunts. This species has been found in England.

DESCRIPTION OF A NEW SHELL.

The Shell of which the three representations are given below was in the possession of Mr R. Weekes, who was uncertain of its locality, but supposed it to be a native of New Holland. It possesses characters different from every Shell with which we are acquainted, and cannot be referred to any of the Lamarckian genera. We, therefore, propose giving it the generic appellation of

FISSILABIA.

Generic Character.—Shell strong, acutely spiral; body about half the length of the spire; aperture nearly circular; outer lip resting on the body; columella interrupted by a fissure, having by its side a tooth-like process.

Specific Character.—*Fissilabia fasciata* (the Banded Slit-lip).—Shell spiral, with seven well-defined volutions, terminating in a somewhat obtuse apex, the body occupying about two-thirds of the Shell; outer lip acute at its edge, but internally flattened for about an eighth of an inch; the columella interrupted by a deep fissure; of a pale yellowish cast; volutions with three bands of interrupted spots of a deep chestnut colour—these are distinctly marked in the aperture, and extend to the margin of the outer lip; the whole external surface is smooth, but not glossy; with obsolete, longitudinal striæ.

This beautiful and elegant Shell is very thick in proportion to its size, and is evidently a Marine species.

The situation which it will occupy in the Lamarckian arrangement is between the genera *Turbinella* and *Pleurotoma*.

ON THE HYBERNATION OF ANIMALS.—NO. II.

We have seen that the lethargy peculiar to hybernating animals is not so profound as to prevent them from being roused by other external impressions besides the application of heat, so there are circumstances which render it probable that they may also be roused by internal sensations, particularly that of hunger. The Dormice lay up a store of provisions in their nests, which they eat during the winter; and they have been observed, when kept in places where they could be watched, to awake, eat, and soon again become lethargic. The Marmots have been observed to come out of their holes early in spring, when the temperature was still lower than when they became lethargic at the commencement of winter. These facts may admit of another explanation, and we only mention them as circumstances which seem to render this opinion very probable.

The circumstances which appear to conduce to this lethargy, in these animals susceptible of it, are partly ascertained and partly conjectural. A certain degree of cold appears to be absolutely necessary to it in animals inhabiting temperate climates. We know too little about the lethargy which appears to be induced in some animals, inhabitants of tropical regions, to enable us to form any conclusions on its nature and character. We have already shown that it is only within a certain range of temperature that this lethargy exhibits itself in the Mammalia of our climate, generally commencing when the temperature is below 50°, and ceasing when it approaches the freezing point. The season of the year, apart from the diminished temperature, does not appear to have any effect in its production. When these animals are supplied with food, and kept in a uniform and temperate atmosphere during the winter, their usual lethargy does not take possession of them; on the contrary, they remain lively and active. Saissy induced this state during summer, by the application of artificial cold, without any injurious effects upon the animal. It has been objected to this experiment, that, at the same time he reduced the temperature, he also prevented the free access of air. This objection must, however, be in a great measure obviated, when we remember that most of these animals, if not all, appear to take precautions against the free admission of air when they are about to fall into their dormant state, and this even has been enumerated as one of the circumstances which favors its production. The Hamster does not become lethargic as long as it is exposed to the free influence of the air. The Marmots stuff the mouths of their holes with earth, hay, or grass, and its removal seems sufficient to rouse them at all times. The Hedgehog and Dormice roll themselves up like a ball in the midst of their nests. The Bats cover each other with their wings. No doubt, some of these precautions are principally to provide against too low a temperature, but they must also, at the same time, impede more or less the free access of air.

A certain degree of fat is also generally believed to favor the production of this lethargy. The hybernating animals generally abound in fat towards the end of autumn, when the dormant state comes on, and they generally come out from their hiding-place exceedingly lean. It is stated, that, in some parts of North America, those Bears that remain lean at the approach of winter migrate southwards, while those that have had an opportunity of fattening themselves become lethargic, and spend the winter in their native districts. It has been supposed, that as the Marmot in a state of domestication never becomes as fat in autumn as those living in a wild state, this explains the circumstance formerly mentioned of some tame Marmots that were observed not to hybernate. Spallanzani, however, states that he has found leaner Dormice as susceptible of the lethargic state from the application of cold as those which were fatter.

Dr Marshall Hall has mentioned some circumstances which seem to show that Bats undergo a daily lethargy, a kind of diurnation. On observing a Bat during summer, he found that it exhibited the permanent characters of lethargy, viz., imperfect respiration, diminished temperature, and the capability of supporting, for a long time, the deprivation of atmospheric air.

J. R.

Petrifaction of Animals by Artificial Means.—A pamphlet has lately issued from the press at Florence, giving an account of some remarkable discoveries by Girolamo Segato (constructor of the Maps of Tuscany, Africa, and Morocco), the principal facts of which are attested by the chief Professors in Florence. The account commences by a statement, that while M. Segato was traversing the deserts of Africa for the purpose of perfecting his Map, he was overtaken, in the valley between the Second Cataract and Mograb, by one of those whirlwinds, or rather sand-spouts, which are not uncommon phenomena in Upper Nubia. After it had passed by, and M. Segato was proceeding in its tract, he observed, in one of the hollows which had been ploughed up by the Spout, some remains of carbonized matter, and on searching still farther, he discovered a body completely charred, both the bones and flesh of which were in good preservation. It immediately occurred to him that the process of Charring could only have been effected by the scorching sand; and that, if the heat of the sand had in this instance effected the complete desiccation and carbonization of animal substances, might it not be possible to effect something similar by artificial means? No sooner had he returned to Italy than he instituted a series of experiments to effect his purpose, and ultimately succeeded in imparting to the limbs and bodies of animals solidity and indestructible durability. So great has been his success, that entire bodies, as well as separate parts, have been preserved; acquiring a firm and compact consistence, which is more decided and obvious according to the hardness or softness of the parts respectively. The skin, muscles, veins, nerves, and even the fat and blood, become consolidated; and what seems still more extraordinary is, that intestines do not require to be removed, and speedily acquire the same durable consistence. Contrary to the Mummies of the Egyptians, and all other modes hitherto practised, the colour, form, and general character remain unchanged; while both limbs and joints continue as flexible and moveable as when alive, and yet they are perfectly free from any smell. After animal bodies have acquired this hardened consistency, they are proof against damp, air, moths, or mites. They may be immersed in water, and allowed to remain for several days without injuring their texture. There is but a very slight diminution of weight; and so far from hairs being lost or injured, they seem more firmly rooted than in the living subject. The skin and feathers of birds and scales of fishes remain unaltered, and insects and worms preserve their natural appearance. These singular facts require no further proof than an examination of the cabinet of Segato, which is rich in specimens preserved by this process. We may mention that he possesses a Canary-bird which was preserved ten years ago, and has been proof against water and the attacks of moths. In the first year after being carbonized it was placed 30, and in the second 40 days under water; and, for a much longer period, was put into a box with a quantity of moths, but was not injured by them. Experiments of a similar kind were made on other animals, attended with the same results. He possesses the hand of a female who died of consumption, which exhibits all the delicate emaciation peculiar to that disease. He has also the hand of a man which is entirely unchanged, and is even flexible at the joints. The most remarkable object in his collection is a table, composed of 214 pieces of animal matter joined together. They look like so many different kinds of stone, and yet are nothing more than pathological portions of human members!

BOTANY AND HORTICULTURE.

The Gingo Tree is a native of Japan, and has not only flowered in the botanic garden of Montpellier, but has also brought to maturity its fruit, the kernel of which, when roasted, has a very agreeable flavour. It will, in all probability, be naturalized in the south of Europe, where the climate appears to be favourable to its growth.

Sugar obtained from Indian Corn.—M. Pallas lately laid before the Académie des Sciences of Paris, a sample of sugar extracted from the stem of this plant. It has been found to contain nearly six per cent. of syrup boiled to forty degrees, a part of which will not crystallize before fructification; but it condenses and acquires more consistency from that period to the state of complete maturity. The time most favourable for obtaining the greatest quantity of sugar is immediately after the fruit is mature, and the time of gathering. The residue, after the extracting of the sugar, is excellent for feeding cattle; or it may be usefully employed in the manufacture of packing paper.

Gigantic Lime Tree.—A gigantic specimen of the Lime or Linden Tree is now to be seen at Ivory, which has reached the unusual stature of 100 feet, while its branches extend over a surface of 245 feet; and these branches commence at a height of not more than 10 feet from the ground. The circumference of the trunk near the ground is 46 feet. It is supposed that this tree is unique in point of magnitude.

ON THE ORIGINAL COUNTRY OF THE CEREALIA, ESPECIALLY OF WHEAT AND BARLEY.—NO. I.

The period when Corn first began to be cultivated for food to Man, marks an important era in the progress of civilization, and in the happiness of the human race. Yet this time, so interesting to the inquirer, is enveloped in the darkness of antiquity; and it is only by a careful comparison of opposite probabilities, that we can arrive at a satisfactory conclusion regarding the time of that Aurora of civilization which marked its introduction.

It is very difficult to determine the native country of the Cerealia, upon principles purely Botanical, because these plants have been cultivated in all civilized countries from time immemorial. Also, we find from the universal experience of the Agriculturist, that Wheat and Barley will perpetuate themselves for two years in our climate after a first tillage, and yet they will die out in the third year. Oats have been observed to grow wild in parts of the woods at Boulogne, which had been occupied by foreign armies, so long as from 1815 to 1819, and then perished. They were also found near the ponds of Auteuil, and along the walls on the road to Neuilly. This same species of Oats (*Avena sativa*) was carried by the Europeans to Rio de la Plata, and there becoming wild, perpetuated itself for more than forty years without any cultivation. This curious fact is stated by M. A. de Saint Hilaire, who resided for six years in that country. Those Botanists, therefore, who fancied that they had discovered, in various places, the native country of the Cerealia, from finding them growing wild, should have remained there a sufficient time to examine carefully whether they continued to grow by spontaneous reproduction for a long course of years.

As we cannot hope for a satisfactory solution of this question from the mere examination of Nature, we must resort to the most ancient traditions. It is requisite to compare the most ancient sculptures with the passages of the Bible. We must contrast the accounts of the origin and migrations of the worship of Ceres, which was probably nothing but the migrations of these plants, with the figures of the spike of corn represented upon the zodiacs in the sign Virgo, and with the grains themselves found in the tombs of Thebes. We may then arrive at a satisfactory result, by applying that rule of criticism proposed by Humboldt, Robert Brown, and other eminent botanists,—that when the native country of a cultivated species is unknown, we must regard that as the probable place of its nativity, where we find indigenous the greatest number of known species belonging to the same genus.

In this way we may circumscribe, within a small zone, the district where the Cerealia must have originally sprung.

The common Wheat (*Triticum hibernum* and *Triticum æstivum*), as well as Barley (*Hordeum vulgare, hexastichon*), are often destroyed by frosts in our climate and on the neighbouring continent. They neither grow in equatorial countries of a medium elevation, nor beyond the tropics at any great height above the level of the sea. From these circumstances, we may infer that the native country of these plants was in the temperate zone, and of no great elevation.

We know positively that they do not reproduce spontaneously, either in the Old or New Continent, or in any place where Europeans have carried their colonies and cultivated this grain, so necessary to the progress of civilization and the happiness of society.

It may also be inferred that the Cerealia do not exist in the wild state in those extensive countries inhabited by tribes of Hunters or Shepherds; for these people would assuredly have changed a precarious and uncertain subsistence for an agreeable food, which yields an abundant return, and which would increase their population, concentrate their power, and ensure the existence as well as happiness of their families.

The Ægyptians, Hebrews, Greeks, and many other nations of Asia and Europe, afford examples of this transition from the Pastoral to the Agricultural state, as soon as they had discovered the Cerealia, or that these had been introduced into their country.

We shall attempt to prove that, according to the most ancient monuments of Ægyptian history, it was at Nysa or Bethsane, in the valley of the Jordan, that Isis and Osiris found the Wheat, the Barley, and the Vine growing wild.

It is proper, in the first place, to ascertain the situation of the city of Nysa. Homer is the most ancient writer who mentions it. "There is a town of Nysa, situate upon a lofty mountain covered with flowering trees, rather farther from Phœnicia than from the waters of Ægypt." This passage, quoted by Diodorus, and four others from the latter writer, fix, with considerable precision, that Nysa lay between the Nile and Phœnicia. Pliny is more precise; he places Nysa or Scythopolis in Palestine, on the borders of Arabia. Stephen of Byzantium relates the same thing; and Josephus informs us that this town of Nysa, called by the Greeks Scythopolis, was in his time styled Bethsane, and was situate in the middle of a plain, beyond the Jordan.

The position of this city is thus laid down by the text of Diodorus, Pliny, Josephus, and Stephanus. It seems also that Nysa, Scythopolis, and Bethsane, are the same city. At the time of Osiris, and even in that of Diodorus, the boundaries of Arabia were as usual very loosely defined on the north and west; and that portion of Palestine, adjoining Arabia, was often included with Syria by one writer, and with the peninsula of Arabia by another. In the ancient history of Java, according to Sir Stamford Raffles, Barley is mentiond to have been imported under the name of *Jawa nusa*. The similarity of these names is striking.

But there is another historical fact which confirms the position of Nysa in the neighbourhood of Palestine. Osiris, or the Ægyptian Bacchus, whom Diodorus and the best informed of the Greeks regarded as the same king, found the Vine growing wild, and entwined round the largest trees near Nysa. It was also in the land of Canaan that Noah discovered the Vine. Moses alludes particularly (Numb. c. xiii.) to the size of the bunches of Grape in the neighbourhood of Hebron; and it is well known that the Vine is a small shrub, inclining in general towards the basin of the Mediterranean. It does not grow wild in Ethiopia, nor in Arabia Proper, nor in Ægypt; but though it has been found in Armenia and Madagascar, these situations are foreign to our present purpose.

Thus, the Sacred Scriptures, the ancient history of the Ægyptians, and Natural History agree upon this important point,—that Agriculture commenced in Palestine. It was here that Wheat, Barley, and the Vine were first cultivated, and the latter was transported to Ægypt by Osiris. These facts follow necessarily from the geographical position of Nysa, determined as above.

It appears, then, that Isis and Osiris discovered Wheat, Barley, and the Vine growing wild in the Valley of the Jordan, that they transported it to Ægypt, demonstrated its utility, and taught its culture.

"The Ægyptian history assures us," says Diodorus, "that Osiris, originally from Nysa, situate in the fertile Arabia, loved Agriculture, and found the Vine in the neighbourhood of Nysa. This shrub was wild, very abundant, and hung generally from trees."—"It was there also," he adds, "that Isis discovered Wheat and Barley, previously growing wild in the country among the other plants unknown to Man." —*Diod. Sic. l.* 1, *c.* 14 *and* 27.

One of the first fruits of this valuable discovery was the cessation of those horrid feasts of Cannibalism, which had hitherto prevailed in Ægypt. Instead of these revolting spectacles, processions with sheafs of Corn and vases filled with Wheat and Barley, served to perpetuate its memory. Diodorus quotes some writers which mention that there existed at Nysa a monument inscribed with Hieroglyphics, which served to perpetuate this discovery of Isis. It bore this inscription, "I am the queen of all this country. I am the wife and sister of Osiris. I taught mortals first to know the use of Corn. I am her who rises in the constellation of the Dog. Rejoice, Ægypt, my nurse."

It is in Palestine, according to Genesis, c. iv. that the Cerealia were discovered, and that Agriculture commenced.

Moses recalls to the memory of the Hebrews this circumstance, which ought to have made the Promised Land still more dear to them.

"For the Lord thy God bringeth thee into a good land, a land of brooks of water, of fountains, a land of Wheat, Barley, and Vines, Fig-trees and Pomegranates, a land of Oil-olive, and Honey,—whose stones are iron, and out of whose hills thou mayest dig out brass (copper)."—Deut. viii. 7.

It is in Palestine that Noah found the Vine (Genesis, ix. 20, 21). It was likewise the country of Bitumen (Genesis, vi. 4). This same Palestine, the land of the Wheat and Barley, is represented in the Bible as the country or situation of the Cedar of Lebanon, of the Balm-tree (*Amyris opobalsamum*), of the Egg-plant (*Solanum melongena*), of the Date-Palm, and of the Sycamore-Fig. It is also the country of the Dromedary, the Jackal, the Deer, the Jerboa, the Lion, the Bear, and the Gazelle. Thus far the Hebrew and Ægyptian histories entirely agree as to the origin of the Cerealia and of the Vine.

We shall now see whether Palestine, according to the most ancient records, unites these conditions. Although the origin of the Cerealia may remain unknown; yet if the country or *habitat* of these different species of indigenous Animals, Vegetables, and Minerals have been correctly stated, one term of the proposition becomes known, and it is then easy to eliminate the remainder.

Anomaly in Grapes.—In a garden at Fernay, a magnificent and solitary cluster of white grapes has been propagated from a vine, which has hitherto produced, and continues to bear, black grapes only.

MINERALOGY.

Extraordinary Application of Gas.—Mr Smith, who gave evidence before the Parliamentary Committee appointed to report upon accidents in mines from gas, in speaking of the coal mines of Nova Scotia, says, "When we first struck the coal at the depth of 180 feet, it was highly charged with water;—the water flew out in all directions with considerable violence;—it produced a kind of mineral fermentation immediately. The outburst of the coal crossed the large river which passed near the coal-pit. We were not aware of the precise outcrop, on account of a strong clay paste, eight to ten yards thick. It is rather difficult to find the outburst of coal, where the clay paste is thickly spread over a country. At the river, the water boiled similarly to that of a steam-engine boiler, with the same kind of rapidity; so that, on putting flame to it on a calm day, it would spread over the river, like what is commonly termed setting the Thames on fire;—it often reminded me of the saying. It is very common for the females, the workmen's wives and daughters, to go down to the river with the washing they have to perform for their families. After digging a hole in the side of the river, about ten or twelves inches deep, they would fill it with pebble stones, and then put a candle to it,—by this means they had plenty of boiling water without further trouble, or the expense of fuel. It would burn for weeks or months, unless put out. I mention this to show how highly charged the coal was with gas. What I am now going to describe may be worth a little attention. There was no extraordinary boiling of the water or rising of the gas, before we cut the coal at the bottom of the pit, more than is usually discernable in a common pond of stagnant water, when a long stick is forced into the mud. As soon as the coal was struck at the depth of 180 feet, it appeared to throw the whole coal mine into a state of regular mineral fermentation. The gas roared as the miner struck the coal with his pick. It would often go off like the report of a pistol, and at times I have seen it burst pieces of coal off the solid wall; so that it could not be a very lightly charged mine under such circumstances. The noise which the gas and water made in issuing from the coal, was like an hundred thousand snakes hissing at each other."

Mercury Mines of Almeyda.—Almeyda is situate in Beira, Portugal, on the Spanish frontier. Its mines are very ancient, for it is recorded by Pliny that the Greeks extracted vermillion from them seven hundred years before the Christian era. The Romans procured from them annually 100,000 livres value of Cinnabar.

Such is their present flourishing condition, that 22,000 quintals of mercury are annually taken from them, and 700 men are constantly employed in mining, 200 in extracting the ore, besides a great number of Muleteers, in conveying the mercury to Seville. The veins are so rich, that although these mines have been worked for ages, the mining has only been extended to the depth of 300 varas, or 800 French metres. They extract the whole vein, which, after being distilled, yields ten per cent. of mercury.

The Mercury produces fatal effects on the industrious miners, most of whom, in the prime of life, present a deadly aspect. These miners possess highly honourable coaracters, and are gentle and discreet in their manners.

GEOLOGY.

Inflammable Gas arising after boring for Salt.—In Jameson's Philosophical Journal it is stated, that from a Salt Mine at Rheine, in Germany, an uninterrupted current of inflammable gas has issued for upwards of sixty years; which is used, not only for light, but also for cooking. In the United States of America currents of inflammable gas are frequently to be met with, issuing from perforations in rocks which have been bored for salt water. The following are some of the instances mentioned in the Transactions of the Philosophical Society of New York:—

In the year 1824, while a company were boring through a rock in Elk Creek, Ohio, at the depth of 24 feet the miners penetrated to a vein of very cold water, somewhat brackish to the taste. At the distance of 116 feet they passed through a rich vein of copper ore, of about three feet in thickness; and at 180 feet they opened a powerful vein of air, which instantly found vent at the top of the shaft or well which they were sinking, and a loud roaring and spouting of water, to the height of 30 feet into the atmosphere. For some distance round this perpendicular jet of water plays a gas, so inflammable in its nature as instantly to take fire whenever a torch is applied to it. The verge of the circumference of this gas is not perceptible; therefore those who are unacquainted with its inflammable quality have found themselves enveloped in flame when attempting to set it on fire. The intervals between the spouting are irregular and uncertain.

Mr Denton mentions, that, "while boring for salt, in the year 1824, about three miles from the village of Sparta, in Tennessee, he hit upon a vein of gas, which in ascending found another vent than the tube, through the natural fissure of a rock in the bed of the Calf-Killer river," forcing a passage through the superincumbent waters, which produced great commotion round the place of escape. A lighted torch being applied, a column of fire, nearly 40 feet high, ascended from nearly the centre of the river, which was about 50 yards wide at that place. Mr Denton met with a similar phenomenon, on the following year, a short distance below the same place. The well was situate on the margin of the river. A bore three inches in diameter was perforated in the limestone rock to the depth of 400 feet; the salt water was forced by the gas through the hole in the rock, in which was placed a tube, the upper end of which was composed of copper, to the extent of 50 feet above the surface of the rock: at the distance of 45 feet a copper faucet was inserted into the wooden tube, and into another of the same kind, standing two feet apart from the first one. The salt water forced up was conducted by the copper faucet into the second tube, from whence it descended 25 feet to a cistern holding 25,000 gallons. While the water was escaping from the first to the second tube, the gas passed up to the top of the first tube—and, upon a lighted candle being applied, it immediately ignited, and flashed up in a flame from 20 to 30 feet in perpendicular height. This place is encircled on three sides by high mountains, which were partly illuminated by this jet of gas, and produced an effect at once magical and sublime.

METEOROLOGY.

Greatest Ascents in the Atmosphere.—On the 16th of December 1831, M. Boussingault, in company with Colonel Hall, ascended Chimborazo, to the height of 19,699 feet, which is the greatest terrestrial elevation yet accomplished. Baron Humboldt was unable to reach a greater height than 19,400 feet. M. Gay Lussac ascended in a balloon from Paris, and obtained an elevation of 22,900 feet. The barometer used by Boussingault fell to 13 inches 8 lines. He found the temperature in the shade was 7.86 (46.6 of Fah.) He conceives it possible for a human being to live in rarified air. It would thus appear, that at a height nearly equal to that of Mount Blanc, where Saussure felt such oppression that he was hardly able to consult his instruments, young females may be seen, in South America, dancing the whole night. During the War of Independence, the celebrated battle of Pichinca was fought at a height little less than that of Mount Rose. Saussure was informed by his guides that they had seen stars in broad day; but Boussingault never observed them, although he reached a much greater altitude.

GENERAL SCIENCE.

Submarine Register Barometer, to be used as an ordinary Deep Sea-Lead.—Mr Payne of the Adelaide Street Gallery of Practical Science, has made and proved the use of an instrument bearing this name. The accuracy with which the mercury rises in descents, and its fall in ascents, has been sufficiently proved by the use of the Barometer; and more recently, by that admirable instrument invented by Mr Adie of Edinburgh, denominated the Sympiesometer, for denoting the heights of mountains and depths of valleys. The invention of Mr Payne, however, differs in many respects from these; but he proposes measuring depths at sea by its means. It is constructed of a tube of glass (or it may be made of iron), rendered tight at top, and the tube then filled with one atmosphere of air or of hydrogen gas. The pressure of the water upon the surface of the mercury in the cistern, is similar to the pressure of the atmosphere upon the surface of the mercury in the Sympiesometer and common Barometer; but in Mr Payne's instrument, the water is prevented from coming in actual contact with the mercury, by the intervention of a piece of fine membrane. The compression of the air in the tube is indicated by a float, somewhat similar to that of the register Thermometer. The glass tube is graduated in atmospheres, and tenths of atmospheres, and also by tables of corrections for temperature, saltness of water, and the depth to which the instrument has sunk,—all of which can be accurately ascertained in fathoms or pounds weight. This Barometer has been graduated by Mr Gordon from 1 to 45 atmospheres, or 247 fathoms, upon the same principle as he graduated the portable gas pressure gauges, which have proved remarkably accurate. By means of this instrument the greatest depth of the ocean may be ascertained with precision.

MISCELLANEOUS.

Fields of Polar Ice.—Of the inanimate productions of Greenland, none, perhaps, excites so much interest and astonishment, in a stranger, as the Ice in its great abundance and variety. The stupendous masses known by the names of Ice-Islands, Floating-Mountains, or Icebergs, common to Davis' Straits, and sometimes met with here, from their height, various forms, and the depth of water in which they ground, are calculated to strike the beholder with wonder; yet the fields of ice, more peculiar to Greenland, are not less astonishing. Their deficiency in elevation is sufficiently compensated by their amazing extent of surface. Some of them have been observed near a hundred miles in length, and more than half that breadth; each consisting of a single sheet of ice, having its surface raised in general four or six feet above the level of the water, and its base depressed to the depth of near twenty feet beneath.

The occasional rapid motion of fields, with the strange effects produced on any opposing substance, exhibited by such immense bodies, is one of the most striking objects this country presents, and is certainly the most terrific. They not unfrequently acquire a rotatory movement, whereby their circumference attains a velocity of several miles per hour. A field thus in motion coming in contact with another at rest, or more especially with a contrary direction of movement, produces a dreadful shock. The weaker field is crushed with an awful noise; sometimes the destruction is mutual; pieces of huge dimensions and weight are not unfrequently piled upon the top, to the height of twenty or thirty feet, whilst doubtless a proportionate quantity is depressed beneath. The view of those stupendous effects *in safety*, exhibits a picture sublimely grand; but where there is danger of being overwhelmed, terror and dismay must be the predominant feelings.

On arriving at the point of collision, between two immense bodies of ice, I discovered that already a prodigious mass of rubbish had been squeezed upon the top, and that the motion had not abated. The fields continued to overlay each other with a majestic motion, producing a noise resembling that of complicated machinery, or distant thunder. The pressure was so immense that numerous fissures were occasioned, and the ice repeatedly rent beneath my feet. In one of the fissures I found the snow on the level to be three and a half feet deep, and the ice upwards of twelve. In one place hummocks had been thrown up to the height of twenty feet from the surface of the field, and at least twenty-five feet from the level of the water; they extended fifty or sixty yards in length, and fifteen in breadth, forming a mass of about two thousand tons in weight. The majestic unvaried movement of the ice, the singular noise with which it was accompanied, the tremendous power exerted, and the wonderful effects produced, were calculated to excite sensations of novelty and grandeur in the mind of even the most careless spectator.

Sometimes these motions of the ice may be accounted for. Fields are disturbed by currents, the wind, or the pressure of other ice against them. Though the set of the current be generally towards the south-west, yet it seems occasionally to vary; the wind forces all ice to leeward with a velocity nearly in the inverse proportion to its depth under water; light ice consequently drives faster than heavy ice, and loose ice than fields; loose ice meeting the side of a field in its course becomes deflected, and its re-action causes a circular motion of the field. Fields may approximate each other from three causes; first, if the lighter ice be to windward, it will, of necessity, be impelled towards the heavier; secondly, as the wind frequently commences blowing on the windward side of the ice, and continues several hours before it is felt a few miles distant to leeward, the field begins to drift before the wind can produce any impression on ice, on its opposite side; and thirdly, which is not an uncommon case, by the two fields being impelled towards each other by winds acting on each from opposite quarters.

The closing of heavy ice, encircling a quantity of bay ice, causes it to run together with such force that it overlaps wherever two sheets meet, until it sometimes attains the thickness of many feet. Drift-ice does not often coalesce with such a pressure as to endanger any ship which may happen to be beset in it; when, however, land opposes its drift, or the ship is a great distance immured amongst it, the pressure is sometimes alarming.—*Scoresby*.

LEARNED SOCIETIES.

Effects of Earthquakes.—At a meeting of the "Geological Society," on the 2d December 1835, a communication was read from Lieutenant Bowers, R.N., stating that he did not observe any change produced on the coast of Chile, or on the relative level of the sea and land; by the earthquake which took place on that coast in November 1822. He was at Valparaiso in 1822, and in February 1823.

A letter was afterwards read from Mr Cumming, who was at Valparaiso during the earthquake, and for some years after, which agreed with the testimony of Lieut. Bowers. Mr Cumming is a collector of shells; and from his frequent and minute investigations, must have noted any variation.

A paper was then read from Mr Parish, Sec. Geological Society, containing a, historical account of the effects produced by earthquake waves on the coast of the Pacific Ocean; from which it appeared that heavy inundations of the ocean accompanied many of the earthquakes which have laid waste the western coast of South America since the year 1590.

Edinburgh: Published for the Proprietors, at their Office, 16, Hanover Street London: Smith, Elder, and Co., 65, Cornhill. Glasgow and the West of Scotland: John Smith and Son, 70, St Vincent Street. Dublin: W. F. Wakeman, 9, D'Olier Street.

THE EDINBURGH PRINTING COMPANY.

THE EDINBURGH

JOURNAL OF NATURAL HISTORY,

AND OF

THE PHYSICAL SCIENCES.

MAY, 1836.

ZOOLOGY.

DESCRIPTION OF THE PLATE—THE GROUSE.

THESE Birds belong to the order Gallinæ, of which our domestic Cock is considered the type. Modern Ornithologists have subdivided the Linnæan genus *Tetrao*. Latham re-established the genus *Perdix*, which was originally instituted by Brisson, and more recently, the Quail and Ptarmigan have each been formed into a distinct genus. The red or common Grouse of Scotland is removed from *Tetrao*, and placed with the Ptarmigan, under the common appellation of *Lagopus*.

Figs. 1 and 2. The Pinnated Grouse, Male and Female (*T. Cupido*).—This is the most remarkable Bird of its genus; it is a native of the United States of America, occupying a tract known by the name of the Brushy Plains of Long Island, in the Queen's County, State of New York, extending for about fifty miles. The soil is a sandy or gravelly loam, covered with trees, shrubs, and small plants. The trees are mostly pitch-pines and white-oaks of inferior size.

This singular species is 19 inches long, and weighs about 3½ pounds. The neck is provided with a pair of supplemental wings, each composed of 18 feathers: the head has a small crest, and a semi-circular comb of orange-yellow extends over each eye. But the most remarkable peculiarity of the Male Bird consists in two curious wrinkled bags of yellow skin situate near the bottom of the neck. When the Bird is at rest these hang loose; but during the breeding season, in particular, they are inflated with air, resemble in colour and magnitude a middle-sized ripe orange, and appear to be formed by an expansion of the gullet with the external skin of the neck. By means of these inflated bags, the Male Bird has the power of uttering a very extraordinary ventriloqual sound, which can be distinctly heard at some miles distant. "It does not strike the ear of a bystander with force, but impresses him with the idea, though produced within a few yards of him, of a voice a mile or two distant. This note is highly characteristic. It is termed *tooting*, from its resemblance to the blowing of a conch or horn, from a remote quarter." During the period of mating, and while the females are occupied in incubation, the males have a practice of assembling, principally by themselves, in some select and central spot where there is very little underwood; and, from the exercises performed there, is called a *scratching place*. The time of meeting is the break of day, and the numbers assembling in one spot are from forty to fifty. When the dawn is past, the ceremony commences by a low tooting by one of the cocks, which is answered by another; they then come forth one by one from the bushes, and strut about like Turkey Cocks, their tail and wings being arranged as represented in the plate, they pass and repass each other, uttering notes of defiance. These are signals for battles, in which they engage with great spirit and fierceness; frequently leaping a foot or two from the ground, uttering a crackling, screaming, and discordant cry.

Figs. 3 and 4. The Spotted Grouse, Male and Female (*T. Canadensis*).—This species inhabits Hudson's Bay during the whole year, where it frequents low grounds; but, in other parts of North America, it is frequently seen on mountains of considerable elevation. The Spotted Grouse has an extensive geographical range, extending from Hudson's Bay as far as the State of New York, where it is frequently a winter visitant. This game is often sent from Nova Scotia and New Brunswick to Boston in a frozen state; as in the North it is known to be kept hanging throughout the winter, and when wanted for use to be taken down and thawed. In winter these Birds feed on spruce, and consequently their flesh is strongly flavoured with that tree. They are very unsuspicious Birds, and are therefore easily approached and killed by the sportsman. They are about 15 inches in length. Frequently they resort to trees, although usually seen on the ground.

Fig. 5. The Sharp-Tailed Grouse- (*T. Phasianellus*).—The disposition of these Birds is very different from that last described, as they are extremely shy, living solitary in pairs during summer, and assembling in packs in autumn, in which state they continue during the winter. Their principal food consists of juniper buds and various sorts of berries; and in winter they eat the tops of evergreens, and sometimes those of birch, alder, and poplar. They are usually seen on the ground, but, if disturbed, resort to the tops of the highest trees.

This beautiful Bird is 16 inches long. It inhabits the southern parts of Hudson's Bay, and is met with on the shores of Lake Superior. In America it is called "the Pheasant."

DESCRIPTION OF THE PLATE—THE SOUI-MANGAS.

This splendid tribe of Birds principally inhabits Africa. Their food consists chiefly of insects, to which some add the nectar of flowers; these they pierce or sip with their tongue. This organ is capable of elongation beyond the bill, and is terminated by a

forked point. During the season of incubation, the plumage of the males shines with the most splendid iridescent and metallic lustre, nearly equal in brilliancy to that of the Humming-Birds. They have an agreeable song, and their disposition is gay and lively. During the rainy season they are subject to a complete change of plumage, which is of the most sombre hues.

Figs. 1 and 2. The Cardinal Soui-Manga (*Cinnyris Cardinalis*).—This Male Bird is remarkable for the harmonious arrangements and beautiful contrast of its colours; the vivid green of its head and neck is subject to reflections of the most beautiful burnished gold, when subjected to a varied play of light; its whole form is very graceful, and the central tail feather being considerably elongated, adds much to the elegance of the bird. Contrary to what is usually the case, the female plumage, although consisting of different colours from that of her mate, is not less beautiful.

Figs. 3 and 4. The Orange Soui-Manga (*C. Osanga*).—The male of this species exhibits a combination of tints of great beauty, the green of the head and neck being subject to the prismatic reflections of the amethyst, ruby, and topaz. The female attire is of a more sober kind, on which account she is less attractive than the male bird.

Figs. 5 and 6. The Red Soui-Manga (*C. Rufa*).—Although a Bird as remarkable for the brilliancy of its attire as any of the preceding, it is less elegant, being divested of the elongated central tail feathers, which add so much to the grace of the others. The more subdued colouring of the female is of a deeply rich tone on the breast and back.

Fig. 7. The Shining Soui-Manga (*C. Famosa*.)—This Bird will yield to none of the feathered tribe for the lustre of its plumage, the whole shining and sparkling by the slightest variation in the play of light, exhibiting colours of the most vivid kinds, which blend and contrast like a galaxy of the most splendid gems. The feathers of the breast present the hues of ruby and gold, bordered with ultra-marine. It is about the size of a Linnet.

Fig. 8. The Black-breasted Soui-Manga (*C. Melanogaster*), is still more elegant in its formation than any of its congeners, and yields to none of them in point of beauty. It is an extremely lively Bird, flitting among the branches of low flowering shrubs with exceeding nimbleness.

Fig. 9. The Fig-eating Soui-Manga (*C. Ficulnea*), although less varied than some others of the genus, is nevertheless beautiful and elegant in its form.

INCONGRUOUS ASSOCIATES.—A gentleman in Teignmouth had a fine large Dog of the Newfoundland breed, a Rabbit, and a bird of the Gull genus, all inhabitants of the same court yard. The friendship between these animals continued for several months; the Rabbit would follow the Dog round the court, attended by the Gull, without betraying the least symptom of fear, as if confident of security in the protection of the noble animal. At the usual time of giving the Dog his meat, his two companions would invariably attend, and the Gull would eat from the same vessel. The Rabbit would join the Dog in all his tricks and gambols. The Dog possessed all its natural fierceness, and would not permit any stranger to enter the court, unless in company with some who had been accustomed to him. He bit several persons, amongst whom was a lady who had been in the habit of feeding him daily, for several months. A workman who was employed on the premises was attacked by him, and with great difficulty escaped by jumping in at a window which had been left open. His affection for the Rabbit never decreased. Since we have been gratified with the sight, the Rabbit escaped into the street, and was worried to death by Dogs. The Dog and Gull remained companions till the owner leaving Teignmouth for his seat at Newry, in Ireland, gave the Dog and Gull to different persons.

A GANDER IN LOVE.—*From the Dumfries Courier.*—"We have the authority of Captain Brown, in his curious work on Dogs, for stating that, brutes though they be, the tender passion is not unknown to the species, whether setter, cur, colly, or mongrel, and why may it not be equally developed in the feathered tribe? Some time since a Gander was located on a farm on the Craigs Barony, and provided with a helpmate that survived the union but a very brief period. The Gander was provided with a wife the second; but meanwhile he had taken a better thought—conceived a devoted attachment to a buxom inhabitant of a cottage, and to this day (1835) gives very unequivocal proofs of it. Though chary of entering the house, evening, noon, and night he stands at the door, and in a moonlight eve serenades his fair one in a fashion peculiarly his own; whenever she appears, he looks so overjoyed, that wings and feet appear all in motion, and the steps of the damsel are actually impeded by his fluttering. The woman confesses that the only way she can get rid of him is by "jouking out o' the house, and whipping hard round the corner;" but once seen, the feathered suitor accompanies her go where she may. If on the harvest rig, there is the Gander by her side; and during the last season, when the farmer conceived it necessary to prolong the labours of his people till after midnight, the Gander accompanied all the motions of the team with which his lady-love was engaged, and watched every sheaf that was placed on it by her hands."

Snakes.—Professor Lugi Metoxa, of Rome, has published an account of some singular experiments made by him upon Snakes. Among others, he endeavoured to ascertain the truth of the assertions of the ancients, respecting the predilection of snakes for music and dancing. In July 1822, about noon, he put into a large box a number of different kinds of snakes, all quite lively, with the exception of some vipers, which were enclosed in a separate box. As soon as they heard the harmonious tones of an organ, all the non-venomous serpents became agitated in an extraordinary manner; they attached themselves to the sides of the box, and made every effort to escape. The *Elaphis* and the *Coluber Æsculapii* turned towards the instrument. The vipers exhibited no symptoms of sensibility. This experiment has been frequently repeated, and with the same results.

Eels.—M. Girardin, Professor of Chemistry at Rouen, while in the act of superintending the digging of a well, the water rushed into it from the neighbouring springs, and contained in it two specimens of small Eels, which have been identified by the celebrated naturalist Dumeril. These must have had a subterranean existence in the springs. At Tours, Eels of various kinds were also brought up by the water of a well in a similar manner.

Horse with Toes on his Feet.—In *Le Globe*, No. 58, we have an account of several papers read in the Academy of Sciences on the 15th August 1827. M. Geoffroy St Hilaire read a Memoir on a Horse which had on his fore-feet three toes, connected by a membrane. This monster is preserved at Lyons, in the private collection of M. Bredin, Director of the Veterinary School of that city. It has been recorded that Julius Cæsar had a favorite horse with toes upon his feet.

BOTANY AND HORTICULTURE.

GENERAL REVIEW OF THE VEGETABLE KINGDOM.—NO. II.

Plants of the Dicotyledonous class within the tropics are frequently conspicuous for the height and circumference of their stems, the richness and variety of their foliage, as well as the bright and finely contrasted colours of their blossom. By the irregularity of their forms, they set off to advantage the arborescent Monocotyledones of the Palm tribe, which have in general the simple sober forms of our columns, of which they were the models. It is towards the Equator that the gigantic climbers, which grow to the length of several hundred yards, are found; as well as those magnificent herbs of the Scitamineæ and Musæ, as tall as the trees of our orchards; with flowers and foliage not less pre-eminent in their dimensions. For instance, the *Corypha umbraculifera*, an East Indian palm, with leaves in the form of an umbrella, and more than six yards across; and the *Aristolochia*, that grows on the river of La Madalena, the flowers of which, according to M. de Humboldt, serve the children for hats. The far greater part of the aromatic plants belong also to the equatorial regions.

By the side of this rich and varied vegetation, that of Europe appears poor and tame. Here the species of trees are few, and all have a port and foliage in which much sameness prevails. Their flowers make so little show, that the generality of people, who think nothing a flower except a coloured corolla, being ignorant of the use and importance of the other parts, believe that most trees have none at all.

The inferiority in the vegetation of our regions will appear in a still stronger light, if we compare the species of the same genera or tribes which grow both in Europe and under the line. In South America, plants of the Fern tribe, with a foliage and fructification not very unlike our common Brake and Polypody, grow like Palms, and have a stalk in the form of a column.

The cold and temperate climates of our quarter of the world abound in dwarf herbaceous turfy *Gramineæ;* hot countries have also many plants of this tribe, but they are on a much larger scale. This difference begins to be perceived even when we reach Italy, where the Millet attains the height of four or five yards. The Bamboos, Panic-grasses, and the Sugar-cane of Asia, Africa, and America, reach the height of eight or nine yards.

It is said that in parts of the East Indies there are antiquated Bamboos, which are real trees, with a haulm of such girth, that a piece divided lengthwise makes two entire canoes.

The herbaceous Monocotyledones of the tropics, such as the *Liliaceæ*, are greatly superior to ours in the beauty of their flowers. The Heaths of the Northern parts of Europe are low bushes, with feeble stems and small bloom; those of the coasts of the Mediterranean have also a small bloom, but their stems are taller and more robust; those of the Cape fascinate by the form, splendor of colour, and size of the corolla. The *Geraniums* of Europe do not approach those of Africa in point of stature or beauty of flower.

All the plants of the Mallow tribe with us are herbaceous; those of hot climates either shrubs or trees. A tribe of so little account in these parts, holds a place among the vegetables of the most note in the equinoctial regions. There it counts among its species the *Baobab* and the *Ceiba*, the colossi of the vegetable creation; besides the "hand-tree" of Mexico, so called from the form and disposition of the stamens of the flower, which represent very tolerably a hand or paw with five fingers.

The *Leguminosæ* or Pulse tribe furnish Europe with many herbaceous species, several shrubs, and one middle sized tree; all of which, however, have leaves composed of but few leaflets. The same tribe in the hot climates of Asia, Africa, and America, teems with lofty trees, graced with leaves of the most delicate texture, divided and subdivided into numberless leaflets, and playing in the wind like plumes.

The *Aroideæ* in Europe never exceed the height of a yard; those of Mexico, the Brazils, and Peru, sometimes tower into the air like the Banana, of which they assume the appearance; at other times lengthening themselves into supple climbers, they mount to the tops of the highest trees.

Differences as strongly marked are exemplified in the *Orchideæ*. In Europe the species are low; their flowers, although equally interesting to the Botanist by the singularity of their structure, as in other regions, are too insignificant to attract the attention of any who do not make plants an object of their study. In the Torrid Zone, the case is quite different in regard to this tribe, the greater portion of which consists of species that excite our wonder by the size and brilliancy of their blossom; and many, as the Vanilla, suspend their long branches covered with a foliage of shining green, and terminated by magnificent garlands of flowers from the summits of trees.

The *Apocyneæ, Boragineæ, Convolvulaceæ*, and many other tribes, are equally examples of contrasts of a like nature. The European Naturalist, whom the ardent thirst of science leads under the Equator, views with ecstacy those fertile regions, which exhibit at every step forms familiar to him, decked in the rich attire bestowed from the hand of a more bountiful and powerful Nature.

There are beauties in a land yet wild and savage, which disappear at the approach of civilization. In Europe the soil abounds chiefly in plants which are of use to Man. Domestic vegetables, by the aid and protection of the cultivator, have so trenched upon the domain of the wilderness, that space is scarcely left for the existence of those for which Man has no call. The primeval forests of the Gauls and Germans have disappeared; our forests are mere formal plantations of large extent. They are intersected in all directions by roads and paths; are explored without difficulty; and the wild animals no longer find safe refuge in them. Generations of trees are renewed in quick succession, on a soil which the industry of the proprietor keeps in constant requisition, and it is mere chance when a single stick is left to end its career by old age. Far in the North there are several forests which still preserve some traces of the primeval vegetation of Europe. In these the Oaks, spared by the axe, acquire an enormous size; while others, worn out by age, fall of themselves, are decomposed, and help unceasingly to augment the surface of the soil covered with high mosses and thick lichens, that preserve a prolific moisture.

None, however, approach in magnificence to the forests which shade the equinoctial regions of Africa and America. One is never satiated with admiring there the endless multitude of vegetables brought into near contact with each other, and mingled promiscuously together; so different among themselves, and often so extraordinary in structure and produce; those enormous trees still exhibiting no symptoms of decay, though their age goes back to a period at but little distance from the last revolution of our globe; those towering Palms, contrasting by their simple forms with all that surrounds them; those extensive climbers; those Rattans, which, knitting together their long and flexible branches by numberless knots and turns, encircle as one group the whole vegetation of these extensive regions. To clear a path through these, neither fire nor axe is sufficient; the one extinguishes for want of circulation in the air, the other is broken or blunted by the hardness of the wood it meets. The soil cannot afford place to the numberless germs which it develops. Each tree disputes with others, which press from all sides, the soil it wants for its existence; the strong stifle the weak; while rising generations obliterate even the slightest trace of destruction and death; vegetation never flags; and the earth, so far from becoming exhausted, acquires new fertility from day to day. Hosts of animals of every kind, Insects, Birds, Quadrupeds, Reptiles, beings as diversified and strange as the vegetation of the place itself, retire under the vast canopy of these ancient thickets as into a citadel, proof against the attack of Man.

Manna of the Desert.—M. Bore, formerly principal gardener and conductor of farming operations to the Pacha of Egypt, has discovered that the Tamarix Mannifera used in medicine grows abundantly about a day's journey from Mount Sinai. He was assured by the Arabs, that after this Manna was purified, it was equal to honey. The Arabs perform this operation by putting it in hot water, and afterwards skimming it. M. Bore gathered some drops himself, which were as large as ordinary sized peas, as they fell from the branches. It was agreeable to the taste, but there was very little gummy or saccharine matter about it. It has been supposed that it was this kind of Manna upon which the Jews subsisted, as mentioned in Scripture, and which some authors have thought was produced from the *Alhagi Maurorum*—a small plant found, according to their account, only in the confines of the desert, where the atmosphere is very humid, which is necessary to the growth of this plant.

Mandrake.—In the vicinity of Uschakan are found two remarkable roots. With one, called *toron*, is made a red colour, which is used in Russia, and the Russian name of which is *morena;* the other, *leschtak* or *manrakor* (mandrake), bears an exact resemblance to the human figure, and is used by us medicinally. It grows pretty large. A Dog is usually employed to draw it out of the ground; for which purpose the earth is first dug from about it, and a Dog being fastened to it by a string, is made to pull till the whole of the root is extracted. The reason of this is a superstitious belief prevalent among the Russian boors, that if a man were to pull up this root, he would infallibly die, either on the spot or in a very short time, and that, when it is drawn out, the moan of a human voice is always heard.

GENERAL SCIENCE.

Scoresby's Experiments on Magnetism.—This gentleman has shown that bars of steel could be rendered highly magnetic by hammering them in a vertical position, with the lower end resting upon a poker or rod of iron. This process, however, he greatly improved by hammering the steel bars between two bars of iron. The steel bars used by him were the eighth part of an inch in diameter.

When only one bar of iron was used, a steel wire, six inches long, lifted a nail weighing 186 grains; but when two bars of iron were used, the wire lifted 326 grains. When the new process was employed with an iron bar eight feet long, a steel wire, six inches long, lifted 669 grains, or four times its own weight.

Mr Scoresby's theory of this process is, that percussion on magnetizable substances in mutual contact inclines them to an equality of condition, in the same manner as all bodies of different temperatures tend to assume the same temperature when in contact. The two great iron bars being made magnetical by position, the interposed bar of steel will, therefore, when thrown into a state of vibration by percussion, receive a portion of their magnetism. In like manner, a magnet, when struck in the air with a piece of flint, or upon a body of inferior magnetic quality, will have its magnetism diminished.

MISCELLANEOUS.
LAWS OF HARMONIOUS COLOURING.

ACCORDING to the theory of Sir Isaac Newton, there are *seven* primary colours, viz. violet, indigo, blue, green, yellow, orange, and red. Artists, however, have long considered that there are only three, namely, red, yellow, and blue; and this opinion has recently been adopted by Sir David Brewster and other philosophers. It is quite certain that, by a combination of these three colours, every other can be made.

"If we look steadily (says Mr Hay *) for a considerable time upon a spot of any given colour, placed on a white or black ground, it will appear surrounded by a border of another colour. And this colour will uniformly be found to be that which makes up the triad; for if the spot be red, the border will be green, which is composed of blue and yellow; if blue, the border will be orange, composed of yellow and red; and if yellow, the border will be purple, being in all cases the complement of the three colours called by artists homogeneous.

"It is well known to all who have studied music, that there are three fundamental notes, viz. C, E, and G, which compose the common chord, or harmonic triad, and that they are the foundation of all harmony. So also there are three fundamental colours, the lowest number capable of uniting in variety, harmony, or system.

"By the combination of any two of these primary colours, a secondary colour of a distinct kind is produced; and as only one absolutely distinct denomination of colour can arise from a combination of the three primaries, the full number of really distinct colours is seven, corresponding to the seven notes in the complete scale of the musician. Each of these colours is capable of forming an archeus, or key, for an arrangement, to which all the other colours introduced must refer subordinately. This reference and subordination to one particular colour, as is the case in regard to the key-note in musical composition, gives a character to the whole.

"This characteristic of an arrangement of colour is generally called its tone; but it appears that this term is more applicable to individual hues, as it is in music to voices and instruments alone. Yet, to avoid obscurity, I shall continue to use it in the sense in which it is generally applied to colouring.

"From the three primary colours, as will be afterwards shown, arise an infinite variety of hues, tints, and shades, so that the colourist, like the musician, notwithstanding the extreme simplicity of the fundamental principles upon which his art is built, has ample scope for the production of originality and beauty, in the various combinations and arrangements of his materials.

"The three homogeneous colours, yellow, red, and blue, have been proved by Field, in the most satisfactory manner, to be in numerical proportional power as follows:— yellow three, red five, and blue eight.

"When these three colours are reflected from any opaque body in these proportions, white is produced. They are then in an active state, but each is neutralised by the relative effect that the others have upon it. When they are absorbed in the same proportions, they are in a passive state, and black is the result.

"From the combination of the primary colours the secondary arise, and are Orange, which is composed of yellow and red, in the proportion of three and five; Purple, which is composed of red and blue, in the proportion of five and eight; and Green, composed of yellow and blue, in the proportion of three and eight. These are called the accidental or contrasting colours to the primaries, with which they produce harmony in opposition, in the same manner in which it is effected in music by accompaniment; the orange with the blue, the purple with the yellow, and the green with the red. They are therefore concords in the musical relation of fourths, neutralising each other at sixteen.

"This neutralising or compensating power, as will be afterwards shown, is the foundation of all agreement and harmony amongst colours, and upon it depends also the brilliancy and force of every composition.

"From the combination of these secondaries arise the tertiaries, which are also three in number, as follow: Olive from the mixture of the purple and green, Citron from the mixture of the green and orange, and Russet from the mixture of the orange and purple. These three colours, however, like the compounds produced by their admixture, may be reckoned under the general denomination of neutral hues, as they are all formed by a mixture of the same ingredients; the three primaries, which always, less or more, neutralise each other. The most neutral of them all being grey, the mean between black and white, as any of the secondaries are between two of the primaries, it may appropriately be termed the seventh colour. These tertiaries, however, stand in the same relation to the secondaries that the secondaries do to the primaries—olive to orange, citron to purple, and russet to green; and their proportion will be found to be in the same accordance, and neutralising each other integrally at 32.

"Out of the tertiaries arise a series of other colours, such as brown, marone, slate, &c. in an incalculable gradation, until they arrive at a perfect neutrality in black, as shown in diagram 2. To all of these the same rules of contrast are equally applicable.

"Besides this relation of contrast in opposition, colours have a relation in series, which is their melody. This melody or harmony of succession is found in all the natural phenomena of colour. Each colour on the prismatic spectrum, and in the rainbow, is melodised by the two compounds which it forms with the other two primaries. For instance, the yellow is melodised by the orange on the one side, and the green on the other; the blue by the green and purple, and the red by the purple and orange. These coincidences can be shown by a diagram where the chromatic scale of the

---

* The Laws of Harmonious Colouring, adapted to interior Decorations, Manufactures, and other useful Purposes, by D. R. Hay, House-painter, Edinburgh, 3d edition. This is a work of great merit, and ought to be in the hands of all persons of taste.

---

colourist is accommodated to the diatonic series of the musician, showing that the concords and discords are also singularly coincident.

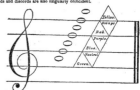

"The following diagram exhibits a general harmony of all the colours of any distinctive character, simple and compound, except the neutral grey. It will be observed, that each limb of this diagram forms a series of hues proceeding from one of the primaries, and predicting a distinct melody or harmony, in succession, of that colour. It will also be seen, that in each of these harmonies, although the primary colour or key-note predominates, the other two primaries enter, in combination, into the arrangement. This, however, is more plainly shown when these three melodies are exhibited separately. There is also shown, upon this diagram, the progress from light to darkness, or from white to black; as also in the nine central divisions, the harmony in succession, and contrast of the primary and secondary colours. The arrangement of this diagram, I trust, will likewise show that all the colours, in their greatest intensity, may be brought together without crudity or harshness:

"In all general arrangements of colours, which are not necessarily confined to any particular leading colour or (to continue the analogy) key-note, it ought to be kept in view, what Nature has pointed out in the most distinct manner in all her colouring, namely, that those cool-toned and neutralised colours which are most agreeable to the eye should predominate, and that vivid and intense colours should, upon all occasions, be used with a sparing hand.

"White is produced by the reflection of all the colours simultaneously in their relative neutralising proportions. Its contrasting colour is black, with which it is co-relative, being the opposite extreme of neutrality. It lies nearest in series to yellow, which may be reckoned its melodising colour. It, however, harmonises in conjunction and opposition with all other colours. Every colour in its series of tints becomes subdued in force proportionally as it approaches towards white. It is the representative of light as black is of darkness; its effect on the eye is therefore cheerful and enlivening.

"Yellow, of the three primary colours, partakes most of the nature of white, being the lightest of all decided colours, and the brightest on the prismatic spectrum. Its contrasting colour is purple, a compound of the other two primaries; its proportional power to which is as three to thirteen, either in quality or intensity. It constitutes, in combination with red, the secondary orange; and, when compounded with blue, it produces the secondary green. These two colours are therefore its melodising hues. It is the most powerful of the positive colours, and consequently the least agreeable to the eye, when unaccompanied, or when predominating in a pure state. Being the lightest of positive colours, it, next to white, forms the most powerful contrast to black. Yellow, of course, forms a component part of all the tertiary or neutral hues, either in predominance or subordination.

"The tertiary, in which it is the archeus or ruling colour, is that called citron, which, being a compound of orange and green, the two secondaries into which yellow enters, has a greater proportion of that colour than either of the other two tertiaries. Citron is of itself a soft and pleasing colour to the eye, and is the lightest of all distinct hues arising out of the treble combination of the primaries. It is very useful as a contrasting colour among low tones of purple and crimson. In tracing yellow still further down in the scale, the next understood colour in which it predominates is the semi-neutral hue brown, a most efficient colour in all the low parts of every warm-toned arrangement.

"In artificial lights pure yellow apparently loses much of its intensity, because it cannot be easily distinguished from white. This occurs from all such lights being less or more of a yellow tone, and consequently diffusing this colour over all objects within their influence; white thereby becoming yellow, and yellow remaining unaltered.

"Orange is the next colour in power; it is a compound of yellow and red, in the proportions of three of yellow to five of red. Between these two colours it appears in the prismatic spectrum, rainbow, and other natural phenomena; they may, therefore, be termed its melodising colours. Its contrasting colour is blue. Orange is the extreme point of warmth in colouring, as blue is of coldness; they, therefore, form the most perfect contrast in this respect, as they do in their numerical proportional power, being eight to eight. In its combination with green it produces the tertiary citron, and with purple the tertiary russet.

"Red is the third in the chromatic series, and second of the primaries. It is the most positive of all colours, holding the middle station between yellow, which is most allied to light, and blue, which is most allied to shade; it is, as Field expresses it, pre-eminent among colours. The hues with which it melodises in series are, of course, orange and purple, being its combinations with the other two primaries. Its contrasting colour is green, a compound of yellow and blue, in the proportion of three yellow to eight blue. Red is decidedly a warm colour, and, to a certain extent, communicates this quality to every hue into which it enters.

"This effect of warmth is most apparent in its combinations with yellow; for in those with blue it becomes more cool and retiring. From the medial situation of red, and from its power in subduing the effect of such colours as enter, in minute proportion, into combination with it, its name is very indiscriminately applied. The first decided hue produced, in its approach towards yellow, is scarlet; and, in its approach towards purple, it produces the most splendid of all hues of this description, crimson.

"The tertiary in which red predominates is russet, a medial hue between purple and orange, and consequently having a double occurrence of red in its composition; therefore, it is the most positive and warm of the neutral colours. It is of great power and value in all the deep parts of any warm-toned arrangement, as a contrasting colour to the deep hues of green, necessarily brought in as relieving colours. The semi-neutral marone is the next understood hue in its descent to black.

"Purple lies next in series to red, of which colour and blue it is composed, in the proportion of five of the former to eight of the latter. In this state of intensity it forms the proper contrasting or neutralising colour to pure yellow. The two primaries of which it is compounded are its melodising colours. Although red be one of its component parts, it is rather a cool colour, and very retiring in effect: being also the darkest of the secondary colours, it bears the nearest relation to black or shade, as its contrasting colour, yellow, does to white or light. From these qualities, purple is a pleasing and agreeable colour to the eye; in this respect it is second only to green. In its combination with green it produces that soft and useful tertiary colour, olive, and with orange, the most powerful of this class, russet.

"Blue is the third of the primary colours, and fifth of the chromatic series. It is, of the primaries, the nearest in relation to shade, as yellow is to light. It is the only absolutely cool colour, and communicates this quality to all hues into the combination of which it enters.

"Green, although the last in the general series which I have adopted, is the medial or second of the secondary colours, being a compound of yellow and blue, in the proportion of three of the former to eight of the latter; the one primary being most allied to light, and the other to shade. Its melodising colours are of course these two primaries, and its contrasting colour the remaining primary, red. As red is the most decided or pre-eminent of the primaries, so green is the most neutral and soft of the secondaries, and the most pleasing and agreeable of all decided hues to the eye. It is also unlike the other two secondaries in this respect—that, in its approximation to either of its component parts, it produces no other distinct denomination of colour; all its hues retaining the same name. Out of the union of green with orange arises the lightest of the tertiary colours, citron; and out of that with purple, the deepest olive, to which it appears particularly allied.

"Green is the natural clothing of the vegetable world, and, in a certain degree of purity, predominates in the same ratio of quantity that red is held subordinate. It is in its greatest intensity and depth when the sun's rays are most powerfully directed to the earth, thereby counteracting the intensity of their reflection, and refreshing the eye by its soft and soothing influence, in that infinite wisdom so conspicuous in all the laws which govern the universe. Green, however, like every other hue in Nature's colouring, seldom appears in vegetation in its primitive purity. Hence the beautiful accordance between the green of the landscape and the blue of the sky, so evidently assisted in both harmony and melody by the intervention of the warm and neutral gray, which prevails intermedially in the distance of the one and the horizon of the other. In its various hues, green, as may naturally be supposed, is a favourite colour in decoration, and would be much more so, were it not that in artificial light its effect is much deteriorated, becoming in most cases dull and heavy.

"Black, as already noticed, is produced by the absorption of the three primaries, and its natural contrast is white. It can only be used in large quantities in arrangements of a cool and sombre character, and ought always to be pure and transparent.

"In the decorative painting, however, of Pompeii and Herculaneum, it was used in much profusion; and in combination with the intense and brilliant colours which accompanied it, produced the most splendid effect."

Edinburgh, February 16, 1836.

Sir,—This last season, I sailed as surgeon in the Sisters, whale ship, and should you deem the following singular phenomenon, which I do not recollect ever having noticed in any work on the Arctic Regions, worthy of insertion in your excellent Journal, you are at liberty to use my name in any manner you think fit.

About midnight, on the 19th September last, being then in latitude 71° N., and about 76° of longitude, it was necessary to warp the ship through some ice, which, I may mention, is done by fixing or striking grapnels, or, as they are called, ice-anchors, into the ice, to which warps are attached, and the ship is then hoven through by the capstan on board the vessel. One of the seamen, in striking the grapnel into the ice for this purpose, a vivid flash of light was emitted at each stroke. I shall not attempt to assign any reason for this singular circumstance, but I may add, that the sea at the same time was beautifully phosphorescent, a phenomenon I frequently observed in those regions; and during the congelation of the ocean and in newly formed ice, even when it has attained the thicknesss of three or four inches, I have observed phosphorescent flashes, evidently produced on the surface, in the substance and below the ice; the temperature at the time being from 18° to 20° of Fahrenheit's scale.

Sea water ice is perfectly fresh until it attains the thickness of a quarter or half an inch, but after that it is quite salt.

To Captain Thomas Brown. (Signed) Nicholas Oliver.

Public Libraries.—"There are at least nine which greatly *exceed* the British Museum Library in extent, those of Berlin, Göttingen, Dresden, Naples, Vienna, Copenhagen, St Petersburgh, Munich, and Paris; and several others of about equal extent, as those of Wolfenbuttel (190,000), Stuttgard (197,000), Madrid (200,000) The extent of the first nine, as nearly as I am able to ascertain it, is as follows:—

	Printed Books.		*MSS.*		*Population.*
Berlin, . .	250,000 vols.		5,000 vols.		248,816
Göttingen, . .	300,000	...	5,000	...	9,594
Dresden, . .	300,000	...	2,700	...	69,000
Naples, . . .	300,000	...	6,000	...	354,000.
Vienna, . .	350,000	...	16,000	...	320,000
Copenhagen, . .	400,000	...	20,000	...	109,000
St Petersburgh, .	400,000	...	16,000	...	320,000
Munich, .	500,000	...	16,000	...	95,718
Paris, . .	700,000	...	80,000	...	890,431
London (Brit. Museum)	220,000	...	22,000	...	1,528,301."

The above is extracted from "A Letter to B. Hawes, Esq., M.P., being Strictures on the Minutes of Evidence taken before the Select Committee on the British Museum," by Edward Edwards, Esq., London, 1836. In reference to our own immediate subject, we can readily bear testimony to the absolute deficiency of the public libraries of Britain in works on all branches of Natural History; and were it not for the monthly draughts of information which we derive from the Continent, answered by drafts of another kind on the Proprietors, the extensive undertaking in which we are engaged would find a speedy termination.

REVIEWS.

The Tower Menagerie, comprising the Natural History of the Animals contained in that establishment. 8vo. Jennings, London. 1829.

The Gardens and Menagerie of the Zoological Society Delineated. 2 vols. 8vo. Tegg, London. 1830.

It may be wondered at by our readers, why we notice at this period two works which have been so long before the public. Our simple reasons are, that they do not seem to be well known and appreciated, and that we consider it a duty to the proprietors and our readers to point out their excellencies.

These two works were of cotemporary projection, and intended to illustrate, from living specimens, the two Menageries whose names they bear. They are, in all respects, twin sisters of exceeding beauty. Both are edited by E. T. Bennet, Esq., the accomplished Vice-Secretary of the Zoological Society; the drawings for both were from the pencil of Mr William Harvey, at the head of his profession as an animal painter, and the engravings for both were executed by Messrs Branstone and Wright, who have brought the art of Wood Engraving to its present high state of excellence. Both are from the justly-celebrated press of Mr Charles Whittingham.

We cannot speak too highly in praise of the graphic and truly scientific manner in which the animals in these three volumes are described. The natural-historical details of the various objects are full of interest, and exhibit proofs of much discrimination and research. They are worthy of the Society under whose auspices they have appeared.

The illustrations are faithful delineations of the different objects, exhibiting a freshness and vigour which at once carry with them a conviction of the accuracy of the character and expression of the animals. They are not merely representations of the animals, but each cut has a pictorial effect, the objects being surrounded with a picturesque and appropriate landscape, giving relief to the figure, and leading the mind to such scenes as are likely to be the accustomed haunts of the beings represented.

Will it be believed, after the high but just character which we have given of these works, they met with such a poor reception, that, within twelve months after their publication, they fell to nearly half their original price?—a libel on British taste. But we are happy to understand they have again risen in value, and we foresee that the day is not far distant when they will bring double their original cost

---

ERRATA.

Animal Kingdom,	page 17,	col. 2,	line 53,	for *always*	read *usually*.
...	... 32,	... 2,	... 74,	... require	... *requires*.
...	... 36,	... 1,	... 26,	... *vower*	... *vomer*.
...	... 39,	... 2,	... 12,	... *Phascogala*	... *Phascogale*.
...	... 40,	... 1,	... 42,	... *Jackalls*	... *Jackals*.
...	... —,	... 2,	... 78,	... *Antilopa*	... *Antilope*.

Edinburgh: Published for the Proprietors, at their Office, 16, Hanover Street. London: Smith, Elder, and Co., 65, Cornhill. Glasgow and the West of Scotland: John Smith and Son, 70, St Vincent Street; and John M'Leod, 20, Argyle Street. Dublin: W. F. Wakeman, 9, D'Olier Street.

THE EDINBURGH
JOURNAL OF NATURAL HISTORY,
AND OF
THE PHYSICAL SCIENCES.

JUNE, 1836.

ZOOLOGY.

DESCRIPTION OF THE PLATE—THE QUAILS.

This genus was instituted by Mr Stephens, in his continuation of " Shaw's General Zoology," for the reception of such of the Partridges as had thick bills. These are only found in North and South America. They frequent the borders of woods, and reside among brushwood, or in plains where the grass is thick and high, or among grain in fields which are cultivated. If disturbed, they fly to trees, where they, perch for safety, and " walk with ease on the branches," according to Audubon; who says they perform occasional migrations from north-west to south-east, usually in the beginning of October, and somewhat in the manner of the Wild Turkey.

Figs. 1 and 2. The Virginian Quail, male and female (*Ortyx Virginiana*). — This handsome bird abounds in the Eastern and Middle States of America, and is to be found in most districts of the Union, where it is called "the Partridge." It emigrates about the beginning of October, at which time, the north-eastern shores of the Ohio are literally covered with them. During these excursions, they frequently fall into the water, and many of them perish, but if they drop at no great distance from the land, they easily reach the shore by swimming, which Audubon affirms they can do " surprisingly."

If Virginian Quails are molested, they take refuge in trees, always resorting to the middle branches; and if they think they are noticed by the sportsman, they erect the feathers on the crown of their head, emit a low note, and escape to another part of the tree, or to a more distant one. When they take to flight without being disturbed, the whole covey pursue the same course; but when frightened they disperse in various directions, and after having alighted call to each other, and are soon congregated by the note of the patriarch-bird of the flock.

The nest of this bird is of a circular form, in which it leaves an aperture not unlike in shape to that of a common oven. It is placed at the side of a thick tuft of grass, and is partly sunk in the ground. The female lays from ten to eighteen pure white eggs; and is assisted by the male in the tedious operation of incubation. They only rear one brood during a year.

Their manner of reposing at night is rather curious, as mentioned by Audubon. He says, " the Partridge rests at night on the ground, either amongst the grass or under a bent log. The individuals which compose the flock form a ring, and moving backwards, approach each other until their bodies are nearly in contact. This arrangement enables the whole covey to take wing when suddenly alarmed, each flying off in a direct course, so as not to interfere with the rest." The flesh of this bird is considered a delicate and agreeable food.

Figs. 3 and 4. The Californian Quail, male and female (*O. Californica*). The first person who noticed these beautiful birds was an Editor of the voyage of the unfortunate La Pérouse, who also figured them in the plates illustrating that work. They are known to assemble in flocks of two or three hundreds in the low woods and plains of California. The flesh is said to be of a fine flavour. These quails are easily tamed, and soon become quite reconciled to a state of captivity. They are birds of an elegant bearing, the crests giving them a fine and striking appearance.

Fig. 5. The Long-tailed Quail (*O. Macroura*). — Nothing is known of the history of this species; if is a native of Mexico.

Fig. 6. Montezuma's Quail (*O. Montezuma*). — This is also a Mexican species, but its locality is not known. Some years ago there was a fine specimen of the bird in the Zoological Gardens of London, which is now dead, and its skin is preserved in the Museum of the Zoological Society.

DESCRIPTION OF THE PLATE—THE KINGLETS.

These very beautiful Birds were formerly ranked among the Wrens, but Cuvier, following Ray, formed a new genus for their reception under the title of *Regulus*. They subsist almost entirely on Insects, in the pursuit of which they exhibit great nimbleness of action. They construct their nest with much neatness. The British species usually suspend them to the extreme branch of a tree, and cover them externally with mosses, selecting the same kind which is on the tree for their purpose.

Figs. 1 and 2. The Ruby-crowned Kinglet (*Regulus Calendulus*). — This beautiful species is a native of North America. They are birds of passage, and visit the United States from the South, about the beginning of April. Their food at this season consists of the blossoms of the Maple Tree, and when these fail, they have recourse to those of the Peach, Apple, and other fruit trees. They eat only the sta-

mens of these flowers, but they also subsist upon the Insects which hover round them. These Birds penetrate far to the North, and even build and incubate in the country around Hudson's Bay. The Ruby-crowned Kinglet is only four inches long, and six in extent of wings.

Fig. 3. Cuvier's Kinglet (*R. Cuvierii*) is a native of Pennsylvania, and was first discovered by Audubon. We know nothing of its history.

Figs. 4 and 5. The American Kinglet (*R. Americanus*). — Wilson and other naturalists confound this Bird with the European Kinglet—fig. 8 of this plate—from which it differs in several essential particulars. The length of the American Regulus is three inches and seven-eighths, while the European species varies from three inches and a half to three inches and three quarters; the bill is also longer, and more dilated at the base, and the crest differs materially.

The American Kinglet is an active, unsuspicious Bird, climbing and hanging occasionally among the branches, and sometimes even on the body of the tree, in search of the larvæ of Insects. It also retires northwards to incubate, and is seldom to be met with in the State of Pennsylvania from May to October; after which it becomes very abundant in orchards, and assists greatly in thinning them of the numerous Insects with which they are infested at this time of the year. It is four inches long, and six in extent.

Fig. 6. Byron's Kinglet (*R. Byronensis*). — This interesting species is a native of Chili, whence it was brought by Lord Byron, who presented it to the British Museum. Its habits are unknown.

Fig. 7. The All-Coloured Kinglet (*R. Omnicolor*) is a native of Brazil, and inhabits the extensive forests which border the Rio-Grande. It is remarkable for the brilliancy of its plumage; and differs from its congeners in being provided with an ample tuft of feathers on the crown of the head; its bill differs, also, in being somewhat straighter.

Fig. 8. The European Kinglet (*R. Cristatus*). This species is to be met with all over Europe, and is plentiful in some parts of this Island. It is a resident with us the whole year round; but Selby records two instances where it migrated; in October 1822 and January 1823. In the latter case, the whole tribe disappeared. This happened a few days prior to the long-continued snow-storm, so severely felt through the Northern Counties of England, and along the eastern parts of Scotland. It is the smallest British Bird, being only three inches and three quarters in length, and seldom exceeds sixty grains in weight. It is commonly known by the name of " The Golden-crested Wren."

MATERNAL AFFECTION OF THE ÆTHIOPIAN SOW. — In chasing the old Sows of this species, with their young ones, Dr Sparmann observed that the heads of the females became suddenly enlarged and more shapeless than they were before. This momentary and wonderful change astonished him so much the more, as, riding hard over a country full of bushes and pits, he had been prevented from giving sufficient attention to the manner in which it was brought about. The whole of the mystery, however, consisted in this: each of the old ones, during its flight, had taken a pig in its mouth; this also readily explained the reason of his surprise, upon finding that all the pigs which he had been chasing along with the old ones, had vanished on a sudden. In this action we find a kind of unanimity among these animals in which they resemble the tame species, and which they have in a greater degree than many others. It is likewise very astonishing, that the pigs should be carried about in this manner, between such large tusks as those of their mother, without being hurt or crying out in the least.

MIGRATION OF WHALES. — It is now well ascertained that all the Whales which frequent the Polar Seas pass annually to the southward, in large bodies, in the months of March and April, about midway between the coasts of Ireland and Newfoundland. From the late report of the Committee of the House of Commons on the public works of Ireland, we learn that Whales appear in great numbers on the western coast of that country in the Spring months, and are totally neglected and unpursued, in consequence of the poverty and want of means of equipment of the people of that coast. This being the case, it is evident that the Polar Seas have been too long and needlessly visited in search of cetaceous animals, at the expense of much loss of property, time, and human life; as Whales may be equally well encountered and captured in the Atlantic Ocean, as in the dangerous northern regions.

THE POISON OF SNAKES. — Sir Thomas Brisbane mentions, that one of the poisonous Snakes which he kept at home, while Governor of New South Wales, bit two of his Pointers, one of which died in three minutes, and the other in about thirty minutes; thus equalling prussic acid in the rapidity of its effects.

ON THE HYBERNATION OF ANIMALS.—NO. III.

We shall now examine more in detail the distinguishing Physiological Conditions of this interesting class of animals, and first shall make some observations on their Respiration.

It has been ascertained by experiment that all the warm-blooded animals, when asleep, consume less oxygen than when awake. In the hybernating animals, during their lethargic state, this is carried to a much greater extent. Mangili carefully observed the fat Dormouse *(Mus Glis)* in this state, and found that its respirations were irregular, with long intervals between them: at one time the animal breathed from 13 to 15 times in succession, then followed an intermission of from 24 to 28 minutes, without any respiration. In another Dormouse *(Mus avellanarius)* he found its respiration only 3 in the minute, though the temperature was as high as 65°. Prunelle placed a Hedgehog, in a state of lethargy, under water for 4 minutes, without injury to the animal. He also found the respirations so imperfect, in a state of lethargy, when the temperature was about 44° or 45° that they could not be reckoned. They became sensible when the temperature was raised to 59°, and in a low temperature they were 3 or 4 per minute, with intervals. Dr Marshall Hall has made the most accurate experiments upon this point, and they fully bear out the statements made by Mangili and Prunelle, that the function of respiration is nearly suspended. He placed a lethargic Bat under an instrument which he has invented for experiments of this kind, and which he has termed a pneumatometer, for 10 hours, and found no perceptible absorption of oxygen. He then roused the animal, and the absorption became immediately apparent. Another lethargic Bat was placed 24 hours under the instrument, and a cubic inch of the oxygen only had disappeared. A Bat placed sixty hours under the instrument, at a temperature varying from 38° to 41°, occasioned an absorption to the extent of 3.8 inches. In a state of activity equal quantities of gas disappeared in less than half that number of minutes.

He retained a lethargic Bat for 16 minutes, and a lethargic Hedgehog for 22½ minutes, under water, without any injurious effects. The same animals, in a state of activity, expire when placed under water for three minutes. Spallanzani enclosed a Marmot and a Bat for four hours in carbonic acid gas without injury.

Great care should be taken, in an experiment of this kind, to observe that the animal is sufficiently lethargic, and that it does not become roused, for immediately the respiration becomes more frequent, more oxygen is consumed within a given time, and if not supplied the animal soon dies. It is only in this manner that we can explain the different results obtained from similar experiments by Prunelle. No doubt these animals, supposing them to remain lethargic, would ultimately perish when confined in an atmosphere deprived of oxygen; but these experiments at least show that when in a state lethargy they consume exceedingly little oxygen. In this respect they resemble the cold-blooded animals, such as the fishes, frogs, &c., which, it is well known, consume a comparatively small quantity of oxygen; and as we proceed with the enumeration of their physiological peculiarities, we shall be surprised to find that they resemble them in almost every particular, affording the strange anomaly of a warm-blooded adult mammiferous animal assuming for a time the physiological condition of a cold-blooded animal. There is an experiment mentioned by Prunelle, which, under this point of view, deserves our attention. It is well known that a warm-blooded animal, when confined in a fixed quantity of air, dies before all the oxygen is exhausted; while, on the other hand, a cold-blooded animal may breathe for a while with impunity in an atmosphere which cannot support the warm-blooded animal, and rarely dies until all the oxygen is exhausted. Prunelle states that he found all the oxygen had been removed from the air, in which a lethargic animal had been confined until it had proved fatal. Sir A. Carlisle was in error when he affirmed that those animals, in the active state, can subsist with a smaller quantity of oxygen than the other Mammalia.

We shall now consider the state of the Circulation of the Blood.

The circulation, like the respiration, in the state of lethargy appears to be very languid. Prunelle states that the pulsations of the heart of the Bat are 200 per minute, in a state of activity, and that they are reduced to 50 or 55, in a state of lethargy. According to the same author, the pulsations of a Marmot, in a state of activity, are 90 in a minute, and are only from 8 to 10 in the minute, and at the same time weak, in the state of lethargy. Marshall Hall found the pulsations of the heart of the Bat 28 in the minute, and regular. The flow of blood in the minuter arteries and veins was slow. Prunelle found the blood in the arteries not florid as it is in a state of activity, but nearly as dark coloured as the blood in the veins, which is exactly what we would expect from the small quantity of oxygen consumed. Buffon states erroneously that the blood of a hybernating animal, in a state of lethargy, will not coagulate.

In the reptiles, the movements of the heart are slower and feebler, and the blood is darker coloured, than in the warm-blooded animals. A state of the circulation, in these respects, which is natural to the cold-blooded animals, would be incompatible with the continuance of life in the warm-blooded animals. During the state of hybernation, however, we find the circulation of the blood in these animals to resemble that in the Reptiles.

With regard to animal heat, numerous observations show that the temperature of the hybernating animals, during their state of activity, is equal to that of the other Mammalia. On the other hand, while in a state of lethargy, their temperature is little elevated above the surrounding media.

The heat of the external surface is generally found to be the same as the surrounding atmosphere, while the internal parts are from 2° to 3° higher. This is a reduction of temperature from which the other adult warm-blooded animals would never recover. Dr Edwards' experiments go to prove that the temperature of the hybernating animals is more easily reduced, by the application of cold in their state of activity, than in the other Mammalia. Marshall Hall has objected to the accuracy of these experiments. We shall not here stop to inquire into the grounds of his objections. Dr Edwards' experiments at least prove that their temperature is more easily reduced, under certain circumstances, than the other Mammalia, and that they resemble the cold-blooded animals and the young of certain of the Mammalia, which are born with their eyes shut, as the kitten and puppy, in admitting of a great reduction of temperature with impunity.

We have already seen that the heart contracts less frequently in a state of lethargy than in a state of activity, and that it circulates dark blood, as in the Reptiles and other cold-blooded animals, and we also find that they resemble them still farther in their tenacity of muscular contractility. It is well known to physiologists that the heart of a Frog, or of a Turtle, or the other muscles of the same animals, will continue to contract, upon the application of a stimulus, for many hours after they have been decapitated, and are apparently dead, and long after the same parts have lost their contractility in the warm-blooded animals. Mangili observed the heart of a Marmot, in a state of lethargy, to beat three hours after decapitation, and after all the principal vessels leading to and from it had been cut through. Upon applying galvanism to the other muscles also, three hours after death, he found them to contract vigorously. In a Marmot, killed in a state of activity, the heart was quiescent fifty minutes after decapitation. The contractility of the muscles, upon the application of galvanism, was scarcely sensible two hours after death. Marshall Hall has confirmed these observations of Mangili.

The other vital functions of Secretion, Digestion, Absorption, are also performed in a very languid manner during the state of lethargy, as in the cold-blooded animals. These animals, in a state of lethargy, throw off very little excrementitious matter. Some do not appear to eat during the whole of their hybernation, and others but sparingly. The fatness of these animals at the commencement of their hybernation, and their leanness at its termination, show that absorption must have been going on, though this, of course, is to a much less extent than would have happened had they remained in a state of activity. Mangili found that a Dormouse *(Myoxus avellanarius)*, weighing 19 oz. 5 grains, lost 2½ oz. of its weight in three months. When killed it had a considerable quantity of fat around its intestines. Prunelle states that two Bats had lost $\frac{1}{32}$ part of their weight in 21 days. Dr Monro found that a Hedgehog, which weighed 13 oz. and 3 drachms on the 25th December, weighed 11 oz. and 7 drachms on the 8th March. This loss was at the rate of 13 grains a-day. According to Mr Cornish, both Bats and Dormice lose from 5 to 7 grains in weight during a fortnight's hybernation.

The astonishing and unexpected fact, that a warm-blooded adult animal should, when placed under certain circumstances, take on the physiological condition of the cold-blooded animals, naturally leads us to inquire if they have any peculiarity in their anatomical structure which accounts for so interesting a phenomenon.

Notwithstanding the labours of Mangili, Otto, Sir A. Carlisle, Prunelle, and others, no satisfactory explanation of the cause of hybernation can be derived from the structure of these animals. It will be unnecessary to point out here the pretended explanations which some of these celebrated men have advanced, as we would require to enter into minute anatomical descriptions, which would be out of place in an account of this kind.

We cannot refrain, however, from stating an interesting particular mentioned by Prunelle and Pallas, and that is, that during hybernation the thymus gland becomes much enlarged. Now when we remember that the thymus gland is much larger during the fœtal existence than in the adult, and that the fœtus is to a great extent a cold-blooded animal, we have here again another point of resemblance between the hybernating animals, in a state of lethargy, and the cold-blooded animals.

J. R.

Speed of the Bull-Frog.—The Bull-Frog (*Rana Catesbeiana of Shaw*) can leap with very great velocity. It is a well known fact, that an American Indian is able to run almost as fast as the best Horse in his swiftest course. In order, therefore, to try how well the Bull-Frogs could leap, some Swedes laid a wager with a young Indian, that he could not overtake one of them, provided the Frog had two leaps before hand. The wager was accepted, and they carried a Bull-Frog, which they had caught in a pond, into a field, and burned his tail. This application stimulated the creature to such a degree, that he made his long leaps across the field with wonderful celerity. The Indian pursued with all his might, and the noise he made in running, added to the fear the poor Frog was probably in of a second burning, made him redouble his efforts, and reached the pond before the Indian could overtake him.

The Water Spider.—The Insects that frequent the waters require predaceous animals to keep them within due limits, as well as those that inhabit the earth, and the Water Spider (*Argyroneta aquatica*) is one of the most remarkable upon whom this office is devolved by her Creator. To this end her instinct instructs her to fabricate a kind of Diving-Bell in the bosom of that element. She usually selects still waters for that purpose. Her house is an oval cocoon, filled with air, and lined with silk, from which threads issue in every direction, and are fastened to the surrounding plants; in this cocoon, which is open below, she watches for her prey, and even appears to pass the winter, when she closes the opening. It is most commonly, yet not always, entirely under the water, but its inhabitant has filled it with air for her respiration, which enables her to live in it. She conveys the air to it in the following manner:—She usually swims upon her back, when her abdomen is enveloped in a bubble of air, and appears like a globe of quicksilver. With this she enters her cocoon, and displacing an equal mass of water, again ascends for a second lading, till she has sufficiently filled her house with it, so as to expel all the water. The males construct similar habitations by the same manœuvres. How these little animals can envelop their abdomen with an air-bubble, and retain it till they enter their cells, is still one of Nature's mysteries that have not been explained. We cannot help, however, admiring and adoring the Wisdom, Power, and Goodness manifested in this singular provision, enabling an animal that breathes the atmospheric air to fill her house with it under water; and has instructed her in a secret art, by which she can clothe part of her body with air as with a garment, which she can put off when it answers her purpose.—*Kirby*.

BOTANY.

BRUGMANSIA ZIPPELII.

This singular plant is nearly allied to the Genus Patma, of which we have already illustrated two species in our former Numbers. Like them it is a parasite, and has hitherto been found only in the Island of Java. It was discovered on the mountain Salax, growing at the height of 1200 to 1500 feet above the level of the sea. This mountain lies in the province of Buitenzorg, on the west of Java, and it almost seems, from the singularity of its vegetable productions, as well as the marked volcanic character of its minerals, to be the favorite shrine both of Flora and Vulcan.

The Genus Brugmansia was constituted by Persoon; but Blume first included this Plant under that denomination. The characters which he assigned to the Genus Brugmansia are the following :—Perianth with one leaf; the crown of the throat interrupted; limb five-parted; segments or partitions twice or thrice cleft; the æstivation valvate induplicate; the central column, subglobore, hollowed above and naked; anthers monadelphous, two-celled, opening by two pores.

This Plant, when it first bursts from the roots of its parent tree, exhibits merely a small tubercle or bud; as it gradually expands, it assumes the different appearances represented above, until finally it acquires the utmost extent of its growth, which is limited in these remarkable parasites to a simple development of their reproductive organs, or mere blowing of the flower. Just before its ultimate expansion it has the following appearance.

The root upon which it grows, or rather blows, belongs to the *Cissus tuberculata* of Blume—a tree very plentiful in the moist woods on the south-west of Mount Salax. The Brugmansia Zippelii is stated to have the property of being remarkably styptical; its specific name Zippelii was given in compliment to the individual who first discovered this curious vegetable production.

THE GUIJANO TEA PLANT.

We are indebted to M. Bonpland for the important information, that South America contains a plant capable of affording a beverage very much resembling the common tea of China. It is found in the neighbourhood of Popayan. The inhabitants of this town make an infusion of its leaves, which have all the properties of Tea, and may be applied to the same purpose. M. Guijano, a distinguished citizen of that place, was the discoverer. Perceiving a great analogy between the leaves of this Melastoma and the common Tea-leaves, he at first thought that his country possessed the real Chinese Tea. He immediately gathered a great quantity of its leaves, and prepared them in the same manner as the Chinese Bohea (*Camellia bohea*). On making the infusion, he at once perceived that the plant under examination was not that of China, but at the same time he ascertained that it could be employed for the same purposes, and would answer equally well in most cases. "We have often drunk with pleasure," says M. Bonpland, "the infusion of the *Melastoma Theezans.* It has the colour of Tea, and is much less astringent, but more aromatic. Many persons would doubtless prefer this drink to Tea, and I think it will be found as useful in most cases. The Melastoma Tea would thrive very well at Toulon, at the Hyères, and other southern countries which enjoy a mild temperature." At a time when our commercial relations with China stand on a very precarious footing, this fact is well worthy the attention of our enterprising countrymen. The Melastoma Theezans is a shrub from 12 to 15 feet high; smooth in all its parts; its leaves are from three to four inches long, oval, and slightly petiolated, of a fine green above, paler below, slightly dentated, and with five nerves. The flowers are white, and exhale during the night a very pleasing odour. They are disposed in a terminal pannicle, are small, numerous, and sessile. The limb of the calyx is membraneous,

with five small short teeth; the petals of the same length as the calyx; the filaments articulated in the middle, compressed, membraneous in the lower part, and charged towards the summit with a very small tubercle; the anthers are wedge-shaped; the ovary almost free; the stigmata flattened. It has a spherical berry, blue when ripe, crowned by the teeth of the calyx, and having three many-seeded cells. Further details regarding this plant may be seen in Bonpland, l. c. p. 17, t. 9.

GEOLOGY.

ON A SINGULAR DETACHED BLOCK OF STONE OCCUPYING THE SUMMIT OF A HILL AT DUNKELD.—Those who are acquainted with the natural history of Cornwall, cannot fail to recollect the theory of Dr Borlase, respecting detached blocks of granite so conspicuous in that country. The Doctor's notions, however, were not exclusively his own, as other antiquaries, over whose judgments the absurdities of the druidical worship seem to have shed their influence, have imagined these and similar appearances to be monuments of that superstition or religious government, all our knowledge of which is comprised in a very few casual hints contained in the Roman historians. The recent increase of attention to natural history, and more particularly to geological investigation, has, however, put to flight all these visions, and left us at no loss to distinguish betwixt the appearances produced by the efforts of art and design, and those which have resulted from the ordinary operations of Nature. Whatever interest, in a historical view, these phenomena may therefore have lost, they have gained a countervailing one in natural objects; in many cases illustrating, either in a curious or useful manner, the changes which time is daily, but slowly, making on the surface of the earth.

In Scotland, as in Cornwall, antiquaries have not been wanting, who were ready to attribute some of these remarkable natural appearances to a Druidical origin; and among these may be enumerated the Rocking Stone in Strathairdle, which has furnished a page to some of the writers of the day. The Rock, of which the following is a sketch, has also been called a Cromlech, which it resembles in the peculiarity of its position :—

But we can have no scruple in admitting it as an example, and a peculiar one at the same time, of those transported stones often found occupying situations so unexpected, as to render an explanation of the course which they have taken a matter of no small difficulty.

This rock occupies the summit of a hill near Dunkeld, in Perthshire, known by the name of Craig-y-barns (the Serrated Rock). Its shape is so irregular, that it cannot be described, but the above cut will supersede the necessity of saying any thing on this part of the subject. The same irregularity renders it difficult to form an accurate notion of its weight, but it probably exceeds fifty tons; a judgment founded on comparing it with other stones of known weights resembling it in shape. The greatest length is twelve feet, and the greatest thickness five, from which circumstances, with the aid of the cut, a sufficiently accurate notion of its form and dimensions may be conveyed.

From the shape of it, it will be seen that the lower flat surface is supported on three loose stones ; and in this circumstance consists that resemblance to a Cromlech,* which has led to the unfounded notion of its druidical and artificial origin. These loose stones also lie at liberty on a flat and solid surface of rock ; a circumstance which adds much to the appearance of artifice.

It is now necessary to remark, that this rock, as well as the supporters, consists of the same materials as the hill on which they rest, which is micaceous schist. It is indeed from this circumstance alone that is derived the proof of its not lying in its native place ; but of its having, on the contrary, been moved to its present position, together with the stones by which it is supported. On examining the direction of the laminæ in all these pieces, it is easy to see that they all lie in different ways, and all different from that of the laminar structure of the solid rock on which they repose. Hence, it is evident that the large block has been placed on the three loose stones which lie on the solid rock ; exhibiting an appearance, it is true, of artifice, as perfect as if it had been the result of the hand of man, though really an accidental operation of Nature.

In accounting for it, however, by natural causes, there seems no reason to doubt that the whole is the result of the accidental fall, or transportation of the larger mass, combined with some posterior circumstances of waste. Originally it had probably been deposited on a bed of loose materials, the smaller of which have disappeared, leaving those three only which were essential to its support, and have been retained by its pressure.

* Cromlechs in British antiquities are huge, broad flat stones, lying upon other stones set upon end. Mr Rowland and Dr Borlase describe them as altars. They have also been considered as originally tombs ; but that, in after times, sacrifices were performed upon them to the heroes deposited within. King Herod is said to have been buried in a monument of this kind in Denmark ; and Mr Wright discovered a skeleton beneath one in Ireland.

The circumstance, in a geological view, most remarkable is, that it now lies on a point, which, if it be not absolutely the highest eminence of the surrounding hill, is yet so nearly at the same elevation with the other summits, that it could not have travelled from any of them to its present place; supposing the surrounding parts to have always been in the state in which they now are. The nearest summit is too little elevated to have permitted a stone of so irregular a form to have moved over the intermediate surface; and that which is higher, is now separated by a hollow or depression, which would equally have prevented its transportation from that point, unless the intermediate ground were restored to an uniform declivity. The integrity of the mass is indeed sufficient to prove that it has not been carried far; and it affords, in fact, a singular example, rather of the results which follow from the degradation of hills, than from the transportation of blocks. Its appearance may probably be explained by imagining that the summit on which it now stands was once higher; and that, in the progress of waste, this mass has now fallen from its original position, on the solid rock on which it now lies; overwhelming in its fall a heap of small materials or rubbish, of which the three supporting stones are the last remains.

We are indebted to Dr Macculloch for the principal part of the above observations.

MISCELLANEOUS.

HAZEL NUTS FOUND IN A SINGULAR STATE AT A GREAT DEPTH.—These nuts were found upon one of the farms at Bonnington, the property of Sir John Hay, Bart., about one mile south from Peebles, in a bog about eight feet below the surface. The top soil was three feet of meadow clay, beneath which was a layer of grayish coloured gravel about four and a half feet thick. The bottom of the bog consisted of a mixture of gray sand and brown moss, with some branches of stumps of trees quite decayed, and the nuts were found near the bottom of this substance. The bog is part of a meadow about 1500 yards long, by about from 300 to 600 feet broad, having a declivity of about one foot in 400.

Upon opening these nuts, it was found that the kernel in all of them had entirely disappeared, though the membrane which enclosed it, and the nut itself, were as entire as if they had been fresh and ripe. By opening the nut carefully, the membrane could be taken out in the form of a perfect bag, without the least opening. The substance of the kernel must, therefore, have escaped through the membrane and the shell in a gaseous form, or must have passed through them when decomposed or dissolved by the water. In some of the nuts that had not arrived at maturity, the bag was very small, and was surrounded, as in the fresh nuts, with the soft fungous substance which had completely resisted decay.

MAGNETISM.—The attention of the "Westminster Medical Society" has lately been directed to the supposed influence of magnetism in the cure of various diseases. This subject was brought forward by Dr Schmidt, when he pointed out the distinction between mineral and universal magnetism, and contended that that fluid acted solely on the nervous system.

Contrary to the opinion entertained by the greatest authorities on this science, he conceived that more magnetic influence existed between the friendly poles than the opposite; but this opinion was successfully overturned by Dr Ritchie and other members, from experiments made at the time.

Dr Schmidt exhibited a very simple instrument for obtaining the magnetic spark. This consisted of a piece of soft iron, round which copper wire was twisted. The extremities were amalgamated with quicksilver, and placed over the poles of a magnet; on one was fixed a plate of copper, and the connexion then being forcibly broken, a vivid magnetic spark was evolved.

ANOMALIES IN THE HEN'S EGG.—In a former Number we described a remarkable instance of *three* shells in the same egg. Since that communication was made, we have been favored with the following observations by an intelligent Correspondent:—

"It is not an uncommon circumstance to find two yolks in a single egg; and it is much rarer to find one egg surrounded with its shell, contained in another egg of a larger size. But I am not aware that there was any case on record, before your notice appeared, of *three* shells being in the same egg. Schurigio has collected many interesting facts relating to anomalies of the Hen's egg, which had been observed from the time of Bartholin to his own. Harvey, Ruisch, Haller, and many others, have also collected a number of similar cases.

"Mery showed to the Academy of Sciences at Paris in 1706, a boiled Hen's egg, in the white of which he found another Egg surrounded with shell, having an internal membrane, and filled with a white substance without yolk.

"Other instances have been recorded by Petit in 1742, and by M. Menière in September 1810."

DISSECTION OF A FEMALE MUMMY.—In the year 1825, Dr Granville dissected a female Mummy before the Royal Society. After depriving the body, by ebullition and maceration, of the bees-wax, myrrh, gum, resin, bitumen, and tannin, with which it had been impregnated and preserved, the parts resembled recent pathological preparations; and though the body must have lived 3000 years ago, Dr Granville was enabled to ascertain the age at which the lady died of ovarian dropsy, and also that she had borne children. Dr Granville gave the dimensions of its various parts, and it is truly singular that these happen to be precisely the same as of the *Venus de Medicis*, whose claims to be considered the most perfect model of the Caucasian female are undoubted.

THE STUDY OF NATURE.—The Mind of man, if it work upon Matter, which is the contemplation of the creatures of God, worketh according to the stuff, and is limited thereby; but if it work upon itself, as the Spider worketh its web, then it is endless, and brings forth indeed cobwebs of learning, admirable for the fineness of thread and work, but of no substance or profit.—*Bacon.*

CAT HOSPITAL AT DAMASCUS.—When M. Baumgarten was at Damascus, he saw there a kind of hospital for Cats; where they were kept in a large house walled round; and it was said that the apartments were quite filled with them. He was told, when he inquired into the origin of this singular institution, that Mahomet, when he once lived there, brought with him a Cat, which he kept in the sleeve of his gown, and carefully fed with his own hands. His followers in this place, therefore, ever afterwards paid a superstitious respect to these animals; and supported them in this manner by public alms, which were very adequate to the purpose.

REVIEWS.

Aperçu d'Histoire Naturelle, ou Observations sur les limites qui separent le Regne Végétal du Regne Animal; par Benj. Gaillon. Boulogne-sur-Mer. 1833. *In* 8vo. *Imprimerie de Leroy-Mabille.*

(Sketches of Natural History, or Observations on the limits which separate the Vegetable from the Animal Kingdom, by Benj. Gaillon, &c.)

THE author of this *Aperçu* has long been engaged in studying the characters of Microscopic beings, and some of the results to which he has arrived exhibit no small degree of perseverance and talent. He has discovered some curious facts relating to the marine body hitherto called after Linnæus the *Conferva comöides*. While observing the mucous filaments of this supposed water-plant, he saw some small yellowish bodies like mere points issue from the filaments. They gradually became oval, and disengaging themselves from these mucous filaments, were finally deposited in immense quantities under the form of a chocolate brown paste, upon the salt water vase which contained them. There they expanded and emitted a globule of small coloured grains, which is evidently their fry. Each of these grains acquired motion and development; the small globular mass gradually extended and ramified itself, and ultimately produced that elongated and plant-like form, which has deceived all Botanists into the belief that it is a plant.

Such is the theory of M. Gaillon, who here proposes to remove the *Conferva comöides* from the vegetable to the confines of the animal kingdom, towards that point which the illustrious Lamarck has represented by the apex of the letter V, where the two branches approach indefinitely near at the base, without ever being confounded.

M. Gaillon proposes to give this new group of animals the name of *Girodella*, and to preserve its specific name *comöides*. He is also convinced, from his Microscopic investigations, that fresh water is still more abundant in productions of this nature than salt. Thus, the bodies which Botanists call *Confervæ*, and included by them in the vegetable kingdom, are, according to M. Gaillon, microscopic animalcules, to which he assigns the name of *Némazoaires*.

These new views have met with much opposition from several Naturalists of eminence; but the observations of MM. Desmazières of Lille and Chauvin of Caen have confirmed the remarks of M. Gaillon. The former of these able observers has declared the pellicules, called *Mycodermes*, which grow on beer, ink, paste, or sour wine, to be composed, like the Némazoaires, of an innumerable quantity of corpuscles, endowed at a certain period with the power of locomotion. M. Chauvin also confirms them by his remarks on the *Conferva zonata*. He observes that the green matter, which fills the interior of its filaments, is at first collected in spheroidal masses, and then forms corpuscles, which grow, burst their tube, escape from the envelop, and disperse themselves over the field of the microscope, where they move about with a swiftness of motion, of which none but an observer can form an adequate idea.

M. Gaillon publishes several elaborate tables of the families and genera of the Némazoaires. He also announces that he will receive, with pleasure, fresh samples of any not included in that table, as well as observations either for or against the animal nature of these bodies; and requests all who make their communications to add microscopic drawings, in order to render their arguments more clear. The last is a gentle hint to non-microscopic Naturalists, which they will readily understand.

WORKS ON NATURAL HISTORY, AND THE PHYSICAL SCIENCES.

Flora Boreali-Americana, or the Botany of the Northern Parts of British North America, by Dr W. J. Hooker, part 7, price 21s.—Botany of Captain Beechey's Voyage, by Dr W. J. Hooker and G. A. W. Arnott, Esq., part 4, 15s.—Icones Filicum; Figures, and Descriptions of Ferns. Fasciculi 1 to 12, by Drs W. J. Hooker and R. K. Greville, L.1, 5s. each plain, or L.2, 2s. each coloured.—Transactions of the Geological Society of London, vol. 3, part 2, L.1, 5s.—Sections and Views illustrative of Geological Phenomena, by H. T. De la Beche, Esq., L.2, 2s.—Virey Philosophie d'Histoire Naturelle, 1 tome.—M. Römer Handbuch der Allgem. Botanik, 1ste Abethl, 1stes Heft, 8vo, 3s. 6d.—A. Breithaupt, Handbuch der Mineralogie, 1ster, bd. 8vo, 12s.—Dr H. F. Link, Propyläen der Naturkunde, 1ster Thl. 8vo, 5s.—D. L. F. Froriep, Notizen aus dem Gebiete der Natur-und Heilkunde, 47ster, bd. 4to, 10s.—Paolo Savi, Studii Geoligici Sulla Toscaña, 8vo.—Dr W. Petermann, Handbuch der Gewächshünde zum Gebraüche bei Vorlesungen, 8vo, 18s.—A. Buchmüller, Handbuch der Chemie für angehende Thierärzte und Oekonomen, 8vo, 7s.—J. A. Buchner, Lehrbuch der analytischen Chemie und Stöchiometrie, Mit 1 Kupf, 14s.—C. C. Person, Elémens de Physique, à l'usage des élèves de Philosophie, 1re partie, 8vo, 4s.—Illustrations of Indian Zoology, by John Edward Gray, parts 19 and 20, L.1, 1s. each.—Plantæ Asiaticæ Rariores, or Descriptions and Figures of Unpublished East India Plants, by D. N. Wallich, L.36.

EDINBURGH: Published for the PROPRIETORS, at their Office, 16, Hanover Street. LONDON: SMITH, ELDER, and Co., 65, Cornhill. GLASGOW and the West of Scotland: JOHN SMITH and SON, 70, St Vincent Street; and JOHN MACLEOD, 20, Argyle Street. DUBLIN: W. F. WAKEMAN, 9, D'Olier Street.

THE EDINBURGH PRINTING COMPANY.

THE EDINBURGH

JOURNAL OF NATURAL HISTORY,

AND OF

THE PHYSICAL SCIENCES.

JULY, 1836.

ZOOLOGY.

DESCRIPTION OF THE PLATE—THE DOGS.

Fig. 1. THE BLOODHOUND.—There was, in early times, a popular belief that this Dog had the instinct to pursue murderers, and, if once put on their scent, that he could trace them with unerring certainty. This exaggeration proceeded from his being able to track fugitives, to follow them through the most secret coverts, and to seize them when found. There is an old law in Scotland, enacting that any person denying a Bloodhound entrance into a house, while in the act of searching for felons, should be treated as an accessory.

The Bloodhound is a large and beautifully-formed Dog; he possesses great strength, and his sense of smell is exquisite. In proof of this, the Hon. Robert Boyle relates the following anecdote:—" A person of quality, to make trial whether a young Bloodhound was well instructed, desired one of his servants to walk to a town, four miles off, and then to a market-town three miles from thence. The Dog, without seeing the man he was to pursue, followed him by the scent to the above-mentioned places, notwithstanding the multitude of market people that went along the same road, and of travellers that had occasion to cross it; and when the Bloodhound came to the cross-market town, he passed through the streets without taking notice of any of the people there, and ceased not till he had gone to the house where the man he sought rested himself, and there he found him in an upper room, to the wonder of those who had accompanied him in the pursuit."

The fame of the English Bloodhound has been deservedly transmitted to posterity by a monument in Basso-relievo, which is said to remain at present in the chimney-piece of the grand ball at the Castle of Montargis, in France. The sculpture, which represents a dog fighting with a champion, is illustrated in a very well known narrative.

Fig. 2. THE STAGHOUND is the largest and most powerful Dog now in general use in Britain for the pleasures of the chase, being employed, as its name implies, for Stag-hunting. It is a mixture of the Bloodhound, old English Southern-hound, and the Foxhound.

Fig. 3. THE FOXHOUND.—No country in Europe can boast of Foxhounds equal in swiftness, strength, and beauty. The reason of this we discover in the attention paid to their breeding, education, and food. The climate also seems congenial to their nature, for, when taken to France or Spain, and other southern countries of Europe, they quickly degenerate, and lose all the admirable qualities they possess in this country. In the words of the Poet of the Chase—

> " In these alone, fair land of liberty,
> Is-bred the perfect hound, in scent and speed
> As yet unrival'd; while in other climes
> Their virtue fails,—a weak degenerate race."

It is a trait in our national character to be excessively attached to hunting, and in no country is the same strict attention paid to the breeding and comfort of hounds as in Britain. The kennel of the Duke of Richmond at Goodwood cost L.19,000; and a thousand guineas have been paid for a pack of hounds.

Fig. 4. THE HARRIER.—This Dog is much smaller than the Foxhound, and is now universally used in Britain for Hare-hunting. It is possessed of great eagerness and perseverance, allowing the Hare but little time to breathe or double; and the keenest sportsmen frequently find it no easy matter to keep up with a good pack of Harriers.

Fig. 5. THE BEAGLE.—This is the smallest of all those Dogs of the Chase which pursue their prey by the scent alone. Its sense of smell is equal to any of those hunting Dogs comprised under the general denomination of Hound. It is a very eager Dog, but its diminutive size renders it incapable of taking the Hare by means of speed, and it succeeds in running down its prey by perseverance alone.

Fig. 6. THE ENGLISH TERRIER is a handsome and sprightly Dog, usually black above, with tan-coloured throat, breast, abdomen, and legs. This Dog, though but small, is very resolute, and a determined enemy to all kinds of game and vermin, in the pursuit of which it evinces an extraordinary and untaught alacrity. Some of the English Terriers will even draw a Badger from its hole. This Dog varies considerably in size and strength, and is to be met with from ten to eighteen inches in height.

Fig. 7. THE SCOTCH TERRIER.—There are several sub-varieties of this Dog in Scotland. Some persons consider that variety with strong and rather long bristly

14

hair and pricked ears to be the model of a real Scotch Terrier, but we believe the form in the plate has greater claims to that title. It is taken from a fine painting of a Terrier well known for his excellent qualities, and once the property of the late Duke of Buccleuch. The most esteemed breed are always of a sandy-brown, varying in intensity.

Fig. 6. THE ISLE-OF-SKYE TERRIER.—This differs from the last in the hair being fully longer, the tail being more villous, the ears being larger and semi-pendulous, the body larger, the legs much shorter, and invariably crooked in front. The most esteemed breed is usually of the same colour as the common Scotch Terrier, although there are exceptions to this rule, and varieties in the colour often appear. Thorough-bred ones may be gray or even black, and recently we have seen a very beautiful and characteristic specimen, the property of Sir John Naesmith of Posso, which was of a pure white. The Isle-of-Skye Terrier is excellent for destroying vermin, to which he is an implacable enemy.

DESCRIPTION OF THE PLATE—THE WARBLERS.

The Warblers are very numerous, and are widely diffused in their Geographical range. They feed chiefly on Insects and their larvæ, and subsist occasionally on fruits. They are migratory in all countries where they are found, passing to the warmer regions during winter, in quest of insect food, and returning to more northern latitudes as the summer advances, for the purpose of incubation. The colours of the plumage, in most of the species, consist of strong and decided contrasts of green, yellow, and black. Nearly the whole species of this extensive genus are Songsters. All those figured on the plate are the size of life.

Fig. 1. The Palm Warbler (Sylvia Palmarum). This lively bird is an inhabitant of St Domingo, and other West India islands, extending its migrations to South Carolina and Philadelphia, in the United States. Its food consists of insects, fruits, and small seeds. It builds its nest on the very top of some lofty palm tree, from which it takes the name of " the Palmist." Its song is limited to five or six notes, and is full, soft, and mellow, although consisting of little variety. There is no difference in the form or hues of the male and female, but the winter plumage is more dull than the summer garb.

Fig. 2. The Blue-Mountain Warbler (S. tigrina). This species inhabits the lofty American range of alpine scenery, whose name it bears, seldom descending from the airy heights and gloomy silence of those dreary fastnesses. It is four inches and three quarters in length, and its song is merely a feeble screech, three or four times repeated. Its food consists chiefly of insects. It is not yet known whether there be any difference of colour between the male and female.

Fig. 3. The Hemlock Warbler (S. parus) was discovered by Wilson in the Great Pine Swamp of Pennsylvania. It is an active and lively bird, climbing and hanging among the twigs like a Titmouse. Its song consists of a few sweet notes, which it never utters while in motion, always singing while in a quiescent state. In pursuit of insects, its usual habit is to commence at the foot of a tree, and hunt, with much spirit and vigour, every branch as it ascends.

Fig. 4. The Autumnal Warbler (S. Autumnalis) visits Pennsylvania in the month of October, and gleans its food principally from such insects as inhabit the willow trees; during which time, the male birds warble out some low, but sweet notes. The size of the bird is four inches and three quarters.

Fig. 5. The Black-throated Green Warbler (S. Arcens) is a very transient visitor of the United States, passing through Pennsylvania, in the end of April and beginning of May, on its way to the north to breed. It is seldom to be met with after the 10th of May. It frequents the high branches and tops of trees, in search of the larvæ of insects that prey on the opening buds. Its song consists only of a few chirruping notes, and-its habits are active and lively.

Fig. 6 and 7. The Maryland Yellow-throat Warbler (S. Marylandica), male and female. This species inhabits the whole United States from Maine to Florida and Louisiana, and abounds in Maryland, Pennsylvania, and New Jersey, frequenting the low swampy thickets, and feeding on insects and their larvæ, which are to be met with among briars, brambles, and alder bushes. It sometimes also visits the cultivated fields of rye, wheat, and barley, ridding the stalks of insects, that might otherwise lay waste the fields. Its song is simple, consisting of the repetition of a twitter, resembling twititee.

Fig. 8. The Kentucky Warbler (S. Formosa) inhabits the country in the United States whose name it bears, and is also to be met with between Nashville and New Orleans, frequenting the solitary and gloomy morasses of those countries, twittering among the high rank grass, which principally covers these desolate regions. In this situation it usually builds its nest, or in the fork of some low shrub. Its notes are loud, and consist of *tweedle* three times repeated. It is said to be a very pugnacious bird, fighting with great violence during the amorous season. The female is destitute of the black under the eye.

Fig. 9. The Yellow-throat Warbler (*S. flavicollis*) remains in the United States for nine months of the year, quitting them during the three winter months. It is principally to be met with among Pine trees, ranging after Insects with much nimbleness, both spirally and perpendicularly, in the manner of a Titmouse. Every three or four minutes it utters its song, which is pretty loud, and somewhat resembling the notes of the Indigo bird.

On the Spine at the extremity of the Lion's Tail.—It has been observed by Homer and many other ancient poets, that the enraged Lion stimulates himself with blows of his tail. Pliny calls the tail the index of the Lion's mind, for, says he, "When the tail is at rest, the animal is quiet, gentle, and seems pleased, which is seldom, however, the case; and anger is much more frequent with him, in the commencement of which he lashes the ground, but as it increases, his sides, as if with the view of rousing it to a higher pitch." Again, we find among the Problemata of Alexander Aphrodisæus the following query:—"Why, since the moving of the tail is, in most animals, a sign of their recognition of friends, does the Lion lash his sides when enraged, and the Bull in the same manner?"

Didymus Alexandrinus, the ancient commentator of Homer, has a note upon that passage of the Iliad where the Lion's rage is mentioned—

"Such the Lion's rage,
* * * * * *
Lash'd by his tail, his heaving sides resound.
He calls up all his rage."—*Book* xx. *line* 199, *&c.*

He asserts, "that the Lion has a black prickle on his tail among the hair, like a horn, when punctured with which, it is still more irritated by the pain." This opinion was by many looked upon as a mere fiction, the more so, as no anatomist, who possessed an opportunity of dissecting a Lion, had hitherto made mention of a spine of this kind. But the matter was put beyond a doubt, some years ago, by Professor Blumenbach, to whom a friend had presented a Lioness, which died a few days previous. He determined to satisfy himself regarding the assertion of the Greek Scholiast. He commenced his dissection, and discovered, on the very tip of the tail, a small dark-coloured prickle, as hard as a piece of horn, and surrounded at its base with an annular fold of the skin; and on cautiously dissecting the hide in this place, he discovered a singular follicle of a glandular appearance, to which the prickle formerly adhered, as represented beneath.

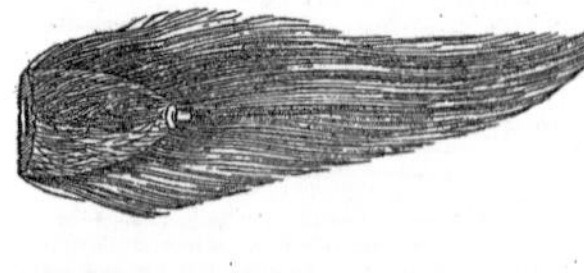

All these parts, however, were so minute, and the little horny apex so buried among the tufted hairs of the tail, that the use assigned by the ancient Scholiasts cannot be regarded as any thing else than imaginary; but the structure of the organ is so elegant, and its form so singular, that it cannot possibly be considered as fortuitous, or what is commonly called a *lusus naturæ*. This simple fact, however, seems to prove that the ancients were better acquainted with some departments of anatomy than could have been supposed.

Tormentors of the Rein-Deer.—Whale Island, during the summer months, is never without three or four families of Laplanders (Fieldfinner), with their herds of Rein-Deer. The causes that induce, nay even compel these people to undertake their long and annual migrations from the interior parts of Lapland to its coast, though they may appear singular, are sufficiently powerful. It is well known from the account of those travellers who have visited Lapland during the summer months, that the interior parts of it, particularly its boundless forests, are so infested by various species of gnats and other insects, that no animal can escape their incessant persecutions. Large fires are kindled, in the smoke of which the cattle hold their heads, to escape the attack of their enemies; and even the natives themselves are compelled to smear their faces with tar, as the only certain protection against their stings. No creature, however, suffers more than the Rein-Deer from the larger species (*Œstrus tarandi*), as it not only torments it incessantly by its sting, but even deposits its egg in the wound it makes in the hide. The poor animal is thus tormented to such a degree, that the Laplander, if he were to remain in the forests during the months of June, July, and August, would run the risk of losing the greater part of his herd, either by actual sickness, or from the Deer fleeing of their own accord to mountainous situations from the gad-fly. From these causes the Laplander is driven from the forests to the mountains that overhang the Norway and Lapland coasts, the elevated situations of which, and the cool breezes from the ocean, are unfavourable to the existence of these troublesome insects, which, though found on the coast, are in far less considerable numbers there, and do not quit the valleys, so that the Deer by ascending the high lands can avoid them.—*De Broke's Travels in Lapland.*

BOTANY AND HORTICULTURE.

BOTANICAL EXCURSIONS IN EGYPT, THE THREE ARABIAS, PALESTINE, AND SYRIA. NO. I.

BY N. BOVE, LATELY CONDUCTOR OF FARMING OPERATIONS AT CAIRO TO HIS HIGHNESS IBRAHIM-PACHA.

Arrival in Lower Egypt—Review of the vegetation and cultivation of the country —Visit to Fayoum.

On the 10th April 1829, I disembarked at Alexandria, and spent six hours in traversing the city and its environs; but the excessive heat of the weather had already parched up most of the annuals. The harvest was now over. I found nothing else to do here but to visit some gardens where I had remarked several trees from different foreign countries, but perfectly naturalized in this place. Among those trees which seemed to me to thrive best were the following:—the Apricot, Quince, Fig, Pomegranate, Orange, Citron, Olive, Mulberry, Peach, and Vine. Some of our common fruit trees, such as the Apple, Pear, Plum, and Cherry, vegetate very feebly in this place. Among the ornamental plants were the common Bead-tree (*Melia Azedarach*), the Sponge-tree (*Acacia Farnesiana*), the Pomegranate with double flowers, the Rose Laurel, the large flowered Jessamine, the Rose with a hundred leaves, the common Carnation, and several varieties of Thyme.

On the 17th April, I set out for Cairo, and remarked on the banks of the Nile the Egyptian Reed (*Arundo Ægyptiaca*), the Egyptian Sugar-cane, the Sea Ambrosia, thickets of the Gum Arabic tree (*Acacia Arabica*), and here and there some isolated tufts of *Acacia Lebbek*. The Cotton tree, the Mulberry, the Maize, and many species and varieties of the Cucurbitaceæ, were cultivated in the open fields.

I arrived at Cairo on the 25th April, and resided there continually during nineteen months. The same plants were cultivated in the fields and gardens as at Alexandria; but I also observed, especially in the gardens of the prince, many species, which seemed to have been introduced from the interior of Africa, and other distant countries.

The luxuriant vegetation of some plants in Egypt is truly surprising, for it is no uncommon thing to see annual shoots which are four or five metres long (from 13 to 16½ feet), especially with the Gleditschia, Cassia fistula, and some others.

The Sycamore Fig (*Ficus Sycomorus*), the trunk of which is from three to four metres broad (10 to 13 feet), and about twenty metres high (65½ feet), may give a still greater notion of the luxuriance of vegetation in Egypt. The branches of this tree, numerous and extensive, form a shade as dense as it is durable; for the leaves are persistent, and always of a beautiful verdure. The fruit is very nearly the size of the common figs, but flattened, and of a yellowish orange colour, approaching to brown. Their taste is more insipid, and not so sweet as that of the common Fig. This tree produces three crops of fruit every year. The latter grows on the branches deprived of leaves, and ends in forming tufts, which might be mistaken at a distance for our Misletoe. To quicken the ripening of the Figs, the Egyptians nip or cut off their upper extremity with a knife. Three or four days afterwards, the fruit acquires a fine golden colour, and a sweet taste, when it is considered as sufficiently ripe. The wood of the Sycamore Fig has the name of being indestructible, or at least of undergoing scarcely any alteration in water. After this maceration, it is susceptible of a very fine polish. The greater number of the wooden amulets of the ancient Egyptians appear to have been formed of it; and at present it is employed with advantage for Hydraulic works.

While visiting the country residences of Ibrahim-Pacha, one of his Inspectors pointed out to me, near the village of Kouba, a stump of St John's Bread Tree (*Ceratonia Siliqua*), which he said had been planted under the reign of a Sultan who governed Egypt three hundred years ago. This Tree had been cut down by the French during their expedition in Egypt. Its roots remained in the earth, without giving any signs of vegetation, until his Highness Ibrahim having cleared the surrounding earth in 1826, and having sunk a well, the moisture occasioned three branches to shoot, which in three years acquired the height of three or four metres (10 to 13 feet), and their base was three decimetres in circumference. Some flower-buds seemed even to show themselves on the branches. Thus, this stump had remained buried in the earth for nearly thirty years, without perishing, and probably without ceasing to increase in size. This fact appears still more surprising than that related by M. Dutrochet, of a species of Pine.—(See Arch. de Botan. 1833, t. 2, p. 231, and Ann. des Scien. Nat. 1833, t. 29, p. 300).

In my Botanical Excursions round Cairo, I found some beautiful and remarkable kinds of Grasses, such as the *Panicum obtusifolium* of Delile, *Poa Ægyptiaca* of Linnæus, and several others. The untilled lands are much infested with the *Poa cynosuroides*. Its roots spread to the depth of more than a metre (about 3 feet). The stalk rises to two metres, and furnishes the material for ropes to the inhabitants, as well as fuel for their ovens, brick fields, and potteries. The *Saccharum cylindricum* serves for the same purposes.

The Lotus (*Nymphæa Lotus*) grows about two leagues from Cairo, in a ditch which contains no water, except when the Nile overflows. In summer this ditch is completely dry, and serves as a public road during seven months of the year, from February to September. At this time, the Nile overflowing fills the ditch, which is covered two months after with Lotus flowers of the purest white.

In the moist sands of the desert, I found among other plants the *Tribulus terrestris*, and *Alhagi Maurorum*, or Manna Tree.

It is in the month of November, after the inundation, that a delightful verdure begins every where to appear. But before this period, the country is completely parched up, and shows the saddest aspect, which contrasts strangely with a cloudless sky. In order to enjoy the splendid prospect which the vegetation occasions, when in its vigour, it must be viewed from some neighbouring height. I witnessed the *ensemble* of this beautiful country, partly wild and partly cultivated, from the mountains Mokadam and Achmar, on the east of the city. An immense horizon, extending over the whole plain of Lower Egypt, presented to the eye villages one beyond the other, surrounded with fruit-trees, and thickets of Dates and the Gum Arabic Tree. Lebbeks and Sycamore Figs were scattered over this plain. Finally, the entire scene was beautifully varied by small sheets of water presenting the dazzling reflections of the solar rays.

Following the desert on the north, I visited the gardens of a village called Madrea, about four leagues from Cairo, where it was pretended that the Balm of Gilead grew (*Amyris Opobalsamum*). This shrub had been naturalized there by one of the Turkish Sultans, who had introduced several plants from the neighbourhood of Mecca, at the time of the conquest of Arabia. It was in vain that I made diligent in-quiries of the Arabs, in my attempts to find this plant, which Linnæus places in Egypt, and he probably received some specimens proceeding from these ancient cultures.

In the month of November, I visited the province of Fayoum, and found here nearly the same plants as formerly stated to grow on the banks of the Nile. Among those few which are not found in Lower Egypt, I observed the magnificent *Asclepias gigantea*, and the *Cyperus alopecuroides*,' which the natives use for fabricating their pretty mats.

At three leagues distance from Medinetta-el-Fayoum, there is a small town called Fedamin, the most ancient in the province, and the environs of which are the best cultivated. It is the only place where the Christians still make wine. The Vines acquire here an enormous size, and their plantation mounts back probably to a very remote age.

The Olives, whose trunks are sometimes more than two metres (4½ feet) in dia-meter, produce three or four thick branches which are about half a metre in diameter, and 5 or 6 metres high. Around them spring up thousands of offsets, which are cultivated at present to be planted out in tufts in the same manner as our Lilacs. These trees appear to have been standing before the era of Mahomet, for since that epoch no plantation has been made in Egypt, excepting those of the reigning princes.

GEOLOGY, AND MINERALOGY.

ACCOUNT OF THE TRAVELLED STONE NEAR CASTLE-STUART.—This mass of stone, of which the following is a representation, is composed of granite, gneiss, quartz, and other rocks of the primitive series, cemented together by a highly indu-rated ferruginous claystone.

It is apparently the very same as the rocks through which the romantic stream of Cawdor cuts its deep and narrow bed, near Cawdor Castle, and no rock of the same kind as the Travelled Stone is found nearer to it than seven or eight miles. Its present situation is on the sands in the little bay near Castle-Stuart, on the Moray Frith; and as it is left entirely dry by every receding tide, it is easily approached over the sands at low water. It is about five feet high at its most ele-vated point, calculating from the surface of the sand, and being to all appearance about one foot imbedded in it. In its horizontal diameters it measures nearly six feet in one way, by nearly seven in the other; its weight being about eight tons.

This large mass of stone is remarkable for having been removed from a situation which it formerly occupied, about 260 yards farther to the E.S.E., by natural means, and in the course of one night, to the position where it now stands. This remarkable circumstance took place on the night between Friday the 19th and Satur-day the 20th February 1799. There had been a long-continued and severe frost; and the greater part of the little bay had been for some time covered with ice, which was probably formed there the more readily, owing to the quantity of fresh water from the stream running near Castle-Stuart emptying itself into this inlet of the sea. The stone was, by means of a projecting ledge all around it, bound fast by a vast sheet of ice, of 18 inches in thickness; and when the influx of the tide took place it was floated in the direction above described, and left in its present situation, the wind having blown with great violence in that direction.

Alexander Macgillivray, of the Sea Mill of Petty, witnessed the fact of the stone being removed, and was the first to discover its absence next morning after it took place. This storm was accompanied by a heavy fall of snow, and as soon as it abated he missed the stone, and perceived that it had been removed much nearer to low water mark, to the position it now occupies. The circumstance soon became generally known in the neighbourhood, and many flocked to the spot on the 20th to see it; among those was Mr Brodie of Brodie, at which time, the hole in which it had been for so many years imbedded still remained to mark distinctly yesterday's site, whilst its tract across the flat oozy sand was very perceptible, extending in a line all the way from its old to its new situation; and an extensive cake of ice still adhered to the stone, and attached to the projecting ledge.

This singular phenomenon will serve as one mode of accounting for the removal of stones to a distance from their original sites, which have long puzzled geologists.

Any person visiting the neighbourhood of Castle-Stuart can see the distance and direction to which this stone has been removed, as the original spot is marked by a wooden post, which was immediately put up as a substitute for the large stone, which served as a *march-stone* between the property of Castle-Stuart, belonging to the Earl of Moray, and the estate of Culloden, belonging to Duncan Forbes, Esq.

Woodside, near Airdrie, 16th March 1836.

SIR,—I have in my possession the wing of a fly, encrusted with calcareous spar, found last summer about twenty-four feet from the surface, and near the bottom of a freestone rock twenty feet in thickness, at Fairybank, parish of Bothwell. It was discovered by a lad while quarrying the rock in one of its horizontal joints, in a spot from which a mass of stone about two tons in weight had just been removed, and where it could not possibly have found its way from the surface. The wing is rather larger than that of the dragon fly; it is of a golden colour, and beautifully mem-braneous. It retains all the freshness of the natural wing, having undergone no petrifying process, and is set, as if by the art of the Lapidary, in the spar. I am not aware of any thing of the kind having hitherto been found in the coal formation, at least I have read of none in Geological works. The relic in my possession must at all events be regarded as a very singular and beautiful monument of Insect existence at a very remote geological epoch; the rock in which it was found being that which lies immediately above the lowest, save one, of our coal strata. I am, &c.

(Signed)   JOHN CRAIG.

To Captain Thomas Brown.

This sketch accompanied the above communication.

SCENERY OF THE VAL DEL BOVE, MOUNT ÆTNA.—Let the reader picture to him-self a large amphitheatre five miles in diameter, and surrounded on three sides by preci-pices from two thousand to three thousand feet in height. If he has beheld the most picturesque scene in the chain of the Pyrenees, the celebrated ' cirque of Gavarnie,' he may form some conception of the magnificent circle of the precipitous rocks, which inclose on three sides the great plain of the Val del Bove. This plain has been deluged by repeated streams of lava, and although it appears almost level when viewed from a distance, it is in fact more uneven than the surface of the most tempestuous sea. Besides the minor irregularities of the lava, the valley is in one part interrupted by a ridge of rocks, two of which, Musara and Capra, are very prominent. It can hardly be said that they

" like giants stand,
To sentinel enchanted land;"

for although, like the Trossachs, they are of gigantic dimensions, and appear almost isolated as seen from many points, yet the stern and severe grandeur of the scenery which they adorn is not such as would be selected by a poet for a vale of enchant-ment. The character of the scene would accord far better with Milton's picture of the infernal world; and if we imagine ourselves to behold in motion, in the darkness of the night, one of those fiery currents, which have so often traversed the great val-ley, we may well recall

" yon dreary plain, forlorn and wild,
The seat of desolation, void of light,
Save what the glimmering of these vivid flames
Cast pale and dreadful."

The face of the precipices is broken, in the most picturesque manner, by the ver-tical walls of lava which traverse them. The masses usually stand out in relief, are exceedingly diversified in form, and often of immense altitude. In the autumn, the black outline may often be seen relieved by clouds of fleecy vapour which settle be-hind them, and do not disperse till mid-day, continuing to fill the valley, while the sun is shining on every other part of Sicily, and on the higher regions of Ætna.

As soon as the vapours begin to rise, the changes of scene are varied in the highest degree; different rocks being unveiled and hid by turns, and the summit of Ætna often breaking through the clouds for a moment with its dazzling snows, and being then as suddenly withdrawn from the view.

An unusual silence prevails, for there are no torrents dashing from the rocks, nor any movement of running water in this valley, such as may almost invariably be heard in mountain regions. Every drop of water that falls from the heavens, or flows from the melting ice and snow, is instantly absorbed by the porous lava; and such is the dearth of springs, that the herdsman is compelled to supply his flocks, during the hot season, from stores of snow laid up in hollows of the mountains during winter.

The stripes of green herbage and forest land, which have here and there escaped the burning lavas, serve by contrast to heighten the desolation of the scene. When I visited the valley, nine years after the eruption of 1819, I saw hundreds of trees, or rather the white skeletons of trees, on the borders of the black lava, the trunks and branches being all leafless, and deprived of their bark by the scorching heat emitted from the melted rock; an image recalling those beautiful lines—

" As when heaven's fire
Hath scath'd the forest oaks, or mountain pines,
With singed top their stately growth, though bare,
Stands on the blasted heath."

Principles of Geology, by Charles Lyell.

IGNES FATUI.—We know that animal substances, in a state of putrefaction, always emit phosphorus, which, taking fire from the contact of the atmosphere, produces light and wandering flames. Such is probably the origin of those *Ignes fatui* which flutter at night over church-yards and fields of battle, and which have given rise to pre-tended apparitions of spirits in churches, where it is the pernicious and superstitious custom to accumulate the remains of the dead. Hydrogen gas is often combined with phosphorus; this mixture is not fit for respiration; it quickly suffocates. There is also a circumstance which seems to have given rise to many histories of spirits and apparitions, namely, the luminous appearance of the inflammable air disengaged from marshes, and composed of hydrogen gas mixed with azotic. The air which inflames on the surface of certain springs, known by the name of *burning fountains*, arises from the presence of hydro-phosphoric gas, or, as it is otherwise termed, phosphoretted hydrogen. One of these springs is met with in the parish of St Bartholomew, in the department of the Isère. The disengagement of inflammable gas during the summer is so considerable, that we frequently see a flame seven feet high; and when travellers first behold it, they imagine that the whole village is on fire.—(Bouvier, Journal de la Médecine Eclairées par les Sciences Physiques, tom. iii. No. 8.)

MISCELLANEOUS.

NO. III.—ANIMAL MAGNETISM.

Since the supposed effects of the Animal Magnetism, then, were not discoverable in those who were incredulous, there was great reason to suspect that the impressions which were produced were the result of a previous expectation of the mind, a mere effect of the Imagination. The Commissioners, therefore, now directed their experiments to a new point, namely, to determine how far the Imagination could influence the sensations, and whether it could be the source of all the phenomena attributed to Magnetism.

The Commissioners had recourse now to a M. Jumelin, who magnetised in the same way with MM. Mesmer and Deslon, except that he made no distinction of the magnetic *poles*. Eight men and two women were operated on by M. Jumelin; but none of them experienced any effect. At length a female servant of Dr Le Roy, who was magnetised in the forehead, but without being touched, said she perceived a sense of heat there. When M. Jumelin moved his hand about, and presented the extremities of his five fingers to her face, she said that she felt as it were a flame moving about; when magnetised at the stomach, she declared that the heat was there; at the back, and the same heat was there: She then affirmed she was hot all over the body, and suffered a headache. Seeing that only one person, out of eleven, had been sensible to the Magnetism, the Commissioners thought that this person was probably possessed of the most mobile imagination. They, therefore, tied a bandage over her eyes, and she was magnetised again; but the effects no longer accorded with the parts to which the Magnetism was directed!

When it was directed successively to the stomach and to the back, the woman only perceived the heat in her head, and a pain in her eyes, and in the left ear! The bandage was removed, and M. Jumelin applied his hands to the hypochondres; she immediately perceived a sense of heat in those parts; and, at the end of a few minutes, said that she was faint, and actually swooned. When she was sufficiently recovered, her eyes were again bandaged. M. Jumelin was then removed to a distance, silence was commanded, and they made the woman believe that she was again magnetised. The effects were now precisely the same, although no one operated, either near her or at a distance; she felt the same heat, particularly in the back and loins, and the same pain in the eyes and ears! At the end of a quarter of an hour, a sign was made to M. Jumelin to magnetise her at the Stomach; he did so, but she felt nothing; he magnetised her back, but without effect; in fact, the heat of the back and loins gradually ceased, and the pains in the head remained! Here, then, was demonstrative evidence of the operation of the Imagination. When the woman saw what was done, the sensations were placed in the parts magnetised; but when she could no longer see, they were referred to the most distant parts, where no Magnetism was directed; and, above all, they were equally felt when she was not magnetised at all, and not felt when she was magnetised, after a little repose, but unknown to herself. The fainting of a nervous woman, when made the subject of a mysterious experiment, and continued in a posture of restraint for a considerable time, is explicable upon natural causes. This experiment also showed, that the distinction of poles was purely chimerical. It was repeated the following day upon a man and a woman, with the same results. Sensations felt when they were not magnetised, could only be the effect of the Imagination: and it was found only necessary to excite and direct the Imagination, by questions, to the parts where the sensations were to be felt, instead of directing the magnetism upon those parts, in order to produce all the effects. A child of five years old was then magnetised; but it felt nothing, except the heat which it had previously contracted in playing.

These experiments were repeated by the Commissioners in various ways, upon many different persons, of all classes, and with the same results, differing only according to the difference of susceptibility in the imaginations of the individuals. They found effects constantly experienced, when no Magnetism was used, and *vice versa* (when the eyes were covered), according to the direction of the patient's attention by questions put to him with address. Now this practice could not lead to any error; since it only deceived their Imagination. For, in truth, when they were not magnetised, their only answer ought to be that they felt nothing.

Some facts communicated to the Commissioners by M. Sigault, an eminent physician at Paris, place the power of the Imagination in a strong light. "Having announced," he says, "in a great house, that I was an adept in the art of Mesmer, I produced considerable effects upon a lady who was there. The voice and serious air which I affected made an impression upon her, which she at first attempted to conceal, but having carried my hand to the region of the heart, I found it palpitating. Her state of oppression indicated also a tightness in the chest, and several other symptoms speedily ensued: The muscles of the face were affected with convulsive twitches, and the eyes rolled; she fell down in a fainting fit, vomited her dinner, and felt herself in a state of incredible weakness and languor. A celebrated artist, who gives lessons in drawing to the children of one of our Princes, complained during several days of a severe headache, which he mentioned to me when we met accidentally on the Pont-Royal. Having persuaded him that I was initiated in the mysteries of Mesmer, almost immediately, by means of a few gestures, I removed his pain to his great astonishment." Dr Sigault justly remarks, that it is probably by such an impression on the mind, that the sight of the dentist removes the toothache when the patient has gone to him for the purpose of having his tooth drawn. He adds, that being one day in the parlour at a convent, a young lady said to him, "You go to M. Mesmer's, I hear." "Yes," he replied, "and I can magnetise you through the grate;" presenting his finger towards her at the same time. She was alarmed, grew faint, and begged him to desist; and, in fact, her emotion was so great, that, had he persisted, he had no doubt that she would have been seized with a fit.

But although the Commissioners were convinced by their experiments, that the *Imagination* was capable of producing different sensations, of occasioning pain, and a scuse of heat, and even actual heat in all parts of the body; and therefore that it contributed much to the effects which were ascribed to *Animal Magnetism;* yet the effects of the latter had been much more considerable, and the derangements of the animal economy, which it excited, much more severe. It was now, therefore, to be ascertained, whether, by influencing the Imagination, convulsions, or the complete *crisis* witnessed at the public treatment, could be produced. In proof of this point, their experiments were not less conclusive, as the following relation of one or two of them will evince. As M. Deslon acknowledged that the complete success of the experiments would depend upon the subjects of them being endowed with sufficient sensibility, he was requested to select some of his patients, who had already proved their susceptibility of the magnetic influence, upon whom the trials might be made.

According to the principles of the Magnetisers, when a tree had been touched by them, and charged with Magnetism, every person who stopped near the tree would feel the effects of this agent, and either fall into a swoon or into convulsions. Accordingly, in Dr Franklin's garden at Passy, an apricot tree was selected, which stood sufficiently distant from the others, and was well adapted for retaining the Magnetism communicated to it. M. Deslon, having brought thither a young patient of twelve years of age, was shown the tree, which he magnetised, while the patient remained in the house under the observation of another person. It was wished that M. Deslon should be absent during the experiment; but he affirmed that it might fail, if he did not direct his looks and his cane towards the tree. The young man was then brought out, with a bandage over his eyes, and successively led to four trees, which were *not magnetised*, and was directed to embrace each during two minutes;—M. Deslon at the same time standing at a considerable distance, and pointing his cane to the tree actually magnetised. At the first tree the young patient, on being questioned, declared that he perspired profusely; he coughed, and expectorated, and said that he felt a pain in the head: he was still about twenty-seven feet from the tree magnetised. At the second tree he found himself giddy, with the headache as before: he was now thirty feet from the magnetised tree. At the third, the headache and giddiness were much increased; he said he believed he was approaching the magnetised tree; but he was still twenty-eight feet from it. At length, when brought to the fourth tree, *not magnetised*, and at a distance of twenty-four feet from that which was, the crisis came on; the young man fell down in a state of insensibility, his limbs became rigid, he was carried to a grass plot, where M. Deslon went to his assistance, and recovered him.

This experiment, then, was altogether adverse to the principle of Magnetism, not negatively, but positively and directly. If the patient, said the Commissioners, had experienced no effects under the tree actually magnetised, it might have been supposed that he was not in a state of sufficient susceptibility; but he fell into the *crisis* under one which was not magnetised; therefore, not from any external physical cause, but solely from the influence of the Imagination. He knew that he was to be carried to the magnetised tree; his imagination was roused, and successively exalted, until, at the fourth tree, it had risen to the pitch necessary to bring on the *crisis*.

The Fire of Saint Elmo is generally considered as an accumulation of electric matter round a point which moves in the air. This fire, then, may be expected to appear frequently at the top of the masts of a vessel sailing along with rapidity. The ancients observed this phenomenon. These fires, when seen in pairs, were called *Castor* and *Pollux;* when the flame was single, it bore the name of *Helen*. The spears of an army often appeared ornamented with these electrical plumes. (Pliny, Hist. Nat. II. cap. 37). A Swedish naturalist, travelling on horseback in snowy weather, saw his fingers, his switch, and the ears of his horse, covered with a fire of this description.—(Forskael, in Bergmann, Géogr. Phys. § 130).

The Glow-worm.—Mr John Murray, in a communication made to the Royal Society, on the luminous matter of the Glow-worm, states some curious facts as the result of his own observation and experiments. His observations tend to show that this light is not connected with the respiration, nor derived from the solar light; that it is not affected by cold, nor by magnetism, nor by submersion in water. Trials of submersion in water, in various temperatures, and in oxygen, are detailed. When a Glow-worm was immersed in carbonic acid gas, it died shining brilliantly; in hydrogen it continued to shine, and did not seem to suffer. Mr Murray infers, that the luminousness is independent, not only of the respiration, but of the volition and vital principle. Some of the luminous matter, obtained in a detached state, was also subjected to various experiments, from which it appears to be a gummo-albuminous substance, mixed with muriate of soda, and sulphate of alumina and potash, and to be composed of spherules. The light is considered to be permanent, its eclipses being caused by the interposition of an opaque medium.

Erratum.—In a part of our first impression—
Journal, page 46, col. 2, lines 49 and 54, *for Bore* read *Bové*.

Edinburgh: Published for the Proprietors, at their Office, 16, Hanover Street. London: Smith, Elder, and Co., 65, Cornhill. Glasgow and the West of Scotland: John Smith and Son, 70, St Vincent Street; and John Macleod, 20, Argyle Street. Dublin: W. F. Wakeman, 9, D'Olier Street.

THE EDINBURGH PRINTING COMPANY.

THE EDINBURGH

JOURNAL OF NATURAL HISTORY,

AND OF

THE PHYSICAL SCIENCES.

AUGUST, 1836.

ZOOLOGY.

DESCRIPTION OF THE PLATE—THE GIRAFFE (*Camelopardalis giraffa*).

It was long a matter of doubt in modern times, whether there existed such an animal as the Giraffe. This doubt was, however, set finally at rest by the zeal and exertions of the indefatigable traveller Levaillant, who saw, chased, overtook, and killed one in the interior of Africa.

The Giraffe is the tallest of quadrupeds, measuring eighteen feet from the hoofs to the tip of the horns. In its native wilds it feeds principally on a tree called *Acacia Giraffæ*. Mr Richard Davis, animal painter to the King, who studied the manners of the young one which arrived in this country in August 1827, says, "In its natural habits, I cannot conclude that the Giraffe is a timid animal, for, when led out by its keepers, the objects which caught its attention did not create the least alarm, but it evinced an ardent desire to approach whatever it saw; no animal was bold enough to come near it. Its docile, gentle disposition, leads it to be friendly, and even playful, with such as are confined with it; a noise will rouse its attention, but not excite its fear.

"I do not think it very choice of its food when out, so that it be green and sweet. It is fond of aromatics; the wood of the bough it also eats; our acacia, and others of the mimosa tribe, it did not prefer; and it never attempted to graze; it seemed a painful and unnatural action when it endeavoured to reach the ground. I have seen it try to do so when excited by an object which curiosity led it to examine; its feet were then two yards apart. It was constantly in motion when the doors of its hovel were open; but it had no sense of stepping over any obstruction, however low.

"It is asserted by travellers, that it resembles the Camel, in having callosities on the breast and thighs, and that it lies on its belly like that animal. There are between the fore-legs and thighs, to the casual observer, what appear to be such, but these are folds of loose skin, which enable it to separate its fore-legs when reaching downwards. Its mode of resting is, like most quadrupeds, on one side; but the operation of lying down is curious and peculiar: I will endeavour to describe it:—

"We will suppose it to be preparing to lie down on the right side; the first action is to drop the fetlock of the right fore-leg, then, on one knee of the left one, to bring down the other knee; it then collects its hind-legs to perform the next movement, the left one being brought rather forward, but wide, until the right hind-leg is advanced between the fore ones"—pretty nearly in the position represented in the second figure of the plate.—" This requires some time to accomplish, during which it is poised with the weight of its head and neck, until it feels that its legs are quite clear and well arranged; it then throws itself on one side and is at ease. When it sleeps, it bends the neck back, and rests the head on the hind quarters."

M. Acerbi, who saw this Giraffe and its companion at Alexandria, differs from Mr Davis in one essential particular. He says, "There are few naturalists who have not contributed to perpetuate the vulgar error, that in eating and drinking from the ground, the Giraffe is compelled to stretch his fore-legs amazingly forwards. Some even assert that he is obliged to kneel down. Of the few animals which fell under my observation, three took their food from the ground without inconvenience; and I am of opinion that when any difficulty exists in this respect, it is the effect of habit, acquired in the progress of domestication."

We hope that these and other discrepant points in the history of this interesting animal will not long remain unsettled, as four beautiful and healthy Giraffes have lately arrived in this country, and are now in the Zoological Gardens, London.

M. Thibaut, who has the charge of the Giraffes in the Gardens, Regent's Park, says they are extremely fond of society, and very sensible; he has observed one of them shed tears when it no longer saw its companions, or the persons who were in the habit of attending it. The Giraffe eats with great delicacy, and takes its food leaf by leaf, collecting them from the trees by means of its tongue. It rejects the thorns, and in this respect differs from the Camel. As the grass on which these Giraffes are now fed is cut for them, they take the upper part only, and chew it until they perceive that the stem is too coarse. Great care is required for their preservation, especially great cleanliness. M. Thibaut says, that he found the flesh excellent eating; the Arabs are very fond of it. On the 15th of August, last year, Thibaut saw the first two Giraffes; a rapid chase, on horses accustomed to the fatigues of the desert, put him and his companions in possession, at the end of three hours, of the larger of the two, the mother of one of these now in his charge. Unable to take her alive, the Arabs killed her with blows of the sabre; and cutting her to pieces, carried the meat to the head-quarters, where it was cooked and eaten.

The Giraffe has its eyes placed on the extreme convex sides of the skull, by which means it can see objects behind it as well as before. This will be observed, as also its manner of lying on the ground, by a back view, which we have given in the most distant figure on the plate.

Some idea will be formed of its mode of laying hold of branches, from the above cut, copied from the excellent paper of Sir E. Home on the Giraffe.

DESCRIPTION OF THE PLATE—THE OWLS.

Of all the birds of prey, Owls are the most useful to man; as their food consists principally of rats, mice, and other vermin which steal abroad, under the cloud of night, to lay waste our corn fields and granaries. Yet it is strange that vulgar prejudice prevails over common sense and daily experience, so that these birds are viewed with hatred, and even dread. A singular appearance, and doleful cry, with their retired and lonely habits, have gained for them a superstitious character which is not likely soon to be wiped away. But their utility demands our kindest regard.

Fig. 1. THE VIRGINIAN HORNED-OWL (*Strix Virginiana*).—This is one of the largest of its tribe, the male bird measuring twenty inches in height, while the female is upwards of two feet. It is to be found in almost every quarter of the United States, and extends its range as far north as Hudson's Bay. It is a bold and noble bird, exhibiting courage equal to that of the Golden Eagle. It builds on high trees, usually fixing its nest on a horizontal branch. A characteristic anecdote of the superstitious notions of our countrymen, in which an Owl of this species was concerned, is related by Dr Richardson. "A party of Scottish Highlanders," says he, "in the service of the Hudson's Bay Company, happened, in a winter journey, to encamp after nightfall in a dense clump of trees, whose dark tops and lofty stems, the growth of more than a century, gave a solemnity to the scene that strongly tended to excite the superstitious feelings of the Highlanders. The effect was heightened by the discovery of a tomb, which, with a natural taste often exhibited by the Indians, had been placed in this secluded spot. Our travellers having finished their supper, were trimming their fire preparatory to retiring to rest, when the slow and dismal notes of the Horned Owl fell on the ear with a startling nearness. None of them being acquainted with the sound, they at once concluded that so unearthly a voice must be the moaning of the spirit of the departed, whose repose they supposed they had disturbed, by inadvertently making a fire of some wood of which his tomb had been constructed. They passed a tedious night of fear, and, with the dawn of day, hastily quitted the ill-omened spot."

Fig. 2. THE LONG-EARED OWL (*S. Otus*).—This species is common both to Europe and America. Like the Great Horned Owl, it also breeds on trees. It is fourteen inches and a half long, and three feet two inches in extent from the tip of one wing to that of the other. In America, Wilson found this species building in the midst of the nests of other birds, and some were even on the same tree; this is identically the same with the habits of this bird in Britain and other places of Europe.

Fig. 3. THE MOTTLED OWL (*S. Nævia*, male), and fig. 4, a young female.—The male and female of this species are considerably different in the colour and markings of their plumage, as well as in the mottling. This has caused authors to describe them as distinct species. There is, beside, a marked difference in the young and adult garbs of both sexes, and this has induced a still further multiplying of specific names. This bird is a native of the Northern States of America, and migrates southward on the approach of winter. It constructs its nest in the hollow of a tree.

DESCRIPTION OF TWO NEW SHELLS.

THE two Shells represented beneath were procured from Orkney, and are now in the Cabinet of William Nicol, Esq., Edinburgh. It is difficult to determine whether they are new species, or only greatly produced varieties of the *Buccinum Anglicanum* and *B. undatum*.

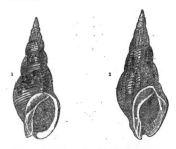

Fig. 1. agrees in all its characters with the *B. Anglicanum* of Lamarck, except in the spire being much longer and more fusiform, the breadth of the body of the shell being only about a third of its length; while in the *B. Anglicanum*, it measures nearly the half of the shell; it is of a reddish brown colour, fasciated and clouded with darker shades of the same colour. If it be really distinct from the *B. Anglicanum*, it might with propriety be distinguished by the name of *B. elongatum*. But until we have seen and examined the animal, the name must remain in abeyance.

Fig. 2. has all the characters and appearance of *B. undatum*, except in its greatly elongated shape, and if the animals really differ, it might be distinguished by the specific name of *acutissimum*.

These Shells are said to be obtained by their adhering to the fishermen's lines in deep sea water; and probably their greatly lengthened shape may be peculiar to this locality.

BOTANY AND HORTICULTURE.

SACCHARINE PLANTS.—The most valuable plants producing Sugar in this country are Beet-Root and Parsnip. The White Beet (*Beta cicla*) is a hardy biennial plant, a native of the Sea Coasts of Spain and Portugal, and introduced into this country in 1570. It was from the roots of this plant that the French and Germans obtained sugar with so much success during the late war, while all their West India colonies were in the hands of the British. The following is the ordinary process of extracting the sugar from this plant:—The roots are reduced to a pulp by pressing them between two rough cylinders; the pulp is then put into bags, and the sap it contains is pressed out. The liquor is then boiled, and the saccharine matter precipitated by quicklime; the liquor is now poured off, and to the residuum is added a solution of sulphuric acid, and again boiled; the lime uniting with the acid, is got rid of by straining; and the liquor is then gently evaporated, or left to granulate slowly, after which it is ready for undergoing the common process of refining raw sugars. The French manufacturers have acquired so much experience in this process, that, from every 100 lbs. of Beet, they extract 12 lbs. of sugar in the short space of twelve hours.

The Parsnip (*Pastinaca sativa*) is next in value to the White Beet as a saccharine root. It is a biennial British plant, common in calcareous soils, and used in England chiefly as a vegetable. One thousand parts of Parsnips contain ninety parts of sugar, nine parts of starch, the rest being water and fibre. An excellent ardent spirit is obtained by distillation from this plant; but the wine manufactured from it, in the opinion of many, possesses a finer flavor, and more nearly approximates to foreign wine than that obtained from any other British produce. The process of manufacturing Parsnip Wine is more clearly and fully described, in an interesting little work entitled "The British Wine-Maker," recently published by Mr W. H. Roberts, than in any other work on the same subject. We may refer generally to Mr Roberts' useful and practical treatise as affording comprehensive and scientific information, while it seems a safe guide in the manufacture of wines from British produce. By the use of the *Saccharometer*—an instrument remarkable for the accuracy of its results—the process, as detailed in the work referred to, is rendered simple and of unfailing success, while, without its aid, wines of uniform quality, from year to year, cannot be otherwise produced. This circumstance arises from the fruits themselves yielding in some years a greater or less proportion of saccharine matter than they do in others, and this difference in quality is accurately determined by the application of the Saccharometer.

RINGING OF WALNUT-TREES.—The Baron de Trehoudi, near Metz, in Lorraine, has successfully introduced into his neighbourhood a practice of ringing Walnut-trees. It is accomplished by abstracting a ring of two inches breadth from the outer bark all around, and then plastering over the part with clay, mixed with moist manure. The Walnut-trees thus treated not only prove more prolific, but the fruit is more early.—*Neill's Horticultural Tour.*

GEOLOGY.

THE FOSSIL ELK OF THE ISLE OF MAN.—In the Royal Museum of the College of Edinburgh there is now the most perfect known specimen of this animal. Its dimensions are given below, but that a more distinct notice may be formed of its great stature, we have placed beside it a human skeleton of six feet, drawn upon the same scale.

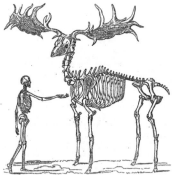

	Ft.	In.
Height to the tip of the process of the first dorsal vertebra, which is the highest point of the trunk,	6	1
Height to the anterior superior angle of the scapula,	5	4
Length from the first dorsal vertebra to the tip of the os coccygis,	5	2
Height to the tip of the right horn,	9	7½
Lateral or horizontal diameter of the thorax, at the widest part, that is at the eleventh rib,	2	0½
Depth of the thorax, from the tip of the process of the eighth dorsal vertebra, to the sternum at the junction of the eighth rib,	2	2

This superb fossil was dug up in the parish of Kirk Ralaff, and secured for our University Museum by the late Duke of Atholl. It was found imbedded in loose shell-marl, associated with numerous branches and roots of trees; over the marl was a bed of sand; above the sand a stratum of peat, principally composed of small branches and decayed leaves; and on the surface of all, the common alluvial soil of the country.

FOSSIL BOTANY.—The researches of M. Adolphe Brongniart into fossil organic remains, have in a great measure led to a knowledge of what must have been the appearance and temperature of the earth, when these fossils were vegetating on its surface; and also how far the various epochs of the existence of these plants accord with those remote Zoological epochs established by modern geologists.

The primitive vegetation, according to M. Brongniart, corresponded in its periods to the three successive formations of soil, from the earliest times, from the Creation down to that of the great Diluvian Change.

To the first period, which is co-etaneous with the simplest and the oldest formation of the globe, and lasted until the occurrence of deposits of coal strata, belong those vegetable bodies, the structure of which is in the highest degree simple. These organic remains are also remarkable for their rarity and the excessive magnitude of their dimensions. With respect to their rarity, compared to those of an analogous character inhabiting the present world, it is asserted, that of the former there are only six distinct families known; whereas of the latter, at least two hundred families exist; and with regard to the magnitude of their dimensions, it will be sufficient to instance the *Fern* trees, which in the actual world, and under the most favorable circumstances, grow to the height of from 20 to 25 feet only, while the same trees rose, in the primitive world, as high as 40 and even 50 feet. Brongniart thinks that the great coal formation which appeared at the termination of this vegetable period is due to the destruction of the plants in question. He arrives at the conclusion *à priori*, as well as from the inspection of the strata in which such plants are found, that life on the surface of the globe began with the vegetable kingdom; that the animals without a vertebrated spine succeeded next; and that probably the oceans contained no fish at the time.

To the second period of Antediluvian vegetation, corresponding to the geological depositions of the *grès*, or freestone, and conchyliferous calcareous stone, which terminated with the clay formation, belong, first, a number of massive and terrestrial plants deposited in those strata, too small to give a decided character to the vegetation of that time; and, secondly, a series of plants found above those same strata, and below the chalk, totally distinct from those which marked the first period, and the remains of which are now to be met with in the calcareous rock of Jura and other districts. The Cryptogamic plants prevailed to a great extent at this epoch, and among the remains of that period, no vestige can be traced of any of our recent Palm trees, nor even of the Di-cotyledonous Plants of our times. During this period, no Mammiferous animal was in existence, whether terrestrial or aquatic. The Reptiles alone constituted the whole range of the existing vertebrated animals, different in kind and dimension from those of the present world, among which may be mentioned those singular beings destined by nature to fly as well as to swim, and since called, from the structure of their parts or entire skeletons found by geologists, Pterodactyles, Plesiosaurus, and Ichthyosaurus.

The third and last period of Antediluvian vegetation, far more interesting than either the first or second, and separated from the latter by the chalk formation, which contains some traces of marine plants, corresponds to the period when the last marine irruptions took place, between the intervals of which those huge animals were propagated, whose vestiges have been distinctly observed, such as the Palæotherium, Anoplotherium, and other genera now lost; and after them the Antediluvian Elephant, the Rhinoceros, and other contemporaneous races. The plants belonging to this period are distributed in two distinct soils—the one resulting from the depositions after marine irruptions, the other from depositions left by fresh water inundations. Their characters are distinct and peculiar; but they, in a great degree, correspond to the plants of the Postdiluvian world. Animalization was at its highest degree of perfection during the third period of Antediluvian vegetation, but *Man formed no part of it.*

Fermenting Ponds in Massachusetts.—A remarkable pond was discovered not long ago in Sharon, Massachusetts, known by the name of Mash-Bog Pond, from which great quantities of lenticular argillaceous oxide of iron and cake-ore are procured. From about the middle of August to some time in September, this pond presents the singular appearance of working or fermenting, as beer does when new.

MISCELLANEOUS.

Effects of Cold on New-Born Infants.—The excessive mortality which prevails among Infants during the first year after birth, has recently led several eminent physicians to investigate this important subject; and the following are the conclusions at which Dr Trevisan arrived, after a most careful investigation:—

1. In Italy, of 100 infants born in the months of December, January, and February, 66 die in the first month, 15 in the course of the year, and 19 only survive.
2. Of 100 infants born in the spring of the year, 48 survive the first year.
3. Of 100 infants born in the summer, 63 survive the first year.
4. Of 100 infants born in the autumn, 58 survive the first year.

The surplus of mortality, during the cold months, is attributed by MM. Milne-Edwards, Fontenelle, Villerme, and Dr Trevisan, exclusively to the practice of exposing the infants to the cold air a few days after birth, in taking them to the church to be baptized; and these distinguished physicians implore the Ecclesiastical authorities to devise some means, consistent with their religious duties, to put a stop to this fatal mistake.

It is not improbable that a great proportion of the numerous deaths which take place in Britain during the first year after birth may be attributable to a similar cause.

Icebergs of the Southern Hemisphere.—In the year 1830, Captain Horsburgh turned his attention to Icebergs which have been met with in the Southern Ocean, and on a strict search through the Journals of the East India Company, he could not meet with any record of the occurrence of these for a whole century previous to 1828, although vessels belonging to the Company, in their passage to and from India, had navigated into the parallels 40°, 41°, and 42° south; while in the years 1828 and 1829, icebergs had occasionally been met with by various vessels, so near the Cape of Good Hope as between the latitude 36° and 39°. The most striking instance mentioned by Captain Horsburgh, is that in which the brig *Eliza* fell in with five icebergs, in 1828, latitude 37° 31′ south, longitude 18° 17′ east of London. They were prodigious masses of ice, from 250 to 300 feet in perpendicular height above the surface of the ocean, and of the shape of church steeples. He accounts for them, by the supposition that a large tract of land exists near the Antarctic Circle, somewhere between the meridian of London and the 20° of east longitude; and attributes their unprecedented descent, during the years 1828 and 1829, to their disruption from the place of their formation by the violent convulsion of an earthquake or volcano. He mentions as a remarkable fact, that icebergs are met with at the same period of the year, namely, April and May, whether in the Northern or Southern Hemisphere, although the seasons are at that time, in each hemisphere, of an opposite character.

Singular Cases of the Effects of Nitrous Oxide.—The following very remarkable cases of the effects of Nitrous Oxide occurred among Professor Silliman's Students, at Yale College. A gentleman about nineteen years of age, of a sanguine temperament, cheerful disposition, and in the most robust health, inhaled the gas, which was prepared and administered in the usual dose and manner. Immediately his feelings were uncommonly elevated, so that (as he expressed it) he could "not refrain from dancing and shouting." To such a degree was he excited, that he was thrown into a frightful delirium, and his exertions became so violent that he sunk to the earth exhausted; and having there remained till he had in some degree recovered his strength, he again rose, only to renew the most convulsive muscular efforts, and the most piercing screams and cries, until, overpowered by the intensity of the paroxysms, he again fell to the ground apparently senseless, and panting vehemently. For the space of two hours these symptoms continued; he was perfectly unconscious of what he was doing, and was in every respect like a maniac. He stated, however, that his feelings vibrated between perfect happiness and the most consummate misery. After the first violent effects had subsided, he was obliged to lie down two or three times, from excessive fatigue, although he was immediately roused upon any person entering the room. The effects remained in a certain degree for two or three days, accompanied by a hoarseness, which he attributed to the exertions made while under the influence of the gas.

The other case was that of a man of mature age, and of a grave character. For nearly two years previous to his taking the gas, his health had been very delicate, and his mind so gloomy and depressed, that he was obliged almost entirely to discontinue his studies. In this case of debility, he inhaled about three quarts of the nitrous oxide. The consequences were, an astonishing invigoration of his whole system, and the most exquisite perception of delight. These were manifested by an uncommon disposition for mirth and pleasantry, and by extraordinary muscular power. The effects of the gas were felt, without diminution, for at least thirty hours, and in a greater or less degree, for more than a week; but the most remarkable effect was *upon the organs of taste.* Before taking the gas, he felt no peculiar choice in the articles of food, but immediately after that event, he manifested a taste for such things only as were *sweet*, and for several days ate nothing but sweet cake. Indeed, this singular taste was carried to such excess, that he used sugar and molasses not only upon his bread and butter, and lighter food, but upon his butcher meat and vegetables; and continued to do so for some considerable time afterwards. His health and spirits recovered their wonted energy, which could only be accounted for from the influence of the nitrous oxide. He afterwards experienced no uncommon exhilaration, but became habitually cheerful, in place of being habitually grave, before he underwent this experiment.

Fall of the Dent du Midi.—On the evening of the 25th of August 1835, a violent storm raged all around the Dent du Midi. On the 26th, between the hours of ten and eleven in the forenoon, a considerable portion of the summit of this mountain was suddenly dislocated from the eastern ridge, and precipitated towards the glacier, on the southern side of which it carried along with it an enormous quantity. This immense mass of stone, earth, and ice, lodged into a deep ravine, which separates the Dent from the Col de Salenfe, and in which flows the torrent of St Barthelemy. At the straitened pass through which this rapid rushes to the smooth and verdant valley of the Rhone, an extensive mass of black slimy mud was noticed, on the surface of which lay fragments of rocks from the dimensions of twelve feet and downwards. It had all the appearance of a stream of lava, and directed its course towards the Rhone, through the Pine forest which mantles a part of the valley, carrying along with it every thing which stood in its course. Large trees were prostrated before it and broken like reeds; large blocks of stone were carried into the Rhone, the waters of which were forced over the opposite bank, and up the channel to a considerable distance. The high road was entirely blocked up by this mass of slime and stone, which rendered necessary the construction of a new road across by means of faggots. Every sort of communication between the upper and lower Valais was prevented for several days, until a temporary bridge could be constructed at the neck of the pass. Dense clouds of dust ascended to a great height for several days after this catastrophe, giving it all the appearance of a phenomenon produced by means of volcanic agency, and a deep valley was hollowed out of the mass.

REVIEWS.

Neue Wirbelthiere, zu der Fauna von Abyssinien gehörig, entdeckt und beschrieben von Dr E. Rüppell. Frankfurt am Main, 1835. Fol. Erste Lieferung—Säugethiere.

(*New Vertebrata, belonging to the Fauna of Abyssinia, discovered and described by Dr E. Rüppell. Frankfort on the Maine, 1835. Fol. First Part—Mammalia.*)

In this important work, the learned author has commenced by describing six new species of Mammalia, being three Monkeys, two Antelopes, and one Goat. The illustrations are got up with great care, and are accompanied by a diagnosis of each species, as well as a circumstantial description; and it is intended that this work shall be annexed to the celebrated Zoological Atlas of the author. Hitherto he has confined himself to the illustration of species previously unknown, and each class of Vertebrata is intended to form a separate series, any one of which may be purchased separately.

The first Monkey is a *Colobus*, to which he assigns the specific name of *Guereza*. It is quick and lively, of a harmless disposition, and lives in small families upon lofty trees. Long hair, white as snow, hangs from the side of its body, which is in every other part, except about the throat and the region around the eyes, of a deep black. Its long tail terminates in a white tuft.

The second species (*Macacus gelada* of Rüppell) is very like the Cynocephalus Hamadryas in the colour and quality of its hair, but in every other respect it is a true Macacus. Its food consists of seeds and roots, and it often makes great devastations among the cultivated fields. It lives in numerous families, among rocky districts covered with low bushes, and is always seen upon the ground.

The descriptions of the Antelopes are not less interesting than the preceding, as well from the striking manner in which allied species are grouped together, as from the important light thrown upon the geographical distribution of the separate groups of species found in the south and west of Africa.

The *Antilope Defassa* of Rüppell attains the size of a full-grown Cow. It lives in small families upon the grassy plains of Western Abyssinia. Its gait is heavy, and its favourite food consists of the leaves and fruit-cups of the *Holcus Sorghum*. In general character, it approaches nearly to the Redunca group of Major Hamilton Smith.

Of the other species, the *Antilope Decula* of Rüppell frequents the bushy mountain-valleys of Abyssinia, and the animals go about in couples. This species may be regarded as the Abyssinian representative of the *A. scripta* of Western Africa, and the *A. sylvatica* of the South. The other species (*Antilope Beisa*) of Rüppell approaches nearly to the South African Oryx, in the entire colour of the body, and in the shape of the horns, which are wanting in the female. It is found on the coast of the Red Sea, and perhaps also in Egypt, at least it was seen by the unfortunate Burckhardt when on his voyage from Schendi to Suakin. It lives in small families among low valleys, with bushes slightly interspersed, and feeds upon grass. When attacked or hardly hunted, it fights courageously, and defends itself with its pointed horns.

The remaining plate represents a Steinbock (*Capra walie, Rüpp.*) Its description is contained in the second part of this valuable work.

Excursions illustrative of the Geology and Natural History of the Environs of Edinburgh, by William Rhind. Second Edition, Royal 18*mo. Edinburgh.* 1836.

In our third number we noticed the first edition of this useful little work, and we now turn to the *second edition* with increased interest. The present contains more than double the matter of the first, and the different localities which the author describes are increased in an equal degree. These important additions to the text have rendered it a most valuable companion in our walks around a district which, our author justly remarks, "may truly be said to constitute the classic fields of Geology. Here Hutton first exercised his active and comprehensive intellect in rearing a system that bids fair to be permanent; and here many a contest has been maintained between the chiefs and heroes of opposing theories." Perhaps there is no situation in Europe where the elements of Geology can be more advantageously studied than in the neighbourhood of Edinburgh; for we have, within the very limited space of a few miles, almost all the varieties of formation, except the Primitive. Certainly no district has been more thoroughly investigated, and by men of the first distinction in this branch of science. Mr Rhind has most judiciously availed himself of all that has been written, while he has himself visited every corner which he describes, and has added many interesting observations,

Mr Rhind's remarks on the Coal Fields of this district are very judicious, and give a clear view of the subject. The Fossil Limestone of Burdiehouse, which excited so much interest among the Geologists during the Meeting of the British Association here, forms the subject of an interesting chapter, and several well engraved woodcuts of all the fossils have been introduced.

The work contains 152 closely printed pages on a small type; a coloured map of the Environs of Edinburgh; two picturesque and geological views of the "Chain of the Pentland Hills and Arthur Seat from the North, and the Pentland range from the South;" besides 54 distinct figures on wood, representing various sections of stratification and rare natural history objects of the district. The price is three shillings and sixpence.

Under Meteorology, we shall insert in our next number an extract from Mr Rhind's work, which will not only interest our readers, but also give them some idea of the contents of the work itself.

OBITUARY.

Captain James Horsburgh, F.R.S. died at his house in London on the 14th May last, in the 74th year of his age.

This distinguished individual and valuable man was a native of Elie, a small town on the East coast of Fife, Scotland, where he sprung from humble, pious, and respectable parents. He commenced his career in the obscure capacity of cabin-boy and cook in a merchant vessel. Through long and unswerving good conduct, he at last was elevated to the command of the *Anna*, East Indiaman, in the year 1802, chartered for Bombay; after reaching which he spent two years in coasting the peninsula of India, visiting Canton and the China Sea, and the islands of the Indian Archipelago, whence he returned to England in 1805.

The Captain possessed a dauntless and enterprising spirit, and a strong natural capacity and inquiring mind, which enabled him to acquire a thorough knowledge of every subject of practical importance and scientific bearing in reference to East Indian Hydrography. It was from his letter written on his arrival from India to the Hon. Henry Cavendish, and published in the Philosophical Transactions, that the scientific world became acquainted with the progress of that astonishing regularity in the rising and falling of the barometer, which so peculiarly distinguishes the tropical regions, and becomes disturbed, or again lost, with an increase of latitude in the station of observation. He also fully developed the nature of the circumstances under which it became manifest, or gradually disappeared, as a ship in her progress alternately crosses the Line, and advances into higher Northern or Southern Latitudes.

Soon after his return he married, and never again went to sea, and left his profession to devote himself to higher objects. His first publication in 1805 was "Memoirs" of his voyage, containing much valuable practical information. He next turned his attention to his great work, the "East India Sailing Directory," which has secured to him the admiration and regard of every maritime nation in the world. It was the result of five years indefatigable research among the archives of the East India Company. In 1810, he was appointed Hydrographer to the Company, in which capacity he constructed the numerous inestimable charts, which appeared in succession from the Hydrographical Office by order of the Court of Directors; which arduous and unremitting labours he continued up till the 18th of April last, when the first symptoms of water in the chest appeared. He lingered until his demise under great bodily suffering, which he endured with fortitude and patience.

The principal writings of Captain Horsburgh, besides his "Directory," are "Atmospherical Register for indicating Storms at Sea," 1816. In 1819 he edited "Mackenzie's Treatise on Marine Surveying," and afterwards his "Compendium of the Winds," "The East India Pilot of general and particular Charts from England to the Cape of Good Hope, Madras, and China;" these were in conjunction with Mr Arrowsmith, and on the largest scale ever published. "Remarks on Icebergs which have been met with in the Southern Hemisphere," published in the Philosophical Transactions for 1830.

LEARNED SOCIETIES.

Ashmolean Society, Oxford.—*May* 20.—The president in the chair.—Dr Buckland communicated to the Society a notice on some very curious recent discoveries of fossil footsteps of unknown quadrupeds, in the new red sandstone of Saxony, and of fossil birds in sandstone of the same formation, in the valley of the Connecticut. The sandstone which bears the impressions of these footsteps is of the same age with that in which, in the year 1828, Dr Duncan discovered the footsteps of land tortoises, and other unknown animals, near Dumfries. In the year 1834, similar tracks of at least four species of quadrupeds were discovered in the sandstone quarries of Hesseberg, near Hildburghausen. Some of these appear to be referable to tortoises, and to a small web-footed reptile. The largest footsteps mark the path of a large quadruped; probably allied to *Marsupialia*, or animals that carry their young in a pouch, like the kangaroo. The name of Cheirotherium has been given to this animal, from a distant resemblance, both of the fore and hind feet, to the human hand. The size of the hind foot was twice as great as that of the fore foot, being usually eight inches long and five inches wide: one was found twelve inches long. These footsteps follow one another in pairs, at intervals of fourteen inches from pair to pair, each pair being on the same straight line. Both large and small steps have the great toes alternately on the right and left side, and bent inwards like a thumb. Each step has the print of five toes. The fore and hind foot are nearly similar in form, though they differ so greatly in size. No bones of any of the animals that made these footsteps have yet been found. Another discovery of fossil footsteps has still more recently been made by Professor Hitchcock, in the new red sandstone of the valley of the Connecticut. In three or four quarries of this sandstone he has ascertained the existence of the tracks of at least seven extinct species of birds, referable, probably, to as many extinct genera. All of these appear in regular succession on the continuous track of an animal in the act of walking or running, with the right and left foot always in their relative proper places. The distance of the intervals between each footstep on the same track is occasionally varied, but to no greater amount than may be explained by the bird having altered its pace. Many tracks are often found crossing one another, and they are sometimes crowded, like impressions of feet in the muddy shores of a pond frequented by ducks or geese. All these fossil footsteps most nearly resemble those of *Grallæ* (waders). The impressions of three toes are usually distinct; that of a fourth, or hind toe, is generally wanting. The most remarkable among these footsteps are those of a gigantic bird, twice the size of an ostrich, whose foot measured fifteen inches in length, exclusive of a large claw measuring two inches. The toes of this bird were large and thick. The most frequent distance of these larger footsteps, from one another, is four feet; sometimes they are six feet asunder. The latter were probably made by the animal when running. There are also tracks of another gigantic bird, having three toes, of a more slender character. These tracks are from fifteen to sixteen inches long, exclusive of a remarkable appendage extending backwards from the heel eight or nine inches, and apparently intended (like a snow-shoe) to sustain the weight of a heavy animal walking on a soft bottom. The impressions of this appendage resemble those of wiry feathers, or coarse bristles, which seem to have sunk into the mud an inch deep;—the toes had sunk much deeper, and round their impressions the mud was raised into a ridge several inches high, like that round the track of an elephant in clay. The length of the step of this bird appears to have been six feet; the footsteps on the five other kinds of tracks are of smaller size, and the smallest indicates a foot but one inch long, and a step from three to five inches. The length of the leg of the African ostrich is about four feet, and that of the foot ten inches. All these tracks appear to have been made on the margin of shallow water, that was subject to changes of level, and in which sediments of sand and mud were alternately deposited. And the length of the legs, which must be inferred from the distance of the footsteps from each other, was well adapted for wading in such situations. Professor Powell afterwards gave a short account of the progress of his researches on Light. Professor Brongniart and Dr Milne-Edwards, of Paris, were present at the meeting.—*Literary Gazette.*

Geological Society.—There was read at the Society a letter from Mr De la Beche, explanatory of the geological position of a collection of fossils from the northern district of Cornwall. He states that, in the grauwacke of Western Somerset, Devon, and Cornwall, natural divisions may be instituted, founded on well-marked characters; but he conceives that the whole of this district belongs to a system older than the Silurian formations of Mr Murchison. Some of these organic remains were procured at Dinas Cove, Padstow Harbour, Trevelga Island (Lower St Columb Porth), and Towan Head, near New Quay, from the slate which is associated with sandstone, conglomerates, and limestone, and which is of the same age with the fossilliferous slate of Tintagel. The other fossils of this collection were procured near Bodmin, by Dr Potts, and in the vicinity of Liskeard. Mr De la Beche also states that there are two distinct evidences, in Somerset, Devon, and Cornwall, of two considerable movements of the land, one to a height of thirty or forty feet above the present level of the sea, and the other to an unascertained depth beneath it, since the production of the existing vegetation of the land, and the Molluscous inhabitants of the neighbouring ocean.

Edinburgh: Published for the Proprietors, at their Office, 16, Hanover Street. London: Smith, Elder, and Co., 65, Cornhill. Glasgow and the West of Scotland: John Smith and Son, 70, St Vincent Street; and John Macleod, 20, Argyle Street. Dublin: W. F. Wakeman, 9, D'Olier Street.

THE EDINBURGH PRINTING COMPANY.

THE EDINBURGH

JOURNAL OF NATURAL HISTORY,

AND OF

THE PHYSICAL SCIENCES.

SEPTEMBER, 1836.

ZOOLOGY.

DESCRIPTION OF THE PLATE—THE CONES.

THE elegance of form, splendid colours, and fine polish exhibited by Shells, have excited admiration in all ages, and procured for them a conspicuous place in Cabinets of Natural History. The Molluscous animals, which once dwelt within these elegant envelopes, perform an important part in the economy of Nature. Their useful as well as noxious qualities render them of high interest. Conchology, however, acquires a higher importance in various points of view. By a knowledge of this branch of Natural History, we are enabled to trace, to a certain extent, the past history of the Earth, in ascertaining the relative antiquity of various strata which lie beneath its surface; for shells are found in general to be the most perfect of all organic fossil remains, and may in truth be termed the "medals of the ancient world." Many kinds of Testaceous animals furnish an excellent and nutritious food, and some tribes supply the table with a delicate luxury. The mother-of-pearl affords materials for ingenuity and art; and the pearl itself, often the rival of the most precious gems, in the estimation of mankind, is the production of testaceous animals. Even the pernicious effects of some tribes demand our attention in studying their history, that we may be the better prepared to avert their depredations. The Snail, in ravaging the garden and the field, marks its progress by the destruction of some of the fairest of the vegetable tribes; and the Ship-worm, the dread of the mariner, appears an insignificant instrument to humble the glory and pride of Man, in demolishing, by its unseen labours, the noblest efforts of his ingenuity and skill.

The Cones are generally more esteemed by collectors than perhaps any other Shells. It is a very extensive Genus, as Lamarck describes 181 recent, and 9 fossil species, besides many varieties; yet the British Seas do not produce a single individual. This genus is confined to the warmer regions of the earth, increasing in numbers as we pass the equator. Many of the species are remarkable for the regularity and beauty of their markings, as well as for the fine colours which they exhibit. They are also prized on account of their elegant form, which is that of a reversed Cone, and turbinated.

Fig. 1. The MATCHLESS CONE. (*Conus Cedonulli.*) This Cone has always been highly esteemed by the curious collector. It was once exceedingly rare, and brought a very high price; as much as one hundred guineas have been paid for a single specimen of the *Cedonulli.* A single variety of this shell, which formerly belonged to the celebrated Naturalist Lyonnet, was valued at three hundred guineas. This shell is subject to considerable variety in the style and character of its markings;— nine are described by Lamarck. The Matchless Cone is a native of the seas of South America.

Fig. 2. The Matchless Cone, a variety.

Fig. 3. Tait's Cone (*C. Taitensis.*)—South American Seas.

Fig. 4. The Diviner's Cone (*C. Augur.*)—Indian Ocean.

Fig. 5 and 6. The Fumigated Cone (*C. Fumigatus.*)—American Seas.

Fig. 7 and 8. The Ornamented Cone (*C. Monile.*)—Indian Ocean.

Fig. 9 and 10. The Hebrew Cone (*C. Hebræus.*)—African and Indian Seas.

Fig. 11. The Tesselated Cone (*C. Tessellatus.*)—Indian Ocean.

Fig. 12. The Marbled Cone (*C. Marmoreus.*)—Indian Seas.

Fig. 13. The Jasper Cone (*C. Betulinus.*)—This cone grows frequently to a large size, sometimes six inches in length. It is a native of the coast of Madagascar.

Fig. 14. The General Cone (*C. Generalis.*)—Indian Ocean.

Fig. 15. The Flea-spot Cone (*C. Pulicarius.*)—Pacific Ocean.

Fig. 16. The Franciscan Cone (*C. Franciscanus.*)—Chinese Seas.

Fig. 17 and 18. The Stone-cutter Cone (*C. Lithoglyphus.*)—Indian Ocean.

Fig. 19. The Music Cone (*C. Musicus.*)—Chinese Seas.

Fig. 20. The Ceylon Cone (*C. Ceylonensis.*)—Coasts of Ceylon and Java.

Fig. 21. The Plated Cone (*C. Lamellosus.*)—Coast of Ceylon.

Fig. 22. The Bridal Cone (*C. Sponsalis.*)—Pacific Ocean.

Fig. 23. The Punctured Cone (*C. Puncturatus.*)—Coasts of New Holland.

Fig. 24. The Geographic Cone (*C. Geographicus.*)—Indian Ocean.

Fig. 25. The Striated Cone (*C. Striatus.*)—Indian Ocean.

DESCRIPTION OF THE PLATE—THE WRENS.

Modern Naturalists have restricted this genus to such birds as have a slender, slightly compressed, curved, and marginated bill, with the nostrils basal, and half covered by a naked membrane; the wings are short and rounded, with the fourth and fifth feathers of equal length, but longer than the others; the tail short, rounded, and erect, and the tarsus the same length as the middle toe.

The birds of this genus are subject to a wide geographical distribution; although there is but one species in Europe, which is common to all its kingdoms.

Fig. 1. The Winter Wren (*Troglodytes Hyemalis*).—This lively little bird is a native of North America, and visits the United States in the month of October, where it generally remains all winter, and migrates to the north in spring, to fulfil the important law of incubation. In its general appearance it has a strong resemblance to the European Wren; and some Naturalists have considered it as of the same species. It sings with great animation while mounted on the point of some branch of a tree. It is by no means a shy bird, as it is to be found in out-houses, yards, and gardens, in different cities of the Union.

Fig. 2. The European Wren (*T. Europæus*).—The Common Wren is a hardy bird, and braves the winter in almost every quarter of Europe, where it enlivens the natives with its sweet and sprightly notes. During this season, it is to be found close to the dwellings of men. It betakes itself to the woods in the summer. The nest is remarkable for its neatness, being of an oval shape, with a hole at the side for an entrance.

Fig. 3. The House Wren (*T. Oedon*).—This species is migratory; it arrives in Pennsylvania about the middle of April, and begins to construct its nest in the second week of May. It is customary for the species to place a small box on the top of a pole in the garden for its reception; and if this be neglected, it will take possession of a hole in the roof or wall, and has even been known to breed in an old hat. The conjugal pair generally hatch two broods in a season,—the one in June, and the other in the end of July.

Fig. 4. Bewick's Wren (*T. Bewickii*).—This bird was discovered by Audubon, near St Francisville, Louisiana. It is also a migratory species; but the breeding station is unknown. It is not known whether it has any song.

Fig. 5. The Marsh Wren (*T. Palustris.*)—The habits of this bird are retired; it arrives in Pennsylvania about the middle of May, or as soon as the water Nymphæ and reeds on the sides of rivers are sufficiently high to shelter it. To such localities it generally limits its excursions. It feeds on insects and their larvæ. Its notes consist of a curious crackling sound, like air bubbles forcing their way through mud or boggy ground when trod upon. Its nest is constructed with admirable neatness and apparent skill.

Fig. 6. The Brown Wren (*T. Furvo.*)—This species is a native of Brazil and Cayenne, and has sometimes been confounded with the House Wren of the United States. It constructs its nest in the low parts of woods. The total length of this bird is four inches.

Fig. 7. The Great Carolina Wren (*T. Ludovicianus*).—The general appearance of this bird at once conveys the idea that it is a Wren, although, on a minute examination, several characters present themselves which render it equivocal; its great size, and larger proportional bill, and other discrepant characters, render it necessary to form a sub-genus for its reception. It is found in Pennsylvania, Virginia, and other parts of the United States, where it is said to build. It is five inches and a quarter long.

Fig. 8. The Long-billed Wren (*T. Longirostris*) is remarkable for the length of its bill, and large size, measuring five inches and three quarters. It inhabits Brazil; but nothing is known of its habits.

A MONGREL DOG.—M. Joannon Navier, Mayor of Cuire, is possessed of a dog, the offspring of a Wolf-Dog and a Jackal. He is of a small size, not exceeding the ordinary dimensions of the Jackal, but so quarrelsome and fierce, that he is the terror of all the neighbouring dogs. He is very voracious, and devours all the ducklings and chickens which he meets with; and is consequently kept always tied up. He is most affectionate to his master, but is not considered a good watch-dog, as he seldom barks. Like all predatory animals, he frequently digs holes in the earth, and is very agile, bounding along the tops of walls with great dexterity when left at liberty. He is subject to frequent changes of his fur, and the last time he cast his coat the under hair was very short; that which covered the thighs was long, and streaked obliquely across, producing a wavy appearance; his tail was long, and finely formed. His ears are like those of his sire—the Wolf-Dog; the conch of which is firm, erect, and pointed backwards; his muzzle is provided with moustachoes, formed of numerous stiff hairs; his eyebrows are prominent, which, with a peculiar expression of the eyes, give him a look of suspicion and ferocity.

BOTANY.

EDIBLE SEA WEED.—The Philippine Islands contribute a very large portion of those Edible Birds-Nests, which are consumed as food in immense quantities by the Chinese, and recently also in Europe. Our particular attention must now be directed to that Edible Sea Weed, which is encountered on the coast of the Philippines, as well as on those of the Bashees, the islands of the Japanese Empire, the Moluccas, and many others, where it serves either as an article of food or for exportation. In the markets of Macao and Canton, one meets with large chests of this dried Sea Weed, which is introduced from Japan. The marine plant constituting this branch of trade is the *Sphærococcus cartilagineus*, variety *setaceus* of Agardh, which is found throughout India in extraordinary plenty. It is eaten by the Salangane or Esculent Swallow (*Hirundo esculenta Lin.*, properly *Cypselus*), for the purpose of employing it in the construction of its nest. The mass, after being changed into a jelly in the stomach of the Bird, is again thrown up, and by its means the nest adheres together. This remarkable Indian Birds-nest comes to China in a raw state, besmeared with dirt and feathers, where it is cleansed in a large warehouse by instruments specially adapted for the purpose, and then consists of nothing else but well-soaked *S. cartilagineus*. On its being thus prepared, it is usual to add a very large quantity of seasoning, when it deserves to be placed in the first rank of Chinese delicacies. The Japanese have long since discovered the method of preparing the substance of this nest in an artificial manner. After having been pulverized, the Sea Weed is boiled down into a thick jelly, and is then drawn out into long threads in the same manner as Vermicelli or Macaroni. It is then brought for sale under the name of *Dschin-schan*. The Dutch call it *Ager-Ager*. These Birds-nests, whether genuine or counterfeit, are used by the Chinese for sauce, which are served up with their meat-dishes; but the Europeans resident in China prefer it in the form of jelly, for which use the *Dschin-schan*, or Ager-Ager, is well adapted. By means of a single boiling, they convert the dried substance into a jelly, which is served up in wine or the juice of fruit. The dried Dschin-schan is sometimes cut into large pieces, and put into thick gravy; it is dissolved in about a minute, and then assumes the form of transparent Vermicelli.

We have dwelt thus at length upon this substance, because there has been so much discussion as to the properties of *Carragheen*, which is nothing else than the dried *Sphærococcus crispus*, found in large quantities on the western and northern coasts of England, and probably resembles the setaceous variety of the *S. cartilagineus*. But we can by no means believe that any other quality can be assigned to the jelly thence obtained, except one purely nutritive, and which does not overload the organs of digestion.—*From the German of Meyen,—Reise um die Erde—(Meyen's Voyage round the Earth.) Bd. II. S. 276.*

RED SNOW PRODUCED BY A FUNGUS.—Mr F. Bauer ascertained that the Red Snow of Baffin's Bay, observed by Captain Ross, was produced by a new species of *Uredo*, which he has termed *Nivalis.* The size of a globule of this fungus he found to be the 1600th part of an inch.

GEOLOGY.

ON THE FOOT-MARKS OF BIRDS, IN NEW RED SANDSTONE.—It has been a matter of some surprise to geologists, how there should be an almost entire absence of Birds among organic remains found in rocks. Till lately, all that have been discovered are those mentioned by Cuvier, consisting of nine or ten specimens found in the tertiary gypsum beds near Paris. All the cases of fossil birds noticed by previous writers are regarded by Cuvier as unworthy of credit. Hence any new discoveries in this interesting branch of Oryctology are valuable to the geologist.

It was mentioned in our last Number, that Professor Hitchcock of Amherst College, North America, had discovered in the new red sandstone of the valley of the Connecticut, in five different places, the foot-prints of at least seven extinct species of birds, which, in all probability, may each represent a distinct genus. All these are imprinted, in regular succession, representing the continuous track of a biped, either walking or running, the right and left feet always appearing in their proper places alternately, as represented in the following figures :

The intervals between the foot-prints are subject to some variation, but not more than may be accounted for, by supposing the animal to have quickened its pace, or the reverse. These foot-prints resemble the tracks which have been left by that order of birds called *Grallæ* or *Waders*, on the muddy margin of a lake. Some have three toes distinctly impressed, as in figures 3, 4, and 5, while others have a fourth toe or hallux, as in figures 1 and 2.

Professor Hitchcock proposes to include all the different impressions discovered by him under the generic term ORNITHICHNITES, signifying *stony bird-tracks*. These he subdivides into 1. *Pachydactyli*, or thick-toed; 2. *Leptodactyli*, or slender-toed. In the former subdivision, the toes are of almost equal thickness through their whole extent, except that they are somewhat tuberous; and they terminate rather abruptly; not, however, without a claw. In the latter, the toes are much narrower, and less thick, with an unequal span.

SUBDIVISION L.—PACHYDACTYLI.

4

O. Giganteus.—The length of the foot, exclusive of the claws, is fifteen inches ; it has three toes, and in one specimen the claw is at least two inches long, and even then a part of it appears to be wanting. In general, however, it is not more than one inch, but seems to be broken off. The whole foot, consequently, is sixteen or seventeen inches! The length of the successive steps, varying from four to six feet! The toes are somewhat tuberculated ; the inner one, in some specimens, distinctly exhibiting two protuberances, and the middle one three, although less obviously. The average thickness of the toes, one inch and one fourth ; and their breadth two inches.

The ordinary step is supposed to be four feet, as most of the foot-prints were that distance apart ; six tracks in succession of this species being found in one spot of this average ; and the greater distance of six feet, seems to indicate a rapid movement of the animal. From the length of step, indicated in the sandstone, the bird must have been about twice the size of an Ostrich, or its head elevated from 12 to 15 feet above the ground. The length of the leg of the African Ostrich is about four feet, and that of the foot ten inches. The Professor says, " Incredible almost as this description may seem, the specimens which I have obtained of this enormous species are, nevertheless, more satisfactory perhaps than any other species. The whole cavity made originally in the mud by the foot of the bird has been filled by a siliceous concretion, differing somewhat from the surrounding rock ; so that the latter may be in a good measure detached, and the former be left standing out very naturally from the rock —presenting, in fact, a petrifaction of the entire foot." The foot-marks of four individuals all pointing in one direction, having been noticed here, shows that they must have moved along nearly together, rendering it probable that this species was gregarious. Found at Mount Tom Quarry.

The other species of this subdivision is the *O. tuberosus*, which is much smaller than the former, being only from seven to eight inches in length ; the length of step measuring from twenty-four to thirty-three inches. Also from Mount Tom. In a quarry to the east of this locality, the prints of another individual were found, which measured only four inches. They agree with those of the *tuberosus*, and may be the young of that bird, but are in the meantime termed *O. dubius.*

SUBDIVISION IL—LEPTODACTYLI.

5

O. ingens.—This bird has also been three-toed ; the foot measuring from fifteen to sixteen inches, exclusive of the hairy appendage, attached to the heel, or hind part of the foot. No visible impressions of the plumlet have been noticed in any of the specimens which were found. The toes are much narrower than in the O. giganteus ; they are quite divaricate, and gradually taper to a point, at a few inches behind the heel. The most perfect specimen exhibits a depression nearly an inch deep, and several inches across ; the anterior slopes to which, in the rear, appear as if large bristles had been impressed upon the mud. This leads to the probability that the bird possessed a sort of knobbed heel covered with wiry feathers, which sunk into the mud when the track was deep. The impression of the bristles extends backwards from the heel at least eight or nine inches, so that the whole length of the foot-print is not less than two feet. The length of the step appears to have been six feet.

The rock on which this species of track appears is composed of a fine blue mud, such as is now common in ponds and estuaries ; and where the bird trod upon it, in some cases, it seems that the mud was crowded upwards, forming a ridge round the track in front several inches in height. " Indeed," Professor Hitchcock remarks, " I hesitate not to say, that the impression made on the mud appears to have been nearly as deep, indicating a pressure nearly as great as if an Elephant had passed over it. I could not persuade myself, until evidence became perfectly irresistible, that I was examining merely the track of a bird."

In the quarry at Horse Race, impressions were discovered exactly similar to those of the *ingens*, but only twelve inches in the length of foot ; and the step measuring from forty to forty-five inches. This the Professor inclines to consider a distinct species, but, in the meantime, has named it a variety of *ingens.*

O. diversus.—Three-toed, with a hairy appendage in the rear ; length of the foot, exclusive of the hairy appendage, from two to six inches ; length of the step varying from eight to twenty-one inches. A great variety of specimens are included under this specific character, from want of definite lines of demarcation. There is a variety named *clarus*, in which the foot-print, exclusive of the hairy appendage, is from four to six inches; the appendage being from two to three inches long; the toes somewhat approximate, accuminate—the inner being shorter than the outer one. Step from eighteen to twenty-five inches. Found in the south-west part of Montague; also at Horse Race.

O. tetradactylus. Fig. 1.—Length of the foot, exclusive of the hind toe, from two and a half to three and a half inches. Toes divaricate; the hind one turned inward, so as to be nearly in the line of the outer toe, prolonged backward. A space, however, usually remains, between the heel and the hind toe, as if the insertion were higher on the leg than the other toes, and its direction obliquely downwards. Length of the step ten or twelve inches.

O. palmatus. Fig. 2.—Four-toed, all directed forward; the fourth toe being short, proceeding from the inner part of the foot; the heel is broad; foot from two and a half to three inches long. Length of step eight inches. Discovered at Horse Race.

O. minimus. Fig. 3.—With three widely-spreading toes, nearly of equal length; feet measuring from half an inch to an inch and a half long; step from three to five inches.

On comparing the descriptions which we have given of the species, it is interesting to observe how the length of the step increases in proportion to the size of the foot; from the *ingens*, having a foot sixteen inches in length, with a stride at least four feet, to the *minimus*, whose foot-print is but one inch long, and its step from three to five inches, which are indicated in the three specimens given in our first cut, drawn to a comparative scale. All these tracks appear to have been made on the margin of shallow water that was subject to changes of level, and in which sediments of sand and mud were alternately deposited. And the length of the legs, which must be inferred from the distance of the foot-prints from each other, was well adapted for wading in such situations.

The sandstone in this valley, where the Ornithichnites occur, extends nearly one hundred miles from New Haven, in Connecticut, to the north line of Massachusetts, varying in width from eight to twenty-four miles. It is divided by one or two ridges of greenstone, protruded through the sandstone, and running nearly north and south. The strata of the sandstone have a general easterly dip, varying from 5 to 30 degrees; so that the lowest or oldest portions of the sandstone lie along the western side of the valley. These lower strata consist, for the most part, of thick layers of red sandstone, not much diversified in appearance. But the upper strata, that is, those on the easterly side of the greenstone ranges, consist of slaty sandstones, red and gray conglomerated sandstones, very coarse conglomerates, shale, and perhaps red marl, with occasional beds of fetid limestone. Indeed, the red sandstone of Hartford is decidedly marly, as it effervesces with acids, and even contains numerous veins of calcspar.

The important discovery of the gigantic foot-prints of the Ornithichnite, which we have first described, is a further proof that there were animals inhabiting the former world of much larger dimensions than any existing races. The Mammoth may be mentioned as the largest of quadrupeds; and the *O. giganteus* seems to have been the largest of birds. In Professor Hitchcock's "Report on the Geology of Massachusetts," published some time ago, he shows that other organic beings, that must have been contemporaries with these immense birds, were their compeers in size; for we find in that work a description of a Sea-Fan (*Gorgonia Jacksoni*), found in the new red sandstone of West Springfield, that has been uncovered, without reaching its limits, eighteen feet in length, and four feet in width! Indeed, the colossal bulk of these birds is in accordance with the early history of organic life in every part of our globe. The much higher temperature that then prevailed seems to have been favourable to a gigantic development of every form of life.

METEOROLOGY.

On the Meteorology of the Neighbourhood of Edinburgh.—The diffusion of heat and moisture over the surface of the earth is a subject full of interest; but it has not yet received that strict attention which its importance demands. We still require a great many local facts and observations to enable us to form general and correct conclusions. The regular prevalence of the equatorial currents, by which a stream of hot and moist air is constantly ascending from the earth's surface, and passing by various modifications to both poles, from whence a drier and colder current flows towards the equator, to occupy the place of the other, seems to be the grand agency employed by nature to warm, and refresh with moisture, every region of the globe. With our insular situation and temperate latitude, these currents are not so constant or so distinctly marked; yet even here we can trace them generally prevailing.

Thus, for eight months in the year, we have either south-west or west winds almost constantly prevailing; while, for the other four months, we have either east or north winds, alternating with north-west and south-west winds.

During our summer and autumn months, the tropical current blows very generally from the southwards: the great summer heat of the northern continent of Europe, by rarifying the air in that direction, causes a current to blow towards it.

This effect is sometimes retarded and interrupted by causes that diminish the heat of the continent—such as long and severe winter snows, and the late thawing of ice. Under these circumstances, we have moist and cold summer weather.

In winter, the low arctic current, blowing towards the tropics, occasionally has the ascendancy, especially when the sun is in the southern ecliptic. And, in spring, an easterly wind prevails generally for six or eight weeks, till the continental land becomes heated, and then the east wind yields to the supremacy of the south and south-west.

With these general principles the local phenomena in the atmosphere in this neighbourhood appear to coincide. And the differences of temperature and moisture between the east and west coasts of Scotland seem very marked.

To-day the summit of Arthur's Seat is enveloped in a dense cloud of vapour, the wind blowing from the southward. This cloud is stationary, and has been so for the last hour; nor does the vapour pass many yards beyond the mountain. The cause of this phenomenon is well understood: the current of wind is loaded with vapour, suspended in it by its high temperature. The temperature of Arthur's Seat is considerably lower: whenever a portion of this current, then, comes in contact with the hill, it is suddenly cooled, deposits its vapour, and each successive portion of the current does the same. The mist does not extend beyond the hill, because the condensing cold is no longer present, and any remaining portions of deposited vapour carried along into the general mass are instantly redissolved.

More diffused fogs are not uncommon in this neighbourhood, especially in the spring months, when there happen to be two currents of air in the atmosphere. An east wind prevails annually here for the greater part of the months of April and May, usually setting in about the middle of April, and continuing, more or less, till the end of May. Blowing over the long tract of north-eastern continent, it is exceedingly chill and deficient in moisture. When it arrives on our insular shores, it gains an accession of temperature, and, with this, its capacity for absorbing moisture is prodigiously increased. This accounts for its chilling and arid effect on the whole vegetable and animal creation, and on the face of the soil generally; it greedily abstracts both their heat and moisture. A cold current of wind, thus sweeping along, and meeting with a warmer mass loaded with moisture, suddenly causes a deposition of vapour; but this cold current is so close to the surface of the earth as to prevent the usual formation of rain—a dense fog usually is the result.

These adverse currents were very perceptible in June 1832, and they appeared not only to cause fogs, but also several tremendous thunder-storms.

For some time previous to the 1st of June, the wind had prevailed from the east. On the 1st and 2d of June, an atmospheric current, passing from south-west to north east, was distinctly indicated by the rapid and continued motions of the clouds in that direction, at a moderate height in the air. On the 3d, the wind was still from the east—the atmosphere was remarkably transparent, and there was a dry scorching heat. In the evening, a dense fog from the eastward suddenly came on, which quickly passed into rain; and, on the morning of the 4th, at seven A.M., loud peals of thunder were heard, accompanied by vivid flashes of lightning. For eight days afterwards the weather continued chill, foggy, and rainy, with wind east-north-east. On the 13th, the weather became milder, the lower stratum of air blew from the eastward, while, above, dense masses of nimbi were seen rolling from south-west to north-east. At seven P.M., a tremendous thunder-storm came on, passing directly over the city from the south-east to north-east. In several cases the report followed the vivid flash instantaneously, with a sharp piercing sound. The electric fluid struck two different parts of the city, doing partial injury. We had then mild cloudy weather, with wind south-west, till the 20th. The morning of that day was clear and cloudless—wind unsteady, easterly; at twelve noon, a thunder-storm in the west passed distantly to the northward of the city. A dense thunder cloud appeared in the western horizon, arched, and tolerably well defined above, while below it was ragged, and constantly joined by numerous accessory clouds. A deep flame-coloured lurid light occupied the whole circular space, from the cloud to the horizon; and the dark accessory clouds shooting along this space, to join the main cloud, had a singular and highly interesting appearance. This lurid light frequently accompanies thunder clouds; whether it be caused by the electric matter, or is the effect of a modification of the sun's light falling on these clouds, is uncertain. A tortuous mass of clouds, something like a tornado, and producing similar effects, is said to have been witnessed to the westward of Edinburgh during this thunder-storm. After this storm, the weather continued throughout the month steady and genial. That these different currents of the atmosphere were connected with the electric phenomena just mentioned, there is every reason for supposing. Most probably, the higher current of heated air, moving from the south, and saturated with aqueous vapour, was positively electrified, and thus coming in contact with the cold, dry, and negatively electrified east wind, would be suddenly condensed, and an equilibrium of the electric fluid would be effected.

Thunder-storms, of any great magnitude, have been rather rare in this neighbourhood for the five years preceding 1831. In June of that year, two violent storms passed over this city. The night previous to the first great storm, flashes of diffused sheet lightning irradiated the heavens for upwards of an hour, accompanied by low peals of thunder. Next morning was calm and serene, but exceedingly sultry. About two P.M., thick masses of clouds began to rise from south-west, and float along the ridge of the Pentland Hills. Gradually the dense mass accumulated over the city—vivid flashes of forked lightning, succeeded instantaneously by a loud report, occurred at short intervals, and rain descended in torrents. The peculiar lurid glare was very conspicuous to the north-eastward, as the storm gradually passed in that direction. Although the lightning thus flashed almost close to the tall spires and houses of the city, yet no injury was done. A vessel, however, was struck in Leith Roads, a situation lower by several hundred feet, and her masts were shattered to pieces. That the lightning did not strike on the higher and exposed parts of the city is wonderful; and, indeed, it is remarkable how very rarely such accidents occur in large towns, while they are so frequent in the open country, and especially where there are trees and water. It, perhaps, may be accounted for thus—that the air overhanging cities is highly rarified from the numerous fires, is free from moisture, and, consequently, can act but as a very imperfect conductor of electricity.

Since 1831, we have annually, about May, June, and July, experienced several severe thunder-storms. In all of these which we have witnessed in this locality, the clouds first make their appearance in dense masses to the southwards, coming along either in the direction of the Lammermoor or Pentland Hills, while the sky to the east and north is generally cloudless and transparent. These electric clouds, becoming larger and denser as they proceed, pass onwards with a slow motion to the north or east.

The exposure of Edinburghshire, on the eastern coast of Scotland, renders it peculiarly subject to easterly winds. On the more inland parts of the country, and especially on the west coast, these winds are not so severely felt; for the space of country which they blow over in their progress serves to mitigate their severity. The climate on the eastern coast, however, is less liable to excess of humidity. The annual average fall of rain is, at Dalkeith, twenty-five inches—Edinburgh, twenty-three; while at Largs, near Glasgow, it is forty-three and a half inches; at Dumfries, thirty-six; at Castle Toward, Argyleshire, fifty inches. This, too, is easily accounted for from local situation. The west wind, charged with moisture, and generally of elevated temperature, on coming to the colder land air, deposits its humidity very rapidly. The soil of Edinburghshire, too, is of a light porous nature, from the quality of the prevailing strata; and its level aspect, removed from the immediate contact of high mountain ranges, tends to preserve it from excess of humidity. The mean annual temperature is 47° 31′. Snow seldom lies for any length of time, from its proximity to the ocean. The winter temperature, then, like most parts of Scotland, is fully milder than that of England; while its summer heat is somewhat less. It has been calculated that the mean annual temperature of Edinburgh is 3° less than that of London.

That the nature of the prevailing strata, and the position of the country as regards mountain ranges, has a very material effect on the climate, is evinced from the country around Aberdeen, and the higher parts of Banffshire. In this part of Scotland the formation is primary, consisting of granite and its accompanying rocks. From the compact, impenetrable nature of this basis, the rain water cannot sink downwards by fissures and porosities, but remains on the surface till it again evaporates, or collects, by numerous superficial rills, into rivers. Thus, the soil is continually moist and cold, and so is the atmosphere around. The long range of Grampian mountains that intersect the north of Scotland also terminate in this quarter; and their summits continually conduct along the atmospheric vapour, so that an additional quantity is deposited on the soil. At Aberdeen, the annual fall of rain during the year 1829 was 28.66 inches, in 1830, 30.60 inches. The mean temperature of the same place, from 1823 to 1830, was 47° 61′.

The country immediately to the north (Morayshire) has a deep sandstone basis, which is light and porous. It is also removed from the line of mountain ridges; and, accordingly, its climate is found to be much drier and warmer. The annual average fall of rain at Fochabers, taking the mean of seven years, is twenty-six inches. Mean annual temperature, at same place, 47°; at Inverness, mean temperature 48°. At Elgin, the mean temperature for 1835 was 47° 6′; the mean fall of rain twenty-four inches.

The northern portion of Edinburghshire lies low, the general range being from one to two hundred feet above the sea level; to the south it gradually rises to five hundred and eight hundred feet. The extreme elevation of the Pentlands is 1879 feet.

The meridian, as calculated at the Observatory on the Calton Hill, is 3° 11′ 4″. W. L.—*Rhind's Excursions, illustrative of the Geology and Natural History of the Environs of Edinburgh.*

Falling or Shooting Stars are appearances every where observed. They are probably the effect of hydrogen gas more or less sulphuretted, for phosphorus is too rapidly inflamed by the contact of the air to be capable of reaching so high an elevation. What seems to prove the hydro-sulphuretted origin of these meteors, is the nature of the circumstances by which they are accompanied. These fires, we are assured, often fall to the ground; and nothing is found at the place of their fall but a fetid glutinous matter of a whitish colour, bordering upon yellow. Now, we know that sulphuretted hydrogen gas holds sulphur in solution; that the hydrogen and the sulphur do not burn at the same moment; that, consequently, the sulphurous part may be precipitated to the earth, whilst the hydrogen, mixed with the oxygen of the air, is kindled by a slight electric spark.

MISCELLANEOUS.

The Hot Blast.—Lately there has been introduced into the great Iron Works of Carron, and other founderies of Scotland, a method of smelting iron by means of heated air, which has produced extraordinary changes in the manufacture of that useful metal, well worthy the attention of English iron smelters, and also the minor founderies which use iron. By the old method, for every ton of iron smelted, it required on an average about eight tons of coal, or a corresponding quantity of coke; while by the "hot blast" *two tons* are sufficient. If it shall be found that iron smelted by this system is not subject to greater deterioration or brittleness than that procured by the old process, the new mode will prove one of the greatest discoveries in modern science, and highly important in a national point of view; for it is said that the actual saving in the quantity of fuel required will be found in practice to be no less than *three hundred* per cent.

Subterraneous Sounds in Granite Rocks.—Humboldt was informed by credible evidences, that subterraneous sounds, like those of an organ, are heard towards sunrise, by those who sleep upon granite rocks on the banks of the Oroonoko. He supposes them to arise from the difference of temperature between the external air, and the air in the narrow and deep crevices of the shelves of rocks. During the day, these crevices are heated to 48 or 50 degrees. The temperature of their surface was often 39 degrees, when that of the air was only 28 degrees. Now, as this difference of temperature will be a maximum about sunrise, the current of air issuing from the crevices will produce sounds which may be modified by its impulse against the elastic films of mica that may project into the crevices. MM. Jomard, Jollois, and Devilliers, heard at sunrise, in a monument of granite, placed at the centre of the spot on which the palace of Karak stands, a noise resembling that of a string breaking.

Inflammable Gas.—When Mr Hughes was travelling in Greece he found, not far from Pollina (the ancient Apollonia) in Albania, a desert place, from the fissures of whose surface an empyreumatic vapour arose, which took fire on the application of a taper, and burnt for some time. From ruins, which he noticed near this place, he inferred that they belonged to that oracle described by Dion Cassius, book xii. p. 45. Mineral pitch abounds in the vicinity. In other sacred places of Greece, as at Delphi and Dodona, where the ignorant were deceived by mineral vapours used in their oracular contrivances, these vapours have now totally disappeared. In the heights of Parnassus, where the remains of the Delphic oracles are found, the celebrated *foramina* (where carbonic acid rose from the fissure of the limestone) have been filled up: and in place of the springs with inflammable gas at Dodona, mentioned in Pliny's Natural History, vol. ii. p. 104, there is found at present near Joannina, along with the remains of the temple, simply a marsh.

Peat Mosses of Holland.—Destitute of coals, and without copse-wood, the Dutch have to depend on their *veener* or peat-mosses for fuel. There are two kinds of these, the higher and the lower. The high mosses afford a layer of what is called gray or dry peat. The upper bed of peat is generally about six feet in thickness; it seems to be composed rather of leaves and stems of reedy plants than of heath, or the plants which commonly accompany heath; and fragments of large branches of trees have sometimes been found in it. Beneath this peat a thin blue clay commonly appears, and which, on the peat being removed, forms arable land. The low mosses afford what are called mud-peats, and when these are taken from the inferior layer of such moss, the excavation speedily becomes covered with water. When the under stratum of moss is formed and contains wood, it is called *derry*. Many trunks of trees occur in it; and these uniformly lie with their heads pointing eastward, showing that the storm or debacle which overwhelmed them had come from the west. Some of the timber, oak in particular, remains sound, so that it can be used in carpentry; but it is of a dark colour, as if stained with ink: thus proving the amazing durability of oak. There is a law in Holland against digging through this derry in the lowest parts of the country, much water being found to ooze in the sand below, and to be repressed by the compact layer of wood moss.

Ignis Fatui, arising from the development of phosphoretted hydrogen, are necessarily soon extinguished; a succession of these fires will therefore appear to the spectator to be one single flame, which moves with rapidity from place to place when we attempt to approach it. The air driven on before us forces the lambent flame to recede. There are other similar fires, which appear to be immoveable when viewed from a particular spot. There was one near Rettwick in Sweden, which was supposed to issue from the mouth of a dragon that kept watch over some hidden treasures. A simple miner ventured to sink a shaft, which discovered a cavern filled with sulphurous pyrites and petroleum, the combustion of which had occasioned the phenomenon. —(Memoires de l'Académie de Stockholm, 1740).

Singular Heat developed in the Fusion of Tin and Platinum.—Mr Fox of Falmouth has found, that a very extraordinary degree of heat is developed by fusing together platinum and tin in the following manner:—If a small piece of Tin-foil be wrapped in a piece of Platinum-foil of the same size, and exposed upon charcoal to the action of the blowpipe, the union of the two metals is indicated by a rapid whirling, and by an extreme brilliancy in the light which is emitted. If the globule thus melted be allowed to drop into a basin of water, it remains for some time red-hot at the bottom; and such is the intensity of the heat, that it melts and carries off the glaze of the basin from the part on which it happens to fall.

Detonating Mud in South America.—Don Carlos del Pozo has discovered in the Llanos of Monac, at the bottom of the Onebrada de Moroturo, a stratum of clayey earth, which inflames spontaneously when slightly moistened, and exposed for a long time to the rays of the tropical sun. The detonation of this muddy substance is very violent. It is of a black colour, soils the fingers, and emits a strong smell of sulphur.

Hot Springs of La Trinchera.—The hot springs of La Trinchera are situate three leagues from Valencia, and form a rivulet, which, in the seasons of the greatest drought, is *two feet deep, and eighteen feet wide.* Their temperature is 93.3 centigrade, from which it appears that they are the hottest in the world, excepting only those of Urijins in Japan, which are asserted to be pure water at the temperature of 100 degrees. Eggs plunged in the Trinchera springs are boiled in four minutes. At the distance of forty feet from them, other springs are found entirely cold. The hot and the cold streams run parallel to each other; and the natives obtain baths of any given temperature, by digging a hole between the two currents.

Coral Reefs of the Pacific Ocean.—These are sometimes very extensive. The inhabitants of Disappointment Islands and those of Duff's Group visit each other by passing over long lines of reefs from island to island, a distance of 600 miles. While on their route, they appear like troops marching upon the surface of the ocean.

LIST OF NEW BOOKS ON NATURAL HISTORY, AND THE PHYSICAL SCIENCES.

Rhind's Excursions, illustrative of the Geology and Natural History of the Environs of Edinburgh, royal 18mo, 3s. 6d. boards; Loudon's Encyclopædia of Plants, new edition, 8vo. half-bound, L.3, 13s. 6d.; Butterfly Collector's Yade-Mecum, 3d edition, 5s. boards; The Medico-Botanical Pocket-Book, by G. Spratt, 10s. 6d. cloth; Outlines of a Journey through Arabia-Petræa, the Edom of the Prophecies, by M. Leon De Laborde, 8vo. with 65 plates and map, 18s.; British Song Birds, by Neville Wood, Esq. foolscap 8vo, 7s.; The Ornithologist's Text-Book, being Reviews of Ornithological Works, published from A. D. 1678 to the present day, with an Appendix, discussing various topics of interest connected with Ornithology, by Neville Wood, Esq. 4s. 6d.; Paley's Natural Theology, with Illustrative Notes by Lord Brougham and Sir Charles Bell, 2 vols. post 8vo, with numerous wood cuts, L.1, 1s.; A Journey across the Andes and down the Amazon from Lima to Para, by Lieut. Wm. Smyth, R.N. with 11 plates and maps, 8vo, 12s.; The Horticultural Magazine and Miscellany of Gardening, by Robert Marnoch, No. 1. 6d.; Narrative of an Ascent to the Summit of Mont Blanc, with 3 illustrations by M. Barry, M.D. 8vo, 4s. cloth; Jacquemont's Journey in India, Thibet, Lahore, and Cashmere, No. 12, price 1s. completing the work; Captain Back's Narrative of the Arctic Land Expedition in 1833, 34, 35, 8vo, 30s. cloth; Loudon's Arboretum et Fruticetum Britannicum, No. 21, price 2s. 6d.; G. P. Deshayes Traité Elementaire de la Conchyliologie, liv. 1, 15 fr. cold.; Ad. Brongniart, Histoire des Vegetaux Fossiles, 1836, liv. 10, 13 fr.

Edinburgh: Published for the Proprietors, at their Office, 16, Hanover Street. London: Smith, Elder, and Co., 65, Cornhill. Glasgow and the West of Scotland: John Smith and Son, 70, St Vincent Street; and John Macleod, 20, Argyle Street. Dublin: W. F. Wakeman, 9, D'Olier Street.

THE EDINBURGH PRINTING COMPANY.

THE EDINBURGH

JOURNAL OF NATURAL HISTORY,

AND OF

THE PHYSICAL SCIENCES.

OCTOBER, 1836.

ZOOLOGY.

DESCRIPTION OF THE PLATE—THE HUMMING BIRDS.

Fig. 1. RIVOLI'S HUMMING BIRD *(Trochilus Rivolii,)*.—This beautiful species is a native of Mexico, and the specimen from which this drawing was made is in the collection of the Duke of Rivoli, who possesses one of the finest private cabinets of birds in Europe. Nothing can exceed the splendid display of colours exhibited by its head and throat, the beryl and ruby vyeing with each other in splendour; and these are beautifully set off by the strong black which surrounds the more brilliant hues.

Fig. 2. THE VIOLET-CROWNED HUMMING BIRD *(T. Stephanioides)* inhabits Chili, and, according to Lesson, sips the nectar of the scarlet Loranthus. It will probably be found hereafter, that this bird feeds upon the insects which prey upon the nectar, rather than upon the nectar itself. It penetrates to the North during winter.

Fig. 3. STOKES' HUMMING BIRD *(T. Stokesii)*. — There are few Humming Birds which surpass this in beauty and elegance of form. The tuft of bright cobalt blue, which decorates its crown, adds great beauty to the bird. It was discovered by Captain King on the island of Juan Fernandes.

Fig. 4. THE NORTHERN HUMMING BIRD, MALE *(T. Colubris)*. Fig. 5. FEMALE; and 6. THE NEST.—This is the Humming Bird of the United States of America, and along with others lately described by Mr Audubon, are the only species of this numerous tribe which migrate so far north. This species arrives in Louisiana about the 10th of April. It is seldom found in the middle districts before the 15th of March, but is beautiful to watch. A person standing in a garden by the side of a common Althæa in bloom, will be surprised to hear the humming of their wings, and then see the birds themselves within a few feet of him; he will be astonished at the rapidity with which the little creatures rise into the air, and are out of sight and hearing the next moment.

> When morning dawns, and the blest sun again
> Lifts his red glories from the eastern main.
> Sips through our woodbines, wet with glittering dews,
> The flower-fed Humming Bird his round pursues;
> Sips, with inserted tube, the honey'd blooms,
> And chirps his gratitude, as round he roams;
> While richest roses, though in crimson drest,
> Shrink from the splendour of his gorgeous breast.
> What heavenly tints in mingling radiance fly!
> Each rapid movement gives a different dye;
> Like scales of burnish'd gold they dazzling show,
> Now sink to shade—now like a furnace glow!

We are assured by Mr Audubon, that this species at least principally lives upon insects, which it seeks out diligently in the nectarium of the flowers, where a great number of insects are always to be found, attracted by the honey. These it rapidly abstracts with its tongue.

Fig. 7. THE CRESTED HUMMING BIRD, MALE *(T. Cristatus)*. Fig. 8. FEMALE. —This pretty little species is a native of the islands of Martinique and Trinité. Its breast is emerald green with iridescent blue reflections. The female is devoid of a crest.

Fig. 9. THE PURPLE HUMMING BIRD *(T. Caligena)*.—Mexico is the native country of this species, where it was first discovered by M. F. Prévost. It ranks among the larger species; and although having no great variety of colours in its plumage, is, however, a very beautiful bird.

Fig. 10. WAGLER'S HUMMING BIRD *(T. Waglerii)*.—The remarkable form of the crest of this species distinguishes it from all others of its tribe; and its elegantly graduated tail gives it a very handsome aspect. It is a native of the warmer parts of Brazil.

Fig. 11. THE HORNED HUMMING BIRD, MALE *(T. Cornutus)*.—This singularly beautiful bird is a native of the elevated Compos-Geraes of Brazil.

Fig. 12. THE HALF-TAILED HUMMING BIRD *(T. Enicurus)* is remarkable on account of having only six quill feathers in its tail. It inhabits Brazil.

DESCRIPTION OF THE PLATE—THE TITMICE.

THIS is a small but sprightly race of Birds, possessed of much courage and strength. Their food consists of seeds, fruit, and insects, and a few of them will eat flesh. Such are their pugnacious habits, that they will assault birds twice their own size; and it is said that they direct their aim chiefly at the eyes. They are very prolific, laying about fifteen eggs at a time. Their voice is harsh and unpleasant.

Fig. 1. THE BLACK-CAPT TITMOUSE *(Parus atricapillus)*.—This is a resident species of the United States of America. In summer they confine themselves to deep and woody recesses, and in winter approach the cultivated fields, and are even seen frequently in farm-yards. They are very sociable in their habits, and are to be met feeding along with the Nuthatch, Creeper, and small Woodpecker. Their principal food consists of the seeds of pines, and those of the sunflower. They also feed on insects.

Fig. 2. THE CANADIAN TITMOUSE *(P. Hudsonicus)* was discovered by Audubon; it extends its range from Canada to the southern limits of the United States. It is a nimble and lively bird, seeking insect food with great perseverance and adroitness.

Fig. 3. THE COLE TITMOUSE *(P. ater)*.—This is a British species, about four inches in length, and its weight only two drachms and a quarter. It abounds in the pine forests of Scotland, feeding on insects and their larva.

Fig. 4. THE CRESTED TITMOUSE *(P. bicolor)*.—This species is found widely diffused over the whole United States. It is a very noisy bird, and remarkable for the variety of its notes; sometimes uttering a squeak like that of a mouse, and at others whistling aloud as if calling a dog, which it will continue for half an hour at a time. It feeds on insects and their larva.

Fig. 5. THE AZURE TITMOUSE *(P. cyanus)*.—This beautiful bird is said by Latham to be a native of India, while Pallas asserts that it is to be found in the north of Europe, extending its range as far as Siberia.

Fig. 6. SMITH'S TITMOUSE *(P. Smithii)*.—This bird was found by Dr Smith in Southern Africa, but nothing is known of its habits.

Figs. 7. and 8. THE PENDULINE TITMOUSE, MALE AND FEMALE *(P. pendulinus)*.—It is an inhabitant of Europe, and is found in Italy, France, Poland, Russia, and Siberia; it frequents marshy grounds, and builds a pendulous nest.

Fig. 9. THE AFRICAN TITMOUSE *(P. Afer)*.—This is one of the largest birds of its tribe, and lives almost exclusively on insect prey, which it searches for among moss and decayed bark of trees with much avidity, pulling off large pieces of bark and moss with apparent ease.

FALCULIA PALLIATA.—A new bird, belonging to the *Passeres*, and allied to the *Upupa*, has been found at Madagascar, by M. Goudot, and forms the type of a new and remarkable genus. The beak is very long, arched, compressed, or flattened, like a blade, and may be compared to that of a small scythe. The nostrils, placed at the base of the beak, and pierced laterally, are not covered by the anterior feathers of the head. The wings, which in length reach the middle of the tail, according to the nomenclature of M. Isidore Geoffroy, belong to the type called by him *eurolitus* —that is, having the fourth and fifth remiges the longest of all. The first, like that of the Hoopoes, is extremely short, and nearly useless in flight. The tail is square, and composed of twelve quills, the externals of which have their stems prolonged, in a very slight degree, beyond the barbs. The feet have three toes directed forwards, and a fourth backwards. All are long, thick, and furnished with curved talons, enlarged at the base by a thick membrane, which has some affinity with that of the Grallæ. The only species now known has the head, the neck, and the under part of the body, white; the back, wings, and tail, of a greenish black, with metallic lights. M. Isidore Geoffroy has named it *Falculia palliata*. It lives on the borders of streams, feeds on small aquatic insects, and the organic remains found in mud.

COCHINEAL.—In the Volume of the "Memoirs of the Academy of Sciences at St Petersburgh," which has just appeared, there is an account given of the Cochineal of Armenia. The insect producing this dye has lately been found to exist in the marshy spots which are scattered through the Valley of Araxa, feeding on the root of the plant called *Poa Pungens*. It is said to differ from the Mexican Cochineal *(Coccus cacti)* in having a greater number of joints in the antennæ, and in its fore feet being shorter, which are adapted for hollowing out the soil. On the termination of the abdomen are numerous filaments, while the Cacti has but two. M. Hamel, the principal author of the above paper, endeavours to establish, that the red colour so often spoken of in Scripture, and also by many ancient writers, was the produce of the Armenian Coccus.

ACCOUNT OF A PRETERNATURAL GROWTH OF THE INCISORY TEETH OF A RABBIT.—In the class RODENTIA of Cuvier, of which the Rabbit is a member, there not unfrequently occurs an extraordinary development of the incisory or cutting teeth. It is a well established point in physiology, that these teeth, like the tusks of the Elephant, are in a constant state of growth, and that they emanate from long roots nearly equal in length to the jaw; they curve backwards under the molar teeth, extending in some instances as far back as the coronoid process. Owing to this beautiful adaptation of Nature to the habits of the tribe, there is a constant, gradual advancement of the interior part of the teeth, to supply that portion worn down by friction, while the animal is feeding or gnawing substances—for which the Rodentia have a strong propensity. Under ordinary circumstances, this gradual increase is so admirably regulated, that the cutting edges of the two pair of incisory teeth uniformly preserve the same relative situation.

The above figure represents the head of a wild Rabbit, which is preserved in the private museum of Mr Robert Frazer, jeweller, 17, St Andrew's Street, Edinburgh; with an extraordinary elongation above the gums, of both the upper and under incisors, the former measuring an inch and five-eighths, and reaching considerably above the nostrils, while the latter is seven-eighths of an inch in length, and very much incurved, so much so, that their points would nearly reach the palate when the mouth is closed. The under incisors are also considerably bent, becoming gradually thinner and more depressed towards their points, where they are divergent, the inner sides being nearly a quarter of an inch apart at the tips. Their ordinary length in the wild Rabbit is about a quarter of an inch. Instances of the same kind have been noticed before, in Plott's History of Staffordshire; in Morton's Natural History of Northamptonshire; and in Loudon's Magazine, where an account of a lusus of this kind is given by the Rev. L. Jenyns, the object of which is preserved in the Museum of the Cambridge Philosophical Society.

Cuvier says, that the prismatic form of the cutting teeth occasions them to grow from the root as fast as they wear away at the edge; and this tendency to increase in length is so powerful, that if one of them be lost or broken, the opposite one in the other jaw having nothing to oppose or communicate, becomes developed to a monstrous extent. The elongation of the teeth in the specimen now before us, could not have been occasioned by any accident of this kind, as the development is excessive in both upper and under teeth, while none of them evince the slightest appearance of having been fractured. Therefore some other mode of accounting for this instance must be sought for, and we would rather attribute it to the unequal action of the jaws, in the under one not being exactly opposed to the upper.

Mr Jenyns supposes that this preternatural elongation may arise either from the food being too soft, or too rapid a secretion of the osseous matter which composes the teeth, or a derangement of the under jaw, produced from a dislocation or some other cause; and to one or other of these he attributes two of the cases which fell under his own observation. In the case of the rabbit's head preserved in the Cambridge Philosophical Society, which, so far as the length and disposition of the lower incisors are concerned, agrees very nearly with our case, he considers that it was occasioned by too rapid a secretion of the osseous matter; and in the second case, both upper and under incisors were preternaturally elongated; "but then," says he, "in this instance there was such an irregularity in their mode of growth, that we may perhaps find a better explanation of the anomaly in some derangement of the jaws, the result either of natural constitution or of accidental injury. Whatever this might have been (for I regret that this rabbit was not preserved, and an examination made of the jaws at the time), the effect was that of causing the lower pair of incisors, when viewed together, to assume the shape and appearance of the letter V, diverging from one another at the surface of the gum, and extending in opposite directions, to the length of nearly an inch and a half. The degree of divergency observed in the upper pair was nearly as great as this in the lower, and their length about the same; but their curvature very much greater; as indeed would naturally result from the greater bend of that portion of the jaw in which these incisers are formed. In this instance, the portion without the gums had completed three parts of an exact circle, and their cutting edges were in close contact with the roof of the mouth."

Both of the rabbits above referred to, when captured, were nearly starved to death from their incapacity to eat their usual food. In that first noticed, the animal seemed to exist solely by means of the small quantity of food which it could nip by the lips at the sides of the mouth, which exhibited marks of its having been used for that purpose. Our specimen has no such appearance. In the second case mentioned by Jenyns, the poor animal was unable to close its mouth, from the curve of the upper incisors pressing upon the tongue. These two were found in the neighbourhood of Cambridge.

Mr Jenyns mentions a third individual, from Lincolnshire, one of whose incisory teeth was still longer than any of the others mentioned, which had grown into the palate, and re-entered that portion of the jaw from whence it sprung, which appeared to have been produced by some local disease, affecting, in the first place, that single tooth, which was much twisted in its direction.

The specimen which has come under our notice was found dead from starvation in the rabbit-warren at Leven Links, Fifeshire, Scotland; and exhibited signs of great emaciation, from its inability to feed.

THE CROW AND RAT.—In the spring of the year 1834, while a person was crossing a field, in an elevated and retired situation, in the parish of Kirk-Marown, Isle of Man, he perceived a Crow flying at a short distance from him which attracted his attention. On account of the unusual noise which it made, and while watching its strange motions, he was able to perceive that it had some object suspended from its bill, which dropped in a few seconds, while the bird was almost flying over him. He immediately hastened towards it, and found it to be a young Rat more than half-grown. It was still alive, but somewhat stunned by the fall. This happened at the time that the young Crows were in the nest, and doubtless was intended for food to them. The particular noise made by the Crow must have been occasioned by the efforts of the Rat to disengage himself from his aërial foe.—(Communicated by a " Young Naturalist," Birmingham, August 1836).

BOTANY.

ON THE SHAMROCK OF IRELAND.—Mr Bicheno has laid before the Linnean Society a paper " on the plant intended by the Shamrock of Ireland;" in which he attempted to prove, by botanical, historical, and etymological evidence, that the original plant was not the white clover which is now employed as the national emblem. He stated, that it would seem a condition at-least suitable, if not necessary, to a national emblem, that it should be something familiar to the people—and familiar, too, at that season when the national feast is celebrated. Thus, the Welsh have given the leek to St David, being a favorite oleraceous herb, and the only green thing they could find on the first of March;—the Scotch, on the other hand, whose feast is in November, have adopted the thistle. The white clover is not fully expanded on St Patrick's day, and wild specimens of it could hardly be obtained at this season. Besides, it was probably, nay, almost certainly, a plant of uncommon occurrence in Ireland during its early history, having been introduced into that country in the middle of the seventeenth century, and made common by cultivation. Several old authors affirm, that the shamrock was eaten by the Irish; one of which, who went over to Ireland in the sixteenth century, says it was eaten, and that it was a sour plant. The name Shamrock is common also to several trefoils, both in the Irish and Gaelic languages. Now, the clover could not have been eaten, and it is not sour. Taking, therefore, all the conditions requisite, they are only found in the Wood-sorrel, oxalis acetosella. It is an early spring plant; it was, and is, abundant in Ireland; it is a trefoil; it is called Sham-roy by the old herbalists, and it is sour; while its beauty might entitle it to the distinction of being the national emblem. The substitution of one for the other has been occasioned by cultivation, which made the Wood-sorrel less plentiful, and the Dutch clover abundant.

MINERALOGY.

THE DIAMOND.—This precious Stone, in its natural state, is of the form of an octahedron. This may be defined a double four-sided pyramid, in which the lateral planes of the one are set on the lateral planes of the other, which will be better understood by the accompanying figure, being a regular octahedron, wherein the triangular faces are equilateral and equiangular, and, of course, the base of the two pyramids is a square. Diamonds are always found in detached crystals, and are more or less well shaped, as they are pure or otherwise; so that they occur in a variety of forms, of which their primitive one is the basis; the faces are frequently curvilinear; they are also subject to the compound crystallization called the macle. The structure is perfectly lamellar, yielding readily to mechanical division parallel to all the planes of the regular octahedron, thus proving that this is the primitive form. Diamond is the hardest of all substances, and its specific gravity 3.5. When heated, it becomes phosphorescent. The general colour of the Diamond is white, but is found of various tints, red like the Ruby, orange like the Hyacinth, blue like the Sapphire, and green like the Emerald; the last of which is most rare, and of the greatest value when it is of a beautiful tint: the rose, blue, and yellow Diamonds are the next in value. Transparency and brilliancy are the natural and ordinary qualities of the Diamond, which exhibits but a single refraction of the rays of light: some, however, are quite opaque. Diamonds are divided into Oriental and Occidental or Brazilian, the former being the most valuable. Boetius de Boot, in his " History of Gems," written in 1609, conjectured that the Diamond was inflammable. Mr Boyle discovered, in 1673, that when exposed to a high temperature, it gave out acrid vapours, in which a part of it was dissipated. Sir Isaac Newton, who composed his work on Optics in 1675, concluded, from its great refracting power, that it must be combustible, and that it might be an unctuous substance coagulated. But the celebrated Averani, in 1695, in presence of the Grand Duke of Tuscany, and several of the most philosophic men of that time, showed, by concentrating the rays of the sun upon it, that the Diamond was exhaled in vapour and disappeared entirely, while other precious Stones only grew softer. Since that period, however, various chemists have burned Diamonds with as much facility as a piece of iron, wire, or wood, by exposing them on a piece of charcoal placed in the flame of a common lamp, or even a candle, and blown with a current of oxygen gas. At a heat less than the melting point of silver, it gradually dissipates, burns, and combines with nearly the same quantity of oxygen, and forms the same quantity of carbonic acid as charcoal.

Dr Murray of Edinburgh has invented the most simple apparatus which we have seen for exhibiting the combustion of the Diamond. The annexed cut gives a representation of it.

A glass globe is filled with oxygen obtained from oxymuriate of potassa over mercury. A portion of the stem of a tobacco pipe, attached to the curved end of a wire fastened to the cork above, carries the Diamond, fixed in a nidus prepared for it. The Diamond is kindled by the oxy-hydrogen blow-pipe, or a stream of oxygen urged over the flame of a spirit-of-wine lamp, and then conveyed into the globe. When the combustion of the Diamond ceases, lime water is passed over the recipient, and the weight of the carbonate of lime formed and precipitated, indicates the quantity of Diamond consumed. Dr Clark exposed a Diamond of six carats, of an amber colour, to the flame of the gas blow-pipe. It became colourless and transparent—after this it became white, and by continuing the heat, it was entirely volatilized in about three minutes. It has also been ascertained by Guyton, Davy, and others, that although Diamonds, whether Oriental or Occidental, are the hardest of all substances, they yet contain nothing more than pure charcoal or carbon. The extreme hardness and transparency of this carbon furnish a problem which has hitherto baffled the efforts of philosophers to solve satisfactorily. Nevertheless, the knowledge of the inflammability of Diamonds has very considerably reduced their mercantile value, and will probably bring it, like that of all other articles, to the common standard of comparative use.

Diamonds were first brought to Europe from the East Indies, where they are found in various parts. The first mine known there is that of Sumbulpour on the river Goual, which falls into the Ganges. A chain of mountains extending from Cape Comorin to Bengal are the most celebrated for producing Diamonds. But the chief of these are in the kingdom of Golconda, furnishing the greatest quantity and most esteemed Diamonds, especially those of Pastall, which are sent to Calcutta, where they are sorted, sealed up in bags, and conveyed to London. In 1770, there were fifty Diamond workings in the kingdom of Visapour; these furnished more than those of Golconda, but were abandoned, being smaller. They are also found in the Island of Borneo. The total annual value of the Diamonds so collected is said to amount to about L.142,500, exclusive of those which are smuggled. The largest Diamond which has hitherto been found, is that called the Mattan Diamond, from Borneo, and particularly described at page 31 of this Journal. The next largest Oriental Diamond is one, the size of a pigeon's egg, of the weight of 193 carats, equal to about *one ounce two pennyweights* Troy. The Empress Catharine II. of Russia offered L.104,166, 13s. 4d. besides an annuity for life of L.1041, 13s. 6d., which was refused; but was afterwards sold to Catharine's favourite, Count Orloff, for the above sum without the annuity, who presented it to the Empress on her birth-day, in 1772. This gem exceeds the famous Pitt Diamond in size, and is reckoned equal in water. It is now in the sceptre of Russia, and is considered the finest of the kind in Europe. The late Queen of Portugal possessed the next largest Diamond in Europe, but we are not aware whether it is still in the family. The Pitt Diamond weighs 136 carats, or nearly an ounce, it was purchased for 2,500,000 livres, and is now the finest of the crown jewels of France.

Brazil Diamonds.—In the year 1728 it was discovered that Diamonds were to be met with in some branches of the river Das Caravilas, at a considerable elevation, and at Serro de Trio, in the province of Mino Geraes, in the Brazils, belonging to Portugal. The Rio de Janeiro fleet brought home at once 1146 ounces of them. Such a considerable supply would have reduced the value of Diamonds, had not the Government of Portugal laid restrictions on the persons who searched for them. The Diamonds are found in such rivers, the courses of which have been more or less altered; and it seems probable that they are washed down by the torrents from the mountains, as they are found in the greatest plenty after violent storms of rain. They are not, however, confined to the beds of rivers alone, as they have been found in cavities and water-courses on some of the highest mountains in the district. The principal work in the present day is at Mandanga, or the Jigitonhona, a very shallow river, which admits of its waters being dammed out, or diverted from their course.

We learn from Mawe's Travels in the Interior of Brazil, that the Diamond mines of that country are situated due north of the mouth of the Rio de Janeiro. The capital of the district is called Tjuco. The face of the country exhibits in all directions a series of gritstone rocks alternating with *micaceous schistus*, in which numerous rounded quartzose pebbles are embedded, giving the whole the appearance of plum-pudding stone.

In the splendid collection of Mr Hewland, there is a Brazilian Diamond embedded in brown iron ore; another, also, in the same substance, is in the possession of M. Schuch, at Lisbon. Eschwege has, in his own Cabinet, a mass of brown iron ore, in which there is a Diamond in a drusy cavity of a green mineral, conjectured to be arseniate of iron. From these facts he infers that the matrix of Brazil Diamond is brown iron ore. The crust which envelopes the rough Diamond is found thicker on those from Brazil than those from the East Indies, and hence they are easily distinguished in their natural state; but as the most skilful lapidaries may be deceived in them after they are cut, they are accordingly of equal value in trade. This equality, however, is only to be understood as relating to small ones, as most of the Occidental Diamonds beyond four or five carats have blemishes, which are seldom found in the Oriental, and in that case the difference is great. Some Mineralogists are of opinion that the latter are harder and more brilliant than the former; but this opinion is not sanctioned by experiment.

North American Diamonds.—A Diamond weighing one carat and a half was found in the autumn of 1835 in the washings of a stream in North Carolina.

Russian Diamond Mines.—When, in the year 1826, Professor Engelhardt undertook a scientific journey into the Uralian Mountains, he remarked that the sands in the neighbourhood of Kouspra, and those of the Platina mines at Nigny-Toura, strikingly resemble the Brazilian sands in which Diamonds are found. Baron Humboldt, during his residence in that country in 1830, confirmed this resemblance; and examinations having been made according to his advice, a young countryman who was employed in washing the auriferous sand, on the grounds of the Countess Polier, discovered a Diamond on the 20th June 1830, which was nothing inferior to those of Brazil; soon after many others were found superior in weight to the first. Thus Russia has added this source of riches to those of which of late years it has obtained in the form of Gold and Platina Mines, from the Ural chain of mountains; and has proved, that what has hitherto been supposed to be the case is not correct, that Diamonds were only found near the line, and none beyond the tropics.

Professor Jameson is of opinion, that Diamonds continue to be formed even at the present day in some alluvial districts of India; as they have been discovered in alluvial beds of clay, not as a secondary deposit, but as an original one; and, says he, "nothing more seems necessary for the formation of the Diamond in such situations, than time, or other favorable circumstances, for allowing portions of the carbonaceous matter in the soil to be reduced to the adamantine state, and afterwards to coalesce, according to the laws of affinity, into the granular and crystallised form—in short, to form Diamond." The Professor also promulgates another theory for the formation of Diamonds; he says, "a direct appeal to the characters of some woods seems to countenance the idea I some years ago suggested to the Wernerian Society, that vegetables may contain carbonaceous matter approaching to the adamantine state. Certain woods which have not the gritty feel of those that contain silica are uncommonly hard, dark coloured, and take a high polish; these, I conjecture, may be somewhat of the adamantine nature. If this should prove to be the case, it would neither be surprising nor unexpected, that such trees may secrete carbon in the adamantine state, which, on being removed from the influence of the living principle of the plant, would, by the power of affinity, form into true Diamonds, just as the silica secreted from the Bamboo takes the form of opal, and that from teak wood the characters of hornstones."

In support of Professor Jameson's theory, we may mention that Dr Hamilton was informed by the workmen, when he visited the Diamond Mines of Parma, in Bengal, that the generation of Diamonds was always going forward, and that they had as much chance of success in searching earth that had been fourteen or fifteen years unexamined, as in digging what had never been disturbed; and, in fact, he saw them digging up earth which had evidently been before examined, as it was lying in irregular heaps as thrown out after examination. These men are so expert at discovering Diamonds, that they never overlook any during their search.

Sir David Brewster states, that Dr Voysey has shown that the matrix of the Diamonds produced in Southern India is the sandstone breccia of the dog slate formation; and that Captain Franklin has found, that in Bundel Kund, the rocky matrix of the Diamond is situated in sandstone, which he imagines to be the same as the new red sandstone of England; that there are at least four hundred feet of that rock below the lowest Diamond beds, and that there are strong indications of coal underlying the whole mass. Sir D. Brewster, from certain cavities observed in Diamond, and from their effects in polarizing light, is led to conjecture, "that the Diamond originates, like amber, from the consolidation of perhaps vegetable matter, which gradually acquires a crystalline form from the influence of time, and the slow action of corpuscular forces."

The usual method of seeking Diamonds is by throwing the stones and rubbish with which they are supposed to be associated into a cistern, full of water, having a cock and plug at the bottom. The lumps are then broken, and the muddy water drawn off till the stones are washed clean. When the sun shines bright, the sand and stones which remain in the cistern are carefully examined. In this business the workmen are so expert, that the smallest crystal cannot escape them.

The cutting and polishing of Diamonds are performed by first cleaving them in the direction of their lamellæ, and then rubbing them against each other, and the powder or dust thus disengaged serves to grind and polish them; these latter operations are performed by the aid of a mill, which turns a wheel of soft iron, sprinkled over with Diamond-dust, mixed with olive oil. The same dust, well ground and diluted with water and vinegar, is used for sawing Diamonds, which is effected by means of an iron or brass wire as fine as a hair.

GEOLOGY.

On the Excavation of Valleys.—There is much diversity of opinion among Geologists on the subject of the excavation of valleys, and of the effects produced by river currents in modifying the form of the solid parts of the earth; and several distinguished men have lately turned their attention to this important and interesting subject. Professor Sedgwick seems to have formed opinions on this subject, which approach near to the true theory. Messrs Lyell and Murchison discussed this subject in a Memoir on certain portions of the volcanic regions of Central France. Their opinions accord with the views of Montrosier, Scrope, and some other writers, who conceive that the existing rivers have, by a long continued erosion, eaten out deep gorges, not only through currents of basaltic lava, which have flowed through the existing valleys, but also through solid rocks of subjacent gneiss. They further seem to prove, that no great denuding wave or mass of water, lifted by supernatural force above its ordinary level, could have assisted in forming such denudations; for the country is still studded with domes of incoherent matter, the remnants of former craters, from which may be traced continuously streams of lava, intersected in the course of the rivers by these deep gorges—the gauges and tests of the erosive power of running water during times comparatively recent.

Mr Conybeare proves that, within the records of history, the river Thames has had no erosive power on the valley, nor produced any effect on the general features of the country through which it flows, and that the propelling force of its waters is not now, and never could have been, adequate to the transport of the boulder stones, which lie scattered on the sides and summits of the chains of hills through which it has found a passage; that much of the water-worn gravel, which has been drifted through the breaches opened in the sinuous line of its channel, is composed of rocks not found within the limits of its basin; and that the form of the country is often the very reverse of that which would have been produced by mere fluviatile erosion, however long continued. Similar facts are supplied by nearly all the greater valleys of England; and on the whole, they point to one conclusion, that the fluviatile erosion, as a mere solitary agent, has produced but small effects in modifying the prominent features of our island: at the same time, they leave untouched all the facts of an opposite kind supported by direct evidence, whether derived from the volcanic districts of Central France, or from any other physical region on the surface of the earth.

The power of mountain torrents in transporting heavy masses of stone is strikingly illustrated in a paper by Mr Culley. He states that a small rivulet descending from the Cheviot Hills along a moderate declivity, carried down, during a single flood, many thousand tons of gravel into the plains below; and that several blocks, from one-half to three quarters of a ton weight each, were propelled two miles in the direction of the stream. Facts similar in kind, but on a scale incomparably greater, must be in the recollection of every one who has seen the Alpine torrents descending into the plains of the north of Italy.

When mountain chains abut in the sea, the laws of degradation are not suspended. At each successive flood, fragments of rock are drifted in the direction of the descending torrents, and rolled beneath the waters. This kind of action is indeed casual and interrupted; but it is aided by another action which is liable to no intermission—the beating of the surf and the grinding of the tidal currents on all the projecting parts of a steep and rocky shore. Under such conditions there are now forming at the bottom of the sea, and at depths perhaps inaccessible, alternating masses of silt and sand, and gravel, which, if ever lifted above the waters, may rival in magnitude some of the conglomerates of our oldest formations.

Professor Sedgwick is of opinion, that the existing drainage of our physical region is a complex result, depending upon many conditions—the time when the region first became dry land—its external form at the time of its first elevation above the sea—and all the successive disturbing forces which have since acted upon its surface. But none of these elements are constant; no wonder, then, that results derived from distant parts of the earth should be so greatly in conflict with each other. In the formation of valleys there is, therefore, little wisdom in attributing every thing to the action of one modifying cause. We know by direct geological evidence, that nearly all the solid portions of the earth were once under the sea, and were lifted to their present elevation, not at one time, but during many distinct periods. This is proved beyond a doubt, by the various marine shells which are found in the strata of the different formations, all of them having existed in the ocean at different epochs of time, and varying in their structure according to the various eras when they existed; the most simply organized being buried in the most ancient beds, and the most complicated in the most recent. We know that elevating forces have not only acted in different places at different times, but with such variations of intensity, that the same formation is in one country horizontal, in another vertical; in one country occupies the plains, in another is only found at the tops of the highest mountains. Now every great irregular elevation of the land (independently of all other results) must have produced, not merely a rush of the retiring waters of the sea, but a destruction of equilibrium among the waters of inland drainage. Effects like these must have been followed by changes in the channels of rivers, by the bursting of lakes, by great debacles, and, in short, by all the vast phenomena of denudation. In comparing distant parts of the earth, we may therefore affirm that the periods of denudation do not belong to one, but to many successive epochs. And by parity of reasoning we may conclude, that the great masses of incoherent matter which lie scattered over so many parts of the surface of the earth, belong also to successive epochs, and partake of the same complexity of formation.

The excavation of valleys seems, therefore, to be a complex result, depending upon all the forces, which, acting on the surface of the earth since it rose above the waters, have fashioned it into its present form. We have old oceanic valleys which were formed at the bottom of the sea in times anterior to the elevation of our continents. Such is the great valley of the Caledonian Canal, which existed nearly in its present form at a period anterior to the conglomerates of the old red sandstone. We have longitudinal valleys formed along the line of junction of two contiguous formations, simply by the elevation of their beds. To this class belong some of the great longitudinal valleys of the Alps. We have other valleys of more complex origin, where the beds through which the waters now pass have been bent and fractured with an inverted dip at the period of their elevation. Such is the valley of Kingsclere described by Dr Buckland. We have valleys of disruption, marking the direction of cracks and fissures, produced by great upheaving forces. Such are some of the great transverse valleys of the Alps. Of valleys of denudation, our island offers a countless number. Some are of simple origin; for example, the dry valleys of the chalk formation, which appear to be swept out by one flood of retiring waters during some period of elevation. Others are of complex origin, and are referable to many periods, and to several independent causes. Lastly, we have valleys of simple erosion: such are some of the deep gorges and river channels in the high regions of Auvergne, excavated solely by the long continued attritions of the rivers which still flow through them.

South American Geology.—M. Orbigny, who explored the countries of Buenos Ayres, Chili, and Peru, mentions some remarkable facts respecting fossil remains being found in those countries at a great height. He found primitive formations in the greater part of Brazil and of the Bande Oriental. The immense basin, which extends from the 25th to the 38th degree of south latitude, was the first place where he detected animal remains in strata, which he reckoned of the tertiary formation. The vegetable remains were below the bones of the Mammiferous animals, which were, in their turn, covered by banks of river shells. The sides of the rivers are high, presenting every facility for observing these superpositions. To the south a primitive chain of mountains separated this geological basin from that of Patagonia. This last is analogous in a certain degree to the basin of Paris, in presenting alternating strata of oysters and freestone with osseous remains, gypsum, and river shells M. Orbigny is decidedly of opinion that the higher plains of the Andes are volcanic at the height of 12,000 feet he discovered marine fossils.

GENERAL SCIENCE.

Transformation of Vegetable Substances into a New Principle.—Bracconot has discovered that vegetable substances produce, when heated with concentrated nitric acid, compounds very different from those afforded when diluted nitric acid is used. Saw-dust, cotton, linen, fecula from the potatoe gum, issoline and soponine, heated with concentrated nitric acid, are transformed into a peculiar mucilaginous substance, called by Bracconot *Xiloidine*. It is transparent, and is reddened by turnsole; cold water coagulates it, and boiling water softens without dissolving it. It is insoluble in alcohol and ammonia; and caustic potash dissolves it with great difficulty. On the other hand, the acids dissolve it in great quantities without altering it; the solutions leave upon bodies a brilliant varnish.

Prospects of the Negro Population in South America, and of the Gradual Extinction of the Original Inhabitants of the New World.—Dr Poeppig, in his account of Chili, has the following important observations on these interesting subjects:—"No country in America enjoys, to such a degree as Chili, the advantages which a state derives from an homogeneous population and the absence of Castes. If this young republic rose more speedily than any of the others, from the anarchy of the revolutionary struggle, and has attained a high degree of civilization and order, with a rapidity of which there is no other example in this continent, it is chiefly indebted for these advantages to the circumstance, that there are extremely few people of colour among its citizens. Those various transitions of one race into the other are here unknown, which strangers find it so difficult to distinguish, and which, in countries like Brazil, must lead, sooner or later, to a dreadful war of extermination, and in Peru and Columbia, will defer to a period indefinitely remote, the establishment of general civilization. * * * If it is a great evil for a state to have two very different races of men for its citizens, the disorder becomes general, and the most dangerous collisions ensue, when, by an unavoidable mixture, races arise which belong to neither party, and in general inherit all the vices of their parents, but very rarely any of their virtues. If the population of Peru consisted of only whites and Indians, the situation of the country would be less hopeless than it must now appear to every calm observer. Destined as they seem by nature herself, to exist on the earth as a race, for a limited period only, the Indians, both in the north and south of this vast continent, in spite of all the measures which humanity dictates, are becoming extinct with equal rapidity, and in a few centuries will leave to the whites the undisputed possession of the country. With the Negroes the case is different; they have found in America a country which is even more congenial to their nature than the land of their origin, so that their numbers are almost every where increasing in a manner calculated to excite the most serious alarm. In the same proportion as they multiply, and the white population is no longer recruited by frequent supplies from the Spanish Peninsula, the people of colour likewise become more numerous. Hated by the dark mother, distrusted by the white father, they look on the former with contempt, on the latter with aversion, which circumstances only suppress, but which is insuperable, as it is founded on a high degree of innate pride. All measures suggested by experience and policy, if not to amalgamate the heterogeneous elements of the population, yet to order them so that they might subsist together without collision, and contribute in common to the preservation of the machine of the State, have proved fruitless * * * *. The late revolutions have made no change in this respect. The hostility, the hatred of the many coloured classes, will continue a constant check to the advancement of the State, full of danger to the individual citizens, and perhaps the ground of the extinction of entire nations. The fate which must, sooner or later, befal the great part of tropical America which is filled with Negro Slaves, which will deluge the fairest provinces of Brazil with blood, and convert them into a desert, where the civilized white man will never again be able to establish himself, may not indeed afflict Peru and Columbia to the same extent, but these countries will always suffer from the evils resulting from the presence of an alien race. If such a country as the United States feels itself checked and impeded by its proportionably less predominant black population; and if there, where the wisdom and power of the government are supported by public spirit, remedial measures are sought in vain; how much greater must be the evil in countries like Peru, where the supine character of the whites favours incessant revolution, where the temporary rulers are not distinguished either for prudence or real patriotism; and the infinitely rude Negro possesses only brutal strength, which makes him doubly dangerous in such countries, where morality is at so low an ebb. He and his half descendant, the Mulatto, joined the White Peruvian to expel the Spaniards, but would soon turn against their former allies, were they not at present kept back by want of moral energy and education. But the Negro and the man of colour, far more energetic than the White Creole, will in time acquire knowledge, and a way of thinking that will place them on a level with the Whites, who do not advance in the same proportion so as to maintain superiority."

Education.—How greatly is Nature neglected in all our Schools! "Ten years," says Erasmus, "have I wasted in reading Cicero." *Decem annos consumpsi in legendo Cicerone.* The Echo replied in Greek, "Ονε (Ass!)

Edinburgh: Published for the Proprietors, at their Office, 16, Hanover Street. London: Smith, Elder, and Co., 65, Cornhill. Glasgow and the West of Scotland: John Smith and Son, 70, St Vincent Street; and John Macleod, 20, Argyle Street. Dublin: W. F. Wakeman, 9, D'Olier Street.

THE EDINBURGH PRINTING COMPANY.

THE EDINBURGH

JOURNAL OF NATURAL HISTORY,

AND OF

THE PHYSICAL SCIENCES.

NOVEMBER, 1836.

ZOOLOGY.

DESCRIPTION OF THE PLATE—THE ANTELOPES.

THE Antelopes are an elegant and active tribe of animals, inhabiting mountainous countries, where they bound among the rocks with so much lightness and elasticity, as to strike the spectator with astonishment. They browse like Goats, and frequently feed on the tender shoots of such trees as are found on their rocky fastnesses. In disposition they are timid and restless, and Nature has bestowed on them long and tendinous legs, peculiarly appropriate to their habits and manners of life. Almost the whole tribe are remarkable for the lively and fine expression of their eyes, and in the East they are considered as the standard of perfection ;—a higher compliment cannot be paid to a female, than to say "She has the eyes of an Antelope."

Fig. 1. THE CHAMOIS (*Antilope rupicapra*).—This extremely active animal is a native of the rocky and mountainous districts of Dauphiné, Piedmont, Savoy, Switzerland, and various parts of Germany. The flesh of this animal is well flavored, on which account it is hunted by the natives of Switzerland. This delicacy of its flesh may arise from its nicety in the choice of food, which consists of the best herbage, the most tender shoots of plants, the flowers, and young buds. It is also very fond of such aromatic herbs as are natives of mountainous districts. It runs along the rocks with great ease and seeming indifference, and leaps from one to another with unerring security.

Fig. 2. DUVAUCEL'S ANTELOPE (*A. Duvaucelii*).—This species is a native of Sumatra, and was discovered by M. Duvaucel. It is somewhat allied to the Cambing-Ootan, but its horns are more recumbent, with longer, larger, and more pointed hoofs; and it differs materially in point of colour, being of an ashy-gray, with a tinge of brown, whereas the former is black.

Fig. 3. THE FOUR-TUFTED ANTELOPE (*A. quadriscopa*) is one of the rarest of the tribe; one specimen only having been seen in Britain, and was exhibited in Exeter 'Change, London. It is a native of Senegal. This and the following species are nearly allied to the Goats.

Fig. 4. THE CAMBING-OOTAN (*A. Sumatrensis*).—The hair on the head, neck, and whole body, is longer and rougher than is common with Antelopes, which gives it much the appearance of an animal of the genus *Capra*; and the character of annulations of the horns adds still further to this peculiar aspect. It inhabits Sumatra.

Fig. 5. THE PRONG-HORNED ANTELOPE (*A. furcifer*) inhabits the New Continent, and its geographical range is further North by several degrees than any other species of its tribe. It is common on the fertile plains which border the Missouri in the United States, and has been observed by Dr Richardson as far North as the 53d parallel, on the north branch of the Saskatchewan. It is a beautifully-formed animal, somewhat larger than the Roebuck.

Fig. 6. THE VLACKTE-STEENBOCK (*A. rufescens*).—This animal is found on the plains of Southern Africa, and was first noticed by Mr Burchell. It is one of the most elegant of the smaller Antelopes, and measures only two feet six inches from the muzzle to the insertion of the tail.

Fig. 7. THE FOUR-HORNED ANTELOPE (*A. quadricornis*).—This is a native of Nepaul, and must not be confounded with the Chickara of Duvaucel. It is remarkable on account of its four horns, a character peculiar to it and the Chickara only, of all the numerous species of their tribe.

Figs. 8 & 9. THE BOMTEBOCK (*A. personata*).—An individual of this rare species belonged to Mr Cross, in whose possession it died. Its skin was purchased by Mr Morgan, and presented to the Museum of the Zoological Society. This was a young animal, and measured only two feet and a half in length ; but when full grown it is said to be little inferior to the Red-Deer in magnitude. It is an inhabitant of the Cape of Good Hope, and has received the appellation of Bomtebock from the colonists.

DESCRIPTION OF THE PLATE—THE THRUSHES.

A striking peculiarity of the Merulidæ is their strong, sweet, mellow, and versatile voice, in which they are perhaps superior to all other Birds. Before the first rays of the sun have appeared above the horizon, this melodious tribe begin to pour forth their varied and delightful notes, which have delighted mankind in all ages, and afforded a theme to many a bard. In this family is classed the Mocking Bird of America, whose extraordinary powers of voice exceed those of all other Birds.

The figures in our plate are all inhabitants of North America, and are represented the size of life.

Fig. 1. THE LITTLE TAWNY THRUSH (*Merula minor*) is found on the banks of the Saskatchewan. It leads a solitary life, and migrates to Pennsylvania in April, where it continues all the summer, employed in incubation and the rearing of its brood.

Fig. 2. THE GOLDEN-CROWNED THRUSH (*M. auricapilla*).—This is also a migratory species, arriving in Pennsylvania in April, and leaving it again late in September. It inhabits the woods, and is frequently to be seen on the ground running in the same manner as a Cock, and moving its tail like the Wagtails. It builds its nest on the ground.

Fig. 3. RICHARDSON'S THRUSH (*M. Richardsonii*).—This richly-coloured Bird was discovered in Nootka Sound, during Captain Cook's third voyage. Richardson found it at Fort Franklin, in latitude 65° 14', in April 1826, and it was also seen by him on the banks of the Saskatchewan. It builds its nest in a bush. Being never observed in the United States, its range to the East is probably bounded by the Rocky Mountains.

Fig. 4. AUDUBON'S THRUSH (*M. Ludoviciana*).—This bird was discovered by Audubon, who says it inhabits the States of Louisiana and Mississippi, where it is found at all seasons in deep and swampy cane brakes. He says its song is equal to that of the Nightingale. It "begins on the upper key, and progressively passes from one to another, until it reaches the bass note ; this last frequently being lost when there is the least agitation in the air." Its nest is built at the root of a tree.

Fig. 5. THE TAWNY THRUSH (*M. Wilsonii*).—This bird arrives at Pennsylvania from the South about the beginning of May, where it remains a week or two, and then passes northward to fulfil the business of incubation. It has also been observed by Richardson to pass as far to the North as the Saskatchewan. Nothing, however, is known of its nest. It frequents alder thickets and dense willow groves.

Fig. 6. THE WATER THRUSH (*M. aquatica*).—This is a shy species, frequenting brooks and the shores of ponds and rivers, where it may be seen wading in search of aquatic insects. It also passes through Pennsylvania for the North early in May, and returns in August.

ASTRACAN SHEEP.—M. Leroux, of Franconville, has a flock of Sheep, presented to him by General Guilleminot, being the produce of a Ram and Ewe, which he brought from Constantinople, and originally introduced from Astracan. At present this small stock consists of two rams and a male lamb, some sheep of a pure race between the first pair, and a considerable number of a mixed breed, between the males and the native sheep of the country. The growth of the wool on these sheep is very rapid, so much so, that it is shorn twice a year, and yielding at each time a greater quantity than any variety of the European sheep, which is only shorn once a year. The wool is long and of a coarse texture, and fit only for the manufacture of common stuffs, the stuffing of mattresses, &c. Its colour is of a silvery gray, and consequently would dye all colours. Underneath the long wool is a thick coating of wool of fine silky or downy texture like that of Cashmere. These animals of the pure breed have the quality of attaining a greater height in a given time, than any other known variety of the sheep ; while the flesh is of a good and pleasant flavour.

OBSTINACY OF A WEASEL (*Mustela vulgaris*).—The following anecdote is communicated by Lieut. John Brown :—" While fishing for Perch on Loch Fitty, a beautiful and picturesque sheet of water, about a mile broad and two long, situate about half way betwixt Kinross and Dunfermline, I observed near the centre of the lake a small object making its way for that part where I stood knee-deep in water. At first it struck me that it was a young Wild Duck (with which this lake abounds), but on its nearer approach, I discovered it to be a full grown Weasel, swimming with his head and back above water in the same manner as a Newfoundland Dog. I called to a friend who fished at some distance from me, and as soon as the Weasel neared us, we commenced an attack on him with our fishing-rods, and he in his turn gave us battle by biting at, and clinging to, the point of our rods, grinning and gnashing his teeth ; and showed a determination to land at the very spot where we stood, in spite of our combined efforts to drive him off, by lashing him with the points of our rods. It was of no avail, for our punishment only increased his ire, and he quitted hold of my friend's rod and made directly towards me, seemingly with an intention of coming to close quarters. He was now too near us to strike at him with our rods, and my friend ran to the shore, seized a piece of paling, and made two or three unsuccessful strokes at him, which served but to increase his fury; and he seemed determined not to relinquish his intentions, except with the loss of his life. As he stood so nobly to the contest, I thought it but humanity to call out "quarter, quarter—allow the brave animal to pass." He landed on *terra firma*, and seeing that hostilities had ceased, shook himself, and trotted quietly off. It was at the broadest part of the lake where he crossed; and from the time when I first saw him till the end of our attack, he must have been little short of three quarters of an hour in the water."

Eupleres Goudotti.—A new genus of Mammalia has been found in Madagascar, by M. Goudot, which M. Doyere, Professor at the College of Henri Quatre, proposes to call *Eupleres*. It is a lively, swift animal, with slender legs, and entirely plantigrade, the sole of the foot being the only part free from hair. It lives on the surface of the ground, is long and thin in the body, and its girth is that of most Insectivora. If any judgment can be formed from its anatomy, its hearing is equal to that of the other Insectivora, and the size of its orbits shows that its sight is likely to be good. The thumb is much the shortest of its five fingers, and all are armed with sharp, thin, and semi-retractile nails. The natives say, that it hollows out the sand, and lives in pits. Flacourt mentioned this animal under the name of Falanou, and thought it to be a Civet, which error has been continued in several works. The animal we now speak of was too young to have completed its dentition, but at present it has six incisors in the upper jaw, two canines, six pointed grinders, and four tuberculous grinders in the under jaw; eight incisors, two canines, with a double root, fitting behind those of the upper jaw like the Mole, four pointed grinders, and six with five tubercles in the lower jaw. M. Doyere gives the specific name in honour of M. Goudot, and calls it *Eupleres Goudotti*.

Introduction of Frogs into Ireland.—It is not generally known that the introduction of Frogs into Ireland is of comparatively recent date. In the 17th number of the Dublin University Magazine, there is a quotation from Donat, who was himself an Irishman, and Bishop of Fesulæ, near Florence, and who, about the year 820, wrote a brief description of Ireland, in which the following passage occurs:—

"Nulla veneta nocent, nec serpens serpit in herba;
Nec conquesta canit garrula rana lacus."

"At this very hour," says the author of the article, "we have neither snakes nor venomous reptiles in this island; and we know, that, for the first time, *frog-spawn* was brought from England in the year 1696, by one of the Fellows of Trinity College, Dublin, and placed in a ditch in the University park, or pleasure ground, from which these very prolific colonists sent out their croaking detachments through the adjacent country, whence their progeny spread from field to field through the whole kingdom." In the Dublin Medical Journal, however, it is stated, "we have learned from good authority, that a recent importation of snakes has been made, and that they are at present multiplying rapidly within a few miles of the tomb of St Patrick."

BOTANY AND HORTICULTURE.

GENERAL REVIEW OF THE VEGETABLE KINGDOM.—NO. III.

North America, under the same degrees of latitude as France and England, and with a colder climate, presents a far richer vegetation. There large trees, such as the *Liriodendron* and *Magnolia*, bear the most superb flowers. Those of many other trees and shrubs vie in beauty with the flowers of the torrid zone; the light waving composite foliage of the *Robinias* and *Gleditschias* are the counterparts of the *Mimosas* of the tropics. The single genus of Oaks comprehends within the United States more species than Europe reckons within the whole amount of its trees.

In the Northern parts of Asia vegetation differs but slightly from that of our own country. We meet with nearly the same genera, and similar types prevail. But in the Southern parts the character of the country is changed. Without water, and swept by scorching winds, the drought is extreme. The carpet of soft verdure, and the refreshing shade of its Northern countries and of Europe, are looked for in vain. Most of the plants have thinly scattered long narrow arid leaves, unscalloped and entire at the edge, and of a gloomy green; several have none at all, or at least such as, instead of leaves, may be truly termed thorns. Yet many of the trees and shrubs have a snowy blossom. Of the former, the largest in those parts belong to the Myrtle tribe, and have a punctured foliage, diffusing an aromatic scent when bruised. There are likewise many shrubs of the pulse tribe with a composite foliage; but the leaflets of the leaves are only evolved on the plants first rising from the seed. As they advance to maturity, the naked footstalks widen into simple lanceolate blades, or become transformed into acicular spines, resembling the leaves of some of the *Asparagi*. In New Holland the *Proteaceæ* abound; and also at the Cape of Good Hope; but the *Liliaceæ*, which decorate the African Promontory so profusely, are, on the contrary, rare in New Holland. It is a fact as notorious as surprising, that no one vegetable belonging to the countries towards the Southern pole produces a single fruit for the food of Man.

There are divers conditions without the performance of which the growth of the different species cannot proceed. An uninterrupted heat is requisite for some; a moment's decrease in it is fatal to them; some withstand a considerable degree of cold while their sap is quiescent, but want a high degree of heat when that is once in motion; some like a moderate temperature, and dread equally the excess of both heat and cold. It is upon the observation of such appearances that the cultivator grounds his practice; he knows that it would be in vain for him to attempt to grow, without shelter, either the Date or Orange beyond the 43d degree of Northern Latitude; that the Olive will do a little beyond; that the Vine is barren beyond the latitude of 50 degrees, or at least never brings its grape to perfection. He is cautious of exposing in a Southern aspect the species whose sap is readily set in motion by the first gleam of warmth; he knows that late frosts destroy them; as in the vineyards round Paris, the plantations there which escape the injuries of frost, are not those which look towards the South, but those that look towards the North. The sap of the latter is set in motion late, and when the heat reaches them the season is already settled, and no risk is run from the inroad of cold.

Late frosts are peculiarly hurtful to the delicate American and Botany Bay plants, which we are attempting to naturalize in Europe. Many of these will bear a very sharp cold in the heart of winter; but no sooner does the spring advance, and a softer air prevail, than their roots begin to elaborate their juices underground, their bark to fill with moisture, their buds to swell and open, and a fall in the temperature, if but for one moment, destroys them.

Local circumstances, such as the elevation of the place, its aspect, the nature as well as dipping of its soil, the proximity of mountains, of forests, of the sea, &c. &c. are all causes of variation of temperature, and must each be attended to, in accounting for the vegetation of any particular district. For instance, the winter is less severe on the Northern coasts of France, than in the interior on the same level—an effect of the vicinity of the ocean. The sea preserves a far more even temperature than the atmosphere, and is constantly at work to maintain some degree of equilibrium in the warmth of the air. In the summer, it carries off a part of the caloric from it—in the winter it gives back a part of that which it contains. It is thus that the mass of water held in the vast basin of the ocean, tempers on its coasts the heat of summer and the cold of winter. For this reason, on the coast of Calvados, the Myrtle, the Fuchsia, the Magnolia, the Pomegranate, the Indian Rose, and many other exotic plants, grow in the open air; but in the department of the Seine, the same plants require shelter. This cause also permits the cultivation of many species in the open ground about London, that near Paris will not thrive without a green-house.

Local circumstances, however, have only a limited influence, and it may be laid down as a general principle, that the cold in the same or nearly the same longitudes is, during winter, in direct proportion to the distance from the Equator. We say during winter, because the length of the days in the summer of the Polar regions sometimes renders the heat even more intense than in our climates; and it is very probable that many of the herbaceous plants of the tropics would succeed in Sweden, Norway, Lapland, and even Spitzbergen, if the frost did not set in too early to admit of their completing the round of the vegetable career.

In proportion as we advance towards the Pole, we are sensible of the change in the appearance of the vegetation. Those species which require a mild and temperate climate, are supplanted by others which seem to delight in cold. The forests fill with Pines, Firs, and Birches—the natural decoration of a Northern land. The Birch of all trees is the one which bears the severity of the climate the longest; but as it approaches the Pole it grows smaller; its trunk dwindles and becomes stunted, and the branches knotty, till at last it ceases to grow at all towards the 70th degree of latitude—the point where Man gives up the cultivation of Corn. Beyond this, shrubs, bushes, and herbaceous plants alone are to be met with. The Wild Thyme, Daphnes, creeping Willows and Brambles, cover the face of the rocks. It is in these cold regions that the berries of the *Rubus arcticus* acquire their delicious flavour and perfume. The Shrubs disappear in their turn. They are succeeded by low herbs, furnished with leaves at the root, from the midst of which rises a short stalk surmounted by small flowers. Such are the Saxifrages, the Primroses, the *Androsaces*, *Aretias*, &c. These pretty plants take up their quarters in the clefts of the rocks, while the Grasses, with their numerous slender leaves, spread themselves over the soil, which they cover as with a rich verdant carpet. The Lichen, which feeds the Rein-deer, sometimes mixes in the turf, sometimes of itself covers vast tracts of country, its white tufts standing in clumps of various forms, looking like hillocks of snow which the sun has not yet dissolved. If we go farther, a naked land, sterile soil, rocks, and eternal snows, are all we find. The last vestiges of vegetation are some pulverulent *Byssi*, and some crustaceous Lichens, which cover the rocks in motley patches.

On the Rapidity of Vegetable Organization.—The vegetable kingdom presents us with innumerable instances, not only of the extraordinary divisibility of matter, but of its activity in the almost incredible rapid development of cellular structure in certain plants. Thus, the *Bovista giganteum* (a species of fungus) has been known to acquire the size of a gourd in one night. Now, supposing with Professor Lindley, that the cellules of this plant are not less than the one-two-hundredth part of an inch in diameter, a plant of the above size will contain no less than 47,000,000,000 cellules; so that, supposing it to have grown in the course of twelve hours, its cellules must have been developed at the rate of nearly 4,000,000,000 per hour, or of more than 96,000,000 in a minute! And, when we consider that every one of these cellules must be composed of innumerable molecules, each of which is composed of others, we are perfectly overwhelmed with the minuteness and number of parts employed in this single production of Nature.

Fibres of the Rose of Sharon.—Mr S. Woodruff recently communicated to Professor Silliman that he had discovered, in the fibres of the bark of the Rose of Sharon, which were in a decayed state, a material very much resembling flax in its texture. The bark on the stalks appeared to be of the earliest growth, and it was separating from the wood; indeed, all of them were so much decayed that it was easy to divest them of the bark. The fibres are strong, and appear much like hemp, but may be divided into fibres as fine as flax. Mr Woodruff twisted a few small cords from this, without attempting any process of preparation, and it was found to be very durable and strong—even although it had not been macerated, and had, besides, the disadvantage of having been exposed to the vicissitudes of the weather, till a natural and gradual decay of its strength may have taken place.

The coat of the Rose of Sharon is much thicker, as well as softer and more silky, than that of hemp; but whether the fibres be sufficiently slender for fabrics of the finest texture remains to be ascertained. The plant is of a robust and healthy character, and is easily grown in a moderately good soil. It is also highly productive in seed, and, being a perennial, might be raised with great facility in a long succession of crops on the same ground, and with much less labour and expense than flax or hemp. Professor Silliman, judging from the specimen sent him, thinks the plant is deserving of the attention of agriculturists.

GEOLOGY.

ON THE PROOFS OF A GRADUAL RISING OF THE LAND IN CERTAIN PARTS OF SWEDEN.—It has been long imagined that the waters of the Baltic, and even the whole Northern Ocean, have been gradually sinking. In 1834, Mr Lyell investigated this interesting subject. On his way to Sweden, he examined the eastern shores of the Danish islands of Moën and Seeland, but neither there nor in Scania could he discover any indication of a recent rising of the land ; nor was there any tradition giving support to such a supposition. The first place he visited, where any elevation of land had been suspected, was Calmar, the fortress of which, built in the year 1030, appeared, on examination, to have had its foundations originally below the level of the sea, although they are now situate nearly two feet above the level of the Baltic. Part of the moat on one side of the castle, which is believed to have been originally filled with water from the sea, is now dry, and the bottom covered with green turf. At Stockholm, the author found many striking geological proofs of a change in the relative level of the sea and land, since the period when the Baltic has been-inhabited by the shells which it now contains. A great abundance of shells of the same species were met with in strata of loam, &c. at various heights, from 30 to 90 feet above the level of the Baltic. They consist chiefly of the Edible Cockle (*Cardium edule*), the *Tellina Baltica*, and the common shore Nerite (*Littorina littorea*), together with portions of the common Muscle (*Mytilus edule*), generally decomposed, but often recognisable by the violet colour which they have imparted to the whole mass. In cutting a canal from Sodertelje to Lake Maelar, several buried vessels were found ; some apparently of great antiquity, from the circumstance of their containing no iron, the planks being fixed together by wooden nails. In another place an anchor was dug up, as also in one spot, some iron nails. The remains of a square wooden house were also discovered at the bottom of an excavation made for the canal, nearly at a level with the sea, but at a depth of sixty-four feet from the surface of the ground. An irregular ring of stones was found on the floor of this hut, having the appearance of a rude fire-place, and within it was a heap of charcoal and charred wood. On the outside of the ring was a heap of unburnt fir wood broken up for fuel ; the dried needles of the fir and the bark of the branches being still preserved. The whole building was enveloped in fine sand.

The author examined minutely certain marks which had at different times been cut artificially in perpendicular rocks washed by the sea in various places, particularly near Oregrund, Gefle, Löfgrund, and Edskösund ; all of which concur in showing that the level of the sea, when compared with the land, has very sensibly sunk. A similar conclusion was deduced from the observations made by the author on the opposite or western side of Sweden, between Udsdervials and Gottenburg, and especially from the indications presented by the islands of Orust, Gulholmen, and Marstrand.

Throughout the paper, a circumstantial account is given of the geological structure and physical features of those parts of the country which the author visited ; and the general result of the comparison he draws of both the eastern and western coasts and their islands with the interior, is highly favorable to the hypothesis of a general rise of the land, every tract having in its turn been first a shoal in the sea, and then, for a time, a portion of the shore. This opinion is strongly corroborated by the testimony of the inhabitants (pilots and fishermen more especially), of the increased extension of the land, and the apparent sinking of the sea. The rate of elevation, however, appears to be very different in different places ; no trace of such a change is found in the south of Scania. In those places where its amount was ascertained with greatest accuracy, it appears to be about three feet in a century. The phenomenon in question having excited increasing interest among the philosophers of Sweden, and especially in the mind of Professor Brezelius, it is to be hoped that the means of accurate determination will be greatly multiplied.

FOSSIL SHELLS ON THE SNOWY MOUNTAINS OF THIBET.—M. Gerard, while making a tour over the snowy mountains of Thibet, picked up some fossil shells on the crest of a pass, elevated 17,000 feet ; and here also were fragments of rocks, bearing the impression of shells, which must have been detached from the contiguous peaks rising far above the elevated level. Generally, however, the rocks formed of these shells are at an altitude of 16,000 feet ; and one cliff was a mile in perpendicular height above the nearest level. M. Gerard further states, " Just before crossing the boundary of Ludak into Bussalier, I was exceedingly gratified by the discovery of a bed of fossil oysters, clinging to the rock as if they had been alive." In whatever point of view we consider the subject, it is sublime to think of millions of organic remains lying at such an extraordinary altitude, and of vast cliffs of rocks formed out of them, frowning over the illimitable and desolate waters, where the ocean once rolled.

THE MOVING BOG OF RANDALSTOWN.—We have already alluded to this rather rare phenomenon, which has been lately witnessed on part of Lord O'Neil's estate in the neighbourhood of Randalstown, on the Ballymena road, and about two miles and a half from the former town. On the 17th September 1835, in the evening, the first movement occurred. A person who was near the ground was surprised to hear a rumbling noise as if under the earth; and immediately after, his surprise was not a little increased, on perceiving a part of the bog move rapidly forward, a distance of a few perches. It then halted, and exhibited a broken rugged appearance, with a soft peaty substance boiling up through the chinks. It remained in this state till the 22d, when it suddenly moved forward, at as quick a rate, covering corn-fields, potatoe-fields, turf-stacks, hay-ricks, &c. ; not a vestige remained to be seen. So sudden and rapid was this movement, that the adjacent mail-coach-road was covered in a few minutes, or rather moments, to a depth of nearly twenty feet. It then

directed its course towards the river Maine, which lay below it ;. and so great was its force, and such the quantity of matter carried along, that the moving mass was forced a considerable way across the river. In consequence of heavy rains at the time, the river found its channel through the matter deposited in its bed, otherwise the water would have been forced back, and immense damage done to the land on the banks. The fish in the river were killed to a great distance. The damage done by the mossy inundation was very considerable. About 150 acres of excellent arable land have been covered and rendered totally useless. Down the middle of this projected matter a channel has been formed through which there is a continual flow of dark peaty substance, over ground which, but a fortnight before, the reapers were at work. A house close by the ground is so far overwhelmed, that only part of the roof is to be seen.

METEOROLOGY.

REMARKABLE PARHELIA.—Lieutenant R. E. Clary, while at Fort Howard, Green Bay, Michigan Ter. in the United States, observed a very singular and interesting phenomenon, on the morning of the 27th of February 1835. It consisted of a large and brilliant halo around the sun, with two parhelia within the circumference, at the extremities of its horizontal diameter, but little inferior in brilliancy to the true sun ; they were accompanied by luminous trains or tails opposite to the sun. Immediately above and beneath the sun, in the circumference of the same circle, there were bright luminous spots of an elliptical form, less intense in brilliancy than the first, but of much greater magnitude. From the superior or more elevated spot, rays, faintly coloured, and slightly curved downwards, appeared to emanate, forming a small portion of an arc of a circle, of less curvature than the halo. Another circle, the plane of which was horizontal, at right angles to, and of greater diameter than the first, with its centre apparently in the zenith, completely surrounded the heavens ; its circumference passed through the sun and two mock suns, the latter being distinctly reflected in the opposite part of the heavens.

About 15° from the zenith, in the direction of the sun, there appeared two faintly luminous arcs of circles, nearly tangent to, and convex towards, each other ; they were but a few degrees in extent.

Two well defined and tolerably brilliant rainbows, situated upon the right and left of the parhelia, with their convexity towards it, completed this rare and interesting appearance.

This phenomenon was first observed a little before 8 o'clock, the lower part of the halo being then about 2° above the horizon, its diameter descending as the altitude of the sun increased ; arrived at its greatest degree of brilliancy and splendour 15 minutes before 10, when it began to decrease, and finally disappeared about 15 minutes before 11 o'clock, the duration of the phenomenon being about three hours.

The morning was extremely cold, the mercury standing at 16° below zero, and the atmosphere was uncommonly clear and serene. In the afternoon it became cloudy, and indicated snow.

The same phenomenon, with some modifications in its appearance, was observed at Fort Winnebago, 113 miles south-west from Fort Howard, which is in latitude 44° 30'.

In the above diagram, the position and appearance of the halo is represented ; as also the horizontal circle, rainbow and parhelia, with their reflections, as they appeared 15 minutes before 10 o'clock :—greater number of reflections, perhaps, than were ever witnessed at one time before, all depending upon the same peculiar state of the atmosphere for their existence.

SHOWER OF FALLING STARS IN RUSSIA.—M. le Comte de Suchteln communicated the following interesting fact to M. Feodorou, which he laid before the " Royal Academy of Sciences" at Paris. " On the 13th November 1832, between three and four o'clock in the morning, the weather calm and the sky serene, the thermometer indicating 55 degrees of Fahrenheit, the heavens appeared to be bespangled by a great number of meteors, which described an arch in the direction of from north-east to south-west. They burst like rockets into innumerable small stars, without producing any perceptible noise, and leaving in the sky a long continued, luminous belt, exhibiting all the varied colours of the rainbow. The light of the moon, however, then in her last quarter, considerably obscured this appearance. At times the heavens appeared as if cleft asunder, and, in the intervening space, there appeared long brilliant whitish bands. At other times flashes of lightning rapidly traversed the arch of heaven, eclipsing the light of the stars, and causing these long and luminous bands of colour to appear. These phenomena continued in succession, without any noise. They were in their greatest splendour between five and six o'clock in the morning. During the same night, and nearly at the same hour, an appearance equally remarkable was witnessed at Hitzkaja-Saschtschita, about seventy-five miles to the south of Orenburg. Two white columns rose from the horizon equidistant from the moon, which at the time had not risen far; about the middle of their height they appeared very brilliant, and considerably curved. Several horizontal bands sprung from this point, the most brilliant of which extended towards the moon, in which they seemed to unite, so that in this way they appeared to form a great H. In the town of Ufa, the seat of the government of the same name, situate 380 miles to the north of Orenburg, a phenomenon similar to that which was observed at Hitzkaja-Saschtschita was noticed, which, however, according to the accounts which have been given, was less brilliant in its appearance."

Remarkable Shower of Hail.—After a violent storm at Clermont, MM. Bouillet and Lecoq found a number of hailstones as large as hens' eggs, and some others as large as those of turkeys. They were all of an ellipsoidal form, and seemed formed of a multitude of needles, united at the extremities of the great axis. They were from eight lines to two inches long. Those needles, on which the fusion had not made much impression, still showed traces of hexagonal prisms, terminated by prisms of six facets. In a second storm, others fell which were not larger than hazel nuts, and these were formed of concentric layers, more or less transparent, rounded, or slightly oval, and possessed a powerful horizontal motion; they were heard to hiss in the air, as if each hailstone rubbed against the other, and their rotation was extremely rapid.

Betwixt the hours of one and two o'clock, on Saturday the 30th April 1836, a heavy shower of hail fell at Edinburgh. Professor Jameson and several members of the Wernerian Society examined the hailstones, and ascertained that they were crystallized in the form of double six-sided pyramids, and at the same time of larger size than usual.

On the Formation of Aerolites, &c.—M. J. L. Ideler has discussed, with great learning, the formation of Fire-balls, and of the Aurora Borealis; and the facts brought forward lead to the following conclusions:—

1. The fall of Aerolites generally takes place in summer, and at the period of the equinoxes; that is, in the season of the most abundant rains.

2. The frequency of this phenomenon diminishes from the equator to the poles, whilst in general the annual quantity of rain diminishes with the mean temperature of localities, allowance being made for the considerable influence of the direction of the winds.

3. The formation of Aerolites in a cloud, having their colour, is analogous to that of rain; as it rains with a clear sky, so in the same manner Aerolites descend unattended with the appearance of clouds.

4. The luminous appearance and the noise resembling thunder, are produced by electricity, which appears in all atmospheric phenomena. The different colours of fire-balls, during their descent, are the effect of the disengagement of different kinds of electricity. It is very likely that Aerolites may fall without being preceded by fire-balls, as it rains very powerfully, without lightning, when the temperature of the aeriform column is below the point of thawing.

5. Aerolites sometimes fall without noise, because the electric explosion has taken place in very elevated regions; there are analogous cases of lightning at the zenith without thunder.

The author, therefore, regards the formation of Aerolites in the atmosphere as the most plausible theory, and recurs to the same idea expressed by Aristotle and Seneca two thousand years ago.

Not satisfied with these observations, M. Ideler adds others in support of his theory. Thus, he quotes certain hail storms, in which the hailstones possessed a metallic nucleus resembling aerolites, is preceded by more or less glimmerings (leceurs) of light, and that the phenomena in question are connected with atmospheric changes, and these again with revolutions which take place within the interior of the earth. The simultaneous fall of meteoric stones in different countries is also in favor of their atmospheric origin, and it often takes place during storms.

M. F. G. Fisher has published in the Memoirs of the Academy of Berlin a Memoir upon the Origin of Aerolites, in which he adopts the foregoing ideas, and supposes that electricity performs an important part in the phenomenon.

Concerning Shooting Stars, M. Ideler endeavours to prove by facts, that they are merely precipitations of animal and vegetable matters disseminated through the atmosphere.

Finally, with respect to the Aurora Borealis, he supposes that the precipitation formed by the dry vapours in the elevated portions of the atmosphere take place in the regions of the magnetic poles, under the form of the Aurora Borealis, from the reason that the ferruginous particles arrange themselves about the pole, in an order similar to that of iron filings around a magnetic oar. Future observations upon terrestrial magnetism will aid in explaining the anomalies of this phenomenon.

The vaporization of all solid and fluid bodies goes on under every degree of temperature. When the maximum of density in the vapours is passed, a precipitate occurs, and clouds, cirri, or mists, are formed, which rest upon the earth, or a concretionary formation takes place. The latter case happens partly from the condensation of clouds, sometimes under a clear sky, sometimes without electric explosion (aerolites), or with the phenomena of electricity (fire-balls); finally, the fall of those bodies takes place in small particles, or agglomerated into masses of a larger size, and analogous to hail.

If such are the phenomena beyond the Polar regions, near the magnetic poles, the precipitates, being attracted, would continually be undergoing an arrangement in a circular series, and thus produce the Aurora Borealis. This kind of precipitation might take place contemporaneously with aqueous precipitation, in which case there would occur rains attended by foreign mixtures.

Professor Gruithuisen has lately been occupied with the origin of Aerolites and Shooting Stars; and supposes he has proved, by mathematical calculations, founded upon physics, that these bodies must necessarily be formed beyond our atmosphere, in the interplanetary space, where the metals and the metaloides, he says, are still held in solution, by means of hydrogen, and where they exist continually for the formation of these opaque bodies.

According to Herschell, the observations of the shooting stars may be useful for the determination of longitude. The height of the meteors seen by M. Quetelet is estimated at from ten to eighteen leagues from the earth, and their motions at from five to eight leagues per second; results which correspond with those of Brandes and other German philosophers.

MISCELLANEOUS.

Acetic Acid.—A most important improvement has recently been introduced into the manufacture of vinegar, which is already extensively practised on the Continent. The introduction of this improvement is chiefly due, we believe, to Mitscherlich. It is founded upon the principle that alcohol, by absorbing oxygen, is changed into acetic acid and water For, two alcohol + four oxygen = one acetic acid + three water (6 H + 4 C + 2 O) + 4 O = 3 H + 4 C + 3 O) + 3 (H + O.)

This oxidation is promoted by the process of fermentation—and, when the fermentation has begun, is much accelerated by the pressure of acetic acid. The oxidation is effected entirely at the expense of the oxygen of the air; to accelerate the process, therefore, by producing as many points of contact as possible between the liquid and the air, the following arrangement is adopted:—A large cask is taken, placed upright, with a stop-cock at the bottom, and a series of holes, half an inch in diameter, bored one in each stave, a few inches above it. It is then nearly filled with chips or shavings of wood, previously steeped in strong vinegar till they are perfectly saturated. Within the upper end of the cask a shallow cylindrical vessel is placed, nearly in contact with the shavings, the bottom of which is perforated with many small holes, each partially stopped with a slender twig, which passes an inch or two beneath the perforated bottom of the cylinder. The alcohol, diluted with eight or nine parts of water, and mixed with the fermenting substances, is now poured into the cylinder, through the bottom of which it trickles, drop by drop, upon the shavings below, becomes oxidized in its passage, and runs out at the stop-cock beneath, *already converted almost entirely into vinegar*. The air rushes in by the holes beneath, and passes out by eight glass tubes, cemented for that purpose, into the bottom of the cylinder—and so rapidly is it deprived of its oxygen, when it escapes above, that it extinguishes a candle. During the process much heat is also developed; so that, from the temperature of 60° (that of the room), the interior cask rises as high as 86° of Fahr. In the proper regulation of this temperature much of the difficulty consists.

A second transmission of the acid, thus obtained, through another smaller cask, finishes the process. The whole is concluded in a few hours; four-and-twenty are considered amply sufficient to convert a given quantity of alcohol into vinegar.

Origin of Amber.—M. T. Aessi says that amber is a resin of the Coniferæ. He has examined particularly that of Castrogiovanni in Sicily, and he cites, though not with perfect confidence, a specimen of amber containing a land shell. M. Graffenauer has given to the Strasbourg Society of Sciences a monograph on amber, which he supposes to have originated in extinct species of trees.

Purification of Water.—In order to precipitate the earths mechanically suspended in water, it is recommended to employ the silicate of potash, gelatinous silica, or phosphoric acid. The last is an excellent reagent for throwing down the oxide of iron, without introducing any foreign principle into the water.

Temperature of Aerolites.—Most philosophers are of opinion that aerolites and meteoric iron are elevated to a high temperature while traversing the atmosphere; nevertheless, there is but little agreement concerning the degree of heat observed in them immediately after their fall. Recently, an experiment of M. Bierley, repeated by M. d'Arcet, has rendered this high temperature doubtful; a bar of iron, heated to whiteness, was held in the current of air from the blowing-machine of a forge—the metal did not cool, but burnt brilliantly, throwing off glowing particles in every direction. The temperature of the iron rather increased than diminished under the influence of the current of air.

African Diamonds.—The Sardinian Consul at Algiers, M. Pelusa, lately purchased from a native three diamonds, which were found in the auriferous sand of the river Gumel, in the province of Constantine.

OBITUARY.

John Pond, Esq.—This celebrated Astronomer died at his house in Greenwich, on Wednesday the 21st September 1836. He was a Fellow of the Royal Society, a Corresponding Member of the French Institute, and an Honorary Member of most of the Astronomical Societies in Europe. During the period of nearly twenty-five years, Mr Pond filled the high and important office of Astronomer Royal, from which a hopeless state of ill-health obliged him last autumn to retire; but his regret at quitting a situation, in the duties of which he had taken so great an interest, was lessened on finding himself succeeded by one of the greatest Mathematicians of the age, Professor Airy—a Philosopher eminently qualified to maintain the character and uphold the dignity of the appointment; to carry forward the improvements introduced by his predecessor, and to extend the boundaries of Astronomical Science.

Errata.—Animal Kingdom, page 64, col. 1, line 6 from the bottom; page 67, col. 1, line 45 from the top; and page 66, col. 1, line 36, *for* Martin *read* Marten.

Edinburgh: Published for the Proprietors, at their Office, 16 Hanover Street.
London: Smith, Elder, and Co., 65, Cornhill. Glasgow and the West of Scotland: John Smith and Son, 70, St Vincent Street; and John Macleod, Argyle Street. Dublin: W F. Wakeman.
The Edinburgh Printing Company.

THE EDINBURGH

JOURNAL OF NATURAL HISTORY,

AND OF

THE PHYSICAL SCIENCES.

DECEMBER, 1836.

ZOOLOGY.

DESCRIPTION OF THE PLATE—THE PTARMIGAN.

MODERN Naturalists have separated the Ptarmigan from the Grouse, under the new generic title of LAGOPUS, or Hairy-legged Grouse. They are wild birds, frequenting and incubating on high and precipitous mountain ranges, particularly in northern latitudes. They are susceptible of considerable variations in their summer and winter plumage, in the latter season being, with the exception of the tail feathers, quite white. They have a wide Geographical range, inhabiting all the islands which lie on the south of Baffin's Bay, and are pretty common on some of the loftiest mountains in the north of Scotland.

Figs. 1, 2, & 3. THE WILLOW PTARMIGAN (*Lagopus Saliceti*).—This is one of the most beautiful species which bear the designation of Game Birds. Its summer plumage is of rich and beautiful dark and pale chestnut, and its winter attire white, with a slight rosy hue. It inhabits the fur countries from the 50th to the 70th degrees of latitude, breeding in the valleys of the Rocky Mountains, and on the Arctic coasts, assembling in vast flocks on the shores of Hudson's Bay during winter. So numerous are these Birds, that Mr Hutchins mentions he has known *ten thousand* to be captured in a season. It is not yet ascertained whether this is identical with the Willow Grouse of Europe, and which is common to Greenland, Iceland, and Scandinavia.

Figs. 4 & 5. THE PTARMIGAN (*L. mutus*), in summer and winter plumage.— These were taken from specimens killed during the Arctic Expedition, under Captain Parry, and seem to differ from the European species, principally in their more diminutive size. The Ptarmigan always keeps near the snow line. It is considered an excellent food, many preferring it to Grouse.

DESCRIPTION OF THE PLATE—THE CHATS.

THIS genus belongs to the group of birds termed *Ampelidæ* by Swainson, and are distinguished by the shortness of the bill and excessive width of the mouth, which gives them the capacity of swallowing large berries, and even moderate-sized fruits. They alight very seldom on the ground, but may be seen constantly moving about among bushes and trees with great rapidity, to which the formation of their feet peculiarly adapts them. Their toes are rather short, more or less united at the base, and the soles broad. They are totally devoid of the nuchal bristles, which protect the mouth of insectivorous birds. This family is almost exclusively confined to America.

Fig. 1. THE YELLOW-THROATED CHAT (*Vireo flavifrons*).—This species is chiefly found in the woods during summer, and utters a slow and plaintive note with but little variation, which it repeats every ten or twelve seconds, and sounds like *preed, preed*, &c. It arrives in the Middle States from the south early in May, and returns with its young in the beginning of September.

Figs. 2 & 3. THE SOLITARY CHAT (*V. solitaria*), Male and Female.—This is a silent and solitary bird, found in Georgia and Philadelphia, and inhabits Louisiana during the spring and summer months, frequenting thick cane-brakes of the alluvial lands contiguous to the Mississippi. It hangs to branches of small berries, feeding upon them as a Titmouse does on buds of trees. The flight of this bird is rather peculiar, consisting of a continued tremor of the wings, as exhibited by birds when they are angry. It is about five inches and a half in length.

Fig. 4. THE PINE-SWAMP CHAT (*V. sphagnosa*).—The favourite haunts of this bird are the deepest and gloomiest pine and hemlock swamps in mountainous regions of the central states of North America. Its place of incubation is still unknown. Its habits are much akin to those of flycatchers, as it seeks after and feeds upon insects with great keenness. It has not been yet ascertained whether it is a bird of song.

Fig. 5. THE YELLOW-BREASTED CHAT (*V. polyglotta*), frequents close thickets of hazel, brambles, and vines, and dense underwood; and when approached utters a scolding note, and seems offended at the intruder. The principal food of this Chat consists of large black beetles and other coleopterous insects. It arrives in Pennsylvania about the first week in May, where it incubates, and returns to the south about the middle of August, the males generally preceding the females by several days. It seems a bird of a wide geographical range, as it has been found in Mexico, Guiana, and Brazil.

Fig. 6. THE RED-EYED CHAT (*V. olivacea*).—This species, like most of its tribe, is a bird of passage, spending its summers in Pennsylvania and the neighbouring

States, and wintering in Jamaica. It builds a hanging nest between two twigs of a young dogwood or other small sapling, seldom more than four or five feet from the ground. Its notes are rapid and lively, consisting of three or four syllables.

Fig. 7. THE WHITE-EYED CHAT (*V. Noveboracensis*), visits Pennsylvania and Georgia about the end of February, and is supposed to winter in Mexico. Its nest is frequently the shape of an inverted cone, suspended by the upper edge of the two sides on the circular bend of a prickly vine. It generally produces two broods in a season. It is five inches and a quarter long, and seven in extent of wing.

Fig. 8. THE WARBLING CHAT (*V. melodia*), arrives in Pennsylvania about the middle of April, and frequents the thick foliage of orchards and high trees. Its voice is very soft and melodious. Its food consists principally of insects and caterpillars, and in its general manners is not unlike the Warblers. It is five inches long, and eight and a half in extent of wing.

ON THE HABITS OF LEECHES, AND TENDENCY OF THE REPTILES OF VALDIVIA AND CHILI TO BECOME VIVIPAROUS.—M. Gay, in a letter to M. de Blainville, dated 5th July 1835, says, "It is a remarkable circumstance, that here all the Leeches exist in the woods, and never in the water; and, indeed, I cannot botanize without having my legs severely punctured by them. They crawl on plants, trunks of trees, and shrubs, never approaching marshes or rivers; and the only one which I have been fortunate enough to discover in such localities is a very small species of 'Branchibelle,' which inhabits the pulmonary cavity of *Avicula Dombeii*; it was while dissecting this molluscous animal that I detected it. I have discovered another species in the neighbourhood of Santiago, which lives on the gills of the Astacus. An equally interesting fact, and which deserves your attention, is the tendency exhibited by reptiles to become viviparous in these southern regions. Almost all those which I have dissected presented this remarkable circumstance. Not only does the harmless Leech of Valdivia give birth to a living progeny, but likewise all the beautiful Iguanas allied to the genus *Leposoma* of M. Spix, and which, on account of their beautiful colours, I have in the meantime termed *Chrysosaurus*. All the species which I have examined, including those which at Santiago deposit eggs, have without exception presented this phenomenon, and I may hence be allowed to generalize. The Batrachians have also furnished me some examples of this description, although in general they are oviparous. Nevertheless, a genus resembling the *Rhinella* of Fitzinger, and of which several species, rather prettily marked, form part of my collection, is constantly viviparous, and therefore increases the proofs of a fact, which is rendered more remarkable by the circumstance that all the examples occur within a radius of two or three leagues only."

The habitat of Leeches, mentioned by M. Gay, is not, however, peculiar to South America, as he seems to imagine; for we find, by an account given in Percival's Ceylon, that they frequent similar localities. "One species of Leech, however," says he, "has left too deep an impression on my mind to be passed over unnoticed. It infests in immense numbers the woods and swampy grounds of Ceylon, particularly in the rainy season, to the great annoyance of every one who passes through them. The Leeches of this species are very small, not much larger than a pin, and are of a dark-red speckled colour. In their motions they do not crawl like a worm, or like Leeches we are accustomed to see in Europe, but keep constantly springing, by first fixing their head on a place, and bringing their tail up to it with a sudden jerk, while at the same time their head is thrown forwards for another hold. In this manner they move so exceeding quickly, that before they are perceived, they contrive to get upon one's clothes, when they immediately endeavour by some aperture to find an entrance to the skin. As soon as they reach it, they begin to draw blood; and as they can effect this even through the light clothing worn in this climate, it is almost impossible to pass through the woods and swamps in rainy weather without being covered with blood. On our way to Candy, in marching through the narrow paths among the woods, we were terribly annoyed by these Leeches; for whenever any of us sat down, or even halted for a moment, we were sure to be immediately attacked by multitudes of them; and before we could get rid of them, our gloves and boots were filled with blood. This was attended with no small danger, for if a soldier were from drunkenness or fatigue to fall asleep on the ground, he must have perished by bleeding to death. On rising in the morning, I have often found my bed-clothes and skin covered with blood in an alarming manner. The Dutch, in their marches into the interior, at different times, lost several of their men; and in our setting out, they told us we should hardly be able to make our way from them."

On the Changes which the Stomach of Crabs undergo during the period of Casting their Shells.—Crabs, it is well known, change their shells at a certain season of the year; and it is a very old opinion that they change their stomachs at the same time, a new stomach being formed around the old, which is digested by the recently developed organ. Baer has proved that the Crab's stomach consists of two coats; one inner, which in every respect may be compared to a callous, horny epidermis, and which is destitute of vitality; and an outer or containing coat, transparent, but sufficiently strong and vascular. The inner coat, as it is well known, consists of various and very curious parts, some resembling bony plates, others compared to teeth. Now, at the period when the Crab changes its skin, it likewise casts the inner coat of the stomach, and on this account this process, analogous to the moulting of birds, and to the renewing of the hair in quadrupeds, is in the Crab attended with very great constitutional disturbance, and a total interruption of the digestive function. Baer relates very accurately the changes which the stomach undergoes preparatory to the casting of its inner coat. It would be beside our present purpose to follow him in this description, however interesting. Some things be mentions are, however, specially worthy of remark. In the first place, the softer parts of the old epidermis, or inner coat of the stomach, are very rapidly digested in the stomach, as soon as it has recovered its functions, and has—which it does quickly—formed a new lining on its inner surface. But there are other harder parts that cannot be readily digested and dissolved, and which are otherwise disposed of. The hard and hollow bones, popularly termed the teeth, are got rid of by being discharged through the external orifice corresponding to the mouth. There are other solid plates of the epidermic portion of the stomach, which are not of a shape calculated to irritate the new and tender epidermis, and consequently they can be retained with impunity, and are destined to perform a new and curious function; for, according to Baer, these plates, for some time preparatory to the act of casting the shells, rapidly increase in weight and in solidity, so as, at the period we are speaking, they may be considered as forming considerable reservoirs of earthy matter, to be gradually dissolved and digested in the newly lined stomach, at the very time earthy matter is required by the animal for the formation of its new shell. These plates are popularly called *Crab-stones*, and when submitted to the digestive process soon lose their roughness, and become smooth and polished before they are entirely dissolved. These Crab-stones are chiefly composed of carbonate of lime; and Baer has proved, by repeated analysis, that the fluid contents of a Crab's stomach contain (at the time these stones are in them) a considerable portion of lime, carbonic acid, and muriatic acid. It is interesting to observe, that the chemical investigations of Dulk render it highly probable, that the chief solvent in the Crab's stomach is the same acid which performs so important a part in human digestion and in dyspepsia, viz. free muriatic acid.—*Dublin Med. Jour.*

Molluscous Animals.—Animals of the Genera Pneumoderma and Hyalea, were only found in the great Ocean, to which they were hitherto supposed exclusively to belong; these have, however, been lately detected in the Mediterranean Sea by Dr Vanbenaden.

BOTANY AND HORTICULTURE.

Cause of the Fragrance on approaching Tropical Lands from Sea.—When vessels begin to near the coasts between the tropics, a delightful fragrance is felt, of which even animals are so sensible, that they become restless, and, appearing to have an instinctive presentiment of the end of their long confinement, not unfrequently leap overboard to reach the shore, which they suppose to be close at hand. "Whoever," says Pœppig, "has made a voyage to the tropical countries of South America or the West Indies, will always remember with pleasure the sensation which he experienced on approaching the land. Perhaps no sense is then so strongly affected as that of smell, especially if the coast is approached in the early hours of a fine summer's morning. On the coast of Cuba, the first land I saw in America, on the 30th of June 1822, all on board were struck with the very strong smell like that of violets, which, as the day grew more warm, either ceased, or was lost amidst a variety of others, which were perceptible as we drew nearer the coast. During a long stay in the interior of this island, I became acquainted with the plant which emits such an intense perfume as to be perceived at the distance of two or three miles. It is of the species Tetracera, and remarkable for bearing leaves so hard, that they are used by the native cabinet-makers and other mechanics for various kinds of work. It is a climbing plant, which reaches the tops of the loftiest trees of the forest, then spreads far around, and in the rainy seasons is covered with innumerable bunches of sweet-smelling flowers, which, however, dispense their perfume during the night only, and are almost without scent in the day time."

On the Duration of the Germinative Power in the Seeds of Plants.—We gave an account of some experiments of this curious subject at page 35 of this Journal, and have now to record a still more extraordinary example of this power. In October 1834, a British tumulus was opened near Malden Castle, by Mr Maclean, who found therein a human skeleton, and a portion of the contents of the stomach, containing a mass of small seeds, which neither the operation of the gastric juice, nor the lapse of probably twenty centuries, had sufficed to destroy. Many of these seeds have been subjected to various careful experiments, to ascertain whether the vital principle was extinct; and we have the satisfaction of announcing, that Professor Lindley has succeeded in producing plants from several of these seeds. These plants have confirmed the opinion expressed by Professor Lindley, on a first inspection of the seeds, that they were those of the *Rubus Idæus*, or common Raspberry. The plants are now very vigorous, have produced much fine fruit this season, and form an object of the greatest curiosity to horticulturists. This highly interesting circumstance proves the Raspberry to be an indigenous plant in this country, growing at a very early period, and then constituting an article of food.

MINERALOGY.

On a Mass of Green Malachite of extraordinary size.—A few months ago, there was met with in the mines at Nischne-Tagilsk, in Russia, a mass of Green Malachite measuring 16.2 French feet in length, 7.5 feet in breadth, and 8.6 feet in height, and weighing about 1300 Russian pounds.

Topaz.—In the vicinity of Villa Rica in Brazil, the Topaz Mines are situated in chlorite-slate, which rests on a sandstone of the primitive class. The Topaz occurs in regular crystals, or in angular masses, or rests on the mineral called Lithomarge, along with rock crystal.

Vitreous Quartzose Tubes.—In various parts of Europe, there have been found long quartzose tubes of a vitreous appearance, and of an inch and a half in diameter, but in length from twenty to thirty feet, and their interior is slaggy and vitreous-like. Several tubes of this description were found at Senner-Heath in Germany, and more recently E. L. Irton, Esq. of Irtonhall, Cumberland, has found on his estate in that county a number of these tubes, agreeing in every particular with those found in Germany. That gentleman presented to the Museum of the Edinburgh University a beautiful series of those found by him. The vitreous tubes of Senner-Heath were found in loose sand. The mottled appearance of the interior of these tubes, together with their form, and the situation in which they were found, favors the inference that they have been produced by lightning.

GEOLOGY AND PHYSICAL GEOGRAPHY.

CHRONOLOGICAL TABLE OF THE MOST IMPORTANT KNOWN ENCROACHMENTS MADE BY THE SEA SINCE THE EIGHTH CENTURY.

BY M. ADRIEN BALBI.

A.D. 800. About this period, the sea carried off a great part of the soil of the island of Heligoland, situate between the mouths of the Weser and the Elbe.

800—900. During the course of this century, many tempests made a considerable change in the coast of Brittany; valleys and villages were swallowed up.

800—950. Violent storms agitated the lakes of Venice, and destroyed the isles of Ammiano and Constanziaco, mentioned in the ancient chronicles.

1044—1309. Terrible irruptions of the Baltic Sea on the coasts of Pomerania, made great ravages, and gave rise to the popular tales of the submersion of the pretended town of Vineta, whose existence is chimerical, notwithstanding the imposing authority of Kant and other learned men.

1106. Old Malamocco, then a very considerable city on the lakes of Venice, was swallowed up by the sea.

1218. A great inundation formed the gulf of Jahde, so named from the little river which watered the fertile country destroyed by this catastrophe.

1219, 1220, 1221, 1246, and 1251. Terrible hurricanes separated from the continent the present isle of Wieringen, and prepared the rupture of the isthmus which united Northern Holland to the county of Staveren, in modern Friesland.

1277, 1278, 1280, 1287. Inundations overwhelmed the fertile canton of Reiderland, destroyed the city of Torum, 50 towns, villages, and monasteries, and formed the Dollart; the Tiam and the Eche, which watered this little country, disappeared.

1282. Violent storms burst the isthmus which joined Northern Holland to Friesland, and formed the Zuyderzee.

1240. An irruption of the sea changed considerably the west coast of Schleswig; many fertile districts were engulphed, and the arm of the sea which separates the isle of Nordstrand from the continent was much enlarged.

1300, 1500, 1649. Violent storms raised three-fourths of the island of Heligoland.

1300. In this year, according to Fortis, the town of Ciparum in Istria was destroyed by the sea.

1303. According to Kant, the sea raised a great part of the island of Rugen, and swallowed up many villages on the coasts of Pomerania.

1337. An inundation carried away 14 villages in the island of Kadzand in Zealand.

1421. An inundation covered the Bergseweld, destroyed 22 villages, and formed the Biesbosch, which extends from Gertruydenberg to the island of Dordrecht.

1475. The sea carried away a considerable tract of land situated at the mouth of the Humber; many villages were destroyed.

1510. The Baltic Sea forced the opening at Frisch-Haff near Pillau, about 3600 yards broad, and 12 to 15 fathoms deep.

1530—1532. The sea engulphed the town of Kortgene in the island of North Beveland in Zealand. In the latter year, it also raised the E. part of the isle of S. Beveland, with many villages, and the towns of Borselen and Remerswalde.

1570. A violent tempest carried off half of the village of Scheveningen, N.E. of the Hague.

1625. The sea detached a part of the peninsula of Dars in Pomerania, and formed the isle of Zingst, N. of Barth.

1634. An irruption of the sea submerged the whole island of Nordstrand; 1338 houses, churches, and towns were destroyed, 6408 persons and 50,000 head of cattle perished. There only remained of this island, previously so fertile and flourishing, three small islets, named Pelworm, Nordstrand, and Lütje-Moor.

1703—1746. In this period, the sea raised the island of Kadzand more than 100 fathoms from its dikes.

1726. A violent tempest changed the *saline* of Arraya, in the province of Cumana, part of Colombia, into a gulf of many leagues in width.

1770—1785. Storms and currents hollowed out a canal between the high and low parts of the island of Heligoland, and transformed this island, so extensive before the eighth century, into two little isles.

"84. A violent tempest formed, according to M. Hoff, the lake of Aboukir in Lower Egypt.

1791—1793. New irruptions of the sea destroyed the dikes and carried away other parts of the island of Nordstrand, already so much reduced.

1803. The sea carried away the ruins of the Priory of Crail in Scotland.

ON THE CONJECTURED BUOYANCY OF BOULDERS AT GREAT DEPTHS IN THE OCEAN.—Travelled Boulder Stones are to be met with in all countries, and in situations where it is extremely difficult to account for their appearance. In addition to the observations which have already appeared in this Journal (see pages 51 and 55), we may remark that Lieutenant W. W. Baddeley, of the Royal Engineers, has communicated to Professor Silliman a curious paper on this interesting subject. He says, "Among the many phenomena which serve to interest and perplex geological students, none are more striking than the formation and position of Boulders, and it is highly probable that no cause, in the first instance, tends more to draw votaries into the labyrinths of this delightful science, than the silent eloquence of these mysterious masses. We who reside on this new and interesting continent (America), are particularly liable to have our conjectures kept alive respecting them, and it appears to me impossible to pass one of these travelled rocks, without feeling the momentary wish that it possessed the power of speech, and the inclination to gratify our curiosity concerning them.

"I wish to call your attention to the applicability of a fact, which, as far as I am aware, has not been pointed out before in any publication, although I doubt not it has occurred to many persons.

"As long as it was maintained that water was incompressible, an opinion originating in the well known Florentine experiment, an augmentation in the specific gravity of sea water, as the result of pressure, could not be rendered applicable to the inquiry. But the effect of experiments in recent times has been to overthrow the Florentine doctrine in this respect, and to render it probable that the density of water, owing to pressure, increases in a ratio proportionate to its depth.

"Now, if this be true, what is the amount of this ratio? Is it sufficient to give to the waters of the ocean, in their greatest depths, a density equal or superior to that of rock? For if this be conceivable, so is it that Pelagian Boulders are now floating about at great depths in all latitudes; and if now, then formerly, an admission which would much lessen the difficulty of accounting for their presence in all climes, and in almost all postures? Thus masses of rock, when once detached from, or reaching the depths of the ocean, might float to the antipodes of the spot where they first commenced their journey, and indeed continue to float until driven by upward currents, or other operative causes, upon the sides and summits of submarine hills and mountains, where they might become deposited, in consequence of their superior relative density compared with that of the water, and they are *now* in, after having undergone the rounding which they would be likely to meet with in the interim from erosion and attrition.

"This conjecture (for it is nothing more) will not be considered absurd, I think, when it is recollected that the compressibility of sea water has been proved by Perkins and others, and is now generally admitted; and that the depth of the ocean, although unknown, must be considerable, some writers making it between two and three miles, while one of them (La Place) considers it to be as high as twenty-five. Bakewell states it to be 'not exceeding ten miles, and more generally admitted not to be over five.' In short, it appears to me that although much is wanted in the way of experiment to determine this interesting question, what is positively known respecting it favors the conjecture now submitted." Lieut. Baddeley proposes the following question, and recommends those who have an opportunity of experimenting to ascertain the relative temperature, density, and saltness of the ocean at various measured depths.

"Given the specific gravity of a boulder (2.6) at the level of the ocean, what would be its specific gravity in relation to the water it is immediately surrounded by, five miles below the surface, and what its ratio of decrease?"

TRAVELLED STONES.—In our last Number, we gave an account of the Travelled Stone of Castle-Stuart. Captain Bayfield has lately laid before the Geological Society an account of the transporting power of the ice peaks framed every winter in extensive shoals on both sides of the River St Lawrence. These shoals are thickly strewed with massive boulder stones, round which the ice froze on all sides; and in the Spring, when the river rises from the melting of the snow, the masses of ice, with these stones still frozen to them, naturally float down the river, and frequently carry these boulders to great distances from their native beds, and are left in situations very remote from any rocks of the same nature. Captain Bayfield also affirms, that icebergs, in which large masses of stone are imbedded in gravel, are annually drifted down the coast of Labrador, through the Straits of Belleisle, and for several hundred miles up the Gulf of St Lawrence. These facts will, in a great measure, account for boulders which are found in situations far apart from formations of a similar kind.

TEMPERATURE OF THE INTERIOR OF THE EARTH.—On Wednesday the 17th August 1835, M. Arago, in delivering a lecture on the theory of the central heat of the earth, related an operation at this time carrying on in Paris, which may be of the highest importance not only to science but to public economy. The municipality have ordered an Artesian well to be pierced near the Barrière des Martyrs; but the men employed, after getting to a depth of 900 feet without finding water, came to a stratum of chalk, so thick, that the undertaking would have been abandoned but for the interference of men of science, who wished it to be continued, with a view to the elucidation of the above theory. According to observations made by means of a thermometer, no doubt remains as to a fact which hitherto it has not been possible to verify with any degree of precision—namely, that the temperature of the earth rises in regular proportion towards the centre; so that, at the 10th degree from the surface, all known matter must be in a state of fusion. At the point to which the perforation in question will have reached, M. Arago expects a spring of water will arise of a sufficient degree of heat to warm public establishments, supply baths, and serve for other purposes.

METEOROLOGY.

EQUINOCTIAL RAINS.—Professor Arago remarks, that navigators occasionally mention rains which fall on their vessels while sailing in equinoctial seas, which led him to think that rain is much more abundant at sea than on land. But this subject still rests merely on conjecture, as it is but seldom that exact measurements are made. Measurements of this kind are not, however, difficult. For example, many measurements were made by Captain Tuckey, in his unfortunate expedition to the River Congo. We know that the discovery ship *Bonite* will be furnished with a small udometer. Its commander should have it placed on the stern of the vessel, in such a situation, that the rain collected by the sails and cordage cannot drop into it.

It would add greatly to the interest of those observations, if navigators would determine the temperature of the rain, and the height from which it falls.

To ascertain the temperature of the rain with precision, it is necessary that the mass of the water should be considerable, relatively to the size of the vessel in which it is contained. A metal udometer is unfavorable for this experiment. It would be greatly preferable to have a large pummel of sand, of some tight stuff, for this purpose, very close in its texture; and the water which runs from the under side of it ought to be received by a glass of small dimensions, containing a little thermometer. The only time at which the elevation of the rain clouds can be ascertained is during the storm; then the number of the seconds which elapse between the appearance of the flash and the time at which the sound reaches the observer, multiplied by 337 metres—the degree of rapidity with which the sound is propagated—indicates the lengths of the hypothenuse of a right-angled triangle, whose vertical side is precisely the height required. This height may be calculated if, by the aid of a reflecting instrument, we estimate the angle formed with the horizon, by the line which, taken from the eye of the observer, terminates in that quarter of the cloud where the lightning first showed itself.

If for an instant we suppose, that rain falls on the vessel whose temperature is lower than that which the clouds must possess, judging from their height and the known rapidity of the decrease of atmospheric heat, every person will understand the place which such a fact would occupy in Meteorology.

If, on the other hand, we imagine that, during a day of hail (for hail does fall in the open sea), the same system of observations had proved that the hailstones were formed in a region where the temperature of the atmosphere was higher than the point at which water congeals, and science would thus be enriched with a valuable result, to which every future theory of hail must necessarily be accommodated.

RAIN IN A PERFECTLY CLEAR SKY.—Some natural phenomena are so extraordinary and improbable in their nature, that many who have witnessed these eccentricities of nature forbear mentioning them, lest they should be ranked as mere visionaries. The rains of the equinoctial regions may be classed among such phenomena.

In intertropical climates rains have frequently been observed when the atmosphere was pure and serene, with the sky of the most beautiful azure! The rain-drops on such occasions do not fall thickly, but they are of much larger dimensions than rain-drops in European climates. The fact is certain; we have the assurance of Baron Humboldt, that he observed occurrences of this kind in the interior of countries: and Captain Beechey observed this phenomenon in the open sea. With respect to the circumstances on which such a singular precipitation of water can depend, we are entirely ignorant of them. In Europe, in calm and clear weather, during the day, sometimes, very small crystals of ice have been observed to fall gently from the atmosphere, their size increasing with every particle of humidity they congeal in their passage towards the earth. Does this approximation put us in the way of obtaining the desired explanation? Have not the large rain-drops of tropical climates been, while in the higher regions of the atmosphere, small particles of ice excessively cold—then in their descent become large pieces of ice by agglomeration; and when lower still, been melted into large drops of water? The object of these conjectures is to exhibit in what point of view the phenomenon may be studied, and to stimulate young travellers in particular, to observe carefully if, during these singular rains, the regions of the atmosphere from which they fall present any traces of halo. If such traces are noticed, however slight they may be, the existence of crystals of ice in the higher regions of the air will be demonstrated.

In almost every country meteorologists are to be met with, but their observations, in too many instances, are made at unsuitable hours, and with instruments which are either inaccurate in themselves, or improperly placed. It seems not to be difficult now to deduce the mean temperature of the day from observations made at any hour: thus a meteorological table, whatever may be the hours noted in it, will be of value by the mere condition that the instruments employed will admit of being compared with the student's barometers and thermometers.

Almost any thermometer will answer the ordinary purposes, but the most useful barometer which we have seen is Adie's sympiesometer, which is used to determine the pressure of the atmosphere, the altitude of any situation above another, or above the level of the sea. To obtain the altitudes from the common barometer, observation tables are required. The sympiesometer gives the altitudes without the use of tables, and only requires a single process of subtraction and multiplication.

DIRECTION OF THE WINDS.—W. C. Redfield, Esq. of New York, makes the following observations on this subject:—"It is deserving of notice, that during some of the coldest periods of winter, in this occasionally serene climate, the predominating winds blow from the south-western or southern quarters of the horizon. This fact appears to be established by the annual reports which have been made to the Regents of the University, and it is believed will become obvious in proportion to the accuracy of our observations. It sufficiently demonstrates, without resorting to other evidence, the fallacy of the notion commonly entertained, that winds are generally rectilinear in their progress, and blow for the most part in right lines over extensive portions of the earth's surface;—an error which appears to remain undisturbed in the minds of some meteorologists."

GENERAL SCIENCE.

On the Use of Nicol's Calcareous Spar Prism in discovering Shoals in the Ocean.—It has been remarked by M. Arago, that the bottom of the sea, or the surface of a shoal *at a given distance from a ship*, is more distinctly seen from its mast-head, or, generally speaking, from a considerable height, than from the deck. He explains this phenomenon on the principle that the reflected light from the surface of the sea, which is always intermixed with that from the bottom, or the shoal, possesses a less and less degree of intensity in proportion as the angle of reflection, reckoned from the surface, is larger. That this reflected light may be entirely removed, when looking into the sea to discover cliffs or shoals, &c., he proposes to observe them by means of a Tourmaline, in which the axis is held horizontally, if possible under a polarizing angle of 37°, reckoning from the surface. The entire and absolute obstruction of the light reflected from the surface of the water cannot possibly take place under a smaller angle than 70°, because it is under this angle alone that it is completely polarized; but under angles of 10° or 12° greater or less than 37°, the number of polarized rays which the Tourmaline can arrest is still so considerable, that the same means of observation cannot fail to be attended with very advantageous results. Poggendorf proposes to use for this important purpose, instead of the Tourmahne, Nicol's *Calc Spar Prism*, because, from its being colourless, it is much better fitted for the purpose. By engaging in such experiments, Arago remarks, that "Mariners will throw a light on a curious question of Photometry; they will probably confer on navigation a means of observation which may prevent many shipwrecks; and by introducing Polarization into the nautical art, they will afford an additional proof of what those individuals expose themselves to, who unceasingly collect experiments and theories without any practical application of them, meeting every remonstrance with a contemptuous *cui bono*." Ere long, we doubt not, ships generally, at least all those vessels specially occupied in geographical and hydrographical researches, will, before leaving port, be provided with Nicol's valuable little instrument.

The Effects of Compressed Air on the Human Body.—Dr Junod has communicated to the Academy of Sciences the results of his experiments with compressed air. In order to operate on the whole person, a large spherical copper receiver is employed, which is entered by an opening in the upper part, and which has a cover with three openings—the first for a thermometer, the second for a barometer or manometer, and a third for a tube of communication between the receiver and the pump. The air in the receiver is perpetually renewed by a cask. When the pressure of the atmosphere is increased one half, the membrane of the tympanum suffers inconvenient pressure, which ceases as gradually as the equilibrium is restored. Respiration is carried on with increased facility; the capacity of the lungs seems to increase; the inspirations are deeper and less frequent. In about eighteen minutes an agreeable warmth is felt in the interior of the thorax. The whole economy seems to acquire increased strength and vitality. The increased density of the air appears also to modify the circulation in a remarkable manner; the pulse is more frequent, it is full, and is reduced with difficulty; the dimensions of the superficial venous vessels diminish, and they are sometimes completely effaced, so that the blood in its return towards the heart follows the direction of the deep veins. The quantity of venous blood contained in the lungs ought then to diminish, and this explains the increased breathing of air. The blood there is then determined in a larger quantity to the arterial system, and especially to the brain. The imagination becomes active, the thoughts are accompanied with a peculiar charm, and some persons are affected with symptoms of intoxication. The power of the muscular system is increased. The weight of the body appears to diminish. When a person is placed in a receiver, and the pressure of the air is diminished one-fourth, the membrane of the tympanum is momentarily distended; the respiration is inconvenienced, the inspirations are short and frequent, and in about fifteen or twenty minutes there is a true dyspnœa. The pulse is full, compressible and frequent; the superficial vessels are turgid. The eyeballs and lips are distended with superabundant fluids, and hemorrhage and tendency to syncope are sometimes induced; the skin is inconveniently hot, and its functions increased in activity; the salivary and venal glands secrete their fluids less abundantly.

REVIEWS.

The Edinburgh New Philosophical Journal, exhibiting a view of the progressive Improvements in the Sciences and Arts, conducted by Professor Jameson, No. 42, July—octavo, 1836. Edinburgh: Adam and Charles Black.

The present Number of this Work contains a great variety of interesting articles, and none more so than that by M. Elie de Beaumont, "on the Temperature of the Earth's Surface during the Tertiary Period;" and also the article by Leopold Von Buch "on Volcanos and Craters of Elevation." The object of this paper is to show, that Craters of Elevation are not Volcanos; that there is a well grounded and important distinction between the two; and that even the cones of Volcanos can be formed only by a sudden elevation, and never by the building up of streams of lava. He considers that *Volcanos* are the constant chimneys, the canals uniting the interior of the earth with the atmosphere, which spread around themselves the phenomena of irruption from Craters that are of small extent, and are only once in operation; while *Craters of Elevation*, on the contrary, are the remains of a great display of powers from within, which are capable of, and actually have raised islands of several square miles in extent, to a considerable height. In a future Number we shall present in detail the views of Von Buch on this highly interesting subject.

The paper by M. Charpentier "on the Glaciers of the Canton of Vallais," contains some remarkable facts respecting the formation and decomposition of Glaciers. One of these we think deserving of notice; namely, that the mass of a Glacier consists of ice, or rather frozen snow, in a pure state, without any mixture of earth or stones. When blocks fall through a fissure to the bottom of the Glacier, they are rolled or pushed forward. If they remain hemmed in between the walls of the fissure, they apppear again after a lapse of a certain period of time on the surface of the Glacier, but at a point farther down the valley than where they fell in. When, however, a block falls quite near the lower end of a Glacier through a fissure to the bottom, and at a time when the Glacier is retiring, it remains nearly at the same point and in the same position which it occupied when it fell.

Bache's paper on the alleged influence of colour on the radiations of non-luminous heat contains a series of important experiments, the results of which are decidedly unfavorable to the specific effect of colour, in determining the radiating powers of bodies; and if the results be admitted as decisive of the radiating powers of the bodies used, they show, that each substance has a specific power not depending upon chemical composition nor upon colour.

Assistant-Surgeon Jameson's notes on the Natural History and Statistics of the Island of Cerigó, in the Mediterranean, are full of new and interesting information; and Dr Graham's Botanical Excursion acquaints us with a number of new localities, in which the rarer plants of our native Flora have been discovered.

This Number contains a very full account of the proceedings of the British Association at Bristol in August 1836; and the Miscellaneous information contained in the Scientific Intelligence is full and varied.

The Edinburgh Philosophical Journal was commenced in June 1819, and has been continued uninterruptedly, by Professor Jameson, up to the present time. It is but justice to say, that no European work, which has appeared during the same series of years, contains so full and varied accounts of the progress of the Sciences and Arts as it does. It is an indispensable companion to the Library of all who study the Physical Sciences, and the advancement of the useful arts.

OBITUARY.

Mr Edward Turner Bennet.—Natural History has met with an incalculable loss by the death of Mr Bennet, who died on Sunday the 21st August, 1836, after a very short illness. He was Secretary to the Zoological Society, which office he filled with much credit to himself, and with great advantage to the Society. The scientific labours which this gentleman undertook, in addition to his official capacity, was to watch and detail the habits of the living Mammalia in the Gardens of the Society, Regent's Park. This task he performed with distinguished ability, which is borne out by his papers in the "Zoological Journal," brought out mainly by his able support and influence. He was also author of those excellent works the "Town Menagerie," and the "Gardens and Menagerie of the Zoological Society," of which we have spoken in high terms, in our notice at page 48 of this Journal.

Mr David Douglas.—It is with extreme regret that we have to announce the death of this indefatigable and excellent young man; and more especially the melancholy accident which led to it. Mr Douglas was engaged by the Horticultural Society of London to travel in various countries as a Naturalist; and has in his Journeys been very successful in discovering many new plants, and various Mammalia and Birds; especially on the Western territories of North America. He was engaged in his scientific pursuits when he met with the fatal accident, which took place on the 12th July 1834. The cause of his death is thus recorded in the Ke Kamu Hawaii, printed at Honolulu, of 26th Nov. 1834:—"From Edward Gurney, an Englishman, we received the following account of the tragical scene: About ten minutes before six o'clock in the morning, Mr Douglas arrived at his house in the Mountain, and wished him to point out the road to Hilo, and to go a short distance with him. Mr Douglas was then alone, but said that his man had gone out the day before (this was probably John, Mr Deill's coloured man). After taking breakfast, Edward accompanied Mr Douglas about three-fourths of a mile, and after directing him on the path, and warning him off the traps, went on about half a mile farther with him. Mr Douglas then dismissed him, after expressing an anxious wish to reach Hilo that evening, thinking that he could find out the way himself. Just before Edward left him, he warned him particularly of three Bullock traps, about two miles and a half a-head, two of them directly on the road, the other on the side. Edward then parted with Mr Douglas, and went back to skin some bullocks which he had previously killed. About eleven o'clock, two natives came in pursuit of him, and said that the European was dead, and that they had found him in a pit in which the bullock was. They mentioned that they were coming up to this pit; one of them observing some of the clothing on the side, exclaimed *lole*, but in a moment afterwards discovered Mr Douglas within the cave, trampled under the feet of the bullock. They went off immediately for Edward, who left his work, ran to the house for a musket and ball, and hide, and on coming up to the pit, found the bullock standing on Mr Douglas's body. Mr Douglas was lying upon his right side. He shot the animal, and after drawing him to the other end of the pit, succeeded in getting out the body. His cane was with him, but the bundle and dog were not. Edward knowing that he had a bundle, asked for it. After a few moments' search, the dog was heard to bark, at a short distance a-head, on the road leading to Hilo. On coming up to the place, he found the dog and the bundle. On further examination, it appeared that Mr Douglas had stopped for a moment and looked at the empty pit, and also at the one in which the bullock had been taken; that after passing up the hill some fifteen fathoms, he laid down his bundle, and went back to the pit in which the bullock was entrapped, and which lay on the side of the pond opposite to that along which the road runs; and that whilst looking in, by making a misstep, or by some other fatal means, he fell into the power of the infuriated animal, who speedily executed the work of death." Thus has perished a Naturalist who has travelled over a wide extent of country, and who, from his zeal, must have collected an immense number of facts which could not fail to have been of much use to science.

Edinburgh: Published for the Proprietors, at their Office, 16, Hanover Street. London: Smith, Elder, and Co., 65, Cornhill. Glasgow and the West of Scotland: John Smith and Son; and John Macleod. Dublin: W. F. Wakeman. Paris: J. B. Balliere, Rue de l'Ecole de Médecine, No. 13 bis.

THE EDINBURGH PRINTING COMPANY.

THE EDINBURGH

JOURNAL OF NATURAL HISTORY,

AND OF

THE PHYSICAL SCIENCES.

JANUARY, 1837.

ZOOLOGY.

DESCRIPTION OF THE PLATE—THE WEASELS.

All the animals of this tribe are carnivorous. Their slender, lengthened, and cylindrical bodies, short legs, and the very free motion allowed in every direction by the loose articulations of their spine, well fit them for pursuing their prey into the deepest recesses. Constituted by Nature to subsist on animals, many of which have great strength and courage, the Weasels possess an undaunted and ferocious disposition. In their predatory character they are inferior to the Cats alone, to which they bear a striking resemblance in many points of their organization. Their disposition is very sanguinary, which frequently impels them to commit the most extensive devastation, simply for the sake of gratifying their excessive thirst for blood. Rats and Mice form a favourite article of their food, and they are very destructive to poultry. All the animals of the Weasel kind are nocturnal, usually remaining the greater part of the day in their retreats asleep, but on the approach of night, they begin to arouse themselves, and prowl about in search of the victims on which they prey.

The Weasels belong to the genus *Mustela* of Linnæus, retained by Cuvier, but from which he has discarded the Ichneumon, and other allied animals. He has formed the remainder into three subgenera, under the titles of *Putorius, Martes,* and *Mephitis,* the last of which embraces the Mephitic Weasel of South America and other fœtid species.

Fig. 1. The Polecat (*Putorius maximus*).—In the winter season the Polecat or Foumart frequents houses, barns, &c. feeding on poultry, eggs, and sometimes milk. It also consumes fish. Bewick relates that one of these, during a severe storm, was traced in the snow from the side of a rivulet to its hole at some distance from it. As it was observed to have made several trips, and as other marks were seen in the snow which could not easily be accounted for, it was thought a matter worthy of some attention; its hole was accordingly examined, the Polecat taken, and eleven fine Eels were discovered to be the fruits of its nocturnal excursions. The Foumart is common all over Europe.

Fig. 2. is a white variety of this animal, commonly called the Ferret, and which many Naturalists have erroneously supposed to be a distinct species.

Fig. 3. Hardwick's Weasel (*P. Hardwickii*).—This animal is a native of the mountains of Nepaul, India. It is of a rapacious disposition, living in forests, and destroying small birds and quadrupeds. This is a large species, measuring from the point of the muzzle to the insertion of the tail two feet two inches, the length of the tail being one foot seven inches and six lines. Its native name is "*Mull-Sānprak*," pronounced with a nasal sound.

Fig. 4. The Common Weasel (*P. vulgaris*).—This animal seldom exceeds nine inches in length from the nose to the tail, the latter being only about two inches and a half in length. It is a bold and active little creature, and makes no scruple of attacking a Hare, which, if it once seizes, it is sure to overcome and destroy; it is more than a match for the strongest Rat.

Fig. 5. The Ermine (*P. erminea*). in its summer fur, and Fig. 6. in its full winter dress. The difference between the Ermine or Stoat and the Weasel is so small, that they have frequently been mistaken for each other when in their summer fur. The Ermine is, however, considerably larger than the Weasel, its length being fully ten inches, and the tail five inches and a half. During winter this animal is wholly of a yellowish white except the tail, which is invariably black at the point. It inhabits Britain, Norway, Lapland, and Russia.

Fig. 7. The Pine Marten (*Martes pinus*) inhabits the North of Europe, Asia, and America, and is also found in some of the more extensive woods of this country. It is said that in its combats with the Wild Cat, it frequently comes off victorious, and instances are mentioned of its killing that animal, although so much larger.

Fig. 8. The White-cheeked Marten (*M. flavigula*).—This species is a native of India, and is one of the largest of its tribe. It is a beautiful animal, but partakes in a small degree of that unpleasant effluvium which is so strong in the Polecat.

Fig. 9. The White-eared Marten (*M. leucotis*).—This animal is twenty inches long, but little is known of its habits, and zoologists are even unacquainted with its native country.

Fig. 10. The Sable (*M. Zibellina*).—This animal, so much in request on account of its valuable and fine fur, is a native of Siberia and Kamschatka. It lives much in trees, and leaps with great activity from branch to branch in pursuit of Squirrels and Birds. In winter it feeds upon berries.

Fig. 11. The Vison (*M. vison*).—America is the native abode of this animal; it is only about fifteen inches in length, exclusive of the tail.

DESCRIPTION OF THE PLATE—THE FINCHES AND BUNTINGS.

The Finches are a numerous and active race, widely dispersed over the world, feeding principally on grain, seeds, and insects. Some of them are remarkable for the melody and variety of their notes, while others are destitute of song, and only utter a chirp.

Fig. 1. The Field Sparrow (*F. pusilla*) is a native of Pennsylvania; it is a migratory species, arriving in that State early in April. It has no song, but a note not unlike the chirp of a cricket. It is the smallest of the American Finches.

Fig. 2. The Swamp Sparrow (*F. palustris*).—This is another summer visitant of Pennsylvania, where it arrives early in April, and frequents the low pine swamps. It roars from two to three broods in a season, returning to the south on the approach of cold weather. It has no song, but a simple chirp.

Fig. 3. The Tree Sparrow (*F. arborea*).—This is a native of the northern parts of America, taking up its winter residence in Pennsylvania, and most of the northern states, where it arrives in the beginning of November, and departs in April. It frequents sheltered hollows, thickets, and hedge-rows. It has a low warbling note.

Fig. 4. The Song Sparrow (*F. melodia*).—This species is very generally diffused over the United States, and is only partially migratory. It commences its song early in the spring, and continues its sweet warblings during the whole summer, and is sometimes even heard in the depth of winter. It frequents the borders of rivers, meadows, and swamps. Its nest is built on the ground.

Fig. 5. The Chipping Sparrow (*F. socialis*).—This is a social American species, inhabiting the city in common with Man in the summer season; but retires to the fields and hedges as the cold approaches, and takes its final departure for the south of America when the frost sets in.

Fig. 6. Henslow's Bunting (*Emberiza Henslovii*).—This new species was first discovered by Audubon, opposite Cincinnati, in the United States. Nothing is known of its history or habits.

Figs. 7 and 8. The Lapland Long-Spurred Bunting (*E. Lapponica*), male and female, inhabit the desolate arctic regions of Europe and America. They are common Birds in Lapland, and have been known to penetrate as far as some of the middle States of America. The Lapland Longspur only 'sings during its aerial flights, at which time it utters a few agreeable and melodious notes.

On the Habits of the Ringdove (*Columba Palumbus*).—Sir,—It is stated in Dr Fleming's History of British Animals, page 47, genus *Columba,* that the Columba Palumbus, or Ringdove of our woods, is easily tamed, but will not breed in confinement. Mr John Robertson, of Rosehall Tea Gardens, Glasgow, has in his aviary a pair of these beautiful Birds. A wooden box was put up in a corner of the aviary last spring, with the intention of giving these Birds an opportunity to breed. The box, however, was not securely fixed, and fell from its position in a short time after, containing a nest and one egg, which was broken to pieces by the fall. The box again being secured, the Cushats again built, and, pursuing the work of incubation, were successful in bringing out and rearing a pair of fine healthy young ones, and are to be seen in Mr Robertson's possession at this time. By inserting the above fact in your useful Journal, you will oblige, &c, (Signed) John Blythe,
30. East Clyde Street,
Glasgow, 28th Nov. 1836.
To the Editor.

The above interesting fact serves to correct a mistake into which most naturalists have fallen, in supposing that the Ringdove will not breed in captivity. This Bird is the largest of the European Columbidæ, and may form a valuable addition to our domestic animals. The first brood should be carefully preserved; for if they can be induced to reproduce, their progeny will in all probability succeed, and be rendered completely domesticated, thus proving a valuable accession to the luxuries of the table. During our residence at Prinlaws House, near Leslie, in Fife, a pair of Ringdoves built and incubated on a spruce fir-tree, in a plantation which bounded our garden, and not more than fifteen yards from the house. Their nest was not more than twenty-five feet from the ground, and close to a garden walk through which the family were constantly passing. Yet this seemed to give the connubial pair no uneasiness, as the female would sit on her nest, with her mate by her side, without attempting to quit the spot on the approach even of several individuals. The first young, which they brought up in the following spring, built a nest in a spruce fir within the garden, and reared their young. Both these nests were occupied every season afterwards during the four years which we remained there. These Birds very frequently alighted and fed in the garden, and even ate occasionally along with the domestic poultry; from which we are of opinion that, if unmolested, the Cushat is not so shy a Bird as is generally imagined, and that it is only from persecution that it retires to the deep recesses of woods for shelter from its numerous enemies.

SINGULAR TENACITY OF LIFE IN AN AQUATIC MOLLUSCOUS ANIMAL.—M. Rang, Member of the Royal Academy of Sciences, Paris, received four young specimens of *Anodonta rubens*, of Lamarck, from Senegal, and although they had been enveloped in cotton for two months, they were still alive; he had learnt that these animals live eight months of the year out of water, upon the ground being suddenly abandoned by the river, and that they remain during six of these months exposed to the ardent heat of an almost vertical sun.

MARINE AND RIVER MOLLUSCA IN THE GULF OF LIVONIA.—So inconsiderable is the saltness of the sea-water in the Gulf of Livonia, that fresh and salt water conchiferous molluscous animals live together promiscuously on the same coasts. M. Freminville informed the Philomatic Society of Paris, that he collected from the same localities species of the fresh water genera, *Anodonta*, *Unio*, and *Cyclas*, which were intermixed with species of the Marine genera, *Tellina*, *Cardium*, and *Venus*.

STEP OF THE CAMEL.—What always struck me, as something extremely romantic and mysterious, was the noiseless step of the Camel, and the spongy nature of his foot. Whatever be the nature of the ground—sand or rock, turf or paved stones—you hear no foot-fall; you see an immense animal approaching you, *still* as a cloud floating on air, and unless he wear a bell, your sense of hearing, acute as it may be, will give you no intimation of his presence.—*Constantinople in 1828, by Charles Macfarlane*

SIVATHERIUM GIGANTEUM.

THIS very large animal forms an important accession to extinct Zoology. In size it greatly surpasses the Rhinoceros, and hence is larger than any known Ruminant in the Class Mammalia to which it belongs; at the same time the forms of structure which the Sivatherium exhibits, render it one of the most remarkable of fossil animals hitherto detected in the more recent strata of the globe.

Of the numerous fossil mammiferous genera previously discovered and established, all were confined to the Edentata and Pachydermata. The fossil species belonging to other orders have all their living analogues upon the earth; and among the Ruminantia, in particular, no remarkable deviation from the existing types has hitherto been discovered. However, the isolated position of the Giraffe and the Camels made it probable that certain genera had become extinct, which formed the connecting links between these and the other genera of the order, as well as between the Ruminantia and the Pachydermata. In the Sivatherium we have a Ruminant of this description, connecting that order with the Pachydermata, and at the same time so marked by individual peculiarities as to be without a living analogue in its own order.

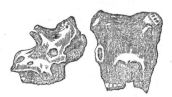

The large cranium of the Sivatherium, of which the above are representations in profile and in front, is a remarkably perfect fossil. When discovered, it was fortunately so completely enveloped in a mass of stone, that although it had long been exposed to be acted upon as a boulder in a water-course, all the more important parts of structure had been preserved. The block might have been passed unnoticed, had not a projecting edge of the teeth, by jutting out in relief, given a promise of some thing valuable concealed beneath. After much labour, the hard crystalline covering of stone was so successfully removed, that the huge head now stands out, with a couple of horns between the orbits, broken only near the tips, and the nasal bones projected in a free arch high above the chaffron. All the molars on both sides of the jaw are present, and singularly perfect. The only mutilation of the fossil is at

the vertex of the cranium, where the plane of the occipital meets that of the brow, and at the muzzle, which is truncated a little way in front of the first molar tooth. The only parts which are still concealed are a portion of the occipital, the zygomatic fossæ on both sides, and the base of the cranium over the sphenoid bone.

The form of the head is so singular and grotesque, that the first glance strikes us with surprise. The prominent features are, 1st. The great size, approaching to that of the Elephant; 2d, The immense development and width of the cranium behind the orbits; 3d, The two divergent osseous cones for horns, starting out from the brow, between the orbits; 4th, The form and direction of the nasal bones, rising with great prominence out of the chaffron, and overhanging the external nostrils in a pointed arch; 5th, The great massiveness, width, and shortness of the face forward from the orbits; 6th, The great angle at which the grinding plane of the molars deviates upwards from the base of the skull.

Viewed in lateral profile, the form and direction of the horns, and the rise and sweep in the bones of the nose, give a character to the head widely differing from that of any other animal. The nose looks something like that of the Rhinoceros; but the resemblance is deceptive, and only owing to the muzzle being truncated. Seen from the front, the head is somewhat wedge-shaped, the greatest width being at the vertex, thence gradually compressed towards the muzzle; with contraction only at two points, behind the orbits and under the molars. The zygomatic arches are almost concealed, and nowise prominent; the brow is broad and flat, and swelling latterly into two convexities; the orbits are wide apart, and have the appearance of being thrown far forward, from the great production of the frontal upwards. There are no spines or ridges; the surface of the cranium is smooth; the lines are in curves, with no angularity. From the vertex to the root of the nose, the plane of the brow is in a straight line, with a slight rise between the horns.

There are six molars on each side of the upper jaw. The third of the series, or last milk molar, has given place to the corresponding prominent tooth, the direction of which, and of the last molar, is well advanced, and indicates the animal to have been more than adult.

The teeth are in every respect those of a Ruminant, with some slight individual peculiarities. The three posterior double molars are composed of two portions or semi-cylinders, each of which incloses, when partially worn down, a double crescent of enamel, the convexity of which is turned inwards. The last molar, which is normal in Ruminants, has no additional complication, like that in the corresponding tooth in the lower jaw. The plane of grinding slopes from the outer margin inwards. The general form is exactly that of an Ox or Camel on a large scale. The ridges of enamel are unequally in relief, and the hollows between them equally scooped. Each semi-cylinder has its outer surface in various sections, formed of three salient knuckles, with two intermediate sinuses, and its inner surface of a simple arch or curve. But there are certain peculiarities by which the teeth differ from those of other Ruminants.

Corresponding to the shortness of jaw, the width of the teeth is much greater in proportion to the length than is usual in this order, the width of the third and fourth molars being to the length as 2.24 and 2.2 to 1.55 and 1.68 inches respectively, and the average width of the whole series being to the length as 2.15 to 1.75 inches. Their form is less prismatic, the base of the shaft swelling out into a bulge or collar, from which the inner surface slopes outward as it rises, so that the coronal becomes somewhat contracted. In the third molar, the width at the coronal is 1.93, at the bulge of the shaft 2.24. The crescentic plates of enamel have a character which distinguishes them from all other known Ruminants; the inner crescent, instead of sweeping in a nearly simple curve, runs zig-zag-wise, in large sinuous flexures, somewhat resembling the form of the *Elasmatherium*. The space occupied by the line of the molars is 9.8 inches.

From the anterior margin of the foramina to the alveolus or first molar tooth,	18.85 inches
From ditto to the truncated extremity of the muzzle,	20.6
From the tip of the nasals to the upper fractured margin of the cranium,	18.0
From ditto to ditto along the curve,	19.0
Width of cranium at the vertex,	22.0
Ditto between the orbits, upper borders,	12.2
Ditto, ditto, lower borders,	16.3
Ditto between the outer surfaces of the horns at their base,	12.5

Among a quantity of bones collected in the neighbourhood of the spot in which the skull was found, there is a fragment of the lower jaw of a very large Ruminant, which is supposed to have belonged to the Sivatherium; and it is even not improbable that it came from the same individual with the head now described. It consists of the hind portion of the right jaw, broken off at the anterior margin of the last molar. The outline of the jaw, in vertical section, is a compressed ellipse, and the outer surface more convex than the inner. The form and relative proportions of the jaw agree very closely with those of the corresponding parts of the Buffalo. The dimensions, compared with those of the Buffalo and Camel, are thus:

	Sivatherium.	Buffalo.	Camel.
Depth of the jaw from the alveolus, last molar,	4.95	2.65	2.70 inches.
Greatest thickness of ditto,	2.3	1.05	1.4
Width of middle of last molar,	1.35	0.64	0.76
Length of posterior, ⅔ of ditto,	2.15	0.95	1.15

No known Ruminant fossil has a jaw of such large size,—the average dimensions above given bring more than double those of a Buffalo, which measured in length of head 19.2 inches, and exceeding those of the corresponding parts of the Rhinoceros.

This gigantic fossil cranium, above described, was discovered near the Markenda river, in one of the small valleys which stretch between the Kyárda-dún and the valley of Pinjór, in the Siválík or sub-Himálayan belt of hills, associated with bones of the fossil Elephant, Mastodon, Rhinoceros, Hippopotamus, &c.

Among a quantity of bones collected from the same neighbourhood with the cranium of the Sivatherium, there are three singularly perfect specimens of the lower portions of the extremities, belonging to three legs of one individual. They greatly exceed the size of any known Ruminant, and there is no other ascertained animal of proportionate size excepting the Sivatherium.

BOTANY.

GENERAL REVIEW OF THE VEGETABLE KINGDOM.—NO. IV.

The principal causes which induce this progression of changes are three; 1st, the excess of duration in the winters, a consequence of the obliquity and disappearance of the solar rays; 2d, the dryness of the air, a consequence of the decrease of heat; 3d, the prolonged action of the light, which illumines the horizon through the whole period of vegetation. It may be proper here briefly to trace the effects of each of these three causes.

1st, It is well known that too great a degree of cold, by congealing the sap, occasions the rupture of the vascular system in plants, and thereby destroys them; but the deleterious action of cold is not confined to purely mechanical results; it has been proved that heat is a stimulus that cannot be dispensed with in vegetation. Many species secrete juices in warmer regions, which are unknown in their economy in colder climates. The Ash yields manna in Calabria, but loses that faculty as it advances towards the north. The Grape, in the south of Europe, abounds in matter of a sweet quality; in the north, it contains an excess of acid. As long as the organic functions, which depend upon the degree or duration of heat, can be carried on, the Ash and the Vine continue to grow; they grow even when those functions are performed incompletely, but their growth is stunted. They finally disappear at that point where the portion of warmth in the atmosphere, though still equal to prevent the freezing of the sap, is no longer able to stimulate their organs or their frame into action. All other vegetables, whose dimension and duration subject them to the full severity of the frost, share the same destiny, at a greater or less distance from the torrid zone, and in proportion as their constitutions require a greater or less degree of heat. So that nothing is found near the Pole but such dwarf shrubs as are sheltered under the snow in winter, or annuals and herbaceous species, endowed with so quick a principle of life, as to rise, flower, and fruit within the space of three months; or some agamous and cryptogamous species, which adapt themselves to all degrees of temperature, and are consequently the last organic forms under which vegetable life is to be descried.

2d, Heat and moisture are highly favorable to the growth of plants. No countries are more abundant in herbaceous vegetables, or better wooded, than Senegal, Guinea, and Cayenne, where both these props of vegetation are in the plenitude of their force. Experiments made with the hygrometer prove that the moisture of the atmosphere increases as we approach the Equator. In hot climates, when the sun sinks below the horizon, the watery exhalations condensed are returned to the earth in the form of dew, that moistens the surface of the foliage, and feeds those vegetables in which the absorbing powers of the parts above ground suffice for their support. Of this number are the succulent plants; the Aloes, the *Cacti*, the *Mesembryanthemums*, some of the Spurges, &c. In these the fibrous root only serves to hold them in their places; and the moisture of the atmosphere is inhaled and retained by the spongy parts above. Thus, in the vast plains that receive the waters from the eastern declivity of the Andes, when the scorching heat of summer has consumed the grasses and other herbaceous kinds which the rainy season had brought forth, we still find some lingering *Cacti*, which, under their dry thorny coats, conceal a cellular system, by which an abundant sap has been imbibed and preserved. But in countries where the atmosphere holds but little moisture in evaporation, either because the soil is wholly destitute of water, or by reason of the coldness of the temperature, we find no plants at all, or such only as are of dry, hard texture. The sands of Africa, watered by no river, are found to be utterly barren. Spitzbergen, Nova-Zembla, Kamschatka, &c. where the influence of the sun is only felt for two months in the year at most, and where, consequently, the air is habitually dry, furnish a very scanty portion of herbaceous only, or some dwarf shrubs, with a narrow leathery foliage. It is true that drought is not in these instances the sole cause of the degenerated state of vegetation, but it would of itself be sufficient to produce it; for it is a fact, that plants acquire height of stem and breadth of foliage only in proportion to the abundance of nutriment which they meet with in the atmosphere, and that nutriment is water reduced into vapour, and held in suspension by the atmosphere.

3d, When vegetables are deprived of light, they extend in length, shoot up pale lank stalks, are of a lax fibre, and of no substance; in short, they spindle themselves out. The way that light acts upon this class of the creation is principally in separating the elementary parts of the water and carbonic acid contained in them, and in extricating the oxygen of the latter. The carbon of the acid, with the hydrogen and oxygen of the water, form the bases of the gums, resins, and oils, which flow in the vessels or fill the cells. These juices nourish the membranes, and induce the woody state in them; and they do this in proportion as the light is stronger, and its action more prolonged. Thus we see that darkness and light have effects directly opposite upon vegetables. Darkness favors the length of their growth, by keeping up the pliancy of their parts; light consolidates them, and stops growth by favoring nutrition. It should follow that a fine race of vegetables, one that unites in due proportion size and strength, depends in part upon the proper reciprocation of nights and days. Now, in the northernmost regions, plants go through all the stages of growth at a time when the sun no longer quits the horizon; and the light, of which they experience the unremitting effect, hardens them before they have time to lengthen. Their growth is quick, but of short duration; they are robust, but under-sized.

The same plants, when transplanted into milder regions, where the atmosphere is moist, and light and darkness follow in regular succession, are seen to lengthen their stems, expand their branches, as well as multiply, dilate, and soften their leaves. They must, however, be endowed with a frame of sufficient pliancy to support their new mode of existence.

Effects of Mountain Heights on Plants.—During the recent interesting tour made by M. Gay among the Cordilleras, he discovered many beautiful and rare specimens of *Baccharis*, *Loasa*, *Alstrœmeria*, and above all, those charming flowers, the *Mutisia*, which exhibit the most singular phenomena. As the tendrils with which these plants are usually furnished would become useless in these cold regions, unprovided with shrubs or bushes, they are changed into real leaves, organs of such great utility to mountain plants. He also remarked, that the plants which are herbaceous in the plains, become here entirely ligneous; and that several trees, especially the *Escaleonia*, instead of assuming that forked appearance which characterises it, becomes stunted, creeping along the rocks, and thus offering less surface to the cold with which the wind is charged in passing over these numerous and immense glaciers. But another observation, which he also made among these cold regions, is still more interesting; it is the form of imbricated leaves which the greater portion of the vegetables assume, among those generally even whose habitual form seems to be entirely contrary to this disposition; thus the leaves of the *Triptilion*, which are so lax and small in the lower regions, become here extremely hard and tough, closely imbricating the stalks, and even the flowers of these beautiful plants. The *Mutisia*, which is nearly devoid of leaves when at the side of mountains, produces at their summit a considerable number. The violets here have not that elegant form, which is observed in those flowers lower down, but possess a conformation altogether different; they have the appearance of a rosette, which may be compared to that of a *Sedum*, with this difference, that the leaves, instead of being almost vertical, are in these alpine violets entirely horizontal. These leaves, which are extremely hard and tough, are round, scabrous, strongly imbricated, and exhibit at the footstalk flowers which are sessile, and of a violet colour somewhat approaching to red. Although M. Gay was very familiar with the genera *Triptilion*, *Escaleonia*, *Mutisia*, and *Viola*, the particular aspect of these alpine species caused him to mistake them entirely, nor did he discover to what genus they belonged, until he studied them after his return.

On the Bedeguar of the Rose.—Mr Jesse makes the following observations on this subject:—"I have often admired a small, round, mossy substance attached to a branch of the Dog-Rose growing in our hedges, and which I was unable to account for until the following circumstance was related to me by an ingenious florist and nurseryman in the King's Road, Chelsea, London, Mr Knight, who informed me, that, having been requested by one of his customers to endeavour to preserve a favorite mulberry tree, which for many years had flourished on her lawn, but which, with the exception of one very large branch, was either dead or decaying, he waited till the sap was ascended, and then banked the branch completely round near its junction with the trunk of the tree. Having filled three sacks with mould, he tied them round that part of the branch which had been barked, and by means of one or two old watering-pots, which were kept filled with water, and placed over the sacks, from which the water gradually distilled, the mould in the sacks was sufficiently moistened for his purpose. Towards the end of the year he examined the sacks, and found them filled with numerous small fibrous roots, which the sap, having no longer the bark for its conductor into the main roots of the tree, had thus expended itself in throwing out. A hole having been prepared near the spot, the branch was sawn off near the sacks, and planted with them, the branch being propped securely. The next summer it flourished and bore fruit, and is still in a thriving state.

"Having heard this fact, I examined the mossy substance on the Dog-Rose, and found that, in consequence of the bark on the branch on which it was found having been removed by some insect, the sap in receding had thrown out roots, which, from the exposure of the air, produced the mossy ball in question, and which was probably made the nest or hybernaculum of some insect. If this mossy substance be examined, the larva of an insect will be found belonging to the genus Cynips."

On the Structure of the Radish Root.—It is well known to most observers, that at the summit of the root of the common radish, at the very base of the stem, or that place which the French call the *collet*, the English the *neck*, is an appendage at first resembling a membraneous sheath, enwrapping the young root, and subsequently, as the root distends, becoming two loose straps hanging down on each side of the root. The nature of this appendage was unknown, until the late L. C. Richard discovered the existence of two modes of germination, called the *exo-rhizal* and *endo-rhizal*, and suggested that the radish was an example of the latter mode, a notion which has been generally admitted by recent writers, notwithstanding the circumstance, that if endo-rhizal, the radish would offer an exception to a very general law, that endo-rhizal germination goes along with the indigenous growth. M. Turpin has lately demonstrated that the supposed fleshy root of the radish belongs to the ascending axis, not to the descending one, and that consequently it belongs to the system of the stem, and not to that of the root. He further asserts, that the tumour, which ultimately becomes the radish, is in the beginning cylindrical, and that its cuticle loses at a very early period the power of distension; in short, that it dies, and separates from the subjacent living matter, just as dead bark separates from liber and young wood in old stems. Now this premature death of the cuticle is connected with the rapid lateral distension of the tumour, the cause of the existence of the two appendages in question, which are nothing more than two straps of dead cuticle, rent asunder by the gradual but rapid distension of the part that they originally ensheathed.

On the Materials of Leghorn and Tuscan Bonnets.—There has been considerable controversy as to the plant from which these bonnets are manufactured. M. Merat, M.D. in a letter to the Editor of the Archives de Botanique, says, "This gramineous plant is a genuine Wheat (*Triticum*), and not a Rye (*Secale*), as M. Chaubard terms it. It is a summer corn, a sort of spelt. It is sown in the spring in barren ground, in a dry soil, and cut down before the expansion of the ear. The entire straw (*Sommites*) is employed, after having first bleached in the dew, and afterwards by a chemical process. There are lands which produce superior straw to others, and of which the middle boles (*centre nœuds*) are longer, a circumstance which is particularly desirable. The same fields in France do not produce such good straw, although the grain answers well."

METEOROLOGY.

OBSERVATIONS ON LIGHTNING, BY M. ARAGO.

M. Fusinieri has been lately studying the effects of lightning under an entirely new point of view.

According to this philosopher, the electrical sparks, seen as they traverse the air from ordinary machines, contain brass in a state of fusion; and incandescent molecules of zinc, when they emanate from a brass conductor, if the sparks issue from a ball of silver, they contain impalpable particles of that metal. In the same way, a globe of gold gives rise to sparks, which contain, during their passage through the air, melted gold, &c. &c.

In the centre of all these sparks the molecules are melted only; but in the circumference, the metallic particles undergo a greater or less degree of combustion, in consequence of their contact with the oxygen of the atmosphere.

When a spark emanating from a globe of gold traverses a silver plate, even of considerable thickness, there is seen on the two surfaces of the plate, at the point where the electric spark entered and emerged, a circular layer of gold, the thickness of which must be very inconsiderable, since the natural volatilization is sufficient to cause it to disappear entirely after a short time. According to M. Fusinieri, these two metallic spots are formed at the expense of the fused gold which the electric spark contains. The deposit on the first face is nothing extraordinary; but by adopting the explanation of the Italian philosopher for the spot on the opposite surface, we are obliged to admit, that the gold disseminated through the spark has passed, at least in part, along with it through the whole thickness of the silver plate! It is unnecessary to add, that a spark issuing from a ball of copper gives rise to the same phenomena.

The spark which emanates from a certain metal, does not merely lose a portion of the molecules with which it was at first impregnated, when it traverses another metal; but it becomes charged with new molecules at the expense of that metal. It is even asserted by M. Fusinieri, that, at each passage of the spark, reciprocal changes are produced between the two metals; that when the spark, for example, quits the silver to pass to the copper, it not only transports a portion of the first metal to the copper, but it likewise transports the copper to the silver! I will not, however, longer insist on these phenomena, and have only now brought them forward with the view of showing that the sparks of our ordinary machines contain ponderable substances.

M. Fusinieri asserts that similar substances exist in lightning, and that in this case also they are in a state of great division, of ignition, and combustion. According to his views, these transported matters are the true cause of the transient smells which always accompany thunder, and also of the pulverulent deposits which remain round the fractures through which the electrical matter has forced a passage. In these deposits, which have been too much neglected by observers, M. Fusinieri has detected metallic iron, in different degrees of oxidation and sulphur. The ferruginous spots left on the walls of houses may be found, when strictly examined, to arise from the iron with which the lightning was charged, at the expense of that which occurs in almost every building; but how are we to explain the sulphurous spots on these same walls, and more especially the ferruginous marks which exhibit themselves in the open field, on trees struck with lightning? M. Fusinieri considers himself warranted to conclude from these experiments, that the atmosphere contains, at every height, or at all events as far as the region of stormy clouds, iron, sulphur, and other substances, on the nature of which chemical analysis has been hitherto silent; that the electrical spark is impregnated with them, and that it conveys them to the surface of the earth, where they form slight deposits round points that have been struck with lightning.

This new method of regarding electrical phenomena unquestionably deserves to be pursued with that accuracy which is suited to the existing state of science. Every one who witnesses the fall of a thunderbolt would render an essential service to science by carefully collecting the black or coloured matter which the electric fluid appears to have deposited at every stage of its progress, when it must have undergone sudden changes in rapidity. A careful chemical analysis of these deposits may lead to unexpected discoveries of very great importance.

Spring on the Peak of the Sarrantina.—M. Darien, who lately made a scientific tour in the kingdom of Asturias, mentions the following singular fact:—"A beautiful spring," says he, "flows from the highest point of the peak of the *Sarrantina*; but as this peak is not commanded by any neighbouring summit, we may necessarily suppose that the other branch of the syphon must be placed at a great distance to the east, to receive, on the flanks of the mountains covered by perpetual snow, and having a much greater elevation, the water which issues from the extremity of the shorter branch, and which an extraordinary accident, or a concealed natural cause, has forced to ascend to the top of a pointed peak."

Annual quantity of Salt raised from the Bowels of the Earth in Europe.—From a careful examination of the most accurate returns, the European Salt Mines and Salt Springs afford annually from twenty-five to thirty millions of hundred weights of Salt!

Effects of Cold upon Ice.—On Lake Champlain, and other American Lakes, and even in narrow rivers, fissures and rents of enormous magnitude are often made in the ice, and are always accompanied with loud reports, like those of cannon. The unwary traveller, who, with his sledge and horses, adventures by night, and sometimes even by day, across the great northern lakes, is frequently swallowed up in the openings which are thus unexpectedly made in the ice. When the weather grows warm again, before the ice melts, the fissures close, and sometimes the edges of them even overlap. At Plattsburg, in the winter of 1819-20, when the thermometer during the night was from 15° to 17° below 0° of Fahrenheit, and during the day from 10° to 12° below it, the reports of the rending of the ice were like that of a six-pounder, and the openings were from 10 to 15 feet wide.

Sepia Colour from Peat.—The stagnant water in Peat-bogs affords, on evaporation, a substance which equals the finest Sepia, and which, on being mixed with a proper proportion of gum or isinglass, may be used as a water colour. It requires no levigation.

Filtration.—When a powder saturated with a fluid, but not in the condition of a paste, is placed in the lower part of an open vase, and another liquid poured upon it, the last liquid permeates the powder, and completely replaces the first without mixing with it. This substitution is independent of the specific gravities of the fluids; thus water drives out alcohol and wine, and alcohol and wine drive out water.

LEARNED SOCIETIES.

Linnæan Society, Nov. 1.—A. Lambert, Esq. V.P., in the chair. Several donations were announced as having been received during the recess, amongst which were presents from the Academy of Sciences, Royal Society of Edinburgh, Asiatic and Geological Societies, &c. Amongst the specimens exhibited was the Spartina polystachya, a species new to the British Flora, discovered last summer growing abundantly on the muddy banks of the river at Southampton, by Dr Broomfield; a cone of the Pinus Sabiniensis, by Mr Warden; and two stems of the true Dahlia, an indigenous plant in the interior of Mexico, which occasionally grows to the height of seventy feet—the present specimen being grown by the Chairman to the height of about seven feet. Mr Gould likewise exhibited a variety of birds from New South Wales, containing several Mellifica, Parrots, and Finches, and a new species of Ptetmorhyneus, distinguished by a beautiful mural band. Mr Richard Taylor, the Secretary, read some remarks by Edward Foster, Esq., on the Euphorbia palustris, a species which was supposed to be a new discovery, but which he proved to have been known in England above two hundred years ago, and described by Major Johnson in an edition of "Gerrard's Herbal;" as also some observations on the "wourale poison," by R. H. Scomburgh, Esq., illustrated by drawings and specimens of the plant, one of which was that seen by M. Martius on the Amazon, in the process of manufacture, and which was proved to be the Strychnos toscifera. The meeting, which was numerously attended, adjourned to November 15th.

Botanical Society of London.—The second meeting of the above Society was held on November 2d, at the Crown and Anchor Tavern, Strand; J. E. Gray, Esq. F.R.S., in the chair. The meeting was numerous, and comprised many noted botanists. After the laws as revised by the Council had been read, confirmed, and ordered to be printed, the Society determined to hold their anniversary meeting on 29th November, being the anniversary of the birth-day of the celebrated English botanist, John Ray. They then determined to hold their next meeting on Thursday, November 17th, at the rooms proposed by the Council at Adelphi Chambers, next door to the Society of Arts. A paper was then read by Mr Daniel Cooper, author of the "Flora Metropolitana," and Curator of the Society, on "The Influence of Light on Plants, and the effects produced by depriving them of it; also the effects of watering them with coloured vegetable infusions when grown in the dark," being a detail of some experiments instituted in the year 1835, together with coloured sketches of the plants after the conclusion of the experiments. Donations of books were received from J. E. Gray, Esq., F.R.S., from G. C. Dennes, from J. Reynolds, Esq. (the Treasurer), and from Dr M'Intyre. Also presents to the Herbarium from D. Cooper, Esq., W. M. Chatterley, Esq., Honorary Secretary, and from Mrs Gawler. The meeting was then dissolved.

Warwickshire Society of Natural History and Archæology.—At a recent meeting of this Society, Professor Buckland stated that he had discovered at Guy's Cliff the remains of an extinct species of animal, which had *never* before been found or mentioned by geologists. The learned Professor said, "He had commenced his studies by collecting fragments of Carisbrook Castle, Corfe Castle, and Warwick Castle; and little did he then dream that he should ever have an opportunity of saying that the stones of Carisbrook Castle contained a species of fresh water fish long extinct; or that, in the distant progress of time, he should have to assert that the Castle, the Collegiate Church, and the town of Warwick, were built upon a stratum utterly unknown to English geologists. Ten years ago he had obtained certain specimens from Guy's Cliff, which he had cherished up among his masses of ignorame, and stored amidst difficulties, in the hope that some ray of light might dissipate the darkness which enshrouded them, and enable him to acquire some accurate information respecting them. Within the last two hours that darkness had been dispelled, and he was able to say that at Guy's Cliff he had discovered an extinct species of animal never before found, and that those portions of rock which were before him on the table, were from a quarry, the name of which had never been uttered in England. Another discovery which he had made was, that the town of Leamington rested on the remains of animals which had existed in other times; and this fact was not hastily acquired, but was founded on strictly logical deductions. It was indeed true, that under the foundations of houses at Leamington (where there had been previously one immense lake) there were to be found the remains of elephants, hippopotamuses, hyænas, tigers, buffaloes, and a string of twenty other animals which he could enumerate."

Geological Society, Nov. 2.—Mr Lyell, President, in the chair. This Society commenced its meetings for the ensuing session. An elaborate paper (the first of a series) was read by Mr Hugh Edwin Strickland, F.G.S., who has recently returned from Asia Minor. The details of the paper were principally confined to the author's observations made during a winter's residence in Smyrna; and two excursions, one into the valleys of the Meander and Cayster, and the other from Constantinople to Smyrna.

Edinburgh: Published for the Proprietors, at their Office, 16, Hanover Street. London: Smith, Elder, and Co., 65, Cornhill. Glasgow and the West of Scotland: John Smith and Son; and John Macleod. Dublin: W. F. Wakeman. Paris: J. B. Bailliere, Rue de l'Ecole de Médecine, No. 13 bis.

THE EDINBURGH PRINTING COMPANY.

THE EDINBURGH

JOURNAL OF NATURAL HISTORY,

AND OF

THE PHYSICAL SCIENCES.

FEBRUARY, 1837.

ZOOLOGY.

DESCRIPTION OF THE PLATE.—THE ELEPHANTS.

THE Elephants constitute a genus belonging to the order *Pachydermata*, and are characterised by their vast size, in which they excel all the terrestrial Mammalia, their long flexible proboscis, forming an instrument of prehension as well as an organ of smell, and their large recurved tusks. Their most distinctive character, however, as Cuvier remarks, is to be found in their grinding teeth, of which the body is composed of a certain number of vertical plates, each formed of bony substance, enveloped with enamel, and connected by a third substance, named cortical. These grinders succeed each other, not vertically, as our second or permanent teeth succeed the first or milk teeth, but from behind forwards, so that in proportion as a tooth is worn down, it is pushed forwards by its successor. An Elephant may thus have one or two teeth on each side of each jaw, according to the period of growth. The first teeth have comparatively few plates. Some individuals are said to change or renew their grinders as many as eight times; but the tusks are not renewed more than once.

Two species only exist at the present day, one belonging to India, the other to Africa; but the bones of another occur, buried in diluvium in many parts of America, as well as the old continent.

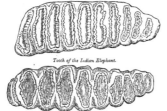

Tooth of the Indian Elephant.

Tooth of the African Elephant.

Fig. 1.—The INDIAN ELEPHANT, *Elephas Indicus*, is distinguished from the African species, by the following external characters:—The head is oblong, the forehead concave, the crown or face of the grinding teeth marked with transverse, parallel, undulating bands, the ears, although very large and pendulous, proportionally smaller than in the next. It has four nails or hooflets on the hind feet. The colour is generally dark brownish-grey, but varies considerably, and sometimes, although very rarely, albino individuals are met with. It inhabits India and the adjacent islands, where it has been domesticated from time immemorial. But it is remarkable that the species has never been propagated in captivity, and all the individuals subjected to man have been caught in the woods and jungles. The use of the ivory of the tusks for handles for knives, musical and mathematical instruments, plates for miniatures, billiard balls, and toys, gives rise to a great annual destruction of these animals, so that it is a matter of wonder that the race has not been extirpated. Although the docility and sagacity of the Elephant has been greatly exaggerated, they are yet probably superior to those of any other animal excepting the Dog.

Fig. 2. The AFRICAN ELEPHANT, *Elephas Africanus*, has the head round, the forehead convex, the ears excessively large, so as sometimes entirely to cover the shoulder, and the crown of the grinding teeth marked by rhomboidal or diamond-shaped ridges, by which they are readily distinguished from those of the Indian species. It has frequently only three nails on the hind feet. The colour is lighter than that of the Indian Elephant, generally yellowish-brown, sometimes reddish. It inhabits Africa, from Senegal to the Cape of Good Hope. The tusks are larger than those of the Indian species, and in the female are nearly of the same size as in the male. Although it is probable that the African Elephant was the species subjected by the Carthaginians, it is not now any where domesticated.

DESCRIPTION OF THE PLATE.—THE TANAGERS.

THE Tanagers are a family of Birds peculiar to America, and belonging to the order *Passeres* of Cuvier, or *Insessores* of many other naturalists. They are in some respects nearly allied to the Finches and Buntings, although by the author just referred to placed among the Dentirostres. The bill is conical, short, pentagonal at the base, its dorsal outline arched. The wings are of moderate length, and sharp, the tail rather short, and even or slightly forked. Many of the species are remarkable for the brilliancy of their colours. In their manners they resemble the Finches, and live upon seeds and berries, as well as insects.

Figs. 1 and 2. The SUMMER RED-BIRD.—*Tanagra æstiva*. The male of this species is of a rich vermilion hue, the female yellowish-brown above, yellow beneath. It inhabits South America and the United States, being found plentifully in the warm season in Florida and the Carolinas, although it seldom proceeds farther north than Boston. It feeds principally on insects, especially coleoptera, which it pursues on the wing, seldom alighting on the ground.

Figs. 3 and 4. The SCARLET TANAGER.—*Tanagra rubra*. The male is equally distinguished for the brilliancy of its plumage as that of the preceding species, while the female is of a plain green tint above, and yellow beneath. It proceeds from South America into the United States, proceeding as far as Canada, feeds on large winged insects and berries, lives chiefly in the depth of the woods, rarely approaching the habitations of man, and returns southward about the end of August.

Fig. 5. The LOUISIANA TANAGER.—*Tanagra Columbiana* of Wilson, *T. Ludoviciana* of Bonaparte. This species was discovered in Louisiana by Lewis and Clark, in their expedition to the Pacific Ocean. Only a few imperfect skins were obtained by these travellers, and little more is known of its history than that it inhabits the extensive prairies between the territories of the Osage and Mandan Indians.

BATS.—Although not many years ago, the number of Bats known to occur in Britain amounted only to six, there are at present described in our more recent works on the Zoology of this country not fewer than sixteen species. The characters of these are briefly given in Jenyns' Manual of British Vertebrate Animals, and more fully in Mr Bell's beautiful work on the British Mammalia, of which six numbers have appeared. Many of the species have been found only in the southern counties of England; and it is somewhat remarkable, that hitherto the species observed in Scotland do not exceed three at most. The common bat of this country is the Pipistrelle, *Vespertilio Pipistrellus* of Gmelin and Desmarest, although, previous to Mr Jenyns' researches, it was considered as the *V. murinus* of Linnæus, a much larger species, which has also been found in England. The Pipistrelle we have determined from species obtained in various parts of the south of Scotland. The only other species which we have hitherto met with, north of the Tweed, is the *Plecotus auritus*, the Long-eared Bat, which in some places is nearly as plentiful as the Pipistrelle. Dr Fleming, besides these, mentions the *V. emarginatus* as having been found in Fife; but it is probable that the information is not to be depended upon, especially as the characters which he gives are vague, and equally applicable to the Pipistrelle, and as he considers the *V. murinus*, which is the Pipistrelle, to be the common Bat, although that has not been found in Scotland. The determination of our Scottish Bats is therefore a desideratum; and we shall feel obliged for the communication of specimens.

PECULIAR AFFECTION OF GALLINULES.—The Water Hen, *Gallinula Chloropus*, and the Corn Crake, *G. Crex*, are both subject to an affection, apparently resulting from terror, which some observers have attributed to epilepsy, and others to a feigned semblance of death. We remember once to have caught a Water Hen, which fell to the ground from fright induced by a shot, although it was not in the least injured, and after carrying it home and keeping it for a night, allowing it to escape next morning, when it flew off with great speed. On another occasion, having fired at and missed a Corn Crake, we were surprised to see it fly directly out to sea and alight in the water, where it floated buoyantly, and seemed indisposed to rise again. Unfortunately for it, a large black-billed Gull, *Larus marinus*, which happened to be flying along the shore, observed it, and descending with a scream, caught it in its bill, and carried it off. We have caught the Corn Crake alive, and have known the same feat performed more than once, in the Hebrides, where that bird is very common. In the sixth number of this Journal, a case of feigned death is related; but we are of opinion, judging from the timidity of this bird, that the phenomena there described could have resulted only from terror.

THE CUCKOO.—In a paper, by Mr E. Blyth, on the species of Birds observed near Tooting, Surrey, published in the Magazine of Natural History for December 1836, the following interesting particulars relative to the Cuckoo are given. It often arrives in the country several weeks before its well known cry is heard. Both its notes are common to the two sexes. The egg is found in a great variety of nests, both of granivorous and insectivorous birds; but, in the inclosed districts, it is most frequent in that of the hedge dunnock, and not rare in the field lark's; whilst, on the open commons, the meadow pipits' and wagtails' nests most usually receive it, and not unfrequently that of the whinchat. Having spared no pains to investigate the economy of this interesting bird, I am now enabled to speak with confidence on most points on which I formerly was doubtful. The species both arrives and departs in flocks, though some of them migrate singly. The adults change the clothing plumage, and also the tail, during their stay with us; but retain the quills to bear them on their migratory journey, the barred markings on the sides of the neck are exhibited in the new feathers which the adult females put forth at the close of the summer, or rather a little after midsummer; and appear, therefore, to be a permanent characteristic of the sex; whereas Mr Selby states that "the female differs in no respect from the male." The young of the year do not change a feather till after they have left us. Unlike many other species with which I am acquainted, the females often continue to lay while in deep moult. They deposit, on the average, five eggs in the course of a season, rarely more; and these, judging from the result of numerous experiments which I have instituted, are not unfrequently ejected by the birds to whose fostering care they are entrusted; hence, I believe, the disproportionate scarcity of their occurrence. A considerable period intervenes between their successive deposition, during the lapse of which the female Cuckoo instinctively watches the proceedings of birds that are carrying about building materials; the time her egg requires to be matured for laying, being at least equal to that which the majority of smaller birds consume in the process of nidification. That she cannot, however, much protract the time of disburthening herself, as has been suggested, when the egg has been received into the oviduct, is sufficiently proved by the fact of her sometimes laying into an unfinished nest. There is no division of her eggs into separate lots, as in the case of birds which incubate their own; a fact which might, indeed, have been suspected *a priori*, seeing that no purpose could be effected by such an arrangement. In point of fact, each separate consecutive egg is analogous to a *lot* in other species. The remarkable deviation of the Cuckoo from the general habit of the feathered race in confiding its progeny to the care of strangers, is dependent on a peculiarity of the vascular system, and particularly on the minuteness of those blood-vessels which supply the parts concerned in the development of the eggs; in consequence of which, each successive egg requires so long a time to become fit for deposition, that they could not be incubated simultaneously. Even when developed, they are, from the same cause, remarkably small for the size of the bird; which, however, adapts them to that of the nests into which they are intruded. I cannot suppose that this peculiar conformation is intended merely for the purpose of retarding the growth of the egg, though this effect is unquestionably occasioned by it; it must rather have reference to some other peculiarity of the bird's habits which we have yet to discover. It is certain that the maternal feelings of the Cuckoo are not quenched, astonishing as this may appear. Mr John E. Gray, of the British Museum, informs us that he has himself seen a Cuckoo, day after day, visit the nest where one of its offspring was being reared, and which it finally enticed away from its foster parents! I had previously heard of analogous cases, but was disposed to regard them as fabulous, until corroborated by so paramount an authority. Since I have proceeded thus far with the Cuckoo's history, I may add that, although caterpillars form its principal subsistence, it feeds likewise on snails and various fruits, and that it is quite true that both sexes devour birds' eggs, as well as callow broods. Of the numerous cases I have now accumulated of the occurrence of the Cuckoo's egg, not one has been met with wherein it could not have been laid into the nest. The egg of the Cuckoo is most commonly found alone, or together with less than the average number of those of the species to which it is confided. I have one instance in which a young Cuckoo was found, half-fledged, in the same nest with two hedge dunnocks of equal age; which fact I can only account for by supposing that the interloper was weak and sickly at the time of its exclusion, and remained so till the instinctive propensity to oust its companions had entirely dissipated, as we know to be the case. It was, however, apparently quite healthy when discovered.

On this communication of Mr Blyth's we have a few remarks to offer. He alleges that in the Cuckoo there is no division of the eggs into separate *lots*, as in the case of birds which incubate their own; but neither is there such a division in these birds; and we are disposed to demand the facts on which Mr Blyth's results are founded, before we can give entire credence to the latter. The organs, moreover, are not smaller than in many other birds, nor differently supplied with blood-vessels. One of the most curious circumstances relative to the organization of the Cuckoo is seen in the stomach, of which the inner coat is generally found stuck full of the hairs of caterpillars, disposed in a circular manner, determined by the action of the muscular coat.

MUSIC OF SNAILS.—In the first number of the Naturalist is the following communication on this subject:—"There is a poetical notion that oysters, amongst other gentle qualities, love minstrelsy; and the fishermen, in some parts, 'sing to charm the spirits of the deep,' as they troll their dredging nets; for

"The Herring loves the merry moonlight;
The Mackerel loves the wind;
But the Oyster loves the dredging song;—
For he comes of gentle kind."

These lines gave rise to a communication from a young lady, which I will send you. Perhaps some of your readers may confirm the supposition of the Snail's musical capabilities. She says:—"One evening I kneeled upon the window seat, when it was nearly dusk, and heard a soft musical sound; not a humming or murmuring, but a truly musical tone. I saw a snail, and, having a desire to annihilate those destroyers of fruit and flowers, took it from the window. I had silenced the music! I recollected what I had heard, and felt a sort of pang." We have heard the same sort of sound emitted under similar circumstances, and believe it to be produced simply by the minute vibrations produced in the glass by the action of the muscular foot of the animal. A similar sound may be produced by bringing the wet finger along a pane. As to Snails, or other molluscous animals, having the faculty of emitting sounds in a manner analogous to that of the higher animals, it is entirely out of the question.

LEPAS ANATIFERA.—The Rev. W. B. Clarke relates, that as one of the Poole pilot-boats was cruising, on the 22d October, in the Channel, the crew picked up, about four leagues from land, a bottle closely corked, on which three congregations of Lepades have rooted themselves. The pedicle or fleshy stalk was fastened to the bottle by a subtance in appearance very like plaster of Paris. He conjectures that it was thrown overboard in the warm latitudes about the Gulf Stream, and that it has been floating across the Atlantic, in the set of the current, towards our shores.

FISHES CAUGHT IN A WHEEL-BARROW.—The same gentleman relates, that on coming out of Cherbourg, on Wednesday, September 21, when off Fort de Querqueville, about seven miles from land, he picked up a wheel-barrow, floating in the tideway, which, by the marks upon it, had evidently been washed from the breakwater, and had been in the water several days. "It was floating bottom upwards; and, on turning it to get it on deck, we found three John Dorys caught between the planks. They were, it is true, extremely small; but their position there showed that small fish are glad to get in (as Paddy said, when he fished under the ditch of Dublin Bridge) '*out of the wet*,' and to shelter themselves from the violence of the surface waters."

IRREGULAR GROWTH OF HORNS AND CLAWS.—The horns of animals have a definite form and direction, according to the different species, and as they increase by the addition of matter to their bases, which is continually secreted, they enlarge indefinitely during the animal's life; but, in their regular state, preserve a direction which prevents them from impeding the actions necessary for the well-being of the individual. Thus, the horn of a ram may attain a very great size, but its curves being regular, it is of no inconvenience to the animal. Should any disturbing cause, however, alter the natural direction of the horn, it assumes an abnormal and frequently fantastic form, and sometimes runs into the face or neck, in which case it requires to be partially cut off. An accidental division of the horn of a young animal will also cause it to grow in a cleft form, which is the case with Sheep that are said to have three or four horns. In these cases there are never, in fact, more than two, but these being split, and running out in a divaricating manner, they seem to be multiplied. The claws of quadrupeds and birds grow in the same manner as horns. Indeed, they are precisely of the same nature. But their growth is checked by pressure and use, the latter of which also tends to wear them down. Thus, although in the natural state, the hooflets of a goat, however old it may be, do not acquire an inconvenient length, because they are constantly worn by being rubbed against the rocks and ground; yet, when the animal is kept in soft pasture ground, or in a house, where it cannot obtain its natural exercise, they often attain a great length, and curve upwards like a Turkish slipper. The claws of confined eagles, and of all cage birds, manifest the same tendency; and in birds of prey the point of the upper mandible sometimes elongates and curves so as to prevent the opening of the mouth. Examples of preternatural growth of the nails of the human subject are common in Medical Museums, but in these cases there is always disease of the secreting organs; yet one may easily imagine to what length the healthy nail would soon grow, when he may see that in a fortnight its free extremity attains a length of more than two-twelfths of an inch.

HYBRID GROUSE.—A bird, presenting characters intermediate between those of the Blackcock and Red Grouse, was lately obtained by Mr Fenton, Preserver of Animals in Edinburgh, and is now in the possession of Mr W. Smellie Watson, of that city. It was much emaciated, but in perfect plumage. The following description was taken from it in its recent state:—In form and proportion it is similar to a female Black Grouse. The bill is of the same form as in that bird. The supraocular membrane resembles that of the Red Grouse, having a thin, free, fringed margin, which is not the case with that of the Blackcock. The feathers are generally oblong, broadly rounded, and have a large tufty plumule. The tail is complete, slightly forked, as in the female Black Grouse, but of only sixteen feathers, as in the Red Grouse. The quills are twenty-six. The tarsi are feathered all round, without a bare space behind. The toes are also feathered a third down, as are the interdigital membranes, and the plumage of these parts is as bushy as in the Red Grouse. They are margined with long linear scales, as in the Black Grouse. The claws are very long, arched, with thin parallel edges, like those of the Red Grouse and Ptarmigan. The bill is brownish-black; the supraocular membrane scarlet; the toes light brown; the claws brownish-black. The upper part of the head is minutely mottled with brownish-red, brownish-black, and grey; the hind neck with a larger proportion of grey. The rest of the neck is black, with a tinge of reddish-purple. On the throat the feathers are margined with white, on the sides of the neck obscurely barred with brownish-red. On the lower parts generally the feathers are black, tipped with white; those of the sides banded with red, of the lower part of the tail black, with a large terminal space. The lower surface of the wing and the axillar feathers are white. The upper parts generally are very minutely undulated with brownish-black and brownish-red, the feathers having very narrow terminal bands of white. The wing feathers and secondary quill-coverts are similar, as are the secondary quills, which are tipped with greyish white. The primary quills, their coverts, and alular feathers, greyish-brown; the outer edges of the primaries mottled with white. A white spot appears at the axilla, but there is not a white band on the wing as in the Black Grouse. The tail is black, the two middle feathers very obscurely mottled with reddish; the eight middle narrowly tipped with white. The tarsal feathers are greyish-white, those on the outer side mottled with red. The length 20½ inches; extent of wings 31. On inspecting the body, the cause of emaciation was obvious. The trachea and bronchi were much inflamed; the left lung perfectly sound, but the right completely gorged with blood. The rectum was greatly dilated and filled with a substance resembling putty, being a deposition of uric acid, which completely obstructed the passage. In form and size this bird resembled a female or a young male Black Grouse, to which it was also similar in its internal organization. In plumage and colouring, it was as exactly intermediate between the two species as could be imagined. Hybrids are alleged to be occasionally produced between the Pheasant and Black Grouse; but we have not seen a specimen, and have reason to think that the birds supposed to be of this character were merely young Blackcocks.

ANECDOTE OF A PARROT.—A reviewer in the Naturalist relates the following anecdote:—" We feel confident that in various instances Parrots have intelligence enough to understand, if not the exact meaning of the words they utter, the *subject* to which they refer. We know an old lady, whose feet were so excessively tender as almost to preclude her from walking, and hence she always went abroad in her Bath chair. She had a favourite Parrot, who, when the tea equipage was placed upon the table, was invariably taken out of his cage by the footman, and placed on the board, as a proper accompaniment to the antiquated china. Poll, no doubt an attentive observer, had long perceived there was ' something rotten in the state of Denmark,' and hence, whenever his mistress failed to dole out what he considered his fair ration, he would, in a threatening manner, exclaim, ' Peck your toes, Madam.' As he sometimes flew from his position, to put his threat into execution, the old lady, to avoid the assault upon her toes, indulged him with a further allowance, which, of course, only led to increased insolence on his part, and the threat of ' Peck your toes, Madam,' was still oftener reiterated. At length, one day Poll having cried ' Wolf,' as he thought, without that attention being paid to the subject which it demanded, proceeded to suit the action to the word with such effect, that the old lady was compelled to scream loudly for help, Poll having administered a dose of toe-pecking that put her in dreadful pain for some days. This was too much to be borne, and the culprit received sentence of transportation. The footman was directed to sell or give him away. Now, though in this case we think it highly probable that the mischievous threat had been taught the bird by the servants, yet Poll must have seen the effect it produced in occasionally increasing his allowance, though doubtless, he did not calculate upon the final *dénoument.*"

BOTANY.

RARE PLANTS FOUND IN JERSEY.—Mr William Christy, jun. F.L.S. &c. has communicated to the public, through the medium of Loudon's Magazine of Natural History, a list of rare plants collected by him in the Island of Jersey, in October 1836. The greater part of the coast, he states, is composed of granite cliffs, too much washed by the sea to admit of much vegetation, except in some of the sheltered bays. The grassy slopes above them afford several very rare plants. The island is remarkably deficient in streams or pools, as well as in salt marshes; nor are there many inland rocks. The greatest number of rare species is afforded by the sandy parts of the shore, and by a great deposit of sand drift in the parish of St Brelade, nearly a mile distant from the sea.

Allium sphaerocephalum. By the road between Beaumont and La Haule.
Asplenium lanceolatum. Common.
Asplenium marinum. Cliffs at Plemont Point and Havre Giffard.
Borago officinalis. Roadsides and orchards.
Carlina acaulis. Common.
Calystegia Soldanella. Sands at Grève de Lecq.
Centranthus rubra. Rocks and walls of Fort Regent.
Crithmum maritimum. La Corbière and Plemont Point.
Cotyledon umbilicus. Common.
Cyperus longus. St Peter's Valley, near St Brelade's Church.
Dianthus prolifer. Near St Ouen's, Quenvais.
Echium violaceum. Quenvais. Mont le Veau.
Elymus arenarius.
Erodium cicutarium, white-flowered. St Aubyn's Bay, Fort Quenvais.
Erodium maritimum. Grève de Lecq.
Erodium moschatum. St Aubyn's Bay, Fort Regent.
Euphorbia Portlandica. Sea shore, all round the island.
Euphorbia paralias. St Aubyn's Bay.
Faeniculum vulgare. Near the sea; not rare.
Glaucium flavum. Quenvais, St Ouen's, Brelade's and St Aubyn's Bays.
Gnaphalium luteo-album. Near Petit Port.
Helianthemum guttatum. Between La Corbière and Noirmont Point.
Hypochaeris glabra. Grève de Lecq.
Iris faetidissima. Sea banks near St Aubyn's Bay.
Juncus acutus. Quenvais, St Ouen's Bay, and Petit Port.
Linum augustifolium. Common.
Lotus angustissimus. Between La Haule and Quenvais.
Matthiola sinuata. St Ouen's Bay, Petit Port, St Aubyn's Bay, Quenvais.
Marrubium vulgare. St Brelade's Churchyard; waste places.
Mentha rotundifolia. Above La Haule.
Mespilus germanica. Hedges between Roselle and Gorey.
Neottia spiralis. Quenvais.
Polycarpon tetraphyllum. Common in Gardens, &c.
Petroselinum sativum. Mont Orgueil Castle.
Scirpus Savii. Between Quenvais and St Brelade's Bay.
Scilla autumnalis. Very common.
Sedum Anglicum. Common.
Silene Anglica. Common.
Silene conica. Quenvais.
Silene nutans. Near St Ouen's, St Aubyn's.

Senebiera didy na. ' St Helier's. St Aubyn's.
Solanum nigrum, with red berries. Quenvais, Petit Port.
Scrophularia Scorodonia. Common.
Statice plantaginea. Quenvais, St Brelade's Bay.
Sibthorpia Europæa. Above La Haule.

GEOLOGY.

ESTIMATE OF THE PROBABLE TEMPERATURE IN EUROPE DURING THE TERTIARY PERIODS.—M. G. P. Deshayes alleges that fossil conchology, studied in a logical manner in its various relations both as to zoology and geology, may become a powerful means of bringing this latter science to perfection. Very numerous observations, repeated upon more than eight thousand species of recent and fossil shells, and more than sixty thousand individuals of all regions, have enabled him to perceive important consequences with regard to an approximate estimate of the temperatures of the geological periods, concerning which man cannot cite his historical annals, since he then had no existence upon the surface of the earth. The animals best adapted to indicate temperature are those which, possessed of but small power of motion, cannot withdraw themselves periodically from the changes of the seasons, and are obliged to sustain all their influence in the places which have given them birth. The greater number of the mollusca and zoophytes are of this kind. At the present day, certain species of the former are found to be peculiar to certain latitudes, some inhabiting the frozen ocean, others the tropical seas; and from the examination of these it is not considered possible eventually to be able to reply to such questions as this : A series of species being given, to point out the climate of the spot from which they have been procured. From comparing all the known species of recent shells, with all those which are brought from the tertiary strata of Europe, the following results have been obtained :—

1. The tertiary strata of Europe contain no species which can be identified with the secondary strata lying beneath them.

2. The tertiary strata are the only ones which contain fossil specimens of existing species.

3. The fossil shells which can be identified with living species, are more numerous in proportion as the strata are more recent, and vice versa.

4. Constant proportions (3, 19, 52 per cent.) in the number of recent species determine the age of the tertiary strata.

5. The tertiary strata are superposed one upon another, and not parallel, as was at first imagined.

6. They ought to be divided into three groups or stages, according to their zoology, the shells which they contain indicating the temperature existing at the period of their deposition.

We are unable here to follow the author in the development of his ideas; but from what he states, it appears to him, that the following conclusions may be drawn :—

1. The first tertiary period took place under an equatorial temperature; and, according to all probability, one many degrees hotter than the present temperature of the equator.

2. During the second period, the beds of which occupy the centre of Europe, the temperature has been similar to that of Senegal and Guinea.

3. The temperature of the third period, at first a little more elevated than ours in the basin of the Mediterranean, has become similar to that which we experience. In the north, the species of the north are fossil; in the south, those of the south.

Thus, since the commencement of the tertiary strata, the temperature has been constantly diminishing. Passing, in our climates, from the equatorial to that which we now enjoy, it is easy to measure the difference. No doubt, naturalists have been able theoretically to conjecture *a priori* these changes of temperature; but it is curious to see their conjectures confirmed by a science which no one had thought of directing towards this end.

METEOROLOGY AND HYDROGRAPHY.

METEORIC STONES.—Several substances of this nature have recently been examined by Professor Berzelius, who has ascertained that meteoric stones are properly speaking minerals, and is of opinion that since they cannot be formed in the atmosphere, which does not contain their component parts, nor be ejected from volcanoes on the earth's surface, as they fall every where, they must be derived from some other body having volcanoes. The nearest to us is the moon, which has gigantic volcanoes, and is destitute of atmosphere to retard the masses ejected. The result of his investigations as to the composition of meteoric substances are the following: Two kinds fall on the earth. One of these is rare, for only three aërolites belonging to it have been remarked, viz. those of Stannern in Moravia, and of Jonzac and Juvenas in France. They contain no metallic iron; the minerals of which they consist are more decidedly crystalline, and magnesia does not form a prevailing component part in them. The second kind includes the other very numerous meteoric stones which have been examined, and which are frequently so similar in appearance, that they might be supposed to have been broken from one piece. They contain ductile metallic iron in variable quantity. It being assumed that these substances come from the moon, we may therefore say that meteoric stones come from only two different volcanoes, of which the one either ejects more abundant masses, or sends them forth in much a direction that they reach us more frequently. Such a state of things corresponds perfectly with the idea that a certain portion of the moon has the earth constantly in the zenith, and that all its ejected masses which are thrown out in a straight line are directed by it towards the earth, whither, however, they do not proceed in a straight direction, since they are also subjected to that motion which they previously possessed as part of the moon. If it is this portion of the moon that sends us meteoric blocks of iron, and if the other parts of the moon do not abound so much in iron, we see a reason why this point should be constantly turned towards the magnetic terrestrial globe. The mineral matter of meteoric stones is composed of the following substances :—

1. *Olivine*, containing magnesia and oxide of iron, and white or greyish, but seldom yellow or green. It is separated by being heated with acids, and dissolving the silica in boiling carbonate of soda. There then remain—2. *Silicates of magnesia, lime, oxide of iron, oxide of manganese, alumina, potash*, and *soda*. 3. *Chromate of iron*, which is always present in both kinds, and is the cause of their greyish-black colour. 4. *Oxide of tin*, containing traces of copper. 5. *Magnetic iron*, probably not contained in all. 6. *Sulphuret of iron*, which exists in all. 7. *Native iron*, not pure, although very ductile, but containing carbon, sulphur, phosphorus, magnesium, manganese, nickel, cobalt, tin, and copper.

Transparency of the Sea on the Coast of Norway.—Nothing, says Sir A. de Capell Brooke, in his Travels in Norway, can be more surprising and beautiful than the singular clearness of the water of the Northern seas. As we passed slowly over the surface, the bottom, which here was in general a white sand, was clearly visible, with its minutest objects, where the depth was from twenty to twenty-five fathoms. During the whole course of the tour I made, nothing appeared to me so extraordinary as the utmost recesses of the deep thus unveiled to the eye. The surface of the ocean was unruffled by the slightest breeze, and the gentle splashing of the oars scarcely disturbed it. Hanging over the gunwale of the boat, with wonder and delight I gazed on the slowly moving scene below. Where the bottom was sandy, the different kinds of asteriæ, echini, and even the smallest shells, appeared at that great depth conspicuous to the eye; and the water seemed, in some measure, to have the effect of a magnifier, by enlarging the objects like a telescope, and bringing them seemingly nearer. Now creeping along, we saw, far beneath, the rugged sides of a mountain rising towards our boat, the base of which was hidden some miles in the great deep below. Though moving on a level surface, it seemed almost as if it were ascending the height under us; and when we passed over its summit, which rose in appearance to within a few feet of our boat, and came again to the descent, which on this side was suddenly perpendicular, and overlooking a watery gulf; as we pushed gently over the last point of it, it seemed almost as if we had thrown ourselves down this precipice, the illusion, from the crystal clearness of the deep, actually producing a sudden start. Now we came again to a plain, and passed slowly over the submarine forests and meadows, which appeared in the expanse below, inhabited, doubtless, by thousands of animals, to which they afford both food and shelter,—animals unknown to Man; and I could sometimes observe large fishes, of singular shape, gliding softly through the watery thickets, unconscious of what was moving above them. As we proceeded, the bottom became no longer visible, its fairy scenes gradually faded to the view, and were lost in the dark green depths of the ocean.

MISCELLANEOUS.

Manufacture of Gun-Flints.—In the last number of the Edinburgh New Philosophical Journal, is a paper on this subject, communicated by Dr James Mitchell, of which the following is a condensed account:—Brandon, in Suffolk, is the only place in England in which gun-flints are made to any considerable extent, and it is said that none are now manufactured in France, or, in fact, in any part of Europe; yet the seventy or eighty men employed could barely make a living by their trade, so much had it been impaired by the cessation of war, and the invention of percussion caps. The masses of flint employed are obtained from a common about a mile south-east of the town, by sinking a shaft about six feet, then proceeding three feet horizontally, and thus alternately until they come to a floor of flint, sometimes going to the depth of about thirty feet. The blocks are handed up from stage to stage, a man being placed about half-way between each to receive them. It is not more than forty years since the present mode of making flints was introduced from France. The workman, called a cracker, who is seated on a chair, has a thick piece of leather strapped to his left thigh, and over it he straps a piece of iron. He takes a large piece of flint-stone, and breaks it into fragments of about two pound weight. Then taking a fragment in his left hand, and applying it to the iron plate on his thigh, he strikes out pieces at short distances from each other. He then strikes with his hammer on the parts of the edge of the flint, which are now separated from the rest, and the effect of the blow, together with the reaction on the plate of iron on his thigh, causes a flake of about three or four inches in length to come off, there being on each side a conchoidal fracture. Of the flakes thus obtained from the mass of flint, some are large and others small. The workman has before him three small casks with the upper end open; into one of them he drops the large flakes, into the second the flakes of a less size, and into the third the flakes of the smallest size. The refuse is thrown into a fourth cask, and is from time to time carried out of doors, and thrown into a heap of rubbish. The three casks with the flakes are intended each of them for a separate workman, who has to finish them into flints, single-barrel flints, double-barrel flints, and pistol flints. The workmen who divide the flakes into flints are called nappers, and one cracker suffices to keep three of them employed. A napper has before him a block, not unlike a butcher's block, upon which a piece of iron is nailed, from which rises a thin piece of iron three inches in length, and only a sixth of an inch in thickness, and brought to a coarse edge. He uses a hammer, which is merely a plate of steel, extending two inches on each side of the handle, and an inch in breadth, and not above a sixth of an inch in thickness. He takes into his left hand one of the flakes, lays it over the little anvil on the block, and with his hammer breaks it into three or four flints. All that he has to do after that, is to see which edge will be best for the flint, and from the other he breaks a little off, and the whole is complete. At Brandon, in the present depressed state of the trade, the best musket-flints, which at one time were sold for two guineas a thousand, now bring no more than from seven to eight shillings.

Luminous Appearance at Sea off the Shetland Islands.—A curious luminous appearance at sea is mentioned in the following extract from a letter to Mr Stevenson, engineer, by the keeper of the Sumburgh Head Lighthouse:—"*Monday, September* 19, 1836.—The herring-boats went out through the night. There came on a severe gale of wind from the north-east, which drove them from their nets, and scarcely any one of them got into their own harbours. Mr Kay's fishermen lost 180 nets, Mr Bruce of Whalsey lost 114, and a great many of the poor men lost the whole of their nets. The fishermen also informed me, that upon the same night, there appeared to them a light, which greatly annoyed them. It appeared to them like a furnace standing in the water, and the beams of the light stood to a great height. It became fainter on the approach of day, and at length vanished away by day-light. It continued for two nights. It stood so near some of the boats that the men thought of cutting from their lines to get out of its way." *Edin. New Philos. Journal, January* 1837.

New Method of Disengaging Heat.—By projecting upon a fire a mixture of water and oily matters in a certain proportion a flame is produced, whose heat is extremely intense. If the water be in excess, the flame languishes; or, if in too small quantity, a smoke is produced. For 1 measure of tar it is necessary to employ $1\frac{1}{2}$ of water. 15 lbs. of oil of turpentine mixed with 15 lbs. of water, and projected upon 25 lbs. of Newcastle coal, produce as much heat as 120 lbs. of this coal.

Poisoning by Arsenic Cured by Hydrated Tritoxide of Iron.—A remarkable case of this description is recorded in the Gaz. Med. de Paris (22d August 1835), by Monod. The subject of it was a hair-dresser, thirty-five years of age, who, in a paroxysm of delirium tremens, swallowed a drachm and a half of white oxide of arsenic. Half an hour afterwards the antidote was given to him, suspended in water, and he drank in twelve hours all the Tritoxide produced by the decomposition of five drachms of the Tritosulphate of Iron. He had no violent colic, and twenty hours afterwards experienced scarcely any uneasiness.

REVIEWS.

British Song Birds; being Popular Descriptions and Anecdotes of the Choristers o the Groves. By Neville Wood, Esq., author of the Ornithologist's Text-Book. London: John W. Parker, West Strand. 1836.

This work contains, as its title announces, popular descriptions of the habits of seventy-three of our smaller birds, all of which, however, cannot with propriety be considered as choristers of the groves. The information afforded appears to be for the most part derived from the author's personal observation, and, although not in general new, is yet communicated in a pleasing style, and in a manner not calculated to mislead the student. In fact, we give full credit to the author when he asserts, that "no one fact is stated, which has not been observed with my own eyes, excepting where other authorities are referred to, which is, in every case, done openly and fully;" and we commend him for his candour, which is favorably contrasted with the disingenuousness of some celebrated ornithologists, who profess to give nothing but their own, while almost every page is full of unacknowledged borrowings. A remarkable difference, too, between the present generation and the last, is the decided preference now given to actual rational information respecting habits and structure, whereas formerly histories of animals were confined to mere technical characters, so brief and unsatisfactory, that in cases of doubt we seldom find them affording any assistance. We believe that our distinguished countryman, Sandy Wilson, the quondam Paisley weaver and packman, was the founder of this modern school of ornithologists, who combine glowing descriptions with accurate details. Mr Wood's "Song Birds," we are happy in being enabled, from having paid considerable attention to the subject, to recommend to all who are interested in this most interesting department of zoology, and more especially to young persons, for the "old chroniclers" are generally too wise to learn, and too knowing to allow merit to any not disposed to flatter themselves. The nomenclature, although strange to those who commenced their studies twenty years ago, is as good as the old; and, in truth, seeing that both are defective, the more innovations that authors make the better, for it will ultimately render it necessary to select unobjectionable names.

Loudon's Magazine of Natural History. New Series. Conducted by Edward Charlesworth, F.G.S. Monthly, 2s. London: Longman, &c.

This popular periodical, which has greatly contributed to the extension of the study of natural history, is particularly valuable as a depository of facts and observations. It is not our intention to offer a formal review of its contents, which are of a very diversified nature; but we take this opportunity of recommending it to those who are devoted to the study of natural history, or who find pleasure in occasionally perusing the remarks and observations made by the students of that science. The new series commences with considerable spirit, and its first number contains several interesting articles. Among others, there is one by Dr Edward Moore, Plymouth, describing a new British fish, the *Peristedion Malarmat* of Lacepede and Cuvier, the *Trigla Cataphracta* of Linnæus. The learned editor gives a description and figure of *Voluta Lamberti*, a fossil shell of the "crag." Mr Bowman endeavours to ascertain the age of the Yew by referring to actual sections of its trunk; Mr Bree offers observations on *Trochilium Crabroniforme*, the Lunar Hornet Sphinx; and Mr Blyth makes some interesting remarks on the Psychological Distinctions between Man and all other animals. The most amusing paper, however, is a review of "The Naturalist," in which that rival periodical is represented to be entirely destitute of all merit, the articles in the first number being, generally speaking, "extremely lame, and written in a style which unavoidably forces upon the reader's recollection the idea of their being got up for the occasion, instead of being the spontaneous productions of those who are engaged hand and heart in the advancement of that science of which they profess to treat." Our opinion of "The Naturalist" is somewhat different; and although we cannot bestow upon it all the praise that we could wish, we must yet do it the justice to recommend it as a performance creditable to its conductors and useful to the public. We shall, however, take another opportunity of discussing the merits of this periodical.

Edinburgh: Published for the Proprietor, at the Office, 16, Hanover Street. London: Smith, Elder, and Co., 65, Cornhill. Glasgow and the West of Scotland: John Smith and Son; and John Macleod. Dublin: W. F. Wakeman. Paris: J. B. Balliere, Rue de l'Ecole de Médecine, No. 13 bis.

THE EDINBURGH PRINTING COMPANY.

THE EDINBURGH
JOURNAL OF NATURAL HISTORY,

AND OF

THE PHYSICAL SCIENCES.

MARCH, 1837.

ZOOLOGY.

DESCRIPTION OF THE PLATE.—THE WOLVES AND FOXES.

The Wolves of the American continent are generally supposed to be of the same species as those of Europe; and Dr Richardson, who has had ample opportunities of observing them, although appearing to consider them as probably distinct, yet, in our present state of knowledge, thinks it preferable to view them as races or varieties. They are very numerous on the sandy plains between the sources of the Saskatchewan and the Missouri, as well as in all the northern regions of America, burrowing like the Fox, and bringing forth their young under ground. So strong a resemblance do they bear to the domestic dogs of the Indians, that a band of them has often been mistaken for the latter, and the howl of both species is so similar, that even an Indian sometimes fails to distinguish the animal by it. Several varieties are met with, characterised by differences in size and colour; of these, two are represented on the Plate.

Fig. 1. The Dusky Wolf, *Canis lupus, occidentalis, nubilus.* This variety is of a greyish-black colour, more or less tinged with brown.

Fig. 2. The .Prairie Wolf, *Canis lupus, occidentalis, latrans,* of a light grey colour, white beneath, may be considered as a variety of the common Wolf of America, although it differs considerably in voice, size, and manners. Dr Richardson states, that on the banks of the Saskatchewan, these animals start from the earth in great numbers, on hearing the report of a gun, and gather around the hunter, in expectation of getting the offal of the animal he has slaughtered. They associate in greater numbers than the grey Wolf, burrow in the open plains, hunt in packs, and excel the other kinds in speed.

Fig. 3. The European Wolf, *Canis lupus,* was formerly generally distributed over the continent of Europe, but, owing to the progress of cultivation, is now confined to the wilder, wooded, or mountainous regions. In Norway, Sweden, and Russia, where they are still plentiful, and hunt the deer and other large animals in packs, they are usually of a light grey colour, and have a woolly fur, like that of the American Wolves; but in Germany and other parts they are generally brown, with coarser and shorter hair; while in the Pyrenees, a variety occurs of a brownish-black colour. The Wolf is, in all probability, the original of the Domestic Dog, of which several races, especially those found with the wild tribes of our own species, so nearly agree with it, that no real distinction can be pointed out between them.

Fig. 4. The Common Fox of Europe, *Canis vulpes,* generally of a yellowish-red colour above, greyish-white beneath, also exhibits considerable diversity in size and colour. Even in Britain, which, from its small size, might not be expected to be favorable to the production of varieties, there may be distinguished the Mastiff Fox, the Hound Fox, of which the figure is a representation, and the Cur Fox, the smallest and darkest of the three. The Common Fox is so well known, on account of his proverbial character of cunning and sagacity, as well as for the amusement which his chase affords to the aristocratical part of the community, that it is unnecessary here to describe his manners.

Fig. 5. The American Foxes exhibit nearly as many varieties as the Wolves, but are generally considered as belonging to a species distinct from that of Europe, and named *Canis fulvus.* The variety represented is the Cross Fox, *C. fulvus, decussatus,* so denominated on account of having a dark line down the back of the neck, crossed by another on the shoulders. Individuals often occur, so similar on the one hand to the Red Fox, and on the other to the Black, that the hunters are in doubt how to name them.

Fig. 6. The Black or Silver Fox, *Canis fulvus, argentatus,* is the rarest and most prized of all the varieties; its fur, on account of its fineness and glossy black colour, more or less tinged with grey or white, fetching six times the price of any other fur produced in North America. It inhabits the same districts as the next variety.

Fig. 7. The Red Fox, *Canis fulvus,* may be considered as the principal, or most common representative of the species, of which the two preceding are varieties. Its size is somewhat less than that of the European Fox, its eyes nearer, its ears shorter, its fur closer and finer, and its colour lighter. . It burrows in summer, takes shelter in winter under a fallen tree, preys on the smaller animals of the rat family, but re-

jects no kind of animal food that comes in its way. It hunts chiefly by night; but is much inferior in speed to the English Fox, and is soon overtaken by a wolf or a mounted horseman. The fur of all the varieties of this species is valuable as an article of commerce, but more especially that of the Black Fox.

DESCRIPTION OF THE PLATE.—THE GNAT-CATCHERS.

The Gnat-catchers have been separated, by Mr Swainson, from the Fly-catchers, between which and the Warblers they seem to be intermediate. The group is peculiar to America, and is composed of small birds, generally bright yellow beneath, and green, blue, or dusky on the upper parts.

Fig. 1. The Hooded Gnat-catcher, *Setophaga mitrata,* is very abundant in the southern states of North America, where it prefers low situations covered with canes or thick underwood, feeding on small winged insects, and returning southwards after the breeding season.

Fig. 2. The Green Black-capt Gnat-catcher, *Setophaga pusilla,* is an inhabitant of the swamps of the southern states, whence it takes its departure early in October.

Fig. 3. The Yellow-tailed Gnat-catcher, *Setophaga Ruticilla,* male; Fig. 4. the Female; and Fig. 5. the Young. This species is found in the interior of the forests, on the borders of swamps and meadows, and in deep wooded glens, in most parts of the United States, which it visits early in summer, to return southward in autumn. It bears a great resemblance to the Redstart of Europe, and exhibits nearly the same habits, feeding on insects, which it pursues in the air, retiring to a tree or bush, and frequently jerking out its tail.

Fig. 6. Selby's Gnat-catcher, *Setophaga Selbii.* Male. This species was named in honour of Mr Selby by its discoverer Mr Audubon, who met with only three individuals, which occurred in Louisiana. Its habits are similar to those of the other species.

Fig. 7. Bonaparte's Gnat-catcher, *Setophaga Bonapartii.* This also was discovered by Mr Audubon, and named in honour of the Prince of Musignano.

Fig. 8. The small Blue-grey Gnat-catcher, *Setophaga cærulea*; and Fig. 9. the Female. The beautiful and diminutive species here represented is a very dexterous fly-catcher, and, moreover, somewhat resembles in its manners the Titmice, with which it occasionally associates. Like all the rest, it is a summer visitant of the United States. .

The Siskin found breeding in Scotland.—Mr Weir of Boghead, a most zealous observer of the habits of birds, has recently favored us with the following account of the breeding of the Siskin in the neighbourhood of Bathgate, in the county of Linlithgow: " About the end of May 1834, as I was returning from Bathgate, I was astonished at seeing, on the parish road between it and my house, a pair of Siskins feeding very greedily on the ripe tops of the dandelion. The head of the male was very dark, and the yellow on its wings uncommonly rich. I followed them for several hundred yards, being exceedingly anxious to discover their nest. In this, however, I did not succeed, as they flew off to a considerable distance, when I lost sight of them. I again and again renewed my search, but without success. A few days after this, two persons who were catching linnets with bird-lime in a small field belonging to me, were struck with the unusual chirping of young birds in a spruce which was planted in the middle of a very strong hawthorn hedge. When they were looking into the tree, in order to discover what kind of birds they were, they immediately flew out of their nest. They appeared to have a resemblance to the female Siskin; but as they were ripe, it was found impossible to secure any of them. The nest was small, built on two of the branches, one side of it resting upon the trunk of the tree, at the height of about five feet and a half from the ground, and within twelve yards of the North Glasgow road. It was one of the best concealed nests I recollect of ever having seen; indeed, had it not been so, it would not have so long eluded the notice of some of our most celebrated nest-hunting youths, who were almost in the daily habit of passing and repassing the place in pursuit of their favorite amusement. The old Siskins, with their young, were seen for two or three weeks afterwards in the immediate neighbourhood. Mr

Macduff Carfrae, bird-preserver in Edinburgh, who came out to pay me a visit, saw the same birds hopping among the branches of some elder trees, about the distance of a quarter of a mile from the place where they were hatched."

Rare Birds recently found in Britain.—In the last number of the Magazine of Zoology and Botany, it is stated, on the authority of P. W. Maclagan, that a Hoopoe, *Upupa Epops*, was shot near Coylton in Ayrshire, on the 16th of October 1836. In the same journal, Mr Albany Hancock gives intimation of the occurrence of the following species: *Falco rufipes*; a male, shot on the Durham coast, between South Shields and Marsden Rocks, in the middle of last October. *Motacilla neglecta*; a male, shot a little to the west of Newcastle, on the 1st of May last. *Larus minutus*; a specimen in first plumage; killed at the mouth of the river Tyne, in September last.

Rare Birds recently found in Ireland.—In the same periodical is an account, by William Thompson, Esq., of Belfast, of two specimens of *Sterna stolida*; which were killed in 1830, between the Tusker light-house, off the coast of Wexford, and Dublin bay: also of a specimen of *Larus Sabinii*, in the plumage of the first year; shot in Belfast bay, on the 18th September 1822, by the late John Montgomery, Esq., of Locust Lodge: and of two others; one shot in Dublin bay, the other in Belfast bay, in September 1834. It appears from the statement given by Mr Thompson, that Bewick's Swan is of more frequent occurrence in Ireland than the common wild Swan; whereas the reverse is the case in England.

Shower of Frogs.—M. Pontus, professor at Cahors, recently addressed to the Academie des Sciences a communication relative to a shower of Frogs. In August 1804, he was travelling in the diligence from Albi to Toulouse, the weather being fine, and the sky without clouds. About four in the afternoon the vehicle stopped a few minutes at La Conseillère, three leagues from Toulouse, to change horses. Just as it was about to set off again, a very thick cloud suddenly covered the horizon, and thunder was heard. The cloud must have been at a very small height, for the drops of rain that fell were very large. It broke upon the road at the distance of a hundred and twenty yards from where they were. Two horsemen, who were coming from Toulouse, and had been exposed to the storm, were obliged to wrap themselves in their cloaks; but were much surprised, and even frightened, at finding themselves assailed by a shower of Frogs. They hastened onwards, and on meeting the diligence, gave an account of their adventure to the passengers. M. Pontus saw that there were still some small Frogs on their cloaks, which they shook off in their presence. The diligence soon reached the place where the cloud had burst, when the passengers became witnesses of the extraordinary phenomenon: the road, and all the fields on both sides of it, were strewn with Frogs, of which the smallest were of the size of at least a cubic inch, the largest nearly two inches, so that they must have been upwards of a month or two old: there were as many as three or four layers of them, and many thousands were crushed by the wheels and the horses' feet. Some of the passengers would have closed the coach-windows to prevent their entering, as they suspected, from their leaps, that they would. The vehicle proceeded for at least a quarter of an hour over this mass of living creatures, the horses trotting all the while.

Guacharo of the Cavern of Caripe.—The Guacharo is a bird which flies abroad in the twilight, and remains concealed by day in some obscure retreat, and in particular, inhabits in great numbers a very deep cave in the Valley of Caripe, in the province of Cumana. Humboldt, who visited this cave in 1799, first made known this remarkable bird, to which he gave the name of *Steatornis Caripensis*, on account of the fat which abounds in it when young, and which the natives apply to the same uses as oil and butter. M. Humboldt's collections having been lost by shipwreck on the coast of Africa, no specimens existed in Europe until 1834, when the Academie des Sciences received from M. Lherminier, a physician at Guadaloupe, a Guacharo preserved in spirits, with a paper relative to its habits and zoological relations. Mess. de Blainville and Geoffroy-Saint-Hilaire drew up a report, which was read on the 6th October 1834, and of which a condensed account is here presented. The body of the Guacharo is not larger than that of a pigeon, its head broad, its mouth very wide, and furnished with long stiff hairs, its bill pretty strong, the upper mandible hooked, with a well-marked tooth on its margin, the lower truncated at the extremity. The nostrils are oval, the eyes of moderate size, the ears rather small, the tongue arrow-shaped and adherent. The wings are very large, the legs short and robust, the toes short, with strong, arched, rather pointed claws, that of the middle toe not toothed as in the Goat-suckers. The tail-feathers are ten. The colour is chestnut red, mixed with brown, and glossed with green, banded and dotted with black, and marked with white spots. M. Lherminier adds, that the sternum is like that of the Goat-suckers, and that, as in them, there is no crop, but only a proventriculus and a gizzard of moderate strength, with two longish cœca at the end of a rather wide and short intestine. He was led to suppose, that as it resembles the Goat-suckers in its nocturnal habits, its form, and the distribution of its colours, it was also similar in its mode of feeding; although Humboldt asserted that the adults are not seen to pursue insects, and that the gizzard of the young is often found filled with nuts. To ascertain this alleged fact, he endeavoured to procure specimens, and at length succeeded in obtaining many young birds, and two old individuals, one of which has been sent stuffed to the Academie, to be deposited in the Museum. The young Guacharos are of the same colour as the old. In most of the individuals examined, the stomach was empty, but in some it contained kernels of fruits, and in none were there any remains of insects; so that Humboldt's opinion has been confirmed by direct observation. M. Bauperthuy, who procured the specimens for M. Lherminier, states that, of ten young birds which he attempted to rear, eight died in the course of the second month, and only two lived to the end of the third. The food which seemed to agree best with them was the fruit of the banana cut into small pieces; but although at first they digested it very well, it at length ceased to afford nutriment to them. In captivity the young birds are dull, and constantly keep their tail raised, and their bill on the ground. When approached they retreat in this position, and present somewhat of the appearance of a toad. When touched, they emit a sharp and very disagreeable cry. During the day, they seek the darkest places and remain quiet, but towards evening they seem to emerge from their usual apathy, and traverse their prison, crying and shaking their wings. The flesh of the old birds is lean and tough, but that of the young fat, tender, and nearly the same in taste as that of a young pigeon. The fat which covers the abdomen is exceedingly abundant, and so fluid that it transudes when handled. When melted over a gentle fire, slightly salted, and then put into a calebash, and well closed, it remained clear and without smell for three months. In taste it is like sheep's fat, but rather more delicate. The seeds found in the gizzards of young birds taken from the nest belonged to various plants, but particularly to one named in the country Mataca. They are round, of the size of a nutmeg, with an aromatic smell, and are rejected entire when they have been deprived of their pericarp, or the fleshy part of the fruit. The Indians who accompanied M. Bauperthuy told him, that the Cave of Caripe is not the only asylum of the Guacharos, but that they are also found in other caverns situated towards the north-east.

The Tarantula.—In a very interesting paper on the Tarantula, *Lycosa Tarentula*, by M. Leon Dufour, published in the Annales des Sciences Naturelles, and of which a translation is given in the number for February 1837 of Loudon's Magazine, are the following particulars relative to its habits:—The Tarantula inhabits dry, barren, uncultivated places, exposed to the sun. It hides itself in burrows of a cylindrical form, often an inch in diameter, sunk more than a foot in the soil, for four or five inches vertical, then horizontal, and again perpendicular. The orifice is ordinarily surmounted by a funnel, sometimes two inches in diameter, rising about an inch above the surface of the soil, composed of fragments of dry wood united by a little clay, and lined with a kind of tapestry, formed of the threads of the animal itself. This funnel not only protects its intrenchment from inundations, and fortifies it against the falling of external bodies, which, swept by the winds, would be likely to close it up, but it also serves as an ambush, by offering to flies and other insects an enticing resting-place. The months of May and June are the most favorable season for seeking it. After various fruitless attempts to procure specimens, M. Dufour succeeded in the following manner:—It occurred to him to take, by way of bait, a stalk surmounted by a spikelet, and to shake it and rub it gently against the opening of the hole. He was not long in perceiving that the attention and desire of the Tarantula were awakened. Tempted by this lure, he advanced, with a slow and irresolute step, towards the spikelet; and on its being drawn back a little, frequently used to throw himself, at one spring, out of his dwelling, the entrance of which was instantly closed. It sometimes happened that, suspecting the snare, or perhaps less pressed by hunger, he held back, immovable, at a little distance from his door, which he did not judge it advisable to pass. When this occurred, M. Dufour, after having observed the direction of the hole, and the position of the spider, drove in the blade of his knife, so as to surprise the creature behind, and cut off his retreat. By employing this mode of capture, he sometimes took so many as fifteen in an hour. We shall conclude our extract with an account of a combat between two Tarantulas.—"In the month of June, one day when I had been successful in the search after the Tarantulas, I chose two full-grown and very vigorous males, which I put together into a large vase, that I might witness the spectacle of a mortal combat. After having many times made the circuit of their arena, in the endeavour to shun each other, they hastened, as at a given signal, to set themselves in a warlike attitude. I saw them, with surprise, taking their distance, and gravely rising upon their hind legs, so as to present to each other the buckler formed by their chests. After looking each other in the face for about two minutes, and, without doubt, provoking each other by glances which I could not discern, I saw them throw themselves upon one another, entwine their legs, and endeavour, in an obstinate struggle, to wound each other with the hooks of their mandibles. Either from fatigue, or by mutual consent, the combat was for awhile suspended: there was a truce for some seconds; and each wrestler, retiring to a little distance, resumed his menacing posture. This circumstance reminded me that, in the singular encounters of cats, there were also suspensions of arms. But the struggle was not long in recommencing, with more fury than before, between our two Tarantulas. One of them, after victory had been a long time doubtful, was at length overthrown, and mortally wounded in the head: he became the prey of the vanquisher, who tore open his skull, and devoured him."

Mode of taking Quails in the Island of Cerigo.—The following notice is extracted from a very interesting account of the Natural History and Statistics of the Island of Cerigo, published in the *Edinburgh New Philosophical Journal*, by the late Robert Jameson, Esq., Assistant Surgeon, 10th Regiment of Foot. The flocks of Quails which appear here in spring and autumn are considerably reduced by various destructive means of the inhabitants; but the most singular is that of finding them by dogs, something similar to a lurcher, and then catching them with hand-nets. Two, or a party of three, go sporting in this way; each net has a mouth somewhat oval, stiffened by a rim of wood two or three feet long, attached to which is a net of a proportionate bulk; to this border is fastened at one end a pole, ten to fourteen feet long; and with such a weapon, a party of these will secure twenty or thirty couples during the day in the following manner:—When the dog makes a point, the party comes up towards the spot in different directions, holding their nets by the ends of the poles; and if the Quails lie so close, as they do in bushes, as to allow the party to touch each other's nets, then the dog is driven in to put them up. On rising, each man strikes at a bird with his extended oval-mouthed net, twisting it in the air to entangle his game, and, when expert, seldom misses.—On their first arrival, the Quails are often so much fatigued as to be taken by the hand, or nets of the simplest construction. In spring they are thin, and scarcely worth the trouble of procuring, while in autumn they are fat, and much prized as delicacies. Great numbers are preserved and fattened for the table, but unless great care is taken, they die quickly. Several experiments have been made here in autumn by private individuals, of several hundreds at a time, but they always died off before the cold weather had fairly commenced.

The Itch Insect, *Acarus scabiei*.—M. A. Gras, an *élève interne* of the Hospital of St Louis, has recently made the following observations on this minute insect:—

The *Acarus scabiei*, or *Sarcoptes hominis*, generally exists on all persons affected with itch who have not commenced a course of medicine. It is almost entirely confined to the hands, where it is found beneath the epidermis, but is also sometimes met with on the feet, in the arm-pits, and other places. It is never found on persons affected with any other cutaneous disease than itch. The insects are all destroyed after the sulfuro-alcaline ointment has been applied; but the patient may not be still cured, for the eruption may remain, unless it be properly treated.

Insects removed from an affected to a sound person, multiply on the skin of the latter, when presently the eruption appears. M. Gras several times communicated the disease in this manner; once, at the desire of Dr Pariset, secretary of the Academie de Medicine, when it produced a sanatory revulsion in a young girl who had fallen into a state of stupor. On the other hand, he several times tried to inoculate himself with the serum of the itch vesicles, but without success. He therefore concludes that the sarcoptes is the sole agent in producing the contagion of itch, which is not contracted unless that animal or its eggs adhere to the skin or clothes of persons coming into contact with those having the disease.

The number of insects on a person has no relation to the extent and intensity of the eruption; for sometimes not more than five or six are found on individuals covered with vesicles and pustules; and again, a hundred have been taken from the hands of a person, who yet had only a few vesicles.

BOTANY.

Alleged Irritability evinced by the Stems of Plants when slit.—Mr Golding Bird has endeavoured, in Loudon's Magazine for February, to explain the retraction observed by the two portions of a slit stem or stalk, on the principle of endosmosis. If a portion of the stem of an herbaceous exogenous plant be divided longitudinally, the division extending to the length of about two inches, the divided portions will instantly separate from each other to the distance of an inch or even more. The same thing occurs if the young shoots of woody stems are slit. Dr H. Johnson of Shrewsbury, who first attempted to explain this phenomenon, attributed it to the vital irritability of the plants, in consequence of its absence in some which are very elastic, and in the woody parts of stems. Mr Bird, however, conceives that it may be explained in another manner. Dutrochet has shown, that when a fluid, as water, or a weak saline solution, is inclosed in an organized membrane, as a piece of bladder, or placed in a glass tube closed by a piece of membrane, and immersed in a solution of sugar, the bladder or glass tube is rapidly emptied; but if, on the contrary, the bladder or tube be filled with syrup, and immersed in distilled water, the reverse takes place, the bladder becoming completely turgid. From these facts he inferred, 1*st*, That when a fluid of low specific gravity, inclosed in an organic membrance, is immersed in one of greater density, the membrane becomes rapidly emptied, in consequence of a current being set up from the lighter to the denser fluid; and, 2*d*, That when a dense fluid is inclosed in a membranous reservoir, and immersed in a fluid of a lower specific gravity, a current is set up, whereby the membrane becomes distended by a supply of fluid from without. The current from within to without, Dutrochet named *endosmose*; that from without to within, *exosmose*. Now, on this principle, rather than on the irritability of the plants, Mr Bird accounts for the following facts observed by him.

1. A piece of the stem of Lamium album, on being longitudinally divided to the extent of three quarters of an inch, immediately exhibited a divergence of its segments, which separated to half an inch. 2. Another piece, similarly treated, showed the same result, fig. 1; and, on being immersed in distilled water, continued to curve in a still greater degree, fig. 2. 3. This piece was then immersed in a weak solution of sugar, when, in the course of an hour, the segments had approached so closely as to touch each other, fig. 3. When replaced in water, the segments again diverged; and so on repeatedly. 4. The upper part of the stem of a young plant, that had stood for twenty-four hours with its lower portion immersed in water mixed with hydrocyanic acid, showed no divergence on being slit. 5. The same piece was placed in distilled water, when, in six hours, the divided portions separated to the extent of half an inch. When removed, and immersed in syrup, it showed a still greater divergence. 6. A piece of the stem of *Stackys palustris*, removed from near the root, was left for some days exposed to the air, and when dry was slit, but did not exhibit the least appearance of divergence. 7. A portion of this dried piece, on being placed in weak syrup, nearly regained its natural stage of turgescence in about twelve hours; when, on being slit at its upper part, it immediately manifested divergence; and which increased on its being immersed in water.

Dr Bird's explanation is to the following effect:—That the property of divergence does not depend upon vital irritability, as Dr Johnson assumed, is shown by experiments 5 and 7, in which the property was restored, by artificial means, to stems deprived of vital influence, by being isolated from the plants bearing them, and submitted to the deleterious influence of a poison, or to desiccation. That it is not caused by elasticity is evident, seeing that the property of divergence is absent in the most elastic parts of plants, as true woody fibre, rattan cane, &c., and present in the most delicate herbaceous plants, as well as in the most inelastic. Dutrochet has shown, that the valves of the pericarp of Impatiens Balsamina, which, when mature, separate from each other, and become considerably curved inwards, are composed of a vesicular tissue, of which the vesicles nearest the external are considerably larger than those nearest the internal face of the valves. The tissue being injected from an accumulation of sap, all the cellules become turgid, and the larger ones occupy a greater space than before. Their complete distension being, however, prevented by the more compact tissue formed by the aggregation of the minuter cells, the whole valve assumes a tendency to curve in such a manner, that the external portion, or that composed of the larger cells, may occupy the convex part of the curve; and this tendency to curve is obeyed as soon as the resistance of the opposite valve is removed by a slight touch or otherwise. This explanation may be applied to the phenomenon in question. The vesicular tissue of the Lamium and other plants is more minute towards the circumference, the internal part being composed of much larger cellules. When a stem possessing this structure is in perfect vigour, its vesicular tissue is injected with sap; the larger central cells become considerably distended, and press upon the smaller cells, which, resisting the pressure, give to the larger cells a tendency to separate, and occupy a greater space in consequence of this distension; their separation being, however, prevented by their intimate lateral organic connection. But when this bond of union is severed, by an incision or otherwise, the segments instantly separate, and curve in such a manner, that the internal portions, or those made up of the larger cellules, form the convex part of the curve, and thus leave more room for their distension; whilst the external portions, formed of a more compact tissue, occupy the concave parts of the curve, in consequence of their not becoming distended so easily or readily as the tissue nearest the axis of the plant.

This explanation is considered to be in perfect accordance with the result of experiment; for, on immersing the piece of stem, when the segments have separated to a certain extent, in water, which is a fluid of lower specific gravity than the sap in the cells, endosmosis ensues, whereby a quantity of water is forced into the already turgid cells, which, consequently, become more turgid, and the curvature increases; and on removing the piece of stem into syrup, a much denser fluid than sap, exosmosis, for reasons already explained, ensues; the cells become emptied, and the separated portions recover their former rectitude; the elasticity of the woody fibre present, also, probably assisting. The reason why poisons prevent the divergence is, that as they destroy the vital energy of the plant, they prevent the cells from becoming distended with sap to an extent sufficient to cause their separation; but, on placing the piece of stem in water, endosmosis occurs, and the divergent property is developed. But when the piece of stem used had been previously dried, the cells were emptied of their sap, and it was necessary to fill them artificially with a tolerably dense fluid, as syrup, before they became sufficiently injected to cause the separation of the segment, on making a longitudinal incision.

Mr Bird, therefore, is decidedly of opinion, that the divergence is the result of a *purely physical action, independent of vitality*; and in this we agree with him, although it is obvious that a much simpler and equally satisfactory explanation of the phenomenon can be given.

The plants found to exhibit it are possessed of a vascular, a fibrous, and a vesicular tissue. The latter occupies the central parts exclusively; while the woody fibres occur towards the exterior, as do the vessels properly so called. This arrangement is represented by the diagram, fig. 4, in which *a*, *a*, indicate the surface of the stem, *b*, *b*, its axis. Now, the woody fibres near *a*, not being extensile, and the cellules at *b*, being dilatable, it will necessarily follow, that when a section of the stem, such as that represented by the figure, is made, the cellules being filled with fluid, or vapour, which readily passes from the one to the other, the larger inner cellules, in consequence of the pressure caused by the unyielding cuticle and herbaceous tissue which compressed the interior when the stem was entire, being now allowed to expand to their full size, or in some degree, while the outer cellules are bound together by the inextensile woody fibres, the curvature outwards of the section is a necessary result, so long as moisture is supplied by the unslit portion of the stem, or communicated from without, this fluid ascending on the principle of endosmose. When, as in experiment 2, the slit stem is immersed in water, the divergence is increased on the same principle; but when, as in experiment 3, a denser fluid is substituted, exosmose takes place in the vesicular tissue, in consequence of which, the pressure is removed from the larger vesicles, which, by their elasticity, regain their original size. In the plant, experiment 4, whose lower parts were destroyed by poison, the upper not having received a supply of fluid to replace the quantity evaporated in twenty-four hours, no divergence could ensue on slitting it; but, as in experiment 5, should it be filled with fluid, it diverges as explained above. For the reasons stated, a dried piece of stem, experiment 6, can of course undergo no change; but if restored to its natural state, as in experiment 7, it exhibits the usual phenomenon. In accordance with this explanation is the fact, that woody stems when slit do not diverge, because all their parts are equally bound together by firm longitudinal fibres, as well as that of purely cellular plants, such as Fucus palmatus, exhibiting no divergence.

Wherefore, we neither agree with Dr Johnson, in considering the divergence or curvature in question as a result of vital contractibility or irritability; nor with Mr Bird, in attributing it solely to the larger vesicles being more readily distended than the smaller.

GEOLOGY.

Fossil Skull of Dinotherium Giganteum.—M. Kaup has addressed to M. Blainville a letter, giving intimation of the discovery of a perfectly-preserved head of *Dinotherium giganteum*, at Eppelstein, by a society, at the head of which is Dr Klippstein, who intends to prefix a geological introduction to the description and figure about to be published by the gentleman above-named. The expense of disengaging the head having been very considerable, the society intend to send it first to Paris, and afterwards to London, to be exhibited to the public. It is probable, however, that it will be purchased by the Academie des Sciences.

FOSSIL BONES DISCOVERED IN A NEW DEPOSIT CONNECTED WITH THE PLASTIC CLAY OF THE BASIN OF PARIS.—M. D'Orbigny, junior, has recently given an account of the discovery of a deposit of a peculiar marine limestone, between the tertiary formation and the chalk, and described numerous organic remains occurring in the lower part of the plastic clay. A trench that had been opened at Basmeudon, at the place called Les Montalets, and especially an opening made at the same point into one of the galléries of the chalk-work of M. Langlois, exposed immediately over the pisolitic limestone, several very interesting strata, of which no notice has until now been taken. The lowest bed is composed of plastic clay and marl, with numerous balls and fragments of chalk and pisolite, torn from the lower deposits, and forming a kind of conglomerate. At the base of this layer are balls, sometimes larger than the human head, of indurated pisolite, with milliolites, and some fibrous nodules of sulphate of strontian. The organic remains found in this bed are:—1. Marine radiaria and shells torn from the chalk by the fresh water which flowed over its surface; viz. *Ananchytes ovata*, *Catillus Cuvieri*, *Ostrea vesicularis*, and *Belemnites mucronatus*. 2. Fresh water shells, *Planorbis*, *Cyclas*, *Paludina lenta*, and *Anodonta*, contemporaneous with the conglomerate. Until now, no species of Anodonta has been found in a fossil state; and M. D'Orbigny believes he has discovered two, which he intends to name *A. Cordierii* and *A. antiqua*. 3. Bones of fishes, which cannot be determined. 4. Reptiles, which no doubt lived in the fresh water which formed the conglomerate; bones of fresh water tortoises, *Trionyx* and *Emys*; several teeth of crocodiles, and of a kind of large saurian animal, nearly allied to the *Mososaurus* or *Monitor* of the Maestricht chalk; also a coprolite, containing small fragments of fishes, and probably belonging to one of the reptiles mentioned. 5. Land quadrupeds carried away by the fresh water. There is a considerable number of teeth, of which two belong to a carnivorous animal of the Otter kind, the rest to pachydermatous animals, viz. to a large species of *Anthracotherium*, to a small species of the same genus, and to *Lophiodons*.

"The presence of these numerous bones of mammifera beneath the plastic clay," says M. D'Orbigny, "appears to me to be of very great interest, as it incontrovertibly demonstrates the existence of these animals at a much older period than is generally supposed. In fact, the only remains of mammifera hitherto found in the lower beds of the Paris deposits, were the jaw of a Lophiodon, discovered by M. Eugene Robert in the *calcaire grossier* of Nanterre, and two fragments of bones, probably also of the Lophiodon, which Cuvier has mentioned as having been taken from the lignite of Laonnais, the age of which is uncertain. These latter facts had already modified the opinion formed by Cuvier as to the depth at which the remains of mammifera might be found in the Paris deposits, and which, he presumed, never went below the gypsum. Now, however, it must be conceded that these animals lived at the period when the first layers of the plastic clay, which is the lowest deposit of the Paris basin, began to be formed. This fact once admitted, it will no longer be difficult to admit, in like manner, some cases which have given rise to much discussion, and which tend to carry still farther back the existence of these animals. One of these is the occurrence of the remains of *Didelphis Bucklandi*, in the oolitic limestone of Stonesfield; another, that of impressions of animals recently observed in the variegated sandstone of Hildburghausen in Saxony, and which some naturalists consider as foot-marks of mammifera or reptiles, while others assert that they are nothing but vegetable impressions; the third and most important refers to the bones of pachydermata, lately found by Professor Hugi in the Portland oolite of Soleure in Switzerland.

From these observations, in connection with those made by myself, may we not conclude, not only that mammiferous animals existed at the commencement of the tertiary period, but even before it, and that further researches will lead to the discovery of a much greater number?

In conclusion, it must be admitted, 1st, That the plastic clay of the neighbourhood of Paris is separated from the chalk by a distinct deposit, which may henceforth bear the name of pisolitic limestone, and which, from its containing only tertiary shells, appears very clearly to be connected with the paleotherian or tertiary epoch, and not with that of the chalk; and, 2dly, That in the lower part of the plastic clay deposit new characters have been found, demonstrating especially that various kinds of mammifera lived at the period when that deposit was formed, and that these mammifera differed greatly from those found in all the upper parts of the Paris formation.

FOSSIL BONES OF THE CAMEL.—M. de Blainville has announced to the Academie des Sciences of Paris, that he has been informed by an individual who has recently arrived from India, that the fossil remains of the Camel have been found in the deposits of the lower ranges of the Himalayan mountains. The remains alluded to consist of a cranium, which was discovered in a very hard sandstone, about two miles from Ramghur, and six from Pingon. This skull, almost entire, appears to have belonged to the single-humped Camel, *Camelus Dromedarius*. There was also found in the same place, the anterior part of the head of an animal intermediate between the *Anoplotherium* and *Paleotherium* of the Paris Basin; together with the tooth of a species of Mastodon, allied to *M. angustidens*.

MISCELLANEOUS.

EXPEDITION FOR EXPLORING CENTRAL AFRICA.—An expedition, under the superintendence of Dr Andrew Smith, fitted out by some individuals resident at the Cape of Good Hope, for the purposes of promoting the objects of science, and of benefiting the commerce of the colonists, has recently penetrated as far as lat 23° 28′. Dr Smith, in concluding his Report, states that the following are the principal results obtained:—

"1. It has put us in possession of much information respecting many tribes even hitherto unknown to us by name; and has enabled us also to extend very considerably our knowledge of those which had previously been visited, by having brought us in immediate connection either with them, or with persons who could furnish information regarding them.

"2. It has enabled us to ascertain the geographical position of many places previously doubtful—to lay down the sources and courses of various rivers which run to the eastward—and otherwise obtain what will considerably add to the utility of our maps of South Africa.

"3. It has enabled us to extend considerably our knowledge of natural history, not only by the discovery of many new and interesting forms in the animal kingdom, but also by additional information in regard to several previously known; and has put us in possession of a splendid collection, which, if disposed of, will in all probability realise a sum more than equal to the expenses which have been incurred.

"4. It has enabled us to ascertain that the Hottentot race is much more extended than has been hitherto believed; and that parties or communities belonging to it inhabit the interior, as far, at least, as the inland lake, which we were told is not less than three weeks' journey to the north of the Tropic of Capricorn.

"5. It has made us aware of the existence of an infinity of misery in the interior, with which we were previously unacquainted—a circumstance, which, in all probability, will lead eventually to the benefit of thousands, who, without some such opportunity of making known their sufferings, might have lived and died even without commiseration.

"6. It has enabled us to establish a good understanding with Umsiligas (a chief), and insure his services and support in the farther attempts which may be made to extend our knowledge of South Africa, which, without his concurrence, could never be well effected from the Cape of Good Hope."

About sixty birds and thirteen quadrupeds, supposed to be new, are characterized in an appendix to the Report. The number of specimens collected is as follows:—180 skins of new or rare Quadrupeds; 3379 skins of new or rare Birds; 3 bands of Snakes, Lizards, &c.; 1 box of Insects: 1 box of Skeletons, &c.; 3 Crocodiles; 2 skeletons of Crocodiles; 23 Tortoises, new or rare; 799 Geological specimens; and 1 package of dried Plants. In botany, however, very little has been done by the party, there having been no person attached to it for the express purpose of collecting and drying plants. It is intended to dispose of the duplicates, and send the most valuable part of the collection to Europe, to be exhibited for the purpose of raising funds to enable the Association to send out other expeditions.

REVIEWS.

The Northern Flora; containing the Wild Plants of the North of Scotland. By Alexander Murray, M. D. Part I. 1836. *Edinburgh: Adam and Charles Black.*

THE object of this work is, to present an account of the native vegetation of the north-eastern parts of Scotland, "which might afford to those resident there the means of acquiring a knowledge of the native plants they may expect to find, without the evident and well-known inconvenience, arising from the extraneous matter, occurring, of necessity, in works of a more general character; while, at the same time, a knowledge of our indigenous species might be imparted to others at a distance, who may be interested in such matters." The principal objection which we have to the Northern Flora is, that the limits of the district or region, whose vegetable productions are described, have not been pointed out, and the range of the Forfarshire hills having been included, apparently for the purpose of taking in Mr Don's discoveries there. We have had occasion, in describing several plants for a projected Flora of Scotland, to refer to Dr Murray's descriptions, which we have invariably found to be accurate and judiciously drawn up. The characters of the genera and species are not elaborated in the usual manner; and yet, in our opinion, they are exceedingly well adapted to the wants of the student, whose aim is to determine a plant unknown to him expeditiously and with certainty. This is certainly the most original of all our local Floras, and indicates much patient research and discriminative acumen. The descriptions are not always so full as we could desire, but the distinctive characters are generally so accurately traced, that the young botanist can scarcely ever be in doubt as to the species in hand, provided his intellect be of ordinary acuteness, although he will find his examinations somewhat laborious from the want of a synopsis of the genera usually prefixed to the classes. We object to the frequently tedious accounts of the real and reputed virtues of the plants, which are quite unnecessary in the present state of science; and it is obvious enough that few people would apply to a Flora for information on this subject, the proper place for which is a treatise on the Materia Medica. The part published does not extend beyond the thirtieth genus of the fifth class. In an appendix are "Notes from the Ancients," by Francis Adams, Esq., more curious than valuable; and "Observations on the agricultural properties of native plants," by the Rev. J. Farquharson, better adapted for the Journal of Agriculture than for the Northern Flora.

The Ornithologist's Guide to Orkney and Shetland. By Robert Dunn, Animal-Preserver, Hull. London. 1837.

WE are much pleased with this amusing and useful little work, which, besides an interesting account of the northern islanders, and of the author's wanderings among them, contains an enumeration of the different species of birds found in Orkney and Shetland, and affords to the intending visitor precise information on subjects of importance to him, such as the expense of boating, lodging, food, and other matters of a like nature. Independently of its interest to the collector, it will also be found useful to the student of British Ornithology, as indicating with certainty the species that occur in these islands..

EDINBURGH: Published for the PROPRIETOR, at the Office, 16, Hanover Street. LONDON: SMITH, ELDER, and Co., 65, Cornhill. GLASGOW and the West of Scotland: JOHN SMITH and SON; and JOHN MACLEOD. DUBLIN: GEORGE YOUNG. PARIS: J. B. BALLIERE, Rue de l'Ecole de Médecine, No. 13 bis.

THE EDINBURGH PRINTING COMPANY.

THE EDINBURGH

JOURNAL OF NATURAL HISTORY,

AND OF

THE PHYSICAL SCIENCES.

APRIL, 1837.

ZOOLOGY.

DESCRIPTION OF THE PLATE.—THE BULLS.

OF the numberless varieties of the Domestic Ox, those peculiar to India and the East coast of Africa, forming a race generally known by the name of Zebu or Indian Ox, are among the most remarkable. They are more especially distinguished by having a large tumour or hump, chiefly composed of fat, on the back between the shoulders. Of this race there are numerous breeds, varying in size from that of our largest Bulls to that of a Mastiff, and dispersed over Southern Asia, the islands of the Indian Archipelago, and the coast of Africa from Abyssinia to the Cape of Good Hope, in which countries it supplies the place of the common Ox, being used as an article of food, as a lactiferous animal, and as a beast of burden. In some parts of India it is also employed for riding, as well as for drawing carriages, and is said to perform a journey of thirty miles a-day. Its flesh, although good, is inferior to that of the European races; but the hump is reckoned a great delicacy. It varies in colour like the other domesticated breeds, the most common tints being ash-grey, cream-colour, or white, but it is often red or brown, and occasionally black. Some of the breeds are horned, others have pendulous or flexible horns, destitute of the core or bony part, and some are entirely hornless.

Fig. 1. Represents an individual of the largest kind.

Fig. 2. One of the smaller horned breeds.

Fig. 3. The smallest race.

Fig. 4. An individual of the hornless variety.

Fig. 5. The American Bison (*Bos Americanus*). This animal attains a much larger size than our domestic breeds, and is peculiar to the American continent, where it occurs in vast herds in the great plains or prairies extending along the Mississippi, Missouri, Arkansas, and Saskatchewan rivers. It has a savage and ferocious aspect, owing chiefly to the enormous size of its head, which it carries low, and partly to the long shaggy hair with which its neck and the fore part of its body are clothed. The shoulders are very high, but the protuberance formed by them does not consist merely of flesh and fat, as in the Zebu, but also of the elongated spinous processes of the dorsal vertebræ. Its flesh is excellent, and affords the principal subsistence of many tribes of Indians. The Bisons are extremely fleet, and although possessed of enormous strength, generally take to flight on the appearance of an enemy, seeking shelter in the forests and swamps. But when wounded they manifest great ferocity, so that it is dangerous for the hunter to show himself, as they are capable of easily overtaking him. Dr Richardson mentions an accident of this kind which occurred while he was residing at Carlton House, one of the Hudson's Bay Company's Stations. "Mr Finnan M'Donald, one of the Company's clerks, was descending the Saskatchewan in a boat, and one evening having pitched his tent for the night, he went out in the dusk to look for game. It had become nearly dark, when he fired at a Bison-bull, which was galloping over a small eminence, and as he was hastening forward to see if his shot had taken effect, the wounded beast made a rush at him. He had the presence of mind to seize the animal by the long hair on its forehead as it struck him on the side with its horn, and being a remarkably tall and powerful man, a struggle ensued, which continued until his wrist was severely sprained, and his arm was rendered powerless; he then fell, and after receiving two or three blows became senseless. Shortly afterwards he was found by his companions lying bathed in blood, being gored in several places, and the Bison was couched beside him, apparently waiting to renew the attack had he shown any signs of life. Mr M'Donald recovered from the immediate effects of the injuries he received, but died a few months afterwards." Individuals of this species are sometimes seen in collections in this country, being exhibited under the classical name of *Bonasus*, to which, of course, they have no title. The weight of an old male is from twelve to fifteen hundred weight. The hair, being remarkably fine and woolly, has been manufactured into coarse cloth, and the dressed skin with the wool adhering forms an excellent blanket, and in Canada often sells for three or four pounds, being there used as a wrapper by persons travelling over the snow in carioli.

DESCRIPTION OF THE PLATE.—THE HUMMING BIRDS.

Fig. 1. The Modest Humming Bird. *Trochilus modestus*, variety. The species of which a variety, partially affected with albinism, is here represented, has obtained its name from the plainness of its colours, which are dull greenish-brown on the upper, and dark grey mixed with green on the lower parts. It lives in the forests of Brazil, but nothing is known of its habits. The variety figured is remarkable for the white patch on the occiput, indicating a tendency to albinism, which is of very rare occurrence in this family of birds.

Fig. 2. The Mango Humming Bird (*Trochilus Mango*). Male. This beautiful species varies exceedingly in colour, according to age and other circumstances. In the adult state, the general tint is a bright green, the throat and middle of the breast black, part of the abdomen pure white. It inhabits Jamaica and other West India Islands, and is said to occur on the Spanish Main, and in Brazil and Guiana.

Fig. 3. The Mango Humming Bird. Young Male. With the upper parts glossed with yellowish-brown.

Fig. 4. The Mango Humming Bird. Young. In this state, although considerably different in colour, and having a large portion of the lower surface white, it is equally beautiful.

Fig. 5. The White-eared Humming Bird (*Trochilus leucotis*). The delicate and beautiful bird to which the name of "white-eared" was somewhat erroneously applied by M. Vieillot, seeing the region of the ears is deep blue, although a white line passes over it, is one of the most common species in Guiana and Brazil, where it is frequently met with in thickets near houses.

Fig. 6. The Blue-throated Humming Bird (*Trochilus Lucifer*). This diminutive species is characterised chiefly by a tuft of scale-like feathers, coloured with the iridescent tints of the Elba iron ore, on the throat. The female is not known. It is a native of Mexico.

Fig. 7. Swainson's Humming Bird (*Trochilus Swainsoni*). M. Lesson, who named this species in honour of a well-known English Naturalist, states that it is extremely rare, and that its native country is Brazil.

ORGANIZATION OF THE ORAN-OUTAN.—M. Geoffroy-St-Hilaire has recently presented to the Académie des Sciences the following observations on this subject:—If we compare the Oran-Outan with Man, we perceive the most remarkable conformity in all their parts. There is not a vessel, nor a nerve, nor a muscular fibre, more or less; but, at the same time, each organic element presents modifications in the length and thickness of the parts. The vertebral axis is comparatively shorter, not from the absence of any of its parts, but on account of their vertical compression. The head is apparently larger, but this more in appearance than in reality. The neck seems wanting, the parts which form it appearing to belong to the hind-head, and to prolong it to the shoulders. This is produced by the following mechanism. In the Oran as in the Bat, the clavicles are extremely long; and to be kept beneath the integuments without occupying too much room, they are directed obliquely, so that their outer extremity has, as it were, ascended towards the skull, and drawn with it a certain number of muscles, which, adding to their thickness, that of the muscles peculiar to the posterior region of the neck, fill up the wide groove formed by the series of spinous processes, which are themselves very large. The action of this strong layer of cervical muscles tends to throw the head backwards. The animal, in consequence of this general modification, must keep its body and head parallel to the trunk of the tree on which it resides, clinging to it by the extremities, and also fixing itself by the hands to the branches which are small enough to be laid hold of. The brain of the young Oran-outan bears a great resemblance to that of a child. The skull might in fact be taken, at an early age, for that of the latter, and the illusion would be almost perfect, were it not for the development of the bones of the face. But it happens, in consequence of its advance in age, that the brain ceases to enlarge, while its case continually increases. The latter becomes thickened, but in an unequal degree, enormous bony ridges appear, and the animal assumes a frightful aspect. When we compare the effects of age in Man and the Oran-outan, the difference is seen to be, that, in the latter, there is a super-development of the osseous, muscular, and tegumentary systems more towards the upper than the lower parts, while the development of the brain is entirely arrested.

QUANTITY OF SEEDS EATEN BY WILD BIRDS.—Rock Pigeons, which are very abundant in the Hebrides, Shetland, and Orkney Islands, live chiefly on grain and the seeds of various plants. We had the curiosity to count the number of seeds contained in the crops of two individuals recently obtained from Shetland. In that of one were 510 grains of barley, while the other contained 1000 grains of the grey oat, besides a few small seeds. Now, supposing there may be 5000 Wild Pigeons in Shetland, or perhaps in Fetlar alone, which feed on grain for six months every year, and fill their crops once a-day, half of them with barley, and half with oats, the number of seeds picked up by them would be 229,500,000 grains of barley, and 450,000,000 grains of oats;—a quantity for which the poor Shetlanders would be very thankful in a season of scarcity like the present. What is the number of Pigeons, wild and tame, in Britain, and how much grain do they pick from the fields and corn-yards? It is probable, that were the quantity of seeds of the cereal plants, which all the granivorous birds in the country devour anunally, accurately known, it would prove much higher than could be imagined; yet by far the greater part could be of no use to Man, were all the birds destroyed, it being irrecoverably dispersed over the fields.

Sternum of Birds.—M. F. Lherminier, a French physician, settled at Gaudaloupe, has recently made extensive observations on the development of the sternum in Birds. Before giving a brief account of his discoveries, we shall shortly describe the bone to which they refer, with the aid of a figure of that of the Black Grouse, *Tetrao Tetrix*:

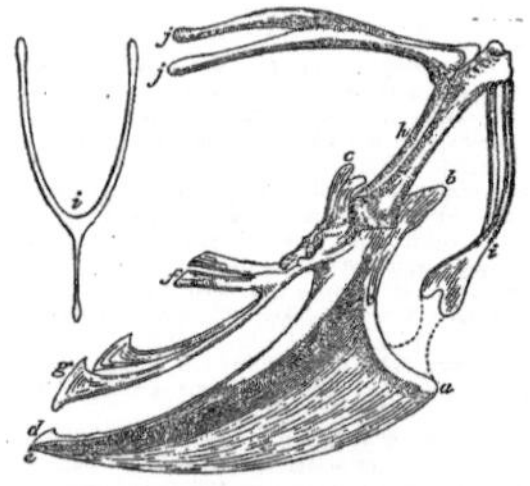

The letters *a, b, c, d, e,* mark the limits of the sternum itself; *h*, indicates the coracoid bones, or clavicles of many authors; *i, i,* the furcula, or anterior clavicles; and *j, j,* the scapulæ. The sternum varies greatly in form in different tribes of birds; the body, *b, c, d, e, f, g,* being sometimes complete, often, as here, with deep vacuities or sinuses, by which it is broken up; and often with holes or notches in its posterior margin. The crest or ridge, *a, e,* is more or less prominent according to the size of the pectoral muscles, which depress the wings, and is therefore, in some degree, indicative of the power of flight of the species. It is wanting in the Ostrich, which is destitute of wings properly so called, that is, organs of flight, and is very highly developed in Eagles, and other birds that fly with great speed.

From the observation hitherto made, only two modes of ossification of the sternum have been admitted; namely, five pieces in the Gallinaceous birds, and two in the Ducks and Ostrich. M. Lherminier, however, considering the diversified forms under which that bone exists in birds, found it difficult to believe that there are only two types for the arrangement of the pieces which enter into its composition; and on extending his researches, came to the conclusion that the number of original pieces is nine; at least, that there are nine distinct bones in the sternum of birds, considered in a general sense, although in no species does the whole number occur at any period of development.

These nine pieces may be considered as belonging to three transverse series, an anterior or *prosternal*, a middle or *mesosternal*, a posterior or *metasternal*. The first row comprehends a median piece, the *prosternum*, and two lateral pieces, the prosternal; the second is composed of a *mesosternum* and two *mesosternals;* the third of a *metcosternum* and two *metasternals*. It is to be remarked, however, that the number of the pieces of each series may sometimes be more than three, and in certain groups of birds, M. Lherminier has seen it amount to six.

The first series, when it exists, is generally confined to a much smaller space than the rest. It is sometimes complete, sometimes reduced to two lateral pieces, and sometimes to the middle piece alone. It serves to support the coracoid bones, *h*, and affords a fixed insertion to the sterno-coraco-clavicular aponeurosis. It is that part denoted by the letters *b, c*.

The second series may, in like manner, be complete, or formed only of the two lateral pieces, but it is never reduced to a single central piece. When the three pieces exist, sometimes the mesosternum forms part of the body of the bone, and is seen at its upper surface; sometimes it belongs to the keel, *a, e,* and occupies its upper part. In the former case, it may be double or single; in the other, it is always a single nucleus. The use of this series is to afford a point of support to the sternal ribs, and to contribute to the development of the crest and body of the sternum. When the first series is absent, the second occupies its place opposite the coracoid bones, which then rest on the mesosternum, or, in its absence, on the mesosternals.

The third series belongs entirely to the body of the bone, and is sometimes complete, sometimes reduced to the two lateral pieces, or to the middle piece alone, which it is sometimes rather difficult to distinguish from the two other central pieces. When these are wanting, the metasternum directly supports the coracoid bones. The metasternals, like the mesosternals, contribute to the support of the ribs, and maintain in a state of tension the fibrous membranes of the sternum, so as to favour the action of the muscles inserted into them.

These pieces exhibit great diversity in their mode of development. Their relative importance, estimated according to the frequency of their occurrence, is as follows:—1. The mesosternals, which are never wanting; 2. the mesosternum; 3. the prosternals; 4. the metasternum; 5. the prosternum and metasternals.

In the Gallinaceous Birds, of which the figure represents the sternum, that bone results from the union of five pieces, there being originally in the chicken five centres of ossification, viz. two lateral pieces of the middle series, and three of the posterior series. The two lateral pieces of the middle row form the part included between *b, c,* and *f;* the middle posterior piece is marked by *a, d, e;* and the two lateral posterior pieces are denoted by *g*.

By researches of this kind, that is, by comparing the parts in Birds of different genera and families, it may be expected that much light will ultimately be thrown on the natural arrangement of Birds, which at present is left in the hands of persons who can do little more than compare skins and stuffed specimens, and who, ludicrously enough, perpetually talk of the "strict analysis" by which they have absolutely fixed the order of Nature.

BRITISH BIRDS.—NO. I.

The Raven.—The habits of many of the most remarkable of our native Birds being very imperfectly described by our most accredited authors, we propose to present those of the species, which, on account of their great size, beauty, utility, or injurious agency, are especially worthy of notice, in a series of papers, commencing with that most sagacious of all birds, the Raven.

This species, which is the largest of the Crow family resident in Britain, with a grave and dignified air, combines much cunning, and in courage is little inferior even to some of the Rapacious Birds. Its body is of an ovate form, rather bulky; the neck strong, and of a moderate length; the head large and oblong; the bill rather long, deep, and nearly straight; the feet of moderate length and ordinary strength. The plumage is compact and highly glossed; the wings long and much rounded; as is the tail. The bill and feet are black; the plumage deep black, with splendent reflections of rich purplish-blue. The length of the male is twenty-six inches, and its extended wings measure fifty-two.

The Raven is a remarkably grave and sedate bird, and, unlike many men who assume an aspect of dignity, is equally noted for sagacity and prudence. It is crafty, vigilant, and shy, so as to be with great difficulty approached, unless in the breeding season, when its affectionate concern for its young, in a great measure, overcomes its habitual dislike to the proximity of Man; a dislike which is the result of prudence more than of mere timidity; for, under particular circumstances, it will not hesitate to make advances which a timorous bird would, no doubt, deem extremely hazardous. It eats from off the same carcase as a Dog, and takes its station close to an Otter devouring its prey, doubtless, because its vigilance and activity suffice to enable it to elude their efforts to inflict injury upon it; and while it yields to the Eagle, it drives away the Hooded Crow and the Gull. It knows the distance, too, at which it is safe from a man armed with a gun, and allows the shepherd and his dogs to come much nearer than the sportsman. It never ventures to attack a man plundering its nest, and rarely pretends to be crippled, in order to draw him away from it, but stands at a distance, looking extremely dejected, or flies over and around him, uttering now and then a stifled croak indicative of anger and anxiety.

When searching for food on the ground, it generally walks with a steady and measured pace; but under excitement, it occasionally leaps, using its wings at the same time, as when driven from carrion by a dog, or when escaping from its fellows with a fragment of flesh or intestine. Its flight is commonly steady and rather slow, and is performed by regularly timed beats of its extended wings; but it can urge its speed to a great degree of rapidity, so as to overtake an Eagle, or even a Hawk, when passing near its nest. In fine weather it often soars to a vast height, in the manner of the birds just mentioned, and floats, as it were, at ease, high over the mountain tops. Some Naturalists having observed Birds thus engaged, have imagined them to be searching for food, and have consequently amused their readers with marvellous accounts of the distances at which the Eagle can spy its prey; but had they patiently watched, they might have found that the quiet soarings of the Raven and the rapacious species have no reference to prey. On the other hand, it may sometimes be observed gliding along, and every now and then shifting its course, in the heaviest gales, when scarcely another bird can be seen abroad. Although there is not much reason for calling it "the tempest-loving Raven," it would be a severe storm indeed that would keep it at home, when a carcase was in view.

In the Hebrides, where this bird is much more abundant than in any other part of Britain, it may be seen either singly or in pairs, searching for food, along the rocky shores, on the sandfords, the sides of the hills, the inland moors, and the mountain tops. It flies at a moderate height, proceeding rather slowly, deviating to either side, sailing at intervals, and seldom uttering any sound. When it has discovered a dead sheep, it alights on a stone, a peat bank, or other eminence, folds up its wings, looks around, and croaks. It then advances nearer, eyes its prey with attention, leaps upon it, and, in a half crouching attitude, examines it. Finding matters as it wished, it croaks aloud, picks out an eye, devours part of the tongue, if that organ be protruded; and, lastly, attacks the abdomen. By this time another Raven has usually come up. They perforate the skin, drag out and swallow portions of the intestines, and continue to feast until satiated or disturbed. Sometimes, especially should it be winter, they are joined by a Black-backed Gull, or even a Herring Gull, which, although at first shy, are allowed to come in for a share of the plunder; but should an Eagle arrive, both they and the Gulls retire to a short distance, the former waiting patiently, the latter walking backwards and forwards, uttering plaintive cries, until the intruder departs. When the carcase is that of a larger animal than a sheep, they do not, however, fly off, although an Eagle, or even a Dog, should arrive. These observations were made by the writer, when lying in wait in little huts constructed for the purpose of shooting Eagles and Ravens from them. The latter were allowed to remain unmolested for hours, that they might attract the former to the carrion; and in this manner, he was enabled to watch their actions when they were perfectly unrestrained.

Although the Raven is omnivorous, its chief food is carrion, by which is here meant, the carcases of Sheep, Horses, Cattle, Deer, and other quadrupeds; Dolphins and Cetaceous animals in general, as well as fishes that have been cast ashore. In autumn it sometimes commits great havoc among barley, and in spring occasionally destroys young lambs. It has also been accused of killing diseased sheep by picking out their eyes. It annoys the housewives by sometimes flying off with young poultry, and especially by breaking and sucking, or rather gobbling eggs, which the ducks or hens may have deposited, as they frequently do, among the herbage

In these islands, should a Horse or a Cow die, as is not unfrequently the case in the beginning of summer, after a severe winter or spring, or should a Grampus or other large Cetaceous animal be cast on the shore, the Ravens speedily assemble, and remain in the neighbourhood until they have devoured it. Whatever may be said by closet-naturalists as to the unrivalled adaptation of the point of the upper mandible of the Rapacious birds for tearing flesh, the bill of the Raven is in practice quite as efficient an instrument. That bird can not only with great ease tear off morsels of flesh, but can pick the smallest shreds from the bones, and rend the intestines in pieces. When engaged upon a large carcase, they conduct themselves very much in the manner of the North American Vultures, as described by Wilson and Audubon. We have seen them thus occupied with a Cow. Some were tearing up the flesh from the external parts, others dragging out the intestines, and two or three had made their way into the cavity of the abdomen.

The Raven sometimes nestles at no great distance from the Eagle, in which case these birds do not molest each other; but in general the former is a determined enemy to the latter, and may often be seen harassing it. Two Ravens attack the Eagle, one hovering above, the other beneath; but without over coming into contact with the object of their dislike, which seems to regard them as more disagreeable than dangerous, and appears to hurry on to avoid being pestered by them. Although they keenly pursue all intruders that seem in any way formidable, they on the other hand allow the Cormorant, the Rock-pigeon, and the black Guillemot, to nestle in their immediate vicinity.

The voice of the Raven is a hoarse croak, resembling the syllables *crock*, *cruck*, or *chrro;* but it also emits a note not unlike the sound of a sudden gulp, or the syllable *cluck*, which it seems to utter when in a sportive mood; for, although ordinarily grave, the Raven sometimes indulges in a frolic, performing somersets and various evolutions in the air, much in the manner of the Rook.

The character of this Bird accords well with the desolate aspect of the rugged glens of the Hebridian moors. He and the Eagle are the fit inhabitants of those grim rocks; the Red Grouse, the Plover, and its page, of those brown and scarred heaths; the Ptarmigan of those craggy and tempest-beaten summits. The red-throated Diver and Merganser, beautiful as they are, fail to give beauty to those pools of dark-brown water, edged with peat banks, and unadorned with silvan verdure. Even the water-lily, with its splendid white flowers, floating on the deep bog, reflects no glory on the surrounding scenery, but selfishly draws all your regards to itself.

The species is also very abundant in the Orkney and Shetland islands. In Sutherland, Ross-shire, and many parts of the county of Inverness, it is also not uncommon. In most of the Highland districts, we have met with it here and there. In the lower parts of the middle division of Scotland, it is of much rarer occurrence; nor is it plentiful even in the higher and more central portions of the southern division, although we have seen it in many places there. In England it is much less frequently met with than in Scotland, although it seems to be generally distributed there also. If we take the whole range of the island as its residence, we must add to its bill of fare many articles not mentioned above, so as to include young Hares and Rabbits; other small quadrupeds, as Rats, Moles, and Mice, young poultry, and the young of other birds, as Pheasants, Grouse, Ducks, and Geese; eggs of all kinds, echini, mollusca, fruit, barley, wheat, and oats; insects, crustacea, grubs, worms, and probably many other articles, besides fish and carrion of all sorts.

In the northern parts of Scotland, the Raven constructs its nest on high cliffs, especially those on the sea-shore; but in the southern parts of the island, where rocks are not so common as tall trees, it is said frequently to nestle in the latter. According to the locality, it begins to repair its nest, or collect materials for forming a new one, as early as from the beginning to the end of February. In the maritime districts, it is generally composed of twigs of heath, dry sea-weeds, grass, wool, and feathers. It is of irregular construction, and very bulky. The eggs are from four to seven, pale-green, with small spots and blotches of greenish-brown and grey, and are about two inches in length. The young are at first of a blackish colour scantily covered with soft, loose, greyish-black down. They are generally abroad by the middle of May. It has been remarked, that when, during incubation, or even when the young have left the nest, one of the old Birds is killed, the survivor soon finds a mate. Ravens, if unmolested, breed in the same spot year after year.

Few Birds are possessed of more estimable qualities than the Raven. His constitution is such as to enable him to brave the fury of the most violent tempests, and to subsist amidst the most intense cold; he is strong enough to repel any bird of his own size, and his spirit is such as to induce him to attack even the Eagle; his affection towards his mate and young is great, although not superior to that manifested by many other birds; in sagacity he is not excelled by any other species; and his power of vision is at least equal to that of most others, not excepting the Birds of prey, for he is generally the first to discover a carcase. To Man, however, he seems to be more injurious than useful, as he is accused of killing weakly sheep, sometimes destroys lambs, and frequently carries off the young and eggs of domestic poultry. For this reason he is generally proscribed, and in many districts a price is set upon his head: but his instinct and reason suffice to keep his race from materially diminishing. He seems to have fewer feathered enemies than most other Birds, for although he may often be seen pursuing Gulls, Hawks, and Eagles, we have never seen any species attacking him, with the exception of the Domestic Cock. It has been alleged, however, that Rooks assail him in defence of their young, and there is nothing incredible in this, for the weakest bird will often in such a case attack the most powerful and rapacious.

Numberless anecdotes are told of the Raven in the domestic state, but as our space is limited, we are unable to present any of them; and have only further to state, that young birds are easily reared, become perfectly tame, may be taught to pronounce words, and form amusing, though occasionally mischievous pets. The species is very widely distributed over the globe, being more or less common in Europe, Asia, and America, but more abundant towards the Arctic regions.

Eagles' Nests.—"An Old Correspondent" of Loudon's Magazine states, that not many years ago, the Eagle had its nest annually in two places on the borders of the counties of Dumfries and Selkirk. One was on a precipice in Eskdale; "the other situation was chosen with much of that touch of reflection that we sometimes observe among birds, as well as others of the lower animals. There is a small rocky islet, almost even with the water, in Loch Skene, which is surrounded with the highest mountains south of the Forth; and, although the side of one of these mountains, that overhangs the lake, is rocky and seemingly inaccessible, the Eagles chose to have their nests on the islet in the loch, because, forsooth, the loch-craig could be approached by ropes from above, while it is almost impossible to convey a boat to the loch, and there never was one there." Now, we would have naturalists to think a little before they state a fact, and be sparing of theory. A boat *has been* conveyed to the loch, and the Eagles of the district are extinct. The side of the mountain "that overhangs the lake" is a rocky slope of less than forty degrees, and certainly does not afford a spot to which a person without a rope might not approach to within ten yards. Eagles often make their nests in very insecure places. We have seen several that could easily be got at by a steady person. One was in a rock not thirty feet high, on the edge of a lake in Harris; another on a smaller rock among the hills of that island; a third at the height of sixty feet in the island of Shellay; a fourth about thirty feet from the top of a rock on the west coast of Harris; a fifth at least five hundred feet of perpendicular height from the sea, but not ten feet from the brink of the precipice, which, however, was surmounted by a very steep and slippery slope; a sixth was on a small island in a lake, in the pass of Miavag, surrounded by precipices, compared with which the rocks of the south of Scotland are contemptible, and on which were many utterly inaccessible spots. The Eagle seems, in fact, to have no precise notions as to either ropes or boats; and it is unnecessary to attribute to her more sagacity than she possesses. She certainly selects what she considers the safest spot, but without regard to its altitude; and, accordingly, her nest may be seen on the gravelly beach of a small lake, or on the summit of a precipice five hundred feet high. In this respect she resembles the Starling, whose nests we have seen in high precipices, on grassy slopes, and in deserted rats' holes, close to the tide-mark.

Rump Gland of Birds.—The uropygial gland of Birds, being one of the few organs accessible to the external-character Ornithologists, has been the subject of much discussion. Its uses are still unknown; although it was at one time confidently asserted to be a reservoir of oil, to be employed by the Bird for anointing its feathers. It is a remarkable fact, however, that the rumpless race of the domestic fowl, which is destitute of this organ, has its feathers not less perfect than those races which are complete in all its parts; and therefore it has been maintained, that the contents of the gland are not applied to the purpose alleged. No person has actually seen a Bird squeeze the oil from the gland, and apply it to the plumage; and it seems quite impossible that many species, whose bills are of great length, or of a form ill-adapted for the purpose, could thus besmear their feathers: the Curlew, for example, the Spoonbill, the Pelican, and the Gannet, whose head and neck, although beyond reach of their bill, is always in as good condition as any other part of their body. It is a well-known fact, that in Birds, when moulting, this gland is greatly enlarged, and in cage-birds often inflames and suppurates in consequence of the increased action which takes place in it. But it has not hitherto been known, that when the moult is completed, or rather, when the feathers of the tail are fully developed, the organ diminishes to such a degree, that frequently little of it can be seen besides its nipple-like prominence. We were led to make this observation by finding it much reduced in a fine Peregrine Falcon, shot in the beginning of winter, and of which the tail seemed complete, while in a Pigeon which had been killed when several of the tail feathers were sprouting, it was extremely large. Since then, we have inspected a multitude of birds, and in *almost* every case where the tail was full grown, have found the rump-gland very small. It therefore, in all probability, has reference to the development of the caudal quills in a manner which we shall explain on a future occasion, and in this respect is analogous to certain other organs in Birds which undergo periodical changes.

Irregular Growth of the Teeth of Rodentia.—Professor Meisner of Basle has recently given some account of the prodigious growth of incisor teeth in some of the *Rodentia*, which he thus accounts for. These teeth, in their normal state, are continually growing in length, slowly rising in height from the alveola, in such proportions as become requisite to compensate for the daily wearing away of their chisel-formed edges. This growth not ceasing during life, he remarks that all such teeth are invariably tubular at their base; and that the same effect is produced not only in the incisor teeth, but in all others whose roots remain unclosed. In animals, such as the Elephant, Babyroussa, Hippopotamus, and Narwhal, where these bony productions serve as a defence, the same observation seems fully to apply; and they sometimes attain an enormous length, no given measure having been ascribed to them for the full period of their maturity, that depending solely upon the duration of the animal's life. In the molar teeth of Hares, Rabbits, the Beaver, and some other Rodentia, this fact holds equally good; but it is not so in the domestic Rat, Mouse, and others, in which the alveola is always closed. He cites the observations of Blumenbach on the monstrous growth of the molar teeth of a Hare, examined by him, and also those of Rudolphi on a similar lusus in a Guinea Pig. We have fully confirmed these observations by an examination of several extraordinary examples of this phenomenon in the matchless Museum of the College of Surgeons. In a Rabbit, we observed the incisor teeth to have grown in a spiral form; in a Hare, also, in which, from their position, they must have occasioned the animal's death, by entering the head, or pressing so firmly upon it, at either side, as to wound the flesh and penetrate in. It thus appears clear, that a beautiful provision of Nature is exhibited in the formation of these teeth; their continual increase enables them to preserve a fine, even, cutting edge, always set to a particular angle with each other, so long as they remain truly in opposition; the motion of gnawing or cutting their food having also the effect of keeping the teeth sharp, by means of their constantly slipping over each other. If, however, by any accident, or malformation of parts, these teeth cease to act against each other, their growth still going on, they form a curved line, extending to an indefinite length during the animal's life, and occasioning, no doubt, in many instances, premature disease and death. So perfect is Nature in all her mechanism, that the slightest deviation from it, by accident or other causes, produces fatal effects.—*The Naturalist.*

BOTANY.

Notes respecting the Value of Specific Characters.—M. Wiegmann, in a letter addressed to the Editor of the Flora (1835, p. 106), communicates some observations made by him on this subject. The results obtained by him give little countenance to those authors who raise the slightest deviation of form to the rank of a species. Certain genera, of which numerous species are cultivated in the gardens, such as *Veronica*, *Verbascum*, *Delphinium*, and *Thalictrum*, are rich in species of which the native country is unknown. The origin of multitudes of nominal species in our catalogues may easily be accounted for by the changes caused by cultivation, and the numerous hybrid forms whose production is favoured by the proximity in which species of the same genus are placed in Botanic gardens. In 1833, the author saw a plant of the onion, *Allium Cepa*, bear a bulb in place of seeds. In the following spring, he put this bulb in the ground, and was astonished to find, that from it sprung up the *Allium proliferum* of Schrader, with a nearly naked, flexuous, and feeble stem, supporting a proliferous umbel, having sterile flowers on long pedicles. He refers to the numerous forms of *Iris* produced by M. Berg, and to the multiplicity of the *Calceolariæ* and other ornamental plants. *Taraxacum palustre* he has found, under cultivation, to assume the appearance of the common species; and seeds of *Myosotis sylvatica* of Ehrart, on being sown in the same spot, furnished five different forms, which authors consider as so many distinct species. Seeds of *Veronica agrestis* gave rise to six different forms. He is of opinion, that the numerous species into which the genus *Rubus* has been divided by Weisse, have been produced in the same manner.

Plants of the North of China.—M. A. Bunge, who resided in China, during 1831, with the Russian missionaries, brought with him on his return a valuable botanical collection, the principal species of which he briefly described in a small work, at the same time communicating specimens to the Botanists, and especially to the herbarium of the Paris Museum. But as among these plants there were several very interesting species, he has thought proper to give more detailed descriptions, and to illustrate some of them by figures. As this work, which was printed at Casan, is not easily procurable, the Editors of the *Annales des Sciences Naturelles* have given an abstract of it in their number for July 1836.

GEOLOGY AND MINERALOGY.

Microscopic Fossil Animals in Tripoli Slate.—M. Ehrenberg has discovered that the friable siliceous slate, commonly called Tripoli, or polishing slate, is entirely composed of remains of Infusory animalcules of the family of Bacilariæ. These remains retain their form so that they can be very distinctly seen with the microscope, and compared with living species. Bog iron-ore is also almost entirely composed of the species named *Gallionella ferruginea*. If a piece of one of these substances be rubbed a little on a glass plate, and the powder thus obtained mixed with water, thousands of animalcules may be seen with a good microscope.

Oxidulous Copper.—Mr G. B. Sowerby, in Loudon's Magazine for March 1837, gives an account of the occurrence of detached cubes of this mineral in Cornwall.—"They vary in dimensions from a quarter to nearly three quarters of an inch: they are of a very dark colour; many of them are nearly complete at all their angles, more particularly the smaller ones. Occasionally two or three are grouped together; some of them are accompanied by a small quantity of green carbonate of copper. Very few are slightly modified, having some of the planes which tend to the rhombic dodecahedron; and one very large crystal, being exactly half an inch long, which is adhering to a like quality, has the planes of the octohedron as its solid angles."

METEOROLOGY.

Luminous Meteor.—Mr Couch, in a paper on Shooting Stars, published in Loudon's Magazine, mentions the following singular phenomenon, which he states to be incapable of explanation on any principle "of our philosophy."—"On July 1, 1832, riding homeward in the evening, whilst the lightning was vivid, and thunder loud and frequent, a ball of light, of about the size of a large orange, met my view on the right, towards the north-west, distant about twenty yards. As if projected from a cannon, it passed straight and rapidly across the lane, close before me, and was instantly lost sight of on the other side. The whole was the work of an instant; the ball, of a steel blue, passing but at the elevation of my head, unattended by any noise or explosion, and evidently unconnected with the lightning that glared around."

Storm of the 29th November 1836.—Mr W. H. White, in his Meteorological Retrospect for 1836, in Loudon's Magazine, gives the following account of this memorable hurricane:—"The violent gusts of wind, chiefly from the N.W., did much damage, leaving many a sad memento behind, both by land and sea. From the most authentic accounts of this gale which have reached me, it appears that it commenced on the 23d, on the eastern shores of North America, off St Lawrence. A ship from Poole fell in with it on the 26th, in lat. 47° N., long. 32° 20′ W., and was thrown on her beam ends. It continued its progress across the Atlantic, and reached the Land's End about 7¾ A.M.; Plymouth, 8½ A.M.; Exeter, 9½ A.M.; Weymouth, 10 A.M.; Poole, 10½ A.M.; Farnham, 12 noon; London, 1½ P.M.; Suffolk coast, 2½ P.M.; and Hamburg, at 6 P.M. Thus the storm travelled at the rate of about 50 miles per hour; but the circular motion of the wind had a velocity of from 120 to 150 miles per hour. The fury of the gale was most felt on the coast of France and Belgium. At Ostend there was scarcely a house which was not unroofed; and so great was the demand for tiles, that they arose from 16 to 30 florins per 1000. The motion of the mercury in the barometer, during the most violent part of the hurricane, attracted great attention. On the morning of the 29th, at 9, the mercury stood at 29.30 in.; it soon afterwards began to sink very rapidly, exhibiting much agitation during the violent gusts of wind for which this hurricane was particularly remarkable, till 12 at noon, when it stood at 28.82 in. At 2 P.M., the barometer had risen to 29.35. soon after which the wind lulled into almost a calm.

MISCELLANEOUS.

Fidelity of the Dog, a Pattern to Man.—In the Rev. Dr Duncan's Sacred Philosophy of the Seasons, recently published, is the following interesting anecdote of Burns: "While yet a school-boy, I enjoyed an opportunity of hearing, in my father's manse, a conversation between the poet Burns and another poet, my near relation, the amiable Blacklock. The subject was, the fidelity of the Dog. Burns took up the question with all the ardour and kindly feeling with which the conversation of that extraordinary man was so remarkably embued. The anecdotes by which it was illustrated have long escaped my memory; but there was one sentiment expressed by Burns, with his own characteristic enthusiasm, which, as it threw a new light into my mind, I shall never forget. 'Man,' said he, 'is the god of the Dog: he knows no other; he can understand no other. And see how he worships him!—with what reverence he crouches at his feet—with what love he fawns upon him—with what dependence he looks up to him—and with what cheerful alacrity he obeys him. His whole soul is wrapt up in his god; all the powers and faculties of his nature are devoted to his service; and these powers and faculties are ennobled by the intercourse. Divines tell us that it ought just to be so with the Christian; but the Dog puts the Christian to shame!' The truth of these remarks, which forcibly struck me at the time, have since been verified by experience; and often have events occurred, which, while they reminded me that 'Man is the god of the Dog,' have forced from me the humiliating confession, that 'the Dog puts the Christian to shame!'"

Uses of the Saliva.—M. Alp Donné has recently published a pamphlet on the saliva, very little light on the nature of which, he remarks, has been thrown since the time of Haller, physiologists being satisfied with attributing to it two uses: 1st, that of moistening the mouth, favouring the motion of the tongue, and facilitating speech and deglutition; 2d, that of penetrating the food, altering its condition, and aiding the solvent action of the gastric juice. M. Donné adds to these uses that of neutralising, by means of the free alkali which it contains, the excess of acid of the gastric juice, in the intervals between digestion. This fact serves to explain a difficulty respecting the gastric juice, which, although examined by the best chemists, has been found to vary much in its composition, being sometimes acid and sometimes alkaline. These differences M. Donné shows to be owing to the circumstances under which it is obtained. Thus, when procured in the morning, it must be but slightly acid or even alkaline, in consequence of being mixed with the saliva which has been swallowed through the night; whereas, when obtained by exciting the stomach of Dogs, it must contain a large proportion of acid, being then pure and unmixed.

Mode of Preventing Beer from becoming Acid.—A patent has been taken out in America, for preserving beer from acidity in hot weather, by Mr Storewell, who gives the following statement:—To every 174 gallons of liquor add one pound of raisins, which are put into it inclosed in a bag, previous to fermentation. The liquor may then be let down at 65°, or as high as 70°. The bag must remain in the vat until the process of fermentation has so far advanced as to produce a white appearance or scum all over the surface of the liquor, which will probably take place in about twenty-four hours. It must then be taken out, and the liquor left until fermentation ceases. The degree of heat in the place where the working vat is situated should not exceed 66°, nor be less than 60°.—*Journ. of Franklin Institut.*

Greatest Waterfall in Europe.—In the mineralogical report of Lapland presented to the Swedish Government, amongst other curious facts, the discovery of a great Waterfall in the River Lulea is particularly mentioned. It is said to be one-eighth of a mile broad, and at its greatest height to fall four hundred feet. It is probable, that the measurements in this document are according to the German standard, a mile being equal to four and a half English miles.

Gas from Resin.—It has been ascertained by experiments, that five cubic feet of gas from resin gave as much light as nine of oil-gas. Respecting the products of combustion, or the purity of the gases, the advantages in favour of gas from resin are incontestable.

Effects of Magnetism on Chronometers.—When Harrison's timekeeper was under trial at Richmond, it did not go as was expected. No one suspected the cause, till Geo. III., who interested himself much about the machine, suggested that it was affected by a magnet which was lying near it. The magnet was removed, and the timekeeper recovered its rate.

Edinburgh: Published for the Proprietor, at the Office, 16, Hanover Street. London: Smith, Elder, and Co., 65, Cornhill. Glasgow and the West of Scotland: John Smith and Son; and John Macleod. Dublin: George Young. Paris: J. B. Balliere, Rue de l'Ecole de Médecine, No. 13 bis.

THE EDINBURGH PRINTING COMPANY.

JOURNAL OF NATURAL HISTORY,

AND OF

THE PHYSICAL SCIENCES.

MAY, 1837.

ZOOLOGY.

DESCRIPTION OF THE PLATE.—THE CERCOPITHECI, OR GUENONS.

THE Long-tailed Apes, to which the French give the name of *Guenons*, have been separated from the rest of the very extensive family of Monkeys, and constitute a genus to which the appellation of *Cercopithecus* is applied. They are of moderate size, have a rather prominent muzzle, with a facial angle of about 60°, cheek-pouches, callosities or bare spaces behind, and a tail of moderate length, or sometimes longer than the body. The species are numerous, live in flocks, and are of a lively disposition. They are easily domesticated, and although petulant, do not manifest the vicious propensities of the Baboons, which are the most intractable and disgusting animals of the family.

Fig. 1. THE GREEN MONKEY OR GUENON (*Cercopithecus Sabæus.*)—This species, which inhabits Senegal and the Cape de Verd Islands, is one of those most frequently imported into Europe. They live in large flocks, ascend the tallest trees with the greatest agility, spring from one branch to another with unerring dexterity, and are at first little frightened by the report of a gun, so that, according to Adanson, specimens may easily be procured. although, on account of their colour, they are not always readily perceived.

Fig. 2. represents a variety of the same species, which some, however, have considered distinct. It is the *Malbrouc* of Buffon, the *Simia Faunus* of Gmelin.

Fig. 3. DIANA MONKEY (*C. Diana.*)—This handsome species, distinguished by the dark colour of its upper parts, which are dotted with white, is also an inhabitant of Senegal, and other parts of Africa.

Fig. 4. THE MONA (*C. Mona.*)—Of a brown colour, with the limbs black, the breast and a great part of the head white, the tail longer than the body. It inhabits various parts of Africa, as well as Arabia, India, and Persia, and is one of the species that agree best with the temperature of our climates. It is easily tamed, and, being naturally timid, is rendered obedient by threats, while it is also capable of considerable attachment, and is not so mischievous as most Monkeys.

Fig. 5. ATYS, OR WHITE MONKEY (*C. Atys.*)—The bare parts being flesh-coloured, and the fur white, this has by some been supposed to be merely an albino variety of some other species.

Fig. 6. SOOTY OR COLLARLESS MANGABEY (*C. fuliginosus.*)—Of a chocolate brown on the upper parts, and said by Buffon to inhabit Madagascar.

Fig. 7. COLLARED MANGABEY (*C. Æthiops.*)—Somewhat similar to the last, but having a white band between the eyes, and another on each side of the head. It is said to inhabit Madagascar and Abyssinia.

DESCRIPTION OF THE PLATE.—THE AGATE-SHELLS.

THE genus *Achatina* belongs to the Trachelipodous order of Lamarck, and to the family of *Colimaces*, or Snails. It is composed of species generally of large or moderate size, of which the ovate or oblong shell, which is usually thin, with an entire, thin-lipped mouth, is for the most part marked with various colours. These shells, in fact, are very beautiful, and several, being also rare, are of comparatively high price. Lamarck supposes them to be terrestrial, but to live in the vicinity of water.

Fig. 1. THE ZEBRA AGATE-SHELL (*Achatina Zebra.*)—Longitudinally striped with red and brown on a white ground, this shell, which is one of the largest of the family, is also one of the most beautiful. It attains a length of six or seven inches, and inhabits Madagascar.

Figs. 2, 3. WHITE-LINED AGATE-SHELL (*A. albo-lineata.*)—This species, which is seldom more than an inch and a half in length, differs in form from the preceding, by having the last turn compressed so as to narrow the mouth. It is found in Martinique.

Fig. 4. ACUTE OR SHARP-POINTED AGATE-SHELL (*A. acuta.*)—Somewhat similar in form and markings to the Zebra-shell. It is found near Sierra Leone.

Fig. 5. PURPLE-MOUTHED AGATE-SHELL (*A. purpurea.*)—This species, which, according to Lister, inhabits Africa and Jamaica, is in much request, on account of the beautiful purple tint with which the inner surface of its mouth is ornamented.

Fig. 6. represents a variety of the same shell, having the spire more variegated, and its turns following an opposite direction, so that the mouth is placed on the other side, an anomaly which sometimes presents itself in shells of this order, as well as in many others, which are then said to be *reversed*.

Fig. 7. CHESTNUT AGATE-SHELL (*A. castanea.*)—The individual represented is paler than usual, the colour being generally reddish-brown. The native country unknown.

Fig. 8. THE FIERY AGATE-SHELL (*A. fulminea.*)—Of an elongated spiral form with red longitudinal bands on a whitish ground. It inhabits St Domingo.

Fig. 9. VARIEGATED AGATE-SHELL (*A. variegata.*)—From the West Indies.

Fig. 10. VIRGIN AGATE-SHELL (*A. virginea.*)—Although common in collections, one of the most beautiful species of the genus, being girt with black and red bands on a white ground, and with the inner lip rose-red. It comes from the West India Islands.

BRITISH BIRDS, NO. II.

THE GOLDEN EAGLE.—Of the various Birds that inhabit this country, perhaps none have attracted more attention than the Eagles, of which two species are indigenous. The Golden or Ring-tailed Eagle (*Aquila Chrysaetus*), although formerly not uncommon in various parts of Britain, is now chiefly met with in the mountainous districts of the middle and northern divisions of Scotland, and in the larger Hebrides, where the species still maintains a rather precarious existence. Excepting the White-tailed Sea Eagle, *Haliaetus Albicilla*, it is the largest of our rapacious birds. As is generally the case among the Raptores, the male is much inferior in size to the female. Several individuals were about two feet nine inches in length, their expanded wings measuring about six feet; the body robust; the neck of moderate length; the head rather large; the wings when closed reaching nearly to the end of the tail, which is rather long, broad, and rounded; the bill is rather short, very deep, compressed with a curved acute tip; and the feet, which are feathered to the lower tarsal joint, are very muscular; the toes strong, united at the base by a short web, and furnished with large, curved, tapering, acute claws, rounded on the sides, and flat beneath; those of the first and second toes being largest. The bill is greyish blue at the base, black at the end, as are the claws; the cere and toes yellow. The general colour of the plumage is dark brown; the hind head and neck light yellowish-brown; the inner and fore sides of the legs and tarsi reddish-brown. The quills are brownish-black, their inner webs irregularly barred with greyish-white; the tail brownish-black towards the end, its proximal part lighter, and irregularly barred or mottled with greyish. The female is generally about three feet two inches in length, with the extended wings measuring about seven feet; the weight varying from ten to twelve pounds. The colours are similar to those of the male, but generally lighter. Young birds have the basal portion of the tail white, that colour being gradually encroached upon by the brown, until the fifth or sixth year, when it entirely disappears.

This beautiful, powerful, and rapacious bird, having very frequently come under our observation both in the wild and captive states, we are enabled to present to our readers a somewhat detailed account of its habits. All Eagles when at rest have a peculiarly clumsy appearance, owing chiefly to the great size of their wings, which they seem to find it difficult to dispose of in a neat and compact manner; but when roused they assume a bold and lively attitude, rendered more imposing by the glare of their full and bright eyes, which are partially overshadowed by the projecting lachrymal bones or eyebrows. When searching the Golden Eagle is more lively than the Sea Eagle, and of more destructive habits; for, although a carrion bird, it frequently seizes grouse, hares, and other small animals, and sometimes attacks even deer and sheep. Great havoc is occasionally made by it among the lambs, before they have attained the age of six weeks; and in consequence of the injury thus inflicted, various methods have been employed for reducing its numbers. Sometimes its nest is assailed from above, by letting down a person upon a rope, who generally succeeds in destroying its contents, whether by removing them, or by lowering among them a bundle of combustible matter with a live coal enclosed. The old birds are shot, by being enticed, by means of a dead Sheep or Horse, to a spot in the immediate vicinity of which a person is concealed under ground, or in a small hut, so covered with heath that it cannot be distinguished from the surrounding surface. More commonly, however, Eagles are trapped, at least on the mainland of Scotland.

The flight of the Golden Eagle is very beautiful. Owing to the great size of its wings, it finds some difficulty in rising from the ground, although it is considerably more active in this respect than the white-tailed Eagle; but when fairly on wing, it proceeds with great ease, and on occasion is capable of urging its speed so as to equal that of most large birds. However, even at its utmost stretch, it is certainly much inferior to that of the Rock Pigeon, the Merlin, and many other species; and the Raven, during the breeding season, finds no difficulty in overtaking an Eagle that may happen to fly near his nest. When searching the hills for food, it flies low, with a motion of the wings resembling that of the Raven, but with occasional sailings and curves, in the manner of many hawks. At times it ascends high into the air, and floats in a circling course over the mountains, until it has discovered some large object; but in tracing grouse and other animals concealed among the herbage, or in hunting for sea-fowls and their young, it does not indulge in those aerial gyrations, which many closet and some field naturalists have supposed to be performed solely for the purpose of enabling it to spy out its prey from afar. In its ordinary flight, it draws its legs close to the body, contracts its neck, and advances by regular flappings of the wings; but when sailing, it extends these organs nearly to their full stretch, curving them at the same time a little upwards at the tips. An Eagle sweeping past in this manner is a most imposing object, the more especially if in the vicinity of its rocky haunts, and still more if the observer be groping his way along the face of a crag, anxiously seeking a point or crevice on which to rest his foot.

Both our native Eagles sometimes ascend to an immense height in fine weather, and float high over the mountain tops for hours together; but certainly not for the purpose of descrying the objects beneath, for no person has ever observed their sudden descent from this sublime station. It is a popular notion, countenanced even by many anatomists and others, who ought to know better, that the Eagle mounts towards the sun in order to enjoy unrestrained the sight of that glorious luminary. They tell us that its eye is peculiarly fitted for this purpose by having a strong semi-opaque nyctitant membrane, by means of which the rays are blunted; but they forget that the common duck, the domestic fowl, and the sparrow, which are not addicted to astronomical investigation, have eyes organized precisely in the same manner.

On the ground, the Golden Eagle, like all others, is extremely awkward; for, owing to its large wings, its great weight, and the form of its toes, which are encumbered with very large curved and pointed claws, it can only walk in a very deliberate manner, or move from place to place by repeated leaps, in performing which it calls in the aid of its wings. Its feet in fact are not adapted for walking; they are most powerful organs of prehension, capable of inflicting mortal injury on any animal not exceeding a Sheep in size. It is with them that it approves its prey of life, and carries it off to its nest or to some convenient place of retreat. With its curved bill it tears off the feathers and hair; separates morsels of the flesh, and even crunches the bones of small animals.

It is seldom that the Golden Eagle ventures under any circumstances to attack a human being. A respectable person in Sutherland relates that two sons of a man of the name of Murray, having robbed an Eagle's nest, were retreating with the young, when one of the parent birds, having returned, made a most determined attack upon them. Although each had a stick, it was with great difficulty that they at length effected their escape, when almost ready to sink under fatigue. The Rev. Mr Inglis, Lochlee, has furnished us with a similar anecdote. The farmer of Glenmark, whose name was Milin, had been out one day with his gun, and coming upon an Eagle's nest, he made a noise, to start her, and have a shot. She was not at home, however, and so Milin, taking off his shoes, began to ascend gun in hand. When about half way up, and in a very critical situation, the Eagle made her appearance, bringing a plentiful supply to the young which she had in her nest. Quick as thought she darted upon the intruder, with a terrific scream. He was clinging to the rock by one hand, with scarcely any footing. Making a desperate effort, however, he reached a ledge, while the Eagle was now so close that he could not shoot at her. A lucky thought struck him: he took off his bonnet and threw it at the Eagle, which immediately flew after it to the foot of the rock. As she was returning to the attack, finding an opportunity of taking a steady aim, he shot her; and, no doubt glad that he had escaped so imminent a danger, made the best of his way down.

The male and the female keep together all the year round, and very probably remain attached for life; but should one of them be killed in the breeding season, the survivor is not long in repairing his loss. This circumstance is not peculiar to Eagles, but has been frequently observed in other birds, more especially those of the Crow family. The Golden Eagle prepares its nest about the beginning of March, choosing a place for it as nearly inaccessible as possible. Although it is often met with on the maritime cliffs of the Hebrides, yet the species has a greater predilection for inland precipices than the Sea Eagle. It is of great size, flat, and formed of sticks, twigs, grass, and other materials. The eggs are generally two; sometimes single, yellowish white, with irregular, pale, purplish dots. The young are fledged about the end of July, and soon after coming abroad, are left to shift for themselves, or are driven off from the haunts of their parents.

The cry of this species is clear and loud, and may be heard in calm weather to the distance of a mile. It resembles the syllable cleuch or queeck, several times repeated; but although in captivity the bird frequently utters it, in the wild state it is less loquacious. When kept a prisoner it is more ferocious than the Sea Eagle, and can scarcely be trusted even by the person who supplies it with food. The capability of existing under long-continued privation has sometimes been exhibited in a wonderful degree by captive Eagles, which have been accidentally neglected for days or even weeks.

Many marvellous tales are told of Eagles, and there is scarcely a parish in Scotland, in which, if tradition be correct, they have not carried off a child. According to popular belief, an Eagle transported one from the island of Harris to Skye, over a space of about twenty miles; but as even more wonderful events are as firmly believed, no confidence can be reposed in such accounts. Although individuals of the species sometimes appear in various parts of England, it is probable that they seldom or never breed in any district of that country. The species has been extirpated from the South of Scotland, as has very nearly been the case with the Sea Eagle; and it is only in the central ranges of the Grampians, or in the wild glens of the northern division, and among the hills of Skye, Rum, Harris, and other islands on the north-west coast, that the Ornithologist has much chance of meeting with it. How few of those who have given detailed histories of this bird have ever seen it, but how much more few they who have enjoyed opportunities of studying its manners! A single fact is of course worth more than a volume of idle imaginings; and however much the above account may yield in interest to those of others, it has the merit of being entirely derived from personal observation.

REMARKS ON THE DIGESTIVE ORGANS OF BIRDS.—Whether the intestinal canal may be employed, in preference to the sternal apparatus, the bill, the feet, or the nervous or respiratory organs, as a basis for the classification of animals, is a question which we do not intend here to discuss; but of its importance in the animal economy, and of the superior facility with which it may be examined, no one having any acquaintance with Comparative Anatomy will entertain a doubt. The accompanying figures represent the digestive organs of the Red Grouse or Ptarmigan (*Lagopus Scoticus*), which being common in our markets during the autumn and winter, can be readily procured for inspection. Its bill is short, somewhat conical, strong, with sharp edges, and a rather rounded point.

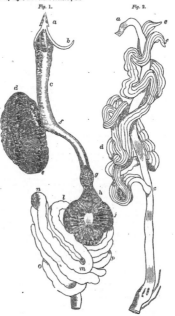

Fig. 1.   Fig. 2.

The Tongue, Fig. 1, a, is short, triangular, pointed, flat above, sagittate and papillate at the base. The Œsophagus, b, c, d, e, f, g, h, is six and a half inches long, narrow, and having a distinct muscular coat composed of transverse or circular fibres, with an internal mucous coat, studded with glandules or crypts, which pour out the clammy fluid with which it is constantly moistened. On the lower part of the neck, it is expanded, or opens into a large membranous sac, of a roundish form, the crop, d, e. On entering the thorax, at f, it is narrowed; but at g appears enlarged, which, however, is owing solely to the thickness of its walls there. This part, named the proventriculus, has, between the outer muscular and inner mucous coat, a great number of oblong hollow glandules, that secrete a peculiar fluid, which is copiously poured out into the cavity of the organ. Here the Œsophagus terminates, and is succeeded by the stomach or gizzard, h, i, j, k. This organ is of a roundish or somewhat rhomboidal, lobulated form. Its outer coat is composed of two very powerful lateral muscles, i, j, of which the fibres converge, become tendinous, and are inserted into a roundish tendinous space on each side; of a lower muscle, k, of the same nature, but less powerful; and of a belt of superior transverse fibres, h. Within this muscular coat is another, of a dense texture, thin, and having a semicartilaginous appearance. The inner coat is a very rough, hard, rugous sac, of which the portions opposite the two lateral muscles, i, j, are thicker and harder than the rest. The intestine comes off on the right side at l, forms a curve, l, m, which returns upon itself, m, n, enclosing the pancreas, and receiving its ducts, together with those bringing the bile from the liver and gall-bladder. The intestine, n, o, p, is then convoluted in the abdomen, and before terminating at q, sends off two processes, or cœca, which are represented by Fig. 2, in which a, b, is a portion of the intestine, c, d, e, f, the two cœca. The intestine has beneath its external or peritoneal filmy covering, a distinct muscular coat, composed of circular fibres, and an internal coat, having its surface villous, or raised into very delicate filamentary processes. The cœca have a similar structure, and are internally villous, but are also marked internally with about seven longitudinal, prominent, white bands. This description, brief and superficial as it is, will afford a more precise idea than a laboured account to the tyro, to whom alone it is addressed, and who, with this beginning, may easily perfect his knowledge by examining the organs of such birds as he can readily procure.

The physiology of the matter, also briefly sketched, is this:—The Red Grouse feeds on the tops of heath, *Calluna vulgaris*, and *Erica cinerea*, the leaves and twigs of *Vaccinium Myrtillus*, and *Empetrum nigrum*, with young heads of cotton-grass, shoots of *Galium saxatile*, carices, grasses, willows, and other plants, as well as berries. While feeding, it walks among the heath, selecting the fresh tips of the twigs, which it breaks off nearly of a size, the largest pieces not exceeding half-an-inch in length. Along with these substances it occasionally picks up a small fragment of white quartz. Before it ceases it has filled the œsophagus and crop—the latter holding a very considerable quantity, as its diameter is nearly three inches. It then reposes among the heath. The crude food is propelled downwards by the successive contraction of the muscular fibres of the œsophagus, until the stomach is filled, when the lateral muscles of that organ, *i*, *j*, alternately contract and relax; and thus agitate the mass of food, pressing it against the horny rugæ of the inner coat, and causing the numerous particles of quartz, a substance harder than glass, to cut or lacerate the fibres, which have been moistened with the copious clammy fluid of the proventriculus, *g*. The crop, *d*, *e*, it must be understood, is merely a recipient, and is destitute of mucous glandules excepting near its mouth; nor do its fluids, which are scanty, exert any solvent power upon the food. But the fluid of the proventriculus, *g*, operates both as a solvent and a diluent. The mass contained in the stomach, being reduced to a coarse pulp, is forced, by the general contraction of the muscles, into the intestines, through the pyloric opening at *l*. In the upper part, or duodenal fold of the intestine, *l*, *m*, *n*, it is further diluted by the pancreatic fluid, and mixed with the bile; and as it gradually passes along the intestine, its nutritious parts are separated, adhere to the surface, and are taken up by the absorbents to be carried into the mass of the blood. As it advances, the woody fibres and fragments of leaves become more perceptible; and as the mass passes the mouth of the cœca, the finer particles are taken up by them, until their entire cavity is filled with a brownish pulpy fluid, which seems to undergo there a second elaboration, and to be absorbed, the residue being returned into the intestine, where it accumulates in the rectum, and is voided in concrete cylindrical fragments.

All the Gallinaceous Birds have a crop, a powerful gizzard, and cœca of which the capacity is about equal to that of the intestine; but these cœcal appendages vary greatly in form and length, being cylindrical in Grouse and Ptarmigans, and oblong towards the end in Pheasants and other species. In the Pigeons there is also a large crop, and a strong gizzard; but the cœca are merely rudimentary, seldom exceeding a third of an inch in length, with a diameter of about two lines. In many Birds the œsophagus has no dilatation or crop, and the stomach exhibits all the gradations, from the powerfully muscular kind or gizzard of the Gallinaceous Birds and Pigeons, to the fibro-membraneous sac of Owls and Hawks. In Birds that feed on fish, the œsophagus is extremely dilated, and the stomach reduced to small size;—in short, the variations exhibited are at least as great in this class as in any other, and the organs in question none but those who are unacquainted with them can possibly consider as of less importance to the scientific Ornithologist than the feathers, or other appendages of the dermal system. If the intestinal canal of any bird be laid before a person who has examined that series of organs, he can instantly determine the group or family of the species to which it belonged.

The Brent Goose.—Having recently had occasion to examine the Cromarty Frith, one of the estuaries of the north-eastern coast of Scotland, we were delighted to find upon it vast flocks of Brent Geese, *Anser Bernicla*. These birds arrive there in the beginning of winter, and remain until the end of spring, keeping generally in the expanded part of the firth, between Invergordon and Cromarty Ferry. On the northern side of that portion of the estuary are very extensive flats of sand and mud, which are exposed when the tide recedes, and then present in many places the appearance of large meadows, being covered with Sea Grass, *Zostera marina*, of which the long, creeping, cylindrical, saccharine roots afford them an abundant supply of food. We have seen probably ten thousand of these Geese on the shoals, together with nearly an equal number of Gulls of various species, but chiefly *Larus canus*, and *Larus ridibundus*. The Brent Geese also frequent the Beauly Frith, but in much smaller numbers. The best account of this species is that by Mr Selby, in his Illustrations of British Ornithology, in which it is mentioned that the gizzard is frequently found "filled with the leaves and stems of a species of grass that grows abundantly in the shallow pools left by the tide," and which may possibly be the Zostera. Mr Selby states that they leave the Northumbrian coast between the end of February and that of March. According to the reports of the "natives," they remain much later in Cromarty Bay; and, on the 22d March, when we saw them there, the flocks had not begun to break up.

Maternal Solicitude of the Partridge.—In the county of Linlithgow, for three years past, the Partridges have been rapidly disappearing. During the month of July in these years, there were several successive days of uncommonly wet weather, in which a vast number of the young lost their lives. Their attachment to their brood is truly astonishing. I have known them, day after day, sitting on their young, until, on account of the rain, they all died; but I never heard or saw such striking instances of their care and tenderness as those which took place last season. I knew of two broods in my neighbourhood, in protecting which both father and mother sacrificed their lives. The one consisted of twelve birds. The male and the female were found sitting close together, with their wings spread out, and six chickens under each. The other consisted of ten birds. The male and the female were found in the same situation, with five birds under each, and all of them dead. So anxious had they been to protect their tender offspring from the rain, that they starved themselves to death, rather than suffer them to run any risk. Mr Mellis, gamekeeper to Sir William Baillie of Polkemmet, told me that he knew several instances of the same kind last season, an occurrence which he never before had observed. How beautifully does it illustrate our Saviour's lamentation over Jerusalem, once the holy and beloved city of God, and the joy of the whole earth, but so soon, alas! to be devoted to irretrievable destruction!—*Boghead*, 21*st March* 1837.—T. D. Weir.

BOTANY.

Pine Forests of Scotland.—The aboriginal woods of Britain have been so much destroyed in consequence of the increase of population, that few remains of them are now to be found beyond the limits of the Highlands of Scotland, and even there it is only in some of the more remote valleys that continuous tracts of sylvan vegetation are met with. The species which are most abundant in those remains of our ancient woods are the Oak, the Birch, and the Pine. Oak-woods, however, are generally reduced to the state of copses, which are periodically cut down for their bark, and for hoops and other articles of domestic and rural economy. Birch forests are still to be seen in many places, especially the glens of the Grampians, and the mountains of the northern division, and as they are not so useful as those of oak and pine, they have a chance of remaining for a much longer period. The pine forests are almost entirely confined to the valleys of the Dee and the Spey, where, however, they have of late years suffered extensively from the axe. In Braemar, trees having a girth of from ten to twelve feet are not uncommon, but most of the larger trunks near the Dee have been cut down, and floated to Aberdeen. In Badenoch and Strathspey, the long ranges of native pine, sombre as they may be in the eyes of the Lowlander or Englishman, afford a delightful feeling to the lover of wild nature, and to the Celt revisiting his native home, after years of absence in a foreign land, must impart a joy more intense than that experienced on contemplating the palm groves of the most sunny climes. In those woods once roamed the wolf and the wild ox, while the stately stag betook itself to the grassy valleys, and the graceful roe browsed among the thickets of alder and willow that skirted the streams. The wild pine of the Highlands has a very different appearance from the trimmed and formal trees of the plantations; its trunk is thicker, shorter, and less rugged; its branches more spreading and covered with a redder bark undisfigured by lichens, and its wood is more resinous and of a redder tint.

On the Wake and Sleep of Plants, by M. Dutrochet.—There are flowers that have only a single wake, which is their expansion, and a single sleep, which immediately precedes the death of the corolla. Of this kind are the flowers of *Mirabilis* and *Convolvulus*. Other flowers present alternations of sleep and waking for several days, such as that of the Dandelion, *Leontodon Taraxacum*.

The flower of *Mirabilis Jalapa* and *M. longiflora* opens its funnel-shaped corolla in the evening, and closes it in the following morning. This flower may be considered as formed by the union of five petals, each of which has its middle nerve. The five nerves which sustain the membraneous tissue of the corolla, in the same manner as the whalebone slips support the cloth of an umbrella, are the sole agents of the movements by which are produced the opening of the corolla, or its state of waking, and its closing or sleep. In the former case, the five nerves curve so as to direct their concavity outwards; in the latter, they bend so as to direct their concavity towards the interior of the flower, and they thus carry with them the membraneous tissue of the corolla as far as the orifice of its tubular canal.

Thus the same nerves, at two different periods, successively perform two opposite motions of incurvation. I have observed, with the microscope, the internal organization of these nerves. They present, at their outer side, a cellular tissue, of which the cells, disposed in longitudinal series, chiefly diminish in size from the inner towards the outer side, so that, during the turgescence of these cells, the tissue which they form must curve in such a manner as to direct its concavity outwards. It is by it, therefore, that the expansion of the corolla, or what is named its wake, must be operated. On the inner side of each nerve there is a fibrous tissue, composed of transparent fibres extremely slender, and intermingled with globules arranged in longitudinal series. This fibrous tissue is situate between a layer of tracheæ or spiral vessels on the one hand, and a layer of superficial cellules filled with air on the other; so that it is placed between two laminæ of pneumatic organs.

On separating, by a longitudinal section, the cellular tissue and the fibrous tissue, of which the nerve is composed, and then immersing it in water, I found that the cellular tissue curved outwards, while the fibrous tissue curved towards the interior of the corolla. These two inverse curvings invariably took place. Thus it is without any doubt the cellular tissue of each nerve that, by its incurvation, produces the wake or opening of the corolla, and the fibrous tissue that, by its incurvation in the opposite direction, effects its sleep or closing.

I separated a nerve from a corolla of Mirabilis which was yet in bud, but near the period of expansion, and on immersing it in water, observed that it curved strongly outwards, thus instantly taking the curve which produces the expansion or wake of the corolla. I then removed it into syrup, when it curved in the opposite direction, or inwards. This proves that, in the second case, there was turgescence of the cellules, the external water then, from the effect of endosmosis, directing itself towards the organic liquid that existed in these cellules; and that, in the second case, there was depletion of the cellules, because their organic liquid, being less dense than the external syrup, then directed itself towards it. It might be thought from this experiment, that the spreading or wake of the corolla being owing to the turgescence of the cellular tissue of its nerves, its shutting or sleep must be owing to the depletion of the same tissue; but it is proved by experiment, that this is not the cause of the shutting or sleep of the corolla. I separated a nerve from a corolla near the period of opening, and immersed it in water. This nerve, which was curved slightly inwards, as is the case in the corolla, when in bud, curved strongly outwards, or in the direction which produces expansion or the state of waking. Endosmosis then determined the turgescence of the cellular tissue, which is the organ of this incurvation. When it had been immersed about six hours, the nerve gave up its outward bend, began to curve inwards, and was soon entirely rolled up.

This succession of phenomena is entirely independent upon the action of light. Thus, the nerve of a corolla of Mirabilis assumes in water the curve which effects the opening of the flower; and again, at the end of some time, takes the bend that produces sleep. If, therefore, as cannot be doubted, it is the turgescence of the cellular tissue of the nerves that produces the incurvation on which depends the wake of the corolla or its expansion, it is to quite a different cause that we must refer the incurvation to which is owing the sleep of the corolla or its closing; for it cannot be admitted that cellular tissue, immersed in water, is emptied. The experiment mentioned above proves that it is the fibrous tissue contained in each nerve of the corolla that is the agent of the inward curvature, or that to which the sleep, or shutting of the corolla, is due. We must, therefore, perceive, that in the nerves of the flower of Mirabilis, the incurvation of waking, or that of which the concavity is directed outwards, and which is owing to the turgescence of the cellular tissue, at first predominates over the incurvation of sleep, or that whose concavity is directed towards the interior of the flower, and which is due to the action of the fibrous tissue; and that afterwards the incurvation of sleep, which depends upon this latter tissue, ultimately proves victorious. The outward incurvation which the cellular tissue affects when the nerve is immersed in water, changes to inward curvation when the nerve is immersed in syrup; which proves that it is endosmosis that acts here. Now, when the nerve, on being placed in water for some hours, has then taken the second curvature, or that of sleep, it does not lose it when removed into syrup. It is not then endosmosis that has occasioned this second incurvation, or that of sleep.—(*To be continued.*)

MISCELLANEOUS.

DESCRIPTION OF THE FRITH OF CROMARTY, AND REMARKS ON ESTUARIES.

The subject of Estuaries being one of great importance, not merely in a geological, but also in an economical point of view, and more especially with relation to the salmon fisheries, we have thought that an account of one of the most remarkable of those of our Eastern coast might prove interesting to our readers.

The Cromarty Frith is an Estuary or Inlet, extending from the northern coast of the Moray Frith, in a slightly curved direction, to the town of Dingwall in the county of Ross, having a length of about 21 miles, a breadth varying from 4 miles in its widest part, to about 4000 feet at Cromarty and Invergordon Ferries; and having an average breadth of a mile and three-fifths, so that its entire superficial area is about 34 square miles. The whole extent of its southern coast is low, the land rising from it gently to form the peninsula of the Black Isle, but having a bank, in several places abrupt, and with an elevation of about 30 feet, excepting at its mouth, where it rises to a greater height. At its upper extremity the shores are nearly flat. Along its northern side, for about a third of its length, they are low, with a bank of small elevation, and not continuous. It is then succeeded by rather steep acclivities, forming the slopes of rounded hills of moderate height, which are themselves the flanks of higher mountains placed farther back. Along the middle third, the mountains recede to the distance of from one to three miles, leaving a gradual slope, which becomes more or less level as it approaches the shores. The ground along the remaining third is very low, but undulated; and the shores are nearly flat, excepting at the mouth of the Frith. There, on both sides, from Cromarty Ferry to the coast line of the open sea, the banks gradually rise, then become rocky and precipitous, and at length are formed of cliffs from one to two hundred feet in height, of a harder nature than those on any other part of its shores. Here, on either side, is a rounded hill or ridge, several hundred feet high. On the south side, the ridge continues some way along the coast of the Moray Frith; but on the north side it soon terminates, by sloping gradually into a nearly level space, extending obliquely from the eastern angle of the bay of Nigg to the sea coast at Shandwick. This elevated ground may have been at one time continuous, as is perhaps indicated by the exact similarity of the rocks on both sides of the very deep channel formed by their disruption. If this were the case, the lake of fresh water, formed by the Conon and other streams, in the basin marked by the steep banks along the present Frith, would no doubt have emptied itself into the sea by the nearly level and very low tract mentioned above as extending from the bay of Nigg to the coast.

A distinct channel, increasing in breadth and depth from the bridge over the Conon to the Moray Frith, exists in the whole length of the basin; while on either side are shoals composed of mud and sand. The depth of this channel, in the lower parts of the Frith, is unusually great, it being at Invergordon Ferry from 60 to 80 feet at low water of spring tides, and at the entrance from 110 to 184. The most remarkable shoals or sand banks are those extending along the bay of Nigg, and at low water filling nearly half the breadth of the space from Invergordon Ferry to Cromarty Ferry, which forms the widest part of the Frith, or that usually called Cromarty Bay. Similar mud-flats occupy a great portion of the other bays on the Frith, and nearly the whole of its upper part. The deposits which thus narrow the channel of the waters, when the tide is out, are formed of sand intermixed with mud, and frequently have fragments of primitive rocks scattered over their surface, together with cooayed shells of various kinds. Along the north shore of Cromarty Ferry is a deposit of pure quartzy sand, forming downs or sand hillocks, covered with sea-bent. Many parts of the shoals in the upper portion of the Frith are composed chiefly of gravel and stones, covered with sea-weeds, while others consist of sand and mud.

The precipitous rocks along the coast of the Moray Frith, and a portion of the narrow channel, extending from it to near the town of Cromarty, are formed of a rock of a very different nature from those on the other parts of the shores of the Cromarty Frith, being of gneiss, and composed of quartz, felspar, mica, chlorite, and hornblende, the quartz and felspar generally predominating. These cliffs and the eminences bounded by them are commonly named the Sutors, although that appellation is applied more particularly to certain fragments or detached masses at their base, projecting from the water, and bearing a fanciful resemblance to cobblers at work. The rocks here are composed of strata, generally thin, which are vertical, or inclined from fifteen to twenty degrees on either side, that is, to the east and west, and often extremely contorted. Their colour is generally red, owing to the predominance of felspar in their composition. At their base are numerous fissures and small caves, some of which are inhabited by Rock Pigeons and Cormorants. Farther up the channel the rocks become of a duller red, having a less crystalline texture, and are chiefly composed of felspar; their stratification is gradually obscured, and the mass at length presents irregular fissures resembling those of an igneous deposit. This rock is succeeded by vertical beds of red conglomerate and sandstone; the former composed of fragments of primitive rocks, chiefly quartz and gneiss; the latter of grains of quartz, cemented by argillaceous matter and oxide of iron, and intermixed with scales of mica. Near the junction of the primitive and secondary rocks, imbedded in the sandstone, and about twelve yards distant from the edge of the conglomerate, is a bed of grey limestone, in thin parallel layers, from one to three or four inches thick, with thinner layers of sandstone interposed. The rocks on the south side are similar; and the limestone bed, which is ten feet thick, runs nearly north and south, and inclines about fifteen degrees to the west, is seen there also. Beyond this the rocks are covered with sand on the north side, and with clay and gravel on the south; but where exposed by quarrying, are found to be sandstone of the old red series, or of that kind named by some primitive sandstone, on account of its resting upon and being formed of the detritres of primitive rocks.

From Invergordon Ferry to Conon bridge the rocks continue of the same nature. The sandstone is generally red, sometimes grey, whitish, or greenish, rather coarse-grained, in some places firm, in others friable, composed of particles of quartz, mica, and chloritic or argillaceous matter, very frequently having pebbles of quartz, gneiss, and granitic rocks interspersed. The conglomerate is composed of pebbles of gneiss, granite, quartz, hornblende-rock, mica slate, and fragments of other primitive rocks, often little rounded, and cemented by argillaceous and arenaceous matter, generally of a deep red colour. Towards Dingwall, on the north side of the Frith, are beds of dark-grey or blackish rocks of the same nature. The strata are nearly horizontal, or incline to the south-west at an angle varying from fifteen to twenty degrees, and, along the shore-line of the southern side, where they sometimes present a fractured face of twenty feet or more in height, have their seams of stratification parallel to the water-line. The same rocks are seen in the bed of the Frith at its margin, but rarely, being almost everywhere covered with sand or mud, and the shores near high-water mark composed of pebbles of quartz, gneiss, and other primitive rocks.

The soil of the surrounding country is a mixture of clay and sand, generally having a large proportion of pebbles; and the subsoil is of a similar nature, with fragments and blocks of primitive rocks.

The vegetation along the shores is composed of the plants common to pasture lands in all parts of Scotland, intermixed with those which are peculiar to maritime stations. There is very little natural wood, excepting on the north side, towards the upper part. In the natural state, the vegetation of the southern side is heathy; but the parts near the northern shores have either been cultivated or are covered with pasture plants. *Ammophila arundinacea* and *Triticum junceum*, which are peculiar to sand, cover the downs above mentioned near the mouth of the Frith. In short, it may be generally observed, without specifying particular plants, that the vegetation is of that mixed nature observed on the sea-coast, along estuaries, or by the mouths of rivers to which the tides have access.

The vegetation of the bed of the Frith is composed of marine Algæ, which, it is well known, are not peculiar to pure salt water, but occur also in the mixed salt and fresh water of estuaries. In the Bay of Nigg, the extensive banks or flats are covered to a great extent with *Zostera marina*, of which the fleshy, saccharine stems, and long grassy leaves, afford abundant food to the vast flocks of Brent Geese, which frequent the Frith in winter. This plant forms submersed meadows, and in winter remains undecayed, although it assumes a duller green. It grows on sand and mud banks and shoals, chiefly in sheltered places, sometimes in salt marshes or ditches along the shore, but is not peculiar to estuaries, as it occurs abundantly on the coasts of the Hebrides and Shetland Islands, in places where there is no intermixture of fresh water. On the Nigg sands, attached to the stones and shells, are also considerable quantities of algæ, of which may be mentioned *Fucus serratus*, *Fucus nodosus*, *Fucus vesiculosus*, *Laminaria saccharina*, *Laminaria digitata*, and *Scytosiphon filum*.

The quantity of algæ in an estuary, or along the shores of the sea, is determined by that of the rocky or stony ground, for these plants adhere only to a solid basis, and do not grow out of mere sand and mud, although a very small point of support, such as a dead shell or a pebble, is sufficient for them. Thus, in the mouth of the Frith of Tay up to Broughty Castle, there are very few sea-weeds, because the bottom and shores are sandy. On the sand and mud flats from thence, several miles above Dundee, scarcely any are to be seen. But wherever rocks or stones occur, either in the bed of the estuary or along its shores, they are found in profusion, and the piers of the harbour of Dundee are densely covered with them nearly up to high-water mark. In the Cromarty Frith these plants are in many places abundant, and some of them are met with to a considerable distance above the mouth of the Conon. On the flats of the Bay of Nigg they are very luxuriant, adhering in large tufts to the stones; along the shores, above Invergordon Ferry, they attain a moderate size; but towards the top of the Frith they gradually diminish, until, in the river itself, they become dwarfish, and at length cease far below Conon Bridge. The predominant species above Invergordon are *Fucus vesiculosus*, *F. nodosus*, and *F. serratus*, which are the plants chiefly used in the manufacture of kelp; but besides these are found *F. canaliculatus*, *Laminaria digitata*, *Ulva umbilicalis*, *U. lactuca*, and many others. Although these and other Algæ are peculiar to salt water, they are yet frequently found in places where fresh water predominates for at least twelve out of every twenty-four hours.—(*To be continued.*)

Edinburgh: Published for the Proprietor, at the Office, 16, Hanover Street. London: Smith, Elder, and Co., 65, Cornhill. Glasgow and the West of Scotland: John Smith and Son; and John Macleod. Dublin: George Young. Paris: J. B. Balliere, Rue de l'Ecole de Médecine, No. 13 bis.

THE EDINBURGH PRINTING COMPANY.

THE EDINBURGH

JOURNAL OF NATURAL HISTORY,

AND OF

THE PHYSICAL SCIENCES.

JUNE, 1837.

ZOOLOGY.

DESCRIPTION OF THE PLATE.—THE MARMOSETS AND TAMARINS.

These generally beautiful and gentle little animals are all inhabitants of South America, where they reside in the midst of the forests, living on vegetable substances. They seldom much exceed a common squirrel in size, and are easily tamed, but being peculiarly delicate, are unable, without much care being bestowed upon them, to withstand the effects of our variable and severe climate. The Marmosets differ very little from the Tamarins, one of the principal external distinctions being derived from the tail, which, although very long in both, is bushy and ringed in the former, while it is slender and uniformly coloured in the latter. The species of both genera have the head of a roundish form, the face flat, the nostrils lateral, the hips clothed with hair; they have no cheek-pouches, and their tail is not prehensile; all their claws are compressed and pointed, excepting those of their great-toes, which toes are however so small, and so little separated from the rest, or capable of being put into opposition to them, that these animals are more bipedal than quadrumanous.

Fig. 1. The Common Marmoset (*Jacchus vulgaris*).—This delicate little creature is characterized by having two large tufts of white hair before the ears, the body grey, transversely banded with dusky, the tail bushy, and ringed in its whole length with blackish-brown and greyish-white. It is pretty generally distributed over South America.

Fig. 2. The Great-eared Tamarin (*Midas rufimanus*).—Of a dusky colour, the hind parts barred with grey, the hands and feet reddish. It inhabits Guiana and other parts of South America.

Fig. 3. The Red-tailed Tamarin (*Midas Œdipus*).—The head and lower parts of this species are whitish, the tail reddish, tipped with dusky.

Fig. 4. The Fair or Silvery Tamarin (*Midas argentatus*).—The face is red, the hair over the whole body of a greyish or silvery white, excepting that of the tail, which is dusky. It inhabits the countries along the Amazons River.

Fig. 5. The Silky Tamarin (*Midas Rosalia*).—This species, of which the fur is of a reddish-colour, the tail with two bands of dusky, is further characterized by the great length of the hair of the head and neck. It occurs in Brazil and Guiana.

Fig. 6. The Negro Tamarin (*Midas Ursulia*).—The fur is black, the back and sides undulated with reddish, the hands and feet black. In other respects it resembles the *M. rufimanus*. Like the other species, it is a native of South America.

DESCRIPTION OF THE PLATE.—THE CROSSBILLS.

The Crossbills constitute a genus of small birds of the Conirostral division of the Insessores, characterized by the peculiar form of their bill, of which the curved and attenuated points of the mandibles are laterally deflected, so as to present the appearance of crossing each other. In other respects, these birds agree with the Bullfinches and Buntings, between which they are in a manner intermediate. They inhabit the pine and larch forests of the northern parts of the European and American Continents, and derive the principal part of their food from the seeds of the cones, the scales of which they separate by a lateral motion of their mandibles. It is in fact probable that this habit, commenced at an early period, when the bill is yet comparatively soft, is the cause of the lateral divarication of the tips of the mandibles. They are irregularly migratory, shifting about in flocks from place to place, according to the abundance or scarcity of food. The plumage varies greatly in colour, that of the young birds being dull, of the females generally not much brighter, and of the males remarkable for its conspicuous, but not generally vivid tints. Only four species are known: The Parrot Crossbill, *Loxia pytiopsittacus*; the Common European Crossbill, *L. Europea*; the American Crossbill, *L. Americana*; and the White-winged Crossbill, *L. leucoptera*. The first is common to Europe and America; the second peculiar to Europe; the third confined to America; the fourth common to both continents. In the plate are represented the American and White-winged species.

Fig. 1. The American Crossbill (*Loxia Americana*). Adult male.—This species is in all respects similar to the Common European Crossbill, excepting in being considerably smaller, and in having the bill more slender, with the tips of the

mandibles much more elongated. The old male has the greater part of the plumage of a dull vermilion tint, more or less mixed with yellow, the wings and tail dusky, their feathers edged with lighter, the back of a darker red, the rump much brighter. It inhabits the woods of Canada, Nova Scotia, and the Northern United States, feeding on the seeds of the pines, larches, and junipers.

Fig. 2. The American Crossbill. Female.—The female is destitute of red, the upper part of the head, the hind neck, and the back, being dull greenish-grey, spotted with dusky, the rump wax-yellow, the lower parts greyish-yellow, the neck, breast, and sides, spotted with dusky.

Fig. 3. The White-winged Crossbill. (*L. leucoptera*.)—This species is about the same size, and has the name general form, but is differently coloured, its principal specific distinction being derived from two white bands crossing the wings, which, as well as the tail, are otherwise deep black. The upper parts are of a light reddish ochre. Its habits are the same as those of the other species.

Fig. 4. The White-winged Crossbill. Young male.—In its second plumage, the male has the upper parts deep-red, largely spotted with black, the rump rose-red, the wings black, with transverse white bands, the tail also black, the lower parts greyish-white.

Fig. 5. The White-winged Crossbill. Female.—The female is greenish-grey, spotted with dusky, above, pale grey, also spotted, beneath, the wings and tail black, but the former with conspicuous white bands.

The Siskin.—Our esteemed correspondent, Mr William Drew, of Paisley, has favored us with the following notice respecting this bird :—" I have remarked the account of the Siskin contained in your Journal of this month. Undoubted as the circumstance of its breeding in Scotland must now be, from the respectable authority of Mr Weir, yet I do not think it amiss so far to corroborate his observations. Early in June 1835, at which time I resided at Inveraray, I went out one morning to fish, and, according to my usual practice, I carried a light gun with me. I was rather surprised, at that season, to see a pair of Siskins among some furze bushes, on the shore of Lochfine, and the birds being close together, I killed both. On dissecting the female, an egg was found ready for exclusion, and I never had any doubt but that the birds were breeding in the neighbourhood, though I did not look for or see the nest. It is very likely that it was on some of the spruces, which were the predominant trees in the place. I subsequently saw and secured a pair of Siskins in the same locality.—*Paisley, 26, Causeyside Street, 3d March 1837.*"

Ventriloquism of the Robin.—The bird endowed with this singular control over his vocal powers, is our favorite and pugnacious little Robin, whom I observed to be as complete an adept in this art as any human ventriloquist could possibly be. While in my garden a few weeks ago, the notes of this bird fell deliciously on my ear, being mellowed, as I believed, by distance. I expected to descry my musician on some distant tree; but, to my great surprise, I perceived him within a few yards of the spot I occupied. I was near enough to observe the alternate contractions and expansions of the breast; but I could not see any motion of the bill.—*Mr Edmunds, in Loudon's Magazine.*

Preservation of Zoological Specimens.—Various substances have been employed in preparing skins of animals to be stuffed as specimens, and with various success. Arsenical soap has of late years been very generally substituted for the powders and solutions formerly used; but this substance, being applied exclusively to the inner surface of the skin, has no other effect than to preserve it; while the hair, the feathers, and especially the down, are left without protection. Now, of what value is a well-preserved skin, from which the feathers have been eaten by moths? We have seen such skins in abundance. The arsenical lathering which they had received did not act in the slightest degree as a preservative from vermin. Others, prepared with alum, corrosive sublimate, and powdered spices, we have seen similarly affected. In short, it is our firm conviction, founded on observation, that nothing applied to the inside of the skin can prevent the hair and plumage from being eaten. As to solutions applied to the outer surface, we have found none that has a permanent effect. Alcohol quickly destroys moths, and so does solution of corrosive sublimate; but neither has a lasting effect. Rectified oil of turpentine acts still more speedily, and even its vapour in a close case destroys insects, but although its action is thus powerful, it is not permanent. Still, we are convinced, it is the best preservative; but, not to injure the plumage, it must be of the best quality, for, if bad, it remains long moist, clogs the feathers, and gathers dust. Preserved birds ought to be kept in close cases, cleaned at least twice a-year, carefully searched for vermin; if much infected, baked in an oven, then slightly impregnated with pure rectified oil of turpentine. Under this treatment skins last long enough; but, judging from appearances, the keepers of museums are a lazy race, the more so if paid by the public; and all naturalists who have visited such places have seen quadrupeds in rags; birds, under which lay small heaps of dust and dead moths, and eggs, of which the membrane had been devoured, and the colours gone; while the minerals, shells, and other comparatively imperishable articles, were carefully laid on cotton, kept perfectly clean, and tended with unusual care. Oil of turpentine we have also found to be the best preservative for dry anatomical preparations, and under its detergent influence, we have seen a collection of muscular and vascular articles, that literally swarmed with Acari and Tineæ, reduced to a state of perfect sanity, although not until they had been repeatedly soaked in it. In short, this oil is invaluable, and greatly surpasses all other substances in use, for where they have failed, we have found it invariably to succeed. To preserve collections of birds' skins in drawers, it is enough to moisten the latter now and then with a little of it by means of a bit of sponge. After a few days, its smell will not nauseate the most delicately organized or clean-fingered inspector of bills, feathers, and claws.

Digestive Organs of Birds.—Having, in a former Number, given a short description of the digestive organs of Birds, illustrated by a sketch of those of the Red Grouse, we now present four figures intended to exhibit the differences on which characters, designative of the various orders, may be founded. These figures are from "A History of British Birds," by W. Macgillivray, newly published, and in which will be found, besides the ordinary technical characters, and an account of the habits, distribution, and relations of the species, various anatomical details, especially with reference to the digestive organs. These figures are representative of the orders Rasores or Gallinaceous Birds, Gemitores or Pigeons, Deglubitores or Conirostral Birds, and Vagatores or Crows and Starlings. It will at once be perceived, that they differ from each other in several particulars.

Fig. 1. 2. 3. 4.

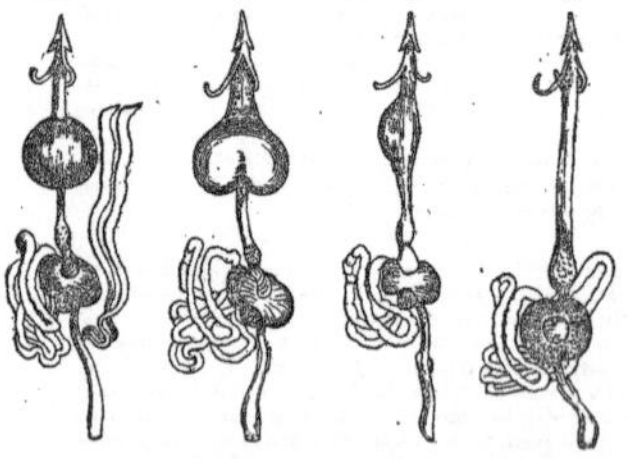

In the *Rasores*, Fig. 1, the œsophagus is narrow, with a very large globular crop, having a small aperture; the proventriculus is bulbiform, with oblong, generally sacculated glandules; the stomach is a powerful gizzard, with very thick lateral muscles, a prominent distinct inferior muscle, and thick rugous epithelium, or inner coat; the intestine of moderate length, and nearly uniform diameter, but having appended two cœca of extreme size, being together of equal capacity with the intestine.

In the *Gemitores*, Fig. 2, the œsophagus dilates immediately with an extremely large transverse crop, internally reticularly rugous in the breeding season, and having a small aperture; the proventriculus is bulbiform, with oblong simple glandules; the stomach a powerful gizzard, with very thick lateral muscles; a prominent distinct inferior muscle, and thick rugous epithelium; the intestine is long and narrow; the cœca reduced to very small cylindrical adnate tubes.

In the *Deglubitores*, Fig. 3, the œsophagus gradually dilates into a membraneous sac of moderate size, lying on the right side, and sometimes curving round the back of the neck; the proventriculus is oblong, with cylindrical glandules; the stomach a powerful gizzard, with very thick lateral muscles, a prominent inferior muscle, and thick rugous epithelium; the intestine short and rather wide; the cœca minute, oblong, adnate.

In the *Vagatores*, Fig. 4, the œsophagus is rather wide, without crop or dilatation; the proventriculus bulbiform, with oblong or cylindrical glandules; the stomach rather large, roundish, more or less compressed, its muscular coat of moderate thickness, not separated into distinct muscles, but composed of uniform fasciculi, inserted into circular tendinous spaces; the epithelium thin, and slightly rugous; the intestine of moderate length and diameter; the cœca very small, cylindrical, and adnate.

All the other orders possess equally distinctive characters, which, when clearly pointed out, will throw considerable light upon the affinities of the different species, and effect changes in classification which will prove utterly subversive of some of our most approved systems.

The Green Wagtail (*Motacilla flaveola*), in Norfolk.—Mr Salmon, in the May number of the Naturalist, gives the following notes respecting this bird:—"The few pairs of spring Oatears, or Green Wagtails, that visit this neighbourhood, resort to the immediate sides of the river, which is bordered by grass lands and uncultivated wastes; it is only in this locality that they are to be seen. I have repeatedly noticed them running upon the weeds on the surface of the water, catching insects, &c. I have found its nest among the ling which grows close to the water. The old birds express considerable anxiety when you approach within the vicinity of their nest, hovering with their peculiar undulating motion whilst on the wing over your head, or alighting upon a bank or any other convenience on the ground, at the same time uttering their note of alarm."

The Green Wagtail, in Mid-Lothian.—In July and the beginning of August, very large flocks of this species are seen in the irrigated meadows to the west of Edinburgh. They are composed of young and old birds, which search for insects and worms on the newly mown grass, as well as on the taller leafy weeds, such as docks. Their flight is similar to that of the Grey Wagtail, that is, performed in long undulations, and they perch occasionally on trees and bushes. A few individuals may be seen in the same place early in summer, but certainly not a fiftieth part of the number that may sometimes be seen in July. None of them remain after the middle of August.

BRITISH BIRDS.—NO. III.

The White-tailed Sea Eagle.—Of this magnificent bird, which is still not uncommon in the North of Scotland, and especially in the Hebrides, the principal distinctive characters are the following:—Its form is robust, the neck of moderate length, the head rather large, the wings very long, the feet strong, the toes stout, the claws large, curved, tapering, acute, rounded above, flat beneath. The bill is nearly as long as the head, very deep, compressed, with a long, curved, acute tip. The plumage is compact and imbricated; the space between the bill and the eye sparsely covered with very small, narrow feathers; those of the neck narrow and pointed, of the back broad, of the belly very soft and downy, of the outer part of the tibiæ elongated. The fourth and fifth quills are longest, and the wings when closed are of equal length with the tail, which is rather short, broad, and rounded. The cere and bill are pale yellow, the iris bright yellow, the tarsi and toes gamboge, the claws blueish black. The general colour of the head, neck, breast, back, and upper wing-coverts, is pale greyish-brown; the lower parts and legs are chocolate-brown; the primary quills blackish-brown, their base and the greater part of the secondaries tinged with ash-grey; the tail white. The length of the male is three feet, of the female three feet two inches, the extent of the wings of the latter seven feet. The colouring varies considerably, young birds being darker. For the first three years, the bill is black or dusky, the iris brown, the feet yellow. The colour of the plumage is dark-brown, mottled with white and light-brown, the tail blackish.

This bird is not remarkable for courage, although its strength is great, and its flight powerful. Its principal nutriment is derived not so much from the produce of its own industry, as from carrion of all kinds; and it is frequently seen sweeping along the sides of the mountains, and hovering over the shores, in search of dead sheep, fish, young birds, and such animals generally as are large enough to furnish a meal, so weak as to make no formidable opposition, or so heavy as to be incapable of very rapid or protracted flight. It sometimes carries off poultry that have straggled to a distance from the house, clutches up young lambs, and, when hard pressed by hunger, is said to attack sheep and even deer; but a fox, an otter, or a seal, it scarcely ever ventures to molest, and a man may carry off its young before its face with little danger of even a scratch. Yet instances have occurred of its manifesting some spirit in defence of its nest.

Being a heavy bird, with very large wings, it finds some difficulty in raising itself on wing, and therefore seldom alights on plain ground for the purpose of resting, but usually settles on a large stone or block, the pinnacle of a rock, or the edge of a precipice. The case is different, however, on an extensive sand, for there it can escape from its enemies without experiencing any danger. When it is fairly on wing, its motions are beautiful in the highest degree. Its wings are extended to their full length, forming an obtuse angle with the back, and as it sweeps along in wide curves, it seems to glide through the air without the least effort, and with very little motion of the wings or tail. The feet are drawn close to the abdomen, and the neck is retracted to such a degree, that the head seems stuck upon the shoulders—a character common to all our rapacious birds. In this manner it searches the hill sides, the moors, and the shores; but in proceeding to a distance, without regarding the intermediate space, it flies in a straight line, always at a great height, and with regular flaps, somewhat in the manner of the Raven. It utters a loud, shrill scream, which in calm weather may be heard at the distance of more than a mile, and emits a noise somewhat resembling the yelping of a Dog, which may be imitated by the syllables *queak, queak*, uttered repeatedly in rapid succession.

It sometimes seizes fishes when they swim close upon the surface, and has frequently been observed watching by a lake or stream for salmon or trouts. When it has young, it provides abundantly for them, and instances have occurred of people obtaining a supply of food, in times of scarcity, by climbing to its nest. On observing a person walking near it, they fly around him at a safe distance, occasionally uttering a savage scream, and allowing their legs to dangle, with outspread talons, as if to intimidate him.

The Raven frequently harasses it, and the Skua and Peregrine Falcon have been seen to pursue it; but it is perfectly safe from the attacks of any bird excepting the Golden Eagle. Like that species, it is often shot from a concealed hut or pit, and deprived of its eggs or young by letting down a person by means of a rope to the nest, which is generally placed in the most inaccessible part of a precipice, although sometimes in no very secure place, and occasionally on the island of a lake.

There seems to be no reason for believing that this Eagle has a very acute sense of smell, especially as it is not aware of the presence of a person lying in wait within fifteen yards of it; but its vision must be very penetrating, otherwise it could not so readily fall upon a carcase on the hills, or a dead fish on the shores, from the great height at which it often soars. In searching for food it usually, however, sweeps along the mountain's sides, at no great height.

In almost every district in the Highlands, stories are told of Eagles that have carried off infants, and it is probable that such an occurrence may have taken place, although the evidence is usually imperfect. Many absurd notions have been entertained respecting their powers and courage: but there is good reason to believe, that the Sea Eagle does not often subject itself to much danger from temerity.

It begins to construct its nest sometime in March, and the young are abroad about the middle of August. The diameter of the nest is about five feet, and it consists of twigs, heath, sea-weed, dry grass, and other materials of a like nature, arranged in a slovenly manner. There are generally two eggs, sometimes only one. They are about the size of those of a goose, but shorter, and of a yellowish-white colour, with a few reddish dots at the large end. The young are fed with fish, carrion, grouse, and young sea birds.

Great numbers of Eagles are annually destroyed in Scotland, on account of the depredations which they commit among the lambs; yet there is little probability of their being extirpated in the northern parts of that country, although very few remain in the south. Dr Bushnan, writing in 1834, doubts whether a pair is to be found wild in Dumfries-shire, or more than a dozen in Galloway. This species is less frequently seen in captivity than the Golden Eagle, probably because its haunts are mere remote from the densely peopled parts of the country. An interesting account is given of a captive Eagle, in Macgillivray's Rapacious Birds of Great Britain, by its owner, Dr Neill, who has kept it since the autumn of 1827.

HABITS OF THE BLUE-BACKED DOVE (*Columba Œnas*).—In "The Naturalist" for May, Mr Salmon gives the following interesting notice on this subject:—"I have lately been asked if I can suggest a better name than *Stock* or *Wood* Pigeon for the *Columba Œnas* of authors. The provincial name in this district is *Sand* Pigeon, which I cannot but consider fully as appropriate as *Bank* Swallow, applied in consequence of the situation the bird selects for nidification. Those who live in woody districts, however, might object to this specific designation, as the bird would then, in all probability, breed in woods; but of this I am not certain. I am inclined to suspect that the species is very local in its distribution in this country during the breeding season, and that it is only towards its autumnal migration that it is seen in very great numbers in the woodlands. I have known an instance of its breeding in the topmost branches of the Scotch Fir, in a similar manner to the Ring Pigeon (*Columba Palumbus*), which is a true arboreal species, and might, with great propriety, be called Wood Pigeon. Indeed, the latter is given by Selby as a provincial name, by which also it is always known in this county. If it be found necessary to make any alteration in the nomenclature of the British Pigeons, I should be disposed to name them thus:—Wood Pigeon, *Columba Palumbus*; Sand Pigeon, *C. arenea*; Rock Pigeon, *C. Livia*." Our mode of nomenclature differs, and we should prefer naming them thus:—*Columba Palumbus*, Ringed Dove, or Wood Pigeon; *C. Livia*, Rock Dove, or White-backed Pigeon; *C. Œnas*, Blue-backed Pigeon; *C. Turtur*, Ring-necked Turtle Dove. But there is no end to alterations of this kind, chiefly because names are imposed at first without consideration, and altered by persons destitute of judgment, or, if possessing that quality, deficient in the necessary knowledge of species.

ARRIVAL OF BIRDS IN THE PRESENT SPRING.—Notwithstanding the severity of the weather, the Wheatear was seen in the King's Park, early in April, and by the end of the month was common. *Hirundo rustica* was seen at Newington on the 29th April. *Cypselus murarius*, *Hirundo rustica*, *Hirundo urbica*, and *Hirundo riparia*, were observed at Canonmills on the 1st May. *Sylvia Trochilus* was shot at Roslyn on the 29th April. *Turdus torquatus* was observed near Swanston on the Pentland Hills on the 4th May; and on the same day, two Cuckoos were seen on Dalmahoy Hill. *Totanus hypoleucos* occurred at Roslyn on the 29th April, but had probably arrived several weeks hefere.

BOTANY.

M. DUTROCHET'S OBSERVATIONS ON THE SLEEP OF PLANTS.—*Continued from last Number.*

THE flowers above described expand and close only once, after which the death of the corolla takes place. I shall now examine the flowers which alternately open and shut for several days in succession, taking the Dandelion, *Leontodon Taraxacum*, as an example. The flower of this plant generally lives ten days and a half, having during that period its expansion in the morning, and its sleep in the evening. Its last sleep takes place in the middle of the day, and is followed by the death of the corollas. In the state of waking, the semiflorets of which this flower is composed curve outwards, producing its expansion; and in sleep curve inwards, effecting its closing. Although these semiflorets are very thin, I have with the microscope observed the internal organization of their nerves, which are very small, and four in number in each. At the inner or upper surface of each of its nerves there is a cellular tissue disposed in lines, of which the cellules are covered with globules, and which is exactly similar to that which I have observed in the nerves of the corolla of *Mirabilis*. At the outer or lower surface of the nerves of the semifloret is a very thin layer of fibrous tissue situate between a layer of tracheæ and a layer of cellules filled with air, and placed superficially. This fibrous tissue, in all respects similar to that which exists in the nerves of the corolla of *Mirabilis*, is in like manner included between two layers of pneumatic organs. It is therefore probable that this fibrous tissue is incurvable by oxygenation, and that the cellular tissue is incurvable by endosmosis, just as is the case with the nerves of the corolla of *Mirabilis*. And in fact, observation shows that the incurvation which produces the wake of the semiflorets of the Dandelion is owing to the impletion of the cellules with fluid to excess; in other words, to endosmosis, and that the incurvation which produces sleep is owing to oxygenation. The semiflorets of the flower of the Dandelion being gathered early in the morning, while they are still in the state of sleep or curved inwards, and being immersed in aërated water, assume the opposite incurvation, or curve outwards. This takes place in darkness as well as in light. If they are immersed in unaërated water, they curve outwards to excess, and so remain. If these semiflorets, thus curved, are placed in syrup, they assume the opposite curvature; and, on being placed in pure water, resume the outward curve. Thus, there can be no doubt that endosmosis produces this action. If semiflorets, in the state of waking, are left for some hours in aërated water, they then assume an inward curve, or that of sleep; and this curve is not altered by transporting them into syrup, which proves that the incurvation of sleep is owing to oxygenation. Thus, the wake and sleep of the semiflorets of the flower of the Dandelion result from the alternately predominating incurvation of an organic tissue incurvable by endosmosis, and of an organic tissue incurvable by oxygenation. The first is undoubtedly the cellular tissue, and the last the fibrous tissue, both contained in the nerves of the semifloret. These two incurvable tissues, according as the one prevails over the other, expand or close the flower.

The causes which render predominant the morning incurvation of the cellular tissue, or agent of waking, are, on the one hand, a more powerful ascent of the sap under the influence of light, which increases the turgescence of this tissue; and, on the other hand, the diminution of the antagonist force of incurvation of the fibrous tissue, which is the agent of sleep, and which takes place at night. In fact, if the semiflorets are gathered in the evening when they have assumed the incurvation of sleep, and immersed in oxygenated water, they there retain without change their incurvation; but if, next morning, there be plucked from the same flower other semiflorets which still retain the incurvation of sleep, and if they be immersed in aërated water, they instantly assume the incurvation outwards, even in the dark. Now, by immersing the semiflorets in water, endosmosis of their cellular tissue is excited, and so its incurvation is effected, producing the state of waking or expansion. If this result does not take place in the evening, it is because the incurvation by oxygenation of the antagonist fibrous tissue is too strong, and cannot be overcome by the incurvation of the cellular tissue. If by immersing next morning in water the semiflorets which have passed the night on the plant, their outward incurvation is produced, this proves that the force of incurvation of the fibrous tissue has diminished, and that, in consequence, this fibrous tissue has lost in the course of the night a part of its oxygenation; so that the cellular tissue, incurvable by endosmosis, which is its antagonist, and which is the agent of the wake or expansion of the flower, then predominates.

Thus the flower which during several days presents the alternations of waking and sleep, is that in which the fibrous tissue, the agent of sleep, loses during the night a part of the oxygen which has been fixed in its interior in the course of the day, and which is the cause of its incurvation; so that this latter having in the morning lost its strength, the cellular tissue, which is incurvable by endosmosis, and is the agent of expansion or waking, becomes again predominant. The sleep of this flower takes place anew in the evening, because the oxygenation of the fibrous tissue, the agent of sleep, gradually increases during the day, and at length renders its incurvation victorious; at the same time, the diminution of the light occasions a diminution of the ascent of the sap, which weakens the turgescence, and consequently the incurvation of the cellular tissue, which is the agent of expansion or waking. These alternations cease only with the death of the corolla. The flowers which have only a single expansion and a single closing, or which wake and sleep only once, are those whose single sleep is immediately followed by the death of the corolla.

The entire flower of the Dandelion, if immersed when expanded in unaërated water, deprived of communication with the atmosphere, remains in the same state without change. If the water is in contact with the atmosphere, it takes up air in solution, and the flower immersed in it assumes the state of sleep in two or three hours.

It may with propriety be remarked, that these experiments confirm what I have said in my Essay on the Respiration of Plants, namely, that vegetables respire like animals, by assimilating oxygen, of which the presence in the organic tissue is as necessary in the one as in the other order of beings.

PINES OF NORTH CALIFORNIA.—The following note is extracted from the "Companion to the Botanical Register," in which a biographical account of the late Mr Douglas is given:—"About an hour's walk from my camp, I met an Indian, who, on perceiving me, instantly strung his bow, placed on his left arm a sleeve of Racoon skin, and stood on the defensive. Being quite satisfied that this conduct was prompted by fear, and not by hostile intentions, the poor fellow having probably never seen such a being as myself before, I laid my gun at my feet on the ground, and waved my hand for him to come to me, which he did slowly, and with great caution. I then made him place his bow and quiver of arrows beside my gun, and striking a light, gave him a smoke out of my own pipe, and a present of a few beads. With my pencil I made a rough sketch of the Cone and Pine-tree which I wanted to obtain, and drew his attention to it, when he instantly pointed to the hills fifteen or twenty miles distant towards the south; and when I expressed my intention of going thither, cheerfully set about accompanying me. At mid-day I reached my long-wished-for Pines, and lost no time in examining them, and endeavouring to collect specimens and seeds. New and strange things seldom fail to make strong impressions, and are therefore frequently overrated; so that, lest I should never again see my friends in England, to inform them verbally of this most beautiful and immensely grand tree, I shall here state the dimensions of the largest I could find among several that had been blown down by the wind. At three feet from the ground its circumference is 57 feet 9 inches; and at 134 feet, 17 feet 5 inches; the extreme length 245 feet. The trunks are uncommonly straight, and the bark remarkably smooth for such large timber, of a whitish or light-brown colour, and yielding a great quantity of bright amber gum. The tallest stems are generally unbranched for two-thirds of the height of the tree; the branches rather pendulous, with cones hanging from their points like sugar loaves in a grocer's shop. These cones, however, are seen only on the loftiest trees, and the putting myself in possession of three of them (all I could obtain), nearly brought my life to a close. As it was impossible either to climb the tree or hew it down, I endeavoured to knock off the cones by firing at them with ball, when the report of my gun brought eight Indians, all of them painted with red earth, armed with bows, arrows, bone-tipped spears, and flint-knives. They appeared any thing but friendly. I endeavoured to explain to them what I wanted, and they seemed satisfied, and sat down to smoke, but presently I perceived one of them string his bow, and another sharpen his flint-knife with a pair of wooden pincers, and suspend it on the wrist of his right hand. Farther testimony of their intention was unnecessary. To save myself by flight was impossible; so, without hesitation, I stepped back about five paces, cocked my gun, drew one of the pistols out of my belt, and holding it in my left hand, and the gun in my right, showed myself determined to fight for my life. As much as possible I endeavoured to preserve my coolness; and thus we stood looking at one another without making any movement or uttering a word for perhaps ten minutes, when at last one, who seemed the leader, gave a sign that they wished for some tobacco. This I signified they should have if they fetched me a quantity of cones. They went off immediately in search of them; and no sooner were they all out of sight, than I picked up my three cones and some twigs of the trees, and made the quickest possible retreat, hurrying back to my camp, which I reached before dusk. Of my three cones, one measures 14½ inches, and the two others are respectively half an inch and an inch shorter, all full of fine seed."

MISCELLANEOUS.

DESCRIPTION OF THE FRITH OF CROMARTY.—NO. II.

On the flats of the Bay of Nigg are extensive beds of the common mussel, *Mytilus edulis*, which the fishers use for bait, together with various other shells, as *Venus pullastra, Solen siliqua, Mya truncata, Lucina radula, Cardium edule, Mactra subtruncatra, Fusus antiquus, Buccinum undatum, Purpura lapillus, Trochus cinerarius, Turritella terebra, Turbo littoreus*, and *T. retusus*. Besides these are observed numerous dead shells of the Great Clam, *Pecten maximus*, the Common Clam, *Pecten opercularis*, as well as most of the above. It is a somewhat curious circumstance, that although shells of the former species occur in the Frith of Forth, as well as in that of Cromarty, living specimens are not found there, while on many parts of the north-western coasts of Scotland they are not uncommon. There is an oyster scalp or bed below Invergordon, near the north shore, and another above it, among the shells found in the upper half of the Frith, are *Buccinum undatum, Purpura lapillus, Turritella terebra, Trochus cinerarius, Patella vulgata, Turbo littoreus, T. retusus, T. rudis, Mytilus edulis, Pecten opercularis, Venus pullastra, V. decussata, Solen siliqua, Mya truncata, Mya arenaria, Cardium edule*, and *Mactra truncata*; but they are not generally found in great numbers, the welks and limpets, in particular, being rare and diminutive.

The fishermen at Cromarty set their lines in the open sea, and do not fish in any part of the Frith, although a few large, generally sickly Cod appear there, with a good number of Codlings, occasionally Haddocks, Skate, and Flounders. Herrings sometimes enter; Sand-eels, *Ammodytes tobianus*, are found in the sands; and Cuddies, the fry of *Gadus carbonarius*, are plentiful in many parts. Fresh-water Trouts are sometimes caught more than half-way down; Eels occur in the upper part, and the Ocellated Blenny is plentiful on the shoals.

Echinus esculentus and *Spatangus cordatus* occur below Invergordon; as do the Spider Crab, the Hermit Crab, and Common Shrimp; but the Lobster and Common Crab are not met with. Several *Asteriæ, Alcyonia*, and *Sertulariæ*, are not uncommon, and the Lug, or Sea-worm, *Lumbricus marinus*, is plentiful in most parts of the Frith.

Seals are numerous at times, and frequently ascend the river as far as the bridge, near which they are sometimes caught by the salmon-fishers. Porpoises also enter the Frith, but are said not to enter the river.

The birds to be seen on the waters, or along the shores in winter, are very numerous. Vast flocks of Brent Geese, *Anser Bernicla*, frequent the shoals of the Bay of Nigg, as already mentioned, and at high-water are seen floating in detached parties over the basin below Invergordon. Long-tailed Ducks, *Herelda glacialis*, are also very abundant there. The Mallard, *Anas Boschas*, and the Golden-eyed Duck, *Clangula chrysophthalma*, are seen in great numbers feeding along the edge of the water, when the tide is out. The other species of common occurrence are, the Common Gull, *Larus canus*; the Black-headed Gull, *L. ridibundus*; the great Black-backed Gull, *L. marinus*; the Guillemot, *Uria Troile*; the Auk, *Alca Torda*; the Red-throated Diver, *Colymbus septentrionalis*; the Cormorant, *Phalacrocorax Carbo*. The Feaser, *Lestris parasiticus*, is of rare occurrence. Along the shores are seen the Redshank, *Totonus Calidris*; the Curlew, *Numenius arquata*; flocks of Ringed Plovers, *Charadrius hiaticula*, and Lapwings, *Vanellus cristatus*, together with the common Heron, *Ardea cinerea*. All these are, properly speaking, sea-birds, excepting the Brent Goose, the wild Duck, the Curlew, the Lapwing, and the Heron, which, however, are as frequently found on or by the sea as on land; while, on the other hand, several of the others are often seen on the land, and on fresh waters, and the Gulls often, and the Red-throated Diver always, breed on the latter.

The general considerations respecting estuaries, a subject of great importance with reference to the salmon-fisheries, which form the second part of this paper, must be deferred till another opportunity. In the meantime, it may be remarked, that the greater number of our estuaries present characters similar to that of the Cromarty Frith, being inlets of the sea entered by a river, and having the same vegetable and animal productions as those enumerated; each, however, having peculiarities of form, depth, and other circumstances. The Beauly Frith, or upper part of the Moray Frith, and the Frith of Tay, in particular, are very analogous to the Cromarty Frith, as will be shown in the sequel.

METEOROLOGY.

Dr Allnat of Wallingford has transmitted, for insertion in this Journal, the following account of a Western Aurora, which appeared on the night of Saturday 18th February last, between the hours of ten and eleven.

From the horizon almost due west a blood-red glow of light arose, shooting upwards to the zenith, where it was joined by another broad sheet of lurid light, which dipped far down into the east, thus forming an irregular arch extending almost continuously across the face of the heavens, from the western to the eastern horizon. The moon, which was high at the time, and the stars, were shining with unusual lustre, and a bitter wind was blowing from the south. There was a kind of nucleus of a brighter hue in the centre of the western or principal aurora, from which emanated streams of light. This nucleus was surrounded by external bands of a deeper and less vivid colour. In the north were also broad insulated patches of dense luminous matter of the same lurid aspect, which occasionally extended themselves so that the moon at one period was overshadowed by them, which caused a halo faintly exhibiting the prismatic colours. There was not a cloud to be seen.

Shortly after eleven these remarkable appearances vanished, when a strong wind sprung up from the south, and the succeeding day was wet and gusty.

A circumstance occurred during the continuance of these phenomena (which fact I do not find noticed in the records of Meteorology), namely, the *sudden rise of the barometer*. It had been sinking on the previous day, and it also rapidly sank after the disappearance of the aurora. This might have been merely the transient effect of natural causes in simultaneous operation; but if it were not, and the barometrical indication on similar occasions be hereafter fully established, it will, I conceive (as do the vibrations of the magnetic needle), lend a collateral and an important aid in determining the real nature of the aurora.—*Wallingford, 12th March* 1837.

REVIEWS.

The Naturalist; illustrative of the Animal, Vegetable, and Mineral Kingdoms. Edited by Neville Wood, Esq.

This useful and entertaining monthly journal, which, after the appearance of its sixth number, underwent what we were apprehensive was nothing less than an entire demolition, has, to our great pleasure, reappeared under a somewhat altered form, under the auspices of the accomplished author of "British Song Birds," and joint editor of the Analyst, a quarterly publication of great merit. The number for April contains, among other papers, an excellent account of the Lemuridæ; A Catalogue of Medicinal Plants found in the neighbourhood of York; Remarks on the sense of Smell in Carrion Birds, by the Rev. F. O. Morris; Observations on the habits of the Fitchet Weasel, by W. R. Scott; together with a variety of interesting miscellaneous notices. The number for May is at least of equal interest. The first paper, by Thomas Allis, on the impropriety of placing the Columbidæ in the order of Rasores, is in all respects to our mind; and the views entertained by that intelligent writer will be found to be similar to those expressed at greater length in Macgillivray's History of British Birds, vol. 1. recently published. The Pigeons certainly differ from the Phasianinæ, Pavoninæ, Perdicinæ, and other families of the Rasores, as much in external as in internal characters and habits, and the only point in which they agree is their feeding on vegetable substances. If this were sufficient to induce an intelligent ornithologist to class them together, why is not the Goose permitted to join the group? In conclusion, Mr Allis remarks:—"To make the order *Rasores* consist of these five families (*Pavonidæ, Tetraonidæ, Cracidæ, Struthionidæ*, and *Columbidæ*), appears to me very incongruous, and to arise more from a desire of adapting them to the exigencies of a preconceived theory, than from any natural affinities observable between them. A circle composed of families differing so greatly in the amount of their aberrations, presents, according to my ideas, chasms so extensive and frightful between the different families of which it is composed, as not at all to accord with the beautiful order of Nature; whereas, if placed in separate orders, we find the extreme species running into each other, and forming one harmonious whole, which cannot be contemplated without feelings of admiration and delight." Right:—"The vagaries of these people" are so absurd, their jargon so ludicrous, their self-sufficiency so amusing, that a reasonable creature, who studies Nature as she is, cannot refrain from associating the most comical ideas with the system advocated by them. "The Naturalist Abroad," by Edwin Lees, is a delightful paper, and Mr Morris's Explanation of the Latin names of British Birds, a very useful one, not to the tyro alone, but to the system-makers, who rarely can spell their names. Various interesting papers follow, in perusing which, we promise the reader much amusement as well as instruction. This eighth number, in fact, is a capital one; and we trust that those which are to come will not disappoint us in fulfilling the expectations to which it gives rise.

Loudon's Magazine of Natural History. New Series. Conducted by Edward Charlesworth, F.G.S.

This long-established, entertaining, and most useful monthly periodical, continues to support its well-earned reputation under the management of its new editor, who has published in it several very interesting papers on Geological subjects. The information which it conveys, although generally of a popular character—that is, intelligible to the reading public at large—is occasionally more strictly scientific or technical, and subjects requiring illustration are very beautifully represented by engravings on wood. The number for May contains a notice of the Teeth of Carcharias megalodon, occurring in the Red Crag of Suffolk, by the editor; A continuation of a Catalogue of the Birds of Devonshire, by Dr E. Moore; Natural Phenomena observed in 1833, by the Rev. W. B. Clarke; On the errors which may arise in computing the relative antiquity of Deposits from the characters of their imbedded Fossils, by H. E. Strickland, Esq.; Observations on the existence of Electric Currents in vegetable structures, by Golding Bird, Esq.; On the enlargement of the Eggs of some marine Mollusca during the period of their hatching, by J. E. Gray, Esq.; Remarks on Generic Nomenclature, by Mr Shuckard; Description of a new genus of British Parasitic Hymenoptera, by Mr Westwood; Observations on the changes of colour in the Fur of Mammalia and in Feathers, by Mr Blyth; besides Reviews, and short Notices on various subjects.

Edinburgh: Published for the Proprietor, at the Office, No. 13, Hill Street. London: Smith, Elder, and Co., 65, Cornhill. Glasgow and the West of Scotland: John Smith and Son; and John Macleod. Dublin: George Young. Paris: J. B. Balliere, Rue de l'Ecole de Médecine, No. 13 bis.

THE EDINBURGH PRINTING COMPANY.

THE EDINBURGH

JOURNAL OF NATURAL HISTORY,

AND OF

THE PHYSICAL SCIENCES.

JULY, 1837.

ZOOLOGY.

DESCRIPTION OF THE PLATE.——THE MAKIS.

THE Lemurs or Makis are quadrumanous animals, which differ from most of the other families of this extensive group in having a pointed muzzle, whence they have been called Fox-monkeys, small ears, and a long unprehensile tail. They have four incisors above, six below, edged canine teeth, and six grinders on each side of either jaw. There is little diversity in their form, and they are said to be very limited in their geographical distribution, being supposed to be peculiar to Madagascar, where they reside in the woods, moving with great agility much in the manner of squirrels, and living on fruit.

Fig. 1. THE WOOLLY MAKI (*Lemur Mongoz*).—As little can be said of the manners of the different species, we shall confine our notices to their characteristic colours. The present is of a brownish-grey colour, with the lower parts lighter, and the throat and sides of the head light red.

Fig. 2. THE RED MAKI (*L. rufus*), of a brownish-red colour, with the muzzle black, and the sides of the head pale greyish-yellow.

Fig. 3. THE WHITE-FRONTED MAKI (*L. albifrons*).— Yellowish-grey, the head whitish, the muzzle black, the bands brownish-red.

Fig. 4. THE RUFFED MAKI (*L. Macoco*).—Variegated with black and white, in large patches.

Fig. 5. THE RUFFED MAKI (*L. Macoco*).—A variety having the body all black, excepting a band across the back, the cheeks and part of the limbs white.

Fig. 6. THE GREY MAKI (*L. griseus*).—This species scarcely differs from the Woolly Maki in colour, but is smaller and has a shorter muzzle.

Fig. 7. THE RING-TAILED LEMUR (*L. Catta*).—Ash-grey, with the tail ringed with black, the head white, the muzzle, a space about the eyes, and a collar black.

DESCRIPTION OF THE PLATE.——THE HUMMING BIRDS.

OF the Humming Birds represented in this plate, little is known beyond their external characters, which present some diversity as to colouring, but with respect to form, proportions, and texture of plumage, are pretty similar.

Fig. 1. THE AZURE-CROWNED HUMMING BIRD (*Trochilus cyanocephalus*).—So named on account of the rich blue of the upper part of its head, is not remarkable for brilliancy of colour. It is a native of Brazil, but its peculiar history is unknown.

Fig. 2. Represents the young of the same species, differing from the adult in having the bill darker, and the green colour of the upper parts more extended over the sides.

Fig. 3. THE BLUE-FRONTED HUMMING BIRD (*T. glaucopis*).—Nearly four and a half inches long, with the top of the head indigo-blue changing to green, the back of a deep gold-green, the wings brown, and the tail steel-blue. It inhabits Brazil.

Fig. 4. TEMMINCK'S HUMMING BIRD (*T. Temminckii*).—Was first described by the celebrated ornithologist, to whom it has since been dedicated by M. Lesson, who states that it " belongs to Brazil, that rich country which yields the diamond, and nourishes the most beautiful birds."

Fig. 5. SAPPHIRE AND EMERALD HUMMING BIRD (*T. bicolor*). Male.—This species nearly three and a half inches long, and distinguished by blue and emerald green colours, is said, by Buffon, to be a native of Guadaloupe, by Sonnini, to occur also in Martinique, and, by Lesson, to have been received from French Guiana.

Fig. 6. The young of this species, nearly of the same colour as the adult.

Fig. 7. CLEMENCE'S HUMMING BIRD (*T. Clemenciæ*).—This beautiful species, which is about five inches in length, and of a bright green colour above, is a native of Mexico.

DESCRIPTION OF THE PLATE.——THE GROUSE.

THE Grouse which inhabit the cold and temperate regions of both continents are characterised by their robust form, full plumage, short, slightly curved, convex bill, concave, rounded wings, and short, stout feet, of which the tarsi are usually entirely or partially covered with feathers, but sometimes bare, and the toes are furnished with lateral pectinated scales, by which especially these birds are distinguished from the Ptarmigans. They exhibit considerable diversity of plumage, some having tufts on the head or neck, and the tail varying in length, as well as in the proportion and form of its feathers. They live upon vegetable substances of various kinds, leaves, twigs, and berries, and afford a generally esteemed article of food. They all nestle on the ground, laying numerous spotted eggs, and their young run about immediately after exclusion.

Fig. 1. THE ROCKY MOUNTAIN SPOTTED GROUSE (*Tetrao Franklinii*).—This bird is by some considered as a variety of *Tetrao Canadensis*, while others hold it to be a distinct species. A certain writer, finding the tail feathers of specimens of the one kind to be emarginate and mucronate, and those of the other plain, looks upon this difference as decisive ; but a little practical knowledge is sufficient to convince us that no dependence can be had on such a circumstance, for feathers which are mucronate when new are often rounded when old, and in other cases the reverse happens. It is plentiful in the Rocky Mountains near the sources of the Columbia River.

Fig. 2. Female.

Fig. 3. THE DUSKY GROUSE (*T. obscurus*). Male.—This species has also been named *T. Richardsonii*, and is stated to be plentiful in the subalpine regions of the Rocky Mountains. The male is greyish-brown above, undulated with dusky, the abdomen whitish.

Fig. 4. Female. The female is more spotted and variegated than the male.

Fig. 5. THE RUFFED GROUSE (*T. Umbellus*). Male.—The beautiful species here represented has a tuft of feathers on the head, and a large ruff on the neck, whence its specific name. It occurs in the British Colonies of North America, and in most parts of the United States, where it is common in the markets.

DESCRIPTION OF THE PLATE.——THE INDRIS, LORIS, GALAGOS, AND TARSIERS.

THESE animals, generally of small size, constitute the last families of the great division of the Quadrumana.

Fig. 1. THE SHORT-TAILED INDRI (*Lichanotus brevicaudatus*).—This is the only authenticated species of the genus, which differs from the Lemurs in having four instead of six lower incisors, and a very short tail. It is of a blackish colour, with the face grey ; has the hind limbs proportionally longer than those of any species of Monkey ; and, according to Sonnerat, is tamed by the natives of Madagascar, and employed in hunting, although in the natural state it is strictly frugivorous.

Fig. 2. THE SLOW LORI (*Stenops tardigradus*).—The Loris have teeth similar to those of the Makis, but with the points of the grinders more acute ; their muzzle is less elongated, and they are destitute of tail. Their eyes are large and approximated, and their tongue is rough. They are nocturnal animals, live upon insects and other small animals, and are said to be very slow in their motions. The present species is of a yellowish-brown colour, with a dusky band along the middle of the back, and is a native of Bengal.

Fig. 3. THE SLENDER LORI (*Stenops gracilis*).—This species is smaller and of a more slender form, with the nose a little raised, the colour nearly similar to that of the last, but without the dorsal line.

Fig. 4. THE SENEGAL GALAGO (*Otolicnus Senegalensis*).—The Galagos have the same kind of teeth as the Loris, and are also insectivorous ; but their tail is very long, and their eyes and ears large. They are natives of Africa, reside on trees, in the holes of which they nestle, and are very active. The present species is about six inches in length, exclusive of the tail, and of a yellowish-brown colour.

Fig. 5. THE LITTLE GALAGO (*Otolicnus Madagascariensis*).—The colour is similar, but the ears are shorter, and the size considerably smaller.

Fig. 6. THE PODGE TARSIER (*Tarsius Spectrum*).—The Tarsiers, so named on account of the great length of their tarsi, are very similar to the Galagos, from which they differ in having the interval between the grinders and incisors filled with several shorter teeth, a shorter muzzle, and still larger eyes. They are equally of nocturnal habits, and live on insects. The present species has the body about six inches long, its tail considerably longer, and terminated by a tuft of hairs. It is said to be a native of Madagascar.

DESCRIPTION OF THE PLATE.——THE SWALLOWS.

THE Swallows constitute an extensive family of birds, remarkable for the elegance of their form, their adaptation for rapid and protracted flight, and their extended migrations. From the liveliness of their manners, their attachment to the habitations of man, and the circumstance of their announcing the return of summer, they are general favourites. This group has been divided into several genera, of which it is unnecessary to speak at present, as the species in the plate, which represent three of these sections, have been referred to the single genus *Hirundo*. These birds are extensively distributed, some species occurring in all the warm and temperate parts of the globe, or, during the summer, extending their migrations to the confines of the frozen regions. Those represented in the plate are all natives of America.

Fig. 1. The Barn Swallow *(Hirundo rufa.)* Male.—This species so closely resembles the *H. rustica* of Europe, that some ornithologists have considered it absolutely the same. It is about seven inches in length, with long pointed wings, and a very deeply forked tail; the forehead and throat reddish-brown, a dusky bar across the fore neck, the upper parts glossy bluish black, and the tail-feathers with a white spot on the inner web. It visits all parts of the United States in the end of spring, nestles in the same manner as our Chimney Swallow, lays five white eggs spotted with reddish-brown, and generally rears two broods in the season.

Fig. 2. The Barn Swallow. Female.—The female has the tail less elongated, the upper parts tinged with green, the rufous colour on the throat and forehead paler.

Fig. 3. The Bank Swallow (*H. riparia.*)—This species, which is much smaller than the last, and of a greyish-brown colour on the upper parts, is common to both continents, its habits in America being precisely similar to those which it exhibits with us. It receives its specific name from its burrowing in sandy banks.

Fig. 4. The White-bellied Swallow (*H. bicolor.*)—The upper parts are of a glossy greenish-blue, the lower white. It has been ascertained by Mr Audubon to winter in Louisiana, and in summer to extend over all parts of the United States.

Fig. 5. The Cliff Swallow (*H. fulva.*)—This species, which is named from its habit of building on the face of rocks, is more variegated in its colours than the preceding, and has the tail abrupt or truncate. It inhabits the southern and western parts of the United States, from whence it returns southward in August.

Fig. 6. The White-collared Swallow *(H. albicollis.)*—Of this species little seems to be known. It is of a uniform blackish colour, with a band of white across the hind part of the neck, and a patch of the same colour on the fore neck; and is said to inhabit Brazil.

Ravages of Field Mice.—The following interesting account of the destructive effects of Field Mice is extracted from Mr Jesse's Gleanings:—" An extraordinary instance of the rapid increase of Mice, and of the injury they sometimes do, occurred a few years ago in the new plantations made by order of the crown in Dean Forest, Glocestershire, and in the New Forest, Hampshire. Soon after the formation of these plantations, a sudden and rapid increase of Mice took place in them, which threatened destruction to the whole of the young plants. Vast numbers of these were killed; the Mice having eaten through the roots of five-year-old oaks and chestnuts, generally just below the surface of the ground. Hollies also, which were five or six feet high, were barked round the bottom; and in some instances the Mice had crawled up the tree, and were seen feeding on the bark of the upper branches. In the report made to government on the subject, it appeared that the roots had been eaten through wherever they obstructed the runs of the Mice. Various plans were devised for their destruction; traps were set, poison laid, and cats turned out; but nothing appeared to lessen their number. It was at last suggested, that if holes were dug, into which the Mice might be enticed to fall, their destruction might be effected. Holes, therefore, were made, about twenty yards asunder, in some of the Dean Forest plantations, being about twelve in each acre of ground. These holes were from eighteen to twenty inches in depth, and two feet one way by one and a half the other; and they were much wider at the bottom than at the top, being excavated or hollowed under, so that the animals when once in could not easily get out again. In these holes at least thirty thousand were caught in the course of three or four months, that number having been counted out and paid for by the proper officers of the forest. It was, however, calculated that a much greater number than these were taken out of the holes, after being caught, by Stoats, Weasels, Kites, Hawks, and Owls; and also by Crows, Magpies, Jays, &c. The Cats, also, which had been turned out, resorted to these to feed upon the Mice; and, in one instance, a Dog was seen greedily eating them. In addition to the quantity above mentioned, a great many Mice were destroyed in traps, by poison, and by animals and birds of prey; so that in Dean Forest alone, the number of those which were killed in various ways could not be calculated at much less than one hundred thousand. In the New Forest, from the weekly reports of the deputy-surveyor of the forest, about the same number were destroyed, allowing the same calculation for those eaten by vermin, &c.; in addition to which, it should be mentioned, that these Mice were found to eat each other when their food fell short in winter. Putting these circumstances together, the total destruction of Mice in the two forests in question would probably amount to more than two hundred thousand."

Matrimonial Union between a Blackbird and a Thrush.—We are indebted to Mr Weir of Boghead for the following notice:—That birds in a state of confinement should be induced, by the solicitations of love, to form matrimonial alliances with other species of the same genus, or such as resemble them most closely in size and habits, when they have not an opportunity of making a choice, is not wonderful; but that they should do so when left at liberty, is of very rare occurrence. Mr Russell of Mosside, my next neighbouring proprietor, and his brother, informed me, that they were surprised at the attention which a male Blackbird paid to a female Thrush about the end of the winter of 1836. They were always in company, and fed together. As the spring approached, their attachment appeared more decided, and they carried on a course of regular flirtation, which ended in matrimony. After a good deal of consultation, they at length determined to build their nest in a bush of heath which hung over what is here called a "farret brae." They had four young ones before the nest was discovered, so cunningly had they concealed it. When they were about half fledged, they were unfortunately carried off by some boys, notwithstanding all the care that had been taken to preserve them.

Snowy Owl shot in England.—On the 13th of February last, a fine male Snowy Owl was shot three miles below Selby-on-the-Moor, where it had been observed by the miller at a mill adjoining, for a day or two previous. The moor is well stocked with Rabbits, and the Owl was most probably preying upon them. It appeared very shy, and when pinioned by the shot was extremely fierce. It was ultimately got into a sac and killed by pressure, when it came into the possession of A. Clapham, Esq. of Potternewton, near Leeds.—*H. Denny, in the Magazine of Zoology and Botany.*

Attachment of the Lark to its Young.—Some mowers shaved off the upper part of the nest of a Sky Lark without injuring the female, which was sitting on its young; still she did not fly away, and the mowers levelled the grass all round without her taking further notice of their proceedings. A young friend of mine, son of the owner of the crop, witnessed all this, and about an hour afterwards went to see if she was safe, when, to his great surprise, he found that she had actually constructed a dome of dry grass over the nest during the interval, leaving an aperture on one side for ingress and egress. My friend immediately hastened to inform me of the circumstance, and I was about to follow him to the spot, but, on his return, he found that some ruffian had, in the meantime, torn open the nest, and made off with the young ones. How disheartening it is for the Naturalist to be so continually annoyed by these callous bird-nesters! I was in hopes, when the brood had left the nest, to have preserved the latter as a most interesting specimen; but, alas! all is, as usual, frustrated. I should add, that the intention of the parents was, obviously, to have preserved their young from the scorching heat of the sun.—*Mr Blyth, in The Naturalist.*

BOTANY.

GENERAL REVIEW OF THE VEGETABLE KINGDOM.—NO. V

Vegetation, in ascending above the level of the sea, undergoes modifications analogons to those which attend its progress from the line to either pole, with this distinction, that in the last case the phenomena succeed by almost imperceptible gradations, while they crowd upon and follow each other in rapid succession on the ascent of mountains. The height of 4000 or 5000 yards in the hottest parts of the globe produces changes as distinct as the 2000 leagues, or more, which lie between the Equator and the polar regions. The three causes of the influence of which we have just spoken, all reappear within this space, viz. a diminution of heat; dryness of air; and protracted duration of light. To these we must add two others, a decrease of depth in the volume of the air, and a scarcity of those substances which abound in carbon, and are produced by the decomposition of organic bodies.

The higher we ascend, the shallower the upper stratum of air becomes; whence the excessive cold at great heights; for it is the action of the atmosphere upon the rays of light which extracts the caloric from them, and we know that the extraction of caloric diminishes in proportion as the mass of air traversed by the rays is shallower; but, on the other hand, the light is purer and more active, just as if caloric was really a simple transmutation of light, as some naturalists have conceived it to be.

The weight of the atmosphere, which at the level of the sea supports a column of mercury 28 inches high, diminishes as we ascend; so that at the elevation of 6000 yards it will only support a column of 13 inches and some lines high. A consequence of this fact is, that the vaporisation of fluids takes place on high mountains at a very low degree of heat. Notwithstanding this, however, the decrease of heat is so great, that the ambient air is very slightly impregnated with moisture.

It is true that heights have not the long days of the polar regions; but they receive the rays of the sun earlier than the plains, and are quitted later by them, so that their nights are shorter than in levels.

In fine, substances containing carbon, the wrecks of organised bodies, are rare on mountains, the rains as well as the waters of the springs dissolving them, and carrying them away as they run off into the valleys.

It cannot be doubted, but these causes united must act powerfully upon vegetation. The slightest degree of heat will cause the plants on mountains to transpire copiously; the severity of the cold, the dryness of the atmosphere, the shortness of the nights, the scarcity of carbon, will impede the enlargement of their leaves, and the growth of their stems; the strength of the light, and the protracted duration of the day, will accelerate the induration of all the parts of their frame.

The course of vegetation on mountains had not escaped the penetration of Tournefort. At the foot of Mount Ararat he had observed the plants which grow in Armenia; a little higher, those of Italy and France; above, those of Sweden; and upon the summits, those of Lapland. Observations of the same kind had been subsequently made on Mount Caucasus, the Alps, the Pyrenees, and other mountains of the old Continent. Every botanist had learned that many of the Alpine plants, that is to say, plants which grow on the various high lands of Europe and Asia, are likewise met with at Spitsbergen, in Nova Zembla, Lapland, and Kamschatka. Swartz had discovered on the mountains of Jamaica, under a still hotter sky, if not plants exactly of the same species with those of our Alpine Phænogamous ones, at least some that were analogous to them; and a great many of the Cryptogamous species precisely the same as our own: For example, *Funaria hygrometrica, Bryum serpyllifolium* and *cespititium, Sphagnum palustre, Dicranum glaucum*, &c. Linnæus, in his own way, had summed up these facts in an axiom. "The different kinds of plants," says he, "show by their stations the perpendicular height of the earth." Yet it was not till lately that any exact survey had been taken of this interesting department of botanical geography.

The first connected series of researches made with the direct intention of ascertaining the progressive succession of plants on mountains, was instituted by M. Ramond. That celebrated individual devoted ten years to the investigation of the entire chain of the Pyrenean mountains; and studied it not only as a geometrician, natural philosopher, and mineralogist, but also as one of the most skilful of botanists; he discovered, with the sagacity that distinguishes him, the stations to which the different species of vegetables belong, and the special circumstances which sometimes cause a derangement in the natural order of their succession. We shall here shortly point out some of the results of his observations.

The Common Oak (*Quercus Robur*) grows in the plains, on a level with the sea;

reaches the slopes of the mountains, and ascends to the height of 1600 yards. It degenerates in proportion as it approaches the point where it ceases to vegetate.

The Beech (*Fagus sylvatica*) makes its first appearance at the height of 600 yards above the sea, and its last at 200 yards above the Oak. The Silver-Fir (*Pinus picea*)..and the Yew (*Taxus communis*), show themselves at 1400 yards, and go on to about 2000. The Scotch-Fir (*Pinus sylvestris*), and Mugho Pine (*Pinus pumilio*), take their stations between the heights of 2000 and 2400 yards.

There the trees stop, and shrubs, with a juiceless foliage, and low or creeping stems, present themselves; these lie hid beneath the snow in the winter. Among them are some of the *Rhododendrons*, *Daphnes*, *Passerinas*, the *Globularia repens*, the two species of *Salix*, *herbacea*, *reticulata*, &c.

Soon after we meet only small herbs with perennial roots, spreading radical leaves, and a naked stalk. These, with the lichens and *Bysri*, arrive at the height of 3000, and even 3400 yards. The first that occur are the *Gentiana campestris*, *Primula villosa*, *Saxifraga longifolia*, and *Aicoon*, &c.; then *Ranunculus alpestris*, *nivalis*, *parnassifolius*, *Arctia alpina*, and finally, *Ranunculus glacialis*, *Saxifraga cæspitosa*, *oppositifolia*, *Androsaces*, and *Groenlandica*. The last brings us to the borders of eternal snow.

Botanists who have explored the Alps, have remarked phenomena perfectly corresponding with those observed by M. Ramond in the Pyrenees. But it was reserved for Messrs Humboldt and Bonpland to demonstrate the succession of modifications in the vegetable structure on the highest mountains yet known, and in one of the hottest and most fertile regions of our globe.

In the equinoctial countries of America, vegetation displays itself to the view of the observer as on the gradually rising steps of an immense amphitheatre, the base of which sinks below the waters of the ocean, while its summit reaches to the foot of the glaciers which crown the Andes, 5000 yards above the level of the sea; show-ing that in America there are vegetables which grow at the height of 1600 or 1800 yards beyond the point where vegetation ceases in the Pyrenees and Alps; a differ-ence that does not depend solely upon latitude, but likewise, according to M. Ramond, upon the breadth, or, if you will, the thickness of the chain of mountains. In chains of but little breadth, such as those of Europe, the air and temperature of the plains have an influence, which is constantly tending to confound the limits of the different kinds of vegetables; but this is not the case in the chain of the Andes, which is from 48 to 60 leagues in breadth. Messieurs Humboldt and Bonpland have had also this advantage in their researches, that as these were made under the Equa-tor, they have been enabled to trace the whole series of modifications which are to be met with between the two extremes of temperature found at the surface of the globe; while other botanists, having explored none but the northern mountains of the old Continent, could only trace the modifications between a mean temperature and extreme cold.

The Plants which belong to dark and humid abodes, such as *Boletus ceratophorus*, and *botrytes*, *Lichen verticillatus*, *Gymnoderma sinuata*, and *Byssus speciosa*, are found on the vaults of caverns and the woodwork of mines, as well in Mexico as in Germany, England, and Italy. Concealed within the bowels of the earth, these less perfect species constitute the last zone of vegetation.

Next come the Plants which belong to fresh-water and salt-water. Of these, a great portion grow without preference in every degree of latitude: the medium in which they exist preserving a more equable temperature than the atmosphere. Duck-weed (*Lemna minor*), and the Greater Reedmace or Cat's-tail (*Typha latifolia*), grow in the marshes both of Asia, Europe, and America. The *Typha latifolia* belongs in common to Jamaica, China, and Bengal. Probably there is no re-gion on the globe where the Gray Bog-Moss (*Sphagnum palustre*) is not to be found. This indifference to climate is still more remarkable in the Sea-plants, such as the *Fuci*, *Lavers*, and *Ceramia*. The Gulf-weed (*Fucus natans*), detaching it-self from the rocks on which it grew, and forming shoals of an immense extent at the surface of the water, obstructs the way of the ships as well towards the poles as un-der the line. On a level with the sea to the height of 1000 yards, we find the palms, the liliaceous plants, the plantain trees, the *Scitamineæ*, the genera *Theophrasta*, *Mussenda*, *Plumeria*, *Casalpinia*, *Hymenæa*, the *Cecropia peltata*, the Balsam of Tolu, the Cospari or Cinchona of Caroæy, with crowds of other species which grow only in a very hot temperature. This is the zone of the palms; a tribe conspicuous for the elegance and grandeur of part of its species, and forming one of the chief ornaments of the scorching plains that lie between the Tropics. Some of them thrive, however, in more temperate regions. The *Cerozylon andicola*, a fine palm, rising 60 yards in height, grows in the Andes at Tolima and Quindio, in the 4° 25′ of northern latitude, setting off at 1860 yards above the sea, and continuing to the height of 2870, an elevation where the atmosphere is at a moderate degree of warmth. Another species has been discovered at the Straits of Magellan, towards the 53° of southern latitude. Two sorts, the fan-palm (*Chamærops humilis*) and date-tree, are even seen to grow on our side of Europe, upon the coasts of the Medi-terranean, and not far from the foot of the Pyrenees, thus advancing their tribe to beneath the 43d degree of northern latitude. But these are the exceptions; the palms in general confining themselves to the hottest parts of the globe, and none being met with towards the polar regions.

The zone of the arborescent ferns and the cinchonas succeeds to that of the Palms and Scitamineæ. The ferns begin at 400 yards, and end at 1600. The cinchonas grow to about 2900 yards high. The oaks begin to appear at 1700 yards. These are deciduous, and, by their periodical evolutions from the bud, remind the European, while wandering in these distant regions, of the mild springs of his native land.

Trees cease to grow at the elevation of 3500 yards, where the shrubs, which be-fore had formed but a small part of the vegetation, take their place, and cover the whole soil.

A good deal lower, at about 2000 yards, the Gentians, Lobelias, Crowfoots, or Ranunculuses, &c. which answer to our Alpine plants, have already begun to show themselves, and keep on from thence to 4100 yards.

At this point, where snow occasionally falls, the grasses, whose numerous species were mingled in the vegetation of the lower steps of the amphitheatre, begin to reign

alone. The oat-grasses (*Avena*), bent-grasses (*Agrostis*), cock's-foot-grasses (*Dactylis*), panic-grasses (*Panicum*), feather-grasses (*Stipa*), *Jarava*, &c. here cover the face of the mountains, and continue their career up to 4600 yards, the point at which Phænogamous vegetation ceases.

THE COFFEE-TREE.

THE COFFEE-TREE (*Coffea Arabica*) seldom exceeds twelve feet in height, and has a slender trunk, sending off long trailing branches at its upper part. The leaves are elliptical, undulated, pointed, smooth, and opposite. The flowers are axillar, sessile, or on very short peduncles, two or three together; the calyx very small, tubular, with five teeth; the corolla funnel-shaped, its limb deeply cut into five ellip-tico-lanceolate, reflexed segments; the stamens five, their filaments short, and at-tached to the tube, their anthers linear; the germen roundish, the style simple, the stigma cleft, its segments reflexed. The fruit is a round fleshy red berry, which contains two seeds, commonly called coffee-beans. The plant thus belongs to *Pen-tandria monogynia* of the Linnæan system, and is referred to the natural order of *Cinchonaceæ*.

It is a native of Arabia and Ethiopia, but is cultivated in various parts of the world. It was first described by Alpinus in 1591, and was planted by the Dutch in Java in 1690, and by the English in Jamaica in 1732. The Mocha coffee, however, is considered superior to all others. It is raised from seedlings, planted out in moist shady places, and watered by artificial rills. When the fruit is mature, cloths are spread under the trees, which are shaken. The berries are afterwards spread on mats, and exposed to the sun until perfectly dry, when the husk is broken with heavy rollers of wood or stone; the kernels are again dried in the sun, and winnowed. It is scarcely necessary to say, that coffee, as used by us, is an infusion of the powdered seeds after they have been roasted.

As an article of food, coffee owes its nutritious quality chiefly to the milk and sugar used in preparing it. Considered medicinally, it counteracts the narcotic effects of opium and spirituous liquors. On strong healthy constitutions it seems to have little effect; but on debilitated or nervous persons it acts injuriously, inducing, when used to excess, headache, giddiness, dimness of sight, and aggravating hysterical affections. It is, therefore, less adapted for sedentary people than for those employed in the open air; but used moderately, as to strength and quantity, along with substantial food, it produces no injurious action. Its effects in promoting digestion and obviating drowsiness are well known; but while, to persons in ordinary health, it forms a wholesome diluent, or a grateful stimulant, neither it nor tea, especially when used in large quantities, ought to be indulged in by those of a nervous temperament, or who are debilitated by disease.

LAMIUM ALBUM AND L. MACULATUM NOT DISTINCT.—At a recent meeting of the Botanical Society of London, Mr Thomas Hancock read a paper on these plants, which he considered as forming only one species, to which he proposed giving Dr Lindley's name, *L. vulgatum*. He had been led to investigate the subject from hav-ing seen many specimens of *L. maculatum* entirely destitute of the longitudinal white patches on the leaves, so particularly insisted on by most authors as its most important specific character, as well as from having several with white flowers, and approaching so closely to *L. album* as to be scarcely distinguishable from it. He stated that Reichenbach, in his figure and description of what he considered to be the true *maculatum*, and Dr Hooker, in his adoption of the same as such, had fallen into an error, and that their plant was doubtless a variety of *L. purpureum*.

MISCELLANEOUS.

ACCOUNT OF AN OSSIFEROUS CAVERN AT YEALMTON BRIDGE, NEAR YEALMTON, DEVONSHIRE.

In the summer of 1835, having casually heard of certain bones having been met with, in the progress of working a limestone quarry at Yealmton Bridge, we undertook to investigate their value, and the circumstances under which they occurred; the results of which inquiry we soon afterwards published, with a feigned signature. The present account regards only the bare facts which obtruded themselves, and is a condensed and corrected relation of what has been before made public in this neighbourhood.

Lime-rock abounds at Yealmton Bridge, and caverns and fissures of various sizes are not unfrequently disclosed to view during its removal for economical purposes. That, on the southern side of the river, at this spot rises to a great height; and before its consumption commenced, its bed projected to the banks of the river. It was in the upright surface of the rock, that the opening or openings of the cave probably formerly existed; but the memory of man can render no account concerning these original entrances. I say entrances, for as there were certain chambers to the cave, each pursuing different directions to the surface, it is reasonable to suppose there were an equal number of apertures; besides which, the fact of the remains of the predatory beasts here discovered, being disposed so that each kind was, generally speaking, separate from the others, seems to point out independent and unconnected movements of these creatures.

A great part of the cavern had been destroyed, and a large quantity of the bones removed, and irrecoverable, at the time we commenced the investigation. The relative positions, directions, and measurements of the remaining cavities, are stated by Captain Mudge as follows:—"Portions of only the eastern and western chambers remained. The former consisted of a descending shaft to the depth of ten feet, which turned at right angles, and again ascended to the surface, both the descent and the ascent being at an angle of 45°. Of the western chamber, a portion remained uninjured. From the present opening, it takes a northerly direction for forty-three feet, the height varying from five to six feet, and the breadth from four to five. It then turns westerly for twenty-five feet, the height varying from five to twelve feet, and the breadth from three and a half to five."

Several deposits, arranged as superimposed strata, occurred in this cavern. The lowest stratum consisted of compact red clay, three feet six inches deep. Above this was found a layer of argillaceous sand in the eastern chamber, and of coarse gravel in the western chamber, the former varying in depth from six to eighteen inches, and becoming broader towards its limit; the latter not exceeding six inches in depth. Over these respectively a bed of stiff white clay, since become red, presented itself, being in depth about two feet six inches. Lastly, above the whole was an accumulation of diluvial clay, three feet six inches, containing pebbles, and the osseous remains. A stalagmitic crust of variable thickness formed an almost general covering to these strata and animal remains. Such were the appearances, and number of deposits, where the space was sufficient and circumstances favorable.

In one direction, where the cave had communication with the surface by means of numerous small, circular, lengthened apertures, an alternation of thin beds of clay and stalagmite was observable; and, contained in the substance of this stalagmite, we discovered the bones of three or four species of Mus, which, however, we have not been able to identify as the remains of existing kinds. One of them is certainly allied to the Water-rat, and another to the common Field-mouse. These facts are mentioned, because we imagine these said alternate deposits of clay and stalagmite to be modern comparatively with those before named, originating in accidental fallings-in of recent soil; and pointing out, at the same time, a difference in the age of these, and of the other animal remains. Besides these, there were other recent exuviæ found in connection with the diluvial clay, namely, certain snail shells, and the bones of a bat, both of which creatures are known to hybernate, and not unfrequently to experience death in such places. There were likewise other relics, of the antiquity of which we are not clearly satisfied, since it is the habit of very many animals to appropriate such cavities for dwellings, to betake themselves during night, during sickness, during winter, or as a resource when pursued, to the hollows and crevices of rocks; and since, by a variety of causes, their bodies, after death, are liable to be found blended with such as are the genuine productions of a former epoch.

The pebbles found in the uppermost stratum are certainly granite, and were, therefore, most likely derived from Dartmoor, where granite universally abounds. Brecciæ, of conglomerates of clay, fragments of rock, bones, pebbles, and stalagmite, coprolitic masses, bodies resembling indurated adipocere, portions of rich spongy fibrous clay, and particles of black mould, were also distributed through the same bed. No hair was found. The bones were in great number and variety, and for the most part it appeared, that besides the separate occurrence of those species which are predaceous, such as the Fox and Hyena, there was, moreover, a separation of the herbivorous from the other kinds; but this may have been accidental, and it must be recollected that this account refers only to such portions of the cavern as remained for our examination. It is, for the same reason, difficult, or impossible, to state the proportions borne by the different kinds to each other; but, if the facts presented by these remaining portions of the cave could be allowed to furnish such a statement, it would be, that the rapacious exceeded in number the other creatures. Very many dozens of Hyenas' teeth were collected; and in one small spot, having an area not greater than four feet, we extracted seven dozens of canine teeth of this animal. Next in frequency of occurrence to the bones of the Hyena and Fox were those of the Horse, Ox, Deer, and Rabbit. After these ranks the Rhinoceros, whilst the bones of the Elephant, Wolf, Pig, Glutton, and Bear, were extremely rare. Phalangeal bones, and a very few others, were all that we found perfect, the rest being in a broken state. The long bones had generally lost their epiphyses, and very many, not excepting those of the Hyena, were marked by teeth of some predatory beast, and evidently show that they had been chipped and gnawed. One or two fragments display on their surfaces scratches resembling those made by the teeth of a Weasel, or animal of that kind. Teeth of very aged animals were found, and there were also bones of young individuals belonging, with the exception of a few of the Hyena, to the herbivorous kinds. The remains of the Elephant are indeed confined to two teeth of a young animal. Some of the bones have been attacked by inflammatory disease, and this occurs among the larger kinds of teeth, which, also, in some instances, are fractured as if they had been submitted to great violence. Some pieces of bone are on one side highly polished, as if they had been subjected to great friction; and Captain Mudge observed a part of the roof of the cavern, which is lower than usual, perfectly smooth and glossy, as though it had been rendered so by the frequent transits of the tenants of the cave.

It is very difficult to determine on the precise number of species of animals found in this cave, since, besides that a very great quantity of the bones had been originally destroyed, our knowledge of fossil osteology is as yet very imperfect, and the broken condition they were found in precludes the possibility of identifying a great many of them, even with the greatest facilities of comparison with other specimens. Add to this, also, that not unfrequently fragments, and even teeth, are met with, which baffle the keenest discrimination, that a degree of uncertainty with respect to date often attaches to some of the animal remains deposited in ossiferous caverns, and that sometimes from a disparity in size, conjoined with a similarity in shape and figure, of some series of teeth, a doubt arises whether there may not have existed several analogous species of such animals. This kind of doubt has unavoidably arisen in the present investigation; but it seems most reasonable to conclude, that there were two or even three species both of Deer and Horse, since there are series of teeth of these genera greatly differing in size.

With regard to the composition of these fossil bones, we ascertained, by means of chemical experiments, that they contained a much less quantity of animal matter than ordinary bones, such as have lain exposed to the atmosphere and rain for some time. They likewise were found to have received into their texture a portion of carbonate of lime, where they had been so situated as to have derived it from the drippings of the cave. These circumstances render them so extremely dry, that they imbibe moisture with great rapidity, and to a great extent, so that if the lips are applied to them, they adhere tenaciously.

Besides the present, there were and are other caverns of a like kind near Plymouth, some of them adjoining this place. The Oreston Caves were well known to the scientific world; another, similar in its nature to that of Yealm Bridge, was discovered near Yealmton, by some workmen, a few years since, and its contents thoroughly destroyed; a third instance is that of a small cavity investigated very recently by a gentleman, on whose property it occurs. It is not far distant from Yealm Bridge, and its contents were precisely similar, except in variety and quantity, to those mentioned in this paper. The bones of the Hyena and Deer were chiefly remarked. A fourth example is Kitley Cave, amongst the pebbles and rubbish of which we found a tooth of the Hyena, and a portion of the head of a Rabbit or Hare, evidently an extinct kind; and there is also a bone of some quadruped firmly fixed among the diluvial pebbles in that part of the cavern which seems to have been choked up with these bodies:—facts which are at variance with the account of the subject of this memoir given by Captain Mudge, in his paper read before the Geological Society, though we are not prepared to say that this cavern had been submitted to the same circumstances as that at Yealm Bridge. Lastly, there is another cavity of small dimensions, occurring, as do all the others, in lime-rock, situated midway between this village and Yealm Bridge, in which, among the pebbles, brecciæ, and mould, we found, after a long search, a single tooth of a Pig, cemented to a portion of indurated clay.

Yealmton, 23d August 1836. JOHN C. BELLAMY.

INFLUENCE OF LIGHT AND DARKNESS ON THE HUMAN BODY.—Dr Allen, in his work on the influence of the atmosphere on the human frame, mentions the following instances:—Baron Humboldt was acquainted with a lady who, at sunset, invariably lost her voice, and did not again receive it till sunrise. Aristotle mentions the case of an innkeeper, who lost her understanding every evening at sunset, but recovered it next morning. When a person has taken too much wine, he becomes much more conscious of the influence of the wine on his brain when the light is removed. He is then no longer able to stand, and the chair or bed on which he may be seems to him to be revolving with rapidity. When he is again placed under the influence of light, all these phenomena cease. In early life, says a German physician, I made a curious experiment of this kind on myself. At a merry breakfast party, I drank a few glasses too many of Malaga wine. It was not until about twelve hours after that, when, in bed, I extinguished my candle, the effect of the wine on my brain became perceptible. Every thing then seemed to move round me in a circle; heat and a feeling of uneasiness came on, and I found it necessary to spring out of my bed. As soon as light was brought every thing became again stationary, and the disagreeable sensations vanished. During the whole day, while under the influence of light, I had been able to follow my usual avocations, without perceiving the slightest symptoms of an unhealthy over-excitement.

ASPIDIUM DUMETORUM.—Professor Don finds that the *Aspidium dumetorum* of Smith is merely an accidental state or variety of *Aspidium dilatatum*, and states that the distinctions derived from the fructification in the English Flora are fallacious, being dependent partly on the age of the frond, and partly on that of the individual plant described.

EDINBURGH: Published for the PROPRIETOR, at the Office, No. 13, Hill Street. LONDON: SMITH, ELDER, and Co., 65, Cornhill. GLASGOW and the West of Scotland: JOHN SMITH and SON; and JOHN MACLEOD. DUBLIN: GEORGE YOUNG. PARIS: J. B. BALLIERE, Rue de l'École de Médecine, No. 13 bis.

THE EDINBURGH PRINTING COMPANY.

THE EDINBURGH

JOURNAL OF NATURAL HISTORY,

AND OF

THE PHYSICAL SCIENCES.

AUGUST, 1837.

ZOOLOGY.

DESCRIPTION OF THE PLATE.—THE WEEPERS.

THE Weeping Monkeys form a section of the Quadrumana characterized by having four incisors above and below, and a canine tooth and six grinders on each side of either jaw; the head round, the forehead somewhat prominent, the muzzle short, the occiput projecting, the ears rounded; and the tail prehensile, but entirely covered with hair. The species greatly resemble each other in their external form, so as to render it difficult to characterize them; and, accordingly, authors have differed as to what ought to be considered as species or varieties. They live in troops in the forests of South America, residing on the trees, and feeding on fruits and insects.

Fig. 1. THE COMMON WEEPER (*Cebus Apella*). This species, which is found in French Guiana, has the upper parts of a dusky-brown, the lower paler, the top of the head, the feet, and the tail, blackish-brown; the outer side of the arms yellowish-brown.

Fig. 2. THE FEARFUL WEEPER, here represented, is generally considered a dark coloured variety of the Common Weeper.

Fig. 3. THE CAPUCHIN WEEPER (*C. Capucinus*) is characterized by having the body of an olivaceous grey, subject, however, to great variation; the top of the head and the hands black; the forehead, cheeks, and outer side of the arms, yellowish-grey. It inhabits Guiana.

Fig. 4. THE WHITE-THROATED WEEPER is nearly a variety of the Capuchin, of a darker colour, but with the forehead, throat, and shoulders, greyish-white.

Fig. 5. THE BEARDED WEEPER (*C. barbatus*) is so named on account of a circle of long whitish hairs surrounding the face. The general colour is a dull greyish-red, varying, however, to brown or grey.

Fig. 6. THE HORNED WEEPER (*C. fatuellus*) is distinguished by two tufts of long hair on the forehead. Its colour is brown, paler beneath; the extremities blackish-brown. It inhabits Guiana.

DESCRIPTION OF THE PLATE.—THE OWLS.

THE Owls are nocturnal predaceous birds, easily recognised by their peculiar form, although closely allied to the falconine family. They are characterized by the enormous size of their head; their flattened face; their very large eyes, which are directed forwards, so as to enable them to see an object at once with both; their hooked bill and claws; and their peculiarly soft and tufty plumage. Their ears are generally extremely large, and furnished with a flap or operculum, but differ greatly in form and extent, sometimes extending from over the eyes to the throat in a semicircle, and sometimes of a roundish form, as in the Falcons. Many species have tufts on the head, which, although vulgarly called ears or horns, have no relation whatever to these organs. Some Owls are diurnal, pursuing their prey in open day; but by far the greater number are nocturnal or crepuscular. The larger species prey on quadrupeds, birds, and fishes; the smaller chiefly on insects. Squirrels, Rats, Mice, and small birds, are favorite articles of food with those of medium size. Their flight is peculiarly noiseless, buoyant, and wavering; and they are extremely light for their size, the great bulk of their body being made up chiefly of feathers, while that of the head is partly owing to the same cause, and partly to the separation of the two tables of the skull by very large cells, which communicate with the organ of hearing. Owls are found in all parts of the globe. They inhabit forests, woods, and rocky places; breed in rocks, towers, old buildings, out-houses, or trees, according to the species. Their eggs are always white, and more rounded than those of any other birds. This great family has been divided into several genera or sections, respecting the limits of which, however, no two authors are agreed. Cuvier considers them as forming only a single genus, which he divides into groups according to their tufts, the size of their ears, the extent of the circle of feathers surrounding their eyes, and other characters. Those represented in the plate belong to his Chevêches, which have no tufts on the head, and are further characterized by having the ears comparatively small.

Fig. 1. TENGMALM'S OWL (*Strix Tengmalmi*). This species is about twelve inches in length; its upper parts are liver-brown, spotted with white, the tail barred with the same colour, the lower parts yellowish-white. It is common to Europe and North America; is strictly nocturnal; builds in trees, laying two white eggs, and feeds on mice and insects.

Fig. 2. THE GREAT CINEREOUS OWL (*S. cinerea*). This is the largest of the American species; and, according to Dr Richardson, inhabits the woody districts of the fur countries; hunts by day, preys on hares and other animals, and nestles on high trees.

Fig. 3. DALHOUSIE'S OWL (*S. Dalhousiana*). This species, one of the smallest with which we are acquainted, is named in honour of Lady Dalhousie, who presented a specimen of it, along with many other birds, to the Museum of the Edinburgh University. It is a native of Canada.

Fig. 4. THE LITTLE OR PASSERINE OWL (*S. passerina*) is nearly of the same size as Tengmalm's Owl, from which it is distinguished by having the lower parts marked with longitudinal brown spots, as well as by other circumstances. It is common to Europe and North America.

REMARKABLE INSTANCE OF SAGACITY IN A DOG.—M. Alphonse De Candolle being last October in the neighbourhood of Aiguesmortes, had occasion to observe a remarkable instance of sagacity in a Dog. The day was hot, and the season unfavorable, on account of the trade winds, which are productive of so much inconvenience on the shores of the Mediterranean. After walking several hours in the desert which separates the town of Aiguesmortes from Carmogne, he arrived with his companions at a plain, where they found some remains of a shipwreck. Of three Dogs which had followed their guide, two accompanied them to this spot. Their black hair imbibed the heat from the sun's rays, and they seemed to find the sand very unpleasant. M. De Candolle sat down on a mat half buried in the sand. One of the Dogs nestled close to a horizontal plank, by way of procuring a little shade, but finding this insufficient, scraped up the sand until it came to the part moistened by the sea-water. It then stretched itself with delight in this fresh and shady bed. The other Dog, however, though of the same breed, never thought what to do, and writhed in the hot sand.

HAIRS IN THE INTESTINES OF HORSES.—M. Maillet of Alfort has proved by numerous dissections, that hairs exist in some parts of the digestive canal in almost all horses. They are found especially in the pylorus and stomach, and are also frequently seen in the colon and cæcum.

BRITISH BIRDS.—NO. IV.

THE SWIFT, *Cypselus murarius*, arrives in Britain from the 20th of April to the beginning of May. It has been stated by Mr Selby, that "it is seldom seen in the northern parts of England before the end of May, or the beginning of June;" but this appears to be a mistake, for in Edinburgh it always comes before the 5th of May, and even in the very severe weather of 1837, it was seen at Newington and Canonmills on the 3d of that month. It is not in general, however, until after the different species of swallow have made their appearance that it presents itself, a few individuals only being seen at first, and the number gradually increasing until at length they become in many places plentiful, and attract attention by their extremely rapid flight and loud screams. The plumage is perfect at the period of its arrival, and it does not moult during its sojourn in this country.

The general form of the Swift is rather full; the body somewhat depressed, the neck very short; the head broad; the bill extremely small, but expanded at the base; the feet remarkably short, but strong; the tarsus anteriorly feathered; the four toes nearly of equal length, and all directed forwards; the claws very strong and curved; the wings are exceedingly long and sickle-shaped; the tail forked. The bill, feet, and eyes, are black; the colour of the plumage is blackish-brown, generally glossed with greenish, the throat whitish. The length of the male is seven and a half inches; the extent of its wings sixteen and a half.

The Swift betakes itself to steeples, high towers, ruinous castles, and abrupt rocks, where it nestles in the holes and crevices. At early dawn, in fine weather, it is to be seen shooting through the air in all directions, with a rapidity scarcely equalled by that of any other bird. Its flight is performed by quick flaps of its long narrow wings, alternating with long glidings or sailings, during which these organs seem motionless, but extended at a moderately open angle. If you watch an individual, you observe it speeding away with quick motions of its wings, which, being raised and depressed over a great range, seem to alternate with each other, although this is not in reality the case, all birds moving their wings synchronously. There it shoots along, turns to the right and left, flutters for a moment, ascends, comes down abruptly, curves and winds in various directions, darts in among its fellows, and is lost to your view. The ease with which it rises, falls, bends to either side, glides in short or long curves, or stops in the midst of its full career, is less astonishing than it ought to be, familiarity in this, as in other instances, producing a disposition to regard as simple what is the result of elaborate mechanism.

It continues searching the air in this manner all day long, when the weather is good; nor does a shower, however heavy, usually induce it to relinquish its pursuit. Even in the midst of heavy thunder-rains, it may often be seen wheeling and diving with unremitted vigour, and in drizzly weather, when the swallows have disappeared, it pursues its avocations, heedless of the damps. On the day on which the accession of her Majesty was proclaimed in Edinburgh, the weather was extremely sultry and oppressive, and a very heavy rain fell in the afternoon, during which I was a little surprised to see the Swifts wheeling joyously over the town at a considerable height. How the insects, of which they were in pursuit, could exist in such a rain, is not less astonishing. In dry and sunny weather, however, it generally rests in the middle of the day, and towards evening is extremely active, filling the air with its shrill and joyous screams.

Its food consists entirely of insects, which it seizes exclusively on wing. Several curious circumstances may be noted with reference to its pursuit of these animals. In rainy or damp coldish weather, the Swifts are to be found flying, at no great height, generally from ten to fifty or sixty yards, frequently in bands of twenty or more, often shooting along the sides of the hedges, descending in curves, and skimming the surface of the grass, wheeling, circling, and performing all sorts of evolutions. On such occasions, they are easily shot, for they often come quite close to the gunner, being altogether heedless of his presence, so intent are they on capturing their prey.

In fine weather, they fly low in the mornings and evenings, and are among the first birds that come abroad, and the latest in retiring to their places of repose; but during the greater part of the day they are to be seen chiefly at a great elevation, apparently that of several hundred yards. Yet I have seen them flying high in rainy weather, when the clouds were separated by long intervals; and, from long observation, I am satisfied that no prognostication of the weather can be based on the flight of Swifts and Swallows. These birds fly high or low, according as their prey is abundant in the higher parts of the air, or near the surface of the ground or woods; and as insects fly lower in the evening and morning, or in damp weather, so the Swifts then descend.

In dry sunny weather they frequently utter a long loud shrill scream, as they pursue their prey; but not in such weather only, for you often hear it before or during rain, especially in the evening. Some have fancied this scream to be an intimation given by the male to his mate that he is at hand, and others that it is caused by the excitement of electricity; but these conjectures are destitute alike of ingenuity and truth. It is not in thundery weather alone that Swifts scream, but often in the clear, dry, and sunny skies, that exhibit no phenomena indicative of a want of electric equilibrium. And, as to the other theory, it suffices to reflect that Swifts scream as frequently over the open fields, at the distance of a mile or more from their resting places, as when wheeling near steeples or towers. The cry of Jackals, Wolves, and Hounds, when in full chase, seems to be analogous to the scream of Swifts under similar circumstances, but the cause and use of either is not satisfactorily ascertained. I have observed, however, that single birds very seldom scream, and that the loudest and most frequent cries are heard when birds are evidently in active and successful pursuit. It is so with Terns, Gulls, and even Gannets; and when you see these birds hovering over the sea, and hear their mingling cries, you may be sure that they have discovered a shoal of fishes, and are enjoying their good fortune. They seem to scream or cry out from pleasure, and thus give intimation to their fellows of the plentiful existence of food. As to the organ of this loud and shrill scream in the Swift, namely, the trachea, it is short, remarkably flattened, and gradually diminishes in diameter to the bifurcation. It has no song or twitter, like the Swallows.

If we suppose that the Swift is destined to feed exclusively on insects as they flutter in air, which is in fact the case, we can be at no loss to trace the reasou of its peculiar form. Its body is light, but moderately stout, and its pectoral muscles are large, otherwise it could not move its wings with the requisite strength and rapidity. The wings are extremely elongated and narrow, because great rapidity of flight is required in the pursuit of animals which themselves fly with speed, and because sudden turns require to be executed in seizing them. A short, broad, concave wing, as that of a Partridge, on being rapidly moved, produces considerable velocity, but is not fitted for either buoyant gliding or quick evolution. For the latter, the surface of the wing must be extended in length and narrowed, and instead of presenting a concavity; must be straight in the horizontal direction. Accordingly, in the Swift, the wing has its humeral articulation peculiarly free, insomuch that, holding one alive in your hand, you at first imagine that its wings have been broken. At the same time, their muscular apparatus is remarkably strong. Then the secondary quills are very short, and the primaries gradually and rapidly elongated, and furnished with very strong, but highly elastic shafts. The tail, although not so long, is similarly constructed, being deeply forked, and so in a manner divided into two pointed and elongated laminæ, similar in some degree to the wings, and aiding their action in executing turns. In seizing its prey, while gliding or fluttering in the air, the bird would be incommoded by any length of neck; that part is, therefore, extremely abbreviated, so that the head seems as if stuck upon the shoulders, as is the case, for a similar reason, in the Cetacea and fishes. A long pointed bill would be of use only to a bird that has objects to pick from the ground or any other surface, or from among soil or foliage. In the present case, the bird, carried with rapidity to its tiny prey, merely requires to open its mouth, which is extremely enlarged, and supplied with an abundant viscid secretion, which immediately entangles the fly that has been caught, and prevents its escape, should the month be opened the next instant. A bird so living has no need of walking, and there being nothing superfluous in nature, its feet are reduced to cramping organs, by which it can cling to any kind of surface when entering its nest, and its gait is merely a hobbling motion, aided by the wings. It cannot rise from a flat surface, as I have ascertained by experiment, but it launches from any little eminence, and if it can spring out horizontally is enabled to fly off, although its usual mode of launching is, like that of the Gannet, by a deep curve.

These two birds are very similar in some points of their organization. Their wings are long and narrow, and their flight is rapid and buoyant; they seize their prey by throwing themselves with velocity upon it; they launch from the rocks in the same manner; and exhibit other points of mutual resemblance; as do the Terns more especially, which, on account of their form and buoyant flight, have received the vulgar appellation of Sea Swallows.

The want of walking feet might be supposed to be somewhat inconvenient on many occasions. Thus, when the bird has its nest to make, it must gather straws and feathers; but so great is its dexterity on wing, that it picks them up with ease as it sweeps along. The nest is placed in the crevice of a wall or rock, in a steeple or tower, in holes under the eaves, or in some such place, at as great a height as possible, and is composed of twigs, straws, and feathers, being bulky, but shallow, and not neatly arranged. The eggs are two or three, of an elongated form, pure white, their average length one inch, their greatest breadth seven and a half twelfths. They are deposited from the beginning to the middle of June, and the young are abroad by the end of July. Only one brood is reared in the season. The Swifts take their departure from the middle to the end of August, thus residing with us only three months and a half.

As the insects on which they live are generally very small, they do not swallow each as it is caught, but collect a number previously to the act of deglutition, for at whatever period they are shot, one generally finds insects in their mouth. When collecting food for their young, they do not return to the nest so frequently as the Swallows, but accumulate a considerable quantity at a time. I have never found any particles of gravel or sand in their gizzards, of which the hard cuticular lining is of a reddish-brown colour, as in most birds that feed on insects, such as Wagtails, Pipits, and Warblers. The insects on which they feed are numerous species of Coleoptera, Ephemeræ, Phryganeæ, and occasionally Libellulæ and Muscæ.

It has been conjectured by some that the viscous saliva of the Swallows is used for agglutinating the pellets of which the outer crust of their nests is composed. I have carefully examined these mud crusts of the Chimney and Window Swallows, and can find in them no appearance of cement. The Sand Swallow, which has no such crust to its nest, has an equally copious viscid saliva. The same is the case with the Swift. But the absurd fancies of ornithological writers are so numerous and so palpable, that it is tiresome to refute them.

The young Swifts are of a dusky colour, at first blind, and almost naked, having merely a few straggling tufts of down. When fully fledged, they are of the same colour as the adults, but of a lighter tint, with the edges of the feathers of the head paler.

Previous to their departure, the Swifts do not collect into large flocks, like the Swallows, but disappear gradually, setting out apparently in small parties, in the same manner as that in which they arrived.

Mortality among Birds.—The following curious statement rests upon the authority of a Lausanne Journal. During the last fortnight great numbers of sick and dead Birds, particularly Thrushes, have been found in the fields of Soleure. An inflammation of the spleen is the cause, and the disease is attributed to some acid exhalations from the earth. All the Sparrows and Finches, it is added, have deserted the infected districts, and in several parts of Switzerland domestic animals have been attacked in a similar way.—*Literary Gazette.*

BOTANY.

Remarks on the Mountain-Ash, or Rowan, and the Apple Tree of Russia, by William Howison, M.D., Lecturer on Botany, Edinburgh.

The Mountain-Ash, or Rowan (*Pyrus Aucuparia*), is distributed over the greater part of the Russian Empire, and towards the northern parts attains great perfection, exhibiting a luxuriant display of blossoms and fruit. It occurs in abundance on the outskirts of the forests, as also around noblemen's and gentlemen's parks or policies, where the tree acquires great size. An open situation appears to be absolutely necessary to its prosperity and increase; for when found in the midst of the forests, which, however, is rarely the case, it is always stunted, and falls into the state of brushwood. In all parts of Russia where I have seen it, it is a natural production, propagating itself by the seeds which are accidentally blown about, and is very seldom artificially planted. During the summer and autumn, it enhances the beauty of the forests by its flowers and fruit, and on that account is esteemed by all classes of Russians, as well as foreigners, in the neighbourhood of their habitations. On the summer holidays or fasts, its branches in full flower are used to ornament the windows of the village Isbac. In a country like Russia, where timber of the finest kinds can be procured in such abundance, and at a trifling price, the wood of the Mountain-Ash, on account of its softness and want of durability, is not applied to any particular use. At times it presents itself amongst the billets of birch and fir sent down from the interior, when, of course, it is consumed along with them as firewood. This tree, however, is by no means destitute of value; for besides affording food to man, it furnishes subsistence, during a considerable part of the year, to the feathered inhabitants of the forests.

The ripe fruit of the Mountain-Ash is used extensively, in a variety of ways, as an article of food. The first application of it to this purpose, which I shall mention, is to form a liquor, which is much esteemed as a stomachic, an agreeable bitter, and for diffusing a glow over the system during the winter months. To make this, take a small cask, two-thirds full of the ripe berries, picked and cleaned; fill it with strong spirits, and allow it to stand in a cold cellar for twelve months. Then run off the spirit, which has become completely impregnated with the colour and flavour of the fruit, and comes away perfectly pure, the macerated berries remaining at the bottom. The spirit or tincture is then bottled. The boors or lower class make use of the Watky, or common fermented spirits of the country, for the above purpose.

while the nobility, the higher orders of the people, and foreigners, employ gin, brandy, rum, or other spirits. A glass or two of this liquor is taken each forenoon during the winter months, and it generally makes its appearance at lunch.

The second application of the fruit of the Rowan Tree as an article of food, is in the form of jelly, jam, or preserve. To make the jelly, put the berries, when ripe and cleanly picked, into a large jar, which is to be placed either in an oven, or in a saucepan of boiling water, until they part with their juice. Strain through a fine sieve, but do not press the berries; weigh the juice, and add to it an equal weight of loaf sugar; boil them together until they acquire a proper consistence. Rowan jelly thus made has a pleasant, slightly bitter taste, and in appearance resembles that made from red currants. It is eaten in considerable quantity with partridges, the different varieties of wild fowl, &c., which are to be had in any quantity at a trifling price, and constitute a daily dish. The jam is made in the same manner as that of the gooseberry, or any other species of fruit, and forms a good remedy in stomach complaints and sore throats. Lastly, the berries, towards the end of the season, when ripe, are collected in great quantities by the boors, for their own consumption, and for that of the nobles on whose estate they live; and are salted, along with various sorts of wild berries, and preserved amongst their winter store in the ice-cellars. During the winter the berries thus kept form a part of their daily meals, and are reckoned antiscorbutic. Far distant, I am afraid, is the period when the fruit of the Mountain-Ash will be applied to any useful purpose by the peasantry of Great Britain, although thus relished and sought after by the more opulent and better fed boors of the interior of Russia.

THE APPLE TREE, *Pyrus Malus*, in all its varieties, attains a state of great perfection, and is found in abundance in the Russian Empire, particularly towards the southern parts, the Crimea and the Ukraine. It presents a healthy and luxuriant appearance, and yields abundant crops. The fruit markets in the great towns, during the autumnal months, are amply supplied with the finest apples of every description; they are handed in abundance through the streets by the itinerant fruit-dealers, and are sold at a cheap rate. They generally attain a great size, and ripen perfectly, as is indicated by the richness of their colour and odour. During the course of the winter they may be had to purchase in the grocery shops in their preserved state.

When the apples are intended to be preserved in their fresh state, those which are in the best condition are carefully picked out and well cleaned by means of a dry cloth. They are then put up in small heaps, or are arranged in regular rows, on the wooden floor of the ice-cellar, with which every Isbac is supplied, for keeping their various sorts of provisions in a fresh state, during the winter and summer months. At regular intervals, any moisture that may have collected upon them is carefully rubbed off; they are frequently turned; those which present the slightest appearance of spoiling are removed for immediate use; and the rest are found to keep in good condition from November until the next crop is ready for use.

I was informed while in Russia that apples may be preserved fresh for a considerable period by another method, which, although it has not been my good fortune to see it practised, I shall now describe, being convinced that it is worthy of a fair trial. A quantity of tight well-made casks or barrels, sufficient to contain the fruit to be preserved, must first be procured. The apples, after being picked out and carefully rubbed with a dry cloth, are to be placed in layers, alternating with others of cranberries, until the casks are completely filled. Cold spring water is then poured in so as to fill up every interval, and, when the process is completed, each cask is secured in a perfect manner by fixing its top on.

Whether the cranberry possesses any preservative power, whether any other kind of small fruit would answer the same purpose, or whether their use might be omitted, I cannot take upon me to determine. I feel inclined, however, to think that it does not exert any such power, and that the preservative or antiseptic quality evinced depends entirely on the water and the consequent exclusion of the air. It is a well known circumstance, that apples which have fallen from the trees in orchards or gardens, and have accidentally got buried under the surface of the earth during the autumn months, have been dug up the following spring or summer in a fresh state; and this, of course, arises from their submersion and consequent protection from the air.

Apples are preserved in great quantity throughout the Russian Empire by another method; they are kept in a dry state. For this purpose, the inferior sorts are generally made use of. They are cut through the apples from top to bottom, into four, six, or eight divisions, according to their size; the seeds with the capsule being generally removed, as is done in this country in preparing apples for pies. The various pieces collected together are dried in the sun, if the heat of the season is sufficient for the purpose; if not, by means of fires; after which they are strung upon twine, formed into bunches, and will remain in good condition for any reasonable length of time. When apples dried in this way are moistened by putting them in warm water at any after period, they regain their natural spongy appearance, and are used by the peasants to make pies during the winter months, and to form a particular kind of drink much consumed in Little Russia.

Preserved in the manner here described, apples brought from the Crimea, Ukraine, or south of Russia, may be had to purchase all the year in the shops of the great towns. In the Island of Mohn, situated at the head of the Baltic, and belonging to Russia, the crab apple only abounds, and of it the boors make a tolerably well tasted cyder.

NEW ESCULENT SEA-WEED.—At a recent meeting of the Medico-Botanical Society, a paper was read by Dr Sigmond, respecting a new esculent sea-weed, which possesses nutritious properties to a much greater extent than Iceland moss, without any of the bitter principle which renders that plant so disagreeable to many. This fucus, of which specimens have lately been brought from Calcutta, is said to contain 54 parts of starch in 100, and to be employed in large quantities by the Chinese, who form with it an agreeable and refreshing jelly. It is abundant in the neighbourhood of Ceylon, and has been much employed by the medical profession in Calcutta. The jelly is said to be quite equal to *blanc mange*.

THE TOBACCO PLANT, *Nicotiana Tabacum*, belongs to *Pentandria Monogynia* of the Linnean system, and to the natural family of *Solaneæ*, or Nightshade Tribe, which are generally narcotic or poisonous, although the Potato, *Solanum tuberosum*, whose farinaceous tubers now afford so important an article of food, ranks among them. The root of the common Tobacco is large, fibrous, and of annual duration: the stem erect, round, hairy towards the upper part, branched, and attains a height of from four to six feet; the leaves alternate, large, ovate, pointed, veined, and clammy; the flowers are disposed in terminal panicles; the calyx is monosepalous, bell-shaped, with five deep-pointed segments; the corolla monopetalous, somewhat funnel-shaped, with a long cylindrical tube, and five acute revolute segments; the stamens five, with in-curved tapering filaments and oblong anthers; the germen oval, with a long slender style, and a round cleft stigma; and the capsule oval, two-celled, containing numerous small roundish seeds.

Tobacco, which is the dried leaf of Nicotiana Tabacum, as well as of several other species of the same genus, was first imported from America into Europe, about the middle of the sixteenth century, by Hernandez de Toledo, who, however, had not the honour of giving it its scientific appellation, which was derived from a Frenchman, Jean Nicot, Lord of Villemain, who being ambassador from Francis II. at the court of Lisbon, sent some to his queen, Catherine de Medicis. The custom of smoking is said to have been introduced into England by Sir Walter Raleigh, who found tobacco cultivated in Trinidad on his first visit to it in 1593. It is believed that the first time the Spaniards saw it smoked was in 1518, at an interview between Grijalva and a cacique of Tabasco, or Tabaco, from which place the plant received its popular name. In spite of the strenuous opposition which was made to its introduction by the princes of Europe, it gradually gained ground, and is now in general use, being extensively cultivated in both hemispheres. The importation of tobacco and snuff into Great Britain, in the course of a recent year, amounted to nearly 17,000 hogsheads. It is related of M. Fagon, physician to Louis XIV., that in the midst of a most energetic speech on the pernicious effects of tobacco, he paused, and deliberately taking his snuff-box from his pocket, refreshed himself with a pinch, to enable him to resume the argument; it may be imagined with how little effect. Another curious fact is related with reference to its cultivation in Virginia, where it was introduced in 1616, under the government of Sir Thomas Dale. The planters at that time being all bachelors, regarded themselves as merely temporary sojourners; but the London Company for the colonization of Virginia sent out a number of respectable young women as wives to the settlers. These ladies were actually sold for a hundred and twenty pounds of tobacco each, that quantity being equivalent to the expenses of the voyage.

Tobacco is extensively cultivated in Cuba, St Domingo, Trinidad, Brazil, Virginia, the Cape of Good Hope, and India. In the United States of America it is raised in the following manner:—The seed is sown in February and March, in beds carefully prepared. In April the young plants are laid out in the fields in rows, at the distance of three feet from each other. Besides being hoed and weeded, they are deprived of the lateral shoots and the tops in order to increase the size of the leaves. When the plants have attained their full size, and the leaves begin to assume a brownish hue, they are cut close to the ground, and left in heaps exposed to the sun for one day. They are then carried to a shed and hung up in pairs on ropes, where they remain until dry, when the leaves are separated and made up into small bundles, which are laid in heaps and covered with a cloth, to favor a slight fermentation, care being taken to prevent occasional exposure to prevent too great heat. They are finally packed in casks for exportation, and undergo a second fermentation, which gives them a dark colour and soft texture.

Along with its narcotic properties, tobacco has a slightly stimulant action, more especially with respect to the digestive organs, on which account it has in certain cases been employed as an emetic and purgative. In obstinate constipation and strangulated hernia, its smoke has sometimes been injected with advantage. But its employment as a medicine is unfrequent, on account of its violent effects, and the uncertainty of its operation, which cannot be restrained within the desired limits. The use of tobacco as a narcotic luxury is now very general in most countries, it being alike relished by the most civilized as by the most savage nations, and by the higher ranks in the former as by the lower. At its first application, it invariably produces powerful narcotic effects, such as giddiness, stupor, and vomiting, and in some cases it has proved a mortal poison. But as, like other narcotics, its effects gradually disappear on its being repeatedly used, although in irritable constitutions it is apt to produce deleterious effects; and it has been remarked that persons who take much snuff, though they seem generally to escape its narcotic power, yet are often affected in the same manner as from the continued use of opium or spirits, and exhibit a loss of memory, and other symptoms of a debilitated state of the nervous system. Dyspepsia and pains in the stomach are also caused by excess in snuffing, as well as in smoking or chewing tobacco. On account of its stimulating property, snuff, by exciting a flow of mucus from the nose, has sometimes been useful in relieving headaches, ophthalmias, and toothaches; but, on the other hand, a ceasing from snuff is apt to give rise to those very disorders, and an inordinate use of it also sometimes produces them. Snuffing at first causes violent sneezing, then an increased flow of mucus, then an agreeable sensation approaching to giddiness, and lastly headache. Smoking and chewing give rise to a copious flow of saliva, then a sensation of giddiness, followed by sickness, vomiting, and headache. These effects, however, are by long use much diminished, although always liable to be induced by excess. Many people seem to be unaware of the real effects of tobacco, and to suppose that it is used merely as a custom, but it is simply on account of its narcotic effect that it is employed. These effects are most powerfully produced by chewing, next by smoking, and in a less degree by snuffing, which, however, is probably the most injurious way of taking tobacco, as it keeps up the excitement continuously, and occasionally produces violent effects. Few people addicted to the use of tobacco are willing to acknowledge its effects. The fact is, they are ashamed to own that they are little better than habitual drunkards. In all ordinary cases tobacco is much more injurious than useful: it causes a great waste of saliva, passes occasionally into the stomach, producing sickness and dyspepsia, stimulates the brain, and causes giddiness, confusion of thought, and impaired memory. In a certain dose it deadens sensibility, and by producing a kind of vertigo or drunkenness, alleviates for a time distressing care. Like wine, ale, porter, and spirits of all kinds, tobacco is productive of much more evil than good; and although its action is apparently less violent than that of spirits, yet it is always deleterious. It is in fact a poison operating slowly but surely on most constitutions, and tending to accelerate the fatuity to which many individuals are otherwise liable in advanced age. Most persons addicted to the use of tobacco are equally fond of the bottle, and the combined use of both produces that most disgusting character, a sot. It is a curious fact, however, that many addicted to both are just the persons who are loudest in denouncing them.

GEOLOGY.

ORGANIC REMAINS IN THE COAL FORMATION AT WARDIE, NEAR NEWHAVEN.—Dr Paterson, in a paper read to the Wernerian Society, and published in the Edinburgh New Philosophical Journal, gives an interesting account of the fossil plants and animals discovered in the strata of slate-clay, bituminous shale, sandstone, and clay-ironstone, exposed on the shore near Newhaven. The plants generally occur in the form of impressions, and are referred to the following species:—*Sphenopteris affinis, S. erythmifolia, S. artemisiafolia, S. furcata, S. elegans, S. Hoeninghausi; Cyclopteris obliqua, C. flabellata, C. trichomanoides, C. reniformis;* undetermined remains of *Calamites; Lepidodendron elegans, L. Sternbergii, L. ramosum, L. aculeatum, L. abovatum, L. appendiculatum, L. selaginoides, L. lycopodioides; Lepidostrobus variabilis, L. ornatus.* "It is now generally admitted," says the author, "that these impressions (*the Lepidostrobi*) must have been made by the reproductive organs of plants, similar in form to the same parts in recent Coniferæ and Lycopodiaceæ. M. Brongniart, from their frequent connection with Lepidodendra, maintains that they belong to that class of plants. *Lepidophylla*, of various sizes, are very common throughout the shale in this place; and frequently attached to them is a small seed-like body, which, when compared with the division of *Lepidostrobus ornatus*, is at once seen to be the same. From this fact, we imagine that Messrs Lindley and Hutton's conjecture of Lepidophylla being referred to some species of Lepidostrobus is correct, and that they are merely the scales or bracteæ of this class of plants. How beautiful, then, must these ancestral members of our vegetation have been, when Lepidodendra waved their luxuriant branches, crowned with Lepidostrobi; and these last being imbricated with Lepidophylla, the whole bearing not a distant resemblance to some of the Coniferæ of the present day!" *Polyporites Bowmannii, Knorria taxina, Sphæria paradoxa, Poacites cocoina, Antholites Pitcairniæ,* undetermined species of Bechera, and *Fucoides Targionii,* "sum up the list of vegetables" hitherto identified by the author.

The fishes discovered are—*Amblypterus striatus, A. nemopterus, A. punctatus; Palæoniscus striolatus; Eurynotus fimbriatus; Acanthodes sulcatus,* and a species of *Pygopterus.* "Coprolites or fœcal balls abound in the slate and clay-ironstone; and here, as in the different coalfields in the middle district of Scotland, contain scales and teeth of fishes."

ELEVATION OF BEACHES BY TIDES.—If the earth were a spheroid of revolution, covered by one uniform ocean, two great tidal waves would follow each other round the globe at a distance of twelve hours. Suppose several high narrow stripes of land were now to encircle the globe, passing through the opposite poles, and dividing the earth's surface into several great unequal oceans, a separate tide would be raised in each. When the tidal wave had reached the farthest shore in one of them, conceive the causes that produce it to cease. Then the wave thus raised would recede to the opposite shore, and continue to oscillate until destroyed by the friction of its bed. But if, instead of ceasing to act, the causes which produce the tide were to reappear at the opposite shore of the ocean, at the very moment when the reflected tide had returned to the place of its origin; then the second tide would act in augmentation of the first, and, if this continued, tides of great height might be produced for ages. The result might be, that the narrow ridge dividing the adjacent oceans would be broken through, and the tidal wave traverse a broader tract than in the former ocean. Let us imagine the new ocean to be just so much broader than the old, that the reflected tide would return to the origin of the tidal movement half a tide later than before; then, instead of two superimposed tides, we should have a tide arising from the subtraction of one from the other. The alterations of the height of the tides on shores so circumstanced might be very small, and this might again continue for ages; thus causing beaches to be raised at very different elevations, without any real alteration in the level either of the sea or land.

If we consider the superposition of derivative tides, similar effects might be found to result; and it deserves inquiry, whether it may not be possible to account for some remarkable and well attested phenomena by such means.

The gradual elevation, during the past century, of one portion of the Swedish coast above the Baltic, is a recognised fact, and has lately been verified by Mr Lyell. It is not probable, from the form and position of that sea, that two tides should reach it distant by exactly half the interval of a tide, and thus produce a very small tide; nor is it likely that, by the gradual but slow erosion of the longer channel, one tide should almost imperceptibly advance upon the other; but it becomes an interesting question to examine whether, in other places, under such peculiar circumstances, it might not be possible that a series of observations of the heights of tides at two distinct periods might give a different position for the mean level of the sea at places so situated.

If we conceive two tides to meet at any point, one of which is twelve hours later than the other, the elevation of the waters will arise from the joint influence of both. Let us suppose, that, from the abrasion of the channel, the later tide arrives each time one hundredth of a second earlier than before. After about 3150 years, the high water of the earlier tide will coincide in point of time with the low water of the later tide; and the difference of height between high and low water will be equal to the difference of the height of the two tides, instead of to their sum, as it was at the first epoch.

If, in such circumstances, the two tides were nearly equal in magnitude, it might happen that, on a coast so circumstanced, there would, at one time, be scarcely any perceptible tide, and yet 3000 years after, the tide might rise thirty to forty feet, or even higher; and this would happen without any change of relative height in the land and water during the intervening time. Possibly this view of the effects which may arise, either from the wearing down of channels, or the filling up of seas through which tides pass, may be applied to explain some of the phenomena of raised beaches, which are of frequent occurrence.—*Babbage's Bridgewater Treatise.*

MISCELLANEOUS.

ST ELMO'S FIRE SEEN IN ORKNEY.—The following account of this phenomenon, extracted from a letter by William Traill, Esq., Kirkwall, is contained in the last number of the Edinburgh New Philosophical Magazine. On Sunday the 19th February last, in a tremendous gale, my large boat sunk, and it was late on Tuesday night before we could get her up and drawn to the shore, after which we had to wait till three o'clock next morning till the tide ebbed from her. She was during this time attached to the shore by an iron chain, about 30 fathoms long, which did not touch the water, when, to my astonishment, I beheld a sheet of blood-red flame, extending along the shore for about 30 fathoms broad and 100 fathoms long, commencing at the chain and stretching along the shore and sea in the direction of the shore, which was E.S.E., the wind being N.N.W. at the time. The flame remained about ten seconds, and occurred four times in about two minutes. Whilst I was wondering not a little, the boatmen, who, to the number of twenty-five or thirty, were sheltering themselves from the weather, came running down apparently alarmed, and asked me if I had ever seen anything like this before. I was about to reply, when I observed their eyes directed upwards, and found they were attracted by a most splendid appearance at the boat. The whole mast was illuminated, and from the iron spike at the summit, a flame of one foot long was pointed to the N.N.W., from which a thunder-cloud was rapidly coming. The cloud approached, which was accompanied by thunder and hail; the flame increased, and followed the course of the cloud till it was immediately above, when it arrived at the length of nearly three feet, after which it rapidly diminished, still pointing to the cloud, as it was borne rapidly on to S.S.E. The whole lasted about four minutes, and had a most splendid appearance.

GEDRITE, A NEW MINERAL.—This mineral was discovered by Count D'Archiea, in the valley of Heas, near Gedre. It scratches glass readily; is scratched by quartz; sp. gr. 3.260. It is very tenacious, and receives the mark of a hammer; colour clove-brown, lustre semi-metallic. Before the blowpipe it fuses into a black enamel, slightly scoriaceous. With borax it fuses into a nearly black glass. It approaches in appearance to anthophyllite. According to Dufrenoy, it consists of silica 38.811, alumina 9.309, protoxide of iron 45.834, magnesia 4.130, lime 0.666, water 2.301, answering to the formula, $5 f. S^2 + M A^2 + A^9$. This approaches to the formula of bronzite.

NEW SPECIES OF RHEA.—It is stated in the Magazine of Zoology and Botany, that Mr Darwin has brought home, among other zoological treasures, specimens of a new species of Rhea, which appears to occupy in Patagonia the place of the formerly known species, from which it is distinguished by being about a fifth less, and by having the tarsi reticulated and feathered below the ancle, or what is commonly called the knee-joint.

EDINBURGH: Published for the PROPRIETOR, at the Office, No. 13, Hill Street. LONDON: SMITH, ELDER, and Co., 65, Cornhill. GLASGOW and the West of Scotland: JOHN SMITH and SON; and JOHN MACLEOD. DUBLIN: GEORGE YOUNG. PARIS: J. B. BALLIERE, Rue de l'Ecole de Médecine, No. 13 bis.

THE EDINBURGH PRINTING COMPANY.

THE EDINBURGH

JOURNAL OF NATURAL HISTORY,

AND OF

THE PHYSICAL SCIENCES.

SEPTEMBER, 1837.

ZOOLOGY.

DESCRIPTION OF THE PLATE.—THE GALEOPITHECI OR FLYING-CATS.

THE Galeopitheci form a small genus of the family of Cheiroptera, which is the first of the order Carnivora, and are especially remarkable for having the skin of the sides extended into a kind of hairy membrane, which stretches from the head to the tail, including the extremities or limbs. Their body is rather elongated and slender, the head small and pointed, the legs strong, the fingers furnished with large hooked claws, the ears short and rather rounded, the tail of moderate length. They are nocturnal animals, live chiefly on fruits, suspend themselves to trees by the hind feet in the same manner as Bats, climb with great agility, and descend from branch to branch with the aid of their expanded membranes. Their native country is the Indian Archipelago, but their manners are very imperfectly known.

Fig. 1. THE COMMON FLYING-CAT (*G. volans*) is the only species of which we have a distinct knowledge. It is about the size of a domestic Cat, light red above and below, inhabits the Moluccas and other Islands of the Indian seas, climbs on trees, and emits a disagreeable odour, like that of the Fox.

Fig. 2. Female. The female is similar to the male, but variegated with grey on the upper parts.

Fig. 3. Young. The young is brownish, variegated with grey and pale reddish, the extremities spotted with whitish.

DESCRIPTION OF THE PLATE.—THE BEE-EATERS. MEROPS.

THE genus Merops is composed of a considerable number of species, remarkable for their vivid colours, slender body, long pointed wings, elongated, tapering, and arched bill, and small feet, of which the anterior toes are united for a great part of their length. They are generally distributed in the warm and temperate regions of the old continent, live on insects, which they pursue in open flight, in the manner of Swallows, and form their nests in holes in the ground, like the Kingfishers, to which they are otherwise allied.

Fig. 1. THE JAVANESE BEE-EATER (*M. Javanicus*) inhabits Java and other Islands of the Indian Archipelago. It is light green above, with a lilac patch on the throat, the ear-coverts black, the secondary quills, tail, and rump, light blue.

Fig. 2. THE SUPERB BEE-EATER (*M. superbus*) is of a beautiful deep red, with the head and rump light blue.

Fig. 3. THE RED-HEADED BEE-EATER (*M. erythrocephalus*). Green above, reddish white beneath, the throat yellow, the upper part of the head bright red. A native of India.

Fig. 4. THE INDIAN BEE-EATER (*M. viridis*). Light green, with the throat blue. It occurs in various parts of India.

Fig. 5. THE RED-WINGED BEE-EATER (*M. erythropterus*). Green above, the throat yellow, a crimson patch on the breast.

Fig. 6. THE BLUE-HEADED BEE-EATER (*M. cæruleocephalus*). Of a beautiful rose colour, the head blue, tinged with green.

NEST OF THE KINGFISHER.—A friend of mine, says R. P. Allington, Swinhope House, Lincolnshire, while fishing in a small trout stream, near Louth, called the Drake, in the early part of June, observed a Kingfisher, with a fish in its mouth, flying several times near his hat with a whirring noise. He watched it until it entered a hole in the bank, the entrance to which was strewed with fish bones. On digging into the hole (which commenced low down in the bank, and ran upwards in a slanting direction for about two feet), he found the nest, containing seven young birds just hatched. The bottom of the nest was excessively thick, and mixed up with small bones of the Stickleback. Its structure, excepting the mixture of fish bones, was not very unlike that of a Thrush. It crumbled to pieces on being touched, and could procure no portion worth preserving. Near the nest was another hole, which had all the appearance of having been the Kingfisher's last year's residence, the bones at the entrance being dry and crumbly; but in this the parent bird again commenced laying, and on opening the nest six eggs were found on the fragments of the structure. They were white and beautifully transparent, showing the yolk through, which gave them an inkish hue at the larger end. I have now in my collection one of the eggs, which, though so transparent, I was surprised to find thicker and stronger than the generality of eggs, and rounder in its form, the circumference being two inches and half, the length eight tenths of an inch.—*Naturalist*, No. 11.

THE PEREGRINE FALCON, *Falco peregrinus*, is the largest British species of its genus, except the Iceland or Jer Falcon, which, however, is of extremely rare occurrence with us, its favorite abode being the arctic regions of both continents. In the olden times it was held in the highest estimation for falconry, for although the species just mentioned was considered superior, the Peregrine was much more easily obtained, and being possessed of great docility, courage, and energy, afforded prime sport to the nobles, who alone were permitted to foster there " generous " birds, which were preserved by legislative enactments. As in other species of this family, the female is much superior in size to the male, and was used for the larger sorts of game, as Herons, Geese and Ducks, while the latter, usually named a Tiercelet, on account of its being often a third less, was flown at Grouse and Partridges. The male generally measures 16 inches in length, while the space between the tips of its extended wings is 36. The corresponding dimensions of the female are 22 and 45 inches.

In form this species is full and robust, its neck rather short, its head large and round. The bill is shortish, thick, and strong, the upper mandible with a trigonal, acute, hooked point, and a strong process or tooth on either side. The legs are robust and short, the tarsi feathered more than half-way down; the toes strong, the second and fourth nearly equal, the first shortest, the third very long; the claws long, tapering, rather compressed, and very acute. The plumage is compact on the upper parts, softer on the lower; the wings very long, pointed, and when closed, reaching to within an inch of the end of the tail, which is slightly rounded. The bill is pale blue, with a bluish-black tip, the cere oil-green, the feet yellow, the claws black. The general colour of the upper parts is dark bluish-grey, the head blackish, the lower parts white, the breast transversely spotted with dusky. In the female the upper parts are tinged with red, the lower are yellowish-white, with larger spots. In young birds the back is blackish-brown, the breast pale yellowish-red, with broad longitudinal spots. In all stages there is a large dark-brown or blackish patch on the side of the head or cheek.

The direct flight of the Peregrine Falcon, which is extremely rapid, is performed by quick beats of the half-extended wings, and is very similar to that of the domestic pigeon. When proceeding in haste from its breeding place or roosting station towards a distant part of the country, it very seldom sails, or moves forward at intervals with extended wings; but when sauntering as it were about its retreats, it employs both modes of flight, as it also does in common with many other hawks, when searching for its prey; yet it is hardly ever seen to float along in circles, as Eagles, Buzzards, and Harriers are wont to do, but performs its short gyrations as if in haste, and the moment an opportunity occurs, comes down upon its prey, either in a curved sweep, or like a stone falling from the air.

Its food consists of birds of moderate size, such as Red Grouse, Partridges, Plovers, Ducks, Auks, Guillemots, and small Gulls, which it pursues in open flight, or pounces upon by perpendicular descent. It also sometimes attacks small quadrupeds, such as Rabbits and Hares. Poultry are very rarely molested by this Falcon, which is by no means so ready to visit the farm-yard or its vicinity as several other species of the family. In raising birds from the water, should they prove too heavy, it sometimes drops them, and sets out in quest of others; yet it is capable of carrying a weight nearly equal to its own, and in the nest of one on the Bass Rock was found a Black Grouse, which had been borne from a distance of several miles.

Although a very shy and vigilant bird, it sometimes ventures to come into more immediate proximity to Man than is prudent. Mr Audubon states, that in America he has seen it fly up, at the report of a gun, and carry off a Teal not thirty steps distant from the sportsman who had killed it; and other authors have made mention of similar displays of audacity.

The cry of the Peregrine is loud and shrill, but is seldom heard, excepting during the earlier part of the breeding season. It nestles on high cliffs, especially those along the coast. The eggs, three or four in number, are of a broadly elliptical form, dull light red, spotted and clouded with deep red, their average length being two inches and a twelfth; their greatest breadth an inch and seven-eighths. The young, which are at first covered with close white down, are able to fly by the middle of July.

In Britain it is generally distributed, but prefers mountainous or rocky situations. It has been seen in the Shetland and Orkney Islands, and is occasionally met with in the Hebrides. The rock of Dumbarton Castle formerly afforded a breeding place to it, as did the Isle of May. It still breeds on the Bass Rock, and there are few rocks of great height or extent on the eastern coast of Scotland on which it is not seen. Many pairs breed in the upper part of Moffatdale in Dumfries-shire. Holyhead, the Great Orme's Head, the Rock of Llandedno in Caernarvonshire, various parts of Devonshire and Cornwall, the Isle of Wight, and many other places, are mentioned as favorite resorts. According to authors, it is found over the whole of North America, in the southern division of that continent, in Australia, at the Cape of Good Hope, and in most countries of Europe.

Although a very beautiful bird, the Peregrine, when perched on a rock, or flying across its face, does not present so imposing an object as one might imagine from viewing a stuffed skin in a museum, for the grey colour of its upper parts blends with the tints of the stone, and its sharp wings and rapid flight have the effect of rendering its apparent size much less than that of the Buzzard. Its mode of flying differs greatly from that of the Buzzards and Eagles, and affords a good example of one of the most remarkable varieties. Birds which have the body bulky, and the wings short and rounded, as the Grouse and Partridge, have a direct flight, not generally rapid; those which have the body full, the wings long and sharp, have also a direct steady flight, but generally rapid, as is the case with the Peregrine and the Rock Pigeon. Birds having the body light and the wings short, as the Gallinules, also fly steadily, but with comparatively little speed. When the wings are large and broad, as in the Heron, the flight is sedate, and capable of being long protracted. When they are very long and rather narrow, it is unsteady and undulating, as in some Owls and Gulls; and when the body is very slender, and the wings extremely long, there is produced that unsteady, bounding, gliding, and buoyant flight, so remarkable in the Terns or Sea-swallows.

Besides the Jer Falcon and Peregrine, there occur in Britain four smaller species of this genus, namely, the Hobby, the Merlin, the Kestrel, and the Red-legged Falcon. Of the latter only five or six individuals have hitherto been met with. The Hobby is also very rare; but the Kestrel is, next to the Sparrow Hawk, the most common of our birds of prey; and the Merlin, the smallest of them, is by no means rare in Scotland, where it breeds on the moors and rocky headlands.

THE MERLIN.—In a paper entitled "Contributions to the Natural History of Ireland," by W. Thomson, Esq., in the June number of the Magazine of Zoology and Botany, is the following notice respecting the Merlin:—"On March the 9th, 1832, when walking on the shore of Belfast Bay, as the tide was flowing, a Merlin, which flew past me, was observed for some time coursing above the uncovered banks, the edge of the waves being the limit to its flight. This at once led me to believe that he was in search of prey, which was confirmed by his giving chase to a large flock of Dunlins (*Tringa variabilis*), in pursuit of which he disappeared. From the oldest of the "shore-shooters" in Belfast Bay, I have heard, that frequently, but chiefly in the autumn, he has seen hawks, which, from his description, were considered to be the Merlin, follow and kill Dunlins on the banks at low water. This the above circumstance, witnessed by myself, tends to corroborate. I am not aware that the Merlins thus resorting to the sea-shore have been before noticed. The weather was mild in such instances." In a work entitled "Descriptions of the Rapacious Birds of Great Britain," by W. MacGillivray, published in 1836, Mr Thomson will find the following paragraph:—"In September 1832, I had an opportunity of observing a Merlin in pursuit of a Sanderling. It came up as I was shooting along the shore at Musselburgh, searched about for some time, flying in various directions at a height of an hundred yards or so, until at length spying the bird of which it had probably been in chase before I observed it, it rushed after it. The Sanderling doubled, and endeavoured to escape by skulking among the stones, but the Merlin kept constantly over it, without, however, attempting to seize it when on the ground. The chase continued for some time, when I observed that the Merlin had perched upon a stone; when, thinking that it had secured its prey, I endeavoured to get within shot of it. On my going up to it, however, it rose, and soon after the Sanderling made its appearance, apparently undecided as to the course it ought to take, on which the Merlin swept towards it in a deep curve and seized it. But now the Hawk became in its turn the pursued party, and on its alighting to feast near one of the bathing-machines, I ran up and shot it from behind the cover."

LARVÆ IN THE HUMAN STOMACH AND INTESTINES.—The following report by Messrs Dumeril and Blainville is the substance of an Essay presented by M. Robineau Desvoidy to the French Academy of Sciences. A woman, aged 57, was affected with dropsy of the abdomen after a mucous fever. On the 3d March 1836, there were administered to her six drops of the oil of *Croton Tiglium*, in three equal doses, each at the interval of an hour. Among the matters ejected by vomiting, four living caterpillars were first observed, and afterwards ten more of the same kind. Two of them were enveloped in paper by the medical attendant, who carried them with him, but afterwards lost them. The rest were sent to one of his friends, who transmitted them to M. Robineau. They were preserved in alcohol, and four of them were submitted to the examination of the reporters.

M. Robineau has accurately determined the similarity of these caterpillars to those of a species of moth designated by Linnæus as the *Pyralis pinguinalis*, but of which the generic name was changed by Fabricius to *Crambus*, and again by Latreille to *Aglossa*, on account of the shortness of the tongue. The history of the development of these insects being well known, M. Robineau supposes that the origin of the larvæ in question ought to be attributed to the deposition of eggs by the parent moth in some fat alimentary substances, and their subsequent introduction into the stomach, where they were hatched and developed.

Being of opinion that *Aglossa pinguinalis* does not live in society, he considers these worms as of a new species, to which he gives the name of *intestinalis*. The reporters, however, are satisfied that they are of the first species, and that the author has probably been deceived by supposing that these caterpillars must have been developed in the duodenum, an opinion which he rests only on the report of the physician, who stated that there was a fixed and excruciating pain in the region corresponding to that part of the intestine. M. Robineau does not believe that the presence of these insects had determined the nature of the disease, but thinks that the mucous fever was a circumstance which favored their development. Supposing that these larvæ were those of a distinct species, living only in society and in the intestinal canal, the author asks by what signs their presence may be discovered and their evacuation procured; but expresses his opinion that in the case in question their expulsion was effected solely by chance.

The reporters then state that the memoir is interesting in a pathological point of view, as it affords an authentic instance of living caterpillars existing in the alimentary canal, and give an account of the notices respecting the species in question, which they have found in various works.

In the Stockholm Transactions for 1837 is a memoir by Rolander on the species of *Pyralis*, of which he has given a full account, accompanied with a plate illustrative of its successive changes. He describes the organization of the caterpillar, and in particular that of its stigmata, which permit it to live a long time and respire in the midst of fat substances, such as butter, suet, lard, and soup; and declares that he has very frequently found them in food, such as broth, cabbage, and pease.

Linnæus, in the second edition of the Fauna Suecica, published in 1761, after repeating Rolander's observation, adds: This caterpillar is very injurious in the human stomach, and may be expelled by an infusion of *Lichen cumatilis*. In Gmelin's edition it is said to be common in houses and kitchens, less frequent in the human stomach, and of all animals that live in others the most pernicious.

De Geer, in his Memoirs, has described and figured the same insect. Lastly, the synonyms have been accurately given by M. Duponchel, who has himself exhibited a perfect figure of it, in his great work on the Lepidoptera of France.

In conclusion, the reporters think that the caterpillars of *Aglossa pinguinalis*, may be introduced from without, along with fat or oily food into the cavity of the human stomach; that they have the faculty of remaining alive there for a certain time; but that this is merely a chance occurrence, and that there is no reason for believing these larvæ to be parasitic.

THE TREE PIPIT.—"We have lately *discovered*," says the Editor of the Naturalist, "that the Tree Pipit (*Anthus arboreus*) is in the constant habit of wagging its tail slowly up and down when perched. This circumstance, which appears to have escaped the notice of all previous writers, and is not recorded in our own *Song Birds*, is not only interesting in itself, but valuable in a systematic point of view. It proves the close relationship of the more arboreal Pipits to the Wagtails, which they also resemble in having short crooked hind-claws. We believe the Meadow Pipit does not wag its tail, or if it does, we have not noticed it." All our Pipits wag their tails, but not slowly up and down, and none more conspicuously so than the Meadow and Dusky Pipits, which are not arboreal.

The Meadow Pipit is a very common bird in most parts of Scotland and England, being met with on moors, in pastures, meadow-land, and cultivated fields. In the haunts of the Grey Ptarmigan, on the stony summits of the central Grampians, in the grassy valleys of the Highland streams, in the fertile plains of the south, and on the downs that border the sea, it is equally at home; but it is more abundant in the green pastures that flank the upland glens, and on the sedgy moors of the interior. There it is seen at all seasons, in small companies, flying about in its peculiar wavering manner, and chirping its weak shrill notes. In winter, however, most of the individuals betake themselves to the lower grounds, many to the sea-shore, where they mingle with the Dusky Pipits. During snow, they search the margins of streams and lakes, frequent unfrozen marshes, and even appear in the stack-yards. Their food consists of insects, pupæ, larvæ, and occasionally small seeds, along with which they pick up particles of gravel, and frequently in the lower districts small bits of coal and other dark-coloured substances. When searching for it, they walk by slow alternate steps, keeping the body close to the ground, in the manner of the Sky Lark, and when alarmed either crouch or spring up, uttering a repetition of their ordinary cheeping note, and fly off to a distance. Like the White and Grey Wagtails, with which they occasionally associate, they vibrate their body, although in general in a much less remarkable degree, but conspicuously when perched on an eminence, or when uttering their notes of alarm. You may sometimes see them perch on a bush or tree, frequently on a wall, a stone, or a rock; but they are essentially ground birds, and while they are employed all day in traversing the meadows and pastures, they repose at night among the dry grass of the moors and the hills, or under the shelter of tufts of heath, furze, or other shrubs. Their ordinary flight is wavering and desultory, but when travelling they fly with speed, in an undulating line. They are not generally very shy, so that they are easily shot, but at the same time they are evidently watchful and suspicious, and fly off when one approaches nearer than thirty yards.

NOTICE OF A KIND OF CORAL BELONGING TO THE GENUS CARYOPHYLLEA FOUND OFF PLYMOUTH.—Several months ago, I found on a dead valve of *Pecten opercularis* a specimen of coral differing from any that I had seen; but from its small size and imperfect state I could not then satisfy myself as to its place in any system. Two weeks since I found five other specimens attached to stones taken up by trawlers from deep water, and by these I was enabled to ascertain the genus. A further examination led me to suspect that it ought to be separated from *Caryophyllea cyathus* of Fleming's British Animals (*Madrepora cyathus* of Ellis's Zoophytes). My largest specimens are half an inch high, of a compressed figure, and measuring at the star four-tenths in the largest diameter, and three-tenths in the shortest. They are generally inversely conical, though the smallest are of nearly the same size at their bases as at the stars. One specimen has the star oval, and presents a curved figure from its base to the smaller end of the star. The two small specimens have the star nearly round. They are generally rough on their exterior, from the attachment of *Serpulæ*, *Flustræ*, and in two instances of *Lepas conoides*. All of them present longitudinal striæ on the outside, derived undoubtedly from the three different kinds of lamellæ of the interior. The margin of the star is observed to project in every instance, and the lamellæ are devoid, for a short extent of their depth, of the external crust. The depth of the disk varies greatly, the centre of this part consisting of small convolutions, or twisted plaits, situate on, or rising from, the basis of the fabric. The smallest kind of lamellæ are sometimes deficient, but the second sized invariably occur between the primary ones. The only circumstance in the history of this species, calculated to afford a specific name of any value, seems to be its great deficiency of stem, and I therefore propose to name it accordingly.

CARYOPHYLLEA SESSILIS.—Primary lamellæ of the star thirteen ; three lamellæ of less size occupying the intervals, and the middle one predominating slightly in height and breadth, and sending off from its base a thin, flexuous, and erect plate or process ; all the lamellæ rough with small tubercles, and more or less plaited on their edges.—J. S. BELLAMY. *Yealmpton, 1st July, 1837.*

NEW SILKWORM.—It is stated that at Rio Janeiro there are several species of Bombyx, the Caterpillars of which inclose themselves in a cocoon, after having spun a thicker and stronger silk than that of the ordinary Silkworm. It has been tried, and found to form a very solid material. A species of mulberry, of which the fruit is small and inedible, grows near the city, and is to be cultivated for feeding the Caterpillars.

BOTANY.

THE ORANGE TREE, CITRUS AURANTIUM.

THIS tree, of which the fruit is so well known in all parts of Europe, belongs to the class *Polyadelphia*, and the order *Icosandria* of the Linnean system, and to the natural family of *Aurantiacea*. It is a native of India, whence it has spread into various parts of the world, and is cultivated in the warmer countries of Europe. It attains a height of from five to fifteen feet, is copiously branched, and has a smooth greyish bark. The leaves are elliptical but rather pointed, smooth, entire, glossy, deep green, with winged or dilated petiols. The flowers are large, white, axillar, and terminal, on short, simple, or branched stalks, with a cup-shaped, five-toothed calyx ; five oblong, concave petals ; about twenty stamens, of which the filaments are united into three or more sets ; a roundish germen, surmounted by a short cylindrical style, and a globular stigma. The fruit, being so well known, requires little description. It is a mass of cellular, pulpy matter, divided by partitions, and containing several seeds towards the centre, the whole inclosed by a thick rhind, externally of an orange yellow, and having numerous glands or crypts containing an essential oil.

Oranges are brought to us chiefly from Spain and Portugal. In England, although orange trees have been long cultivated as ornamental greenhouse plants, they never produce fruit equal to that imported from the countries just mentioned. There are two principal varieties, the China or sweet orange, and the Seville, both of which are in common use. The latter, however, is the kind chiefly employed medicinally, its juice being an agreeable acid liquor, which is of considerable use in febrile and inflammatory disorders, as it quenches the thirst, promotes the secretions, and diminishes the action of the arterial system. The yellow rhind, which is an agreeable aromatic bitter, has been employed with success as a stomachic, as well as in intermittent fevers. It abounds with a volatile aromatic oil, which, with the bitter contained in it, gives it a pungent and harsh taste. This part is extensively prepared in Scotland as a conserve under the name of marmalade. The flowers and leaves, which are also beset with minute glands secreting an essential oil, have been employed as a remedy against epilepsy and other convulsive disorders, but have fallen into disuse.

The lemon, the lime, and the citron, are species of the same family, and also natives of India. The production of the common orange, Dr Lindley remarks, is enormous. A single tree at St Michael's has been known to produce 20,000 oranges fit for packing, exclusively of the damaged fruit and the waste, which may be calculated at one third more. Mr Audubon found it apparently wild in Florida. " What-ever its original country may be supposed to be, the plant," he says, " is to all appearance indigenous in many parts of that country, not merely in the neighbourhood of plantations, but in the wildest districts. Nothing can be more gladdening to the traveller, when passing through the uninhabited woods of East Florida, than the wild orange groves which he sometimes meets with. As I approached them, the rich perfume of the blossoms, the golden hue of the fruits that hung on every twig, and lay scattered on the ground, and the deep green of the glossy leaves, never failed to produce the most pleasing effect on my mind. Not a branch has suffered from the pruning knife, and the graceful form of the trees retains the elegance it received from nature. Raising their tops into the open air, they allow the uppermost blossoms and fruits to receive the unbroken rays of the sun, which one might be tempted to think

are conveyed from flower to flower, and from fruit to fruit, so rich and balmy are all. The pulp of these fruits quenches your thirst at once, and the very air you breathe in such a place refreshes and reinvigorates you. I have passed through groves of these orange trees fully a mile in extent. Their occurrence is a sure indication of good land, which, in the south-eastern portions of that country, is rather scarce. The Seminole Indians and poorer squatters feed their horses on oranges, which these animals seem to eat with much relish. The immediate vicinity of a wild orange grove is of some importance to the planters, who have the fruit collected and squeezed in a bone mill. The juice is barrelled and sent to different markets, being in request as an ingredient in cooling drinks. The straight young shoots are cut and shipped in bundles, to be used as walking sticks."

THE MELON IN BOKHARA.—There are two distinct species of Melons, which the people class into hot and dry. The first ripens in June, and is the common musk or scented Melon of India, and not superior in flavor ; the other ripens in July, and is the true Melon of Toorkistan. In appearance it is not unlike a water-melon, and comes to maturity after being seven months in the ground. It is much larger than the common sort, and generally of an oval shape, exceeding two and three feet in circumference. Some are much larger, and those which ripen in the Autumn have exceeded four feet. One has a notion that what is large cannot be delicate or high-flavored ; but no fruit can be more luscious than the Melon of Bokhara. I always looked upon the Melon as an inferior fruit till I went to that country ; nor do I believe their flavor will be credited by any one who has not tasted them. The Melons of India, Cambool, and even Persia, bear no comparison with them ; nor even the celebrated fruit of Ispahan itself. The pulp is rather hard, about two inches thick, and is sweet to the very skin ; which, with the inhabitants, is the great proof of superiority. A kind of molasses is extracted from these Melons, which might be easily converted into sugar. There are various kinds of Melons ; the best is named *Kochoolue*, and has a green and yellow-coloured skin ; another is called *ak nubat*, which means white sugar-candy ; it is yellow, and exceedingly rich. The winter Melon is of a dark green colour, called *Kara-Koobuk*, and said to surpass all the others. Bokhara appears to be the native country of the Melon, having a dry climate, sandy soil, and great facilities for irrigation. Melons may be purchased in Bokhara throughout the year, and are preserved by merely hanging them up apart from one another ; for which those of the winter crop are best suited. The water-melons of Bokhara are good, and attain also an enormous size ; twenty people may partake of one ; and two of them, it is said, form sometimes a load for a donkey.—*Burnes' Travels into Bokhara.*

GEOLOGY.

FOSSIL REMAINS OF MONKEYS.—M. de Blainville has reported to the Academy of Sciences, in his own name and that of *Messrs* Dumeril and Flourens, some observations upon the fossil bones discovered in the Commune of Sausan near Auch, by M. Lartet. After discussing all the facts relative to the discovery of the fossil bones of Monkeys, he goes on to say :—

" Thus, till quite lately, it was certain that no trace had been discovered left by any animal of the Monkey tribe, in even those beds which lay nearest the surface of the earth, nor even in the alluvial strata, when M. Lartet announced to the Academy, in a letter read at the scientific meeting of the 17th of last January, that he had just found in the numerous and curious assemblage of fossil bones, discovered by him in the environs of Auch, the lower jaw of an Ape, properly so called, one of the grinders of a Marmoset (sapajou), and the anterior extremity of the lower jaw of an animal of the family of Makis."

The singular interest of so unexpected a discovery, the co-existence in the same deposit, on the one hand, of the bones of the Rhinoceros, the Dinotherium, the Mastodon, the Stag, and the Antelope ; and on the other, bones of Quadrumana of Asia, America, and Madagascar, caused the correctness of his determination to be questioned. A second letter, containing a detailed description of the lower jaw of the Ape, accompanied by a figure, might have established beyond doubt the truth of a part of M. Lartet's announcement. Nevertheless, to prove not only that it was certainly an Ape which was under consideration, but moreover a Gibbon (a group of Quadrumana which are scarcely known, except in the large islands of the Indian Archipelago), more than a representation was necessary. M. Lartet, in consequence, has sent the bone itself, as well as all those which he has thought referable to the Quadrumana.

The jaw attributed to the Gibbon is an almost complete lower one, in which only the terminal parts of the rami are wanting, and it is provided with all its teeth. The total number of teeth is 16 ; that is to say, 4 incisive, 2 canine, 4 false grinders, and 6 true ones. It is the dentary formula of Man, and of all the Apes of the old continent.

The incisive teeth are equal in size, almost vertical, and ranged in a transverse line ; the canine teeth are short, vertical, and would meet without going beyond each other ; the first false grinder is not at all inclined backward from the pressure of the upper canine, and is, on the contrary, quite vertical, as in Man ; the grinders have their crown armed with blunt tubercles disposed in oblique pairs. By all these characters it is easy to recognise the jaw in question as belonging to one of the Quadrumana, to an Ape properly so called, and to one high in the series.

" Now," says M. de Blainville, " as the Gibbons are certainly the group of Apes which ought to follow immediately after that of the Orangs, if, indeed, they be distinct from each other, we see that M. Lartet is very near the truth, so much the more as the true grinders have tolerably distinct the fifth tubercle characteristic of these teeth among the Gibbons. Yet as this disposition is not certainly so well indicated in the fossil Ape as in the living Gibbons that we are acquainted with, and as besides this it offers a much more evident peculiarity in the proportion of the last grinder, which comes very near to that existing in the Semnopitheci, and even in the Baboons, it seems decisive that the fossil Ape should form a small separate section, unless we can refer it to the Colobi, which in South Africa seem to represent the Semnopitheci of India. The other fragments, which M. Lartet supposes, it is true, to have belonged to Quadrumana, have appeared to us to be referable to other groups."

"Though we are at present unable by any possibility to admit the extraordinary fact of the assemblage in one locality of fossil remains belonging to animals so rigorously limited in their geological boundaries as the true Apes, the Marmosets, and the Makis; yet the discovery of fossil bones belonging indubitably, as M. Lartet has clearly seen and pointed out, to an Ape more nearly related to the Gibbons, which are limited to the farthest parts of Asia, than to any other living species, does not the less remain to be considered as one of the most fortunate and unlooked for discoveries which has been made in Palæontology of late years; and we propose, in consequence, that the Academy should continue to M. Lartet the encouragement which it has begun to afford him, in order to facilitate his researches."—*L'Hermes, as quoted in Loudon's Magazine.*

Effect of Forests on the Size of River Currents.—The following observations, translated from *L'Echo du Monde Savant*, we extract from Loudon's Gardener's Magazine. M. Bousingault, in a memoir recently presented to the Academy of Sciences at Paris, has endeavoured to show the effects which the clearing away of forests has upon the force and abundance of the river streams in a country. He thinks that the current of water diminishes in proportion as the clearings extend; and was led to take this view from observations made in America, especially in the lake of Valencia in Venezuela, which has no outlet. This lake, in fact, diminished in depth as fast as the forests were grubbed up; but as soon as, on account of political troubles, the grubbing up ceased, the waters began to assume their primitive level. Similar results have been furnished by the lake of Ubate, in New Grenada, and even by those of Switzerland.

M. Bousingault also thinks that clearing away the forests has a direct tendency to diminish the quantity of rain. In the provinces of San Buenaventura, of Choco, and of Esmeralda, which are situate to the south of Panama, and where rains are almost continual, the soil is covered with thick forests, whilst towards Paita, beyond Tumbez, the forests have disappeared, and rains may be said to be unknown. This want of rain is in like manner observed in all the country near the desert of Sechara, and even to Lima; yet these two countries enjoy the same temperature, they present nearly the same surface, and have a like position relatively to the mountains.

M. Arago remarks that a contrary result has been observed at Viviers, in the department of Ardèche, where the quantity of rain fallen has augmented since the clearing of forests from the country.

On the other hand, M. Devèze of Chabriol has come to the conclusion, from the examination of several historical documents, that in the department of Cantal, in the environs of Saint Fleur, there has been an abatement of temperature since the disappearance of the forests. For example, from the records of the 13th and 14th centuries, it is proved that at this period the vine was cultivated on the slope of the hill of Saint Fleur, and this culture will not succeed at present. The chestnut has also disappeared from many of the cantons where it formerly flourished; and many villages situated near the summits of mountains have been abandoned. It is also remarked that, in this country, many streams have been dried up in consequence of the clearing of the forests.

DOLOMITES.

On the Phenomena of Dolomisation, and the Transformation of Rocks in General, by M. Theodore Virlet.—The general question of the Transformation of Rocks is one of the newest and most important inquiries of Geology, the solution of which ought to furnish us with the means of making rapid progress, in the study of the composition of rocks, and lead to the elucidation of a multitude of isolated facts, hitherto considered as inexplicable.

While describing to the Geological Society of France, some years ago, the modifications which presented themselves in a bed of hematitic iron, which I had an opportunity of inspecting, near Sargans, in the Canton of St Gall, Switzerland, I was led by the recollection of numerous analogical facts falling under my notice, and mentioned in my account of the Geology of Greece, to regard the phenomena of the transformation of rocks under two different points of view, and to separate the modified rocks into two very distinct classes.

1. Such as have been modified, whether by the prolonged action of heat, or by that of electro-chemical agents, or by both of these causes united, which have altered combinations or primitive arrangement of the molecules in relation to each other.

2. Rocks which have been modified by chemical actions and reactions, with the assistance of foreign agents (such as the gases), which acted directly upon them, and changed their primitive nature. It is in this class the Dolomite, or Magnesian Limestone, should be arranged.

The first manner of regarding the modification of rocks, which was originally suggested by me, serves to explain how certain beds, placed in the midst of other beds, may be more modified than the latter, or may even undergo a complete modification, without the others, whether they were in contact, or even formed the lower part of the same deposit, experiencing any sensible change in their original state, and that without any of the beds being confounded with each other. The opinion which I advance on this subject results as much from my own observations as from the manner in which I regard the first sandy deposits, as being formed at a period when the waters began to condense on the surface of the earth, and although many may consider it as somewhat heretical, I have no doubt that it will soon be admitted by all accurate observers; namely, that all stratified rocks, without excepting gneiss, the mica slates, or clay-slates, &c. have been originally rocks of sediment, formed by mechanical aggregation, and that they have acquired the crystalline characters which now distinguish them by a series of modifications, which they have undergone posteriorly to their being deposited.

It is conceived, on the contrary, according to the second kind of modification of rocks, that in the greater number of cases, all the beds are confounded in such a manner as to present a single mass without distinct stratification; such, for example, as Dolomite, certain deposits of sandstone and clay transformed into jaspers, or trachytes, or porphyries, and other rocks which I have had occasion to enumerate, for the chemical agents, by penetrating across a certain number of beds, or even the entire mass, have separated a part of the elements of the original rocks and substituted others, or else have formed new combinations, and finally united the whole mass of the deposit. It is to these considerations that I wish chiefly to direct attention, as they have reference to the phenomena of Dolomisation.

I do not dispute, that there are Dolomites which should be called *primitive*, whatever may be their geological age; that is to say, which were the result of simultaneous deposits of carbonate of lime and of magnesia, for magnesia was at least as abundant in nature as lime, particularly at the time when the old deposits were formed. These *primitive* Dolomites, however, always present a distinctive character in being regularly stratified, like the other rocks to which they may be found subordinate; while the Dolomites of which I now speak, and which I shall designate *Dolomites of Transmutation* (such as are described by M. de Buch as occurring among the Alps, and many others which I could mention), are without stratification, presenting irregular masses, combined with other characters, which individuals accustomed to observe modified rocks can seldom mistake. No one who has visited the Dolomites of the Alps can entertain any doubt of the reality of the phenomenon of Dolomisation, however difficult the explanation of it may at first appear, since chemistry teaches us that carbonate of magnesia is not volatile, or that it is decomposed at a red heat, an objection that has not been urged by M. Thenard. It was in fact these considerations that caused me to be among the first to publish my doubts on the subject, at a time when no one undertook to ascertain, by chemical analysis, that the parts of the deposit which had not been modified were not equally magnesian, that is, did not form beds of *primitive Dolomite*, a circumstance which would have reduced the phenomenon of the change of Limestones into Dolomite to a simple phenomenon of modified crystallization, analogous to that, for example, which has determined the change of the compact Jura Limestone of Carrara, and that of the compact chalks of some parts of the Pyrenees, into granular Limestones or Statuary Marble. One of my friends, M. Des Genevez, who possessed a very extensive knowledge of chemistry, and whose early scientific works afford so much reason to lament his premature death, has unhappily been lost to the sciences before publishing the results of his chemical researches on Dolomisation. These, he has many times assured me, had demonstrated to him that there existed an insensible passage from beds of unaltered carbonate of lime to dolomite or double carbonate of lime and magnesia. Thus, the transformation of certain calcareous rocks into dolomite, posteriorly to their formation, appears to me to be a well established phenomenon, and requires, in my opinion, only to be properly explained, in order to be admitted by all.

Who does not know how many facts, perhaps among the most difficult to comprehend previously, have already been explained by the excellent researches of M. Becquerel, in electrical chemistry, and the important labours of M. Fournet, regarding the formation of veins? Numerous other facts, although not yet fully explained, have been brought forward and admitted without dispute. For example, I have proved that the Emery of Naxos comes from veins, and consequently had been formed, like the greater number of specular iron ores, by means of volatilization and sublimation; yet the corundum and oxide of iron, the mixture of which constitutes Emery, are not more volatile than the carbonate of magnesia, which forms the subject of dispute.

Since our chemical knowledge, then, does not always enable us to explain the phenomena whose existence we can prove, does it follow that we ought to call them in question? Has Nature no mode of acting which surpasses our knowledge? And could she not proceed, for instance, by means of double chemical decomposition? On this supposition the phenomena will admit of easy explanation. It is well known that all the muriates are volatile, or at least susceptible of sublimation. Magnesia might then easily reach the state of a muriate, and occasion the formation of a soluble hydrochlorate of lime, which would be carried off by the infiltration of water; while the magnesia, on the contrary, would be combined with that portion of the carbonic acid set at liberty, and would thus serve to form the double carbonate of magnesia and of lime, which constitutes dolomite, properly so called. In this there is certainly nothing inadmissible or contrary to reason, inasmuch as the hydrochloric acid gas is one of the gases most frequently disengaged from volcanoes; and the muriates ought to have been disengaged more abundantly in former times, if we admit, with geologists of the modern school, that the immense deposits of rock-salt which exist in saliferous formations, are deposited by volatilization in the midst of the strata which they penetrate.

I am therefore of opinion, that the modifications of rocks of the second class may henceforth be explained by means of double decomposition—a process which has enabled one of my friends, M. Aimé, to produce in the laboratory crystallized specular iron ore, analogous to that of the island of Elba, as well as pure iron equally well crystallized—a substance hitherto unknown to mineralogists; whence I conclude that the time is not perhaps far distant when we shall be able to produce with ease all the species of precious stones, without even excepting the diamond.

Fossil Footsteps in Sandstone and Greywacke.—Professor Hitchcock has discovered in the valley of the Connecticut River, the imprints of what he considers fourteen new species. Some bear so near a resemblance to the feet of living Saurians, that they have been denominated *Sauriodichnites*. The Professor says, "I have no certain evidence as yet that any of these impressions were made by four-footed animals, although, in respect to two or three species, I have strong suspicions that such was the fact. I have sometimes thought they might have been made by Pterodactyles, yet they have in general fewer toes than those described by Cuvier and Buckland. Within a few weeks past, I have found on the flag-stones, in the city of New York, some marks, which I suspect were made by the feet of a didactylous quadruped, which, like the Marsupialia, moved by leaps. The rock is slaty greywacke, from the banks of the Hudson, between Albany and the Highlands."—*Silliman's Journal.*

Edinburgh: Published for the Proprietor, at the Office, No. 13, Hill Street. London: Smith, Elder, and Co., 65, Cornhill: Glasgow and the West of Scotland: John Smith and Son; and John Macleod. Dublin: George Young. Paris: J. B. Balliere, Rue de l'Ecole de Médecine, No. 13 bis.

The Edinburgh Printing Company.

THE EDINBURGH

JOURNAL OF NATURAL HISTORY,

AND OF

THE PHYSICAL SCIENCES.

OCTOBER, 1837

ZOOLOGY.

DESCRIPTION OF THE PLATE.—BATS OF THE GENERA PLECOTUS AND NYCTICEIUS.

THE Bats constitute the first family of the order Carnivora, and are peculiarly distinguished by the elongation of their anterior extremities, which are expanded, and, by the intervention of a delicate membrane between the fingers, converted into wings. This membrane also includes the hind legs, of which, however, the toes are not similarly developed, and extends to the tail. In consequence of this organization, they are technically named Cheiroptera. These animals live chiefly on insects, are of nocturnal or crepuscular habits, fly with a fluttering kind of motion, betake themselves by day to caves, crevices, and retreats of a similar nature, and in cold climates become torpid during a great part of the winter. Some of the larger tropical species are frugivorous. They have been arranged under a great number of genera, of which two are represented in the plate.

The Long-eared Bats, forming the genus *Plecotus*, are characterized chiefly by the great size of their ears, which are united by a membranous flap extending across the head.

Fig. 1, 2, 3. THE COMMON LONG-EARED BAT (*P. auritus*) is of frequent occurrence in this country, as well as in various parts of the continent. It is easily distinguished by the great size of its ears, which are nearly as long as the body.

Fig. 4. THE BARBASTELLE (*P. Barbastellus*) is very similar, but has the ears much smaller.

Fig. 5. THE TIMOR LONG-EARED BAT (*P. Timoriensis*) is about the same size and colour, with the ears intermediate between those of the two former species. The peculiar characters of these animals requiring minute description, will be subsequently given at length.

A group of Bats, belonging chiefly to North America, and having only two incisors in the upper jaw, with short ears, has been separated under the name of *Nycticeius*. To this genus belong the following species.

Fig. 6. THE NEW YORK DWARF-EARED BAT (*N. Noveboracensis*).

Fig. 7. THE BOURBON BAT (*N. Borbonicus*.) From the Island of Bourbon.

Fig. 8. THE CAYENNE or ROUGH-TAILED BAT (*N. lasiurus*). From Cayenne.

DESCRIPTION OF THE PLATE.—THE DOGS, LYCAONS, AND FENNECS.

SEVERAL species of the genus *Canis* have already been figured and described, so that it is unnecessary here to present any general character.

Fig. 1. THE GREY WOLF (*Canis Occidentalis*) is peculiar to the arctic regions of North America, but differs little from the European Wolf.

Fig. 2. THE ARCTIC FOX (*C. Lagopus*), which, as its name implies, is a native of the circumpolar regions of Europe and America, is of small size, with a very soft and close fur, which in winter becomes white, its summer tint being greyish-brown.

Fig. 3. THE CORSAC (*C. Corsac*) is very similar to the Common Fox in form, but of a duller tint.

Fig. 4. THE JACKAL (*C. aureus*) is distinguished by its short bushy tail. It varies in colour, but is generally of a greyish-brown tint, the limbs reddish externally. Animals apparently of this species are distributed over a great portion of Asia and Africa; they hunt in packs; and from the ease with which they are domesticated, and their form, are supposed by some to be the original of the domestic dog.

Fig. 5. THE CAPE JACKAL (*C. Mesomelas*) is similar to the Jackal, but has the tail long, and as the pupil is elliptical, it is referred to the section of Foxes.

Fig. 6. BURCHELL'S LYCAON (*L. Tricolor*) occurs in Southern Africa, and is in a manner intermediate between the Dogs and Hyænas, having more of the general appearance of the latter. It varies in colour, being yellowish-red, or ochraceous, and marked with white spots circled with dusky.

Fig. 7. BRUCE'S FENNEC (*Megalotis Brucii*). A small group of little animals similar in most respects to the Dogs and Foxes, but especially distinguished by their very large ears, constitute the genus *Megalotis*. That first described by Bruce is of a light grey colour tinged with brown, and inhabits Nubia, burrowing in the sand.

Fig. 8. SMITH'S FENNEC (*M. Smithii*) is very similar, with the tail more bushy.

Fig. 9. LALAND'S FENNEC (*M. Lalandii*) has the ears much shorter. It is somewhat smaller than the Common Fox, and occurs in Southern Africa.

DESCRIPTION OF THE PLATE.—THE HUMMING BIRDS.

Fig. 1. THE TOPAZ-THROATED HUMMING BIRD (*Trochilus Pella*). Adult male. This, although the most common species, is one of the most splendidly coloured, being of a brilliant ruby tint, varying to dusky red, the head velvet black, the throat emerald green, changing to gold yellow. The male is distinguished, moreover, by two very long dusky feathers in the tail. This species is plentiful in Guiana and the neighbourhood of Cayenne.

Fig. 2. Represents a variety, in which the body is mottled with white, and which is remarkable in manifesting a tendency to albinism, in a family of birds almost exempted from that change.

Fig. 3. Female. The female wants the long tail-feathers, and is of a deep green colour, with the throat yellowish-red.

Fig. 4. THE VIOLET-EARED HUMMING BIRD (*T. auritus*) is so named on account of two tufts of feathers on the sides of the head, of which one is of a violet purple colour, while the other is emerald green. The upper parts are of a gilded green tint, the lower pure white. The middle tail feathers are bluish-black, the rest white. It is one of the most common species of Guiana and Brazil.

Fig. 5. The female resembles the male, but differs in wanting the green and purple tufts on the auricular region, and in having the white of the breast and abdomen mixed with numerous brown or dusky spots.

Fig. 6. THE SAPPHIRINE HUMMING BIRD (*T. sapphirinus*). Female. Green above and on the sides, violet purple on the fore part of the neck. Inhabits Brazil.

DESCRIPTION OF THE PLATE.—THE MACACOS AND MAGOTS.

THE MACACOS (*Macacus*), as characterized by Cuvier, have molar teeth furnished with a fifth tubercle, like the *Semnopitheci*, and callosities and cheek-pouches, like the Guenons; but their limbs are thicker and shorter than those of the former, and their muzzle more prominent, and their superciliary ridge more bulging than those of the latter. Although rather docile in early life, they ultimately become intractable; their tail, which varies in length in the different species, is not prehensile. Most of them are natives of India.

Fig. 1. THE OUANDEROU (*M. Silenus*). Male. This species, which is distinguished by having the head encircled with a kind of beard of a whitish colour, the forehead black, and the body and limbs of a dusky tint, inhabits the forests of Ceylon, feeding on fruits, leaves, and buds, and refusing to associate with other species.

Fig. 2. THE CHINESE-CAPPED MACACO (*M. Sinicus*) is so named on account of the manner in which the hairs on the upper part of the head spread out in all directions. It is yellowish-red on the upper parts, white beneath, and is found in Ceylon and Bengal.

Fig. 3. THE HARE-LIPPED MACACO (*M. Cynomolgus*), distinguishable by an erect tuft of hair on the top of the head, has the upper parts of a greyish hue, the lower whitish, but varies in colour. It inhabits Africa, and is said to commit great havoc on the cultivated fields, which it attacks in large flocks.

Fig. 4. THE RHESUS (*M. Rhesus*) has a very short tail, and is generally of a reddish tint, the fore-parts grey. It is said by Cuvier to inhabit Ceylon.

Fig. 5. THE PIG-TAILED MACACO (*M. nemestrinus*) is of a reddish colour, with the lower parts whitish, a dark line from the head along the back, and a short slender tail somewhat curved like that of a pig.

The MAGOTS (*Inuus*) are very similar to the Macacos, differing chiefly in having the tail reduced to a mere knob.

Fig. 6. THE BARBARY MAGOT (*I. Magotus*) is a native of Barbary, and remarkable for being the only species of quadrumanous animal that occurs in any part of Europe, it having been naturalized on the Rock of Gibraltar. It bears our northern climates better than any other species. The colour is generally light-greyish-brown.

DESCRIPTION OF THE PLATE.—THE HOOPOES AND PROMEROPS.

THE genus *Upupa* is characterized by a very long, slender, slightly arched bill, triangular at its base, convex above, compressed, without notch, and somewhat blunt; open, elliptical nostrils; an even tail, composed of ten feathers; and wings of moderate length, with the fourth and fifth quills longest. The species are remarkable for the longitudinal crest of elongated feathers with which the head is ornamented. They live on insects, are migratory, nestle in hollow trees and fissures of rocks, and moult only once in the year.

Fig. 1. The Common Hoopoe (*Upupa Epops*). This species, which inhabits all parts of Africa, and migrates into most of the countries of Europe, occasionally appearing in Britain, is of a reddish colour, with a broad band of black on the back. The wings and tail are also of that colour, the former with five white bands, the latter with one. The crest, which is erectile, is red, the feathers tipped with black, and some of them having a white band. It lives in woods and thickets in the vicinity of low and moist ground; feeds on beetles and other insects, larvæ, ants, and young frogs; and nestles in the holes of trees.

Fig. 2. The Lesser Hoopoe (*U. minor*) resembles the Common, but is considerably smaller, and wants the white band on the crest. It occurs in India and Africa.

Fig. 3. The Madagascar or Cape Hoopoe (*U. Capensis*) differs so much in the form of its bill, that M. Temminck has referred it to the genus *Pastor*, and M. Lesson thinks it might be with propriety placed in the genus *Fregilus*. Its crest is composed of very slender feathers, some of which are curved forwards over the nostrils.

Fig. 4. The Striped Promerops (*Epimachus fuscus*). The genus *Epimachus* is composed of species which agree in having a long slender hill, more curved and compressed than that of the Hoopoes, from which they differ also in wanting the crest of elongated feathers on the head, while, on the other hand, some of them have the feathers of the sides greatly developed. The species here represented is one of the smallest, and inhabits Africa.

BRITISH BIRDS, NO. VI.—THE BLACK GROUSE.

The genus *Tetrao* of authors, comprehending birds of the Gallinaceous order, characterized by a full habit of body, short, concave, rounded wings, legs of moderate size, and having the tarsus completely or partially covered with feathers, a small head, and bare superciliary crests, has by some been divided into two genera, one including the Grouse properly so called, which have the toes pectinated with elongated lateral scales, the other including the Ptarmigans, of which the toes are destitute of these lateral scales and covered with feathers. There are three species of this genus resident in Britain, of which one belongs to the first or Grouse section, the rest to the second, or that including the Ptarmigans. They are popularly named the Black Grouse, the Red Grouse, and the Ptarmigan.

The Black Grouse, *Tetrao Tetrix*, is nearly of the size of the domestic fowl. The male differs greatly from the female in colour, and in the form of its tail, which is lyrate, or has the outer feathers longer and curved outwards. The general colour of the plumage is black, that of the neck and back glossed with deep blue; the lower wing-coverts, lower tail-coverts, and bases of the secondary quills, white. The female has the tail slightly forked, its lateral feathers straight; the general colour is yellowish-red, spotted and undulated with brownish-black.

In its internal organization this species exhibits the peculiar characters distinctive of the gallinaceous birds in a remarkable degree of development. The œsophagus is dilated into a very large membraneous crop, which covers the fore part of the neck; it then contracts, and enlarges into a proventriculus of moderate size. The stomach is a very muscular gizzard, lined with a thick and dense rugous membrane. The intestine is of considerable length, and has two extremely elongated cœcal appendages, furnished internally with longitudinal ridges, and in which the residuum of the food receives a second elaboration.

It feeds on fresh twigs of *Erica cinerea*, *Calluna vulgaris*, *Vaccinium Myrtillus*, willows, and other shrubs, as well as on berries and leaves of various plants, gradually filling its crop, which is capable of containing a globular mass from three and a half to four inches in diameter. The food varies according to the season, although heath twigs always form the principal part. In spring the tops of eriophora, carices, blades of grass, willow-catkins, and buds of trees; in summer, leaves of various shrubs; and in winter, juicy twigs of all kinds, are found in the crop.

The Black Grouse, then, is a phytophagous bird, which, feeding on substances containing comparatively little nourishment, introduces a large quantity at a time, like a ruminating quadruped, and gradually triturates it in the gizzard, with the aid of particles of white quartz, which it picks up as required. In the gizzard the mass, acted upon by the fluid abundantly secreted by the proventricular glands, is reduced to a pulpy mass, which in the duodenum is further diluted by the pancreatic juice, and mixed with the bile. The nutritious parts are absorbed as it passes along, and the cœcal appendages subject it to a further elaboration.

In searching for food it frequents the lower grounds of the less cultivated districts, keeping for the most part in the vicinity of woods or thickets, to which it retreats for shelter or protection. Sometimes it makes an excursion into the stubble fields, or even attacks the standing corn. It walks and runs among the herbage with considerable agility, perches adroitly on trees, and may often, especially in spring, be seen on the turf tops of the low walls inclosing plantations. Its flight is heavy, direct, and of moderate velocity, but is capable of being protracted to a great distance. This species, however, does not generally wander far from its ordinary haunts, which are the lower slopes of hills covered with coppice, interspersed with heath, rank grass, and ferns, or valleys flanked by rocky and wooded ranges. In such situations, it is plentiful in many parts of the northern and middle divisions of Scotland, in several districts in the southern division, and in the north of England; but is rare in other portions of that country.

In autumn it falls an easy prey to the sportsman, but in winter and the early part of spring it is shy and difficult to be procured. As an article of food it ranks high, the flesh being whiter than that of the Red Grouse, and the males weighing from three to four pounds. Its natural enemies are Foxes, Polecats, Eagles, and Falcons. Vipers are said to destroy its eggs and young, as do Ravens, Hooded Crows, and Carrion Crows.

The males keep by themselves in autumn and winter, and towards the middle of spring join the females, fighting with each other when they meet. At this season, the red space over the eye assumes a deeper tint, and the bird manifests increased activity and vigour; but when the excitement is over, the males appear fatigued and emaciated, meet together without manifesting animosity, and seem intent on recruiting their diminished energies.

The female forms a rather inartificial nest of dry grass, in which she deposits from five to ten eggs, of a regular oval shape, generally two inches long, with a yellowish ground colour, irregularly spotted and dotted with brownish-red. As the nests are usually placed in low situations, they are frequently partially or entirely inundated in very wet seasons.

The young, which are at first covered with close, fine down, are able to run about the moment after they leave the egg.

Hybrids are sometimes, though very rarely, produced between this species and the Red Grouse, as well as the Pheasant.

Formerly another species of this genus, the Wood Grouse, or Capercailly, *Tetrao Urogallus*, occurred in the northern parts of Scotland, but has been extirpated. The last individual recorded to have been seen was killed in 1769, in Strathglass, to the north of Inverness. This species, which is about equal in size to a Turkey, is still abundant in Sweden and Norway, whence it is imported in winter, when it may frequently be obtained in the London markets.

BOTANY.

Directions for the Preservation of Sea Plants. By J. S. Drummond, M.D. Abridged from the Magazine of Zoology and Botany, No. VIII.—The first object to be attended to in preserving marine plants, is to have them washed perfectly clean before spreading. It is a good practice to cleanse them before leaving the shore either in the sea, or in a rocky pool. All foreign bodies, as fragments of decayed sea-weeds, sand, gravel, or portions of the softened surface of the rock on which the specimens may have grown, together with the smaller testacea, and corallines, must be carefully removed. On getting home, it is further necessary to prepare each specimen by examining it in fresh or sea water in a white dish or plate, so that every foreign substance may be detected and removed. The next thing to be attended to is the quality of the paper on which the specimens are to be spread. A great error is generally committed in using it thin and of inferior quality. Indeed, much of the beauty of many species depends on the goodness of the paper, just as a print or drawing appears better or worse, as it is executed on paper of a good or an inferior kind. Some species contract so much in drying as to pucker the edges of the paper, if it be not sufficiently thick. That which has been found best is a thick music paper, closely resembling that used for drawing, and of which the sheet divides into four leaves about the size of royal octavo.

Whatever pains may have been taken in cleaning the recent specimens, it is often found, when spreading them, that some foreign particles continue attached, and for the removal of these, a pair of dissecting forceps, and a camel-hair pencil of middle size, will be found very convenient. A silver probe, with a blunt and a sharp end, is the most convenient instrument for spreading out and separating branches from each other, but any thing with a rigid point, such as a large needle, or the handle of the camel-hair pencil, will answer. A large white dish serves for spreading the specimens in, and all that is further necessary is a quantity of drying papers, and some sheets of blotting paper, with three or four flat pieces of deal board. Nothing answers better for drying than old newspapers, each divided into eight parts, but it is necessary to have a large supply of these.

The beautiful and common *Plocamium coccineum* is one of the most easily preserved species, and may be taken as an example of the mode of proceeding with most of the others. The steps to be pursued are as follows:—

1. The specimen is to be perfectly well cleaned.
2. A dinner dish to be filled about two thirds with clean fresh water.
3. The paper on which the specimen is to be spread to be immersed in the water in the dish.
4. The specimen to be then placed on the paper, and spread out by means of the probe and camel-hair pencil.
5. The paper with the specimen on it to be then slowly withdrawn from the dish, sliding it over its edge.
6. The paper with the specimen adhering to it to be held up by one corner for a minute or two, to drain off the water.
7. To be then laid on a paper or cloth upon a table, and the superfluous water still remaining to be removed by repeated pressure of blotting paper upon the specimen, beginning this operation at the edges, and gradually encroaching towards the centre, till the whole can be pressed upon without danger of any part adhering to the blotting paper, which probably would be the case were the latter applied at once to the whole specimen.
8. The specimen then to be laid on a couple of drying papers placed on a carpet or a table; two more papers to be laid over it, and then the piece of board, on which latter a few books are to be put, to give the necessary pressure.
9. These papers to be changed every half hour or oftener, till the specimen is sufficiently dry. A number of specimens with drying papers interposed, may be pressed at once under the same board.

Destruction of the Vine by Insects.—The Académie des Sciences having lately appointed a commission to repair to Argenteuil, and examine into the disasters caused there by a species of insect, which destroys the vines, MM. Dumeril and St Hilaire visited the place, and went over an extent of ground of a league in length, and half a league in breadth. All the vines which covered this place were entirely destitute of leaves and grapes, and nothing was to be seen but vine-props supporting blackened stalks almost in a state of atrophy, owing to several sorts of insects, among which the first place must be assigned to the Pyralis, of which M. Dumeril presented the Academy with specimens of the Eggs, Caterpillar, Chrysalis, and Butterfly. The presence of all these metamorphoses, and above all the deposit of eggs on the leaves of the vine, indicate that the Pyralis is continually reproduced, and that in consequence it would be difficult to destroy it by fire, as is generally done with other insects. M. Dumeril gave notice of his intention of concerting with his colleagues, MM. St Hilaire and Dumas, on the means of preventing the return of this scourge.

THE MULBERRY TREE.—MORUS NIGRA.

THIS tree, which is a native of Italy, but is cultivated in many parts of Europe, belongs to the class *Monœcia* and the order *Tetrandria* of the Linnæan system, and to the natural family of *Artocarpeæ*, the Broad-fruit tribe. It does not attain a considerable height, but sends off numerous crooked branches, and is covered with rough brown bark. The leaves are numerous, cordate, acute, serrate, veined, rough, bright green, and placed on short stalks. The male flowers are disposed in close roundish catkins, each floret with four oval, concave leaflets, and four stamens. The calyx of the female flower is divided into four obtuse persistent segments; the germen is roundish; and there are two rough styles, with simple stigmas. The fruit is a large succulent berry, composed of a number of roundish bodies, each containing an oval seed, and affixed to a common receptacle.

The fruit, having an agreeable subacid taste, is esteemed as an article of food, and is used medicinally for allaying thirst. The bark has an acrid bitter taste, and has been successfully used as an anthelmintic, particularly in cases of tapeworm. But the principal use to which this tree is applied is the very important one of rearing Silkworms, which feed on its leaves. In Europe, however, the leaves of the white mulberry, *Morus alba*, and in China, those of *Morus tartarica*, are preferred. The fruit of the black mulberry is often made into a syrup or rob, in the same manner as currants; a kind of wine is also made from it, and its juice is sometimes employed to give a deeper colour to red wine. The wood is yellow, pretty hard, and well adapted for turning.

The white mulberry is a native of Asia, and is cultivated extensively in various parts of Europe, for the purpose of rearing Silkworms. It answers in almost every variety of soil, but thrives best in light mould. It is raised from root-shoots, cuttings, or seed, or by grafts, and may be stripped of its leaves two or even three times every summer. Sheep are fond of the leaves, and fatten on them, and in Spain especially are thus fed during part of the winter. Birds also prefer the fruit of the white mulberry to most others, and it is said that Thrushes and Blackbirds fed on it are considered a great delicacy; but it is much inferior to that of the black mulberry, being nearly insipid, and therefore is not, like it, used as food by Man.

MISCELLANEOUS.

Catalogue of Land and Fresh-water Shells found in the vicinity of Plymouth, with Remarks on their Habitats, &c. By J. C. Bellamy, Esq., Yealmton.

The arrangement and nomenclature of Turton's Manual are adopted.

BIVALVE SHELLS.

1. CYCLAS PUSILLA.—In stagnant pools, running water, and spots flooded by rivers. We have a shell here, found in the same situations, of a rust colour, but differing so little in shape as to be entitled to consideration only as a variety of *C. Pusilla*.

2. UNIO MARGARITIFERUS.—In rapid rivers. Dead and occasionally living specimens are found in the sand-pits dug in the banks. These specimens are always numerous according to the quantity of rain which falls in the autumn and winter; the virulence of the stream, being thereby increased, drives the shells into these catch-pits. A person assured me that Crows watch on the banks of rivers for these shell-fish, and fly with them into the air, then drop them on some rock, in order to break the shell, and descend to devour the contents. Crows are reported to destroy, and feed on the Swan-mussel in the same manner.

3. LIMACELLUS PARVUS.—Found in *Limax maximus*.

4. L. OBLIQUUS.—In the common Field-slug. Possibly *L. unguiculus* and *L. variegatus* are to be found here, but they have not come under my notice.

5. TESTACELLUS HALIOTIDEUS.—I have one specimen of this shell, and it seems as yet to have been found only here and at Bideford.

6. VITRINA PELLUCIDA.—Among damp moss. Dead shells are very frequent under stones. I think I have found *V. elongata*, or at least elongated specimens of *V. pellucida*.

7. HELIX ASPERSA.—Common in hedges and fields.

8. H. HORTENSIS.—In hedges and gardens. Not common.

9. H. NEMORALIS.—Common in hedges and gardens. In winter I find this and the two foregoing species either lying loose in warm hedges, with their apertures sealed up, or cemented firmly to stones or old trees. A few remain quiescent, without any epiphragm, and stir out for food when the weather is tolerably mild. These species also will, on receiving an injury to the mouth, secrete a thin lid of the same description, and continue quiescent in this state until the fracture has been repaired.

10. H. RUFESCENS.—Common in heaps of stones and among rubbish, and generally where there is much moisture and shade. Its colour varies.

11. H. SERICEA.—Common in damp and shaded situations, especially where dead leaves abound.

12. H. VIRGATA.—Very common in dry situations. In some places, particularly elevated fields, they are so numerous as to be a pest to the agriculturist. This shell and *Bulimus fasciatus* are liable from exposure to become denuded of the outer layer, and bleached.

13. H. CAPERATA.—Common in dry situations. In old limestone quarries, large and beautiful specimens are found crawling on the rock facing southwardly.

14. H. SPINULOSA.—In woods among leaves, and in heaps of stones. They seem to retire in winter into mould formed by decayed moss on rocks, or into other sheltered situations. Not common.

15. H. FUSCA.—In woods among moss. Rare.

16. H. NITENS.—Very common among wet moss and under stones. Large specimens are usually found dead. The finest I ever collected, and which were fully three-fourths of an inch in breadth, were lying on rock sloping down to the sea.

17. H. ALIARIA.—In the same situations as the last. I have as yet found it only in my own garden. Full-grown shells would certainly be with difficulty distinguished from *H. nitens*, were it not for their odour; but the fry have a remarkably glossy aspect, and a considerable depth at the aperture, while the young of *H. nitens* retain the characteristic bend and narrowness of mouth observable in older shells. Some of this species seem not to emit the scent until immersed in hot water.

18. H. HISPIDA.—Not common. In old walls and under stones. Some occur without hairs.

19. H. CRYSTALLINA.—Common in wet hedges among moss.

20. H. RADIATA.—Common under stones, wood, &c., generally selecting damp places. I have found the white variety.

21. H. RUPESTRIS.—Common on dry walls and old buildings. I have noticed them on the walls of churches. In summer, when a shower falls, they will come out in numbers from their recesses, and appear greatly refreshed and enlivened.

22. H. FULVA.—This rare species is found with us among leaves, and under stones, especially where there is a little moisture. Specimens vary in shape.

23. H. PULCHELLA.—Rare. The smooth variety only has come under my notice. I have found it under stones. Mr T. Colley found a colony of the rough sort in a very dry wall at Trematon Castle in Cornwall. According to Dr Turton's experience, this last variety is found only in damp places.

24. CLAUSILIA BIGOSA.—Abundant in old walls.

25. BULIMUS OBSCURUS.—Rather rare. Chiefly found in old dry hedges. I lately ascertained that many of this kind retire in winter to the crevices of rocks, and in spring come forth besmeared with dirt derived from their habitations.

26. B. LUBRICUS.—Found pretty commonly under stones, and in wet hedges among moss.

27. B. FASCIATUS.—In fields bordering on the sea.

28. BALÆA FRAGILIS.—Not common. In moss attached to the trunks of trees, in old walls, and on rocks. I have occasionally found it in the cylindrical holes bored by insects in rotten stumps of trees.

29. SUCCINEA OBLONGA.—In pools and streams attached to stones and aquatic plants. In corroboration of Dr Turton's view of the difference of the species from *S. amphibia*, I may mention that it is found here without any admixture of the last named species.

30. CARYCHIUM MINIMUM.—Common among wet mosses, leaves, and stones. I have found great numbers of dead shells in summer among the moss, attached to rocks on hills. The heat of summer destroying the moss, the shells are in their turn sacrificed.

31. PUPA UMBILICATA.—Found with *Clausilia rugosa*, and equally common. It occurs amongst rubbish in wet situations.

32. P. MARGINATA.—Chiefly in spots near the sea, under stones. Not common.

33. P. EDENTULA.—Two specimens of this shell have occurred to me in a heap of stones. These belong to the "more elongated and cylindrical" variety.

34. VERTIGO PYGMÆA.—Found by Mr T. Colley under stones in a damp situation, at Broisand.

35. V. SEXDENTATA.—Found by Mr Colley with the last, and also by him in a similar situation, a very short distance from Plymouth.

UNIVALVE SHELLS.

36. PLANORBIS VORTEX.—Common in pools and streams.

37. P. ALBUS.—Common in pools, attached to plants and stones.

38. P. IMBRICATUS.—Rather local. In a standing pool near my residence I find them in plenty attached to plants.

39. P. SPIRORBIS.—Found by Mr Colley in pools. Rare.

40. LIMNÆUS PEREGER.—Very common in pools, streams, and rivers.

41. L. FOSSARIUS.—Rather common in streams and pools.

42. PHYSA FONTINALIS.—Not common in ponds, streams, and rivers, attached to stones, sticks, and other objects.

43. PALUDINA STAGNORUM.—Very rare in pools and streams. I have never procured more than two specimens.

44. ANCYLUS FLUVIATILIS.—Common on stones and plants, in pools, streams, and rivers.

A dealer in shells living here sold me specimens of *Physa rivalis*, and some of *Clausilia bidens*, saying he had collected them in this neighbourhood. They have never been found by me or any other collector with whom I am acquainted, and I therefore feel great doubts on the subject. Mr Colley has a series of shells, seemingly *Helix lucida*, collected in this neighbourhood, but I am not as yet sure of their proper appellation.

I shall proceed shortly, with the assistance of Mr Colley, to make out a list of the salt-water shells of this neighbourhood.

Falling Stars in the Mauritius.—M. Louis Robert, who has long resided in the Mauritius, and devoted his attention to meteorology and astronomy, has addressed to the Institute of France an extract from his Journal relative to shooting stars. The coincidence between the appearance of these luminous bodies in an island of the torrid zone, and those which have been observed at the same time over a great part of Europe, is not without interest. Towards three o'clock in the morning of the 13th November 1832, during calm and somewhat cloudy weather, there were seen from all parts of the heavens, where there were no clouds, and especially towards the zenith, at some degrees to the south, a great variety of shooting stars, which traversed the heavens in all directions. The number was so great that it was impossible to count them. Their courses were not in straight lines, as those of shooting stars generally are, but they described all sorts of curves in the sky. The phenomenon was at its height about four in the morning. A little before the rising of the sun, but few of these meteors were to be seen. The mercury of the barometer was at its usual height, and the thermometer of Reaumur was two degrees lower than for some days preceding.

Artificial Formation of Crystals.—M. Gaudin has lately obtained microscopic crystals of some of the insoluble salts in great perfection, by means of a process which he believes to be applicable to those salts furnishing crystals of all sizes. The method consists in placing certain solutions in an artificial atmosphere; for example, placing under the same bell-glass a capsule containing carbonate of ammonia moistened, and another open glass vessel containing a weak solution of any soluble salt of lime, or baryta, strontia, lead, &c.; in a few hours we find on the sides of this vessel crystals of the carbonate of these bases. Sulphate of baryta has been obtained in crystals by placing under the same bell-glass a vessel containing strong hydrochloric acid, and in another open glass vessel, water, sulphate of lime, and carbonate of baryta. The solution of a pure salt of lime gives crystals under the rhomboidal form with its principal modification, whilst the solution of baryta gives simultaneously crystals of sulphate and carbonate of that base. The author has obtained sulphate of tin from an atmosphere of sulphur vapour. Its appearance resembles that of snow.

Remarks on Varieties.

No two individuals of any species of animal or vegetable are exactly similar. Examine two Elephants, two Chaffinches, two Oysters, two Oaks, two specimens of any one kind of Moss or Lichen; examine the individuals composing each pair at one time, and placed under apparently similar conditions of existence, and it will be found that differences exist between them. These differences, however, are of a very unimportant character; they affect the organization and economy of the being in the most trifling manner. Nature, for some reason, has thought fit that no two of her productions should correspond, not only through the aggregate of species, but also through the aggregate of individuals constituting a species.

Besides these minor differences between individuals generally, there are others of a higher degree occurring to a great proportion of species in all the tribes and orders of living beings. To these extended degrees of aberration from regular specific character and appearance, the term "varieties" has been arbitrarily attached by systematists; and in consequence of there being no line of demarcation, natural or artificial, between these and the minor differences first named, contrarieties of opinion and waverings of individual judgment commonly attend inquiries relative to specimens being deserving of the epithet "variety," just as disputes arise whether highly extended varieties may induce suspicions of their title to rank as species.

This term "variety" has also been in use in Mineralogy and Geology; but surely its application here must rest on different principles from those which gave it introduction into Zoology and Botany; for, as in the case of animals and plants, the laws of organization and vitality are undoubtedly concerned in the production of varieties, and as the boundaries of species are absolutely fixed by nature, one kind having no alliance with another, so, in the case of inorganic bodies, we witness precisely opposite facts, and a "variety" seems to be an addition to, or reduction from, the characters of the true species, or an assumption of a form of crystallization of the substance, not usual to the species. The number and degrees of these varieties are infinite, and the commixture of species seems to be in a great measure casual, and under no restrictions.

Some persons have imagined that in the original state of the globe no varieties existed; but, by arguing the importance of these in the economy of nature, I presume a refutation of this notion is effected. Yet I think it quite certain that varieties have increased numerically, the additional instances chiefly occurring among those tribes which, by their organization, allow of variation upon the interposition or presence of some circumstance not usually in operation on the species, but having the power of modifying its structure or aspect, either for a time or during the life of the individuals so affected. It is also highly probable that these new forms of difference may sometimes be permanent, and transmit their peculiarities of structure or appearance to their progeny, with or without the aid of those circumstances which originally induced these varieties.

These are the two forms of variety mentioned by authors, "accidental" and "permanent;" and besides their ordinary occurrence, they may, as I have just said, be produced afresh, or additional instances may be originated.

In prosecuting our inquiry into the present subject, there is one fact which especially excites attention, namely, the rarity of "varieties" in the upper classes of animals, and their great frequency in the lower departments, and still farther increase in numbers in the vegetable world. Indeed, with certain interruptions to the observance of this rule, it seems that varieties gradually augment in frequency as we descend in the scale of organization. Farther than this it may be remarked, that amongst the lowest of vegetable productions, no determinate configuration or general aspect of a species is observed, so that there is no standard whereby to judge of the extent of deviation, as can be done in higher classes, but with the exception of certain marks ascertained to be the specific characters, the deviations between specimens are interminable, all is variety. We must also here notice, by the way, that this unrestrained disposition for variety in the lowest tribes of plants furnishes a proof amongst others that there is a gradual transit from the vegetable to the mineral world, for it happens, as before stated, that the varieties in the species of minerals, or the grades of difference between kinds, are infinite, so that we here find evidence respecting the chain of creation being continued by resemblances between the lowest forms of vegetation and mineral substances. Some of the zoophytes also which curiously simulate the lowest grades of vegetation in their mode of growth and other circumstances, are likewise perfectly unrestricted in the forms and aspects they assume.

In taking a view of the whole chain of being, from the most highly to the most simply organized, the interruptions to the rule of increase of varieties are frequent, and in some cases remarkable. Instances occur in most of the families of animals and plants, in which no tendency to variation can be traced: whilst, in direct contradiction to our rule, there are animals high in the scale of being suffering numerous variations in the species, and again among plants instances can be adduced, where little or no variety is detected. In the Mammalia varieties are seldom met with. Those generally observed are such as affect only the colour and size of the individual, and these varieties are seldom found to be permanent or capable of being transmitted to future races. There are instances, however, and some of them among British quadrupeds, of varieties in colour, size, and figure, being permanent. Such are the white variety of the Mole, and the three varieties of the Fox. Accidental varieties of colour occur also in the Common Rat, the Domestic Mouse, the Hare, and others. Domestication, however, occasions very considerable variations of structure in quadrupeds, and these capable of being perpetuated. Besides, accidental variations are of very frequent occurrence. But *domesticated* and *cultivated* varieties form a subject quite apart in argument from that under consideration.

Among birds varieties are rather more frequent than among Mammalia, and the slight differences between all individual specimens of each species is a circumstance much more observable here than in the preceding class. I apprehend that permanent varieties seldom occur, but that those usually observed are of the accidental kind, consisting simply in an alteration of the colouring matter of the feathers,—such are the White Starling, the White Rook, the Cream-coloured Blackbird, the Black Bullfinch, and so forth,—or in deviations from the ordinary number of tail-feathers, or in the addition of a tuft or crest. Domestication causes surprising differences in plumage and structure.

In the class of reptiles the slight variations observed between individuals are still more considerable than in birds. Varieties occur in respect of size and colour; and of these an instance is found in our common Lizard, of which three or even more varieties are enumerated. In fishes the differences between individual specimens are very general and considerable; and varieties in regard to colour and size, as well of the entire body as of parts, are frequently remarked. There is also one form of variety in this class deserving of separate notice. Some kinds of fish are not symmetrical in their parts, such, for instance, as the Flounder, which has its eyes placed on the right side of the head. Now, in such cases, varieties occur wherein the lateral arrangement of organs is reversed, the eyes and their contingent structures appearing on the left side. It seems also probable that this sinistral variety is sometimes, or perhaps always, permanent; but it cannot be a matter of much moment to determine whether a variety is constantly being reproduced lineally from the original stock, or first parents or originators of that variety, or whether it is continually being produced as an accidental form of variation from parents of the ordinary configuration and appearance.

In examining the class of Molluscous animals, we find the differences between specimens of still greater degree. Varieties are likewise numerous, and are carried to a greater extent than in the classes we have yet viewed. Colour and size in particular are subjects of variation; so are relative proportion and arrangement of parts. These all form accidental or permanent varieties, and I believe that sometimes the same form of variation occurs both accidentally and permanently. In the case of Univalve shells, reversed specimens are sometimes noticed, such as the sinistral variety of *Helix Pomatia*. But, besides these, there are instances of varieties in shells, where the individual must suffer much more considerable alteration or deviation from the ordinary growth of the species. Such are the varieties in the genus Chiton having seven or even six valves, eight being the regular number. So also we frequently see varieties in spiral shells, where one or more whorls are superadded to the usual number. Lister records a variety of *Lymnea stagnalis* having the tentacula or feelers branched; and I have seen a specimen of *Helix umbilicata*, collected by Mr T. Colley of Plymouth, in which the last whorl, instead of completing its gyration, protruded diagonally from the shell, forming a projection nearly a quarter of an inch long. By such circumstances it appears that great latitude of variation obtains among the animals of this class, and that these varieties are by no means confined to colour and size, instances of which abound in all collections of shells, but extend likewise to differences affecting more decidedly the structure and anatomical arrangement of the animal.

In the remaining tribes of animals the differences between specimens belonging to the same kind are very considerable, insomuch that in the majority of species it is necessary to disregard external aspect and general appearance in distinguishing species, and to concentrate the attention on those parts likely to afford a clue to correct and final opinion. Moreover, we may observe that these differences are so great as to be equivalent to varieties in higher classes, and they seem to merge by insensible degrees into positive varieties amongst themselves. In the Asteriadæ we notice an interesting form of variety, consisting of the reduction from the ordinary number of arms, or of additional developments of the same. This is seen in two or more of our native Asteriæ.—*(To be continued.)*

Edinburgh: Published for the Proprietor, at the Office, No. 13, Hill Street. London: Smith, Elder, and Co., 65. Cornhill. Glasgow and the West of Scotland: John Smith and Son; and John Macleod. Dublin: George Young. Paris: J. B. Balliere, Rue de l'Ecole de Médecine, No. 13 bis.

The Edinburgh Printing Company.

THE EDINBURGH

JOURNAL OF NATURAL HISTORY,

AND OF

THE PHYSICAL SCIENCES.

NOVEMBER, 1837.

ZOOLOGY.

DESCRIPTION OF THE PLATE.—THE CIVETS AND GENETS.

The Civets and Genets form together a family of the Carnivora, characterised by their elongated form, conical and pointed head, short limbs, and long tapering tail. Their tongue is covered with pointed stiff papillæ, their claws are retracted when they walk, and in the vicinity of their anus is a deep bag into which is secreted an unctuous matter which is highly odorous. There is very little essential difference between the two genera, the Civets merely having the anal bag deeper, and the pupil round, whereas in the Genets the bag is reduced to a slight depression, and the pupil is a vertical slit.

Fig. 1. The Common Civet (*Viverra Civetta*) is of a greyish or brownish-grey colour, irregularly banded and spotted with black; the tail is shorter than the body, and ringed with dusky; there are two black bands on each side of the neck; and along the middle of the neck and back is a kind of mane. This species, from which was obtained the civet of commerce, formerly much employed in perfumery, is a native of various parts of Africa.

Fig. 2. The Grey-headed Civet (*V. Poliocephala*), characterised by the dark brown colour of the body and limbs, the yellowish-white of the throat, and the bluish-grey tint of the head, is also a native of Africa.

Fig. 3. The Zibet (*V. Zibetta*), which inhabits India, is very similar to the Common Civet, being of a light-grey colour, spotted with black, the tail ringed, and the neck patched with the same; it is destitute of mane, and has a white stripe down the back.

Fig. 4. The Luwak (*V. Musanga*), greyish-brown all over.

Fig. 5. The Javanese Civet (*V. Rasse*), brownish-yellow, longitudinally banded and spotted with black, inhabits Java.

Fig. 6. The Common Genet (*Genetta vulgaris*) is grey, or brownish-grey, spotted with black, the tail banded, and as long as the body. This species is found in the southern parts of Europe, in Africa and Asia, and exhibits numerous varieties.

Fig. 7. Variety of the Common Genet from Malacca.

Fig. 8. The Fossane Genet (*G. Fossa*), reddish-grey with reddish-brown spots, and four longitudinal lines of black on the back, inhabits Madagascar.

Fig. 9. The Banded Genet (*G. fasciata*), brownish-yellow, with longitudinal brown bands.

Fig. 10. Slender Genet (*G. gracilis*), clear reddish-yellow, with transverse brown bands and spots, the tail annulated with brown; inhabits Java.

DESCRIPTION OF THE PLATE.—THE EAGLES.

The order of Rapacious Birds is divided into two sections, one containing the diurnal, the other the nocturnal species. The former are further subdivided into two families, the Vulturine and Falconine. Of the latter family the Eagles constitute a genus, characterised by their great size, their strong, straightish, compressed bill, of which the tip of the upper mandible is decurved and trigonal, its edges festooned. The wings are very large and rounded, and the feet extremely muscular, with strong toes, and long, curved, very acute claws.

Fig. 1. The Golden Eagle (*Aquila Chrysaetos*) inhabits various parts of Europe, Asia, Africa, and America. The female, as in most species of this order, is much larger than the male. When full-grown, it is of a dark-brown colour, the tail uniform with the back, the hind head and neck yellowish-brown, the cere and toes bright yellow, the bill and claws bluish-black. This species has already been pretty fully described in our 20th Number.

Fig. 2. The Golden Eagle when young, has the greater part of the tail white, and in this state was formerly considered as a distinct species, and named the Ring-tailed Eagle.

Fig. 3. The Plaintive Eagle (*A. Nævia*) is much smaller than the Golden Eagle, and inhabits various parts of Europe, especially the southern and eastern, and is also common in Egypt. It is of a nearly uniform glossy reddish-brown colour; the tail dusky, with a terminal light red band; the bill and claws are bluish-black; the cere and toes yellow. This species feeds on hares, rabbits, other small quadrupeds, birds of various kinds, and in summer large insects.

Fig. 4. The Young of the Plaintive Eagle differs from the adult bird in having numerous large oval white spots on the wing-coverts.

BRITISH BIRDS.—NO. VII.—THE RED GROUSE AND PTARMIGAN.

The Ptarmigans are distinguished from the Grouse by having the tarsi and toes feathered, and by the want of the lateral pectiniform scales appended to the toes of the latter. In most other respects, however, the genera *Tetrao* and *Lagopus* are so similar, that it may well be doubted whether they really ought to be considered as distinct. It has been stated, that only one species of Grouse, which was described in our last Number, occurs in Britain, while in that country there are two species of Ptarmigan, the Red Grouse of authors, or, more properly, the Brown Ptarmigan, and the Common or Grey Ptarmigan.

The Red Grouse (*Lagopus Scoticus*) is peculiar to Great Britain and Ireland, not having hitherto been observed in any other part of the world. It is a strong full-bodied bird, the male about sixteen inches in length, its extended wings measuring twenty-seven on an average. The colour varies a little with the seasons. In winter the adult male is chestnut brown, inclining to red on the neck; on the body variegated with black; on the breast blackish, with many of the feathers tipped with white. The general colour of the female is yellowish-red, spotted and varied with black. In summer the male is chestnut brown, minutely varied and spotted with black; the head and neck also varied; the breast darker and more obscurely varied. The female is yellowish-red, spotted and varied with black; most of the feathers on the upper parts tipped with yellowish-white. The superciliary membranes or crests, which are larger in the male, are vermilion, the bill brownish-black, the iris hazel, the claws blackish-brown, greyish-yellow at the end.

Although this bird occurs in Ireland and England, it is more abundant in Scotland, where it is met with on the heaths, from the level of the sea to the height of about two thousand feet. The low sandy heaths of the eastern parts of the middle division appear to be less favorable to it than the more moist peaty tracts of the western and northern districts, where the shrubs on which it feeds attain a greater size. In the central regions of the Grampians it is equally abundant as on the moors of the Hebrides; and on the hilly ranges of the south, the Pentlands, the Lammermuir, and the mountains of Peebles, Dumfries, and Selkirk, it is still plentiful.

It is pleasant to hear the bold challenge of the cock at early dawn on the wild moor, remote from human habitation. You may fancy it to resemble the syllables *go, go, go, go back, go back, go back*, although the Highlanders, naturally imagining the bird to speak Gaelic, interpret it as signifying *co, co, co, mo-chlaidh, mo-chlaidh*, that is, who, who, (goes there,) my sword, my sword.

The food of this species consists chiefly of the tops of heath, *Calluna vulgaris*, *Erica cinerea*, and the leaves and twigs of other shrubs and herbaceous plants, as has already been stated in No. xx, where a sketch of the digestive organs has been presented. On ordinary occasions, the species does not fly much, but keeps concealed among the heath, seldom choosing to rise unless its enemy comes very near. On the approach of danger, it lies close to the ground, when, being of a colour not contrasting strongly with that of the plants around, it is with difficulty perceived by rapacious birds, among which its principal enemies are the Golden Eagle, the Peregrine Falcon, the Common Buzzard, and the Henharrier. The quadrupeds which occasionally prey upon it are the Polecat, the Pine Martin, the Fox, and perhaps the Ermine. When traced by a dog, it either runs to some distance, or squats at once. On such occasions the male is generally the first to rise. He erects himself among the heath, stretches out his neck, utters a loud cackle, and flies off, followed by the female and young, affording by their straight-forward, heavy, though strong, flight, an easy mark to a good shot. Young birds often allow a person to come within a few yards or even feet before they fly off, and even the old males, unless previously harassed, rise within shooting distance.

In a district where there is choice of situation, the Red Grouse prefer the slopes of hills not exceeding two thousand feet high; but they are to be found on the lowest and most level peat-bogs, especially if there are large tufts of heath surrounded by banks. Those which in summer and autumn reside on the heights usually descend in winter; but even during that season, individuals may be found in their highest range, which is bounded less by actual elevation than by the disappearance or scantiness of heath.

This species generally flies low and heavily, moving its wings rapidly, with a whirring noise, and proceeding in a direct course without undulations. Occasionally, when at full speed, and especially when descending parallel to a declivity, it sails at intervals, that is, proceeds for a short time with expanded and apparently motionless wings. Its flight is strong, often protracted to a considerable distance, and capable of being urged to a surprising degree of velocity, when the birds have been pursued by a hawk.

Although the haunts of the Red Grouse are the heathery moors, it has sometimes been found in stubble-fields, or among corn, bordering on uncultivated tracts; and when it finds an opportunity of feeding on oats, it does not scruple to avail itself of it. Unlike the Black Grouse, it is seldom or never met with in woods.

The male is not polygamous, nor does he at any time desert his mate. When incubation is over, and the young run about, they are tended by both parents, the female manifesting great anxiety for their safety, and feigning lameness to induce a person who has approached them to follow her. When surprised on the nest, she flies with a low undecided flight to a short distance, and runs off among the heath. The young are soon able to fly, and the flock keeps together until the end of autumn, unless scattered and thinned by sportsmen and vagabonds. Towards the beginning of winter, several flocks often unite and keep together, forming what are called packs. They are then generally more shy, and continue so until the beginning of spring, when they separate and pair, without manifesting any remarkable animosity; for although the cocks may occasionally fight, they have not those regular periodical battles described by authors as enacted by many species of the Grouse genus.

The nest is found in the midst of the heath, in a shallow cavity, and formed of bits of twigs, grass, and sometimes a few feathers. The eggs are from eight to twelve, or even more, generally an inch and seven-twelfths in length, an inch and three-twelfths across, of a regular oval form, yellowish-white, pale yellowish-grey, or brownish-yellow, thickly clouded, blotched, and dotted with blackish and amber brown.

The young leave the nest soon after they are freed from the shell. They are at first covered with a fine close down of a pale yellowish-grey tint, mottled beneath with pale brown, patched above with deep brown, the top of the head chestnut, margined with darker.

As an article of food, the Red Grouse is highly esteemed. The flesh is very dark coloured, and has a peculiar somewhat bitter taste, which by some is considered as extremely pleasant. The species is capable of living in a state of domestication, and then feeds on grain, bread, potatoes, and other substances, although it always prefers its natural food. A few instances have been known of its breeding in captivity; but, from its habits, it does not seem probable that it could be trained in subjection, like the domestic fowl.

The Grey Ptarmigan (*Lagopus cinereus*) is the only other species of this genus that occurs in Britain, where it is now confined to the summits of the higher mountains of the middle and northern divisions of Scotland. It resembles the Red Grouse in form and proportions, the male measuring about fifteen inches in length, and about twenty-five inches between the points of its extended wings. In winter the male is white, with a black band from the bill to the eye, the tail-feathers greyish-black, barred and tipped with white, the shafts of the primaries brown. The female in winter is also white, the feathers between the bill and the eye black at the base only, the tail-feathers brownish-black, based and tipped with white. In spring both sexes are white, mottled with dark grey, and yellow feathers, which are barred with black; the wings, lower parts, and tail, as in winter. In summer the head, neck, upper parts and sides, are spotted and barred with yellow and brownish-black; the wings, lower parts, and tail, as in winter. In autumn the plumage of the upper parts and sides is finely barred with greyish-white and greyish-black; the head, neck, and sides, retaining the yellow summer feathers longest; the wings, lower parts, and tail, as in winter. The young are spotted and barred with yellow and dark brown; the wings white, the shafts of the primaries dusky; the tail brownish-black, the middle feathers barred with yellow and dark grey.

This beautiful bird is met with in flocks on the bare and weather-beaten summits of the Grampians, and other high mountains of the North, where they reside from the beginning of spring to the close of autumn, seldom descending into the heathy tracts, unless in winter when the ground is covered with snow. Its food consists of various plants, chiefly of a shrubby nature; twigs and leaves of *Calluna vulgaris*, *Erica cinerea*, *Empetrum nigrum*, *Vaccinium myrtillus*, *V. vitis-idæa*, *Salix herbacea*, and others, being in fact similar to that of the Red Grouse. Its habits are also similar, but it generally allows a much nearer approach, and seems in no degree aware of danger from the proximity of man, remaining squatted on the ground, or on a stone until you almost trample upon it. When in packs, however, it is more shy, although even then it seldom flies off until one comes within shooting distance.

While feeding, the Ptarmigans run and walk among the lichen-crusted and crumbling fragments of rock, from which it is very difficult to distinguish them when they remain motionless, as they invariably do should a person be in sight. Indeed, unless you are directed to a particular spot by their strange low croaking cry, which is not very unlike that of a frog, you may pass through a flock without observing a single individual. When squatted, however, they utter no sound, their object being to conceal themselves; and if you discover the one from which the cry has proceeded, you generally find him on the top of a stone, ready to spring off the moment you show an indication of hostility. If you throw a stone at him, he rises, utters his call, and is immediately joined by all the individuals around, which, to your surprise, you see spring up one by one from the bare ground. They generally fly off in a loose body, with a direct and moderately rapid flight, resembling, but lighter than, that of the Red Grouse, and settle on a distant part of the mountain, or betake themselves to one of the neighbouring summits, perhaps more than a mile distant.

Early in spring they separate and pair. The nest is a slight hollow, scantily strewn with a few twigs and stalks, or blades of grass. The eggs are of a regular oval form, about an inch and seven-twelfths in length, an inch and from one to two-twelfths across, of a white, yellowish-white, or reddish colour, blotched and spotted with dark brown, the markings larger than those of the Red Grouse. The young run about immediately after leaving the shell, and from the commencement are so nimble and expert at concealing themselves, that a person who has accidentally fallen in with a flock vary seldom succeeds in capturing one. They are at first covered with a light yellowish-grey down, patched on the back with brown, and having on the top of the head a light chestnut mark, edged with darker. When fledged, they are very similar to the young of the Red Grouse, but banded and spotted with brighter reddish-yellow.

The flesh of the Grey Ptarmigan is not so dark as that of the Red Grouse, nor quite so bitter, although it has the same flavor. A very considerable quantity is annually killed, the bird being held in estimation not only as an article of food, but, when stuffed, as a domestic ornament.

Hooded Crow.—The Hooded Crow, or *Hoody*, is the Carrion Crow of Scotland; but whether identical with the English bird of that name, is doubtful. It is not migratory, like the Carrion Crow of England, but is found at all seasons on most of the headlands and rocky shores of the Highland lochs and Western and Northern Islands. In these places it does not breed in trees, as in England; for there are no trees; but it makes its nest in precipitous crags. It is the habit of this bird, as of the Eagle, that one pair appropriate to themselves a breeding-place, and drive away all intruders. If one of the mates be killed, another very speedily appears and takes its place. In like manner, two or three pairs of *Hoodies* appropriate a district of coast or an island, the numbers being always limited to the means of subsistence. These do not fight among themselves, but prevent strangers from encroaching on their feeding-grounds. The boldness, rapacity, and cunning of the Hooded Crow are very remarkable. Two gentlemen, who were lately on a visit to Barra Head Lighthouse, observed to the very intelligent chief light-keeper there, that a fine-looking Domestic Cock had lost the feathers of his neck. "That," said he, "is the consequence of fighting with the Hoody Crows in defence of his hens, which the Hoodies would kill and devour." He added, that even the lighthouse Dog was not a match for the Hoodies; but on him they practised cunning bearing the stamp of reasoning. When the Dog had got a bone, and was couching with it between his paws, one Hoody was observed to come in front, and another to approach behind: the one in front of the Dog manœuvred impudently, till the indignant cur, losing temper, left his bone and made a spring forward at the presumptuous bird; at this moment, the accomplice Crow from behind instantly struck in and flew off in triumph with the prize!—[For the above notice, from the pen of Dr Neill, we are indebted to Mr Fraser.]

Carnivorous Geese.—The carcase of a horse lying near the Gas Works, by the side of the river at Musselburgh, at present affords an object of great attraction to the domestic Geese. These animals, which we saw crowded about it to the number of nearly twenty, have nibbled off nearly all the flesh from the back and ribs. It is remarkable, however, that they have not succeeded in making any thing of the intestines, their soft and blunt bill not enabling them to tear asunder so tough a substance. Indeed, so ill adapted is that instrument for picking the bones, that they have only been able to tear off the muscular fibres, leaving all the tendinous and ligamentous parts adherent. We are aware that domestic Ducks and even Geese will eat almost any thing that comes in their way, and have seen them devour fish, suet, potatoes, blood, mice, &c.; but do not recollect to have seen the Goose recorded as a carrion bird.

Remarks on Varieties of the Fox observed in Scotland. By W. MacGillivray.
(Read to the Wernerian Natural History Society.)

The object of the following remarks is not to furnish anything new respecting the organization and habits of an animal so well known as the Fox, but to point out differences observed in individuals living in a limited range of country, with the view of drawing attention to peculiarities generally neglected by those who study nature from books.

It appears to me, that there are four races or varieties of the Fox in Scotland.

The first of these, the *Hound* Fox, is tall, slender in the limbs, with a very attenuated muzzle; a bright reddish-yellow fur; the lower parts of the body greyish-white; the tail greyish-yellow, with long black hairs scattered towards its extremity; and about three inches of its extremity white.

The second, the *Cur* Fox, is similar to the first variety in colouring, but is smaller, with the body deeper, and the legs shorter. These two races seem to pass into each other, and can scarcely be distinguished except in the extremes.

The third, the *Dog* Fox, is compact in form, with comparatively short limbs; the head rather broad, the muzzle pointed; the fur deep red; the lower parts brownish-red; the tail yellowish-grey, darkened with black hairs, and having the tip of the same colour.

The fourth, the *Mastiff* Fox, is larger and stronger; its limbs more robust; the head much broader; a dull, greyish-yellow fur, profusely interspersed with white hairs; the tail dusky, with long black hairs scattered over it, and a small white tip.

The first and second varieties are of common occurrence in the lower districts of the country. The third is very rare, and individuals of it being considered inferior specimens of the species, it is not to be seen in collections. The fourth kind seems peculiar to the Highland districts.

The first variety approaches in its general form to the most characteristic variety of the Shepherd's Dog. The head is of ordinary size, with a long, attenuated muzzle; the ears roundish, erect, with a slit at the lower part of the outer margin; the eyes of moderate size; the mouth opening to beneath the middle of the eye; the snout small, the nostrils terminal, with a lateral slit; the upper lip grooved; the neck of moderate length, but strong; the body rather long and compact; the limbs longish and rather slender; the tail reaching to the feet.

On the fore feet are five toes, the inner small and raised from the ground; on the hind feet four toes; the claws slightly arched, compressed, bluntish, grooved beneath, near the end; those of the hind feet broadest.

$$\text{Incisores } \frac{6}{6} \quad \text{Canini } \frac{2}{2} \quad \text{Molares } \frac{12}{14}$$

Incisors of the upper jaw incurvate, rounded, with two small lateral processes, excepting the two larger side ones, which are caniniform. Canini tapering, incurvate, long, distant from the incisors. Laniarii 4; 1st small, compressed, obtuse, with a posterior shoulder; 2d and 3d nearly equal, compressed, tapering, with two shoulders; 4th much larger, oblique, with an anterior tubercle, a central point, and a posterior edge. Of the two molares, the anterior is much larger; both are similar in form, transverse, with two external tubercles, one internal, and a projecting margin.

In the lower jaw the incisors are smaller, rounded, with an external lateral process, the two outer considerably larger. Canini close to the incisors, smaller than those above, and more curved. Of the laniarii the first is very small, rounded, with a posterior shoulder; the rest nearly equal, with anterior and posterior shoulders, the last one with two posterior. The next tooth, which is the largest, combines the two forms, its anterior part being sharp-edged with two points, the posterior, lower, with four unequal tubercles. The first grinder has four tubercles, and the last, which is very small, has three.

The fur is rather soft, pretty close, longish, and consists of a fine woolly covering, and long hairs of a stiffish texture; the tail bushy; the hair of the limbs and face short; the soles covered with hair, excepting the tuberculous eminences on which they have been worn.

The general colour is yellowish-red, pure on the sides of the neck and chest, and lower sides of the abdomen, mixed with white hairs on the shoulders, thighs, upper flank, sacral region, and tail. The snout and lips are blackish; the inner surface and edges of the ears whitish, outer part of the ear pure black. Tail pale greyish-yellow, the tips of the hairs brownish-black, its extremity white to the length of about three inches. Lower part of the cheeks, fore part of the neck, lower surface of the body, inner thighs, pudic region, a narrow line down the fore part of the leg, and the hind and inner part of the fore legs, greyish-white. Fore part of all the feet, brownish-black, that colour extending from the toes to near the cubital and knee-joints; hair of soles deep red; claws light brown; mystachial bristles black.

The male and the female are precisely similar in external appearance; nor does there seem to be much difference in size.

Dimensions of two individuals:

	Male.		Female.	
	FT.	IN.	FT.	IN.
From the tip of the nose to the root of the tail..................	2	6	2	4½
Length of the tail (not including the hair).........................	1	4	1	3½
Total length (including the hair)......................................	4	0	3	10
Height at the shoulder } (following the flexure of the limbs) {	1	6	1	6
Height at the rump }	1	7½	1	7
From the nose to the occiput..	0	7	0	7

The second variety is similar to the first, but smaller, with the limbs shorter, the colour of a deeper red, and having a larger mixture of white hairs on the parts mentioned.

Of the Dog Fox the following is a description taken from a male shot in Peeblesshire, in December 1835. The general colour is yellowish-red, the forehead, shoulders, and haunches with greyish-white hairs interspersed. There is a dark patch between the mystachial bristles and the eye, and the lower jaw is dusky. Edge of upper lip, base of lower, throat, and fore neck, with a short band down each humerus anteriorly and internally, and another on the middle of the breast, greyish-white. The lower parts of the body are brownish-red; the axillæ and groins light red; the scrotal hairs yellowish-white; outer surface of ear black, inner pale greyish-yellow. Legs dark brownish-red, fore part black from the toes up to the cubital and knee-joints; hind part of the hips deep red; tail yellowish-grey, the tips of the long hairs brownish-black, the extremity of the same colour; claws yellowish-grey.

	FT.	IN.
From the tip of the nose to the root of the tail..................	2	3
Length of tail	1	3
Total length (hairs of tail included)	3	8
From the nose to the occiput...	0	6½

A skin and skull of a Fox procured in Sutherland by Sir William Jardine, who favored me with the inspection of them, belonged decidedly to the fourth variety, the Mastiff Fox. The skull greatly exceeds that of the other kinds in the width of its zygomatic arches, and the size of the cerebral cavity. The fur is longer and coarser, of a duller red, and much intermixed with white hairs; the lower parts light grey, and the extremity of the tail white for about three inches. Respecting this variety I am unable to furnish any further particulars, not having had an opportunity of late of examining a recent individual.

I have reason to believe that similar varieties exist in other Mammalia inhabiting this country. For example, the Red Deer of Lewis and Harris are much smaller and of a dunner colour, than those of the mainland. Of the Weasel, there is a variety much inferior in size to the common; and of the Water-Rat I am aware of three, the common brown, the black, and an intermediate kind. The latter, as well as the black, may be a distinct species, it being considered as such in the neighbourhood of Keith, where it is said to lay up a store of potatoes for winter provender.

Whether the different kinds of Fox above described breed together, or remain distinct, is as yet unknown, nor are we in possession of any positive knowledge as to their peculiar habits.

THE SISKIN BREEDS IN SCOTLAND.—The following notice, interesting as affording evidence of a fact hitherto doubted, was handed to us by Mr Carfrae, preserver of animals, Edinburgh. The Siskin is a common bird in all the high parts of Aberdeenshire, which abound in fir woods. They build generally near the extremities of the branches of tall fir trees, or near the summit of the tree. Sometimes the nest is found in plantations of young fir wood. In one instance I met with a nest not three feet from the ground. I visited it every day until four or five eggs were deposited. During incubation the female showed no fear at my approach. On bringing my hand close to the nest, she showed some inclination to pugnacity, tried to frighten me away with her open bill, following my hand round and round when I attempted to touch her. At last she seized a firm hold of my finger, and held fast. I visited her almost every day, and could with perfect confidence stroke down her back. As last she would only look anxiously round to my finger, without making any attack on me. The nest was formed of small twigs of birch or heath outside, and neatly lined with hair.— *J. M. Brown, Abergeldy.*

BOTANY.

THE WATER TREFOIL, OR BOGBEAN.—MENYANTHES TRIFOLIATA.

THIS plant, which is generally distributed in Britain, growing abundantly in marshy places, peat-bogs, and by the sides of lakes and pools, is one of the most beautiful of our native species. It belongs to *Pentandria monogynia* of the Linnæan system, and to the natural family of *Gentianeæ*. The root is perennial, long, creeping, jointed, and sends out numerous verticillate white fibres. The leaves are alternate, petiolate, ternate, the leaflets obovate, thick, smooth, and deep green. The flower stalk rises to the height of from six to ten inches, and supports an oblong or conical raceme of numerous very beautiful flowers, which are pentapetalous; the corolla previous to expansion rose-coloured externally, afterwards reddish-white, the petals on the inner surface covered with numerous fringe-like white filaments.

All parts of this plant are extremely bitter, and in some countries it is used as a substitute for hops in the preparation of ale. The root, although almost equally bitter, Linnæus informs us, is dried and powdered by the poorer people in Lapland, to be made, with a little meal, into a coarse unpalatable bread. In this country the plant is not applied to any use, if we may except its occasional employment in some parts of the north of Scotland, as a purgative for calves. Formerly it was much employed in various chronic diseases, as scurvy, dropsy, jaundice, asthma, and gout, the paroxysms of the latter of which complaints it was supposed to keep off, but at present it is neglected on account of the preference given to gentian and other bitters. There can be no doubt, however, that, as an astringent and stomachic, it is equally powerful with many exotic plants.

MISCELLANEOUS.

ON VARIETIES.—NO. II.

(Continued from Page 116.)

AMONGST articulated animals (and I believe in most cases where any organ or part is greatly multiplied), we observe instances of the number of legs being indeterminate, such as *Julus indicus*. Another instance is found in *Aphrodita aculeata*. We also observe that extraordinary form of variety noticed above in certain fishes and in shells, consisting of a transfer of certain structures ordinarily peculiar to one side of the animal to the opposite side. This is seen in our common Lobster, in which the knobbed claw and the serrated claw exchange sides as it were capriciously. I say capriciously, because, in this instance, unlike those remarked in fishes and shells, there is no predominating arrangement of these organs.

But besides these more remarkable forms of variety in the lower grades of animals, there are others less notable, generally dispersed. The Hercules Beetle is said to vary so extensively in size, as to have raised doubts occasionally whether these variations were not entitled to specific consideration. The Tarantula has several varieties differing in regard to the quality of the hair. Varieties in regard to colour are of frequent occurrence. All the Echinidæ are disposed to vary extensively in this respect; more than thirty varieties as to colour have been observed in *Echinus angulosus*. Many species vary also largely in structure and shape, particularly in their spines. Thus there are four varieties of *Echinus cidaris*, all determined by the form of the spines.

In the extreme grades of animal organization, in which the aspect at least of vegetables is assumed, we find the differences between individuals in the same species carried to their highest pitch. Between these and absolute varieties no line can any way be drawn, although something must be allowed for our very imperfect knowledge of these plant-like animals. Yet authors have been pleased to enumerate varieties occasionally, and in these very considerable amounts of difference are set forth. Thus in *Corallina rubens* three varieties occur, distinguished by structural peculiarities :— See Fleming's British Animals, p. 514. *Halichondria ramosa* (Flem. p. 523) has several varieties, in all of which the structure differs greatly ; but I apprehend that in viewing the subject of varieties in these tribes, we should keep in view their near relationship to vegetables, and that in consequence the varieties observed in them will partake of the same affinity. It might, therefore, be reasonably supposed that in simulating the aspect of plants, they would put on those forms of variation which peculiarly distinguish the vegetable world, consistently, however, with their actual animal nature. This is actually the fact. They observe no fixed or precise dimensions, although there is a limit to their growth. In numerous instances the figure of the species varies largely. Frequently the shape of some of their parts suffers great variation; very often specimens of one kind present a variety in regard to colour; and, lastly, in some, as before noticed, no adherence to any dimension or figure (as a whole or as regards parts), within limits which are allotted to other zoophytes, can be detected, so as to furnish a clue to what might be the prevailing appearance. We shall afterwards see that these forms of variety in the zoophytes are represented in various departments of the vegetable kingdom.

It must be borne in mind, that although there is a most evident gradual transition from animals to vegetables, the traces of analogy are soon lost in the predominating characteristics of the latter class, and that there are as many differences between extreme classes, or departments of vegetables, as are found between extreme classes of animals; so that it might reasonably be inferred, that a review of the classes of vegetables of the same nature, such as that we have given to the tribes of animals, would be attended with advantage. Yet, it must also be recollected, that throughout the vegetable kingdom, our knowledge of organization, and consequently our knowledge of the limitations of species, is extremely circumscribed, insomuch, that the science of Botany, as far as regards the determination of species and varieties, is involved in opinion and conjecture. In a review, therefore, of the above named kind, we should be constantly impeded by a want of fixed rules and examples, and we shall content ourselves with an enumeration of such ascertained and determined forms of variety as are met with in this class.

In the first place, one rule must be stated relative to plants in general, that consistently with an increased proneness to vary in proportion as the subject is low in the scale of organization, they display the most extended series of variations, a great portion of which assuming too large an amount of difference from the standard structure and appearance, to be regarded as ordinary specimens, have been formally arranged by authors as varieties; yet, as would naturally be expected, disputes have occurred, first regarding the number of those specimens entitled to be ranked as varieties, and, secondly, regarding what should be deemed varieties and what species. The vegetable world, then, is a continued scene of variation, and the rule of fixing *varieties*, technically so termed, is in the highest degree arbitrary and artificial. The most extensive mode of variation among plants is that of size and dimensions. The same plant is liable from circumstances, such as soil or station, to suffer alterations, in this respect, of considerable amount, so that it has frequently been matter of dispute among botanists, whether the most remarkable of these should be esteemed species or varieties. An alteration in size is most commonly accompanied by alterations of other kinds, especially variations in the shape of the whole or of parts, but these also occur independently of a variety in size. The mode of growth and the habits of plants are subject to variety. A plant commonly erect may be found prostrate, and *vice versa*. Colour is a mode of variation particularly observable in plants. We here refer more especially to the flower. Frequently there are several varieties in regard of colour in one species. Varieties which affect the structure or organization of the species are extremely common. This is illustrated in a most ample manner by the hair or pubescence, a species very frequently having varieties with this kind of clothing variously disposed, and in various quantities. The number of styles is not constant in some species of plants; in *Arenaria rubella*, for instance, they vary from three to five. In *Acer Pseudoplatanus* the cotyledons are said to vary from one to four. In those species where the number of any given organ is small, and so far uniform as to furnish a specific character, varieties in respect of such number are sometimes met with. Thus *Erigeron alpinus*, which usually has but one flower or anthodium on each stem, has two varieties, in one of which the number of flowers ranges from one to three. *Orchis bifolia* is occasionally furnished with a third leaf. With the exception of such instances as these, flowers, leaves, &c., are numerically unrestricted and undefined.

As I have before stated, the lowest of vegetable forms enjoy a perfect freedom in regard of size and shape, excepting that a maximum is appointed for the first, and that the shape is at all times in some degree characteristic of the species respectively. I once more repeat, that in this second kingdom of organized beings, varieties obtain to their utmost extent, and that it is an arbitrary proceeding to appoint certain of these variations to be entitled "varieties" in books. In my remarks, therefore, I have not confined myself to these last, but have thought fit to view the subject of varieties in this class as a whole. It remains to be said, that accidental and permanent varieties are both met with amongst plants, the former by far the most numerous, and oftentimes found to be constantly liable to reproduction. Permanent varieties are very frequent, as might easily be supposed from the lasting influence exercised over vegetable productions by the soil and medium in which they are placed, and these circumstances so prone to alteration within even confined limits. Besides this, vegetables are in very numerous cases endowed with the power of originating forms dissimilar to the stock producing them. *Draba verna*, which ordinarily displays a flat pouch, has a variety with a swollen pouch; *Poa alpina* has a variety in which the spikelets are densely crowded together; *Lolium perenne* exhibits great modification in its spike; *Artemisia maritima* has two permanent varieties both found growing in the same localities, one distinguished by a drooping, the other by an erect raceme. These differences seem so far permanent and inherent in the plant, that they are occasionally seen growing from the same root. We shall return at intervals to the subject of the two kinds of varieties mentioned in this paper.

Having, in the course of the preceding remarks, enumerated and instanced the several kinds of variation occurring in animals and vegetables, we shall in the next place inquire into the causes and uses of varieties. The uses are proximate and ultimate; that is to say, either dependent in an evident way on surrounding influences or circumstances, or produced by some agency, and for some reasons, respecting which we cannot be exactly certain. It is of course easy to refer all occurrences and all productions to the great source of beauty and variety in nature; but it is the business of the naturalist to search for intermediate causes, even though the search be likely to prove fruitless, and though we should be in danger of setting that down as a cause which is perfectly inadequate to the office, whilst the true cause remains hidden, and unadapted to human comprehension. If, however, no cause can be discerned or surmised, it is equally the business of the philosopher to ascribe the circumstance to the superintending agency of God; at least until further discoveries have been made, very many circumstances in connexion with our subject must be disposed of in this way.

Varieties in respect of size are caused by soil, station, climate, and food. This might easily be conceived, for it is not reasonable to suppose that species disposed to vary should, when their individuals are exposed to adequate causes, continue unaltered in their dimensions. Some species, however, are acted on only by one or more causes, and refuse to be influenced by others though exposed to them. The tendency to variation is not displayed alike by all kinds, not even by species nearly allied. These facts seem of themselves to show that some intention or design, in regard to varieties, pervades nature. Size is perhaps of all kinds of variety the most prevalent or general, and climate, in co-operation (for the most part) with food, the most usual cause of such diversities. One of the most interesting phenomena in connexion with our subject is that of some natural productions attaining their greatest magnitude, luxuriance, and vigour, as they approach the tropics. This, of course, can only be observed in cases where the geographic range is very extensive. On a small scale, it cannot be expected to be perceptible, unless the stations occupied should present great dissimilarities in climate within confined limits. It might not be easy to cite cases of variety (technically so named) in support of these assertions, but the circumstance of great variety in size occurring to species is quite sufficient for our purpose, and of this diversity numerous instances can be furnished. The Lion and the Tiger, whose geographic ranges are considerable, attain dimensions near the equator far greater than those which they reach at their furthest limits. The Slow-worm, *Anguis fragilis*, attains the length of three feet in countries farther south, and where the heat is more considerable, while with us its usual length is one foot. But besides this rule, as regards increase of size, as the species approaches the equator or an increased temperature, there is another law, which clearly shows that each species has its peculiar station at which it enjoys the full vigour of its growth and endowments without regard to heat. The Sea Ear, *Haliotis gigantea*, in the Polar Seas attains the length of six or seven inches; but this size, as well as the number of individuals, diminishes as we recede southwards, so that here is a reverse position to that before stated. So also is it with certain kinds of plants. Some are found to degenerate in size as they get into a colder latitude, or are found elevated into the cold atmosphere of a mountain, whilst with others, whose growth is in perfection on the summits of mountains, or in a latitude equally cold, a degeneracy in size can be observed as the species recede from such spots. But, on the other hand, there are instances, as before said, of species resisting almost entirely influences affecting others so freely as occasionally to raise doubts whether they might not be entitled to specific rank when observed under the extreme of variation to which they have been submitted. The White or Barn Owl inhabits the whole of Europe and America, and yet through this vast extent it exhibits remarkably slight differences. The Wolf also, found from the torrid zone to high northern latitudes, differs chiefly in the colour of its fur only.

We occasionally see varieties induced in plants by their being inundated through some accidental cause, in which case the medium (which is equivalent to soil) effects an alteration in their stature or size, besides sometimes producing structural differences; but I have before stated that plants, beyond all other organised beings, are liable to vary.

Upon the same principle, that species suffer diminution in size and deterioration generally in their structure and functions, as they recede from the spot at which their maximum of size and utmost luxuriance are attained; they are found in very many cases to be affected almost in a similar way by the same description of causes acting in the immediate vicinity perhaps of specimens which have attained to their extreme size and full development of parts and qualities. This remark applies to the lower classes, especially such as are denied the power of locomotion, but coming into existence in an unfavorable soil or station, or where the natural pabulum is furnished sparingly, constitute, in the language of naturalists, "dwarf," "impoverished," "starved," or "depauperated" specimens. Such also not unfrequently appear in books ranked as varieties technically. On the same principle, moreover, in these tribes suffering variation so freely, "gigantic" specimens (elevated occasionally to the rank of varieties) are met with, dependent for their extravagant luxuriance on some peculiarly appropriate site, or peculiarly adapted soil or medium.

Artesian Wells.—At present more than thirty Artesian wells are in progress in the departments of France. That at the Abbatoir, near the Barrière de Grenelle, becomes every day more interesting. The boring has already reached the depth of 1360 feet. This well was commenced on the 30th December 1833, and during the 1200 days which have elapsed since then, the works have been directed by M. Mulot, jun., and have not been discontinued for a single day. From this it appears that the average progress has been upwards of thirteen inches per diem. It appears that the administration are about to make an engagement with M. Mulot to bore to the depth of 1800 feet, if water be not previously met with. A well begun at Dresden had obtained in October last an abundant supply of water at the depth of 840 feet. This source, having a temperature of 68° Fahr., furnishes a supply of 14 gallons of good water per second. They penetrated through 62 feet of sand and gravel; 810 feet of marl and chalk; 43 feet of pure marl; and 22 feet of greyish freestone. Admitting the above temperature to be that of the strata at this depth, and comparing it with the mean temperature at the surface of the earth at Dresden [48° Fahr.], we find a uniform increase of temperature of 1.20° Cels. for every 100 feet, or, 1° for every 78 feet of depth; but this increase being greater than that actually observed in boring, we must conclude that the water of this well comes from a greater depth.—*L'Echo du Monde Savant.*

Edinburgh: Published for the Proprietor, at the Office, No. 13, Hill Street. London: Smith, Elder, and Co., 65, Cornhill. Glasgow and the West of Scotland: John Smith and Son; and John Macleod. Dublin: George Young. Paris: J. B. Balliere, Rue de l'Ecole de Médecine, No. 13 bis.

THE EDINBURGH PRINTING COMPANY.

THE EDINBURGH
JOURNAL OF NATURAL HISTORY,

AND OF

THE PHYSICAL SCIENCES.

DECEMBER, 1837.

ZOOLOGY.

DESCRIPTION OF THE PLATE.—THE ORANG OUTANGS, PYGMIES, AND GIBBONS.

THE genera of the extensive order of Quadrumana represented in this plate, are those which exhibit in their conformation the nearest approach to the human species. They have four incisors in each jaw, a more or less prominent nose and elongated arms.

PITHECI.—MEN-OF-THE-WOODS, OR ORANG OUTANGS.

ONLY one species of this genus is known. It is that which presents the greatest resemblance to Man; and is characterized by a large and rounded head, a narrow flattened nose, with expanded nostrils, a large projecting mouth, reddish shaggy hair, and other features, which it is unnecessary to particularize here, as a full description will be subsequently presented.

Fig. 1. THE RED ORANG OUTANG (*P. Satyrus*). It inhabits Cochinchina, Malacca, and the islands of the Indian Archipelago, and is said to attain a height of six or more feet. The individuals brought to Europe have been young, and have never attained maturity. An individual described by Dr Abel did not practise the grimaces of other Apes or Monkeys, but exhibited a gravity approaching to melancholy, and a mildness superior to that of almost any other animal.

Fig. 2. THE FEMALE differs from the male chiefly in being less hairy.

TROGLODYTES.—THE PYGMIES.

OF this genus also, only a single species is known. It bears a close resemblance to the Orang Outang, from which it differs in having a prominent superciliary ridge.

Fig. 3. THE CHIMPANZEE (*T. niger*), or Black Orang Outang, inhabits some parts of Africa, and especially the coasts of Angola and Congo. It lives in troops, and is said to be very intelligent; but little is known respecting its habits; and all the individuals brought to Europe alive have been young.

HYLOBATES.—THE GIBBONS.

THESE animals are remarkable for the extreme elongation of their arms, and are further distinguished from the preceding genera by having callosities on their buttocks.

Fig. 4. THE BLACK GIBBON (*H. Lar*), is characterized by its black fur, and the circle of grey hairs which surrounds its face. It inhabits Coromandel.

Fig. 5. THE FEMALE differs in wanting the grey hairs on the face.

DESCRIPTION OF THE PLATE.—THE HARRIERS, AND SNAKE-EATERS.

THE HARRIERS.—CIRCUS.

THE HARRIERS constitute a genus of the order *Accipitres* or *Raptores*, intermediate in some measure between the Hawks, properly so called, and the Owls. They are generally of a slender form, with elongated wings, tail, and tarsi, their toes rather short, their claws slender and moderately curved. The aperture of their ears is very large, and their head is surrounded with a kind of ruff resembling that of the Owls, but less distinct. They prey on small quadrupeds, birds, reptiles, and insects, and generally attack birds while on the ground, although their flight is rapid and bouyant.

Fig. 1. THE LONG-LEGGED HARRIER (*Circus Acoli*) is of a light greyish-blue colour above, and on the fore neck, the lower parts white, transversely barred with black, the cere rich orange, and the feet yellow. It inhabits Africa.

Fig. 2. THE BUSON HARRIER (*C. Buson*) is of a brownish red, barred with blackish-brown beneath, the quills and tail dusky, the latter barred with white spots.

Fig. 3. THE INDIAN HARRIER (*C. melanoleucus*) has the upper parts and fore neck of dusky brown, the lower parts, smaller-wing coverts, and terminal portion of the secondaries and their coverts, with the tail, white.

Fig. 4. THE HEN HARRIER (*C. cyaneus*) is light blue in the adult state, with the rump white; but the female and the young, of which the latter is here represented, are brownish-red with dusky markings. In this state it was formerly considered as a distinct species, under the name of the Ringtail.

THE SNAKE-EATERS.—SERPENTARIUS.

Fig. 5. THE SECRETARY OR AFRICAN SNAKE-EATER (*Serpentarius Secretarius*). This is the only species of the genus, which, on account of its long legs and other circumstances, has by many systematists been referred to the order *Gralla* or waders. Its true place, however, is among the *Accipitres*, as is shown by the form of its bill, its claws, and its habits, which are essentially those of a rapacious bird. It preys especially on snakes, which it attacks with its wings and feet, trampling them down, and raising them in the air to allow them to drop on the ground. It inhabits various parts of Africa, and is particularly abundant at the Cape of Good Hope. The name secretary has been given to it on account of the occipital crest suggesting the idea of a clerk with his pen stuck behind the ear.

BRITISH BIRDS.—NO. VIII.—PARTRIDGES.

THE generic name of Partridges was formerly applied to a vast number of birds, which, in the present improved state of ornithology, have been arranged into several distinct genera. In many respects they are very similar to the Grouse and Ptarmigans, from which they differ, however, in having the tarsi always bare, and sometimes furnished with a tubercle behind, and in generally having a bare space behind or about the eye, while they want the coloured superciliary membranes. Properly speaking, only two species are indigenous in Britain, the Common or Grey Partridge, and the Common Quail; but the Red Partridge, which is common in France, and the Virginian Quail, which is peculiar to America, have been introduced, and are spreading in some counties of England. To the genus *Perdix*, characterized by a short, strong, convex bill, a bare space behind the eye, short strong tarsi anteriorly covered with two rows of scutella, and in the male bearing a blunt tubercle behind, marginate toes, strong compressed and arched claws, wings having the first quill longer than the seventh, and a short rounded tail, belong the Red and Grey Partridges. The Virginian Colin is referred to the genus *Ortyx*, having a short and stronger bill, rounded wings, and tail of moderate length. The Quail belongs to the genus *Coturnix*, of which the species are small, have the head entirely feathered, and the tail extremely short.

THE RED PARTRIDGE, *Perdix rubra*, frequently called the Guernsey or French Partridge, is somewhat larger and more robust than the common species, which it resembles in form and proportions, but from which it differs greatly in colour. The bill, the naked space about the eye, and the feet, are bright red, as are the irides. The upper parts are reddish-brown tinged with grey; the forehead ash-grey; the throat and cheeks white; a black band extends from the bill to the eye, and thence down the neck, becoming broader on its fore part, which is spotted with the same colour; the lower parts are ash-grey and light red, the sides transversely banded with ash-grey, white, black, and red. The male has large flat tubercles on the tarsus. This species is said to occur in various parts of Asia and Africa, and to be plentiful in Spain, Portugal, Italy, and the South of France, where it inhabits the low grounds, feeding on seeds, grain, and insects. In England, where it is now not uncommon in some places, it is said to prefer waste heathy ground to corn fields, and to afford less sport than the common species, as it runs before the Dogs, the individuals composing a covey dispersing, and rising one after another. It is accused, moreover, of driving off the Common Partridge, which, it is feared, may in time be extirpated by it, as the Black Rat has been by the larger and more mischievous brown species.

THE GREY PARTRIDGE, *Perdix cinerea*, has the bill and feet bluish-grey; the upper parts minutely mottled with ash-grey, yellowish-brown, brownish-black, and brownish-red; the scapulars and wing-coverts darker, with longitudinal whitish streaks; the forehead, cheeks, and throat, light red; the neck ash-grey, minutely undulated with black; the sides broadly banded with brownish-red, of which there is a large patch on the breast. This well-known bird is generally distributed in Britain, being found in all the lower parts of England and Scotland, with the exception of some of the wilder districts of the latter country. Although not peculiar to cultivated land, it thrives best in those parts that are most extensively covered with crops, among which it finds comparative security during a considerable part of the year. It is of rare occurrence in the narrow valleys of the moorlands, and in the heaths is seldom seen unless in the immediate vicinity of corn fields. It is fond of rambling, however, into waste or pasture grounds, which are covered with long grass, furze, or broom; but it does not often enter woods, and never perches on trees. It runs with surprising speed when alarmed or in pursuit of its companions, squats when apprehensive of danger, and in flying rises obliquely to some height, and then moves off in a direct course, rapidly flapping its wings, which produce a whirring sound. Its food consists of tender blades of grass or corn, grain of all kinds, seeds of various plants, insects and larvæ. These substances, first lodged in the crop, are subsequently ground in the powerful gizzard, with the aid of numerous particles of quartz. Partridges feed principally in the morning and towards night, betake themselves during the middle of the day to places covered with shrubs or ferns, or bask under the hedges. In the evening, before betaking themselves to rest, they are often heard in the fields uttering their harsh sharp cry, apparently for the purpose of apprising each other of their position, so that the stragglers may come up, or the female join her mate. They repose at night on the ground, generally in an open and comparatively bare place. During winter they keep together in coveys, seeking their food among the stubble; but early in spring they separate, and by the beginning of March are generally paired, although the eggs are not laid until June. The place selected for their nest is various, it being found in corn and grass fields, in pastures, among shrubs, by hedges, sometimes even by road-sides. It is merely a slight hollow scraped in the soil, with a few straws, and contains from ten to fifteen eggs, of a pale greenish-brown colour, averaging an inch and a half in length. Although the male takes no part in incubation, he remains in the neighbourhood of the nest, and on apprehension of danger to it comes up and endeavours to entice from it the person who may have approached too near for its safety. The young ase led about by both parents, who manifest the greatest anxiety for their welfare. The female cowers over them in wet or cold weather, protects them in the same manner at night, and evinces the most striking marks of maternal tenderness in all her actions. The principal food of the young is insects and larvæ, and especially those of ants. During autumn and winter the brood remain with their parents, and towards the commencement of the latter season, several families unite into a pack. Partridge-shooting is a favorite diversion; but, notwithstanding the vast numbers annually killed, these birds seem to be rather increasing than diminishing; so that, as an article of food, they are not beyond the reach of the middle classes of society. The Partridge thrives in a state of captivity when properly fed, but refuses to breed in that condition. As may be expected in a bird apparently so dependent upon the labours of man for shelter and subsistence, it varies considerably in size, it being found that in the valleys of the hilly and little cultivated districts it usually attains a smaller size than in the rich plains, where it finds an abundant supply of nutritious food.

The Virginian Colin, *Ortyx Virginiana*, is much smaller than the Common Partridge, and has the upper parts brownish-red, variegated with black; the throat and a broad band over the eye, white; the loral space, and a broad band passing down the neck and crossing it in front, black; the lower parts greyish-white, undulated with black. The habits of this species have been admirably described by Wilson and Audubon, the latter of whom states, that it "has been introduced into various parts of Europe, but is not much liked there, being of such pugnacious habits as to drive off the Grey Partridge, which is considered a better bird for the table."

The Common Quail, *Coturnix dactylisonans*, is a very beautiful little bird, not exceeding eight inches in length, and having the upper parts variegated with reddish-grey and brownish-black, and marked with whitish longitudinal pointed streaks, of which are three bands on the head. It is generally distributed over Europe, and a great part of Asia and Africa. In the former region it is migratory, arriving in the beginning of summer, and departing in September, generally in vast straggling flocks. Their arrival in England takes place in the middle of May. They never appear with us in great numbers, but coming quietly, like the Corn Crake, spread over the country unobserved, and are pretty generally distributed, although nowhere plentiful, and in the northern counties very rare. It is seldom that they are now met with in Scotland. The food of the Quail consists of seeds, herbage, and occasionally insects; and its haunts are chiefly the cultivated fields and pastures, where it continues during the season, never entering the woods or perching. The nest is a slight hollow, with some dry blades, on which it deposits its numerous eggs, sometimes amounting to twenty, and of a regular oval form, reddish-yellow, marked all over with brown spots and blotches. The males are extremely pugnacious, and when they meet each other in their haunts engage in desperate combats. They utter a loud shrill cry, composed of several notes, which have been considered as constituting a kind of song, in consequence of which these birds are often kept in cages on the Continent.

CURIOUS HABITS OF A DOMESTIC PIGEON.

The following account of a Pigeon, exhibiting in a very remarkable degree the social instinct misapplied, has been transmitted to us by the Rev. Mr Adam of Peebles. "About fourteen years ago," says the owner of the bird, in a letter to the gentleman just mentioned, "the right wing of the Pigeon which you saw in our house was broken by a shot, which was the means of his coming into our possession. After recovering of his wound, he showed his courage by defeating a hawk, who had the audacity to attack him, as he sat in the sole of a window. On one occasion he was sent with his mate—for we took care to furnish him with one—to the house of a gentleman in our neighbourhood; and while there, he gave many proofs of his superiority as a bird of courage, at least when opposed to others of his own species; and this superiority he maintained until his companion fell a prey to the Rats. Poor Poodle, (as we call him,) disconsolate at the lose of his spouse, now left this place for his former habitation, no doubt thinking that a change of scene might do him good; but, unable to fly, he was obliged to walk a distance of somewhat more than a quarter of a mile. Wayworn and bedaubed with mud, he trudged into the room, and emitted his wonted coo-roo as he took possession of his old castle, to wit, that part of the floor in a corner, on which stands an old table, and into which he will not allow Dog or Cat to enter, or even to approach, without a blow of his wing or bill.

"Next spring, finding no mate, he attached himself to a stocking-foot stuffed with straw, round which he built a nest. The year following he took up with a Rabbit that used to run about the house. This animal in its pranks, with a kind of half wicked and half sportive design, would sometimes destroy Poodle's nest. This was no doubt a great annoyance to the Pigeon, for it generally cost him the labour of a whole week to repair the injury done. It was curious to see how he proceeded in this operation: having lifted a piece of twig, and placed in the position in which he wished it to rest upon those intended to be placed under it, he perhaps found it too large, on which he would not attempt to shorten it, for some kind of intuitive knowledge seemed to assure him, that the attempt on his part would be vain; but laid aside the long twig for future use, and had recourse to one of smaller dimensions. While this Rabbit was his associate, he used to remind it at night that it was time to retire to rest; if disobeyed, he gave the intimation in another form, went out from his retirement, and compelled compliance. During the day, if the Rabbit was on the floor, he used to come out and attempt to decorate it, in which occupation he took great pleasure, especially in trimming its long ears. The Rabbit would sit still all the while, unless the Pigeon became rude, when a battle would take place. At length the Rabbit was killed, and Poodle for some time had no mate; but thinking it better to have a partner of any sort than none at all, he attached himself to the dog, who allowed him to perch on his back, and use any liberty short of inflicting pain. He always, however, keeps possession of his castle, which the hens, the cat, and the dog, sometimes seem inclined to enter, but from which, though age has damped his fire, he succeeds in repelling them.

"Another curious circumstance is, that he seems fond of knowing all that is going on. If a person with whom he is acquainted calls, he is sure to regard him with very particular attention; and as a proof of his inquisitive disposition, it may be stated, that for some time there was a hole in the floor, at which he would place himself, frequently remaining half an hour, and giving good heed to what was going on in the apartment beneath. Again, let it be supposed that he is sitting on the outside of the window, and that he perceives preparations going on for breakfast or dinner, he manifests great joy by half flying and half tumbling down to the ground, and makes his appearance by coming up stairs. If he be asleep in his castle at supper time, and any of the family say "Poodle, why don't you come for your share?" he replies in his own way, and presently comes forth."

These remarks, says our correspondent, are from the pen of an eye-witness; and the family in which so many creatures so different in nature dwell in amity, may well be regarded as an amiable one. The dove is an emblem of peace, and if ever peace dwelt on earth since the fall of man, it is among the members of that family, in the midst of which Poodle has had the good fortune to find an asylum.

OBSERVATIONS ON THE JERBOAS.—BY M. F. CUVIER.

Particular circumstances having for some time directed the attention of the Author to the Glires, he has anxiously availed himself of every opportunity of extending his knowledge of them, in order to be able to apply to the numerous species of which that great family is composed, those precise rules of the natural method, without which confusion and obscurity must prevail in the study of animals. The results of his new observations on the Jerboas have recently been communicated to the Academy of Sciences.

After much consideration, naturalists had agreed in forming the genus Jerboa of Glires remarkable, more especially for the great size of their hind feet; for having three toes on these feet, which alone in walking or leaping are placed on the ground, and are articulated to a single metatarsal bone; for their broad head, short muzzle, large eyes, and short tail. These Glires, however, are subdivided into several sections, characterized by the absence or the number of the rudimentary toes on the hind feet. M. Lichtenstein, to whom we are indebted for an excellent memoir on the Jerboas, published at Berlin in 1828, has formed three divisions of these animals: 1st, those which have only three toes in the normal state on the hind feet; 2d, those which have an additional rudimentary toe on these feet; 3d, those which have two rudimentary toes on them. Further, the form of the teeth of the species of this last division was known, and was attributed to the teeth of all the species of the genus.

The possession of skulls of several species of the first division, that which is characterized by three toes only on the hind feet, has shown M. Cuvier, that these species are not distinguished solely by the number of the toes, from those which have five on the hind feet, but that they are moreover characterized by the form of the molar teeth, and by the structure of several of the parts of the head. Thus, while in these latter, the true molares present numerous irregular folds of enamel, these folds are in the others reduced to a single one on each of the lateral faces of these teeth. On the other hand, if the general structure of the head is the same in all the animals, and is characterized by the size of the skull, the shortness of the muzzle, and especially the great size of the infraorbitary foramen, yet there are differences in these circumstances sufficient to mark the different divisions. Thus, the species which have three hind toes are remarkable for the great breadth of the head, and the capacity of the skull, and this breadth is in part occasioned by the enormous development of the petrous portion of the temporal bone, and the breadth of the zygomatic arch. On the contrary, the species which are furnished with five toes have the capacity of the skull much reduced, all the parts of the ear are of moderate dimensions, and those which compose the zygomatic arch are narrow, so as to present surfaces of small extent to the muscles which arise from them.

From these observations M. Cuvier concludes, that the species of Jerboa which have three toes ought to be generically distinguished from those which have five; and as among these latter are the species which Pallas names Allactagas, he proposes giving them that name, and allowing the others to retain that of Jerboas.

The observations are followed by the description of a new species of Allactaga, a native of Barbary, to which M. Cuvier gives the name of *Allactaga arundinis*, from the account of its manners given by Shaw, who appears to have known it, although he has very imperfectly described it.

Beside the Jerboas, are pretty generally placed in systems some Glires of small size, with long hind legs, terminated, like the anterior, by five more or less developed toes. These animals are collectively designated by the names of *Gerboides*, *Gerbillus*, and *Meriones*, and several of them have been considered as true Jerboas.

M. Cuvier enters into a historical and critical examination of this genus, and of all the species which have been referred to it, under one or other of the common names mentioned. This leads him, after having shown all that is known of the organization of these animals, and consequently of their generic characters, to distinguish the species which have been referred to this genus, without belonging to it; those which have been referred to it, on more or less doubtful grounds; and lastly, those which really belong to it, and which, from twenty-one are reduced to six species:—1. Olivier's Gerbillus, which does not differ from the *Meriones quadrimaculatus* of M. Ehrenberg; 2. M. Geoffroy's Jerboa of the pyramids, to which perhaps may be referred M. Ruppell's *Meriones robustus*; 3. The Gerbille Pygargue of M. Cuvier, or *Meriones Gerbillus* of M. Ruppell; 4. Mr Gray's *Gerbillus Africanus*, which is not distinguished from that named *Gerbillus Schlegelii* by M. Smutz; 5. The Indian Gerbillus, for the knowledge of which we are indebted to General Hardwick; 6. The Jird, or *Mus Meridianus* of Pallas. M. Cuvier concludes his long investigation by extending and rectifying the characters of the first five species just mentioned, and describing three new species:—The short-tailed Gerbillus, which is found at the Cape of Good Hope, and appears to occur also in the Indian Peninsula; the short-eared Gerbillus, which also comes from India, and Burton's Gerbillus, from Senaar.

The result of M. Cuvier's observations as to the relations of the Gerbilli to the other Glires is, that these animals have no connection with the Jerboas, but are intimately allied to the Rats and Mice. The memoir is accompanied with numerous figures which represent Burton's Gerbillus, and the heads and teeth of eight of the nine species, which are ascertained to belong to the genus.—*Annales des Sciences Naturelles.*

BOTANY.

THE OLIVE TREE. OLEA EUROPÆA.

This celebrated plant belongs to *Diandria Monogynia* of the Linnæan system, and to the natural order of *Oleineæ.* It usually attains a height of about twenty feet, and sends off numerous long branches, with opposite, lanceolate, narrow, entire bright-green leaves, of which the lower surface is whitish. The flowers are numerous, small, white, and disposed in clusters in the axils of the leaves. The calyx is tubular, with four small, erect, deciduous segments ; the corolla funnel-shaped, having a short tube, and form semiovate segments. The filaments are tapering, and crowned with erect anthers, the germen round, the style short, the stigma cleft. The fruit is an elliptical drupa, containing a nut of the same form.

The Olive tree is a native of the south of Europe, and flowers in July and August. Its varieties are numerous, and distinguished by the form of the leaves, and the shape, size, and colour of the fruit. " It has been celebrated from the earliest ages, and is the second tree, with which we are acquainted, which is mentioned in the sacred writings. It must have been known before the Flood, as the Dove returned to Noah in the Ark with a leaf of it in her mouth. There can be little doubt of this incident having been the origin of the olive's being considered the emblem of peace. This tree must have been very extensively cultivated in Judea, to have furnished the vast quantities of oil which were used in the sacrifices and service of the temple ; besides its general consumption as an article of food.

" Olive trees sometimes attain a great age. There is an Olive tree in the environs of Villa Franca, near Nice, the lowest extremity of the trunk of which, next the surface, measures about thirty-eight feet, and three feet and a half above the surface, nineteen feet above the circumference. One of its main branches is six feet and a half in circumference, and the trunk itself eight feet and a half in height. This is both the oldest and largest olive tree in that part of the country, and though fast decaying, still retains much of its stately appearance. The celebrated Olive tree of Pescio, which has hitherto been considered the most ancient in Italy, and is stated by Marchentini to be seven hundred years old, is much younger than this wonder of Nice. There are records now extant, which show that as far back as the year 1516 the latter was accounted the oldest in those parts. In 1818, it bore upwards of two hundred weight of oil, and in earlier days, in good years, more than three hundred and fifty."

The Spanish and Provence olives are pickled, and in this state are to many extremely grateful. They are prepared from the green unripe fruit, which is repeatedly steeped in water, with the addition of quicklime or some alkaline salt, which quickly extracts their bitterness. They are then washed, and preserved in a pickle of common salt and water, to which an aromatic is sometimes added.

The principal consumption of Olives is in the preparation of the common sallad oil, or olive oil, which is obtained by grinding and pressing them when thoroughly ripe. The finer oil issues first by gentle pressure, and the inferior kinds are obtained by beating the residuum, and pressing it more strongly. The best olive oil is of a pale clear amber colour, of a mild taste, and without smell. It becomes rancid when long kept, and congeals at 38° Fahr. In some shape oil forms a considerable part of our food, and is very nutritious, but, as it does not readily unite with the contents of the

stomach, it does not agree with some persons, and is often brought up by eructation. As a medicine it is supposed to correct acrimony, and to relax the fibres, or lubricate the parts ; hence it has been recommended in coughs, catarrhs, erosions, nephritic cases, spasms, colics, constipations, and worm cases. Externally it is a useful application to bites and stings of various poisonous animals, burns, tumours, and other affections. The application of a drop of it to the bite of a bug removes the smarting sensation in less than a minute.

M. DUTROCHET'S OBSERVATIONS ON THE SLEEP OF PLANTS.

(Continued from Page 100.)

While reflecting on this singular phenomenon, I was led to think that it is not without reason that Nature has lavished respiratory organs on the fibrous tissue, which is situated between two layers of hollow organs filled with air. As it was not by being filled with liquid that the fibrous tissue took its active state of curvature, it might be by impletion with oxygen. If this supposition were correct, the nerve, which, on being placed in aerated water, there assumed, first the outward, and subsequently the inward curvature, would, on being immersed in unaerated water, unvaryingly preserve its first outward curvature, or that of sleep, which is due to the endosmosis of the cellules of the cellular tissue ; it would thus never present the inward curvature, or that of sleep, which I considered to be owing to the oxygenation of the fibrous tissue.

I ought first to observe, that when any thin part of a vegetable is immersed in unaerated water, the latter quickly dissolves the air contained in the pneumatic organs of this vegetable part, and takes its place, so that no respiratory oxygen remains in it. My supposition was justified by experiment. A nerve of a flower of Mirabilis, on being immersed in unaerated water, assumed and retained, without change, its inward curvature. An expanded flower which, on being entirely immersed in aerated water, three assumes at the end of some hours the state of closing or sleep, does not assume this state in unaerated water, but there unvaryingly retains its expanded state.

It might perhaps be thought, that the air contained in the pneumatic organs of the nerves of the corolla, would act in virtue of its elasticity to produce the incurvation of sleep, and not in virtue of the chemical action of the oxygen which it contains ; whence it would happen, that the incurvation of sleep would not take place on immersing the corolla in unaerated water, which dissolves the air contained in the pneumatic organs, and takes its place. But this is not the case, for experiment has proved to me, that the air never returns into the pneumatic organs that have been filled with water, in parts of vegetables which continue to remain submersed. Now, this does not prevent a corolla of Mirabilis from assuming the state of sleep after two or three days, when the unaerated water in which it is plunged in an expanded state, is left to be aerated by contact with the atmospheric air. It is, therefore, without doubt, by the chemical action of the oxygen dissolved in the water, that the fibrous tissue acquires the force of incurvation which produces the state of sleep. Thus, in flowers of the genus Mirabilis, waking and sleep—in other words, the expansion and closing of the corolla, result from the alternately predominating action of two organic tissues situated in the nerves of the corolla, and which tend to curve in opposite directions, viz :—1st, A cellular tissue which tends to curve towards the exterior of the flower, from being filled with liquid to excess, or by endosmosis ; 2dly, A fibrous tissue, which tends to curve towards the interior of the flower, by oxygenation.

The corolla of Convolvulus purpureus exhibits precisely the same phenomena as that of the different species of Mirabilis, with reference to the mechanism which effects its opening and shutting, the internal structure of its nerves being exactly similar. The flower of Mirabilis opens in the evening, and closes in the morning ; that of Convolvulus purpureus opens toward midnight, and does not close until the next evening. Thus, these two flowers are equally nocturnal as to the hour of their expansion. If the flower of the Convolvulus remain open during the day, while that of the Mirabilis shuts in the morning, this depends in a great measure upon the circumstance that the former is much slower than the latter in oxygenating the fibrous tissue of its nerves under the influence of light and heat.

The flowers above mentioned have only a single wake and a single sleep, followed by the death of the corolla. I now proceed to examine such as for several days present an alternation of waking and sleep, and take the Dandelion as an example. The flower of this plant generally lives two days and a half, showing during this time a state of waking in the fore part of the day, and of sleep in the evening. On the third day, the last sleep takes place in the middle of the day, and is followed by the death of the corolla. In the state of waking, the semiflorets of which this flower is composed curve outwards, producing its expansion ; in sleep, they curve inwards, effecting its closing. Although these semiflorets are very thin, I have been able with the microscope to observe the internal organization of their nerves, which are very small, and four in number in each semifloret. At the internal or upper surface of each of its nerves, there is a linear cellular tissue, the cellules of which are covered with globules, and which is perfectly similar to what I have observed in the nerves of the corollas of the genus Mirabilis. At the outer or lower surface of the nerves of the semifloret is a very thin layer of fibrous tissue situated between a plate of trachem and a plate of cellules filled with air and placed superficially. This fibrous tissue, which is perfectly similar to that which exists in the nerves of the corolla of Mirabilis, is in like manner inclosed between two layers of pneumatic organs, and it thence becomes probable that it is more incurvable by oxygenation, and that the cellular tissue is incurvable by endosmosis, as is the case with the nerves of the corolla of Mirabilis. In fact, experiment proves that the incurvation which produces the state of waking in the semiflorets of the Dandelion is owing to an excessive impletion with fluid, that is, to endosmose, and that the incurvation which produces sleep is owing to oxygenation. The semiflorets of the flower of the Dandelion being gathered early in the morning, when they still retain the incurvation of sleep, and being immersed in aerated water, assume in it the contrary incurvation, which is that of sleep. This takes place in darkness as well as in light. If they are immersed in unaerated water, they assume an exaggerated curvature of sleep, and unvaryingly retain this curvature. If these semiflorets thus curved outwards are transferred to syrup, they curve in the opposite direction; and if replaced in pure water, resume the outward curvature. Thus, there is no doubt that all this is owing to endosmose. If we allow the semiflorets which are in the state of waking to remain some hours in aerated water, they assume the incurvation which is that of sleep, and this incurvation is not destroyed by transporting the semiflorets thus curved into syrup, which proves that this incurvation of sleep is not owing to endosmose. As this incurvation of sleep does not take place in unaerated water, this proves that it is owing to oxygenation. Thus the wake and sleep of the semiflorets of the flower of the Dandelion result from the alternately predominant incurvation of an organic tissue incurvable by endosmose, and of an organic tissue incurvable by oxygenation. The first is undoubtedly the cellular tissue, and the second the fibrous tissue, both contained in the nerves of the semifloret. These two incurvable tissues, alternately prevailing the one over the other, open or close the flower.

The causes which effect the prevalence of the morning incurvation of the cellular tissue, the agent of the state of waking, are, on the one hand, a more powerful ascent of the sap under the influence of light, which increases the turgescence of that tissue, and, on the other hand, the diminution of the antagonist force of incurvation of the fibrous tissue, the agent of sleep, and which takes place during the night. In fact, if the semiflorets be gathered in the evening, when they have assumed the curvature of sleep, and immersed in aerated water, they there retain without change their curve; but if there be taken next morning from the same flower, other semiflorets still retaining the curvature of sleep, and they be immersed in aerated water, they instantly resume the opposite incurvation, even in the dark. Now, by the immersion of the semiflorets in water, endosmose of their cellular tissue is excited, and in consequence the curvature of sleep is produced. If this result has not taken place in the evening, it is because the incurvation by oxygenation of the antagonist fibrous tissue is too strong, and cannot be overcome by the incurvation of the cellular tissue. If next morning, by immersing in water the semiflorets which have passed the night on the plant, their curvature of sleep be produced, this proves that the force of incurvation of the fibrous tissue has diminished, and that in consequence this fibrous tissue has lost a part of its oxygenation during the night; so that the cellular tissue incurvable by endosmosis, which is its antagonist, and which is the agent of the state of waking, then prevails.

Thus the flower, which for several days presents the alternations of waking and sleep, is that in which the fibrous tissue, the agent of sleep, loses during the night a part of the oxygen which has been fixed in its interior during the day, and which is the cause of its incurvation; so that the latter having in the morning lost some of its power, the cellular tissue by endosmose, the agent of the state of waking, becomes again predominant. The sleep of this flower takes place again in the evening, because the oxygenation of the fibrous tissue, the agent of sleep, gradually increases during the day, which renders its incurvation predominant; at the same time the diminution of light occasions the diminution of the sap, which weakens the turgescence, and consequently the incurvation of the cellular tissue, the agent of waking. These alternations cease only on the death of the corolla. The flowers which present only a single wake and a single sleep, are those of which the single sleep is immediately followed by the death of the corolla.

The entire flower of the Dandelion immersed in its expanded state in unaerated water, deprived of communication with the atmosphere, unchangingly retains this state of waking. If the water is in contact with the atmosphere, it dissolves a portion of the latter, and the flower which is immersed assumes the state of sleep at the end of two or three hours.

It may be remarked, that these experiments confirm what I have said in my paper on the Respiration of Plants, namely, that vegetables respire like animals, by assimilating oxygen, the presence of which in the organism is as necessary in the one as in the other set of beings.

MISCELLANEOUS.

ON VARIETIES.—NO. III.

(*Continued from Page* 120.)

Varieties in colour are attributable to much the same causes as those producing variations in size. This, however, is a subject more difficult of investigation, so little being known of the *modus operandi* of these causes, and of the laws which regulate their influence. Observations also are not likely to disclose much to us. Some persons may be content with an explanation of this kind of varieties, derived from the idea of nature ordaining them as so many additions to the beautiful variations every where present in the universe; but if we have recourse to such an explanation in the majority of cases, in the absence of immediate causes, there are instances which, plainly contributing to the welfare of the object, or being, evince a benevolent design in the Creator. Such an instance is seen in the common Trout, which, according to the observation of anglers, conforms in colour to that of the river in which it is found, that is, to the tint of the water or of its bed. It is also seen that plants, and other stationary productions, are influenced in their colour by local circumstances. The brilliancy of their colours is deteriorated or destroyed, according to the degree of obstruction offered by surrounding bodies to the rays of the sun. Soil likewise affects the colour of plants, though in a mode entirely unknown to us. Some localities produce certain coloured varieties of a given species of plant, and it is very probable that many "accidental" varieties are attributable to the same cause.

Food is thought in some cases to produce varieties in certain animals, and this is not improbable in the case of such as are found scattered and exposed to differences in the kinds of nourishment appointed for them. Among the facts available in support of the idea of variety in colour being an ordination of nature to increase the beauty and interest of a species, it may be urged that a given colour is always liable to give rise to modifications of the same in all its gradations, and these both as permanent and accidental varieties; but then there is one rule among plants which seems to place the question on a different footing, namely, that all blue flowers are prone to produce white varieties, that is, varieties destitute of colouring matter. In cases where two, three, or more colours appear together, and are diffused in patches over the individual, we sometimes find no regularity adopted as to the size of such patches or spots, or their figure, or relative position. Now, these are irregularities with the causes of which we are not likely to become acquainted. The great tendency to variation in colour of animals and plants is abundantly exemplified in domesticated and cultivated species.

As the causes that operate in producing varieties in size and colour are so obscure, it cannot be surprising that an investigation of the same kind relative to varieties affecting still more decidedly the species, and inducing differences in structure appreciable for the most part by the naked eye, should be equally unsatisfactory in results. Indeed, if the peculiarities distinguishing species cannot be satisfactorily explained in the majority of cases, by reference to causes presumed to be in operation, those which distinguish varieties must be also in the same predicament. The hairs and prickles of plants are a kind of structure very prone to variation, and it seems that they increase in number in dry situations, and diminish in such as are wet. But instances in which causes can be traced are extremely rare, and little else can be done than to enumerate varieties of this description. In many such we can discover without difficulty that although varieties are in themselves departures from natural structure, they are nevertheless produced under the influence or guidance of a law of formation. Thus, in the case of *Julus indicus*, the additional members are produced by pairs, it being in contradiction to suppose that a single leg could be formed by vessels, or the secreting organs, which had previously formed two members at a time, the disposition or arrangement of these secreting vessels being such that two structures instead of one must as it were necessarily ensue. So, in the production of the parts of plants, if the usual mode is by pairs, triplets, &c., the varieties will display the same arrangement. Yet this rule is by no means invariably abided by.

In addition to the forms of variety yet noticed, there are others which it is necessary to enumerate, though their causes are quite inappreciable. Plants usually odoriferous are liable to become scentless under certain circumstances, and it is thought that the rays of the sun are in some cases essential to the production of this quality. Scent is of course dependant on the presence or condition of some structure, and it would therefore be requisite to inquire what this structure is in each case, and what circumstances are liable to vary its formation and its qualities. The taste or flavor also, both of animals and plants, is liable to slight differences, dependant most probably on soil, situation, and food. The Oyster, for instance, varies in taste according to the river from which it has been procured. The Salmon, the Trout, and the Perch, also differ in this respect. Peppermint varies in flavor according as the soil suits its constitution, and the same is well known to be the case with regard to Grapes, Melons, and many other kinds of fruit.

The song and notes of birds, especially of those which are not confined to a monotonous song, or to one particular key, is in some instances not a little different from that proper to the species. Besides, minor differences are frequently observed in individuals. On this subject see some remarks in the "Journal of a Naturalist." The habits or actions of animals, too, exhibit some individual peculiarities of the same amount, there remaining, however, those distinguishing features which serve to characterize the animal, and separate it from every other. In short, all that I contend for is this:—That, as there exist between species certain differences of a greater or less extent, by which they are respectively known and characterized; so there exist between the individuals of a species differences of a greater or less degree forming a line of demarcation, or peculiarities characterizing and separating one from another. To my apprehension, nature has been desirous of fixing these bounds, and impressing on individual productions marks which cannot be effaced, thereby preserving the same rule as regards the component parts of species, as is noticed between the specific components of the organic world. These last-named varieties are well seen in domestic animals, and in cultivated specimens of plants, though I by no means appeal to these in support of my opinion.

Much of the difficulty attached to an inquiry as to the uses of varieties would be banished by admitting that, besides nature appointing a certain series of differences between individuals of a species, she has also in very many cases appointed uses, separate or additional, for varieties occurring to species. Unless we admit that some particular office, or function, is set apart for very many varieties we meet with, I do not see how we are to account for their differences being so great. I conceive that in many instances, not only are the amounts of difference assumed as varieties, as great as those between species, but that their uses are as peculiarly important in the great scheme of the universe, as those of a vast number of species belonging to the same class or tribe as the varieties. It is not pretended, however, that these uses can be defined, and it is but an inference to say varieties execute functions different from ordinary specimens. In the case of animals and plants which we employ as food, or for economical purposes, we are led to the conviction that here varieties are indispensable, because, although some maintain existence under unaltered forms when dispersed over a large extent of country, and exposed to great diversity of circumstances, surrounding media, and conditions are ever differing, and the surface of the earth is composed of heterogeneous materials, different kinds of soil being frequently found in adjacency within very confined limits, and consequently the occurrence of such species would be very circumscribed, unless by the means of variation they could adapt themselves to such diversifications. In the vegetable world we may instance wheat, barley, maize, other kinds of grain, and various species of grass. Now, if these did not conform to the different soils, man would be debarred from participating in the benefits they confer, excepting in those particular localities whichh appened to agree with them. And so, if certain heights were ordained for trees, they could only be found in those soils and situations which allowed them to attain those heights.—*To be continued.*

Edinburgh: Published for the Proprietor, at the Office, No. 13, Hill Street.
London: Smith, Elder, and Co., 65, Cornhill. Glasgow and the West of Scotland: John Smith and Son; and John Macleod. Dublin: George Young. Paris: J. B. Balliere, Rue de l'Ecole de Médecine, No. 13 bis.
THE EDINBURGH PRINTING COMPANY.

THE EDINBURGH

JOURNAL OF NATURAL HISTORY,

AND OF

THE PHYSICAL SCIENCES.

JANUARY, 1838.

ZOOLOGY.

DESCRIPTION OF THE PLATE.—THE PROBOSCIS AND SOLEMN APES.

Fig. 1. THE PROBOSCIS MONKEY of Pennant (*Nasalis larvatus*), distinguished from all others by the elongation of its nose, is the only species of the genus to which it belongs. It is of a reddish fawn colour, lives in troops in the forest, and inhabits the Island of Borneo.

Fig. 2. THE DOUC (*Semnopithecus Nemœus*) belongs to a section of Monkeys which have thirty-two teeth, a facial angle of about 45°, a round head, flat nose, ears of moderate size, very long limbs, short thumbs, and a very long and slender tail. This species, which inhabits Cochin China and Madagascar, has its fur diversified with brilliant colours.

Fig. 3. THE ENTELLUS (*S. Entellus*), of a yellowish white colour, with the face and hands black, inhabits various parts of India.

Fig. 4. THE CIMEPAYE (*S. melalophus*) has its upper parts yellowish red, the lower whitish, and the face blue. It inhabits the island of Sumatra.

Fig. 5. The CRESTED APE (*S. comatus*), gray above, dingy white beneath, with a tuft of black hair on the head, inhabits Sumatra.

Fig. 6. The Tchincou (*S. Maurus*), or Negro Monkey, so named on account of its black colour, is found in the island of Java.

DESCRIPTION OF THE PLATE.—THE FISHER-EAGLES.

THOSE Eagles which are characterized by having the bill nearly as long as the head, very high, and compressed, with naked tarsi, and very large toes, of which the claws are remarkably strong, have been separated from the rest, to constitute a genus, to which the name of *Haliaetus* is given. They are mostly maritime or la-custrine, and feed in a great measure on fish, which they frequently clutch as they swim at the surface, but which they also obtain by depriving other birds of the pro-duce of their industry.

Fig. 1. THE WHITE-HEADED FISHER-EAGLE (*H. leucocephalus*) may be briefly described as having the head, the neck, the hind part of the back, and the tail white, the rest of the plumage being brown; the bill, cere, iris, and feet yellow, the claws black. It is about the size of our White-tailed Fisher-Eagle, *H. Albicilla*, measur-ing nearly three feet in length, and seven feet between the tips of the wings. It is not uncommon along the coasts, and on the lakes and great rivers of the United States of America. "During spring and summer," says an enthusiastic ornithologist of that country, "no sooner does the Fish-Hawk make its appearance along our Atlantic shores, or ascend our numerous and large rivers, than the Eagle follows it, and, like a selfish op-pressor, robs it, of the hard earned fruits of its labour. Perched on some tall summit, in view of the ocean, or of some water-course, he watches every motion of the osprey, while on the wing. When the latter rises from the water, with a fish in its grasp, forth rushes the Eagle in pursuit. He mounts above the Fish-Hawk, and threatens it by ac-tions well understood, when the latter, fearing perhaps that its life is in danger, drops its prey. In an instant, the Eagle, accurately estimating the rapid descent of the fish, closes its wings, follows it with the swiftness of thought, and the next moment grasps it. The prize is carried off in silence to the woods, and assists in feeding the ever-hungry brood of the Eagle. It now and then, however, procures fish for itself, by pursuing them into the shallows of small creeks. It does not confine itself to this kind of food, but greedily devours young pigs, lambs, fawns, poultry, and the putrid flesh of carcasses of every description, driving off the vultures and carrion-crows, or the dogs, and keeping a whole party at defiance until it is satiated."

Fig. 2. THE WHITE-HEADED FISHER-EAGLE. Young. In its first plumage this species is of a uniform chocolate brown on the head, neck, and upper parts, the tail dusky, with brownish-white patches.

Fig. 3. THE VOCIFEROUS FISHER-EAGLE (*H. Vocifer*), which is of smaller size than the preceding, has the head, neck, and tail white, the back and wings dark brown, the breast white, spotted with dusky, and the abdomen and tibial feathers brownish-red, inhabits various parts of Africa.

BRITISH BIRDS.—THE RINGED DOVE OR CUSHAT.

THE Pigeons or Doves form a very natural family, perfectly distinct from the Gal-linaceous birds, to which they exhibit little more affinity than is indicated by their very large crop, which, however, is of a different form, and their granivorous habits. They

vary much in form, some having the body full, others slender, while the tail is very short, moderate, or greatly elongated. In all, however, the head is small, oblong, compressed, with the fore part rounded. The bill is especially characterized by hav-ing the nasal membrane bare, generally scurfy, fleshy, and tumid, with the narrow longitudinal nostrils placed under its anterior margin. It varies in size, but the upper mandible has its ridge always obliterated at the base, by the encroachment of the nasal membranes, and its extremity horny, arched or convex, more or less compressed, with a blunt thin-edged point. The œsophagus expands into a large crop, consisting as it were of two united membranous sacs; the stomach is a powerful gizzard, of a some-what rhomboidal form; the intestine is long and slender; and the cœca are very small and cylindrical. The tarsi are generally short and stout, either scutellate or feathered; the foot is of that kind equally adapted for walking or perching, having three toes be-fore, and one behind, the latter on the same level, and shorter than the lateral, the claws short, and moderately arched. The plumage is various, but the feathers have always the tube very short, and the shaft thick at the base. The wings are large, more or less pointed, and the tail is even, rounded, or graduated. Several genera, some of them founded on slight differences, have been instituted in this family. The four species which occur in Britain belong to the genus *Columba*, which may be cha-racterized as follows: The bill rather short and slender, the upper mandible tumid and scurfy at the base, horny, arcuato-declinate, convex above, compressed, thin-edged towards the end; the wings long, broad, rather pointed, the second and third quills longest; the tail of moderate length, or rather long, straight, even, or slightly round-ed, of twelve broad feathers.

THE RINGED DOVE OR CUSHAT (*Columba Palumbus*) is distinguished by having the plumage of the upper parts greyish blue; the wings and scapulars tinged with brown; the hind part and sides of the neck bright-green and purplish red, with two cream-coloured patches; the fore part of the neck and breast light reddish purple; a white patch on the wing, including the four outer secondary coverts. It is the largest of our native species, measuring about seventeen inches in length, and about thirty between the tips of the wings. It is a strong bird of its size, elegantly formed, and agreeably, although not gaudily, coloured. Its food consists of seeds of the culti-vated cereal grasses, wheat, barley, and oats, as well as of leguminous plants, beans and pease, the field mustard and charlock, leaves of the turnip clover, and other plants, beech-nuts, acorns, and other vegetable substances. It is generally distributed, being found in all the more or less wooded districts of England and Scotland; but it pre-fers cultivated tracts, and is not found in the bare and treeless parts of the country. In winter it appears in large flocks, sometimes amounting to many hundreds, when the individuals of a district congregate in some favorable locality, although in ordi-nary circumstances it is not so decidedly gregarious as the Rock-Dove. Its flight is strong and rapid, being performed by quick beats of the half-extended wings, with oc-casional intermissions, its pinions sounding as it glides along. When it has espied a place likely to afford a supply of food, it alights abruptly, and usually stands for a short time to look about, after which it commences its search. It walks in the man-ner of the Domestic Pigeon, that is, with short and quick steps, moving its head gently backwards and forwards. The flock disperses and spreads over the field, it being seldom that two or three individuals keep close together, and they generally take care not to approach the enclosing walls or hedges, so that it is difficult to shoot them on the ground. In the time of snow or hard frost, they frequent turnip fields, and are more easily approached; but in general they are very suspicious and vigilant, ever ready to fly off, on the slightest appearance of danger. Frequently, however, in the woods, one may surprise them within shooting distance; and by waiting for their arrival at their roosting places in winter, considerable execution may occasionally be done among them. As the flesh of the pigeon affords a sufficiently palatable article of food, it is abundant in our markets in winter and spring, but generally brings a low price, from sixpence to a shilling.

Soon after sunset, the Cushats betake themselves to their roosting places in the woods, and before settling, usually wheel round the spot selected. Should they be disturbed, they fly off to a short distance and return; but if repeatedly molested, they betake themselves to a distant station. In severe weather they sometimes per-form partial migrations, but in general are stationary, not finding it necessary to ex-tend their range in the cultivated and sheltered districts, where turnips may always be had during the winter.

In fine weather they bask in the sun on dry banks, or in the open fields, rubbing themselves, as it were burrowing, in the sand or soil, and throwing it about with their wings, as if washing in water, which they do like most birds. In drinking, they

32

immerse the bill to the base, and take a long draught; a circumstance in which Pigeons differ from Gallinaceous birds. Nor does the Cushat, like them, scrape up the earth with its feet, while searching for food.

Early in spring the Cushats make preparations for rearing their young. Their courtship is conducted much in the same manner as that of the Domestic Pigeon, the male strutting with elevated head, protruded breast, and quick step, round the female, or, if on a branch, performing various movements and often turning round, as he utters his murmuring love-notes. At times he rises in the air, produces a smart noise by striking the points of his wings against each other, descends, rises again, and thus continues to gambol in the presence of his mate. The cooing of this species may be imitated by pronouncing the syllables coo-roo-coo-coo, the two last protracted. It is softer, deeper, and more plaintive than that of the Rock-Dove.

The nests are placed on a branch, or in the fork of an oak, beech, fir, or other suitable tree, more especially a fir or pine; and in the latter case, often only a few feet from the ground. Sometimes a nest may be seen in a holly or hawthorn bush, or in a hedge, but in general a thick wood is preferred. It is composed of twigs loosely put together, in a circular form, flat above, and varying in thickness from two to four inches. The eggs are always two, of a regular oval or sometimes elliptical form, pure white, an inch and seven-twelfths in length. It appears that two or even three broods are reared in the year.

The young are at first rather scantily covered with a yellowish down, and when fledged are of the same colour as the adult, but duller and tinged with brown, the white spots on the neck and the changing tints of that part being awanting. The colours are perfected at the first moult, the only change that they afterwards undergo being to a somewhat deeper and purer tint.

The domestication of this species has often been attempted, but almost always without success; for although individuals become perfectly tame in confinement, they embrace the first opportunity of regaining their freedom. An instance, however, has recently occurred of its breeding in captivity, as appears from the following statement made by Mr Allis of York, in the Naturalist of last month. "I have this year succeeded in breeding the Ring Pigeon in confinement. I took the old birds from the nest in the autumn of last year. This year they bred a pair of young, which have now passed through the first moult, and are not distinguished from the old birds."

The Ringed Dove is generally distributed over Europe, but is more abundant in the southern parts.

On the Geographical Distribution of Animals, with an Account of the Species that inhabit the South of Devon. By J. C. Bellamy. Yealmpton.

Great exertions have been made, in every department of science, to determine principles and ascertain laws. The successful prosecution of this subject must depend, in a great measure, upon acquaintance with detail; and since this species of knowledge has recently received great accessions, philosophers have been guided by a reasonable hope, that a renewed inquiry after principles would be attended with proportionately important results. It is to be feared, however, that the difficulties of the investigation have been very often insuperable, and have hitherto prevented us from acquiring any satisfactory knowledge, particularly as concerns the geographical distribution of animals. Indeed, as to primary or first causes, the reasons of the institution of laws, or the occurrence of facts in connexion with this interesting subject, we know absolutely nothing. It has been customary, until very lately, to confound together primary or general laws, and secondary or partial laws. By primary laws, I understand those ordinances constituting the plan, system, or method, according to which the whole animal kingdom is arranged or distributed over the surface of the globe; it being opposed to reason, and at variance with all scientific considerations, to suppose that living beings have been placed on the earth promiscuously, indiscriminately, and without regard of order and adaptation.

Since the greatest benefits and most important uses of natural science depend upon the determination of principles and of general conclusions, the labour bestowed on the study should centre on this great object; and in the enumeration of facts, we should be careful to inquire, as we proceed, what effect that detail has on admitted doctrines, or what influence it might have in establishing or disclosing new views and theories. In accordance with this idea, I shall combine a recognition of laws with a statement of facts in the present paper; first mentioning the primary and secondary laws by which the distribution of animals is governed; and then entering upon the detail of zoological geography, as observed in the south of Devon; availing myself, in this second part of the subject, of every occasion to advance the knowledge of the higher department of principles and general results. It will be needful to remember, that this subject of the Geography of Animals is intimately associated with several others of great interest, and more particularly with migration and the "polity of nature;" and that these have, equally with the present question, laws, and general considerations connected with them, which are by no means to be confounded with those we are now about to state.

On inquiring how far the dispersion of animals is affected by, or connected with, the relative temperature of the earth, according to distance from the extreme points of heat and cold, we find that the animal creation is greatly accumulated within the tropics, and that it gradually diminishes in extent as we recede toward the poles. Throughout these vast spaces, however, numberless exceptions and deviations from this general ordination occur, in consequence of the influence of secondary or partial laws, as will be shown in the sequel; but the fact of the tropical countries being the great seat of animal creation, the temperate regions possessing fewer, and the polar districts the least number, is incontrovertible.

But not only do the intertropical regions contain the largest proportion of animals, both as regards species, and as regards the number of individuals, but they are characterized also by giving place to the most highly organized creatures of the whole series; while the temperate climates, and polar regions, are respectively characterized by animals having less and less of this endowment. Corresponding latitudes will therefore be found to agree in their animal productions, in so far as they will present to view creatures possessing similar degrees of organic endowment. These statements, however, although defensible in a general way, are greatly qualified by secondary influences, as will subsequently appear.

The laws which we have here stated will receive elucidation by reference to the Fauna of continental mountains, where, on a small but similar scale, as regards temperature, we see the progressive advancement of numbers, and of organization, from the summit to the base, though the occurrence of highly and of lowly organized beings in both extremes forms a partial impediment to the principle. It would almost appear that we had arrived at the knowledge of one of the primary causes of natural phenomena, in finding such definite results in connexion with heat and cold; but there are too many and too palpable exceptions to this rule to allow of such a conclusion. If animals were governed, in their dispersion, by heat and cold, except in a secondary and partial way, quadrupeds and birds, belonging to the same tribes as are found within the tropics, would not be present in polar or alpine situations.

There is no portion of our globe, even the most desolate, which is not at times visited by certain of the higher animals, and also permanently inhabited by certain kinds of insects and inferior creatures. It thus appears, that, besides the lower orders of animals being more numerous than the higher classes, they are likewise more generally dispersed. The resources of the lowest tribes are, in all probability, so obscure and occult as to be not only unknown, but even inconceivable to our minds; on the other hand, wherever vegetation attains to any tolerable degree of perfection in its various forms, there a whole series of animal productions presents itself. If the extent of the Flora and Fauna of any given country be examined into, the above result will infallibly be arrived at.

One primary law on which the distribution of animals depends, having a pretty general influence, and which seems indeed altogether in unison with the aggregate of our zoological knowledge, is the gradual failure in number of individuals of a given species as we recede from the point which, from their comparative plenty there, we presume to be their principal seat. Together with this numerical failure, we see likewise, as might easily be conceived, a failure, or deterioration in size, in qualities, in colour, and in all other endowments. To so great an extent is this occasionally carried, that naturalists are frequently at variance in their decisions on the species, some considering such specimens as deteriorations, others viewing them as separate species, or at least as formal varieties. It is seen that, independently of distance from the seat of luxuriant growth, and great numerical increase, specimens having all the appearance of such as are found at the very verge of the geographical range of a species, are constantly detected within short distances of, or absolutely in the metropolis itself. There are unquestionably two sets of causes in operative influence on animals in regard to distribution, and it is of great moment to refer the phenomena connected therewith to the right sources. The primary causes are quite unknown to us, and are likely to continue so, though it would appear that the various parts of the organized creation being ordained to counterbalance each other; that, as the laws of dependence pervade the world of living beings in all its parts, any determination or regulation, such as the one mentioned, namely, the diminution of numbers and deterioration in size and qualities of individuals, provided it were general, and observed in all classes and species, need not excite surprise.

The secondary causes appear to be temperature, food, situation, and the hostility of other species. The influence of these appears to be very considerable, and though we cannot be altogether warranted in attributing the above-named circumstance of diminution of number and deterioration in size, &c. to these causes, however plausible it might seem to do so, they are undoubtedly the agents that cause deteriorations generally. These secondary causes become indeed of the greatest moment in investigating the zoology of a given district. Primary laws can be seen and estimated only by reference to zoology as a whole, by taking into our view the phenomena exhibited by the entire series of animals; secondary causes must be appreciated by examining the phenomena of animal geography on smaller scales. It is then we see temperature, food, situation, and other circumstances, operating to the production of certain modifications in the distribution of animals, within a comparatively small compass, while the primary laws influencing their situation on the earth are uninterrupted, and, as it were, overrule the others. In confirmation of the supposition that the gradual lessening of numbers, and gradual deterioration in size and other qualities of individuals of a species, as we recede from their metropolis, depend on a primary law, we see the same rule applied to entire tribes and classes of animals in numerous instances. If secondary causes, such as food or climate, determined the limits of species, it would not be found that the verge of the range of one species was the principal seat of another possessing similar endowments and organization, and feeding for the most part similarly. The reasons or causes then of this peculiar law, or ordinance of Nature, are hidden from us. The great seat of the feline tribe is in the tropical regions, and we see the species there found gradually diminishing in number of individuals as we advance northward. We see also that the individuals situated at the outskirts of this great metropolis of rapacious creatures, are diminished in their size and bodily vigour, and that their ferocity has suffered decrease. The place of this tropical series is now supplied by a new set, and a third still more northwardly may without exaggeration or difficulty be detected, each undergoing within its own limits the same gradual diminution and deterioration. Eventually, if we compare the contents of the two opposite points with regard to this genus, the difference becomes remarkable: we find the species few, the individuals also few, their size small, and their vigour and ferocity greatly reduced at their northern limit, while at the point where we commenced these features are totally reversed. The principal seat of the cetaceous animals is in the arctic and antarctic seas, different species occurring at these two extremes; they gradually diminish in number as we enter the temperate regions; and are at their minimum in the equatorial seas. The Turtle and Tortoises are chiefly inhabitants of the warm latitudes; yet they extend sparingly northwards, and even in England a few stray individuals have been captured, the Hawk's-bill Turtle in the Severn, in Orkney, and in Zetland, and the Leathern Tortoise in Cornwall. Peron says, "the seat of the Phasianellæ is at Maria Island; all traces of them are lost at King George's Sound, after passing through insensible gradations." The Corals and other zoophytes, so plentiful and luxuriant in tropical seas, are replaced in temperate climes by (for the most part) a new series of less size and less luxuriant aspect; yet the tropical genera and species still linger with us in spare quantities. In England the Madrepores are reduced to two or three species, of rare occurrence and diminutive size; the Gorgoniæ are also comparatively small; and the *Gorgonia flabellum* of the tropics has been two or three times found on the Cornish and Leith shores. Instances of this kind might readily be multiplied.

It appears, from the foregoing remarks, that however desirable it might be, in all inquiries like the present, to determine general or primary laws, and that however obvious it may be that such determinations should be the principal aim of naturalists, yet the subject we are inquiring into will scarcely allow us to proceed farther than the discovery of secondary influences. But since these are liable to have an undue importance ascribed to them, and to be viewed as primary laws, it would be right, as far as possible, to ascertain their absolute weight, and to see how far they modify those more general influences.

It remains to be observed, that there are a certain number of facts in the Geography of Animals which do not appear to come under any law; nor are they explainable by, or referable to, any cause of which we have knowledge. No species of animal is cosmopolitan, but the extent of geographic range of species varies very largely. The greatness of this extent, however, is, except in a comparatively few cases, so ordered that there are certain divisions of the globe inhabited by races of animals peculiar to them, these races defining, as it were, by their limits, the zoological divisions of the earth. Thus, with a few exceptions (not considering those cases in the northern parts, where the two continents join or approximate), the Fauna of America is peculiar to it; and the same may be said of Australia, with the difference only of still fewer exceptions being present. Now, besides the exceptions to this rule of exclusive Faunas, the ranges of animals within their zoological divisions is frequently very extensive; whilst, on the other hand, the limits of very many are extremely circumscribed, sometimes a small spot of land, or a single river, being the extent of the habitat; in all of which cases, no clue to the cause of such peculiarities can be discovered. Indeed, unless we at once confess that animals occupy stations on the earth assigned to them by the will of the Creator, and determined only by a Providence and an Omnipotence perfectly inscrutable by us, we must be content to believe that animal distribution depends on circumstances connected with the constitution of the species, of the nature of which we are ignorant, and likely to continue so. It is not enough to point out instances of evident adaptation of animals to the circumstances which surround them, or to show that their peculiar food is found around them; for it might easily be demonstrated, that, in numerous cases, the same circumstances and the same food abound where the animals never come, and where, if brought by man, they readily become naturalized. It is not enough to say, that, in the instances where no adaptation is manifest, it nevertheless must exist, because we see that these animals are invariably found in one particular kind of situation; for, although some do certainly confine themselves as thus stated, yet they are frequently peculiar to one region or spot, and denied to others equally suited to their existence; besides which, there are kinds that, in seeming opposition to the whole analogy of zoological science, occupy a range of country or of abode, including opposite kinds of circumstances and situations, and these instances occurring too in the same tribe or family, where, as above stated, adaptation was in some species proved by the uniform character of the abode.

With respect to migration also, the causes are not always obvious, for birds of precisely similar endowments and character observe different habits, some migrating, and some being fixed; whilst at times the migrating species will for the winter remain with us, and seem to live as well as our common residents, so that the reason of migration is not always clear. The Llama and Vicugna, and the Sapajous, are peculiar to America; the Ornithorhynchus, Kangaroo, and Wombat, to New Holland. The Jay inhabits equally almost every country of Europe, and the immediately adjoining Asiatic countries, but extends no farther. The Barn Owl inhabits Europe, America, and part of Asia, and to my own knowledge some of the South Sea Islands. The Peregrine Falcon inhabits Europe, America, and Australia. The Blue Jay is confined to North America, the *Leptocephalus morisii* to the southern shores of England; and the *Physa alba* to the River Towyn of North Wales. The Sapajous, or prehensile-tailed Monkeys, are certainly well adapted to the forests of America; but are they less suited to the forests of other countries? The Nightingale certainly finds its peculiar insect food in those countries, and counties in England, to which it now resorts, the climate also being congenial to its feelings and habits; yet though it is found in Sweden and Germany, it is absent from Scotland and Northumberland, and though it is found in the middle and some southern counties of England, it is not seen in Devon or Cornwall. The Great Bustard is found enjoying a distribution latitudinally, whereas the same climate, situation, and food, could be obtained to the north and south of this zone. The *Achatina acicula*, though found in some limestone districts of England, has never been seen in the south of Devon, where lime abounds. The Swallow never migrates to America or China, though the food and climate there would suit its constitution. The Hedge Warbler is stationary, while the Blackcap migrates. Lastly, the Willow Wren, which ordinarily migrates, will yet at times remain with us through the winter.

The first of those secondary causes or influences, ranking as laws of geographic distribution of animals, which we shall mention, is *Climate*, a term which includes a consideration of temperature, of seasons, of winds usually prevalent, of the dryness or humidity of the air, of rains, drought, continued cold or heat, &c. It deserves notice that the presence of mountains, rivers, seas, barren spots, the quality of soil, the degree of cultivation, and the clearness or cloudiness of the sky, have all some influence in forming the climate, and in consequence the Fauna of a country. Secondary laws seem to act and re-act largely on each other, so that no one of them appears to have a separate or unmixed influence on animal distribution. The influence of climate on the distribution of animals may at once be seen by considering the vast increase of living beings as we approach the equatorial regions from the poles; but then it must be clearly understood, that this effect of climate, or rather of heat, is observed only in a very general way, and that, owing to a great variety of causes, some quite incomprehensible, others connected with food, situation, &c. The interruptions of this rule of increase are both numerous and important; still, on the whole, heat may be considered one of the secondary causes that influence the geography of animals. The alternations of the seasons, which, besides bringing an alteration of temperature, induce considerable difference on the food of animals, have a decided influence on their situations, causing a variety of movements termed Migrations. These changes of place are more immediately dependent on temperature and the state of the atmosphere, than on food, or other causes. Winds frequently affect the Fauna of a country, by driving aquatic animals to land, or by putting migrating animals from their destined courses. The state of the atmosphere as regards dryness or humidity, together with a continuance of rains or drought, will affect the general nature of the climate, and thereby the vegetable produce and animals of the country in which such conditions occur. Unusually hot or fine summers are most likely the causes of our receiving certain birds from the southern parts of Europe at that season. Long continuance of wet and cold at the time of the autumnal migration, will influence the period of departure of perhaps all our summer visitants.

Geographical situation, relations, and arrangement of a country, have considerable power over the extent and nature of a Fauna. The adjacency of an ocean, or large river in connexion with the sea, implies of course the presence of marine productions; the intersection of a country by smaller rivers and inland waters, will afford fluviatile and lacustrine animals; mountains and hills are the resorts of a variety of creatures; heaths and uncultivated spots have their peculiar animals; cultivated land, by originating a large proportion and variety of plants and trees, invite thither a great variety of passerine and other birds, either in search of insects in connexion with the vegetation, or for the purpose of feeding on the various seeds; together with the passerine birds are found the Climbers, they being insectivorous; and, lastly, certain of the Hawks, or other predatory birds allured thither by the presence of smaller species.

Such is the usual ornithology of many of our wooded districts in Devon, and notwithstanding that we owe much to our hills and heaths, perhaps our geographical sitiou and our relations to other countries, and, above all, the extent of our woods and cultivated ground, may be considered as more generally influential in determining our species of birds than any other secondary cause. Certainly Devon and Cornwall are two of the mildest counties in England, and in conformity with that character, the Stone Curlew has been known, according to Montagu, to remain all winter with us at the Start, the most southern point of land in England, except the Lizard in Cornwall. So also the Chiff-chaff was observed by the same eminent naturalist to stay the winter with us near his house at Kingsbridge.

Storms and other phenomena of weather referable to the head Climate are, as above said, and as will in the sequel be illustrated, of considerable consequence in forming and influencing a Fauna; but Climate is very much dependent on situation, arrangement, and other local circumstances of a country, especially adjacency of sea, which renders the temperature of all countries bordering on it mild and agreeable, provided the prevalent winds are in a direction from it. If Devon were not situated in connexion with the sea, of course no mildness of climate, or storms, or phenomena of that kind, could confer on us those marine products so conspicuous in our Fauna; and if our situation were not at the southern limits of the island, and opposite to the southern states of Europe, we should necessarily have none of those animals which by accident, or the invitations of unusually fine weather, cross over to experience the gentle warmth of our summer, or else are driven by the violence of equinoctial storms on our coasts; or, lastly, in the case of autumnal migrants, are enticed to stay the winter with us by reason of our southernmost locality, together with agreeably genial warmth.

The Flying-fish has occurred to the north of us, in the Bristol Channel, possibly under the influence of equinoctial gales; yet our situation must be taken into account rather than this phenomenon of our climate. The *Hippocampus vulgaris* and *Echineis Remora* have both been captured on our shores, yet situation, equally with, or probably more than climate, should be regarded as the cause. Many of our birds are influenced in their visits hither, and in their stay with us in winter, as well as in many peculiarities of movements exhibited by them, by our climate; but the abundance of wood and shelter, and the diversification of the surface of our country, will alone supply some explanation of the vast number of terrestrial birds found with us. Consequently, not only must we be compelled on most occasions to consider these two causes of distribution, climate and geographic position, with the other physical conditions of a country, in connexion, but to reflect that the latter influence is of the two the more powerful and extensive in operation.

If it were demanded of us to state the general nature and qualities of the climate of South Devon, we should say it was characterized by equality of temperature and humidity of atmosphere. Our summers are short, generally fervent, and attended by long droughts; our autumns are particularly rainy; our winters stormy, and sometimes very cold and lengthened; our springs chilly, unsettled, deceptive, and on the whole characterized by frequent intervals of gentle warmth of short continuance, between the long-continued rains, the protracted blasts, and blighting winds, ordinarily prevalent. Vegetation having made several unavailing efforts in these intervals, and having received frequent checks and blights, is at length permitted to put forth its energies in May. Occasionally this month is with us unusually dry and fine, so as to be productive of calamitous consequences both as regards the feeding of cattle, and the crops of grass and corn; for at this period vegetation makes its greatest efforts, and requires a supply of moisture to proceed with, in defect of which the harvests are rendered late and scanty, and cabbages and other garden produce are greatly injured. All this happened to us last year. Occasionally also, and such was our lot in the present year, May is uncommonly unpropitious, and vegetation makes no decided advances till June. On 24th March, snow fell and lay three or four inches deep, and ice formed in the estuary of the Yealm half an inch in thickness. On the 2d April, snow again fell; on the 11th, a fresh deposit took place, and in some spots remained two weeks. There were no leaves on the trees till after the first week in May! In the end of March, frosts even entered our hot-houses, and destroyed the young grapes.

According to my remarks, the arrival of spring birds of passage is deferred in accordance with the weather experienced in that season. They arrived late in the present year, or, at all events, were not seen or heard till after their usual periods. It is much easier to state facts than to assign reasons for facts; and so in the present case it will be found rather speculative to trace out the sources of these characters of our climate, and which it is requisite to do, because these sources will be found in the other physical conditions of the country, and themselves influential on the geography of animals, proving, as before said, that these secondary causes do not act independently of each other.—(*To be continued.*)

THE SPARROWHAWK.—Every person who attends to the habits of birds, must have seen this Hawk pursued by a troop of Swallows or other small birds, which hover over and behind it,, for what purpose is not well known. One fine evening last month, just after sunset, we observed one flying at a moderate rate over the fields, with a Pied Wagtail close upon it, and loudly uttering cries of alarm. It is probable that the Hawk had carried off its companion, although we did not perceive any thing in its talons, which, by flitting about, distract their attention, and thus prevent them from singling out an individual; and we do not remember to have ever before seen a single bird have the audacity to follow its dreaded enemy.

BOTANY.

THE CINNAMON TREE—LAURUS CINNAMOMUM.

THE natural family *Laurineæ* contains many highly interesting plants, which are strikingly similar in their properties, and perhaps none is more important than the present species, which grows in great abundance in the island of Ceylon. It has also been introduced into several parts of South America, and many of the West India islands. The tree grows to the height of 20 or 30 feet, and the diameter varies from 12 to 18 inches. The bark of the trunk is ash-coloured and scabrous, while the inner bark, which is of a reddish colour, is the cinnamon of commerce. The bark of the young shoots is smooth, and often beautifully-variegated. The leaves are opposite, elliptical, pointed at both ends, entire, three-nerved, and of a bright green colour. The flowers are arranged in the form of a paniculated umbel, and are of a greenish white colour; corolla of six acute, oval, and spreading petals, of which the three outer are broader than the others; stamens nine in number, thus referring it to the Linnæan class *Enneandria*, while the single pistil indicates the order *Monogynia*; filaments flattened, and arranged in three sets, shorter than the corolla, and at the base of each inner petal two small round glands are situated; the germen is oblong; stigma angular and depressed. The fruit is of a deep blue colour, being a pulpy pericarp containing an oblong nut, and the whole somewhat resembles a small olive. The cinnamon of commerce is the inner bark; camphor may be obtained from the root; a highly flavored oil is distilled from the leaves; the cassia buds of the shops are supposed to be the receptacle of the seed; and in Ceylon, candles are made from the kernel. The cinnamon generally arrives at perfection when the tree has attained the age of about six years, and only the small shoots are stripped of their bark, which, being made into bundles, is sent to Europe. Independently of its well known use as a condiment, it is also employed in medicine. The fragrant smell, the pungent and glowing taste, as well as all its properties, depend on an essential oil, of a whitish yellow colour, which is obtained by distillation from the bark. Cinnamon, being a most grateful aromatic, is stimulating, stomachic, and tonic, but is seldom used alone, being commonly exhibited along with other remedies which are more powerful. The oil has a very pungent taste, and is a powerful stimulant, being sometimes employed in cramp of the stomach, paralysis of the tongue, and in toothache. But its principal use is to cover the taste of other disagreeable drugs. A watery infusion of the bark is often advantageously given to check vomiting, and relieve nausea. Cinnamon, however, is more prized by the cook than by the physician.

VICTORIA REGINA.—Mr R. H. Schomburgk has recently communicated to the Botanical and Geographical Societies of London, an account of a remarkable species of plant belonging to the natural family of Nymphæaceæ, which he discovered on the 1st January 1837, on the river Berbice, in British Guiana. The leaves are from five to six feet in diameter, orbicular, flat, with a raised margin from three to five inches high; their upper surface bright green, the lower crimson, with eight principal nerves, nearly an inch high, and numerous subordinate branches, and beset with prickles. The young leaf is convolute, and expands but slowly, rising at first above the surface, but when expanded falling upon it. The stem of the flower is an inch thick near the calyx, and, like the leaf-stalk, is covered with prickles. The calyx has four sepals, each upwards of seven inches in length, and three in breadth, reddish-brown, and prickly externally, white on the inner side. The petals are very numerous, those next to the sepals larger, fleshy, and of a white colour, their general form oblong; the inner petals gradually smaller, bright red, and passing into the stamens. The flowers measure from twelve inches to nearly two feet in diameter, and finally assume a pink colour all over. The fruit is globular, prickly, many-celled, with numerous seeds. It appears, however, from a note by Dr Weissenborn, in Loudon's Magazine, that it is the *Euryale Amazonica* of Dr Pöppig; and even if it were new, its name, if it is necessarily to confer honour on our Queen, or to be honoured by her, ought to be Victoriana at least, agreeably to all precedent. A coloured figure will be seen in the Number for December 1837 of the Magazine of Zoology and Botany.

MISCELLANEOUS.

REMARKABLE CAVERNS IN BRAZIL.—Dr Lund, a Danish traveller, now in Brazil, has discovered in the mountain chains between the Rio Francisco and the Rio das Valhas, a great number of caverns, among which, Sappa nova de Marquiné, in the Sierra de Marquiné, is one of the most remarkable. The mountain consists of clay-slate, flinty-slate, and transition-limestone, in which last is the cavern described. Its total length, from north to south, is 1440 feet, the height being from 30 to 40 feet, and the breadth from 50 to 60. It is separated by masses of stalactite into twelve divisions, of which only three were known before Dr Lund explored them. The others, especially the innermost, were of such extraordinary beauty, that his attendants fell on their knees, and expressed the greatest astonishment. On the river Valhas, the banks of which the traveller afterwards traversed, the vegetation assumes a peculiar character. The inhabitants call the forests *catuigas*, or white forests. They form a thicket of thorny trees and bushes, interwoven with parasitical plants of the same nature. The leaves fall in August, and, from the beginning of September till the rainy season, the catuigas are as bare as European forests in winter. On this excursion Dr Lund had an opportunity of examining nineteen caverns, all of which confirmed his opinion of their geological formation. He has collected many remarkable particulars respecting the circumstances which must have taken place in a great inundation, as well as respecting its effects, and convinced himself, by several indications, that its course in South America was from north to south. In three of the nineteen caverns which he explored, he found petrifactions of quadrupeds, which he had not discovered in the Marquiné cavern, viz. *Cervus rufis*, *Calogenys paca*, *Cavia aperia*, six Bats, four Mice, *Lepus Brazilliensis*, and an Owl. In the first mentioned cavern he found two species of ruminating animals, far larger than those now living in Brazil, and a Megatherium, of the size of an Elephant.—*Literary Gazette.*

EARTHQUAKE IN CROATIA.—The following is an extract from a letter, dated Agram, October 15.—" We have lately witnessed an extraordinary phenomenon. Since the 1st of October, loud rumblings have been heard, proceeding as if from under our feet; the affrighted cattle *were* hurrying in all directions, the wild animals entered even into the very streets of the city, and the birds of prey settled on our roofs, and allowed themselves to be taken without resistance. The would-be-wise and fortune-tellers, of whom we have an overabundance, predicted the end of the world, or, at the very least, some great revolution of nature. On the 6th of October, about three o'clock, a loud noise similar to a discharge of artillery was heard, and the earth trembled. The alarm was now general, and people quitted their houses, and fled to the open country. The bells rung of their own accord, and many houses were overturned. These reports continued at intervals of half an hour, or an hour, till the evening; during the night they occurred at longer intervals, and the trembling of the earth was less powerful. On the morning of the 7th, two reports were heard, and the motion of the earth then ceased altogether. The air became cooler, and a north wind began to blow. The barometer was at 26° 4' 10", and the thermometer at 7° above zero. Fortunately no lives were lost by the falling of the houses, but three women and two children have died from fright, and more than sixty persons are suffering seriously from fear and exposure to the weather. Letters from different parts of the country announce that the noise was heard and the shocks felt throughout the extent of Croatia, and that much damage has been done and many lives lost."—*Athenæum.*

CORROSIVE QUALITY OF EARTHY SUBSTANCES.—M. C. Moritz, a German, who is now travelling in the north of South America, and whose letters are successively publishing in the Berlinische Nachrichten, passed near Guigue, over tracts from which the lake has retired,—the former extent of it being indicated by strata of little petrified fresh-water shells of the same species, which are still living in the lake. The dust which arises from the pulverization of these shell-beds has a corrosive quality, and, when brought into contact with the human skin, causes a very disagreeable burning sensation, from which M. Moritz had much to suffer, and which he ascribes to the remains of the hellcites. The inhabitants of the district use to say, *El aroma de la laguna pica*. It appears, however, that this quality of the dust is to be accounted for by the admixture of saline particles, as M. Michel Chevalier was annoyed in the same manner in 1835, when travelling in Mexico over those beds of ancient lakes, which, by the laceration of their high banks through earthquakes, have been changed into dry savannahs. These lakes appear to have been salt, as there are many extensive deposits of rock-salt, and as the ground is in many places so impregnated with muriate of soda, that it is altogether unfit for cultivation.—*W. Weissenborn, in Loudon's Magazine.*

EDINBURGH: Published for the PROPRIETOR, at the Office, No. 13, Hill Street. LONDON: SMITH, ELDER, and Co., 65, Cornhill. GLASGOW and the West of Scotland; JOHN SMITH and SON; and JOHN MACLEOD. DUBLIN: GEORGE YOUNG. PARIS: J. B. BALLIÈRE, Rue de l'Ecole de Médecine, No. 13 bis.

THE EDINBURGH PRINTING COMPANY.

· THE EDINBURGH

JOURNAL OF NATURAL HISTORY,

AND OF

THE PHYSICAL SCIENCES.

FEBRUARY, 1838.

ZOOLOGY.

DESCRIPTION OF THE PLATE.——THE BEARS.

The genus *Ursus* is composed of animals, generally of large size, which are for the most part frugivorous, although some of them feed also on flesh, and all of them readily become carnivorous when their ordinary food cannot be obtained. They belong to the Plantigrade Tribe of the Carnivora, characterized by walking on the entire soles of their feet, and by having five toes on the posterior as well as the anterior extremities. The Bears are of an apparently unwieldy form, with thick and muscular limbs, a very short tail, rounded ears, and a rather sharp muzzle. They pass the winter in a semitorpid state, without food, in cavities dug by themselves, and lined with leaves.

Fig. 1. 2. The Polar Bear (*Ursus maritimus*). The size and ferocity of this species was greatly exaggerated by the older navigators, one of whom alleges that the skin of a White Bear killed by him and his companions measured twenty-three feet in length, while another states that the Bears frequently seized on the seamen, carried them off with the greatest ease, and leisurely devoured them within sight of the survivors. It appears, however, that the ordinary length does not exceed from seven to eight feet, and that, although carnivorous, it is not essentially a predatory animal, but prefers the carcases of whales and fishes. It also lies in wait for seals at the openings of the ice, and seizes them with great dexterity. Dr Richardson states that, in autumn, it searches the shores for berries and other vegetable substances. The Polar Bear is of a more elongated form than the species which live in the forests, and the outline of its forehead and muzzle is straight, without the indentation between the eyes observed in them. Its neck is nearly twice as long as the head, its ears are very short, and its feet are larger than those of the Brown Bear. The fur is yellowish white, short on the head, and woolly on the back, long and woolly on the sides, belly, hinder parts, and legs. It inhabits the shores and icebergs of the polar seas, rarely extending its range beyond the arctic circle. The females retire to their winter quarters in December, and come abroad with their young in March or April.

Fig. 3. 4. The Brown Bear of Europe (*U. Arctos*). This species formerly inhabited the whole of Europe, as far south as the Alps and Pyrenees; but it has long been extirpated in the interior of France, in Holland, Germany, and the British Islands. It is plentiful in Siberia, extends as far eastward as Kamtschatka and Japan, and occurs more sparingly in North America. Its usual length is about four feet, its height two and a half, and it is covered with a long, soft, woolly fur, which, in young individuals, is of a deep brown, while, in the older, it has a mixture of yellowish-grey and reddish-brown. During the summer and autumn it lives chiefly on roots, leaves, and fruits, occasionally also on small quadrupeds; and at the approach of winter betakes itself to the hollow of a tree, or a crevice in a rock, or a retreat constructed by itself of branches of trees, and lined with moss, subsisting during the winter by the absorption of the fat which it has accumulated. Towards the middle of spring it again comes abroad, and at this period, being urged by hunger, is more dangerous than at any other, although it never attacks man unless provoked.

Fig. 5. The Black Bear of America (*U. Americanus*). The American Black Bear differs from the Brown Bear of Europe, in having the forehead less elevated, and the fur, which is of a glossy black, composed of soft smooth straight hairs. It is abundant in Canada, the Rocky mountains, and most of the thinly peopled districts of the United States. Its food consists chiefly of vegetable substances, but it also preys on animals of various species, sometimes makes great havoc among pigs, and now and then seizes a calf or a sheep. The females always retire to their winter quarters when the frost sets in, but, in the southern districts, the males often remain abroad all winter. The flesh of this species, as well as that of the European, is considered good food, and that of the bear hams are said to be excellent.

DESCRIPTION OF THE PLATE.——THE MUFFLED COCKATOOS AND PARROTS.

The Parrots, which Linnæus considered as constituting a single genus, have latterly been constituted into a family, under the name of Psittacidæ, the various subdivisions of which it is unnecessary here to describe, as the subject will be resumed in another part of our work. These birds may be briefly characterized by their short feet, of which the toes are flattened beneath, two of them only directed forward, their compact body, large rounded head, and short strong bill, of which the upper mandible is hooked, while the lower is truncate at the extremity. Their wings are generally short, and their tail exhibits a great variety of forms, being sometimes short and even, sometimes graduated and of extreme length. They are all frugivorous, and reside in the

33

warmer regions of the globe, climbing among the branches by means of their feet and bill. Their plumage is ornamented with the most gaudy colours, and as they are easily kept in captivity, and may generally be taught to articulate words, they are favourite domestic pets.

Fig. 1. The Rose Muffled Cockatoo (*Calyptorhynchus Eos*) has the head, neck, and all the lower parts, rose coloured, the rest of the plumage grey. It inhabits Malacca.

Fig. 2. The Racket-tailed Parrot (*Psittacus setarius*) is so named on account of the elongated shafts of the two middle feathers. It is of a bright green colour, with a red spot on the head, and a band of red and yellow across the hind neck.

Fig. 3. Hurt's Parrot (*P. Hentii*). Light green, the head yellow, the edge of the wing and the tail feathers bright red, the latter tipped with green.

Fig. 4. Peeters' Parrot (*P. Pretrii*). Light green, the forehead and sides of the head, with the edge of the wing, the alæna and primary coverts carmine red.

Fig. 5. The Mitred Parrot (*P. mitratus*). Bright green, with the upper part of the head rose coloured.

BIRDS OBSERVED ON THE FRITH OF FORTH IN DECEMBER 1837.

On the 27th December, accompanied by a young friend, I crossed by the ferry-boat from Trinity to Kirkaldy. The water was exceedingly smooth, for although fleecy clouds drifted rapidly eastward, it was calm below, and the sun emerging from the clouds gave promise of a fine day. At first, no birds were to be seen, excepting one or two wandering Gulls, which hovered and wheeled in silence over the water; but when we had proceeded about a mile, small parties of Tarrocks, *Larus ridibundus*, advanced, screaming, and now and then dipping in pursuit of their prey. Farther on, numerous little bands of Guillemots and Auks, *Uria Troile* and *Alca Torda*, were seen flying up the Frith, or floating on the water. These birds, which are precisely similar in their manners, and differ very little in appearance, float lightly, with erect necks, dive with rapidity, partially opening their wings as they plunge headlong into the water, and fly in strings, at the height of two or three feet from the surface, with a direct and rapid motion, simultaneously inclining themselves alternately to either side. Six large Cormorants, *Phalacrocorax Carbo*, with sedately flapping wings and long outstretched necks, presented an interesting sight, as they flew past in a line, almost touching the smooth water. Small groups of Red-throated Divers, *Colymbus Septentrionalis*, composed of from two to four or five individuals, now and then shot past in rapid flight, and scarcely at a greater height than the Guillemots, although on many occasions I have seen them fly at a great elevation, especially in rough weather. A few Great Northern Divers, *Colymbus glacialis*, also made their appearance, flying precisely in the same manner as the red-throated species, but with somewhat less rapid motions of the wings. A single Great Black-backed Gull, *Larus marinus*, sailed quietly along at a considerable height, and now numerous groups of the Black-headed Gulls, at this season of the year however unhooded, danced buoyantly and gaily at the distance of a few yards from the water, often wheeling, and occasionally stooping to pick up some small fish.

The tide was rising, and almost all the birds were advancing in the same direction toward the entrance of the inner Frith or estuary. Passing Kinghorn, and entering the Bay of Kirkaldy, we found the number of Guillemots and Auks diminished, while the Gulls had disappeared; but here vast numbers of Velvet Ducks, *Oidemia fusca*, were dispersed over the waters in groups of from two to fifteen or twenty. In a flock that rose before us, however, I counted thirty-eight individuals. These birds, on account of their black colour, and large size, have a remarkable appearance, which is rendered still more so when they are on wing, as then the white patch across that organ becomes exceedingly conspicuous. They swim lightly, and fly with moderate speed, at the height of three or four feet. In rising from the water, they ascend very gradually, striking it with their wings along a distance of two or three yards, and in alighting they settle as it were upon their hinder part, and then fall forward. Interspersed among them in smaller numbers were groups of the Black Duck, *Oidemia nigra*, a species very similar, but inferior in size, and destitute of white on the wing. It exhibited precisely the same modes of flying and swimming. A shot fired by a person on board at the ducks, started from a rock off Seafield Tower, a large flock of Turnstones, *Strepsilas Interpres*, and from another in its vicinity, a smaller flock of what seemed to be Dunlins, *Tringa alpina*. Finally, on approaching the harbour of Kirkaldy, we saw a single beautiful Long-tailed Duck, *Anas glacialis*.

Having walked to Queensferry, we there went on board the steamer from Newhaven to Stirling, at eleven o'clock next day. The surface of the Frith was exceedingly smooth, and from Queensferry to Charleston was sprinkled with Guillemots, Auks, Divers, and Gulls. The latter, however, were chiefly congregated, to an extent that one could hardly have conceived, along the northern shores, over the eddies of which they hovered in pursuit of the young herrings that had been stationary there for several weeks. The number seen at one glance along the coast could not be less than a hundred thousand. The different species were easily distinguishable. Possibly nine-tenths of the individuals belonged to the Black-headed kind, *Larus ridibundus*, a most inappropriate name by the by, as its cries bear no resemblance to laughter. The young birds of this species were comparatively few, and did not generally keep apart, though sometimes small groups of them might be seen; of the remaining tenth, one half belonged to the Common Gull, *Larus canus*, of which there seemed to be more young than old birds. The other half was composed of Herring Gulls, *Larus argentatus*, young and old, Smaller Black-backed Gulls, *Larus fuscus*, and Greater Black-backed Gulls, *Larus marinus*, the latter in very small numbers. Two Feasers were seen, both young birds, of a dusky colour, with the wings mottled with whitish, the tail even, and therefore probably *Lestris pomarinus*. They did not attack the Gulls, but fished for themselves, picking up the small herrings from on wing.

The Guillemots, which were very numerous, but kept chiefly in the open part of the Frith, sometimes rose as the vessel approached them, and ran, as it were, along the surface in a straight line, flapping their wings all the while, to the distance of a hundred yards or more, although most of them dived when we came near them. The Red-throated Divers, which, although similar in colour to the Guillemots, were easily distinguished by their superior size, and the comparatively greater length of their necks, also frequently rose and splashed along to the distance of from two to four hundred paces. Although they fly with great speed when fairly on wing, they are heavy birds, and, in removing to a short distance, apparently do not think it necessary to rise into the air as a Gull or Tern would do, but shoot out in a straight line, striking the water with their wings and feet, the latter, in particular, throwing it to a distance behind.

Above Charleston, the Gulls and Divers disappeared from the open water, and the flocks of the former seen along the bays were not more numerous than they would be in ordinary cases. At Bo'ness, on the southern side of the Frith, not a single bird of any description was to be seen. Near Alloa, however, we observed a flock of smaller Black-backed Gulls, *Larus fuscus*, composed of thirty-eight individuals; and farther up many birds of the same species, with a few of *Larus canus*, *Larus argentatus*, and *Larus ridibundus*, were seen floating here and there in the bays. In the narrow part of the estuary, of which the flat margins are secured by low embankments, and sometimes fringed with very tall reeds, *Arundo Phragmites*, a Heron, a flock of Knots, *Tringa cinerea*, several Mallards, *Anas Boschas*, and three Red breasted Mergansers, *Mergus serrator*, were observed.

On the whole, few sights could be more interesting to the ornithologist than that of the vast number of sea-birds which are at present collected in the Frith, and especially in that part of it above Queensferry, to which doubtless they have been attracted by the great shoal of small Herrings which have sought refuge there, and which are caught in great quantities by the fishermen, and sold in the neighbouring towns as an article of food.—W.M.G.

ON THE GEOGRAPHICAL DISTRIBUTION OF ANIMALS.—NO. II.

The great source of the general equability and mildness of our climate is certainly our connexion with the sea. The humidity of the air is referable also to the adjacency of the ocean, the sea winds conveying with them the continued exhalations from its surface; but it depends likewise on the presence of our hills, which are great accumulators of vapour, and attractors of the lower clouds. A great influence is also exerted by our inland waters, which exhale considerable quantities of moisture. Lastly, the great abundance of trees, and of vegetation generally, must have the effect of condensing a large quantity of vapour, and of collecting a great quantity of rain, and subsequently yielding it to the atmosphere. But the great alterations effected in the appearance of our country by such extensive plantation and culture of various kinds of vegetable produce, influence not only indirectly our Fauna through the medium of climate, but also act directly on animals by accommodating a larger number than could otherwise find subsistence with us. This adoption of new residences by animals is a fact so generally allowed that I need not here insist on it.

I have stated that equability and mildness of climate influence our Fauna; but does humidity also? I am not aware that it does. The two principal seasons in which our Fauna is rendered extensive are, on the one hand, winter, when cold and storms cause a great variety of waders and water-fowl to seek our shores, and, on the other, summer, when the sun approaches us for a short space of time, and when humidity is in a great measure obviated by the absence of sea winds. At all other seasons but summer, our climate is rendered colder than it would otherwise be, by the great abundance of our wood, which forming considerable shelter, keeps off the benign influence of the sun's rays. Perhaps also, when our summers are not very hot, the earth does not imbibe so much caloric as to allow of its giving off any subsequently, to temper the severity of the winter's cold. This is in some measure borne out by our experience of the last year. On the other hand, a compact soil, such as is ours, generally has the quality of imbibing the fervour of the summer's sun, without restoring it to the air so readily as do loose sandy soils, so that a scorching atmosphere is to a great extent prevented. On the whole, our summers may be stated as being generally moderately hot, thereby preserving the character of uniformity to our climate, other circumstances above named having only minor degrees of power. Other alterations in the surface of a country besides planting and culture affect the number, condition, and situation of animals; many are exterminated by man's interference, some are thereby excited to unusual multiplication, many are restricted in their numbers, or in their range, or in both; and some are encouraged to disperse locally or generally. The alterations here referred to comprise tillage, draining, irrigation, fencing, building, &c. As regards Devonshire, we find the number of quadrupeds diminished by the advances of agriculture and civilization. Many kinds of birds also have been thereby lost to us, though possibly some species have been gained, and certainly the number of individuals has in many cases been increased, and a great variety of alterations effected in geographic position. The number of our Insects and Molluscous tribes has certainly been increased, and their geographic limits and positions considerably interfered with, contracted, or enlarged. Many instances in point will be afforded in the sequel.

Very little needs be said on the subject of food as affecting the geography of animals. By the polity of nature, the vegetable world and the series of animals are intimately blended and connected: an extensive Flora will for the most part imply a large proportion of animals, and so likewise the weaker creatures draw to them the carnivorous tribes. In our county we find an extensive Flora, and our woods are numerous and deep; the series of animals also is found very perfect, and the parts of it would, if not subjected to our interferences and persecutions, be relatively proportionate. Food influences the migration of animals to and from this country, subject, however, to the higher influence of weather. In order to comprehend the relative proportions of influence exerted by these two causes, we must suppose adverse and propitious cases in point. If food is plentiful, and the weather intemperate, a summer bird of passage will forthwith undertake its journey, if near the usual period for migrating; if food is scarce, and the weather fine, it will also depart; if both circumstances are adverse, it will hasten its departure still more decidedly; and if provision be in plenty, and the weather fine, its stay will be prolonged. Similar remarks might be made relative to winter migrants. Food has considerable influence in determining the other kinds of migrations besides the vernal and autumnal. It also causes a variety of unusual movements in animals, as will appear in detail. Food is known to determine with precision the habitats of many kinds, though, as shown under the head of Primary Laws, it has not that amount of power supposed by some, and many situations producing the required pabulum, yet do not produce the animals so dependent, and where, if removed, they thrive well. Our *Bulimus fasciatus* and *Helix virgata*, which seemingly need maritime localities from some preference of food, do not occur universally all along the coast, though the vegetation is seemingly uniform in character, but they are collected together in parties at certain spots.

Man conducts a warfare against certain animals, which he finds or supposes to be prejudicial to his interests. In some cases, as where by our agricultural operations, &c. animals have been permitted to multiply more than their natural enemies would have allowed, our destruction of their superfluous number is justifiable; and likewise, in the case of such creatures, whose lives and actions are incompatible with our security and actions, extirpation is demanded. Many of those animals, however, consigned to unlimited destruction, form important links in the chain of creation, and, in consequence, their deficiency will cause alterations of various kinds in the proceedings of those other creatures with whom they were associated in the general scheme and polity of nature. The general destruction of our rapacious birds by gamekeepers and others, must permit a vast accumulation of those species of smaller animals on which they feed, and, in consequence, a more general and unnatural dispersion. A consideration, therefore, of the operations of man, whether as respects his agricultural or his other improvements and refinements, or whether as regards his hostilities to the animal creation, is worthy of some regard in framing an estimate of the causes in active influence on the positions, ranges, and migrations of our native animals of Devon.

It is fairly to be presumed, that our acquaintance with secondary causes is as yet very imperfect; at least we are still unaware of the reasons of a very great number of phenomena which, from being peculiar to certain tribes, or to certain species, cannot possibly be referable to general laws; and until some light has been thrown on these circumstances, our knowledge on this head must be deemed incomplete.

Upon the whole, it will be found difficult to render an inquiry into circumstances and detail interesting, though the chief objects and questions we shall have in view will be—firstly, to determine or illustrate general laws, if the limits of such an investigation should by possibility admit of it; secondly, to determine and illustrate secondary laws, or causes of geographic distribution, by attention to the following circumstances:—Comparisons of the phenomena of this district with others of similar extent, and similar or dissimilar aspect, and contingencies; irregular distributions; the occurrence of varieties, and other modifications, as provisions to suit local circumstances; the times selected for migration (of the various kinds), with an attention to the causes possibly influencing these movements; modifications of habits; peculiarities in the Zoology of the whole of the spot selected, or in any of its parts; remarkable deficiencies in its Zoology; general, numerical, and other results. Lastly, it will be right to make mention of a variety of other circumstances, though no explanation of their occurrence can be given, since records of unexplained facts may serve to invite notice and inquiry into causes.

There are two methods of considering the Natural History of a given spot, each having its peculiar advantages. We may investigate it with regard only to those phenomena and circumstances properly and peculiarly its own, irrespectively of all interferences and additions by the inroads of agriculture, planting, &c.; or we may consider it in its present state under all its alterations. The first mode has the advantage of being the more natural, and it is also calculated to display a great number of remarkable features of the spot. The second plan has the advantage of setting things in their present light, and serves also to show to what extent animals are influenced by our proceedings, how far their distributions are modified by our advances and operations; the instincts guiding them in their defences against our intrusions; and, lastly, it serves to demonstrate an important principle in the economy of a great proportion of creatures, namely, that of adaptation to variety in surrounding conditions, a quality almost unrestricted in the human species, but enjoyed in a more limited degree by animals, but without which their lives would be dependent upon the slightest alterations in surrounding circumstances, and without which we should not have been enabled to subject any of them to our uses and pleasures. I believe it will be expedient to examine the Zoology of this district with reference both to original and to existing features, giving to each its respective value, and peculiar considerations.

A Cuckoo seen in Morayshire in December.—Mr John Barclay of Caloots, near Elgin, has transmitted to us the following very interesting account of the appearance of a Cuckoo in his neighbourhood:—" The winter, so far as it is gone, has been remarkably mild in Morayshire, and I suppose all over the kingdom. The last ten days have been like April or May. On Friday last (29th December) my neighbour, Mr James Geddes, at Maft, told me he had seen a Cuckoo on the paling about his garden, and not only seen it, but heard it utter its well-known cry. I see Bewick mentions that they have been found in this country in a torpid state in winter; and I remember an old man in the parish of Auchterless, Aberdeenshire, who said he had found one among whins which he was cutting, and that on being brought into his house, after some sleep it uttered the peculiar note, cuckoo, and escaped. But the fact I have mentioned is something new, so far as I know. I enclose a note from Mr Geddes, whose word may be relied on. He also mentions the Black Snail being out at another proof of the mildness of the season. I have not seen any; but a shoot of a rose on the outside of my window, looking to the Lossie, has grown upwards of five inches since the beginning of November." Mr Geddes's note to Mr Barclay is as follows:—" Maft, January 2, 1838.—My dear Sir,—I was not at home when your note came yesterday. The Cuckoo perched upon the railing of my garden on Wednesday last. One of my boys, taking it for the common hawk that flies about at this season, fired at it, when it flew down to the nearest wood, about a hundred yards, and perched on a tree, and immediately began its call " cuckoo," which it continued for five minutes, leaving it no longer doubtful what kind of bird it was. I find the large Black Snail has been seen out lately by others as well as myself. I only saw one about the beginning of last week. I have never before observed them out till May, and even them at the beginning of a course of fine weather only.—I remain, yours truly, J. Geddes."

The Pied Wagtail.—Mr Gould has recently offered some observations, in London's Magazine, respecting our Pied Wagtail, which he considers to differ from the Motacilla alba of Linnæus. " While engaged upon this tribe of birds during the course of my work on the Birds of Europe, I was surprised to find that the sprightly and Pied Wagtail, so abundant in our islands at all seasons, could not be referred to any described species, and that it was equally as limited in its habitat (as M. flava); for, besides the British Islands, Norway and Sweden are the only parts of Europe whence I have been able to procure examples identical with our bird, whose place in the temperate portions of Europe is supplied by a nearly allied, but distinct, species, the true M. alba of Linnæus; which, although abundant in France, particularly in the neighbourhood of Calais, has never yet been discovered on the opposite shore of Kent, or in any part of England. As, therefore, our bird, which has always been considered as identical with the M. alba, proves to be a distinct species, I have named it after my friend W. Yarrell, Esq., as a just tribute to his varied talents as a naturalist.

" The characters by which these two species may be readily distinguished are as follows:—The Pied Wagtail of England (M. Yarrellii) is somewhat more robust in form, and, in its full summer dress, has the whole of the head, chest, and back, of a full deep jet black; while in the M. alba, at the same period, the throat and back alone are of this colour, the back and the rest of the upper surface being of a light ash-grey. In winter, the two species more nearly assimilate in their colouring; and this circumstance has, doubtless, been the cause of their hitherto being considered as identical; the black back of M. Yarrellii being grey at this season, although never so light as in M. alba. An additional evidence of their being distinct (but which has, doubtless, contributed to the confusion) is, that the female of M. Yarrellii never has the back black, as in the male; this part, even in summer, being dark-grey, in which respect it closely resembles the other species."

Cossus Ligniperda.—The larva of this insect is capable of living to a surprising length of time without sustenance or food, and also without preventing its coming to maturity or the perfect imago, as the following circumstances will corroborate.—On the 13th of June 1836, I took an excursion for the purpose of procuring a few specimens of the Cossus, in company with an intelligent naturalist, in the neighbourhood of Nottingham, where they so greatly abound among the willow trees. I was fortunate in procuring a dozen chrysalids. I also found abundance of the larva, in all its stages, one of which was in its second year, and this I put into a tin box, the lid being perforated by small holes to admit air. On my return home, I placed the box in a situation which for six months afterwards escaped my recollection, when I again laid my hand upon it, and, on opening it, to my surprise the Caterpillar was not only alive and healthy, but to all appearance larger than when I first saw it. I afterwards removed it into a large box, which could not be perforated, and watched its progress very closely for a considerable length of time, when I put in a small quantity of saw-dust, for the purpose of allowing it to spin itself a cocoon or nest, respecting which, at the time, it appeared careless. But about the beginning of May 1837, it commenced operations, which it completed in a few days, since which it remained until the 17th of July, when it emerged into the perfect moth (which I placed in my cabinet), thus remaining without food or support in the larva state ten months, and two in the chrysalis. The box was placed in a very warm situation, in a cupboard near the fire.—R. H. Cowlishaw.—The Naturalist.

Species of the Genus Mustela.—The Prince of Musignano, Charles L. Bonaparte, has recently communicated to the Magazine of Natural History, a notice respecting the genus Mustela of authors, which he has divided into four genera: Zorilla, Martes, Putorius, and Mustela, the latter including the small slender-tailed species, such as the Ermine and Weasel. The following are the species of this genus, as he has restricted it.

1. Mustela Erminea. Linn. Europe.
2. M. Cicognanii. Bonap. North America.
3. M. Boccamela. Bonap. Sardinia.
4. M. vulgaris. Linn. Europe.
5. M. Richardsonii. Bonap. (M. Erminea, Richardson, Fauna, Bor. Amer.) North America.
6. M. longicauda. Bonap. (M. Erminea, Richardson, Fauna, Bor. Amer.) North America.
7. M. frenata. Licht. Mexico.

Skull of the Guadaloupe Human Skeleton.—In the number for July 1837 of Silliman's Journal, is a description of the skull of the fossil skeleton found in Guadaloupe, and deposited in the British Museum. It was procured on the spot by M. L'Herminière, and has been placed in the collection of the Literary and Philosophical Society of South Carolina. Dr Moultrie, Professor of Physiology in the Medical College of that State, having examined the fragments with the view of ascertaining whether it belonged, as was supposed, to an individual of the Carib race, or American variety, in so much that were it possible to detach them from their incrustation, the vacancies might be filled with the corresponding parts taken from the head of the Peruvian.

BOTANY.

THE MISSELTOE.—VISCUM ALBUM.

This singular and celebrated plant, which belongs to Tetrandria monogynia of the Linnæan system, and to the natural family of Loranthea, is parasitical, growing mostly on Apple-trees, but also on the Pear, Hawthorn, Service, Oak, Hazel, Maple, Ash, Elm, and many others. In Worcestershire and Herefordshire it is very common in orchards, but in the northern counties it is seldom met with. It is an evergreen shrub, insinuating its radical fibres into the wood of the trees on which it grows. Its branches are numerous, regularly dichotomous, smooth and yellowish green; its leaves oblong, entire, striated, opposite, on short stalks; the flowers small, axillar, in close spikes; the calyx of the male flower divided into four ovate equal segments; the anthers four, attached to the calyx, which in the female flower is divided into four small ovate leaves, and placed upon the oblong, three-edged germen, which is surmounted by a blunt and somewhat notched stigma; the fruit, a globular, white, smooth one-celled berry, containing a fleshy, heart-shaped seed. It is supposed to be propagated by the Missel Thrush and Fieldfare, which are said to eat the berries, of which the seed passes through

them unchanged, and adheres to the branches of trees where it germinates. There is no proof of this, however, and it has been observed that the roots are always inserted on the under side of the branches, but this again is accounted for by the action of rain. Withering states that sheep eat it very greedily, and that it is frequently cut off the trees for them in hard weather. Birdlime is sometimes made from the berries, whence the saying, Turdus suum malum cacat. Their pulp is so slimy and tenacious, that if they are rubbed on the smooth bark of almost any tree, they will adhere and produce plants the following winter. It was formerly in great repute as a remedy in epilepsy, but is now entirely disregarded. Boyle mentions a case of this disease in which it was employed with remarkable success. Colbach strongly recommends it in various convulsive disorders, and other authors consider it as tonic and astringent. If we add to this that the Druids attributed to it the most astonishing virtues, we have all that can be said in its favour. At the present day, it is not employed in medicine, and attracts attention merely on account of its peculiar habit and aspect; for even its use as birdlime has been superseded by that of the bark of the holly.

Hybrid Ferns.—A triumph has been obtained by M. Martens, the professor of chemistry at the university of Louvain, and Dr L'Herminier, over those who assert that no hybrid plant can be produced where no stamina exist. The former shook the fronds of Gymnogramma calomelanos and G. chrysophylla, reciprocally over each other, at the time when the fructification was fully developed, and thus produced a new plant which is to be called G. Martensii. It is worthy of remark, that the hybrid plant bids fair to be easily propagated in our greenhouses, while the parents constantly languish and die. While M. Martens was making his experiment at Louvain in Belgium, Dr L'Herminier watched the same process taking place naturally in the woods and savannahs of Guadaloupe, and sent some dried fronds, in excellent preservation, of the hybrid to M. Bory St Vincent.—Athenæum

REVIEWS.

A History of British Birds. By William Yarrell, F. L. S. Secretary to the Zoological Society. *Illustrated by a Woodcut of each species, and numerous Vignettes.* London: John Van Voorst.

The author of this work, already so well known to the public as a naturalist, cannot fail to increase his reputation, and add to the advancement of ornithological science, by the publication of a series of descriptions of the Birds of Britain, so judiciously drawn up, and so beautifully illustrated. In the prospectus, we are informed that it will be completed in two volumes, 8vo, and will contain a greater number of British Birds than has yet been included in any work on the same subject. Four parts have made their appearance, so that we are enabled to form an estimate of its value, and compare it with other works on the same subject at present in the course of being published. The figures are generally good, and in all cases beautifully engraved. Of those most to our mind, as exhibiting the natural aspect and attitude of the birds, we may mention the Gyr-falcon, the Kestrel, the Hen Harrier, the Snowy Owl, the Great Grey Shrike, the Red-backed Shrike, the Wood-chat Shrike, the Flycatchers, the Dipper, and the Fieldfare. They are all, in our opinion, much superior to the figures of Selby's Illustrations, which, indeed, we have always considered as neither accurate in a scientific point of view, nor creditable as works of art. But, if we were disposed to criticise them, we should find some faults here and there, although possibly those very faults might by others be considered as beauties. Thus, the Golden Eagle has the bill too large; the white-tailed Eagle is not well-shaped, the body being too full and rounded, the bill not deep enough, and the tail so represented that one cannot decide as to which of its surfaces is exhibited; the Merlin, the Hobby, and the red-footed Falcon, are clearly skins stuffed with tow, and not living birds; the brooding Buzzard, at p. 80, has the bill at least three times as large as it ought to be; the Marsh Harrier is flat and stiff; and the Goshawk is more like a Buzzard than it ought to be. Nevertheless, the cuts seem to us to be very beautiful, and we know none else in which fewer faults exist. But neither the merits nor the defects of this department belong to Mr Yarrell, whose descriptions we conceive to be uniformly good, and to contain much interesting information, obtained, however, less from personal observation than we had anticipated or could wish. The distribution of the species, not only in the different parts of Britain, but also in foreign countries, is carefully traced; their habits are given in detail, although sometimes rather briefly; their nests and eggs, as well as their external appearance, are described in a succinct manner; and occasionally some particulars relative to their organization are given and illustrated by vignettes. However, that the reader may be enabled to judge for himself as to the execution of this part of the work, we lay before him Mr Yarrell's description of one of our winter visitors.

"The Fieldfare is a well-known migratory Thrush that comes to us from the north, and is one of the latest, if not the last, species that makes its annual and regular winter visit to Great Britain and the North of Ireland. It seldom appears before the beginning of November, depending on the temperature of the season, and frequently later than that, arriving here in large flocks in search of food; and if the weather continues open and mild, spreading themselves over pasture-lands to look for worms, slugs, the larvæ of insects, and any other soft-bodied animals of that sort; but, on the occurrence of snow or frost, they betake themselves to the hedges, and feed greedily on haws, and various other berries. At this time they are much sought after by youthful gunners, who find them shy and difficult to approach; the whole flock taking wing and keeping together, settle by scores on some distant tree, from whence, if again disturbed, they wheel off in a body as before. Should the weather become very severe, the Fieldfares leave us to go farther south, and are again seen on their return. They are known to go as far to the south and to the east as Minorca, Smyrna, and Syria. The Fieldfare does not return to its breeding-ground till late in the season. I have known them shot on the 12th of May, and others have been seen much later. White of Selborne says, that one particular season they remained till the beginning of June; and he asks, why do they not breed in the Highlands? Some instances have occurred of the Fieldfare breeding in this country; and Pennant, or the editor rather of the last edition of the British Zoology, mentions two instances that came to his knowledge. More recently, a nest has been found in Kent, and others in Yorkshire and Scotland; but in Orkney and Shetland, according to the observations of Mr Dunn, it is only seen on its passage to and from other countries. Mr W. C. Hewitson, whose zeal in the cause of Natural History induced him to visit Norway a few summers since, in the hope of obtaining many rare specimens for illustration in his excellent work on the Eggs of British Birds, thus describes the nesting habits of the Fieldfare:—After a long ramble through some very thick woods, 'our attention was attracted by the harsh cries of several birds, which we at first supposed must be shrikes, but which afterwards proved to be Fieldfares, anxiously watching over their newly-established dwellings. We were soon delighted by the discovery of several of their nests, and were surprised to find them (so contrary to the habits of other species of the genus *Turdus* with which we are acquainted) breeding in society. Their nests were at various heights from the ground, from four feet to thirty or forty feet, or upwards, mixed with old ones of the preceding year: They were, for the most part, placed against the trunk of the spruce fir; some were, however, at a considerable distance from it, upon the upper surface, and towards the smaller end of the thicker branches. They resembled most nearly those of the Ring Ouzel. The outside is composed of sticks, and coarse grass and weeds gathered wet, matted together with a small quantity of clay, and lined with a thick bed of fine grass: None of them yet contained more than three eggs, although we afterwards found that five was more commonly the number than four, and that even six was very frequent. They are very similar to those of the Blackbird, and even more so to the Ring Ouzel. The Fieldfare is the most abundant bird in Norway, and is generally diffused over that part which we visited, building, as already noticed, in society, two hundred nests or more being frequently seen within a very small space.' The eggs are light blue, mottled all over with spots of dark red-brown; the length one inch three lines, the breadth ten lines. William Christy, Esq., junior, who, with a party of naturalists, visited Norway in the summer of 1836, says, on the mountains called the Dovrefield, Fieldfares were rearing their young; they were just able to fly about on the 6th of August. The call-note of the Fieldfare is harsh, but its song is soft and melodious. In confinement it soon becomes reconciled, and sings agreeably. At night, when at large, it frequents evergreens and thick plantations; but, unlike its congeners, it has frequently been known to roost on the ground among fern, heath, or furze, on bushy commons. This bird is well known in Sweden, Russia, and Siberia, where it is found only in summer; in Poland, Prussia, and Austria, it remains the whole year; but in France, and the still more southern countries of Europe, it is only a winter visitor, extending its migration in that season, as before stated, to Minorca, Smyrna, and Syria."

It will be seen from the above, that there is in Mr Yarrel's History more careful and judicious gleaning from other works than original matter. To the student, however, this is rather a recommendation than otherwise; and, in acknowledging his sources of information, the author evinces a degree of honesty which we in vain look for in the writings of many other candidates for ornithological renown.

The Natural History of the Birds of Great Britain and Ireland. Part I. Birds of Prey. Illustrated by thirty-six Plates. By Sir William Jardine, Bart. F.R.S.E., F.L.S. &c. Edinburgh: W. H. Lizars.

This work, which professes to be, not *a* history, but *The* History of British and Irish Birds, differs in some essential respects from Mr Yarrell's. In the first place, the Figures of the birds described are engraved on copper, and coloured, so as to resemble nature in more respects than one. Secondly, the peculiarities of the bills, feet, and wings, are illustrated by woodcuts. In the third place, the descriptions are more original, and their style bolder, although not so uniformly correct. Of the plates, several are excellent, some good, and others very poor. In the first class may be ranked those of the Orange-legged Falcon, the Golden Eagle, the Honey Buzzard, the Marsh Harrier, and the White Owl; in the second, the Jer Falcon, the Hobby, the Kestrel, and the Swallow-tailed Nauclerus; in the third about half of the whole number. Thus, for example, the Eagle Owl has the head much too small, the markings on the breast not sufficiently minute, and the feathers on the sides and tibiæ stiff, in place of being peculiarly soft and almost downy. Thus, also, the Long-eared Owl is washed with blue, of which it in reality has no traces; the Tawny Owl has light blue eyes, whereas they are in nature so dark as at a little distance to seem black; the Kite has the bill too elongated; the Goshawk is too clumsy and awkward; the female Sparrowhawk is much more like the species last mentioned; the grey tints generally are converted into blues; and the eggs seem to us remarkably ugly. As to the woodcuts, we can only say that their execution is in all respects exceedingly poor. In this volume are described as British, one vulture, *Neophron Percnopterus*, of which a single specimen was killed in Somersetshire in 1825; twenty Falconidæ, among which are included the Swallow-tailed Nauclerus, of which two specimens have been killed, the one in Argyleshire, the other in Yorkshire, and the Crested Spizaetus, which is added to the British Fauna, on the authority of Mr Wingate, animal preserver at Newcastle-upon-Tyne, who received a skin, in a fresh state, from Aberdeen; and eight owls, among which is the Hawk Owl, of which an individual is recorded to have been captured "a few miles off the coast of Cornwall, on board a collier, in a very exhausted state."

In an Introduction is given a sketch of the progress of ornithological science in Britain, together with some remarks on the changes that have taken place in the number and distribution of our native species. The characters of the order *Raptores*, or Birds of Prey, and of the genera and species of which it is composed, are briefly sketched, and illustrated by woodcuts. In the descriptions of the species, the author evidently strives hard to imitate Wilson, Buffon, Audubon, and others, who have succeeded in combining popular narrative with scientific information; but his attempts are generally unsuccessful, and often ludicrous, although, on the whole, the histories of the species may be recommended as generally correct, however ungracefully managed. If they be compared with the notes to Wilson's Ornithology and White's Selborne, it will be found that he has improved considerably, although he is still far from being an adept in composition. As an example of his manner, we here give a portion of his history of one of the most celebrated of our native birds.

"The Golden Eagle has ever been associated with majesty or nobility; in ancient mythology, an Eagle was alone thought worthy to bear the thunder of Jove. By rude and savage nations he is combined with courage and independence. The young Indian warrior glories in his Eagle's plume as the most distinguished ornament with which he can adorn himself. The dress of the Highland chieftain is incomplete without this badge of high degree. And if, by the trammels of system (which, nevertheless, is indispensable, when the number of objects to be arranged exceeds eight thousand), we are forced to place him in an aberrant or less honourable situation, yet, when met with on his native mountains, free and uncontrolled, we cannot refuse the tribute which has been rendered to him by our predecessors.

"In England and the south of Scotland the Golden Eagle may be accounted rare, very few districts of the former being adapted to its disposition, or suitable for breeding places. Some parts of Derbyshire are recorded as having possessed eyries, in the mountainous parts of Wales there are others, and the precipices of Cumberland and Westmoreland also boasted of them. Upon the wild ranges of the Scottish border, one or two pairs used to breed, but their nest has not been known for twenty years, though a straggler in winter sometimes is yet seen amidst their defiles. It is not until we really enter the Highlands of Scotland by one of the grand and romantic passes, that the noble bird can be said occasionally to occur, and it is not until we reach the very centre of their "wildness," that he can be frequently seen. But the species must be gradually, though surely decreasing, for such is the depredation committed among the flocks during the season of lambing, and which is the time when a large supply of food is required by the parent birds for their young, that every device is employed, and expense incurred by rewards, for their destruction."

Edinburgh: Published for the Proprietor, at the Office, No. 13, Hill Street. London: Smith, Elder, and Co., 65, Cornhill. Glasgow and the West of Scotland: John Smith and Son; and John Macleod. Dublin: George Young. Paris: J. B. Balliere, Rue de l'Ecole de Médecine, No. 13 bis.
THE EDINBURGH PRINTING COMPANY.

THE EDINBURGH

JOURNAL OF NATURAL HISTORY,

AND OF

THE PHYSICAL SCIENCES.

MARCH, 1838.

ZOOLOGY.

DESCRIPTION OF THE PLATE.—THE PANDAS AND COATIS.

The genus *Ailurus* is composed of a single species of carnivorous quadruped, allied to the Civets and Martins on the one hand, and to the Benturongs and Coatis on the other, being in some measure plantigrade like the latter, but in form more resembling the former.

Fig. 1. The Shining Panda (*A. refulgens*), of a light red colour above, paler on the sides, the face white, and the lower parts and legs dusky. It inhabits South America.

The Benturongs have been separated from the genus Viverra, to constitute a small group characterized by their semipalmated feet, plantigrade mode of walking, and long spirally twisted tail, which, however, is not prehensile. They inhabit the Indian Peninsula, and in habits resemble the Civets, feeding on small quadrupeds, birds, and other animals.

Fig. 2. The White-fronted Benturong (*Ictides albifrons*) is characterized by having the fur long, and formed of a mixture of black and white hairs, the forehead whitish, the tail black, and a patch of the same colour including the eye.

Fig. 3. The Black Benturong (*I. niger*), the Musanga of Raffles, is of a black colour all over, excepting the face, which is whitish.

The Coatis, which by Brisson were referred to the genus Ursus, and by Linnæus to Viverra, are now separated to constitute the genus Nasua, so named on account of the prolongation of their nose, which is also possessed of considerable mobility. They are small, plantigrade, nocturnal animals, with an elongated tapering head, limbs of moderate size, and a long tail. They feed on animal substances, climbing on trees like the Opossums, and inhabit South America.

Fig. 4. The Brown Coati (*N. narica*), of a dusky-brown colour, the tail ringed with dark brown, and with three white spots near the eye.

Fig. 5. Female Brown Coati.

Fig. 6. The Red Coati (*N. rufa*). This species differs from the Brown, in having its fur of a light brownish-red colour, the bands on the tail lighter, and the muzzle greyish-black, with white spots round the eye.

DESCRIPTION OF THE PLATE.—THE PROMEROPSES.

The genus *Epimachus* is referred by Cuvier to the family of the Tenuirostres, or slender-billed birds, and is nearly allied to the Hoopoes, having the bill longer than the head, arched, and tapering ; the mandibles pointed ; the tarsi short, the outer toe united at the base to the middle one. The plumage is glossed or velvety, the tail elongated, and the wings long and rounded, the fourth and fifth feathers longest. These birds, which inhabit Africa and New Guinea, are supposed to live chiefly upon insects.

Fig. 1. The Red-billed Promerops (*Epimachus erythrorhynchus*), which has the bill and feet red, the plumage black, but lightly glossed with green and purple, the tail very long, with all the feathers, excepting the two middle, marked near the end, with an oval white spot. It inhabits New Guinea.

Fig. 2. The Superb Promerops (*E. superbus*) resembles the Birds of Paradise in the extraordinary development of its hypochondrial feathers, and in the texture of its plumage, of which the colour is black, with purple and green reflections : the scapulars sickle-shaped, purplish-black on the inner web, greenish-yellow on the edges and tip. The tail is extremely elongated, so that the bird, although its body is not much larger than that of a Jay, measures four feet in length. It inhabits Southern Africa.

Fig. 3. The Blue Promerops (*E. cœruleus*) is of the same form as the red-billed, but of a greyish-blue colour all over.

BRITISH BIRDS.—NO. X.—THE ROCK DOVE.

Having in a former Number described the Ring Dove or Cushat, we now present an account of the Rock Dove, which is the original of our Domestic Pigeons, and is of frequent occurrence on many of our rocky coasts. The plumage of this species is light greyish-blue, the lower parts being as deeply coloured as the upper. The middle of the neck all round is splendent with green, its lower part with purplish-red. The back and the upper part of the sides, from near the shoulders to near the tail, are pure white, as are the lower wing-coverts and axillaries. The primaries and their coverts

34

are brownish-grey on the outer web, the former dusky towards the end, as are the outer secondaries. There are two broad bars of black on the wing, one extending over the six inner secondary quills, the other over the secondary coverts, the outer two excepted. The tail has a broad terminal band of black, and the outer web of the lateral feathers is white. The downy part of the feathers is greyish-white, excepting on the white part of the back, where it is pure white. The horny part of the bill is brownish-black, as is the anterior half of its tumid portion, of which the rest is white ; the iris is bright yellowish-red ; the bare space around the eye flesh-coloured ; the tarsi and toes carmine-purple ; the claws greyish-black. The length is fourteen inches, and the extended wings measure twenty-seven. The female is a little smaller, and has the shining colours on the neck less extended.

This beautiful bird occurs in great abundance in the outer Hebrides, in Skye, Mull, and other islands on the north-west coast of Scotland, in Shetland and Orkney, and here and there along the rocky shores of all parts of the northern division of Britain, as well as on some of the southern. At early dawn, the Pigeons may be seen issuing from their retreats in straggling parties, which soon take a determinate direction, and meeting with others by the way, proceed in a loose body along the shores until they reach the cultivated parts of the country, where they settle in large flocks, diligently seeking for grains of barley and oats, pods of the charlock, seeds of the wild mustard, polygons, and other plants, together with several species of small shell-snails, which abound in the sandy pastures. When they have young, they necessarily make several trips to the course of the day, but from the end of autumn to the beginning of summer, they continue all day in the fields. In winter they collect into flocks, sometimes composed of several hundred individuals ; and, as at this season they are anxious to make the best use of the short period of day-light, they may easily be approached by a person acquainted with the useful art of creeping and skulking. In general, however, they are rather shy, and very seldom allow a person to advance openly within sixty or seventy yards. After the corn has ripened they soon become plump, and continue in good condition until the middle of winter, or even later, should the season prove mild ; but in spring and summer they are lean and tough. Their flesh is superior to that of the Wood Pigeon, but not equal to that of the Golden Plover, a bird which is equally abundant in the northern and western islands.

The manners of the Rock Doves are similar to those of our Domestic Pigeons. When searching for food, they walk about with great celerity, moving the head backwards and forwards very prettily at each step. In windy weather, they usually move in a direction more or less opposite to the blast, and keep their body nearer to the ground than when it is calm, the whole flock going together. When startled, they rise suddenly, and by striking the ground with their wings produce a crackling noise. Their flight is very similar to that of the Ringed and Golden Plovers, birds which in form approach very nearly to the Pigeons ; when at full speed, they fly with great celerity, the air whistling against their pinions. They usually alight abruptly when the place is open and clear, and if very hungry immediately commence their search ; but on other occasions they fly over the field in circles, gradually descend, and on alighting stand and look round them for a few moments. The notes of this species resemble the syllables coo-roo-coo, quickly repeated, the last prolonged. Its nuptials are celebrated with much cooing and circumambulation on the part of the male, and the first eggs are laid as early as the beginning of April, several broods being reared in the season. The nest is rudely constructed of straws, grass, various other plants, and frequently a few feathers ; the eggs, like those of all other Pigeons, are two in number, smooth, and pure white. The young, which are at first covered with loose yellow down, are fed by their parents, who, applying their bill to that of the nestlings, force up the food from their crops, so as to be within reach of the bill of the young, which all the while flaps its wings, and utters a low cheeping note, indicative of its eagerness to have its wants supplied. When fledged, they are of the same colour as the old birds, the head and neck, however, being of a dull purplish-blue, without the bright green and purple tints of the old, and the wings tinged with brown. At the first moult they acquire their full colour.

In Shetland, according to the Rev. Mr James Barclay, they are met with in considerable numbers, and are especially abundant in the parishes of Sandwick and Dunrossness, as well as in Fetlar, these places being the most extensively cultivated. They flock twice in the year, first in August or September, and again in winter. One pair when tamed generally breed four times in the season ; and so great is their fecundity, that they are frequently seen sitting on eggs long before the former brood is able to leave the nest, so that the parent bird has at the same time young birds and eggs to take care of. Mr Smith, of the same country, has also transmitted us specimens, and states that in Fetlar they are seen in large flocks in the winter and spring months, when they frequent barn-yards, especially should the ground be covered with snow. The crops of Mr Smith's specimens were completely filled up to the mouth: that of one with a mixture of barley and oats, together with a considerable number of eggs of snails, and some fragments of the pods of charlock; that of another, with oats, a few seeds of polygona, and fragments of charlock; and that of the third with oats alone. The number of oat seeds in the crop of the second amounted to 1000 and odds, and the barley seeds in that of a specimen sent by Mr Barclay were 510. From these facts it may be imagined what a quantity of seeds is annually devoured by all the Pigeons, wild and tame, in Britain.

After a long continuance of snow, these birds become extremely emaciated, for they scarcely find any thing to eat on the shores, and they do not appear to betake themselves to turnip fields, like the Wood Pigeons. Indeed, in most of the districts in which they abound, turnips are very rarely cultivated. From ten to twenty have sometimes been killed at a shot, when they have settled on corn-stacks, and it is not uncommon to obtain as many as four or five even on stubble or newly sown fields, for they often move very close together. The Peregrine Falcon and the Sparrowhawk seem to be the only feathered enemies of this species, whose power of flight, however, is such as to render it little liable to persecution even from them, while its rocky haunts exempt it from the attacks of rapacious quadrupeds. It is easily tamed when procured from the nest, and readily breeds with the Domestic Pigeon. On the other hand, individuals of the latter often fly off to the rocks, and either form colonies apart, or mingle with the Wild Pigeons. In its truly wild state, the Rock Dove presents no remarkable variations of colour, and the variously coloured individuals sometimes seen among the rocks are emancipated slaves, or their descendants.

Another British species, the Stock Dove, is very similar to the present, but differs in having no white on the rump, and in its habits and distribution, it being entirely unknown in the northern parts of the country; but of this and the Turtle Dove, we shall present the principal characters on another occasion.

Ornithology of Australia.—Mr Gould, whose splendid illustrations of the birds of Europe, and other ornithological works, have gained him a well merited celebrity, intends to leave England, this spring, for New Holland, where he proposes to remain two years, for the purpose of studying the habits of the numerous species of birds occurring in that country. Having already commenced a work on them, he will carry out with him the materials necessary for continuing its publication. The outlines will be drawn by himself, and the lithography will be executed, as formerly, by Mrs Gould, who will accompany him for that purpose.

Longevity of Birds.—Dr Weissenborn gives an account, in the last number of the Magazine of Natural History, of a Nightingale which had been caught in its adult state, and had lived in captivity nearly thirty years. He also states that a German paper, the Nürnberger Correspondent, (October 1837,) mentions that an Amsterdam merchant has been in possession of a grey Parrot for the last thirty-two years, after a relation had had the same bird forty-one years, so that its age is now seventy-three years, besides the time it may have lived previously to its transportation to Europe. It is now in a state of marasmus, and its powers of vision and memory are gone. Till sixty, it regularly moulted once a year, and the last time, the red feathers in its tail were exchanged for yellow ones.

Diseases of Birds.—The pathology of birds has hitherto received little attention. Domesticated individuals, as might be supposed, seem to be subject to more diseases than those in a state of freedom. We recently examined a parrot, of which the lungs were in a state of atrophy, and had moreover undergone a change similar to that exhibited in the human species when affected with melanosis. The internal or cuticular lining of the gizzard, which, in its natural state, is hard and dense, resembling the skin of the heel, was enormously thickened, and presented the appearance of a mass of transparent jelly.

Mr Shaw's Experiments on the Development of the Fry of the Salmon.—At a recent meeting of the Royal Society of Edinburgh, Mr John Shaw presented a very interesting account of experiments on the development of the young of the Salmon, which has been published in the Edinburgh New Philosophical Journal, and of which the following is an abstract. Mr Shaw had formerly stated his opinion that the Parr is the young of the Salmon, but it had been objected to his observations, that they were made upon ova taken from the bed of the Nith, which might have belonged to another species. The ponds in which these recent experiments have been made are three in number, two feet deep, thickly imbedded with gravel, and supplied by a small stream in which larvæ of insects abound. The waste water from them is conducted by wooden pipes secured by grating, and any accidental overflow is prevented by embankments two feet high.

With a bag-net extended on an iron hoop five feet in diameter, he caught, on the 4th January 1837, two Salmon, a male and a female, which were engaged in depositing their spawn; and having drawn them ashore, placed them successively in a trench on the beach made for the purpose, pressed out a quantity of the ova and melt, allowing the latter to pass down the stream, so as to be mingled with the former, and then transferred the spawn to a basin, and deposited it in a stream connected with a pond previously formed for its reception. The temperature of this stream was 39°, of the river Nith, from which the Salmon were taken, 33°, and of the atmosphere 36°. On the 23d February, fifty days after impregnation, the embryo fish was distinctly visible to the naked eye, and moved feebly in the egg. The temperature of the stream was 36°, and of the atmosphere 38°. On the 28th April, a hundred and fourteen days after impregnation, the young were excluded from the eggs. At this period the little fish has a very peculiar aspect; the head is large, the entire length is five-eighths of an inch, and the colour pale blue or pink; the bag attached to the abdomen conical, and of a beautiful transparent red, so as to be easily distinguishable at the bottom of the water, even when the fish itself can with difficulty be observed. A slightly indented fringe extends from the dorsal and anal fins, to the termination of the tail. On the 24th May, twenty-seven days after being hatched, the young fish had consumed the yolk; but in a few days afterwards, the whole of this family, with the exception of one individual, was found dead at the bottom of the pond, a circumstance which, having more than once occurred before, Mr Shaw attributes to a deposition of mud.

To show the effect of increased temperature in hastening the development of the infant fish, he relates an experiment made upon a few of the same ova. On the 20th April, a hundred and six days after impregnation, finding these ova unhatched, and the temperature of the stream being 41°, he took four of them and placed them in a tumbler of water, covering the bottom with fine gravel, in which he imbedded the ova. He then suspended the tumbler from the top of his bed-room window, above which he placed a large jar, from which a stream of pure water was directed into the tumbler, the overflowings of which were carried out at the window along a wooden channel. The average temperature of the room was 47°, that of the water 45°; but during the night it was considerably increased, and the young fish in the tumbler were hatched in thirty-six hours, whereas those remaining in the stream did not hatch till the 28th of April, a difference of nearly seven days.

At this stage they are so very transparent, that their viscera are distinctly visible. Their pectoral fin is continually in rapid motion, even when they are otherwise in a state of perfect repose. On the 24th May, thirty-nine days after their birth, the fish in the tumbler were completely divested of the yolk, and the characteristic bars of the Parr had become visible. At this time they measured nearly one inch in length. They were returned to the pond, where they perished with the rest of the family.

In another experiment, the ova were procured in the same manner, and deposited in the stream entering a pond, on the 27th January 1837, the temperature of the stream being 40°, and that of the river 36°. On the 21st March, fifty-four days after impregnation, the embryo fish was visible to the naked eye. On the 7th May, a hundred and one days after impregnation, the little fish had burst the shell, and were to be found amongst the shingle of the stream. The temperature of the water was 43°, of the atmosphere 45°. When two months old, the fry presents in miniature the proportions of a mature fish; at the age of four months, the characteristic marks of the Parr are distinctly visible; when six months old, it is only three inches and a quarter in length, but its approximation to the features of the parent fish is more striking, and on comparing it with the Parr in the river, no marked difference can be observed. The whole of this family, as well as another family in a separate pond, were in December in perfect health, and fed freely on worms and larvæ, with which they had been supplied during the summer.

On comparing the Parr taken from the river at a corresponding period with those taken from the pond, they were found to be uniformly of a darker colour, which Mr Shaw attributes to the more impure or muddy quality of the river water.

In the course of Mr Shaw's visits to the experimental pond, he had often observed, that, while the little fish remained stationary in any particular part, they were always found to be of a colour corresponding to that of the bottom, and when they removed to any part of a different colour, that, after resting on it for a few minutes, they gradually assumed a corresponding hue. Wishing, therefore, to prove the fact of this assimilation by actual experiment, he procured two earthenware basins, one nearly white inside, the other nearly black, placed a living fish in each, and kept up in them a constant supply of fresh water. The fishes, which at first were of their natural colour, had not remained in the basins more than four minutes till each had gradually assumed a colour nearly approaching to that of the respective basins. He next took the fish out of the white basin and placed it in the black one, and the fish which was in the black basin he placed in the white, and the results were the same. He next placed both fishes in one basin, when the contrast for a short time was exceedingly striking. Exclusion of the light by means of a mat produced a dark colour, which gradually disappeared when the mat was removed.

Mr Shaw, in conclusion, considers that he has now succeeded in establishing the fact, that the young Salmon does not proceed to the sea the same year in which it is hatched, and hopes that his experiments will therefore be admitted as beneficial both in a scientific and economic point of view. The belief that the Salmon migrates the same year it is hatched, he observes, has created an indiscriminate slaughter of that fish, at an age when it especially requires the protection of the legislature. There being no fish in our rivers that takes the fly more readily, the destruction of the fry by juvenile anglers is incalculably great.

Although the author of these experiments may not have succeeded in convincing naturalists that the Parr and the fry of the Salmon are identical, he is persuaded that they are so. But the apparent maturity of the sexual organs of the male, and the immaturity of those of the female Parr, are perplexing circumstances, which cannot be reconciled with any of the proposed theories respecting that fish. He suggests the hypothesis of an analogy between the female Salmon and the queen Bee, and imagines that the former like the latter may have the aid of a plurality of males in propagating her species. At the same time these male Parrs attend the female partly for the purpose of devouring the ova which descend with the stream.

In conclusion, Mr Shaw, in our opinion unnecessarily, disclaims all pretension to scientific attainments, wishes to be considered only as an honest inquirer after truth, possessing facilities of observation peculiarly advantageous, and states that he intends to continue his investigations. If all the naturalists who have so confidently spoken on this interesting and important subject, had bestowed as much care upon it as Mr Shaw has done, their observations might have proved useful in place of being so contradictory and vague that no reliance whatever can be placed upon them.

The Bernacle Shell.—The animal represented by the accompanying figure belongs to the Pedunculate Cirripeda, which, agreeably to Lamarck's definition, are characterized by having their body supported by a tubular coriaceous contractile peduncle, of which the base is fixed to marine bodies. The plurivalve shell of these animals is considered by the author just mentioned as analogous to the operculum of the sessile cirripeda, of which the external crust or calcareous envelope does not exist in the pedunculate species. The genus *Anatifa*, to which our animal belongs, is defined as having the body covered by a shell, placed upon a tubular, flexible, tendinous, and contractile stalk ; with numerous ciliated and jointed tentaculary bodies attached to the respiratory organs ; the shell compressed, of five unequal contiguous valves, the lower lateral valves larger. Several species are known, of which the most common is that here figured.

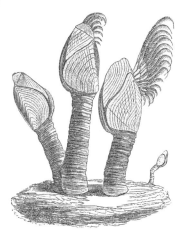

Anatifa lævis, the Smooth Barnacle Shell.—In this species the shell is compressed and comparatively smooth, the striæ on its surface not being prominent as in another species, named on that account *A. striata*. This shell is composed of five pieces, two lateral, and one dorsal ; the lower lateral pieces much larger than the upper, and the dorsal very narrow and keeled. The pieces are connected by membrane, unless in front towards the top, where the tentacula project. The body of the animal is of an oblong compressed form, curved and enveloped in a soft skin or mantle. Appended to the branchiæ or respiratory organs are ten pairs of long, slender, tapering, curved, and ciliated tentacula, five pairs on each side. At their base are two shorter pairs, and several small flattened and curved organs, equally ciliated, but connected with the mouth. The shell is about an inch and a quarter in length, and is supported upon a peduncle varying in length from a few inches to a foot or even two feet, and in diameter from a quarter of an inch to half an inch. This peduncle is of a reddish colour, and has an external, transversely corrugated, elastic, cuticular envelope, which at its base is attached to a piece of wood, floating or fixed in the sea, and at its upper part to the shell. Internally it consists of a cylinder of longitudinal, whitish, muscular fibres, separated by a large quantity of aqueous fluid, and at the base adhering to the cuticle, which covers the wood, while at the top of the peduncle they are continuous with the body of the animal. The contraction of these muscular fibres of course shortens the peduncle, and throws its envelope into strong transverse wrinkles. These animals adhere often in great numbers to timber of any kind floating in the ocean, to logs, portions of wrecks, casks, and the bottoms of vessels, and are frequently thrown ashore on the western coast of Britain. Gulls of different species, but particularly the Great Black-backed and Herring Gulls, sometimes eat the animal as well as its fleshy peduncle. A foolish notion was once generally entertained respecting this animal, and is still prevalent among uneducated persons, namely, that the Barnacle or Brent Goose, *Anser Bernicla* and *A. Brenta*, originate from it. The ciliated tentaculary organs, which have some slight resemblance to feathers, were the cause of this misconception, having been taken for the sprouting quills of the young bird.

BOTANY.

Victoria Regia.—On the 22d January was read to the Royal Geographical Society of London, a note upon this plant by Professor Lindley. " I have great satisfaction in stating to the Society, that some specimens of the flowers of this extraordinary plant, which have lately been received from M. Schomburgh, completely confirm the statement of that traveller in all essential particulars, and at the same time establish the new genus Victoria upon the most complete evidence. The most startling circumstance named by M. Schomburgh was, that the flowers measured fifteen inches in diameter ; one of the specimens now received measures fourteen inches in diameter, although its petals have rotted off in consequence of the bad manner in which they have been prepared. With respect to the genus, it has been already mentioned in the Journal of the Geographical Society, at my request, that although *Victoria* is possibly the same as the *Euryale amazonia* of Pœppig, yet it is, in my opinion, quite distinct from the latter genus. I am not aware that any one in this country, of any botanical reputation, has called this opinion in question ; and therefore it may appear unnecessary to notice it further. But Professor Pœppig is so good a naturalist, that it is due to him to state upon what grounds I consider him to be wrong in the genus to which he referred the plant. Euryale is an East Indian water plant, with very large floating leaves, sometimes as much as four feet in diameter, light purple underneath, and then reticulated with numerous very large prominent veins. It is, however, covered with sharp prickles on the under side of the leaves, the leaf-stalks, flowerstalks, and calyx. In these particulars it agrees with Victoria, but in little else. Victoria has the inner petals rigid, and curved inwards over the stamina, into which they gradually pass ; in Euryale there is no transition of this kind. In Victoria there is a double row of horn-like sterile stamens, curving over the stigmas, and adhering firmly to their back ; Euryale has no such structure. In Victoria there are thirty-six large, uniform, compressed fleshy stigmas ; in lieu of this very singular character, Euryale has only the margin of a cup, with six, seven, or eight curvatures. Victoria has twenty-six cells to the ovary ; Euryale only from six to eight. And, finally, to say nothing of minor distinctions, the ripe fruit of Victoria lies at the bottom of a regularly truncated cup, which stands high above the water, while the flower of Euryale sinks into the water after flowering, and the fruit, when ripe, is invested with the decayed remains of the calyx and corolla. These facts will, I think, confirm my original statement, that, notwithstanding the prickles of the leaves and stalks, the genus Victoria is more closely allied to Nymphæa than to Euryale, and will, I hope, set at rest all future ingenious speculations upon the first of these genera being untenable."

GEOLOGY.

Remarks on the Erratic Blocks of the Jura, by M. Agassiz.—Having spent several months in the neighbourhood of the Alps, for the purpose of studying the glaciers, and of examining the observations of MM. Venetz and Charpentier on the great moraines found at a distance from existing glaciers, he was not less struck by the polished appearance presented by the rocks on which they had moved. The flanks of the valley of the Rhone are entirely polished, even to the borders of the Lake of Geneva, more than a day's journey from the glaciers, wherever the rock has been hard enough to withstand the action of the weather. On seeing this phenomenon, evidently produced by the glaciers, which, when they retired, left on their edges the concentric masses of rolled blocks named *moraines*, he remembered that the northern slope of Jura, which faces the Alps, presents similar polished surface, termed *lunes*, and on his return to Neufchatel, he hastened to examine them more attentively. He found that they have no connection with the stratification of the rocks, or with the direction of the chain of mountains, and that they extend along the whole surface, following its undulations, and presenting an uninterrupted smooth surface wherever the rock has been recently exposed. The furrows which often traverse these surfaces never occur in the direction of the slope of the mountain, but are oblique, so that they could not have been caused by currents of water ; and the minute markings on the surface are generally parallel to these furrows. It appeared evident that this polishing had been produced by ice ; it was found to occupy an extent of more than twenty leagues to the east and west of Neufchatel, although no one had taken notice of it. As the erratic blocks of the Jura are found resting on these polished surfaces, the question naturally occurs, whether, as M. Charpentier once supposed, the glaciers had extended to the Jura, pushing before them blocks of Alpine rocks, and polishing the surface on which they moved. Such an idea is confuted by the fact that the erratic blocks of the Jura are angular, while those of the moraines are always rounded, as are also those of existing glaciers. If, therefore, the erratic blocks in question had been pushed by glaciers to so great a distance, they ought to be much more rounded than those of the moraines, which ought also to have been the case had they been transported by currents of water. Had the latter been the agent, it is impossible to account for the lakes between the Alps and Jura not having been filled up, especially as it can be demonstrated that they existed previously to the blocks. Nor can the phenomena be accounted for by the supposition of the angular blocks having been transported on floats of ice moved by currents of water : for the erratic blocks of Jura do not rest immediately on the polished surfaces, but on a bed of rolled pebbles several feet in thickness, the larger at the surface, and the smaller, often as fine as sand, at the bottom. Now, had the transportation been effected by currents, the reverse would have been the case. The existence of fine sand on the surface of the polished rocks, also proves that no powerful cause has acted on the surface of Jura, since the period at which these Alpine rocks were transported ; and the fine lines on the polished surfaces are no doubt due to this sand, which, however, could not have been moved by a current of water, for neither torrents nor lakes, when charged with sand, produce any thing analogous on the same rocks. M. Agassiz is of opinion that most of the phenomena attributed to great diluvial currents, and especially those recently described by M. Selfstroem, have been produced by ice.

MISCELLANEOUS.

Pearl-Fishing in Arabia.—The Pearl-bank extends from Sharja to Biddulph's Group. The bottom is of shelly sand and broken coral, and the depths vary from five to fifteen fathoms. The right of fishing on the bank is common, but altercations between rival tribes are not unfrequent. Should the presence of a vessel of war prevent them from settling these disputes on the spot, they are generally decided upon the island where they land to open their oysters. In order to check such quarrels, which, if permitted, would lead to general confusion, two government vessels are usually cruising on the bank. The boats are of various sizes, and of various construction, according from ten to fifty tons. During one season it is computed that the island of Bahrein furnishes, of all sizes, three thousand five hundred ; the Persian coast, one hundred ; and the space between Bahrein and the entrance of the Gulf, including the Pirate Coast, seven hundred. The value of the pearls obtained at these several ports is estimated at forty lacs of dollars, or four hundred thousand pounds. Their boats carry a crew varying from eight to forty men, and the number of mariners thus employed at the height of the season is rather above thirty thousand. None receive any definite wages, but each has a share of the profits upon the whole. A small tax is also levied on each boat by the Sheikh of the port to which it belongs. During this they live on dates and fish, of which the latter are numerous and good, and to such meagre diet our small presents of rice were a most welcome addition. Where polypi abound, they envelope themselves in a white garment, but in general, with the exception of a cloth around their waist, they are perfectly naked. When about to proceed to business, they divide themselves into two parties, one of which remains in the boat to hand up the others who are engaged in diving. The latter having provided themselves with a small basket, jump overboard, and place their feet on a stone, to which a line is attached: Upon a given signal, this is let go, and they sink with it to the bottom. When the oysters are thickly clustered, eight or ten may be procured at each descent, the line is then jerked, and the person stationed in the boat hauls the diver up with as much rapidity as possible. The period during which they can remain under water has been much overrated; one minute is the average, and I never knew them, but on one occassion, to exceed a minute and a half.

Accidents do not very frequently occur from Sharks, but the Saw-fish, *Pristis antiquorum* of Linnæus, is much dreaded. Instances were related to me where the divers had been completely cut in two by these monsters; which attain, in the Persian Gulf, a far larger size than in any other part of the world where I have met with them. As the character of this fish may not be familiar to the general reader, I will add a few words in the way of description. They are of an oblong rounded form, their head being somewhat flattened on the fore part, and tapering more abruptly towards the tail. They usually measure from thirteen to fifteen feet in length, being covered with a coriaceous skin, of a dark colour above, but white beneath. The terrific weapon from whence they derive their name is a flat projecting snout, six feet in length, four inches in breadth, armed on either side with spines resembling the teeth of a Shark. Diving is considered very detrimental to health, and without doubt it shortens the life of those who much practise it. In order to aid the retention of the breath, the diver places a piece of elastic horn over his nostrils, which binds them closely together. He does not enter the boat each time he rises to the surface, ropes being attached to the side, to which he clings, until he has obtained breath for another attempt. As soon as the fishermen have filled their boats, they proceed to some of the islands with which the bank is studded, and there, with masts, oars, and sails, construct tents. They estimate the unopened Oysters at two dollars a hundred.—*Wellsted's Travels in Arabia.*

Periodical Thunder-storms at Constantinople.—In a letter to M. Arago Admiral Roussin states, that on the 10th August a violent thunder-storm occurred at Constantinople, and that about the same period one generally occurs yearly. It commenced in the south, and then veered to the north, approaching the Black Sea, its ordinary domain. It lasted from one o'clock in the morning to daybreak, and was accompanied by furious rain. The lightning struck three points near each other at Pera, at the Danish embassy; and at that of Spain, it broke the doors, and burnt the carpets and curtains, but did not touch the glass. The other bolt fell on a small Greek vessel, broke its mast, killed a man, and injured another. He says that at other times thunder-storms are by no means frequent.

LEARNED SOCIETIES.

Zoological Society of London.—At the meeting of the 9th January 1838, Thomas Bell, Esq., F.R.S., in the chair, Mr Gould exhibited a collection of Australian birds, and laid on the table descriptions and characters of the whole for publication in the Society's Transactions. He also described forty species of birds from Mr Darwin's collection, about a third of which were new. Mr Gray described a small marsupial quadruped from Van Dieman's Land, which he named *Perameles fascialis.* A communication was read from Captain Harris, descriptive of a new Antelope found by him in South Africa, about the size of a Horse, black above, white beneath, with the horns beautifully curved. It inhabits the great mountain range in the country of Mataveld, and is gregarious in small families.

Jan. 23d.—Richard Owen, Esq., F.R.S., in the chair. Mr Ogilby gave an account of several new species of Mammalia contained in the collection made by Captain Alexander in South Africa; and Mr Gould described several new birds from the same collection. Among the rapacious species were an Eagle with a white breast and red tail, two small Falcons scarcely larger than a Sparrow, and a very small Owl; among the insessores a Jay which perches on the horns of the Rhinoceros, and which the hunters of that animal anxiously watch, as when they see it fly up, they know that the object of their pursuit is alarmed. Mr Ogilby described a new species of Galago from Madagascar, in which the fore-finger was of the same size as the other, and partially opposable to them. Mr Owen also read some observations on the anatomy of the Giraffe, from dissections of the animals which died at the Regent's Park and Surrey Zoological Gardens.

PROCEEDINGS OF THE ROYAL SOCIETY OF EDINBURGH.

The first ordinary meeting of the 55th Session of the Society took place on Monday, 4th December 1837, Sir T. M. Brisbane, Bart., President, in the chair. A communication on the Food of the Vendace, Herring, and Salmon, by John Stark, Esq., was read. The author, after some preliminary observations, stated, that fishes, like the Vendace, residing in lakes, and feeding on animal food, must necessarily subsist on the small aquatic animals found in these lakes; that there is no reasonable analogy between the Vendace and the Herring, because they live in different mediums, the one in salt, the other in fresh water; that their food cannot therefore be the same; that writers on natural history state the animalcules which are found in the stomach of the Vandace, and the other minute animals found in lakes, to form the food of fresh-water fishes generally; and that Leeuwenhoeck had even figured the identical animal found in the stomach of the Vendace in 1833 more than 130 years before. As to the Herring, the author stated that its food was known and described from personal observation by Neucrantz, previous to 1654, by Leeuwenhoeck in 1696, by Fabricius in 1781, by Muller in 1785, by Bloch about the same period, by Lacepede and Latreille in 1798, by Scoresby in 1820, and by many others; and that what has been stated by all authors on the subject is corroborated by the examination of the stomach and intestinal canal of that fish, and by specimens on the table of the Society. With regard to the food and reproduction of the Salmon, he gave an account of the observations of the most esteemed authors, and exhibited preparations by Dr Parnell, confirming their statements. He noticed the valuable evidence taken before a committee of Parliament in 1824 and 1825, which corroborated the statements of systematic writers; and gave an abstract of the evidence as to the period of the ascent of the Salmon in the different rivers.

Dr Barry made some verbal remarks on the physiology of *Proteus anguinus.*

REVIEW.

Illustrations of Ornithology. By Sir William Jardine, Bart., F.R.S.E., F.L.S., M.W.S., &c.; and Prideaux John Selby, Esq., F.R.S.E., F.L.S., M.W.S., &c. 3 vols. Edinburgh, Daniel Lizars; London, Longman, Rees, Orme, Brown, and Green, and S. Highley; Dublin, Hodges and M'Arthur.

Some writers have affected to make a distinction between naturalists, whom they arrange primarily into professional and amateur; but this arrangement is altogether artificial being dichotomous, otherwise we should have referred the authors of this work to the latter section, seeing that they assume the pen and the pencil more from a benevolent desire to benefit the human race, or a wish to distinguish themselves, or a pure love of their favorite science, than from any necessity of taking it up as a trade or profession. As it is, however, we must simply view them as working ornithologists, painters, and engravers, for not only have they elaborated the descriptions of the objects which they have selected, but they have also represented them *more pictorum et sculptorum.* Their purpose has been to advance the progress of science by describing new and rare or remarkable objects; and their performance, besides in some measure accomplishing this object, will remain a monument of praiseworthy perseverance. The figures, although not generally characterized by a perfect semblance of living birds, are yet in all cases sufficiently accurate to enable a person, by comparing his specimens with them, to satisfy himself as to their specific identity. Some of them are very well executed; for example, *Pteroglossus maculatus, Cursorius bicinctus, Aegotheles lunulatus, Sitta castaneo-ventris, Gallus Bankiva, Tropidorhynchus corniculatus, Emberiza erythroptera, Trogon Reinwardtii, Corythaix Buffonii, Passer Indicus, Trichoglossus hæmatodus, Squatarola rubecula, Platyarchus Stanleyii, Halcyon Macleayii,* which, with several others, are all that could be desired; but generally the forms and attitudes are ungraceful; the head in many of the smaller species is not nearly large enough, and the curvature of the neck is often absurd, as in *Ptiloris paradiseus,* female, and *Tachyphonus Vigorsii.* The descriptions are generally brief and confined to the external form and plumage, the internal organization of animals not forming, in the estimation of our best naturalists, an object of sufficient importance to attract attention. As to the manners or habits of the species represented, it was not to be expected that persons residing in England could know much about them, and thus the display of erudition made by the authors consists of remarks as to affinities of species and genera, without much that could be of interest to the philosophical ornithologist, who combines in his notion of the science the internal and external structure, the habits, relations, and distribution. Yet, to those who can afford to purchase it, the work will be useful, chiefly because it will afford them the means of naming the skins represented there, should they happen to possess them or see them in a museum. In several cases, a synopsis of all the known species of a genus is given, and in general all is done that might reasonably be expected in the matter of mere dried skins, the animating principle of which has not been taken into consideration. The specific characters are given in Latin, but as we think unnecessarily, and had the learned authors employed their vernacular tongue, they might have avoided the many grammatical errors and inelegancies, which many schoolboys might detect. Had a lad in the Rector's Class of our High School written such a version as the following, he ought to have been whipped:—*Accentor supra fusco-cinereo, dorso strigis fuscis vario, gula alba fusco-maculata; tectricibus alarum nigris, apicibus albis, infra cano-rufescente maculato, rectricibus lateralibus apicibus rufo albidis.* However, this is, after all, nothing very remarkable, for one of our first naturalists in many cases fails to give the correct orthography of the very names which he himself invents, and our ornithologists, from Montagu downwards, have too often manifested a contempt of the rules of composition.

Finally, the work seems to us creditable to its authors, as indicative of great perseverance and love of approbation, as well as of science, so called at least; but it exhibits no quality beyond what hundreds of individuals might manifest, had they the means of collecting skins, comparing them with published figures and descriptions, and giving them to the world, not caring much whether the expenses should be covered or not. Mere descriptions of bills, feet, and feathers, are easily manufactured, for in this branch of science a short apprenticeship is sufficient to make a tolerable journeyman; and as to the affinities and analogies, which one might suppose to require a little judgment, nothing more is needed than to obtain a knowledge of the slang employed by the founders of a system, and carry out their views, taking care to overlook every circumstance unfavorable to the adopted theory, and to abuse all who place truth and nature in opposition to it. What we especially admire the authors for is, the dexterity they have attained in transferring the delineations of their pencil to copper; although we conjecture that there are not many journeymen engravers who would feel disposed to boast of such performances as theirs. In short, we give them all the credit due to laborious and well-meaning cultivators of ornithology, and recommend to our readers to patronize them accordingly. If they do not attain the summit of Parnassus, it is not for want of scrambling.

Edinburgh: Published for the Proprietor, at the Office, No. 13, Hill Street. London: Smith, Elder, and Co., 65, Cornhill. Glasgow and the West of Scotland: John Smith and Son; and John Macleod. Dublin: George Young. Paris: J. B. Balliere, Rue de l'Ecole de Médecine, No. 13 bis.

THE EDINBURGH PRINTING COMPANY.

THE EDINBURGH

JOURNAL OF NATURAL HISTORY,

AND OF

THE PHYSICAL SCIENCES.

APRIL, 1838.

ZOOLOGY.

DESCRIPTION OF THE PLATE.—THE OSTRICHES.

In the system of Cuvier the Ostriches and Cassowaries constitute the family of Brevipennes, belonging to the order of the Grallæ. They are characterized by the shortness of their wings, which prevents them from flying ; by a form of bill and granivorous habits which indicate an affinity to the Gallinaceous birds ; and by the great length and muscularity of their legs, which approximate them to the other Grallæ. They are destitute of the hind toe, as well as of the crest of the sternum ; their furcula is imperfect, and their pelvis complete. The Ostriches, of which two species are known, have the bill horizontally depressed, of moderate length, with the point blunt ; the tongue short and rounded ; their œsophagus dilated into a large crop ; their stomach a powerful gizzard ; their intestines large, and their cœca long. Their eyes are large, and their eyelids furnished with lashes. Their plumage is soft, the feathers are in a manner double, the plumage being equal in length to the other part, and the wings, instead of quills, are ornamented with gracefully undulating plumes, of which the filaments are disunited, while the tail is similarly decorated.

Fig. 1. The Ostrich properly so called (Struthio Camelus), male. This is the largest bird known, its height sometimes exceeding seven feet. The male is generally black, mixed with white, of which latter colour are the beautiful plumes of the wings and tail, which are in great request as ornamental articles of dress. Ostriches live in large flocks in the sandy deserts of Africa and Arabia. Although incapable of flying, they run with extreme rapidity, so as, for a time, to outstrip the swiftest steed ; and their legs, which are of vast size, are furnished with only two toes, a rudimentary inner toe, however, having been recently detected beneath the skin. Their flesh is considered good when the birds are young, but otherwise it is extremely tough, although it was considered a delicacy by the Romans. The eggs, which are about twenty times the size of those of our domestic fowl, are of an elliptical form, with a very thick, yellowish shell, which is sometimes used as a cup. Their number, according to Vaillant, is about ten, but so many as forty have been found deposited together. They are laid on the sand, and are hatched by the heat of the sun, although the female sometimes sits upon them at night; but, in the more temperate regions, and particularly near the Cape of Good Hope, they are deposited in a pit about three feet in diameter, and the female sits regularly upon them. The young are tended with the greatest care. The food of this bird consists of vegetable substances. Like other granivorous and graminivorous birds, it swallows fragments of stone to aid its trituration, and, in captivity, is not nice as to the articles selected for this purpose, picking up pieces of iron, copper, lead, &c. This circumstance has been attributed to the peculiar obtuseness of its taste, but with little reason, for in granivorous birds, which swallow their food without mastication, the tongue is an organ of prehension and deglutition, but not of taste, so that in this respect the Ostrich does not differ from other birds.

Fig. 2. Female. The female is generally of a grey colour, and is somewhat inferior to the male in size.

DESCRIPTION OF THE PLATE.—THE LORIES AND PARRAKEETS.

The numerous family of the Psittacidæ is divided into several sections or families, representatives of two of which are here figured.

Fig. 1. The Sparkling Lory (Lorius scintillatus) is so named on account of the bright yellow streaks on the plumage of its breast and sides. It is from New Guinea.

Fig. 2. The Tiriba Parrakeet (Psittacula cruentata), having the top of the head dark brown, the plumage generally green, with patches of bright red, yellow, and purplish blue, was discovered in Brazil by Prince Maximilian of Neuwied.

Fig. 3. The Coquette (Ps. pleocentis), male. This species was found in New Guinea, by the Dutch naturalists Macklot and Müller, who visited that country in the Triton corvette.

Fig. 4. The Coquette, female.

Fig. 5. The Slender Parrakeet (Ps. euteles) was discovered by the same naturalists in the island of Timor.

Fig. 6. The Iris Parrakeet (Ps. Iris) was also found by them in the same island, where it appeared to inhabit the woods near the shores in great numbers.

35

SYMPATHETIC AFFECTION AMONGST ROOKS.—During a severe winter, when the snow had lain long and deep upon the ground, the feathered tribes were reduced to the point of starvation. One morning the Strathendry Crows had fixed on some barley, a little to the east of the steading, and had nearly uncovered the stack to get at the grain. To save the barley, one of the men took his gun, and contrived to get within range ; but the moment he raised his head, the sentinel on duty sounded the alarm, and the man fired into the dense cloud, as they floated off the stack. Amongst the wounded, one had lost the extremity of his right wing by the joint. Thus disabled, he was soon secured, and given in charge of a servant's wife, who had shown herself an adept in training birds and cats, to try what we could elicit of the innate dispositions and mental faculties of the Crow.

Daily were this person's birds and cats to be seen feeding, in perfect harmony, from the same dish ; and I have frequently seen a cat pretty sharply admonished by a blackbird, when overstepping the bounds of good manners while feeding. No cages were wanted for her birds, though both doors and windows stood open as occasion required. Her feathered and feline family went and came as pleased themselves. If any were absent at feeding time, she went to the door and called them in Gaelic, as she said she never could make bird or cat obey her in English.

The first lesson given to the poor mutilated crow was to place him on her knee, while yet starving, with a hand over his shoulders to prevent his stirring, but allowing him full freedom to look down on Larks, Linnets, Blackbirds, and Cats, forming a circle round the feeding dish. He seemed to eye the cats with suspicion, but " hunger tames the Tiger." Stretching his neck towards the provisions, and indicating a desire to raise his wings, he was allowed to go down. He shyed at first, walked round them for a time, but at length struck in, and made a hearty meal. From this time he felt quite at home, seemed to study the rules of the house, and kept his place in the crowd, or before the fire, where he often lay quite at his ease, along with the cats, enjoying himself on the warm hearth. It was quite amusing to see with what familiarity he would stalk about, with all the strut and dignity of a lord of the manor, mixing inquisitively among the servants, and eyeing all their motions, like an attentive superintendent, returning to head-quarters when called, or marching off to dine with the servants in the bothy. The servants dubbed him " Captain." The most remarkable part of his history, however, remains to be told.

Experience had taught him that his confidence was not misplaced. His gratitude for the protection and ease which he enjoyed was evinced in his filial familiarity and obedience. One day, however, he was observed evidently watching an opportunity to carry off a piece of boiled potato, which he at last accomplished, and walked with it to the door, as if to hide it, for some future occasion. This he was observed to repeat as often as he found opportunity, as he conceived, unobserved. The circumstance was mentioned to me, and I resolved to watch him out of doors. To prevent detection within, he left the house, and I soon after observed him casting his eyes about him, pause, and then march off towards the bothy,

" Looking round wi' canny care,
Lest boggles catch'd him unaware,"

walk cautiously forward, drive his beak into a cold boiled potato, march away in double quick time, crouching as if afraid of being detected in the act of thieving, and, turning the nearest stack, disappear. Something prevented my following him at the moment, but he soon reappeared for a further supply, took the same route, and again disappeared. I immediately took a cold potato in my hand and followed. On rounding the stack, to my astonishment I found him in the act of feeding another disabled crow. I cautiously neared them. The stranger shyed, lifting his wings to fly off. Captain, however, remained undisturbed. I held out my potato. Captain came with evident satisfaction, took a portion of it from my hand, while the stranger, who halted at a safe distance, was looking on. Captain then walked with the potato towards his friend, who met him, and in the course of feeding I observed, with a painful sensation, that the poor stranger had lost his bill, and consequently was incapable of helping himself, although food had lain before him. What free-masonry passed between them I know not ; but on Captain's returning for the remaining portion of my potato, the stranger followed in his rear, with all the familiarity and confident bearing of an old acquaintance. However much my admiration was excited on this first interview with the stranger, I was still more astonished to see him walk side by side with his preserver into the servants' bothy, while the servants were at dinner, without shying, or betraying the least symptom of fear. When they had been fed sumptuously, Captain marched him over to head-quarters, and introduced him to his mistress, cats, and comrades, by whom he was "most graciously received." There he remained an inmate, under the title of Nebby.

Nebby could fly as well as ever, and took frequent flights round, for intelligence or amusement. I have often seen him on returning alight, and implore his mistress for water, by gently moving his wings and holding up his head. On her sitting down, Nebby was immediately on her knee, to receive it; and it was given him by dipping the finger, and dropping it from the point into his throat.

One Sunday, some idle blackguard boys, from some of the mills down the water, carried off both Captain and Nebby. I made every inquiry, but never recovered them.—GAVIN INGLIS.

REMARKABLE ELONGATION OF THE BILL OF A ROOK.—The accompanying figure represents the head and bill of a Rook, shot near Dalkeith in the beginning of March, and submitted to our inspection by Mr Carfrae, preserver of animals in Edinburgh. The bird was in every other respect of the ordinary appearance of the species, and was in good condition, notwithstanding the extreme severity of the season, and although one might imagine it impossible for it to pick up its food with a bill so constructed. The malformation consists of an elongation of the upper mandible, of which the point extends beyond that of the lower an inch and a quarter, being curved gently downwards, grooved beneath, and having a breadth of about a quarter of an inch.

We have seen instances of a like elongation in other birds, although not to the same extent; and in the Rook itself a case occurred in which the upper mandible was not only elongated, but deflected laterally. On the other hand, we have seen the lower mandible abbreviated, and imperfect, in as much as its two sides did not meet at the commissure, which presented a vacant space. In birds kept in captivity not only the bill, but also the claws, frequently become extremely elongated; and the same circumstance is occasionally observed in the hoofs of ruminating animals. In such cases the elongation is easily accounted for by the want of the ordinary action of the animal, which would tend to wear down the extremities of the horny parts, or repress their extension by the pressure applied to them. The case is somewhat analogous to the occasional elongation of the incisors of the glires; and we would humbly suggest to the quinarian systematists, that the Rook, and consequently the Crows in general, are analogous in their own circle to the Hares! Indeed, they have often seized upon less palpable analogies to support their peculiar views.

PISCATORY HABITS OF THE CARRION CROW.—"Stobo-hope, Peebles-shire, 22d January 1838.—In a former communication, which you have inserted in your History of British Birds, I told you of two Carrion Crows that for upwards of twenty years have inhabited the ground of which I have charge. In this glen is a small meadow irrigated from the stream which runs along its bottom. On the brink of this principal head I noticed my old friends busy eating, and thinking it might be some part of a sheep, I made towards them, and found them standing with their heads toward each other, pecking with all their might at something that lay between them. In their eagerness, they sometimes tossed it athwart, to obtain mouthfuls. When I went near, one of them carried off the remainder of the feast in his bill. I found neither wool nor feathers, however, bones nor entrails, nothing, in short, but a little blood on the snow; but, on a more minute inspection, I observed a small trail, as if something had been pulled out of the water, and the marks of some drops of water that had been splashed out and sunk among the snow. I could make no more of it, and so left it, carelessly thinking it might have been a water-mouse which had seized in a fit of desperate hunger. A day or two after, I fell in with Sir James Montgomery's man, who has charge of the watered meadows, ditches, and drains, and told him of the circumstance, when he assured me that he had oftener than once surprised the Carrion Crows devouring fish, taken in the meadows, and that one time they had eaten all but the bones of the head. The reason why this happens in a meadow is as follows. To irrigate a meadow rightly, the water must be taken off at times, and on the occurrence of certain changes in the state of the air. During the time that it is flooded, small fish or trouts sail down the principal lead, then distribute themselves along the small canals where sustenance for them abounds. When the water is instantaneously let off, the poor trouts can find protection nowhere from the Heron, who diligently searches all the sinks and shallows of the half-dried pool. But I had no conception of the Carrion Crows taking and feeding on fish; this I thought had been a prerogative of fowls and other animals whose structure adapted them for searching in and under water. I had no doubt, when I considered the omnivorous nature of this Crow, however, that if it found a dead fish, it would readily eat it up; but that it would plunge into the water, and seize a fish swimming deep, I could not have supposed; yet this had certainly been done on this occasion, for the water out of which they had dragged the fish was rather more than a foot deep, and on taking a second look of the place, I found that no protection could be afforded to the trouts, as there were neither stones nor hanging banks.—WM. HOGG."

ON THE GEOGRAPHICAL DISTRIBUTION OF ANIMALS.—NO. III.

(Continued from page 130.)

CONTINUING our subject, we propose to include in our survey that portion of Devon-shire which extends from the heart of Dartmoor to the southern coast, and which is included between the rivers Exe and Tamar respectively on the east and west. There is thus presented to the view an extent of country having in its northern part the character of sterility, and in its southern that of fertility. Now, as we have already had occasion to observe, the animal part of the creation is almost entirely dependent on the vegetable world; whilst the vegetable kingdom, in its turn, is dependent on inorganic matter. Accordingly, if it appears that the northern division of our limits is not calculated, from the nature of its superficial soil, to maintain vegetable life, except in a limited degree; so also, it is obvious that the animal productions of this spot must be likewise restricted. The central districts of Dartmoor present to the eye a series of hills of great size, covered with detached blocks of granite. On the summits of many of these hills are found swamps, and even pools of great depth, and between them streams pass on for future coalescence; and where the surface is level for a sufficient space, the drainings of the country rest, and form morasses and lakes. Altogether, Dartmoor and its vicinity present a large proportion of water, since it appears that five principal rivers, twenty-four secondary streams, fifteen brooks, two lakes, and seven heads, are found on it.

The Flora of this wild district consists, with but few exceptions, of the lower tribes, such as mosses, ferns, lichens, &c., and of such plants as are peculiar to marshes and other collections of water. The soil cannot possibly support many of the higher orders; but the beauty, variety, and luxuriance of those vegetable forms which mantle the rude blocks of granite, spring from the spongy soil of the bogs and marshes, raise themselves into notice above the stream, or maintaining their existence in the body of the current, attached to some fixed point, move in conformity to its undulations, are sufficient to attract the notice of even the incurious. In this sterile spot the most common creature excites regard, and those which are peculiarly its own cannot fail to be contemplated with much interest.

The Quadrupeds of Dartmoor, though now reduced to a small number, were formerly pretty numerous. The following are recorded as its ancient inhabitants: the Wolf, the Brown Bear, the Boar, the Wild Ox, the Red Deer, and the Wild Cat. The Wolf appears to have become extinct on Dartmoor about the close of the reign of Elizabeth. It was a pure native of our country, and required great exertions for its removal. The Bear seems to have been extirpated in the eleventh century; and unless its food consisted chiefly of vegetable productions, it is difficult to understand how its existence could have been maintained. The Boar and Wild Ox have been taken under the protection of man, and the date of extirpation of the wild stock is not recorded. That noble animal the Red Deer was, until within the last fifty years, pretty common in the remote wooded districts of the county. Its race, too, has undergone extirpation in a very gradual manner. "Sometimes, but rarely, one has been perceived near Ashburton;" and it is not more than three years since I saw an account in a paper of the chase of one near that town, which had unfortunately been spied in some coppice. It is quite reasonable to suppose that both the Wild Cat and the Goat were natives, or rather frequenters, of this district, so perfectly congenial in its aspect to their natures. All these were most likely found in less degree throughout the woods and wilds that lie to the south of Dartmoor; but by increase of population and agriculture they were no doubt soon removed from these spots, and eventually their limits restricted to the Moor itself. But here also they suffered extermination soon after the king took part in the more noble field amusements, and when punishments were inflicted for interference with the game.

At present, the following are the Quadrupeds found in the south of Devon, inclusive of the Moor, though there are some of them which are unfitted to live in spots so barren and wild as are the more remote and central portions of this district.

1. Barbastelle Bat. *Vespertilio Barbastellus*.
2. Horse-Shoe Bat. *Rhinolophus Ferrum-equinum*.
3. Common or Pipistrelle Bat. *Vespertilio pipistrellus*.
4. Great Bat. *Vespertilio Noctula*.
5. Long-eared Bat. *Plecotus auritus*.
6. Smaller Horse-shoe Bat. *Rhinolophus Hipposideros*.
7. Hedgehog. *Erinaceus Europæus*.
8. Common Shrew. *Sorex araneus*.
9. Water Shrew. *Sorex aquaticus*.
10. Mole. *Talpa Europæa*.
11. Badger. *Meles Taxus*.
12. Fox. *Canis vulpes*.
13. Weasel. *Mustela vulgaris*.
14. Stoat. *Mustela erminea*.
15. Polecat. *Mustela putorius*.
16. Common Martin. *Mustela foina*.
17. Pine Martin. *Mustela Martes*.
18. Otter. *Lutra vulgaris*.
19. Common Mouse. *Mus domesticus*.
20. Field Mouse. *Mus sylvaticus*.
21. Harvest Mouse. *Mus messorius*.
22. Black Rat. *Mus Rattus*.
23. Brown Rat. *Mus decumanus*.
24. Squirrel. *Sciurus vulgaris*.
25. Hare. *Lepus timidus*.
26. Rabbit. *Lepus Cuniculus*.
27. Dormouse. *Myoxus glis*.
28. Water Vole. *Arvicola amphibia*.
29. Field Vole. *Arvicola agrestis*.
30. Fallow Deer. *Cervus Dama*.

We have here a list of thirty terrestrial mammalia. The extent of this catalogue must needs excite surprise, for the present state of Dartmoor would by no means lead to the belief of its supporting more than two or three quadrupeds of the smaller kind; and, indeed, upon inquiry into facts, we find that the barren open portions are frequented only by the Rabbit, Mole, Weasel, and perhaps the Stoat. How then can this region have maintained those large animals recorded as extirpated? There is undeniable evidence that the central department of what we ordinarily term Dartmoor was in former years a forest, and that it was set apart for the king's use as a royal chase. With this explanation difficulties vanish. At the present period the woods and plantations in the immediate vicinity of the Moor harbour the same quadrupeds as those found in the southern districts. The Martin and Polecat, however, are now more peculiarly frequenters of the deep woods, remote from cultivated parts. The former, indeed, I have not known to be captured anywhere but in the noble woods at Buckland-in-the-Moor; but it is reported to be also found in the woods at Lydford. I am told that the Yellow species (*M. Martes*) is found with the other. The Otter is said to confine itself to the river Dart, probably from its superior size and depth, and from its being better supplied with fish. The fur of specimens from hence is said to have an admixture of white hairs. The Mole is a creature by no means limited to cultivated districts, as appears by its occurrence on a barren hill, of very considerable height, in the immediate neighbourhood of the Moor. The most numerous and characteristic species, however, is, the Rabbit, which, as will subsequently be seen, draws thither a variety of rapacious birds that otherwise would not find food in such a district.

The Ornithology of Dartmoor is in many respects interesting. The Rabbits which abound there draw numerous species of rapacious birds to it. The Raven, *Corvus Corax*, Carrion Crow, *C. Corone*, and Hooded Crow, *C. Cornix*, likewise traverse in their wanderings these wild spots. The Ring Ouzel, *Turdus torquatus*, frequents many of the rocky and rapid streams, in parties, nestling, to my knowledge, in hollows of the rocks. The Water Ouzel, *Cinclus aquaticus*, is a frequenter of similar situations, being a great lover of solitude; and I am not aware that either of these birds has been traced, except sparingly, beyond the barren portions of the Moor. The Titlark, *Anthus pratensis*, Stonechat, *Saxicola Rubicola*, Whinchat, *Saxicola Rubetra*, and Grasshopper Warbler, *Sylvia Locustella*, are found occupying their respective stations on the heaths and stony fields; and the Wheatear, *Saxicola Œnanthe*, and Reed Warbler, *Sylvia Phragmitis*, are reported to make their abode within the limits of the more barren parts of the district. It has been told me that Cuckoos, *Cuculus cunorus*, are at times seen haunting rocky spots on the borders of Dartmoor; and this may be true enough, although it is possible that my informant may have mistaken the Nightjar, *Caprimulgus europæus*, for it, as they are not very dissimilar in appearance, and since I well know that Nightjars are found on the borders of the Moor in large wooded inclosures, but particularly at Buckland-in-the-Moor, where the oaks have attained a great size. From these woods they usually select positions in the adjoining commons or brakes for nestling. The Great Bustard, *Otis Tarda*, which formerly frequented the Moor, has, I fear, been extirpated. In times past also, no doubt, the Crane, *Ardea Grus*, frequented Dartmoor. One was shot, in 1826, on the borders, and there is a hill in the heart of the Moor, having on its summit a pool of great size, called Cranmere Pool, a name signifying the abode of Cranes, as though these birds had been in the habit of resorting thither, as is the practice of some other birds at present. The Thick-kneed Plover, *Œdicnemus crepitans*, frequents the downs and wastes; and it would seem that they wintered with us, and were driven to inclosed lands, as in severe winters they have been brought to Plymouth market, where I have myself seen them, though rarely. In the summer months I have seen Curlews, *Numenius Arquata*, on the marshy grounds, where indeed they breed, but I presume the numbers killed on the coast in severe winters must be derived principally from the northern counties. It has been proved that many individuals of the Snipe family breed on Dartmoor, but it is only of late years that this fact has been observed. The same observation applies to the Duck, Wigeon, and Teal, according to report, though I can answer only for the Wild Duck, *Anas Boschas*, which unquestionably breeds in several spots on the Moor, besides in those situations in our cultivated grounds where care has been taken to protect it. The swamps of the Moor are also the breeding-places of many individuals of several species of wading birds found in the autumnal and winter months in our cultivated lands and shores. The Lapwing, *Vanellus cristatus*, descends in flocks in winter. I have noticed them arriving in vast quantities in December yearly, on the high grounds bordering Bigbury Bay. It has been found also that the Golden Plover, *Charadrius pluvialis*, Grey Plover, *Charadrius Helveticus*, and Dunlin, *Tringa cinclus*, breed with us, and there is great reason to suppose that very many other similar birds do so likewise. The Coot, *Fulica atra*, Water Hen, *Gallinula chloropus*, and Water Rail, *Rallus aquaticus*, are well known to breed not only in the marshes of Dartmoor, but also plentifully in very many swampy woods, and other secluded watery spots not far from Plymouth, dispersing from these retreats as soon as the cold sets in. The Black Grouse, *Tetrao Tetrix*, is sparingly dispersed over the moors, and in winter roams with its progeny over the woods and cultivated parts of the country, being occasionally shot, and brought to market in the severer months.

Besides the above named birds, there are others recorded to have been observed sparingly on or in the vicinity of the Moor. The Honey Buzzard, *Pernis apivorus*, Sea or Cinereous Eagle, *Haliaetus Albicilla*, Golden Eagle, *Aquila Chrysaetos*, Goshawk, *Astur palumbarius*, Kite, *Milvus vulgaris*, Little Owl, *Strix passerina*, Short-eared Owl, *Strix brachyotus*, Nutcracker, *Nucifraga Caryocatactes*, Greater Spotted Woodpecker, *Picus medius*, Lesser Spotted Woodpecker, *Picus minor*, Rose Ouzel, *Turdus roseus*, Wryneck, *Jynx torquilla*, Crossbill, *Loxia curvirostra*, Hawfinch, *Coccothraustis vulgaris*, Hoopoe, *Upupa Epops*, Snow Bunting, *Plectrophanes nivalis*, Turtle Dove, *Columba turtur*, Quail, *Coturnix dactylisonans*, Little Bustard, *Otis tetrax*, Great Snipe, *Scolopax major*, Barker, *Totanus fuscus*, Spotted Rail, *Gallinula Porzana*, and Little Rail, *Gallinula minuta*, are some deserving notice. But it is to the deep and unfrequented woods before named, as bordering the Moor, that we are principally indebted for these rarities; and it cannot be altogether surprising that these spots, so secluded, and so generally calculated to be the abodes of the feathered tribes, should contain within them objects so precious to the naturalist. They who have read Vaillant's Travels in Africa will, I think, agree with me in the remark, that the transition from these woods to the sterile tracts of the Moor contiguous, where even in summer little else can be seen save the Curlew flying from the summit of one Torr to another, and by its harsh note adding to the dreariness of the scene, the Stonechat, the Ring Ouzel, or perchance a Buzzard hovering aloft, is not very unlike the sudden changes experienced by that adventurer, and which he so touchingly describes; at one time surrounded by hundreds of beautiful birds, enlivening by their actions and notes the thick groves; then situated in a trackless desert, and guided only on his way by the harsh note of a duck, flying at a great height in quest of some rock which might happily contain water in its basins.

The arid and remote portions of the Moor are frequented by only a few birds, not found (or but rarely and at certain times) in the southern and cultivated districts. The Eagles and birds of that kind are generally, however, partial to remote spots, or restricted to them by our interference. The Golden Eagle, if still a Devon bird, must be accounted as in some measure peculiar to Dartmoor, though most of this kind roam to immense distances at certain periods. The Sea Eagle, too, has been seen both on Dartmoor and frequenting cultivated land, and has likewise been captured at the Eddystone. The Goshawk, Kite, and Honey Buzzard, may be considered almost confined to Dartmoor; the latter, however, has been noticed at Slapton Ley. The Short-eared Owl has been killed both on Dartmoor and Exmoor, but seems to be very rare, and to confine itself to open and remote spots. The Dipper and Ring Ouzel are both lovers of the Dartmoor solitudes. The Black Grouse, Little Bustard, Thick-kneed Plover, and the various waders before named as most likely breeding on the moors, together with the Great Bustard and Crane (if still resident in the county), must be considered in some degree peculiar to Dartmoor, though they are all constrained, on the occasion of severe weather, to seek shelter and food in the cultivated parts.

The woods bordering Dartmoor are well adapted to shelter a variety of birds of the rarer kinds; but yet none that I know of are limited to them, though the Turtle Dove is more frequent in these situations than in the southern parts; and several species of the rarer Hawks are mostly obtained from thence.

I am not aware that anything need be said relative to the amphibia of Dartmoor, excepting that the Lizard, *Lacerta agilis*, and Viper, *Coluber Berus*, are both found pretty commonly on the downs and other dry situations, as indeed they are throughout the whole county. Nothing can be said relative to the Ichthyology, nor anything on the Conchology, unless it be that *Helix trochilus* has been noticed close to the Moor. Having but slight acquaintance with Entomology, I can only say that the lists of the Dartmoor insects are very extensive.

When we come to examine the Fauna of the central districts of South Devon, we find considerable alterations in its character, besides its extent being greatly increased. The diversification in the surface of the country, together with every variety of soil, and vegetable produce, is no doubt one great reason of this circumstance, while another cause is our being situated at the southern limits of the island, by which means we are more likely than other counties to partake of the ornithology of the continent, and likewise to receive a variety of birds which migrate from northern counties or kingdoms.

Reverting to the mammalia, we have still a few remarks to make. The Stoat has been known to assume in winter the white clothing ordinarily supposed to be peculiar to northern latitudes, or at most to some of our northern counties, and I have myself seen a specimen particoloured. Our woods and thickets are so numerous, and sometimes so little frequented, that the Hedgehog, Badger, Fox, Hare, Squirrel, and Dormouse, are all found pretty abundantly, and the Shrew, Weasel, and various kinds of mice, frequent in plenty our fields and hedges. The cream-coloured Mole is found with us, but I believe only in one locality. White Rats have been captured at times in rabbit-grounds, where they are known to resort for the sake of the young rabbits. I am not quite sure respecting the Water Shrew, but believe I have taken it while a boy in the stagnant waters of the entrenchment round Davenport, and it is generally believed to be a common animal, though very shy in its nature. In the space of ten years, I have not seen above three or four specimens of Polecat in this neighbourhood, though I have been much in the habit of inquiring on such subjects among gamekeepers. On the other hand, the Stoat and Weasel are plentiful throughout the county. I have good evidence that the Harvest Mouse is found in Cornwall; and, from reports, there is great reason to believe it is tolerably common in Devon. I very lately captured the Mouse figured in Shaw's Miscellany as a rough-haired variety of the Meadow Mouse, or Field Mole, but, in all probability, a distinct kind. The Black Rat is very scarce with us, if found here at all, and I have never yet noticed it. A white Hare was last year seen in a wood near my house during the winter months, and, though that season was unusually severe, it is certainly difficult to understand why certain individuals of this animal, and of the Stoat, should assume this change, while all others of the species remain unaltered.

In all cultivated districts abounding in wood, and productive of an extensive Flora, as is the south of Devon, we necessarily meet with a great variety of birds belonging to the Passerine order. In the central portions of South Devon, now under consideration, there are but few birds besides the Passeres observed, and these, excepting half a dozen species at the most, may be regarded as peculiar to their central part. The Ring Ouzel and Water Ouzel may to a certain extent be considered as Dartmoor birds, though some, no doubt, have been seen beyond those precincts, and a pair of Water Ouzels build yearly in a fish-house not far from my house, and very close to a flour-mill, a saw-mill, and several houses besides, being within fifty yards of the main road. The Rock Lark, *Anthus petrosus*, is a frequenter only of the sea-coast; as is the Red-legged Crow, *Fregilus Graculus*, though I see they are both marked as Dartmoor Birds in an ornithological list of that place by Dr Tucker, a good naturalist. The Raven, Hooded-Crow, and Nutcracker, are likewise birds which can scarcely be claimed for the ornithology of a cultivated tract, though the two first roam at times over every variety of country. The Nutcracker has only twice been killed in Devon, according to a late authority, and then not within the limits we are now treating of, though Dr Tucker's list, prefixed to Canington's Dartmoor, has authorized our naming it among the birds of that spot.

The entire number of birds, wholly or partially inhabiting the cultivated, wooded, and well watered part of Devon we are now speaking of, may be estimated at about 134, allowing about 14 to be Accipitres, 5 Scansores, 5 Gallinæ, 8 Palmipedes, about 22 Grallæ, and 80 Passeres, the total being more than one-half of the whole South Devon list, which may be computed at 237, or 242 Devon Birds, the remainder being made up by those few birds more peculiarly belonging to the Moors, and by the great variety of those birds and web-footed birds furnished by our coasts, and hereafter to be noticed. But, giving all due weight to these last-named sources in swelling our ornithological list, that part of it which we are now more especially examining is found, by comparison with ornithological lists of well wooded and well cultivated counties, to be unusually extensive. The number of birds in Oxfordshire does not exceed 120, inclusive of several kinds of web-footed birds (besides the more common sorts of Ducks, &c., which, we may presume, would form part of most ornithological lists), which occasionally roam inland, or are driven thither by stress of weather, instanced by the Herring Gull, and Leach's Petrel, three kinds of Waders which are at times detected in inland localities, the Ibis, Phalarope, and Greenshank; so that, more properly, not above 106 land and fresh-water birds are found in Oxfordshire, making a difference of 28 species between the two counties. This difference we shall examine into, as it will serve to illustrate in some measure the peculiarities, and more remarkable features of the ornithology of the cultivated parts of Devon.

First, with regard to the Accipitres, the deficiencies consist of these birds (not taking into account certain species which I shall have to mention in considering the third portion of the south of Devon):—the Rough-legged Falcon, *Buteo Lagopus*, Ash-coloured Falcon, *Circus cinerascens*, Moor Buzzard, *Circus æruginosus*, Gyrfalcon, *Falco islandicus*, Great-eared Owl, *Bubo maximus*, and Little Owl, *Strix passerina*. In examining into such subjects, we may perceive that, with respect to some species, the reasons of absence are evident enough, while, with respect to others, the causes of restriction of limits, and occasional or vacillating appearance are quite unknown. Speaking generally, the cause of our possessing so many species of this tribe, indeed, all the British species, except the *Scops Aldrovandi*, a straggler, and the Snowy Owl, *Surnia Nyctea*, whose range does not extend so far southward, seems to be the rocky and mountainous character of our county, together with so great an abundance of uncultivated land interspersed, and a vast number of woods of great depth, and removed from the neighbourhood of man. The only species of the Accipitres, however, which may be accounted at all common in Devon, are the Kestrel, Sparrowhawk, Common Buzzard, Moor Buzzard, White Owl, and Brown Owl. In former years the Kite was a common bird in this country, but, at the present day, it is particularly scarce, furnishing an illustration of the uncertainty of the geographical position of rapacious birds, and a proof likewise that we are inadequate to fathom very many of the phenomena of animal dispersion, for we know no reasons why that species should abandon us, and according to Dr Pulteney, it is very frequent in Dorset, and by an authority of my acquaintance, it is not very uncommon in Oxfordshire. But this uncertainty of position is not confined to rapacious birds.

When we turn to the Passeres of Oxfordshire, we do not find more than 12 deficiencies, allowing 80 to be the number observed in the cultivated parts of Devon, so that it is not in this department that the chief part of the difference is found They are the Mealy Redpole, Golden Oriole, Pied Fly-catcher, Dartford Warbler, Bearded Titmouse, Cirl Bunting, Twite, Siskin, Bee-eater, Lesser Redpole, Wood Lark, and Reed Warbler; the last three being doubtful; and allowing the whole to be deficient, there is but one which is not a rarity in the British Isles. I have considered the Ring Ouzel and Water Ouzel to be almost wholly moorland birds in Devon; but although the latter is not met with, so far as I know, in Oxfordshire, the former has been sparingly observed. But although we seem to possess almost every rarity not ranked in the ornithological lists of other countries, there is one bird found in Oxfordshire and not with us, the Mountain Sparrow, *Passer montanus*. I have likewise noticed the following birds as common in that county, which with us are scarce: the Redstart, Grasshopper Warbler, and Lesser White-throat. The Pettychaps is likewise much more frequent in Oxfordshire than in Devon. The Nightingale is common in Oxfordshire, but in Devon has not been noticed farther west than Exeter, though in Dorset it is plentiful, hereby defining very accurately its south-western limits in England. It was once only heard at Kingsbridge by Montagu. Now, in all these instances of remarkable dissimilarity between the two counties, no explanation whatever can be given of the preference for localities so observed. The distribution of some animals is latitudinal, that of others longitudinal; some species inhabit a district of a rounded or irregular form, some are found regularly dispersed through the whole of a natural geographical division of the globe, some are found inhabiting one division, and only partly one or more of the others; some inhabit one or more countries, omitting, or refusing to dwell in, certain spots in that range, whether large or small; some are uncertain in their stations, are continually changing their position, or remain an indefinite period, and then disappear *in toto*, or return after the lapse of some time; in all which phenomena little can be detected of secondary causes exercising a decided influence, and yet I cannot but think we are largely indebted to these causes for the great variety of natural productions of which we boast, allowing all necessary weight to those unknown primary laws of dispersion under which very many species appear with us, and many are denied to us. But depending upon all these causes combined, we are enabled to rank as Devon birds a very large number of rarities, perhaps more than any other county in Great Britain. Perhaps the whole number of British birds may be considered as 274, and upon inquiry where our deficiencies of 37 chiefly occur, it is found that they consist in part of stragglers and rarities, and in part of birds whose limits of distribution are confined to the Northern Isles, Scotland, or to certain counties to the north of us; or, lastly, whose geographical position may with greater propriety be referred to the north than to the south of Devon. In the Passeres, now under consideration, our list is defective in the Pine Grosbeak, Crested Tit, and Tree Sparrow, the first two being limited to the Scottish pine forests, and the last to the central counties of England.

There are certain passerine birds which are irregular visitors of Great Britain, such as the Bee-eater, Oriole, Rose Ouzel, and Hoopoe, which have several times been found with us, and which, properly speaking, have their station in Africa, but migrate into Europe yearly, and at times pass over to the British Isles, so that this offers some explanation of the fact of so many of the British specimens having been captured in the south, and especially in Devon. Another class of irregular visitants seem to arrive from opposite sources. The Nutcracker is in our list an instance of this, as the northern countries appear to be its true station, the cause of its coming hither not being evident. A third series of irregular emigrants, consisting of the Pied Flycatcher, Bohemian Chatterer, Crossbill, and Greater Grosbeak, are in all probability derived from the Continental states. The Bearded Titmouse, *Parus biarmicus*, has been noticed only near Thorverton and Dacolise. The Cirl Bunting, *Emberiza Cirlus*, is, as regards Devonshire, confined to its southern parts adjoining the sea, frequenting furze-brakes. It has been noticed also in Cornwall, and seems for the most part limited to the southern portions of the kingdom, where they enjoy a climate more in accordance with that experienced by the bulk of this species on the continent, its chief station. The Reed Warbler is found sparingly with us, though it has not been noticed in Wilts, Somerset, and Dorset. The Dartford Warbler is said to have been common near Plymouth in former years, but is now scarce and local. The Brambling and Snow Bunting are chiefly observed in winter. The Great Shrike, *Lanius excubitor*, Lesser Redpole, *Linaria minor*, Mealy Redpole, *Linaria canescens*, Twite, *Linaria montium*, and Siskin, *Carduelis spinus*, are all noticed only sparingly and casually.

Wild Swans.—The unusual severity of the winter, although it has driven many individuals of the Duck family farther to the south than usual, has not caused the appearance in Scotland of any rare visitants belonging to other families. A great number of Swans has been obtained in different parts of the country; but among those which have come to Edinburgh, we have not observed specimens of any other than the common species. The windpipe of this bird, from its great size, and the facility with which it may be made to emit a sound resembling the ordinary cry of the bird, is one of the best that can be selected to show that the bronchi and lower larynx are the parts in which the cries of birds are produced. On blowing into those of the Swan, one may, by relaxing the membranes in a sufficient degree, produce a sound exactly similar to that emitted by the bird while alive, and almost as loud. If the trachea be cut across, the sound is still produced, but is then considerably more acute. This proves that the cry is not produced in the upper larynx, as is the case in the Mammalia, the glottis merely performing the office of dividing the current of air, so as to form it into distinct notes.

REVIEWS.

Annals of Natural History; or Magazine of Zoology, Botany, and Geology. Conducted by Sir W. Jardine, Bart., P. G. Selby, Esq., Dr Johnston, Sir W. J. Hooker, and Richard Taylor, F.L.S. London. R. and J. E. Taylor.

The Magazine of Zoology and Botany, which, notwithstanding its excellence, and its being replete with "scientific papers and facts, unadorned and truthful," as its editors state, and although, in their opinion, it was conducted "in a manner which they believe has been acknowledged to stand high in the estimation of those who were inclined to dip below the surface of the subjects which others pretend to study and admire," ceased with the twelfth number, in February 1838, and has been incorporated with Hooker's Companion to the Botanical Magazine, and metamorphosed into the "Annals of Natural History," of which "a number will be published on the first of every month, containing from five to six sheets, with plates, coloured or uncoloured, according to circumstances." "No. 1. New Series, price 2s. 6d." has accordingly appeared. As to the boast of dipping below the surface, we are not aware that either its former or its present editors have exhibited any diving propensities, with the exception of Dr Johnston, who, as a gentleman, a scholar, and a naturalist, stands high in our estimation. The "Annals" contain an account of a new Ocillatoria by Dr Drummond; Remarks on the Germination of Limnanthemum lacunosum, by Dr Grisebach; a continuation of Mr Thompson's excellent and highly interesting contributions to the Natural History of Ireland; an account of some new species of quadrupeds and shells, by Mr J. E. Gray; Prodromus of a Monograph of the Radiata and Echinodermata, by Dr Agassiz; continuation of Dr Johnston's Miscellanea Zoologica, composed of "the Scottish Mollusca nudibranchia," which, being more than a description of the external form, and written in the author's usual elegant and perspicuous manner, cannot fail to be interesting as well as instructive; and, lastly, some straggling "Information respecting Botanical Travellers, showing how Mr Cuming collected 1150 species of plants in the island of Luyon, secured about 60 species of Orchideæ, and 125 Fungi, &c. Then come Bibliographical Notices, Proceedings of Societies, and Miscellanea. Some additional information is given "respecting the splendid Nymphæaceous plant discovered by Dr Schomburg in the River Berbice. A new genus has been formed of it, which has been dedicated to our young Queen. It is the *Nymphæa Victoria* of its discoverer; *Victoria Regina* of Mr Gray; *Victoria regia* of Dr Lindley," &c.

Edinburgh: Published for the Proprietor, at the Office, No. 13, Hill Street. London: Smith, Elder, and Co., 65, Cornhill. Glasgow and the West of Scotland: John Smith and Son; and John Macleod. Dublin: George Young. Paris: J. B. Balliere, Rue de l'Ecole de Médecine, No. 13 bis.

THE EDINBURGH PRINTING COMPANY.

THE EDINBURGH

JOURNAL OF NATURAL HISTORY,

AND OF

THE PHYSICAL SCIENCES.

MAY, 1838.

ZOOLOGY.

DESCRIPTION OF THE PLATE.—THE CAT TRIBE CONTINUED.

In the Number for January 2, 1836, is an account of several species of this genus, which is characterized by the ferocity, agility, and insidious nature of its different species. Among the larger of these is the Jaguar, which on the New Continent may be considered as analogous to the Tiger of the old.

Fig. 1. THE JAGUAR (Felis Onca). Although this animal is the most powerful of the American Cats, it is inferior in size and strength to the Tiger, and so closely resembles in form, as well as in the distribution of its colours, the Leopard, that the two species have been confounded by many naturalists. The Jaguar, however, is much superior in size to that animal, individuals often measuring nearly five feet from the nose to the insertion of the tail. The body is much thicker, the limbs shorter, and the tail inferior in length, so that its tip scarcely trails on the ground. Both animals are of a yellowish-red colour, marked with circular dusky rings, which in the Leopard are smaller, and composed of rather widely separated spots, whereas in the Jaguar they rather form continuous rings. This species appears to be generally distributed over South America, but does not extend northward beyond the isthmus of Panama. Although not uncommon in the parts remote from cultivation, it has disappeared in the neighbourhood of the cities, and is daily becoming more scarce, as its predacious habits, together with the value of its skin, render it an object of pursuit. The Jaguar, like other species of this genus, lies in wait for its prey, and springs upon it unawares, bearing it down by its weight and great strength, and instantly depriving it of life, when it devours a portion of it on the spot, or carries it off to the thickets. It very rarely ventures to attack a man, unless urged by extreme hunger, and, notwithstanding its great vigour, is considered as cowardly. It climbs trees with facility, swims with almost equal ease, and although generally nocturnal, sometimes hunts by day. Its strength is so great, that it can drag so large an animal as a horse or a cow to a considerable distance.

Fig. 2. THE JAGUAR. Female. The female differs from the male chiefly in being somewhat smaller, and in having the colours less bright.

Fig. 3. THE PERSIAN LYNX (Felis Caracal). The Lynxes are distinguished from the other Cats by having the ears pointed, with a tuft of hairs at their extremity. The present species, which is of a yellowish-red colour, and the tail reaching to the heels only, inhabits various parts of Asia and Africa. The Persian variety here represented is marked with small dusky spots.

Fig. 4. THE COLOCOLO (F. Colocolo), which inhabits South America, is of small size, with the head short, broad, and flattened, the body greyish-white, with longitudinal streaks of dusky and red, and the tail beautifully ringed with the same.

DESCRIPTION OF THE PLATE.—THE SHORT-TAILED CROWS OR BREVES.

The genus Pitta of Temminck is composed of birds generally of very splendid plumage, of a singularly abbreviated form, arising chiefly from the shortness of the tail, whence the name Breves applied to them by the French. They are in a manner intermediate between the Crow and Thrush families, having the bill of the former with the feet of the latter. According to Mr Swainson, they have the gradually curved bill of the true thrushes, but much stronger; the predominant colour of their plumage is green, the sides of the head and the wings being generally variegated with vivid blue; and all are confined to New Holland, and the neighbouring isles of the Indian seas.

Fig. 1. THE GIANT BREVE (Pitta cyanoptera) is one of the largest species, but inferior in beauty to many of the rest, although coloured with green, blue, yellow, and black. It was discovered in Sumatra by Messrs Diard and Duvaucel.

Fig. 2. MACKLOT'S BREVE (P. Macklotii) has the head reddish-brown, the throat dusky, the fore-neck light blue, the lower parts dull-red, and the upper green.

Fig. 3. THE NOISY BREVE (P. strepitans) is a very beautiful species, having the upper part of the head brown, the throat and upper part of the neck, with a patch on the breast, black, the fore-neck, breast, and sides, yellow, the abdomen red, the back green, and the wing-covets and secondaries blue. It inhabits New Holland.

Fig. 4. THE GRENADINE BREVE (P. granatina) is also splendidly coloured with red, brown, green, and blue; as is

Fig. 5. THE RED-BREASTED BREVE (P. erythrogastra), which inhabits the Philippine Islands, has the head brown, the neck azure blue, the back green, and the wings blue, with two white spots.

36

ON THE GEOGRAPHICAL DISTRIBUTION OF ANIMALS.—NO. IV.

(Continued from page 140.)

THE remark which we made relative to the Accipitres as being prone to roam, of individuals of one species being stationed in localities of very different natures, of their partaking of a variety of food, and of the same species being found in various quarters of the globe, applies, though less forcibly, to the family of the Pies, and to the Shrikes. The appetite of the Raven extends to every sort of animal food: it has of necessity a roaming disposition, and it is spread over the globe from the northern countries to the Cape of Good Hope. The Crow and Magpie also enjoy a very extensive range, and the latter is found in America. Those Pies, however, which are in some degree granivorous, or less decidedly carnivorous, are not so widely dispersed, or of so roaming a disposition: the Rook, the Jackdaw, the Jay, the Hooded Crow, and the Chough. The Great Shrike, Lanius Excubitor, has a range almost equal to that of the Raven, while the Flusher, Lanius Collurio, which is chiefly insectivorous, is confined to Europe, or extends at most to Egypt. Among land birds, therefore, the Accipitres and rapacious or carnivorous Passeres, enjoy the most extensive distribution. The Kingfisher, a piscivorous bird, is also very widely dispersed, being common to Europe, Asia, and Africa. The Waders and Web-footed birds are, however, more extensively distributed than land birds, more especially the former, as we may have occasion to observe hereafter. Fish enjoy a very wide range, but quadrupeds seem to characterize the natural divisions of the earth, at least in a great measure. The other tribes are likewise in some degree characteristic of different quarters of the world.

Proceeding in our comparison of the cultivated parts of Devon with Oxfordshire, we come next to the Scansores, in which division we do not possess more species than are met with in Oxfordshire, or most other counties. There are two stragglers, the Black Woodpecker, Picus martius, and the Hairy Woodpecker, Picus villosus, which have not been noticed with us. The Lesser Spotted Woodpecker, Picus minor, and Wryneck, Yunx Torquilla, are rare with us, and seemingly prefer some stations to others.

In the Gallinæ, we have the advantage of Oxfordshire, in possessing the Stock Dove, Columba Œnas, a bird which appears here in large flocks in winter, but is not noticed at other times. The Rock Dove, C. Livia, is not known to exist on our coasts in any considerable numbers. The Quail is rather scarce with us: it is found principally in October, at the time of departure, but a few stay with us through the winter.

The order of Waders, Grallæ, comprehends some which reside in swampy situations, or in connexion with inland rivers and lakes, besides being principally composed of those usually termed shore birds (which likewise at times resort to rivers and lakes, and for the most part breed in fens or retired swampy spots), and a few other species which are by no means water birds, and are found in very different situations from the other Grallæ. Accordingly, some birds of this order find a place in the Oxfordshire list, and may with propriety be enumerated amongst our own birds of the South Hams. Of these birds some breed on Dartmoor, and appear in autumn and winter in the cultivated lands, as before noticed with respect to the Golden Plover, the Grey Plover, and the Lapwing. Some individuals of the Woodcock and Snipe also breed in retired moorland situations, and appear with the main body of those migrators in the cultivated parts, on the occurrence of the first cold. The Curlew and Dunlin (or some of them) breed on the moors, and pass over to the shores and rocks in winter; while very many species may be regarded as common to the cultivated parts and to the coasts, rendering it in some degree questionable which situation should claim them. The Grallæ, however, generally are shore birds. I enumerate about twenty-two Waders as frequenters of the cultivated districts of South Devon, of which thirteen are found in Oxfordshire, the remainder consisting of the Grey Plover, the Great Snipe, Olivaceous Gallinule, Spotted Gallinule, Dottrel, Little Bustard, Little Gallinule, Ruff, and Green Sandpiper, all of which are, as in the preceding instances of birds found here and not in Oxfordshire, rarities, or at least uncommon birds. Nor does that county claim any species not found with us, as may readily be imagined from the fact that there are but two British birds belonging to that division of the Grallæ which we may term inland and fluviatile Waders. Sabine's Snipe, Scolopax Sabini, and the Courser, Cursorius isabellinus, have been seen in England but twice or thrice. The Thick-kneed Plover, however, which in our cultivated districts is scarce, is in Oxfordshire common, though probably this depends on that county possessing such a noble and extensive forest, and other uncultivated tracts. In Oxfordshire the Golden Plover appears only in winter, whereas here it breeds on our moors, and appears in the southern parts afterwards. Considering the very confined limits usually observed by the Ruff, it is surprising that any of them should have been noticed in Devon, however rarely. One Devon specimen was shot in December 1808, on that fertile source of aquatic birds, Slapton Ley, which may serve perhaps in some degree to illustrate the eligibility of our county for such birds, and species from every class of birds are known occasionally to remain in England through the winter, whilst the main body observe the accustomed migration; and, on the other hand, some also remain to breed if the position is found eligible, whilst the bulk of such species retire to other counties or to other countries, where the rearing of young is conducted with certain concealment. These may possibly be cases of altered character of species, but in respect of the rarity of the Ruff in this country, it should be taken into consideration that though it is now restricted to certain of the eastern counties, it may possibly in former years, when cultivation had made no great advances, have extended generally over a much greater number, our own included. Slapton Ley is a lake of rather large size, situated towards the verge of the southern coast of Devon. By a wise ordination of Nature, the birds peculiarly termed marine, or rather some of the species, upon the occasion of want in the winter months, divide their search after food between the ocean and inland waters; besides which, some individuals of these species betake themselves wholly to the waves, or wholly to inland water, roaming for the entire season from spot to spot, or keeping constant to one locality; conclusive proofs against the notion of instinct being a defined, constrained, and very limited mental operation in brutes, a doctrine adverse also to the history of most species in which instinct is detected. Upon the occurrence then of cold, Slapton Ley adds to its visitors a great variety of shore-birds and pelagic fowls, so that this piece of water alone sufficiently shows the arbitrary nature of making the coast a geographical limit for the marine birds. In consequence of our maritime situation, and the freedom with which marine birds pass the limits of the coast, it becomes difficult to state precisely which of the Grallæ should be enumerated as belonging to the cultivated districts, and I have allowed myself to be guided in some measure by the Oxfordshire list, because that county is not maritime, but it is very likely that many birds of the Heron kind which I shall rank as shore-birds are equally entitled to be considered as inland Waders. It should be mentioned, that not only do the shore and marine birds pass the limits of the coast, and obtain food inland, but the land and fresh-water Grallæ are very often in winter, especially if it be severe, found exploring the shores for provender. The Heron, Coot, Gallinule, Lapwing, Golden Plover, Grey Plover, and I believe some other kinds, act thus. I am not aware of any species of the Waders we are now speaking of quite peculiar to South Devon, but the Little Gallinule deserves notice, as having been first discovered here; only three specimens are known, two of these being Devon birds, and the other obtained from the river Ware.

The last tribe of birds belonging to Oxfordshire, and to the cultivated districts of South Devon, consists of those few web-footed birds which roam to inland waters, meadows, stubbles, &c., or from our own coasts, or which breed and abide wholly or partially in such situations. In order to state the case as near the truth as possible, I have allowed the same number of these birds to South Devon as are found to occur in Oxfordshire; though, from our adjacency to the sea, those species which in an inland locality would be stationary, with us change their situation, and again some marine birds of our coast at times repair inland, which in counties remote from the coast are never seen. The following are the species referred to:—The Little Grebe, the Wild Duck, Teal, Wigeon, Grey Lag Goose, Common Gull, Great Black-backed Gull, and Golden-eye. The Little Grebe breeds in all the fish-ponds and small lakes in the county, and in winter very many are seen busy diving in all our inlets. The several kinds of Ducks frequent the marshes and large ponds during winter, some of the Wild Duck, as before stated, breeding with us. The Wild Goose is a frequenter of meadows, marshes, and stubbles, besides the sea coast, during the winter. And, lastly, the two Gulls, and perhaps other birds also, particularly the Red-legged Gull, make excursions in winter to inland waters and marshes after food, and some of them are known to abide during the breeding season at certain ponds, and other collections of water in the south of this county. We here bring to a close our comparison between the ornithology of Oxfordshire, and that of the South Hams of Devon. We have found a difference of twenty-eight species, or, excluding the doubtful deficiencies in the Passeres, a difference of twenty-five. It is of course always desirable to trace out the principal features in such deficiencies, and upon examination it is seen, that the proportion of deficiencies in the Accipitres and Grallæ is equal. It is here, then, that our county excels in the ornithological department, at least such is the fair conclusion by comparison with the products of another county, possessing a tolerably extensive ornithological Fauna, and considering the respective histories of the species in which Oxfordshire is deficient, our advantage cannot be set down to maritime position. On the contrary, I believe the true reasons may be stated as follows:—*1st*, There are certain European birds, whose chief situation is in France, Italy, Holland, and other adjoining countries, but whose range extends to the southern portions of Great Britain, just as the range of others is found to extend from Europe to the northern shores of Africa. There are other species also whose principal station is in Africa, and which migrate yearly into Europe, reaching in small numbers the British Isles; and since the number of animals generally diminishes northward, the proportion of birds resident in, or migrating to, the southern shores of Britain, will be greater than that of the northern parts. *2dly*, Montagu states as his opinion, that in the autumnal migration of the long, soft-billed Waders (and I suppose other kinds also) from their northern haunts to the southern portions of Europe, they experience in their transit across the north sea, equinoctial gales which gradually drive them to the southern parts of England, so that hence we are more likely than northern counties to have rarities conferred on us. *3dly*, This county contains almost every kind of retreat for the various sorts of birds: it is mountainous, well-wooded, and well-watered. *4thly*, In the retreat of those birds which come to this country from southern latitudes to their winter residences, the southern coasts of England offer them a resting place previous to their departure, so that here we see a larger number both of individuals and of species, than the generality of other counties; besides which, many, upon finding suitable abodes and mild climate, are induced to abide with us through the winter. Lastly, the occurrence of some species in this county, which have never been noted, or but sparingly, in others, must be set down either to accidental causes, or to influences of which we are ignorant. Since the reasons of our ornithology being so extensive are so various, the species of which we boast as supernumeraries, or as being found here, whilst they are but very sparingly scattered over the rest of England, cannot be exclusively Grallæ, or Accipitres, although, as before hinted, it is in these classes that the greatest preponderance is observed; neither must we be altogether guided by the ornithology of any one county, such as that selected for comparison, in arriving at conclusions respecting the peculiarities of our own, or respecting the ornithology of the southern counties of England generally. As we proceeded, we pointed out instances which illustrate the various reasons here given for the extent of our ornithological list. Other cases might be cited, but as our knowledge of their geographical ranges is imperfect, they could be mentioned only with great hesitation.

A few words on the remaining tribes of animals will close our consideration of this division of our subject. With respect to the Reptiles of South Devon but little can be said, as they do not differ materially either in number or in geographical position from those of most other counties. The Nimble Lizard, *Lacerta agilis*, is found in plenty on our heaths and commons, but is not confined to these spots, as I have taken it in gardens, and I have likewise once seen it on the stump of an old tree in a willow-ground. The Dumfries-shire Snake, Warty Eft, Edible Frog, and Natter Jack Toad, do not occur in Devon, so far as my knowledge extends.

It is not in my power to add any thing on the Fresh-water Fishes, or on the Insects of Devon. A list of Land and Fresh-water Shells of this county has already appeared in this Journal, so that it is here only necessary to remark how singularly well defined are the limits of some species of these tribes, so much so, as to induce the belief that these limits in some cases correspond with the boundaries of soils, or with the geographical distribution of certain vegetable productions on which they feed, though possibly the facts of these cases may after all be quite as inexplicable as the limits of dispersion observed by the Nightingale, and some other birds. At the least, however, it will be allowed, that comparisons between the geographical limits of the various productions, mineral, vegetable, and animal, of a given district, would in all probability lend considerable assistance in the determination of the causes of a great number of now unexplained facts; for the three kingdoms of nature are not merely associated, but intimately connected, and since secondary causes have so great influence on animal and vegetable dispersion, we may not unfrequently discover the cause of distribution of these to rest with the qualities of the soil on which they are maintained, and still more frequently the phenomena of animal distribution to depend on the nature of certain vegetable products occurring on the spot, or diffused over those tracts to which certain animal forms are peculiar. Again, the selection of particular food by certain of the carnivorous animals, will, in a great number of instances, determine with exactness the limits of such species.

THE CROSSBILL A PERMANENT RESIDENT IN SCOTLAND.—The Crossbill, *Loxia curvirostra*, is not found in great numbers in any part of Scotland. It remains with us during the whole year, and may be met with in small flocks of from eight to twenty or more individuals, among the pines in the midland and higher districts of the country. They feed most eagerly on the seeds extracted from the cones of the Scotch fir, larch, and spruce, and whilst thus occupied, they keep very quiet, and can be discovered only by the chuckling noise made by their tearing open the cones with their powerful and curious beaks, and the occasional dropping of the cones they have rifled. They always move from one part of the forest to another at a signal given by one of the party that acts as leader, and is stationed on the summit of the tree. They manifest their desire to move by uttering a sharp loud note; and when the watchman observes any symptoms of impatience among those below among the branches, he takes the lead in uttering his shrill note in a louder and more rapid manner than the others for a few seconds, and on his taking wing the others instantly let go the cones on which they have been operating, and accompany him, flying in a compact body, and uttering their note as they fly along. They often take long flights, and frequently return to a neighbouring tree after making a few circuits. When feeding on the low branches of a tree, it is surprising how little fear they exhibit, even when approached so closely as to be almost within reach of the hand. Having slightly wounded one in the wing, I carried it home and placed it in a cage, with a quantity of larch cones, which it immediately attacked, without showing any symptoms of fear, and after helping itself most plenteously, died shortly after, which I attributed to my neglect of supplying it with water to drink. I have often endeavoured to find out the nests of these birds in the *usual season*, but never succeeded; and was surprised at last on discovering that their broods are all on the wing before their neighbours of the woods have set about preparing their nests. I was attracted one day in the end of February, or early in March, during a heavy snow-storm, by the peculiar chirping of nestlings in the act of feeding; and, on ascending the tree, found five or six young Crossbills, almost fully feathered, and quite vigorous, notwithstanding the severity of the weather, snugly huddled together in a nest composed of small sticks externally, and well lined with matted wool. In mild seasons I suppose they breed, even in this country, during the month of January. They are not entirely destitute of song. I have often been delighted with a concert of these birds perched on the sunny side of a tree. Their song is not loud, but pleasingly varied, a good deal resembling that of the Water Ouzel, when you chance to overhear his gentle warbling, as he sits on a ledge of ice, in a frosty winter day.—J. M. BROWN.

A few statements like the above, by persons who have had opportunities of seeing the Crossbills in their native forests on the Dee and Spey, would settle definitively the questions so much agitated respecting the breeding and dispersion of those birds, which, on appearing in the lower parts of the country, are supposed to come from the Continent. The Siskin in all probability breeds in the same districts, as it has been seen there in flocks very early in August. The Snow-Bunting has also been found in summer on the higher Grampians, such as Benmucdui, Bennabuird, and Lochnagar.

Many parts of the mountainous and more remote districts of Scotland have as yet merely been cursorily visited, and even some birds known to occur in them, such as the Crested Tit, have never been subjected to minute examination as to their habits and distribution.

DIGESTIVE ORGANS OF BIRDS.—NO. II.

In the Number for May 1837, p. 94. was presented an outline of the digestive organs of the Red Grouse, *Tetrao Scoticus*, accompanied with a short description. The subject being of great importance, both in an anatomical point of view, and in reference to the classification of birds, the intestinal canal of a bird of another family is here figured and described. It has been seen that the Grouse, and Gallinaceous birds in general, have an extra-thoracic dilatation of the œsophagus, or in other words a crop; a powerful muscular gizzard, lined with a dense and tough epidermis; intestines of great length; and cæcal appendages usually of a capacity equal to that of the intestine. In the birds of prey there is a different arrangement, which will be shown by the digestive organs of the Golden Eagle, *Aquila Chrysaetus*, here represented in outline.

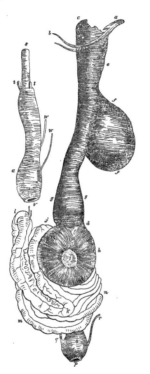

In this bird, which is carnivorous, the tongue, *a*, is short, emarginate, fleshy above, horny beneath towards the end, and papillate behind. There are numerous crypts along the sides of the tongue, beneath it and anteriorly to the glottis or aperture of the windpipe, which pour forth a viscid fluid, analogous to saliva. The hyoid bones, *b, b*, are seen in their natural situation. The œsophagus, *c, d*, is of great width, and on the fore-part of the neck, previously to its entering the thorax, is dilated into an extremely extensible sac, or crop, which has not a narrow aperture like the crop of the Gallinaceous birds, but is simply a dilatation of the œsophagus, having the same structure as it. Behind the crop, where it enters between the coracoid bones, the œsophagus is a little contracted; and at its lower part enlarges to form the proventriculus, *g, g*. At its upper part, the œsophagus has a thin layer of longitudinal muscular fibres, some of which extend over the crop. In its whole length, it is encircled by transverse fibres, forming a distinct muscular coat. Its inner or mucous coat is soft, smooth, and perforated by the apertures of numerous mucous crypts, the secretion from which is copiously spread over it. This inner or mucous coat is less elastic than the muscular, and when the œsophagus is empty, is thrown into longitudinal folds.

In the proventriculus, *g, g*, is a broad belt of cylindrical glandules, having a central cavity, with thick walls, and a spongy or villous inner surface. Their inner extremity is narrowed, and they open on the mucous membrane by an inconspicuous aperture, placed in the centre of a small rounded eminence. The fluid secreted by these glandules, and copiously poured out so as to cover the entire surface of the proventriculus, as well as part of that of the stomach, is of a greyish-white colour, clammy, and when cold having the consistence of slightly coagulated albumen.

The stomach, *h*, which commences at the lower edge of the proventriculus, is of a roundish form, a little compressed. Its outer or muscular coat is thin compared with that of birds of other families, but yet of considerable thickness, and is composed of fibres arranged in fasciculi, which are broader in their middle, or along the edges of the organ, and are inserted into two thin tendinous spaces, *i*, one on each side. At its upper part, the fibres diverge. Within the muscular coat is a thin layer composed of whitish interlaced fibrous tissue. The inner coat is of a softish homogeneous texture, elevated into strong rugæ running in various directions. It is considered as analogous to the cuticle, and somewhat resembles that on the heel in man; for which reason it is named the cuticular lining, or epithelium.

At its commencement, the œsophagus is placed directly in front when not distended, but when filled inclines immediately to the right side, on which the crop lies. At the lower part of the neck it inclines to the left side, passes into the thorax in the centre, when the trachea, which had run along the left side, comes in front, and bifurcates over it. The stomach occupies the middle and left side of the abdomen, and when distended fills a very large portion of its cavity.

The intestine is short, and of very small diameter, in proportion to that of the œsophagus. The pyloric orifice of the stomach has a kind of knobbed valve. The duodenum, or first curve of the intestine, *j, k, l*, is of much greater capacity than the succeeding portion. It is accompanied by the pancreas, which lies in its curvature, and after receiving its ducts, terminates at *l*, immediately under the right lobe of the liver, where it receives the biliary ducts. The intestine then diminishes in capacity, *m, n*, and is convoluted, lying chiefly on the right side. Over the stomach this contracted portion terminates, and the rectum commences, which is much dilated, runs along the sacrum, enlarges at the end, to form the cloaca, *o*, which receives the urine by the two ureters, *q, r*, and ends at the anus, *p*. Besides the peritoneal covering, the intestine has a muscular, and an inner or mucous coat. The former is thickest in the duodenal portion, the inner surface of which is delicately villous. On the rest of the small intestine are long slender villi, which, towards its lower part, become more sparse. The inner surface of the rectum is furnished with numerous mucous crypts.

The cæca, in the Golden Eagle, *t, t*, are very different in capacity from those of a Gallinaceous bird, being mere mucous sacs, scarcely more than a quarter of an inch in length. They are seen to occupy the same place, namely, the commencement of the rectum, *t, v*. The smaller figure shows the termination of the small intestine, *s*; the cæca, *t, t*; the rectum, *t, v*; the cloaca, *u*; the anus, *v*; and the ureters, *w, w*.

The Golden Eagle is a bird organized for rapine; its frame is firm and compact; its long and broad wings enable it to fly with great speed; its legs are furnished with strong muscles, and its claws are curved, tapering, and capable of being thrust into the vitals of its victims, which it can also carry off in its talons when of moderate size. But it does not feed exclusively on animals which it has itself captured, for it often devours carrion of all kinds. Keeping its prey down with its feet, it deprives it in a rude manner of part of its hair or feathers, and then tears up and swallows fragments of the flesh. Its bill, being hooked at the point, and sharp-edged, is obviously well adapted for this use. If the subject be large it fills the stomach, and then the crop and œsophagus, these parts being capable of holding more than two pounds of flesh. The crop is merely a recipient for the food, which is found in it quite unaltered. When the crop, stomach, and intermediate part are found filled, the solvent action is first perceived in the proventricular space; and it is probable that the secretion from its glandules effects the solution of the food in all species, for in those of which the inner coat of the stomach is thick and horny, there can be no effusion from it. The mass of flesh, mixed with feathers, hair, and bones, being in the stomach reduced to a soft pulp, the nutritious parts pass into the intestine by the pylorus, which rejects the indigestible substances. The hair and feathers are, by the contraction of the muscular fibres, thrust into the œsophagus, the muscular fibres of which contract in succession, so that the pellet is ejected. In the duodenum the pulpy mass of the food is further diluted by the pancreatic fluid, assumes a homogeneous appearance, and is of a light red colour. On being mixed with the bile, it assumes a yellowish or greenish tint, and deposits the chyle on the surface of the intestine, whence it is absorbed. The refuse enters the rectum, where it is diluted by the urine, and it finally ejected in a semi-fluid state, of a dull green colour, mixed with flakes of white, being projected to a considerable distance, so that the feathers be not soiled.

The digestive organs of Vultures are similar to those of Eagles and Hawks; but various modifications are observed in those of the different species. In the Falconine species generally, the intestinal canal is about three times the whole length of the bird. Owls differ from Hawks in having no crop; but their cæca are much larger, and even approach in size to those of some of the Gallinaceous Birds.

On comparing the figure and description here presented with those of the Red Grouse formerly offered, it will be seen that the digestive organs of the Rasores and Raptores differ in several essential respects. Similar differences occur in all the really natural groups of birds, such as the Gemitores or Pigeons, the Deglubitores or Conirostral tribe, the Cantatores or Songsters, the Parrots, Snipes, Divers, and Ducks.

PROLIFICACY OF THE BLACKBIRD.—A pair of these birds built four successive nests last season upon the island in St James's Park, and succeeded in rearing seventeen young ones; the three first broods consisting of five each, and the last of two only. There cannot be the least doubt as to the identity of the female, as she is well known to the person who attends them, and so tame as to take food from his hand while sitting on the eggs. There were, moreover, no other individuals of the same species near the place. Another isolated pair which I knew of raised, unmolested, three broods in a garden near my residence, so that the Blackbird would appear to raise as many young as the Partridge, which produces only one brood in a season.—*Edward Blyth*, in "*The Naturalist*."

GEOLOGY.

Depth of the Frozen Soil of Siberia.—In a paper on this subject, communicated to the Geographical Society by Professor Baer of Petersburgh, the author stated that it has long been ascertained that the soil of Siberia, over a great extent of country, is never entirely free of ice, the summer heat producing only a partial thaw at the surface. Gmelin the elder mentions that the ground at Yakutzk, in Lat. 62½° N., Long. 130° E., was found to be frozen at a depth of 91 feet, and that the people were compelled to desist in sinking a well. Many similar facts had been collected by travellers, but they were not generally credited; and even in 1825, Von Buch rejected them as erroneous. Yet they have recently been corroborated by Erman and Humboldt. Within these few years, a merchant, of the name of Schargin, having attempted to sink a well at Yakutzk, was about to abandon the project, when Admiral Wrangel persuaded him to continue his operations. The whole stratum of ice was at length perforated, and, at the distance of 382 feet from the surface, the soil was very loose, and the temperature ½° Ream. (31° Fahr.) Near the surface it was much lower, and had increased as follows:—at some feet from the surface, 6° R.; at 77 feet, 5°; at 119 feet, 4°; and at 217 feet, 2°; at 305 feet, 1½°; at 350 feet, ½°; at 382 feet, ½°. As the soil had become loose at 350 feet, and as the aperture of the well was eight feet square, and the work carried on partly during the winter, when the column of cold air must have lowered the temperature, it is probable that the spot at which the thermometer marked the freezing point was at that depth. This vast thickness of frozen ground indicates that Siberia must have been for a long period in the same physical condition as at present. Although the extent of this layer of ground ice is not determined, yet enough is known to show that it occupies a vast space. Humboldt found the soil frozen at the depth of six feet at Bosgolowsk, near the Ural, in 60° N. Lat. Near Beresow, Erman found the temperature at a depth of 23 feet, still + 1°.6, (35½° F.); but, in 1821, a body was disinterred which had been buried 92 years, and it showed no signs of decomposition, the earth around having been frozen. It has long been known that, at Obdonk, in N. Lat. 68°, the ground is always frozen. Near Tobolsk, no ice is found in the soil, but, as we proceed eastward, the ground ice advances farther south. From the discussion which followed the reading of this paper, it seemed to be generally considered that the experiment at Yakutzk had not been made with sufficient care to authorize the belief that the frost penetrates to so great a depth as 350 feet; but that the statements of Arago and Von Buch, and of individuals in this country, had been fully borne out by M. Schargin's observations, and almost exactly in the same ratio as hitherto found. Captain Back stated, that in the cold regions of North America, even in the height of summer, he had never found the ground thawed more than four feet below the surface; but that experiments on the subject were wanting before correct information could be obtained.

Raised Beaches at Coquimbo.—In a letter lately received from Mr Pentland, Her Majesty's Consul-General in Bolivia, and dated Tacna, Peru, 3d September 1837, he says:—I had occasion to observe at Coquimbo raised Beaches on a very large scale, and attaining an elevation of 400 to 500 feet above the present sea-level. They consist of beds of sea sand, alternating with others, exclusively formed of large oysters, and in general capped by a mass of boulders and gravel, some of the former weighing several tons, and covered with parasitic marine mollusca. It is this modern marine deposit which forms the parallel roads spoken of by Captain Hall, and referred to by Lyell, and has evidently been raised at a very modern period, many of the shells preserving their brilliant colours. The vicinity of Coquimbo is composed of a transition granite, with masses of porphyry in veins, and both contain the rich metallic veins of that celebrated metalliferous district. The Andes in this neighbourhood are very high, and if the position of the peak seen from the port is accurately laid down in the map (and which I had no time to verify), its elevation must exceed 20,000 feet, according to some zenith distances I took of it. The country around the town from which I now write is an arid sandy desert, without a trace of vegetation. It is covered with loose sand of the new red sandstone series, through which the quartziferous trachyte rises near Tacna, and continues to form a band at the base of the Andes.—*Edinburgh New Philosophical Journal.*

Large pair of Fossil Horns found in Essex.—Though the geological feature of the county of Essex may not be of that highly interesting character exhibited by the mining districts of England; and though the facts respecting the physical history of our planet are not developed in such quick succession here as they are found to be in some other localities, owing in a great measure to the mineral properties of our tertiary beds not being of that quality to warrant extensive excavations; still as the constant action of the sea upon the blue-clay cliffs of our coast washes those cliffs down, and brings to our notice the fossils which they have so long concealed, or if, by any of our artificial removals of soil, those relics are brought to light, the facts are as worthy of being recorded, as if the organic remains were of still higher antiquity. I am induced to make these observations by my having just received from a friend the two bony portions (commonly called the core, or slug of the horn) of the interior of a pair of very large horns of the Ox, lately discovered at Clacton, on the Essex coast, about ten miles south of Walton. They were found in a mass of drift sand, overlying the London clay, and in consequence of the cliffs slipping down, they were disentombed. Many portions of the skull were found with them, and the *os frontis* was attached to them. By measurement, I find them to be three feet long on the outer curve, from the base to their tips; they are curved about eight inches, and are eighteen inches in circumference at the base. The diameter of each horn at the base is six inches in one direction, and five inches in the other. They have not the fluted character so conspicuous in the horn which I found three years ago at Copford, but they have the punctures so common to the bony portions of horns in general. In both instances the remains of the Elephant were found associated with those of the Ox. At Capford, a vertebra and the cuneiform bone of the right fore-foot of the former was discovered, subsequently to those of the Ox above mentioned. And with the horns recently discovered at Clacton was found a perfect Elephant's tooth, eleven inches in length, and three inches wide upon the grinding surface, and eight inches in depth.—*John Brown, in Magazine of Natural History.*

MISCELLANEOUS.

DESCRIPTION OF THE FRITH OF CROMARTY, AND REMARKS ON ESTUARIES.

(Continued from page 96.)

The term Estuary is generally, although not in popular language, applied to that part of a river which the tide ascends. It has also been defined a space traversed by a river, and into which the tide enters. By geological writers and others, it is frequently employed to designate a basin, partially or entirely filled with sea-water, more or less mixed with fresh-water, and in which are deposited quantities of sand and mud carried into it by a river or rivers during floods. Indeed, this latter seems to be the general acceptation of the term. The words frith and estuary are to a certain extent synonymous, for all the inlets in Scotland called friths, viz. the Solway Frith, the Frith of Clyde, the Frith of Forth, the Frith of Tay, the Beauly Frith, and the Cromarty and Dornoch Friths, are in some part of their extent estuaries, that is, receive large rivers, which deposit in them mud and sand. Most of them, in fact, are actually named after their principal rivers, and have thus evidently been considered as either formed or greatly influenced by them. It is therefore clear that the 'Frith' of Tay is thus but another word for the estuary of that river. Why the general principle of popular nomenclature should thus have been deviated from in a few cases is not apparent. What, agreeably to it, ought to have been named the Ness Frith or Beauly Frith, has been named the Moray Frith, probably because of its great extent, and of its being bounded on one side by that province; while the Cromarty and Dornoch Friths have taken their names from the towns so called. In the parts of Scotland where the Gaelic language has prevailed, the inlets are named lochs, a term which is also applied to bodies of fresh water. But very few of these sea-lochs receive rivers of considerable size, and almost all of them are named from circumstances having no connection with their streams.

Of the larger rivers of Scotland there are very few that have an estuary of the most simple kind, or, in other words, which open directly into the sea, without the intervention of a basin common to the river and the ocean. Such, however, are the Dee, the Don, and the Ythan, in Aberdeenshire, of which the estuary or tide-way is not two miles in length. Were rivers generally of this kind, there could be no disputes concerning their estuaries. The highest tide mark might then be considered as the upper, and the boundary of the lowest tide as the lower limit.

As an example of the second kind of estuary mentioned above, or that formed by a space traversed by a river, and into which the tide enters, may be mentioned the basin of Montrose, which is about eight miles in circumference, and is left dry at low water excepting the channel of the South Esk, by which it is traversed. There can be no dispute as to the propriety of considering this basin an estuary.

The estuaries of the Forth, the Tay, the Clyde, and the Beauly, are of the third or most common kind, being basins, from a great part of which the sea never recedes. They vary in extent, some being longer and broader, and their channels are more or less deep, according to the original nature of the ground, or the force of the currents by which it is traversed. But no rule can possibly be laid down for the determination of an estuary of this kind, which is neither a river, nor yet the sea, but a compound of both. The medium level of the sea at high and low water, which has been proposed for determining the limit by which a river is separated from the sea, is obviously a merely arbitrary character, which may answer for defining the mouth of the river properly so called, but which can separate it from the sea only when no basin intervenes, when the sea itself is in fact the basin, into which the river pours its waters. On the contrary, when there exists an intermediate basin or estuary, common to the sea and the river, the line in question may form a boundary between the river properly so called and the estuary, but it can have no reference to the lower limit of the latter, which must be determined by other means.

(To be continued.)

LITERARY NOTICES.

Works preparing for Publication. (Longman and Co.)—The Rev. L. Vernon Harcourt (son of the Archbishop of York) has in the press a work on the "Doctrine of the Deluge." His object is to vindicate the Scriptural History of the Deluge from the doubts which have been recently thrown upon it by geological speculations. This the author has endeavoured to accomplish by showing, upon the testimony of a long list of ancient and modern authors, that since the era of that catastrophe a set of religionists never ceased to exist, whose opinions and usages were founded upon a veneration of the Ark as the preserver of their race. In 2 vols. 8vo.—Mr Westwood's "Popular Introduction to the Modern Classification of Insects," which has been so long announced for publication, is at length in the press, and will be published in Monthly Parts; the first will appear on the 1st of June. The author has for eight years been employed upon it, collecting materials, from the Continental as well as British Museums. It will be illustrated with many thousand figures engraved on wood. The author has paid very minute attention to the Natural History of the Transformations of Insects, and confidently hopes that there will be found much new and interesting matter in his work. It is intended to form a sequel to the popular work of Messrs Kirby and Spence. 1 vol. 8vo.

—Essays in Natural History. By Charles Waterton, Esq. With a view of Walton Hall, and an Autobiography of the Author. 1 vol. fcap. 8vo.

The second volume of Mr MacGillivray's History of British Birds, including the *Cantatores, Reptatores, Scansores, Volitatores*, and *Excursores*, being all the remaining land birds, the *Raptores* excepted, will be published in autumn.

Edinburgh: Published for the Proprietor, at the Office, No. 13, Hill Street. London: Smith, Elder, and Co., 65, Cornhill. Glasgow and the West of Scotland: John Smith and Son; and John Macleod. Dublin: George Young. Paris: J. B. Balliere, Rue de l'Ecole de Médecine, No. 13 bis.

THE EDINBURGH PRINTING COMPANY.

THE EDINBURGH

JOURNAL OF NATURAL HISTORY,

AND OF

THE PHYSICAL SCIENCES.

JUNE, 1838.

ZOOLOGY.

DESCRIPTION OF THE PLATE.—THE BABOONS.

The Cynocephali constitute a well marked section of the great tribe of the Simiæ, being characterized by having a muzzle somewhat like that of the Dog, whence their name, terminal nostrils, large canine teeth, flattened and angular ears, cheek-pouches, longitudinal streaks on the face, large callosities on the buttocks, limbs of nearly equal length, and a moderately long tail. They attain a considerable size, are active and vigorous, excitable, ferocious, and indomitable.

Fig. 1. The Guinea Baboon (*Cynocephalus Sphinx*), which inhabits the coast of Guinea, is of a light yellowish-brown colour, with the face black, and the cartilages of the nostrils surpassing the jaws.

Fig. 2. Represents a young female of the same species.

Fig. 3. The Little Baboon (*C. Babouin*) is a native of Southern Africa. It has the fur of a greenish-yellow tint, the face of a livid flesh-colour, the cartilage of the nostrils of the same length as the upper lip.

Fig. 4. The Chacma (*C. porcarius*) is of a dusky green colour above, with the face dusky, and the circumference of the eyes pale. The male, which has the hairs of the neck long, forming a kind of mane, is extremely ferocious. This species inhabits the Cape of Good Hope.

Fig. 5. The Dog-Faced Baboon (*C. Hamadryas*) is said to occur in the neighbourhood of Moca on the Persian Gulf, and in Arabia. Its general colour is ash-grey, the face flesh-coloured, and the hands black; the hair on the neck and fore part of the body elongated. It exhibits the same ferocity as the other species, becoming unmanageable when old, although at first somewhat docile.

Fig. 6. Represents a young female of the same species.

DESCRIPTION OF THE PLATE.—THE VULTURES.

The order of Rapacious Birds may be divided into three great sections, the Vultures, Hawks, and Owls. The former birds are always of large size, and are easily distinguished from the others by their having the head and part of the neck bare. Their bill varies, being in some very strong, in others slender, but the point of the upper mandible is always curved and acute. The feet are strong, but the claws are not so large or curved as in Eagles and Hawks. The wings are very large, being not only long, but of great breadth. The flight of these birds is accordingly powerful and sustained. They often ascend to a vast height, and float as it were in circles. Their sight is extremely penetrating, but their smell, although by some alleged to be very acute, is by others denied to be in any way remarkable. They feed on dead animals of all kinds, and are extremely useful in the warmer regions of the globe, where they chiefly abound, in often disposing of putrid carcases which would otherwise infect the air with noxious effluvia. Their feet are not formed for pouncing upon live prey, or for carrying off objects in their talons; and, therefore, they generally remain satisfied with dead prey, which they gorge in an excessive degree whenever an opportunity occurs; they are for this purpose furnished with a large crop; their stomach is membranous, like that of the Eagles; their intestine of moderate length, and their cœca extremely small.

The Vultures, properly so called, constitute a genus of this group, peculiar to the Old Continent, and characterized by having the bill strong, straight at the base, convex above; the nostrils large and transverse, the head and neck destitute of feathers, but sparsely covered with a very short down; a collar of large feathers at the lower part of the neck; the wings very large, the first and sixth quill nearly equal, the fourth longest. Two species occur in the south of Europe, the rest in the warmer parts of Asia. and in Africa.

Fig. 1. The Aquiline Vulture (*Vultus Ægyptius*) inhabits all the northern parts of Africa.

Fig. 2. The Crested Vulture (*V. galericulatus*) is a native of India.

Fig. 3. The Pondichery Vulture (*V. Pondicerianus*) also inhabits India.

Fig. 4. The Indian Vulture (*V. Indicus*) belongs, as its name implies, to the same region.

87

BRITISH BIRDS.—THE CUCKOO.

The Cuckoo arrives in the south of England about the 15th of April, in that of Scotland towards the end of the same month, and in the northern parts of Britain soon after the beginning of May. The periods, however, vary considerably according to the character of the season, and as the birds do not always announce their return by emitting their well known-cry, they may sometimes be met with at a time when their presence is not suspected. There seems to be hardly any part of the country which they do not visit; for while some remain in the southern counties, others settle in the most remote islands of the north, and although they are met with in the most cultivated districts, they also frequent the valleys of the wildest of our hilly and mountainous tracts. Perhaps the most favourite resorts of the species are the parks and plantations bordered with fields and pasture-grounds, or the woods and thickets of the upland glens; but on the rocky hills of the most treeless regions, and the bleak moors or fern-clad slopes of the interior, it is found often in great numbers, although never in flocks, for if gregarious during its migrations, as some suppose, it manifests no social disposition during its residence.

In the maritime Highlands and Hebrides, after the time of the arrival of the Wheatear, every one is on the look-out for the Cuckoo. Both birds are great favourites with the Celts, the latter more especially, but both may be the harbingers of evil as well as of good; for should the Wheatear be first seen on a stone, or the Cuckoo first heard by one who has not broken his fast, some misfortune may be expected, whereas, should the reverse be the case, the individual will prosper all that year. Such at least is the popular creed in the North. The Saxons of the south have no such fancy, and the lover of nature, Saxon or Celt, gladly hails the bird of summer.

" Cuckoo! Cuckoo! O welcome, welcome notes!
    Fields, woods, and waves rejoice
    In that recovered voice,
As on the wind its fluty music floats,
        At that elixir strain,
    My youth resumes its reign,
And life's first spring comes blossoming again."

Early in the mild mornings of May and June, and towards the close of day, he who wanders along the wooded valleys will be sure to be greeted by the ever-pleasing cry of the Cuckoo, unvaried though it be, as the bird perched on a rock, or lichen-crusted crag, or balancing itself on the branch of some tall tree, coos aloud to its mate. You hear nothing but the same *hu hu*, or, if you please so to syllable it, *coo coo*, repeated at short intervals, but if you attend better you will find that these two loud and mellow notes are preceded by a kind of churring or chuckling sound, which, if you creep up unseen, you will find to consist of a low and guttural inflection of the voice, during which the throat seems distended. But the Cuckoo, ever vigilant and shy, has observed you, and flies off, followed by two small birds, which, by their mode of flying and incessant cheeping notes, you know to be Meadow Pipits. They keep pace with it, and when it alights on the grassy bank, they alight too, and take their stand in its vicinity. You have heard that Cuckoos lay their egg in the nest of a Pipit or other small bird, and you at first suppose these to be its foster-parents; but this is not a young Cuckoo, but an old grey male, just arrived from Africa mayhap. They attend it, fly after it, stand beside it, and seem to be concerned about it, to be distrustful of it, to watch its motions, and to indicate their dislike to it by their continued cries. But Cuckoos are not always followed by Pipits, for often you may see them glide among the trees without any attendants.

The flight of the Cuckoo is swift, gliding, even, rapid on occasion, generally sedate, usually at no great height. In the hilly parts it may be seen skimming along the ground, alighting on a stone or crag, balancing itself, throwing up its tail, depressing its wings, and then perhaps emitting its notes. In woody districts it glides among the trees, perches on their boughs, and makes occasional excursions into the thickets around. On the ground it is not often seen, unless when cooing, and then it walks in a constrained manner; but on trees it alights with facility, clings to the twigs with firmness, glides among the foliage, and by the aid of its tenacious grasp and ample tail, throws itself into various and always graceful postures, as it searches for its prey. Its food consists of coleopterous, lepidopterous, and dipterous insects, in procuring which it must visit a variety of places; and very much of hairy caterpillars, which it picks from among the grass and heath, where, however, it cannot search by walking, like the Plover or Curlew, as its feet are too short, and its toes misplaced for such a purpose. Yet it can hobble round a bush to pick the worms from it, as well as cling to its twigs, and is thus greatly beneficial to the gooseberry bushes in large gardens in the neighbourhood of its haunts. The inner membrane of its stomach is often found stuck full of the hairs of caterpillars, so as to present a very curious appearance, the hairs being disposed in a circular manner, showing the rotatory motion of the contents of the organ. The remains of its food are ejected in pellets, as in Hawks and Owls. It has been conjectured that it occasionally feeds on eggs, especially those of the small birds in the nests of which it deposits its own; but I am not aware of its having been caught in the act. It has also been accused of eating young birds, but no one has found bones or feathers in its stomach.

The most remarkable trait in the character of the Cuckoo is its confiding the charge of hatching its eggs, and rearing its young, to some other bird, always much smaller than itself. The species on which it thus imposes its progeny is generally the Meadow Pipit, *Anthus pratensis*. In Scotland its egg has not been found in the nest of any other bird, but in England it has been discovered in those of various species:—the Hedge Chanter, White Wagtail, Sky Lark, Nightingale, Garden Warbler, and others. The egg is small in proportion to the size of the bird, being generally not much larger than that of its foster-parent, its average length being ten and a half-twelfths of an inch, its greatest diameter eight-twelfths, its colour white, greyish-white, or reddish-white, speckled with ash-grey or greyish-brown. Various conjectures have been hazarded as to the cause of this disproportionately small size of the eggs. If we say that as the Cuckoo is physically constrained to deposit its egg in the nest of some small bird of the insectivorous kind, its egg must be nearly of the size of those of its dupe, we may state a truth, but we afford no explanation of the phenomenon Why should it be so constrained; why does it not form a nest, hatch its eggs, and rear its young? Because, as some say, it leaves its summer residence only in July, and as it remains but two months or a little more with us, it could not leave its young in a sufficiently advanced state to shift for themselves. By why should it thus hurry away? has it not abundance of food? does it not go at the very time when insects are most abundant? If it dreads the cold of autumn, is not that of April or even May much greater; and if its tender young find enough of heat until September, how is it so much more sensitive? It has been alleged, conjecturally, that the ovary is much less plentifully supplied with blood than that of other birds of similar size, and therefore the eggs are not developed; but, in fact, no difference exists between the Cuckoo and Jay in this respect; and if there were, although the smallness of the eggs might be accounted for in so far, how is it necessary that they should be so small?—In short, all that we know about the matter is just this: the Cuckoo arrives in the end of Spring, and departs in July; it forms no nest; but deposits its eggs singly in the nests of various small birds, which hatch them, and rear the young. The latter remain until September. The great size of the stomach, the dilatation to which it is subject from the nature of its food, which chiefly consists of hairy caterpillars; the necessity of being ever in motion to procure a sufficiency of this food, which in proportion to its bulk contains very little nutriment, and other fanciful notions, have also been adduced as reasons why the Cuckoo does not nestle and hatch its eggs. All our Owls have stomachs equally large, and swallow as much hair, yet they all rear their young. According to Wilson, the Carolina Cuckoo has the stomach capable of great distension, and covered internally with hair, so that in all probability it is precisely the same as that of our Cuckoo; yet it also hatches its own eggs.

It appears from the observations of various persons, that the Cuckoo, having found a nest, watches for the absence of its owner, then deposits its egg, and flies off; that, in general, the nest in which it places its egg contains none or few eggs; that the owners of the nest sometimes eject the intruded egg; and that in a few instances two Cuckoo's eggs have been found in the same nest. It is also stated that the Cuckoo, on depositing its egg in a nest already containing eggs, sometimes carries off one or more of them; but frequently nests have been found containing the ordinary number of eggs along with that of the Cuckoo. Pipits or other small birds, finding a Cuckoo at or near their nest, manifest alarm, anxiety, and hatred towards it, just as they would toward a Jay or other suspected bird.

It was known to the ancients that this bird leaves its egg to be hatched by another, but they mingled truth with fable, believing that the young devoured not only those of its foster-parents, but finally the latter themselves. The manner in which the young Cuckoo's fellow-lodgers disappear from the nest is perhaps as marvellous as any thing else in the history of this strange bird. A pair of Pipits, Wagtails, or Hedge Chanters, would find it a sufficient task to provide their own young with food, and probably would be unable to satisfy, in addition, the incessant cravings of the young Cuckoo, which grows very rapidly; and as it soon completely fills the nest, would crush to death or suffocate its feeble fellow-lodgers. The young Cuckoo, as if in order to obtain sufficient nourishment, and prevent the protracted misery of its foster-brethren, ejects them from the nest, and their parents, unable to replace them, or failing to recognize them, leave them to perish. The exclusive occupation of the nest by the young Cuckoo was first satisfactorily accounted for by Dr Jenner, the discoverer of vaccination, who, in the Philosophical Transactions for 1788, states, that having found a nest of the Hedge-Sparrow, containing a Cuckoo's egg and three of the Hedge-Sparrow's, but on the day following a young Cuckoo and a young Hedge-Sparrow, two of the eggs having disappeared, he "saw the young Cuckoo, though so lately hatched, in the act of turning out the young Hedge-Sparrow. The little animal, with the assistance of its rump and wings, contrived to get the bird upon its back, and making a lodgment for its burden by elevating its elbows, clambered backwards with it up the side of the nest."—"It remained in this situation for a short time, feeling about with the extremities of its wings, as if to be convinced whether the business was properly executed, and then dropped into the nest again. With these, the extremities of its wings," he continues, "I have often seen it examine, as it were, the egg and nestling before it began its operations; and the nice sensibilities which these parts seem to possess seemed sufficiently to compensate the want of sight, which as yet it was destitute of. I afterwards put in an egg, and this, by a similar process, was conveyed to the edge of the nest and thrown out. These experiments I have since repeated several times, in different nests, and have always found the young Cuckoo disposed to act in the same manner." He then states that its shape is well adapted for this purpose, as its back is very broad, with a depression in the middle, which is not filled up until it is about twelve days old. When two Cuckoos' eggs happen to be deposited in the same nest, a severe contest takes place between the newly-fledged young, and continues until the weaker is ejected.

These observations have been verified by Montagu, who having taken a young Cuckoo, five or six days old, to his house, "frequently saw it throw out a young Swallow for four or five days after. This singular action was performed by insinuating itself under the Swallow, and with its rump forcing it out of the nest with a sort of jerk. Sometimes, indeed, it failed after much struggle, by reason of the strength of the Swallow, which was nearly full-feathered; but after a small respite from the seeming fatigue, it renewed its efforts, and seemed continually restless till it succeeded. At the end of the fifth day this disposition ceased, and it suffered the Swallow to remain in the nest unmolested." Mr Blackwall also having taken a young Cuckoo that was hatched in a Tit-Lark's nest seven days after the old birds had quitted the neighbourhood, had an opportunity of observing the same phenomenon. "The nestling, while in my possession, turned both young birds and eggs out of its nest, in which I had placed them for the purpose, and gave me an opportunity of contemplating at leisure the whole process of this astonishing proceeding, so minutely and accurately described by Dr Jenner. I observed that this bird, though so young, threw itself backwards with considerable force when any thing touched it unexpectedly."

Beyond this, there is nothing very marvellous in the history of the young bird, which, carefully fed by its foster-parents, who no doubt believe it to be their own progeny, grows apace. It appears that very many species of birds, having hatched the eggs of others, consider the produce to be really their own; and that many also, without having incubated, will adopt a helpless youngling and feed it. It is not more wonderful that the Pipits or Wagtails should harbour no suspicion of the alien character of the large bird that fills their nest, than that a hen should continue to perform a motherly part toward the ducklings which manifest the difference of their nature by gladly betaking themselves to the water, of which she has a salutary dread. While the young Cuckoo remains in the nest, it is plentifully supplied with food by its friends, who having their parental feelings excited by its continued demands, cheerfully labour in its behalf. When it can fly, and has left the nest, they continue to provide for and protect it to the best of their power, and this conduct of theirs seems the more strange that it contrasts with that of other little birds, even of the same species, but especially Swallows, which fly after and endeavour to molest it. The young are extremely voracious, and when threatened or seized assume an air of boldness, ruffle their feathers, and show a disposition to fight. When confined they may be fed with raw meat, but seldom survive the winter.

The old birds cease to emit their notes about the end of June, and are never seen beyond the middle of July. The young disappear about the beginning of September. This species is dispersed over the whole of the Continent, extending to Norway and Lapland, from which, and all the northern and middle regions, it departs early in the season. It is also found in Africa, as far as its southern extremity. The old birds arrive in full plumage, and depart without having moulted. The young also take their departure previously to moulting, which, as in the old birds, takes place in winter.

A curious instance of a young Cuckoo's being fed by a Thrush is related by the Bishop of Norwich. "It was taken from the nest of a Hedge-Sparrow, and a few days afterwards, a young Thrush, scarcely fledged, was put into the same cage. The latter could feed itself, but the Cuckoo, its companion, was obliged to be fed with a quill; in a short time, however, the Thrush took upon itself the task of feeding its fellow prisoner, and continued so to do with the utmost care, bestowing every possible attention, and manifesting the greatest anxiety to satisfy its continual cravings for food. The following," he continues, "is a still more extraordinary instance, corroborating the above, and for the truth of which we can vouch in every particular. A young Thrush, just able to feed itself, had been placed in a cage; a short time afterwards, a young Cuckoo, which could not feed itself, was introduced into the same cage, a large wicker one, and for some time it was with much difficulty fed; at length, however, it was observed that the young Thrush was employed in feeding it, the Cuckoo opening its mouth and sitting on the outer perch, and making the Thrush hop down to fetch food up. One day, when it was thus expecting its food in this way, the Thrush seeing a worm put into the cage, could not resist the temptation of eating it, upon which the Cuckoo immediately descended from its perch, and attacking the Thrush, literally tore one of its eyes quite out, and then hopped back: the poor Thrush felt itself obliged to take up some food in the lacerated state it was in. The eye healed in course of time, and the Thrush continued its occupation as before, till the Cuckoo was full grown."

A case of a like nature, but referring to the Cow-Bunting, a small bird, whose mode of propagation is similar to that of the Grey Cuckoo, is related by Wilson, in his American Ornithology. Having taken from the nest of a Maryland Yellow-Throat a young male Cow-Bunting, he "placed it in the same cage with a Red Bird, *Loxia Cardinalis*, who, at first, and for several minutes after, examined it closely, and seemingly with great curiosity. It soon became clamorous for food, and from that moment the Red Bird seemed to adopt it as his own, feeding it with all the assiduity and tenderness of the most affectionate nurse. When he found that the grasshopper which he had brought it was too large for it to swallow, he took the insect from it, broke it in small portions, chewed them a little to soften them, and, with all the gentleness and delicacy imaginable, put them separately into its mouth. He often spent several minutes in looking at and examining it all over, and in picking off any particles of dirt that be observed on its plumage." But this assumption of the office of a nurse has been manifested by many birds of the orders Cantatores, Deglubitores, and Vagatores, with regard to helpless individuals, not only of their own, but of other species; insomuch that it would seem to result from the excitement of the parental instinct affected by the solicitations of the destitute orphan.

In form, as well as colour, the Cuckoo is one of the most elegant of our native birds. Its entire length is fourteen inches, the extent of its wings twenty-three, the length of the bill ten-twelfths. The wings and tail are very long, the former straight and pointed, the latter broad and graduated. The plumage is soft and blended. The mouth wide, the bill shorter than the head, compressed toward the end, pointed, and somewhat arched. The feet are small, and zygodactyle, that is, having the outer, as well as the inner toe, directed backwards. The bill is greyish-black, at the base orange, the iris also orange, and the bare margins of the eyelids gamboge-yellow, the feet orange, the claws ochre-yellow. The upper parts are bluish-grey, more or less tinged with green; the quills dusky, the inner webs of the primaries marked with oblong transverse white bands; the tail greyish-black, glossed with green, the feathers tipped, and along the shafts and inner edges spotted with white. The throat and fore-part of the neck are light ash-grey; the breast and sides white, transversely barred with brownish-black. The young has the upper parts variegated with light red and dusky.

GIZZARD OF THE GALLINACEOUS BIRDS.

We have in two preceding Numbers given a short account of the digestive organs of the Gallinaceous and Rapacious Birds. The structure of the stomach in the former, however, requires more particular notice, as it has frequently been misrepresented. The accompanying figures are taken from the gizzard of the Black Grouse, *Tetrao Tetrix.*

Fig. 2.   Fig. 1.

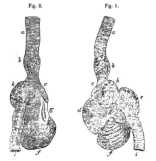

Fig. 1. represents the lower part of the œsophagus, *a* ; the proventriculus, *b*, which may be considered as part of the stomach, it being composed of a great number of hollow glandular bodies or crypts, which secrete an abundant clammy fluid of a greyish colour. This fluid is carried along with the food into the cavity of the stomach, and is by many supposed to be a solvent of the alimentary mass, analogous to the gastric juice in Man and the Mammalia. There is reason, however, for doubting this, and it is probable that the real gastric juice is exhaled by the minute vessels upon the inner surface of the stomach, in the space marked *c*, of which the cuticular lining is soft. The mass of food, lubricated by the mucous fluid of the œsophagus and proventriculus, being deposited in the stomach, is there ground to a pulp by the action of the two lateral or digastric muscles, *d*, *e*, the fibres of which converge toward two tendons, one of which is seen at *g*. The lower part of the gizzard is furnished with a muscle, *f*, less thick, but still powerful, which, after trituration is completed, probably aids in propelling the mass towards the intestine, *i*, while it is prevented from escaping into the œsophagus, *a*, *b*, by the contraction of the superior muscle, *e*. The internal coat of the stomach is thick, tough, elastic, and rugous, and has two much thickened portions opposite the muscles, *c*, *d*, which perform as it were the office of millstones. There is a muscular dilatation or sac at *h*, from which the intestine proceeds. The view here given is that of the parts seen from above. Fig. 2. shows the same parts viewed from the right side,—*a*, the œsophagus; *b*, the proventriculus; *c*, the superior or anterior muscle; *d*, one of the lateral muscles; *f*, the inferior muscle; *g*, the tendon; *h*, the sac from which proceeds the intestine, *i*.

The birds in which this extremely muscular stomach is found are very numerous. All those belonging to the Gallinaceous order, the Pigeons, most of the Gallatores, and the Swans, Geese, Ducks, and Mergansers, exhibit it in its highest degree of development. A less muscular stomach, although formed on the same plan, is seen in the order Passeres or Insessores of authors, but there it presents various modifications. Many, but not all birds, which have a powerful gizzard, are also furnished with a crop, or dilatation of the œsophagus, as is the case more especially with the Rasores and Pigeons; the former of which have extremely large cœca, while in the latter these organs are reduced nearly to the minimum. On the modifications thus presented might be founded an arrangement of birds, which would at least have the merit of placing together those of similar habits.

ON THE GEOGRAPHICAL DISTRIBUTION OF ANIMALS.—NO. V.

(*Continued from page* 142.)

We now pass on to consider the third division of the subject, the shores and marine productions of South Devon, subjects in a great measure apart from those first discussed; and, although we found it convenient to consider the Fauna of Dartmoor, and the uncultivated districts separately, the productions of those spots are not so distinct or peculiar as those of the shores and sea will be found to be. Divisions of this kind, arbitrary as they are, will nevertheless be found useful, because, as the qualities of the elements and constitution of the climate in each may be expected to differ, it might be also ascertained, that the tribes of creatures inhabiting or frequenting them were more or less classified or brought together by reason of those influences. It is further convenient for the purpose of comparison with other spots of similar or dissimilar natures. We must, in the first place, say a few words on the Mammalia.

The chief abode of our Bats is toward the coast, because they principally reside during their daily and diurnal quiescence in the caves and fissures of our limestone rocks, and these occur for the most part toward and on the coast. All our rarer species have been chiefly noticed in caverns on the coast. Thus Montagu found the two species of Horse-shoe Bat in Kent's Hole at Torbay, and one of these has been captured near Plymouth. The Barbastelle has occurred at Kingsbridge, and the Great Bat I have seen several times affecting quarries round Plymouth. The two common sorts seem (perhaps because more plentiful) to resort as freely to old buildings, hollow trees, &c., as to cavities of rocks; but, I believe there is no decided preference shown by these animals for natural cavities of the earth; hollow trees, &c., being the appointed substitutes, where and when caves cannot be obtained. It is the same precisely with the Barn Owl, to which hollow and wild trees form a natural resort, and caves a natural succedaneum; but both with Bats and the Barn Owl, old buildings, such as barns and sheds, are to be considered in some respects as unnatural habitations, because the creation of animals was anterior not only to the erection of such buildings, but to the adoption of this island as a residence by our race. It has, however, been most wisely directed that the instinctive faculties of brutes should not be so definite and so limited in their operation, as to preclude all departure from their more peculiar habits and actions, and hence it has been found, that the construction and situation of the nests of birds have at times varied remarkably from ordinary rule; hence birds have been enabled to sustain themselves on novel food, when driven by stress of weather, or other adverse circumstances, to countries which they have never before seen; hence, when detained by weakness, or the allurement of climate, and supply of food, from making their accustomed migration, they can support themselves against unaccustomed impressions; and hence, amongst a great variety of other instances not only with birds, but with other kinds of creatures, we are enabled to domesticate them, and to cause, by our interference, extraordinary alterations in character and variations, or rather improvements in instinctive powers. With respect, however, to the adoption of our buildings as places of resort, and situations for nestling by birds and other animals, I believe too little has been said by writers, because such irregularities have been made to pass current in their mode of expression, and doubtlessly in the mind of many readers, as the natural habits of such species: whereas an useful lesson might be thence drawn, relative to the faculties and mental operations of the beings around us. With us the natural nestling places of the Jackdaw are on the sea-cliffs, and rocky eminences in general are its natural abodes wherever found. In defect of these, however, Nature has prompted it to make use of hollow trees and rabbit holes in some instances; but towers, and ruined buildings, are such faithful imitations of its native cliffs, that this bird is diffused very generally through our country. With the Wheatear, old walls and heaps of stones answer the purpose of rocks, both for obtaining food and for nestling. This bird chiefly resides during the breeding time on our coasts, and though it frequents quarries and old walls, these situations are never far removed from the sea-side. At other seasons, it either quits us, or frequents fallows, &c., for food. The natural breeding-places of the Martin, with us, are the cliffs, but owing to similarity of position, and that extraordinary dependence on man observed in this and other species, it usually builds against houses, and frequents our neighbourhood. It is remarkable, however, that in the Swift and Swallow so few instances should be on record of their building in other situations besides houses, out-buildings, and churches. I have known a pair of Swallows build a nest in some ivy under a bridge not far from my house, and this too in a loose state, so that it was constantly swinging about with the least current of air, but I am not aware of another instance in this neighbourhood of either of these birds affecting other than the above-named situations; but, in all likelihood, their natural abodes are like the Martin's and other species known to build similarly, such as the Cliff Swallow and Esculent Swallow. It would be easy to cite other cases of adopted nestling-places of the same class, but these are here named because the birds in question seem to belong almost wholly to that portion of South Devon into the productions of which we are now inquiring. The whole character and history of species, and consequently the philosophy of Zoology in general, can only be arrived at by tracing the habits, physical endowments, &c., of animals in each different locality they inhabit; and hence this is one cogent reason for prosecuting the natural history of districts, and comparing and combining the same.

Now, the Otter is an illustration of this. It is usually thought to be a fluviatile animal only; but in Devonshire it is both fluviatile and marine, quite as many residing on the coast, and fishing to a short distance off the land, as on the banks of our rivers. In the former case, they take possession of small hollows in the rocks, and are yearly hunted in these situations near Plymouth. It might be expected that the Seal would find a place among our Devon animals, but I know of no instance of its capture here; though, from its occurrence in Cornwall, we might reasonably expect that it would be found with us also.

The Cetaceous Mammalia of our coast deserve more than usual notice in this place, because these animals have never been properly examined by British naturalists, and because those which have been seen on our coasts have as yet received very limited notice, at least in respect to the number of species. In the first place, it is highly probable that the distribution of this class has never yet been completely understood, and that, so far from their being scarce in the British seas, as usually supposed, they are tolerably common even now, and were doubtless more so formerly. I have a most respectable authority for stating that, in the middle of July 1836, several Whales were seen between the Azores and the Land's End. The crew of the ship in which he took his passage indeed captured one, and one was noticed by him immediately off the Land's End. The species, however, he could not tell. A Whale was found dead off Penzance not long since. Amongst the "rarer fish of Cornwall" is mentioned the Blower or Fin Fish, *Balæna Physalus*, Linn. I have seen a tooth of a Whale thrown ashore a few miles from Plymouth, which proves to belong to the *Physeter Microps*, L. In October 1831, a Sharp-lipped Whale, *Balæna Boops*, L., found dead off the Eddystone, was towed into Plymouth Sound. In 1814, a specimen of *Delphinus Tursio* was captured about five miles up the Dart. The Grampus is found in the Channel, and has been often seen off the coasts of Cornwall; and the Porpoise is common on all our coasts in herds, and frequently enters our estuaries in pursuit of fish. We have here, therefore, an account of six species, or probably more, found off the coasts of Devon and Cornwall.

We have now to conclude the subject of the Birds of Devon by speaking of those on the shores and sea. With regard to the Accipitres, we enumerate the following as more or less peculiar to the coasts:—The Peregrine Falcon, Red-legged Falcon, Osprey, Sea Eagle, and Kastrel. In this tribe of birds we perceive that, besides their disposition to roam and change residence, there are remarkable examples of diversity of habitat in the same species. Thus, while the Kestrel is found pretty commonly in inland counties, with us it is almost wholly confined to the coasts. The Red-legged Falcon, *Falco rufipes*, has only been once obtained here, and then in the Channel. I must here state, that I owe my acquaintance with this last fact, and also a great many others relative to the rarer birds of Devon, to the "Catalogue of the Birds of Devonshire," published by Dr Moore of Plymouth in the Magazine of Natural History. The Peregrine Falcon and Osprey are not very uncommon; but the Sea Eagle is decidedly rare. A few Passerine birds must be mentioned in connexion with the coasts. The Rock Pipit, *Anthus petrosus*, is stationary, its food seeming to consist of the smaller marine insects. The Chough is likewise a coast bird, but is not stationary on the Devon shores, visiting us only in small numbers towards Winter. Both these birds occur in the list of Dartmoor birds prefixed to Carrington's poem of that name. They seem to me to be the only true Passerine birds of our coast; at least, alterations have taken place in the geographical situation of the Wheatear, Jackdaw, and Martin, as before named; in addition to which, these do not confine themselves to one abode, and are, moreover, observed to affect inland rocks equally with shores. Their appetites are not limited to shore productions, and though the Chough is in some other countries an inland bird, it is not so in England, save through accident or necessity. The Raven at times builds on our cliffs. The Crow is noticed very frequently in Autumn and Winter, examining the rejectamenta of the tide. The King-fisher migrates partially to the sea-side in October; and those which are found there remain till Spring. This is another instance of diversity of action in individuals of one species according to situation. It is certainly an ordination of Nature to allow of more extensive dispersion of the whole of the species during the season of greatest want; for by this arrangement of their appetites the removal of some portion of the species to the estuaries and coasts permits a very general though very slight change in the position of all the members of the kind, and at the same time gives to each a more extensive range for capture of prey during the time of necessity. Mr Knapp observes that the universality of the Robin, that is to say, its general dispersion, is remarkable. This is confirmed by my notice of it, though rarely, on the furze of our cliffs, far removed from farms or villages. How remarkably does the history of one species differ from that of another, when thoroughly investigated, and how evident is it that the completion of these histories is essential to the development of those various plans instituted by the Creator for the perfection of his wondrous scheme of nature. There is one of the Gallinaceous birds recorded by Polwhele as a frequenter of the Devon coasts, the Rock Dove, *Columba Livia*, though he refers it exclusively to Combe Martin and Lundy Island on the north of Devon. It is, however, now understood to be an inhabitant of the southern shores, occurring rarely in the caves.

We calculate that there are fifty-six Waders belonging to South Devon, and since twenty-two of these may be regarded as belonging to the terrestrial and fluviatile portion of this class, there is a preponderance in favour of the marine portion. With respect to very many of this class, it is quite impossible, as before remarked, to assign to them unequivocally an inland or marine station. They are either so imperfectly known that their preference is merely suspected by their accidental occurrence in one of these situations on the occasions when seen, or they have been found to resort equally to both kinds of habitats; even in the case of those species which breed on our moors, and appear in Autumn and Winter on the coasts, allowing that they do not at those seasons visit lakes and rivers also, it must cause a doubt on the mind whether it would be natural to refer them to the inland or to the marine class exclusively. In fixing the numbers, however, as above, at twenty-two for the cultivated districts, and the remainder, thirty-four, for the uncultivated parts and the shores, I have been guided thus: I first selected those respecting whose station no doubt could be felt, and then classified the rest by ascertaining, as far as possible, which parts of South Devon they evinced the most preference for, or in the case of the rarer birds by considering that to be their station where each had by good accident been observed, unless it seemed to me that this station was at variance with the general character of the bird, and assumed only by mere casualty. Of the thirty-four, then, I have reckoned two as belonging to Dartmoor, and thirty-two as shore birds. Then, besides the two which are peculiar to the moor, namely, the Great Bustard and Crane, more than two dozen other species have been noticed on that spot, very many of which breed there. Some of these I have considered as belonging to the cultivated parts, and some to the shores, according to the bias they exhibited. Again, of the twenty-two Grallæ of the cultivated parts, seventeen are also at times visitors of the coast. And, lastly, of the thirty-two shore birds about to be named more particularly, only one-half have exclusively been seen on the coasts, and at the mouths of rivers.

REVIEWS.

REMARKS ON THE NATURAL HISTORY PERIODICALS.

The *Magazine of Natural History* for May contains several interesting papers. The first, on the Influence of Man in modifying the Zoological Features of the Globe, by Dr W. Weissenborn, being part of a series, is devoted to the Zubr, *Bos Urus*, which anciently inhabited the whole tract between the Baltic and Haemus, whilst the Black Sea and the Steppes of Russia confined it on the east, and the cold hindered it from penetrating farther in a north-easterly direction, but which "is now restricted to a single habitat, the wild and swampy forest of Bialowieza, in Lithuania." After reviewing and comparing the statements made by authors from Aristotle downwards, he comes to the result, "that the *Urus* and *Bison* are the same animal; and that we have for the creature now called in the systems *Bos Urus* four sets of synonyms, viz. one probably of barbarous origin, but employed by Aristotle, *Bonasus*, *Monassus*: One, it seems also of barbarous origin, *Bison*, *Visen*, *Wisont*, *Wisant*, &c.; one derived from the same root as *Taurus*, (the syllable *or*, or *ur*, of the primitive language, from which the Greek 'ορος, and the very word *origin*, are derived, and which conveys the idea of what is ancient and grand), *Urus*, Gall. and Lat. *Our*, *Auer*, *Ur*, *Auerochs*, Germ., *Ureox*, Engl., *Tur*, Pol., *Tor*, Russ., *Tyr*, Dan.; and one of Sclavonian origin, *Zimbr*, Mold., and *Zubr*, Lith."—"We may, therefore, with much probability, consider the zubrs of the forest of Bialowieza as the only survivors of a species which was formerly found, in great numbers, in the vast swampy forests of the whole of Middle Europe, and perhaps Great Britain, whilst no other bovine animal inhabited the same tract within the historical times." The next paper is by Mr Blyth, Curator to the Ornithological Society, and is entitled "Outlines of a new Arrangement of Insessorial Birds." This arrangement is professedly formed on the basis of anatomy and physiology, and he takes occasion to show some of the more palpable defects of the "preposterous mode of classifying" usually denominated the quinary system. The class of birds appears to him "resolvable into three primary divisions, which might be respectively styled—*Insessores*, *Gressores*, and *Natatores*, that is, Perchers, Walkers, and Swimmers. The class of mammalia might be divided precisely in the same manner, namely, into Perchers, as Monkeys and Squirrels; Walkers, as Horses, Oxen, and Tigers; and Swimmers, as Seals, Shrews, and Whales. If into a single "subclass" are admitted Perchers, Parrots, Swallows, and Woodpeckers, as well as Falcons, Owls, and Pigeons, we see no reason why into another should not be admitted Baboons and Oposums, Mice and Elephants. Mr Blyth's Insessores then include all birds that do not belong to his Gressores and Natatores, for it is impossible to define them otherwise. His Gressores include the Gallinaceous birds and Waders. His Natatores we presume are the palmipede birds. We can scarcely admit that all the Insessores, including Vultures, Falcons, Owls, Crows, Shrikes, Thrushes, Warblers, Cuckoos, Parrots, Swallows, Woodpeckers, Creepers, and King-fishers, are modelled on a single anatomical type; but when the different orders and families are explained, we shall be able to form some idea of the system. In the meantime, it is pleasant to us to observe that from our own small beginnings in the physiological method of arranging birds, there is thus presented a prospect of the science's being rescued from the hands of mere collectors of dry and arsenicated skins. In the third article Mr Newman combats the idea of the antennæ of insects being analogous to the ears of higher animals. "Further Observations on Rules for Nomenclature," by Mr Ogilby, are amusing and clever, but the subject, having been so much discussed of late, and the parties being as pugnacious as ever, sober folks would rather hear of something else.

The *Naturalist* for the same month commences with a pleasantly written paper by Mr R. Adie on the physical power of insects as labourers, and on their architecture. The next paper affords an explanation of a peculiar mechanism in the trachea of Birds, which appears to have been overlooked, and which consists of an arrangement of the rings, so managed that one side of a ring slips over the sides of its two neighbours, while the other side of the same ring slips under their sides. Mr H. C. Watson then corrects an error in the Meteorological Journal of the Royal Society, regarding the minimum of the thermometer in January last. Mr Rylands favours the young entomologists with "farther hints," and earnestly cautions "the student against suffering a passion for collecting to choke the desire of investigating nature." From this results the misfortune which Mr Swainson justly regrets, "that nearly all naturalists are more bent upon increasing the contents of their cabinets, than on studying the economy of those living objects which are perpetually crossing their path." Mr Swainson we presume classes himself among the small number of persons who study nature more than the contents of their cabinets; yet we are not assured that the gentleman may not be decrying his besetting sin. A long list of flowering plants for May occupies four pages and a fifth; the remaining four-fifths being allotted to the "British Swans," of which absolutely nothing is said by "G. L. Lister, Game-dealer," &c. Mr T. B. Hall deals more hardly with Mr Yarrel's "British Birds" than we should feel disposed to deal were we to criticise the engravings. Proceedings of societies, with reports of lectures and various notices, occupy the remainder.

The *Annals of Natural History* contain "Observations on the Coregoni of Loch Lomond," by Dr Parnell; an account of a journey to and residence of nearly six months in the organ mountains in Brazil, with remarks on their vegetation, by Mr George Gardner; continuation of Mr Thompson's Contributions to the Natural History of Ireland; on a new English species of Urtica, namely, *U. Dodartii*, by Mr Babington; and other articles which we have not room to notice.

EDINBURGH: Published for the PROPRIETOR, at the Office, No. 13, Hill Street. LONDON: SMITH, ELDER, and Co., 65, Cornhill. GLASGOW and the West of Scotland: JOHN SMITH and SON; and JOHN MACLEOD. DUBLIN: GEORGE YOUNG. PARIS: J. B. BALLIERE, Rue de l'Ecole de Médecine, No. 13 bis.

THE EDINBURGH PRINTING COMPANY.

THE EDINBURGH

JOURNAL OF NATURAL HISTORY,

AND OF

THE PHYSICAL SCIENCES.

J U L Y, 1838.

ZOOLOGY.

DESCRIPTION OF THE PLATE.—THE EURYLAIMES.

THE genus *Eurylaimus*, first instituted by Dr Horsfield, is composed of birds remarkable for the great size of their bill, of which the upper mandible has its dorsal outline convex, the sides sloping, the tip small and deflected, the notches very small ; the lower mandible dilated at the base ; the bristles generally of moderate size. The tarsi are short, the toes rather small ; the claws compressed, moderately curved, and acute. The plumage is compact, often splendidly coloured. They inhabit the tropical regions of Asia, and are considered as belonging to the family of Flycatchers.

Fig. 1. THE COATDOM EURYLAIME (*Eurylaimus Corydon*), of a brownish-black colour, with the bill blood-red, the middle of the back orange-red, the throat whitish, and a white band on the primaries and tail feathers.

Fig. 2. THE GREAT-BILLED EURYLAIME (*E. nasutus*), which has the upper parts, the chin, and a band on the fore-neck, bluish-black, the rest of the lower parts and the rump deep red, and the scapulars white, inhabits the Indian Islands.

Fig. 3. HORSFIELD'S EURYLAIME (*E. Horsfieldii*), originally made known by Dr Horsfield under the name of *E. Javanicus*, is bluish-black above, vinaceous beneath, the wings and tail patched with yellow. It occurs in Java.

Fig. 4. HORSFIELD'S EURYLAIME. Female. Dusky, spotted with yellow above ; yellowish-white beneath.

Fig. 5. THE HOODED EURYLAIME (*E. cucullatus*), crested, the head and upper parts black, the breast rose-coloured, a white ring on the neck, the scapulars yellow. It is nearly allied to the last species, and inhabits the same country.

DESCRIPTION OF THE PLATE.—THE THROSH-SHRIKES.

THE Thrush-shrikes, constituting the genus *Laniarius*, are distinguished by having the bill less curved at the tip than in the Shrikes, and approaching somewhat in form to that of the Thrushes. They inhabit Africa and India, and are in general remarkable for the beauty of their plumage.

Fig. 1. THE BACBAKIRI THRUSH-SHRIKE (*Laniarius Bacbakiri*), greenish-yellow above, ochraceous beneath, with the head ash-grey, and a semi-lunar black band on the fore-neck.

Fig. 2. The Female is deep green above, paler beneath.

Fig. 3. THE GREEN THRUSH-SHRIKE (*L. viridis*), green above and on the sides ; the throat and breast bright red, a black semi-lunar band on the fore-neck.

Fig. 4. THE WHITE-THROATED THRUSH-SHRIKE (*L. albicollis*), green above, yellow beneath. with the head black, and the throat white.

Fig. 5. THE OLIVE THRUSH-SHRIKE (*L. olivaceus*), green above, yellow beneath, with a black band on the side of the head and neck.

Fig. 6. THE BARBARY THRUSH-SHRIKE (*L. Barbarus*), black above, deep red beneath, the head and neck yellow.

THE AUROCHS OR ZUBR, BOS URUS.—Dr Weissenborn states that this animal, which formerly inhabited a great part of Europe, is now confined to the marshy forest of Bialowicza, where it is protected by very strict laws. The size of the animal in its present confined abode is much less than it was in former times, when it was uncontrolled in its range. A six-year-old male measured seven French feet from the crown of the head to the *tuber ischii*, and its height at the withers was four feet nine inches ; but in 1795, an individual thirteen feet long, and seven feet high, was killed near Friedrichsburg, and another killed in Bialowicza in 1752 weighed 1450 lb. The game-keepers say it lives to the age of fifty years and upwards. Its strength is enormous, and it easily defends itself against the Wolf or Bear, an old Zubr being a match for four of the former. It runs with great speed, but has little bottom, swims with agility, is exceedingly shy, and is incapable of being tamed. Its food consists of the bark of trees, lichens, and various herbaceous plants. Its voice is a short grunt. " On the 19th of October 1836, a hunt was held with a view of furnishing some of the museums in Prussia with specimens ; and Prince Dolgorukow, the governor-general of the province of Balystock, who presided at it, caused it to be conducted with more than usual solemnity. Two thousand drivers and marksmen were assembled, besides an immense number of spectators whom curiosity had drawn to the spot, where a balcony of fir trees had been erected for the prince and his suite. Thither different herds of Zubrs were driven, to be fired at, and the flesh of the largest bull was dressed, to give additional interest to the concluding act of the party, a plentiful dinner."

38

BRITISH BIRDS.—THE GOATSUCKER.

THE Goatsucker or Nightjar arrives in this country from the middle to the end of May, being among the latest of our summer visitants, and departs about the end of September. It is generally distributed, but is nowhere very common, and in many large tracts is not met with. Dr Edward Moore states, that it is " common about the South Hams of Devonshire, where they frequent orchards," and my friend Mr Barclay has met with it near Elgin. It is chiefly found on furzy commons, wild bushy heaths, and broken hilly ground covered with ferns, especially in the neighbourhood of thickets and woods. It is rarer in Scotland than in England, which is the reverse with the Cuckoo, a bird in some respects similar in its habits, and which appears to be much more plentiful in the wild valleys of the north, than in the cultivated plains of the south.

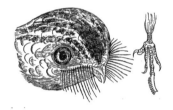

The bill of this bird is extremely small and feeble, the mouth excessively wide, the palate, flat, covered with a smooth membrane, which is transparent, as are in some measure the bones. The tongue is very small and triangular. The gullet is rather wide, the stomach large, round. membraneous, its muscular coat being composed of fasciculi, as in the Owls and Cuckoos, the intestine short, the cœcal appendages small. The eyes are very large, as are the apertures of the ears. The head is very large, depressed, and flattened above. The feet are extremely small ; the tarsus very short, anteriorly feathered except at its lower part ; the first toe very short, slender, and directed inwards, the second and fourth about equal, the third much longer, the anterior toes connected by membranes as far as the second joint ; the claws very small, arched, and compressed ; that of the middle toe is proportionably longer, and has its inner convex edge expanded and pectinated, being cut in two by parallel slits. The plumage is full and blended ; the wings very long and narrow ; the tail very long, often broad, rounded feathers. The bill and claws are dusky, the feet flesh-coloured ; the general colour of the upper parts is ash-grey, minutely dotted and undulated with dusky, and variegated with brownish-black and pale yellowish-red, the head and back being marked with elongated spots of the latter colour. On the inner webs of three of the primaries is a large roundish white spot ; the two lateral tail-feathers have also a large patch of white at the end. The lower parts are transversely barred with dull reddish-yellow and dusky ; and on the throat are some white feathers. The length is eleven inches, and the extended wings measure twenty-three.

This unfortunate bird has a strong claim on our sympathy, on account of the manner in which all its actions and habits have been misrepresented. The ancients accused it of milking goats, and thus it received the names of Caprimulgus and Goatsucker, which it retains to the present day. Then it was alleged to be so awkward as to be obliged to fly with its mouth wide open, and so slovenly as to need an instrument on its foot with which to cleanse its chops. Lastly, so malignant were its traducers, as to hint that it could not see like other creatures, but was obliged to gape widely, and then turn its eyes downward to look through the roof of its mouth, for which purpose that part was made thin and transparent.

The substances which I have found in the stomach of this bird were remains of coleopterous insects of many species, lepidoptera, and sometimes larvæ. I have seen the inner surface slightly bristled with the hairs of Caterpillars, as in the Cuckoo. Towards evening the Goatsucker may be seen skimming along the edges of woods with a light and buoyant flight, winding in varied curves, in the manner of a Swallow, but with less velocity, and by its noiseless motions also reminding the observer of the Owls. As it proceeds, it now and then emits a shrill squeaking cry. It is seldom that more than one or two individuals are seen at a time; but Montagu remarks that he observed "in Scotland eight or ten on wing together in the dusk of the evening, skimming over the surface of the ground in all directions, like the Swallow, in pursuit of insects."

During the day it generally rests on the ground, among furze or fern, or on the branch or bough of a tree, on which it reposes in a direction parallel to its axis. This arises from the disposition of the toes, which is such, that it cannot securely grasp a branch in the ordinary way. When disturbed while on the ground, it flies off with a wavering buoyant flight, and generally alights on a tree, if there be one in the neighbourhood.

The eggs, which are two in number, broadly elliptical, whitish and clouded with ash-grey and brown, are deposited on the bare ground, among furze, heath, or fern. The young are densely covered with long whitish down. During the breeding season, according to Montagu, "it makes a singular noise, like the sound of a large spinning wheel, and which it is observed to utter perched, with the head lowermost."

The serrature of the middle claw of this and the other species has elicited various conjectures as to the use of so curious a structure. Several persons have supposed or imagined it to be for the purpose of enabling the bird to clear away from between the bristles that fringe its mouth, the fragments of wings or other parts of lepidopterous insects, which, by adhering, have clogged them. This at first sight seems a remarkably plausible account of the matter, but a very little reflection, with a slight inspection of the parts, will suffice to show its futility. The bristles are large, strong, and placed at some distance from each other. The teeth of the claw are extremely thin, and very close, being separated only by mere chinks. The claw then cannot act as a comb, because one of the bristles is as broad at the base as two or three of the teeth, so that it cannot enter between them; and although it tapers away toward the end, yet even there it is too wide to be insinuated. But, although the claw may not act as a comb, it may be said that its serrated edge will more readily than a continuous edge catch hold of any thing stuck between the bristles. This is likely enough; but then the species of the genus Podargus or strong-billed Goatsuckers have similar bristles, but are destitute of clefts on the claw. Gannets, Herons, and other birds, that have no bristles, have yet a serrated claw. Therefore, the serrature is not intended for the purpose of clearing the bristles. Yet it may be quite true that the Goatsucker uses its claw to produce that effect; but it is not less true that Parrots, Finches, and other birds, having no such serrature, employ their claws for scratching the parts about the head. And so another reason must be sought for.

The young Goatsucker has at first no serrature on its claw, any more than the young Gannet. One fully fledged, and shot about the end of September, now before me, has the toe scarcely half the length of that of an old bird, and with only five teeth, the old bird having ten. The chinks in the young bird's claw are less deep than those in that of the old bird. A young fledged Gannet shews the same circumstance. All birds whose middle claw is serrated, have that claw elongated, and furnished with a very thin edge. It therefore appears that the serration is produced by the splitting of the edge of the claw, after the bird has used it, but whether in consequence of the pressure caused by standing or grasping can only be conjectured.

The Flycatchers, and other birds of the same family, which have strong bristles, intended for the same use, have not serrated claws; yet if their bristles become clogged, they no doubt will clean them in the same manner.

It appears that the use of the serratures is not that of clearing the bristles of the scales of lepidopterous insects, because fish eating birds without any bristles have similar serratures; but there is no reason for doubting that Goatsuckers brush away adherent matter with their claws, just as other birds do, the domestic fowl, for example.

Another supposition is, that the serrature enables the Goatsucker to hold more securely a large insect which it has caught with its foot. And observers have stated that they have been pretty sure of having seen that bird, when flying, raise its foot to its mouth, as if, in the manner of a Parrot, to carry to it an insect. It may be so; but as yet no one has quite satisfactorily seen a Goatsucker catch a moth or a beetle with its foot; and this cannot be the use of the serrature, for the Gannet and Heron, which do not seize their prey with their feet, have serrated claws.

The notion of a bird's flying with open mouth, for the purpose of seizing its prey, is preposterous. It has been alleged that Swifts and Swallows do so; but I have satisfactorily ascertained that they do not; and there is no reason to suppose that Goatsuckers are so awkward as to require to keep their jaws constantly wide open lest their prey should escape them.

But the most absurd notion of all is that expressed by Mr Selby as follows. "The membrane that lines the inside of the mouth is very thin and transparent, particularly opposite to the posterior part of the eye, which organ is pretty clearly discernible through the membrane. As the mouth opens to such great lateral extent, it has been suggested that the bird may possibly be capable of turning the eye in its socket, so far as to look through this almost transparent veil in a straightforward direction, when the mouth is extended in its nocturnal flights. I have consequently directed my attention to this point, but as yet without any satisfactory result." Indeed, it was unworthy of exercising the observation of so sagacious an ornithologist. How desperately imaginative must those persons be, who, not content with allowing a bird to seize its prey like other birds, by opening its bill when it comes up to it, must represent it as flying about with its mouth wide open, and instead of using its eyes as all other birds use them, turning them round, to the imminent danger of separating the optic nerve, so as to spy moths and beetles through a window in the palate!

OBSERVATIONS ON RABIES IN DOGS, OXEN, HORSES, PIGS, AND SHEEP. BY DR WAGNER.

The following is a condensed account of some important observations made by Dr Wagner on the diseases commonly called Rabies or Madness in domestic animals.

1. *The Dog.*—Dread of water is not always present, nor is the habit of gnawing wood, straw, or hides, or snapping in the air, a sure symptom. An appearance of dejection and shyness, drooping of the head and tail, flashing of the eye, foaming at the mouth, refusal of food and drink, are suspicious, but not peculiar to the disease. The only decisive indication of it is a dog's running about and recklessly attacking men and animals, especially those of its own species. In a rabid state, some dogs will even leap over high fences, in order to reach dogs or cats which they discover to be on the other side, thus showing that smell, sight, and hearing, are still unimpaired. Others will sneak along a wall, or run forward in a straight line, attempting to bite whatever they meet with. Some again never move from one spot, but gnaw at every thing within reach, snap at vacancy, and ultimately refuse food and drink, or if they attempt to partake of either, appearing incapable of deglutition. In such cases, paralysis of the loins and posterior extremities appears to have taken place at the outset. The moment a dog evinces any appearance of illness, he ought to be locked up, or fastened to a strong chain. When it begins to gnaw wood, to show a dull eye, or snap at animals, to bark hoarsely, to attempt to run away or break its chain, to eat and drink in a snapping manner, to appear alternately lively and sulky, to disregard its master's call, to growl and snarl at well known persons, it ought to be killed, for no doubt can remain of its being rabid.

2. *In Oxen.*—The disease occurs in horned cattle in two distinct forms. In the first, the animal loses its appetite, eats and drinks by fits only, appears at times as if suddenly stupified, recedes from the manger, forgets to chew, then chews on, listens keenly to any noise, notices every object, but continues to obey its keeper. Borborygmi, however, are heard, and sometimes there is slight straining or disposition to tenesmus. In the open air, the animal ceases to graze, seems as if lost, and strays from the herd, but generally allows itself to be led back. The symptoms increase: hunger and thirst are not experienced, the eyes flash at times, without seeming inflamed, the animal seldom lows, but when it does, either emits a hoarse noise or a clear and strong cry. The borborygmi increase, the animal licks some parts of its body, especially one of the feet, where probably it has been bitten, until it is excoriated and bleeds. Paralysis of the loins ensues, the animal remains lying, can only rise with great difficulty, and totters on its hind-legs; an increased straining is remarked, and is followed by hard but ultimately thin evacuations; it moves its head from side to side, licks at cloth or fur, which it by degrees gets between its teeth and tears. The secretion of milk, which has been diminishing, now ceases. On the sixth or seventh day, the animal is unable to stand, refuses food, being unable to swallow; the straining at the rectum continues; but a dread of water is never observed. Sometime between the sixth and ninth day, it sinks on one side, commonly the left, the head stretched backwards, but the eye still lively and uninflamed. The trunk continues motionless, but the legs are languidly agitated, until the animal dies. In this form of the disease there is little danger in approaching the animal. The other form commences in the same manner. In the stall, the animals recede still more from the manger, and will try to break the rope with which they are fastened. Their bellowing is not frequent, but prolonged, and its tone clear; they scratch the ground with their fore-hoofs so violently as to throw the dung to the roof, and with their hind-feet kick at any one who approaches. The paroxysms are periodical, and in their intervals little is to be observed, except a reluctance to feed or drink. About the fourth day they will snap every kind of fastening during the paroxysm, attack and gore all who approach them, rage about in the stable, and gnaw the cribs and other objects, until paralysis of the limbs supervenes, when they fall, and ultimately lie on one side. On the seventh, eighth, or at most the ninth day, they die. On dissection, nothing unusual is to be discovered, excepting extreme distension of the gall-bladder.

3. *In Horses.*—The symptoms are at first unwonted activity, kicking and biting of their fellows, refusal of food and drink, frequent nodding or jerking of the head, and a fiery eye. On the second or third day the symptoms have greatly increased in degree, the animals become furious, and strive to bite and kick, so that it becomes necessary to destroy them.

4. *In Pigs.*—Only one instance occurred to the author. A fattening boar had been bitten by a mad dog four days previously to his seeing it, but it already raged with such violence that he could only observe it through the crevices of the stye. It gnashed its teeth, leaped against the sides of the stye, and threatened by its violence to burst through them. A butcher, who killed it with a hatchet, cut it up, contrary to orders, and exposed it to sale, but no mischief resulted.

5. *In Sheep.*—Five sheep, which were bitten by a mad dog, all became affected after the lapse of a few weeks, left off grazing, and dispersed the flock by indiscriminate attempts to butt and mount upon their fellows. They were all killed save one, which, being confined in a stall, kept quietly staring at the wall, but on hearing the slightest noise turned to the direction whence it proceeded, and jumped up against the sides of the stall. It neither ate nor drank, but the proprietor declined to wait the natural termination of the disease.

Country people frequently hold fast the tongue of the rabid cattle with one hand, while they thrust the other into the throat to endeavour to force nourishment into the stomach; yet, although the hands and arms could scarcely on all such occasions be entirely free from injury, and the people neglected to wash their hands, the author never witnessed any evil effects to result. However, he knew a farmer who died of rabies merely from having washed out the wound inflicted on a pig by the bite of a mad dog. He has frequently known the milk of rabid animals to be taken without detriment; and in two instances the flesh of rabid oxen, which was clandestinely eaten, proved innoxious. But, at a period when many cattle perished of rabies, the instances of canine madness became unusually numerous. He concludes with remarking, that having witnessed many instances where the bite of decidedly rabid animals has produced little or no effect upon the human subject, although the remedies employed were merely such as were suggested by superstition, he infers that in the human species a predisposition to hydrophobia rarely exists.

ON THE GEOGRAPHICAL DISTRIBUTION OF ANIMALS.

(Concluded from p. 149.)

DEVONSHIRE boasts of all the English marine Grallæ except seven, one of which, the Red Coot-foot, does not extend in range beyond Orkney, and the rest are either stragglers or great rarities. Among the rarer Grallæ of our shores, the following deserve enumeration: the Great White Heron, Purple Heron, Little White Heron, Freckled Heron, Night Heron, White Stork, Black Stork, Little Bittern, Spoonbill, Ibis, Brown Snipe, Pigmy Curlew, Temminck's Sandpiper, Little Stint, Sanderling, Phalarope, Greenshank, Wood Sandpiper, Stilt, Avocet, and Spotted Redshank; and if the records of the occurrence of these birds in England be consulted, it will be seen that not only are they of extreme rarity, but that some of them have occurred more frequently in Devon than elsewhere. Some of our other Grallæ also are tolerably common, while in other parts they are scarce. What then are the reasons of our possessing so very many rarities, and of having so many individuals of the less rare birds of this division of the Grallæ? Is it because our climate is so genial and uniform, because our shores and harbours are so suitable for their sustenance and retreat; on account of our southern situation; or from all these causes combined? The Oyster-catcher seems partial to the rocks and portions of the coast far out towards the open sea, where it occurs in small parties in autumn and winter. The Ringed Plover has been supposed to repair to other countries on the occurrence of the winter's cold; but in Devon, I am quite sure, it resides on the coasts of our estuaries through the winter. The common Godwit, *Limosa rufa*, has been regarded as a species arriving in this country in autumn, and departing in spring; yet, besides having received this species during winter, I have had it on the 10th of May (1830), and, according to Dr Moore, four out of a brood were shot on the estuary of the Tamar in June 1828.

The web-footed birds of Devon are very numerous, compared with those of other maritime counties; and yet, if we regard Devon relatively to its marine ornithology alone, the greatest number of deficiencies will be found amongst the present tribe; for whilst we are deficient in seven of the marine Grallæ, or nine Waders altogether, out of 39, or 65 altogether, we are wanting in 13 of the Palmipedes out of 67, or 75 inclusive of those allowed as inland birds. This deficit of 13 makes up the number 37 formerly mentioned as the amount in which Devonshire was wanting in respect of the whole of the British Birds. These deficient species are either rarities, or such as are limited to the northern isles; so that we may fairly state that we possess all the English Birds any way common, and which are not by the laws of their physical distribution confined to more northern abodes, and further that we have a very large proportion of those rare birds which occur, however sparingly, in nearly all parts of our island, or on the other hand, have been noticed a very few times in the whole of the country. The following are the rarer birds of this class observed with us:—Crested and Red-necked Grebes, Black-throated Diver, Black Guillemot, Little Auk, Puffin, Cinereous Shearwater, Fork-tailed Petrel, Skua (one shot in September 1831 off the Eddystone), Arctic Jager, Glaucous and Little Gulls, Sandwich, Arctic, Lesser, and Black Terns, Goosander, Merganser, Ferruginous, Eider, Scoter, Velvet, Golden-eye, Harlequin, Long-tailed, Garganey, Pintail, Gadwall, and Shoveller Ducks; Bean, White-fronted, Red-breasted, Beruacle, and Brent Geese, and Wild Swan. The occurrence of so many of the rarer pelagic birds on our coasts seems to allow of readier explanation than that of rarities from amongst other tribes, because we find that the former are bestowed on us in most profusion and very frequently only, on the occasion of storms, or of very severe cold. A long-continued or violent squall from the south-west has afforded us specimens of the Little Auk, the two Petrels, and the Phalarope. A severe winter confers on us specimens of the Wild Swan, Bean, White-fronted, and Brent Geese; Pintail, Golden-eye, Pochard, and other species of Ducks. The object of nestling has likewise influence with some; but winter is the season in which by far the greater number of the web-footed species appear here; and not only do many of those reared in more northern countries migrate thus far south, but others come from their breeding places in our own island. Thus, the Gannet repairs to us in winter from Lundy Island on the north of Devon, and the numbers are greatly increased if the season be severe. The Shieldrake breeds in Braunton Burrows in the north of the county, and many individuals appear here in winter. The Red-legged Gull is abundant at that season, having come from some of the northern English counties, where it breeds; it is scarce in summer, but one was shot near Plymouth in the first week of March 1835. The Herring Gull breeds here, and after that its numbers are lessened on our coasts until the next spring. The Shearwater and the Puffin breed on Lundy Island, and are seen off our shores, and in the channel, chiefly in winter. A great many of the water birds, however, occur on our coasts without our being able to assign any cause for their appearance, as they have arrived independent of storms or severe cold. The Little Gull has been obtained in England seven times, and five of the birds were from Devon. The rarer kinds of Tern mentioned above have occurred without assignable causes, besides many other birds; but possibly the security of our bays and harbours, together with the mildness of climate, may be the attractions to such of them as lead a wandering life.

The geographical distribution of fishes is so very imperfectly known, that but very little can be offered respecting it. About 150 species have been recorded as British, and of these about 30 have hitherto been taken only on the coasts of Devon and Cornwall. About 60 species have come under my observation, or are known to occur on the coast. Many approach our shores at fixed periods, and of these none is more interesting than the Pilchard, on account of its numbers and its importance as an article of winter food to the poorer inhabitants. It appears in August, and generally remains till the end of September. Its comparative scarcity with us for the last few years is remarkable.

Our subject gradually loses interest as we descend to the lower tribes. Of the Mollusca we possess a considerable number. They seem to predominate, however, in those portions of the class which characterize bold, rocky shores; and possibly the same remark holds good relatively to other tribes of marine animals. It is likely that this will account for the same animal being frequently found on shores of the same character, though far distant. Thus, we very often meet with shells stated to have been taken on the coasts of Devon and Shetland, so that although the geography of shells is so very intricate, it may yet be detected that situation has much control over it. The British Marine Mollusca at present may be computed at 497, of which number nearly 60 are recorded as peculiar to Devon. But, as in the case of Ichthyology, much may be set down to the want of attention to the science in other parts, so it appears that in Conchology we have been peculiarly fortunate in the possession of such observers as Montagu and Turton.

Since the time of Montagu, but little has been done in illustration of the remaining marine tribes of Devon, the Radiata, of which, however, at the present time we can boast of more than twenty species peculiar to the Devon shores; and I have certainly no doubt that this number might be greatly increased by diligent and keen research. The sponges of our coast in particular require illustration; and I am of opinion that not more than three-fourths of them have been named. We see frequent instances, among the lower tribes, of species peculiar to the southern shores of England, representing, by their general similarity, species peculiar to more northern stations. But altogether these facts require elucidation and careful consideration; and no naturalist should deem such matters unworthy of his attention, upon finding that detail of this kind is not merely essential to the development of the laws of animal geography, but that it is requisite to complete the history of species, and to become acquainted with many of the laws regulating their existence and their actions, and in general operation upon the entire series of living beings.

We have shewn that our county yields to none in the importance of its Fauna, and this chiefly because of the peculiar eligibilities of its physical conditions. I do not know, therefore, that it would be of use to inquire into the relative proportions borne between the various tribes, or to institute a comparison between the animals of the south of Devon, and those of any given spot of the same extent. An answer to the former question will be found only by tracing the dependence observed in nature from the inorganic kingdom, to the highest conditions of organization. A competent reply, indeed, cannot be given till our knowledge of the laws of life shall be greatly increased, and till we recognise, as the denizens of our country, hundreds of creatures which have as yet escaped notice. A comparison of our animals with those of any other district would avail nothing, since any enumeration of species does not imply a knowledge of the conditions which influence their situation, or control their limits. In framing a list of animals inhabiting a given spot or country, it would be very right to apply the principles of zoological geography. To speak of the frequency or scarcity of animals independent of these, is to betray ignorance to those who can judge, and to perpetuate error to those who would learn. It is not questionable but that one half of the lists published have been formed without regard to the circumstances causing the residence or visits of animals; nor is it doubtful, but that in asserting that an animal is scarce or frequent, the authors of these lists have overlooked the fact, that it is one scarce or frequent in the country as a whole, or scarce or frequent in many countries conjointly. Supposing a bird stated in books, and known to be found generally as a common inhabitant of England, to be recognised as such in any provincial list; it is clear that no knowledge is thus communicated; and supposing that in any other district it was found not quite so common, and yet reported as common in the ornithological catalogue of that district, it is obvious that the truth is kept back. In short, the manner of these communications is altogether far too general, and deficient in the necessary precision. If such terms as "common," "scarce," and the like, be not used relatively, and if there be not precision used in referring to the occurrence of animals, but little information is imparted by these lists; and I think also that their value would be doubled, if, to the bare intimation of the frequency or scarcity of occurrence relatively to the aggregate of each animal respectively, there were added the causes in operation, and a reference to the conditions by which the number is controlled.—J. C. B., Yealmpton, October 5, 1837.

BOTANY.

TEA-TREE.—In a letter from the Abbé Voisin to M. Stanislas Julien, we find a statement which proves that the Tea-tree may be cultivated in our northern climates. He has resided twelve years in China, near the frontiers of Thibet, in which country all the species of tea are successfully cultivated in the plains, as well as on the mountains; although the degree of cold there much exceeds that of our winters, and the snow never melts before the end of April. Twenty-four treatises concerning tea have been composed in Chinese, from the seventeenth century to the present time, and which contain all the requisite instructions for the culture and preparation of this plant, and will be translated by M. Voisin, if required in Europe.—*Athenæum*.

VANILLA.—M. Charles Morren has succeeded in raising the Vanilla in France, and making it produce fruit. He placed the plants in coke, strewn with the remains of rotten willow wood. In this situation the same plant bears fruit only every alternate year, and twelve months elapse between the fecundation of the flower and the maturity of the fruit.—*Ibid*.

CUSCUTA EPILINUM—A BRITISH PLANT.—Mr Bowman states in the Magazine of Natural History, that having, in 1836, gathered a species of *Cuscuta* upon flax, in a field near Ellesmere, Shropshire, and forwarded specimens to Sir W. J. Hooker, the latter found it to be the *Cuscuta Epilinum* of Weihe. It "may be distinguished from *C. Europæa*, by its simple not branched habit, and by its very pale capitula or heads, which are without any of the rosy tinge of the latter. These heads consist of fewer flowers than in the latter species (about five), and these are large, fleshy, and succulent. Both the heads and their component flowers are more decidedly sitting than in *C. Europæa*; the heads are subtended by a membranous, obovate, reflexed bractea, of a reddish-brown colour; but there is no bractea under each individual flower. The calyx is large and spreading, its segments thick and deltoid, almost as long as the corolla, which has very acute segments and a globose tube, even before the enlargement of the germen. The filaments are also very short and very acute; beneath each of which, at the base of the corolla, is inserted a broad membranous scale, whose jagged tips do not reach so high as the insertion of the stamens. The capsule is globose, two-celled, the cells two-seeded, and the seeds subtriquetrous from compression as they swell, covered with chaffy granulations and deeply pitted. It is supposed to grow exclusively on flax, and to have often been mistaken for *C. Europæa*.

GEOLOGY.

Submarine Volcano.—On the 25th of last November, the captain and passengers of the brig Cæsar, from Havre, on passing the bank of Bahama, saw an enormous fire, which increased till it had tinged the whole of the sky and part of the horizon. It was kept in sight for four hours, and could only be accounted for as proceeding from a submarine volcano. On the 3d of January, the captain of the Sylphide, also from Havre, being on the same spot, found the sea disturbed, and whitish in colour, which he attributed to the same cause.

Natural Soda Fountain.—An American Missionary, Mr Spalding, gives an account of a natural soda fountain which he and his party passed, on their route across the Rocky Mountains, at the distance of three days' journey from Fort Hall. One of the three openings of this fountain "is about fifteen feet in diameter, with no discovered bottom. About twelve feet below the surface are two large globes, on either side of this opening, from which the effervescence seems to rise. However, a stone cast in, after a few minutes throws the whole fountain into violent agitation. Another of the openings, about four inches in diameter, is through an elevated rock, from which the water spouts at intervals of about forty seconds. The water, in all its properties, is equal to any artificial fountain, and is constantly foaming and sparkling. Those who visit this fountain drink large quantities of the water with good effect to health.

CHEMISTRY.

Gold.—The medical properties of gold have lately occupied the attention of M. Legrand; and he is of opinion that this metal, reduced to an impalpable powder, its metallic oxides, and the perchloruret of gold and sodium, possess in a very high degree the property of restoring vital strength, and of increasing the activity of the organs of digestion and nutrition.

Lead.—Chemists have long turned their attention towards the different combinations of water and acetic acid, with oxide of lead, and which are so valuable to medicine, the arts, and analysis. M. Payer has discovered a new acetate, and an equally new combination of water and protoxide of lead. In the course of his researches, he has been able to explain several phenomena hitherto unknown, and which are highly interesting in the matter of analysis.

REVIEWS.

Yarrell's History of British Birds. London: Van Voorst.

It has been alleged, that, in the recent numbers of this work, a great falling off in the execution of the engravings has been observed; but, on carefully reviewing the whole, and comparing the figures with each other, we are unable to perceive either deficiency or improvement in those that have lately appeared. As a series, we cannot but consider them as extremely beautiful, although, as in all other productions of art, deficiencies may be pointed out. In form, and especially with regard to the bill, they are, in our opinion, generally superior even to Mr Gould's much larger figures; and there is scarcely any of them that a person moderately acquainted with the species could fail to recognize. The heads of some, however, as of the Black Redstart, and especially the Grasshopper Warbler, are obviously much too small. The information afforded in the descriptive part continues to be of that condensed and carefully elaborated character which it has all along presented.

Essays on Natural History, chiefly Ornithology. By Charles Waterton, Esq., Author of "Wanderings in South America." With an Autobiography of the Author. London: Longman.

This is merely a reprint of various articles that have appeared in Mr Loudon's Magazine of Natural History, some of which, however, are highly interesting, as affording accurate information respecting the habits of our native birds, written in a very pleasing, although diffuse and discursive style, of which the following, painful as the subject is, may be taken as a good example:—"Sad and mournful is the fate which awaits this harmless songster (the Chaffinch) in Belgium, and in Holland, and in other kingdoms of the continent. In your visit to the towns in these countries, you see it outside the window, a lonely prisoner in a wooden cage, which is scarcely large enough to allow it to turn round upon its perch. It no longer enjoys the light of day. Its eyes have been seared with a red-hot iron, in order to increase its powers of song, which, unfortunately for the cause of humanity, are supposed to be heightened and prolonged far beyond their ordinary duration by this barbarous process. Poor Chaffinches, poor choristers, poor little sufferers! my heart aches as I pass along the streets, and listen to your plaintive notes. At all hours of the day we may hear these hapless captives singing (as far as we can judge) in apparent ecstacy. I would fain hope that these pretty prisoners, so woe-begone and so steeped in sorrow to the eye of him who knows their sad story, may have no recollection of those days when they poured forth their wild notes in the woods, free as air, 'the happiest of the happy.' Did they remember the hour when the hand of man so cruelly deprived them both of liberty and eyesight, we should say that they would pine in anguish, and sink down at last, a certain prey to grief and melancholy. At Aix-la-Chapelle may be seen a dozen or fourteen of these blind songsters, hung out in cages at a public-house, not far from the cathedral. They sing incessantly, for months after those in liberty have ceased to warble; and they seem to vie with each other which can carol in the loudest strain. There is something in song so closely connected with the overflowings of a joyous heart, that when we hear it, we immediately fancy we can see both mirth and pleasure joining in the party. Would, indeed, that both these were the constant attendants on this much to be pitied group of captive choristers! How the song of birds is involved in mystery! mystery probably never to be explained. Whilst sauntering up and down the Continent in the blooming month of May, we hear the frequent warbling of the Chaffinch; and then we fancy that he is singing solely to beguile the incubation of his female, sitting on her nest on a bush close at hand. But on returning to the town, we notice another little Chaffinch, often in some wretched alley, a prisoner with the loss of both its eyes, and singing, nevertheless, as if its little throat would burst. Does this blind captive pour forth its melody in order to soothe its sorrows? Has Omnipotence kindly endowed the Chaffinch with vocal faculties, which at one time may be employed to support it in distress, and at another time to add to its social enjoyments? What answer shall we make? We know not what to say. But be it as it will, I would not put out the eyes of the poor Chaffinch, though by doing so I might render its melody ten times sweeter than that of the sweet Nightingale itself. O that the potentate, in whose dominions this little bird is doomed to such a cruel fate, would pass an edict to forbid the perpetration of the barbarous deed! Then would I exclaim, O king of men, thy act is worthy of a royal heart. 'That kind Being, who is a friend to the friendless, will recompense thee for this.'" Truly, Mr Waterton, thou art a kind-hearted man; but how, seeing thou abhorrest the cruelty of burning the eyes of birds, hast thou aimed so many red-hot shafts at a brother ornithologist? Let the reader compare the above with some criticisms contained in the Essays, and he will be led at least to suppose that "birds of a feather" are not always apt to be friendly. The autobiography is not the least interesting part of the work.

A Geographical and Comparative List of the Birds of Europe and North America. By Charles Lucien Buonaparte, Prince of Musignano.

This Pamphlet of 67 pages, containing merely a list of the names of the birds of two portions of the two great continents, the booksellers value at five shillings. The species found in Europe are 503, those in North America 471. As to the genera, it is useless to speak of them, as these imaginary groups, formed on no principle whatever, have existence merely in the realm of fancy. Wilson and Audubon, the principal discoverers of American species, have scarcely a single name of all that they imposed left to them. A great man in science erects an edifice, beautifully proportioned, as he thinks, and the next giant in science who comes to the building ground, just quietly sweeps it away to make room for a new pagoda; picking up, however, most of the stones, on which, when aptly disposed, he inscribes such magic characters as Linn. Ill. Vieill. Briss. Dum. Brehm. Sav. Cuv. Vig. Bechst. Boie. Tem. Lath., with a profusion of Nob, Nobs. In this catalogue the genera are reduced to the minimum size; and the ornithology of Europe and America is thus "brought up to the present level of the science." The species which were generally understood to be common to both continents, such as Falco Peregrinus, Corvus Corax, Corvus Pica, are greatly reduced in number, the American Peregrine being named Falco Anatum, the Raven Corvus Catatotl, the Magpie Pica Hudsonica; but as distinctive characters are not given, we are unable to imagine the grounds on which some of these distinctions are made. The author professes to adopt always the specific name used by Linnæus, and then presently converts Falco *Buteo* into Buteo *vulgaris*, Falco *Milvus* into Milvus *regalis*, Strix *Otus* into Otus *vulgaris*, &c. However, he certainly does retain a large proportion of the Linnæan specific names, for which he merits the thanks of the ornithologists. The figures of the works of Audubon and Gould are "quoted as types of the species under consideration;" but while the latter is characterized as "the most beautiful work on ornithology that has ever appeared in this or any other country, of the former it is stated merely that the merit of M. Audubon's work yields only to the size of his book." What this phrase is intended to signify, we cannot comprehend; but we will maintain against prince or peasant, that Mr Audubon's drawings of birds, being taken generally from entire individuals, and executed in a peculiar manner, cannot be with truth placed on a level with those of Mr Gould, of which the forms are very frequently defective, and the parts disproportioned. At the same time, Mr Gould's work is truly beautiful; but Mr Audubon's is magnificent, not merely in size, but in character. Both works, however, are less useful to the world than a good octavo volume of plain letterpress would be, for they are beyond the reach of ordinary students. To those who already know the birds of Europe and America, the prince's List will be useful, but to the student it is a sealed book. A comparative Catalogue, having ordinal, generic, and specific characters, is a desideratum.

The Naturalist's Library. Vol. X. Flycatchers. By William Swainson, Esq., with a Memoir of Baron Haller.

The Memoir is certainly the best written and not the least interesting part of the book. The Muscicapidæ, sectioned as usual into fives, and analogised into Raptorial, Rasorial, Grallatorial, Natatorial, and Insessorial, will be found by those who have adopted this fanciful system to be treated in a very interesting manner. Even they who have other views will find it advantageous to give this essay a careful perusal, as the elucidation of affinities of any kind must be useful. But the work is obviously intended, not for the student, so much as for the adept. The thirty-two plates illustrative of the text are remarkably well executed, the drawing, engraving, and colouring, being all of a superior character, and the resemblance of the figures to their originals being found by us perfectly satisfactory, although, it is true, we have only prepared skins of some of the species to compare with them, and we suppose the author can have had nothing better. The decorations are also much to our taste, the scenery introduced being beautiful and appropriate. The illustrations of this volume, in short, are among the best of the series; and as pleasing pictures, are worth the price even to those who can understand little of the descriptions. The conductor and publisher certainly deserve praise for their exertions in rendering the series of essays not merely amusing, but more or less instructive, by employing individuals of acknowledged talent, such as the author of the present volume, of whose extensive knowledge and ingenuity, however much we may differ from him as to the most useful method of examining birds, we have, without presuming to pass judgment upon a master in science, a very high opinion.

Edinburgh: Published for the Proprietor, at the Office, No. 13, Hill Street. London: Smith, Elder, and Co., 65, Cornhill. Glasgow and the West of Scotland: John Smith and Son; and John Macleod. Dublin: George Young. Paris: J. B. Balliere, Rue de l'Ecole de Médecine, No. 13 bis.

The Edinburgh Printing Company.

THE EDINBURGH

JOURNAL OF NATURAL HISTORY,

AND OF

THE PHYSICAL SCIENCES.

AUGUST, 1838.

ZOOLOGY.

DESCRIPTION OF THE PLATE.—THE BABOONS AND MANDRILLS.

The Baboons, or *Cynocephali*, are generally large and ferocious species of Monkeys, having cheek-pouches, callosities on the hips, an elongated and truncate muzzle, and a tail varying in length, but usually rather long. The only species here represented is,

Fig. 1. The Little Baboon (*Cynocephalus Babouin*). A variety of a more uniform green colour, with the face dusky. It inhabits Western Africa.

The Mandrills have the muzzle more elongated than any other species of this family. Its extremity is truncate, as in the Baboons, the nostrils being terminal; their tail is extremely short and erect; and they have callosities behind. In character they are similar to the Baboons, being ferocious, irritable, and malicious. They are all natives of Africa.

Fig. 2. The Great Mandrill (*Papio Maimon*). Male. The adult males of this species are of a greyish-brown colour above; white beneath; their callosities blue, as is the face, the latter being obliquely banded with white, and the extremity of the muzzle more or less tinged with red. The hair on the head is elongated, that of the cheeks being yellowish-brown, and there is a short yellowish beard on the chin. It is said to attain nearly the size of a Man, and to be an object of dread to the Negroes on account of its ferocity.

Fig. 3. The Great Mandrill. Young.

Fig. 4. The Drill (*P. leucophæus*). Male. This species is very similar to the former, being about the same size. greyish-brown, tinged with green above, white beneath, but the face black. It inhabits Guinea and the Gold Coast.

Fig. 5. The Drill. Female.

Fig. 6. The Drill. Young.

DESCRIPTION OF THE PLATE.—THE PIGEONS.

The Pigeons, which were included by Linnæus in the single genus Columba, have recently been arranged under several genera, and by some ornithologists have even been formed into a separate order. The position of this group has been a subject of much dispute, some referring it to the Gallinaceous order, others to the Passerine, to both of which it in fact presents affinities. Cuvier places the Pigeons at the end of the Gallinaceous Birds, and considers them as forming only a single genus, which may be divided into three subgenera: the *Columbi-Gallinæ*; *Columbæ*, or common Pigeons, including Turtles; and the *Vinagines*, or thick-billed Pigeons. The species represented in this plate belong to the second of these subgenera.

Fig. 1. The Speckled Pigeon (*Columba scripta*), having the sides of the head marked with dusky bands on a whitish-ground, the upper parts brown, the inner secondaries and their coverts bronzed. Is a native of New Holland.

Fig. 2. The Mantled Pigeon (*C. lacernulata*), with the head and lower parts bluish-grey, the throat yellow, the abdominal feathers and lower tail-coverts brightred, occurs in Java.

Fig. 3. The Grey-headed Pigeon (*C. capistrata*). The upper part of the head ash-grey, the throat cream-coloured, the hind neck and back light-red. Java.

Fig. 4. The Double-crested Pigeon (*C. dilopha*), so named on account of the disposition of the elongated feathers on its head, of which the posterior are crimson, is from New Holland.

Fig. 5. The Red-collared Pigeon (*C. humeralis*), with a half-collar of red on the hind neck, the upper parts light-brown, the tail graduated. New Holland.

Dentition of the Walrus.—Individuals of this remarkable animal present so great differences in the number of their teeth, some of which are always wanting in adults, that authors have given different statements on the subject, some representing it as having four, others as five grinders, on each side in either jaw, and all being in error as to the incisors. "The Morses," says Cuvier, "resemble the Seals in their limbs and the general form of the body, but differ much in the head and teeth. Their

39

lower jaw has no incisors or canine teeth. The grinders have all the form of short and obliquely truncated cylinders. There are four on each side above and below; but at a certain age, two of the upper fall out. Between the two canine teeth are moreover two incisors resembling the grinders, and which most authors have not recognised as incisors, although they are inserted in the intermaxillary bone; and between them are, in young individuals, two others which are small and pointed."

Having recently examined a number of skulls, I have come to a different conclusion. The normal dentition of this animal is shown by the skull of a young individual in the Museum of the Edinburgh College of Surgeons. In the upper jaw there are on each side *three* incisors, the first or inner extremely small, the second a little larger. The third or outer disproportionately large, being equal to the largest grinders. The socket of this tooth is placed in the intermaxillary bone, but, towards its mouth, is partly formed by the maxillary. The small incisors have deep conical sockets. The

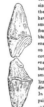

canine tooth is displaced, being thrust outwards, and of enormous size, the lateral incisor is on the level of its anterior margin, and the first grinder is opposite its middle. There are five grinders, having conical blunt sockets, and consequently single roots; the first smaller than the lateral incisor, the second and third largest, the fourth much smaller, the fifth very small; all shortly conical and blunt, with enamel only at the tip. The canine tooth is also at first enamelled at the tip. In the lower jaw are two very small incisors on each side; the canine tooth is wanting; five grinders, with single conical compressed roots, and short compressed conical crowns, enamelled at the point; the first, second, and third nearly equal, but the latter a little larger, the fourth much smaller, the fifth very small. The tusks are compressed, conical, directed downwards, a little curved backwards, from one to two feet long, somewhat diverging, but when very long converging again towards the point. In old individuals the incisors are obliterated, excepting the lateral pair in the upper jaw. The fifth grinders in both jaws are also obliterated, although not always, and sometimes the fourth in one or both. One of the most remarkable circumstances with regard to the teeth of this animal is, that although the grinders are so placed as to meet at the points only, the outer surface of the lower not falling within the line of the outer so as to meet the inner surface, nor the reverse, in no possible case, yet the upper and lower grinders are always worn so as to present an oblique flattened surface on their inner side, which of course cannot be produced by the rubbing of the teeth against each other. Indeed, the jaw is incapable of lateral motion on account of the form of its articulation, and the manner in which it is jammed in between the tusks. The food of the Walrus is said by some to consist of fuci, by others of fish, or of both; but the mastication of neither seems capable of producing the peculiar kind of flattening described.

A Honey Buzzard killed in Scotland.—Having been permitted, through the kindness of Mr Carfrae, to examine a recently killed individual of the Honey Buzzard or Pern, *Pernis apivorus*, a species of which I have seen only one other specimen obtained in Scotland, which was shot at Drumlanrig, and is now in the Museum of the University of Edinburgh, I embrace this opportunity of presenting a full description of it. This bird, a male, and apparently an old individual, is of a rather slender, elongated form, with the body moderately full, the neck of ordinary length or rather short, the head ovato-oblong. The bill, although slender compared with that of other birds of this order, is rather stout; the aperture of the mouth is wide, and extends to beneath the anterior angle of the eye; the cere large; the upper mandible with its outline as far as the edge of the cere convexo-declinate, then curved in the third of a circle, the sides convex, the edges soft to beneath the anterior angle of the nostrils, then hard, direct, and sharp, the tip slender, descending, acute; the lower mandible comparatively small, with the back broad, the sides rounded, the edges as in the upper, the tip rounded. The gape-line is arched from the base. Nostrils oblongo-linear, large, and oblique. Upper mandible a little concave, lower broadly channelled, with a median prominent line. Tongue deeply concave above, with the sides nearly parallel, the tip rounded but emarginate. The legs are short; the tarsus anteriorly covered with feathers half way down, on the rest of its extent with angular scales. The toes are of moderate size; the first stouter, the second next, the fourth least so; the first with four large scales above, the second with three, the third with four, the fourth with three. The claws are long, rather slender, tapering, arcuate; the first and second strongest, the third longest, with a thin inner edge, the second next in length, the fourth smallest.

The plumage is compact. The feathers on the fore part of the head and cheeks are ovate, compact, and small, especially on the loral space and around the eye. The feathers in general are ovate, curved, with a large downy plumule. On the lower parts they are nearly as compact as on the upper. The wings are long, very broad, and extend to two inches and a half from the end of the tail. There are twenty-three quills; the outer six are separated at the end when the wing is extended, and have the inner web cut out towards the end; all the rest are broad and rounded, with a minute tip. The tail is long, a little emarginate and rounded at the end, the feathers broad. The first quill is short, being two and seven-twelfths inches shorter than the second, which is eleven lines shorter than the third, this latter being the longest; the fourth is only a line shorter, and the other primaries gradually diminish. The middle tail feathers are three lines shorter than the third, which exceeds the lateral by ten-twelfths.

The cere is pale yellow at the base, but dusky in the greater part of its extent. The bill is black; the base of the lower mandible flesh-coloured; as is the mouth internally, excepting the mandibles and the horny part on the back of the tongue, which are black. The margins of the eyelids are also black, the iris yellow, the tarsi and toes orange, the claws black. The loral space and anterior part of the forehead are brownish-grey; the head reddish-brown; the rest of the upper parts umber-brown. The feathers generally are darker on the shaft and towards the end. The primary coverts and primary quills are blackish-brown at the end, and in the rest of their extent have generally on both webs three bands of dark-brown on a lighter ground; the inner webs white, except at the end, where they are light-brown mottled with darker. The outer quill has, however, only a single dark band, reduced to two spots, the second and third two bands, also reduced to spots; on the secondaries the dark bands are reduced to two, and gradually approximate inwards. The tips of the tail-feathers are brownish-white; there is then a broad band of brownish-black, and a dusky space with seven indistinct darker bands, between which and the base are three large blackish bands; the upper tail-coverts are light umber. The throat is light reddish-brown; the rest of the lower parts umber, each feather with the shaft and a portion near it dusky. The feathers of the legs are lighter, as are the lower tail-coverts, which have two bands of white towards the base. The concealed and downy part of the whole plumage is white, that colour appearing externally on the hind neck and head when the feathers are raised, as it there extends over more than half their length. The lower wing-coverts are umber brown.

Length to end of tail twenty-four and one-fourth inches; extent of wings fifty-two; wing from flexure sixteen and three-fourths; tail eleven and one-half; bill along the ridge one and four-twelfths; tarsus one and eleven-twelfths; first toe eleven-twelfths, its claw one; third toe one and nine-twelfths, its claw one and three-twelfths.

This species was formerly referred to the genus *Buteo*, but has recently been considered as, with some others, constituting a distinct genus, to which the name of *Pernis* was given by Cuvier. In some respects, Pernis is more nearly allied to *Milvus* than to *Buteo*, especially in the shortness and thickness of the tarsi, the slender moderately curved claws, the very long wings and tail. *Pernis apivorus* is in fact intermediate between *Buteo vulgaris* and *Milvus regalis*, and its habits are in correspondence with this circumstance, for both the birds mentioned are insectivorous, and I have found even earth-worms in the stomach of our Common Buzzard. The Honey Pern, although fond of the larvæ of Bees and Wasps, cannot live exclusively on them; but as yet, from its very rare occurrence in this country, little is known of its habits. The individual above described was shot in the county of Stirling, and affords the third authentic instance of the occurrence of the species in Scotland.

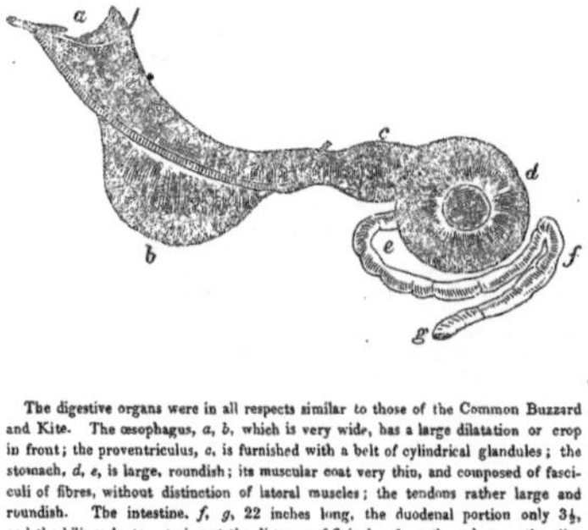

The digestive organs were in all respects similar to those of the Common Buzzard and Kite. The œsophagus, *a*, *b*, which is very wide, has a large dilatation or crop in front; the proventriculus, *c*, is furnished with a belt of cylindrical glandules; the stomach, *d*, *e*, is large, roundish; its muscular coat very thin, and composed of fasciculi of fibres, without distinction of lateral muscles; the tendons rather large and roundish. The intestine, *f*, *g*, 22 inches long, the duodenal portion only $3\frac{1}{2}$, and the biliary ducts entering at the distance of 3 inches from the pylorus; the diameter varies from five-twelfths to two-twelfths; there are no cœcal appendages; and the cloaca is 2 inches long, and of an elliptical form. The stomach was filled with fragments of bees and numerous larvæ, among which no honey or wax was found.

"Besides various specimens obtained in Suffolk, Norfolk, and along the eastern coast as far north as Northumberland, the Honey Buzzard," says Mr Yarrell, "has been killed in several western counties, including Dorsetshire, Devonshire, and Worcestershire. Dr Heysham considered it very rare in Cumberland, and had only met with one specimen. Mr Thomson of Belfast has recorded one example killed in the north of Ireland. According to Linnæus, Brunnich, Muller, and Pennant, the Honey Buzzard inhabits Denmark, Norway, Sweden, and Russia. From thence southward, it is found in Germany, France, Italy, and the south of Europe generally. M. Temminck considers it very rare in Holland. It is said to be a native of Eastern climes; and Mr Gould states that he has seen it in collections of skins from India.

BOTANY.

Scottish Oaks.—Although the Oak does not attain in Scotland dimensions equal to those of English trees of the species, and in the native forests of that country is generally of diminutive size, yet many individuals have been marked as entitled to some consideration. Of these the following list is presented in Mr Loudon's splendid Arboretum Britannicum, recently published. In the neighbourhood of Edinburgh, there is an oak in Dalmeny Park 70 feet high, with a trunk 15 feet 9 inches in circumference, diameter of the head 96 feet; another oak, 70 feet high, has a trunk only 6 feet 5 inches in circumference, but carries nearly that thickness to the height of 30 feet before it throws out branches. At Barnton Hall is an oak 80 feet high, with a trunk 11 feet in circumference, and a head 82 feet in diameter; the trunk is sound, and without branches to the height of 20 feet, but the head is stag-horned and much decayed. At Hopetoun House is a growing tree 75 feet high, with a trunk 11 feet in circumference. At Melville Castle is an oak 70 feet high, with a trunk 18 feet in girt, at 4 feet from the ground, and a head 90 feet in diameter. South of Edinburgh. In Ayrshire, at Kilkerran, is an oak 50 feet high; the girt of the trunk is 12 feet 6 inches, and the diameter of the head is 90 feet. In Haddingtonshire, at Yester, is an oak 89 feet high, with a trunk 12 feet in girt, and a head 70 feet in diameter. In Lanarkshire, at Bothwell Castle, is an oak 59 feet high, with a trunk 14 feet in circumference, and a head 98 feet in diameter. In Roxburghshire, at Minto, are several oaks about 200 years old, which are 70 feet high, the girt of the trunk about 12 feet, and the diameter of the head 63 feet. North of Edinburgh. In Aberdeenshire, at Fintray House, are four oaks with trunks varying from 5 feet 6 inches to 5 feet 10 inches in circumference. The oak does not ripen its acorns, and rarely its young wood, in this county. In Banffshire, at Gordon Castle, is an oak 66 feet high, with a trunk about 10 feet in girt, and a head 66 feet in diameter. In Cromarty, at Coul, there is an oak 162 years old, which is 80 feet high, the circumference of the trunk 12 feet, and diameter of the head 60 feet. In Fifeshire, at Donibristle Park, is one 70 feet high, with a trunk about 11 feet in girt, and 40 feet clear of branches, diameter of the head 45 feet. At Largo is an oak 100 feet high, with a trunk 9 feet 6 inches in circumference, and 35 feet clear of branches, and a head 53 feet in diameter. In Forfarshire there is an oak on the estate of Lord Gray, at Gray House, which was 68 feet high, the circumference of the trunk 17 feet 6 inches, and the diameter of the head 90 feet, when it was measured in June 1836, by Mr Robertson, his lordship's gardener. The same oak, when measured in 1821, was, we are informed by Mr Robertson, then only 16 feet in circumference, and consequently it has gained 18 inches since that period. It is *Quercus pedunculata*, and is in great health and vigour. In Perthshire, at Taymouth, is a growing oak 45 feet high, with a trunk 14 feet in girt, and a head 72 feet in diameter. The tree stands in the park in a loamy soil, on a dry subsoil, and is about 100 years old. In Ross-shire, at Brahan Castle, is an old oak 80 feet high, with a long straight trunk 12 feet in circumference, and a head 90 feet in diameter. In Stirlingshire, at Blair-Drummond, is a growing oak 120 years old, 86 feet high, with a trunk 20 feet in the bole, and 14 feet in circumference, diameter of the head 60 feet. There are many fine oaks at Blair-Drummond from 15 feet to 50 feet in the bole, but no other is quite so much in circumference. In Callender Park, *Q. sessiliflora* is 50 feet high, the circumference of the trunk 15 feet 6 inches, and diameter of the head 56 feet. In Sutherland, at Dunrobin Castle, is an oak 10 feet high, the diameter of the head 47 feet, and the girt of the trunk about 11 feet.

Destruction of Mosses in Pastures.—Mr Thomas Bishop, Methven Castle, has recently made experiments with the view of ascertaining the best method of destroying mosses growing in pasture land. The results of these experiments are communicated in the last number of the Transactions of the Highland and Agricultural Society of Scotland. "The growth of musci will always be found rapidly to take place when rye-grasses are the only grass-seeds sown, excepting on the very best land, as it is not the nature of that kind of grass to form a close sward. The production of them is also much encouraged in older pastures, by eating the grass too bare in August, and the early part of September, as also by taking a crop of meadow-hay where the ground is not much trodden upon, or has been under water in irrigation or otherwise. With the view to destroy musci and invigorate the better grasses, top-dressings have been used and found very expensive, and their effects as evanescent. Experiments have been made here for that purpose with bone-dust, put on in spring, at twenty bushels per acre, price 2s. 6d. per bushel. Lime and soot to the same amount of cost. Common salt was also tried to a fifth part of less expense, and found very efficient, although fears were entertained, for a time, that the grass would suffer. The soot in the following season gave evident indications of its tendency to extirpate the musci, but not so powerfully as the salt; the lime and bone-dust used had little or no effect. Liquid manure collected in a tank, under cover, from the feeding and cow byres, and washings of the dung-yard, have also been used, but not for the length of time to warrant a decisive opinion of its tendency to destroy musci, although their beneficial effects on grass land is unquestionable. The last and most efficient remedy for the prevention and destruction of musci, and easiest to have recourse to, when the ground has not become altogether exhausted, or in an over damp state, is to allow a great portion of the summer's grass to remain unconsumed on the ground, until the following winter, when the barer it is eaten before the new growth of spring, the finer will be the following summer's grass. But should the repetition of this treatment fail to extirpate the musci, it will be more profitable to put the grass lands under a rotation of crops, and sow them out anew with a mixture of grass seeds suited to the soil and climate. A little more manure than would have sufficed for a top-dressing will thus be repaid with a richer and much more enduring herbage."

Botany of the Neighbourhood of the Lake of Thun.—In a communication from Col. P. J. Brown of Eichenbühl, Switzerland, read to the Botanical Society of Edinburgh, it was stated that the Lake of Thun, having an elevation of about 1900 feet above the sea, and the surrounding country being much intersected by hills or long ridges, the vegetation assumes a subalpine character on the pastures about 1800 feet above the lake, comprising *Trollius europæus, Hieracium aureum, Tussilago alpina*, &c. The following was given as an approximation to the species usually met with at different altitudes on the surrounding mountains. Between 2000 and 3000 feet, *Arenaria verna* and *ciliata, Dryas octopetala, Cotoneaster vulgaris, Hieracium villosum*, &c. Between 3000 and 4000 feet, *Silene acaulis, Cerastium alpinum, Phaca astragalina Oxytropis uralensis, Saxifraga oppositifolia, Hieracium aurantiacum, Arbutus alpina, Ajuga alpina, Orchis pallens, Carex atrata*, &c. Above 4000 feet, *Gnaphalium alpinum* and *Leontopodium, Petrocallis pyrenaica, Draba tomentosa* and *stellata, Androsace bryoides*, &c.

GEOLOGY.

Fossil Bones of Yealm Bridge Cave.—Mr Bellamy, to whom we are indebted for an account of his discovery of this cavern, has recently favoured us with the following additional remarks on the subject.

Notwithstanding that the bones from the cave I was so fortunate to discover betray no symptoms of altered character, so far as external aspect and superficial examination go, being to appearance in no way different from bones which have been exhumed after a few years' interment, it is natural to inquire whether, in their intimate composition, alterations cannot be detected whereby some information may be added to the chemistry of bone, and to the inquiry concerning the operation of those circumstances under which these bodies were placed, as well as the period during which they have lain entombed. So far as respects the hard earthy portion of bones, it seems that when excluded from the action of air, they will remain for ages unaltered; though when I exposed some of my specimens for two or three days and nights to the action of the weather, in a rainy season, they soon split and cracked, showing proofs of incipient disintegration. Even without this exposure to the weather, the whole of my specimens suffered so much from contact of the air, as to oblige me to use a kind of varnish for their protection. But this decomposition must be regarded as the effect of unusually lengthened age; for in general we see that bones of ancient date do not suffer in this way on being brought to light. Besides the durability of their external figure, these bones had lost much of their moisture, and had imbibed in lieu the droppings from the cavern, which convey much of the calcareous matter of the rock; this latter circumstance in particular determining their absorbent quality noticed before. With respect to the animal part of their composition, I shall have to offer a modified opinion. That it is somewhat lessened in quantity in all bones of this class and date, I have no doubt; but I also think that there are differences dependent on the bone selected for examination, on its being entire or fractured, on its situation in the cave, and so forth. If you experiment on a small fragment, the result will seem to be very different from what it is when experimenting on an entire bone, at least so far as regards the ordinary method of employing muriatic acid to dissolve the earths, and exhibit the remaining animal portion. Thus, when by experiment I endeavoured to determine the comparative difference between these and bones of ordinary occurrence, I found that after selecting a fragment of fossil bone and a piece that had lain exposed on the high-road for some months, each weighing half a drachm, and putting them in maceration in glass vessels, violent and rapid escape of carbonic acid gas, due no doubt to the imbibed calcareous matter, proceeded directly on the fossil piece being immersed; gradual corrosion, or rather gradual removal of the earthy portions, was soon evident, and in the space of seven hours nothing remained of the original fragment but a small spongy, flocculent mass, or pellicle, weighing eleven grains. On the contrary, the other fragment gave off slowly and deliberately gaseous matter; the process of removal of the earths was not finished for a very long time, and at the end, the original form of the immersed piece was retained; it was soft, fibrous, flexible, and elastic, and weighed eighteen grains. But again my brother finds that in exhibiting the animal form of bones, no external difference is observable between specimens of this kind derived on the one hand from a phalangeal bone of the fossil Hyæna, and on the other from any common bone. The first *seems* to exhibit the fact as well as the other; but I decidedly think there is a diminished quantity of the albuminous substance, though it may not appear so. Mr Martyn, in his Treatise on Fossils, speaks cautiously on the subject of the chemistry of fossil bones by saying, through Professor Playfair, that they *often* contain a portion of gelatin (or rather, by recent examination, albumen) in their composition, particularly in their interior, the surface only having undergone a change. In other instances, he adds, the gelatin (albumen) is *wholly* displaced, while a greater proportion of carbonic acid than that which existed in it originally is found united with the calcareous matter. Those in the Rock of Gibraltar seem to have been so circumstanced. The question, therefore, respecting the composition of fossil bones must be answered in a cautious and qualified manner, and with reference to the conditions of each particular case.

The occurrence of the bones of Mice in the substance of some of the stalagmite, is a circumstance to which too much attention cannot be directed in framing a theoretic statement of the age of the contents of this cavern, because, as before mentioned, they are not identically the same species with those now in existence, and from their peculiar position seem not to have been contemporary with the other animals. Yet in consideration that other specimens of these remains were discovered in the diluvial clay, it may not be unfair to suppose that those found in stalagmite were so impacted previously to the catastrophe which effaced the whole series of Antediluvian creatures, during indeed that period in which the Hyænas and other predatory beasts employed this cavern, or its compartments, as their dens. There is the same analogy or resemblance of these Mice to the present kind, as subsists between the other Antediluvian creatures, and their representatives of this day,—the same remarkable affinity. Looking also to the wise provisions of Nature in regard of food, we see that these small creatures would hold a decided relation to the predaceous habits of the animal I have ventured to designate Glutton, from its evident similarity to our *Mustela Gulo*.

I perceive that two animals named in my manuscript as having been identified among the rest, have been omitted in the printed account. The reason of this I am well aware of: it was thought that some mistake had arisen in naming the specimens, particularly perhaps the *Sheep*, this animal seeming to be inconsistent with an undomesticated or unreclaimed state. Of this I have been warned in a private letter from Dr Buckland; but what precludes our supposing the existence of a wild animal *allied* to our *Oves Aries*? and what shall forbid the conclusion, when by comparison it is seen that *the analogy is most perfect*? Messrs Clift and Owen, of the College of Surgeons, have through Colonel Mudge, in his paper before alluded to, confirmed my original statement made in 1835 by the test of comparison with recent teeth. With regard to the *Duck*, I can only say I have diligently compared the specimens with recent skeletons in my museum, and feel fully justified in renewing the assertion. Dr Buckland mentions "a bird" in his list of the animals of the Kirkdale Cave, and Colonel Mudge mentions in his account of the Yealm Bridge Cave (1836) "a bird of considerable size."

I cannot avoid mentioning, in conclusion, that Colonel Mudge's account of the strata found in the cave is incorrect. The red clay was unquestionably the lowermost stratum, not the argillaceous sand. From want of specimens, I presume, he has also omitted to insert the names of the Deer, Pig, Glutton, and Field Mouse.

Wealden Strata at Elgin.—At a meeting of the Geological Society, on the 25th April, a communication from Mr Malcolmson was read, giving an account of the occurrence of Wealden Strata at Linksfield, near Elgin. The country around consists principally of old red sandstone, but at Linksfield, about a mile south of the town, that formation is overlaid by a series of beds, formerly considered to be lias, but ascertained by their organic remains to belong to a fresh-water deposit of the age of the Wealden of England. The succession of the strata in descending order is as follows:—1. Blue clay with thin beds of compact shelly limestone. 2. Bands of limestone and clay. 3. Blackish clay. 4. Compact grey limestone with shells. 5. Green clay. 6. Red sandy marl, enclosing rolled pebbles of granite, gneiss, &c., also angular fragments of old red sandstone. Among the fossils are *Cyclas media*, a common shell in the fresh-water strata of Essex; an *Avicula*, which occurs in the lower purbeck beds at Swanwich, also remains of fishes, and great abundance of a new species of Cypris. Strata equivalent to the Wealden of England were discovered in the Isle of Skye by Mr Murchison, in 1827. The Rev. G. Gordon has recently found the Linksfield fossils at Lhanbryde, three miles to the eastward of that locality; also a Pinna, considered by Mr James Sowerby as closely resembling a species belonging to the Portland sand; and, in making the canal by which a great part of the Loch of Spynie was drained, fossils were found belonging to the coral rag and the lias of England. Mr Malcolmson, therefore, hopes that many of the formations above the old red sandstone, hitherto undetected in that part of the kingdom, will be discovered. He also announced, that Mr Martin of Elgin has found in the old red sandstone of that neighbourhood, among other remains of fishes, scales identified with those of the old red sandstone of Clashbennie.

CHEMISTRY.

Dr Traill's Indelible Ink.—On the 19th February last, Dr Traill read to the Royal Society of Edinburgh a paper on the composition of a new ink, which, by resisting chemical deletion, promises to diminish the chances of the successful falsification of bills, deeds, and other documents. The author was led to the investigation of this subject by its connection with that branch of medical jurisprudence which treats of the prevention and detection of forgery. It is well known that common writing ink may be totally effaced from paper by certain chemical agents, and that several others so impair its colour, that the characters traced with it become illegible. To the first class of chemical agents belong chlorine, and substances containing it, as well as oxalic acid; to the second diluted solutions, or the vapours of the mineral acids, and of the caustic alkalies. These agents were applied to written specimens of a great number of different inks, and the degree of resistance to their effects was considered as the criterion of the durability of each. These views engaged the author in an extensive series of experiments on coloured metallic preparations, suspended in different vehicles, the results of which were not satisfactory. He then attempted the composition of a carbonaceous liquid which should possess the qualities of good writing ink. The inks used by the ancients were carbonaceous, and have admirably resisted the effects of time; but the author found that the specimens of writing on the Herculaneum and Egyptian papyri were effaced by washing with water; and on forming inks after the descriptions of Vitruvius, Dioscorides, and Pliny, he found that they did not flow freely from the pen, and did not resist water. The carbonaceous inks with resinous vehicles, rendered fluid by essential oils, though they resisted water and chemical agents, had the disadvantages of not flowing freely from the pen, and of spreading on the paper, so as to produce unseemly lines. Solution of caoutchouc in coal-naphtha, and in a fragrant essential oil, lately imported from South America, under the name of *aceite de sassafras* (the natural produce of a supposed Laurus), were subject to the same objections.

The author tried various animal and vegetable fluids as vehicles of the carbon, without obtaining the desired result, until he found, in a *solution of the gluten of wheat in pyrolignous acid*, a fluid capable of readily uniting with carbon into an ink possessing the qualities of a good, durable writing ink. To prepare this ink, he directs gluten of wheat to be separated from the starch as completely as possible, by the usual process, and when recent to be dissolved in pyrolignous acid with the aid of heat.

This forms a saponaceous fluid, which is to be tempered with water until the acid has the usual strength of vinegar. He grinds each ounce of this fluid with from eight to ten grains of the best lamp black, and one and a half grain of indigo. The advantages of this ink are:—1. It is formed of cheap materials; 2. is easily made; 3. has a good colour; 4. flows freely from the pen; 5. dries quickly; 6. when dry is not removable by friction; 7 is not affected by soaking in water. Lastly, slips of paper written on by this ink have remained immersed in solutions of various chemical agents, capable of immediately effacing or impairing common ink, for seventy-two hours, without change, unless the solutions be so concentrated as to injure the texture of the paper.

Dr Traill offers this composition as a writing ink, to be used on paper, for the drawing out of bills, deeds, wills, or wherever it is important to prevent the alteration of sums or signatures, as well as for handing down to posterity public records, in a less perishable material than common ink.

OBITUARY.

On the 11th of May, in the 80th year of his age, died in London, Thomas Andrew Knight, Esq., of Downton Castle, in Herefordshire, the President of the Horticultural Society of London. A correspondent in the *Athenæum* gives the following account of his life.

Mr Knight was born at Wornsley Grange, near Hereford, on the 10th of October 1758. He was the youngest son of the Rev. Thomas Knight, a clergyman of the Church of England, whose father had amassed a large fortune as an iron-master, at the time when iron-works were first established at Colebrook Dale. When Mr Knight was three years old, he lost his father, and his education was in consequence so much neglected, that at the age of nine years he was unable to write, and scarcely able to read. He was then sent to school at Ludlow, whence he was removed to Chiswick, and afterwards entered at Balicl College, Oxford. It was in the idle days of his childhood, when he could derive no assistance from books, that his active mind was first directed to the contemplation of the phenomena of vegetable life; and he then acquired that fixed habit of thinking and judging for himself, which laid the foundation of his reputation as an original observer and experimentalist. He used to relate an anecdote of his childhood, which marks the strong original tendency of his mind to observation and reflection. Seeing the gardener one day planting beans in the ground, he asked him why he buried those bits of wood. Being told that they would grow into bean plants and bear other beans, he watched the event, and finding that it happened as the gardener had foretold, he determined to plant his pocket knife, in the expectation of its also growing and bearing other knives. When he saw that this did not take place, he set himself to consider the cause of the difference in the two cases, and thus was led to occupy his earliest thoughts with those attempts at tracing the vital phenomena of plants to their causes, upon which he eventually constructed so brilliant a reputation.

It was about the year 1795 that Mr Knight began to be publicly known as a vegetable physiologist. In that year he laid before the Royal Society his celebrated paper upon the inheritance of disease among fruit-trees, and the propagation of debility by grafting. This was succeeded by accounts of experimental researches into vegetable fecundation, the ascent and descent of sap in trees, the phenomena of germination, the influence of light upon leaves, and a great variety of similar subjects. In all these researches, the originality of the experiments was very remarkable, and the care with which the results were given was so great, that the most captious of subsequent writers have admitted the accuracy of the facts produced by Mr Knight, however much they may have differed from him in the conclusions which they draw from them.

The great object which Mr Knight set before himself, and which he pursued through his long life with undeviating steadiness of purpose, was utility. Mere curious speculations seem to have engaged his attention but little; it was only when facts had some great practical bearing that he applied seriously to investigate the phenomena connected with them. For this reason, to improve the races of domesticated plants, to establish important points of cultivation upon sound physiological reasoning, to increase the amount of food which may be procured from a given space of land, all of them subjects closely connected with the welfare of his country, are more especially the topic of the numerous papers communicated by him to various societies, especially the Horticultural, in the chair of which he succeeded his friend Sir Joseph Banks. Whoever calls to mind what gardens were only twenty years ago, and what they now are, must be sensible of the extraordinary improvement which has taken place in the art of horticulture during that period. Tuis change is unquestionably traceable in a more evident manner to the practice and writings of Mr Knight, than to all other causes combined. Alterations first suggested by himself, or by the principles which he explained in a popular manner, small at first, increasing by degrees, have insensibly led, in the art of gardening, to the most extensive improvements, the real origin of which has already, as always happens in such cases, been forgotten, except by those who are familiar with the career of Mr Knight, and who know that it is to him that they are owing. Of domesticated fruits or culinary vegetables there is not a race that has not been ameliorated under his direction, or immediate and personal superintendence; and if henceforward the English yeoman can command the garden luxuries that were once confined to the great and wealthy, it is to Mr Knight, far more than to any other person, that the gratitude of the country is due.

The feelings thus evinced in the tendency of his scientific pursuits were extended to the offices of private life. Never was there a man possessed of greater kindness and benevolence, and whose loss has been more severely felt, not only by his immediate family, but by his numerous tenantry and dependants. And yet, notwithstanding the tenderness of his affections for those around him, when it pleased Heaven to visit him, some years since, with the heaviest calamity that could befall a father, in the sudden death of an only and much-loved son, Mr Knight's philosophy was fully equal to sustain him in his trial.

Mr Knight's political opinions were as free from prejudices as his scientific views; his whole heart was with the liberal party, of which he was all his life a strenuous supporter. It is no exaggeration to add, that great as is the loss sustained by his country and his friends, it will be equally difficult to fill his vacancy in science. No living man now before the world can be said to rank with him in that particular branch of science to which his life was devoted.

Died, at his house in Ridley Place, Newcastle, on the 5th of May, aged 69, Nathaniel John Winch, Esq., greatly respected. Mr Winch was well known as an excellent botanist. He was author of an "Essay on the Geographical Distribution of Plants through the Counties of Northumberland, Cumberland, and Durham," which has passed through two editions; also of "Observations on the Geology of Northumberland and Durham, 4to, 1814;" and of a very elaborate "Flora of Northumberland and Durham," printed in the Transactions of the Natural History Society of Northumberland, Durham, and Newcastle-upon-Tyne. He has bequeathed the whole of his very extensive Herbarium and his Library of Natural History to the Linnæan Society, of which he was a member, and has left a legacy of L.200 to the Newcastle Infirmary, to which he acted as secretary for 21 years.

MISCELLANEOUS.

Manganese Mine at Grandholm, near Aberdeen.—We are informed, says the Editor of the Edinburgh New Philosophical Journal, by Dr Fleming, that the ore occurs in a rock of mica-slate, which stretches in a northerly direction, and has an easterly dip, varying from 30° to 50° and upwards. In some places it is thin, slaty, even, or waved, while at other parts thin beds of gneiss and granite make their appearance. When the mine was opened upwards of twenty years ago, an excavation, or opencast, was made across the stretch of the strata, at the eastern termination of which a mass of felspar porphyry makes its appearance, the relations of which are not seen at present. The ore, which is the Grey Manganese ore, or Hydrous Binoxide of Manganese, occurs in irregular thin beds, rounded concretions, or anastomosing films in the rock, accompanied by small quantities of sulphate of barytes. As the working has but recently commenced, little more than a dozen of tons of the ore have been obtained; but as the undertaking is in hands possessing abundance of capital and enterprise, Messrs Cookson, of Newcastle, the mine will have a fair trial, and it is hoped that a new branch of trade will thus be added to those already so successfully carried on at Aberdeen.

Phosphorescency of the Ocean.—The naturalists of *La Bonita*, in her late voyage round the globe, have made many observations respecting marine phosphorescence, which are thus reported to the French Academy of Sciences. Numerous observations made upon phosphorescent water, by means of reagents, of filtration, boiling, simple examination, and with the help of the microscope, have led us to the following conclusions:—The phosphorescent property of sea-water is not inherent in the nature of this liquid, but is essentially owing to the presence of organized beings. The animals which produce the phosphorescence belong to different classes. In the first rank we find the minute species of crustacea which swarm in the sea, but especially a very small species having two valves, which possess this remarkable property in the highest degree. All these species have been collected, and are carefully preserved in alcohol. Many mollusca, principally small Pelagic Cephalopoda, Biphora, &c., also many zoophytes, among which we remark Diphyes, Medusæ, &c., possess the phosphorescent property. Lastly, in certain localities we also find, on the surface of the ocean, very small yellowish bodies, which are, nevertheless, extremely phosphorescent. We encountered these small bodies in immense abundance, when landing at the Sandwich Isles, and in crossing from that archipelago to the Marianne Islands. We met with them in such vast quantities in the Straits of Malacca and on the coasts of Pulo Penang, that the whole surface over a great extent seemed as if covered by a thick yellowish dust. These small bodies have been examined with the microscope; but although they have been for a long time submitted to our notice, we have never been able to detect the slightest movements connected with them. At the same time, the experiments we have made on them, through the means of various reagents, lead us to the conclusion that they are organized and living bodies. They appeared somewhat different as taken at the Sandwich Islands, and in the Straits of Malacca. The former were globular and transparent, with a yellowish point in the centre, the latter were rather oval, with a depression in the centre, so that they were somewhat kidney-shaped; they also were entirely yellowish.

In all the animals which possess phosphorescence, the property has appeared to us to depend upon a particular principle, probably a secretion. Some of them, as the small phosphorescent crustacea, can distinctly emit it in streams, especially when irritated. Others did not appear to possess the power of emitting this matter, and in them it was developed only in certain circumstances, as when they struck a body, or were irritated. In others again, as in the Cephalopoda, and in some Pteropoda, the phenomenon showed itself in a nearly passive way: the phosphorescent matter in their nucleus, or other parts of their bodies, shone constantly and uniformly as long as the animal was in the enjoyment of life, and along with this disappeared their light. Lastly, in the yellowish corpuscles above described, the phosphorescent matter shines almost uniformly, but if brought into contact with any reagent, their lustre is first increased, and then insensibly vanishes away. The phosphorescent matter which we collected on the sides of the vessel was yellowish, slightly viscous, and very soluble in the water, which it rendered luminous at the moment it was projected by the animal. —*Comptes Rendus*, 5 *Avril. Ed. New Philos. Journal.*

Edinburgh: Published for the Proprietor, at the Office, No. 13, Hill Street. London: Smith, Elder, and Co., 65, Cornhill. Glasgow and the West of Scotland: John Smith and Son; and John Macleod. Dublin: George Young. Paris: J. B. Balliere, Rue de l'Ecole de Médecine, No. 13 bis.

THE EDINBURGH PRINTING COMPANY.

THE EDINBURGH

JOURNAL OF NATURAL HISTORY,

AND OF

THE PHYSICAL SCIENCES.

SEPTEMBER, 1838.

ZOOLOGY.

DESCRIPTION OF THE PLATE.—THE HOGS.

In the order of Pachydermata, the Hogs constitute a family distinguished by having the feet cleft, with two hoofed toes, and two additional smaller ones behind, which do not reach the ground ; as well as a moveable prolongation of the muzzle, supported upon a bone, and which enables them to dig into the soil in search of roots. This family is composed of the *Hogs* properly so called, the Ethiopian Hogs, and the Peccaris.

Fig. 1. THE CHINESE HOG (*Sus Scrofa, var. Sinensis*) is a small variety of the Common Hog, with the ears erect, the upper parts blackish, the lower white.

Fig. 2. THE CAPE HOG (*S. capensis*) is similar to the Chinese in form, and generally of a black colour.

Fig. 3. THE BABYRUSSA (*S. Indicus*) is distinguished by the elongation of the tusks of both jaws, which are curved backwards. Its ears are short, its feet comparatively long, its colour brownish-grey. It inhabits the Indian Islands.

Fig. 4. THE BABYRUSSA. Female. Of a brownish colour, with the tusks not projecting.

Fig. 5. ÆLIAN'S HOG (*Phacochærus Æliani*). The Phacochœri are distinguished from the Hogs by their grinders, which are composed of cylinders connected by cortical matter. Their tusks are of enormous size and spread outwards. This species, of which the head is furnished on each side with a large fleshy lobe, is peculiar to Africa.

Fig. 6. THE COLLARED PECCARI (*Dicotyles torquatus*). The Peccaris, which are peculiar to South America, resemble our Common Hog in form, but have no outer toe on the hind feet, and their tusks are not protruded. The present species has a band of white on the shoulder and lower part of the neck.

Fig. 7. THE WHITE LIPPED PECCARI (*D. labiatus*) is brownish-black, with the lower lip and the throat white.

DESCRIPTION OF THE PLATE.—THE JAYS.

THE JAYS constitute a genus of birds very nearly allied to the Crows, from which they differ chiefly in having the bill less strong, the tail more elongated, their colours more gaudy, and their habits approximating to those of the Titmice, to which also they are nearly allied. They reside chiefly in woods and thickets, occasionally betaking themselves to the open country in the neighbourhood, and feed on fruits, seeds, insects, worms, sometimes eggs and young of other birds, as well as small quadrupeds, and carrion. They are, in short, as omnivorous as the Crows. They are generally dispersed, and some of the species are spread over a vast extent of country. The species represented on the plate are American, with the exception of the first, the Common Jay, which belongs to Europe and Asia.

Fig. 1. THE EUROPEAN JAY (*Garrulus Glandarius*). This bird, which is remarkable for the beautiful blue barred patch on the wings, although its colours are not otherwise fine, is not uncommon in many parts of Britain and the Continent of Europe. Like several other species, it imitates the notes of birds and the cries of quadrupeds.

Fig. 2. THE FLORIDA JAY (*G. Floridanus*), although a native of Florida, as its name implies, occurs also in Louisiana, Kentucky, and other parts of the United States.

Fig. 3. THE BLUE JAY (*G. cristatus*) is the most extensively distributed and best known of the North American Jays. It is with the description of this species that our countryman Wilson commences his celebrated "American Ornithology ;" and its manners have with equal felicity been described by his successor, our esteemed friend, the enthusiastic Audubon.

Fig. 4. STELLER'S JAY (*G. Stelleri*) is peculiar to the north-western coast of America, and was first described by Lathan from a specimen obtained, on Cook's Expedition, at Nootka Sound.

Fig. 5. THE CANADIAN JAY (*G. canadensis*) inhabits the northern parts of the United States, and the British settlements in North America.

40

BRITISH BIRDS.—THE KINGFISHER.

THE Kingfisher, although one of the least elegantly formed of our native birds, is among the most distinguished for the beauty of its plumage, which is such as at once to recall to mind the splendour of the feathered denizens of the tropics. Its large body, short and thick neck, disproportionately long bill, diminutive feet, and abbreviated tail, give it a peculiar appearance, so that the least observant cannot mistake it for any other bird. The bill is considerably longer than the head, straight, rather slender, higher than broad in its whole length, four-sided, its outlines almost straight, and its tip pointed. The very short tarsi are roundish, and destitute of defined scales ; the first toe shorter than the second, the third slightly longer than the fourth ; the claws arched, slender, compressed, and acute. The plumage is soft and blended ; the feathers generally long, especially on the hind neck and rump ; of an oblong form, without plumules. The wings are rather short, but very broad, the secondary quills being of great length ; the tail very short, a little rounded, of twelve rather narrow, rounded feathers. The upper mandible is dark-brown, as are the margins and tip of the lower, the remaining part being pale orange. The tarsi and toes are orange-red, the claws dark-brown. The upper part of the head is dull green, each feather with a transverse bar of light greenish-blue near the end ; the hind neck, sides of the back, scapulars, and wing-coverts, are of a similar dull green, tinged with purple in a different light, the latter feathers tipped with light blue. The middle of the back, the rump, and tail-coverts, are of a beautiful glossy light blue, the tail of a duller purplish-blue. The quills are brown, with the outer webs dull green. A band of yellowish-red from the nostril to the eye ; the loral space dusky ; behind the eye a similar yellowish-red band ; below which, and extending from the lower mandible, is a band of greenish-blue, terminating behind in a yellowish-white patch. The throat is of the latter colour, and the rest of the lower parts yellowish-red, of a richer tint anteriorly. The length is 7¼ inches. The female is somewhat smaller, but similar in the colours, the tints being only a little less bright.

Let us now imagine ourselves on the banks of the Esk, the woods resuming their green mantle, and the little birds chanting their summer songs. From afar comes the murmur of the waterfall, swelling and dying away at intervals, as the air becomes still, or the warm breezes sweep along the birchen thickets, and ruffle the bosom of the pebble-paved pool, margined with alders and willows. On the flowery bank of the stream, beside his hole, the Water-Rat nibbles the tender blades ; and on that round white stone in the rapid is perched the Dipper, ever welcome to the sight, with his dusky mantle and snowy breast. Slowly along the pale blue sky sail the white fleecy clouds ; as the lark, springing from the field, flutters in ecstacy over his happy mate, crouched upon her eggs under the shade of the long grass, assured that no rambling urchin shall invade her sanctuary. But see, perched on the stump of a decayed wil low jutting out from the bank, stands a Kingfisher, still and silent, and ever watchful. Let us creep a little nearer, that we may observe him to more advantage. He cautious, for he is shy, and seeks not the admiration which his beauty naturally excites. There he is grasping the splint with his tiny red feet, his bright blue back glistening in the sunshine, his ruddy breast reflected from the pool beneath, his long dagger-like bill pointed downwards, and his eye intent on the minnows that swarm among the roots of the old tree that project into the water from the crumbling bank. He stoops, opens his wings a little, shoots downwards, plunges headlong into the water, reappears in a moment, flutters, sweeps off in a curved line, wheels round, and returns to his post. The minnow in his bill he beats against the decayed stump until it is dead, then tossing up his head swallows it, and resumes his ordinary posture, as if nothing bad happened. Swarms of insects flutter and gambol around, but he heeds them not. A painted butterfly at length comes up, fluttering in its desultory flight, and as it hovers over the hyacinths, unsuspicious of danger, the Kingfisher springs from his perch, and seizing it returns to his post. There, swift as the barbed arrow, darting straight forward, on rapidly moving pinions, gleams his mate, who alights on a stone far up the stream, for she has seen us, and is not desirous of our company. He presently follows, and our watch being ended, we may saunter a while along the grassy slopes, inhaling the fragrance of the primrose, and listening to the joyous notes of the Blackbird, that from the summit of yon tall tree pours forth his soul in music.

It is chiefly by the still pools of rivers and brooks that the Kingfisher is met with. Although not plentiful in any part of this country, nor any where gregarious, it is generally dispersed in England, and occurs in the southern and part of the middle division of Scotland, but has not, I believe, been met with beyond Inverness, for the Kingfishers, so called, of the North, are merely Dippers. It remains with us all the year, shifting its station on the streams, and in summer selecting some place having a steep bank, in a hole in which it deposits its eggs. Mr Henry Turner states, in Loudon's Magazine of Natural History, Vol. IV. p. 450, that at Bury St Edmund's in Suffolk, "some boys watched an old one into a hole in the bank of the river Lark, and attempted to capture it on its exit; but without success in this case. They then with a crooked stick pulled out a portion of the nest, consisting of a few feathers, old dried roots, and hay. I subsequently examined the hole," he continues; "it was in a low meadow 300 yards east of Northgate Street, and on the bank of a small stream. The entrance to the hole was about three feet from the water, and one foot beneath the level of the meadow. Hole nine inches in diameter, and about five feet in length, straight, and somewhat larger at the end than at the entrance." Various other individuals have alleged that the nest is formed of dry grass, roots, and feathers; but Montagu gives a different account of it.

"The hole chosen to build in is always ascending, and generally two or three feet in the bank; at the end is scooped a hollow, at the bottom of which is a quantity of small fish-bones, nearly half an inch thick, mixed in with the earth. This is undoubtedly the castings of the parent birds, and not of the young, for we have found it even before they have eggs, and have every reason to believe both male and female go to that spot for no other purpose than to yield this matter for some time before the female begins to lay, and then they dry it by the heat of their bodies, as they are frequently known to continue in the hole for hours long before they have eggs. On this disgorged matter the female lays to the number of seven eggs, which are perfectly white and transparent, of a short oval form, weighing about one dram. The hole in which they breed is by no means fouled by the castings; but before the young are able to fly it becomes extremely fetid by the fæces of the brood, which is of a very watery nature, and cannot be carried away by the parent birds, as is common with most of the smaller species. In defect of which, instinct has taught them to have the entrance to their habitation ascending, by which means the filthy matter runs off, and may frequently be seen on the outside. We never could observe the old birds with any thing in their bills when they went in to feed their young; from which it may be concluded they eject from their stomach for that purpose."

This account of the nest, however, is very improbable; and accordingly Mr Rennie, in his edition of Montagu's Dictionary, doubts its accuracy. "In the bank of a stream at Lee in Kent, we have been acquainted with one of those nests in the same hole for several successive summers, but so far from the exuviæ of fish-bones ejected, as is done by all birds of prey, being dried on purpose to form the nest, they are scattered about the floor of the hole in all directions, from its entrance to its termination, without the least order or working up with the earth, and all moist or fetid. That the eggs may by accident be laid upon portions of these fish-bones is highly probable, as the floor is so thickly strewed with them, that no vacant spot might be found, but they assuredly are not by design built up into a nest. The hole is from two to four feet long, sloping upwards, narrow at the entrance, but widening in the interior, in order, perhaps, to give the birds room to turn, and for the same apparent reason the eggs are not placed at the extremity. I am not a little sceptical as to its sometimes selecting the old hole of a Water-Rat, which is the deadly enemy to its eggs and young; but it seems to indicate a dislike to the labour of digging. It frequents the same hole for a series of years, and will not abandon it, though the nest be repeatedly plundered of the eggs or young."

The question as to its nestling in a Water-Rat's hole can be decided only by observation. Its bill is certainly adapted for digging into earth or sand, but its feet, one might suppose, would prove very inadequate instruments for scraping out the debris along a tunnel of three or four feet. On the other hand, its hole is often at a greater height from the water than we ever find that of the Water-Rat; in one case it has been seen twelve feet above it; and all accounts agree in describing it as straight and sloping upwards, whereas the holes of Water-Rats are usually tortuous. It is possible enough that sometimes the Kingfisher may take possession of a Water-Rat's hole, or even that of a Common Rat or Mole, and enlarge it, as the Starling has been known to do, and that it may also dig a hole for itself, like the Bank Swallow. At all events, we have certain evidence that the American Kingfisher, *Alcedo Alcyon*, digs its hole. Mr Audubon states that "the male and female, after having fixed upon a proper spot, are seen clinging to the bank of the stream in the manner of Woodpeckers. Their long and stout bills are set to work, and as soon as the hole has acquired a certain depth, one of the birds enters it, and scratches out the sand, earth, or clay, with its feet, striking meanwhile with its bill to extend the depth. The other bird all the while appears to cheer the labourer, and urge it to continue its exertions; and, when the latter is fatigued, takes its place. Thus, by the co-operation of both, the hole is dug to the depth of four, five, or sometimes six feet, in a horizontal direction, at times not more than eighteen inches below the surface of the ground, at others eight or ten feet."

In the second volume of the Naturalist, p. 274, Mr Allington gives the following account of it:—"A friend of mine, while fishing on a small trout stream, near Louth, called the Crake, in the early part of June, observed a Kingfisher with a fish in its mouth, flying several times near his hat with a whirling noise. He watched it until it entered a hole in the bank, the entrance to which was strewed with fish-bones. On digging into the hole (which commenced low down in the bank, and ran upwards in a slanting direction for about two feet), he found the nest, containing seven young birds just hatched. The bottom of the nest was excessively thick, and mixed up with small bones of the Stickle-back. Its structure, excepting the mixture of fish-bones, was not very unlike that of a Thrush. It crumbled to pieces on being touched, and I could procure no portion worth preserving. Near the nest was another hole, which had all the appearance of having been the Kingfisher's last residence, the bones at the entrance being dry and crumbling; but in this the parent bird again commenced laying, and on opening the nest six eggs were found on the fragments of the structure. They were white, and beautifully transparent, showing the yolk through, which gave them a pinkish hue at the larger end. I have now in my collection one of the eggs, which, though so transparent, I was surprised to find thicker and stronger than the generality of eggs, and rounder in its form, the circumference being two inches and a half, the length eight-tenths of an inch."

The flight of the Kingfisher is direct and rapid, performed by quick beats of the wings, and very similar to that of the Dipper, which it, however, excels in speed. The movements of the wings are indeed so rapid that one can scarcely perceive them, and the flight of this bird, the Dipper, Auks, Guillemots, and other short-winged birds, might induce the closet-naturalists to revise their opinions as to flight, founded merely upon the length and breadth of wings; for a long wing is not always so well adapted for speed as a short one, and a Guillemot can easily outstrip a Gull. Its feet are not adapted for walking or hopping, and therefore it takes its stand on a stone, a stump, a rail, or a branch overhanging the water, waits with patience, and when a minnow or a stickle-bat comes near the surface, darts upon it and secures it. In like manner it sallies forth in pursuit of the larger insects. Although very shy, insomuch that one can very seldom get within shot of it when perched, it does not shun the vicinity of human habitations, but, on the contrary, often breeds at no great distance from them. It does not associate with any other birds, and it is seldom that even two of its own species are seen together. Being highly prized by collectors and others, it is much harassed, and although nowhere plentiful, may be obtained in almost any district to the south of the Forth and Clyde. In some places they leave the larger streams in autumn, and betake themselves to the brooks, so that a person not aware of their habits in this respect might suppose them to be migratory. Even in the more northern parts, however, they remain all the year, and many individuals have been shot near Edinburgh in December and January.

ON THE MOVEMENTS OF THE MOTACILLÆ IN SOUTH DEVON.

BY J. C. BELLAMY, ESQ. YEALMPTON.

The Grey Wagtail, *Motacilla Boarula*, visits us without deviation yearly in the month of September, and remains until the end of March or the first of April, frequenting rivers, brooks, spring-heads, and the sea-coast. Some circumstance determines a slight irregularity of a few days, both in their arrival and departure,—most probably it is their food; but in respect of number there is apparently little difference. They seem to come in a body, and attract immediate attention by their tameness, and the briskness of their motions. Their retreat, however, is accomplished in a straggling manner: suddenly we lose the bulk of the party, and hear only the twit of a solitary bird or so, on the bank of a river, or at a spring-head; in a few days these are gone, and we see no more of them for a season. But though I am not able to bear testimony to the fact, this species has been known to remain all summer and breed. In 1831 I saw one frequenting a pond near Tavistock on the second of September. This may have been an unusually early arrival, or one of a pair that had stayed through the season. Their chief resorts are rivers and streams, but some repair to the sea-coast, and fare with the Pied Wagtail. In hard weather they seem all to frequent the roads, and seek support from the droppings of cattle, frost seeming to cause a general retirement of the insects on which they feed, or sealing down the soil and stones beneath which they harbour. I believe that if any precise dates for their arrival and departure could be ventured on, they would be September 15th and April 8th; yet in 1835 I saw a flock arrive at a small village on the sea-coast on August 13th. In their retreat also, in the spring preceding, I observed an unusual tardiness, and they disappeared gradually. No phenomena that we know of can enlighten us respecting these irregularities, any more than concerning the cause of their migrations: we see that the Pied Wagtail haunts the same situations, feeds similarly, and is content to remain with us the year through; but some impulse carries the Grey hundreds of miles northward to rear its young. It is now clearly made out that in the spring our flocks retire to the northern counties, it being there a stationary bird also; but, independently of Selby's authority for this, a paper which I possess, written by a naturalist living at Kendal, tallies so well in its account of the transits of this bird there with its movements in this county, as to have led me to suspect the nature of their retreat before I read Selby's statement. This gentleman, Mr Gough, thus writes:—"The Grey Wagtail is a partial migrator; a few remain about the town through the winter, and these are joined by great numbers from the south in March, when they all retire to the rugged banks of the river Muit to spend the season of incubation." A remarkable feature in the habits and economy of birds is their adaptation of appetite to a variety of food, both as regards one season or time, and as regards various seasons, in which we frequently notice a change in their food. But few of the class confine themselves to one particular species of food, whereas a very large proportion partake of a variety, but still not similar in character. Thus, the Grey Wagtail searches out various insects, and is content to feed on such as are found on the shore, which necessarily must differ widely from those inhabiting the sides of rivers, or the streamlets. Very many birds, again, have appetites still more accommodating, and will devour food quite incongruous. This portion of the economy of the Grey Wagtail permits the extension of the species much more than would otherwise be effected, and we see also that it even protects the species to a great extent from death, for if it could not on emergency betake itself to the food afforded by the roads, when frost deprives it of more genial supplies, it must necessarily be the victim of want. We conclude also, that it is this principle of accommodation in the appetites and digestive powers of birds, and other creatures, which fits and enables them to live in the midst of alterations in their ordinary provender, effected by the operations of man, and which permits us to avail ourselves of their services in a domesticated or reclaimed state, without much trouble or inconvenience, their appetites shortly becoming adapted to an unaccustomed diet. And so, in their habits and actions, we must not fail to note a principle of accommodation of the same description. A little reflection must bring to our minds a thousand alterations in the face of Nature, wherever man has fixed his abode, or extended his domains; and as, on the one hand, we might *a priori* imagine the actions and habits of animals to be as undeviating and determinate on all points, as are the fundamental laws of their organizations and constitutions, so, on the other, we find a corresponding alteration and conformity of action in them, to suit our intrusions on their territories, our planting, our tillage, our building, and all our various operations on and perversions of Nature. In civilized and cultivated territories scarce an animal moves but it encounters alterations of our making; and, though the lower tribes can experience but slight impediments, and can have to adapt themselves thereto only in a very minor degree, yet the higher tribes must certainly employ some portion of thought at times to overcome these hinderances; and, as before said, if instinct were so confined and constrained a power, as usually conceived, these alterations in Nature would infallibly disarrange all their proceedings. Judging by the analogy of a vast number of instances of departure from accustomed actions, and by the anomalies of individual cases, as contrasted with the proceedings of species taken in the aggregate, we conclude that the instances of the Grey Wagtail's breeding in Devon are determined purely by choice, and are not dependent on any human causes or interferences, and that these are also cases showing that instinct is not so very constrained a faculty, but involves a certain portion of thought and volition. If the instincts implanted in the Grey Wagtail were of a definite, precise, constrained, and unalterable nature, they would necessarily pervade every member of the species; and so far from any pair of birds choosing to stay the summer with us, while all their fellows were preparing to migrate, no inducements of food ever so great, nor even any accidents or ailments impairing their bodily power, would prevent their essaying a flight ever so short and feeble; in fact, they would be compelled to exert themselves to the very utmost, and to sacrifice every feeling to this one object.

The Yellow Wagtail, *Motacilla flava* of Ray and other English writers, offers an illustration of the same diversity of operations among individuals of one species. This summer Wagtail arrives here about the very time the Grey Wagtail leaves us, and it also quits us about the period the winter species comes. Still stragglers are seen on to November, according to some remarks I made in 1831, frequenting both the coast and inland stations; and in October 1833, and other years, they have been noticed haunting the beaches near Plymouth, so that, without contending for their stay through the whole winter, I have reason to infer that, like the Grey sort, they are occasionally induced to act differently from the aggregate of their kind. Facts of the same nature as those I have above recorded, relative to these two birds, are also named by Mr Markwich in Linnæan Transactions, I. 126. The Yellow Wagtails congregate in August and September, and abide for several days on the beaches and shores, feeding among the sea-weed. They likewise affect open fields. The number collected at these times is disproportioned to our summer stock, so that probably this species approaches the southern parts of the kingdom previously to departure.

The Pied Wagtail, *Motacilla alba* of most English writers, appears to be a bird of more enduring constitution than the other kinds, because it suffers the alternations of our seasons without removing to other situations. It is resident with us all the year. Its actions, however, clearly indicate the possession of powers of accommodating itself to circumstances of necessity; not that the species acts in concert, or that the movements and operations of the individuals are simultaneous and uniformly similar; on the contrary, each bird seems intent on its own peculiar interests, and its having been ordained that the appetite of this species of bird should not be restricted or very limited in capacity, some individuals are found to diet on the sea-shore, whereby greater space is allowed to other individuals to procure food. In summer, however, when the supply of food is so ample for the generality of creatures, the number of Wagtails haunting the beaches is very small, whereas, toward winter, they augment greatly. Although I have said that the individuals appear to have separate and exclusive interests, yet it seems that some portion of the kind congregate and depart, we know not whither, some thinking they remove to other countries, and some contending that they merely take up fresh quarters in this same kingdom. However, this removal in August must act beneficially towards the remainder, both of that species and also towards the Grey sort. During summer they may be found distributed by the sides of rivers and ponds, on roads and in gardens, besides being also on the shores and inlets, as before said. In June I have seen them both in that situation, and in my own garden, and before the house on the road, searching for insects. About September, they are more particularly noticed arriving in the vicinity of houses and stable-yards. From that time, on through the winter, they obtrude themselves greatly in gardens, where they pick up the insects disturbed by the spade of the gardener, which had secreted themselves, and been wrapt in their winter's sleep, or in temporary torpor.

British Shrews.—The Reverend Leonard Jenyns, in a paper published in the last number of the Annals of Natural History, presents the following synopsis of British Shrews:—

Sorex, Linn.—Two middle incisors much produced; the upper ones curved, with a spur behind more or less prolonged; the lower ones almost horizontal; lateral incisors or false grinders small, from 3 to 5 above, 2 below, on each side; true grinders 4 above, 3 below, on each side; fur short and soft; snout attenuated; tail long.

Sect. I.—*Amphisorex*, Duvernoy.—Middle incisors in the lower jaw with the edge denticulated; the upper ones forked, the spur behind being prolonged to a level with the point in front; the lateral incisors which follow in the upper jaw five in number, and diminishing gradually in size from the first to the last; all the teeth more or less coloured at their tips.

1. *S. rusticus*, Jen., *Common Shrew*.—Snout and feet slender; tail moderately stout, nearly cylindrical, not attenuated at the tip, well clothed with hairs, which are very divergent in the young state, and never closely appressed. Appears principally to frequent dry situations, gardens, hedgebanks, &c.

2. *S. tetragonurus*, Herm., *Square-tailed Shrew*.—Snout broader than in the last species; feet, fore especially, much larger; tail slender, more quadrangular at all ages, and slightly attenuated at the tip; clothed with closely appressed hairs in the young state, in age nearly naked. More attached to marshy districts than the last species, though not confined to them.

Sect. II. *Hydrosorex*, Duv.—Middle incisors in the lower jaw with an entire edge; the upper ones notched, or with the spur appearing as a point behind; the lateral incisors which follow in the upper jaw four in number; the first two equal, the third somewhat smaller, the fourth rudimentary; the tips of all the teeth a little coloured.

3. *S. fodiens*, Gmel., *Water Shrew*.—Deep brownish-black above; nearly white beneath; the two colours distinctly separated on the sides; feet and tail ciliated with white hairs. Marshes and banks of ditches; but it is occasionally met with at a distance from water.

4. *S. ciliatus*, Sowerby, *Ciliated Shrew*.—Black above; greyish-black beneath; throat yellowish-ash; feet and tail strongly ciliated with greyish hairs. Found in the same situations as the preceding.

Horns of the Arnee.—A skull with the horns of an old bull of this species, which inhabits the plains and jungles of the upper parts of India, has lately been presented to the Edinburgh College of Surgeons by Dr James Smith. The length of the skull is 2 feet, and the forehead, which is slightly convex, measures $10\frac{1}{2}$ inches across, opposite to the middle of the horns. In young animals, the horns rise obliquely, but in very old ones, as in this individual, they spread out nearly horizontally, curve upwards and backwards, and have the points slightly incurvate. They are compressed and somewhat triangular, flattened in front, with numerous transverse broad rugæ, excepting towards the point, which is more rounded. The length of each horn, measured on the outer edge, is 4 feet 2 inches, its breadth at the base 7 inches, its circumference there 18 inches, and the space between the tips of the horns is 5 feet.

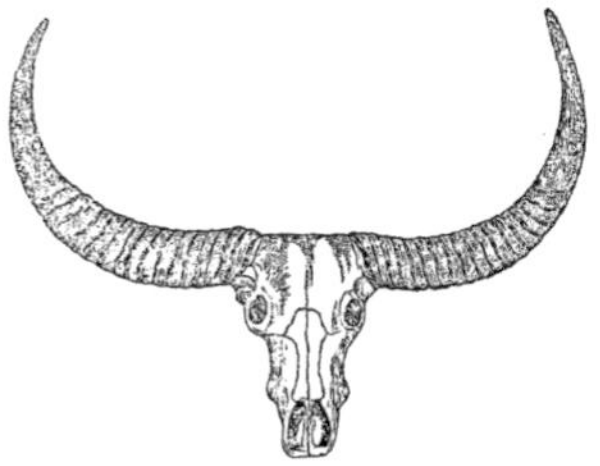

The Goatsucker.—Of the birds which abound in the Maremme of Patria, says Colonel Maceroni, in his Memoirs, in which I frequently used to pass weeks together, many varieties of the Bittern and of the Goatsucker abound. Of the Bitterns, one very small variety is remarkable for its Mephistopheles-like appearance, and accounts for the frequent appearance of its representation upon Egyptian obelisks and monuments; its body is not so large as a Pigeon, but the beak is a foot long, and the usual attitude of the bird mounted on its equally long legs, is with the sharp sword-like beak pointed vertically upwards. Of Caprimulgi, I have shot a very great variety; the most remarkable of which is one that has the exact representation of a white moth depicted on each side of the expanded tail of the bird, so that when flying about in the night, the moths, which constitute its principal food, seeing the moths upon the tail, come fluttering round their devourer, instead of avoiding him. I forget whether it is in this same specimen of the Goatsucker or in another, that I have remarked a very curious arrangement for the purpose of enabling the bird to see into its own mouth when extended wide open; and their mouths are enormous, so that they cannot easily miss their flying prey. Upon opening the mouth of this kind of Goatsucker, which they can do so as to place each half of their beaks at a straight line with the other, a large portion of the skin beneath the eyes becomes stretched and quite free of feathers; is as transparent as glass, so as to allow the eyes to see directly through it into the mouth, or further on in that direction. The operations of the Caprimulgi being carried on during the darkness of the night, have not been so much noticed by naturalists as they deserve.—This extract affords a good example of the ridiculous fancies which authors indulge in when they choose to imagine themselves physiologists.

Malay Albino.—On landing at Gressik I was struck by the singular appearance of a Malay lad, an albino, standing under the shade of a tree on the river-bank. His skin was of a reddish-white, with blotches here and there, and thinly covered with short white hairs. The ears were small and contracted; the iris of a very light vascular blue; the lids red, and fringed with short white lashes; the eyebrows scanty, and of the same colour; the pupil much contracted from the light. On calling him to come near he appeared to be ashamed. He evinced an extreme sensibility to the stimulus of light, from which he almost constantly kept his eyes guarded by shading them with his hands. He told me he could see better than his neighbours in imperfect darkness, and best by moonlight, like the "moon-eyed" albinoes of the Isthmus of Darien. He is morbidly sensitive to heat; for this reason, and on account of the superstitious respect with which the Malays regard him, he is seldom employed by his friends in out-door labour, although by no means deficient in physical strength. The credulous Malays imagine that the Genii have some furtive share in the production of such curiosities, though this they tell as a great secret. To this day the tomb of his grandfather, who was also an albino, is held sacred by the natives, and vows are made at it. Both his parents were of the usual colour. His sister is an albino like himself. Albinos I believe are not common on the Peninsula, nor are there any tribes of them, as according to Voltaire, existing in the midst of Africa. In the only two instances I recollect observing, the eyes were of a very light blue; the cuticle roughish, and of a rosy blush, very different from that of the two African albinos seen and described by Voltaire, and quoted by Lawrence.—*Journal of Asiatic Society of Bengal.*

METEOROLOGY.

Observations on the Effects of a remarkable Storm or Tornado. By Walter R. Johnson, A.M. Abridged from Vol. VII. of the Journal of the Academy of Natural Sciences of Philadelphia.

Considered as a meteorological phenomenon, the calamity which, on the 19th of June 1835, desolated a part of the city of New Brunswick in New Jersey, is worthy of the most attentive investigation. All accounts concur in representing the air of the morning, and, indeed, of the whole day up to the time of the tornado, as unusually sultry. At four o'clock the sun was still unobscured at Princeton; but within half an hour, a cloud from the north-west had reached that place, and a shower of rain, accompanied by a brisk wind from the south-west, had commenced. Before five o clock the rain had ceased, and the air was less oppressive. The evening continued tranquil until ten o'clock, when another shower of rain fell, accompanied by some wind; but within half an hour the sky was once more cloudless, and the wind began to rise with much force from the west or north-west. Between eleven and twelve a sensible depression of the dew-point was noticed, as indicated by the action of the air on the lungs, as well as on the surface of the body. From twelve at night to five the next morning the wind was boisterous; and a great change in the state of the atmosphere had obviously taken place. An electrical machine, which it had on the day previous been found impossible to excite, was, at nine or ten o'clock, A.M., able to yield sparks an inch and a half or two inches long, between balls three-fourths of an inch in diameter, a sure indication of an increased distance between the dew-point and the temperature.

Intelligence of the occurrence at New Brunswick having been received during the forenoon, it was resolved to visit the spot, and endeavour to ascertain such facts as might explain the mode of action of the tornado. On arriving within six miles of the city, we were informed that it had been seen about a mile and a half to the north-east, and that the dense black cloud was, by the junior observers, conceived to be filled with crows; an appearance afterwards explained by the fact that shingles, boards, &c., had been carried upward by the tempest from buildings destroyed in that vicinity. On reaching the height about half a mile from the dense portion of the city, the first buildings which had been damaged were passed. A barn had been completely demolished, and most of the lighter materials scattered to a great distance. The house was not thrown down, but left leaning with no part of the roof remaining, except some of the rafters; and the fact here witnessed was repeatedly observed in the town below, where several houses within the path of the tornado were deprived of their shingles, and the ribs which had held them to the rafters, but the latter still continuing partially or entirely undisturbed. In a few cases, in which the ridge of a building lay in a northerly and southerly position, the eastern slope of roof was observed to be removed, or at least stripped of its shingles, while the western slope remained entire. Many buildings were likewise observed with holes in their roofs, whether shingled or tiled, but otherwise not much damaged, unless by the demolition of windows. These appearances clearly demonstrated the *upward* tendency of the forces by which they were produced, while the half unroofed houses, already mentioned, prove that the result out of all the forces in action at the moment was not in a perpendicular to the horizon, but inclined to the east. Such a force would apply to the western slope of the roof some counteracting tendency, or relieve it from some portion of the upward pressure. Had there been no other facts to shew the powerful rushing of currents upward, the above would, it is conceived, have been sufficient to settle the question, but taken in connection with the circumstance that roofs so removed were carried to a great height, and their fragments distributed over a large extent along the subsequent path of the storm—that beds and other furniture were taken out of the upper stories of unroofed houses—that persons were lifted from their feet, or dashed against walls; and that, in one instance, a lad, of eight or nine years old, was carried, upward and onward with the wind, a distance of several hundred yards; and, particularly, that he afterwards descended in safety, being prevented from a violent fall by the upward forces, within the range of which he still continued. In connection with these and similar facts, it seems impossible to doubt that the greatest violence of action was in an upward and easterly direction.

The next point to which attention was called by the appearances around, was the manner in which this upward current had been supplied from below; and for the solution of this question, it was necessary to compare objects throughout the whole breadth of the track left by the storm. A peach orchard on the slope of the hill descending to the town, gave the first indication in regard to this matter, but the larger fruit and ornamental trees, in some gardens in the neighbourhood, together with an inspection of the forest on the east side of the river, shewed conclusively that on the extreme borders of the track, the forces were nearly or quite at right angles to its general direction. Uprooted trees along the southern border lay with their tops towards the north, those on the northern border to the south; thus pointing to a common object in the central line of the current. From the outer edges, however, towards this central line, the trees were observed on both sides to have a gradually increasing inclination towards the east, and in the middle to be entirely in *that*, as a general direction. None were seen with the tops from the centre of the path. A frame-building, which was on the southerly part of the track, was unroofed, and the remaining part of the structure, with its contents, removed bodily three or four feet to the northward. All the herbage, shrubs, and trees, in its immediate vicinity, and the trees in a garden, were found lying with their heads in a northerly or north-easterly direction. A stone near the river, and on the northern border of the path of the tornado, was lifted from its foundation about four or five feet towards the south. A row of poplars, which had been prostrated in the lower part of the city, and on the northern part of the path, fell southward. Another evidence of lateral inward currents was found in the appearance of many forest trees, east of the river, which, though too far removed from the central line of the path to be uprooted, were still so much within the range of the lateral forces as to have their outside limbs, or those most remote from the central line, broken off by the effect of cross strain; while no similar fracture was seen on limbs turned towards the centre of the path. This result will be easily understood when we consider the well known difference between breaking a limb by cross strain, and that of drawing it asunder by simple longitudinal tension.

Another fact, indicative of the direction of currents from the sides inward, was noticed on the plain east from the Raritan, where the shingle and fragments of boards lay with their longitudinal direction generally towards the point to which the storm was moving. Many of them were found far beyond the belt of ground on which the violence of the wind had been exerted. Their position may be explained by referring to the three forces in action at the moment they reached the ground:—First, the force of gravity, which, if the air had been motionless, and the bodies descending perpendicularly, would probably, from the unequal density of the parts of the several masses, have caused most of them to descend endwise; and then the position, subsequently taken by them respectively, would have been a matter of indifference, and we might have expected to find them lying promiscuously. But, second, they were, while in the air, moving onwards with the storm in an easterly direction, and when the lower end struck the ground, the composition of this force with gravity would naturally have thrown the centre of gravity over to the east, and we should have expected to find the lighter end of every piece of timber in that direction. But, third, if a current of wind were encountered near the ground, running towards the centre of the path, we should, on the north side of the path, expect to find the lighter ends of each piece directed to the south-east, and on the south side to the north-east; precisely what appeared to be the case, so far as could be judged from the general appearance of the masses.

The next set of facts observed had reference to the course of the materials projected upwards after they had arrived at a considerable elevation. All accounts agree that the appearance of the cloud was that of a funnel or inverted cone, with the apex resting on the ground. The falling rafters, scantlings, and other parts of the ruined buildings, generally indicated that they were, subsequently to the upward action, carried outward by the gradual enlargement of the current into which they had been drawn. The shingles and boards, just described, were cases in point, being found far beyond the trail of the tornado as marked upon the surface. Rafters, which penetrated buildings south of the track, entered them on the north side, and in a direction inclining to the south-east. Their descent in some instances was with great violence, contrary to what happened in the ravage of the upward motions; where a lad, already referred to, was deposited in safety after an aerial journey of one-fourth of a mile. A window-frame and brick wall were, in one instance, penetrated by a rafter twenty feet in length, eight inches wide, and from four to six inches thick. In the passage of the storm from the city to the opposite bank of the Raritan, no indications were, of course, left to mark its action upon the waters; but it was stated that it had laid the bed of the channel bare, and it is probable that had it traversed a great extent of water, it would have assumed the character, as it certainly had the form, of a water-spout. The upper edge of the opposite bank was marked by two well-defined stripes, each from ten to twenty feet wide, and one hundred or more feet asunder. Here, it was supposed, must have been the outer edge of the aerial trunk, or funnel, through which the air rushed upwards, and as the tornado, in its onward movement, advanced against the bank, the air coming in on every side to fill up the partial vacuum would exert the greatest force at the moment when it changed its horizontal for a vertical motion. The surface of the ground beyond this point seemed, in some places, to have been raised, as if the air beneath, by its sudden rarefaction, had thrown up small portions of the soil, which was rather dry and porous; and it is perhaps worth consideration, whether this cause may not, in this and similar occurrences, have facilitated the overturning of trees themselves.

In conclusion, it may be remarked, that the directions and intensities of the forces in this occurrence, together with the hygrometric states of the air, preceding and following the meteor, and the inverted conical form of the moving column, as confirmed by several witnesses, not less than the fall of hail, and the distribution of fragments of materials beyond the path of the ground current, seem most satisfactorily accounted for, on the supposition that a disturbance of atmospheric equilibrium results from a deposition of moisture in the higher regions of the atmosphere giving out a great amount of latent heat, which, in turn, expands the cold dry air above the forming cloud, and creates an ascending movement; the expansion of pure air by an addition of heat, being in such cases much greater than the contraction of this atmospheric mixture by a condensation of its moisture. In this effect is, of course, involved the well-known principle, that the capacity of air for heat is augmented as its volume expands, but the increase of capacity for heat being less rapid than the supply of heat from aqueous depositions, an ascending current is maintained with a force due to the difference of these two causes. The origin of this view of the subject, with which the writer had been made acquainted previously to the examination above detailed, is due to Mr J. P. Espy.

Fall of Meteoric Stones in Brazil.—On the 11th December 1836, about half-past eleven o'clock, P.M., the sky clear, and the wind south-west, a fire-ball, of uncommon size and brilliancy, appeared over the village of Macoa, at the entrance of the river Assu. It immediately burst with a loud crackling noise, and a shower of stones fell within a circle of ten leagues. They fell through several houses, and buried themselves some feet deep in the sand, but they did not occasion any further damage than killing and maiming a few oxen. The weight of those picked up varied from one to eighty pounds.—*Poggendorf's Annalen.*

Edinburgh: Published for the Proprietor, at the Office, No. 13, Hill Street.
London: Smith, Elder, and Co., 65, Cornhill. Glasgow and the West of Scotland: John Smith and Son; and John Macleod. Dublin: George Young. Paris: J. B. Balliere, Rue de l'Ecole de Médecine, No. 13 bis.
THE EDINBURGH PRINTING COMPANY.

THE EDINBURGH

JOURNAL OF NATURAL HISTORY,

AND OF

THE PHYSICAL SCIENCES.

OCTOBER, 1838.

ZOOLOGY.

DESCRIPTION OF THE PLATE.—THE CATS.

The characters of the extensive family of predaceous quadrupeds usually designated by the name of Cats have already been given.

Fig. 1. The Black Leopard (*Felis Leopardus, var.*) is a melanitic variety similar to that of the Jaguar.

Fig. 2. The Maned Hunting Leopard (*F. jubata*), which is of a light yellowish-red colour, covered with small roundish black spots not confluent, is a native of various parts of Asia, and has been tamed so far as to be sometimes used in hunting.

Fig. 3. The Red-eared Lynx (*F. caligata*), of a light-red colour, and with the tail of moderate length and banded with dark-grey, belongs to the section in which the ears are pointed and terminated by a tuft of hairs.

Fig. 4. The Serval (*F. Serval*) is very similar in form to the last, but wants the tufts on the ears, and is spotted with black. It inhabits Southern Africa.

Fig. 5. The Nepaul Cat (*F. torquata*) bears a great resemblance in colour and markings to the Wild Cat of Europe.

Fig. 6. The Collared Cat (*F. armillata*), of the same general form as the Serval, is marked with longitudinal elongated bands of light red, margined with black.

DESCRIPTION OF THE PLATE.—THE SHRIKES.

The Shrikes, which are small birds, generally of rapacious habits, with the bill strong, and having a prominent tooth or projection on each side of the upper mandible, near the decurved tip, are generally distributed over the globe; but certain groups are in a great measure confined to particular regions, and differences of form, combined with geographical distribution, indicate the propriety of a more extended division of this family than that adopted by the older systematic writers. The species of which the bill is strong, with the point curved, constitute the genus *Lanius* of Cuvier, while those whose bill approaches to that of the Thrushes are referred by Vieillot to the genus *Laniarius*.

Fig. 1. The Buff-breasted Shrike (*Lanius icterus*), olivaceous above, and yellow beneath.

Fig. 2. The Cruel Shrike (*L. pendris*), bluish-grey above, white beneath, with the fore neck black.

Fig. 3. The Supercilious Shrike (*L. supercilious*), light red above, whitish beneath, with a white frontal band extending over the eyes.

Fig. 4. The Whiskered Shrike (*L. mystaceus*) is one of the handsomest species of the genus, and moreover remarkable for the vivid red of its lower parts and tail.

Fig. 5. The Boubou (*L. Æthiopicus*), dusky above, and white beneath.

WILD CATTLE OF CHILLINGHAM PARK.

The following account of the Wild Cattle kept at Chillingham, and which are supposed to be the descendants of a species which formerly inhabited this country, is extracted from a letter addressed by Lord Tankerville to Mr Hindmarsh, who read a paper on these animals at the recent meeting of the British Association. His lordship supposes that " they were the ancient breed of the island, inclosed long since within the boundary of the park," but states that he is not in possession of any documents respecting them, or the period at which the park was first inclosed. " They have pre-eminently all the characteristics of wild animals, with some peculiarities that are sometimes very curious and amusing. They hide their young, feed in the night, basking or sleeping during the day; they are fierce when pressed, but, generally speaking, very timorous, moving off on the appearance of any one, even at a great distance. Yet this varies very much in different seasons of the year, according to the manner in which they are approached. In summer, I have been for several weeks at a time without getting a sight of them, they, on the slightest appearance of any one, retiring into a wood, which serves them as a sanctuary. On the other hand, in winter, when coming down for food into the inner park, and being in contact with the people, they will let you almost come among them, particularly if on horseback. But then they have also a thousand peculiarities. They will be feeding sometimes quietly, when, if any one appear suddenly near them, particularly coming down the wind, they will be struck with a sudden panic, and gallop off, running one after another, and never stopping till they get into their sanctuary. It is observable of them as of Red Deer, that they have a peculiar faculty of taking advantage of the irregularities of the ground, so that, on being disturbed,

41

they may traverse the whole park, and yet you hardly get a sight of them. Their usual mode of retreat is to get up slowly, set off in a walk, then a trot, and seldom begin to gallop till they have put the ground between you and them in the manner that I have described. In form they are beautifully shaped, short legs, straight back, horns of a very fine texture, thin skin, so that some of the bulls appear of a cream colour; and they have a cry more like that of a wild beast than that of ordinary cattle. With all the marks of high breeding, they have also some of its defects. They are bad breeders, and are much subject to the rush, a complaint common to animals bred in and in, which is unquestionably the case with these as long as we have any account of them. When they come down into the lower part of the park, which they do at stated hours, they move like a regiment of cavalry in single files, the bulls leading the van, as in retreat it is the bulls that bring up the rear. Lord Ossulston was witness to a curious way in which they took possession, as it were, of a new pasture recently opened to them. It was in the evening about sunset; they began by lining the front of a small wood, which seemed quite alive with them, when all of a sudden they made a dash forward altogether in a line, and charging close by him across the plain, they then spread out, and after a little time began feeding. Of their tenacity of life, the following is an instance. An old bull being to be killed, one of the keepers had proceeded to separate him from the rest of the herd, which were feeding in the outer park. This the bull resenting, and having been frustrated in several attempts to join them by the keeper's interposing (the latter doing it incautiously), the bull made a rush at him and got him down; he then tossed him three several times, and afterwards knelt down upon him and broke several of his ribs. There being no other person present but a boy, the only assistance that could be given him was by letting loose a deer hound belonging to Lord Ossulston, who immediately attacked the bull, and by biting his heels drew him off the man, and eventually saved his life. The bull, however, never left the keeper, but kept continually watching and returning to him, giving him a toss from time to time. In this state of things, and while the dog, with singular sagacity and courage, was holding the bull at bay, a messenger came up to the castle, when all the gentlemen came out with their rifles, and commenced a fire upon the bull, principally by a steady good marksman, from behind a fence at the distance of twenty-five yards; but it was not till six or seven balls had actually entered the head of the animal, one of them passing in at the eye, that he at last fell. During the whole time he never flinched nor changed his ground, merely shaking his head as he received the several shots. Many more stories might be told of hair-breadth escapes, accidents of sundry kinds, and an endless variety of peculiar habits observable in these animals, as more or less in all animals existing in a wild state: but I think I have recapitulated all that my memory suggests to me as most deserving of notice."

THE BLACK-TAILED GANNET.—SULA MELANURA. TEMM.

A specimen of this supposed species, which, however, seems to us to be merely a variety of the Common Gannet, is in the Museum of the University of Edinburgh. It was caught on the Bass Rock, in May 1831, and was submitted to our inspection while it was recent. The notes taken at the time are as follows. Apparently adult. The principal differences between this and the common are these:—

The bill is shorter, and at the base thicker.

The space from the eye to the base of the bill is shorter.

The black line of bare skin is extended down the neck to more than half its length, whereas in the common it is only about two inches long.

The secondary quills are dark purplish-brown; white in the common; but dark brown in the second year.

The tail is dark brown; white in the common; but dark brown in the second year.

Length 36 inches; extent of wings 64; bill along the ridge 3¾, along the edge of lower mandible 5½; tarsus 2⅝; middle toe 4½; primaries 10; secondaries 26; tail-feathers 12.

The size and proportions are thus nearly the same as those of the Common Gannet; from which it differs, as stated, more especially in having the bill shorter and thicker, and the naked longitudinal band on the throat more extended. Should these differences occur in other individuals resembling the adult of the common kind, but with the secondary quills and tail dark brown, the Black-tailed Gannet may be in reality a distinct species. A few individuals have been seen on the Bass Rock, but the people who rent it do not consider them of a different kind. This notice, it is supposed, is the first account of the occurrence of the black-tailed variety or species of Gannet in Scotland.

HARES OF THE UNITED STATES AND CANADA.

OUR highly esteemed friend, the Rev. Dr Bachman, of Charleston in South Carolina, has recently published, in the Journal of the Academy of Natural Sciences of Philadelphia, an elaborate monograph on the different species of *Lepus* inhabiting the United States and Canada, of which we here present a very brief abstract.

The Hares constitute the genus *Lepus*, of the order *Glires* or *Rodentia*, characterized by having four small teeth placed behind the two upper incisors, two incisors below, six grinders on each side in the upper and five in the lower jaw, in all twenty-eight; by having the ears and eyes large; five toes to the fore feet, four to those behind, with claws slightly arched; the interior of the cheeks covered with hair, as well as the soles of the feet; the tail short; the mammæ from six to ten.

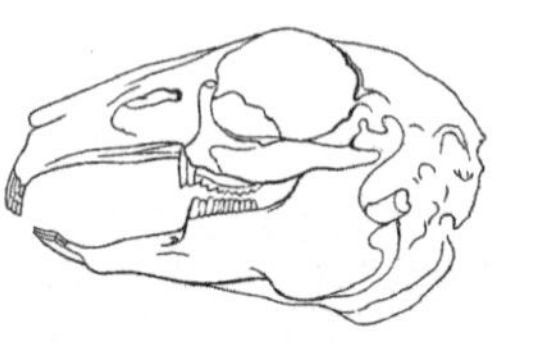

1. *Lepus glacialis* (Leach). Polar Hare. Larger than *L. Virginianus*; colour in winter white; hair of a uniform white to the roots; in summer of a light-grey above; ears black.

This species is the largest at present known in North America. It has a wide range in the northern portions of America, having been found on both sides of Baffin's Bay, on the Baren Grounds, as far north as the country has yet been explored; in the North Georgian Islands; and in Newfoundland and Nova Scotia. It shelters itself among large stones, or in the crevices of rocks, feeds in winter on the Labrador Tea plant and the berries of the Alpine Arbutus, produces several young at a birth, and affords excellent eating, its flesh having a finer flavour than that of any other American Hare.

2. *L. Virginianus* (Harlan). Northern Hare. Larger than the American Hare, less than the Polar; white in winter, the roots of the hairs blue, then yellowish-fawn, tipped with white; in summer, reddish-brown above, white beneath; ears a little shorter than the head.

This species is plentiful in the Northern States, in Canada, in the fur countries as far north as lat. 64° 30′, and has been seen on the Columbia River, the plains of the Missouri, and in the eastern states as far south as Virginia. It is supposed to be the fleetest species known, and Lewis and Clarke ascertained by measurement that it could leap 21 feet at a bound. It confines itself to the densest forests or to prairies overgrown with tall grass, feeds principally at night on grasses, bark, leaves, and buds of shrubs and trees, but as an article of food is inferior to the other species.

3. *L. aquaticus*. Swamp Hare. Larger than the American Hare, being nearly the size of the Northern Hare; tail, ears, and head, long; feet long, narrow, less covered with hair than those of the American Hare; general colour nearly black above, white beneath.

This species, now first described, is numerous in all the swamps of the western parts of Alabama, and still more abundant in the State of Mississippi and in the lower parts of Louisiana. It differs from all the other species in frequenting marshy places, occasionally resorting to the water, and swimming with ease.

4. *L. Americanus* (Harlan). American Hare. In summer the fur above yellowish-brown, on the outer surface of the fore and hind legs rufous; beneath, and under surface of the tail, white. In winter the upper surface considerably lighter, being grey or greyish-white, intermixed with black and brown hairs.

It is very widely diffused in the United States, and is very prolific, producing from five to seven at a birth. Its principal enemies, besides Man, are the Ermine, Canada Lynx, Wild or Rufous Cat, Foxes, and various birds of prey. It is fond of frequenting plantations, conceals itself by day in a brush-heap or a tuft of grass, and often commits havoc in the gardens.

5. *L. palustris* (Bachman). Marsh Hare. Smaller than the American Hare; ears much shorter than the head; eyes rather small; tail very short; feet small, thinly clothed with hair; upper surface yellowish-brown; lower grey.

It confines itself to marshy places near the coast, in Carolina, Georgia, and Florida; produces from five to seven young at a birth, and is superior to the American Hare as an article of food.

6. *L. Nuttallii* (Bachman). Nuttall's Little Hare. Very small; tail of moderate length; general colour above, a mixture of light-buff and dark-brown; beneath, light yellowish-grey; ears broad and rounded; lower surface of tail white.

Mr Nuttall, who discovered it, states that it is met with to the west of the Rocky Mountains, inhabiting thickets by the banks of several small streams which flow into the Shoshonee and Columbia rivers.

7. *L. campestris* (Nuttall). Prairie Hare. In summer the head, neck, back, shoulder, and outer parts of the legs and thighs, lead colour; the belly, breast, and inner parts of the legs and thighs, white. In winter it is pure white, except the black and reddish-brown of the ears, which never change.

It occurs in the northern and western parts, especially on the plains to the west of the Rocky Mountains.

8. *Lepus (Lagomys) princeps* (Richardson). Little Chief Hare. Head short and thick; ears rounded; legs short; no tail; colour above blackish-brown; beneath greyish-fawn.

It inhabits the Rocky Mountains from lat. 52° to 60°.

Dr Bachman concludes with the following general observations:—"The *Lepus glacialis* is an inhabitant of the polar regions, having been seen as far north as our discoveries have as yet extended, ranging across the whole of the northern part of our continent, and found as far south, at least, as Newfoundland, where it is abundant, occupying a range of more than 30° of latitude. There are, no doubt, wide intermediate spaces, where it is not found. Professor M'Culloch, residing at Picton, Nova Scotia, has not observed it in his vicinity. It generally avoids swampy situations, preferring high grounds, and is often seen along the sides of hills, and in rocky situations. The *Lepus Virginianus*, according to Richardson, is found as far north as lat. 68°, and is known to exist as far south as the mountains of Pennsylvania. The *Lepus Americanus* is believed to exist in all the settled parts of the United States. The *Lepus aquaticus* has not yet been discovered to the east of the State of Alabama, nor to the west of the Mississippi river, or the south of New Orleans, although it will probably be found to extend considerably beyond these limits. The *Lepus palustris* has not yet been seen north of the maritime districts of South Carolina, nor to the west of Georgia, but extends to the southernmost portions of East Florida. The *Lepus Nuttallii* seems peculiar to the district west of the Rocky Mountains. The *Lagomys princeps* has been found from lat. 42° to 60°, and probably extends along the whole range of the Rocky Mountains, especially in the most elevated regions.

"These different species, although they have a strong general resemblance during a period of the year, may yet, by a little attention, be easily distinguished from each other. *L. glacialis* may be known by its black ears in summer, and by its hair being snowy white, even to the roots, in winter. *L. Virginianus* may be recognised in summer by its reddish-brown colour, and in winter, by its fur being white at the tips, and plumbeous at the base. *L. aquaticus* may be distinguished from the Northern Hare by its never becoming white in winter; from the American Hare, by its larger size, by its fur becoming black in winter, instead of whiter, as is the case with the former, and by its aquatic habits; and from the Marsh Hare, by its being one-third larger, by its much longer head, ears, and tail, and by its swiftness of foot. *L. palustris* may at any time be distinguished by its short tail, which is never white beneath, by its small hind feet, resembling in this respect the Cavy, and by its aquatic habits. *L. Nuttallii* may be known by its diminutive size, and differs from *Lagomys princeps* by the presence of a tail. *L. campestris* may be distinguished from the Polar Hare, by the fur on the back never becoming pure white to the roots, in winter."

GEOLOGY.

ON THE SLATE OF SOUTH DEVON. BY J. C. BELLAMY, ESQ. YEALMPTON.

SLATE, in its varied appearances, occupies the generality of South Devon southward of Dartmoor, a granitic tract, being, as it were, diffused in all directions, the other strata, lime trap, &c., being intruded between its courses and hills. Under the term slate I here include every kind of rock popularly so called, and in consequence, for the sake of convenience, confound together rocks both "primitive and transition." However perfect systems and tables as set forth in books may seem to be, and however desirable it might be for me to present definite terms and divisions of the strata under examination, I am so clearly satisfied of the immatured condition of this science, simply from the fact that local phenomena have not yet been narrowly investigated and compared, that I think it highly probable a few years will once more remodify existing opinions and classifications, and that, consequently, it can be no great outrage to consider provisionally all our rocks of a slaty nature under one head, though I do not mean to exclude conjecture and classification altogether. Thus, while under the term slate I comprise mica-slate, gneiss-slate, clay-slate, roofing slate, dunstone, greywacke-slate, flinty-slate, and greywacke, with perhaps some other minor kinds, I believe it would be unphilosophical to disregard the principle of separation between the fossiliferous and the nonfossiliferous rocks, or the principle of arrangement derived from the occasional alternation, and intimate blendings of certain strata, together with the truly natural association of deposits by the occurrence of the same description of animal remains, and other structures in their substance. But who, in the midst of conflicting statements of authors, will undertake to explain what is clay-slate, and what is not, particularly as this is generally ranked amongst the primitive (nonfossiliferous) rocks, and with us contains indubitable proofs of being fossiliferous, besides being apparently in union with a mass of other slate containing no such demonstrations? In a popular work now before me, giving a succinct statement of strata, primitive formations are described as being destitute of petrifactions; then the rocks so classed under this head are briefly described, and amongst them stands "clay-slate." It is added, "the clay-slates of Switzerland are celebrated for their impressions of fishes. Mont Pilate consists of thin laminæ, and in almost every plate is impressed a fish!" Thus it would be gratuitous and presumptuous in me to draw a line of demarcation between the clay-slate and the other kinds of slate, supposing a natural distinction to exist, which I am not sure is the case, for, as respects a distinction founded on the occurrence of fossils, the dunstone and greywacke contain no fossils, and are, it is well known, in immediate connection with patches of fossiliferous slate, and consequently the hills of slate connected with the fossiliferous slate above named may, for aught I see, be equally regarded as coeval in its deposition. If, therefore, it were demanded of me to state what I regarded as clay-slate, and to draw a line of demarcation between it and other kinds, I should say, either that the separation must be dependent on the presence or absence of organic relics, presuming on the possibility of making a separation by that means, or that the entire mass of our slate not evincing the characters of mica, or of gneiss-slate, on the one hand, or betraying none of the characters of flinty-slate or of greywacke-slate on the other; in short, all that rock which amongst amateurs and persons loosely informed on the matter is termed clay-slate, and which I have above recognised in a general way as being fossiliferous, must receive this appellation. From what I have just stated, it might be gathered that the interruptions to the courses of the different varieties of slate were not sudden, but rather the reverse. This circumstance, indeed, is remarkable amongst the features of this rock considered as a whole. Mica-slate and gneiss-slate, however, hold a station quite apart from the others, and are certainly altogether distinct. On the other hand, clay-slate (used in a limited sense), roofing-slate, dunstone, and greywacke-slate, are observed to pass gradually into one another, and to reciprocate each other's qualities. Again, greywacke-slate, flinty-slate, Lydian-stone, (?) and greywacke, graduate into one another. Roofing-slate is found sparingly, and in small patches. It obviously passes into the general mass of clay-slate, of which, notwithstanding its containing organic remains, I have above surmised that it may in propriety be ranked apart. The general body of clay-slate assumes a great variety of aspects, which are manifestly graduated, sometimes approximating the typical roofing-slate, and sometimes degenerating into a loose brown debris, or becoming indented, or closely impacted, constituting "dunstone;" not unfrequently also running into decided greywacke, either in small patches, or even in extensive beds. Altogether, these last named rocks form the generality of our hills which are round-backed. In situations where the slate is in tolerably-sized fragments, fossils are found in plenty. Greywacke, flinty-slate, and Lydian-stone? are found in one situation, in close approximation to greywacke-slate, this last being itself the staple rock of that spot, and passing with great freedom into greywacke, and rather less so into the others. It and the flinty-slate are fossiliferous. Although I have mentioned certain kinds of slate as being fossiliferous, I believe that no fixed rule can be laid down on this subject, for the same kind of slate will in one place exhibit these remaining, and in another be destitute of them. Greywacke-slate is most constant in this respect. The rubbly and loose kinds of clay-slate, and the best kinds of roofing-slate, seldom contain them. The generality of our slate, that, namely, to which I presume the term clay-slate may be applied, is found assuming for the most part the appearance of small and loosely joined fragments of a light-grey colour, sometimes blueish-grey, much tinged with iron, and frequently intersected by seams of quartz. At intervals, these hills of clay-slate give good quarries of roofing-slate, of a bright lead colour, in which fossils are freely distributed. These quarries are in general soon worked out, the stone soon becoming dense with greywacke, or degenerating into a rubbly state, or a dunstone, which is an indurated form of clay-slate. The course and dip of our slate vary greatly, even at times within a small space. I have understood that near Exeter is a bed of slate horizontal in its course. This, it was thought, was a rare circumstance, but in this neighbourhood it is far from being an uncommon appearance, though only observed in small patches. From a level it varies to a close approximation to verticality. It faces usually from about south-west to about south-east, but likewise at times looks southwardly, eastwardly, and westwardly, or thereabouts; but though it is thus prone to variation, even repeatedly within a small distance, I am not aware that it ever faces northwardly of east and west. Its solidity varies sometimes within very confined limits, and this often depends on the sudden presence of greywacke, which is seen to pervade the slate with great freedom in some spots. At times there is reason to believe that access of air determines the decomposition of slate, though in the cases where this is supposed to have happened, I should consider from its brown and powdery appearance, that some peculiarity of chemical composition had existed previously. Generally, however, the loose rubbly condition in which we observe a deal of our slate to exist, cannot have been owing, at least in the first instance, and principally, to atmospheric influences; for in some places, as at the slate quarries, this presumed decomposition shews itself throughout the whole depth to which the rock has been worked—at one spot the rubbly and loose state of the rock gradually disappearing, and the substance by degrees getting closer, until it is fairly of the character of dunstone, or at the least of ordinary slate. Indeed, this transition from loose to indurated slate is exemplified most freely in this neighbourhood, and seems to be inconsistent with a supposition of decay from the agency of air. The oxide of iron occasionally stains our slate in fantastic shapes. It seems also to occur in greatest plenty in the immediate neighbourhood of fossils, and in their composition. I occasionally find fragments of slate in the quarries on which are marked dark, broad, concentric lines, in the manner of the layers of wood in the boles of trees, of which appearances I can offer no explanation. There is one fact in the history of our slate worthy of notice, and worthy of being further investigated with a view to some explanation. This is the bendings or reflexions in the slope of the rock observed in very many places. In one spot in particular I have observed it to be bent four times in the depth of as many feet, so as to assume a perfectly zig-zag appearance. It must be recollected that this occurrence is not limited to loose and disconnected kinds of slate, but may be seen on larger scales in good solid sorts also, though the repeated reflexions may not be here so remarkable or perfect. At Crab-Tree, near Plymouth, there is a hill of slate, which towards its summit turns off at an obtuse angle, which gives this rock a peculiar aspect. Mr Hennah, in a work before named, records an instance where the slate seems as if thrust up by some violent impulse from below, between two beds of lime. I am of opinion that several similar instances, as regards this appearance of the rock, can be shown in our neighbourhood, though as to the cause of this, as well as of the reflexions above specified, I can offer no satisfactory notion.

Greywacke-slate assumes a very different appearance from all others. It is harder, and not so extensively tabular, is of a dark lead colour, and passes very freely, and very suddenly, into greywacke, at numerous points. Viewing it as it appears on the coasts, it displays itself in large flat cakes, having thin and rounded edges between which other cakes are in their turns inserted. It passes also with freedom into flinty-slate, and also I suspect into Lydian-stone, but here I want the opinion of some competent judge. In the flinty-slate are found the turbinolias formerly alluded to. Beneath the water these are dislodged by the action of that element from their bed, and thrown up in more or less perfection of form on the beach, a fact corroborative of the general rule of fossils exceeding in density the matrix in which they occur. Dr Leach reports that at Buckfast in this county the lime is seen covering the slate, and that they both at this spot contain a quantity of flinty matter. Now, it is singular that close to the place where I reside, there is seen a small quantity of lime overlaid by the slate, and both containing flinty matter!—a fact seemingly confirming amongst others before named, the contemporaneous deposition of these strata, or at least their close relationship with respect to the convulsions which have disturbed the crust of our globe. The circumstance of strata at their points of junction being more or less cemented, blended together, and intermixed, appears to me to have attracted too little notice, for though in some cases their admixture seems rationally accounted for by the previous well-founded conjecture, that their deposition was coeval, there are other instances in which we have no ground for imagining that their depositions were coëtaneous. Lime and slate are most freely intermixed at their points of junction; and we see the same thing occurring between the slate and sandstone in spots where these meet, the latter frequently being wedged into the other; and in some places there being a union of qualities. But then, there are other reasons for imagining the contemporaneousness in the deposit of these strata. On the other hand, where slate abuts against granite or trap, we find that in some spots the former is cemented freely to the latter. This is seen in numerous blocks scattered profusely through the valley at Joybridge, but I am not sensible of any cases where there is a thorough intermixture, as seen between slate and limestone.

Amongst the varied relations assumed by slate in its before named general diffusion, there is one which seems to indicate that even the oldest (allowing for argument that they are of different ages) kinds of slate were not contemporary with granite, for this rock not only extends into contact with granite at our north, but even spreads between the hills of that primitive formation, assuming, I presume, in its run a lengthened three-sided figure. Whatever be the actual relations between lime and slate (or at least certain sorts of the latter), and whatever may be the depth to which the lime as a whole, or in some parts, descends, I must not omit to furnish an additional proof of the intimate connexion subsisting between these two strata, namely, that "Lime appears in masses on the north, south, and west sides of the Dart, insulated in schistus." (Carrington's Dartmoor, Preface.)

The most common fossils in our slate are the encrinites, of which there seem to be several species, and all different from those in the limestone. Perhaps the next kind most generally diffused is a sort of coral, or rather, more properly certain species of corallites, for the specimens differ very greatly. One which I have is thick, separate, and fan-shaped; another is fixed in relief, and of a loose powdery texture; while a third sort is like the last alæform, and appears as a mere impression in roofing-slate. They are all small in size. The turbinolias seem limited to the flinty-slate at Bovisand, but are there most plentiful. My brother procured from the slate at Crab-tree a species of "trilobus." Madrepores are found, though rarely, in slate. They seem to be very similar to, if not identical with, those so common in lime. In the flinty-slate at Bovisand, Miss Hook has found two anomalous specimens. One has the appearance of a Turbo, and the other bears a considerable similarity to a crab; but these are highly ambiguous. There are also various other fossils, of which I can render no definite account.

Decomposed and comminuted slate from old buildings, is used with benefit as a manure, for besides acting mechanically in separating tenacious clots of soil with which we may combine it, its composition shews that it is no despicable ingredient, of a productive soil. Shoemakers use what I conceive to be Lydian stone as "lap-stones," they being found on the shore of the required rounded shape, and possessing great density and tenacity.

VARIETIES.

Instinct of Animals.—The following curious statement is copied from a letter which the late William Fischbein, the well known animal painter, wrote in explanation of one of his most beautiful coloured sketches, now in possession of M. Meyer, of Hildburghausen. It represents five little red mice, in the presence of a young cat. "That instinct is an inherent or innate quality of animals, is clearly proved by experience. The cat possesses the instinct of catching and eating mice, and the mouse that of shunning the cat as its most dangerous enemy. Once, in Rome, I happened to open a drawer which I seldom had occasion to use, when I saw a mouse jumping out of it, and found among the papers a nest with five young mice, naked and blind, and of a pale flesh colour. I placed them on a table, handled them, &c., and they evinced no symptoms of fright, nor any inclination to get away, but only appeared eager to approach each other for the sake of warmth. There happened to be in the house a very young cat, which had never tasted anything but milk. I placed it near the little mice, by way of experiment, but to my astonishment it did not even look at them, nor perceive them, even when I turned its eyes in the proper direction, until at last, when I had repeatedly approached its nose to the mice, it suddenly caught a scent which made it tremble with desire. The propensity became more and more violent, and the cat smelled at the mice, touching them with its nose, when all at once, the pale-coloured like creatures became suffused with blood, and began to make great exertions to get out of the way of imminent danger, whilst the cat as eagerly followed them."—*Dr Weissenborn, in Mag. of Nat. Hist.*

The Vampire Bat of South America is often the cause of much trouble by biting the Horses on their withers. The injury is generally not so much owing to the loss of blood, as to the inflammation which the pressure of the saddle afterwards produces. The whole circumstance has lately been doubted in England; I was therefore fortunate in being present when one was actually caught on a horse's back. We were bivouacking late one evening near Coquimbo in Chile, when my servant, noticing that one of the horses was very restive, went to see what was the matter, and fancying that he could distinguish something, suddenly put his hand on the beast's withers and secured the Vampire. In the morning, the spot where the bite had been inflicted was easily distinguished, being slightly swollen and bloody. The third day afterwards we rode the horse without any ill effects. Before the introduction of the domesticated quadrupeds, this Vampire Bat probably preyed on the Guanaco or Vicugna, for these, together with the Puma and Man, were the only terrestrial Mammalia of large size which formerly inhabited the Northern part of Chile.—*Darwin, in the Zoology of the Voyage of H. M. S. Beagle.*

Preservation of Corn.—Thinking that all granaries are but imperfect shelter from the vicissitudes of weather, General Demarcay has made use of an ice-house situated on his estate, but no longer used as such. Its depth was sufficient to render it impervious to atmospherical changes, and he lined it with wooden planks so as to form a large case, but which was at some little distance from the bottom and the sides of the ice-house, so that it was not liable to the damp of the surrounding earth, and allowed a free circulation of air around its exterior. The corn was placed so as to fill the case to within a yard of the top; three layers of loose planks were placed at a third of a yard distance between each, and the roof was then formed of thatch, and in a conical shape. The experiment has lasted twelve years, and been constantly attended with satisfactory results. The same grain has remained there for three years, without the slightest alteration, and, what is remarkable, some newly-thrashed corn which had been completely wetted while it was measured in the open air, having been placed in the case, was three weeks afterwards found to be as dry and glossy as flax-seed.—*Athenæum.*

CHANGES OF COLOUR IN BIRDS.—In an excellent paper on the Hares of the United States and Canada, by the Rev. Dr Bachman of Charleston, S. C., are the following observations on this subject:—"In birds, as well as in quadrupeds, there are some species in which great changes of colour take place from the young to the adult state, and also at different seasons of the year; the question is often asked, whether this is effected by a gradual change of feathers, or of hair, or by a sudden moult in birds, and a shedding of the hair in animals. I am inclined to the belief, from many experiments on animals and birds in captivity, that Nature, in effecting these changes, does not proceed with uniformity in all the species. I have observed in the Yellow-crowned Warbler, *Sylvia coronata*, that the change occurs by a sudden moulting in spring; and in a very few days it changes its homely dress, with which it came to us in autumn, for its bright nuptial livery of spring and summer. The *Sylvia petechia* is, during autumn and winter, so nearly of the colour of the *S. coronata*, that they cannot, without careful examination, be known from each other. In spring, this bird becomes olive-green on the back, beneath yellow; the head at the same time becomes of a bright chestnut-bay colour. In this state of plumage, Bonaparte has described it as a different species, *S. palmarum*. In this bird also the change is effected, as I have observed on an examination of more than fifty specimens, by a moulting of the feathers from the 1st to the 15th of March, in Carolina; and it, in the course of two weeks, receives its bright summer plumage. On the other hand, there are many birds in which these changes take place slowly, and by a somewhat different process: the gradual fading or brightening of the feathers immediately after a moult. Thus I have seen the young of the Whooping Crane, *Grus Canadensis*, undergoing its mutations. A pair of these birds, when nearly two years of age, as they increased in size, began gradually to change their colours. I was in the habit of visiting them every two or three days during four or five months. They had just moulted; many of their new feathers were still sheathed. I could, at every renewed visit, perceive that the plumage was continually becoming lighter, till at last they were nearly pure white; and I was gratified to find my previous conjectures correct, that the birds I had so anxiously watched were Whooping Cranes, *Grus Americanus*. A male of the Summer Duck, *Anas Sponsa*, which I have preserved in an aviary for some years, loses its brilliant plumage at the time of moulting in summer, and continues for several weeks afterwards in the plain livery of the female, when the feathers begin to brighten gradually; from day to day the change continues, until, after the expiration of a few weeks, the bird assumes all its bright and beautiful plumage.

All our birds that assume one set of colours during six months of the year, and different ones during the other six months, appear to moult in spring as well as in summer. My examinations have extended to the *Sylvia coronata*, *Fringilla savanna*, *Fringilla Pennsylvanica*, *Sylvia petechia*, *Charadrius Helveticus*, *Strepsilas interpres*, and several species of *Sterna* and *Larus*.

ORNITHOLOGICAL LOGIC.—In his remarks on Gould's Birds of Europe, Mr Wood, the Editor of the Naturalist, has the following note:—"Red Ptarmigan, *Lagopus Britannicus*, Tétras rouge, Fr. The British ornithologist feels a peculiar interest in this species, as it has never been known to occur out of the British Islands; but since, as every sportsman knows, it is so abundant on the heaths of Cumberland, Westmoreland, Yorkshire, Wales, and many parts of Ireland, surely we cannot encourage the monopoly of our northern brethren by continuing the name *L. Scoticus*." The original name of *Tetrao Scoticus* was not given to the bird in question by the "northern brethren;" but having been imposed and received as genuine, and being moreover correct as the species is a native of Scotland, although also of England and Ireland, it must be retained, the species however being referred to its proper genus, *Lagopus*. Is *Pandicilla Suecica*, so named by Mr Wood, as well as others, peculiar to Sweden; or *Plectrophanes Lapponica* to Lapland; or *Sterna Anglica* to England; or are a hundred and fifty other birds named after particular sections of the world peculiar to those spots? If not, let the catalogue be revised, and we will consent to adopt the proposed alteration in the name of our Ptarmigan; but, in the meantime let it be honoured with the name of a country abounding in that heather in which it delights to dwell.

HYBERNATION OF SWALLOWS.—M. Isidore Geoffroy, in his zoological instructions drawn up for the new scientific expedition to the north, calls upon naturalists to observe any facts which they may meet with concerning the hybernation of Swallows. In consequence of this, M. Dutrochet communicates to him, that he found two of these birds in a state of torpor, in a recess, formed in the wall of a building. On being warmed by the hand they flew away, proving thereby that Swallows are occasionally capable of wintering in a northern climate.

GALVANIC TELEGRAPH.—The highly scientific mode of making instantaneous telegraphic communications by galvanic power, which has so long been considered attainable, has already been put to the most decided test on the London and Birmingham railway, under the direction of Professor Winston and Mr Stephenson, the engineer to the Company. Four copper wires, acted upon at each end of the line at pleasure, by the agency of very simple galvanic communicators, have been laid down on the line of the London and Birmingham railroad, to the extent of twenty-five miles. They are inclosed in a strong covering of hemp, and each terminus is attached, to a diagram, on which the twenty-four letters of the alphabet are engraved in relative positions, with which the wires communicate, by means of moveable keys, and indicate the terms of the communication. The gentlemen are fully satisfied that communications to almost any extent may be made instantaneously by the agency of galvanism.—*Quarterly Journal of Agriculture.*

ATHOLL FOREST.—It is said that there are 7000 Red Deer in the Athol Forest, and the number is not over-rated, if the extent of ground, of which they have the undisturbed possession, be any criterion. The Roe Deer also are numerous in the different plantations of the country. The Fox, the Wild Cat, the Marten, the Pole-cat, the Weasel, and the Alpine Hare, are common. The Rabbit, the Squirrel, and the Rat, have lately made their way into the country, and have increased so rapidly as to become troublesome and destructive.—*New Statistical Account of Scotland.*

REVIEWS.

A History of the British Zoophytes. By GEORGE JOHNSTON, M.D. W. H. Lizars, Edinburgh; S. Highley, London; W. Curry, Jun. and Co., Dublin.

THIS important work is recommended to the lovers of Natural History, not less by the very pleasing manner in which it is written, than by the scrupulous accuracy of description which it displays. An accomplished scholar, as well as an acute observer, the author has so illustrated what is technical by the choicest selection of what is common to the affections of all persons of cultivated faculties, that his work, while it affords a clear view of a class of animals hitherto little studied on account of the want of sufficient guides, is calculated to interest even those who satisfy themselves with a vague contemplation of the beauties of Nature, or derive pleasure from the familiar exposition of her wonders.

The first part is devoted to a history of Zoophytology, and the structure, physiology, and classification of Zoophytes. Restricting the term Zoophyte to those creatures "which in their form, or most remarkable characters, recall the appearance of a vegetable or its leading properties," he defines the class thus.—Animals avertebrate, inarticulate, soft, irritable and contractile, without a vascular or separate respiratory or nervous system; mouth superior, central, circular, edentulous, surrounded by tubular, or more commonly by filiform tentacula; alimentary canal variable; where there is an intestine the anus opens near the mouth; asexual; gemmiparous; aquatic. The individuals (*Polypes*) of a few families are separate and perfect in themselves, but the greater number of Zoophytes are compound beings, viz. each Zoophyte consists of an indefinite number of individuals or polypes, organically connected and placed in a calcareous, horny or membranous case or cells, forming, by their aggregation, corals or plant-like *Polypidoms*. The arrangement which he adopts, or rather elaborates, is as follows.

Subclass I. RADIATED ZOOPHYTES. Body contractile in every part, symmetrical; mouth and anus one; gemmiparous and oviparous.

Order I. *Hydroida.* Polypes compound, the mouth encircled with roughish filiform tentacula; stomach without proper parietes; reproductive gemmules pullulating from the body and naked, &c.

Order II. *Asteroida.* Polypes compound, the mouth encircled with eight fringed tentacula; stomach membranous, with dependent vasculiform appendages; reproductive gemmules produced interiorly, &c.

Order III. *Helianthoida.* Polypes single, free or permanently attached, fleshy, naked or encrusted with a calcareous Polypidom, the upper surface of which is crossed with radiating lamellæ; mouth encircled with tubulous tentacula; stomach membranous, plaited, oviparous, &c.

Subclass II. MOLLUSCAN ZOOPHYTES. Body non-contractile, and non-symmetrical; mouth and anus separate; gemmiparous and oviparous.

Order IV. *Ascidioida.* Polypes aggregate, the mouth encircled with filiform ciliated retractile tentacula: a distinct stomach, with a curved intestine terminating in an anus near the mouth; ova internal. Polypidoms horny and fistulous, or calcareous, membranous, &c.

The second part of the work, necessarily more extended, is composed of a history of all the British species, including characters of the orders, families, genera, and species. Besides the technical descriptions, we are favoured with numerous and often highly interesting historical notices, referring to the structure and physiology of the animals described, their uses, discoverers or describers, and other points. Sometimes quaint, and frequently antiquarian, although never obscure or antiquated, some of our author's scientific friends, if consistent with themselves, must deem him an affected imitator. We, on the contrary, are delighted with those peculiarities of style and manner which contrast so strongly with the dry didactic details, and awkward attempts at freedom of speech, of persons who do not always succeed in spelling their names, much less in constructing a sentence according to rule. It is pleasant, too, to be reminded of "the charming Hippothoa and rosy-armed Hipponoe," Nereids of old Hesiod; or of Drayton's

> "Antheia, of the flowers that hath the general charge,
> And Syrinx of the reeds, that grow upon the marge."

The descriptions are illustrated by forty-four plates, and numerous engravings on wood, for all which the drawings were made by the author's accomplished consort, "by Mrs Johnston, who is herself the engraver of fourteen of them. The naturalist who may have attempted similar illustrations will appreciate the labour, perseverance, and skill that has been bestowed upon them, and will not harshly censure any errors of detail which a minute criticism may discover." The discovery of these errors we leave to the envious, and conclude our brief notice of a work that deserves all commendation, by expressing our wish that it may quickly be dispersed over the land, and inspire in those who peruse it some of that spirit of unaffected piety which has dictated many of its most beautiful passages.

EDINBURGH: Published for the PROPRIETOR, at the Office, No. 13, Hill Street. LONDON: SMITH, ELDER, and Co., 65, Cornhill. GLASGOW and the West of Scotland: JOHN SMITH and SON; and JOHN MACLEOD. DUBLIN: GEORGE YOUNG. PARIS: J. B. BALLIERE, Rue de l'Ecole de Médecine, No. 13 bis.

THE EDINBURGH PRINTING COMPANY.

JOURNAL OF NATURAL HISTORY,

AND OF

THE PHYSICAL SCIENCES.

NOVEMBER, 1838.

ZOOLOGY.

DESCRIPTION OF THE PLATE.—THE DEER.

The Ruminantia, so named on account of the singular faculty which they possess, of subjecting to mastication a second time the food which they have hastily bruised and deposited in the stomach, may be primarily divided into two groups : those destitute of horns, as the Camels, Lamas, and Musks ; and the horned species, which again arrange themselves into two distinct families or sections, those with caducous bony horns or antlers, and those with bony persistent horns of the same texture as hoofs, and moulded upon a bony core. The antlered Ruminantia constitute the family of *Cervidæ* or Deer, of which there are numerous species, some of which are found in all the continents and large islands, excepting New Holland. The Deer are generally remarkable for elegance of form, and rapidity of motion. Their body is of moderate bulk, their limbs long and slender, their tail very short, their eyes and ears large ; and the males or Stags are furnished with branched horns or antlers, which are shed and renewed annually. The female of one species, the Rein Deer, is also horned. These appendages, varying in size and form, afford good distinctive characters for the species, taken in conjunction with the stature, colour, and other circumstances. The Stags represented in the accompanying plate belong to a section characterized by having the branches of the horns round, and generally tapering, not flattened.

Fig. 1. The Wapiti Deer (*Cervus Canadensis*) is much larger than the Stag of Europe, but very similar in form. It differs in not having the horns forked or branched at the extremity, as well as in other particulars. The northern parts of America are inhabited by this species, whose range is daily becoming more limited on account of the increase of population.

Fig. 2. The Red Deer (*C. Elaphus*). This beautiful and stately animal, once generally distributed over Europe, still occurs in a wild state in many of the hilly and wooded districts of that continent, as well as of the northern and temperate parts of Asia. In Britain it is confined to some districts of the Highlands of Scotland, where it is fostered by the land proprietors.

Fig. 3. The Bengal Deer (*C. Bengalensis*) is about the same size as our Red Deer, and inhabits various parts of India.

Fig. 4. The Porcine Deer (*C. porcinus*) is a small species, so named from its comparatively large body and short legs, giving it somewhat of the aspect of a hog. Its horns are very slender, with two very small antlers, one of which is placed near the summit.

Fig. 5. The Timor Deer (*C. Timorensis*), of larger size, is somewhat similar in form and colour, and has also two short branches on the horns, which are proportionally stronger.

DESCRIPTION OF THE PLATE.—THE PIGEON.

The extensive family of Pigeons, *Columbidæ*, characterized by a more or less tumid soft space at the base of the bill, a large double crop or dilatation of the œsophagus, a strong gizzard, and two very small cœca, together with other circumstances, such as their laying two white eggs, emitting a cooing or murmuring sound, and feeding their young by disgorging the food partially macerated in their crop, and mixed with a milky secretion, is divided into several groups or genera, to one of which, *Columba*, with slender bill, the birds represented in the plate belong.

Fig. 1. The Dwarf Pigeon (*Columba nana*) is a very beautiful small species, in the colouring of which green and yellow predominate.

Fig. 2. The Pearly Pigeon (*C. perlata*) is also green on the upper parts, with the fore-neck red, and the wing-coverts marked with oblong reddish spots.

Fig. 3. The Red-Headed Pigeon (*C. ruficeps*), light red, with the feathers of the hind-neck edged with green.

Fig. 4. The Banded Pigeon (*C. leptogrammica*), having the upper parts light-brown and green, transversely banded with dusky.

Fig. 5. The Ash-coloured Pigeon (*C. cinerea*), female ; and

Fig. 6. The Female of the Diadem Pigeon (*C. Diadema*).

There is little of particular interest known of these beautiful birds, which do not require to be technically described here, as they will be the subject of extended description in a subsequent part.

BRITISH BIRDS.—THE CORN CRAKE.

The Corn Crake, which, although it seldom comes under the observation of unprofessional admirers of Nature, is yet familiarly known by its cry, insomuch, that to most people it is "*cœt et præterea nihil*," is a small bird intermediate in size between the Quail and the Partridge, having the body much compressed, the neck rather long and slender, the head small and oblong. The bill is somewhat shorter than the head, stout, tapering, and much compressed ; the tongue rather short, fleshy, emarginate, and papillate at the base, flattened above, and pointed. The œsophagus is six inches long, of nearly uniform diameter and narrow, the stomach roundish, compressed, large, with strong lateral muscles ; the intestines twenty-five inches long, and of moderate width ; the cœca three and a quarter, being large, and, like those of the Gallinules, approaching in form to the cœca of the Gallinaceous birds. The eyes are bare for nearly half an inch ; the tarsus is of moderate length, compressed, but stout ; the toes very long, slender, and compressed ; the first very small ; the claws of moderate length, slender, and slightly arched. The plumage is blended ; the wings short, concave, and rounded ; the first quill much shorter than the second, which is slightly longer than the third ; the tail extremely short, arched, much rounded, of twelve very weak feathers. The bill is light brown, the lower mandible whitish at the end ; the iris light hazel, the feet bluish flesh-colour. The upper parts are light yellowish-brown, each feather marked with an oblong central spot of brownish-black, and laterally tinged with grey ; the wing-coverts are light red, some of them imperfectly barred with white ; a broad band of ash-grey passes over and behind the eye and ear, and the cheeks are tinged with the same ; the face, fore-part and sides of the neck, are light yellowish-brown, tinged with grey ; the sides and breast barred with light red and white ; the lower wing-coverts and axillar feathers light red ; the chin and abdomen brownish-white ; the quills and primary coverts light brown, the outer webs tinged with light red ; the edge of the wing, and outer web of the first alular feather and first quill, reddish-white ; the inner secondaries, and the tail-feathers, like the back. The length is 10¾ inches, the extended wings measure 16, the bill is eleven-twelfths, the tarsus 1½, the middle toe 1½, its claw three-twelfths.

Were we to betake ourselves on some beautiful summer morning to one of the pastures that skirt the sandy shores of the remote Hebrides, anticipating the rising of the sun, and listening as we proceed in the grey twilight to the cries of the distant gulls, and the loud crash of the little wavelet, whose fall on the beach produces a louder noise than the rush of the mighty billow would do in a storm, we should not fail to see as well as to hear the Corn Crake. Here let us crouch behind the turf wall, in view of that thicket of iris, and watch. There already, dimly seen, one is quietly walking along the grassy ridge, lifting high foot after foot, and sometimes stooping as if to pick up something. Now it stops, stands in a crouching posture, but on unbent legs, and commences its curious but monotonous song. Another is observed threading its way among the short grass of the adjoining piece of meadow land. The ruddy streaks in the east betoken the sun's approach to the horizon. There, along tide-mark, some dark-coloured bird approaches ; it perceives us, wheels round, and comes up, announcing itself by its croak as the Hooded Crow. The Crakes seem to understand the warning, and immediately betake themselves to the thicket, whence we can easily start them. Yet they sometimes allow you to come within a yard or two before they rise, and so closely do they sit, that I have once or twice seen a small pointer, which I had trained to bird-nesting, spring upon and seize one.

The Corn Crake visits us early in May. It may seem strange that a bird apparently so ill adapted for continued flight, should yet be capable of performing the long journeys necessary for its annual visits. Its ordinary haunts are fields of corn and grass, and in the less cultivated parts of the country the large patches of flags and other tall herbaceous plants, which occur in moist places. It runs with great celerity, so much so, that I think a man could hardly overtake it, and it seems extremely averse from flying, for it seldom rises until one gets quite close to it. When it has started, it flies heavily, with considerable speed, allowing its legs to hang, and soon alights. In an oat field in Harris, I once shot at a Rail that suddenly rose among my feet, when, apparently not having been hitten, it flew off in a direct course to the sea, about four hundred yards distant, where, to my surprise, it alighted and floated motionless, sitting lightly on the water, like a Coot or Gallinule. Soon after, a Black-backed Gull coming up, spied it, and, uttering a loud chuckle of delight, descended with rapidity, and carried it off in its bill. In this case, I think the bird was so frightened, although not hurt, that it entirely lost its presence of mind, as the Water Hen sometimes does under similar circumstances.

At all times of the day, but more especially in the early morning, and towards twilight, it utters its singular and well-known cry, resembling the syllables *crek, crek*, repeated at short intervals, and often continuing for many minutes, probably a quarter of an hour or more, if the bird is not disturbed. It has the reputation of being an expert ventriloquist, and whether or not it deserves that title, it is certain that one is very apt to be mistaken as to the spot in which the bird is when he listens to its cry, which is at one time loud, at another low, now seems to indicate a close proximity, now a remote position, and even appears to come from various directions. I have heard the Thrush and the Robin so sing, close at hand, that I imagined them to be far away, and it is probable that other birds have the same faculty, which seems to depend upon the elongation or contraction of the trachea. When uttering its cry, the Corn Crake usually remains still, standing with its neck considerably drawn in. I have watched it so employed through a hole in a wall. But I have also often seen it walk leisurely along at the time. As to its neck being "stretched, perpendicularly upwards," as alleged by Mr Selby, it may perhaps sometimes be so, but not usually. At the period when the nights are shortest, I have heard it commence its cry so early as one in the morning.

Although not gaudily attired, the Corn Crake is richly coloured, and when observed in its wild haunts, has an appearance of great elegance. It moves in a graceful manner, and when proceeding leisurely, walks with what might be called sedateness, lifting its feet rather high, jerking up its short tail, and bending its neck backwards and forwards at every step, like the Water Hen. If alarmed, either while walking, or while uttering its cry, it instantly ceases, stands still or crouches, and if it judges it expedient, starts off, throwing its neck out and its body forward. It is not gregarious at any period of its residence with us, although in favourable situations, such as extensive meadows, many individuals may sometimes be found not far from each other. Its food consists of worms, testaceous mollusca, and insects, especially lepidoptera.

Soon after its arrival, it begins to form its nest, which is composed of a few straws, laid in a slight hollow, among corn, or herbage of any kind. The eggs, which are of an elongated oval form, and of a light or cream-colour, pale greyish-yellow, patched, spotted, and dotted with umber or brownish-red and light purplish blue or grey, are generally about ten, or from eight to twelve. In colour they bear a remarkable resemblance to those of the Missel Thrush. Their average length is two inches and one-twelfth; their greatest breadth an inch and five-twelfths. The young are at first covered with long hair-like down of a blackish colour, and leave the nest immediately after they burst the shell, to follow their mother among the grass or corn. When only a few days old, they run with amazing celerity, and scatter about, so that when one falls in with a flock, it is very difficult to catch more than one or two of them. On such an occasion, I have seen the old bird come up and run about in great distress.

Towards the middle of July, the Crake ceases to utter its cry, and one might suppose that it then leaves the country; but the period of its departure is protracted to the beginning of September. I have seen young birds in the end of that month, and instances of their having been shot in winter have occurred in various parts of the country.

The flesh of this bird is white, and affords delicate eating; but this sort of game is not easily obtained in the more highly cultivated tracts. In the Hebrides and West Highlands, however, few birds are more common, insomuch that there is hardly a patch of yellow iris or meadow-sweet, of the nettle, dock, or other tall weed, in which a Crake or two may not be found. Its cry may be so successfully imitated by drawing an edged stick along the teeth of a comb, or a thin piece of bone along another which has been notched by a saw, that by this artifice the bird will sometimes be induced to come up. Pennant and Montagu state that on its first arrival it is very lean, but before its departure becomes excessively fat. I have never, however, seen any great difference in this respect, birds obtained early in the season being in as good condition as afterwards.

The young, when fully fledged, differ from the old birds chiefly in wanting the bluish-grey markings on the head and neck. Individuals are often seen so late as the end of September, and a very few instances of their having been shot in winter have occurred.

THE PARROT GROSBEAK.

Having in our possession specimens of this very rare bird, which has not yet been minutely described, we have thought it may prove interesting to our ornithological readers to present an account of it. The genus *Psittirostra* is composed of a single species, about the size of *Loxia curvirostra*, which it resembles in form and proportions, differing, however, in the bill, which approximates more to that of *Corythus Enucleator*. It is precisely intermediate between the genera *Loxia* and *Corythus*.

The bill is short, and very robust; the upper mandible has its dorsal line curved from the base into the fourth of a circle, the ridge narrow, the nasal sinus short and rounded, the sides convex and much declined, the edges sharp, overlapping, inclinate at the base, decurved, destitute of notch, the tip descending much beyond that of the lower mandible, trigonal, and acute; the lower mandible with the angle rather short and rounded, the dorsal line ascending and very convex, the sides of the crura erect, towards the end convex, the edges sharp and somewhat inclinate, the tip broad and rounded, but with a narrowed slightly ascending point.

The nostrils are basal, small, half closed by a horny operculum, and partially concealed by short feathers. Aperture of eyes of moderate size; eyelids feathered. Aperture of ear rather large, and roundish.

The head is large, and roundish-ovate; the neck short; the body moderately stout. The feet are rather short, and of moderate strength; the tarsus short, compressed, with seven scutella in front, and two lateral plates meeting behind so as to form a sharp edge; toes four, moderate, compressed; the first stout, the lateral about equal, the third much longer; the third and fourth united at the base; claws rather large, arched, much compressed, very acute, laterally grooved.

Plumage soft and blended; the feathers ovate and rounded; those on the fore part of the head very short. Wings of moderate length, broad, somewhat rounded; primary quills nine, the first being obsolete, the second longer than the sixth, the third longest, but scarcely exceeding the fourth, the fifth considerably shorter; secondary quills ten, broad and rounded, the inner short. Tail rather short, emarginate, of twelve moderately firm, obliquely rounded feathers.

This genus differs from *Corythus* in having the bill larger and higher, with the upper mandible more elongated, and the lower more convex and rounded; from *Loxia* in not having the tips compressed and laterally divaricate. It was instituted by Temminck, from *Loxia psittacea* of Gmelin and Latham, and is appropriately named, the bill having a great resemblance to that of a Parrot.

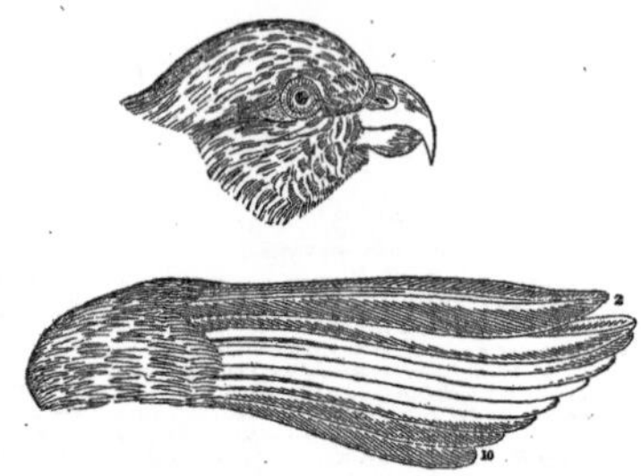

Psittirostra psittacea. Parroquet Parrotbill.

Male.—The bill and feet are light brown. The head and neck yellow; the upper parts olive-green, tinged with grey; the wings and tail blackish-brown; the first row of small coverts, and the secondary coverts, tipped with whitish, the quills and tail-feathers margined with light greenish-grey. The throat is greyish-white, the sides of the neck and body light brownish-grey, the middle of the breast and abdomen yellow; the lower wing-coverts and tail-coverts white.

Length to end of tail 7 inches; bill along the ridge 7½ lines, along the edge of lower mandible 7½ lines; wing from cubitus 3¾ inches; tail 2¼; tarsus 10½ lines; hind toe 4½ lines, its claw 5 lines; middle toe 8 lines, its claw 4 lines.

Female.—The bill and feet are as in the male; all the upper parts are olive-green tinged with grey; the throat whitish; the cheeks, sides of the neck and body, light grey; the other parts as in the male. The size is somewhat smaller.

M. Temminck having stated that the first quill is wanting, Mr Swainson, not having a specimen, remarks that this requires explanation. The Parrotbill is not singular in this respect, for all those passerine birds of which the outer quill is as long as the next or nearly so, are equally destitute of the first quill, these species having only nine primaries. I am not aware of this fact having hitherto been observed.

GEOLOGY.

Lake of Arendsee.—Near Arendsee, in the circle of Magdeburg, there is a remarkable lake of the considerable extent of about a German square mile, or about eighteen English square miles. It has been formed in a flat country, within the historical times, probably by the superficial strata sinking into an immense cavern excavated by subterraneous currents of water. According to Almonius, this event appears to have taken place about a thousand years ago. The lake was considered as unfathomable, and within the memory of man it never had been frozen, the great depth of its water preventing the latter from taking a sufficiently low temperature through that severity and duration of frost which the winters of Northern Germany commonly present. Last winter, however, this rare phenomenon did occur, long after the greatest rivers had been covered with a solid crust; and after having spent its free caloric in large masses of vapour, which for many days hovered over its surface and banks, the morning of the 31st of January exhibited it all covered with one smooth and polished plate of ice. The thickness of the latter was nine inches, and in a few places not above four or five inches. This was a convenient opportunity for taking accurate measurements of the depth of the lake, and it was then first ascertained that the opinion of its being unfathomable is unfounded. The general depth does not exceed 157 feet, only near the ruins of an old convent, at a distance of 400 steps from the bank, it was found as deep as 161 feet, which may be taken for its greatest depth. Beginning from the south bank, at a place where a large piece of ground sunk in 1685, the depth increased within distances of 400 steps each, at the following rate: —42½ feet, 87, 116, 137, 157. Among the many remarkable phenomena presented by this lake, the one that it throws out *yellow amber* is, perhaps, the most striking. This substance is only found on its eastern bank, and the more violently the west winds blow, the more yellow amber is there collected. The size of the fragments does not, however, generally exceed that of a French bean. As the whole tract of Magdeburg to the Baltic Sea is pretty uniform, we may conclude that in one of its strata it contains an almost continuous bed of yellow amber, which on the shores of the Baltic is exposed in a great part of its length, whereas near Arendsee it has been accidentally opened by the sinking of the ground. Many petrifactions of wood and other substances are likewise thrown out. Innumerable fish, as Eels, Pike, Tench, Perch, &c., inhabits its waters. The fishery is, however, comparatively little productive, on account of the great depth of the lake. Pikes of the enormous weight of 50lb., and Eels of 15lb., are not unfrequently caught.

DEER HUNTING IN THE TEXAS.

We have been favoured by a friend, who has recently visited Texas, with the following notice respecting one of its principal islands, and the habits of a species of Deer found there.

It has seldom happened to me to see a more interesting island than that named Galveston, which lies in the inner curvature of the Bay of Mexico. It is flat, somewhat sandy, in places cut up by numerous diverging bayous, between which are marshes of a brackish nature. Its sandy shores produce perhaps the best Oysters in the world, while around it, in the deep waters of Galveston Bay, are found numerous species of fish, not less pleasing to the palate of the epicure, than interesting to the curious ichthyologist. There too are Prawns of finer quality than any found along the Atlantic Coast of North America.

This island, besides being celebrated for Oysters, Prawns, and fine fish, in which respect, however, it does not differ from many others in the same country, is still more so on account of its having been the great place of rendezvous of the ruffians commanded by the famous Pirate Lafitte, a man who, twenty years ago, was looked upon with terror even by the superior authorities of the countries around. There was a time, indeed, when this reprobate had entire possession of Galveston Island, on which he erected his fortresses, raised his numerous huts, and could conceal within the harbour the whole of his curious fleet in the most perfect state of security, for as I can well assure you, the entrance of Galveston Harbour or Bay is most intricate and difficult of access. However, this atrocious villain was at last secured, and, I believe, suffered the punishment which his vile deeds so much merited.

If ever there was a spot where game might be found in abundance, it is Galveston Island. There the common American Deer, *Cervus virginianus*, or perhaps a species not yet pourtrayed or described, exists, not as Deer are seen in the parks of England, by hundreds, but by thousands. Leave your grassy couch at dawn, see that your rifle is in good order, that your pouch contains fifty bullets, and your horn a sufficient quantity of powder, be a good marksman, calm, steady, and perfectly adept in the work of destruction, and then tell me, when you return towards the setting of the sun, if you have not procured more venison than would suffice for a hundred of the most hungry and voracious flesh-eaters.

Methinks now I am on that loved spot, that I have risen a little before the appearance of the orb of day, that, clothed like a true hunter, I stand in the midst of my companions, who are similarly equipped. We are landed from our vessel at the distance of some eight or ten miles from the Town of Galveston, say, if you choose, at the "Eagle's Nest." Then here we all are, quite ready for a frolic. Our guns are in trim, our sailors, as sailors generally are, gay and sprightly, and towards yon cover of small bushes we proceed with some caution. The shooters take their stations at pretty regular distances, under the lee of the thicket, and at such spots as we conceive to be those at which the game will probably issue. Our sailors, now well trained, for we have no dogs, enter the bush windward, and now their loud shrieks announce to us their willingness to perform their duty faithfully. "Halloo! Halloo! Halloo!" these sounds come pleasantly on our ears, however much terror they may inspire in the timid deer. Up starts,—not a Stag, but a Galveston Owl. "Quite a stranger!" quoth the sportsman in advance; but, as it proves somewhat of a curiosity, in the way of feathers and flesh, no sooner has it come within range of my "Long Tom," than aim I take, pull the slick trigger, and down to the earth falls the bird. Think not that such an incident tends to mar the main sport. No; Deer, such at least as those of Galveston Island, on such occasions shrink into their form, lay down their ears, and perhaps, for aught I know, indulge in the thought of their being quite safe. But hark!—"Watch! Watch! Watch!" Ah! how delightful the sound! The Deer is a-coming; the guns are all instantly cocked, and every shooter has crouched close down. There!—the antlered, smooth, and sleek-skinned Buck emerges, bounding violently at first, but now advancing timorously, for you see it crouching almost to the ground as it enters the plain; its tail nevertheless goes this way and that way, and again and again waves, like the flag of some party willing to come to a parley. But, alas for thee, thou forlorn Buck! I have heard the report of a gun. I see the beautiful animal stagger; his tongue is out, his eye becomes dim. There he lies at our mercy extended on the ground! The shouts of the sailors are repeated. All the Deer in the thicket are now truly alarmed. Out and in again bound some of the most timorous; now they are hard pressed by the tars, and many are forced to take to the open plain; but before they have sped fifty yards toward another cover, the agonised bleatings of several are heard amidst the noise of the shots.

"Enough for a week's food," says one. "Agreed," say all. And now the sailors are called. "Bring your oars, brave fellows;" these most useful implements have been used in beating the bushes. The Deer are now fastened so as to swing by their four feet, three or perhaps four on the same oar. Two sturdy fellows shoulder the bending sticks; others follow, laden in the same manner; and toward the sandy shore we return.

Did you ever see a Deer skinned in a trice, and cooked in sixty minutes? I have. It is done thus. Imagine first that the game and all have reached the inner margin of the bay, and that the venison "in the skin" lies on the ground. Our encampment is prepared, by which I mean you to understand, that four oars have been put up, and the sail of our craft attached to them, so as to keep the sun's rays from parching our eyes. A fire has been kindled, to chase away the myriads of mosquitoes, now fully as hungry as ourselves. The Deer are dragged to the water, and there skinned *sans ceremonie*; that is to say, by cutting and tearing off their hide in a manner which an Edinburgh "flesher" would hold most unbutcherly. Then the carcase is washed in the brine, partially salted, if you please to call it so, and hung in front of the augmented blaze. There let it roast a while. "Turn it again, John." "Aye, aye, sir;" and so it fizzes and drips. Each hunter, who, in the meantime, has collected along the shore his hundred or so of Oysters, now watches them as they gasp in the agonies of death, hissing and opening their shells wider and wider, in the midst of the burning embers. "Done, sir, quite done!" "Thank you, John." Now, every man, his own carver, cuts and slashes according to his own taste; the juicy meat allays our hunger, the Oysters follow as a dessert; and should we apprehend an apoplectic attack, we guard against it by washing all down with a glass of grog.

My opinion, that the Deer which are so abundant on Galveston Island, as well as in all the maritime parts of the Texas, is different from the Virginian Deer, is founded on the following facts. The latter, which is common and well-known, not only in Virginia, but throughout the whole of our Atlantic states, from Maine to South Carolina, is a much larger animal. It differs also materially in its general colour at all seasons, being more red in summer, and of a darker grey in winter. Its average weight, when in good condition, may be estimated at about a hundred and fifty pounds. Its antlers are much larger, more bent forwards, and considerably less palmated than those of the Texian or Floridan species. In its movements, the Virginian Deer, though extremely swift and graceful, differs from the Deer of Galveston Island, in exhibiting a certain *je-ne-sçai-quoi* in its bound, which resembles in this respect that of the Roe-Buck of Scotland and of Europe generally. The Galveston Deer is also wont to congregate in immense numbers at all times, so much so, indeed, that I and my companions have passed through spaces on Galveston Island where their tracts were as thick as those in a sheep-fold. In spring and summer, the general colour of the pile is in some individuals almost white, and in winter red, instead of dark-grey or brown. Whether this may be owing to the influence of climate or not I cannot positively say, although I think not. At Galveston, their principal enemy, besides Man, is a small wolf, probably of a species yet unknown to Naturalists, which hunts in packs, and has been known to chase the Deer even into the town.

THE HYDRA, OR FRESH-WATER POLYPUS.

Leeuwenhoek discovered the Hydra in 1703, and the uncommon way its young are produced; and an anonymous correspondent of the Royal Society made the same discovery in England about the same time, but it excited no particular notice until Trembley made known its wonderful properties, about the year 1744. These were so contrary to all former experience, and so repugnant to every established notion of animal life, that the scientific world were amazed; and while the more cautious among Naturalists set themselves to verify what it was difficult to believe, there were many who looked upon the alleged facts as impossible fancies. The discoveries of Trembley were, however, speedily confirmed; and we are now so familiar with the outlines of the history of the Fresh-water Polype, and its marvellous reproductive powers, that we can scarcely appreciate the vividness of the sensation felt when it was all novel and strange; when the leading men of our learned societies were daily experimenting upon these poor worms, and transmitting them to one another from distant countries, by careful posts, and as most precious gifts, and when even ambassadors interested themselves in sending early intelligence of the engrossing theme to their respective courts.

The Hydræ are found in fresh, and, perhaps, also in salt waters, but the former species only have been examined with care, and are the objects of the following remarks. They prefer slowly running or almost still water, and fasten to the leaves and stalks of submerged plants by their base, which seems to act as a sucker. The body is exceedingly contractile, and hence liable to many changes of form. When contracted it is like a tubercle, a minute top, or button, and when extended it becomes a narrow cylinder, being ten or twelve times longer at one time than at another, the tentacula suffering changes in their length and diameter equal to those of the body. "It can lengthen out or shorten its arms, without extending or contracting its body; and can do the same by the body, without altering the length of its arms; both, however, are usually moved together, at the same time and in the same direction." The whole creature is apparently homogeneous, composed of minute pellucid grains cohering by means of a transparent jelly, for, even with a high magnifier, no defined organization of vessels and fibres can be detected. On the point opposite the base, and in the centre of the tentacula, we observe an aperture or mouth which leads into a larger cavity, excavated as it were in the midst of the jelly, and from which a narrow canal is continned down to the sucker. When contracted, and also when fully extended, the body appears smooth and even, but "in its middle degree of extension," the sides seem to be minutely crenulated, an effect probably of a wrinkling of the surface, although from this appearance Baker has concluded that the Hydra is annulose, or made up of a number of rings capable of being folded together or evolved, and hence in some measure its extraordinary ability of extending and contracting its parts. That this view of the Hydra's structure is erroneous, Trembley has proved; and the explanation it afforded of the animal's contractility was obviously unsatisfactory, for it was never pretended that such an anatomy could be detected in the tentacula, which, however, are equally or more contractile. These organs encircle the mouth, and radiate in a star-like fashion, but they seem to originate a little under the lip, for, the mouth is often protruded like a kind of small snout; they are cylindrical, linear, or very slightly tapered, hollow, and roughened, at short and regular intervals, with whorls of tubercles which, under the microscope, form a very beautiful and interesting object, and I have thought when viewing them, that every little tubercle might be a cup or sucker similar to those which garnish the arms of the Cuttle-fish. Trembley has shown us that this is a deception, and that there is really no exactness in the comparison. The tentacula are amazingly extensible, from a line or less to one, or, as in *Hydra fusca*, to more than eight inches; and "another extraordinary circumstance is, that a Polype can extend an arm in any part of its whole length, without doing so throughout, and can swell or lessen its diameter, either at the root, at the extremity, in the middle, or where it pleases; which occasions a great variety of appearances, making it sometimes terminate with a sharp point, and at other times blunt, knobbed, and thickest at the end, with the figure of a bobbin. We naturally inquire how this wonderful extension is made,—by what power a part without muscularity is drawn out until it exceeds by twenty or even by forty times the original length? The dissections of Trembley have proved beyond all doubt that the body is a hollow cylinder or bowel, and that the tentacula are tubular, and have a free communication with its cavity; and in this structure, combined with the loose granular composition of the animal, we find an answer to the question. Water flows, let us say by suction, into the stomach, through the oral aperture, whence it is forced by the *vis a tergo*, or drawn by capillary attraction, into the canals of the tentacula, and its current outwards is sufficient to push before it the soft yielding material of which they are composed, until at last the resistance of the living parts suffices to arrest the tiny flood, or the tube has become too fine in its bore for the admission of water attenuated to its smallest possible stream,—how inconceivably slender may indeed be imagined, but there is no thread fine enough to equal it, seeing that the tentacula of *Hydra fusca* in tension can be compared to nothing grosser than the scarce visible filament of the gossamer's web."

The Hydra, though usually found attached, can nevertheless move from place to place, which it does either by gliding with imperceptible slowness on the base, or by stretching out the body and tentacula to the utmost, fixing the latter, and then contracting the body towards the point of fixture, loosening at the same time its hold with the base; and by reversing these actions it can retrograde. Its ordinary position seems to be pendant or nearly horizontal, hanging from some floating weed or leaf, or stretching from its sides. In a glass of water the creature will crawl up the sides of the vessel to the surface, and hang from it, sometimes with the base, and sometimes with the tentacula, downwards; and again it will lay itself along horizontally. Its locomotion is always very slow, and the disposition of the zoophyte is evidently sedentary; but the contractions and mutations of the body itself are sufficiently vivacious. While in seizing and mastering its prey it is surprisingly nimble; seizing a worm, to use the comparison of Baker, "with as much eagerness as a cat catches a mouse." It is dull, and does not expand freely, in the dark, but enjoys light, and hence undoubtedly the reason why we generally find the Hydra near the surface and in shallow water.

The Hydræ are very voracious, feeding only on living animals, but when necessary they can sustain a fast of many weeks without other loss than what a paler colour may indicate. Small larvæ, worms, and entomostracous insects, seem to be the favourite food, and to entrap these, they expand the tentacula to the utmost, and spread them in every direction, moving them gently in the water to increase their chances, and when a worm, &c., touches any part of them, it is immediately seized, carried to the mouth by these flexible and contractile organs, and forced into the stomach. "'Tis a fine entertainment," says Baker, "to behold the dexterity of a Polype in the mastering its prey, and observe with what art it evades and overcomes the superior strength or agility thereof. Many times, by way of experiment, I have put a large worm to the very extremity of a single arm, which has instantly fastened on it with its little invisible claspers. Then it has afforded me inexpressible pleasure to see the Polype poising and balancing the worm with no less seeming caution and judgment than a skilful angler shows when he perceives a heavy fish at the end of a single hair-line, and fears it should break away. Contracting the arm that holds it, by very slow degrees, he brings it within the reach of his other arms, which eagerly clasping round it, and the danger of losing it being over, all the former caution and gentleness is laid aside, and it is pulled to the Polype's mouth with a surprising violence." Sometimes it happens that two Polypes will seize upon the same worm, when a struggle for the prey ensues, in which the strongest gains of course the victory; or each Polype begins quietly to swallow his portion, and continues to gulp down his half until the mouths of the pair near, and come at last into actual contact. The rest which now ensues appears to prove that they are sensible of their outward position, from which they are frequently liberated by the opportune break of the worm, when each obtains his share; but should the prey prove too tough, woe to the unready! The more resolute dilates the mouth to the requisite extent, and deliberately swallows his opponent, sometimes partially, so as, however, to compel the discharge of the bait, while at other times, the entire Polype is engulphed! But a Polype is no fitting food for a Polype, and his capacity of endurance saves him from this living tomb, for after a time, when the worm is sucked out of him, the sufferer is disgorged with no other loss than his dinner. This fact is the more remarkable when it is contrasted with the fate which awaits the worms on which they feed. No sooner are these laid hold upon than they evince every symptom of painful suffering, but their violent contortions are momentary, and a certain death suddenly follows their capture. How this effect is produced is mere matter of conjecture. Worms, in ordinary circumstances, are most tenacious of life even under severe wounds, and hence one is inclined to suppose that there must be something eminently poisonous in the Hydra's grasp, as it is impossible to believe, with Baker, that this soft toothless creature can bite and inject a venom into the wound it gives. "I have sometimes," says Baker, "forced a worm from a Polype the instant it has been bitten, (at the expense of breaking off the Polype's arms,) and have always observed it to die very soon afterwards, without one single instance of recovery." To the Entomostraca, however, its touch is not equally fatal, for I have repeatedly seen Cyprides and Daphniæ entangled in the tentacula, and arrested for some considerable time, escape even from the very lips of the mouth, and swim about afterwards unharmed; perhaps their shell may protect them from the poisonous excretion. The grosser parts of the food, after some hours' digestion, are again ejected by the mouth; but, as already mentioned, the stomach is furnished with what, in one sense, may be called an intestine, to which, according to Trembley and Baker, there is an outlet in the centre of the base, and the latter asserts that he has, "several times, seen the dung of the Polype, in little round pellets, discharged at this outlet or anus."—*Dr Johnston's British Zoophytes.*

MISCELLANEOUS.

Blindworm.—Mr Ryland has the following notice respecting this animal in the last number of the Naturalist. "I am led to suppose from Mr Salmon's remarks, that a prejudice against these inoffensive reptiles exists in his neighbourhood, similar to the belief prevalent in Lancashire. The notion of the lower orders here respecting this Snake is curious, and will be best shown by a conversation I once had with a turf-cutter:—'Well, my man, you have other kinds of Snakes here (Woolston Moss) besides the Viper?' 'Aye, sir, we sometimes light on Blindworms and Edders.' 'Indeed!' 'Aye: but the Blindworms are the worst, and desperate hard to kill. Whoy, if you were to cut one into half a dozen pieces, they 'ud join together again!' A medical friend informs me that a belief in the power of separated parts of this Snake reuniting is prevalent amongst the lower classes in Scotland. How can it have originated?"

In Scotland, the "lower classes" imagine the Blindworm to be as poisonous as the Viper or Adder, and generally do not distinguish the one species from the other. Knowing that Serpents have often bitten (or, as they think, stung) persons, as well as cattle, and as they are thus dangerous, they naturally on falling in with one kill it, and to make sure work, usually cut it into pieces with a knife, it being a prevalent belief among them, that a Serpent, if merely bruised, will or may come alive again. Their abhorrence of the animal, and their apprehension of the possibility of reviving, even leads them to separate the pieces to a distance; but the belief of a reunion of the parts, if left in proximity, is merely the result of fear carried to excess. The Blindworm is a very harmless creature, but, like many other innocuous Snakes, participates in the bad character of those which are venomous. I knew a gentleman in Harris, who, when a child, carried home a great number of them in his lap, to the great terror of his friends, as well as his own, after he came to be informed of their nature. A very slight blow with a stick is sufficient to disrupt this animal, and individuals are often seen with part of the tail wanting, in which respect it resembles the Lizards.

Magnificent Aurora Borealis.—Happening to be at Loch Achray on Sunday the 16th October, I was gratified by the sight of a more splendid display of the Aurora than I have ever witnessed. About nine o'clock, having gone out with a friend to enjoy the pure air, and the stillness of a most lovely night, I observed an arch extending from N.E. to S.W., its highest part being about 20° below the zenith. At nine, the arch broke up, and numberless streams of light, issuing from the horizon, shot upwards, and converged in a point a few degrees to the east of the zenith. At one time these streams rose from every part of the horizon, being, however, broader and brighter in some places. From behind Benvenue, in particular, rose a vast sheet of yellowish light, which changed into dim red and purple, and showed the outlines of the mountain as distinctly as they could have been seen on the clearest day, while its black mass, in which no feature could be distinguished, added greatly to the sublimity of the phenomenon. To the eastward, in the direction of the arch, which had disappeared, sheets of light shot across the sky, some apparently proceeding from the N.W., others meeting them from the S.E. This continued some time; the sheets then broadened, and finally were converted into thin streamers, shooting towards the zenith. The magnificence of the spectacle then gradually diminished, and the Aurora assumed a more ordinary appearance. Next day, the weather was fine, as also on Tuesday; but on Wednesday it rained at Glasgow from nine in the morning until late at night. This I mention, because it has been alleged that the more brilliant the Aurora, the sooner is it succeeded by rain or a tempest. In this case, after the most gorgeous exhibition imaginable, no rain fell for sixty hours.

Musk Deer.—The Musk Deer, *Moschus moschiferus*, is found in Thibet, Western Tartary, and the adjacent provinces of China. It is a species of Antelope, about the size of a moderate Hog, with a small head, a round hind quarter, delicate limbs, and without a tail. From the upper jaw two long curved tusks project downwards. Long bristles, not unlike the quills of a Porcupine, cover the body. The musk is a secretion, formed in a little bag resembling a wen, near the navel, which is found only upon the males. Its flesh is eatable. In Thibet it is considered the property of the state, and can be hunted only by permission of government. It delights in the coldest regions of the mountains, and bids defiance to the pursuit of man. In taking the musk from the animal, when still alive, it is necessary to bind up the bag instantly, for otherwise it is absorbed in the flesh, and retains no smell.—*Gutzlaff's China.*

Chinese Cormorants and Quails.—The Fishing Cormorant, which is trained to dive and catch the unwary fish, proves very useful. To prevent it from swallowing its prey, an iron ring is put around its neck, so that it is obliged to deliver its quota to its owner. It is as well trained as the Falcon in Europe, and seldom fails to return to its master, who rewards its fidelity by feeding it with the offals of the fish it has caught.—Quails, which are to be met with in great quantities in the north, are greatly valued by the Chinese, on account of their fighting qualities. They carry them about in a bag, which hangs from their girdle; treat them with great care, and blow occasionally a reed, to rouse their fierceness. When the bird is duly washed, which is done very carefully, they put him under a sieve with his antagonist, and strew a little millet on the ground, so as to stimulate the envy of the two Quails. They very soon commence a fight, and the owner of the victor wins the prize. Good fighting Quails sell at an enormous price, and are much in request.—*Ibidem.*

Polygonum Tinctorium.—This plant is now cultivated with success in the experimental gardens of M. Vilmorin, near Paris. Attention was called to it by M. Jaume St Hilaire, in consequence of its being used in China for dyeing a deep blue. M. Chevreul has examined it, and ascertained that it owes its properties to the true Indigotine, of which it yields a greater proportion than the *Isatis tinctoria.*—*Athenæum.*

Prognostics of Weather.—If the sound of the rapid or cataract descend with the stream, it foretells such rainy weather to be at hand as will swell the brook or river to its margin; whereas, if the sound ascend along the stream, and die away in the distance, it is an omen of the continuance of dry weather. If, during a storm of frost and snow in winter, the Ptarmigan, the hardiest among the feathered tribes of the Grampians, be repeatedly heard in the face of the mountain, an additional fall of snow may soon be expected.—*Rev. Robert Macdonald, in Stat. Account.*

Edinburgh: Published for the Proprietor, at the Office, No. 13, Hill Street. London: Smith, Elder, and Co., 65, Cornhill. Glasgow and the West of Scotland: John Smith and Son; and John Macleod. Dublin: George Young. Paris: J. B. Balliere, Rue de l'Ecole de Médecine, No. 13 bis.

THE EDINBURGH PRINTING COMPANY.

THE EDINBURGH

JOURNAL OF NATURAL HISTORY,

AND OF

THE PHYSICAL SCIENCES.

DECEMBER, 1838.

ZOOLOGY.

DESCRIPTION OF THE PLATE.—THE HOWLERS.

THE Monkeys of America are distinguished from those of the Old Continent by having four additional grinders, and by being destitute of cheek-pouches as well as callosities. Some of them have an elongated prehensile tail, and of these the first group is that composed of the *Howlers*, constituting the genus *Mycetes* of Illiger. They are so named on account of the frightful noises which they are enabled, by a peculiar inflation of the hyoid bone and larynx, to emit.

Fig. 1. THE RED OR ROYAL HOWLER (*Mycetes Seniculus*), which is of the size of a Fox, and somewhat similar in colour, inhabits the forests of Guiana.

Fig. 2. THE URSINE HOWLER (*M. ursinus*) is very similar to the last, and occurs in the same country.

Fig. 3. THE GOLDEN-TAILED HOWLER (*M. chrysurus*) is also similar in size and colour, but with the upper part of the body and the terminal portion of the tail bright yellow.

Fig. 4. THE BLACK HOWLER (*M. niger*), of a blackish colour, paler beneath, inhabits Brazil.

Fig. 5. The female is of a yellowish-grey tint.

Fig. 6. Represents a variety of the same species of a yellow colour.

DESCRIPTION OF THE PLATE.—THE SHRIKES.

THE Shrikes are in a manner intermediate between the Thrushes and the Falcons, resembling the former in their general appearance and in the form of their feet, while their notched and curved upper mandible indicates their affinity to the latter, which they also imitate in their predatory habits, although they also feed on insects and fruits. Most of the species represented in the plate are such as may be called typical.

Fig. 1. THE CAPE SHRIKE (*Lanius capensis*), variegated with white and black, inhabits southern Africa.

Fig. 2. THE SHARP-TAILED SHRIKE (*L. pyrrhonotus*), with an elongated and graduated tail, and variegated with ash-grey, light red, white and brownish-black, is from the same country.

Fig. 3. THE RED-BACKED SHRIKE (*L. Collurio*) is pretty similar to the last, but smaller and with a shorter tail. It is a regular summer visitant in most parts of Europe, and is not uncommon in some portions of England, chiefly frequenting copses and commons.

Fig. 4. THE BLUE SHRIKE (*L. bicolor*) has the upper parts bright blue, the lower white.

Fig. 5. THE GREAT CINEREOUS SHRIKE (*L. Excubitor*) is common to Europe and North America, although some ornithologists have alleged that the Shrike of the latter country differs from ours. It is a rare visitant in Britain, coming late in autumn and departing in spring.

Fig. 6. THE COLLARED SHRIKE (*L. collaris*), which inhabits Africa, is described by Vaillant as transfixing its prey on a thorn, which is a habit common to many other species.

Fig. 7. THE RUFOUS SHRIKE (*L. rufus*) has been seen in England, where, however, it is extremely rare, although it occurs in summer in many parts of the continent.

BRITISH BIRDS.—THE CURLEW.

WITH the history of the Curlew might be connected, and not inappropriately, much of the wild scenery of our land, for during the breeding season, its retreats are the barren heath and the mountain side. Let it be now the middle of October. We are traversing the mud flat that extends from the village of Cramond to near Queensferry, on the southern shore of the Frith of Forth. Many Gulls are scattered over the sands, small flocks of Ducks are swimming in the river, straggling bands of Terns hover and scream along the edge of the water, here and there may be seen a solitary Gannet gliding past, and far out at sea are some dusky birds, which may be Cormorants or Red-throated Divers. On that island is a vast assemblage of small birds, probably Dunlins; farther on are some black and white Waders, which we may conjecture to be Oyster-catchers; and lastly, scattered over the miry flat, are very many grey-backed, long-legged, long-necked, and long-billed stragglers, the very birds of which we are in search. They observe us, one utters a loud shrill cry, to which an-

43

other responds, and presently all are on the wing. Mark how they fly, at a moderate height, with contracted neck, outstretched bill, feet folded back, wide-spread wings moved in regular time. Away they speed, one screaming now and then, and alarming the gulls and other birds in their course; nor do they stop until, arriving at a suitable spot a quarter of a mile off, they perform a few circling evolutions, and alight by the margin of the sea, into which some of them wade, while the rest disperse over the sand. All that we can see or say of them here is, that at this season they have arrived on the sea shore, where they frequent the beaches, searching for food in the same way as the Godwits, Longshanks, and Sandpipers, but in what precise manner they procure it, or of what it consists, remains to be discovered. To see these vigilant and suspicious birds at hand, we must find some place resorted to by them, in which we may draw near without being perceived. Let us then betake ourselves to the Island of Harris.

Here is a low tract of sandy pasture, with a shallow pool upon it, and extending along a large ford or expanse of sand, covered by the tide, and laid bare when it recedes. Many Curlews and Golden Plovers, a few Ringed Dotterels, two or three Mallards, and, doubtless, hundreds of Snipes, are dispersed over the plashy ground. That old turf cattle-fold will enable us to approach the birds unseen, unless some of the Curlews should happen to fly over head and discover us, when they will be sure to sound an alarm. Now crawl this way, and see that the muzzle of your gun is not filled with sand. From this slap in the wall, cautiously raising our heads until we can bring an eye to bear on them, we may observe their motions. There, twenty paces off, stalks an old Curlew, cunning and sagacious, yet not aware of our proximity. He has heard, or fancied that he has heard, some unusual sound, and there he moves slowly, with raised head, and ear attent; but some appearance in the soft sand has attracted his notice, and, forgetting his fears, he thrusts, or rather works his bill into it, and, extracting something which he swallows, withdraws it, and proceeds, looking carefully around. Now from the surface he picks up a Snail, and that small bird named *Helix ericetorum*, which, raising his head, and moving it rapidly backwards and forwards, at the same time slightly opening and closing his mandibles, he gradually brings within reach of his tongue, when he swallows it. There he has dragged a worm from the sand, and again has obtained a small crab or insect. But now two others have come up; they are within range; let me fire. There they lie, two dead; the other, with broken wing, runs off loudly screaming. Curlews, Plovers, Redshanks, Dotterels, Ducks, and Snipes, all rise, and move to a distance corresponding to their fears, the Curlews flying out of sight, the Snipes coming back to the same spot, and the Plovers alighting about two hundred yards off.

The Curlew is extremely shy and suspicious, so that at this season, unless by some stratagem or accident, one can very seldom obtain a shot at it. In Harris I once shot three from a fold in the manner described above. On another occasion, having a musket with large shot, I let fly at one feeding in a field as I was passing along, hit it in the wing, and on measuring the distance found it to be seventy-five yards. In the Hebrides it is a common saying, that to kill seven Curlews is enough for a lifetime; but one, by lying among the rocks on a point frequented by them, might, I doubt not, shoot as many in less than a week. This method, however, I have never tried, it being much more pleasant to be moving about than lying jammed into the crevice of a cliff. When alarmed, they spread out their wings, run rapidly forward some paces, and, springing into the air, uttering their loud cries, fly off at a rapid rate. When looking for food, they generally walk sedately, unlike the Redshank, which is continually running, stooping, or vibrating, but sometimes run, and that with great celerity. Dry pastures, moist ground, and shallow pools, are equally frequented by them, and they may be seen wading in the water up to the tarsal joint.

Towards the end of March they generally leave the shores, where they have resided in flocks from September, and separating in pairs, betake themselves to the interior, where, in the higher and less frequented moors, they deposit their eggs and rear their young.

It is now the beginning of May. The sunny banks are covered with primroses, the golden catkins of the willow fringe the brooks, while the spikes of the cotton-grass ornament the moss-clad moor. Let us ascend the long glen, and, wandering on the heathy slopes, listen to the clear but melancholy whistle of the Plover, the bleating of the Snipe, and the loud scream of the Curlew. Here is a bog, interspersed with tufts of heath, among which is a profusion of Myrica Gale. Some Lapwings are coming up, gliding and flapping their long broad wings; a black-breasted Plover has stationed himself on the top of that mound of green moss, and a Ring Ouzel has just sprung from the furze on the brae. See, what is that? A Hare has sprung from among our feet! No, a Curlew, fluttering along the ground, wounded, unable to escape. Run. She has been sitting; here is the nest in a hollow under shelter of two tufts of heath and a stunted willow. It is composed of dry grass, apparently *Eriophora, Eleocharis palustris, Scirpus cæspitosus*, some twigs of heath, and, perhaps, portions of other plants, not very neatly disposed. It is very shallow, and internally about a foot in diameter. The eggs are four, pyriform, excessively large, three inches long, an inch and ten-twelfths across, light olive or dull yellowish-brown, or pale greenish-grey, blotched and spotted with umber brown, the markings crowded on the larger end. They vary considerably in size and form, some being only two inches and three-quarters in length. Those in the nest before us are of the largest size, very darkly coloured, and so little contrasting with the surrounding objects, that unless the bird had sprung up among our feet, we should scarcely have observed them.

Far up on the hill-side you hear the loud cry of the Curlew, which is presently responded to from the opposite slope; in another place a bird commences a series of modulated cries, and, springing up, performs a curved flight, flapping its wings and screaming as it proceeds. Presently the whole glen is vocal, but not with sweet sounds, like those of the Mavis and the Merle. But it is vain to pursue the birds, for these are the males, and at this season you will find them fully as shy as they were in winter on the sea-shore. Some weeks hence, when the young are abroad, the females, and even the males, will flutter around you, if you approach the spot where their unfledged brood lie concealed among the herbage, and will attempt, by feigning distress, to lead you into a vain pursuit.

Like all the other birds of this genus, the young are covered with long, stiffish down, and run about presently after exclusion from the egg, squatting to conceal themselves from their enemies. Up to the age of three weeks they are still unfeathered; their forehead, throat, and under surface, yellowish-grey, their upper parts of the same colour, with patches of dark brown; the bill not longer than the head. That organ gradually elongates, as the feathers sprout, and by the end of about seven weeks they are able to fly.

At this season, old and young feed on insects, larvæ, and worms. The latter are very fat, but the former are not in good condition until the middle of autumn, about which period the Curlews unite into small flocks, gradually disperse, and betake themselves to the shores. Their flesh is delicate and well flavoured, and they are not unfrequently to be seen in our markets. I am not aware of any difference produced in the quality of their flesh as an article of food by their change of residence.

Montagu has given, in the Supplement to his Ornithological Dictionary, a very interesting account of a tame bird of this species. " One which was shot in the wing, was turned amongst aquatic birds, and was at first so extremely shy, that he was obliged to be crammed with meat for a day or two, when he began to eat worms; but as this was precarious food, he was tempted to eat bread and milk like Ruffs. To induce this substitution, worms were put into a mess of bread mixed with milk, and it was curious to observe how cautiously he avoided the mixture, by carrying every worm to the pond, and well washing it previous to swallowing. In the course of a few days this new diet did not appear unpalatable to him, and in a little more than a week he became partial to it, and, from being exceedingly poor and emaciated, got plump and in high health. In the course of a month or six weeks, this bird became excessively tame, and would follow a person across the menagerie for a bit of bread, or a small fish, of which he was remarkably fond. But he became almost omnivorous; fish, water-lizards, small frogs, insects of every kind that were not too large to swallow, and (in defect of other food) barley with the Ducks was not rejected. This very great favourite was at last killed by a Rat (as it was suspected), after a short life of two years in confinement; but he had in that time fully satisfied our inquiries into his natural habits."

An adult male Curlew measures 25 inches in length, 42 from the tip of one wing to that of the other. The body is ovate and rather full, the legs long and slender, the neck also long, the head rather small, the bill extremely long, measuring six inches; the tibia bare at its lower end, the tarsus reticulated, the toes rather short, slender, three before, one behind. The throat is very narrow; the œsophagus very long and rather slender; the proventriculus oblong; the stomach a large and powerful gizzard; the intestine long, of moderate width; the cœca rather slender, cylindrical, 4½ inches long. The plumage is moderately full, soft, and blended; the wings very long, narrow, pointed, the first primary longest; the tail rather short and rounded.

The bill is black, the base of the lower mandible and the basal margins of the upper flesh-coloured. The general colour of the upper parts and neck is light greyish-yellow tinged with red, each feather with a central blackish-brown streak; the scapulars with serriform yellowish-red spots on the edges; the primaries deep brown, the first five quills unspotted on the outer web, the rest with serriform white spots on the outer, and all with similar large spots on the inner web; the back white, with narrow longitudinal black marks; the upper tail-coverts barred with black; the tail white, with twelve brownish-black bands; the breast, sides, and abdomen, white; the first with lanceolate spots, the second with spots and bars; the last tail-coverts with narrow lanceolate spots.

DIGESTIVE ORGANS OF THE SNAKE-BIRD.

In the recently published volume of Mr Audubon's Ornithological Biography, in which are contained numerous descriptions of the digestive and respiratory organs of birds, is the following account of those of the Anhinga or Snake-bird, one of the most interesting inhabitants of the lakes and marshes of the Southern States of North America. In external appearance and habits, the Snake-bird is very nearly allied to the Cormorants. The structure of the feet is essentially the same in both genera, as is that of the wings and tail, the latter, however, being more elongated in the Anhinga, in correspondence with the neck. If one might suppose a small Cormorant elongated and attenuated, with the feet rather enlarged but shortened, the head diminished in size, and the bill formed on the model of that of a Heron, being destitute of the dis-

tinct ridge and curved unguis, he would form a pretty correct notion of this bird. Not only is the bill like that of a Heron, but the vertebræ of the neck are very similar to those of that family, and form the same abrupt curvatures between the seventh and eighth vertebræ. But all the other bones are those of the Cormorants and Pelicans. The sternum in particular is almost precisely similar to that of the Crested Cormorant, so that, without entering very minutely into its description, no differences could be pointed out.

The lower mandible has a distinct oblique joint at about a third of its length, enabling it to be expanded to the extent of an inch and a half. The pouch, which is small, is constructed in the same manner as that of the Pelicans and Cormorants; its muscular fibres running from the lower edge of the mandible downwards and backwards, and a slender muscle passing from the anterior part of the hyoid bones to the junction of the crura of the mandible. The tongue is reduced to a mere oblong knob, one and a half twelfth of an inch in length, and a half twelfth in height.

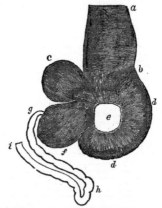

The œsophagus, a, b, is seventeen inches long, exceedingly delicate and dilatable, with external longitudinal fibres, the transverse fibres becoming stronger toward the lower part. Its diameter, when moderately dilated, is one and three-fourths inch at the top, one inch farther down, at its entrance into the thorax nine-twelfths, and finally one and a half inch; but it may be dilated to a much greater extent. The proventricular glands, instead of forming a belt at the lower part of the œsophagus, are placed on the right side in the form of a globular sac, c, about an inch in diameter, communicating with the œsophagus, b, and stomach, d. For two inches of the lower part of the œsophagus, a, b, or at that part usually occupied by the proventriculus, the transverse muscular fibres are enlarged, and form an abrupt margin beneath; on the inner surface there are four irregular series of large apertures of gastric glandules or crypts. The proventriculus itself, c, is composed of large crypts of irregular form, with very wide apertures, and covered externally with muscular fibres. The stomach, d, d, is roundish, about an inch and three-quarters in diameter, with two roundish tendinous spaces, e, and fasciculi of muscular fibres; its inner coat thin, soft, and smooth. It opens by an aperture a quarter of an inch in diameter into a small sac, f, precisely similar to that of the Pelican, which has a muscular coat, with a soft, even, internal membrane, like that of the stomach. The pylorus has a diameter of two-twelfths, is closed by a semilunar valve or flap, and is surrounded by a disk of radiating rugæ three-fourths of an inch in diameter. The intestine is three feet four inches long, its average diameter two and a quarter twelfths, but only one-twelfth at its junction with the rectum, which is three and a half inches long, three-twelfths in diameter. There are no cœca properly so called, but a small rounded termination of the rectum, two-twelfths in length, as in the Herons.

Many other very curious modifications of the digestive organs of birds are contained in this volume, which thus affords subjects for rumination to the system-makers. It is the only example, excepting the first volume of the History of British Birds, and the Description of the Rapacious Birds of Great Britain, of an ornithological work in the English language, containing an account of these organs in the species described. Hitherto the characters of birds have been derived from the colours of the plumage; but an era has commenced in which systematic ornithology is likely to be improved in a degree of which we can at present form no adequate conception. Since the publication of the works mentioned above, several individuals have adopted the ideas expressed in them, but, as in other instances of a like nature, carefully abstain from making any allusion to the source of their ideas.

ATTRACTIVE QUALITIES OF THE STARLING.—The Starling becomes wonderfully familiar in the house. As docile and cunning as a Dog, he is always gay, wakeful, soon knows all the inhabitants of the house, remarks their motions and air, and adapts himself to their humours. In his solemn tottering step, he appears to go stupidly forward; but nothing escapes his eye. He learns to pronounce words without having his tongue cut, which proves the uselessness of this cruel operation. He repeats correctly the airs which are taught him, as does also the female, imitates the cries of men and animals, and the songs of all the birds in the room with him. It must be owned that his acquirements are very uncertain; he forgets as fast as he learns, or he mixes up the old and new in utter confusion; therefore, if it is wished to teach him an air, or to pronounce some words clearly and distinctly, it is absolutely necessary to separate him from other birds and animals, in a room where he can hear nothing. Not only are the young susceptible of these instructions, the oldest even show the most astonishing docility.—*Bechstein's Cage Birds.*

THE following letter from a celebrated Naturalist to the Editor will no doubt prove interesting to our readers.

On Board the Crusader, Cote Blanche, 18*th April* 1837.

MY DEAR FRIEND,—Being just now snugly anchored in a bay, the description of which may prove agreeable to you, I sit down to give you an account of what I have been doing since I last wrote to you.

After visiting "Rabbit Island," on which, as I have already told you, not a single Rabbit or Hare is to be seen, we made our way between it and Friskey Point, by a narrow and somewhat difficult channel leading to the bay in which I now write. The shores around us are entirely formed of a bank, from twenty to thirty feet high, and composed of concrete shells of various kinds, among which the Common Oyster, however, predominates. This bank, which at present looks as if bleached by the sunshine and rains of centuries, is so white that it might well form a guiding line to the vessels which navigate this bay even in the darkest nights. The bay, however, is so shallow, that it is rarely entered by vessels larger than schooners of about seventy tons burthen, which visit its shores to take in the sugars and cottons grown in the neighbouring country.

The "Crusader" is a somewhat curious craft, small, snug withal, and considerably roguish-looking. She has not fewer than four "grunters" on her fore deck, her sails are of pure white cotton, and although she bears the lively flag of our country at the peak, her being painted purely black gives her the aspect, not merely of a smuggler, but of a pirate. But here she is, at the entrance of the canal of a sugar plantation, and close to another craft, much the worse for wear, and, for aught I know to the contrary, belonging to the captain alone, who, I would almost venture to assert, belongs to no country at all.

It is now four weeks since a razor came in contact with my chin. All my companions are equally hircine; or, if you please, hirsute. As to our clothing, were you to see us at this moment, you would be ready to exclaim, "What vagabonds these fellows are!" Coats and trousers plastered with mire, shirts no longer white, guns exhibiting the appearance of being in constant use, and all sorts of accoutrements that pertain to determined hunters, complete our *tout ensemble.* But, as I have said, here we are, and on shore must go. "Man the gig," quoth our captain. In a trice the gig is manned. One after another, for there are five or six of us, we swing ourselves into the after-sheets. The word is given, the oars are plied, and now we once more are on terra firma.

The crossing of large bays, cumbered with shallow bars and banks of oyster-shells, is always to me extremely disagreeable, and more especially when all these bars and banks do not contain a single living specimen of that most delectable shell-fish. Nay, I am assured by our pilot, who is no youngster, that ever since he first visited this extensive waste, not an oyster has been procured in these parts. But now in single file, like culprits or hungry travellers, we proceed along the margin of the canal. Ah, my dear friend, would that you were here just now to see the Snipes innumerable, the Blackbirds, the Gallinules, and the Curlews that surround us;—that you could listen, as I now do, to the delightful notes of the Mocking-bird, pouring forth his soul in melody as the glorious orb of day is fast descending toward the western horizon;—that you could watch the light gambols of the Night Hawk, or gaze on the Great Herons, which, after spreading their broad wings, croak aloud as if doubtful regarding the purpose of our visit to these shores! Ah! how well do I know you would enjoy all this; but, alas! we are more than four thousand miles apart.

Hark! what's that? Nothing but a parcel of men coming to greet us. Here they are, seven or eight Negroes. Who lives here, my good fellows? Major Gordy, massa. Well, now shew us the way to the house. Yes, gentlemen, come along. So we follow our swarthy guides.

The plantations here are of great value, both on account of their proximity to the Gulf of Mexico, and the excellence of the soil, which, as in other parts of Louisiana, is composed of a fat, black mould. The Indian corn was at least six feet high, and looked most beautiful. As we approached the mansion of Major Gordy, I observed that it had a pleasant aspect, and was furnished with a fine garden, and a yard well stocked with cattle, together with a good number of horses and mules, just let loose from labour. A mill for grinding corn and making sugar particularly drew our notice to it, as the Crusader happened just then to be destitute of both articles; and as I saw some women milking the cows, my heart fairly leaped with joy, and the hope that ere long we might procure a full bowl of the delightful and salubrious beverage. The short twilight of our southern latitudes had now almost involved every object in that dim obscurity so congenial to most living creatures after the toils of the day, as allowing them to enjoy that placid quiet which is required to restore their faded energies.

Near the entrance of the mansion stood an elderly man, of tall stature and firm aspect, leaning on what I would call a desperate long gun. As I approached this Cote Blanche planter, I thought that something not so very friendly as I could have wished was expressed in his countenance. As he rested his heavy frame on his monstrous rifle, he neither moved his head, nor held out his hand to me, until I presented mine to him, saying, "My good sir, how do you do?" His answer was a rather suspicious look at me and my companions; but notwithstanding, and probably because he was on his own ground, he asked us what was our wish, and then desired us to walk in.

Cote Blanche Bay, you must be informed, has for a number of years been infested by a set of rascally piratical vagabonds, who have committed extensive depredations, in consequence of which, a few years ago, a United States' revenue cutter was sent to protect the coast. I have no doubt that the major took us, to a man, as members of the gang who had more than once visited, not his house, but his plantation, on which they had played many wanton and atrocious freaks.

We now, however, had entered the house. Candles were lighted, and we at once came face to face, as it were. It curiously happened that our captain was without his uniform, and fully as rough-looking otherwise as any of us. I was, however, much pleased to see that the major himself was not much superior to us in respect of apparel; nor had his razor been employed for many days. I happened to have about me some unequivocal credentials, from the head departments of the United States, which, on my observing that some degree of suspicion still remained, I placed in his hands. He read them, spoke kindly to us, promised to forward our letters to the nearest post town, and invited us to consider his dwelling as our own. From that moment until we returned to our vessel, we were all as comfortable and merry as men can be when distant from their own dear homes.

Next morning we received from Major Gordy a barrel of sugar, another of corn meal, some pails of milk, and a quantity of newly made butter, together with potatoes and other needful articles—and all this without our being allowed so much as to offer him the least recompence. The day after, we returned to breakfast by invitation, and found in the house several strangers, armed with rifles and double-barrelled guns. After we had been introduced to all around, we seated ourselves, and made a vigorous attack upon our host's eggs and bacon, coffee, tea, and milk. As this important business was proceeding, I was delighted to hear the following anecdotes, which I hope you, my dear friend, will relish as much as I did.

"Gentlemen," said our host, straightening himself in his arm-chair, "I am considerably suspicious as regards the strangers who happen to anchor within the range of my dominions. Indeed, gentlemen, I must acknowledge that even after you returned on board last night, I sent off some of my men in various directions, to let my neighbours know that a strange craft had anchored near the landing-place; and here, gentlemen, are those neighbours of mine; but as it happens that the name of the gentleman who calls himself a "Naturalist" is well known to some of them, I now feel quite satisfied as respects the purposes you have in view. But let me tell you what happened to me some years ago.

"Such a shark-looking craft as the one you call the Crusader happened to drop its anchor abreast of my landing-place, about dusk one evening, and as I guessed that the fellows on board were not better than they should be, I watched their motions for a while from my back piazza. But nothing happened that night. Next morning, however, I heard the firing of guns down the meadows where my cattle and hogs were in the habit of feeding. So I took my rifle, walked towards the spot, and soon found, sure enough, that the rascals had killed a fine ox and several hogs, which they were dragging to the shore. Indeed, gentlemen, I saw the yawl crammed with the spoils of my plantation. Well, I took as good an aim as I could at the nearest man, and cracked away, but without hitting. At the report of my gun the fellows all took to their heels, and on getting on board hoisted sail and went off. I have never heard of them since. Well, gentlemen, about the same hour next morning, a black-looking barge, hardly as large as your Crusader, came to, off the very some spot, and although I watched it and every one on board nearly the whole night, and it was a beautiful moon-shiny one, not a soul of them came on shore until morning. Then, however, I saw some bustle on board. Several men got off in a very small affair, which was fastened a-stern of the large boat. I saw them land, and deliberately walk towards the meadows. No sooner had they reached the wettest part, and that is where my hogs generally root for food, than crack, crack, crack, went off their guns in all directions. You may well suppose how vexed I was at all this, and conceive how soon I mustered my men with clubs, and armed myself with my rifle. On reaching the ground, think, gentlemen, what were my thoughts, when I saw the fellows all advancing towards me and my people, as if they were the honestest men in the world. I was so mad when they came close up, that I had a great mind to shoot the one in front, for he looked for all the world as if he cared not a pin for any one. However, I did not shoot, but asked him why he was shooting my hogs? 'Hogs! goodman, you are quite mistaken; we are shooting snipes until we come in contact with the rascally pirates who infest the coast, and lay waste your plantations. My name, my good Sir, is Captain ———, of the United States' Navy; and these are some of my men. Will you come on board, and breakfast with us on your own snipes?'" No wonder that the Major, having been subject to the visits of these marauders, should have taken us in the dusk, armed as we were, and withal not having precisely the aspect of sober citizens, for persons not quite so good as we should be. But I must now conclude, and in my next you shall hear something of the result of my expedition into the marshes.

HYDROGRAPHY.

REMARKS ON THE COLOUR OF THE OCEAN, BY M. ARAGO.

MR SCORESBY compares the colour of the polar seas to *ultramarine blue;* M. Cortaz considers the finest *indigo* or *sky-blue,* as that of the Mediterranean; Captain Tuckey characterizes the waters of the Atlantic in the equinoctial regions as *bright azure;* and Sir Humphrey Davy gives *bright blue,* as the tint reflected by the pure water obtained from melted snow or ice. Sky-blue, then, more or less intense, or mixed with smaller or greater quantities of white light, would appear to have been always the proper tint of the ocean. But is there any deception in this?

The waters of the sea are often impregnated with foreign matter. For example, the extensive green bands, so peculiarly striking in the polar regions, contain myriads of medusæ of a yellowish tint, which, combined with the blue colour of the ocean, produce green. Near Cape Palmas, on the coast of Guinea, Captain Tuckey's vessel appeared to move in milk; that appearance arising from multitudes of animals floating on the surface, and so colouring the natural colour of the water. The belts of carmine which several navigators have traversed arise from the same cause. Sir Humphrey Davy states that in Switzerland, when the tint of the lakes passes from blue to green, it is because the waters are impregnated with vegetable substances. Lastly, near the mouths of large rivers, the sea has often a brownish colour, arising from the mud and other earthy matters held in suspension. It is necessary, then, to insist on the colours produced by foreign matter mixed with the water, so that they may not be confounded with those to be described.

The sky-blue tint of the sea is modified, and sometimes even entirely changed, in places where the water is not very deep. This is because the light reflected by the bottom reaches the eye, mixed with the natural light of the water. The effect of this superposition may be calculated by the laws of optics; but we must add to our acquaintance with the nature of the two commingled tints, what is more difficult to be ascertained, that of their comparative intensities. Thus, a bottom of *yellow* sand reflecting but lightly, gives the sea a green tint, because yellow mixed with blue, as is well known, produces green. Now, without changing the shades, if you replace the dull yellow by a bright yellow, the slight blue of pure water will scarcely produce a lively light green, and the sea will appear *yellow*. In the Bay of Loango, the waters are always *deep red*, insomuch that they are said to be mixed with blood; and Captain Tuckey ascertained that the bottom is intensely red. Let us substitute for this bright red bottom one of the same colour, but dull or reflecting slightly, and the waters will then appear *orange*, or even perhaps *yellow*.

An objection, which at first sight appears important, is made to this mode of regarding the subject:—a bottom of white sand, it is said, ought not to alter the tint of the sea, for if white weakens the colours with which it mixes, it yet does not change the hue. But there is a ready answer to this objection: for how can we be certain that the sand at the bottom is white? Is it not in the open day, after we have brought up a portion, and exposed it to the *white* light of the sun or the clouds? Is the sand in the same condition when beneath the water? If in the open air you were to illuminate it with red, green, or blue light, it would appear red, green, or blue. We have then still to inquire what colour strikes it at the bottom of the water.

Water is in the condition of many other bodies which philosophers have very deeply studied, and which possess two kinds of colours; a certain colour which is *transmitted*, and another, quite different, which is *reflected*. Water appears of a blue colour by reflection, and some imagine it to be green by transmission. Thus, water disperses in all directions, after having *blued* it, a portion of the white light which went to illuminate it. This dispersed light constitutes the *proper colour* of liquids. As to the other *irregularly transmitted rays*, their passage through the water makes them green, and this the more intensely the thicker the traversed mass.

This being admitted, we may return to the case of a not very deep sea, with a bottom of white sand. This sand receives the light only through a stratum of water. The light then is green when it strikes the bottom, and it is with this tint that it is reflected; and in the second traject which the luminous rays make through the same liquid in returning from the sand to the open air, this green tint sometimes so predominates as to prevail over the blue. This, then, perhaps may be the whole secret, which is to the practical navigator, in the time of calm, a certain and invaluable index of great depths. We say *in time of calm*, for when the ocean is agitated, the waves suitably elevated may in fact convey to the eye so large a quantity of *transmitted or green rays*, that the reflected blue rays shall be entirely masked.

Let us imagine a triangular prism placed in the open air horizontally before an observer, somewhat lower than he is. This prism cannot by refraction conduct to the eye any ray coming directly from the atmosphere. On the contrary, the anterior face of the prism will throw towards the observer a reflected atmospheric pencil, a great part of which, it is true, would pass over his head. This portion would require to be bent in its course, to be refracted from above downwards to reach the eye. A second prism, placed like the first, but nearer to the observer, would produce this effect.

From these few words of explanation, the reader has already, no doubt, made the assimilation which must lead to the conclusion towards which we are tending. The waves of the ocean are a kind of prism; no wave is ever solitary; the continuous waves advance nearly in parallel directions. When two waves, then, approach a vessel, a portion of the light which the anterior face of the second wave reflects *traverses the first*, is there refracted from above downwards, and thus arrives at the observer placed on the deck. Again, then, we see transmitted light, light consequently *made green*, reach the eye at the same time with the common blue tints; but these are the phenomena of great depths over white sand produced without deep water, and the green colour of the sea arises from the predominance of the transmitted over the reflected colour.

We have now hastily traced the imperfect outlines of a theory of the colours of the ocean, that voyagers may thereby be directed in the investigations which they may have occasion to make on the subject. The examination of circumstances which may oppose this theory will suggest to them experiments, or at least observations, which otherwise they probably would not have thought of. Thus, every one will understand that the prism waves ought not to produce the same effects under differences in the direction of their propagation, and some variation in the colour of the sea will be expected under a change of wind. This phenomenon is apparent on the Swiss lakes. Is it so also on the open ocean?

Some persons persist in assigning an important influence to the blue of the sky in the production of that of the sea. It appears to us that this idea may be subjected to a decisive proof, thus. The blue rays of the atmosphere do not return from the water to the eye till after they have been regularly reflected. If the angle of refraction equals 37°, they are polarised. A piece of tourmaline will then completely eliminate them, and thus the blue of the sea will be seen without any extraneous mixture.

In order to get rid, as much as possible, of the influence of reflection, when examining the colours of the ocean, some able navigators have recommended that we should always examine it through the aperture of the ship's rudder. In this way the water in some points of view exhibits beautiful *violet* tints; but we may easily satisfy ourselves that these tints are merely the effect of contrast, and that they proceed from atmospheric light feebly reflected in an almost perpendicular direction, and coloured by their approximation to the transmitted green colours which almost invariably surround the rudder.

To those who may wish to develope this attempt to explain the colours of the sea, or to those who may desire to refute it, and substitute a better, it will be necessary to begin with investigating the colour of the water when seen by transmission *with the aid of diffused light*. Those who will recall to their mind the pre-eminently green hue of the cut edge of a crystal glass, even when the latter is only illuminated in front and perpendicularly, will be aware of the importance of this remark. The following appears a very simple means of obtaining a satisfactory conclusion.

I shall suppose the observer to be furnished with one of those large hollow prisms which philosophers are in the habit of using when they study the refraction of liquids. To render our apprehensions more precise, let us make the refracting angle 45°. We shall then suppose that the prism is partially immersed in water, so that the edge of its refracting angle is downwards, and that one of its faces, that which is farthest from the observer, shall be vertical; whence it will result as a necessary consequence, that the other face will be inclined to the horizon at an angle of 45°. Under these circumstances, the light which moves horizontally in the water at a fraction of an inch below the surface, that which constitutes its *edge colour*, if the expression may be used, will strike perpendicularly the vertical side of the prism; it will penetrate into the interior of the instrument, traverse the small quantity of air which it encloses, reach the second plate, and then be reflected vertically upwards. In looking upon this inclined surface, the observer may then judge of the proper colour which the water has by refraction, quite as well as if his eye were in the liquid. In this form the experiment is so simple and easy, and requires so little time, that we shall venture to request the Academy to recommend to our voyagers to repeat it as often as possible, not only in sea-water, but also in that of lakes and rivers. When science shall be enriched with all these observations, we shall no longer run the risk of constructing theories sooner or later to be contradicted by facts.—*Comptes Rendus.*

MISCELLANEOUS.

Toes of the African Ostrich.—It had been alleged some time ago, that Dr Riley of Bristol had discovered a rudimentary toe in the Ostrich, in addition to the two which it requires no minute investigation to discover; but Mr Thomas Allis of York, having carefully examined a specimen for the express purpose of satisfying himself on this point, asserts, "that whatever rudimentary toe may have been discovered in Dr Riley's Ostrich, there was certainly no third toe" in the one examined by him. He further corrects a mis-statement of Cuvier regarding the number of phalanges in the toes of the Ostrich, having found them to be four and five, in place of four and four.

Blindworm.—Mr Rylands, in a letter to the Editor of the Naturalist, mentions his having captured, on Woolston Moss, a specimen of Anguis agreeing, in almost every respect, with Pennant's description of *A. Eryx*. Its description is as follows:—Length, 14¾ inches; belly of a bluish lead-colour, with a few scattered white spots, which become more continuous and regular under the tail; the remainder of the body is greyish-brown, with three longitudinal dark lines, one extending from the head along the back (becoming indistinct towards its termination) to the point of the tail, the others broader, and extending the entire length of the sides; scales, &c., as in *A. fragilis*. In conclusion, he remarks that further observations being necessary to settle the question, he is desirous of obtaining such.

There is in my collection a specimen of the Blindworm, found in Aberdeenshire, and which I kept alive for some time. Its tail had been partially broken off and cicatrized, so that its length cannot be ascertained; but what remains is 10 inches long, and the lost part must have been three or four more. The common Blindworm is described as being greyish-brown above, bluish-black beneath, with several parallel rows of small dark spots along the back, and a dusky band on each side; and the variety or supposed species named *Eryx* is said by Pennant to have the belly of a bluish lead-colour, marked with small white spots irregularly disposed; the rest of the body greyish-brown; with three longitudinal dusky lines, one extending from the head along the back to the point of the tail, the others broader, and extending the whole length of the sides. My specimen, which was found on a moor in Buchan, agrees sufficiently with this alleged *Anguis Eryx*, and yet I have reason to believe is nothing but the common Blindworm. It is of a light greyish-brown above, with a central undulated narrow line, and on either side three still more slender lines of black; then on each side is a broader black line or band, below which are five interrupted or dotted lines of black; and the belly or lower surface is bluish-black in its whole length, but with six paler lines, and numerous whitish spots beneath the tail. Such variations cannot be considered as specific. Sometimes the Slow-worm has the upper parts without lines at all, sometimes with a single line, and sometimes with seven lines; the lower parts either uniformly bluish-black, or with a broad band of the same colour as the back, and on each side a central dark band. The number of lines is, I think, fourteen on each side, twenty-eight in all, that being the number of series of scales. Some specimens are destitute of white spots on the lower part of the abdomen and tail, but in most individuals they are more or less apparent. Young, or at least small individuals, generally have the lines more distinct.

Edinburgh: Published for the Proprietor, at the Office, No. 13, Hill Street. London: Smith, Elder, and Co., 65, Cornhill. Glasgow and the West of Scotland: John Smith and Son; and John Macleod. Dublin: George Young. Paris: J. B. Bailliere, Rue de l'Ecole de Médecine, No. 13 bis.

THE EDINBURGH PRINTING COMPANY.

THE EDINBURGH

JOURNAL OF NATURAL HISTORY,

AND OF

THE PHYSICAL SCIENCES.

JANUARY, 1839.

ZOOLOGY.

DESCRIPTION OF THE PLATE.—THE DEER.

The species here represented belong to the section having flattened antlers.

Fig. 1. The Moose-Deer, or Elk (*Cervus Alces*), a young individual in winter.

Fig. 2. The Moose-Deer. The same in summer.

This species, which inhabits the northern parts of both continents, is as large as a Horse, and stands very high in proportion to its length. Its form is inelegant, its muzzle being enlarged and tumid, its shoulders prominent, its hind quarters lower, its tail extremely short, and its limbs strong and greatly elongated. The antlers are at first simple, as in Fig. 2, but in the fifth year assume an expanded triangular shape, with projecting digitations on the outer edge, and sometimes increase to a large size. They are vigilant and timid, but when wounded will attack a Man. Their flesh is esteemed excellent food.

Fig. 3. The Rein-Deer (*C. Tarandus*). Male.

Fig. 4. The Rein-Deer. Young female.

The Rein-Deer has attracted more notice than perhaps any other species, on account of the uses to which it is applied by the inhabitants of the northern parts of Europe, who have reduced it to a state of perfect subjugation. These animals, in fact, constitute the chief riches of the Laplanders who keep them in herds, subsist on their milk and flesh, and employ them for drawing their sledges. They are of the size of the Red-Deer, but are less elegantly formed, with stouter and shorter limbs. This species occurs in the arctic parts of America, where it is known by the name of Caribou.

Fig. 5. The Fallow-Deer (*C. Dama*). Male of the black or dark-coloured variety.

Fig. 6. The Fallow-Deer. Female.

Fig. 7. The Fallow-Deer. Young.

The Fallow-Deer, which is inferior in size to the Red-Deer, and exhibits great variations in colour, is kept in a semi-domesticated state in many parts of Europe, and especially in Britain, where it forms a conspicuous ornament to the parks of the Nobility and others. The light-coloured or common variety is usually spotted with whitish: but the dark-brown, here represented, is for the most part destitute of such markings.

DESCRIPTION OF THE PLATE.—THE HORNBILLS.

The Hornbills, principally distinguished by their enormous bill, generally having a large projection on the upper mandible, are considered by Cuvier as forming a genus of the Syndactylous Family of the Passerina. In the form and size of the bill they seem allied to the Toucans, while in their general aspect they resemble the Crows, and in their short, syndactylous feet, approach the Bee-eaters and Kingfishers. The tongue is extremely small, so as to be incapable of being used as an organ of prehension. They are said to be omnivorous, like the Crows, and are peculiar to Africa and India.

Fig. 1. The Mourning Hornbill (*Buceros anthracicus*). Black, with the tail-feathers, except the middle, white in the terminal half; the bill yellow.

Fig. 2. The White-breasted Hornbill (*B. convexus*). The head, neck, and upper parts, black; the lower parts and tail yellow.

Fig. 3. The Yellow-billed Hornbill (*B. gracilis*). Black, with the terminal parts of the tail white.

Fig. 4. The Red-billed Hornbill (*B. sulcatus*), with the bill red, the upper parts of the body bluish-black, the tail white, with a terminal black band. Inhabits the Philippine Islands.

Fig. 5. The Langur Hornbill (*B. galeritus*). Dusky, with the tail whitish, terminated with black.

Fig. 6. The Red-necked Hornbill (*B. ruficollis*). The head and neck light red, the other parts bluish-black, except the tail, which is white.

44

BRITISH BIRDS.—THE WATER-HEN.

The Water-Hen, or Green-footed Gallinule, when seen running along the banks of a stream or pool, invariably calls to mind the idea of a young domestic fowl, its form and attitudes being extremely similar. The body, although much compressed, is rather full anteriorly: the neck of moderate length, the head oblong, compressed, and rather small. The bill, which does not exceed the head in length, is rather stout, tapering, and much compressed. The feet are large; the tibia muscular, its lower part bare, the tarsus of moderate length, large, compressed, anteriorly covered with broad, curved scutella; the toes very long, slender, compressed, the first, however, very small; the claws long, slender, slightly arched, compressed, acute. The plumage is blended, soft, glossy above; the wings short, concave, and rounded; the tail very short, arched, much rounded, of twelve weak, narrow, rounded feathers.

In the end of autumn, when the moult has been completed, the bill is greenish-yellow beyond the nostrils, the basal part, and frontal-plate, crimson-red, the latter somewhat paler. The iris, which is very narrow, seems red at a little distance, but is composed of three rings, the outer hazel, the middle dusky, the inner bright red. The feet are dull green, with a ring of bright red above-the tibio-tarsal joint; the claws dusky, the head, neck, and lower parts, are of a dark greyish-blue, the abdomen tinged with pale-grey, and the uppermost hypochondrial feathers, which are very long, have a longitudinal band of white on the outer web. The back and smaller wing-coverts are of a deep olive-brown. The quills, alula, and primary coverts, are dark brown, the secondary coverts, the same tinged with olive-brown, the first quill and first alular feather with the outer edge white, of which colour also is the edge of the wing. The tail is blackish-brown; the proximal under tail-coverts white, and a tuft of feathers under the middle of the latter deep black. The length is thirteen inches, and the extended wings measure twenty-two.

The Water-Hen is found in all parts of England and Scotland that are adapted to its nature. It frequents marshy places, pools, lakes, still streams, mill-dams, and even ditches, where it searches for food chiefly among the reeds and other aquatic plants along the shores. It swims with great ease and elegance, sitting lightly on the water, with its neck erect, and its tail obliquely raised. It dives with equal facility, and in travelling among the reeds, sedges, and other aquatic plants, makes its way with surprising ease, owing to the compression of its body, and its elongated toes. When surprised in a narrow stream or ditch, it usually dives, and conceals itself among the plants or beneath the banks, often remaining for a long time submersed, with nothing but the bill above the water. I have seen it thus betake itself to the margin, when on my going up to the spot, thinking the motion among the grass had been produced by a Water-Rat, it sprung up from under the water, and flew away. On other occasions, I have traced it under the overhanging earth, in a hole among the stones, and behind a waterfall. When disturbed in a large pool or lake, it either swims out to the open water, or betakes itself to the reeds or sedges, among which it remains concealed until the danger is over; and from its hiding place it is not easily scared, for as its power of flight is not of a high order, it prefers the asylum of the water.

In swimming, it moves its neck backward and forward, as a Pigeon does when walking, a circumstance which becomes remarkable in this, as in some other birds, when compared with the Swans, Geese, and Ducks, which keep the head steady while advancing on the water. In general, it is not so ready on being disturbed to betake itself to the open water as the Coot, but prefers skulking along the shores. When a shot is fired at one, and has not hit, it often flies off, but often also keeps steadily swimming on. Being one evening with a friend at Seaton Marsh, on the Dun, near Aberdeen, I started a Water-Hen, and let fly after it, on which it alighted at a very short distance, and concealed itself. My companion, however, having discovered it, took it up, but could perceive no injury that it had sustained. We carried it home, and having satisfied ourselves with observing its form and attitudes, took it back next day and let it loose, when it flew directly off to a great distance. It had evidently been paralyzed by terror, as was the case with the Corn Crake already mentioned. I have seen another when swimming right down the wind, after a shot had been fired at it, raise up the hind part of its body, and spread out its tail like a fan, which thus answered the purpose of a sail, and would have carried the bird on at a good rate, even if it had not made use of its oars.

It often perches on the stumps or trunks of willows growing in the water or hanging over it, or rests on a tuft or turf, where it may be seen standing on one foot, with its neck drawn in. Its ordinary position when reposing resembles that of the Heron, the body being oblique, the legs straight, the head retracted; and in walking it raises its feet high, probably to prevent its long toes from being entangled. Early in the morning, often even at any time in the day, if it suspects no danger, it makes excursions into the fields or pastures adjoining its watery retreat, and walks along precisely with the air of a domestic fowl searching for food. It is extremely vigilant when on shore, and on the least alarm runs off with great speed, throwing its body forward, and stretching out its neck. Its flight is heavy, straight, performed by regular flappings, and very similar to that of the Corn Crake. When flying over a short space, it allows its legs to dangle, and when alighting on the water, enters it at a very low angle, splashing it up with its wings, as is the manner of the Coot and most species of the Duck tribe. In rising, also, it moves a considerable way before fairly quitting the water, which it strikes with its wings, like the Gannet and most aquatic birds.

It is curious to observe with what facility the Water-Hen makes its escape, in circumstances in which one might at first suppose it impossible for it to get off in security. Thus, you may come upon one feeding in a narrow ditch filled with water. It instantly dives, or flies off a short way, and when you run up to the place where it has just alighted, and think you are sure of it, you find no traces of its existence. Watch as long as you please, no bird makes its appearance; it has sunk, and concealed itself somewhere along the margin, and there it will remain, with nothing but its bill above the surface, until you have departed, for it would require an eye sharper than that of a Lynx to discover it. Although, when accustomed to the molestation of Man, it is very vigilant, easily alarmed, and always prepared for flight, it is less wary in remote and unfrequented places. In some of the rushy lakes of the islands of Harris and North Uist, I have found it easier to get within shooting distance than in the mill-dams and streams of the lower districts of Scotland, where, should it observe you, even at a great distance, it is sure to be off instantly, and by the time you get to the place, it has concealed itself.

From the middle of April to the beginning of May, when vegetation has made some progress, but in the northern and more exposed parts of the country not until the middle of that month, the Water-Hen commences the construction of its nest, which it places in the midst of a tuft of rushes or sedges, or fixes among reeds, or builds on a sedgy spot close to the water, or even sometimes on the trunk of a decayed tree or fallen willow. It is bulky, and composed of blades of reeds, grasses, fragments of decayed rushes or flags, and other aquatic plants. The eggs, which sometimes amount to eight or even ten, vary in form from regular ovate to nearly elliptical, and have a pale dull brownish-grey or greyish-yellow ground, with irregularly dispersed spots and dots of a deep brown colour, varying in size from the smallest perceptible by the naked eye to a diameter of nearly a quarter of an inch. Their average length is an inch and three-quarters, their breadth an inch and a quarter. The young, which are at first covered with long stiffish black down, leave the nest soon after they are hatched, and follow their mother. The sight of a flock is interesting, especially if you come suddenly upon it, for then the young scatter about in all directions, dive and conceal themselves, the old bird in the meanwhile lingering and displaying the greatest anxiety, until her brood is safe, when she too dives, and is no more to be seen.

The flesh of this bird is white, and in autumn and the beginning of winter, when there is a layer of fat under the skin, affords good eating, not much inferior to that of the Partridge.

TO THE EDITOR OF THE EDINBURGH JOURNAL OF NATURAL HISTORY.

Paisley, 4th *December* 1838.

Sir,—I sometime ago procured the skin of a bird, among some specimens said to have been brought from Australia, in every respect resembling our Common Grey Wagtail, *Motacilla Boarula*, except in dimensions, and a slight variation as to colour, the latter of no importance. I am aware of birds analogous to the Wagtails being found in India (*Enicurus*), and Mr Swainson has recorded a new Wagtail, *Motacilla gularis*, from the west coast of Africa; but I had no knowledge of the range of the true Wagtails extending to Australia. That they are supposed *not* to extend to that province may indeed be inferred from what is said by Mr Caley, as quoted by Mr Swainson in his recent work on the Muscicapidæ in relation to the "Dishwasher Fantail, *Seisura volitans*."—"I have often," says Mr Caley, "considered it, when I witnessed its manners, to be the Wagtail of the colony;" and Mr Swainson adds, "such it truly is."

Assuming it as a fact, therefore, that the present bird is from Australia, and supposing it were identical with the European species, the circumstance of its being so remotely distributed would of itself be worthy of remark. But after a careful examination, and comparison of the bird with nine specimens of the common kind, I have arrived at the conclusion that it is a distinct species, and probably an immature individual. I draw the conclusion entirely from the disparity which exists in the proportions of the two birds. The following table shows their relative dimensions:—

	European Bird.	Australian Bird.
Total length	7½ inches.	6⅛ inches,
Gape	¾	¾
Front rather less than	½	½
Wings	3½	3⅛
Tail from base	4	3½
Tail beyond wings	2¾	2
Tarsus	¾	¾
Circumference	3¾	2¾

It is thus seen that the measurements all differ, except in the cases of the tarsus and bill, showing the Australian bird to be much the smaller of the two,—differences sufficient to constitute it a distinct species.

The plumage of the specimen is evidently that of an adult bird considerably worn. The colours may be said to agree with those of the female of the common species, only they are duller. The scapulars do not reach the tips of the wings by half an inch, while in the common species they cover the tips when closed. The nail of the hind toe is rather larger than in the common kind.

The bird was brought by a sailor from New Holland, and was disposed of to Mr Wm. Small, Animal Preserver, Paisley, along with other small birds of that region, such as *Rhipidura flabellifera, Sw.*, specimens of *Cinnyris, Melliphaga, Zosterops, Arachnothera*, &c. I see, on the whole, no reason to doubt the habitat.

I remain, &c. William Drew.

Motacilla lugubris, M. Boarula, and *M. flava*, are mentioned by Temminck in his Catalogue of European Birds found in Japan, and we have a specimen of the latter species from Southern India. If the principal difference between Mr Drew's Motacilla and British specimens of *Boarula* be merely the length of the elongated secondary quills and tail-feathers, it may perhaps be found that these parts have been worn.

Bittern.—The body of a Bittern, *Ardea Stellaris* of Linnæus, having been sent to the editor from Whittingham in East Lothian, where it was shot in the middle of December last, an opportunity has been afforded of examining the digestive organs of a bird now extremely rare in Scotland. As in all other birds of this tribe that have been examined by the writer,—about fifteen species,—the œsophagus, *a*, *b*, is very wide and extremely thin, although the external transverse and internal longitudinal fibres are distinct. It is seventeen inches in length, at the distance of three inches from the commencement contracts to eight-twelfths, then enlarges to one inch two-twelfths, and so continues until it enters the thorax, when it enlarges to one inch ten-twelfths. The proventriculus is of the same width. The stomach, c, *d*, is a comparatively small, roundish, compressed sac, 1½ inch in length, somewhat less in breadth, with a roundish pyloric sac, from the upper part of which the intestine comes off. Its muscular coat is thin, its inner coat soft and smooth. The intestine, *d*, e, is 6 feet 7½ inches long, from 2½ twelfths to 1½ twelfths in diameter; the rectum ½ inch in diameter, and its cœcal extremity about a quarter of an inch long. There are no cœca, properly so called, in this, more than in any other member of the family.

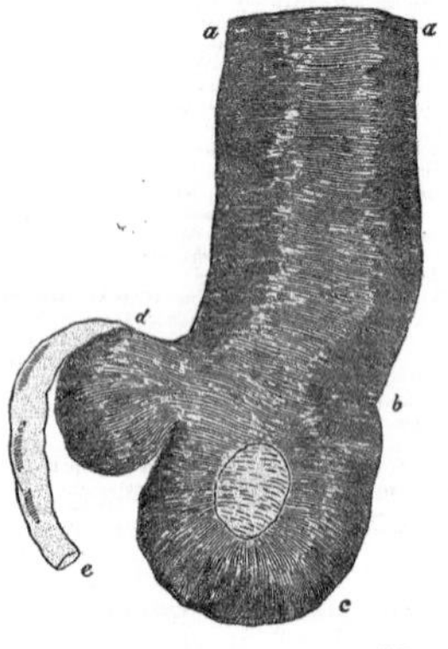

Wild Ox of Scotland.—Dr Knox has published in the Agricultural Journal a paper on the Wild Ox, in which he gives an account of Cuvier's ideas as to the different species of Bos, figures of the skeleton and skull of a specimen of the Duke of Hamilton's white cattle, a historical view of the subject, and some curious remarks as to the incapacity of persons calling themselves Naturalists. The principal conclusion seems to be the following:—"In the absence of historic facts, it would be rare to offer any conjectures as to the origin or history of the wild cattle of Britain. They may be a distinct species from all others, and the present improved short-horned breed may be derived from them; they do not seem to me to have been the original breed of cattle of this country, but are probably an Italian breed introduced by the Romans." However this may be, it is clear that the skeleton figured presents characters sufficient to constitute a new genus, for no other ruminating animal known to us has the ribs articulated in the manner represented. After all, the matter has been left just as it was, and as it will remain, until some sufficiently educated and industrious person shall undertake its investigation.

ON SLEEP. BY J. C. BELLAMY, ESQ., YEALMPTON.

Sleep is enjoyed by animals in varied proportions. In some it takes place for a certain space of time daily, in others, a state of quiescence is remarked after the lapse of indefinite periods of activity. Some enjoy it daily during the continuance of light, others during the interval of darkness; others, again, observe no regularity in this respect, but take a certain portion of rest, with frequent interruptions, during the twenty-four hours, all these variations of habit being in accordance with the respective varieties of their economies, and in particular with the different natures of their food. But some partake of a large proportion of sleep daily, while others are content with a much less quantity, and some, also, enjoy but a limited amount after protracted activity. And here we are deficient in reasons, for this quantity of sleep does by no means appear proportioned to the quantity of exercise sustained by the respective animals, sleep or rest being ordinarily and rationally regarded as a compensation to the animal powers for previous activity. The Swift and Swallow, which take both incessant and violent exercise during the long days of summer, take only the same portion of sleep as the Sparrow or the Robin, which move only to the adjoining gardens and homesteads throughout the day; the Stormy Petrel, which continues on the wing, and is ever changing the direction of its flight throughout the tempest, and is for the most part in action at other times, seems to take but a very small quantity of compensating rest and sleep. Birds of prey, which for the most part are not obligated to use much exertion in the capture of their quarry, seem nevertheless to devote a large proportion of time to repose. The Alligator, according to report, indulges much in rest, although its actions are limited in comparison with very many other creatures feeding and living similarly. Very many kinds of insects whose spheres of activity are very circumscribed, apparently set apart a considerable deal of time to sleep, or at least inertion. Sleep is enjoyed by creatures with different degrees of perfection, some being awakened with the slightest noises, others allowing themselves to be approached, and even touched or injured, before their faculties are resuscitated; "the Rook," as Mr Knapp says, "seems rather to rest than sleep. Pass what time I may in a road adjoining a rookery in this parish, I inevitably hear many of them disturbing the quiet of the night, as if impatient to see the dawning of the next day." The Alligator, it seems, may be approached quite near while it sleeps, and from its colour, dimensions, and figure, very many persons have thus been exposed to imminent peril. Perhaps, generally speaking, this unsusceptibility to noise and intrusion, during repose, may be proportioned to the powers and courage of the animal, for, though the Rook roosts high, and ought to be free from the apprehension of danger, yet a variety of the weaker kinds of creatures need to be wary and watchful, and surely are so. The Sedge-bird gives notice of this wakefulness, by uttering its peculiar notes frequently on the mere passage of a person on a neighbouring road, independently of any further interruption, such as throwing a stone into the bush where it roosts, as Mr White and others have stated. If sleep be perfect and healthy, the whole nervous system during this interval of rest is passive and inert; thought and all mental action, together with sensation, are suspended; if the two former are continued, one part only of the nervous system is at rest, if they are continued, we see no demonstration of them, and if their action is supposed to be partial, of what use would such imperfect action be? But there are few persons so foolish as to conceive that dreams and other demonstrations of continued nervous operation occur in a true state of nature, and consequently it seems to follow, that these are only concomitants of disease, or at least of depraved condition of system, which is the same thing as disease, only in a less degree. Still the nervous system, while we have seen that it enjoys its repose in varied extents, and in varied degrees of perfection, in accordance, no doubt (though we are not able to trace these connexions on all occasions), with the habits and economies of creatures, is found to be more susceptible of derangement and lesion in some domesticated animals than in others, for it is in the domesticated state that disease nearly always occurs. We do not draw this conclusion from the instance of our own species, because man seems to take pains, by unsuited diet and depraved actions, to induce disease, and this has been perpetuated through a succession of generations; but the Dog is remarkably the subject of dreams, while the Horse is seldom, if ever, so affected, but altogether the nervous system is that most usually disordered and disorganized in domesticated and confined animals; witness Palsy, Epilepsy, Mania, Hydrophobia, Deafness, &c. In ourselves there are certain peculiarities of our dreams which, to weak-minded persons, suggest a degree of importance respecting them, which it may be clearly shown they have no right to. In the first place, it is not fair that any support should be granted to such an idea, from the fact of dreams having in primitive ages been made use of as divine instruments and intimations, because we have no grounds whatever for supposing that in latter ages such manifestations have been vouchsafed to any members of our race, or that we have been favoured with such immediate and appreciable interferences and demonstrations. Science is not called on to refute the errors of vulgar superstition, or the chimerical delusions of religious fanatics. It is said, that if one should dream a dream several times repeated, it is sure to be eventually verified; but who will stand forth to corroborate this assertion? And without staying to lay stress on certain points which furnish objections to the validity of such a belief, may not such occurrences be safely referred to the mind of the dreamer having dwelt for some time on the subject of the dream, or on something analogous thereto, and we shall presently show that the mind is very apt in us to perform its offices in a partial, imperfect, and incoherent manner during sleep, or rather our imperfect repose. Perhaps the most important of the presumed supports to the prophetic nature of dreams, or a portion of them, is drawn from the extraordinary dream of the celebrated Mr Newton, and other cases of the same character; but without entering into an inquiry of the particulars of these, which would lead me further from my present argument than I should wish, I will content myself by observing, that since our sleep is so frequently unsound, especially where the mind is in an habitually disturbed and excited state, as in the case of Newton, and since it does appear, as I shall presently again state, that unsound sleep is compatible with a greater or less development of the imaginative, and even of the reasoning powers also, so it is not more wonderful that in such kind of sleep, partaken of by persons of a disturbed and nervous habit, the mind should foresee events connected essentially with its immediate character, than that the same prophetic indulgence of the mind, during a waking state, should actually receive a verification; "for," as Mr Mayo observes, "the mind in dreaming moves in its accustomed channels." Still, though I am disposed, from the general analogy of the argument, to exclude all credit in the importance and prophetic character of dreams, I feel that it would be rash and indecorous to deny the possibility that in such cases as Mr Newton's a divine influence may have been interposed.

It appears to me that true natural sleep excludes at once consciousness and thought, that consciousness is capable of being brought into action on the application of stimuli, or excitants of the nervous system, with various degrees of ease in different individuals, just as we have seen animals, more or less readily aroused from sleep on the intrusion of noise, contact, &c. We usually find that on the removal of the intrusions to our dormant consciousness, the latter again withdraws itself to rest; at other times, according to the condition of the health of persons, these intrusions constitute the basis of some imaginative flights in their subsequent imperfect sleep. The movement of the limbs, in order to attain more convenient postures, together with other momentary demonstrations of consciousness, serve only to show incidental interruptions to our repose, induced by painful impressions, but yet so transitory and incomplete, as seldom to leave any trace of it on the memory. With regard to mental operations during sleep (imperfect sleep), they are excited by various causes, and are in themselves different in their natures; some of the causes of dreams will sufficiently account for the character they evince; a disordered state of stomach from repletion, ill-adapted food, &c., will draw forth the imagination to luxurious repasts or revellings, in the sense of taste; the receipt of injuries or calumnies will often suggest dreams of a vindictive or disputative character; the study of some abstruse or important subject, on which we may be intently bent, will occasionally be found the subject of our dreams; in all of which cases we usually find that the mind occupies itself upon them in a desultory and incoherent manner, probably combining objects of importance with those of an opposite nature, associating the ludicrous with the grave, or bringing some unconnected and indifferent materials into opposition with a connected train of thought on the realities of some preceding day, or on some future probabilities: On the contrary, at times, though rarely, the mind will act as perfectly as though the dreamer were awake, and the senses themselves receiving no actual impressions by which the mind is ordinarily obstructed during its investigations, the processes of the imagination and of reasoning are conducted with more than usual speed, perspicuity, method, and perfection. It has happened to me when my mind was occupied in retaining a deal of relative anatomy, than which nothing can be more arduous, to have dreamed with the utmost accuracy and with great profit, the whole of what I had been engaged in learning on previous days. I have also at times been enabled in my sleep to repeat the various steps of an argument or theory, respecting which I had been interested. At seasons when the mind is occupied for many successive days in the contemplation of some given subject, and in looking forward to future results, it cannot be deemed extraordinary, that on the same consecutive nights there should be a repetition of certain prospective images, or that these fancies of our sleep being the reflected images of our previous waking thoughts, when reason guided our apprehensions, should on some future occasion receive their realization.

The condition of persons who think or recapitulate correctly in what is ordinarily termed their sleep, is not essentially different from their state while in a reverie, for, in the latter, consciousness, sensation, and motion (or the consciousness of it), are in a great measure suspended, and it seems that in this species of repose the mind acts while consciousness, sense, and the muscular fibre, are recruiting their powers. It is, in my estimation, an error to attribute real, perfect, and healthy sleep to persons who are in the habit of dreaming incoherencies, and much more so to allow it to those in whom the mental operations are skilfully carried on, because, if sleep be a destined interval between the seasons of mental and bodily exertion, it is certainly not attained by those who are observed to use more or less muscular efforts in what is termed their sleep, nor by those in whom ratiocination, exertion of memory, or excited interest, is remarked; neither by fair inference is it enjoyed by those who dream incoherently, or indulge their fancies vaguely and loosely in the blandishments of sense. The mind very commonly in our sleep occupies itself on the subjects which last engaged our attention previously to our repose, and if we fall asleep with the desire to awake at an hour earlier than has been our habit, that object is kept in view throughout the night, and we wake at the time desired. On one occasion I was a somnambulist; in my dream I fancied I had been called on by a knock at the door to let some one into the house; I threw back the bed-clothes, opened my door, avoided, by means of my sight, objects that might impede me, descended a flight of stairs, holding by the bannisters, unbolted the street door, and returned with care to my bedroom, these facts having been brought to my remembrance by finding the door unsecured in the morning. Now, in this and the preceding cases, it seems contradictory to suppose the occurrence of actual sleep, for, besides the evidence they bear of mere partial repose, it is seen, that where the person rests with the intention of being engaged the next day on an occupation of importance, where the dream is unusually exciting, where there is a wish to awake sooner than usual, and where somnambulism occurs, the sleeper is, *cæteris paribus*, awakened with unusual ease; moreover, many persons dream but seldom, and can then attribute the circumstance to some evident cause; lastly, many partake of a sleep in which they have some very indistinct perceptions of passing occurrences; in all these instances, the conclusion of sleep being void of consciousness and thought in a healthy state of body, and connected with the same by reason of disease or moral excitement, being apparently unavoidable. Dreams have reference to past facts, impressions, and mental prospects of the future; they are mostly reflected images of realities we had either experienced, or wrought up by our fancies in a waking state. They who advocate the prophetic nature of dreams must acquaint us whether they were realized to them in a strict and satisfactory manner with relation to the dreams. They must say also why all dreams are not prophetic, why so large a proportion consist of incoherent materials, and especially they must explain why dreams, seemingly important, and quite consistent in detail, are never verified.

If the above named opinions relative to sleep and dreaming are correct, those of Mayo, Elliotson, and others, must be wrong. Mr Rennie, in "Habits of Birds," after allowing that the Redbreast, Restart, and other birds, occasionally sing at night, mentions a supposititious case of a Dunnock singing at ten P.M. *while asleep*. That caged birds may do thus, is what I will allow as readily as Mr Rennie, but is it not most probable that this Dunnock was awakened by the "cold and frosty night" (such are Mr R.'s words), in the same way as torpid animals are awakened by intense cold?

J. C. Bellamy.

CULTURE OF TOBACCO IN INDIA.

In an excellent paper on the Agriculture of Hindostan in the Quarterly Journal of Agriculture, the following account is given of this plant, which forms so widely dispersed an article of luxury.

"Two varieties of the *Nicotiana Tabacum* have been cultivated, the Orinoco, and the sweet-scented Tobacco. They are very similar in appearance, and differ principally in the form of their leaves, those of the latter being broader and shorter. The leaves are light green, and grow alternately at intervals of two or three inches on the stalk, which is strong and erect, attaining a height of from six to nine feet. The lower leaves grow to the extent of twenty inches, and they decrease in size as they ascend, those at top being about ten inches long and five broad. The young and tender leaves are rather smooth, and of a deep green colour, but become yellowish and rougher as the season advances. The extremities of the stalks are surmounted by clusters of flowers, which are yellow externally, and are within of a delicate red. As these fade, kidney-shaped capsules succeed, each containing on an average a thousand seeds. A single plant has been known to produce many hundreds of capsules, so that its produce may be from three to five hundred thousand seeds.

Although Tobacco be very generally cultivated throughout the East, and more especially in the northern and western provinces of India, and although no plant of European production is in such general request there, its consumption being almost universal among the inhabitants, yet it is in many respects inferior to that of American growth. This, however, we feel assured is much more attributable to the want of skill in the curers, than to the soil or climate of Hindostan, and may doubtless be remedied by superior care and attention on the part of cultivators.

The Javanese annals inform us, that the smoking of Tobacco was introduced into that island at the very commencement of the seventeenth century. The plant is, however, true to its more northern origin, and will not arrive at perfection in the hot plains and valleys, unless it has been first raised from seed upon the cooler mountainous tracts. Two distinct classes of husbandmen are therefore required in its culture. The seedlings reared on the hills are sold to the lowland farmers, under whose care it comes to maturity, and seeds again resold by them to their hill friends. Unless this plan be adopted, the plants are sure to degenerate.

No other proof need be mentioned of the extraordinary fertility of the valleys of Java, than the circumstance that two crops are taken annually from the finer lands, the one of Rice and the other of Tobacco, after which the fields are again laid under water, and sown with rice. This submersion is the only dressing that the soil receives, and its only relief an occasional half year's fallow. The Tobacco seedlings are transplanted in June, and the crop is reaped before the setting in of the rains, in October and November.

Tobacco requires a rich soil, and will not thrive on inferior lands; and it is only on this account that particular districts enjoy superior reputation for its successful culture. The ground is prepared for its reception by being broken up with hoes, and the clods worked until sufficiently pulverized. The mould is then drawn with the hoe round the projected leg of the labourer, until it form a mound reaching to his knee. The leg is then withdrawn, and he proceeds to form another and another mound, until the whole ground has been gone over. These hillocks are made to run in lines, three or four feet apart in every direction.

The plants, which have been reared from the seed with great care and diligence, being tender, and unable to stand much cold, thereby requiring a covering of mats at the least threatening of frost, are taken out of the ground, when it has been sufficiently moistened by rain, when the fourth leaf has sprouted, and the fifth is just appearing, and conveyed to the field in baskets, one plant being dropped at every hillock; while a person immediately follows, who, with his finger, makes a hole in the centre, and places the plant in it, carefully firming the ground with his hand, and in such a manner as not to hurt the leaves, which are easily injured.

If there be much moisture in the soil, it will be evident, in a very few days, that the sprouts have taken. Such as have not are withdrawn, and their places supplied by others, a favourable shower being waited for.

No crop requires greater care and attention, through the whole course of its growth, than Tobacco. The plants must be continually weeded and earthed up, and any dead leaves which may be discovered about them must be carefully shred away and removed. When the plants have attained the height of two feet, the flower-branches appear, and the operation of topping must then be gone through, lest too much of the nourishment should go to the flower and seed, depriving the leaves of their strength. It is also necessary to remove from time to time all suckers or superfluous shoots; and this is done by the finger and thumb-nail. Grubs and other destructive insects must also be carefully removed as they appear. Some cultivators keep flocks of Turkeys for this purpose, the Caterpillars being especial favourites with these birds.

The disease termed *firing* is a blight which takes place from long continued drought; the plants withering and dying out in some particular spots. When ripening, the plants change their colour from a dark to a yellowish-green, and the fibres of the leaf thicken and become more prominent. The time for cutting and gathering in requires an experienced eye; but this should not be unnecessarily deferred, the touch of frost producing entire destruction.

The plant is cut near the ground, and if the stem be thick, it is then split down the middle, that the leaves may be more exposed to the air in the process of curing. In this state the plants are left on the ground, until sufficiently exposed to the drying rays of the sun. After this they are gathered into a barn, whose sides are constructed so as to admit a partial circulation of air. Poles, four feet apart, are stretched horizontally across from wall to wall, and these again intersected with what are called tobacco-sticks, on which the leaves are hung to be cured. This is done by suspending the plants by the split which has been made in the field at cutting time. The air is generally sufficient for the purpose of curing; but if the weather is unusually moist, small fires of bark, or rotten wood, are used in different parts of the barn, care being taken that the heat is not too great, as something resembling the disease of firing may be thus induced.

In four or five weeks from the time that firing has commenced, the Tobacco is said to be in case; that is, the leaves are tough and elastic, and covered with a glossy moisture. The first damp day is then chosen for stripping the leaves from the stalks, which are distributed into three heaps, according to their qualities. Small bundles are then tied together at the thickest ends, by means of a leaf tightly wrapped round them. This fasciculus is termed a *hand*, and is about the thickness of a thumb.

The little bundles are then thrown together on a platform, to undergo the operation of sweating, a process in which a slight degree of fermentation is created. When this has subsided, the leaves again acquire their elasticity, and the Tobacco is considered fit for shipment.

SALMON-FISHING IN THE TAY.

In the Statistical Report of the parish of Balmerino, in Fifeshire, the Rev. John Thomson makes the following statement:—"The Salmon-fishery, once so productive here, is now very inconsiderable, being confined exclusively to the *toot-net* method of capturing the fish. The net employed is from 50 to 80 yards long, and differs little from the common seine or sweep-net. Instead of being constantly kept in motion, as the latter is, it is attached to a boat at anchor, and only hauled when the tootsman, who watches in the boat, observes a fish to strike the net. It is totally unsuited to estuary fishing.

"The number of boats connected with the fisheries in this parish amounts to seven, and they keep employed 14 men during the open season. The fisheries belong to Mr Wedderburn of Birkhill, Mrs B. Morison of Naughton, and Mr Stuart of Balmerino, and are rented at about L.50 per annum.

"In the year 1797, the stake-net was introduced in the Solway Frith (where it is at present legal!) and in the Frith of Tay. This engine was constructed by driving strong stakes in a row, from the shore towards low water mark, and nearly at right angles to the tide. On these stakes were stretched nets with open meshes, three inches from knot to knot, or twelve inches round. Thus a wall of open netting was constructed, sufficient to intercept the large fish, but through which the water, and all small fishes, could pass freely. In this wall of netting were placed courts or labyrinths. The Salmon, influenced in their movements by the tide, met this netted wall, and, seeking along for an opening through which to pass, entered these courts, where they were detained and taken out at low water. By means of these nets, great quantities of Salmon, of the first quality, were caught, and exported in boxes packed with ice to the London market. The proprietors of the river fisheries now got alarmed for their monopoly, and, taking advantage of the antiquated statutes referred to above, some of them passed by the Scottish legislature more than four hundred years ago, and long before it was ever dreamt of that Salmon-fishing could be successfully carried on in the friths or seas, applied for an interdict, and ultimately succeeded in prohibiting the use of such fixed machinery: the courts having held, that, as these statutes had never been repealed, any infringement of their provisions must be held illegal. The effect of this decision has been greatly to deteriorate the estuary fishery; while the proprietors of the river fishings enjoy a monopoly price in the market.

"In framing these regulations, the object of the Scottish legislature must have been to protect the public interests and prevent injury to the fishery generally; and this is the only ground upon which they can yet be defended. Experience, however, has proved that this defence of them cannot be maintained. At the early period when these regulations were framed, little was known regarding the Salmon; but from what is now known of its habits and history, it is perfectly well ascertained, that while in the friths and in the sea (where alone stake-nets can be used), none but fish in the best condition are caught; in the rivers, Salmon of inferior quality, and often in a foul and unwholesome state, are taken,—that it is in the *rivers*, and in these alone, that any injury can be done to the spawn, or salmon fry, by the heavy ground ropes which are drawn across the spawn-beds; that the stake-nets do not interrupt or interfere with the river fishings, as it has been proved that the quantity caught in the river, since the stake-nets were abolished, does not exceed the quantity caught when these were in use;—that, in short, the effect of the prohibition has been to secure a monopoly price to the river fishers, while the myriads of Salmon which escape from the rivers and find their way to the sea, and which might become a valuable article of commerce, and be made available as a rich and nutritious article of food to the public, at a moderate price, are totally lost, or only abandoned as a prey to the monsters of the deep.

"The extent of the loss in the Tay generally, in consequence of the suppression of these nets, has been estimated at from 200 to 300 tons, or from 20,000 to 30,000 head of Salmon annually. The whole estuary does not now produce above 3000 fish. (Evidence before a Committee of the House of Commons, 1827.) In this parish alone the loss may be estimated in rent at L.1000 to L.1200 annually to the different proprietors; and about L.1000 in the shape of wages. Other evils also have resulted. The aged females and others have been deprived of an excellent and healthy employment in the working of nets, while a hardy and expert race of seamen were regularly trained through means of their connection with the fishery. Let us hope that an enlightened and paternal legislature, under whose revision the fishery laws are again to be brought, will speedily remove the present oppressive restrictions upon this lucrative branch of industry, by which not individual proprietors alone, but the community at large, must be so extensively benefited, and in the decision of which question they have so deep an interest."

Edinburgh: Published for the Proprietor, at the Office, No. 13, Hill Street. London: Smith, Elder, and Co., 65, Cornhill. Glasgow and the West of Scotland: John Smith and Son; and John Macleod. Dublin: George Young. Paris: J. B. Bailliere, Rue de l'Ecole de Médecine, No. 13 bis.

THE EDINBURGH PRINTING COMPANY.

INDEX TO THE JOURNAL.

Aborigines of South America, prospects of, 68
Acarus scabiei, the, 86
Acetic acid, improvement in the manufacture, 72
Adders at Methven, 2; the bite of, 6
Address, our, to the public, 1
Aërolites, fall of in Brazils, 160; motion of; 72; temperature of, 72; nature of, 83
Africa, allusion to some of the animals of South, 136
Agate shells, our plate of, described, 93
Aglossa intestinalis noticed, 110
Ailurus, (Pandas,) plate of, described, 133
Albino, a Malay, described, 159
Amber, facts concerning, 24, 72; the yellow, of Lake Arendsee, 166
Ambergris, account of, 24
Anatifa, genus of, described, 134
Andes, Ulloa's ascent of, 31
Animal magnetism, 20, 32, 52, 56
Animalcules of snow, 21
Annals of natural history noticed, 140, 148
Anser bernicla, in Cromarty and Beauly Frith, 95
Ant, anecdote of the, 25
Antelopes, great, plate of, described, 69
Anthus arboreus, habits of, 110
Apes, proboscis and solemn, plate of, described, 125
Apple tree, the, in Russia, 106
Aquila Chrysaetus described, 93
Ardea stellaris, see Bittern
Arendsee, Lake of, described, 166
Arnee, the horns of, described, 159
Arsenic, antidote for, 84
Artesian wells, notice concerning, 120
Ascents, great, into the atmosphere, 44
Aspidium dumetorum, 104
Ass, peculiarity in giving milk, 29
Audubon, letter from, on board of the Crusader, 171
Aurochs, notice of, 149
Aurora borealis, 100, 168; Sir John Ross' views of, 24
Australian ornithology, by Gould, 134, 136

Baboons, plates of, described, 145, 153
Bachsthanne, Dr, description of the hares of North America, 162
Baloon excursion, 9
Barley, native country of, 43
Barometer, submarine register, 44
Bats, plate of, described, 113; the Scottish, 81
Beaches, see Sea Beaches
Beagle, the, described, 53
Bears, our plate of, described, 129
Bee-eaters, our plate of, described, 109
Beer, preventive against acidity of, 92
Bees, culture of, 9; anecdote of, 37
Beet-white, Beta cicla, sugar from, 58
Beetles, plate of, described, 37
Bellamy, Mr J. C., contributions of, 104, 111, 115, 126, 130, 138, 140, 147, 151, 155, 156, 162, 174
——— George, Wanderings in New S. Wales, noticed, 27
——— J. W., his fishes of Ceylon noticed, 42
Bernacle shell described, 134
Bill of birds, its frequent elongation, 138
Birds, diseases of, 134; mortality of, 105; quantity of seed devoured by, 90; social habits of, 9
Bittern, the, noticed; digestive organs described, 174
Blackbird, breeding with the thrush, 102; prolificacy of, 143

Blind-worm, notices of, 168, 172
Blood-hound, the, described, 53
Body, human, effect of compressed air on the, 76
Bog-bean, the, described, 119
Bogs, moving, notice of, 31, 71
Bone caves in New Holland, 10
Borneo, notice concerning the inhabitants of, 16
Bos Urus, notice of, 149
Botany, fossil, Brongniart's researches in, 58; of Lower Egypt, 54
Boulder, at summit of Dunkeld hill, 51; near Castle Stuart, 55
Boulders, buoyancy of at great depths, 75; upborne by icebergs, 75
Brent Goose in Cromarty Frith, 95
Breves, our plate of, described, 141
Bridgewater treatise, Mr Kirby's, noticed, 4
British birds, Mr MacGillivray's descriptions of, 90, 93, 98, 105, 109, 114, 117, 121, 125, 133, 145, 149, 157, 165, 169, 173
Brugmansia Zippelii, the plant described, 51
Buceros, see Hornbills
Buccinum acutissimum, 58; Anglicanum, described, 58
Bulls, our plate of, described, 89
Buntings, plate of, described, 89
Bonaparte, Prince Charles L., his comparative list of European and American birds, 152
Butterflies, our plate of, described, 13; feasts upon, in New Holland, 18

Camels, plate of, description of, 25; pace of, 9; step of, 78
Canary, anecdote of the, 29
Caprimulgus, the, described, 149
Cargueroes, or men-carriers of the Andes, noticed, 40
Caryophyllea sessilis, coral, described, 110
Cat tribe, plates of, described, 19, 21, 144, 161; domestic cat, anecdote of, 26; Mahomedan cat-hospital, 52
Caverns, remarkable, in Brazils, 128
Cebus, plate of, described, 105
Celery, new mode of blanching, 3
Cercopitheci, plate of, described, 105
Chaffinch, Mr Waterton on the, 152
Chats, plate of, described, 73
Chillingham Park, Mr Hindmarsh's account of wild cattle of, 161
Chimpansee, figure of, described, 121
Cinnamon tree, described, 128
Circus, plate of, described, 121
Civets, plate of, described, 117
Claws, irregular growth of, 92
Coal, discovery of in Greece, 10; in Mount Lebanon, 23
Coatis, plate of, described, 133
Cochineal of Armenia, described, 65
Cockatoos, plate of, described, 129
Cod, the fecundity of the, 33
Coffee tree, description of the, 103
Cold, effects of, on new-born infants, 102
Colouring, Mr Hay's remarks on harmonious, 47
Colour, change of, in birds, 164; in plants, 38; sepia colour obtained from peat water, 80
Columba palumbus, described, 125
Comets, account of, 4; notice of Halley's, 11, 24
Cones, conus, plate of, described, 61
Coral found near Plymouth, 110
Cormorant, notice of the Chinese, 168

Corn, effectual mode of preserving, 163
Corn-crake, described, 165, 81
Coniguerina volcano, eruption of, 15
Cossus ligniperda, endurance of the larva without food, 131
Coturnix dactylisonons, see Common Quail
Cowdie tree, the timber of the, 10
Crabs, changes in the stomac of, when casting their shell, 74; crab-stones, 74
Cromarty Frith, Mr MacGillivray's description of, 96, 100, 144
Crossbills, plate of, described, 97; resident in Scotland, 142
Crow, anecdotes of, 37, 66; anecdote of the hooded, 118; of the carrion, 118; piscatory habits of the carrion, 138; short-tailed crows described, 141
Crystals, formation of artificial, 116
Cuckoo, the, described, 145; seen in November in Morayshire, 131
Curlew, the, described, 169
Cuscuta epilimum, a British plant, 151
Cushet, the, described, 125
Cuvier, the Baron, Statue of, in Montbelliard, his native town, 16
———, M. Fr., his observation on the Jerboas, 122
Cynocephalus, plate of, described, 145
Cypselus murarius, described, 105

Darkness, influence of, on human body, 104
Davy's, Sir H., Safety Lamp, objections to, 12
Dayaks of Borneo, description of these aborigines, 27
Deer, plates of, described, 6, 165, 73. Deer hunting in Texas, described, 167
Deluge, Mr Harcourt's Treatise on the Doctrine of, alluded to, 144
Dent du Midi, fall of, 59
Devonshire, South, Mr Bellamy on the fauna of, 126, 138, 140, 147, 151; on the slates of, 162
Dew-butter, the so called, 31
Diamond, the, described, 66; facts regarding, 72; the Matton, 31
Digestive organs of birds, described, 94, 98, 143, 147, 170
Diseases of birds, remarks on the, 134
Divers, the powers of, 16
Dodo, the authenticity of the, 150
Dogs, plates of, described, 41, 53, 113; anecdotes of, 43, 92, 105; a hybrid between the dog and jackal, 61; hydrophobia in the, 150
Dolomites, phenomena connected with, 112
Douglass, Mr David, obituary notice of, 76
Dove, habits of the blue-backed, 99; the ring-dove, described, 125; the rock-dove, described, 133
Dreams, Dr Gregory's notes on, 20
Drew's, Mr, remarks on the Australian wagtail, 174
Dromedary, plate of, described, 25
Duhn's Ornithological Guide, notice of, 88
Dutrochet's Observations on the Sleep of Plants, 95, 99, 123

Eagles, plate of, described, 117; fisher-eagles, plate of, described, 125; the golden-eagle, described, 93; digestive organs described, 143; white-tailed sea-eagle, described, 98; nests, the localities of their, 91
Eaine, an aërolite called inflammable snow, 4
Earth, temperature of, 23; of the interior, 75

Earthquake at Conception, described, 10 ; at Chichester, 15 ; in Croatia, 128
Earthy substances, corrosive quality of some, 128
Edible birds' nests, 62
Edible rocks, 3
Edinburgh, meteorology of the neighbourhood of, from Mr Rhind's Excursions, 63
Edinburgh New Philosophical Journal. notice of, 76
Egg, hen's, anomalous structure of, 27,.52 ; preserved fresh for 300 years, 15
Egypt, remarks on the vegetation of, 54
Electricity, effects of, on plants, 38
Elephants, plate of, described, 84 ; anecdote of,' 36 ; their love of sweatmeats, 15 ; cause of their occasional rage in confinement, 9.
Elevation of the land in Sweden,.71
———— of sea beaches by the tide, 118
Elk, fossil, described, 58
Erratic blocks of Jura, Agassiz on the, 135.. See Boulders
Estuaries, Mr MacGillivray's remarks on, 96, 144
Euphrates Expedition, notice of, 8
Eupleres Goudotti, a new Insectivorous animal of Madagascar, 70
Eurylaimes, plate of, described, 149
Expedition for exploring South Africa, results of, 88

FAIRY-RINGS, opinion of the cause of, 18
Falco peregrinus, described, 109
Falcula palliata of M. Is. Geoffroy, described, 65
Fallow-deer, plate of, described,.173
Feigning death, animals, 39, 81·
Felis, caracal, described, 101 ; colocolo. 141 ; leo, 21 ; onca, 141
Fennecs, plate of,' described, 113·
Ferns, hybrid, 131·
Finches, plate of,' described, 77
Fire-fly, notice of the, 37
Fishes of Ceylon, Mr Bennett's account of, 12·
Fishes, eruption of living, from a volcano, 26.
Fissilabi., a new shell, 42
Flint, singular sections of Kentish, 19
Floating islands, notice of, 15
Flying-cats, plate of, described, 53
Flying-fish, account of, 17
Flood, remarkable of the Garonne, 7
Foot-marks on rocks, instances of, 15, 60, 62
Foot of the fly, mechanism of, 30
Forests, effects of, on rains, 112 ; pine forests of Scotland, 95
Forth, filth of, birds observed in, 129
Fossil bones. in Paris Basin, 88 ; fossil camel in the Himalayas, 88 ; remains in the coal formation at Wardie, near Newhaven, ·108 ; remains of monkeys, 111 ; horns found· in Essex, 144 ; bones in Devonshire, 155 ; oak in Aberdeen harbour, 7 ; ferns, 23
Foxes, plate' of, described, 83 ; Scottish varieties, 118 ; Arctic, plate of, described, 113
Fox-hound, described, 53
Fragrance, cause of the, of tropical countries, 74
Friendship among animals, 45
Frogs, anecdote of, 50 ; alleged shower of, 50
Frozen soil, depth of, in Siberia, 144
Fruit trees, management of, 10
Fuci, the, used as food, 107

GUILLON'S Aperçu d'Histoire Naturelle, notice of, 52
Galagoes, plate of, described, 101·
Galeopithecus, plate of, described, 109
Galvanic·telegraph, noticed, 164
Gannet, black-tailed, described; 161
Garrulus, plate of, described, 157
Gas, inflammable, natural evolution of, 43, 44; 64.
Gedrite, a new mineral, 108.
Geese sometimes.carnivorous, 118
Genets, plate of, described, 117
Geographical distribution of animals, 126, 130, 138, 141, 147, 151.
Germinative power of seeds, 35, 52, 74·
Gibbons, plate of, described, 121·
Gingo tree, introduced into Europe, 42·
Giraffe, plate of, described, 57
Gizzard of gallinaceous birds, described, 147
Glow-worm, the, noticed, 56
Gnat-catcher, plate of, described, 85
Gnats, tormenting power of, 54
Goat-sucker, the, described, 149 ; Col. Maceroni's opinions of the, 54
Gold, sources of, 36 ; veins of, in North Carolina ; composition of native, 3 ; coinage, notice of the, 4 ; medical virtues of, 152
Golden eagle, description of the, 93
Goose, anecdote of the, 45
Grouse, plate of, described, 45, 101 ; the black, described, 114 ; the red, 117 ; hybrid with the black-cock, 82

Guscharo of the cavern of Caripe, in Cumana, 86 ·
Gundaloupe, human skeleton of, 131
Guenons, plate of, described, 93
Guyana tea plant, 51
Gulf Stream, account of the, 10
Gun flints, manufacture of, 84
Gunpowder, history and manufacture of, 12

HAILSTONES, the shape and size of, 72
Hair in the intestine of the horse, 165.
Haliaetus, plate of, described, 125
Hares, the, of North.America, described, 162
Harrier, the, described, 53
Harriers, plate of, described, 121
Hay,.Mr D. R., notice of his work on colouring, 47
Hedgehog, the, said to be proof against poison, 32
Heifer, which yielded milk, 2
Herring, Mr Shaw on the food of, 196
Hibiscus·mutabilis, change of colour in the flowers of, 26
Hogs, plate of, described, 157
Honey buzzard, described,. 153 ; killed in Scotland, 153
Hoopoes, plate of, described, 113 ; observed in Scotland, 86
Horsburgh, Capt. James, biographical notice of, 60
Horn, irregular growth of, 82
Hornbills, plate of, described, 173
Horny excrescences in men, 36
Horse, rabies in the, 150 ; toes on the feet of the, 46
Hot blast, the, in ironworks, 64.
Hot springs, notices respecting, 39, 64.
Howison, Dr, on the mountain-ash, and apple-tree in Russia, 106
Howlers, plate of, described, 169·
Humboldt, Baron Wm. de, his death, 4 ·
Humming birds, plates of, described, 29, 65, 89, 101, 113.
Hybernation of animals, 34, 42, 50 ; instances in swallows, 164 ·
Hybrid animals, 82 ; plants, 131·
Hydra, described, 167
Hydrophobia as occurring in the lower animals, 150
Hylobates, plate of, described, 121

ICE affected by cold, 80 ; striking light with steel, 48·
Icebergs of Southern hemisphere, 59
Ictides, plate of, described, 133
Igneo fatui, instances of, 55, 64
Indian death blast, effects of, 4
Indris; plate of, described, 101
Infusoria, fossil in Tripoli Slate, 92
Ink, indelible, Dr Traill's, 155
Insects in India, numbers of, 2
Instinct, animal, 163
Ireland, introduction of frogs and snakes into, 70.
Irritability in the stems of plants, 87
Itch, insect, noticed, 66 .

JACKAL, plate of, described, 113
Jaguar, described, 141
Jardine's, Sir Wm.. History of British Birds, Vol. I., noticed, 132 ; illustration of ornithology, noticed, 136
Jays, plate of, described, 157 .
Jerboas, M. Fr. Cuvier's observations on, 122.
Jersey, the rare plants of, described, 83

KINGLETS, plate of, described, 49
King-fisher, the, described, 157 ; its nest, 109·
Kirby's Bridgewater treatise, noticed, 4·
Knowle-Park beech, dimensions of the,.14
Knight, Mr Thos.. A., biographical·notice of, 156
Krubut, Rafflesia Arnoldii, described; 14

LACERTO GRECO, mechanism of foot of, 30
Lagopus Scoticus, described, 117 ; cinereus, described, 118
Lakes, filling up of North American, 38
Lamium album, and maculatum, not distinct, 103
Lanius, plate of, described, 161, 169
Lark, maternal affection of, 102
Laurus cinnamomum, described, 128 ; minutus observed in·Britain, 86 ; sabinii found in Ireland, 86
Larva in the human stomach, 110
Lead, the salts of, 152
Leeches, the habits.of, 73
Leopard, (Felis Leopardus,) described, 21 ; the black, described, 161 ; the hunting, 161
Lepus, species of, in North America, 162
Libraries, public, number of books in some of the principal, 45
Life, tenacity of, in animals, 78
Light, influence of, in human body, 104·
Lightning, Fusinieri's views of, 80
Lion's tail, the spine at the point of the, 54
Longevity, case of, 16 ; in birds, 134

Loris, plates of, described, 101, 137
Loxia curvirostra, habitat of, 148
Luminosity of the sea, example of, 84
Lycaons, plate of, described, 113
Lynx, the, described, 141

MACACOES, plate of, described, 113
Magazine of Natural History, Charlesworth's, noticed, 64, 100, 148
Magnetism, Mr Scoresby's experiments on, 46
Maggots, plate of, described, 113
Makis, plate of, described, 101
Man, extinct race of, in South America, 32
Manna, on the, of the desert, 46
Mandrake root, in Russia, 46
Mandrils, plate of, described, 153
Manganese in Aberdeenshire, 156
Marble, the, of the Western Islands, 14
Marmosets, plate of, described, 97
Maternal affection of animals, 26, 49, 95
Melon, the, of Bochara, 111
Menyanthes trifoliata, described, 119
Mercurial mine of Almeyda, 43
Merlin, habits of the, 109
Merops, plate of, described, 109
Meteors, appearance of, 7, 11, 92 ; near London, 10
Mexico, inundations in, 31
Mice, field, ravages of, 102
Migration of birds, on the, 6
Military, mortality of the, in the British Colonies, 7
Mines, greatest depth of, 14
Misseltoe, the, described, 121
Monkey. See Apes
Moon, mountains, &c. in the, 36·
Moose deer, plate of the, described, 173
Morus nigra, described, 115
Mosses, destruction of, in pastures, 154
Motacilla neglecta, observed in Britain, 86 ; Yarrell[illegible], Mr Gould's description of, 131 ; boarula, described. ed by Mr Drew, 174 ; motacillæ of South Devon, 158
Mountain-ash of Russia, 106
Mulberry-tree, description of, 115
Mule, anecdote of the, 13
Mummy, examination of a, 52
Murray's, Dr, Northern flora, noticed, 88
Musk deer, notice of the, 188
Mustella Genus, Prince Musignano's division of, 131
Mycetes, plate of, described, 169

NATURALIST, the, notice of, 100, 148
Naturalist's Library, Sir Wm. Jardine's, notice of, 152
Niagara, fall of table-rock at, 10
Nichol's spar-prism for assisting to discover shoals in the ocean, 76
Nightjar, the, described, 149
Nitrous oxid, its effects on human frame, 59

OAKS, dimensions of Scottish, 154
Ocean, Arago on the colour of, 171
Ocelos, (Felis pardalis,) described, 21
Olive-tree, the, described, 123 ·
Orang-outang, plate of, described, 121 ; organization of, 89
Orange-tree, the, described, 111 ; productiveness of, 3
Orioles, plate of, described, 111
Oryx Virginiana, described. 122
Ossiferous caverns, near Yealmton, Devonshire, 104·
Ostriches, plate of, described, 137 ; rudimentary toe of (?) 172
Owls, plates of, described, 57, 105 ; snowy, shot in England, 102
Ox, hydrophobia in the, rabies, 150 ; wild, in Chillingham Park, 161 ; in Scotland, 174

PANDOS, plate of, described, 133
Parhelia, remarkable at Fort Howard, 71
Parrakeets, plate of, described, 137
Parrots, plate of,·described, 129 ; anecdote of, 83
Parrot, grossbeak, described, 166
Parsnip, (Pastinaca sativa), sugar from, 58
Partridge, the, described, 121 ; anecdote of, 95
Pearl fishing in Arabia, the, 135
Peat mosses in Holland, the, described, 64
Peccaries, plate of, described, 157
Peregrine falcon, described, 109
Pernis apivorus, described, 153
Petrifaction, artificial, of animals, 42
Phosphorescence of the ocean, 27, 156
Pig, rabies in the, 150
Pigeons, plates of, description of, 17, 153, 165 ; curious habits of, 122
Pines in North Carolina, 99
Pipit, the tree, habits of, 110
Pithci, plate of, described, 121

Pitta, plate of genus, described, 141
Plan of this Journal, 1
Plants, geographical distribution of, 26; age of, 38
Platina found in Siberia, 7
Pointer, the English, 41; the Spanish, 41
Polar ice, fields of, described, 44
Pole-cat, description of the, 77
Polygonum tinctorium, notice of, 168
Polypus, fresh water, described, 167
Pond, obituary notice of Mr, 72
Prognostics of weather, 168
Promerops, plates of, described, 113, 133
Pteterostra psittacea, described, 166
Ptarmigans, plate of, described, 73; described, 117
Puma, (felis concola,) described, 21

Quails, plate of, described, 49; the common, described, 122; the Chinese, notice of, 168; mode of capture in Cerigo, 86
Quartzose tubes, remarkable instances of, 74
Quinary system in Zoology, remarks on, 148

Rabies, as occurring in the lower animals, 150
Rafflesia Arnoldi, described, 14
—— Patma, described, 30
Rains, equinoxial, described, 75; diminished in some parts of France, 16; during a perfectly clear sky, 75
Raven, the, described, 90
—— anecdote of, 30
Rein deer, plate of, described, 173
Rhea, new species of, 108
Rhind's, Mr, Geological and Natural History Excursions round Edinburgh, 1st edition, noticed, 12; 2d edition, 60, 63
Rhinoceros, description of, 5
Ring-dove, the habits of, 77
Roberts', Mr, safety lamp for miners, 12
Rook, a domesticated, 137
Rump gland in birds, uses of, 91
Rüppel, Dr E. vertebrate fauna of Abyssinia, notice of the, 59

St Elmo's fire, instances of, 56, 108
Saccharine matter, plants which yield, 58
Safety lamp, Sir H. Davy's, 12; Mr Roberts', 12
Saliva, the uses of the, 92
Salmon fry, Mr Shaw on the development of, 134; do. on the food of, 136
Salmon, number fished in the Tay, 176
Scallop shell, pecten nebulosus, described, 9
Sea-beach, raised at Coquimbo, 144
Sea, account of its more remarkable encroachments upon the land, 74; its transparency, 84; on the colour of, 26
Sea plants, the best mode of preserving, 114
Sea weed, sparococcus cartilegineus, edible, 62; valuable one from Ceylon, 107
Secretary, the bird, plate of, described, 121
Seeds of plants, see Germination of
Seine, river, quantity of water in the basin of the, 3
Selby, P. J. Esq., illustrations of ornithology, noticed, 136
Sepia-colour, fine, procured from water of peat, 80
Serpentarius, plate of, described, 121
Setters, English, 41
Shamrock of Ireland, what plant is really the, 66
Shaw, Mr J., on the development of salmon fry, 134; on the food of vendace, herrings, and salmon, 136
Shells, the, in the neighbourhood of Plymouth, described, 115
Sheep, rabies in the, 150; the Astracan, 69

Short's popular observatory, noticed, 16
Shrews, British, Mr Jenyns' synopsis of, 159
Shrikes, plates of, described, 161, 169
Shoals in the ocean, help to the discovery of, 76
Silk-worm, on the food of the, 14; new species, 111
Siskin, its breeding in Scotland, 85, 97, 119
Sivatherium giganteum, described, 78
Skuts of South Devon, described, 162
Sleep, Mr Bellamy on, 174
Sleep of plants, on the, 95, 99, 123
Snakes, on the poison of, 49; on their predilection for music, 46
Snake-eaters, plate of, described, 121
—— bird, digestive organs of the, described, 170
Snow, red, 62; late fall of, 32
Soda, natural fountain of, 152
Somnolency, extraordinary cases of, 40
Sounds, subterranean, in granite rocks, 64
Soui mangas, plate of, described, 45
Spaniel, King Charles', described, 41
Sparrows, various species, described, 77; anecdote concerning, 37
Specific characters, value of, 92
Spider, the water, her dwelling under water, 50
Springer dog, the, described, 41
Squirrels, plate of, described, 33
Stag-hound, the, described, 53
Stars, fixed, computed distance of, 39
—— falling or shooting, noticed, 64, 71, 116. See too Aerolites
Starling, the attractive qualities of, 170
Sterna stolida, observed in Ireland, 86
Sternum, the, of birds, described, 90
Storm of November 29, 1836, described, 92; a remarkable one in New Jersey, 160
Strix, plates of, described, 57, 105
Straw for bonnet-making, kinds of, 79
Study of Natural History, advantages of the, 5
Submarine vessel proposed, 16
Sugar from Indian corn, 42
Suia melanura, the, described, 161
Sus. See Hog
Swainson's volume of fly-catchers, noticed, 152
Swallows, plate of, described, 101; the torpidity of, 2
Swans, wild, notice of, 140
Swift, the, described, 105

Tail-gland in birds, Reamuer's account of, 25
Tallipot-tree, immense size of its leaves, 14
Tamarins, plate of, described, 97
Tanagers, plate of, described, 81
Tarantula, habits of, noticed, 86
Tarsiers, plate of, described, 101
Tatty, the, of India, described, 27
Tea-tree, may be cultivated in temperate climates, 151
Teeth, irregular growth of, in rodentia, 91; preternatural growth of the incisive in the rabbit, 66
Temperature of the earth, 23; of Europe, during the tertiary period, 83
Terriers, different species, described, 53
Tetrao, plates of, described, 45, 101; T. tetrix, described, 114; T. urogallus, 114
Thermal springs. See Hot Springs
Thrushes, (merula,) plate of, described, 69
Thrush-shrikes, plate of, described, 149
Thunder storms, periodic, at Constantinople, 136
Tiger, (felis tigria,) described, 21
Titmice, plate of, described, 65
Tivoli, notice of the Cascade near, 10
Tobacco, the, plant, described, 107; culture of, in India, 176

Tornado, Mr Johnson's account of the New Jersey, 160
Tower, Menagerie, by E. T. Bennett, noticed, 48
Traill's, Dr, indelible ink, 155;
Transformation of rocks, doctrine of, explained, 112
Travelled rocks. See Boulders
Trees, ages of, 7; gigantic, 42
Trogons, plate of, described, 1
Troglodytes, plate of, described, 121
Tulip, enormous price of new variety of, 14

Underwood, T. R., obituary of, 8
Upupa, plate of the, described, 113; U. Epops, observed in Scotland, 86
Uropygial gland in birds, uses of, 91
Ursus, plate of genus, described, 129

Val del Bove, Mount Etna, scenery of, 55
Valleys, on the excavation of, 67
Vampire of South America, 163
Van Dieman's Land, the aborigines of, 16
Vanilla, raised in France, 151
Varieties, Mr MacGillivray's remarks on, in animals and plants, 116, 119, 124
Vegetable Kingdom, general view of, 34, 46, 70, 79, 102
Vegetable substances, transformation of, into xiliodine, 68
Vendace, Mr Shaw on the food of, 131
Ventriloquism in the robin, 97
Vesuvius, eruption, 10
Victoria regina, the plant, described, 128, 135
Vine, facts regarding the, 43; destruction of, by insects, 114
Viscum album, described, 131
Volcano, submarine, 152
—— mud, of Grobogan, 35
Voyages of discovery, some, noticed, 12

Wagtails, the, of South Devon, described, 158
Wagtail, the pied, Mr Gould on, 131; the green, in Norfolk, 98
Walnut trees, the ringing of, 58
Walrus, normal dentition of, described, 153
Warblers, plate of, described, 53
Wasps, African, anecdote of, 26
Waterfall of River Lutea, the greatest in Europe, 92
Water-trefoil, described, 119
Water-hen, described, 173; apparently feigns death, 81
Waterton's Essays on Natural History, chiefly ornithology, noticed, 152
Wealden strata, the, observed near Elgin, 155
Weasels, plate of, described, 77; anecdote of one, 69
Weepers, plate of, described, 105
Westwood's introduction to the classification of insects, noticed, 144
Wier, T. D., Esq., contributions to the Journal, 95, 102
Whales, migration of some species, 49
Wheat, on the native country of, 43
Whidah Buntings, plate of, described, 37
Winch, Nat. John, Esq., biographical notice of, 156
Winds, on the direction of, 75
Wolves, plates of, described, 85; the grey, 113
Wood, Mr Neville, his British Song Birds, noticed, 84
Woodpeckers, plate of, described, 41
Worms in the eye of the perch, 3
Wren, anecdote of, 6

Yarrell's History of British Birds, noticed, 132; 2d edition, noticed, 152

LIST OF THE WOODCUTS.

1. Figure of the Mexican Deity, Vitzliputzli, dressed with a helmet, in the form of a Golden Couroucou, ... 2
2. Representation of the Clouded Scallop, Pecten nebulosus, ... 9
3. Fig. 1. An Egg of the small Tortoise-shell Butterfly, (Vanessa urticæ.) Fig. 2. Eggs of the large Tortoise-shell Butterfly, (Vanessa polychloris.) Fig. 3. Caterpillar of the Purple Emperor Butterfly, (Aptera Iris.) Fig. 4. Larva of the Goat Moth, (Cossus ligniperda,) 13
4. The full-blown flower of the Krubut, the great flower of Sumatra, ... 14
5. Fig. 1. First appearance of the Krubut, in form of a round knob. Fig. 2. The female flower. Fig. 3. One of the anthers, somewhat magnified, ... 14
6. Fig. 1. 2. 3. Appearance of Portraits in fractures of flint nodules, ... 19
7. Representation of the Dodo, taken from the "Exotica" of Clusius, published in 1605, ... 22
8.   do.   do.   Travels of Herbert in Africa, Asia, &c., published 1634, ... 22
9.   do.   do.   Bontius' Historia Naturalis et Med. Indiæ Orient., 1698, ... 22
10.   do.   do.   in the British Museum, copied into Edwards' Birds, 1760, ... 22
11. Remarkable structure of a Hen's Egg, ... 27
12. Engraving of the Rafflesia Patma, from Dr Blume, ... 30
13. Three figures, showing the successive development of the flower, ... 30
14. Indication of a Volcano on the dark part of the Moon, ... 36
15. Figures 1. and 2. Representations of what are called the bristles in a Porcupine Boy, exhibited to the Royal Society; that is, of one affected with a peculiar disease of the skin, ... 37
16. Representation of his whole hand, ... 39
17. Three figures representing a new shell designated Fissilabia, ... 42
18. Diagram exhibiting the melody of colours, according to Mr D. R. Hay, 47
19. Diagram exhibiting a general harmony of all the colours of any distinctive character, ... 47
20. Gradual expansion of the Brugmansia Zippelii from the first small tubercle to its full expansion, ... 51
21. The same just before its ultimate expansion, ... 51
22. The Rocking-stone of Strathairdle; a Boulder on the summit of Craigy-barns, near Dunkeld, ... 51
23. The Spine at the extremity of the Eion's tail, ... 54
24. The Boulder or travelled stone near Castle Stuart on the Moray Frith, 55
25. A Fossil regarded as the wing of a fly, encrusted with calcareous spar, 55
26. Mode in which the Giraffe lays hold of branches with its tongue, ... 57
27. Figures 1. 2. Regarded as representations of Buccinum Anglicanum, and B. Undatum, ... 58
28. Representation of the Fossil Elk of the Isle of Man, in the Royal Museum of the Edinburgh University, ... 58
29. Foot-marks of Birds in new Red Sandstone. Figs. 1. and 2. Four-toed indicated. Fig. 3. Three, ... 62
30. Example of Professor Hitchcock's Pachydactali, or thick-toed stony bird-tracks, ... 62
31. Do.   do.   Leptodactyli, or slender-toed bird-tracks, 62
32. Extraordinary elongation upwards of the upper and lower Incisors in a wild Rabbit, ... 66
33. Natural shape of the Diamond represented, ... 66
34. Dr Murray's apparatus for exhibiting the Combustion of the Diamond, 67
35. Diagram of a remarkable Halo, Horizontal Circle, Rainbow, and Parhelia of the Sun, ... 71
36. Profile and front views of the Cranium of the Fossil Sivatherium, ... 78
37. Fig. 1. Teeth of the Indian Elephant. Fig. 2. Teeth of the African, 82
38. Figs. 1, 2. 3. 4. Pieces of the stem of the Lamium album, illustrating Mr Bird's views of their retraction on the principle of endosmosis, ... 87
39. The Sternum of the Black Grouse, Tetrao tetrix, ... 90
40. Figs. 1. 2. The Digestive Organs of the Red Grouse or Ptarmigan, (Lagopus Scoticus,) 94
41. Figs. 1. 2. 3. 4. Represent respectively the Digestive Organs of Gallinaceous Birds, Pigeons, Conirostral Birds, and Vagatores, or Crows and Starlings, ... 98
42. Representation of the Coffee tree, (Coffea Arabica,) 103
43. Leaf and flower of the Orange tree, (Citrus aurantium,) 111
44. Leaf, flower, and fruit of the Mulberry tree, (Morus nigra,) 115
45. Leaf and flower of the Water Trefoil or Bogbean, (Menyanthes trifoliata,) 119
46. Leaf, flower, and fruit of the Olive tree, (Olea Europæa,) 123
47. Leaf and flower of the Cinnation tree, (Laurus cinnamomum,) 128
48. Stem, leaves, flower, and fruit of the Misseltoe, (Viscum album,) 131
49. The smooth Bernacle shell, (Anatifa lævis,) 135
50. Remarkable elongation of the Bill of the Rook, ... 138
51. The Digestive Organs of the Golden Eagle, (Aquila chrysaetus,) 143
52. Figs. 1. and 2. Represent the Gizzard of the Black Grouse, (Tetrao tetrix,) 147
53. Figs. 1. and 2. Head and Foot of the Goatsucker or Nightjar, ... 149
54. Two figures illustrating the Dentition of the Walrus, ... 153
55. The Digestive Organs of the Honey Buzzard, (Pernis apivorus,) 154
56. Cranium and Horns of the Arnee Bull, ... 159
57. Cranium of the Polar Hare, (Lepus glacialis,) 162
58. Head and Wing of the Parroquet Parrotbill, ... 165
59. Digestive Organs of the Snake-bird, ... 170
60. Digestive Organs of the Bittern, (Ardea stellaris,) 174

DIRECTIONS TO THE BINDER.

The Binder will place the Titles, &c. in the following order :—

I. TITLE OF THE JOURNAL.

II. THE EDINBURGH JOURNAL OF NATURAL HISTORY, AND OF THE PHYSICAL SCIENCES.

III. INDEX OF THE JOURNAL, with the LIST OF WOODCUTS.

IV. TITLE OF THE ANIMAL KINGDOM, with the PREFACE.

V. THE ANIMAL KINGDOM OF THE BARON CUVIER.

VI. LIST OF THE PLATES.

VII. The PLATES, arranged according to the List.

THE

ANIMAL KINGDOM

OF THE

BARON CUVIER,

ENLARGED AND ADAPTED TO THE PRESENT STATE OF

ZOOLOGICAL SCIENCE.

ILLUSTRATED AFTER THE ORIGINAL DRAWINGS

OF

AUDEBERT, BARABAND, CRAMER, D'ORBIGNY,

EDWARDS, GEOFFROY-ST-HILAIRE, GILPIN, HUET, MARÉCHAL, NITSCHMANN, OUDART,

TITIAN, R. PEALE, PRÊTRE, REDOUTE, REINOLD, A. RIDER,

SAVIGNY, WERNER, A. WILSON, &c. &c.

EDINBURGH:

PUBLISHED FOR THE PROPRIETOR, 13, HILL STREET.

GLASGOW AND THE WEST OF SCOTLAND: JOHN SMITH AND SON, 70, ST VINCENT STREET; AND
JOHN M'LEOD, 20, ARGYLE STREET. ABERDEEN: A. BROWN AND CO.

MANCHESTER: AINSWORTH AND SONS, 107, GREAT ANCOATS STREET. LEEDS: THOMAS AINSWORTH, JUNIOR,
50, QUEEN'S PLACE, COBURG STREET.

DUBLIN: GEORGE YOUNG, 9, SUFFOLK STREET.

PARIS: J. B. BALLIÈRE, RUE DE L'ÉCOLE-DE-MÉDECINE, 13, (BIS.)

LONDON: SMITH, ELDER, & CO., 65, CORNHILL.

M.DCCC.XXXIX.
5

PREFACE.

THE importance of Natural History, and that of Zoology in particular, being fully appreciated by all liberal minds, no apology is necessary on our part for presenting to the public a New Edition of the " Règne Ani-, mal ; " with additions and alterations, corresponding to the more advanced state of the Science, since the publication of that celebrated Work. Indeed, the absence of any accurate or complete SPECIES ANIMALIUM, or SYSTEMATIC CATALOGUE OF ANIMALS, either in our own or any other language, will, it is hoped, render the present undertaking of the highest interest to the Scientific Zoologist.

The outline which the BARON CUVIER so ably sketched, as explanatory of his views of arrangement, he chiefly exemplified by noticing such animals as were best known to him, while the more rare, though not less interesting, species were merely alluded to by referring to the most expensive works on the subject. It will, accordingly, be our care to fill up his outline with descriptions of all the species which have ever been noticed in any work of authority, and these will either be drawn directly from the living animals, or by direct reference to the original works in our possession, wherein they were first described. Those valuable acquisitions which have been made to the science by the discoveries of AUDUBON, BÉLANGER, DIARD and DUVAU-CEL, D'ORBIGNY, EHRENBERG, HARLAN, HORSFIELD, LESSON and GARNOT, LE VAILLANT, LICHTENSTEIN, The Prince MAXIMILIAN of Wied Neuwied, The Prince of MUSIGNANO, QUOY and GAIMARD, RICHARDSON, RÜPPEL, SPIX, SMITH, VIEILLOT, ALEXANDER WILSON, &c. &c., together with the accurate materials supplied by the able Monographs of AUDOUIN, BRONGNIART, FREDERIC CUVIER, DE BLAINVILLE, MILNE-EDWARDS, E. and I. GEOFFROY-ST-HILAIRE, The BARON VON HUMBOLDT, GOULD, GRAY, LAMARCK, LATREILLE, SAVIGNY, SWAINSON, TEMMINCK, VALENCIENNES, VIGORS, and others, will find their proper places in the system ; while such new species as may be discovered during the progress of this publication will be described in supplementary sheets, paged so as to indicate their natural position. Throughout the work the utmost care will be taken to clear the system of all those doubtful and imaginary species, with which the inexperience of compilers has hitherto crowded our systematic catalogues. The introductory portion, containing the generalities of the Animal Kingdom, will be enlarged by a selection of the most approved physiological views, chiefly from the other writings of the BARON CUVIER, as well as those of the most distinguished British and Continental Writers.

But " a well-executed design," to use the words of M. TEMMINCK, " is always more valuable than the most minute description, especially in those classes of animals where the species are very numerous, and the characters of which are so difficult to define by words ; " and it is accordingly our intention to illustrate this work in a manner worthy of the merits of its illustrious author. The original designs, which are widely scattered throughout numerous foreign works in all civilized languages, have been collected together at considerable expense, and most of these will be here introduced, with appropriate English names, for the first time, to the British Student, with scrupulous accuracy and at a moderate price, the whole arranged in generic groups, and with appropriate backgrounds. Original Drawings, made by our own Artists, after such undescribed living animals as we are able to meet with, will likewise appear.

THE

ANIMAL KINGDOM.

INTRODUCTION.

SECT. I.—NATURAL HISTORY IN GENERAL; AND METHODS OF ARRANGEMENT.

Nature—Division of the Physical Sciences—Natural History—General Principles—Conditions of Existence, or Final Causes—Observation—Classification—Artificial or Natural—Subordination of Characters.

We deem it necessary to commence our work by clearly defining the object of Natural History, and by establishing a precise line of demarcation, so as to separate this science from others, to which it is nearly allied. This is the more requisite, as confused and indistinct notions on the subject very generally prevail.

Various significations have been applied to the term *Nature*, in our language as well as in most others;—sometimes it denotes the qualities of a being derived from original constitution, as distinguished from those acquired by art;—sometimes it signifies the vast concourse of beings composing the universe;—and sometimes the laws which govern those beings. It is especially in this last sense that we are accustomed to personify Nature; and, from a proper feeling of respect, to use this term for the name of its divine Author. Physics, or Physical Science, may consider nature in any of these three points of view.

The name of the Supreme Being, which never ought to be pronounced without emotion, could not be introduced into philosophical discussions upon every occasion, without a violation of decorum. Accordingly, it has become an established practice to use the milder term *Nature*, as an appellation of similar import. This is done without any intention of deifying the powers of nature. On the contrary, the best and wisest natural philosophers agree with Dr Clarke in considering " that there is no such thing as what we commonly call the course of nature, or the power of nature. The course of nature, truly and properly speaking, is nothing else than the will of God, producing certain effects in a continued, regular, constant, and uniform manner, which course or manner of acting being in every moment perfectly arbitrary, is as easy to be altered at any time as to be preserved. So that all those things which we commonly say are the effects of the natural powers of matter and laws of motion, of gravitation, attraction, and the like, are indeed (if we will speak strictly and properly) the effects of God's acting upon matter, *continually* and *every moment*, either immediately by himself, or mediately by some created intelligent being." In these and following passages, the term *law* is used in a metaphorical sense. An ordinary law of civil society is addressed by an intelligent legislator to persons, capable both of understanding the meaning of the law, and of regulating their actions accordingly. But, when we use the phrases *law* of vegetable life, *law* of gravitation, it is evident that the word is used in a sense widely different from the former. Nature, that is, the Supreme Being, not only prescribes the law, but executes it; a *law of nature* being nothing more than that particular regular mode of acting which the Deity has prescribed to himself.

Physical Science is either general or particular.

When we deduce effects from causes, and, by explaining the various phenomena of nature, obtain the power of applying the materials she presents to purposes useful to mankind, it is termed *General Physics*, or *Natural Philosophy*; but, when we consider the various objects presented by Nature, simply for the purpose of obtaining a knowledge of their order, their arrangement, and the disposition of their parts, without referring effects to their causes, it is termed *Particular Physics*, or *Natural History*. It thus appears, that while Natural Philosophy is the ultimate object of science, Natural History is the source whence all science must necessarily arise. The former is unrivalled for sublimity of ideas and depth of investigation, the latter for variety of character and interest in its details. While Natural Philosophy is best fitted to occupy the mind in its severer moments, Natural History affords an agreeable relief, by the general beauty of its objects, the elegance of their forms, the richness of their colouring, the singularity of their habits and instincts, and the exquisite adaptation of all their parts. Both alike lead to form elevated and enlightened conceptions of the power and beneficence of the Creator.

General Physics examines, in an abstract manner, each of the properties of those moveable and extended bodies, to which we apply the general term *matter*.

That branch called *Mechanics* considers the particles of matter as collected in masses, and deduces mathematically, from a very small number of experiments, the laws of equilibrium, of motion, and of its communication. Its several divisions take the names of Statics, Dynamics, Hydrostatics, Hydrodynamics, Aerostatics, &c. according to the nature of the bodies, the motions of which are under examination.[*] Optics considers solely the peculiar vibrations of light; but in this science various phenomena, ascertained entirely by experiment, are daily becoming more numerous.

Chemistry, the other division of General Physics, explains the laws, by which the elementary particles of bodies act on each other, at indefinitely small distances; the combinations or decompositions resulting from the affinity of their ultimate elements; and the manner in which the operation of affinity is modified by various circumstances, capable of increasing or diminishing its action. Being chiefly an experimental science, it cannot, on that account, be classed with others more exclusively mathematical.

The theories of heat and electricity may belong almost equally to Mechanics or to Chemistry, according to the point of view in which each of them is considered.

The mode of proceeding adopted in all the departments of General Physics is, to consider, either mentally or experimentally, only a small number of the properties of bodies at once, in order to reduce them to the greatest attainable simplicity; then, to calculate or discover the effects resulting therefrom; and finally, to generalize and incorporate the laws of these properties so as to form series of theorems; and, if possible, to resolve them into one universal principle, which will serve as a general expression for them all.

Particular Physics, or Natural History (for both of these terms are used indiscriminately), may [be extended so as to] include the particular application of the laws, ascertained by the different branches of General Physics, to the numerous and diversified created beings existing in nature, in order to explain the phenomena which each exhibits. When used in this extensive signification, it also includes Astronomy; but this latter science, being fully elucidated by the light of Mechanics alone, is entirely subservient to its laws, and employs methods of investigation, too different from those admitted by Natural History, to be [extensively] cultivated by the same persons. It is usual, also, to include Meteorology among the branches of General Physics, and to confine Natural History to objects which do not admit of rigorous mathematical investigation, or precise measurement in all their parts.

Geology ranks next to Astronomy for the sublimity and depth of its investigations, and ought, logically, to be classed with Natural Philosophy. But the science is based upon so vast a mass of historical detail, and is still so much in its infancy, that it will long continue to be arranged with the branches of Particular Physics. It contrasts with Astronomy in this respect, that while the last-mentioned science leads us to explore the infinity of space, Geology unfolds the secrets of the infinity of past time. In the one, the *present place* of man is considered but as a point in the vast regions of space; in the other, the *present time* but as an instant in the middle of two infinities—time past and time to come.

Natural History properly considers only the inorganic bodies called Minerals, and the various kinds of living beings [called Animals and Plants,] almost all of which are under the influence of laws, more or less unconnected with those of motion, of chemical affinity, and of various others,

---

[*] In this, and in similar passages, we have thought it more agreeable to received usage to transpose the terms Mechanics and Dynamics from the position in which they stand in the original, the former being, as we conceive, the more general term.—*Translator.*

analysed in the several departments of General Physics. We ought, in treating of Natural History, to employ precisely the same methods as in the General Sciences; and, therefore, we endeavour to adopt them, whenever the subjects under examination become sufficiently simple to permit that mode of investigation. But as this is seldom practicable, there arises, hence, an essential difference between the General Sciences and Natural History. For, in the former, the phenomena are examined under circumstances completely within the reach of the inquirer, who arrives, by analysis, at general laws; while, in the latter, they are removed, by unalterable conditions, beyond his control. In vain, he attempts to disengage them from the influence of general laws, already ascertained. He cannot reduce the problem to its elements, and, like the experimental philosopher, withdraw successively each condition; but he must reason upon all its conditions at once, and only arrive, by conjecture, at the probable result of such an analysis. Let him seek to ascertain, by direct experiment, any one of the numerous phenomena essential to the life of an animal, though but slightly elevated in the scale of being,

"And ere he touch the vital spark—'tis fled."

Thus, it appears that, while Mechanics has become a science chiefly of CALCULATION, and Chemistry of EXPERIMENT, Natural History will long remain, in most of its departments, a science wholly of OBSERVATION.

The latter part of this remark must, however, be restricted to the early stages of Natural History; because, in its more matured condition, it becomes a science of demonstration. Every branch of physics has one leading object in view, and that is, the discovery of the ultimate laws of Nature. Philosophy regards this as of primary importance; while utility is held only as of secondary rank. Science, in its most comprehensive sense, is a superstructure founded on facts, or acquired by experience; and hence, in its early stages, we consider it as entirely limited to observation: but when we have learned to generalize, and find that truths agree in their several relations, we have arrived at the demonstrative part of the science. It is not, therefore, from a mere knowledge of correct nomenclature, or from a capacity to recognize at sight a natural object, that we are entitled to apply the name of *scientific* knowledge to Natural History; but only when we have succeeded, by observation, in deducing the laws which regulate these objects, in their relations to surrounding beings.

These three terms, Calculation, Observation, and Experiment, express, with sufficient accuracy, the manner of cultivating the several branches of Physical Science; but, by exhibiting among them very different degrees of certainty, they indicate, at the same time, the ultimate point to which Chemistry and Natural History ought to tend, in order to rise nearer to perfection. Calculation, in a manner, sways Nature; it determines the phenomena more exactly than can be done by observation alone: Experiment obliges Nature to unveil: Observation watches when she is refractory, and seeks to surprise her.

Natural History employs with advantage, on many occasions, a principle of reasoning peculiar to itself, termed *the conditions of existence*, or, more commonly, *final causes*. As nothing can exist except it contains within itself all the conditions which render existence possible, it is evident, that there ought to be such a mutual adaptation of the various parts of each being among themselves, and such an accommodation of their structure to the circumstances of surrounding beings, as to render possible the existence of the whole. The analysis of these conditions often leads to the discovery of general laws, with a clearness of demonstration, surpassed only by the evidence of direct experiment or calculation.

It was by the knowledge of this principle, that the celebrated Dr William Harvey was enabled to discover the circulation of the blood in Man. The Honourable Robert Boyle relates his conversation with Dr Harvey on this subject, in the following words:—"I remember, that when I asked our famous Harvey, in the only discourse I had with him (which was but a little while before he died,) what were the things which induced him to think of a circulation of the blood? he answered me, that when he took notice that the valves in the veins of so many parts of the body were so placed, that they gave free passage to the blood towards the heart, but opposed the passage of the venal blood the contrary way, he was invited to think, that so provident a cause as Nature had not placed so many valves without design: and no design seemed more probable than that, since the blood could not well (because of the interposing valves,) be sent by the veins to the limbs, it should be sent through the arteries and return through the veins, whose valves did not oppose its course that way." It is evident from this, and many other similar instances, that, in examining the subjects of Natural History, we shall best advance the science, by considering attentively the uses and ends designed by Nature in their formation, and the functions which their organs are destined to perform. This manner of investigation has been objected to by some philosophers, among whom is Des Cartes, as being a presumptuous attempt on the part of human reason, far above its powers, to penetrate into the secret designs of the Creator. The following passage, extracted from the works of Mr Boyle above quoted, forms a satisfactory answer to this objection:—"Suppose that a countryman, being in a clear day brought into the garden of some famous mathematician, should see there one of the curious gnomonic instruments, that show at once the place of the sun in the zodiac, his declination from the equator, the day of the month, the length of the day, &c. &c., it would indeed be presumptuous in him, being unacquainted both with the mathematical disciplines, and the several intentions of the artist, to pretend or think himself able to discover *all the ends* for which so curious and elaborate a piece was framed: but when he sees it furnished with a style, with horary lines and numbers, and, in short, with all the requisites of a sun-dial, and manifestly perceives the shadow to mark from time to time the hour of the day, it would be no more a presumption than an error in him to conclude, that (whatever other uses the instrument was fit or was designed for,) it is a sun-dial, and was meant to show the hour of the day." The whole science of Natural History teems with instances, showing the successful application of the general principle called the conditions of existence. Thus, when we see an animal, possessed of a capacious stomach, long intestines, and a massive structure, we may safely infer that it is *herbivorous*, or feeding on vegetables, slow in its movements, and of timid and gentle habits. On the contrary, when we find an animal with short intestines, straight stomach, and armed with weapons of offence, we immediately conclude it to be *carnivorous*, or feeding upon flesh, and of a fierce and active disposition.

It is further observed by the author, in his Lectures on Comparative Anatomy, that the construction of the alimentary canal determines, in a manner perfectly absolute, the kind of food on which the animal is nourished. For, if the animal did not possess, in its senses and organs of motion, the means of distinguishing the kinds of aliment suited to its nature, it is obvious that it could not exist. An animal, therefore, which can only digest flesh, must, to preserve its species, have the power of discovering its prey, of pursuing, of seizing, of overcoming, and of tearing it in pieces. It is necessary, then, that the animal should have a penetrating eye, a quick smell, a swift motion, address and strength in the jaws and talons. Agreeably to this necessity, a sharp tooth, fitted for cutting flesh, is never co-existent in the same species with a hoof covered with horn, which can only support the animal, but cannot grasp any thing: hence the law, according to which all hoofed animals are herbivorous, and also those still more detailed laws, which are but corollaries to the first, that hoofs indicate molar teeth or grinders with flat crowns, a very long alimentary canal, with a capacious and multiplied stomach.

It is only after having exhausted all the laws of general physics, and the conditions of existence, that we are compelled to resort to the simple laws of observation. The most effectual mode of deducing these is by comparison; by observing the same body successively in the various positions in which it is placed by Nature; and by comparing different bodies with each other, until we obtain a knowledge of some constant relations between their structure and the phenomena exhibited by them. These various bodies thus form a species of experiments, performed entirely by Nature's hand, where different parts of each are supplied or abstracted, as we would desire to treat them in our laboratories; and the results of these additions or abstractions are presented to us spontaneously. We are thus enabled to deduce the invariable laws influencing these relations, and to apply them in a manner, similar to the laws determined by general physics. Could we but incorporate these laws of observation with the general laws of physics, either directly or by means of the principle called the conditions of existence, the system of natural science would be complete, and the mutual influence of all beings would be perceived throughout the whole. To approach this great end the efforts of naturalists should be steadily directed.

All researches of this kind presuppose that we have the means of distinguishing with certainty, and of describing to others with accuracy, the objects under investigation; otherwise, we shall be continually liable to fall into confusion, amidst the innumerable beings which surround us. Natural History ought, therefore, to have for its basis, what has been technically termed *a system of nature*, or a methodical and extensive catalogue, arranged with divisions and subdivisions, in which all beings shall bear suitable names and distinct *characters*.* That we may always be able to discover the character of any particular being from knowing its name, or the name from knowing its specific character, we must found this peculiar description upon some essential or permanent properties of the being. We must not derive the character from habits, or colour alone, as these properties are ever liable to be modified by external circumstances, but from INTERNAL ORGANIZATION OF COMPOSITION.

When Natural History was in its infancy, the objects were few and easily remembered. Systems of classification were either neglected as unnecessary, or confined only to those general divisions and subdivisions, which it was impossible to overlook. But ever since the days of Aristotle, A.C. 330, when Alexander the Great had increased the number of known species of animals by some of the productions of the conquered East, the necessity of a precise system of classification has been universally admitted; and now, the progress of geographical knowledge has enlarged the bounds of the science to so vast an extent, and disclosed a variety so inconceivable of forms hitherto unknown, that the naturalist would, without classification, be overwhelmed with endless details. Yet the different kinds of animals are daily becoming more numerous by the contributions of enlightened travellers. In 1750, the number of distinct species of insects was estimated at 20,000, and now it cannot be less than 100,000. And when it is considered how small a part of the globe has been

* The word *character*, in Natural History, denotes that peculiar description of an object which distinguishes it from all others. Thus we say, the character of *man* is, "Teeth of three kinds, posterior extremities furnished with feet, anterior with hands, &c. &c."—*Translator.*

carefully examined, when there are vast tracts in the interior of Asia, Africa, America, and the isles of the Southern Ocean, which have never been trod by civilized man, while many portions, even of Europe, are but superficially explored, and when the depths of the vast ocean present insuperable barriers to investigation, we may reasonably expect, that the whole number of species will be found to be very much greater. Improved microscopes have disclosed myriads of animalcules previously unknown, and almost every fluid contains an enormous variety of distinct forms, many of them peculiar to each kind of liquid. All this apparent chaos is by the art of the naturalist reduced to a beautiful system, and immediately one universal principle of order may be traced throughout the whole.

Scarcely any object in nature is so peculiar in its formation, as to be at once defined by any single trait in its character. We are almost always under the necessity of combining many of these peculiarities, in order to distinguish an object from others to which it is nearly allied; especially when these allied objects possess some, though not all, of its peculiarities, or when these peculiarities are united to other properties of a different character. The more numerous the objects are which have to be distinguished, the more it becomes necessary to multiply the terms of their several characters; so that, without some contrivance, they would become descriptions of inconvenient length. To remove this objection, divisions and subdivisions are employed. A certain number of allied species are collected together into one group, and it then becomes necessary only to express, for their respective characters, the points wherein they differ, which, according to the above supposition, form but a small part of their description. The whole group is termed a *genus*. The same difficulty would be experienced in distinguishing the genera from each other, if we did not repeat the operation, by grouping the allied genera to form an *order*; and then assembling the allied orders to constitute a *class*. Subdivisions intermediate to these are established when necessary. This aggregation of the inferior, is termed a *system* or *method*. It may be compared, in some respects, to a dictionary, wherein the properties of things are an index to their names, being the reverse of ordinary dictionaries, in which the names are given, as an index to their meanings or properties.

Thus it appears, that a collection of individuals of the same form

constitute	a species,
Of species	a genus,
Of genera	an order,
Of orders	a class,
And of classes	a kingdom.

To explain this arrangement more clearly, we shall take an example from the Animal Kingdom; suppose, the horse. This animal belongs to the class *Mammalia*, containing all which suckle their young; to the order *Pachydermata*, or thick-skinned animals, such as the elephant, boar, and rhinoceros; and to the genus *Equus*, composed of animals with solid hoofs, as the ass and zebra. From these allied species it is finally distinguished by the term *caballus*. Thus, the scientific name of the horse is *Equus caballus*, terms derived from its genus and species. But, as different naturalists often give different names to the same animal, it becomes necessary to add to these the name of the naturalist who first introduced the generic and specific names. In the above example, we therefore write *Equus caballus*, LINN. for the celebrated naturalist Linnæus. In the following pages, we shall give an extensive list of the various synonymes, or names belonging to the same animals, for facilitating reference to other works on the same subject.

Such is the method indispensably required, in framing the arrangement of the almost unbounded objects of Natural History.

We need scarcely caution our readers against the errors of the Realists, once the cause of so much contention in the schools. The individuals alone, or more properly the particles composing each individual, have a real existence in nature, while species, genera, &c. are but general words, invented by man, to express certain points of resemblance, which he perceives among their properties.

There are two different principles observed in the formation of systems of arrangement, according as they are intended to be *artificial* or *natural*. The design of an artificial system is to enable the student to find *the name* of an object, whose properties are known, and to this alone its utility is, in general, confined. Thus, Linnæus arranged plants, chiefly according to the number and situation of the stamens and pistils contained in their flowers. But, being founded on the comparison of only one single organ, the artificial method conveys no general knowledge of other properties, and frequently separates objects which ought never to be disjoined. It is altogether different with a *natural* method. Its divisions are not founded upon the consideration of a single organ, but are derived from characters presented by all the parts of the object. Accordingly, the objects are disposed in such a manner, that each bears a greater affinity to that which immediately precedes and follows it, than to any other.

When this method, therefore, is good, it is not confined to a mere list of names. If the subdivisions have not been selected arbitrarily, but rest upon real and permanent relations, and upon the essential points of resemblance in objects, the natural method is the means of reducing the properties of beings to general laws, of expressing them with brevity, and of fixing them permanently on the memory. To produce these results, objects must be assiduously compared under the guidance of another general principle, necessarily proceeding from that of the conditions of existence formerly explained, called *the subordination of characters*, which

we shall here briefly elucidate. The several parts of a being having a mutual adaptation, there are certain constitutional arrangements which are incompatible with others; again, there are some with which they are inseparably connected. When, therefore, certain peculiarities belong to an object, we may calculate with facility what can, and what cannot, co-exist with them. We, accordingly, distinguish by the terms *important* or *leading characters*, those parts, properties, or constitutional arrangements, having the greatest number of these relations of inconsistency, or of necessary co-existence; or, in other words, which exercise upon the whole being the most marked influence. Others of minor importance are termed *subordinate characters*. The superiority of characters is sometimes determined in a satisfactory manner, by considering the nature of the organs described in the character. When this is impracticable, we must resort to simple observation; and, from the nature of a character, must infer such to be the most decided as are found the least liable to vary, when traced through a long series of beings, differing in degrees of resemblance. For this reason, we should select for the grand divisions, those characters which are at once important and permanent; and may reserve, with propriety, the subordinate and variable characters for the minor subdivisions of our system.

There can be but one complete system, and that is, the natural method. Here species of the same genus, order, or class, resemble each other more than they do the species of any other corresponding division; the place of each object is decided by its relation to surrounding beings; [and the whole arrangement forms a type of that beauteous system of nature which, "changed thro' all, thro' all remains the same." Even Linnæus, who framed the best artificial system ever presented to the world, observes, in his Philosoph. Bot. § 77, that natural historians should regard the natural method of arrangement as the ultimate aim of their labours.

In a word, the natural method is the very soul of Natural History.

> " Unerring nature, still divinely bright,
> One clear, unchanged and universal light."]

SECT. II.—OF LIVING BEINGS, AND OF ORGANIZATION IN GENERAL.

Life—Its definition—Death—Organization—Generation—Spontaneous Generation —Reproduction—Species—Varieties—Permanence of Species—Pre-existence of Germs.

LIFE, being the most important of all the properties of created existence, stands first in the scale of characters. It has always been considered the most general principle of division; and, by universal consent, natural objects have been arranged into two immense divisions, ORGANIC beings [comprising animals and plants], and INORGANIC beings [comprising minerals.]

The word *Life* is used under two significations which are often confounded. It may be applied merely as a general term to express, with brevity, the various phenomena peculiar to living beings; or it may signify the *cause* of these phenomena. It is in the latter sense that the terms *vital principle*, or *principle of life*, are employed; being, in this respect, perfectly analogous to the terms gravity, heat, attraction, and electricity, which are used in the general sciences under a twofold signification,—the one physical,—the other metaphysical. But, it is with the phenomena alone, or the physical sense of these terms, that Natural Philosophy has any concern. The knowledge of causes is removed far beyond the reach of human reason; and, by neglecting to discriminate between these two senses, ancient philosophers before Lord Bacon, and too many modern ones since his time, have fallen into endless discussions, and obscured the light of real science. Yet, it is difficult, upon a subject so interesting as life, in which we all feel deeply concerned, to restrain curiosity within the bounds of reason and philosophy. A recent anonymous writer asks, " Who has not put to himself the question ' What is life?' Who would not receive a clear and just solution of the inquiry, with a feeling of interest, far beyond that afforded by the successful result of any ordinary scientific investigation? We can comprehend part of the mechanism by which life acts; we *feel* its result. We see that mechanism to be so delicate, so complicated, so fragile, so easily set wrong, while our interest is so deep that it should act well, and permanently well, that the exquisiteness of adjustment, the skill of contrivance, and the completeness with which the intended result is secured—all subjects of distinct and interesting investigation—only increase the earnestness of our wish, that we could see beyond the mechanism, and understand that, which it is permitted us to know only by examining its phenomena.

" We do not commonly consider *how much* is given us in life,—the daily enjoyment of the boon renders us insensible to the variety and plenitude of its richness. We shall become more sensible of it upon contemplating the various tissues of organic particles that have been formed; the number of properties that are attached to each; the number of organs that are constituted by their aggregation and arrangement; the number of functions that are exercised by those organs; and the number of adjustments by which all are combined, harmonized, and made effectual to the production of one grand result. It is then we perceive, how many things must exist, how many relations must be established, how many actions must be performed, how many combinations of actions must be secured, before there can be sensation and motion, thought and happiness."

Many attempts have been made to account for the vital principle, but hitherto all these have proved abortive. It is possible, that various functions of the animal frame may hereafter be discovered to proceed from mechanical or from chemical laws; but, we believe, that the ultimate springs of the phenomena of life will ever remain concealed from human knowledge.

In order to form a just idea of the essential conditions of life, we must first examine those beings, which are the most simple in the scale of creation; and we shall readily perceive that these vital conditions consist, in a power possessed by certain bodies, for a period of time only, of existing in a determinate form; of continually drawing into their composition a part of the surrounding substances; and of returning back, to the influence of the general laws of matter, certain portions of their own materials.

These phenomena are exhibited by the *conferva rivularis*, a small bundle of green filaments, finer than hair, found in rivulets and stagnant pools. Being without root or leaves, it is simply attached by a broad surface to the margin of the water. While life exists, it increases in size and weight, throws out filaments like branches, assimilates the particles of water, and of other inorganic substances around it, into vegetable matter, and lays them down in an oblong cellular form. In animals and plants, nutrition is the effect of an internal power; their growth is a development from within. In minerals, on the contrary, growth goes on by the external deposition of successive strata or layers; whilst organised bodies, by means of their vital power, grow and increase by the assimilation of different substances. The *stalactite*, once supposed to be an exception, is now proved to be subject to the ordinary laws of inorganic matter.

Thus life may be compared to a whirlpool of variable rapidity and intricacy, drawing in particles of the same kind, and always in the same direction; but where the same individual particles are alternately entering and departing. The form of living bodies seems, therefore, to be more essentially their own, than the matter of which they are composed.

The matter forming the bones of animals has been ascertained to undergo a very considerable change in a few days; and from this fact the probability of a corresponding change in the other parts of the frame is inferred. The very singular rapidity with which this change is effected was accidentally discovered. Certain animals were fed with madder (*rubia tinctorum*), a plant cultivated for its red dye; and in twenty-four hours all their bones were found to be deeply tinged with its colour. On continuing the same food, the colour became very deep; but upon leaving it off, the colour was completely removed in a very few days. By alternately changing the food, the bones were found to be marked with concentric rings of the red dye, according to the number of times that the change was made. These phenomena, so far surpassing any thing that could have been anticipated, are well calculated to convey an idea of the extraordinary rapidity with which the particles of the animal frame are removed, while the form remains without any apparent alteration.

While this movement continues, the body wherein it takes place LIVES; when it entirely ceases, the body DIES. After death, the elements which compose the living frame, being surrendered to the influence of the ordinary chemical affinities, begin to separate; and the dissolution of the once living body speedily follows. It was, therefore, by the vital movement, that dissolution had been previously arrested, and that the elements of organized bodies were preserved in a state of temporary union. All bodies cease to live after a certain period of time, the duration of which is fixed for each species. Death appears to be a necessary effect of life; and the very exercise of the vital power gradually alters the structure of the body, so as to render its longer existence impossible. The frame undergoes a regular and continual change, as long as life remains. Its bulk first increases in certain proportions, and to certain limits, fixed for each species, and for the several organs of each individual; and then, in the course of time, many of its parts become more dense or solid. This last change appears to be the immediate cause of natural death.

If different living bodies be examined with attention, we shall find them to be composed of an organic structure, which is obviously essential to such a whirlpool, as that to which we have already compared the vital action. There must not only be solid particles to maintain the forms of their bodies, but fluids to communicate the motion. They are, therefore, composed of a tissue of network, or of solid fibres and thin plates (or laminæ,) which contain the fluids in their interstices. It is among the fluid particles that the motion is most continuous and extensive. Foreign substances penetrate into the innermost parts of the body, and incorporate with it. They nourish the solids by interposing their particles; and, in detaching from the body its former parts, which have now become superfluous, traverse the pores of the living frame, and finally exhale under a liquid or gaseous form. During their course, the foreign substances enter into the composition of the solid framework, containing the fluids; and, by contracting, communicate a part of their motion to the liquid particles within them.

This mutual action of solids and liquids—this transition of particles from the one form to the other, presupposes a great chemical affinity in their elementary constituents; and we accordingly find, that the solid parts of organized bodies are composed chiefly of such elements as are capable of being readily converted into liquids or gases. The solids would also require to be endowed with considerable powers of bending and expanding, in order to facilitate the mutual action and reaction between the solids and the fluids; and hence, this is found to be a very general characteristic of the solid parts of organized bodies. This structure, common to all living bodies—this porous or spongy texture, whose fibres or laminæ, ever varying in flexibility, intercept liquids, ever varying in quantity—constitutes what has been termed *organisation;* and, from the definition we have already given of the term life, it necessarily follows that none but organized bodies are capable of enjoying life. Thus we see, that organization results from a great number of arrangements, all of which are essential conditions of life; and hence it follows, that if living bodies be endowed with the power of altering even one of these conditions, to such an extent as to obstruct or arrest any of the partial movements, composing the general action, they must possess within themselves the seeds of their own destruction.

Every organized body, besides the ordinary properties of its texture, possesses a form peculiar to its species; and this applies, not merely to its external arrangement in general, but even to the details of its internal structure. From this form is derived the particular direction of each of its partial movements; upon it depends the degree of intricacy in the general motion; and, in fact, it is this which constitutes the body a species, and makes it what it is.

Life is always attended by organization, just as the motion of a clock ever accompanies the clock itself; and this is true, whether we use the terms in a general signification, or in their application to each particular being. We never find life, except in beings completely organized and formed to enjoy it; and natural philosophers have never yet discovered matter, either in the act of organizing itself, or of being organized, by any external cause whatever. The elements forming, in succession, part of the body, and the particles attracted into its substance, are acted upon by LIFE, in direct opposition to the ordinary chemical affinities. It is impossible, therefore, to ascribe to the chemical affinities those phenomena, which are the result of the vital principle; and there are no other powers, except those of life, capable of re-uniting particles formerly separated.

The birth of organized beings is, therefore, the greatest mystery of organic arrangements, and indeed of all nature. We see organized bodies *develop* themselves, but they never *form* themselves; on the contrary, in all those cases where we have been able to trace them to their source, they are found to derive their origin from a being of similar form, but previously developed; that is, from a *parent.* The offspring is termed a *germ,* as long as it participates in the life of its parent, and before it has an independent existence of its own. In various species differences are found to exist in the place where the germ is attached to its parent; and also, in the occasional cause which detaches it, and gives it a separate existence; but, it is a rule which holds universally, without one single exception, that the progeny must have originally formed part of a being like itself. The separation of the germ is termed *generation.*

Many ancient, and some more recent philosophers, believed that certain organized beings could be produced without parents; and this opinion, though now completely exploded among the learned by the most convincing experiments, still maintains its ground with the ignorant. It originated, as most errors do, from hasty and inaccurate observation. Virgil gravely attempts, in a very elegant passage of the Georgics, to

<blockquote>
Explain
The great discovery of the Arcadian swain;
How art creates, and can at will restore
Swarms from the slaughter'd bull's corrupted gore.
</blockquote>

And Kircher, who lived in the seventeenth century, gives a recipe to make snakes; which, however, he does not appear to have tried.

In Scotland, the country people still believe that the hair-worm (*Gordius aquaticus, Linn.*) can be formed artificially by placing a horse's hair in water; and this unfounded opinion is, we understand, generally diffused throughout the kingdom.

The mites in cheese, the blight on plants, and the maggots in meat, seem at first sight to favour the belief in spontaneous generation; but in all these cases the insects have been demonstrated to proceed from eggs, deposited instinctively by the parent, upon a substance capable of affording nutriment to her young. The popular mistakes on this subject are generally, however, concerning the lower tribes of animals. But the ancients taught that even man could be produced without a parent. The newly-formed earth was supposed to have been originally covered with a green down, like that on young birds; and, soon afterwards, men, like mushrooms, rose from the ground. Lucretius (A. C. 60) relates, that even in his time, when the earth was supposed to be too old for generation, "many animals were concreted out of mud by showers and sunshine."

Every organized being produces others resembling it. Without this provision, all species would become extinct, since death is the necessary consequence of the continued action of life. Certain animals possess the power of reproducing some of their parts, after these have been removed. This power is termed REPRODUCTION, and it is found in various degrees of perfection, according to the species.

In general, this power of renovating mutilated parts is found to exist most perfectly in the lower species of organized beings. The head of the snail (*Limax, Linn.*) may be cut off, and the whole organ, including its elegant telescopic eyes, will be reproduced. The claws, feet, and feelers (or *antennæ*) of crabs and lobsters, as well as the limbs of spiders, when amputated, are completely restored by the fresh growth of new organs. When accident deprives a shark of its teeth, they are replaced with facility. If the fins of fishes be cut, they will reunite, and the rays themselves will be reproduced, provided only the small parts at their bases are left. The eyes of lizards, though possessed of an intricate apparatus of coats and humours, if removed, will be replaced by new eyes equal to the former. Even man and the higher animals possess the same power, only restricted within narrower limits. Injuries to various parts of our frame are speedily repaired, and the wounds heal. The effect of injury to a living bone is curious. A new bone is produced round the old one; which finally dies, and is absorbed or discharged. The new bone, which at first was spongy in its texture, and irregularly formed, assumes, in a few years, its natural dimensions, and all appearance of change is completely removed. Thus we see the bountiful provision of Nature, and the effect of that principle of *reproduction*, which restores most of the organs of the body to their natural form and action, when deranged by injury or by disease.

Organized beings are developed with greater or less rapidity and perfection, according as they are placed in favorable or unfavorable circumstances. Heat, the quantity or quality of their nutriment, and other causes, exercise considerable influence over them; and this influence may extend over the whole frame, or be confined only to certain organs. Hence, it follows, that the resemblance between the progeny and its parents can never be perfectly exact. These minor differences among organized beings are called *varieties*.

The different kinds of dog (*Canis familiaris, Linn.*), of horse (*Equus caballus, Linn.*), of sheep (*Ovis'aries, Desm.*), are all varieties of the same species, and are produced by merely accidental causes, such as domestication, climate, &c. By cultivation, the sloe has been transformed into the plum, and the crab-tree into the apple-tree. The cauliflower and red cabbage, though apparently very different plants, are descended from the same parents,—the wild Brassica oleracea,—a weed growing near the sea. Mr Herbert relates, in the Horticultural Transactions, that he succeeded in raising, from the natural seed of a highly-manured red cowslip, a primrose, a cowslip, oxlips of the usual and other colours, a black polyanthus, a hose-in-hose cowslip, and a natural primrose, bearing its flower on a polyanthus stalk;—all these are instances of varieties, depending upon soil and situation.

There is, however, no real ground for supposing that *all* the differences observable in organized beings are the result of accidental circumstances. Every thing hitherto advanced in favour of this opinion is purely conjectural. On the contrary, experience clearly shows, that, in the actual state of the globe, species vary only within very narrow limits; and, as far as past researches have extended, these limits are found to have been in ancient times the same as at present.

The French naturalists, who visited Egypt with Bonaparte, found the bodies of the crocodile, the ibis, the dog, the cat, the bull, and the ape, which had been embalmed three thousand years ago by the Egyptians as objects of veneration, to be perfectly identical with the living species now seen in that country, even to the minutest bones and the smallest portions of their skins. The common wheat, the fruits, seeds, and other parts of twenty different species of plants, were also discovered, some of them from closed vessels in the sepulchres of the kings; and they resembled in every respect the plants now growing in the East. The human mummies, also, exactly corresponded with the men of the present day.

We are, therefore, compelled to admit that certain forms have been regularly transmitted to us from the first origin of things, without having transgressed the limits assigned to them, [except in a slight degree, when modified by certain accidental circumstances.] All beings, derived from the same original form, are said to constitute a *species*; and the *varieties* are, as has been stated, the accidental subdivisions of species.

Generation appears to be the only means of ascertaining the limits by which varieties are circumscribed; and we may therefore define a SPECIES to be—a group or assemblage of individuals, descended, one from another, or from common parents, or from others resembling them, as much as they resemble each other. However rigorous this definition may appear, its application in practice to particular individuals is involved in many difficulties, especially when we are unable to make the necessary experiments.

In conclusion, we shall repeat, that all living bodies 'are endowed with the functions of absorption [by which they draw in foreign substances]; of assimilation [by which they convert them into organized matter]; of exhalation [by which they surrender their superfluous materials]; of development [by which their parts increase in size and density]; and of generation [by which they continue the form of their species.] Birth and death are universal limits to their existence: the essential character of their structure consists in a cellular tissue or network, capable of contracting; containing in its meshes fluids or gases, ever in motion: and the bases of their chemical composition are substances, easily convertible into liquids or gases; or, into proximate principles, having great affinity for each other. Fixed forms, transmitted by generation, distinguish their species, determine the arrangement of the secondary functions assigned to each, and point out the part they are destined to perform on the great stage of the universe. These organized forms can neither produce themselves, nor change their characters. Life is never found separated from organization; and, whenever the vital spark bursts into a flame, its progress is attended by a beautifully-organized body. The impenetrable mystery of the pre-existence of germs alike defies observations the most delicate, and meditations the most profound.

We trace an individual to its parents, and these again to their parents. After a few generations the clue is lost, and in vain we inquire, Whence arose the first animal of the species? and what produced the first germs from which have descended the innumerable tribes of animals and plants, that we see in constant succession rising around us? Whence did the species MAN arise? Philosophical inquiry fails to lead us through the labyrinth; and we feel the force of the same principle which inspired Adam, when he says, with Milton,

> " Thou sun, fair light,
> And thou enlightened earth, so fresh and gay,
> Ye hills and dales, ye rivers, woods, and plains,
> And ye that live and move, fair creatures, tell,
> Tell if you saw, how came I thus, how here,
> Not of myself?"

SECT. III.—DIVISION OF ORGANIZED BEINGS INTO ANIMALS AND PLANTS.

Animals and Plants—Irritability—Animals possess Intestinal Canals—Circulating System.—their Chemical Composition—Respiration.

LIVING or organized beings have been subdivided by universal consent, from the earliest ages, into ANIMALS endowed with sensation and motion, and into PLANTS destitute of both, and reduced to the simple powers of vegetation.

Some plants retract their leaves when touched; and all direct their roots towards moisture, and their flowers or leaves towards air and light. Certain parts of plants even exhibit vibrations, unassignable to any external cause. Yet, these different movements, when attentively examined, are found to possess too little resemblance to the motions of animals, to authorize us in considering them as proofs of perception and of volition.

They seem to proceed from a power, possessed in general by all living substances, of contracting and expanding when stimulated,—a power to which the name of *irritability* has been assigned. The fibres composing the heart of animals alternately expand and contract, altogether independent of the will of the animal; and thick hair will grow on the skins of some animals, when removed into a cold climate. As we neither ascribe volition nor sensation to the heart or to the hair, so we cannot attribute these qualities to the heliotrope, to the sun-flower, or to the sensitive plant. The nice distinction of character must be cautiously observed, between sensation and mere irritability: like the higher powers of reason and instinct, they are

" For ever separate, yet for ever near."

The power of voluntary motion in animals necessarily requires corresponding adaptations, even in those organs simply vegetative. Animals cannot, like plants, derive nourishment from the earth by roots; and hence they must contain within themselves a supply of aliment, and carry the reservoir with them. From this circumstance is derived the first trait in the character of animals. They must possess an intestinal canal, from which the nutritive fluid may penetrate, by a species of internal roots, through pores and vessels into all parts of the body. The organization of this cavity, and of the parts connected with it, ought to vary according to the nature of the aliments, and the transformations necessary to supply the juices proper to be absorbed; whilst the atmosphere and the earth have only to present to vegetables the juices already prepared, when they are immediately absorbed.

Animal bodies, having thus to perform more numerous and varied functions than plants, ought to possess a much more complicated organization; and, in consequence of their several parts having the power of changing their position relatively to each other, it becomes necessary that the motion of the fluids should be produced by internal causes, and not be altogether dependent on the external influences of heat and of the atmosphere. This is the reason that animals are endowed with a *circulating system*, or organs for circulating their fluids, being the second characteristic peculiar to animals. It is not so essential, however, as the digestive system, for it is not found in the more simple species.

The complicated functions of animals require organized systems, which would be superfluous in vegetables; such as, the muscular system for voluntary motion, and the nerves for sensation. It was also necessary that the fluids should be more numerous and varied in animals, and possessed of a more complicated chemical composition than in plants, in order to facilitate the action of these two systematic arrangements. Therefore, another essential element was introduced into the composition of animals, of which plants, excepting some few tribes, are generally deprived; and while plants usually contain only three elements, oxygen, hydrogen, and carbon, animals add to these a fourth, namely, azote or nitrogen. This difference in chemical composition forms the third trait in the character of animals.

Plants derive their nourishment from the soil and atmosphere, and thence obtain water, composed of oxygen and hydrogen; also, carbonic acid, which is a compound of carbon and oxygen; while the atmosphere yields an unlimited supply of air, composed of oxygen and nitrogen [with a slight mixture of carbonic acid.] From these materials, the supplies necessary to preserve their own composition unaltered, are obtained; and, while hydrogen and carbon [with a certain portion of oxygen] are retained, they exhale the superfluous oxygen [untainted.] The nitrogen, on the contrary, is [either absorbed in very small quantities, or] altogether rejected. Such is the theory of vegetable composition; in which one of the most essential parts of the process, namely, the exhalation of oxygen, can only be performed by the assistance of light.

When plants are deprived of light, an opposite effect ensues. Instead of giving off oxygen gas, and absorbing carbonic acid, the reverse takes place; and carbonic acid is disengaged, while oxygen is absorbed. The effect of plants upon the air is, therefore, to increase its purity during day-light, but to deteriorate its quality during the darkness of night.

Animals require for their nutriment, directly or indirectly, the same substances which enter into the composition of vegetables, namely, hydrogen, carbon, [and a certain portion of oxygen.] But, in addition to these, it is essential, for the preservation of their peculiar constitution, that they accumulate a much larger portion of nitrogen, and disengage any excess of hydrogen, and especially any superfluity of carbon: This is performed by respiration, or breathing, in which process the oxygen contained in the atmosphere combines with the [excess of] hydrogen and carbon in the blood; with the former of these, it forms watery vapour, and with the latter carbonic acid. The nitrogen, to whatever part of the system it may penetrate, seems chiefly (though not altogether) to remain there.

The quantity of nitrogen retained in the system varies with the seasons, being greater in summer, and less in winter. The degree of variation is different for animals of different species: in some it is very small in quantity, while in others it is equal to their entire bulk.

The effects produced upon the atmosphere by plants and animals, are of an opposite kind; the former decompose water and carbonic acid, while the latter reproduce them. Respiration forms the fourth characteristic of animals, and is the most distinguishing function of the animal frame; namely, that which forms its essential difference from all other beings, and in a manner constitutes it an animal. So important is its influence over the whole body, that we shall presently be able to show, that animals perform the functions of their nature with greater or less perfection, according as their respiration is more or less perfect.

Thus we perceive that animals are distinguished from plants by the following characteristics:—1st, They are possessed of an intestinal canal; 2dly, Of a circulating system; 3dly, Nitrogen enters largely into their composition; and finally, They are endowed with organs adapted for respiration.

SECT. IV.—THE ORGANIC FORMS OF THE ANIMAL BODY, AND THE PRINCIPAL CHEMICAL ELEMENTS OF ITS COMPOSITION.

Cellular Tissue—Membranes—Gelatine—Medullary Substance—Muscular Fibre—Fibrin—Blood—Albumen—Secretion—Nutrition.

A POROUS tissue of network, and at least three chemical elements (carbon, hydrogen, and oxygen), are essential to all living bodies, while a fourth element (nitrogen) may be almost considered peculiar to animals. We shall now proceed to describe the various kinds of meshes, of which the network is composed, and the different combinations into which these four elements are found to enter.

There are three kinds of organized principles, or forms of network; the *cellular tissue,* the *medullary substance,* or marrow, and the *muscular fibre.* To each of these forms is attached a peculiar combination of chemical elements, as well as a particular function.

The *cellular substance* is composed of an indefinite number of small laminæ, without any apparent arrangement, crossing so as to form very small cells, communicating with each other. It may be compared to a species of sponge, similar in form to the entire body; while all other animal particles either occupy its cells, or traverse its substance. It possesses the property of contracting indefinitely when the causes, which preserved it in a state of extension, are removed. This power retains the body within the limits, and in the form, assigned to it by Nature.

The cellular substance, or tissue, enters into the composition of every part, forming regular series of cells. We find it equally in the brain, the eye, and the nerves, only somewhat finer in its texture than in the bones and muscles. Its cells move with facility, and accommodate themselves to the motions of the body, being moistened, at the points of contact with the adjacent cells, by a liquid, which lubricates them like the synovia, or oily fluid of joints, so as to facilitate their motion.

When the cellular substance is compressed into compact planes, it forms laminæ of various extents, called *membranes.* These membranes, when united into cylindrical tubes, more or less ramified, receive the name of *vessels.* The filaments, called *fibres,* are entirely composed of cellular substance; and the bones are nothing more than cellular substance, rendered hard by the deposition of earthy particles.

The general matter of which the cellular substance is composed, consists in the proximate principle or combination, called *gelatine;* the distinguishing character of which is, that it can be dissolved by boiling water, and, upon cooling, takes the form of a tremulous jelly.

Gelatine, when analyzed by Gay-Lussac and Thenard, was found to contain in 100 parts, by weight—carbon, 48; hydrogen, 8; oxygen, 27; and nitrogen, 17; very nearly.

The *medullary substance* cannot be resolved into any simpler organic structure. It appears to the eye as a soft whitish pulpy matter, composed of an infinite number of very minute globules. No peculiar motions can be observed in it; but it possesses that most wonderful of all properties, the power of transmitting to the mind the impressions made on the external organs of sense, and of rendering the muscles subservient to the determinations of the will. The brain and spinal marrow are almost entirely composed of medullary substance; and the nerves, which are distributed through all the organs capable of sensation, are, in respect to their composition, nothing but bundles (or fasciculi) of this substance.

The *muscular,* or fleshy *fibre,* is composed of a particular kind of filaments, having the peculiar property, during life, of contracting or folding themselves up, when touched or injured by any external body; or when acted upon, through the medium of the nerves, by the will.

The muscles are the immediate organs of voluntary motion, and are composed entirely of bundles of fleshy fibres. All the membranes and vessels, which are required to exercise any compressive force, are armed with these fibres. They are always united intimately with the nervous filaments, or threads; but certain muscles are observed to execute motions, altogether independent of the will, especially in the exercise of functions possessed in common with plants. Thus, although the will is frequently the cause of muscular motion, yet its power is neither general nor uniform in its action.

Fleshy fibre has, for the basis of its composition, a particular principle, named *fibrin,* which is [nearly] insoluble in boiling water, and seems naturally to assume a filamentous arrangement.

It consists of white solid fibres, inodorous and insipid. When analysed by Gay-Lussac and Thenard, 100 parts were found to contain about 53 parts of carbon, 7 of hydrogen, 20 of oxygen, and 20 of nitrogen.

The nutritive fluid, or blood, when recently extracted from the circulating vessels, may not only be ultimately resolved, for the most part, into the general elements of the animal body, carbon, hydrogen, oxygen, and nitrogen; but it already contains fibrin and gelatine, prepared to contract their substance, and to assume respectively the forms of filaments or of membranes, according to circumstances, whenever a slight repose enables them to exhibit this tendency. In addition to these, the blood contains another proximate principle, called *albumen,* [composed very nearly of 53 parts of carbon, 7 of hydrogen, 24 of oxygen, and 16 of nitrogen.] Its character is to coagulate in boiling water, [like the white of eggs, composed almost entirely of albumen.] We also find in the blood nearly all the other elements, which enter into the composition of each animal body in small quantities; such as, the lime and phosphorus deposited in the bones of the higher animals; the iron, which seems essential to the colour of the blood and other parts; and the fat, or animal oil, placed in the cellular tissue to render it flexible. In fact, all the solids and fluids of the animal body are composed of chemical elements contained in the blood. It is only by possessing some elements, of which the others are deprived, or by a difference in the proportions in which they combine, that [in general] they can be distinguished. From this it appears that it only requires, for their formation in the body, to abstract the entire, or a part, of one or more elements of the blood; or, in a few cases, to add a foreign element, procured from another source.

Some substances, differing very much in character, seem, however, to possess nearly the same chemical composition; we must therefore consider the peculiar arrangement of the particles as an essential distinction among animal fluids and solids, as well as their composition, and the proportions of their elements.

We might, without impropriety, assign the term *secretion* to denote the various operations by which the blood nourishes and renovates the solid and fluid parts of the body. But, we shall restrict the term to the production of *fluids* only; while, we shall apply the term *nutrition*, to signify the production and deposition of the materials, necessary for the growth and maintenance of the *solids*. To each solid organ, and to every fluid, is assigned that peculiar composition which is suited to its place in the system; and, by the renovating power of the blood, their composition is preserved during health, and the continual waste repaired. Thus, by affording continual supplies of nutriment, the blood would undergo a perpetual deterioration, were it not restored by the new matter obtained from the digestion of the food; by respiration, which relieves it of the superfluous carbon and hydrogen; by perspiration, and various other means, which deprive it of any excess of other principles.

These continual changes in the chemical composition of the several parts, are as essential to the vital action, as the visible motions of the old particles, and the constant influx of new ones: indeed, they seem to be the final object for which the latter motions were designed.

SECT. V.—ACTIVE FORCES OF THE ANIMAL BODY.

Muscular Fibre—Nerves—Hypothesis of a Nervous Fluid.

THE muscular fibre is not confined, in its functions, to be merely the organ of voluntary motion. We have shown, that it is one of the most powerful agents employed by Nature, in effecting such necessary motions and transference of particles in the bodies of animals, as are possessed by them in common with vegetables. Thus, the muscular fibres of the intestines produce the peristaltic motion, which renders these canals pervious to the aliment; and the muscular fibres of the heart, with the arteries, are the agents in the circulation of the blood; and thus, ultimately, of all the secretions.

The Will contracts certain portions of the muscular fibre through the medium of the nerves. Certain other fibres, such as those to which we have just alluded, are independent of the Will, and yet are animated by nerves extending through them. We may therefore conclude, from analogy, that these nerves are the causes of their involuntary contraction.

The nerves are composed of several distinct filaments, resembling each other in every respect; and they appear to be formed of the same soft pulpy material, commonly called marrow, or medullary substance, surrounded by a cellular membrane. The filaments are again enveloped in a tube of this membrane, forming a continued nerve, extending from the brain to various parts of the muscles and skin. Yet the functions of the several filaments of the same nerve are very different. One filament is designed for voluntary muscular motion, another for sensation, and a third for involuntary motion. Sir Charles Bell, to whom we owe this remarkable discovery, divides all the nervous filaments of the body into four general systems; namely, of voluntary motion, of sensation, of respiration, and of involuntary motion. The last of these performs the functions of nutrition, growth, and ultimately, of decay. Besides these, there are nerves destined to particular functions of sensation; such as sight, smell, and hearing.

When the sensitive filament of a nerve is injured in any part of its course, pain and not motion is the result; and the pain is referred by the animal to that part of the skin where the remote extremity of the filament is distributed into minute fibres. A patient, whose leg has been amputated, will feel a pain, which long-continued habit has taught him to refer to the extremity of the toes; when, in reality, the injury has been inflicted upon that portion of the nervous filament which terminates at the stump.

In the remainder of this section, our author proposes to explain the phenomena of the nerves upon the hypothesis of a *nervous fluid*, acted upon by certain chemical affinities. We are aware that several, almost insurmountable, objections may be urged against this theory, and indeed against every other which attempts to explain the complicated functions of life. Yet, if an hypothesis correspond pretty accurately with observed facts, it may have its uses, by fixing the phenomena in the memory, provided we always recollect, that it is but an hypothesis, to be modified as knowledge extends. Thus the phenomena of heat are referred to the imponderable fluid *caloric*; of light, to the vibrations of a highly elastic medium; of electricity, to the electric fluid;—none of which can be demonstrated to have a real existence in nature. But, in adopting an hypothesis, we must never forget that it is a temporary, not a final, theory ;—a motive for seeking further analogies, or, as Dr Thomas Brown rightly observes, " a reason for making one experiment rather than another."

HYPOTHESIS OF A NERVOUS FLUID.

Every contraction, and, in general, every change in the dimensions of inorganic matter, is occasioned by a change of chemical composition; either, by the absolute addition or abstraction of some solid matter, or by the flux or reflux of an imponderable fluid, such as caloric. In this way the most violent convulsions of nature arise, such as explosions, conflagrations, &c.

It is therefore probable that the nerve acts similarly upon the muscular fibre, by means of an imponderable fluid, especially as it has been proved that the impulse is not mechanical.

The medullary matter of the entire nervous system is formed throughout of the same material; and, blood-vessels accompanying all its ramifications, it is thus enabled to exercise, in every part, the functions belonging to its nature.

All the animal fluids being secreted from the blood, there is every reason to infer that the nervous fluid is derived from a similar source, and that the medullary substance is the agent in the secretion. On the other hand, it is certain that the medullary substance is the sole conductor of the nervous fluid; all the other organic elements are non-conductors, and arrest it, as glass opposes the progress of the electric fluid.

All the external causes, capable of producing sensation, or of occasioning contractions in the muscular fibre, are chemical agents, possessing a power of decomposing, such as light, caloric, salts, odorous vapours, &c. It is therefore extremely probable, that these causes act in a chemical manner upon the nervous fluid, by altering its composition; and this view appears to be confirmed by the fact, that the action of the nerves is enfeebled by long continuance, as if the nervous fluid required a supply of new materials to restore its composition, and enable it to undergo a further alteration.

An external organ of sense may be compared to a kind of sieve, which only permits those agents to pass through it, and act upon the nerve, that it is fitted to receive at that place; but it often accumulates the nervous fluid so as greatly to increase its effect. Thus, the tongue has spongy papillæ, which imbibe saline solutions; the ear is furnished with a gelatinous pulp, violently agitated by the sonorous vibrations of the air; and the eye is supplied with transparent lenses, which concentrate the rays of light.

Those substances which have obtained the name of *irritants*, from their power of occasioning contractions in the muscular fibre, probably exercise this action through the medium of the nerves; and they influence them in the same manner as the Will does, that is, by affecting the nervous fluid, in the manner necessary to alter the dimensions of the muscular fibre under its influence. Yet the Will is not concerned in producing these effects; often the mind is totally unconscious of their action. Even when the muscles are separated from the body, they are susceptible of being irritated, so long as that portion of the nerve, which accompanies them, retains its power of acting. In this case, the phenomena are totally removed from the influence of the Will. The state of the nervous fluid is altered by muscular irritation, as well as by sensation and voluntary motion: there exists, therefore, the same necessity for restoring its original composition. Irritants occasion those movements and transferences of particles necessary to the functions possessed in common by animals and plants; thus, the aliment stimulates the intestine; the blood irritates the heart. These motions are all performed independent of the influence of the Will, and, in general, while health continues, without the consciousness of the animal. To effect these objects, the nerves which produce the motion have, in most cases, an arrangement entirely different from those affected by sensation or controlled by the Will.

The nervous functions, by which we mean sensation and muscular irritability, are exercised with more or less vigour upon every point, in proportion as the nervous fluid is more or less abundant there; and as this fluid is produced by secretion, its quantity ought to depend jointly upon the quantity of the medullary matter secreting it, and upon the supplies of blood received from this medullary substance. In animals possessing a circulating system, the blood is distributed to all parts of the body, through the arteries, by means of their irritability and the action of the heart.

If these arteries be irritated in an unusual degree, they act more forcibly and propel a greater quantity of blood; the nervous fluid becoming more abundant, increases the local sensibility; and, reacting upon the irritability of the arteries, carries their mutual action to a high degree. This is called nervous excitement, or *orgasm*; when it becomes painful and permanent, it is termed *inflammation*.

This mutual influence of the nerves and muscular fibres, whether in the intestinal or arterial systems, is the true source of those involuntary actions, common both to animals and plants.

Each internal organ is susceptible of irritation only from its peculiar irritant, to which it is in a manner especially adapted, just as an external sense can be affected only by its particular objects. Thus mercury irritates the salivary glands, and cantharides, the *vesica*. These agents have been called *specifics*.

As the nervous system is continuous and of uniform structure, local irritations, and frequently repeated sensations, fatigue it throughout the whole extent; so that any function, when excessively exercised, may enfeeble all the others. Thus, too much food impedes the action of the intellectual powers, and long protracted study impairs the powers of digestion.

An excessive local irritation may affect the whole body, just as if all the vital energies were concentrated upon one single point.. But a second irritation, in another place, will diminish the first, or, as it has been called, *determine* the first into another part; such is the effect of blisters, laxatives, and other counter-irritants.

We have thus shown, in the above brief sketch, that it is possible to account for all the phenomena of physical life; if we merely assume hypothetically the existence of a nervous fluid, possessed of certain properties, which are deduced from generalizing the phenomena of the vital system.

SECT. VI.—THE ORGANS OF ANIMAL BODIES, THEIR APPROPRIATE FUNCTIONS, WITH THEIR VARIOUS DEGREES OF COMPLICATION.

Sensation—Touch—Taste—Smell—Sight—Hearing—Muscular Sense—Head—Brain—Voluntary Motion—Muscles—Bones—Tendons—Ligaments—Nutrition—Stomach—Gastric Juice—Chyme—Chyle—Lacteals—Arteries—Veins—Lymphatics—Respiration—Lungs—Gills—Trachea—Capillary Vessels—Secretory Glands—Generation.

AFTER having considered the organic elements of the animal body, the chemical elements of its composition, and the active forces which prevail in it, nothing now remains to complete a general view of the animal system, excepting a summary account of the several functions of which life is composed, with a description of their appropriate organs.

The functions of the animal body may be divided into two classes—the animal functions, which are peculiar to animals—and the vital or vegetative functions; common to animals and plants. The former comprise sensation and voluntary motion, the latter nutrition and generation.

We shall commence with SENSATION, which resides in the nervous system.

The sense of touch is the most extensively diffused of all the external senses. It is seated in the skin, a membrane enveloping the entire body, and traversed in every part by nerves. Their extreme fibres are expanded at the surface of the skin into minute *papillæ*, or small projecting filaments, where they are protected by the outer skin, and by other insensible coverings, such as hair or scales.

The degree of perfection in which different animals possess this sense varies considerably; but its exercise, in a high degree, is always accompanied by certain conditions. The organ must be supplied with numerous nerves and papillæ, under a very fine cuticle; with a soft cellular substance, like a cushion; and with a hard resisting base. It must also be endowed with a considerable degree of flexibility, as a close contact with the surfaces of bodies is indispensable. Most animals are possessed of some particular organ, in which the sense of touch is developed in a high degree. In the hand of man, and particularly at the extremities of the fingers, we find all the necessary requisites of this sense, combined in their most perfect form. The proboscis, or trunk, of the elephant seems to rank next to the human hand; and, among the higher orders of animals, either the snout or the lip if endowed with much sensibility. This quality is particularly observable in the nose of the tapir, and of the hog, in the lips of the mole, and in the upper lip of the rhinoceros. The seal, and animals of the cat kind, such as the lion and tiger, have whiskers, possessed, near their roots, of considerable delicacy, which renders them of important use to these animals as feelers. Certain species of monkeys have delicate prehensile tails, which they use with surprising agility. In birds, the nerves of touch seem chiefly developed in the feet and toes, and most of the aquatic species are endowed with bills of considerable feeling. Serpents use their slender tongues as instruments of touch; and the great flexibility of

their bodies renders them well adapted for the exercise of this sense. The snouts of fishes have some nicety; but, with this exception, these animals seem nearly destitute of delicate sensation. Insects feel chiefly by means of their antennæ; and the several tribes of annelida, actiniæ, and polypi, by their tentacula. Several animals are covered with a dense integument, in many of their parts, which are thus wholly unfitted for this sense. The thick hides of the elephant and rhinoceros, the feathers of birds, the scales, horny coverings, and shells of the lower animals, are evidently inconsistent with the necessary conditions of touch. Bats are enabled to fly in the darkest places, by the extreme acuteness of their tactual nerves.

Taste and smell are merely more delicate modifications of this sense, for the exercise of which the membranes of the tongue and nostrils are specially organized.

In most of the lower animals the sense of taste is very imperfect, or it is altogether wanting. The tongue of man is supplied with numerous papillæ, of a conical form and spongy texture, projecting in a manner visible even to the naked eye. Taste seems in him to attain its most perfect state; and he not only enjoys the natural varieties of an omnivorous animal, but also a number of acquired tastes, which other species are wholly denied. The tongues of birds, of reptiles, and of fishes, are often covered with a hard and horny cuticle, which renders them altogether unfit for the delicate exercise of this sense. Many animals swallow their food without mastication; and they must be thereby effectually deprived of the enjoyments of taste, as a certain degree of contact between the food and the organ is essentially necessary for its exercise.

The sense of smell resides in an organ, rendered susceptible by the extreme delicacy and extent of its ever humid surface.

Very minute particles of an odoriferous substance are darted forth in every direction, and are received upon the extensive and complex membrane, which lines the internal parts of the nasal cavity. Matter is thus perceived, when in a state of great subdivision, with a degree of acuteness far surpassing any of the other senses. The extreme minuteness of these particles may be inferred from the fact, that musk, and many other substances, will exhale odour for several years, and yet no loss in their weight can be detected, even by the most delicate balances. Carnivorous animals, in general, possess a more acute sense of smell than those living upon vegetable food; and the structure of their nasal cavities is consequently much more intricate. This power was obviously given to facilitate the discovery of their food. In man the sense of smell seems best adapted for vegetable effluvia. A dog, though surpassing him in detecting the most minute effluvium of another animal, will derive no pleasure from the finest vegetable odours. M. Audubon is of opinion that birds of prey are not endowed with an acute sense of smell. The degree in which this sense is enjoyed by the lower tribes of animals has not yet been completely determined, but it is observed to exist in bees and snails.

The beauty of the eye, and the unbounded sphere which it exposes to observation, give to the sense of sight a decided pre-eminence. Light, when emitted from the sun or any luminous body, strikes upon the external covering of the eyeball. By means of the crystalline lens, it is then refracted or bent from its original direction to a focal point, from which the rays of light are again distributed on the expanded extremity of the optic nerve, prepared to receive them. The size of the eyes in quadrupeds, and the intensity of their vision, bear a constant relation to the nature of their food. Herbivorous animals, such as the elephant and the rhinoceros, have very small eyes in comparison with their entire bulk. The eyes of the whale, when viewed singly, are very large; but they seem disproportionately small, if we contrast them with the enormous mass of their entire body. But quadrupeds and birds feeding on flesh, require powers of vision of very great intensity. In these animals we accordingly find the organ large, and highly developed, so as commonly to impart a peculiar expression of ferocity to their countenances. The animals which are the objects of pursuit are frequently supplied with acute vision, thereby enabling them to escape or avoid danger; and this is particularly exemplified in the squirrel, the rat, the deer, and the hare. Animals which burrow under ground, as the mole and the shrew-mouse, have, in general, exceedingly small eyes; while in some they have been found nearly wanting, as in the blind rat (*Mus typhlus, Linn.*) The cat, the lemur, and other animals which pursue their prey during the night, are peculiarly adapted, by the construction of their eyes, for acutely perceiving objects, when illuminated by a very small quantity of light. The eyes of reptiles and fish are accommodated to the medium in which they reside. The chameleon can move one eye with rapidity, and in various directions, while the other remains fixed. Reptiles residing generally in the water, also fish, and the cetacea, such as the dolphin and seal, have their eyes covered with a dense skin, and the lens is more convex than in other animals. The arachnides, or spiders, possess generally eight eyes, arranged upon the upper part of the head in a symmetrical form; and there are not less than twenty-eight in the common millepede (*Julus terrestris.*) The insect tribes enjoy great variety and intensity in their visual organs; but the precise limits of this sense among the lowest animals in the scale of creation is not yet clearly ascertained.

The organ of hearing is excited by vibrations or undulations of air, of water, or of some solid medium, recurring at intervals, with different degrees of frequency. These impulses are received upon the tympanum or ear-drum; thence they are communicated to the acoustic nerve, and are finally transmitted to the brain. When the vibrations are not performed in equal times, or do not occur more frequently than seven or eight in a second, there is heard merely a *noise*. But when they rise much above this velocity, a very low or grave musical note is first heard. By an increase of velocity, the note becomes higher or more acute, and the ear is finally capable of perceiving sounds resulting from 31,000 impulses in a second. There is a regular gradation among animals of important use to the perfection of the organ of hearing, but none of them can rival the delicacy with which the practised ear in man perceives minute changes of tone, alterations in the quality or expression of sound, and varieties in its intensity and loudness. Feeble and timid quadrupeds generally have their ears directed backwards, to warn them of approaching danger; while, in the predaceous tribes, the ears are placed forwards, to aid in discovering their prey. Animals, though seldom susceptible of musical notes, sometimes exhibit an aversion for the low or grave sounds. This is remarkably the case with the lion. In bats, the sense of hearing is surprisingly acute.

L'organe de la génération est doué d'un sixième sens, qui est dans sa peau intérieure.

Perhaps a greater claim to the right of being termed the *sixth* sense, may be established in favor of that feeling of resistance, or *muscular* sense, by which we perceive the degree of force exercised by particular muscles. The mouth and lips of a new-born animal are directed by this sense to their proper function; and the adult would be in danger of a fall while engaged in walking, leaping, or other active exercises, if he were for a moment unconscious of the present state of the muscles appropriated to those actions. Shooting, bounding, and taking aim, presuppose a consciousness of the degree of muscular exertion sufficient to produce a certain effect; and instances are not wanting of its surprising accuracy. Thus, the Indian fresh water fish called the *Chætodon rostratus*, will hit an insect with a small drop of water at a distance of several feet, and the encumbered insect speedily falls an easy prey. When the elephant is annoyed by flies, he will discharge a large quantity of water upon the part attacked, with sufficient accuracy and force to dislodge them. The deadly spring of the lion and tiger exhibits the instantaneous result of the muscular sense in its most tremendous form. Dr Yellowley mentions the case of a woman who was afflicted with the disease called *anæsthesia*, where the muscular sense of her hands was lost, although the muscular power remained. On turning her eyes aside, she used to drop glasses, plates, &c., which were held in safety as long as another sense supplied the place of the lost one.

The stomach and intestines are possessed of certain peculiar sensations which declare the state of these viscera; and indeed every part of the body is susceptible of sensations, more or less painful, when affected by accident or disease.

Pain teaches an animal to avoid hurtful objects, and is wisely given as a safeguard to his frame; accordingly, its seat is mostly at the surface. The deep parts of the body have but little sensation, as it would there be only a useless encumbrance. The animal is continually warned, by uneasy sensations, to change his posture frequently, to avoid high degrees of heat, and, in general, to accommodate his frame to surrounding circumstances.

Many animals are defective both as to ears and nostrils, several are destitute of eyes, and some are reduced solely to the sense of touch, which is never wanting.

In the higher animals, impressions made upon the external organs of sense are transmitted by the nerves to the brain and spinal marrow, which form the central masses of the nervous system. The elevation of an animal in the scale of creation may [frequently] be determined by the volume of its brain, and the degree in which the power of sensation is concentrated there. Animals of a lower grade have the medullary masses much dispersed; and in the more simple genera, all trace of nerves seems to be lost in the general substance of the body. That part which contains the brain and principal organs of sense is called the *head*.

We now proceed to consider the second animal function—namely VOLUNTARY MOTION.

When the animal wills to move, in consequence of a sensation upon an external organ, or any other cause, the motion is transmitted to the muscles by means of the nerves.

This power of originating motion, residing in the nervous system of living animals, is one of the most wonderful properties of their nature. Every machine, however complicated or varied in its structure, can only be set in motion by some external power already existing in nature, or produced by art, whether it be the expansive force of steam, the descent of weights, the action of running water, or the recoil of a spring. No perpetual motion can ever be preserved by any arrangement of the parts of a machine among themselves; they must rest ultimately upon a prime mover. But the exquisite arrangement of the animal frame surpasses, in this respect, the highest mechanical skill. The mind wills—the muscle contracts. How much soever we may desire to unravel the mystery, the process is inexplicable, and seems for ever removed beyond the reach of human ingenuity. The only fact hitherto ascertained is, that if the nerve be separated, seriously injured, or even tightly compressed, the motion of the muscle will not follow the volition of the mind.

The muscles are bundles of fleshy fibres, by the contraction of which the animal body performs all its motions. The extension and lengthening of the limbs are equally the result of muscular contraction with their bending and drawing in. They are arranged in number and in direction to suit the motions which each animal is destined to perform; and when it becomes necessary to execute these motions with vigour, the muscles are inserted upon hard parts, which are so articulated, one over the other, as to constitute them so many levers. These parts, in the vertebrated animals, are called bones. They are situate internally, and are formed of a gelatinous mass [of cellular substance, the pores of which are] penetrated by particles of phosphate of lime. In some of the lower tribes of animals, such as the Mollusca, the Crustacea, and the Insects, these hard parts are external, and composed either of calcareous or of horny sub-

stances, called shells, crusts, or scales, all of which are secreted between the skin, and the epidermis or cuticle.

A considerable difference is found between the chemical composition of the bones belonging to the higher orders of animals and the external coverings of crustaceous animals. Human bones, when analyzed by Berzelius, were found to contain in 100 parts nearly as follows: of animal matters, (being chiefly gelatine, cartilage, and marrow) 34 parts; of phosphate of lime, 51 parts; of carbonate of lime, 11 parts; of fluate of lime, 2 parts; of phosphate of magnesia, 1 part; and of soda, muriate of soda and water, 1 part. Here the principal ingredient is *phosphate* of lime; but in the hard parts of crustaceous animals, such as crabs and lobsters, the *carbonate* of lime is considerably in excess. The shells of the mollusca, such as muscles and oysters, are almost entirely formed of the carbonate of lime. On the contrary, the horny coverings of insects contain a very minute portion of earthy matter, and are mostly composed of animal substances. The same proximate elements enter into the composition of horns, nails, and hoofs, being gelatine, with a membranous substance, resembling the white of eggs boiled hard. The scales of fish are composed of layers of membrane alternately with those of phosphate of lime, which arrangement is the cause of their brilliancy; but the scales of serpents contain no phosphate of lime, and very much resemble, in their constitution, the horny coverings of insects.

The fleshy fibres are inserted upon the hard parts, by means of other fibres of a gelatinous nature, called *tendons*, which seem to be a continuation of the first.

These tendons exercise the same office as straps or ropes in ordinary machinery, when it is required to transfer motion from one part to another. By this means a moving power can be exercised, in a spot where its immediate presence would be highly inconvenient. Thus, the hand is moved by tendons communicating with muscles, fixed at a considerable distance upon the arm; and the velocity and delicacy of its movements are not obstructed by their presence. Often these tendons are strapped down by cross cords, and pass along grooves in the bones, or through a pulley formed by a ligament. By these mechanical contrivances, the direction in which the muscular power acts may be changed; the forces of different muscles are compounded, and altered in intensity; and the velocity of the resulting motions modified according as circumstances may require. This arrangement also permits the accumulation of force upon one point; for a great number of muscular fibres are employed to contract one tendon, in the same manner as several horses may be employed to draw the same rope.

The peculiar shape observable in the articulated surfaces of the hard parts confine the motions of the tendons within certain limits, and they are still further restrained by cords or envelopes, usually called *ligaments*, attached to the sides of the articulations. Animals become enabled to execute the innumerable motions involved in the exercise of walking or leaping, flying or swimming, according as the bony and muscular appendages are adapted for these various motions; and also, according to the relative forms and proportions which the limbs, in consequence, bear among themselves.

NUTRITION, which we shall now explain, forms the first of the vegetative functions.

The muscular fibres connected with digestion and circulation are not influenced by the will, but, on the contrary, as we have already explained, their principal arrangements and subdivisions appear to be specially intended to render the animal completely unconscious of their exercise. It is only when the mind is disturbed by violent passions, or paroxysms, that its influence is extended beyond the ordinary limits, and that it agitates these functions common to vegetable life. Sometimes, when the organs are diseased, their exercise is accompanied by sensation; but, in ordinary cases, digestion and circulation are performed without the consciousness of the animal.

The aliment is first masticated, that is, minutely divided by the jaws and teeth, or sucked in, when taken by the animal in a liquid form. It is then swallowed entirely by the muscular action of the back parts of the mouth and throat, and deposited in the first portions of the alimentary canal, which are usually expanded into one or more stomachs, where the food is penetrated and dissolved by corrosive juices.

This gastric juice possesses the very remarkable property of dissolving most animal and vegetable matters, when deprived of life, and some mineral substances. It more especially acts upon such as yield nutriment to the animal, and are adapted to its general habits and formation. When recently procured from the stomach of a healthy animal, it appears as a clear mucilaginous fluid, slightly salt to the taste. Substances, when undergoing fermentation or putrefaction, are immediately checked in their action by the gastric juice, and are formed by its corrosive influence into a new fluid, possessed of entirely different properties, called *chyme*. But most mineral substances are indigestible. Certain tribes of savages, as the Otomacs, will, however, swallow daily large quantities of earth to allay the cravings of hunger. But this substance does not appear to be digested; it merely acts mechanically in distending the stomach.

The higher region of the alimentary canal is occupied by the stomach, which receives the food conveyed to it through the œsophagus or gullet. The form and structure of the stomach bears a constant relation to the nature of the food. In herbivorous animals, it is composed of a complicated system of reservoirs, where, by a slow and intricate process, the small quantity of nutriment contained in vegetable matter is abstracted and conveyed into the system. In carnivorous animals, the stomach is comparatively simple; and a supply of abundant nourishment is readily procured from animal food.

After passing through the stomach, the food is received into the remaining part of the canal, where it is acted upon by other juices destined to complete its preparation.

The chyme formed in the stomach having passed into the intestine, comes in contact with the bile and the pancreatic juice. An immediate change takes place. The chyme acquires the yellow colour and bitter taste of bile, and at length divides into two portions; the one, a white tenacious liquid called chyle, and the other, a yellow pulp.

The coats of the intestinal canal are supplied with pores, which imbibe that portion of the alimentary mass adapted for the nutrition of the body [being the chyle], while the useless residue is finally conveyed away and ejected.

The canal in which this first function of nutrition is performed, appears to be a continuation of the skin, and it is composed, in a similar manner, of laminæ. Even the surrounding fibres are analogous to those adhering to the internal surface of the skin, and called the fleshy pannicle. A mucous secretion takes place throughout this canal, which seems to have some connexion with the perspiration from the surface of the skin; for, when the latter is suppressed, the former becomes more abundant. The skin exercises a power of absorption very much resembling that possessed by the intestines.

The whole length of the intestinal canal is much greater in herbivorous, than in carnivorous animals.

It is only in the very lowest tribes of animals that the same orifice is applied to the double purpose of receiving fresh supplies of aliment, and of ejecting the substances unfitted for nutrition. Their intestines assume the appearance of a sack with only one entrance. But in a far greater number of animals, having the intestinal canal supplied with two orifices, the nutritive juice [or chyle] is absorbed through the coats of the intestines, and immediately diffused [by the lacteals] through all the pores of the body. This arrangement appears to belong to the entire class of insects.

If we commence from the arachnides [or spiders] and the worms, and then examine all animals higher in the scale of creation, it will be found that the nutritive fluid circulates through a system of cylindrical vessels; and that it only supplies the several parts requiring nourishment by means of their ramified extremities [or lacteals], through which the nutriment is deposited in the places requiring sustenance. These vessels, which distribute the nutritive fluid or blood to all parts of the body, receive the name of arteries. Those, on the contrary, are called veins, which restore the blood to the centre of the circulating system. This motion of the nutritive fluid is sometimes performed simply in one circle; often there are two circular motions, and even three, if we include that of the vena-porta [which collects the blood of the intestines, and conveys it to the liver.] The velocity of its motion is frequently assisted by certain fleshy organs called hearts, which are placed at some one centre of circulation, often at both.

In the vertebrated and red-blooded animals, the nutritive fluid, or chyle, leaves the intestines either white or transparent; and is conveyed into the venous system, by means of particular vessels called lacteals, where it mixes with the blood. Other vessels similar to the lacteals, and composing with them one arrangement, called the lymphatic system, convey into the venous system those nutritive particles which have either escaped the lacteals, or have been absorbed through the cuticle or outer skin.

Before the blood is fitted to renovate the substance of the several parts of the body, it must receive, from the surrounding element, through the medium of respiration, that modification which we have already noticed. One part of the vessels belonging to those animals, which possess a circulating system, is destined to convey the blood to certain organs, where it is distributed over a large extent of surface, in order that the action of the surrounding element may be the more energetic. When the animal is adapted for breathing the air, this organ is hollow, and called lungs. But when the animal only breathes [the air dissolved] in water, the organ projects, and is called branchiæ, or gills. Certain organs of motion are always arranged so as to draw the surrounding element either within or upon the organ of respiration.

In animals which do not possess a circulating system, the air penetrates into every part of the body, through elastic vessels called tracheæ; or else water acts upon them, either by penetrating, in a similar manner, through vessels, or simply by being absorbed through the surface of the skin.

In Man, respiration is performed by means of the pressure and elastic force of the air,

which rushes into the lungs, where a vacuum would otherwise have been formed by the elevation of the ribs, and the depression of the diaphragm. Muscular force then expels the air, after the necessary purification of the blood existing in the lungs has been performed; and the same actions are again repeated. The blood, which was of a dark purple colour, while slowly travelling from all parts of the body to the heart, has no sooner been purified by yielding its excess of carbonic acid to the surrounding air, and by absorbing oxygen, than its colour changes into a bright vermilion.

In Birds, it was necessary to combine lungs of small bulk with an extensive aeration of the blood; and, accordingly, the blood not only passes into the lungs, but through them into capacious air cells; from which, by the action of the chest, it is again expelled. The lungs thus act twice upon the same portion of air.

The change of the tadpole into the frog is accompanied by extraordinary alterations in its respiratory organs, which will be more fully explained hereafter. In the first, or tadpole state, the organs are branchial, and in the frog they are pulmonary. The arrangements are striking and singular.

All respiration must be either aquatic or atmospheric. In the former case, the respiration is said to be cutaneous or branchial, according as it is performed through the skin or through gills. On the other hand, atmospheric respiration may be either tracheal or pulmonary, according as it is performed through the air-tubes called tracheæ, or by means of lungs.

After the blood has been purified by respiration, it is fitted to restore the composition of all parts of the body, and to execute the function of nutrition properly so called. The wonderful property, possessed by the blood, of decomposing itself so as to leave precisely, at each point, those particular kinds of particles which are there most wanted, constitutes the mysterious essence of vegetative life. We lose all traces of the secret process by which the restoration of the solids is performed, after having arrived at the ramified extremities of the arterial canals. But in the preparation of fluids we are able to trace appropriate organs, at once varied and complicated. Sometimes the minute extremities of the vessels are simply distributed over extended surfaces, from which the liquid exudes; and sometimes the liquid runs from the bottom of minute cavities. But the more general arrangement is, that the extremities of the arteries, before changing into veins, form particular vessels called capillary, which produce the requisite fluid at the exact point of union between these two kinds of vessels. The blood-vessels, by interlacing with the capillary vessels which we have just described, form certain bodies called conglomerate or secretory glands.

With all animals destitute of a circulation, and especially with Insects, the nutritive fluid bathes the solid parts of the body; and each of them imbibes those particles necessary for its sustenance. If it become requisite that any particular fluid should be secreted, capillary vessels, adapted for this purpose, and floating in the nutritive fluid, imbibe, through their pores, the elements necessary for the composition of the fluid to be secreted.

It is thus that the blood continually renovates all the component parts of the body, and repairs the incessant loss of its particles, resulting necessarily from the continued exercise of the vital functions. The general idea which we are able to form of this process is sufficiently distinct, although the details of the operations performed at each particular point are involved in obscurity, from our ignorance of the precise chemical composition of each part, and our consequent inability to determine the exact conditions necessary for their reproduction.

In addition to the secretory glands necessary for performing a part in the internal economy of the system [such as the liver and the pancreas], there are others which secrete fluids destined to be rejected, either as being superfluous, or for some purpose useful to the animal. Of the latter we may mention the black fluid secreted by the Cuttle fish [with which, when pursued, he obscures the water to cover his retreat], and the purple matter of several Mollusca.

The function of GENERATION is involved in much greater obscurity and difficulty than that of simple secretion; and this difficulty attaches chiefly to the production of the germ. We have already explained the insuperable difficulties attending the pre-existence of germs; yet, if once we assume their existence, no particular difficulty remains attached to generation [which is not equally applicable to ordinary secretion.] While the germ adheres to the mother, it is nourished as if it formed a part of her own body; but when the germ detaches itself, it possesses a distinct life of its own, essentially similar to that of an adult animal.

The form of the germ, in its passage through the several progressive states of development, successively termed the embryo, the fœtus, and, finally, the new-born animal, never exactly resembles that of the parent; and the difference is often so very great that the change has received the name of metamorphosis. Thus, no person could ever anticipate that the caterpillar would finally be transformed into the butterfly, until he had either observed or been informed of the fact.

These remarkable changes are not peculiar to Insects, for all living beings are more or less metamorphosed during the period of their growth; that is to say, they lose certain parts altogether, and develop others which were formerly less considerable. Thus, the antennæ, the wings, and all the parts of the butterfly, were concealed under the skin of the caterpillar; and, when the insect cast off its skin, the jaws, the feet, and other organs, which belong not to the butterfly, ceased to form a part of its body. Again, the feet of the frog are inclosed within the skin of the tadpole; and the tadpole, in order to become a frog, loses its tail, mouth, and gills or branchiæ.

Even the infant, before its birth, at that period, and during its progress to maturity, undergoes several metamorphoses. In the earlier periods of development, the embryo corresponds, in some of its parts, with certain of the lower animals. At first, it seems destitute of a neck, and the heart is situate in the place where a neck afterwards appears, an arrangement which is found to exist permanently in fish. There is also a striking resemblance between the lower extremity of the vertebral column in the embryo, and the tail of the fish. About the end of the fifth month, it is covered all over with a yellowish white silk, like the down of a young duck, which entirely disappears in six or seven weeks. The limbs are formed under the skin, and reaching it, gradually shoot out into their permanent position; yet, even when fully developed in other respects, the shoulders and thighs are still concealed under the skin. In this respect, the embryo resembles the horse and other animals, which have the shoulders and thighs permanently enveloped under a thick covering of muscle. The fingers, when first formed, are surrounded by a skin, which entirely covers them, like the mitten-gloves used for an infant. This covering is gradually absorbed, when it takes the form of a duck's web, and finally disappears. M. Tieddeman and M. Serres, have shown that the brain of the fœtus, in the highest class of animals, assumes in succession the various forms which belong to Fishes, Reptiles, and Birds, before it acquires those additions and modifications which are peculiar to the Mammalia. "If you examine the brain of the Mammalia," says M. Serres, " at an early stage of uterine life, you perceive the cerebral hemispheres consolidated, as in Fish, in two vesicles isolated one from the other; at a later period, you see them affect the configuration of the cerebral hemispheres of Reptiles; still later again, they present you with the forms of those of Birds; finally, they acquire, at the era of birth, and sometimes later, the permanent forms which the adult Mammalia present."

As the infant grows towards manhood it loses, at a certain age, the thymus gland; by degrees it acquires hair, teeth, and beard; the relative size of its organs changes; the body increases at a much greater rate than the head, and the head more rapidly than the internal part of the ear.

Le lieu où les germes se montrent, l'assemblage de ces germes se nomme l'*ovaire;* le canal, par où les germes une fois détachés se rendent au dehors, l'*oviductus;* la cavité où ils sont obligés, dans plusieurs espèces, de séjourner un temps plus ou moins long avant de naître, la *matrice* ou l'*utérus;* l'orifice extérieur par lequel ils sortent, la *vulve.* Quand il y a des sexes, le sexe mâle est celui qui féconde; le sexe femelle celui dans lequel les germes paraissent. La liqueur fécondante se nomme *sperme;* les glandes qui la séparent du sang, *testicules;* et, quand il faut qu'elle soit introduite dans le corps de la femelle, l'organe qui l'y porte s'appelle *verge.*

§Ect. VII.—A BRIEF NOTICE OF THE INTELLECTUAL FUNCTIONS OF ANIMALS.

Mind—Matter—Sensation—Illusions—Perception—Memory—Association of Ideas —Abstraction—Judgment—Faculties of Man and other animals compared—Instinct—Connexion between the Brain and Intellectual Faculties.

We have already explained, when treating of the nervous system, that before the mind can perceive an object, an impression must be made upon an organ of sense, either immediately, or through some material medium; and that this impression must be transmitted through the nerves to the brain.

But the manner in which sensation, and its consequent perception are produced, is a mystery impenetrable to the human understanding; and, since philosophy is unable to *prove** the existence of matter, it is only hazarding a gratuitous hypothesis to attempt to explain mind by materialism, [or by analogies borrowed from the qualities of matter. The consideration of the Physiology of the Human Mind, or Metaphysics, forms the subject of another science.] But it is the province of the naturalist to ascertain the conditions of the body attendant on sensation,—to trace the extreme gradations of intellect in all living beings,—to investigate the precise point of perfection attainable by each animal,—and, finally, to ascertain whether there be not certain modifications of the intellectual powers, occasioned by the peculiar organization of each species, or by the momentary state of each individual body.

It has been already explained, that, to enable the mind to perceive,

* First truths do not admit of proof; they are assumed. We cannot prove the existence of mind, but we are conscious of its existence; and we cannot prove the existence of matter, for we perceive it.

there must be an uninterrupted communication of nerves between the external organ of sense, and the central masses of the nervous system. The mind is, therefore, conscious only of some impression made upon these central masses. It follows, then, that the mind may be conscious of real sensations, without any corresponding affection of the external organ; and these may be produced either in the nervous chain of communication, or in the central masses themselves. This is the origin of dreams, and visions, and of several casual sensations.

The various kinds of spectral illusions proceed from impressions, which, being made on the retina, are thence communicated to the brain, and are referred by the mind to an object in actual existence. "When the eye or the head receives a sudden blow, a bright flash of light shoots from the eyeball. In the act of sneezing, gleams of light are emitted from each eye, both during the inhalation of the air, and during its subsequent protrusion; and in blowing air violently through the nostrils, two patches of light appear above the axis of the eye and in front of it, while other two luminous spots unite into one, and appear as it were about the point of the nose, when the eyeballs are directed to it. In a state of indisposition, the phosphorescence of the retina appears in new and more alarming forms. When the stomach is under a temporary derangement, accompanied by headache, the pressure of the blood-vessels upon the retina shows itself, in total darkness, by a faint blue light floating before the eye, varying in its shape, and passing away at one side. The blue light increases in intensity—becomes green and then yellow, and sometimes rises to red; all these colours being frequently seen at once; or the mass of light shades off into darkness. When we consider the variety of distinct forms which, in a state of perfect health, the imagination can conjure up when looking into a burning fire, or upon an irregularly shaded surface, it is easy to conceive how the masses of coloured light which float before the eye may be moulded, by the same power, into those fantastic and unnatural shapes which so often haunt the couch of the invalid, even when the mind retains its energy, and is conscious of the illusion under which it labours. In other cases, temporary blindness is produced by pressure upon the optic nerve, or upon the retina; and under the excitation of fever or delirium, when the physical cause which produces spectral forms is at its height, there is superadded a powerful influence of the mind, which imparts a new character to the phantasms of the senses."*

Many circumstances render it extremely probable, that the pictures drawn in the mind by memory, or created by imagination, do not merely exist "in the mind's eye," but are actually figured on the retina. During health, and in ordinary cases, these images are faint, and are easily distinguished from the sensations resulting from real perception. It is only when the body is affected by certain diseases, or during sleep, that the impressions on the retina appear to proceed from objects in actual existence.

Several instances might be brought forward to illustrate the illusions of the senses. By the well-known experiment of making a galvanic circuit through the tongue, a piece of zinc and one of silver, there is produced a pungent metallic taste, in the same manner as would have followed the real application of a rapid substance. Thus it may be seen that, if we communicate an impression to the nerve on its passage to the central mass, the mind will be affected in the same manner as if the impression had been made on the external organ.

By the terms *central masses,* we understand a certain portion of the nervous system, which is always more circumscribed as the animal is more perfectly constructed. In Man it is exclusively a limited portion of the brain. On the contrary, in Reptiles the central mass may include either the brain, the entire marrow, or any portion of them taken separately; so that the absence of the entire brain does not deprive them of sensation. The extension of the term, when applied to lower classes of animals, is much greater, as their sensitive power is still more widely diffused.

We are hitherto completely ignorant of the nature of the changes which take place in the nerves and brain during perception, and of the manner in which the process is carried on. Analogies derived from matter, sensible species, images, and vibrations, obscure rather than explain this mysterious subject.

A certain state of mind follows a certain impression upon an external organ. We refer the cause of the sensation to some external object. This constitutes *perception;* and the mind is said to form an *idea* of the object. By a necessary law involved in the constituion of the mind, all the ideas of material objects are in time and in space.

When an impression has once been made through the medullary masses upon the mind, it possesses the power of recalling the impression after the exciting cause has been removed. This is *memory,* a faculty which varies much with the age and health of the individual.

During childhood, and in youth, the memory is very vivid. Accordingly, this period of life is most favorable to the acquisition of knowledge, especially of those subjects involving a great extent of detail, such as languages, geography, civil history, and natural history. The memory fails with increasing years.

Vivid perceptions and sensations are easily conceived; but the memory of a former mental impression is in general more faint.

Certain diseases, such as apoplexy, destroy the memory, either entirely or partially. A disordered state of the stomach will deprive the mind of the power of following a continued train of deep thought. This is also the case in the first stages of fevers.

* Letters on Natural Magic, by Sir David Brewster.

Blows and other injuries of the head will often affect the memory in a manner altogether incredible and surprising; and similar effects are sometimes produced by a high degree of nervous excitement.

Ideas which resemble [which contrast], or which were produced at the same time [or in the same place], have the power of recalling each other. This is termed the *association of ideas.* The order, the extent, and the quickness in which this power is exercised, constitutes the perfection of the memory.

Every object presents itself to the memory with all its qualities, and all the ideas associated therewith. The understanding possesses the power of separating these associated ideas from the objects, and of combining all the properties resembling each other in different objects under one general idea. This power of generalization, by which an object is imagined to be divested of certain properties, which in reality are never found separate, is termed *Abstraction.*

The power of abstraction appears to belong exclusively to Man; who, by the invention of general terms, is enabled to reason concerning entire classes of objects and events, and to arrive at general conclusions, comprehending a multitude of particular truths.

Every sensation being more or less agreeable, or disagreeable, experience and repeated trials readily point out the movements necessary to procure the one, or to avoid the other. The understanding thence deduces general rules for the direction of *the Will* relatively to pleasure and pain.

An agreeable sensation may produce unpleasant consequences; and the foresight of these consequences may react upon the first sensation, and thus produce certain modifications of the abstract rules framed by the understanding. This is *prudence* or *self-love.*

The lower animals seem influenced only by their present or very recent sensations, and they invariably yield to the impulse of the moment. Man alone appears able to form the general idea of *happiness,* and, by taking a comprehensive view of things, to lay down a plan for the regulation of his future conduct, and the attainment of his favourite objects.

But an inseparable barrier is placed between man and inferior intelligences, by the power of perceiving those qualities of actions which are termed *right* and *wrong,* and the emotions which attend their perception. The supremacy of conscience, and its claim to be considered an original faculty of the mind, are clearly pointed out by Bishop Butler. "Virtue," he elsewhere observes, "is that which all ages and all countries have made profession of in public—it is that which every man you meet puts on the show of—it is that which the primary and fundamental laws of all civil constitutions over the face of the earth make it their business and endeavour to enforce the practice of upon mankind, such as justice, veracity, or a regard for the common good."

By applying terms to express our general ideas, we obtain certain formulæ or rules, which are easily adapted to particular cases. This is *judgment* or *reasoning* [which may be either *intuitive* or *deductive.*]

When original sensations and associations forcibly recur to the memory [the mind possesses the power of combining and arranging them, to form a new creation of its own], this is called *imagination,* and it may be accompanied by agreeable or painful associations.

Man being endowed with superior privileges, possesses the faculty of connecting his general ideas with particular signs. These are more or less arbitrary, easily fixed in the memory, and serve to suggest the general ideas, which they were intended to represent. We apply the term *symbols* to designate these signs when associated with our general ideas, and they form a *language* when collectively arranged. Language may be addressed either to the ear or to the eye; in the former case it is termed *speech,* in the latter, *hieroglyphics.* Writing is a series of images, by which the elementary sounds are represented to the eye [under the form of *letters.*] By combining them [into *words*], the compound sounds of which speech is composed are readily suggested. Writing is therefore an indirect representation of our thoughts.

This power of representing general ideas by particular signs or symbols, which are arbitrarily associated with them, enables us to retain an immense number of distinct ideas in the memory, and to recal them with facility. Innumerable materials are thus readily supplied to the reasoning faculty and to the imagination. The experience of individuals is also communicated by written signs to the whole human species, and by this means the foundation is laid for their indefinite improvement in knowledge through the course of ages.

The art of printing, by multiplying copies, has ensured the permanence of knowledge, and has afforded a powerful aid to the intellectual progress of the species.

This capacity for indefinite improvement forms one of the distinguishing characters of human intelligence.

The most perfect animals are infinitely below Man, in respect to the degrees of their intellectual faculties; but it is nevertheless certain that their understandings perform operations of the same kind. They move in consequence of sensations received; they are susceptible of lasting affections; and they acquire by experience a certain knowledge of external things, sufficient to regulate their motions, by actually foreseeing their consequences, and independently of immediate pain and pleasure. When domesticated, they feel their subordination. They know that the being who punishes them may refrain from doing so if he will, and they assume before him a supplicating air, when conscious of guilt, or fearful of his anger. The society of man either corrupts or improves them. They are susceptible of emulation and of jealousy; and, though possessed among themselves of a natural language, capable of expressing the sensations of the moment, they acquire from man a knowledge of the much more complicated language through which he makes known his pleasure, and urges them to execute it.

We perceive, in fact, a certain degree of reason in the higher animals, and consequences resulting from its use and abuse, similar to those observed in Man. The degree of their intelligence is not far different from that possessed by the infant mind, before it has learned to speak. But, in proportion as we descend in the scale of creation to animals far below man in organization, these faculties become more languid; and, in the lowest classes, they are reduced to certain motions obscurely indicating some kind of sensation, and the desire of avoiding pain. The degrees of intellect between these extremes are infinite.

Dogs, cats, horses, birds, and other animals, may have their original faculties modified by personal experience; and they are accordingly trained to the performance of those extraordinary feats, which in all countries form a favourite amusement of the people. "By experience," says Mr Hume, "animals become acquainted with the more obvious properties of external objects; and gradually, from their birth, treasure up a knowledge of the nature of fire, water, stones, earth, heights, depths, &c. The ignorance and inexperience of the young are here plainly distinguishable from the cunning and sagacity of the old, who have learned by long observation to avoid what hurt them, and pursue what gave ease and pleasure. A horse that has been accustomed to the field, becomes acquainted with the proper height he can leap, and will never attempt what exceeds his force and ability. An old greyhound will trust the more fatiguing part of the chase to the younger, and will place himself so as to meet the hare in her doubles; nor are the conjectures which he forms on this occasion founded on any thing but his observation and experience. This is still more evident from the effects of discipline and education on animals, who, by the proper application of rewards and punishments, may be taught any course of action the most contrary to their natural instincts and propensities. Is it not experience which renders a dog apprehensive of pain, when you menace him, or lift up the whip to beat him? Is it not experience which makes him answer to his name, and infer, from such an arbitrary sound, that you mean him rather than any of his fellows, and intend to call him when you pronounce it in a certain manner and with a certain accent?"

The,e exists, however, in a great number of animals, a faculty different from intelligence, called *instinct.* This power causes them to perform certain actions necessary to the preservation of the species, but often altogether removed from the apparent wants of the individual. These are often so very complicated and refined, that it is impossible to suppose them the result of foresight, without admitting a degree of intelligence in the species performing them, infinitely superior to what they exhibit in other respects. The actions proceeding from instinct are still less the effect of imitation, for the individuals executing them have sometimes never seen them performed by others. The degree of instinct is by no means proportioned to the general intelligence of the species; but it is in those animals which, in their other actions, manifest the utmost stupidity, that instinct appears most singular, most scientific, and most disinterested. It is so much the property of each entire species, that all individuals exercise it in precisely the same manner, without ever attaining to higher degrees of cultivation.

"Every other animal, but Man, from the first outset of the species and of the individual, is equal to his task; proceeds in the shortest way to the attainment of his purpose, and neither mistakes the end nor the means by which it is to be obtained: In what he performs, we often justly admire the ingenuity of the contrivance and the completeness of the work. But it is the ingenuity of the species, not of the individual; or rather it is the wisdom of God, not the deliberate effect of invention or choice, which the created being is fitted to employ for himself. His task is prescribed, and his manner of performing it secured. Observe the animals most remarkable for a happy choice of materials, and for the curious execution of their works. The bird, how unvaried in the choice of the matter she employs in the structure, or in the situation, she has chosen for her eyrie or nest! Insects, most exquisitely artful in the execution of their little works, for the accommodation of their swarms, and the lodgement of their stores; how accomplished in their first and least-experienced attempts! Nature appears to have given to the other animals a specific direction to the means they are to employ, without any rational conception of the end for which they are to employ them."

Thus, the working bees, from the creation of the world, have always constructed edifices of great ingenuity, upon principles deduced from the highest branches of geometry, for the purpose of lodging and nourishing a posterity which is not even their own.

"It is a curious mathematical problem," observes Dr. Reid, "at what precise angle three planes which compose the bottom of a cell in a honey-comb ought to meet, in order to make the greatest saving, or the least expense, of material and labour; and this is the very angle in which the three planes in the bottom of a cell do actually meet. Shall we ask here, who taught the bee the properties of solids, and to resolve the problems of *maxima* and *minima*? We need not say that bees know none of these things. They work most geometrically, somewhat like a child, who, by turning the handle of an organ, makes good music, without any knowledge of music. The art is not in the child, but in HIM who made the organ. In like manner, when a bee makes its comb so geometrically, the geometry is not in the bee, but in that Great Geometrician who made the bee, and made all things in number, weight, and measure."

The solitary bees and wasps construct very complicated nests for the reception of their eggs. From each egg there proceeds a worm which has never seen its mother, which knows not the structure of the prison enclosing it; and yet, after it has undergone its metamorphosis, will construct another nest, precisely similar, to contain its own egg.

No satisfactory explanation can be given of the phenomena of instinct, except we admit that these animals possess some innate and constant internal power, which determines them to act, in the same manner as when they are influenced by ordinary and accidental sensations. Instinct haunts them like a perpetual reverie or vision; and all the actions proceeding therefrom may be compared to those of a man walking in his sleep.

Instinct has been wisely bestowed upon animals by the Creator, to supply the defects of their understandings, the want of bodily force or fecundity; and thus the continuation of each species is secured to the proper extent.

There is no visible mark, in the conformation of an animal, by which we can ascertain the degree of instinct which it possesses. But so far as observation has hitherto extended, the degree of intelligence seems proportioned to the relative size of the brain, and especially of its hemispheres.

Without venturing to decide upon this point, we must remark that the latter assertion has been controverted by many recent observers, especially by Dr Herbert Mayo, in his valuable Outlines of Human Physiology. "It does not appear," he remarks, "that an increase in the absolute weight of the brain confers a superiority in mental endowments. Were this the case, the intellects of the whale and of the elephant should excel the rational nature of man. Neither does the relative weight of the brain to the whole body appear the measure of mental superiority. The weight of the human brain is but one thirty-fifth part, while that of a canary bird is one-fourteenth part. Nor in conjunction with parity of form, and structure even, does this relation appear of any value. The eagle is probably as sagacious as the canary bird; but the weight of the brain is but one two-hundred-and-sixtieth part of its entire weight.

"We may next inquire," he proceeds, "whether an increasing number and complication in the parts of the brain is essentially connected with improved mental functions. The first instances which occur to the mind are in favour of the affirmative of this supposition. It may be inferred, from their docility and surprising capability of receiving instruction, that birds have higher mental endowments than fish; and accordingly, in place of the nodules of the fishes' brain, which are scarcely more than tubercles to originate nerves, birds possess an ample cerebrum and cerebellum (or lobes of their brain). But in pursuing this argument, if we compare, on the other hand, the brain of birds with those of alligators and tortoises, we find no striking difference or physical superiority in the former over the latter; yet in mental development, the tortoise and alligator are probably much nearer to fish than to birds. The *instantia crucis* (or decisive experiment), however, upon this question, is found in the comparison of the brain of the cetaceous mammalia (such as whales or dolphins), with the human brain on the one hand, and with that of fish on the opposite.

"The cetaceous mammalia," he observes, "have brains which, besides being of large size, are nearly as complicated as those of human beings; they might therefore be expected, if the opinion which I am combating were true, to manifest a remarkable and distinguishing degree of sagacity. Endowed with a brain approaching nearly in complexity and relative size to that of man, the dolphin should resemble in his habits one of the transformed personages in eastern fable, who continued to betray, under a brute disguise, his human endowments. Something there should be, very marked in his deportment, which should stamp his essential diversity from the fishes, in whose general mould he is cast. His habits too, not shunning human society, render him especially open to observation; and the class of men who have the constant opportunity of watching his gambols in the sea, are famed for their credulity, and delight to believe in the mermaid, the sea-snake, and the kraken. Yet the mariner sees nothing in the porpoise or the dolphin but a fish, nor distinguishes him, except by his unwieldy bulk, from the shoal of herrings he pursues. The dolphin shows, in truth, no sagacity or instinct above the carp, or the trout, or the salmon. It is probable even that the latter, which have but the poorest rudiment of a brain, greatly exceed him in cunning and sagacity. I am afraid that the instance which I have last adduced is sufficient to overthrow most of the received opinions respecting the relation of the size, shape, and organization of the brain to mental development; nor is it easy to find a resting-place for conjecture upon this subject."

SECT. VIII.—ON THE CLASSIFICATION OF THE ANIMAL KINGDOM.

General Distribution of the Animal Kingdom into Four Great Divisions.
1. *Vertebrata*—2. *Mollusca*—3. *Articulata*—4. *Radiata*.

AFTER the observations which have already been made concerning systems of classification in general, we have now to ascertain those leading characters of animals, upon which we must found the primary divisions of the Animal Kingdom. It is evident that these must be derived from the animal functions, that is, from sensation and motion; for not only do these functions constitute them animals, but they point out the rank which they hold in the animal world.

Observation confirms the correctness of this reasoning, by showing that their development and intricacy of structure correspond in degree with those of the organs performing the vegetative functions.

The heart and the organs of circulation form a kind of centre for the vegetative functions, in the same manner as the brain and the trunk of the nervous system are the centres of the animal functions; for we see these two systems become gradually more imperfect, and finally disappear together. In the very lowest classes of animals, where nerves can no longer be discovered, all traces of muscular fibres are obliterated, and the organs of digestion are simply excavated in the uniform mass of their bodies. The vascular system [or systematic arrangement of vessels] in insects, disappears even before the nervous; but, in general, the medullary masses are dispersed in a degree corresponding to the agents of muscular motion. A spinal marrow, on which are various knots or ganglions, representing so many brains, corresponds exactly to a body divided into numerous annular [or ring-like] segments, supported upon pairs of limbs, distributed along its entire length.

This general agreement in the construction of animal bodies, resulting from the arrangement of their organs of motion, the distribution of the nervous masses, and the energy of the circulating system, ought, then, to form the basis of the primary divisions of the Animal Kingdom. We shall now proceed to examine what the characters are, which ought to succeed immediately to the above, and give rise to the first subdivisions.

If we divest ourselves of the popular prejudices in favor of long established divisions, and consider the Animal Kingdom upon the principles already laid down, without reference to the size of the animals, their utility, the greater or less knowledge we may have of them, or to any of these accidental circumstances, but solely in reference to their organization and general nature, we shall find that there are four principal forms, or (if we may use the expression) four general plans, upon which all animals appear to have been modelled. The minor subdivisions, by whatever titles they may be ornamented by Naturalists, are merely slight modifications of these great divisions, founded upon the greater development or addition of some parts, while the general plan remains essentially the same.

1. VERTEBRATA—*Vertebrated Animals.*

In the first of these forms, which is that of Man, and of the animals most resembling him, the brain and the principal trunk of the nervous system are enveloped in a bony covering, composed of the cranium [or skull], and the vertebræ [or bones of the neck, back, and loins.] To the sides of this medial column are attached the ribs, and the bones of the limbs, forming collectively the framework of the body. The muscles, in general, enclose the bones which they set in motion, and the viscera are contained within the head and trunk.

Animals possessed of this form are called Vertebrated Animals (*Animalia vertebrata*), [from their possessing a *vertebral* column, or spine.]

They are all supplied with red blood, a muscular heart, a mouth with two jaws, one being placed either above or before the other, distinct organs of sight, hearing, smell, and taste, in the cavities of the face, and never more than four limbs. The sexes are always separate, and the general distribution of the medullary masses, with the principal branches of the nervous system, are nearly the same in all.

Upon examining attentively each of the parts of this extensive division of animals, we shall always discover some analogy among them, even in species apparently the most removed from each other; and the leading features of one uniform plan may be traced from man to the lowest of the fishes.

The following are examples of Vertebrated Animals: Man, quadrupeds, whales, birds, serpents, frogs, tortoises, herrings, carps, &c.

2. MOLLUSCA—*Molluscous Animals.*

In the second form of animals we find no skeleton. The muscles are attached solely to the skin, which forms a soft envelope, capable of con-

14   GENERAL REVIEW OF LIVING BEINGS.

tracting in various ways. In many species earthy laminæ or plates, called shells, are secreted from the skin, and their position and manner of production are analogous to those of the mucous bodies. The nervous system is placed within this covering along with the viscera; and the former is composed of numerous scattered masses, connected by nervous filaments. The largest of these masses are placed upon the œsophagus, or gullet, and are distinguished by the term *brain*. Of the four senses which are confined to particular organs, we can discover traces only of taste and of sight, but the latter is very often found wanting. In only one family, however, there are exhibited the organs of hearing. We always find a complete circulating system, and particular organs for respiration. The functions of digestion and of secretion are performed in a manner very nearly as complicated as in the vertebrated animals.

Animals possessed of this second form are called Molluscous Animals (*Animalia mollusca*), [from the Latin, *mollis*, soft.]

Although the general plan adopted in the organization of their external parts is not so uniform as in the vertebrated animals, yet, in so far as regards the internal structure and functions, there is at least an equal degree of mutual resemblance.

The cuttle-fish, oyster, slug, and garden-snail, are familiar instances of this class of animals.

3. ARTICULATA—*Articulated Animals.*

The third form is that which may be observed in Insects and Worms. Their nervous system consists of two long cords, extending the entire length of the intestinal canal, and dilated at intervals by various knots, or ganglions. The first of these knots, placed upon the œsophagus or gullet, and called the brain, is scarcely larger than any of the others, which may be found arranged along the intestinal canal. It communicates with the other ganglions by means of small filaments, or threads, which encircle the œsophagus like a necklace. The covering of their body is divided into a certain number of ring-like segments, by transverse folds, having their integuments sometimes hard, sometimes soft, but always with the muscles attached to the interior of the envelope. Their bodies have frequently articulated limbs attached to the sides, but they are also very frequently without any.

We shall assign the term Articulated Animals (*Animalia articulata*) to denote this numerous division, in which we first observe the transition from the circulating system in cylindrical vessels of the higher animals, to a mere nutrition, by imbibing or sucking in the alimentary substances; and the corresponding transition, from respiration through particular organs, to one performed by means of tracheæ, or air cells, dispersed throughout the body. The senses most strongly marked among them are those of taste and sight. One single family exhibits the organ of hearing. The jaws of the Articulated Animals are always lateral, but sometimes they are altogether wanting.

As instances of this form, we may mention the earth-worm, leech, crabs, lobsters, spiders, beetles, grasshoppers, and flies. From the circumstance of their coverings, or limbs, being divided, or jointed, they derive the name of "articulated," from the Latin *articulus*, a little joint.

4. RADIATA—*Radiated Animals.*

To the fourth and last form, which includes all the animals commonly called Zoophytes, may be assigned the name of Radiated Animals (*Animalia radiata*.) In all the other classes the organs of motion and of sensation are arranged symmetrically on both sides of a medial line or axis; while the front and back are quite dissimilar. In this class, on the contrary, the organs of motion and of sensation are arranged like rays around a centre; and this is the case even when there are but two series, for then both faces are similar. They approach nearly to the uniform structure of plants; and we do not always perceive very distinct traces of a nervous system, nor of distinct organs for sensation. In some we can scarcely find any signs of a circulation. Their organs for respiration are almost always arranged on the external surface of their bodies. The greater number possess, for intestines, a simple bag or sac, with but one entrance; and the lowest families exhibit nothing but a kind of uniform pulp, endowed only with motion and sensation.

The following are instances of this singular class of animals:—The sea-nettle, polypus, hydra, coral, and sponge. The name zoophyte is derived from two Greek words, ζωον (zoon), an animal; Φυτον (phyton), a plant; while that of *radiata*, evidently points out the *radiated* or *ray-like* arrangement of their parts.

" Before my time," says the Baron Cuvier in a note to his first edition, " modern naturalists divided all Invertebrated Animals into two classes—Insects and Worms. I was the first who attacked this view of the subject, and proposed another division, in a paper read before the Society of Natural History at Paris, the 21st Floreal, year iii.

(or 10th May 1795), and which was afterwards printed in the " *Decade Philosophique*." In this paper, I pointed out the characters and limits of the Mollusca, the Crustacea, the Insects, the Worms, the Echinodermata, and the Zoophytes. The red-blooded worms, or Annelides, were not distinguished until a later period, in a paper read before the Institute, on the 11th Nivose, year x. (or 31st December 1801.) I afterwards distributed these several classes into three grand divisions, analogous to that of the Animalia Vertebrata, in a paper read before the Institute in July 1812, and afterwards published in the *Annales du mus. d'Histoire Nat. tome xix.*"

SECT. IX.—GENERAL REVIEW OF LIVING BEINGS.

Life—Animals and Plants—Definition of an Animal.

WHEN we contemplate the face of the earth, we perceive it to be covered with living beings. Animals and plants are to be found in every corner of the globe, with the exception of the poles, where perpetual frosts and the long darkness of winter render the land incapable of supporting them; and where, to use the words of the poet, " Life itself goes out." We even find the remains of living bodies at enormous depths below the surface, in spots which once formed the beds of running streams, or the bottom of a mighty ocean, from which situations they have been elevated by the ordinary laws of volcanic agency. The mould forming the surface of the earth is composed of the remains of generations which are now no more: it serves to maintain the growth of living plants, and, through them, of all living animals. In the atmosphere surrounding the globe, every thing is fitted for life: light and heat bring organized bodies into existence; the air, covering the earth in every direction to the depth of many leagues, continually exchanges its particles with those of living bodies. Finally, water, which passes incessantly from the sea to the clouds, and from the clouds to the sea, is another element essential to Life.

Life is one of those mysterious and unknown secondary causes, to which we assign a certain series of observed phenomena, possessing mutual relations, and succeeding each other in a constant order. It is true that we are completely ignorant of the link which unites these phenomena, but we are sensible that a connexion must exist; and this conviction is sufficient to induce us to assign to them one general name, which is used in two senses: first, as the sign of a particular principle; and, secondly, as indicating the totality of the phenomena which have given rise to its adoption.

As the human body, the bodies of the other animals, and of plants, appear to resist, during a certain time, the laws which govern inanimate bodies, and even to act on all around them in a manner opposed altogether to those laws, we employ the terms *Life* and *Vital Principle* to designate these apparent exceptions to general laws. It is, therefore, by determining exactly in what these exceptions consist, that we shall be able to understand clearly the meaning of these terms. For this purpose, let us consider living bodies in their active and passive relations to the rest of nature.

For example, let us contemplate a female in the prime of youth and health. The elegant form, the graceful flexibility of motion, the gentle warmth, the cheeks crimsoned with the blushes of beauty, the brilliant eyes sparkling with the fire of genius, or animated with the sallies of wit, seem united to form a most fascinating being. A moment's is sufficient to destroy the illusion. Motion and sense often cease without any apparent cause. The body loses its heat, the muscles become flat, and the angular prominences of the bones appear; the cornea of the eye loses its brightness, and the eyes sink. These are, however, but the preludes of changes still more horrible. The neck and abdomen become discoloured, the cuticle separates from the skin, which becomes successively blue, green, and black. The corpse slowly dissolves, a part combining with the atmosphere, a part reduced to the liquid state, and a part mouldering in the earth. In a word, after a few short days there remain only a small number of earthy and saline principles. The other elements are dispersed in air and water, prepared again to enter into new combinations, and to become the constituent particles, perhaps, of another human body.

It is evident that this separation is the natural effect of the action of the air, heat, and moisture; in a word, of external matter upon the dead animal body; and that its cause is to be found in the elective attraction of these different agents for the elements of which the body is composed. That body, however, was equally surrounded by those agents while living, their affinities for its molecules were the same, and the latter would have yielded in the same manner during life, had not their cohesion been preserved by a power superior to those affinities, and which never ceased to act until the moment of death.

All living beings are found to possess one common character, whatever differences may prevail among them. They are all born from bodies similar to themselves, and grow by attracting the surrounding particles which they assimilate with their substance. All are formed with different parts, which we call *organs*, and from which they derive the appellation of *organized* beings. These organs united together form a whole, which is a perfect unity in respect to form, duration, and the phenomena it exhibits; and, as one of these properties cannot be abstracted from the rest without annihilating the whole, a living being receives the name of *individual*. Each being possesses a degree of heat, differing in different beings, and, to a certain point, independent of surrounding bodies. They all resist the laws of affinity which sway the mineral kingdom, and the compositions which they form are submitted to laws different from those influencing the mixtures of the chemist. They all absorb something from without, and transform it within; and all exhale certain principles, the product of the vital action. All reproduce other and similar beings, by the same actions by which they were themselves produced. All exist for a time, variable for each individual, but nearly the same for the same species, when in the wild state of nature. After this active individual existence; they all cease to live; and, finally, their bodies are dissipated into their more simple elements, according to the universal laws of Inorganic Chemistry.

Thus every living being forms, by its unity, a little world within itself; yet this little world cannot remain isolated from the universe without. In Life, there is always a bond of mutual dependence among the organs—a universal concourse and agreement of actions. Every part corresponds with the whole, and the whole with the universe.

If, then, we wish to distinguish a living body from another organized body, but without life, we have only to ascertain whether it continue to interchange particles with the soil, or gaseous fluids, which surround it; or, on the contrary, whether it maintain no active or efficacious relations with the universe. Again, if we wish to distinguish an organized body, which has ceased to live, from a mineral, we have only to ascertain whether the particles are otherwise united than by the ordinary molecular attractions, and whether the free action of the elements is about to annihilate it either by destruction or putrefaction.

The division of Living Beings into Animals and Plants has been already explained. The former, being of a complex nature, are provided with an internal cavity which receives their aliment, and are endowed with sense and spontaneous motion. Directed by instinct, they are alike capable of avoiding injury, and of pursuing their natural good. The latter, fixed to the earth by their roots, and deprived of the faculties of sensation and motion, are placed by Nature in situations fitted to supply their wants. The materials necessary for their sustenance are absorbed directly, without instinct or motion, and are abundantly supplied without either preparation or complicated labour. Animals, endowed with the distinctions of sex, both of which sometimes co-exist in the same individual, but more frequently in separate individuals of the same species, preserve these distinctions during the whole period of their lives. Almost all plants, on the contrary, have the two sexes united in the same being; and the distinctive characters of sex are lost and renewed every year. Again, the internal structure of animals is more complicated than that of plants: it is *internally* that the great functions of life are performed. With plants, on the contrary, the principal organs are placed on the surface; and their functions are mostly performed *externally*. As soon as an animal is born, its organs are exhibited: they require nothing but development and growth to form a perfect animal; and, if we except certain metamorphoses, the external form of the adult is already sketched. The vegetable, born from a seed, develops its organs successively; first the root, then the stalk, leaves, and flowers;—and when the flowers have bloomed, they die; the rest of the organs perish, the whole ceases to live, or sometimes only the stalk, or perhaps only the leaves. Not a year elapses but each flower is destroyed or renewed, partially or entirely. Thus, the two classes of beings possess in common the powers of nutrition and of reproduction. The animal has, however, something more than the vegetable, and enjoys the higher powers of sensation and voluntary motion. The animal alone possesses nerves, muscles, blood, and some kind of stomach. One at least of these organs is always visible; and, as the nerves and muscles are intermittent in their action, and incapable of maintaining a long-continued exercise without repose, animals possess a new distinctive mark in that periodical sleep to which they are at intervals subjected.

To a person who has considered Life only in Man, or in those higher animals which most resemble him, it appears almost superfluous to explain the essential difference between an animal and a plant. If there existed upon the face of the earth only such animals as Birds, Fishes, or Quadrupeds, there would then be no occasion to enlarge so fully upon the distinctions in their functions: the line drawn by the hand of Nature would suffice. We should readily be preserved from error on this point by their senses, their voluntary motion, the symmetry and complexity of their structure, but, above all, by the instinct which directs their actions. Then we might say with Linnæus, "Vegetabilia crescunt et vivunt; Animalia crescunt, vivunt et sentiunt," (Vegetables grow and live; Animals grow, live, and feel); and this definition would be as accurate as it is brief. We should not be obliged to separate Corals, Polypi, Insects, Crustacea, and Symmetrical Shells, from the Vegetable Kingdom.

But such is not the case. All animals do not exhibit the distinctive marks of complicated structure and voluntary motion. This may be easily inferred from the fact, that Tournefort, a man of great talents, and an able naturalist, actually formed nine genera in the seventeenth family of his Botanical system with those Polypi which were known to him and to his learned contemporaries. At a later period, Trembley hesitated for a long time before he could determine whether the Hydra was an animal or a plant; and the experiments which he performed to determine the question have been admired by all the philosophers of his time. The dexterous manipulations of Trembley are the more remarkable, as Peyssonel had previously observed that minute animals inhabit the different compartments of the corals. This discovery was extended by Ellis and Solander to all kinds of Polypi; while Donati, Réaumur, and B. de Jussieu, brought the subject prominently forward in their public lectures and writings. The question, however, still remained in an unsatisfactory state, and attracted the attention of the distinguished naturalists of the eighteenth century. Buffon proposed to establish an intermediate class between animals and plants. Linnæus adopted this suggestion, although it proceeded from Buffon; and rendered the distinction permanent by the title of Zoophytes, or Animated Plants. The celebrated Pallas followed Linnæus; Cuvier adopted the word and the distinction; while Lamarck rejected them both.

These doubts and differences of opinion among enlightened men could only have proceeded from the obscurity of the subject. One cause of the obscurity arose from the false direction which their studies had unfortunately taken. Confining themselves to their cabinets, Naturalists remained too far from Nature. They had found solid bodies—Corals, Sponges, Alcyonia, Polypi, of innumerable shapes, sometimes covered with soft and moveable bodies, and sometimes without them. Instead of considering the soft body as the artificer of the solid mass, they believed that the latter produced the former; and as the solid masses were observed to grow and vegetate, they were hastily considered to be plants, while the soft bodies were regarded as the flowers of these extraordinary vegetables. The error was further confirmed by the circumstance, that at the particular period when these Polypi reproduce other beings of the same species, their bodies are covered with little buds and shoots, which bear a great resemblance to certain flowers, the structure of which cannot be very distinctly perceived. But when these supposed flowers were observed to be endowed with spontaneous motion, and that they were possessed of sensation, a great difficulty arose; and the name of Zoanthes, or *animated flowers*, was assigned to them.

It has now, however, been completely ascertained that the Polypi themselves fabricate these solid apparent vegetables, which serve for their abodes. They secrete them in very nearly the same manner as the Mollusca form their shells; the Teredo its testaceous tube; the Lobster its crustaceous envelope; the Tortoise its shield; the Fishes their scales; Insects their elytra or wing-cases; Birds their plumage; the Armadillo his scaly covering; the Whales their horny laminæ; Quadrupeds their skins and organs of defence; and Man, his hair, nails, and cuticle. In all these beings there are to be found some parts which vegetate; and if it were necessary to class with plants all beings which are found to vegetate in any of their parts, we ought, consistently, to include all the animals just named with the Zoophytes or animated plants of Linnæus and Pallas.

The following are the characters by which we may always ascertain whether a living being, organized, growing, drawing in nutriment, possessing an internal temperature peculiar to itself, and reproducing its kind, be an Animal or a Plant.

If it be irritable to the touch, and move spontaneously to satisfy its wants,—if it be not deeply rooted in the soil, but only adhere to the surface,—if its body be provided with a central cavity,—if it putrify after death,—if it give out the ammoniacal odour of burnt horn,—and finally, if in its chemical composition there be found an excess of azote over carbon,—then we may be certain that it is an Animal. But if, on the contrary, the doubtful being under examination enjoy no lasting or spontaneous power of motion,—if it be destitute of an internal cavity,—if it be deeply inserted in the soil,—if, when detached, it speedily fade and die,—if, when dead, it merely ferment, but do not putrify,—if it burn without the odour of a burnt quill or horn,—and if its residue be very considerable and chiefly carbon,—then we may venture to declare it to be a Plant.

These characters are sufficient, and can, in general, be easily ascertained. In this enumeration, no allusion has been made to sensation as a distinctive mark of the two classes of living beings; because, in the lowest classes of animals, where alone any difficulty can arise, it is only from the property of irritability that we can infer sensation. The phenomena of reproduction have likewise not been alluded to, because it is in the lowest animals, which we are the most likely to confound with plants, that this power is still involved in great obscurity, or altogether unknown. It is not, as we might at first sight suppose, the most perfect, or, to speak more correctly, the most complicated plants that are likely to be mistaken for animals. A moment's reflection will readily show how utterly impossible it is to confound a plant, bearing leaves and flowers, with any animal whatever. But it is otherwise with the less characterized beings; and the Animal and Vegetable Kingdoms may be compared to two mighty pyramids, which touch each other by their bases, while their opposite vertices diverge to two infinitely remote points in either direction.

We have thus shown how extremely difficult it is to characterize the essential differences of animals and plants in one short definition. Even Cuvier himself, who spent twenty years of his life in examining the organization of animals, from the simple Polypus up to Man, has carefully abstained from proposing any such definition.

This difficulty increases in proportion to the number of animals under examination. It does not consist in ascertaining the characters appropriated to particular animals, but in selecting such a trait as shall be common to them all. We know that none but animals are possessed of a brain, nerves, muscles, heart, lungs, stomach, or skeleton. We know that they alone move, digest, respire; that they alone have blood, and seem to feel;—but the point is to ascertain which of these characters remains throughout the vast chain of beings, and which of them can be traced in the last link as well as in the first. We see the lungs disappear, then successively the glands, the brain, the skeleton, the heart, the gills, the blood, the nerves, the muscles, and finally, even the vessels; while in the lowest animals of all, we can scarcely ascertain whether they possess a digestive cavity or a stomach. However, as we find this last-mentioned organ in almost all animals, and as it can be clearly observed even in those which have no other externally visible organ, we may reasonably conclude that it is to be found in all; and, if we fail to discover a stomach in many, we should rather suppose our failure to proceed from want of skill, or from want of sufficient delicacy in our senses, arising probably from the excessive minuteness of the beings under examination. We shall, therefore, assume that all animals possess a stomach, and that they digest; we may infer that they are all possessed of sensation; but it is absolutely certain that they all, and they alone, permanently possess the power of voluntary motion.

If, therefore, we may venture to propose a definition which shall be generally applicable to all animals, we should define them to be *Living Beings having stomachs.* The stomach is, in fact, the great essential spring of every animated being. Nerves and muscles, organs of sensation and motion, appear indeed to be of a higher and more elevated character than the organ of digestion. Yet would these golden wheels of animated nature be inert and motionless, if they were not influenced by this prime-mover, formed of a coarser, but more energetic material, which supplies the fuel to their fires, and enables them to maintain undiminished the original vigour of their motions.

SECT. X.—GENERAL REVIEW OF LIVING BEINGS CONTINUED.

Their Unity and Perfection—Symmetry—Mutual Dependence—Classification of Living Beings.

All living beings are organized; that is, they are composed of different organs, each performing its separate function, and in its own peculiar manner. These organs collectively form a whole, perfect in each being; and the aggregation of those actions compose all that we are permitted to know of Life. Without the healthy state of the body, life cannot exist; yet the organs remain after life has ceased. We behold a body, which has just been deserted by the breath of life; we perceive an exquisite machine, where nothing seems defective; the wheel-work remains entire, but it wants the propelling hand of the workman. We may admire the sublime mechanism of that mighty Being who formed it, but the moving power ever escapes our research.

The greater number of living beings possess numerous organs, and a complicated structure. When the functions are various, the structure becomes intricate; but there exists a regular gradation—a well-marked hierarchy of functions, as well as of organs. All living bodies absorb nutriment, and reproduce their species; all animals move spontaneously at least some of their parts; many visibly respire; Man thinks. But it is evident that the first order of these functions is Nutrition—the other phenomena always presuppose this one. Let us, then, examine the subject of Nutrition, and we shall assuredly commence at the first link in the vital chain.

The greater number of Plants have a root fixed in the earth, a stem which shoots into the air, and directs itself towards the light. This stem bears leaves, branches, and flowers: these flowers, of various degrees of complication, produce fruits or seeds, destined to form a succession of beings, similar to those which have produced them. If we desire to ascertain which of these organs is essential to the existence of the plant, and, with this view, we successively abstract these several parts; if we cut off the fruits and seeds, the remainder of the plant rests uninjured. The stalk may lose its leaves without perishing; it may be cut, and the roots will continue to live and absorb in their ordinary manner—nay, the root will often even reproduce parts similar to those of which it has been deprived. The root is, therefore, the most important part of the Plant, and by it principally the whole vegetable is nourished.

Something similar to this may be observed in Animals. We see an animal of a very complicated structure. A bony skeleton, nerves, organs of sensation, a brain, muscles for motion, a heart for circulating the blood, lungs for absorbing air, a stomach in which the nutriment is deposited and prepared, glands for secreting the humours, arrangements for continuing the species, a general covering for protecting the whole, and limbs for changing its situation;—all these organs, and many more, compose its substance. In beings of this degree of complication, it is impossible to assign to each organ its proper degree of importance, because we cannot abstract any without injuring them all; and many cannot even be touched without subverting the entire fabric. But this separation, which we should in vain attempt to perform, Nature has herself realized in the long chain of animated existence. In descending from the viviparous Quadrupeds to the Birds, from the Birds to the Reptiles and Fishes, and passing from the Birds and Fishes, by the Mollusca and Insects, to the Worms and Polypi, we see these living machines become more simple, until at length we find, in the lowest orders, nothing but that first principle indispensable to all animals. The whole body of the Polypus forms, in fact, nothing but one entire stomach, without any other perceptible organ; and this alone is essential to the existence of a being so extremely simple.

We may thus conclude, that as the root is the first and essential element of the plant, so the stomach is the foundation of animal organization. Nature confirms this principle throughout all her works. She has created vegetables which are composed entirely of one vast root, and has formed animals of a simple gelatinous mass, containing only one enormous stomach. All the functions, are, however, of an extreme simplicity in bodies so homogeneous. In order that a vegetable may exist, composed entirely of a root, it is necessary that the substances proper to be absorbed should surround this root; it must be attached to a soil, composed of mould, and saturated with moisture, or to another plant; and these conditions are sufficient for its individual existence. As it produces no flowers, the species can only be continued by off-sets, buds, artificial or natural divisions of the root; and it is chiefly in this way that such bodies are propagated. But that an animal—Polypus or Worm—composed of one entire stomach, may exist, different arrangements are requisite. The stomach is placed internally; therefore, it is evident that the food must be carried into it. The animal must be able to move towards the food, and to draw it, by certain partial movements, within the cavity. In order to seek its food, it must feel and perceive; while a certain degree of instinct must exist, that it may adopt these movements in proportion to its wants. Thus, from one fundamental arrangement, there arises a being, perfect though simple, but which, though simplest of its kind, already appears complicated.

We have styled the Polypus a simple being, because it is composed of one entire stomach. Although it moves, and must feel, we can perceive no muscles, brain, or nerves; it possesses powers, while the instruments remain concealed. Yet the Polypus must be considered as a perfect being, because to it is assigned all the conditions necessary for its continued existence: it is in this respect as perfect as a Bird, or as one of the Mammalia. It is true that the animal possesses neither a heart nor lungs, no vessels or glands; but it has no occasion for them. When the body is one entire stomach, and when the animal is perfectly simple and homogeneous throughout, it is evident that these structures would be superfluous. Organs are only necessary when circulation and respiration are confined to particular parts. Every portion of the animal can draw from the alimentary canal that part of the nutriment necessary for its sustenance: it can breathe and assimilate these particles into its proper substance. But when the animal is not possessed of this perfect homogeneity throughout, it then becomes necessary that it should have a proper stomach to receive the nutriment, a heart to distribute it along with the blood into all the organs, and gills or lungs to purify this nutriment by exposure to the air. Unity of action is a first principle in life; and, in the higher orders of animals, it is the heart and the lungs which produce this unity in respect to nutrition, in the same manner as the brain realizes the unity of sensation.

Organization may exist without life, as living bodies are subject to death; but whoever says *Life*, also says *Organization*. Buffon was therefore guilty of a pleonasm, when he defined animals to be—*Bodies, living and organized.*

This organization of living bodies is regulated by certain fixed rules, which have received the name of laws from their constancy and universality. We have just spoken of the perfection and unity observable in all living bodies. The latter, however, is not absolute. Animals possessing a complicated structure are in truth *individuals;* but with plants and with the lower animals, individuals cannot be strictly said to exist, at least in the sense in which the term is understood in regard to Man and the higher animals. It is true that, as long as the several organs remain untouched, they enjoy one common life, and form one perfect and consistent being; but it is not impossible to abstract and prune away some parts, without interrupting the life of the being thus mutilated. We know well that a plant can be deprived of its flowers, leaves, and branches: there may remain nothing but a divided root, with a mutilated stem; and even this vestige of a living being will not cease to enjoy life. Nay, frequently many of the detached parts will themselves become new beings, when placed under circumstances favorable to their development. A branch and a leaf are sometimes adequate to form a vegetable similar, in all its parts, to the being whence they were derived. Upon this fact rest the whole theory and practice of slips and layers. The same thing is found with certain animals. A naked Polypus, when cut into several pieces, forms so many new and perfect Polypi, which continue to live in exactly the same manner as their original stock. Many of the rays of an Asterias, or Sea-star, may be detached without destroying the animal. The heads of slugs may be cut off, and the animals survive, even without any apparent diminution of their vigour. But what seems still more astonishing, some of the vertebrated animals themselves may be similarly mutilated without being instantly deprived of life. Tortoises and Salamanders, which have been decapitated, will still maintain their existence for a considerable time. The Emperor Commodus used to amuse himself with knocking off the heads of Ostriches while running round the Circus at Rome; and we are told by the historians of the times that they still continued their course. This singular power is even perceptible in the newly-born animals of the class Mammalia, which preserve their existence for a very short period, even when similarly injured. Still, however, these are but exceptions to a general law prevailing throughout the Mammalia, the Birds, and even among animals less complex and less elevated in the scale of creation. With these we in general find, that the extirpation of any important organ is incompatible with life. Sudden-death speedily follows such an operation. They are only capable of supporting the amputation of a limb or appendage; they can only endure a superficial wound, or injury. There exists, among all the Vertebrated Animals, a perfect dependence among their primary organs. If one of these be taken away, the remainder of the body ceases to live. If one of them be sick or wounded, the injury affects the other parts. There are five important organs, the integrity of which is absolutely essential to the continued existence of an animal possessing them; these are the heart, the brain, the organs of respiration, the spinal marrow, and the stomach. When these are once associated in a living animal, their co-existence is indispensable; and any serious division or decapitation of a body, provided with these five organs, is speedily mortal.

The parts of a plant are less united and more independent of each other; while the destruction of a part does not lead to the annihilation of the whole, because plants are nearly homogeneous. The portions remaining are provided with the same organs as the entire being. Precisely the same cause enables those lower animals to exist, which are formed but of one simple stomach. They possess no special and circumscribed organs; each of their divided segments partakes of an equal degree of complexity with the whole. But it is evident that a different result ought to be observed among the higher animals, where the functions necessary to their existence are isolated in special and circumscribed organs. With them the existence of the *individual* rests upon the exact mutual relation of the varied pieces composing the entire body.

In fact, it is a general rule, which prevails throughout the entire Animal Kingdom, that the organs essential to life are concentrated and intimately united in an animal, according to its elevation in the scale of creation, or, in other words, according as its structure is more or less complex. The variety and intricacy of the wheel-work requires a greater concentration of the moving power.

The symmetrical forms observable in all Living Beings are surprising. In regard to the roots of plants, and the branches of large trees, we observe that a great irregularity generally prevails. But this is owing rather to inequalities of the soil, and to varieties in the intensity of light, than to any natural disposition to irregularity in the plants themselves. The soil is not composed of uniform materials, and the roots always direct their fibres toward those parts which are most easily moved and yield the most abundant nutriment. The leaves and buds, again, are delicately sensible to nice degrees of light. We accordingly observe that the Coniferæ, such as the Pine and Fir, being resinous, and ever-green trees, upon which these powers have least influence, present the most regular and symmetrical forms.

The regular arrangement among plants is no where found in greater perfection than among the Labiatæ. We do not here allude to their flowers, which are not so very remarkable in this respect, but to their square stems, their opposite leaves, their branches, and their peduncles. In most of these plants, each leaf, taken separately, is arranged with regularity. But none even of those can compare with the beautiful symmetry observable in the leaves of the Sensitive Plant, the Acacias, and the Firs. In by far the greater number of plants, we find the utmost exactness in the distances between the several divisions of the calyx and corolla,—the flower-cup, and the flower itself; in the dimensions of each stamen, of each pistil; in every compartment of the ovarium, and of the fruit. With the exception of certain flowers analogous to those of the Acacias, of the Labiatæ, of the Orchideæ, and some others, the irregularities which many occasionally present are due to the abortion of certain parts, to their adherence, or to their transmutation into other forms.

Ascending to the Animal Kingdom, and arriving at the Polypi, those lowest of animated beings, we already find the same symmetrical arrangements. Their cilia, their tentacula, or little arms, these appendages of mere animated sacs, are disposed with regularity, around that single orifice, which we dignify by the name of *mouth*. It is only in those calcareous and arborescent masses which they form and inhabit, and which compose by their aggregation, rocks, islands, and rudimentary continents, that we fail to observe this regular arrangement. We may recognise the same order in the starry rays of the Euryalia, and in the spinous compartments of the Echini, or Sea-urchins. In respect to Insects, the symmetry is exquisite. We find the same quality in many Mollusca, but most particularly in their shells, and in the crustaceous envelopes of Crabs and Lobsters.

It is, however, in the higher or Vertebrated Animals that symmetry is brought to its greatest degree. Their bones, their nerves, their organs of sense, their brain, their muscles, their glands, their gills or lungs, are all arranged in lateral pairs, when their number is even; or they are placed in the exact central axis of the body, when their number is odd. We must admit, however, that it is externally we can best trace this correspondence, for the internal organs are not thus arranged. In this respect the contrast is altogether surprising: in vain we seek for symmetry in the disposition of the intestines, the liver, or the heart.

This physiological arrangement is ably illustrated by the excellent Dr Paley. "The regularity of the animal structure," he observes, "is rendered remarkable by the three following considerations:—First, the limbs, *separately* taken, have not this co-relation of parts, but the contrary of it. A knife taken down the chine, cuts the human body into two parts, externally equal and alike; you cannot draw a straight line which will not divide a hand, a foot, the leg, the thigh, the check, the eye, the ear, into two parts equal and alike. Those parts which are placed upon the middle or partition line of the body, or which traverse that line, as the nose, the tongue, the lips, may be so divided, or, more properly speaking, are double organs; but other parts cannot. This shows that the correspondency which we have been describing does not arise by any necessity in the nature of the subject; for, if necessary, it would be universal; whereas, it is observed only in the system or assemblage: it is not true of the separate parts; that is to say, it is found where it conduces to beauty or utility; it is not found where it would subsist at the expense of both. The two wings of a bird always correspond; the two sides of a feather frequently do not. In centipedes, millepedes, and the whole tribe of Insects, no two legs on the same side are alike; yet there is the most exact parity between the legs opposite to one another. The next circumstance to be remarked is, that, whilst the cavities of the body are so configurated as *externally* to exhibit the most exact correspondency of the opposite sides, the contents of these cavities have no such correspondency. A line drawn down the middle of the breast, divides the thorax into two sides exactly similar; yet these two sides enclose very different contents. The heart lies on the left side, a lobe of the lungs on the right, balancing each other neither in size nor shape. The same thing holds of the abdomen. The liver lies on the right side, without any similar viscus opposed to it on the left. The spleen indeed is situate over against the liver, but agreeing with the liver neither in bulk nor form. There is no equi-pollency between these. The stomach is a vessel both irregular in its shape and oblique in its position. The foldings and doublings of the intestines do not present a parity of sides. Yet that symmetry which depends upon the co-relation of the sides, is externally preserved throughout the whole trunk; and is the more remarkable in the lower part of it, as the integuments are soft; and the shape, consequently, is not, as the thorax is by its ribs, reduced by natural stays. It is evident, therefore, that the external proportion does not arise from any equality in the shape or pressure of the internal contents. What is it indeed but a correction of inequalities?—an adjustment, by mutual compensation, of anomalous forms into a regular congeries?—the effect, in a word, of artful, and, if we might be permitted so to speak, of studied collocation? Similar also to this, is a third observation; that an internal inequality in the feeding vessels is so managed, as to produce no inequality of parts which were intended to correspond. The right arm answers accurately to the left, both in size and shape; but the arterial branches, which supply the two arms, do not go off from their trunk, in a pair, in the same manner, at the same place, or at the same angle, under which want of similitude it is very difficult to conceive how the same quantity of blood should be pushed through each artery: yet the result is right;—in the two limbs which are nourished by them, we perceive no difference of supply, no effects of excess or deficiency. Concerning the difference of manner, in which the subclavian and carotid arteries, upon the different sides of the body, separate themselves from the aorta, Cheselden seems to have thought, that the advantage which the left gains by going off at an angle much more acute than the right, is made up to the right, by their going off together in one branch. It is very possible that this may be the compensating contrivance: and if it be so, how curious—how hydrostatical!"

Many animals form singular and remarkable exceptions to this general law of symmetry. The Mollusca have generally their digestive orifices, as well as the distinctive characters of sex, placed on one side of the body, and that is usually the right side. Flat fishes swim on one side; both their eyes are placed on that which is turned uppermost, and this again is almost always the right side. Even in those animals which are most beautifully arranged, one side of the entire body surpasses the other in strength, energy, and activity, and this stronger half of the body is almost always the right side. We can observe this circumstance among the Crustacea; we see it in the side-walk of the Crab; and remarkably so in the Pagurus Bernhardus, or Hermit Crab, where the right forceps is larger than the left. We even see it in the larger birds, and the feathers of the right wing are always stronger and of a better quality than those of the left. The same inequality can be traced among the Mammalia, and in none of them more so than in Man, who is, perhaps, less ambi-dextrous than any other animal. With him the superiority of the right hand over the left is not altogether the effect of habit, but is founded in nature. In walking, it is the right leg and foot that give the greater impulse to the body; in hopping or leaping, every schoolboy, who is not naturally left-handed, uses his right leg in preference to the left. Diseases of the right are more acute than those of the left side. When a person wishes to examine an object most minutely, he looks at it with one eye, and that is almost always the right eye. Whether it be not a consequence of that more general law, that a concentration of vital force in one organ is followed by a diminution of vigour in others adjacent to it, and that the presence of the heart at the left side deprives that entire division of the body of the vigour enjoyed by the right side, we shall not at present venture to determine.

We have now shown that one general plan can be traced throughout the whole of Living Beings; that analogies, sufficiently precise, may be observed throughout the Animal and Vegetable Kingdoms; and that in every portion of created existence, we find a degree of unity and perfection, a mutual dependence among their parts, and the most exquisite symmetry in their forms. We shall now proceed to trace the analogy perceptible in the essential functions of all these beings. Whether we examine the arrangements for the continuation of the several species, the manner in which that constant ingress and egress of particles, constituting nutrition, is fulfilled, the temperature belonging to each class of beings, or that necessity which compels every one of them to come in immediate contact with pure air, the results are the same for all. It is only the details in the workmanship of the great artificer that vary, but the same divine hand is perceptible throughout the whole. Thus, all Living Beings require nutriment, but animals alone receive the food into central cavities, and digest it. To all Living Beings air is equally essential—all absorb it and respire; but the instruments of respiration are infinitely diversified in the several classes of living beings. Man and other Mammalia, Birds, and Reptiles, breathe through lungs; the Fishes, on the contrary, the Crustacea, and the Mollusca, respire through gills or branchiæ. Insects, again, perform this function through tracheæ, or minute holes, with which their surfaces are perforated; while many Worms and Polypi appear only to absorb air through the pores of their skin, with which they are every where covered. Plants breathe through their leaves; and many of them, deprived even of leaves, only perform this function through the pores of the epidermis which covers their substance.

Again, in respect to the arrangements for continuing the several species, we observe the same general design, while the means are ever various. How different do we see this function in the Mammalia—those viviparous animals, where the young, already active and nearly perfect, immediately commence, from the moment of their separation from the parent, those instincts and actions, which can be terminated only by death; in the numerous class of oviparous, and in the ovo-viviparous animals! Again, how immense the chasm between all these animals just alluded to, and the lower beings which are destitute of any distinctions of sex!—how different is the function performed by the Polypi, without sexes, without germs, producing their kind only by buds or off-sets!—and then, again, another mighty chasm between these and plants, continuing their species by hermaphrodite flowers, or else by flowers of distinct sexes! Nor even here does variety cease to exist, for many are cryptogamous, or apparently destitute of any means for continuing their species, except by certain minute and almost evanescent sporules or reproductive corpuscles.

In all functions we trace this analogy in the end, and diversity in the means; and nowhere in a greater degree than in the functions peculiar to animals. They all appear to feel, yet many possess no other organ of sensation than the skin. In very many we find no brain, and in others not even can a vestige of nerves be traced. It is evident that they all move spontaneously, yet in many we can find no visible marks of muscles or organs of motion. We shall, however, not enlarge at present much further on this point, which will be illustrated hereafter in every page of The Animal Kingdom, but merely allude to the analogy observable among the Vertebrated Animals.

The analogy among the functions and organs of these animals is so remarkable, and the attention which has in consequence been paid to them so great, that we are exceedingly apt to form limited and erroneous views of the other parts of the animal world;—we expect to find in the lower animals the same parts, the same functions, which are plainly observable in them. Deeply impressed with their structure and functions, we can scarcely bring ourselves to imagine any living being without circulating fluids, a heart, blood, or vessels. So prejudiced are we in favour of the arrangements observable in the higher animals, that we can scarcely imagine any sensitive being without nerves, or any creature capable of moving without muscles. Tournefort even admitted plants to have muscles; nay, further, he actually described them. At the present day, there is little probability of our falling into a similar error; yet we are all naturally disposed, on observing a great analogy in the functions of all animals, to suppose them to be identical in their structure.

We have said that the analogy among the Vertebrated Animals is very remarkable. They are all possessed of a spinal column, composed of numerous vertebræ. Within this solid column is lodged the spinal marrow, and it carries at one extremity a well-defended bony case or head, which contains the brain. In all these beings we find a heart, red blood, lungs, or gills; in all, the organs of the five senses are seen in greater or less degrees of perfection: we find nerves, muscles, a digestive canal, more or less complicated, a liver and pancreas, with evident arrangements for continuing the species. With the exception perhaps of one species, they all have their mouths disposed horizontally; and when they have limbs, these are always four in number. This similarity prevails throughout their structure and functions. It is true that their surfaces vary remarkably according to their several destinations, while the organs of motion differ greatly as they may be designed for swimming, flying, or walking. The organs of respiration vary according as they are intended to breathe in water or in the air. But these differences in external arrangement do not prevent us from tracing the most exact analogy among them all. If we take all the organs, one by one, and compare them separately in any two vertebrated animals, we shall find the most exact equivalents in the two beings; the analogy will be found perfect in all the essential circumstances; it is only the details which are observed to differ. The fish at first sight appears to have neither neck nor thorax: but on inspecting it more attentively, we find it to be possessed of all the series of vertebræ; and that the different pieces of its thorax are concentrated near the cranium, with which they are almost confounded. M. Geoffroy has illustrated this curious organization of the Fishes in a philosophical and truly interesting manner. There is, however, one very remarkable distinction between these aquatic vertebrata and the aerial vertebrata, in the organs of voice, of which the former are completely deprived.

The principles, which must form the bases of a natural system of classification, have been already explained. A knowledge of internal organization, with the laws of the subordination and co-existence of functions, will alone lead us to this result.

Every function presupposes another function. Thus, when we see a being apparently moving voluntarily when irritated by any stimulant, we infer that it feels. We, therefore, conclude that voluntary motion presupposes sensation. Again, Life is temporary in its action; it therefore presupposes the reproduction of individuals with the extinction, and perhaps also the creation, of new species. We also conclude that circulation presupposes respiration; because, wherever we find a heart, we also meet with lungs, just in the same manner as we invariably find nerves wherever we can discover muscles. In fact, Life is but an aggregation of phenomena produced by organs connected and governed by these laws of co-existence.

But in forming a system of classification, the difficulty consists in detecting the law of subordination existing among the various combinations of these instruments of Life. Reflection upon the final cause or design of the functions will often lead us to detect these laws; but there are innumerable relations which no discernment could detect, without the nicest dissection of the bodies, or the most arduous observation of the habits of the animals when in their native elements. The anatomist in his laboratory, and the "out-of-door" Naturalist, who haunts the wilds of Nature, must unite their labours before we can form a satisfactory system of Classification.

After examining the internal structure of every known animal, it has been found that some of them have vertebræ, and others have none: this is a fundamental fact. Again, on examining further, it is found that all those having vertebræ are also possessed of a spinal marrow and a complicated brain; that they have always four organs of sense, of various degrees of perfection, with horizontal jaws placed in the head; and that they have never more than four limbs, and always red blood. On the contrary, when the Invertebrated animals are examined, they are never found to possess either a brain or spinal marrow; their senses are not so distinctly marked, their blood is white, or not so red, and they all have more than four limbs, or none whatever. Proceeding further, when the Vertebrated animals are more closely examined, some of them are found to continue their species by eggs—they are oviparous; others, on the contrary, produce their young alive—they are viviparous. The latter are found to be alone possessed of mammæ, for suckling their young, and hence they are called Mammalia.

Whenever, therefore, we find an animal with a bony skeleton, we know that it must either belong to the Mammalia, or to one of the three classes of oviparous Vertebrata. If it have feathers and lungs, it is a Bird; if it have lungs and no feathers, it is a Reptile; if it have gills and not lungs, it is a Fish. On looking further into the details of the structure, there are found other varieties, yet ever co-existing with certain essential differences. We are thus enabled to assign precisely the rank of an animal from knowing the smallest part of one of these essential organs; and we can even discover the most curious relations between these differences in the structure of animals, and their habits or instincts. All the Carnassiers, or beasts of prey, for example, have the digestive canal more simple, shorter, less powerful, and consequently their body more slender; on the contrary, they have the canine teeth, or parts analogous to them, much longer, stronger, better armed, and moved by muscles of great energy. Birds of prey have the talons of their claws more fitted for tearing, the beak strong and hooked. The Lion, and all others of the Cat genus, are similarly armed with formidable retractile claws, with alternate and sharp teeth, and with a solid jaw-bone, moved by powerful muscles. These fundamental characters are in a manner reflected throughout the whole structure, in such a manner that, upon examining a process or projection in one of the teeth of a Carnivorous Quadruped, or the condyle of its jaw-bone, we can describe the remainder of its frame-work, and write the history of its habits. In the same manner, we can form an estimate of the force with which a bird flies, by examining the formation of its sternum or breast-bone, to which the muscles of the wings are attached. Whenever we find those two small bones, called Marsupial, in the

pelvis of an animal, we may be certain that its young are produced before their time; that they are received and protected in a ventral pouch or bag. Finally, we know that the Ruminantia, or ruminating animals, all have a cloven hoof; that they all have four stomachs, and no incisive teeth in the upper jaw; and that all which carry antlers or horns on their front, have no canine teeth in the upper jaw. The history of the Animal Kingdom offers many facts analogous to these, which will be pointed out in the course of this work.

But we must remark, that all the organs of each being have the most perfect agreement among themselves. Never does Nature unite among them characters of an opposite kind: we never find the teeth and jaw-bone of the Carnassier, with the cloven foot of an herbivorous quadruped. The poets, painters, and statuaries of former times, loved to blend these distinctive characters into imaginary and fantastic forms. Deceived by their fertile imaginations, they knew not the laws regulating their co-existence. Sometimes we see enormous wings that no muscle can move; sometimes the heads of many animals of different species, united to a trunk which belongs to one of them, or perhaps to a different animal. Nature disdains to present the discordant characters of the Cerberus, Demon, or Angels, of our painters and our poets. One universal harmony characterises all her works, and every part of her perfect mechanism corresponds to the whole.

These, then, are the principles of our Classification, founded on the comparative importance of the organs, their constancy, and the laws of their subordination.

A STOMACH represents the Animal Kingdom, and a ROOT the Vegetable Kingdom. As these can exist isolated from every other part, we must seek for other organs, to form the secondary divisions in the two Kingdoms.

With Animals, we must first examine whether they are Vertebrated; and in that case, whether they are Viviparous or Oviparous; that is, whether they have mammæ or not. If they have none, we must next inquire whether they breathe through lungs or gills; and we may further examine whether they are or are not carnivorous, whether they fly, walk, swim, or crawl.

If, on the contrary, the animals under examination be without vertebræ, we examine the general arrangement of their body, their movements, whether they breathe through branchiæ, tracheæ, or simply through the skin; whether they have one or more hearts, or none whatever; whether they have wings, feet, antennæ, or tentacula; whether they have testaceous coverings, shells, or elytra; or whether they have nerves, nervous cords, swelling into knots, or an imperfect brain: we may investigate their intestines, or their metamorphoses. In this way, we are conducted by degrees from those first great divisions, which overwhelm us by their magnitude, into the more circumscribed groups of genera and species.

THE ANIMAL KINGDOM,

CONTAINING LIVING BEINGS WITH STOMACHS, ENDOWED WITH SENSATION AND VOLUNTARY MOTION.

Divisions.		Classes.
I. VERTEBRATA	Animals with a bony skeleton, consisting of a cranium, spinal column, and generally also of limbs; the muscles attached to the skeleton; distinct organs of sight, hearing, smell, and taste, in the cavities of the face; never more than four limbs; sexes separate; blood always red.	1. MAMMALIA. 2. AVES. 3. REPTILIA. 4. PISCES.
II. MOLLUSCA	Animals without a skeleton, the muscles being attached to the skin; body almost always covered with a mantle, which is either membraneous, fleshy, or secreting a shell; nervous system composed of scattered masses, or ganglions, connected by filaments; with distinct organs of digestion, circulation, and respiration; never with five senses, and generally without sight and hearing; blood white or bluish; sexes separate; hermaphrodites, perfect or reciprocal; oviparous or viviparous; eggs sometimes without shells.	1. CEPHALOPODA. 2. PTEROPODA. 3. GASTEROPODA. 4. ACEPHALA. 5. BRACHIOPODA. 6. CIRRHOPODA.
III. ARTICULATA	Animals without a skeleton, divided into a number of ring-like segments, having their integuments sometimes hard, sometimes soft, and the muscles always attached to the envelope; with or without limbs; respiring through tracheæ or air-vessels, sometimes through branchiæ; nervous system composed of two long cords, swelling at intervals into knots or ganglions.	1. ANNELIDES. 2. CRUSTACEA. 3. ARACHNIDES. 4. INSECTA.
IV. RADIATA	Animals having the organs of sensation and motion, arranged around a common axis in two or more rays, or in two or more lines extending from one extremity to the other; approaching nearly to the uniform structure of Plants. No circulation in vessels; nervous system obscure.	1. ECHINODERMATA. 2. ENTOZOA. 3. ACALEPHÆ. 4. POLYPI. 5. INFUSORIA.

THE VEGETABLE KINGDOM,

CONTAINING LIVING BEINGS WITH ROOTS, WITHOUT SENSATION OR VOLUNTARY MOTION.

Divisions.		Classes.
I. A-COTYLEDONES	Agamous, or rather cryptogamous Plants, without stamens or pistils.	1. APHYLLÆ. 2. FOLIACEÆ.
II. MONO-COTYLEDONES	Plants, having the embryo with only one cotyledon perianth simple, consisting of a calyx only; floral organs generally three, or multiples of three; nerves of the leaves generally longitudinal; stem composed of cellular tissue, with scattered vascular fasciculi.	1. HYPOGYNIA. 2. PERIGYNIA. 3. EPIGYNIA.

III. DI-COTYLEDONES	Plants, having their embryo with two cotyledons, excepting the Coniferæ, where there are often from three to ten verticillate cotyledons; all the parts of the stem disposed in concentric layers; flowers generally with a calyx and corolla, the parts of which are usually five, or some multiple of five; nerves of the leaves generally ramified.	1. MONO-CHLAMYDEÆ.
		2. DI-CHLAMYDEÆ.
		  a. COROLLIFLORÆ.
		  b. CALYCIFLORÆ.
		  c. THALAMIFLORÆ.

THE ANIMAL KINGDOM.

DIVISION I.—VERTEBRATA, SUBDIVIDED INTO FOUR CLASSES.

1. MAMMALIA ...... { Man and Beasts, with warm blood; heart with two ventricles; females suckling their young with milk, secreted in breasts or mammæ; viviparous, excepting the Monotremata, which are either oviparous or ovo-viviparous.

2. AVES.............. { Birds, with warm blood; heart with two ventricles; no mammæ; oviparous; body covered with feathers, and organized for flight.

3. REPTILIA......... { Reptiles, with cold blood; heart with one ventricle; having lungs, or sometimes only gills or branchiæ; oviparous, or ovo-viviparous; generally amphibious.

4. PISCES............. { Fishes, with cold blood; heart with one ventricle; no lungs, but breathing by branchiæ; generally oviparous; body organized for swimming.

DIVISION II.—MOLLUSCA, SUBDIVIDED INTO SIX CLASSES.

1. CEPHALOPODA... { Cuttle-fishes, having the mantle furnished with a shell, and united under the body, forming a muscular sac; head connected with the mouth of the sac, and crowned with long and strong fleshy limbs, for walking on, and seizing their prey; with two large eyes; and two gills placed in the sac. Sexes separate.

2. PTEROPODA.......... Marine animals without feet; with two fins, placed one on each side of the mouth; head distinct; hermaphrodites.

3. GASTEROPODA ... { Snails or Slugs, and Limpets, with a distinct head; crawling on a fleshy disc; very seldom with fins; generally with a shell; tentacula from two to six.

4. ACEPHALA ...... { Aquatic animals, generally with a bivalve or multivalve shell; without an apparent head or limbs; mouth concealed between the folds or in the bottom of the mantle; hermaphrodites; branchiæ external; incapable of locomotion.

5. BRACHIOPODA.... { Marine Animals, without a head; having two fleshy arms, furnished with numerous filaments; bivalve shells; incapable of locomotion.

6. CIRRHOPODA..... { Barnacles, inclosed in a multivalve shell; with numerous articulated limbs or cirrhi, disposed in pairs; incapable of locomotion. General structure approaching to the articulated animals.

DIVISION III.—ARTICULATA, SUBDIVIDED INTO FOUR CLASSES.

1. ANNELIDES .... { Worms, generally with red blood; without limbs; usually hermaphrodites, perfect or reciprocal; body soft; more or less elongated, and divided into numerous segments; circulation double, with one or more hearts or fleshy ventricles; respiring generally through branchiæ; sometimes dwelling within membraneous, horny, or calcareous tubes.

2. CRUSTACEA .... { Marine Animals, with a crustaceous envelope, having articulated limbs attached to the sides of the body; blood white; always with articulated antennæ or feelers in front of the head, and generally four in number; distinct organs of circulation; respiring through branchiæ.

3. ARACHNIDES....... { Spiders, with the head and breast united in a single piece, and with the principal viscera situate in a distinct abdomen, behind the thorax; without antennæ; oviparous.

4. INSECTA.............. Insects, divided into three distinct parts, the head, thorax, and abdomen; always with two antennæ, and six feet.

DIVISION IV.—RADIATA, SUBDIVIDED INTO FIVE CLASSES.

1. ECHINODERMATA { Sea-hedgehogs, and Sea-stars, with distinct viscera and organs of respiration; with a partial circulation; often with a kind of skeleton, armed with points or moveable spines; destitute of head, eyes, and articulated feet; nervous system indistinct; organs of motion extremely imperfect.

2. ENTOZOA............ { Intestinal Worms, with no distinct organs of circulation or respiration; body generally elongated, and organs arranged longitudinally; without head, eyes, or feet.

3. ACALEPHÆ........... Medusæ, or Sea-nettles, without organs for circulation or respiration; with only one entrance to the stomach.

4. POLYPI............. { Small Gelatinous Animals, with only one entrance to the stomach, surrounded with tentacula; generally adhering together and forming compound animals.

5. INFUSORIA.... ... { Animalcules, or Minute Microscopic Animals, found in fluids, or vegetable infusions. As their internal structure is but little known from their extreme smallness, this class will probably be found hereafter to contain animals which ought to be placed in some of the higher divisions.

SECT. XI.—GENERAL REVIEW OF LIVING BEINGS CONTINUED.

Subordination of Characters—Imaginary Chain of Beings—Circular Hypotheses.

THE preceding Tables exhibit the primary and secondary divisions of Living Beings. In the course of this work, we shall describe the organs and functions from which their characters are derived.   At present let us consider somewhat further the laws of their subordination and co-existence.

Upon investigating the internal structure of the entire Animal Kingdom, certain beings are discovered, consisting of a stomach isolated from every other organ, without visible nerves or muscles, without a heart or vessels, and destitute of a brain and organs of sense.   We are, therefore, led to consider the stomach as the most essential character.   The most variable organs must be regarded as of the least importance; and we thence conclude that the nerves, muscles, heart, lungs, and brain, are subordinate characters.

But on investigating the more complex animals provided with all the organs just enumerated, and upon studying the gradual progress of their development, it is found that the heart is the first formed of the organs, or at least it is the first visible organ, and that one in which the vital action is most evident. Upon examining the structure of monstrous beings, we observe that the heart can exist without the other organs much oftener and more perfectly than they can exist without the heart. Again, when we observe an animal already brought to light, and increased in magnitude, we see the organs of sense, the brain, and the greater number of muscles, suspend their functions in a periodical sleep; we see the lungs themselves sometimes cease to act for a short space of time; while the heart continues to beat as long as life exists. For all these

reasons, the heart appears to be the most important organ among the higher animals.

It must be admitted, however, that many difficulties prevent us from determining precisely which of the five organs, essential to the life of a vertebrated animal, is the most important, when we see the animal healthy, full grown, perfectly formed, each organ exactly performing all its functions, and the entire being in the full exercise of all its powers.   It has been already explained that the whole of the organs presuppose a stomach which nourishes them. The lungs and gills cannot exist without the brain; the brain in its turn requires the action of the heart; and the heart itself cannot perform its functions without the aid of the spinal marrow and of the lungs, which are ruled by the brain. All the organs form a mutually-connecting bond of union. It is true that if we examine in detail any one of the subordinate organs, it appears to have more need of the blood than of the nerves, and that it can exist longer without the action of the brain than of the heart. But if we contemplate any part of the complicated wheelwork essential to Life, it is found to be reciprocally connected, and this even in the most varied and intricate manner. Yet, when we see the heart commencing to beat before the stomach and lungs are in action,—when it is observed to throb during the absence of respiration, or after it has altogether ceased,—when we see that mutilations of the brain do not always produce instantaneous death, while the destruction of the spinal marrow speedily causes the heart to cease its movements,—we have sufficient grounds for supposing that the circulation of the blood is the primary essential condition of existence among any of the higher animals. For these reasons, in arranging the numerous subjects of the Animal Kingdom, the spinal marrow, which appears to govern the action of the heart, must be considered as the primary organ of the body; and as this delicate system of nervous matter requires for its protection a bony column of vertebræ, it is necessary to assume the existence or absence of a vertebrated column as the foundation of our primary divisions of the Animal Kingdom.

We shall defer the greater part of our observations upon systems in general, until we come to treat of the history of Zoology. At present our remarks will be confined to the theory of one universal chain of existence, which may not improperly be termed the theory of the straight line; and we believe it to have as little foundation in Nature as the corresponding theories of the circle.

On contemplating the long chain of organized beings, we observe them to become complicated by degrees, without sudden breaks and transitions. The lowest have nothing but a simple root;—the highest possess an exceedingly complicated brain. In passing from one extreme to the other, we first find imperfect plants, or we should rather say, plants of very simple structure; some of which are composed of an umbrella-shaped covering attached to a root,—that essential organ of every plant, excepting perhaps the Krubut; others apparently consist but of simple leaves; and some have only pediculated flowers without leaves. On the other hand, we find plants composed at once of a root, leaves, stem, and flowers; while the flowers either simply present only an ovarium, stamens, and pistils—organs essential to the production of seed—or, besides these indispensable organs, they also exhibit petals and a calyx, more or less complicated.

In the Animal Kingdom, the successive gradations in the complication of structure are much more numerous. To the stomach, which we have already mentioned as composing the most simple of animals, we see added, in succession, various appendages, moveable tentacula, and afterwards some rudimentary appearances of vessels filled with white blood. Continuing our observations further, we begin to perceive some scattered nervous filaments, and then some colourless muscular fibres. Soon after, we find that the digestive canal becomes more complicated; instead of one orifice, we now find two; at length we arrive at an elongated and convoluted intestine. In yet higher gradations than these, we perceive lungs, tracheæ, and gills, with complicated muscles, destined to move particular members, connected by joints of an elaborate mechanism. Further upwards, we perceive hearts of a simple construction, evident organs of sensation, distinct arrangements already of a complicated character, for continuing the species, ganglions or knots of nerves, with a marrow dilated at one extremity. Finally, we arrive at a vertebrated column, perfect senses, a spinal marrow enclosed in a bony tube, a skull, and, to crown the whole, a beautifully-organized brain.

However perfect this chain of animated existence may at first sight appear, we must admit that many objections may be made to its details. The transitions are often harsh, and by no means always very obvious, from one link to another; and if it be true that, whatever fractures a link,

"Tenth or ten-thousandth breaks the chain alike,"

we fear that the advocates for one uninterrupted chain of existence, from the minutest conferva or lichen, to the throne of the Eternal, will find many chasms which cannot be united even in the most fertile imagination.

In attempting to trace this supposed chain between the lowest animals and the Vertebrata, we find that the progressive development of the organs of nutrition and of sensation greatly disturb its uniformity. The organs of sensation and motion have already arrived at a great degree of perfection in animals, in which we can find no heart, no evident circulation, or observable respiration. On the contrary, in other beings, an opposite result may be observed; and while some have a heart already manifest, with vessels and complicated respiratory organs, we find that the organs of sensation have but a very slight degree of development.

If we be desirous of forming in the imagination one of these universal chains of existence, we have but to assign to the mountain rocks, or to the filaments of the asbestos, the faculty of absorbing nourishment and of growing, and we have formed the idea of a being resembling a plant, which possesses two orders of functions,—the one essential to the preservation of the individual, and the other necessary for the continuation of the species. To these two subordinate, yet well-defined functions, let us add the powers of voluntary motion and sensation,—let us add a central cavity for digesting the aliment; and we thus produce an animal of the lowest possible degree. To this moving, sensitive, and instinctive mass, let us join numerous nerves traversing every part, senses of a complex form circumscribed in special organs, a central brain, the instrument of perception and volition; let us add to these, muscles for obeying the determinations of the Will, with a skeleton for affording a support to the muscles, and firmness to the whole fabric, and an animal is constructed of the highest order and of the most complicated form. On the summit of this series of superior beings, let us place Man—a being remarkable for the vertical situation of his body, the volume of his brain, the perfect adjustment of his senses; for his Prudence, Curiosity, and Wisdom; for the energy of his Will, the lights of his Reason, and the sublimity of his Genius.

Many philosophers, but especially Donati and Charles Bonnet, have ingeniously attempted to arrange all the bodies in nature, in a manner similar to what we have here attempted to explain, according to the progressive analogies which they offer to the observer. They have endeavoured to pass by insensible gradations from one natural production to another, just as in the rainbow or solar spectrum we arrive, by unperceived transitions, from colour to colour, from the violet ray to the blue, from the blue to the indigo, from this to the green, yellow, orange, and red, and finally, from the red, by a new circle, round again to the violet. The philosophers whom we have just named thought that every thing in nature formed one long chain, without break or interruption, and Bonnet illustrated his views in the following manner:—

He thought that the tales, the slates, the schists, but especially the amyanthus, formed a natural and easy transition from the Mineral to the Vegetable Kingdom. Again, the Sensitive Plant, as well as many species of Algæ and Fuci, formed a natural link between plants and the most simple kinds of animals. After that, a thousand different shades and nice transitions presented themselves in the Animal Kingdom. If certain species of simple Polypi form the connecting link between the two great Kingdoms of organized Nature, they serve, at the same time, to unite the Infusoria, those microscopic inhabitants of fluids, with the Acalephæ, Sea-nettles, or Medusæ. Again, these last-mentioned animals conduct us gradually to the Worms and Mollusca, on the one hand, and, for different reasons, to the Insects, the Arachnides or Spiders, and the Crustacea, on the other. Proceeding further, we are led from the aquatic Worms to the Mollusca, by means of the Hirudineæ or Leeches, and from the Mollusca to the Reptiles, by the Limax or Snail. The Reptiles, in their turn, form the bond of connexion with the Fishes, by means of the tadpoles, the young of the Frog, in the same manner as the Insects, by another circle, merge successively into the Worms, Mollusca, and Reptiles, by their Larvæ and Caterpillars. Water-serpents are not very different from Eels. The Fishes are related to the Birds by means of the Flying Fish, the Trigla or Gurnard, and the Exocetus; and, finally, the Birds are linked to the Mammalia by the Ornithorynchus in one sense, and by the Bats and Flying Squirrels in another.

Many analogies of a similar kind are traced by the ingenious Bonnet. Thus, the Palmipedes, or Web-Footed Birds, are said to lead us, by a gentle transition, to the Fishes, just as the Penguins and Ostriches merge gradually into the Mammalia. We are conducted from the Mammalia to the Fishes by the Otters and Whales, to the Reptiles by the Seals, and to the Birds by the Bats and the spiny Echidnæ. The transition is not abrupt, according to Bonnet, from the Monkey to Man, and Man himself is formed after the image of his Creator. He adds, with his usual elegance, "Un seul être est placé hors de la chaine, et c'est celui qui l'a créée." (One being alone is placed without the chain, and that is—The Creator.)

We should not have dwelt thus at length upon these analogies, many of which are altogether imaginary, were we not fully persuaded that even these imperfect comparisons are useful in giving a general idea of living beings to persons ignorant altogether of Zoology, and consequently are appropriate for these Introductory pages.

But if Naturalists have failed in attempting to resolve the intricate dispositions of Nature by the straight line, they are equally at fault in proposing Circular theories. In vain do they attempt the solution of problems, which even the highest geometry cannot resolve, by the simple theories of the straight line, and circle. Seduced by an excessive love of simplicity, they depart from those physiological views which should form the basis of a sound system of classification.

Mr W. S. Macleay was the first proposer of the circular system. He thought that the several kingdoms of Nature, as well as their various subdivisions, returned into themselves, and may therefore be represented by circles. He considered the number Five as the basis of this system. Each circle formed precisely five groups; each of these composed other five, and so on, until we arrived at the extreme limit of the system. The proximate circles were thought to be connected by the intervention of lesser groups, to which the term *osculant* was assigned; and relations of analogy were pointed out between certain corresponding points in the circumferences of contiguous circles. We must admit that this theory has been applied with some degree of success to two of the branches of Natural History—Ornithology and Entomology; and the reason of this evidently arises from the great number of objects included in these branches, which gives an unusual facility to the circular theorist.

The objects of Natural History are infinite in number; that is to say, their number is so vast that no individual, however industrious, can possibly, within the usual period of a lifetime, comprehend their various phenomena and relations. Again, these phenomena themselves are innumerable; the connexion of their properties is absolutely overwhelming, by their intricacy and the closeness of their approximation. If, then, we are willing to form a Circular theory, the basis of which is intended to be three, five, seven, or any other number, arbitrarily assumed, we have but to take some one leading group; and, casting about for some other leading group which can join on to this first one, and a third on to the second, we must necessarily fall in with some other leading group which will join on to the first, and thus a circle will be formed. We have said necessarily, because, according to the common theory of probabilities, the number of objects being infinite, and the number of groups, and the relations of groups, also infinite, we must necessarily, without the aid of any very fertile imagination, fall in with some leading property which will conduct us to the spot whence we started.

This capability of arrangement in circles is not exclusively a law of Nature, as the advocates of the circular theories would lead us to suppose. Works of Art may be arranged in a similar manner. The merchant may arrange his goods, or the librarian his books, in circles, according to the most approved principles. Commencing with folios bound in morocco, and passing through all the gradations of binding, size, and colour, he might be easily conducted, by these and other relations, to the unbound folio, stitched in red cloth, which would lead him, by a nice transition, back again to his original starting place; and if any difficulty attached to this arrangement, it might easily be remedied by the invention of groups normal or aberrant.

We are apt to imagine, on falling in by accident with any of the recent works proposing Circular theories, that we have mistaken treatises on Geometry or Mechanics for volumes of Natural History. Considering internal organization and laws of co-existence, as subjects irrelevant to Natural History, they substitute (what a distinguished circular theorist of the present day rightly terms) the "wheels within wheels" of a fertile imagination. They may not be unaptly compared to the Ptolemaic system of Astronomy; and like it, could only be tolerated in the infancy of science.

"With cycle on epicycle,—orb on orb,"

they almost call from us the just, though somewhat startling, observation of Alphonso X. king of Castile.

The combinations of properties among natural objects are so numerous, that many beings must necessarily have the same parts, and there must always be a great number presenting very slight differences. On comparing those resembling each other, it is easy to form series, which will appear to descend gradually from a primitive type. These considerations have accordingly given rise to the formation of a Scale of Being, and to Circular Theories; the object of the former being to exhibit the whole in one series, commencing with the most complicated, and ending with the most simple organization,—and that of the latter to form two series, which, like two semi-circles, described with the same radius, shall exactly fit and correspond at their extremities. In each, the mind is led from one link to another by insensible shades, almost without perceiving any interval.

On considering each organ separately, and following it through all the species of one class, we observe that its progression, within certain limits, is preserved with a singular regularity. The organ, or some vestige of it, is to be found even in species where it is no longer of any apparent use, except to prove that Nature strictly adheres to the law of doing nothing by sudden transitions. Yet, the organs do not all follow the same order of gradation. One part is found absolutely perfect in a certain animal, while another part is in its most simple form. Again, on examining a different animal, the relative complication of the two organs is absolutely reversed. If, therefore, we were to class different species according to each organ taken separately, we should be under the necessity of forming as many series as we should have regulating organs. Thus, to make a general scale of complication, it would be essential to calculate the precise degree of complication resulting from each combination, which is far from being practicable.

As long as the great central springs remain the same, and while we confine ourselves to the same combinations of the principal organs, these gentle shades of an insensible gradation are found to prevail. All the animals of each of the primary divisions seem formed on a common plan, which serves as the basis of all their minute external modifications. But the moment that we direct our attention from one principal group to another, wherein different leading combinations take place, the scene directly changes. There is no longer any resemblance, and an interval, or marked transition, is obvious to every one. Thus, it is impossible to find in the whole Animal Kingdom any two beings which sufficiently resemble each other to serve as a link between the Vertebrated and Invertebrated animals.

The Creator never outsteps the bounds which he has prescribed to himself in the laws of the conditions of existence. Ever adhering to the small number of combinations that are possible, Nature seems to delight in varying the arrangement and structure of the accessory parts. There appears in them no necessity for a particular form or arrangement, while it frequently happens, that particular forms and dispositions are created without any apparent views of utility. It seems only sufficient for their existence that they should be possible, that is to say, that they do not disturb or destroy the harmony of the whole. These varieties augment in number, in proportion as we turn our attention from the leading and essential organs to those which are less important; and when we finally arrive at the external surface of the body, where the laws of external Nature require that the least essential organs, and those least liable to injury, should be placed, we find the number of varieties absolutely infinite. The labours of naturalists have not yet succeeded in tracing all their differences, and newly-discovered species are continually rising, as it were, into existence. Yet not even is a bone varied in its surfaces, in its curvatures, or in its eminences, without subjecting the other bones to corresponding variations.

THE FIRST GREAT DIVISION OF THE ANIMAL KINGDOM.

ANIMALIA VERTEBRATA—VERTEBRATED ANIMALS.

HAVING A BONY SKELETON, CONSISTING OF A CRANIUM, SPINAL COLUMN, AND GENERALLY ALSO OF LIMBS; THE MUSCLES ATTACHED TO THE SKELETON, DISTINCT ORGANS OF SIGHT, HEARING, SMELL, AND TASTE, IN THE CAVITIES OF THE FACE; NEVER MORE THAN FOUR LIMBS; SEXES SEPARATE; BLOOD ALWAYS RED.

THE bodies and limbs of the Vertebrated Animals are sustained by a solid framework or skeleton, composed of separate pieces joined together, and moveable upon each other. This enables them to execute their movements with vigour and precision; while the solid support afforded by the bones permits them to attain a considerable size. It is, consequently, in this division that the largest animals are found.

There are three important purposes answered by the bones of Vertebrated Animals: a solid framework is supplied to the softer parts; bases are furnished upon which the muscles are fixed; and a protection from external injury is afforded to the vital parts, and to the central masses of the nervous system.

A limit to the size of animals is fixed by the force of gravity at the earth's surface. Were animals to exceed this limit, they would fall to pieces by the weight of their limbs; and a certain degree of cohesion, constituting rigidity, is inconsistent with animal motion. This forms a complete refutation of the idle tales of crakens, giants, and other monsters.

The nervous system of the Vertebrata is more concentrated, and its central masses are of greater size than in other animals. Their sensations also are more vivid and prolonged, and they possess superior intelligence and capacity for improvement.

The bodies of the Vertebrata are always composed of a head, a trunk, and [generally] also of limbs.

The head is formed of the cranium or skull, and of the face. The former contains the brain, and in the latter are placed the receptacles for the organs of sense.

The human skull is composed of eight bones, one frontal, two parietal, one occipital, two temporal, one sphenoid, and one æthmoid. The frontal bone (or os frontis), forms the entire covering of the forehead, with the upper orbit of the socket for the eye, and extends towards the temples. In the infant, this bone was originally two, which have gradually coalesced into one. Two parietal bones (ossa parietalia) form the sides and upper part of the skull. The occiput, or hinder bone of the head (os occipitis), forms the base of the skull. There are two temporal bones (ossa temporalia), one on each side; and one sphenoid bone (sphenoides.) The principal part of the sphenoid bone, and the whole of the æthmoid (æthmoides), are placed in the internal part of the skull; they support the brain, and allow the nerves to pass through their irregular grooves and cavities.

In the skull of the fœtus, during the earlier months of its existence, cartilage is substituted for bone. By degrees portions of the cartilage are absorbed, and earthy particles are as gradually deposited in their places from certain points, which are thence called centres of ossification.

The form, which was rudely defined at first, advances by gradual steps toward perfection, and by definite and fixed laws. A certain system of ossifying centres belongs to each separate bone. The fibres proceeding from centres of the same bone, have a natural affinity for each other, and generally refuse to unite with the fibres of other bones. This, however, sometimes takes place, as in the Os frontis, when the two bones are said to form an anchylosis.

As these bones of the skull are destined to protect the brain, they are formed in a manner peculiarly adapted for that purpose. Each bone is composed of two plates; the internal layer (or tabula vitrea) is dense, hard, and well adapted to resist a violent blow; the external layer is fibrous and tough, and fitted to check the vibration of the internal part. It has been rightly compared to a soldier's helmet, lined with leather and ornamented with hair.

The edges of the bones, where they meet together, are beautifully dove-tailed, forming sutures, which give the whole skull unity and strength. These irregular lines of junction are formed by the fibres of opposite centres, which continue to secrete the bony particles, and to insinuate themselves until finally stopped by their mutually increased resistance.

Fourteen bones compose the human face. These are usually grouped together under the names of upper and lower jaws. They consist of two superior maxillary, two malar, two nasal, two palate, two lachrymal, two turbinated, one vomer, and one inferior maxillary bone. Of these, the most remarkable is the lachrymal, which consists of a delicate bony scale, as thin as paper, and containing the passage which conveys the tears from the eye into the nasal cavity.

The relative size and arrangement of the bones of the face and skull vary considerably in different tribes of animals.

The head is always placed at the anterior extremity of the vertebral column, and opposite to the tail. It is divided into three parts, which may vary in their relative proportions, but are never wanting. These divisions are, first, the Cranium, containing the brain, in the partitions of which are situate the cavities of the internal ear, and frequently a part of those of the nose; secondly, the Face, terminated below by the upper jaw, and containing the orbits of the eyes, with the nasal cavities; and thirdly, the lower jaw.

The trunk is sustained by the spine of the back and by the ribs.

The spine is composed of vertebræ. Each vertebra is, in general, moveable on the adjoining ones. The first carries the head; and through each vertebra there is an annular or ring-like part, all of which collectively form a canal. In this canal is placed that portion of the medullary substance called the spinal marrow, from which the nerves proceed.

In the spinal column of adult animals, great strength is combined with a considerable degree of general flexure. The arrangement of the solid matter of which it is composed, is admirably calculated to give lightness to the whole fabric. It is neither collected into one solid mass, nor generally diffused as in sponge. But the column is hollow; and, by the laws of Mechanics, it has been clearly ascertained that while the height and the quantity of matter of two pillars remain the same, that column is the stronger which is, to a certain degree, of a hollow construction.

The cartilages between the several vertebræ yield considerably to the pressure of the body, after remaining for a long time in the vertical position, and expand after repose. Hence arises the very striking phenomenon, that a man is considerably taller in the morning after a night's rest, than in the evening after the fatigues of the day. The long absence of pressure upon the cartilages of the vertebræ causes them to expand. This difference of height has been observed only in the human species; they are the only creatures who walk erect, and throw the pressure of the upper parts of the body upon the back bone. Thus we find no difference in the height of horses, even after the longest day's journey.

Among young people this difference is more observable than in those who are more aged. Persons of very laborious habits sink rather less than those of sedentary habits; and when the height is once lost, it cannot be restored for that day, not even by the use of the cold bath; and it can be alone regained by a night's repose in a horizontal position. [*]

* Philosophical Transactions, No. 383, p. 87;

The Reverend Mr. Wasso found that several persons, enlisted as soldiers in the morning, had been discharged for want of height, on their being measured again before the officers in the evening. On this occasion, he measured several other persons at different times, and found that the variation, in many cases, was not less than an inch. This gentleman observed, from his own personal experience, that on fixing a bar of iron, which he just reached with his head on getting out of bed in the morning, he would nearly want half an inch of his height in an bour or less, if he employed the time in rolling his garden walks, or any other exercise of the laborious kind. He also observed that the height was very suddenly diminished by riding; and, what appeared remarkable, he found that on sitting almost motionless in his study for several hours he often lost a whole inch of his height.

It appears evident that this change is occasioned merely by the back bone, from the circumstance that it arises in persons when they sit as well as when they stand.

The structure of this part of the body is beautifully adapted to the several purposes for which it was intended in the several tribes of animals, and in none more so than in Man. By the thickness and shortness of the several vertebræ, and by the intervening cartilages, with the bony projections, it is adapted for motions peculiar to itself. Had the several vertebræ been of any considerable length, the articulations must have inclined at a large angle upon their innermost edges, and the spinal marrow within the bones would have been continually liable to injury. Again, if the cartilages had been entirely wanting, it would have been as useless as if it were but one bone, and then the body would have been rendered incapable of bending, and would have remained for ever in an erect posture. The remarkable differences among the cartilages of the three kinds of vertebræ are other singular instances of exquisite arrangement.

The vertebræ of the back require but little motion; and the cartilages in that part of the column are small and thin when compared with those of the loins. In this latter part, the motion is much greater, and being placed lowest, it also supports a greater weight. It therefore follows, that during the period of the day in which we are actively engaged in the several duties of life, until we dispose ourselves for rest, the cartilages of the spine will become more close and compact from the pressure they sustain; and consequently, the whole spine, which alone supports the body, will become shorter. On the contrary, when this superior weight is entirely removed, by placing the body in the horizontal position, as it always is when we are in bed, the compressed cartilages will begin to enlarge themselves, until they gradually recover their expanded state.

As the cartilages between the several vertebræ are twenty-four in number, and as every one of these is pressed in our daily employments, the aggregate of their several expansions cannot be supposed less than an inch. This pressure is occasioned by the weight of the body alone upon the spinal column; it must therefore be much greater in persons constantly employed in carrying burdens. That the compression and expansion of the cartilages in older people is less than in younger, is a necessary consequence of the cartilages growing harder in the course of time, and becoming less capable of compression. It also follows, that old persons must lose some part of their former height, from the cartilages shrinking into a smaller compass as they grow bony; and this shortening, or "growing downwards," is not imaginary, as persons commonly suppose, but a real phenomenon.

In general, the spine is prolonged into a tail, extending beyond the lower extremities.

Man possesses a kind of rudimental tail, in the os *coccygis*, terminating the spinal column. Among Mammalia, the Ternate Bats alone are destitute of this rudimental tail; it is wanting altogether in the Batrachia or Frog tribe.

The ribs may be compared to semi-circular hoops, which protect the sides of the cavity of the trunk. They are articulated at one extremity to the vertebræ, and generally the other end is fixed to the sternum or breast bone; but frequently they do not enclose the entire trunk, and there are some species in which they are scarcely visible.

The vertebræ are never entirely wanting, although their number is exceedingly variable. Those which sustain the ribs are called *dorsal;* those between the dorsal and the head are termed *cervical;* those below the dorsal, *lumbar;* those connected with the pelvis, or hinder extremity, *sacral* or *pelvic;* and those forming the tail, *coccygeal* or *caudal.* The ribs are wanting in Frogs, also in Rays, Sharks, and a great number of cartilaginous Fishes. It is obvious that in the animals without ribs, the distinction of the three first kinds of vertebræ cannot take place, and that the distinction of the three last disappears in those having no hinder limbs, or where their limbs are not attached to the spine.

The ribs which proceed from the vertebræ, and join the sternum or breast bone, are called true ribs; those which do not extend so far are termed false ribs. In Quadrupeds these false ribs are always behind; in Birds they are both before and behind. This distinction ceases to prevail in animals that have no sternum. The sternum is absent in Serpents and Fishes, unless we give that name to denote the anterior part of the bony girdle, which supports the pectoral fins, or anterior extremities of the Fishes. Several Fishes have no very apparent neck.

The Vertebrata never have more than two pair of limbs; but these are sometimes altogether wanting, or only one pair of them is deficient. The forms of their limbs vary according to the movements which they are destined to execute. The fore or anterior limbs may be so organized as to perform the office of hands, of feet, of wings, or of fins; the hinder may be either feet or fins.

The limbs are totally wanting in Serpents and in some Fishes. The fore limbs are absent in one species of Lizard; while the hinder limbs are not found in the Apodes order of Fishes, that is to say, in those having no ventral fins; and also in the Cetaceous tribes of Mammalia. No Vertebrated animal ever has more than four limbs, unless we include in the number the kind of wing which belongs to the Flying Dragon (*Draco volans, Linn.*), a little animal next to our Lizard.

The Limbs, when perfect, are divided into four parts. The fore limbs comprise the Shoulder, the *Arm*, the Fore *Arm*, and the Hand; the hinder limbs contain the Hip, Thigh, Leg, and Foot. These distinctions do not hold among those Fishes having their limbs consisting only of bony rays; that is to say, of bones constituted like a fan, and articulated to parts corresponding with the shoulder or hip: yet even some analogy may be found between these parts and the divisions in the limbs of the higher animals.

The shoulder consists of the Scapula, placed against the back, and the Clavicle attached to the Sternum. The last is wanting in some Quadrupeds, as well as in the Cetacea, as will be explained hereafter; but it is double in Birds, Tortoises, Frogs, and many Lizards. The Scapula is never wanting when the fore limb exists. The Arm is formed of one bone only; and the fore arm is almost always formed of two. Even when the fore arm has but one bone, there generally appears a furrow, or other vestige, of its ordinary construction. The Hand varies with respect to the number of its bones; but those which exist in it always form the Wrist or Carpus, the body of the hand or Metacarpus, and the Fingers. This organization prevails even in Birds, which have their fingers enveloped in a skin covered with feathers. It likewise prevails in the Cetacea, in which the whole of the fore limb is reduced to the figure of an oar or fin.

The parts of the skeleton are usually disposed with a strict regard to symmetry; so that the halves of the body formed by a longitudinal section are exact counterparts of each other. In one family of Fishes only, called Pleuronectes, including Soles, Plaice, Turbots, and other Flat Fishes, the head is so formed that the two eyes and two nostrils are on the same side, but the symmetry is preserved in the remainder of the skeleton.

Each class and each order of animals have particular characters relative to their skeleton; consisting in the general form of their trunks and limbs, and in the number of their different parts. These particulars will be explained in the future pages of the Animal Kingdom. We may, however, remark here, that though an animal of one class may have some resemblance to those of another, in the form of its parts, and the use it makes of them, that resemblance is external, and affects the skeleton only in its proportions, but neither in the number nor in the arrangement of the bones. The Bats, for example, appear to have wings; but an attentive examination demonstrates that they are real hands, the fingers of which are merely somewhat lengthened. In the same manner, though the Dolphins appear to have fins all of one piece, we find under the skin all the bones that compose the fore limbs of the other Mammalia, only shortened and rendered almost immoveable. The wings of the Penguin, which likewise resemble fins in one piece, contain internally the same bones as those of other Birds.

The skeletons of the Vertebrated animals determine their most important forms and proportions. Retaining a general resemblance throughout all the classes of this division, they do not differ so much as their external figures, while they preserve a remarkable uniformity, which would not be always anticipated from the aspect of the parts they sustain. This property cannot be observed among the Invertebrated animals, because their hard parts are placed externally, and must therefore have the same forms as the animals themselves.

In general the bones are joined or articulated together, thus forming one connected frame; but some exceptions to this law are to be found. The bones supporting the tongue in Mammalia and Birds are not connected with the other bones except by soft parts, though in Fishes they are articulated to the rest of the skeleton. Again, the entire of the fore extremities in quadrupeds destitute of clavicles are attached to the remainder of the skeleton by muscles only; but in the others they are united to the sternum by single clavicles, or collar bones. Among Birds this union is effected by double clavicles. In the Fishes, a bony girdle connects the bones of the fore fins strongly with the spine. The skeletons of the hinder fins of the Fishes, on the contrary, are usually isolated, and fixed only in the muscles; while, in the other three classes, the hinder extremities are always attached strongly to the rest of the skeleton by means of the pelvis, or are wanting altogether.

It has already been explained that the bones of the Vertebrata are composed principally of phosphate of lime and of animal matters, such as gelatine, cartilage, and marrow. The quantity of the calcareous phosphate increases in the bones with age; the gelatinous substance, on the contrary, appears most abundant in proportion as it is examined near to the period of birth. The bones of the fœtus, in the earlier period of existence, consist merely of cartilage or indurated jelly; for cartilages resolve almost entirely into jelly when subjected to the action of boiling water. In the very young embryo there is no real cartilage; in its place we observe a substance which has all the appearance, and even the semi-fluidity, of ordinary jelly. It has already assumed a determinate shape, and is covered by a membrane, which afterwards forms the external covering of the bones, or periosteum. The flat bones, again, have the appearance of simple membranes during the first stages of ossification. Those which are intended to move on each other exhibit visible articulations, although the periosteum passes from one to the other, and envelopes the whole in one common sheath; while those connected only by sutures, such as the bones of the skull, form a continued whole, in which nothing indicates that these sutures will one day exist.

The phosphate of lime, which gives consistency and strength to the bones, is deposited in this gelatinous basis; this deposition does not proceed irregularly, but by laws, fixed and determinate for each bone. During ossification, we first observe fibres developing themselves separately; and these are shortly succeeded by new and smaller fibres, extending in every direction, and uniting the former into one uniform mass.

The surfaces of bones are generally formed of close and compact fibres, which are placed parallel to each other in the long bones, but diverge like radii, from centres in the flat bones. These fibres proceed from certain systems of points called centres of ossification. Each long bone has usually three systems of this kind; one towards its middle, where a series of ossifying points surround it like a ring, having the bony fibres extended in a direction parallel to the axis; and another principal centre at each extremity of the bone, sometimes accompanied with several subordinate points. When the three bony pieces, formed by the successive extension of these three centres of ossification, have even approached so as to be in contact with each other, they remain for a long time unconsolidated, and there appears between them a quantity of matter purely gelatinous, capable of being dissolved by boiling water. The extremities, while separated, are called *epiphyses*, while that in the body of the bone is termed *diaphysis*. In the flat bones, the centres of ossification may be compared to suns, of which the bony fibres are the rays, rendered visible through the semi-transparent cartilage by their opaque whiteness. These centres vary their appearances in different bones: in the round bones they resemble small grains or nuclei, but in the angular bones they assume a great variety of forms and positions.

When the fibres of one centre have advanced so far as to come every where in contact with those next to them, the bones are then only separated by sutures, which may afterwards be more or less promptly effaced. Some of these fibres turn aside to the right and left, and thus produce the appearance of lattice-work; while new strata, placed above and below the former, cause the texture of the bone to assume a lamellated aspect.

We are in the habit of considering as single bones, all those of which the different parts ossify and unite in youth, as the vertebræ, the occipital and frontal bones; while we consider those that do not form a union with the neighbouring bones until an advanced period of life, as distinct. Thus the frontal bone, which sometimes remains separate from the parietal bones to a very old age, is regarded as a distinct bone; while at the same time it is composed of two parts, which frequently remain separate until the age of thirty or forty.

Ossification is not found to proceed with an equal rapidity, whether we consider it in each kind of animal, or in the different bones of the same animal. In Man, and all other Mammalia, we observe that the bones of the internal ear are not only first ossified, but that they surpass all others in density, and in the quantity of calcareous phosphate they contain. Thus, the bone of the cavity of the tympanum in the Cetacea, but particularly in the Whale and the Cachalot, is superior to marble in hardness and density. Its section appears equally homogeneous, and exhibits no vestige of fibres, lattice-work, or vessels. On the contrary, other bones are very slow in acquiring the consistency they ultimately possess. The epiphyses, for example, do not ossify until long after the bodies of the bones to which they belong. Finally, there are some cartilages, which, in certain classes of animals, never admit a quantity of calcareous phosphate sufficient to render them completely bony; such as the cartilages of the ribs, and the larynx. It is certain, therefore, that there are several cartilages which are never converted into bones, although there is no bone which did not formerly exist in the state of cartilage; yet, there is a general tendency in all gelatinous parts to receive calcareous matter, as the tendons, and several white parts, ossify with greater facility than the others. The same differences which exist in this respect between the several bones of the same species, are also found to exist between species and species, on comparing the entire skeleton.

We not only find that the bones of an animal are slow in arriving at the degree of hardness which belongs to them, in proportion to the period of the growth of the animal; but we further know, that there are some animals in which ossification is never complete, and whose skeletons are always cartilaginous. This is the case with Sharks, Rays, Sturgeons, and all those Fishes which are thence called cartilaginous, or Chondropterygii. Although the bones of the other Fishes, and of many Reptiles, attain a greater degree of hardness, they still, however, preserve much more flexibility, and retain a far greater proportion of the gelatinous substance, than the bones of animals having warm blood. They grow, therefore, during the whole period of their existence; because it is cartilage only that can grow. On the contrary, when once the bones have attained their proper degree of hardness, their dimensions cannot alter; and the animal can only increase in thickness. At this period, the animal economy commences a retrograde movement, and the first steps are made towards old age and decrepitude.

Animals differ greatly in respect to the texture of the bones, and the cavities of various kinds formed within them, as well as in the rapidity of ossification, and in the proportions which the constituent parts of bone bear to each other. In Man, the internal texture of the bones is very fine. The laminæ of their spongy substance are small and close; and where this texture is most unlike lattice-work, it exhibits long and delicate fibres. In Quadrupeds, the texture of the bones is in general coarser; in the Cetacea, it is more loose, the cells are larger, and the laminæ which form them much broader. It is easy to distinguish their external fibres, which in the jaws and ribs of Whales and Cachalots may be rendered as evident, by long maceration in water, as the fibres of half-decayed wood. With respect to size, however, they seem to bear no relation to the magnitude of the animal to which they belong. The bones of Birds are of a slender, firm, and elastic nature, and seem formed of laminæ soldered, the one over the other. The bones of Reptiles and Fishes are in general more homogeneous, and the calcareous particles seem more uniformly distributed throughout the gelatinous substance. This observation appears the more striking as we approach the cartilaginous Fishes, in which the gelatinous substance completely overcomes, and appears to conceal, the phosphate of lime.

Several animals have no large medullary cavities even in their long bones. There are none in the Cetacea and Seals. Caldesi and Cuvier have long remarked the same thing in respect to the Tortoise. The Crocodile, however, has these cavities very distinct.

In some bones we find other cavities, called *sinuses*, which contain no marrow. They all communicate, more or less directly, with the exterior of the body. Man has sinuses in the frontal, sphenoid, and maxillary bones of the skull. In several Mammalia, these sinuses extend much farther backwards, and penetrate through a great part of the body of the cranium. In the Hog they proceed as far as the occiput; and it is these which swell so singularly the cranium of the Elephant, and which superficial observers are apt to mistake for an extraordinary development of the brain. In Oxen, Goats, and Sheep, these sinuses extend even into the centres of the horns. The Gazelles are the only animals with hollow horns, having the nucleus of their horns solid or spongy without any large cavity.

Other sinuses exist in the temporal bones; these communicate with the cavity of the tympanum. These are particularly extensive in Birds, and occupy as much space as the nasal sinuses do in Quadrupeds. They produce the same effect on the cranium of the Owl, as the other kinds of sinus produce on that of the Elephant.

The growth of the horns of the Deer present singular phenomena, which will be described in their proper place.

The blood of the Vertebrata is always red; and it appears to possess a peculiar composition, fitted to preserve that energy of sensation and muscular vigour observable in this division of animals. These properties, however, greatly depend upon the degree of perfection in which respiration is performed; and this circumstance gives rise to the subdivision of the Vertebrata into four classes [Mammalia, Birds, Reptiles, and Fishes.]

The external senses are always five in number; and they reside in two eyes, two ears, two nostrils, the integuments of the tongue, and those of the whole body. Certain species seem [at first sight] destitute of eyes.

But this is only apparent, as all the Vertebrata have two eyes composed of the same essential parts as those of Man. The only apparent exceptions are the *Mus typhlus*, or Blind Rat, where the eyes are concealed under the skin; and the Fish called *Cobitis anableps*, where the same eye, having two pupils, appears double.

The nerves reach the marrow through the holes of the vertebræ and of the skull; and they all appear to become incorporated with it. After interlacing its filaments, the marrow expands itself in forming the various lobes of which the brain is composed, and terminates in two arched masses called hemispheres; the volume of which is [sometimes] proportioned to the degree of intelligence possessed by the animal.

Most of the nerves of the trunk and of the limbs arise from the spinal marrow, whence they are distributed, and through their means sensation and motion are transmitted to the several parts of the body. When any portion of the spinal marrow is compressed or divided, all those parts to which nerves are transmitted, arising from the portion of the spinal column situate below the part compressed, are immediately paralyzed and deprived of sensation and motion. In some species remarkably tenacious of life, a remnant of consciousness will remain, even after the entire separation of the brain and spinal cord. If the head of a serpent be removed, and shortly afterwards the skin of its tail be punctured by a sharp instrument, the headless trunk will turn instinctively to defend the part attacked.

The internal structure of the spinal cord is well exhibited on making a section of the spine of a Carp. Its spinal cord is composed of six columns running parallel, and arranged symmetrically round a central canal filled generally with fluid. The pair of columns situate behind are appropriated for sensation, and the two in front for voluntary motion. Fibres are transmitted from each side; these unite and form nerves, yet the fibres still retain their characteristic functions. The nerves of sensation have small ganglions or knots near the central column; and the whole are formed into one system by the sympathetic nerve, which is parallel to the spine, and passes near it on each side.

There are four principal portions of the Human Brain—1, the Medulla oblongata; 2, the Cerebellum; 3, the Cerebrum; and 4, the Optic tubercles. Besides these, there are many intricate parts, which we shall not attempt to describe in this outline.

1. The Medulla Oblongata is that part of the brain which forms the immediate continuation of the spinal cord. It appears to contain the portion most essential to life and consciousness. Every other part of the brain may be successively taken away, at every mutilation the sphere of vital action may be diminished; yet, if the *Medulla oblongata* be preserved entire, nay, even if that minute portion near the origin of the fifth and eighth nerves be uninjured, the animal will still exhibit marks of instinct and sensation. Upon removing the cerebrum, cerebellum, and tubercles of a living hedge-hog, M. Magendie found that the animal was rendered blind; yet it remained acutely sensible to smell, taste, or bodily punctures; and its powers of locomotion were unimpaired. Infants are sometimes born *acephalous*, that is, without a brain, and possessed only of the *Medulla oblongata*, nerves, and spinal cord. When they survive for a short time after birth, which does not often happen, they always possess sensation and motion. Yet their brain only resembles that of the Mollusca, such as the snail or the oyster. One infant, described by Mr Lawrence, survived four days. "The brain and cranium of this infant were deficient, and the basis of the latter was covered by the common integuments, except over the *foramen magnum* (or that hole in the skull which communicates with the vertebral column), where there existed a soft tumour, about equal in size to the end of the thumb. The smooth membrane covering this was connected at its circumference to the skin. The child, as is generally the case in such instances, was perfectly formed in all its other parts, and had attained its full size. It moved briskly at first, but remained quiet afterwards, except when the tumour was pressed, which occasioned general convulsions. It breathed naturally, and was not observed to be deficient in warmth, until its powers declined. From a fear of alarming the mother, no attempt was made to see whether it would take the breast; but a little food was given it by the hand." From these instances, it appears that the brain is not essentially necessary to the performance of the ordinary instinctive functions of an animal; but that it is the *Medulla oblongata* which forms the essential organ. If the head of a tortoise or frog be cut off, sensation will remain either in the separated head, or in the headless trunk, according as the section has been made above or below this vital part. It is a curious fact, that the usual effects of an emetic are prevented by pressure on the *Medulla oblongata*, or vomiting is instantly arrested, if it has already commenced.

2. The Cerebellum, or little brain, is the tubercle or tubercles arising from the expansion of the *hinder* portions of the spinal marrow. It would therefore appear to be more especially designed for sensation. In Man, the *Cerebellum* is composed of two large hemispheres, connected together by the vermiform processes, so as to form one structure, being composed of a white nervous substance, enveloped in a uniform covering of gray matter. When this part of the brain is injured, the animal is affected with a kind of giddiness, by which it appears to be hurried forwards. In attempting

to check itself in this imaginary career, it actually effects a movement in the opposite direction. Dr Mayo describes with accuracy the phenomena attending injuries to the Cerebellum in various animals. ' The removal of the Cerebellum in fish produces no further immediate effect than that of weakening the animal; and frogs, from which this organ is removed, show an indisposition to move, unless irritated or placed in water, when the movements, though less lively than before, are not observed to be otherwise affected. In Birds and in Mammalia, more important results ensue upon the injury or removal of the Cerebellum, which, it may be remarked, does not appear to be sensible to pain from mechanical lesion. If the Cerebellum be wounded upon one side, the animal appears to be generally weakened upon the same side: if the wound be deep, the body upon the injured side is rendered paralytic. If, in a rabbit, the upper and middle portion of the Cerebellum be removed, the hind legs are observed to be spread, the fore legs are extended forwards in a state of rigidity; the whole attitude is that of preparation for moving backward, or throwing itself over. After a short time, the animal beats the ground with its fore paws, the hind legs not moving, and urges itself backwards. The flight and walk of pigeons are not affected by the removal of the upper part of the Cerebellum. After a deep section has been made, the bird totters, falls on its breast, rises again, and is in continual agitation. A deeper section still causes it to walk and fly backwards. After the entire removal of the Cerebellum, the bird, when irritated, walks as usual: when thrown into the air, it moves its wings regularly, and lights upon its feet. M. Magendie mentions the case of a young woman, who was affected with a nervous malady, that forced her to run rapidly backwards, disregarding every peril. If, in a _rabbit_ a section through the middle portion of the Cerebellum be made, in the median plane, the eyes of the animal are observed to be in extraordinary agitation, and as if starting from their sockets: the animal inclines towards one side, then is suddenly thrown towards the opposite, as if unable to balance itself with precision: its fore legs are rigidly extended forwards, as if it were in the act of receding. If a vertical section of the Cerebellum be made, leaving one-fourth of the whole adhering to the crus (or shank) of the right side, and three-fourths to the left, the animal rolls over and over incessantly, turning itself towards the injured side. The right eye is directed downwards and forwards, the left eye upwards and backwards. On making a similar section upon the left side the animal stops, and the eyes resume their natural direction."

3. The Cerebrum, or larger lobes of the brain, arise from the expanded portions of the anterior or fore columns of the spinal marrow. When certain portions of the Cerebrum are divided in various animals, they spring suddenly forwards, and continue to advance steadily in a straight line. Even when opposed by some obstacle, they continue to preserve the attitudes of one advancing.

4. The Optic Tubercles give rise to the optic nerves, and are formed by a production of the central columns of the spinal marrow. Upon injuring this part of the brain, blindness immediately follows.

The relative arrangement of these several parts of the brain differs considerably among the several classes of animals. In Fish, they are arranged nearly in a straight line, while in the Mammalia and other higher tribes, they are disposed in a more complicated manner. The relative proportions also vary. In the higher animals, the cerebral hemispheres are much larger, in proportion to the tubercles, than in fishes.

The Vertebrated animals have always two jaws. The lower jaw possesses the greatest power of motion, and may be either raised or depressed. The upper one is, in general, entirely fixed.

The Upper jaw is immoveable in Man, in Quadrupeds, and in some Reptiles, as the Tortoise and Crocodile; but it is more or less moveable in Birds, Serpents, and Fishes. The lower jaw is always moveable in the Crocodile, although the contrary has been asserted.

Both jaws are almost always armed with teeth. These are excrescences of a peculiar nature, nearly resembling bone in their chemical composition, but which grow by the deposition of matter in certain sacs. Yet one entire class of Vertebrated animals (Birds), have their jaws covered with a horny substance; and among the Reptiles, one entire genus (Tortoises) are similarly supplied.

The teeth are used by the various classes of Vertebrated animals for different purposes. In general they are for masticating the food; often for weapons of defence; in some they are employed for digging and for seeking out the food; and in others, they seem designed for no other purpose than for defending the eyes, as in the Phaeochœrus Æthiopicus, or African boar.

The formation of teeth proceeds upon a plan entirely different from that employed by Nature in the deposition of bone. It will be recollected that the bones are pervaded in every direction by vessels which nourish, renovate, and absorb their particles; but the teeth, on the contrary, are almost entirely destitute of vessels. When once deposited, they remain in a certain degree unchanged; and hence, when once destroyed, they cannot be renewed. The foundations of the teeth are laid before birth, and each tooth is formed in a small sac, by the deposition of earthy matter. They are covered by the enamel, which is the hardest animal substance in nature, and will even strike fire from steel.

The structure of the teeth in graminivorous animals is peculiar. A grinder is composed of several distinct teeth, each tooth being covered with its own enamel, and the whole united together by a kind of cement.

The intestinal canal of the Vertebrated animals proceeds from one extremity of the body to the other, undergoing various bendings, with several expansions and contractions. It possesses subsidiary organs; and receives various secretions, having a digesting power. Some of these are seated in the mouth, and called the saliva; others, formed only in the intestines, bear several names. The two most important secretions are the Juice formed by the gland called the Pancreas, and the Bile, which is produced from another very large gland called the Liver.

After the food has been digested, it passes into the alimentary canal. That portion fitted for nutrition, called the chyle, is absorbed by particular vessels called Lacteals, and transported into the veins. After the several parts of the body have been nourished by the blood, the nutritious particles remaining unabsorbed are also introduced into the veins by vessels analogous to the Lacteals, and forming with them an arrangement called the Lymphatic system.

The veins bring back to the heart the blood that has served to nourish all parts of the body, and which has just been supplied with chyle and lymph. But before it is in a proper state again to be transported by the arteries throughout the body, it is obliged to pass wholly, or in part, through the organ of respiration. In the three highest classes [Mammalia, Birds, and Reptiles] the respiration is pulmonary, or performed through Lungs, consisting of an assemblage of small cells through which the air penetrates. In the Fishes alone, and in certain Reptiles during the first periods of their existence, the respiratory organ consists of Gills, composed of a series of thin plates between which the water flows.

In all the Vertebrated Animals, the blood supplying the liver with materials for the bile, is derived from that venous blood which has partly circulated in the coats of the intestines, and partly in a particular organ named the Spleen. After being collected in a canal termed the Vena-portœ, this blood is again subdivided at the liver.

Tous ces animaux ont aussi une sécrétion particulière, qui est celle de l'urine, et qui se fait dans deux grosses glandes attachées aux côtés de l'épine du dos, et appelées reins: la liqueur que ces glandes produisent, séjourne le plus souvent dans un réservoir appelé la vessie. Les sexes sont séparés; la femelle a toujours un ou deux ovaires, d'où les œufs se détachent au moment de la conception. Le mâle les féconde par la liqueur séminale; mais le mode de cette fécondation varie beaucoup. Dans la plupart des genres des trois premières classes, elle exige une intromission de la liqueur; dans quelques reptiles, et dans la plupart des poissons, elle se fait quand les œufs sont déjà pondus.

DIVISION OF THE VERTEBRATED ANIMALS INTO FOUR CLASSES.

1. Mammalia (Man and Beasts).—2. Aves (Birds).—3. Reptilia (Reptiles).—4. Pisces (Fishes).

We have just explained the several points in which all the Vertebrated Animals resemble each other. There are, however, certain differences, which give rise to their separation into four large subdivisions or classes. These are characterized by the particular manner in which their motions are performed, or by the degree of their energy or vigour; and these again depend upon the quantity of their respiration. The muscular fibres possess a greater or less degree of irritability and general energy, according as the respiratory organs are more or less perfect.

There are two conditions which determine the quantity of Respiration. The first is, the relative quantity of blood supplied to the respiratory organ in a given time; and the second is, the relative quantity of oxygen, entering into the composition of the surrounding fluid. The quantity of blood, purified by respiration, depends upon the arrangement of the organs adapted for respiration and for circulation.

The organ of circulation [or heart] may be either double or single. The entire blood, arriving from all parts of the body through the veins, may be obliged to circulate through the respiratory organ, before being again distributed by means of the arteries; this occurs when the heart is double: or a part only of the blood returning from the body may be obliged to traverse the respiratory organ, while the remainder returns through the body, without having been submitted to the action of respiration—which is the case when the heart is single. The latter arrangement occurs in Reptiles; and the quantity of their respiration, with all the qualities depending thereon, is determined by the quantity of the blood transmitted to the lungs at each pulsation.

The circulation of the warm-blooded animals, possessed of a double heart [being Mammalia and Birds] is performed in the following manner:—The blood is propelled from the left ventricle into the aorta, and thence diffused throughout the body by means of the arteries. It passes through the minute capillary vessels at the extremities of the arteries into the veins; from which it slowly collects in the vena cava, and is deposited in the right auricle of the heart, and thence removed into the right ventricle. It is then propelled by the right ventricle through the pulmonary artery into the lungs, where it receives the action of the air; and being restored by the pulmonary veins back again to the left auricle of the heart, it passes thence into the left ventricle, prepared again to resume its circuitous course.

The force exercised by the heart upon the blood, and the velocity of the circulation, may be understood from the following observations of the learned Dr Paley:—"There is provided in the central part of the body a hollow muscle, invested with spiral fibres, running in both directions, the layers intersecting one another: in some animals, however, appearing to be circular rather than spiral. By the contraction of these fibres, the sides of the muscular cavities are necessarily squeezed together, so as to force out from them any fluid which they may at that time contain; by the relaxation of the same fibres, the cavities are in their turn dilated, and, of course, prepared to admit every fluid which may be poured into them. Into these cavities are inserted the great trunks, both of the arteries which carry out the blood, and of the veins which bring it back. This is a general account of the apparatus; and the simplest idea of its action is, that, by each contraction, a portion of blood is forced as by a syringe into the arteries; and at each dilatation, an equal portion is received from the veins. This produces at each pulse a motion and change in the mass of the blood, to the amount of what the cavity contains, which, in a full-grown human heart, I understand is about an ounce, or two table-spoons full. How quickly these changes succeed one another, and by this succession, how sufficient they are to support a stream or circulation throughout the system, may be understood by the following computation:—Each ventricle will at least contain one ounce of blood. The heart contracts four thousand times in one hour; from which it follows, that there pass through the heart, every hour, four thousand ounces, or three hundred and fifty pounds of blood. Now, the whole mass of blood is said to be about twenty five pounds; so that a quantity of blood, equal to the whole mass of blood, passes through the heart fourteen times in one hour; which is about once in every four minutes. Consider what an affair this is when we come to very large animals. The aorta of a whale is larger in the bore than the main pipe of the water-works at London bridge; and the water roaring in its passage through that pipe is inferior, in impetus and velocity, to the blood gushing from the whale's heart. Hear Dr Hunter's account of the dissection of a whale:—The aorta measured a foot in diameter. Ten or fifteen gallons of blood are thrown out of the heart at a stroke, with an immense velocity, through a tube of a foot in diameter. The whole idea fills the mind with wonder."

The circulation through a single heart may be seen in the frog. The heart is composed of only one ventricle, and of one auricle. From the ventricle the blood is propelled through two divisions of the aorta, finally terminating in one large branch, and is thence transported through the ramified extremities of the arteries throughout the body. Returning by the vena cava, it is again carried to the auricle, and thence restored to the ventricle. But during its passage a *part* only of the blood was transported to the lungs through the pulmonary arteries, and again brought back, through the pulmonary veins, after having been purified. This partial aeration of the blood imparts to the Batrachia or frogs a cold and sluggish character.

The Fishes have a double circulation, but their respiratory organ is formed for breathing through the medium of water; and their blood receives the action only of that portion of oxygen which is dissolved or mixed in the water. From this circumstance it follows that the degree of their respiration is still less than that of the Reptiles.

The gills of Fishes are situate at each side of the throat, and immediately adjoining the heart. Dr Monro is of opinion, that they present an extent of surface to the action of the water equal to that of the entire human body. The fibres resemble the teeth of an exceedingly fine comb, and they are covered with minute protuberances, resembling the pile of velvet, while innumerable blood-vessels distribute their delicate fibres over the entire surface. The distribution of these vessels on the folds and divisions of the gills forms one of the most minute and delicate arrangements in the animal economy. By means of these organs, the Fish is enabled to absorb the oxygen dissolved in the water; and after yielding this substance, the water is discharged through the branchial openings. The Fishes form a contrast with animals of the other divisions, in this respect, that they do not inspire by the same opening through which they expire.

In the Mammalia, the circulation is double, and the aerial respiration is simple, that is to say, it is performed only in the lungs. The quantity of their respiration is therefore greatly superior to that of Reptiles, in consequence of the form of the heart, or circulating organ, and also to that of the Fishes, from the nature of the surrounding element.

The quantity of respiration possessed by Birds is yet greater than that of the Mammalia, because they not only have a double circulation, with a direct aerial respiration, but they also breathe through many other cavities besides the lungs. The air penetrating into the cells distributed all over the body, acts upon the branches of the aorta, or arteries of the body, as well as upon the ramifications of the pulmonary artery.

From these circumstances are derived the four kinds of motion for which each of the four classes of Vertebrated animals seems particularly designed.

1. The Quadrupeds, in which the quantity of respiration is moderate, are generally formed for walking, for running, and for developing these motions with vigour and precision.

2. The Birds, wherein respiration is much more perfect, possess that muscular vigour and that lightness of construction necessary for flight.

3. The Reptiles, endowed with a more feeble respiration, are condemned to crawl upon the earth, and many of them pass a part of their life in a continued state of torpor.

4. The Fishes, in order to execute their less vigorous motions, require to be supported in a fluid of nearly the same specific gravity with their own bodies.

All the other organic arrangements proper to each of these four classes, and especially those which are connected with motion and with external sensation, bear a necessary relation to these essential characters.

MOTIONS OF THE VERTEBRATED ANIMALS.

Walking—Leaping—Running—Trotting—Galloping—Climbing—Flying—Darting—Paddling—Diving—Swimming.

To perform all the different kinds of progressive motion which are enjoyed by Man and the lower animals, it is necessary that a certain velocity should be communicated, in one particular direction, to the *centre of gravity* of the animal body, or that point in the body around which all the parts balance and remain at rest. A certain number of joints must exist, capable of a greater or less degree of flexure. Their relative position must be so adapted that it may be comparatively easy to extend them on the side to which the centre of gravity is made to incline, and difficult on the opposite side, so that the general movement may tend in the former direction.

The mechanical part of Animal motion may be understood from the following illustration:—If we imagine a spring divided into two branches, one of which rests upon a firm resisting base, and then suppose that the branches are compressed by some external force, their elasticity will cause them to recede as soon as the compressing force has been removed, and the two branches will be inclined at the same angle to each other as they were before the compression. But as that branch which rests upon the basis is unable to overcome its resistance, the movement takes place wholly in the opposite direction, and the centre of gravity of the spring is forced from the resisting body with more or less velocity. Accordingly, in any animal, while the muscles (*flexors*), which bend the part employed in effecting the movement, represent the external compressing force of the spring, and those muscles (*extensors*), which stretch it out, correspond to the elasticity that makes the branches of the spring fly asunder, the ground supporting the animal, or fluid in which it moves, forms the resisting basis.

In *Walking*, the centre of gravity is alternately moved by one part of the extremities and supported by the other, the body never being completely separated from the ground. It differs essentially from *Leaping*, where the entire body is projected into the air; and from *Running*, which consists of a number of short leaps.

In general, it is less painful to walk than to stand, because the same muscles are not continued in action for so long a period; and it is much easier to counteract those unsteady motions which occur in walking by contrary and alternate actions than it is to prevent them entirely. Thus it follows, that though all animals which stand erect on two legs, such as Man and Birds, can also walk on two legs, yet many moving in an upright position with sufficient ease, cannot stand on two feet for any time without very great fatigue and exertion.

When Man intends to walk on level ground, he first extends one foot. His body then rests equally upon both legs, the advanced leg making an obtuse angle with the tarsus or instep, and the other an acute angle. As the ground does not yield to the point of the foot, the heel and the remainder of the leg must necessarily be raised, otherwise the heel could not be extended. The pelvis and trunk are consequently thrown upwards, forwards, and somewhat in a lateral direction. In this manner they move round the fixed foot as a centre, with a radius consisting of the leg belonging to that foot, which, during the movement, continually diminishes the angle formed with the tarsus. The leg communicating this impulse is then thrown forward and rests its foot upon the ground; while the other, which now forms an acute angle with its foot, has the heel extended in its turn, and in like manner makes the pelvis and trunk turn round upon the former leg. The centre of gravity is thus carried forward by these movements at each progressive step, inclining, however, at the same time to the right and left alternately, so as to be supported by each leg in its turn. It will also be seen that each leg, immediately on extending its heel, bends and rises, in order to its being moved forward,—extends in order to rest its foot upon the ground,—turns upon this foot as on a fixed centre, so as to support the weight of the body,—and then extends its heel again in order to transfer this weight to the other leg.

In this manner, each leg supports the body in its turn; but it is also necessary that the extensors of the thigh and knee should be brought into action, to prevent their articulations from giving way; and this motion is followed by a corresponding action of the flexors of the same articulations. It will be observed, that the three principal articulations of each leg are situate in opposite directions to each other, that the foot should be raised by their flexion immediately over the place which it occupied during their extension. It would otherwise be impossible to bend them without throwing the foot backwards or forwards.

In consequence of the impossibility of regulating the undulatory motion, in a manner perfectly equal on both sides, a man cannot walk in a straight line with his eyes shut; nor could he even preserve a uniform direction, did he not correct these deviations by the sense of sight.

In descending a stair-case, or in walking down an inclined plane, the advanced leg is placed lower than that remaining behind; and the body would fall upon it with a fatiguing and dangerous jerk, were it not carefully checked by the extensors of the hip. By this means, the body is compelled to descend gradually; but the muscles of the loins soon become fatigued by the exertion.

On the contrary, in ascending a stair-case, or an inclined plane, it is requisite at each step, not only to transport the body horizontally, as on a level surface, but also to bear it up against its own weight, by means of the extensors belonging to the knee of the advanced leg, and to the heel of the leg remaining behind. The knee and calf of the leg are therefore fatigued in ascending. A mechanical advantage is gained by leaning the body forward in ascending, because the lever, by which its weight retards the motion of the knee, is thereby shortened in equal proportion. A fatigue, similar to that produced by the action of ascending, is occasioned by walking with a very wide step. As the legs are thereby placed considerably apart, the body sinks lower at the moment of their separation; and as it is necessary to raise the body proportionally, when turning alternately on each leg, the fatigue is consequently greater.

Man is not compelled to swing his arms greatly to assist his walking, except when confined to a very narrow path from which he cannot depart, and then he employs every means to correct the unsteady motion of the body. Apes always require the assistance of their arms in walking; and such as have these extremities longest, like the Orang Outang (*Pithecus Satyrus*), and the Long-armed Monkey or Gibbon (*Pithecus lar*), use them with the greatest advantage.

Among Quadrupeds, the action of walking is performed in the following manner:—The articulations of the hind-legs are first bent slightly, and extended in order to carry the body forward; in which movement the extensors of the knee and heel particularly contribute. The breast is thrown forward by this movement, the fore-legs incline backwards, and the animal would certainly fall, did it not instantly throw its fore-legs forward in order to support itself. The trunk is drawn upon the fore-legs, which are now fixed in this position, and the action of the hind-legs is again repeated. But it must be observed that, in the action of walking, these movements are not performed at the same instant, by the legs of each pair; for, in that case, the animal would necessarily be completely suspended for a moment over the ground. Its motion would then no longer be a *walk*, but a succession of leaps, particularly denominated a *full-gallop*. On the contrary, each step is executed by two legs only, one belonging to the fore pair, and the other to the hind pair. When the motion is performed by the legs on the same side, it is called an *amble*—when by legs on opposite sides, a *pace*.

During the amble, the body being alternately supported by two legs of the same side, is obliged to balance itself to the right and left, in order to avoid falling, and the right fore-foot moves to sustain the body, urged onwards by the right hind-foot. It is this balancing movement which renders the amble of the Horse and Ass so agreeable to invalids.

In the *pace*, the body is supported alternately by two legs placed in a diagonal manner. The right fore-leg is advanced to sustain the body, thrown forward by the extension of the left hind-foot; and at the same instant the latter bends in order to its being moved forward. While these are raised, the right hind-foot begins to extend itself, and the moment they touch the ground, the left fore-foot moves forward to support the impulse of the right foot, which again moves forward.

Quadrupeds having the fore-feet longer than the hinder, as may be observed in the Giraffe, or Camelopard, possess the chief strength of their body in the fore-legs, and accordingly the principal impulse is given by extending the fore-foot. The Sloths, and all animals which like them have the fore-legs greatly disproportioned to the others, drag themselves onwards with a laborious and tedious movement, by first extending the anterior legs, and then bending them so as to draw the body onwards; and with the Sloths, this difficult motion is further increased by the imperfect articulations and general feebleness of the hinder-legs.

The legs of the Mammiferous Quadrupeds move forwards and backwards in planes nearly parallel to the spine, and not far from the middle plane of the body upon which the weight operates. On the contrary, in the Oviparous Quadrupeds, the thighs are directed outwards, while the bendings of the limbs take place in planes perpendicular to the spine. In the latter case, the weight of the body acts with a much longer lever in opposing the extension of the knee. These animals, therefore, have the knees always bent, and the belly drags upon the ground between the legs. For this reason they have received the name of Reptiles.

The short leaps of the Hares, Rats, and particularly of the Jerboas, are occasioned by the great length of their hinder as compared to the fore-legs. Indeed, their fore-legs are so short, that had they not the precaution to make this prancing movement, these animals would be thrown down by each impulse of the hind-feet. It is only in ascending a hill, that they can be said to walk at all. Their movement on level ground is performed by a succession of short leaps; and when they attempt to walk slowly upon level ground, they are compelled to move themselves by the fore-feet, and merely to drag the hinder pair after them. We may observe the latter movement in the Rabbit, and still more distinctly in the Frog.

The Otters, Beavers, Water Tortoises or Turtles, and other quadrupeds designed for swimming, have the hinder-legs placed very far apart to facilitate the motion. They are, therefore, impelled laterally, the line of motion becomes crooked, and the trunk is urged onwards from side to side.

In Leaping, the body rises entirely from the earth, darts into the air, and remains suspended for a momentary period, depending for its duration on the force of projection. This movement is performed by the sudden extension of all the inferior articulations, after they have undergone an unusual degree of flexion. Their rapid extension gives a violent shock to the bones composing the articulations. The impulse is then communicated to the centre of gravity of the animal's body, and it is projected with a determined velocity depending on its weight. A leaping body is, therefore, a projectile which gradually loses the acquired velocity by which it ascended, its motion being continually retarded, and finally destroyed by the force of gravity exercised by the earth. We are therefore enabled to ascertain the curve described by a leaping body in the air, with the time and place of its descent, when the projectile force and the force of gravity are given, and allowance made for the resistance of the air.

All the animals which leap best have the hinder-legs and thighs much longer and thicker than the anterior—the projectile force, and consequently the extent of the leap, being regulated by the proportional length of the muscles. The surprising leaps of the Kangaroo, Jerboa, and Frog, are plainly owing to this cause.

The smaller animals leap much farther than the larger, in proportion to their size. This must follow obviously, if it be considered that when the projectile force impressed on two bodies is in proportion to their different magnitudes, their velocity will be equal, and that the extent of the space through which they pass depends entirely upon their respective velocities. The leaps of small and large animals are therefore nearly equal.

Man and Birds are the only animals capable of leaping vertically or hopping, because they alone have the trunk placed directly over the legs, and the direction of a leap depends upon the situation of the centre of gravity, in respect to the member by which the impulse is given. They are also capable of leaping forward, by impressing a greater degree of force on the rotatory motion of the thigh than on that of the leg; or they may even leap backwards, by making an opposite exertion. On the contrary, Quadrupeds can only leap forwards.

Running differs from walking, only in the body being projected forward at each step, and in the hinder-foot being raised before the anterior foot touches the ground. It consists, in fact, of a series of low leaps performed by each leg. As the acquired velocity is preserved, and augmented at each bound by the new velocity thereby added to it, running is more rapid than the quickest walking step. An animal cannot, therefore, stop itself instantaneously when running, though a stop may be made at each step in walking. In leaping forward, a previous run is advantageous, because it adds the momentum acquired during the run to that obtained from the leap itself; but a vertical leap or hop would be entirely prevented by a run, or at least considerably diminished. For this reason, a horse in full gallop, preparing to leap, retards his velocity before making the spring. In running, an animal inclines its body forwards, that the centre of gravity may be in a proper situation for receiving an impulse in that direction from the hinder-leg. It is also requisite to move the fore-leg rapidly forward to guard against falling. Were any obstacle to intervene, so as to prevent this leg from reaching the ground in time to support the body, a fall would be the consequence. It also follows, that interruptions of this kind are more dangerous in running than in walking, on account of the greater momentum of the body, and for the same reason they occur more frequently. Man never varies his manner of running, except in taking longer or shorter steps, or in giving to his body a greater or less degree of velocity; but Quadrupeds vary their mode of running, by the different order in which they raise each foot, or bring it to the ground.

The feet diagonally opposite rise simultaneously in the *trot*, and fall at once, each pair alternately, but in such a manner, that for a moment all the four feet are off the ground. The sound of the animal's steps are therefore heard two and two in succession, and a regular motion is produced.

The Dog, Hare, and many other quadrupeds, can only run in the manner particularly denominated the *full gallop*, which is the most rapid motion of the Horse. These animals raise the anterior feet at each step; the body is projected forwards by the extension of the hind-feet; the two fore-feet descend at the same time, and are followed by the two hind-feet also descending together. By this means, the steps of the horse are heard by two beats at a time, differing in this respect from the *common gallop*, where the two fore-feet are lifted unequally, and fall one after another; and from those other varieties of the gallop, where the horse's footsteps are heard by a series of three or four beats, from the hinder-feet falling to the ground either both together or one after the other.

Many animals leap by organs different from feet, but they all agree in this respect, that the movement is occasioned by the sudden extension of several articulations. Serpents leap by folding their bodies into several undulations, which are unbent at the same instant, according to the degree of velocity which they wish to impart to their bodies. Only a few genera are assisted in this motion by the scales of the belly, which they are able to elevate and depress at pleasure. Some Fishes leap to the tops of cataracts by bending their bodies strongly, and then unbending them suddenly, so that they rise with an elastic and powerful spring.

Several animals, which in reality leap, have been improperly said to fly. The Flying Lemurs, Flying Squirrels, and Flying Phalangers, have membranes between the feet, but their toes are not elongated. These membranes serve to support them for some time in the air, and enable them to take great leaps in descending; but the membrane acts merely as a parachute, as these animals cannot raise themselves in the air. In the same manner, the Flying Dragon, a small lizard found in the East Indies, supports itself for some moments during a leap, by a membrane sustained by a few bony rays, articulated to the spine of the back.

Man and various other animals possess the power of seizing objects, by surrounding and grasping them with their fingers. For this reason, it is necessary that the fingers should be separate, free, flexible, and of a certain length. Man has such fingers on his hand only; but Apes and some other kinds of animals have them both on the hands and feet; hence they are termed *quadrumana*, or four-handed.

Man surpasses all other animals in the delicate operations which his hand is capable of performing. The Apes and Lemurs alone possess with him a thumb opposable to the other fingers, and forming with them a kind of forceps. They are consequently the only animals capable of holding moveable objects in a single hand. But it is indispensable to perfect prehension that they should have the power of rotating the hand upon the fore-arm, and the bones of the shoulder must be placed so as to prevent the scapula or shoulder-blade from being thrown forwards.

The Squirrel, Opossum, Rat, and other animals, possess fingers sufficiently small and flexible to enable them to take up objects, but they can only hold them by the assistance of both paws. Dogs and Cats, which have the toes shorter, and besides are under the necessity of resting on their fore-feet, can retain their hold of substances solely by fixing them upon the ground with their paws. Those animals having the toes united and drawn together under the skin, or enveloped in horny hoofs, are incapable of exercising any prehensile power.

Climbing is greatly facilitated by a power of seizing and grasping firmly. Man is but an indifferent climber, because he can only grasp with his hands. His feet are chiefly adapted for supporting the body, and afford but an imperfect means of elevating it by the extension of the knees and heels. The arms form, therefore, the chief means of drawing the body upwards in climbing.

Monkeys, and other Quadrumana, are the best climbers. They can seize equally well with their four extremities; and the position of their hind-feet is still more favorable to this action, as the soles are turned inwards, instead of being directed outwards. Ant-eaters and Sloths have a considerable protuberance on the heel, which nearly accomplishes the same end; and with the Opossums and Phalangers climbing is assisted by a thumb almost always directed backwards, and forming a kind of heel very powerful in its operation. The animals just mentioned, as well as most of the Quadrumana, are assisted in climbing by their tail, which is capable of seizing bodies almost as powerfully as a hand. For this purpose, additional force is imparted to the common muscles of the tail.

The sharp and hooked claws found in animals of the Cat kind, enable them to climb with facility. Their nails are retained between the toes, with the points elevated by two elastic ligaments, altogether independent of the will of the animal. When they wish to use the nails, either for tearing their prey or seizing moveable objects, the nails are protruded by the muscle, which moves the last phalanx of the toe on the preceding one.

In the Sloth, the ligaments are differently disposed, and the nails being naturally inflected, must be raised when the animal wishes to use them. As the toes are of an inconvenient form, being composed of two phalanges, one of which is very short, and the other entirely covered with the nail, while the metacarpal bones are ossified together, and immoveable, the Sloths perform their movements with constraint and difficulty.

The climbing birds are enabled by their claws to fix themselves to the inequalities in the bark of trees; and they perform this action by the assistance chiefly of the hinder-toes, which are used in supporting and preventing them from falling. The greater number of these genera have two hinder-toes, but the Creepers and Nut-hatches have only one. The Woodpeckers, as well as the Creepers, are assisted in climbing by the quills of the tail, which are stiff and capable of being fixed firmly into the inequalities of surfaces.

Some birds can raise the food to their mouth by means of the one foot, while they stand upon the other. Parrots have their toes conveniently disposed for this purpose, and also the Owls. Without this provision, the latter would frequently fall whenever they attempted to peck, in consequence of the great weight of the head, and the corresponding elevation of their centre of gravity. But as most birds require both feet in order to stand firmly, they seldom use them for holding substances except during flight, when the feet are disengaged. The Cormorant and Pelican will sometimes swim with the one foot and carry some substance in the other; and the Wading Birds frequently stand for a long time on the one leg, which they are enabled to do without much difficulty, while they hold a stone, or some heavy substance, in the other as a counterpoise.

Among the Reptiles, the Chameleon seems to possess adaptations for climbing, enjoyed by no other animals except the Quadrumana. With a prehensile tail, and hands resembling forceps, he exhibits a degree of agility unusual with the Reptiles.

Flying and swimming are leaps taking place in fluids, and the motion is produced by the resistance which the fluid makes to the surface of the wings or fins, when moved by the animal with great rapidity. Leaping, however, takes place on a fixed surface, possessing the power of resistance, from its magnitude and firmness. If we suppose the ground to be either soft or elastic, leaping may still be performed; but there arises a diminution in the velocity of the leap, proportional to the resiliency of the support. It is necessary, therefore, that the moving power should be increased in proportion, to produce an equal momentum by the extent of the vibrating surfaces, and by the rapidity of their vibrations. The velocity with which the wings or fins must be used, depends on the rarity of the medium in which they move. It is less in water, greater in air near the earth's surface, and increases as the animal ascends into the higher regions of the atmosphere. Birds cannot therefore fly above a certain height, dependent on the strength of their muscles; and they are capable of rising to a greater height when the barometer is high than when it is low. The muscles moving the wings and fins, but especially the former, require a force vastly superior to that necessary to produce a simple leap upon a firm surface.

As a flying or swimming body is entirely surrounded by the medium in which it is placed, it experiences an equal resistance in front as well as from behind on striking the fluid. An animal would be incapable of advancing, if it did not possess the power of greatly diminishing the surface of its wing or fin, immediately after having struck the fluid.

Flying and swimming are sometimes performed by the same animal; but the former is executed most perfectly by Birds, and the latter by Fishes. Some birds never fly. The Ostriches, Auks, and Penguins, are possessed of small rudimental wings, but they seem to have them but for the purpose of conforming more nearly in external resemblance to other birds. Some Mammalia can fly, although they have no wings. The Bats possess a membraneous expansion, extending to the feet and to both sides of the tail, but supported chiefly by the humerus, the fore-arm, and the four fingers. These bones being greatly elongated, serve to support the membraneous surface, which is of firmness and extent sufficient to raise and maintain these animals in the air, when acted upon by the powerful muscles of the breast.

The first motion of a bird in attempting to fly is an ordinary leap with the feet. Accordingly, those birds having the wings very large and the feet very short, as we observe in the Booby and the Martens, commence their flight with great difficulty, as they cannot leap sufficiently high to obtain the space necessary for the extension of their wings.

In flying, the resistance of the air is in proportion to the mass struck at one time. On this account the short-winged birds must repeat their vibrations very frequently; they are therefore soon fatigued, and unable to continue their flight for a long time. When a bird attempts to fly, the humerus is first elevated, and then the entire wing, which had hitherto remained folded; while, at the same time, it is extended in a horizontal direction by means of the fore-arm and the last division of the wing, corresponding to the foot of Quadrupeds. After the wing has thus acquired all the superficial extent which it is capable of attaining, the bird suddenly depresses it, until it forms an acute angle with the vertical plane of the body, subtending the ground. The air resists this motion, which is performed with great rapidity, and produces a reaction of part of the force upon the body of the bird. Its centre of gravity then rises in the same manner as in other leaps. The wings may be compared to a lever, of which the pectoral muscles are the moving power, and the body of the bird the weight; while the air resisting, by its inertia, the action of the expanded wing, is the fulcrum. The impulse being once given, the bird refolds the wings by bending their joints, elevates them again, and gives a new stroke to the air. The force of gravitation diminishes the velocity which the body thus acquires in ascending, in the same manner as it affects every other projectile. There consequently occurs a moment in which the bird neither ascends nor descends. If it seize this moment precisely, and give a new stroke to the air, its body will acquire a new velocity, which will carry it as far as that obtained by the first impulse; it will then rise in a continuous manner, and with a uniform velocity. If the second impulse of the wing commence before the impulse arising from the first has been lost, the bird will ascend with an accelerated motion; or, if the bird do not vibrate its wings at the exact moment when the ascending velocity is lost, it will begin to descend with great rapidity. Yet the bird may keep itself always at the same height by a series of equal vibrations, if a point be seized in the fall so situate that the velocity which would have been acquired in descending, and the small space there would have been to reascend, reciprocally compensate and destroy each other; but if it once allow itself to descend to the point from which it departed, it can only rise by a much stronger exertion of the wings.

In descending, the bird has only to repeat less frequently the vibrations of its wings. The *darting* of Birds of Prey is occasioned by their suppressing the vibrations of the wings altogether, when the bird, being continually acted upon by the force of gravity, falls with an accelerated velocity. When a bird in descending suddenly breaks its fall, it is called a recover. The resistance of the air then increases in proportion to the square of the velocity, and the bird rises again.

The preceding remarks apply only to flight when made in a vertical direction, either in ascending or descending. The Quails, Larks, and other birds which are observed to fly upwards in a straight line, have the wings placed entirely horizontal, but in the greater number of birds the wing is inclined, and turned backwards. This inclination may be further increased at the will of the bird. It is greatly assisted by the length of the quills, which enable the resistance of the air to act on their extremities with a mechanical advantage, while they are the more elevated by it, from their fixed points being placed at the base. By this arrangement, birds are enabled to advance in a horizontal, as well as in a vertical direction, by a series of oblique curves.

The oblique motion upwards in flight may be resolved into two other motions, the one in a horizontal direction, independent of the force of gravity, and the other in a vertical direction, opposite to that power. In flying horizontally, the bird rises in an oblique direction, and does not make a second movement of the wings until it is on the point of descending below the line of the intended direction of its flight. These partial movements, therefore, will not take place in a straight line, but in a series of curves, nearly approaching to the straight line, and in which the horizontal motion greatly prevails over the vertical. In ascending obliquely, the wings move with greater rapidity; in descending obliquely, their vibrations are less frequent; and both of these motions are performed by a series of curves.

Some birds cannot sufficiently diminish the obliquity of their wings, and with them the horizontal motion is always very considerable. When the wind blows strongly in the same direction with the flight of these birds, they are carried to a very considerable distance out of their intended path. For this reason, those birds of prey which the falconers term *noble*, are under the necessity of flying against the wind, when they wish to rise perpendicularly upwards. The anterior quills of their wings being extremely long, and their extremities pressing closely upon each other, the horizontal motion with them is proportionally greater than that of other birds. On the contrary, with the *ignoble* birds the quills of the wings are separated at their extremities, and permit the air to pass between them, which renders the wing less capable of assuming the oblique position.

Deviations from the rectilineal path to the right and left, are chiefly occasioned by the unequal vibrations of the wings. When the left wing vibrates the more frequently, or with the greater force, the left side moves more rapidly, and the body necessarily turns to the right. The rapid movement or greater force of the right wing produces a corresponding turn to the left. The difficulty of suddenly turning increases with the velocity of flight; and this arises partly from the inertia of the body, which perseveres in its rectilineal course, and partly from the increased difficulty of making the one wing to surpass the other in its velocity. For this reason, birds of rapid flight make great circuits in turning. Some will turn on the side, and make use of the tail as a rudder, when they wish to change the horizontal direction.

The tail of birds, when expanded, serves to sustain the hinder part of their body. When depressed during flight, the resistance of the air forms an obstacle which raises the hinder part of the body and depresses the anterior; upon turning the tail upwards, a contrary effect is produced.

As all Birds do not fly, so all Fishes do not swim, yet there are many Birds which perform two motions, resembling those more particularly belonging to each class. Aquatic Birds are improperly said to *swim*; and the poet, describing the swan "sailing with the breeze," is perhaps not aware that his term is philosophically correct.

The bodies of Aquatic Birds are naturally lighter than water, from the great quantity of air which they contain within the abdomen, and from the feathers, which are oily and impervious to moisture. They precisely resemble a boat, and have no further occasion for the feet than as oars for moving forwards. As the fore part of the body is completely sustained by the water, the legs are situate farther backwards than those of other birds, that their effect may be more direct, as their presence farther in advance would be superfluous. The legs and thighs are short, that the resistance of the water to the muscles may be as slight as possible. The tarsus, or instep, is compressed for cutting the water, while the toes are very much expanded, or even united by a membrane, in order to form an oar of greater breadth, and capable of acting upon a greater surface of water; and when the bird inflects its foot in order to give a new stroke to the water, it closes the toes upon each other to diminish the resistance of the fluid.

In diving, these birds are obliged to compress the breast with much force, in order to expel the air which it contains. The neck is then elongated, that the body may acquire an inclination forwards, while, by striking the feet upwards, it is forced downwards. The Swan, and some other aquatic birds, spread their wings to the wind, and use them as sails to diminish the labour of paddling.

The body of the Fish is rendered of the same specific gravity as the water, by means of the air-bag or swimming-bladder. By the assistance of this organ, it can raise or depress its body in the water. When this air-bag is burst, the fish remains always on its back, and is unable to ascend. In an ordinary state, the fish is able to compress the air-bag precisely in the degree sufficient to enable its body to remain in equilibrium with the water, and to retain it in the same horizontal plane. It compresses the air-bag in a greater degree when it wishes to descend, and dilates this organ when about to ascend. This compressive force is accomplished by the lateral muscles of the body, which tend to contract the bladder by elongating it. In this manner, though the extent of its surface remains the same, the capacity can be diminished, since it is further removed from a spherical form; and it is well known that the sphere possesses the greatest solid content of all bodies of equal surface. Some Fishes are capable of having their air-bags so much dilated by heat, that when they remain for a long time on the surface of the water, exposed to the burning heat of a tropical sun, they cannot compress the bag in a degree sufficient to enable them to descend again.

When the fish is in equilibrium with the water, and wishes to advance, its tail is bent in two different directions, resembling the letter S, by means of its strong and complicated lateral muscles. It then augments the surface of its tail to the utmost extent, by means of the dorsal, anal, and caudal fins. The tail is then extended with great velocity, the resistance of the fluid serves as a solid substance, a part of this velocity is imparted to the fluid, and the body of the fish is propelled onwards by the remainder of that velocity, diminished of course by the resistance of the fluid before the fish. But this is not great, because the force with which it advances is much less than that employed to extend the tail; and also, when the tail returns to a right line, the fish presents to the fluid only the thickness of its body, which is by no means very considerable.

It is necessary that the fish should bend its tail again to give a second stroke to the water. This motion, however, is directly contrary to the direction of the power by which the tail was extended, and produces in the fluid an equal resistance in the opposite direction. This resistance would be equally powerful, and would completely counteract the progressive motion of the Fish, if the surface of the body continued the same as before. But the anal and dorsal fins are then laid down upon the body, while the caudal fin becomes folded and narrow. Again, the curvature of the tail takes place very slowly, while its extension is violent and sudden. On returning to the right line, the tail is bent a second time; but this takes place precisely in the opposite direction, and the impulse resulting therefrom has an equal obliquity only on the opposite side of that imparted by the first stroke. By this means, the course of the body is rendered straight; and, by striking the water more on the one side than on the other, the fish is enabled to move to the right or left, and to turn round horizontally.

It does not appear that the pectoral and ventral fins are of very much use in the progression of Fishes. They seem intended to aid in preserving them in a state of rest, or in equilibrium with the fluid, and they are extended whenever it becomes necessary to correct the vacillations of the body. They are used also in the slight turnings of their progressive motion, and in preventing themselves from falling on one side in swimming. Perhaps in those tribes of fishes where these fins are unusually large, they have some other uses, which a more accurate acquaintance with the habits of these fishes would enable us to describe.

Fishes without air-bags experience a much greater difficulty in changing their elevation in the water. The greater part remain always at the bottom, unless the disposition of their body enables them to strike the water from above downwards with great force. The Rays *(Raiæ)* elevate themselves with their large pectoral fins, which are very properly termed wings, as these fishes use them in raising themselves, precisely in the same manner as the bird elevates itself in the air by means of its wings. Unlike the other fishes, the Flat Fishes *(Pleuronectes)* are compelled to swim in an oblique position with the back on the one side and the belly on the other, in consequence of their eyes being placed on the same side of the head. In swimming, they accordingly strike the water from above downwards. As both the Rays and Pleuronectes are unable to strike the water conveniently to the right and left, they are compelled to make a succession of leaps in order to impart a horizontal direction to the whole of their motions. The tail is struck downwards with great force, which elevates them slightly, and this force, combined with that of gravity, brings them back to the horizontal line after describing a curve. They depart from this line by a new leap, in a manner similar to the flight of Birds.

The same means are employed by the Whale and other Cetacea. It must be observed, that the bodies of these Mammalia are organized for swimming as perfectly as those of Fishes; but they differ from them in this respect, that the efforts of the tail are made principally in a vertical direction. The use of the air-bag is supplied by the lungs, which are compressed or dilated at will by the action of the diaphragm and the muscles between the ribs. Serpents, and many of the Invertebrated Animals having long bodies and no fins, swim in the same manner as Fishes, by suddenly inflecting their bodies.

The Quadrupeds, Aquatic Birds, and Reptiles, swim by means of their feet, which propel them onwards in precisely the same manner as a boat is moved by oars.

When the oar is in a state of rest, it forms two angles with the side of the boat, and these may either correspond or be unequal. The boatman moves the oar so as to render the anterior angle more obtuse, and the hinder one more acute. If the water did not resist by its inertia, the boat could not change its place; but as its resistance obstructs the motion of the oar, the angle in question widens by the progressive motion of the boat. When once the impulse has been given, the boatman draws back his oar or turns its edge, that it may not interrupt the motion, and then repeats the same operation.

The above description of the mechanical process in rowing, is directly applicable to the animals just named, if we only consider their feet as oars, and their bodies as so many boats. The Seals, Morses, and other Amphibia, swim the most perfectly of all the Mammalia; while they resemble the Cetacea and Fishes in the form of their bodies more nearly than any other animals of the same class. In many Quadrupeds the mechanical power of the feet in paddling is greatly increased by membranes between the toes, as may be observed in the Otter and Beaver, but in general they swim simply by the action of the four limbs. Of these, the hinder serve to urge the body onwards, and the anterior to sustain the fore-part of the body, which is the heaviest.

As the weight of the head in Man is greater in proportion to the size of the body than almost any other of the Mammalia, he experiences greater difficulty in supporting his head when swimming than most of these animals, and he alone, of all Mammalia, seems incapable of swimming naturally, and without repeated trials.

ANALOGIES OF THE VERTEBRATED ANIMALS.

Identity of their Construction—Natural Scale of Animal Organization—Laws of Monstrous Development.

On stating, for the first time, to a person ignorant of Zoology, that all the Vertebrated Animals are analogous to each other, he is apt to reject the assertion as paradoxical. He may ask, what analogy can exist between a Serpent and a Mammiferous Quadruped, or between a Frog and a Bird? If we answer, that they are composed of the same elements, of the same tissues, however different they may externally appear, and that they are all possessed of the same or nearly corresponding properties, he may reply, that this is natural to all bodies possessed equally of life.

It may be necessary to explain that, when all the Vertebrated Animals are stated to be analogous, it is meant that they are composed of the same constituent materials, and that, to a certain point, each similar organ is formed of the same number of pieces. By this it is not meant that these animals are all possessed rigorously of the same number of organs, or of parts of an organ; on the contrary, many are altogether deprived of certain parts which are found well developed in other species. It is evident that the Boas, destitute of limbs, cannot resemble in this respect the Fishes provided with fins, or still less the Mammalia, possessed of four limbs. It is only asserted, that when we investigate the corresponding organs of these animals, we find them to be composed of the same materials, and of the same constituent pieces.

It is unnecessary at present to allude particularly to the internal organs of animals, such as the muscles, arteries, and viscera, as every one is aware that a certain degree of analogy prevails in their internal parts or entrails. It is only necessary to point out some of the leading organs of the animal body, to be assured that there exist undoubted points of analogy in their structure. In attempting to trace these analogous parts, it is convenient to select young animals, or rather the unborn embryos, in preference to the adult animals, because the identity and similarity of the original and essential outline become gradually disguised by those characters, which afterwards form the distinctions of genera and species.

The external coverings of the Vertebrated Animals appear at first sight destitute of every pretension to analogy. We are struck with the remarkable differences between the scales of the Fish, and the feathers of the Bird, or between the shield of the Armadillo, and the glittering and delicate skin of the Eel. Yet, if we examine the first periods of the existence of these animals, they are all equally covered with a soft and thin membrane,—a simple or naked skin; and, on tracing the subsequent changes which this envelope undergoes, we are led to perceive productions analogous to that same epidermis or cuticle, which invariably forms the external covering of each animal. Whatever changes the outward garb of the animal may undergo, whether it become hairy, scaly, feathered, or covered with a shield, these are merely subsequent modifications which the epidermis or outer skin undergoes in the course of its development.

In the skeleton, these analogies appear still more evident. We see in all the Vertebrated Animals a central column of bones, piled one over the other, called vertebræ; in all, certain projections or processes arise from these vertebræ; and among the greater number, these projections are sufficiently lengthened out in front of the body to form ribs, and to constitute the walls of a cavity destined to contain the heart and the lungs. All these peculiarities appear greatly diversified in the different species of full-grown and perfect animals: but, if we look back to their origin, and investigate at the moment that ossification begins, we find that the first constituent elements are identical—that the points of ossification are the same in all. This appears still more evident in the structure of the heads of animals. It seems difficult to conceive that the heads of the Crocodile and of the Sparrow are composed of precisely the same number of pieces as those of Man, or any other of the Mammalia. If we were only to compare the skulls of these several animals, in their permanent adult states, we should be compelled to admit that their differences were more numerous than their analogies, and they would then appear to differ as much in their details as in their general form when taken as a whole. But upon examining these several heads at the period when they first begin to ossify, they exhibit the most perfect similarity in the number of pieces of which they are thus originally composed.

It would doubtless form a deeply interesting object of inquiry, to ascertain the cause of this remarkable contrast in the results, while the materials are identical, and whence it arises that so great an analogy among the elements finally produces such evident differences in the final structure, and in its progress towards completion. The cause, however, is involved in impenetrable obscurity; such is the constitution of Nature, and we are only left to ascertain the facts, without venturing to speculate upon the cause.

The essential pieces of the skull are at first of the same number in all the Vertebrated Animals; but they have neither the same form nor size. A certain part, which in one species is excessively reduced in size, takes an extremely large volume in another; and this part, which in some is so greatly developed, occasions the neighbouring parts to become abortive. There are some heads in which the vomer and the bones of the nose are as large as the frontal bone; yet these bones do not on this account lose their distinctive characters. Another cause of variety in the ultimate development, while there is a perfect similarity in the primitive elements, arises from this circumstance, that the different portions of these bones remain isolated in certain species, while in others the greater number of scattered pieces become grouped together, and coalesce into one bone. These modifications are particularly realized in the bones of the Skull; as it was especially necessary that this solid covering should be modified so as to suit the peculiarities of the brain and the other essential organs, which it was destined to enclose and to protect.

This instance is not solitary; the other pieces of the skeleton are similarly modified in their arrangement, to correspond with the adjustment of the more important organs. In the Fishes, for example, we find that the bones of the breast follow the respiratory organs into their appropriate situation, and are grouped along with them in the vicinity of the head. Without disturbing the order of Nature, it was impossible that the air, when dissolved in water, could retain that elasticity belonging to its gaseous form; it is unable to rush by its pressure within the body of the fish, and to seek out, as it were, the blood in the respiratory organ, for the purpose of purifying it. The gills are accordingly placed in the vicinity of the jaws; and it is only a consequence of these remarkable laws of analogy, that they should be accompanied in their places by the bones fitted for sustaining them, and by the muscles which enable them to act. These analogies, so obvious in the structure of the bones, may also be traced in the nervous system of the Vertebrated Animals.

It has been already observed, that the elements composing the bodies of the animals belonging to the First Great Division are analogous; but the analogy exists most perfectly only up to a certain period of animal development, and this period is not the same in all, but varies according to the degree of complication in their structures. A young unborn Mammiferous quadruped resembles a Human embryo, which is less grown than itself; but at the same time it resembles the embryo of a Bird in a higher state of advancement, that of a Reptile in a still more perfect condition, and that of a Fish, when completely finished, and perhaps already born. Thus, the classes of Vertebrated Animals are perfectly analogous only at ages differing for each class. Fishes and Reptiles may only resemble Birds and Mammalia which are much younger than themselves. These last-mentioned animals, on the contrary, continue to grow, and to increase in complication at a period of time long after the Reptiles and Fishes have ceased to develop themselves. We thus see the cause of the greater complication of structure among the higher animals. They are all identical in their first formation, but the Reptiles continue their growth for a longer time than the Fishes, the Birds longer than both Reptiles and Fishes, and Man and other Mammalia longer than these three classes of Birds, Reptiles, and Fishes.

Since all the Vertebrata can thus be brought to one common type, and to a base nearly identical for all classes, it follows that each higher animal undergoes revolutions analogous to those, observable in the whole series of animals inferior to it. One single Vertebrated animal of the highest order, will hence exhibit in those transitory states of existence, which pass from its first origin to its final completion, all those characters which are permanent arrangements in the lower classes of Vertebrated Animals. It also arises, that the first ages of the human embryo form analogous types to the other animals; it first resembles a Fish, then successively a Reptile, Bird, and finally a Mammiferous animal. Hence it appears, that in studying the development of the human embryo, we are reading a minute, and, in its leading characters, an accurate description of the physiology belonging to the higher classes of animated nature. It has been often admired that so small and delicate an organ as the retina of the human eye can exactly represent a distant and intricate scene with a precision which the most perfect human artist would in vain attempt to imitate; and shall we not equally admire that minute and exquisite disposition of things, by which the most complicated characters belonging to the distant inhabitants of the sea, of the land, and of the air, are thus brought together and transcribed in miniature, so that a complete history of a large portion of the Animal Kingdom is traced within the shell of an egg, or the membranes of a human embryo!

This remarkable disposition of Nature requires further elucidation.

The skin of the Human embryo precisely resembles that of certain Reptiles, or even of the Medusæ and Polypi, by its softness, its perfect nakedness, as well as its simplicity. In the earliest age, the anterior and central opening of the abdomen corresponds to a disposition observable in the Oyster and some other Mollusca, which have the mantle divided during the whole period of their life. At this time, likewise, the muscles are colourless, soft, gelatinous, and destitute of tendons, as we may see among the lower animals, such as the Worms. The bones of the human fœtus are nearly rounded, as we find in adult animals of the lower grades. The same bone, which is designed to form afterwards a whole, perfect in itself, is at this early period broken up into as many separate points of ossification as we can find divisions, permanently separate among the Mammalia and Birds, but especially among the adult Reptiles and Fishes. This singular correspondence of the temporary state of the human bones with the permanent state of the same bones in the other Vertebrata, is particularly observable in the occipital and sphenoid bones, in the upper jaw and temporal bones, in many bones of the face, and also in the sternum, or breast-bone. The breast-bone, for example, consists almost always of nine pieces in the first ages of the human embryo, and nine pieces are to be found in the Tortoise during the whole course of its life.. The upper jaw-bone is at first composed of five pieces in the human fœtus, and the Crocodile continually preserves these five pieces isolated. We might bring forward many other parallel instances of analogy among ages and classes, by entering more minutely into the details of the skeleton. It is only in consequence of this law that the bones of inferior animals are more numerous than those of animals more elevated than themselves, and that the animal skeleton exhibits a greater number of bony pieces, as we approach the first periods of ossification. Indeed, there is almost as much difference between the skeletons of the human fœtus and that of the full-grown Man, as between those of the adult Reptile, and of a Mammiferous Quadruped in its embryo state.

We shall find that these analogies, between the permanent states of the inferior classes, and the transitory states of embryos of the higher animals, can be traced equally in the heart and circulating organs, in the lungs and respiratory organs, in the nervous system, and in the organs of sense, as well as in the arrangements finally destined to continue the several species.

The heart of the human fœtus is first composed of one single cavity. It afterwards divides into four, which form momentarily a communication with each other, but speedily separate, in such a manner that the two cavities of the left side end in having no direct communication with the cavities on the right. This sketch of the progress of the human heart, which is equally applicable to that of the other Mammalia and of Birds, indicates a new analogy between the very young human embryo, still destitute of a heart, and the Worms, which never have a heart. Again, when the heart of the human embryo has only one ventricle, it resembles that of the Arachnides and Crustacea; afterwards, the heart with two cavities corresponds to those of the Fishes and Batrachian Reptiles. Finally, when there are but three cavities, the two auricles having coalesced into one, it resembles the hearts of the Tortoises and Serpents; and when the partition of the auricles is pierced by the hole termed the *foramen ovale*, the heart of the human fœtus bears a striking analogy to a permanent arrangement found in the Seals.

In following out this comparison of the progress of the human embryo, with the permanent arrangement of the adult animals, we see the venous blood of the human fœtus communicate primitively with the arterial blood, which is a natural arrangement in all the inferior animals, beginning from the Birds. The digestive canal is at first short and simple, as we find in animals of the lowest degree. The liver is originally composed of little comportments, as may be seen continually in the Crustacea; afterwards, it resembles that of the Mollusca, in being formed of lobes slightly united. The spleen and thymus gland are always absent among the Invertebrated Animals, and the latter is wanting even in the Fishes. These organs are developed very late in the embryo of Man and other Mammalia. The same remark applies to the breast-bone. It is wanting in many Reptiles and in all the Fishes, and it is very late in making its appearance in the embryo of the larger animals. In general, we find that those organs, of which the lower animals are altogether destitute, are the slowest in being developed in the human fœtus. On the contrary, the greater number of the organs which exist only temporarily in the human fœtus are the first to appear; and thus the gills, which in our species appear in its very earliest state, soon vanish; and that kind of tail, which may be seen in the human embryo when forty days old, does not exist longer than the fiftieth day.

The kidneys of the fœtus among the Mammalia are always very large, which arrangement is found to remain continually with the Fishes. They are at first lobed and of an unequal surface in the human embryo, nearly resembling that form which may be seen in the adult Fishes, Birds, and in many Reptiles and Mammalia. The subrenal capsules are at first very large in the human embryo, a disposition which exists in the Apes, and in several adult Rodentia, such as the Squirrel and Mouse. There appears also to exist a *cloaca* in the embryo of Man and of the Mammalia, as may be invariably found during the whole lives of the oviparous animals, and in the Monotremata of M. Geoffroy-Saint-Hilaire.

We see corresponding analogies in the instruments of respiration, and in the manner in which this function is performed. The Birds commence by breathing through means of the membraneous filaments of the allantois, just as the Polypi respire through the skin. The fœtus of Man, at first, has gills resembling those of the Fishes; and the Batrachian Reptiles, before they have lungs fit for acting, breathe through gills like the Crustacea; Birds breathe through simple membranes before they respire through lungs; and Man receives, through the placenta, blood already purified, as long as his lungs cannot gain access to the air of the atmosphere. There are even some animals, such as certain Reptiles during their metamorphoses, which successively present all these different modes of respiration, and which thus resemble in turn all the different classes of animals, excepting the Insects, breathing through tracheæ. These Reptiles respire, at first, through a naked skin like the Polypi; afterwards, through external gills, like the Crustacea and Annelides; after that again, through internal gills, like the Fishes; and finally, through lungs, in the same manner as the other animals of their own and superior classes.

But these analogies among the transitory states of the fœtus and the permanent forms of the lower animals, when in a state of perfection, are in no respect more evident than in the details of the nervous system. Thus the human fœtus, which we have selected as forming the standard of comparison, has at first certain parts of the brain *(corpora quadrigemina)* exactly similar to those belonging to the perfect Reptiles and Fishes. These organs were originally hollow, lobed, not quadruple, but double only, and placed merely on the surface of the enkephalon; yet they finally assume the permanent form of the Reptiles and Fishes. The Mammalia are the only animals in which these tubercles become quadruple, and they are the only ones where they become solidified by the obliteration of the central cavity. Besides the above-mentioned analogy in regard to the nervous system, the human embryo presents many other peculiarities corresponding to those of the lower animals. For example, the hemispheres of the brain are possessed at first of only a small volume, and they are rounded in a manner resembling the adult Fishes. Certain parts of its brain *(corpora callosa)* are so divided that they appear at first sight to be absent, in the same manner as with the Birds of every age.

The spinal marrow of the human embryo exhibits a central cavity; and a similar arrangement is found in all perfect animals of classes inferior to the Mammalia, while even the lateral nervous cords are originally so much isolated, as to give to the entire structure the appearance which it always preserves in the Articulated animals. Besides this, the marrow of the human embryo originally occupies the entire length of the vertebral canal, as in the other animals, but it is only in the third month that it ascends as high as the loins. Finally, its entire nervous system exhibits several analogies to the permanent dispositions in the other classes of Vertebrata, such as analogies of consistency, volume, greater or less numerous subdivisions, and even of faculties.

In regard to the organs of sense, analogies of a similar order may be traced. The mouth of the human fœtus is at first without lips, as in the Vertebrated Animals of the inferior classes. Its palate is then divided, and the mouth, on that account, communicates directly with the nasal fossæ, as in the Reptiles and Birds. The tongue is originally small as we find in the Fishes. The nose and ear present no external projection, in which respect they again resemble these organs in the Cetacea, and in the great oviparous animals. Again, the eye appears at first without an eyelid, as it is always found in Insects, the Crustacea, the Mollusca, the Fishes, and certain Reptiles.

Finally, the general form of the human embryo does not exhibit a less degree of analogy with the perfect states of other animals. The head is at first so slightly developed, as to give to the body the appearance of an Invertebrated Animal. It resembles a Fish or a Reptile by the absence of limbs; and the caudal or tail-like appendage, which we have already mentioned, gives it for a short time the aspect of an ordinary quadruped.

From what has been said, it appears that the embryos of the higher animals exhibit in the course of their growth the greater part of the characteristic peculiarities of all classes of animals, and present the fleeting models of almost all the different kinds of organization. . The rudiments of Man thus form a reduced, yet striking, image of the entire Animal Kingdom.

We must admit, however, that these resemblances which have just been enumerated are far from establishing among all beings a perfect and absolute identity, whether we consider different animals in their states of perfection, or upon comparing one animal at the different ages of its progress, and with other animals of different yet inferior species. Every animal preserves continually, and throughout every age, certain well-marked characters peculiar to its species. These differences are so striking, even in those which appear to resemble each other most nearly, that it becomes impossible for us to conclude, that they may all be arranged in one graduated scale, every where complete and continuous; or that they all possess one common framework, one visible and identical basis, with the same number of essential organs, possessing the same natural characters. Still less can they be supposed to be derived by gradual metamorphoses and complications from one common stock, from binary, or even from ternary types. These analogies being always partial, and frequently vague, can by no means warrant the adoption either of the universal chain of existence proposed by Bonnet, or the transmutation of species, and the successive filiation of Demaillet and Lamarck. · Still less will they admit the adoption of the organic identity proposed by certain French and German writers.

If animals resemble each other universally, it is only in the great phenomena of existence, of which we have already treated. When, however, we descend to the instruments producing these phenomena, we are surprised to find the most striking difference, instead of resemblance, which can be considered as perfect. One animal seemingly superior to another in some of its organs, is often evidently inferior to the same animal in other details of its structure. We sometimes find that two animals which resemble each other entirely in respect to one set of organs, are sometimes so dissimilar in their conformation, that it is impossible to blend them together even in imagination. Finally, there are some organs of which certain entire classes of animals are altogether deprived, and yet they may exhibit, in two other classes of animals apparently allied to each other, the most discordant characters. Indeed, it is certain that there is not a single organ which does not vary from one genus or from one family to another; but at the same time, there is not an organ, except perhaps the stomach, which can be found in all families and in all genera. We shall see numerous proofs of this in the details of the Animal Kingdom.

The real cause of analogies in the different ages and species must be found in this,—that *they are all constructed upon models evidently analogous*. The same Divine Artificer formed them all. We find the same *style* in every page of the great book of Nature; and we every where see the most evident affinities both in the essential organs of animals, and in the phenomena which they exhibit.

From what has already been said, we may easily perceive the cause of those wonderful productions, those *Lusus Naturæ*, or monstrous births, which in all ages have astonished and alarmed the ignorant. These evidently arise from a retardation in the growth of some organs. An animal, though remaining incomplete in regard to one organ, may yet continue to grow in all other respects, and the disproportion of the organs may thus proceed to the most shocking disparity. Another consequence of this law is, that the abortive and imperfect organs of an animal must resemble the usual organs of the same animal in an earlier state of its existence, or those of a lower animal in an adult form. Though all the organs may have been originally perfect, yet if one ceases to develop itself, while all the others continue their progress, the monstrosity of the final result is the necessary consequence.

It follows also, that what is a deformity in one animal, may be a constant and permanent character of another. The monstrosity of a higher animal may bear a decided analogy to the regular form of one of another grade. For example, it is evident that one of the Mammalia, born without hair, is analogous to the lowest classes of animals having the skin naked. If the skin be scaly, it brings the Quadrupeds down to the level of the Fishes and the Ophidian Reptiles. The Mammalia and Man often have a divided palate, like the Birds, the Fishes, and many Reptiles. The occasional absence of the teeth in Mammalia may be explained on the same principle. The monstrous development of the liver is found naturally in Birds, Fishes, and also in some Cetacea and Reptiles. To want the tongue, or to have it forked—to have the limbs abortive, or altogether wanting, though monstrous developments, are natural and constant arrangements in other species.

The following are the most remarkable of the laws of Monstrosity, or, as Lord Bacon calls them, " the laws which govern the sports of Nature:"—

1. Monstrosities are always found more frequently in females than in males. The cause of this is to be found in the fact, that the male sex is a more advanced state of development than the female. All embryos, of whatever sex they may ultimately be, are at one time females. In infancy, too, the female character predominates. Young boys preserve for many years the smooth chin, the narrow larynx, the silver voice, and the rounded limbs of the young female. The young birds of both sexes have at first the same plumage as their mothers, and moult at the same time that they do.. The same thing holds with all the other characteristics of the male sex.. The mane of the Lion, the crests, spurs, and other ornaments of the male Birds, the antlers of Stags, the horns of Cattle, the vivid colours and powerful energy which belong to the males of different species, are all characters slow in developing themselves.

2. Monstrosities never exceed certain limits, and deformities have their fixed laws. Thus, when there are supernumerary fingers, they are always equally disposed.

3. They always preserve a degree of symmetry, even among the most shocking irregularities. A double monster seems to result from the same law, which occasions both sides of the body to be symmetrical.

4. The absence or excessive smallness of one organ is always followed by the extreme development of another.

5. Monstrosities are more frequently found in the left side than in the right, because the left side is always the more feeble and imperfect.

6. Deformities generally go together in pairs. Thus, the presence of supernumerary fingers and the division of the palate; the excessive smallness of the lungs, with a great development of the liver; and these dispositions, which usually co-exist in monsters, are found naturally in many animals.

7. The most diversified organs of the animal series are the most liable to monstrosity in the higher animals.

8. Monsters always have some of their organs below their age, and consequently below their class, but never above them. A monstrous Bird or Quadruped often has its organs analogous to those of a Reptile or of a Fish; but a Fish or Reptile never has those of a Bird. This rule is general and constant.

Several more laws might be mentioned, but these seem sufficient for our present purpose. We shall now proceed to consider the Animal Kingdom more in detail, and accordingly commence with the Mammalia.

THE FIRST CLASS OF THE VERTEBRATED ANIMALS.

THE MAMMALIA—MAN AND BEASTS.

WITH WARM BLOOD; HEART WITH TWO VENTRICLES; FEMALES SUCKLING THEIR YOUNG WITH MILK SECRETED IN BREASTS OR MAMMÆ; VIVIPAROUS, EXCEPTING THE MONOTREMATA, WHICH ARE EITHER OVIPAROUS OR OVO-VIVIPAROUS.

THE Mammalia should be placed at the head of the Animal Kingdom, not only because it is the class to which Man himself belongs, but because it surpasses all others in the enjoyment of more numerous faculties, of more delicate sensations, of a greater variety of motions, and where all these properties are combined so as to form beings of greater intelligence, fruitful in resources, less the slaves of instinct, and more susceptible of improvement.

· This class possesses characteristics peculiar to itself, in its VIVIPAROUS generation [the young being born alive], in the manner by which the fœtus [or embryo] is nourished in the womb, by means of the placenta, and in the MAMMÆ, or breasts, by which the young are suckled.

On the contrary, the other classes are OVIPAROUS [or produced from eggs previously laid by the parent]; and if we contrast them generally with the Mammalia, we shall find that they possess numerous points of resemblance among themselves, which clearly exhibit a special plan of organization in the general system of the Vertebrated animals.

As the degree of their respiration is moderate, the Mammalia are in general adapted for walking upon the ground, but at the same time their movements are performed with vigour, and in a continuous manner. For this reason, the articulations of their skeletons have very precise forms, which determine the direction and extent of their motions with precision.

Some of them can, however, raise themselves in the air by means of elongated limbs, connected by extensible membranes; others have their limbs so much shortened that they cannot move easily except in the water. But this circumstance by no means deprives these last-mentioned animals of the general characters of the class.

This variety in the character of their locomotion requires a corresponding difference in the organs by which their movements are effected; and, consequently, the skeletons exhibit many important peculiarities when compared with those of Man, and with each other. Yet, whatever differences may be found to prevail, one general character can be discovered throughout the whole. The same bones which compose the arm and hand of Man, or of the Monkey tribe, are found equally in those genera which use their fore extremities for walking, swimming, or flying; the proportions of the bones being only altered, or some of the parts being obliterated, or changed. We find them equally in the fore-leg of the Deer, in the wing of the Bat, and in the fin of the Whale. One order of animals having an abdominal pouch for containing the young, are supplied with two additional bones, called *marsupial*, moved by appropriate muscles.

The interior composition in the bones of the Mammalia is generally the same; but their texture varies, being denser and closer in the smaller tribes, but very fibrous, and loosely arranged in the Cetacea. The texture of the bones in the sea-beasts is particularly adapted for locomotion in water, being aided by the lightness of their structure, and by having all the cells of their bones filled with an oily fluid.

The joints or articulations by which the bones of the Mammalia are connected together greatly influence the habits and general economy of these animals. Some of the joints are only capable of performing an imperfect motion, while a greater latitude is assigned to others. Appropriate names have been assigned to these several kinds of articulations, and they have been further arranged in divisions and sub-divisions. Some kinds of articulation are observed to be altogether peculiar to certain classes of Mammalia; while it frequently occurs that the corresponding bones will vary in their mode of articulation for different Mammalia. Other bones again, which are separated in one genus, will be intimately and closely united together in another.

The first class of articulation *(Diarthrosis)* contains the free and perfect joint. The opposite surfaces of the bones are distinct and well defined, being covered with a polished and smooth cartilage. A continuation of the periosteum or external covering of each bone passes onwards from one to the other, forming a kind of capsule, or little bag, and permitting nothing either to pass within or to escape from the cavity. But the periosteum does not cover the articular cartilages, it merely forms the external covering of the entire joint. Other ligaments frequently strengthen this covering either externally or within the capsule, and thus will limit the motions of the bones more perfectly than the capsule alone could have done. The interval within the capsule is occupied by an oily fluid, or else each bone is covered by a smooth and polished cartilage. Sometimes a piece of cartilage will be found; perhaps also some peculiar gland, or other solid body.

The direction and extent of motion possessed by the bones of the Mammalia is dependent upon the rigidity or number of the ligaments, as well as upon the form and depressions found in the articulated surfaces.

When a bone is articulated to another by one of its extremities, it admits only of two species of motion, torsion or flexion.

The torsion, or twisting, takes place when the bone is capable of moving round its own axis, or else round an imaginary axis passing through the articulation. This kind of motion can only be found when the articulated surfaces are plain or spherical, and the latter are alone capable of motion in every direction. Flexion or bending takes place when the extremity of the bone, farthest from the joint, approaches the bone which is fixed.

The different kinds of torsion have been assigned different names; there is the hinge-joint *(Ginglimus)*, the ball and socket-joint *(Enarthrosis* or *Arthrodia)*, and the rotating joint *(Trochoides)*.

The manner in which the head is attached to the trunk, the lower jaw to the head, and the several parts of the limbs to each other, differs in the several classes of animals. The head of the Mammalia is united by a hinge-joint to the neck; in the Birds it is connected by a ball and socket-joint. Even among the Mammalia themselves the articulations are found to differ. In Man, the radius of the fore-arm is connected by a ball and socket-joint with the humerus, at its one extremity, and it rotates upon the other. But in the Rodentia, and many Pachydermata, the radius is connected by a hinge-joint with the humerus, and is immoveable at its other extremity. In some species these bones are even completely united.

The second class of articulation *(Synarthrosis)* admits of no motion whatever; it is said to form a *suture* when two flat bones join each other by the edges; to be *squamous*, when the thin edge of the one bone covers that of the other; *denticular*, when the edges are notched and indented together; and *harmonic*, when they simply touch each other. We find, in the bones of the face and head of Man, instances of these different kinds of articulation. The manner in which the bones of the skulls belonging to the several Mammalia are joined together, bears a great resemblance to that observable in Man; and we find, in all, that they have a tendency to ossify as their ages increase. A variety of this kind of articulation is found in the teeth of Man and Quadrupeds. These are inserted like wedges into the cavities of other bones. To this style of connexion the name of *Gomphosis* has been assigned.

There is no instance found in the human skeleton corresponding to that singular kind of articulation observed in the nails of the Cats. These are inserted into small cavities in the last phalanges of the toes, and at the same time they receive a pivot, or eminence of the phalanx, into a small cavity of their own, prepared to receive it. This curious contrivance is also found in other quadrupeds with powerful claws. We also find it in the tusk of the Morse, where a small pivot is observed to project from the basis of the *alveoli*.

The third kind of articulation *(Amphiarthrosis)* admits only of a slight and restricted motion. This is not occasioned by the form of the bones, which are perhaps perfectly adapted for free motion, but by the cartilages and ligaments which are placed between the bones forming the articulations, and uniting firmly with them. The vertebræ of the back exhibit this restricted motion; but the bones of the pelvis are joined in such a manner as scarcely to permit any motion whatever. The bones of the wrist and instep are considered, by some anatomists, to belong to this class; for, though they appear to be provided with a few smooth articulated surfaces, yet they are confined so greatly by the surrounding ligaments, that they move upon each other with great difficulty and through a very narrow space.

In the second and third kinds of articulation, the edges or surfaces of the bones either come immediately into contact, or else they are bound together by a substance which attaches itself throughout their entire parts of connexion. Also, the periosteum is continued from one bone to the other, and is more intimately connected at the place of their junction than at any other part. In this respect they differ wholly from the kind first described.

The muscles which set these bones into motion are as various as the movements destined to be performed. Those composing the trunk of the Elephant are unrivalled for the union of strength, variety, and delicacy. As the snout of most animals is incapable of performing any considerable motion, we are naturally struck with astonishment at seeing an organ, which appears at first sight to be merely a prolongation of the snout, performing all the offices of the human hand. This delicacy is owing to the immense number of minute muscles which are arranged in various directions, and thus enable the animal to execute the various movements of the organ. Upon making a transverse section of the proboscis, counting the number of short muscles, and then allowing the breadth of a line for the succeeding ones, which is considerably more than their thickness, some estimate may be formed of the number of muscles composing the trunk of the Elephant; and, upon adding these to the number of bundles comprising the horizontal layers, they are found to amount to between thirty and forty thousand. The snout of the Tapir is formed on a similar principle, with an additional muscle, corresponding to that which raises the upper lip of the horse.

In Monkeys, the muscle which frowns (*corrugator supercilii*) is large, and is frequently used, but without expressing the feelings indicated by that action in Man. In most quadrupeds the muscles, moving the external ear, are more strongly developed than in Man. They are thus enabled to give a great variety of attitudes to that organ, which enables them to collect sounds in every direction. Most quadrupeds, after an agreeable sensation, will erect their ears, and depress them when displeased, in the same manner as the Horse is in the habit of doing. Those animals which possess the power of rolling themselves up, as the Hedgehog, have a number of curious muscles for that purpose. The muscles of the tail are generally strongly developed; especially in the Kangaroos, which use them for standing and leaping, and in the Monkeys having prehensile tails.

A certain degree of similarity prevails between the muscles of Man and those of all the other Mammalia; but this resemblance is, upon the whole, greatest between Man and the Quadrumana. But we cannot fail to remark the small development in the calf of the leg, and in the buttock, among the Apes, as these muscles are intimately connected with the upright posture peculiar to Man, and the beauty of the human form. Many muscles, however, are found exclusively in Quadrupeds. Thus the fleshy pannicle *(panniculus carnosus)*, that sub-cutaneous covering of the body, is of very great size in the Hedgehog, Armadillo, Porcupine, and all animals possessing the power of rolling themselves up. This cutaneous expansion is even found in the Cetaceous tribes; and the inhabitants of the Aluntian Islands are said to fabricate a thread of great delicacy from the tendinous fibres of this muscle, procured from the Whale.

Certain muscles are distinguished for their very great strength. Thus the Horse is enabled to kick backwards with very great force, owing to the great development of the Gluteus medius and Gemellus muscles. The Mole is enabled to burrow under ground, and to throw up the earth, by the great magnitude of the *pectoralis major*, *latissimus dorsi*, and *teres major* muscles.

There is also a curious arrangement in the muscles belonging to the epiglottis of the several Mammalia. In the motion by which we elevate and depress the hyoid bone and the larynx, the muscles acting on the bones, and other hard parts, may be compared to ropes drawing a resisting object in a certain direction. Innumerable muscles of a complicated form may be seen in the tails of these Apes having that organ prehensile. It is said that no less than two hundred and eighty muscles were discovered by Mery in the prehensile tail of a Cercopithecus.

Although the manner in which the bones are articulated determines the motions which they are capable of performing, yet it is by the number and direction of the muscles attached to the bones that the motions performed by each bone are fixed.

The muscles are attached to the bones by tendons. The fibres of which the tendon is composed are of a closer and denser texture than those of the muscles, and of a silvery whiteness. Being penetrated by fewer vessels and no nerves, its substance seems altogether gelatinous. It possesses neither irritability nor sensibility, and forms the passive link, by means of which the muscle acts upon the bone. Portions of tendon are found both inside and on the surface of several muscles; and even those tendons, by which the muscles are inserted to the bones, penetrate a certain length into their fleshy substance, where they are interlaced in various manners. The term *aponeurosis* has been applied to those tendons which are broad and thin. The tendons have a great affinity for phosphate of lime, which they often absorb with facility when their action is frequently repeated, and when employed to execute violent motions. This frequently occurs with the Jerboas and other animals, which constantly leap with their hinder limbs. It is considered probable that all the elementary muscles exercise an equal force at the moment of their contraction; but the degree of force with which a muscle can be exercised greatly depends upon the manner in which its fibres are disposed, and the situation of the muscle itself, in respect to the bone or part it has to move. We therefore cannot estimate the force of a muscle by its mass, or by the number of fibres of which it is composed, but must also consider the composition of the muscle and the method of its insertion.

The muscles are either *simple or compound*. In the simple muscles all the fibres have the same disposition. The most usual are the *ventriform*, having all the fibres nearly parallel, and forming a long bundle of a round form. The fleshy parts swell in the middle, forming the belly of the muscle, and becoming smaller at each extremity, where they terminate in tendons. Another kind of simple muscle consists of those that are flat, and have parallel fibres, forming a sort of fleshy membrane, which is terminated by aponeuroses or tendinous membranes, instead of ending in small tendons. Both these kinds sometimes have tendons or aponeuroses in their middle, or in other points of their bodies. It is obvious that, in either, the total action of the muscle is equal to the sum of all the particular actions of the fibres; and that, if the action experiences any mechanical disadvantage, it arises from the mode of insertion, and not from the composition of the muscle. This is not the case with the two other kinds of simple muscles, the *radiated* and *penniform.*

The *radiated* muscles are those which have their fibres disposed like radii of a circle, and which proceed from a base more or less extended, while they incline towards each other, and are inserted in a small tendon.

The *penniform* muscles have their fibres disposed in two rows, uniting in a middle line, and forming angles more or less acute, so that they resemble in some degree the arrangement of the feathers in a quill. The tendon forms the continuation of this middle line.

It may be easily perceived that, in the two last-mentioned kinds of muscles, the total or resulting force is less than the sum of the component forces; and that, if we take successively the lengths of every two fibres, which unite in producing one angle, as the measures of their individual forces, the diagonal of the ultimate parallelogram which may be formed thereon will represent the entire resultant, in quantity and direction, belonging to the fibres of the whole muscle.

When several muscles unite in one common tendon, the result is called a *compound* muscle. These muscles may be similar in their nature, but sometimes they are formed of very different kinds, such as the radiated and the ventriform uniting to form one compound muscle. We may, then, estimate the particular action of each according to the preceding observations, and the total action can then be estimated according to the degree of their inclination. Other muscles, again, are styled *complicated:* these may have only one belly with divided tendons; or they may have several fleshy parts, wherein the tendons are interlaced in several ways.

The absolute force of the muscles is determined from these several dispositions; but it is their insertion which determines their real effect. The muscular insertions may be referred to eight distinct classes:—1st, the fleshy envelope; 2d, the sphincter or ring; 3d, the curtain; 4th, the rotatory; 5th, the rope; 6th, the lever of the first kind; 7th, the lever of the second; and 8th, of the third kind. The first four have a striking similarity, in their being all formed of a girdle, or portion of a girdle, which contracts upon the surrounding parts.

1. The diaphragm and abdominal muscles are instances of the fleshy envelope. Being destined to compress the soft parts contained in a certain cavity, they envelope that cavity in every direction, in the form of membranes or bands. When all the fibres act simultaneously, it is for excretory purposes; but they usually act alternately, and then the effect is to enlarge one cavity and to diminish the other. Thus, at each inspiration the abdomen becomes wider and shorter, while the contrary happens on each expiration. The heart, arteries, and intestines, have muscles of this kind; and the muscles moving the tongue in Man and Beasts must also be referred to this class.

2. The sphincter muscles are calculated to widen or contract some soft aperture. Some of them surround the orifice like rings, and others are inserted in a manner, more or less directly, upon the edges of the opening. If the muscle be uniformly distributed around the orifice, it always preserves its figure, and is dilated or contracted always in the same manner. But when these muscles have different directions, and make different angles with the edges they have to move, the form of the aperture is very variable, as we may see in the lips of Man. No animal possesses so great a mobility of this part, and none can therefore possess so expressive a physiognomy.

3. The curtain muscle is seen in the eyelids of Man and other Mammalia. When these muscles are placed in the body of the membrane, which is destined to cover some other parts, their structure is such as we have just described; but when they are situate externally, they have the form of very complicated pullies, as will be explained when we come to treat of the Eye in Birds.

4. The rotatory motion of the muscles may be seen in the means by which the globular mass of the eye is rolled and supported on every side.

5. The rope muscle has already been alluded to, in speaking of the larynx, and may be regarded as the most advantageous form in which a muscle can be applied.

6, 7, 8. When a bone intended to be moved is articulated at any particular point, it cannot be elevated or depressed in a direct line, but must be considered as a lever having its fulcrum in the articulation. The bone forms a lever of the first kind when the articulation is between the two extremities, and the muscles are placed at one of them, as we may observe in the muscles attached to the olecranon and heel-bone. But the most usual case is when the articulation is at one of the extremities of the bone; and then the most favorable position for the muscle is when it rises from another bone parallel to that which it has to move, or which forms with it only a very small angle. This is the case with the muscles between the ribs (*intercostales*), and several others. Yet these muscles possess a degree of obliquity which considerably diminishes their power. The muscles closing the mouth of Man may also be compared to those just mentioned with respect to their small obliquity; but they are inserted much nearer to the point of support than the former, a circumstance which also considerably diminishes their force.

The most usual kind of insertion is where a muscle attached to one bone is inserted into another, which last is articulated either mediately or immediately with the first, and may be extended until they both form a line, or inflected so as frequently to make a very small angle. This mode of insertion appears to be the most disadvantageous of all in respect to mere force, on account of the obliquity of the insertion when the moving bone is extended, and also on account of its proximity to the fulcrum. The first inconvenience is partly corrected by the heads of the bones. Their articular extremities are usually enlarged, so that the tendons of the muscle, by turning round a convexity, in order to be inserted below it, form more obtuse angles with the lever, or body of the bone, than would be practicable if the head did not exist. By this means the obliquity of their insertion is diminished, and rendered less variable.

The proximity of the fulcrum was necessary to prevent the members from being monstrously large in the state of flexion, but particularly for producing a prompt and complete flexion. As the muscular fibre loses only a determinate fraction of its length by contraction, if the muscle were inserted at a greater distance from the joint, the moveable bone would only be approximated to the other by a small angular quantity. On the contrary, by inserting it near the apex of the angle, a very small contraction occasions a considerable approximation. Velocity is gained in proportion as the space through which the muscle acts is diminished. In this manner, muscles of this kind exercise a power which surpasses all imagination.

There are many instances of muscles inserted at a considerable distance from the fulcrum, especially in the short bones, which must be completely inflected. The vertebræ and phalanges of the fingers are in this situation. Muscles extended from the one to the other of these bones would not have produced a sufficient degree of flexion. In the phalanges, the fingers would have been two thick. It was also necessary that the tendons of these muscles should be attached to the bones over which they pass. If this were otherwise, it would happen that, whenever the phalanges were bent so as to form an arc, the muscles with their tendons would remain in a straight line, and form its cord. We may hence perceive the necessity of the annular ligaments, the sheaths and perforations. The last-mentioned arrangement occurs solely in the flexions of the fingers and toes of Man, Quadrupeds, and some other animals, and consists in the muscles which have to extend farthest being placed near to the bones, while their tendons, perforating those of the muscles, are inserted at a shorter distance, and lie over the first. When there are only three phalanges, there is but one perforation. The muscles moving the tail in the Quadrupeds are placed at a great distance from it; but their long and slender tendons are inclosed in sheaths, which they do not leave excepting immediately opposite the points into which they are to be inserted.

The whole of the Mammalia have the upper jaw fixed to the skull; and the lower one is composed of only two pieces, articulated to the temporal bone, by a projecting part [called the condyloid process.]

By the elongation of the condyles, which fit into the zygomatic process of the temporal bone, this joint is nearly restricted to the motions of a hinge, alternately raising and depressing, while the lateral motion is only just sufficient for the grinding of the food.

There is a single or double bone, found in most Mammalia, called the inter-maxillary bone, but of which Man is entirely destitute. In these animals the upper jaw-bones do not touch each other under the nose, nor do they contain all the teeth, but the inter-maxillary bone is wedged in between the former, and contains the incisive teeth of those animals possessing them. The size of this bone varies surprisingly in the several orders and genera of Mammalia, being small in the Walrus and many Carnassiers, but large in the Beaver, Marmot, Hippopotamus, and Cachalot, but especially in the Wombat. In the Ornithorynchus it is constructed of two pieces in the form of hooks. This bone is seen to exist in animals altogether destitute of teeth, and is also found in such Ruminantia as have no incisive teeth in the upper jaw. Some anatomists have doubted whether the upper jaw-bones and inter-maxillary bones are not the same, and that the latter is merely the anterior or incisive portion of the former. The latter opinion appears to be the more probable, as the division is found in the human fœtus, while, in some quadrupeds, the two bones are frequently seen to coalesce. The lower jaw surpasses all other bones in the variety of its forms among the different Mammalia. It possesses very strong projections on the under side in the Wombat; and we may remark in the *Cercopithecus Beelzebub*, and other Brazilian Monkeys, a remarkable lateral development of the bone, which assists the larynx in the emission of that extraordinary deafening sound peculiar to these animals. In the Ornithorynchus, the anterior part of the lower jaw is shaped like a shovel.

An intimate relation may be observed between the kind of food with which an animal is nourished, and the motions performed by its lower jaw; and these again are greatly influenced by the form of its condyles. Thus, Mammalia living on vegetables possess a power of moving their lower jaws from side to side, so as to produce that grinding effect necessary for pulverizing and dividing grain, and for bruising grass. These animals are in this way able to move their lower jaw in almost every direction, by the form of the condyle, and of the cavity to which it is articulated. On the contrary, with the Carnassiers, we find that the lower jaw is altogether incapable of any other motion than simply downwards and upwards, being destitute of that lateral grinding motion attendant on mastication in its most perfect form. Thus, while the teeth of the Herbivorous quadrupeds may be compared to the stones of a mill, the movements of the teeth, or rather tusks, in the Carnivorous quadrupeds greatly resemble the dividing motion of scissors.

The neck consists of seven vertebræ, one species excepted [the three-toed sloth] which has nine.

A great variety is found in the number of their vertebræ, excepting those of the neck. In the Cetacea, where the neck is very short, the bodies of the Cervical Vertebræ are extremely thin, and form by *anchylosis* one bone; so that the original number of vertebræ, with their processes, can scarcely be perceived. In Quadrupeds having long and flexible necks, such as the Camel, and Camelopard, the spinous processes of the vertebræ of the neck are small, or they are nearly obliterated. A peculiar substance of great strength, called the *ligamentum nuchæ*, is attached to the necks of the larger quadrupeds. By means of this elastic body, the great weight of the head is supported. In the Elephant, it is of a very great size. The short-necked Cattle have double transverse processes, and in the bodies of the Cervical Vertebræ, both of Ruminating animals and Horses, there is a longitudinal ridge running along the front. With Carnivorous animals, the *ligamentum nuchæ* is small; and as the pendent position of their heads require strong muscles for their support, the Cervical Vertebræ have their transverse processes very large and flat, both in the front and back, and thus afford places of attachment for the muscles of the neck, as well as for those which contribute to open their mouths.

The length of the neck does not depend upon the number of the cervical vertebræ; for, as we have already observed, this is nearly always the same in most quadrupeds. In general, we find the length of the neck to be such, that, when it is added to the head, their united lengths are exactly equal to the height of the animal from the ground. Were this otherwise arranged, quadrupeds could not easily have reached either the herbs on which they feed, or the water they must drink. The bulk of the head, in all those animals where this rule is observed, is very nearly in an inverse proportion to the length of the neck, else the muscles would be unable to elevate the head. This rule, however, is not adhered to in such animals as lift their food to the mouth by means of hands, or of feet constructed somewhat similar to hands. Neither do we find it in the Elephant, where the proboscis is substituted for hands, nor is it to be found in the Cetacea, which obtain their food in water: the latter possess the shortest necks of all the Mammalia. It appears singular that the number of cervical vertebræ should remain constantly the same, although the necks of different animals differ so very widely in length. In Man, we sometimes observe the vertebræ of the back and loins to vary from their usual number, but never those of the neck.

The Dorsal Vertebræ are very large and long in all quadrupeds with long necks and ponderous heads, especially in the Horse, Camel, Elephant, and Camelopard, which arrangement seems necessary to afford a place of attachment for the *ligamentum nuchæ*.

The number of Lumbar Vertebræ varies exceedingly in different Mammalia; and when the length of the body is remarkably great, it is usually occasioned by a greater or less number of these vertebræ of the loins. Their motion is more or less restrained in nearly all quadrupeds; and this is effected by the exterior side of each posterior articular process being directed backwards, so that the anterior articular process of the next vertebra falls between two prominences. The shape of the body in animals, whether slender, short, or thick, is chiefly determined by the length of the loins, and this again depends upon the number of the lumbar vertebræ.

The anterior ribs are attached in front by cartilaginous portions to a sternum or breast-bone, composed of a certain number of vertical pieces.

Some of the ribs are attached only by the hinder extremity to the spine, and are called *false* or *shaking ribs*, to distinguish them from the *true*, which are united to the sternum. The ribs of the Mammalia vary greatly in number. In no instance is the total number less than twelve, being the number in the human skeleton. The horse has 18 ribs, being 8 true and 10 false; the Elephant has 20 ribs, being 7 true and 13 false. The strength of the spinal column, and its consequent ability to sustain great weights, depend very much upon the size of the ribs, and upon the figure which they give to the rest of the body; accordingly, we find that in the large herbivorous quadrupeds, which are usually employed as beasts of burthen, the ribs are thick and broad. Those quadrupeds which have no clavicles have less curvature upon their sides than the others. Being never required to use the anterior extremity as a prehensile member, the chest is narrowed and flattened upon the sides, especially towards the sternum, whereas Mammalia with clavicles have their chest shaped nearer to the human form.

The ribs are remarkably strong and compactly set in all quadrupeds destined to roll themselves up when attacked by other animals. In all Mammalia, they have only a very limited motion upwards and downwards, and their articulations are strengthened by a great number of ligaments. There are capsules at each articulated extremity of the ribs, which retain them upon the bodies and transverse processes of the vertebræ. They are further secured by means of two ligaments, the one being inserted into the transverse process of the superior vertebra on the *inside*, and the other into the lower articulating projection of the same vertebra, but on the *outside*. By this means the cavity of the chest is rendered secure, as well as by the capsule which unites the other extremity to the prolonged cartilage of the breast. There is also a ligamentous expansion between the ribs, connecting the lower edge of the one rib to the upper edge of the next.

In all animals, excepting perhaps the Marmot, the thorax or chest is narrower than in Man, and deeper from the spine to the breast. This peculiarity arises from the greater length of their breast-bone, and the less-marked flexure of their ribs. The Camelopard, and other animals having very long legs, possess the keel-like form of the chest in a remarkable degree; this is especially observable in the Deer tribe.

The Sternum, or Breast-bone, differs generally from that of man, in being composed of a greater number of pieces, and in being rounder and narrower. It is also longer in proportion to the rest of the body. In the Mole, the sternum is remarkably thick and strong. To enable this animal to excavate the earth for the admission of its body, the anterior portion of the clavicle is compressed upon the sides, so as to give it the form of a ploughshare. It projects beyond the line of the first rib, and thus enables the animal to burrow with singular rapidity.

The anterior extremity of the Mammalia commences in a shoulder-blade, which is not articulated, but merely suspended in the flesh; often resting on the sternum, by means of an intermediate bone, called a clavicle.

The anterior extremities, or fore-limbs, often appear to differ widely from each other upon superficially comparing the external forms in different species with each other and with Man; thus, the Dolphin and Whale seem to possess fins, and the Bat wings. But this difference is more apparent than real. Upon examining their internal structure, we find that the fore-limb consists always of four component parts, the Scapula or shoulder, the Humerus or arm, the Fore-Arm, and the Hand. In the fins of the Cetacea we perceive all these bones flattened in their form, and scarcely capable of moving upon each other, while the wing of the Bat is really a hand, with its fingers excessively elongated. There are also found a class of limbs intermediate between the fore-foot of the quadruped and the pectoral fins of the Cetacea. The Otter, Seal, and Walrus, have their bones covered with a web-like integument, adapted for the purpose of swimming. Their limbs are much more freely developed than in the Cetacea, and possess a greater freedom of motion, so that they form an intermediate transition from the one structure to the other.

The Scapula, or shoulder-blade, is found in all Mammalia. In general, the edge of this bone, next to the spine, is rounded, and the posterior angle is thus rendered blunt. The shape of the Scapula depends on the presence or absence of a clavicle; the acromion not being so prominent when this clavicle is absent; and then there is another process called the *recurrent process*, pointing backwards almost perpendicularly to the spine. The posterior angle is also most elongated in those species having complicated motions of the anterior limbs. In animals having only the rudiments of clavicles, or none, the acromion process is nearly deficient.

The Clavicle, or Collar-bone, is not found in species which employ their anterior

limbs only for progressive motion. In the Mole, the clavicle is of an extraordinary thickness, being nearly square, and slightly greater in breadth than in length. In the Bat, it is very large and strong. Indeed the Clavicle is found in all Mammalia which use their fore-limbs for burrowing, like the Mole; for raking the ground, like the Hedgehog and Ant-eater; for climbing, like the Sloth; or for holding objects, like the Beaver and Squirrel. In the other Mammalia, we often find in its place a smaller bone called the *Os claviculare*, analogous to the true clavicle, but merely connected to the muscles. This arrangement is observed in most Carnassiers, and in many Rodentia.

Sir Charles Bell observes, that "Animals which fly, or dig, or climb, as bats, moles, porcupines, squirrels, ant-eaters, armadilloes, and sloths, have the clavicle; for in them a lateral or outward motion is required. There is also a certain degree of freedom in the anterior extremity of the cat, dog, marten, and bear; they strike with the paw, and rotate the wrist more or less extensively; and they have therefore a clavicle, though an imperfect one. In some of these, even in the Lion, the bone which has the place of the clavicle is very imperfect indeed; and if attached to the shoulder, it does not extend to the sternum, it is concealed in the flesh, and is like the mere rudiments of the bone. But however imperfect, it marks a correspondence in the bones of the shoulder to those of the arm and paw, and the extent of motion enjoyed. When the Bear stands up, we perceive by his ungainly attitude and the motion of his paws, that there must be a wide difference in the bones of his upper extremity from those of the Ruminant or Solipede. He can take the keeper's hat from his head and hold it; he can hug an animal to death. The Ant-bear especially, as he is deficient in teeth, possesses extraordinary powers of hugging with his great paws: and, although harmless in disposition, he can squeeze his enemy, the Jaguar, to death. These actions, and the power of climbing, result from the structure of the shoulder, or from possessing a collar-bone, however imperfect. Although the clavicle is perfect in man, thereby corresponding with the extent and freedom of the motion of his hand, it is strongest and longest, comparatively, in the animals which dig or fly, as in the Mole and the Bat."

It follows from these observations, that animals possessing a clavicle, and thus having the power of hugging, are unable to bear a severe shock on their fore-limbs without running the risk of fracturing the collar-bone. "If we observe the bones of the anterior extremity of the horse," continues the same eminent anatomist, "we shall see that the scapula is oblique to the chest; the humerus oblique to the scapula; and the bones of the fore-arm at an angle with the humerus. Were these bones connected together in a straight line, end to end, the shock of alighting would be conveyed through a solid column, and the bones of the foot, or the joints, would suffer from the concussion. When the rider is thrown forwards on his hands, and more certainly when he is pitched on his shoulder, the collar-bone is broken, because in man this bone forms a link of connection between the shoulder and the trunk, so as to receive the whole shock; and the same would happen in the horse, the stag, and all quadrupeds of great strength and swiftness, were not the scapulæ sustained by muscles, and not by bone, and did not the bones recoil and fold up."

The varieties of form observed in the clavicles are very great. In the Two-toed Sloth they have the form of a rib; in the Mole they are nearly cubical. The clavicles are very long in the Bat, but in the Quadrumana they greatly resemble the same bone in Man. The clavicle is not found in the Cetacea, Ruminantia, or Solipeda, and generally it is deficient in all long-legged quadrupeds, having a keel-shaped chest.

The anterior extremity of the Mammalia is continued by an arm, a fore-arm, and a hand.

The Arm-bone, or Humerus, varies considerably in the elevation of its processes, and in its length and breadth. In quadrupeds, the Humerus is much shorter comparatively than in the human subject. It was this circumstance which has led the ancient anatomists, and persons ignorant of Comparative Anatomy, to consider quadrupeds as having the elbow turned forwards. That part which is usually considered the knee of a Horse, corresponds to the human wrist, the arm-bone being concealed within the muscles of the shoulder. The Humerus is very long in the Bat, but very short and thick in the Mole.

The human fore-arm is composed of two distinct bones, the Ulna and the Radius; the former swings with a hinge-like motion upon the elbow, and the latter gives the wrist and hand a rotatory motion. Accordingly, in all animals which use their fore-arms, like Man, for other purposes than walking, both of these bones are distinctly developed; but, in the true quadrupeds the Ulna diminishes in size, is altogether absent, or becomes a mere appendage to the Radius, which is then the principal bone of the fore-arm. In cattle, the Ulna is immoveably united to the Radius throughout its entire length, becomes finally ossified, and may then be considered but as a single bone. They thus form a perfect hinge-like joint, which does not admit of any rotatory motion, and having the pulley placed on the end of the Humerus. These quadrupeds have therefore their anterior extremities always in a state of pronation, that is, the back of the wrist is always turned upwards.

The general arrangement of the bones in the anterior limbs, though the same throughout the entire class, yet changes surprisingly in its details with the different orders and genera. In the Bat the radius is nearly deficient, being reduced merely to a slender and sharp-pointed rudiment. Its thumb is short, and furnished with a hooked nail, while the phalanges of the four fingers have no nails, but are extremely long and thin, almost like the spines of a fish, while the membraneous, or wing-like expansion, is extended upon them. Again, in the Mole, we see a striking difference in the form of the anterior extremity. The bone of its fore-arm is thin in the middle, but surprisingly expanded at either extremity; and a peculiar bone, called the *falciform* bone, is found at the extremity of the radius. Its paw is shaped like a shovel, the phalanges of the fingers are supplied with *sesamoid* bones and numerous processes, which increase the angle of insertion for the tendons, and facilitate the great muscular motion required by this little animal. The Flying Squirrel possesses a peculiar arrangement for enabling it to spring from great heights. This consists in a sharp-pointed bone at the outer edge of its Wrist, connected to that part by means of two smaller round bones.

In general, the Radius forms the principal bone in the fore-arm of the Mammalia, while the Ulna is a slender and small bone, which is frequently consolidated with the radius, and terminates in a point before reaching the wrist. There are only a few genera which possess the power of rotating the wrist freely, so as at one time to present the palm of the hand downwards (pronation), and at another time the palm of the hand upwards (supination). This power diminishes in proportion as the fore extremities are used for progression, and for supporting the body in standing. While in these positions, the fore extremity is always in a state of pronation. The radius and ulna are flattened in the Cetacea and Seal; and in the Elephant the lower extremity of the ulna is larger than that of the radius, a conformation which is peculiar to that animal.

The Hand is formed of two rows of bones called the Carpus or Wrist; of one row called Metacarpus; and of Fingers, each composed of two or three bones, called Phalanges.

The forms of the wrist and remote extremities vary with the delicacy of the organs of touch. It is only in few genera that a Hand, properly so called, is found; but when it exists it is always much less complete, and therefore less useful than that of Man. The mechanism of the human hand is exceedingly curious, and admirably adapted for the various purposes of life. Anaxagoras is said to have maintained that Man owes all his wisdom and superiority over the other animals to the use of the hand; but Galen's view of the matter was rather more philosophical. According to him, Man is not the wisest creature because he has hands; but he had hands given him because he was the wisest creature, for it was not hands that taught him the arts, but Reason. The great superiority of the human hand over that of any other animal, arises from the circumstance that his thumb is of a great size and strength, and can be brought in opposition to the fingers. It thus gives him the power of holding whatever he seizes; and were it not for the thumb, various arts and manufactures of civilized life would either remain unexecuted, or would require the awkward concurrence of both hands. Albinus calls the thumb a second hand. "Manus parvæ major adjutrix." The want of the thumb, and the absence of fingers of any great length, compel the Squirrel, Rat, and Opossum, to hold objects in both paws. The Cat and Dog, which are obliged to use their fore-paws for progression, only hold objects by fixing them on the ground. The Solipeda and Ruminantia cannot hold objects at all. On comparing the hand of the Apes with that of Man, which it most resembles, we cannot fail to remark the smallness of their thumbs in proportion to the length of the fingers. These are slender and very long, while the thumb is weak, small, and short. In the Cercopitheci, the thumb is concealed under the integuments, and their mode of seizing food and bringing it to the mouth differs but slightly from that employed by the Squirrel. The fore-foot of the Horse is terminated by a single bone, called the coffin bone, corresponding in some degree to the third phalanx of the human finger, as the pastern, to which it is united, is analogous to the first phalanx. There are also two short and immoveable bones placed behind and on each side of the coffin bone, called splint bones.

If we except the Cetacea, all the Mammalia have the first part of their hinder extremity attached to the spine, and formed into a Pelvis or Basin, which, during youth, is formed of three pairs of Bones, the Ileum, attached to the Spine; the Pubis, which forms the anterior; and the Ischium, forming the hinder part of the Pelvis.

All these bones are generally more narrow and elongated in the quadrupeds than in Man. In no instance do they form a basin like the human Pelvis; while frequently the distinction between the large and small Pelvis does not exist, and the cavity often looks obliquely upwards, towards the Spine. One class of Mammalia, possessing abdominal pouches, have two small Marsupial bones placed on the anterior part of the Pubis. These are of a flat and elongated form, and serve to support the abdominal pouch of the female; yet in some species they are also found in the male.

At the point of union among the three bones of the Pelvis, is placed the cavity, to which the Thigh-bone or Femur is articulated. To this last bone is attached the Leg, composed of two bones, the Tibia or Shin-bone, and the Fibula.

The Femur is remarkably short in quadrupeds having a long Metatarsus, as in the Horse, and in common cattle; and the bone is so enveloped with muscle, that the part usually called the thigh is really the leg. The Fibula, in many quadrupeds, bears a striking analogy to the Ulna of the fore-arm, from its declining in size, and becoming united, by anchylosis, with the Tibia, or else appearing merely in a rudimental form.

This extremity of the Mammalia terminates in a Foot composed of parts analogous to those of the hand, namely, a Tarsus, Metatarsus, and Toes.

These bones are altogether wanting in the Whale, Dolphin, and other Sea Beasts. They have no pelvis, properly so called, as the Ischia are absent; yet we find two small isolated bones which may be compared to the ordinary Pubis.

The head of the Mammalia is always articulated by two condyles upon their atlas or first vertebra.

This name of Atlas is assigned to the first vertebra, because it sustains the globe of the head. The second vertebra is called the Dentata, or Axis, because it has a tooth-like process, or axis upon which the first turns. "I challenge any man," says Dr Paley, "to produce in the joints and pivots of the most complicated or the most flexible machine that was ever contrived, a construction more artificial, or more evidently artificial, than that which is seen in the vertebræ of the Human neck. Two things were to be done. The head was to have the power of bending forward and backward, as in the act of nodding, stooping, looking upwards or downwards, and at the same time, of turning itself round upon the body to a certain extent, the quadrant

we will say, or rather perhaps a hundred and twenty degrees of a circle. For these two purposes, two distinct contrivances are employed: First, the head rests immediately upon the uppermost of the vertebræ, and is united to it by a hinge-joint, upon which joint the head plays freely forward and backward, as far either way as is necessary, or as the ligaments allow, which was the first thing required. But then the rotatory motion is improvided for. Therefore, secondly, to make the head capable of this, a further mechanism is introduced—not between the head and the uppermost bone of the neck, where the hinge is, but between that bone and the bone next underneath it; it is a mechanism resembling a tenor and mortice. The second, or uppermost bone but one, has what anatomists call a process, viz. a projection somewhat similar, in size and shape, to a tooth: which tooth, entering a corresponding hole or socket in the bone above it, forms a pivot, or axle, upon which that upper bone, together with the head which it supports, turns freely in a circle, and as far in the circle as the attached muscles permit the head to turn. Thus are both motions perfect without interfering with each other. When we nod the head, we use the hinge-joint, which lies between the head and the first bone of the neck. When we turn the head round, we use the tenor and mortice, which runs between the first bone of the neck and the second. We may add, that it was on another account also expedient that the motion of the head, backward and forward, should be performed upon the upper surface of the first vertebra; for, if the first vertebra itself had bent forward, it would have brought the spinal marrow, at the very beginning of its course, upon the point of the tooth."

The Brain of the Mammalia is always composed of two hemispheres, united by a medullary layer called the Callous body (Corpus Callorum), inclosing two ventricles, and enveloping the four pairs of Tubercles, or eminences called the Striated Bodies (Corpora Striata), the Beds of the Optic Nerves (Thalami Nervorum Opticorum), the Nates, and the Testes. Between the Beds of the Optic Nerve is placed the third ventricle, which communicates with the fourth, situate under the Cerebellum. The crura of the Cerebellum always form, under the Medulla Oblongata, a transverse eminence called the Bridge of Varolius (Pons Varolii).

The Brain in the Monkey tribe is rather flatter in the superior surface of its hemispheres than in Man; but in Quadrupeds it is very considerably flatter. In the Dolphin, and other Sea-Beasts, the Brain has a different shape from that of the other Mammalia, being rounded in every part, while its greatest diameter is across. There are no olfactory nerves in the cetaceous animals, while those of quadrupeds are of an enormous size, especially in the larger herbivorous tribes.

The proportion which the size of the Brain bears to that of the entire Body, varies greatly for different Mammalia. Even in the same individual it will change with the degree of fitness, or with the age of the animal. As these circumstances cannot be supposed to affect the powers of the mind very materially, we may naturally inquire how the relative size of the brain, and of the entire body, can be assumed as the measure of intelligence in an animal. To enable the student to form his own conclusions on this subject, we annex the following TABLE, showing the proportion that the size of the whole body bears to that of the Brain in several animals:—

The Squirrel Monkey (Callithrix sciureus),	as 22 to 1
Capuchin Monkey (Cebus capucinus),	... 23 ... 1
Striated Monkey (Jacchus vulgaris),	... 28 ... 1
Field Mouse (Arvicola vulgaris),	... 31 ... 1
MAN, according as he is young or old,	as 22, 25, 30, and 36 ... 1
The Mole (Talpa Europæa),	as 35 ... 1
Coaita Monkey (Ateles paniscus),	... 41 ... 1
Mouse (Mus musculus),	... 43 ... 1
Varied Monkey (Cercopithecus mona),	... 44 ... 1
Gibbon (Hilobates lar),	... 48 ... 1
Collared Mangabey Monkey (Cercopithecus Æthiops),	... 48 ... 1
Rat (Mus decumanus),	... 76 ... 1
Ruffed Lemur (Lemur Macaco),	... 84 ... 1
Porpoise (Delphinus phocæna),	... 93 ... 1
Great Bat (Vespertilio Noctula),	... 96 ... 1
Dolphin (Delphinus delphis),	as 25, 36, 66, and 102 ... 1
Great Baboon (Papio Maimon),	as 104 ... 1
Barbary Ape (Innus magotus),	... 105 ... 1
Ferret (Mustela furo),	... 138 ... 1
Rabbit (Lepus cuniculus),	as 140, and 152 ... 1
Cat (Felis catus),	as 82, 94, and 156 ... 1
Hedgehog (Erinaceus Europæus),	... as 168 ... 1
Fox (Canis vulpes),	... 205 ... 1
Calf (Bos taurus junior),	... 219 ... 1
Hare (Lepus timidus),	... 228 ... 1
Wolf (Canis lupus),	... 230 ... 1
Panther (Felis pardus),	... 247 ... 1
Ass (Equus asinus),	... 254 ... 1
Bear (Ursus arctos),	... 262 ... 1
Beaver (Castor fiber),	... 290 ... 1
Sheep (Ovis aries),	... as 192, and 351 ... 1
Marten (Viverra martes),	... as 365 ... 1
Dog (Canis familiaris),	as 47, 50, 57, 154, 161, and 365 ... 1
Horse (Equus caballus),	... as 400 ... 1
Domestic Hog (Sus scropha),	... 412 ... 1
Elephant (Elephas Indicus),	... 500 ... 1
Wild Boar (Sus scropha),	... 672 ... 1
Ox (Bos taurus),	... 860 ... 1

From the above Table it would appear that the Brain is proportionably largest in the smaller animals. Man is surpassed in this respect only by a small number of

Mammalia, and these are lean and meagre. The Rodentia generally possess the largest proportional Brain, and the Pachydermata the smallest. It is very difficult, if not impossible, to arrive at these results with any great degree of accuracy, because the weight of the brain generally remains the same, while that of the body will vary considerably according as an animal is lean or fat.

The proportion which the Cerebrum bears to the Cerebellum is, in

The Squirrel Monkey (*Callithrix sciureus*), . .	as 14 to 1
Man, ...	... 9 ... 1
The Ox (*Bos taurus*), ...	... 9 ... 1
Dog (*Canis familiaris*), ...	... 8 ... 1
Wild Boar (*Sus scropha*), ...	... 7 ... 1
Horse (*Equus caballus*), ...	... 7 ... 1
Cat (*Felis catus*), ...	... 6 ... 1
Hare (*Lepus timidus*), ...	... 6 ... 1
Sheep (*Ovis aries*), ...	... 5 ... 1
Mouse (*Mus musculus*), ...	... 2 ... 1

"It is a common opinion," observes Dr Herbert Mayo, "that the front of the brain is the seat of the intellectual faculties; yet in Monkeys and in Man the back part of the brain is that which has the largest relative size. The sheep, on the other hand, has an ample front to its brain, a large intellectual region, according to the phrenological theory, while its instinct of attachment to its young has a poor locality in its moderate posterior cerebral lobe. Has nothing then been discovered to mark an essential superiority in the brain of Man? The question must, I believe, be answered in the negative. No physical condition, distinguishing the human brain from that of animals, and therefore fitting it to co-operate with a rational soul, has as yet been ascertained, or even plausibly conjectured to exist."

Physiologists have been led, in all ages, by that marked superiority of mental power which Man possesses above the other animals, to seek in the structure of their brains for some corresponding difference. It was long supposed that Man has the largest brain in comparison to his body; but the above Tables show that he is surpassed by several Quadrumana, and by the Mouse.

There is another point of comparison which seems to approach nearer to their actual comparative intelligence, which was first proposed by Sömmering. By comparing the quantity of the brain with that of the nerves arising from it, we ascertain more accurately the degree in which its purer intelligence excels its mere animal nature. "Let us divide the brain into two parts; that which is immediately connected with the sensorial extremities of the nerves, which receives the impressions, and is therefore devoted to the purposes of animal existence. The second division will include the rest of the brain, which may be considered as connecting the functions of the nerves with the faculties of the mind. In proportion, then, as any animal possesses a larger share of the latter and more noble part—that is, in proportion as the organ of reflexion exceeds that of the external senses—may we expect to find the powers of the mind more vigorous and more clearly developed. In this point of view Man is decidedly pre-eminent; here he excels all other animals which have hitherto been investigated." Sömmering found that the brain of Man never weighed less than 2 lb. 5½ oz., while that of the Horse never exceeded 1 lb. 4 oz. in weight. But the nerves arising from the brain of the Horse were at least ten times larger than those in Man.

However ingenious this theory may be, it is not found to hold good in every instance; and even if proved, it would still leave the nature of the union between Mind and Matter as mysterious and as incomprehensible as ever.

The nerves of the Mammalia bear a striking resemblance in their disposition to those of Man, with the exception of the olfactory nerves, which are large and hollow processes of the anterior lobes of the cerebrum, the cavities of which communicate with the lateral ventricles of the brain.

Their eyes, invariably placed in their orbits, and preserved by two eyelids and the vestige of a third, have their crystalline humour preserved by the ciliary process, and sclerotic coat, composed of simple cellular substance.

The eye-ball of Man is nearly globular, and about one inch in diameter. It is defended externally by a white opaque membrane, having the density of tanned leather, called the *Sclerotica*, which surrounds every part, excepting the small circular portion in front called the *Cornea*, and a small perforation behind to admit the optic nerve into the brain. The Cornea is perfectly transparent and possesses great sensibility; yet at the same time it is so tough, as to offer a powerful resistance to external injury. Within the Cornea, and immediately in contact with it, is placed a small quantity of pellucid fluid called the *Aqueous* humour, which occupies the external visible portion of the Eye. The *Iris* is situate behind the Cornea and the Aqueous humour. It consists of a circular membrane, perforated by a small hole in the centre called the *Pupil*. The colour of the Eye resides in this membrane, and both its structure and functions are very remarkable. It is formed of two layers of fibres; in the one, they are arranged like rays from the inner to the outer margin; and in the other, they form concentric circles. By the action of the radiated fibres, the pupil is contracted; but it is dilated by the action of the circular fibres. These delicate motions are executed in a manner which human ingenuity in vain attempts to equal. The pupil instantly contracts when exposed to a strong light; but when the light is deficient, it dilates readily, in order to admit as great a number of rays as possible. Behind the pupil, a fluid in the shape of a double convex lens is placed, called the *Crystalline* humour, formed of denser materials than any other liquid in the human Eye. The remaining portion is filled up with the *Vitreous* humour, which is the most plentiful of all its fluids. The inside of the Sclerotica being lined with a coloured viscid secretion called *Pigmentum nigrum*, or black Pigment, and the inner surface of the Iris, with a dark brown pigment, any scattered rays of light which would otherwise render vision obscure are intercepted. These pigments are therefore of the same utility with the black paint which lines the interior of telescopes and other optical instruments. The Retina, or expanded extremity of the optic nerve, is the immediate seat of sensation. Upon this delicate membrane, an inverted picture of the external scene is exactly delineated; and though scarcely half an inch in diameter, it contains the forms, positions, and colours of the most distant objects, without confusion or irregularity.

The black pigment of the Eye is wanting in that variety of the Human race called Albinos, and this deficiency is connected with the want of colouring principle in the hair. The Rabbit, Mouse, and Horse, are sometimes found to possess this peculiarity when they are said to be *glass-eyed*. They are capable of transmitting it to their posterity, and thus forming a breed of white animals. The Ferret's eye is naturally destitute of the black pigment.

The immediate object of vision is colour, and to this alone its function was originally confined. By habit and by comparison with the other senses, but especially with that of touch, it acquires new powers; and the coloured canvas of original perception is embodied into a real scene, in which the distance, magnitude, and figures of objects may be, in general, instantaneously discovered. This remarkable fact was fully established by Mr Cheselden, who couched the eyes of a young gentleman born blind, and ascertained the effects produced on the mind by the first exercise of the power of vision.

By certain exquisite contrivances, the Eye is so constructed as to correct the defects which Opticians experience in their own artificial instruments. But the consideration of the Spherical, Paralactic, and Chromatic Aberrations of light, belongs more properly to the science of Optics.

In the Mammalia, the eye is composed of the same coats and humours as in Man, being only slightly modified and adapted to surrounding circumstances. The eyes of the Cetaceous tribes more nearly resemble those of Fishes, as they are flattened on the anterior side, and adapted to the dense medium in which they reside. By the exquisite arrangement of the sclerotic coat and cornea in the Seal tribe, these animals are able to adapt their vision to the two different media of air and water; and they are enabled to shorten or elongate the axis of the eye according to circumstances. The eye of the Mus typhlus, or Blind Rat, is not larger than a poppy seed, and is altogether covered with hair, so that the animal can scarcely perceive the difference between light and darkness. The eyes of the Mole are so minute that most persons imagine them to be entirely absent. The pigment at the back of the eye receives the name of the *Tapetum lucidum*; it is of different colours in different animals.

In the Monkey it is	dark-coloured.
Ox, ...	green.
Sheep, ...	pale yellow.
Rabbit, Hare, and Hog, ...	brown.
Lion, Bear, Cat, and Dolphin, . .	pale golden yellow.
Dog, Wolf, and Badger, ...	pure white, fringed with blue.
Horse, Goat, Stag, and Buffalo, . .	silvery blue, changing to violet.

It has been supposed that the Tapetum enables these animals to see more distinctly in the dark.

In the Ear a cavity is always found called the Tympanum or Ear-drum, which communicates with the hinder part of the mouth by a canal called the Eustachian tube. This cavity is closed externally by a membrane, termed the *membrana tympani*, and contains within it a chain of four bones, called the Malleus, or Hammer Bone; the Incus, or Anvil; the Os Orbiculare, or Lenticular Bone; and the Stapes, or Stirrup. The Stapes rests upon a vestibulum or central porch, which leads on the one side to three semi-circular canals, and on the other to a Cochlea, or spiral canal, which communicates by one extremity with the vestibulum, and by the other with the Tympanum.

The Vestibule, Cochlea, and semi-circular canals, collectively form the *Labyrinth*. This is the essential part of the organ; and no obstruction or removal of the external parts can altogether destroy the sense of hearing as long as the Labyrinth remains uninjured. The precise uses of these several parts are not yet fully understood.

The Mammalia do not appear to be capable of distinguishing musical sounds with that nicety, or of deriving that extent of enjoyment which has been bestowed on Man. The Dog may be trained to distinguish one particular tune from another; but the extent of his pleasure or acquirements seems greatly limited. But Man, besides being able to perceive differences between acute and grave tones, is capable of distinguishing four or five hundred varieties in their quality and intensity. A flute, hautboy, and violincello may all sound the same note, and yet the peculiar quality of each may be readily perceived. An attentive ear will observe some differences although twenty human voices sound the same note, and with equal strength; nay, even the same voice may be varied in many ways, by sickness or health, youth or age, leanness or fatness, good or bad humour; and the same words spoken by foreigners and natives, or even by the inhabitants of different provinces in the same nation, will be readily distinguished.

To ascertain the effect of the high and low notes of the piano-forte upon the Elephant and Lion, Sir Everard Home procured one of Broadwood's piano-fortes to make the experiment at the Menagerie in Exeter Change, London. The Elephant was first tried. His attention was scarcely attracted by the high notes; but when the low ones were played, he brought his broad ears forward, remained evidently listening, and made use of sounds rather expressive of satisfaction than otherwise. The full sound of the French horn produced the same effect. But the Lion was much more forcibly affected. When the high notes were played he remained silent and motionless, but listening with deep attention. No sooner, however, were the low notes sounded than he sprang up, endeavoured to break loose, lashed his tail, and appeared to be enraged and furious, so much so, as to alarm the female spectators. This was accompanied with the deepest yells, which ceased with the music.

The Cranium, or skull of the Mammalia, may be subdivided into three compartments; the anterior portion, containing the two frontal bones and the Æthmoid; the central portion, being the Parietal bones and the Sphenoid; and the hinder-portion, being the occipital Bone. Among the occipital, the two parietal, and the sphenoid, are interposed the Temporal bones, of which one portion properly belongs to the face.

The bones found in the skulls of the Mammalia frequently differ in number from those of Man. In some, the sutures which are always observable in the human Cranium are obliterated at an early period of life, and two, three, or more bones, are consolidated into one. In other species, some bones which become consolidated into one in Man, remain during their entire lives as separate pieces. In the Elephant, all the sutures of the skull soon become united into one solid piece.

The occipital bone is divided into four portions, during the first or Fœtal period of life. The body of the sphenoid bone is then composed of two middle parts, which are themselves subdivided, so as to form three pairs of lateral wings. The temporal bone is composed of three portions; one of these serves to complete the Cranium; another to close the labyrinth of the Ear; while the third forms the walls or parietes of the Ear-drum. These bony portions are multiplied to a still greater extent in the first age of the Embryo; they coalesce more or less rapidly according to the species; and the bones themselves finally unite into one in the adult animal.

The face in the Mammalia is essentially formed by the two maxillary bones, between which passes the canal of the nostrils. In front of them are placed the two intermaxillary bones, and behind the two palate bones; between them descends the single projecting plate of the æthmoid bone, called the *vomer*, and upon the entrances to the nasal canal are situate the bones distinguished by the proper term *nasal*. To the external parts of its entrance are found the inferior turbinated bones; the superior turbinated bones, on the contrary, belong to the æthmoid bone, and are placed behind and above.

To this complicated arrangement of the bones of the nose in the Mammalia, these animals owe their superiority over man in receiving impressions of odoriferous effluvia. The inferior and superior turbinated bones are greatly subdivided and convoluted. The obvious design of this arrangement, is to extend the surface of the pituitary membrane which is spread over them; and the extent of this surface is always found to bear a constant relation to the acuteness of the sense of smell.

The frontal sinuses, and in general the sinuses of all the bones in the neighbourhood of the nasal cavity, are very large, which has led several eminent physiologists to consider them as subservient to the organ of smelling; others consider these cavities to be merely reservoirs for containing a watery fluid, which lubricates the parts where this sense more especially resides.

The Cetaceous tribes do not possess the sense of smell, nor have they any organ which appears capable of exercising it. The two canals which correspond to the nostrils are used by the Whale tribes for transmitting air to and from the lungs. They do not respire through the mouth, and the nostrils are placed on the top of the head. By this arrangement they can swallow their food and keep the mouth in water, without interrupting their respiration.

The Jugal, or cheek-bone, unites the maxillary to the temporal bone, and often to the frontal. The lachrymal bone occupies the internal angle of the orbit, and sometimes a part of the cheek. In the embryo, all these subdivisions are much more numerous. The tongue in the Mammalia is always fleshy, and attached to the hyoid bone (*os hyoides*); it is composed of several pieces, and suspended to the cranium by ligaments.

It is generally supposed that the sense of taste resides exclusively in the tongue; but this is not strictly correct. Some substances will excite particular tastes on passing over the inside of the lips and fauces. Blumenbach mentions that he had seen a man, in other respects well formed, who was born without a tongue; yet he could distinguish very readily the tastes of solutions of salt, sugar, and aloes, when rubbed on his palate, and would express the taste of each in writing. The tongue of the other Mammalia differs always from that of Man. In the Monkey tribes it is longer and thinner. The entire Cat genus have horny integuments surrounding the conical papillæ, which are on the middle of the tongue. These are small hooks or claws, sharp-pointed, and inflected backwards; so that when any of the larger animals of this genus employ the tongue in licking the human hand, they tear off the skin.

There does not appear to be any conical papillæ on the tongues of the Cetacea. Cuvier was unable to discover them, even with a glass, upon the tongues of the Dolphin and Porpoise; and John Hunter compared the tongue of the large Whales to a feather-bed. The worm *(lytta)* of the dog's tongue is a tendinous bundle of fibres, running length-wise under the tongue. Casserius thought that it assisted dogs in lapping up fluids. We need scarcely observe that the practice of cutting out the worm as a preventive of Hydrophobia, though sanctioned by Pliny, is an old prejudice long since exploded. The Edentata, such as the Ant-eater and Manis, possess a long worm-like tongue, which is apparently used for no other purpose than for taking up the food.

Their Lungs, two in number, are divided into lobes, composed of an infinite number of small cells: they are always inclosed loosely in a cavity formed by the ribs and the diaphragm, and lined by the pleura.

The number of lobes in the lungs often varies in individuals of the same species, but in general they are more numerous than in the human species. A due proportion is always observed between the size of the lungs and that of the animal, although the external form of the chest would lead an observer to arrive at an opposite conclusion. The convexity of the diaphragm is not considerable, and the thorax is proportionately wide in species having a short chest; but, when the thorax is long, the diaphragm projects far into the chest, and the thorax is narrowed. Thus, in the rhinoceros, elephant, and horse, the diaphragm passes up into the thorax, and permits the viscera to lie within the margin of the ribs.

The Mammalia respire in a manner exactly resembling Man. Atmospheric air rushes into the cells of the lungs through the windpipe the instant after birth; it is expelled and replaced by fresh air, and the action continues as long as life remains. Although the muscles which enlarge the chest were to act with unlimited force, no air could enter the lungs at each attempt at inspiration, if they were of a firm and inelastic texture. A vacuum would, on the contrary, be formed between the pleura pulmonalis or external covering of the lungs, and the pleura costalis or internal lining of the ribs. But the lungs are highly elastic and free in their motion, so that atmospheric air rushes into and dilates the cells, exactly in proportion to the expansion of the area of the chest. When any cause prevents the air from rushing into the lungs, death by suffocation or *asphyxia* is occasioned. On examination, the lungs are found collapsed, as during expiration; the right cavities of the heart, and the veins leading to them, are filled with dark blood, while the left cavities of the heart and the arteries are nearly empty. In animals of the first class, which are hanged, death is occasioned by strangulation, and not by apoplexy, as is frequently supposed. This was proved by Gregory, who opened the windpipe of a dog, and passed a noose round his neck above the wound. The animal, when hanged, continued to live, and to breathe through the small aperture; but he died when the rope was attached below the wound. M. Richerand asserts that a respectable surgeon in the Austrian army had informed him that he once saved the life of a soldier by performing the operation of opening the windpipe, a few hours before his execution. The soldier, feigning to be dead, was cut down, delivered over to the surgeon for anatomization, and thus finally escaped.

The glottis through which the external air rushes into the lungs, is so small that it may be readily obstructed when the epiglottis rises during the act of swallowing, and the substance swallowed may stop up the mouth of the larynx. Anacreon, the celebrated poet, was in this manner suffocated by a grape seed, and Gilbert, also a poet, met his death in a similar manner.

The organ of voice, in the Mammalia, is always at the superior extremity of the Trachæa or windpipe;—a fleshy prolongation, called the velum palati or palate-curtain, establishes a direct communication between the Larynx and the back part of the nostrils.

"The human voice," says Sir Charles Bell, "commences in the Larynx, but reverberates downwards into the Trachea, and even into the chest, whilst it may be directed with different effects into the cavities of the head, mouth, and throat. The organ of voice is neither, strictly speaking, a stringed instrument, nor a drum, nor a pipe, nor a horn, but it is all these together; and we will not be surprised at this complication, if we consider that the human voice is capable of every possible sound,—that it can imitate the voice of every beast and bird,—that it is more perfect than any musical instrument hitherto invented;—and, in addition to every variety of musical note, it is capable of all combinations, in articulate language, to be heard in the different nations of the earth. The essential and primary parts of the organ are the *vocal cords*, or thyro-arytænoid ligaments. The membrane lining the larynx is reflected over these ligaments, so as to be drawn by them in their motions; and this is what is meant when it is said the organ is like a drum, for these membranes must vibrate in the air. The muscles of the arytænoid cartilages draw tight the vocal cords and their attached membranes, and thus give them a certain tension; and the air being expelled forcibly from the chest at the same time, they cause a vibration of these ligaments and membranes. This vibration is communicated to the stream of air, and sound is produced. This sound may reverberate along all the passages from the lungs to the nostrils; but unless there be a certain vibration in these cords of the larynx, there is no vocalization of the breath. For example, a man in whispering articulates the sounds of the mere breath, without the breath being vocalized and made audible by the vibrations in the larynx. In singing, the vocalized breath is given out uninterruptedly through the passages, the rising notes in the gamut being produced, first, by the narrowing of the glottis, and secondly, by the rising of the larynx towards the base of the skull. In the graver notes, the larynx is drawn down, and the lips protruded; and in the higher notes the larynx is elevated to the utmost and the lips retracted." The various sounds emitted by different animals, to which we assign the terms roar, bray, howl, purr, scream, whistle, bark, grunt, snort, and hiss, are all caused by peculiarities in the construction of their vocal organs, which will be explained hereafter.

As the Mammalia [generally] reside on the surface of the earth, where they are exposed to moderate variations of temperature, their covering of hair is but moderately thick; and in many of the animals inhabiting warm countries this integument is generally deficient. The Cetacea, which live entirely in the water, are, however, the only species wherein it is altogether wanting.

The abdominal cavity of the Mammalia is hung round with a membrane called the Peritoneum, and their intestinal canal is suspended to a fold of this peritoneum, called the Mesentery, containing numerous conglobated glands, in which the lacteal vessels are ramified. Another production of the peritoneum, called the Epiploon, hangs before and beneath the Intestines.

The uses of these several parts are precisely the same in the other Mammalia as in Man; but their form and extent depend upon the convolutions and length of the intestinal canal; and therefore its reflexions, which form the omentum and the envelopes of the intestinal canal differ greatly among the several quadrupeds. There are lateral omenta in some of the quadrupeds, which hybernate, such as the Polish and Alpine Marmots, in addition to the usual omenta of other quadrupeds. They arise from the loins, cover the sides of the abdomen, and advance nearly to its centre. These processes of the Peritoneum become loaded with fat, about the period that the animals remain torpid, and the fat is entirely expended during the time of their hybernation. The use of these lateral omenta is sufficiently obvious; yet it is very singular that other species which sleep during the winter, and are nearly allied to those just mentioned, such as the Garden Dormouse (*Myoxus nitela*) and the common Dormouse (*Myoxus avellanarius*) are destitute of them.

L'urine, retenue pendant quelque temps dans une vessie, sort, dans les deux sexes, à un très petit nombre d'exceptions près, par les orifices de la génération.

In all the Mammalia [with the sole exception of the Monotremata] the generation is essentially viviparous. Immediately after conception, the fœtus descends into the Womb, surrounded with its membranes, of which the exterior is called the *chorion*, and the interior *amnios*. It is fixed to the sides of this cavity by one or more folds of vessels called the *placenta*, which establish the communication with the mother, from whom it derives its nourishment, and probably also its oxygenation. In the earlier periods of gestation, the fœtus of the Mammalia possesses a small vessel, analogous to that which contains the yolk of the Oviparous animals, and receiving supplies from the vessels of the mesentery in a similar manner.

Ils ont aussi une autre vessie extérieure, que l'on a nommée allantoide, et qui communique avec celle de l'urine, par un canal appelé l'ouraque.

La conception exige toujours un accouplement effectif, où le sperme du mâle soit lancé dans la matrice de la femelle.

The young are nourished for some time after their birth by MILK, a fluid peculiar to this class, and produced by MAMMÆ, or Breasts. This secretion commences at the moment of birth, and continues as long afterwards as the young may require. It is from these Mammæ that the class has obtained its name of MAMMIFÈRES, or MAMMALIA. This being a characteristic peculiar to the animals composing this class, serves to distinguish them more precisely from the remaining classes than any other external character. It remains, however, still doubtful whether the Monotremata possess mammæ or not.

Meckel could find no traces of Mammæ in the male Ornythorynchus, but thought he perceived them in the female. " I detected, on the right side of the abdominal muscles," he observes, "a small round mass, which at first bore the appearance of a portion of intestine accidentally pushed into this situation. I was satisfied that this gland was a true Mamma, an opinion which was more forcibly impressed upon my mind from its structure and situation, from its marked development in the female, and the want of it in the male, or at least its existence in so minute a degree as to have hitherto eluded the closest examination." Oken and De Blainville asserted, *à priori*, and without having ever examined a female Ornithorynchus, that its Mammæ must exist, and would no doubt be discovered hereafter, on account of the very numerous analogies which this animal presents to the other Mammalia. Sir Everard Home describes the Mammæ of the Ornithorynchus in the Philosophical Transactions for 1802.

On the other hand, M. Geoffroy considers that these organs are not real Mammæ, but are analogous only to the lateral glands of the Muscardin (*Myoxus avellanarius*.) Again, the Ornithorynchus is either oviparous or ovo-viviparous, which properties are always connected with the absence of Mammæ, and its bill evidently appears unfitted for sucking; so that, upon the whole, it must still be considered as doubtful whether these organs really perform the functions of Mammæ.

Although the Mammæ are always found, with the above exception, in the females, yet the males of many species are destitute of them, as the Hamster (*Cricetus vulgaris*), and the *Lemur mongoz*, while in some others, as the Horse, they are found in an unusual situation. The Mammæ are frequently less numerous in the male than in the female. Milk has often been secreted in the breasts of Men, as well as of other male animals, such as the Goat, Ox, Dog, Cat, and Hare. Blumenbach describes a he-goat which it was necessary to milk every other day for the space of a year. It is very common to find milk in the breasts of newly-born children of both sexes; and the same circumstance has likewise been observed in the calf and foal.

In the Cetacea and Marsupialia the Mammæ do not project so as to form udders or breasts, but they lie flat under the skin. In general the Mammæ are very observable only during the period of suckling, at which time they are largely distended with milk, except in those animals having them placed upon the chest, when they possess that graceful and delicate form observable in the human female of the Caucasian race during the bloom of youth. It is very difficult to discover them in the Marsupial animals, except at the period when the young are actually contained in the abdominal pouch of the female. The number, as well as the position of the Mammæ, varies greatly in different animals. It would appear that there are frequently twice as many teats as the number of young usually produced by each animal. Yet this rule is not without several exceptions, among which may be included the Guinea-pig (*Cavia cobaia*), and the Domestic Sow. Indeed it is among the domesticated races that these exceptions are chiefly found. Thus, according to Buffon, the number of the Sow vary from ten to twelve; of the Cow from four to six; of the Rat from eight to ten. The Mare and Ewe may have from two to four, while the Ferret sometimes has three on the right side, and four on the left. From these examples we may readily perceive that no fixed law is observed in the number of the mammæ.

10

1. *Bimana*—2. *Quadrumana*—3. *Carnassiers*—4. *Rodentia*—5. *Edentata*—6. *Marsupialia*—7. *Pachydermata*—8. *Ruminantia*—9. *Cetacea*.

THOSE variable characters, which establish the essential differences of the Mammalia among themselves, are derived jointly from the organs of touch and from those of mastication. The forms of the hands or feet chiefly determine the degree of their agility and dexterity, while those of their teeth not only correspond to the nature of their aliments, but draw along with them innumerable other distinctions, relative to the digestive organs, and even to the intellectual functions.

The degree of perfection in the organ of touch is estimated by the number of the fingers, their capability of motion, and the extent in which their extremities are enveloped in a nail or hoof.

A HOOF which entirely surrounds that extremity of the finger nearest to the ground, blunts its sense of touch, and renders it incapable of grasping an object.

The opposite character is found in the NAIL, composed of a single layer, which covers the one side only of the extremity, and leaves to the other the utmost sensibility of touch.

The nature of their ordinary food is determined by the form of the MOLAR or CHEEK TEETH, and this always corresponds to the mode in which the jaws are articulated. In order to cut flesh, the Molars must be serrated, or saw-like, and the jaws united in the manner of scissors, which can only open and shut. On the contrary, in order to crush grains, it is necessary that they should have Molars with flat crowns, and jaws capable of moving horizontally. It is also requisite that the crown of these teeth should possess that kind of inequality which the millstone acquires, that its substance should be of different degrees of hardness, and that some of its parts should wear away more rapidly than others.

All animals with Hoofs [hence called UNGULATED] must of necessity be herbivorous, that is, possessed of Molar teeth with flat crowns, because the structure of their feet prevents them from seizing a living prey.

It is different with those animals said to be UNGUICULATED, from their possessing Nails. They are susceptible of several varieties, and may partake of different species of food; but they differ still more from each other in the extent of motion possessed by the fingers, and the delicacy of their touch. There is one characteristic which exercises a mighty influence on the degree of their address and means of industry—that is, the power of opposing the thumb to the other fingers, for the purpose of seizing small objects, which constitutes it a HAND, properly so called. It is in Man, whose fore-extremity is entirely free, and capable of being employed in seizing, grasping, or holding, that this power reaches its limit of perfection.

These different combinations, which determine rigorously the nature of the different Mammalia, have given rise to their subdivision into the following orders:—

THE UNGUICULATED MAMMALIA.

1. BIMANA.—Man alone possesses hands solely at his fore-extremities, and at the same time is privileged in many other respects, so as to entitle him to the first place among the unguiculated animals; his lower extremities alone support his body in a vertical position.

2. QUADRUMANA.—The order next to Man possesses hands at all the four extremities.

3. CARNASSIERS.—The third order has not the thumb free and opposable to the other anterior extremities.

All the animals of the above orders possess three kinds of teeth, namely, Molars, Canines, and Incisors.

4. RODENTIA.—The fourth order differs but slightly in the structure of the fingers from the Carnassiers, but it wants the Canine Teeth, and the Incisors are disposed in front for the peculiar kind of mastication, termed Gnawing.

5. EDENTATA.—Next follow those animals having the fingers very much confined, and deeply sunk into large nails, which are often very crooked. They also have the imperfection of wanting Incisors. Some also want the Canines, and others have no teeth at all.

6. MARSUPIALIA.—This distribution of the Unguiculated animals would have been perfect, and might form a chain of some regularity, if New Holland [and America] had not furnished us with a small collateral chain, composed of animals WITH POUCHES. All these genera resemble each other in the whole character of their organization, yet some of them correspond to the Carnassiers by the structure of their teeth, and the

nature of their food; others agree with the Rodentia in these particulars, and others again with the Edentata.

THE UNGULATED MAMMALIA.

The animals with Hoofs are less numerous, and at the same time less various in their structure.

7. PACHYDERMATA, or Jumenta, comprise all the hoofed animals which do not ruminate. The Elephant, though included in this class, would properly form a class of itself, which is allied to the Rodentia by some remote analogies.

8. RUMINANTIA.—The Ruminating animals form a very well-marked order, from their cloven feet, their four stomachs, and the absence of true Incisors in the upper jaw.

THE SEA-BEASTS.

9. CETACEA.—Finally, we arrive at the Mammalia altogether destitute of hinder extremities. From their partaking of the form of the Fishes, and their aquatic life, we should be led to constitute them a separate class, did not the remainder of their economy resemble the Mammalia in every respect. These are the Fishes with warm blood of the ancients [the Sea-Beasts of the present day], which unite the strength of the other Mammalia to the advantage of being sustained by the watery element. It is accordingly in this class that the most gigantic animals are found.

The characters upon which these orders are founded will be seen more clearly in the following Analytical Table:—

DIVISION OF THE CLASS MAMMALIA INTO NINE ORDERS.

CLASS I.—MAMMALIA.

CONTAINING MAN AND BEASTS, WITH WARM BLOOD; HEART WITH TWO VENTRICLES; FEMALES SUCKLING THEIR YOUNG WITH MILK, SECRETED IN BREASTS OR MAMMÆ; VIVIPAROUS, EXCEPTING THE MONOTREMATA, WHICH ARE EITHER OVIPAROUS OR OVO-VIVIPAROUS.

							Orders
			With three kinds of teeth,	With two hands,			1. BIMANA.
	With nails or claws,	Without Marsupial bones,		With four hands,			2. QUADRUMANA.
				Without hands,			3. CARNASSIERS.
Limbs Four,			Without canine teeth,				4. RODENTIA.
			Without incisors,				5. EDENTATA.
		With Marsupial bones,					6. MARSUPIALIA.
	With hoofs,	With less than four stomachs,					7. PACHYDERMATA.
		With four stomachs,					8. RUMINANTIA.
Limbs Two,							9. CETACEA.

GENERAL REVIEW OF THE MAMMALIA.

External relations of the Mammalia to the other Classes, and to each other.— Usage of the terms Mammalia, Beast, Quadruped, Bimanous, Quadrumanous, and Cetaceous.—Further subdivision of the Mammalia into Families and Tribes.

IN the preceding outlines, the internal organization of the Mammalia, and the leading principles of their classification, have been briefly explained. We shall now proceed to consider, in a general manner, their external relations to the remaining classes of animals and to each other.

In those superficial characters, which strike the observer most forcibly at first sight, the Mammalia present many traits which are to be found equally in the other Classes, a fact which is not sufficiently adverted to in ordinary discourse. Thus, by the term Beast or Quadruped, it is usual to understand an animal covered with hair, and having four feet; and whenever a Bird or a Fish is referred to, the feathers of the former and the scales of the latter offer themselves readily to the imagination. Yet these external characters by no means serve to distinguish the several classes of Vertebrated Animals. The property of having four feet, which is possessed by a large and important portion of the Mammalia, is not confined solely to them. Many oviparous animals belonging to the Third Class *(Reptilia)* possess the same characteristic; and in this respect the four-footed Beasts of the earth, which approach Man so nearly in their other characters, and occupy so high a place in the economy of Nature, are not superior to the Lizards and Frogs. Again, the Armadilloes *(Dasypus)*, instead of being covered with hair, are armed with a solid covering like the Tortoises, or even like the Crustacea. The animals of the genus Manis are covered with scales not very different from those of the Fishes, and the same structure is found in the tail of the Beaver *(Castor Fiber.)* The Porcupines *(Hystrix)*, and the Hedgehogs *(Erinaceus)*, are covered with a species of sharp quills, without feathery fibres on the extremity, but having the tube very like that of Birds. The Cetacea, or Sea-beasts, resemble the Fishes so forcibly in their external forms, that the uninformed portion of mankind persist in calling them Fishes in opposition to the universal decision of Naturalists. The Whale, Dolphin, Grampus, and other animals of this order, have nothing in common with the Fish, except the circumstances of their living in the same element, in being destitute of hair, and in possessing that external form necessary for rapid motion in a fluid of considerable density. Yet the term Whale-fishery will long preserve its usage among that numerous class of persons, who are apt to reject the critical observations of Naturalists, from their apparent over-refinement.

Nature appears to evade, by the variety of her combinations, those obvious divisions which a superficial examination would lead us to form; and the Mammalia approach to the Birds, the Reptiles, the Fishes, and even the Crustacea, in the character of their external covering. This variety in the superficial appearance establishes clearly the necessity of seeking, in their internal organization, for the principles of classification. It has often been stated, that whole Error lies on the surface, Truth must be sought deeply in the hidden parts; and this assertion, which is only made *metaphorically* in reference to moral subjects, is *literally* true in Natural History.

The Birds share their quills with the Hedgehogs and Porcupines; and their long bills destitute of teeth, with their tongue, are imitated by the trunk and tongue of the Ant-eaters *(Myrmecophaga.)* The Reptiles are not alone armed with a solid covering. The Fishes share their scales with the Beaver and Manis, and their fins with the Seals *(Phoca)*, the Morse *(Trichechus)*, the Manatus, and the true Cetacea. The Birds have their powers of flight assigned also to the Bat; the crawling

of the Reptiles and Eels is imitated in some degree by the slow movements of the Sloth *(Bradypus)*; and the Fishes share their powers of swimming with most Mammalia, but more especially with the tribe Amphibia, and order Cetacea.

As the meanings of the terms Beast, Bird, Fish, and Quadruped, are established by popular usage alone, they are necessarily destitute of that precision which should characterize the language of science. The term *Mammalia*, which has been generally adopted by Naturalists, is much more wide in its signification then that of *Quadruped*; it agrees more nearly with the word *Beast* than perhaps any other term, although not exactly, as the latter term excludes Man, and the Cetacea are not always understood by the vulgar to be really Sea-beasts. The term *Quadruped* is still more improperly considered as synonymous with *Mammalia*, with which, however, it is often confounded. In the last-mentioned class Man is included as well as the Cetacea, although *he* is a *Biped*, and *they* are altogether destitute of hinder limbs. The Ape tribes are possessed of four hands, and properly *Quadrumanous*. Even of those animals which are, strictly speaking, Quadrupeds, from their walking habitually on four feet, many either frequent the water or are capable of supporting themselves in the air. The Seals and other Amphibia, although Mammalia, cannot properly be styled Quadrupeds, and the same observation applies to the Bats.

The true Quadrupeds live exclusively on the land; they may be said to divide it with Man, whose Nature they approach more nearly than that of the Birds, Reptiles, or Fishes. But we must observe that the term quadruped strictly supposes that the animal walks on four feet. If it be destitute of feet like the Manatus and true Cetacea, if it be supplied only with arms and hands like the Ape, or if it possess wings like the Bat, the term Quadruped ceases to be applicable. Man is the only Biped and Bimanous animal, because he alone possesses two feet and two hands; the Manatus is only Bimanous, and the Bat is a Biped, while the Ape is Quadrumanous or four-handed. The Jerboa *(Dipus)*, and Kangaroos *(Macropus)*, cannot properly be styled Quadrupeds, because they can walk only on their hind-feet, in consequence of the fore-limbs being too short and weak. The signification of the term Quadruped is further restricted by removing all those animals which are able to use their fore-paws as a substitute for hands, such as the Bears *(Ursus)*, the Marmots *(Arctomys)*, the Coatis *(Nasua)*, the Agoutis *(Dasyprocta)*, the Squirrels *(Sciurus)*, and the Rats *(Mus)*; and those last-mentioned animals form a kind of intermediate class between the Quadrupeds and the Quadrumanous tribes. The term Quadruped is thus applicable only to one half of the Mammalia: it is totally inapplicable to at least one quarter, and is not strictly applicable, in its full extent, to the remainder.

The Quadrumana fill up the link which would separate the form of Man from that of the Quadrupeds. Those animals, with true clavicles, form another subordinate link between the Quadrumana and Quadrupeds; while the Bipeds with wings lead us to the Birds. None of the vague terms of ordinary discourse correspond exactly with those nice distinctions which the philosophical student loves to trace in the works of Nature.

As it is the leading design of classification to assist the memory by a clear and lucid arrangement of Natural objects, it frequently becomes necessary to multiply subdivisions in a few orders, which would be altogether superfluous in the remainder. By this contrivance we are enabled to arrive at general views in every department of Nature, and to remember a vast mass of phenomena not otherwise attainable. These subordinate divisions are termed *Families* or *Tribes*, and are determined by some general resemblance prevailing throughout that whole department, or else by some particular character possessed by all the individuals included therein. The selection of these characters is more or less arbitrary. No general rules can be given for their institution, and they must depend chiefly on the skill of the Naturalist. Yet they are not altogether capricious, as will be readily seen hereafter. The general style of the objects under examination must be seized at a glance; and the groups must be strictly natural, or they will defeat the end for which they were instituted.

In the First Order (*Bimana*) Man alone is included; and it admits, therefore, of no further subdivision. The Second (*Quadrumana*) comprises the Monkeys, Baboons, Sapajous, Sagoins, Ouistitis, and Makis—animals which form a decidedly natural group, and all partake more or less of the same physical peculiarities.

It is different with the Third Order (*Carnassiers*), being those Mammalia, without Marsupial bones, which have three kinds of teeth, and are destitute of hands. Among these we find the Bat, the Mole, the Bear, the Cat, and the Seal; all which animals differ greatly in the subordinate characters of their structure, and consequently in their habits and external appearance. All the Carnassiers, as their name denotes, subsist either partially or entirely upon animal food. But some of them possess a remarkable fold of skin, which connects the sides of the neck with all the limbs, and the fingers of the anterior pair. This singular membrane confers upon the group the power of flight, exercises a remarkable influence over their general habits and structure, and hence we distinguish the first family, *Cheiroptera*.

Of the remaining Carnassiers, some have their molar teeth with conical crowns; their habits are subterranean or nocturnal, and they feed on Insects. These Carnassiers form the second family, *Insectivora*.

We are thus left only with those Carnassiers which are destitute of a membrane fitted for flight, and whose molar teeth are destitute of conical crowns. To these negative characters they join the positive one of being more decidedly Carnassiers, or of living more exclusively on flesh; for which reason they compose the third and last family, *Carnivora*. But this numerous and interesting family admits of further subdivision into Tribes.

The *Plantigrada* walk on the entire soles of their feet.

The *Digitigrada* walk on the ends of their toes.

The *Amphibia* have their feet furnished with webs, which adapt them for an aquatic life.

The Fourth Order (*Rodentia*) is a very natural division, and does not require to be subdivided, unless we were to consider the presence or absence of perfect clavicles as a sufficient ground for the institution of two tribes founded on this distinction.

The Fifth Order (*Edentata*) would remain undivided, did not the extraordinary peculiarities of the Sloths (*Bradypus*) authorise their separation from the ordinary Edentata, the former tribe being marked by its very long and crooked claws.

The Sixth Order (*Marsupialia*), among which we propose to include the Monotremata, forms a division of animals possessing marsupial bones, but at the same time partaking of the characters of many of the preceding orders in general structure and habits. Their anomalous dentition renders any classification, founded upon this character, liable to some objections. It is, therefore, not without some hesitation that we venture to propose an arrangement, founded on *the presence or absence of incisors and canines in the lower jaw*.

The first tribe (*Didelphida*) has both incisors and canines in the lower jaw, and includes the genera *Didelphis*, *Thylacinus*, *Phascogale*, *Dasyurus*, and *Perameles*, all of which are more or less carnivorous.

The second tribe (*Macropoda*) have incisors, but the canines are either wanting altogether in the lower jaw, or else are very small. They live chiefly on fruits or herbs. In this tribe we propose to include the genera *Phalangista*, *Petaurus*, *Potorous*, *Macropus*, *Lipurus*, and *Phascolomys*.

The third tribe coincides exactly with the *Monotremata* of M. Geoffroy St Hilaire, being destitute both of incisors and canines, and containing the two genera *Echidna* and *Ornithorynchus*.

Arriving at the Mammalia with Hoofs, we find that, in the Seventh Order (*Pachydermata*), it is necessary to distinguish the remarkable proboscis of the Elephant—a character which establishes his claim to a separate tribe (*Proboscidea*), if not to a separate order. The solid hoof peculiar to the genus *Equus* also gives rise to the formation of a tribe of *Solipeda*, leaving the remaining genera to form a natural group of *Pachydermata*, or thick-skinned Mammalia.

The *Ruminantia* or Eighth Order exhibit, in their four stomachs, and indeed in their entire conformation, that close resemblance which would render any intermediate divisions at present superfluous.

The Last Order (*Cetacea*) admits of further subdivision into the *Herbivora*, destitute of spiracles on the top of their head, and destined, by their dentition and general construction, to feed on marine vegetables; and the true *Cetacea*, with spiracles on the top of the head.

These subdivisions, and the leading characters on which they are founded, are shown in the following Table, with a few examples of each family and tribe, to enable the student to fix them more easily in the memory:—

SUBDIVISION OF THE ORDERS OF THE CLASS MAMMALIA INTO FAMILIES AND TRIBES.

Orders.	Families.	Tribes.	Examples.
1. BIMANA,			Man.
2. QUADRUMANA,			Monkeys, Lemurs.
3. CARNASSIERS,	With a fold of skin connecting the sides of the neck, with all the limbs, and the fingers of the anterior pair, — 1. CHEIROPTERA,		Bats, Flying Cats.
	Without a fold of skin as above, — Molar teeth, with conical crowns, — 2. INSECTIVORA,		Moles, Hedgehogs, Shrews.
	Molar teeth, without conical crowns — 3. CARNIVORA, — Feet without Webs, — Walking on the entire soles of the feet, — 1. PLANTIGRADA,		Bears, Badgers.
	Walking on the toes, — 2. DIGITIGRADA,		Weasels, Dogs, Foxes, Cats.
	Feet with Webs, — 3. AMPHIBIA,		Seals, Walrus.
4. RODENTIA,			Squirrels, Mice, Hares.
5. EDENTATA,	Nails long and bent, — 1. TARDIGRADA,		Sloths.
	Nails short, — 2. EDENTATA (proper),		Ant-eaters, Armadilloes.
6. MARSUPIALIA,	Lower jaw — With incisors and canines, — 1. DIDELPHIDA,		Opossums.
	With incisors, but the canines wanting or very small, — 2. MACROPODA,		Kangaroos.
	Without incisors or canines, — 3. MONOTREMATA,		Ornithorynchus or Duck-bill.
7. PACHYDERMATA,	With a proboscis, — 1. PROBOSCIDEA,		Elephants.
	Without a proboscis, — With two or four hoofs on each foot, — 2. PACHYDERMATA (proper),		Rhinoceroses, Hogs.
	With only one hoof on each foot, — 3. SOLIPEDA,		Horse.
8. RUMINANTIA,			Camels, Deer, Sheep, Oxen.
9. CETACEA,	Without spiracles on the top of the head, — 1. HERBIVORA,		Dugong.
	With spiracles on the top of the head, — 2. CETACEA (proper),		Dolphins, Whales.

GENERAL REVIEW OF THE MAMMALIA CONTINUED.

Some popular and external characters of the preceding Tribes and Families.

We shall defer our observations upon the first order BIMANA, until we come to treat of the physical history of Man. At present it is necessary to add a brief review of the superficial and external characters of those tribes and families which we have enumerated above.

2. QUADRUMANA.—Next to Man, but at a considerable distance below him, we find the numerous tribe of Apes, from the Orang-Outang to the Sagoins, all possessed of hands on their hinder extremities, and if we except a few genera, also a

thumb free and opposable. The latter characteristic gives them the utmost facility of climbing trees, and of grasping the branches. Accordingly, the Apes feed in general upon fruits and nuts. Some of the American species, such as the Sapajous, are capable of hanging to the trees by means of a prehensile tail, which twines around the branch, and enables them to swing with the head downwards. These animals, as well as the Sagoins, are distinguished from the Apes of the old continent by a nose so broad and flat, that both nostrils can be seen on either side. Among the Apes of the old Continent, we find the genera Macacus, Inuus, Hilobates, Cercopithecus, the ferocious Mandrills (*Papio*), and several others. All these animals live in the forests of tropical countries, where they form numerous bodies. They compose a tribe remarkable for its great resemblance to Man, and the natural propensity to imitate his actions—qualities which are combined with some degree of skill and intelligence, a singular liveliness of disposition, and innate fondness for mischief. It is said that they form regularly organized bodies in their native forests, and establish among themselves a kind of rude police for pillaging or guarding the fruits of the different districts. The females carry their young in their arms like the Negresses, and are often observed to kiss them tenderly, and frequently to beat or bite them as a punishment. This close resemblance between the Ape and the human species when in the savage state, will hereafter become the subject of our consideration.

The Makis, a branch of the Quadrumana, are diminutive like the Apes, whom they resemble greatly, both in manners and disposition, but are at once distinguished by their pointed muzzle. They live mostly on Insects, and are marked by meagre and elongated fingers and arms.

3. Carnassiers.—After this family we find the Bats, which bear a near relation to the preceding, both in general conformation, and in having their mammæ placed on the chest. Their wing-like arms, and their elongated fingers attached to a membraneous expansion, impart the power of flying, or rather of supporting themselves by a rapid succession of vaulting movements. They are all of a hideous aspect. The young cling to their mother, who gives them the breast even when flying. Unable, by the delicacy of their eyes, to endure the full blaze of day, they appear only at night, when they vault rather than fly after the insects which form their prey. In the warmer climates there are enormous Bats, which live also on the fruits of trees. All these animals hang by their thumb-nails in the depths of obscure grottoes, caverns, and other retreats, and in our climates they are observed to become torpid during the winter.

The remaining tribes of Carnassiers follow the preceding races, and constitute two numerous, as well as interesting families. Among the Insectivora, we find the Hedgehogs (*Erinaceus*), the Shrews (*Sorex*), and the Moles (*Talpa*). The paws of the last are not very unlike hands in their general appearance, and are used for digging in the earth, as well as for climbing and raising food to the mouth. All the genera of Insectivora are fond of darkness and retirement. Among the Carnivora we find the family Plantigrada, which walk with the entire soles of the feet upon the ground. In this first tribe of the Carnivora, we have the Bears (*Ursus*), the Badger (*Meles*), and a few others—mostly animals of a surly and savage disposition, retiring during the winter into caverns and other obscure retreats. Among the Digitigrada, forming the second family of the Carnivora, we find the Weasels (*Mustela*), the Otters (*Lutra*), the Civets (*Viverra*), possessed of a fine and glossy fur, a long and slender form, and a light step. Concealing themselves among crevices, they steal slyly upon their prey, whose blood they suck with delight. The animals of the Genus Mephitis exhale a most insupportable odour. The Otters frequent the borders of streams or the sea shore, and seek their prey in the water. Among the more ferocious Digitigrada, we find the Wild Dog, Wolves, Foxes, and Jackalls (*Canis*), with the Hyæna—animals with a keen sense of smell, hunting together in packs, and overcoming by force of numbers the most powerful beasts of prey. They present a haughty demeanour, an elevated head, and are eminent at once for brilliancy of instinct, and for sanguinary courage. In the last quality they are only surpassed by the Lion, the Tigers, Leopards, Panthers, and the Lynx, all forming part of an extensive genus (*Felis*), of which our domestic Cat is considered the type. These animals are enabled by their strong and retractile claws to climb with facility. Their head is round, their eyes glitter in the darkness of night, their tongue is roughened like a file, and their teeth are exceedingly powerful. They wait for their prey; with a sudden spring they dash it to the ground, and enjoy with ecstasy the flesh yet throbbing from the breast of their victim, and the blood still warm from its heart. Among the Amphibia, which conclude the long list of Carnassiers, we find the Seals (*Phocæ*) and the Morse (*Trichechus*), animals which greatly resemble the Cetacea in external form and habits, but differ decidedly from them in the structure of the teeth.

4. Rodentia.—The family of Gnawers follow the Carnassiers, from whom they are separated by very distinct characters. We find in them two long cutting teeth, in the front of each jaw, no canines, but molars, and intestines of great capacity. These timid animals, destined for the most part to gnaw vegetable substances, do not eat flesh except under extraordinary circumstances. Their hinder legs, and the entire hinder part of their body, is stronger than the fore, and they run and leap rapidly. Their muzzle is more or less arched, their eyes project, yet their sight is not acute, but this deficiency is compensated by the fineness of their hearing. Such is the general character of the Dormice (*Myoxus*), the Marmots (*Arctomys*), the Hamsters (*Cricetus*), the Field Mice (*Arvicola*), animals which become torpid during the winter season; also the Squirrels (*Sciurus*), the numerous tribes of Rats and Mice (*Mus*), the Hares (*Lepus*), the Guinea-Pigs (*Cavia*), the blind Rat (*Spalax*), the Beavers (*Castor*), the Porcupines (*Hystrix*), and some others. We find the most singular habits and instincts among these animals. The industry of the Beavers is known to all, and the sagacity and skill with which they fell trees by means of their powerful teeth, draw them across rivers, form dikes with their hand-like paws, and construct cabins above the water, where they amass stores of bark for their maintenance. Their large and flat tail serves them as a trowel to work up the earth with which they form the walls of their singular masonry. Every one is familiar with the nimbleness of the Squirrels (*Sciurus*); but all are not aware of that instinct which leads them to peel off a piece of bark from a tree, when about to cross a brook. Mounting this frail boat, with their bushy tail stretched like a sail before the breeze, they gain the opposite bank in safety. In the genus Pteromys, we find the Flying Squirrels provided with a membrane extending from the fore to the hind feet, so that, by stretching out their limbs, they form a parachute, which assists their great leaps from tree to tree. Many species of Rats (*Mus*), and Field Mice (*Arvicola*), live in the earth, united together in social bands, where they amass magazines of provisions, and construct warm retreats of hay and moss, for their protection during the winter. The Hamsters (*Cricetus*), have large cheeks, which they fill with provisions, and transport the contents to their subterranean retreats. There are other Rodentia, such as the Lemmings (*Lemmus*), which emigrate every year, according to the seasons, to gather the provisions which Nature has scattered over different countries. Certain species of economical Mice in Tartary, collect so large a quantity of nutritious roots, that the people of that inhospitable climate avail themselves of the supply afforded by their nests as a provision for themselves during the winter. In the East Indies and in Africa, we find Jerboas (*Dipus*), a kind of Rat with hind-legs of so great a length, that they are supported almost in a vertical position, and are enabled, like grasshoppers, to leap continually, and to an immense distance.

5. Edentata.—The first tribe (*Tardigrada*) of the Edentata, is composed of those singular animals found in America, called Sloths (*Bradypus*), from the excessive slowness of their movements. They present some slight resemblance to the Monkey tribes in their general form, and in the circumstance of their mammæ being pectoral, but they are destitute of the front teeth, and instead of hands, exhibit large crooked nails fitted for climbing on trees. They live entirely upon leaves, and lead a life which we would consider melancholy, uttering the most lamentable cries, and moving themselves with great apparent difficulty. Of the remaining tribe composing the true Edentata, we find that nature has provided some compensating protection for the want of canine and incisive teeth, either by bestowing on them scales, placed one over the other, resembling the sepals or flower-cup-leaves of the artichoke, as may be found in the Manis, or else a bony cuirass of moveable pieces, seen in the Armadilloes (*Dasypus*). The Ant-eaters (*Myrmecophaga*), are supplied only with hard and tufted hair, but being altogether destitute of teeth, they are furnished with an elongated and viscid tongue, which they extend into the nests of Ants, and these insects, adhering to it in numbers, are speedily brought into their mouth.

6. Marsupialia.—Commencing with our first tribe, Didelphida, we find in America, the Opossums (*Didelphis*), and in New Holland, the Dasyurus—animals which are more remarkable than perhaps any of the preceding. In form, they somewhat resemble a very small Fox, with a long tail, naked and flexible at will, while their fore-paws approach the form of hands. But the singular part of their structure is the abdominal pouch found in the females, which serves as a double womb. The young are produced before their time, and attach themselves immoveably to the teats of the mother, which are placed within the pouch. After their second birth, they retreat into this natural pocket, which protects them from the severity of the cold. When arrived at a more advanced age, they climb upon the back of their mother, and, by means of a long and flexible tail, they hook themselves to her tail or limbs, so that she is able to carry them when alarmed, in this manner, and can run or climb trees with considerable speed when pursued. This tribe is naturally carnivorous, feeding on birds and other small prey. The second tribe, Macropoda, of which the Kangaroo (*Macropus*) may be considered the type, contains several remarkable animals, some of which are almost as large as our sheep. They have strong and large hind-limbs, with a long and stiff tail, collectively forming a tripod, upon which they stand, or rather sit, securely. In this position they usually remain, for their fore-paws are very short, and are used only as hands. Instead of walking, they move nimbly by a succession of elevated leaps; but as their feeble progeny are unable to follow their mother in this rapid movement, a kind and benevolent Nature has bestowed upon them an abdominal pouch, like the Opossums, to transport their young ones. These species are of a mild disposition, are easily tamed, and possess that timidity which we find in most animals living exclusively upon vegetable food. The last tribe (*Monotremata*) are also found in New Holland. These quadrupeds are covered either with smooth or bristly hair; but instead of jaws, they exhibit the singular anomaly of a beak exactly resembling that of a Duck, with reproductive organs like the Birds. These curious animals frequent the water, and burrow under ground.

Nearly all the genera of which we have spoken in the preceding outline have Clavicles, or collar-bones, which enable them to use their fore-feet for other purposes than walking. They can seize various objects; their fingers are separate and furnished with nails, which distinguishes them from the Ungulated or Fissipede classes. The former are also, in general, more expert and intelligent than the species of which we shall now treat; for the Ungulated animals being less free in the motions of their limbs, have also less skill and intelligence. The Ungulated Mammalia are mostly polygamous. That fond affection for their offspring which is found in the Unguiculated classes, is almost unnecessary with them, as their young are more precocious, that is, they arrive sooner at the full exercise of their faculties than the progeny of the Unguiculated Mammalia.

7. Pachydermata.—In the first rank of Ungulated Mammalia, we find the Elephant distinguished by the superiority of his intelligence from the proper Pachydermata. These last mentioned animals are, on the contrary, very rude and unintelligent. They are covered with thinly scattered bristles rather than hair. The form of their bodies is clumsy and inelegant. They are fond of wallowing in the mire, and of frequenting the water, or low and moist grounds, where they live on coarse food, such as stalks and roots. Their sight is not acute, but their sense of smelling is very fine. Under the skin, we usually find a thick layer of lard, which renders them but slightly sensitive, except towards the nose and mouth. We next find the interesting family of the Solipeda, so called from their feet being enveloped in a single hoof, such as the Horse, Ass, Zebra, and some other animals, all of the genus Equus, which are equally fitted for running rapidly, or for the transportation of burdens.

8. Ruminantia.—Arriving at the ruminating animals, we here find the genera of the Camel and Dromedary (*Camelus*), the Lama and Vicugna (*Auchenia*), the Musk (*Moschus*), the Elk, Rein-deer, Stag, Fallow-deer (*Cervus*), the gigantic Camelopard (*Camelopardalis*), the beautiful Gazelles or Antelopes (*Antilopa*), the Goats (*Capra*), the Sheep (*Ovis*), the Buffalo, Musk-Ox, and common Ox (*Bos*), and many others of great interest. All these animals are readily marked by their cloven feet, that is, their feet divided into two hoofs, by the horns which most of them possess, and by the want of front teeth in the upper jaw. Those ruminating animals naturally without horns, like the Camels, Vicugna, and Musk, find an equivalent in the canine teeth of their upper-jaw. Among the Deer, the horns are branched, and fall each year after the rutting season, when their warlike ardour is over. In the other genera, the horns are hollow, and fit firmly into a bony receptacle, which prevents them from falling. All these animals feed on grass or leaves; they have four stomachs, and ruminate, that is, they restore their food a second time

into the mouth for undergoing a final mastication. The females are easily tamed; they yield an abundance of milk, and instead of fat are supplied with suet. The males, which are less numerous in each species, are consequently polygamous, and the females produce only one or two young ones, which are able to walk from their birth. The mammæ are always placed near the abdomen; the flesh forms a healthy food. Every one is acquainted with the immense advantages which Man derives from the domestication of these genera, with the fleeces and skins of the Sheep, Goat, Vicugna, and with the leather yielded by the skins of all the animals of this tribe. Without the Rein-deer, the Polar regions would be uninhabitable by the Laplander and many other nations. Without the Horse and Ox, agriculture would be impossible, and nations could no longer exist in their present state of civilization. The Arab in vain might attempt to traverse the Deserts without the aid of the Camel.

9. CETACEA.—Finally, we arrive at the Cetacea, whose limbs are formed into oars or fins. They all live upon the water rather than in it, for they can only breathe atmospheric air, and may be drowned by too long an immersion in the water. The Herbivora, a tribe of aquatic Mammalia, are analogous in many respects to the Amphibia. Among them we find the Manatus or Sea-cow, and the Dugong (*Halicore*), animals which have probably given rise to the accounts of Tritons, Sirens, and Mermen. The Cetacea proper are more peculiarly aquatic than the Amphibia, for they are never found to rest upon the ground. The female usually produces one or two young ones alive in the water, where she gives them the breast, watches over them, and supports them when fatigued upon her back and sides.

GENERAL REVIEW OF THE MAMMALIA CONTINUED.

Relations to the other Classes—Gradual degeneracy of form—Fitness for their several stations.

FROM the brief outline which has been attempted in the preceding sections, it may be easily understood that the other Mammalia approximate very closely to Man in their general nature, and more especially that portion of them which compose the Viviparous Quadrupeds. These form unquestionably the most important portion of the Animal Kingdom, from the similarity of their external shapes to our own, the superiority of their instincts above all other animals, the meek submission of some to the force of domestication, and the determined hostility of others to any modification of their original habits.

Their marked resemblance to ourselves naturally leads us to view the Quadrupeds, and indeed all the terrestrial Mammalia, with feelings of interest, which the other classes of Vertebrated Animals must in vain attempt to claim. It is true that we admire the delicacy and lightness exhibited in the forms of most Birds, the general warmth of their temperament, their liveliness, and perpetual motion, which we are apt to compare to their own airy medium. The Fishes, on the other hand, are naturally a stupid race, without animation or sensibility; and, like the ocean which serves as their dwelling, preserve nearly the same temperature at all times. Their exertions are seen only in efforts to swim, or to satisfy their most pressing necessities; and that scaly covering which surrounds many of their bodies blunts their sense of touch, and renders them more or less insensible to external impressions. On the contrary, the Quadruped preserves a middle station between the heights of the atmosphere and the abysses of the ocean. He seems to share with Man the sovereignty of the earth, and, like him, to exhibit an intermediate character. He neither possesses the ardour and vivacity of the Bird, the stupidity of the Fish, nor the heavy apathy of the Reptile. But, fixed to the firm and dry land, the Quadruped has received a certain degree of solidity and firmness of structure. His walk has not that rapidity which characterizes the flight of Birds, or that nimbleness which we observe in the movements of the Fish; yet his motions do not partake of the laborious dulness observed in the Tortoise, and other Reptiles. His moderate swiftness permits the muscles to act with greater vigour, and allows his faculties time to expand. Indeed, without considering Man, the other Mammalia contain beings the most susceptible of intelligence on the face of the globe.

We have already pointed out the leading characters which belong to Man, as well as all the other Mammalia; yet we must observe, that throughout their entire orders, from the Bimana to the Cetacea, we may easily trace a kind of gradual departure from the external form fitted for Man. Whether we consider the Monkey tribe in their external appearance, or in their internal organs, we find the closest resemblance. The skeleton, the muscles, and all the internal organs, even to the ramifications of the smallest vessels, present a degree of similarity to Man at once startling and mortifying. In fact, the Apes, though forming many distinct genera and species of themselves, seem but a rough sketch of human degradation. The same shades of deviation can be seen in descending from the Quadrumana to the Cheiroptera, the Carnassiers, the Tardigrada, and indeed throughout all the series. We must, however, recollect, that the most important organs, such as those which are the essential attendants of their internal functions, never change materially. They are identical in all the Mammalia, and fulfil their uses in nearly the same manner. It is only externally and superficially that this degeneracy of form exhibits itself. Thus, for example, the hand of Man may be recognized in that of the Ape. In the Makis it already begins to appear deformed, and continues to deteriorate through the Hedgehogs, the Moles, and Bears. It becomes a paw, when we arrive at the Dogs. Afterwards the nails exhibit, in the Sloth, the transition to the solid hoofs of the Sheep, Stag, and Ox, and terminate in the uniform hoof of the Solipeda. Finally, we find in the Whales and Dolphins no other vestige than a stump rudely fashioned as an oar. Yet, if we open the skin of this part, we still find the principal bones of the hand and arm, but in that rudimental form which serves but to mark the wideness of their separation from the perfectly-developed hand of Man. This law of degeneracy is, however, by no means invariably adhered to, and we have intentionally passed over several genera which exhibit marked and decisive exceptions to its generality

From the Quadrumana to the Cetacea we observe a decided contrast to Man in the elongation of their muzzles, their general tendency towards the earth, and the violence of their passions, unrestrained like his by the voice of Reason and Conscience. It is probable that their enjoyments of sense are more vivid than those of Man; they always yield to the present impulse, and are susceptible only in a slight degree of intellectual improvement.

It is for action and not for reflection that the Beasts of the earth are designed. Their limbs are more robust than those of Man; and this natural vigour is further improved in the wild races, especially the Carnivora, by continual exercise. Their constant activity increases this muscular vigour, their bodies are more healthy, and become more capable of resisting external injury or the inclemencies of the seasons. Nearly the same kind of contrast which we remark in our own species, between the vigorous and thickly-set Mechanic and the delicate and lively Female, may be observed between a wild animal of the forest and a robust Man. In proportion as the external qualities of the body are improved, sensibility and delicacy of feeling diminish; and it would almost appear to be the necessary result of civilization and refinement that the muscular vigour of our species should diminish, and that their liability to disease should increase.

However inferior the other Mammalia may be to Man in intellect, they are of all animals the best able to understand his commands. The Birds are not capable of holding these intimate relations to ourselves: for, whatever degree of intelligence may be attributed to the domesticated Parrot or tame Canary-bird, these are greatly surpassed by the superior instincts of the Dog, the Beaver, and the Elephant. Still less are we capable of forming modes of connexion with the Reptiles and Fishes, while the Mollusca and lower divisions of animated Nature form other natural societies in which the influence of Man cannot be felt. In short, his power becomes extensive in proportion as the animal approximates to his Nature. We can teach the Insect, the Fish, or the Reptile, absolutely nothing; our influence increases over the Birds; but the other Mammalia are capable of considerable instruction. They are not mere automata, but possess a certain degree of perfectibility. Indeed, the instincts of the Mammalia seem to establish an intermediate intelligence between the Human Soul and that mere animal existence enjoyed by the other divisions, whose whole lives are absorbed in seeking their food or continuing their species.

Every animal must necessarily be fitted for the station in which it is placed by Nature. For if, by any accidental or natural event, an animal be placed in a situation for which it is unfitted, it will either perish absolutely, or else its original constitution will be modified so as to correspond accurately with its new condition. Thus the animals of the torrid zone are supplied with a very slight coat of hair, as we see in the Barbary Dog, and the Apes; while under the frigid zone they exhibit the warmest and thickest furs, as in the Sable (*Mustela zibellina*), the Bears, and the Arctic Fox (*Canis lagopus*).

This adaptation to surrounding circumstances is found equally in their senses, their means of defence, the greater or less swiftness of their movements, and the ferocity or mildness of their dispositions.

Though all the Mammalia possess five senses, they do not enjoy each sense with the same degree of intensity. Those species which dwell in the mountains, like the Chamois (*Antilope rupicapra*), and the Ibex (*Capra ibex*), whose flight is rapid, and which lead a wandering life, are far-sighted; on the contrary, the heavy animals dwelling in the valleys, like the Hogs and Rhinoceroses, are near-sighted. Those again whose eyes are too delicate for the full blaze of daylight, come from their dens only at night, or in the twilight, like the Bats, or else conceal themselves in the earth like the Armadilloes and Hedgehogs. The more timid and feeble races make a greater use of their ears than of their eyes; the Hare, the Rabbit, the Jerboa, the Mouse, and other Rodentia, raise their ears at the slightest noise, preparatory to flight; but the more powerful and courageous races, such as the Lion, the Tiger, the Lynx, and other Cats, endowed with a keen and piercing sight even at night, have their ears small and their hearing indistinct. Thus the feebleness of one sense is made up by the perfection of another, just as in Man, when accident deprives him of sight, the sense of hearing becomes more acute. The power of smell, in the Mammalia, always refers to their proper food or to their own species. A dog, which finds no pleasure from the scent of the Tuberose or the Carnation, will discover the female of his own species, or the carcase of another animal, at an immense distance. With the Carnivorous animals the sense of taste becomes a fierce and sanguinary appetite; with the Herbivorous tribes it possesses an equal sensibility in distinguishing the nutritious plant from the poisonous weed.

The same adaptation to their wants and enjoyments is found in the general form of their limbs. "In some," observes Goldsmith, "they are made for strength only, and to support a vast unwieldy frame, without much flexibility or beautiful proportion. Thus the legs of the Elephant and Rhinoceros resemble pillars; were they made smaller they would be unfit to support the body; were they endowed with greater flexibility or swiftness, they would be needless, as they do not pursue other animals for food; and, conscious of their own superior strength, there are none that they deign to avoid. Deer, Hares, and other creatures that are to find safety in flight, have their legs made entirely for speed; they are slender and nervous. Were it not for this advantage, every carnivorous animal would soon make them a prey, and their races would be entirely extinguished. But in their present state of nature, the means of safety are rather superior to those of offence, and the pursuing animal must owe success only to patience, perseverance, and industry. The feet of some that live upon fish alone are made for swimming. The toes of these animals are joined together with membranes, being web-footed like a goose or duck, by which they swim with great rapidity. Those animals that lead a life of hostility, and live upon others, have their feet armed with sharp claws, which some can sheath and unsheath at will. Those, on the contrary, who lead peaceful lives, have generally hoofs, which serve as weapons of defence, and which in all are better fitted for traversing extensive tracts of rugged country, than the claw-foot of their pursuers."

In obedience to the same universal law of adaptation, we find that the Armadilloes and Manis, which are destitute of teeth, find a counterbalancing defence in their horny cuirass or scales. In the Porcupine and Hedgehog, which are in other respects both feeble and defenceless, Nature has converted the ordinary hair of the other quadrupeds into a forest of pointed darts; and these animals, rolling themselves into a spiny ball on the approach of danger, are invincible to all other species. The herbivorous tribes do not possess strong teeth or hooked claws, but many of them have the head armed with powerful horns. The timid Rodentia either seek with instinctive industry to hide themselves under ground, like the Marmot, the Rabbit, and the Rat, or they leap with agility, like the Squirrel, from tree to tree, or else, like the Jerboas and Cape Rat, they avoid their pursuers by wide and frequent springs resembling Grasshoppers; again, the Vicugna and Llama have no means of defence, yet when attacked they dart upon their enemies an acrimonious and disgusting saliva. The Pole-cats and the Mephitis exhale, when pursued, odours so execrable, that they compel their most irritated enemy to desist in his pursuit. Some animals, like the Howling Sapajous (*Ateles* and *Lagothrix*), attempt to terrify their enemies by the most frightful howls; others avoid them by climbing trees, by darting into their subterranean retreats, by vaulting, by leaping, by plunging into the water, by distracting their pursuer with a host of ingenious devices and precautions, or by the construction of fortified dens or impenetrable recesses.

Besides these means of defence, the smaller species are more productive, both in number and frequency, than the larger species; they are also more robust, lively, and active, in proportion. Before an Elephant or a Whale can turn round once, a Muscardin or a Mouse will have made a hundred movements. The smallness of their limbs gives more unity and solidity to their bodies. Their shorter muscles contract more easily, and more forcibly, than in these larger and more unwieldy machines. Were an animal to exist three or four hundred feet in length, and of a proportionate thickness, it would lie gasping on the earth overwhelmed with its weight, and would become the easy prey of all other animals, even of the most feeble.

Thus we find that the Mammalia are fitted in every respect for the stations which they occupy, and that a bountiful Nature provides, by means of their complicated relations, for one continual scene of activity and enjoyment.

GENERAL REVIEW OF THE MAMMALIA CONTINUED.

Their Food—Carnivorous Tribes—Final causes of their Mutual Destruction—Herbivorous Tribes.

The surface of the earth, clothed with verdure, is the inexhaustible source whence Man and Animals derive in common their subsistence. Every animated being lives *ultimately* upon vegetables, and vegetables are maintained by the *debris* or remains of every thing which has lived and vegetated. A perpetual round of existence is thus maintained. Without Death there could be no Life, and it is only by annihilating other beings that animals are able to support themselves, and to continue their species; they must either feed on vegetables or upon other animals. Yet Nature, like an indulgent mother, has fixed limits to this apparently indiscriminate destruction. The carnivorous and voracious individuals are reduced to a small number, while she has largely multiplied both the species and individuals which are herbivorous. Man too has greatly assisted in exterminating, or confining within narrow limits, the predaceous species, and in establishing the more peaceful tribes. Among the Marine genera, although some are herbivorous, yet the greater number are nearly equally voracious. These devour their own and different species without ever appearing to exterminate each other, because their fecundity is as great as the destruction; and nearly all this mutual consumption acts as a new incentive to reproduction.

Man stands foremost among the carnivorous tribes. Being the predominant species, he exercises over the other Mammalia the privileges of a master. He has chosen those which please his taste, and forms them into humble dependants. By causing them to multiply more rapidly than unassisted Nature would have done, they have given rise to numerous flocks; and from the care bestowed in their production, he acquires a natural right of immolating them to satisfy his wants. This power, however, extends much farther than his necessities would require; for, independent of those species which he has subdued and can dispose according to his pleasure, he carries on a war of extermination against the wild Beasts, the Birds, and the Fishes. He does not even confine himself to the climate which he inhabits, but seeks for new delicacies in the remoter parts of the globe. Nature seems scarcely adequate to supply this continual demand for variety, and Man alone may be said to consume more animal food than all the other Mammalia taken together.

Next to Man, the carnivorous beasts possess the most destructive habits, and are at once the enemies of their fellow-animals, and the rivals of Man. Having the same appetites and the same fondness for animal food, they are under the necessity of disputing with him the possession of their prey; and in the first ages of human society these formed one of the most formidable checks to civilization. Even at the present time, in civilized Europe, it is by the utmost vigilance alone that he can preserve his flocks and poultry from the ravages of the Wolf, the Fox, the Ferret, and the Weasel.

Man thus carries on a continual war against the carnivorous animals, which he either pursues for pleasure or for safety. However superior to him in bodily strength or swiftness, the most powerful fall ready victims to the union of numbers, the superior powers of his mind, and especially to that peculiar art with which he avails himself of the inert materials of Nature as instruments of destruction. No race of animals can resist the agency of gunpowder; the Whale falls before the harpoon; the Elephant and Lion cannot evade the pit-fall and the snare. The largest animals receive death or captivity at his hands, as certainly as the smallest; and Man can confine the limits, or even exterminate every animal which comes within the sphere of his influence.

All animals, whether of the same or of different species, are naturally in a state of warfare. It is chiefly in the tribes more particularly styled carnivorous, that this war proceeds to open hostilities; yet there is a silent and a secret opposition of interest, even among the most peaceful tribes. As their numbers continually increase, food becomes scarce, disease thins their numbers, and the remainder fall a ready prey to the stronger and fiercer animals. Like plants, they destroy each other as effectually by the mere occupancy of space as they could have done by the fiercest conflicts. The rising generation soon repairs the loss occasioned by the latter, but nothing can extend the numbers of a species beyond the limits marked by Nature in the quantity of its food.

This universal war of species is an established law of Nature, and, however startling it may appear at first sight, is advantageous on the whole. Violent deaths are as necessary to the proper regulation of Nature as natural deaths. The latter preserve the perpetual bloom of youth over the face of the earth; the former assist in maintaining the correct balance among the numbers of different species, and in restraining their exuberance within the proper limits.

To illustrate this important law of Nature, let us consider for a moment some one of the inferior species, which serve for food to the higher classes. The Herrings offer themselves, at certain seasons, in myriads to our fishermen; and, after nourishing the Birds which sport on the surface of the ocean, as well as the predaceous tribes frequenting its abysses, form the principal support of many nations of Europe during a considerable part of the year. The destruction which takes place among these Fishes is overwhelming; yet the consequences would be tremendous if their fecundity were not thus restrained. They would soon cover the surface of the sea, their numbers would then destroy each other, for want of sufficient nourishment their fecundity would diminish, and famine and disease would produce the same results which other animals now effect. But their undevoured bodies would taint the atmosphere, perhaps the ocean itself, and the putrid miasmata arising therefrom would carry disease and death into all species of animated beings, as well as their own. Thus the false sensibility which seeks to restrain the mutual destruction of animals would effectually ensure the entire annihilation of them all. As Nature is at present constituted, Life is the consequence of Death; were it otherwise disposed, one universal death-like stillness would pervade the face of Nature. As in the animal frame, the continued action of the vital power necessarily occasions death, so in the frame of Animated Nature the continuance of reproduction must be followed by a corresponding destruction. The same observations which have been made upon the Herrings are applicable to every other species, and hence we may fairly infer that there exists an absolute necessity for the mutual destruction of animals. The futility of that philosophy of the Brahmins which condemns the use of animal food, is sufficiently obvious. From being founded in Nature it is a legitimate usage, and absolutely essential to the well-being of the whole.

By taking a general view of the constitution of Nature, we are enabled to explain those apparent incongruities which strike the observer at first sight. We then discover that "these scattered evils are lost in the blaze of superabundant goodness, as the spots on the disk of the sun fade before the splendour of his rays." In every department of Nature we find

"All partial evil—universal good."

In these wars of the animals, Nature has provided that each creature should meet its death in the easiest possible manner. There is a certain spot in the spinal marrow where the two ascending main nerves that form the great brain cross one another, and if this spot be injured, death is the immediate consequence. This fact is well known to Huntsmen and Butchers. The latter plunges his knife into the neck of the Ox at that exact spot, the animal immediately drops, and ceases to live after a few convulsions. On the same principle the Huntsman cuts through the neck of his game.

The Carnivorous Animals always seize their prey by the neck, and bite through this part. In the same manner the Hound kills the Hare, and the Bird of prey its quarry. The Pole-cat also destroys its prey at a single spring. Dr Gall locked up a Pole-cat for some time, during which he fed it on bones till its teeth were blunted. While in this state, it was unable to kill the Rabbits placed in its kennel with the same despatch as formerly; but when they had again grown sharp, Gall observed that, on the very first leap it made on the Rabbit, it cut the little animal's neck on that very spot with a sharp fang, and instantaneous death ensued. He observed the same thing at a hawking party of the Emperor Joseph the Second. As soon as the Hawk had reached the Hare, it would immediately cut through that part of her neck with its bill.

Yet Nature seems to have stamped a character of marked ferocity upon most Carnivorous tribes. The Cat torments the captive Mouse, and seems to take delight in its convulsive struggles to escape. The Tigress or female Leopard brings her prey still palpitating to her den, and gives the first lessons of ferocity to her progeny.

That sentiment of humanity towards our own species, imparted by Nature for the proper regulation of social intercourse, is transferred by us to the more intelligent and sensitive animals—in other words, to those which most nearly resemble ourselves. We cut and eat a live Oyster without the slightest commiseration, because it does not exhibit external signs of sensation, nor does it raise a cry of suffering when the fatal moment arrives: yet few, whose feelings have not been blunted by early habit, can bear to immolate a Lamb. A wise Providence has thus protected the higher animals from the gratuitous infliction of pain on the part of Man. However necessary the trade of Butcher or of Executioner may be to society, it always appears in some degree odious. The Brahmins have carried this sentiment to a ridiculous extent. They permit the most disgusting Insects to frequent their houses, their food, and their persons, without destroying them; and the Mussulmans have erected hospitals for the accommodation of infirm Dogs.

However odious the Carnivorous Animals may appear to us in the exercise of their legitimate calling, our sense of retributive justice is satisfied in knowing that most of them undergo the same fate which they have inflicted on others. "Every dog has his day," according to the proverb. The proper counterpoise and equilibrium of animals could not be established without them, and their own final fate shows the general system of reciprocity and that balance of good and evil prevailing throughout the Animal Kingdom.

The chief benevolent emotions which the Carnivorous tribes present are seen in their casual attachment to the females of their own species, the regard of the mother for her young, and that occasional language of signs by which they communicate their wants or their passions. If a Lion or a Tiger meet his mate at an unfavourable time, they both become furious, and a conflict often fatal to one or to both is the result. The circumstance of their both living by the chase renders them natural enemies to each other. This singular combination of love and hatred is wisely given by Nature to assist in preventing the too rapid increase of the more destructive animals.

It is the organization of the Carnivora—the possession of teeth, of claws, of short and narrow intestines—that imposes the office of Nature's executioners upon these animals by an imperative necessity. The sharp teeth of the Leopard or Panther might attempt in vain to grind plants: and even when we compel these animals to swallow bread and other purely vegetable substances, the gastric juice of their stomach is unable to dissolve them. On the contrary, the Lamb and the light Gazelle would refuse animal food with disgust. Their teeth are not formed for tearing, and their entire economy is adapted to a vegetable diet. It is thus that we find, in the organization of the animal, the reasons for all its actions.

This exquisite relation of all the parts of an animal to each other, enables the Naturalist to describe the whole creature on seeing only a part. Thus, from knowing the size of a tooth, we can judge of the height of the animal which bore it; by the shape of the tooth we can tell whether it be carnivorous or herbivorous. Thence follow the general structure of the body, not only of the stomach and viscera, but also the form of their paws, of claws with the one, or of hoofs with the other, the liveliness of their passions, as well as the habits which attend this kind of life and constitution.

Besides the claws and teeth, which form the offensive arms of the Carnivora, they are endowed with superior strength, agility, cruelty, and treachery. The source of these qualities must be sought in the nature of their food—in the superior organization of flesh and blood. The herbivorous tribes want offensive arms in general, yet they are seldom of a timid or peaceful disposition. They love to unite together in social bands, to pasture on the plains or by the mountain side, or else to hoard the common fruits of their industry. The carnivorous tribes, like tyrants, are unfitted for society by their ferocious and domineering tempers; they dread the rivalship of their own species, and the natural attachment of the sexes is with them but a momentary passion. They can endure hunger much longer than the herbivorous tribes, whose food is always spread out before them; and this power of fasting is necessary to animals obliged by their structure to overpower their prey by violence, to run them down by perseverance, or to surprise them by stratagem. They can fast for several weeks, but as their necessities increase they become bolder and more ferocious. The Wolf, with an appetite sharpened by famine, becomes an intrepid and formidable enemy. He then invades the villages, breaks into the stables during the daytime, and even ventures to contend with Man. But when he has found an abundance of nourishment, he gorges himself for several days; and, with an admirable sagacity, conceals tbe remainder under ground as a provision for future want.

This continual use of animal food, and the high state of organization at which all the solid and fluid parts of their bodies have arrived, renders their flesh at once unpalatable and unwholesome. Their excretions are all fetid, and the slightest check to the vital activity brings on a rapid decay. On the contrary, the vegetable nutriment of the herbivorous tribes imparts to their flesh a high degree of delicacy. Their milk is sweet, agreeable, and nutritious. Thus the herbivorous tribes yield an abundance of nourishment to Man, while he rejects with disgust the flesh of those which are carnivorous.

The natural antipathy of some of the carnivorous animals for each other, proceeds from their rivalship in the chase. It is thus that the Lion, Tiger, Panther, or Bear, permits no poachers upon his hunting grounds. These despots of the Animal Kingdom allow few intruders to share their authority, and clear the forest of all those petty tyrants, which prey only upon small game; and which, like the inferior *noblesse* of the middle ages, oppressed the lower ranks, and diminished the population.

"It is not among the larger animals of the forest alone," says Goldsmith with his usual elegance, "that these hostilities are carried on; there is a minuter and a still more treacherous contest between the lower ranks of Quadrupeds. The Panther hunts for the Sheep and the Goat; the Catamountain for the Hare or the Rabbit; and the Wild Cat for the Squirrel or the Mouse. In proportion as each carnivorous animal wants strength, it uses all the assistance of patience, assiduity, and cunning. However, the arts of these to pursue are not so great as the tricks of their prey to escape, so that the power of destruction in one class is inferior to the power of safety in the other. Were this otherwise, the forest would soon be dispeopled of the feebler races of animals, and beasts of prey themselves would want, at one time, that subsistence which they lavishly destroyed at another.

"Few wild animals seek their prey in the daytime; they are then generally deterred by their fears of Man, in the inhabited countries, and by the excessive heat of the sun in those extensive forests that lie towards the south, and in which they reign the undisputed tyrants. As soon, therefore, as the morning appears, the carnivorous animals retire to their dens; and the Elephant, the Horse, the Deer, and all the Hare kinds, those inoffensive tenants of the plain, make their appearance. But again, at night-fall the state of hostility begins, the whole forest then echoes with a variety of different howlings. Nothing can be more terrible than an African landscape at the close of evening; the deep-toned roarings of the Lion, the shrill yellings of the Tiger, the Jackal pursuing by the scent, and barking like a dog, the Hyæna with a note peculiarly solitary and dreadful, but, above all, the hissing of the various kinds of Serpents that then begin their call, and, as I am assured, make a much louder symphony than the Birds in our groves in a morning.

"Beasts of prey seldom devour each other; nor can any thing but the greatest degree of hunger compel them to it. What they chiefly seek after is the Deer, or the Goat; those harmless creatures that seem made to embellish Nature. These are either pursued or surprised, and afford the most agreeable repast to their destroyers. The most usual method, with even the fiercest animals, is to hide and crouch near some path frequented by their prey, or some water where they come to drink, and seize them at once with a bound. The Lion and the Tiger leap twenty feet at a spring; and this, rather than their swiftness or strength, is what they have most to depend on for a supply. There is scarcely one of the Deer or Hare kind that is not very easily capable of escaping them by its swiftness; so that, whenever any of these fall a prey, it must be owing to their own inattention.

"But there is another class of the carnivorous kind that hunt by the scent, and which it is more difficult to escape. It is remarkable that all animals of this kind pursue in a pack, and encourage each other by their mutual cries. The Jackal, Syagush, the Wolf, and the Dog, are of this kind; they pursue with patience rather than swiftness; their prey flies at first, and leaves them for miles behind, but they keep on with a constant steady pace, and excite each other by a general spirit of industry and emulation, till at last they share the common plunder. But it too often happens that the larger beasts of prey, when they hear a cry of this kind begin, pursue the pack, and, when they have hunted down the animal, come in and monopolize the spoil. This has given rise to the report of the Jackal's being the Lion's provider; when the reality is, that the Jackal hunts for itself, and the Lion is an unwelcome intruder upon the fruit of his toil."

It is in barren and unfrequented districts that the carnivorous animals are most fierce and sanguinary, because their prey is scarce, and the possession of it is continually disputed by a host of famished rivals. From these continued scenes of violence their character acquires an unusual ferocity. The Bear of the Alps is a formidable and dangerous animal to the traveller. But the beasts that frequent the plains or fertile valleys find their food more easily, and when found it is less disputed. Their character being thus softened down by the comforts of life, loses that high degree of courage and asperity which distinguishes the mountain races.

The carnivorous animals associate in troops only for the convenience of a combined attack; on the other hand, the herds of herbivorous animals seem intended only for their mutual defence. Placing the young ones in the centre, and the females in the rear, the males advance to the front, united in a phalanx, and presenting their horns to the enemy, repel his attack with vigour, and generally with success.

Most of the Frugivorous tribes, such as the Apes, the Makis, and the Loris, ramble about in numerous troops, for the purpose of pillaging the fruits of a district. Like expert marauders, they establish a regular order of pillage. They place sentinels in advance, and, forming a chain, pass the fruit from hand to hand. Upon the slightest alarm being given by the sentinels, the whole troop retreats to the woods or mountains, carrying off as much as they can hold in their hands and cheek-pouches.

GENERAL REVIEW OF THE MAMMALIA CONTINUED.

Domesticated Animals are not Slaves.—Methods of Taming Wild Animals—Influence of Mild Treatment—Hunger—Sweetmeats—Caresses—Chastisement—Their Occasional Revenge.

Animals, whether domesticated or in a wild state, always preserve their real characters, and act in a manner suited to their situation.

The absolute submission which we are in the habit of requiring from our domestic animals, and that kind of tyranny which we exercise over them, have given rise to the belief that they are really Slaves. It is commonly supposed that our superior power *compels* them to resign their natural fondness for independence, to yield implicit obedience to our will, and to perform those offices for which they are adapted by their organization, intelligence, and instincts. We are in the habit of attributing to our own influence the submission obtained from these animals; *we* are imagined to be the source of those instincts developed under domestication, and to have commanded obedience, just as our superior power maintains them in captivity.

This conclusion is, however, altogether fallacious. Judging from appearances only, we have confounded two things totally distinct in their nature, namely, Domestication and Slavery. Domestication is a state of freedom, and hence the difference between the human Slave and the domesticated Animal is as great as that between Slavery and Liberty.

The domesticated animal makes use of its natural faculties within the limits marked out by its situation, in a manner exactly similar to the wild animal in the woods. Being never urged to act except by external causes, or by internal instincts, as soon as its will has conformed itself to the constraints of its situation, it makes no further sacrifice. The animal, in fact, is not in reality in a different situation from what it would have been if left to itself. It lives, without constraint, in society with Man, because doubtless it was naturally a sociable animal. It conforms itself to the will of Man, within certain limits, because its herd would have had a leader in the strongest or most active animal, to whom submission would have been naturally paid. If a Dog is by our care rendered a good courser, it is because he was a hunter by nature, and we have only developed one of his original qualities. The same rule is observed in all the different qualities which we impart to our domestic animals. They perform nothing which is not agreeable to their nature; in doing so, they only fulfil the original purposes for which they were formed; they never acquire different qualities, and thus enjoy, under Man, a perfect state of liberty.

It is true that Man possesses an immense power over the domestic animals, and one which he often abuses. Yet he usually develops qualities natural to the animal; hence it acquires a degree of improvement unattainable in its original state, and thus its condition really becomes ameliorated. Thus we may see the immense difference between Slavery and Domestication. The Slave is not only a social being, with the power of willing, but he is naturally a free being, whose mind cannot confine itself spontaneously to the situation in which he is placed. He knows his condition, considers its consequences, and feels its oppression. The natural power which he possesses of reflecting upon his situation, shows it to him in all its degradation. He feels that he is in chains, that he cannot use his natural free-will, and that he is a degradod being. On the contrary, a domestic animal satisfies all its wants; hence it lives in a state of Nature, and is conformable to the situation in which it is placed.

The Slave who is compelled to renounce his free-will is far from being in the same condition. He holds the same rank in the Moral as a mutilated being or a monster in the Physical World.

The essential difference between these two states is further seen in the opposite means which are employed to enforce them. A Man can only be reduced to Slavery, and maintained in it, by violence; because it is the very nature of Liberty to be unrestrained. An animal can only be domesticated by kindness. Its will exists, and shows itself only in its wants, and it can only be acted upon through its necessities, either by satisfying or enfeebling them.

Violence is altogether useless in disposing a wild animal to obedience. As it has a natural aversion to Man, from his being of a different species, it runs away, if at large, upon the first impression of fear which he occasions, and if captive, maintains a determined hatred towards him. It is only by restoring confidence that it can ever be rendered familiar to him, and this can be effected only by kind treatment. It is thus that the social instinct of the animal becomes gradually developed, and its natural feeling of distrust of every thing which is new or strange becomes proportionately weaker.

The methods to be adopted in taming an animal are as various as the creatures themselves. Each process must be adapted to the peculiar likings of the animal.

To satisfy its natural wants is one method which, in the course of time, brings on its entire submission, especially when applied to a very young animal. The habit of receiving its food constantly from our hands renders it familiar, and finally it becomes attached. But, except when very long continued, the attachments thus formed are but slight. The benefits which the animal thus procures it could have obtained of itself, had it been allowed to fulfil its natural disposition. As soon as we attempt to bend it to any particular service it runs away, and quickly returns to its original independence. It is therefore necessary, not only to satisfy its original and natural wants, but, by creating in it new wants and enjoyments, to render the society of Man absolutely necessary to its existence.

Hunger is one of the most powerful means of taming animals. As the extent of a benefit conferred is always in proportion to the want of the person relieved, so the gratitude of the animal is more profound according as the food given to it was the more necessary. This method is applicable to all the Mammalia. It gives rise to a feeling of affection on the part of the animal, and at the same time produces a physical debility which reacts upon and enfeebles its Will. It is thus that the education of Horses begins, when they have passed their first years in a wild state. On being first caught, a very small quantity of food is given to them, and at very long intervals of time. They hence become gradually familiarized to their keeper, and acquire a certain degree of affection for him, which he readily turns to his own advantage, and thus confirms his power.

If to the influence of hunger that of delicious food be added, the empire of kindness becomes greatly extended; and this power arrives to a degree perfectly astonishing, if we can succeed in pleasing the palate of animals by any kind of confectionery or cookery, in a higher degree than could have been done by the best food attainable in their wild state. In fact, it is chiefly by means of dainties, especially of sweetmeats, that the herbivorous animals can be induced to go through those wonderful feats and exercises which may often be witnessed in the Circus.

These delicacies influence the will of the animal to such a degree, that starvation and physical deprivations become no longer necessary. In a short time it acquires a high degree of affection for those who contribute to its enjoyment, and willingly performs whatever they may require.

But the services rendered by animals do not always proceed from so selfish an origin. Caresses are one chief means of gaining their affection, which cannot be termed sensual, because these are addressed to no particular sense. Their fondness for caresses is altogether an acquired taste. No wild animal requires them from others of its own species. Even among our domestic animals, we see the young rejoice at the approach of their mother, the male and female happy in each other's society, individuals accustomed to live together pleased to meet again after having been separated; but these feelings are always accompanied with much reserve, and they never extend to reciprocal caresses. It is from Man alone that they receive them, and their attachment to him increases with the strength of the acquired taste for them. The pleasure of animals may be further heightened by a soft tone of voice, or even by touching their mammæ.

All animals are not sensible to caresses in an equal degree. The Ruminating animals seem but slightly influenced by them; the Horse, on the contrary, enjoys them with ecstasy; it is the same with many Pachydermata, and especially with the Elephant. The Cat is not indifferent to them; sometimes even it seeks them with ardour; but it is unquestionably in the Dog that the influence of caresses produces the most marked results; and what is remarkable, all the other varieties of the genus Canis, which have been hitherto observed, share this quality with him. M. F. Cuvier mentions, that in the *Menagerie du Roi* there was a She-Wolf, upon whom caresses with the hand and voice produced so powerful an effect as almost to amount to a state of delirium, and her joy was exhibited as much by cries as by movements. A Jackal from Senegal was similarly influenced when treated in the same manner; and a common Fox was so forcibly affected by them, that it was necessary to abstain from all demonstrations of this kind, as the result might have been fatal to the animal. It will be interesting to know that all these animals were females.

It may be doubted whether we should consider the chanting of airs, or the sound of bells, amongst those artificial pleasures by means of which animals are gratified and captivated. The songs of the Camel-driver are perhaps only the simple signs by which the Camels learn to mend or slacken their pace.

That animals may continue to perform those acts of docility which we require from them, caresses must follow as well as precede their performance. The constraint employed in urging them to act would, if too long continued, have an injurious effect. It is then only by repeating their delicacies or caresses that the calmness and confidence is restored.

When once familiarity and confidence prevail on the part of the animal by means of kind treatment, and that habit has rendered the society of Man necessary to it, we may then venture to use higher degrees of constraint, and even to inflict punishment. But our means of severity are very limited. We can only use blows, with certain precautions to prevent the chastised animal from running away. Punishment always produces the same effect; it changes the disposition which we wish to suppress into fear. By the association of ideas, the former impression yields, or is entirely merged into the latter.

It is always dangerous to carry the punishment of an animal to excess. Violent fear may either totally intimidate the animal and render it for ever useless, or else it drives it to despair, and it becomes altogether ferocious and unmanageable. A Horse, naturally timid, if corrected imprudently, plunges in the madness of its fright, along with its rider, into the deepest abyss. The Spaniel when kindly treated is intelligent, docile to its master, and in every respect fitted for sporting; yet if an undue severity has been used in its education, it is undecided, hasty, or cringing.

When once the severity of punishment has passed a certain limit, which varies in species and even in individuals, the animal begins to resist. In a moment, the instinct of self-preservation awakes in all its force. Thus we often see our domestic animals, and even the Dog itself, revolt against ill usage, and inflict the most cruel punishment on its perpetrators.

Many instances might be adduced of vengeance inflicted by the domestic animals, especially by the Horse, against those who have ill-treated them, as well as the hatred shown by these animals, and the very long time for which this feeling of aversion has preserved its force. The cases are numerous and well known; and although they have long demonstrated *that brutality is not the way to obtain obedience from animals*, these creatures still continue to be treated as if it were unnecessary to court their compliance. One example of an Elephant may be mentioned here, which happened under the eyes of M. F. Cuvier.

This animal had been entrusted, when three or four years old, to a young man who took charge of it, and had trained it to perform various feats for the amusement of the public. It paid implicit obedience, and seemed to feel a tender affection for him. It not only yielded, without a moment's hesitation, to all his commands, but seemed absolutely unhappy without him. It rejected the attentions of any other persons, and even ate its food with sadness, when given by the hand of any other individual. While this young man remained under the eyes of his father, who owned the Elephant, he always treated it properly; but when it was transferred to the *Menagerie du Roi*, and that the young man was thus left to himself, his attentions diminished, the wants of the animal were neglected, and in a moment of drunkenness he went so far as to strike the Elephant. The poor animal immediately lost its habitual gaiety; it became so sad and dejected that it was supposed to be unwell. It, however, still obeyed the keeper, but no longer performed its exercises with the same alacrity as heretofore. Signs of impatience were sometimes shown, and then suppressed, as if two opposite feelings were in secret conflict; the animal became less disposed to obey, which increased the discontent of its leader. It was in vain that orders were given to the young man on no account to strike the Elephant, whose former docility could only be restored by the kindest treatment. Vexed at having lost his authority, and in not being able to exhibit the feats of the Elephant with his former success, his irritation increased, and one day he struck the animal with so much brutality that it became excited to the utmost pitch of fury, and uttered such a yell that its terrified keeper, who heard it for the first time, was glad to escape its vengeance. Never afterwards would the animal permit him to approach; even at the sight of him it became enraged, and all attempts to manage it were unsuccessful. It became wholly untractable, and no longer could be induced to perform for the amusement of the public.

It thus appears that kindness on our part is absolutely necessary to dispose animals to obedience, and that interest as well as humanity agree in pointing out the same course for the proper management of their instincts.

GENERAL REVIEW OF THE MAMMALIA CONTINUED.

Taming of the Mammalia—Forced Watches—Castration—Susceptibility of different Tribes to the influence of Domestication—Formation of Domestic Races—Relation to the Social Instincts.

The different methods of taming wild animals pointed out in the preceding section, are completely applicable only to those animals which are susceptible of affection and of fear. When animals feel a certain degree of attachment for kindness received, or when they dread the repetition of punishment, it is sufficient merely to recal these emotions to produce an immediate effect upon their Will.

It often happens, however, from the peculiar nature of individuals or of species, that certain habits or likings have acquired so powerful an influence, that no other emotion can maintain the ascendancy. With animals of this character, neither kindness nor punishment has any effect; and if persisted in, they would tend but to increase their constitutional bias. It is only by acting immediately upon their Will, so as to weaken the force of the ruling passion, that they can be rendered susceptible of gratitude or of fear. With tempers so refractory as these, the only means of domestication hitherto discovered are Forced Watches and Castration.

Without proceeding to actual mutilation, it appears that of all methods Forced Watches exercise the most powerful influence in enfeebling the Will of an animal, and in disposing it to obedience, especially when united to a prudent combination of rewards and punishments. Animals may be prevented from sleeping by applying the whip more or less frequently, or still more effectually by a loud reverberating noise, such as that of a Drum or Trumpet, which must be varied so as to avoid the effect of uniformity. By keeping them long without food, and then feeding them slightly during their usual time of sleep, the same effect may also be produced.

This method is applicable to all animals and to both sexes, although it does not always produce the same result. The other method, that of Castration, applies to male individuals solely, and is absolutely necessary only with certain Ruminating animals, but chiefly with the Bull.

All the other animal appetites, which were given for the preservation of the individual, such as Hunger, Thirst, and the desire of Sleep, when opposed, lead to an immediate physical debility on the part of the animal. Those passions, on the contrary, which were given for the continuation of the species, increase in proportion to the obstacles presented to their gratification. Hence it is only by depriving them of the organs from which these passions derive their source that we can bring them under our power.

In fact, the Bull, the Ram, and other Ruminants, can be domesticated solely after having undergone this mutilation. We may thus perceive the error of holding out the Ox and the Sheep as models of patience and submission. So far from this being really the case, the Bull and the Ram can only be used for propagation; and we have merely succeeded in domesticating the females of these races.

This operation is not necessary for Horses, although such as have undergone it are generally more tractable than the others. The Dog loses by castration his entire vigour and activity; and this appears to be its usual effect upon other Carnassiers, for we find that the domestic Cat is affected in the same manner as the Dog.

It thus appears from the preceding observations, that we can only obtain an authority over animals by means of their natural wants and propensities, by giving them a new direction, by developing, or else by annihilating them altogether.

The very small number of animals which we have hitherto succeeded in rendering practically useful, when compared with the total number of species, renders it extremely probable that we have not yet carried the art of domestication to its extreme limit, and that hereafter we shall discover the means of training new species to our use, as well as more perfect methods of educating the old.

It may easily be gathered from what has been already said, that the arts of taming present very different results when applied to animals of different species. There can be no comparison, for example, between the Dog and the Buffalo. The former is devoted in his attachments, submissive, and grateful; the latter wants docility, and indeed every benevolent affection. Between these two extremes, we may range in their order of susceptibility, the Elephant, the Hog, the Horse, the Ass, the Dromedary, the Camel, the Lamas, the Reindeer, the Stag, the Ram, and the Bull. We shall defer the further investigation of the peculiar characters of all these animals, until we come to describe the animals themselves. At present it is necessary merely to take a rapid glance over the several tribes of Mammalia, in reference to their different susceptibilities for domestication.

It might have been expected that the Apes of the Old Continent, which combine a high degree of intelligence to a structure the most favorable for the development of all their qualities, would have presented conditions well adapted for Training; yet no male adult Ape has yet been induced to submit to Man, however kindly he may be treated. We allude here to the Genera *Cercopithecus*, *Macacus*, and *Cynocephalus*; for the Orangs, with the Genera *Hilobates* and *Semnopithecus*, are still too little known to assert any thing positive concerning them. But in regard to the Genera first mentioned, their sensations are so vivid, their natural distrust so great, and all their emotions so violent, that they cannot be brought to observe any degree of order, or to habituate themselves to any given situation. Nothing can satisfy their wants, which alter with every change of circumstances, and even with the movements of the keeper round their cage. For which reason we can never expect any kind feelings on their part. At the time when they are rendering the most affectionate returns, they are ready in a moment to tear their master to pieces; and this does not seem to proceed from any premeditated treachery, but all their faults arise from the excessive unsteadiness of their tempers.

Yet it would appear that by great severity, and by keeping them almost continually in torture, they can be made to go through certain exercises. It is thus that the inhabitants of Sumatra succeed in training the Maimon (*Macacus nemestrinus*) to climb trees when ordered, and to gather fruits; but these arts always perish with the individual. As this kind of training is conducted solely by force, it cannot be regarded as a real domestication. It is by the same means that we see some of these animals, and especially the Magots (*Macacus inuus*), learn to obey their master, to leap with skill and precision, and to perform those astonishing dances for which they are so well adapted by their organization and natural dexterity. But being subdued by force alone, they are ever ready to run away; and in warm climates where they can obtain food, and do not require shelter, they are never known to return.

The American Apes with prehensile tails, such as the Ateles and Sapajous, are much more tractable, as they combine a great fondness for caresses and some attachment, to a high degree of intelligence and social instinct. With the Lemurs, there are so many difficulties in training from their excessive timidity, that all attempts of this kind must prove abortive.

This last observation is also applicable to the Insectivora, which, in addition to other difficulties, are possessed of an organization and limbs little favorable for training.

All the Carnassiers of solitary habits, such as the Lion, the Panther, the Martins, the Civets, the Wolves, and the Bears, are easily accessible to kindness, but fear has no power over them. While at large, they keep at a distance from all danger; and when confined, ill treatment only serves to enrage them. But if you satisfy all their wants as soon as these become urgent, if they receive nothing but kindness at your hands, and if no sound of your voice or motion of your limbs be threatening, soon will these powerful animals show the satisfaction which they feel at your approach, and give the most unequivocal proofs of their affection. Often the apparent mildness of the Ape is followed by some treacherous act, but the external signs of a Carnassier never deceive. If he be inclined to do mischief, every look and gesture betrays his intention, and it is the same when he is mildly disposed. Lions, Panthers, and Tigers, after having been tamed, may even be harnessed to a carriage, and they will readily obey their keepers. Wolves trained for the chase have been known faithfully to follow the pack of Hounds to which they belong. Every one has witnessed the feats which Bears may be induced to execute. Yet we have not succeeded in bringing these races to perform any actual service. Had this been effected, their superior strength would doubtless have rendered them valuable acquisitions.

The Seals are sociable animals, and in addition are gifted with surprising sagacity. They seem of all the Carnassiers the most susceptible of Kindness, and may easily be induced to perform any thing that their structure permits. Among the Rodentia, the Beaver, Marmots, Squirrels, Dormice, and Hares, are so little gifted with intelligence, that when we say *they feel*, the whole of their acquirements are summed up in one word. It is true that they may be made to go through certain exercises, because they are attracted by pleasure and avoid pain. But none of these animals will distinguish their keeper from any other person, however attentive he may be to them; and in this respect the social are not different from the solitary species. This seems to proceed from the excessive weakness of their memories.

Passing onwards to the Tapirs, the Pecaris, the Cony (*Hyrax*), the Zebra, and other Pachydermata or Solipeda, we find animals associating together in herds, grateful for kindness received, and afraid of punishment, capable of distinguishing their keeper, and often becoming very strongly attached to him.

This is also the case, to a certain degree, with the Ruminantia, but chiefly with the females; for, without any exception, the males of this tribe are possessed of an excessive brutality, which punishment only increases, and kindness fails to improve.

All that has here been adduced only shows the different means which may be adopted in taming these animals, and in attaching them to our persons. Something more than this is required to produce actual Domestication, for it may be seen that animals may be made to feel the influence of Man, and yet they may not necessarily become domesticated.

Had we been compelled, with each generation of animals, to begin anew the process of taming, we should not, properly speaking, ever have had domestic animals. At least their domestication would not have had its full effect, and the important consequences to the civilization of the human race would not have followed in its train. Such would have been the result, had there not existed a most important general law, which is found also to prevail in every department of animated life;—that the changes undergone by the first tame animals did not die along with them, but were transmitted to their offspring.

It is a well-known fact, that the young of all animals bear a great resemblance to their parents. This fact is equally true in regard to the human race, and seems not merely to be confined to their physical qualities, but to extend also to their moral and intellectual capabilities. Yet there are certain subordinate points in which animals depart from their original type, and these arise from the circumstances under which they have lived—such as the quantity or quality of their food, confinement, shelter from the inclemencies of the weather, the attentions or punishments of Man. It therefore follows, that those qualities which parents may transmit to their young are capable of being influenced by accidental circumstances, and hence we are able to modify animals and their descendants within certain limits, or, in other words, to form domestic races. Thus we have given rise to numerous varieties of the Horse and the Dog. Each breed or race possesses some qualities which adapt it for certain purposes in preference to any other race, and these are transmitted to its descendants as long as a course of opposite circumstances do not arise to disturb the effects of the former. For these reasons, we are obliged to adopt various means to preserve the purity of the several races, or else to obtain, by the crossing of races, new or intermediate qualities to those already formed.

We may also observe, that those races which are the most domesticated, and the most attached to Man, are precisely such as have received the action of the greater number of means for attaching them to his person. The Dogs, for example, upon whom, whether male or female, caresses have so powerful an effect, are undoubtedly the most domesticated of all animals; while the Bull, which is only attached to us through its food, and whose females are alone subjected to us, is certainly the least domesticated. This difference between the Dog and the Bull is farther increased by the difference in the fecundity of the two species. The Dog submits to our influence a much greater number of generations, in a given time, than the Bull. We are, of course, ignorant of the circumstances which induced the Dog to attach itself to Man, at the commencement, and also the manner in which he was reduced to his present state of submission; but every thing leads us to believe that his original disposition must have been exceedingly favorable to domestication. From the great facility with which the Elephant is tamed, we may conclude that if he were induced to breed in captivity, a race of domestic Elephants might be formed, rivalling the Dog in submission and attachment. Hitherto this has not been fairly tried; no attention has been paid, until very lately, to the breeding of captive Elephants; and even in those warm countries where their services are most necessary, wild Elephants are caught and tamed, while no efforts are made to transmit these acquired characters to their descendants.

An excessive fondness for society seems, however, to be another quality necessary to form a true state of domestication, besides that power just explained of transmitting to posterity their acquired instincts. There was originally a natural sociability of disposition in all the domesticated animals which assisted our efforts. Had they all resembled the Wolf, the Fox, and the Hyæna, in their fondness for solitude,—had they always avoided the presence of their own species,—it is difficult to suppose that we ever could have been successful in our attempts. Perhaps, indeed, we might have succeeded, by long continued perseverance through a course of generations, in forming a race, domesticated to a certain point only, which would acquire a habit of living along with us, until our luxuries would become almost necessary to it, as has been done in the case of the domestic Cat; but the difference between this sullen state of a mere toleration of Man and a real domestication is very great. We may also rest assured, that, had not these animals originally presented some striking partiality for the society of the human race, the attempt to domesticate them would never have been made. It thus appears evident that the possession of great intelligence, of a general mildness of character, and a susceptibility to rewards or punishments, are insufficient of themselves to produce domestication. Without dispositions naturally social, the animals now domesticated never could have been induced to attach themselves to Man, and to place themselves under his protection.

There are many social animals which cannot be domesticated; but it is an observation which holds true without one single exception, that all the domesticated animals form troops or herds, more or less numerous, whether they are observed in the wild state, or whether we consider only those portions of them which, being left to themselves, have reverted to their original state of wildness. On the other hand, no solitary species, however easily it may be tamed, has ever given rise to domestic races.

Whenever we succeed by kindness to attach an animal, naturally sociable, to our persons, we merely induce it to transfer, for our own advantage, that allegiance which it would naturally have paid to other animals of its own species. The habit of living with us becomes to it a necessary of life; and the Sheep, which has been brought up by our care, follows its keeper just as it would have followed the flock in which it was born. Our superior intelligence soon destroys all equality between ourselves and these animals; our Will guides them in the same manner as the strongest Stallion of the herd would have become the chief, and be followed by all the weaker individuals of which his herd is composed. The submission with which animals obey us is not greater than what they would have yielded to their natural leader in the field. It is true that our power is greater than his, because our means of persuasion are more numerous, and we are able to suppress the greater number of those wants which, in the wild state, would have estranged them from their leader. It may be said that the Hack-Horse which has passed from hand to hand, and been owned by numerous masters, so that all its natural attachments are weakened, if not altogether effaced, appears to have the same degree of docility to every person, and to be in a manner obedient to the entire human race; and we must admit that this case has no corresponding situation in the wild herd. But this objection will have no weight when we consider, that when an animal, whether isolated or in a herd, has had only one master, it is to him alone that he yields obedience and pays his allegiance. Every other person is disowned or even treated as an enemy, just as a strange animal or the member of another society would be in a wild herd. The Elephant allows himself to be guided solely by the Mohout whom he has adopted. Even the Dog, when brought up in solitude with his master, is fierce to all other Men; and every one is aware of the danger of intruding among a herd of Cows, in pastures which are but little frequented, without being accompanied by the Herdsman.

Thus, every animal which acknowledges Man as the chief of his herd is domesticated. The converse is equally true, as Man could not enter into such a society without immediately becoming the chief.

From these observations it will readily appear, that in domesticating the inferior animals, Man has only become a member of that society which these animals form among themselves, and the authority he has acquired rests solely upon the superiority of his intelligence.

GENERAL REVIEW OF THE MAMMALIA CONTINUED.

Mammalia can be accurately studied only in Captivity—Popular errors from considering Mammalia when in the Wild State—Importance of Menageries or Zoological Gardens—Several Species of Mammalia now wild are capable of being domesticated.

It is commonly, but erroneously supposed, that the character of the Mammalia can only be studied when those animals are at liberty in their native haunts, and that in a state of confinement, we can learn little of their real nature. Pining under the restraints of confinement, they are supposed to offer for our observation nothing but a series of artificial actions, totally unfit to convey those accurate notions which we should acquire upon seeing them at liberty. According to Buffon, the confinements of slavery are opposite to that state of Nature best fitted to exercise and develop all their faculties. "L'animal sauvage," he observes, "n'obéissant qu'à la Nature, ne connaît d'autres lois que celles du besoin et de la liberté."—(Tom. iv. p. 169.)

This popular error, as to the method of studying the characters of animals, has been repeated on the authority of Buffon by most subsequent writers, and tends very materially to retard the progress of scientific zoology. Its source may be traced, partly in the prevailing notion that domestication is a state of slavery, an error which we have already exposed, and partly in those visionary views of a pristine state of native innocence and simplicity in which Man is placed by the imaginations of our poets. From a natural association of ideas, these views are transferred to those animals which most resemble him. Their mistake upon this important point might have been avoided, if Naturalists had considered that when an animal is at large, it by no means enjoys that fancied independence which is usually connected with our ideas of a state of Nature. Its natural character is as liable to be modified by the irresistible pressure of those circumstances in which it is placed, as it would have been in the iron cage or paddock of its keeper. A wild animal prowling about with unresisted sovereignty in the midst of forests and uninhabited deserts, is very different from the same animal living at large in a thickly peopled country. Its character further changes with the plenty or scarcity of its food—the sudden or gradual variations of temperature—the numbers or vicinity of its own species—the strength and courage of its rivals, and a thousand other circumstances. The same animal when made captive is still further modified; it will scarcely be recognized if we succeed in taming it, and still less so if it become susceptible of a true domestication. But whatever modifications the animal may undergo, it always exhibits certain natural instincts and dispositions which are peculiarly its own. Its condition may be altered, but the original Nature remains the same under all circumstances. If new influential causes come into operation, they produce corresponding effects, but these are always relative to the faculties of the animal presenting them. Hence we consider these successive modifications which it undergoes, however numerous or varied, merely as the means of adaptation employed by Nature for bringing it into harmonious correspondence with the several changes of situation; and we are thus enabled to deduce its real Nature from a proper comparison of the phenomena it presents with the conditions under which they arose.

That this is the only method of ascertaining the characters of Mammalia with accuracy, may be farther seen upon considering fully the situation of an animal in the wildest state of independence which can be imagined. Let us take a Ruminating animal for example, whose wants are more easily satisfied than those of a carnivorous quadruped, and place him in the middle of the rich savannahs of South America, in the company of animals which are less able perhaps than those of any other country to disturb his repose. This surely would appear at first sight a situation the most favorable for the development of his natural propensities.

As long as all the wants of the Ruminant are satisfied, he remains at rest on the soft couch, which accident or his own choice has assigned to him. He sleeps sound in the consciousness of security, and when Hunger urges him to action, he finds his food spread out before him. If it be Thirst that troubles him, the neighbouring brook suffices to quench it. Thus his life passes on with a perfect sameness, alike uninteresting to the philosophical observer, until the rutting season arrives. Then urged onwards by a blind fury, he seeks the female. Bellowing in the ardour of his pursuit, he follows her traces, kills her if she resist and cannot fly, and either remains the conqueror, or becomes the victim of those rivals whom he encounters on the road. If successful, he is enfeebled by the violence of his passions, his ardour cools, and he returns to his retreat in search of a repose, which to him has now become necessary. There he remains following the same round of animal existence, until the anniversary of the putting season again urges him on in his temporary career of madness.

If we now consider the life of a Carnassier in the wild state, there is but little to add to the uniform picture which we have here attempted to represent. Instead of pasturing in the savannah, this animal springs upon his prey in the jungle, or else pursues it in the desert. He is thus compelled to make use of other qualities beyond those which a mere vegetable diet would have required. Sleep is perhaps equally necessary to him, and probably as long in its duration as that of the Ruminant. All the difference we find between them is, that the nature of the food with the Carnassier demands the exercise of a greater degree of cunning, sagacity, and strength, more caution in ensuring his own individual safety, or, if a female, also that of her youthful progeny.

Now, we ask is there any thing in the course of life followed by these animals, which cannot be learnt equally well when they are in a state of captivity?

If we succeed in taking both the Ruminant and Carnassier alive, and transport them into some Zoological Garden, we no longer find their Nature stupified with that dull inactivity which has here been exposed. We can now place them in situations much more complicated than any they could have experienced in the wild state. We can vary these situations, we can multiply their wants, or increase the dangers to which they are exposed. It is then that we observe their natural dispositions developing themselves, that we find new propensities arise, new resources expand, and an entirely different view of their Nature arises gradually before us. Then we begin to perceive that the state in which animals are placed by the hand of Nature on the earth, is not the most favorable for the development of their faculties. That constant equilibrium of forces which prevails among all animal societies, gives to the most powerful a preponderating influence over the weaker, which never allows the latter liberty to act. It is only when the industry of Man intervenes, that animals acquire the power of developing their faculties. When the overpowering forces, to which they are subject in the wild state, are restrained, or diminished in their action, we are able to discover the natural instincts and propensities of the animal, and arrive at definite as well as varied results.

The older Naturalists have fallen into many important errors from considering animals only in the wild state, and the characters which they have given to most Mammalia are in consequence imperfect, and in many instances altogether erroneous. The illustrious Buffon, to whom we owe so many glowing descriptions of their characters, adorning with the charms of his eloquence subjects hitherto confined only to the severer studies of the learned, gives many striking instances of these mistakes, which, of course, have been repeated after him by most popular compilers.

"The Lion," he tells us, "unites with a high degree of fierceness, courage, and strength, the more admirable qualities of nobleness, clemency, and magnanimity. Often he forgets that he is the sovereign, that is, the strongest of all animals—and walking with a gentle step, he does not deign to attack Man except when provoked to the combat. He neither quickens his step nor flies, and never pursues the inferior animals except when urged by hunger." Again, on referring to his description of the Tiger, we find the same eminent Naturalist observing, "That the Tiger presents a compound of meanness and ferocity; he is cruel without justice, that is, without necessity. He seems always thirsty for blood, although his hunger may be satisfied with flesh. His fury knows no other intermission than the time spent in ambush for his prey. He seizes and devours a second prey with the same fury, which seems to have been only exercised, and not glutted, in the blood of the first."

These differences which Buffon describes, probably on the authority of travellers, could only have arisen from the different circumstances under which these animals had lived. It is one of the facts which the institution of properly-regulated Zoological Gardens or Menageries has disclosed, that *the Lion and the Tiger have very nearly the same natural dispositions.* When placed in the same circumstances, they constantly present the same phenomena. The one is tamed with the same facility as the other; they have the same attachment to their keeper, they make the same acknowledgments for kindness received, and their hatred or passion is excited by the same causes. Their sports and gambols bear the same resemblance as their fears and desires. They both seize their prey with the same eagerness, and defend it with the same fury. In a word, if we abstract the differences of their form, they seem to be absolutely the same animal, so close do their characters correspond in every respect.

Again, there is the Hyæna. Every one has heard of "the untameable Hyæna, that fierce beast which," according to the Showman, "was never tamed since the memory of Man." Its name has been long considered as the emblem of the most determined ferocity and cruelty. Buffon, and the most eminent Naturalists, lend their names as authorities for the assertion. Yet when we come to submit the Hyæna to the experiments of the Menagerie, its character yields to the influence of science, like the most untractable earth before the galvanic pile of the chemist. We then find that *the Hyæna is a most tameable animal.* When treated with kindness, it comes like a Dog to the feet of its master, receives his caresses with pleasure, and takes its food mildly from his hand.

These are only a few of the instances which might be brought forward to show the immense importance of Menageries, or Zoological Gardens, to the proper knowledge of the characters of animals. When at large in the desert, it is with the utmost difficulty that we can ascertain the real condition of an animal so as fully to appreciate the influence of surrounding circumstances. On the contrary, when confined in the Experimental Garden, we possess the means of successively abstracting those forces which in its former state constrained and overpowered its natural propensities. We may further submit it to new combinations of influential forces, and we may thence deduce those general laws by which all the productions of Nature are equally swayed.

What, we may inquire, would be the present state of Natural Philosophy, if mankind had confined themselves only to those phenomena which Nature spontaneously presents, if they had not invented complicated instruments and apparatus for the purpose of placing the forces of Nature in new and untried conditions? To suppose that animals, when captive, exhibit actions of a different Nature to those performed in the wild state, would be to assign to Man the absurd power of altering the Nature of animals, of creating in them other dispositions than such as were assigned to them by their Maker; in other words, of subverting the laws of created existence. No person can suppose that the Chemist, however he may vary his experiments, can create a single particle of matter, or alter any one of the laws of inorganic substances. In a similar manner, the Zoologist, however he may vary the condition of the animal under examination, arrives by analysis only at those particular and general laws which Nature has assigned to the animal, but its original constitution remains the same under all circumstances. As long as our observations of animals are confined to those at liberty, this important branch of Natural History will contain nothing but a crude collection of isolated facts often at variance with each other, because they are united by no connecting link. The observers, perhaps, are guided by no sound views of science, or the facts they record are accidental, or arise from local causes, while fantastical hypotheses are formed of the nature of animals, derived from their views regarding the nature of Man. When, however, the captive animals of a Zoological Garden are submitted to a rational course of experiments, that branch of Natural History which considers the actions of animals and their causes, becomes elevated to the rank of a science from the richness and variety of the general truths which it unfolds.*

It is only necessary here to point out, in a few words, some of the important facts which have been brought to light by a properly-regulated course of experimental inquiry into the nature of animals, and to exhibit a few of the subjects which still remain open for inquiry.

For a long time it was imagined that the moral perfection of Man depended upon the development of his organs, and if this error has at length been abandoned by all except a few popular theorists and their followers, it still holds its sway in regard to animals. Those animals which enjoyed the greatest delicacy of sense, with pliable limbs well adapted for rapid motion, ought, according to this theory, to be the most intelligent; and the Monkeys, as well as many Carnassiers, seem at first sight to confirm the rule. But the examination into the intelligence of many species of Seals (*Phoca*) has demonstrated the important truth, that *the intelligent powers of animals are not in proportion to the perfection of their organs.* Of all Mammalia the Seal seems at first sight least fitted by its structure for intelligence. Its limbs are modified into oars or fins, it has no external ears, its eyes are adapted for vision in the dense medium of water, and hence it can see very imperfectly in the air; its nostrils are only opened when the animal breathes; and its body is covered all over with a thick layer of blubber, which deprives it of the exercise of touch, except at the places where the whiskers are inserted. Yet the Seal equals, if not surpasses, the Dog in its susceptibility for attachment, in docility, and in the brilliancy of its instinct. This fact demonstrates that the most exact acquaintance with the organic characters of animals is but an imperfect kind of knowledge, if we are ignorant of the inward principle which animates and guides their external frames. There is another striking instance of the importance of studying animals in captivity. It was always supposed from examining animals only in the wild state, that their intellects were developed in the same manner as those of Man. A young animal born with faculties still in the bud, seemed, during the ardour of its youth, to exhibit vivacity rather than strength; and it was thought that its intellect became matured, as in Man, with increasing years. This prejudice has been altogether overthrown by the examination of animals in captivity. It is there found that in the first ages of youth their intellects arrive at the full development, and that *young animals are beyond all comparison more intelligent than their parents.* It is clear that this fact never could have been ascertained with wild animals, because it was necessary to follow them throughout all the stages of their growth. There were many precautions requisite to ensure the success of the inquiry. All animals were not proper for the investigation; those of very limited capacity presented no apparent result; those modified by domestication could not be relied on; and the Carnassiers being under the continual necessity of using their faculties for subsistence, had their original nature so much altered, through the experience of the individual, as to be unfit for the experiment. It was necessary to confine the inquiry to the Apes, which have been most favored with intelligence, and yet whose existence does not depend upon the use they make of it, as the forests of their native climates yield a continual supply of abundant nourishment.

The fact that young animals are more intelligent than their parents, marks an important difference between the nature of Man and that of the Brutes. While human nature is capable of an indefinite improvement in the lapse of time, the nature of the brute blazes forth at once in its greatest brilliancy. The latter, by the continual decay of its original powers, points out that eternal rest to which it will soon be consigned; while the aspiring mind of Man sees, in the gradual perfectibility of his Nature, a glimpse of the immortal existence beyond the grave, on which his hopes love to repose.

These are not the only kind of truths to which experimental inquiries into the Nature of animals lead us. They also give much important information regarding their instincts, those necessary actions to which they are blindly urged by a superior power.

While the examination of the Beavers was confined to those in their wild state, it was remarked that such only as lived together in society, and in uninhabited countries, ever constructed habitations, while the solitary individuals encountered sometimes in densely peopled countries, retired into the natural cavities of the rocks on the banks of lakes and rivers. Buffon says that these animals are not urged to work and to build by that inward instinct, or physical necessity, which guides the Ants or the Bees; but that they act *par choix*, that is, from understanding the design and utility of their work, and that their industry ceases when the presence of Man inspires them with the dread of his power. Of all previous writers upon the Nature of animals, Buffon had probably the most just and elevated ideas concerning them, yet upon this point he fell into a serious error, which subsequent experimental inquiry has not failed to discover. It is found that when one of the solitary Beavers is placed in a convenient situation, when he is supplied with the proper materials for his edifice, such as earth, wood, and stone, neither his solitude nor the presence of Man has any effect upon his industry; he still continues to build. Had Buffon submitted one of these animals to experimental inquiry, he would have regarded the huts and dikes of the social Beavers not as "the result of combined projects founded on the reason and convenience of their ends—of natural talents perfected by repose," but he would have regarded them, as they really are, the result of an industry purely mechanical, as the object or gratification of an internal want wholly instinctive. Numerous experiments made with several of the solitary Beavers, taken from the banks of the Isère, the Rhone, and the Danube, have demonstrated that they are always naturally disposed to build, although they already may have a commodious habitation, and no apparent advantage could result from their labour, except that of blindly satisfying an instinct which they are, in a manner, forced to obey.

We shall only allude here to one more error which the examination of captive animals has completely served to expose. The belief that the herbivorous animals are of dispositions milder, more tractable, and more affectionate than the Carnivora, has infected the works of nearly all our popular writers on Natural History. It has exercised an important influence on philosophical and religious systems, upon the received views of the Nature of Man, or of the effect of food upon the moral development of his Mind, upon the laws of nations, and even upon their poetry. The dark-eyed Gazelle has become the emblem of mildness as well as of beauty, and it has been the same with the Hind and other animals with large eyes and a light or timid step. On the other hand, the Tiger, the Panther, the Hyæna, and the Wolf, are held up as glaring instances of a brutal ferocity as well as cruelty, fitted only to inspire us with hatred and detestation.

But upon a minute and close examination, upon becoming in a manner *personally acquainted* with them—a state of things which can only happen in a Menagerie—we are compelled absolutely to reverse these epithets; in a word, to assign to the Herbivorous tribes those ideas of brutality with which we had been previously taught to regard the Carnassiers. In fact, all the adult Ruminating animals, but especially the males, are rude and brutal in their manners; they can neither be soothed by good treatment, nor attached by caresses. If they have intelligence sufficient to know the hand that feeds them—a circumstance not always the case—they owe him no attachment. The requisite attentions of their keeper are performed only with the necessary precautions to ensure his own safety. The moment he ceases to intimidate them, they are ready for an attack. A secret sentiment urges them to regard every animal as an enemy which is not of their own species. We have seen that it is altogether different, even with those animals which feed most exclusively upon flesh. While the former are of a low and narrow capacity, the Carnivora are equally remarkable for the extent, refinement, and activity of their intelligence. So true is it, even with animals, that the development of their intellectual powers is more favorable than otherwise to the advance of those nobler qualities which attract our regard and esteem.

The importance of Zoological Gardens is not confined merely to the acquisition of scientific truths; they may lead to practical results of the utmost importance to society. There are numerous animals whose powers of becoming domesticated have not yet been fairly tried, and even in the present state of our knowledge, it is more than probable that not a few will be rendered practically available, and become to the next generation as familiar as Steam Boats and Gas Lights are to us.

Upon applying the principles already explained to different tribes of animals, we shall be able to point out some species which may hereafter become domesticated.

Beginning with the Apes, we find qualities highly favorable for domestication, such as the social instinct and great intelligence; but these are entirely counteracted by their excessive irascibility, violence, and fickleness of disposition, which render them altogether incapable of yielding submission. Hence they are entirely excluded from the list of animals with whom Man could associate. A like exclusion must be given to the American tribes of Quadrumana, to the Makis, and to the Insectivora, for the feebleness of their bodies would render them useless to Man, whatever susceptibility their dispositions might possess.

But with regard to the Seals, it seems altogether surprising that Fishermen have not made use of their instincts, or taught them to assist in fishing, in the same manner as the Hunter has brought up the Dog to aid in the chase.

We may pass over the intervening tribes of Rodentia, Edentata, and Marsupialia. The feebleness of their bodies, and their limited intelligence, disqualify them from sharing our labors. It is different with the Pachydermata, as most animals of this order have already been domesticated, or are fit to become so.

The Tapir (*Tapir Americanus*), it is to be regretted, is still in an unreclaimed

* At a time when London, Liverpool, and Dublin, have their own Zoological Gardens, increasing daily in wealth and importance, it seems singular that Scotland should be so far behind her neighbours in this branch of science; that the study of animals should be left to the generous munificence of private individuals, although the establishment of a public Zoological Garden, in Edinburgh, offers a reasonable prospect of remuneration, when considered even in the light of a mere commercial speculation, and without any reference to its important effects in elevating the public taste.

state. Being very much larger than the Wild Boar, and at the same time much more docile, it would yield domestic races of far greater value than the common Pig, and of a different quality. Yet, as it presents but few means of defence, this valuable animal is gradually becoming more scarce in America, where it is in great demand for the delicious flavor of its flesh as an article of food. It is probable that this important race, if not previously domesticated, will become totally extinct as America becomes more peopled.

All the different kinds of Solipeda might be rendered, with care, as domestic as the Horse or the Ass. The training and breeding of the Zebra (*Equus Zebra*), the Quagga (*E. Couagga*), the Dauw (*E. Montanus*), the Dzhiggtai (*E. Hemionus*), would be a useful labor to society, and probably a lucrative undertaking for the projectors.

Nearly all the Ruminantia are social animals, living together in troops, and thus most of the species of this numerous tribe are fitted by their Nature to become domesticated. There is at present one species, perhaps two, which are now only partially domesticated in South America, and are nearly unknown in our climates. This is much to be regretted, as they would yield fleeces of great fineness, and at the same time be useful as beasts of burthen. The Alpaca (*Auchenia paco*), and the Vicugna (*Auchenia vicugna*), are more than twice as large as the largest races of our Sheep. The qualities of their fleece are very different from those of the ordinary wool, and might be made into stuffs possessing an intermediate quality between wool and silk. This would certainly give rise to a new branch of industry, and serve to extend the commerce of our nation.

It has often been objected to the domestication of animals inhabiting warm countries, that the difference of climate would form with us an insurmountable difficulty. This error might have been avoided, if the objectors had been more aware of the resources of Nature in adapting animals to differences of temperature, as well as of our extensive influence over all living beings. In reference to the Alpaca and Vicugna this difficulty could not exist, and the objection resolves itself into mere ignorance of their habitat; for these animals reside only on the very temperate parts of the Andes, in Peru and Chile. It is not even applicable to the Tapir, although originally from the warmest climates.

The nature of domestication has now been fully explained. We have seen that its foundation exists in the natural disposition of animals to live together in herds or troops, and to form mutual attachments;—that it can only be induced by kindness, chiefly by augmenting their wants and afterwards satisfying them. Yet, by these means, we could only produce domestic individuals and not domestic races, if we were not aided by one of the most general laws of living beings,—the power of transmitting their organic and intellectual modifications to their posterity. This is one of the most remarkable phenomena of Nature, and well worthy of profound attention. That an accidental modification of the body should become a permanent alteration of form is extraordinary, but that a passing desire or habit should become, in the course of time, an original instinct, is without doubt altogether astonishing.

We have also seen the importance of studying animals in captivity, as connected with the progress of Zoological science. However the study of wild animals may serve to point out the part they have to play on the great theatre of Nature, it totally fails to discover their faculties and dispositions: we must resort to captive animals for this information. If it were true that animals must be examined when at liberty in order to ascertain their Nature, then the advancement of this branch of science is hopeless, as the difficulties of studying wild animals are so great as to be equivalent in practice to an impossibility. When at liberty they view with distrust every person whom they do not know, and either fly from or attack all who molest them. Again, animals could not be examined in savage and remote countries with which we are altogether unacquainted. The mere circumstance of pursuing an animal alters its original condition, and even then its natural state is as much disturbed as if it were really in captivity.

If it be true that the state of an animal, in whatever part of the earth it may be placed, is the natural consequence of the faculties and instincts imparted by its Creator, it follows, that if we have ascertained the latter, we may predict the former. As soon as we know exactly the general faculties and dispositions of the species, it is easy to state how it will act in every situation in which it can be placed. It becomes no longer necessary for us to follow the animal into the details of its existence, to visit the country of its residence, to find it out, and to hunt it down. Having once ascertained its Nature by Analysis, we can then apply the principles thus established Synthetically to every other possible case. This is the way in which all the sciences proceed, and Zoology can be properly cultivated only when it follows a similar course.

Under whatever view the subject may be considered, we must arrive at one conclusion—that the examination of animals in Menageries or Zoological Gardens is, of all methods, the best for studying and knowing them, as they ought to be investigated by the lover of Nature.

GENERAL REVIEW OF THE MAMMALIA CONTINUED.

Analysis of the principles which guide the actions of Animals—Intelligence and Instinct—Effect of Habit in transforming the character of Actions—Intelligence of the higher Animals compared with that of Man.

Before entering upon an analysis of the inward principles which determine the actions of animals, it must, in the first place, be recollected, that our knowledge of the intellects or sensations of animals will rest ultimately upon the consciousness of what passes in our own minds. It is only by examining that internal light which we possess within ourselves that we can arrive at any satisfactory conclusion. We compare our own actions with theirs; we are conscious of the internal cause which incites us to act, and we infer a similar cause in the animal. Should the Creator have bestowed a faculty to animals altogether different from those we are conscious of possessing, it must remain concealed for ever from our thoughts. The boundaries of our own intellectual world form the limits of our knowledge regarding the causes which produce the actions of animals.

Some of the principles which urge Mammalia to act are evidently of the simplest kind. The cries of an Infant when in pain, or in want of assistance,—the determination of a newly-born animal to the breast and the action of sucking,—the flight of a young animal when influenced by fear, although it has had no experience of danger,—its resistance when we attempt to seize it,—the attention of an animal just born to the cries of its mother, are all actions of this kind. Whether simple or complex, they arise previous to all experience; and have been regarded, by the common consent of all Naturalists, as purely Instinctive. They proceed from an irresistible and uniform internal power, which leads invariably to the same course of action.

But all the actions of animals are not of this uniform description. The Dog obeys and does not fly from the whip which his master raises to chastise him. He seeks for the object which has been pointed out, instead of remaining indifferent to the order he has received. If he be confined in a cage with wooden bars, he is agitated with rage and attempts to destroy them; but if they are made of iron, he lies down resigned to his confinement. All these actions are Intelligent; and it is the very nature of this Intelligence that it is capable of being modified by experience, and of conforming itself to the variable circumstances which incite it to act.

Other instances of Intelligence may be mentioned. When a Horse has to choose between two roads, of which one is known to him, he always takes the latter, however long the period of time since he may have travelled thereon. The Dog leaps before his master, and loads him with caresses, when he sees him preparing to go out, and wishes to accompany him. The same animal confines the flock, which has been entrusted to his care, within the precise limits marked out by his master. The Wolf attacks his prey openly and by force when in the recesses of the forest; but, if he be in the neighbourhood of a village, he approaches it cautiously, and attacks it by surprise.

All these actions are evidently Intelligent, and not Instinctive. The slightest circumstance would have induced the Horse to take the road which he had not previously travelled. If the Dog, by his disobedience, had offended his master, instead of leaping before him with joy, he would crouch and tremble at his feet. We also know that he acquires the remarkable talent of guarding the flock entirely from a previous education, and in being trained expressly for that purpose.

On the contrary, it is the common character of Instinctive actions to be fixed and invariable; to be constantly produced by the same causes and the same conditions. We accordingly consider the following actions as Instinctive;—the Dog, when he hides under ground the remains of his meal;—the Horse and Reindeer, when they remove with their hoofs the snow which covers the earth, to expose the food of which they are in want;—the Cows, when they come together in a circle, upon the approach of an enemy, with their heads and horns in the circumference, and their calves in the centre;—the Beavers, when they build cabins and construct dikes,—when they cut the wood necessary for their edifices, and repair the ravages which time or an enemy has occasioned to their buildings;—the Rabbit, when it excavates its burrow;—the Bird, when it constructs its nest. All these actions, and many others, are presented to us with a certain degree of uniformity, essentially the same in all its more important particulars. The Dog hides his food with the same blind Instinct, although its superabundant supply renders such a precaution unnecessary. The Horse or Reindeer that uncovers the grass or moss concealed under the snow, does the same thing when he sees the snow for the first time, and prior to all experience; he acts in the same manner after a meal as when oppressed with hunger. The Beaver builds in all situations, under the closest confinement as well as when in the enjoyment of the greatest liberty; when in the possession of the most comfortable abode, as well as when in want of all shelter. The Cows, which exhibit so much ingenuity in defending their young when in a herd, do not change their plan of defence though surprised in a small party, and when this method becomes wholly insufficient. The Rabbit which takes so much pains to burrow its retreat, knows not how to conceal it, or to adapt its construction to the changes of the seasons, to the circumstances of the place, or to the nature of its enemies. The lower classes of animals present instances still more striking and extraordinary of the blindness of their Instincts.

Upon considering all the Instinctive actions of animals, we find that these are of a nature very different from their Intelligent acts. Instincts are exercised or exhibit themselves only at certain periods; they are always of a limited number in each species, but they go on increasing greatly in number and importance among the lower classes of animals, generally in proportion as their organization differs from that of the human race.

Numerous instances might be adduced in support of these views. We at once perceive an immense difference between those intelligent actions which have already been enumerated, alike remarkable for their complexity, and those involuntary Instincts, always of great simplicity, which are occasioned by fear, passion, desire, or hunger. The latter seem purely organic,—that is, they result from the direct influence of a superior Power,—while, to the former, Intelligence appears indispensable. It is also only at certain periods, and for a limited time, that these animals seek their females, prepare their abodes, or construct their nests. The Dog, the Horse, and the Bull present few actions which can be regarded as instinctive; yet their lives pass on with considerable activity. Their intelligent actions nearly fill up their entire course, and are sufficient for the numerous situations in which they are placed. We can perceive a trace, among the Mammalia, of that diminution of Intelligence which results from the prevalence of the Instinctive acts. The Dog presents a great number of Intelligent actions, and only a small number of Instincts. The Bull, on the contrary, leads an active life within very narrow limits; and, though his Instincts are not positively numerous, they become relatively so, when compared with the very small number of his intelligent actions.

But the marked difference between Intelligence and Instinct becomes still more striking when we extend our views beyond the limits of the Mammalia, and consider the entire Animal Kingdom. It is then we shall perceive that the Quadrumana and Carnassiers, which stand at the head of the list, may almost be styled Intellectual, if compared to animals of the lower divisions, whose entire existence appears to be swayed by a uniform and constant force. Indeed, were not all analogies between mental and material phenomena altogether inadmissible, we would be disposed to compare the Instincts of the inferior animals to those inert powers of Nature which form the prime movers of our own machinery. The most complicated Instinct of the Dog, requiring for its fulfilment the concurrence of the greatest number of intelligent acts, appears absolutely nothing in comparison to actions of this nature, which may be seen in the Invertebrated animals, but chiefly among the Insects. In the Dog and other Mammalia of the higher orders, that Instinct which urges them to store up provisions for future want, shows itself only in a few isolated acts. Among the Insects, on the contrary, their entire existence, however varied it may appear, seems composed only of one single invariable action, from which nothing external can divert them, and to which they seem invincibly urged by a superior power. None of the Mammalia exhibit in any of their actions such a combination of sagacity, foresight, and skill, as might have been inferred from the industry of the Bee, did we not see in its actions, proofs of the existence of a Mind not its own. Were Man incapable of receiving evidence of a Creative Power in his own constitution, he must read it in that principle which urges these lower animals to perform a complicated course of actions, continuing for days and months,—ever directed to one end, and that end invariably the same. He must perceive that Wisdom is not the sole property of Man, when he finds profound combinations, calculations of the greatest complexity, and the most ingenious views, urging these lower animals to work with a degree of perfection which all his learning and experience, accumulated for a long course of ages, can scarcely equal.

Although we have said that Intelligent actions may be varied at the pleasure of the animal, while the Instinctive actions are irresistible, these assertions must be understood with some qualification. While performing an Instinctive action, the animal always preserves the power of using its senses, and of exercising its Intelligence to the degree natural to its species, and employs both in the manner most favorable for the execution of that Instinctive action to which it is actuated.

An animal is capable of exercising its Intelligence in a degree inversely proportional to the force of its Instinct. As the Instinctive wants become urgent, its Intelligence appears more fettered. There is no comparison, for example, in the degree of Instinctive force between the Hamster (*Cricetus*), which stores up magazines of provisions for the winter, and the Dog who hides his superabundant food. Nothing could divert the Hamster from its purpose; the slightest circumstance would cause the Dog to neglect that precaution.

Having pointed out the difference between Intelligent and Instinctive actions, we now come to draw the probable line of demarcation between the Intellectual powers of Man, and the Intelligence of the lower animals.

We evidently perceive that animals, especially the higher classes, have the power of Attention; that their senses receive impressions analogous to those we are conscious of experiencing in ourselves; that their ideas follow each other in a certain regular order constituting a train of thought; that a former idea can be recalled; that their ideas are variously associated; and that they can form some conclusions. This seems to be the extent of their powers. We are, of course, reduced merely to conjecture the intensity or qualities of their sensations or perceptions, and are therefore unable to point out those qualities of bodies which can be perceived by Man alone.

There is, however, one curious circumstance which may be noticed in regard to the sense of Hearing. With animals it is a Sensation and not a Perception; in other words, they are unable to refer sounds to an external cause. If a wild Bull or Horse feel himself struck violently, he makes no mistake as to the cause. He rushes immediately at the person who has inflicted the blow, even when struck only with a stone or other projectile, just as the Wild-Boar rushes upon the Hunter, whose rifle-ball has struck him. But when captive animals, in course of taming, are tormented by a Drum or Trumpet to prevent them from sleeping, they have no perception either of the instrument from which the sound proceeds, or of the person who plays it. They suffer passively, as if by some internal injury, where the cause of the evil is within themselves. It is curious that their head and ears are notwithstanding directed instantaneously to the precise quarter whence the sound proceeds. It is different with the sensation of Colour. The Bull rushes at a piece of red cloth, in the same manner as he would have done at an assailant; from which we may infer, that when the Horse and Bull are unable to refer a Sound to its proper cause, it is owing less to the distance which separates them from the instrument, than to the peculiar nature of their sense of Hearing.

In other respects, they generally seem to have the same senses as ourselves, and to perceive analogous qualities in bodies. Their motions result from the qualities of their sensations; they attempt to fly, to defend themselves, to seize, or to attack, according as they are moved by pleasure or pain.

Being capable of forming certain relations to Man of a benevolent or malevolent character, they acquire a marked affection for those who treat them well, and a determined hatred to their tormentors. Some species form an attachment for each other solely from the habit of living together for some time, and frequently their mutual hatred arises from mere caprice.

These dispositions presuppose Memory, and at least some confused knowledge of the relations of those qualities which distinguish one person from another. They exhibit the internal affection of the moment by external signs, which are in general very like those employed by Man for the same purpose.

The Mammalia acquire from experience a certain knowledge of natural objects,—of those which are safe or dangerous; they avoid the latter in consequence of this experience, and of that memory from which it is derived; without being determined by an Instinctive Attraction or Repulsion. This experience enables them to infer the consequence of their own conduct, when domesticated. They know that a certain action will be punished by their master, and that a contrary one will be rewarded. Their final determination does not proceed from any internal attraction, but often in direct opposition to some very powerful Instinct, and from the sole knowledge of the reward or punishment which will follow. This knowledge, besides memory, also presupposes a power of reasoning from analogy, or of inferring that similar causes produce similar effects. Knowing well the power of their master, that he can either punish them or not, they assume before him a supplicating air, on perceiving him to be angry.

Their passions and emotions react upon their involuntary functions in precisely the same manner as with Man. Surprise stops their respiration; they tremble with Fear; Terror throws them into a cold perspiration; and Love agitates their frames.

They may be corrupted or improved by Domestication. Habits of ease or luxury create in them artificial wants unknown in the fields or woods. Education may fit them for actions for which they are not adapted by their structure. By proper training they may be rendered docile, mild, and active; or, if improperly managed, they may become more obstinate, passionate, stubborn, or lazy, than Nature had formed them.

Race Horses give evident proofs that they are actuated by Emulation, and Dogs dispute with each other for the caresses of their master. The Jealousy of the latter does not merely relate to the possession of their food or other enjoyments wholly physical, but also to the benevolent affections.

The natural language of the Mammalia enables them to explain to each other the wants or sensations of the moment, and, in their intercourse with Man, they understand that more complicated language by which he makes known his commands. Not only do the young know the cry of their mother when she gives notice of approaching danger, but they comprehend a number of artificial words used by Man, and act in consequence. We have been acquainted with a gentleman who spoke to his Dog only in the French language. The animal would go home, or leave the room at his master's command if announced in the common phrase, but would remain undecided and look into his face with eyes of inquiry when the order was given in an unusual style or language. Some species and genera have very great powers of Imitation.

There can be no doubt that all the lower animals, without exception, are unconscious of their existence, and incapable of reflecting upon their own condition. They cannot turn their thoughts within themselves, and consider what it is that they see, feel, think, and perform. The acts of their Minds, like the movements of their bodies, are the mere result of external causes or of internal Instincts. They cannot form the notion of Liberty, for this can only be acquired by Reflection. For the same reason, they are not Moral and accountable creatures.

Indeed, it is chiefly to the want of Abstraction—of that power by which Man forms general ideas, and arrives at general conclusions, that the inferiority of the lower animals may be attributed.

Many philosophers, and especially Condillac, imagined that animals can reflect; and they founded their views upon those invariable actions which we have regarded as Instinctive. But there appears to be an evident contradiction, in attributing a constant and seemingly necessary action to a power such as reflection, which presupposes Liberty. It is also evident, that if the Dog concealed his food from really foreseeing the chance of future want, or, in other words, from reflecting upon his former necessity, and considering the probability of its recurrence, he would not have confined himself merely to set aside a supply of provisions, but would have taken means to provide shelter, a manger—in a word, to procure a supply for satisfying *all* his wants. This method of reasoning is conclusive against our regarding Instinctive actions, which are always partial and limited in their application, as the result of Reflection.

Other philosophers, imagining that the power of Reflection was usually proportioned to the force of their original desires, thought that the Instincts of animals might still be owing to Reflection, as they might depend for their exercise upon the force of the wants or likings of the animal. But if this were the case, the results of Reflection would be seen in their most trifling attachments or likings, as well as in their most pressing wants; and this does not agree with observed facts. It is certain, that, for animals generally to satisfy their appetite for food is of the greatest importance, and exercises the most powerful influence over each individual. Also it appears much more indispensable for their existence that it should be satisfied than the want of shelter. Yet we see many animals continuing to dig burrows and retreats, or, according to this theory, appearing to foresee the necessity of providing shelter, and yet the same animals do not foresee that which ought to be most urgent with them, the necessity of laying by a supply of food for future use.

All other attempts to explain the actions of animals in a general manner, and without admitting particular faculties, are equally objectionable, and the same thing may be observed in reference to Instinctive actions.

To avoid the contradictions which have been here pointed out, some philosophers have thought that the Instinctive actions of animals proceeded from some peculiar form of the Brain. When stated in the above simple form, this theory presents many difficulties. Its partisans have pointed out what they conceive to be the particular forms which manifest Instincts, and have collected many striking coincidences in support of their views. We find, it is true, in the structure of their brains certain forms which are in some manner connected with the Intelligent functions; but their experiments have not yet demonstrated these functions, and the extreme difficulty of such a task will long render it almost impossible. In the meantime, it is useless to occupy ourselves with suppositions, which are ever liable to be overturned by future inquiries. Natural History justly excludes all consideration of that question, once so much agitated in the schools—Whether animals have Souls.

There is a class of phenomena very different from the preceding, which are founded on more certain analogies, and throw some light upon the connexion between Intelligence and Instinct. We allude to the remarkable effects of Habit in transforming an Intelligent into an Instinctive action.

When a Man, who has perhaps studied with great care some treatise on the Art of Horsemanship, attempts for the first time to mount a Horse, none of his attitudes or movements, notwithstanding all his science, are what they ought to be. His body falls behind or before when it should be in a vertical position. His limbs are shaken when they ought to remain motionless. In a word, there exists no harmony between his own motions and those of the Horse. At first, it is only by a great effort of the Mind, that he can perform any one of the requisite movements; afterwards he is able to execute two motions at once, and at length he performs them all simultaneously. But after a little practice, the effort of the mind becomes less necessary, he ceases to pay any attention to the act. In a word, it is no longer an Intelligent action, but it becomes Instinctive. If the Horse make a movement in opposition to the Bridle, instantly the motion of the Horse communicates to the body of its rider that counter-motion proper to restore the equilibrium. This is done also without the slightest reflection, like the winking of the eye, or the turn of the head when a blow is threatened. The power of Reflection would arrive too late for the desired purpose, and it is therefore supplied by an action purely Instinctive.

These differ from the other kind of Instinctive actions only in their origin, being acquired by Habit, instead of being imparted by Nature.

The transformation of Intelligence into Instinct is capable among many animals of being transmitted to their descendants. We are assured that when Rabbits are kept for a long course of years in places where they are unable to burrow, they give birth to races which are not naturally disposed to form these subterranean retreats. Leroy informs us that young Foxes, born in countries which are thickly peopled, show, even before quitting their kennel, a much greater degree of cunning and prudence than those living in countries uninhabited by Man, and where they have few enemies to fear, or to avoid. Certain races of Dogs are naturally adapted for Hunting, to which they betake themselves Instinctively, and without any training, while other kinds of Dogs, such as the Bull-dog, do not course naturally. More instances of occasional habits being transformed into permanent Instincts, and transmitted as such to their descendants, might here be enumerated. But having treated of animals as individuals, we must now view them in those societies which they form among themselves, and trace the various forms of the Social Instinct.

GENERAL REVIEW OF THE MAMMALIA CONTINUED.

The Social Instinct—Occasional attachments of the Unsociable species—Social habits of a herd or troop of wild animals—All the domestic animals are naturally sociable.

It has long been admitted, that the love of Society is an original principle of the Human Mind, one which precedes all reflection and all knowledge. "Man," says Montesquieu, "is born in society, and there he remains." A kind of Instinct overpowers him, and prevails as well in the most savage hordes as among the most cultivated nations. "Send Man to the solitude of the desert, he is a plant torn from his roots. The form remains, it is true, but every faculty decays; the human personage and character alike cease to exist."

This Instinctive feeling is equally prevalent among the lower animals as with Man, and it is equally an original principle. It can neither be regarded as an intelligent phenomenon, nor as the result of habit. We find but slight traces of the love of Society in many species which possess a very high degree of intelligence, and it would seem that the greatest as well as most remarkable instances of it are found in the lower classes of animals, such as the Insects.

The proofs that it is not the result of habit are equally convincing. If it proceeded from Education, or from the influence of parents over their young, these causes would act with equal force upon all animals which are dependent for the same period of time upon their parents. The Bears, which tend and bring up their young equally as long as the Dogs, and with the same tenderness and solicitude, ought then to be equally sociable. Yet the Bear is essentially a solitary animal. Further, we find with all Mammalia that the influence of Habit never can prevail over the Instinct of Nature, that the Social Instinct lies dormant in some species of animals, although they have never had an opportunity of exercising it, and that it disappears in these species which are not intended for a permanent state of society, although the utmost care be taken to preserve its existence. The uniform experience of the Menagerie proves that Mammalia, naturally sociable, are always ready to form attachments, although brought up from their earliest moments in solitude. M. F. Cuvier observed this fact on comparing Dogs with very ferocious Wolves, both kinds being reared in the strictest solitude. The Social Instinct showed itself in the Dogs as soon as they had recovered their liberty. On the other hand, young Stags, which, in the first years of their lives, formed herds and lived together in society, separated permanently from each other when they had arrived at the age of puberty. The early habits and the social Instinct of the last mentioned animals became equally effaced by time.

A great number of Mammalia form connexions which are necessarily of a very temporary kind, as they are founded only upon those appetites which disappear when gratified. With these species, the males and females seek each other's society only during the rutting season. The female attends her young with the utmost affection, and defends their lives at the peril of her own, from the moment of their birth; and this affectionate regard continues on her part as long as her breasts are capable of secreting milk. Also the attachment is reciprocal on the part of the offspring, as long as they require her to provide for their wants. But as soon as the rutting season is over, the males absent themselves; as soon as the breasts of the mother cease to secrete milk, and the young can procure their own food, all attachment is at an end; the temporary union is dissolved, and they live apart in the most perfect state of estrangement. Then the slight trace of social habits which they had contracted in their infancy is effaced, each animal becomes a solitary individual, and shifts for itself. The wants of the one animal form the principal obstacle to a neighbour in satisfying its own. This opposition of interest brings on enmity and war, which is the habitual state of all solitary animals in respect to their own species. Among them, force is the only law. The feebler animal avoids the stronger, and dies in its turn if it cannot find a still more feeble animal to destroy, or another solitude to inhabit. This is the order of things which prevails among all the species of the Cat tribe, among the Martins, the Hyænas, and the Bears. We find it in all those animals which have no other wants than such as tend either to the preservation of the individual, or to the continuance of the species; and these passions, so far from being the cause of the Social Instinct, as some philosophers have imagined, are manifestly opposed to all social intercourse.

The examples which have just been adduced are afforded by animals presenting habits of the most solitary kind. Nature, however, does not pass at once to the truly Social species. Other animals possess the Social instinct in different degrees, and modified more or less by other instincts. We find traces of a permanent union in that attachment which exists between the Wolf and his female, even beyond the rutting time. The same pair appear to be remotely connected to each other during their entire lifetime, although they only form an intimate union during the rutting season. At other times they go about alone, each seeking its own food, and if they be sometimes seen to hunt in company, it is more from their having the same pursuit than from any Social feeling. The influence of these casual meetings in developing the Social instinct is but slight, and hence the Wolf easily bears a complete separation from his companion.

The Roe-bucks (*Cervus Capreolus*) are an instance of a higher degree of Social Instinct, but still not in its full extent. Among these animals, the male and female continue always together; they share the same retreat, feed in the same pastures, and run the same chances of fortune. If one of them perish the other does not survive, unless it can find consolation in another individual of a different sex, and in the same solitary condition. But the affection of these animals for each other is exclusive; they do not differ from the solitary animals in respect to their young, and they always separate from the latter as soon as their presence is no longer indispensable for their preservation.

In a union such as that last described, the mutual influence of the two individuals is very limited. There is neither rivalry nor subordination, and they acquire habits of the most perfect harmony.

It is different with those numerous animals which are naturally sociable, and live together in bands or troops, although their individual interests are often at variance. We here see the Social principle in its full extent, and to a degree which may almost be compared to human societies. It is not merely confined to the union of individuals in a family; but it holds together numerous families, and maintains peace among hundreds of individuals of either sex and of every age. All the individuals of the same herd know each other, and are mutually attached according to the relations which circumstances and their individual qualities have established among them. Harmony reigns until some foreign cause appears. But this mutual feeling exists only for the other individuals of the same herd. A stranger is at first refused admittance, almost always they fall upon him as an enemy, and ill-treatment usually compels him to fly. On the other hand, with every isolated individual the love of society is a want which must be satisfied. At first he follows them at a distance, he gradually approaches, and for the sake of being admitted, gives up his natural will to that point where the instinct of self-preservation urges him either to defend himself or to fly.

We may be able to exhibit the state of society which prevails in the troops of wild animals by tracing the life of a young individual from its birth until it becomes the leader of the herd. From the moment that it ceases to be nourished entirely by milk, and to venture out of the ring under the conduct of its mother, it begins to recognise the surrounding localities, its food, and the features of the other members of the herd. The mutual relations of the latter had been previously determined by the incidents of their growth and education, and these same causes exercise a similar influence upon the young animal, whose progress we are now tracing. It can neither fight with the older animals to establish its superiority, nor can it avoid them by deserting the herd; it is not strong enough for the former course, and is prevented from the latter by the Social Instinct. Submission is therefore its only resort. Whenever any circumstance arises which places its interest in opposition to that of the others, the most feeble animal is of course sacrificed. In this case, it must either yield to necessity or escape by stratagem. This is, in fact, the situation of all young Mammalia in the middle of the herd. They are very soon taught both the extent of their liberty and of their strength. If they be in a troop of Carnivora, such as the Wild Dogs, when a prey has been hunted down, each individual shares in it, according to the authority which he has been able to exercise over the others. The young animals are left to feed only upon the remains of the feast, or upon such bits as they can carry off by stealth. They attempt to appropriate a few fragments, or to slip in behind the others, happy if they can escape the blows which the older animals deal unsparingly. In this manner, they feed largely if the prey be abundant, but if scarce they die of hunger. By this continual exercise of authority which the older animals sway over the younger, the obedience of the latter is confirmed, and becomes with them a fixed habit.

Meantime, the young animals increase in age and strength. Other things being the same, they do not prevail in a battle with those which have preceded them by one or two years, but they are more nimble and vigorous than the younger individuals which have have just made their appearance. If force alone prevailed in these societies, the last-mentioned animals would have to yield obedience to the former, but this does not usually happen. The relations which custom has established are preserved, and if the society be conducted by a leader, it is the oldest animal which possesses the greatest power. His authority, which originated in force, preserves itself by that habit of obedience which the others have acquired through time. His influence obtains a kind of moral force, founded as much upon confidence as on fear. It ceases to be contested; and it is only those single or combined authorities which are in course of being established that meet with opposition, which they always encounter when the object is one that does not admit of being shared. A joint or combined authority of several animals requires only an equality of strength, aided by a strong social instinct, and the habit of living together in the same course of life. The wild and sociable animals do not fight except when excited by some violent passion; and if we except the cases where they have to defend their lives, the possession of their females, or the latter the safety of her young, they feel no mutual rivalry. The absolute superiority of one animal over another is asserted only when it becomes impossible to share the object: then contests begin, especially during the rutting season. Indeed, it is usually the female, by the marked preference which she bestows upon the younger and more vigorous males, whom she singles out with an extraordinary sagacity, that first induces the latter to assume their proper place, and break through that constraint and obedience which habit had confirmed.

There are probably some herds of animals where age alone acquires the force of authority. That such a state of society should exist, it only requires that all their wants should be satisfied, which perhaps takes place among those herds of Ruminants, living in the rich prairies of Africa and America, which Man has not yet subdued to his use. Their food being always plentiful, could not be with them a matter of rivalry, and but for the females, their lives would pass on in the most profound peace. A contrary effect would be produced where the interests of individuals are at variance; and when food continues long to be scarce, the entire breaking up of the society is the necessary consequence.

Hitherto we have supposed all the individuals composing the herd to be of the same natural disposition. But this is not strictly the case; some are more violent in their passions than others, or their wants are more urgent. One animal is naturally mild and peaceable; another is timid; a third perhaps is brave or passionate; peevish or obstinate; and then the former state of things is disturbed. However, these accidental causes soon produce their effects, each animal obtains its proper influence, and acquires its natural level. New-comers gradually become accustomed to yield obedience to those whom they find invested with the command, until their own turn to command arrives, which happens when all the others are newer associates than themselves, or that they are the oldest animals of the society.

The Social Instinct does not only show itself in the mutual affection which the animals of the same herd bear for each other. It is also to be seen in that sentiment of hatred and estrangement which they bear to every unknown individual. Two herds never unite willingly; and if circumstances compel them to do so, violent battles are the result. The males single out the males, the females attack their own sex; but if a single strange individual, especially when of another species, happen to be thrown by chance in the middle of a herd, he can only escape certain death by a timely retreat.

From this it follows, that the tract of country occupied by one herd, is in a manner inviolable to all neighbouring herds. It belongs to them by a kind of right. If they be Carnivorous, it is there they find their prey, if Herbivorous, their pasture. No other tribe in ordinary circumstances ventures to transgress its limits. Any pressing danger, such as a great famine, will, however, serve to break through this natural state of things; and the instinct of self-preservation then bursts through all restraints. But in ordinary cases, the right of property, as well as its effects, may be seen not only in the sociable herds, but even in the solitary species. Every one of the latter considers the place where he has established his residence, the retreat he has prepared, as well as the district upon which he obtains his food, as exclusively his own. The Lion permits no other Lion to reside in his neighbourhood. Two Wolves never inhabit the same district together, except in those countries where they are merely stragglers, and are every moment liable to be hunted to death. The same thing happens with the birds of prey. The Eagle from her eyrie extends her dominion over the immense tract of country within the range of her piercing sight.

The state of society above described prevails in every animal association, if we make allowance for their specific characters, such as the peculiar instincts, likings, or faculties which distinguish any one species from the others. But troops or herds of different species present characteristics which are peculiarly their own, and which serve to modify the Social Instinct. These will fall to be described in their proper place.

In all societies where the natural wants of the individual become very urgent, force prevails in a considerable degree. Among the Carnassiers, which are often pinched by the severest famine, the authority of the leader is much more liable to change than among the Ruminants. The same thing happens with Birds, whose wants and rivalries always proceed to the utmost pitch of fury. On the other hand, particular likings and special instincts may serve to augment and perfect the social sentiment. Many animals unite to the instinct of society that of combining for their mutual defence. Some dig capacious burrows; others erect solid habitations; and it is to the social instinct developed in a high degree that we owe the domestic animals.

By separating a sociable animal from the herd to which it naturally belongs, we may acquire some idea of the force of that instinct which urges it to forsake solitude, and live with its own species. Many instances might be adduced illustrative of the powerful affection which results from that sentiment. A Cow, a Goat, or a Sheep, when separate from its flock, feels an uneasiness often fatal to its life. M. F. Cuvier had a Corsican Ewe (*Ovis musimon* of *Pallas*) which fell into a declining state, and could only be restored to health by replacing her in the same pen with her companions. Travellers are usually well aware of the danger in meeting a herd of Wild Horses. Without the greatest precautions, they run the risk of losing all their own Horses; for the latter, although domesticated, cannot resist the force of that Instinct which urges them to join the wild herd that surrounds and neighs to them.

The two following anecdotes, illustrative of the force of the Social Instinct, are related on the authority of M. F. Cuvier. A Lioness had lost the Dog with which she had been brought up, and for the purpose of exhibiting these animals together in the public Menagerie, she was supplied with another, with whom, in a little time, she consented to associate. She had not appeared to suffer much from the loss of her former companion, as her affection for it was very slight. In a little time the Lioness died. The surviving Dog exhibited a very different result. He refused to leave the den where they had been confined. He continued two days to take his food, but his illness increased. The third day he refused all nourishment, and he died of exhaustion on the seventh.

The other instance was that of a Roebuck. It was very young, and had been captured during the spring in a forest. A young Lady, who took charge of it during the summer, became a companion, from whom it could not be separated. It followed her every where, and was as little afraid when she was present, as it was wild and fierce, when she happened to be away. On the approach of winter, it became inconvenient to leave the Roebuck in the same place where it had been brought up; and besides it had been unwell. But as the young lady was going to town, it was thought best to bring it up also, and to place it in a neighbouring garden, with a young Goat for its companion. On the first day, it would not rise upon its legs, or move from its place. On the second it began to take a little food; whether it would have continued is doubtful. However, its mistress visited it on the third day, and it returned with ardour all the caresses which she bestowed; but as soon as she left the animal, it lay down and rose no more.

Our domestic animals have always given striking instances of this deep and exclusive affection, which causes them to die of grief when separated from the object of their attachment. This obviously arises from the circumstance explained in a former section, that all the truly domestic animals are sociable in a high degree.

Without the social instinct, it is impossible to conceive how these animals could ever have become domesticated, if we consider the very early stage of human civilization when they became so. It is true, that by means of kind treatment, continued for a long course of generations, we can succeed in habituating some unsociable animals to live along with us; but this is far from being a real domestication. If Man had been originally placed in an unhospitable climate, the necessity of providing for his daily wants would not have allowed him much leisure for experiments on domestication, and had he been placed in one of those fertile countries where every thing is spontaneously bestowed, he would scarcely have subjected himself to a painful and continual industry for so remote an object. Indeed, no savage nation has been found with animals which themselves alone have rendered domestic. On the other hand, we have in the Cat an evident proof that unsociable animals do not naturally become domestic. It lives amongst us, accepts our protection, receives our kindness, but does not yield an equal submission or exchange of favors with the really domestic species. If Time alone were sufficient to produce domestication, the Cat would be as domestic as the Dog, the Cow, or the Horse.

Confidence, among the animals, is always paid to strength. The one succeeds the other; but it is chiefly by the former that authority is maintained. Nature affords numerous proofs of this. We are informed by travellers of undoubted credit, that the wild Horses have a chief, the bravest of the herd, who walks always at their head, whom they follow with devotion, and who gives the signal either for flight or for battle, according to the estimate which he forms of the strength of his enemies, or the extent of the danger. But if by accident he falls, the herd, being without a leader, disperses; each individual flies at random; some attempt to join other herds, and many fall victims to their indecision and mistakes. The same thing may be seen among our domestic animals; and the shepherd, in relation to his flock, is nothing else than the strongest individual of the herd, whose strength they have experienced, and in whose skill they repose the greatest confidence.

When strange animals first meet together, they are in the same relative situation as a wild animal in the presence of Man. When once the society of Man has become necessary, it is tamed. This fact is shown by the manner in which wild Elephants are captured.

The domestic Elephants, obedient to the Man who conducts them, are in the same state of hostility and estrangement, when placed in the presence of a wild Elephant, as every herd is in respect to an isolated stranger or the member of another herd. The wild Elephant, on the contrary, is urged invincibly by the social instinct to approach the other individuals of his species, and to submit to them within certain limits.

Elephants, like other sociable animals, can be made to use this influence to catch the wild individuals. Some tame Elephants, usually females, are conducted in the immediate neighbourhood of the retreats where the wild animals have established themselves. If they can find near the wild herd any individual, who is compelled by the others to keep at a distance, being perhaps driven out of the herd by some more powerful leader than himself, and thus compelled to live in solitude, so as to do violence to his social principle, the wanderer does not fail to find out the domestic Elephants, and to approach them. The masters of the latter, who are not far distant, run and tie the wild Elephant with ropes, being protected by the tame Elephants which belong to them: and if the stranger attempt to make the least resistance, the tame Elephants compel him, by blows of their trunk and tusks, to submit himself to be led away into captivity.

A most striking instance of the inefficiency of mere force, in comparison to that confidence which is established by time, was often exhibited in the *Ménagerie du Roi*. When the Moors of Barbary catch a young Lion, they are in the habit of bringing it up with a young Dog. These two animals become mutually attached, but especially the Dog to the Lion. As the former grows faster than the latter, it arrives sooner at its adult state, that is to say, at that time of life when carnivorous animals acquire strength and courage. From this difference of growth, it follows that the Dog maintains so great an ascendancy over the Lion as completely to direct the physical strength of the latter, and he always preserves this power, especially if the Lion be of a mild and quiet temper.

Other animals afford instances that muscular strength does not always acquire the ascendancy; courage and perseverance are also means of obtaining the command. M. F. Cuvier had a Cashmere Goat, which was placed in company with three other Goats, each of which was at least twice as large and as strong as the former. Yet the little Goat managed to get the mastery over the others, although in fighting he lost one of his horns, and thus was deprived of the advantage which the others enjoyed of striking both to the right and left. But his fury and obstinacy were so great, that he ended by obtaining, through dint of perseverance, an authority as complete as if it had proceeded from an undoubted superiority of physical strength. Two of the Goats which he had subdued followed their little master every where, and when separated from him could not rest until he was restored to them.

Buffon relates a fact, authenticated in a letter from M. Dumourtier, which shows how much the influence of animals over each other is increased by time. "The paternal authority among the Rabbits is much respected. I observe that all my Rabbits pay a great respect to their grandfather, whom I can easily recognise by his gray hair. His family have greatly increased. Those who have become fathers in their turn still preserve their submission to him. Whenever his sons fight together, whether for the females or for the possession of the food, the old grandfather, when he hears the noise runs at full speed to the spot. As soon as they perceive him, order is immediately restored; and if he catches any of them fighting, he separates them, and makes an example of the refractory animals by immediate punishment. I may mention another instance of his influence over his posterity. For having always accustomed them to go into their holes on blowing a whistle, whenever I give the signal, however distant they may be, I observe the old grandfather to place himself at their head, and though he arrives first at the holes, he makes all the others defile in before him, and is always the last to go in himself."

It cannot be said that this authority on the one hand, and submission on the other, are instinctive. They depend upon accidental and variable causes. They often present opposite features in the same individual; and even the slightest change in the external appearance of the animals is sufficient to dissolve all harmony between them. A trifling circumstance of this kind would cause them not to know each other, and to recommence their battles. If two Rams which have long lived together in the most perfect harmony be shorn, they look at each other with fury, and rush together with such violence, that unless separated, they will fight until one of them either flies or remains dead on the spot. A boy belonging to the *Ménagerie du Roi* nearly lost his life only from changing his dress. He had acquired an absolute authority over a Bison from North America. His command alone was sufficient to make this powerful animal go in or out of his stable, and the mere presence of the Boy made him tremble. One day having obtained from the tailor a new suit of clothes, a little different in its colour and shape from that which he habitually wore, he went into the stable to perform some service for the animal, when the latter, having looked at him attentively for some time, made a sudden attack; and the young lad would certainly have been killed, if he had not had the agility to leap over the gate of the ward into which he had so imprudently entered. Having thus escaped, and suspecting the cause of this unexpected attack, he resumed his usual clothes. The animal immediately recognized him, and regained his former fear and docility.

Force, however, exercises a very important influence in all animal societies. We even see its influence in places where it might be expected that Nature would oppose some obstacle. In a flock of Goats, the She-Goat exhibits a remarkable care for her young, and is ready to defend them with her life from the attack of any stranger. But if one of her kids receive blows from any of the other Goats of her own flock, she shows no opposition to their violence, and takes no notice of the cries of her young one, provided that they proceed only from the blows of the other members of her society.

Cunning is so often the attendant of weakness, that we may readily expect all the young animals of a herd will possess a great share of it. M. F. Cuvier observed a remarkable instance of this in the conduct of a young Rhesus Monkey towards his mother. Although she treated him in the most affectionate manner while he was suckling, she would never allow him to eat any thing. He could obtain nothing except by stealth; and even after he had filled his pouches, she would compel him to disgorge. In this way, the skill and cunning of the young Monkey became developed in a surprising degree. He used to watch the moment for seizing his food, when his mother was about to turn her head or eyes, and he always anticipated her movements with remarkable accuracy.

It may readily be expected that a herd will separate when famine prevails. Then each animal is attentive only to its own preservation. Some species and individuals even devour each other if driven to extremities. This takes place among the Rats (*Mus*), and also, it has been said, among the Field Mice (*Arvicola*). A dissolution of the society likewise occurs, when one of the Instincts essential to its existence cannot be exercised. In densely-peopled countries, the Beavers, instead of constructing habitations, lead a solitary life in the natural excavations of the rocks on the banks of lakes or rivers.

These several facts entirely confirm the correctness of those general views which have here been laid down regarding the Social Instincts of animals; and M. F. Cuvier has contributed more, by his talents and industry, to expose the character and manners of the Mammalia, than perhaps any other Naturalist.

The preceding observations serve to show that there results from the instinctive union of several individuals in herds or troops, a certain mutual dependence, which passes into a habit, and becomes a necessary of life. The authority of one animal over the other originates in force, but when once established, it is maintained by confidence, until passions more powerful than the social instinct arise, and snatch the authority from the chief, to vest it in a stronger and more courageous individual. It is in these mingled states of peace and war that the greater number of animal societies pass their existence, and they are dissolved when the instinct of self-preservation becomes more powerful in each individual than the Social Instinct.

Societies of this description have nothing either Intellectual or Moral in their constitution. We view, with mingled feelings of astonishment and admiration, a state of things in which authority is maintained without force, where harmony exists without the influence of Reason, and a variety of opposite wants and desires, without discord or dissension. We can ascribe this solely to the great First Cause of all things. The animals themselves take no active part in it, and are, under this view of the subject, but the blind and passive instruments in the hand of an invisible and all-powerful Being.

When Societies of Men approach this passive state, they bear a great resemblance in character to animal societies. It is sad to think that human nature can exist in such a state of degradation; yet the accounts of enlightened travellers inform us that the savages of New Holland, for example, lead nearly the same kind of animal life, where those faculties, which distinguish Man from the other Mammalia, have scarcely received any development.

It is only when the activity of Man is roused, that the mere animal societies, which we have here described, assume a new appearance. Phenomena of habit then become phenomena of conscience. The same action which was formerly produced by mere likings or necessities, now results from the light of Reason. The authority of the strong and the submission of the weak become ennobled by the feeling of Duty. Thus Society, which among the other Mammalia is purely Instinctive, is transformed with civilized Man into an Intellectual and Moral condition.

GENERAL REVIEW OF THE MAMMALIA CONTINUED.

Tame Races have become wild—Alterations and Development of their Instincts and Intelligence under Domestication—Sensibility—Imitation—Sympathy—Incapacity to distinguish between Justice and Injustice.

THE Intelligent Powers and Instincts of the Mammalia, which have formed the subject of the preceding sections, may be made to undergo various modifications and alterations. There is a certain degree of perfectibility connected with each animal nature, and the changes induced may either affect individuals only, or be also capable of being transmitted permanently to their posterity. Hence arise the peculiar Instincts and Intelligence of the different races or varieties of a given species.

When the animals of an uninhabited country first encounter Man, they exhibit no fear of his power, nor do they seem apprehensive of danger. The early navigators of the South Seas often allude to this innocent confidence of the Mammalia and Birds. Dr Richardson found the wild Sheep of the Rocky Mountains exhibiting that simplicity of character so often remarked in the domestic animals; and, in the retired parts of the mountains, where the hunters seldom penetrate, he had no difficulty in approaching them. He adds, "where they have been often fired at, they are exceedingly wild; they alarm their companions on the approach of danger by a hissing noise, and scale the rocks with a speed and agility that baffles pursuit." But the young of all Mammalia, which have been much exposed to persecution, exhibit an Instinctive fear for strangers; and this acquired Instinct, perpetuated by generation, may be induced as well by any of the larger and fiercer Carnassiers as by Man. Thus Danger, whether proceeding from Man or other animals, may perform the converse of Domestication, and render those races wholly wild which had originally been tame.

It has been often observed, that a certain resemblance exists between the characters of some classes of Men, and of the animals with whom they habitually associate; for example, between the Drover and his Oxen, the Shepherd and his Sheep, the Muleteer and his Mules, the Arab and his Steed. This modification is usually supposed to have been undergone solely by the Man; but this is not strictly correct, as the characters of the animals themselves insensibly approach that of their master and companion.[1]

[1] Ælian has long ago observed the curious fact, that the domestic animals, and especially the Dog, acquire the faults and good qualities of the society to which they belong. "The Molossian Dog," he remarks, "is the bravest, while that of Caramania, like the people of that nation, is the most ferocious and the least susceptible of Domestication." On comparing English Horses and Dogs with those of French origin, M. Dureau de La Malle observed certain well-marked national peculiarities which confirm the truth of Ælian's remark. In this way the habits and manners of the domestic animals may form an index to the civilization of a great nation. Even among the different grades of society in the same country, we find the animals adopting the peculiarities of their masters, and acquiring traces of their vices as well as virtues. The Dog which becomes so dainty when brought up in a Lady's chamber, is ferocious with the Butcher, submissive in the poor man's cabin, or thieving and cringing with the beggar. When standing at the Nobleman's lodge, he even adopts the tone and manners of the great man's porter. M. Edwards tells us that he has often seen Dogs, educated by weak females, become excessively timid, and that this timidity was transmitted to their offspring. A Terrier-dog, born in the house of M. de La Malle, and treated like a spoiled child by a kind-hearted woman, who amused herself with speaking to it all day, had its sensibility brought at six months old to such a state, that when its mistress caressed the Cat, or pretended to scold the little animal, its large eyes would fill with tears, and it would end by crying like an infant.

In the wild and savage state, the lower animals and Man are possessed of much less sensibility than when domesticated or civilized. They also retain a much greater physical power in resisting pain, and can endure without complaints the pangs of sickness, of deadly wounds, and all the evils arising from their original constitution or their want of civilization. The fortitude with which the savages of North America and of New Zealand endure torments is well-known. According to Azara, the Charruas, a savage race of Paraguay, do not utter a complaint even while under the knives of their enemies. This feebleness and want of fortitude in civilized nations has many points of analogy among the Wolves, Foxes, and proper Dogs, when they are placed in similar circumstances. The domestic Dog raises a most hideous yelping if a person tread on his paws, pinch his ears, or give him a whipping; but if the wild Dog, the Fox, and the Wolf be wounded or taken in a trap, they suffer the sharpest pangs without uttering a cry, and expire without groans, in the midst of the most cruel torments. The observed habits of the Dingo, or wild Dog of New Holland, perhaps the wildest of the species, bear the same relation to the domestic Dog in the scale of sensibility.

The ancient Greeks and Romans endured pain more patiently than the moderns. The Turks have nearly the same fortitude, and the differences among the tenets of Paganism, Islamism, and our own, are not the sole cause; for the colonists of Africa, America, and New Holland, and the sturdy peasants of our own country, endure pain more patiently, and with less complaints, than the inhabitants of towns. By this circumstance alone, we might almost be able to determine the degree of civilization among the different classes of society, and it will be found to vary usually in the inverse proportion of their capability to endure pain. M. de La Malle says that the English, of all European nations, take the greatest care to preserve themselves from sickness; that they have the greatest dread of pain; and show the least fortitude and firmness when the necessity for enduring the pain is not absolute. This is certainly a high compliment to the civilization of our nation, although made at the expense of our fortitude. The cause is strictly physiological. As the nervous system of Man becomes more susceptible to refined and vivid impressions, it acquires greater irritability; and when the imagination, with the powers of reflection and foresight, are highly developed, we may readily expect that the intensity of the pain will be increased. Habit with the Savage resigns him to pain; the civilized Man either discovers a remedy or roars out with anguish.

Some Instincts do not exhibit themselves until the animals have attained a certain age. When gnawing a bone, the Dog does not know until two months old how to hold it down steadily with his paws, and he is ten or twelve months old before he hides his superfluous food. The latter Instinct is also seen, according to Azara, in the Puma (*Felis Concolor*), and in several other wild Animals of Paraguay.

Many of the domestic animals, but especially the Dog, express their Contempt or Aversion for any object by rolling themselves over it. If they find the carcass of a Mole, a Shrew (*Sorex*), or other Insectivora, they immediately roll over it, which they never do upon the carcass of a Ruminant or Solipede, of which they are very fond. M. de La Malle had two Spaniels, which devoured with pleasure the bones of Woodcock and Snipe; but when he threw into their mouths the gizzards of these Birds, which had a very strong marshy flavour, they rejected them with well expressed signs of disgust, and when the gizzards fell upon the floor, they immediately rolled over them. On attempting to urge them by commands and threats to eat the gizzards, they smelt them and rolled over them as before, nor could they be diverted from this Instinctive action either by the presence or injunctions of their master. This experiment was repeated several times, and always with the same result.

A remarkable instance of the force of Imitation is related by M. de La Malle of his Dog, named *Fox*. This gentleman had a male kitten, aged six months, when the Scotch Terrier, *Fox*, then two months old, was given to him. It was of that variety with long and rough hair, with straight ears directed forwards, which attaches itself to Horses, and is used for Fox-hunting. This Dog, when two years old, had never been out of the house where he was allowed to run at large; he had never seen other Dogs, and had received his education solely from the three daughters of the porter, and from the Cat. The latter was the companion of his sports, and was with him continually; hence these animals had acquired a singular affection for each other. The Dog had adopted the mildness and timidity of the females, who took charge of him; but the Cat, being older than *Fox*, was his master in point of muscular force, and the Dog showed, in a marked manner, the influence of his preceptor. He bounded like the Cat, and rolled a ball or a mouse with his fore-paws in the same manner. He even licked his paw, and rubbed it over his ear just as he had observed in his instructor. The imitation was striking; it might have been expected that, in this state of isolation, the Dog being the more intelligent animal, would have acquired the greater influence over his companion; but the contrary happened. This circumstance is easily explained from the power of Imitation being greater in the Dog than in the Cat. But although *Fox* had showed such an attachment to his friend, it was not powerful enough to overcome his aversion to the species. If a strange Cat presented itself in the garden, *Fox* immediately put it to the rout. The Cat also manifested his hatred to a strange Dog, which M. de La Malle brought for the first time into the house. The visitor could not be taught to endure the caresses of *Fox*, but exhibited the utmost astonishment and aversion for his unnatural and Feline accomplishments. We are informed that M. Audouin had a Dog, which died in the year 1831, and had acquired all the manners of a Cat, particularly that one of licking his paw and passing it over the ears.

It is well known that Dogs can open a latched door, and ring the bell for the porter; this proves the facility with which they imitate the actions of Men. Many Cats are known to leap upon the bell-rope when they wish to have the room-door opened. M. de La Malle had another Dog, which was brought to Paris when eight years old. On the day of its arrival, it went out of the house, but being fatigued it wished to return, and barked at the entrance for a long time without effect. At length a stranger rapped at the door by raising the knocker. The Dog observed the action, and came in along with him. That same day M. de La Malle saw it come in six times by raising the knocker with its paws. It must be observed, that there were no knockers at that gentleman's country-seat, where the Dog had been brought up from its birth; and also that it had not previously been absent from home.

Signor Bennati,—a learned physiologist of Milan, who has written a curious memoir on the mechanism of the voice during singing, and received the favorable notice of the Baron Cuvier in May 1830,—had a Water-Spaniel, which always came near the Piano-forte whenever the S. Bennati struck the chords, and seemed to show a taste for music. The learned Doctor, himself a skilful musician, was then studying the merits of Dr Gall's system of Phrenology, and accordingly searched the Dog's cranium very carefully for the bump of music, but without the slightest success. Not discouraged by this important circumstance, he tried to teach the gamut to the Dog. He began with the Piano, but failed; he then tried the Violincello, the Flute, and the Clarionet, also without success. At length he recollected that Dogs usually bark when a Bell is rung, and, therefore, concluded that Bells exert a peculiar action upon the Acoustic nerve of Dogs. He procured seven diatonic Bells; and, by making them vibrate with the bow of a Violin, succeeded in making the quadruped-musician sing the gamut very correctly after nine days' lessons. He even brought the musical education of the Water-Spaniel as far as to make him sing an accompaniment in thirds to his own voice, which is one of very considerable power.

These several facts show that we are still very far from being able to point out limits to the intelligence of animals possessed of this remarkable faculty of imitation; and, at the same time, they serve to exhibit the influence of a rational course of education upon domestic races so intelligent and so capable of improvement as our Dogs.

The intelligent powers of the domestic animals are thus capable of undergoing a much greater degree of development than is commonly imagined; and this improvement is not confined merely to the faculty of Imitation, but extends also to the other powers of Memory, Judgment, and Reasoning. The facts observed during the training of Pointers, Setters, Shepherd Dogs, and Water-Spaniels, are evident proofs of that development of intelligence which increases with time, and may be induced by the care and skill of their instructor.

One instance of intelligence in a Dog belonging to M. de La Malle may be mentioned here, as it shows that the animal, judging from the impressions of its senses, combined their relations, and drew a just conclusion from the appearances and facts which he had observed:—"I reside," says that gentleman, "when in the country, in a tolerably large Château, with a great number of windows, as well in the dwelling-house itself as in the offices. The Spaniel, named *Pyramus*, to which I allude, sleeps in an open niche in the wall at the end of a very large court-yard, and I am in the habit of introducing him into my room during the night. This animal always finds

some food in my room, and a fire in the winter; he is, therefore, fond of his master, for Dogs, as well as Men, love society. I usually rise at midnight during winter, for I then retire to rest at five o'clock in the evening. As soon as I have risen and have lighted my lamp, I hear the Dog *Pyramus* under my window whining and howling gently. If I delay in opening the window, his cries become louder, with an occasional bark to give me notice of his presence. On opening the window, and on telling him that I am going to let him in, he is silent; but if I forget my promise, or am long in performing it, he begins in about half an hour his plaintive howls and barking. I have often observed him by moonlight, and when there is no light in the room, sitting with his eyes fixed on my window, but always remaining silent, and neither expressing his wishes by cries nor any other sound. From these facts I draw the following conclusions:—1st, That the Dog, by means of the sense of sight, combined the appearance of the light with the idea of his master, and of the agreeable things he was in the habit of getting from him; 2d, The absence of light indicated that his master either slept or was absent, and that then his cries would be superfluous. I may add, that my room is on an upper story, and though the Dog cannot get at it, except by a staircase and a long gallery with many turnings, yet this animal never mistakes the position of my window, although it is exactly the same as twelve others in front. And whether there be a light in my room or not, he regularly places himself at the same hour under my window, always silent when he perceives no light in the room, but calling me and asking to be let in whenever he observes the light."

It would be going too far to assert that well-educated Dogs can acquire notions either of Delicacy or Decency; but there cannot be the least doubt that they possess powers of Memory, Reflection, Judgment, and Association of Ideas. They can even combine Relations, and draw just inferences from the notions received directly from the sense of sight. Many of their perceptions of sight are also acquired. If a pup of two or three months old be called from an upper story when lying in a court, he knows not how to direct his eyes in the direction of the sound which strikes his ear. He must first learn to combine the relations of these two senses, which in this respect have an intimate connexion. But when once he has by chance directed his eyes to the quarter whence the sound proceeds, he treasures up this fact ascertained by experience; the result is fixed in his Memory, and he does not again make the same mistake.

The Domestic Pig, which is brought up with us only for the market, appears, when confined in its stye, to be excessively stupid and devoid of intelligence. Yet education, and the habit of living in the society of Man, develop his social character, and he exhibits some amiable qualities. At the town of Brives-la-Galliarde, in France, Pigs are domesticated like Dogs, and live in society with the inhabitants. They go up-stairs even in houses of three stories high, and often sleep in the same room with their masters. From this treatment they have acquired singular habits of cleanliness, which are further improved by their mistresses taking them usually twice a-day to the river to be washed and rubbed. While undergoing this operation, it is curious to see them going voluntarily to the water, and turning themselves first on one side, then on the other, and then on their back, to assist her; and M. de La Malle has seen them thank their mistress after their own fashion, when it was all over, by licking her hand. The Irish Pigs have long been remarked for their intelligence, and this is evidently owing to their living so much in the houses of the lower classes, and associating with the children. There is, however, no instance on record of their having been guilty of an equal degree of gratitude with their cleaner brethren of Brives-la-Galliarde.

The Intelligence of the Elephant is capable of undergoing a very considerable degree of development under Domestication. An Elephant, at the *Jardin du Roi*, was brought to understand the meaning of several words. When his guide said "*En arrière*," without elevating his voice or making the slightest gesture, the animal backed immediately. A remarkable instance of foresight was observed in the War-Elephants of Cochin China, which is related here on the authority of an intelligent traveller, and an eye-witness of the circumstance. Seventy Elephants were ranged against a Tiger; and one of them, urged on by his Mohout or guide, advanced to the attack. The Tiger waited until the Elephant was in the act of striking with his tusks, and making a sudden spring, alighted on the neck of the Elephant, with his hind paws inserted on the animal's trunk. The Elephant was wounded, and fled; but all the other Elephants who witnessed this conflict profited by the inexperience of their companion, and advanced against the Tiger with their trunks rolled up under their throats in the most careful manner, thus showing a degree of observation, foresight, and judgment, which might not have been expected in so large and heavy an animal.

We shall only add here a few instances which serve to prove that the Domestic Dogs, from their living in society with Man, have acquired the power of reflection, of combining means to an end, and of foreseeing difficulties in their execution. They also learn the meanings of many artificial words. They communicate their ideas to each other by means of natural signs, and assign to each the part necessary to be performed, in a combined plan of action—qualities which require operations of their minds but little inferior to the results of the Human Intellect. We shall even show that they form plans when hunting by themselves, which exactly resemble those ingenious devices invented by Man, and practised by him in the Art of War.

M. de Puymaurin, a *député* or Member of the French Parliament, had a female Water-Spaniel, whose education had been very carefully attended to, and it accordingly showed extraordinary intelligence. During the occupation of Paris by the allied armies in 1814, General Stewart, who lodged in M. Puymaurin's house at Toulouse, remarked that the Dog would take nothing that was offered to it with the left hand, and he tried to deceive the animal by crossing his arms, and even by exciting its appetite by some marked difference in the quality of the food held in each hand, but without effect. Being determined to subject the Dog to a very peculiar experiment, he requested one of his Aides-de-camp (Colonel Cameron), whose right arm had been amputated, to offer the Dog some food. The Dog approached, and without noticing the hand containing the food, rose upon its hind-legs, and applied its nose to the place where the Colonel's right arm ought to have been, as if to be sure that there was no deception, and being satisfied that the Colonel had only one arm, it passed on to his left hand and took the food. This fact was reported to M. de La Malle by M. Auguste de Puymaurin, the son of the *député*, on whose authority it is inserted here.

A well-trained Dog can often be brought to understand the meanings of words, even though spoken without the slightest gesture or alteration of tone. M. Edwards has been heard to mention an anecdote of a Dog, which was in the habit of seeking and bringing back Gloves. If in the course of conversation, when the Dog would appear to be paying no attention to what was going on, any mention was made of his talents, and the word Gloves (*Gants*) happened to be used, the Dog was off immediately seeking out for them; and when they were found, he again resumed his former position of careless listener to the conversation. Another Dog, which belonged to an aunt of M. Audouin, was excited in the same way when Gingerbread cakes were alluded to, of which he was very fond. If this word (*Gimblettes*) happened to be mentioned in the course of conversation, and without any peculiar emphasis, he was excited and ran to the cupboard where the cakes were shut up. This experiment was often repeated before several people, who would not at first believe the statement.

M. de La Malle informs us, that one of his neighbours, the Count de Fontenay, was engaged in some agricultural speculations relating to the breeding of the Merino Sheep, jointly with the Marquis des Feugerets, whose property was situate about two leagues from his own. The Count had a very fine Pointer, possessed of great intelligence, and as he had educated this Dog himself, it almost seemed to anticipate his wishes. One day he had an urgent message to communicate to his neighbour, and as no one was at hand to whom it could be entrusted, it occurred to him to try whether the Dog would carry it. Accordingly he fastened the letter to *Soliman*'s collar, and told him carelessly, and without expecting him to obey the command, "Carry that to Feugerets!" (Porte cela aux Feugerets.) The Dog did as he was desired, and would permit no one to touch the letter except the Marquis. "I have seen this Dog," says M. de La Malle, "for four or five years acting as messenger between these two Châteaux with a remarkable quickness and fidelity. When the Dog delivers the letter, he goes to the kitchen to be fed. As soon as he has had his meal, he sits down before the window of the Marquis des Feugerets' study, and barks at intervals, to show that he is ready to take back the answer. On the letter being attached to his collar, he sets off and brings it to the Count his master."

It has been proved, beyond the possibility of doubt, that the property of pointing and setting game, which some races of Dogs are made to acquire by feeding them well, and then exercising a certain degree of constraint and punishment, is transmitted unaltered to their descendants. M. Magendie, happening to hear that there was a race of Dogs in England which brought back game naturally, procured two adult Retrievers. These animals produced a female Retriever, which always remained under M. Magendie's immediate inspection, and though it had received no instruction, it stopped and brought back game, from the very first day that it was led to the field, and this it did with a degree of steadiness fully equal to those Dogs which have learned this art solely under the stern discipline of the whip and collar.

When the Spaniards discovered America, they introduced Dogs as auxiliaries in their military expeditions against the Indians. Columbus first employed them for that purpose, and we are informed in his own Memoirs, that at his first conflict with the Indians, his army consisted of 200 foot soldiers, 20 horsemen, and 20 dogs. These Dogs were employed in the conquest of several parts of the New World, especially in Mexico and New Grenada, wherever the resistance of the Indians was prolonged. We are informed by M. Roulin, that this race is still preserved pure on the Plateau of Santa Fé, where it is used for Stag hunting. This it performs with an extreme ardour, and still uses the same mode of attack, which must have rendered it so formidable to the Indians. It consists in seizing the animal by the abdomen, and then overturning it by a sudden jerk, which is given at the moment, when the weight of its body is thrown upon the fore-legs. Sometimes the weight of the animal thus overturned is six times that of the Dog.

Without receiving any previous education, the Dogs of pure breed, naturalized in South America, bring to the chase certain dispositions which the newly-introduced coursing Dogs, though of a superior European breed, have not yet acquired. Thus, the American Dogs never attack a Stag in front in the middle of its course, and even when the latter comes towards a Dog without perceiving him, the sagacious animal swerves to one side, and waits his opportunity to attack it in flank. A foreign Dog, who is unaccustomed to these precautions, is often left dead on the spot, from having the vertebræ of his neck dislocated by the violence of the shock.

Among the poor people inhabiting the banks of the Magdalena, this Dog has degenerated, partly from the cross of another breed, and partly from the want of sufficient food. Even in this degenerate race, a new Instinct seems to become hereditary. It has been long used exclusively in hunting the White-lipped Peccari (*Dicotyles labiatus*). The art of the Dog consists in moderating his ardour, and in not attacking any particular animal, but thus keeping the entire herd in check. The very first time that these Dogs are brought to the chase, they show their knowledge of this art, which has been transmitted to them by their parents. A Dog of a different breed rushes into the midst of the herd, is surrounded, and no matter how great his strength may be, he is devoured in an instant.

Those instances, where different varieties of the Dog unite their several talents while hunting, and form one combined plan of operations, are perhaps still more striking than any of the preceding. "I had at one time," says M. de La Malle, "two sporting Dogs, the one an excellent Pointer with a very smooth skin, and of remarkable beauty and intelligence. The other was a Spaniel, with long and thick hair, but which had not been taught to point, and only coursed in the woods like a Harrier. My Château is situate on a level spot of ground opposite to a copse-wood filled with hares and rabbits. When sitting at my window, I have observed these two Dogs, which were at large in the yard, approach and make signs to each other, and first glancing at me as if to see whether I offered any obstacle to their wishes, slip away very gently, then quicken their pace when they were a little distance from my sight, and finally dart off at full speed when they thought I could neither see them nor order them back. Surprised at this mysterious manœuvre, I followed them, and witnessed a singular sight. The Pointer, who seemed to be the leader of the enterprise, had sent the Spaniel out to beat the bushes, and give tongue at the opposite extremity of the brushwood. As to himself, he made with slow steps the circuit of the wood, by following it along the border, and I observed him stop before a passage much frequented by the rabbits, and there point. I continued at a distance to observe how this intrigue was going to end. At length, I heard the Spaniel, which had started a hare, drive it with much tongue towards the place where his companion was lying in ambush, and the moment that the hare came out of the passage to gain the fields, the latter darted upon it, and brought it towards me with an air of triumph. I have seen these two Dogs repeat the same manœuvre, and in the same manner, more than a hundred times; and this conformity has convinced me that it was not accidental, but the result of a concerted agreement and combined plan of operations arranged beforehand."

Leroy was of opinion that Wolves do the same thing; but he founded his conclusions solely upon the traces of their foot-marks left on the snow or mud. The same thing has been said by Hunters respecting Foxes, but the truth of it is very doubtful. Indeed, these wild animals which hunt during the night, especially when timid, are so difficult to observe, that these assertions require further confirmation, especially when made of animals known to be of solitary habits.

The fact that the domestic Dogs often combine their different talents to execute one manœuvre, is further corroborated by M. Louis Châteaubriand, nephew of the celebrated writer, who has witnessed the same thing between two Harriers and a Pointer. It is clear, that whatever differences there may be between this contrivance of the Dogs, and the ambuscade of a skilful general who hides his forces in the woods or copses, and sends a small body of troops with orders to fall back before the enemy, and draw them on towards the defiles, they both agree in being an ambuscade—a trick played upon the credulity of the enemy, and require the same operations of the Mind to direct them both.

The workings of Sympathy among the domestic animals are very striking, and the observations of M. de La Malle on this point have served to lead the way towards a better acquaintance with its powerful influence. Having been educated in the country during the earlier part of his life, he had amused himself with imitating the cries of many wild and domestic animals, and from habit he acquired a skill so great, as to deceive the animals themselves. In this way, by expressing after their own manner the external signs of Pain, Anger, or Desire, he could excite the same passions in them, and call forth at pleasure the external signs of those passions. By numerous experiments, he found that the imitation of the sound always produced a sympathetic effect, and he thus succeeded with Dogs, Cats, Asses, Cocks, Hens, &c., in producing the same results as a good comic or tragic actor upon an assembled audience, and in making the house cry or laugh, according as his voice and gestures excited the emotions of Grief or Joy.

By yawning, and at the same time imitating the sound which accompanies the yawn, Dogs may be made to yawn at pleasure; but to succeed in the experiment, the animals must be lying quiet for some time. If they be moving or in the field, their attention will be otherwise engaged. When several Dogs are lying down together, the first that yawns makes all the others follow, except those which have their attention occupied about any passing matter.

In all ages, impostors have not been wanting, who have pretended to know and to translate several words in the language of the Mammalia, the Birds, and even of the Insects. There is, however, a natural language, or language of signs, which can be interpreted, and under this point of view, the animals may be considered as savages, who are visited by civilized men for the first time, and of whose language they are wholly ignorant. It is at first necessary to invent a language of signs for communicating their ideas before a vocabulary has been formed of the most essential words. We are told by M. de La Malle, that in most instances he has completely deceived these animals; and at other times, when the imitation was less accurate, they have perceived the failure, and either treated it with contempt, or received it with an expression of ironic gaiety, as if they understood the joke, but at the same time were not duped by it. These experiments were repeated so often by that able observer, and so frequently produced the same results, that he is of opinion we may interpret the language of signs and the symbols of the passions among Dogs as accurately as we can the cries and gestures of the human race. While entering the house, he one day imitated the cries of fighting Dogs with such accuracy, that his Dog, who was very much attached to him, darted out and bit him in the leg. At the first word the animal perceived his mistake, and threw himself howling on the ground, and asking pardon for the offence in the most affecting manner. Sometimes when behind a screen, on imitating the gentle cries of the female, the Dogs were immediately excited, raised their ears, howled, and gave the usual indications on the approach of the female.

Numerous other instances might here be adduced of an extraordinary development of intelligence in the domestic races, and many of these are so striking, that they have led some philosophers to assign to the animals certain qualities which are properly Moral. Public exhibitions have been made of the extraordinary abilities of some individuals, and interest has not failed to exaggerate their talents beyond all reasonable bounds.

In the year 1830, there were two Poodle-Dogs or Water-Spaniels, called *Fido* and *Bianco*, which were exhibited at Paris as the most learned individuals of their race. They were said to be able to spell in different languages any word they heard pronounced, to tell the name of the reigning sovereign, or to name the card which a visitor had selected. These feats would doubtless have established their claim to the possession of intellectual and moral qualities, but on examination, it was found that their intelligence was even more limited than that of many others of their species, and that all their learning resolved itself into a small matter. It was suspected that their master had some private sign which the Dogs understood, but although the exhibition was attended by several eminent Naturalists, it was long before they could detect it. Those who have not seen the exhibition must be informed, that all the letters of the alphabet, or all the playing cards, were arranged in a circle round the

Dog, at a tolerable distance from each other, and that the Dog kept continually moving round the circle. When giving an answer he carried successively to his master all the letters which composed the proper word or phrase, and he did the same thing with the card on which a visitor thought. The manœuvre, as far as the Dog was concerned, resolved itself merely to bringing objects. We can easily suppose that the animal, passing slowly from letter to letter, and touching each piece with his muzzle, would go on until his master would make the sign which meant "*Fetch it.*" The Dog would immediately seize that particular letter and bring it to his master, and thus entire sentences might be formed without any understanding of their meaning on the part of the Dog. It was curious to find out the sign which the master had invented for the Dog, and it was considered by M. Feuillet, librarian to the Institute, and several other members of the *Académie des Sciences*, to consist in a gentle tick of his nail; for it was observed that the master usually kept one of his hands covered by the other, or behind his back, or else in his pocket, to conceal the motion. On listening very attentively, they heard this sound every time that the Dog passed by the letter necessary to form the required word. It would be important to know the plan of instruction adopted by the master in teaching the Dog to obey a sound almost imperceptible to an unpractised ear; but this it was his interest to conceal with the utmost care.

An instance is related by M. Arago, the celebrated astronomer, which induced him to believe that Dogs know the difference between justice and injustice (du juste et de l'injuste), yet we cannot help thinking that the facts do not quite warrant the inference drawn by that eminent Natural Philosopher. Several years ago, when about four leagues from Montpellier, he was detained by a storm in an indifferent country inn. Nothing could be had for dinner except a single fowl, and this was ordered to be placed on the spit. The spit was attached to a large hollow wheel into which the Dogs were made to enter, and to turn it round by their weight and motion. One of the Dogs was in the kitchen, and the innkeeper attempted to seize him; but the animal hid himself, showed his teeth, and refused to go into the wheel at the command of his master. M. Arago, surprised at this, inquired the cause; he was answered, "The Dog knows it is his comrade's turn." The "comrade" was sent for at his request, and as soon as the animal arrived, at the first sign made by the cook, he entered the wheel and turned it round for about ten minutes. Wishing now to try the first Dog, the philosopher stopped the wheel, and ordered them to put the animal in which had formerly refused. This Dog being now convinced, according to M. Arago, that his turn had arrived, at once entered the wheel, and continued there till the fowl was roasted.

A similar anecdote is related of four black Mâtin Dogs which turned the wheel in the Jesuits' College of La Flèche. These Dogs, it was said, always knew their turn of service, and invariably revolted, as if against an evident injustice, whenever they were ordered into the wheel out of their proper course. M. du Petit-Thouars, who passed La Flèche in 1767, after the Jesuits were expelled, had this story told of the College Dogs by several inhabitants, who had witnessed the fact.

We must, however, refuse to assign moral qualities, or the faculty of discerning right from wrong, to these lower animals, if the above phenomena admit of being explained on the common principles of habit, or the association of ideas. When a Dog has long been accustomed to perform a disagreeable office every fourth day, he will come in the course of time to know his particular day of service, and a more frequent demand will naturally excite his resentment. To assign moral qualities to the lower animals is a very popular error, and we are inclined to suspect that these eminent philosophers allowed themselves to be deceived by the ambiguous language of the country people. It is a common circumstance to see a Dog, after having committed some fault for which he has often been punished, enter the room slowly and sneak into a corner; some one then observes, "He knows he has done wrong." In this case, the Dog only infers that the same action will be followed by the same punishment; or, in other words, he reasons from analogy. But this is far from entitling him to the rank of a moral being, or one capable of distinguishing right from wrong; and it is plainly unphilosophical to multiply the Mental powers of animals, if the ordinary operations of intelligence are sufficient to explain all the facts hitherto authenticated.

GENERAL REVIEW OF THE MAMMALIA CONTINUED.

Individual modifications are seldom transmitted to posterity—Connate modifications generally are transmitted—Varieties in the external forms of Wild and Domesticated Mammalia.

ALTHOUGH the original and constitutional powers of animals remain the same under all circumstances, we have now seen that they may undergo some very considerable modifications. It has also been shown that, in many instances, these acquired characters are transmitted to posterity. The latter conclusion has been doubted by several Naturalists, who are inclined to consider these acquired characters as resulting either from the principle of Imitation, the influence of situation, or a combination of these causes. It has been asserted that no modification, induced after birth in the intelligent powers or instincts of an animal, is capable of being transmitted to its offspring, and that all transmitted powers must be connate, that is, imparted with the vital existence. With the human race there must always be an extreme difficulty in ascertaining how far those hereditary characters, so often observed to prevail in particular families, are the result of physical constitution or of Imitation and situation; but it is abundantly evident that the acquired properties of the lower animals are sometimes transmitted to posterity. The races of Rabbits which wholly lose their Instinct of burrowing from confinement;—the young Foxes of thickly-peopled countries which inherit the acquired cunning of their parents;—the young Pointers which set game naturally and without any previous instruction;—and the excessive timidity of young Lap-Dogs, whose parents have been long in the company of timid females;—are instances of this fact. Further, we have, in the docility of those races which have long been domesticated, an evident proof that the modifications of the earlier-tamed animals were transmitted to posterity, while the Apes of Sumatra, which have not this transmitting power, are wholly incapable of yielding domestic races.

It must, however, be carefully observed, that animals do not transmit to their posterity *all* their acquired modifications. On the contrary, it is only a *very small* number of acquired habits or alterations which are thus transmitted to their descendants. Nearly all those arts which a well-trained Dog has acquired perish with the individual, and the process of education has to be recommenced with the pup. The reverse takes place with the other kind of modifications which are properly connate, or born with the animal; for the latter are very frequently inherited by their descendants, and the instances of their not being so transmitted are by no means numerous. Thus, the artificial modifications of Instincts and Intelligence *sometimes* become hereditary; but connate modifications *usually* are so. The latter is remarkably the case in Man. "We see," says Dugald Stewart, "one race, for a succession of generations, is distinguished by a genius for the abstract sciences, while it is deficient in vivacity, in imagination, and in taste; another is no less distinguished for wit, gaiety, and fancy, while it appears incapable of patient attention or of profound research." We have many remarkable instances of the transmission of connate varieties, in the intellectual qualities of the several races of Dogs. No care or education will induce those qualities in the stupid Greyhound, which we find in the intelligent and docile Water-Spaniel.

Some species have an inherent tendency to produce these connate modifications of intelligence; but unless care is taken to preserve them free from foreign admixture, they soon become blended in the mass of average talent usually found in the species.

Hence connate modifications of Intellect may be either *original* or *transmitted*. The first stupid Greyhound, as well as the first intelligent Water-Spaniel, might have been descended from a pair of average talent; and these qualities, which were *original* connate modifications when they first appeared, have become *transmitted* connate modifications in our present races.

But the changes which the Mammalia are capable of undergoing are not confined solely to their Intelligent powers. We see that their external forms are also modified, and that this sometimes happens to a degree so great, as to render it a matter of considerable difficulty to ascertain whether the animals, so altered by circumstances, belonged originally to the same or to different species. These changes are induced by causes which may affect the individual only, or they may also alter the Offspring. The same law is observed in the changes of external form as in the modifications of Intelligence and Instinct; and while the individual varieties which are induced after birth are but rarely transmitted to posterity, the connate varieties give rise to most of those permanent alterations which distinguish the several races.

"We see no instance of connate variety," observes Dr Prichard, "however trifling, which does not manifest a tendency to become hereditary and permanent in the race. White animals with red eyes produce offspring resembling themselves, and the stock will retain its character permanently as long as no intermixture is suffered to take place. The progeny of black animals have the sable hue of their parents. On this account, black Rams are always killed in this country, and never suffered to remain with the flocks. In other countries black Sheep are preferred, and are bred up, while the white, when that variety springs up, are destroyed; accordingly, the general colour of the flocks is black. All the other varieties, as is well known, have a tendency to hereditary transmission. We may observe, that the disposition to variation is more frequently shown in some species than in others, and requires the agency of less powerful causes to excite it into action. The tendency to hereditary descent also is different, both among the animal and vegetable species. For in some species of the latter class, varieties are observed to reappear in the plants produced from the seed, and to continue constantly in the stock, resembling in this particular the nature of animal varieties. On the other hand, some species of animals approach to the capricious character of the vegetable kinds, and the variations which arise in them evince little tendency to become permanent."

It may be proper to recall to the memory of the student, the definition already given of the term *Species*. All animals are said to be of the same species, when they are descended one from another, or from common parents, or from others resembling them as much as they resemble each other. On the other hand, all those differences which are found among animals, sprung from the same original stock, are termed *Varieties*.

The most superficial characters are always the most liable to variation:—The colour of the hair depends greatly on the quantity of the light,—the thickness of the fur upon the degree of heat,—the size and corpulence of the animal upon the quantity of food, jointly with the degrees of temperature and moisture. Often the variations of colour are wholly connate, and can be traced to no external cause.

Among the wild animals, these variations are very much limited by the natural propensities of the animals themselves, as they never willingly ramble to a great distance from those localities where they find a convenient supply of all the things necessary for the maintenance of their species. Their migrations are therefore limited by the circumstances which unite all these conditions. Thus, the Wolf and the Fox inhabit every latitude from the frigid to the torrid zones, yet they scarcely undergo any alteration, through all the changes of climate, except a greater or less degree of beauty and richness in their fur. The variations are still less among those wild animals, especially Carnassiers, which are confined within a small geographical range. A thicker mane forms the sole difference between the Hyæna of Persia and that of Morocco. These variations would be confined within still narrower limits, if the wild animals were at liberty to choose their own localities. But from the earliest ages, they have been hunted by Man, or by the more formidable Carnassiers, and exiled from their native haunts. Some have been driven into unfavorable situations; those possessed of a sufficient flexibility of form have extended their range to remote distances, while the others were left no other retreat but the pathless deserts bordering on their native country. There are no species which, like Man, have established themselves in every country. A great number are confined to the tropical parts of Asia and Africa, others solely to the warm districts of America. Some are seen only in the Arctic regions; another animal world opens before the voyager towards the South pole. Many islands have their own peculiar creation; and even some chains of mountains and valleys are seen to abound with animals unknown to every other part of the globe. Among all these species, we find, when all other conditions remain the same, that those confined to a small district exhibit the smallest and least important variations.

The Herbivorous Mammalia are much more liable to be influenced by climate than the Carnassiers, because their food is more variable, both in quantity and quality. For this reason, the Elephants in one forest will be larger than in another. Their tusks will be longer in those places where the food yields a greater quantity of the matter necessary to enter into the composition of ivory. The same thing will happen to the Rein Deer, and to the Stags, in respect to their horns. However, the nature of their food serves to confine the range of the Herbivorous tribes within still narrower limits than that of the Carnassiers.

Thus the entire influence of the climate and food upon wild animals is by no means very great. Some variations are, however, due to another cause. When the same male always continues attached to the same female, as happens with the Roe-bucks, the young exhibit that uniform resemblance to each other, and to the parents, which demonstrates the fidelity of their attachment. It is evident, that where the same female attaches herself to several males, as happens with the common Hinds (*Cervus elaphus*), the varieties must be greatly increased. We must also expect to find considerable variations in those smaller species which are very productive. Females bearing five or six young ones at a birth, and producing perhaps three times a year by different males, must greatly augment the number of these varieties.

Among the wild animals, any connate varieties which may arise from local causes, are soon blended by a continual intercourse with the original race, and in a few generations they wholly disappear. Hence we find in the herds of wild animals characters of marked uniformity, which cannot be discovered among those domestic races where the care of Man has intervened, and rendered the varieties permanent.

Nature has placed a fixed barrier to those variations, which might have arisen from the union of males and females of different species. We usually find that species, nearly allied to each other in zoological characters, bear a mutual aversion of the most marked and decisive kind. It requires the greatest degree of ingenuity and constraint, on the part of Man, to deceive the animals so far as to form these unnatural unions between different species; and when the Hybrids thence produced are themselves productive, which seldom happens, they do not continue so for many generations. Even this partial fecundity would not probably have existed, without the continuation of that care by which the first union had been induced. We never find in the woods any animals of a character intermediate to the Hare and the Rabbit, to the Weasel and the Polecat, or to the Stag and the Fallow Deer.

There are, however, distinct species which are capable of producing by their union fertile individuals. The offspring of the Dog or his female, with the female or male Wolf or Fox, are prolific. The same thing occurs with the offspring of the He-Goat and Ewe. The Mules produced between the Ass and Mare are sometimes prolific, and fertile races may be produced from the unions of several Birds of distinct species. Yet these are merely exceptions to the more general law, that the unions of animals of different species are either wholly unproductive, or the offspring is a Hybrid, and incapable of procreation.

There is a tendency among most animals, whether wild or domesticated, to pass occasionally into White varieties, distinguished by the term Albinos. Their hair is remarkably soft, and perfectly white, the iris of the eye is of a bright red colour, and the sight is acutely sensible to light. These varieties are therefore crepuscular, that is, they appear only during twilight or moonshine. In dark woods, old cathedrals, and obscure subterranean retreats, the common brown Mouse acquires, from the absence of light, the red eyes and white hair of the Albino variety. This property becomes hereditary, and thus races may be formed, as has happened in the case of the common Ferret, which is probably only a variety of the Polecat (*Mustela Putorius*). The offspring of two brown Mice are white when the old animals are retained in absolute darkness. These Albino varieties have also been seen in the Bactrian Camel, the Elephant, the Beaver, and in a very great number of other animals.

"By far the most powerful cause of the evolution of varieties in the Animal Kingdom," remarks Dr Prichard, "is Domestication. To be convinced of the truth of this fact, we need only look at the phenomena which surround us on every side. In all our stocks of domesticated animals, we see profuse and infinite variety, and in the races of wild animals, from which they originally descended, we find a uniform colour and figure for the most part to prevail. Domestication is to animals what cultivation is to vegetables, and the former probably differs from the natural state of the one class of beings in the same circumstances, which distinguish the latter from the natural condition of the other class. The most apparent of these is the abundant supply of the peculiar stimuli of each kind. Animals in a wild state procure a simple and unvaried food in precarious and deficient quantities, and are exposed to the inclemencies of the seasons. Their young are produced in similar circumstances to the state of seedlings which spring uncultivated in a poor soil. But in the improved state, all the stimuli of various food, of warmth, &c. are afforded in abundance, and the consequence is a luxuriant growth, the evolution of varieties, and the exhibition of all the perfections of which each species is capable."

Hence it is in Man and Domestic Animals that varieties are most numerous and perplexing. At present, we shall confine our remarks to the changes experienced by the Domestic Animals, leaving the more interesting and difficult consideration of the varieties in the Human Race to a more advanced stage of this work. Indeed, it is only after a careful study of the limits of variation amongst the Domestic Animals, that we shall be competent to consider the varieties of the Human Race in an unprejudiced and impartial manner.

In those species which have experienced an imperfect domestication, the variation is but slight, and it is due chiefly to climate. Thus, we have but few varieties of the Cat, and their modifications are by no means considerable. Some have a softer fur, the colours of others are perhaps more vivid, or their size is greater. These are the limits of their variations, and they are further confined by the habits of intercourse which the tame Cats preserve with the wild individuals of their species, established in their immediate neighbourhood. In this way, all the tame Cats bear a marked resemblance to the wild Cat of pure breed. The colour of the latter is uniformly gray, with blackish longitudinal bands, while its fur is somewhat rough. When tamed, in some climates, its colours become vivid or the fur grows smooth; in other countries the colours are softened down and become blended together, and the fur grows rough. When the European Cats are removed to the warmer parts of Africa, their form does not change with the climate. Their variations of colour prevail chiefly in Anatolia, Spain, and Persia, where distinct races have been formed. It is said, that there exists in China a variety of the Domestic Cat, with pendant ears, and in the Isle of Man there is a distinct race destitute of tails. These form the extreme limits of variation in a species, which serves to mark the transition from the Wild to the Domestic Mammalia.

We may naturally expect to find a greater degree of variation in the Domesticated Herbivorous Mammalia, which have been transported into every climate for our use, and to whom we allot various portions of food and labour. Their modifications are, however, merely superficial. A greater or less degree of size; longer or shorter horns, or the want of them altogether; or perhaps a lump of fat on the shoulder, or near the tail, are the general limits of their variations. The colour also is ever variable, and often without any assignable cause.

The Goat has experienced many changes of colour, and in Spain it has lost its horns. The different qualities of fleece found in the Cashmere, Thibet, and Angora Goats, are well known in commerce. It is, however, the last of these varieties, which appears, from its pendant ears, to have departed the most widely from the original type. The usual and well-known influence of the climate of Anatolia, joined to the long domestication of the species, in a nation civilized at an early period, has produced this variety by the long-continued action of these causes. Buffon remarked that the Angora Goats born in France were losing those long and pendant ears which characterize the Syrian variety, and it was expected that in a few generations they would acquire ears and fleece resembling those of the common Goats of that country; but, according to Blumenbach, the Angora variety continues permanent when the animals are removed to other climates.

Among the Sheep, the variations are numerous, but these chiefly refer to the fleece, which it has been the constant care of Man to alter and improve. In some districts the sheep are always black, and very often a white Ram and Ewe will produce a black lamb. In other places they may be brown, spotted, reddish, or even yellowish. These varieties of colour are still more accidental than the other differences, which arise among races from alterations of food and climate. The limit of variation among the Sheep may be seen in those enormous accumulations of fat, which swell the tails of a race found in some parts of Africa and Asia. Pallas, who saw the Sheep of the Kirguis, a tribe of Siberia, describes them as being more fat and deformed than any he had ever seen. They are taller than a young calf, very heavy, and somewhat resemble the Indian Sheep in their proportions. Their heads are much swollen, with large pendant ears, and the lower lip extending far beyond the upper. The greater number have one or two bunches covered with hair, which hang down from their neck. In the place where in other Sheep a tail is usually found, there is a round and large protuberance of fat, with scarcely any wool beneath, and these protuberances often weigh from thirty to forty pounds, and yield from twenty to thirty pounds of suet. These peculiarities continue permanent wherever they may be removed. The native Sheep of Ethiopia are covered with coarse hair, and those of Thibet with very fine wool. The Ankon variety of Sheep from Connecticut have the fore-legs bent like an elbow; and this deformity, which is usually communicated to their descendants, as well as the general shortness of all the legs, was at one time much cultivated, from the Sheep being thence unable to climb the fences.

We have already seen, that the Bull can scarcely be considered a domestic animal, when viewed in reference to its disposition; yet its colour varies equally with its more domesticated female. Innumerable varieties of the Cow are distinguished by Graziers; and France alone reckons at least sixteen varieties, deriving their names from the Provinces which they inhabit. The possession of udders of enormous size, and the property of giving milk all the year round, are qualities acquired by Domestication. We are informed by Pennant, that the American Bison is covered in the winter with a long shaggy fleece, which hangs over his neck, and partakes somewhat of the nature of wool; but that in summer he is almost naked. In the island of Celebes there is found a variety of the Buffalo, not larger than our common Sheep.

The Ass undergoes, from domestication, several changes in the colour and quality of its hair; but the wild Asses, inhabiting the country of the Calmucks in immense numbers, resemble each other precisely in all those particulars which are observed to vary among the domesticated species. In other respects, they only differ from the Domestic Asses, in being of a greater size and beauty. The wild Asses inhabiting the deserts of Barbary are uniformly gray, and are said even to outstrip the Horse in speed of foot.

The joint influence of Climate and Domestication seems to have a greater power over the Horse, and its varieties are accordingly almost infinite in number. In some the head is small and slender, the nostrils are wide and easily moved, the ears are slim and directed forwards, and the eyes are lively. In other varieties we have a complete contrast in all these particulars. The head is heavy, the nostrils are narrow and close, the ears are large and directed backwards, and the eyes have a marked expression of dulness. An equal degree of variety is found in their colours, which may be black, bay, brown, or white, or any combination of these shades. In Ceylon there is said to be a variety of the Horse which is not more than thirty inches high; in other climates the Horse is nearly as large as the Bactrian Camel. In bulk it sometimes rivals the Ox, and often it emulates the lightness of the Stag. We are informed, by John Hunter, that all foals are usually of the same colour, and that, though the hair may vary as they become older, still the skin remains the same, being no darker in black than in white Horses, which is contrary to what is observed in most species. There is an exception in cream-coloured Horses, which have the skin of the same hue as the hair. In size, colour, and form, as well as the quality of the hair, they may differ greatly, but always within certain well-defined limits. We may perhaps consider the Calmuck variety, with very long, thick, and white hair, as the greatest deviation from the original type.

With the Domestic Hog, the extreme points of variation must be placed in its soft and pendant ears, the smallness of its tusks, and the union of its hoofs. This animal appears to depart most widely from its original form, when domesticated in the warmest countries; and then the variations in size and shape are innumerable, and its colour usually changes to white. That variety of the Hog with undivided Hoofs, observed long ago by Aristotle, is sometimes found in England; and in Normandy there is said to be a race of Hogs with the fore legs much shorter than the hinder.

It is, however, in the Dogs that varieties are most striking and important. Man has transported these animals into every part of the globe, and the extent of his power is seen in the extraordinary differences of their forms. As their unions may be regulated entirely by the will of their master, and as the connate varieties of individuals are readily transmitted to their offspring, we find among them singular deviations from the original type. Not only is their colour infinitely various, but their hair becomes more or less abundant, and sometimes it is wanting altogether. The height of some Dogs is five times as great as that of others, and their bulk may, therefore, be more than one hundred fold. These are not the only differences. They seem to be acutely susceptible to all those circumstances which affect the growth of the different parts of the body. The forms of their nose, ears, and tail; the relative height of their limbs; the progressive development of their brain; and the form of their skulls, are alike affected by these sources of variation. Sometimes the head is slim, the muzzle slender, and the forehead flat. Often the face is fore-shortened, and the forehead projects. Indeed, the differences between the French Mâtin Dog and the Water-Spaniel, between the Greyhound and the Bull Dog, are more strongly marked than those among many wild animals of the same genus, but of different species. It is unquestionably in the varieties of the Dog that we see the highest degree of deviation yet ascertained to prevail among the individuals of any species throughout the entire Animal Kingdom.

Of all the characters which the domesticated animals possess, it may be observed that the colour of the hair and skin is the most liable to variation. Being placed externally, this part of the body is exposed more than any other to the influence of outward causes of change. The hair of the different Dogs exhibits this natural versatility in a remarkable degree with respect to colour, quality, length, and arrangement. In cold climates the Dogs have usually two kinds of hair; the one, being short, fine, and woolly, immediately covers the skin, while, in the other, the hairs are long and silky. It is the latter kind of hair which imparts the coloured appearance to the animal. In tropical climates, the fine and warm woolly hair becomes obliterated, and at length wholly disappears. The same thing happens in our houses, when Dogs are protected from the changes of weather and the severity of winter. The skin of the Barbary Dog is naked and oily; the Bull Dog, the Mastiff, the Greyhound, and the Carlin, have the hair short and smooth. It becomes longer in the Shepherd Dog, the Wild Dog of New Holland, the French Mâtin, and the Iceland Dog; it is very long in the Wolf Dog, the Spaniels, and especially in the French Bichon, where it sometimes reaches nearly to the ground. Again, if the hair be viewed in respect to its quality, we find at least as many shades of difference. The Shepherd Dog, Wolf Dog, and the Griffon, have coarse hair, while in the French Bichon, some Water-Spaniels, and the Great Dog of the Pyrennees, it is both silky and soft. In some it is straight and smooth, in others woolly and curly. Many races have the body clothed entirely with long hair, while on the head and limbs it is perfectly thin and smooth. Others, on the contrary, have the bead and neck furnished with a mane, and the remainder of the body is covered with short hair. The Wolf Dog is an instance of the former kind, and the Lion Dog of the latter. In these respects, we find in the hair of Dogs all those variations of quality and quantity which can be found in the several genera and species of the Mammalia. Their colours may be white, a deeper or paler brown, fawn, or black. Some Dogs are seen entirely of one of these colours, but most frequently the tints are distributed irregularly in spots, which may be either large or small. Sometimes these spots have a tendency to become symmetrical; at other times the longer hairs are of a different colour from the short ones, and then the joint effect of the two colours produces different shades, according as the white, black, fawn, or brown, predominates. Thus, we may see Dogs with hair apparently resembling that of the Wolf, and upon a closer examination this is found to proceed from the mixture of white, fawn, and black hairs; or more rarely the general effect may produce a gray slate colour. These colours are not connected with any particular variety, nor does it necessarily follow that Dogs of different colours must be further distinguished by the forms of their heads, the quality of their hair, or the proportions of their bodies. It usually happens, in all these cases, that when care is taken always to unite individuals of the same colour, form, and size, that the race perpetuates itself. It is from the constant union of individuals having the same, or nearly the same, colour that the Danish Dog, Greyhound, Bull Dog, and Mastiff, are fawn, the Shepherd Dogs are black, the Wolf Dogs white, and the Gallic Hound, the Braques, Bassets, and Spaniels, have black spots upon a white ground; but when this precaution is not observed, the colours of the above-mentioned Dogs will be modified in proportion to their degree of admixture with other races. However, the connate modifications of colour, as well as the more important ones of shape and size, usually end in becoming hereditary, when they are not counteracted by some neutralizing cause.

Thus we find, upon the whole, that the more important variations of the Mammalia may be ranged under the following heads:—

1. The skull and face may be shorter or longer, broader or higher; the forehead may be elevated as in the Wild Boar, or depressed as in the domestic Hog. Thus, the head of the Neapolitan Horse differs remarkably from that of Hungary and Transylvania in the shortness and breadth of its lower jaw-bone. Camper also remarked, that the lachrymal depressions (*jovea lachrymales*), which can be clearly observed in the Wild Bull, had disappeared by degeneration in the domestic Ox.

2. The general figure and proportion of the limbs may be altered to a most remarkable extent. We see striking instances of this variation on comparing the Syrian and Arabian Horses with those of the North of Germany and the Shetland Isles.

3. In stature there may be a singular disparity; thus, the Hogs transported to the Island of Cuba acquire a size nearly double that of the common European Pig; and a very considerable growth takes place among the Wild Cattle of Paraguay.

4. The texture and quality of the Hair may vary from the soft wool of the Thibet Sheep to the dense and almost rigid hair of the Ethiopian variety. In Normandy, the bristles of the common Pig lose all their stiffness. But the most singular instances of variation in the hair are effected by the climate of Anatolia, where this cause equally affects different species of Mammalia, and transforms the short fur of our Cats and Rabbits, as well as the wool of our Sheep, into the long and silky fleece of the Angora varieties.

Lastly, the colour of the Hair may vary from black to white through all the shades of brown or red. The fleeces of Angora often assume a silvery whiteness. Indeed, we have only to look around us to see innumerable instances of diversity in the colours of our domestic animals.

GENERAL REVIEW OF THE MAMMALIA CONTINUED.

Permanence of Species—Difficulties in distinguishing between Species and Varieties.

As the variations which have arisen among the Domestic animals, and especially in the Dog, appear very considerable, it has been thought probable by many Naturalists that our different races of Dogs have descended from several distinct species. In this way the difficulties of explaining the causes of their variations are wholly avoided, rather than resolved. It remains to be shown, under this view of the subject, from how many species the domestic Dogs have descended. No Naturalist could propose to establish a species for every distinct race, which are upwards of fifty in number. Still less, as all these races are capable of forming crosses with each other, could they institute a species for every combination in pairs; nor could they extend it to those secondary and tertiary crosses which might be formed among their posterity, both with each other and with the original races, thus rendering the number of species absolutely infinite. It has been rather attempted to limit the sources of the several races of Dogs to a small number, marked by important differences. Yet it becomes equally impossible to point out the particular stocks from which these races have descended, or the variations which must be regarded as important, without falling under objections of another kind.

There are several considerations which clearly establish the important fact that Species have a real existence in Nature;—that certain forms have been assigned to each animal from the origin of things; and that although the animals are liable to diverge from their primitive forms, they always possess a preservative tendency, a *nisus formativus*, and are ever ready to revert to the original type when the external causes of change are removed. The entire Animal Kingdom is divided into a number of distinct species, each of which perpetuates its own form, without ever transgressing certain limits, or acquiring the characters of another species.

In all the varieties observed to arise among the Mammalia, the form of the bones preserves a remarkable stability, which would not always be expected from the appearance of the external parts. The Baron Cuvier compared the skulls of Foxes from the North of Europe and from Egypt with those of France and with each other, yet he found no other differences than such as might distinguish one individual from another. The antlers of the Rein-Deer and Stags often vary in size, and the same may happen with the tusks of the Elephant; but two individuals of any one of these species, however dissimilar they may be in size, do not exhibit the slightest difference in the number of their teeth, or the articulations of their smallest bones. This is also observed with the domestic Cattle, which may be destitute of horns, or have them of variable length, and yet they possess an exact correspondence in all the other parts of the skeleton.

Thus the forms of the bones in general vary but little; while their modes of connexion, their articulations, and the form of the great molar teeth, remain constantly the same in each species. The divided Hoof of the Hog sometimes becomes consolidated, and this may be regarded as the extreme limit of variation among the bones of our domestic herbivorous animals.

The variations of the bones in the different kinds of Dog have undergone a special examination by M. Frederic Cuvier, performed at the request of his brother upon the specimens at the *Museum d'Histoire Naturelle*. To enter fully into the details of this investigation would at present be out of place. It will suffice here to mention, that a general correspondence in all the parts of their skeletons was found to exist, and at the same time some important variations, especially in the degree of elevation of the frontal sinuses. The teeth were always of the same number and general form; sometimes an additional false molar, or a tubercle, was observed on one side or on the other. It is well known that all Dogs have five toes on the fore-feet, and only four on the hinder, while there is a slight trace of a rudimental fifth toe in the hinder metatarsal bone, which, however, shows no appearance on the outside. These toes being of unequal length, usually preserve the same relations in all the races; but sometimes a fifth toe exhibits itself on the internal surface of the hinder-feet. It is, however, generally very short and imperfect, and this last is the maximum of variation found in the skeletons of all the races of Dogs.

It thus appears abundantly evident, that animals now possess certain characters which remain permanent, and resist all modifications, whether arising from climate or domestication, or from a natural tendency to run into connate varieties.

Time, however, it has been said, may effect a perceptible modification in the entire characters of species. Fossil remains, and other Geological monuments, appear to show that millions of years have elapsed since the first species of animals inhabited the earth, and it is asked, may they not have undergone many modifications during the interval?

It is evident that we can only ascertain the effect which a very long time will produce, by comparing it with the change actually observed to have taken place during a shorter period. MM. Cuvier and Geoffroy-Saint-Hilaire sought out the most ancient documents which Egypt could afford, for the purpose of solving this question, so important to the Naturalist, and essential to a knowledge of the past history of our globe. M. Cuvier examined with great care the ancient Ægyptian obelisks transported to Rome, and found a perfect resemblance between the general form of the animals engraved thereon, and the common species of our own day. M. Geoffroy collected as many mummies of the lower animals and of Man as he could find, and was led to form a similar conclusion. These monuments must have been from 2000 to 3000 years old. "For a long time," says M. Lacépède in the report which he made upon these objects in common with MM. Cuvier and Lamarck,—"For a long time philosophers have been anxious to know whether species change their forms during the course of ages. This question, apparently trivial, is yet essential to the history of the globe, and to the solution of a thousand other questions not far removed from the gravest objects of human veneration. Never were we in a better condition to decide the question upon a great number of remarkable species, and for a long period of years. The superstition of the ancient Ægyptians would almost seem to have been inspired by Nature, for the purpose of bequeathing to us a monument of her history. A people of fantastical opinions, by embalming with so much care the brute beings, objects of their stupid adoration, have left in their sacred grottoes complete cabinets of zoology. The climate has united with the art of embalming to preserve these bodies from all corruption, and we have now the means of ascertaining with our own eyes what was the condition of these animals 3000 years ago. One can scarcely restrain the raptures of Imagination upon seeing at the present day an animal preserved, with the smallest bone and hair perfectly distinguishable, which had 3000 years since in Thebes or Memphis its priests and altars. But without wandering into all the subjects to which these associations give rise, we shall confine ourselves to noticing the simple fact—that these animals perfectly resemble those of the present day."

Although the bones of a species do not vary to any extent, yet the identity of osteological characters is not alone sufficient to establish an identity of species; and some species which possess a most exact similarity of structure are held, by the general consent of Naturalists, to be of different species. It is almost impossible to distinguish the skeleton of the Wolf from that of the Wild Dog of New Holland. Their teeth are the same; the vertebræ of the tail are equal in number; the feet have the same number of toes; and the bones of the head exhibit the same relations, except that the orbital fossæ are slightly larger in the Wolf. The same thing occurs in the Wolf of Canada, which is smaller than the common Wolf, and larger than the Dog of New Holland. The Jackal also resembles the Wolf-Dog very closely, especially in the form of the head. There is likewise a most exact similarity in respect to the organs of Sense among the New Holland Dog, the Canadian Wolf, and the Jackal. Again, the quality and arrangement of the hair exhibit no essential differences, for they all may have either woolly or silky hair, according as they have been naturalized in cold or temperate countries. In fact, they only differ in colour. Yet all these genera merely vary from white to brown or black, and excepting the Black Wolf, which has the hair of a uniform colour, the others have hairs of fawn, black, or white so mingled together, that it is difficult to set down any colour as peculiar to either species, and which will not pass by insensible shades into another.

What, then, it may naturally be inquired, forms the distinction between a species and a mere variety; and how are we to ascertain those permanent characters which were assigned to our domestic animals at the origin of things?

If these questions be considered in a purely abstract form, no difficulty can arise, as we have only to include in the same species all those animals, whose differences of external form and garb can be traced to some acknowledged causes of variation; while the animals whose differences cannot be thus explained, must be held to belong to separate species. "Where two races of animals are distinguished by any undeviating marks in such a way that they never will, under any circumstances, pass into each other, or that the progeny of either can never acquire the characters of the other, they are of distinct species, and it matters not how wide or how narrow be the line of discrimination, provided that it never be broken in upon." But when we come to apply these abstract rules to the realities of Nature, we find that they are not always sufficient to distinguish the mere variety from the genuine species.

The difficulty is further increased by the circumstance that many varieties or races of some species differ more decidedly among themselves, than the species of certain genera, where the objects are very numerous. Again, the greater part of our acquisitions are imported from remote and barbarous countries. "A large proportion," as Mr Lyell observes, "have never even been seen alive by scientific inquirers. Instead of having specimens of the young, the adult, and the aged individuals of each sex, and possessing means of investigating the anatomical structure, the peculiar habits and instincts of each, what is usually the state of our information? A single specimen, perhaps, of a dried plant, or a stuffed bird or quadruped; a shell without the soft parts of the animal; an insect in one stage of its numerous transformations;—these are the scanty and imperfect data which the Naturalist possesses. Such information may enable us to separate species which stand at a considerable distance from each other; but we have no right to expect any thing but difficulty and ambiguity, if we attempt from such imperfect opportunities to obtain distinctive marks for defining the characters of species which are closely related. When our data are so defective, the most acute Naturalist must expect to be sometimes at fault, and, like a novice, to overlook essential points of difference, or pass unconsciously from one species to another."

Buffon established the criterion for the determination of species in the power of producing, by their union, races equally fertile with themselves, and this rule seemed to be confirmed by the experiments of John Hunter. They were of opinion, that "if a male and female produce an offspring which is prolific, the tribes to which the parents respectively belong are hence proved not to be specifically different, and whatever diversities may happen to characterize them, are in this case to be looked upon as examples of variation. But if the third animal be unprolific, it is to be concluded that the races from which it is descended are originally separate, or of distinct kinds. The fact that most hybrid animals are wholly unprolific, would appear to be a provision for the attainment of this desirable end, and for maintaining the order and variety of Nature. For if such had not been the condition of these intermediate animals, we have reason to believe that all the primitive distinctions would have been long ago totally effaced; a universal confusion of species must have ensued, and there would not be at this day one pure and unmixed species left in existence. The Naturalists above mentioned, inferring, from the apparent utility of this law, that it must universally prevail, obtain by means of it a ready method of determining on identity and diversity of species."

It is very clear that if two animals are prevented by any great disparity of organization or disposition from uniting, that the criterion of generation holds good to a certain extent. The Bull and the Goat, for example, would at once be pronounced to be distinct species. This rule may enable us to assert that two animals are not of the same species, but it does not always serve to discriminate between nearly-allied species. Hence it seems rather to be the first rude attempt at forming a criterion, than one which serves to mark out nice distinctions. The crosses among the Dog, the Wolf, and the Jackal;—between the Goat and the Sheep;—the Horse and the Ass;—the Lion and the Tiger, with the occasional appearance of fertile Hybrids in many, and the possibility of its occurrence in them all, show that the converse often fails. Although animals which do not generate together belong to distinct species, yet it is not true that distinct species must not generate together, nor does it follow that their progeny must always be sterile.

The determination of species by the property of producing fertile races, had previously been restricted by Frisch to such as generate together of their own accord, "von Natur mit einander gatten." Those artificial unions brought about by restraint, artifice, or domestication, were wholly excluded by him. But this restriction renders the rule useless in practice for determining those points where difficulties may chiefly be expected to arise. It is in respect to Man and the domestic animals, or with animals brought from distant and uncivilized countries, that a rule is most required to distinguish the species from the mere variety. Blumenbach inquires, "When will it come to pass that all nearly-allied animals shall be brought together from remote countries, so as to submit them to the requisite experiments,—for example, whether the Chimpansé (*Troglodytes niger*) from the Angola Coast, will form a fertile race with the Orang Outang (*Pithecus Satyrus*) from Borneo?" This is a desideratum which the general establishment of Zoological Gardens alone can supply, but in the meantime we must seek some other criterion, which shall be applicable to Man and the domesticated animals, for the determination of species.

It is here that difficulties arise in drawing the line between the species and the variety. Tillesius considers that several distinct species are confounded under the name of Jackal or Chacal (*Canis aureus*), while both Pallas and Guldenstaedt regarded the Jackal of Caucasus as the original source whence our domestic Dogs are descended. Others again thought that the different kinds of Dog have diverged from the Shepherd's Dog, while some considered them all but as degenerations from the Hyæna, the Wolf, or the Fox.

Thus it is precisely in those places where a fixed rule is most required that the breeding principle wholly fails, and we may seek in vain for any other. Blumenbach could propose none, but referred the determination of species to Analogy and Probability. "Fere desperem," he observes, "posse aliunde quam ex ANALOGIA et verisimilitudine notionem speciei in Zoologiæ studio depromi." (I may almost despair of being able to derive the idea of species in the study of Zoology from any other source than analogy and probability). Two races of animals which possess a general resemblance, and differ only in those respects which have been observed to vary, and can be traced to some well-known causes of variation, must at once be admitted to belong to the same species; but however near their general appearance may be, if they exhibit any difference which, in all our experience of the Animal Kingdom, has never been known to exist as a variety, they must be set down as distinct species. The proper determination of species rests, therefore, upon the knowledge of an immense number of facts, and forms one of the most difficult, as it is one of the most important, subjects to which the Naturalist can direct his attention.

Thus Blumenbach considers the Ferret to be merely a variety of the Pole-Cat (*Mustela putorius*), not because they generate together, for perhaps the experiment may not yet have been made, but because the former is white with red eyes; and, from that well-known rule of analogy, that the same effects must be referred to the same causes, its origin is the same with those Albino varieties produced daily among the domesticated Mammalia. Again, the Indian Elephant (*Elephas Indicus*) differs remarkably from that of Africa (*E. Africanus*) in the number and form of its molar teeth. Whether these animals will engender together it is perhaps difficult to determine; but on examining every specimen which reaches this country, the same difference is found to exist. Further, we know of no analogous instance of variety in the formation of the molar teeth among wild or domestic animals. We, therefore, do not hesitate to set down these two Elephants as distinct species.

There are other difficulties arising from the want of accurate information; and these, in the present state of the science, occur but too frequently. For example, the skin of an animal arrives from the Cape of Good Hope. At the first glance it appears, perhaps, to be a specimen of the common Cape Otter (*Lutra capensis*), and this opinion may be farther confirmed on examining the structure of its teeth. The colour of the breast and throat may seem of a purer white, and to be more extensive than usual, but this is a characteristic which might belong to a mere variety. On looking at the feet, we are much surprised at finding all the toes without nails, excepting on the second and third of the hinder-feet, where only a rude vestige of a nail can be observed. A Carnassier without claws would seem an anomaly in creation. To suppose a being, compelled by its structure to live on animal food, and yet to be refused by Nature the weapons fitted for seizing its prey, disturbs our ideas of final causes, and we delight to trace order and regularity in the works of creation. The specimen must then be imperfect. It belongs to an old individual;—perhaps the claws may have dropped off through age or disease. We set it down, therefore, as a mero variety of the *Lutra capensis*. Some years afterwards, young indi.

viduals are brought with the nails entire; the first view then seems abundantly confirmed; perhaps even we applaud our own sagacity, and our extensive knowledge of final causes,—of those ends and uses for which the Creator designed the various parts of the animal world. Finally, another young individual is imported with all the characters of the original specimen, thus proving it not only to have been a distinct species, but entitling it to the rank of a separate genus—the *Aonyx*, or Nail-less Otter of M. Lesson: and one more instance is afforded of the inexhaustible variety in the works of Nature. When we find that even the possession of Claws is not always indispensable to the subsistence of the Carnassier, we may thence derive the salutary caution, not to confide too implicitly in analogical reasoning, if we wish to form correct views regarding new or unknown natural objects.

GENERAL REVIEW OF THE MAMMALIA CONTINUED.

Supposed Degeneration of Species—Theory of Original Stocks proposed by Linnæus and Buffon—Lamarck's Theory of the Transition of Species.

SINCE the supporters of the permanent characters of Species thus find it difficult to fix any very definite rule for determining them, and as the characters themselves are often seen to run into innumerable varieties, two very different theories have been proposed. Linnæus and Buffon asserted that only a small number of stocks were originally created, from which all the existing species have degenerated and diverged, from the influence of climate, food, and domestication, aided by a promiscuous intercourse, which has been limited only by their progeny ceasing to produce fertile races. On the other hand, Lamarck considered that the form of the body, and all the characters of species, were the consequence of the habits, the manner of living, and other circumstances, which have, in the course of time, given rise to the form of each species. Further, that Man, and each higher animal, has originally arisen from some lower Division of the Animal Kingdom, by the gradual transition of the characters of one species into another, but always from the lower to the higher, with the transmission of such commuted characters to their posterity. These theories both agree in denying the fixed character of species. That of Linnæus and Buffon would remove the character of durability from the species to fix it in some original stock, the type of the Genus, the Family, the Tribe, or perhaps even of the Order. That of Lamarck would overturn the permanent character of all forms. The first asserts the degeneration, the second the gradual development and perfectibility, of species. While the one reposes chiefly on the phenomena of Variation, the other rests upon those general analogies among species, which have led Bonnet to form his universal chain of existence, and later writers their circular theories.

It is to Linnæus that we must assign the merit of relieving Systematic Botany from those accidental varieties which spring up daily in our gardens, and had been improperly raised by Tournefort and other former Botanists to the rank of species. But the zeal of this great Naturalist in bringing down Varieties from their undue elevation, led him to conjecture that many of those Plants which had been discovered since the time of Tournefort might have been produced, during the intervening period, by the intermixture of species. From the impregnation of one kind of Plant with the pollen of another, he was induced, not only to suspect that Nature now produced new species by this means, but that, even at the origin of things, there had been created only a certain number of simple genera, the continual crossing of which has given rise to the immense number of species at present known. This hypothesis, which originated from the consideration of Plants, was afterwards extended by Linnæus even to Animals, and however plausible it may at first sight appear from contemplating those races, by which Nature has so infinitely varied some species in different parts of the globe, it seems, on a further consideration, to be wholly untenable. Contrivance and ingenuity, on the part of Man, are always seen to be necessary to bring about the production of a Hybrid or cross between two different species. There is further an impossibility of perpetuating these crosses as species or distinct races, arising either from their absolute or relative want of fecundity, or from that degeneration and deterioration to which their issue is subject. They always require the assistance of one of their primitive stocks, to prevent the new race from becoming wholly extinct. Further, in those genera and classes where the objects are very numerous, we often see two or more species formed evidently upon the same model, which may be more or less varied, yet they always remain distinct from each other. Examples of this are not wanting from the Quadrumana and Cheiroptera to the lowest species of Zoophytes. We also see that those peculiarities which serve to characterize the several species, genera, or even natural families, continue to exist without there ever appearing before our eyes new links between allied species. For nearly two centuries, Animals and Plants have been observed with great care, yet there has not been one authenticated instance of a distinct and constant species, which has yet been proved to be of modern origin. Finally, those fossil Shells and Bones found in earthy strata, deposited during the earlier ages of animal life, exhibit the same variety, not only of those forms which are found at the present day, but also of many others now wholly extinct. These facts are opposed by a mere probability or conjecture, and we are hence compelled to consider species, although very nearly resembling each other, to have been so formed at the origin of things.

Buffon has carried these views regarding the Degeneration of species among animals to a much greater extent than Linnæus did in respect to Plants. After reducing the numerous races of domestic animals to certain original stocks, he grouped the allied species of quadrupeds into races or natural families. Assuming certain species to be the primitive stocks from which the numerous allied species at present existing have descended, he thence attempted to explain their degeneration, partly by their close affinities, but chiefly by those causes which are sufficient to vary the domestic animals. He thought that species, such as now are commonly admitted, did not formerly exist, and that we must seek for their characters in those natural groups which have served to form genera or families. The degeneration of species, according to Buffon, was one which preceded all history, and formed the most ancient of their changes. It appeared to arise in each family, or in each of those genera under which nearly-allied species are usually comprised. Only a few isolated kinds, he remarked, formed, like Man, at once the species and the genus. The Elephant, Rhinoceros, Hippopotamus, and Camelopard, according to him, composed simple genera and species which were continued in a direct line, and without any collateral branches; while all the others appeared to form families, in which a chief and common stock might generally be observed, from which there seemed to proceed different offsets, increasing in number according as the individuals in each species were smaller and more fertile. Buffon on these principles reduced all the species of quadrupeds then known to thirty-eight families. He admits that this state of Nature has not come down to us, but is, on the contrary, the remnant of a former state of things, and that we can only acquire a knowledge of it "by inductions and relations nearly as fugitive as the time, which seems to have obliterated all traces of its existence."

Notwithstanding the opinion which M. F. Cuvier has hazarded upon this theory, "that it even now presents an appearance of the greatest probability," it is one to which we can by no means subscribe. After making due allowance for the influence of climate, food, and the numerous accidents to which all the individuals are subject, these causes are wholly insufficient, however long we may suppose them to operate, to change the entire forms of animals, their proportions, and even their internal structure, to such a degree as this hypothesis would require. We see that those domestic animals which Man has transported to the most opposite climates, have only changed the quality of their hair or their colour. The influence of pasture can only alter the height, the proportion of the horns, or perhaps add some lumps of fat to the body. But a small number of generations spent on another soil are sufficient to overturn whatever this race may have acquired during ages of cultivation. Again, if we consider those species, whether Mammalia or Birds, which are most populous in individuals, and at the same time the most fertile, the entire of their observed variations are by no means great. Some species which are very populous are nearly exempt from varieties, while others, though less fertile, vary much. Thus the common Mouse and the Mulot (*Mus Sylvaticus*) are perhaps as populous as any species; yet their variations are rare, and an infinite number of instances might be brought forward among the Fishes and Insects. There are even species very nearly allied to each other, and almost equally distributed in opposite climates, of which the one has run into a great number of varieties, while the others every where preserve a uniform resemblance to each other. The Polish Marmot (*Arctomys bobac*), and the Siberian Marmot (*Spermophilus citillus*), are striking instances that a vegetable diet does not give rise to greater varieties than animal food. The Polish Marmot lives only on vegetables without ever touching animal substances, yet it remains unvaried, according to Pallas, from Poland to the banks of the Lena. On the contrary, the Siberian Marmot, which is as carnivorous as the Surmulot (*Mus decumanus*), has undergone many important variations of size, colour, and proportion in the same latitudes, and under similar circumstances.

Although the influence of Domestication has a much more powerful tendency to occasion variation, than all those reverses and changes which the wild species can experience, yet Man has not succeeded in altering the Nature of any one of these animals, so as to form a new, distinct, and permanent species. The Horse and the Ass, in their transitions from the wild to the domesticated state, have undergone less variation than some other wild species, which climate alone has been sufficient to modify. The Bactrian Camel and the Dromedary retain their natural forms in the few countries where they are naturalized. Buffon considered the humps of the Camel to have been occasioned by the long habit of carrying burdens; but the wild Camels of Thibet and China have the same humps and callosities as their domesticated brethren. Pallas has correctly observed, that he might as well have regarded the follicule of the Musk, and the dorsal gland of the Peccari, as abscesses arising from disease. The Ass is more harshly treated than the Camel, the Alpacas are as much accustomed to carry burdens, yet they are without humps. The Horse and the Ass have not acquired callosities on those places where they have so long been exposed to the friction of the saddle and harness.

Climate and Food, however long we may suppose these causes to operate, are wholly unable to account for the existence of the numerous species of animals which cover the face of the globe. We see that the preservation of the Races among our domestic animals, and the improvement of the breeds, depend chiefly upon the peculiarities of the individuals selected to propagate. Graziers have long since laid down those rules by which the domestic animals, and especially the Horse, can be rendered larger, more beautiful, or more vigorous than they would have been if left uncultivated. But it is only by continued care that the purity of the breeds can be preserved, and they ever exhibit an inclination to resume the characters of the wild animals. We thus see that the tendency of the offspring to retain the characteristics of its parent is powerful enough to counteract all those causes which may modify the external forms of animals. Thus the introduction of Rams of a good breed corrects the fleeces of the worst flocks in a single generation, and even in the least favorable climates. The Angora Goat has imparted his silky fleece to the Swedish flocks, and they maintain this character for several generations. In Russia also, the Stallions with a frizzled and crisp hair, impart to their foals a similar coat and of the same colour. The wild, as well as domestic animals, also tend continually to maintain their primitive forms in opposition to all the influences of climate and food, which are wholly insufficient to induce this supposed degeneration and degradation of species. Whenever some accidental connate deformity or partial excrescence becomes hereditary, as sometimes happens, the natural liberty of intercourse soon re-establishes the original form, and it is only by interfering with their unions that we can succeed in rendering permanent the accidental varieties of our domestic animals. In the wild state also, the females are led instinctively to prefer the most courageous of the males, the most perfect, and the most masculine of their species. The males, likewise, instinctively prefer the most beautiful of the females, and thus they both tend to transmit to their offspring the most perfect form of their species.

Since Nature then has placed an instinctive mutual aversion in animals of different species,—since she has rendered Hybrids either sterile or weak and imperfect,—if allied animals distributed in remote parts of the globe are found to be incapable of yielding fertile races, we have presumptive evidence that this supposed degeneration of species cannot have existed, and we derive from the known insufficiency of the present causes of change a positive ground for inferring their descent from distinct original types. The mere circumstance of our being able to induce by art and contrivance a fertile union between two species, is not sufficient to counteract this evidence, when we see that these same species preserve themselves distinct in the wild state, and continually maintain certain well-defined peculiarities.

Pallas was led to infer that some of our domestic animals, such as the Sheep, the Goat, and the Dog, are factitious beings, not proceeding from any permanent origin, but from the union of several distinct species, such as the Dog from the Wolf, the Fox, and Jackal; the Sheep from the Mouflon and Siberian Argali; and the common Goat from the Persian and Caucasian Goats with the Ibex. We know that these animals have given rise to fertile hybrids; and hence it becomes impossible to say how far their varieties may be owing to foreign contamination, or to the occurrence of connate varieties in the original species. It is, however, useless to indulge in conjecture where data are defective; but from analogy we might infer that a very small part of their varieties have been owing to foreign admixture.

It now remains for us to notice the theory of the successive Transition of Species proposed by M. Lamarck. According to him, the habits and manners of life assigned to each animal do not follow from any original form peculiar to its species: but that, on the contrary, the form of each species is the result of its habits, its manner of life, and other influential causes, which, in the course of time, have constituted the shape of the body and the parts of the animal. With new forms, new faculties have been acquired; and thus gradually Nature has produced the animals as we now see them.

We must in justice remark, that this theory has been censured in this country with undue severity, from its appearing at first sight to dispense with the agency of a First Cause in the creation of the several species of animals. But in reality, a creative power is as indispensable in *maintaining* the successive transition of forms, as in originally *creating* them. Lamarck himself was well aware of this, for he observes, " When I see that Nature has placed the source of all the actions of animals, of all their faculties, from the most simple to those which constitute instinct, industry, and finally reasoning—in their *wants*, which alone establish and direct their *habits*; ought I not to acknowledge in this power of Nature, that is to say, in the existing order of things, the execution of the will of its Sublime Author, who has imparted them the power?"

As an illustration of this supposed transition of species, we shall show M. Lamarck's method of explaining upon his theory, how it comes to pass that some Mammalia can fly. A very ancient race of common Squirrels had long amused themselves with leaping from tree to tree, and thence had acquired a habit of extending their limbs like a parachute. From frequent repetitions of this act, the skin of their sides became gradually enlarged, in course of time, and a loose membrane extending from the fore to the hind feet, embraced a large volume of air, and broke the force of their fall. In a word, they acquired the characters of the Flying Squirrels (*Pteromys*). These animals, however, were still without membranes between their fingers.

But a race of Squirrels of much higher antiquity, after undergoing the preceding metamorphosis, had acquired a habit of taking still longer leaps than the former. Accordingly the skin of their sides became more ample, uniting not only the fore and hind legs, but even the tail with the hinder feet, as well as the fingers with each other. These now form our Flying Lemurs (*Galeopithecus*).

There was, however, a third race of Squirrels vastly more ancient than any of these, which had contracted a habit, in the course of time, of extending not only their limbs, but also their fingers. From this habit, long preserved and become inveterate, they not only acquired lateral members, but an extraordinary elongation of the fingers of the anterior limbs, with large intermediate membranes, so that at length they constituted those singular wings which we find in the Bats (*Vespertilio*).

" So great is the power of habit," observes M. Lamarck, " that it singularly affects even the conformation of the corporeal parts, that it imparts to those animals which have contracted certain habits through a long course of ages, certain faculties which other animals of different habits do not enjoy."

Upon this theory, it was requisite that the higher orders of animals should be regarded as of the greatest antiquity, a longer time being necessary for their transition from those simple forms, which were supposed to have been first created. " I have no doubt," proceeds Lamarck, " that all the Mammalia have originally sprung from the ocean, and that the latter is the true cradle of the whole Animal Kingdom. In fact, we still see that the least perfect animals are not only the most numerous, but that they either live solely in the water, or in those very moist places, where Nature has performed, and continues to perform, under favorable circumstances, her direct or spontaneous generations; and there, in the first place, she gives rise to the most simple animalcules, from which have proceeded all the animal creation."—(Philosophie Zoologique, tom. 2, p. 456).

We must remark, that there has never yet been, within the historical era, a well authenticated fact of any animal of one species having acquired organs, or faculties belonging to another; nor are any species known to have lost any of their senses or powers to make way for new ones. It must further be observed that, while we have never found any of these transitions in circumstances within the sphere of our investigations, Lamarck places them precisely in those, where they cannot be proved or disproved by direct observation. Where did these transitions begin? In the abysses of the Ocean, where Man has never penetrated, and where myriads of beings lie concealed from his observation, perhaps for ever. What animals owe their origin to spontaneous generation? Animalcules, a class of beings the most remote from our observation, and whose forms can only be traced through the deceptive medium of the microscope. When did these transitions occur? Before the historical era, in those remote and inaccessible ages, whose existence is alone attested by the organic remains imbedded within the surface of the earth.

But, observes Lamarck, " there is a very good reason why we do not see those changes successively performed, which have diversified the known animals, and brought them to their present state. We see them only when they are finished, and not when undergoing the change; and we very naturally infer that they always have remained as we see them." This is a prejudice. " If the average duration in the life of each generation of Men were only a second, and if there be a pendulum mounted and in motion, each generation would consider this pendulum really to be at rest, never having seen it change in the course of their lives. The observations of thirty generations would not demonstrate any thing positive concerning the vibrations of this instrument."

We may remark, that our sole means of judging of unknown objects is by comparing them with others which are known, and that it is unphilosophical to found a theory of what occurs, or has occurred, in remote and inaccessible parts of the creation, in direct opposition to what is seen to happen within our own sphere of observation. The earth appears to be at rest, if it be compared with objects on its surface; and we reason correctly, for, *in respect to them*, it is at rest. But on referring it to the Solar System, we at once perceive it to be in motion. Again, if we compare the entire Solar System with the more remote heavenly bodies, analogy would lead us to expect that our system may be in motion towards the Fixed Stars, and that these Stars themselves may only be fixed, relative to our own limited means of observation. To suppose the Fixed Stars to be really motionless, would be as great a violation of analogical reasoning, as those theories inflict which deny the permanent characters of species. All the sciences adopt this mode of reasoning when the contemplated object is inaccessible to direct experiment or observation. On looking abroad into Nature, the Chemist finds every thing in a state of composition. He nowhere discovers pure oxygen, chlorine, calcium, or potassium, because nearly all the unions which simple substances were capable of forming spontaneously have already occurred. The Naturalist is disposed to imagine that something similar to this may have taken place among the species of animals and plants; but the Chemist analyzes the compounds of these substances himself, and he sees their combinations going on before his eyes. The Naturalist cannot bring forward one single instance of the degeneration or transition of species from one form to another.

The weak point of the Lamarckian doctrine, in the absence of positive proof, is a violation of one of the first rules of analogy. Mr Lyell correctly remarks, in his recent criticism on this subject, that " no positive fact is cited to exemplify the substitution of some *entirely new sense*, faculty, or organ, in the room of some other rendered useless. All the instances adduced go only to prove, that the dimensions and strength of members, and the perfection of certain attributes, may, in a long course of generations, be lessened and enfeebled by disuse; or, on the contrary, be matured and augmented by active exertion, just as we know that the power of scent is feeble in the Greyhound, while its swiftness of pace and its acuteness of sight are remarkable;—that the Harrier and Staghound, on the contrary, are comparatively slow in their movements, but excel in the sense of smelling. It is evident, that if some well authenticated facts could have been adduced to establish one complete step in the progress of transformation, such as the appearance in individuals descended from a common stock, of a sense or organ entirely new, and a complete disappearance of some other enjoyed by their progenitors, that time alone might then be supposed sufficient to bring about any amount of metamorphosis. The gratuitous assumption, therefore, of a point so vital to the theory of transmutation, was unpardonable on the part of its advocate."

We have now seen that some Mammalia are capable of undergoing a very considerable variation, not only in their Instincts and Intelligence, but also in their external forms;—that the variations which each individual can be made to undergo by the circumstances in which it is placed are but very rarely transmitted to posterity, while connate modifications usually end in becoming hereditary;—and that there are certain limits beyond which no species has been observed to vary; so that we are fully entitled to conclude, that a certain form was assigned to each species at the origin of things.

GENERAL REVIEW OF THE MAMMALIA CONTINUED.

Forms to which the Domestic Animals have reverted on becoming wild.—Their modifications during the Historical era.

If it be true that the numerous varieties of the Cow, the Horse, the Dog, and other Domestic Mammalia, are the effects of the slow and continued influence of certain causes, which, in the first instance, induce a departure from the primitive type in the evolution of connate varieties, and afterwards transmit these variations to posterity, giving rise to their several distinct races;—it ought to follow, that in all these artificial beings, whose characters Man has for a time rendered permanent, there should be a continual tendency, when left to their own resources, to assume the form of the original type. On allowing the domestic animals to run wild,—on permitting them to substitute the wandering habits and precarious subsistence of mountains and forests for the uniform and regular diet of the stable, we ought to find that their acquired characters disappear, that all the individuals bear that marked resemblance to each other, which will serve to indicate both the identity of their species, and the original form from which the races have diverged.

The experiments confirming the truth of this conclusion have long been performed on the largest scale in the immense continent of America. It is well known that the Europeans, on the first discovery of the New World, sought in vain for any vestige of that animal creation to which they had so long been familiarized. Those useful animals, without whose aid, in the first instance, the civilization of Man might have been indefinitely retarded, had to be transported to America to supply the immediate necessities of the earlier colonists. Soon, however, the accidental flight of some animals to the woods, hastened probably by the abundant supply of food, and a favorable climate, which, in increasing their fecundity, rendered a vigilant care of them superfluous, a large proportion became absolutely wild, and the establishment of wild individuals in the immediate neighbourhood of the tame herds, soon exerted a direct modifying influence over the latter. Hence, in America, we may see performed, on a magnificent scale, the converse of that gradual modification which the domestic animals underwent in their original transition from the wild state; and may further compare those half domesticated herds, acknowledging only a partial submission to Man, with the humble individuals of their own species, which still yield him a patient and implicit obedience.

It is evident that careful observations should be multiplied over the whole continent of America, in order to render this investigation complete; but we owe to M. Roulin the merit of having traced some changes in a portion of this vast country. That learned physician, during his residence in Colombia for six years, has collected a number of interesting facts which were communicated to the Royal Academy of Science at Paris in the year 1828. These observations were made in New Grenada and a part of Venezuela, from the 3d to the 10th degree of North Latitude, and from the 70th to the 80th degree of West Longitude. However limited this tract of country may appear, it offered unusual facilities for observation, being traversed throughout its entire course by the great Cordilleras of the Andes, which are here divided into three principal chains; so that, within the distance of a few leagues, the same living animals were investigated, though resident in one district, where the medium temperature is only 50° Fahrenheit, and in another where it varies from 77° to 86°.

The Mammalia transported from Europe to America were the Hog, the Horse, the Ass, the Sheep, the Goat, the Cow, the Dog, and the Cat. It becomes important to ascertain whether these animals retain the forms acquired in Europe, or whether they have undergone any considerable change. By carefully comparing these phenomena with the circumstances under which they have arisen, much light may be thrown upon those modifications which probably attended the transition of these animals from the wild to the domestic state.

The first Hogs brought to America were introduced by Columbus, and became established at Saint Domingo in November 1493, being the year which followed its discovery. During the following years they were successively carried into all those places where the Spaniards attempted to fix themselves, and, in the period of about half a century, they might be found wild from the 25th degree of North, to the 40th of South Latitude. In no place do their important changes appear to have been effected by climate, and they have reproduced every where with the same facility as in Europe.

Most of the pork consumed in New Grenada comes from the warmest valleys, where the Hogs are bred in large numbers, from their maintenance costing but little. During some seasons they are even supported wholly by wild fruits, and especially by those of the several species of Palms. From roaming constantly in the woods, the Hog has lost in this district all traces of his former domestication. His ears are straight and erect, his head has widened and become elevated in the upper region. The colour has again become constant, being entirely black. The young Pigs have several fawn-coloured stripes, like the European Wild Boar in its youth, and upon a ground of the same colour. Such are the Hogs brought to Bogota from the valleys of Tocayma, Cunday, and Melgar. Their hair is scanty, and on this account they bear a striking resemblance to the Wild Boar of Europe, from a year to eighteen months old. This deficiency of hair is not, however, peculiar to the Hogs of Grenada, but is also experienced by the common Wild Boar of Europe. M. Roulin observed an instance of it in France, at a farm near Fougères, where seven or eight of these animals were brought up together. One of them, being about two years old, had been fed in a stable from the beginning of spring, with the intention of fattening it for the market. Though the animal had not been closely confined in this place, the good feeding of the stable was sufficient to induce it to remain at home. Its hair had almost wholly fallen off from the effects of the heat, and it exhibited a most perfect resemblance to the Hogs of Melgar above described, except that the two horizontal stripes on the sides of the muzzle were more decidedly marked, and gave it a stronger expression of ferocity. The Hogs of the Paramos, which are mountains at least 8,200 feet above the level of the sea, approach much more nearly in appearance to the Wild Boar of the European forests, from the thickness of their hair, which has even become frizzled. Beneath, in some individuals, it has been observed to assume a woolly appearance. The Hog of these elevated regions is, however, small and stunted, from the want of sufficient food, and the continued action of an excessive cold. In some sultry districts, the Hog is not black like those above described, but red, like the Peccari, during its youth. At Melgar, and in the other places above mentioned, instances have been known where the Hogs are not entirely black; but these are comparatively rare. There is a variety called Cinchados, or *girthed*, because they have a large white band underneath, which usually unites on the back, and always preserves a uniform breadth; and the young individuals of this variety bear the same stripes as those of the pure black breed. The only Hogs in Colombia which resemble the common Pigs of Europe, have been imported within the last twenty-five years, and these do not come direct from Europe, but from the United States of America; and it must be recollected that in the neighbourhood of New York, where this race has long been domesticated, it experienced the influence of a climate very nearly the same as our own.

The Horse has become wild in several districts of Colombia, especially in the plains of San Martin, among the sources of the Meta, the Rio Negro, and the Umadea, where small troops of Chestnut Horses may be observed. Their limited numbers, the narrow range to which they are confined, and the immediate neighbourhood of the inhabited districts, have prevented them from acquiring those peculiarities which Azara has related of the Wild Horses of Paraguay. They go about in small squadrons, composed of an old Horse, five or six Mares, with some Foals, and one squadron is completely isolated from another. Instead of approaching the caravans to entice the domestic Horses, they run away on the first appearance of a Man, and do not stop their flight until he is out of sight. Their movements are graceful, especially those of the leader, but their forms, though not heavy, are wanting in elegance.

In the Hatos des Llanos, the Horses are almost wholly left to themselves. The herds are assembled together at intervals to prevent them from becoming absolutely wild, to extract the larvæ of the Gad-fly, and to mark the Foals with a red-hot iron. From this independent kind of life, they begin to acquire the uniform colour of the savage races. The Chestnut bay is not merely the prevalent colour, but it is very nearly the only one. Something similar to this has probably happened in Spain with the Wild Horses (*cavallos cerreros*) which wander in the mountains; for in the Spanish proverbs, the Horse is often noticed by the name of *el bayo* (the Bay), as well as the Ass by the term *rucio* (Gray).

In the small Hatos which are found on the plateau of the Cordilleras, the effects of domestication are more perceptible. The colours of the Horses become more various; there is also a greater difference in their height; and while many are more diminutive than the average of the species, only a few surpass the medium size. As long as they live continually in the fields, their hair is tolerably thick and long, but a few months' residence in the stable is sufficient to render it short and glossy. It is customary to cross this breed of Horses with the races from the warm valleys, especially with those of Cauca. On some properties where this precaution has not been attended to, the Horses have become perceptibly smaller, though the pastures have long been celebrated for their richness. The hair has grown to such a degree as to render their appearance absolutely deformed. In respect to the useful qualities, this breed has lost but little; and the Horses belonging to one canton are even celebrated for their swiftness.

When a Horse is brought from the Llanos de San Martin, or from Casanare, to the plateau of Bogota, he must be kept in the stable until he is accustomed to the climate. If allowed to run loose at once into the fields, he grows thin, contracts a cutaneous disease, and often dies in a few months. The pace which is commonly preferred in the saddle-horses is the amble; this they are made to acquire early, and the greatest care is taken not to allow them, when mounted, to take any other pace. In a short time, the limbs of these Horses usually become stiff; and then, if otherwise of a good form, they are allowed to run in the Hatos as Stallions. From them a race has descended, in which the amble is with the adults the natural pace. These Horses are called *aguilillas*; and they form a remarkable instance of the transmission of acquired habits from the parent to his offspring.

The Ass has undergone very few alterations in its form or habits in all the provinces visited by M. Roulin. At Bogota it is very common, being there used for transporting building materials; but being badly taken care of, and exposed to the inclemencies of the weather, without receiving sufficient nourishment, the race has become small and pitiful. It is covered with very long and uncombed hair. Deformed individuals are often seen, not only among the adults, which are loaded prematurely, and before they have acquired sufficient strength, but also among the Foals at their birth. Perhaps the latter circumstance may arise from the ill treatment of the dams during the period of gestation.

In the low and warm provinces this animal is less neglected, as it is required for the production of Mules. Being well fed, at least in these districts, it becomes larger and stronger; its hair also is shorter and more polished. In no province, however, has the Ass reverted to its wild state.

The Sheep was originally transported to the New World from Spain; and the earliest importation appears not to have been the Merino variety, but another, which the Spaniards call *de lana burda y basta* (with wool coarse and rude). It is very common on the Cordilleras, at an elevation of 3,300 to 8,200 feet.

In no place do the Sheep appear to have escaped from the protection of Man, and hence we find that their manners have undergone scarcely any change; nor can any alteration in their forms be observed, except a slight diminution of stature. Within the limits above defined, the Sheep propagate readily, and almost without requiring any care; but the reverse happens in the hotter districts. It appears that in the plains of Meta it is very difficult to rear Lambs; and no Sheep are to be seen from the river to the foot of the Cordilleras, although their skin is very much in demand to make a kind of parchment, and that its price is as high as the hide of an Ox. In the valley which separates the most eastern chain from the central, they may perhaps be sometimes seen, but always in small numbers. The females are not very fruitful, and the Lambs are difficult to rear.

There is one very curious phenomenon exhibited by the Sheep of this district. The fleece grows upon the Lambs in the same manner as in most temperate climates, provided they are sheared as soon as it has arrived at a certain degree of thickness, in which case the wool grows again, and continues to observe the same order. But if the favorable period for stripping the animal of its fleece be allowed to pass, the wool thickens and becomes matted together, it detaches itself in flakes, and finally leaves behind—not, as we might expect, a growing fleece, or a naked and diseased skin—but, a short, glossy, and compact hair, exactly resembling that of the common Goat in the same climate.

Although the Goat is evidently best fitted for a mountainous region, it seems to thrive better in the low and sultry valleys than in the more elevated regions of the Cordilleras. In the former districts it multiplies rapidly, generally bearing two young at a birth, often three, but never six, as some have been pleased to assert. Its height is diminutive, but in other respects its form has greatly improved. Its body is more slender, the shape of its head is more elegant, more pleasingly disposed, and usually less overloaded with horns. The agility of this animal, and its taste for climbing and leaping, are also singularly increased. In the public square of a village, M. Roulin has often seen them leaping more than four feet upwards to the mouldings on the pilasters of the church. The projecting place on which their feet rested was not three square inches; yet in this position, so difficult to preserve, they remained for hours together, without any other apparent object than that of warming themselves in the direct solar rays, as well as in those reflected from below. These Goats are covered with short hair, very glossy and thick; and although they may be seen to possess all the shades of colour, yet the most common is fawn, with a brown stripe on the back, and black symmetrical marks upon the face. The She-Goats of Europe strikingly exhibit the influence of domestication in causing a great enlargement of the udders; for this acquired character has entirely disappeared in the She-Goats of America.

The establishment of the larger Cattle in America must be dated, like that of the Hogs, from the second voyage of Columbus to St Domingo. In the latter place they multiplied rapidly, and the island soon became the nursery from which these animals were transferred to different points on the coast of the Mainland, and thence to the interior of the continent. Although these numerous exportations must have diminished their numbers considerably, yet we are informed by Oviedo, that within twenty-seven years from the discovery of St Domingo, herds of 4000 head of Cattle might frequently be encountered, and that there were even some containing at least 8000. In the year 1587, the number of hides exported from this island alone amounted to 35,444, while 64,340 appear to have issued from the ports of New Spain. This was the sixty-fifth year after the capture of Mexico, before which event the Spaniards were entirely occupied in warfare, and it strikingly evinces the extreme rapidity with which these animals will increase their numbers when placed under favorable circumstances.

While the Cattle were in small numbers, and grouped around the habitations of their masters, they succeeded equally well almost every where; but as soon as their numbers became greatly increased, it was discovered that in certain districts they could not exist without the assistance of Man. Unless they were able to find a certain quantity of Salt, either in the substance of the plants which formed their food, or in the streams which in some districts acquire a brackish taste from the saline particles contained in the soil, it was found to be absolutely necessary to furnish it to them directly. If this precaution were not attended to, they became stunted and poor; many of the females ceased to be fruitful, and the herds rapidly disappeared. Even in those districts where the Cattle can exist without this assistance, it has been found advantageous to distribute salt at stated intervals to the herd. This is one principal means of attaching them to a particular spot; and so great is the avidity with which they take this substance, after being for some time deprived of it, that when it has been distributed to them two or three times at the same place, they are seen running from all quarters to the spot as soon as they hear the horns which the herdsmen sound before making the distribution.

If, however, the country yields a sufficient supply of salt, and if the herdsmen neglect to assemble the Cattle from time to time, they become in a very few years wholly wild. This has happened at two places to M. Roulin's knowledge, the one in the province of San Martin, in a property belonging to the Jesuits, at the time when this religious order was expelled, the other in the province of Mariquita at Paramo de Santa Isabel, in consequence of the abandonment of some works where the natives washed for Gold. In the latter place, the Cattle have not remained in the districts where they were originally placed, but have mounted the heights of the Cordilleras to seek the region of the Grasses, and there live in a temperature almost uniform of 48° to 50° Fahrenheit. To this spot the peasants of the villages Mendez, Piedras, and some others situate in the plains, sometimes come to hunt them. They drive with knotted cords small divisions of the herd towards the places where snares have been previously prepared. Whenever they obtain possession of one of these animals, it is often impossible to conduct it alive from among the mountains. This does not arise from the resistance which the captive makes, for after a little time its violence begins to diminish; but when the animal begins to perceive the futility of its efforts to escape, it is often seized with so great a tremor over the whole body, that it falls to the ground; to make it rise becomes impossible, and it dies in a few hours. The want of salt to preserve the meat, the distance from any inhabited district, and the difficult nature of the roads, prevent the hunters from deriving any other advantage from the slain animal than the portion which they can consume upon the spot. These disadvantages render the hunting of wild Cattle by no means frequent; and the hunters always run the risk of being surprised by the snow, which often falls in these elevated regions. When the snow lasts many days, these unfortunate men, accustomed to the continual warmth of the adjoining valleys, are sure to perish. If, however, they are so fortunate as to bring one of these animals from the mountains, it is not difficult to tame; this is effected by confining it near to the farm, by supplying it regularly with salt, and habituating it to the sight of Men. M. Roulin never had an opportunity of seeing one of these animals alive, but he tasted the flesh of a wild Calf which had been killed on the evening of his arrival. Its flavour did not in any respect differ from that of the common domestic Calf; the hide was remarkably thick, in other respects of the usual size; the hair was long, thick, and rough. In the province of San Martin he, however, saw a wild Bull of a chestnut colour, pasturing in the Llanos in the midst of the domestic cattle. The Wild Bulls pass the morning in the woods which cover the base of the Cordilleras, and do not appear in the Savannah until about two hours after noon, when they come out to feed. As soon as they perceive a Man, they hasten to regain the forest at a full gallop.

The hide of the Wild Bull does not appear to differ in any respect from that of the domestic Cattle which inhabit the same districts. In both they are much heavier than the hides of the Cattle brought up on the plateau of Bogota, and the latter yield in this respect, as well as in respect to the thickness of their hair, to the wild Cattle of Paramo de Santa Isabel.

In the warmest parts of the provinces of Mariquita and Neyba, there are some herds of horned Cattle with their hair extremely scanty and fine; they are given, by antiphrasis, the epithet of *Pelones*. This variety is transmitted to their descendants, but no care is taken to preserve the breed, as the Pelones are unable to bear the cold of the elevated regions of the Cordilleras, where the cattle intended for consumption or exportation must remain for some time to fatten. There is also another variety of Cattle in this district called *Calungos*, having the skin entirely naked like the Barbary Dog. As these animals are more feeble and delicate, it is usual to kill them before they are old enough for breeding. These never appear in the cold districts.

In Europe, where the milk of the Cow forms a very important article of rural economy, it is usual to milk her continually from the moment of the birth of her first Calf until she ceases to be fertile. This practice, continually repeated upon all these animals for a long series of generations, has had the effect of producing permanent alterations in the species. The udders have acquired an extraordinary size, and the milk continues to be secreted even after the Calf has been removed. In Columbia, however, the introduction of a new rural system, the abundance of cattle in proportion to the number of inhabitants, their dispersion in pastures of very great extent, and a number of other circumstances which need not here be detailed, have counteracted this effect of domestication. The organization and function of the udder soon resume their original state when freed from the long-continued influence of habit. At present if a Cow of Columbia be intended to yield milk for the dairy, the first care must be to preserve the Calf; it is allowed to remain along with its mother for the entire day, during which she is permitted to suckle it. They can be separated only at night, and the milk secreted during the interval of their separation alone becomes available for economical purposes, and accordingly it is abstracted every morning. If the Calf happen to die, the secretion of the milk is immediately stopped.

In America, the Cat has scarcely undergone any alteration, except in its having no period of the year corresponding to the rutting season. This peculiarity, which might naturally be expected in a climate always equal, exists also with the Hog, the Bull, the Horse, the Ass, and the Dog. Although Kids and Lambs are born all the year round, yet there are two periods of the year, Christmas and Whitsunday, when the number of births is greatly increased.

These particulars, furnished by M. Roulin, though necessarily defective on many interesting points, enable us to draw several important conclusions, which serve to throw a light upon the past history of our domestic animals, and directly also upon the philosophy of species in Zoology. However, the extreme difficulty of distinguishing those phenomena which are due solely to domestication from those belonging to food, climate, and situation, lead us naturally to inquire whether the records of ancient History can yield us any information regarding the progress of variation among our domestic animals. If we can discover in these writings any traces of their gradual deviation from the form of the wild races, we have an additional evidence in support of those views which have here been laid down.

Unfortunately the notices of the domestic animals in the writings of the ancients are neither numerous nor full; yet however scanty, they possess a peculiar interest and importance. It is true that we find those civil and military events which attend the rise and fall of great Empires, or the establishment of different religions, and other historical events concerning the Human Race, recorded with a scrupulous attention; but the ancients may be said to have wholly neglected the minor histories of the farm-yard and stable. The gradual modifications of their domestic animals presented none of those brilliant events and striking positions which compose the ordinary pages of history, but, moving onwards with a silent and almost insensible step, they escaped the notice of their contemporaries. Man, on becoming civilized, soon forgot, with characteristical ingratitude, those early companions of his labours, without whose aid his own progress might have been indefinitely retarded. It is only now by examining the aggregate of their changes through a long course of years, that we are led to perceive the extent of their modifications, and can fully appreciate the importance of their contributions towards the wealth and happiness of society.

Modern Naturalists have commonly supposed that the native country of our domestic Mammalia cannot be ascertained; yet it would appear that these animals were all living in a wild state in Europe at the time of Aristotle. This great Naturalist himself attests the fact, and mentions the Horse, Bull, Hog, Sheep, Goat, and Dog, as familiar instances. We are also informed by Pliny, after having alluded to the intercourse between domestic Pigs and the wild Boar, that there were no domesticated animals in his time, which could not also be found in a wild state. (In omnibus animalibus placidum ejusdem invenitur et ferum).

The concordance of these two passages is striking, and they prove that in the 450 years which elapsed from the time of Aristotle to that of Pliny, the domestic animals had not been widely distributed over the globe, nor had they undergone much variation. Indeed, as long as wild animals reside in the immediate vicinity of the tame herds, it is certain that the domestication of the latter will be exceedingly imperfect. The continual intercourse of the wild with the tame animals, and the contagious example of herds running wild in neighbouring mountains and deserts, must have diverted the captive animals from those domestic habits to which the restraints of Man would otherwise have reduced them. In this respect, the partially domesticated races would have resembled those Indians of the United States of America, which are taken from their tribes during their infancy, and educated in the midst of towns, both in the religion and manners of the Europeans. At the age of twenty or thirty years, if they happen to encounter in the woods a tribe of Hunters of their own nation, so hereditary are their propensities, that they at once reject their former peaceable life, with all its advantages, moral and intellectual, and plunge without reflection into the savage and adventurous life of their ancestors.

Varro appears to confirm the opinion of the Greek and Oriental Philosophers, that the Sheep, in consequence of its superior docility and mildness, was the first animal which became domesticated. "The Sheep," says he, "is not only of a very peaceful nature, but it is the animal most fitted to supply the wants of Man, since it yields not only milk and cheese for food, but also its wool and skin for clothing." "In several countries," continues Varro, "there still exist in the wild state some of the animals which we have rendered domestic. In Phrygia and Lycaonia, many flocks of Wild Sheep are to be found. The Wild Goat exists in Samothrace, and there are several in Italy, in the mountains adjoining Fiscellum (now Monte della Sibilla, near Abbruzo), and Tetrica (near the most elevated point of the Appenines in the Upper March of Ancona). In respect to the Hog, every one knows that he is descended from the Boar, which is found wild in all countries. There are still a great number of Wild Bulls in Dardania, Mysia, and Thrace; there are Wild Asses in Phrygia and in Lycaonia, and Wild Horses in some parts of Hither Spain." (Re Rustic, II, I, 4–6).

This passage of Varro fully corroborates the testimonies of Aristotle and Pliny, and his evidence is important, as we know that Varro himself travelled through all the countries where he places these wild animals. Modern researches have verified a part of Varro's declaration, and recognize the original localities of the Ass. These are the mountains of Taurus and lower Curdistan, separating Persia from Afghanistan. Here it still exists in the wild state, and the pursuit of this Solipede has long been one of the chief amusements of the Persian Kings.

Buffon and other modern Naturalists differ in opinion from Varro and the Oriental Philosophers, regarding the priority of domestication with the Sheep. The Dog, according to Buffon, was the first animal which Man acquired for his use; and it was by the assistance of this animal that he was able to seize and subdue all the other species necessary to supply the wants of an infant society. This opinion rests chiefly upon the extreme facility with which wild animals of the Dog Genus are tamed, arising from their great sociability and their power of imitation. Azara mentions an Aguarachay of Paraguay (*Canis cinereo-argenteus*) which became as tame as a Dog, but ate up all the fowls. Yet the opinion of ancient writers regarding the prior domestication of the Sheep seems to be by far the more probable. The Sheep lives habitually in large flocks; the mildness of this animal, its simplicity, and disposition to follow its companions even to certain destruction, must have rendered it an easy prey to the savage in those first ages which followed the creation of Man. Its utility for food and clothing must have been evident. On the contrary, the Wild Dog lives in troops; he is a Carnassier, fierce, and daring; he unites with his fellows to form a combined plan of attack and defence. He is as strong and more to be dreaded than the Wolf. No use could be made of his skin, of his flesh, or the milk of the female. Hence it is not very probable that the savage would have at once foreseen all the future advantages which he would derive from associating the Dog in his labours to reduce and subdue the other animals. Even if he could have entertained this project, the difficulties and dangers with which it was beset would have diverted him from the enterprise. In this case, we must admit, that the more simple and natural idea would first present itself to his mind.

It may easily be imagined, that in those early ages, when the globe was less peopled than it is at present, the great work of Domestication must have been slowly and gradually accomplished. The remarkable property which these animals possess of transmitting their acquired qualities to their descendants, and of perpetuating modifications of form, colour, and even of intelligence, render their races singularly capable of improvement. The several races of Men are far less capable of undergoing this relative improvement than the domestic animals, which receive his influence in innumerable ways. Yet we are not without some striking instances of the transmission of acquired properties even in Man. Among the Negro children of Sierra Leona, the offspring of the Negroes, who have long been liberated, and who are born in the colony, possess an immense relative superiority of intelligence over the children of Negroes which have recently been emancipated from their slavery. Their parents inhabit the same country; but the older liberated Negroes have commenced a moral and intellectual education, while the more recent Slaves have long endured a savage and degraded existence. It has, however, never been attempted to bring the Human race, like the Domestic animals, to a greater physiological perfection, by always uniting individuals, remarkable for the beauty of their forms, the goodness of their temperament, and the extent of their intellectual faculties. Absolute monarchs might, in the course of a long dynasty, have made this curious experiment, and endeavoured to promote the good of their subjects, by improving the breed of their own ministers. Hence Man, considered as a race, that is, in reference to his physiological qualities, is much less capable of improvement than the domestic animals.

In consequence of this remarkable property of transmitting acquired faculties to posterity, the notices of the ancients, which date back perhaps from twenty to twenty-five centuries, however meagre, become peculiarly important and interesting.

Wants, dangers, and necessities, develop the more violent and fiercer passions of animals; the suppression of these exciting causes improves the milder and more useful qualities. From the descriptions of Aristotle, the passions of the domestic animals were formerly much more violent than they are at the present day.

The progress of domestication, as recorded by the ancients, in respect to the Horse, the Ass, the Dog, and the Cow, presents many interesting facts. With the Dziggtai (*Equus hemionus*), domestication seems to have made a retrograde movement.

Herodotus (iv. 52) informs us that Horses existed in the wild state on the banks of the Hypanis (now the Dniester). These Horses, he adds, were white. Further, also, that in Thrace, the Pæonians of lake Prusias fed their Horses and beasts of burthen with fish instead of hay. Strabo says that the Wild Horses were to be found in India, on the Alps, in Iberia, among the Celtiberians, and finally in Caucasus, where the intensity of the cold had given them thick coats of hair. The last remark is confirmed by modern observations on the Norwegian and Lapland Horses, which have a thick and woolly hair like the fleece of our Sheep. Pliny says that the North contains herds of Wild Horses. Strabo relates, on the authority of Megasthenes, that the greater number of our domestic animals were wild in India. Ælian makes the same remark for the interior of India.

Since Wild Horses thus existed in great numbers on several parts of the Old Continent, the progress of domestication must have been very slow in all those places where they came in contact with the tame herds. Azara observed, that the Wild Horses which live at liberty in the plains of Paraguay, in herds consisting of many thousand individuals, have an instinctive habit of seducing the domestic Horses. As soon as they perceive one, says this able Naturalist, even at the distance of two leagues, they form into an uninterrupted column, and approach at full gallop to entice him. They either surround him on every quarter, or merely come along side; they caress him by neighing gently, and always end in carrying him off never to return, without his offering them the slightest resistance. The inhabitants of that country hunt the Wild Horses very keenly, to drive them away from their own studs, for, without this precaution, the Wild Horses would seduce away all the tame herds. Gerbillon notices the Wild Horses in the desert of Chamo in nearly the same terms.

This fact may serve to explain one of the causes that in ancient times the herds of Wild Horses disappeared very rapidly when the population increased. According to the accounts of those Missionaries who were best acquainted with China, Wild Horses are still to be found in Western Tartary and in the territory of Kalkas. They live in large troops in the neighbourhood of Ha-mi, and appear to resemble the common Horses. Grosier, in his Description of China, mentions that if they meet a domestic Horse, they surround him on all sides, and, urging him onwards, draw him to their forests of Saghatur.

A passage of Xenophon (περὶ ἱππικῆς, III.) alludes to this characteristic of the Wild Horses, so forcibly described by Azara and the Chinese Missionaries. His remark serves to show that, at the period of 450 years before the Christian era, the domestication of the Horse was still recent, and had not yet overcome this primitive instinct. In speaking of a Horse broken in by the groom, Xenophon observes, "It is proper to ascertain whether, when mounted, he will willingly separate from other Horses, or whether, when passing them at a short distance, he does not attempt to join them." Another observation of Xenophon, "One can teach nothing to a Horse by word of mouth" (Ibid. VIII.), shows how imperfect their domestication must have been in his time. We have so many proofs and examples to the contrary, as to render an allusion to them only necessary at present.

The modern Wild Horse, as described by Pallas, has his tail and mane very long and thick. He carries his ears depressed backwards, like a domestic Horse of the present day when preparing to bite. Xenophon and Varro describe a Stallion, the model of a War-horse, in words nearly synonymous to those used by Pallas in describing the Wild Horse of the Russian Steppes (juba, cauda, crebra, subcrispa, auribus applicatis). We have here an evident proof that the Domestic Horse, in the last century of the Roman Republic, still retained the characters now peculiar to the Wild Horses of the old continent.

It must be observed that Herodotus describes the Wild Horses to be white (λευκοι), while the dark bay has become the prevalent colour of the Wild Horses in America. Naturalists have generally concluded that the latter was the primitive colour of the species. This difference between the primitive hues of the Old and New World is supposed by some to be owing to the excessive cold of the climate in some parts of the former, where it has been supposed that the temperature might act upon the Solipeda and Ruminantia in the same manner as it is known to do upon Hares, Rabbits, and other Rodentia. But Leo Africanus and Marmol relate that the Wild Horses of Africa are small, and either white or ash-coloured. Pallas also informs us that the Wild Horses which inhabit the country between the Jaik and the Volga are fawn, red, or dun-coloured. Aristotle attributes the changes in the colour of the hair of Mammalia, as well as in the feathers of Birds, jointly to the cold and the influence of the water. The streams of Psychus, near to Chalcis in Thrace, according to him, caused the White Ewes to produce Black Rams. In the neighbourhood of Antandros, he states that there are two rivers, one of which causes the lambs to be white and the other black. We must remember that Aristotle belonged to Stagyra, and that he here mentions a fact which, it is probable, had fallen under his own observation. The same remark is made by Varro, Pliny, Ælian, and by Anatolius (Hippiatric. p. 59). It would be interesting to verify their declaration by observations made on the spot, as it seems to be rather of doubtful authority.

The progress of education with the Horse, and the influence of domestication during 1800 years, are seen in the development of his paces both in number and permanence. The natural paces of the Horse are the walk, the trot, and the gallop; those which he has acquired from education, for the purpose of combining swiftness of pace with comfort to the rider, are the amble, the *pas relevé*, and the *aubin* of French authors.

The *pas relevé* consists in raising two feet on the same side, not at once as in the amble, but successively. It is a close trot which beats the ground, as in the walk, at four successive times. In the *aubin*, the Horse gallops with the fore feet and trots with the hinder. The Greeks and Romans had induced neither the pas relevé nor the aubin. That pace which they call *tolutarii*, and which the Lexicons give as synonymous with εὔδρομος, is evidently the amble, and seems to have been induced during the last century of the Roman Republic. It is described by Varro, Pliny, Nonius, and Vegetius, in a manner which leaves no doubt that the amble *(tolutarem ambulaturam)* was produced by training (*traditur arte*). The race at that period had not been so long domesticated, that this property should have been transformed from an artificial acquirement into a permanent quality. It must then have been in the interval of time which has elapsed since the days of Pliny and Varro, that the amble, the pas relevé or trot with four beats, and the aubin, where the Horse gallops with the fore limbs and trots with the hinder, all of which are wholly artificial, had become natural paces, and were transmitted as such to posterity. At the present day, these acquired paces are as permanent as the properties of pointing and bringing back game with the Setter Dogs and Retrievers. M. de la Malle has remarked more than a hundred times in the pastures of Normandy, that the Foals descended from a sire and dam endowed with the pas relevé, or even where the sire alone possessed this quality, have exhibited this artificial movement in the meadow before receiving the slightest education, or even leaving the side of their dam.

As we might readily expect, the ancients were acquainted with very few varieties of the Horse. Only two distinct races, the Thessalian and African, can be traced on those ancient monuments which have reached our times. There are, however, two intermediate varieties, the Sicilian and Apulian races, formed probably from crosses between the Thessalian and the Wild Horse of Italy, and between the Italian and the African races. The descriptions of authors agree precisely with the representations on the statues, the basso-relievos, and the medals, at least in respect to the two primitive races. We have the Thessalian Horse faithfully represented on the Parthenon, in the equestrian statues and basso-relievos of the Greeks, and even on the columns of Trajan and other Roman sculptures, where this variety is always adopted as the type of the heroic Horse. The African race is seen on the medals of Carthage and on a medal of Mauritania, supposed to be a Juba (Catalogue de M. Mionnet, t. vi. Nos. 5 and 6). In the time of Oppian, who was contemporary with Septimius Severus, the races of the Horse had greatly increased in number, and he accordingly enumerates fourteen varieties. The Persian Horse of the age of the Achæmenides is figured on the monuments of Persepolis. At the present day, in consequence of the continual crossing of these races during twenty centuries of domestication, and the joint influence of climate and food, this species, so useful to Man, has been transformed into varieties almost innumerable.

The Horse is now reared under domestication with greater facility than formerly. The foal, according to Varro, was suckled by its mother until the age of two years; —we separate them at six months. At three years old the young Horse was exercised, and when he perspired, was rubbed over with oil. If the weather were cold, fires were lighted in the stables. The modern Horses do not require these minute attentions even in our less congenial climate.

The Ass, being less useful than the Horse, has been more neglected by Man, and consequently his physical and intelligent powers are not so highly developed. Yet there are some interesting conclusions which may be drawn from an attentive comparison of his ancient and modern history, and may serve to clear up some obscure points as to the causes which have served to retard the progress of his domestication. The imbecility of the Ass, and his imperfect education, may partly be owing to the circumstance that the domestic species were continually united with the wild animals during many centuries. This practice was one chief cause of the slow progress of domestication among the ancients. Indeed, their rural system of large commons allowed a liberty almost absolute to their herds. These animals passed the spring in the valleys, the summer on the mountains, and the winter on the plains. It was, therefore, impossible to prevent the wild individuals, which then existed on several points of the globe, from accidentally uniting with several domesticated individuals of their own species or genus. This may probably explain the fable of the Mares of Bœtica, said to have been fecundated by the West wind. Wild Horses were very numerous in Spain, and the ignorant herdsman, seeing products formed with whose origin he was unacquainted, easily resolved the problem by referring it to a miracle. With the Ass, however, it was the constant practice of the Romans, according to Varro and Pliny, to select the Wild Asses (*onagri*) as Stallions. Luitprand, Bishop of Cremona, who wrote in 968, mentions that the domestic Asses of Cremona differed but slightly, in his time, from the Wild Asses of Asia Minor.

The attempts of the ancients to produce Hybrids or crosses between different species were so common, that they had proper names to denote the Hybrids between the Dog and the Wolf, as well as those between the Sheep and the Goat. They also had names for the cross between the Pig and the Wild Boar, and between the Sheep and the Mouflon. To obtain a fine race of Mules, the Romans united the Mare with the Wild Ass. Columella remarks, "that the Mule, the immediate descendant of the Wild Ass, remains wild, difficult to tame, and slender like its father; but that the Stallion of this species is more useful in the second generation than in the first. For when a Mare is united with an Ass descended from a Wild Ass and a domesticated female, the savage nature of the Mule appears to have been softened down by the influence of time, and the product of this union combines the beauty of form and the mildness of its sire, with the courage and swiftness of its grandsire." This important observation of Columella strikingly exemplifies the influence of domestication, as well as the transmission of certain physical and intelligent qualities in the course of generations, and is the more valuable, as we may at the present day search Europe in vain for a Wild Ass to repeat this interesting experiment.

Those Chapters in the writings of Varro, Pliny, and Columella, which treat of the production of Mules, contain minute directions as to the precautions which were necessary in their days to bring about an unnatural union between different species. The Ass, intended ultimately to propagate, had to be taken from its mother the moment it was born, and placed under a Mare without its perceiving the change. The Mare, on the other hand, had to be deceived by keeping her in the dark, and her own foal had also to be removed. She would then suckle the Ass' foal intended for propagating, and treat it as if it were her own offspring. In this way, the foal selected to be a Stallion formed an attachment to Mares from its infancy. It had to be constantly introduced into the society of Mares even while yet at the breast, that it might be habituated to their approach at the earliest age. The above mentioned authors go on to describe that l'accouplement doit se faire dans un lieu étroit, fermé, obscur, avec une jument liée, qui a déjà porté, et dont les désirs ont été d'avance irrités par un âne commun qui les éveille sans les satisfaires.

These precautions clearly show that domestication had not yet induced that kind of depravity which is its consequence, nor had it yet been sufficient to corrupt the manners of the Ass and Horse as at the present day; for we know that these Hybrid unions, formed between different species, can now be procured without the necessity of resorting to the slightest artifice. It must, however, be recollected, that such unions can only arise among domestic animals of nearly-allied species, or between animals of which one sex at least is domesticated.

The Ruminantia, it has been already explained, are those over whom domestication has had the least influence. Yet among the Romans, it was found necessary to employ only the most robust and powerful men, of a loud and menacing voice, to conduct their Herds of Oxen. Before yoking an Ox for the first time to the plough, it was requisite to tie him strongly to his manger, to put the yoke on his neck, to enfeeble him for four days by hunger and forced watches, and then to coax him with cakes, salt, and wine. At the present day these precautions are wholly superfluous; and in any of our modern farms, a girl of fifteen years of age can induce the strongest Bull to obey her commands, although he may have lived for many years at large in the meadows.

There is a singular fact recorded by the ancients respecting the food of the Ox, which was long considered to be of doubtful authority. Ælian and Athenæus have related, on the authority of Zenothemis, that in a lake of Pæonia, certain Fishes were produced, which the Oxen ate with as much pleasure they would have eaten hay, provided the Fishes were presented living and palpitating. When dead, the Oxen would not touch them. The singularity of this assertion, which would serve to break down the usual distinctions between the digestive functions of Herbivorous and Carnivorous animals, has, however, been removed by modern writers, several of whom relate, that in the cold countries of the North of Europe bordering on the sea the Oxen and Horses are fed on Fish. In respect to Horses, there can be no doubt as to the fact, for the Horses which were brought in 1788 from Iceland to France, by M. de Calonne, had no other food than Fish on the passage, as well as during their stay at the port of Dunkirk. M. du Petit-Thouars, who was garrisoned at the latter place, reported this fact to M. de la Malle, on whose authority it is inserted here. Torfæus (Hist. Norveg.) relates the same fact for the Norwegian Horses.

The more recent experiments of M. Magendie have fully confirmed this omnivorous property of the Domesticated Animals; and it is perhaps one of the most curious consequences of their association with Man. Wild Animals appear, however, to possess this quality to a certain extent. M. Roulin reports that the Martin (taira) of Columbia will eat bananas and green maize, as well as Quadrupeds, Reptiles, Birds, and Insects. M. de la Malle has known a Polecat to devour pears, peaches, apricots, grapes, and other fruits of our garden trees, besides its ordinary animal food. These facts appear fully to verify the observations of Ælian and Zenothemis.

Another consequence of domestication, in modern times, may be remarked in the permanent secretion of milk with the Cows, Ewes, and She-Goats. The wild races only suckle their young during the interval necessary to habituate the digestive organs of their progeny to other food. We have already seen that the domestic species, transported into the New World, have lost this property of their ancestors in acquiring their independence, and only preserve their milk as long as the calves and kids are kept along with their dams. We have a further proof of the imperfect domestication of the Ass, in the circumstance that the secretion of milk in the female Ass does not remain permanent, but continues only during the time that the foal remains with its dam.

An interesting passage of Aristotle appears to show, that one of the most important consequences of domestication, the permanent secretion of milk, which is at present maintained by an irritation of the Mammæ almost mechanical, was first induced by a stimulus procured from some plants of the Nettle family (*Urticeæ*). He adds, in reference to the She-Goats, that even when they have not been fecundated, it was customary to rub their udders with Nettles so violently as to excite pain. At first milk was drawn mixed with blood, then a quantity of purulent matter, and finally a milk as pure, as healthy, and in a quantity as copious, as that rendered by a She-Goat which had just produced.

The progress of domestication may, however, be seen more especially in the Dog, who has in all ages been the companion, the guardian, and we almost say the intimate friend of Man. Being possessed of a superior genius, and habituated to the society of his master, domestication has been truly wonderful in developing his natural capacity. The ancients were acquainted with but few varieties of the Dog, as far as we can gather from the descriptions of authors, and the figures on the monuments of antiquity. They had the Watch Dog, the Coursing Dog, the Shepherd's Dog, and the little Maltese Dog, supposed to have somewhat resembled the French Bichon. The intelligent qualities of these varieties had been but slightly developed; and the ancient Greeks and Romans were wholly unacquainted with those Dogs which set game, such as the Pointers and Spaniels, upon whom a modern education can produce results so surprising. The Water Spaniel or Poodle Dog, whose fame is now widely spread for the constancy of his attachment and the extent of his acquirements, was wholly unknown to them.

Aristotle and Xenophon have expressly declared, that animals can be made to understand *nothing* by word of mouth. Those who have witnessed the intelligence and dexterity of the modern Poodle Dog, will be able readily to appreciate the influence of domestication when continued through a long period of time. These animals can be induced, at the word of command, to ring the bell, or perform many of the ordinary duties of a servant, such as to shut and open the door, or deliver a letter. A black Poodle belonging to Robert Wilkie, Esq. of Ladythorn, in the county of Northumberland, would feign all the agonies of death in a very correct manner. When commanded to die, he rolled over on one side, stretched himself at full length, and moved his hinder legs with a convulsive motion, first slowly and afterwards quickly, as if in extreme pain. After putting his head and body in motion with these affected convulsions, he would then stretch out all his limbs, or lie on his back with the legs turned upwards as if he had expired, and remain motionless until the word of his master restored him again to his customary animation.

These instances, and numerous others, which need not here be produced, clearly establish the important fact, that the education of the domestic animals has always followed a gradual progress, which may be either slow or rapid, according to circumstances. We may thence further expect, that future ages, by bestowing more care as well as skill, and being aided by the influence of a longer period of time, may develop the intelligent powers of our domestic animals in a still higher degree.

It appears, that with one species at least domestication has gone retrograde. The Dziggtai (*Equus Hemionus*) of Mongolia was once domesticated in Syria. Aristotle declares (Nat. Animal. vi. 30), that "in Syria animals are to be found called Hemionus, a species resembling the mule in appearance, but being in reality different from it. These Hemionus are swifter than Mules. They produce among themselves a constant race. Some animals, which still remain in Phrygia, where they were introduced in the time of Pharnaces, the father of Pharnabazus, prove the truth of this assertion. Three animals now remain out of nine."

Although the later writers among the ancients generally confound the Hemionus with the different kinds of Mule, yet Aristotle carefully distinguishes them. There was the Mule (*οὐρεος, mulus*), or the Hybrid between the Ass and Mare; the Bardeau (*ἴννος, hinnus*), or the Hybrid between the Horse and female Ass; and a cross of the second degree (*γίννος, hinnulus*), between the Mule and the Mare. From all of these Aristotle separates "the Hemionus (*ἡμίονος*), which is not at all of the same species as the Mule, notwithstanding its resemblance, since they propagate together, and continue their race." Theophrastus confirms this remark of Aristotle; and more modern writers, such as Constantine Porphyrogenitus, and Eustathius in his Commentary on the Iliad, remark, that the Hemionus was formerly domesticated in that part of Asia Minor called Paphlagonia.

Pallas has recently identified the Hemionus of Aristotle with the modern Dziggtai of Siberia. It is probable that this species may have been brought to Syria by some of the Tartar hordes, and that it remained there in domestication until the era of Aristotle; for after this time all notice of it disappears from the writings of the ancients, and its place is supplied by the Horse and Mule. In certain parts of Central Asia, the Dziggtai is said to be domesticated at the present day.

Thus, upon considering the domestic animals in reference to those phenomena which have attended their return from the domestic to the wild state, and upon investigating the records of antiquity, we are led to form several important conclusions which it may be proper here to recapitulate.

In the first place, we find that the numerous variations of the domestic animals, in respect to the colour and quality of the hair, are brought back by a state of liberty to a uniformity almost invariable. In the New World the common colour of the hair is a chestnut bay for the Horse, a dark gray for the Ass, and black for the Hog. In the Old Continent it seems to be gray for the Ass as in America, but a different colour for the Horse, which here becomes white. We are hence entitled to infer, that all shades which diverge from these primitive hues are the evident consequences of domestication. This discrepancy between the original colours of the Horse in the Old and New Continents is not, however, without an analogous instance. The Ox, on becoming wild in South America, appears, from the observations of M. Roulin, to have reverted to a chestnut brown, while in Britain we know that the wild breed of the Ox, now exterminated, was entirely white, excepting a slight tinge of red on the ear, and a black muzzle. Further, we find that the domestic animals on becoming wild reacquire other properties corresponding to their independent mode of life. The ears of the Hog are diminished, and his skull is enlarged; the speed and agility of the Horse are increased; the courage of the Ass reappears especially among the Stallions; and the petulance of the Goat seems to be augmented with the ease and agility of his movements. We also find that the permanent secretion of milk in the Cow and She-Goat is an acquired property of domestication. In conducting these inquiries, it often becomes difficult to distinguish those changes which are entirely attributable to the loss of properties formerly acquired by domestication, from those new changes induced by climate, food, and other physical conditions under which the animals are placed. It is to some accidental influence of this kind, that we must ascribe the difference in the primitive hues of the Horse and Ox, which in America are chestnut bay and chestnut brown; while in the Old Continent white is the original colour. Yet, after making due allowance for the joint or separate influence of food and climate, and after comparing the several races with each other, and with the circumstances in which they are placed, we are compelled to admit the general principle, that habits of independence occasion the wild races to revert continually towards a primitive form and colour, which can be no other than those from which they have diverged in the course of ages.

In the second place, upon examining the writings and monuments of antiquity, we find that all our domestic animals have existed throughout Europe in the wild state. Most of them have undergone modifications dependent on the antiquity of their domestication. This progress can be traced in the Horse which has undergone perceptible changes during the interval of seventeen centuries from the age of Pliny to the present time. The pace of the pas-relevé has been acquired by our Horses since the time of the Romans, and this quality is now transmitted to posterity. We further perceive that while the ancients were acquainted with only four varieties of the Horse, and but few of the Dog, the variations of these animals at the present day are absolutely innumerable. The influence of domestication in developing the milder and more useful qualities of the Horse, the Ox, and the Dog, as well as in perfecting their intelligence, may be clearly traced. There also exists a tendency to break down the original distinctions between the carnivorous and herbivorous animals, by inducing a kind of omnivorous habit, especially when these animals are reduced to extremities.

The dense fleece of the Sheep and the barking of the Dog have been considered to be the acquired results of Domestication. Hereafter we shall investigate the grounds upon which these opinions seem chiefly to repose.

Every where we are struck with that general tendency of the Mammalia, and indeed of all living beings, to preserve the forms impressed upon them at the moment of their creation. As soon as the industry of the Horticulturist, or the skilful precautions of the Grazier and Veterinarian, are suspended, both Plants and Animals alike feel the influence of this *atavism*, which leads them to revert to the forms of their remotest ancestors. The vegetable resumes its rustic garb, or the bitter and useless secretions of its wild condition, the animal loses some of the most important and valuable of its properties. Both alike revert to a uniform type in their external and internal characters. Animated beings are soon stripped of those rich attributes which they had derived from the cultivation of the soil, or from civilization, the abundance of nutritious food, a careful shelter from the inclemencies of the weather, or their habits of intercourse with the superior genius of Man, but above all, from his care in regulating their unions among themselves. A bountiful Nature is ever ready to substitute qualities, which bear relation only to the wants of the animal, and the part it should perform on the great stage of created existence, for those other properties, which doubtless were imparted only for the purpose of administering to the wants and necessities of Man.

GENERAL REVIEW OF THE MAMMALIA CONTINUED.

Recapitulation—Relations which the dimensions of the Mammalia bear to the peculiarities of their organization, and the stations they are designed to occupy—Occasional Difference of Size between the Sexes.

THAT original types have been impressed upon species at the moment of their creation, seems then to be one of the most general and important laws of Animated Nature. If the preceding observations have any force, the conclusions in which Lamarck and other experienced Naturalists have inferred the perpetual variation of species, and the indefinite extent of their modifications, during the course of ages, become wholly inadmissible. Great as the variations of Animals and Plants may appear upon a superficial consideration, they seem, upon a more cautious investigation, to be in reality confined within certain very narrow and well defined limits. The care of the Horticulturist can modify the secretions of a plant, and the relative magnitude of its parts; he can obtain an extraordinary development of one part, at the expense of another; he can transform the stamens into petals, and occasion a single flower to become double; he can impart a delicious flavor to the fruit; or lead to the development of fleshy and tuberculous roots by suppressing the branches, shortening the stalks, or diminishing the flowers. Availing himself of a corresponding law of Nature, the Grazier can modify the general functions of Nutrition and Generation among Animals. Among the Ruminantia, the cares of Man may lead to the alteration of particular secretions; their milk may be rendered permanent, or their hair fine and silky. By regulating the temperature to which they are exposed, or the quality of their food, hy the annihilation of other organic functions, Man can not only succeed in modifying individuals, but Nature lightens the labours of his posterity, in transmitting their acquired properties to future generations. But on abandoning these artificial products to their own mutual action, the original equilibrium of their functions restores itself. The balance of animal forces either becomes rectified in the individuals, or their posterity undergo a course of regeneration in resuming their original habits, or perhaps in uniting with the wild individuals of their own species.

From this invincible tendency of each species to resume its original form, we are led to regard all the variations of Animals and Plants but as the vibrations of a pendulum, which continues to oscillate around a fixed and determinate axis. The original type is continued by generation, according to constant laws, and the innumerable disturbing causes to which it is exposed, whether internal or external, are insufficient to subvert this harmony of parts. The inherent disposition of each specific type reappears after all our attempts to annihilate it, and it is in Natural History as in Morals,

Naturam expellas furcâ, tamen usque recurret.

Nature further prevents all permanent confusion of species by the instinctive aversion of allied animals. Even when Hybrid or adulterous unions do arise, the Mules are usually sterile, and these Mongrel products appear to stand in relation to surrounding beings as something unnatural and monstrous.

The tendency of a species to produce the same form on the one hand, and the causes of deviation on the other, compose two opposite and counter-balancing forces, by the mutual reaction of which each separate force becomes modified, and from their combined action there thence proceed effects which may be regarded as the resultant of the two forces.

The peculiarities of a race are the more decidedly marked according as it is more ancient. Among the domestic animals, there can be no doubt that a great number of these individuals belong to races whose origin dates back from a very remote period of antiquity. Those races, on the contrary, which are known to be of more recent origin, preserve their peculiarities with greater difficulty, and always tend to revert to the forms of those more ancient types, to the crossing or modification of which they owe their own existence. Instances of this law occur daily under our eyes, and indeed are matter of notoriety to gardeners and breeders of cattle. They are, however, most perceptible in the Dog, where there often appears, from the crossing of races, a new variety, which, however, is found to possess a short and fleeting existence, the common lot of all types of modern origin. These considerations would lead us to assign a very high antiquity to the period of the first appearance of the most permanent races.

The attempts of some recent German and French philosophers to explain the immense diversity of animals and plants upon Physiological principles, and without an appeal to an original and specific creation, appear to us to be wholly unsatisfactory. As well might they attempt to resolve by Mechanical principles how it happens that one time-piece shows the day of the month, another only the hour of the day, while a third will point the minutes and seconds,—differences which can only be explained by the intention and design of the Horologian.

Instead of speculating in these inaccessible regions, the Natural Historian endeavours to trace the relations of created beings with each other, and with the general laws of inanimate Nature, rather than to indulge in conjecture upon the physical causes of their diversity.

All created beings must necessarily be formed in direct correspondence with each other, and the places they are destined to inhabit. In the same manner, as particular organs are adapted to particular purposes, so must the general dimensions of the body correspond accurately with surrounding circumstances. We see the eye exhibit different relations in respect to light, according as it is intended to see to a small or remote distance, or through the medium of air or of water. The ear again is organized relatively to the vibrations of the air, to melodious sounds, as in many Birds; the nose to odoriferous effluvia, to animal odours in the Carnassier, to vegetable in the Herbivorous tribes; the organs of mastication and suction to the nature of the food; the arms and means of defence for the preservation of each species, and the destruction of its prey.

It is evident, therefore, that if we can trace design and correspondence between particular organs and functions, a certain general equilibrium of functions and organs must also exist, and each species or original type must possess that general form, dimensions, and duration, which will enable it to continue its existence for a limited time, and perform its part among created beings.

On comparing all the Mammalia among themselves, we are at once struck with the remarkable differences in their dimensions, which present a greater amount of variation in these respects than perhaps in any other class of animals. The Minute Shrew *(Sorex exilis)* is the smallest of known Mammalia, and measures only one eight hundredth part of the length of the Basque Whale *(Balæna boops)*, while in bulk it is only about one part in half a million. In other words, assuming the dimensions of the smallest species of the Mammalia = 1, then the length of the largest animal of this class is = 800, and its bulk is = 500,000.

This great disparity of size chiefly arises, it must be observed, between Mammalia differing considerably from each other in their external organization. If we compare together animals of the same order only, the discrepancy in their dimensions is brought within much narrower limits. The sizes of these animals approximate still more closely if we descend to tribes, families, or genera; and so invariably does Nature preserve this relation between the bulk of an animal and its external characters, that if we find two congenerous species, which present remarkable differences in size, we may be almost certain that there will also be found important differences in some of those organs which commonly serve to supply the generic characters. Among the Quadrumana, for example, the Apes form a most natural family, and one in which the general height remains tolerably constant. If we except the Orangs and the Cynocephala, which are the largest, and the Ouistitis, whose size is much less than the others, and which almost stand out as an isolated group from the remaining genera in respect to their external characters, we shall find that the remainder differ but slightly in their dimensions. Among all the Cynocephala, the length of their head and body remains uniformly constant, being a little more than two feet. Again, the Ouistitis compose a genus very numerous in species; and yet, when considered by themselves, they present a result much more remarkable. On comparing together all the known species of this genus, and then taking the mean of their dimensions, M. Isidore Geoffroy found that the largest species only exceeded this mean by an inch and a quarter, while the smallest remained below the average by an equal quantity. If, however, the Ouistitis be subdivided into those three sections usually received by Naturalists, then the dimensions of the species belonging to each section will differ from the medium magnitude proper to that section only by six lines one way or another.

Among the Bats the genus Vespertilio, as formerly established, would seem at first sight to present a striking exception to this rule, in respect to the Roussettes, in which the distance between their expanded wings is four feet, while the smaller species of this country scarcely measure as many inches. But upon investigating the characters upon which the genus had been instituted, M. Geoffroy observed several important points of difference between the organization of the larger and the smaller Bats, which have ultimately led to their being separated into distinct subdivisions and genera. Many other apparent exceptions have also vanished in a similar manner on being examined more minutely, and the consequence has been, that several new Genera have been adopted by the general consent of Naturalists. Indeed, it may be stated generally, that wherever there have existed striking differences of size between nearly allied species, Naturalists have always felt the necessity of establishing sub-genera, or groups in which these remarkable anomalies are made to disappear.

Since it thus appears that the sizes of the Mammalia bear a determinate relation to those external characters which usually serve to determine the Genera or higher divisions, it will follow that their magnitude must bear a corresponding relation to the conditions of their existence; such as the element in which they move, their mode of life, their food, their climate, and their situation. In all these respects we may expect to find that their sizes will be so apportioned by Nature, as to bring them into harmonious correspondence with the circumstances of their condition.

In reference to the element in which they move, we find that all those Mammalia which dwell in the bosom of the ocean acquire the largest dimensions. The different species of Whale (*Balæna*), of Cachalot (*Physeter*), and of Dolphin (*Delphinus*), attain a bulk to which few other Mammalia can compare. Even among those groups of Mammalia where some genera commonly frequent the water, and others live habitually on the land, we find that the former attain to a magnitude much greater than that of the latter. Thus, among the Carnassiers, no species reaches so great a size as the Sea-Horse or Morse (*Trichechus rosmarus*). Again, among the numerous animals composing the Genus Mustela of Linnæus, the Weasels, Martins, and other terrestrial species, are much smaller than the Otters. Even among the Otters (*Lutra*), it is precisely those species which are the most essentially aquatic, such as the Brazilian Otter (*L. Braziliensis*), and the Sea-Otter (*L. lutris*), which attain the greatest dimensions. The same observation is also applicable to the Rodentia and Insectivora. The Beaver is larger than the Mouse, and the Water Shrew (*Sorex fodiens*) of greater magnitude than the common Shrew (*S. araneus*), and for a similar reason. This adaptation of Nature seems obviously intended to accommodate their bodies to the density of the medium in which they more habitually reside. A greater bulk, by displacing a larger quantity of water, renders them more buoyant, and leaves the muscular force of their limbs more unfettered to execute the movements proper to each animal. Again, the force of gravity at the earth's surface being counteracted in a more sensible degree by the reaction of the denser fluid, enables some of the aquatic species to attain a bulk which would be impossible in a land animal.

On the other hand, those Mammalia which live more exclusively in the air, such as the Bats, or upon trees, like the greater part of the Monkeys, never attain any very considerable dimensions. The agility which their situation requires would have been inconsistent with a heavy form, and the dimensions of the trees in which most Quadrumana fix their abodes, necessarily confine their magnitude within very narrow limits. Among the remaining Mammalia which commonly live on the surface of the land, and may thence be more particularly denominated terrestrial Mammalia, we find that their average bulk maintains a size intermediate to these lighter forms of the aerial Mammalia on the one hand, and those belonging to the more ponderous inhabitants of the ocean on the other. The relative magnitude depends, however, with the terrestrial Mammalia, upon certain other conditions; for, while we find some animals among them which only yield in magnitude to the aquatic tribes, we at the same time discover in this division the very smallest animals of the entire class without exception. This rule, therefore, does not hold so accurately in respect to those animals which dwell habitually on the ground.

If, however, we investigate the terrestrial and aerial Mammalia in reference to the nature of their food, we find certain constant relations established by the Creator between the quantity of food necessary for their maintenance and that which is supplied to them; in other words, between the demand of their stomachs and the supply of food sufficient to sustain them. The largest of all terrestrial Mammalia are the Herbivorous animals, such as the Elephant, Hippopotamus, and Rhinoceros, because the grosser kinds of vegetable food are supplied in immense quantities throughout many parts of the globe, but especially in the tropical climates. The warmer regions of the globe yield, in a dense and luxurious vegetation, an ample supply of nourishment to these ponderous frames; and supply those succulent plants, soft stems, and leaves, the want of which renders the regions around the poles wholly incapable of supporting the larger terrestrial Mammalia. In the group of Herbivorous Mammalia, while we have the colossal magnitude of the Elephant on the one hand, we have on the other the Java Musk (*Moschus meminna*), scarcely the size of a Rabbit.

The Carnivorous Mammalia compose a group which forms, after the preceding, a series of an inferior order. Among them we have the Lion and the Tiger for the *maximum*, and find a *minimum* limit probably in the Ermine (*Mustela erminea*). Although these animals find an abundant supply of nourishment among the inferior tribes of every denomination, it is neither so ample nor so constant as to permit them to attain any very great dimensions. An unwieldy bulk would ill correspond with that activity which their predatory habits seem necessarily to require.

The Frugivorous tribes of Mammalia form a third group confined within much narrower dimensions. On the one hand we have the Orangs, and on the other the smallest species of Roussette Bats (*Pteropus*). These species are accordingly confined to those more favored regions of the earth, where fruits are to be found throughout all seasons of the year.

Lastly, we see among those Mammalia which feed exclusively on Insects a further instance of the relation which the nature of the food bears to the average bulk of species. While we have the Ant-eaters (*Myrmecophaga* and *Orycteropus*), whose length does not exceed four feet, we have, in the minute Shrews, some of the smallest of known Quadrupeds.

Thus, upon considering all the Mammalia, in a general point of view, in reference to their comparative dimensions, we are led to perceive that there always exists a relation between the bulk of the animals composing an entire group and the conditions of their diet. The more capacious animals feed upon these substances which are found most abundantly on the face of the globe. Those of smaller size usually attain to dimensions proportional either to the magnitude of the animals upon which they are destined to feed, or to the nature of those vegetable substances to which their digestive organs are adapted. Every where we perceive a most exact correspondence between the quantity of nourishment which their constitution requires, and that which is bestowed by the hand of Nature. This method of apportioning to each animal, by an equitable division, its share of the produce of the earth, is surely one of the most admirable and beneficent arrangements of the Creator.

Although the subject of the Geographical distribution of the Mammalia will hereafter receive our most attentive consideration, it may be proper to remark at present, that the dimensions of animals always bear a certain relation to the magnitude of the regions in which they reside. It has long been remarked, that islands which are either very small or much isolated contain very few Quadrupeds, and those only of small dimensions, while some are even wholly destitute of Mammalia. Indeed, the largest animals of this class are found only upon the continents, in the largest islands, or upon those smaller islands which are so near to the larger continents as to be intimately affected by their proximity. Even in respect to the continents themselves, the Mammalia belonging to the Old Continent, which is the larger, surpass in dimensions those of the American, which is the smaller. The Mammalia of New Holland come next in magnitude, then those of Madagascar, Britain, and the lesser islands. A similar law may be traced among the aquatic Mammalia, for those which inhabit the Ocean greatly surpass in bulk the species that frequent the Rivers. The largest of the latter do not exhibit Mammalia which can be compared in any degree with the Morse and the Basque Whale.

Thus, whether we investigate the Land or the Sea, every where we perceive that the dimensions of the Mammalia are proportioned to the magnitude of the regions which they are destined to inhabit. If the Southern hemisphere be compared, as a whole, with the Northern, omitting Africa and those islands which are traversed by the Equator, as the animals must be nearly the same on both sides of this line, we shall find that the Southern hemisphere will contain Mammalia whose size is generally less than that of the corresponding animals in the North. This is, however, only a particular case of that more general law already explained, for the Southern hemisphere contains at most only small continents or large islands.

Differences of latitude and climate lead to many important corresponding differences in the sizes of animals, but these do not admit of being expressed in a general law, but must be noticed in detail. It most commonly happens that genera and species arrive at their *maximum* of size in the hottest regions of the globe, and descend to their *minimum* in the coldest. There are, however, some Mammalia, such as the Bears, which have their *maximum* in the Polar Regions, while their congeners of the tropics are greatly inferior in bulk and strength. But there is no instance in any one genus where the largest species are found alone in temperate or moderately warm climates; and the same observation is equally applicable to the several individuals of the same species.

It commonly happens, when individuals belonging to the same species inhabit both the mountains and the valleys, that the inhabitants of the mountains will be the smaller of the two groups. This probably arises from exposure to cold, and a scanty supply of food. Here again we find that correspondence between the dimensions of the animals and those active habits which a mountain residence demands. A heavy and unwieldy form would have been unsuited to the difficulties of those almost inaccessible heights, where the mountain races are often compelled to gather a scanty and precarious subsistence.

Among the domestic animals, *individual* variations of height happen very rarely, and are commonly confined within narrow limits, while, on the contrary, the variations in the dimensions of *races* are sometimes both very numerous and remarkable. With some domestic animals the primitive height of the wild species is preserved, or it has been very slightly modified. In these instances all the races have the same height, or differ very slightly; and whenever they are found to vary from the height of the wild races, it is always to become a little smaller.

There are some species of domestic animals, such as the Dog and the Horse, which present some races of very large dimensions, and others, on the contrary, are very small. When, however, the medium height of all the races is ascertained, it is found to differ but slightly from the height of the original type, as deduced from measurement or from reasoning. Thus the ordinary height of those species which vary but slightly, as well as the medium height of all the races of those species which vary much, approach very nearly to the dimensions of the height belonging to the primitive type. In other words, species have varied but slightly in their average dimensions from the time when they were first domesticated.

There are certain lesser variations in dimensions, which depend upon the greater or less care which the individuals receive on the part of Man. Those species which have experienced a slight diminution, belong to such as have been generally neglected or badly nourished.

The predominance of the Male over the Female in dimensions is much more general, and more strongly marked among the Mammalia than among the Birds, or in any other class. There is a remarkable disproportion between the size of the Bull and that of the Cow; of the Ram and the Ewe; and of the He and She Goat. This superiority of volume is not, however, essential to the male sex; and far from being general, it is confined to a very small proportion of the Animal Kingdom. If this superiority of size were characteristic of the sex, the loss of the reproductive organs during early youth ought to prevent it. The effect is, however, precisely the contrary, for castration, which brings the constitution of the male near to that of the female, is highly favorable to the growth of the former; these organs were therefore rather an obstacle to their development.

This disparity seems to arise from the fact, that the nutritive powers of the female are, in the Mammalia, expended upon their offspring. The Cow, whose weight is, perhaps, not one half that of the Bull, lives in a continual state of gestation or lactation; and it is the same with the Ewe and She-Goat. Every Grazier is aware, that the young females cease to grow as soon as they begin to produce; and that yielding milk is still more prejudicial to their growth than gestation.

If we compare the Ruminantia and Herbivorous Cetacea with the Pachydermata, we shall find, that the young of the former consume much more milk than the young of the latter; and, accordingly, it is in the two classes first mentioned that the predominance of the male over the female in size is most decidedly marked. Indeed, the Pig is the only Pachydermata which is very prolific, and it is precisely in this Genus that the male more sensibly exceeds the female in dimensions. Of all the Rodentia, the Rats are the most fertile; and the predominance of the male Rat over his female is more apparent than in any other of the Rodentia.

Those Mammalia which live upon Insects and Fruits, such as the Cheiroptera and Insectivora, do not exhibit the same difference of magnitude between the sexes as the proper Carnivora. The females of the former, from their situation, find an easy and abundant prey either in the larvæ of Insects which hatch around them, or in the Fruits which fall and ripen near their retreats. The female of the latter is obliged, on the other hand, to pursue an alert and nimble prey, which often eludes her pursuit. Her young ones are not deficient in number, and she consequently loses a large quantity of nutritive power. From these causes, the male, who is always at large, and lives for himself alone, is wholly exempt; and hence the female of the Bat, the Hedgehog, and the Mole, is at least as large as the male, while the Lioness is smaller than the Lion.

Among the Marsupialia, where the females produce an embryo, or rudimentary fœtus, which always travels about with its mother, and cannot keep her confined to a spot remote from her food, we find that the female is at least as large as the male.

With the Edentata and Tardigrada, the female is usually larger than the male. It is especially remarkable in the Ant-eaters, where the female, by the aid of her long tongue, an organ usually more developed in the female than in the male, enables her to catch the Ants, her prey, with a superior nimbleness and agility.

Among the domestic animals, whenever it happens that the female is made to work like the male, and that she is not compelled to submit to a continuous and depressing lactation, she does not yield to him in size. The She-Ass is as large as the male; the Mare as the Horse; and the Dog is not larger than his female. In these cases, Man provides equally for their wants and necessities.

Some Naturalists have considered the Polecat (*Mustela foina*) to be a domesticated variety of the Martin (*Mustela martes*). In the former, the sexes are of an equal size, while the male is greater than the female in the latter. With the common Hare (*Lepus timidus*) the male is not so bulky as the female; on the contrary, with the Rabbit (*L. cuniculus*) the male is the larger of the two. This evidently may be traced to the superior fecundity of the latter species.

We may easily see how the dimensions of animals should depend so much on the quantity and quality of their food, since all substances do not contribute an equal quantity of nutriment. Vegetable substances, which are mucilaginous and herbaceous, contribute much more powerfully towards the development of animals than those which are fibrous and of an animal nature. These are more favorable than acid substances; and the latter again surpass those which are saccharine. Thus among all the Mammalia it is the Herbivorous and proper Cetacea, the Pachydermata, and the Ruminantia, which attain greater dimensions than the Carnassiers, and these again than the Quadrumana and Edentata. The same thing may also be traced among the Birds, for the Waders (*Grallæ*) and the Web-footed Birds (*Palmipedes*) become larger than the Birds of Prey (*Accipitres*), the latter are in their turn larger than the Thrushes (*Turdus*), and these again than the Humming Birds (*Trochilus*).

This advantage in respect to dimensions, to which a plentiful supply of food contributes, is unfavorable to reproduction, and hence acts ultimately against the species; for the difficulties of procuring a sufficient supply of food are always greater in the larger than in the smaller species. Large species are hence comparatively rare upon the earth, except where human industry has ministered to the insufficiency of their own resources. The smaller races of Goats and Sheep might maintain themselves without assistance in our temperate climates, but it would not be possible to preserve the larger races of these animals; and with still greater reason of the Horse, the Cow, or the Ass.

The superiority of the male over the female ought then to be more apparent in the larger than in the smaller species. With our Oxen and Sheep the difference is greater among the larger than among the smaller races. It is greater in the Rat than in the Mouse. This inequality between the sexes would have been still greater in the largest species, if the deficiency of nutrition sustained by the female did not become progressively less according as there exists a progressive diminution of fecundity. Among those domestic animals, whose females supply us continually with milk, the inequality becomes enormous in the largest species, because frequent milking is still more unfavorable to development than a very great fecundity. Good Cows fatten during gestation, and become lean when milking commences, whatever may be the quality or quantity of their food.

The primitive cause of this inequality of size between the sexes seems to show a tendency to return to an equilibrium; and we may thence infer that there formerly existed a greater disparity between the males and females of the Pachydermata than at present, when we find this disparity still existing among the Amphibia, which are more productive than the Elephant, the Rhinoceros, and the Hippopotamus.

When the capacity of reproduction is extinguished, species arrive at their end. We may infer that it is chiefly to the feeble powers of reproduction among the Pachydermata that we find so many fossil species belonging to this order which have no living analogues. Species, like individuals, decline and die, when they have attained the limit of their dimensions.

From what has been said, it may easily be inferred that in those orders of animals where the male is usually monogamous, and shares with the female the care of her progeny, he is not in general susceptible of that superior development beyond the female, as where he is polygamous.

We have now seen that when two or more species of Mammalia resemble each other perfectly in their generic characters, their height is the same, or but slightly different. Those families, genera, or species, which inhabit the bosom of the ocean, or pass a part of their lives in the water, arrive at a large size comparatively to the other families, genera, and species of the same group; and the increment of their dimensions is the greater, all other things being the same, in proportion as their organization renders them more essentially aquatic. The genera with wings, or which live in trees, on the contrary, never attain to any but very small dimensions. Those Mammalia which are purely terrestrial, may be arranged in series according to their dimensions, very large in the first, less in the second, and so on, that is, into herbivorous, carnivorous, frugivorous, and insectivorous. In other words, there always exists an exact co-relation between the volume of the animals and the volume or quantity of organized beings which they are destined, by the formation of their digestive organs, to consume.

It has also been shown that there exists a constant relation between the height of the Mammalia and the extent of the places where they live; the largest species inhabit the oceans, continents, or large islands; the smaller reside in rivers or small islands. Even the Mammalia of a more extensive continent surpass in dimensions their analogues of a less extensive continent, and the Mammalia of the Northern Hemisphere are larger than the corresponding animals of the Southern. In general, also, though not always, the height of Mammalia resident in the mountains is inferior to that of the analogous animals residing in the plains.

The preceding observations are true without exception in reference to the Mammalia, but when applied to lower classes of animated Nature, they gradually lose their general correctness, and are finally lost when we arrive at the lowest classes of all, in an infinity of exceptions. We, however, always find that when other circumstances remain the same, the variations of size observable in any one class are always confined within narrower limits in proportion as that class is more natural.

We have also seen that the size of the body depends upon the quantity of nutritive particles which it is capable of retaining. The female would always attain a larger size than the male, as she is endowed with a greater power of absorbing nutriment, did she not experience the influence of certain counteracting causes which do not act upon the male. She has to submit to a severe lactation, in many cases to a frequent parturition, and is often compelled to undergo privations of food in her cares for her offspring. As these causes do not affect the male, there hence arise inequalities in their dimensions, or relations of volume, and these differences of size between the sexes, when they exist, depend upon the intensity of these causes. It must, however, be admitted, that sometimes the relative sizes of the sexes seem to be *inexplicable* by any of the causes just enumerated, and in these cases we must infer that differences in the bulk of the sexes have been originally impressed upon the species at the moment of their creation.

GENERAL REVIEW OF THE MAMMALIA CONTINUED.

Phenomena of Nutrition among the Mammalia—Manner of obtaining their Food.

THAT Animals can only be nourished by substances which have once lived, in other words, by other Animals or by Plants, we have already had occasion to explain. Some persons have hastily concluded, from imperfect observations, that animals can nourish their bodies with inert mineral substances. This opinion is, however, erroneous. That yellow earth which the famished Wolves have been seen to swallow in their rage, serves but to deceive the intensity of their hunger. The Mollusca do not devour the fragments of rocks or old wood which they destroy or perforate, nor do Birds digest those hard mineral substances which are sometimes found broken or pulverized in their gizzards.

The food of the Mammalia is very various, since it takes in all other animals and vegetables. There always exists a certain correspondence between the degree in which the organs of an animal are complicated, and the nature of the food which it consumes; for it has been generally remarked, that the simplest beings always require the simplest food. The Tiger feeds on a living prey, the Wolf upon carcasses, the Otter upon fish, the Hog upon roots and animal food, the Myrmecophaga upon ants, the Apes upon fruits, the Rodentia and Ruminantia upon simple herbs, Man and the Bear upon almost any thing.

Thus we may remark among the Mammalia the greatest diversity in their tastes for food. While in some the nutriment wholly belongs to the Animal Kingdom, in others it is as entirely confined to the Vegetable. Others again, whose nourishment is mixed, seem to avail themselves at once, or indifferently, of the produce of either Kingdom. These varieties are usually expressed by the terms *sarcophagous* or *carnivorous*, *phytophagous* or *herbivorous*, *polyphagous* or *omnivorous*.

These first differences of animals in respect to their food, presuppose or draw along with them certain other differences in respect to their organs. A carnivorous animal has more teeth than an herbivorous one; these teeth are more unequal, more fitted for tearing, and more trenchant like a saw. His jaws are more free, more powerful, and moved by larger and more vigorous muscles; his stomach is not so large, and its sides are thinner; his intestines are shorter, and proceed within a slimmer form through a less capacious abdomen. His limbs are also differently disposed than in the herbivorous animals, as the voracious instincts require instruments at once fitted for agility and destruction.

Many carnivorous animals will only devour a living prey, which leads them to continual battles, aggressions, and carnage. Others, again, avoid the dangers of these murderous combats, by feasting on carcasses recently dead, or even already putrified. The Lion kills all that he eats, but the Hyæna will extract his meal from the charnel-house or tomb. There are also some Carnivorous Mammalia which confine themselves to sucking the blood of other animals; some species of Weasels and Bats are examples of this, especially the Polecat (*Mustela putorius*) and the Vampyre Bat (*Vampirus spectrum*). The Opossums (*Didelphis*) are nourished almost wholly upon the eggs of other animals. The Ant-eaters (*Myrmecophaga*), being destitute of teeth, feed upon Ants and other Insects which adhere to their glutinous tongue.

The Herbivorous animals, on the contrary, as has already been observed, have their jaws less powerful, moved by more feeble muscles, armed with teeth fitted for grinding rather than biting or tearing. Their limbs are less disposed for aggression, but in return, their stomach is more capacious, its sides are more muscular and thick, and sometimes it is multiplied and complex, their intestines are larger and longer, while their forms are more massive. Those animals which Ruminate (*Ruminantia*), that is, whose food returns a second time to the mouth through the œsophagus, to be again chewed, after having already remained in the stomach, generally have horns on their foreheads, and want the incisive teeth in the upper jaw. They all have four stomachs, or rather one stomach subdivided into four cavities. These divisions are disposed in the following manner. The first is the Paunch (*ventriculus*); this is the largest, and occupies almost the entire left side of the abdomen. The second stomach or Honey-comb (*reticulum*), being the smallest of the four cavities, is placed on the right and before the Paunch. Still further to the right, and almost behind the liver, is the third stomach or *feuillet* (*omasum*), which communicates by a small opening with the fourth stomach or *caillette* (*abomasum*). The last is analogous to that single one found in most other Mammalia, and it communicates with the duodenum or intestine, by a kind of pyloric opening. The separation between the first and second stomach is not very strongly marked; but the others are divided from each other by well defined contractions which prevent any confusion among them. The œsophagus is inserted on the right side of the Paunch, and a kind of prolonged gutter causes it to communicate with the second and third stomachs.

When the food has just been masticated and is swallowed for the first time, it is introduced into the first stomach or paunch; afterwards into the second; and it is only after the food has been submitted to the action of these organs, after it has been impregnated with the juices which are secreted there, and has been softened, that it reascends through the œsophagus into the mouth, in order to undergo a new trituration more perfect than the former. When swallowed this second time, the food is placed in the third stomach, without having gone near the Paunch and Honey-comb. The very young Ruminantia, which are still fed alone upon the milk of their mother, have not yet obtained this ruminating power; but the fluid which they suck passes at once into the last stomach, just as happens among the adult animals after they have ruminated.

Several Cetacea have stomachs nearly as complicated as the Ruminantia. The Dolphin (*Delphinus delphis*) and the Porpoise (*D. phocæna*), for example, have, for a stomach, four cavities placed in a row, one after the other. There exists, also, between the first three cavities, a kind of short canal, forming a narrow passage, by means of which the communication is established from one to the other. Yet none of these animals have been observed to ruminate.

Many Rodentia have their stomach divided into several cavities by contractions; some appear to have two stomachs, but this last arrangement is found more particularly among the Marsupialia, and especially in the Kangaroo-Rat (*Hypsiprymnus White*). Those herbivorous animals which do not ruminate, commonly have the œsophagus inserted towards the centre of the stomach. The latter organ is disposed in a manner to prolong the stay of the food on the same side as the spleen, being the left, and the orifice of the pylorus is very narrow. We may remark here, that the Rodentia usually have two incisive teeth in each jaw, isolated from all the other teeth, and their hinder limbs being almost always longer than the fore, dispose them naturally to leap.

The digestive organs both of Man and the Quadrumana hold a medium station between the Herbivorous and Carnivorous animals. Man has all the kinds of teeth, the trenchant or tearing teeth, like the Carnivora, and the molar or grinding teeth, like the Herbivorous animals, without any very sensible inequalities. His lower jaw moves in all directions; horizontally, like those animals which live upon herbs, and perpendicularly, like the Carnassiers. His stomach is single, but tolerably large, and the sides are of a medium thickness. The rest of his organs hold a mean between those two divisions of Mammalia just mentioned.

The Bears (*Ursus*) and the Badgers (*Meles*), which appear specially organized for being carnivorous, will, however, eat, almost indifferently, all kinds of food, drawn indiscriminately from the two kingdoms of organized beings. But generally speaking, it must be considered a rare occurrence to find a carnivorous animal feeding on vegetables, or an herbivorous animal eating animal substances. This is only remarked when they are urged by an extreme famine, or where they have long been domesticated, and have thence acquired the omnivorous propensity of Man himself. Thus the famished Dog eats bread, and sometimes even vegetables. Cats, when deprived of all nourishment, have been known to devour, in the extremity of their hunger, even the flaxy fibres of a rope. The Rats, also, although organized in every respect for a vegetable diet, will sometimes eat animal substances. It has been remarked, that when flesh is placed in the stomach of a Horse, it remains there without undergoing any alteration, but he has been known to eat fish; and Goats have been seen to devour animal substances, which they manage somehow to digest.

In respect to drinking, the Carnivorous animals, whose digestion is more rapid, generally experience a less urgent necessity for water than the Herbivorous tribes. It has been proved by Marcorelle, that when all the other conditions continue the same, they can remain without water most easily when fed upon fat and oily aliments. The Camel and Dromedary are exceptions; for, though herbivorous, they can remain without drinking for many days longer than any other animal.

The manner in which the Mammalia drink varies much. Man swallows liquids in the same manner as solids, but he drinks also by suction. The carnivorous animals lap up liquids, and they could be made to die of thirst by keeping their trachæa open externally, which would take away from them the power of sucking up the liquid. The Bear bites the water like a fruit or any other solid aliment, and neither laps nor sucks. Most of the Herbivorous animals drink by suction, and to make them perish with thirst it is only sufficient to paralyze their tongue. It has been said that Man is the only animal who drinks without being thirsty.

There are some animals which are seldom observed to drink; these are chiefly carnivorous. We have already noticed that the Camel and Dromedary can remain several days without drinking, and this abstinence from liquids does not appear to make them suffer. There are, however, in the stomachs of these animals certain separate cavities, which seem to be intended to keep the fluids in reserve. In other respects, animals appear to fatten in proportion as they drink less, up to a certain limit, after which a too great abstinence from water makes them fall off in bulk. The Horse and the Ass, according to Aristotle, form an exception to this rule, on account of the enormous quantity of fleshy and often dry herbs with which these animals fill their stomachs almost without intermission. It is customary to suppress the drink of Pigs and other domestic animals gradually when it is wished to fatten them.

Marcorelle, a member of the ancient *Académie des Sciences*, made several experiments to ascertain the effect of drinks upon the bulk of the body. He passed two entire months without drinking water, wine, or any other fluid; and he lost, during that interval, five pounds and a half of his entire weight; he at first weighed 120 pounds. After this he resumed his usual diet, eating the same things as before, but adding to it wine, either pure or diluted with water. During six days of this altered regimen he recovered six pounds of his substance, that is, one eleventh part more than what he had lost. He observed also, that vegetables, of all food, were the most liable to excite his wish for drink. We are not, however, to infer from this, that drinking much is favorable to *enbonpoint*; for the contrary is nearer the truth. Too much drink fatigues the stomach, and weakens the digestion of the food. Tea, and other hot drinks, hasten the digestion; but it is rather to hurry it on, than to accelerate it beneficially, for such drinks occasion the aliment to pass through the pyloric duct before being sufficiently chymified. In respect to alcohol and other exciting fluids, spicy, salt, or acid, these fluids favor the production of chyle, and, in one sense, they occasion a more abundant flowing of the intestinal juices, gastric, pancreatic, and biliary. But alcoholic drinks ultimately impede nutrition by the excitement which they cause in all the organs, in unnaturally quickening the pulsations of the heart They are also injurious from interrupting the sleep, which they render either short or troubled.

Whatever may be the nature of the food, or in whatever animal it may be deposited, the alimentary mass accumulated in its stomach usually remains there one or more hours before any considerable alteration can be perceived. After that, the aliments begin to soften, to change their colour, and often their smell. With the exception of those grains which are entirely covered by an insoluble epidermis, the change begins on the surface of the alimentary substances; and it must be remarked, that the properties of the chyme differ exceedingly according to the kind of food from which it results. Those herbs which have been triturated and twice masticated by the Ruminants give a different stomachic product than animal food, or the grains of the Cerealia. It may also be observed, that the digestion of animal substances is more rapid than that of a vegetable diet. Accordingly, the Carnivorous animals digest their food more rapidly than the Herbivorous; and Man, with the Plantigrada, which are omnivorous, digest meat more rapidly than leguminous plants or fruits, It often happens with these animals, that vegetable food will traverse the entire intestinal canal, without having lost its natural and distinctive qualities, or being in any way altered, which happens very rarely with animal substances.

The aliment, when softened and digested by the stomach, being now *chyme*, forms a mass nearly homogeneous, and of a different colour in different animals, but usually grayish in our species, and in several others, of a sharp odour and taste, and, from its acidity, turning all vegetable blues into red.

In the greater number of Mammalia, it is the stomach alone which performs this first part of digestion—the formation of chyme, and the intestines do not proceed to transform the chyme into chyle, until the moment when the stomach has completed this preliminary process. But there are some animals among whom the relative functions of the stomach and intestines are not confined to these definite functions. The Horse, for example, which has a very narrow stomach, and yet will eat without intermission for several hours, cannot retain the food in his confined stomach for a sufficient length of time to be completely chymified. Accordingly, the pyloric duct of the Horse remains continually open, and the food passes onwards into the intestine without interruption, although not completely chymified; and the process is completed by the intestines, without this confusion of functions being in any way prejudicial to digestion.

The time which the food takes to chymify varies much from one animal to another. The Mammalia require only a few hours to perform this function, and the same thing happens with the Birds of Prey; but the Serpents and the greater part of the Reptiles require entire days and weeks to digest a single meal. The period of time which the same animal takes for digestion is quicker or slower, according as the state of its health is more or less perfect, according as the food is more or less abundant, or is in a greater or less state of minute division; and also, according to the peculiar nature of the substance to be digested. The aliment is finally submitted to the action of the intestinal juices, and the surrounding organs continually maintain a degree of heat, which is always very near to 104° Fahrenheit.

The minute division of the food by mastication is a process indispensable to a rapid digestion. While the teeth and jaws are performing their function, the salivary glands of the mouth continue to secrete a fluid in very considerable quantity. In the Ruminantia, and generally in all animals living on a vegetable diet, these salivary glands are very large, while in the carnivorous tribes they are very small.

The entire length of the alimentary canal differs materially among the Mammalia. Its length, with the Ruminantia, is twenty-seven times that of the body, while it is only from three to five times their length among the Carnivora. In omnivorous animals, such as Man, its length is intermediate to these, being six or seven times that of the body.

Several experiments have been made by Sir Astley Cooper to ascertain the digestive power of the Dog. He found that this animal digested pork more easily than mutton, the latter more rapidly than veal, and beef with greater difficulty than any of the others. He found that fish and cheese were easily digested by the Dog, and boiled veal more readily than roast. The fat of meat seemed more digestible than cheese, codfish dissolved more readily than beef, and beef than potato. The order of digestion for the different parts of the same kind of animal food was fat, muscle, skin, cartilage, tendon, and bone, the last being the least digestible.

Young Dogs, when they have acquired strength, and are in good health, can digest bones; and what is remarkable, Spallanzani has observed that the gastric juice of their stomachs made an impression even upon the enamel of teeth. Boerhaave asserts the contrary, but the observations of Spallanzani have been confirmed. This power of dissolving bone is not peculiar to the Mammalia, but is also possessed by some animals of the other classes. Thus the Falcons, Eagles, and Crows, usually refuse bones; but when introduced into their stomachs, with proper precautions, these refractory substances are digested. Serpents and Adders also digest them, as has been remarked by Spallanzani. Only the smaller bones, however, possessing the least solidity, are dissolved entirely and rapidly; the harder bones require to be minutely divided in order to be softened and dissolved, otherwise they merely undergo a small loss of substance. It must be observed, that before digesting they pass into a cartilaginous state, and resemble indurated gelatine, as if they had been submitted to the action of nitric acid.

The Ruminantia, like the granivorous Birds, can digest herbs and grains, only when these substances have been previously divided, mashed, or ground. When entire herbs and solid grains are introduced into their stomachs, whether uncovered, or inclosed in linen, or perforated tubes, these substances undergo no digestion; they are merely moistened or softened, and this is the extent of their modification. The same result is obtained even when they are moistened with saliva. On the contrary, if bags or tubes of mashed herbs or grains be introduced into the stomachs of the Ruminants, the digestion is then performed in a few hours. These experiments have been made upon Oxen, Sheep, and many other Ruminantia, and they present similar results for all this order of Mammalia. They have also been made upon the Horse, and with the same result, although he does not ruminate.

Many physiologists have attempted, with various success, to effect artificial digestions out of the body, by extracting the gastric juices from the stomachs of different animals, and afterwards mixing them with the food. Spallanzani, by these experiments with the gastric juices of different animals, obtained several important results. When cold, the gastric juice produced scarcely any effect; it merely opposed putrefaction, but did not exercise its dissolving and digestive power until it was raised to its proper temperature. It did not act upon grains and herbs until they were ground, mashed, and impregnated with saliva. The gastric juice of Man softened and seemed to digest beef in about thirty-six hours, when raised to a temperature equal to that of the stomach. He also observed, that the gastric juice of one species often acted upon a great variety of substances, and yet it did not always act upon substances which could be dissolved by the gastric juice of another species.

From these instances of artificial digestion, we may readily expect that the stomach, being lubricated by the gastric juice, will continue to digest after the death of the animal, and even it has been said to digest itself. Hunter first noticed the fact, that the gastric juice will act upon the sides of the stomach after death, and to this cause he attributed the erosions and perforations which are sometimes found in the stomachs of human subjects. Spallanzani made several experiments upon Dogs and Cats with a similar result. He caused the animals intended for trial to fast for a long time, and then to be fed immediately before being killed. Their bodies were placed in stoves which preserved their natural temperature, and in the course of a few hours he found that the food in their stomachs was sensibly digested.

The chyme, after being slowly formed on the surface of that alimentary mass which the stomach contains, accumulates as it forms near the pylorus or intestinal opening of the stomach. It is raised from this situation into the narrow pyloric entrance, by the increased action of those gentle and almost insensible contractions which it had already experienced. The more violent contractions which are necessary to expel the chyme from the cavity of the stomach, usually originate in the duodenum or small intestine, from which they are transmitted to the pylorus, and thence gradually to the entire stomach. But this ascending movement of the chyme, during which some persons experience much uneasiness, is immediately converted into another movement in the opposite direction; and it is by means of this reaction that the chyme finally traverses the pylorus, which is opened for that purpose, and by the same cause. The movement is repeated, and continues as often as the quantity of chyme newly formed in the stomach requires to be removed. In this way, the duodenum is gradually filled in small quantities at each operation by the food chymified in the stomach. Afterwards the course of the chyme is very slow in the intestines, and the same observation applies to the intestinal movements generally, except when excited by disease or mental emotion.

The fulness of the duodenum favors the secretion of the bile and pancreatic juices as well as of mucus and the intestinal juices. All these new fluids being mixed with the chyme immediately change its nature. It almost wholly loses its acidity; its colour changes from gray to yellow, and it becomes bitter to the taste. This bitterness in some animals extends even to the stomachic product itself, especially in Birds, because the bile often penetrates through the pyloric entrance into the stomach. If the aliments contain fat or oily matters, these substances pass into the intestine without having undergone any alteration in the stomach, but the bile uniting with them, forms a kind of soap, which is easily dissolved. With the exception of herbs, and these fat and oily substances, the duodenum permits all bodies, which the stomach has not previously digested, to pass without alteration. The same aliments produce a similar chyme in animals of the same species when in health, and the changes which the chyme afterwards undergoes are equally the same. But the chyme produced from animal food is thicker and more viscous than that produced from vegetables; it is reddish, and does not curdle milk. Vegetable chyme, on the contrary, is almost fluid; it has a yellowish tinge, and curdles milk. Further, the chyme furnished by vegetables is less rich in nutritious matter, and it wants that albuminous substance, which is found in chyme resulting from animal food. Different kinds of gas are disengaged during the process, but their nature varies according to the species of animal, its age and state of health, the kind of food which is used, and especially according to the part of the intestinal canal from which they proceed.

In a short time after the chyme has descended into the duodenum, and undergone the action of the bile, it divides into two parts. The one is a fluid termed *chyle*, which is the part destined to nourish the animal; the other is more solid and coarse, and less homogeneous, and, being the useless residue of the aliment, is finally rejected.

This separation of the chyle, and even its formation, appear to be more especially due to the influence of the bile; at least it is certain that digestion is always imperfect, and that the chyle is either deficient or in small quantity, when the bile cannot mix with the chyme prepared by the stomach.

Among the greater part of the Mammalia, the time which the chyle takes to form, after the chyme has passed from the stomach into the duodenum, varies from two to four hours. This function is much slower among the Fishes, and still more so with the Reptiles. Well formed chyle has sometimes been found in the white vessels which adjoin the stomach, and it has been said in consequence, that the stomach can form chyle. But this has only been seen in certain animals, whose bile frequently mingles with the gastric juice. The Dog appears to be an instance; but it does not seem to be of frequent occurrence among the Mammalia generally.

"Les excrémens," observes M. Isidore Bourdon, "séparés du chyle qui les surnage et dont l'absorption s'opère dans le haut de l'intestin, perdent peu-à-peu, à mesure qu'ils descendent vers les gros intestins, la fluidité qu'ils avaient dans le milieu de l'intestin grêle. Le mucus des gros intestins en favorise la marche vers l'anus, mais les loges que présentent ces conduits de distance en distance, en prolongent le séjour et en accroissent la consistance. C'est par l'action des fibres musculeuses des intestins que les excrémens son peu-à-peu poussés vers l'anus, et c'est par les muscles abdominaux qu'ils sont finalement rejetés hors du corps. Cette expulsion résulte d'un mécanisme assez compliqué où la glotte, au moins chez les Mamifères, joue un rôle important. Le rejet des matières fécales est beaucoup plus facile chez les animaux ovipares et dans l'Ornithorrhynque; et cette différence résulte de ce que ces animaux ayant un cloaque, leurs urines s'amassent dans ce lieu aussi bien que les excrémens, qu'elles délayent. Les excrémens diffèrent pour chaque espèce d'animal; mais la plus grande différence s'observe surtout entre les carnivores et les herbivores. Le même animal, s'il est omnivore, a des excrémens très différens, suivant qu'il use d'alimens végétaux ou d'alimens tirés de l'autre règne. Les excrémens provenant d'une nourriture animale ont la propriété de faire cailler le lait, et il n'existe rien de semblable pour les fécès des alimens végetaux. C'est absolument le contraire de ce que nous avons dit pour le chyme des carnivores et des herbivores."

Bordeu has made several interesting remarks on this subject; the curiosity of an ingenious mind having overcome the natural disgust towards a study so repulsive. The researches of Prout are more precise than those of Bordeu.

Complicated fluids are digested as well as solids, but a large part of them are absorbed directly by the stomach. Water, alcohol, and other simple fluids, contribute towards nutrition, chiefly by imparting their fluidity to some of the animal bodies.

Those movements of the intestines by means of which they are traversed by the stomachic product, are termed *peristaltic*. There are, however, other movements which are directed from below, upwards, in the contrary direction, and are hence termed *anti-peristaltic*. These movements produce various phenomena, such as Regurgitation, Rumination, and Vomiting. Among the Mammalia no animal vomits more readily than the Cat; there are, however, few Vertebrated animals which do not possess the power of vomiting. The Horse cannot vomit in ordinary cases, because the situation of the cardia opposes an obstacle to the return of the food towards the œsophagus.

It is ascertained that the chyle separates itself from the alimentary mass, after it has remained for some time in the distended cavity of the duodenum, and that in a short time the bile and pancreatic juices act upon it, although we are ignorant of the precise nature of the operation. With respect to the characters and quality of the chyle, this substance is very plentiful in the Mammalia alone. The chyle is always of an opaque white, which has caused it to be compared to milk. When taken from the body, and left to itself, it separates into two portions; the one is serous and saline, while the other is fibrous, in this respect nearly resembling the blood when similarly treated. If placed in an inert vessel, the chyle usually acquires a reddish tinge, which appears to be owing to the action of the oxygen in the air; and a thick kind of cream forms on the surface. This resemblance of the chyme to milk has led some physiologists to consider the one as the product of the other. One thing is certain, that nothing tends more towards the abundant secretion of milk, in the female, than that plentiful production of chyle resulting from the abundant supply of nutritious food. In other respects, the nature of the food with which the animal is nourished greatly influences the chyle resulting from digestion. Different substances do not produce the same kind of chyle; fat matters produce a chyle which is white, and more opaque than that yielded by substances which are not fat. The chyme never acquires the hue of any colouring matter which may have been introduced into the intestine; and it is even with difficulty that it can be made to acquire an odour.

The chyle, when once separated from the chyme, of which it may be regarded as the extract, floats on the surface of this matter, and accumulates by small rivulets in the mucous valvules, with which the interior of the smaller intestines are supplied. It remains in this place for a few instants, when it is absorbed by the small vessels, which serve to bring it into the mass of the blood. It is impossible to say in precisely what manner, or by what force or mechanism, this absorption of the chyle is effected, and to indulge in conjecture would but lead to certain error. On carefully examining, with the naked eye, the interior of the intestine, at the moment when the chyle is formed, there may be seen, at the surface of the intestinal membrane, small eminences like spongioles, which appear to erect themselves, and become filled with the fluid. On compressing these spongioles or small projections, the chyle exudes; and when they are examined with the microscope, we may perceive them to be ramified with innumerable small vessels, and their surfaces perforated with minute pores like the point of a needle. These pores are conjectured to be the commencement of the white vessels or lacteals, which carry off the chyle, and that by their means the chyle is gradually pumped out or absorbed at the surface of the intestine. We are entirely ignorant of the nature of that power by which this absorption is effected, but it has been ascertained from experiment, that the chyle penetrates into the lymphatic vessels of the intestine, and traverses the glands of the mesentery; that it is conveyed by proper vessels to the thoracic duct, through which it is finally carried into the blood. Once united to the blood, the chyle experiences the propelling force of the heart, traverses the organs of respiration, and comes in contact with the air always existing in the lungs. We may perceive many points of resemblance between the chyle and blood, in the spontaneous separation of their parts, in the fibrine which they both contain, and in their being similarly affected by oxygen, which colours them both red; and we are fully entitled to conclude that this fluid, arising from the digested aliments, is actually changed into blood during its passage through the organs of respiration, for on leaving the lungs, the chyle has lost all those characters which formerly distinguished it from blood.

As the blood is continually undergoing waste in its contributions towards the formation of the several secretions, as well as the reparation of the organs, this loss must be supplied by the aliment, without which life soon becomes extinct. The digested product of the food being altered in its properties, and completely animalized, finds its way into all the organs of the body, which it renovates and repairs. Thus, an identification of new matter with the former substance of the animal body is finally effected, and this process constitutes the essential part of the function of nutrition. All portions of the animal frame undergo continual changes of dimension, form, and structure, from the first period of their formation, until the body is finally subjected to the ordinary laws of inanimate substances. A part of the elements of which they are composed is incessantly dissipated in various ways, such as by respiration, perspiration, friction, and many others. These losses in the human frame amount to as much as several pounds weight of substance in the course of the twenty-four hours. Without an adequate supply of nutriment, the strength of the animal soon becomes reduced, its bulk diminished, and it finally perishes. There appears to exist a constant internal action, by which all the organs appear to be continually worn away and destroyed, only again to repair themselves, when supplied, through the food, with the proper elements for their composition.

Such are the leading facts hitherto ascertained, relative to the obscure function of Nutrition. The necessity for a supply of food is felt by all animals; yet it is not experienced in an equal degree by all species, nor by animals of the same species, nor even by the same animal when placed in different circumstances. This appetite for food is heightened by youth, fatigue, long-continued want of sleep, by violent passions when the paroxysm has passed, by convalescence after a long illness, by a dry and cold air, and the influence of climates and seasons. On the other hand, old age, prolonged sleep, hybernation, perfect repose, and hot baths, diminish the necessity for food. With the human species, luxurious habits lead to a loss of appetite, while it is heightened by labour; and thus Hunger, which declines the invitation of the opulent epicure, comes an unwelcome guest into the hovels of the destitute.

In general, the carnivorous animals endure a long-continued fast with less inconvenience than the herbivorous. This remark must not be confined to the Mammalia, for it extends to the Birds of Prey, especially to the Eagle, to Serpents, and Spiders, all which animals can remain a very long time without food, and do not appear to suffer from their continued abstinence. On this account they are in general of a more meagre habit of body than such animals as live either on herbs or fruits. There are many instances on record of old Men, but more especially of Women, who have lived for several weeks, some say months, without food. A mad enthusiast who imagined himself to be Christ in person, remained, it is said, during the forty days of Lent without using any food whatever; but confined himself, without swallowing any thing, merely to washing his mouth with water or wine. These instances are not, however, always very well authenticated; and it would be difficult to prove, in this case, that the fanatic did not actually swallow some of the fluid. Moisture, darkness, and repose, tend to diminish the usual effects of abstinence. A dog has remained alive under these circumstances for nearly fifty days without food. Persons of a vivid imagination, as well as frantic madmen, have in general a digestion extremely energetic, and they sometimes consume enormous quantities of food. Idiots also are frequently tormented with a devouring hunger. Next to Sleep, which wholly suppresses this appetite for the time, nothing tends more to drive away Hunger than the long-continued exercise of deep thought.

This appetite for food, which Man is enabled to confine within the bounds of Reason and Temperance, becomes in the lower animals one of the leading principles of action. Indeed, if we except the reproductive principle, and the principle of self-preservation from external danger, there are no others which approach in violence to the appetite for food, especially when heightened by abstinence. To obtain a sufficient supply of nourishment, is the great end, to which a large proportion of the instincts of each animal bear an immediate reference; and we commonly find, that those animals which possess the greatest facility in obtaining a subsistence, have the greater number of enemies to avoid. Such instincts as lead immediately to self-preservation from external danger, are more developed in the Herbivorous animals, than those other kind of instincts which relate more especially to their maintenance; and it is among the Carnivorous animals, whose existence depends solely upon their skilful exertions, that we find the most ingenious devices to deceive and destroy their prey.

The Quadrumana, especially the Monkeys, find an easy maintenance in the fruits of those warm countries, where alone they have fixed their abodes. Secure on the tops of trees, they have few other enemies to avoid than the Serpent tribes, which infest the lower branches. If we except those marauding parties, which they are sometimes compelled to form, in a great measure they are relieved from the cares which harass most other animals. But the Lemurs, being chiefly nocturnal, prey upon the small Birds and Insects while sleeping upon the branches. The Loris, favored by the darkness, steals upon its reposing victim, with a step so noiseless and excessively slow, that it is enabled to secure its prey with as much certainty as those Carnivora which depend for subsistence upon the extreme rapidity of their movements.

Some of the Cheiroptera, such as the Roussette Bats, feed almost wholly on fruits; the remainder pursue the Moths and Gnats which fly about during the summer evenings. A few in South America venture to suck the blood of Man, and of the larger quadrupeds, but their bites are neither deep nor dangerous. During the day, and in winter, they hang securely suspended by their thumb-nails to the roofs of caverns, and other obscure retreats. The Galeopitheci, or Flying Cats, by means of their membranes, extended like a parachute, dart from the tops of trees, by parabolic leaps, upon the small birds reposing on the lower branches.

The Insectivora, as their name denotes, feed chiefly on Insects; to these they add Worms, Snails, and tender roots. Some of these animals, such as the Mole, seek out their prey beneath the surface, by long mining operations; others, as the *Scalops Canadensis*, or Aquatic Shrew, add to their subterranean habits a mode of life almost subaqueous.

The Plantigrada, though omnivorous, differ in their tastes; some, as the Bears, are partial to a vegetable diet, while others, like the Glutton (*Gulo arcticus*), prefer animal food. The latter devours enormous quantities of flesh, and when urged by famine, conceals itself among the lower branches of a tree, from which it watches for an opportunity to leap upon the back of some quadruped passing beneath, whose blood it continues to suck, until exhaustion compels the larger animal to yield to its more cunning enemy.

The numerous genera of Carnivora are compelled, by the sagacity of their prey, and their more exclusive propensity for animal food, to resort to many ingenious devices for obtaining it. With the greater number of these animals, the principle of destruction is so strong, that they will destroy every living animal within their reach, although their hunger may be completely satisfied, so well are they disposed to execute the office of Nature's executioners, in curbing the excessive fecundity of the smaller tribes. Animals of the Genus Felis, such as the Lion, Tiger, and Leopard, never attempt to run down their prey by swiftness. Their sense of smell being somewhat obtuse, they rather seek to conceal themselves in a thicket near those places where the herbivorous animals come to drink, and spring upon their prey by one, or at most two or three bounds. If unsuccessful, which seldom happens, they retreat to their covert, or remove to a more favorable spot. On the other hand, the Genus Canis, such as the Jackal and Wolf, are skilful in tracking their game, which they run down by perseverance, or overcome by force of numbers. These animals, with the Hyæna, do not refuse carcasses, though in the last stage of decay, and disinter human bodies from the sands of the African deserts, or the cemeteries of the East. The Adives collect during the night, like bands of robbers, around the tents of the Moors or the Bedouin Arabs, who remain in momentary expectation of an attack from these ferocious brigands. The "Jackals' shriek," which is re-echoed by the distant hills, their voracity, and formidable numbers, strike the wanderer with terror; and when once accustomed to human flesh, they cannot enjoy any other. They will assemble at night to the number of two or three hundred, for the purpose of attacking caravans. At their frightful clamour, the Antelope and other herbivorous animals are roused from their coverts, and take to flight, when they fall, perhaps, into the ambuscade of some Lion or Leopard, while the band of Adives witness the success of this other brigand with jealous eyes, and are left only to dispute the mangled remains of the feast.

The Amphibia feed chiefly on fish, which they always devour in the water; though some species seem capable of living occasionally on Fuci.

Those instincts of the Rodentia which refer to their self-preservation from external danger, are more remarkable than any others. No animals are so skilful in forming subterranean retreats, which are usually executed by the combined labour of an entire settlement. One individual props up the earth which threatens to fall, another divides a large cavity into apartments, and a third forms a water-proof roof, with a layer of clay, to preserve the entire dwelling from the rain. One apartment is destined for the nursery, another for the granary. Here these animals amass, during the latter part of the autumn, a plentiful supply of provisions, and they find, on waking in the spring from their long winter sleep, that maintenance which would otherwise have completely failed them, until the returning autumnal fruits and grains again permitted them to amass another hoard. The Squirrels accumulate hazel-nuts, or the cones of the pine; the Dormouse gathers acorns and kernels; the Marmot seeks for different roots; and many species of Rats select bulbous roots in particular. Other species penetrate into our granaries and storehouses, composing a kind of vermin, which nothing can entirely extirpate.

Among the Edentata, the Tardigrada or Sloths feed chiefly on the leaves of trees, while the proper Edentata, such as the Armadillo, prefer insects and carcasses, though they all seem on an emergency to be likewise capable of digesting vegetable food. None of these animals ruminate. The Sloth is enabled to endure a long-continued fast without inconvenience, and it never drinks, being supplied only with Vegetable fluids. Prevented by its singular organization from any rapid movement, the Sloth devours every soft part of the tree within its reach, commencing with the leaves, and following on with the buds, tender shoots, and bark, until the whole tree is left entirely bare. Here the animal remains motionless and without eating for many days, until extreme hunger finally compels it to seek for food. Rolling itself in a ball, and falling heavily from the branches upon the ground, it crawls with measured pace to the nearest tree. The Armadillo burrows under ground into the numerous Ant-hills of South America, and the larger species frequent the neighbourhood of burying-grounds in great numbers. By means of subterraneous excavations, they invade the graves of the inhabitants, unless carefully protected by boards from their incursions.

The three tribes of Marsupialia present a great variety in their tastes for food. Among the Didelphida we find a strong partiality for animal food of every kind, yet they do not refuse fruits. Some live chiefly on the eggs of Birds, or on Crabs and Insects. Others devour carcasses, and even venture to make unwelcome visits into the houses of the Americans in search of food. The Macropoda, on the other hand, live almost wholly upon herbs or fruits. Of the Monotremata, the one Genus (*Echidna*) appears to feed, like the Hedgehog, upon land Insects and fruits; the other (*Ornithorynchus*) upon aquatic Insects, Worms, and Mollusca.

The Pachydermata being in general of great bulk, are obliged, by their organization, to feed chiefly on vegetables, while some of the smaller species, such as the Hog, seem almost omnivorous. The larger species find their food either among the trees of the forest, or in the marshes bordering on large rivers; the smaller generally seek with their snouts for the coarser kinds of fruit which fall from the trees, and lie concealed beneath the surface of the soil. All the Solipeda are essentially herbivorous.

Among the Ruminantia, the taste for food is wholly limited to the vegetable kingdom. The Camels, whose callous feet are well adapted to the sandy soil of Arabia, find in these Deserts a scanty herbage of prickly trees or shrubs; for this purpose their gums and tongue are almost cartilaginous, as a protection against the spinous processes of their food. The Rein-Deer, which is the sole sustenance of the Laplanders, Samoiedes, and Jakutes, scratches the snow for a supply of Lichens and Mosses, which is sufficient for his support. On the other hand, in the sultry plains of Ethiopia, the colossal Camelopard pastures on the foliage of the highest trees. The Ox and the larger Cattle feed on the rich herbage of the plains; some of the smaller, such as the Sheep and Goats, are satisfied with the more stunted plants of mountain regions.

The Herbivorous Cetacea feed in numerous herds on the marine vegetables accumulated at the mouths of rivers, as well as on the terrestrial herbs which float down the streams. Some, however, confine themselves to Fuci. The proper Cetacea are chiefly carnivorous, preying upon Fishes and Mollusca. Some, as the Dolphin, do not refuse vegetable substances; while others, as the Grampus (*Delphinus gladiator*) and Narwhal (*Monodon monoceros*), carry on a deadly warfare against the very largest Fishes, and even upon their own order. Combining together in troops, they do not hesitate to attack the great Whale, apparently for the sole purpose of devouring his tongue, for which the Narwhals seem to have a great partiality, leaving the remainder of his enormous body as a prey for epicures of a lower grade.

In general, animals of the class Mammalia seek their food separately, or in company, in which cases each individual labors for himself alone. It is only in a few species, such as the Beaver, Hamster, and Economic Mouse, all of which construct dwellings of great complexity, that each individual assists in accumulating a common board. In this arrangement we see one of the simplest states of society, where there exists community of goods, without any permanent division of labour. It is in Man alone that the Mechanic becomes distinguished from the Agriculturist.

Thus the Mammalia derive their subsistence from all inferior classes of living beings, as well as from their own; and hence they exert a very great influence in regulating the numbers of all other animals, and in establishing a universal equilibrium among living beings in general. The earth, destitute of herbivorous animals, would soon be covered with a rank and dense vegetation. A few luxuriant species of herbs would wholly engross each Botanical province, and annihilate all others. Hence the herbivorous animals are requisite to curb the exuberance of the Vegetable Kingdom; but as the herbivorous animals themselves would multiply, in their turn, to an inconvenient degree, even so far as to devour all plants to their very roots, the carnivorous animals are created to restrain the excessive multiplication of the herbivorous tribes, and thus become the indirect, yet necessary, allies of the Vegetable Kingdom.

In respect to the kind of food which is most suited to each animal, and the relative facility with which different substances are digested, these are questions which apply chiefly to Man. Each wild animal only uses that kind of food which is best suited to it; and its aliments are consequently very much restrained in their number. But Man is omnivorous, every kind of aliment can be rendered suitable to him, and he does not scruple to avail himself even of those which are most prejudicial to his health.

In all species of wild animals, and in some which have been domesticated, we perceive a most remarkable caution in avoiding such kinds of food as are deleterious to them. Nature commonly imparts a special instinct to each animal, in those cases where the ordinary processes of knowledge by experience would arrive too late to ensure the desired effect. "In looking at a pastured field," says Dr Fleming, "we observe that there are some plants which are left untouched, while others are cropped to the ground. But as the tastes of animals in this respect are exceedingly various, we observe that what is left untouched by one species is greedily devoured by another. What is eaten by the Goat, for example, with avidity, and with impunity by the Horse or Sheep, as the Water Hemlock (*Cicuta virosa*), is certain poison to the Cow. Hence it has been called Water-Cowbane, and we have heard a Fifeshire farmer, with a sigh, which intimated his experience of its effects, call it 'deathen.'" Cantharides, if taken by the Dog in a very small quantity, produce convulsions and death; yet the Hedgehog, being chiefly insectivorous, devours with impunity these poisonous insects.

Domestication exercises a certain influence over those instincts which lead animals to discriminate between nutritious and poisonous food, for some species are observed to lose, when long domesticated, that instinctive aversion to deleterious substances, so necessary for their preservation in the wild state. Dr Fleming remarks that Cows, which have been kept within doors during the winter, and supported chiefly on dry food, when turned out to pasture in the spring, devour indiscriminately every green herb, and frequently suffer for their indiscretion. Linnæus relates in his Lachesis Lapponica, that when he visited Tornea, the inhabitants complained of a distemper which killed multitudes of their cattle, especially during spring, when turned out into a meadow in the neighbourhood. He soon traced the disorder to the Water Hemlock which grew plentifully in the place, and which the cattle did not know how to avoid. In the Orkney Islands, the Fox-glove becomes fatal to the Goslins, when first turned out into the hills to pasture. It is probable that, in a wild state, this instinct remains unimpaired, and directs them invariably to avoid those substances which are unsuited to their digestive organs.

Civilized Man appears to have lost, in a greater degree than any other animal, this power of discriminating between noxious and nutritious food. By means of the art of Cookery, which he alone knows how to employ, numerous substances, though in their natural state they may be nauseous to the taste or even poisonous, are rendered highly nutritious; and thus the original properties of substances become disguised or neutralized in endless variety. Habit soon modifies his taste; and Man, being now left to the suggestions of Reason, is denied that Instinctive power of discrimination which the wild animals so largely enjoy.

Many interesting facts relative to the comparative facility with which different substances are digested, have been elicited from numerous experiments made on Man and the other Mammalia.

Milk being a fluid peculiar to the Mammalia, is, of all substances, the most nutritious to them. This proceeds from its containing the three ingredients essential to a perfect regimen. "All other matters appropriated by animals as food," observes Dr Prout, "exist for themselves, or for the use of the vegetable or animal of which they form a constituent part. But Milk is designed and prepared by Nature expressly as food; and it is the *only material*, throughout the range of organization, that is so prepared. In Milk, therefore, we ought to expect to find a model of what an alimentary substance ought to be—a kind of prototype, as it were, of nutritious materials in general. Now, every sort of Milk that is known is a mixture of three staminal principles; that is to say, Milk always contains a *saccharine* principle (sugar), a *butyraceous* or *oily* principle (butter), and a *caseous*, or, strictly speaking, an *albuminous* principle (cheese). Though in the milk of different animals these three principles exist in endless modified forms, and in very different proportions, yet none of the three is at present known to be entirely wanting in the milk of any animal."

It has been remarked, that the following kinds of aliment are the most digestible for Man:—beef, mutton, veal, lamb, and chicken; fresh eggs when half boiled, the milk of the Cow, Mare, Ass, Camel, and Goat; several kinds of Fish, when seasoned only with salt and parsley, but if used with oil or dripping, they are less digestible. Those vegetable substances easiest to digest are spinage, celery (chiefly the root), young asparagus, hop-buds, the placenta of artichokes, the boiled pulp of fruits with stones or pippins, especially if they be sweet and aromatic; the farinaceous seeds of the Cereal plants, wheat, rice, peas, &c.; bread on the day after it is baked, but especially stale bread, and chiefly white bread; turnips; new potatoes; and gum-arable.

The following substances are less digestible:—the flesh of pork, the different kinds of raw salad, cabbages, beet, onions, carrots, horse-radish, warm bread, figs, pastry, fried fish, and seasonings with vinegar or oil. The stomach can attack these substances but imperfectly; and that digestion which it is unable to accomplish is finished in the intestine.

Finally, we may mention as the most indigestible substances, the tendinous and cartilaginous parts, and especially the membranes of beef, pork, veal, fowls, &c.; bones, even when minutely divided; fat and oily substances; the white of egg hardened by heat; mushrooms; truffles; oily seeds, such as walnuts, almonds, pistachia nuts; the pippins of raisins, apples, &c.; olives; cocoa; the different oils; raisins; grape-skins; the epidermis, or outer skin of different seeds and fruits; the skins of peas; the bark of different trees; and many emulsive and ligneous grains. These last-mentioned seeds undergo so little change from the action of the stomach, that they germinate without difficulty on leaving the intestine. In this way many Plants are disseminated from one country to another.

There are several substances which serve to facilitate digestion, when mixed in small quantities with the food. The Ruminantia cannot exist without a supply of salt; and Man experiences, with advantage, the moderate use of spices, wine, liqueurs, cheese, sugar, and some bitter substances, particularly the products of the Cashew nut.

Numerous other substances are in an eminent degree prejudicial to digestion, and produce a more marked effect; such as the acids, Peruvian bark when taken after a repast, and the several emetics and poisons, in however small a quantity they may be used. Sedentary habits, excessive mental exertion, or violent emotions, also disturb or retard the function of digestion. Water, particularly when warm, if taken in large quantities after a meal, occasions the aliments to leave the stomach before they are digested.

It is by means of a well-regulated regimen, that Man and the domesticated animals are brought to the state of the highest possible health. Race-Horses, Greyhounds, and Fighting-Cocks, as well as Boxers, Racers, and other Athletæ, acquire by this means an extraordinary increase of physical force, and are enabled to continue their exertions for a very long time. This training of Men to athletic exercises produces surprising improvements in their external appearance. Their appetite is improved by this means, and digestion rendered more perfect. Giddiness of the head, after violent exertion, never occurs. The skin becomes clear, smooth, and well-coloured, and the veins are seen distinctly through it. The bones get harder and rougher; they thence become less liable to injury from blows and exercise, while the shape is improved. But the most important effects of training are upon the lungs, which acquire a free and powerful respiration, without which no animal can long maintain a vigorous action. The mental powers are also said to become improved; the attention is more ready and the perceptions more acute. These important effects are produced by temperance without abstemiousness, and regular exercise in the open air. "By these processes," says Sir John Sinclair, "the nature of the human frame is totally changed, and in the space of two or three months, the form, the character, and the powers of the body, are completely altered from gross to lean, from weakness to vigorous health, and from a breathless and bloated carcass to one active and untiring. Thus the very same individual, who but a few months before became giddy and breathless on the least exertion, has his health not only improved, but is enabled to run thirty miles with the fleetness of a Greyhound; or, in a shortness of time hardly to be credited, to walk above a hundred; or, varying the object in view, to excel in wrestling, or to challenge a professed boxer. The mind also becomes more courageous, corporeal sufferings are borne with patience; a command of temper, and a presence of mind, are also acquired and preserved undisturbed amidst pain and danger." It appears that these important results of training are produced by the most simple means, which every man may practise to a certain extent; general ill health might thus be commonly prevented, and many diseases wholly removed.

It will be seen, from the preceding outline, that by far the greater number of Mammalia exist upon more than one kind of food; and even in those species which are more especially restricted to an animal or vegetable diet, a certain degree of variation from their ordinary habits is allowed to them, by means of which they can subsist in unusual situations. Thus the Squirrel will sometimes devour Birds, and the Marten and Pole Cat can subsist upon fruits. Domestication tends greatly to produce this omnivorous habit, yet there are some instincts connected with the food of animals which it fails to overcome. The Dog continues to hide his food, though fed regularly and plentifully; and civilized Man pursues the wild game with alacrity, although hunting has long ceased to be necessary to his subsistence.

When the numbers of herbivorous animals are not kept down by other tribes, or when the carnivorous species fail in finding their prey, food begins to fail, and no resource remains to the famished animals but Migration from their native haunts. Excessive changes of temperature may be the ultimate causes of these migrations, by occasioning the destruction of those insects or plants from which the animals derived their maintenance. The Mammalia are, however, in general sufficiently protected by their covering of hair or blubber from the changes of the seasons. They seek for shelter beneath the surface of the earth or sea, perhaps they sleep or hybernate. But when animals are threatened with famine, either by a season excessively favorable to their multiplication, or any other cause, and their provisions in consequence suddenly become scarce, a simultaneous movement is the certain consequence. Migrations occurring in spring seem to owe their origin chiefly to this scarcity of provisions arising from an excessive population. Dr Richardson informs us that the Black Bears of America migrate from Canada into the United States in very severe winters; but in milder seasons when they have been well fed, they remain and hybernate in the North. Among some of the sociable Mammalia, the force of hunger, the confidence arising from the example of their fellows, and the excitement of the Social impulse, urge even the feebler and more timid animals to attempt migrations on the greatest scale of magnitude, and fraught with the highest danger to themselves. The common Squirrels, compelled by a scarcity of provisions to desert their abodes, migrate from Lapland into lower latitudes in amazing numbers. Onwards they travel in a direct line, nor do rocks, forests, the deepest ravines, or the broadest waters, disturb the invariability or impetuosity of their course. Numbers are drowned in passing large firths and rivers, or fall a prey to their numerous enemies. The Lemmings of Norway and Sweden often pour down in myriads from the mountains of the North and devastate the country. They move generally in lines, about three feet from each other, and exactly parallel. The general direction of their march lies from north-west to south-east, and they pass directly onwards through rivers and lakes. When stacks of hay or corn interrupt their passage, they gnaw through them instead of passing round. Pennant relates, that the Rats of Kamtschatka becoming too numerous at the commencement of Spring, proceed in great bodies westward, swimming over rivers, lakes, and arms of the seas. Many are drowned or destroyed by Water-fowl or Fish. As soon as they have crossed the River Penchim, at the head of the Gulf of the same name, they turn southward, and reach the rivers Judoma and Ochot by the middle of July, a district surprisingly distant from their point of departure.

Mr Lyell has correctly observed, that the large Herbivorous animals which are gregarious, can never remain long in a confined region, as they consume so much vegetable food. The immense herds of Bisons, which often blacken the surface, in the great valley of the Mississippi, near the banks of that river and its tributaries, are continually shifting their quarters, followed by Wolves, which prey on the rear. "It is no exaggeration," says Mr James, "to assert, that in one place, on the banks of the Platte, at least ten thousand Bisons burst on our sight in an instant. In the morning we again sought the living picture, but upon all the plain, which last evening was so teeming with noble animals, not one remained." Vast troops of Dziggtai, which inhabit the mountainous deserts of Great Tartary, feed during the summer in the tracts East and North of Lake Aral. In the autumn they collect in herds of hundreds, and even thousands, and direct their course towards the North of India, and often to Persia. Bands of two or three hundred Quaggas are sometimes seen to migrate from the tropical plains of Southern Africa to the vicinity of the Maleleveen river. During their migrations they are followed by Lions, who slaughter them nightly. Myriads of Springboks or Cape Antelopes pour down like a deluge upon the cultivated regions near the Cape, when the stagnant pools of the immense deserts south of the Orange River dry up, which often happens after intervals of three or four years. The havoc committed by them resembles that of the African Locusts; and so crowded are the herds, that the Lion has been seen to walk in the midst of the compressed phalanx with only as much room between him and his victims, as the fears of those immediately around could procure by pressing outwards.

There are certain secluded spots in the neighbourhood of Melville Island, which are visited annually by herds of Musk-Oxen and Rein-Deer; during the short summer of the arctic regions, various plants put forth their leaves and flowers the moment the snow is off the ground, forming a carpet, spangled with the most lively colours, and these animals travel over immense distances of dreary and desolate regions, to graze undisturbed in these luxuriant pastures.

Mammalia which frequent the ocean, like the Whales and Seals, or the air, like the Bats, possess unusual facilities for executing these periodical migrations. The Whales of the Northern Seas are known to desert one tract of sea and visit another at a very remote distance. The Seals, according to Krantz, retire from the coasts of Greenland, in July, return again in September, and depart again in March, to return in June. They proceed in great droves northwards, directing their course where the sea is most free from ice. This migration of the Seals must, however, proceed from some other object than a mere search for food, as they are observed to be very fat when they set out on this expedition, and very lean when they come home again. The Great Bat (*Vespertilio noctula*) visits England during the summer, but retires in winter to Italy, where it hybernates.

The daring manner in which Land animals attempt to cross large tracts of water is an immediate consequence of the urgency of their wants. "Rivers and narrow firths," says Mr Lyell, "can seldom interfere with their progress, for the greater part of them swim well, and few are without this power when urged by danger and pressing want. Thus, among Beasts of Prey, the Tiger is seen swimming about the islands and creeks in the Delta of the Ganges, and the Jaguar traverses with ease the largest streams in South America. The Bear, and also the Bison, stem the current of the Mississippi. To the Elephant in particular, the power of crossing rivers is essential in a wild state, for the quantity of food which a herd of these animals consumes, renders it necessary that they should be constantly moving from place to place. The Elephant crosses the stream in two ways. If the bed of the river be hard, and the water not of too great depth, he fords it; but when he crosses great rivers, such as the Ganges and the Niger, the Elephant swims deep, so deep that the end of his trunk only is out of the water—for it is a matter of indifference to him whether his body be completely immersed, provided he can bring the tip of his trunk to the surface, so as to breathe the external air. Animals of the Deer kind frequently take to the water, especially in the rutting season, when the Stags are seen swimming about in search of the Does, especially in the Canadian lakes; and in some countries where there are islands near the sea-shore, they fearlessly enter the sea and swim to them. In hunting excursions in North America, the Elk of that country is frequently pursued for great distances through the water."

Without this power of shifting their quarters, a far greater number of animals would have become extinct than has occurred under their present constitution. The mutual action and reaction of species is the necessary consequence of these general laws of Nutrition, by which all Living Beings are governed. Individuals maintain their existence for days or years—species for centuries and ages. Each arrives at its termination when its resources wholly fail, from the influence of surrounding causes of change, and it is to their mutual struggles for subsistence we owe that equilibrium of animal forces which is found to prevail in all parts of the globe. In every place it is decreed that the demand for food shall bear a determinate ratio to the supply, and Nature never hesitates to deal indiscriminate destruction on all individuals or species which transgress this law.

It is evident from the mutual dependance of animals upon each other and upon plants, that the creation of certain species has preceded that of others in the order of time. Vegetables must have become numerous upon the earth before the Frugivorous tribes made their appearance; while the Herbivorous animals must have multiplied upon the earth, and become widely distributed, previous to the institution of predaceous types. The phenomena of nutrition thus clearly point out that the creative power has been exerted successively, and probably at remote periods of time—a conclusion which is fully confirmed by the investigation of Fossil Remains.

The mutual reaction of Animals upon each other, and upon Plants, follows necessarily from the limited duration which is allotted to the existence of individuals and species. Had Living Beings not been subject to Death, there would have been no reproduction; the checks to reproduction would not have existed; in a word, there would have been no activity, no prey to pursue, no enemies to avoid—no mutual reaction, in short, Life would lose that stamp of animation which marks its phenomena so distinctly from those of Inorganic Nature. The liability of Animals to Death is thus the ultimate cause of their greatest enjoyments and sufferings.

"The law of universal mortality," observes Dr Buckland, "being the established condition on which it has pleased the Creator to give being to every creature upon earth, it is a dispensation of kindness to make the end of life to each individual as easy as possible. The most easy death is proverbially that which is least expected; and though, for Moral reasons peculiar to our own species, we deprecate the *sudden* termination of our mortal life, yet, in the case of every inferior animal, such a termination of existence is obviously the most desirable. The pains of sickness and decrepitude of age are the usual precursors of death, resulting from gradual decay; these, in the human race alone, are susceptible of alleviation from internal sources of hope and consolation; and give exercise to some of the highest charities and most tender sympathies of human nature. But throughout the whole creation of inferior animals no such sympathies exist: there is no affection or regard for the feeble and aged; no alleviating care to relieve the sick; and the extension of life through lingering stages of decay and old age would to each individual be a scene of protracted misery. Under such a system, the natural world would present a mass of daily suffering, bearing a large proportion to the total amount of animal enjoyment. By the existing dispensations of sudden destruction and rapid succession, the feeble and disabled are speedily relieved from suffering, and the world is at all times crowded with myriads of sentient and happy beings; and though to many individuals their allotted share of life be often short, it is usually a period of uninterrupted gratification; whilst the momentary pain of sudden and unexpected death is an evil infinitely small, in comparison with the enjoyments of which it is the termination.

"To the mind which looks not to general results in the economy of Nature, the earth may seem to present a scene of perpetual warfare and incessant carnage; but the more enlarged view, while it regards individuals in their conjoint relations to the general benefit of their own species, and that of other species with which they are associated in the great family of Nature, resolves each apparent case of individual evil into an example of subserviency to universal good.

"The appointment of death by the agency of Carnivora, as the ordinary termination of animal existence, appears therefore in its main results to be a dispensation of benevolence; it deducts much from the aggregate amount of the pain of universal death; it abridges and almost annihilates throughout the brute creation the misery of disease and accidental injuries, and lingering decay; and imposes such salutary restraint upon excessive increase of numbers, that the supply of food maintains perpetually a due ratio to the demand. The result is, that the surface of the land and depths of the waters are ever crowded with myriads of animated beings, the pleasures of whose life are co-extensive with its duration; and which, throughout the little day of existence that is allotted to them, fulfil with joy the functions for which they were created. Life to each individual is a scene of continued feasting in a region of plenty; and when unexpected death arrests its course, it repays with small interest the large debt which it has contracted to the common fund of animal Nutrition, from whence the materials of its body have been derived. Thus the great drama of universal life is perpetually sustained; and though the individual actors undergo continual change, the same parts are ever filled by another and another generation; renewing the face of the earth, and the bosom of the deep, with endless succession of life and happiness."

GENERAL REVIEW OF THE MAMMALIA CONTINUED.

The internal functions of the Mammalia in harmony with the revolutions of the Earth, and the laws of inanimate Nature—General Relations to Light, Heat, and Electricity.

It has already been shown, that the dimensions and forces of animals bear a certain determinate relation to the circumstances of their conditions;—that the Creator has organized them so as to correspond accurately with their intended habitations. When an aquatic animal removes permanently to the air or earth, it receives an organization suited to that change. The Frog is assigned the characters of a Fish while in its Tadpole state, and acquires those of a Reptile when it is designed also to reside upon the land. But this correspondence of animals to the circumstances of their condition is not confined merely to the media, whether air or water, in which they are intended to move; for their forces also bear a determinate relation to the earth, considered Mechanically as a mass of matter, or Astronomically in its relation to the other bodies of the Solar System.

The dependance of all animal motions upon the attractive force of the Earth is sufficiently obvious. Each animal body is acted upon by Gravity, in proportion to its mass; in other words, it possesses weight; and in order that animals may exercise the power of moving, it is necessary that their forces shall bear a certain relation in excess to that of gravity, otherwise no motion could follow. Their forces must also be proportioned to the resistance which gravity offers to their exertions, or else animals would lose their balance, their motions would proceed by jerks; at one time they would endanger their own safety by an excessive rapidity, at another by an excessive slowness; in all they would be devoid of grace, energy, and convenience. Animals would come into collision with other animals, or with harder substances than their own bodies; and this globe of Earth, like a machine out of order, would soon lie in a state of inactivity and disorganization.

Mr Whewell remarks, that if the force of gravity were increased in any considerable proportion at the surface of the earth, all the swiftness, and strength, and grace of animal motions must disappear. If, for instance, the earth were as large as Jupiter, gravity would be eleven times what it is; the lightness of the Fawn, the speed of the Hare, the spring of the Tiger, could not exist with the existing muscular powers of those animals; for Man to lift himself upright, or to crawl from place to place, would be a labour slower and more painful than the motion of the Sloth. The density and pressure of the air, too, would be increased to an intolerable extent, and the operation of respiration and others which depend upon these mechanical properties, would be rendered laborious, ineffectual, and probably impossible. If, on the other hand, the force of gravity were much lessened, inconveniences of an opposite kind would occur. The air would be too thin to breathe; the weight of our bodies, and of all the substances surrounding us, would become too slight to resist the perpetually-occurring causes of derangement and unsteadiness: we should feel a want of ballast in all our movements. Things would not be where we placed them, but would slide away with the slightest push. We should have a difficulty in standing or walking, something like what we have on ship-board when the deck is inclined; and we should stagger helplessly through an atmosphere thinner than that which oppresses the respiration of the traveller on the tops of the highest mountains.

The force of gravity depends upon the mass or quantity of matter in the Earth. For any reason that we can discover, this globe might have been as large as Saturn or Jupiter, its mean density might have been that of cork or of gold, in any of which cases the force of gravity would have been very different from what it is at present; and we can easily imagine, that if every thing were seven times as heavy, or one-seventh lighter than it actually is, animals could not exist in their present state. The Moon and Planets all differ in size and density from the Earth, and from each other; and in general, the smaller seem to be nearest to the Sun, for our imperfect gaze fails to discover the lesser bodies which probably exist in the outward regions of the Solar System. For this reason, the inhabitants of other globes must be so different from ours, as to render it almost impossible for us to form any conception of their nature.

Thus the Earth is not only the common source whence Animals and Plants derive their subsistence, but it is the common source of all animal motions, not only in the reaction offered by its inertia, but in the looseness of that invisible tie which connects our bodies to its surface. Our relations to the earth are not even confined to the surface, but extend to those remote depths which the miner and geologist contemplates only in imagination; and every particle of matter towards the centre exercises an influence in proportion to its magnitude. The intimate relation of the Earth to the inhabitants of its surface, has led all ages to regard it as the common mother of all; and nations have been proud to consider themselves αὐτόχθονες (*autochthones*), or sprung from "the dust of the earth,"—their native soil.

It is, however, in the great phenomena of Astronomy, in those revolutions of the heavenly bodies which have served to mark the epochs of time, that we perceive the more astonishing, because more unexpected, correspondence between these remote phenomena, and the periodical functions of organized beings.

The diurnal revolutions of the globe are always performed in the same time, being that which elapses between the appearance of a star on the meridian until it again returns to the same meridian. This regular and constant movement, constituting a sidereal day, forms the *unit* or measure of time, and gives rise to the periodical changes of Day and Night. All animals and plants, which decorate the surface of the earth, partake in this revolution around its axis, and to this phenomenon all functions of animals, depending upon the presence or absence of Light, such as Sleeping and Waking, Hunger, states of Exertion or Repose, bear an immediate reference. The internal clock-work of the animal frame has been made to run for twenty-four hours, when the same states of the animal frame succeed each other in the same order, and in exact conformity with the revolutions of the globe.

Besides this diurnal period of the animal clock-work regulated by the diurnal revolution of twenty-four hours, there also exist periodical functions referring to divisions of weeks and months, as the epochs of menstruation, also the incubations of Birds, which may endure for two or three weeks, and the gestations or internal incubations of the Mammalia, extending from three weeks to nine or eleven months.

The year, or period of the Earth's revolution round the Sun, is the most important astronomical phenomenon in reference to organized beings. It is felt through every portion of animated Nature; it measures the great epochs of their existence, and forms the limit of duration to a multitude of animals of the Class of Insects in particular, and of Plants. All their functions are distributed in reference to the periods of the year. The annual species are born in the Spring, the Summer becomes the period of their puberty and reproduction, their fruits or productions appear in Autumn, and they die on the approach of Winter. Man, and the other persistent beings, from the Mammalia to Trees and Herbs, experience more or less the influence of the seasons over their physiological functions.

Spring, being the morning of the year, is favorable to births and bodily growth; it is in fact the period of youth, expansion, and gaiety. Experience proves that the human frame then undergoes, like Plants, its highest degree of growth and development.

Summer, analogous to mid-day, is the season of heat, ardour, strength, and the highest development of the faculties. It corresponds with the age of puberty, and the impetuosity of the passions. The rutting period, with most animals, happens towards the summer solstice.

Autumn is the evening of the year. Plants then yield their fruits, they afterwards become ligneous and dry, and finally fade away. Animals, after performing the act of generation, cast their hair, skin, or feathers, and undergo that moulting which strips them of their more gaudy attire. This is the epoch when the faculties become concentrated, a period of melancholy and sadness. Vegetation ceases, and plants in general lose their foliage.

Winter, the cold night of the year, renders the vegetable world dormant, and especially the cold-blooded animals. It is the season of repose, of nutrition, and internal repair, preparatory to future action. Animal bodies become inert, moist, and phlegmatic. Life is rendered at this period almost stationary and nullified; it remains either in a state of concentration, or in absolute torpidity.

Thus, besides the nychthemeral periods, or diurnal revolutions, which regulate the daily functions of existence, the crises of maladies, the hours of repast or excretion, we have monthly periods of gestation and incubation, menstruation, rutting, and moulting, corresponding to the flux and reflux of the tides, to the periodical winds of the tropics, and the revolutions of the Moon. Again, the annual periods fix a limit to the lives of all annual and biennial species, and determine the periods of their growth, the metamorphoses of Insects, with the phenomena of reproduction and decay among most animals.

Thus the revolutions of our globe, and its relative situation to the heavenly bodies, maintain the circles of our existence in equilibrium with them. A philosophical Astrology may read our lives and destinies in the stars, which move in their curvilinear orbits by the same force that urges all Living Beings onwards in their physiological periods. Time, measured by the successive revolutions of our planet, draws onwards all the generations of Plants and Animals which decorate its surface; it marks the fatal hour to each individual, as it brings round the periods of love and the necessities of nutrition. The fœtus of animals and the fruits of vegetables arrive to maturity at the appointed period. Each species of Mammalia has its fixed time of gestation, sufficient for the proper elaboration of the fœtus, which period may, however, sometimes vary by a few days in proportion to differences of food, temperature, or the season of the year. Minerals, on the other hand, are only moved by general impulses, without each of them partaking in a special activity. With them no period of time marks out their duration, whilst, with us, each pulsation of the heart, and every second of time, urges us onwards in the vital career, without the possibility of avoiding or retreating.

The well-being of most animals is intimately connected with the degree in which they conform all their habits and functions to the periods of day and night. This is most remarkable in Man, who in all ages has his fixed periods of the day and night for food and repose, and this regular circle of actions has a direct reference to his internal constitution, and independent of mere external stimuli. "In the voyages recently made into high northern latitudes, where the Sun did not rise for three months, the crews of the ships were made to adhere, with the utmost punctuality, to the habit of retiring to rest at nine, and rising a quarter before six, and they enjoyed, under circumstances the most trying, a state of salubrity quite remarkable. This shows that, according to the common constitution of such Men, the cycle of twenty-four hours is very commodious, though not imposed on them by external circumstances." Some Men are naturally *nyctalopes*, or night-eyed, such are Albinos and white Negroes, Dondos or Blafards, which cannot endure the full blaze of day-light.

Among the Quadrumana, several Howling Apes (*Mycetes seneculus* and *Beelzebul*) are either nocturnal or at least crepuscular, preferring the twilight of the morning and evening for their time of feeding and exertion. It is the same with certain Makis, who have thence derived their name of *Lemur*, from their haunting the twilight like the shades (lemures) of the departed.

The Cheiroptera or Bats, especially the Genus Noctilio, with the Galeopitheci or Flying-Cats; the Insectivora, such as the Hedgehogs, Shrews, and Moles; the Plantigrada, such as the Bears and Badgers; also the entire genus of Cats; the Weasels, Polecats, and many Opossums, are strictly nocturnal; and this quality seems eminently appropriate to all those Carnivora or Marsupialia, which watch for their prey, and endeavour to surprise them while sleeping. Like assassins and brigands of our own species, they bury themselves in silence and obscurity to render their blows the more deadly.

We find many crepuscular or semi-nocturnal species among the Rodentia, which move in obscurity through fear of their enemies. Thus, the Rats, Dormice, and Hares, come from their retreats in the evening, or very early in the morning.

The Edentata, such as the Armadillos, Anteaters, and Manis, are also nocturnal, through timidity and their want of offensive arms.

The Pachydermata and Ruminantia, on the contrary, feed only during the day. With all these animals, the remaining part of the twenty-four hours is devoted to sleep and repose. Crepuscular feeders sleep partly during the night and partly during the day, while diurnal sleepers are nocturnal feeders, but in all cases the same round of functions succeeds in equal periods of twenty-four hours.

The time selected by each animal for its period of food or repose is capable of undergoing much modification, from the presence or absence of particular stimuli, such as the different states of the air, the states of electricity, moisture, and heat, at the several periods of the day and night. The presence or absence of Light and Heat seems chiefly to regulate the periods of activity and repose in all animals. The Day being warmer than the Night, tends to establish in some of the bodies a movement towards the surface,—a period of waste and destruction of force,—the Night with these is devoted to a reparation and accumulation of energy.

The dependance of certain functions of animals upon the presence or absence of Light, becomes more perfect when any particular formation of their visual organs specially marks them out for enduring Light of different degrees of intensity. When animals run into white varieties, as may often be observed in white Negroes, Albinos, white Rabbits, Mice, Dogs, Cats, Pigeons, and many others, their eyes are commonly red; and these organs then become so acutely sensible to light, that they are unable to support the full blaze of day; but at the same time, they can see much more clearly in twilight than individuals which have not experienced this degeneration.

The cause of this extreme sensibility in their visual organs is completely ascertained. If we examine the inside of the Sclerotica and Iris, which commonly form the obscure chamber of the eye, we shall find in the leucose individuals, that these membranes are deprived of the black or brown pigment which is designed to defend the eye from the rays of Light, except at the transparent aperture of the pupil. The retina being thus insufficiently defended against the luminous rays, becomes easily dazzled in bright daylight, but receives a sufficient number of rays in the twilight for the animals to see clearly. An opposite effect is produced in black or brown individuals, such as the Negroes, in whom this pigment (*pigmentum nigrum*) which lines the interior of the sclerotica and iris, defends it perfectly from the entrance of luminous rays, excepting at the proper aperture of the pupil. For this reason, Negroes, and generally all individuals with black eyes, can easily support the full blaze of sunshine, while the blue, gray, or ash-coloured eyes of the fairer inhabitants of Europe are so tender, from the intensity of the tropical sun, that they require to be defended by coloured glasses, else they become affected with ophthalmia.

Men and the lower animals, with very white skins and light hair, are thus destitute of that brown or black colouring matter, which not only lines the sclerotica and iris, but also impregnates the mucous tissue under the skin, and, passing onwards, tinges the hair or fleece of different colours. Black or chestnut hair usually accompanies an iris more or less brown. It thus follows, that black and dark-brown animals can endure the blaze of day, and that the light-brown or white animals, which are naturally better adapted for the cold and polar regions of the earth, become the most proper to see during the twilight or at night. All the nocturnal animals are further capable of dilating their pupils largely in the dark, in order to receive a larger pencil of rays than the diurnal animals; the latter, on the contrary, are compelled to close the pupil to avoid being dazzled. The inhabitants of the polar regions possess this power of dilating and contracting the pupil in a remarkable degree, as they experience, at one season of the year, the dazzling reflections of the snow, and at another, the long twilights and Auroræ Boreales of winter.

When deprived more or less of this pigment, Man and the lower animals have a very sensitive skin, the fibres of which are delicate and slim, while their hair is light, fine, and silky. These individuals are easily overcome by the heat of the day and the intensity of its light; they soon become exhausted during the day, and find the feeble rays of the night better proportioned to the delicacy of their temperament. Hence, hey transform the Day into a period of repose, the Night into one of activity and exertion.

We have a further confirmation of the accuracy of these observations in the fact, that with the greater part of the nocturnal animals, the colouring pigment of the skin is less vivid than in the diurnal races. It may generally be remarked, that animals with nocturnal eyes are clad in a mournful and dingy vestment of gray or ash-colour, striped with black or spotted, not only among the Mammalia and Birds of Night, but even in the Insects, when we compare them to the allied diurnal species. We see a remarkable contrast between the tints of the diurnal Butterflies and those of the Moths and Sphinges. The Owls are sad and sombre birds when placed against the Parroquets or Humming Birds, glittering in the brilliant sun of the torrid zone. Many animals of the Cat kind, the Lemurs and Bats, cannot compare in these respects with the gayer quadrupeds.

Nocturnal animals further possess the peculiar quality of advancing to surprise their prey with a noiseless step. The almost imperceptible flight of Nocturnal Birds of prey is well known to proceed from the soft feathers of their wings; and the same effect is produced by the wings of the Bats, and the nocturnal Butterflies. The crepuscular Sphinges alone produce a humming noise by the vibration of their wings; but they suck the nectar of flowers, and the Bombyx and Cossus do not take animal food. All other nocturnal animals are, for the most part, carnivorous, attack their prey by surprise—and Nature inspires them with the same instinct as the cowardly assassin, who does not dare to face his enemy in the blaze of day. In return, however, they are often impregnated with fetid odours, which serve to announce their approach.

It is thus to the peculiar constitution of their bodies that the nocturnal animals owe their property of sleeping during the day and waking at night, a peculiarity so opposite to other beings. An analogous state exists also in the Vegetable kingdom. Some flowers appear to close and languish during the heat of the day. The sun acts too vividly upon the frail texture of certain petals, and occasions the sap and nutritious fluids which fill their laminæ to evaporate too freely; but during the freshness of night the sap and juices being less dissipated, accumulate in the tissues of these plants, their canals are dilated, and the flowers and leaves expand. Their tender organs of reproduction would soon become dessicated by the heat of the sun; hence the plant withdraws them from its influence, and displays them only before the pale light of the moon. Diurnal flowers also have, with animals, a more solid tissue than the nocturnal. The former require to be stimulated by light and heat, that their reproductive organs may develop themselves, while the more tender nocturnal flowers resemble those animals of the night which shield their eyes from the dazzling influence of the day-light. Further, the sexual organs of Nocturnal Plants fade more rapidly than the diurnal: their monopetalous and polypetalous corollæ are of a texture extremely frail, generally blanched and etiolated. Their evanescent perfume is exhaled only at night; it fails during the day.

In general, all nocturnal beings, whether Mammalia, Birds, Reptiles, Fishes, Insects, or Plants, present sombre and tarnished hues, while the diurnal species, under the fiery influence of the sun, assume garments of dazzling brilliancy. Light thus stimulates to activity, and is one of the principal causes of the development in organized beings of animal and vegetable poisons and perfumes, when acting upon special constitutions. Darkness benefits those acrid, venomous, faded, and inert plants, whose juices, feebly elaborated, require its shelter: it is favorable to the development of animals generally at their birth—to the larvæ of insects in their dark asylums—to mushrooms and lichens, the mysterious product of evanescent sporules, in the depths of forests, and the hollows of caverns—and in general to all feeble and imperfect organizations. The moisture and coldness of night are further unfavorable to the waste of living bodies, and hence it is the period when they experience the highest degree of growth and vegetation, provided that the cold be not too intense. In fact, the internal functions of nutrition and repair are performed more intensely during the repose of night, and the absence of external stimuli. Then the organs grow, and become replenished with nutritious fluids. Thus the mushrooms are mostly the offspring of the night, or multiply in the secret obscurity of subterranean excavations.

Each day, like each season, distributes to every living being some portion of heat, light, and nutriment, and measures the rhythm of their functions of waking or sleeping, nutrition or excretion. When these are maintained in harmony with the movements of the globe, health and regularity are alike maintained. We find the influence of the periodic return of day and night in places where its presence could scarcely have been anticipated. According to the researches of Messieurs Burch, Quetelet, and Villermé, the mortality of the human species increases towards sunrise, diminishes towards sun-set, and almost no deaths happen at mid-day. Further, it is observed that births occur most frequently during the night, and deaths during the day. Births and deaths are generally most numerous between the hours of three and six in the morning, and least numerous from three to six in the evening, corresponding to the maximum and minimum of temperature. This prevails as well in the seasons of the year as in the hours of the day. The greater number of births and deaths occur in the most stimulating periods of the day and year, being at six in the morning, and in the month of April. Those periods, on the contrary, when stimuli begin to fail, are most deficient in births and deaths, such as three o'clock in the afternoon, and in the month of August. The paroxysms of fevers, and the pains of an approaching *accouchement*, usually begin in the evening, while the crisis or result arrives towards the morning.

Thus there exists a remarkable correspondence between the structure of animals and plants, and that periodical order of light and darkness resulting from the rotation of the earth around its axis. Although this succession of functions depends partially on the presence of the external stimuli of light and heat, yet there appears to be a diurnal period belonging to the constitution both of animals and vegetables, and this structure corresponds with the astronomical day. The power of accommodation possessed by living beings in this respect, is not sufficient to allow us to suppose that the periods of the day and night could be very greatly lengthened or shortened without causing their ultimate destruction. "We may be tolerably certain," says Mr Whewell, "that a constantly recurring period of forty-eight hours would be too long for one day of employment, and one period of sleep, with our present faculties; and all whose bodies and minds are tolerably active will probably agree that, independently of habit, a perpetual alternation of eight hours up, and four in bed, would employ the human powers less advantageously and agreeably than an alternation of sixteen and eight. A creature which could employ the full energies of his body and mind uninterruptedly for nine months, and then take a single sleep of three months, would not be a man. When, therefore, we have subtracted from the daily cycle of the employment of men and animals, that which is to be set down to the account of habits acquired, and that which is occasioned by extraneous causes, there still remains a periodical character, and a period of a certain length, which coincides with, or at any rate easily accommodates itself to, the duration of the earth's revolution. We can very easily conceive the Earth to revolve on her axis faster or slower than she does, and thus the days to be longer or shorter than they are, without supposing any other change to take place. There is no apparent reason why this globe should turn on its axis just three hundred and sixty-five times while it describes its orbit round the sun. The revolutions of the other planets, as far as we know them, do not appear to follow any rule by which they are connected with the distance from the Sun. Mercury, Venus, and Mars, have days nearly the length of ours. Jupiter and Saturn revolve in about ten hours each. For any thing we can discover, the Earth might have revolved in this or any other smaller period, or we might have had, without mechanical inconvenience, much longer days than we have. But the terrestrial day, and consequently the length of the cycle of light and darkness, being what it is, we find various parts of the constitution, both of animals and vegetables, which have a periodical character corresponding to the diurnal succession of external conditions; and we find the length of the period, as it exists in their constitution, coincides with the length of the natural day."

The want of colour, or *Albinism*, in animal and vegetable bodies, when they are said to be *leucose* or *white*, has its proximate cause in the original want, or the diminished secretion, of the coloured layer of mucous net-work placed immediately under the epidermis, or outer skin of animals. With Plants, this is owing to the inert secretion of the green matter, or *chromule*, and its ceasing to colour the cuticular tissues. In all species, softness and moisture are the results of this albinism or whiteness. Its ultimate cause is the want of vital energy, arising either from the prolonged absence of the influence of Light upon the organic structure, or from the intensity of a long-continued cold. Its effects may be either absolute and total, or merely partial and local, even among the white varieties of animals and plants. Its general tendency is to effeminate all beings.

Accidental albinism may arise from old age, or the want of a continued renewal of this coloured layer, which communicates its hue to the Hair, Feathers, or Scales. It may be even induced before old age by disease, or by the absence of the usual supplies of nutriment, or, among animals, by the violence of fear or any sudden emotion, which may serve to withdraw from the exterior of the body its secretions, and render the skin pale, or the hair white. There is also an accidental albinism from the mechanical injury of the mucous pigment, arising from the bruising or tearing of the skin, and on these spots, white hair or feathers will arise in the place of coloured appendages.

An opposite state of deep blackness, or *Melanism*, when the surface is said to be *melanose* or *black*, arises from the superabundance of the mucous subcutaneous tissue in animals and plants, in which carbon exudes towards the exterior. Such are Negroes, and all black or dark-brown animals, lurid and venomous plants, as the Solaneæ. This state of the skin is well fitted for skies, resplendent with Light and Heat. It is attended in individuals with dryness, rigidity, and shortness of stature.

Excessive cold, combined with the absence of Light, serves to drive the nutritive and repairing juices far from the skin. This kind of albinism is especially remarked in animals inhabiting the highest mountains and the polar regions, where they become white in winter and coloured in summer. The large species of the Porcupine exhibit these alternate annulations of white and brown, which are due to the alternations of summer and winter. A similar effect might be produced on live Sparrows, by plucking the feathers, and rubbing their naked bodies with Spirit-of-wine. The feathers which then succeed remain white, because the alcohol prevents the secretion of the colouring subcutaneous matter, in the same manner as an excessive cold. A corresponding effect may be produced by similar means upon the Mammalia.

The colours of animals are intimately connected with the latitude of the place as well as with the changes of the seasons, and seem, in general, delicately sensible to the external stimuli of Light and Heat. In Mammalia and Birds, we find that the hotter regions of the globe, as well as the summer months, are favorable to deep and bright colours.

Sir John Leslie's experiments on colour, as affecting the radiation and absorption of Heat, afford the best explanation of the final causes of these changes. "The rate at which bodies cool is greatly influenced by their colour. The surface which reflects heat most readily suffers it to escape but slowly by radiation. Reflection takes place most readily in objects of a white colour, and from such, consequently, heat will radiate with difficulty. If we suppose two animals, the one black, and the other white, placed in a higher temperature than that of their own body, the heat will enter the one that is black with the greatest rapidity, and elevate its temperature considerably above the other. These differences are observable in wearing black and light coloured clothes during a hot day. When, on the other hand, these animals are placed in a situation, the temperature of which is considerably lower than their own, the black animal will give out its heat by radiation to every surrounding object colder than itself, and speedily have its temperature reduced; while the white animal will part with its heat by radiation at a much slower rate. The change of colour in the dress of animals is therefore suited to regulate their temperature by the radiation or absorption of caloric. While it is requisite that the temperature of some species should be preserved as equally as possible, the cooling effects of winter are likewise resisted by an additional quantity of heat being generated in the system. An increase in the quantity of clothing takes place, to prevent that heat being dissipated by communication with the cold objects around, and the dress changes to a white colour to prevent its loss by radiation. In summer, the pernicious increase of temperature is prevented by a diminished secretion of heat, or the secretion of cold, increased perspiration, the *casting* of a portion of the winter covering, and by a superior intensity of colour in the remainder giving it a greater radiating power. The last character would, in the sunshine, by absorbing heat, prove a source of great inconvenience, were its effects not counterbalanced by other arrangements, and by the opportunity of frequenting the refreshing shade, or bathing in the stream." Animals become light or gray in old age, and thus the too great dissipation of heat in their systems is prevented.

If it were possible for any one to doubt the fact that the functions of animals and plants correspond with the movements of the terrestrial globe, he would find a convincing proof in the influence of the seasons, upon the casting of hair among the Mammalia, the moulting of Birds, the changes of skin among the lower animals, and the defoliation of Plants.

In the Spring, all Nature is living and vegetating, expanding and developing its productions; the earth is clothed with verdure, the animals are dressed in their nuptial garbs, and their amours commence. The cause of this external expansion of all beings originates in the circumstance that their functions, long oppressed by the cold of winter, have acquired a superabundance of juices, sap, and nourishment, which only await the favorable moment of external heat to expand. Their germs are developed with extreme vigour. In the human race, there is at this season a determination towards the skin, eruptive maladies become more prevalent, and exanthemata sometimes appear as though budding were not exclusively confined to the vegetable kingdom. The Hair, Feathers, Horns, Scales, and Epidermis of animals, as well as the Leaves, Flowers, and Fruit of vegetables, which have grown and expanded during the Spring, assume their most glittering hues, if not during Summer, at least during the six months of the year, when in our climate the sun is most above the horizon. But at the approach of the autumnal equinox, living bodies, whether Animals or Plants, are exhausted by the excessive action of their vital forces during summer, and their functions become less vigorous, in proportion as Light and Heat diminish with the enfeebled rays of the Sun. Those external parts, produced in the preceding Spring, cease to receive nutriment from the body; they have further arrived at the full period of their growth, and are incapable of receiving any more. Hence they dry up or fade away; others become detached and fall. Sooner or later at this season, we invariably witness the casting of hair, feathers, scales, horns, and epidermis, as well as the dropping of flowers, leaves, and fruit, when each Living Being enters into a kind of autumnal concentration preparatory to the rigour of winter.

In the southern hemisphere, our winter being then its summer, and reciprocally, the periods of casting or moulting are in each year opposite to ours.

In the Torrid Zone, the Sun passes twice a year from the one tropic to the other, so that it produces, to a certain extent, two winters and two summers. The winter is there the season of continual rains; it also determines twice a year the casting or moulting period of animals and plants, and doubles the number of the rutting seasons. From these circumstances, it arises that Living Beings experience a two-fold waste of vital force, and live in these warmer regions at a faster rate than elsewhere. They are continually producing or wasting; new flowers spring up next to the fruit; new leaves replace the old and faded; the Bird prepares its nest of eggs, and sings new carols within the hearing of its brood of six months old; and the Quadrupeds consume in a continual state of generation, gestation, and lactation.

In colder countries, and on the summits of elevated mountains, there exists another kind of change in the feathers of Birds and the hair of Mammalia, which arrives at the period of winter. The white robe, the symbol of chastity and sexual indifference, is particularly fitted to these cold regions, in the same manner as the brilliant robe of summer is in correspondence with the full vigour and activity of the reproductive system. Thus the Hare of the Alps (*Lepus variabilis*) and the Ermine, as well as a great number of other Mammalia, with an immense multitude of Birds of the Northern Regions, especially the Waders (*Grallæ*) and Web-footed Birds (*Palmipedes*), which are covered in the summer season with hair or feathers of brown and more brilliant hues, acquire a pale gray or uniform white during the winter.

It has been considered by some, that the white garb of arctic animals serves to protect them from their enemies, by assimilating their colour with that of the snow. But Nature, pursuing a fair system of reciprocity, imparts the same colour to the beasts and birds of prey, so that, in reality, this provision is less effectual than has been commonly supposed.

Dr Fleming has made the following observations on the cause of the change of colour in those quadrupeds, which, like the Alpine Hare and Ermine, become white in winter:—"It has been commonly supposed that these Mammalia cast their hair twice in the course of the year; at harvest, when they part with their summer dress, and in spring, when they throw off their winter fur. This opinion, however, does not appear to be supported by any direct observations, nor is it countenanced by analogical reasoning. If we attend to the mode in which the human hair becomes gray as we advance in years, it will not be difficult to perceive that the change is not produced by the growth of new hair of a white colour, but by a change in the colour of the old hair. Hence there will be found some hairs pale towards the middle, and white towards the extremity, while the base is of a dark colour. Now, in ordinary cases, the hair of the human head, unlike that of several of the inferior animals, is always dark at the base, and still continues so during the change to gray; hence we are disposed to conclude from analogy, that the change of colour, in those animals which become white in winter, is effected, not by a renewal of the hair, but by a change in the colour of the secretions of the *rete-mucosum*, by which the hair is nourished, or perhaps by that secretion of the colouring matter being diminished or totally suspended." An Ermine shot by Dr Fleming in May 1814, in a garb intermediate to its summer and winter dress, confirmed this view of the subject. In all the under parts of its body, the white colour had nearly disappeared, in exchange for the primrose-yellow, the ordinary tinge of these parts in summer. The upper parts had not fully acquired their ordinary summer colour, which is a deep yellowish-brown. There were still several white spots, and not a few with a tinge of yellow. Upon examining those white and yellow spots, not a trace of interspersed new short brown hair could be discerned. This would certainly not have been the case if the change of colour is effected by a change of fur. Besides, while some parts of the fur on the back had acquired their proper colour, even in those parts numerous hairs could be observed of a wax-yellow, and in all the intermediate stages from yellowish-brown, through yellow to white. These observations leave little room to doubt that the change of colour takes place in the old hair, and that the change from white to brown passes through yellow. If this conclusion be not admitted, then we must suppose that this animal casts its hair at least seven times in the year. In spring it must produce primrose-yellow hair, then hair of a wax-yellow, and lastly of a yellowish-brown. The same process must be gone through in autumn, only reversed, and with the addition of a suit of white. The absurdity of this supposition is too apparent to be further exposed. Thus the hair, as long as it remains connected with the body, participates in the general life of the system, and is influenced in respect to its colour by the secretions of the mucous net-work of the skin.

There exists a general tendency in all living bodies to develop themselves from within, outwards. This evolution of living bodies is the ultimate cause of those changes which the external surface of their bodies undergoes during the several periods of their existence, and determines the variations in the quantity of their clothing: the proximate causes of change are the external stimuli of light and heat. As each appendage of the animal body is endowed with a vital power peculiar to itself, it must have its peculiar periods of youth, perfection, decay, and death. When any organic portion of the body is completely dead, it separates and falls, because a living substance cannot co-exist with a dead one. The casting of hair, and all other kinds of moulting or external change, is nothing more than the natural death of a certain portion of an animal body, in consequence of the development of other parts interior to it, and this kind of function is regulated by fixed laws.

The external parts of animals and plants which are renewed each year are of two kinds. They may have a peculiar organic conformation, as we find in hair, horns, teeth, feathers, and leaves, or they may have a simple structure, scaly or foliaceous, as we find in the epidermis or outer skin, shells, and membranes.

The changes of all these bodies partake more or less of the same general character. A tree may be considered as a body composed of an infinite number of germs which are successively developed. Besides the fruit produced each year, it pushes forth an immense number of leaves, which extract their nutriment from the sap, expand, and arrive at their full growth. Then having received all the nutriment which the ariolæ of their tissue can maintain, they dry up, become yellow or brown, the leaf ceases to extract the sap, and dies of old age. The vessels of the petiole are broken by this drying up and obstruction, and the leaf falls. This is observed generally in autumn with the trees of our climates, and happens also in the ever-green trees, the only difference being, that in these the new leaves are repaired as fast as the old ones fall, so that the tree is not at any one time completely destitute of verdure.

The same thing happens with the feathers of Birds, and the hair of the Mammalia. The bulbous root of the hair is penetrated by a blood-vessel, and that portion of nutriment and growth necessary to its development is thence communicated to the shaft. When the root dries, and the canal ceases to admit nutriment, the hair falls; the nutriment finds its way to other bulbs, the germs of hairs yet in embryo, concealed beneath the epidermis; and a new coat of hair succeeds to the former. Thus the hair is a kind of plant, which has its bulb or root, and its shaft or stalk, composed of long sheathy tunics, one within another, like the tubes of a telescope.

The casting of hair among the Mammalia arrives at different seasons, according to the peculiar constitution of each animal in reference to Heat, and in general its degree bears an immediate reference to the temperature of the district, whether arising from the season of the year, the latitude of the place, or the degree of elevation. "In the warmer regions," says Dr Fleming, "it is requisite to have the temperature of the body diminished, in the colder regions, the very opposite object is aimed at. In the former case the hair or feathers are thinly spread out, while in the latter, they form a close and continuous covering. In the Dogs of Guinea, and in the African and Indian Sheep, the fur is so very thin that they may be almost denominated naked. In the Siberian Dog and Iceland Sheep, on the other hand, the body is protected by a thicker and longer covering. The clothing of animals, living in cold countries, is not only different from that of the animals of warm regions in its quantity, but in its arrangement. If we examine the covering of Swine of warm countries, we find it consisting of bristles or hair of the same form and texture; while the same animals, which live in colder districts, possess not only common bristles or strong hair, but a fine frizzled wool next the skin, over which the long hairs project. Between the Swine of the South of England and the Scottish Highlands, such differences may be observed. Similar appearances present themselves among the Sheep of warm and cold countries. The fleece of those of England consists entirely of wool, while the Sheep of Shetland and Iceland possess a fleece, containing, besides the wool, a number of long hairs, which give to it, when on the back of the animal, the appearance of being very coarse. The living races of Rhinoceros and Elephant, inhabitants of the warm regions, have scarcely any hair upon their bodies; while those which formerly lived in the Northern plains of Europe, the entire carcasses of which have been preserved in the ice of Siberia, were covered with fur similar to the Iceland Sheep, consisting of a thick covering of short-frizzled wool, protected by long coarse hairs. These species, now extinct, possessed clothing, suiting them to the climate where they lived, and where they became at last enveloped in ice. Had they been transported by any accident from a warmer region, they would have exhibited in the thinness of their covering, unequivocal marks of the climate in which they were reared. By means of this arrangement, in reference to the quantity of clothing, individuals of the same species can maintain life comfortably, in climates which differ considerably in their average annual temperature. By the same arrangements, the individuals residing in a particular district are able to provide against the varying temperature of the seasons. The covering is diminished during winter, and increased in summer, as may be witnessed in many of our domestic quadrupeds. Previous to winter, the hair is increased in quality and length. This increase bears a constant ratio to the temperature; so that, when the temperature decreases with the elevation, we find the Cattle and Horses, living on farms near the level of the sea, covered with a shorter and thinner fur than those which inhabit districts of a higher level. Cattle and Horses, housed during the winter, have shorter and thinner hair than those which live constantly in the open air. The hair is likewise shorter and thinner in a mild, than during a severe winter."

The approach of the hot seasons of each year, by occasioning the development of new hair, transfers to them that nutrition which the former coat was in the habit of receiving. Hence, as the summer advances, the hair falls off, and the animal becomes sleek; and the warm covering of winter is exchanged for a lighter and more commodious garb. The Sheep in our climates casts its fleece before the end of June, and the Mole about the end of May. The time when the wild animals, whose furs are used in commerce, acquire their winter coats, corresponds with the hunting season. "During the summer months the fur is thin and short, and is scarcely ever an object of pursuit; while, during the winter, it possesses in perfection all its valuable qualities. When the beginning of winter is remarkable for its mildness, the fur is longer in *ripening*, as the animal stands in no need of the additional quantity for a covering; but as soon as the rigours of the season commence, the fleece speedily increases in the quantity and length of the hair. This increase is sometimes very rapid in the Hare and the Rabbit, the skins of which are seldom ripe in the fur until there is a fall of snow, or a few days of frosty weather; the growth of hair in such instances being dependent on the temperature of the atmosphere. In the northern islands of Scotland, where the *shears* are never used, the inhabitants watch the time when the fleece of their Sheep is ready to fall, and pull it off with their fingers. The long hairs, which likewise form a part of the covering, remain for several weeks, as they are not ripe for casting at the same time as the fine wool. The operation of pulling off the wool, provincially called *rooing*, is represented by some writers, more humane than well-informed, as a painful process to the animal. That it is not even disagreeable, is evident from the quiet manner in which the Sheep lie during the pulling, and from the ease with which the fleece separates from the skin."

The shedding of those antlers, which are produced each year on the Stags, and other Deer, may be explained on the same principles as the shedding of hair, and other external appendages. As long as the bony protuberances on the forehead of the Stag continue to absorb the nutritive fluids holding phosphate of lime in solution, and permit them to penetrate abundantly into the parts yet soft and gelatinous, the horns grow in the form of antlers of various shapes. But when these horns, being completely filled with phosphate of lime, refuse to admit any more, the latter accumulates in a lump at the root of the horns, and obstructs the nutritive canals. These soon die, and the passage from within outwards, being thus interrupted, the antler dies and falls like the withered leaf or the dead feather.

In a similar manner we may explain the shedding of the milk teeth in the Human infant and the other Mammalia. The germs of the second teeth, before they appear externally, exist at the root of the gums, in the form of small capsules, which receive their nutriment from the blood-vessels of the maxillary arteries, and their sensation from the dentary nerves. When the first teeth have attained their full growth, and cease to admit any more nutriment, the latter is diverted to the other germs of teeth situate below. The second teeth, having thus acquired more force, expel the others and assume their place.

From these instances, it may be seen that the shedding of teeth, horns, hair, feathers, or scales, is the same phenomenon of organization; and that these bodies resemble leaves, or rather those parasitical animals and plants, which draw their nutriment from a body larger than themselves, on which they grow or live. So exact is this comparison, that the Hair and Nails maintain a separate Life on the corpse of Man or other Mammalia, and continue their growth until the dead body, by being entirely decomposed, ceases to supply them with a nutritious lymph.

Thus the moults and changes which living bodies undergo at the surface, in different periods of their existence, depend upon the general fact that organized bodies develop themselves continually from within outwards, so that the matter composing them never remains the same. The nutritive particles derived from the food, after being assimilated to our bodies, and incorporated into our proper substance, are ever transitory, and tend to undergo decomposition and waste at the surface, so that as fast as the internal organs are repaired, the vital force impels the nutrition towards the exterior, where it is decomposed and finally rejected. Each portion of the individual participates in the general nutrition; but besides this general life which the organs enjoy in common with the entire frame, each organ partakes of a special living power, which can maintain itself distinct from the whole, or even occasion a growth at the expense of the other parts. Hence each animal appendage has its special birth, age, and limited duration, besides those which it derives from the entire body, as we find in the organs of generation, the teeth, hair, feathers, and the leaves of plants. These appendages, though developed a long time after the birth of the individual, perish notwithstanding before it, and various external germs develop themselves successively. Thus, the special vital forces of particular parts possess a much shorter duration than the general Life of the body. Further, these productions which succeed each other, whether hair, feathers, or teeth, may neither have the same form nor colour. The radical leaves of Plants often have forms and colours very different from those of the branches and floral peduncles. The feathers of the winter plumage are more downy and thick than those of summer, or the nuptial period of Birds. The second teeth of the Mammalia have very different roots from the former; an old Stag receives a more formidable defensive weapon than the Fawn whose first horns are beginning to shoot. Thus, Nature has implanted in animals and plants different kinds of germs, appropriate to the several epochs of Life, as well as the external circumstances of their situation, and even in reference to their relative situation in respect to the heavenly bodies. The rich variety which we find in these arrangements at once demonstrates the admirable economy of Nature, which operates incessantly in evolving or developing, according to fixed and determinate laws. Every one is compelled to acknowledge that organized bodies are formed in exact correspondence with the physical agents which surround them, otherwise the harmony and concourse of all portions of Nature could not subsist. Living bodies are not only formed in direct co-relation to Air, Food, and Moisture, but also with the laws of Light, excepting, perhaps, in certain subterranean animals and plants, and require the influence of a moderate Heat.

Electricity, and that form of Electricity which we commonly term Magnetism, may also contribute towards the vital action in certain circumstances. So close is the co-relation of Electricity with the vital power, that several later writers have confounded the one agent with the other. Although it is impossible to admit that Life is the same as Electricity, yet their intimate connexion is undoubted. Animal bodies are in this respect delicate electro-vital machines, and acutely sensible to the electrical state of surrounding bodies.

Animal electricity has been shown by Mr Faraday to be identical with all the others, only that it resides in those imperfect conductors which compose the animal tissues, in the same way as Voltaic Electricity penetrates into the metallic substance, and the ordinary Electricity exists at the surfaces of bodies. In fact, all these Electricities may be converted into each other, and Magnetism itself is only a particular form of Electricity.

There is no phenomenon among the Mammalia which can compare in intensity to the electric batteries of certain Fishes. The presence of Electricity is, however, demonstrated in various ways. The Hair and Skin of Man, when heated, have been accompanied, under certain circumstances, by remarkable electric and luminous sparks. Hales and Bellingeri have shown the different states of Electricity in the humours of the Human Body. Friction can draw electrical sparks from the Fur of the Cat and several other Mammalia, chiefly carnivorous. The same thing has been found with the plumage of certain Birds, as the Parrots.

It was conjectured by Humboldt, and confirmed by the experiments of MM. Prévost, Dumas, and Edwards, that every muscular contraction, and every act of the Will or volition, is accompanied by a kind of electrical discharge of the nerves which animate it, and that the nerves serve to deposit and distribute an electro-vital fluid.

The scintillations and corruscations which emanate from the eyes of certain Mammalia when in the dark, are phenomena of a very different kind from that general phosphorescence which prevails over the entire bodies of many Fishes, Molluscs, Crustacea, and Zoophytes. They seem to depend upon a certain state of the nervous expansion of the retina, when the animal is under the influence of rage, love, hunger, or any violent emotion, especially in the more furious species of Carnivora. This property of the retina is not peculiar to the Mammalia, but is also found among the Mollusca, Arachnides, and Crustacea. The enormous eyes of certain Cephalopoda, as the Cuttle-fishes (*Sepia*), appear luminous in the middle of the Ocean, and terrify the Fishes, their prey. The eight eyes of the Tarantula Spider, a voracious and nocturnal species, are also luminous, according to M. Léon Dufour. We also find this property among several Saurian Reptiles, such as the genera Anolius and Gecko, whose eyes scintillate in the darkness of night, and the same assertion has been made regarding the Alligators, which are thus said to frighten their prey. The ancients have related many fables concerning the piercing looks of the Basilisk Serpents, and modern authors have given credit to the fascinating powers of the Rattlesnake. Certain credulous believers in Animal Magnetism have also attributed the most terrific effects to the glances of the Toads, and have illustrated their credulity by examples. When carefully considered, these examples only prove a nervous state of the imagination when under the influence of fear. Such effects may be induced in sensitive frames by the approach of a hideous or dangerous object, and may be observed in the lower animals where these involuntary sensations occasion them even to tremble and faint. The effect which the Pointer Dog produces upon the Partridge is a striking instance. But the greater part of the Carnivora being nocturnal in their habits, such as the Cat genus, as the Lions, Lynx, Ounce; the Dog genus, as the Wolves, Foxes; the Martens, and probably also the Bats, with the Nocturnal Birds of Prey, have luminous eyes in the dark, whether during the night, or during the day, when confined in a dark chamber. These animals then dilate their pupils, so that the expanded surface of the retina at the back of the eye shines vividly, and illuminates the external chamber of that organ. Light is thus projected from within upon those objects on which the animal fixes its gaze, that one can distinguish them very well at the distance of more than a foot and a half. This emanation appears clearly to proceed from the expanded extremity of the optic nerve. It lasts nearly for a minute at the pleasure of the animal, or even involuntarily when under the influence of violent emotions. Certain Apes, as the *Nocthora trivirgata*, and the Howlers (*Mycetes*), have nocturnal eyes possessed of this radiating property. Inflammation, in various diseases of the Eye, gives to some Men a temporary power of seeing in the dark, or of emitting luminous rays. The eye when rubbed, or when it receives a blow, becomes dazzled by the sudden influx of blood into this organ, and not only do scintillations appear, but there is a luminous emission when these animals are enraged, like that proceeding from an electrical discharge.

The luminosity of the retina does not proceed from a simple reflection of those scattered rays of Light which may chance to fall upon the Eye, as Tréviranus and Benj. Prévost have considered probable, for these phenomena can be observed in the most perfect darkness. The Cat, when irritated, darts forth fiery rays of light; its eyes sparkle intermittently when enraged, according to Esser and Rengger; and Gruithuisen remarked that the rays acquired a greenish tinge when the animal was caressed. Dogs, when enraged, impart to their ocular radiations a tinge sometimes yellowish and sometimes blueish. Their corruscations vary with individuals, but the luminous emanations appear to be most brilliant in animals with black or ash-coloured hair. They neither proceed from the crystalline nor vitreous humours of the Eye, for these can be altogether removed without destroying their luminous property, which only then acquires a more greenish hue. But on wounding the optic nerve, or on scraping the retina, the radiation becomes extinct; thus proving that it proceeds neither from the cornea or uvea, or indeed from any of the transparent portions of the Eye.

GENERAL REVIEW OF THE MAMMALIA CONTINUED.

Phenomena of Sleep—the Hybernation of some Species.

THE phenomena of Sleep bear an immediate relation to the most general laws of Nature, and form an important illustration of the fact, that the periodical motions of the animal economy are in direct correspondence with the movements of our planetary system, and especially with our situation relative to that Sun which regulates the periods of the day and year. "All our wants reappear," says Cabanis (Rapports du Physique et du Moral de l'Homme), "and all our functions execute themselves, in fixed and iso-chronous periods. The duration of the functions is the same in each period; the same appetites or the same wants have the hours marked for their return; and it commonly happens, when these wants are not immediately satisfied, they diminish and disappear for a certain time, only to return again with the greater force and importunity at the next succeeding period, which ought to produce a return of the impression. This character of periodicity is particularly remarkable in the returns and duration of sleep, which commonly reappears during each astronomical day at the same hour; continues nearly for the same period of time; and according as it is regular in its periods, slumber is the more easy, while the repose which follows is the more salutary and refreshing."

There are two principal states of vital activity, of which all animals partake in different degrees. When the vital excitement exists to its full extent, the animal is said to be *awake*; when the functions of life are suspended, either wholly or partially, it is said to be *asleep*. From this waking state, when life exists in all its plenitude, there may be many degrees of its diminution, called *Reverie, Delirium, Dreaming, Sleep, Torpor, Stupor, Asphyxia, Lethargy*, according to their intensity, of which states the last is but one degree removed from absolute death.

The principal occasions on which these states of vital repose naturally present themselves to our observation are, *1st*, When the body and mind of an animal languish either from the return of their period of natural repose, or through excessive exertion. *2d*, When the cold of winter, or perhaps also the heat of summer, acting on special constitutions, suspends the animal functions of life either partially or entirely. The former phenomenon appears daily, while the latter is of annual recurrence.

The first of these occasions, or Sleep properly so called, differs from Death, with which it has often been compared, in the circumstance that all the involuntary functions of life continue their action uninterruptedly. It may be recollected that animals have two kinds of vital functions; the one vegetative and internal, which continue, with the exception of generation, to exert themselves during the entire existence of each individual, and the other purely animal, which refer to external objects. The former being essential to their existence, are never suspended; the latter are intermittent in their action. If the heart ceased to propel the blood through the arteries, if the lungs ceased to respire air, if the functions of nutrition and secretion were discontinued, or if they depended upon the mere Will of the animal, life would soon become extinct. But all these internal actions are involuntary, and hence it is only the external and purely animal functions of life which can have their periods of action and repose, of waking and sleeping. These latter actions are therefore less essential to life than the former. An animal, when profoundly asleep, is reduced to a state very analogous to that of a plant. Though dormant, he is still a living being, for he continues to perform the functions of nutrition and secretion even more perfectly than when awake; but he is destitute both of sensation and motion, and must awake before he can fully resume these functions peculiar to animals. Thus animals have two states of existence, waking and sleeping, while a plant has only one. The state of the latter may, however, be more or less active, according to the different degrees of heat or light to which it is exposed. There can be no difference with plants between the activity of the internal and external functions, and they always appear to be plunged in a state of repose more or less profound. Many of the lower animals, such as an Oyster or a Zoophyte, when considered superficially, appear to exist in a continued state of torpor, rather than to possess an active life, because they maintain but few relations to external objects, and hence are commonly considered rather as vegetating than as living, although Plants are possessed of life as well as animals, but only in different degrees.

Sleep, in fact, consists in the suspension of the organs of sense and voluntary motion, qualities which chiefly serve to distinguish animals from plants. All the voluntary muscles repose completely, and the sleeping animal maintains no active relations with external objects. The functions belonging to vegetable life continue their existence, but the consciousness of existence is lost. The heart and the lungs continue to act without interruption, while the organs of thought and sensation possess but a temporary action. It is thus precisely those organs which are the most intimately connected with the Mind, namely, the organs of thought and sensation, which most require repose, and the human Soul, though immortal, when entirely separated from the Body, cannot now maintain its consciousness uninterruptedly for twenty-four hours together.

Night, or the absence of light, is favorable to the sleep of all animals not naturally of nocturnal habits. Silence, repose, the absence of noise, and in general every thing which interrupts the relations of the animal with external objects, are favorable to sleep. As long as the purely animal functions continue to be stimulated, they maintain their action, until an excess of action produces a contrary effect. A violent exertion of the body, profound thought, or any powerful sensation, disposes for sleep. Often the fatigue of a single sense brings on the sleep of all the senses, through that intimate connexion maintained among all the parts of the body. The monotonous murmurs of a brook, the howling of a forest, bad music, protracted reading, bad verses, or a long lecture on an uninteresting subject, gradually fatigue the sense of hearing or sight, and lead the vital forces of these organs to seek in sleep for an accession of energy, and the repose of the entire animal functions speedily follows.

The inclination to Sleep is announced by a slowness of motion, by languor of the Attention and Will, and by the gradual stupefaction of the senses. But the different kinds of functions are suspended in a certain order of succession, according to their nature and relative importance. The muscles which move the arms and legs are relaxed and cease to act before those which sustain the head, and the latter before those which support the spine. When the sense of sight is first suspended by the falling of the eyelids, the other senses still maintain their action. The sense of Smelling is obliterated before the taste; Hearing after smelling; and Touch last of all. Even during the most profound sleep, the sense of touch continues to suggest different movements and changes of position, when the long duration of the same posture renders it disagreeable. At length animal exertion is at an end; the muscles, excepting those of Circulation and Respiration, cease to act; and the body sinks down, obedient to the ever-acting force of Gravitation.

These phenomena of Sleep are very analogous to that insensibility of particular organs, during our waking moments, when the Attention is fully engaged. A profound Mathematician, when absorbed in a calculation, neither sees, hears, nor feels; all the functions are asleep except the organ of thought. Other Men, like mere machines moved by habit, perform the same operation a thousand times with their hands, while the thinking principle remains buried in a profound lethargy.

At the precise moment when the Mind loses its consciousness, there results a general relaxation of all the muscles. If the body be at rest and in health, this sudden change in its state of obedience to the Will is attended with no marked result. But if the body be fatigued, or in an uneasy posture, or if the joints or muscles be painful, this first result of Sleep has the effect of removing it entirely. Hence arises the difficulty of sleeping in a sitting posture, or during an attack of gout or rheumatism. The pain which the sudden starting of the muscles occasions is often so great in these diseases, that Sleep can only be induced by strong doses of opium or some other narcotic. It also follows from this relaxation of the muscles, that the limbs become bent during Sleep, and that a substance grasped firmly in the hand, falls at the instant when consciousness is lost.

During Sleep the character of the Respiration is altered; it becomes less frequent and deeper. The heart also beats more slowly, but the pulse is stronger. The Heat of the surrounding air, when imperfectly renewed, tends, however, to increase its movements.

The heat of the body is not naturally higher during Sleep; on the contrary, a diminished respiration tends to lower the temperature. It usually happens, however, from external circumstances, that there is an apparent rise of temperature, from the body being surrounded by imperfect conductors of heat, and from the circumambient air being but slowly renewed.

As the stomach is a muscular organ, and as the passage of the food through the pylorus depends upon the rapidity with which the almost insensible contractions of the Stomach are performed, it follows that Sleep retards digestion, while, at the same time, it renders it more complete. This slowness of digestion is further increased by the state of rest in which the body remains, as nothing tends more to excite a rapid digestion than the gentle motion of the limbs, or of the entire body. The same phenomena take place in the intestines, where the aliments remain almost inert in the several portions of the alimentary canal. However, the slowness of this movement favors the formation of chyle, and renders its absorption more complete.

Absorption is very active during Sleep, and the danger of slumbering in noxious air hence becomes very great. Travellers are usually advised to avoid sleeping in marshy situations, such as the Pontine Marshes of Italy, especially in the warm season of the year. Perspiration also is performed more easily, because the pores remain open during the state of muscular relaxation. "Une nutrition plus efficace, la réparation graduelle des forces qui en résulte, et aussi la réplétion de la vessie, toutes ces choses réveillent en nous, durant le sommeil, des idées de jouissance et des souvenirs de volupté."

As all the senses do not fall asleep at the same time, so they differ in the order in which they awake. Taste and Smell commonly resume their functions last of all. The sense of Sight is roused with greater difficulty than that of Hearing. An unexpected noise will often awaken a Somnambulist from his lethargy, upon whom the strongest rays of light will have had no previous effect, although his eyes continue open. Touch, as it was the last sense to become dormant, so it appears to be the most easily roused. The same person who cannot be awakened by very loud noises, will rise instantly on being gently tickled on the soles of the feet. Often the mere approach of the respiration of another will be sufficient to rouse the soundest sleeper.

The positions which the Mammalia assume during Sleep are very various. The young animal sleeps with its limbs gathered together, in a posture most resembling that of the fœtus in the womb. This situation is very favorable to the renewal of the animal forces, by permitting the relaxation of all the articulations, and in preserving the heat of the more sensitive parts. For the latter reason, the Dog and Cat sleep with their bodies formed into a circle. Some Mammalia sleep in the open air, while others retire to caves and sheltered places. Many repose without any covering, while others prepare a bed of some imperfectly-conducting substance, to preserve the temperature of their bodies, which would otherwise fall during Sleep below the natural standard. It is usually on the right side that Man reposes. This posture favors the action both of the heart and stomach, as the vibrations of the former would reverberate through the body from the reaction of the substance upon which it reposes, and the latter would be compressed by the weight of the liver. After sleep, all the organs, being refreshed, repaired, and completely nourished, acquire a greater size; thus Man and other animals which commonly hold the spine more or less erect, are taller in the morning than in the evening after the fatigues of the day.

Sleep is not always profound; some of the animal functions continue to act; ideas succeed each other, and the animal is said to dream. The power of dreaming is falsely ascribed to Man alone; other Mammalia dream likewise, because they are capable of thought, and possess a certain degree of intelligence. Sometimes the Dog is observed to howl, struggle, and perspire copiously. Moving his tail and limbs rapidly, he pursues the Hare in imagination, and, on the point of seizing it, closes his teeth and lips as if in the act of dyeing them in blood. Some Birds are also known to dream, as the Parrots. Those animals which are most easily excited dream more frequently than the others; thus the Horse is more liable to dream than the Bull. According to Chabert, this phenomenon among Cattle is observed only in the Bull, the Ram, or in Cows which are suckling.

It is possible to protract the usual period of sleep by an unusual excitement; but if the stimulus be long continued its effect goes off, and then nothing can prevent sleep as long as the health continues good. In fact, sleep, once in the twenty-four hours, is as essential to the existence of the Mammalia as the momentary respiration of fresh air. The most unfavorable conditions for sleep cannot prevent its approach. Coachmen slumber on their coaches and couriers on their horses, while soldiers fall asleep on the field of battle, amidst all the noise of artillery and the tumult of war. During the retreat of Sir John Moore, several of the British soldiers were reported to have fallen asleep upon the march, and yet they continued walking onwards. The most violent passions and excitement of the mind cannot preserve even powerful minds from sleep; thus Alexander the Great slept on the field of Arbela, and Napoleon upon that of Austerlitz. Even stripes and torture cannot keep off sleep, as criminals have been known to slumber on the rack. Noises which serve at first to drive away sleep, soon become indispensable to its existence; thus a stage-coach stopping to change horses, wakes all the passengers. The proprietor of an iron forge, who slept close to the din of hammers, forges, and blast furnaces, would awake if there was any interruption to them during the night; and a sick miller, who had his mill stopped on this account, passed sleepless nights until the mill resumed its usual noise. Homer, in the Iliad, elegantly represents sleep as overcoming all men, and even the gods, excepting Jupiter alone.

The length of time passed in sleep is not the same for all men; it varies in different individuals and at different ages; but nothing can be determined from the time past in sleep, relative to the strength or energy of the functions of the body or mind. From six to nine hours is the average proportion, yet the Roman Emperor Caligula slept only three hours. Frederic of Prussia and Dr John Hunter consumed only four or five hours in repose; while the great Scipio slept during eight. A rich and lazy citizen will slumber from ten to twelve hours daily. It is during infancy that sleep is longest and most profound. Women also sleep longer than men, and young men longer than old. Sleep is driven away during convalescence after a long sickness, by a continued fasting, and the abuse of coffee. The sleepless nights of old age are almost proverbial. It would appear that carnivorous animals sleep in general longer than the herbivorous, as the superior activity of the muscles and senses of the former seem more especially to require repair. Satiated with their prey, they are obliged to seek repose to digest those very substantial matters which compose their aliment.

In general, it may be stated, that during sleep the internal functions predominate over those relating to the exterior of the body. Every thing which tends to interrupt the relations of the external with surrounding objects serves to induce sleep. On the contrary, the existence of external stimuli tend to expel it, until at length they lose their effect by long-continued exercise.

From this it ought to follow that excessive cold, which benumbs the external powers, ought to occasion sleep. When exposed to the action of a low temperature, animals experience an irresistible desire to sleep, which soon terminates in death. Of this there are frequent examples in the inhospitable climates of the north, Siberia, Lapland, and Kamschatka, or on the tops of high mountains, as the glaciers of Switzerland. Dr Solander and party nearly lost their lives from this cause among the hills of Terra del Fuego. Surprised by an excessive cold, he was with difficulty prevented by his companions from yielding to this impulse of nature, although knowing well the consequences of sleeping.—(See Captain Cook's First Voyage). Travellers on horseback are peculiarly liable to be overcome by this propensity to sleep, when the cold is very intense, in which case they are sure to be frozen to death.

There prevails among many Mammalia a singular internal modification, which cannot be explained by any cause more general than itself, but must be referred to some unknown original constitution. We refer to that state of torpidity commonly called Hybernation, into which some animals fall during a part of the autumn and in winter, but from which they escape early in spring. Although we are wholly ignorant of the cause of this winter sleep, the effects and design are well known. It seems obviously intended to preserve the animals in situations where they could not have maintained their existence, from the impossibility of finding an adequate supply of food. Accordingly, all the active functions of life are suspended, where their exercise would be incompatible with more general laws.

At a more or less advanced period of the autumn, depending on the degree in which the temperature is lowered, animals possessed of this peculiar constitution seek to shelter themselves from the cold and wind, by retiring into holes excavated in the ground, walls, trees, or among the bushes. These retreats they line carefully with grass, green leaves, moss, and other bad conductors of heat. Hybernation occurs among several of the Mammalia, as in the Fat Dormouse (*Myoxus glis*), the Garden Dormouse (*M. nitela*), the Common Dormouse (*M. avellanarius*), the Hedgehog (*Erinaceus Europæus*), the Bats, the Alpine Marmot (*Arctomys marmota*), the Hamster (*Cricetus vulgaris*), the Jumping Mouse of Canada (*Meriones nemoralis*), and some others. Animals with cold blood hybernate as well as some of the Mammalia. Many Reptiles become torpid in cold climates, as well as some Insects, Mollusca, and Worms; but in general the degree of their lethargy is much less profound than that of the hybernating Mammalia. They pass their time of hybernation without food, but are not always deprived of sensation and motion, even at the freezing point.

"It is highly important," observes Dr Marshall Hall, "to distinguish that kind of torpor which may be produced by cold in any animal from true hybernation, which is a property peculiar to a few species. The former is attended by a benumbed state of the sentient nerves, and a stiffened condition of the muscles; it is the direct and immediate effect of cold, and even in the hybernating animal is of an injurious and fatal tendency; in the latter, the sensibility and motility are unimpaired, the phenomena are produced through the medium of sleep, and the effect and object are the preservation of life. Striking as these differences are, it is certain that the distinction has not always been made. In all the experiments which have been made with artificial temperatures especially, it is obvious that this distinction has been neglected. True hybernation is induced by temperatures only moderately low. All hybernating animals avoid exposure to extreme cold. They seek some secure retreat, make themselves nests or burrows, or congregate in clusters; and if the season prove unusually severe, or if their retreat be not well chosen, and they be exposed, in consequence, to excessive cold, many become benumbed, stiffen, and die. To induce true hybernation it is quite necessary to avoid extreme cold, otherwise we produce the benumbed and stiffened condition to which the true torpor or torpidity may be appropriated. I have even observed that methods which secure moderation in temperature lead to hybernation. Hedgehogs supplied with hay or straw, and Dormice supplied with cotton wool, make themselves nests, and become lethargic; when others, to which these materials are denied, and which are consequently more exposed to cold, remain in a state of activity. In these cases warmth, or moderate cold, actually concur to produce hybernation."

The kind of retreat which each animal prepares varies with the species. The Bats, besides hybernating in holes, also hang suspended in grottoes and caverns, where the temperature is milder than in the open air. Other hybernating animals are satisfied with bringing their head nearer to their lower extremities, so as to present a surface of less extent to the cold. On discovering them in their retreats, they are found rolled up, cold to the touch, motionless, stiff, their eyes closed, their respiration slow, interrupted, scarcely perceptible, or none at all. Their insensibility is sometimes such that they may be moved, rolled about, and shaken in every possible way, without being disturbed from their torpor.

During the spring or summer, when these animals enjoy their full activity, they possess an elevated temperature, which may vary according to the species and individuals, between 95° and $98\frac{1}{2}^{\circ}$ Fahrenheit, and consequently is between those limits which characterize the animals with warm blood. When examined in the autumn, with the view of ascertaining the changes which they then undergo, it is observed that their temperature falls rapidly as the cold season advances. Their respiration gradually becomes impeded, their movements are less rapid, and their appetite diminishes; but these animals still maintain the use of their senses and the power of locomotion. This intermediate state, between the full possession of life in all its vigour, and absolute torpidity, may continue for one or two months.

The degree of external temperature at which they become absolutely torpid varies with the species, and even according to the individual. In general, the disposition to hybernate follows in a descending scale of temperature, of which the following is the order:—The Bat, the Hedgehog, the Dormouse, the Marmot, and the Hamster; the comparison has, we believe, not yet been instituted for the other species. Although there is no precise degree at which these animals lose the faculties of sensation and motion, it has been remarked that the Bats become torpid between the temperatures of 50° and 44½° of Fahrenheit; the Hedgehog about 44½°; the fat Dormouse at 41°. The Marmot and Hamster cannot become torpid except at a temperature considerably below the freezing point; and they further require that their respiration should be impeded by diminishing, or altogether preventing, the accession of fresh air in the boxes or holes where they are confined.

Absolute torpidity can only be said to belong to these animals when their temperature has been fully reduced, and their respiration diminished, so that they at length come to be wholly deprived of sensation and voluntary motion. Hybernation is, however, susceptible of different degrees, which are characterized by the number of inspirations made in a given time; the absence of all respiratory movement marks in this case the highest degree of torpidity. All the species do not partake of it in the same degree. The Bats experience a very slight lethargy. On the other hand, the Marmot undergoes the most profound torpidity.

The pulsations of the heart and arteries become greatly enfeebled during torpidity. In the active state of the Hamster, the heart makes 150 pulsations in a minute, while in its torpidity it beats only 15 times. Bats, during summer, have about 100 pulsations in a minute. When they begin to grow torpid they have only 60; and as their lethargy increases, the action of the heart is so feeble that only 14 beats have been distinctly counted, and these were at unequal intervals. Dormice breathe so rapidly when they are awake, that it is scarcely possible to count their pulse; but as soon as they begin to grow torpid, 68 pulsations may be counted in a minute, 31 when they are half torpid, and only 20, 19, or even 16, when their torpor is not so great as to render the action of the heart wholly imperceptible.—(Reeve's Essay on Torpidity.)

The results of the recent researches of Dr Marshall Hall regarding the sensibility of hybernating animals are at variance with those of preceding observers. According to him, the slightest touch applied to one of the spines of the Hedgehog immediately rouses it to draw a deep inspiration; the merest shake induces a few inspirations in the Bat, and the slightest disturbance is felt, as appears from its effect in inducing motion in the animal. In fact, he considers the sensibility of these animals during hybernation to be in the same condition as in ordinary sleep. On the other hand, according to MM. Prunelle, Spallanzani, Mangili, Legallois, and Edwards, the strongest stimuli, with the exception of heat, make no impression upon them. Marmots are not roused from their torpid state by an electric spark strong enough to give a smart sensation to the hand, and a shock from a Leyden phial excited them only for a short time, as Spallanzani relates in his experiments made upon them jointly with Volta. They are insensible to the pricking of the feet and nose, and remain motionless and apparently dead. Bats are insensible also to every kind of stimulus except heat, or to a stream of air blown upon them, which affects their sensations powerfully. Wounds have been inflicted, and their limbs broken, without the mutilated animals exhibiting any external signs of pain.

The internal temperature of these animals during their lethargic sleep chiefly depends upon that of the external air, yet it is usually from 5½° to 7° more elevated than the latter. Hence their temperature is very variable; it may descend to 37½° without changing the state of the animal; but the internal temperature cannot go below 32°, the freezing point of water, without either waking the animal or occasioning its death.

There exists, therefore, a degree of external cold, which is incompatible with the torpidity or life of these animals. Species which most easily become torpid, such as the Bats, Hedgehog, and Dormice, cannot support an external temperature of 14°, and a heat of 50° to 54° likewise awakens them.

They may also be aroused by different mechanical means, such as by shaking them either gently or violently, according to the depth of their torpidity, without it being necessary at the same time to change the degree of external temperature. But if capable of resuming their activity in this way, they cannot long maintain it without the aid of a gentle heat. On being roused from their torpidity, they present all the phenomena of waking from ordinary sleep. When the torpid Hedgehog is touched it coils itself up more forcibly than before. The Dormouse unfolds itself when similarly treated, and the Bat moves variously. There is no stiffness nor lameness in their movements, and the Bat even flies about with great activity, although exhaustion and death are the certain consequences. Dr Marshall Hall thinks that those physiologists who assert the contrary have mistaken the phenomena of torpor from cold for a true Hybernation.

From what has been said, it is evident that the repose of the hybernating Mammalia is neither uniform nor constant in its duration. As it is influenced by the changes of the atmosphere, it may be continuous or interrupted according to the variations of the weather, or the precautions which the animals have taken to shelter themselves from the sudden changes of temperature, as well as their individual susceptibility.

When these animals are very liable to be awakened, either from their constitution or habits, they instinctively take the precaution of amassing stores of provisions to supply their future wants. The Hedgehog, for example, has been seen to form several separate stores, and to resort to them at different periods of the torpid season. Its traces have also been observed upon the snow.

When torpid animals are suddenly and frequently awakened, their respiration becomes heightened, and death soon follows. "All those Bats which were sent me from distant parts of the country," says Dr Marshall Hall, "died. The continual excitement from the motion of the coach keeping them in a state of respiration, the animal perished. One Bat had, on its arrival, been roused so as to fly about. Being left quiet, it relapsed into a state of hybernation. The excitement being again repeated the next day, it again flew about the room; on the succeeding day it was found dead." We may thus see one reason of the precautions which these animals take to preserve themselves from being suddenly disturbed or excited. They select sheltered spots, such as burrows or deep caverns, at once secure from their enemies as from the inclemencies of the weather. The Common Bat (*Vespertilio murinus*) hangs itself by the claws of the hinder feet, with the head downwards, while the Horse-shoe Bat (*Rhinolophus ferrum-equinum*) spreads its wings to protect and embrace its companions. Many other animals form nests, and some congregate together The Hedgehog and Dormouse roll themselves in a ball;—all which dispositions are evidently intended to preserve them from being disturbed by a low temperature.

There is no external character by which the hybernating animals can be distinguished from the others. Though some species belong to the same genus, such as the several Dormice, yet this phenomenon is also found in the Bats belonging to a family, separated by a wide interval, and Comparative Anatomists have sought in vain in the internal structure of these animals for an organization peculiar to them.

It might, however, be expected, that their organization would approach to that which these animals possessed when in their embryo state; and this actually happens to a certain extent. A large quantity of fat is lodged in different parts of the body, but especially in the appendages of the peritoneum, which are always more numerous and extensive than in other species. The sub-renal capsules, of whose use we are ignorant, but which we know are more developed in the fœtus than in the adult, are stronger and obtain some growth in these animals of which we are now treating. It is the same with the thymus gland and its appendages; that is to say, those granulous organs which are found to surround the necks of the torpid animals, such as the Marmots, Dormice, and Bats, and may even extend between both shoulders, as in the Bat, according to the observations of M. Jacobson of Copenhagen. These are nearly all the peculiarities yet observed in the hybernating animals, and they are very far from explaining the causes of this singular phenomenon.

Hybernating animals are not found in all orders of the Mammalia. None of the Quadrumana become torpid, probably because none of them are designed to inhabit a cold climate. The order of Carnassiers, on the contrary, contains several, especially among such as reside in cold countries; several are also found among the Rodentia, but the remaining orders do not contain any torpid animals.

We cannot by any means agree with M. Edwards in thinking that no species of animal is condemned by its nature to hybernate, and that the state of hybernation depends upon external circumstances, so that we can make it come and go by regulating the conditions under which animals are placed. On the contrary, we find that the nearer hybernating animals are permitted to approach to their natural mode of life in a domesticated state, the more they are disposed to follow their natural habits. Thus, torpidity seems perfectly congenial to the nature of the Marmot, and if any animal can be said to be naturally torpid it is this. Although it can live during the whole winter without becoming torpid, it by no means follows from this that its tendency to become torpid is artificial. We could no more compel other animals to become torpid on the approach of winter, to whom such a state was unnatural, than we could assign to ourselves a new organ of perception. The circumstance that we have not yet succeeded in referring the phenomena of hybernation to any cause more general than itself, only proves that the ultimate cause is complicated and obscure, but ought not to lead us to doubt of its existence.

In the preceding observations, reference has been made to those species only of whose hybernation there can be no doubt. Some species of Bears and Badgers, however, undergo a kind of lethargic sleep, termed *quiescence*. This state differs materially from that of ordinary hybernation, as the females bring forth their young during their interval of retirement. The common Bear (*Ursus arctos*) is always loaded with fat in the autumn, when he retires to a den previously lined with branches and soft moss. Here he sleeps but little if the winter should be mild, and licks his fore-paws and soles of the feet continually during the intervals of repose; but when the winter is severe, he sleeps much. This state of quiescence cannot be studied with the same facility as that of hybernation, as these animals never become quiescent when in confinement, but remain as much awake during the winter as in the spring and summer.

Some writers maintain that the number of animals susceptible of hybernation is very great; others are inclined to extend this supposition to all, even to Man himself. Thus, Addison mentions an Englishman who underwent a lethargic sleep from the 5th to the 11th of August annually. Sheep in Iceland have been known to live under the snow; and instances often occur in Cumberland, Westmoreland, and the Highlands of Scotland, of sheep existing for four or five weeks under drifts of snow, where they can procure little or no food, and must, it is supposed, have become torpid. Persons have been known to continue asleep from seven to fourteen days, and some much longer, apparently from the influence of fear, anxiety, or other causes which tend to weaken the vital powers. Yet these and other instances are far from establishing the fact of torpidity, when we are unable to induce that state in any of these animals, under circumstances which would be certain to bring it on in those predisposed to hybernate.

Had Man not been exempt from that unknown law of Nature which compels certain of the lower animals to become torpid, we should find it exemplified in all those cases where men have been exposed to cold, and no allusion to such a fact is made in the history of the human species. Yet Gmelin measured a natural cold of 120° below zero, at Jenislisk, lat. 58° N., long. 110° E., in the year 1735; and Pallas, in 1772, found the temperature at Krasnajorsk, lat. 56° N., and 110° E., to be 80° below zero, so that a mass of quicksilver exposed to the air was frozen and became malleable. The Greenlanders go about with very light clothing, and the Norwegian peasants work during the winter with their bosoms bare, or roll themselves in the snow.

It would appear that there are certain animals which experience a corresponding state, that cannot properly be termed hybernation, as it happens during the hottest months of summer, and in tropical climates. The Tenrecs, or Madagascar Hedgehogs (*Centenes ecaudatus*), are asserted by Bruguière to undergo this summer torpidity or æstivation, but his statement has more recently been called in question by Mr Telfair, in an account of that animal. Humboldt, however, has observed this remarkable state in the hottest parts of South America, in certain Reptiles which pass a part of the year buried in the earth, and do not leave their state of torpor until the rainy season drives them from their retreats. The singular state of torpidity, induced by the excitement of a high temperature, may be considered as analogous to that daily sleep of Man and the lower animals during the hottest parts of the day, and called the *sieste* in Spain and Italy.

Neither cold nor heat can thus be said to be the cause of hybernation and æstivation, although they are auxiliaries to those states. In the same manner as it is neither cold nor heat that compels us to sleep every night, but a constitutional fatigue, and the necessity of repairing our forces; so in these annual sleeps the absence and presence of heat are merely proximate causes. Nature has wisely constituted these animals thus; for it is probable that their feeble constitutions would have otherwise failed to resist the cold, or they would perhaps have been unable to procure an adequate supply of food. When the period of torpidity is over, the rutting season usually commences, and these animals then enter anew upon the functions of life with renovated faculties.

GÊNERAL REVIEW OF THE MAMMALIA CONTINUED.

Phenomena of Reproduction—Growth—Duration of Life.

SCARCELY have the Mammalia attained the full period of their growth, when another order of functions make their appearance. Their animal forces then acquire a new direction, and Life, which was formerly confined to the development of the individual, now applies itself towards the continuation of the species.

The time when the phenomena of reproduction first exhibit themselves, is termed *Puberty.* Then the reproductive organs, which previously were but slightly apparent, acquire a remarkable development, and in some species obtain certain external characters which remain during the whole course of their lives. Infancy is the period comprised between birth and puberty. It is during the time preceding puberty that the growth of the body chiefly takes place, although it may continue for some time afterwards. The length of the period of infancy bears to that of life a certain relation, which may be regarded as almost constant. Buffon remarks, that in our climates, for the largest animals, it is about the one-seventh part of their entire life.

At the age of puberty, the Mammalia assume the characters of maturity. Their height attains its greatest limit, and the distinctive marks of each animal become bold and well-defined. The physiognomy assumes a more animated expression; their voice becomes hoarser or stronger, and the fur handsomer: while the vivacity of their movements marks the impetuosity of those passions which animate them at this epoch. The male becomes distinguished from the female by colours which are commonly darker or browner, and in many species by certain definite external characters. Thus, some male Apes acquire a beard and a coat of long hair; the Lion obtains a mane; and the Stags and Roebucks are armed with branching horns, of which the females are nearly always deprived. The He-Goats and Rams are at once distinguished from the females by their horns, their masculine gait, and combative disposition. This superiority in the males is most marked among the Ruminantia, which are commonly polygamous, and where each male having several of the other sex to keep in subjection, it becomes necessary to assign him a physical superiority, unnecessary in the monogamous species, where the sexes are always more equal in strength.

Puberty constantly exhibits itself much sooner in females than in males, although the reproductive power remains longer with the latter than with the former. In our climates Man attains this condition at the age of fifteen or sixteen, and Woman at that of fourteen or fifteen; in warmer climates it exhibits itself at the age of twelve to fourteen in the former, and at ten to twelve in the latter. With most of the other Mammalia, excepting the domestic animals, we are still ignorant of the precise periods when puberty commences. Dogs are capable of reproducing at the age of nine or ten months; Cats from a year to eighteen months. A Lioness of the "Ménagerie" at Paris was six years old when she exhibited these phenomena for the first time. Rabbits can procreate at the age of five or six months; Hares a little later; and Guinea-Pigs at five or six weeks. Sheep show signs of puberty when one year old; Rams, He-Goats, and Stags, at eighteen months. Horses produce at two years and a half, and Mares a little sooner. Camels, according to the ancients, at three years; Wolves at two years; Cows at eighteen months; Bulls six months later; the She-Ass from eighteen to twenty months; and the Ass at two years. It is, however, the interest of the Grazier to prevent the domestic animals from procreating before they have attained their full growth, otherwise the deterioration of the races is sure to follow.

There are certain seasons of the year when most Mammalia become susceptible of the instincts of reproduction. This is termed the *rutting* season, during which the usual character of the animals is totally changed, especially of the males. The most timid animals, being excited by the abundance of food and the internal suggestions of instinct, acquire a degree of courage and even fury, which urges them on in a career of madness, which can be compared only to the habitual ferocity of the most formidable species. The females also, at this period, lay aside their habitual reserve, and are seen to provoke the males by biting, teasing, and following them everywhere.

Some Mammalia in our countries, as well as in those of the south, whether males or females, remain always in a state adapted for procreation, after having once attained the age of Puberty. With the exception of the Monkeys, this happens only to those species, which either receive an abundant nourishment from Man, or else obtain a plentiful supply by plundering his stores. Of the first kind are the Dog, Cat, Rabbit, Guinea-Pig, Hog, Bull, Buffalo, Horse, and Ass; in the second division must be placed the common Rat, the common Mouse, the Wood Mouse, the Economic Mouse, and the Hamster. With the animals of the Menagerie, the change of climate which they experience, and the constraints of confinement, occasion them to undergo certain deviations from their natural period of rutting. Among these may particularly be enumerated, the Lion and other Cats from warm climates, the Cape Ichneumon (*Mangusta grisea*), and the Cape Genet (*Paradoxurus typus*), the Ichneumon of Egypt, the Gnu Antelope, and the Zebra; also the Axis Deer from the banks of the Ganges, and the Kangaroo of New Holland.

It commonly happens, however, with the Mammalia, when they have not been modified by domestication or confinement, that each has a peculiar season when the phenomena of the rut more especially present themselves. Thus, winter is the rutting time of the Wild Cat and Martens of Europe; of the Wolf, from December to February; of the Jackal and Corsac Fox (*Canis Corsac*), only in winter. The Arctic Fox (*Canis lagopus*) is in season at the end of February; the Bear, in summer; the Hedgehog, at the end of winter; and the Hare in February or March. The Beavers seek the females in the beginning of January; the American Ondatra and the Common Squirrel, in spring. The Dromedaries seem to be more in season about the month of January than at any other time. Camels begin about the middle of November, and end at the commencement of February. The month of September is the chief time for the Sheep and Goats, although the males of both species are always fit for procreation. In the Stags of our countries, the Roebuck, and other Deer, the rutting season succeeds to the period when the horns are renewed, that is in November, and after this time the horns fall. The Rein-Deer are in the same case.

Thus the season of the rut varies with the species; but it is always so arranged in reference to the term of gestation, that the young may make their appearance at a favorable season of the year, when the heat of summer will serve to aid their growth, and assist in developing their forces. At this season, also, a luxurious vegetation supplies the herbivorous animals with abundance of food, which favors the secretion of milk, and ensures its continued supply.

The external signs of the rutting period vary greatly with the several species. In those which are capable of procreating at all seasons, such as Man, the Monkeys, Dogs, Cats, and Horses, no particular sign is observed. It is different with the Rodentia. "Dàns la plupart des Rongeurs (Rodentia), les testicules, ordinairement petits et comme cachés dans l'abdomen, prennent un volume très considérable et deviennent fort apparens. C'est en particulier, ce qu'on remarque dans les Rats, les Surmulots etc. où ces parties font, à cette époque, une saillie très-remarquable à la base de la queue, et donnent au corps une figure pointue vers cette extrémité." At this period the Elephants secrete, on the side of the head behind the ears, a brownish fluid, which proceeds from glands situaté under the skin. The Bactrian Camel diffuses a most disagreeable odour at this season. At first he undergoes a violent perspiration, which lasts for fifteen days; then a blackish and viscous fluid exudes from the neck, not through any particular opening, but merely from the pores of the skin, so that the Persians are obliged to cut his hair very close. In the Dromedary, also, the male presents at this particular season a similar phenomenon. We find, likewise, that all those odoriferous Mammalia, which are supplied with pouches from whence the odours emanate, emit their perfumes at this time with unusual force. In the greater number of animals belonging to the Deer Genus, and in several Antelopes, the larynx or windpipe of the male projects considerably; and it cannot be doubted that the change of tone which his voice undergoes is owing to this cause.

It usually happens that the females exhibit the external signs of the rutting season in a milder and more subdued form than the males. "Alors seulement," observes M. Desmarest, "les organes externes de la génération se tuméfient légèrement, s'entr'ouvrent, et sont continuellement humectés par un fluide plus ou moins visqueux, qui, chez les jumens, où il est particulièrement abondant a reçu le nom d'hippomanès. Neaumoins, la tuméfaction et la rongeur excessive des fesses de certaines femelles de singes doivent être considérées comme un signe du rut, et sans nul doute aussi, les écoulemens sanguins qui ont lieu à des epoques régulières et plus ou moins rapprochées, mais fixes chez celles de quelqaes espèces." These appearances may present themselves in the females at intervals more or less considerable in the course of the year.

Among the Mammalia, and indeed in all living beings, the period of puberty and reproduction is one of energy and strength; all their affections become more ardent, and their wants irresistible. The term *rut*, from *ruere*, to rush headlong, serves to express the fury which transports these lower animals.

In furias ignesque ruunt, amor omnibus idem.

Alike ferocious and untameable, they are susceptible at this period neither of fear nor any other passion, and seem deaf even to the calls of hunger or sleep. The Bull forsakes the meadows, and rambles everywhere in search of his mate. The forests resound with the howling of contesting Wolves, and the Lion, with a deafening roar, defies his rivals to the combat. We may easily perceive the final cause of these contests among the lower animals during the rutting season. Nature ever sacrifices the interests of individuals towards the perfection of species. The most vigorous males always possess the most formidable weapons of attack and defence, while the more effeminate individuals exhibit their feebleness at once in their horns and their want of courage. It is especially among the polygamous races where these combats of the rutting season are more conspicuously observable, because each male fights for several females. In the monogamous species, on the contrary, where the numbers of the sexes are nearly equal, these battles seldom occur. Again, in the Carnivora, when the number of the males surpasses that of the females, duels become both frequent and sanguinary. The Seals (*Phoca*) are perhaps more polygamous than any other of the Mammalia. Each maintains a kind of seraglio or family, composed, perhaps, of one hundred and twenty females, which he defends from the approach of any other male, with the utmost jealousy and rage. Other species, less faithful or more complaisant, pass from conquest to conquest, and pay their court to all the beauties of the neighbourhood.

The duration of this season varies with different species; but, in general, among the wild animals it ceases as soon as the females have been fecundated. With most of the latter, the external signs of the rut immediately disappear; the females resume their usual reserve, and repel with rudeness the approaches of the male. There are exceptions, in the Monkeys, the Mare, and in our own species. The female Rabbit is likewise an exception, though only an apparent one; as from the peculiar formation of the matrix, she is susceptible of a twofold impregnation, or superfetation.

"Le mode d'accouplement varie peu dans les Mammifères; en général cet acte a lieu comme dans nos espèces domestiques d'Europe. On avait dit, que ceux de ces animaux qui, dans l'état ordinaire, ont la verge dirigée en arrière, comme les Rhinocéros, les Chameaux, les Lamas, les Dromadaires, etc., s'accouploient en arrière; mais il n'en est rien; dans l'érection, leur verge reprend sa direction en avant, et le coït a lieu comme à l'ordinaire, mais, à la vérité, avec plus de difficulté. Les singes seuls s'accouplent à la manière de l'homme, mais c'est à tort qu'on a prétendu que l'éléphant en faisoit de même. Selon Buffon, le mâle et la femelle du hérisson ne peuvent s'accoupler comme les autres quadrupèdes; il faut qu'ils soient face à face, debout ou couchés. Les rats s'accouplent en se mettant debout, ventre contre ventre. Dans beaucoup d'espèces chez lesquelles, comme les chats, les lions, les gerboises, etc., le gland du mâle est muni de pointes cornées plus ou moins longues, et quelquefois dirigées en arrière, l'accouplement est très-douloureux. Chez les chiens il dure fort long-temps, ce qui est dû à une conformation particulière du vagin de la femelle. Dans d'autres, comme dans l'espèce du taureau, il est terminé en quelques secondes. Tantôt la femelle se tient debout, tantôt elle s'accroupit sur ses deux jambes antérieures; tantôt le mâle se maintient à l'aide de ses deux mêmes membres de devant, ou saisit la peau du cou de la femelle avec ses dents, etc. On a cru long-temps qu'un sixième ongle surnuméraire, qui se trouve au côté interne des pieds de derrière des Echidnés et des Ornithorhinques étoit destiné á faciliter l'accouplement; mais on a découvert depuis peu que ces organes avoient un tout autre objet, et que c'etoient des armes empoisonnées dont ces animaux se servoient contre leurs ennemis; et c'est ce que M. de Blainville a confirmé. (Desmarest in Nouv. Dict. des Sciences Nat.)

In some species of domesticated animals, especially in the Dog, copulation is maintained for a long time after the emission of the fecundating fluid; whilst among the greater part of the Birds, especially in the Gallinæ, the union is instantaneously dissolved. It is always dangerous in the former cases to force a separation, which is sometimes attempted, although opposed by the peculiar organization of the sexual organs; the intention of Nature apparently being, by this extraordinary prolongation of the union, to render conception more certain. "L'étroite conjonction parôit destinée, dans le principe, à produire l'irritation nécessaire à l'émission de la semence; et le plaisir qui en résulte est le ressort qui détermine le plus puissamment les animaux à la propagation, quoiqu'elle paroisse douloureuse d'abord dans quelques espèces, comme dans le genre chat, dont les femelles poussent souvent alors des cris aigus. Dans quelques oiseaux polygames aussi, comme les Faisans, l'accouplement parôit être un acte plus violent que voluptueux; car on voit les femelles redouter l'approche du mâle, qui fait usage de sa force pour les y contraindre." (Yvart.)

The older females of each species exhibit an attachment for the males, at an earlier season of the year than those of a less advanced age. After conception, as has already been observed, the females, in general, repel the approaches of the male. In all cases where the races are peculiarly ferocious, as in the Lion, Tiger, Panther, and other large Cats, the females are the first to solicit the approaches of the male. Had this not been the case, it is difficult to conceive in what manner their races could have been continued. In species of a milder disposition, the males endeavour to please the other sex, and often exhibit a strongly-marked feeling of jealousy towards their own. The Monkeys remain attached to one or two females, rarely to more. Their union seems to be a kind of marriage; they require fidelity, are exceedingly jealous, and severely punish their female companions, who are well-disposed to coquetry, on finding them in company with other males.

The phenomena of generation must, however, be considered in another point of view, at once singular and surprising. It seems now beyond a doubt, that the power enjoyed by male animals of continuing their species depends upon the presence in the spermatic fluid of a certain kind of animalcules, which have thence received the name of *Zoospermata*. The testicle, or secreting organ, is well known to be the means by which the ovaria are fecundated; and it is remarkable, that within this organ alone Zoospermata have been hitherto observed.

All animals, before arriving at the age of puberty, are incapable of reproduction, and they are accordingly wholly destitute of Zoospermata. M. Dumas informs us that he has made a considerable number of experiments upon young animals, and that he found all of them to be destitute of Zoospermata. He particularizes the young of the Rabbit, Calves, Foals, young Asses, Guinea-pigs only a few months old, Mice of the same age, a great number of Norwegian Rats, with Pullets, young Ducks, and even Frogs. The fluid extracted from their organs contained the same kind of irregular globules which are found in the testicles of the Mule; but it was wholly deprived of moving bodies, and nothing was found which could in any way approach to the peculiar form of those animalcules proper to fertile animals.

It is well known that animals become sterile at a certain period of life, varying with the species. To ascertain whether the presence of these animalcules is essential to fecundity, it becomes necessary to investigate whether very old animals possess them or not. This, to a certain extent, has been done. M. Dumas examined a Stallion aged twenty-five years, and which had been incapacitated through age for about four or five years, as well as some Dogs of a very advanced period of life. Their sexual organs were perfectly healthy, yet he found them to be destitute of animalcules, and the fluid within them resembled in every respect that of the young individuals already mentioned.

These facts serve to establish the importance of animalcules, and appear to show that their presence is essential to the fecundating power of animals.

With the view of setting this interesting point beyond a doubt, many experiments have been instituted. The spermatic fluid of a Dog was placed in two silver capsules in equal quantities. In one of these, a metallic rod, polished at its extremity, was plunged, in such a manner, that the rod and capsule might be placed in communication with the two surfaces of a Leyden phial, strongly charged. An electric spark was then made to pass through the fluid, but not at its surface. After a few discharges, the animalcules became entirely motionless, while the other capsule, which had not been electrified, was animated with them as completely as it had been previously to the experiments, which did not last for five minutes.

The result of this and other experiments, which cannot be detailed here, appear to show that these animals possess irritability, and are destitute of the muscular system of other animals. It is, however, certain that all male animals hitherto examined, possess spermatic animalcules when in a state of puberty. Young individuals, as well as those that are aged, exhibit no traces of them, and even it is remarked, that Birds are destitute of spermatic animalcules, except at those particular periods of the year which Nature has fixed for their procreation. The domestic Cock and Pigeon, being fertile all the year round, are, of course, exceptions.

These spermatic animalcules exist within the testicle in a state of complete perfection; they are transmitted to the deferential canals, without undergoing any alteration during the transition. Neither their motion nor their form is influenced by the mixture of fluids from the other glands, so that, on being emitted, they appear in the same state as when in the spermatic vessels themselves. The spontaneous movement of the animalcules is intimately connected with the physiological state of the individual which supplies them. Each species possesses a species of animalcule, of a form peculiar to itself, and no two species hitherto examined have the same kind of animalcules, although they always remain the same in the same species. The electric spark kills them, but they are not affected by the galvanic current, even when used in a degree of intensity sufficient to decompose water and the salts which it contains.

Whatever opinion may be held as to the part performed by these animalcules, in the function of generation, it cannot be doubted that they exist solely in the essential part of the generative organ, in all animals which do not reproduce by buds or offshoots. On the other hand, they are wholly wanting in animals incapable of generation, and their presence in the seminal fluid may thence be assumed as the index of its fecundating power.

With the exception of a few solitary instances of superfetation, it is always found among the Mammalia, that the fecundation of the male refers only to one birth, being that which next immediately follows the union of the sexes. On the contrary, with the Birds, a single union may influence several successive broods. Thus, the domestic Fowl will produce fertile eggs at tolerably remote periods of time, after having once received the influence of the male. A young and vigorous Cock is adequate for fifteen fowls, and serves to fecundate all the eggs they may produce during twenty days. Hence one male may be sufficient to give existence in a single day to three hundred chickens.

The phenomenon of gestation can only be observed in the Mammalia and other viviparous animals. The term *Gestation*, from the Latin *gestare*, to carry, denotes the period of time which elapses between conception and birth. Among Birds and all other oviparous animals, a real gestation cannot exist, because the eggs detach themselves from the ovaries, pass along the oviducts, and are deposited as soon as they are formed. With these animals, gestation becomes superseded in general by incubation, to which it is greatly analogous, and the former function may thus be considered as little else than an internal incubation. The apparent design of Nature, in both cases, is to favour the gradual development of the embryo or fœtus—the first rudiment of the new animal resulting from conception. It is also observed, that the rapidity of growth in the fœtus, whether during the gestation of the viviparous animals, or the incubation of the oviparous, always diminishes in proportion as the fœtus approaches the time appointed by Nature for its birth.

The length of the period of gestation, like that of incubation, varies greatly among the several genera and species. It further obtains certain accidental variations, which appear to depend upon the age of the mother, her state of health, an increase or diminution in the velocity of the circulation, the quantity or quality of the food, and all those causes, derived from the influence of climate, soil, shelter, and the different kinds of treatment which these animals receive from Man. The period of gestation may also be either shortened or prolonged, according to the temperature which prevails during that interval. It is a matter of common observation among graziers, that two cows, though fecundated on the same day, will yet produce at an interval of several weeks. The variation among sheep under similar circumstances amounts to a few days, but in general, this difference among domestic animals of the same species may extend as far as twenty days.

It commonly happens in all those species, where the individuals take a long time in arriving at their full growth, that the period of gestation is considerably prolonged; and the converse is equally true; for in all those species which are very precocious, the time of gestation is extremely short. This rule is not, however, without many exceptions. Thus, the Goat and Sheep are capable of reproducing at the age of two years, and have commonly attained their full growth at this period, while their ordinary time of gestation is about five months. On the other hand, a Lioness at the *Ménagerie du Muséum*, in 1801 and 1802, seemed unfit for procreation before the age of two years, and yet she produced after a gestation of 108 days only, or rather more than three months and a half.

The duration of gestation seems further to depend upon the comparative volume of the species; this rule, however, is by no means invariably preserved. Thus, the Ass and Zebra, though less in volume than the Cow and Buffalo, employ less time in performing this function than the latter species.

It hence appears that the duration of gestation varies in different animals, and the empirical laws deduced from multiplied observations are not without many exceptions. By combining, however, the general organization of the Mammalia, with the time necessary for each animal to arrive at its full growth, as well as with the comparative bulk of the females, it is possible to obtain a general and definite result; while the characteristic thus obtained may, with propriety, be added to those which commonly serve to distinguish the leading groups of Mammalia. Thus in Man, nine months is the well-known period of gestation. Among the Quadrumana, it is also nine months for the larger species, but only seven for the smaller. In the Carnassiers, gestation endures six months with the Bear; 108 days with the Lion; nine weeks with the Arctic Fox (*Cunis lagopus*); from 55 to 56 days with the Cat; the same period for the Martens and Weasels; from 62 to 63 days with the Dog; and nine months with the Morse. Those Mammalia which experience the shortest term of gestation are unquestionably the animals belonging to the Order Marsupialia. Among the large Kangaroos, for example, the young are scarcely more than an inch in length, when they first attach themselves to the breasts of their mother, although the full-grown animal is at least five feet in height. Gestation is also of short duration in the Rodentia, being only four months in the Beaver, one of the largest animals of this order. It is still less in the smaller Rodentia, being from 30 to 40 days in the

Hares and Rabbits; 31 days in the Dormice; four weeks in the Squirrels and Rats; and three weeks in the Guinea Pig. Among the Pachydermata, gestation is of much longer duration; it endures with the Elephant from 22 to 23 months; it lasts from 11 to 12 months in the Horse and Ass; in the Zebra for a year and some days; in the Tapir, from 10 to 11 months; in the Hog and Boar for four months. Further, it endures among the Ruminantia, for twelve months in the Dromedary; for nine months in the female Buffalo and Cow; for eight months and some days in the females of the common Red Deer (*Cervus elaphus*), the Fallow Deer (*C. dama*), and the Rein Deer (*C. Tarandus*); five months and a half for the Roebuck (*C. capreolus*); five months for the Goat, the Sheep, the Mouflon, and several Antelopes. We are hitherto without any positive information regarding the period of gestation among the Cetacea.

It is evident that the number of births appropriate to each species will mainly depend upon the average length of each period of gestation. On this account, the larger species do not produce every year, especially when a long term of lactation also intervenes. The smallest species, on the contrary, multiply most prodigiously, and it may be generally stated that, if we except the Rabbit and Hog, both the number of births, and the number of young ones at each birth, are in general more considerable in proportion as the size of the animal is less. The Guinea Pig can produce every two months; the Hamsters, the Rats, the Mice, the Field-mice (*Arvicola*), and the Shrews, do not produce less than three or four litters in the course of the spring, summer, and autumn. With respect to the exceptions above stated, it may be inferred that the abundance of food which those animals obtain from Man has modified their nature; for we find that in all the wild species which approach nearest to these domestic races, the number of young produced at each birth is always less, as may be remarked in the Hare, producing only from three to four young at a birth, and the female of the Boar from three to eight.

The number of young in each litter also bears an immediate reference to the length of gestation. At each birth, Man and the Quadrumana commonly produce only one, very rarely two or more, and the Cheiroptera bear two. Among the Carnassiers, the Tiger produces one; the Lion, three or four; the Cat, four or five; the White Bear, two; the Brown Bear, from one to three; the Wolf, the Fox, and the Adive, from four to five; the Arctic Fox, from five to seven; the Badger, from three to four; the Mole, from four to five; and the Seals, one or two. Among the Marsupialia, the Opossums produce from eight to ten, but the Kangaroos only one or two. Of the Rodentia, the Beaver bears two or three at a birth; the Rabbit from four to eight; the Hamster from five to six; the common Rat, the Mouse, and the brown Rat, or Surmulot, from eight to ten. The Agouti bears four, according to Laborde, or only two, according to Buffon and d'Azara. The garden Dormouse produces five or six young ones at a birth; the common Dormouse three or four; the Guinea-pig from seven to ten; the common Squirrel from three to five; and the Marmot three or four. Among the Edentata, the Sloths produce only one, as also the Ant-eaters, while the Armadilloes bear four at each of their births, which occur pretty frequently. With the exception of the Pig, the Pachydermata produce but few young at a time; thus, the Elephants, Rhinoceroses, Hippopotamus, Tapir, and all the Horse genus, have only one; the Peccari has two, while the female Pig will bear as many as twelve, and even twenty. All the Ruminantia produce two or more, excepting the largest species, which have only one. The Cetacea produce, in general, but one young one at each birth. It most commonly happens that the first and last litter of each animal are deficient in number, and often also in strength.

It thus appears, that the largest and most formidable species are far less fruitful than the smallest and weakest. Not only are the former longer in arriving at their age of puberty, but their periods of gestation and lactation are prolonged, and the number of young at each birth is, in general, less. Thus, while the Tiger produces only one Cub at a time, the Wild Cat will bear four or five. "In this manner, the lower tribes become extremely numerous; and, but for this surprising fecundity, from their natural weakness, they would quickly be extirpated. The breed of Mice, for instance, would have long since been blotted from the earth, were the Mouse as slow in production as the Elephant. But it has been wisely provided, that such animals as can make but little resistance, should at least have a means of repairing the destruction, which they must often suffer, by their quick reproduction; that they should increase even among enemies, and multiply under the hand of the destroyer. On the other hand, it has as wisely been ordered by Providence, that the larger kind should produce but slowly; otherwise, as they require proportional supplies from Nature, they would quickly consume their own store; and, of consequence, many of them would soon perish through want, so that life would thus be given without the necessary means of subsistence. In a word, Providence has most wisely balanced the strength of the great against the weakness of the little. Since it was necessary that some should be great and others mean, since it was expedient that some should live upon others, it has assisted the weakness of one, by granting it fruitfulness; and diminished the number of the other by infecundity."

Thus in general, it would appear that the fecundity of animals is greater in proportion as they are more liable to perish from external causes. Insects, plants, and the smaller species of Mammalia, which cannot escape from danger, are exceedingly fertile, because Nature diminishes the chances of death by those of life, so that the species may exist continuously. The number of young at a birth thus serves as an index to the probable perils of each species, as well as the voracity of its enemies.

Among all the monogamous species, such as several Rodentia, Bats, and Moles, a kind of family is established during the interval, necessary for the support of the young, when the father and mother divide the cares of their family between them. A mutual tenderness seems to prevail in these little societies, which are connected by affection alone. Each animal shares in the common labour, and each partakes in its share of the produce. These societies, among many Rodentia, are almost as intimate as those of Man, whose articulate language is here supplied by a natural system, where cries and gestures supply the place of words. But when once the young have become strong enough to maintain themselves without the aid of their parents, they become estranged, and form other family connexions. They cease to recognize each other after a certain time, and become wholly indifferent to the nearest ties of blood. They even do not scruple to contract alliances with their nearest relations of the other sex. It is seldom, however, that the young show any partiality of this kind towards their older relations. On the contrary, this anomalous feeling nearly always exhibits itself in the attachment of the older for the younger animals.

With polygamous species, such as the Ram, Goat, and Bull, there exists no attachment of relationship on the part of the male. Possessing several females, he has no affection for his young, and the mother alone takes charge of them, during their unprotected state of infancy. In these species, the mother bears a much less number of young at a time than the monogamous kinds, and hence the female suffices to nourish and protect them. The polygamous species being generally also herbivorous, and the young being capable of walking from the moment of their birth, they are sooner able to subsist without the aid of their parents than the Carnassiers. The latter, accordingly, are monogamous in general. Their young are often born with their eyes closed, and with imperfect senses.

It may be remarked, that although the Herbivorous animals produce, in general, only one or two young ones at a birth, this limited power of production is compensated by the greater number of females which the males are capable of fecundating; thus, a single Bull or Ram is sufficient for a flock of twenty Cows or Ewes. But the Carnivorous animals, being chiefly confined to one female, produce a more numerous race. It thus results in animals, as well as in man, that fruitfulness is the common attendant of monogamy and chastity.

Animals of different species are destitute of the power of producing fertile races. Not only are the individuals themselves naturally averse to unite, but there is found a great variety in the forms of their organs of generation, and in their different periods of gestation, while the Hybrid produce, or Mule, is, in general, unfruitful. These adulterous unions can hence only take place between animals which are very nearly allied to each other, as between the Horse and Ass, the Buffalo and Cow, the Bison and Zebu, the Camel and Dromedary, all the combinations in pairs among the Wolf, Fox, Jackal, and Dog; between the Ram and Goat, the Hare and the Rabbit. These animals are capable of mutually fecundating each other, whenever man can succeed in overcoming their natural antipathy. But there are certain disparities of organization, which wholly prevent the Dog from uniting with the Cat, the Bull with the Mare, the Ass with the Cow, although some examples of the last kind of union have been imagined. The unions between the larger species of Apes and the human species, as well as their fabulous product, are not authentic, as some have credulously supposed.

Animals which produce more than one young one at a birth usually bear an even number. This proceeds from the circumstance that each ovary supplies its contingent of ovaria to be fecundated. In the same manner, Nature assigns an even number of Mammæ to these viviparous quadrupeds. Human twins are most commonly both males or both females, although sometimes they are male and female, but these last happen more rarely. Four at a birth is very uncommon in the human species.

This gemelliparous property is often peculiar to particular families. Twin brothers are often the fathers of twins at several successive births, and in one case, a second marriage having taken place, the latter wife produced twins likewise. In this kind of generation, it is probable that the impregnation of the two ovaries happens at the same moment, especially as we know that animals, habitually multiparous, only require a single union, although doubtless superfetation may also be induced by subsequent unions.

The young of nearly all Mammalia are born with their eyes closed, and do not open them for several days. The mother cuts the umbilical cord with her teeth, and, even without being carnivorous, devours the membranes or after-birth, as in the Cow, the Sheep, and many others.

As soon as the young are born, their mother takes a peculiar care of them, until they are sufficiently strong to find a maintenance without her aid. The female Rabbit prepares a bed of fur for her litter, which she tears from the under part of her body a few days before producing. The She-Bear collects hay and other soft substances in her retreat for a similar purpose. "Whatever be the natural disposition of animals at other times, they all acquire new courage, when they consider themselves as defending their young. No terrors can then drive them from the post of duty; the mildest begin to exert their little force, and resist the most formidable enemy. Where resistance is hopeless, they then incur every danger, in order to rescue their young by flight, and retard their own expedition, by providing for their little ones. When the female Opossum, an animal of America, is pursued, she instantly takes her young into a false belly, with which Nature has supplied her, and carries them off, or dies in the endeavour. I have been lately assured," continues Goldsmith, "of a She-Fox, which, when hunted, took her Cub in her mouth, and ran for several miles without quitting it, until at last she was forced to leave it behind, upon the approach of a Mastiff, as she ran through a farmer's yard. But if, at this period, the mildest animals acquire new fierceness, how formidable must those be that subsist by rapine! At such times, no obstacles can stop their ravage, and no threats can terrify; the Lioness then seems more hardy than even the Lion himself. She attacks Men and Beasts indiscriminately, and carries all she can overcome reeking to her Cubs, which she thus early accustoms to slaughter. Milk, in the Carnivorous animals, is much more sparing than in others; and it may be for this reason, that all such carry home their prey alive, that, in feeding their young, its blood may supply the deficiencies of Nature, and serve instead of that milk, with which they are so sparingly supplied. The choice of situation in bringing forth is also very remarkable. In most of the rapacious kinds, the female takes the utmost precautions to hide the place of her retreat from the male, who, otherwise, when pressed by hunger, would be apt to devour her Cubs. She seldom, therefore, strays far from the den, and never approaches it while he is in view, nor visits him again, till her young are capable of providing for themselves. Such animals as are of tender constitutions, take the utmost care to provide a place of warmth, as well as safety, for their young. Some dig holes in the ground; some choose the hollow of a tree; and all the amphibious kinds bring up their young near the water, and accustom them betimes to their proper element. The rapacious kinds bring forth in the thickest woods."

The young are at first nourished entirely by the Milk secreted from the Mammæ or breasts of their mother. Each mamma is a conglomerate gland, covered with a tenacious cellular tissue: it is formed of rounded grains, separated from each other by fat, and surrounded by spongy and cellular tissues. In the midst of this gland, a number of lactiferous canals cross each other, being semi-transparent, susceptible of dilatation, and re-uniting in several leading branches towards the nipple. Besides this general conformation, there are several thoracic, epigastric, or hypogastric arteries, independently of numerous lymphatic vessels, which carry their ramifications throughout these organs. They are also very numerously supplied with nerves, for their sensibility is very great. The nipple, which is only covered by a mucous tissue, with a very fine skin and epidermis, is delicately sensible to the slightest touch. "Elle est formée d'un tissu vasculaire particulier qui jouit de la propriété d'entrer en une veritable érection analogue à celle de la verge et du clitoris; car ces organes ont beaucoup de sympathie entre eux. Elle reçoit de sang et devient rouge et très-sensible alors. Les conduits s'ouvrent et sont prêts à faire jaillir le lait de même que le sperme est éjaculé par les canaux excréteurs des vésicules seminales. En effet, il y a une grande ressemblance entre l'action de la glande mammaire et celle des organes de la génération."

The mammæ may be placed, according to the species, on the breast, the groin, or the abdomen. Their number is often relative to that of the young. In the larger species, which have only one or two young at each birth, there are usually but two mammæ, whether pectoral or ventral. With the species of medium size, there are most commonly eight; although some may have as many as fourteen.

The Carnivorous animals most commonly have from six to ten placed longitudinally under the abdomen; the Opossums and Kangaroos have four to eight, fixed within a fold of skin, or inguinal purse, within which the young lodge securely. The Elephant, as also the Quadrumana, usually have two upon the breast, as in Man. The female Hog has from ten to twelve, and the Ruminantia, whose milk seems to be more substantial than that of any other domestic animals, have generally two to four mammæ. These numbers point out the maximum limit to the number of young, which each female, when in a healthy state, is capable of nourishing without inconvenience. Among the gregarious tribes of Mammalia, the young recognise their mother with surprising accuracy by the sound of her voice, or by the smell, in the midst of the most numerous flock. Those young possessed of the greatest vigour will, however, take milk from several mothers, at the expense of the weakest, which are thus deprived of a portion of the food intended for them by Nature. Some unnatural mothers drive their young away on first approaching their udder, without exhibiting the slightest compassion for the unprotected state of their offspring, which are thus in general left to perish.

With the greater number of Mammalia, the young take and leave the breast according to their wants; but it is different with the Marsupialia, the young of which attach themselves so forcibly to the mammæ, that they would rather permit themselves to be decapitated than leave the nipple. They remain continually in this position until their bodies become entirely covered with hair, and they possess strength sufficient to gambol around their mother. Among most species of this singular class of Mammalia, the skin of the abdomen forms a purse or pocket containing the mammæ, and to which cavity the young resort for refuge, even after the time when they cease to derive their sustenance from their mother's milk alone. Only two species of Mammalia, the Ornithorhynchus and Echidna, are without any apparent mammæ; but many interesting questions regarding their habits, and especially the cares which they bestow upon their young, still remain unsolved.

It has been said that the young Elephant sucks with its trunk. This, however, is an error, as it makes use of its mouth, in nearly the same manner as other Mammalia.

The time of suckling varies with the period of gestation, as well as with the time necessary for the growth of the young. Thus it is prolonged as long as the ninth or tenth month in Man, the Horse, and the greater part of the larger quadrupeds, while it is very short with the Rodentia, which have in each year a considerable number of births. With the Guinea-pig, which is the most fertile of known Mammalia, the period of lactation terminates in about twelve or fifteen days.

After having fed their young during the days immediately succeeding to the period of birth, entirely with the milk of their mammæ, the females of the Carnivorous animals take themselves to the chase, and bring home to their young different kinds of prey, so as gradually to accustom them to the use of a more solid food. At this time they seem to lose their natural ferocity, and gambol with their young; but on being attacked, they are only thereby rendered the more formidable. After having tried every possible means to place their family in a place of security, they fight with the most determined obstinacy and courage. The particular history of each species exhibits, in general, many interesting details relative to the care which the female takes of her young, until they are sufficiently strong to provide for themselves. As soon, however, as they have attained this period, the mothers are often seen suddenly to change their feeling towards their progeny, and drive away, with the greatest obstinacy, the same young ones which had so long been the continual objects of their warmest attachment. This is particularly observable in all those species which experience a rut at a particular period of the year, and also most remarkably among the larger Carnivorous animals, who would soon become pinched for want, if too many were permitted to reside together in the same district.

It is commonly during the interval which elapses between the termination of lactation and the commencement of puberty, that the first or milk teeth are replaced by others. This only happens to those species which have simple teeth, fixed by true roots. It begins with the incisors, and ends with the molars, while it often happens that the latter are not changed until long after the age of puberty. The Hog never loses its first teeth, as they do not fall, but always continue growing. In certain other quadrupeds, the teeth continue to grow during the whole course of their lives, such as the incisors of the Rodentia, the compound molars of some animals of the same order, and those of the Elephants. The same property is observed in the teeth of the Kangaroos as well as in the Elephants, but with this difference, that the molars are developed from the back of each jaw forwards, and do not grow out of the gums as in most other Mammalia. There are, however, numerous variations in these respects among the several genera and species, as well as in the forms, which the teeth present, according to the respective ages of the animals. Those Mammalia which change their teeth, and especially the Carnassiers, experience at this critical period the most painful nervous affections, which often prove fatal.

In general, the term of life among the Mammalia is in direct proportion to the time which they severally take in arriving at their full growth, exclusive of the period of gestation. Buffon calculated, from many observations, that they lived seven times the period of growth; but it is very often only six times this period.

Among the most remarkable exceptions to the above rule, we find Man, with whom the average duration of life is far less than that of other species, relative to his time of growth. As he does not attain his full size until about the age of twenty years, his life ought to average a duration of 120 to 140 years. Several individuals have attained these ages, and some have even passed them; but of those few who survive the first years of infancy, by far the greater number do not pass beyond the ages of seventy or eighty. This anomaly to the rule of Buffon is due to a multitude of circumstances, which it would be premature to detail at present; such as the mode of life, the abundance and excess of food, the want of temperance, and other results of an imperfect and misdirected civilization.

For the same reason, the relation which the period of growth bears to the whole term of life, is not without many exceptions among the domestic animals. On the one hand, they receive the influence of a superabundant nourishment, and on the other, are more frequently preserved from those excesses to which this abundance might have given rise. Hence, the duration of life is often prolonged among the domestic animals beyond the term already specified.

The growth of the Horse being commonly completed in about four or five years, it lives twenty-five or even thirty-five, provided the natural term of its existence has not been shortened, as happens too frequently by ill treatment of every kind, by violent fatigues, as well as the want of attention and suitable nourishment. This animal presents, notwithstanding, several instances of remarkable longevity, and some individuals have been known to attain the advanced ages of sixty and even seventy years.

As the Ass takes nearly as long as the Horse in reaching its full growth, the duration of its life ought to be nearly the same; yet it often breaks down before that period through injuries or neglect, which it receives most undeservedly from all quarters. It is observed that animals, naturally disposed to chastity, live longer than those of different propensities. The Mule and Bardeau are usually unable to procreate, and accordingly they live longer than either the Horse or Ass. Very frequently Mules die at the age of forty, and one has been known to attain the age of eighty years.

The Bull takes about two or three years in growing, and the natural period of its life terminates at fifteen to twenty years. The Buffalo approaches the former very nearly in both of these respects; yet it appears to take a little longer time in reaching its full growth, and hence lives to a more advanced age. The Sheep has nearly the same period of growth, and also a corresponding period of life. The Goat approaches to the same terms, both in respect to its growth, and the duration of its existence; yet the extreme attachment of these two last-mentioned species to sexual propensities serves to abridge the ordinary period of their lives, in those few cases where Man does not terminate their existence suddenly for his own advantage.

The Hog being two years in attaining its full development, may reach the age of fifteen or twenty years, if not fattened before the term of puberty, as is most commonly done, though some old Boars have been known to pass far beyond the above-mentioned terms.

We may thus perceive that the relation of the period of growth to the duration of life does not remain constant among the domestic animals. It is, however, more precise with the wild Mammalia. The Lion lives twenty-five years according to Buffon, though several Lions of the Tower Menagerie of London lived in confinement to the extraordinary ages of sixty-three and seventy years, on the authority of Shaw. The Mococo (*Lemur catta*) lives at least twenty years, the Rabbit eight or nine; the Hare seven; the Mouse only a short time. The Elephant, it is said, lives for two hundred years; the Bear thirty; and the Wolf fifteen or twenty.

Further, the Dog usually lives fourteen years, though the lives of some individuals have been prolonged to twenty; the Cat lives nine or ten years, and the Dromedary forty or fifty.

Nothing positive is known regarding the ages to which the Seals and the Cetacea respectively attain; it is, however, probable, from their near approximation to the Fishes, in external characters, that they resemble them also in the average duration of life; in other words, they live to a very advanced age. This presumption is further confirmed with the Seals, by the fact that they take a very long time in growing.

GENERAL REVIEW OF THE MAMMALIA CONTINUED.

The Structure of Teeth—their growth—the phenomena of successive dentition—their varieties of form.

The teeth among the Mammalia are always found upon the jaws or maxillary bones, which is far from being the case with the lower classes of vertebrated animals—the Reptiles and Fishes. Though useful auxiliaries of digestion, they are by no means essential to that function; for some animals are wholly destitute of teeth, and in others, they are far removed both from the mouth and the intestinal canal. Their existence is not exclusively proper to the vertebrated animals, nor are they always confined in them to the bones of the mouth.

The teeth of animals may be defined as bodies, generally hard or of a calcareous appearance, produced by the secretion of special organs, fortifying the anterior parts of the alimentary canal, and by the assistance of which, the greater part of these animals seize, retain, or divide the food with which they are nourished, while some employ them further as weapons of offence and defence. The teeth of the Mammalia, with which alone we are at present concerned, may further be restricted by the circumstance already alluded to, that they are only found upon the margins of the maxillary bones.

It was for a long time supposed that the teeth were bones, that they were produced in the same manner, and had a similar structure. This view of the subject has been wholly abandoned, since the publication of the admirable treatise of the Baron Cuvier upon the grinders of the Elephants, in the Annales du Muséum d'Histoire Naturelle, tome viii, in the year 1806. Although the differences between teeth and bones appear to be very numerous and essential, there seems, however, to be a considerable analogy between them, especially when considered in a point of view purely anatomical. When physiologically considered, they possess many peculiarities in common with the horns, nails, and hair.

At first, the constituent matters of the teeth and bones are precisely the same; and if we revert to the first formation of these bodies, it appears that they are equally secreted and deposited by proper vessels. Under this point of view, the teeth may be considered as bones, the vessels of which are united in a single mass, and deposit the osseous matter around them; while the bones may, on the other hand, be viewed as teeth, within which the minute subdivisions of the vessels cause this matter to circulate in every direction.

At their origin, and during the greater part of their existence, the teeth are composed of a secreting organ and a secreted substance. The former, or secreting organ, is always concealed in the lower part of the tooth, or in the interior; and when entirely formed, consists of three, or at least two other organs. It is essentially composed of vessels and nerves, which communicate directly with the remainder of the organization. The latter, or secreted substance, is merely deposited outside the first. It is composed of a greater or less number of different substances, and being deprived of all vessels and nerves, bears no necessary or immediate connexion with the other organs.

The secreted substance is of a calcareous appearance, and composed of two parts; the one external, called the *crown (fust* or *couronne)* of the tooth; the other being more or less concealed, is termed the *root (racine)*. The intermediate point is distinguished by the appellation *neck (collet)*.

The crown of the tooth may be composed of different kinds of matter, deposited one over the other. In the most complicated kinds of teeth, three of these may be obtained by a mechanical analysis. The central part is termed the *ivory;* the second the enamel, and the most external part the *cortex*. These three substances are found combined in four different ways. Some teeth are composed of ivory, enamel, and cortex; others only of ivory and enamel. Some, again, are formed of ivory and cortex, the enamel being wanting; others of ivory alone, this last being never observed to be deficient except in those Mammalia which are wholly destitute of teeth.

The root may be real or apparent. In the first case, it is formed of ivory alone, as in Man, the Carnassiers, and the Ruminantia; or of ivory and cortex, as in the Cachalots. In the second case, the root is merely a continuation of the crown, and has all the characters of the latter. Such are the roots of all tusks properly so called, the incisive teeth of all the Rodentia; the molars of Hares, of Guinea-pigs, and of the Cabiais (*Hydrochærus*).

The secreting organ of the tooth or dentary capsule, according to M. Frederic Cuvier, appears to be dependant on, or produced by, the nerves and maxillary vessels. It is not, however, without relation to the contiguous parts, being even united to the gums; but much less than some authors have imagined. It is certain that the secreting organ of the second teeth, for a long time after its formation, is altogether independent of these parts, and it is only subsequently that it becomes united to the gums.

The dentary capsule corresponds, both in its structure and functions, with the substances or materials of which the teeth are composed, in such a way, that it is more simple in teeth formed of one substance alone, than in those composed of two or three. It is the same with its forms, as well as its growth, in relation to the forms and growth of the teeth, the one always being the consequence of the others. The most complicated kind of dentary capsule, being that observed whenever the teeth are composed of three substances, is itself formed of three very distinct secreting organs. The central one, called the *bulb*, produces the ivory; the second, under the form of a membrane, secretes the enamel, and may thence be termed the *enamellating membrane;* and the third, which surrounds all the other parts, produces the cortex or external ivory. The last may be termed the *external membrane*.

The bulb which secretes the ivory by its external surface, appears to be entirely composed of nerves and vessels. Several arterial trunks, which extend from the one extremity to the other, are ramified infinitely before arriving at its extremities, where their divisions sometimes form tufts and fringes of an almost imperceptible degree of fineness. This part of the teeth may be studied with the greatest facility when they first begin to form; it is then found to be naturally injected, and is not exposed to injury during the abstraction of those bony portions in which the teeth are enclosed, while a very slight degree of maceration is sufficient to extract the bulb from the coating of ivory by which it is surrounded. It seems to be homogeneous throughout, and always has the same shape as the tooth will ultimately have. In fact, it is the mould upon which the tooth is modelled.

The enamellating membrane produces the enamel by its internal surface. It surrounds the bulb entirely, and follows all its circuitous outlines, thus possessing the same form, except at the base of the bulb, corresponding to the neck of the tooth where it abuts and terminates. M. F. Cuvier was unable to detect any vessels in this membrane. It is transparent and brittle when thick and about to deposit enamel; but it soon softens, becomes of a milky whiteness and great elasticity. Finally, it ends by disappearing altogether, when it has no longer any function to discharge, that is, when the external membrane, by depositing the cortex, resumes its place. The transparency of this membrane, its extreme thinness thereafter, and its final obliteration, in those teeth where the ivory is formed, have been the cause that many Naturalists have failed to observe it. But it may be seen very easily upon the parts contiguous to the molars of the Ruminantia, and especially on the hinder ones, at the moment when these animals are born; and, if once remarked here, it becomes easy to detect it upon all teeth possessed of enamel.

The external membrane, like the bulb, is of a nature essentially vascular, and may be considered as an external bulb. It is homogeneous in respect to its intimate structure; but its two faces have not always the same forms, nor do they perform the same functions. It deposits the cortex by its internal surface, and follows all the contortions of the tooth. In the compound teeth it juts outwards, whenever they present any hollows. The parts which line these cavities are not merely membranes, at least when the cortical matter is about to be deposited, for they then have the same thickness as these cavities, and this gives them all the appearance of bulbs. Before the above period, it is sufficiently thin upon the surface of the compound teeth, and this observation is applicable to most teeth. But it may be presumed that the external membrane is always of a great degree of thickness in the capsules of those teeth where the ivory has to be covered with a great thickness of cortex, as may probably happen in the molar teeth of the Cachalots. Its external surface is always simple, being merely the protecting and uniform envelope of the entire dentary organ, and its form when complete is always more or less spherical. It is pierced at its summit by the evolution of the tooth; but its margins are attached to the gums, and become in some measure a continuation of them.

These three parts, composing the dentary capsule, are intimately united, and become confounded together towards the inferior part of this organ, at the point where the vessels and principal nerves are introduced, at least from the time when the roots begin to develop themselves, and to become distinguished from the crown. It appears that all the three parts originate from this point, and likewise all the essential vessels which traverse and nourish them pass from thence. Their other portions are from the very commencement entirely independent of each other. The external membrane may be raised without occasioning the slightest injury to the enamellating membrane, which detaches itself without effort from the layers of enamel just deposited; and the bulb may be separated from its cones of ivory like a blade from its scabbard; or if the cones be broken, it may be disengaged and displayed without being destroyed, or in any way injured.

This capsule, however, is not entirely formed before the teeth are secreted, in those at least which have roots. It develops itself successively, and in proportion as the different parts are formed, beginning from the summit of the crown, and finishing by the extremity of the root.

The bulb and enamellating membrane seem to deposit simultaneously the matters which they respectively secrete; and the first molecule of ivory receives the first molecule of enamel. It is only at a later period, that the external membrane deposits the cortex, being at the time when the crown is already formed, and when the bulb, as well as the enamellating membrane, cease to deposit matter in this part of the tooth, for these secreting organs have still to give birth to the roots.

The above detailed analysis of the most complicated kind of dentary capsule, enables us to pass rapidly over those destined solely to secrete the ivory and enamel, or the ivory and cortex, or the ivory alone, and being consequently of a more simple structure.

Those capsules intended to form teeth composed of ivory and enamel alone, are not, on that account, deprived of the external membrane, but this body always appears to be thinner, instead of being thick, as in the preceding kind of teeth, when about to deposit the cortex. It is raised with difficulty and by shreds, and seems only to be intended to protect the function of dentition; it accordingly envelopes the organ in every part. The enamellating membrane presents itself in these capsules with all the general characters which have been assigned to it. The bulb does not differ from that belonging to teeth composed of three substances.

With respect to those teeth which are composed of ivory and cortex, such as the molars of Cachalots, we also find the external membrane in them to be of a certain thickness, in addition to the bulb, which is never wanting.

Having thus shown that the dentary capsule of the most complicated kind of teeth, produces three distinct and different substances which can be accurately separated from each other, it now remains for us to consider the secreted bodies themselves, composing, as they do, the proper substance of the teeth.

The ivory forms the essential and fundamental part of the tooth. As it covers the organ by which it is secreted, it is deposited from without, inwards, and does not appear to be absolutely identical in all kinds of teeth. In some, as the tusks of the Elephants for example, it is deposited by concentric beds, in such a manner that they are composed of cones, the one encasing another, and being numerous in proportion to the length of the tusk. This conical appearance is especially shown in fossil tusks, as the cones themselves do not appear to have been separated artificially. Other teeth have a more homogeneous kind of ivory, but the differences of texture which this substance presents are very numerous. These tusks of the Elephants show on their transverse sections a number of segments of circles regularly disposed, which intersect each other, and form a waving mark, by which the true ivory may always be recognised. The teeth of Man, the Quadrumana, and Carnassiers, possess an ivory of a silky appearance, apparently composed of fibres. Those of the Cetacea, the tusks of the Hippopotamus and others, have their ivory simple, and of the most uniform texture; those of the Rat-Moles (*Bathyergus*) seem formed of longitudinal and parallel fibres, like those of a rush. These characters arise doubtless from the peculiar structure of the bulbs which secrete these different kinds of ivory; yet their essential differences have not been determined by experiment, but will probably be ascertained hereafter, when these bulbs are submitted to a more minute investigation.

This central part, being the most important and considerable portion of the crown of the teeth, is chiefly formed of a very compact gelatinous substance. The calcareous matter which gives it the external appearance, is merely deposited in the meshes of this substance, and composes only the smallest portion. It may be abstracted by means of a small quantity of dilute acid, and the gelatine remains pure, and of the same form as the ivory. This calcareous matter, the only part of the tooth really destitute of life, is a phosphate of lime.

The Enamel is deposited in a manner contrary to the ivory, being from within, outwards, and always immediately over the latter; this it appears to do by a kind of crystallizing process. On being examined upon a section of a tooth, it is found to have the appearance of brilliant needles, perpendicular to the surface of the ivory. The ivory and enamel do not form one body, although they are united together very closely, for the enamel can be detached from the ivory without injuring the latter, and reciprocally. But the essential distinction between them consists in the circumstance that the enamel does not possess gelatine for its base; for, although it contains some traces of that substance, they are always very minute in quantity. The enamel, on the other hand, is essentially composed of fluate of lime, which contributes its stony character, and imparts a degree of hardness superior to that of any other portion of the teeth, and indeed of any animal substance.

The Cortex, like the Enamel, is deposited from within, outwards; but it cannot be discovered upon teeth possessed of enamel until the latter is entirely formed. M. F. Cuvier is of opinion that, in teeth composed of ivory and cortex alone, it is deposited over the ivory like the enamel. The intimate nature of cortex is absolutely the same as that of ivory, on which account it might with propriety be termed the *external ivory*. Gelatine forms its principal base, and phosphate of lime is deposited between the meshes of that substance. The cortex is found in layers more or less thick. It is of an extreme thinness on the projecting surfaces of the molars in the Ruminantia, but is much thicker in the hollows found on the summits of their crowns. It is observed, however, to possess a still greater thickness in the crowns of the teeth belonging to Cachalots. In this place it equals the ivory in quantity and thickness; for the whiter substance, which surrounds the central part of these teeth, is not enamel, as some Naturalists have supposed, but a true external ivory.

It commonly happens that the cortex contains nothing but gelatine and phosphate of lime. In some cases, however, it contains some colouring matter in addition to these, as may be seen in the teeth of several Ruminantia, and in the incisors of the Beavers, Pacas, Agoutis, Porcupines, and some others. The colour of the anterior part of these teeth depends upon a very delicate layer of true cortex, as M. F. Cuvier ascertained by many careful experiments. The colour becomes brown only on that part of the tooth which projects from the gums, while the portion within them is of a dark green. It has been said that this colour is owing to the presence of iron, and that the change which it undergoes from the contact of the air is a true oxidation.

The above details regarding the structure of the dentary capsule, which produces the teeth, as well as the composition and structure of the teeth themselves, have been hitherto demonstrated upon a very small number of Mammalia, and they are applied only by analogy to the remainder. In fact, the teeth of Man, of some Carnassiers, Rodentia, and Ruminantia, with the Solipeda and the Indian Elephant, have alone been studied in respect to their dentary capsules, and the substances of which the teeth are composed. It is probable that a special investigation of teeth belonging to other Mammalia may lead to the restriction or extension of some of the preceding observations.

The above remarks explain to a certain extent the manner in which the crown of the tooth is formed. As the dentary bulb is the mould of the crown, and as the matter which it secretes is deposited upon its surface, the crown cannot fail to exhibit the same projections, hollows, and angles—in a word, to have the same identical figure; but there is nothing in the structure of this bulb which can explain the form of the roots.

By the term *root* is commonly understood, that part of the tooth contained within the gums; but it is essential, as has already been explained, to distinguish those insertions which differ from the crown neither in structure nor form, from the roots properly so called, which begin from the neck of the tooth, and diminish gradually, until they terminate in a point more or less obtuse, and more or less irregular. The first are not real roots, but are formed merely by the prolongation of the crown within the gums.

When the time at which the true roots have to be formed has arrived, the enamellating membrane ceases to maintain its activity, and even becomes wholly obliterated. The bulb and the external membrane alone continue to grow and to produce roots, which usually correspond, in number and situation, with the principal tubercles of the crown, and appear to be numerous in proportion to the number of leading branches which the maxillary arteries transmit into the bulb. It seems probable that these vessels and their branches form an inferior prolongation of the bulb, as soon as the crown has been deposited; or, in other words, that the bulb continues to develop itself under their influence, which is restricted to the points immediately surrounding them. Under this point of view, the roots of the teeth may be regarded as the evanescent crowns of the same teeth, reduced to a rudimentary state; for we can easily see how they might be continued, if the vascular system did not become obliterated. In fact, those teeth, where the capsule never ceases to reproduce the crown as fast as it wears away, and which are consequently destitute of true roots, only become such in consequence of the undiminished vitality of their bulb, which continually maintains its vigour and activity as at the commencement. Thus we see that teeth possessed of roots, obtain them at periods of their existence more or less advanced. Among the herbivorous animals, the Horse for example, the vitality of the bulb continues for several years, while it ceases in a very short time with the Carnassiers. In this respect, the Mammalia offer a great variety of examples.

There are several circumstances which serve to confirm the accuracy of these views. As long as the dentary capsule is wholly occupied in depositing the crown, we see, at the precise point where the membranes composing it reunite and become confounded, a uniform disc, supplied with an immense number of vessels, which distinguish it readily from all the adjacent parts. It is from this surface that the capsule continues to grow uniformly, until the crown has acquired its entire height. At the latter period, however, it undergoes a total change; the isolated portions of the vessels disappear, and those which remain compose little circles, more or less numerous, and distinct from each other. From these circles the roots grow; during which operation, the external membrane detaches itself from the bulb on all the intermediate points of the partial circles. The crown is then terminated by the deposition of ivory between the roots and beneath both the crown and the bulb; further, as this deposition takes place from different points of the circumference of the tooth, it is at the internal surface of the roots that it reunites. The little circles continue to diminish; sometimes they divide after a certain growth has taken place in the root, causing them to appear more or less forked; and they end in disappearing gradually, so as to occasion all the roots to terminate in a point or thin layer. By this growth, the bulb, now reduced within very narrow dimensions, remains inclosed within the crown, and the roots are found to be pierced through their entire length by those vessels and nerves which formed them; thus connecting them with the bulb on the one hand, and on the other with the maxillary vessels and arteries.

The first traces of the dentary capsule can be discovered in the fœtus, it is said, during the earliest days of its life. There can be no doubt, however, that the teeth are in a great measure formed at the period of birth in a large portion of the Mammalia, and the young animals are even compelled to use them before the period of their lactation has entirely terminated. Physiologists are not, however, agreed as to the nature of the process carried on within the jaws, in those parts which are traversed by the teeth, before leaving the gums. Some have supposed, that there exists a natural passage, leading from the capsule, out of the gums; and it is imagined, that this cavity is enlarged by the expansive force of the tooth, aided by the elasticity of the adjacent parts. Others have conceived, that the tooth tears everything which opposes its passage; and they have even attributed to this cause, some of the accidents which occasionally accompany the dentition of young animals.

The former of these views is opposed by the observed phenomena of the second dentition, where another set of teeth is developed immediately beneath the first, in such a way, that the second cannot appear before the first have fallen. No such natural passage has been observed; and it ought not to be presumed before adequate proof, that Nature has employed two different methods of evolving these organs. It should rather be inferred, that if the second teeth are able to surmount the obstacles presented to their growth by the first teeth immediately above them, these will also be able to overcome the resistance of the membranes and cartilages, when they are required to leave the jaws, to satisfy the new wants of the young animal. It further appears, that teeth of the most complicated form, having their crowns terminated by many tubercles, and having between them many intervals of considerable depth, obtrude themselves, by the summits of their tubercles, on several points at the same time, beyond the gums; yet the gums still continue to occupy the intervals which separate their tubercles.

The hypothesis of a violent tearing is still less admissible than that of a natural passage. During the time that the teeth are growing, not the slightest trace of such a phenomenon can be observed; and analogy does not appear to justify this second supposition. Nature appears to have a surer and more effectual means than these mechanical hypotheses would lead us to infer; for the present is, in reality, only a particular case of a very general law, of which it forms one of the most exact applications.

There is no truth in Physiology better established than this, that the nutritive power of any organic part is enfeebled, when it receives the continued mechanical action of any foreign body whatever, and the nutrition of the part may even be wholly interrupted, if this action acquire a certain degree of intensity. It seems, that in the perpetual interchange of particles which constitutes life, the new molecules become incapable of replacing the former, whenever a foreign body compresses the parts from which the others have escaped. It may be said, either that the place of the first bodies has ceased to be occupied, or that the assimilating force, which ought to have supplied new molecules, has ceased to act. The consequence is, that the part becomes obliterated; and the molecules, which should have nourished it, are dissipated, or go to supply the adjacent parts.

There can be little doubt that the development of the teeth is a phenomenon of this description. When the crown of a tooth begins to be formed, and still more, before this period, all that part of the gum, which is intended to be opened for its passage, is thick and filled with vessels. As the tooth grows, this part becomes smaller, and the time at length arrives, when it consists of nothing more than a compact and dry skin, which soon disappears in order to allow a free passage to the tooth. This view of the subject is, however, incapable of explaining how it happens that the pressure of the teeth is exerted contrary to the gums, rather than in the opposite direction. Although the tooth begins at first to form only on the side next to the crown, this circumstance does not completely account for the fact that the tooth tends exclusively to emerge on this side. The reaction of a tooth growing in the direction of its root, is equal to its action in the direction of the crown; and if the degree of firmness possessed by the adjacent parts be regarded in this question, instead of piercing the gums, the teeth ought to descend on the side where the roots are afterwards found; for the inferior parts of the capsule and its bulb would offer much less resistance than the denser structure of the gums. It is therefore probable that we ought to attribute the natural direction of the teeth to some special impulse which the circulation impresses upon the dentary organ, as well as to the mere growth of the capsule by its interior part. The addition of matter to the inner extremity of the crown is far from being sufficient of itself to explain this phenomenon. The pressure of the gum upon the teeth would even be sufficient wholly to arrest it, and it is, on the contrary, the life of the gum which would then have to be suspended. While the teeth are growing, the vital action of their capsules is raised to an intense degree, the blood is directed towards them with great force, their irritation becomes extreme, and hence probably result the fatal consequences which frequently occur to young animals during the period of their dentition.

The protrusion of the teeth from the sockets, in consequence of their secretion and growth, is not the sole movement which these organs present. Other changes succeed, the object of which being the mastication of the food, is rather more obvious than the causes which produce them.

Among these may be considered, in the first place, the secondary movement of the crown in teeth with distinct roots, after they have emerged from the sockets. The capsules of all these teeth being entirely inclosed within the jaws, have their lower parts, which correspond with the neck of the tooth, much below the dentary margin of these bones; but when the teeth are entirely formed, the neck is on a level with this same border—that is to say, the inner part of the crown, which in some manner has been formed in the bottom of the jaws, ultimately finds itself on a level with their exterior margin. This protrusion of the crown appears to be owing, at first, to the growth of that part of the capsule which is about to give birth to the roots—a growth which does not make its appearance until after the formation of the crown. Subsequently to this, it must be attributed to some special impulse of the circulation, which maintains itself in a high degree as long as the dentary capsule preserves its secreting power. Further, at this particular period of its growth, the gum no longer opposes any resistance to the growth of the teeth.

There is seen in the molars of the Horse a second kind of movement, which they probably possess in common with all other herbivorous animals as well as the Ruminantia. It consists in the continued obtrusion of their teeth, even when completely formed, and opposed by others in the opposite jaw, against which they act during mastication. This movement was fully demonstrated by Tenon (Académie des Sciences, an. 6), who, however, did not investigate its cause. It may be regarded as a continuous ossification of the jaws,—an operation which only ceases with the life of the animal. In fact, a third movement of these teeth exhibits this ossification tending continually to expel the teeth from their sockets; and this occurs when a tooth, not being opposed by others, is pushed out of the jaws. As no force then opposes the continued secretion of bone, the sockets become filled, and the teeth are driven from the place which they occupied, as if they were foreign bodies. This movement, which is prejudicial to most animals, has one advantage for those which are obliged to wear out their teeth in grinding their food; for, although the wearing out of the teeth in these Mammalia is often very unequal, the dentary organs do not on that account remain uneven at their summits; and the consequence is, that the grinding of the food may be continued to the most advanced periods of life.

There is a secondary movement in the incisors or front teeth of the Rodentia wholly opposite to the preceding, and still more difficult to explain. That part of the tooth which corresponds in situation with the root, is placed much less towards the front of the bones containing them, among young animals, than among the old. These teeth continually fall back at their extremity where the bulb is placed, in proportion as the animal grows, while they advance forward by the other extremity. M. Frederic Cuvier, who observed this singular phenomenon in the Rabbit and Guinea-Pig, supposes that the bulb continues to grow by its hinder part, being influenced by the nerves and vessels which thence derive their life; and this phenomenon appears to be common to all teeth approaching to tusks in their general character.

Another problem connected with the incisors of the Rodentia is much less difficult to solve,—we mean their curvature, and the peculiar curves which they affect. To produce an arched tooth, it is sufficient that the capsule be arched; but if the curvature of the capsule remain always the same, these teeth, which can grow indefinitely when no obstacle arrests their course, would present in this case the form of a regular circle, of which frequent examples are found. Instead, however, of this curve, the incisors of the Rodentia exhibit one nearly approaching to a spiral, where the first portions of the teeth are inclosed in those that follow. It is necessary, therefore, that the capsule producing the teeth should change its curvature, and that it should approach continually towards the right line, as these animals advance in age, up to a certain point, perhaps, when it ceases to be further modified. We may also remark, that these changes of curvature are the same in the incisors of both jaws; for these teeth, at all periods of life, preserve among themselves the same relations.

The appearance of the teeth beyond the gums usually commences, among the Mammalia, with the period when the milk begins to be insufficient for the nourishment of the young animal; but it very rarely happens that they are all developed at the same time. In this respect, great differences are found among them; and Nature, in most cases, fails to impart at one time all the teeth necessary for the use of each animal. These are very few, we may almost say, no Mammalia, where some of these organs are not renewed; that is to say, that certain kinds of teeth fall, and are reproduced, or rather replaced, once or oftener, by the successive growth of other teeth beneath, behind, or before the former.

These first teeth, which give place to new ones, are distinguished by the term *milk teeth*, or teeth of the first dentition; those which succeed are termed the *second teeth*. But these terms, founded upon what has been observed in the human species, ought not to be taken in the strict sense when applied to other Mammalia; for among these, it will be seen that the milk teeth may fall before birth, or a long time after the adult age. To avoid the mistakes to which the ambiguity of these terms might give rise, it will be proper to employ the terms *first*, *second*, and *third* teeth, to denote the order of their appearance.

This department of Natural History, which shows the succession of the teeth, their mutual influence, the coincidence of their appearance with other parts, and with the new wants of each animal, as well as the relations of form and number between the teeth of different dentitions, has, unfortunately, been much neglected, and it is only now beginning to receive that attention which it merits. In a Zoological point of view, a knowledge of the dentary system in different ages is almost indispensable; but we are still without a series of drawings, showing the teeth of young Mammalia, corresponding to the valuable lithographic sketches of M. Frederic Cuvier (Sur les Dents des Mammifères), made from adult specimens.

Before explaining the few particulars that have hitherto been ascertained on this subject, it will be proper to premise a few words respecting the different kinds of teeth, as well as to explain the system of notation which we intend to use in describing them.

The teeth of the Mammalia emerge solely from the inter-maxillary and maxillary bones. The incisors, or front teeth, make their appearance first; and these may be followed either by the canines or molars. The last are subdivided into false molars, carnassier molars, and tuberculous molars: while the tuberculous molars themselves may be further distinguished by their having simple, compound, or proper tubercles.

(1.) The Incisor, or Cutting Teeth (*Incisores* or *Primores*), are somewhat broad and long, with their margins often parallel, and cut away obliquely at their free extremity. We see them so, for example, in the front teeth of Man; but this form is still more strongly marked in the Rodentia or Gnawers. This term *Incisor* ought properly to be applied only to those teeth which have a form especially fitted for cutting; but it has been extended not only to all such as are found in the incisive or inter-maxillary bone, but even to those opposed to them in the lower jaw, although the latter often have neither the form nor use of true Incisors.

(2.) The Canines (*Laniarii*), or Tearing Teeth, have the general form of those teeth which appear most prominently in the Dogs. They are longer than all the others, and always have a single root and a single point to the crown. As these teeth are usually placed in the upper and lower jaw, immediately behind the Incisors, the term Canine has been extended to all teeth which appear to occupy this place. They are likewise called *Corner* teeth; and, from the chief use to which they are applied, have obtained the name of *Laniarii*, from *laniare*, to tear.

(3.) As the names of most parts of animals are derived from the corresponding parts in Man and the Ruminantia, which were principally dissected by the ancient Anatomists, the term *Molar* (*Molares*) is correctly applied only to those which act, as we may remark in the Ruminantia, almost like a mill-stone—(in Latin *mola*, a mill). Hence, we understand Molar teeth to be compound, semi-compound, or even simple teeth, having the crown broad and flat, with broken projections and small eminences corresponding to each other in both jaws. Afterwards, however, the term has come to be applied indiscriminately to all teeth situate behind the Canines, and occupying the entire inner extremities of the dentary lines, although they sometimes possess no other character of molars than the place they occupy. Thus, in the Cats, where these teeth are trenchant, and correspond in each jaw, so as, in fact, to act in the same manner as true Incisors, they are not on that account deprived of the common appellation of Molars.

Hence it becomes necessary further to distinguish the different kinds of Molars from each other. In many Mammalia, the Molars differ greatly both in size and form, and have on that account been divided into False and True Molars. By *false* Molars, being most commonly the anterior ones, we understand such as are small and pointed. *True* Molars are considered to be thicker and larger, with their crowns studded with several points, or altogether flat. In the Carnivorous animals, there is found a very large Molar, which more especially fulfils the tearing purposes of these animals; and this tooth is further distinguished by the terms *carnassier Molar*.

The importance of possessing a good system of Nomenclature for the teeth becomes sufficiently obvious, as soon as the necessity of defining clearly the different kinds and combinations of teeth, both in respect to their forms and relative position, has been made apparent.

In stating the dentary systems of animals, two methods of notation have hitherto been employed; for example, the adult teeth of the human species, being eight incisors, four canines, eight false molars, and twelve tuberculous molars, have long been represented thus:

$$\text{Incisors } \tfrac{4}{4};\ \text{Canines } \tfrac{1-1}{1-1};\ \text{Molars } \tfrac{5-5}{5-5} = 32.$$

In his work on the Teeth of the Mammalia, M. Frederic Cuvier expresses the same thing, under the following form:

32 Teeth.	16 upper.	4 Incisors.
		2 Canines.
		10 Molars.
	16 under.	4 Incisors.
		2 Canines.
		10 Molars.

The former expression, besides its inconvenience from the smallness of the figures, does not represent the nature of the Molars. The latter is not compact. In both, the teeth on each side are confounded together unnecessarily.

As a new system of notation, which will combine the advantages of brevity and clearness, yet remains to be proposed, we venture to suggest the following, which possesses, in our opinion, some of the most essential requisites. Let M represent any Molar tooth; C a Canine; F a false Molar; and C′ a carnassier Molar. Let a number annexed to an explanatory letter denote that there are as many teeth of the kind, represented by the letter, as there are units in the number. Further, let a number without an explanatory letter denote an Incisor, or front tooth, and a Molar, unless otherwise expressed, be always understood to be tuberculous. Then, adopting the ordinary signification of the Algebraical symbols, the dentary system of the Adult Man will be conveniently represented as follows:

$$^{2}\left|\frac{2 + C + (2\,F + 3)\,M}{2 + C + (2\,F + 3)\,M}\right. = \frac{16}{16} = 32.$$

where the numerator denotes the number, nature, and relative position of the teeth on one side of the upper jaw, and the denominator of those on one side of the lower. The small figure in the corner indicates that each expression must be doubled to represent both sides of each jaw, and the vertical line on the left hand shows the medial axis of the body, passing in the middle of the front teeth.

To avoid repetitions, only one side of either jaw is alluded to in referring to particular teeth; and what is said of one side must be understood of the other, which precisely resembles the first in all its relations. It is always customary to count from the anterior extremity of all the parts which bear these organs. Thus, the first Incisor among the Mammalia is that tooth found nearest to the suture, by which the inter-maxillary bones are united. This suture is represented in the formulæ by the vertical line. All extraordinary cases are excepted, such as the appearance of teeth before birth or in extreme old age, while the ordinary and most natural process of development is always understood to be meant, unless otherwise expressed.

In the human species, the first dentition generally takes place from the sixteenth or eighteenth month to the age of two years or two years and a half, and it usually commences with the lower jaw. The first Incisors precede all the others, and these are followed by the second; so that, towards the end of the first year, all the incisors are developed. The first tooth which pierces the gums after the incisors is a molar; and it is subsequent to this that the Canine, though placed before it, makes its appearance; finally, the first dentition is terminated by a second Molar. It must be remarked that Molars, and not false Molars, immediately follow the Canine; and this is contrary to what is observed in the final dentition of the human species. The general law, of which this is only a particular case, will come afterwards to be explained.

The formula for the Milk-teeth, or first teeth of the Infant, is therefore

$$ {}^{2}\left|\frac{2 + C + 2\ M}{2 + C + 2\ M}\right. = \frac{10}{10} = 20 $$

All the teeth of the first dentition fall exactly in the same order as they have appeared. The Incisors and Canines are replaced by Teeth of the same nature as themselves, only stronger and larger than the first. On the contrary, the two first Molars are replaced by false Molars only. This operation is finished towards the twelfth year, and the first tuberculous molar, being the third in the order of time, shows itself about the sixteenth or eighteenth year. This tooth is larger than any which preceded it, or even than those which follow. Finally, the last of these teeth, commonly termed *wisdom teeth*, and which may not improperly be styled a third dentition, make their appearance a few years later, though sometimes they are delayed as long as the thirtieth year.

All the teeth of the second dentition are formed, according to the important researches of M. Serres, by the vessels and nerves of a special-dentary canal, developed beneath the first, and which replaces it as soon as the first set of teeth begins to fall. We may thence conjecture that, at the time when the other Mammalia change their teeth, an analogous phenomenon takes place.

When the teeth of the first dentition fall, it is observed that the greater part of them have lost their roots, and that the lower part of the crown is tinged black, and covered with asperities which seem to be the effect of a species of corrosion. It will be proper, however, to postpone further notice of this curious phenomenon until the subject of successive dentition has been concluded.

The Apes of both continents present nearly the same phenomenon as the human species, in respect to their first and second dentition. The Makis and Insectivora have not yet been studied in these respects, but some Carnivora have been satisfactorily examined, and especially the Dogs and Cats, in which two dentitions are recognised.

The first dentition of the Cats consists in the upper jaw of three incisors, one Canine, one rudimentary false Molar, one carnassier, and one small tuberculous Molar; in the lower jaw of three incisors, one Canine, one false Molar, and one carnassier. Or,

$$ {}^{2}\left|\frac{3 + C + (F + C' + 1)\ M}{3 + C + (F + C')\ M}\right. = \frac{14}{12} = 26 $$

In the second dentition of the Cats, the incisors and canines are replaced without any important changes, by teeth similar to them in nature. It is also the same with the first two false Molars; but the carnassiers are replaced by the second false Molars, and both are developed immediately after the others, so that, from being second Molars in the first dentition, the carnassiers pass onwards to be third Molars in the second; that is to say, in the upper jaw, the carnassier Molar has taken the place of the tuberculous Molar, which in the second dentition appears in the fourth and last place, while the carnassier of the lower jaw is developed in the place where no tooth was found in the first dentition. This will be readily understood on comparing the following formula of the second and adult dentition of the Cats with the preceding.

$$ {}^{2}\left|\frac{3 + C + (2\ F + C' + 1)\ M}{3 + C + (2\ F + C')\ M}\right. = \frac{16}{14} = 30 $$

In the dentition of the Dogs we find phenomena very analogous to the above. On completing their first dentition, they have in the upper jaw three incisors, one Canine, one false Molar, one carnassier, and one large tuberculous Molar; in the lower jaw, three incisors, one Canine, two false Molars, and one carnassier. The formula for the first dentition of the Dogs is therefore

$$ {}^{2}\left|\frac{3 + C + (F + C' + 1)\ M}{3 + C + (2\ F + C')\ M}\right. = \frac{14}{14} = 28 $$

In the same manner as among the Cats, we find that the incisors and canines of the Dogs are renewed, at the second dentition, in both jaws, without any important change. Immediately after the canine, in the upper jaw, there appears a rudimentary false molar in a spot where no tooth had previously existed. The false molar of the first dentition is replaced by a tooth similar to itself; the carnassier by a third false molar, and the tuberculous molar by a carnassier. Finally, a tuberculous molar, with a second or smaller one, appears after the carnassier. In the lower jaw, as in the upper, there occurs a rudimentary false molar after the canine. The two false molars of the first dentition are replaced by teeth which resemble them, and the carnassier by another false molar. Next follows the new carnassier, and immediately behind it, one large tuberculous molar, and one rudimentary, in a place where no tooth had appeared during the first dentition. These changes appear obvious to the sight, on contrasting the following formula of the adult dentition in the Dogs with that last given.

$$ {}^{2}\left|\frac{3 + C + (3\ F + C' + 2)\ M}{3 + C + (4\ F + C' + 2)\ M}\right. = \frac{20}{22} = 42. $$

It follows from this, that in the second dentition of the Dogs and Cats, the teeth are not only more numerous, but the carnassier molars are placed at a much greater distance from the canine teeth than in the first dentition.

This observation is equally applicable to all Carnassiers, and it is not difficult to perceive the design of Nature, in thus altering the position of these teeth, which are most important to animals feeding almost exclusively on flesh. To render the action of these teeth always powerful, they are brought nearer to the fulcrum or hinge of the jaw; and thus the effect which the jaw would otherwise produce from its growth, in diminishing their power, is effectually counteracted.

As the Rodentia do not possess different kinds of molars, they cannot present these changes which we observe in the Carnassiers. Excepting the Cabiais (*Hydrochærus*), their teeth of the second dentition are developed immediately under those of the first, and the latter entirely resemble the former. The Cabiais, on the contrary, possess a peculiar mode of dentition, in common with the Elephants and Ethiopian Hogs (*Phacochærus*).

It has not yet been ascertained, whether the incisors of the Rodentia fall, and are replaced. The Baron Cuvier has proved, that all species of Rodentia which have only three molars, possess only a single dentition; and that there is a second dentition only among those species which have more than these three teeth; that is to say, all molars surpassing three in number, and placed before them in the jaws. He has further made the singular observation, that the teeth of the first dentition fall, in the Guinea Pigs, while these animals are yet in the womb of their mother. In species of the Hare genus (*Lepus*), the first teeth fall a few days after birth, and this phenomenon is found even in those rudimentary incisors, which are known among all animals of this genus to develop themselves behind the principal incisors.

We proceed to the Pachydermata, as the Edentata have not yet offered any important observation in this branch of the subject. The first dentition of the Hippopotamus consists of two incisors and one canine in each jaw, of three false molars and three tuberculous molars in the upper jaw, and of two false molars and three tuberculous molars in the lower. Or,

$$ {}^{2}\left|\frac{2 + C + (3\ F + 3)\ M}{2 + C + (2\ F + 3)\ M}\right. = \frac{18}{16} = 34 $$

In the second dentition, the incisors and canines of both jaws experience no change. The first of the three false molars in the upper jaw falls, and is not replaced; the two others are replaced by teeth the same as themselves, and a false molar succeeds to the first real molar. But, at the same time, another molar is developed at the extremity, so that the number of these real molars remains always the same, notwithstanding the fall of the first real molar. In the lower jaw, the first false molar falls without re-appearing; the two following are replaced by teeth of the same kind, and then, as in the other jaw, the last molar appears. The second dentary system of the Hippopotamus may, therefore, be represented thus;

$$ {}^{2}\left|\frac{2 + C + (4\ F + 3)\ M}{2 + C + (2\ F + 4)\ M}\right. = \frac{20}{18} = 38 $$

We may apply the same observation which has already been made regarding the Carnassiers to the Hippopotamus; and it is, probably, for a similar reason, that the first real molar of the first dentition is replaced by a false molar in the second.

The Ethiopian Hogs (*Phacochærus*) exhibit a new mode of change, which they possess in common with the Cabiais (*Hydrochærus*). Their last molar possesses a movement from behind towards the front, so that, when entirely grown, the two small teeth which preceded it have disappeared, and it alone occupies the jaws.

The Elephants have a mode of dentition resembling the Cabiais and Phacochœrus. Their molars begin to show themselves by the fore part, and continue to advance from behind forwards; from which it follows, that these animals have at first only one molar in each jaw, afterwards two, then only one, then two again, and so on. It appears that this movement is the consequence of the growth of eight teeth. The first, which occurs soon after birth, has not fallen when the second makes its appearance. At the age of two years, the latter remains alone; and this continues until the appearance of the third, which remains alone until the sixth year. At nine years of age, it also disappears to give place to the fourth, and so on. It should be noticed, that these teeth always appear at first by their fore part, which, on that account, is much sooner worn out than the hinder.

Passing to the Horses, we return to the mode of dentition first described. The teeth of the second dentition develop themselves immediately under such of the first as are intended to fall; that is to say, under all the incisors, the canines, and the first three molars. The only particulars requiring notice at present are, that the first teeth are narrower than those which succeed, and that the last molars appear as soon as the first have fallen.

The Ruminantia present analogous phenomena. All the incisors and canines of the first dentition give place to teeth of the same nature as themselves; and of the six molars found in each jaw, the first three are replaced by the same kind of teeth, being only less complicated. It is then, also, that the last molars, which are very complicated, present themselves; and we may thus perceive the same design of Nature, already noticed, in respect to the Carnassiers and others.

Among all these animals, the greater part of the teeth, at the time of falling, present nearly the same appearance as those of Man. Their roots have disappeared; and from the irregularities in these teeth at their innermost surface, it might be imagined that they had been corroded: and further, that being composed of different substances, some of them were less accessible than others to the action of the corrosive substance. As spots or black stains are perceived on the whole extent of this surface, they might appear to exhibit manifest traces of a kind of corrosion, and forcibly recall the appearance of caries, or decay of the teeth, as has often been remarked.

Many hypotheses have been proposed to explain this singular phenomenon. Yet it is not easy to perceive how a corrosive fluid, if it really existed, could spare the adjoining parts, and especially the teeth of the second dentition. The mechanical action of the second teeth upon the first has also been supposed, and of all attempts to explain the phenomena, this is undoubtedly the most unfortunate. One tooth cannot wear out another without wearing itself out likewise; and the second teeth are always in their most unblemished state at the time when the first have fallen: finally, this has been attributed to a power of absorption; and the last opinion is most generally adopted. It is not unlikely that caries, or decay of the teeth, may be produced by a similar cause.

From the observations which have already been made, on the complication of the dentary capsules, on the variety of substances of which the teeth are composed, on the care which Nature takes in supplying the places of those teeth destined to fall, on the different situations which they occupy, and the names they have received, will

lead the student to perceive the importance of these organs, and the different functions which they are destined to fulfil. But a much more enlarged view is acquired by
studying the different forms of the teeth, in their relations to each other, as well as
to the nature and habits of the animals possessing them. It remains for us to give a
rapid outline of this part of the subject.

On placing before our view the teeth of all known Mammalia, we soon perceive
that they admit of being classified under a small number of different forms. With
some of them, as we have already remarked, there is no difference between the root,
or rather, the parts inserted within the bones which bear the teeth, and the crown,
or part beyond these bones. Teeth of this nature have no real roots, in the proper
acceptation of the word; that is to say, the crown is continued inwards as far as the
dentary capsule, which never produces anything but the crown, as long as it remains
free and active—a circumstance occurring with some animals during the whole course
of their lives. Among others, on the contrary, the roots are very distinct from the
crown; they may be either simple or complex, but do not, in general, exhibit in
their forms that constancy of character which is always to be recognised in the
forms of the crown. This circumstance arises naturally from the different manner
in which each of them is formed.

Restricting, therefore, our view of the teeth solely to their crowns, we find that
there exist three principal forms among them. These may be almost infinitely modified, and some transformed into others, in such a manner, that it becomes impossible rigorously to determine the precise point where the one form passes into the
other. This manner of classifying the teeth must be regarded as nothing more than
a purely artificial method, for enabling us to speak of their forms without too much
obscurity and confusion, by restraining, within their proper limits, the observations
necessary to be made. The crowns of all teeth may be regarded as conical, trenchant, or tuberculous.

(1.) The Conical Teeth vary in form from the cylinder, more or less compressed,
terminated by a point, more or less obtuse, to the oval or ellipse. Some are straight,
some angular, others curved. Those of an elliptical, or oval form, are the least
common, and are observed among the Cachalots. The conical teeth, which comprise
the Canines of the Carnassiers, the tusks of the Elephants, Hippopotamus, &c. are the
most numerous. Finally, the cylindrical may be seen in the molars of such Edentata
as are possessed of teeth. Among the Conical teeth, only two kinds of composition
can be observed. Some are formed only of ivory and cortex, such as the Molars
of the Cachalots; for although the external part of these teeth possesses a whiter
tinge than the central, it is not formed of enamel, as some have thought. Both
substances are ivory in reality, and it is the same with the tusks of the Elephants.
Others are covered with enamel, such as the Canines of the Carnassiers, and many
others. In this class of conical teeth are found by far the greater number of those
destitute of roots, such as nearly all tusks; and of those, wherein the root is distinct
from the crown, only a small number have been observed with many roots, as the
Canines of the Moles, for example.

(2.) The Trenchant or Cutting Teeth may be presented under a simple or compound form. Among the former may be placed the Incisors of the Rodentia, which
belong as much to the first division as to the second, the Incisors of the Quadrumana, the Carnassiers, the Ruminantia, and others. In this division we may
place the false and carnassier molars of the Carnivorous animals. There are, however, many among the former which approach nearer to conical than to trenchant
teeth. All teeth of this class are composed of ivory and enamel, though some also
have cortex. These last are the incisors of the Rodentia, which present the singular
anomaly of having enamel only on their anterior surface. They are with simple or
multiple roots; and those of the Rodentia alone are possessed of the same peculiarity
as tusks, in having no roots properly so called.

(3.) The Tuberculous Teeth present the greatest variety of form, and are all
Molars. The simple tuberculous Molars are those of the Quadrumana, the hind-
most molars of some Carnassiers, the grinders of Squirrels and Rats, those of
the Babyroussa, or Indian Hog. The proper tuberculous Molars are found
in the Insectivora, &c. The compound tuberculous molars belong to a great number of Rodentia, such as the Beavers, the Pacas (Coelogenys), the Agoutis, the
Hares, the Guinea Pigs, and others. The simple tuberculous molars are always formed
of ivory and enamel, while they are all possessed of several roots. This observation
is equally applicable to the proper tuberculous molars.

Among the compound tuberculous molars there are perhaps none which do not
possess the cortex, in addition to the ivory and enamel. Some of these teeth are
found with several roots, as in the Beavers, Elephants, Horses, and Ruminantia;
and without roots, as in the Hares, the Cabiais, the Lagomys, and other Rodentia.

The uses which animals make of these different forms of teeth are exceedingly
various. To some they are powerful arms, by means of which they attack their prey,
or any enemy that threatens them, or else defend themselves when attacked. In
others they seem rather to be intended to retain a prey, which has already been seized.
Some kinds are used for dividing the food like pincers; others for cutting it like
scissors. Again, we find another class of teeth which grind like the stones of a mill,
or which triturate their food, like jagged pestles fitting into mortars as jagged as themselves. Sometimes they crush by a single jerk, or pressure. All these forms and
different modes of action find their final object in the ever-varied substances, which
may serve for the nourishment of animals. The kind of food which each animal requires is determined by its nature; this again regulates the influence which it exercises upon other beings, and determines its station in the scale of creation.

The different kinds of teeth are found combined together in different manners. In
many Carnassiers, we find conical, trenchant, and tuberculous teeth, all united in the
same individual. Among the greater number of the Ruminantia, we can discover
only the trenchant and tuberculous teeth. The conical teeth alone are found in some
Edentata, and in the Cachalots and Dolphins, while only the trenchant and conical
teeth are found in the common Seal.

In fact, we find in almost all Mammalia at least some of these forms—simple teeth,
semi-compound, or compound, with one or more roots—conical, compressed, pointed,
with flat crowns, tuberculous or trenchant. At present it is unnecessary further to
enumerate the different possible combinations of teeth. A general idea of the subject may be obtained from the following

SYNOPSIS OF THE MAMMALIA, EXHIBITING AN OUTLINE OF THE NATURE, FORM, AND POSITION OF THEIR TEETH.

Instances.

Teeth
- Calcareous
  - in both jaws
    - of various forms
      - completely lining the margins of both jaws
        - of three kinds not very strongly defined (Anomalous)
          - molars simple
            - tubercles blunt .... Man, Orang-Outang.
            - pointed ...... Lemurs, Flying-Cats.
            - very sharp ... Hedgehogs, Shrews, Moles.
          - molars compound .... Anoplotherium (fossil).
        - of three kinds, well defined (Normal) .... Apes generally, the Ouistitis, nearly all CARNASSIERS, Hippopotamus, Hog, Tapir, Opossums.
      - not completely lining the margins of both jaws, leaving a vacant space
        - in the middle, both above and below ...
          - really ... RODENTIA, Asiatic Rhinoceros.
          - apparently ... Horse, Kangaroo-Rat.
        - in the front ...
          - both above and below .... Sloths, Morse.
          - below, and in the middle above .... Elephants, Mastodon (fossil), Dugong (adult), Manatus (young).
          - above, and in the middle below .... RUMINANTIA.
    - of one form only
      - all Molars .... Armadilloes, Orycteropus, Megatherium (fossil), African Rhinoceros, Manatus (adult).
      - all Conical or Canines .... Dolphins generally.
  - in one jaw only
    - the upper .... Narwhal, some Dolphins.
    - the lower .... Cachalots?
- Horny
  - in both jaws .... Ornithorhynchus.
  - in the upper jaw only .... Whales.
- Wanting .... Manis, Ant-eaters, Echidna, some Dolphins?

These varied forms and positions, and even the number of the several kinds of teeth,
often afford the best specific characters for determining the Mammalia; in all cases
they offer the surest characteristics of the genera, and even of other divisions of a higher
order. In this way, one of the Mammalia may be immediately recognised by a simple
inspection of its teeth; and, reciprocally, we may determine the nature of the animal
to which a single isolated tooth has belonged. The importance of this study towards
the knowledge of fossil animals, as well as for establishing the generic groups of fossils, has been forcibly illustrated in the celebrated work of the Baron Cuvier, on the

Fossil Bones of Quadrupeds (Sur les Ossemens Fossiles), and by MM. Frederic
Cuvier and Illiger. The teeth may, indeed, be regarded as one of the most important subjects in Zoology, and one of the most certain marks for ascertaining the nature of animals, and the relations established among them. They are, in fact, the
foundations of the science; and, hence, should occupy an important place in any
system of classification, as they serve as an index to the order of facts and their relations; and, hence, may be considered as indispensable to the existence of the whole
science.

GENERAL REVIEW OF THE MAMMALIA CONTINUED.

The Structure of Skin, Hair, Horns, Nails, Scales, and other integuments.

THE entire surface of all organized bodies is terminated by an envelope of a peculiar nature, varying in thickness according to the species of animal or plant, or the different parts which it covers. In animals, this integument commonly receives the name of *skin*, and seems to be essentially the same in all the Vertebrated animals, the external differences being merely owing to the development of certain additional parts. One of its surfaces is always intimately united to the body of the animal or plant; while the other, remaining unattached, bears immediate and various relations to the surrounding bodies.

In the Mammalia generally, the skin is composed of four substances, more or less distinct, and varying in their properties. The most external is termed the *epidermis or cuticle;* the second from the surface is the *mucous tissue* or *rete mucosum;* the third is the *papillary* or *nervous substance;* and the fourth, or innermost, forms the true skin, *chorion*, *cutis*, or *dermis*. These successive layers may be of greater or less thickness, and some of them may not always be present in the several species of animals belonging to this class.

The epidermis or cuticle is the most universal of all the layers, being found on the bark of trees, the stalks of herbs, the petals of flowers, the pellicle of the fruit,—as well as upon the entire surface of all animals. It appears to be an intermediate substance between horn and true skin, being nothing more than a thin membrane, formed by the hardening and drying up of the most superficial layer of the mucous tissue immediately beneath it, and of the albuminous fluids with which the latter is impregnated. It does not possess life in common with the other animal tissues, being merely composed of a greater or less number of inanimate layers placed one over the other. This cuticle is not confined to the surface of the body, but extends into its several apertures, protecting them, as well as all the nerves of the body, from a prejudicial contact with the media of air and water, to which they are continually exposed. The consistency of the cuticle varies with the nature of the circumjacent fluid; thus, it is observed to be dry and almost horny in animals living permanently in the air, while it is viscous and mucous in the aquatic species. The cuticle appears to be folded in a variety of ways, among those Mammalia, which remain continually exposed to the drying influence of the air. Sometimes these folds take the form of circles, wrinkles, or spiral curves, corresponding to the elevations and depressions of the skin, or that part of it called the mucous tissue. The thickness of the cuticle becomes considerable whenever a part of the body is exposed to a continuous friction; for example, upon the sole of the foot, the palm of the hand, and other parts used for holding or grasping, such as the prehensile tails of some American apes. The holes through which the hairs protrude may be perceived in the furrows of the cuticle. These appear to be conical elongations, forced outwards by the hairs, to which they serve as rudimentary sheaths.

The epidermis is very thin in Man, excepting on those parts which cover the palm of the hand and the sole of the foot. Yet it may be considerably hardened, and even changed into a substance nearly approaching to horn in consistency, either by friction, long exposure to a dry air, or to certain chemical agents, while the sense of touch becomes deadened in consequence, and almost wholly obliterated. We see frequent instances of this in the hands of hard labourers, of blacksmiths, dyers, or in those natives of Africa who walk barefooted upon burning sands. On the back of the human hand the furrows of the cuticle exhibit angular figures of various forms; on the palm they assume the appearance of parallel and elongated lines; while under the extremities of the toes they take the form of arcs of circles, curves of different kinds, and especially some very remarkable close and symmetrical spirals.

Among the other Mammalia, the cuticle, being always thinner in proportion as the hairs which protect it are more compact, is found to exhibit nearly the same appearance as in Man. The epidermis, covering the wings of the Bat, is very thin, and possesses furrows of many angles, very similar to those seen on the back of the human hand. This integument is thin in the Porcupine, and not very distinct from the other strata of the skin, which, in these animals, is always gelatinous. In the tails of the Beaver, Rats, Ondatra, and others, the epidermis is remarkably dry and scaly, as well as upon the surface of those scales which cover the body of the Manis and the Armadillo. Where the skin is very thick and deeply furrowed, as in the Elephant, Rhinoceros, and Hippopotamus, the epidermis is likewise thick, being covered with small plates, which sink into the several furrows, and may be separated like scales. The soles of their feet exhibit a remarkable structure in respect to the cuticle, being divided externally by deep depressions, nearly circular, with six or eight surfaces more or less regular, each of which contains an infinite number of small polygons of great irregularity. The entire surface of the skin thus acquires the appearance of shagreen. When separated from the foot, the epidermis exhibits elevated lines upon its external surface, corresponding to the furrows of the greater polygons, as well as smaller ones, corresponding to the lesser polygons. This arrangement gives it the appearance of net-work *in relievo*, of a pretty regular design, and resembling lace with large points. The Cetacea are covered with a very smooth epidermis without any remarkable fold, and are always moistened with a mucous oily secretion, which prevents the surfaces of these animals from becoming macerated by the action of the water.

The rete mucosum, or mucous tissue, is situate immediately between the epidermis and the villous surface of the skin. It is not membraneous, but forms a mucous layer, the colour of which varies in different species and races of animals, and sometimes in different sexes and individuals, or even in the parts of the same animal. This apparent colour of the surface depends upon that of the mucous tissue; for the epidermis when removed is almost transparent, and the cutis or true skin is also destitute of colour.

The villous or papillary surface of the skin is placed between the cutis and mucous tissue, and immediately beneath the latter. It does not possess the membraneous structure of the epidermis, but is a surface produced by the aggregation and approximation of a number of minute papillæ or small tubercles of various shapes, and formed apparently by the external extremities of the cutaneous nerves. The figures of these nerves are exceedingly various, but their structure is nearly the same. They are easily exhibited on being macerated in water for some days. Each tubercle may then be observed to consist of a bundle of minute fibres, united at their base, like the hairs of a pencil. The fibres of the centre are sometimes longer than those of the circumference, and then the papilla assumes the form of a cone; often they are of the same length, in which case it appears flat. As the sense of touch resides more particularly in these papillæ, they are accordingly found in the greatest number, and most conspicuously, on the tongue, the lips, and at the extremities of the fingers. In Man, the papillæ on the soles of the feet and the palms of the hands are particularly remarkable, as they are placed close together in a compact manner, and distributed in lines corresponding to the external grooves of the epidermis already noticed. The papillæ under the nails present a villous surface, the minute and compact fibres of which are all directed obliquely towards the extremity of the fingers. The minute fibres of the lips are disposed in the same manner, but are still longer, closer, and more delicate. In the other Mammalia the same rule is constantly observed, and the papillæ are always more developed in proportion as the parts to which they belong are employed in touch. Thus, the nervous papillæ are very visible on the snout of the Moles, the Shrews, and the Hogs, where they form tufts consisting of very close fibres. They may also be remarked on the proboscis of the Elephant, and very distinctly on the tail of the Cayenne Opossum, and it is probable that they exist in the same manner in all Mammalia with prehensile tails. Cuvier was unable to detect them on the skin of the Dolphin and Porpoise.

The cutis or true skin is situate most internally. Its structure has been developed by anatomists in a very distinct manner, by certain modes of preparation, and especially through maceration in water. They have demonstrated that it is composed of a tissue of gelatinous fibres, crossing each other in every direction, and so interwoven that the substance may be compared to felt. Among these fibres may be observed a great number of fine ramifications of nerves, as well as arterial, venous, and lymphatic vessels. The organization of the cutis is such, that the fibres composing it are capable of elongation and extension in every direction, and we may easily perceive that these qualities were necessary to give the surfaces of animals the power of evading the mechanical action of other bodies. These properties of elongation and extension, possessed so remarkably by the skins of the Mammalia, have enabled manufacturers to apply them to different purposes, where strength and flexibility are necessary, or where great friction has to be sustained; and the process of inducing these requisites constitutes the art of the currier. The fibres are further approximated or separated, to form the leather and adapt it for different uses, and this again is the foundation of the arts of the tanner, skinner, parchment and morocco-maker. The cutis in Man is from a line to a line and a half in thickness in certain parts of the body. From maceration, as well as the process used by skinners, we perceive that the fibres which enter into its composition are long, fine, and very solid, but united in a lax manner. In the Mammalia generally, the cutis is thickest on the dorsal, and thinnest on the ventral region. It is also much thinner on birds than on the Mammalia.

The obvious intention of Nature, in providing Animals with a skin or epidermis, was to protect them from an injurious contact with surrounding bodies. For some purposes this covering is insufficient, in which case other appendages are added, differing in form and consistency, and suited to their several purposes. These integuments have received the names of Hair, Horns, Nails, Scales, &c.

Hairs are filaments of a horny substance, more especially intended to cover the skin of the Mammalia. One extremity of each hair is implanted in the cutis, and sometimes penetrates even as far as the muscular layer beneath. This extremity is enlarged into a bulb, more or less thick, sometimes containing a small drop of blood, the whole being inclosed in a membraneous sheath. When the hair is young this cell is large, and its size diminishes in proportion as the hair grows older. If punctured during its earlier stage, the blood flows, and it becomes soft and flaccid.

That entire portion of the Hair placed externally to the skin, is termed the Shaft. It forms a very elongated cone, the free extremity being the apex. The hairs grow from their base, and hence are finer in young Animals than in the old; for a similar reason, they appear to augment in number when cut, though in fact their extremities only are increased in diameter. When the nails rise out of the skin, they carry with them a small portion of the epidermis, which forms a kind of sheath around their base; this becomes gradually detached, under the appearance of transparent and whitish scales.

Some Animals have the hair in some parts of their bodies more or less developed at the time of their birth. In other parts no hair appears until a more advanced period of life.

Linnæus remarks that "Mammalia have hairs, Birds have feathers, and Fishes have scales." These assertions, as we have already remarked, are true only in a general manner, for many Mammalia either want hairs altogether, or are furnished with a very small number. This fact did not escape the observation of Linnæus himself, who alludes to it in another part of his *Systema Naturæ*. Some species, such as those of the genus Manis, are in fact covered with true scales, and others, like the Cetacea, have a naked skin. These exceptions are, however, more apparent than real in respect to the Manis, in which the scales are little else than compound hairs. According to M. Blainville, the last remark is equally applicable to the Cetacea, where the hairs, becoming blended together, unite in forming a kind of crust or general envelope. This celebrated Naturalist is even of opinion that the term *Pilifères*, or hair-bearers, might form an advantageous substitute for that of *Mammifères*, an observation with which we can by no means agree, as true hairs are also found on many Birds, so that the term *Hair-bearers* would apply equally to them.

In general, the Mammalia have two kinds of hair; the one bristly, more or less stiff, and external; and the other woolly, very fine, soft to the touch, and commonly hidden beneath the stronger hairs. The domestic species of Sheep form a remarkable exception to this observation, on account of the abundance and length of the woolly hairs, and at the same time of the almost total disappearance of the coarse hairs.

Animals of cold countries approach towards this peculiarity of the Sheep. In warm regions, on the contrary, the bristly hairs are more strongly developed, and the woolly hairs become almost wholly wanting. The quantity, or rather the proportional abundance of the latter, is generally in the inverse ratio of the temperature, while that of the former is directly proportional to it, or nearly so.

The climate has a great influence on the nature of the hair, especially among the domestic animals. It becomes long and rigid in cold regions, as we remark in the Siberian Dog and Iceland Ram. Syria and Spain produce an opposite effect, and these more favoured climates produce a corresponding change in the hair, which may become tufted, fine, and silky. These qualities may be remarked in the Spanish Sheep, the Maltese Dogs, as well as in the Goats, Cats, and Rabbits of Angora.

The coarser hairs bear a greater predominance in certain parts of the body, especially among the Males, such as in the cervical region of the Lion and Horse, where they form a mane, and on the tail in many species. Other animals are covered in every part of the body by very long hairs; such, in particular, is the Bear of India (*Ursus labiatus*), in which the hairs are almost every where from seven to nine inches in length, and in some places even a foot long.

In some species the fur is mixed, and in others it is sometimes composed entirely of spines, more or less abundant, and of various structure; such are the Hedgehogs, the Tenrecs, the genus Echimys, the Porcupines, and many others. These spines or prickles are usually pointed, as their name indicates, and each is composed in general of a single hair. A great development of the muscles of the skin is always found in all species armed with spines, and this arrangement is especially remarkable in the Hedgehog. It may also be observed, on comparing different species, that the spines are arranged in small and regular groups, the disposition of which is peculiar in each species.

It becomes difficult to examine the structure of the hairs on the human body, because they are slender; but the bristles of the Hogs or the whiskers of the Cats are better adapted for this kind of inquiry. On examining the bristle of a Wild Boar with the microscope, we observe that it is grooved throughout the whole of its length by about twenty furrows, formed by an equal number of filaments, and the union of which constitutes the surface of the hair. In the middle of the bristle there are two canals, which contain a humour called the pith or marrow. The filaments of the hair separate on being dried, as may be remarked in the bristles of brushes, where the cavities may be observed to be empty, and a few laminæ cross each other in different directions.

The hairs of the Elk and Musk, with the spines of the Hedgehog, the Tenrec, and the Porcupine, are not altogether similar; their surfaces being covered with a horny lamina, varying in thickness, and a few furrows only can be observed. Internally they contain a white spongy substance, which appears at first sight strongly to resemble the pith of the Elder-tree (*Sambucus nigra*).

The colour of the hair appears partly to depend upon that of the mucous tissue; for where animals have differently coloured spots upon the hair, these usually indicate corresponding colours below them in the skin. Even in the human species, many striking relations of this kind may be remarked. Thus, Negroes in general have the head black. Persons with red hair often have the skin freckled or covered with reddish spots, while black hair usually accompanies a dark complexion.

The external colours of Animals depending on those of their respective mucous tissues are exceedingly various. Among the Mammalia, it is very seldom that they appear of a vivid hue. On some species of Mandrills, the nose and hips are bright red, violet, and carmine. The mucous tissue is also pure white on their cheeks; and of a beautiful silvery whiteness on the bellies of the Cetacea. In the Mammalia generally, the mucous tissue imparts its hue to the hair and nails. It is often observed to be coloured within the cavities of organs, into which it has been prolonged, such as on the palate, the tongue, the ear, the conjunctive and nasal membranes of the Quadrumana, Dogs, Ruminantia, and Cetacea. The mucous tissue appears to be thickest in the class last mentioned. On the backs of the Dolphins and Porpoises it is very thick, and of a deep black.

The colours of the Mammalia have not that metallic lustre which characterises a large number of the genera of Birds, there being one solitary exception to this observation in the brilliant lustre of the Chrysochloris, or Cape Mole. We do not find among these animals the dazzling brightness of the Parroquets, the Tanagers, or the Flamingos, nor can we discover anything analogous to those ornamental appendages which adorn the plumage of many Birds. There is another peculiarity in the colours of the Mammalia, that they are in general much paler and fainter beneath than on the back or flanks. This may be observed not only in the true Quadrupeds, but even in those species, such as the Kangaroos, which more or less are in the continual habit of maintaining an upright position. Yet, without enumerating all those species which are entirely of one colour, such as the Coaita (*Ateles paniscus*), the Polar Bear (*Ursus maritimus*), we find some exceptions among the Rodentia, and especially in the order of Carnassiers, such as the Hamster, the Gluttons, the Badger, and some other species, while many even have the belly absolutely black. In particular, we may notice a Carnassier recently described, for the first time, by M. Frederic Cuvier, who has assigned to it the specific name of Panda.

The colours of the males among the Mammalia are most commonly the same as those of the females, excepting perhaps that the shades of the latter are not quite so deep. In this respect they differ remarkably from the greater number of Birds, in which the colours of the female differ almost wholly from those of the male. However, all the other circumstances which influence the colours of Birds act equally upon those of the Mammalia, although most commonly in a different manner. Age, for example, varies the colours of the fur only in a small number of species, as among the Stags, the Tapirs, and the Lion, all of which are clothed at their birth in a kind of livery, or peculiar arrangement of colours. Their coats, instead of being uniform, as in the adults, are at first ornamented with spots, regularly disposed, and analogous in their arrangement to those observable in the adult animals of other species belonging to the same genus. Thus the spots of the young Fawns are white, similar to those of the adult Axis; while they are black in the Lioness' cubs, in the same manner as we see in most adult Cats. This very remarkable relation between the system of colouring belonging to the young individuals of one species, and that observed in other species of the same genus, may also be traced among the Birds. But the young of the Mammalia differ from those of Birds in this respect, that—while the plumage of the latter, most commonly resembling that of their mothers, is duller in its hues than at a more advanced period of life—the livery of the young Mammalia is, on the contrary, an ornament which they gradually lose as their years advance, until they finally resign the spotted and agreeable garb of youth for one of a more simple and uniform character.

The colouring matter of the hairs resides in the horny part of their substance, and not in the pith, which is commonly white. We can observe this structure most conveniently in the spines of the Porcupine, from their unusual magnitude. Some hairs are coloured differently in several parts of their length, while the colours themselves may be infinitely various, both in quality and intensity. In general, the hairs of the Mammalia are round, and this form is observed more especially in the hair of the head or mane. On the tail of the Hippopotamus, as well as on the body of the Great Ant-eater, and especially on the Ornithorhynchus, they assume a flattened appearance; and in several species of the Ruminantia, especially the Musk (*Moschus moschiferus*), the hairs appear as if they had been crimped. In some varieties of Goats, Cats, and others, the hairs are fine, long, and silky; they appear both crisped and frizzled in the Rams. From their great thickness, stiffness, and elevated position in the Hogs, Hedgehogs, and Porcupines, they have received the name of *Spines* in the two last-mentioned animals, and of *Bristles* in the first.

All Mammalia possess a certain quantity of hair, without excepting the Cetacea, in general destitute of this covering. Man is covered in almost every part of the body by scattered hairs, although they are not easily perceived in some places from their excessive fineness. Those of the head and beard are the longest; those of the axilla and pubes are next in length. On the interior of the nose and ears they are shorter, and on the remaining parts of the body they appear of a still more diminutive length. Contrary to the arrangement in the other Mammalia, the hair is longer on the breast and abdomen than on the back. There is never any hair on the palms of the hands or on the soles of the feet.

Among the Quadrumana, the true Apes have the hair of the head in general of the same length as that of the body. The hairs which cover the fore-arm point upwards towards the elbow, instead of being directed towards the hand, as we may see in the Orang-Outang and some other species. The buttocks are callous in a great many Quadrumana, and entirely deprived of hair.

In most species of Cheiroptera, a few scattered hairs only can be seen on the membranes of the wings, the nose, and the ears. One species of Bat (*Vespertilio lasiurus, Linn.*) has also a few upon the tail. The remainder of the body is covered by short, fine, and villous hair, as may be seen in the Flying-cats (*Galeopithecus*), and other animals of their order.

The spines of the Hedgehog are found only upon the head and back; the limbs and lower parts of the body are covered with stiff bristles. In these respects, the Tenrecs resemble the common Hedgehog. In some species the spines and bristles are mixed together indiscriminately.

The hair is fine, short, and close in the Moles and Shrews, so that their skins seem as soft as velvet to the touch.

Among the Carnivora the hair varies considerably. There are two kinds of hair in the Weasels, Sables, Ermines, Martins, and others; the one being very fine, thickset, intermixed, and placed close to the skin; while the other, which is longer and stiffer, alone appears at the surface. These two kinds constitute the finer furs. The amphibious Mammalia have short, rigid, and very close hair.

The arrangement found in the Carnivora may also be remarked among the finer haired Rodentia. The spines of the Porcupine are more slender, short and flexible on the head, neck, and belly, than on the back. There are about ten or twelve placed upon the tail, and resembling the tubes of quills, truncated at the free extremity. A rustling sound is emitted from these spines when the animal moves its skin.

Among the Edentata there is found a considerable diversity in the quality of the hair. It is broad and flat in the great Ant-eater (*Myrmecophaga jubata*), and has a longitudinal furrow in both surfaces, so that each hair presents the appearance of a dried blade of grass. The two-teed Ant-eaters are covered, on the contrary, with very fine wool. Several have hard and sharp-edged scales placed one over another, like the tiles of a house, as we see an example in the Manis. Others are covered with prickles, like the spinous Ant-eater (*Echidna*). The Armadilloes (*Dasypus*) have, in addition to the scales or osseous bands, which cover the back and head in regular compartments, some scattered hairs, which are short and rigid like those of the Elephant. These hairs drop off, however, as the animal advances in age.

The Hogs, of all Pachydermata, have the greatest quantity of hairs, which in them are called bristles. These are scattered and frequently bifid at their free extremity. The other genera of this family are comparatively almost destitute of hair. It is in general short in the Solipeda, excepting on the mane and tail, where it receives the name of Horse-hair (*crines*).

The Bulls, Deer, Antelopes, and Giraffe, have short hair in general. In the Camels it is very fine and soft, and remarkably so in the Lamas. All of them may have callosities, which are destitute of hair, on the knees and breast. Goats' hair is long and fine, extending to a pointed beard under the chin. The hair of Sheep is long, and readily distinguished by that crisp and frizzled appearance, well known in the wool of commerce.

The Hair of all these animals, when submitted to a chemical analysis, yields nearly the same results, whether it be examined under the form of wool, bristles, spines, or scales. On being subjected to the action of Heat, in open vessels, it fuses or liquifies at first by swelling up. It subsequently emits a white flame, and resolves into a black carbon, the incineration of which is very difficult. Hair yields, on distillation, a reddish liquor, containing prussiate of ammonia, and another salt of an ammoniacal base, combined with a peculiar animal acid, which Berthollet has named Zoonate of Ammonia. The carbonaceous residue at the bottom of the still is light, and contains carbon and phosphate of lime. The Hair does not dissolve completely in boiling water, a mucilaginous matter, which is the pith or medulla, being separated from it. Caustic alkalis and some acids dissolve it entirely. Sulphur, silica, iron, and manganese, may be traced in the hair.

There are certain prolongations of a horny substance, which grow upon the heads of some species of Mammalia, especially the Ruminantia. These also appear on several other parts of animals.

The term *Horn*, as applied in the arts, would exclude the excrescences of the Stag, Deer, Rein-deer, Elk, and others, which consist rather of a bony substance, distinguished by the term *antler*. The horns of this division of Ruminants are true bones and composed like them of a cartilaginous matter, within the meshes of which, particles of phosphate of lime are deposited, constituting a kind of earthy salt, commonly known as earth of bones or hartshorn.

These antlers, in their perfect state, are true bones both in their texture and elements, the external part being hard, compact, and fibrous; the internal spongy, but very solid. There are no large cells, medullary cavities, or sinuses. The bases of these antlers adhere to the frontal bone, forming one body with it, in such a way as to render it impossible to point out, at certain ages, the limits between them. The skin which covers the forehead does not extend farther. It is surrounded by a denticulated bony substance, called the *burr*. Neither skin nor periosteum covers this substance or the rest of the antler. Furrows more or less deep, which are the vestiges of vessels distributed along their surfaces while they were yet soft, are alone to be traced on the exterior. These hard and naked horns remain only for one year on the head of the Stag. The period of their fall is varied according to the species; but when near, there appears, on sawing them longitudinally, a reddish mark of separation between them and the supporting eminence of the frontal bone. This mark becomes gradually more apparent, and the bony particles at length lose their adhesion at that part. A very slight shock then makes the antlers drop off at that period, and two or three days commonly intervene between the fall of the one antler and that of the other.

The eminence of the frontal bone resembles, at that time, a bone broken or sawed through transversely, and its spongy texture is exposed. The skin of the forehead soon, however, covers it; and when the horns are again about to shoot, tubercles arise, which remain covered by a production of the same skin until their perfect size has been attained. During the whole of this operation, the tubercles are soft and cartilaginous. Under the skin a true periosteum is found, in which vessels, sometimes of great size, are distributed, and penetrate the mass of cartilage in every direction. The cartilage ossifies gradually, and, passing through the same stages as the bones of the fœtus, ends in becoming a perfect bone. During this time, the burr at the base of the horn penetrates the indentations through which the vessels pass, and also develops itself. The indentations by their growth confine the vessels, and finally obstruct them. Then the skin and periosteum of the bones wither, die, and fall off. The bones, now become bare, in a short time fall off, only to be renewed by others, but always of a larger size than the first. The antlers of the Stag are subject to diseases equally with other bones, and of the same kind. In some the calcareous matter is extravasated, and forms different exostoses; in others, on the contrary, it is found in too small a quantity, and the bones continue porous, light, and without consistence.

The true horns, such as those of the Bull, Ram, Goat, and Chamois, are formed upon processes of bone, and differ materially from antlers in this respect, that they grow at their root or base, and bear a great analogy to the other integuments.

This view is established by investigating the manner in which the horns of the Calf exhibit themselves. In the third month of conception, while the fœtus of the Cow is still inclosed in the membrane, the cartilaginous frontal bone presents no mark of the horns which it afterwards bears. It becomes partly ossified towards the seventh month, and presents in its two portions the small tubercle, which appears to be produced by the elevation of the osseous laminæ. These bony tumours soon after appear externally, and raise the skin in proportion as the tumour grows. At last it becomes horny as it elongates, and forms a kind of sheath which covers externally the process of the frontal bone. There are numerous branches of blood-vessels, which serve to nourish the bony part, and are placed between this sheath and the frontal bone.

Thus, the horns are merely solid, hard, elastic, and insensible sheaths, which protect the osseous prolongation of the frontal bone. These sheaths, generally of a conical figure, are broadest at the base, or extremity from which they grow. The curvatures assumed by the horns are different in the several species; they also present different channels or transverse furrows, depending on the age of the animal; and these denote, in a very certain manner according to the species, the number of years it has lived. The horns grow in precisely the same manner as the nails of animals and the beaks of birds; that is to say, from the bone which serves as a base there exudes a gelatinous matter, which takes the form of the horn, and hardens on coming in contact with the air. M. Vauquelin has found this gluten, or animal mucus, to be of precisely the same chemical nature as that found in hair.

The texture of the horns appears to be much the same in the Ram, Goat, Antelope, and Bull. They consist of fibres of a substance very analogous to hair, and appear to be agglutinated in a very solid manner. These fibres in the two genera first mentioned are short, and covered by superincumbent layers like the tiles of a house. In the last two they are longer, more compact, and form elongated cones, the one being incased within the other.

It appears that the horns of the Rhinoceros differ materially from those of the Ruminantia. They have no osseous part, and are not placed upon the frontal bone, but upon the ridge of the nose. They are formed, however, of the same substance, and we observe fibres analogous to hairs more distinctly in this animal than perhaps in any other. The base of the horn presents externally an infinite number of rigid hairs, which seem to separate from the mass, and render that part rough, like a brush, to the touch. When sawed transversely, and examined with a magnifying glass, we perceive a multitude of pores, seeming to indicate the intervals resulting from the union of the agglutinated hairs. When divided longitudinally, there are numerous parallel and longitudinal furrows, which demonstrate a similar structure. This kind of horn is attached solely to the skin. The horns of the Two-horned Rhinoceros (*Rhinoceros bicornis*) always appear in some degree moveable. When fixed, as in the Indian Rhinoceros (*R. Indicus*), there is a thick mucus interposed between its base and the bone over which it is situate.

The whale-bone, which lines the interior of the upper jaw in the Whales, also consists of hair united into laminæ.

The colour of horns depends, like that of the hairs, on the colour of their mucous tissue. Heat softens bone, and even fuses it; hence this agent is largely employed in manufacturing them into different articles.

It appears from these observations, that true horns differ essentially from the bony prolongations called *antlers* in the Deer. The latter increase at their extremity, and are covered with skin during their growth. They fall off when their growth is completed, and are replaced by others. The true horns are developed at the base, are covered with skin, and remain permanently.

The Nails of the Mammalia form, with the preceding, useful arms of attack, or necessary shields against external injury. These horny prolongations are generally equal in number to that of the fingers and toes, whose extremities they serve to arm and protect. Their form depends upon that of the last phalanx of each finger or toe, and they bear the same relation to these phalanges, as the hollow horns to the processes of the frontal bone which they cover.

The nails seem to be incased in a fold of the skin, the portion covered by the latter being called the root of the nail. They grow by this part precisely in the same manner as hairs, but the opposite extremity wears away by friction, from the various uses to which animals apply their nails. Accordingly, they are observed to grow exceedingly long in animals that are confined, and have few opportunities of motion. No part of a nail is sensible, except that which adheres to the skin, and the free extremity may be cut or broken without occasioning the slightest pain. The colour of the nails depends upon that of the mucous tissue.

The human nails appear in the third month of conception; and their development takes place nearly in the same manner as in the common horns. They appear at first like a kind of cartilage, which gradually acquires a proper consistency. Almost all animals have the nails formed, in some degree, at the time of their birth. The nails of Man, as well as the greater part of the Unguiculated Mammalia, appear to be formed of extremely thin layers, placed one upon another. The external laminæ are larger than those of the inferior surface; on which account we do not readily perceive this kind of imbrication which actually takes place. When diseased, however, or upon making a transverse section of the nail, after it has become completely dried, this structure becomes evident. Often we observe striæ, or very fine longitudinal and parallel lines, apparently resulting from the manner in which this part is moulded upon the laminæ beneath it.

Nails are generally wanting in animals which do not employ their extremities either in walking or grasping, as we may remark in the Cetacea. When analyzed chemically, the nails afford nearly the same results as hair, to which they bear considerable analogy, both in structure and their mode of growth.

Hoofs surround the phalanx entirely, in which respect they differ completely from nails. They are neither pointed nor cutting at the extremity, and both surfaces meet to form a round and blunt edge. Their interior is rendered remarkable by deep and regular furrows, which receive projecting laminæ not observable in nails. In the Elephant and Rhinoceros these furrows are very strongly marked. They are also conspicuous in the Horse, but do not appear very prominently in the Ruminantia. A layer of mucous matter may always be observed between the nails and the soft parts of the phalanx. In hoofs there is found a soft substance abounding in nerves, which serve to maintain a certain degree of sensibility in these parts.

Scales may be regarded as very flat horns, in the same way as hairs admit of being considered very slender horns. They bear a great resemblance to hairs, feathers, horns, and nails, both in their mode of growth and use, as well as in respect to their chemical analysis. Only a very few species of Mammalia possess scales on some parts of their body, and in Birds they are found on the feet alone. Reptiles and Fishes are, on the contrary, almost wholly covered by them.

The term *scales* is applied to a variety of substances of very different natures. In general, they consist of laminæ or small plates of a substance which may either be horny or bony. The scales of the animals belonging to the Genus Manis consist of a kind of flat nails of a horny substance, but of considerable thickness; their anterior third, which is bevelled and sharp edged, is free, while they adhere to the skin by the other portion. The external surface is channelled longitudinally, particularly in the Long-tailed Manis (*Manis tetradactyla*), in which animal they usually terminate in three points. They are furrowed transversely on the side next the skin, and appear to be formed of imbricated laminæ. In the Armadilloes, the scales consist of small compartments of a calcareous substance, covered with a thick, smooth, and apparently varnished epidermis. The scales covering the tail of the Beaver consist of thin laminæ of a horny substance, similar to those on the feet of Birds. The tails of Rats, Opossums, and most animals with prehensile tails, are covered with scales of the same nature.

GENERAL REVIEW OF THE MAMMALIA CONTINUED.

Their Organs of Voice.

MAN alone of all Mammalia possesses the exclusive privilege of uttering articulate sounds, to which the great flexibility of his tongue and lips, as well as the general form of his mouth, alike contribute. The power of communicating his ideas by artificial words, forms a means of communication of the greatest value; and there are no signs capable of being employed with the same convenience for this purpose, or which could be perceptible at so great a distance, or in so many directions. This faculty of speech, joined to the perfection of the hand in Man, contributes largely to his power.

The other Mammalia can express their wishes by cries alone; yet these natural signs are themselves subject to many modifications. Although incapable of communicating any complicated idea, they at least serve to express the passions by which the lower animals are agitated. Thus, we can readily distinguish the cry of rage, when an animal threatens his prey; that of distress in the unfortunate victim; the amorous cry of one sex for the other; the rallying cry of the female, when she wishes to assemble her young ones around her; the bursts of resentment; and in some species, even a cry of gratitude towards Man.

The larynx of the Apes greatly resembles that of Man in many respects; yet in those species which approach most nearly to him in organization, there are found membraneous sacs of a greater or less extent, where the air becomes stifled, and hence they are rendered incapable of producing any other than hoarse and inarticulate sounds. The large Apes of Africa, such as the Mandrills, merely emit a low sound, resembling *aou, aou*, pronounced in the throat, and never very loud. Animals of the genera Cercopithecus and Macacus utter very sharp and disagreeable cries. The Apes of the South American Continent, belonging to the genus Mycetes, have received from travellers the appellation of Howlers. These unite together in troops, and each Ape utters a continuous howl, which can be heard for at least a mile in every direction. For this purpose, they are supplied with a laryngeal apparatus of a remarkable appearance. Their lower jaw has its branches very distant from each other and much elevated, for the purpose of receiving a bony drum in the interval which separates them. The extremity of this organ is marked externally by a line across the skin of the throat, it being nothing more than the hyoid bone greatly enlarged and rendered hollow. This drum is the chief cause, in the Howlers, of the frightful noises which are re-echoed continually through the forests of those remote regions. The Sapajous also, from South America, merely utter a soft sonorous cry, for which reason they have been called *Weeping Monkies* by some travellers.

The Varied Lemur makes, it is said, a great noise in the woods of Madagascar; however, the other species of this genus only emit a low and continuous grunt when in confinement.

The Bats, as well as the Shrews and other Insectivora, have no other voice than a very sharp and feeble cry.

The sound of the Bear is a growl, or loud murmur, often intermingled with the gnashing of his teeth. That of the Kinkajou (*Cercoleptes*), when alone at night, resembles the barking of a very small Dog, and always beginning with a kind of sneeze. On being injured, its cry changes to the note of a young Pigeon. When threatened, it hisses like a Goose, and if irritated, utters a noisy and confused cry. The Badgers and Weasels, whether Martens or Polecats, always walk in silence, until they meet with some injury, when they emit a sharp and hoarse cry, expressive of resentment or pain.

The yelping of the Fox consists of a kind of barking howl, which forms the proper cry of that animal. It is chiefly in winter, especially during frost and snow, that the Foxes yelp, at which season their cry becomes further distinguished by a sharper and more elevated tone of voice. The howling of Wolves, and of many other Carnivorous animals, is a mournful and prolonged cry, which they utter either when pressed by famine or transported with desire. It is more especially during the night that the Wolves howl, and remarkably so in the long nights of winter. The Dog, on losing his master, utters a melancholy and plaintive noise, which is in fact a kind of howl. The barking of this animal is better known than the cry of any other species. An attentive ear can distinguish the bark of pleasure on again meeting his master; the bark of pain when wounded or struck; and that of joy when playing with other animals of his own kind. Those varieties commonly called Watch-dogs are generally silent during the day, but bark more especially at night. When a pack of hounds are in pursuit of any animal, it is easy to tell, from the different modulations of their cries, at what point the hunt has arrived.

The sound of the Hyæna, when at large in the woods, has been compared by some Naturalists to that of a Man when vomiting with great difficulty. In the Menagerie, however, this animal emits a cry of that kind only when irritated, at which time it is also accompanied by a sound expressive of resentment, very similar to that used by other Carnassiers on the same occasion. In the mewing, or rather the *miauling* of the Cats, it is very easy to distinguish the attracting cries of the females; the cries of pain with which they repel the approaches of the males; and the low and soft sounds with which they invite their young ones to follow them. The males, during their nocturnal combats, interrupt the silence of the midnight hour with their noisy sounds, preceded or followed by smothered hisses and growls more or less prolonged. When these animals are at their ease, especially when warm and after a repast, they produce a continual noise or *purr*, similar to that of a spinning-wheel when in motion, which sound is due to a peculiar formation of the larynx.

The roar of the Lion and Tiger is a frightful and imposing cry. The former consists of prolonged and rather grave sounds, intermingled with others of a sharper tone, and a kind of tremor, and is susceptible of numerous variations, according to the age, sex, as well as the passions which animate the Lions. All African travellers agree in representing the deep terror which suddenly seizes the other animals, and especially Horses, when, in the depths of the forests, they hear this dreadful cry around them. The roar of the Tiger is also very powerful and continuous for four or five minutes; that of the female is more plaintive, less interrupted, and of a much longer duration. According to Azara, the roar of the Jaguar may be represented by the words *houa, houa*. There is something plaintive in it, and, at the same time, strong and grave, like the lowing of a Bull; while that of the Panthers, with which the Jaguar has often been confounded, resembles the noise of a saw.

The voice of the adult Seals may be compared to that of a hoarse Dog. The cry of the young is much clearer, approaching more nearly to the mewing of the Cat. On separating the young from their mothers, the former *miaule* continually; while the old ones bark furiously at those who molest them, and use every exertion by biting to avenge the loss of their offspring. When attacked by Man, the Morses low in a most dreadful manner.

All the Marsupialia, with which we are acquainted, appear to possess merely a low and feeble voice; but the sounds emitted by many of them still remain unknown.

The Rodentia in general emit no other sound than a whistle. The common Squirrel, as well as the Palm Squirrel, has a strong voice; and these animals emit a kind of murmur, or grumble of discontent, with the mouth shut, on being irritated. The Striated Squirrel, and the Pteromys or Flying Squirrel, when unmolested, appear to have no kind of voice, but in other circumstances, they utter a cry like the scream of a Rat. The Marmot and Bobac, when playing or being caressed, have the voice and murmur of a young Dog; but on being irritated or frightened, they utter a whistle so shrill and piercing as to offend the ear. According to Erxleben, the sound of the Hamster is a kind of bark. The Campagnol Water-rat, when pursued so that he cannot escape, makes a cry resembling a snore. The Guinea-pigs have a grunt similar to that of a young sucking Pig, and a kind of purring noise when pleased; their cry expressive of pain is very acute. The Agoutis and Pacas have also a grunt like that of a Pig, while the Cabiai brays like an Ass. The voice of the Rat-moles is merely a snore. The Ondatras make a kind of groaning, which is especially remarkable in the females. The Siberian Jerboas, when irritated, have a cry similar to that of a young Dog just born, and sometimes a kind of snore, whilst the Common Jerboa raises a sharp cry. The Hares and Rabbits pass their lives in silence, and cry only when wounded or tormented; on these occasions, the former of these amimals emits a sound which bears some resemblance to the human voice. The note of the Alpine Hare (*Lagomys Alpinus*) is a simple and acute whistle, much resembling the cry of a young bird; while that of the Calling Hare (*L. pusillus*), another species of the same genus, is a very strong and grave note, somewhat similar to that of a Quail. This voice is composed of simple sounds, but repeated at equal intervals, three, four, and often six times, and it can often be distinguished at the distance of half-a-mile, although the animal which produces it is of very small dimensions. Finally, the Porcupine has a grunt similar to that of a Pig, from which circumstance it has in part derived its name.

Among the Edentata, one animal has received the name of *Ai* from its cry; in the same manner as one of the Rodentia has been assigned the name of *Aye-aye*. The Unau or Two-toed Sloth, which belongs, like the Ai, to the genus Bradypus, cries very rarely. Its note is brief, and is never repeated twice consecutively. Although plaintive, it bears, however, no resemblance to that of the Ai. It is not known whether the Ant-eaters, the animals belonging to the genera Manis, Orycteropus, Echidna, and Ornithorhynchus, have any peculiarity in their voice.

The ordinary cry of the Elephant is a grunt, which he changes into a whistling noise when irritated. The appellation of River-horse, applied to the Hippopotamus, derives its origin from the neighing noise made by this animal; its cry of pain is a kind of lowing, which bears much analogy to that of the Buffalo. The Rhinoceroses emit a grunting noise, somewhat resembling that of the wild Boar and domestic Pig; that is to say, a series of hoarse, short, and rough sounds, following each other at short intervals of time. The Tapirs have no other cry than a sharp whistle, which might not have been expected from an animal of its magnitude. The Daman has only a very feeble cry, and of short duration.

The Camels and Dromedaries are commonly silent, if we except the period of the rutting season, at which time they emit a very disagreeable rattling noise in the throat. The voice of the Lamas appears to be a gentle moaning like the word *he-em*, pronounced in the tone of voice of a complaining female; and some interval of time usually elapses before the animals repeat this sound. The Stags during the rutting season emit their rough bellow; it is a hoarse and disagreeable sound, which reaches to a great distance. The Antelopes, the Goats, and Sheep, bleat; and it is remarked, that the bleating of the Rams is stronger and graver than that of the young, the Ewes, or the Wethers. It has also been observed, that the bleating of the Goat is shorter than that of the Sheep; and that all these animals, when domesticated, are most clamorous on leaving the stable for the fields, and in the evening on their return. The greater part of the animals belonging to the Bull genus low or bellow; that is to say, they emit a very grave, powerful, prolonged sound. There is one exception, however, in the Yak, which grunts like a Pig. On this account it has been called the Grunting Bull or Cow (*Bos grunniens*). The Zebu, which is usually considered as a simple variety of the common Bull, grunts likewise, which circumstance, according to M. Desmarest, renders it probable that it should be regarded rather as a variety of the Yak. The voice of the Buffalo is a frightful bellow, much more forcible and clamorous than that of the Bull.

The neighing of the Horse is emitted whenever he experiences any vivid sensation, or is animated by some passion. He utters this cry alike when inspired by courage, pride, or desire; and neighs in the battle as if courting danger; or during the race, as though he defied his rivals to the contest. The braying of the Ass is a well known discordant sound, which it emits when pressed by any want, or inspired by passion.

Finally, the Cetacea, on happening to run aground, signify their consciousness of danger by uttering a very strong howl, with but few repetitions.

The voice of the Mammalia is not always confined to the simple purpose of expressing externally the interior or moral state of the individual; it is also occasionally employed as a means of preserving the species. While the Marmots are feeding on herbs in some elevated meadow of the Alps, one of them placed on a rock surveys in security all the others; and this advanced sentinel gives notice to his companions on the approach of an enemy by a loud whistle, which is the signal for immediate flight. In the same way the Wild Horses collect together in a dense troop, as soon as some of them, on being apprised of the danger, give notice of their fear by a peculiar neigh. The voice also serves to attract the sexes during the rutting season. At this period it usually changes its nature, as we may remark in the Stags, whose throats swell, and impart a graver tone to their cries. The peculiar sounds of the carnivorous animals further serve as a warning note to the feebler quadrupeds, and may thus contribute indirectly towards the preservation of the latter.

It can be readily perceived from the preceding observations, that the greater part of the Mammalia are generally incapable of producing any other sounds than such as are either noisy or disagreeable. Yet after having once heard each particular noise, we can easily recognise these animals by their respective cries. The variety of sounds is so great, that it often becomes very difficult, if not impossible, to convey, by any form of words, an adequate idea of the peculiar sensations which many of them occasion.

The Mammalia differ considerably from Birds, in respect to their organs of voice. In the latter, the several varieties of sound are produced in a more simple manner; and, from the organs of voice in the Birds approaching very near, in the principles of their structure, to several well-known musical instruments, they are, at the same time, more fully understood. But the cries of the Mammalia are most commonly of a very complicated and discordant nature; so that we may attempt in vain to imitate them by any mechanical means.

In ordinary language, we understand by the term *voice*, those sounds which animals produce in expelling the air from their lungs, through the opening of the glottis. From the above definition, it follows, that animals with lungs, being the first three Classes of Vertebrated Animals, the Mammalia, Birds, and Reptiles, can alone enjoy this power.

The voice, being formed of vibrations communicated to the air, consists, like all other sounds, of three orders of properties, perfectly distinct from each other: (1.) The *tone*, or the different degrees of depth and acuteness, which depend on the slowness or rapidity of the vibrations; (2.) The *intensity*, or different degrees of loudness, regulated by the extent of the vibrations; and, (3.) The *quality*, which depends upon a variety of circumstances hitherto undetermined, relative either to the internal structure, the substance, or the figure of the sonorous body. Man alone being capable of speech, becomes susceptible of a fourth order of modifications, which we represent by the letters of the alphabet. These may be further divided into two sub-orders—the one, relative to the principal sounds, which we represent by *vowels* and the other dependant upon their mode of articulation, and distinguished by the *consonants*. Those circumstances which give rise to the several qualities of tone, and the articulate words of Man, are still involved in great obscurity, although the investigations of De Kempelin in 1791, and the more recent experiments of Messrs Willis and Wheatstone, demonstrate that it is not impossible to imitate the sounds of the human voice by certain mechanical contrivances.

In respect, however, to the tone and intensity of sounds, the theory has long been well understood. We know that the rapidity of the vibrations of cords is inversely proportional to the length of the latter, and directly proportional to the degree of their tension. It is also ascertained, that a cord producing a tone will give at the same time others corresponding to the aliquot parts of its length, such as the half, the third, or the quarter; and the sounds thus formed are termed harmonic tones or chords. The vibrations of the entire cord coincide with the smaller but more rapid vibrations of the aliquot parts, and the sounds thence resulting are found to be harmonious, or agreeable to the ear. We further know that wind instruments of music can produce sounds corresponding to their total length, at the same time that they emit others relative to the lengths of their aliquot parts; and that it merely requires some apparently very slight cause, whether with cords or wind instruments of music, to occasion one of these partial or harmonic tones to prevail over the whole or fundamental tone. It has also been remarked, in respect to the tubes of wind instruments, that their form, in most cases, does not affect the tone. If the extremity opposite to the embouchure be closed, they produce a sound corresponding to a tube of twice their length, but when it is only partially closed, as in the chimney or funnel-pipes of the organ, the tone is always more grave than if it had been open, but less so than when entirely closed. Wind instruments of music can emit no sound by simply blowing into the tube. There must be at the entrance of the tube some sonorous body, that is to say, a thin plane capable of vibrating, or at least of breaking the current of air against its edge. This condition is absolutely essential to the production of sound, properly so called.

The organ of voice, being found only in animals possessing lungs, always consists of the canal formed by the bronchial tubes, the trachea, and the mouth; in other words, of an irregular tube to which the lungs act as bellows. The planes capable of breaking the air and producing a true sound may, however, be placed in different positions relative to the length of the tube. The entire portion comprised between the vesicles of the lungs and these vibrating planes, which have received the name of *glottis*, may be considered as nothing more than the nozzle of the bellows. That portion of the tube placed beyond these planes, being the *larynx*, must alone be regarded as the sonorous instrument, whose length and other circumstances serve to influence and modify the voice.

Many Birds are found to possess, in the interior of their bronchial tubes, small planes, being a kind of rudimentary glottis; but all of them have a complete one at the point where their bronchial tubes unite in forming the trachea. On this account we should thence regard the trachea itself in all Birds as a true musical instrument. In the Mammalia and Reptiles, on the contrary, no glottis is found except at the upper extremity of the trachea, where it enters the mouth. We must, therefore, consider the mouth in these animals as the real instrument of music, and the trachea with them is merely a wind-pipe or *porte-vent*.

It thus appears that the voice of animals is formed by the air which is discharged from the lungs by the muscles of expiration; that it traverses the bronchial tubes, and sometimes also the trachea; and arrives at a contracted portion, edged with two thin and flexible planes, called the glottis, where the sound is really produced. It then traverses a second tube, consisting either of the trachea and mouth or of the mouth solely, where it receives the last modifications, from the length, the form, and the differences in the complication of these cavities. Finally, it passes between the lips, which may be more or less opened or differently formed.

The possible intensity of the voice depends upon the proportional volume of the lungs and aerial cavities, and hence results the extraordinary volume of voice possessed by most Birds. The facility of modulating the voice during singing depends upon the facility of motion possessed by the muscles which contract the lungs. That portion of the trachea or the bronchial tubes, situate within the glottis, cannot influence the quality of the sound, excepting, perhaps, that the proportion of its diameter, in respect to the glottis, may influence the possible velocity of the air in its passage. The glottis itself affects the sound like the embouchure of a wind instrument of music, while that portion of the canal situate externally to it acts like the tube of the instrument, inasmuch as its several lengths determine the respective fundamental tones which the animal can assume; while the glottis, by its tension and the shape of its

orifice, occasions the several harmonic tones of the fundamental note belonging to each particular length. The external opening may lastly be compared to the remote extremity of the organ-tube, which may be more or less closed. Upon the facility with which an animal can vary these three conditions depends the extent and flexibility of its voice.

Those modifications which we represent by the letters of the alphabet, are formed in the mouth, and depend upon the greater or less power of motion possessed by the tongue, and especially by the lips, to which circumstances Man owes the superior power of speech. Some animals which seem to possess considerable flexibility in their organs of voice have certain additional parts, nullifying the advantageous form of the others, such as cavities, in which the air is obliged to circulate after leaving the glottis.

In respect to the Mammalia generally, we are far from having a complete knowledge of their manner of producing those disagreeable and complicated noises which our musical instruments fail to imitate. A few general facts have, however, been ascertained. Thus, the interval of the fibrous, and more or less sharp, cords of the larynx, placed on the upper extremity of the trachea, and called the vocal cords, is the place where the sound is formed; while the size, freedom, and tension of these cords influence the sound at its very origin. As the entire trachea serves merely as a wind-pipe, it varies little in its form. The rings are scarcely ever complete, but leave behind them a simple, membraneous band.

The sound produced by the vocal cords, or inferior ligaments of the glottis, may be modified,

1. By the form and dimensions of the passage opened for it as it traverses the remaining parts of the larynx;

2. By the resounding or dispersion of the sound in the cavities contiguous to the larynx, such as the ventricles of the glottis, the furrows and pouches which sometimes communicate with it, or the pouches which occasionally open in front of the larynx;

And, lastly, By the form and dimensions of the double passage furnished by the mouth and nostrils, or by the different positions of the tongue and lips.

It would be inexpedient to pursue this subject in detail at present, and our further observations upon those modifications which influence the voice will be found under the specific descriptions.

GENERAL REVIEW OF THE MAMMALIA CONTINUED.

Anomalous adaptations for motion and prehension.—Special organizations for digging, flying, and swimming.

ALTHOUGH some Mammalia possess the power of plunging in the water, of elevating themselves in the air, or of burrowing under ground, these seem rather to be anomalous states of existence. The normal or proper state of their organization fits them more especially for terrestrial animals.

Among those destined to reside upon the earth, Man alone has his fore extremities adapted solely for prehension, and his hinder limbs for maintaining the body in an erect posture. We shall merely remark at present, that he owes this vertical position chiefly to the size of the soles of his feet; the largeness of the muscles belonging to the legs and thighs; the breadth of the pelvis; the position of the head upon the neck; and the shortness of the arms when compared with the length of the legs. It follows from these arrangements, that his forward movement consists simply in the successive position of the lower extremities one before the other and in parallel lines; while running differs but little from walking, except in the greater rapidity of the action.

It is different in the true Quadrupeds, where the extremities are very nearly of equal length. In these we have a variety of modes of progression; such as the walk or pace, where the two diagonal feet, either the right fore and the left hind foot, or the left fore and the right hind foot, act successively, but in such a way that the advance of the fore foot is almost instantaneously followed by the advance of the hind foot on the opposite side, so that the four feet are raised and set down one after another. In a more rapid pace, being the trot, the two feet diagonally opposite rise and are set down at the same instant. The canter or common gallop, the full gallop, and the amble, have been already explained (see page 26), as well as the *aubin*, or Spanish amble, and the *pas relevé* of French authors (see page 63).

The greater number of Quadrupeds, when they wish to advance slowly, go at a simple walk. The trot is the proper motion of certain species, such as the Horse, the Fox, some races of Dogs, the Bear, and of the Elephant when hurried. The amble is the natural pace of certain races of the Horse and Dog, and invariably so with the Hyæna, which circumstance gives a singular and striking appearance to the gait of that animal. The canter, and especially the full gallop, are used by most land Mammalia when hotly pursued, or when they are in pursuit of any prey.

Those Mammalia having the fore extremities much longer than the hinder, are prevented from walking upright like Man, or on the four feet like the Quadrupeds. The position of their body is therefore oblique, and their speed on a flat surface being very limited, they are more disposed to climb trees than to use any other kind of exercise. There are other Mammalia, such as the Jerboas, the Gerbils, the Kangaroo, and the Kangaroo-Rat, where the hinder are very considerably longer than the fore legs, which seem by their excessive shortness to depart as much in defect from the ordinary size as the others do in excess. These animals walk with great difficulty. The Kangaroos in particular make use of their tail, which is of considerable strength and size, either to counterpoise the weight of the fore extremities, or to assist in raising the body, while the fore paws touch the ground. When they run, or rather jump, for such is their real motion, the hinder feet alone act, and propel the body to a considerable distance in advance. The tail, however, follows to their assistance, and the moment the feet touch the ground, it extends and forms with the two metatarsi a kind of tripod, which maintains their upright position, and enables the animal to execute a new leap. As another instance of this kind of movement, we may adduce the Jerboas, animals about the size of a Rat, with their hinder feet much longer in proportion than those of the Kangaroo. As the soles of their feet are very small, the metatarsi elevated, with the tail very long and hairy at its extremity, we perceive a striking difference in their mode of progression from that of the Kangaroos. At each leap the Jerboas fall upon their small fore feet, but they raise themselves very quickly by means of a long and heavy tail, serving as a counterpoise, and preventing the body from falling backwards, which accident happens whenever the tail has been lost.

Among the Rodentia, and especially with the Hares, where this disproportion between the hinder and fore feet is not excessive, the animals usually run at a full gallop; but they avoid descending the declivities of the mountains at this pace, as the forcible action of their hinder legs, aided by the inequality of the ground, would occasion them to perform a somerset. In ascending, they possess, on the contrary, a considerable advantage over an ordinary quadruped, for the gradual rising of the ground counteracts the inequality in the length of the legs, and brings their body more nearly into the horizontal position.

In the Martens, the Pole Cats, and the Ferrets, which have their feet small, and placed very far apart, the advance is always effected by a full gallop, the back being curved into an arch, in order to bring the two extremities nearer to each other, and to prevent the abdomen from trailing on the ground.

Among the terrestrial Mammalia, a great many species are obliged to seek their food among the branches of trees. The greater number of these have their thumbs opposable to the other fingers, either on the four feet, as in the Apes and Makis, or on the hinder feet alone, as in several Marsupialia. This arrangement gives to these animals the power of seizing the branches with facility, and of applying the entire palm of the hand to the inequalities of surfaces. Other Mammalia, being deprived of moveable fingers, make use of their long nails as hooks, which fulfil the same purpose, although rather in an imperfect manner, such as we find in the Sloths and Ant-eaters.

Finally, among the climbing species we find many which make use of their tail as a fifth limb, on account of the surprising flexibility which it possesses, as well as its power of rolling itself round the branches, and even of picking up very small and slender bodies. This tail, termed prehensile (*Cauda prehensilis*), belongs more especially to some American Apes, to the Opossums, the Phalangers of the islands in the Indian Archipelago, to the Brazilian Porcupine (*Hystrix prehensilis*), and a few others.

All those Mammalia, which, like Man and the Apes, have the anterior extremities supplied with hands, possess at the same time complete clavicles, and one hand alone is sufficient to seize the food and bring it to the mouth. It is different with all the other animals destitute of thumbs, and whose fingers, being moveable all together, and in one direction, are armed with crooked nails. In these animals, such as the Squirrels and Rats, the assistance of both hands is necessary to hold an object. Hence, while feeding, they are in the habit of sitting down, which enables them to use both of the anterior extremities with freedom. Those Carnassiers which have rudimentary clavicles, and whose paws are well furnished with claws, are obliged to hold their prey between the two fore paws, and tear it with their teeth. The Cats alone are able to carry with one paw to their mouth a very small portion of food by means of their sharp claws.

Some Mammalia dig in the earth with great facility, and among them the extremities are specially adapted for this purpose. The nails of the fore feet among the digging Mammalia are commonly very strong and short, so that these animals are rarely swift runners, excepting perhaps the Jerboas, which hollow out their dwellings in the loose sands of the steppes. The Mole, the Scalops, and the Chrysochloris, are especially remarkable for the strength of their hands, where all the fingers are united together, and form a kind of shovel or spade, very well adapted for removing the earth. In these animals the abdomen trails upon the ground.

In general, we find that all the anomalous points in the organization of these animals adapted for digging, may be reduced to two particulars—to modifications in the anterior extremities, and proportional differences in the organs of sense. The first traces of these anomalies may be found in such Mammalia as burrow in the earth to shelter themselves and their young ones; but they are most clearly seen in those animals which scarcely ever come from beneath the surface of the ground, but are continually in the habit of seeking a subterranean subsistence, whether vegetable or animal.

The first and most striking examples may be observed among the insectivorous Carnassiers, such as the Mole, the Chrysochloris, and a few neighbouring genera. Among these animals, we find that the anterior limbs are proportionably much more developed than the hinder, which are very slender. The thorax forming the point of insertion for the muscles of the shoulder, acquires a peculiar solidity by the speedy ossification of the cartilages of the ribs, the pieces of the sternum, and by the existence of a kind of breastwork or medial crest. The clavicle being very strong and solid, projects much in front. The shoulder-blade is very broad, with the acromion and coracoid processes distinctly pronounced, and indicating the presence of muscles of a proportional size. The shoulder loses much of its usual length to acquire a breadth so considerable, as almost to make it square. The two bones of the fore arm, though less strong, are yet tolerably so; and, finally, the hand, which is shaped like a shovel, is very short and broad, terminated by five very thick and cutting nails, and is still further augmented by a sharp bone as a supernumerary, which occupies the cubital margin. It may be added to this description, that the entire limb, which bears, at its insertion, upon the lateral parts of the neck, is turned beneath and behind in such a way, that the hand acts constantly in those directions.

There is another modification in these animals, which properly belongs to the organs of motion. The nose or muzzle is modified to be an instrument fit for excavating or digging, at the same time that it serves as a finder through the delicacy of its olfactory organ; hence result the ossification of the cartilage separating the nostrils, the great development of the muscles of the nose, and, finally, the great force of the extensor muscles of the head.

In respect to the organs of sense, we may readily expect that the hearing will be developed to an extent proportionably greater than in the other Mammalia, by means of a larger drum, a very short auditive canal, a very wide orifice, and no kind of concha. On the contrary, the sense of sight continually diminishes, until it ends by scarcely existing except in a rudimentary form, as is found in many of the Rodentia, and especially in the Zemni (*Spalax typhlus*). In those species where the anterior limbs are a little less adapted for burrowing, the eyes disappear almost wholly, and the organ of hearing appears, on the contrary, to develop itself in an inverse proportion to the former sense.

In respect to the Mammalia capable of flight, these are of two kinds. Some have simply the skin of the sides extended behind the fore feet and before the hinder, as we see in the Flying Cats (*Galeopithecus*), the Squirrels, and the Flying Squirrels (*Pteromys*). The effect of these expansions of the skin merely consists in preventing the body, when darting from a very elevated situation, the top of a tree for example, from falling too heavily, and hence their function confines itself to that of a parachute. Other Mammalia, on the contrary, do really fly, that is to say, they can elevate themselves by means of the movements of their anterior extremities, which are prodigiously developed and furnished with long fingers, united together by expansions of a very fine skin; these are the Bats.

The first tendency towards the anomalous organization for flight is seen in those Mammalia which are disposed to seek their food in trees, and hence are constituted so as to climb with facility. In fact, there is no Mammiferous animal, before sustaining itself in the air, which is not obliged to elevate itself to a height more or less considerable. The latter movement may be effected by a mere digital compression of the fore and hind feet, as we see in the Apes; or by embracing the tree, as in the Bear; or, finally, by means of hooks; such as we see in the Cats, the Sloths, and even in the Squirrels. The first traces of that disposition by which these animals are enabled to leave their elevated position, and maintain themselves wholly in the air, is found in some Rodentia nearly allied to the Squirrels, being the Pteromys or Flying Squirrels, which have on their flanks a fold of skin, as already described, and likewise in the Phalangista or Flying Opossums. A still more perfect state of this organization is found in some species of Carnassiers nearly allied to the Apes, such as the Galeopithecus. In these animals, the fore extremity is considerably elongated. The fore arm has acquired that peculiar ginglymoidal arrangement, which has been assigned to the Cetacea for a different purpose; and, as may be easily supposed, the principal moving, and especially the pectoral, muscles are very largely developed. On the other hand, the hinder limbs have diminished very sensibly, and also the tail, which is almost wanting, in order to throw the centre of gravity of the body within the axis of the fore extremities. The most perfect form of this anomaly is, however, seen in the Bats.

The aquatic Mammalia are of two kinds; some frequent the margins of lakes, rivers, or streams, into which they plunge occasionally to seize their prey, or to find the aquatic plants on which they feed. Their extremities are either wholly webbed, as in the palmated feet of the Ornithorhynchus, in which the expansion of the skin is that of the palm or sole of the foot, and not merely of the fingers, as we find in the Otters and Cabiais. These animals have webs upon all the four feet, being different from the Beavers and Cheironectes, whose hinder extremities alone are webbed, while the skin extending between them is entire. In the Desmans, the feet are only semi-palmated, and in the Hydromys, they are palmated for two-thirds of their length.

The Ondatra, whose mode of life is so analogous to that of the Beaver, has its hinder feet fitted for swimming; but instead of having the feet united by a membrane, each of them is bordered on the right and left by a row of elongated, stiff, and serrated hairs, which cross each other by the points with those of the adjoining fingers, thus forming a surface capable of offering sufficient resistance to the water.

Among the marine Mammalia, the Seals and Morses deserve to be noticed in the first place, because they are supplied with all the four extremities, of which the anterior have the fingers united, and armed with claws. In some species of Seals, being the Otaries of Peron, the skin on the tip of each finger is prolonged into a long and narrow strap, forming a band. The hinder feet, placed entirely at the extremity of the body, also have apparent fingers, but these are united together by the skin. Others, such as the Lamantins and Dugongs, are destitute of the hinder extremities; but all the fingers of their limbs are invested by a thick skin, on which vestiges of nails are to be found. Finally, the proper Cetacea, which also have the hinder extremities wanting, depart still more widely from the quadrupeds, as we here fail to discover the slightest trace of nails. These differences in the organization of the extremities occasion corresponding variations in their mode of life. The Seals are very agile when in the water, and execute a number of evolutions and movements in consequence also of the extreme flexibility of their vertebral column; but they walk with much apparent difficulty on the land. The Lamantins, Dugongs, and Stelleres (*Rytina*), have still greater difficulty in leaving the water, and remain like inert masses whenever they chance to run aground upon a bank. The proper Cetacea do not voluntarily leave the sea, where they swim with a prodigious velocity, by means of the movements of their tail and fins. In consequence of the former being flattened from above downwards, and not compressed from right to left, this organ moves chiefly in the vertical direction, instead of horizontally, as we find in the greater part of the Fishes.

On comparing all aquatic Mammalia with each other, and with the remaining animals of their class, we find that the organic modifications which have been experienced by those Mammalia destined to reside more or less in the water, consist essentially in the following particulars:—The Hairs which, in the amphibious animals, are observed to be short and exceedingly numerous, terminate in the Cetacea by forming an agglutinated or universal envelope, as we see in the Lamantins and Whales. The external Ear diminishes in size, until it ends, in certain Cetacea, by disappearing almost wholly, so that it is scarcely possible to discover any vestige of the concha. The crystalline lens of the Eye is observed gradually to become more convex, and to approach insensibly towards the spherical form, which it acquires most completely in species residing constantly in the water. We may further remark the absence of lachrymal ducts and pores. It may also be noticed, that the olfactory system tends gradually to diminish.

In the organs of locomotion, it is observed that the bones lose their medullary cavity, and become spongy throughout their whole extent in the Cetacea. But the most remarkable part of their conformation consists in the general fish-like shape which the true Cetacea exhibit. Their bodies usually ending in a point both before and behind, and expanded in the middle, approach nearly in form to the solid of least possible resistance. Already we remark a tendency towards this form in the Otters, and still more so in the Seals, although these animals possess four complete limbs. Their resemblance in shape to the Fishes is, moreover, most perfect in the Lamantins and proper Cetacea, where there exist only a few slight rudiments of a pelvis, and where the vertebral column is terminated by a powerful and broad tail, formed externally by a large horizontal expansion, sometimes bifurcated and acting as a fin.

GENERAL REVIEW OF THE MAMMALIA CONTINUED.

Relations of the Mammalia to Man—Their injuries and depredations—Economical purposes to which their products are applied—Management of the domestic Mammalia, relative to Station, Soil, and Climate.

As the other Mammalia are influenced by the same wants and necessities as Man, as they are under an equal necessity with him of seeking their own preservation, of finding their food, and reproducing their kind, it will often happen that these instincts will run counter, and appear in opposition or competition with the corresponding wants of Man. Hence result those injuries to his person and property which some of them inflict.

The Tiger, the Panther, and Jaguar, are the principal carnivorous animals which venture to attack our species by open force, in the forests of those warm countries which they inhabit. It has often been said and repeated, that the Lion is generous towards Man; but his generosity is that of a Cat, and there are few who would be disposed voluntarily to place their persons within his reach. The Wolves, Hyænas, and Bears, do not attack Man except when pressed with hunger, or when they have young ones, which they think it necessary to defend. The greater part of the remaining Carnassiers confine themselves in all cases to a defensive combat. Certain foreign Bats are said, probably with some exaggeration, to be capable of inflicting death on a sleeping person, by opening a vein, and then sucking the blood, through means of certain horny papillæ with which their tongue is supplied. In our own country, and on the continent of Europe, it sometimes, though very rarely, happens that a Weasel or Pole-cat insinuates itself into the cradle of a newly-born infant, and sucks its blood, so as to occasion death. It is well known that the domestic Pig, which devours flesh with avidity, often occasions similar accidents. In general, we find that the Carnassiers attack women and children, with whose feebleness or tenderness of flesh they seem to be instinctively acquainted, in preference to the adult Man.

Some of the larger herbivorous animals, such as the Buffalos, in the neighbourhood of the Cape of Good Hope, and certain Elephants of a savage disposition, and known in the East Indies by the name of *Grondahs*, as well as the Rhinoceros, do not hesitate to attack Man, if they happen to meet him in their road, and speedily put him to death, by trampling him under foot.

Certain Mammalia maintain a continual war against those flocks of herbivorous animals, or domestic fowls, which Man has tamed for his own use. The Wolves on the continent of Europe roam continually round the parks where the Sheep are assembled, and carry off all stragglers. The Foxes, Weasels, and Polecats, introduce themselves into the farm-yards, where they destroy the fowls, and devour their eggs. Many other species of Martens, as well as the Opossums of North America, and the Dasyures of New Holland, make depredations similar to the preceding. Even Horses are attacked by Wolves in mountainous and woody regions, in preference to other animals, against which enemies they have no other defence than their heels; but by an admirable instinct of self-preservation, they collect together in a circle with their heads towards the centre, and the hinder feet in the circumference. The Oxen of Africa are sometimes surprised by the Lions. After killing them by biting through the hinder part of the neck, the Lions transport their victims to their retreats, with a degree of ease, which serves to exemplify the extraordinary force of these animals. The herds of Paraguay are likewise diminished in number by similar attacks from the Jaguars.

But the Carnassiers are far from inflicting so severe an injury on Man as the herbivorous animals, and especially those belonging to the smallest species, which attack the seed when under ground, the harvest on its surface, or the hoards of the granary and store-house. The wild Deer, Goats, Hares, and Rabbits of Europe, devour the corn while in leaf; Rats, Field-Rats, and Hamsters, devour it while in grain, and the last-mentioned species, not content with destroying its share, amasses a store for future use, which may be estimated, on an average, at about a bushel for each individual. The common and garden Dormouse attack the fruits of our garden trees; the Rats and Mice destroy our provisions of every description; and the Moles, while seeking their food, consisting of Earthworms, Insects, and their larvæ, plough the surface of our meadows. On the continent of Europe, the wild Boars, whose destruction, except by a privileged few, is prohibited in some countries by game-laws of doubtful justice, advance in innumerable troops to attack the stores of potatoes, which perhaps are the only resource of the injured cultivator of the soil. The Otters plunder the fish-ponds of the Continent, and the Water-Rat, by living on the fry of the fishes, prevents their increase. Our own country is happily exempted, by its insulated situation, from many of these depredations.

Each kind of animal, which lays waste our own territory or that of our continental neighbours, finds analogous plunderers in more remote or foreign countries. In Africa the Apes, descending in immense numbers, fill their cheek-pouches with grains of maize, and when molested, take to flight, their hands being filled with as many ears of this plant as they can conveniently carry. In America the Agoutis, Cabiais, and Cobayes, execute the same kind of devastations which, in our country, are inflicted by the Hares and Rabbits. A herd of forty or fifty Elephants soon causes every vestige of cultivation to disappear from an entire canton of the East Indies. In the North of Europe, thousands of Lemmings, descending in a body from the mountains of Norway and Lapland, direct their route in a straight line towards the south, and may almost be said to destroy every plant in their way. These legions of the northern hive are soon followed by another plague. Innumerable Foxes, which at first followed the Lemmings, and lived at their expense, are soon obliged to change their prey; and, after the complete annihilation of the Lemmings has been effected, to regale themselves on the fowls of the farm-yard, or some of the smaller quadrupeds.

However great these injuries already enumerated may appear, they are fully compensated by advantages of a more solid kind, derived from the use of that small number of species which Man has succeeded in taming for economical purposes.

The flesh of the herbivorous Mammalia forms a well known article of food. Those animals chiefly used in European countries for this purpose are the Bull, the Hog, the Sheep, the Goat, the Hare, the Rabbit, the Stag or Red Deer, the Fallow Deer, the Roe-buck, the Chamois Antelope, the Ibex, the Squirrel, the Dormouse, and a few others. The Laplanders feed chiefly on the flesh of the Rein Deer; the Canadians upon that of the Wapiti (*Cervus Canadensis*), and of the Elk (*C. alces*). The Negroes eat that of the Elephant, the Rhinoceros, the Hippopotamus, the Manis, and several Apes. The Americans of New Spain do not refuse the flesh of the Armadilloes; the inhabitants of Chile esteem that of the Lama and Vicugna; and the Arabs eat the flesh of the Horse and Dromedary. With the inhabitants of the Indian Archipelago, the flesh of the Roussette Bats forms a daily article of food.

Certain species have been interdicted in some countries by laws dictated either by superstition or convenience. The flesh of the Hog is held in horror by the orthodox Turk and Jew, while the Bull is the object of veneration among the Brahmins.

Particular parts of some of the larger Mammalia are highly esteemed by certain nations. Thus, the foot of the Elephant is generally considered by the Negroes of Africa as a delicious article of food; while the Dutch sailors, who pass half their lives in the midst of the polar ice, look upon the tongue of the Whales as a very delicate morsel.

Seals, Dogs, and Otters, are the only carnivorous animals forming articles of food. The first two are in common use among the inhabitants of Kamtchatska, and the last, from its fishy flavour, is in great request among the monks of Catholic countries during Lent.

A few of those Mammalia which Man has succeeded in taming and subjugating completely to his use, appear to have been distributed, like him, from the more elevated parts of Central Asia, to all other points of the globe. These species are the Horse, the Ass, the Sheep, and the Goat. The Dog and the Rein Deer are proper to the climates of the North; the Buffalo and Elephant to those regions of India situate at the foot of the elevated mountains of Central Asia. The Lama is peculiar to the New Continent. Our race of Bulls, usually considered to be peculiar to Europe, is, with very great probability, identical with the Zebu, or Humped Bull of India; and, according to M. Desmarest, it does not differ specifically from the Yak, or Grunting Bull of Thibet and the frontiers of China.

Two species of Mammalia, in particular, have served to assist in subjugating the remainder; being the Horse and the Dog. Many others are employed in hunting, such as the Chetah or Hunting-Leopard, and the Ferret; or for fishing, as the Otters.

Others have been destined to carry burthens of greater or less weight; of these we may particularize the Bull, the Camel, the Dromedary, the Yak, the Horse, the Elephant, the Ass, and the Lama. Many have been harnessed to carriages of various forms, such as the Dog, the Rein Deer, the Horse, and the Bull; or have served for riding, as the Elephant, the Horse, the Ass, and the Bull.

To some, Man has confided the care of his property. The Dog seems as it were consecrated for this purpose; while the Cat, the Guinea-Pig, and, it is said, also the Ichneumon of Egypt, have been destined to defend his provisions from the attacks of the smaller parasitical species.

The art of medicine has derived many useful materials from this class of animals. Without noticing the ridiculous properties which have been assigned to the excrements of Dogs and Rats (*Album Græcum* and *Album nigrum*), it is generally admitted that the flesh of the Calf possesses relaxing properties; the empyreumatic oil, procured by the distillation of horns, is used on several occasions, and especially as a vermifuge. The blood of the Ibex, Goat, or Chamois Antelope, was once considered as a useful medicine during attacks of pleurisy.

The organic products of the Mammalia have, from the earliest ages, furnished the materials for a variety of useful arts. The horns and bones are used for combs, boxes, button-moulds, the handles of cutting instruments, and the innumerable products of the toy manufacturer. The longer hairs, such as those of the mane or tail, are used for fishing lines, or coarse stuffs; the shorter or finer hairs, such as wool, the hair of the Goat, the Rabbit, and Cat, when dressed and spun, enter into the fabrication of a multitude of different tissues; the bristles, or large and stiff hairs, serve for brushes; and the finer hairs of other species for pencils. The skin, prepared in divers manners, furnishes the soles or upper leathers of shoes, the materials for gloves, harnesses, portmanteaus, the roofs of coaches, bottles, &c. &c.; and the Korekis form boat-sails with the skin of the Rein-Deer.

The blood of the Mammalia serves to clarify liquids, and is particularly useful to sugar-refiners, and in the manufacture of Prussian blue. The tendons are used for thread by the Samoiedes, the Laplanders, and the Greenlanders. The fat, more or less liquid, according to the species from which it is taken, may be used under the form of lamp-oil, lard, or suet. The marrow of the bones forms the basis of many kinds of pomatum. The intestines, after having been washed, dried, and twisted, compose the strings of some musical instruments; and the gall serves for extracting grease from stuffs, or for laying on colours over a greasy surface.

The distillation of flesh and bones yields many chemical products of great utility in the arts, such as ammonia, phosphorus, &c. Finally, the manure of the Herbivorous animals may with justice be said to be the *primum mobile* of Agriculture. It restores to the soil those principles which are annually extracted from it, and thus conduces eminently to its fertility. Indeed, it has always been found, that the success of agricultural enterprises depends in general upon the extent to which these animals are multiplied and improved.

The acknowledged importance of the domestic animals, in an economical point of view, has led agriculturists in all ages to pay peculiar attention to them, in order to bring up, feed, dress, treat, and shelter these animals in such a way, as to draw from them the greatest possible amount of benefit in the most economical manner; and, by multiplying and improving the breeds, to render them proper for their several destinations.

Without entering into those details, which belong to Agriculture rather than Zoology, it may be proper here to point out some of the important advantages which result to the cultivator himself, as well as to society at large, by the successful cultivation of the domestic animals.

In respect to the former, unless the cultivator can easily and economically procure a sufficient supply of those manures, of which he is in almost daily want, the produce of the soil will in general be feeble and uncertain. Destitute of these animals, the agriculturist would be deprived of the principal articles of his daily consumption. In fact, rural establishments would want that activity which renders them at once agreeable and useful.

When considered as objects of public utility, the domestic animals possess many claims to our regard. The cultivation of the Cereal plants, which contribute so largely to our maintenance, deservedly occupies the first consideration, and those animals, which are raised for alimentary purposes, possess at least a second claim to our regard, since they tend directly, on the one hand, to increase the former by their manures, and indirectly to economize their consumption. In equal bulks of animal and vegetable food, nearly twice the quantity of nutritive particles is contained in the former as in the latter; and a pound of meat will in general be as nutritious as two pounds of bread.

The numerous and important advantages which Man derives from the domestic animals, have led all nations from their earliest origin to regard them with the most scrupulous attention. On referring back to the first ages of which we have any authentic records, we see the chiefs of tribes, the patriarchs and first sovereigns, paying a special attention to the management of cattle, and founding on this solid basis, not only their own prosperity, but that of their contemporaries and descendants. The sacred books and the most ancient historians furnish repeated examples of these facts, which are too well known to be repeated here. At this remote period, when Man, just formed by the Divine Power, entered upon the dawn of his civilization, the domestic animals were considered not only as the most firm support of Agriculture, but they yielded the most valuable materials for Commerce. As the principal wealth of the times consisted in domestic animals, these naturally became the first medium of exchange between nations. Cattle were therefore the first money that existed, as they were the first article which possessed exchangeable value. We have a confirmation of this fact in the circumstance, that the first acknowledged representatives of mercantile value, the earliest metallic money which passed current, was decorated with an image of these animals, indicating that it maintained an equal value. They were also the earliest offerings presented by most nations to their deities; and the ancient Ægyptians worshipped the Bull, *Apis*, with the highest veneration. If we turn to that nation, which has left us the most extensive and important written monuments of its experience in the different branches of rural economy, we shall find the ancient Romans applying themselves with remarkable zeal to the training and management of cattle. We have an evidence of this fact in the term *jumenta*, which they applied generally to all kinds of cattle, derived from *juvare*, to help. There is also the term *pecunia*, money, from which we have derived our English adjective *pecuniary*, and the Latin term *peculium*, from which we have derived our *peculation*, alike derived from *pecus*, which the Romans applied to cattle in general. Cato the elder, the first of their Agriculturists who has transmitted his precepts to our times, on being asked by some persons to point out that particular branch of rural speculation which should command their first attention, if they wished to acquire wealth in the quickest possible manner, is said to have replied, "Manage your cattle well;" and on being again asked, what was the next best object of their attention, if they wished to derive only a tolerable return for their labour, he replied, "Manage your cattle tolerably well." In the countries of civilized Europe, we find that a large portion of the wealth of their inhabitants consists in cattle; and we can commonly form a good notion of their respective degrees of agricultural prosperity, as well as of the comforts of the cultivator, by noticing the number and quality of the domestic animals.

Since the proper management of Cattle is thus an undoubted and inexhaustible source of wealth, it may be interesting here to trace the principal rules of conduct, which should form the guide of the Agriculturist and Grazier. These remarks will be equally valuable to the Naturalist, as they are more or less applicable to all terrestrial Mammalia. The chief points to which our remarks are confined are the influence of station, soil, climate, food, exercise, lodging, dressing, as well as the application of these to the several purposes for which the animals are finally intended.

The station in which the domestic animals are maintained may be low or elevated, dry or moist; and these four qualities impart corresponding properties to the animals which receive their influence. Peculiarities of station are often combined together in pairs; thus, an elevated station is often dry, while a low station is usually damp.

When an elevated station is dry, it is generally more healthy than a low one when damp. The air in the former is lighter, keener, and more pure; and communicates its bracing qualities to such animals as are continually exposed to its influence. The vegetable nutriment which it yields is more scanty, but it is, at the same time, more substantial, and rather imparts force and energy to the animals which are fed thereon than volume of body. This kind of soil is best adapted for the Goat, the Sheep, and the greater part of the Ruminantia, which select it naturally when allowed to run at large.

A low soil, when damp, appears to be unfavourable to most constitutions. The air seems overloaded with heterogeneous miasmata; it is, therefore, less healthy, and being of greater specific gravity, communicates a corresponding dulness of motion to those animals which are habitually exposed to its influence. The excessive moisture constantly surrounding them relaxes the fibres of their bodies, elongates their membranes, extends their limbs, and renders the whole animal more massive, ponderous, and slow in its movements. Vegetable food is more abundant in these situations, but it is more watery, and less nutritious; it loses in quality what it gains in quantity, and induces corpulence rather than energy; and while it promotes an increase of size, diminishes strength. Poisonous plants appear to be more abundant in these situations than in stations of an opposite kind, and animals are not only exposed to the prevailing miasmata, but are, as it were, in a continual bath of vapours. This kind of station is best adapted for the Buffalo, the Bull, and the Hog.

The middle point between both extremes, as in many other matters, seems to be the most favorable to a large number of domestic animals. It is best suited, in particular, when accompanied by a proper degree of heat, to the Horse, the Ass, the Dog, the Cat, the Rabbit, and the Hare. Plains are especially adapted for the Solipeda, in which stations they are more at liberty to exercise their limbs.

We may perceive from these general observations, that there necessarily exists a most intimate relation between the nature of the places inhabited by the domestic animals, and the general aspect of these animals. The prevalent character of each race appears greatly to depend upon that of its station. Thus, we see in low and moist districts, that these animals exhibit an aqueous temperament; their flesh appears soft, and the animals themselves acquire a certain degree of apathy and stupidity. Upon an elevated and dry soil, they possess, on the contrary, a certain degree of fineness of structure; their flesh is delicate and muscular; their girth slender; their movements rapid; and they assume an increased sensibility corresponding to their agility and vigour.

The quality of the soil likewise exercises a considerable influence upon the average bulk of animals, and it has been observed, by a law which is equally applicable to Plants, that those frequenting elevated mountains, whether granitic or schistous, and silicious soils when dry and arid, are smaller than those frequenting calcareous plains, luxuriant in herbage, as well as low and moist countries. In the latter case, their fibres are soft and better supplied with nutriment; the meshes of their animal tissues remain more lax, and acquire a greater degree of extension than in the preceding case, where the fibre continues dry and short. On this account, in low and moist soils, and in fertile valleys, the same races of domestic animals exhibit greater bulk and corpulency than upon a dry, elevated, stony, and steril soil. It is also from this cause that the Horses, Bulls, and Sheep of Holland, Belgium, and the rich pastures of Switzerland and France, become more bulky than animals of the same species brought up in the Alps, the Pyrenees, the Appenines, and all rough and mountainous situations.

It thus appears unquestionable, that the nature of the soil exerts a powerful influence over the constitution of the domestic animals, and that it demands the most scrupulous attention on the part of the rural Economist.

Climate acts in a powerful manner upon the physical constitution of animals, and demands at least an equal share of attention. By the terms difference of climate, we commonly include, in a general manner, all those conditions of the atmosphere which occasion a greater degree of heat and moisture to prevail in one place rather than in another; and it may be easily imagined, that if the nature of the media, in which animals habitually reside, exercises an important influence over them, they will also be influenced by the temperature and moisture of the climate. They are more susceptible than Man to the immediate influence of changes of temperature, from being continually exposed to the inclemency of the air, and seem acutely sensible of great and sudden changes of the atmosphere. We even observe them foretelling and announcing an approaching change of the weather by various premonitory signs.

As the climate may be either hot or cold, dry or moist, each of these conditions induces very different results in respect to their reproduction, constitution, amelioration, and, in general, all the vital functions of the domestic animals.

Heat being one of the most powerful stimuli of the vital reproductive powers, seems conducive both to fertility and growth, especially when accompanied by moisture; cold, on the contrary, is generally injurious. We remark that Nature develops all her treasures of fertility in the ardent climates of the South, while the icy regions of the North are generally less peopled, more uniform and inanimate in their general aspect. Melancholy solitudes replace, in these desolate regions, the most active and well-marked scenes of animation, which, however, are less permanent, and pass more rapidly away.

It thus appears, that the active force of heat, which bears an intimate relation to that of light, exalts the intensity of all the faculties and properties, and gives them the fullest energy which they are capable of acquiring. By the same law which assigns to the plants of the South more exquisite flavours, aromata, essential oils, perfumes, and colours, than to those of the North, we find the animals of warm countries also exhibiting a greater richness and variety in their hues, more vivacity and energy of character, more activity and strength in all their parts. Everything proclaims in Nature the beneficial influence of warmth over reproduction, as well as upon the form and qualities of its productions.

It appears, however, that heat, while it augments the energy of the vital powers, contributes a more diminutive growth to the organs of the different functions, probably because the moisture which contributes much to this development is less abundant, and because the solids of the animal body bear a greater ratio to the fluids, which are more or less dissipated by heat. Climates of dry and warm character render their fibres rigid, slim, moveable, and irritable, and they become deprived of that moisture, which had lessened their sensibility by softening them. Thus, we constantly observe that Horses, Bulls, Sheep, Goats, Dogs, and other domestic animals, are proportionably smaller, but more vivid, ardent, and active in warm countries, than animals of the same species in colder regions, provided always that the cold be not too intense. An excess of cold is, however, still more injurious to growth. The largest races of cattle are found in temperate climates, which are moderately cold and moist. A moderate degree of cold, by giving density and elasticity to the animal fibre, when influenced by an adequate supply of moisture, becomes at once favorable to the growth and multiplication of the species.

We may also remark, that the influence of climate upon the reproduction of animals imported from foreign countries merits a high degree of attention from the Agriculturist. In the same manner that Vegetables, when transplanted from a burning to a cold climate, multiply but seldom and with difficulty in the ordinary way, we remark that animals imported from a very warm to a very cold country often become unfruitful. It has frequently been noticed, that Arabian Mares, when brought to Britain under different circumstances, either become unfruitful, or yield feeble and unprofitable results. The Stallions of many races are sometimes in the same situation, even when transported to a much shorter distance. M. Yvart remarks, that the Asses of Tuscany and Spain are not always productive in France, or in countries lying farther to the northward; and it is well known, that in all the Northern countries of Europe, animals of this species yield products greatly inferior in appearance to those of the South. The other domestic animals present us with results, which may be regarded as equivalent to the preceding, after making due allowances for the differences between the climate of their residence and that of their original country. Thus, we may remark that the Sheep and Bull seem rather to deteriorate on removing from the North to the South of Europe; now these animals appear to have belonged originally to countries where a cold and moist atmosphere was more prevalent, than one of an opposite character. On the contrary, as we have just observed, a different result is obtained in respect to the Horse and Ass, which were originally natives of the South.

That degeneration of individuals, so frequently remarked in animals and plants, results inevitably from their being imperfectly acclimated, and many of the diseases with which they are afflicted proceed from a similar cause. In the Southern countries of Europe, the insensible exhalations which transpire from the surface of the skin are usually considerable, while the contrary takes place in its more Northern regions. Hence, in importing animals from the South to the North, due care should be taken to overcome their constitutional habit in a gradual manner. There is a constant determination of all useless or hurtful matters towards the skin in warm climates; while, in cold countries, transpiration is counteracted, arrested, or suspended, and always modified in a greater or less degree. Hence, unless their change of situation is cautiously effected, animals become predisposed to several cutaneous disorders, to obstructions, enlargement of the liver, and other maladies of this nature. Again, when animals are suddenly transported from the North to the South, and without the necessary precautions, the consequences are not less dangerous than those already enumerated, as the excretory functions of the skin are less energetic in cold than in warm countries, the internal functions possess a greater relative energy; and, on removing them to the South, their constitutional habit becomes modified. The insensible transpiration of the skin necessarily becomes greater, and the active forces of the system tend towards the surface,—a change which may occasion many dangerous maladies, such as putrid fevers. The only effectual way of counteracting these serious inconveniences, is by adopting a system proper to all the circumstances of the locality, according to the principles laid down in our best Medical treatises.

Climate exercises an important influence over many of those characteristics, which commonly serve to distinguish one species from another; and it is highly probable, that many animals, which are commonly considered by Naturalists as belonging to allied species, may in reality be nothing more than permanent races, descended from the same original stock, and preserved distinct solely by the influence of climate. The usual characters of animals, when long exposed to dry and warm climates, may be stated in general terms to be the following:—Their skin is thin, supple, and oily; their hair scanty and fine; their limbs long; the tendinous parts distinct; their horns hard, dry, and brittle; the hoof contracted; the feet narrow and sound; the muscles dry and but slightly fat; and their temperament rather sanguineous than lymphatic. The circulation of the blood becomes accelerated; they possess much ardour, energy, and courage; while the several parts of their bodies seldom acquire very voluminous proportions. On the contrary, animals exposed to a cold and moist climate, along with more strongly marked proportions, have their skin thicker, harder, and dryer; their hair longer, coarser, and more bushy; their extremities shorter, with the tendons less strongly pronounced; the horns softer and more spongy; the feet larger, broader, more flattened, and less compact; the muscles stronger, closer, and well supplied with fat. Their temperament is rather lymphatic than sanguineous; their circulation is slower; they possess less physical and mental energy, and may almost be said to consist wholly of matter, as they are visibly deficient in ardour, energy, and courage. The animals of temperate climates occupy in all respects a mean between these two extremes.

Animals have, as well as vegetables, their natural habitations and stations, to which they should be approximated as much as possible in the state of domestication; and it is always dangerous to separate them from these localities without the greatest caution. Nature often places insurmountable obstacles to their migrations, by depriving them, as we have already seen, of the power of reproducing any where except in their native countries. The study of habitations and stations is therefore of the highest importance in the management of the domestic animals.

By the term *habitation*, we commonly understand the climate which each animal prefers, because it is best adapted to its organization; and by *station*, that particular place which each of them chooses in the same country and under the same climate, from its finding more resources in that locality for living and satisfying all the conditions of its organization.

Thus, the habitation of the Reindeer appears to be irrevocably fixed to the frozen countries adjoining the North Pole, where this animal has long been domesticated, and yields the most important services. After the many unsuccessful trials which have been made, it may be considered as almost impossible to render it acclimated in the temperate plains of Europe. Perhaps it might succeed, with the proper precautions, on the summits of our coldest mountains. Again, the natural station of the Rabbit is on a sandy and dry soil; that of the Sheep and Goat in dry and elevated regions; the Buffalo and Bull delight in low and moist situations. These animals cannot be separated entirely and suddenly from their natural stations, without exposing them to inconveniences more or less serious. In all attempts at acclimating foreign animals it is, therefore, as important to study their natural station as their habitation.

Wherever the same temperature prevails, and in whatever latitude, it is generally possible to find some spots where animals may be imported with success, where they will multiply like plants in analogous situations. It appears also, that those animals which Nature has placed in the temperate climates, may extend themselves insensibly towards the opposite extremes of heat and cold; for, as Pallas has judiciously observed, all our domestic animals of the North and South are found wild and apparently native, in the temperate regions of Central Asia.

It has long been remarked that those animals, as well as plants, which have their natural station in dry and elevated countries, are analogous to the living productions of cold countries; and that those species which delight most in low and moist grounds approach more nearly in general character to the productions of the South. This serves to indicate that it is commonly more advantageous to attempt the acclimation of animals from warm countries in low localities, whilst those of the North are most easily naturalized in dry and elevated regions, and it is always useful in practice to study these analogies by attending to the natural disposition, whether low or elevated, which a cold or warm country is capable of affording. It seems probable, also, that individuals will be more easily acclimated in places which form the natural stations of congenerous species, than of those greatly removed from them, for the same dispositions and qualities are usually found to exist in animals belonging to the different species of the same genus. The chances of a successful acclimation are further increased by the adoption of a similar, or at least a kind of food analogous to that which they would have received in their native country; and, in some instances, this is indispensable to their existence. Thus, we often see Birds, directed by the migratory instinct, resorting to localities where they can find that kind of food which is necessary to their existence, and of which they have been deprived by the severity of the climate.

It follows from the above observations, that whenever animals are imported from a country which is very hot or very cold, very dry or very moist, to one which is less so; and that it becomes desirable to maintain them in a state of health, so that they may continue their species by generation, and in general maintain the healthy exercise of all their functions, it becomes necessary to observe the following precautions:—1st, To approximate them by a convenient and suitable position to their original and natural situation; and, 2dly, To avoid all sudden transitions with the greatest caution, so as to acclimate them gradually. The climate, as we have already remarked, exercises a most direct and powerful influence upon the physical and intelligent powers of all animals as well as upon their offspring; and hence we may readily anticipate alterations more or less sensible and permanent, on transporting them suddenly, and without the suitable precautions, to remote distances, or perhaps to situations of an opposite kind to those whence they were abstracted. The effects become more apparent when their transportation is effected from the warm to the cold climate. It may be added, that it is frequently more advantageous to remove animals which are still young, because, from their being more pliant at this age, they habituate themselves readily to the change, and in the end endure the unfavorable circumstances to which they are exposed. A very sudden and powerful change is, however, better endured by the adult animal, whose frame being more matured, is better capable of resisting the shock.

With the domestic animals, we commonly find that temperate climates, where they are exposed but little to sudden changes of the atmosphere, are in general those which agree best with their natures, and where they are least subject to deformity and disease. In these situations they also become more mild and tractable, as their natures assume the general aspect of the climate, while they seem to acquire a certain degree of rudeness and asperity from the contagious influence of an unhospitable region.

The particular kind of food which animals receive when domesticated, exercises a most marked influence over their physical and intellectual constitutions; and unquestionably forms one of the most important branches of their management.

As the attachment of each species to any particular kind of food is regulated by its internal organization, it will often happen that a description of food which is greedily sought after by one animal is rejected by another, and if taken by the latter may even become poisonous, of which numerous instances are to be found. Animals, when unconfined, have the advantage of removing from place to place, often to very great distances, as well as with great rapidity, and, guided by an unerring instinct, are seldom deceived in the choice of their food. In their domesticated condition, they are, on the contrary, confined within very narrow limits, and, being entirely submitted to the absolute dominion of Man, who is not always guided by the views of an enlightened economy, are reduced to the necessity of appeasing their hunger with the food presented to them by his hand. The contrivances of Art are often at variance with the instincts of Nature, and the most fatal accidents, not always attributed to this cause, are the consequences of a violation of her laws.

The character and habitual dispositions of the domestic animals are influenced, in the most direct and well-marked manner, by the general description, and even the particular variety, of their food. An animal feeding solely upon grass, especially when very watery, is usually dull, slow, and possesses but little activity and vigour, although it may obtain a certain degree of *embonpoint*; another feeding on grass nearly ready for cutting, and deprived of its excess of moisture, acquires more force, and a genuine plumpness. If its nature admit of the use of fruit, these qualities become still more apparent, or, if grain be made choice of, its energy is greatly improved. Finally, an animal feeding upon flesh excels the others in its agility, and its animal forces assume the highest degree of energy, or even ferocity. We have frequent instances of the truth of these remarks in all the omnivorous animals, whose characters undergo a considerable metamorphosis, according to the kind of food on which they may happen to be maintained. Let two Dogs be made the subject of an experiment, the one being fed constantly on flesh, and the other on bread, and we shall soon be able to distinguish the former by the superiority of its energy, courage, strength, and ferocity. The carnivorous animals are thus more robust and active than the herbivorous, because flesh is more nutritive and sustaining than a vegetable diet.

It has been correctly remarked by Buffon, that the influence of food is greatest, and produces the most sensible effects, upon herbivorous animals. Animals feeding exclusively upon flesh are much less liable to be influenced by their food than by climate, and the several other circumstances, whether favorable or unfavorable, of the situation in which they are placed.

It is chiefly by a proper selection of their food, that we can succeed in rendering animals, when domesticated, more fruitful than they would naturally have been. By the same means, their flesh can also be rendered more tender, savoury, and delicate. It is more especially during the early periods of their youth, that an abundant and well chosen food is deserving of the highest attention, for the slightest negligence in this respect may produce unfavorable consequences upon their general health, whilst, by an opposite course, we may even succeed, to a certain extent, in correcting an original and constitutional weakness, accelerate the period of puberty, or promote their growth as well as strength. It is even possible, by taking advantage of accidental connate varieties, to perpetuate new races of great value, and render them capable of transmitting these properties undiminished in utility to their posterity. These facts are not always attended to by rural economists, nor is a sufficient degree of attention paid to the kind of food given to young animals. This exercises an important influence over their physical and intelligent dispositions, and an undue parsimony in the distribution of their food, or an injudicious choice in its quality, may be regarded as the vice of a false economy, which deteriorates the qualities of the most valuable species or races. either by diminishing their fecundity, or preventing the development of their most valuable qualities. It is an admitted fact, that, in early life, the preponderating function is that of nutrition, while in the adult the reproductive function prevails. Considerable differences in the height and proportions of individuals are induced by the abundance, the nature, and the quality of their food; and it is to a superiority in these respects, that the domestic animals are generally larger and more prolific than the same wild species, which are not so well nourished.

It may be useful to practical economists to know, that, in general, small animals eat more in proportion to their size than large ones; and for the same reason their vital energy is greater.

The quantity of food necessary to the maintenance of the domestic animals, is in the direct ratio of the loss of substance which they experience from various causes. For this reason, all those which labour much, and all species naturally exposed to violent exercise, stand in need of food in proportion to the degree in which their muscular strength is exerted. All animals whose movements are slow, and labour light, require but little food, as their loss of force is inconsiderable; and those, again, who pass their winter in a state of torpidity, may remain for a very long time without food, as their loss of strength during this time is still less. An elevated temperature, by diminishing the force of the digestive organs, and by moderating the movements of the body, renders less food necessary than a low temperature. Hence, we may diminish their allowances, with propriety, during the warmest seasons of the year.

It has already been noticed, that the distinction commonly made between herbivorous and carnivorous animals is by no means constant. This fact has been advantageously applied by rural economists in various ways. Thus the young of herbivorous animals, shortly after their birth, are frequently supplied, when very feeble, with fresh eggs. The same nutritious food is likewise given occasionally aux étalons avant la monte, and, it has been stated, with beneficial results; likewise, also, to race-horses, with marked success.

We are assured by M. Yvart that, in Auvergne, fat soups are given to cattle, especially when sick or enfeebled, for the purpose of invigorating them. The same practice is observed in some parts of North America, where the country-people mix, in winter, fat broth with the vegetables given to their cattle, in order to render them more capable of resisting the severity of the weather. These broths have long been considered efficacious by the veterinary practitioners of our own country, in restoring Horses which had been enfeebled through long illness. It is said by Peall to be a common practice in some parts of India, to mix animal substances with the grain given to feeble horses, and to boil the mixture into a sort of paste, which soon brings them into good condition, and restores their vigour. Pallas tells us that the Russian boors make use of the dried flesh of the Hamster reduced to powder, and mixed with oats; that this occasions their Horses to acquire a sudden and extraordinary degree of *embonpoint*. Anderson relates, in his History of Iceland, that the inhabitants feed their Horses with dried fishes when the cold is very intense; and that these animals are extremely vigorous, although small. We also know that in the Feröe Islands, the Orkneys, the Western Islands, and in Norway, where the climate is still very cold, this practice is also adopted; and it is not uncommon even in some very warm countries, as in the kingdom of Maskat, in Arabia Felix, near the Straits of Ormuz, one of the most fertile parts of Arabia. Fish and other animal substances are there given to Horses in the cold season, as well as in times of scarcity.

The milk of Cows, fed in this manner, has a disagreeable flavour, while the flesh of such animals as are killed for the table is not pleasant. In general it acquires the flavour, whether good or bad, of the substances on which the animals had been nourished, and for this purpose, therefore, vegetable substances are always preferable. Thus, the flesh of the Carnassiers, whether true Carnivora, or merely Insectivora, of Ant-eaters, &c., is disgusting; and in the same manner, the flesh of Birds is always agreeable in proportion as they feed more exclusively upon vegetables. Animal substances being easily susceptible of putrefaction, impart to the flesh of those which are fed upon them an alkaline and ammoniacal odour. The corrupt Fish, sometimes given to the domestic animals of the North, contributes greatly to their inferiority; and it is well known that Fish in general imparts less muscular vigour and energy than the flesh of Quadrupeds.

The habitual use of animal food renders the herbivorous animals less docile, more untractable, and even dangerous in some cases, as many facts have demonstrated. Cases are quoted of Horses, fed in this manner, having devoured their own masters. It is not probable that these animals could long exist on such a diet, without inconvenience, from their internal organization being greatly different from that of the carnivorous animals, especially in respect to the Ruminants. Still it is abundantly demonstrated, that animal substances can be administered with advantage, especially in cases of scarcity of their ordinary food, or of weakness, whilst the carnivorous animals have an indispensable necessity of living upon flesh, in order to derive a sufficient nourishment, and to maintain that kind of life for which Nature has intended them.

The food given to the domestic animals may either be composed of entire and unprepared substances, such as Nature spontaneously presents, or it may be divided and prepared in various manners; while its good qualities are susceptible of being improved in several ways, according to the object which is had in view. It may consist of plants either green or dried, whole or divided, moist or dry, raw or boiled, fermented or the reverse, sweet or sour, plain or seasoned with different substances; and, according as it is given to them in these different states, the results obtained are very different.

The mechanical division of boiled food, whether green or dried, facilitates the several acts of masticating, swallowing, and also of ruminating when it occurs; hence, by a necessary consequence, their digestion being more perfectly performed, an equal weight of food becomes more profitable, and this mode of preparation should always be adopted, except when the food is consumed on the field. For this purpose several useful instruments have been invented, such as turnip-cutting machines, choppers, mills, and many others more or less ingenious, which divide quickly, economically, and completely, the different kinds of food, whether roots, seeds, or forage. Every farmer who feels any interest in the improvement of his domestic animals, should be provided with one or other of these instruments, and he will not fail, sooner or later, to be completely indemnified for the additional expenses they may occasion.

Green food is in general more profitable to these animals, especially when it is intended to fatten them, than such as is either faded or dry: for, independently of the loss of nutritive principles which it experiences more or less while drying, it is digested more easily, rapidly, and completely, in the former than in the latter case.

For the same reasons, food which has been moistened and softened after being dried is usually more profitable than when given under a hard or dry form. Seeds especially, when broken or reduced to flour, or even made into a paste or broth, are more quickly assimilated into the animal substance than when entire. Hence, they are nearly every where reduced to a state of minute division before being given to animals in course of fattening, and numerous experiments have clearly established their comparative superiority over those which have not undergone this process.

The boiling of their food, by performing or facilitating its division, is one of the best means known of promoting digestion, and even of increasing the quantity as well as quality of the alimentary substances which undergo this process.

This advantageous result appears to originate in part from the circumstance that the molecules of the alimentary substance are separated by the coction which they undergo, and thus present a greater surface to the influence of the gastric juice, and partly from the influence of the water wherein they are immersed, as well as of the high temperature to which they are exposed, augmenting their nutritive powers. The water seems actually to become solid as in the making of bread, by entering into union with them, or by imparting its hydrogen, which afterwards becoming united to carbon, may contribute towards the formation of fat. These facts have been established by a great number of experiments made here and elsewhere, with roots, grains, and even with raw and boiled hay or grass, used for fattening the domestic animals. Potatoes and Jerusalem artichokes, which, in their raw state, are either cared for but little by the cattle, or unprofitable, acquire by boiling new properties which render them extremely advantageous after having undergone this operation. Indeed, the general practice of boiling the food cannot be too strongly recommended, especially when the low price of fuel, and the other circumstances of the locality, allow it to be performed conveniently and economically. It is also proper to administer it to the cattle while still warm, if possible, for the reason that it appears to be more agreeable to them when given in that state, and that it invigorates and refreshes them more quickly than when allowed to cool after boiling.

As a confirmation of the correctness of these views, regarding the superiority of boiled over raw food in the fattening of cattle, we have only to consider for a moment what actually takes place every day before our eyes in respect to Man. We here see how greatly substances which have been submitted to the action of heat, such as bread, meat, soups, broths, and other articles, surpass those used in their natural state. A small quantity of wheat, maize, barley, or rice, well boiled and eaten warm with a little milk, gains in nutritive matter an immense superiority over the same quantity of these substances, if eaten without this preparation. The same remark is applicable to all kinds of grain.

It may be noticed here, that the food intended for cattle can be conveniently and economically boiled by steam, by putting it into a common barrel, cased with iron, and having at its base a grating of the same metal, with the bars tolerably close. After filling it with the roots intended to be boiled, it is exposed to the vapour of boiling water arising from a cauldron placed upon an economical furnace. This arrangement permits the food to be boiled cheaply and in a very short time. Care must, however, be taken that the base of the barrel fits accurately into the upper rim of the cauldron, and that it has at its top a moveable cover so as to permit the roots to be easily placed there and withdrawn. There must also be a small hole in the cover to allow a part of the vapour to escape when it has reached the top.

The addition of some coarse provender, such as chopped straw, to boiled roots, is admitted to be advantageous; probably because it renders the mastication of these substances more complete, and serves also as a kind of ballast, which should always bear a certain proportion to the nutriment, properly so called.

Fermentation, which may be regarded as a sort of cooking afforded spontaneously by Nature, adds greatly to the nutritive qualities of the substances which undergo this process. It has long been recommended to allow the barley, intended to fatten cattle, to germinate, and this may be regarded as the first step in the process of fermentation, which the grain undergoes when used for making beer. By this means, the saccharine principle becomes more fully developed, while the food is unquestionably made more digestible and nutritious. Hence cattle-dealers seek with avidity, and employ with great advantage, the residue of breweries, distilleries, and starch manufactories. A part of the grain thus prepared, or its refuse, is used largely for feeding cattle in Belgium, Alsace, and generally in the immediate neighbourhood of all large manufacturing towns. The nutritive properties of the food are further augmented by rendering it sour, or at least, it tends in this state to render the digestive function more energetic. Hence, the farinaceous substances used for food, especially when it is intended to fatten the cattle, are made in a great number of places to undergo the acetous fermentation. Indeed, all the modes of preparation already enumerated are but little useful to animals destined for hard labour. Seasoning renders the food more agreeable to their taste, more digestible, and therefore more profitable. Common salt is probably the most powerful and useful of all substances for this purpose, and hence it is employed almost every where with advantage. It sharpens the appetite, excites to drink, facilitates digestion, renders the flesh of animals intended for the table of a superior quality, and either promotes or supplies the acidity induced by the second stage of fermentation. All Mammalia seek salt with as much avidity in their wild state as in that of domestication, and show a degree of pleasure, which is a sure index of its utility when mixed with their food, and of its power of correcting the hurtful qualities of their aliment when it happens by some accident to have become vitiated.

In addition to those precautions, which are essential to the proper selection and preparation of food for the domestic animals, it is of great importance to regulate the rations or quantity of food distributed to them at intervals, in order that they may be rendered as profitable as possible. The quantity of food ought always to be in proportion to their age, state of health, the violence of their exercise, and final destination, always observing, at the same time, the general principle, that the quantity of the food must be more considerable when it is less substantial, as any diminution of its nutritive qualities can only be compensated by a proportional increase of its quantity. It is always impossible to determine, in a fixed and positive manner, how much of each kind of food an animal should consume in a given time, because this depends upon a great number of circumstances relative to its species, its race or breed, the peculiar constitution of the individual, its employment, as well as its age and state of health. The daily allowances further change with the very variable nature of their food, the different ways in which it is administered, the state of the atmosphere, the season of the year, and several other circumstances, all of which should be taken into consideration before we can determine their proper daily rations with any degree of accuracy. Hence result the various and contradictory opinions emitted on the subject by most writers who have attempted to fix quantities. Some have laid down as a principle, that certain domestic animals will daily consume the third part of their weight of watery food, such as turnips, beet-root, or green clover; while others have fixed for the same animals a fourth part of their weight of cabbages, carrots, and parsnips, and a fifth or a sixth of beet-root, potatoes, and Jerusalem artichokes. There must be, however, a great variation according to the different circumstances just enumerated. It appears to us that all these matters should be regulated by particular and individual trials, and be left wholly to experience. This is of more real use than the futile attempts made in most practical books to fix quantities, and which only serve to demonstrate the real ignorance of the persons attempting to enforce them. Physiologists, and all who have studied this matter properly, know very well, that although there are certain well ascertained general laws which regulate the entire animal economy, each individual possesses a peculiar constitution, or *idiosyncrasy;* which more or less serves to modify these laws. Hence we frequently find a disparity of effects resulting from the same apparent or real cause, and these variations show themselves in the quantity of food which animals consume, as well as on a great many other occasions, the explanation of which can only be obtained on the principles already explained.

Along with the really nutritive food, there must always be mixed a certain quantity of ballast, that is, of some coarse and slightly nutritious food, otherwise the sides of the stomach, as well as the intestines, will not be sufficiently distended and stimulated, so as to perform completely the functions for which Nature intended them. Unless this condition is rigorously attended to, the digestion, elaboration, and assimilation of the nutritive juices, will always be incomplete even in healthy and well constituted animals. It is therefore a very important error to overload the stomachs of these animals with any very nutritious food unmixed, even when it is exclusively intended to fatten them.

In respect to the distribution of their food, it is only necessary to notice one excellent maxim, *Good food, a little at a time, and often;* they should be allowed to eat quietly and slowly in order that they may digest the largest quantity of food in the shortest possible time. Regular intervals of feeding should be observed, with occasional fasting, which serves to appetize them, and give an impulse to their digestive organs. They should not, however, be allowed to grow impatient, which occasions a loss of animal force and nutrition. Digestion never proceeds rapidly as long as the animal continues eating. It is only when sufficiently filled that the circulation becomes accelerated, the temperature of the body more elevated, and digestion proceeds with its greatest activity. All these phenomena succeed in the course of a few hours, after which the temperature of the body falls, the respiration becomes moderate, and hunger returns. It is only at this time that more food should be given, in small rations at a time; and when treated in this manner, the animal consumes less, and derives more benefit from its food.

To alternate and vary the kind of food used is always necessary, because the continual use of the same aliment does not sharpen the appetite so well as a judicious selection and rotation. A variety of food serves to stimulate the digestive organs, and prevent that disgust which the same diet continued too long always occasions by its uniformity. Care should be taken, in respect to these changes of food, to avoid a sudden alteration of diet, especially from green to dry food, or *vice versa,* for these are always more or less prejudicial. It is also very important not to overload the stomachs of labouring animals, immediately before they set out to their work, as is too frequently done, for this often occasions indigestion, or at least renders it imperfect or laborious. From want of food or other circumstances, these animals are often obliged to submit to a long fast, which they are always better able to endure in proportion as their food has been the more substantial.

There are some domestic animals, such as the Camel and the Ass, which are remarkable for their frugality, as well as their capacity of remaining long without food. There are also some races of other animals which are equally celebrated for these qualities; and when they do not originate in some constitutional defect, or from ill health, and when it is not effected at the expense of their other useful properties, this forms a powerful inducement for propagating some races in preference to others. The Mule is an instance of the above, as well as some of the improved breeds. A quantity of barley, equal to about one feed, is sufficient, according to the report of travellers, for the daily maintenance of an Arabian saddle Horse, after a long journey in the deserts; while a European Horse performing the same service would have consumed, in the same time, a much larger quantity of barley, besides a considerable bulk of hay and straw. The remarkable frugality of the former, although doubtless owing to an original constitution improved by habit, is partly due to a difference in the nutritive qualities of the food, as well as to the climate. If animals of the South consume, in general, a smaller quantity of food than those of the North, this is in part due to the circumstance, that the food is much more nutritious in the former than in the latter case, and also that it possesses a greater specific gravity. It may not be improper in this place to notice a remarkable error almost universally adopted in this country, of giving out corn, which is the most substantial part of their food, by *measure* instead of by *weight,* as it has been ascertained by many trials, that the quantity of really nutritive matter may vary in bulk by nearly one-half, according to the quality of the corn.

As the most useful and important of our domestic animals are herbivorous, it may be advantageous briefly to notice here the general qualities of the several vegetable substances which usually form the basis of their diet.

The substances principally used for this purpose are, *1st,* Grass, either fresh, or under the form of Hay. *2d,* The Straw of the Cereal plants. *3d,* Leaves or Stalks. *4th,* Roots or tubers. *5th,* Seeds, Grains, or Fruits. Each of these subjects admits of being treated somewhat in detail.

Grass is the most natural food of the herbivorous animals, and is often sufficient to restore feeble animals to a good condition when they have fallen off, upon any other kind of diet. This food is not, however, adapted for hard-working animals.

The best kind of green food is fine, substantial, not very watery or faded, and should not have grown in a shady situation; it is usually found upon natural or artificial meadow-land. The Natural families of the Gramineæ and Leguminosæ are the most abundant in important Plants. In the former we may notice the Meadow-grasses (*Poa*), Fescue-grasses (*Festuca*), Fox-tail-grasses (*Alopecurus*), Oat-grasses (*Avena*), Cat's-tail-grasses (*Phleum*), Bent-grasses (*Agrostis*), Canary-grasses (*Phalaris*), Wheat-grasses (*Triticum*), the Barleys (*Hordeum*), Hair-grasses (*Aira*), Soft-grasses (*Holcus*), Dog's-tail-grasses (*Cynosurus*), Quaking-grasses (*Briza*), Millet-grasses (*Milium*), and a few other genera. Of the Leguminosæ, the following are the most remarkable:—The Medicks or Lucerns (*Medicago*), the Trefoils (*Trifolium*), Saintfoin (*Onobrychis*), the Melilots (*Melilotus*), the Vetches (*Vicia*), the Tares (*Ervum*), the Milk-Vetches (*Astragalus*), and the Bird's-foot Trefoil (*Lotus*). There are some plants, which not only have the property of exciting a more abundant secretion of milk in those females which are fed thereon, but also render it of an excellent quality; such are the roots of the Parsnip or Carrot, and the stalks of the Maize; while others, such as the Garlics (*Allium*), actually impart a disagreeable odour, or other unfavorable qualities. Each domestic animal shows a marked predilection in favor of some plants, and either refuses certain others altogether, or feeds upon them only when compelled by a scarcity of food, as Linnæus and several of his followers have long ago remarked. Not only do they derive pleasure from particular parts of certain plants in preference to the remainder, but the different states of vegetation in which each of them is found, as well as the different situations and nature of the soil on which the plants grow, contribute still more strongly in determining their choice. With a very small number of exceptions, we find in general that when plants are in their flowering state, or one which nearly approaches to it, they are most nutritious. At this time, their nutritive particles are diffused abundantly and equally throughout the whole plant, and they hold a middle state between the aqueous condition which is too relaxing, or not sufficiently nutritious, and the ligneous condition, which renders difficult the functions of mastication, deglutition, and digestion. In general, also, medium qualities of the soil, as well as intermediate stations, should be preferred for pasture grounds.

After numerous comparative trials made at Upsal in Sweden upon the common plants of the meadows, fields, and other pasture lands, it was found, by M. Hesselgreen, that the plants used by each species of domestic animals vary greatly in number. His results are represented in the following table:—

Of 575 Plants,	the Goat eats	449,	and refuses 126
528	the Sheep	387,	141
494	the Bull	276,	218
474	the Horse	262,	212
243	the Pig	72,	171

This serves to indicate that the Goat is the least delicate in his taste, and can eat without inconvenience a great number of plants hurtful to other species. The Sheep feeds upon nearly three-fourths of all the plants it encounters; the Oxen and Horses refuse nearly one-half, while the Hog can eat the leaves and roots comparatively of a very small number of species. The above results are, however, very incomplete, and must be considered merely as approximations.

Subsequently to the investigations of M. Hesselgreen, M. Yvart examined nearly seven hundred of the most common plants of France, or those capable of being naturalized there, and as his inquiry appears to have been conducted with much care, it may be interesting to compare his results with those already given of the Swedish investigator.

	Goat.	*Sheep.*	*Bull.*	*Horse.*	*Hog.*
Can eat	547	408	311	268	86
Is very fond of	28	81	121	113	36
Sometimes eats	32	33	70	39	23
Takes in all	607	522	502	420	145
Refuses	83	133	183	235	169
Total Plants examined . . .	690	655	685	655	314

Many plants are wholly refused by all animals. Among the principal of these growing in marshy places we may notice the following: The Common Butterwort (*Pinguicula vulgaris*), Common-hooded Milfoil (*Utricularia vulgaris*), Forget-me-Not (*Myosotis palustris*), Perfoliate Pond-Weed (*Potamogeton perfoliatum*), Long-leaved Cowbane (*Cicuta virosa*), the Long-leaved Sun-Dew (*Drosera longifolia*), the Round-leaved Sun-Dew (*D. rotundifolia*), Water-Pepper (*Polygonum Hydropiper*), Sweetflag (*Acorus calamus*), Water Crowfoot (*Ranunculus aquatilis*); Great Spearwort (*R. lingua*), and Water Milfoil (*Myriophyllum spicatum*).

There are several other plants which either grow in somewhat moist pastures or in the shade, and are likewise refused by all cattle. These are the Common Thorn-apple (*Datura Stramonium*), Common Henbane (*Hyoscyamus niger*), Black-berried Nightshade (*Solanum nigrum*), Dwarf-Elder (*Sambucus Ebulus*), Mountain Dryas (*Dryas octopetala*), Black Horehound (*Ballota nigra*), Common White Horehound (*Marrubium vulgare*), Impatient Lady's Smock (*Cardamine impatiens*), Common Celandine (*Chelidonium majus*), and the Blue Erigeron (*Erigeron acre*). It must be noticed, however, that many of these plants, when very young, are sometimes cropped by the cattle without inconvenience, while some even of the most nutritious plants are refused when in grain, from their perfume being too strongly diffused. After the animals have endured a long continued fast, their discrimination in these respects is not so nice; and the climate may occasion some further differences. Thus, the young sprouts of the Wolf's-Bane and Hemlock become esculent even for Man in the North of Europe, where their deleterious properties are not sufficiently developed to become hurtful.

Some plants are often eaten by the cattle while green and fresh, and yet are generally refused by them if offered in a dry or faded state. These are Cock's-Comb (*Rhinanthus crista-galli*), the Horse-Tails (*Equisetum*), the Bedstraws (*Galium*)—which spoil the Hay, and the Common Buckbean (*Menyanthes trifoliata*); while others, such as the Crowfoots (*Ranunculus*), and Swallow-Worts (*Asclepias*), lose their injurious properties when dried, and in that state are eaten by the cattle without inconvenience. Others serve as seasoning, such as the Garlics (*Allium*), and the Docks (*Rumex*), either of which may be used occasionally as a stimulant or corrective; while the Cotton-Grasses (*Eriophorum*), and some others, become hurtful from their hairs, which serve as a nucleus to those dangerous ægagropiles or concretions, sometimes found in the first stomach of the domestic Ruminants.

There are also a great number of plants eaten without inconvenience by the Goat, and even greedily sought after by that animal, while they are refused by all other cattle. The principal are the Common Mare's-tail (*Hippuris vulgaris*), Common Prickly-Seed (*Echinospermum Lappula*), the Greater Water-Plantain (*Alisma Plantago*)—highly detrimental to all other domestic animals, the Wood Anemone (*Anemone nemoralis*), that of the meadows (*A. pratensis*), the Spring Anemone (*A. vernalis*), Celery-leaved Crowfoot (*Ranunculus sceleratus*), the Knotty-rooted Figwort (*Scrophularia nodosa*), and Tamepoison (*Asclepias vincetoxicum*), of which it is extremely fond. The last mentioned plant can be eaten by the Horse, only after it has been killed by the frost. To these we may add, the Small Water-Wort (*Elatine Hydropiper*), Box-leaved Andromeda (*Andromeda calyculata*), Biting Stonecrop (*Sedum acre*), Snapdragon (*Antirrhinum linaria*), Stinking Camomile (*Anthemis cotula*), Black-berried Bryony (*Bryonia alba*), Marsh Lousewort (*Pedicularis palustris*), that of the woods (*P. sylvatica*), Hemp Agrimony (*Eupatorium cannabinum*), the Annual Mercury (*Mercurialis annua*), which is poisonous to all other animals, according to Ray and Linnæus, the Corn Horsetail (*Equisetum arvense*), that of the marshes (*E. palustre*), and the Male Polypody (*Polypodium filix mas*).

Some plants are eaten solely by the Hog, and it is often only their roots that are sought after. The chief of these plants are the Common Cyclamen (*C. Europæum*, Common Asarabacca (*Asarum Europæum*), the White Water Lily (*Nymphæa alba*), and the yellow (*N. lutea*), for which the Horse exhibits a marked aversion, the Water Soldier (*Stratiotes aloides*), Sea Wrack-Grass (*Zostera marina*), and Maiden Hair (*Asplenium trichomanes*).

A few plants are very much sought after by all cattle, and almost with equal avidity. These are the Common Millet-Grass (*Millium effusum*), Meadow Soft-Grass (*Holcus lanatus*), Annual Meadow-Grass (*Poa annua*), Oats, Barley, and Wheat, the Carrot and Parsnip, the Great Round-leaved Willow (*Salix caprea*), the Norwegian Cinquefoil (*Potentilla Norvegica*). Also, the Creeping Trefoil, the Common Lucerne, and Sainfoin. But many of these plants must be in different states, in order to be liked equally by the several species of cattle.

On considering the entire vegetable kingdom in a general manner, we find that scarcely any Acotyledonous plants are fitted for the maintenance of cattle. Indeed, if we exclude the Grasses, nearly all of which may be used for this purpose, we find but few even among the Monocotyledonous plants. It is unquestionably in the Dicotyledonous class that the greatest number of useful materials for this purpose are to be found.

The following natural families are arranged according to the order of their utility for food to cattle:—The Gramineæ, Leguminosæ, Cruciferæ, Rosaceæ, Amentaceæ, Umbelliferæ, Cucurbitaceæ, and Polygoneæ.

The best Hay is afforded by the more elevated meadows, and its quality depends greatly upon the care with which it has been dried. In this article quality is much more important than quantity; for a stone of good Hay, well selected, and carefully dried, affords more nutriment than several stones of coarse or ill prepared material,—a matter to which sufficient attention is not always paid. The exposure to the sun or air, during its making into Hay, always occasions grass to lose some portion, more or less considerable, of its nutritive substance, which is evaporated along with the watery matter. New Hay often occasions indigestion, and it should not be given to cattle for several months after being made, at which time it is entirely deprived of its uncombined aqueous substance.

Straw should be considered rather as a useful kind of ballast proper to be mixed with the really nutritive food of the domestic animals, than as a substantial nourishment. The best quality is fine, white, short, and massive. It is often advantageous to have it chopped and even moistened.

Dried leaves, as well as the small branches of a great number of trees, shrubs, and bushes, may sometimes form a useful substitute for straw or hay, when the latter cannot be easily procured. The Elm, the Mulberry tree, the Ash, the Hornbeam (*Carpinus betulus*), the Lime trees (*Tilia*), the Common Maple and Sycamore (*Acer*), the Common Acacia (*Robinia pseudacacia*), the Willows, the Poplars, the Birches, Beeches, Planetrees, Chestnuts, Oaks, Dogwood (*Cornus*), Hazel (*Corylus*), Furze (*Ulex*), and the Vine, are frequently used for this purpose on the Continent, in places where they happen to be plentiful. The same substances, if given in their green state, may also replace the newly-mown grass of the meadows; but they should always be administered with caution, and with a due attention to their effects, which vary according to the species, as well as in their several states of vegetation. The green leaves of a tolerably large number of vegetables are annually cultivated on a large scale, either as food for Man or for Cattle; such are the leaves of the Maize, Beet-root, Cabbage, Carrot, Parsnip, Potato, and some others, all of which may be used for this purpose in many cases with advantage.

Roots, or rather their tuberous appendages, which are often very large and voluminous, such as those of the Parsnip, Carrot, Beet-root, Potato, Jerusalem Artichoke, and Turnip, are frequently superior to any of the substances already mentioned as a daily article of food for cattle, and many comparative trials have clearly shown that they are in general much more profitable.

Seeds, grains, or fruits, contain, of all the parts of a plant, the largest quantity of nutritive substance under the smallest bulk. They ought to be given judiciously and sparingly to cattle, from their being in general very costly, and there are some other inconveniences to which their frequent use may give rise. Sometimes they are ground, broken, or prepared in different ways in order to render them more digestible and economical. The principal seeds used for the food of the domestic animals are also, in great part, furnished by the useful families of the Gramineæ and Leguminosæ. Other farinaceous fruits, procured from some of the remaining families, are occasionally added to these; such as the Buckwheat (*Polygonum Fagopyrum*), the Chestnut, Horse-chestnut, and Acorn, as well as the oleaginous seeds of cruciferous plants, especially of some varieties of the Cabbage, and Gold-of-pleasure (*Camelina*). To these may be added, the seeds of Flax, Hemp, some species of Poppy; also Beech-nuts and Walnuts, or rather their refuse, and some other fruits less common or important. Most of these, however, have the inconvenience of imparting to the flesh of the animals fed thereon an odour and taste by no means agreeable. In respect to the bran or husk of grains, it is nourishing only when it contains some flour mixed therewith, for the outer rind itself is not only destitute of nutriment, but very indigestible, and often injurious.

Herbivorous animals are wholly overcome by famine, while carnivorous animals are more easily vanquished by an excess of their food. Long-continued hunger exasperates the latter, and renders them furious, while many striking instances are known of the most ferocious animals being wholly tamed by an abundant supply of food, united to other precautions.

Exercise, to which but little attention is commonly paid, is a subject requiring almost an equal degree of consideration with that of their food. By this term we commonly understand the amount of bodily motion necessary to maintain the proper circulation of the fluids, and to impart that degree of activity which the natural condition of their body requires. Exercise is the contrary of repose, and without either of these, the animal machine would soon be destroyed. It greatly assists the insensible transpiration, the most abundant of all the secretions, and keeps off a number of diseases depending upon the superabundance of the fluids, their impurity, or stagnation, enlargements or obstructions of the viscera. Far from diminishing the animal forces, it reanimates them; a languishing appetite is restored, and the consequences of exercise are reflected throughout the entire vital economy.

The influence of exercise upon fecundity and longevity are not less remarkable. Very fat animals are often unfruitful, while a long-continued repose frequently leads to obesity, which again induces impotence, and often death. These are not the only consequences of a continual want of exercise. Their limbs are deprived of that play and spring necessary in preserving all the parts of the body in their state of health.

Exercise should, however, be regular and moderate. Very violent labour may affect all the organs, and render the stature diminutive; hence all excess in this respect should be avoided if possible, especially during the growing period of life.

The domestic animals also require much attention in respect to their lodgings. When in their wild state, they are constantly in the open air; in their domesticated condition they are often abstracted from it. This essential difference necessarily affects the conditions of their existence; and, in proportion as they are brought nearer to their natural state, their health becomes improved, while an opposite course of treatment may be attended with the most fatal consequences.

Our most useful domestic animals are often confined in narrow stables, which are perhaps rather injurious than beneficial, from their vicious construction or pernicious arrangement. A knowledge of these defects has suggested to some rural economists the idea of exposing their cattle continually in the open air; but these persons do not perceive, that in avoiding one error, they fall, as frequently happens, into another not less important. When in their wild state, animals are always *in the open air*, it is true; but it does not follow that they are continually *exposed to the weather*, which is a very different circumstance of their condition. In their native haunts, they always endeavour to withdraw themselves from excesses of every kind, whether of heat or cold, moisture or dryness, as well as from storms, tempests, violent winds, or the attacks of their enemies, while they are free to change, whenever they please, either their place or position. It therefore becomes a serious error, through inattention to these circumstances, to expose domestic animals to the inclemencies of the weather, without the slightest shelter, for the mere purpose of avoiding the common disadvantages of a stable. We have often seen flocks of Sheep shut up in narrow parks, exposed in winter to the frost, in summer to the burning heat of the sun, in spring and autumn to excessive moisture, and in all seasons to the sudden changes of the atmosphere, and consequently to the most sudden alterations of temperature. The natural consequences of this mismanagement have invariably followed, while their undue mortality and impoverished condition fully demonstrated, that animals exposed to all kinds of weather are far from being in that state of nature to which it was intended to reduce them. Their amelioration, their prosperity, and even their existence, are compromised as much by this injudicious treatment, as they would have been by the most confined, uncleanly, and ill-constructed stable.

It is no doubt true, that we should endeavour to bring animals to a real state of nature, and place the enjoyments of liberty and fresh air as much as possible within their reach. Yet this can only be done effectually by giving them the power, by having some enclosed space where they may be free to move, and with a sufficient number of retreats or sheltered spots, to which they can resort at those times when there is more real inconvenience and disadvantage in being without than within. This is the only legitimate way in which we can approximate the domestic animals to their natural condition.

To dress their coats occasionally, and clear the entire surface of their bodies from all impurities, by the aid of suitable instruments, are attentions imperatively required by the state of domestication, and are apparently indispensable to the health of the most useful animals which have submitted to the empire of Man. As the skin of these animals is perforated by an infinite number of pores, or orifices of the smaller arteries adjoining the epidermis, there exhales continually during the healthy state of the animals an excrementitial vapour, which has been considered in Man to surpass in quantity all the other evacuations taken together. This important function, known by the name of *insensible transpiration*, is indispensable to the well-being of all the domestic animals, and it cannot be arrested, or even suspended or modified, without being attended with injury to the system. When this function is performed regularly and suitably, it clears the skin, maintains it in the supple state fitted for the play of all the organs, and smooths as well as nourishes the hair, which then looks sleek and glossy. When, however, by any cause, this passage for the superfluous humours has been interrupted, they either flow back towards the centre, or become fixed in the exterior. In either case the vital functions are disturbed, and a great number of dangerous maladies are the consequence.

These accidental derangements of their natural order may be observed among all animals, but they are more frequent and acquire greater intensity with the domestic quadrupeds, especially such as labour severely or are in course of fattening, from their being more exposed to the causes whence they arise. Being often obliged to remain stationary for a considerable length of time, and most frequently in narrow and confined places, exposed to a continual and abundant dust, with the exhalations arising from their food or other matters, various foreign substances fix themselves upon the skin, and if daily care be not taken to remove all these obstructions to the insensible transpiration, their general health becomes seriously affected, and thus the improvement of the breed may be retarded or their fecundity diminished.

The particular purposes for which each race or species may be intended is one of the points to which the Economist pays special attention. A general distinction, depending on their different adaptations and the variety of their products, is made among all animals intended to be improved artificially. Some, for example, such as the Horse, the Ass, and sometimes the Mules proceeding from them, are chiefly used in Europe for carriage, draught, or speed, while their economical products during life, or afterwards, amount to a small matter, being merely the hide, hoofs, tendons, and the oil abstracted from the marrow of their bones. The Ox and Buffalo again are specially used for draught or the table, while their females chiefly yield milk and its modifications. The Sheep and Goats present us with the three-fold tribute of their fleece, milk, and flesh; and different uses are made of the horns with which

some of the animals are armed, as well as of their skin. The Hog yields little else besides his flesh and bristles, and sometimes his skin. The Rabbit only imparts its flesh and fur; and we esteem the Dog and Cat rather for intellectual than physical qualities.

From the variety of these products, and the different kinds of service which they render, a particular attention is commonly paid to such points as contribute more especially towards the several advantageous results. Thus, their size, weight, volume, tendency to grow fat, smallness of the bones compared with the other parts of the body, the abundance and fineness of their flesh, are qualities which are particularly esteemed in species or races intended to be fattened. The relative volume of the most useful parts compared with those of less value is another point of importance. Intestines of small size are in this case to be desired, as well as small bones, with a fine and supple skin. An abundant cellular tissue, when the accumulation of fat is an object, becomes an essential point; and a broad back with the dorsal and lumbar muscles strongly developed is no uncertain promise of a large quantity of delicate beef-steak. Another and a very different set of qualities are esteemed in animals intended for draught. The size, weight, and massiveness of the body, the breadth of their base, the thickness of the loins, and the force of the bony skeleton, are essential characters in all animals for draught and burthen; while an ordinary Saddle-Horse should be rather active than heavy. Animals specially intended for laborious occupations should have a broad chest, the fore quarters elevated, and the hind quarters neatly made, large, well sloped off and proportioned. For ordinary draught or light work the characters should be less strongly marked, and the general form disposed for agility. The Race-Horse requires much suppleness in all his limbs; and the form best adapted for this purpose consists of a low front, a broad chest, a body rather elongated than shortened, with a great deal of freedom in all his limbs.

In the Ox and Buffalo, muscular force and largeness of the extremities are esteemed, together with suppleness of the skin, and all the qualities for fattening already enumerated.

With the Cow and female Buffalo, as well as in Ewes and she-Goats, the first objects to be considered are the development of the udders, the size of the lacteal vessels and mammary veins, as well as the fineness and suppleness of skin.

In the Sheep, a long and abundant fleece, free from all blemish, united to a suitable height and form, and a great aptitude to fatten quickly, are the most desirable qualities.

The Goat is esteemed for a long, fine, and silky hair, when united to lightness, agility, docility, and other qualities already enumerated, especially in reference to its milk, which is often used.

In the Hog, we esteem an excessive voracity, supple and abundant bristles, with a disposition rather tranquil than wild, and especially a disposition to grow fat quickly.

An abundant and fine fur, large size, powers of reproduction well pronounced, and a tendency to fatten, are valuable qualities with the Rabbit.

In all domestic animals, their liveliness and vigour, with the complete development of all their organs, are the surest guarantees for their strength, energy, and courage, and these qualities ought always to fix the attention of the rural Economist.

The head of young animals, as well as their organs of nutrition and digestion, are usually more voluminous in proportion than the other parts of the body, because the growth of the individual is at this age the principal object of Nature. Their bones are rather cartilaginous than solid. Their skin, whose absorbing power is stronger than at a more advanced age, is a loose and thin tissue. Their blood is not deeply coloured; their fat is white and spongy, with little consistency or flavour, and is most prevalent towards the exterior. Their muscles are softer, and more watery than in mature age, and their flesh is consequently more tender; but, at the same time, more insipid, as well as less nutritious and juicy than in the adult.

With old animals, on the other hand, the solids of the system predominate, and the nutritious parts are lost faster than they are repaired. Not only are the bones very hard, but the cartilages are often ossified; the skin is coriaceous, and adheres forcibly to the subjacent parts; the colour of the flesh is deep; the fat often of a bright yellow, thick, viscous, and more prevalent at the centre than at the circumference; the muscles are shortened and dried up; the flesh is consequently but little nutritious, and difficult of digestion.

It follows from these observations, that the most nutritious, savoury, and substantial meat, without being hard or indigestible, is procured from animals at the medium age, between the two extremities of life.

[We have now considered the Mammalia under most of those general points of view which appear necessary to render the consideration of Species intelligible to the general reader, as well as to impart a sufficient degree of interest to their details. The Geographical distribution of the Mammalia over the surface of the globe, the consideration of the remains of animals now found only in the Fossil state, as well as the causes which have led to their extinction—all subjects of great and general interest, require to be postponed, until we have gone over the numerous species of this Class; for, without a previous knowledge of species and their differences, the importance of these branches of the science cannot be fully appreciated. We also think it advisable fully to develop the system of arrangement adopted in the "Regne-Animal," with the additions and improvements suggested since the publication of the last Edition of that work, before entering upon the History of this branch of the Science, being fully convinced that an intimate acquaintance with some one system, at least, is absolutely necessary for the proper understanding of the several systems which have been proposed from time to time for the arrangement of the objects comprised in this Class of the Animal World. Previous to the consideration of Genera and Species, it will, however, be proper to define some of the most important terms used in describing them.]

GLOSSOLOGY OF THE MAMMALIA,

BEING AN EXPLANATION OF THE PRINCIPAL TECHNICAL WORDS USED IN

MASTOZOOLOGY.

ABBREVIATIONS.—LAT. LATIN—GR. GREEK—FR. FRENCH—GERM. GERMAN.

(1.) GLOSSOLOGY, from γλῶσσα (*glôssa*), tongue or language, and λογος (*logos*), a discourse,—supplies the explanation of the technical terms belonging to any Art or Science. It corresponds with the French " *Terminologie*," and with the German " *Kunstwörter*."

(2.) MASTOZOOLOGY, from μαστος (*mastos*), the breast, ζῶον (*sôon*), an animal, and λογος,—is the science which treats of the Mammalia or mammiferous animals. The corresponding French word " *Mammalogie*," being derived partly from the Greek and partly from the Latin, is inadmissible. M. Desmarest has suggested the term " *Mastologie*," which, however, is more limited in its signification than our term, originally proposed by M. de Blainville.

(3.) THE MAMMALIA, *Lat.* Mammalia, *Fr.* Mammifères, *Germ.* Säugthiere,—corresponds to the English terms " MAN and BEASTS" taken together. These are the MASTOZOA of M. de Blainville, the MAMMALIA of Linnæus, Erxleben, and others. To distinguish them from the four-footed Reptiles which are oviparous, they have been styled " VIVIPAROUS QUADRUPEDS," with much impropriety, as the Cetacea want the hinder limbs (see page 38). Some recent writers have attempted to introduce the barbarous term MAMMALS.

I.—THE SKELETON IN GENERAL.

(4.) THE SKELETON, *Lat.* Sceleton, *Fr.* Le Squelette, *Germ.* Geripp, Knochengerüst,—is the bony frame-work of the body, destined to protect the nervous system and other vital parts, and serving as a point of support to the organs of active motion. It is divided into the *head*, *trunk*, and *extremities*.

(5.) THE HEAD, *Lat.* Caput, seu Cranium, *Fr.* La tête, *Germ.* Kopf, Schädel,—forms the *anterior* portion of the skeleton [the *superior* in Man], containing the brain and the principal organs of sense.

(6.) THE SKULL, *Lat.* Calvaria, seu Cranium, *Fr.* Le crâne, *Germ.* Hirnschädel,—is the upper and hinder part of the Head, especially intended to contain the Brain. Its volume varies relatively to that of the head, and is by some thought to be proportioned to the degree of intelligence. The exceptions are, however, very numerous.

(7.) THE FACE, *Lat.* Facies (*maxilla*), *Fr.* La face; *Germ.* Gesicht,—forms the anterior part of the Head in Man. It contains the organs of sight, smell, and taste. M. Desmarest is wrong in considering the ear as a part of the face. The length and size of the face are chiefly determined by the dimensions of the organs of the senses, and the degree of intelligence is very often in the inverse ratio of this development. The face is said to be *flat* and *perpendicular* in Man, the Orang-Outang (*Pithecus satyrus*), and the Sloths; sometimes it is *prolonged* into a sort of tube, as in the Echidna, or into a muzzle, as we find in the Dog and most other Mammalia.

(8.) THE JAWS, *Lat.* Maxillæ, Mandibulæ, *Fr.* Les mâchoires, *Germ.* Kiefer, Kinnladen,—composing the upper and under parts of the face, are united together by an articulation, and form the mouth.

(9.) THE UPPER JAW, Lat. Maxilla, seu Mandibula superior, *Fr.* La mâchoire supérieure, *Germ.* Oberkiefer,—is composed of two maxillary bones, and generally of an incisive bone.

(10.) THE INTERMAXILLARY, or INCISIVE BONE, *Lat.* Os intermaxillare, seu incisivum, *Fr.* Les intermaxillaires, præmaxillaires ou incisifs, *Germ.* Zwischenkieferbein,—is a simple or compound bone belonging to the upper jaw, and supporting the incisive or front teeth. It varies in size, being wanting in the adult Man, although found in the human fœtus.

(11.) THE PALATE, *Lat.* Palatum, *Fr.* Les palatins, *Germ.* Gaumen, —is the lower surface of the upper jaw, and forms the roof of the mouth.

(12.) THE LOWER JAW, *Lat.* Mandibula, seu Maxilla inferior, *Fr.* La mâchoire inférieure, *Germ.* Unterkiefer,—which chiefly determines the form of the face, is sometimes arched in front, as in Man; or its two branches meet in front at a more or less acute angle, as in most other Mammalia. Sometimes its branches do not form a single bone, but are separated at their point of contact, as in the Rodentia; at other times it presents an inferior point. The sides of the lower jaw terminate behind in two elevated portions called the *ascending rami or branches* [*Fr.* les branches

montantes], on which are placed the *condyles*, or articulations with the cranium, in the *glenoid cavity*. The condyles are sometimes *transverse*, as in the Carnassiers, or *longitudinal*, as in the Rodentia, and remarkably so in the Ruminantia. The form of the glenoid cavity corresponds to that of the condyles, though in certain genera, as in the Ant-eaters, they disappear altogether.

(13.) THE CHIN, *Lat.* Mentum, *Fr.* Le Menton, *Germ.* Kinn,—forms the anterior and lower margin of the under jaw. It may be more or less *prominent* or *concealed*.

(14.) THE TEETH, *Lat.* Dentes, *Fr.* Les Dents, *Germ.* Zähne,—are small and very hard bones of the mouth, inserted either in the jaws or the palate, and having a free or projecting extremity. Animals are said to be *edentulous*, *Fr.* Edentés, *Germ.* Zahnlose, when the teeth are wanting.

(15.) THE TRUNK, *Lat.* Truncus, *Fr.* Le Tronc, *Germ.* Leib,—is composed of the *spinal column*, the *ribs*, and the *sternum*.

(16.) THE SPINAL COLUMN, *Lat.* Spina dorsi, Gr. 'Ραχις (Rhachis), *Fr.* La Colonne vertébrale, *Germ.* Rükkgrat, Wirbelsäule,—is formed by the union of the small bones of the back, composing a continuous tube of a triangular or circular form for the protection of the spinal marrow. It is divided into several regions.

(17.) The small bones composing the spinal column are called VERTEBRÆ, *Lat.* Vertebræ, *Gr.* Σπόνδυλοι (Spondyli), *Fr.* Les Vertebres, *Germ.* Wirbelbeine, Rükkenwirbel.

(18.) THE RIBS, *Lat.* Costæ, *Fr.* Les Côtes, *Germ.* Rippen,—are elongated bones inserted on the sides of the vertebræ, and converging at their other extremities. They are said to be *sternal* or *true ribs* when they extend as far as the sternum, and are articulated to it by means of a cartilage. The *asternal* or *false ribs* are much shorter than the former, and placed further behind [below in Man].

(19.) THE STERNUM or BREASTBONE, *Lat.* Sternum, *Fr.* Le Sternum, *Germ.* Brustbein,—which may be either simple or compound, is placed between the inferior [anterior in Man] summits of the true ribs. It varies in size in different species, being very large in the Cheiroptera and Moles.

(20.) THE CHEST or THORAX, *Lat.* Thorax, *Fr.* La Cavité thoracique, *Germ.* Brustkasten,—is the anterior cavity of the trunk containing the heart and lungs, and bounded above [behind in Man] by the spinal column, on the sides by the ribs, and beneath [before in Man] by the sternum. Man, the Quadrumana, many Cheiroptera, and the Manatus, have their mammæ placed on this region.

(21.) THE SACRUM, *Lat.* Os Sacrum, *Fr.* L'Os Sacré, *Germ.* Kreuzbein, —consists of those vertebræ adjacent to the tail and connected with other bones. The spinal marrow most commonly terminates here, and the sacrum may either be simple or composed of several bones.

(22.) THE HAUNCH or PELVIS, *Lat.* Pelvis, *Fr.* Le Bassin, *Germ.* Bekken,—serves to protect the hinder part [the lower in Man] of the abdominal cavity. It consists of the sacrum and two ossa innominata, in which are articulations for inserting the bones of the hinder limbs.

(23.) THE OS COCCYGIS or CAUDAL VERTEBRÆ, *Lat.* Os Coccygis seu Vertebræ caudales, *Fr.* L'Os Coccygien ou caudal, *Germ.* Schwanzbein, Schwanzwirbel,—formed of one or more bones annexed to the extremity of the sacrum, serve to support the tail when it happens to be present.

(24.) THE CERVICAL VERTEBRÆ, *Lat.* Vertebræ collares, *Fr.* Les Vertèbres cervicales, *Germ.* Halswirbel,—are the vertebræ placed between the head and the chest.

(25.) THE DORSAL VERTEBRÆ, *Lat.* Vertebræ Pectorales seu dorsales, *Fr.* Les Vertèbres dorsales, *Germ.* Brustwirbel oder Rükkenwirbel, —are the vertebræ of the chest supporting the ribs.

(26.) THE LUMBAR VERTEBRÆ, *Lat.* Vertebræ lumbales, *Fr.* Les Vertèbres lombaires, *Germ.* Lendenwirbel,—are the vertebræ between the chest and the sacrum.

(27.) THE LIMBS or EXTREMITIES, *Lat.* Artus seu Extremitates, *Fr.* Les Membres ou les Extrémités, *Germ.* Gieldmassen,—are articulated bones on each side, connected in pairs with the thorax and pelvis. In general, they are four in number, as in the greater part of the Mammalia, thence called Quadrupeds. Some Mammalia have only two limbs, as the Cetacea, and then the place of the hinder limbs is occupied by a single bone enveloped in the flesh.

(28.) The ANTERIOR or PECTORAL LIMBS, *Lat.* Artus pectorales seu antici. *Fr.* Les Membres antérieurs, *Germ.* Brustgliedmassen, Vordergliedmassen,—are the extremities attached to the thorax, each consisting of a *shoulder-blade*, *arm*, *fore-arm*, and a *hand*, or more commonly a *foot*. These have no distinct articulations with the trunk, but are wholly isolated, except when the collar-bones happen to be present.

(29.) The HINDER or POSTERIOR LIMBS, *Lat.* Artus abdominales seu postici, *Fr.* Les Membres postérieurs, *Germ.* Bauchgliedmassen, Hintergliedmassen,—are the extremities articulated to the pelvis, each consisting of a *thigh*, a *leg*, and most commonly a *foot*, or sometimes a *hand*.

(30.) The SHOULDER, *Lat.* Humerus, *Fr.* L'Epaule, *Germ.* Schulterglied.—is the first articulation or joint of the anterior limbs, comprehending the *shoulder-blade* and *collar-bone*.

(31.) The SCAPULA or SHOULDER-BLADE, *Lat.* Scapula, *Fr.* L'Omoplate, *Germ.* Schulter-blatt,—is the broad and flat bone of the shoulder, placed on the chest towards the spinal column.

(32.) The CLAVICLE or COLLAR-BONE, *Lat.* Clavicula, *Fr.* La Clavicule, *Germ.* Schlüsselbein,—is the other bone of the shoulder, situate between the shoulder-blade and the breast-bone. It is said to be *perfect* in all animals which can raise their fore-limbs to the mouth. In most of the Carnassiers and Rodentia it is *imperfect*; while it is wholly *wanting* in all animals specially intended for walking and running.

(33.) The HUMERUS or SHOULDER-BONE, *Lat.* Brachium seu Os Humeri, *Fr.* L'Humérus, *Germ.* Ober-Arm,—is the second articulation or joint of the fore-limb inserted upon the shoulder-blade, and bearing the fore-arm at its other end.

(34.) The FORE-ARM, *Lat.* Anti-Brachium, *Fr.* L'Avant-Bras, *Germ.* Unter-Arm,—is the third joint of the fore-limb, supported by the shoulder-bone, and articulated to the wrist at the other end. Sometimes it is *simple*, or almost consisting of only one bone, as in the Ruminantia and Solipeda; and sometimes it is *double*. In the latter case, it consists of the *ulna* and *radius*, which may be *free*, and capable of moving one over the other, as in Man and the Apes; or *fixed*, as in the greater part of the Carnassiers and Rodentia.

(35.) The ULNA, *Lat.* Ulna, *Fr.* Le Cubitus, *Germ.* Ellenbogenbein,—is the primary bone of the fore-arm, articulated by a hinge-joint to the shoulder-bone, and having a process or projection at its hinder and upper extremity, forming THE BONE OF THE ELBOW, *Lat.* Olecranon, *Fr.* L'Olécrane, *Germ.* Ellenbogenhökker.

(36.) The RADIUS, *Lat.* Radius, *Fr.* Le Radius, *Germ.* Speiche, Spindel,—the remaining bone of the fore-arm, is sometimes reduced to the rudimental state, and forms merely an apophysis of the ulna, as in the Solipeda and Ruminantia.

(37.) The HAND [in Man], THE FORE HAND [in the Apes], THE FORE PAW, or FORE FOOT [in Quadrupeds], THE FIN [in the Cetacea], *Lat.* Manus seu Pes anticus, *Fr.* La Main, *Germ.* Hand oder Vorderfuss,—consists of all the remaining articulations of the fore-limb taken together, being the *carpus*, *metacarpus*, and *phalanges*.

(38.) The CARPUS or WRIST-BONES, *Lat.* Carpus, *Fr.* Les Os carpiens, *Germ.* Oberhand,—is the basal joint of the hand, nearest to the Ulna, and consisting of several small bones, usually disposed in two rows. They never exceed nine, nor are less than five in number.

(39.) The METACARPUS, *Lat.* Metacarpus, *Fr.* Les Métacarpiens, *Germ.* Mittelhand,—is the joint contained between the wrist and the finger-bones. The number of metacarpal bones is variable, as they usually correspond with the fingers, though sometimes they merely represent a rudimentary finger. There are five of these bones in the hands of Man, the Apes, and the greater part of the Carnassiers, and four in the Hippopotamus and Hogs. There are three in the Horse, a principal one called the CANNON or SHANK-BONE (*Fr.* Le canon), and two rudimentary called the SPLENT-BONES (*Fr.* Les péronés). The Ruminantia have two metacarpal bones united into a single Cannon bone.

The total length of the Carpus and Metacarpus varies in an inverse sense to that of the humerus. Thus, animals with a very short humerus have very long cannon bones.

(40.) The FINGERS or TOES, *Lat.* Digiti, *Fr.* Les Doigts ou Orteils, *Germ.* Finger, Zehen,—usually articulated, form the apex of the hand or fore-foot.

(41.) The PHALANGES, *Lat.* Phalanges, *Fr.* Les phalanges, *Germ.* Fingerglieder,—are the articulations of each finger. Among the quadrupeds, every finger, excepting the thumb, has three phalanges, of which the last supports the nail or hoof. The thumb has only two phalanges, and is often wanting. In the Cetacea, the fingers are formed of a considerable number of flattened phalanges, united together by cartilages, so as to form a kind of fin.

(42.) The LAST PHALANX, *Lat.* Rhizonychium. *Fr.* Le dernier phalange, *Germ.* Klauenglied, Nagelglied,—bears the nail or hoof, and varies in its form and dimensions, according to the figure and disposition of its horny covering.

(43.) The FEMUR or THIGH-BONE, *Lat.* Femur, *Fr.* Le fémur, *Germ.* Schenkel (Hüfte),—is united to the pelvis, and forms the first articulation of the hinder limb. It corresponds to the Humerus of an anterior extremity.

(44.) The SHIN-BONE or TIBIA, *Lat.* Tibia, *Fr.* Le tibia, *Germ.* Schiene (Schenkel).—is the second articulation of a hinder limb. It is supported by the femur, and articulated to a foot or hand at its other extremity.

(45.) The FIBULA, *Lat.* Fibula, *Fr.* Le péroné, *Germ.* Wadenbein, —is a long bone, sometimes added to the Tibia, though often wanting.

The Tibia and Fibula correspond to the Ulna and Radius of the Fore-arm, and present the same variations. Sometimes these bones are very distinct, and moveable one over the other, as in the Apes and Makis. Sometimes they are distinct, though but slightly moveable. Most commonly the Fibula is the mere rudiment of a bone. In most Mammalia, the Tibia and Fibula together form the LEG. In the Horse and other digitigrade Quadrupeds, this is improperly styled THE THIGH.

(46.) The FOOT [in Man]. THE HINDER-HAND [in the Apes], THE HINDER-PAW or HINDER-FOOT [in Quadrupeds], THE HAND [in the Opossums], *Lat.* Pes, *Fr.* Le pied, *Germ.* Fuss, Hinter-fuss,—includes all the remaining articulations of a hinder-limb, being the *Tarsus*, *Metatarsus*, and *Phalanges*.

(47.) The TARSUS or INSTEP, *Lat.* Tarsus, *Fr.* Le tarse, *Germ.* Oberfuss, Fusswurzel,—the basal joint of the foot, nearest to the Tibia, consists of several bones, never exceeding seven in number. In the Horse, this is called THE HOCK.

(48.) The METATARSUS, *Lat.* Metatarsus, *Fr.* Le métatarse, *Germ.* Mittelfuss,—is the second joint of the foot, between the Tarsus and Phalanges. It never consists of more than five bones corresponding to the toes, or of less than two, which, however, may sometimes become united together, so as to form one bone. It consists, in the Horse, of a CANNON or SHANK-BONE, and SPLENT-BONES, as in the Metacarpus.

(49.) The TOES and PHALANGES may be compared to the corresponding parts of the anterior limb. See (40.) (41.) and (42.)

(50.) The ELBOW, *Lat.* Cubitus, *Fr.* Le coude, *Germ.* Ellenbogen,—is at the junction of the Humerus and Fore-arm.

(51.) The WRIST, *Lat.* Flexura, *Fr.* Le poignet, *Germ.* Handbeuge,—is at the junction of the fore-arm and carpus. In the Horse, and other digitigrade quadrupeds, this is very improperly called THE KNEE.

(52.) The KNEE, *Lat.* Genu, *Fr.* Le genou. *Germ.* Knie,—is at the union of the Femur and Tibia. In the Horse, it receives the name of the STIFLE JOINT.

(53.) The PATELLA or KNEE-PAN, *Lat.* Patella seu rotula, *Fr.* La rotule, *Germ.* Kniescheibe,—is a small isolated bone in front of the knee.

(54.) The HAM, *Lat.* Poplites, *Fr.* Le jarret, *Germ.* Kniekehle,—is the hollow part at the back of the knee, in Man and the Apes. This part is concealed in the Horse, and all digitigrade quadrupeds.

(55.) The ANKLE-JOINT, *Lat.* Suffrago, *Fr.* L'articulation de la cheville, *Germ.* Fussbeuge, Hakkengelenk,—is at the union of the Tibia and Tarsus. This is the HOCK-JOINT of HINDER-KNEE of the Horse.

(56.) The HEEL, *Lat.* Calcaneus, Talus. seu Calx, *Fr.* Le Talon, *Germ.* Hakken, Ferse,—is the hindermost point of the Tarsus. In the Horse, this is called the POINT OF THE HOCK.

(57.) The ANKLE, *Lat.* Malleolus, *Fr.* La cheville, *Germ.* Knöchel,—is the inner process or projection at the end of the Tibia.

(58.) The PASTERN or FETLOCK-JOINT, *Fr.* L'articulation du fanon, ou le boulet, *Germ.* Hüfhaaregelenk, in digitigrade quadrupeds,—is the joint at the extremity of the metatarsus in the hinder-leg, or of the metacarpus in the fore-leg.

(59.) The UPPER PASTERN, *Lat.* Os Suffraginis, *Fr.* Le pâturon. *Germ.* Fessel,—in digitigrade quadrupeds, is the second bone from the hoof, adjoining the fetlock, corresponding to the first phalanx in Man.

(60.) The LOWER PASTERN or CORONET-BONE, *Lat.* Os coronæ, *Fr.* La couronne, *Germ.* Krone,—in digitigrade quadrupeds, is the bone next to the hoof, corresponding to the second phalanx in Man.

(61.) The COFFIN-BONE, *Lat.* Os pedis, *Fr.* L'Os du sabot, *Germ.* Hüfbein,—is the bone of the hoof, analogous to the last phalanx of the finger in Man.

II.—THE HEAD IN GENERAL.

(62.) The VERTEX, or TOP OF THE HEAD, *Lat.* Vertex, *Fr.* Le vertex. *Germ.* Scheitel,—is the highest portion of the skull, in a line drawn between the ears, perpendicularly upwards.

(63.) The SINCIPUT, or FORE-PART OF THE HEAD, *Lat.* Sinciput. *Fr.* Le sinciput, *Germ.* Vorderkopf,—is that portion of the head reaching from the vertex to the eyes.

(64.) THE OCCIPUT, or HINDER-PART OF THE HEAD, *Lat.* Occiput, *Fr.* L'occiput, *Germ.* Hinterkopf,—is that portion of the skull extending from the vertex backwards to the cervical vertebræ.

(65.) THE FACE, *Lat.* Vultus, *Fr.* La face, *Germ.* Antlitz,—placed at the anterior part of the skull, contains most of the organs of sense. See (7.)

(66.) THE FOREHEAD, *Lat.* Frons, *Fr.* Le front, *Germ.* Stirn,—is that portion of the Sinciput, extending from the eyes to the anterior margin of the vertex. It is said to be

(67.) PROPORTIONATE, *Lat.* Proportionata, *Fr.* Proportionné, *Germ.* Ebenmässige stirn,—when it occupies a third part of the length of the face ;

(68.) HIGH, *Lat.* Alta, *Fr.* Haut, *Germ.* Hohe, lange,—when it is longer than the third-part ; and

(69.) LOW, *Lat.* Brevis, *Fr.* Bas, *Germ.* Kurze,—when shorter. The Forehead is *very open* in Man and some Apes.

(70.) THE MUZZLE, *Lat.* Rostrum, *Fr.* Le museau, *Germ.* Schnauze, —is the prolongation of the face.

(71.) THE FACIAL ANGLE, *Lat.* Angulus facialis, *Fr.* L'angle facial, *Germ.* Gesichtswinkel,—is the angle, more or less acute, formed between two imaginary lines, the one drawn from the external hole of the ear to the extreme point of the upper-jaw next to the teeth, and the other from the latter point as a tangent to the most prominent part of the forehead. This angle is seldom measured except in Man and the Apes. In the former it varies from 90° to 70°, and in the latter from 65° to 30°. Of all Mammalia, the Orang-Ouang has, next to Man, the most open facial angle, and the Great Ant-eater (*Myrmecophaga jubata*), the most acute.

(72.) THE TEMPLES, *Lat.* Tempora, *Fr.* Les tempes, *Germ.* Schläfen,— are the portions of the head on each side of the forehead, situate above a line drawn from the eye to the ear.

(73.) THE CHEEK, *Lat.* Bucca, *Fr.* La joue, *Germ.* Bakke,—is that portion of the face extending from the corners of the nose and mouth to the ear.

(74.) THE UPPER-CHEEK, *Lat.* Gena, *Fr.* La joue supérieure, *Germ.* Wange,—is that portion of the cheek between the eye and the ear, immediately covering the zygomatic arch.

(75.) THE UNDER-CHIN, *Lat.* Ingluvies, *Fr.* La partie inférieure du menton, *Germ.* Unterkinn,—is that portion of the lower jaw between the external margin of its branches and the throat.

(76.) THE UNDER-CHEEK, *Lat.* Mala, *Fr.* La partie inférieure de la joue, *Germ.* Kinnbakke,—is the hinder-part of the lower jaw, extending beneath a line drawn from the corner of the mouth to the ear, and thence downwards to the lower margin of the face.

(77.) THE PAROTID REGION, *Lat.* Regio parotica, *Fr.* La région parotique, *Germ.* Ohrengegend,—is the part of the head round the ear.

(78.) THE OPHTHALMIC REGION, *Lat.* Regio ophthalmica, *Fr.* La région ophthalmique, *Germ.* Augengegend,—is the region around the eyes.

(79.) THE NASAL REGION, *Lat.* Regio nasalis, *Fr.* La région nasale, *Germ.* Nasengegend,—is that portion of the face around the nose.

(80.) THE ORAL REGION, *Lat.* Regio oris, *Fr.* La région orale, *Germ.* Mundgegend,—is the part round the mouth.

(81.) THE SUPERCILIARY RIDGES, *Lat.* Cristæ superciliares, *Fr.* Les crètes sourcilières,—are projections of the frontal bone, placed horizontally over the orbits of the eyes. These are found in certain Apes.

(82.) THE SAGITTAL RIDGES, *Lat.* Cristæ sagittales, *Fr.* Les crètes sagittales,—are found on the top of the head at the upper part of the parietal bone when single, or at the junction of the parietal bones. They are found particularly among the Carnassiers.

(83.) THE OCCIPITAL RIDGES, *Lat.* Cristæ occipitales, *Fr.* Les crètes occipitales,—are placed transversely on the occipital bone; and form the point of attachment for the muscles which raise the head, as well as for the cervical ligament in the Apes, Carnassiers, Ruminantia, Solipeda, and others.

In respect to its form and size the HEAD may be

(84.) ROUND, *Lat.* rotundum, *Fr.* arrondie, as in Man and most Apes ;

(85.) LENGTHENED, *Lat.* elongatum, *Fr.* alongée, as in the Horse ;

(86.) GREATLY LENGTHENED, *Lat.* prælongum, *Fr.* très alongée, as in the Great Ant-eater ;

(87.) PYRAMIDAL, as in the Howling Apes ;

(88.) VERY LARGE, *Lat.* prægrande, *Fr.* démesurément grosse, as in the Whales, Cachalots, and Elephants ;

(89.) MIDDLE-SIZED, *Lat.* medium, *Fr.* moyenne, as in the Dog ;

(90.) SMALL, *Lat.* parvum, *Fr.* petite, as in the Ai ;

(91.) FLATTENED, *Fr.* aplatie, and

(92.) DUCK-BILLED, *Fr.* En bec de canard, as in the Ornithorhynchus.

(93.) THE ORBITAL FOSSÆ, *Lat.* Fossæ orbitales, *Fr.* Les fosses orbitaires,—are the cavities in the skull, for the reception of the eyes. They may be

(94.) ANTERIOR, *Fr.* antérieures, as in Man and the Apes ;

(95.) LATERAL, *Fr.* latérales, as in the Rodentia.

(96.) THE TEMPORAL FOSSÆ, *Lat.* Fossæ temporales, *Fr.* Les fosses temporales,—are the depressions of the temples. Sometimes they are

(97.) DISTINCT, *Fr.* distinctes,—when they are separated from the orbital fossæ, as in Man and the Apes ;

(98.) COMMUNICATING, *Fr.* communiquans,—when they are united to the orbital fossæ by the bottom of the latter, as in the Horse.

(99.) MARGINED, *Fr.* marginées,—when the orbital and temporal fossæ have a common margin, as in the Carnassiers and Rodentia.

(100.) THE NASAL FOSSÆ, *Lat.* Fossæ nasales, *Fr.* Les fosses nasales, —are the holes in the skull, corresponding to the apertures of the nose.

III.—THE ORGAN OF HEARING.

(101.) THE EAR, *Lat.* Auris, *Fr.* L'oreille, *Germ.* Ohr,—is the organ of hearing. See pages 8 and 35.

(102.) THE HOLE OF THE EAR, *Lat.* Meatus auditorius externus, *Fr.* Le conduit auditif externe, *Germ.* Gehörgang,—is the tube of the ear opening externally.

(103.) THE AURICLE or EXTERNAL EAR, *Lat.* Auricula, seu Concha, *Fr.* La conque externe, *Germ.* Ausseres Ohr, Ohrmuschel,—is a hollow cartilaginous cavity, for conveying the vibrations of the air into the internal ear. It is wholly wanting in the Cetacea, the Seals, the Rat-Moles, Common Moles, and some others.

(104.) THE HELIX, *Lat.* Heligma, seu Helix, *Fr.* L'hélix, *Germ.* Ohrleiste,—is the outer and hinder margin of the Ear, usually convoluted.

(105.) THE ANTHELIX, *Lat.* Anthelix, *Fr.* L'anthelix, *Germ.* Gegenleiste,—is the inner margin of the Ear, running almost parallel with the helix.

(106.) THE TRAGUS, *Lat.* Tragus, *Fr.* L'oreillon ou le tragus, *Germ.* Ohr Ekke,—is the projection at the anterior margin, immediately before the hole of the ear. It assumes an enormous size in some Bats.

(107.) THE ANTITRAGUS, *Lat.* Antitragus, *Fr.* L'anti-tragus, *Germ.* Gegen Ekke,—is the hinder process of the ear, opposite the Tragus.

(108.) THE LOBE, *Lat.* Lobulus, *Fr.* Le lobule, *Germ.* Ohrläppchen, —is the lowest part of the ear below the Tragus.

The AURICLE is said to be

(109.) OPERCULATED, *Lat.* Auricula operculata, *Fr.* L'oreille operculée, *Germ.* Gedekkeltes Ohr,—when the tragus lines the ear, so that it appears to be double ;

(110.) MARGINATED, *Lat.* Auricula marginata, *Fr.* L'oreille rebordée, *Germ.* Gerandetes Ohr,—when it is supplied with a convoluted helix ;

(111.) RUDIMENTARY, *Lat.* Auricula abscondita, *Fr.* L'oreille rudimentaire, *Germ.* Verstekkte Ohren,—when it is almost concealed, as in the Marmot.

(112.) ROUNDED, *Fr.* arrondie, and

(113.) APPLIED TO THE HEAD, *Fr.* appliquée contre la tête,—as in Man, and such of the Apes as most resemble him ;

(114.) ANGULAR, *Fr.* anguleuse,—as in the Macacos and Baboons ;

(115.) HORN-SHAPED, *Fr.* En cornet,—with the opening in front, and the base enlarged, as in Cats, Dogs, and Weasels ;

(116.) PEDUNCULATED, *Lat.* Auricula pedonculée,—in the form of an elongated horn, with a kind of branch, which gives its great mobility, as in the Ruminantia, the Horse, and Rhinoceroses.

(117.) THE AURICULAR OPERCULUM, *Lat.* Operculum auriculare, *Fr.* L'oreillon, *Germ.* Ohrdekkel. This term is applied to the tragus, when it is elongated so as almost to cover the auricular cavity.

IV.—THE ORGAN OF VISION.

(118.) THE EYE, *Lat.* Oculus, *Fr.* L'œil, pl. les yeux, *Germ.* Auge, —is the organ of vision.

(119.) THE EYE-BALL, *Lat.* Bulbus oculi, *Fr.* Le globe de l'œil, *Germ.* Aug Apfel,—is the body of the eye, more or less globular, composed of membranes and humours.

(120.) THE CONJUNCTIVE MEMBRANE, *Lat.* Tunica conjunctiva seu adnata, *Fr.* La conjonctive, *Germ.* Verbindende augenhaut,—is the anterior membrane of the eye-ball, being a continuation of the skin of the eye-lids.

(121.) THE EYE-LIDS, *Lat.* Palpebræ, *Fr.* Les paupières, *Germ.* Augenlieder,—are moveable cutaneous coverings, enveloping the whole or a part only of the eye.

(122.) THE SCLEROTICA, or WHITE OF THE EYE, *Lat.* Tunica sclerotica, *Fr.* La sclérotique, *Germ.* Weisse Augenhaut,—is a firm and white membrane, covering the Eye-ball, and seen partially on its external surface.

(123.) THE CORNEA, *Lat.* Tunica cornea, *Fr.* La cornée, *Germ.* Hornhaut,—is that transparent anterior membrane or coat of the eye filling the circular aperture of the sclerotica.

(124.) THE IRIS, *Lat.* Iris, seu Tunica iridea, *Fr.* L'iris, *Germ.*

Regenbogenhaut,—is the coloured circle of the eye, seen through the Cornea It varies in colour, from light-blue to yellow, or deep orange, and is most commonly of a deep yellow or brown colour.

(125.) THE PUPIL, *Lat.* Pupilla, *Fr.* La pupille, *Germ.* Sehe,—is the dark central disc of the eye, surrounded by the iris. When fully dilated it is most commonly round.

(126.) THE ORBIT, *Lat.* Orbita, *Fr.* L'orbite, *Germ.* Augenhöle,—is the external margin of the cavity of the skull, destined to contain and protect the eye-ball.

(127.) THE APERTURE OF THE EYE, *Lat.* Apertura oculi, *Fr.* L'ouverture de l'œil, *Germ.* Augen Offnung,—is the space occupied by the eye-ball, and appearing externally when the eye-lids are drawn back.

(128.) THE ANGLES OF THE EYES, *Lat.* Canthi oculorum, *Fr.* Les angles des yeux, *Germ.* Augenwinkel,—are the corners formed on each side by the joining of the eyelids.

(129.) THE INTERNAL OR NASAL ANGLE, *Lat.* Canthus nasalis, *Fr.* L'angle intèrieur ou nasal, *Germ.* Nasenwinkel,—is the inner corner of the eye nearest to the nose.

(130.) THE EXTERNAL OR TEMPORAL ANGLE, *Lat.* Canthus temporalis, *Fr.* L'angle extérieur, *Germ.* Schläfenwinkel,—is the outer corner of the eye nearest to the ear.

(131.) THE NICTITATING MEMBRANE, *Lat.* Membrana nictitans, *Fr.* La troisième paupière, *Germ.* Blinzhaut,—is a cutaneous covering of the eye placed at the nasal angle, and capable of covering it like a curtain.

(132.) THE LACHRYMAL FOSSA, *Lat.* Fossa lacrymalis, *Fr.* La fosse lachrymale, *Germ.* Thränengrube.—is the dilated upper extremity of a duct, in the nasal angle of the eye, for conveying the tears from the eye to the nose.

(133.) THE SUB-ORBITAL SINUS, *Lat.* Sinus suborbitalis, *Fr.* Le Larmier,—is a naked furrow beneath the eye, secreting a peculiar humour, as in the Antelopes.

The EYES are said to be

(134.) RUDIMENTARY, *Fr.* Rudimentaires,—when they are not visible externally, and some minute traces alone can be discovered beneath the skin, as in the Blind-rat (*Spalax typhlus*);

(135.) APPARENT, *Fr.* Apparens,—when they are visible externally, as most commonly happens in the other Mammalia.

In respect to their size, the EYES may be

(136.) VERY LARGE, *Fr.* Très-grands,—in many nocturnal animals, and several aquatic species, as the Galagos, Hares, Flying-Squirrels, Seals, and Others;

(137.) MEDIUM SIZE, *Fr.* mediocres ou moyens,—as in most terrestrial quadrupeds; or

(138.) SMALL or VERY SMALL, *Fr.* Petits ou très-petits,—as in subterraneous species, such as the Moles, and Cape-Moles (*Bathyergus*), or in some nocturnal species, as the Bats.

The EYES may vary in their relative position, and are said to be

(139.) ANTERIOR, *Fr.* Antèrieurs,—when they are directed in front, and more or less approaching to each other, so that their visual axes are nearly parallel, as in Man and the Apes; or

(140.) LATERAL, *Fr.* Latéraux,—when they are widely separate, and placed on the sides of the head, as in the Hares and other Rodentia, and generally in most herbivorous animals, where the eyes have nearly the same visual axis.

The CORNEA may have different degrees of projection. It is

(141.) VERY CONVEX, *Fr.* Très bombé,—in the nocturnal species, as the Galagos, also in the Tapir and Hare;

(142.) ORDINARY, *Fr.* Ordinaire,—as in most diurnal Mammalia; and

(143.) FLAT, *Fr.* Plat.—as in species which are habitually immersed in water, such as the Seals and Cetacea.

(144.) THE CARPET OF THE EYE, *Lat.* Tapetum lucidum, *Fr.* Le tapis,—is a portion of the choroid coat, situate at the bottom of the eye, opposite to the point where the optic nerve enters. It is variously coloured in different Mammalia. See page 35.

V.—THE ORGAN OF SMELLING.

(145.) THE NOSE, *Lat.* Nasus, *Fr.* Le nez, *Germ.* Nase,—is the organ of smelling. See pages 8 and 36.

(146.) THE NOSTRILS, *Lat.* Nares, *Fr.* Les narines, *Germ.* Nasenlöcher,—are the two external orifices of the nose for admitting the air. These are said to be

(147.) CLOSE, *Fr.* Peu ouvertes,—when they consist of simple clefts but slightly open, as in the Apes, some Carnassiers, and Rodentia;

(148.) CAVERNOUS, *Fr.* Caverneuses,—when they open into large cavities, as in the Horse, the Ass, and Hippopotamus;

(149.) SPIRAL, *Fr.* En spirale,—when they are convoluted, as in the Makis;

(150.) OPERCULATED, *Fr.* Operculées—when they are closed by a lid, as in some Rats, especially of the genus Nycteris;

(151.) OBSERATE, *Lat.* Obseratæ,—when the nostrils can be closed by muscles at the will of the animal, as in the Seals.

(152.) THE SPIRACLES, *Lat.* Spiracula, *Fr.* Les évents, *Germ.* Luftlöcher, in *Dutch* Lugtstippen,—are nostrils united together, and placed at the top of the head, as in the Whales, through which these animals discharge the enormous quantity of water swallowed while pursuing their prey.

(153.) THE PARTITION OF THE NOSTRILS, *Lat.* Dissepimentum seu septum narium, *Fr.* La division des narines, *Germ.* Nasenscheidewand,—is found in most Mammalia.

(154.) THE EXTERNAL NOSE, *Lat.* Nasus externus, *Fr.* Le nez extérieur, *Germ.* Aussere Nase,—is the external part of the face containing the nostrils.

(155.) THE BULB OF THE NOSTRIL, *Lat.* Pterygium, Rima, *Fr.* La bulbe du nez, *Germ.* Nasenflügel,—is the expanded part on each side of the external wall of the nose.

(156.) THE SNOUT, *Lat.* Rhinarium, *Fr.* Le museau, *Germ.* Nasenkuppe,—is the extreme part of the nose, distinguished by a smooth granular and moistened surface, on the sides of which the nostrils are commonly placed.

(157.) THE CHILOMA, *Lat.* Chiloma, *Fr.* Le mufle, *Germ.* Maul,—forms the projecting muzzle of some Ruminants. It is wanting in the Sheep, Goats, Camels, some Stags, the Musk-Ox, and a few others; in most of the remainder it swells outwards, and comprises the upper lip and the part of the nose immediately adjoining.

The NOSE, which is usually placed in the middle of the face, may be

(158.) PROMINENT, *Lat.* Prominulus, *Fr.* Proéminent, *Germ.* vorragende,—when it projects beyond the upper lip, as in Man, and the Proboscis Monkey (*Nasalis larvatus*), in which cases the nostrils are *inferior*;

(159.) FLATTENED, *Lat.* Impressus, *Fr.* Camus, *Germ.* Gepletschte,—when it is depressed within the upper lip, and begins to resemble a muzzle, as in most of the Apes;

(160.) SPREADING, *Lat.* Repandus, *Fr.* Repandu, *Germ.* Verbreitete,—when its extremity is broader than the rest of the nose;

(161.) POINTED, *Fr.* Pointu,—when the head is narrowed in front, so as to make the nose entirely terminal, as in the Makis, Moles, and Ant-eaters;

(162.) SHORT, *Lat.* Abbreviatus, *Fr.* Court, *Germ.* Kürze,—when the head is not prolonged, so that the nose scarcely projects, as in the Cats, and most Rodentia;

(163.) TUBULAR, *Lat.* Tubulosus, *Germ.* Röhrige,—when the nose is elongate, inclosed on every side, and having no perceptible bulb;

(164.) HOOKED, *Lat.* Resimus, *Fr.* Crochu, *Germ.* Ramsnase, Umgebogne Nase,—when it is curved downwards, so that the ridge of the nose forms an arch;

(165.) TURNED-UP, *Lat.* Simus, *Germ.* Stülpnase, Aufgebogne Nase, —when it is curved upwards, so that the upper ridge appears hollow;

(166.) SIMPLE, *Lat.* Simplex, *Fr.* Simple, *Germ.* Einfache,—when the nose is destitute of any remarkable appendage or sinuosity, as in most Mammalia; and

(167.) COMPLICATED, *Lat.* Complex, *Fr.* Compliqué,—when it is ornamented with naked membranes, more or less developed.

(168.) THE NASAL APPENDAGE, *Lat.* Prosthema, *Fr.* La feuille membraneuse, *Germ.* Nasen-Ansatz,—this term is applied to the leaf-like membrane superadded to the nose, found in many Genera of Bats. The nasal appendage is said to be

(169.) FOLIATED, *Lat.* Foliatum, *Fr.* Folliculée, *Germ.* Geblätterte,—when it is shaped like a simple leaf;

(170.) CORDATE, *Lat.* Cordatum, *Fr.* En forme de cœur, *Germ.* Geherzte.—when in the form of a heart;

(171.) FUNNEL-SHAPED, *Lat.* Infundibuliforme, *Fr.* Infundibulifere, *Germ.* Trichterförmige—when it resembles the funnel of a chimney;

(172.) HASTATE, *Lat.* Hastatum, *Fr.* En forme de fer-de-lance, *Germ.* Spiessförmige,—when it assumes the form of a lance;

(173.) LYRATE, *Fr.* En forme de lyre,—when it is shaped like a lyre;

(174.) CRISTATE, *Lat.* Cristatum, *Germ.* Kammrandige,—when the margins of the nostrils are surrounded with small folds or crests; and

(175.) STELLATED, *Fr.* En forme d'étoile,—when the nostrils are surmounted with a membrane in the form of a star.

The NOSE is said to be

(176.) PROBOSCIDEAL, *Lat.* Proboscideus, *Fr.* En forme de trompe, *Germ.* Rüsselförmige,—when it extends slightly beyond the point of the jaw and is moveable, as in the Coatis (*Nasua*);

(177.) THE PROBOSCIS, *Lat.* Proboscis, *Fr.* La trompe, *Germ.* Rüssel, —is a very long and moveable muzzle, as in the Elephants and Tapirs.

(178.) THE CHANFRIN, *Fr.* Le chanfrein,—is the upper part of the nose, comprised between the forehead and the nostrils. It is observed to be arched upwards in the Sheep; curved in an opposite direction in

27

the Goats; armed with one or two horns in the Rhinoceros, and furrowed longitudinally, as in the Bats of the Genus Nycteris.

VI.—THE MOUTH AND ORGANS OF TASTE.

(179.) The Cavity of the Mouth, *Lat.* Cavum oris, *Fr.* La cavité de la bouche, *Germ.* Mundhöhle,—is the hollow place formed by the jaws and cheeks, commonly divided into three portions; the *superior*, between the tongue and the palate; the *inferior*, between the tongue and the lower jaw; and the *anterior*, between the tongue and the teeth or lips.

(180.) The Lips, *Lat.* Labia, *Fr.* Les lèvres, *Germ.* Lippen,—are the extremities of the skin, upon the external margins of the jaws, distinguished into the *upper* and *lower lip.*

(181.) The Corners of the Mouth, *Lat.* Anguli oris, *Fr.* Les angles de la bouche, *Germ.* Mundwinkel,—are the angles, formed at the points of union of the lips.

(182.) The opening of the Jaws, *Lat.* Rictus, *Fr.* La gueule, *Germ.* Mundöffnung,—is the distance of one jaw from the other when the mouth is distended.

(183.) The Cheek-pouches, *Lat.* Sacculi buccales, Buccæ saccatæ, seu Thesauri, *Fr.* Les abajoues, *Germ.* Bakkentaschen,—are cutaneous sacs in each cheek within the cavity of the mouth, and fitted for holding food.

(184.) The Gums, *Lat.* Tomia, *Fr.* Les Gencives, *Germ.* Ladenranden,—are the margins of the jaws within the mouth, adapted for mastication, and on which the teeth are most commonly placed.

(185.) The Alveolar Cavities, *Lat.* Alveoli, *Fr.* Les fosses alvéolaires, *Germ.* Zahnhöhle,—are the depressions of the jaws, into which the teeth are inserted.

(186.) The Tongue, *Lat.* Lingua, *Fr.* La langue, *Germ.* Zunge,—the principal organ of taste, supplied with nervous papillæ (see p. 36), is most commonly fleshy and flexible. It may be

(187.) Medium-sized, Oval, and Flat, *Fr.* Médiocre, ovale, et a-platie,—as in Man, the Apes, and many other Mammalia ;

(188.) Long, and very thin, *Fr.* Longue et très mince,—as in most Carnassiers, especially the Dogs and Cats ;

(189.) Long and thick, *Fr.* Longue et épaisse,—as in the Horse and Ruminantia, in which animals it serves to pluck the herbage ;

(190.) Very long and vermiform, *Fr.* Très-longue et vermiforme,—as in the Ant-eaters, Armadilloes, and Orycteropus:

In respect to its movements, the Tongue may be

(191.) Extensible, *Fr.* Extensible,—in a greater or less degree, as in Man, the Apes, the Carnassiers, Rodentia, Pachydermata, and Ruminantia ;

(192.) Very extensible, or Protractile, *Fr.* Très extensible, ou protractile,—as in the Ant-eaters, Orycteropus, Armadilloes, and the Bats of the Genus Glossophaga;

(193.) Fixed, *Fr.* fixée,—by the entire of its lower surface, as in the Cetacea.

The surface of the Tongue may be

(194.) Smooth, *Fr.* Douce,—when the papillæ, with which it is covered, are fine and soft, as in Man, the Apes, Dogs, Ant-eaters, Cetacea, and many others ;

(195.) Rough, *Fr.* Rude,—when the papillæ are horny, and have their points directed backwards, as in the Cats, Civets, Opossums, and the Bats of the Genera Phyllostoma and Pteropus ;

(196.) Scaly, *Fr.* Ecailleuse,—when its sides are protected with large scales, having two or three points terminating in an angle, as in the Porcupines ;

(197.) Funnel-shaped,—when its point terminates in a disc, shaped like a cupping-glass, as in the Bats of the Genus Glossophaga; and

(198.) Furrowed, *Fr.* Sillonnée,—when its upper surface is marked with a longitudinal furrow.

VII.—THE TEETH,

IN RESPECT TO THEIR TEXTURE, PARTS, AND FORM.

(199.) The Ivory, *Lat.* Substantia ossea, *Fr.* L'ivoire, *Germ.* Knochenmasse,—is the central or bony part of the teeth, usually constituting the principal part of its substance.

(200.) The Proper Ivory, *Lat.* Ebur, *Fr.* L'ivoire proprement dit, *Germ.* Elfenbein,—is the same bony substance, when composed of conical layers, as in the Tusks of the Elephant.

(201.) The Enamel, *Lat.* Substantia vitrea, *Fr.* L'émail, *Germ.* Schmelz,—is the white, hard, and dense substance covering the teeth externally, or intersecting them internally.

(202.) The Cortex, *Lat.* Indumentum corticale, *Germ.* Zahnkütt,—is a substance of less density, covering the enamel, or connecting its interstices in compound teeth.

(203.) The Whale-bone, *Lat.* Elasmia, *Fr.* Les fanons ou barbes, *Germ.* Barten,—horny laminæ in place of teeth, hanging transversely from the sides of the palate.

(204.) The Root of a Tooth, *Lat.* Radix, *Fr.* La racine, *Germ.* Zahnwurzel,—is the lower part of a tooth placed within the alveolar cavity and the gum, and most commonly destitute of enamel.

In respect to its Root, a Tooth may be

(205.) Mono-rhizal, Di-rhizal, Tri-rhizal, or Poly-rhizal,—according as it has one, two, three, or many roots ;

(206.) Cœlorhizal,—when its root is hollow ; or

(207.) Stereorhizal,—when its root is solid.

(208.) The Crown of a Tooth, *Lat.* Corona dentis, *Fr.* La couronne, *Germ.* Zahnkrone,—is the part of the tooth beyond the alveolar cavity and gum, serving for mastication.

(209.) The Neck of a Tooth, *Lat.* Collum dentis, *Fr.* Le Collet, *Germ.* Zahnkranz,—is the interval more or less distinct, separating the crown from the root of a tooth.

(210.) A Simple Tooth, *Lat.* Dens obductus, *Fr.* Dent simple, *Germ.* Uberlegter Zahn, Einfacher Zahn,—has its ivory entirely covered with enamel, but only on the exterior. It consists of a Root, a Neck, and a Crown, and being always of a determinate form, ceases to grow after having left the bulb whence it originated.

(211.) A Compound Tooth, *Lat.* Dens lamellosus, *Fr.* Dent composée, *Germ.* Blättriger Zahn,—has its ivory intersected with folds of enamel in every direction, so that it appears to be formed of perpendicular laminæ.

(212.) A Complicated Tooth, *Lat.* Dens complicatus, *Fr.* Dent demi-composée, *Germ.* Schmelzfaltiger Zahn,—along with a simple root, has its crown more or less intersected with folds of enamel, but not so far as to separate the tooth into laminæ.

(213.) The Ridge of a Tooth, *Lat.* Machæris, *Fr.* Le bord tranchant, *Germ.* Schmelzleiste,—is the sharp external line of a fold of enamel, emerging from the crowns of a compound or complicated tooth.

(214.) A Fibrous Tooth, *Lat.* Dens fibrosus, *Fr.* Dent fibreuse, *Germ.* Fasriger Zahn,—is composed of fibres or longitudinal tubes, resembling the stalk of a reed.

(215.) The Cusps or Points, *Lat.* Cuspis, *Fr.* Les points, *Germ.* Zakke,—are the sharp points on the crown of a tooth.

(216.) The Tubercles, *Lat.* Tuberculum, *Fr.* Les tubercules, *Germ.* Zahnhökker,—are the small blunt planes on the crown of a tooth.

A Tooth is said to be

(217.) Tearing, *Lat.* Sectorius, *Fr.* Déchirante, *Germ.* Reiss Zahn,—when it terminates in a sharp point cutting unequally ;

(218.) Conical, *Fr.* Conique,—when it varies in form from the Cylinder to the Oval or Ellipse (see page 88) ;

(219.) Trenchant, *Lat.* Incisorius, *Fr.* Tranchante, *Germ.* Schneide Zahn,—when it terminates in a sharp edge cutting equally (see page 88) ;

(220.) Chisel-shaped, *Lat.* Cestriformis, *Fr.* En ciseau-à-tailler, *Germ.* Meisselförmiger Zahn,—when a long and narrow trenchant tooth terminates in a thin edge, hollowed or scooped out on one side ;

(221.) Wedge-shaped, *Lat.* Acutatus, *Fr.* En biseau, *Germ.* Zuge-schärfter,—when a trenchant tooth is cut obliquely off at its extremity, as in the upper Incisors of most Rodentia, and the lower Incisors of some.

(222.) The Apex, *Lat.* Scalprum, *Fr.* Le coin, *Germ.* Schneide,—is the point of the crown in a tearing, trenchant, chisel-shaped, or wedge-shaped tooth.

A Tooth is said to be

(223.) Unicuspidate, Bicuspidate, Tricuspidate, or Multicuspidate,—according as its crown ends in one, two, three, or many cusps ;

(224.) Tuberculous, *Lat.* Tuberculatus, *Fr.* Tuberculeuse, *Germ.* Hökkriger,—when it is furnished with Tubercles or small blunt planes on its crown (see page 88) ;

(225.) Cuspidate, *Lat.* Cuspidatus, *Germ.* Zakkiger,—when its crown is supplied with many cusps ;

(226.) Ridged, *Lat.* Rugosus, *Fr.* à collines transverses, *Germ.* Runzliger,—when its crown is armed with several elevated ridges ;

(227.) Plane, *Lat.* Lævis, inermis, *Fr.* Plane, *Germ.* Glatter,—when its crown is smooth, and without cusps, tubercles, or ridges ;

(228.) Growing Indefinitely, *Lat.* Auctus, *Fr.* Poussant, *Germ.* Erweiterter,—when the crown continues to be pushed outwards on the sides, before, or behind, from its interior part, during the entire life of the animal, as in the Incisors of the Rodentia.

Teeth are said to be

(229.) Homogeneous, *Lat.* Homogenei, *Germ.* Gleichartige Zähne,—when they all resemble each other in form and texture ; and

(230.) Heterogeneous, *Lat.* Heterogenei, *Germ.* Ungleichartige Zähne,—when they differ in form and texture.

VIII.—THE TEETH,

IN RESPECT TO THEIR INSERTION AND POSITION.

A Tooth is said to be

(231.) Separable, *Lat.* Injunctus, *Fr.* Séparable, *Germ.* Eingekeilter Zahn,—when it has a distinct root inserted into an alveolar cavity of the jaw, and capable of being separated from it, as in the teeth of most Mammalia;

(232.) Inseparable or Innate, *Lat.* Innatus, *Fr.* Inséparable, *Germ.* Eingewachsner Zahn,—when the tooth is inserted in its alveolar cavity, in such a way that it appears to be a continuous process or excrescence of the jaw-bone, so that it cannot be separated from it without fracture, as in the Molars of the Orycteropus or Cape Ant-eater;

(233.) Imposed, *Lat.* Impositus, *Fr.* Imposée, *Germ.* Eingefleischter Zahn,—when the tooth is merely attached to the gum, and has no alveolar cavity of its own;

(234.) Adherent, *Lat.* Agglutinatus, *Fr.* Attachée, *Germ.* Angehefteter Zahn,—when a tooth is attached to the jaw or palate, without any distinct root, and solely by an intermediate membrane, as in the Ornithorhynchus.

(235.) Maxillary Teeth, *Lat.* Dentes maxillares, *Fr.* Les dents maxillaires, *Germ.* Ladenzähne,—are inserted on the jaws.

(236.) Palatine Teeth, *Lat.* Dentes palatini, *Fr.* Les dents palatines, *Germ.* Gaumenzähne,—are inserted on the palate.

(237.) The Incisors or Fore-teeth, *Lat.* Dentes primores, *Fr.* Les Incisives, *Germ.* Vorderzähne,—are maxillary teeth in the front of the mouth. These are distinguished into the *Upper Incisors*, placed, except in Man, upon the intermaxillary bone, and the *Lower Incisors* opposite to the former. (See page 86.)

(238.) The Molar Teeth or Molars, *Lat.* Dentes molares, *Fr.* Les dents molaires, *Germ.* Bakkenzähne,—are maxillary teeth placed far within the mouth, upon the hinder margins of the jaws. (See page 86.)

(239.) The Canine Teeth or Canines, *Lat.* Dentes laniarii, angulares, seu canini, *Fr.* Les dents canines, *Germ.* Ekkzähne,—are simple maxillary teeth, placed at the sides of the front teeth and near the corners of the mouth, whence they are sometimes called *corner teeth*. They are always pointed, and of a conical form. When the jaws are closed, the canine teeth of the upper jaw always fall *behind* those of the lower. (See page 86.)

The Molar Teeth are said to be

(240.) Continuous, *Lat.* Molares continui, *Fr.* Les Molaires complètes, *Germ.* Anschliessende Bakkenzähne,—when they immediately adjoin the Canines, or, when these are wanting, the Incisors;

(241.) Abrupt, *Lat.* Molares abrupti, *Fr.* Les Molaires incomplètes, *Germ.* Abgesetzte Bakkenzähne,—when they are separated from the Canines or Incisors by a broad space, although continuous to each other.

(242.) The Interval, *Lat.* Diastema, *Fr.* L'espace vide, *Germ.* Zahnlükke,—is the large vacant space between the Incisors or the Canines, and the Molars, as in the Horse and Bull.

The Teeth, in respect to each other, are further said to be

(243.) Approximated, *Lat.* Approximati, *Germ.* Gedrängtstehende,—when they stand close together;

(244.) Divided, *Lat.* Discreti, *Germ.* Vereinzelte, *Germ.*—when there are interstices between them; and a Tooth is

(245.) Remote, *Lat.* Dimotus, *Germ.* Weggerükkter,—when separated by a broad interstice from the others of its own kind.

The Teeth of the one jaw, in respect to those of the other jaw, are said to be

(246.) Opposite, *Lat.* Oppositi, *Germ.* Entgegengesetzte,—when the crowns of the upper teeth are opposite those of the lower;

(247.) Congruent, *Lat.* Congrui, *Germ.* Dekkende,—when each crown of every individual tooth in the upper or under jaw is opposite to the corresponding crowns of the opposite teeth;

(248.) Obverse, *Lat.* Obversi, *Germ.* Abgeschrägte,—when the oblique crowns of the one jaw are fitted into the corresponding oblique crowns of the other;

(249.) Alternate, *Lat.* Alternantes, *Germ.* Wechselständige,—when the crown of a tooth in either jaw occupies the space between two teeth of the opposite jaw;

(250.) Inclined, *Lat.* Acclinati, *Germ.* Ubergreifende,—when the sides of the teeth in the one jaw cover the sides of the teeth in the opposite jaw, and this they may do either *externally* or *internally*.

The Teeth, in respect to the Jaws and Lips, may be

(251.) Erect, *Lat.* Erecti, *Germ.* Aufrechte,—when they are placed vertically in the gum;

(252.) Procumbent, *Lat.* Procumbentes, *Fr.* Proclives, *Germ.* Liegende,—when they lie more or less horizontally upon the gum, as in the lower Incisors of the Makis and Kangaroos;

(253.) Oblique, *Lat.* Obliqui, *Germ.* Schräge,—when they are joined to the gum, so as to form an obtuse angle with it;

(254.) Transverse, *Lat.* Transversi, *Germ.* Queerzähne,—when they are so placed in the gums as to stand inwards beyond the remaining teeth of the same series;

(255.) Inclosed, *Lat.* Inclusi, *Germ.* Bedekkte,—when they are completely covered by the jaws and lips on the mouth being shut;

(256.) Projecting, *Lat.* Exserti, *Germ.* Freie, vorragende Zähne,—when they appear externally although the mouth be closed.

IX.—THE TEETH,

IN RESPECT TO THEIR POSITION AND FORM JOINTLY.

The Teeth are said to be

(257.) Anomalous, *Fr.* Anomales,—when the forms of the three kinds of Teeth, Incisors, Canines, and Molars, are not very distinctly pronounced, as in Man and the Orangs;

(258.) Normal, *Fr.* Normales,—when the differences among the Incisors, Canines, and Molars, are strongly marked, as in the Carnassiers and Hogs.

The Molar Teeth have received the names of

(259.) Grinders, *Lat.* Dentes tritores, *Fr.* Les Machelières, *Germ.* Mahlzahne,—when they have broad crowns, which may be tuberculous, cuspidate, ridged, or plane;

(260.) Carnassiers, *Lat.* Laniarii ambigui, *Fr.* Les Carnassières, *Germ.* Zweideutige Ekkzahne,—when they are strong and lobed, compressed and cutting on their margins, so that, excepting from their position, it would remain doubtful whether they should be set down as Molars, Canines, or Incisors;

(261.) False Molars, *Lat.* Molares incurrentes, *Fr.* Les Fausses Molaires, *Germ.* Ubergehende Bakkenzahne,—when they are placed anteriorly, and are somewhat conical, like the canine teeth;

(262.) Tuberculous Molars, *Lat.* Molares tuberculati, *Fr.* Les Molaires tuberculeuses, *Germ.* Hökkriger Bakkenzahne,—when they present tubercles or blunt excrescences, and belong to an animal which also has Carnassier Molars.

The Canine Teeth assume the name of

(263.) Tusks, *Lat.* Dentes falcati, *Fr.* Les défenses, *Germ.* Fangzahne,—when they project strongly from the mouth. The Tusks are *curved downwards*, as in the Morse ; *directed laterally*, as in the Boar and Ethiopian Hog ; or *curved upwards*, as in the Indian Hog.

The Incisive Teeth are said to be

(264.) Cleft, *Lat.* Pectinati, *Fr.* Pectinées,—when their margins exhibit deep scissures, as in the Flying-Cats;

(265.) Bilobed or Trilobed, *Fr.* Bilobées ou Trilobées.—when they have one or two furrows upon their edge, as in some Bats and young Dogs ;

(266.) Bifurcated, *Fr.* Bifurquées.—when they are in the form of a fork with two prongs, as in some Sea-Lions (*Otaria*);

(267.) Spoon-shaped. *Lat.* Cochleari-formes, *Fr.* En cuiller,—when they are flattened, rounded, and slightly hollowed at their internal surface, as in the lower jaws of the Genus Condylura;

(268.) Awl-shaped. *Lat.* Subulati, *Fr.* En alène,—when they end gradually in a sharp point. as in the lower Incisors of most Rodentia ;

(269.) Cylindrical and Truncated, *Fr.* Cylindriques et tronquées,—as in the Wombat (*Phascolomys*).

The Incisive Teeth sometimes become Tusks, see (263), when they may be

(270.) Straight, *Fr.* Droites,—as in the Narwhal; or

(271.) Curved upwards, *Fr.* Arquées en en-haut,—as in the Elephant.

(272.) A Supernumerary Tooth, *Lat.* Dens accessorius, *Germ.* Nebenzahn,—is a minute homogeneous Molar, superadded either before or behind the other Molars.

(273.) Rudimentary Teeth, *Lat.* Dentes spurii, *Fr.* Dents rudimentaires, *Germ.* Unächt Zahne,—are small deciduous teeth placed before the Molars.

X.—THE NECK.

(274.) The Neck, *Lat.* Collum, *Fr.* Le cou, *Germ.* Hals,—is the intermediate portion between the Head and Trunk, and covering the Cervical Vertebræ. (See page 32.)

(275.) The Cervix or Back of the Neck, *Lat.* Cervix, *Fr.* Le cou supérieur, *Germ.* Hinterhals,—is the upper side of the Neck (the hinder in Man), extending from the Occiput to the first Dorsal Vertebra.

(276.) The Nucha or Nape of the Neck, *Lat.* Nucha, *Fr.* La nuque, *Germ.* Genikk.—is that part of the Cervix next to the Occiput.

(277.) The Auchenium or Lower part of the Neck, *Lat.* Auchenium, *Fr.* Le cou postérieur, *Germ.* Nakken,—is the region of the Cervix below the Nucha.

(278.) The Throat, *Lat.* Guttur, *Fr.* La gorge, *Germ.* Vorderhals,—is the lower region of the Neck (the fore in Man), extending from the Under-chin (75) downwards to the Breast.

(279.) The Gullet, *Lat.* Gula, *Fr.* L'encolure, *Germ.* Kehle,—is the region of the Throat next to the Under-chin.

(280.) The Jugulum or Lower Part of the Throat, *Lat.* Jugulum, *Fr.* La gorge inférieure, *Germ.* Gürgel,—is the region of the Throat between the Gullet and the Breast.

(281.) The Side of the Throat, *Lat.* Parauchenium, *Fr.* Le cou latéral, *Germ.* Halsseite,—is the part of the Neck on each side between the Cervix and Throat.

(282.) The Pit of the Throat, *Lat.* Fossa jugularis, *Fr.* La fosse jugulaire, *Germ.* Gurgelgrube,—is the hollow part before the breast-bone at the base of the Jugulum.

(283.) The Hyoid Bone, *Lat.* Os hyoides, *Fr.* L'os hyoïde, *Germ.* Zungenbeine,—which serves to support the tongue, sometimes appears externally, when it assumes the form of a drum, as in the Howling Apes.

(284.) The Collar, *Lat.* Torques, *Fr.* Le Collet, *Germ.* Ringkragen,—is a coloured ring surrounding the neck.

XI.—THE TRUNK.

(285.) The Trunk, *Lat.* Truncus, *Fr.* Le tronc, *Germ.* Rumpf,—is the primary part of the body containing the viscera and alimentary canal, and bearing the head and neck, the limbs, and frequently the tail.

(286.) The Upper Region of the Trunk [the hinder region in Man], *Lat.* Notæum seu Pars supina, *Fr.* Le dos, ou la partie supérieure du tronc, *Germ.* Rückenseite,—is the entire upper part of the body, extending along the spinal column from the Nucha to the Anus.

(287.) The Gastric or Lower Region of the Trunk [the fore region in Man]. *Lat.* Gastræum, seu Pars prona, *Fr.* La partie inférieure, *Germ.* Bauchseite,—is the entire lower or sternal part of the body, extending from the Gullet to the Anus.

(288.) The Front Region of the Body [the superior in Man], *Lat.* Stethiæum, *Fr.* La partie antérieure, *Germ.* Vordertheil, Vordergeschlepp, —is the entire of the front or thoracical portion of the body.

(289.) The Hinder Region of the Body [the inferior in Man], *Lat.* Uræum, *Fr.* La partie postérieure, *Germ.* Hintertheil, Hintergeschlepp, —is the entire posterior or inferior portion between the Thorax and the Anus, including the abdominal cavity.

(290.) The Dorsal Region, *Lat.* Dorsum, *Fr.* La région dorsale, *Germ.* Rukken,—is the middle part of the Upper Region, resting upon the spinal column.

(291.) The Interscapular Region, *Lat.* Interscapulium, *Fr.* La région inter-scapulaire, *Germ.* Vorderrukken,—is the fore-part of the dorsal region, situate between the Scapulæ, and opposite to the breast. In the Horse, it receives the name of the Withers, *Fr.* Le garrot.

(292.) The Small of the Back, *Lat.* Tergum, *Fr.* Défaut des côtes, *Germ.* Hinterükken,—is the hinder part of the dorsal region, next to the interscapular region.

(293.) The Crupper, *Lat.* Prymna, *Fr.* La Croupe, *Germ.* Kreuz,— is the hindermost part of the dorsal region, opposite to the insertion of the thighs.

(294.) The Uropygium, or Root of the Tail, *Lat.* Uropygium, *Fr.* L'uropygium, *Germ.* Steiss, Schwanzgegend,—is the hindermost part of the trunk, immediately above the anus.

(295.) The Pectoral Region, *Lat.* Pectus, Præcordia, *Fr.* La poitrine, *Germ.* Brust,—is the anterior part of the thorax [the superior in Man], immediately covering the sternum and ribs, and having its Sternal Region longitudinally in the centre.

(296.) The Abdomen, *Lat.* Abdomen, *Fr.* L'abdomen, *Germ.* Bauch, —is the hinder part of the belly [the lower in Man], between the thorax and the anus.

(297.) The Navel, *Lat.* Umbilicus, *Fr.* Le nombril, *Germ.* Nabel, —is the external vestige of the umbilical cord, placed usually near the middle of the abdomen.

(298.) The Umbilical Region, *Lat.* Regio umbilicalis, *Germ.* Nabelgegend,—is the region around the navel.

(299.) The Epigastric Region, *Lat.* Epigastrium, Scrobiculus cordis, *Germ.* Oberbauch, Vorderbauch,—is that portion of the belly next to the breast.

(300.) The Groin, or Inguinal Region, *Lat.* Inguina, Sumen, *Fr.* L'aine, *Germ.* Unterbauch, Hinterbauch,—is the extreme hinder part of the belly [the lower in Man], next to the anus, and between the lower extremities.

(301.) The Perinæum, *Lat.* Perinæum, *Germ.* Damm,—is the narrow isthmus between the organs of generation and the anus.

(302.) The Humeral Region, or Side, *Lat.* Armus, *Fr.* La région humerale, *Germ.* Schultergegend,—is the region of the shoulder on the lateral part of the thorax.

(303.) The Hypochondriac Region, *Lat.* Hypochondria, *Germ.* Weichen,—is the lateral region of the trunk, between the thorax and the loins.

(304.) The Loins, or Lumbar Region, *Lat.* Lumbi, Coxa, Regio lumbaris, *Fr.* Les reins, *Germ.* Hüftengegend,—consist of the hinder-most part of the hypochondriac region, around the insertion of the thighs.

XII.—THE TAIL.

(305.) The Tail, *Lat.* Cauda, *Fr.* La queue, *Germ.* Schwanz,—formed by the vertebræ of the os coccygis projecting beyond the trunk, is not found in all Mammalia. It is wanting in the Orangs, some Bats, the Rats, Moles, Cabiais, &c., or its place is occupied by a mere tubercle, as in the Magot.

The Tail is said to be

(306.) Very Long, *Lat.* Cauda longissima seu elongata, *Fr.* La queue extrêmement longue, *Germ.* Sehr langer Schwanz,—when it is longer than the body, as in the Guenons and Makis;

(307.) Medium Length, *Lat.* Mediocris, *Fr.* Médiocre, *Germ.* Mittellanger,—when it is scarcely shorter than the trunk;

(308.) Short, very Short, and Abrupt, *Lat.* Brevis, brevissima, abrupta, *Germ.* Kurzer, sehr kurzer, abgekürzter,—when it is shorter than the thigh, and most commonly only the stump of a tail;

(309.) Annular, *Lat.* Annulata, *Fr.* Annulaire, *Germ.* Geringelter,— when the skin of the tail is divided by rings;

(310.) Loricate or Shielded, *Lat.* Loricata, *Fr.* Plaquée, *Germ.* Gepanzerter, when it is covered with a bony case;

(311.) Voluble or Rolling, *Lat.* Volubilis, *Fr.* S'enroulante, *Germ.* Wikkelschwanz, when the tail is very long, capable of being rolled around the animal, and of continuing in that position;

(312.) Prehensile, *Lat.* Prehensilis, *Fr.* Prénante, *Germ.* Greifschwanz,—is a rolling tail, with the under part of its apex usually smooth, and supplied with a soft skin fitted for touching and holding;

(313.) Loose, *Lat.* Laxa, *Fr.* Libre, *Germ.* Schlaffer,—when it is neither voluble nor prehensile;

(314.) Bushy, *Lat.* Comosa, Jubata, *Fr.* Touffue, *Germ.* Buschiger, —when it is ornamented to its base with long and pendulous hair;

(315.) Tufted, *Lat.* Floccosa, *Fr.* Floconneuse, *Germ.* Gequnastter, —when its apex is ornamented with a tuft of long hair;

(316.) Distichous, *Lat.* Disticha, *Fr.* Distique, *Germ.* Zweizeiliger, —when the tail is covered with long hair, arranged in two series diverging from the centre.

The Tail may be

(317.) Thick, Oval, and Flat, *Fr.* Epaisse, Ovale, et Aplatie, as in the Beavers and Ornithorhynchus;

(318.) Square, *Fr.* Carrée,—as in some Shrews;

(319.) Triangular and Robust, *Fr.* Triangulaire et Robuste,—as in the Kangaroos.

(320.) The Stump of the Tail, *Lat.* Stirps caudæ, *Fr.* Le tronc de la queue, *Germ.* Schwanzrube,—is the body of the tail when considered without the hair.

(321.) The Switch-hair of the Tail, *Lat.* Coma, *Fr.* Les longues poils de la queue, *Germ.* Schweif,—are the longer tail-hairs considered by themselves.

XIII.—THE MAMMÆ, &c. &c.

(322.) The Mammæ, *Lat.* Mammæ, *Fr.* Les Mamelles, *Germ.* Euter, Brüste,—consist of a number of glands secreting milk, and placed symmetrically in a more or less considerable number on each side of the lower part of the body.

(323.) The Teats or Nipples, *Lat.* Papillæ, *Fr.* Les trayons, *Germ.* Slugwarzen,—the excretory ducts of the milk, are placed upon the Mammæ.

(324.) The Areola, *Lat.* Areola, *Fr.* L'aréole, *Germ.* Hofe,—is the circle surrounding the nipple, and frequently coloured.

The Mammæ are said to be

(325.) Pectoral, *Lat.* Pectorales, *Fr.* Pectorales, *Germ.* Brust-Euter,—when they are placed upon the breast;

(326.) Abdominal, *Lat.* Abdominales, *Fr.* Abdominales, *Germ.* Bauch-Euter,—when they are placed upon the belly;

(327.) Inguinal, *Lat.* Inguinales, *Fr.* Inguinales, *Germ.* Schaam-Euter,—when they are placed upon the groin between the thighs.

(328.) The Abdominal or Marsupial Pouch, *Lat.* Mastotheca seu Marsupium abdominale, *Fr.* La Poche marsupiale, *Germ.* Zitzensakk, —is a large fold of skin, placed in front of the belly, and capable of being closed, so as to form a bag or pouch, containing the mammæ, and some times also several young in the embryo state.

The Mammæ are said to be

(329.) Exposed, *Lat.* Apertæ, *Fr.* Ouvertes, *Germ.* Unbedekkt,— when they are not covered by an abdominal pouch.

(330.) The Anus, *Lat.* Anus, *Fr.* L'anus, *Germ.* After,—is the external opening of the rectum, placed under the tail for excretory purposes.

(331.) THE ANAL POUCH, *Lat.* Rima odorifera, Saccus analis, *Fr.* Les follicules anales, *Germ.* Riechetide Hautfalte, Aftertasche,—is a sac placed between the tail and the anus, or between the latter and the organs of genenation, and emitting an odoriferous secretion.

(332.) THE PENIS, *Lat.* Penis, *Fr.* La verge, *Germ.* Ruthe,—is the male organ of generation placed upon the groin. It is said to be

(333.) ADNATE, *Lat.* Adnatus, *Germ.* Angewachsne,—when the basal part is concealed by the skin of the abdomen, and its apex alone is free towards the umbilical region.

ANIMALS are said to be

(334.) RETROMINGENT, *Lat.* Animalia retromingentia, *Germ.* Rükkwärtsharnende Thiere,—when the penis is directed backwards.

(335.) THE VULVA, *Lat.* Vulva, *Fr.* La vulve, *Germ.* Wurf,—is the female organ of generation, placed upon the perinæum.

(336.) THE CLOACA, *Lat.* Cloaca, *Fr.* Le cloaque, *Germ.* Kloake,—is the common outlet for the intestinal canal and the organs of generation, found in the Monotremata.

XIV.—THE LIMBS.

(337.) THE LIMBS, *Lat.* Artus, *Fr.* Les membres, *Germ.* Gliedmassen,—are the articulated extremities, fitted for walking, and usually having their apices furnished with fingers or toes, and nails, claws, or hoofs. In general the limbs are of equal length; sometimes, however, the fore-limbs are the longer, as in the Gibbons and Sloths; and sometimes the hinder limbs exceed the former, as in the Kangaroos, and most Rodentia.

THE LIMBS are said to be

(338.) RETRACTILE, *Lat.* Retracti, Obvoluti. *Germ.* Eingezogne,—when their articulations are very short, so that the basal joints are buried in the flesh, and the terminal almost hidden in the fur.

(339.) PINNIFORM, *Lat.* Artus pinniformes, *Fr.* En forme de nageoires, *Germ.* Flossenartige Gliedmassen,—when the pectoral limbs are so immersed in the trunk, and covered with skin, that their articulations are only discoverable by anatomization. They are then called

(340.) THE PECTORAL FINS, *Lat.* Pinnæ pectorales, *Fr.* Les Nageoires pectorales, *Germ.* Brustfinnen,—from their resembling the fins of a fish, both in general form and use.

(341.) THE CAUDAL LIMBS, *Lat.* Pedes compedes, *Fr.* Les Nageoires caudales, *Germ.* Verwachsne Beine,—are hinder-limbs placed horizontally like a tail, so that their articulations can only be perceived anatomically.

(342.) THE FOOT,—is the extreme part of a limb, adapted for walking. See (37.) and (46.)

(343.) THE SOLE, *Lat.* Planta, *Fr.* La plante du pied, *Germ.* Sohle,—is the inferior side of the foot from the wrist or ancle-joint to the extremities of the fingers or toes.

(344.) THE FINGERS or TOES,—form the apex of a limb. See (40.) Their number varies from one to five on the several extremities.

Thus, there are on each limb in

Man	5	Above and 5	Below.
The Bears, Elephant	5	Before and 5	Behind.
Most RODENTIA	4		5
Pecaris, Cabiais, Agoutis	4		3
Hippopotamus, Suricate	4		4
Two-toed Ant-eater	2		4
Sloths	2 or 3		4
Rhinoceroses	3		3
RUMINANTIA	2		2
SOLIPEDA	1		1

A FOOT is said to be

(345.) MONO-DACTYLOUS, DI-DACTYLOUS, TRI-DACTYLOUS, TETRA-DAC-TYLOUS, or PENTA-DACTYLOUS,—when it has one, two, three, four, or five toes or fingers;

(346.) ADACTYLOUS, *Lat.* Adactylus, Mutilatus, *Fr.* Adactyle, *Germ.* Ohnzehiges,—when the toes or fingers are wanting.

(347.) THE AXILLA or ARM-PIT, *Lat.* Axilla, *Fr.* L'aisselle, *Germ.* Achsel,—is the hollow under the fore-limb at its insertion with the thorax.

(348.) THE THIGH, *Lat.* Clunis, *Fr.* La jambe, *Germ.* Keule,—is the femur, together with its fleshy covering.

(349.) THE BUTTOCKS, *Lat.* Nates, *Fr.* Les fesses, *Germ.* Gesäss,—are the hinder [or lower] sides of the thighs, and frequently prominent.

(350.) THE CALLOSITIES OF THE BUTTOCKS, *Lat.* Tylia, Natis calvæ, *Fr.* Les callosités, *Germ.* Gesäss-schwiele,—are hard and smooth portions of the buttocks, frequently coloured, found in some Apes.

(351.) THE CALF, *Lat.* Sura, *Fr.* Gras de la jambe, *Germ.* Wade,—is the swelling muscle behind the upper part of the tibia.

XV.—THE FEET OR HANDS,
IN RESPECT TO THEIR PARTS AND APPENDAGES.

(352.) THE THUMB or GREAT-TOE, *Lat.* Pollex, Hallux, Digitus primus, *Fr.* Le pouce, *Germ.* Daumen, Innenzehe,—is the innermost finger or toe, usually distinguished from the others by its situation, and by being always the shortest and thickest.

(353.) THE INDEX or FIRST FINGER, *Lat.* Digitus index seu secundus. *Fr.* Le doigt indicateur, *Germ..* Zeigefinger, Die zweite zehe,—is the finger next to the thumb.

(354.) THE SECOND or MIDDLE FINGER, *Lat.* Digitus tertius, *Fr.* Le medius, *Germ.* Die dritte zehe,—is the finger next to the index.

(355.) THE THIRD or RING FINGER, *Lat.* Digitus quartus, *Fr.* L'annulaire, *Germ.* Die vierte zehe,—is the third finger from the thumb.

(356.) THE FOURTH or LITTLE FINGER, *Lat.* Digitus quintus, *Fr.* Le petit doigt, *Germ.* Die funfte zehe,—is the fourth finger from the thumb. The same phraseology is applied to the Toes.

THE THUMB is said to be

(357.) RUDIMENTARY, *Lat.* Verruca hallucaris, *Fr.* Le pouce rudimentaire, *Germ.* Daumenspur,—when it scarcely emerges from the skin;

(358.) OPPOSABLE,—when it is capable of being applied to the fingers. This may be upon the fore-limbs only, as in Man, or on all the limbs, as in the Apes, or only on the hinder limbs, as in the Opossums.

THE FINGER or TOE is said to be

(359.) INSISTENT, *Lat.* Insistens, *Germ.* Auftretend,—when it touches the earth while the animal rests upon its feet;

(360.) ELEVATED, *Lat.* Amotus, *Germ.* Hinaufgerükkte,—when it is inserted so high, that it does not touch the ground while the animal is walking.

(361.) THE GLOVE, *Lat.* Podotheca. *Germ.* Fuss-scheide,—is the fur covering the entire foot or hand.

(362.) THE FINGER-GLOVE, *Lat.* Dactylotheca, *Germ.* Zehenscheide,—is the part of the glove covering each separate finger.

(363.) THE FINGER-BALLS, and HEEL-BALLS, *Lat.* Tylari, *Germ.* Zehenballen, Hakkenballen,—are the naked and callous parts under the fingers and the heels.

(364.) THE NAIL, *Lat.* Lamna, Unguis, *Fr.* L'ongle, *Germ.* Nagel,—is the broad and flat horny surface, covering in a greater or less degree the upper side of the last phalanx.

(365.) THE CLAW, *Lat.* Falcula, Unguis falcularis, *Fr.* Griffe, *Germ.* Kralle,—is an elongated, compressed, and rounded nail.

A CLAW is said to be

(366.) RETRACTILE, *Lat.* Falcula vaginata, *Fr.* Rétractile, *Germ* Gescheidete Kralle,—when it can be drawn within a proper sheath upon the last phalanx;

(367.) TEGULAR, *Lat.* Tegularis, *Germ.* Kuppennagel,—when it approaches to a Nail in form.

(368.) THE HOOF, *Lat.* Ungula. *Fr.* Le sabot, *Germ.* Huf,—is a horny covering, enveloping the point of the phalanx on every side.

(369.) THE SOLE OF THE HOOF, *Lat.* Solea, *Fr.* La sole, *Germ.* Hutsohle,—is its entire under surface, including the hollow part.

(370.) THE CORONET, *Lat.* Coronamen, *Fr.* La couronne, *Germ.* Hufkranz,—is the upper margin of the hoof, where it presses the finger or foot.

(371.) THE FROG, *Fr.* La fourchette,—is an elevated portion in the form of a V, sometimes found behind the middle of the sole of a hoof.

(372.) THE HAND, *Lat.* Manus, *Fr.* La main, *Germ.* Hand,—is the extreme part of a limb, having its thumb free and opposable, and covered with a flat nail. Sometimes this term is applied to a foot, when its toes are very long and much separated from each other.

(373.) THE PAW, *Lat.* Palma, *Fr.* La patte, *Germ.* Tatze,—is the broad part, consisting of the carpus and metacarpus in the fore-limb, or the tarsus and metatarsus in the hinder.

(374.) THE PALM, *Lat.* Vola, *Fr.* La paume, *Germ.* Handhöhlung,—is the flat inner surface of a hand.

XVI.—THE FEET OR HANDS,
IN RESPECT TO THEIR FORM AND USE.

THEIR FINGERS or TOES may be

(375.) DIVIDED, *Lat.* Fissi, *Fr.* Séparés, *Germ.* Gespaltne,—when they are not connected by any intermediate membrane; or

(376.) HALF-DIVIDED, *Lat.* Semi-fissi, *Fr.* Demiséparés, *Germ.* Halb-gespaltne,—when they are only partially connected.

ANIMALS are said to be

(377.) FISSIPEDE, *Lat.* Animalia Fissipeda, *Fr.* Les fissipèdes, *Germ.* Spaltflüssige Thiere,—when their fingers or toes are unconnected by a membrane.

THE FINGERS or TOES are said to be

28

(378.) UNITED, *Lat.* Coadunati, *Fr.* Réunis, *Germ.* Verwachsne,—when they are connected by no membrane, and yet adhere so closely together that they are contained in the same finger-glove (362).

THE FEET are said to be

(379.) PALMATED, *Lat.* Palmati, *Fr.* Palmés, *Germ.* Schwimmfüsse,—when the fingers or toes are connected together by a membrane reaching nearly to their extremities, so as to be fitted for swimming;

(380.) SEMI-PALMATED, *Lat.* Semipalmati, *Fr.* Demipalmés, *Germ.* Halbe Schwimmfüsse,—when the membrane between the fingers extends to about one half of their length;

(381.) PINNATED, *Lat.* Lomatini, *Fr.* Pinnés, *Germ.* Gesäumte Füsse, —when the fingers are supplied with membranes only on their sides.

(382.) THE WEB, *Lat.* Palama, *Fr.* La membrane pour la natation, *Germ.* Schwimmhaut,—is the membrane belonging to a palmated or semipalmated foot.

THE FEET are said to be

(383.) CHEIROPTEROUS, *Lat.* Chiropteri, Volatiles, *Fr.* Cheiroptères, *Germ.* Flugbeine,—when the fingers of the fore-feet are excessively elongated and supplied with a light membrane, so as to adapt them for flight.

(384.) DERMOPTEROUS, *Lat.* Dermopteri, *Fr.* Dermoptères, *Germ.* Flatterbeine,—when the fore and hinder feet are connected together by a membrane, which is merely an expansion of the skin of the trunk, and often extending before and behind the limbs.

(385.) THE EXTENSIBLE MEMBRANE or WING, *Lat.* Patagium, *Fr.* La membrane extensile, ou L'aile, *Germ.* Flughaut,—is the membraneous appendage of a dermopterous or cheiropterous foot.

It is said to be

(386.) DIGITAL, *Lat.* Digitale, *Germ.* Zehen-Flughaut,—when the membrane extends between the elongated fingers of the fore-limb;

(387.) CERVICAL, *Lat.* Collare, *Germ.* Halsfittig,—when it extends between the neck and the expanded fore-limb;

(388.) LUMBAR, *Lat.* Lumbare, *Germ.* Seitenfittig,—when the membrane between the fore and hinder limb proceeds from the sides of the trunk;

(389.) INTERFEMORAL, *Lat.* Interfemorale, Anale, *Germ.* Steissfittig, —when the membrane extends behind the thighs;

(390.) HAIRY, *Lat.* Pelliceum, *Fr.* Pileuse, *Germ.* Flugfell,—when the membrane is thickly covered with hair;

(391.) NAKED, *Lat.* Membranaceum, *Fr.* Nue, *Germ.* Flughaut,—when it is light and destitute of hair.

THE HINDER FEET are said to be

(392.) SALTATORIAL or LEAPING, *Lat.* Saltatorii, *Fr.* Propres à sauter, *Germ.* Springbeine,—when they are nearly twice as long and strong as the fore-feet, and fitted for leaping;

(393.) AMBULATORIAL FEET, *Lat.* Pedes ambulatorii, *Fr.* Les pieds propres à marcher, *Germ.* Gangbeine,—fitted for walking, and such as are not palmated, cheiropterous, dermopterous, or saltatorial, but may be either fissipede or united;

(394.) FOSSORIAL FEET, *Lat.* Pedes fossorii, *Fr.* Les pieds propres à fouiller la terre, *Germ.* Grabfüsse,—are very broad, and armed with strong FOSSORIAL NAILS, *Lat.* Ungues fossorii, *Germ.* Grabklauen,—so as to be fitted for digging.

THE FEET are said to be

(395.) PLANTIGRADE, *Lat.* Plantigradi, *Fr.* Plantigrades, *Germ.* Sohlenschreitende Beine,—when they are destitute of hair as far as the heel, from the animal placing the entire sole of the foot upon the ground when walking;

(396.) DIGITIGRADE, *Lat.* Digitigradi, *Fr.* Digitigrades, *Germ.* Zehenschreitende Beine,—when they are covered with hair almost to the apices of the toes, which alone touch the ground when the animal is walking;

(397.) BISULCATE or CLOVEN, *Lat.* Bisulci, *Fr.* Pieds fourchus, *Germ.* Spalthufige,—when the toes are only two in number, insistent (359), and ungulated or hoofed;

(398.) SUB-BISULCATE, *Lat.* Subbisulci, *Fr.* Pieds demi-fourchus, *Germ.* Kerbhufige,—when the two toes are almost united, and their apices alone are free and covered with hoofs;

(399.) SOLIDUNGULATE, *Lat.* Solidunguli, *Fr.* Solipèdes, *Germ.* Einhufige,—when the foot consists of a single finger, covered by a single hoof;

(400.) UNGUICULATE,—when the toes are furnished with nails;

(401.) UNGULATE,—when they are protected by hoofs;

(402.) TRIUNGULATE, QUADRIUNGULATE, or MULTUNGULATE,—when they are protected by three, four, or more hoofs.

(403.) THE SIDE-HOOFS, *Lat.* Ungulæ succenturiatæ, *Fr.* Onglons surnumeraires, *Germ.* Nebenhufe,—are the hoofs of elevated toes (360), found in bisulcate and mulungulate feet.

XVII.—THE SKIN.

(404.) THE SKIN, *Lat.* Cutis, *Fr.* La peau, *Germ.* Haut.—is the general integument of the entire body, composed of several layers. See page 89.

It is said to be

(405.) LOOSE, *Lat.* Laxa, *Fr.* Lâche, *Germ.* Schlotternde,—when it hangs down and forms folds.

(406.) THE DEW-LAP, *Lat.* Palearia, *Fr.* Le fanon, *Germ.* Wamme, —is a loose skin hanging from the neck and fore-part of the breast, as in the Bull.

(407.) THE HUMP, *Lat.* Tophus, Gibber, *Fr.* La bosse, *Germ.* Höcker,—is a broad and swelling projection, formed of fat under the skin, as in the Indian Bull.

(408.) THE CALLOSITY, *Lat.* Callus, *Fr.* Le callosité, *Germ.* Schwiele, —is a naked skin protected by a hard and horny epidermis, upon which some animals rest upon the ground, as the Camels and Dromedaries. It is also found on the palms of the hands, the soles of the feet, and the thighs of some Apes.

(409.) A WART, *Lat.* Verruca, *Fr.* Une Verrue, *Germ.* Warze,—is a small, hard, and round tumour.

(410.) THE SCALES, *Lat.* Squamæ, *Fr.* Ecailles, *Germ.* Schuppen,— are flat, horny, or bony parts, inserted on the skin, and frequently imbricated, or arranged one over the other, like the tiles of a house.

(411.) THE SHIELD, SHELL, or COAT of MAIL, *Lat.* Lorica, Clypeus, Testa, *Fr.* Le bouclier, ou la plaque, *Germ.* Panzer, Schild, Schale,— is a horny or bony case covering the trunk and tail, or the most part of them, as in the Armadilloes.

It is said to be

(412.) AREOLATED, *Lat.* Scutulata, *Fr.* Plaqué, *Germ.* Getäfelter, —when the surface of the shield is covered by regular partitions, and a shining epidermis.

(413.) THE AREOLA of a SHIELD, *Lat.* Scutulum, Assula, *Fr.* Le compartiment d'un bouclier, *Germ.* Feld, Schildchen,—is each partition of an areolated shield.

(414.) THE BANDS, *Lat.* Cingula, Zonæ, *Fr.* Les bandes transversales, *Germ.* Gürtel,—are the distinct divisions of a transverse shield, and moveable by means of the skin.

(415.) THE TAIL-RINGS, *Lat.* Gyri, annuli, *Fr.* Les anneaux d'écailles, *Germ.* Schwanz-ringel,—are the separate divisions of a shielded tail.

(416.) THE DORSAL FIN, *Lat.* Pinna dorsalis, *Fr.* La nageoire dorsale, *Germ.* Rükkenfinne,—is a fin-like cutaneous process placed on the back, and sustained by small bones.

(417.) THE CAUDAL FIN, *Lat.* Pedalium, Pinna analis, *Fr.* La nageoire caudale. *Germ.* Schwanzfinne,—is a cutaneous process placed horizontally in the shape of a fin upon the apex of the tail.

THE SKIN is said to be

(418.) APPLIED, *Lat.* Applicata, *Fr.* Appliquée,—when it exactly fits upon the body, as in the Deer and Antelopes;

(419.) VERRUCOSE, *Lat.* Verrucosa, *Fr.* Verruqueuse,—when it is covered with small naked eminences, or warts;

(420.) SCALY, *Lat.* Squamosa, *Fr.* Ecailleuse,—when the epidermis is folded over in such a way as to resemble the scales of a fish, as on the tail of the Beaver;

(421.) NAKED, *Lat.* Nuda, *Fr.* Nue,—when it is destitute of hair or any other integument;

(422.) THICK, *Lat.* Densa, *Fr.* Epaisse,—as in the Elephant, Rhinoceros, and others; for which reason these animals are said to be PACHYDERMATOUS, or THICK-SKINNED, *Lat.* Pachyderma;

(423.) ROUGH, *Lat.* Asperata, *Fr.* Rugueuse,—as in the animals already mentioned; and

(424.) CALLOUS, *Lat.* Callosa, *Fr.* Calleuse,—when it is hard to the touch.

XVIII.—THE HAIRS.

(425.) THE HAIRS, *Lat.* Pili, *Fr.* Les poils, *Germ.* Haare,—are horny filaments of various shapes (see page 89), covering the body externally in various degrees. They may be fine like silk, as in the Chinchilla, or coarse like hay, as in some Ruminantia.

(426.) THE BRISTLES, *Lat.* Setæ, *Fr.* Les soies, *Germ.* Borsten,— are hard and rigid hairs frequently divided at the points, as in the Hog.

(427.) THE PRICKLES or SPINES, *Lat.* Aculei, Spinæ, *Fr.* Les piquans ou les épines, *Germ.* Stacheln,—are very strong, hard, and rigid hairs, frequently ending in a sharp point. Sometimes they are nearly *conical* and *of medium length*, as in the Echidnas; *very long* and *bulging in the middle*, as in the Porcupine; *flattened* like the blade of a sword, as in the Echimys and some spiny Rats; *alone*, as in the Hedgehogs, Coendou, and Spiny Echidna; or *mixed* with hair, as in the Silky Echidna and Canada Porcupine.

THE BODY is said to be

(428.) HAIRY or COVERED, *Lat.* Pilosum, Vestitum, *Fr.* Pileuse, *Germ.* Behaart,—when it is furnished with Hair;

(429.) HAIRLESS, *Lat.* Depilis, *Fr.* Nue, *Germ.* Haarlos,—when it is destitute of hair; as upon all callosities; on most part of the face in Man and some Apes; on the Chiloma (157) of most Ruminants; and on the snout of all Carnassiers and Rodentia.

(430.) THE FUR, *Lat.* Vellus, *Fr.* La fourrure, *Germ.* Pelz,—consists of the hairs of the entire body or a part, considered together.

(431.) THE FELT, *Lat.* Pellis, *Fr.* Le feutre, *Germ.* Fell,—is the hairy skin and fur, considered together.

(432.) THE WOOLLY FUR, *Lat.* Codarium, *Germ.* Wollpelz.—consists. of the finer and softer hairs of the fur, amongst which the longer hairs are usually intermixed.

(433.) THE WOOL, *Lat.* Lana, *Fr.* La laine, *Germ.* Wolle,—consists of long, fine, frizzled, and curly hair, as in the Sheep.

(434.) THE HAIR OF THE HEAD, *Lat.* Capilli, Crines, Coma, *Fr.* Les Cheveux, *Germ.* Haupthaare,—are the long hairs proceeding from the skin of the skull.

(435.) THE CREST, *Lat.* Caprona, Antiæ, Crista, *Fr.* L'aigrette, *Germ.* Stirnschopf,—consists of long hairs, proceeding upwards and backwards from the top of the head, and it may either be *Stellated* (455) or *Vorticillated* (456).

(436.) THE BEARD, *Lat.* Barba, aruncus, *Fr.* La barbe. *Germ.* Bart, —consists of long hairs hanging from the chin, as in the Goat and Bison.

(437.) THE WHISKER, *Lat.* Mystax, barba malaris, *Fr.* La moustache, *Germ.* Bakkenbart,—is the long hair covering the under-cheek (76).

(438.) THE MOUSTACHIO, *Lat.* Mastax, *Fr.* La moustache, *Germ.* Knebelbart,—is the beard of the upper lip. It is very long in the nocturnal or aquatic Carnassiers, as the Cats and Seals. It can scarcely be observed in the Kangaroos and Ruminantia, and is wholly wanting in the Whales and Dolphins.

(439.) THE STIFF HAIRS, *Lat.* Vibrissæ, *Fr.* Les faisceaux de poils, *Germ.* Schnurrhaare,—are long elongated bristles of great strength upon the nose, and some other parts of the face, and sometimes also upon the inner sides of the fore-limbs. These gigantic moustachios serve as weapons of defence to the Manatus.

(440.) THE EYE-BROWS, *Lat.* Supercilia, *Fr.* Les sourcils. *Germ.* Augenbraune,—are transverse series of hairs placed at the lower part of the forehead and above the eyes.

(441.) THE EYE-LASHES, *Lat.* Cilia, *Fr.* Les cils, *Germ.* Wimpern.— are long hairs proceeding from the eyelids.

(442.) THE MANE, *Lat.* Juba, *Fr.* La crinière, *Germ.* Mähne,—consists of long hairs adorning the ridge of the back, sometimes extending from the occiput to the end of the tail, as in the Civet and Zibet; sometimes it extends no lower than the shoulders, where it becomes mixed with the long hairs of the withers, as in the Lion; and sometimes its hairs have their points turned towards the head, as in the *Antelope leucoryx*.

(443.) THE TAIL-TUFT, *Lat.* Floccus, *Fr.* Le Flocon, *Germ.* Quaste, —consists of long loose hair covering the extremity of the tail, as in the Lion, the Ass, and some Apes.

(444.) THE HAIR-BUNCH, *Lat.* Scopa, *Fr.* La brosse, *Germ.* Haarbüschel, —is a bunch of long and loose hair hanging from the wrist (51), as in some Antelopes, or from any other part of the body.

(445.) THE PENCIL, *Lat.* Penicillus, *Fr.* Le pinceau, *Germ.* Pinsel,— is a very small bundle of rigid hair placed on the top of any part.

(446.) A LOCK, *Lat.* Cincinnus, *Fr.* Une touffe, *Germ.* Lokke,—is each small bundle of hair involved at its apex.

(447.) THE PUBES, *Lat.* Pubes, *Fr.* Le pubis, *Germ.* Schaamhaar,— is the hairs placed upon the groin.

(448.) THE DOWN OR MILK HAIR, *Lat.* Lanugo, *Fr.* Le duvet, *Germ.* Milch-haar,—is the soft and fine hair of young animals, not yet changed, and frequently variegated with different colours.

(449.) THE HAIR-SEAM OF SUTURE, *Lat.* Sutura, *Germ.* Haarnaht,— is the line formed by the points of the hairs of the fur converging together.

THE HAIRS are said to be

(450.) SMOOTH, *Lat.* Pili incumbentes, *Fr.* Les poils couchés ou lisses, *Germ.* Anliegende haare,—when they are pressed close to the skin along their entire length;

(451.) ERECT, *Lat.* Erecti, *Fr.* Droits, *Germ.* Aufrechts,—when they are placed almost vertically upon the skin;

(452.) KNOTTED, *Lat.* Tomentosi, *Fr.* Noueux, *Germ.* Filzige,—when they are entangled together and interwoven, so that they can be extricated only with difficulty;

(453.) SILKY, *Lat.* Sericei, *Fr.* La bourre, *Germ.* Seidenhaare,—when they are soft, short, and shining, as in the Makis;

(454.) VILLOUS, *Lat.* Villosi, *Fr.* Zottige,—when they are long, thin, and straight;

(455.) STELLATED, *Lat.* Stella, *Germ.* Haarstern—when the hair proceeds from a centre like the radii of a circle, as on the head of the Hair-tipped Macaco (*Macacus cynomolgus*);

(456.) VORTICILLATED, *Lat.* Vortex, *Germ.* Haar-wirbel,—when the hairs run from the circumference to the centre in bent radii, as on the head of the Chinese-capped Macaco (*M. Sinicus*);

(457.) DISTICHOUS, *Lat.* Varicula. *Germ.* Scheitelung,—when the hair of the fur is arranged in two series, diverging in opposite directions;

(458.) FISTULOUS, *Fr.* Fistuleux. *Germ.* Fistelartig,—when it is shaped like a reed or quill, as in the hairs beneath the body of the Porcupine.

XIX.—THE HORNS.

(459.) THE HORNS, *Lat.* Cornua, *Fr.* Les cornes. *Germ.* Horner,— are hard processes, composed either of horn or bone. These may proceed either from the nose, when they are said to be *nasal*, or from the forehead, when they are termed *frontal*. According to the substance of which they are composed and their form, they may be either *osseous* or *corneous, solid* or *hollow.*

(460.) THE PROPER HORNS, *Lat.* Cornua vaginantia, *Fr.* Les cornes creuses, *Germ.* Scheidenhörner,—are hollow bodies or sheaths covering osseous processes of the frontal or nasal bones, found in several Ruminants, such as the Bulls, Goats, Sheep, and Antelopes.

(461.) THE HORN-BASE or CORE, *Lat.* Embolus. *Germ.* Hornzapfen,— is the bony process of the frontal or nasal bone, covered by a proper horn.

(462.) THE ANTLERS, *Lat.* Ceras, cornu. *Fr.* Les bois. *Germ.* Ge. weihe,—consist of two solid, frontal, and corneous horns, which fall and are renewed every year, as in the Deer. They are called the ATTIRE by hunters, and are said to be

(463.) BRANCHING, *Lat.* Cerata ramosa, *Fr.* Les bois branchus, *Germ.* Astige Geweihe,—when they emit branches.

(464.) THE BEAM OF MAIN-STEM, *Lat.* Caulis, *Fr.* Merrain, *Germ.* Stange, —is the principal trunk of a branching antler.

(465.) THE BURR, *Lat.* Stephanium, *Fr.* Le Noyau. *Germ.* Krone, —is a granulated prominence at the base of the beam, and covering the horn-base (461).

(466.) THE BRANCHES OF SNAGS, *Lat.* Rami, *Fr.* Les andouillers, *Germ.* Zinken, Enden,—are the processes of the beam.

(467.) THE BROW ANTLER, *Lat.* Propugnaculum, Amynter, *Germ.* Augensprosse,—is the foremost and lowest branch of a beam.

THE ANTLERS may be

(468.) PALMATED, *Lat.* Cerata palmata, *Germ.* Schaufelförmige Geweihe.—when the branching antlers are dilated at their extremity like a shovel;

(469.) DECIDUOUS OF ANNUAL, *Lat.* Decidua, annua, *Fr.* Caducs, *Germ.* Wechselnde, Abfallende,—when they fall and are renewed every year.

THE HORNS and ANTLERS may be

(470.) PENDANT, *Lat.* Prona. *Germ.* Vorgelegte,—when they hang downwards above the muzzle;

(471.) REFLECTED, *Lat.* Reclinata. *Fr.* Recourbées, *Germ.* Rükkegelegte,—when they are turned backwards, as in the *Capra Mambrica;*

(472.) INCURVATED, *Lat.* Camuræ, *Fr.* Courbés en dedans, *Germ.* Eingebogne,—when they are bent inwards;

(473.) DIVERGENT, *Lat.* Vara, *Fr.* Divergens, *Germ.* Auswärtsgebogne.—when they are bent outwards, as in the *Cervus Dama;*

(474.) REDUNCATE, *Lat.* Redunca, *Fr.* Courbés en avant, *Germ.* Hakige,—when they are curved forwards, as in the *Antilope Tragocampus;*

(475.) LYRATE, *Lat.* Lyrata, *Fr.* En lyre ou lyroides, *Germ.* Leierförmige.—when they are so bent as to exhibit the form of an ancient lyre, when viewed behind, before, or on the side;

(476.) ROUND, *Lat.* Teretia, *Fr.* Arrondies,—as in the Stags;

(477.) ANNULATED, *Lat.* Annulata, *Fr.* Annelées,—covered with rings, as in the Gazelles.

(478.) BIFID, *Fr.* Bifourqués,—divided into two prongs at the point, as in the Roebuck;

(479.) TRIFID, *Fr.* Trifourqués,—divided into three prongs, as in the *Cervus pygargus;*

(480.) UNCINATE, *Fr.* Recourbés à leur extremité,—bent at the points;

(481.) SPIRAL, *Fr.* En spirale,—as in the *Ovis sterpsiceros;*

(482.) TRIANGULAR.—as in the *Capra depressa;*

(483.) CARINATED, *Fr.* Garnies de petites cannelures,—with furrows, as in the *Capra Ægagrus;*

(484.) KNOTTED, *Fr.* Moniliforme,—as in the *Capra Ibex.*

(485.) THE PANNICLE, *Lat.* Pannicula, *Germ.* Bast,—is the rough skin covering deciduous horns when they have just begun to appear.

(486.) THE PRICKET OF DAG, *Lat.* Patialus, *Fr.* Dague, *Germ.* Spiess, —is the simple antler of a young animal, before it begins to have branches.

(487.) A SPITTER, *Lat.* Subulo, *Fr.* Daguet, *Germ.* Spiessern,—is a young animal, having only prickets or dags upon its forehead.

ABDOMEN, 296
Abdominal Pouch, 328
Abrupt, 241
Adactylous, 346
Adherent, 234
Adnate, 333
Alternate, 249
Alveolar Cavities, 185
Ambulatorial, 393
Anal Pouch, 331
Ankle, 57
—— joint, 55
Annular, 309
Anomalous, 257
Anthelix, 105
Antitragus, 107 ;
Antlers, 462
Anus, 330
Apex, 222
Applied, 418
Areola, 324, 413
Areolated, 412
Arm-pit, 347
Attire, 462
Auchenium, 277
Auricle, 103, 109 to 116
Auricular operculum,' 117
Axilla, 347

Back, Small of the, 293
Bands, 414
Beam, 464
Beard, 436
Bifurcated, 266
Bisulcate, 397
Body, Front Region of the, 288
—— Hinder Region of the, 289
Branch'2, 466
Breast-bone, 19
Bristles, 426
Brow-antler, 467
Burr, 465
Buttock, 349

Calf, 351
Callosity, 408
—— of the Buttock, 350
Callous, 424
Canine Teeth or Canines, 239
Cannon-bone, 39, 48
Carinated, 463
Carnassiers, 260
Carpus, 38
Caudal Fin, 417
—— Limbs, 341
Cervical, 387
Cervix, 275
Chanfrin, 178
Cheek, 73
—— pouches, 183
Cheiropterous, 383
Chest, 20
Chiloma,' 157
Chin, 13
Clavicle, 32
Claw, 365
Cloaca, 336
Coat of Mail, 411
Coffin-bone, 61
Collar, 284
Collar-bone, 32
Congruent, 247
Conjunctive Membrane, 120
Cordate, 170
Cornea, 123, 141 to 143
Coronet, 370
Coronet-bone, 60
Cortex, 202
Crest, 435
Cristate, 174
Crown of a Tooth, 208
Crupper, 293
Cusps of a Tooth, 215

Dag, 466
Deciduous, 469
Dermopterous, 384
Dew-lap, 406
Didactylous, 345

Digital, 386
Digitigrade, 396
Distichous, 316, 457
Dorsal Fin, 416
—— Region, 290
Down, 448

Ear, III. 101
—— Hole of the, 102
—— external, 103
Elbow, 50
Elevated, 360
Enamel, 201
Epigastric Region, 299
Extensible Membrane, 385
Extremities, 27
Eye, IV. 118, 134 to 140
—— angles of the, 128
—— aperture of the, 127
—— carpet of the, 144
—— external or temporal angle of the, 130
—— internal or nasal angle of the, 129
Eye-ball, 119
Eye-brows, 440
Eye-lashes, 441
Eye-lids, 121

Face, 7, 65
Facial Angle, 71
False Molars, 261
Feet, XV. XVI.
Fels, 431
Femur, 43
Fetlock-joint, 58
Fibula, 45
Fin, 37
—— dorsal, 416
—— caudal, 417
Finger-balls, 363
——glove, 362
Fingers, 40, 344, 352 to 356
Fissipede, 377
Fistulous, 458
Foliated, 169
Foot, 46, 342
Fore-arm, 36
Fore-foot, hand, or paw, 37
Fore-head, 66 to 69
Fore-teeth, 237, 264 to 271
Fossorial, 394
Frog, 371
Fur, 430

Gastric Region, 287
Glossology, 1
Glove, 361
Great-toe, 352
Grinders, 259
Groin, 360
Gullet, 279
Gums, 184 .

Hairs, XVIII. 425
—— of the head, 434
Hairbunch, 444
Hairscum, or suture, 449
Ham, 54
Hand, 37, 46, 372
—— hinder, 46
Hastate, 172
Haunch, 22
Head, II. 5, 84 to 92
—— Fore part of the, 63
—— Hinder part of the, 64
—— Top of the, 62
Heel, 56
Helix, 104
Heterogeneous, 230
Hock, 47
—— point of the, 56
—— joint, 55
Homogeneous, 229
Hoof, 368
Horns, XIX. 459
—— proper, 460

Hornbase, 461
Humeral Region, 302
Humerus, 33
Hump, 407
Hyoid Bone, 283
Hypochondrias Region, 303

Imposed, 233
Incisors, 237, 264 to 271
Index, 353
Inguinal Region, 300
Innate, 232
Insistent, 359
Instep, 47
Intermemoral, 389
Intermaxillary Bone, 10
Interscapular Region, 291
Interval, 242
Iris, 134
Ivory, 199
—— proper, 200

Jaw, upper, 9
—— lower, 12
Jaws, 8
—— opening of the, 182
Jugulum, 280

Knee, 52
—— of a Horse, 51
—— hinder, 55
Kneepan, 53

Lachrymal Fossa, 132
Limbs, XIV. 27, 337
—— anterior, or pectoral, 28
—— hinder, or posterior, 29
Lips, 180
Lobe, 106
Lock, 446
Loins, 304
Loricata, 310
Lumbar, 386
—— Region, 304

Main-stem, 464
Mamma, 322, 325 to 327 to 329
Mammalia, 3
Mane, 442
Marsupial Pouch, 328
Masto-zoology, 2
Maxillary Teeth, 235
Metacarpus, 39
Metatarsus, 48
Milk-hair, 448
Monodactylous, 345
Molars, or Molar Teeth, 238, 240, 241
Moustachio, 438
Mouth, VI. 179
—— corners of the, 181
Multungulate, 402
Muzzle, 70

Nail, 364
Nape of the neck, 276
Nasal appendage, 168 to 175
—— Region, 79
—— Fossæ, 100
Naval, 297
Neck, X. 274
—— Back of the, 275
—— of a Tooth, 209
—— Lower part of the, 277
Nictitating Membrane, 131
Nipples, 323
Normal, 258
Nose, V. 145, 156 to 167, 176
—— external, 154
Nostrils, 146 to 151
—— Bulb of the, 155
—— Partition of the, 153
Nucha, 276

Obserate, 151

Obverse, 248
Occipital ridges, 63
Occiput, 64
Operculated, 109, 150
Ophthalmic Region, 78
Opposable, 358
Oral Region, 80
Orbit, 125
Orbital Fossæ, 93 to 95
Os Coccygis, 23

Pachydermatous, 422
Palate, 11
Palatine Teeth, 236
Palm, 374
Palmated, 379, 468
Panniole, 485
Parotid Region, 77
Pastern, 58
—— upper, 59
—— lower, 60
Patella, 53
Paw, 373
—— hinder, 46
Pectoral Fins, 340
—— Region, 295
Pedunculated, 116
Pelvis, 22
Pencil, 445
Penis, 332
Pentadactylous, 345
Perinæum, 301
Phalanges, 41
Phalanx, last, 42
Pinnated, 381
Pisiform, 359
Plantigrade, 395
Points of a Tooth, 215
Prehensile, 312;
Pricket, 466
Prickles, 427
Proboscis, 177
Procumbent, 252
Protractile, 192
Pubes, 447
Pupil, 135

Quadriangulate, 402

Radius, 36
Redunecate, 474
Retractile, 338, 366
Retrosdingent, 334 .
Ribs, 18
Ridge of a Tooth, 213
Roots of a Tooth, 204 to 207
Rudimentary, 134, 357

Sacrum, 21
Sagittal Ridges, 82
Saltatorial, 392
Scales, 410
Scapula, 31
Sclerotica, 122
Semi-palmated, 380
Shank-bone, 39, 48
Shell, 411
Shield, 411
Shin-bone, 44
Shoulder, 30
Shoulder-blade, 31
—— bone, 33
Sincipat, 63
Skeleton, I. 4
Skin, XVII. 404
Skull, 6
Snags, 466
Snout, 156
Sole, 343
—— of a Hoof, 369
Solidungulate, 399
Spinal Column, 16
Spines, 427
Spiracles, 152
Spitter, 467
Splent-bones, 39, 48
Stellated, 175, 455

Sternum, 19
Stiff Hairs, 439
Stifle-joint, 52
Sub-bisulcate, 398
Sub-orbital Sinus, 133
Superciliary Ridges, 81
Switch-hair of the Tail, 321

Tail, XII. 305 to 319
—— Root of the, 294 ,
—— Stump of the, 320
Tall rings, 415
Tailtuft, 443
Tarsus, 47
Teats, 333
Teeth, VII. VIII. IX. 14, 229 to 230, 243 to 258
Tegular, 367
Temples, 72
Temporal Fossæ, 96 to 99
Tetradactylous, 345
Thigh, 348
—— bone, 43
Thorax, 20
Throat, 278
—— Lower part of the, 280
—— Pit of the, 282
—— Side of the, 281
Thumb, 352
Tibia, 44
Toes, 40, 344
Tongue, 186 to 198
Tooth, 217 to 221, 223 to 228, 231 to 234
—— compound, 211
—— complicated, 212
—— fibrous, 214
—— rudimentary, 273
—— simple, 210
—— supernumerary, 272
Tragus, 106
Transverse, 254
Trenchant, 219
Tridactylous, 345 .
Triangulate, 402
Truncated, 269
Trunk, XI. 285, 15
—— Upper Region of the, 286
—— Lower ———— 287
Tubercles, 216
Tuberculous Molars, 262
Tusks, 263

Ulna, 35
Umbilical Region, 298
Uncinate, 480
Under-chin, 75
Under-cheek, 76
Unguiculate, 400
Unguiate, 401
Upper-cheek, 74
Uropygium, 294

Verrucose, 419
Vertebra, 17
—— caudal, 23
—— cervical, 24
—— dorsal, 25
—— lumbar, 26
Vertex, 62
Villous, 454
Voluble, 311
Vorticillated, 456
Vulva, 335

Wart, 409
Web, 382
Whale-bone, 203
Whisker, 437
Wing, 385
Wool, 433
Woolly-fur, 432
Wrist, 51
Wrist-bones, 38

ORDER I.—BIMANA.

MAMMALIA WITH FOUR DISTINCT UNGUICULATED LIMBS; WITH THREE KINDS OF TEETH; AND OPPOSABLE
THUMBS ON THE PECTORAL LIMBS ALONE.

SYNONYMS.

PRIMATES (in part).—Linnæus and others.
BIMANES.—Duméril, Cuvier, and others.
BIMANA.— Hamilton Smith and others.
ERECTA (Aufrecte Säugthiere).—Illiger.

Man forms but a single Genus, and this Genus is the only one of its Order.

GENUS.—HOMO. MAN.

SYNONYMS.

HOMO.—Erxleben, Illiger, Fischer, and others.
HOMO SAPIENS.—Linnæus and others.
L'HOMME.—Cuvier, Buffon, Duméril, Desmarest, and others.
MENSCH.—Tiedemann and others.
MAN is excluded from the Zoological systems of Pennant, Brisson, and others.

GENERIC CHARACTERS.

THE TEETH continuous, approximated, erect, and nearly of equal length; the FORE-TEETH eight, trenchant, the upper not inserted in an intermaxillary bone; the CANINES four, pointed, unicuspidate; the MOLARS twenty, grinders, tuberculous; eight being FALSE MOLARS, bicuspidate; the remaining twelve, TRUE MOLARS, and quadri-cuspidate. The DENTAL FORMULA for the Adult is therefore

$$\frac{2 + C + (2 F + 3) M}{2 + C + (2 F + 3) M} \frac{16}{16} = 32$$

or the Infant it is

$$\frac{2 + C + 2 M}{2 + C + 2 M} \frac{10}{10} = 20$$

THE PECTORAL LIMBS, with pentadactylous HANDS, eminently disposed for prehension.

THE LOWER LIMBS, with pentadactylous and plantigrade FEET.
THE NAILS flat and feeble, on all the fingers and toes.
THE HEAD, with its FACIAL ANGLE from 70° to 90°. THE CHIN prominent. THE ORBITAL and TEMPORAL FOSSÆ distinct.
THE CLAVICLES perfect.
THE MAMMÆ two, and pectoral. THE PENIS free.
THE SKIN naked except on the skull, also on the chin, axillæ, and pubes of adults.
THE TAIL wanting.
WALKS erect.
FEEDS on many animal and vegetable substances (Polyphagous).
INHABITS all parts of the globe, except the regions immediately surrounding the Poles (Cosmopolite).

As the physical history of Man is directly interesting to ourselves, and ought to form the standard of comparison to which that of other animals is to be referred, it will be proper to consider the subject more in detail.

The several points in which the organic arrangements of Man are peculiar to himself will be briefly contrasted with those possessed by him in common with the other Mammalia. We shall point out the advantages which these peculiarities give him above the other species, and indicate the natural order in which his individual and social faculties have developed themselves. Finally, we shall enumerate the several races of Man and their distinguishing characteristics.

THE PECULIAR CONFORMATION OF MAN.

THE foot of Man is very different from the hinder-hand of an Ape. It is broad, the leg bears vertically upon it, and the heel expands beneath. Its toes are short, and can scarcely bend; the great-toe, longer and thicker than any of the others, is placed on the same plane, and cannot be opposed to them. This kind of foot is therefore proper for supporting the body in an erect position, but can be of no use either for seizing or climbing. Further, the hands of Man are not adapted for walking, and he is therefore truly Bipedal and Bimanous—qualities which are possessed by no other animal.

In fact, the entire frame of Man is disposed for an upright posture. His feet, as we have just seen, supply him with a broader base than those of any other Mammiferous animal. The *extensors* or muscles which maintain the leg and thigh in a state of extension are extremely vigorous, whence result the projections of the calf and buttock. The *flexors* of the leg are attached very high, so as to permit the knee to be completely extended, and thus occasion the calf to appear more prominent. The pelvis is broad, and serves to separate the thighs and feet, giving to the body a pyramidal form highly

29

favorable to equilibrium. The necks of the thigh-bones form, with the bodies of these bones, an angle, which increases the separation of the feet, and enlarges the base of the entire body. Finally, the head, by this upright posture, is balanced upon the trunk, from its articulation being placed above the centre of gravity of the entire mass.

When a Man makes the attempt, he cannot walk conveniently on all-fours. His feet being short and inflexible, and his thigh too long, the knees are thrown against the ground. His broad shoulders and arms, placed far apart from the medial line, are but ill adapted for sustaining the fore part of the body. The rhomboid muscle which, like a girth in quadrupeds, suspends the trunk between the shoulder-blades, is smaller in Man than among any of them. His head is heavier on account of the size of the brain, and the smallness of the sinuses or cavities in the bones of the skull, yet the means of supporting it are most feeble; for Man has not the cervical ligament of quadrupeds, and the vertebræ of his neck are not disposed so as to prevent it from bending forwards. At the very most, he can but sustain his head on a level with the spine, and then his eyes and mouth are directed downwards to the earth, so that he cannot see before him. On the contrary, the arrangement of these organs is perfect when he returns to the upright posture.

The arteries which serve to convey the blood to the brain are not subdivided, as in many quadrupeds, and the blood necessary for so large an organ would be poured into it with too great a velocity, so that frequent apoplexies would be the consequence of his persisting in a horizontal position.

Man is, therefore, formed for resting on his feet only in an upright posture. He thus preserves the entire freedom of his hands for the arts and occupations of life, and his organs of sense are placed in the most favorable position for receiving external impressions.

His hands, which derive so many advantages from their freedom of movement, are not less favored in respect to their structure. The thumbs, which are longer in proportion than those of the Apes, impart in consequence a greater facility for holding small objects; while every finger, excepting the third or ring-finger, is capable of moving separately—a peculiarity which cannot be found in any other animal, not even in the Apes. As the nails protect one side only of the extremity of each finger, they supply a point of attachment to the organs of touch, without depriving them of any portion of their delicacy. Again, the arms which support these hands possess a solid point of attachment in their broad shoulder-blades, their strong clavicles, and the general disposition of the shoulder-joint.

Man is thus highly favored in respect to his fitness for dexterous or skilful movements, but these qualities have been assigned at the expense of his strength. His speed in running is much less than that of other animals of the same size. He is likewise without offensive arms; his jaws are flat, his canine teeth do not project, and his nails are not crooked; while his body, destitute of hair on the back and sides, is absolutely unprotected from the inclemencies of the atmosphere. Finally, he is a longer period of time than any other animal in acquiring that degree of strength necessary to enable him to provide for his own maintenance and defence.

This natural feebleness has, however, one important advantage, that of compelling him to resort to the resources within himself, and particularly to that intelligence which has been assigned to him in a supereminent degree.

No quadruped approaches to Man, in respect to the size and the number of convolutions in the hemispheres of his brain, that is to say, in that portion of the organ which serves as the principal instrument to his intellectual operations, and the hinder part of that organ even extends backwards so as to cover the cerebellum. The very shape of the skull proclaims the magnitude of the brain, while the smallness of his face announces how little that part of the nervous system, influenced by the external senses, is predominant in the human species.

The external senses of Man are all of medium power, yet they are at the same time of great delicacy, and in due proportion to each other. His eyes are directed forwards, and he cannot see on both sides at once like many quadrupeds, yet their position imparts more unity to the results of his vision, and serves to direct his attention more especially to sensations of this kind. The eye-ball and iris

can vary their dimensions but slightly, and this confines the sphere of his vision to a limited distance and a determinate intensity of light. His external ear cannot move to any great extent, and its small size scarcely augments the intensity of sounds; yet he is better able than any other animal to appreciate minute differences of tone. His nostrils, more complicated than those of the Apes, are less so than in any other genus, yet he appears to be the only animal whose sense of smell is sufficiently delicate to be affected by disagreeable odours. This delicacy of smell would lead us to expect a corresponding delicacy in his organs of taste, and Man must possess considerable advantages in this respect at least over those animals having the tongue covered with scales. Finally, the fineness of his touch results from the thinness of the skin, the absence of all insensible parts, as well as the form of a hand, better adapted than that of any other animal for accommodating itself to the minute inequalities of surfaces.

Man enjoys a peculiar pre-eminence in respect to his organs of voice. He alone of all Mammalia can produce articulate sounds, to which the form of his mouth, and the great flexibility of his lips, alike contribute. This means of communicating his ideas is to him of the greatest value, the various modifications of sound being employed most conveniently for this purpose, as they may be perceived at greater distances and in more directions at the same time than any other signs.

It would appear that even the position of the heart and the larger vessels is suited to the upright posture. The heart is placed obliquely upon the diaphragm, and its point directed to the left side, which arrangement requires a disposition of the aorta different from that in most quadrupeds.

Man seems formed for feeding chiefly upon fruits, roots, and other succulent parts of plants. His hands enable him to gather them with ease; while his short and comparatively weak jaws, his canine teeth not projecting beyond the line of the remaining teeth, and his tuberculous molars would permit him neither to pasture upon grass, nor to devour flesh, did he not prepare his food by a culinary process. Once, however, in possession of fire, and a knowledge of those arts which have enabled him to seize or kill at a distance the other animals, all living beings can be made to contribute to his maintenance. This circumstance has further enabled him to increase the numbers of his species without any apparent limit.

The digestive organs of Man correspond with those of mastication. His stomach is simple; the intestinal canal of medium length; the larger intestines well defined; the cœcum short, thick, and augmented with a narrow appendage; the liver is divided only into two lobes and a lobule; and the epiploon hangs before the intestines even as far as the pelvis.

To render this abridged statement of the anatomical structure of Man more complete, it may be sufficient to add, that he has thirty-two vertebræ, seven of which are cervical, twelve dorsal, five lumbar, five sacral, and three coccygeal. Seven pairs of ribs are united to the sternum by cartilaginous appendages, and are termed *true ribs*, the five following pairs are called *false ribs*. The cranium of the adult has eight bones, namely, one occipital, forming the base of the skull, two temporal, two parietal, one frontal, the æthmoid, and the sphenoid. The bones of the face are fourteen in number; two maxillary; two jugal, each of which serves to connect the temporal and maxillary bones by a kind of bridge, called the zygomatic arch; two nasal; two palatine behind the palate; a vomer between the nostrils; two turbinated bones within the nostrils; two lachrymal at the nasal angles of the eyes; and the single bone of the lower jaw. Each jaw contains sixteen teeth, four trenchant incisors in front, one pointed canine at each corner, and ten molars with tuberculous crowns, five on each side, making in all thirty-two teeth. His shoulder-blade has at the end of its spine or projecting crest, a process called the *acromion*, to which the clavicle is attached, and, below its articulation, there is a point called the coracoid process, for the attachment of several muscles. The radius turns completely upon the ulna, on account of its peculiar mode of articulation with the humerus. The carpus has eight bones, four being in each row. the tarsus has seven; the remaining bones of the hands and feet may be easily counted according to the number of the fingers.

By means of his industry, Man commonly enjoys the advantages of a regular diet. With him a uniform attachment to the other sex supplies the place of that periodical rut observed in many other animals. The male organ of generation is not sustained by a bony axis; the prepuce is not attached to the abdomen; and the penis is therefore pendulous. Numerous large veins serve to lead the blood of the testicles back into the general circulation, and contribute towards the moderation of his desires. The matrix of the female is a simple and oval cavity; her mammæ, two in number, are placed upon the breast, and correspond to that facility with which she holds the infant upon her arms.

On comparing the purely corporeal properties of Man with those of the other Mammiferous animals, we thus find that he presents only a few slight differences, insufficient to separate him from their class. Among these, his upright posture is at once the most remarkable and important; for while the quadruped carries the trunk of his body nearly parallel to the ground on four supports, Man rears an upright column erected upon a narrow yet firm basis.

Those who are disposed to consider the anatomical structure of Man with attention, will readily appreciate the doubtful veracity of the accounts of savages found wandering on all-fours in the woods of Europe. Linnæus erroneously considered these as forming a distinct variety of the human race, under the name of *Homo ferus tetrapus* (Linn. Syst. Nat. ed. Gmel. p. 21), of which he enumerates several examples. They were probably only the descendants of some unfortunate outcasts of civilized society, abandoned by their parents during the early years of infancy. Indeed, the property of walking upon four feet appears to be so incompatible with the human organization, that we may safely consider the narrations of Tulpius, Connor, and Camerarius, to be erroneous in this particular. All the more modern and best authenticated accounts of these savages represent them in every case as walking erect. Among these narrations we shall select, as being most entitled to credit, that of the young boy of Aveyron, who resided for a long time in Paris. at the Asylum for the Deaf and Dumb, under the care of the celebrated Sicard. His history has been detailed by MM. Bonnaterre and Virey, from the latter of whom we have obtained the following particulars.

During the year 1795, a naked child was observed searching for acorns and roots in the woods of La Caune, Department of Tarn. He fled at the approach of strangers, but was taken, and afterwards escaped. Fifteen months afterwards he was retaken by three hunters, although he had climbed a tree, and was conducted to La Caune. He again escaped, and lived at large for six months, exposed to the cold of one of the most rigorous of winters, until at length he was compelled by hunger to enter one of the houses in the outskirts of the town of Saint-Sernin, with only a slight remnant of his former garments. He was offered some potatoes, which he devoured raw, as well as chestnuts and acorns. He refused every other kind of food, such as flesh, raw or boiled, bread, apples, pears, grapes, nuts, or oranges, smelling them carefully before tasting them. As he uttered no articulate sound, he was supposed to be naturally dumb. He seemed terrified, and had apparently no other design than that of eating and then fleeing again to the woods. He could scarcely endure clothes, refused to sleep in bed, and seemed devoid of every feeling of decency or cleanliness, qualities which appear to be peculiar to the highly-civilized Man.

This lad was seen by M. Virey at Paris, at the age of eleven or twelve years. He was then strong and well formed for his age, and though his new mode of life had rendered him rather fat and unwieldy, he was still able to run very fast. He walked erect, balancing himself with his arms, and remained nearly all day upon the ground in a sitting posture, eating continually while awake, and sleeping immediately afterwards. His skin, which appeared brown and dirty when he was first taken, became white after being washed. His nails were very long, and his face almost concealed by long flaxen ringlets. On being taken to Paris, he was attacked with the small-pox in a mild form, from which he soon recovered, having refused to take anything during the entire course of the malady. He appeared at times to have spasmodic movements, as if he had been very much frightened. His teeth were nearly bare to their alveolæ, and, being of a careless temper, liking nothing but eating and sleeping, he had grown rather corpulent. All his movements were hasty but sure. He could not swim, and did not usually climb trees, unless compelled by the approach of danger. Once he leaped from the second story of a house in order to flee to the woods. His hands were by no means callous or hard; and his fingers were surprisingly flexible. Although he appeared not to dread the most extreme cold or heat, yet he seemed to prefer the cool shade in summer and the fire in winter. His skin was covered with many scars and marks of burns. When he perspired, he strewed dust upon his skin, not liking the moisture. Though fond of sleeping, his slumbers were never profound; and when at rest, he gathered himself up like a ball, rocking himself by way of assisting his slumbers. He hated children of his own age, yet he was not ill-natured, and never attempted any mischief. Though innocent and foolish, he could not be considered as imbecile; his character was very mild, but he would not endure contradiction. He was frank in his manners, and excessively selfish, though simple and confined in his notions.

This savage continued always upon his guard, and showed his fondness for solitude by seeming annoyed at the presence of strangers. He had not learned to throw stones; and, without being actually timorous, did not exhibit courage superior to other children of his own age. When enraged, he raised a blustering cry, or a murmur in the throat, and had, when first taken, some natural signs of resentment, fear, and other passions. He had no defect in his organs of speech, but his excessive want of attention, and ignorance of the vernacular language, rendered him careless on the subject: at length he became able to comprehend many things, but without attempting to speak himself. This young Aveyronese was very much disposed to steal fruit and other articles of food, but attached no value to anything except it contributed towards his immediate natural wants. He continued in a half-savage state, without ever learning to speak, although much trouble had been taken to instruct him.

Many other stories are quoted in the foreign journals of savages found in Hungary and elsewhere, but they offer nothing remarkable. The authentic instances of girls found in the wild state are perfectly analogous to that of the boy of Aveyron, with the exception that these females exhibited some marks of modesty, in which he was wholly deficient.

It is therefore absurd to maintain, with Moscati and other writers, that Man is formed for walking on all-fours, a position to which the objections already stated are insurmountable obstacles. This supposition must be classed with that equally probable one of Adrian Spigel, who attributed the habits of reflection peculiar to Man to the size of the muscles upon which he usually sits, and the consequent ease of that posture.

The form of the pelvis is one of the most important consequences of our upright position. Its direction is oblique in the Apes and quadrupeds, and the os coccygis, which turns inwards, and is wholly concealed with us, usually appears externally with them in the form of a tail. This obliquity of the os coccygis is one of the chief causes of the difficulty in human parturition. "Aussi la direction du vagin, chez les femelles d'animaux, est parallèle à l'axe des vertèbres sacrées; ces femelles accouchent et urinent en arrière; les mâles s'accouplent aussi à elles par derrière (Venus præpostera); il n'en est pas ainsi des singes, et surtout de la femme, dont la station, plus ou moins rapprochée de la perpendiculaire, ramène en devant l'ouverture du vagin. La direction du canal utéro-vaginal est, en ce cas, oblique de devant en arrière, d'où il suit que l'écoulement des urines, des menstrues, a lieu en devant, de même que l'accouplement (Venus antica), et le part est plus laborieux." This inconvenience would not occur, had the human species been supplied with a tail and went on all-fours, as certain credulous travellers have related. The other animals are never afflicted with inguinal hernia, which is the occasional consequence of the downward pressure of the intestines, and the upright position of Man.

The suspensory muscle of the eye-lid is not found in the human species, as it was not the intention of Nature that Man should keep his eyes directed towards the earth, like the Ruminating animals. He is also destitute of the *paniculus carnosus*, or sub-cutaneous muscle, the *pancreas Asellii*, the *corpus Highmorii*, the hepato-cystic ducts, the nictitating membrane, and the incisive fossa behind the upper teeth. The cervical ligament also has not been assigned to us, as our upright position would render it useless.

The great size of the human head and the weakness of the arms are obstacles which prevent Man from swimming naturally like the quadrupeds. The infant, different from the young of other animals, sinks with its head foremost, and even the adult will swim more easily with his back downwards. It is evident, therefore, that our species was never intended for an aquatic or amphibious life, as some have supposed. The accounts of Mermen and Mermaids found in various places, either resolve themselves into absolute imposture, or the objects erroneously described were merely Seals or Lamantines. They are generally described with palmated hands, like a duck, short arms, flat noses, the figure of a beast. a body terminated with two paws or a forked tail, the skin covered with scattered hairs of a gray or brown colour,—characters which agree sufficiently with those of a Seal, to warrant the above conjecture.

THE PHYSICAL DEVELOPMENT OF MAN.

BIRTH AND INFANCY:

THE number of offspring in the human species is usually one at each parturition, and twins do not occur more frequently than once in five hundred accouchements. Births, in which a greater number is produced, occur still more rarely. Gestation continues for nine months. A fœtus of one month old is usually an inch in length; of two months, two inches and a quarter; of three months, five inches; of five months, six or seven inches; of seven months, eleven inches; of eight months, fourteen inches; of nine months, eighteen inches. Infants born at a less period than seven months very seldom survive.

The milk-teeth begin to appear some months after birth, commencing by those in front. There are twenty at the age of two years (see the Dentary Formula), all of which fall successively towards the seventh year, to be replaced by others. Of the twelve back molars which do not fall, four appear at the age of. four and a half, four at nine years, and the last four do not sometimes appear until near the twentieth year.

The fœtus grows more and more in proportion as it approaches the period of birth:—the infant, on the contrary, grows at a less rapid rate as it advances in years. At the period of birth, it is more than a quarter of its ultimate height; it has reached the half at the age of two years and a half; and the three quarters at nine or ten years.

The first cries of the newly-born infant indicate the uneasiness of its change from the one mode of existence to the other. On its first appearance it is washed in tepid water, and dried; its umbilical cord is tied, and cut above the ligature. The women in savage countries bite this cord through with their teeth, and do not always tie it, yet hæmorrhage is not always the consequence of this neglect, and the Hottentot women do not even remove the slight mucous fluid left upon the skin. Among several nations of the North, it is customary to plunge the newly-born infants into cold water, or even to roll them in the snow. This practice was anciently adopted by the Scotch, Irish, Helvetians, and Germans. In our own times, it is still practised by the Morlachs, Icelanders, Siberians, and several others; and, though it sometimes hardens the cellular tissue, renders it violet-coloured, and causes them to perish, yet it has the effect of accustoming the survivors to the cold at an early age, and induces a more robust habit of body. At the moment of birth an important change is effected in the circulation of the blood. The air rushes into the lungs, and the blood which fills these viscera returns to the heart by the vena portæ, and is distributed throughout the body by the aorta and its branches. Before this period the blood passed immediately from the right to the left ventricle of the heart. The infant at birth is still cartilaginous. Its limbs are small; its flesh soft, gelatinous, and moist. Its vessels' are large and wide; the brain considerable; the belly distended; the cellular tissue surrounding its. organs loose, spongy, and filled with lymph; its glands swollen and filled with watery fluid. A milky fluid can sometimes be squeezed from its breasts during the first days after birth.' Its eyes are dull, wrinkled, and covered with a slight membrane (tunica Halleri), which prevents the too violent action of the light upon these still delicate organs. The ears are closed by a mucous fluid, which prevents the admission of loud sounds. The pituitary sinuses of the nose are obstructed by a viscous humour, so as to be incapable of smell; the skin is too soft to convey any sensations of touch; and the tongue can scarcely taste anything. The use of the senses is acquired gradually; and instinct alone directs the infant mechanically towards its mother's breast, and instructs it to suck.

At birth, the infant is about twenty inches in length, and weighs from six to ten pounds. The first milk of the mother, or colostrum, is serous and laxative, and serves to clear the intestines of the meconium, or blackish fluid, which is discharged during the first day after birth; yet the infant is commonly delayed for twelve hours before it is permitted to be fed. Nature has wisely adapted the qualities of the mother's first milk to the wants of the infant. The milk of nurses is much less suitable, being too old and substantial, but it is more serous in the mother in proportion as it approaches the period of birth. At all times, the milk of a stranger is not so well adapted to the temperament of the infant as that of its own mother, and the milk of any other animal is still more objectionable. There is nothing so judicious in these matters as a scrupulous attention to the suggestions of Nature.

The newly-born infant sleeps almost continually, and requires the breast every time it wakes. The Negro infant clings to the long breast of its mother, and holds itself so firmly to her back, that she can attend to her other labours without the trouble of supporting it upon her arms. Towards the fourteenth day the infant begins to smile and recognize those

who approach it; but it does not attempt to speak until about the tenth or twelfth month. The words most easily pronounced are composed of labial consonants, such as papa, mamma, baba, and these have, therefore, served in most languages to denote the same objects.

For the first three months after birth, no other food should be given to the infant than the milk of its mother; afterwards, several nutritions substances, easy of digestion, may also be employed. The natural period for discontinuing the milk is commonly on the appearance of the first teeth. The incisive teeth, eight in number; being four in the front of each jaw, appear at the age of eight or ten months. Their growth is painful, and is announced by fever and inflammation. At this critical period, very little food should be given to the infant. The order in which the teeth exhibit themselves has already been explained (See pages 86, 87): Infants are sometimes born with their front teeth; but these examples are very rare.

The hair of newly-born infants is always more or less fair in the European variety, but in other races of Man it is already quite black. The same remark applies to the iris; and the colour both of the hair and eyes becomes deeper as they advance in years. The infants of Negroes, and persons of dark complexions are born rather of a lighter tint, but they become gradually darker and darker, although they are not exposed to the rays of the sun. The growth of girls is usually more rapid than that of boys.

PUBERTY.

IN our climates, the human species exhibits the first signs of puberty from the ages of twelve to fourteen in girls, and from fifteen to seventeen in boys; but these periods vary all over the globe. They seem to depend upon the temperature of the climate, the quantity and quality of the food, the general purity of morals, the temperament of the individuals, their employments, as well as the peculiar constitution of the races to which they belong.

Heat, as is well known, tends to increase the activity of the vital power in all organized bodies; and ought, consequently, by hastening their growth, to bring the period of puberty nearer to that of birth. An inhabitant of Finland or Iceland is scarcely marriageable at eighteen years of age, and some are even as old as twenty-two before they become so, from their exposure to the excessive cold of their climate; the women also are not marriageable until seventeen or even nineteen. On the other hand, in Hindoostan, Persia, and Arabia, the males are capable of marrying at the age of thirteen or fourteen, while girls are often mothers at ten or twelve. Intermediate climates may accelerate or retard the puberty of the people exposed to their influence.

Among the white races of Europe, these variations are very considerable, especially in respect to the females. Thus, in Saxony, Thuringia, and Upper Germany, the women are not marriageable before fifteen even in the towns, and they are still slower in countries lying farther to the north, or in elevated places, where they are sometimes delayed until the ages of twenty to twenty-four. For this reason, the females in the islands of the North, the Orkneys, and the Western Islands, preserve their fecundity to an advanced age. In Ireland, sixteen appears to be the usual period; in France, it is fourteen or even thirteen in the southern departments or the large towns, where various causes combine, and induce a greater degree of precociousness. In Italy, the women are formed at twelve years of age; and it is the same in Spain; while at Cadiz, marriages frequently take place at that age. In Minorca, eleven is the period of puberty; and at Smyrna, mothers have been seen at eleven or twelve. The Persian women are marriageable at nine or ten; and nearly the same thing takes place at Cairo. The Berber women are often mothers at eleven. according to Shaw; also the Agows of Abyssinia, according to Bruce. From nine to ten, they appear to be marriageable at Senegal, according to Adanson. The age of ten years is the usual marriageable period, not only in Arabia, but also in most parts of Africa.

Many instances are quoted of an equally great precociousness being observed on the Malabar Coast, where the females are married at eight to ten years, and become mothers soon afterwards. In the Deccan, according to Thévenot, women have been known to have children at eight years of age. Paxman has seen married children from four to six years of age; but it is impossible to believe that these could have reached the period of puberty. In fact, a common custom prevails in India, for the inhabitants to betroth and even to marry their children together, for which reason, girls have been known in Java and Hindoostan to be mothers at ten years of age.

These instances are not, however, very general, and remarkable exceptions of the same kind are often observed in the temperate regions of Europe. Thus, Haller saw Swiss girls of twelve years, and Smellie, English girls of the same age, exhibiting the usual signs of puberty. In Belgium and Switzerland, girls have produced at nine years of age; but nothing general should be inferred from these solitary cases.

When individuals of the Negro races are removed to more temperate climates than Africa, such as Europe or North America, they arrive at puberty sooner than the White races, the difference in this respect commonly amounting to a year. This serves to indicate that their races are naturally more precocious than ours. The same remark is also applicable to the Mogul races. On the authority of several travellers, the females of Siam, Golconda, China, and Japan, are marriageable at eleven or twelve years; but even in climates much colder than our own, individuals of the same races still continue to be more precocious. A Calmuck, or a Mogul woman of Siberia, in a climate as cold as that of Sweden, is marriageable at thirteen, while a Swedish woman would not be so under fifteen or sixteen. The Samoiede and Lapland women present the usual appearances of puberty at the age of eleven or twelve, and the males at twelve or thirteen, while the English, German, and French, become marriageable only at a much later period. We thus see that each human race possesses a peculiar aptitude in this respect, that the individuals of one race naturally become formed at an earlier period than those of another, and this they do notwithstanding the various differences of climate, food, and temperament.

Females who become marriageable at an early age soon lose their powers of conception. From the ages of thirty to thirty-five years, the women of India are accounted old, according to Paxman. The Javanese women do not conceive beyond thirty years, and even in Persia, according to Chardin, many women do not produce after twenty-seven years. The Siamese women are apparent exceptions to this observation, for they are said to have children at the age of forty, although they attain to puberty at a very early period. Upon the whole, it may be stated generally, that the usual period for the commencement of puberty in females varies, under the burning sun of the Tropics, from nine to twelve years, and terminates about thirty, though sometimes extending as late as forty years. On the other hand, the Samoiede women, though marriageable at an early age, continue their powers unimpaired to forty-one. In our own country, its termination may be stated at the ages of forty to forty-five years.

Puberty is represented by external signs. These appear (in France) from the ages of ten to twelve in girls, and from twelve to sixteen in boys. In warm countries they are sooner observable; and either sex is rarely capable of procreation before that period.

Among other well-known signs of puberty, the deepening of the voice in the male, and the expansion of the breast in the female, are the most obvious and constant.

MATURITY.

After attaining the age of puberty, the body soon acquires its maximum of height. Some young men do not grow after fourteen or fifteen years of age; while others continue as long as twenty-two or twenty-three years. Nearly all of them, during this time, are of a slender make, their thighs and legs small, with the muscular parts not so perfectly developed as they ultimately become.

At the age of eighteen years the youth usually ceases to grow. The adult man rarely exceeds the height of six feet, and is seldom found below five feet. Women are usually some inches less.

The stature of individuals is subject to great variations. Dwarfs have been known scarcely more than two feet high, while we are not without authentic accounts of giants nearly nine feet in height. Contrary to what is observed among the domestic animals, the medium height of one human race is but little different from that of another. The height of women is less variable than that of men. Among all nations of great stature, they are very considerably smaller than the men, while the difference in the relative height of the sexes varies but slightly among the races of diminutive stature. The height of the smallest dwarf bears to that of the largest giant the ratio of 1 to 4 very nearly; and, supposing them to be equally well-proportioned, their ratio of bulk will therefore be as 1 to 64. On the other hand, the medium height of the smallest races is to that of the largest only as 1 to 1½; consequently, their ratio of bulk will be very nearly as 1 to 3½.

The following tables, deduced by M. Isidore Geoffroy-St-Hilaire, from a number of scattered observations, published by different authors, serve to exhibit the average amount of the hereditary variations of height.

NATIONS REMARKABLE FOR THEIR GREAT STATURE.

	INHABITING	CLIMATE.	HEIGHT. FEET. IN.	AUTHORITIES.
Patagonians, ...	45° to 50° S. lat.	Rather cold.	6 4.7	La Giraudais, Malaspina.
Do.	do.	do.	6 2.6	Commerson, De Gennes.
Inhabitants of the Navigators' Islands...........................	14° S. lat.	Warm.	6 2.6	La Pérouse.
Carribees, ...	8° to 10° S. lat.	Very warm.	6 1 5	Humboldt.
Patagonians, ...	45° to 50° S. lat	Rather cold.	6 0.5	Bougainville.
Mbayas, ...	20° to 21° S. lat.	Warm.	6 0.5	Azara.
New Zealanders, ...	35° to 45° S. lat.	Rather warm.	5 11.4	Garnot & Lesson.
Otaheitan Chiefs, ...	17° S. lat.	Very warm.	5 10 3	do.
Marquesas, ...	10° S. lat.	do.	5 10 3	Marchand.
Patagonians, ...	45° to 50° S. lat.	Rather cold.	5 9.	Cook, Wallis.

NATIONS REMARKABLE FOR THEIR SMALL STATURE.

	INHABITING	CLIMATE.	HEIGHT. FEET. IN.	AUTHORITIES.
New Holland, ...	35° S. lat.	Warm.	5 2.9	Quoy & Gaimard.
Inhabitants of Vanikoro, ...	12° S. lat.	do.	5 2.3	do.
Orotchys Tartars, ...	51° N. lat.	do.	5 1.8	La Pérouse.
Kamtschatkadules, ...	50° to 60° N. lat.	Very cold.	5 1.8	do.
Papous of Offuck, ...	0° to 1° S. lat.	Very warm.	4 10.6	Garnot & Lesson.
Different European and Asiatic Nations near the Arctic Circle, ...	60° to 75° N. lat.	Very cold.	{ 5 1.8 { to { 4 9.6	{ Krusenstern, La Pérouse, { Regnard, Depaw.
Esquimaux, ...	70° N. lat.	do.	4 3.2	Hearn, Depaw.
Boschismans, ...	30° S. lat.	Rather warm.	4 3.	Barrow, Péron.

The differences in the statements regarding the average height of the Patagonians may partly be explained by the circumstance, that these tribes are of migratory habits, and partly through the observers paying too much attention to the individual variations of stature.

It has long been remarked, that the nations of smallest stature abound more especially in the northern hemisphere, and towards its most northern extremity. The second of the preceding tables shows this fact clearly; but it also exhibits some exceptions. On the other hand, the tallest nations are almost confined to the Southern Hemisphere, where they form two series, the one continental, from the Carribees to the Straits of Magellan, and the other insular, extending from the Marquesas to New Zealand, throughout the islands of the Southern Ocean. This peculiarity, though often interrupted, commences about 8° 10' of S. latitude, and terminates towards the 50th degree. In the Southern Hemisphere we meet, however, with many nations, whose medium height, without being very

small, is still above the average; and reciprocally, there are many in the Northern Hemisphere of very considerable height.

Upon comparing the geographical position of nations, whose stature is very great, with that of the very diminutive races, we are led to notice a very curious, and apparently paradoxical result. Nations of very small stature are always found in the immediate vicinity of the very tallest inhabitants of the whole globe; and, reciprocally, there are nations of considerable stature dwelling in the neighbourhood of the very smallest. Thus, in the Southern Hemisphere, the island of Terra del Fuego, though separated from Patagonia merely by the Straits of Magellan, and only at a short distance from the Navigators' Islands, is inhabited by a very diminutive and-ill-made race of men. It is the same in the Northern Hemisphere, where the inhabitants of Sweden and Finland, though bordering upon Lapland, are rather above the middle stature. Thus the influence of climate upon the height of the human races appears unquestionable, al-

30

though often modified and even wholly counteracted by other causes. Excessive cold tends to arrest the development of the human frame; while, on the contrary, a moderate degree of cold is favorable to it. We find that the nations of the coldest climates in Europe, Asia, and America, such as the Laplanders, Samoiedes, and Esquimaux, are of small stature, and likewise the people of Terra del Fuego in the Southern Hemisphere.

Again, we see that the inhabitants of all countries, which, in reference to the temperate parts of Europe, we should term *rather cold*, are of very considerable dimensions. The Swedes, Finlanders, Saxons, the inhabitants of the Ukraine, and many other nations of Europe, Asia, and North America, are instances in the Northern Hemisphere, and we have the Patagonians in the Southern. Again, in our own island, while the people of the south and centre of England are of ordinary stature, the inhabitants of the Border counties and of several districts in the Lowlands of Scotland are in general of very considerable stature. Further, in the Highlands of Scotland, where the cold is severe, the stature falls rather below the average. The comparative moisture of the several localities may be another cause of these variations, and when united with those of temperature, serve to account for some of the most remarkable differences.

The elevation of a country is another cause of these variations. In tropical climates, the inhabitants of the several regions of elevated mountains present an epitome of those differences of stature which we trace throughout the several climates of the globe. Nations dwelling upon slightly elevated plateaus are in general tall and robust; while men only of small stature are found in the neighbourhood of mountain fastnesses, which are as desert as the Polar regions, and covered, like them, with eternal snows. In the mountains of temperate, and especially of cold climates, the height of the people dwelling on plateaus, even but slightly elevated, diminishes considerably, in consequence of the more marked differences of temperature. These relations are not, however, invariable; the mountaineers of Puy-de-Dome, and especially the Swiss, are, in some rich cantons, according to M. Villermé, not only of middle stature, but rather above the average standard.

From the various circumstances already noticed, it follows that under every isothermal line, except in the immediate neighbourhood of the poles, nations are to be found of very great stature, others very small, and others again of medium size. Even in the same regions, or in countries apparently identical in physical character, we find races of very different degrees of stature. Thus, the Hottentots, in the immediate neighbourhood of the Caffres, but unquestionably belonging to another race, are much smaller; and, what is still more remarkable, we may find in several islands, such as the Friendly, the Society, and Sandwich Islands, two classes of men very unequal in height. "In the Sandwich Islands," observes M. Gaimard, "the population is divided into two very distinct classes, being the chiefs and the populace. The first enjoy a more abundant diet, consume more animal food, are never compelled to labour excessively, and intermarry together; they are consequently tall, strong, and well made. The others possess no land, and cannot always obtain good food; they are generally of inferior height and strength."

The causes which M. Gaimard assigns for these variations are fully confirmed by the recent observations of M. Villermé upon the average height of the population in France. He found, as Haller and other physiologists had previously conjectured, that the human stature is always greater, other circumstances remaining the same, when the country is rich and fertile; that, where there are good clothing, lodging, and especially wholesome food, it improves, and diminishes where difficulties, fatigue, and privations are experienced during infancy and in early youth. From these facts M. Villermé concludes, that the hardships experienced by most mountaineers form one of the causes which have hitherto retarded their growth, and this observation may further be extended to the inhabitants of the arctic regions, where they receive the two-fold influence of cold and want.

The differences found in the heights of the several nations of Africa cannot be explained by any of the above causes. They serve to show that there must have been some original difference of stature in the primitive types of the several races; as well as to demonstrate the tendency of the human species, in common with the domestic animals, to transmit the connate varieties of races to their posterity.

It is, at the least, very improbable, that the average stature of the human species could have sensibly diminished through the lapse of ages. Antiquity believed in the existence of whole nations of Giants; but it also credited the existence of Pygmies, Troglodytes, or Myrmidons. Many travellers, and especially Péron, show, that savages, far from being stronger than individuals of the more civilized races, are usually feebler. Man, on becoming civilized, has, therefore, lost nothing of his original strength.

A man is longer in arriving at his full growth than a woman, the latter being usually as completely formed at twenty years as a man at thirty. Every part of the form in either sex, to use the words of Buffon, announces the superiority of the human species over all living creatures. Man

maintains his body erect and elevated; his attitude is that of command; his countenance is directed towards the heavens; and presents an august face on which is impressed the character of dignity; the image of his soul is painted in his physiognomy, and the excellence of his nature penetrates through the material organs which surround it, animating the features of his face with a divine expression. His majestic carriage, his firm and resolute step, announce the nobleness of his rank. He touches the earth solely by his most remote extremities, and seems to regard it at a disdainful distance. His arms are not given to him merely as pillars to support the mass of his body; his hand is not permitted to tread on the ground, or to lose, by a continuous friction, that fineness of touch, of which it is the chief organ. His arm and hand are reserved for nobler purposes,—to execute the suggestions of his will, and be subservient to the various circumstances of life.

All the features of the face remain in a state of calm repose while the mind is tranquil;—their proportion, connexion, and harmony, serve to indicate the tranquillity which reigns within. When the mind is agitated, the human face becomes a living tablet, upon which the passions are transcribed with delicacy and energy; where every expression of the mind is represented by a corresponding trait, every mental process by a characteristic, the vivid impression of which often serves to betray the intended action, and represent externally the image of our secret thoughts. The body of a well-made man, according to Buffon, ought to be rather square, the muscles well-expressed, the form of the limbs well-defined and the features strongly marked. When contrasted with the female, we find him of a taller stature, larger and firmer muscles, a browner skin, a larger brain, the bones more robust, the voice deeper, the chest broader, the hairs more numerous and of a deeper tint.

In woman, every thing is more rounded, the lines are softer, and the features more delicate. "To man," says Buffon, "belong strength and majesty, while grace and beauty form the embellishments of the other sex." The hair of her head is longer, finer, and more flexible; her skin lighter and more delicate, her limbs more graceful, the pelvis broader, the thighs thicker, and the limbs smaller. In the man, the upper parts of the body, such as the chest, the shoulders, and the head, indicate strength and power; the capacity of his cranium is considerable, and contains three or four ounces of brain more than the skull of the female, according to the experiments of M. Virey; but his haunches, his pelvis generally, and thighs, are narrower and thinner than hers. The upper part of a man is, therefore, broader than the lower, so that he somewhat resembles a reversed pyramid. In the woman, on the contrary, the head, shoulders, and chest, are small and narrow, while the pelvis and adjacent parts are broad and large, for which reason her body appears to converge upwards towards a point, like an erect pyramid. This difference in their form corresponds to the appropriate functions of either sex. The man being destined by nature for labour, is formed, rather for the employment of his physical energies, in making provision for the maintenance of that family of which he is the chief; while the other sex, to whom the business of reproduction more especially belongs, requires a more capacious pelvis to fulfil the conditions of parturition. The trunk of the female is longer in proportion than that of the man; her lumbar region is more extended, her neck thinner and longer; while her legs, thighs, and arms, are shorter. From these circumstances result her more slender form, as well as the elegance, lightness, and ease, of all her motions.

There are many circumstances in the constitution of woman analogous to the characteristics of infancy in both sexes, and serving to indicate that her organization is not so highly matured as that of Man. Her bones are smaller and thinner; her cellular tissue more spongy and humid, imparting a roundness and plumpness to her form, and increasing the flexibility of her whole frame. Her pulse is weaker and more rapid; her skin is smooth, and almost deprived of scattered hairs, as well as of a beard, excepting after the age of parturition has passed, when the hair begins to grow plentifully upon the chin. It often happens that women have a smaller number of molar teeth than men, so that it may be said, with truth, that the wisdom teeth of many women never appear at all. They in general eat less than the other sex, preferring soft and saccharine food; while the man, being more energetic and vigorous, is instinctively led to prefer the more substantial and stimulating qualities of animal substances.

The beauty of the fair sex varies greatly all over the globe. In the north of Europe, the women are found more frequently than the men with light hair and eyes, and their dazzling whiteness often degenerates into insipidity. All the southern women are brown, and more or less striking; but the most beautiful of the sex, according to our notions of beauty, are found in the temperate parts of Europe and Asia. The centre of Spanish beauty seems to lie towards Cadiz or Andalusia, while the most agreeable Portuguese women are in the neighbourhood of Guimaraens. Women of great beauty are also seen in many parts of Italy and the adjacent islands; in particular, the Sicilian and Neapolitan ladies, descended from ancient Greek colonies, are accounted exceedingly beautiful. The Albanian women are well made; the females of the island of Chio appear charming.

Those of the Ægæan Archipelago are very white, lively, and agreeable; like all Greeks, their eyes are very large and of great beauty.

But the most enchanting models formed by the hand of Nature are allowed to be the Circassians, the Cashmerians, the Georgians, the Mingrelians, and in general all those from Gurgistan, Imerita, and the neighbourhood of the Caucasian chain of mountains. All travellers agree upon this point; and the beauties of these nations being largely exported for slaves, are exclusively reserved, by the laws of Mahomedan countries, for the Faithful alone, while Jews and Christians are not permitted to purchase them in the markets of the Turkish empire. According to the most recent observations, the Lesghian women surpass all the others in beauty. Their manners are not, however, the most unobjectionable. In the regions inhabited by this beautiful race, scarcely an ugly countenance can be seen in either sex. Constancy is there as rarely to be found among the women as jealousy among the men. It is a remarkable fact, that this handsome race is immediately surrounded by the ugliest inhabitants of the earth, the hideous Calmucks and Nogais Tartars, with flat noses, high cheek-bones, their eyes far apart, their skins deeply sun-burnt, and of a dark brown colour. Yet the climate, the soil, the habits and mode of life in both, are the same; but the races are very different. The Calmuck women are not less frightful than their husbands. Imagine a mouth almost reaching to the ears; a skin of the colour of soot; oblique eyes, not very dissimilar to those of a goat; a nose so flat that the holes of the nostrils are alone visible; the lips and cheeks projecting and elevated; the hair stiff, black, and coarse as a horse's mane; a small stature and meagre limbs; flaccid mammæ hanging down like sacks of tanned leather, surmounted with a jet black nipple,—and we have the picture of a Calmuck beauty. The young Circassian female is a complete contrast in every particular. With a skin of the utmost delicacy and whiteness, she possesses fair and flowing tresses, blue eyes, a gently-swelling bosom, a slender and flexible figure,—qualities which are combined with a lightness of step, a softness of voice, and an expression of the eye, which renders her the most charming of women, in respect at least to her external form. We must not, however, expect to find in her either the polished education, or that propriety of conduct which belongs exclusively to the most civilized nations. It is indeed chiefly from intermarrying with the females of Cashmere, Circassia, and other nations inhabiting the ancient Colchis, that the higher classes among the Persians are distinguished from the lower. The descendants of the ancient Guebres, or Parsees, of the sect of Zoroaster, who were forbidden, like the Jews, to marry out of their own caste, still continue brown and very ugly.

The expressive physiognomy and brilliant complexions of the English ladies are admired by foreigners. The general elegance of the bust, probably derived from their Norman ancestors, finds an exact counterpart at the present day in Belgium, Normandy, and Switzerland. Their hair and eyes are most commonly light, sometimes the former is even red, especially towards the northern part of the island. In Scotland, the bones of the cheeks and ancles appear rather too prominently, and the general complexion is paler, approaching more nearly to that of the Dutch ladies. The latter are remarkable for an excessive *embonpoint*.

Among the Germans, the Saxon ladies bear the palm of beauty. Scarcely an ugly face can be seen in the territory of Hildesheim, and the charming complexions of its inhabitants have given rise to the German proverb, that " The pretty women spring from the earth like flowers." Although the Austrian ladies are not in general ugly, the Hungarians far exceed them in beauty; but among all the German nations, a tendency towards *embonpoint* is rather common.

The Polish women are said to possess all the whiteness, as well as the coldness, of their native snows. This observation must, however, be considered as rather hyperbolical, for nearly all the females of Sclavonian origin are lively and ardent, though doubtless their countenances want expression. The Russian ladies were recently in the habit of daubing their faces with a thick paint, while the abuse of the vapour-bath soon deprived them of their attractions. Their forms are masculine, and like most Sclavonian women, their dispositions are energetic and passionate. The Albanian females are more agreeable than the Morlachians. The skins of the latter are very much sun-burnt, while long, pendant mammæ, surmounted by black nipples, are often exhibited to the traveller's eye. In the extreme North of Europe, in Denmark, and Sweden, the women are almost always of a pale blonde, with blue eyes, and complexions often fading into a deadly whiteness. They usually have very large families, especially on the borders of the Baltic Sea.

The most beautiful French women are found towards Avignon, Marseilles, and the ancient Provence, which was formerly peopled by a Greek colony of Phocians. Further to the north, the Cauchoises, the Picardes, and the Belgians, are the prettiest, their skins being of a dazzling whiteness. There is, however, less elegance in their motions, as well as delicacy in their forms. The Parisian ladies are more distinguished for the finished elegance of their manners than for beauty. In Brittany, the ancient Armorica, the women are generally too thick in the limbs. The

prettiest Portuguese ladies have rather long necks; those of the Castilians are excessively short. The more attractive beauties of Italy are found in Tuscany, about Florence, Sienna, or even Venice; but in Lombardy and the neighbourhood of the Alps, their forms, being more voluminous and massive, are less attractive.

In those regions of Asia situate on this side of the Ganges, and peopled, like Europe, by the same white race, we still observe some beautiful features in the females. The Persian women, born under a fertile and temperate climate, are generally very agreeable. The women of Turkey are pretty, for the most part. " Even among the lower classes in the East, every woman," says Belon, " has a fresh and blooming countenance, with a white skin as soft as velvet," probably owing to the frequent use of the hot-bath. They destroy the hair on every part, excepting the eyebrows and head, with the *Rusma*, a depilatory substance made of lime and orpiment; while they tinge their nails and fingers red with *Henna* (*Lawsonia inermis, Linn.*) From the excessive inactivity of their harems, their countenances acquire, according to the Turkish expression, the roundness of the full moon. An unusual roundity of form is here considered as the highest beauty; so that, according to Volney, beauty is estimated by the quintal. As in Egypt, their mammæ are of enormous size. Yet nothing can be more monotonous than the physiognomy of all Turkish women, owing to their faces being always covered. Indeed, so much are they attached to this practice, that some of the poorer women, who can afford only a partial clothing, prefer exposing any part of their body rather than the face. Their countenances thence become wholly destitute of every expression.

The Arab women, although tolerably agreeable in their extreme youth, and remarkable at all times for large black and brilliant eyes, which their poets compare to those of the Gazelle, disfigure themselves by passing a large ring through the cartilage of the nostril; also by designs engraved upon the skin with the point of a needle, and dyed of various colours. The Hindoo women place a similar ring in the left nostril. The heat of the sun dries up and browns the Bedouin and Hindoo females. They sometimes paint the forehead and cheeks blue, and the nails always red.

Nearly the same observations are applicable to the Moorish and Berber women of the white races; their features are considered tolerably regular. Those who never leave the harems and towns preserve a very white appearance, according to Bruce and Poiret. They are even violated or blanched like plants which vegetate in obscurity.

In Malabar, Bengal, Lahore, Benares, all Hindoostan, and Mongolia, the women seem agreeable in general, but small, yellow, and slender; partly from the heat of the climate, which enervates them, and partly from marrying excessively young, at ten or twelve years of age, before their constitutions can be completely formed. The continual transpiration which they experience from the surface of the skin renders its appearance always fresh, and this is increased by the use of perfumed oil of cocoa. The latter is also copiously applied to the hair; and they make frequent application of some depilatory substance. It is stated that the jaws of the women of Malabar are very narrow, that their legs are long in proportion to the body, and their ears placed very high. All the women of the East, according to many travellers, have the pelvis very broad, a defect which the Armenian and Jewish dealers in female beauty endeavour to remedy by tight bandages.

The females of the White races, according to the notions prevalent in our climates, engross the whole beauty of the sex. We must, however, add a few words regarding the Yellow and Black beauties,—such, at least, as they appear in the eyes of those who have learned by habit to get rid of the prejudices of colour.

In Asia, the yellow ladies of Golconda and Visapour are much prized; their features are lively and attractive. Those of Guzerat are olive-coloured; but paler than the men, who are more tanned by the heat of the sun. It is said that the prettiest Chinese beauties come from the province of Nan-king, and Nan-chou its capital.

Even the Negresses are not without their degrees of beauty. In the region of the East, they bear their proportionate prices, especially the younger females. According to the reports of the slave-merchants, no black beauty is ever imported from countries where the waters are bad, or the soil steril. The black women from the shores of the Red Sea are much esteemed by the Persians, who import a great number annually. The East Indians are also partial to the Caffre girls, who are entirely black, and exported in large numbers from Mozambique. " Les femmes Kamtchadales et Samoïedes ont, dit-on, les parties de la génération très-larges. On sait que plusieurs Hottentotes ont les grandes lèvres du vagin longues et pendantes comme le fanon du bœuf, et quelquefois découpées en festons; mais elles n'ont point ce prétendu sablier de peau qu'on leur attribuoit; les femmes des Houzouanas portent vers la croupe un coussin de graisse qui ressemble à un cul postiche."

Many attempts have been made to estimate the total number of human beings on the surface of our globe; but relative to this subject nothing has hitherto appeared but conjectures of great vagueness. While some would consider them as amounting to a thousand millions, others, with greater probability, estimate their number as low as six hundred millions. This principally arises from our ignorance regarding the amount of the population in the great empires of Asia, the innumerable states of the interior of Africa, the vast territories of America, and in New Holland. M. Malte-Brun makes the following approximations, which are probably very far from the truth—

> Europe,......................170 millions.
> Asia,............................320 to 340.
> Africa,...................... 70
> America,.................... 45 to 46.
> South Sea Islands........ 20
>
> Total,   625.to 646 millions.

The population of China by some accounts is said to amount to 333 millions, while others make it as low as 19½ millions.

In respect to the proportion which the number of the one sex bears to that of the other, our information is more precise. More males than females are always born in the civilized countries of Europe; yet the number of females exceeds that of the males. In England and Wales, and in Sweden and Finland, the male births are to the female as 100 to 96; in France as 100 to 95. The ratio of 25 : 24 is approached very nearly in all other places where observations have hitherto been made. Thus, at Petersburg, 100 male infants appear to 95 females; at Paris, 100 males to 96 females. By one enumeration made in France, during the ministry of M. Chaptal, the returns were 100 males to 95 females; and once for Paris they were 100 males to 97 females. Süssmilch assures us, that 100 males are born for 93 females in North America. In New Spain, according to Humboldt, 100 males to 97 females.

This law of Nature will be more satisfactorily illustrated by the following Table, showing

THE PROPORTION OF THE SEXES AT BIRTH, .

In	During the	Preceding	Males.	Females.	Authorities.
England and Wales,	.	.	29 years	1800	3,285,188	3,150,922	Population Abstracts.
Sweden and Finland,	.	.	20 .	1795	1,006,420	965,000	Wargentin.
France,	.	.	.	3 ..	1802	110,312	105,287	La Place.
Scotland,	.	.	.	29 ..	1800	67,353	62,636	Population Abstracts.
Carlisle,	.	.	.	18 .	1796	2,400	2,271	Heysham.
Montpellier,	.	.	.	21 .	1792	12,919	12,145	Mourgue.
Stockholm,	.	.	.	9 .	1763	12,015	11,706	Wargentin.

The mortality among male infants is, however, greater than that of females; hence it results that their numbers become equal about the 15th year. Men are always exposed to greater dangers than women, owing to various circumstances peculiar to their sex, such as their removal to the colonies, military and naval service, unhealthy arts and dangerous trades, accidents, and the temptations to excesses of all kinds. The women, even in our climates, are therefore always more numerous than the men. From the observations of Kersseboom, Deparcieux, and others, it further appears, that women live longer than men in the ratio of 18 to 17, when once they have passed the more critical ages. More boys die than girls, and more men than women, nearly in the ratio of 10 to 9, at Paris, London, and elsewhere. In 1778, the female population of France exceeded the male by one-sixteenth part; in 1763, Wargentin remarked one-fifteenth in Sweden. In Venice, during the year 1811, there were 10 women to 9 men; and, by more recent observations, 9 women to 8 men in Paris.

In warm countries, notwithstanding the misrepresentations of prejudice, we find the number of females very considerably greater than that of men. Kempfer relates that at Meaco, a large city of Japan, there are 6 women to every 5 men; and the same proportion was noticed at Quito by Ulloa. M. Labillardière found nearly 11 women to 10 men in the south of New Holland. Among the Guaranis, in America, there are about 14 women to 13 men, according to Azara. Major Pike found a much greater proportion of women in several of the savage tribes of New Mexico; there being in some of them 7 women to 6 men, or even 3 women to 2 men; and among the Sioux, the surprising proportion of 2 women to 1 man. In the large towns of Mexico, according to Humboldt, there are 5 women to 4 men.

This numerical excess of women is more especially remarkable on the Coast of Guinea, and in the different Islands of the East Indies, such as Java; likewise on the Malabar Coast, in Bengal, and at Bantam. It may arise partly from the traffic of Negroes, which exists in Africa, and partly from the commerce and navigation of the East removing a large portion of the male population. It is still, however, very probable that their number is considerably greater, as the accounts of nearly all travellers agree, although their statements are not in general founded upon any precise enumeration. It is asserted that there are at Cairo 7 women to 6 men, 6 to 5 in the East Indies, 5 to 4, or even 4 to 3, in the different regions of Southern Asia.

According to the estimates of some eminent Statisticians, in a district of country where 10,000 infants are born annually, we must expect to find 295,022 inhabitants of both sexes. Of these 93,003 are children below the age of 15, and 202,019 above that age. Of the latter individuals, in the most civilized country of Europe, there will not be more than 23,250 marriages, the average duration of which may be estimated at 21 years, 5812 widows, and 4359 widowers.

The number of births always exceeds that of deaths among the most civilized nations, but the proportion varies according to local circumstances.

OLD AGE AND DEATH.

Scarcely has the body attained its full height, than it begins to increase in thickness, and the fat accumulates in its cellular tissue. The different vessels gradually become obstructed; the solids grow rigid; and, after a life of greater or less duration, more or less agitated, more or less painful, old age follows, bringing in its train decay, decrepitude, and death. Those who live beyond one hundred years are rare exceptions; by far the greater number perish before that term, by disease, by accidents, or merely by old age itself.

The systematic order in which the human race are snatched away by death, is one of the most remarkable phenomena of Nature. Nothing appears, at first sight, to be more uncertain than the life of a single individual, because the estimate is derived from a limited experience of the mortality among a few private acquaintances. Yet the Tables of observation obtained by recording the numbers of individuals who die at every period of life from infancy to extreme old age, indicate that a remarkable regularity actually exists. Once in possession of a knowledge of the law of mortality prevalent in a given country, or among a given class of individuals, we are not only able to form general estimations, but to calculate the nicest shades of risk or adventure.

All tables of observation exhibit the precarious life of the infant in a striking point of view. The expectation of human life increases gradually until the ages of six, seven, or eight, which may therefore be considered as the safest period of life, while it diminishes to the most advanced ages.

The following Table exhibits the mortality experienced among the inhabitants of our crowded metropolis.

THE MORTALITY IN LONDON,
OBSERVED AMONG 190,565 INHABITANTS, FROM 1811 TO 1820.

Ages.		No. Living.	Deaths.	Ages.		No. Living.	Deaths.	Ages.		No. Living.	Deaths.
From 0 to 2		190,565	52,970	From 60 to 70		35,589	15,888	From 103 to 104		13	4
2	5	137,595	18,772	70	80	19,701	12,247	104	105	9	1
5	10	118,823	7,848	80	90	7,454	6,210	105	108	8	2
10	20	110,975	6,363	90	100	1,244	1,205	108	109	6	2
20	30	104,612	13,600	100	101	39	16	109	111	4	2
30	40	91,012	17,916	101	102	23	5	111	113	2	1
40	50	73,096	19,668	102	103	18	5	113	114	1	1
50	60	53,428	17,839								

On comparing this Table with Dr Heysham's, showing the mortality at Carlisle, or with the returns of the Registry Commission for the kingdom of Sweden, we find that the inhabitants of London are, with the sole exception of the centenaries, subject to a greater mortality than the residentes in small towns, villages, or the open country. This arises chiefly from the vices, unhealthy occupations, sedentary habits, and the want of cleanliness, so remarkable among the lower classes of the people, who in all large cities form the great mass of the population.

The apparent difference is, however, greatly diminished, if we make allowance for the circumstance, that the hospitals are almost always established in towns, and many of the sick brought to them from the country. Out of 21,000 deaths recorded in Paris, nearly 7000 took place at the hospitals. Although the inhabitants of the country enjoy a purer air, a more sober and regular life, it cannot be denied that rural employments subject them to many hardships, so that it may be fairly questioned whether the real disadvantages of cities are so great as a comparison of Tables of Mortality would indicate.

From the Returns obtained in the whole kingdom of Sweden and Finland, by the Tabell-verket, or Registry Commission, during the 50 years preceding 1805, the expectation of human life was found to vary as in the following Table:—

THE EXPECTATION OF HUMAN LIFE,
Deduced from the Observed Mortality in Sweden and Finland, during the 50 years from 1755 to 1805.

At Birth,	36.64 years.	Age 25	35.46 years.	Age 50	18.71 years.	Age 75	5.95 years.
Age 5	48.24 .	30	32.00 .	55	15.66 .	80	4.43 .
10	46.28 .	35	28.64 .	60	12.69 .	85	3.43 .
15	42.69 .	40	25.20 .	65	10.07 .	90	2.48 .
20	38.98 .	45	21.98 .	70	7.74 .		

These investigations were commenced under the direction of M. Wargentin, and continued by MM. Nicander and Leyonmarck.

We here see that the probable expectation of the life of an Infant just born, is rather more than that of a person aged 25 years : that the best period of life is midway between the ages of 5 and 10, and that the value of life does not diminish in the direct ratio of the years which gradually roll away, a circumstance not usually attended to by ordinary calculators. Although the expectation of human life diminishes with increasing years, it always decreases in a smaller ratio. Thus, at 65 years of age, the expectation of life is 10 years, while at 75 it is still 6 years. The probabilities of death during these 10 years being already decided and converted into certainty, the remaining probabilities after 75 can alone affect the result.

We are still in want of accurate returns of the Mortality prevalent at the several ages over the entire Kingdom of Great Britain and Ireland. In this respect, the government of Sweden has set an example to the rest of Europe, well worthy of imitation.

In the absence of more accurate Observations, Dr Thomas Young formed a hypothetical Table of all the observations made in Great Britain previous to the year 1824 ; but, from the influence of the London Bills of Mortality, and the want of returns from other parts of the kingdom, the mortality of this country would appear by his Table to be very considerably greater than that of Sweden and Finland. However, until some system of observations, embracing the entire kingdom, is set on foot by our government, we must remain in comparative ignorance of the probabilities of life at the several ages as far as regards our own country.

THE EXPECTATION OF HUMAN LIFE,
According to the Mean of all the Observations in Great Britain, previous to the year 1824.

Age. At Birth.	Years.	Age.	Years.	Age.	Years	Age.	Years.	Age.	Years.	Age.	Years.	Age.	Years.	Age.	Years.
At Birth.	30 0	15	37.6	30	27.8	44	20.1	58	13.1	72	7.2	86	3.2	100	2.4
1	36.8	16	36.8	31	27.3	45	19.5	59	12.6	73	6.8	87	3.0	101	2.1
2	40.6	17	36.0	32	26.7	46	19.0	60	12.1	74	6.5	88	2.8	102	1.8
3	42.5	18	35.3	33	26.1	47	18.5	61	11.7	75	6.1	89	2.6	103	1.8
4	43.5	19	34.6	34	25 6	48	18.0	62	11.2	76	5 8	90	2.5	104	2.4
5	43.9	20	33.9	35	25.0	49	17.5	63	10.8	77	5.5	91	2.4	105	3.0
6	43.8	21	33.2	36	24.4	50	16.9	64	10.3	78	5.2	92	2.5	106	3.5
7	43.6	22	32.5	37	23.9	51	16.4	65	9.9	79	4.9	93	2.5	107	2.8
8	43.1	23	31.9	38	23.3	52	15.9	66	9.5	80	4.6	94	2.5	108	2.5
9	42.6	24	31.3	39	22.8	53	15.4	67	9.1	81	4.3	95	2.7	109	2.0
10	41.9	25	30 7	40	22.2	54	15.0	68	8.7	82	4.1	96	2.9	110	1.5
11	41.2	26	30.2	41	21.7	55	14.5	69	8.3	83	3.9	97	2.9	111	1.0
12	40.5	27	29.6	42	21 2	56	14.0	70	7.9	84	3.6	98	2.9	112	0.5
13	39.8	28	29.0	43	20.6	57	13.5	71	7.5	85	3.4	99	3.1	113	0.0
14	38.3	29	28.4												

From the limited number of survivors who attained the age of 90, the results in the above table above that age cannot be relied on.

The Tables of Observation made at Carlisle by Dr Heysham approach very nearly to those procured from the whole kingdom of Sweden, and probably represent the average Mortality among the upper and middle classes of society more correctly than any other in this country. We there find, out of 10,000 infants at birth, that one-fourth die before they attain the age of 3 years, and one-third before the age of 6 years. Further, that only one arrives at

The age of 41	out of		2 Infants.
...	62	...	3 ...
...	69	...	4 ...
...	73	...	5 ...
...	75	...	6 ...
...	80	...	10 ...
...	85	...	22 ...
...	90	...	70 ...
...	95	...	333 ...
..	100	...	1,111 ...
...	104	...	10,000 ..

The annual mortality observed to prevail among the entire population of civilized countries, varies according to local circumstances, more espe-

cially in respect to their state of peace or war, plenty or scarcity. In Sweden and Finland, one male out of 33½, and one female out of 39, died annually, during the 20 years preceding 1795. In the whole population of England, there died annually in the 10 years preceding 1810

One out of 43½ males,
and One out of 48 females.

Owing to the sudden changes of temperature, the beginnings of spring and autumn, about the time of the equinoxes, are the most unfavorable periods of the year ; and in tropical climates, where there are only two seasons, the most dangerous periods are about the times when these seasons change.

By Dr Heysham's Observations at Carlisle, it appeared that the intensity of mortality there was least in the month of August for both sexes, and at its maximum for females in the month of May ; and for males, as well as the whole population without distinction of sex, in October. Dr Short's Observations, collected at Derby, Chester, York, Lancaster, and other parts of England, indicate that the maximum mortality occurs in April, and its minimum in August; so as fully to confirm the popular opinion that settled weather is healthy, and frequent transitions unhealthy, especially sudden changes from heat to cold, and the contrary.—[Milne's Treatise on the Valuation of Annuities and Assurances.]

According to the London Bills of Mortality, during the ten years preceding 1810, the numbers cut off by certain well-known diseases were as follows :—Consumption, 43,905 ; Fevers of all kinds, 16,204 ; Old Age,

14,749; Small-pox, 12,534; Dropsy, 8023; Asthma, 5780; Measles, 5780; Asthma, 5472; Apoplexy, 2995; Childbirth, 1942; Palsy, 1165; Gout, 808; Miscarriage, 24. Total, 113,601.

The principal fatal diseases which occurred among the inhabitants of London, consisting of a population of 1,178,374 persons, included within the Bills of Mortality during the year 1831, are indicated in the following Table :—

CLASSES OF DISEASES.	SPECIFIC DISEASES.	DIED.	TOTAL DEATHS BY EACH CLASS OF DISEASES.	RATE PER CENT.
Pectoral Complaints, . . .	Consumption, .	4807		
	Inflammation of lungs and pleura,	16		
	Asthma, . . .	1061	7865	.667
	Hooping-cough, .	1732		
	Disease of heart, .	127		
	Hydrothorax, . .	122		
Inflammation, ...	Inflammation, . .	2812		
	Abscess, . . .	161	3280	.278
	Mortification, . .	307		
Convulsions, ...	. . . .	2980	2980	.253
Age and Debility, ...	. . . .	2677	2677	.227
Miscellaneous Diseases, . . .	. . . .	2439	2439	.207
Diseases of the Brain, . . .	Apoplexy, . . .	485		
	Paralysis, . . .	246		
	Insanity, . . .	226	1864	.158
	Epilepsy, . . .	54		
	Dropsy in brain, .	853		
Eruptive Fevers, ...	Scarlet fever, . .	143		
	Small-pox, . .	563	1544	.132
	Measles, . . .	750		
	Erysipelas, . . .	88		
Fevers, ...	Intermittent, . .	36		
	Typhus, . . .	223	1224	.104
	Common, . . .	965		
Still-born, ...	. . . . . .	898	898	.076
Diseases of Liver, . . .	Disease of liver, .	296	340	.029
	Jaundice, . . .	44		
Diseases of Bowels, . . .	Dysentery, . .	11		
	Diarrhœa, . . .	33		
	Inflammation of bowels & stomach,	138	230	.019
	Cholera, . . .	48		
TOTAL,		25,341	25,341	2.150

Every country has its local maladies, which select their victims, and serve to diminish the probabilities of life. In the metropolis nearly one-fifth of the annual deaths are occasioned by consumption. Scurvy and diseases of the lungs are common in the North of Europe; in the Southern parts, acute fevers are prevalent. Under the tropics, pestilential fevers prevail during the periods of greatest heat, and dysentery in the rainy season. We find the plague in Egypt, Syria, and Turkey, the yellow fever in America. The nature of the soil and climate, the food used by the inhabitants, the qualities of the air which they habitually respire, customs more or less injurious to health, and other circumstances not fully understood, are the exciting causes of certain morbid states of the body.

Diseases are termed Endemic, when they belong to a particular country, as the Cholera on the Banks of the Ganges. Endemics again may be either Sporadic or Epidemic. In the former case, they attack only a few individuals here and there at a time, as the Asiatic Cholera has done in this country since 1832;—in the latter, they attack a very great number of the inhabitants of any country at the same time, as the Cholera of 1832, or the Influenza of the winter of 1836-7. An epidemic is sometimes nothing more than an ordinary sporadic disease, which has become general in its attacks, such as the common continued fever of this country, and thus any sporadic disease may become epidemic. Sometimes it is a foreign malady supposed to be introduced by contagion, or produced in the country by some unknown influence, as the plague, and more recently the Asiatic Cholera; or it may be a new disease altogether, as happened in Paris during 1828-1829, and of which the cause remains unknown.

In many cases, the local circumstances of a country appear to be the evident causes of certain endemic maladies; but others result from a complication of different causes, such as the Plica Polonica, a disease of the hair, by which it becomes long and coarse, matted and glued in inextricable tangles, and peculiar to Poland, Tartary, and Lithuania, towards the autumn. Baldness and epilepsy are frequent in the islands of the Ægæan Archipelago; St Vitus' dance (Chorea) in the territory of Suabia; and Tarantulism, a kind of spasmodic affection, common in Italy, where it was erroneously attributed to the bite of the Tarantula Spider. It seems difficult to assign any satisfactory reason why Dogs scarcely ever go mad in Mexico and Manilla, although hydrophobia is common on the Coromandel Coast, according to Legentil; or why the plague does not spread from Egypt towards the East Indies, Tonquin, and thence to China, while it

tends continually towards the west. St Petersburg and the Feroe Islands are almost exempted from intermittent fevers; and quartan agues are scarcely to be found in Scotland, peculiarities which are probably owing to the dryness and keenness of the air.

When the climate of a country is affected by the general cultivation of the soil, these results a corresponding change in the endemic maladies of the locality. In proportion as the ancient forests of Pennsylvania were cut down, the inflammatory fevers, then very common, disappeared, and were replaced by bilious agues, according to Dr Rush. The climate of France and Germany was formerly much colder and moister than at present, owing to the dense forests with which the country was covered, as well as the pastoral and almost savage lives of the inhabitants; and gave rise to other endemical affections than those observed at the present day.

The inhabitants of all marshy situations, without exception, where the standing waters continually give rise to putrid exhalations, are subject to intermittent fevers, especially to tertian and quartan agues. These maladies are more or less dangerous, according to the heat of the climate and the season of the year. Tertian agues, which may be mild in the spring, often become continuous in summer, malignant towards the autumnal equinox, while winter again renders them chronic, and deprives them of their virulence. The simple tertian ague of Amsterdam assumes, under the burning sun of Batavia, the aspect of an intermittent of the most dangerous character.

Endemics may also be attributed to the nature of the food and drink. Nearly all maritime nations, who live continually on fish, appear to be subject to frequent diseases of the skin. These affections prevail in the Indian Archipelago, in the Antilles, as well as in the Western Isles, Iceland, and especially round the Baltic Sea; likewise in Friesland, Scotland, Ireland, Brittany, and generally among all classes whose occupations are chiefly those of fishing and the coasting trade. Certain fishes, such as the Sharks (Squalus), Skate (Raia), and others, especially towards the spawning season, actually induce the cutaneous eruptions. On the Indian Seas, a great number of these maladies are produced by several Branchiostegous Fishes, such as Diodons and Tetradons; also, in the neighbourhood of the Caspian Sea, and the rivers of Northern Asia, from the abuse of Caviar and other unhealthy preparations of Fish.

Some kinds of vegetable diet are also the cause of endemical maladies. The coarse bread used by the inhabitants of Westphalia, and the Sarrain, Buckwheat, or Black corn (*Polygonum Fagopyrum*), on which the poorer inhabitants of Sologne, in the Orleanois, are almost wholly maintained, occasion pains in the joints and other diseases. In like manner, the glutinous dishes of polenta, macaroni, millet-broth, and new chestnuts, produce different glandular enlargements and other endemical maladies, in every place where the populace feed too exclusively upon these substances. The abuse of acid wines on the Rhine, and other districts of Germany, as well as the constant use of cider, dispose towards gout and colic.

According to Forster and other observers, we can only attribute the sloughing ulcers found among the inhabitants of several Islands in the South Seas to the acrid drinks, which they prepare from the roots of a species of Pepper. Several other diseases must be considered rather as resulting from a particular kind of diet prevalent in one particular place, than from any other cause. Of this kind are the flabby and leucophlegmatic condition of those nations which subsist chiefly on milk, butter, and cheese, as in Friesland, in the Alps, and all places where much cattle are maintained. The dysentery and diarrhœas, so fatal in very warm or tropical climates, proceed rather from indigestible substances, the abuse of fruits and spirituous liquors; for we find that these diseases can often be avoided by abstaining from the excesses which led to their prevalence.

Hippocrates has long ago remarked, in his Treatise on Air, Water, and Soils, that the local circumstances of each territory predispose the human constitution to particular maladies, or relieve it from diseases of an opposite kind. At the present day, we find the dull inhabitants of the banks of the Phasis equally subject to disorders of the lymphatic system with the Sauromates of the Palus Mæotis; we may contrast, with equal justice, the mild and timid Asiatic with the robust and courageous European, or the corpulent inhabitant of a fertile valley, with the energetic and nervous mountaineer. We likewise find that in low and humid grounds, where the air is stagnant, or exposed to the influence of warm and moist winds from the south and west, as in Holland, putrid and eruptive diseases become very prevalent. Broken-down constitutions are often affected by vertigo, deafness, catarrhal ophthalmias, difficulty of breathing, coughs, lethargy, apoplexy, catarrhs, &c.

On the other hand, in dry and northern exposures in elevated regions, agitated by winds from the north and east, such as the upper Auvergne, the Vivarais, or at Marseilles, Montpellier, or Grenoble, the inhabitants are much exposed to inflammatory diseases, active hæmorrhages, a strong disposition to acute maladies, to inflammations, pneumonia, rheumatism, and acute ophthalmias. Diseases of the chest are common among the inhabitants of cold and mountainous countries.

The two characters of a locality just enumerated, give rise to endemical affections of an opposite kind. In low, moist, and tolerably warm places, the constant humidity habitually relaxes the frame. Diseases here assume a chronic character, with imperfect crises, and humoral degenerations, inducing a precocious old age, among the most of the inhabitants. Elevated regions, on the other hand, bring the body into a state of vigour and energy.

From these endemical dispositions, it follows that strangers often remain exempt from the diseases prevalent among the natives of a country, or, on the contrary, the same circumstances, which have become, through habit, essential to the health of the inhabitants of a district, occasion the illness of a stranger. The water of the Seine often causes Diarrhœa to every one, except the Parisians, who are accustomed to drink it. The Cretin of the Valois loses his stupid appearance in the arid and stimulating heights of the surrounding mountains, while the mountaineer is less affected with Hæmorrhages and acute affections by descending into the denser and moister air of the neighbouring valleys.

Hence it follows, that maladies, like plants, do not disseminate themselves equally in all regions. The miliary fever, frequent in Normandy, is almost unknown in the other provinces of France. Aphthæ, common in Holland, are scarcely ever to be found in Vienne. The carbuncle, common in the south of France, can scarcely be encountered in the north. For analogous reasons, it may be said that the peculiar nature of each country serves to modify the types of the several diseases of the human race. A pleurisy, for example, will be different in intensity in a mountainous locality and in deep valleys. On this account, however exact the descriptions of diseases may be made by physicians, they always exhibit varieties in different climates, which have not been elsewhere remarked.

The Laplanders, according to Schœffer and Linnæus, are subject to inflammations of the head and lungs, and especially of the eyes, in consequence of their being exposed to smoke and dust, as well as to the glare of the sun upon the snow; also to mortification of the extremities from the cold. The frequent use of the milk of the Rein-deer and smoked flesh, often occasions *pyrosis*, and violent colics, followed by ptyalism. They are also disposed to vermes, and are singularly liable to spasmodic affections. They are never affected with plague, acute fever, or agues.

In Norway and Sweden, in some parts of Finland, of Russia and Denmark, however, agues, paralysis, gout, dropsy, and rheumatism, are prevalent, according as the country is more or less moist or cold. But the dryer and elevated regions of Iceland, Norway, and Sweden, are salubrious, and the inhabitants attain a remarkable degree of longevity.

The Muscovites, Cossacks, and Tartars of Kasan, inhabit more healthy countries: with the exception of affections of the chest, caused by cold, they experience few maladies, and have good appetites. Sometimes they experience a morbid hunger during severe frosts. Intermittent fevers prevail on the banks of the Volga, the Don, and other large rivers.

We know that the peculiar affection of the hair, termed the *Plica Polonica*, is endemic in Poland, Lithuania, Transylvania, and Silesia, and sometimes is found in Alsace, Switzerland, and the Low Countries. It is said to prevail more especially among Jews and Christians, of uncleanly and intemperate habits, but especially among the former. This disease, according to many writers, was introduced originally from Tartary in the Ukraine. It is often accompanied by a general affection of the lymphatic system, and other diseases. Some instances are not wanting among the uncleanly Fakirs of Hindoostan.

In Hungary, the inhabitants are sometimes affected with pestilential fevers, accompanied by purple or miliary eruptions. In Thrace, Macedonia, and Turkey in Europe, we find many acute fevers, affections of the brain, and dysenteries. It is well known that the plague often prevails in Constantinople, and extends its ravages among the Turks. From the frequent use of the hot bath and opiates, their constitutions assume less energy than those of Europeans in other climates.

In Germany, purple and miliary fevers are very prevalent towards Leipsic; in Misnia, these affections are frequently complicated with smallpox and measles.

We find a great number of endemical affections in Britain. Tuberculous consumption and catarrh are very frequent maladies in this country. The counties of Essex, Cambridge, and Lincolnshire, were once very subject to intermittent fevers, owing to their marshes, though lastly they have greatly diminished from the general drainage of these districts. Common continued fever is very prevalent, especially towards autumn, and in our large manufacturing districts, hooping-cough, measles, and scarlatina, occur almost universally among children.

In France, calculous disorders prevail in the Barrois and the wine country, which some would attribute to the nature of the waters, but are more probably owing to the wine. In the moist territory of Languedoc, children become subject to the disease called *la sarretie*, a kind of locked-jaw, and *crinons*, *maclous*, or sub-cutaneous vermes, found likewise in the north of Europe.

The Swiss are often troubled with nostalgia, or an excessive longing for their native land, when in foreign countries. The districts of Vaud, Faucigny, Maurienne, and especially the Valais, are subject to cretinism, bronchocele, glandular swellings, accompanied by cachexia, dropsy, and idiocy. During the greatest heats of summer, the inhabitants of these deep valleys are also afflicted with inflammations of the brain and coups-de-soleil.

In Italy, diseases vary according to the localities. The maladies endemical in marshy countries increase towards Mantua, the lagunes of Venice, the marshes of Pisa, and especially during autumn, in the *aria cattiva* of the Pontine marshes, near Rome. Towards Naples, there are often to be seen red spots upon the skin, being a kind of urticaria or nettle-rash. The Greeks are often afflicted with ordinary leprosy, attended with alopecia, or a falling away of the hair from the entire body.

In the moist gorges of the mountains in the Asturias, there prevails a peculiar scorbutic leprosy, called *mal de la rosa*, described by M. Thierry in the Journal Medicale.

The elevated plains of Tartary in Central Asia maintain a great number of wandering nations, whose disorders can scarcely be termed endemical, as these people continually change their place of residence. Some Siberians are subject at birth to an occasional relaxation of the muscles of the upper eye-lid, occasioning a temporary blindness, like the young of many quadrupeds. Southern Asia exhibits most of the endemics peculiar to tropical climates; the hepatic and nervous systems become highly excited, and lead to corresponding diseases. In Asia Minor, besides the plague and many affections of the lymphatic system, such as leprosy and elephantiasis, there prevail spasmodic affections, and especially the *cholera morbus*, which is also frequent in Batavia. At Ceylon, ascites and tympanites are very prevalent, especially during the rainy season. Among all these nations, the nervous system is excited by a kind of habitual irritation from the heat, which gives rise to a corresponding debility in the muscular system, and feebleness of the digestive organs, the vital energy being determined towards the surface of the body. According to Duhalde, there prevails a peculiar kind of erysipelas among those Chinese who work in varnish, and the Asiatics generally are often afflicted with a kind of eruptive disease or *pemphigus*, from being exposed to the heat of the sun.

Many of the diseases found in Asia prevail also in Africa, but modified by the peculiar conditions of the climate and the races of men. The plague appears to be endemical in Egypt; it ceases at the period of the greatest heat, when the Nile rises, and the Northern or Etesian winds blow. When, however, the winds blow from the deserts, during the fifty days following Easter, they raise whirlwinds of a hot and fine sand, which occasion frequent ophthalmias. In Cairo, there are an immense number of blind individuals, and at least one half of the inhabitants are afflicted with diseases of the eye. While the plague prevails, other diseases cease, and particularly intermittent fevers. Diseases of the skin are very common in this country.

Mungo Park found numerous goitres and frequent swellings of the sub-maxillary glands, in different regions of Bambarra, along the river Niger. In the island of St Thomas, the inhabitants are affected with a species of elephantiasis, to such a degree, that Buffon mistook them for a new variety of the human species. There is a dry and burning wind, called Har-mattan, blowing from the north-east, which traverses the Sahara, loaded with a reddish vapour, or rather a fine and hot sand, which dries up the vegetation, chops the lips, occasions ophthalmias, but at the same time produces very salutary effects upon the system, so that, when it arrives at the marshy districts of Africa, it immediately drives off fevers, dropsy, and other diseases.

Some disorders are confined to the Negro race. The yaws, a disease in which elevated red blotches appear upon the skin, are so peculiar to this race, that they do not attack the Europeans in the colonies of America, although apparently under the same circumstances. The inhabitants of the African deserts, who feed upon locusts, are subject, according to Drake, to the morbus pediculosus, of which disease numbers die before the age of forty. The western coasts of Africa are more unhealthy than the eastern, from the trade-winds blowing from the east, and becoming heated as they traverse the continent.

The extensive hemisphere of America comprises a vast number of different climates, and is liable to an immense number of endemical affections. At its northern extremity, such as Labrador and Hudson's Bay, and on the western coasts at Nootka Sound, few diseases are to be found except such as arise from the excessive cold. The descendants of the French and English, who have settled in Canada, have acquired the same hardy constitution as the Swedes.

Intermittent fevers prevail greatly in the United States, from the marshy nature of the country. The frequent changes, as well as the humidity of the atmosphere, occasion catarrhal affections, inflammations of the pleura and lungs, with phthisis. Connecticut is more healthy. Louisiana is much subject to spasmodic affections and opisthotonos, a form of tetanus.

Mexico, and indeed all equinoxial America, is moister, and covered more densely with forests than Africa. It is chiefly towards Vera Cruz, and its fatal coasts, that the yellow fever has long been known. According to M. Humboldt, the ancient Mexicans, or Toltecs, had experienced this malady before the arrival of the Spaniards. It has spread rapidly throughout all the Spanish colonies and elsewhere, as at New York, St Domingo, Porto-Bello, where it bears the name of the black vomit, or vomito prieto. This fatal disease is chiefly endemical on the marshy banks of rivers, and towards the end of autumn. It principally attacks Europeans, and seems to spare the Negroes. Some ports of Spain and Italy are not exempt from this malady. Dropsy is very common on the coasts of Mexico. In all these warm climates, tetanic affections very frequently follow ordinary wounds, causing sudden death, which has often been erroneously ascribed to the poison of the woorara and upas.

During the rainy seasons in Jamaica, acute fevers and colics are the most common maladies, followed by paralysis. Many African maladies prevail here, especially among the Negroes, with whom they are imported. Most Europeans, on passing under the Tropics, experience a kind of feverish delirium called calenture, the effect of the heat, and which goes off by vomiting. On arriving at the colonies, they fall into a state of extreme debility and languor. Afterwards, the abuse of strong liquors, fruit, and other habits unsuited to the climate, occasion them to become afflicted with dysenteries, diarrhœas, and boils. The Brazilians are exposed to frequent ulcerations of the feet, called biahos, produced by a species of Flea (Pulex penetrans, Linn.), which penetrates into the flesh. A red Insect (Ixodes nigra, Latr.) occurring in the Savannahs of Martinique, occasions much inconvenience to the Negroes.

Most writers consider Syphilis as imported from the New Continent, and it cannot be denied that it was found in Peru; but the warm climate and the vegetable diet of the inhabitants render it less dangerous than in our own climate. In some of the South Sea Islands, this disease, introduced by Europeans, cures itself without medical treatment. The elevated plateaux of the Andes are very healthy, and contain many centenaries.

The influence of the various occupations of civilized life upon the human frame is a subject of equal interest with that of the prevalence of diseases in particular localities. Under this head we may include seden-

tary habits; want of ventilation; insufficient exercise of particular parts of the body; exposure to cold; over-exertion; the excessive use of particular parts; unnatural or constrained positions; exposure to heat, moisture, and the noxious fumes or minute particles of animal, vegetable, or mineral substances.

Sedentary habits, when continued for a long time, and without those occasional relaxations necessary to the health of the system, are certain to shorten life. Persons of these habits soon become afflicted with stomach complaints, and various organic diseases, in many cases arising from the pressure on the sternum and lower part of the stomach. Want of exercise is peculiarly fatal to the young, among whom it is as necessary as food to the development of the several structures of the body. The effect of confinement is strikingly observable on comparing the crowded inhabitants of a manufacturing town with a body of active agriculturists. It not only stunts the growth but produces deformity, and depresses the mental powers. Mr Owen states that, in his factory at New Lanark, the children were frequently deformed in their limbs, their growth was stunted, and they were incapable of making any progress in the first rudiments of education. The evil effects of confinement are greatly increased by excesses of any kind, such as too much food, or the use of ardent spirits. On the other hand, too limited a supply of food is almost equally prejudicial. Literary men in general suffer in an especial manner from the want of bodily exercise, on account of the disproportionate manner in which their mental powers are over-strained. In various classes of artizans, such as tailors, weavers, jewellers, engravers, and watchmakers, the effects of long confinement are especially observable, and in the several classes of writing-clerks. Tailors are particularly subject to curvatures of the spine, to inflammations of the stomach, bowels, and liver. In all the instances just enumerated, the muscular and nervous systems acquire an unnatural degree of subserviency to the lymphatic system.

Want of ventilation is especially prejudicial in factories, where children and adults are congregated together in vast numbers and in over-heated apartments. The air becomes tainted with an excess of carbonic acid and animal effluvia, while oxygen is supplied in quantities too small to purify the blood during respiration. When the rooms are not heated in the usual manner by a common fire-place, but by pipes of warm air or by steam, the ill effects of a want of ventilation become greatly increased.

Exposure to cold is one chief cause of many diseases. Indeed, it may almost be said, that one-half of the deaths and two-thirds of the diseases that occur among the children of the poor, are more or less caused by cold. Numbers fall victims, during the winter and spring, to their want of sufficient clothing. A brief or moderate exposure to cold, during perfect health, acts as a useful stimulus to the vital action; but a very intense or long continued abstraction of heat acts as a direct impediment to its exertion.

Exposure to heat is chiefly injurious from the subsequent transition to cold. The perspiration is suddenly checked, and the consequences often become fatal. Rheumatism, asthma, catarrhs, and inflammation of the lungs, are the results of sudden exposure to cold, without the precaution of warm clothing. Bakers, brewers, sugar-refiners, forgers, and glass-blowers, are particularly liable from their occupations to be affected with these diseases.

Moisture, in itself, does not appear to be positively injurious, except in so far as it lowers the temperature of the body by evaporation. As the sea is usually warmer than the air during winter, it happens that, in cases of shipwreck at this season, an almost total immersion in the water is less injurious than sitting in wet clothes exposed to the cold air, and the rapid reduction of temperature during the time they are drying. Cullen records an instance of shipwreck, where the persons who lived longest were almost totally immersed in salt water; while the consequences were fatal to those who were exposed to the freezing influence of the wind only, or to the wind, assisted by the evaporation from wet clothes. Immersion for a long period in salt water is not so injurious as in fresh, from the former being more stimulating. The use of spirituous liquors during shipwreck increases the danger, as it raises the temperature of the body for a short time, only to render it more sensibly affected by the subsequent cold. It is not unfrequently followed by apoplexy. Inattention to change wet clothes, whether from rain or perspiration merely after severe labour, is very injurious. This is the exciting cause of most of the diseases found among fishermen, water-carriers, fullers, and washerwomen.

The animal effluvia of candle-manufactories, slaughter-houses, and dissecting-rooms, are generally unhealthy, although no positive diseases can be assigned to them. In most of these occupations, where persons are much exposed to animal effluvia, there are certain causes serving to counteract the ill effects that would otherwise follow. Tanners are preserved, by the tan and lime, from the injurious consequence of exposure to animal matter and moisture. M. Patissier observes, that butchers and catgut manufacturers are free from phthisis, while glue and size boilers are comparatively healthy.

The particles of vegetable matter in a minute state of division, found floating in the atmosphere of corn-mills, when inhaled by the work-people, bring on asthma, indigestion, and frequently consumption. Millers are consequently pale and sickly in their appearance, and their lives usually short. The heated and sulphureous vapours arising from kilns render maltsters liable to many diseases. Snuff-makers, from being exposed to the dust of the tobacco, are often affected with diseases of the lungs and head, partly from its mechanical effects, and partly from its narcotic influence.

Mineral particles in general are peculiarly noxious. The fumes of mercury become speedily fatal to the workmen in quicksilver mines, to gilders, and glass-platers. M. Jussieu states (Mémoires de l'Académie des Sciences, 1719), that by proper precautions, by care in changing the dress, and by minute attention to cleanliness, the free workmen in the mines of Almaden escaped disease for a long time; while the slaves who could not afford a change of clothing, and took their food in the mines without ablution, speedily became diseased in the throat and lungs. Lead occasions paralysis and colic. M. Merat (Traité de la Colique Métallique, Paris, 1812) states, that out of 279 cases of colic in the hospital of La Charité, at Paris, in the years 1776 and 1811, the numbers were—Painters 148 ; Plumbers 28 ; Potters 16 ; Porcelain-makers 15 ; Lapidaries 12 ; Colour-grinders 9; Glass-blowers 3 ; Glaziers 2 ; Toymen 2 ; Shoemakers 2 ; Printer 1 ; Lead-miner 1 ; Shot-manufacturer 1. Of the remainder (36) there were 17 belonging to trades connected with copper. The same writer recommends the artisans never to take their meals in the workshop, or without ablution, and in general to preserve great cleanliness. The acid vapours of chemical works frequently occasion inflammation of the throat, and the most corpulent person is speedily reduced to a small size. Chlorine is similarly injurious, but at the same time acts as a disinfectant. The chemical manufactory of Belfast was preserved by the Chlorine fumes from the effects of the epidemic that ravaged Ireland for the three years preceding 1819.

Particles of matter acting mechanically on the lungs are perhaps of all others the most certainly fatal to the artisan. By irritating the bronchial surface, pulmonary diseases are speedily induced. These causes act principally among needle and steel-fork pointers, dry grinders, and sandstone cutters. These unfortunate victims of their industry seldom live above the age of 40, while the greater number die at the ages of 30 and 35. Philanthropists of every description have long attempted, by various contrivances, to remove these evils ; but the ignorance, perverse habits, and blind fatuity of the workmen themselves, are the principal obstacles to their success. Mr Abrahams of Sheffield proposed magnetic masks to intercept the minute particles of steel ; M. D'Arcet invented the *fourneau d'appel ;* Dr Johnstone the damp crape, and Dr Gosse the sponge. But the carelessness of the workman mars the good intentions of the philosopher ; so strong is the influence of habit and the recklessness consequent on the certainty of a short career.

TEMPERAMENTS.

Besides the marked differences observable among Mankind in respect to age and sex, there are others arising from the relative energy of the different functions of the human body, while in a state of health, and occasioning that peculiar aspect and physiognomy termed the *temperament,* which strikes an observer at the first glance. This word, *temperament,* must not be confounded in its signification with *constitution ;* for one individual may be of a robust constitution, and another frail in the extreme, although both are of the same temperament.

Some modern writers enumerate as many as seven temperaments ; the more ancient authors admit only four. We shall describe the sanguineous, the bilious, the lymphatic, and the nervous ; to which may be added the subordinate temperaments, called the athletic and melancholic, making six in number.

1. THE SANGUINEOUS TEMPERAMENT is characterized by the predominant activity of the heart and blood vessels. Externally, it is marked by rosy cheeks, an animated countenance, and all those physical characters which are so accurately represented in the superb statues of Antinous and the Apollo Belvidere. Its moral character is exhibited in the lives of Alcibiades and Marcus Antonius. The Duc de Richelieu is a striking instance of the sanguineous temperament among the moderns.

These peculiarities constitute the *Muscular* or *Athletic,* when men of a sanguineous temperament devote themselves to the habitual exercise of their physical strength, and the entire frame undergoes a corresponding modification. The head becomes small, the shoulders broad, the chest large, the haunches solid, and the intervals of the muscles deeply marked. Of this acquired temperament, we find an excellent model in the statue of the Farnese Hercules.

2. THE BILIOUS TEMPERAMENT is characterized by a brown skin, inclining towards yellow ; moderate fulness and firmness of body ; the muscles well-defined ; and the forms harshly expressed. It is chiefly among men of this temperament that we find those splendid virtues and enormous crimes, which have been at once the admiration and terror of the world. Alexander the Great, Julius Cæsar, Marcus Brutus, Mahomet, Charles XII. of Sweden, the Czar Peter the Great, Oliver Cromwell, Sixtus V., and Cardinal Richelieu, are commonly cited as examples.

Whenever the bilious temperament is attended by a morbid obstruction of the abdominal viscera, or derangement in the nervous functions, the skin acquires a deeper hue, the aspect becomes uneasy and gloomy, and it assumes the characters of the *atrabilious* or *melancholic* temperament of the ancients. Louis XI., Tiberius, Rousseau, Tasso, Pascal, Gilbert, and Zimmerman, are its models.

3. THE LYMPHATIC TEMPERAMENT arises from the undue proportion of the fluids over the solids, and is chiefly marked by the form becoming rounded and without expression, all the vital actions more or less languid, the countenance pale, the memory treacherous, the attention interrupted, the pulse weak and slow. Such Men are but little fitted for business, and never produce any of those great characters, which occupy an eminent place in the moral history of the human race.

4. THE NERVOUS TEMPERAMENT is marked by the predominance of the nervous or sensitive system, over the muscular or motive. This excessive sensibility of the organs is rarely natural or primitive, but is most commonly the acquired result of a life too sedentary and inactive, in which the mental powers have attained a great development. Voltaire and Frederick the Great of Prussia are illustrious instances of the nervous temperament. (See Richerand, Elémens de Physiologie, 10° Edit. 1833.)

It is seldom that we find all the particulars of any temperament united in the same individual, and every person is born with peculiarities of his own, which constitute the *idiosyncrasy* of that individual. The sanguine temperament is, however, directly opposed to the melancholic, bilious, and lymphatic; although it may happen, that an individual, sanguineous in early youth, becomes melancholic with advancing years.

According to M. Thomas (Physiologie des Tempéramens et des Constitutions, 1826),the human temperaments depend upon the relative proportions in the cavities of the cranium, the thorax, and the abdomen. He enumerates,

1. THE MIXED, or just proportion of these cavities, constituting the Apollo Belvidere, or complete physical Man. This corresponds with the sanguineous temperament.

2. THE CRANIAN, or relative predominance of the cranium over the thorax and abdomen. This is the bilious temperament of other writers.

3. THE THORACIC, or relative predominance of the thorax over the cranium and abdomen, forming the athletic or muscular temperament.

4. THE ABDOMINAL, or predominance of the abdomen over the cranium and thorax. Here the pelvis is broad, the cellular tissue widely distributed, as in the Venus de Medicis ; accordingly, this temperament is usually found in the female sex.

5. THE CRANIO-THORACIC, or predominance of the cranium and thorax over the abdomen. It is directly opposed to the abdominal. When this temperament is highly developed, the muscles are hard and well pronounced, the cellular tissue is rare in all parts of the body. This temperament seems merely to be a modification of the nervous already described.

6. THE CRANIO-ABDOMINAL, or relative preponderance of the cranium and abdomen over the thorax. This form of the nervous temperament, most commonly found in females, is directly opposed to the thoracic.

7. THE THORACO-ABDOMINAL, easily recognised by the predominance of the face over the cranium, and directly opposed to the cranian, is more widely distributed over Asia, Africa, and America, producing, when excessive, imbecility of mind and idiocy.

It can scarcely be denied, that the temperaments exercise a considerable influence over the moral character of the individual. Although we admit that every virtue and every vice in all its degrees may be distinctly exhibited in the several temperaments ; yet there are certain general facts which mark the natural tendency of the moral sentiments to follow corresponding states of the body.

If, for instance, we find the lungs of great extent, the chest capacious, and the heart of considerable size, attended by a high degree of animal heat, and a very active state of the vital functions ; with a muscular fibre and a cellular tissue of medium consistency; we shall also find the moral character mild and amiable, generally amorous, light, inconstant, and volatile.

Again, if we find in addition to these a large hepatic system, and copious secretions of bile, with corresponding powers of procreation, the animal heat becomes higher, the circulation obtains greater rapidity, and the vessels acquire a size still larger than in the former instance. Violent dispositions of mind are the result, with a character of great energy, ambition, magnanimity, intrigue, or cruelty.

On the other hand, if we remark a high degree of softness in the muscular fibre, a feebleness in the nervous system, attended with slight activity of the abdominal and thoracic viscera, the prevalent states of mind may be safely predicted to be mildness, want of energy, indolence, idleness, and an almost total inactivity of the mental powers.

Those states of mind, which habitually belong to an individual of high

32

intellectual and moral attainments, of intense habits of application, whether in business or study, are the usual consequences of a nervous temperament, when attended by a considerable firmness of the muscular fibre. Persons of delicate constitution, whose nervous system is highly developed, never acquire this moral character. A morbid sentimentalism occupies the place of energy of thought and action, the individual becomes timorous, undecided, and often excessively prone to superstitious observances.

The athletic temperament is most frequently accompanied by a want of sensibility, of intellectual capacity, and even of real vital energy, requiring some great stimulating cause before it will exert its enormous physical power. The melancholic temperament seems to belong to a kind of mental pathology. (See the Table Analytique of M. Le Comte De Tracy, prefixed, in some editions, to Cabanis, sur les Rapports du Physique et du Moral de l'Homme.)

THE INTELLECTUAL AND MORAL DEVELOPMENT OF MAN.

THE infant requires the assistance of its mother for a much longer time than the nourishment of her breast, and its nature being at the same time susceptible of an intellectual as well as of a physical education, a durable attachment arises between them. From the circumstance that the sexes are not very unequal in number, and, from the difficulty of supporting more than one wife, where wealth does not supply the means, we may infer that monogamy is the natural bond of union for our species. The father consequently assumes a share in the education of his offspring, a circumstance not peculiar to the human race, but common to all other species where this kind of union is observed to prevail. The length of this education permits him to have other children during the interval, and hence the perpetuity of the matrimonial union seems to have a real foundation in Nature. Thus, the great length of the period of infancy gives rise to the paternal influence, and, indirectly, to all the subordination of society, as the young people who form new families will preserve for their parents the same relative feelings of respect, which have so long been experienced under their mild sway.

The natural disposition of Man to social labour has multiplied, without any apparent limit, those advantages which he would otherwise have obtained from his personal skill and intelligence. It has enabled him to tame or repel the other animals, to preserve himself from the inclemencies of the most rigorous climates, and to extend his species over the face of the entire globe.

Man does not appear to be swayed by any principle, which can be compared [in intensity] to the instinct of the lower animals, to that constant industry produced by an internal irresistible impulse. His knowledge is the entire result of his sensations and observations, or of those of his predecessors. The accumulated experience of ages, transmitted by word of mouth, improved by meditation, and applied to the various purposes of necessity, or the enjoyments of life, has given rise to the Useful Arts. Speech and writing, by preserving the knowledge already acquired, seem to be sources of an indefinite improvement of the species. It is thus that he obtains all his science, and becomes entitled to occupy an important place in the economy of Nature.

The Intellectual and Moral Development of the human species has proceeded, however, by very distinct gradations.

The first hordes, reduced to live by the chase, by fishing, or on wild fruits, compelled to devote their whole time in obtaining a scanty subsistence, could not multiply to any great extent without exhausting their resources, and consequently were incapable of making any important progress. Their arts were confined to the construction of huts and canoes ; to covering themselves with skins, or to the fabrication of arrows and nets. They made no physical observations, excepting, perhaps, on those more obvious stars which served to guide them in their wanderings, or on a few natural objects whose properties were immediately useful. They domesticated no animal excepting the Dog, because it seemed to be disposed by Nature to the same predatory mode of life.

As soon, however, as Man succeeded in taming the larger herbivorous animals, he found a more secure means of subsistence in the possession of numerous herds, and a certain degree of leisure, which enabled him to extend his knowledge. Some degree of industry was bestowed upon the fabrication of dwellings and clothes, the value of property became known, and consequently commerce, wealth, and the inequality of conditions,—at once the incentives to the noblest emulation as well as to the basest passions. Yet the necessity of seeking new pasturage, and of migrating according to the seasons, still retained his civilization within narrow limits.

It was only since the invention of Agriculture, and the conse-

quent division of the soil among hereditary proprietors, that Man has really succeeded in multiplying the numbers of his species to a high degree, and carried to a great extent his Science and the Useful Arts. By means of Agriculture, the manual labour of a part only of the members of society can produce a sufficient quantity of food to maintain the whole. A sufficient degree of leisure is thus left for other pursuits which are less necessary ; while, at the same time, the hope of securing, by industry, a comfortable subsistence for each individual and his posterity, has given a new stimulus to emulation. The invention of a circulating medium for representing exchangeable values, raises this emulation to the highest degree. By facilitating the means of intercourse, it has at once rendered capital more independent and susceptible of a greater increase ; while, by a necessary consequence, it has augmented the vices of luxury and the fury of ambition.

In all stages of social progress, the natural propensity of Mankind to reduce every thing to general laws, and to find out the causes of phenomena, has given rise to men of philosophical minds, who have added new ideas to the mass of the previously-acquired knowledge. As long as the great body of the people continued unenlightened, they have sought, by exaggerating their merit, and disguising their limited knowledge under the propagation of superstitious notions, to make their personal abilities the means of ruling over others.

A more incurable evil is the abuse of physical force. At the present day, Man is the only species capable of contesting with Man ; and he is almost the only species which is continually at war with his fellows. Savages dispute the possession of their forests, the wandering shepherds their pastures, and both classes make irruptions as often as they are able, in the territories of the neighbouring Agriculturists, to carry off without trouble the fruits of a labour not their own. Civilized nations themselves, far from being satisfied with their share of the enjoyments of life, fight for mere objects of national vanity, or for the monopoly of commerce. From these circumstances arise the necessity of Governments to direct the national wars, and to suppress, or reduce to regular forms of law, the quarrels of private individuals.

Circumstances, more or less favorable, have retarded the social progress of Mankind within certain limits, or have served to promote its development.

The frozen climates in the North of both continents, and the impenetrable forests of America, are still inhabited by savage hunters or fishermen.

The immense plains of sand or salt in the centres of Asia and Africa are covered by pastoral tribes and innumerable herds of cattle. These half civilized hordes assemble together at intervals, on the call of some enthusiastic chieftain, and fall upon the cultivated countries which surround them, where they establish themselves, and eventually become civilized, only to be subdued by other shepherds in their turn. This is the cause of that despotism which, in all ages, has served to crush the rising germs of industry and science in the delightful climates of Persia, India, and China.

Mild climates, soils naturally well-watered, and rich in vegetation, are the appropriate cradles of Agriculture and Civilization. Wherever their geographical situation shelters them from the irruptions of barbarians, all kinds of talent are naturally excited. Such were Greece and Italy, in the early days of Europe, and such at the present day is all this happy portion of the globe.

There seem, however, to be other intrinsic obstacles which serve to arrest the progress of some races of Mankind, even in the midst of circumstances, apparently the most favorable to their improvement.

VARIETIES OF THE HUMAN SPECIES.

ALTHOUGH Mankind appear to compose a single species, [partly] from the circumstance that individuals of all the races are capable of producing a fertile progeny, [and partly from other considerations], they present certain hereditary peculiarities which constitute what are termed *Races*.

THREE of these appear to be eminently distinct from each other; namely, the White, or CAUCASIANS ; the Yellow, or MONGOLIANS ; and the Black, or NEGROES.

We shall distinguish by the term NORMAL, those varieties of Mankind which admit of being readily referred to one or other of the preceding types. The remainder, or ANOMALOUS Races, will be arranged under SIX divisions. These may be termed the MALAYANS, the POLYNESIANS, the AUSTRALASIANS, the TASMANIANS, the HYPERBOREANS, and the AMERICANS.

Though obviously distinct from each other, the characters of the Anomalous races approach more or less nearly to those of some or all of the Normal races.

NORMAL RACES.

Syn. LES RACES ÉMINEMMENT DISTINCTES.—Cuv.[1] Reg. Anim. I. 80.
VARIÉTÉS DE RACES BIEN CARACTÉRISÉES.—Desm.[2] Mam. 47.

I. CAUCASIANS.

Syn. RACE CAUCASIQUE.—Cuv. Reg. Anim. I. 80.—Desm. Mam. 47.
HOMO SAPIENS, EUROPÆUS, VAR. β. Lin.[3] Gmel. I. 22.—VAR. ζ. Erxl.[4]
2.—VAR. α. Albin Tab. Oss. Hum. (fide Fischer).—VAR. CAUCASICA. Blumenb.[5] Handb. et Abbild.[6]
HOMO JAPETICUS.—Fisch.[7] Syn. Mam. 2.
CELTO-SCYTH-ARABES.—Desmoul.[8] Tab.
EUROPAEL.—Camp.[9] Gesichtsz.

LA CAUCASIQUE, OU ARABE-EUROPÉENNE.—Dum.[10] Zool. Anal. 6.
RACE BLANCHE, OU CAUCASIENNE (in part)—Less.[11] Mam. 24.
PREMIÈRE RACE. BLANCHE.—Virey,[12] Hist. Nat. du Gen. Hum. I. 438, et Nouv. Dict. d'Hist. Nat. art. Homme.[13]
RACE EUROPÉENE, OU CAUCASIQUE.—De Lacepède, in Dict. des Sc. Nat. art. Homme.[14]

The Caucasian variety, to which we belong, may readily be distinguished by the beauty of the oval form of the head. Some of the groups belonging to this division have constituted nations of the highest comparative civilization, and which have most frequently exercised dominion over the remainder. They may vary in the tint of the skin and colour of the hair, and have been [rather improperly] termed *Caucasians*, from the circumstance that the early traditions of nations would refer [some of] their tribes to the group of mountains between the Black and Caspian Seas, from which they [are conjectured by several writers to have] emigrated in all directions. The inhabitants of the Caucasus itself, such as the Circassians and Georgians, are still accounted at the present day among the most beautiful people of the globe.

The leading branches of these races may be distinguished by the analogy [and affinity] of their languages.

Whenever two languages have a general resemblance in their grammatical structure, and when a great number of the roots or elements are common to both, they are said to be *allied* to each other. Thus, the Hebrew, the Chaldee, the Syriac, and the Geez or Ethiopic, have a natural affinity. Again, there are others which are wholly distinct in their vocabularies, with few words in common, yet they bear a striking resemblance to each other in their grammatical structure, such as the monosyllabic languages of the Chinese, Tibetans, Siamese, &c. These may be termed *analogous* languages.[15]

By combining a philological inquiry into the affinity or analogy of languages, with a careful examination into the physical diversities of nations, we are enabled to classify the several tribes of Men under appropriate subdivisions.

All the Caucasian races may be reduced to five principal sub-varieties, (A.) *Homo Caucasicus*, or Caucasians Proper; (B.) *H. Iapeticus*, or

Iapetans : (C.) *H. Celticus*, or Celts ; (D.) *H. Semiticus*, or Aramæans ; (E.) *H. Scythicus*, or Scythians. Minor differences of form and language give rise to further subdivisions into groups or families of nations.

(A.) HOMO CAUCASICUS.—CAUCASIANS PROPER.

Syn. HOMO JAPETICUS, α a CAUCASICUS.—Fisch. Syn. Mam. 2.
CAUCASIENNE.—Desmoul. Tab.
RACE CAUCASIQUE (ORIENTALE).—Bory.[16] Ess. Zool. I. 110.
1° SOUCHE EUROPÉENE, 1° TIGE CAUCASIQUE.—Broc.[17] Ess. 28.
RACES GRECQUES ET PÉLASGIQUES (in part).—Malte-Brun,[18] Géog. Univ.
Icon. Blumenb.[19] Dec. Cran. III. t. 21. (Skull of a Georgian female.)

The various tribes known by the names of Georgians, Imeritians, Mingrelians, Abassians, Tscherkessians or Circassians, and Lesghians, have long been celebrated for the extreme regularity and general beauty of their features. They inhabit the mountain chains of the Caucasus and the adjacent valleys, situate in the immediate neighbourhood of the Black and Caspian Seas, between the 41° and 45° of N. latitude. From the earliest ages, these regions have been the abode of numerous tribes, all of whom, excepting the Ossetes, according to M. Julius Klaproth, speak languages, the idioms of which are wholly distinct from those of all other known tongues.

These circumstances, combined with the characteristic physiognomy of the Caucasians Proper, entitle them to be regarded as indigenous and primitive tribes of great antiquity.

The Georgian races, as we are informed by Chardin,[20] are the most beautiful in the East, and we may even say, in the world. During the twelfth century, numerous poetical and historical works were composed in their own peculiar language.[21] Their women are not so white as the Circassians, nor are their figures quite so graceful, yet they possess great beauty, and scarcely an ugly countenance can be found in all the country. They are tall, well made, extremely slight round the waist, and of a most

---

[1] CUV. REG. ANIM.—Le Règne Animal distribué d'après son organisation. Par M. Le Baron Cuvier. Paris, 1829. Nouvelle Edition.
[2] DESM. MAM.—Mammalogie, ou Description des Espèces de Mammifères. Par M. A. G. Desmarest. Paris, 1820.
[3] LIN. GMEL.—Caroli A. Linné. Systema Naturæ per Regna Tria Naturæ. Curâ Io. Frid. Gmelin. Lugduni, 1789.
[4] ERXL.—Io. Christ. Polyc. Erxleben, Systema Regni Animalis. Classis I. Mammalia. Lipsiæ, 1777.
[5] BLUMENB. HANDB.—Handbuch der Naturgeschichte, von J. F. Blumenbach. Gött. 1821.
[6] BLUMENB. ABBILD.—Abbildungen Naturhistorischer Gegenstände, von J. F. Blumenbach. Gött. 1797.
[7] FISCH. SYN. MAM.—Synopsis Mammalium. Auctore J. B. Fischer. Stuttgardtiæ, 1829.
[8] DESMOUL. TAB.—Tableau Général, Physique, et Géographique des Espèces et des Races du Genre Humain. Par A. Desmoulins.
[9] CAMP. GESICHTSZ.—Ueber die Verschiedenheit der Gesichtszüge von P. Camper, Berl. 1792 (fide Fischer).
[10] DUM. ZOOL. ANAL.—Zoologie Analytique, ou Méthode Naturelle de Classification des Animaux. Par A. M. Constant Duméril, Paris, 1806.
[11] LESS. MAM.—Manuel de Mammalogie. Par René-Primevere Lesson, Paris, 1827.
[12] VIREY, HIST. NAT. DU GEN. HUM.—Histoire Naturelle du Genre Humain. Par J. J. Virey, Paris, 1824.
[13] NOUV. DICT. D'HIST. NAT.—Nouveau Dictionnaire d'Histoire Naturelle appliquée aux arts. Par une Société de Naturalistes et d'Agriculteurs, Paris, Deterville, 1817.
[14] DICE. DES SC. NAT.—Dictionnaire des Sciences Naturelles. Par plusieurs Professeurs du Jardin du Roi, et des principales Ecoles de Paris, Strasbourg, et Paris, 1821.
[15] Abstract of a Comparative Review of Philological and Physical Researches. By J. C. Pritchard, M.D., in Reports of the British Association for 1832.
[16] BORY ESS. ZOOL.—L'Homme, Essai Zoologique sur le Genre Humain. Par M. Bory de Saint-Vincent, Paris, 1836, 34 Edit.
[17] BROC. ESS.—Essai sur les Races Humaines. Par P. P. Broc. Paris, 1836.
[18] MALTE-BRUN, GÉOG. UNIV.—Géographie Universelle. Par M. Malte-Brun, Paris, 1816.
[19] BLUMB. DEC. CRAM.—Decas Craniorum. Ed. J. F. Blumenbach, Gött. 1790-1820.
[20] Les Voyages de Jo. Chardin, vol. I. p. 171, Amsterdam, 1735.
[21] Eugene, Annales des Voyages, XII. p. 86, 90 (fide Malte-Brun).

pleasing expression of countenance. The Men are likewise very hand-some, have good abilities, and might excel in the sciences and useful arts, did not a defective education render them very ignorant and vicious.[1] There is perhaps no country where libertinism and dissipation prevail to a greater extent than in Georgia. The skull of a Georgian female, who died at Moscow, after having been taken by the Russians in one of their wars with Turkey, is figured by Blumenbach in his Decades of Skulls, and would strike the most careless observer by the noble expansion of its frontal region, and the general symmetry and elegance of its proportions. The Imeritians speak a Georgian dialect.

In Mingrelia, the women are equally beautiful, but perfidious. When love or hatred happens to be the ruling passion, a Mingrelian female is equal to any action, however atrocious. The men are as remarkable for their immorality, and theft and assassination are common occurrences. They exchange wives without the slightest scruple, and are not particular as to the degrees of their consanguinity to spouses, which are commonly two or three in number, or to concubines, in general as numerous as their means will allow. Husbands have here but little jealousy, and a gallant Lothario, when convicted, is compelled to atone for his offence, according to Chardin, by paying a pig to the injured husband; and it is not uncommon, he adds, for the pig to be eaten together by the three parties interested in the affair. Large families are anxiously desired in this country, for the sordid purpose of selling the miserable progeny as slaves. The master disposes of his servants, the brother of his sister, and the father of his children, without the slightest compunction. On this account slaves are very cheap. Their prices average as follows :—

Handsome girls, aged from 13 to 18 years........20 crowns.
Men from 25 to 40 years of age....................15   ...
Married women.......................................12   ...
Men above 40 years of age......................... 8 to 10
Children.. 3 or 4

The physiognomy of the Abassians is very remarkable,—an oval face, a head very much compressed on the sides, a short chin, large nose, and hair of a deep chestnut colour, form its usual traits.

The Circassian nobles speak a language peculiar to themselves, and dif-ferent from the vernacular language of their country. They are of a robust make, with a small foot, and strong wrist. The females are delicate, pleas-ing, and graceful in their forms; their skins white, with black or brown hair. It is chiefly the remarkable cleanliness of their persons which renders them so attractive to Europeans, for they are often surpassed in regularity of form and features by some of the neighbouring tribes. The Leshgian women, in particular, rival them in respect to personal attractions, as well as in courage. The dialects of the latter tribes are very numerous, and have some affinity to the language spoken by the inhabitants of Finland.

To this sketch of the more important tribes, belonging to the Proper Caucasian races, we may add, that the medium height of the men is about 5 feet 8 inches, their temperament usually sanguineous and bilious.[2] Their hair is most commonly black, fine, shining, and very much curled ; the nose straight; the shape of the face perfectly oval, and the facial angle vary-ing from 85° to 90°. The women are occasionally subject to an excessive rotundity of form. Their mouth is small, their bust most graceful, and skin perfectly white. Their eyebrows, excessively narrow, have been com-pared to the gently-curved filaments of silk. Such are the peculiarities of the Caucasian females, whose beauty is so celebrated in the East. They serve to ornament the harems of the Mahometans from the centre of Asia to the kingdom of Morocco.[3]

(B.) HOMO IAPETICUS.—IAPETANS.

Syn. LE RAMEAU INDIEN, GERMAIN, ET PÉLASGIQUE.—Cuv. Reg. Anim. I. 81.—Less. Mam. 24 (in part).

THE Indian, German, and Pelasgian branch is much more widely distributed than the remainder, and became subdivided at an earlier age. We are still able, however, to recognise innumerable affinities among its four principal languages,—the Sanscrit, at present the sacred language of the Hindoos, and the parent of most of the dia-lects of Hindoostan ; the ancient language of the Pelasgi, the com-mon mother of the Greek, the Latin, and several others now extinct, also of all our languages in the South of Europe ; the Gothic, Teu-tonic, or Tudesque, from which are derived the languages of the North and North-west of Europe, such as the German, the Dutch, the English, the Danish, the Swedish, and their dialects ; finally, the languages called the Sclavonian, from which are derived the Russian,

the Polish, the Bohemian, the Wend, and other dialects of the North-East of Europe.

The nations of this powerful and important branch of the Cau-casian race have raised philosophy, the sciences, and the arts, to their present advanced state, and for more than thirty centuries have been the depositors and guardians of human knowledge.

The ancient Persians have the same origin as the Hindoos, and their descendants at the present day bear the most striking marks of their affinity with the nations of Europe.

All these nations may be termed *Iapetans*, not from any fancied descent from Japhet, son of Noah, but rather from Iapetus, the father of Pro-metheus, whose daring exploits are celebrated in the legendary history o remote antiquity.

> Audax Iapeti genus,
> Ignem fraude mala gentibus intulit.
>   Hor. lib. 1. Od. 3.

This illustrious branch of the Human Race would be justly entitled, from moral and political considerations alone, to occupy the first place in our classification. Some of its subdivisions emulate the Proper Caucasians in personal beauty, and the facial angle approaches nearly to 90°. As in them the face is oval, the forehead open, the nose straight, or nearly so ; the eye-brows more or less arched ; the eye-lashes of medium length ; the mouth middle-sized ; the beard long ; and the ears closely applied to the head. Their hair, generally fine, and even silky, varies from black and deep chestnut to a blonde, approaching to white. A complexion more or less vivid relieves the excessive paleness of the face, and betrays the passions of the moment by the changes of its colour. This ruddiness may, on the one hand, become degenerate in individuals, who are etiolated by confinement, or, on the other, it may merge into a deep brown when exposed to the excessive heats of a tropi-cal sun. In every part of the globe, the skin of the Iapetans preserves its primitive whiteness, when protected from the direct solar rays.

With the exception of the Indo-Persians, all the nations of this sub-variety are essentially monogamous. Polytheism was their primitive re-ligion, with some vague notions regarding the immortality of the soul. Christianity and its numerous modifications are now professed by nearly all of them excepting the Indo-Persians, who conform themselves to other creeds more congenial to the prejudices of a degraded people.

Industrious, patriotic, and brave, with a taste for the Sciences, the Fine and Useful Arts—in a word, endowed with talents of the highest order, the Iapetan races have produced, without exception, all those great geniuses who have astonished and enlightened the world.

1. PELAGIUS.—PELASGIANS.

Syn. HOMO JAPETICUS b. PELAGIUS.—Fisch. Syn. Mam. 2.
LES PÉLAGES.—Cuv. Reg. Anim. I. 81.
RACE PÉLAGE (MÉRIDIONALE).—Bory Ess. Zool. I. 114.
RACES GRECQUES ET PÉLAGIQUES (in part).—Malte-Brun, Geog. Univ.
ETRUSCO-PÉLASGE.—Desmoul. Tab. Hum.

Icon. Blumenb. Dec. Cran. IV. t. 32. (Skull of a Roman Prætorian Soldier): VI. t. 51. (Skull of an ancient Greek).

From the earliest ages, the Pelasgi were divided into two distinct branches, the Proper Pelasgians and the Etruscans. Among the former were included the Phrygians, Lydians, Carians, Trojans, Thracians, Il-lyrians, and the aborigines of Greece. Among the latter we find the native Italian nations.

Next to the Proper Caucasians in the beauty of their features, we may observe their models at the present day in the statues of the Jupiter Olym-pus, the Apollo Belvidere, and the Venus de Medicis. The characters are the following ; medium height about five feet seven inches ; the hair fine, brown or chestnut, rarely blonde, and often of surprising length. The foot is larger and the leg thicker at its base than agrees with our ideas of beauty ; the nose is perfectly straight, and in the same line with the fore-head, without the slightest depression at the point of junction. The eyes, remarkable for their size, have often been compared to those of an Ox (βοῶπις.) Their temperament is most commonly sanguineous and bilious.

Though nearly extinct, or lost among their numerous alliances with the neighbouring races, the traits of the pure Pelasgian race may still be found in a few Roman and Grecian Ladies. They have, however, en-tirely disappeared among the great mass of the people who now inhabit the Ægean Archipelago, Turkey in Europe, Italy, and Sicily,—regions once the exclusive abodes of this interesting race. To them we owe the

---

[1] BUFF. HIST. NAT.—Histoire Naturelle générale et particulière, Paris, 1750, tome III. p. 433, 434. Par M. De Buffon.
[2] Bory Ess. Zool. I. 110.
[3] Bory Ess. Zool.—Also, Sir R. K. Porter's Travels in Georgia, Persia, &c.

introduction of the Cereal plants, the cultivation of the Olive, and they appear to have first domesticated the Bull. Having received the knowledge of letters and the art of writing from the Phœnicians, belonging to the Aramean race, their first poets became their historians, and philosophy, born on the banks of the Nile, was developed by the genius of a Socrates, an Aristotle, or a Cicero. Attached to their native soil, the Pelasgians seldom ventured upon maritime expeditions, except when impelled by views of ambition or self-defence. The mighty empires of Greece and Rome were the results of their genius and enterprise.

Among their descendants at the present day we may include the modern Greeks, who speak the Romaika language; the Albanians or Schypetars, whose dialect exhibits some traces of a Celtic origin; the Wallachians or Roumouni, partly blended with the Sclavonians; and perhaps also the great Ceito-Latin nations, whose languages, such as the Italian, French, Spanish, and Provençal, though partly of Celtic origin, are now the principal vernacular languages of the south of Europe.

2. GERMANICUS.—GERMANS.

Syn. RACE GOTHICO-GERMANIQUE.—Malte-Brun, Geog. Univ.
RACE GERMANIQUE (BORÉALE).—Bory Ess. Zool. I. 129.

This is the tallest of all the Iapetan races, its medium height being from five feet ten inches to five feet eleven inches, and men are sometimes found of six feet and a half in height. Their temperament is usually lymphatic; the complexion often animated; the skin of a dazzling whiteness, sometimes approaching to albinism; the eyes usually blue;[1] the hair very fine, straight, of a golden blonde, and becoming grey only at a very advanced age. Their frames are well-proportioned, the tissues soft, and overloaded with fat. The men are robust, brave, accustomed to fatigue, but often passionately fond of fermented liquors. The women are tall, of strong make, fine complexion, and of remarkable *embonpoint*. " La plupart," says Col. Bory de Saint-Vincent, " répandent une odeur qu' il est difficile de qualifier, mais qui rappelle celle de la chair des animaux fraîchement dépecés; elles sont rarement nubiles avant seize à dix sept ans, passent pour avoir certaines voies fort larges, accouchent conséquemment avec plus de facilité que les femmes de la race Celtique, et n' ont en général que peu de ce qui, chez ces dernières, garnit en abondance certaines parties du corps que doivent cacher les ajustemens."

Two great branches of Germans, the *Teutonic* and *Sclavonian*, became distinguished at a remote age.

α. TEUTONICUS.—TEUTONIC RACES.

Syn. VAR. TEUTONE.—Bory Ess. Zool. I. 132.
H. JAPETICUS æ GERMANICUS.—Fisch. Syn. Mam. 2.

The Teutonic races exhibit the German physical characters already enumerated in their greatest purity. Naturally gay, and of a jovial temper, they are fond of good cheer and spirituous liquors. With much frankness and loyalty, they are brave, warlike, capable of executing the most daring enterprises, determined enemies to slavery, and very punctilious as to points of honour. To them we owe the practice of duelling; and their females have always enjoyed the highest degree of influence and consideration.

Under the name of Cimbri, the Teutonic race penetrated into Scandinavia; and the Suenones, afterwards the Goths, who descended towards the south of Europe, upon the feeble remnants of the Roman empire, have left in Gaul, Italy, and Spain, numerous traces of their invasions. Other tribes under the names of Saxons, Danes, and Normans, ravaged the coasts of the British Isles and Gaul, then inhabited by Celts and Romans, and even extended their incursions to the territories of Italy and Greece, where they formed permanent settlements. Further to the north they became the Borusci, and under the name of Norwegians and Swedes, extended their domains to Iceland, Norway, and the regions of the Arctic circle. Of their modern languages, we need only particularize the English, the German, the Dutch, the Flemish, the Swensk or Swedish, Dansk or Danish, the Frieslandic, Icelandic, Norse, Dalska, and the innumerable Germanic dialects of central Europe.

The love of liberty has always been a ruling passion among the Teutonic races, and this feeling has extended alike to their political and religious institutions. " Their opinions were not blindly received from priests, nor was their liberty of action fettered by chiefs."[2] Nearly all the

nations of Teutonic origin profess the doctrines of the Reformed Church as being more congenial to the freedom of their opinions: The Anglo-American colonists, and their descendants, preserve undiminished the haughty and unyielding spirit of their Saxon ancestors.

A. SLAVONICUS.—SCLAVONIAN RACES.

Syn. VAR. SCLAVONE.—Bory Ess. Zool. I. 135.
H. JAPETICUS æ SLAVONICUS.—Fisch. Syn. Mam. 3.
RACE SCLAVONNE.—Malte-Brun, Geog. Univ.
Icon. Blumenb. Dec. Cran. III. t. 22. (Skull of a Lithuanian.)

The traits and manners of the genuine Sclavonian are still forcibly impressed upon the Muscovite Russians, the Polanders, Lithuanians, and Bohemians. With eyes commonly brown, and the colour of the hair rather dark, their cheek-bones are somewhat prominent, the nose often slightly turned upwards, their eyes piercing, their voice strong and coarse. They are in general of an elevated stature, sometimes middle-sized; their step is masculine, and their temperament usually bilious or lymphatic. Capable of enduring the greatest fatigue, hospitable, and brave, they are, at the same time, ignorant, idle, and cunning. The women do not receive the same high consideration and respect as among the Teutonic races, and they have still preserved that habit of sitting with crossed legs, indicative of an Asiatic origin. Indeed, to use the words of Gibbon, " they seem to unite the manners of the Asiatic barbarians with the figure and complexion of the ancient inhabitants of Europe."[3]

Some of their tribes, such as the Cossacks, maintain at the present day the same habits of plunder as the Scythians their neighbours, with whom they are daily becoming more and more blended. For a long time the wandering Bohemians of Western Europe have preserved the predatory habits as well as the language of their Sclavonian ancestors. A branch of the same race, emigrating into the regions occupied almost exclusively by the Teutonic tribes, has established itself on the Elbe, where it still maintains unaltered the characteristics of the genuine Sclavonian, in the little state of Bohemia.

The two great branches of Germans, with a few tribes of genuine Scythians, though really of *Oriental* origin, are celebrated in the history of the middle ages as the *Northern* nations, and as such, became the scourge of the Roman Empire. Scandinavia, so long censured as the great " Northern hive,"[4] has been fully " vindicated" of the charge by a modern writer.

3. INDO-PERSICUS.—INDO-PERSIANS.

LES ANCIENS PERSES ET LES INDIENS.—Cuv. Reg. Anim. 1. 82.

A certain degree of resemblance in the physical traits of the Persians and Indians, the vicinity of their geographical stations, and, above all, the remarkable affinity, which the Sanscrit and the Zend bear to each other, and to the Greek, the Latin, the Teutonic, the Gothic, and the Icelandic, both in their roots and inflexions,[5] have led us to place the Hindoos and Medo-Persians as subordinate branches of the Iapetan races.

α. INDICUS.—HINDOO RACES.

Syn. H. INDICUS.—Fisch. Syn. Mam. 3.—Bory Ess. Zool. I. 225.
INDOUES.—Desmoul. Tab.—Cuv. Reg. Anim. I. 81.
Icon. Blumenb. Dec. Cran. VI. 53. (Skull of a young Bengalese Iadian.)

The Hindoos, or descendants of the ancient Indians,[6] though now mingled with many foreign races, occupy the finest and most extensive regions of the vast peninsula extending to the south of the Himalayan mountains. In respect to stature, they are very considerably below the average, being about five feet six inches. The traits of their physiognomy most forcibly resemble those of the other Iapetans, and we should almost be led to place them in the same subdivision as the Germans, were it not for their colour, which is a very dark yellow, tending towards a bronze, and always with a slight olive tinge. The nose is aquiline, and never flat; the mouth of middle size, with the teeth placed vertically in the gum; the lips thin and coloured; the chin round, and usually marked with a slight dimple; while the skin betrays, by its sudden paleness, the emotions of the individual. The eyes are large and round, with a yellowish cornea; the eye-lashes very long; the eye-brows narrow and arched; the hair straight, long, and very black; the ears well made, and of medium length; the beard scanty. Their legs and feet are extremely elegant, especially among the women. The latter are very short in the body, and elongated in the

---

[1] Tacitus (De Moribus Germanorum. Sec. 4.) notices the strong family likeness of the Ancient Germans.
[2] Sir James Mackintosh's History of England, vol. 1, c. 1, in Dr Lardner's Cabinet Cyclopædia.
[3] Gibbon's Decline and Fall of the Roman Empire, c. 42.
[4] The " *Officina Gentium,*" and " *Vagina Nationum,*" of Jornandes, c. 4. The opinion that the Goths were of Scandinavian origin has been too hastily adopted by Gibbon and Montesquieu.
[5] Adelung's Mithridates, and Klaproth's Asis Polyglotta, passim. Also, Paul de St Bartholomé Dissert. de Antiq. et Affin. Linguarum Zend., Sanser., et German. (fide Malte-Brun).
[6] M. Broc arranges the Hindoos, with great impropriety, among the Mongolians, with whom they have little affinity, excepting in colour.

33

limbs, without being thin; their shoulders in just proportion, and the bust nearly hemispherical—" Elles n'ont presque pas de poil au pubis, mais il y est ordinairement très-dur ; elles accouchent avec une prodigieuse facilité, passent pour très-lascives, et font connaitre leur penchant à la volupté par la variété de mouvemens et d'attitudes qu'elles savent prendre avec tant de souplesse dans ces dancés qui les ont rendues des Bayadères célèbres." (Dict. Class. d'Hist. Nat. art. Homme.)[1] The puberty of the females is very precocious, as they often become mothers at the age of nine or ten years, but lose this power of production at thirty. Few instances of longevity occur among the Hindoos.

The Sanscrit, or dead language in which their sacred books were composed, is written with fifty-two letters, and many thousands of abbreviating syllables. The Pracrit or common dialects of the country are very numerous. From the earliest ages the Hindoo nation has been subdivided into four Tchadi or castes, consisting of the Brahmins, or sacred order ; the Kshatriyas or soldiers, sometimes called the Rajas or Rajepootras, princes or sovereigns ; the Vaisjas, or shepherds and agriculturists ; and the Sudras or labourers. This division, and the reputed dishonour of a marriage out of each caste, have tended greatly to preserve the primitive traits of the genuine Hindoo from foreign intermixture. Every thing serves to prove the extreme antiquity of their civilization; yet so strong is the influence of their religious and political prejudices, that they have preserved themselves almost stationary like the Chinese, though exposed to a constant intercourse with Europeans. The simplicity, mildness, and docility of their character, with their ignorance of the art of war, have led them to submit to a handful of Europeans, allured to their coasts merely by views of commercial advantage. Industrious and sedentary in their habits, they leave the rich commerce of their country to Arabs, Jews, Malays, hinese, and especially to Europeans. Rice, seasoned with pepper and other stimulants, forms their habitual food. The use of Elephants in war, as well as for domestic purposes, was first introduced by the Hindoos, and adopted afterwards by the Carthaginians in Northern Africa, Spain, and Italy.

The Hindoos believe in the transmigration of human souls into the bodies of other animals, and consequently never embalm their dead. All animals are more or less the objects of veneration. The extravagant mythology of the Hindoos would almost defy all powers of analysis ; and the existence of a Tçimurti or Indian Trinity, consisting of Brahma, the Creator,—Vishnou, the Preserver,—and Siva, the Destroyer, bears a prominent place among Oriental dogmas. They practise no barbarous mutilations of the body; but from the earliest ages, purification in the Ganges has been considered a sacred duty. The self-immolation of widows on the funeral piles of their husbands is now discouraged; and an enlightened government endeavours by humane regulations to lighten the terrors of that superstition, which once led thousands to sacrifice themselves under the car-wheels of Juggernaut.[3]

β. PERSICUS.—MEDO-PERSIAN RACES.

Syn. LES ANCIENS PERSES.—Cuv. Reg. Anim. I. 82.
Icon. Blumenb, Dec. Cran. V. t. 35. (Skull of a Persian ;) V. t. 41. (Skull of an Armenian.)

From the earliest ages, until the subjugation of Persia by the Arabs, that country has been the abode of a distinct indigenous race, composed of different nations, and speaking languages in many respects allied to the Germanic dialects. As might be expected from their frequent alliances with the females of other nations, and especially with those of the Proper Caucasians, their physical traits approach more nearly to the European nations, than to the Hindoo branch of the Indo-Persians. Like them, however, the Persians and Armenians have a tinge of yellow and even of olive, but not so dark a skin. The hair is black, the nose aquiline, and the countenance oval, with an elevated forehead. The females are of middle stature, with long black hair, large and dark eyes ; the nose small, and the feet narrow. To these we may add small feet and hands, slender shape, a soft skin, the neck long, and the breast of moderate proportion, long eye-lashes, arched eye-brows, a slight rosy complexion, and we have a portrait of the Persian beauty of the present day. The man of rank and wealth is often distinguished by a portly rotundity of form.

A most extraordinary resemblance may be traced between the Persian and German languages, in their roots, their inflexions, and the forms of their syntax. This affinity may be traced likewise in many Gothic words belonging to the English, Danish, or Icelandic languages, and especially in the latter.[2] Thus the English word door, corresponds with the German thür, the Danish dor, and the Persian dar. The most ancient dialect of the Persians is the Zend, in which the sacred books called the Zend-Avesta, of great antiquity, are written. The ancient Parsee has been preserved in the Shah-Naameh, or Historical Work of Ferdoosi, and in the Ayer-Akbeir. The modern Parsee, the Deri, and the Pehlevi, are now the prevailing dialects of the race. The ancient philosophy of Zerdusht or Zoroaster, once the prevailing religion, and tenaciously held by the persecuted Guebres or Fire-worshippers, has now given place to the Mahometanism of the sect of Ali. The Persians are said to possess that politeness, and versatility of mind, which characterizes a neighbouring nation, and they have been not unaptly termed the Frenchmen of Asia. Many of the arts have made great progress, but their science is almost nominal.

The Armenians must be placed in the same division as the Persians. Their language, according to Adelung, possesses much analogy to others of the Iapetan races.

C. HOMO CELTICUS.—CELTS.

Syn. RACE CELTIQUE.—Bory Ess. Zool. I. 120.—Desmoul. Tab. Hum.—Bose. Ess. 30.
H. JAPETICUS c. CELTICUS.—Fisch. Syn. Mam. 2.
RACES OCCIDENTALES DE L'EUROPE.—Malte-Brun, Geog. Univ.

The Iapetan races were preceded in Europe by the Celtic tribes, who came originally from the North [or more probably from the East],[4] and were once widely distributed ; and by the Cantabrians, who emigrated from Africa to Spain. The former were soon confined to the most western extremities of Europe, and the latter have now become blended among the numerous nations whose posterity inhabits the Spanish Peninsula at the present day.

The languages spoken by the Celts, in so far as they have reached our times, appear to bear some affinity to those of the Iapetans, an affinity which, according to some writers, amounts almost to an absolute identity. The Celtic dialects may be arranged under two distinct divisions, each of which is unintelligible to men of the other division. The Gaelic dialects are spoken in the Highlands of Scotland, in Ireland, and the Isle of Man; and the Cymbric dialects in Wales, Cornwall, and Lower Brittany. Both branches unquestionably belong to the same family of nations. The dialects spoken in the Basque provinces of Spain are commonly referred to this sub-division.

The Celts, Kelts, or Gauls, were the aborigines or primitive inhabitants of Western Europe, and once extended from Ireland as far as the Danube. Before they adopted an agricultural mode of life, they threatened Spain, Greece, and Italy, with their migratory bands, and succeeded in forming permanent settlements in many places, even as far as Asia Minor, where they gave their name to the province of Galatia.

They differed from the Germans in obeying an order of Druids, who united the sacerdotal and political functions, and practised human sacrifices, with other barbarous rites. Their skulls are of an unusual thickness ;[5] the forehead rather protuberant on the sides ; the nose is not rectilineal, but more or less marked by a depression between the eyes. Their hair is of no great length, but thickly furnished, of a deep chestnut, or brown colour, and of tolerable fineness. Their eyes are not so large and prominent as in the Proper Caucasians and Pelasgians, and generally black or brown, sometimes grey. In respect to stature, they are rather taller than the two races just mentioned, their medium height being about five feet nine inches. Their body is well proportioned, robust, and more plentifully covered with hair than that of almost any other race.

All the nations on the left bank of the Rhine were anciently[6] of Celtic origin, and at the present day, nearly three-fourths of the French population exhibit at least some considerable proportion of Celtic characters, though greatly mixed with those of the Pelasgians and Germans, and occasionally with the traits of some Arannan races.

---

[1] DICT. CLASS. D'HIST. NAT.—Dictionnaire Classique d'Histoire Naturelle. Paris, 1825. The article " Homme," by M. Bory de St Vincent, forms the basis of his " Essai Zoologique," already quoted.

[2] Malte-Brun, Geog. Univ., tom. III., lib. 50, consists of an excellent chapter on the Moral and Political State of the Hindoos.

[3] See M. Balbi, Atlas Ethnographique du Globe, Paris, 1827, wherein the several nations of the world are arranged according to the analogy of the idioms and roots of their languages, as well as their manners and customs.

[4] Prichard, on the Eastern Origin of the Celtic Nations.

[5] M. Latour d'Auvergne (Origines Gauloises, Hamburgh, 1801) considers the thickness of the skull as a special distinction of the Celtic races, and to be found at the present day among the Lower Bretons. We have examined the skulls of several ancient Druids from the Hebrides, deposited in the Museum of the Phrenological Society of Edinburgh, and found them to be in general of very considerable thickness. One skull, in particular, from the Monastery of Iona, and apparently of great antiquity, was remarkable in this respect.

[6] Cæsar de Bello Gallico, lib. 1.

p. HOMO SEMITICUS.—ARAMEANS.

Syn.   HOMO ARABICUS.—Bory, Ess. Zool. I. 162.—Fisch. Syn. Mam. 3.
LE RAMEAU ARAMÉEN, OU DE SYRIE.—Cuv. Reg. Anim. I. 81.—Less. Mam. 24.

The Aramean or Syrian branch of the Human Species is stationed to the south of the regions inhabited by the Proper Caucasians. It has produced the Assyrians; the Chaldeans; the ever-unconquerable Arabs, who attempted under Mahomet to render themselves masters of the world; the Phœnicians, the Jews, the Abyssinians, and other Arabian colonies; and probably also the Egyptians. It is from this branch, ever prone to the dissemination of mystical doctrines, that the most widely-prevalent systems of religion have derived their origin. Science and literature have sometimes flourished among them, but always under a repulsive form or a figurative style.

1. ARABICUS.—ARABIANS.

Syn.   H. ARABICUS, (b.) ADAMICUS.—Fisch. Syn. Mam. 3.
RACE ADAMIQUE (ORIENTALE).—Bory, Ess. Zool. I. 179.
RACE ARABE.—Malte-Brun, Geog. Univ.—Desmoul. Tab.
Icon.   Blumenb. Dec. Cran. I. t. 1; IV. t. 31, and VI. t. 52. (Skulls of Ægyptian Mummies.) IV. t. 34, and Ill. t. 28. (Skulls of Jews.)

In the primary division of the Arameans, we include all nations and tribes who speak dialects nearly resembling the Hebrew. Of these, we may enumerate the Old Syriac, spoken by the tribes in the neighbourhood of Damascus and Mount Libanus; the Hamyarite and Koreishite dialects of the Arabic, the latter of which is consecrated by being the language of the Koran; and the Coptic, a relic of the ancient Ægyptian, with its dialects, the Memphitic, Saidic, and perhaps also the Bashmooric.[1] The Hebrew and its dialect the Chaldee are well known as the original language of the more ancient division of our Sacred Scriptures.

The Arab race, which the conquests of Mahomet have distributed over an extensive territory, is distinguished by an oval but elongated face, elevated forehead and prominent chin; the nose well marked, and in general aquiline; the eyes black or deep brown, of large size, and of a peculiar expression in the females, which has often led to their being compared to those of the Gazelle. Their eye-brows are tolerably thick and arched, their lips narrow, and the mouth agreeable. Their hair, which is black, smooth, seldom curly, and rather coarse, usually grows to an extraordinary length; and the women wear it in plaits, which commonly hang as low as the ancles. The Arab is rather of an elevated stature; while on the contrary the Arab female is diminutive—a disparity between the sexes which seems to be one of the characteristics of their race. The Arab women being exceedingly precocious soon lose their powers of procreation, while the men preserve their youthful vigour to an advanced age. This physical peculiarity has rendered polygamy from the earliest ages[2] the prevailing custom of the Arabic races. Circumcision—a rite once sanctioned by Religion—appears to have tended greatly to preserve these races from alliances with foreign nations. The modern Jews, dispersed in all parts where a lucrative commerce prevails, have maintained to the present day many of the sacred institutions of their ancestors, and the same abhorrence of swine's flesh,[3] with other branches of the Aramean race. By a kind of over-refinement, the Arabian devotee, according to Sonnini, occasionally extends the practice of circumcision further than antiquity would have authorised.[4]

The modern Arab has been not improperly termed the *brown* variety of the White races. On the burning sands of Abyssinia his skin acquires a deep tint, but becomes almost eviolated in the cool mountain valleys of the same territory, while the females of rank, confined to their tents, are of a dazzling, and often an insipid whiteness. The brown tint of their skin seems, therefore, to be rather an accidental result of climate than a peculiarity of race.[5]

The Bedouin Arab of the desert preserves the same predatory habits as in the days of Jacob or Moses; and by the assistance of the Dromedary and the Horse, with some cattle, he is enabled to maintain the nomadic life in a country of peculiar sterility. All his senses acquire from habit an extraordinary acuteness, in detecting the traces of an enemy in the sand, or the footsteps of a lost Camel among thousands of other impressions. A

stranger is usually received as an enemy; yet some instances of a romantic hospitality are not unknown.[6] The genuine Arab possesses a poetical imagination, which often degenerates into extravagance. Religious enthusiasm and fanaticism seem in all ages to have been congenial to his disposition.

Many tribes, devoting themselves to agriculture, and especially to commerce, exhibit a peculiar tact and penetration in matters relating to their own pecuniary advantage.

The splendid monuments of ancient Ægypt attest the early period at which the arts were cultivated by these races, but the writings of the Arabian physicians in the middle ages exhibit the extreme limit of their scientific attainments.

We may recognise the Arabian features in the inhabitants of a large portion of the east coast of Africa, in the Comora Islands, Socotora, and the north of Madagascar, though often blended with the traits of the genuine Moor, the Ethiopian, and the Caffre.

2. ATLANTICUS.—ATLANTIC RACES.

Syn.   H. ARABICUS (a.) ATLANTICUS.—Fisch. Syn. Mam. 3.
RACE ATLANTIQUE (OCCIDENTALE).—Bory, Ess. Zool. I. 174.
Icon.   Blumenb. Dec. Cran. V. t. 42. (Head of the Mummy of an ancient Guanche from Teneriffe.)

The caverns of the Peak of Teneriffe have preserved the remains of an ancient branch of the Aramean race, which the philological researches of MM. Hornemann and Marsden have proved to be allied to that of the Be,be,s or Shêlooks, who inhabit a large portion of the north of Africa. To this race, whose early civilization appears undoubted, we must refer the ancient Phœnicians, the Numidians, and Carthaginians, and probably also the Getulians and Garamantes. From the invasions of Greeks and Romans, Goths and Vandals, Turks and Arabs, the Atlantic races of the north of Africa exhibit many characteristics of the other white races, such as fine hair, tending towards chestnut, and even blonde, and in the mountain districts some families are almost white. In general, their nose is less aquiline than among the other Arameans, and they are not usually of so dark a colour, but approaching towards an olive tint.

The modern Moors inherit the maritime genius of their Phœnician ancestors, and have long been celebrated for their piracies; in which respect they differ remarkably from some nations of the Arabian branch, and especially from the Ægyptians, whose aversion to the sea is of great antiquity. The Moorish females would be accounted beautiful in any country. Those of the kingdom of Tripoli, though next to the Ægyptians, differ considerably from them, and are much taller. Like many of the Arab females, they tattoo their faces, especially upon the cheeks and chin. Red hair, as in Turkey, is much esteemed, and the locks of their children are often dyed with vermillion. The practice of rubbing the hair of the eyelids with plumbago is very general.[7]

E. HOMO SCYTHICUS.—SCYTHIANS.

Syn.   LE RAMEAU SCYTHE ET TARTARE.—Cuv. Reg. Anim. I. 82.
HOMO SCYTHICUS (in part).—Bory, Ess. Zool. I. 236.

The Scythian and Tartar branch of the human race, established from the earliest ages towards the north and north-east of the ancient continent, has preserved its nomadic life in the immense plains of those regions; from whence it has never descended upon the more favoured regions of the other branches, but with views of plunder and devastation. The Turks, who overthrew the empire of the Greeks, and subverted in Europe the miserable remnant of the Greek empire, belonged to this populous race. The Finns and Hungarians are Tartar nations, who may be almost said to have lost their way among the Sclavonian and Teutonic Races. The North and East of the Caspian Sea, apparently their original country, contains other nations of the same origin, and who speak similar languages, although intermixed with a considerable number of small tribes, differing from each other in descent as well as language. The Tartar nations have remained for a greater period unblended in all the regions which extend from the mouths of the Danube as far as the banks of the Irtisch, whence they have long threatened the

---

[1] Professor Vater in Adelung's Mithridates.
[2] Genesis, and the Books of Samuel, passim.
[3] Pliny (Nat. Hist., lib. VIII.) imagines that the Hog cannot live in Arabia.
[4] Un développement excessif des nymphes a rendu presque nécessaire une opération analogue à la circoncision chez les hommes. Consult on this subject Sonnini, Voyage en Egypte; also Virey, Hist. Nat. du Gen. Hum., tome I.
[5] Niebuhr's Description de l'Arabie, and Burckhardt's Travels in Arabia, passim.
[6] Volney (Voyage en Syrie), Niebuhr, and Burckhardt (Notes on the Bedouins and Wahabys), give many interesting details regarding the manners and customs of the wandering Arab.
[7] Buff. Hist. Nat., tome III. 430.

Russian Empire, though ultimately subdued by its power. In this subdivision we may place those Ancient Scythians, who at a remote period made irruptions into Upper Asia ; and the Parthians, who subverted the Asiatic dominion of the Greeks and Romans. Owing to the conquests of the Mongolians, numerous alliances have been contracted with that race, and many evident traces of Mongolian blood may be recognised, especially among the inhabitants of Lesser Tartary.

I. OTHMANICUS.—OTTOMAN TURKS.

Syn. LES TURCK.—Cuv. Reg. Anim. I. 82.
        RACE TURQUE (in part).—Desmoul. Tab.
Icon. Blumenb. Dec. Cran. I. t. 2 (Skull of a Turk).

The origin of that tribe of Scythians, who, in modern times, have obtained so much celebrity under Othman and his successors, is involved in considerable obscurity. From time immemorial they appear to have wandered in the plains near the Southern bank of the Oxus, until the feebleness of their neighbours induced them to desolate with their inroads the fertile territories of Greece, Asia Minor, Syria, and Egypt.[1] Their language bears at the present day no inconsiderable affinity to the dialects of Tartary, though mixed with many Persian and Arabic words, derived from an intercourse with the nations whom their valour had subdued.

Possessed of a noble but harsh physiognomy, the features of the Ottoman Turks nearly approximate to those of a European. Being rather tall, robust, and well-made, with tawny complexions, and black or dark-brown hair, it is with some difficulty that we can trace any resemblance to those Mongolian races, with whom their own historians, and some other continental writers, would confound them. A large and flowing dress, thick rolls of turban upon the head, and long mustachios, impart an imposing aspect to a deportment of no ordinary gravity.[2] Polygamy forms the luxury of the wealthy, but the poorer Mussulman is glad to escape from the trouble and extravagance of an extensive Haram. Indolent, proud, and ignorant, the Turks are averse to business, and passionately devoted to the immoderate use of coffee, opium, and tobacco. Their customs and manners, their religious bigotry, the feebleness of their political institutions, and the steadiness with which they have opposed all attempts to improve their condition, mark them out as Scythians in the strictest sense of the term.

2. FINNICUS.—FINNS.

Syn. RACE FINNOISE.—Malte-Brun, Geog. Univ.—Desmoul. Tab.

The Finnic or Tchoude nations, surrounded by races of different extraction, may be recognised by their language, customs, and physiognomy. The Finlanders, Biarmians, Ehstes, Livonians, Wotiaks, Woguls, Tchouvashes, Tcheremisses, and Ostiaks, tribes of the Russian Empire; the Magyars, Ougres,[3] Ungres, or Hungarians, who form a large portion of the population of modern Hungary ; and the Laplanders, may be included with propriety in this subdivision.

The Finnic races, widely dispersed throughout the provinces of Russia, are marked by a sallow complexion, prominent cheek bones, red or yellow hair, a large occipital region, and scanty beard. In the Russian language they are termed *Tchoudes*, which signifies *strangers*. A few relics of the mythology and history of these obscure nations are preserved in the Saga of Saint Olaf, and in some Scandinavian and Russian monuments.

The Finns are possessed of a literature and mythology peculiar to themselves. The Finnic language is understood by their neighbours the Ehstes or Esthonians, who still retain under all the degradation of slavery the yellow hair and other characteristic features of their race. Their poetry is metrical and full of alliteration. The Wotiaks are a weak and ugly race, mostly with red or yellow hair and scanty beard. The religious notions of these tribes are peculiar; and the wicked, after death, are placed for ever in cauldrons of burning pitch, while the Tchouvashes change them into skeletons, which roam eternally in a frozen desert. The Ostiaks pay a peculiar veneration to the Great Bear, and swear allegi-

ance to every new Russian Czar upon a skin of that animal, or upon an axe with which a bear has been killed.

The Hunns or Magyars, though of Finnic origin, are connected with those tribes of Tartars, who, under the name of Turks, devastated Europe during the middle ages. The Hungarian peasant in general is of a robust and energetic constitution, but of moderate stature, and still retains many of his Tartar habits. The modern traveller may recognise in his peculiar costume many points of resemblance to that of most nomadic Scythians of Central Asia.

The remarkable affinity of the Hungarian language to that of the Laplanders led Sainovicz to consider them as identical ;[3] and M. Klaproth has proved the intimate relation of the former to that of the Ostiaks.[4]

The Laplanders belong unquestionably to the Finnic or Tchoude race. They are of small stature, with dark brown complexion, black hair, high cheek bones, broad face, and pointed chins. The men are thickly set and active, their beards scanty ; the women are robust, and produce with slight inconvenience. Naturally of a roaming disposition, the Laplanders reside in tents made of cloth or the skin of the Rein-deer ; and seldom remain for any considerable time in the same locality. From time immemorial they have been maintained by numerous herds of Rein-deer, which almost compose their entire means of sustenance. Polytheism, wherein every object in Nature is changed into a Deity, and a universal idolatry, in which the elements are typified, form the bases of their religious opinions. Their superstition is extreme, and many tribes, who even profess Christianity, still preserve their ancient idols, their magical drums, and certain knots with which they allay the frequent storms of those desolate regions.

3. TARTARUS.—TARTARS.

Syn. LES PEUPLES TARTARES.—Cuv. Reg. Anim. I. 82.
        RACE TURQUE (in part).—Desmoul. Tab.
Icon. Blumenb. Dec. Cran. II. t. 12. (Skull of a Tartar of Kasan.)

The Tartar, Tatar, or Turkish nations of Central Asia, who are probably identical with the Asiatic Scythians of the Ancient Greeks, must be carefully distinguished from the Mongolians. Their European cast of countenance, though tending slightly towards a yellow, their long beards, curly hair, and slender figures, mark them out as belonging to the White races. The Moor is not more different from the Negro, than the real Tartar from the genuine Mongolian.

The pastoral life has been in all ages the favorite mode of existence of the Tartar races, who roam undisturbed, over the extensive plains between the Beloor mountains, and the basins of Lake Aral and the Caspian Seas. With extensive herds of cattle, with a numerous body of horse, which the Tartar can manage with skill, and those habits of continual motion which are essential to the practice of successful war, the tribes of the Scythians have always been prepared to advance upon unknown territories, wherever they expected to find a powerless enemy or a plentiful subsistence. Led by their Khans, the Tartars are well known at various periods of history as having acted an important part in several mixed emigrations of Tartar and Mongolian nations. On these occasions, both were included under one common appellation, and they may be recognised at one period as the Euthalites, Nephthalites, or White Hunns ; at another, as the Turks of Transoxiana ; again as the Hunns of Attila ; as the Abares in the sixth century ; and finally, under Chingis Khan and Timoorlane they assisted in devastating Persia, India, and Western Asia, for more than two centuries ; while the most fertile countries of Europe have been struck with terror at the approach of these martial shepherds.[5]

Their invincible courage, overwhelming numbers, and rapid conquests, have given a military aspect to the pastoral mode of life, which our poets are in the habit of adorning with the attributes of peace and innocence. The flocks and herds which accompany the Tartar during his inroads supply him with milk and flesh, and he feeds indiscriminately on animals which have died by accident or disease. Horse-flesh is particularly esteemed as an article of food, but the animals themselves are in general too valuable to be used for this purpose except on an emergency. The ignorance of these nations is extreme, and it is seldom that even a Tartar Khan can either read or write.

AT the conclusion of this enumeration of the Caucasian races, but without referring them to any particular section, we shall place the Zigeunes, Zingani, Tchinganes, Atchingans or Gypsies, who appear to have wandered from time immemorial in most parts of Europe, but especially in Turkey,

1 Von Hammer, Geschichte des Osmanischen Reiches, I'. Band.
2 Lady Mary Montague's Letters, Volney's Voyage en Syrie, Olivier's Voyages, and Malte-Brun, Geog. Univ.
3 Sainovicz, Demonstratio Idioma Hungarorum et Laponum idem esse, Copenhagen, 1770.
4 Klaproth, Asia Polyglotta.
5 Gibbon (Decline and Fall of the Roman Empire, chap. 26, 34, and 64) describes with elegance and fidelity the manners of the pastoral nations, the conquests of Attila and his Hunns, and the devastations of the united bands of Mongolians and Tartars in the thirteenth and fourteenth centuries.

Wallachia, Moldavia, Hungary, and Transylvania. " They have wandered through the world, in every region, and among every people they have continued equally unchanged by the lapse of time, the variation of climate, and the force of example. In the neighbourhood of civilized life, they continue barbarous, and near cities and settled inhabitants, they live in tents and holes in the earth, or wander from place to place like fugitives or vagabonds." The skull of a genuine Gypsy is figured in Blumenb. Dec. Cran. II. 11.

II. MONGOLIANS.

Syn. , RACE MONGOLIQUE.—Cuv. Reg. Anim. I. 82.—Desm. Mam. 47.
LA MONGOLE (in part).—Dum. Zool. Anal. 7.
H. SCYTHICUS (in part).—Fisch. Syn. Mam. 4.
RACE JAUNE, ou OLIVATRE.—Virey, Hist. Nat. du G. Hum. I. 457.

RACE JAUNE, ou MONGOLIENNE (in part).—Less. Mam. 25.
HOMO SAPIENS, VARIUS, VAR. β.—Lin. Gmel. I. 23.—TATARUS. β.
Erxl. I.—VAR. MONGOLICA, Blumenb. Hand. et Abbild.

THE Mongolians may be recognised by their prominent cheek-bones, a flat face, narrow and oblique eyes, straight and black hair, a scanty beard [if we except the Ainoos, or hairy-men of the Kurile Islands], and an olive complexion. They have formed empires in China and Japan, and sometimes extended their conquests beyond the Great Desert; but their civilization has always remained stationary.

The geographical station of the Mongolian races commences at the east of the Tartar branch of the Caucasians, and extends to the Pacific Ocean. Some of their branches, still nomadic, roam over the Great Desert, under the names of Calmucks or Kalkas. Three times their ancestors, under Attila, Chingis, and Timoorlane, spread far and wide the terror of their name. The Chinese appear to have been civilized at an earlier period than any other of these races, and indeed we may almost say, than any nation in the world. The Mantchoos, forming a third branch, have recently conquered China, which they still continue to govern. The Japanese, the Coreans, and nearly all the hordes which extend to the north-east of Siberia, under the dominion of the Russians, may be referred to this division, as well as the natives of the Marianne Islands, the Carolinas, and the adjacent islands of that Archipelago. If we except some of the Chinese literati, the Mongolian nations are generally devoted to the different sects of Buddhism, or the religion of Fo.

The origin of these races appears [according to the conjectures of some writers] to have been in the mountains of Altai, in the same manner as our White races have been assigned to the Caucasian mountains; but it has not been found possible to follow out with equal ease the [supposed] affiliation of their several branches. The history of these nomadic races is as fugitive as their settlements; and the records of the Chinese, confined within the limits of their empire, give but a few short and unconnected notices of the neighbouring nations.

The affinities of their languages are also too little known to guide us through the labyrinth. The dialects used to the north of the peninsula beyond the Ganges, as well as in Thibet, bear some relations to the Chinese language, at least in their nature, being in some respects monosyllabic, while the nations who speak them are not without some traits of personal resemblance to the other Mongolians.

A. HOMO CALMUCCUS.—CALMUCKS, & MONGOLIANS PROPER.

Syn. LES CALMOUQUES.—Cuv. Reg. Anim. I. 83.
Icon. Blumenb. Dec. Cran. I. t. 5. (Skull of a Calmuck); also II. t. 14.

The languages of the Calmucks or Kalkas, the true Mongolians, and the Booriaits of Lake Baikal, bear a considerable degree of resemblance to each other, but are wholly distinct from those of the Tartar races, in their vocabularies as well as in the forms of their syntax.[1] One important circumstance proves them to be allied in some degree to the dialects of China and Thibet, being the frequent recurrence of monosyllabic words. The Calmuck language is highly poetical, abounding in romances and epic compositions of considerable beauty.

The Calmuck is of a middle stature, with the internal angles of the eye

directed downwards towards the nose, the eye-brows black and narrow, the interior ends of the arches low, the nose flat and broad at the point, the cheek-bones prominent, the head and face very round. Their complexion is a brownish-yellow, differing in intensity according to the sexes and individuals. The acuteness of their senses is much celebrated, but not more so than their ugliness, which is described by travellers as being something terrible. The men shave their heads, with the exception of a small tuft, which is allowed to grow sufficiently long to form a lock of considerable length on either side of the face. The women allow their hair to hang in two braids over the shoulders, but without shaving any part of the head.

The proper Mongolians have thick lips, short chins, scanty beards, large and prominent ears, flat noses, and oblique eyes, nearly resembling those of the true Calmucks. Their language is but little known. Polygamy, though permitted by law, is among them, as in other places, rather uncommon. The religion of all these benighted tribes is that of the Dalai Lama, in which the people are held under the entire subjection of priests and jugglers.

B. HOMO SINICUS.—CHINESE.

Syn. LES CHINOIS.—Cuv. Reg. Anim. I. 83.
H. SCYTHICUS β. SINICUS.—Fisch. Syn Mam. 5.
H. SINICUS.—Bory, Ess. Zool. I. 249.
INDO-SINIQUE.—Desmoul. Tab.

Icon. Blumenb. Dec. Cran III. t. 23 V. t. 44. (Skulls of Chinese).
CHINESE.—Griff.[2] Anim. King. (Head of a fur-dealer).

Under the general term Chinese we shall include the inhabitants of China Proper, Corea, Tonquin, Thibet, Cochin-China, Siam, and the natives of the Birmese Empire. The Japanese, also, may probably be referred to this head, until a more minute acquaintance with their language and history shall fully establish their claim to a distinct subdivision.

Of these nations, the Birmese and Siamese are the tallest, being commonly from five feet six inches to about five feet nine inches in height. They strikingly exhibit the ordinary features common to all the Mongolian races, such as prominent cheek-bones, with oblique and narrow eyes; they also have scanty beards. Their ears are very broad, and project outwards, so as to be entirely visible in front. On the beautiful porcelain wares of China and Japan, their physical traits are often depicted with precision. The intensity of their complexions varies in the several races, being darker among the Cochin-Chinese and Siamese, and brown-olive among the Tonquinese. Their hair, which has nearly the coarseness of horse-hair, grows within an inch of the eyebrows. The Chinese ladies of rank, confined to their abodes by the barbarous treatment which their feet undergo in early youth, as well as by the forms of etiquette, often exhibit an etiolated appearance corresponding to that of our European ladies; yet there is always something about them, to use the words of M. Bory, which forcibly recalls the idea of *suet*. In general, the Chinese races have an oily skin, a yellowish-green complexion, passing towards a brown, according to climate, and even becoming dark below the 20th parallel of north latitude, from their alliances with the Malay races. It is however remarkable, that the most northern Chinese have the darkest complexions of the whole.

The greater number of the races of Chinese descent have the utmost horror of intermarrying with strangers, whom indeed they generally regard with aversion. To preserve themselves from the aggressions of foreigners, they constructed their celebrated wall, which serves to protect the northern frontier of their empire. Agriculture forms their chief

---

[1] Bergmann. Nomadische Streifereinen unter den Kalmuken, I. p. 125. (Fide Malte-Brun).
[2] GRIFF. ANIM. KING.—The Animal Kingdom. By Edward Griffith, and others. London, 1827.
34

means of sustenance. " The staff of life in China," observes Sir John Barrow,[1] " is rice, and it is the chief article of produce in the middle and southern provinces. The grain requires little or no manure; age after age the same piece of ground yields its annual crop, and some of them two crops a year. In the culture of rice, water answers every purpose." All the details of agriculture are prescribed by the laws; national festivals are consecrated in honour of the art, and the Emperor of China ostensibly professes to be the first agricultural labourer of this vast territory. A sea-faring life is held in abhorrence by the greater number of the natives, and those Chinese who leave their country for purposes of trade, only do so in defiance of the laws, for, if recognised, they are not permitted to return without molestation.

Silk is the material most commonly used for their larger garments, and cotton dresses are by no means so common as among the Hindoos. Being scarcely ever addicted to the abuse of spirituous liquors, tea forms the favorite beverage of the Chinese; and perfumes are highly esteemed. Their soldiers have little courage; and their dress and appearance are said to be " most unmilitary, better suited for the stage than the field of battle; their paper helmets, wadded gowns, quilted petticoats, and clumsy satin boots, being but ill adapted for the purposes of war." The people, however, are very industrious and skilful in business. Clever artizans in nearly all the useful and elegant articles of life were common in China, at a time when Europe was sunk in profound barbarism. Their history, and the peculiarities of their monosyllabic language, may be traced to a remote period of antiquity; yet their civilization has long remained stationary, and the most trivial actions of individuals are, here regulated by the forms of law.

The Emperor and his court have long adhered to the ancient religion as originally taught by Kong-fu-tse, commonly called Confucius, but the present Mantchoo dynasty exhibits a very marked leaning towards Buddhism, or the religion of Fo.

The traditions of the Japanese tend to show that they were originally a Chinese colony, yet their language is wholly different; few Chinese terms can be recognised; the words are not monosyllabic; while the syntax and conjugations possess a distinct and original character. There is a peculiarity in the eye of a Japanese, which indicates a slight resemblance to the Mantchoo; it is oblong, sunken, and narrow, so as to appear as if constantly winking; the eyelids form a deep furrow, and the eyebrows appear higher than we find generally in other nations of this race. The head of a Japanese is commonly large; the hair thick, black, and glossy; the neck short, and the nose broad.

The native inhabitants of Pegu, Ava, and Aracan, do not differ greatly from the proper Chinese in their physical traits, but are merely a little darker in their complexions. It is said that the Aracanese admire a broad and flat forehead, and with this view continue to flatten the foreheads of their infants soon after birth. Their nostrils are broad and open, their eyes small and lively, and their ears hang down to the shoulders. The females are not very dark, and their ears are fully as long as those of the male population.

C. HOMO SERICUS.—MANTCHOOS.

Syn. Les Mantchoux.—Cuv. Reg. Anim. 1. 83.—Lesson, Mam. 25.
Icon. Blumenb. Dec. Cran. II. t. 16. (Skull of a Tongoos), and III. t. 23. (Skull of a Daourian).

According to M. Langlès,[2] the language of the Mantchoos is wholly different to that of the true Mongolians, the Chinese, or any of the Tartar tribes, and bears some remote affinities to the languages of Europe.

Most of the leading families of Mantchoo extraction have settled in China, where they have adopted the common dress of the country. For these reasons they are gradually becoming incorporated with the Chinese, whom their arms have subdued. Their religion has some affinity with Shamanism. They are more robust in their figure than the Chinese, but their countenances have less expression. The feet of the genuine Mantchoo women are not cramped and rendered useless by bandages.

The several wandering tribes of hunters, known by the name of Tongooses or Œvœns,[3] who roam over the barren wastes situate to the eastward of the Sea of Baikal, may be placed in this subdivision.

D. HOMO KURILIANUS.—AINOOS.

Syn. Kourilienne.—Desmoul. Tab.
Icon. Desmoul. Hist.[4] Nat. des R. Hum. pl 5 and 6.

The islands extending in a chain from Japan to Kamtchatka are inhabited by a race of men peculiar in language and appearance, called by the Japanese Mo-sins, or Hairy men, though they style themselves Ainoos, according to Krusenstern.[5] They are taller than the Japanese, and of a more robust frame, with very thick and black beards, and the hair of the head also black and somewhat frizzled. The forehead rises rather squarely upwards, the nose is straight, and nearly on a level with the forehead, as in the Celtic, Iapetan, or Aramean races, only shorter and thicker. Their complexion is deep brown, approaching towards black; the beard and eyebrows so thick as almost to conceal the face; the whole body covered with hair; and, if we may credit the accounts of the Russians, a child of five years old, found, in 1806, at the bay of Mordwinoff, had his body already covered in this remarkable manner.

Their height is about five feet seven inches in the neighbourhood of Jesso; their limbs well proportioned. The women appear to European eyes far uglier than the men; their complexion is equally dark, their lips painted blue, and their hands tattooed. M. Desmoulins (Hist. Nat. des R. Hum. pl. 6) represents an Ainoos family, from a singular Japanese design, which, though rude, is executed with some spirit. The mother of the family appears to be suckling a young pig, a common practice in the island of Jesso, where the females rear young bears, dogs, or pigs, in this manner, and confine them when old enough in cages until they are sufficiently fat for killing.

The language of the Ainoos is said to bear no affinity to the Japanese, the Kamtchatskadale, or the Mantchoo, and, as far as hitherto known, seems to be very different from any other.

III. NEGROES.

Syn. Race Nègre.—Cuv. Reg. Anim. I. 80.
       Race Noire ou Mélanienne (in part).—Less. Mam. 26.
       Race Ethiopienne ou Nègre.—Desm. Mam. 47.—Dum. Zool. Anal.

       Homo Sapiens, Afer δ.—Linn. Gmel. 1. 23.—r. Erxl. 2.—Var. Æthiopica.—Blumenb. Hand. et Abbild.

The Negro races are confined to the region south of Mount Atlas. Their complexions are black, their hair woolly, the cranium compressed, and the nose flattened, while their projecting muzzle and thick lips indicate a near approach to the characters of the Monkey tribes. The hordes composing this division have always remained in a barbarous state.

A. HOMO ÆTHIOPICUS.—ETHIOPIANS.

Syn. Ethiopien.—Less. Mam. 26.—Desmoul. Tab.
       H. Æthiopicus.—Bory, Ess. Zool. II. 29.
Icon. Blumenb. Dec. Cran. I. t. 6, 7, and 8. II. t. 17, 18, 19.

The traits of the genuine Ethiopian are so very different from those of

any of the races already described, as to strike an observer at the first glance. But independently of the nature of his woolly hair, the excessive darkness of his entire skin, and the clear tones of his voice, there are some striking anatomical differences, which, according to some writers, would be accounted specific, if recognised in any other animal than Man. The cranium is narrowed in front, flattened on the top, and becomes rounded in the occipital region; the sutures are very close, the bones of the nose considerably flattened; the incisive teeth inserted obliquely; while the skeleton of the entire body surpasses that of all other races in whiteness.[6] The bones of the pelvis are larger, especially in the females, and the thighs and legs possess a certain degree of curvature, so as to impart a bow-legged appearance to the best made Negro.

The following may be briefly stated as the characters of the true Ethiopian :—The skin of the entire body black, excepting the palms of the

[1] Encyclopædia Britannica. Art. China. By Sir John Barrow. Also, Barrow's Travels in China; and Narrative of a Journey in the Interior of China. By Dr Clarke Abel. London, 1818.
[2] Langlès, Alphabet Mantchoo. (Fide Malte-Brun.)
[3] See Malte-Brun. Geog. Univ. And the Synoptical Table at the end of the XLth division.
[4] Desmoul. Hist. Nat. des R. Hum.—Histoire Naturelle des Races Humaines. Par A. Desmoulins. Paris, 1826.
[5] Krusenstern's Voyage, II. p. 7.
[6] This remark has been made by M. Bory Saint-Vincent; we have not had an opportunity of verifying his observation.

hands and the soles of the feet; the forehead depressed; the hair woolly and curly, forming a true cap on the top of the head, and leaving the inferior margin almost regular; the eyes large and prominent; the sclerotica tinged with yellow; the nose flattened; the cheek-bones projecting; the lips very thick and protruding forward; the chin depressed; the ears long and directed laterally; the occiput thrown far behind. The shoulder-blades are longer and more pointed than those of any of the Caucasian races; the anterior limbs are singularly elongated; the lower limbs are thin, the thighs flattened, especially in the internal parts; the calf of the leg is very small, and placed at a short distance from the cavity at the back of the knee, the heel projects far behind, and the sole is excessively flattened.

It has been shown by M. Soemmering, that the brain of the Ethiopian is comparatively more confined than ours, the cranium being always smaller in proportion as the face projects forward; while, on the other hand, the nerves which issue from it are larger. The colour of his blood is evidently deeper, as well as that of the muscles, the bile, and generally of all the humours of his body. His perspiration is also more ammoniacal, and taints linen with a very disagreeable odour. The breasts of the females hang very low, and from the earliest marriageable age assume a long, pendant, and pear-like form, which permits them to suckle their infants over the shoulder. "Elles ont aussi le vagin en tout temps large et proportionné au membre viril du mâle, souvent énorme, mais à-peu-près incapable d'une érection complète. La grande facilité avec laquelle conséquemment les Négresses accouchent dès l'âge de onze à douze ans où elles sont définitivement réglées, dégénère en inconvénient, et nulles femmes ne sont plus sujettes à l'avortement. Dans le fœtus, la tête n'est pas aussi grosse proportionnellement qu'elle l'est dans les autres espèces; aussi la fontanelle du nouveau-né est très peu considérable et presque fermée dès la naissance. les os du crâne ne devant pas jouer les uns vers les autres, quand il est question de la délivrance."

Every thing in the constitution of the proper Negro denotes an approximation towards the animal, where the mere physical impressions predominate over, and often almost extinguish, the moral or intellectual. His sight is piercing, his sense of smelling extremely subtle, and his hearing very sensitive to musical impressions. The Ethiopian is sensual in his tastes, gluttonous, and excessively amorous. Every other variety of the human species is excelled by his race in agility, dexterity, and indeed in all those imitative qualities which more or less depend upon the animal frame. He is an adept in dancing and swimming, but seldom rides. He performs the most striking feats of address, climbs, and leaps with an agility surpassed only by the Apes. While dancing, the Negro agitates every part of the body, and seems indefatigable. He can distinguish a remote object, which the gaze of a European could only reach by the aid of a telescope; and can detect the presence of a serpent, or hunt an animal by the scent. The slightest noise does not escape his ears, and the fugitive Negro slave frequently evades the pursuit of his master through the superior delicacy of his organs of sense. His touch is surprisingly acute; but being thus keenly attentive to the impressions of mere sense, his reflecting powers are but little exercised. The dread of the most cruel punishments, and even of death itself, does not prevent the Ethiopian from abandoning himself to the passion of the moment; and even when writhing under the lash of the overseer, the sound of the tam-tam, or some other execrable music, will inspire him with fortitude, or bind up the feelings of the past in forgetfulness. The monotonous notes of some dull chant, picked up by chance, are sufficient to support him under the most violent fatigues. A moment of pleasure is sufficient to obliterate the remembrance of a year of pain. Ever devoted to the affections and feelings of the moment, the past and the future are nothing in his eyes; his griefs are fugitive, and as he follows the suggestions of sense rather than the dictates of reason, he is extreme in all things,—like a lamb when oppressed.—like a tiger when power is placed in his own hands. In a moment of vengeance, he does not hesitate to massacre one of his wives, or dash his infants upon the earth. Nothing is more terrible than his despair, or more sublime than the devotedness of his attachments.[1]

The genuine Negroes are subject to certain disorders, such as the yaws, which they do not communicate to the other varieties. The small-pox is very dangerous in them, and commonly appears before the age of fourteen; but after that period, we are assured that the Negroes are comparatively free from it. Though exceedingly muscular, the prevalent temperament is lymphatic; the pulse appears to be more accelerated than among the White races.[2]

Whatever opinion may be adopted regarding the cause of the intellectual and social inferiority which the Ethiopians exhibit, when contrasted with the Caucasian races, or even with several of the Mongolians; whether we consider it as proceeding from their original constitution, or from the unculkivated state in which their faculties have always been suffered to remain,—the fact cannot be denied. In general, the Ethiopian is lazy, without the least foresight, deriving no knowledge from the experience of the past, with few wants but such as Nature can readily satisfy in a tropical climate; and thus he vegetates in a condition which cannot be termed absolutely savage, although without the smallest pretensions to be accounted as a genuine civilization.

The true Negroes are generally divided into petty tribes or small nations, despotically governed by chiefs of the most sanguinary dispositions, nearly always at war with each other, for the purpose of making prisoners, to satisfy the avidity of the execrable slave-dealer. Some of these tribes, according to their geographical position, live on fish; others devote themselves to trade; and others, again, cultivate some kinds of pulse or grain, or lead the nomadic life of shepherds. Without any established form of religious belief, they attribute supernatural powers to any remarkable object which strikes their attention. Some adore a Serpent, or any other animal, some a Baobab, or any large tree, according to the peculiar form of Fetishism, which happens to be countenanced in the locality.[3]

Polygamy, in the widest sense of the word, seems to be practised throughout the entire races; and seldom do sentiments either of modesty or humanity penetrate into their savage breasts. Blood flows unheeded, and the most inhuman torments and mutilations are inflicted on their vanquished enemies; such as tearing off the lower jaw, or some of the limbs, as a trophy. Whole tribes wander entirely naked, armed with bows and arrows, or wooden javelins pointed with iron; and it is only in the European colonies that they consent to wear the langouti, or blue girdle. Some of the tribes, who have become more civilized or degraded by their commerce with Europeans, adopt the cotton cloths or stuffs of foreign manufacture. The beautiful natural productions of the vast and almost unknown continent of Africa are exchanged for ardent spirits, gunpowder, iron, or trinkets. "Les Ethiopiennes passent pour très lascives ou plutôt elles paraissent ignorer qu'on puisse repousser les solicitations d'un homme, surtout lorsqu'il est blanc. Cependant il est quelques nations Nègres où une sorte d'état social ordonne la fidélité des femmes envers les maris. et où l'on punit l'adultère, en enterrant tout vifs les deux coupables."

The Negroes do not appear to be equally long-lived with the individuals belonging to the other races of mankind; they become decrepid at sixty years of age, even when they enjoy, in a free country, the utmost extent of that domestic liberty which they seem to be capable of enjoying. Their woolly hair is observed to become grey sooner than among the White races.

Before the avarice of Europeans had transported the Ethiopian races to the New World, Africa was their exclusive abode; here they continue to occupy a vast tract of coast, extending along the Gulf of Guinea, from the river Senegal. and the 16th or 17th degree of north latitude, as far south as the parallel of Saint Helena, that is to say, to about the 16th or 17th degree of south latitude. The Ethiopian races scarcely appear beyond the Tropics, but probably extend far into the interior of the continent. They are rarely to be met with on the eastern coast, which is inhabited by races of Men differing from them in many respects. Towards the west, we find the Foulahs, on the banks of the River Gambia, already slightly mingled with the Moors; the Ghiolofs or Yalofs, a very black, tall, and robust race; the Sousous of Sierra Leone; the Mandings of the Grain Coast; the Ashantees of the Gold Coast, whose warlike and dangerous character is much celebrated; the Negroes of the Coast of Ardra and Benin, where the greater number of slaves are at present procured; the inhabitants of the Coast of Gabon, which are shunned by Europeans; and the nations of Loango, Congo, Angola, and Benguela, more or less civilized from their intercourse with the Portuguese for several centuries.

B. HOMO CAFFRARIUS.—CAFFRES.

Syn. H. CAFER.—Bory, Ess. Zool. II. 86.

CAFRE.—Less. Mam. 27.

KAFFERN.—Lichtenst.[4] Reise. II.

EURO-AFRICAINE.—Desmoul. Tab.

Icon. Péron, Voy.[5] pl. 54. (A native of the Mosambique Coast.)

The term Caffre, or *Infidel*, originally applied by the Mussulmans to designate all Negroes who refused to submit to the rite of circumcision.

---

[1] Virey, Hist. Nat. du G. Hum. tome II.

[2] Bory, Ess. Zool. II. p. 33.

[3] Journal of an Expedition to explore the Course and Termination of the Niger. By Richard and John Lander. London, 1832.—Also Annales des Voyages. passim.

[4] LICHTENST. REISE.—Reise nach dem südlichen Afrika von H. Lichtenstein. Berlin, 1811.

[5] PÉRON, VOY.—Voyage de découvertes aux Terres Australes, fait pas ordre du gouvernement sur les Corvettes le Géographe, le Naturaliste, et la Goëlette le Casuarina, pendant les années 1800 à 1804. Rédigé par François Péron, et continué par M. Louis de Freycinet.—Atlas par MM. Lesueur et Petit. Paris, 1824.

has been latterly confined to a race of Negroes, occupying the country to the south-east of Africa, called Caffraria, which extends from the Cape of Good Hope to Monomotapa.

The races inhabiting this extensive district differ alike from the Ethiopians, the Hottentots, and the Aramean races, which adjoin them. The skull of the Caffre exhibits an elevated arch, like that of the European ; the nose, far from being flattened, approaches to the aquiline form. He has the thick lips of the Ethiopian, and the high cheek-bones of the Hottentot ; his curly hair is less woolly than that of the Negro, and his beard stronger than that of the Hottentot. In general he is tall and well-made ; with the skin not quite so black as in the Negro, and is usually in the habit of painting his face and entire body with a red ochre. The height of the females contrasts forcibly with that of the males, for scarcely do they attain the stature of a European female, though in other respects they are equally well-formed. The limbs of a well-made Caffre present that rounded and graceful contour which we admire in the antique statues of the Pelasgian races ; his countenance is mild and lively. The Caffre girls are highly esteemed for their beauty, and form an important branch of a disgraceful export trade. The clothes of the Caffres are made of skins, and their ornaments consist of ivory or copper rings, which they carry on the left arm or in the ears. Cattle form their principal wealth, although the cultivation of the soil, performed exclusively by the females, yields no inconsiderable portion of their sustenance.[1]

All the Caffres are very warlike and active, fond of long journeys, either to visit their friends, or merely from the restless desire of change.

The Betjouanas have already exhibited some rude approaches towards civilization. Their countenances are intelligent ; their memory retentive ; and they exhibit no small degree of inquisitiveness during their intercourse with strangers. Their priests, the chief of whom is second only to the king, preside over certain religious ceremonies, such as the circumcision of the male infants, the consecration of cattle, and predictions of the future. They are unacquainted with the art of writing: their arithmetic is confined to addition ; they count on their fingers, and have no signs of a decimal notation. The form of their houses distinguishes them most advantageously from the other nations of Southern Africa, and some considerable towns are occasionally to be found, with a population of several thousands.

The Koussas have a decided attachment for a pastoral life ; yet they do not hesitate to take up arms in defence of their country, and have successfully resisted the attacks of Europeans. The Maroutzas and Makinis manufacture the dresses, ornaments, arms, and domestic implements of the other tribes. Hence the Caffre races would appear to have advanced further in civilization than any of the Ethiopian races, though none of them have yielded their faith to the exertions of the Christian missionaries. Some Caffre families have emigrated to the southern extremity of the Island of Madagascar.

The language of the Caffres is sonorous, rich in vowels and aspirations, with very few of those harsh guttural sounds which render the Hottentot dialect so disagreeable to foreigners.

C. HOMO CAPENSIS.—HOTTENTOTS & BUSHMEN.

Syn. HOTTENTOT.—Less. Mam. 27.
      H. HOTTENTOTUS.—Bory, Ess. Zool. II. 113.
      AUSTRO-AFRICAINE.—Desmoul. Tab.

Icon. Blumenb. Dec. Cran. V. t. 55. (Skull of a Bushman Hottentot.)
       Geoff. et F. Cuv.[2] Hist. Mam. (Femme Boschesmanne) ; and Péron,
       Voy. pl. 57.

The Hottentots of Southern Africa present the widest divergence in their physical traits, as well as in their anatomical characters, from the White races of Europe ; and in many respects assume the characteristics of the Orangs, and larger Apes. As we find in the Genus Macaous, the bones of the nose are united, according to Lichtenstein, into a single scaly lamina, flattened, and much broader than in any other human skull. The olecranon cavity of the humerus is also pierced with a hole ; and the front teeth with their alveolæ are oblique.

The complexion of the Hottentot is more or less brown or yellowish-brown, but never black. His head is small, the cheek-bones very prominent, the eyes sunk, and the sclerotica pure white ; the face, very broad above, ends in a point, the nose is flat, the lips thick, and the teeth very white. He is well made and tall, with small hands and feet in proportion to the rest of his body. The hair is black, curly, or woolly ; but in many tribes, instead of covering the surface of the scalp, it is collected into small tufts, at certain distances from each other, resembling the pencils of a stiff shoe-brush, only curled and twisted into hard and round lumps. When allowed to grow it forms small tassels like the fringe of a curtain. The Hottentot is almost destitute of beard, his ears are directed backwards, and the concha is so small that no part of that organ is visible in front. The form of the foot is very different from that of an Ethiopian or Caffre, and the impression of a Hottentot foot on the sand in consequence is readily recognised.

The Boschismans, or Bushmen, called Saabs by some of the native tribes, seem to have been separated from the proper Hottentots at a very remote era. These races having been hunted like wild beasts by the colonists, and driven to the deserts, exist only by hunting or plunder. They live in caverns, and clothe themselves with the skins of animals killed in the chase, while their industry is confined to the fabrication of poisoned arrows and fishing-nets. They exist in the extreme of wretchedness ; their meagre limbs and famished appearances betray the privations to which they are reduced. They remain without leaders, without property, or even a social tie, excepting the transient passion of the moment.[3]

We must here notice the existence of two singular anomalies in the organization of the female Boschismans; these are the extraordinary size of their haunches, and that remarkable prolongation of the sexual organ, vulgarly called the apron.

In respect to the first of these peculiarities, Le Vaillant assures us that he saw it in a girl of three years old.[4] The large projection of her haunches consisted in a fleshy and adipose tissue, oscillating at every movement of the body like a tremulous jelly. The mother while walking occasionally places her infant upon this protuberance, and Le Vaillant observed one female running while her child stood upon the haunch, like the groom behind a cabriolet.

The female Boschisman, exhibited in Paris during the year 1815, measured about nineteen inches across the haunches, while the hips projected full seven inches, a peculiarity which was afterwards found to proceed from a large mass of fat placed immediately under the skin.[5]

According to Péron, of the apron of the Boschisman female has nothing in common with the ordinary sexual organs as observed in the females of other nations ; while, according to the Parisian anatomists, the peculiarity of the female already alluded to consisted merely in an unusual development of the nymphæ. The labia were slightly pronounced, and intercepted an oval of 4.26 inches in length. From the upper angle there descended between them a half-cylindrical eminence about 1.6 inches in length, and .53 in thickness, the lower extremity of which enlarging became forked, and was prolonged into two fleshy petals, about 2.66 inches in length, and 1.06 inches in breadth ; each of them rounded on the summit; their bases enlarging and descending along the internal margin of the great labium, on its side, and changing into a fleshy crest which terminated at the lower angle of the labium. On raising these two appendages, they formed together the figure of a heart, with long and narrow lobes, the centre of which was occupied by the vulva. It may be readily seen on comparing this description with the analogous parts in the European female, that the two fleshy lobes which form the apron are formed at the upper part by the clitoris and the summits of the nymphæ, while in all the remainder of their extent they consist merely in an excessive growth of the nymphæ. This view of the subject is confirmed by the fact that the length of the nymphæ varies greatly even in Europe, and in general becomes more considerable in warm climates, and that the Negro and Abyssinian females are sometimes incommoded to such a degree as to be compelled to extirpate them by fire or amputation. The amount of this development observes no constant law among the Boschismans. Blumenbach states that he is in possession of drawings of the organs which were 8.5 inches in length, and they vary frequently in respect to their form.

This excrescence is evidently not the result of art, as all the Boschisman females possess it in their earliest youth ; and the female already mentioned concealed it so carefully, and as though it were a deformity, that its existence was not even suspected until after her death.

The countenance of this Boschisman female presented an odious combination of the Mongolian and Negro features. Her muzzle projected still more than in the Negro, and the face was more flattened than in the Calmuck; while the bones of her nose were smaller than in either. Her breasts hung downwards in large masses, obliquely terminated by a blackish areola, about 4.3 inches in diameter, furrowed with radiated striæ, in the centre of which appeared a flat and almost obliterated nipple. The

1 Lichtenst. Reise II.
2 GEOFF. ET F. CUV. HIST. MAM.—Histoire Naturelle des Mammifères. Par M. Geoffroy-Sainte-Hilaire ; et par M. Frederic Cuvier. Paris, V.Y.
3 Barrow's Narrative of a Journey amongst the Boushouanas ; and Campbell's Second Journey.
4 LE VAILL. VOY.—Voyage dans l'Intérieur de l'Afrique, par le Cap de Bonne-Espérance.—1° Voyage, Paris, 1790.—2° Voyage, 1795.
5 Geoff. et F. Cuv. Hist. Mam. Femme Bosch. p. 5.

general colour of her skin was a brownish yellow, almost as deep as that of her face. Her movements had something hasty and capricious in them, like those of an Ape. She had the same habit of pushing out her lips, as may be observed in the Orang-Outang. Her height was about four feet nine inches, which appears, according to the report of her country-men, to be much above the usual size.

The tribe of Boschismans, which is probably identical with that of the Houzouanas, described by M. Le Vaillant, has its facial angle no greater than 75°, and approaches nearer to the brute in this respect as it does in many others, than most other branches of the human species.

The Hottentots and Boschismans occupy the extreme Southern point of Africa, but do not extend within the tropics; they are spread around the basin of the Orange River under the names of Gonaquois, Namaquois, Coranas, Boschismans, and Houzouanas.

The Caffres, as well as the Colonists of the Cape, wage a war of extermination against the unfortunate Hottentots, who wander about clothed in skins, besmeared with black or red grease, in the midst of their cattle, which form their whole possessions.

ANOMALOUS RACES.

Syn. Variétés de Races moins distinctes. — Desm. Mam. 47.

The southern portion of the peninsula, situate immediately beyond the Ganges, is inhabited by the Malays (*H. Malayensis*), who approximate more nearly to the Hindoos in their physical characters than to the Mongolians. Their race and language are widely dispersed throughout the coasts of nearly all the islands of the Indian Archipelago.

The innumerable small islands of the Southern Ocean are also peopled by a handsome race (*H. Polynesius*), apparently resembling the Hindoos, while their language bears many [remote] analogies to the Malayan.

But the interior districts of the larger islands, and more especially their wildest territories, are inhabited by another race (*H. Australasicus*), with dark complexions, and approaching more nearly to the Negro in form. These islanders, who live in a state of extreme barbarity, are commonly termed Alfooroos [and may be subdivided into two branches, the Proper Australians (*H. Australis*) and the Oceanic Negroes (*H. Melaninus*)].

Again, we find upon the coasts of New Guinea and the neighbouring islands, a race commonly termed Papoos (*H. Papuensis*), and nearly resembling the Caffres of the Eastern Coast of Africa. The inhabitants of New Holland (*H. Australius*) are regarded as Alfooroos, while those of Van Diemen's land should rather be considered as Papoos [or Tasmanians].[1]

It is not very easy to refer either the Malays or Papoos to any of the three Normal Races, for the former merge by insensible gradations, on the one hand, into the Hindoos of the Caucasian races, and on the other, into the Chinese of the Mongolian race, so that we can scarcely point out any characteristics sufficiently marked to distinguish them. We are not yet in possession either of figures or descriptions sufficiently exact to decide whether the Papoos [as some have conjectured] are not merely Negroes, who have anciently lost their way upon the Indian Seas.

The tribes (*H. Hyperboreus*) who inhabit the northern extremity of both continents, such as the Samoids[2] and the Esquimaux, belong, according to some writers, to the Mongolian race; according to others, they are merely some degenerate off-sets of the Scythian or Tartar branch of the Caucasian races. The Americans themselves (*H. Americanus*) have not hitherto been clearly referred to any one of the races found in the Ancient Continent, yet they have not those precise and constant characters which would permit us to elevate them to the rank of a Normal race. Their reddish copper-coloured complexion is not a sufficient one, while their hair, generally black, and their scanty beards, would lead us to include them among the Mongolian races, if their harshly-defined countenances, their nose as prominent as ours, their large and projecting eyes, did not oppose this arrangement, and assimilate them more nearly to the European forms of countenance. Their languages are as innumerable as their tribes, and all attempts to demonstrate a satisfactory analogy between the latter and the dialects of the Old World have hitherto failed.[3]

A. HOMO MALAYENSIS — MALAYS.

Syn. Les Malais.—Cuv. Reg. Anim. I. 83. - Less Mam. I. 24.
Race Malaie.—Desm. Mam. 47 (in part).
H. Neptunianus, Race Malaise (Orientale).—Bory, Ess. Zool. I. 281.
Malaise ou Oceanique.—Desmoul. Tab.
H. Sapiens. var. Malayana (in part).—Blumenb. Hand.
Icon. Blumenb. Dec. Cran. IV. t. 39. (Skull of a Javanese.)
Péron Voy. pl. 38 to 43.

The Malays were first recognised as a distinct race during the twelfth century, when some of their tribes, emigrating from Menang-Kabou in Sumatra, founded Singhapoura, and established the principal seat of their power at Jolor, in the peninsula of Malacca. This warlike and commercial people have rendered themselves masters of the sea coast of most of the islands in the Indian Archipelago, and have either expatriated the original inhabitants, or driven them to the mountains of the interior, where the Alfooroos long existed, unknown to European navigators.

From a continual intercourse with the Moors of the Red Sea they have acquired the Mahometan Religion, with many Arabian customs, while other traditions and customs are evidently derived from their neighbours the Hindoos. The inhabitants of Java, mixed with Arabic blood, have long composed several powerful and populous states. In the neighbouring Islands of Borneo, Celebes, Tidore, Ternati, Sumatra, and Sooloo, we find the same race established, though often modified in some degree from an extensive intercourse with Europeans and Chinese. But the Malayan race may be found in its greatest purity in the Islands of Guebay, Oby, Gilolo, Floris, Lombok, and Bali.

In all the governments of the Malays, we find the despotic form of the Hindoos universally adopted, where the person of the Rajah is held in the profoundest veneration. Capable of the blackest perfidy and duplicity, and with an ardent thirst for revenge, the Malays are as celebrated in the East for their treachery, as some Atlantic nations once were for their " Punic faith." The annals of Malayan history are one continued record of assassination and treason, whilst in all ages piracy has been, with a large portion of the population, one favourite mode of acquiring a livelihood. Professing the tenets of Islamism, the Malays adopt polygamy, and other precepts of the Koran, but modified by several Hindoo dogmas. Their chiefs are richly clad according to the Hindoo manner, while the lower orders go about entirely naked, with the exception of a narrow piece of cloth round the waist. In general, the Malay is sensual and dissolute in the extreme, passionately devoted to intoxicating liquors and opium; and, above all, to the practice of chewing the betel-nut. This drug appears almost peculiar to the Malayan race, in whose territory the materials are plentifully to be found. The Pinang (*Areca*), the Pepper, and occasionally also the Cashew, with some calcareous earth, form its ingredients.

---

[1] For a further knowledge of the different races who people the Islands of India and the Pacific Ocean, consult the Dissertation of MM. Lesson and Garnot, in the " *Zoologie du Voyage de la Coquille,*" p. 1–113; and for the languages of the Asiatic nations and their mutual relations, see the " *Asia Polyglotta*" of M. Klaproth.— *Note of the Baron Cuvier.*

[2] M. Cuvier includes the Laplanders in this enumeration (Reg. Anim. p. 84) ; and M. Garnot (Voyage de la Coquille, tome I. p. 512) is disposed to consider a part only as Hyperboreans. We have been induced (see before, p. 132), from the affinity of their language to the Finnic or Tchoude dialects, to place them in the Scythian branch of the Caucasians.

[3] On the subject of the Americans, besides the Voyage of M. Von Humboldt, so rich in important documents, consult the Dissertations of Vater and Mitchell.— *Note of the Baron Cuvier.*

35

The Malayan race, resident in the immediate vicinity of the equator, is seldom found beyond the 92d and 132d meridian of Eastern longitude ; yet they can be traced to the eastern coast of the Island of Madagascar, though sometimes partially mingled with the Moors. They may be said to form the entire population of the shores of the Indian Archipelago, the Sonda and Molucca Islands ; from whence they have been distributed to some of the Philippine Islands, and as far as the Island of New Guinea, on the north of which they have formed some permanent settlements. The Malays are even to be found at Waijoo, at the Isles of Aroo, and in Dampier's Straits.

The physical traits of the Malays are as characteristic as their manners and customs. In general they are of medium size, robust and well proportioned, their complexions of a yellowish copper-colour, slightly mixed with orange. M. Bory remarks that their mucous membranes have a deep violet tinge. The females, every where subjected to a jealous surveillance, are in general of diminutive stature, well rounded, their breasts voluminous, their hair very coarse and black, their mouth wide, and their teeth might be accounted beautiful, if they were not blackened and corroded by the immoderate use of the betel. Both sexes are violent in their passions, irritable, treacherous, capable of the grossest deceit, submissive and crouching to the yoke of the strongest, barbarous and merciless to their enemies or their slaves.[1]

The Malayan language is spoken throughout all the islands, with slight local variations. It is mild, harmonious, and simple in its rules ; full of oriental terms of expression, and abounding in figures of speech.[2] Their religion and knowledge being derived from the Arabs, the inhabitants of Malacca have adopted the characters of the Arabic, with the practice of writing from right to left, while those of Java and Sumatra write like the Europeans, from left to right[3]

B. HOMO POLYNESIUS.—POLYNESIANS.

Syn. RAMEAU OCÉANIEN.—Less. Mam. 25.
      H. NEPTUNIANUS OCÉANIQUE (OCCIDENTALE).— Bory, Ess. Zool. I. 298.
      RACE JAUNE DU GRAND OCÉAN.—Quoy et Gaim. Zool. de l'Astr.* I.
         18 (in part).

Icon. Blumenb. Dec. Cran. III. t. 26. (Skull of an Otaheitan) ; V. t. 50.
         (Skull of a Marquesan.)
      Cook and King, Voy.[5] Pl. XI. XVIII. XXIII. LXIII. and LXIV.
      Langsd. Reise.[6] I. •. 7 and 8.
      Kotzeb. Voy.[7] II. fig. and III. fig. tit.
      Quoy et Gaim. Zool. de l'Astr. pl. I. and fig. 4 and 5 of pl. 2.

The Polynesian or Oceanic variety of the human race is far superior to the remaining population of the Southern Ocean in the beauty and symmetry of its proportions. In general, the South Sea Islanders are of an elevated stature, with their muscles well formed, a well formed and elevated cranium, and an expressive physiognomy, varying from placid timidity to warlike ferocity. Their eyes are large and protected by dense eyelashes. The colour of their complexion is a clear yellow, deeper in those natives who are compelled to seek for subsistence among the coral reefs, and much fainter among the females. Their noses are broad and flat, the nostrils widely-dilated, the mouth large, the lips thick, the teeth very white and beautiful, and the external ears remarkably small. The beauty of the women, though somewhat exaggerated by the earlier navigators, is not inconsiderable. Their eyes are large, their teeth of the purest enamel, their skins soft and smooth, their hair long and black, and tastefully arranged over breasts of the most perfectly hemispherical form. In other respects they may be termed ugly, having, like the men, large mouths and flat noses. The colour of their complexions is almost white, their stature short, their forms corpulent. The inhabitants of Mendocia and Rotooma, according to Krusenstern,[8] are the most comely ; next to these we may place the Otaheitans, the Sandwich Islanders, and those of Tonga. At New Zealand, the beauty of the females declines, while the males are more robust and athletic than any others of the same race.

The greatest analogy may be traced in the manners and customs of these islanders, though separated from each other by an immense expanse of ocean, and the identity of the race has been demonstrated upon the greater portion of the islands situate to the south-east of the Indian Archipelago and Australia. In fact, all the volcanic and coral islands of this ocean within the Southern Temperate Zone are peopled by the Polynesians, while they appear to have sent only a single colony to the northward, which occupies the Sandwich Islands. The entire Archipelago of the Carolinas, on the contrary, with the Philippine and Marianne Islands, is peopled by a totally different race. The Polynesians are thus widely distributed in the Friendly and Society Islands. A branch has extended to the Isles of Mandana, Washington, Mangea, Rotooma, Lady Penryn, Sauvage, Tonga, and New Zealand. About one half of the population of the Fidjee and Navigators' Islands belongs to this race, which does not extend, according to MM. Lesson and Garnot, beyond the Island of Rotooma in that direction.

Both sexes of the Polynesian variety clothe themselves in the most graceful manner with long flowing robes, wherever the variations of temperature require this covering. The chiefs alone possess the prerogative of wearing the tipouta, a garment which bears much resemblance to the poncho of the Araucanos in South America. The New Zealanders, placed beyond the tropics, have adopted garments suited to their climate, consisting in an ingenious fabric, formed of the silky fibres of the Phormium. All these islanders agree in possessing a singular taste for head dresses. Those of Otaheite and Sandwich crown themselves with flowers, while those of the Marquesas and Washington, Rotooma and the Fidjee, attach a superstitious value to the teeth of the Cachalot. These ornaments are replaced in New Zealand by plumes of feathers. Throughout the entire islands, the practice of tattooing the skin is widely practised, either to distinguish the different ranks of the people, or merely for ornamental or superstitious purposes. The inhabitants of the Pomotoo Islands cover their entire bodies with these designs ; in Otaheite and Tonga, they are more limited and simple ; while in the Sandwich Isles and New Zealand, the entire countenance is covered with devices, arranged in a symmetrical and highly expressive form. The women of New Zealand and the Marquesas tattoo the internal angle of the eyes, the angles of the mouth, and often also the chin. In general, the tattooing of the South Sea islands is composed of circles and semicircles, opposed or bordered by notches somewhat resembling the never-ending circle of the Hindoo mythology.

The same domestic habits may be traced throughout the entire race. Their food is cooked in subterranean ovens by means of heated stones ; the leaves of plants are used for culinary purposes ; the Bread-fruit (Artocarpus incisa), the Cocoa, and the Taro, are boiled for food. They all drink the Kava or Ava, the juice of a species of Pepper, which intoxicates or refreshes. Before the arrival of Europeans, the women were excluded from all entertainments. Their dwellings are adapted to the circumstances of the locality. In some places, such as the Society Islands, Tonga, Mangea, the Marquesas, and Rotooma, the houses are large and capacious, serving for several families, without closed walls, and built nearly on the same plan. In other places, such as New Zealand, where each tribe is continually at war with its neighbours, and where the tempests are violent and prolonged, their hippahs-are almost inaccessible, surrounded with palisades ; while the narrow buildings, sunk in pits almost level with the ground, and sufficient to contain only two or three persons, are entered on all-fours.

The language of the Polynesians, though apparently simple, and rich in Oriental figures, is directly opposed to the genius of the pure Malayan. All navigators agree in remarking the singular affinity which prevails throughout the dialects of the great Southern Ocean. An Otaheitan can be understood in the Marquesas, the latter at Sandwich, and the native of these last islands in New Zealand.[9]

Our limits do not permit us to notice at present the religious opinions prevalent among these tribes, their human sacrifices, their Morais, or the occasional cannibalism of some nations.[10]

¹ Sir Thomas Stamford Raffles' History of Java. London, 1817.
² Wm. Marsden's History of Sumatra. London, 1811. Also Grammar and Dictionary of the Malayan Language. London, 1812.
³ Crawford and Leyden's Memoirs, in the Asiatic Society's Transactions.
⁴ QUOY ET GAIM. ZOOL. DE L'ASTR.—Voyage de la Corvette l'Astrolabe, exécuté par ordre du Roi, pendant les années 1826 à 1829, sous le commandement de M. Jules Dumont D'Urville.— Zoologie, par MM. Quoy et Gaimard. Paris, 1830, et suiv.
⁵ COOK AND KING, VOY.—A Voyage to the Pacific Ocean, undertaken by the command of his Majesty, for making discoveries in the Northern hemisphere, performed under the direction of Captains Cook, Clerke, and Gore, in his Majesty's ships the Resolution and Discovery, in the years 1776 to 1780. Vols. I. and II. by Captain James Cook ; III. by Captain James King. London, 1785.
⁶ LANGSD. REISE.—Bemerkungen auf einer Reise um die Welt, von G. H. von Langsdorff. Frankfort, 1812.
⁷ KOTZEB. VOY.—Otto Von Kotzebue. A Voyage of Discovery into the South Seas and Behring's Straits, undertaken in the years 1815–1818, in the ship Kurick, (in German.) Translated by H. E. Lloyd. London, 1821. Also, a new Voyage round the World in 1823 to 1826. London, 1830.
⁸ KRUSENST. VOY.—A. J. Krusenstern—Voyage round the World, (in Russ.) Petersburg, 1809, and atlas in fol. 1813. (Translated into English by R. B. Hoppner. London, 1813.)
⁹ Otaheitan Grammar, published by the Missionaries at Otaheite, 1823. A Grammar and Vocabulary of the Language of New Zealand, 1820.
¹⁰ See the Voyages of Cook, Bougainville, Vancouver, Carteret, Turnbull, Mariner, Wallis, Krusenstern, La Pérouse, Langsdorff, Lisianskoi, &c. &c.—Also Forster, in the 2d Voyage of Cook.

To this subdivision of the Human race, some writers would refer the Dayaks of Borneo,[1] as well as the Ancient Peruvians and Mexicans.[2]

C. HOMO AUSTRALASICUS.—AUSTRALASIANS.

Syn. ALFOUROUS.—Cuv. Reg. Anim. I. 84.

These barbarous races, known by the name of Alfoorous, may be subdivided into two branches, the Proper Australians or New Hollanders (*H. Australius*), and the Oceanic Negroes (*H. Melaninus*).

1. AUSTRALIUS.—AUSTRALIANS.

Syn. H. POLYNESIUS —Fisch. Syn. Mam. 8.
   H. AUSTRALASICUS.—Bory, Ess. Zool. I. 318.
   AUSTRALIENS.—Less. et Garn. Zool. de la Coq. I. 106.
   AUSTRALASIENNE.—Desmoul. Tab.
   ALFOUROUS-AUSTRALIEN.—Less. Mam. 28.

Icon. Blumenb. Dec. Cran. III. t. 27, and IV. t. 40. (Skulls of New Hollanders.)
   Péron Voy. pl. 20 to 28.
   Griff. Anim. King. (New Hollander.)
   Quoy et Gaim. Zool. de l'Astr. pl. 5.

The indigenous tribes of New Holland, who exhibit a marked resemblance to each other, according to the statements of navigators,[3] have already been noticed for their ignorance, their wretchedness, as well as their moral and intellectual debasement. Their tribes are not numerous, have little communication with each other, and are sunk into a state of almost hopeless barbarism.

The natives of New South Wales live in the rocks and thickets surrounding the European settlements, without adopting any of the manners of civilized life, excepting, perhaps, the taste for intoxicating liquors. Modesty seems wholly foreign to the race, and even in the midst of a populous European colony, they are not particular in adopting any kind of covering. The most unrestrained liberty appears indispensable to the existence of the genuine Australian, who preserves his independence in the rocky fastnesses of the neighbourhood, reclining near a wood fire, and protected from the wind merely by a few branches, or a large piece of bark torn from the Eucalyptus.

The stature of the Australians is commonly below the average. Many tribes have meagre limbs, apparently of disproportionate length; while a few individuals, on the contrary, have the same parts long and well proportioned: Their hair is not woolly, but coarse, very black, and plentiful, usually worn loose and disordered, but most commonly short and collected into curly masses.

The face is flat; the nose very broad, with the nostrils almost transverse; the lips thick; the mouth widely cleft; the teeth slightly inclined outwards, but of the purest enamel; and the external ears very large. Their eyes are half-closed through the laxity of the upper eye-lids, a circumstance which imparts to their savage countenances a physiognomy peculiarly repulsive. The colour of their skins commonly assumes a darkish tint, varying in intensity, but never becoming absolutely black. The external which appears to separate their forms from that of the Pelasgian races appears immense to our eyes.

Marriages are concluded by force. At a certain period of life an incisive tooth is extracted from every man, and a phalanx amputated from the finger of every female. Their head and breast are usually covered with some red colouring matter, and this ornament is of the highest importance in all their *coroboris*, or great ceremonies. The habit of painting the nose and cheeks by the same rude means is also common, with the addition of white rays along the forehead and temples. On the arms and the sides of the thorax they raise the same conical tubercles, which are practised largely by the Negro race. Numerous families insert rounded sticks, from four to six inches in length, into the partition of the nostrils, a practice which imparts a savage aspect to their physiognomy. Finally, the garments of this race never extend beyond a rude coat of Kanguroo skin thrown over the shoulders, or a forest style of filaments weaved into a coarse kind of net work.

These tribes are superstitious to an excessive degree; jugglers are encouraged, and witchcraft punished. Their differences are decided by a kind of duel, consisting of equal numbers, or equal arms; and the judges of the field decide the rules of the combat. Their offensive arms consist of a kind of javelin, a wooden sabre, a club or woodah, while the shield alone is used for defence. The bow and arrows are wholly unknown in the entire continent of Australia.

The inhabitants of King George's Sound are subject to an intense cold during winter, and cover themselves with large mantles made of Kanguroo skins. Those in the neighbourhood of Sydney and Bathurst prepare the skins of the Petaurista, while the New Hollanders within the tropics live in a state of absolute nudity. Their ornaments consist of small collars made of stubble. Their dwellings around Port Jackson are made of the branches or bark of trees; elsewhere they seem to consist in a sort of nests formed of interwoven branches, and covered with bark.

The care which they bestow upon their tombs serves to indicate their belief in a future state. In general, it has been observed that they burn their dead, and inter the ashes with a religious veneration. Their industry is confined to the construction of lines for hunting and fishing, the product of which is devoured on the spot, after being roasted over a wood fire, which they always carry along with them. The women are treated with contempt, compelled to the most laborious occupations, such as carrying their utensils and children, while the man walks about leisurely with nothing but a javelin in his hand. They also prepare the food, of which they are only permitted to eat the fragments rejected by their masters. The pungent root called dingona is gathered and prepared by the women for their own use, and only eaten by the men in times of scarcity, when the chase is unsuccessful. Their canoes vary with the tribes; near Port Jackson these are formed of a solid piece of the bark of the Eucalyptus joined tightly together at each extremity. They have some rude ideas of drawing and music, and their dancing consists in an awkward imitation of the leaps of the Kanguroo.

Their languages, differing in every tribe, are nearly unknown to Europeans.[4]

2. MELANINUS.—OCEANIC NEGROES.

Syn. H. MELANINUS.—Bory, Ess. Zool. II. 104.
   ALFOUROUS-ENDAMÈNES.—Less. et Garn. Zool. de la Coq. I. 102.
   NÈGRE OCÉANIENNE.—Desmoul. Tab.

Icon. Less. et Garn.[5] Zool. de la Coq. t. 1. (Crânes d'Alfourous-Endamènes).
   Labillardière, Voy.[6] pl. VII. and VIII.
   Quoy et Gaim. Zool. de l'Astr. pl. 3.

The indigenous population in the Islands of the Indian Archipelago appears to have consisted of a black race, which has been extirpated in some islands, and in other places driven to the mountains of the interior, as the ancient history of Malacca confirms. This nation, with black skins, coarse, black, and straight hair, live in places inaccessible to the other races, and are known by various appellations. The central plateau of the Molucca islands is now occupied by the Alfonroos or Haraforas, and of the Philippines by *los Indios* of the Spaniards. At Mindanao, they are styled *los Negros del monte*; at Madagascar they are the *Vinzimbers*, or native inhabitants of that island, and at New Guinea, they are styled *Endamenes*.

The Oceanic Negroes live in the most savage and miserable manner. Always at war with their neighbours, they are wholly occupied in avoiding the ambuscades and pit-falls laid for their destruction. It is therefore with difficulty that they can be examined by Europeans, who visit only the coasts. The Papoos represent their enemies of the mountains as ferocious, cruel, and vindictive, without the knowledge of any art, while their entire lives are occupied in obtaining a scanty subsistence in the forests. M. Lesson considers this description as exaggerated by the hatred of the Papoos. Those seen by him had a repulsive expression of countenance, their noses flattened, their cheek-bones prominent, their eyes large, their teeth inclining outwards, their limbs long and meagre, their hair black, thick, coarse, and straight, but of no great length, and their beards very coarse and thick. An expression of extreme stupidity was impressed upon their features, perhaps owing to the individuals examined being in slavery. These Oceanic Negroes, whose complexions were of a dirty brown nearly approaching to black, went entirely naked. They had incisions upon the arms and chest, and carried a small stick about six inches long in the partition of the nose. Their countenances were ferocious, and their movements capricious. The Southern Coast of New Guinea is probably inhabited by the Endamenes of the interior.

[1] Less. et Garn. Zool. de la Coq. tome I. p. 46.
[2] Fisch. Syn. Mam. p. 4.
[3] See the Voyages of Phillips, Collins, White, D'Entrecasteaux, Péron, Flinders, Grant, King, &c. &c.
[4] Journal of two Expeditions into the Interior of New South Wales, undertaken by order of the British Government, in the years 1817-18. By John Oxley. London, 1820.
[5] Less. et Garn.—Zool. de la Coq.—Voyage autour du monde, exécuté par ordre du Roi. Sur la corvette de sa Majesté la Coquille, pendant les années 1822 à 1825, par M. L. I. Duperrey. Zoologie, par MM Lesson et Garnot. Paris, 1826, et suiv.
[6] Labillardière, Voyage à la recherche de La Pérouse. Paris, an. VIII.

D. HOMO PAPUENSIS.—PAPOOS & TASMANIANS.

Syn. H. NEPTUNIANUS, PAPOUE (intermédiaire)—Bory. Ess. Zool. I. 303.
    PAPOUAS et PAPOUS.—Quoy et Gaim.[1] Zool. de l'Uran. p. 1.—Less. et
    Garn. Zool. de la Coq. 1. 84.
    PAPOUE.—Desmoul. Tab.

Icon. Quoy et Gaim. Voy. de l'Astr. pl. 4, fig. 4 and 5. (Natives of New
    Guinea), and Voy. de l'Uran. pl. I. and II.
    Péron, Voy. pl. 4 to 8. (Natives of Van Diemen's Land.)

The races who inhabit the shores of the islands of Waijoo, Salwatty, Gammen, and Balenta, and all the northern coast of New Guinea, from Point Sabelo to Cape Dory, are known by the name of Papoos. Their hair and the general colour of the skin hold medium characters between those of the Malays and Oceanic Negroes, from the intermixture of whom they have in all probability originated.   Their existence as a distinct race, though noticed by Dampier in 1699, was not fully recognised until the first voyage of MM. Quoy and Gaimard.

The Papoos are in general of medium stature, and tolerably well-made, though some individuals are seen with feeble and meagre limbs.   The colour of the skin is not black, but rather a deep brown, midway between the tints of the Malay and Oceanic Negro.   Their hair is very black, neither straight nor curled, but woolly, tolerably fine, and frizzled, which gives to their head an extraordinary voluminous appearance, especially when they neglect turning it up behind.   The beard is scanty, but very black on the upper lip; the pupil of the eye is of the same colour. Although the nose is slightly flattened, the lips thick, and the cheek-bones prominent, their physiognomy is not disagreeable.   Hated by the other races, as being a kind of hybrids, they live with the adjoining tribes in a continuous and permanent state of distrust, and wander about armed with a bow and two or three large quivers filled with arrows. Suspicion, hatred, and all the passions which naturally arise from their situation, are forcibly depicted on their countenances; and, as with all the other black races, the instinctive faculties exhibit a marked prevalence over the moral or intellectual   The women, with a few exceptions, are unusually ugly, and compelled by their despotic masters to perform all the offices of slavery.

The Papoos of the Bouka Islands, New Britain, and Port Praslin, wear no garments of any kind. The natives of Dorery and the north of New Guinea are exceptions to this custom, and procure cotton fabrics, dyed blue or red by the Malays, in exchange for birds of paradise, tortoise-shell, or slaves.   These nations are in the habit of covering the chest and shoulders with elevated and papillated cicatrices, arranged in curved or straight lines, according to some regular pattern.

Some tribes of New Guinea, Waijoo and Bouka, give their hair that singular frizzled appearance, which has been regarded by some as characteristic of the Papoo race.   Other tribes, however, such as those of Rony in New Guinea, of New Britain, and New Ireland, permit their hair to fall upon the shoulders in tangled and flowing masses.

In general, the Papoos are fond of daubing the hair and face with a composition of red ochre and grease, variegating the breast and face with transverse bands of earth of coral. Different from the Polynesians, the practice of tattooing is but sparingly adopted, and they confine themselves to tracing a few scattered lines upon the arms or the lips of their females. Ornaments of every kind are anxiously sought, and worn indiscriminately on the head, breasts, and arms. Bracelets of a dazzling whiteness, made with great skill and beautifully polished, are frequently to be seen; but these are chiefly procured merely from the larger extremity of those enormous Cones which are plentifully found in the surrounding seas.   They pierce the nostril occasionally for insertion some small ornament. The custom of chewing the betel, with areca and lime, has been partially introduced from their intercourse with the Malays.

The Papoos of Dorery and Waijoo have a peculiar taste for carving idols, which they place on their tombs, in particular parts of their cabins, as well as on the prows of their canoes.   Their religion is a pure Fetishism, though a few traces of the Mahometan rites may be noticed.

Their barbarous and gustural dialects wholly differing from tribe to tribe, are as unintelligible to each other as they are to foreigners.

To this division once belonged the aborigines of Van Diemen's Land, now said to be exterminated by the colonists. Their contrast to the New Hollanders already described is singular. Péron remarks that in passing from Van Diemen's Land to New Holland, one is at once struck by the extraordinary difference between them. They have absolutely

nothing in common, whether in their manners, customs, arts, instruments for hunting or fishing, dwellings, canoes, arms, the form of the skull, the proportions of the face, or their language.   This remarkable dissimilarity prevails likewise in their colour. The indigenous inhabitants of Van Die-men's Land are much browner than those of New Holland; the former have short, woolly, and curly hair; among the latter it is straight, long, and smooth.

The many points of resemblance which these proper Tasmanians of Van Diemen's Land bear to the Papoos of New Guinea, have induced us to include them in the same division. They have the same habit of painting their hair with a red ferruginous earth; of elevating small cicatrices upon the skin; of cooking their food upon wood fires; of living un-covered upon the ground without shelter along side of large fires; of fabricating elegant baskets with twigs of trees; and of fashioning small ornaments, especially a kind of ear-ring called *rodray*.   They are polygamous, and construct conical buildings over the tombs of their deceased relations.

Though exposed to a rigorous climate, they seldom build cabins for themselves, but merely raise a temporary shelter from the winds with the bark of a tree.

Their language differs so remarkably from the barbarous and innumerable dialects of Australia, that M. Labillardière at once declared them to be of a different origin.[2]

E. HOMO HYPERBOREUS.—HYPERBOREANS.

Syn. LES HABITANS DU NORD DES DEUX CONTINENTS.—Cuv. Reg. Anim. I.
    85.
    H. SCYTHICUS χ. HYPERBOREUS.—Fisch. Syn. Mam. 5.
    H. HYPERBOREUS (in part)—Bory, Ess. Zool. I. 262.
    HYPERBORÉEN ou ESKIMAU (in part)—Less. Mam. 25.
    HYPÉRBORÉENNE.—Desmoul. Tab.

Icon. Cook and King, Voy. pl. LXXV. and LXXVI. (Natives of Kamts-
    chatka.) Pl. XXXVIII. and XL. (Natives of Nootka Sound.) Pl.
    XLVI. and XLVII. (Natives of Fr. William's Sound.) Pl. XLVIII.
    and XLIX. (Natives of Oonalashka.)
    Desmoul. Hist. des R. Hum. pl. 1 and 2.

The obscure tribes which inhabit the northern extremity of the old continent under the names of Samoids, Yookaghirs, Koriaks, Tchooktches or Kamtchatdales, and under the name of Esquimaux, have wandered in North America as far to the southward as Nootka Sound, Labrador, and the frontiers of Canada, belong to the Hyperborean races.[3]

The inhabitants of these desolate regions are of very small stature, their medium height being about four feet nine inches. Their persons are short and thickly made, with short legs, a large and flat head, the lower part of the face projecting greatly, the mouth wide, the ears large, and the beard very scanty. Their eyes are small, black, and angular; the skin olive-coloured and shining with grease. Their hair, black and bristly, is arranged, however, with much care.   The men have a very harsh voice, nearly resembling that of the Ethiopian Negroes. The women, apparently very ugly to European eyes, are nearly of the same height as the men, and comparatively more muscular. Their soft, pendant, and pear-shaped mammæ are very long from their earliest youth, and can be thrown over the shoulder to suckle their infants, which are usually carried on the back. The areolæ are large, the nipples long, wrinkled, and black as coal.   They arrive very late at the period of puberty. " Absolument glabres, excepté sur la tête, elles accouchent avec une extrême facilité, ce que tient à une telle dilatation de certaines voies, qu'on a dit qu'elles élargissaient artificiellement ces parties en y portant sans cesse enfoncée une énorme cheville en bois."[4]

The complexions of all the Hyperborean races are much darker than those of the nations of Europe and Central Asia, and are generally blacker in proportion as they approach the Pole.   It is not rare to find tribes living near the seventieth degree of latitude, who are deeper in tint than the Hottentots at the opposite extremity of the Old Continent, and almost as dark as the Ethiopians of its equatorial regions.

These tribes, though ever wandering from place to place, feel that strong attachment for their native wastes which incapacitates them from existing in the temperate regions of the globe. Clad in furs from head to foot, the Hyperborean fishes for his subsistence, or maintains extensive herds of Rein-deer. The Dog, after undergoing castration, shares his labours or draws his sledge. On the borders of the Northern Ocean, he

[1] QUOY ET GAIM. ZOOL. DE L'URAN.—Voyage autour de la monde sur la corvette l'Uranie et la Physicienne, par M. Freycinet; Zoologie, par MM. Quoy et Gaimard. Paris. 1824, et suivr.

[2] For a further account of this race, consult Péron's Voyage, and Labillardière (Voyage à la recherche de la Pérouse), passim.

[3] E. Sabine in the Journal of Science, XIII. ; Captain Parry's Journal of a Third Voyage; Captain Sir John Ross' Voyage of Discovery ; and Scoresby's Arctic Regions.

[4] Bory, Ess. Zool. I. 262.

is skilful in surprising the inhabitants of the deep, and is not even wanting in the art of taking the larger Cetacea. The Hyperboreans prefer the blubber of these animals to any other food; they delight in the oil thence procured, and drink whatever escapes being consumed in their lamps during the long night of a Polar winter. Besides the flesh of the animals killed in the chase, they partake of the smoke-dried carcases of Dogs, Rein-deer, or Fish, all of which are most esteemed when putrid or dried rather than when fresh. They make a kind of bread, which no stomach but their own appears capable of digesting, with the roasted and pounded extremities of Lichens, such as the Iceland Moss (*Cetraria Islandica*) and the Rein-deer Cenomyce (*Cenomyce rangeferina*), mixed with the bark of young Birches and Pines reduced to a coarse flour. Salt is seldom used by them, and they have little taste for spiritous or strong drinks, while they delight in oil and milk. A few tribes, such as the Kamtchatdales, are acquainted, however, with the art of extracting an intoxicating beer from a species of Mushroom usually accounted poisonous (*Agaricus acris*), which they drink under the name of *Machomor* to an excess which often occasions death. They build neither towns nor villages, and can hardly be said to exist in society. In their scattered huts, half-buried in the earth, the members of a polygamous family promiscuously reside, in the midst of smoke and the confusion of their domestic animals, where the idea of modesty never enters. Next to the Hottentot in the uncleanliness of his person, the Hyperborean brings along with him a most insupportable odour.

F. HOMO AMERICANUS.—AMERICAN INDIANS.

Syn. LES AMÉRICAINS.—Cuv. Reg. Anim. I. 84.
RACE AMÉRICAINE.—Desm. Mam. 48.—Less. Mam. 25.
HOMO SAPIENS, AMERICANUS.—α. Lin. Gmel. I. 22.—VAR. AMERICANA.—Blumb. Hand. et Abbild.

The enumeration which has been attempted in the preceding pages of the several races of mankind would now be complete, had not the genius of Columbus, at the latter end of the 15th century, disclosed to Europeans the wonders of a western continent, which by a singular injustice now bears the name of another. The aborigines of every part of that immense region exhibit, with slight exceptions, the same physical characters; and their innumerable dialects have those singular affinities and analogies, which fully establish the claims of the American Indians to be considered a distinct and original race. There are, however, important differences among them, which have led to their being subdivided into North American Indians (*Borealis*), and South American Indians (*Australis*).

I. BOREALIS.—NORTH AMERICAN INDIANS.

Syn. HOMO COLOMBICUS.—Bory. Ess. Zool. II. 1.—Fisch. Syn. Mam. 6.
COLOMBIENS.—Desmoul. Tab.
Icon. Blumb. Dec. Cran. I. t. 9, 10. II. t. 20. IV. t. 38.
North American.—Griff. Anim. King. I.

In this subdivision we include all those scattered tribes of Red-Indians whom the progress of civilization is gradually confining within narrow limits, but who once peopled the Canadas, the territory of the United States, the eastern part of Mexico, the Antilles or West India Islands, the mountain chain of the Andes, Terra Firma, and the Guianas; from Cumana to the Equator—under the names of Hurons, Sioux, Cherokees, Chippeways, Iroquois, Arkansas, Illinois, Apalaches, Chicacas, Mohicans, Oneidas, Carribees, Mexicans or Aztecs, Peruvians, &c. &c.

The North American Indians, wherever they have not been confounded with European or African blood, are tall, robust, and well made, stronger and more active than is usual with savages. Their limbs are not meagre like the Australian, but well proportioned. The skull is of an agreeable oval shape, but with the forehead flattened to so remarkable a degree, as induced the more early writers to imagine this depression of the forehead to be owing to the skilful application of bandages or planks to the surface of the head. Although it may be true that some tribes assist the natural peculiarity of the races, yet it is found among tribes who use no such art, such as the Mexicans, and being usually considered as a great beauty by the latter, the Aztec gods and heroes are represented with an extreme depression of the forehead amounting to exaggeration.[1] The nose of the North American Indian is long, well-defined, and aquiline.[1] The mouth is moderately cleft, the teeth are placed vertically in the gums, and the lips are similar to those of a European. Their eyes are brown; their hair commonly black, straight, coarse, and glossy; of medium length, and

seldom reaching beyond the shoulders. It is said never to become grey with advancing years. The men have naturally but a scanty beard, and in general pluck the hair very carefully wherever it appears. The colour of their skin is reddish, approaching to the hue of copper. The females, condemned to the severest drudgery, and almost reduced to the condition of beasts of burthen, are not without their charms in some of the mountain districts, though elsewhere they are in general of low stature, with high cheek-bones, prominent eyes, and flat bosoms; and their breasts, though often well made, are in general somewhat pendulous and flat. Puberty commences at an early age.

The Indians, who wander over the extensive country situate between the Pacific Ocean and the Alleghany Mountains, in a state of savage independence, seem destined soon to undergo an inevitable extermination.[2] The warlike tribes of the Sioux, who traffic largely with the Anglo-Americans in furs, appear to have emigrated from the north-west. They use symbolical writing, like the Mexicans. The Chippeways are of a more pacific character, but passionately devoted to the abuse of spiritous liquors, which the avidity of the fur-dealer supplies in large quantities. Hieroglyphics are likewise in use with this tribe. The Menomenies have a fine expressive countenance, much intelligence, and a patriarchal simplicity of manners. Some considerable progress has been made in agriculture and domestic manufactures by the Cherokees, who employ Negro slaves to execute the most laborious parts of their employments. Among the numerous tribes who wander between the sources of the Missouri and the frontiers of Mexico, we find a great diversity of language, manners, and customs; yet there are some points in which they exhibit a remarkable similarity to each other. They wander from place to place, occasionally building huts or permanent lodges for hunting the Buffalo, the flesh of which forms their principal sustenance. Many nations go almost naked, though in general their dress consists of a robe of Buffalo skin, attached to the shoulders; an apron covering the waist and middle; and a rude form of boots called *mocassins*, attached to the legs. The females wear a cloak like the males, with an under garment of elk or deer-skin, reaching down to the knees. Feathers are worn on the heads of the chiefs, while various rude ornaments and showy garments distinguish their days of state. It is considered highly ornamental to paint the face red and black; their bodies are also painted during their warlike expeditions. Some tribes bore their noses, and wear different kinds of pendants; others slit their ears and load the helices with brass wire, so that the extremities almost drag on the shoulder. Horses are extensively used by the Indians of the plains westward of the Mississippi, but are seldom found to the eastward, where the difficult nature of the country renders it impassable to Horses.[3]

The North American Indian is reserved and circumspect in all his words and actions, and nothing ever induces him to betray the emotion of the moment. His thirst for vengeance is excessive, and cannot be eradicated.[4] The same peculiarities of character extend to the Mexican Indians, who are melancholy, grave, and taciturn, when they are not under the influence of intoxicating liquors. Even the children of the Indians at the age of four or five years display a remarkable contrast to the Whites at these early ages, and delight to throw an air of mystery and reserve over their most trifling observations. But when once the state of repose, in which the Indian habitually indulges, is disturbed, the transition to a violent and ungovernable state of agitation is at once both sudden and terrific.[5]

The Mexicans and Peruvians had made considerable progress in the arts and sciences, before the barbarous persecutions of the Spaniards plunged them into a state of poverty and degradation. They had a most correct knowledge of the true length of the year, and the construction of the calendar; they framed geographical maps of their own country; they constructed woods, canals, and enormous pyramids, the sides of which were accurately directed to the four cardinal points.[1] Their civil and military hierarchies and their feudal systems presented a complicated and intricate form, indicative of no very modern origin. They cultivated no other grain than the maize (*Zea*), and they knew no preparation of milk, although the females of two species of the Ox might have contributed an abundant supply. The Mexicans still preserve a peculiar taste for painting and sculpture, as well as for flowers, with which they delight in ornamenting their persons and dwellings.

Some wild tribes of Mexican hunters still preserve a savage independence. The Apaches are a warlike and industrious nation, dreaded by the Spaniards, who are compelled to oppose a large force to their depredations.

Much ingenuity has been employed in tracing affinities and analogies between the languages of North America and the north-east coast of Asia, but without leading to any satisfactory result.

[1] A. Von Humboldt, Vues des Cordillieres et Monumens des Peuples indigènes de l'Amérique, 2 vols. folio, 1811.
[2] Major Pike's Travels.
[3] Hodgson's Letters from North America.
[4] Lewis and Clarke's Travels. These officers penetrated for the first time across the continent of America to the Pacific Ocean.
[5] Godman's American Natural History. Philadelphia, 1826. Vol. I. p. 21.

36

2. AUSTRALIS.—SOUTH AMERICAN INDIANS.

Syn. H. AMERICANUS.—Bory, Ess. Zool. II. 17.—Fisch. Syn. Mam. 6.
AMÉRICAINE.—Desmoul. Tab.

Icon. Blumenb. Dec. Cran. V. t. 46, 47, and 48, VI. t. 58.
Pr. Max. Reis.[1] pl. 2 and 3 (Pusis Indians), pl. 7 (Patachos Indians of the Rio de Prado), pl. 10 and 17 (Boticudos Indians).

The races of South American Indians, differing materially from those already described, are still imperfectly known. They are widely dispersed over the extensive plains of the Amazon and Oronoko, Brazil, Paraguay, and part of Chile, under the names of Omaguas, Guaranas, Coroads, Atures, Otomacs, Boticudos, Charruas, Chaynas, &c. &c.

These races, with few exceptions, have their skulls oval-shaped, of a disproportionately large size, sunk between the shoulders, flattened on the vertex, with a broad forehead, but depressed to an extreme degree; the superciliary ridges very much elevated; the cheek-bones prominent; the eyes small and sunken; the nose flattened with expanded nostrils; the lips thick; the mouth wide, with the teeth placed almost vertically in the gums. The hue of their skin is neither black, yellow, nor copper-coloured, but a dark brown. Their hair is black, straight, and of the coarseness of horse hair.

Their stature is rather below the middle size. Hunting, and the use of a few nutritious roots, supply the simple sustenance of tribes whose intelligence is excessively limited. They are destitute of religion, and what is more remarkable, are said to be almost devoid of superstition. The bow and the arrow, the javelin and the club, are their only arms. Their aversion to the North American Indians is most decided, and they carry on a continual war with them at all points where they come in mutual contact.

Of the races situate to the East of the Andes and La Plata, we may notice the Omaguas, who assist the natural depression of their foreheads by artificial means; the Guaranas, Coroados, and Atures, obscure tribes, who are daily diminishing in numbers. The Ottomacs include ants, gum, and, what is more remarkable, a kind of potter's clay, in their list of edible substances. This earth is kneaded into balls of a few inches in diameter, and roasted before a slow fire. When analyzed by Vauquelin, it was found to consist of 50 parts per cent. of silica, 40 of alumina, 4 of magnesia, 1 of oxide of iron, besides water. It is procured only from particular beds, and all kinds are not equally pleasing to their palates. Fish, lizards, fern roots, and other animal and vegetable substances, are also used for food, when they can be procured, but under any circumstances a ball of clay usually concludes the repast. The Puris, a ferocious tribe, are said to roast and eat the prisoners taken in war. All these tribes are of a very deep-brown colour. The Buticudos, of a light-brown, sometimes approaching almost to white, adopt the singular custom of inserting a very large and round block of wood into the lobe of each ear, and in the under lip, so as to give the entire countenance a most singular and characteristic appearance. The Charruas, a warlike tribe of Paraguay, have successfully resisted the attacks of Europeans; they are of a deep-brown complexion approaching to black.

At the southern extremity of the South American continent, we find several wandering tribes, very elevated in stature, whom M. Bory has raised to the rank of a distinct species, under the name of *Homo Patagonus*. They are known to navigators under the names of Patagonians, Puélches, Araucans, or Tehuetlets. Exaggeration and the love of the marvellous have elevated the stature of the Patagonians to seven feet and upwards; yet it seems pretty well established that six feet, four inches, is no very extraordinary height among this people.

In the adjoining island of Tierra del Fuego, we find the Pescherais, a diminutive and stunted race, scarcely taller than the Hyperboreans of the Northern Hemisphere.

SUB-VARIETIES OF THE HUMAN RACE.

All the preceding races of men are capable of producing by their union a fertile progeny, which possesses intermediate characters between those of its parents. These combinations give rise to most of the variations of features which may be observed in all quarters of the globe, but especially among the European colonies. In those ordinary instances of inter-marriage, where the differences between the characters of the parents are inconsiderable, the variety thence resulting has no particular name assigned to it; but in alliances where the characters are very remote, the progeny has been distinguished by various denominations.

The MULATTO proceeds from the Caucasian or White Race and the Negro, well known in the European colonies, as forming a dangerous caste called *men of colour*, or *petit blancs*, despised by the Whites of pure race as being of inferior blood, and detested by the Negroes, from their pretending to usurp the authority of Whites, without possessing a legitimate title. In his physical features, the Mulatto holds an interme-

dine station between his parents, both of whom he resembles in colour and form, in his half-curly hair, his muzzle slightly projecting, as well as in his moral and intellectual qualities. When Mulattoes intermarry, their posterity resembles themselves, and form a race called *Casque*, probably a corruption of the word *caste*. In general, they are well-made and robust, violent in their passions, talkative and volatile. In the East Indies there is a race of oriental Mulattoes, called *Bouganese*, the progeny of the Hindoo and Negro. They are browner and more meagre than the Mulattoes of European descent.

The MESTIZO, generally of feeble constitution, is the result of a union between the American Indian (*H. Americanus*) and the European.

The ZAMBI or LOBOS is the descendant of the African and American Indian. These individuals are of a dark-brown copper-colour, very muscular and robust. In Mexico they are called *Chino*.

The TEKO is the descendant of the Chinese and Malay.

The BASTER is the progeny of the Caucasian and Hottentot (*H. Capensis*). His skin is of the colour of dried citron. In respect to his intellectual character, he partakes more of the European than of the Hottentot, being braver and more energetic than the latter. The prominence of the cheek-bones continues in these unions for several generations. An excellent figure of this sub-variety is given in the Atlas to Péron's Voyage, pl. 55.

The BLACK BASTER results from the union of the Negro and Hottentot. This race is superior in stature to the common Baster; and the black complexion of the Negro is mitigated by the olive tint of the Hottentot. A female Hottentot, according to Le Vaillant (Première Voyage), is much more fertile when unised to a White or Negro than to one of her own race.

The above are the principal unions of the first degree to which particular names have been assigned. But, as each of these sub-varieties may combine with another, and with the original races, these again with their progeny, and so on *ad infinitum*; there hence arise combinations of the second, third, fourth, and fifth degrees, after which they cease to have particular denominations.

Commencing with the unions of the second degree, the TERCEROON or MORISCO is the progeny of the Caucasian and Mulatto. Sometimes this race is incorrectly termed Quadroon.

The GRIFFE or CABER is the descendant of the Negro and Mulatto. This is sometimes called Zambo.

The QUATRALVI or CASTISSE proceeds from the American Mestizo and the Caucasian or White.

The ZAMBAIGI descends from the American and the Zambi.

The ZAMBO PRIETO results from the union of the Negro and the Zambi.

The TRESALVE is the progeny of the American and the Mestizo.

The DARK MULATTOES result from the American and the Mulatto.

To ascertain the purity of these several combinations of the second degree, let us put W for White, B for Black, and A for American; we then find

The Terceroon or Morisco	$\frac{1}{2}\left(W + \dfrac{W+B}{2}\right) = \frac{3}{4}W + \frac{1}{4}B$
The Griffe or Caber	$\frac{1}{2}\left(B + \dfrac{W+B}{2}\right) = \frac{1}{4}W + \frac{3}{4}B$
The Quatralvi or Castisse	$\frac{1}{2}\left(W + \dfrac{W+A}{2}\right) = \frac{3}{4}W + \frac{1}{4}A$
The Zambaigi	$\frac{1}{2}\left(A + \dfrac{A+B}{2}\right) = \frac{3}{4}A + \frac{1}{4}B$
The Zambo Prieto	$\frac{1}{2}\left(B + \dfrac{B+A}{2}\right) = \frac{3}{4}B + \frac{1}{4}A$
The Tresalve	$\frac{1}{2}\left(A + \dfrac{A+W}{2}\right) = \frac{3}{4}A + \frac{1}{4}W$
The Dark Mulatto	$\frac{1}{2}\left(A + \dfrac{W+B}{2}\right) = \frac{1}{2}A + \frac{1}{4}W + \frac{1}{4}B$

In the combinations of the third order,

The QUADROON proceeds from the White and the Terceroon. These are sometimes incorrectly termed Albinos.

The OCTAVOON descends from the White and Quatralvi.

The SALTATRAS is the progeny of the Mulatto and Terceroon.

The COYOTE proceeds from the Mestizo and Terceroon.

The GIVEROS descends from the Zambi and the Griffe.

The CAMBUJO is the result of the Mulatto and Zambaigi.

In respect to the purity of their blood, we find

The Quadroon	$\frac{1}{2}\left(W + \dfrac{3W+B}{4}\right) = \frac{7}{8}W + \frac{1}{8}B$
The Octavoon	$\frac{1}{2}\left(W + \dfrac{3W+A}{4}\right) = \frac{7}{8}W + \frac{1}{8}A$

[1] PR. MAX. REIS.—Reise nach Brasilien in den Jahren 1815 bis 1817, von Maximilian, Prinz zu Wied-Neuwied. Frankfurt, 1821.

The Saltatras $\cdot \frac{1}{2}\left(\dfrac{W+B}{2}+\dfrac{3W+B}{4}\right)=\frac{5}{8}W+\frac{3}{8}B$

The Coyote $\cdot \frac{1}{2}\left(\dfrac{W+A}{2}+\dfrac{3W+B}{4}\right)=\frac{5}{8}W+\frac{1}{4}A+\frac{1}{8}B$

The Giveros $\cdot \frac{1}{2}\left(\dfrac{B+A}{2}+\dfrac{W+3B}{4}\right)=\frac{5}{8}B+\frac{1}{4}A+\frac{1}{8}W$

The Cambujo $\cdot \frac{1}{2}\left(\dfrac{W+B}{2}+\dfrac{3A+B}{4}\right)=\frac{3}{8}B+\frac{3}{8}A+\frac{1}{4}W$

The combinations of the fourth order, which have been distinguished by particular names, are the following:

The QUINTEROON descends from the White race and the Quadroon.
The PUCHUELAS is the progeny of the White and the Octavoon.
The HARNIZOS comes from the White and the Coyote.
The ALBARASSADOS is the descendant of the Mulatto and the Cambujo.
Proceeding, in the same manner as before, to calculate the purity of descent, we find

The Quinteroon $\cdot \frac{1}{2}\left(W+\dfrac{7W+B}{8}\right)=\frac{15}{16}W+\frac{1}{16}B$

The Puchuelas $\cdot \frac{1}{2}\left(W+\dfrac{7W+A}{8}\right)=\frac{15}{16}W+\frac{1}{16}A$

The Harnizos $\cdot \frac{1}{2}\left(W+\dfrac{5W+2A+B}{8}\right)=\frac{13}{16}W+\frac{1}{8}A+\frac{1}{16}B$

The Albarassados $\frac{1}{2}\left(\dfrac{W+B}{2}+\dfrac{3B+3A+2W}{8}\right)$
$=\frac{7}{16}B+\frac{6}{16}W+\frac{3}{16}A$

The only combination of the fifth order which we shall notice is The BARZINOS, proceeding from the White and the Albarassados, giving for its formula

$\frac{1}{2}\left(W+\dfrac{7B+6W+3A}{16}\right)=\frac{11}{16}W+\frac{7}{32}B+\frac{3}{32}A$

The remaining sub-varieties are not noticed by writers, as they present nothing remarkable. The combinations may proceed absolutely without limit, each preserving more or less the original character of its race, in proportion to its affinity with the several primitive stocks. It will be remarked that the preceding terms are mostly derived from the Spanish and Portuguese languages, because the several castes were first observed in the colonies of those nations.

The individuals resulting from alliances between remote races are in general vigorous and robust, and fully confirm the observations made by Buffon and others, that the crossing of races tends to perfect the individual. It is a singular fact, that the form of the head in the offspring is always more like that of the father than of the mother, and it is not unworthy of notice, that the same observation holds good in reference to mules, or the sterile progeny of different species.

To prevent the deterioration of the individual in the human race, it is unnecessary to resort to remote alliances between foreign races. A European, married to another of a foreign country or of a different family, may produce individuals as well-formed as those resulting from the union of a European and Negro. In consequence of the intercourse of families, the national characters of the primitive races are daily becoming more and more obliterated. The migrations of nations from the north to the south, the revolutions of empires, conquest and colonization, have further tended to blend the human species together. Thus, the Turkish and Persian blood becomes improved by their alliances with the Mingrelians, Circassians, and other Proper Caucasians. In the East Indies, the intercourse of the Europeans with the Hindoos has given rise to a sub-variety of the White races, called *Half-caste* or *Mêtis*. These are as troublesome in the East as the Mulattoes in the West Indies, and their consanguinity to the pure European is distinguished by several gradations.

The CASTISSE, the progeny of a White and a Half-caste Hindoo, is therefore three-fourths European and one-fourth Hindoo.
The POSTISSE, the descendant of the European and Castisse, is seven-eighth parts European and one-eighth Hindoo.

In proportion as these combinations are multiplied, all the great distinctive characters of the races become gradually effaced, and blended into each other, so that no definite characteristic remains.

The Terceroons and Quadroons, the progeny of the Mulatto and the White, are more or less of dark complexions. The females have their mucous membranes of a white tinge; the Quadroons preserve the dark scrotum of the Negro. It is remarkable that the black tinge maintains itself longer in the nutritive and generative organs than in any other parts of the body.

In the preceding observations we have taken no notice of the Creole, this variety being entirely the result of climate. When a European marries another European, and settles between the tropics, his offspring is termed Creole, and the same name is applied to the offspring of Negro

parents, born in the East or West Indies. In fact, this word, derived from *creare*, to generate, is applied indiscriminately to all persons born in the Indies, and even to the lower animals. It is, however, most commonly limited to the progeny of Europeans.

In the constitution of the Creole, we may trace the influence of a warm climate upon the human frame. He is in general well-made, of good stature, rather of meagre limbs, but of delicate frame. His passions are violent, naturally haughty, imperious, and accustomed to exercise a despotic rule over a host of slaves.

The female Creoles of the tropical regions of the globe are very liable to abortion, and yield little milk, for which reasons, their offspring is usually nursed by Negress slaves.

The Anglo-Americans and other inhabitants of the temperate parts of North America do not materially differ in constitution from other Europeans.

FICTITIOUS RACES.

Inaccurate observation and the love of the marvellous have given rise to imaginary races, which on further examination have proved to arise from individual malformation, disease, or the mistakes of early navigators.

The ALBINO, called *Blafard* on the continent of Europe; *Bedas*, *Chacrelas*, or *Kakerlaks*, in India; *White-Negro* or *Dondos*, in Africa; *Darien* in America, is an individual malformation or degeneration in the colouring matter of the skin and hair, usually dying with the individual, but sometimes becomes hereditary, and is transmitted to their offspring. It presents the same characters in whatever race it appears, and is found likewise among the lower animals. (See before, pages 74 and 75.)

The human Albinos are of a feeble constitution, the skin of a dull white, the eyes weak with the iris red, and the hair of a pale yellow. They are most commonly found, or at least are most remarked, among the races of dark complexions. At Java, they are reported to form a wandering and proscribed race, roaming in the woods under the name of *Chacrelas*. Labillardière observed an Albino female of Malay descent upon one of the Friendly Islands. The Albinos of Ceylon, called *Bedas* or *Bedos*, appear to belong to the Hindoo race. They are also found among the Papoos; and have been seen, but very seldom, among the Hyperboreans. A White Negress from Madagascar was observed by M. Bory de St Vincent; she had two children, the one by a White, the other by a Negro. Each of these children presented intermediate characters between its parents, having the usual traits of the father combined with the Albino features and white hair of the mother. Albinos are reported to be common in the woods of the Isle of France. They are also common on the continent of Africa. In America, the most remarkable are the Dariens, who reside in the isthmus connecting the northern and southern portions of that continent.

The CRETINS, who are found in the mountain gorges of the Valois, have been improperly raised to the rank of a variety of the human race. They are usually found among the Japetic or Celtic races. Imbecile in mind, with a goitre disfiguring the anterior part of the neck, where the glands are materially altered, a yellowish skin, languishing manner, and extreme bodily weakness, these degenerate individuals drag on a wretched existence in some mountainous regions, such as the Pyrenees, the Alps, Styria, and the Carpathian chain. Sometimes they are born from well-formed parents, and sometimes they compose small families, generally hidden in the obscure recesses of the valleys. In most places they are looked upon with disgust; a healthy mountaineer would disdain to contract a matrimonial alliance with one of these unfortunate individuals; but in the Valois they are regarded with a superstitious veneration. It is said that Cretins are found in the Uralian mountains, the Himalayan, and the Andes. They have also been remarked in the heights of Sumatra.

It is almost unnecessary to repeat here the numerous fables with which credulous travellers have crowded their narratives. The *Quimos*, of the mountains of Madagascar, were represented as a variety, only three feet and a half in height, with long arms, the form of an Ape, a white and shrivelled skin, defending themselves with great courage, &c. &c. The *Men with tails*, who have been found in the Indian Ocean, and especially in the Island of Formosa, have dwindled down into ordinary apes before more accurate observations. A race of Malay women is described by Struys, *with beards* as long as their husbands'. The diminutive Africans, also, who live on Grasshoppers, as mentioned by Drake, until the age of forty, and in their turn are devoured by worms, must be placed in the same list. We no longer discover the Pygmies and Troglodytes of antiquity, who fought with the Cranes. One of the fathers of the Church gravely repeats a conversation which he had with a Centaur in Africa, where he saw men without a head, and one eye in the middle of the breast. Raleigh found the same kind of Cyclops in South America, where also the Amazons of the ancients have been resuscitated. Races have been mentioned with a single leg and thigh supporting their bodies like a column. Tritons or Sea-men, and Mer-maids or Sea-women, have likewise been taken in Holland, and taught to sew with great precision. It is humorous to find in works, otherwise of some pretensions. such fables as that of the Hindoo race of Saint Thomas, with flat thighs, said to be found in the Island of Ceylon. or the six-fingered nations of the human race. We must not, however, confound the well-authenticated accounts of Porcupine-men with the above fictitious narrations.

CLASSIFICATION OF RACES.

Linnæus (A.D. 1766) was the first systematic writer who ventured to include Man as a member of the Animal Kingdom. He established the order Primates, consisting of four genera ; 1. Homo ; 2. Simia ; 3. Lemur ; and, 4. Vespertilio. The genus Homo, which he characterized by the brief phrase, " Nosce te ipsum," consisted of the *Homo sapiens*, and the *Homo ferus*, the latter founded on a few accidental instances of juvenile outcasts, while the former, subject to variation. *culturâ et loco*, was subdivided into five races ; α Americanus ; β Europæus ; γ Asiaticus ; δ Afer ; and ε Monstrosus ; the last being composed of all the defective individuals observed among the remainder. He avoided the error of those subsequent writers, who consider the races of men as so many distinct species. Yet his classification was exceedingly arbitrary, and in attempting to apportion the human race among the four divisions of an antiquated geography, he blended together a number of races, very different in their physical characters, and failed to notice the inhabitants of many extensive regions of the globe which cannot be referred with propriety to any of the principal continents.

Buffon (A.D. 1766), in the excellent treatise, " Sur les Variétés dans l'espèce Humaine," with his usual disregard to systematic arrangement, did not propose any natural subdivision of the races. He collected the results scattered over the innumerable voyages and travels of his day, and discriminated with caution among the mass of errors and contradictions with which their writings abounded. Subsequent travellers have added more precise information for correcting and completing the valuable treatise of Buffon, which even now may be read with pleasure and advantage. Already the critical eye of Buffon distinguished the Malay from the other Asiatics, and the Tartars from the Chinese. He admitted the physical differences of the Hyperborean races, distinguished the Hottentots from the other Africans, and acknowledged the unity of the Ethiopians.

Blumenbach (A.D. 1797) admitted five varieties of the human species ; 1. Caucasica ; 2. Mongolica ; 3. Æthiopica ; 4. Americana ; 5. Malayana. These are little more than the old division of Linnæus, with the substitution of the Malayan variety in place of the H. Monstrosus, Linn. Our chief objection to this arrangement consists in the obvious impropriety of including the Americans and Malays, whose characters are not very decisive, in the same rank with the Caucasians, Mongolians, and Ethiopians. The Malayan division has now become insufficient to contain the numerous and varied races of the Southern Ocean.

Duméril (A.D. 1806) instituted the order Bimanes, which was a most decided improvement upon the order Primates of Linnæus, who placed intellectual Man in the same order with the Apes and Bats. He subdivided the human race into six varieties ; 1. La Caucasique, or Arabe-Européenne ; 2. L'Hyperboréenne ; 3. La Mongole ; 4. L'Américaine ; 5. La Malaie ; 6. L'Ethiopienne, or Nègre. His arrangement coincides pretty nearly with that of Blumenbach, with the manifest improvement of separating the Hyperboreans from the Mongolians.

The Baron Cuvier, in the first edition of the " Règne Animal" (A.D. 1816), admitted only three principal varieties ; 1. Blanche ou Caucasique ; 2. Jaune ou Mongolique ; and, 3. Noir ou Ethiopique ; at the same time remarking, that he did not know to which of the above to refer the Malays, Papoos, or Americains.

In the same year M. Malte-Brun published his enumeration of the human races, but without attracting much attention from systematic writers. He distinguished sixteen races. 1. Polaire. 2. Finnoise. 3. Sclavonne. 4. Gothico-germanique. 5. Occidentales de l'Europe. 6. Grecques et Pélagiques. 7. Arabe. 8. Tartare et Mongole. 9. Indienne. 10. Malaie. 11. Noire de l'Océan Pacifique. 12. Basanée du Grand-Océan. 13. Maure. 14. Nègre. 15. De l'Afrique Orientale. 16. D'Amérique.

M. Virey, in the Nouv. Dict. d'Hist. Nat., article Homme, which appeared A.D. 1817, was the first naturalist who ventured, in defiance of the received opinions, to divide mankind into two species, characterized by the magnitude of the facial angle.

His classification was as follows :

Genre Humaine.

1° Espèce, Angle facial de 85° à 90°.

1. Race Blanche,	{ Arabe-Indienne. { Celtique et Caucasienne.
2. Race Basanée,	{ Chinoise. { Kalmouke-Mongole. { Lapone-Ostiaque.
3. Race Cuivreuse,	Américaine ou Caraïbe.

2°. Espèce, Angle facial de 75° à 80°.

4. Race Brune-foncée,	Malaie ou Indienne.
5. Race Noire,	{ Cafres. { Nègres.
6. Race Noirâtre,	{ Hottentots. { Papous.

This enumeration, though more complete than those of any of his predecessors, with the exception of Malte-Brun, is still liable to many objections. Characters derived from the magnitude of the facial angle are too variable in Man to constitute a specific difference, and we may seek in vain for any other.

M. Desmarest, in his Mammalogie (A.D. 1820), adopting the hint of the Baron Cuvier, made the distinction of which all other writers appear not to have observed the importance. The three great races, which we have termed Normal, and the Anomalous, or more indistinct varieties of the human species, are here distinctly pointed out. He divided Man, the only species of the Genus Homo, as follows :

  † Variétés de races bien caractérisées.
    A . Race Caucasique.
    B . Mongolique.
    C . Ethiopienne ou Nègre.
  †† Variétés de races moins distinctes.
    D . Race Malaie.
    E . des Papous.
    F . Américaine.

This arrangement coincides exactly with that in the " Règne Animal." Hitherto the important differences between the Malays, Polynesians, and Australasians, passed unnoticed, and M. Desmarest failed to adopt the Hyperboreans of Duméril.

This was the state of the science (A.D. 1825) when the ingenious treatise of M. Bory de St Vincent appeared in the Dict. Class. d'Hist. Nat. art. Homme. Omitting to notice the distinction proposed by Cuvier and Desmarest between the Normal and Anomalous races, and persevering in the error begun by M. Virey, he incautiously distributed the Human race into no less than fourteen distinct species. Yet he has the merit of subdividing the White races with much accuracy, though the leading features had been previously laid down by Malte-Brun, and of distinguishing most of the varieties in the islands of the Southern Ocean.

His divisions are as follows :

  † Léiotriques.
    * Ancient continent.
      1. Japétique.—H. Japeticus.
       +. Gens Togata.
        1°. Caucasica. (Orientale).
        2°. Pélage (Méridionale).
       + + Gens Bracata.
        3°. Celtique (Occidentale).
        4°. Germanique (Boréale).
        α Teutone.
        β Sclavone.
      2. Arabique.—H. Arabicus.
       1°. Atlantique (Occidentale).
       2°. Adamique (Orientale).
      3. Hindoue.—H. Indicus.
      4. Scythique.—H. Scythicus.
      5. Sinique.—H. Sinicus.
      6. Hyperboréenne.—H. Hyperboreus.
      7. Neptunienne.—H. Neptunianus.
       1° Malaise (Orientale).
       2° Océanique (Occidentale).
       3° Papoue (Intermédiaire).
      8. Australasienne (H. Australasicus).
    ** Nouveau Continent.
      9. Colombique (H. Columbicus).
      10. Américaine (H. Americanus).
      11. Patagone (H. Patagonus).
  †† Oulotriques.
    12. Ethiopienne (H. Æthiopicus).
    13. Cafer (H. Cafer).
    14. Mélanienne (H. Melaninus).
    15. Hottentote (H. Hottentotus).
  ‡‡‡ Monstrueux.
    α. Crétins.
    β. Albinos.

M. Desmoulins (A. D. 1826) published his Tableau général du Genre Humain, in which the number of species in the Human Genus was further augmented to sixteen. They were as follows :

1. Scythique,	{ 1. Indo-Germaine. { 2. Finnoise. { 3. Turque.

2. Caucasienne.

3. Semitique. . . . . { 1. Arabe.
2. Etrusco-Pélasge
3. Celtique.

4. Atlantique.
5. Indoue.

6. Mongolique. . . . . { 1. Indo-Sinique.
2. Mongole.
3. Hyperboréenne.

7. Kourilienne.
8. Ethiopienné.
9. Euro-Africaine.

10. Austro-Africaine. . . { 1. Hottentote.
2. Houzouânas ou Boschismane.

11. Malaise ou Océanique, with 5 subdivisions.
12. Papoue.
13. Nègre Océanienne, with 4 subdivisions.
14. Australasienne.
15. Colombienne, with 2 subdivisions.
16. Americaine, with 5 subdivisions.

In this classification the analogies of language are singularly violated in many instances, and subdivisions carried to a greater extent than the actual state of our knowledge seems to warrant.

Another writer on this subject is M. Lesson, who, in his Manuel de Mammalogie (A.D. 1827), has proposed the following arrangement :

1°. Race Blanche ou Caucasienne.
   1. Araméen.
   2. Indien, Germain, et Pelasgique.
   3. Scythe et Tartare.
      1. Malais.
      2. Océanien.
2°. Race Jaune ou Mongolienne.
   1. Mantchoux.
   2. Sinique.
   3. Hyperboréen ou Eskimau.
   4. Américain.
   5. Mongol-Pelagien ou Carolin.
3°. Race Noire ou Mélanienne.
   1. Ethiopien.
   2. Cafre.
   3. Hottentot.
   4. Papou.
   5. Tasmanien.
   6. Alfourous-Endamêne.
   7. Alfourous-Australien.

Here M. Lesson has attempted to reduce all the races of Mankind to three principal divisions. chiefly characterized by colour. The Malays, though of a yellowish copper-colour, are placed among the White races. The Americans, though copper-coloured, are referred to the Yellow races; and the Hottentots, though brown, or yellowish-brown, are placed among the Black. Colour, we have always conceived, is one of the worst characteristics of races, and any classification which does not admit the distinction between the Normal and Anomalous races—which does not assume the analogies and affinities of language as the guide in the minor subdivisions of races—or which looks to mere varieties of colour, without attending to those of form as well as of intellectual character, is liable to insurmountable objections.

The latest writer whose observations have yet reached us is M. J. B. Fischer. His arrangement (A.D. 1829), though in many respects an improvement on those of his predecessors, is still faulty. Besides several minor defects, such as wholly omitting to distinguish the Tartar, Finnish, and Ottoman races, he places the South American Indian and the South Sea Islander in the same rank with the Caucasian, the Mongolian, and the Ethiopian. We shall, however, by exhibiting his arrangement, enable our readers to form their own conclusions respecting its merits.

1. Homo Japeticus.
   α. a. Caucasicus (Orientalis).
      b. Pelagius (Meridionalis).
      c. Celticus (Occidentalis).
      d. Germanicus (Borealis).
      e. Slavonicus (Intermedius).
   β. Arabicus.
      a. Atlanticus (Occidentalis).
      b. Adamicus (Orientalis).
   γ. Indicus.

2. H. Neptunianus.
   δ. Occidentalis.
   γ. Papuensis.
3. H. Scythicus.
   β. Sinicus.
   γ. Hyperboreus.
4. H. Americanus.
   β. Patagonus.
5. H. Columbicus.
6. H. Æthiopicus.
   β. Caffer.
   γ. Melanoides.
   δ. Hottentottus.
7. H. Polynesius.

The classification of the human races is a subject which has long been totally neglected in our own country. All the systems which we have explained emanate either from France or Germany; while in this island, Man is either wholly omitted in our works on Natural History, as though he were something foreign to the Animal Kingdom, or the views promulgated date as far back as the writings of Blumenbach; so that, in our most modern systematic treatises,[1] we still find the human race divided into five varieties. Our own arrangement, which we subjoin to facilitate a comparison with those of preceding writers, is, we believe, the first yet presented to the British reader. It is hoped that the English terms by which we have designated some continental divisions, may not be supposed identical with the common geographical names. Many races of North American Indians are found to the southward of the Isthmus of Darien ; the Finns are not confined to Finnland, nor are the Arabians to be found only in Arabia. We have retained the Hottentots and Bushmen among the Negro races provisionally only, as it is very questionable whether they should not be transferred to the Anomalous Races, where they might be placed between the Papoos and Hyperboreans.

NORMAL RACES.

I. CAUCASIANS.
   A. H. Caucasicus.—Caucasians Proper
   B. H. Iapeticus.—Iapetans.
      1. Pelagius.—Pelasgians.
      2. Germanicus.—Germans.
         α. Teutonicus.—Teutonic Races.
         β. Slavonicus.—Sclavonian Races.
      3. Indo-Persicus.—Indo-Persians.
         α. Indicus.—Hindoo Races.
         β. Persicus.—Medo-Persian Races.
   C. H. Celticus.—Celts.
   D. H. Semiticus.—Arameans.
      1. Arabicus.—Arabians.
      2. Atlanticus.—Atlantic Races.
   E. H. Scythicus.—Scythians.
      1. Othmanicus.—Ottoman Turks.
      2. Finnicus.—Finns.
      3. Tartaricus.—Tartars.
II. MONGOLIANS.
   A. H. Calmuccus.—Calmucks and Mongolians Proper.
   B. H. Sinicus.—Chinese.
   C. H. Sericus.—Mantchoos
   D. H. Kurilianus.—Ainoos.
III. NEGROES.
   A. H. Æthiopicus.—Ethiopians.
   B. H. Caffrarius.—Caffres.
   C. H. Capensis.—Hottentots and Bushmen.

ANOMALOUS RACES.

   A. H. Malayensis.—Malays.
   B. H. Polynesius.—Polynesians.
   C. H. Australasicus.—Australasians.
      1. Australius.—Australians.
      2. Melaninus.—Oceanic Negroes.
   D. H. Papuensis.—Papoos and Tasmanians.
   E. H. Hyperboreus.—Hyperboreans.
   F. H. Americanus.—American Indians.
      1. Borealis.—North American Indians.
      2. Australis.—South American Indians

---

[1] Such as that of Major Hamilton Smith, in Griffith's Anim. King. vol. 5. Some Naturalists commence their systems with the *Simiadæ* or *Monkeys*, which are thus made to stand at the head of the Mammiferous Animals, as well as of the entire Animal Creation.

ORDER II.—QUADRUMANA.

MAMMALIA. WITH FOUR DISTINCT UNGUICULATED LIMBS, WITH THREE KINDS OF TEETH, AND FOUR HANDS.

SYNONYMS.

PRIMATES (in part).—Linn. Gmel. I. 21.—Fisch. Syn. Mam. I.
QUADRUMANES.—Cuv. Reg. Anim. I. 85.—Geoff. Ann. Mus.[1] XIX. 85.—
    Temm. Mon. Mam.[2] I. pref. pag. 13 (excluding the Genus Galeopi-
    thèque). Q. ou TETRACHIRES.—Dum. Zool. Anal. 9.
QUADRUMANA.—Ham. Smith, Syn.[3] p. 4.
QUADRUMANA (Vierhander). Voigt, Thierr.[4] I. 73.
POLLICATA (Daumenfüsser).—Illig. Prodr.[5] 66,
VIERHANDIGE.—Schinz, Thierr.[6] I. 94.

CHARACTERS OF THE ORDER.

THE TEETH consisting of Incisores, Canines, and Molars.
THE PECTORAL LIMBS generally pentadactylous, sometimes only tetradactylous.
THE HINDER LIMBS always pentadactylous.
THE FINGERS with Nails or Claws.
THE HEAD with its facial angle varying from 30° to 65°. THE EYES directed forwards. THE ORBITAL
    and TEMPORAL FOSSÆ distinct. THE CLAVICLES perfect.
THE MAMMÆ usually two, sometimes four, always pectoral. THE PENIS free, and with a Scrotum.
LIVE mostly in trees, where they climb with great facility.
FEED on fruits, roots, and insects.
INHABIT the tropical parts of the entire globe, rarely extending far beyond.

INDEPENDENTLY of the anatomical details already enumerated, which distinguish this order from that of Man, it presents a remarkable difference from our species in the conformation of the lower extremities. The thumbs of the hinder limbs are free and opposable to the fingers, which are long and flexible like those of the fore-hand. Hence all the species of this order climb trees with great ease, while it is only with the utmost difficulty that they can hold themselves upright, or walk in an erect position. The soles of the lower limbs then rest only upon the outer margins, and their narrow pelvis is very unfavourable to equilibrium.

In respect to their intestines, they are tolerably similar to our species; their eyes are directed forwards; the mammæ are pectoral; the penis hangs freely; the brain has three lobes on each side, the hinder of which covers the cerebellum; and the temporal fossæ are separated from the orbits by a bony partition. In other respects they gradually degenerate from the form proper to Man, in exhibiting a muzzle more or less elongated, a tail, and a walk more exclusively quadrupedal. Yet the freedom of their fore-arms, and the complicated form of their hands, enable them to execute many actions and gestures similar to those of Man.

For a long time they have been divided into two genera; now in some measure become two small families, the APES and MAKIS, by the continual accession of new species. Between these it is necessary to place a third, the OUISTITIS, which cannot be referred with propriety to either of the preceding.

FAMILY I. SIMIA.—APES.

SYNONYMS.

SIMIA.—Linn. Gmel. I. 26 (in part).
SINGES.—Cuv. Reg. Anim. I. 86.—Desm. Mam. 48 (in part).—Geoff. Ann. Mus. XIX. 86.
QUADRUMANA (Vierhander) in part.—Illig. Prodr. 67.
APE.—Shaw,[7] Gen. Zool. I. 1 (in part).
AFFEN.—Schinz, Thierr. AFFE.—Voigt, Thierr.

CHARACTERS OF THE FAMILY.

GENERAL FORM approaching to that of Man.
NAILS flat on all the fingers, and of the same form, excepting that of the thumb, which is the flattest.

To this family belong all the Quadrumana with four straight incisive teeth in each jaw, and with flat nails on all their fingers,—characters which assimilate them to Man more than to the succeeding families. Their molar teeth also, like ours, have only blunt tubercles, and they feed essentially on fruits; but their canine teeth passing beyond the others, supply them with an offensive weapon, which is wanting in our species, and requires a vacant space in the opposite jaw to receive the projecting canine, when the mouth is shut.

Buffon subdivided the Apes into five tribes: 1. *Les Singes propres*, without tails; 2. *Papions*, with a short tail; 3. *Guenons*, with a long tail, and callosities on the buttocks; 4. *Sapajous*, with

[1] ANN. MUS.—Annales du Muséum d'Histoire Naturelle, par les Professeurs de cet établissement, Paris, 1802-1813. By the terms GEOFF. ANN. MUS. we quote the several memoirs by M. Geoffroy-Saint-Hilaire in the distinguished work just mentioned.
[2] TEMM. MON. MAM.—Monographies de Mammalogie, ou description de quelques genres de Mammifères, dont les espèces ont été observées dans les differens Musées de l'Europe, Paris, 1827 et seq.
[3] HAM. SMITH, SYN.—Synopsis of the species of the class Mammalia, by Major Charles Hamilton Smith, forming vol. V. of Griff. Anim. King., London, 1827.
[4] VOIGT, THIERR.—Das Thierreich, vom Baron von Cuvier, von F. S. Voigt, Leipzig, 1831.
[5] ILLIG. PRODR.—Caroli Illigeri, Prodromus Systematis Mammalium et Avium, Berolini, 1811.
[6] SCHINZ, THIERR.—Das Thierreich, von dem Herrn Ritter von Cuvier, aus dem Franzosischen von H. R. Schinz, Stuttgart und Tubingen, 1821.
[7] SHAW, GEN. ZOOL.—General Zoology, or Systematic Natural History, by George Shaw, M.D., London, 1800 et seq.

a long and prehensile tail, without callosities ; 5. *Sagouins*, with a long tail, not prehensile, and without callosities. Erxleben adopted this division, and translated their names by the words Simia, Papio, Cercopithecus, Cebus, and Callithrix. Thus, the last two terms, which were used by the ancients to designate the Apes of Africa and the East Indies, were transferred to the Apes of America. It has since become necessary to suppress the genus *Papions*, founded solely on the shortness of the tail, because it broke too

much the natural affinities of species. All the others have been subdivided ; and it has been requisite to remove out of the division the Ouistitis, formerly included among the *Sagouins*, as they do not correspond accurately with the characters of the remaining Monkeys.

The Apes may be divided, according as the number of their molar teeth is 20 or 24, into two principal tribes, which again require to be subdivided into several genera.

TRIBE I. CATARRHINA.—APES OF THE OLD CONTINENT.

SYNONYMS.

CATARRHINI (κατα, *kata*, down, ῾ριεα, *rrhina*, nostrils).—Geoff. Ann. Mus. XIX. 86.
LES SINGES PROPREMENT DITS, OU DE L'ANCIENT CONTINENT.—Cuv. Reg. Anim. 1. 87.

CHARACTERS OF THE TRIBE.

THE DENTAL FORMULA the same as in Man (see before, page 113).
THE NOSTRILS separated by a narrow partition, and opening beneath the nose.
THE TAIL never prehensile, sometimes wanting.
INHABIT the Old Continent.

The Apes of the Old Continent have the same number of molar teeth as Man ; but they differ from each other by many characters which have furnished the distinctions to the following genera and species.

GENUS I. PITHECUS.—MEN-OF-THE-WOODS.

Syn. ORANGS PROPREMENT DITS.—Cuv. Reg. Anim. I. 87.—F. Cuv. in Dict.
des Sc. Nat. XXXVI. 27.
ORANG (Pithecus).—Geoff. Ann. Mus. XIX.—Isid. Geoff.[1] in Bélang.
Voy. p. 22.
SIMIA (in part).—Linn. Gmel.—Illig. Prodr.—Fisch. Syn. Mam.

These are the only Apes of the Old Continent entirely without callosities on the buttocks. The hyoid bone, the liver, and cœcum, resemble those of Man. Their nose does not project, they have no cheek-pouches, nor any vestige of a tail.
This genus comprises but a single species, the Orang-Outang, of which the young alone has yet been carefully examined by Naturalists.

1. PITHECUS SATYRUS.—ORANG-OUTANG.

THE YOUNG.

Syn. L'ORANG-OUTANG.—Cuv. Reg. Anim. I. 87.—F. Cuv. in Dict. des
Sc. Nat. XXXVI. 281.—F. Cuv. et Geoff. Hist. Mam.—Isid. Geoff.
in Bélang. Voy. p. 23.
ORAN-OTAN.—Shaw, Gen. Zool. I. 3.
MAN-OF-THE-WOODS.—Edw.[2] Glean. I. pl. 213.
GREAT APE.—Penn.[3] Quad. pl. 36.
JOCKO.—Buff. Hist. Nat. suppl. VII.
DER EIGENTLICHE ORANG-UTANG.—Voigt, Thierr. I. 74.
DER ORANG-UTANG (WALDMENSCH).—Schinz, Thierr. I. 98.
SIMIA SATYRUS.—Linn. Gmel. I. 26.—Erxl. p. 6.—Blumenb. Hand. et
Abbild.—Fisch. Syn. Mam. p. 9.—Kuhl,[4] Beitr.—Tiles.[5] Naturh. Russ.
PITHECUS SATYRUS (Orang roux).—Geoff. Ann. Mus. XIX. 87.—Desm.
Mam. 50.—Less. Mam. 30.—(Orang-Outang).—Ham. Smith, Syn.
p. 5 (Red or Asiatic Orang-Outang).—Jard. Syn.[6] p. 204.
SIMIA AGRIAS.—Schreb.[7] Saugth.
SIMIA ABELII.—Fisch. Syn. Mam. p. 10.

PONGO ABELII.—Less. Mam. 31.
Icon.[8] LE JOCKO.—Audeb.[9] Sing.
L'ORANG-OUTANG Femelle.—F. Cuv. et Geoff. Hist. Mam.
ORANG-OUTANG.—Abel. Chin.[10]
Donovan,[11] Nat. Rep. pl. 58 and 59.
Vosm.[12] Descr. pl. 14 and 15.
Camp.[13] Kort Berigbt.
RED OR ASIATIC ORANG-OUTANG.—Wils.[14] Illustr. pl. 5, fig. 1 and 3.

THE ADULT (very probably).

Syn. GROOTE ORANG-OUTANG OF OOST-INDISCHE PONGO.—Wurmb.[15] in Ver-
handl. van het Batav. Genootsb. II. p. 245.
PONGO VURMBII.—Geoff. Ann. Mus. XIX. 89.—P. WURMBII.—Desm.
Mam. 52.—Less. Mam. 32.
SINGE DE WURMB.—Audeb. Sing. p. 21.
PAPIO WURMBII.—Latreille in Buff.[16] ed. Sonn. XXXVI. p. 296.
SIMIA WURMBII.—Fisch. Syn. Mam. p. 32.
LE PONGO.—F. Cuv. in Dict des Sc. Nat. XXXVI. 295.
Icon. PONGO (squelette).—Audeb. Sing. ADEL pl. 2.

SPECIFIC CHARACTERS.

THE MUZZLE short in the young, very long in the adult.
THE FOREHEAD elevated in the young, greatly depressed in the adult.
THE PECTORAL LIMBS very long, reaching as low as the ancles.
THE EXTERNAL EAR medium size like that of Man.
THE HANDS narrow, and the FINGERS elongated.
THE TAIL, CHEEK-POUCHES, and CALLOSITIES, all wanting.
THE HAIR scanty, of a brownish red in the young, black in the adult.
INHABITS Cochin-China, the Peninsula of Malacca, and especially the Island of Borneo.

The Orangs, properly so called, have their arms sufficiently long to reach the ground when they stand upright, while their thighs, on the contrary, are very short.

---

[1] ISID. GEOFF. IN BÉLANG. VOY.—Voyage aux Indes-Orientales, pendant les années 1825 à 1829, par M. Charles Bélanger.—Mammifères par M. Isidore Geoffroy-St.-Hilaire, Paris. 1834.
[2] EDW. GLEAN.—Gleanings of Natural History, by G. Edwards, London, 1758 et seq.
[3] PENN. QUADR.—History of Quadrupeds, by Th. Pennant, London, 1793.
[4] KUHL, BEITR.—Beiträge zur Zoologie und vergleichenden Anatomie von J. H. Kuhl, Franckfurt. 1820 et seq.
[5] TILES. NATURH. RUSS.—Naturhistorische Früchte der ersten Kaiserlichen Russischen Weltumseglung, von W. G. Tilesius, Petersburg. 1813.
[6] JARD. SYN.—Synopsis of the Simiadæ, at the conclusion of the Naturalist's Library, vol. I. Monkeys, by Sir William Jardine, Bart., Edinburgh, 1833.
[7] SCHREB. SAUGTH.—Die Säugthiere in Abbildungen nach der Natur mit Beschreibungen, von J. C. D. Schreber, Erlangen, 1775 et seq.
[8] The only good figure of the Orang-Outang for a long time was that of Vosmaer, made after a specimen kept at the Hague. That of Buffon (Suppl. VII. pl. I.) is faulty at all points ; that of Allamand (Buff. d'Holl. XV. pl. XI.) is a little better, and has been copied in Schreber, pl. II. B. That of Camper, copied in Schreber, pl. II. C. does not want precision, but it is easy to see that it was not drawn after the living animal. Bontius, Med. Ind. 84, presents a creature of his own imagination, although Linnæus made it the type of his Troglodytes (Amœn. Acad. VI. pl. I. § 1). There are some tolerably good figures in Griff. Anim. King., and in Kruzenstern's Voyage, pl. XCIV. and XCV., but all taken from young specimens.—*Note of the Baron Cuvier.*
[9] AUDEB. SING.—Histoire Naturelle des Singes et des Makis, par J. B. Audebert, Paris, An. 8 (1799-1800).
[10] Narrative of a Journey in the Interior of China, and of a Voyage to and from that country, in the years 1816 and 1817, by Dr Clarke Abel. London, 1818.
[11] DONOVAN, NAT. REP.—The Naturalist's Repository, or Monthly Miscellany of Exotic Natural History, by Edward Donovan, London, 1821 et seq.
[12] VOSM. DESCR.—Description de l'espèce de Singe, &c. nommé Orang-Outang, par A. Vosmaer, Amsterdam, 1778.
[13] CAMP. KORT BERIGHT.—Kort beright wegens de Ontleding van verschiedene Orang-Utangs, van P. Camper, Amsterdam, 1778.
[14] WILS. ILLUSTR. ZOOL.—Illustrations of Zoology, by James Wilson, Edinburgh, 1831. The synonym *Red,* applied to the Orang-Outang, is at least premature, inasmuch as the adult appears to become black. This appellation is also adopted in Jardine's Naturalist's Library, and in Jard. Syn.
[15] Wurmb's Paper in Verhandlingen van het Bataviaasch Genootschap der Konsten en Wetenschapen, Batavia, 1792 et seq.
[16] BUFF. ED. SONN.—Histoire Naturelle, par M. Le Clerc de Buffon. Nouvelle Edition, par C. S. Sonnini, Paris, An. 7 and 8 (1798-1800).

Orang is a word in the Malayan language signifying *reasonable creature*, and applied equally to Man, the Orang-Outang, and the Elephant. *Outang* [or rather *Utan*, according to Mr Marsden] signifies *wild*, or *belonging to the woods*, and thus the earlier navigators, translating these words, gave the animal the name of *Man-of-the-Woods*.

The Orang-Outang, of all animals, is the most similar to Man in respect to the shape of its head, the size of the forehead, and the volume of its brain. Yet the exaggerated expressions of some writers on those points seem to originate in the circumstance that only young individuals have been examined by them, and every thing leads us to believe that their muzzle becomes more elongated with advancing years.

The body of the Orang-Outang is covered with thick red hair, the face is blueish, and the thumbs of the hinder limbs are very short in comparison to the remaining fingers. Its lips are susceptible of a singular elongation, and enjoy a great facility of motion.

The fore-hands of the Orang-Outang are shaped exactly like those of Man, excepting that the thumb is very short, reaching only to the first joint of the index or first-finger. The hinder-hands have likewise five fingers, but the thumb is placed much lower than in the human species, and in its ordinary position, instead of lying parallel to the fingers, it forms a right angle with them. The fingers of the hinder-hands have the same structure as those of the fore-hand, and are equally free in their movements, and all, without exception, have nails. The calves of the legs as well as the buttocks are but slightly prominent.

It is highly probable that the Orang-Outang has the same number of teeth as Man; yet this cannot be stated positively, until the adult has been carefully examined. All the descriptions hitherto given of its teeth apply merely to the milk or first teeth of the young animal. If it be true that the Pongo is really the adult Orang-Outang, then they are precisely of the same number as in the human species. In the upper jaw, the first incisor is very broad and wedge-shaped; the second terminates likewise in a line, but its inner surface is inclined towards the first incisor, by the action of the very large canine tooth next to it. This canine, which is separated from the second incisor by a small interval, is very long, strong, thick, and hollow beneath, from the action of the opposite tooth. The first and second false molars are divided by a longitudinal furrow into two portions, worn off obliquely, and presenting two blunt tubercles, one on the inner margin, and the other on the outer. The three real molars which follow are nearly of equal size, their crowns flat, and apparently worn down by constant use. In the lower jaw, the two incisors are equal in size, but the first terminates in a straight line, and the second in an oblique one, beginning towards the first incisor, and ending towards the canine. The latter is very strong, long, and sharp, yet somewhat less than the canine of the upper jaw, and its shape is rendered triangular by an elevated crest on its internal surface, partly formed by the action of the canine and of the first incisor of the opposite jaw. The first false molar is cut off obliquely before and behind, forming a very thick conical tubercle, presenting an oblique plane to the upper canine, and strongly supported, against which the animal can exercise great force in cutting and tearing. The other molars resemble those of the upper jaw.

The neck of the Orang-Outang is very short; the tongue smooth; the nose wholly flattened at the base, and on a level with the rest of the face at this part, but projecting slightly at its extremity; the nostrils open beneath, as in all other Apes of the Old Continent. The eyes resemble those of other Apes; the iris is brown, and the ears are exactly similar to those of our own species; the nails are black.

Almost the whole body is covered with hair, which is darker and thicker in some parts than in others. The colour of the skin is generally a blueish slate; but the ears, the circles around the eyes and mouth, and the inside of all the four hands, tend towards a flesh copper-colour. The hair of the head, fore-arm, and legs, is of a deeper red than elsewhere. It is thickest on the head, back, and upper part of the arm; but very scanty on the belly, and still more so on the face. The upper-lip, nose, and palms of all the four hands, are the only parts entirely naked. The hair of the whole body is rather woolly, and of the same nature throughout; on the fore-arm it points upwards towards the elbow, while on the upper-arm the points are directed downwards. The skin, especially that of the face, is thick and shrivelled, while beneath the neck it hangs so loosely as to appear like a gôitre, when the animal lies on its side.

Camper discovered, and accurately described, the two membraneous sacs which communicate with the ventricles of the glottis of this animal, and render its voice hoarse; but he was wrong in stating that the nails are always wanting on the thumbs of the hinder hands.

The Orang-Outang is entirely formed for climbing and residing in trees, as it walks with great difficulty. When ascending a tree, it seizes the trunk by all the four hands, and uses neither arms nor thighs as a Man would do in similar circumstances. It passes easily from tree to tree, when the branches touch, so that in its native tropical forests there can be but little occasion to come near the ground. In general, all its movements are slow, and seem to be executed with pain, when performed on a flat surface. It first rests the fore-hands firmly on the ground, raises itself on its long arms, and throws the body forwards by passing the hinder-hands between the arms, and carrying them beyond the fore-hands; then, resting on its hinder-hands, it advances the upper part of the body, supports itself again upon the wrists of the fore-arm, and repeats the same movements. It is only when supported by its fore-hand that it ventures to walk on the hinder-hands; sometimes it rests upon the palms, but most usually upon their external margins only, as though it wished to preserve the fingers from all contact with the ground. Whenever it ventures to rest upon the entire palm, it holds the last two phalanges of the fingers curved up, excepting the thumb, which remains open and at a distance. When resting, it sits down on the thighs, with the legs crossed, according to the Oriental custom. It reposes indiscriminately on the back or the sides, drawing up the legs towards the body and crossing the arms on the breast. When about to sleep, it is fond of being well covered, and for this purpose makes use of every kind of clothing placed within its reach.

When young, such only as it has hitherto been seen in Europe, it is rather a gentle animal, easily tamed, and readily becoming attached. From its conformation it is capable of imitating a great number of our actions; but its intelligence does not appear to hold so high a rank as has been reported, or even much to surpass that of the Dog.

The Orang-Outang makes use of its fore-hands in the same manner in general as we do, and it seems only to require experience to be able to do so in nearly all instances. It often raises its food to the mouth by means of its fingers, but sometimes also seizes them with its long lips, and drinks by sucking up, as all animals with long lips usually do. It makes use of the sense of smell to distinguish the nature of its food, and seems to trust greatly to this sense on all occasions. Fruit, pulse, eggs, milk, and meas, are eaten by the Orang-Outang indiscriminately. It is very fond of bread, coffee, and oranges, and on one occasion swallowed the contents of an inkstand without experiencing any injury. It seems to have no stated hour for its meals, but, like a child, is ready for its food at all times. Both the sense of sight and that of hearing are very acute. Music produces no effect upon this animal.

When molested, the Orang-Outang strikes its opponent with the hand, and attempts to bite; but these actions appear in general to proceed rather from impatience than ill-nature, as it is commonly mild, affectionate, and very fond of society. It delights in being caressed; gives real kisses to the object of its attachment; and seems fond of sucking the fingers of the persons who approach, yet it never sucks its own. Whenever it is very anxious for any thing, its cry is sharp and guttural. Then all its signs are very expressive: it inclines the head forward to show its disapprobation; pouts when its wants are not immediately satisfied; and when in a passion, cries out very loudly, rolling itself on the ground, and its neck, at the same time, swelling out in a singular manner.

The history of the Orang-Outang has been rendered very obscure by the earlier authors, from its having been confounded with the remainder of the larger Apes, and especially with the Chimpansee. After submitting their writings to a severe criticism, it has been found that the Orang-Outang inhabits only the most eastern portions of the Old World—such as Malacca and Cochin-China, and especially the great island of Borneo, whence it has been transported, though rarely, to Java.

There is so strong a resemblance between the Orang-Outang and an Ape of Borneo, which is yet only known by its skeleton, and by the name of Pongo,[1] in the proportions of all its parts and the disposition of the fossæ and sutures of its skull, that we can

---

[1] Auseb. Singes, pl. anat. II.—This name of *Pongo*, corrupted from that of *Boggo*, which is given in Africa to the Chimpansee and Mandrill, was applied by Buffon to a pretended large species of Orang-Outang, which was nothing more than the product of his own imagination. Wurmb, a naturalist of Batavia, applied it to the animal described by him for the first time, of which Buffon had not the smallest idea (See the Memoirs of the Batavian Society, tom. ii. p. 245). The idea that it might be nothing more than the adult Orang-Outang occurred to me on seeing the head of a common Orang, with its muzzle projecting much more forward than that of the young individuals hitherto described, and I made my views public in a Memoir read at the *Académie des Sciences* in 1818.—MM. Tilesius and Rudolphi appear also to have made the same conjecture. See the Memoirs of the Academy of Berlin for 1824, p. 131.—*Note of the Baron Cuvier.*

scarcely help believing it to be the adult, either of the Orang-Outang, or at least of some species nearly allied to it; although the great projection of its muzzle, the smallness of its cranium, and the height of the branches of its lower jaw, might perhaps lead to a different conclusion. The length of its arms, and of the apophyses of its cervical vertebræ, with the swelling of the bone of its heel, are favourable to the upright position and the facility of walking on two feet. This is the largest of all the Apes, and a most formidable animal, approaching to Man in height.

Mr J. Harwood (Trans. Linn. xv. p. 471) describes the feet of an Orang-Outang, fifteen inches in length, which dimensions appear to announce a very considerable height. He would have inferred that the Pongo was the adult Orang-Outang, had not the skeleton of the Pongo in the Royal College of Surgeons, London, exhibited one lumbar vertebra more than the skeletons of the Orang-Outang. This objection appears, however, to have no real weight, as the same variation has been observed more than once in the human species.

The Pongo, or adult Orang-Outang, is very rare in Borneo, where it bears the character of having great strength and ferocity. The only specimen hitherto obtained defended itself vigorously with large branches of trees; so that it became impossible to take it alive.

On the skull of the Pongo there is a singular ridge of bone, passing from the occiput to the vertex, and there dividing into two branches, extending towards the external sides of the orbits. Two other lateral crests divide the occiput into equal portions, and reach as far as the auricular fossæ.

The excessive length of its arms show, that when the adult stands on all the four hands, its body must assume a diagonal position, nearly approaching to the perpendicular. In this attitude, the enormous projection of its muzzle requires a considerable muscular power to sustain its weight, and it is doubtless for this purpose that the skeleton possesses those enormous cervical apophyses, whose length is not equalled in any known Mammiferous animal. The height of the skeleton in the Paris Museum is 4 French feet, or about 4 ft. 3 in. English, from the top of the head to the palms of the hinder-hands. Many scattered notices of this animal, in its young state, are interspersed among the British and Foreign Journals.[1]

GENUS II. TROGLODYTES.—PYGMIES.

Syn.   CHIMPANSÉS.—Cuv. Reg. Anim. I. 89.
  TROGLODYTES.—Geoff. Ann. Mus.—Desm. Mam.
  SIMIA (in part).—Linn. Gmel.—Illig. Prodr.—Fisch. Syn. Mam.—
    Temm. Mon. Mam.

The arms of the other Orangs [the Chimpansés of the Baron Cuvier] reach only as low as the knees. These animals have no forehead, and their cranium curves backwards immediately from the ridge of the eye-brows.

Like the preceding, this genus comprises only a single species.

I. TROGLODYTES NIGER.—CHIMPANSEE.[2]

Syn.   LE CHIMPANSÉ.—Cuv. Reg. Anim. I. 89.
  ORANG-OUTANG, HOMO SYLVESTRIS, or PYGMY.—Tyson,[3] Anat. Pyg.
  PONGO.—Buff. Hist. Nat. Suppl. VII.
  DER SCHIMPANSEE.—Voigt, Thierr. I. 76.—Schinz, Thierr. I. 99.
  SIMIA TROGLODYTES.—Linn. Gmel. I. 26.—Blumenb. Handb. et Ab-
    bild.—F. Cuv. in Dict. des Sc. Nat. XXXVI. p. 285.—Kuhl, Beitr.
    —Fisch. Syn. Mam.
  TROGLODYTES NIGER (Troglodyte Chimpansé).—Geoff. Ann. Mus. XIX.
    87.—Desm. Mam. 49.—Less. Mam. 29.—Isid. Geoff. in Bélang.
    Voy. 21.
  SIMIA PYGMÆUS and S. SATYRUS.—Schreb. Säügth.

TROGLODYTES LEUCOPRYMNUS.—Less.[4] Illustr. Zool.
Icon.   LE PONGO.—Audeb. Sing.
  Tyson, Anat. Pyg. pl. 1.
  Less. Illustr. Zool. pl. 32 (var. à Coccix blanc).
  BLACK ORANG OF AFRICA.—Wils. Illustr. Zool. pl. 5, fig. 2.

SPECIFIC CHARACTERS.

THE MUZZLE short.   THE FOREHEAD very low.   THE SUPERCILIARY RIDGES prominent.

THE EXTERNAL EARS very large, but of human form.   THE NOSE flat.

THE PECTORAL LIMBS reaching down to the knees.

THE HANDS broad, pentadactylous.   THE FINGERS of medium length. THE NAILS very flat, as in Man.

THE TAIL and CHEEK-POUCHES wanting.

THE CALLOSITIES slightly developed.

THE HAIR black, long on the back, and scanty elsewhere.

INHABITS the coasts of Angola and some other parts of Africa.

The Chimpansee is covered with black or brown hair, scanty in front. If we may credit the reports of travellers, it approaches or surpasses the stature of Man; but we have as yet seen no specimen in Europe which would indicate so great a size.

It inhabits Guinea and Congo, lives in troops, constructs huts of boughs, arms itself with stones and clubs, using them in repelling Mén and Elephants from their dwellings; pursues the Negresses [most probably a fable], and sometimes carries them off to the woods, &c.

Naturalists have long been in the habit of confounding this species with the Orang-Outang. When domesticated, it is sufficiently docile to be dressed, to walk, to sit, and to eat according to our manner.

This animal, like the preceding, is chiefly organized for climbing trees. Owing to the great strength of the four fingers of its pectoral limbs, it can swing upon them for hours without inconvenience. It walks with difficulty on all-fours, clenching the fingers, and resting upon the knuckles, so as to avoid placing the palms upon the ground. It very rarely assumes the erect attitude, though it can run nimbly on the hinder-limbs for a short distance. During this movement, it assists the equilibrium of the body by placing the fore-hands upon the thighs.

The hair is usually black, upon a skin of a light yellow. Occasionally, a few scattered white hairs appear in various parts of the body, especially near the uropygium, sometimes forming a patch upon the buttocks.[5] On the back of the thighs and on the fore-arms, the points of the hair are directed upwards, while they point downwards in every other part of the body where they happen to be present. There is no hair on the palms of all the hands, and the abdomen is almost naked.

The canines in all the young specimens hitherto examined scarcely project beyond the line of the other teeth, to which they are continuous and approximated, as in Man. The dentition of the adult is unknown.

For a long time it has been supposed, that callosities did not exist in the Chimpansee, yet they have lately been detected in a rudimentary state by M. Isidore Geoffroy-St-Hilaire, and thus it may be remarked with truth, that no Ape of the Old Continent, excepting the Orang-Outang, is. wholly destitute of callosities.[6]

The Chimpansee, when young, and residing in its native regions, is active and cheerful, but soon grows languid and dull, on being transported to our ungenial climate. Here it delights in warm clothing, rolling itself carefully in a blanket on retiring to rest. Its cry presents little variation; sometimes it emits a kind of howl or loud barking noise, when irritated; at other times, it cries like a petted child; or utters a sound like *hem*, pronounced in a grave tone, especially on being presented with sweetmeats. The habits of a female specimen, bought, by Captain Payne, from a native trader from the banks of the Gaboon, are thus described by Dr Traill.

[1] See, in particular, RICHARD OWEN, On the Osteology of the Chimpanzee and Orang-Utan, in the Transactions of the Zoological Society of London, vol. I.;—J. HARWOOD, An Account of a Pair of Hinder-Hands of an Orang-Outang, in the Transactions of the Linnæan Society. XV.;—JEFFRIES' Account of the Dissection of a Simia Satyrus, in the Philosophical Magazine, LVII.;—FRED. CUVIER, Description d'un Orang-Outang, in the Annales du Muséum, XVI.;—TIEDEMANN, Das Gehirn des Ourang-Outangs, in the Zeitschrift für Physiologie, II.;—and RUDOLPHI, Ueber den Orang-Utang und Beweis dass derselbe ein junger Pongo sey, in the Abhandlungen der Königl. Preussischen Akademie der Wissenschaften zu Berlin, for 1824. Also, the Mémoire sur les Orangs-Outangs, by MM. Cuvier and Geoffroy-St-Hilaire, published in the Magasin Encyclopédique, III., wherein the genus *Orang* was first proposed.

[2] This is the *Quojas Morou* or *Angola Satyr* of Tulpius, who gives a bad figure of it (Obs. Med. p. 271), which is represented much better by Tyson (Anat. of a Pygmy, pl. 1), copied by Schreber, pl. 1, B. Scotin has given a tolerable figure, copied in Amœn. Acad. VI. pl. 1, fig. 3, and in Schreb. I. C. A specimen kept by Buffon, preserved in the Museum, is represented rather indifferently in the Hist. Nat. XIV. 1, under the name of Jocko. The same specimen is figured much better in Lecat (Traité du Mouvement Musc. pl. 1. fig. 1) under the name of Quimpesé. It is the same given by Audebert, under the name of Pongo, but after the stuffed specimen merely.—*Note of the Baron Cuvier.*

[3] TYSON, ANAT. PYG.—The Anatomy of a Pygmy, by Dr Edward Tyson, London. 1699.

[4] LESS. ILLUSTR. ZOOL.—Illustrations de Zoologie, ou Recueil de Figures d'Animaux, par R. P. Lesson. Paris, 1831.

[5] One of these traits of albinism, more than usually developed, led M. Lesson to raise his *Troglodytes leucoprymnus* to the rank of a distinct species.

[6] The Baron Cuvier places the Orang-Outang and the Chimpansee together in the genus *Orang*, characterised chiefly by the absence of callosities. This genus must now be suppressed. See the Dict. Class. d'Hist. Nat. XV. 447.—M. J. B. Fischer confirms the preceding observation, by remarking, " Nates etiam in hac specie esse callosas, nonnullæ observatæ sunt. Nates etiam in hac specie esse callosas, neperrime innotuit."

38

The Chimpansee " ate readily every sort of vegetable food; but at first did not appear to relish flesh, though it seemed to take pleasure in sucking the leg-bone of a fowl. At that time it did not relish wine, but afterwards seemed to like it, though it never could endure ardents spirits. It once stole a bottle of wine, which it uncorked with its teeth, and began to drink. It shewed a predilection for coffee, and was immoderately fond of sweet articles of food. It learned to feed itself with a spoon, to drink out of a glass, and shewed a general disposition to imitate the actions of men. It was attracted by bright metals, seemed to take pride in clothing, and often put a cocked hat on its head. It was dirty in its habits, and never was known to wash itself. It was afraid of fire-arms; and, on the whole, appeared a timid animal."[1]

GENUS III. HYLOBATES.—GIBBONS.

Syn. Les Gibbons.—Cuv. Reg. Anim. I. 90.
  Hylobates (ὕλη, *hyle,* wood, βατις, *bates,* wandering).—Illig. Prodr. 67.—Kuhl, Beitr.—Temm. Mon. Mam.—Isid. Geoff. in Bélang. Voy.
  Simia (in part).—Linn. Gmel.—Ertl.—Fisch. Syn. Mam.
  Pithecus (Orang) in part.—Geoff. An. Mus. XIX.—Desm. Mam.

GENERIC CHARACTERS.

The Muzzle short. The Superciliary Ridges prominent. The Ears medium size.

The Pectoral Limbs excessively long, reaching to the hinder-hands.
The Hands pentadactylous. The Fingers long and narrow.
The Nails of the thumbs flat; the remainder convex and semi-cylindrical.
The Cheek-pouches and Tail wanting.
The Callosities always present, and more or less prominent.
The Hair very dense.
Inhabit Sumatra, Java, the Sonda, and Molucca Islands, Borneo, and some parts of Hindoostan.

The Gibbons must be distinguished from the species already described. They have the long arms of the Orang-Outang, and the low forehead of the Chimpansee, and the callosities of the Guenons; but they differ from the latter in being destitute of a tail or cheek-pouches. They are found only in the most secluded parts of India and the Eastern Archipelago.

This genus contains five species.[2] Two have long been known, but specimens of the remainder have only very recently been sent to Europe for the first time, by MM. Diard and Duvaucel.

The Gibbons compose a most natural and well-defined group in the order of Quadrumanous animals. They resemble the preceding genera in having no tail or cheek-pouches, while they preserve the rudimentary forehead of the Chimpansee; yet the existence of well-defined callosities exhibits a descent in the scale of organized being, connecting them more nearly with the lower genera of Apes. The most striking peculiarity in the countenances of the Gibbons arises from their having the chanfrin, or ridge of the nose, concave, which in the Chimpansee is convex and very prominent. They are covered by a dense coat of hair, having the same direction on the fore-arm as in the succeeding genera.

The dentition of the Gibbons has been examined in three species (2, 4, and 5). In the upper jaw, the first incisor terminates in a straight line, slanting obliquely inwards; the second, smaller than the first, slopes towards the canine. The latter is very long, greater in breadth than in thickness, trenchant on its external margin, having two longitudinal furrows on its internal surface, separated by a projecting crest, the hinder furrow broader and deeper than the anterior one. The second false molar is larger than the first, both with two blunt tubercles, the one on the internal margin being smaller than that on the external. The three true molars go on increasing in size from the first to the third. Each is composed of four tubercles, two of equal size on the external margin, and two on the internal, the hinder tubercle being much smaller than the one before it. These tubercles are formed by furrows dividing the tooth into unequal portions. In the lower jaw, the first incisor is small, terminated by a straight line; the second is rounded on its internal surface, and terminates in a point. The canine is more square than in the upper jaw, and terminates behind by a heel; its internal surface having the two furrows and crest as in the opposite jaw. The first false molar placed obliquely has only a single point; the second has two, one within, the other without. Three real molars follow, increasing in size. Their surfaces have five tubercles, two in front, and three behind, arranged triangularly. Here we find the first instance of this kind of molar.

1. HYLOBATES ALBIMANUS.—WHITE-HANDED GIBBON.

Syn. Le Gibbon noir.—Cuv. Reg. Anim. I. 90.
  Le Gibbon aux mains blanches.—Geoff. Cours.[3] Le;. 7, p. 33.
  Simia albimana.—Vigors and Horsfield in Zool. Jour. XIII. 107.
  Hylobates albimanus.—Isid. Geoff. in Bélang. Voy. p. 29.
  Simia Lar.—Linn. Gmel. I. 27.—Fisch. Syn. Mam. Suppl. p. 534.
  Simia longimana.—Schreb. Säügth.
  Long-armed Ape.—Penn. Quadr. 99.—Shaw, Gen. Zool. I. 12.
  Gibbon.—Ham. Smith, Syn.
  Common Gibbon and White-handed Gibbon.—Jard. Syn.
Icon. Le Gibbon.—Audeb. Sing.
  Le Grand Gibbon.—Buff. Hist. Nat. XIV. pl. 2.

SPECIFIC CHARACTERS.

Hair whitish on the hands; a whitish circle round the face; black or dusky brown elsewhere.
Callosities small.
Inhabits the East Indies.

This animal is covered with thick dark hair, and its face is surrounded with a whitish circle.

But its more important distinction from the other Gibbons consists in the whitish patches of hair upon the backs of all the hands, from which its specific name is derived.

The arms of this Gibbon are so long, that, when seated, it can place the elbows upon the ground, and, resting its head between its hands, goes quietly to sleep. It always moves in an erect posture, even when resting on all the four hands, as the arm is nearly as long as the body and legs taken together. Its eyes are large but sunken; its ears naked; its face flat, of a deep tan-colour, and rather resembling that of Man.

The manners of this Gibbon are mild and quiet. Its movements are not precipitate like those of the lower Apes, and it eats with gentleness its food, consisting of bread, fruits, or nuts. It seems very delicate, avoiding cold and moisture, and does not long survive an absence from its native country. The height of the adult is between three and four feet. It is found more particularly on the Coromandel Coast, Malacca, and the Molucca Islands.

2. HYLOBATES VARIEGATUS.—VARIED GIBBON.

Syn. Le Gibbon brun.—Cuv. Reg. Anim. I. 90.
  Le Wouwou (Hyiobates agilis).—F. Cuv. Dict. des Sc. Nat. XXXVI. 288.
  Pithecus variegatus.—Geoff. Ann. Mus. XIX. 88.—Desm. Mam. 51.
  Simia variegata.—Fisch. Syn. Mam. 11.
  Hylobates variegatus.—Kuhl, Beitr. 6.
  Little Gibbon and Active Gibbon.—Ham. Smith, Syn., and Jard. Syn.
Icon. Vouvou. Mâle, fem., et petit.—Geoff. et F. Cuv. Hist. Mam.
  Petit Gibbon.—Buff. Hist. Nat. XIV. pl. 3.

SPECIFIC CHARACTERS.
THE ADULT MALE.

Hair on the back, loins, thighs, and hinder part of the head, yellow or clear brown; round the face greyish white; elsewhere brown.
Callosities small.

THE ADULT FEMALE.

Hair whitish only on the eyebrows; otherwise resembling the male.

THE YOUNG.

Hair of a uniform clear yellow.
Inhabits Sumatra.

Their agility is extreme; they live in pairs; and the Malayan name Wouwou is derived from their cry.

The height of the Varied Gibbon, when erect, is about three feet. Its hair is of the same nature throughout, dense and apparently woolly. Its thighs, which are much shorter than the arms, are very much turned outwards. The fingers of the hinder-hands are short, the thumbs long and capable of bending backwards; in the fore-hands, the fingers are long and the thumb very short. The nose has this peculiarity, that the nostrils open upon the sides; so as almost to form an exception to the character of the tribe Catarrhina.

The manner in which the colours of these Gibbons vary with the sex and age of the individual, makes a minute description of their tinus useless, if not impossible. They live rather in isolated couples than in fami-

---

[1] Observations on the Anatomy of the Orang-Outang (Chimpansee), by Dr Thomas Stewart Traill, in the Memoirs of the Wernerian Natural History Society, vol. III. Edinburgh, 1817-20.
[2] The Baron Cuvier (Reg. Anim.) admits only four species, but he confounds the *Ounko* of Fred. Cuvier with the *Gibbon noir*, Temminck (Mon. Mam.) admits four as being well known, and one as doubtful. Desmarest (Mam. et Suppl.) describes five species, of which the *H. variegatus* and *H. agilis* are identical. Fischer (Syn. Mam.) and Isid. Geoff. (Bélang. Voy.) enumerate five. Sir William Jardine (Naturalist's Library and Syn.) describes *seven* species, two of which are purely nominal.
[3] Geoff. Couns.—Cours de l'Histoire Naturelle des Mammifères, par M. Geoffroy-St-Hilaire. Paris, 1829.

lies, and are the rarest of all the Sumatra Gibbons. Their agility is almost that of a bird, and perceiving the approach of danger at an immense distance, they immediately wake flight. Climbing rapidly to the tops of the trees, the Varied Gibbon seizes the most flexible branch, and balancing itself two or three times before making its spring, clears a distance of fourteen or fifteen yards, several times in succession, without showing any signs of fatigue. When in confinement, though still active, it exhibits no signs of this extraordinary muscular power. Its abilities are not very considerable, yet it seems susceptible of some slight education, is inquisitive, familiar, sometimes gay, but always greedy. Its forehead is very low, and its larynx destitute of any membraneous sac.

3. HYLOBATES LEUCISCUS.—ASH-GREY GIBBON.

Syn. Le Gibbon cendré.—Cuv. Reg. Anim. I. 90.
Pithecus leuciscus.—Geoff, Ann. Mus. XIX.—Desm. Mam.
Simia leucisca.—Fisch. Syn. Mam.
Hylobates leuciscus.—Kuhl, Beitr. 6.—F. Cuv. Dict. des Sc. Nat. XXXVI. 289.—Geoff. Cours, Le . 7.
Die aschgraue Gibbon.—Voigt, Thierr. I. 77.
White Gibbon.—Shaw, Gen. Zool. I. 12.
The Wow-wow.!—Ham. Smith, Syn.—Jard. Syn.
Icon. Le Moloch.—Audeb. Sing.
Simia leucisca.—Schreb. Saügth. pl. 3, B.

SPECIFIC CHARACTERS.

Hair soft and woolly, of a uniform ash-grey; the face black, or dark grey; the circle round the face clear grey.
Callosities very large.
Inhabits the Molucca and Sonda Islands.

This species lives among the reeds, and climbs the highest stems of the bamboo, balancing itself on them by means of its long arms. This is also called Wouwou by the natives.

The black face of the Ash-Grey Gibbon contrasts forcibly with the colour of the hair on the rest of the body. Its height is rather more than three feet. The habits of this Gibbon are little known, our knowledge of it resting merely upon two specimens in the Paris Museum, and a few observations made by Camper upon the living animal.

4. HYLOBATES RAFFLESII.—RAFFLES' GIBBON.

Syn. Le Gibbon Ounko, Hylobates Rafflei.—Geoff. Cours. Leç. 7.
Simia Rafflesii.—Fisch. Syn. Mam. Suppl.
Pithecus Lar.—Geoff. Ann. Mus. XIX.
Simia Lar Ungka-etam.—Raffles in the Linn. Trans. XIII. 242.
Simia concolor.—Fisch. Syn. Mam.
Hylobates Rafflei.—Isid. Geoff. in Bélang. Voy.
Simia Hoolock.—Harlan in the Transactions of the American Philosophical Society.
The Hoolock (H. Hoolock).—Jard. Syn.
Icon. Ounko, mâle et fem.—Geoff. et F. Cuv. Hist. Mam.
Simia concolor.—Harlan, Journ. Acad. Nat. Sc. Philad. V. pl. 9.

SPECIFIC CHARACTERS.

THE MALE.

Hair black, changing to brown according to the angle at which the light is reflected, eyebrows white, and cheeks grey.
Callosities small.

THE FEMALE.

Hair of the eyebrows clear grey, and cheeks black, elsewhere resembling the male.

Inhabits Sumatra, and the territory of Assam in British India.

This animal, called *Ungka-etam* by the Malays, is smaller than the Varied Gibbon, which it resembles in most other respects, excepting colour. It is confounded with the White-Handed Gibbon by the Baron Cuvier, as well as by its first describer, Sir Thomas' Stamford Raffles.[2] It has been stated that the females of this species have the fingers united as in the Syndactylous Siamang ; such, however, is not the case, as we are assured by M. Isidore Geoffroy-St.-Hilaire.

To this species we must assign the *Simia concolor* and *Simia Hoolock* of Dr Harlan, which some Naturalists would consider as distinct species.[3]

Dr Burrough thus describes the habits of the latter animal. The Hoolocks, he observes, " walk erect, and, when placed upon a floor, or in an open field, balance themselves very prettily, by raising their arms over their head, and slightly bending their arm at the wrist and elbow, and then run tolerably fast, rocking from side to side ; and if urged to greater speed, they let fall their hands to the ground, assist themselves forward, rather jumping than running, still keeping the body, however, nearly erect. If they succeed in making their way to a grove of trees, then they swing with such astonishing rapidity from branch to branch, and from tree to tree, that they are soon lost in the jungle or forest." To these particulars he adds among others, that the principal food of the animal was the banana, that it was fond of spiders and flies, but disliked flesh ; that its temper was mild, and its cry loud and shrill, consisting of whoo-whoo-whoo, repeated for five or ten minutes without intermission.

5. HYLOBATES SYNDACTYLUS.—SYNDACTYLOUS GIBBON.

Syn. Le Siamang.—Cuv. Reg. Anim. I. 90.
Pithecus syndactylus.—Desm. Mam. Suppl. 53].
Simia syndactyla.—Raffles in the Linn. Trans. XIII.—Fisch. Syn. Mam.
Hylobates syndactylus.—Isid. Geoff. in Bélang. Voy.
The Siamang.—Ham. Smith, Syn.—Jard. Syn.—Stark, Elem.[4]
Icon. Siamang.—F. Cuv. et Geoff. Hist. Mam.
Simia syndactyla.—Horsf. Zool. Jav.[5]

SPECIFIC CHARACTERS.

THE MALE.

The Hinder Hands with the first and second phalanges of the index and middle fingers united by the integuments.
The Throat with a large naked space beneath.
Callosities small.
The Hair generally black, reddish upon the eyebrows and chin.

THE FEMALE.

The Hinder Hands with the first phalanx only of the index and middle fingers united ; otherwise resembling the male.

Inhabits Sumatra.

We must now notice the Syndactylous Gibbon, as being the most remarkable species of this genus, from its having the first and second fingers of the hinder hands united together by a narrow membrane reaching the entire length of the first phalanx [in the females, and to the end of the second phalanx in the males]. The animals of this species live together in numerous troops, which are conducted by brave and vigilant leaders, making the forests resound with their deafening cries at the rising and setting of the sun. The larynx is supplied with a membraneous sac.

These animals are very common in the forests of Sumatra. They are slow in their motions, dull and stupid ; they climb without security, and leap without agility, so that, on surprising them, they are easily captured, yet they can hear a noise at the distance of a mile from the object, when they immediately take fright and abscond. If found upon the ground apart from trees, they are sure to be taken. At first they attempt to flee, but their body being too high and heavy for their short and meagre thighs, inclines forward, and their arms performing the office of stilts, they advance forward by jerks like a lame old man, whom fear has compelled to make a great effort. However numerous the troop, a wounded Syndactylous Gibbon is deserted by his companions, except it be a young one, when the mother, who carries, or stands near her progeny, falls with it, and raising a hideous yell, throws herself with open mouth and extended arms upon the assailant. The care which the mothers bestow on their young is very remarkable, and M. Duvaucel, by means of proper precautions, has seen them carefully washing their progeny in the river, regardless of the cries of the young ones. This species often falls an easy prey to the Tiger, through the paralyzing influence of fear.

In respect to their intelligence, the Syndactylous Gibbons stand nearly the lowest of the Monkey tribe. They are almost equally insensible to kind or unkind treatment. Hatred, as well as gratitude, are alike strangers to these animated machines. Mostly in a crouching posture, rolled up

---

[1] This practice of adopting the barbarous and vernacular names of animals as specific, is certain to plunge the science into inextricable confusion. Thus the name *Wouwou* is applied by the Malays indiscriminately to the Varied and Ash-grey Gibbons, and the appellation *Ounko*, both to Raffles' and the White-handed Gibbon. Savages are not likely to care much about specific distinctions, and all such terms as Ounko, Wowwow, Siamang, Hoolock, should be suppressed, wherever their use is likely to be attended with confusion.
[2] Descriptive Catalogue of a Zoological Collection, made on account of the Hon. East India Company, in the Island of Sumatra and its vicinity, by Sir Thomas S. Raffles, in the Transactions of the Linnæan Society, vol. XIII. London, 1818-22.
[3] After an attentive examination of the characters assigned to these animals, we can find nothing which cannot be stated with equal correctness of the Gibbon of Raffles, excepting the *alleged* absence of callosities, which has probably arisen from want of care in the stuffing of the specimens.
[4] Stark, Elem.—Elements of Natural History, by John Stark. Edinburgh, 1828.
[5] Horsf. Zool. Jav.—Zoological Researches in Java and the neighbouring Islands, by Dr Thomas Horsfield. London, 1825.

in its long arms, with the head resting between its knees, the Syndactylous Gibbon utters a disagreeable cry, like that of a Turkey, and apparently without any motive. When in confinement it takes its food with
indifference, raises it to the mouth without eagerness, and allows it to be
taken away without astonishment. The manner of drinking consists in
plunging the fingers in water, and then sucking them.

The forehead is almost wholly wanting, the eyes are sunk in their
orbits, the nose is broad and flat. The nostrils, placed likewise in this
species on the sides of the nose, are very large; the mouth opens the
whole extent of the jaws, and the cheeks are buried under the projecting
cheek-bones. A large naked sac, oily, and flabby like a goitre, hangs
under the throat: the hair is glossy, soft, long, and thick, of a deep
black, except on the eyebrows and chin, where it is reddish. The thighs
being arched inwards are always bent. The guttural sac of this animal
extends and swells largely, occasioning the peculiarity of its cry. It is
rather above three and a half feet in height.

All the Apes of the Old Continent, which now remain to be described, have the liver divided into several lobes: the cœcum large,
short, and without appendage; and the hyoid bone shaped like a
buckler.

GENUS IV. CERCÓPITHECUS.[1]—GUENONS.

Syn. Les Guenons.—Cuv. Reg. Anim. I. 91.
     Cercopithecus (in part).—Briss.[2] Reg. Anim. 193.—Erxt. 22.—Illig.
         Prodr. 68.—Temm. Mon. Mam.
     Cercocebus (in part) et Cercopithecus.—Geoff. Ann. Mus. XIX.
         —Desm. Mam.
     Simia (in part).—Linn. Gmel.—Fisch. Syn. Mam.

GENERIC CHARACTERS.

The Muzzle slightly elongated. The Stomach round, and of medium
size.
The Body and Limbs slender.
The Thumbs of the anterior hands short.
The Nails of the thumbs flat, the remainder semi-cylindrical.
The Hands pentadactylous.
The Tail long.
The Callosities and Cheek-pouches always present.
The Hair plentiful.
The Last Molar of the lower jaw with four tubercles only.
Inhabit Africa.

With a muzzle projecting about 60°, the Guenons have cheek-
pouches, a tail, and callosities, while the last true molar in the lower
jaw, like ,the other two, has only four tubercles. The species are
very numerous, varying greatly in size and colour. These animals are abundant in Africa, where they live in troops, and make
great havoc in gardens and cultivated fields. They may be easily
tamed.

This genus contains fifteen species, the reality of which cannot reason.
ably be doubted.[3] Three others, noticed by writers of good authority,
are apparently referable to this genus, but the information given is as yet
insufficient fully to establish their claims.

The dentition of the Guenons has been verified upon all the species.
In the upper jaw, the first incisor is twice as broad as the second; the
latter is narrow, and does not rise to the level of the first. The canine
is very long, sharp, and trenchant on its hinder part, and a small interval separates it from the incisors. The first false molar, which touches the canine,
presents externally a conical point, and an oblique plane,
swelling in the middle, and circumscribed at its lower part by a project.
ing border. The second false molar is larger than the first, and has the
same form, except that the internal border is so much elevated, as to ap.
pear almost like a tubercle. The three real molars are nearly of equal
size, and composed of four similar tubercles, arising from a horizontal
and a transverse furrow, intersecting each other at right angles, and di.
viding the tooth into four equal parts. In the lower jaw, the first incisor,
though smaller than the corresponding tooth in the opposite jaw, is still
larger than the second incisor. It terminates in a straight line, while the
second incisor is sloped off towards the canine. The latter mentioned

tooth is not so strong as the canine of the opposite jaw, but is sharp,
rounded, and terminated at its base behind by a very prominent heel, di.
vided by a slight groove into two lobules. The first false molar presents
no conical point, but is remarkable for an inclined plane extending an.
teriorly and externally, much longer than the other, upon which the in.
ternal and flat part of the opposite canine glides by a movement exactly
similar to that exhibited by the carnassier teeth of the carnivorous animals, being in fact the same as the action of a pair of scissors. The
second false molar exhibits a conical tubercle in front, and a circular depression towards the middle of the hinder part. The three following
molars increasing gradually in size from the first to the third, exactly resemble the molars of the opposite jaw.

The Guenons, in respect to their organization, seem to hold a medium
station among all the Apes of the Old Continent. The head is tolerably
round, although the muzzle projects, and their facial angle is about 50°.
Their ears are of medium size, and similar in form to those of man.
The nose is flat, their forms light and slender, their tail and limbs elongated, but not so much so as in the Solemn Apes (*Semnopithecus*), while,
on the contrary, the thumbs of their anterior hands, though short, are
longer than those of the latter. The callosities of all the species are very
strong, and their cheek-pouches well marked. Their teeth have very
prominent tubercles, and are not worn down by detrition as in the Solemn
Apes. This arises from the circumstance that the Guenons live principally upon fruits and roots, while the Solemn Apes feed chiefly upon
leaves. The stomach of the Guenons is round and of medium size.
There is no important variation in the colours of the sexes.

These animals are evidently formed for residing on trees, which are at
once their abode and place of refuge. On being alarmed, they instantly
take flight, and leaping rapidly from bough to bough, soon disappear.
The leap is their habitual pace, for they can walk on two limbs only with
considerable difficulty, and they are equally unadapted for making any
rapid progress upon four. Hence they never willingly adopt these
paces, excepting for very short distances, or when they are not hurried.

Their cheek-pouches, which are large, serve as magazines for depositing their food. Numerous troops disperse themselves in the fields
and gardens near their native forests, pillage them of fruit, and, filling
their cheek-pouches, retreat on the slightest alarm to their inaccessible
abodes in the woods.

There is something hasty and capricious in the manners of the Guenons, which strikes an observer at the first glance. Nothing can fix their
attention for any length of time to one object. The dread of continual
torture serves to command it for a moment; and in a few rare instances
they have been known to become attached by kind treatment. Their
curiosity is very great; but, when apparently occupied in attentively examining some object, the slightest circumstance is sufficient to divert
their attention, and the object in their hands is instantly allowed to fall
to the ground. It is interesting to remark the rapidity and caprice with
which they change every moment their temper and occupations.

Although there are many species of Apes in Africa, yet it is remarked
by travellers that they do not mix promiscuously, but that each occupies
a separate district.

All the Guenons yet known are of African origin. One species is said
to come from Bengal, but this is most probably an error.

1. CERCOPITHECUS RUBER.—RED GUENON.

Syn. Le Patas.—Cuv. Reg. Anim. I. 91.
     Simia rubra.—Linn. Gmel. I. 42.—Fisch. Syn. Mam. 24.
     Cercopithecus ruber.—Geoff. Ann. Mus. XIX. 96.—Desm. Mam. 59.
     Red Monkey.—Penn. Quadr. 208.—Shaw, Gen. Zool. I. 49.

Icon. Patas mâle adulte.—Geoff. et F. Cuv. Hist. Mam.
     Le Patas à queue courte.—Audeb. Sing.
     Patas à bandeau noir femelle.—Geoff. et F. Cuv. Hist. Mam.
     Buffon, Hist. Nat. XIV. pl. 25 and 26.

SPECIFIC CHARACTERS.

The Hair of a bright yellowish-red above, white beneath, a black or a
white band over the eye.
Inhabits Senegal.

The Red Guenon is remarkable for the brilliancy of its coat, which is
of so bright a red on the upper part, that it appears as if exaggerated by
the hand of a painter. There are two varieties of this species, the one

---

[1] Cercopithecus (κερκος, *kickos*, tail, and πιθηξ. *pithēx*, an Ape), Apes with tails, a name in use among the ancient Greeks.—*Note of the Baron Cuvier.*
[2] Briss. Reg. Anim.—Le Règne Animal, divisé en IX. classes, par M. Brisson. Paris, 1756.
[3] The Baron Cuvier, in the second edition of the *Règne Animal,* enumerates only thirteen species. Two others have since been established, making the number as stated in the text. The catalogues of most other systematic writers run very wide of the mark. Temminck (Mon. Mam.) asserts that there are 19 or 20, but does not specify them. Geoffroy (Ann. Mus.) distributed the species of this genus among two genera (Cercopithecus and Cercocebus), and his example was followed by Desmarest. But the institution of the genus Semnopithecus by Fred. Cuvier occasioned the latter (Cercocebus) to be suppressed in all works of any authority, and reduced the number of species in the genus Cercopithecus from 20 or 21 to 13. However, the Catalogues of British Zoologists still exhibit the old division Cercocebus, which rests upon no real basis, in addition to the new genus Semnopithecus, thus multiplying sub-divisions without any adequate reason. Major Hamilton Smith, Sir William Jardine, and Mr Stark, have followed M. Desmarest pretty closely, merely omitting C. maurus. Of their 18 species, at least five are fictitious or referable to other genera. These nominal species are C. auratus, latibarbatus, pileatus, albocinereus, and Atys.

with a black, the other with a white band over the eyes. They are not so capricious as the other Guenons. The damage occasioned by these animals in the cultivated fields of Senegal, at the seasons when the millet and other grain become ripe, is incalculable. Forty or fifty Guenons assemble together. One Guenon mounts upon a tree as an outpost, listening and watching on every side, while the remainder are plundering. As soon as he perceives any one approaching, the sentinel cries out like an enraged person to give notice to the remainder, who start off with their booty on hearing the signal, and leap with prodigious agility from tree to tree. The females, who carry their young clinging against the abdomen, run off with the rest, with the same agility as if they had no burthen to carry. These animals do not agitate their jaws, when displeased, like the other Guenons, and they walk more frequently upon all the four hands than on two only. They are from a foot and a half to two feet in length from the point of the muzzle to the insertion of the tail; and the tail is not so long as the body and head taken together.

The body is slender. The head medium size; with the cranium slightly lengthened and flattened upon the vertex; the forehead projecting above the orbits of the eyes, and above the upper part of the nose. The face is flesh-coloured, the nose covered with short black hair; the eyes sunken; a black or white band passing over the eye, resembling a prolonged eyebrow; the hair very plentiful upon the cheeks, forming cheek-tufts; the ears naked. Its bright yellowish-red hair, which extends over the forehead, the vertex, occiput, upper part of the neck, the back, sides, crupper, sometimes only the upper part of the tail, and sometimes the whole tail, and the thigh, are not without some mixture of black and grey, proceeding from the circumstance that many of the hairs are black on the points and elsewhere grey. The red hair becomes paler upon the outside of the arm, the fore-arm and leg; while it finally tends, upon the cheeks, tip of the muzzle, neck, lower part of the neck, arm-pits, the inner surfaces of all the limbs, the breast and the belly, to a white, mixed in several points with yellow, pale-red, and grey. The hair throughout is generally rough and glossy. The nails are black, the palms of the hands brown.

Mr Bennett remarks, that a specimen in the Gardens of the Zoological Society of London was " lively and active, but somewhat irascible if disturbed or handled. It was, however, too young to be dangerous. When pleased it danced on all-fours in a peculiar and measured step, which was far from being ungraceful; although after a time it became ludicrous from its regular monotony."

2. CERCOPITHECUS ÆTHIOPS.—COLLARED MANGABEY GUENON.

Syn. LE MANGABEY À COLLIER.—Cuv. Reg. Anim. I. 91.
CERCOCEBUS ÆTHIOPS.—Geoff. Ann. Mus. XIX.
CERCOPITHECUS ÆTHIOPS.—Desm. Mam. 62.
SIMIA ÆTHIOPS.—Fisch. Syn. Mam. 23.—Linn. Gmel. I. 38.
WHITE EYELID MONKEY.—Penn. Syn. p. 114.
Icon. Mangabey à collier.—Geoff. et F. Cuv. Hist. Mam.
Le Mangabey var. A.—Audeb. Sing.
Buff. Hist. Nat. XIV. pl. 33.

SPECIFIC CHARACTERS.

THE HAIR slate-grey above, whitish beneath,[1] also on the temples, and on the back of the neck; bright chestnut-brown on the top of the head; a greyish band beneath the ear.

THE EYELIDS white.

INHABITS the Western coast of Africa.

Buffon says that this animal comes from Madagascar, while Hasselquist assigns it to Abyssinia. The fact is, as we are assured by Sonnerat, there are no Apes at all in Madagascar.

It possesses so many intimate relations to the following species, (3) that Buffon and Pennant confounded them together under the common names of Mangabey and White Eyelid Monkey. Yet their differences, slight as these may be considered by some, are found to be so invariably the same in numerous individuals, that we cannot hesitate to pronounce them distinct species. Their variations, it will be remarked, are chiefly in the colours of the head and neck.

The height of the Collared Mangabey Guenon is about a foot and a half, being rather less than that of the species next to be described. Its hair is very long and soft to the touch. The first incisor of the upper jaw being very broad, renders its grin at once obvious and peculiar. It may be readily distinguished from the other White-eyelid Monkey by the bright chestnut-brown on the upper surface of its head, and the collar of pure white crossing the fore-part of its neck, and including the large bushy cheek-tufts, which extend backwards, beneath and behind the ears.

3. CERCOPITHECUS FULIGINOSUS.—COLLARLESS MANGABEY GUENON.

Syn. LE MANGABEY SANS COLLIER.—Cuv. Reg. Anim. I. 91.
CERCOCEBUS FULIGINOSUS.—Geoff. Ann. Mus. XIX.
SIMIA FULIGINOSA.—Fisch. Syn. Mam. 24.
CERCOCEBUS CYNOMOLGUS.—Geoff. Ann. Mus. XIX.
WHITE EYELID MONKEY.—Penn. Quadr. p. 204.—Shaw. Gen. Zool. I.
Icon. LE MANGABEY.—Audeb. Sing.—Buff. Hist. Nat. XIV. pl. 32.
MANGABEY femelle.—Geoff. Hist. Mam.

SPECIFIC CHARACTERS.

THE HAIR uniform slate-grey above; whitish beneath; black on the backs of the hands.

THE EYE-LIDS white. THE EARS violet-grey.

INHABITS Congo and the Gold Coast.

Buffon thought that this animal came from Madagascar, and believed it to be a variety of the preceding.

Though one of the most common species of Guenons, it was a long time before its native country was indicated with precision. Continually in motion, it exhibits in captivity the most grotesque attitudes. The males are chiefly remarkable for their agility, and enliven their motions by a singular grin approaching to a laugh, at the same time showing their incisors, which are always very large. They are constantly in the habit of holding the tail turned forwards upon the back, and not elevated in a semicircular form, as in most of the other Guenons. The length from the muzzle to the insertion of the tail is about two feet; the height from the shoulders to the palm about 1½ feet. The females are usually more tranquil, and fonder of caresses than the males. " A l'époque du rut, c'est-à-dire chaque mois, elles éprouvent aux parties génitales un gonflement considérable, qui, près de l'anus est très-large, et qui, après s'être rétréci tout-à-coup, descend vers la vulve et l'entoure. Alors on voit paraître une véritable menstruation."

When in captivity, their docility is considerable. Audebert notices one individual which danced on the tight rope, holding a balance-pole in its hands; took up a book, placed it on a table, and turned over the leaves with much ease, making grimaces at it as though it contained some provoking intelligence. The same writer significantly remarks, " On sent que le fouet du maître jouoit ici un grand rôle."

The muzzle is thick and projecting; the circle round the eyes prominent. The face varies in colour, sometimes being of a deep flesh-coloured tint; sometimes blackish on the fore part of the muzzle, and the remainder copper-coloured. Above the eyelids there is constantly a white band in the form of a crescent, very striking; there are coarse hairs on each side of the nose, and others stiff and bristly on the lower part of the forehead just above the nose. The ears are naked, violet-coloured, without margin, and slightly folded back at their extremities. The hairs of the cheek-tufts are directed backwards, whitish, with a grey band. The hair of the entire upper part of the body as well as of the tail is a slate-grey, with a slight tinge of yellow upon the head; that of the throat, breast, belly, and the interior of the limbs, of a greyish-white. The extremities of the limbs, from the fore-arm in front, and from the heel behind, are of a deep black. The tips of the fingers are very thick, especially of the thumb; and the nails are flat.

4. CERCOPITHECUS SABÆUS.—GREEN GUENON.

Syn. LE CALLITRICHE.[2]—Cuv. Reg. Anim. I. 91.
CERCOCEBUS SABÆUS.—Geoff. Ann. Mus. XIX.
CERCOPITHECUS SABÆUS.—Desm. Mam. p. 61.
SIMIA SABÆA.—Linn. Gmel. p. 32.—Fisch. Syn. Mam. 21.
GREEN MONKEY.—Penn. Syn. 113.—Quadr. p. 203.—Shaw. Gen. Zool. I. 42
Icon. Callitriche mâle.—Geoff. et F. Cuv. Hist. Mam.
Le Callitriche.—Audeb. Sing.—Ménag. du Mus.[3]—Buff. Hist. Nat. XIV. pl. 37.

SPECIFIC CHARACTERS.

THE HAIR yellowish-green above, tending to a grey upon the limbs; whitish beneath; the cheek-tufts and the tip of the tail yellowish.

THE FACE black. THE SCROTUM greenish; surrounded with yellow hair.

INHABITS Senegal.

The Green Guenon possesses many points of resemblance to the Malbrouck (5). It is one of the most beautiful of the Monkey tribe, its hair being disposed in alternate rings of black and yellow, which, by combining,

[1] By the term *above*, we understand all the superior and exterior parts of the body, such as the shoulders, back, sides, arms, and fore-arms, thighs and legs, feet and tail; by the term *beneath*, all the inferior and interior parts of the body, such as the neck, breast, belly, and the inner surfaces of all the limbs.
[2] The name of Callithrix is assigned by Pliny (l. VIII. c. 54) to an Ape of Ethiopia, furnished with a beard and a bushy tail, probably the Ouanderou. Buffon applied it arbitrarily to the above species.—*Note of the Baron Cuvier.*
[3] MÉNAG. DU MUS.—La Ménagerie du Museum National d'Histoire Naturelle, par les Citoyens Lacépède et Cuvier. Paris.—An. X.—(1801.)

39

give it, when viewed at a certain distance, the appearance of a bright
green. This tint, so unusual among Mammiferous animals, probably induced Buffon to apply the term Callithrix (signifying *beautiful hair*) to
this animal.

We know little regarding the habits of the Green Guenons when in the
wild state, excepting the short notice of Adanson. They keep together
in troops, most frequently on trees, maintain a profound silence, and
would pass unnoticed, were it not for the branches which they break and
throw down on the passenger. In places where Man does not often
penetrate, they do not fear him; and, though some fall down mortally
wounded, the remainder do not take flight on that account.

In confinement they are not easily tamed. One individual in the Paris
Menagerie always continued ferocious. " Les femmes lui causaient une
fureur d'une autre espèce, qu'il témoignait de la manière la plus brutale."
His voice was a kind of growl, commencing in a grave and ending in an
acute tone, resembling that of the Baboons, but not so loud. F. Cuvier
describes it by the syllable *grou.* Most frequently it remained seated,
with the eyes closed. Its colour became deeper in winter, and the hair
of the breast and belly fell off in large quantities during summer, so as to
leave them almost bare.

Its body, though slender, approaches in form to that of the Malbrouck
Guenon. The head is pyramidal, the muzzle elongated; the upper part
of the orbits low in front, projecting greatly just above the nose; the
ears large, rounder than in the Malbrouck Guenon. The hair on the
upper part of the body of a yellowish-green, as already described; the
external surface of the limbs more grey, the yellow hairs having nearly
disappeared. The lower parts of the body, and the internal surfaces of
the limbs, beneath the cheek, throat, and neck, are of a yellowish white.
The colour of the back is continued on the upper side of the tail to its
extremity, which is ornamented by a long pencil of yellow hairs. The
face, ears, and hands, are black; the cheek-tufts yellowish, with the
hairs directed backwards, and dispersed so as to form a kind of ruff. The
skin of the scrotum is greenish, and surrounded with yellow hairs.

5. CERCOPITHECUS FAUNUS.—MALBROUCK GUENON.

Syn. LE MALBROUC.—Cuv. Reg. Anim. I. 92.
       CERCOPITHECUS CYNOSURUS.—Geoff. Ann. Mus. XIX.
       DOG-TAILED MONKEY.—Shaw, Gen. Zool. I. 1, p. 32.
       SIMIA FAUNUS.[1]—Linn. Gmel. I. p. 31.
       SIMIA CYNOSUROS.—Fisch. Syn. Mam. 22.
Icon. Le Callitriche var. A.—Audeb. Sing.—Buff. Hist. Nat. XIV. pl. 29,
         copied in Schreb. Säügth. pl. XIV. C.
       Simia Cynosuros.—Scopoli, Deliciæ Floræ, et Faunæ Insubr. I. pl. 19.

SPECIFIC CHARACTERS.

THE HAIR greenish-grey above, ash-coloured beneath and on the
limbs; no yellow on the tail; a white band above the eye; tufts of
white hair on each side of the face.

THE FACE black, flesh-coloured round the eyes.

THE SCROTUM bright ultramarine.

THE CALLOSITIES bright red.

INHABITS

This animal is said to inhabit the forests of Bengal. If this be true,
which has not yet been satisfactorily proved, it forms a remarkable exception to the other Guenons, which are all confined to Africa. Its agility
is extreme, but it seldom permits its voice to be heard, which is at best but
a feeble or sharp cry, or a low growl. The males when young are tolerably
docile, but become exceedingly unmanageable when arrived at the adult
period of life; the females continue mild, and seem susceptible of some
attachment. The irritability of the males, though excessive, is always
tempered by a certain degree of caution; they are fond of attacking an
enemy from behind with their teeth and nails, darting off immediately
before he can turn, but not losing sight of him so as to prepare for a new
sally. They use their hands with much address, seizing the smallest objects between the thumb and first finger, notwithstanding the shortness of
the former. The rind of fruits and roots is carefully peeled off with their
teeth, and they smell over every object before tasting it. They drink by
supping up.

Their ears are similar to ours, but without the helices. " Les mâles
paraissent toujours disposés à l'accouplement. Les femelles ont l'ouverture du vagin très simple, avec un clitoris fort petit." In general, it
would appear that these Guenons are less disposed to breed in confinement than the Apes of most other genera. The remarkable colour of the
scrotum is the most striking characteristic of the Malbrouck Guenon.

Its body, unlike the other Guenons, is strong and muscular; its head
tolerably large, and pyramidal; its muzzle projecting; and its lips very

extensible. The upper part of the body is generally of a greenish grey,
resulting from the mixture of alternate hairs with yellow and black tips;
the lower parts of the body and beneath the tail are grey from the mixture of hairs tipped with white and black. All the hairs are, however, grey at their base. The hairs on the sides of the cheeks are very
long, and directed backwards, forming very prominent cheek-tufts; the
muzzle is black, flesh-coloured around the eyes, the latter character being
more perceptible in the young than in the adults; the ears and palms of
the hands are black; the callosities very red, especially at periodical seasons; the scrotum very voluminous, of a bright ultramarine, within which
the penis is almost concealed.

6. CERCOPITHECUS ERYTHROPYGUS.[2]—VERVET GUENON.

Syn. LE VERVET.—Cuv. Reg. Anim. I. 92.
       CERCOPITHECUS PYGERYTHRÆUS.—Desm. Mam. Suppl. p. 534.
       SIMIA PYGERYTHRA.—Fisch. Syn. Mam.
       CERCOPITHECUS POUILLOU.—Desmoulin in Dict. Clas. d'Hist. Nat. VII. 568.
Icon. VERVET MÂLE.—Geoff. et F. Cuv. Hist. Mam.

SPECIFIC CHARACTERS.

THE HAIR greyish-green above, white beneath, red around the anus,
black on the point of the tail.

THE SCROTUM greenish, surrounded by white hair.

INHABITS the Cape of Good Hope.

This animal differs from the Malbrouck Guenon in having its
scrotum [greenish] surrounded with white hair, and red hair near
the anus.

It possesses many points of general resemblance to the Guenons with
greenish hair, already described (4) and (5): M. Lalande brought several specimens to Europe from the Cape of Good Hope, in the forests
of which he found them very plentiful, without meeting with any of the
others, which tends to confirm the idea that it forms a distinct species.
It is found only in the woods remote from the colony.

Its face and ears are black, flesh-coloured around the eyes, a very prominent band of white hair across the forehead, the cheek-tufts white,
the upper part of the body greyish-green, changing into a grey upon the
limbs; the scrotum of a very brilliant green; the arms surrounded with
hairs of a deep-red; all the hands black from the joints of the heels and
wrists, and the tip of the tail black.

7. CERCOPITHECUS GRISEO-VIRIDIS.—GRIVET GUENON.

Syn. LE GRIVET.—Cuv. Reg. Anim. I. 92.
       CERCOPITHECUS GRISEO-VIRIDIS.—Desm. Mam. p. 61.
       SIMIA SUBVIRIDIS.—F. Cuv. in Dict. des Sc. Nat. XX. p. 26.—Fisch.
         Syn. Mam.
Icon. Grivet mâle.—Geoff. et F. Cuv. Hist. Mam.

SPECIFIC CHARACTERS.

THE HAIR greenish above, excepting the limbs and tail, which are
grey; white beneath, also the tufts of the cheeks.

THE SCROTUM bright green, surrounded with bright yellow hair.

INHABITS Nubia.

The green scrotum surrounded with yellow hair serves to distinguish this animal from the Malbrouck Guenon.

It possesses a strong general resemblance to the other Green Guenons
in size and general proportions. According to Cailliaud, it is found in
the forests of Nubia.

The upper parts of its body are of a dingy green, resulting from annulated hairs of dark grey and bright yellow; the hairs on the thighs are
similar, but with very little yellow; the hair on the backs of the hands
marked with alternate rings of grey and white. The cheek-tufts, as well
as a band over the eyes, are white; the face, the ears, and the palms of
all the hands, of a violet-black; the circle round the eyes flesh-coloured.
There are a few scattered hairs, like bristles, on the superciliary ridge
between the eyes.

It has the savage disposition of all the larger kinds of Guenons, and
bears that strong specific affinity to the Malbrouck and Green Guenons
which seems to indicate the transition from the one form to the other.
It resembles the Malbrouck in the general colours of the hair, but differs from it in the shape of the head, which is not so round; in the colour of the scrotum, which is of a bright green instead of ultramarine; and
in that of the hairs surrounding these parts being white in the Malbrouck,
and bright yellow in the Grivet. This appellation seems wholly accidental and arbitrary, being the name to which the individual in the Paris
Menagerie used to answer when called.

There is a strong general affinity among the Green, Malbrouck, Vervet,
and Grivet Guenons, which would constitute them a distinct group.

---

[1] The *Cercopithecus barbatus* of Clusius, which Linnæus quotes as an example of his *Faunus*, is rather an *Ooanderoo* than a *Malbrouck.*—*Note of the Baron Cuvier.*
[2] Erythropygus, from ἐρυθρός, red, and πυγή, anus.

8. CERCOPITHECUS TALAPOIN.—TALAPOIN GUENON.

Syn. Le Talapoin.—Cuv. Reg. Anim. I. 92.
    Cercopithecus Talapoin.—Geoff. Ann. Mus. XIX. p. 93.—Desm.
      Mam. p. 56.
    Simia Talapoin.—Linn. Gmel. I. p. 35.—Fisch. Syn. Mam. 21.
    Talapoin Monkey.—Penn. Syn. 114.—Shaw, Gen. Zool. I. 1, p. 46.
Icon. Melarine, jeune fem.—Geoff. et F. Cuv. Hist. Mam.
    Buff. Hist. Nat. XIV. pl. 40.

SPECIFIC CHARACTERS.

The Hair greenish above, whitish beneath, the tufts of the cheeks whitish.

The Nose black, in the middle of a flesh-coloured face.

Inhabits

This animal comes probably from Africa, though its precise locality has not yet been verified. It belongs to the same group of Guenons with those now about to be described.

The head is round, and the muzzle projects but slightly; the ears are large, round, and naked; the nose, the ears, and the palms of all the hands, are black; the circle round the eyes and the tip of the lips flesh-coloured. The hair on the cheeks, temples, forehead, the top of the head, the occiput, above and on the sides of the neck, the back, loins, crupper, the sides of the breast and belly, as well as the external surfaces of the limbs and the backs of the hands, are covered with a mixture of yellowish green and black, each hair being dark-grey through a great part of its length from the root, afterwards greenish-yellow, and terminating in black. The lower jaw, the inferior surface of the neck, throat, breast, belly, armpits, and the inner surfaces of the limbs, are whitish, with some slight tinges of yellow; the tail beneath is of an ash-grey. The nails of all the thumbs are round and flat.

9. CERCOPITHECUS MONA.—VARIED GUENON.

Syn. La Mone.—Cuv. Reg. Anim. I. 92.
    Cercopithecus Mona.—Geoff. Ann. Mus. XIX. 95.—Desm. Mam. p. 58.
    Simia Mona.—Linn. Gmel. 34.—Fisch. Syn. Mam.
    Varied Ape.—Penn. Quadr. and Syn.
    Varied Monkey.—Shaw, Gen. Zool. I. 17.
Icon. La Mone.—Audeb. Sing.
    Mone mâle.—Geoff. et F. Cuv. Hist. Mam.
    Simia Mona.—Schreb. Saügth. pl. 15.
    Simia Monacha.—Ibid. pl. 15. B.
    Buff. Hist. Nat. XIV. pl. 36.

SPECIFIC CHARACTERS.

The Hair of the body brown; the limbs and tail black; the belly and inside of the arms white; a black band on the forehead; the top of the head greenish-yellow; the tufts of the cheeks straw-coloured; a white spot on each side near the insertion of the tail.

The Ears and Hands flesh-coloured.

Inhabits Africa.

This Guenon, according to Buffon, F. Cuvier, and others, is playful, gentle, and affectionate; yet an adult specimen, preserved in the Gardens of the Zoological Society of London, deserved no such good character, but showed, on the contrary, a temper as capricious and savage as any of its tribe.[1]

The name Mona appears to be of Arabic origin, and is applied by the Moors of Barbary to all Apes with long tails. This Guenon is about a foot and a half in length; and seems to thrive well in our climates.

The head is small and round; the muzzle thick and short; the eyelids, nose, and lips, naked and flesh-coloured; the intervals of the eyes blueish. The top of the head is of a bright greenish-yellow, resulting from the intermixture of hairs which are wholly black at the points, afterwards of a greenish-yellow beneath the black, and finally of an ash-colour near their roots. The back and sides are of a bright brown, speckled with black; above the limbs, thighs, and the top of the tail, of a pure slate-grey passing into black. The breast, belly, and the inner surfaces of the limbs, are of a dazzling white. The cheek-tufts are straw-coloured, mixed with black points; a black marginal band commencing from the centre of the forehead extends on each side to the ear, and is thence prolonged down the shoulders and fore-arms. Two very white spots appear on each side of the tail above the thighs. The hair surrounding the callosities is reddish; the tail is black, and arched forwards over the back; the palms of all the hands are naked and brown, the nails short, black, and flattish.

The Varied, Spotted, Moustache, Vaulting, Winking, and Diadem Guenons, constitute a group of small and agreeable Monkeys.

10. CERCOPITHECUS DIANA.—SPOTTED GUENON.

Syn. Le Rolowal.—Cuv. Reg. Anim. I. 92.
    Exquima[2].—Marcgr. Brasil, p. 227.
    Cercopithecus Diana.—Geoff. Ann. Mus XIX. 96.—Desm. Mam. p. 60.
    Simia Diana.—Linn. Gmel. I. 32.—Fisch. Syn. Mam. 19.
    Simia Rolowal—Fisch. Syn. Mam.
    Spotted or Diana Monkey and Palatine Monkey.—Shaw, Gen. Zool. I. 37 and 38.—Penn. Syn. and Quadr.
Icon. La Diane.—Audeb. Sing.
    Buff. Hist. Nat. Suppl. VII. pl. 20.

SPECIFIC CHARACTERS.

The Hair dark slate-grey, spotted with white above; white beneath; the crupper of a purplish-red; inside the thighs orange.

The Face surrounded with white. The Beard whitish, long, and scanty.

Inhabits Guinea.

Linnæus assigned the name of *Diana* to this Guenon, from the fancied resemblance of the crescent-shaped hairs ornamenting its brow to the ancient representations of that goddess. Mr Bennett describes the specimen in the Gardens of the Zoological Society of London as "one of the most graceful and good-tempered of its tribe. Like the greater number of them, its disposition is more mild and pliant in youth than after it has attained its full maturity. It is fond of being caressed, and nods and grins with peculiar expression when pleased; but, after a certain age it becomes more sedate, and seldom indulges in those antics." Gardens and Menagerie, Vol. i. p. 36.

The body is rather slender; the head elongated; the face triangular and black; the ears rather small and round; the hair on the top of the head short and black, with a border formed of stiffer hairs than the rest, in which some are pure white. The cheek-tufts rather long; a white pointed beard, about two inches in length, and scanty, appears behind a small brownish black spot at the tip of the chin. The sides of the head and neck as far as the ear, the breast, and the interior surfaces of the limbs, are white; the hinder part of the head and neck, the shoulders, sides, the external surface of the arms, and the upper parts of the thighs, are covered with dark hairs, annulated with yellowish-white, which gives them a greenish tinge. A purplish-red patch, in the form of an isosceles triangle, beginning about two-thirds the length of the back, extends down to the loins for a base; the anterior parts are black, as well as all the hinder limbs, with the exception of the fore part of the thigh, which is divided by a narrow and oblique band of white hairs reaching from the base of the tail to the knee. The insides of the thighs are orange, the circle round the callosities white.

11. CERCOPITHECUS CEPHUS.—MOUSTACHE GUENON.

Syn. Le Moustac.—Cuv. Reg. Anim. I. 92.
    Cercopithecus Cephus.—Geoff. Ann. Mus. XIX. 20.—Desm. Mam. 57.
    Simia Cephus.—Fisch. Syn. Mam.
    Moustache Monkey.—Penn. Syn. and Quadr.-Shaw, Gen. Zool. I. 1, 44.
Icon. Le Moustac.—Audeb. Sing.
    Moustac mâle.—Geoff. et F. Cuv. Hist. Mam.
    Buff. Hist. Nat. XIV. pl. 39.

SPECIFIC CHARACTERS.

The Hair greenish-brown above, greenish-grey upon the limbs, with a slight tinge of yellow; dark-grey beneath.

The Face blueish-black, tending to black near the lips.

The Tail grey near the insertion; elsewhere orange-red.

The Upper Lip with a white moustachio.

Inhabits Guinea.

The habits of this species, so remarkable for the singular ornament on the upper lip, are not far different from others of its congeners. It is mild in captivity; and specimens vary greatly in size.

The body is rather slender; the head round; the muzzle slightly elongated; the nose projecting at its origin between the eyes; the face of a blueish black; the upper lip with a white moustachio; the circle of the mouth covered with black hairs; the upper part of the head and body, and the external surface of the limbs, are of a brown, speckled with green, resulting from the manner in which the hairs of these two colours are interspersed; a white spot before each ear, and near the eyes; the hair greyish-brown at the base of the fore-limbs, darkening towards their extremities; the hinder hands not so deep as the others; beneath the chin a dirty white, blending into the dark grey beneath the belly; the internal surface of the arms and thighs of a uniform grey; the tail brown at its base, blending into an orange-red, which is the colour of its latter portion.

---

[1] The Gardens and Menagerie of the Zoological Society Delineated. By E. T. Bennett. London, 1835, vol. i. p. 40.
[2] The figure in Marcgravius connected with the description of the *Exquima* is that of an Ouarine Howler, while the figure of the Exquima refers to the description of the Ouarine or Guariba Howler. This transposition has occasioned many errors in their synonyms.—*Note of the Baron Cuvier.*

12. CERCOPITHECUS PETAURISTA.—VAULTING GUENON.

Syn. L'ASCAGNE.—Cuv. Reg. Anim. I. 93.
　　CERCOPITHECUS PETAURISTA.—Geoff. Ann. Mus. XIX. 95.—Desm. Mam. p. 59.
　　SIMIA PETAURISTA.—Linn. Gmel. I. 35.—Fisch. Syn. Mam. 18.
　　VAULTING MONKEY.—Shaw, Gen. Zool. I. 1, 51.
Icon. Ascagne, femelle.—Geoff. et F. Cuv. Hist. Mam.
　　LE BLANC-NEZ.—Audeb. Sing.—Ménag. du Mus.
　　L'Ascagne.—Audeb. Sing. (var.)

SPECIFIC CHARACTERS.

THE HAIR greyish or greenish-brown above; grey beneath; the upper part of the nose black; the lower white; a white tuft before each ear.
THE FACE covered with black hair; sometimes naked and violet blue; always flat.
THE UPPER LIP with a black moustachio.
INHABITS the Coast of Guinea.

This interesting little animal is one of the smallest and most docile species of the genus. It is at once recognised by the white patch on its nose, consisting of smooth, short, and closely set hairs. It is lively, and active in its manners, and generally good-tempered in its disposition. A specimen in the Gardens of the Zoological Society of London was by no means familiar, appearing particularly anxious to conceal its face, crying out and kicking with all its might when handled for the purpose of examination.

The head is round; the ears large; the hair of the forehead and cheeks rather short. The top of the head, the upper part of the neck, the back, the sides, and the external surfaces of the limbs, are covered with hair of a dark brown, sometimes with yellow and grey intermixed; the lips naked and brown, covered with scattered hairs; the root of the nose between the eyes black; a white spot upon the nose, formed of very thick and short hairs, cut off horizontally on its upper part, and bordered beneath by the nostrils. The under part of the neck and the sides of the head are white, sometimes singed with yellowish, which is prolonged towards the breast and belly, where a reddish tinge sometimes prevails. The inner surfaces of the limbs are of a whitish-grey. The hinder limbs and hands are more grey than those before, and without any greenish tinge. The tail is of a dirty white beneath, separated by a line from the greenish-brown of the upper part; the anterior hands are black. Its usual length is from ten to twelve inches, and that of the tail is usually 15 to 18 inches, though it commonly wants some of its vertebræ.

There is a variety of this species with the face naked and of a blueish-violet, figured by Audebert under the name of L'Ascagne.

13. CERCOPITHECUS NICTITANS.—WINKING GUENON.

Syn. LE HOCHEUR.—Cuv. Reg. Anim. I. 93.
　　CERCOPITHECUS NICTITANS.—Geoff. Ann. Mus. XIX. 95.-Desm. Mam. 58.
　　SIMIA NICTITANS.—Linn. Gmel. I. 33.—Fisch. Syn. Mam. 18.
　　WINKING MONKEY.—Penn. Quadr. and Shaw, Gen. Zool.
Icon. Le Hocheur.—Audeb. Sing.
　　Hocheur.—Geoff. et F. Cuv. Hist. Mam.
　　GUENON À NEZ BLANC PRO-EMINENT.—Buff. Hist. Nat. Suppl. VII. pl. 18.

SPECIFIC CHARACTERS.

THE HAIR black or brown, speckled with white.
THE NOSE prominent and white, in the middle of a blueish-black face.
INHABITS Guinea.

In respect to its form, proportions, habits, and disposition, this Guenon approximates nearly to the Spotted Guenon (10) already described.
The Winking Guenon is about eighteen inches in length, and the tail twenty-eight. The naked parts of the face and ears are blueish-black, the eyelids flesh-coloured, the hands entirely black, and the skin of the body white, slightly tinged with black. On the naked parts of the face only a few isolated dark hairs are to be seen. But the nose is entirely covered with short and thick hair, black between the eyes, and of a fine white throughout the rest of its length. The head and upper parts of the body, as well as the cheek-tufts, are black, speckled with yellowish hairs. The sides are black, speckled with white; the breast brownish-white. The neck, limbs, and tail, are entirely black, and a black line separates the cheek-tufts from the rest of the face. On the lower jaw, on the inner surface of the thighs, and under the arms, the hair is grey. These colours are formed for the most part by hairs which are grey at their base, and annulated with black and yellow, or black and white through the remainder of their length.

These last five species (9, 10, 11, 12, and 13) are small, and prettily varied in their colours. They are very mild in their manners, and are common in Guinea.

14. CERCOPITHECUS DIADEMATUS.—DIADEM GUENON.

Syn. LE GUENON À DIADÈME.—Isid. Geoff. in Belang. Voy.
　　SIMIA LEUCAMPYX.—Fisch. Syn. Mam.
　　SIMIA DIANA.—Desmoulins in Dict. Clas. d'Hist. Nat. VII. 565.
Icon. Diane femelle.—Geoff. et F. Cuv. Hist. Mam.

SPECIFIC CHARACTERS.

THE HAIR of the body and cheeks greenish-grey, speckled with black; a whitish band like a crescent across the forehead; the tail speckled with white; elsewhere black.
INHABITS the Western Coast of Africa.

This Guenon was figured and described by M. Fred. Cuvier as a variety of the C. Diana. It is, however, very different from the Spotted Guenon (10), as may be readily perceived on comparing their descriptions.

A specimen of this animal lived for many years at the Paris Menagerie. On its first arrival, the upper part of the neck, shoulders, arms, fore-arms, neck, breast, belly, and tail, were uniformly black, but not of so deep a tint as in the lower parts of the body; the back and sides were speckled with black and white, the hairs having small alternate rings of black and white. The cheek-tufts were speckled also with black and white, and a slight tinge of yellow might be remarked in the white band, shaped like an inverted crescent upon the forehead, just above the eyes. Yellow hairs could be discovered only under the callosities, and very few in number. The entire face was of a violet colour, of which the blue predominated upon the cheeks, while the red seemed concentrated upon the muzzle and eyelids. The hands were entirely black, and the eyes of a brownish-yellow. In the course of a few years, the general distribution of these colours did not change, but the white rings of the hairs on the back became yellow, and this colour had increased upon the cheek-tufts. The hair which covered the internal surface of the thighs became varied with grey and white rings, which gave these parts a mild grey appearance, and the hairs of the tail were covered with similar rings, but the grey had almost become black. The entire coat of the animal was very thick on the upper part, and thin beneath, where the skin, as well as on the remainder of the body, had a violet tinge.

15. CERCOPITHECUS PYRRHONOTUS.—EHRENBERG'S GUENON.

Syn. et Icon. LE NISNAS mâle.—Valenciennes in F. Cuv. et Geoff. Hist. Mam.
　　DER NISNAS.—Ehrenb. Symb.[1] I. pl. X.

SPECIFIC CHARACTERS.

THE HAIR bright red above; white beneath, and on the outside of the limbs.
THE FACE entirely black.
THE SCROTUM bright green.
INHABITS Abyssinia.

This new species was recently discovered by M. Ehrenberg in Abyssinia, and brought to the Royal Prussian Menagerie, near Potsdam. It is known to the inhabitants of Darfoor by the name of Nisnas, and has many relations to the Red Guenon. It appears, however, to be more robust, its muzzle is broader and more obtuse, the tail longer, and the face entirely black, while the Red Guenon (1) is black only on the nose.

It is of a fine brick-red colour upon the body, on the arms, the anterior part of the thighs, and on the tail. This hue becomes feeble, and passes into a straw-colour on the occiput; the forehead is rather of a deeper red than the back; the cheeks white; the naked portion of the face is blackish; the fore-arms, legs, and the hinder parts of the thighs, of a pure white; the palms of all the hands black.

DOUBTFUL SPECIES.

1. C. BARBATUS of Clusius, Exol. p. 371, described by Linn. (Syst. Nat. I. p. 36), is an obscure species, by some referred to C. Diana. It is merely said to be black and brown above, white beneath, the beard white and ending in a point; the tail ending in a tuft.

2. C. HIRCINA.—Shaw, Gen. Zool. I. p. 58. This is the Goat Monkey of Pennant, Quadr. p. 212, described with a blue naked face ribbed obliquely, a long beard like that of a Goat; the whole body and limbs deep brown; the tail long, from a specimen said to be in the British Museum.

3. C. JOHNII.—Fisch. Syn. Mam. p. 25. The hair of a shining black, and bristly, the head greyish-brown. The hairs of the head very spinous, giving the animal a peculiar aspect. It is said to come from Tellichery, in the East Indies. (John. Beschreib. einger Affenarten in Neuen Schriften der Gesellsch. Naturf. Freunde I. p. 215.)

---

[1] EHRENB. SYMB.—Symbolæ Physicæ, seu Icones et Descriptiones Corporum Naturalium ex Itineribus per Africam Borealem et Asiam Occidentalem, à C. G. Ehrenberg. Berolini, 1828.

1. C. AURATUS (Geoff. Ann. Mus. XIX.) is undoubtedly the same as the Semnopithecus Pyrrhus of Dr Horsfield.

2. C. LATIBARBATUS (Geoff. Ann. Mus. XIX.) is identical with the Semnopithecus leucoprymnus of Otto. This is the Purple-faced Monkey of Pennant and Shaw; probably also the Broad-toothed Baboon of the latter.

3. C. PILEATUS (Geoff. Ann. Mus. XIX.—Desm. Mam. and others) is identical with the C. Talapoin of Buffon. The colours of the specimen described by Geoffroy had become altered by being preserved for a long time in alcohol.

4. C. ALBO-CINEREUS (Desm. Mam. Suppl.) We are assured by-M. Isidore Geoffroy that no animal was brought from India by MM. Diard and Duvaucel, answering to the description of Desmarest, nor does any Guenon from that country exist in the collections of the Paris Museum.

5. C. ATYS, figured by Audebert under the name of L'Atys, is undoubtedly an Albino variety of some other species, of Macacus cynomolgus according to Temminck (Mon. Mam.), or of Semnopithecus auratus according to Isidore Geoffroy.

6. C. FUSILLUS (Desmoul. in Dict. Class. d'Hist. Nat. Art. Guenon) rests upon three very young specimens, which have since been proved to be merely the young of the C. erythropygus, or Grivet Guenon.

7. LA GUENON COURONNÉE of Buffon (Hist. Nat. Suppl. VII. pl. 16) is a Macacus, probably M. Sinicus.

GENUS V. NASALIS.—PROBOSCIS-APES.

Syn. NASALIS (NASIQUE).—Geoff. Ann. Mus. XIX.—Isid. Geoff. in Bélang. Voy.

SEMNOPITHÈQUES (in part).—Cuv. Reg. Anim.—Temm. Mon. Mam.

SIMIA (in part).—Fisch. Syn. Mam.

The Proboscis-Apes are included by some Naturalists [among whom is the Baron Cuvier] with the Solemn-Apes; yet when we consider that, besides their remarkable nasal prominence, they differ from the latter in several important points of their organization, it appears advisable to place them in a group intermediate to the Guenons and Soleman-Apes.

As yet we are acquainted only with one species.[1]

1. NASALIS LARVATUS.—KAHAU PROBOSCIS-APE.

Syn. LE NASIQUE OU KAHAU.—Cuv. Reg. Anim. I. 94.

NASALIS LARVATUS.—Geoff. Ann. Mus. XIX.—Isid. Geoff. in Bélang. Voy.

CERCOPITHECUS NASICUS.—Lacepède.—Desm. Mam.

SEMNOPITHECUS NASICUS—F. Cuv. Mam.

Icon. SIMIA NASICA.—Schreb. Säügth. pl. 10 B. and 10 C.

GUENON À LONG NEZ.—Buff. Hist. Nat. Suppl. VII. pl. 11 and 12.

PROBOSCIS-MONKEY.—Penn. Quadr. pl. 104 and 105.

LE KAHAU.—Audeb. Sing.

SPECIFIC CHARACTERS.

THE MUZZLE very short. THE FOREHEAD rather prominent.

THE NOSE very broad, excessively elongated, pierced beneath with two enormous nostrils.

THE HANDS pentadactylous; the anterior long, thumbs short.

THE NAILS rather flat, broad and thick on the hinder thumbs.

THE TAIL longer than the body.

THE CALLOSITIES and CHEEK-POUCHES always present.

THE HAIR abundant; yellowish passing to a clear red on the breast, neck, and arms; reddish on the back, and upper part of the head.

INHABITS Borneo, and perhaps also Cochin-China.

This Ape lives in Borneo in numerous troops, which assemble in the morning and evening on the branches of large trees, near the margins of rivers. *Kahau* is its cry. Its nose is excessively long, and projecting in the form of a sloped spatula.

This characteristic nasal prominence distinguishes the Proboscis-Ape from every other mammiferous animal. The nose is between four and five idches in length, narrow at its extremity, and in the middle there is a furrow, which appears to divide it into two lobes. The nostrils, separated by a narrow septum, are large, and open horizontally. They are placed at the very extremity of the nose, which is very much elongated in front, so that they do not adjoin the upper lip. The entire face, as well

as the nose, is wholly destitute of hair, and the skin is of a dark brown, blended with blue and red. The head is round, covered on the top, behind, and on the sides, with a short tufted hair of a reddish-brown. The ears, almost hidden under the hair, are naked, thin, broad, round, and blackish, with a visible slope on their margin. The forehead is low, the eyes rather large, and remote from each other, without eyebrows, and destitute of eyelashes beneath, while the latter are tolerably long on the upper eye-lid. The mouth is large, and furnished with strong canine teeth; but we are without any minute description of the dentition. The body is massive, and covered with a reddish-brown, more or less deep upon the back and sides, tending towards orange-red on the breast. It is of a yellow mixed with grey on the abdomen, thighs, and arms. On the chin, and over the neck and shoulders, the hair is much longer than on any other part of the body, and contrasts remarkably with the dark and naked skin of the face. The tail is very long, furnished with short yellowish hairs; the hands and feet are naked within, and covered externally with short yellowish hairs, mixed with grey. All the nails are black, those of the thumbs are flat, the remainder convex.[2]

The colour of these animals varies with their age. Their height is rather above three feet.

1. N. INCURVUS (Vigors and Horsfield, Zoological Journal, No. 13) rests upon the examination of a single specimen, which would appear to be merely a young individual of N. larvatus.

GENUS VI. COLOBUS.[3]—THUMBLESS-APES.

Syn. CERTAINES GUENONS SANS POUCES.[4]—Cuv. Rég. Anim. 1. 93.

COLOBUS.—Illig. Prodr.—Geoff. Ann. Mus.

LES COLOBES.—Temm. Mon. Mam.

SIMIA (in part).—Linn. Gmel.—Fisch. Syn. Mam.

GENERIC CHARACTERS.

THE MUZZLE short. THE FACE naked.

THE ANTERIOR HANDS tetradactylous, the THUMBS being wanting.

THE HINDER-HANDS pentadactylous; the thumbs placed very remote from the fingers.

THE TAIL long and slender, with a tuft at the end.

THE CALLOSITIES and CHEEK-POUCHES always present.

THE LAST MOLAR of the lower jaw with five tubercles.

INHABITS Africa.

This genus, now consisting of three species,[5] is admitted by several Naturalists with some doubt. In respect to the form of their cranium, and the characters of their dentition, they exactly resemble the Solemn-Apes; and the peculiarity of their anterior hands merely arises from the want of the last phalanx and nail of the thumb, which is very short in the genus Semnopithecus. They bear the same relation to the Apes of the Old World as most of the Ateles do to the other Apes of America.

1. COLOBUS COMOSUS.—ROYAL THUMBLESS-APE.

Syn. COLOBUS POLYCOMOS.—Geoff. Ann. Mus. XIX.—Desm. Mam.

SIMIA COMOSA.—Shaw, Gen. Zool.

ATELES COMATUS.—Geoff. Ann. Mus. VII. 273.

SIMIA POLYCOMUS.—Fisch. Syn. Mam.

Icon. FULL-BOTTOMED MONKEY.—Penn. Quadr. pl. 46.

GUENON À CAMAIL.—Buff. Hist. Nat. Suppl. VII. pl. 17.

Schreb. Säügth. pl. 10 D.

SPECIFIC CHARACTERS.

THE HAIR of the head light yellow mixed with black, very long, and hanging down upon the back and shoulders; elsewhere black.

THE TAIL white.

INHABITS Sierra Leone.

This Thumbless-Ape is known to the Negroes of Sierra Leone by the name of *King of the Monkeys*, apparently from the beauty of its coat, and the singular head of hair, resembling a large periwig or a diadem, according to the views of the observer. This hair is much esteemed for various purposes, chiefly ornamental. The face is black; the body and limbs are furnished with a very short and shining hair of a beautiful black, contrasting remarkably with the colour of the crest, which is long and bushy, and of a yellowish tinge mixed with black, and still more so with the tail, of

-----

[1] A second species proposed by Vigors and Horsfield in the Zoological Journal, No. 13, is not admitted by other Naturalists.

[2] See an excellent description of the Proboscis-Ape by the Baron Wurmb, in the Verhandl. van het Batav. Genootsch. III. p. 146 (Memoirs of the Batavian Society).

[3] Colobus, from κολοβος, mutilated.

[4] Pennant describes certain thumbless Guenons (Simia polycomos and Simia ferrugines), of which Illiger has formed his genus Colobus, but not having yet been able to see them, I avoid noticing them in the text. M. Temminck assures us that they resemble the genus Semnopithecus in respect to their cranium and teeth.—*Note of the Baron Cuvier*

[5] Geoffroy-St.-Hilaire (Ann. Mus. XIX.) was only aware of the two species described by Pennant. Kuhl (Beitr.) admitted three, one of which is altogether nominal. Desmarest followed Kuhl, but doubted the reality of Colobus ferrugineus. Temminck confirmed the existence of C. ferrugineus; and finally, Ruppell discovered a third species, the description of which is here presented for the first time to the British reader.

a snowy whiteness, and terminating in a large tuft of hair. This animal is rather more than three feet in height when standing erect. · Its limbs are very slender.

2. COLOBUS FERRUGINEUS.—BAY THUMBLESS-APE.

Syn.   SIMIA FERRUGINEA.—Shaw, Gen. Zool.—Fisch. Syn. Mam.
        COLOBUS FERRUGINOSUS.—Geoff. Ann. Mus. XIX.—Desm. Mam.
        COLOBUS TEMMINCKII.—Kuhl Beitr.—Desm. Mam.
        BAY MONKEY.—Penn. Quadr.

Icon.

SPECIFIC CHARACTERS.

THE HAIR dark bay on the back, light bay beneath and on the cheeks, black on the top of the head, and on the limbs.
THE TAIL black.
INHABITS Sierra Leone.

This animal, first described by Pennant along with the preceding, was conjectured by Buffon, Lacepède, and Desmarest, to be merely a variety of the Royal Thumbless-Ape (1); its specific reality has, however, been recently proved by M. Temminck, as well as its identity with Colobus Temminckii.

3. COLOBUS RUPPELII.—MANTLED THUMBLESS-APE.

Syn. et Icon.   COLOBUS GUEREZA.—Rupp. Neue Wirbelth.[1]

SPECIFIC CHARACTERS.

THE HAIR of the body, face, top and hinder part of the head and limbs, black; the chin, neck. side of the head, and margin of the forehead, white; a mantle of long white hair, hanging from the shoulders, sides, and crupper, covering the thorax, loins, and thighs.
THE TAIL with the first half black, ending in a long white tuft.
THE CALLOSITIES black, edged with white.
INHABITS Abyssinia.

The Mantled Colobus[2] is found in small families on the loftiest trees usually in the neighbourhood of running water. It is active, lively, and taciturn, generally of a harmless disposition, not inflicting those depredations upon the cultivated fields so common among the other Apes. Its food consists of fruits, grain, and insects; during the whole day, it is occupied in seeking its food, at night it sleeps on the trees. The agility of this Ape is great. Ruppell witnessed downward leaps of forty feet in height The Thumbless-Ape of Abyssinia is found only in the low grounds of the Provinces of Godjam, Kulla, and especially in Damot. The natives of the last named place hunt these animals regularly at stated periods, and the singular mantle is considered among them as a mark of distinction, and worn as an ornament upon their leathern bucklers. Guereza is the name by which this Ape is known to the Abyssinians.
The face, eyes, top of the head, neck, the interscapular region, shoulders, breast, abdomen, the first half of the tail, the limbs, and feet, are of a beautiful velvet black. The edge of the forehead, the temporal region, the side of the neck, chin, and throat, are of a snowy whiteness, as is also the singular mantle composed of long silky hairs, extending from the shoulders and sides of the body upon the chest, abdomen, and haunches. There is likewise a white margin round the black callosities of the buttocks. The hinder half of the tail is very bushy and white. Each hair is marked with several grey rings, which gives it a silvery grey appearance. On the hands and face, white hairs are mixed upon a dark ground. The hair above the head is long and soft to the touch; the white hair on the sides of the body, forming the mantle, is more than a foot in length. The callosities, the soles of the feet, and nails, are black; each nail is rather long, convex, and compressed. The colours of the sexes or of the young do not vary; but in the young females, the hair of their mantle is rather shorter. The length of the adult from the point of the nose to the base of the tail is two feet and a half; and the tail is as long as the body.

IMAGINARY SPECIES.

1. C. TEMMINCKII (Geoff. Ann. Mus. XIX.,.and Desm. Mam.), resting upon a single specimen in Bullock's Museum, now in the possession of M. Temminck, is identical with C. ferruginea according to, the latter.

GENUS VII. SEMNOPITHECUS.[3]—SOLEMN-APES.

Syn.   LES SEMNOPITHÈQUES.—F. Cuv.[4] Dents des Mam. p. 14.—Cuv. Reg. Anim. I. 93.
        SEMNOPITHECUS.—Desm. Mam. Suppl.—Isid. Geoff. in Bélang. Voy.
        LASIOPYGA, and CERCOPITHECUS (in part).—Illig. Prodr.—Desm. Mam.
        PYGATHRIX and CERCOPITHECUS (in part).—Geoff. Ann. Mus. XIX.
        SIMIA (in part).—Linn. Gmel. I.—Fisch. Syn. Mam.

GENERIC CHARACTERS.

THE MUZZLE very short.   THE NOSE scarcely projecting.
THE LIMBS long.   THE BODY slender and elongated.
THE TAIL very long.
THE HANDS pentadactylous; the anterior narrow and very long, with the anterior thumbs very short.
THE CALLOSITIES always present.
THE CHEEK-POUCHES rudimentary or altogether wanting.
THE HAIR very long and abundant.
THE LAST MOLAR of the lower jaw with five tubercles.
INHABITS the East Indies.

The Solemn-Apes differ from the Guenons in having a small additional tubercle in the last molar tooth of the lower jaw. They are peculiar to the Oriental countries, while their elongated limbs, and especially their very long tail, give them a singular air. Their muzzle scarcely projects more than in the Gibbons, and they are equally provided with callosities. Further, they appear to be almost destitute of cheek-pouches. Their larynx is supplied with a sac.
Eleven species[5] compose this natural group, first instituted by M. Frederic Cuvier, after a careful examination of the Entellus Solemn-Ape (2).
The dentition of all the species has not yet been carefully verified. That of the Negro Solemn-Ape (5) exhibits the following peculiarities: In the upper-jaw, the first two incisors are nearly of the same size and form. The canine following them immediately afterwards is slightly longer, terminating in a point, and presenting on its internal border a strong worn-down surface, which renders its margins trenchant in some degree. The first and second false molars usually exhibit a point on their external and an oblique plane on their internal surface. The three following molars are each composed of four tubercles formed by a very deep transverse furrow, and a longitudinal furrow which is less deep than the former, and cuts it at right angles. These three teeth are nearly of the same size. In the lower jaw, the two incisors are similar to, though slightly broader than, those of the opposite jaw. The canine is pointed and slightly stronger than the opposite one, and also presents a single oblique plane on its internal surface. The first false molar is usually composed of a single obtuse point, though sometimes we may remark a small heel behind the point. The second false molar resembles the first, its crown being merely somewhat flatter. Of the two real molars which follow, the first is the smaller, and both of them are composed of four tubercles, resembling those in the opposite jaw, already described. Lastly, the third molar, which is the largest, besides its four tubercles, has a fifth, in the form of a heel at the hindermost part.
The Solemn-Apes are remarkable for mildness of disposition, great intelligence, and a slowness of motion quite opposed to the vivacity and petulance of the Guenons. It is in India, and chiefly in the islands of the Eastern Archipelago, that these animals are found in great numbers. They are treated by the natives with a kind of religious veneration, which they probably owe to the mildness of their manners and the gravity of their deportment. Some of the animals composing this genus have been for a long time confounded with the Guenons; but the most of them are only very recently discovered. The anatomical investigations of Dr A. W. Otto (Nov. Act. Acad. Cur. XII.)[6] have proved that in one species at least (9) the stomach is more than three times as large as in the Guenons, and that it differs from theirs equally in its structure, its form, and volume. The left portion forms a broad cavity, while the right is narrow, and convoluted so as perfectly to resemble an intestine, and the entire organ is so very considerable, that its whole curvature measures not less than two feet and three or four inches. It further resembles an intestine from being fixed by two well marked muscular bands, one placed along each portion, and as these bands are much narrower than the stomach itself, the walls of that organ usually expand and form, as in the colon, an

---

[1] RUPP. NEUE WIRBELTH.—Neue Wirbelthiere zu der Fauna von Abyssinien gehörig, von Dr Edward·Ruppell. Frankfurt am Main, 1837.
[2] Lulolphius (Æthiop. I. c. 10) notices this Ape under the title of " Animalculum e genere eorum, quæ Hollandi Sanguinem vocant." His indifferent figure is erroneously referred by Erxleben (Syst. p. 57) to the Hapale Jacchus.—Salt appears to have seen a fragment of the Skin (Travels in Abyssinia, Appendix, p. 41) ; and Bennett mistook it for the Colobus comatus.
[3] Semnopithecus, from σεμνός, semnos, solemn, and πίθηξ, pithēx, an ape.
[4] F. Cuv. DENTS DES MAM.—Des dents des Mammifères considérées comme caractères Zoologiques, par M. F. Cuvier. Strasbourg et Paris. 1825.
[5] Only six species, in addition to Nasalis larvatus, were known to the Baron Cuvier. Temminck admits the number stated in the text. The lists of British Systematic writers are very defective in respect to this genus, in no instance extending beyond six species.
[6] Nov. ACT. ACAD. CUR.—Nova Acta Physico-Medica Academiæ Cæsareæ Leopoldino-Carolinæ Naturæ Curiosorum—1757 et seq. Dr Otto's Memoir will be found in Vol. XII. published in 1825.

uninterrupted course of spacious compartments, tied by small muscular fibres, which gradually lose themselves transversely between the muscular bands.

Other species of Semnopithecus have more recently been dissected by Professor Duvernoy of Strasbourg, who finds their stomachs equally voluminous and remarkable in their form, though slightly different from those already described.

These animals are said to feed chiefly on leaves; a kind of provision for which the structure of their stomach appears specially adapted, in some respects approaching to that observed in the Ruminantia. Their cheek-pouches are so small, that they can hardly be said to exist.

1. SEMNOPITHECUS NEMÆUS.—COCHIN-CHINA SOLEMN-APE.

Syn. Le Douc.—Cuv. Reg. Anim. I. p. 93.
      Pygathrix Nemæus.—Geoff. Ann. Mus. XIX.
      Lasiopyga Nemæus.—Illig. Prodr.—Desm. Mam.
      Simia Nemæus.—Linn. Gmel. I. 34.
      Semnopithecus Nemæus.—F. Cuv. Mam.—Geoff. Cours.—Isid. Geoff. in Bélang. Voy. .
      Cochin-China Monkey.—Penn. Quadr.—Shaw, Gen. Zool.
Icon. Le Douc.—Audeb. Sing.
      Douc femelle.—F. Cuv. et Geoff. Hist. Mam.
      Buff. Hist. Nat. XIV. pl. 41.

SPECIFIC CHARACTERS.

THE HAIR of the body, upper parts of the head and arms, of a slate-grey, faintly dotted with black; the fore-arms, throat, tail, and a triangular space on the rump, pure white; the thighs, fingers, and backs of the hands, black; the thighs and tarsi of a bright red; the face light orange; a red and black collar more or less complete round the neck; tufts of yellowish or whitish hairs on the cheeks.

INHABITS Cochin-China.

This species, known long previously to any of the remainder, is remarkable for the bright and varied colours of its hair.

M. Diard brought several specimens from Cochin-China to Europe, of different ages and sexes, thereby proving that they do not undergo any considerable variation, and at the same time correcting the error of Buffon and Daubenton, which refused callosities to this species.[1] The length of the adult is about two feet three inches, exclusive of the tail; but unfortunately we are as yet wholly unacquainted with its habits and manners.

2. SEMNOPITHECUS ENTELLUS.—ENTELLUS SOLEMN-APE.

Syn. L'Entelle.—Cuv. Reg. Anim. I. p. 94.
      Cercopithecus Entellus.—Geoff. Ann. Mus. XIX.—Desm. Mam.
      Semnopithecus Entellus.—F. Cuv. Mam.—Geoff. Cours.—Isid. Geoff. in Bélang. Voy.
      Simia Entellus.—Dufresne, in Bulletin de la Société Philomatique for 1797.—Fisch. Syn. Mam.
Icon. L'Entelle (young).—Audeb. Sing.
      Entelle mâle (young).—L'Entelle vieux (adult).—F. Cuv. et Geoff. Hist. Mam.

SPECIFIC CHARACTERS.

THE ADULT.

HAIR of the body yellowish-grey, mixed with black hairs on the back and limbs; straw-yellow, approaching to orange, on the sides; black hairs on the eyebrows directed prominently forwards; the tail almost black.

THE YOUNG.

HAIR of the body nearly white, interspersed with black and yellowish hairs; a white beard directed forwards.

THE FACE and HANDS naked. The SKIN blueish-black.

INHABITS Hindoostan.

This species, along with some others, is especially venerated in the religion of the Brahmins.

It bears with them the name of Houlman, and holds a very respectable place among at least thirty thousand divinities. Towards the end of the wet season it becomes very plentiful in Bengal. The pious votaries of Bramah permit their gardens to be wasted, and their tables stripped

before their eyes, by herds of Entellus Monkeys, while the visits of the latter, though doubtless inconvenient, are always regarded as a great honour. " From the respect in which the Entellus Monkeys are held by the natives, it appears that, whatever ravages they may commit, the latter dare not venture to destroy them, and only endeavour to scare them away by their cries. Emboldened by this impunity, the Monkeys come down from the woods in large herds, and take possession of the produce of the husbandman's toil with as little ceremony as though it had been collected for their use; for, with a degree of taste which does them credit, they prefer the cultivated fruits of the orchard to the wild ones of their native forests. Figs, cocoa-nuts, apples, pears, and even cabbages and potatoes, form their favourite spoil. The numbers in which they assemble render it impossible for the sufferer to drive them away without some more efficient means than he is willing to employ."[2]

It will be observed, that a considerable difference exists between the young and the adult in the colour of the hair. In addition to the variations already noticed, the cranium undergoes considerable depression with increasing years; it ceases to have any forehead, and the profile view exhibits merely the arc of a large circle, so greatly do its cerebral contents diminish. These organic changes are followed by a corresponding variation in the intellectual character. The young Entellus exhibits an astonishing degree of penetration in perceiving the qualities of objects, a great susceptibility to kind usage, and an invincible propensity to obtain by cunning whatever he is unable to acquire by force. In the adult mildness and apathy resume the place of his former intelligence. The old Entellus is fond of solitude, slow in his movements, alike incapable either of planning or executing any device to obtain his object.

The Entellus Solemn-Ape, as well as the Simpai (3) and the Negro (5), possesses a sub-guttural pouch, which communicates with the larynx. Its cheek-pouches, if not altogether wanting, are at least very slightly developed. The cœcum is long and capacious; the liver composed of unequal lobes; the right lung has four lobes, while the left has only three. Owing to the great length of its limbs, and especially of the hinder, and the general proportions of the body, it appears well adapted for making prodigious leaps.

The specimen brought by Thunberg died on its passage homewards; and those lately in the Paris and London Zoological Gardens being unable to endure the rigour of our climate, have not long survived.

3. SEMNOPITHECUS MELALOPHUS.—SIMPAI SOLEMN-APE.

Syn. Le Cimepaye.—Cuv. Reg. Anim. I. 94.
      Simia Melalophus (Simpai).—Raffles, in Linn. Trans. XIII. 245.— Fisch. Syn. Mam.
      Semnopithecus Melalophus.—Desm. Mam. Suppl.
Icon. Cimepaye (young).—F. Cuv. et Geoff. Hist. Mam.

SPECIFIC CHARACTERS.

THE HAIR of a very bright reddish-brown above; whitish beneath; a crest with black hairs intermixed on the upper and hinder parts of the head.

THE FACE and EARS blueish. THE HANDS black.

INHABITS Sumatra.

The Solemn-Ape, called Simpai by the Malays, from its cry, was first described by our distinguished countryman Sir Thomas S. Raffles, from specimens procured for him in the woods near Bencoolen by MM. Diard and Duvaucel. It is not less remarkable for its colours than for the peculiar shape of the face. The hair is very long, silky, and of a brilliant reddish-brown on the back, sides, neck, tail, the outer surfaces of its limbs, the backs of the hands, the forehead, and cheeks. The chest, abdomen, and the inner surfaces of the limbs, are whitish; a circle, or rather a crest, of black hairs intermixed with brown cover the upper and hinder parts of the head, and a few scattered black hairs may also be seen along the back and upon the shoulders. The face is blueish as low as the upper lip; both lips and chin are flesh-coloured. The eyes are brown, the ears blueish, like the face; the hands are black beneath, and so are the callosities. The hairs of the cheeks, directed backwards, form cheek-tufts; the abdomen is almost naked; and the hair on the inner surfaces of the limbs is very scanty when compared with that on the remainder of the body. The length of the animal, exclusive of the tail, is about one foot seven inches, and the tail is long and tapering, exceeding thirty inches. The line of the face may be noticed as singularly straight and perpendicular, at least in the young.

[1] The genus Lasiopyga (from λασιος, lasion, hairy, and πυγη, pugé, anus) was instituted by Illiger, to contain the Cochin-China Solemn-Ape, which Buffon had stated to be destitute of callosities, from the examination of a specimen altered in the stuffing. M. Diard having sent several Cochin-China Apes to the Natural History Museum of Paris, it is now certain that they have callosities. Hence the genus Lasiopyga of Illiger, founded upon this error, must be suppressed.—Note of the Baron Cuvier.
The above observation applies equally to the genus Pygathrix of Geoffroy (Ann. Mus.) Sir William Jardine preserves the fictitious genus Lasiopyga throughout his volume on the Monkeys in the Naturalist's Library, published in 1833; the error bad, however, been corrected by Fred. Cuvier (Hist. Mam. art. Douc.) as early as 1820, and again noticed by the Baron Cuvier in the Second Edition of the Règne Animal, which appeared in 1829.
[2] E. T. Bennett after Thunberg (Travels in Europe, Asia, and Africa, Upsal, 1793), and Wolf (Residence in Ceylon, Berlin, 1782).

4. SEMNOPITHECUS COMATUS.—CRESTED SOLEMN-APE.

Syn.   LE CROO.[1]—Cuv. Reg. Anim. I. 94.
   SIMIA CRISTATA (CHINGKAU).—Raffles, in Linn. Trans. XIII. 244.
   PRESBYTIS MITRATA.—Eschscholz, in Kotzeb. Voy. III. p. 353.
   SIMIA COMATA.—Fisch. Syn. Mam.

Icon.   Croo.—F. Cuv. et Geoff. Hist. Mam.

SPECIFIC CHARACTERS.
THE ADULT.

THE HAIR of the body iron-grey above; white beneath, and along the under surface of the tail; long black hairs, forming an elevated crest, on the top of the head.

THE YOUNG.

THE HAIR of a reddish fawn colour.

INHABITS Sumatra.

The Crested Solemn-Ape, which is identical, according to Temminck, with the Presbytis mitrata described in Kotzebue's Voyages, occurs frequently in the forests near Bencoolen, in the Island of Sumatra. It is about two feet in length, and fourteen inches in height, when standing on its four hands; the tail is nearly two feet and a half. The hairs of the crest are long, and diverge round the face. The colour of the young animal contrasts remarkably with that of the adult, being of a reddish fawn.

There is a variety of this species with the hair light grey, or whitish, called Chingkau Puti by the natives.

5. SEMNOPITHECUS MAURUS.—NEGRO SOLEMN-APE.

Syn.   LE TCHINCOU.—Cuv. Reg. Anim. I. 94.
   SIMIA MAURA (LOTONG).[2]—Raffles, in Linn. Trans. XIII.—Linn. Gmel. I. 39.—Fisch. Syn. Mam.
   CERCOPITHECUS MAURUS.—Geoff. Ann. Mus. XIX.—Desm. Mam.
   SEMNOPITHECUS PRUINOSUS.—Desm. Mam. Suppl.
   NEGRO MONKEY.—Penn. Quadr.—Shaw, Gen. Zool.

Icon.   Tchincou.—F. Cuv. et Geoff. Hist. Mam.
   SEMNOPITHECUS MAURUS.—Horsf. Zool. Jav.
   BLACK MONKEY.—Edw. Glean. pl. 311.

SPECIFIC CHARACTERS.
THE ADULT.

THE HAIR black, sometimes with a white spot beneath the origin of the tail.

THE YOUNG.

THE HAIR entirely reddish-brown; afterwards varied with black spots.

INHABITS Sumatra and Java.

This animal has long been known under the name of LA GUENON MAURE, by which Buffon distinguished it (Suppl. VII.), and was supposed by him to have come from Guinea. The hair, which is uniformly black in the adults, is very scanty on all the inferior parts of the body, and especially on the abdomen. The ears and face are naked, excepting the lips and sides of the mouth, where some white hairs may be observed; the colour of the skin is blueish. The hair on the hands is scanty, the skin of the hands and callosities black. The iris of the eye is of a bright orange-yellow. The hair diverges from the crown of the head, so as to project over the forehead in front, and to form a kind of crest behind. According to Raffles, it is not easily tamed.

The young, instead of being black, are at first reddish-brown, and it is not until they have cast their hair that they gradually assume the dark hue of the adult.

6. SEMNOPITHECUS FLAVIMANUS.—YELLOW-HANDED SOLEMN-APE.

Syn.   LE SEMNOPITHÈQUE AUX MAINS JAUNES.—Isid. Geoff. in Bélang. Voy.
Icon.   SEMNOPITHECUS FLAVIMANUS.—Less.[3] Cent. Zool. pl. 40.

SPECIFIC CHARACTERS.

THE HAIR reddish-brown above, white beneath; a tuft of long grey hairs above, and on the back of the head; hands of a clear yellow.

INHABITS Sumatra.

In form this Solemn-Ape approaches nearly to the Simpai (3), but is sufficiently well characterized by the colour of the hands. The upper part of the body is of a clear reddish-brown in which black hairs are intermixed; the dark hairs are much less abundant on the sides, and consequently the reddish tint becomes much more pure. The inner surface of the arms is of the same whitish tint as the under part of the body; and the under surface of the tail is white throughout the first quarter of its length, and afterwards reddish, which is the colour of the entire upper surface. The external regions of the hinder limbs, of the fore-arms and hands, are of a fine deep gold-yellow, inclining to red upon the thighs and fore-arms, and very pale upon the fingers. The internal surface of the limbs, beneath the body and head, as well as the very long hairs which cross the hinder surface of the cheeks, are white.

This animal is of the same dimensions as the Simpai, but its tail is longer. It has the same remarkable tuft on the head as S. melalophus and S. comatus, but the forehead, and sides of the head, as far as the ears, are covered with hairs of the ordinary length, of a bright gold.yellow, inclining to red. The hairs in the middle of the crest are, on the contrary, very long. and form a kind of compressed cap. In S. melalophus the crest is black, while it is of a dirty white in S. flavimanus, with the exception of the anterior part, which is blackish.

The face appears to be blackish, the eyelids white, and the nails brownish.

The specimen, first described by Isidore-Geoffroy and figured by Lesson, was sent from Sumatra by Diard and Duvaucel.

7. SEMNOPITHECUS FASCICULARIS.—KRA SOLEMN-APE.

Syn.   SIMIA FASCICULARIS (KRA).—Raffles, in Linn. Trans. XIII. 246.
Icon.

SPECIFIC CHARACTERS.

THE HAIR of the back and upper part of the head reddish-brown; the tail and sides of the body grey; lighter beneath, and on the limbs.

THE FACE brown, covered with short grey hairs.

INHABITS Sumatra and the Malay Islands.

These animals, for our knowledge of which we are indebted to Sir T. Stamford Raffles, occur very frequently in the forests of the Malay Islands in large companies. Their name is derived from their cry; they are not easily tamed.

The body is about twenty inches in length, and the tail rather more. The cheeks are furnished with light grey tufts much longer than the beard. The eyelids, particularly the upper ones, are white; the eyes are brown, the eyebrows prominent, and the muzzle projecting. The nose is prominent between the eyes and flat at its point, where the nostrils open obliquely some way above the lip. The ears are rather round, and pointed obtusely behind.

A whiter variety, with a reddish shade on the back, is distinguished by the natives.

A smaller animal, probably the young of the Kra, is called Kra Buku by the natives. It agrees in most respects with the Kra, but is not more than a foot in length, and occurs very commonly. The head has very little hair on the temples, and it wants the circle round the face.

8. SEMNOPITHECUS CUCULLATUS.—HOODED SOLEMN-APE.

Syn.   LE SEMNOPITHÈQUE À CAPUCHON.—Isid. Geoff. in Bélang. Voy.
Icon.   Bélang. Voy. pl. 1. (Mammifères.)

SPECIFIC CHARACTERS.

THE HAIR of the body dark-brown; of the limbs and tail, black; of the head, light brown.

THE TAIL very long.

INHABITS Hindoostan.

This Ape was discovered by M. Leschenault de la Tour, in the mountains of the Ghauts. Subsequently M. Bélanger found several individuals in the western Ghauts, and Dussumier brought some specimens from Bombay to the Paris Museum.

The upper part and sides of the head, as well as the throat, are yellowish-brown, and, by their clear tint, contrast remarkably with the remainder of the hair, which is dark-brown on the flanks, loins, and thighs; blackish on the medial line of the back, and on the thighs, legs, and arms; while the fore-arms, all the hands, and the tail, are pure black. Beneath the body, and on the internal surface of the arms and thighs, the hair is scanty. The nails are black. The face is mostly naked, as in the other Solemn-Apes, and surrounded by a circle of black bristles, stiff, and tolerably long. On the sides of the face, these bristles are not numerous, and point outwards; while on the forehead the bristles are very abundant, and are more or less directed upwards. This arrangement is found in other Solemn-Apes, and remarkably so in the Entellus. The ears are covered with black hairs, and strike the eye prominently in the middle of the light-brown hairs of the remainder of the head. The length of this animal is about two feet, and the tail is slightly shorter.

---

[1] The name *Croo* is given by the Malays indiscriminately to certain Apes of Sumatra belonging to the genera Macacus and Semnopithecus, from their cries. It is written *Erro* by Desmarest, and *Orro* by Desmoulins. but both of these appear to be typographical errors.
   [2] There is some variation in respect to these Malayan names. Raffles calls the S. comatus *Chingkau*, and the S. maurus, *Lotong.*—Note of the Baron Cuvier. We must remark, however, that Raffles always gives the Malayan characters, and he is certainly the best authority.
   [3] LESS. CENT. ZOOL.—Centurie Zoologique, ou Choix d'Animaux Rares. Par R. P. Lesson. Paris, 1830.

9. SEMNOPITHECUS LEUCOPRYMNUS.—OTTO'S SOLEMN-APE.

Syn. LE SEMNOPITHÉQUE AUX FESSES BLANCHES.—Isid. Geoff. in Bélang. Voy.
CERCOPITHECUS LATIBARBATUS.— Geoff. Ann. Mus. XIX.—Desm. Mam.

Icon. CERCOPITHECUS (?) LEUCOPRYMNUS.—Otto in Nov. Act. Acad. Cur. XII. pl. 46 bis.—pl. 47 (skull and stomach).
PURPLE-FACED MONKEY.—Penn. Quadr. pl. 43, fig. 2.

SPECIFIC CHARACTERS.

THE HAIR blackish above; dark brown beneath; top of the head and neck brown; throat, under part of the neck, and hinder parts of the cheeks, yellowish-grey; a triangular whitish patch behind; tail whitish.
INHABITS Ceylon.

The animals belonging to this species are said by Pennant to be very harmless, feeding on leaves, or buds of trees, and soon becoming tame. We are indebted to Dr Otto for an excellent figure, accompanied by a most minute description of the anatomical structure, and especially of the remarkable peculiarities of the stomach, which have already been noticed in our general observations upon the Genus Semnopithecus.

The length of the animal is about one foot, eight inches, and of the tail about a foot and a half; the forehead is broad, and the snout projects but slightly, the facial angle being rather more than 60°. The fingers and toes are remarkably slender, and the abdomen appears of very small dimensions. The upper part of the head and neck is of a deep brown; the body and limbs black; the internal surface of the limbs and the under part of the body passing to a blackish brown; the throat, under part of the neck, and hinder part of the cheeks, are covered with long hair of a yellowish-grey; the tail is whitish in the adult. A large triangular patch of greyish-white, commencing on the medial line of the back, about four inches above the origin of the tail, covers the entire of the buttocks, and the upper part of the thighs.

A young specimen, brought from Ceylon by M. Leschenault de la Tour, has lately been added to the Paris Museum.

10. SEMNOPITHECUS VELLEROSUS.—LONG-HAIRED SOLEMN-APE.

Syn. LE SEMNOPITHÉQUE À FOURRURE.—Isid. Geoff. in Bélang. Voy.
Icon.

SPECIFIC CHARACTERS.

THE HAIR of the body black, and very long on the back and sides; throat, side of the head, and tail, yellowish-white; a large grey spot on the buttocks, and on each side near the origin of the tail; the tail whitish.
INHABITS

This species, described by Isidore Geoffroy, is of the same size as the Cochin-China Solemn-Ape (1), but very nearly allied both in form and colour to Otto's Solemn-Ape (9). The hair of the limbs and tail is rather short, that of the head is slightly longer, but on the upper part of the body and on the sides it attains the unusual length of five, six, and seven inches. All these hairs are smooth, recumbent, and directed backwards; those beneath the body, on the contrary, are slightly frizzled, and disposed very irregularly.

The body, limbs, and upper part of the head, are of a brilliant black. The throat and lower part of the neck, on the contrary, are covered with hair of a dirty white; but on each side, on the hinder and internal part of the thigh and on the buttocks, we find a large spot of clear grey, passing to a yellow, round the callosities. The hairs composing this spot are mostly of a greyish-white, but a great many black hairs are interspersed. The tail is entirely white.

The above description was obtained from a skin, purchased by Delalande in the Brazils, where it had in all probability been transported from the East Indian Archipelago. Its mutilated state did not permit the colours of the face, hands, fore-arms, and lower part of the legs, to be accurately described.

11. SEMNOPITHECUS AURATUS.—GOLDEN SOLEMN-APE.

Syn. CERCOPITHECUS AURATUS.—Geoff. Ann. Mus. XIX.—Desm. Mam.
SIMIA AURATA.—Fisch. Syn. Mam.
Icon. SEMNOPITHECUS PYRRHUS.—Horsf. Zool. Jav.

SPECIFIC CHARACTERS.

THE HAIR of a uniform golden-yellow above, paler beneath.
INHABITS Java and the Molucca Islands.

This animal, called *Lutung* by the Javanese, agrees with the Negro Solemn-Ape (5) in all respects excepting colour. The tint extending over the upper parts of the animal, and over the exterior of the limbs, is essentially different from the fulvous tint in the young of the *S. Maurus* before the change of colour to black takes place. The hair is long, soft, and silky; reddish-brown, with a beautiful golden gloss on the back, head, tail, and extremities, varying slightly in its degree of intensity as it approaches the sides and forehead; beneath and along the interior of the extremities it is pale yellowish, with a golden lustre. The long, shaggy and thickly disposed hair, which covers the upper parts, is separated by a regular boundary stretching along the hypochondriac region, from the hair on the abdomen, which is very thickly disposed, curled, silky, and of a very delicate texture.

The specimen described by Geoffroy-St.-Hilaire under the name of *Cercopithecus auratus*, appears to have over the knee-pan a black spot, which is wanting in the Javanese specimen illustrated by Dr Horsfield.

IMAGINARY SPECIES.

1. S. EDWARDSII (Fisch. Syn. Mam.), derived from the Middle-sized Black Monkey, figured in Edwards' Gleanings, pl. 311, is probably identical with *S. Maurus*.

2. S. FULVO-GRISEUS (Desmoulin, in Dict. Class. d'Hist. Nat.) is founded, according to Isid. Geoffroy, upon two specimens, one of which is a young *S. leucoprymnus*, and the other probably a *S. comatus*.

GENUS VIII. MACACUS.—MACACOS.

Syn. LES MACAQUES et LES MAGOTS.—Cuv. Reg. Anim. I. 94, 96.
MACACUS.—Lacépède.— Desm. Mam.—Isid. Geoff. in Bélang. Voy.
CERCOCEBUS (in part), INUUS, PAPIO (in part).—Geoff. Ann. Mus. XIX.
SIMIA (in part).—Linn. Gmel.— Fisch. Syn. Mam.
GUENON (in part).—Temm. Mon. Mam.

GENERIC CHARACTERS.

THE MUZZLE large and rather elongated. The FACIAL ANGLE about 40°. The NOSE but slightly projecting.

THE LIMBS robust, of medium length. The BODY rather short and thick.

THE ANTERIOR THUMBS short. The NAILS of the thumbs flat, the remainder cylindrical.

THE TAIL, varying in length, sometimes replaced by a simple tubercle.

THE CHEEK-POUCHES and CALLOSITIES always present.

THE HAIR generally abundant on the fore part of the body.

THE LAST MOLAR of the upper jaw, with five, and of the lower jaw, with six tubercles.

INHABIT the East Indies, North of Africa, and the Rock of Gibraltar.

The Macacos' resemble the Solemn-Apes in having additional tubercles to their last molars, and the Guenons in their callosities and cheek-pouches. Their limbs are thicker and shorter than those of the first, their muzzle more prominent, and the superciliary ridges more elevated than in either. Though tolerably docile in early youth, they become intractable with age. They all have a sac which communicates with the larynx under the thyroid cartilage, and is filled with air when they cry. The tail hangs down, and takes no part in their movements. They produce at an early age, but are not completely adult until the ages of four or five years. The period of their gestation lasts about seven months, and the females often have, during the rutting season, " des énormes gonflemens aux parties postérieures."[2] The greater part of these animals are peculiar to the East Indies.

This group has been instituted to contain such Apes as have their characters intermediate to those of the genera Cercopithecus and Cynocephalus. It is by no means a well defined or rigorous division, but blends insensibly into the characters of these adjacent groups. The facial angle in some species becomes as low as 30°, in others above 40°. The muzzle is shorter than that of the Cynocephali, and longer than that of the Cercopitheci, yet the differences, which are considerable with some species, become nearly evanescent in others. In respect to habits and disposition, they are intimately connected with both genera, some of the species being nearly as fierce, destructive, intractable, and lascivious as the Baboons, while others, again, have the volatility and caprice of the Guenons.

The characters of their dentition do not differ materially from those of the Guenons, already described, excepting in respect to the last molars. In the upper jaw, the last molar is terminated by a very small unequal tubercle, accompanied by several small dentations at its external surface. The canine is rounded, and not flattened at its internal surface, and the

---

[1] The name *Macaco* is applied indiscriminately to all Monkeys by the Negroes on the coast of Guinea, and the slaves of the West Indian colonies. Marcgravius describes a species with *nares elatas bifidas*, and these vague words, adopted solely after him, have remained as one of the characteristics of the *Macaque* of Buffon, although nothing of the kind is to be seen.—*Note of the Baron Cuvier.*

[2] It was this circumstance which led Ælian to remark that Apes were to be seen in India, afflicted with *prolapsus uteri.*—*Note of the Baron Cuvier.*

41

external surface exhibits a strongly marked depression. In the lower jaw, the heel of the last molar is composed of two tubercles, the external one, equal in size to the one in front of it, and the internal much smaller. These particulars were drawn from the examination of a M. Sinicus, and they appear to belong equally to all the other species of the genera Macacus and Cynocephalus.

The Macacus may be arranged in three sections, depending on the length of the tail; the Long-tailed Macacos (Cercocebes); the Short-tailed Macacos (Maimons); and the Tail-less Macacos (Magots).

(A.) LONG-TAILED MACACOS. (CERCOCEBES.)

The first section contains five species. These approach nearly to the Guenons, and many authors have placed them in the same group; others have formed the genus Cercocebus by uniting them to some Guenons with longer muzzles, and arranged them immediately after the Guenons. Indeed, the Long-tailed Macacos bear much resemblance to the Guenons in their general form. Their muzzle is shorter than in the other sections, the brain more voluminous, the body not so clumsy and massive, the tail is as long or longer than the body, and some are rather mild in their dispositions.

1. MACACUS SINICUS.—CHINESE-BONNETED MACACO.

Syn.   LE BONNET CHINOIS.—Cuv. Reg. Anim. 1. 95.
       SIMIA SINICA.—Linn. Gmel. 1. 34.—Fisch. Syn. Mam.
       CERCOCEBUS SINICUS.—Geoff. Ann. Mus. XIX. 98.
       MACACUS SINICUS.—Desm. Mam.—Isid. Geoff. in Bélang. Voy.

Icon. Le Bonnet Chinois.—Audeb. Sing.—Buff. Hist. Nat., Suppl. VII. pl. 16.
       Bonnet Chinois.—F. Cuv. et Geoff. Hist. Mam.

SPECIFIC CHARACTERS.

THE HAIR of a bright gold-yellow above, white beneath; the face flesh-coloured; the hair on the top of the head arranged in rays, forming a kind of bonnet.

INHABITS Bengal and Ceylon.

The Chinese-Bonneted Monkey, so called from the manner in which the hair diverges on the top of the head, displays the usual disposition of the Guenons, exhibiting, when in confinement, a mixture of playfulness and malice extremely amusing. It seems to share the religious veneration of the Hindoos in common with many other Apes.

All the upper parts of the body are of a brilliant gold-yellow, resulting from hairs which are grey at the base, but covered with rings of black and yellow through the rest of their length, in which, however, the yellow rings predominate. The tail is slightly browner; the cheek-tufts, the inner surface of the limbs, the under part of the neck, the breast, and abdomen, are whitish; the hands, feet, and ears, blackish, and its face flesh-coloured; the under-lip only is imagined with black. The eyes are brown. The hair of the head appears to hang down in long tufts, rather than to compose a compact bonnet. In the young, this ornament is more divided, and exactly resembles the hair of the next species, to which it is very nearly allied.

2. MACACUS RADIATUS.—RADIATED MACACO.

Syn.   LE TOQUE.—Cuv. Reg. Anim. 1. 95.
       CERCOCEBUS RADIATUL.—Geoff. Ann. Mus. XIX. 98.
       MACACUS RADIATUS.—Desm. Mam. 64.—Isid. Geoff. in Bélang. Voy.
       SIMIA RADIATA.—Fisch. Syn. Mam.

Icon. LE BONNET CHINOIS.—Buff. Hist. Nat. XIV. pl. 30?
       Toque Mâle.—F. Cuv. et Geoff. Hist. Mam.

SPECIFIC CHARACTERS.

THE HAIR of the head and body greenish-brown above, white beneath; the limbs grey externally; the tail brownish or blackish above, white beneath.

INHABITS the East Indies.

This animal, which occurs most frequently on the Malabar Coast, is remarkable for the singular form of its head and muzzle, and in these respects differs remarkably from all other known Macacos. These have the muzzle thick and clumsy, while in the Radiated Macaco it is thin and narrow, and the forehead is naked and full of wrinkles. It is further characterized " par la forme du gland de la verge. Chez les autres Macaques cet organe est simplement pyriforme; chez les Toques, il se compose de trois parties distinctes; l'antérieure, qui est en forme de poire, et la postérieure, formée de deux bourlets epais; de sorte que, dans l'érection la coupe longitudinale de ce gland présenterait la figure d'une feuille à trois lobes, les deux latéraux arrondis, et le moyen allongé."

The hair is silky, and of a greenish-grey, owing to the hairs, which are grey at their innermost half, being divided throughout the remainder of

their length by rings of black and dirty yellow. The diverging hairs on the upper part of the head are not very long, but their radiated form is constant in all the species hitherto examined. The skin of the hands has a violet tinge, that of the face and all other naked parts is flesh-coloured.

3. MACACUS CYNOMOLGUS.—COMMON MACACO.

Syn.   LE MACAQUE DE BUFFON.—Cuv. Reg. Anim. 1. 95.
       CERCOCEBUS CYNOMOLGUS.—Geoff. Ann. Mus. XIX.
       MACACUS CYNOMOLGUS.—Desm. Mam. 65.
       SIMIA CYNOMOLGUS.—Linn. Gmel. 1. 31.—Fisch. Syn. Mam.
       MACACUS IRUS.—F. Cuv. in Mém. Mus.[1] IV.

Icon. Macaque mâle, femelle adulte, jeune mâle, tête de femelle d'un jour.
       —F. Cuv. et Geoff. Hist. Mam.
       Buff. Hist. Nat. XIV. pl. 20.—Schreb. Säugth. pl. 13.

SPECIFIC CHARACTERS.

THE HAIR greenish-brown, dotted with black above, whitish beneath; the tail blackish above, ash-coloured beneath.

INHABITS Sumatra and Java.

Though wholly unacquainted with the peculiarities of the Common Macaco in its wild state, we are fortunately in possession of many important facts, obtained by M. Frederic Cuvier, from the examination of several living specimens of different ages and sexes, in the Paris Menagerie. He was thus enabled to record the several phenomena of reproduction, as well as the changes which the young undergo from birth to their mature age.

The adult is rather more than one foot nine inches in length, and the tail about one foot eight inches. Its entire form is heavy and clumsy, especially in the fore part of the body. The head is broad, flattened at the vertex, and very strong in proportion; the muzzle is short and obtuse, the nose flat, and a strong crest, advancing over the eyelids, covers the eyes. The fingers are united by a membrane as low as the second phalanx. The colours of its coat arise from the intermixture of golden-yellow hair, with black, over a greyish ground, from the combination of which it presents a general tint of pale greenish-brown; all the inferior parts are of a light grey, as well as the inner surface of the limbs. The tail is blackish, the feet entirely black, the face flesh-coloured and almost naked. Between the eyes there is a spot much whiter than any of the surrounding parts. The cheek-tufts are composed of short greenish hairs. The head is destitute of any crest or hairy appendage, and the hairs lie flat with their points directed backwards, those of the cheeks are grey, scanty, and directed forwards. The iris is brown. The parts of generation are flesh-coloured, the gland pear-shaped, and the scrotum remarkably large. The canine teeth are long and very strong.

The female is considerably smaller than the male, being only about fifteen inches in length; her form is more compact. The head is smaller, and the superciliary crest, which entirely covers the eyelids, is not nearly so prominent. The canine teeth are small, and do not pass beyond the incisors, and this peculiarity is common to all the females of the genus. The face is surrounded with long grey and straight hairs, giving it a bristly appearance, of which the male is wholly deprived. The hair on the top of the head is directed towards the medial line, and forms rather an elevated crest, extending from the top of the forehead to the occiput. In other respects the female entirely resembles the male.

At birth, the head of the young Macaco is rather long in front, compared to its dimensions from right to left; the muzzle projects, but the forehead is straight. The skin is flesh-coloured, excepting between the interval of the eyes, where it is white. All the hairs are black, and appear in greatest plenty on the upper parts of the body; the under parts are nearly naked. The hair at the end of the tail appears long, and terminates in a tuft. On the top of the head it extends from the medial line, pointing obliquely backwards, and, finally uniting in the occiput, forms a kind of crest. It has two pectoral mammæ, the callosities are prominent, but are not yet become hard.

In the course of the first year, the muzzle gradually lengthens, and the head becomes narrower, the superciliary crest being still wanting. The incisives appear, and the first canine begins to protrude in the lower jaw. After the first casting of the hair, the greenish hair of the adult succeeds, excepting on the fore-part of the top of the head, and the face is not yet surrounded with those thick hairs which afterwards appear. All its proportions resemble those of the adult, and the interval separating the eyes is always white. In the third year, the young male very much resembles the adult female.

The Common Macaco is one of the most untractable animals of the genus, and yields with difficulty to the ordinary methods of taming. It rests either on all the four hands, or upon the callosities, and eats in either

<hr>

[1] MAM. MUS.—Mémoires du Muséum d'Histoire Naturelle de Paris, par les Professeurs de cet établissement.   Paris, 1816, et seq.

of these positions by raising the food to the mouth with the fore-hands, or by seizing it with the mouth itself. Before swallowing it always fills the cheek-pouches, and it drinks by sucking up. When retiring to rest, it sleeps on the side, with the limbs folded up, and the head between the legs, or else in a sitting posture, with the back curved and the head resting upon the breast. Its voice is a hoarse cry, becoming very loud when the animal is enraged, but when pleased it emits a soft kind of whistle. The period of gestation in the female is seven months, " et ses parties de la génération ne paraissent point entourées, à l'époque du rut, de ces exubérances si remarquables, et quelquefois si monstrueuses chez d'autres espèces de Macaques, de Babouins, et même de Guenons. Ils s'accouplent chaque jour trois ou quatre fois, à la manière à-peu-près de tous les quadrupèdes. Pour cet effet, le mâle empoignait la femelle aux talons, avec les mains de des pieds de derrière, et aux épaules, avec ses mains antérieures, et l'accouplement ne durait que deux ou trois secondes."

VAR. AYGULA.—EGRET MACACO.
Syn. L'AIGRETTE.—Cuv. Reg. Anim. I. 95.
CERCOCEBUS AYGULA.—Geoff. Ann. Mus. XIX.
Icon. Buff. Hist. Nat. XIV. pl. 21.—Schreb. Säügth. pl. 22.
L'Aigrette.—Audeb. Sing.

The Egret Macaco (L'Aigrette), figured in Buffon, XIV. pl. 21, appears to be merely a variety distinguished by a bunch of long hair on the top of the head.

There are two other varieties, the one with long and thick hair of a deeper green, the other marked with black on several parts of the body.

4. MACACUS AUREUS.—TAWNY MACACO.
Syn. TAWNY MONKEY.—Penn. Quadr. I. 211.
LE MACAQUE ROUX-DORÉ.—Isid. Geoff. in Bélang. Voy.
Icon. Bélang. Voy. pl. 2. (Mammifères.)

SPECIFIC CHARACTERS.

THE HAIR of the body orange-red, dotted with black above, tawny on the sides, greyish beneath.
INHABITS Bengal, Pegu, Java, and Sumatra.

This animal was first noticed by Pennant, from a living specimen in Brookes' exhibition ; lately, skins have been sent from the East Indies to the Paris Museum by Leschenault, Renaud, Duvaucel, and Diard, procured from the localities above-mentioned. It is very common in the markets of Calcutta.

In every respect excepting colour it greatly resembles the Common Macaco, the greenish tint of the former being replaced by red. The upper part of the head and body is covered with hairs, grey at their base, with the points annulated with black and red, forming by their combination an orange-red, dotted with black. The limbs are greyish externally, and white on their internal surface ; the under part of the body and under surface of the tail are likewise white. The flanks are tawny, the red blending insensibly into grey. The cheeks are covered behind with long white hairs directed backwards. The eyelids are white, and are separated on the medial line by some black hairs. Finally, there is usually found beneath the chin a bunch of red hair pointing downwards.
This animal is merely said by Pennant to be " very ill-natured."

5. MACACUS MONTANUS.—MOUNTAIN MACACO.
Syn. et Icon. MACACUS GELADA.—Rupp. Neue Wirbelth. pl. 2. (Säügthiere.)

SPECIFIC CHARACTERS.

THE HAIR very long on the back and hinder part of the head; body deep reddish-brown above, blending into light wood-brown[1] on the head, neck, sides, limbs, and tail ; beneath the body, the fore-arms, and all the hands, dark-brown.
THE TAIL ending in a tuft.
THE FACE and CALLOSITIES naked and blackish grey.
THE THROAT and UPPER PART OF THE BREAST each with a naked space in front.

This well characterized Macaco was discovered by Ruppell in the elevated mountain chain of the Abyssinian provinces of Haremat, Simen, and Godjam, at a height of about 8000 feet above the level of the sea. It is found in numerous families throughout those rocky regions which are overgrown with bushes, but always upon the ground, differing in this respect very remarkably from most other Quadrumanous animals. Seeds, roots, and the young buds of plants, form its usual food, which it seeks for in large companies. Its devastations upon the cultivated fields of the natives are very frequently experienced ; by night, it remains concealed in the holes and clefts of the rocks. When attacked, it emitted a loud hoarse bark, and attempted to defend itself.

In the adult male, the hinder part of the head, the cheek-tufts in front,

the parotid region, neck, and back, are densely covered with hair about ten inches in length, giving the animal the appearance of being covered with a mantle, which hangs down over the neck and arms. The hair of the forehead, ears, and neck, the cheek-tufts, as well as the hinder-legs and tail, are hazel colour or wood-brown ; that of the sides and along the back is a deep reddish-brown blending into hazel. On the fore part of the neck and breast, there are two large naked spots, flesh-coloured and angular, with their corners directed to each other, somewhat resembling the form of an hour-glass. These naked spots are marked with rings, in which grey and white hairs are scantily dispersed. The naked callosities of the buttocks are closely approximated to each other, their colour being blackish-grey. The nails are black, long, and arched; those of the fore-hands being much larger than on the hinder. The tail is long, very hairy, and terminates in a thick tuft. The entire animal has a very massive appearance, owing to the thick and long hair in which it is enveloped. It carries its body bent rather backwards, but in a horizontal position ; and holds the tail curved upwards near the root, but with the tuft hanging vertically downwards. The hair round the face stands erect, giving the animal a wild and formidable appearance, especially when it shows its teeth. It is known to the inhabitants by the name of *Guereza*.

The young males have the hair of the neck much shorter and more plentiful, and the deep reddish-brown colour appears over the whole ; as the animal increases in size, it gradually acquires the light hazel tint. The adult female is altogether as deeply coloured as the young male. The adult male is about 3 feet 2 inches in length.

(B.) SHORT-TAILED MACACOS. (MAIMONS.)

Some species of Macacos are distinguished by having a short tail.

In this section, which contains four species, the muzzle becomes more elongated, and the tail, which is always less than the body, in some of the species is excessively short. The Rhesus and Pig-tailed Macacos form the types of the subdivision.

6. MACACUS SILENUS.—WANDEROO MACACO.
Syn. LE MACAQUE À CRINIÈRE.—Cuv. Reg. Anim. I. 94.
PAPIO SILENUS.—Geoff. Ann. Mus. XIX. 102.
MACACUS SILENUS.—Desm. Mam.—Isid. Geoff. in Bélang. Voy.
LION-TAILED MONKEY.—Penn. Quadr.
SIMIA SILENUS.—Linn. Gmel.—S. LEONINA.—Linn.
Icon. L'OUANDEROU.—Audeb. Sing.
Ouanderou femelle.—F. Cuv. et Geoff. Hist. Mam.
Buff. Hist. Nat. XIV. pl. 18.—Schreb. Säügth. pl. 11.

SPECIFIC CHARACTERS.

THE HAIR black; a large greyish crest and white beard surround the head; the under part of the body white ; the tail ending in a tuft.
INHABITS Ceylon and Hindoostan.

This animal, first noticed in Knox's Ceylon, is called Nil-Bandar by the natives. The adult is exceedingly ferocious, but the young appears susceptible of some education. It frequents the woods.
The abdomen, the breast, and the circle round the head, are white ; the remainder of the body is of a fine black. The hairs are generally long, especially round the head, where a greyish crest appears on each side of the forehead, uniting in a white beard on the chin, and extending backwards over the cheeks. The tail terminates in a tuft. The callosities are reddish, but the face and hands black.
The earlier accounts of this animal are, as usual, full of fable and exaggeration. According to Father Vincent Muria, " All the other Monkeys pay such profound respect, that they submit and humiliate themselves in his presence, as though they were capable of appreciating his superiority and pre-eminence. The princes and great lords hold him in much estimation, because he is endowed above every other with gravity, capacity, and the appearance of wisdom. He is easily trained to the performance of a variety of ceremonies, grimaces, and affected courtesies, all which he accomplishes in so serious a manner, and to such perfection, that it is the most wonderful thing to see them acted with so much exactness by an irrational animal." (Vincent Maria fide Bennett.)—Robert Knox, in his Historical Relation of Ceylon, tells us, with more probability, that " They do but little mischief, keeping in the woods, eating only leaves and buds of trees; but when they are catched they will eat any thing." The specimen in the Museum of the Zoological Society of London being young, " was extremely active, and occasionally very troublesome, but at the same time a perfectly good-tempered fellow. He was very strong, and had his teeth been full-grown, would in all probability have proved a dangerous animal." (E. T. Bennett, Gardens and Menagerie of the Zoological Society.)

---

[1] Wood-Brown—Colour of the hazel-nut,—No. 105. Werner's Nomenclature of Colours. By Patrick Syme. Edinburgh, 1821.

7. MACACUS RHESUS.—RHESUS MACACO.

Syn.   LE RHÉSUS.—Cuv. Reg. Anim. I. 96.
     INUUS RHESUS.—Geoff. Anim. Mus. XIX.
     MACACUS RHESUS.—Desm. Mam.
     MACACUS ERYTHRÆUS.—F. Cuv.—Isid. Geoff. in Bélang. Voy.

Icon.   LE RHÉSUS.—Audeb. Sing.
     LE PATAS À QUEUE COURTE.[1]—Audeb. Sing.
     Rhésus mâle adulte, Rhésus femelle âgé de 49 jours, Rhésus femelle 4
     face brune, Maimon femelle.—F. Cuv. et Geoff. Hist. Mam.
     Buff.[2] Hist Nat., Suppl. VII. pl. 14.

SPECIFIC CHARACTERS.

THE HAIR greenish-grey above, passing to bright yellow on the loins
and thighs; whitish beneath.
THE FACE flesh-coloured.
INHABITS Bengal.

The Rhesus Macacos frequent the forests on the banks of the Ganges
in large numbers. Encouraged by the reluctance of the Hindoos to de.
stroy animals, they carry their depredations to the very suburbs of the
cities. Their disposition appears most untractable. During extreme
youth, a certain degree of familiarity may be encouraged with impunity;
but they soon become mischievous, and age renders them ferocious.
Their ferocity is the more dangerous from its being combined with con.
siderable foresight and intelligence.

All the upper parts of the body are of a fine greenish-grey, resulting
from hairs which are entirely grey at their base and throughout great part
of their length, while their points are either black or yellow. This yellow
becomes paler on the arms and legs, so as to render these parts almost
grey; while it assumes a brighter hue upon the loins and thighs. The throat,
neck, breast, abdomen, and the internal surfaces of all the limbs, are white.
The tail is greenish above and grey beneath. The skin of the face, the
ears, and hands, is of a clear copper-colour, and destitute of hairs. The
thighs appear of a bright reddish-yellow, extending upon the crupper over
the origin of the tail. The hairs are very fine and silky; plentiful on the
upper parts of the body, but scanty beneath. An extreme flaccidity may
be remarked in all animals of this species, and the young have those hanging
folds of skin on each side of the throat, which in other animals are seen
only in old age.

In the female, the thighs are of a bright red; and this colour, which is
entirely owing to the blood, appears on the legs, and upwards upon the
crupper, towards the insertion of the tail, especially during the rutting sea.
son. At this period the nipples are rose-coloured.

The males differ from the females only in having their cheek-tufts more
bushy, their proportions more massive, their height greater, and their ca.
nines stronger.

8. MACACUS NEMESTRINUS.—PIG-TAILED MACACO.

Syn.   LE MAIMON.—Cuv. Reg. Anim. I. 96.
     INUUS NEMESTRINUS.—Geoff. Ann. Mus. XIX.
     SIMIA NEMESTRINA.—Linn. Gmel. I.
     MACACUS NEMESTRINUS.—Desm. Mam.—Isid. Geoff. in Bélang. Voy.

Icon.   Le Maimon.—Audeb. Sing.
     SINGE À QUEUE DE COCHON, mâle, adulte.
     SIMIA PLATIPYGOS.—Schreb. Säügth. pl. 5, B,
     PIG-TAILED MONKEY.—Edw. Glean. pl. 214 (young).
     Buff. Hist. Nat. XIV. pl. 19.

SPECIFIC CHARACTERS.

THE HAIR greenish-brown above; with a black band on the top of the
head, extending along the back to the tail; whitish beneath.
THE TAIL very short and curved backwards.
INHABITS Java and Sumatra.

These animals have the same manners and disposition as the preceding.
Their general colour is of a deep greenish-brown, proceeding from grey
hairs annulated with black and yellow. The top of the head, for the
breadth of two or three fingers, is black, and this shade extends along the
neck, back, and tail, but gradually acquiring the greenish-brown tinge.
The latter colour covers the shoulders, becoming more yellow upon the
fore-arm. The thighs are likewise green, but with a mixture of grey;
the cheeks, the under part of the chin and neck, the breast, abdomen,
under surface of the tail, and the internal surfaces of all the limbs, are
white or flesh-coloured. Before the ears, at their base, and on the cheeks

beneath the eyes, there are some blackish hairs, and behind the ears they
are entirely black. The dark brown face is almost naked from the eyes
to the mouth, with the exception of a few long and black hairs. The
ears and the palms of the hands are naked, and of the same colour as the
face, so are also the callosities. The males and females resemble each
other in colour, and the young are of a brighter yellow than the adults.

9. MACACUS MAURUS.—URSINE MACACO.

Syn.   LE MACAQUE DE L'INDE.—Cuv. Reg. Anim. I. 96.
     MACACUS ARCTOIDES.—Isid. Geoff. in Bélang. Voy.
     MACACUS MAURUS.—F. Cuv

Icon.   MACAQUE DE L'INDE.—F. Cuv. et Geoff. Hist. Mam,
     Isid. Geoff.[3] Etud. Zool. pl. 11. (var.)

SPECIFIC CHARACTERS.

THE HAIR of a uniform dark brown.
THE FACE, EARS, and HANDS, black.
THE TAIL very short.
INHABITS the East Indies.

The characters of this species are obvious and decisive, forming a gra.
dual transition into the third division, or Tailless Macacos. It rests, how.
ever, merely upon a single specimen, sent by M. Alfred Duvaucel from
the East Indies, and has by some been considered identical with M. niger
(11). The specific name *Maurus* is derived from the colour of the face
and hands, and that of *Ursine* from the resemblance of its hair to that of
the Brown Bear (Ursus Arctos).

There is a variety of this species, sent by M. Diard from Cochin-China,
and figured by M. Isidore Geoffroy (Etud. Zool. pl. 11), under the name
of *Le Macaque Ursin*, which appears to differ from the above, in having
the nose alone black, with the face and hands flesh-coloured. Its dark
brown hair is also scantily dotted with clear red. The specimen was
about two feet ten inches in length.

(C.) TAILLESS MACACOS. (MAGOTS.)

These are merely Macacos, in which a small tubercle supplies
the place of a tail.

10. MACACUS INUUS.—BARBARY MACACO.

Syn.   LE MAGOT COMMUN.—Cuv. Reg. Anim. I. 96.
     INUUS ECAUDATUS.—Geoff. Ann. Mus. XIX.
     MACACUS INUUS.—Desm. Mam.—Isid. Geoff. in Bélang. Voy.
     SIMIA SYLVANUS, SIMIA INUUS.—Linn. Gmel. I. 27, 28.

Icon.   LE MAGOT.—Audeb. Sing.
     Magot mâle.—F. Cuv. et Geoff. Hist. Mam.
     Buff.[4] Hist. Nat. XIV. pl. 7 and 8.

SPECIFIC CHARACTERS.

THE HAIR of a bright greyish-yellow.
THE TAIL tuberculous.
INHABITS the North of Africa and the Rock of Gibraltar.

Of all Apes, the present species seems best capable of enduring
our climate. Originally from Barbary, it has become naturalized
upon the most inaccessible parts of the Rock of Gibraltar.

In size the Barbary Macaco never exceeds a middle-sized Dog. The
top and sides of the head, the cheeks, neck, shoulders, the corresponding
parts of the back, and the fore part of the anterior limbs, are of a bright
gold-yellow, mixed with a few black hairs. The other parts of the body
are of a greyish-yellow. Each hair is of a dark grey at its base, and an.
nulated with yellow and grey throughout the rest of its length. The face
and ears are wholly naked, and of a bright flesh-colour. The tips of the
ears are covered with long hairs, the hands are blackish and well furnished
with hair. The cheek-tufts are thick and directed backwards. All parts
of the body are covered with hair, having the points directed downwards,
excepting on the fore-arm, where they point upwards. The entire coat
is very long, dense, and uniform; for which reason, these Monkeys become
best able to resist the cold during the winter.

The Barbary Ape walks habitually on all its four hands, but without
ease, as it is more especially organized for climbing. It sleeps either on
the side, or in a sitting posture, with the head between its hind-legs. It
raises the food to its mouth by the hand, or seizes it with the lips; and

---

[1] The two individuals which served as models to the designs of Audebert are in the Paris Museum. I have examined them, and they belong to the same species.—*Note of the Baron Cuvier.*

[2] The *Macaque à queue courte*, figured in Buff. Hist. Nat. Suppl. VII. pl. 13 (Simia erythræa of Schreber), appears to be a Common Macaco (M. cynomolgus), the tail of which has been cut.—*Note of the Baron Cuvier.*

[3] ISID. GEOFF. ÉTUD. ZOOL.—Etudes Zoologiques, par M. Isid. Geoffroy-Saint-Hilaire. Paris, 1832.

[4] The *Pithèque* of Buffon, Suppl. VII. pl. 4 and 5, is merely a young Barbary Macaco. His *Petit Cynocephale*, pl. 6, and the Cynocephalus major and minor of Prosper Alpin, are likewise of this species.—*Note of the Baron Cuvier.*

[5] Πίθηκος is the Greek name for Apes in general; and the Ape, whose anatomy is given by Galen, is nothing else than a Barbary Macaco, although Camper considered it to be an Orang-Outang. M. De Blainville has remarked this error, and I have confirmed his observations by comparing with the two species every thing referred by Galen to his Pithecus.—*Note of the Baron Cuvier.*

smells every suspected object very carefully. Almost every kind of food can be given it. In the wild state, fruits and leaves form its habitual diet; in confinement it eats fruit, bread, and boiled vegetables, especially carrots and potatoes. It drinks by sucking in. When enraged, its jaws are agitated with astonishing rapidity, its movements become violent, and it emits a loud and hoarse cry, which becomes rather mild when the passion subsides. Its strong canine teeth, and its thick and long, though flat nails, are capable of inflicting severe wounds. A natural fondness for society induces it to adopt any little animals which may be placed with it : these are carried about, loaded with caresses, and cannot be taken away without putting the Barbary Macaco in a violent passion. These animals have the highest affection for their young, which they preserve in a state of great cleanliness. Their geographical range does not extend eastward beyond Egypt.

11. MACACUS NIGER.—BLACK MACACO.

Syn. Simia Nigra.—Cuv. Reg. Anim. I. 98.—Fisch. Syn. Mam.
　　Cynocephalus Niger.—Desm. Mam. Suppl. pl. 534.—Isid. Geoff. in Bélang. Voy.
　　Macacus Niger (The Black Ape).—Bennett, Gard. Zool. Soc. I. 189.

Icon. Quoy et Gaim. Voy. de l'Astr. Mammifères, pl. 6,—pl. 7 (anatomical). Gray,[1] Spicil. Zool. pl. 1, fig. 2.

SPECIFIC CHARACTERS.

The Hair entirely black ; a tuft of long hair forming a crest on the top of the head.

The Tail tuberculous.

Inhabits the Molucca Islands.

The present species is arranged by the Baron Cuvier, Desmarest, and others, among the Baboons (Cynocephalus) ; but it wants the terminal nostrils of that Genus, and ought, therefore, to be considered as a true Macacus ; at the same time it indicates the passage towards the Baboons, to which in other respects it bears a near affinity.

The hair on all parts of the body is of a pure and shining black, very long and woolly ; that of the cheeks shorter, blacker, and more dense, forming cheek-tufts, and the entire upper part of the head is ornamented with a crest of hair which is very long towards the occiput. The callosities are of a bright red. The face is broad and prominent, narrowed at the nostrils, and abruptly truncated, with the nostrils placed obliquely on the upper surface. The projection of the muzzle is not disagreeable to the eye ; and its countenance, unlike that of the Baboons, is agreeable and intelligent.

DOUBTFUL SPECIES.

1. M. carbonarius, figured in F. Cuv. and Geoff. Hist. Mam. under the name of Le Macaque à face noire, appears to differ from the M. cynomolgus merely in having a black face.

2. M. speciosus, illustrated in the work just mentioned under the name of Le Macaque à face rouge, differs from M. cynomolgus only in having its tail considerably shorter, and a brighter reddish tint upon the face.

3. M. arctoïdes (Isid. Geoff. in Bélang Voy.) seems to be a variety of the M. Maurus of Fred. Cuvier.

GENUS IX. CYNOCEPHALUS.[2] BABOONS.

Syn. Les Cynocéphales et les Mandrills.—Cuv. Reg. Anim. I. 97 and 98.
　　Cynocephalus.—Briss. Reg. Anim. (in part).—Illig. Prodr.—Desm. Mam.—Isid. Geoff. in Bélang. Voy.
　　Papio (in part).—Briss. Reg. Anim.—Erxl.—Geoff. Ann. Mus. XIX. 101.
　　Simia (in part).—Linn. Gmel.—Fisch. Syn. Mam.
　　Cynocéphale.—Temm. Mon. Mam.

GENERIC CHARACTERS.

The Muzzle much elongated, and truncated at the extremity. The Facial Angle from 30° to 35°. The Nostrils terminal.

The Limbs very robust, and nearly of equal length.

The Cheek-pouches and Callosities always present.

The Hair wanting about the callosities.

The Last Molar of the upper jaw with five, and of the lower jaw with six, tubercles.

Inhabit Africa and some adjacent parts of Asia.

The Baboons have the teeth, cheek-pouches, and callosities of the preceding genus, but in addition to these characters, their muzzle is very much elongated, and truncated at the extremity, where

the nostrils are placed. The latter circumstance occasions their physiognomy to resemble that of a Dog, rather than of the other Apes. Their tail varies in length.

These, in general, are large, ferocious, and dangerous Apes ; most of them are found in Africa.

From the particulars above mentioned, it will readily be perceived that the Baboons approach more nearly in their characters to the lower orders of Mammiferous animals than to any other Apes of the Old World. With them the vertical position is rendered still more difficult to maintain, and their habitual mode of progression is consequently on all the four hands. The forests are not their favourite places of resort ; in general they either prefer the mountains, or localities interspersed with hillocks, rocks, and brushwood. Notwithstanding the clumsiness of their forms, they climb trees with much ease, and exhibit no small agility in their leaps. Travellers notice, probably with some exaggeration, the danger which females, residing in the neighbourhood of the Baboons, undergo from the ferocity of the males. Negresses are said to have been forcibly carried off by the Cynocephali, and even to have lived very happily with them for several years, while the animals, detaining them in caverns, supplied them regularly with provisions, &c. These statements derive their probability from the fact that adult Baboons exhibit the most frantic and outrageous gestures at the sight of a woman, especially a very young one. When visited by the latter in the menageries, they rush against the bars of their cage, shake them with all their might, accompanying this violence by the most terrific cries or disgusting gestures, which are heightened still more if the object of their regard happens to be accompanied by a male of her own species.

All these animals attain a considerable size, which is nearly that of a Wolf. When attacked, they defend themselves vigorously ; but, though ferocious, do not usually attack others, except at a distance, by throwing branches of trees, or menacing them by their cries. Their dispositions present a singular compound of passion and ferocity, blended with much intelligence and cunning. In a few seconds, they pass from one extreme to another, from indifference to the most violent passion, without any apparent cause for the sudden change. Their fury, when in confinement, is capable of rising to a pitch sufficient to occasion death itself. M. F. Cuvier tells us that he has seen several expire from the consequences of their passion.

Being without the elevated hinder limbs of most other Apes of the Old World, they walk on all the four hands with greater ease, though far from equalling, in this respect, the true quadrupeds. Their movements on the ground are always constrained, their walk slow, and their run is merely a kind of trot or shuffling gallop. Rarely they stand in an erect posture, and advance only a few steps in this manner. During their extreme youth, the agility with which they climb trees is remarkable ; in old age, they usually continue resting on their callosities. Their chief food consists of fruits and roots, with the tender leaves and young shoots of certain plants. When about to eat, they always commence with filling their cheek-pouches, and drink the fluid, by sucking inwards, like other animals with long and moveable lips. These animals are said by M. F. Cuvier to be " très-lascifs, toujours disposés à l'accouplement, et bien différents des autres animaux, les femelles reçoivent les mâles même après la conception. Celles-ci, lorsqu'elles ne sont pas pleines, entrent tous les mois en rut ; et cet état se manifesté par un gonflement considérable, causé par l'accumulation du sang dans les organes génitaux, et les parties qui les avoisinent et il est accompagné d'une véritable menstruation." Their growth is slow, and they do not become completely adult until the eighth or tenth year. The females are smaller and milder than the males.

The Baboons compose a very natural group of Quadrumanous animals, consisting of six well-authenticated species, which may be recognised at a single glance by their terminal nostrils. They admit of being arranged under two sections, distinguished by the length of their tails ; the Proper Baboons (Papions), with tails nearly as long as the body ; and the Mandrills, with a very short tail. The dentition of all these animals perfectly resembles that of the Macacos already described. Their tail rises upwards near its base, and the remainder, if any, not being susceptible of muscular motion, falls perpendicularly downwards.

(A.) Proper Baboons. (Papions.)

Syn. Les Cynocéphales.—Cuv. Reg. Anim. I. 97.

The Proper Baboons cannot be regarded as generically distinct from the Mandrills ; at the same time, they are destitute of those singularly bright coloured markings, which distinguish the latter from all other Mammiferous animals, if we except the greenish scrotum found in the Grivet, Vervet, and Green Guenons, with the bright ultramarine of the Malbrouck.

---

[1] Gray, Spicil. Zool.—Spicilegia Zoologica, or Original Figures and Short Systematic Descriptions of new and unfigured Animals. By John Edward Gray.

[2] Cynocephalus (from κυων, *kuōn*, dog, and κεφαλή, *kephalē*, head) is a term well known to the ancients, and this animal occupies a prominent place in the symbolical figures of the Ægyptians, where it represents Thoth or Mercury.—*Note of the Baron Cuvier.*

42

1. CYNOCEPHALUS SPHINX.—GUINEA BABOON.

Syn. LE PAPION.—Cuv. Reg. Anim. l. 97.
PAPIO SPHINX.—Geoff. Ann. Mus. XIX.
SIMIA SPHINX.—Linn. Gmel. I.—Fisch. Syn. Mam.
CYNOCEPHALUS PAPIO.—Desm. Mam.

Icon. Le Papion.—Le Papion, var. A.—Audeb. Sing.
Le Papion mâle, Papion femelle très jeune.—F. Cuv. et Geoff. Hist. Mam.
SIMIA CYNOCEPHALUS.—Brongn.[1] in Journ. d'Hist. Nat. l. pl. 21.
Le Papion.—Buff, Hist. Nat. XIV. pl. 13.

SPECIFIC CHARACTERS.

THE HAIR yellow, tending more or less towards brown, the cheek-tufts yellow.

THE FACE black. THE TAIL long.[2]

INHABITS Africa.

Specimens of this animal vary in size, probably owing to some difference in their ages. The adult is ferocious, and of brutal manners.

The face, the ears, and the palms of all the hands, are entirely black; the upper eyelids white. Its general colour is a yellowish-brown, resulting from hairs covered alternately with small rings of black and clear brownish-yellow, so that the animal, when viewed very closely, appears speckled with these colours. The cheek-tufts are yellowish, and directed backwards. The hair on the back of the neck is longer than on any other part of the body; the inner surfaces of the legs and thighs are scantily covered, as well as the lower part of the belly, beneath the neck, and on the breast; the bases of the hairs are usually grey. The females and young do not differ from the adults in colour, but much so in form. They are not so robust, and their muzzles are much less elongated.

This species may at once be distinguished from the following, by the cartilage of its nostrils projecting forwards beyond the other parts of the muzzle. It is found on the coast of Guinea, also in the Island of Meroe, and rarely in Senaar, according to Calliaud.

2. CYNOCEPHALUS BABOUIN.—LITTLE BABOON.

Syn. LE BABOUIN.—Cuv. Reg. Anim. l. 97.
PAPIO CYNOCEPHALUS.—Geoff. Ann. Mus. XIX.
SIMIA CYNOCEPHALOS.—Linn. Gmel. I.—(S. cynocephala).—Fisch. Syn. Mam.
CYNOCEPHALUS BABOUIN.—Desm. Mam.—C. ANTIQUORUM.—Schinz. Thierr. l.

Icon. Babouin mâle.—F. Cuv. et Geoff. Hist. Mam.—F. Cuv. in Mem. du Mus. pl. 19.
LE PETIT PAPION.—Buff. Hist. Nat. XIV. pl. 14.

SPECIFIC CHARACTERS.

THE HAIR greenish-yellow above, clear yellow beneath, cheek-tufts whitish.

THE FACE flesh-coloured. THE TAIL medium length.

INHABITS Northern Africa.

This species is nearly allied to the preceding.

It occupied an important place in the Theogony of the ancient Ægyptians, and was worshipped at Hermopolis, where a celebrated temple stood in its honour. It is probably this species, which appears represented so frequently among the hieroglyphics of that singular people. The upper parts of the body are of a pretty uniform greenish-yellow, and this results from hairs covered with large yellow rings alternating with small black ones, so that the former predominate, and the greenish tint ensues. All the lower parts of the body are of a paler yellow than the upper, and the tufts of hair on each side of the face are whitish. The young are of the same general colour as the adults, but beneath they are of a dirty white. Their muzzle is not so prominent, and the colour of the thighs, instead of being red, is deep brown. The nostrils of the adult, placed at the extremity of the muzzle, are separated above by a very well-marked groove, and the lateral cartilages do not advance as far forward as the central. The tail, elevated at its origin, soon hangs downwards, and terminates at the ham, or hinder part of the knee. The face is of a bright flesh colour, rather paler around the eyes.

VAR. ANUBIS.—ANUBIS BABOON.

Syn. SIMIA ANUBIS.—Fisch. Syn. Mam.
Icon. Anubis.—F. Cuv. et Geoff. Hist. Mam.

The Anubis Baboon differs from the Little Baboon already described in being of a deeper green, its muzzle more elongated, and the cranium flatter.

3. CYNOCEPHALUS PORCARIUS.—CHACMA BABOON.

Syn.[3] LE PAPION NOIR.—Cuv. Reg. Anim. l. 97.
PAPIO PORCARIUS.—Geoff. Ann. Mus. XIX.
SIMIA PORCARIA.—Linn. Gmel. I.—Fisch. Syn. Mam.
SIMIA SPHINGIOLA.—Herm.[4] Obs. Zool. l. p. 2.—Fisch. Syn. Mam.
CYNOCEPHALUS PORCARIUS.—Desm. Mam.
URSINE BABOON.—Penn. Quadr. l. No. 104.

Icon. Le Papion var. B.—Audeb. Sing. (incorrect).
CHACMA mâle très vieux.—Tête d'un très jeune individu.—F. Cuv. et Geoff. Hist. Mam.
Simia porcaria.—Bodd.[5] Abhandl. in Naturf. XXII. pl. 1 and 2.
LONG-NOSED MONKEY.—Penn. Quadr. l. No. 111.
LA GUENON À FACE ALONGÉE.—Buff. Hist. Nat. Suppl. VII. pl. 15.
SINGE NOIR.—Le Vaill. Voy. II.

SPECIFIC CHARACTERS.

THE HAIR black, with a yellowish or greenish tinge; the cheek-tufts grey.

THE FACE and HANDS black. THE TAIL very long, with a tuft at the end.

INHABITS the Cape of Good Hope.

This species resembles the preceding in form and manners. The adult has a long mane, and the tail, which terminates in a tuft, extends as low as the heel.

Several animals of this species, called Choak-kama or Chacma by the Hottentots, were preserved for a long time in the Paris Menagerie. An incident occurred with a male specimen, brought by Captain Baudin from the Cape of Good Hope along with its female, serving forcibly to illustrate the peculiar disposition of these animals. Though rather mild when first imported, it soon lost its docility. Having one day escaped from its cage, the keeper imprudently threatened it with a stick; in an instant the animal flew upon him, and inflicted three deep wounds upon his thigh with its strong canine teeth, which penetrated as far as the femur, so that the life of the man was for a long time rendered precarious. To induce the animal to return to its cage, the following stratagem was adopted. The keeper had a daughter, who often fed the Baboon, and for whom it had at all times exhibited a powerful attachment. She placed herself at the side of the cage, opposite to the door at which the animal was to enter, and a man was made to approach the young woman as if he were about to caress her. The moment the Baboon perceived this movement, it raised a frightful cry, and throwing itself in the fury of its jealousy upon the individual, rushed into the cage, which was instantly closed from behind.

The greenish-black hue of its coat proceeds from hairs which are grey at the base, and otherwise black, excepting some rings of yellow more or less dingy. The face, ears, and the palms of the hands, are naked, and few hairs are to be seen on the internal surfaces of the arms and thighs. The fingers, especially of the hinder hands, are covered with short, coarse, and black hairs; the tail terminates in a strong black tuft; and the neck is furnished with long hairs, forming a mane, which is wanting in the female. The skin of the hands, face, and ears, is of a violet-black; but the circle round the eyes has a paler tint, and the upper eyelid is white as in the Mangabey Guenon. The nostrils are separated above by a deep furrow, the upper and anterior portion of the head is wholly flat, and the callosities are very small.

4. CYNOCEPHALUS HAMADRYAS.—DOG-FACED BABOON.

Syn. LE TARTARIN DE BELON.[6]—Cuv. Reg. Anim. l. 98.
PAPIO HAMADRYAS.—Geoff. Ann. Mus. XIX.
SIMIA HAMADRYAS.—Linn. Gmel. I.—Fisch. Syn. Mam.
CYNOCEPHALUS HAMADRYAS.—Desm. Mam.

Icon. TARTARIN mâle.—Tartarin femelle jeune.—F. Cuv. et Geoff. Hist. Mam.
Cyn. Hamadryas, male, female, and young.—Ehrenb. Symb. pl. 11.
DOG-FACED MONKEY.—Penn. Quadr. I. pl. 43, fig. 1.
SINGE DE MOCO.[7]—Buff. Suppl. VII. pl. 10.

---

[1] M. Brongniart's Paper on Simia Cynocephalus, in the JOURN. D'HIST. NAT. (Journal d'Histoire Naturelle, par MM. Lamarck, Bruguières, Olivier, Haüy, et Lepelletier, Paris. 1792).

[2] Those figures which represent the tail short, as in Buff. Hist. Nat. XIV. pl. 13 and 14, are made after mutilated specimens. M. Brongniart has represented the former with some precision, but under the improper name of Simia Cynocephalus. His figure is copied in Schreber, pl. 13 B.—*Note of the Baron Cuvier.*

[3] All these fictitious species have arisen from the greater or less state of preservation of the specimens, or differences of age.—*Note of the Baron Cuvier.*

[4] J. Herrmanni Observationes Zoologica, Edidit F. Hammer, Argent. 1804.

[5] Boddaërt Abhandlung über den Affen mit dem Schweinskopfe (Der Naturforscher, Halle, 1774 und 1804).

[6] See fol. 101 of L'Histoire de la Nature des Oiseaux, avec leurs Descriptions, &c., par P. de Belon du Mans. Paris, 1555.

[7] This has been copied in Schreber, and the colouring is faulty.—*Note of the Baron Cuvier.*

SPECIFIC CHARACTERS.

THE MALE.

THE HAIR of a blueish-grey, very long on the neck and cheeks, forming a mane.

THE TAIL long and tufted.

THE FEMALE AND YOUNG.

THE HAIR of a dark greenish-grey. No mane. THE FACE flesh-coloured.

INHABITS Arabia, Persia, and Ethiopia.

This animal is likewise exceedingly brutal and ferocious.

The blueish-grey colour of its fur has a slight tinge of green, resulting from rings which are alternately black and yellowish-grey. The hinder parts of the body are paler than the anterior, the fore-limbs almost black, the cheek-tufts and abdomen whitish. The face, ears, and hands, are of a deep-brown, the buttocks red, as in all the adult Baboons. A well-marked groove separates the nostrils above. The hair of the mane is about six or seven inches in length, commencing from the neck, and covering all the anterior parts of the body. The abdomen and inner surface of the thighs are scantily covered, and the tail terminates in a small tuft.

The female differs from the male in having no mane, and her hair is of a dark greenish-grey. The young males resemble the adult females.

M. Ehrenberg saw several wild individuals of this species in Arabia. They live together in small families.[1]

(B.) MANDRILLS. (LES MANDRILLS, Cuv.)

Of all Apes, the Mandrills have the longest muzzle (about 30°); their tail is very short; they are likewise very brutal and ferocious. The nose does not differ from that in the preceding section.

In this section, two species alone are now admitted by Naturalists.

5. CYNOCEPHALUS MORMON.—VARIEGATED BABOON.

Syn. LE MANDRILL.—Cuv. Reg. Anim. I. 98.
PAPIO MORMON.—Geoff. Ann. Mus. XIX.
SIMIA MORMON ET MAIMON.—Linn. Gmel. I.—(S. MAIMON.)—Fisch. Syn. Mam.
CYNOCEPHALUS MORMON.—Desm. Mam.

Icon. Le Mandrill.—Audeb. Sing.—Ménag. du Mus.
Mandrill mâle vieux.—Mandrill mâle jeune.—F. Cuv. et Geoff. Hist. Mam.
LE MANDRILL, BOGGO, CHORAS.—Buff. Hist. Nat. XIV. pl. 16, 17.—Suppl. VII. pl. 9.
GREAT BABOON.—Penn. Quadr. I. pl. 40 and 41.—RIBBED-NOSE BABOON (young).
VARIEGATED BABOON.—Shaw, Gen. Zool. I. pl. 10.

SPECIFIC CHARACTERS.

THE MALE.

THE HAIR greyish-brown, olive-brown above, white beneath, a citron yellow tuft under the chin.

THE CHEEKS with bright purple stripes. THE BUTTOCKS bright purple. THE NOSE bright scarlet, likewise the CALLOSITIES.

THE FEMALE AND YOUNG.

THE HAIR nearly the same as the male, no tuft under the chin.

THE CHEEKS with a few small blue stripes.

THE FACE otherwise black.

INHABITS Guinea.

In the adult males, we find the nose red, especially at the point, where it becomes scarlet, and this circumstance has very erroneously occasioned the adult to be set down as a distinct species[2] from the young. The parts of generation and the circle of the anus are of the same colour, and the buttocks are of a bright purple, so that one can scarcely imagine an animal so hideous and extraordinary. It attains nearly the size of Man, and the Negroes of Guinea hold it in much dread. Many traits of its history have been confounded with those of the Chimpansee, and consequently of the Orang-Outang.

The adult Variegated Baboon (S. Mormon, Linn. and Choras, Buff.) presents a singular and revolting combination of characters, in which the

peculiarities of Man are brought into close and degrading approximation to those of the brute. Its bright and yellow eyes are deeply sunk in beneath a low forehead, and approximate so nearly to each other, as to impart a peculiar air of ferocity to the countenance. Its muzzle is enormous, terminating in a round and flat surface of a bright scarlet, continually moistened by a disgusting mucus. The cheeks are very prominent, and furrowed with longitudinal ridges of a bright blue, changing into violet. A narrow stripe of crimson, running along the centre of the nose, divides the face into two parts, giving it a wounded or lacerated appearance. Its hair is of a greyish-brown, spotted with yellowish-brown, especially near the head; white on the breast and abdomen, as well as inside the thighs, on the neck, and behind the ears. The tail being very short, is continually erect, disclosing the brilliant hues of red and blue already noticed. Its limbs are exceedingly muscular, and its strength is much superior to that of Man.

The hair owes its colour to alternate rings of black and yellow, forming by their combination a greenish-brown, common to many Apes, but of a darker hue in the Variegated Baboon than in any others. A white band commences from each ear, and passing upwards is interrupted at the vertex. The skin round the eyes is of a violet-brown. The hairs on each side of the head, being very long, unite together on the summit, forming a kind of crest, the centre of which is sometimes elevated into a pointed tuft.

The colours of Mammiferous animals are in general dull and tarnished, Nature having sparingly bestowed upon them those vivid hues, which she has so liberally distributed among the feathered tribes. By a singular exception, however, the Mandrills, when arrived at mature age, exhibit such dazzling combinations of red and blue, as may vie with the colours of the brightest birds. These brilliant reflections do not proceed from the hairs, but from the skin itself; yet they are not inherent in its substance, but depend upon the vital energy of the animal. Before the adult period of life, they are scarcely seen, and they become dull and tarnished whenever the animal is indisposed.

The young Variegated Baboon (S. Maimon, Linn.), before the appearance of its canine teeth, has a broad and short head, with the body rather thick; the face is bluish, with only the sides of each jaw blue and furrowed, while the thighs present no particular colour. As soon as the canines begin to appear, the body and limbs become longer, while the physiognomy gradually acquires the characters of the adult.

The females always remain of smaller size than the males, and their skin does not acquire any bright or vivid colours, while the nose never becomes entirely red. At the rutting period, which occurs every month, the vulva is surrounded by a monstrous protuberance, resulting from the accumulation of blood in those parts, and generally assuming a spherical form. When the rutting period is over, this spherical protuberance gradually diminishes, but reappears in about twenty-five or thirty days.

It is singular that no animals, excepting the Baboons, distinguish women in a crowd from the other sex; at the same time, no species gives more striking marks of this singular partiality than the Variegated Baboon. It marks out the youngest ladies, however disguised by the fashion of the day, invites them by voice and gesture, and there can be little doubt that if unconfined, the danger of the fair ones would be very imminent. May not these have been the Satyrs of antiquity? George and Frederic Cuvier are of opinion that they were wholly unknown to the ancients.

Originally from Africa, and especially from the regions near the Gulf of Guinea, they do not appear to have extended as far south as the Cape of Good Hope.

Camper and Vicq-d'Azyr have described the membraneous sac, which communicates with the larynx, and serves to render their voice hoarse. They emit a kind of growl, which may be expressed by the syllables aoo-aoo.

6. CYNOCEPHALUS LEUCOPHÆUS.—DRILL BABOON.

Syn. LE DRILL.—Cuv. Reg. Anim. I. 99.
SIMIA LEUCOPHÆA.—F. Cuv. in Ann. Mus. IX.—Fisch. Syn. Mam.
CYNOCEPHALUS LEUCOPHÆUS.—Desm. Mam.

Icon. Drill mâle, Drill très vieux, Drill très jeune, Drill femelle.—F. Cuv. et Geoff. Hist. Mam.—Ann. Mus. IX. pl. 37 (young).

SPECIFIC CHARACTERS.

THE MALE.

THE HAIR yellowish-grey; white beneath, and on the side of the head.

---

[1] To this place the Baron Cuvier has referred his Simia nigra in the following words :—" We must distinguish from the other Cynocephali, a species entirely black and without any tail (S. nigra, Cuv.), but the head of which resembles the others." The last portion of this sentence does not appear to have been stated with M. Cuvier's usual precision, as the nostrils are not terminal. We have accordingly followed Bennett in considering this animal as a true Macacus (see Macacus niger already described). Isid. Geoffroy forms this animal into a third section, under the name of Cynopitheques, Cynocéphales-Magots, or Tailless Baboons, analogous to the Magots or Tailless Macacos of the preceding genus.

[2] We have ourselves seen, along with M. Geoffroy, two or three Mandrills, or S. maimon, change into Choras, or S. mormon, in the Menagerie of the Museum. The tuft of hair which is added to the characters of the Mormon is often found also in the Maimon.—Note of the Baron Cuvier.

The Face black. · The Under-jaw bright red.

The Tail very short and slender; bright red near the base.

THE FEMALE AND YOUNG.

The Ha'r greenish-grey above; white beneath, on the lower jaw, and on the side of the head; cheek-tufts brownish.' The Face black.

The Tail very short and slender, greyish-white.

Inhabits Africa.

This animal was first described by M. Fred. Cuvier, who has given us a very complete account of the species in all the different stages of its growth. It bears many points of resemblance to the Variegated Baboon already described, and seems to share the same propensities. The upper parts of the body are, however, of a deeper green, and there is more white on the other parts. It wants the deep blue stripes upon the face, as well as the scarlet and violet-blue colouring of the buttocks.

DOUBTFUL SPECIES.

1. The Crested Baboon (Simia cristata of Fisch.), from Africa, is described by Pennant (Quadr. No· 101) from a faded specimen in the

Leverian Museum. Shaw (Zool. I. p. 26) copies Pennant's description, but erroneously assigns it to India.

2. The Long-nosed Monkey (Penn. Quadr. No. 111), the Guenon à museau alongé of Buff. Hist. Nat. Suppl. VII., and the Simia nasuta of Shaw, is probably identical with the Chacma Baboon already described. Also, the Prude Monkey of Penn. Quadr. No. 111.

3. The Wood Baboon, Cinereous Baboon, and Yellow Baboon (Penn. Quadr. Nos. 95, 96, and 97), are identical either with Cynocephalus mormon, or C. leucophæus. As to which of these should claim the preference, it is impossible to decide, for the bright colours of the Mandrills fade after death, and Pennant's descriptions are merely derived from skins belonging to the Leverian Museum.

IMAGINARY SPECIES.

1. Le Babouin chevelu (Papio comatus), Geoff. Ann. Mus. XIX. is identical with Cynocephalus porcarius.

To these we shall add,

2. The Little Baboon of Pennant, Simia Apedia (Linn. Gmel; I. 28), which is probably the young of some species of Macacus.

ADDITION TO
GENUS I. PITHECUS.—MEN-OF-THE-WOODS.
(See pages 147—149.)

Before concluding the Natural History of the Apes of the Old Continent, it will be necessary briefly to notice the important acquisition made to Zoological science by the recent discovery of the adult Orang-Outang,[1] as well as the highly probable existence of two species of Pithecus in the Island of Sumatra.

1. PITHECUS SATYRUS.—ORANG-OUTANG.

THE ADULT.

Add. Syn.   Orang de Wurmb.—Geoff. Cours. d'Hist. Nat. I* leç. 31.

Icon.   Temm. Mon. Mam. II. pl. 41 (old male).—Pl. 42 (old female).—Pl. 43 (views of the head).—Pl. 44 (young).—Pl. 45 and 46 (skeleton).

SPECIFIC CHARACTERS.
THE ADULT MALE AND FEMALE.

The Muzzle very prominent; a large protuberance on each cheek, in the male only. The Forehead much depressed.

The Pectoral Limbs very long, reaching as low as the ancles.

The External Ear, Eyes, Mouth, and Nose, small.

The Hands narrow, the Fingers long; the Thumbs of the hinderhands most commonly without nails.

The Tail, Cheek-pouches, and Callosities, all wanting.

The Hair long and scanty, of a deep chestnut brown; a long and pointed beard, yellowish-red, in the male only; the eyelashes wanting.

THE YOUNG.

The Muzzle prominent; no protuberance on the cheeks. The Forehead elevated. The Hair of the same colour as the adult; no beard; otherwise resembling the adult.

Inhabits Borneo and Sumatra.

The adult Orang-Outang, which is now ascertained to be identical with the Pongo of the Baron Van-Wurmb, is the largest of all known quadruhumanous animals, approaching very nearly to Man in stature. An old male in the possession of M. Temminck is above four feet three inches in height; but one in the possession of the Dutch Scientific Expedition, established at Banjarmassing, in the Island of Borneo, has attained the extraordinary stature of five feet seven inches. The head of the adult male is of singular dimensions; the cheeks are prolonged laterally, and bear a very prominent swelling on each side, in the form of a crescent. These fleshy protuberances give a deformed appearance to his face, while the excessive prolongation of the muzzle, and thickness of the lips, above which its very diminutive nose appears engrafted, combine in rendering its countenance one of the most hideous in the entire range of the Animal Kingdom. Its protuberances are nearly six inches in length, and about two inches in thickness. · They resemble those excrescences which are found in certain species of Hogs, and in all the known species of Phacochœrus; their texture consists of an adipose substance, hard to the touch, disposed in a very abundant cellular tissue. Nothing is yet known of the functions to which this peculiar organization may be subservient. It is only developed in the male, when very nearly adult, probably about

the age of eight or ten years, and there is no appearance of it in the females.

The forehead of the adult male is almost wholly naked. The orbits are prominent, the eyes one-third smaller than those of Man. It has no eyelashes, but its diminutive eyelids are surrounded by a few stiff hairs. The nose is depressed, blending into the growth on the cheeks, and projecting only at the point, on the sides of which the nostrils open. These are separated by a partition, extending beyond their lower margin, and blending into the thick upper-lip; the latter, as well as the lower lip, is very thick and fleshy. The lower jaw terminates in a very broad chin, truncated, and projecting beyond the upper jaw; it bears in the male a long and pointed beard. The mouth is a horizontal cleft, very small in proportion to the height of the animal. All these parts are nearly destitute of hair, excepting a few scattered ones, of a yellowish-red, on the temporal ridges. The lateral parts of the lips are supplied with a kind of moustachio, arising at the angles of the nose, and extending to the angle of the mouth. The ears are small, and formed like those of Man, with a fixed lobe. The hinder part of the head is of a roundish form; all the hairs with which it is covered proceed from a common centre, and are disposed in rays; the vertex stretching far behind terminates in a depressed occiput.

All parts of the trunk are heavy, massive, and destitute of elegance, owing to the extreme size of the haunches and the volume of the abdomen. The breast is almost naked; the hairs become more abundant along the sternum to the abdomen, where they are neither so long nor dense as to cover the skin, which may be perceived throughout. The back, as far as the haunches is still less hairy, but the sides of the body are abundantly supplied with long hairs, which fall down upon those with which the legs are covered. The fore-limbs are very considerably longer than the hinder; they nearly touch the ground, when the animal stands erect; and the fore-arm especially is of considerable length. These parts are very hairy, but less so towards the hands and fingers, where they are very short. All the hairs of the fore-arm point towards the elbow, where they unite with those of the humerus, and end in a point. The fingers, as well as the metacarpus and metatarsus, are much longer than those of Man, and hence the thumb is placed at a considerable distance behind. The hinder-thumbs are rather short, perfectly opposable to the other fingers, and forming, with the index of the hinder-hands, a semicircle. This organization plainly indicates that the Orang-Outang is not adapted for walking on two feet, but that it is wholly organized for climbing trees. Its movements on the ground are constrained, either when erect or on all the four hands.

M. Temminck has examined six individuals of different ages killed in the wild state, without being able to find the slightest indication of a nail upon the thumb of the hinder feet, and the skin covering the last phalanx of the thumb is not even harder than in any other place. A seventh specimen, which lived for several years in captivity, had the thumb of the right hinder-hand without a nail, but a perfect nail appeared upon the left. Two other skeletons, in the collection of the King of Holland, had nails on all their thumbs, and these also died in captivity. The much agitated question may therefore be fairly considered as decided, and the thumbs of the hinder-hands are wanting in the normal state of the Orang-Outang. All the other fingers are furnished with black nails, longer and more curved than those of Man. The same relative length prevails

---

[1] Our account of the Orang-Outang was written in June 1838, and we have now (Oct. 1838) just received the excellent description of M. Temminck, with lithographic views of the adult.

among the fore-fingers as in Man, but the index of the hinder-hands is invariably the longest of all. and the other fingers diminish gradually to the fourth finger, which is the shortest. The naked parts of the hands present the same arrangement of the papillæ of the skin in concentric curves, and as the papillæ on the tips of the fingers are very fine, we may infer that the organ of touch is extremely delicate in the Orang-Ou-tang.

All the naked parts of the body and head, excepting the orbits and the lips, are of a blueish slate grey. The hair is uniformly throughout of a deep chestnut brown, more or less glossy, but the beard and moustachios are of a yellowish-red. There is no difference in the colours of the male and female, even in their different periods of age. The young of the year, those of five, six, or eight years, do not vary in this respect from the full-grown adults; but there is a slight difference in the quantity of hair, the young being more plentifully furnished than the adult.

None of the individuals in the possession of M. Temminck have true callosities; the epidermis being merely hardened by the frequent sitting posture. The teeth of the old animals, especially the males, are much used by detrition, so that their original structure and the crown have to-tally disappeared. The canines of the males are much stronger than those of the females, which are regularly straight and conical; while those of the males are very strong compared with the other teeth, and their direc-tion outwards is strongly marked.

On comparing the skeleton of the adult male with that of Man, several modifications were noticed in its structure. The seven vertebræ of the neck form a column as long as that of Man, but their spinous processes, commencing with that of the axis, are vastly longer, though but slightly forked at their extremity. The hole for the passage of the spinal cord is much narrower than in Man. The variation in the magnitude of the fa-cial angle, according to the ages of individuals, is very remarkable.

In the skulls of two very old males, it was from . 35° to 37°
In a female nearly adult, ... 38°
In a female at the period when the last molar was appearing, 40°
In an individual of less size, . . . 48°
In two individuals about 1 ft. 11 in. high, . 52°
In a very young specimen about 1 ft. 6 in. high, . 65°

· 2. PITHECUS ABELII.—RED ORANG.

THE YOUNG.

Syn. ORANG ROUX.—Temm. Mon. Mam. II. 136.
JEUNE ORANG-OUTANG.—Marion de Procé in Ann. des Sc. Nat. V. (2d series), p. 313.
Icon. ORANG-OUTANG.—Abel. Chin. p. 318.

The existence of a species of Red Orang in the Islands of Sumatra and Borneo is rendered extremely probable, by the discovery that the young Orang-Outang is of the same dark chestnut brown colour as the adult. Hitherto, Dr Abel's Orang-Outang, as well as the notice of M. Marion de Procé, have been considered as referring to the young Orang-Outang; but they now appear to belong to the young of an unknown species, which may be called the Red Orang.

It differs from the young of the Orang-Outang in being covered with long red hair; that of the head extends in front upon the forehead, pro-ducing the appearance of a periwig. It has long eyelashes, and its muzzle is not prominent.

ADDITION TO

GENUS VIII. sp. 8.—MACACUS NEMESTRINUS,—PIG-TAILED MACACO.

(See page 164.)

Add. Syn. SIMIA CARPOLEGUS.—Raffles, in Linn. Tran.. XIII. 243.

The Pig-tailed Macaco, called Bruh by the Malays, is very common in the neighbourhood of Bencoolen. Of the three varieties found there, the Bruh-setopong is the largest, the most docile, and most intelli-gent. It is much esteemed by the inhabitants, who train it to ascend trees for the purpose of gathering cocoa-nuts, in which service it is very expert. When sent to gather this fruit, it selects the ripe nuts with great judgment, and pulls no more than it is ordered. Its height is about two feet when sitting. The other varieties, called Bruh-selasi and Bruh-puti, are of a darker colour, more intractible, and less intelligent.

TRIBE II.—PLATYRRHINA.—APES OF AMERICA.

SYNONYMS.

PLATYRRHINI (πλατὺς, *platās*, broad, 'ρινα, *rrhina*, nostrils).—Geoff. Ann. Mus. XIX. 104.
LES SINGES DU NOUVEAU CONTINENT.—Cuv. Reg. Anim. I. 99.
CEBUS.—Fisch. Syn. Mam. 37.

CHARACTERS OF THE TRIBE.

THE DENTAL FORMULA, $\frac{2 + C + (3\ F + 3)\ M}{2 + C + (3\ F + 3)\ M} \frac{18}{18} = 36$

THE NOSTRILS separated by a broad partition, and opening on the sides of the nose, the genus Eriodes excepted.
THE CALLOSITIES and CHEEK-POUCHES always wanting.
THE TAIL always long, sometimes prehensile.
INHABIT America.

THE Apes of the New Continent have four molar teeth more than the others, making thirty-six teeth in all; they have long tails, no cheek-pouches; the buttocks hairy, and without callosities; the nostrils [usually] pierced in the sides of the nose, and not beneath. All the larger Quadrumanous animals of America belong to this division; their great intestines are less inflated; their cœcum longer and thinner than in the Apes of the Old Continent.

These Quadrumanous animals form a natural group, wholly distinct from those hitherto described. Buffon was the first to notice the re-markable difference in their characters, which would almost seem to evince that they belonged to different creations. It will be recollected that the Apes of the Old World have, with the single exception of the Orangs, their buttocks destitute of hair, while natural and inherent cal-losities cover those parts; they most commonly have cheek-pouches for holding their provisions; and the partition of their nostrils is narrow, and opens beneath the nose as in Man. All these characters are want-ing in the Apes of America. The partition of their nostrils is, with a few exceptions, very thick; the nostrils open on the sides of the nose and not beneath; their buttocks are entirely covered with hair, and they have no callosities. They are wholly desitute of cheek-pouches, and they differ not only specifically from the Quadrumanous animals of the Old Conti-nent, but generically, and these primary variations in the characters, which their generality renders highly remarkable, draw along with them a number of subsidiary differences, rendering the subdivision at once natural and satisfactory.

Although in the normal and perfect state of these Apes of America, we find six molar teeth on each side and in each jaw, it occasionally hap-pens, as well as with those of the other Continent, that a less number is found either when young individuals have not acquired their full com-plement of teeth, or when individual specimens have lost some of them through old age. Hence it will sometimes happen that only five molars will be found, as in the Apes of America. M. Geoffroy has, however, noticed on one occasion the existence of seven molars on each side of the upper jaw, in a very old Cebus Apella; and the same number of molars have been found by M. Isidore Geoffroy in both jaws, but only on one side, of an Ateles pentadactylus.

Buffon first proposed the subdivision of the Apes of America into two sub-tribes, the SAPAJOOS and SAGOINS, according as their tails are, or are not, prehensile. Subsequently M. Spix separated them into GYMNURI, or Naked-tails, wherein the extremity of the tail is naked and callous beneath, and TRICHURI, or Hairy-tails, where it is entirely covered with hair. Were it not for the genus Cebus, or Weepers, whose tails are at once hairy and pre-hensile, these two divisions would coincide; the Sapajoos and Gymnuri being otherwise prehensile and naked, while the Sagoins and Trichuri are not prehensile, and are covered with hair over the entire surface of the tail.

The prehensile nature of the tail, upon which Buffon's subdivision is founded, is at once a striking and singular adaptation of an organ, which in most other animals is either rudimentary, or hangs uselessly downwards. It becomes among the Sapajoos in some respects a fifth hand, by which the animal can seize distant objects without moving its body, or hang sus-pended from the branches of a tree, even after life is extinct.

43

SUB-TRIBE I.—CATECHURA.[1]—SAPAJOOS.

SYNONYMS.

HÉLOPITHECI (Helopithèques).—Geoff. Ann. Mus. XIX. 105.
LES SAPAJOUS.—Buff. Hist. Nat.—Cuv. Reg. Anim.
CEBUS.[2]—Erxl. p. 44.

CHARACTERS OF THE SUB-TRIBE.

THE TAIL prehensile, and long.

Some of the American Apes have their tails prehensile ; that is to say, its extremity is capable of rolling itself with sufficient force round other bodies, so as to seize them like a hand. These are more particularly styled Sapajoos.

GENUS I. MYCETES.[3]—HOWLERS.

Syn. LES ALOUATTES.—Cuv. Reg. Anim. I. 99.
MYCETES.—Illig. Prodr. 70.—Kuhl. Beitr.—Desm. Mam.
STENTOR.—Geoff. Ann. Mus. XIX. 107.
CEBUS (in part).—Fisch. Syn. Mam.—Erxl. p. 44.
SIMIA (in part).—Linn. Gmel. L.

GENERIC CHARACTERS.

THE HEAD pyramidal. THE FACE oblique. THE FACIAL ANGLE 30°.
THE HYOID BONE cavernous, capacious, and appearing externally.
THE TAIL naked beneath the point.
THE HANDS pentadactylous. THE NAILS short and convex.
INHABIT South America.

AT the head of the Sapajoos we may place the Howlers, which are distinguished by their pyramidal head. Their upper jaw descends much lower than the cranium, while the ascending branches of the lower jaw are much elevated, for receiving a bony drum, formed by a vesicular expansion of the hyoid bone, which communicates with the larynx, and gives an enormous volume and terrific tone to their voices. Hence their name of Singes hurleuses, Brullaffe, or Howlers. The prehensile portion of their tail is naked beneath.

This genus contains several species, the distinctive characters of which are not very definitively fixed, as the colour of the hair, upon which it is founded, varies according to the differences of age and sex.

These animals are at once distinguished from all other Apes by the expansion of the throat, and by their terrific howl, resembling the grunting of a herd of swine. Though monogamous, they are found in troops of fifteen or twenty, and fill the air at the rising and setting of the sun with their mournful howls, which may be heard to a very remote distance. The concert usually commences with the note of a single Howler, whose example is speedily followed by all the remainder. These Apes are by no means nimble, but heavy, stupid, and lazy. They feed chiefly on leaves. When perceived by the hunters in their inaccessible retreats among the dense foliage of lofty trees, surrounded by rocks and rivers, they do not fly with the agility of the other Apes, nor do they take flight to any distance, but moving slowly, and howling piteously, they climb higher towards the tops of the trees. The females carry their young clinging to the back or under the belly. As the Howlers are in general very large and fat, they are in great request among the Colonists and Indians, who use them as food. They are dressed with the skins on, well singed, and roasted before a fire. Bouilli à la singe is likewise accounted very palatable ; but the resemblance of the animal to a human child, especially about the head, gives the dish a revolting appearance. This kind of food is easily procured, as the Howlers are at once discovered by their cries, and the slowness with which they take to flight commonly exposes the entire troop to certain death.

This genus is natural, well defined, and characterized by having its limbs of medium length, and all terminated by five fingers, the anterior thumb being half as long as the first finger, very confined in its movements and scarcely opposable, but especially by the remarkable form of the skull and hyoid bone. The skull is pyramidal, and shaped in such a manner, that when it is made to rest upon the dentary margins of the upper jaw, that is to say, when the plane of the palate is held horizontally, the occipital foramen is on a level with the upper part of the orbits. The position of the occipital foramen is likewise singular ; it recedes back-

wards, and instead of being placed at the base of the skull, is perpendicular to it. The lower jaw is excessively developed, especially in its branches, which are so extensive, as to equal the entire skull in extent of surface. They form two deep partitions, containing between them a large cavity, in which is deposited a hyoid bone modified in a remarkable manner. The body of that bone is transformed into an osseous chest, with very thin and elastic sides, presenting a large opening behind, on the sides of which are articulated two pairs of horns, forming the half of an ellipsoid, when they have attained their full growth. This chest is about two inches and a half in each diameter, and almost square. In consequence of this enormous growth, the hyoid bone extends beneath the lower jaw, and forms a projection, covered externally and concealed by a long and thick beard. The precise manner in which this apparatus influences the sound, so as to produce a volume so enormous, has not yet been distinctly explained. The larynx does not differ from that of the Weepers (Cebus), except by the presence of two membraneous sacs, into which the ventricles open, and communicate with the hyoid bone.

The females of the Howlers, as well as those of other American Apes, do not appear to be subject to the "écoulement périodique;" they produce only one young one at a time, and this they carry on the back. They appear to have much affection for their young. D'Azara seems to consider them as polygamous, but Spix positively assures us that they are not so. They are domesticated with difficulty, and, as far as we know, have never yet been seen alive in Europe.

At the present date, only four species are plainly recognizable ; but this number has been more than doubled by the vague indications of several German and French writers.

1. MYCETES SENICULUS.—ROYAL HOWLER.

Syn. L'ALOUATTE ROUSSE.—Cuv. Reg. Anim. I. 99.
STENTOR SENICULUS.—Geoff. Ann. Mus. XIX.
MYCETES SENICULUS.—Kuhl. Beitr.—Desm. Mam.—Latr.
CEBUS SENICULUS.—Erxl. p. 46.—Fisch. Syn. Mam.
SIMIA SENICULUS.—Linn. Gmel. — (Mono Colorado). — Humb.[4] Obs. Zool. p. 342 and 354.
ROYAL MONKEY.—Penn. Quadr. No. 132, a.
Icon. L'Alouate.—Audeb. Sing.
Buff. Hist. Nat. Suppl. VII. pl. 25.

SPECIFIC CHARACTERS.

THE HAIR of the head, arms, hands, and tail, deep brownish-red ; elsewhere, bright yellowish-red ; a beard long and bushy.
THE FACE naked and black.
INHABITS French Guiana, Carthagena, and the banks of the Magdalena.

The Royal Howler, about the size of a large Fox, of a bright yellowish-red, deeper on the head, [limbs], and tail, often comes to us from the woods of Guiana, where it lives in troops.

Its food consists of leaves rather than fruits, and its cry is composed of short and hoarse sounds, proceeding from the depth of the throat, resembling the grunting of a Hog, but infinitely louder. When first captured, this animal is very savage ; but when brought up in captivity, we are told that it loses its voice, becomes melancholy, and does not long survive.

The face of the Royal Howler is black, and naked with the exception of a few scattered red hairs, some scanty black hairs for eyelashes, and a few on the lips. The hair of the forehead is deep brownish-red, very short and thick, and pointing backwards, while the hair of the occi-

---

[1] Catechura—from κατέχω, katechō, to hold fast, and ουρά, oura, a tail.
[2] Cebus, or Cepus (Κῆπος)—names of an Ethiopian Ape, which appears, from Ælian's description (XVII. c. 8), to have been the Red Guenon (Cercopithecus ruber).—Note of the Baron Cuvier.
[3] Mycetes—from μυκητης, mukētēs, howling.
[4] HUMB. OBS. ZOOL.—Recueil d'Observations de Zoologie et d'Anatomie Comparée, faites dans l'Océan Atlantique, dans l'Intérieur du Nouveau Continent, et dans la Mer du Sud, pendant les années 1799 à 1803, par Al. de Humboldt et A. Bonpland. Paris, 1811.

put points forwards, occasioning a very small tuft at the vertex. On the temples and cheeks it is of the same colour, but very long; the beard is very broad and bushy. The hair of the arms from the shoulder to the elbow, as well as of the back, sides, breast, and abdomen, is bright yellowish-red; that of the fore-arms, hands, thighs, legs, and tail, of a very deep brownish-red. The tail is as long as the body and head taken together.

2. MYCETES URSINUS.—URSINE HOWLER.

Syn. L'ALOUATE OURSON.—Cuv. Reg. Anim. I. 99.
STENTOR URSINUS.—Geoff. Ann. Mus. XIX.
MYCETES URSINUS.—Kuhl. Beitr.—Desm. Mam.—Pr. Max.[1] Beitr. II. 48.
CEBUS URSINUS.—Fisch. Syn. Mam.
SIMIA URSINA (Araguato de Caracas).—Humb. Obs. Zool. p. 329 and 355.
Icon. Mycetes ursinus.[2]—Pr. Max. Abbild.[3]
Humb. Obs. Zool. pl. 30.

SPECIFIC CHARACTERS.

THE HAIR reddish-brown throughout, scanty on the abdomen; the beard strong and thick.
THE FACE naked and blackish.
THE naked portion of the TAIL black.
INHABITS Brazil.

The Ursine Howler does not differ greatly from the preceding species.

It abounds in the primitive forests of Brazil, where it interrupts the silence of night by its stunning howl, resembling the sound of a drum. The natives regard it as very excellent game, especially in winter, when it is very fat. Being of a very mild nature, it is easily tamed when taken young; but the slowness of its movements, and the disagreeable monotony of its howl, must render it by no means a very agreeable domestic animal. It appears to inhabit a great portion of South America.

The hair, which is of a uniform reddish-brown, is much darker in the young, than approaches in the adult to a rusty red, or reddish-brown. M. Humboldt, who frequently observed the females carrying their young upon the shoulders, did not remark any difference in the colours of the sexes. They are found in immense numbers, sometimes as many as forty being seen on a single tree. The leaves of trees, rather than the fruit, appear to be their habitual food. When domesticated, they are more steady than most other Monkeys, and apparently of a less delicate constitution.

VAR. FUSCUS.—BROWN HOWLER.

Syn. STENTOR FUSCUS.—Geoff. Ann. Mus. XIX.
MYCETES FUSCUS.—Desm. Mam.—Kuhl. Beitr.
Icon. Spix,[4] Sim. et Vespert. Bras. pl. 30. (Mycetes fuscus mas.)

This is considered by some to be the Ouarine of Buffon, but it undergoes much variation. Though generally of a dark chestnut brown, with the back and head passing to a bright chestnut, and the points of the hairs golden yellow, it seems scarcely distinguishable from the Ursine Howler already described.

There appear to be some other species, which are of a black, brown, or paler colour. In one of them this pale tint is ascertained to belong to the female [and young].

3. MYCETES CHRYSURUS.—GOLDEN-TAILED HOWLER.

Syn. Stentor chrysurus (L'Hurleur à queue dorée).—Isid. Geoff. Mém. Mus. XVII.
Icon. Isid. Geoff. Etud. Zool. pl. 7 (Mammifères).

SPECIFIC CHARACTERS.

THE HAIR of the back, sides, and hinder half of the tail, bright golden yellow; elsewhere dark chestnut brown.
THE FACE almost naked.
INHABITS Columbia.

The Golden-tailed has long been confounded with the Ursine Howler. It occurs frequently on the banks of the Magdalena, where it is known by the name of Araguato, which term is applied indiscriminately to several different species of Monkeys, all agreeing, however, in having a beard.

Like most other Monkeys, it lives in troops. M. Roulin informs us, that when a troop of these Howlers is passing from one tree to another, all the individuals composing it act in a manner precisely similar to each other, as in the school-boys' game of "follow-the-leader;" they leap successively to the same points, and place their hands in the same positions, as if each individual were obliged to imitate the motions of the animal preceding it.

On the hinder half of the tail, and on the upper surface of the body, from the shoulders to the insertion of the tail, the hair is of a very brilliant golden yellow; on the rest of the tail it is of a light chestnut brown; while on the remainder of the body, head, and limbs, it is of a very dark chestnut brown, especially on the limbs, where it merges into a violaceous tint. The face is almost wholly naked, but less so than in the Royal Howler.

4. MYCETES NIGER.—BLACK HOWLER.

Syn. STENTOR NIGER.—Geoff. Ann. Mus. XIX.—Desm. Mam.
CEBUS CARAYA.—Fisch. Syn. Mam.
SIMIA CARAYA.—Humb. Obs. Zool. p. 355.
MYCETES NIGER.—Pr. Max. Beitr. II. 66.
LE CARAYA.—D'Azara,[5] Quadr. Parag. II. p. 208.
Icon. MYCETES BARBATUS.[6]—Spix, Sim. et Vespert. Bras. pl. 32 (male).—Ib. pl. 33 (fem. and young).
Mycetes niger.—Pr. Max. Abbild. (fem.)

SPECIFIC CHARACTERS.

THE MALE.

THE HAIR soft and long; entirely black, tending to reddish on the breast; the beard very long.

THE FEMALE AND YOUNG.

THE HAIR light greyish-yellow.
INHABITS Brazil.

This Howler, called *Bugiu* by the Brazilians, is found plentifully in the interior of Minas Geraes and Bahia, among the low forests, distinguished by the name of Catinga. They live much retired, but always in numerous troops. In some parts of Bahia, the males are becoming very rare, owing to their being much hunted for their elegant black fur, which is used for ornamenting hats and saddles. They have the sagacity, when attacked, of sheltering themselves behind trunks and branches; and unless surprised by a sudden shot, they place themselves in such a position, that their bodies when deprived of life cannot fall to the ground. On this account the hunter loses the greater part of his game, as the trees which they frequent are almost inaccessible. The flesh of this species is preferred by the Portuguese and South American Indians to that of Ducks and several other animals.

The Black Howler is generally very corpulent. Its hair is plentiful above, but very scanty on the interior surface of the entire body, excepting on the centre of the breast, where there is a tuft of black hairs found in both sexes. The hair of the body, limbs, tail, and beard, is of a shining black in the adult male; on the top of the head and back of the neck it tends slightly to a brown, and to a greyish-white on the fingers. The hairs lie flat on the back and tail, are directed forwards on the top of the head; while they are short, straight, and directed backwards, on the forehead. The entire face is surrounded by a very dense beard; but is naked on the forehead, beneath the eyes, on the lips and chin, with a few scattered black, stiff, and short hairs, mixed with others of considerable length. The ears are round, very distinct, and slightly hairy behind. The tail is nearly as long as the body, very thick, with about a fourth part of its under surface callous, and sloping gradually to a point. The nails are rather long, black, slightly curved, and the thumb of the hinder-hands flattened. Its voice resembles the croaking of a Frog. The female has the same characters, excepting that her body is less corpulent; the back, and sometimes the upper surface of the tail, is blackish; the remainder of the body of a greyish-yellow, the beard less dense, shorter, divided near the throat, and of a greyish or reddish-yellow. The forehead is broader, and marked on the sides and at the centre of the forehead by dark brown lines; the tail is less thick and callous beneath for nearly one-half of its length. The young have the same characters as the females, excepting that the dark hues become deeper with their age.

[1] PR. MAX. BEITR.—Beiträge zur Naturgeschichte von Brasilien, von Maximilian, Prinzen zu Wied. Weimar, 1825.

[2] Regarding the Mycetes ursinus of Prince Maximilian, the Baron Cuvier remarks, " It appears to be much browner than the ursinus of M. Geoffroy, and to approach more nearly to the M. fuscus or the M. discolor of Spix, pl. 30 and 34. It is this last which appears to be the Stentor fuscus of Geoff." We do not coincide in this opinion of the Baron. It appears to us that Spix, pl. 34 (M. discolor), is the Cebus Belzebul of Erxleben, and merely a variety of the Stentor niger of Geoffroy; while Spix, pl. 30 (M. fuscus), is identical with the S. fuscus of Geoffroy, and a variety of the M. ursinus. See the Synonyms in the text.

[3] PR. MAX. ABBILD.—Abbildungen zur Naturgeschichte Brasiliens, herausgegeben von Maximilian, Prinzen von Wied. Weimar, 1824—1831.

[4] SPIX, SIM. ET VESPERT. BRAS.—Simiarum et Vespertilionum Brasiliensium Species Novæ, ou Histoire Naturelle des Espèces Nouvelles de Singes et de Chauve Souris, observées et recueillies pendant le voyage dans l'Intérieur du Brésil, exécuté par ordre de S. M. Le Roi de Bavière, dans les années 1817 à 1820, publiée par Joan de Spix. Monachii, 1823.

[5] D'AZARA, QUADR. PARAG.—Essais sur l'Histoire Naturelle des Quadrupèdes de la Province du Paraguay, par Don Félix d'Azara, écrits depuis 1783 jusqu'en 1796, traduits par M. L. E. Moreau-Saint-Méry, Paris, 1801.

[6] Marcgravius (Bras. 227) describes a species of Howler entirely black and bearded. The figure will be found at p. 228 of that work, but under the erroneous name of Ezquima. It seems to be the Mycetes barbatus of Spix, pl. 32. The female, pl. 33, is of a pale yellowish-grey, and the male will be found to be the Mycetes niger of Kuhl and Prince Maximilian of Wied Neuwied. The Caraya of Azara, said to be black, with the breast and belly dark red, and the female of which is brownish, will probably belong to this species.—Note of the Baron Cuvier.

M. Spix, while on a hunting expedition, observed a female of this species, which had been wounded, continuing to carry a young one on her back, until fainting through loss of blood, she employed her dying efforts in throwing her young one upon the adjoining branches to a place of safety.

VAR. STRAMINEUS.—STRAW-COLOURED HOWLER.

Syn.   STENTOR STRAMINEUS.[1]—Geoff. Ann. Mus. XIX.
      SIMIA STRAMINEA (ARABATA).—Humb. Obs. Zool. p. 355.
Icon.   MYCETES STRAMINEUS.—Spix, Sim. et Vespert. Bras. pl. 31.

This animal is in all probability a mere variety of age or sex of the Black Howler, from which we can find no essential ground for specific distinction. It was found by M. Spix in the forests between the Rio Negro and Solimaens towards Peru. Its hair is of a uniform straw colour, tending in some places towards orange yellow. It is much smaller than the Black Howler, and is probably a young male of the first or second year.

VAR. RUFIMANUS.—RED-HANDED HOWLER.

Syn.   MYCETES RUFIMANUS.[2]—Desm. Mam.—Kuhl. Beitr.
      CEBUS BELZEBUL.—Erxl. p. 44.—Fisch. Syn. Mam.
      SIMIA GUARIBA.—Humb. Obs. Zool. 355.
      SIMIA BEELZEBUL.—Linn. Gmel. I. 355.
      PREACHER MONKEY.—Penn. Quadr. I. No. 132.
Icon.   MYCETES DISCOLOR.—Spix, Sim. et Vespert. Bras. pl. 34.

The Red-handed variety blackish, with the latter half of its tail and its hands reddish, is smaller than the Black Howler, and approaches still nearer to this type than the straw-coloured variety already described. It probably represents a male about to assume the characteristics of the adult.

DOUBTFUL SPECIES.

1. M. FLAVICAUDATUS, THE YELLOW-STRIPED HOWLER (Kuhl Beitr. and Desm. Mam.), is in all probability a variety of some of the species already described, perhaps the young of M. chrysurus. It was found on the banks of the Amazon, where it is known by the name of Choro. The face is yellowish-brown, and scantily supplied with hair; the body is dark-brown; and the tail, shorter than the body, has a yellow stripe on each side. M. Humboldt distinguishes it under the name of Simia flavicauda.—(Choro).—Obs. Zool. p. 343 and 355. It is the Stentor flavicaudatus of Geoff. Ann. Mus. XIX.

All the Sapajous which now remain to be described (Les Sapajous ordinaires, *Cuv.*) have the head flat; the muzzle slightly projecting, and the facial angle about 60°. Some of them have the thumbs of the fore-hands wholly or partially concealed beneath the skin, while the prehensile portion of their tail is naked beneath.[3]

GENUS II. ATELES.—SPIDER-MONKEYS.

Syn.   LES ATÈLES (in part.)—Cuv. Reg. Anim. I. 101.
     ATELES (in part.)—Geoff. Ann. Mus. XIX. 105.—Spix, Sim. et Vespert. Bras.
     LES ATÈLES.—Isid. Geoff. Mém. Mus. XVII.
     SIMIA (in part.)—Linn. Gmel. I.

SPECIFIC CHARACTERS.

THE HEAD rounded.—The Facial Angle about 60°.—THE FACE perpendicular.—THE EARS large and naked.

THE HYOID BONE slightly cavernous, not appearing externally.

THE TAIL naked beneath the point.

THE FORE-HANDS tetradactylous, the thumb being wanting or rudimentary.—THE HINDER HANDS pentadactylous.—THE NAILS wide, and semicylindrical.

THE LIMBS long and thin.

INHABIT South America.

All these animals come from Guiana and Brazil; their fore-hands are very long and thin; while their mode of progression is singularly slow. They bear a remarkable similarity to Man, in the disposition of some of their muscles, and they alone of all animals have the biceps cruralis constructed as in the human species.

The Spider-Monkeys are generally mild, timid, melancholy, lazy, and very slow in their movements, so as always to appear in pain or unwell; yet, when there is any occasion for exertion, they can exhibit much agility, and clear a very considerable space at a single leap. They live in troops, on the elevated branches of trees, and feed chiefly on fruits. We are likewise assured that they eat Roots, Insects, Mollusca, and small Fish, that they seek for Oysters at low-tide, and break the shells between two stones. Some add that they use the point of the tail as a bait for Crabs and small Fish. Dacosta and Damper relate that, when the Spider-Monkeys are desirous of passing across a river, or from one tree to another, without touching the ground, they take fast hold of each other by the tail, so as to form a kind of chain, which is made to oscillate, until the lowermost Monkey has been swung sufficiently near to the object, which it seizes, and then draws after it all the others. The tail, besides its most ordinary use, that of rendering their position secure by grasping the branch of a tree, is employed by them for several purposes. It serves to seize objects at a distance, without obliging them to move the body, or even the eyes, the sense of touch being so admirably developed in its callous portion as to render the co-operation of another sense on most occasions unnecessary. Sometimes they roll the tail round the body, so as to protect themselves from the cold, to which they are very sensitive, or two individuals clinging closely together, roll their tail round each other. It has been ascertained that they sometimes use the tail for carrying food to the mouth; but for this purpose they usually employ their hands, which, though wanting the thumb, and of a disagreeable form, owing to their great length and narrowness, are far from being awkward. The genus is widely dispersed throughout South America, and contains several species, which are nearly allied, and greatly resemble each other in the colours of their hair. Like the Orang-Outang, they walk with great difficulty, and when on all the hands, they close the latter, and place the outer surfaces upon the ground. When sitting down on the haunches, they sometimes draw the hinder part of the body forwards by fixing the fore limbs upon the ground, and using them like a pair of crutches.

The cerebral cavity is rounded and voluminous, and forms nearly two-thirds of the entire skull. Their orbits, broad and deep, are remarkable in the adults for a kind of crest, appearing on the superior and exterior portion of their circumference. The lower jaw is rather deep, and its branches broad, but not so much as in the Howlers. The hyoid bone resembles that of several Apes of the Old Continent, such as the Guenons and Baboons. It is analogous to that of the Howlers, but smaller, and does not impart any volume to their voice. The Spider-Monkeys and succeeding genera emit a mild and sonorous whine, resembling the fluted cry of some birds.

The molar teeth in both jaws are small, with their crowns irregularly rounded. The upper incisors are very unequal in size, the first being much longer and broader than the second; in the lower jaw, the incisors, on the contrary, are equal in size, but are considerably larger than the molars. Their nails are wide and semicylindrical; the ears large and naked. The nostrils are of an elongated shape, situate at a distance from each other and wholly lateral, being exactly placed on the sides of the nose. The clitoris is excessively long, so that the sexes are distinguished with difficulty. M. Isidore Geoffroy found it to be two inches and a half in length in a female of Ateles Brissonii. The tail being much longer than the body, is naked beneath for a third part of its length from the point. Their hair is silky and generally long, as in the Howlers. The forehead is covered with scanty hairs, which are directed, at least partially, from the front backwards. All the other hairs of the head are very long, and point from behind forwards, so that at the places where the points meet a kind of crest or tuft, more or less distinctly pronounced, is formed, the disposition of which varies according to the species. These latter characters serve to distinguish a Spider-Monkey from the succeeding genera, at the first glance, without submitting them to any detailed examination.

The Genus ATELES, as originally proposed by Geoffroy-St-Hilaire (Ann. Mus. VII.), contained five species, one of which (Le Camail) belongs more properly to the genus Colobus of the Old World. This error M. Geoffroy was one of the first to acknowledge, and he accordingly substituted (Ann. Mus. XIX.) the Chuva of Humboldt in its place, leaving the total number of species five as before; but the discovery by Prince Maximilian of a new species with a very small thumb, subsequently induced Desmarest (Mam. p. 72) to separate the genus into two sections

[1] The Alouatte couleur de paille, Stentor stramineus, Geoff. and Mycetes stramineus of Spix, pl. 31, of a yellowish-grey, appears from its skull to differ in the species, but it may be merely a female of one of the preceding. We may easily comprehend, regarding the Howlers generally, that if their characters possess so little certainty, their synonyms must have still less.—*Note of the Baron Cuvier.*

[2] Marcgravius (Bras. 236) speaks of a black *Guariba* with brown bands. This animal Spix refers to his Seniculus niger (see the Mém. de Munic for 1813, p. 333).—It is the Mycetes rufimanus of Kuhl.—*Note of the Baron Cuvier.*

[3] The Baron Cuvier here proceeds to remark, that " These Sapajoos have been formed into the genus Ateles by M. Geoffroy (Ann. Mus. VII. 260). Two species (Ateles pentadactylus and Esiodes tuberifer) have been separated from the remainder by M. Spix, to form the genus Brachyteles, and serving to connect the genera Ateles and Lagothrix together. The remaining Ateles to which M. Spix reserves the name of Coäïta, Buff. are entirely deficient of any apparent thumb on the fore-hands." For this arrangement, which has justly been considered as wholly artificial, we have ventured to substitute another, apparently better suited to the present state of the science.

characterized by the presence or absence of this rudimentary appendage. Spix, carrying these views still farther, proposed the genus BRACHYTELES, and the Baron Cuvier, adopting them in the Second Edition of the "Règne Animal," assigned two species to the genus Brachyteles, and retained five in the genus Ateles. The excellent Memoir of M. Isidore Geoffroy (Mém. Mus. XVII. published in 1828) clearly demonstrated, however, that this arrangement should be regarded as purely artificial, violating as it does the most natural analogies. Three additional species have likewise been discovered since the Baron Cuvier published his last Edition, so that the genus Ateles would now contain ten species, were not three of them assigned to M. Isidore Geoffroy's new genus ERIODES, which will doubtless be universally adopted.

The term Ateles, from *ατελής, imperfect,* is, strictly speaking, inapplicable, as one species has the thumb, although rudimentary. It is, however, a well characterized division, being in the New World analogous to the Semnopitheci and Colobi of the Old. The Ateles have the same slowness, gravity, and mildness; their head, likewise, is round; their limbs long and thin; the abdomen voluminous, and the tail long. They are essentially destined to live on trees. When on the ground, their movements are excessively awkward; they drag themselves along, rather than walk, and, instead of resting the fingers or the soles of the feet upon the ground, so as to be either digitigrade or plantigrade, they rest on the inner side of their fore-hands, and the outer margin of the hinder. These uncouth crawling gestures have led to their being called *Spider-Monkeys;* but they atone for their awkwardness upon the ground by their agility when on trees. They run along the smallest branches with the greatest activity and address, and leap from tree to tree, though separated by a considerable interval; for, as they live chiefly on fruit, they have no occasion even to come to the ground except when they require water. They assist each other in danger, and attack a stranger by throwing small branches at him, or even their own excrements. When attacked by hunters, and one of them has been wounded, the remainder fly to the tops of the trees and raise the most lamentable cries, while the injured animal, placing its hands on the wound, watches the flowing of the blood until it loses consciousness and dies. It then commonly remains suspended by the tail; for this organ has the property of closing itself at the extremity, though it remains extended throughout the rest of its length.

These animals are easily tamed, while kindness and attention render them very affectionate. It has even been stated, that they learn to assist in different domestic offices; but this requires confirmation.

1. ATELES PANISCUS.—COAITA SPIDER-MONKEY.

Syn. LE COAITA.—Cuv. Reg. Anim. I. 101.
ATELES PANISCUS.—Geoff. Ann. Mus. VII. and XIX.—Desm. Mam.
SIMIA PANISCUS.—Linn. Gmel. I. 36.—Humb. Obs. Zool. p. 353.
FOUR-FINGERED MONKEY.—Penn. Quadr. I. No. 133.
Icon. Coaita femelle,—Geoff. et F. Cuv. Hist. Mam.
Le Coaita.—Audeb. Sing.
Buff. Hist. Nat. XV. pl. 1.

SPECIFIC CHARACTERS.

THE HAIR entirely black. THE FACE naked and flesh-coloured.
THE FORE-HANDS tetradactylous, the thumb being wanting.
INHABITS Guiana and Brazil.

This animal is wholly covered with black hair, like the following species, but is absolutely without a visible thumb on the fore-hand; its face is flesh-coloured.

Though long known as a distinct species, its history has always been more or less confounded with that of other Monkeys. It certainly does at first sight appear paradoxical, that an animal apparently organized for rapid motion, with long and slender limbs, and a tail capable of acting like a fifth hand, should move with slowness and constraint; yet, when upon the ground, its arms and legs seem to move with pain to the animal, and to require the influence of some urgent motive for action. When on a tree, however, we are assured by Pennant and others, that the activity of this animal is very great. It does not appear to be desitute of intelligence; it is mild and affectionate to its keeper. At all times the tail is firmly rolled round any object within its reach, as if to protect the individual from an accidental fall. Audebert tells us, that he saw one specimen raising straw and hay to its mouth with the tail, in nearly the same manner as an Elephant uses its trunk. These animals are said to be found in numerous troops in the woods of Guiana and Brazil, suspended from the branches of trees.

The entire body of the Coaita Spider-Monkey is covered with black, glossy, long, and coarse hairs, rather scanty beneath, and without the slightest trace of woolly hairs. The face, as well as the skin of the body, is of a brownish flesh-coloured tint, and the hands are black. Its ears resemble ours, but they want the lobe; the abdomen is of very great capacity, and seems to announce the presence of voluminous intestines. The cry of this animal is acute and plaintive. The mammæ of the female, placed

beneath the arm-pits, are marked by a black nipple. The clitoris is enormously developed, being nearly two inches in length.

2. ATELES PENTADACTYLUS.—FIVE-FINGERED SPIDER-MONKEY.

Syn. LE CHAMEK.—Cuv. Reg. Anim. I. 100.
ATELES PENTADACTYLUS.—Geoff. Ann. Mus. VII. and XIX.
SIMIA CHAMEK.—Humb. Obs. Zool. p. 353.
ATELES SUB-PENTADACTYLUS.—Desm. Mam.
Icon.

SPECIFIC CHARACTERS.

THE HAIR entirely black. THE FACE naked.
THE FORE-HANDS with a rudimentary thumb.
INHABITS Guiana, Brazil, and Peru.

The Five-fingered Spider-Monkey differs from the preceding in having the thumb slightly apparent; yet this consists only of a single phalanx, and wants the nail. The hair is wholly black.

So great is the resemblance between this species and the Coaita in form and colour, that they were long regarded as identical. Their skulls are, however, very different. That of the Five-fingered Spider-Monkey is broader, shorter, flatter towards the suture of the parietal bones, and more expanded towards the temples. The lower jaw-bone is likewise proportionably larger; the inferior margin is straight, while it is vaulted in the Coaita; and the ascending branches are so extensive, that we might almost imagine that they served, as in the Howlers, to support a hyoid bone of unusual magnitude. The thumb differs greatly in the two species. In the Coaita Spider-Monkey, the bone of the metacarpus is, at most, only half the length of the adjoining bone, and the terminal phalanx is so small, that it forms but a fifth part of the length of the preceding; these two bones are slender in proportion, so that they are lost in the common integuments, without permitting the slightest trace of them to appear externally. In the Five-fingered Spider-Monkey, the same bones are found, their chief difference consisting in their thickness, but the bone of the metacarpus is not quite so long. The first and only phalanx is still less so, being about a third of its length; but it is much broader, especially near the extremity. This phalanx, detaching itself wholly from the second bone of the metacarpus, constitutes the thumb of the Chamek. It is very short, and wholly wants the second phalanx as well as the nail, which terminates the fingers of most other Monkeys.

The hair, like that of the Coaita, is coarse, rough, dry, and of a deep black.

3. ATELES ATER.—BLACK-FACED SPIDER-MONKEY.

Syn. LE CAYOU.—Cuv. Reg. Anim. I. 101.
ATELES ATER.—F. Cuv.
ATÈLE COAITA DE CAYENNE.—Geoff. Ann. Mus. XIII.—Desm. Mam.
Icon. Cayou.—Geoff. & F. Cuv. Hist. Mam.

SPECIFIC CHARACTERS.

THE HAIR and FACE very black.
THE FORE-HANDS tetradactylous.
INHABITS South America.

This animal has its face black, like the remainder of its body.

The hairs are long and silky, but rather dry and coarse, like those of the Coaita; they are as long on the head and tail as on the remainder of the body, where they point in the usual direction, from the front backwards; while on the head their points are directed forwards. The skin is black throughout, the pupil of the eye brown, and the organs of generation flesh-coloured, The ear is oval, and the antihelix remarkable for its large size.

4. ATELES MARGINATUS.—STRIPE-FACED SPIDER-MONKEY.

Syn. LE COAITA À FACE BORDÉE.—Cuv. Reg. Anim. I. 101.
ATELES MARGINATUS.—Geoff. Ann. Mus. XIII. and XIX.
SIMIA MARGINATA (CHUVA).—Humb. Obs. Zool. p. 354 and 340.
Icon. COAITA À FRONT BLANC, femelle,—F. Cuv. et Geoff. Hist. Mam.
Ann. Mus. XIII. pl. 10.

SPECIFIC CHARACTERS.

THE HAIR black, a margin of yellowish hairs round the face of the male; whitish in the female.
THE FACE flesh-coloured. THE FORE-HANDS tetradactylous.
INHABITS Brazil.

These Monkeys occur frequently on the banks of the Rio Santiago and Amazon. M. Humboldt was informed by the Indians, that they live in numerous bands apart from the Marimondas (Ateles Brissonii). The male is rather ill-tempered, the female mild and intelligent.

When seated, they raise the tail perpendicularly upwards, and roll the point into a spiral curve. Their physiognomy bears a very striking resemblance to that of a Negro.

The Stripe-faced Spider-Monkey is entirely covered with long, black, and silky hairs, much scantier on the lower parts of the body than on the upper. The internal surfaces of the hands and the naked portion of the tail are violet-coloured. The hairs of the back, sides, thighs, legs, arms, and tail, are directed in the usual manner; those of the head point from the front backwards, while the hairs of the forehead rise almost erect upwards, and form, when opposed to the others, a kind of crest, so that the yellowish or whitish ornament of the forehead is the first object which attracts the attention of the observer. The hair of the fore-arm near the elbow is directed backwards.

5. ATELES BRISSONII.—BRISSON'S SPIDER-MONKEY.

Syn. Le Coaita à ventre blanc.—Cuv. Reg. Anim. 1. 101.
ATELES BELZEBUTH.—Geoff. Ann. Mus. VII. and XIX.
SIMIA BELZEBUTH (MARIMONDA).—Humb. Obs. Zool. p. 325 and 253.
CEBUS BRISSONII.—Fisch. Syn. Mam.
Le Belzébut.—Briss. Reg. Anim. p. 211.
Icon. Belzebuth (young).—Geoff. et F. Cuv. Hist. Mam.—Ann. Mus. VII. pl. 16.

SPECIFIC CHARACTERS.

THE HAIR dark brown above; yellowish-white beneath; changing into yellowish-red on the abdomen.
THE FACE violet-black; flesh-coloured round the eyes. THE FOREHANDS tetradactylous.
INHABITS Guiana.

This animal is widely dispersed throughout Spanish Guiana, where it is known by the name of Marimonda. Its hair is dark brown, very long and shining on the upper part of the back. The hair of the occiput and vertex is directed forwards, while that of the forehead points backwards, thereby forming a tuft on the top of the head, and contributing towards its extreme ugliness. The face is naked and black, while the tips of its very extensible lips, and the points of its nose, are of a reddish-white. The mouth is surrounded with stiff grey hairs; the neck and chin are almost naked; the eyes brown, and furnished with long black eyelashes. The abdomen, the interior of the thighs, and the inferior surface of the tail, are covered with yellowish-red hair, the points of which, when strongly illuminated, give a slight metallic reflection.

The Marimonda of the banks of the Orinoco is excessively slow in its movements, of a melancholy, mild, and timid disposition. Through excess of timidity it is very apt to bite even those who are attending to its wants; and it announces its approaching passion by making a grimace and raising the cry *co-oh*. Of all prehensile tails, that of Brisson's Spider-Monkey exhibits the greatest perfection. It can raise even a straw, and fully equals the trunk of the Elephant, so that, as Humboldt remarks, it seems as if the very eyes of this Monkey were placed at the end of its tail. Without turning its head, this little animal can introduce its tail into the smallest holes for the object of its search. This organ, however, is not observed to raise food to the mouth, that office being always fulfilled by the hands. When collected together in large numbers, they interlace their limbs and tails in the most grotesque forms. Their attitudes announce the greatest apathy and indolence, while the joints of their limbs are so flexible, that they almost appear dislocated. When exposed to the burning heat of the sun, they throw the head backwards, their eyes are directed upwards, and, folding their arms on the back, they remain motionless in this extraordinary position for hours together.

In young specimens, the hair is of a dirty white beneath, and greyish black above.

6. ATELES HYBRIDUS.—MONGREL SPIDER-MONKEY.

Syn. L'Atèle métis (MONO-ZAMBO).—Isid. Geoff. Mém. Mus. VII.
Icon. ATELES HYBRIDUS.—Isid. Geoff. Etud. Zool. pl. 1 (Mammifères).

SPECIFIC CHARACTERS.

THE HAIR clear brown above; yellowish on the thighs; white beneath; a white circular patch on the forehead.
THE FACE dark brown. THE FORE-HANDS tetradactylous.
INHABITS Columbia.

The present species differs remarkably in the general colour of the hair from its congeners, which are either black or dark brown. It is at once known by the white patch placed in the centre of its forehead. Beneath the head, the body, and along the tail, as far as its callous portion, the Mongrel Spider-Monkey is of a dirty white; the upper parts are generally of a clear brown, which passes into a pure brown on the head, the fore-limbs, thighs, and beneath the tail, and into a well defined pure yel-

low on the thighs, as well as on the sides of the tail, and a part of the lower extremity.

This animal is very common in the Valley of the Magdalena, where it is known by the name of Marimonda, a name common to several other Monkeys. It is likewise termed Zambo or Mono-Zambo, from the resemblance which its colour bears to that of the Zambo, or descendant of the African and American Indian. The Mongrel Spider-Monkeys live in troops of twelve or fifteen individuals, and the traveller through the woods is informed of their presence by the noise which they make in throwing themselves from one branch to another. The females appear much attached to their young, which they carry on their backs from place to place. An old female, embarrassed by her young one, had a considerable leap to make; M. Roulin saw an old male place himself on the extremity of the branch, and make it oscillate so as to bring it to the level of the female, who took advantage of the proper moment to effect her passage. Their attention to the young is shared by both sexes.

This species rests upon several adult females and one young male in the Paris Museum. M. Roulin admits its reality, and assigns it to Columbia.

7. ATELES MELANOCHIR.—BLACK-HANDED SPIDER-MONKEY.

Syn. ATELES MELANOCHIR.—Desm. Mam.
Icon. ATÈLE MELANOCHIR (femelle).—Geoff. et F. Cuv. Hist. Mam.

SPECIFIC CHARACTERS.

THE HAIR on the top of the head, and the outside of the arms, legs, and tail, black; whitish beneath; elsewhere grey tinged with yellow.
THE FACE flesh-coloured round the eyes and mouth, elsewhere black.
THE FORE-HANDS tetradactylous.
INHABITS Peru.

Of the present animal little is known. It seems to have all the characteristics of the other Spider-Monkeys; such as being mild, affectionate, sociable, and excessively slow in its movements.

The head, limbs, and tail, are covered with black hairs above; the inner surface of the arm and fore-arm, down to the fore-hands, is white, as well as the inner surfaces of the legs and thighs, the inferior surfaces of the neck, breast, and abdomen, the sides of the thighs, and the inferior surface of the tail; the shoulders are of a yellowish-grey, and the remainder of the upper parts of the body, as well as the cheek-tufts, are of a pure grey. All the hands, and the naked portion of the tail, are black, likewise the cheeks and the lower half of the nose; but the circles round the eyes and mouth are flesh-coloured. The hair is entirely composed of silky filaments; those on the black as well as the white parts of the body are of a uniform shade throughout, while those on the grey parts are annulated with black and white, more or less mixed with yellow.

DOUBTFUL SPECIES.

1. THE BAY SPIDER-MONKEY (Ateles fuliginosus), is known to us only by the description in Kuhl, Beitr. p. 26, taken from an individual in the Paris Museum. From the silence of the French Naturalists, and the want of any figure of this species, we are inclined to place it here until an opportunity occurs for examining a perfect specimen.

2. GEOFFROY'S SPIDER-MONKEY (Ateles Geoffroy), appears from Kuhl's description (Beitr. p. 26) to be identical with the Ateles Melanochir of Desmarest.

GENUS III. ERIODES.[1]—WOOLLY-MONKEYS.

Syn. LES ERIODES.—Isid. Geoff. Mém. Mus. XVII. 138.

GENERIC CHARACTERS.

THE HEAD rounded. THE FACIAL ANGLE about 60°. THE EARS small and hairy.
THE NOSTRILS separated by a very narrow partition, and almost opening beneath the nose.
THE TAIL naked beneath the point.
THE FORE-HANDS tetradactylous or pentadactylous. THE HINDER-HANDS pentadactylous. THE NAILS compressed, excepting those of the hinder thumbs, which are large and flat.
THE LIMBS long and thin. THE HAIR very woolly.
INHABIT South America.

These Monkeys bear a general resemblance to the Spider-Monkeys already described, in their long and meagre limbs, but differ in several important points of their organization. Externally, they may be at once distinguished by the woolly nature of their hair, their short hairy ears, the compressed form of their nails, as well as by the remarkable anomaly in the disposition of their nostrils. In the last particular, they may be considered as holding a medium rank between the Apes of the Old Continent

---

[1] Eriodes, from ἐριώδης, *woolly.* The term *Wolliffe* (Woolly-Monkeys), given by Kuhl to the genus Lagothrix, seems to be more applicable to the present.

and those of America, being in fact more nearly allied to the former than to the latter. Their nostrils are rounded, very nearly approximate to each other, and are rather inferior than lateral, owing to the extreme thinness of the partition of their nostrils. The nails bear more resemblance to those of some Carnivorous animals, such as the Dogs, than of the Spider-Monkeys; they are compressed, and may be described as formed of two laminæ surmounted by a blunted crest. The nails on the thumbs of the hinder-hands are, however, an exception to this rule, being broad and flat as in Man. Their ears are small, and for the most part covered entirely with hair.

The molar teeth of the Eriodes are generally very large and quadrangular. Their incisive teeth, in both jaws, are arranged nearly in a straight line; they are very small, of equal length, and much less than the molars, by which characters the dentition of the Woolly-Monkeys may be distinguished from that of all other Sapajous, excepting the Howlers.

Their hair is soft to the touch, woolly and very short. That of the head, still shorter than on the body and tail, is directed backwards; and this arrangement, being opposite to that of the Spider-Monkeys, gives to their physiognomy an aspect wholly different.

The Woolly-Monkeys live together in troops, among the branches of trees; they leap with much agility, and greatly resemble the preceding or succeeding genera, in their manners, as far as the latter have been ascertained.

1. ERIODES HEMIDACTYLUS.—DWARF-THUMBED WOOLLY-MONKEY.

Syn.   ERIODES HEMIDACTYLUS.—Isid. Geoff.
       ATELES HYPOXANTHUS.—Desm. Mam.—Less. Mam.—Kuhl, Beitr.
Icon. Mém. Mus. XVII. pl. 22.

SPECIFIC CHARACTERS.

THE HAIR yellowish fawn-colour, tending to black upon the back; the tail and hands of a brighter yellow.

THE FORE-HANDS pentadactylous; the THUMB being very short, with a small compressed NAIL.

THE FACE flesh-coloured, spotted with grey.

INHABITS Brazil.

This species was discovered by Delalande in 1816, but has always been confounded with the Ateles hypoxanthus of Prince Maximilian. It may, however, be at once distinguished from its having the thumb of the fore-hand unguiculated, very narrow, short, and scarcely reaching to the origin of the first finger, so as to be wholly useless to the animal for any practical purposes. Its hair is in general yellowish-fawn, assuming a darker tint upon the back. The hands and tail are of a purer yellow than the remainder of the body. There is a naked space at the base of the tail and near the anus, surrounded by hair of a ferruginous red. The face, which is completely naked in the neighbourhood of the eyes, appears to be spotted with grey over a flesh-coloured ground. The habits of this animal are unknown.

2. ERIODES ARACHNOIDES.—THUMBLESS WOOLLY-MONKEY.

Syn.   LE COAITA FAUVE.—Cuv. Reg. Anim. I. 101.
       ATELES ARACHNOIDES.—Geoff. Ann. Mus. VII., XIII., and XIX. Desm. Mam.
       ERIODES ARACHNOIDES.—Isid. Geoff. Mém. Mus. XVII.
       SIMIA ARACHNOIDES.—Humb. Obs. Zool. p. 354.
Icon. Ann. Mus. XIII. pl. 9.

SPECIFIC CHARACTERS.

THE HAIR is yellowish fawn-colour above; yellowish-white beneath; reddish on the outside of the limbs and beneath the tail; the eyelashes long and black.

THE FACE flesh-coloured.—THE FORE-HANDS tetradactylous.

INHABITS Brazil.

This animal is easily distinguished from all other Woolly-Monkeys known at present, by the absence of any external appearance of a thumb on the fore-hands. Its hair is generally of a clear yellowish-fawn colour, passing into reddish-grey upon the head, and into bright red beneath the tail, upon the hands, and especially near the hinder wrists or heels. Some specimens are, however, sometimes found of a uniform clear yellow. The hair of the ears is of a deeper chesnut tinge, that of the forehead approaches to white, set off by a row of long, stiff, and black eyelashes, with which the forehead is bordered. Its hair has this peculiarity, that it gives out a fawn-coloured tint, when rough and bristled as it usually appears, but passes into a chesnut brown when perfectly smooth. This proceeds from the circumstance that the tips of the hair are of a fawn-colour, while their points being of a deeper brown near the head and ears, occasion the latter tint to predominate on those parts of the body.

We are at present ignorant of its habits and manners. It comes from Brazil, where it bears the name of *Macaco vernello*. Several specimens exist in the Paris Museum.

3. ERIODES TUBERIFER.—TUBEROUS WOOLLY-MONKEY.

Syn.   LE MIRIKI (typographical error for MIRIKI).—Cuv. Reg. Anim. I. 100.
       ERIODES TUBERIFER.—Isid. Geoff. Mém. Mus. XVII.
Icon. BRACHYTELES MACROTARSUS (fem').—Spix, Sim. Vespert. Bras. pl. 27.
       ATELES HYPOXANTHUS (male).—Pr. Max. Abbild.

SPECIFIC CHARACTERS.

THE HAIR yellowish-grey; the base of the tail and anal region yellowish-red.

THE FACE flesh-coloured.

THE FORE-HANDS with a rudimentary thumb, rarely bearing a NAIL.

INHABITS Brazil.

The Tuberous Woolly-Monkey, known to the inhabitants of Brazil by the names of *Mono* and *Miriki*, is the largest Quadrumanous animal of South America. It inhabits the lofty primeval forests of the interior, in those desert regions overrun with wood, which are seldom visited by Man.

This animal may be easily distinguished from the preceding, by having its thumbs rudimentary, and appearing externally under the form of simple tubercles, which, according to the observations of M. Spix, always want the nail, though the contrary is asserted by others. The hair on the upper part of the body is short, and rather thick, of a yellowish fawn-colour, as in the rest of its congeners, blending into a fiery red towards the roots of the hair, behind the thighs and legs, on the fingers, and beneath the tail. The hair of the head is rather darker, but becomes lighter about the face. The latter is of an oblong form, naked, flesh-coloured, slightly tinged with grey, and bears long, black, and stiff hairs on the margin of the forehead and eyelashes. The cranium is broad and arched; the ears prominent, truncated on the margin, hairy, of a deep brown beneath and behind; while the lower jaw slightly ascends at the inferior angle.

These large Monkeys are very plentiful in the maritime provinces from St Paul to Bahia. They travel about in troops during the day, and make the air resound with their loud cries. On perceiving a hunter, they ascend quickly to the tops of the highest trees, and leap swiftly and silently from branch to branch, until they are lost in the gloom of their impenetrable forests.

GENUS IV.—LAGOTHRIX.—GLUTTONOUS MONKEYS.

Syn.   LES LAGOTHRIX.—Cuv. Reg. Anim. I. 101.
       LAGOTHRIX.—Geoff. Ann. Mus. XIX.—Desm. Mam.
       GASTRIMARGUS.—Spix, Sim. et Vespert. Bras.

GENERIC CHARACTERS.

THE HEAD rounded. THE FACIAL ANGLE about 50°. THE EARS very small.

THE TAIL naked beneath the point.

THE HANDS pentadactylous.

THE NAILS slightly compressed.

THE LIMBS of moderate length.—THE HAIR rather woolly.

INHABITS South America.

These animals have the head round, like that of the Spider-Monkeys, their thumbs are developed as in the Howlers, and the tail is partly naked, as in both. They all come from the interior of South America, and are said to be singularly gluttonous in their habits.

The genus Lagothrix, containing two species only, was instituted by M. Geoffroy-Saint-Hilaire, and may be distinguished from any of the preceding by its having the limbs much shorter, and especially by the fore-hands being pentadactylous, as in the Howlers and Weepers (*Cebus*), the latter of which it greatly resembles in the general proportions of the body. The fingers are of moderate length, and the first or index is even short. The nails of the anterior hands are slightly compressed, not excepting even those of the thumbs, and in respect to their shape, are intermediate to those of the Ateles and Eriodes. On the hinder-fingers, excepting the thumb, the nails are still more compressed, and similar to those of the Eriodes, especially in respect to the three last fingers. The head in the Gluttonous Monkeys is rounded, and their hair soft to the touch, very fine, and almost as woolly as in the Eriodes; but their incisive teeth and nostrils resemble those of the Ateles. Their facial angle is 50°, and their ears are very small.

It is to Humboldt that we are indebted for the discovery of these animals, which still remain but little known, whether in their organization or manners. We are merely informed that they live together in numerous troops, and appear to be very mild in their disposition; they often stand erect upon their hinder hands.   Spix, having found them in Brazil, adds, that they have a harsh, disagreeable voice, and that they are very gluttonous, for which he gives them the name of Gastrimargus: but the rule of priority warrants us in preferring the name given by Geoffroy.

1. LAGOTHRIX CANUS.—GREENISH GLUTTONOUS MONKEY.

Syn.   Le Grison.—Cuv. Reg. Anim. I. 101.
         Lagothrix canus.—Geoff. Ann. Mus. XIX.—Desm. Mam.
         Simia cana.—Humb. Obs. Zool. p. 354.
Icon.   Gastrimargus olivaceus.¹—Spix, Sim. et Vespert. Bras. pl. 28.

SPECIFIC CHARACTERS.

The Hair grey olive above ; black or dark brown beneath ; very black and frizzled on the head.   The Face and Hands black.
Inhabits the banks of the Amazon.

These animals are known to the Brazilians by the name of *Barigudos*, from their singular gluttony.   They acquire a great size, and were found in troops, occupying the highest trees in the neighbourhood of the great river Solimoëns.   When once tamed, they become very familiar, and, approaching the dinner table, sit down patiently on their haunches, waiting for boiled meat or oranges, of which they are very fond, though the latter often prove fatal to them.   One specimen was brought over as far as the Azores, but died there from the cold.   Towards the month of November, they are seen in great numbers carrying their young ones on the back or belly.
This species is olive-coloured throughout, and greatly resembles a Negro in the face, by the dark, short, thick, and woolly hair of the head ; by its hue, which is very black ; by the remarkable whiteness of the teeth, and its flat, short, and depressed nose.   Its hair is extremely short and thick ; grey olive above, and black or dark brown beneath ; on the head, on all the hands, the inner surfaces of the limbs, and beneath the tail, it is entirely black ; each hair on the back is of a dirty white, annulated with light yellowish-grey, pointed with black.   The tail is longer than the body, very strong, hairy, and naked beneath for about a quarter of its length from the point.   The hair under the body is rather black.   The face is somewhat squared, and the parts round the eyes are naked.   Before the ears and behind the cheeks, the hair is turned back so as to form a kind of dark cowl.   The ears are extremely short, truncated, and covered with slightly brown hairs.   The incisive teeth are almost square ; the canines are strong ; the anterior hands slightly elongated ; the nails black or grey, long, nearly triangular, and slightly curved ; those on the thumbs of the hinder-hands are almost flat.
The female and young are distinguished by having the cowl not so black as in the male.

2. LAGOTHRIX HUMBOLDTII.—HUMBOLDT'S GLUTTONOUS MONKEY.

Syn.   Le Caparo.—Cuv. Reg. Anim. I. 101.
         Lagothrix Humboldtii.—Geoff. Ann. Mus. XIX.—Desm. Mam.
         Simia lagothricha (Caparro).—Humb. Obs. Zool. p. 321 and 354.
Icon.   Gastrimargus infumatus (fem.).—Spix, Sim. et Vespert. Bras. pl. 29.

SPECIFIC CHARACTERS.

The Hair deep brown, tipped with black.
The Face and Hands black.
Inhabits the banks of the Rio Guaviare and Rio Iça.

Though greatly resembling the preceding species in many respects, this animal may at once be distinguished by its neck and body being more slender.   Its face is not so square, nor its lower canines so long.   The head and entire body are of a deep brown, approaching in many places to black, especially on the breast and hands.   All the hairs of the body are directed backwards, and are mostly black on their points.   The cowl on the head nearly resembles that of the species just described.   All the nails are black, triangular, and shorter than in the former ; that of the longest finger of the fore-hand is not compressed.
Some of the females have the back and head of a paler brown, approaching to grey ; sometimes the hair becomes wholly white, with the points black, so as to produce the effect of a uniform grey.

IMAGINARY SPECIES.

1. Lagothrix infumatus (Le Lagothriche enfumée) of Isid. Geoffroy, Dict. Class. d'Hist. Nat. art. Sapajou, founded upon the Gastrimargus infumatus of Spix, is absolutely identical with Humboldt's Gluttonous Monkey, on the authority of Temminck (Mon. Mam.)

GENUS V.   CEBUS.—WEEPERS.

Syn.   Les Autres Sapajous.—Cuv. Reg. Anim. I. 102.
         Cebus (Sajou).—Geoff. Ann. Mus. XIX.—Desm. Mam.
         Cebus (in part).—Erxl.—Fisch. Syn. Mam.
         Simia (in part).—Linn. Gmel. I.

GENERIC CHARACTERS.

The Head round.   The Muzzle short.   The Forehead slightly projecting.   The Facial Angle about 60°.
The Body and Limbs medium size.
The Tail prehensile, and entirely covered with hair.
The Hands pentadactylous.   The Nails compressed.
Inhabit South America.

The remaining Sapajoos have the head round, and the thumb distinct, while at the same time the tail is wholly covered with hairs, although it still continues to be prehensile.   Their species are much more numerous, and almost as difficult to characterize distinctly as those of the Howlers.
The Weeper Monkeys are of a mild disposition, their movements quick and lively ; they are very readily tamed ; and it is from their little fluted cry that they derive the name of *Singes Pleureurs*, or Weepers.
The genus Cebus appears to occupy the same station in the New World as Cercopithecus does in the Old, each being in an eminent degree the type of its tribe.   All the animals belonging to this division come to us from Guiana and Brazil, where they live in troops, on the elevated branches of trees.   They feed chiefly on fruits, but willingly devour Insects, Mollusca, and Annelides, or even sometimes meat.   They are believed to be monogamous.   The females usually produce a single young one at a time, which they carry about with them on the back, and treat with the most affectionate attention.   Many instances are known of their producing in confinement in Europe.   Some of them are noticed by travellers under the name of *Singes musqués*, or Musky Monkeys, in consequence of the strong odour of musk which they emit, especially during the rutting season.   They make their little fluted cry on all ordinary occasions, but when agitated by passion, whether jealousy, fear, or joy, their voice becomes strong, approaching to a noisy bark.   These animals are of great agility and intelligence, very quick, and always in motion, yet docile and easily educated.   M. Isidore Geoffroy noticed one individual, after many unsuccessful attempts to break a nut, first with its teeth, and then on the wood work of its cage, expertly making use of a bar of iron for that purpose.   They have not the same volatile character as the Guenons of Africa, but resemble them in the indelicacy of their behaviour.   They require to be kept very warm in our climates, and are extremely liable to diseases of the chest, when exposed to cold or moisture.   These Monkeys are rather common in most large cities.
The limbs of the Cebus Monkeys are strong, powerful, and elongated ; consequently, they leap with remarkable agility.   Their anterior thumbs are rather short, not very free in their motions, and but slightly opposable to the fingers, as is the case with the Howlers and Gluttonous Monkeys.   Their nails are cylindrical and somewhat flat ; the tail is nearly the length of the body ; sometimes it is wholly covered with long hairs, at other times the hairs on the terminal portion are excessively short, from its friction against other bodies, but it never exhibits a true callosity.   M. F. Cuvier notices the existence of a slight callosity at the extremity of the tail of the Cebus hypoleucus, but this is at variance with our observations upon an individual of that species.   The hyoid bone has its central portion enlarged, but it does not appear externally ; the head is rather round, the face broad and short, and the eyes very large and approximated to each other.   The opening of the nostrils is broad, but rather narrow from above downwards.   The dentary margins are almost parallel to each jaw ; the molar teeth of medium size, and six in number on each side of both jaws, as in all the other Sapajoos.   In one solitary instance M. Geoffroy found seven molars on each side of the upper jaw, in a very old individual of his *Cebus variegatus*, which we consider identical with C. xanthosternus of Prince Maximilian.   The incisive teeth are arranged almost in a straight line, the first incisor being the larger in the upper jaw, while the second is the larger in the lower ; the canines are very strong in all old individuals.   The cerebral cavity is very voluminous ; it is broad, and at the same time extends far from the back forwards ; the occipital hole is situate directly under the base of the skull.   The tail, being entirely covered with hair, is an organ of motion, but not of touch.   The males have an external scrotum, and the glans resembles an inverted pyramid, the base being outermost.

¹ According to the Baron Cuvier, the Gastrimargus olivaceus of Spix is identical with Lagothrix Humboldtii of Geoffroy, while G. infumatus is the same as L. canus. These synonyms, we think, should be transposed, as in the text.   It must be admitted, however, that the characters given by Geoffroy do not correspond very accurately with the representations of Spix.

There is no genus of Mammiferous animals, wherein the species are so difficult to characterize as the present. The earlier writers admitted but few species: Brisson recognised three, Linnæus four, Gmelin six, Buffon two, and the Baron Cuvier thought that possibly there might only be one. A more accurate acquaintance with these animals serves to announce that their species are fully as numerous as the Guenons of the Old World; that there exists much constancy in their distinctive characters; that the observed variations in the colours of their hair are inconsiderable; and that the differences in their external markings must be ascribed rather to the internal organization of each species and its influence upon their colouring, than to external and accidental circumstances.

The lists of systematic writers are in general very incorrect, and the catalogue of doubtful species more numerous than might be desired. We think, however, that twelve species may be safely admitted in the present state of our knowledge, without much risk of ultimate error.[1]

(A.) PROPER WEEPERS. (SAJOUS.)

Some of the Weeper Monkeys have the hair on the forehead of a uniform length.

1. CEBUS APELLA.—COMMON BROWN WEEPER.

Syn. LE SAJOU.—Cuv. Reg. Anim. I. 102.
    CEBUS APELLA.—Erxl.—Geoff. Ann. Mus. XIX.—Desm. Mam.
    SIMIA APELLA.—Linn. Gmel. I.—Humb. Obs. Zool. p. 355.
    CAPUCIN MONKEY.—Penn. Quadr. No. 185.
Icon. LE SAJOU BRUN.—Buff. Hist. Nat. XV. pl. 4.
    Le Sajou.—Audeb. Sing.—Mém. Mus. (male and female).
    SAI (var.)—F. Cuv. et Geoff. Hist. Mam.

SPECIFIC CHARACTERS.

THE HAIR black on the head and hands; brown on the back, breast, and belly; yellowish-white on the arms, shoulders, and sides of the face and forehead; a black band along the cheek.

THE FACE dark brown.

INHABITS Guiana.

It was doubtless from confounding this species with several of its congeners, that the Baron Cuvier and others were led to suppose that it undergoes much variation. We have seen many specimens of the Common Brown Weeper, and have found them to bear so minute a resemblance to each other, as to leave no doubt as to their specific distinctness from all those so-called varieties. The " Sajou" of Audebert accurately represents this species; and it must be noticed that his figure does not differ more from the " Sai" of M. Fréderic Cuvier, than might have been anticipated from the circumstance that the latter is drawn after the living animal, and the former from a stuffed specimen.

The characteristic colours of this species already mentioned result from hairs, which are, throughout their entire length, of the same colour as the points. Each cheek is divided into two parts by a dark band, arising near the anterior margin of the ear, from the black cowl on the top of the head, and passing in a curved line downwards to the chin, so as to meet the corresponding band of the other cheek. All the hairs are silky and soft, occasionally assuming a woolly appearance. The skin of the hands and face, as well as of the parts covered with hair, is dark brown.

2. CEBUS GRISEUS.—GREYISH WEEPER.

Syn. CEBUS GRISEUS.—Desm. Mam.—Fisch. Syn. Mam.
Icon. LE SAJOU GRIS.—Buff. Hist. Nat. XV. pl. 5.
    LE SAJOU mâle.—F. Cuv. et Geoff. Hist. Mam.

SPECIFIC CHARACTERS.

THE HAIR yellowish-brown above; lighter beneath; a black cowl on the top of the head; the cheeks, breast, shoulders, and fore-arms, white.

THE FACE and EARS flesh-coloured. THE HANDS dark violet.

INHABITS

Of this animal there is little particular to record, so much is the character of the genus likewise that of the species. We are not certain of its native country, which may probably be French Guiana. All its hairs are

silky, long, thick, and always grey at their base, whatever may be their colour at the points. The hinder part of the head, the neck, back, sides, the hinder part of the thighs, and the upper surface of the tail, are yellowish-brown; the abdomen and lower part of the thighs are of the same colour, but paler, and the under surface of the tail is of a dirty yellow; on the top of the head there is a black patch or cowl. The fore-part and sides of the head, the upper-part of the arms, the anterior surface of the fore-arm, the neck, and breast, are white. The face and ears are flesh-coloured, the hands and feet of a dark violet, and this colour prevails upon the scrotum; all these parts are nearly destitute of hair.

3. CEBUS CAPUCINUS.—CAPUCHIN WEEPER.

Syn. LE SAI.—Cuv. Reg. Anim. I. 102.
    CEBUS CAPUCINUS.—Erxl.—Geoff. Ann. Mus. XIX.—Desm. Mam.
    SIMIA CAPUCINA.—Linn. Gmel. I.—Humb. Obs. Zool. p. 355.
    WEEPER MONKEY.—Penn. Quadr. I. No. 136.
Icon. Le Sai.—Buff. Hist. Nat. XV. pl. 8.—Audeb. Sing.
    SAJOU BRUN (femelle).—F. Cuv. et Geoff. Hist. Mam.

SPECIFIC CHARACTERS.

THE HAIR dusky brown, with golden reflections; lighter on the face, shoulders, and fore-arms; a black cowl on the top of the head, ending in a point on the centre of the forehead.

THE FACE and HANDS violet-grey.

INHABITS Guiana.

This animal has its muzzle remarkably thick and short; its entire body is covered with soft and silky hairs, which are dusky brown throughout the greater part of their length, but are bright golden yellow at the points, communicating a greenish tinge to the animal, when viewed by a transmitted light, and emitting bright golden reflections, if seen obliquely. The fore-arms, the fore-part of the shoulders, the cheeks, and the temples as far as the ears and the sides of the forehead, are of a lighter brown approaching to yellow; the top of the head is black, in the form of a cowl, which is prolonged over the eyes, and ends in a point about the centre of the forehead. The dorsal line is darker than the other parts of the body; the hair is light, and very scanty beneath. The skin on all the naked portions of the body is of a violet-grey.

4. CEBUS HYPOLEUCUS.—WHITE-THROATED WEEPER.

Syn. CEBUS HYPOLEUCUS.—Geoff. Ann. Mus. XIX.—Desm. Mam.
    SIMIA HYPOLEUCA (CARIBLANCO).—Humb. Obs. Zool. p. 337 and 356.
Icon. SAI À GORGE BLANCHE mâle.—F. Cuv. et Geoff. Hist. Mam.
    Sai à gorge blanche var. A.—Audeb. Sing.
    Buff. Hist. Nat. XV. pl. 9.

SPECIFIC CHARACTERS.

THE HAIR on the arms and sides of the head white; on the neck and breast yellowish-white; elsewhere black.

THE FACE, FOREHEAD, and EARS, flesh-coloured. THE HANDS, and other parts of the body, violet-grey.

INHABITS the Banks of the Magdalena.

The White-throated Weeper is one of the species most commonly met with in Europe, where it is brought from Brazil or Guiana. Humboldt found several individuals near the Rio Sinu, and was informed by the Indians that they lived together in numerous troops, wholly distinct from the Common Brown Werpers (C. Apella).

The entire face, forehead, and ears, are naked, and of a pure flesh colour. The hands, and all the inferior parts of the body, likewise naked, are of a violet-grey. The hair is white on the sides of the cheeks, and on the arms from the elbows to the shoulders; on the neck and breast they are yellowish, but elsewhere of a very deep black. M. Frédéric Cuvier remarks, that " the White-throated Weeper alone has the point of its tail naked, but not by any means to the extent of the Coäita Spider-Monkey, for example. This character," he adds, " is proper to the species, and may serve to distinguish it from all other Sapajous." The tails of all the Weepers are liable to lose their hair at the point from friction, and even sometimes they may become callous; but that this is not a char-

---

[1] It will be readily perceived that we here depart widely from the views of the Baron Cuvier regarding the species of this genus. His observations, sufficiently vague, have now become, through the progress of science, in many respects obsolete; they are as follows:—

" The *Sajous* and *Sais* differ so much from brown to yellowish and whitish, that one would be tempted to constitute them into so many distinct species, if we had not also the intermediate varieties. Such are the Simia trepida, syrichta, lugubris, and flavia of Linnæus and Schreber, as well as some of those which are distinguished by Geoffroy (Ann. Mus. XIX. 111 and 112). Spix, again, has multiplied these species to a still greater extent, and on very insufficient grounds. We should approximate to the Sajou (C. Apella, Linn.) the C. robustus of Prince Maximilian, which even appears to us nothing more than an adult *Sajou*. The Macrocephalus of Spix, pl. 1, does not appear to belong to a different species. We approximate to the *Sai* (Simia Capucina, Linn.), the Sai à gorge blanche of Buffon (C. hypoleucus); the C. ishninosus of Spix, pl. 2; the C. xanthosternos of Prince Maximilian, or C. xanthocephalus of Spix, pl. 3; and the C. cuculatus. Id. pl. 6. We would rather be disposed to regard as distinct species, the Sajou à pieds dorés of Fred. Cuvier; the Sajou brun or C. unicolor of Spix, pl. 4; and the Simia flava of Schreber, pl. 31, B. from which the C. gracilis of Spix, pl. 5, appears to differ merely in the stuffing; but we are still in want of numerous observations made in the localities where these animals inhabit, before we can be satisfied that the species are not arbitrarily determined."

" After the Sajou corms (Simia fatuellus, Linn. Gmel.) should come the C. cirrifer of Geoff. and a Cebus of the same name of Prince Maximilian, but which is different, C. cristatus of Fred. Cuvier."

acteristic of the species, we have had an opportunity of ascertaining from a living specimen, in the collection now forming for the Zoological Society of Edinburgh. The tail of our Cebus hypoleucus is entirely covered with hair, and no trace of a callosity is visible on any part.

All the movements of this animal are exceedingly brisk and lively. It is also very gentle and intelligent; its eye too is quick, so that it watches all one's gestures, and appears to comprehend them; nay, it almost seems to read one's thoughts.

5. CEBUS CHRYSOPUS.—GOLDEN-LIMBED WEEPER.

Syn.   CEBUS CHRYSOPES.—F. Cuv.
Icon. SAJOU-À PIEDS DORÉS.—F. Cuv. et Geoff. Hist. Mam.

SPECIFIC CHARACTERS.

THE HAIR dusky brown above, white beneath; a white circle round the face; bright golden yellow upon the limbs.
THE FACE and EARS brown. THE HANDS bluish-grey.
THE TAIL bushy; dusky brown towards the base, elsewhere yellowish-white.

INHABITS

The height and proportions of this Weeper, for a knowledge of which we are indebted to M. Frédéric Cuvier, do not differ materially from those already described. Its head is very large and round; the face is naked, and rather brown, and surrounded by a broad circle of white hairs, which covering the forehead and cheeks as far as the ears on both sides, meet beneath the lower jaw. The remainder of the head is of a dusky brown, and this colour extends along the upper part of the back; the sides of the body and shoulders are somewhat lighter, while the neck, breast, belly, and the inner surfaces of the thighs, are white. The tail near its origin is, on the upper surface, of the same colour as the back, elsewhere it is yellowish-white. All the limbs are of a bright golden yellow. The skin of the fingers and palms of all the hands is bluish, and the ears are of the same colour as the face. The hairs are thick, soft, and very silky, on all the upper parts of the body, but more scanty beneath, especially on the abdomen; the tail is very bushy.

M. Cuvier informs us that he had an opportunity of studying the habits of this animal only for a short time; and that its dispositions served to correspond with those of the other Weepers. It exhibited the same petulance and caprice: its voice was something like a slight whistle, during its joy; and when alarmed or in wrath, its cry was sharp and rough.

6. CEBUS ALBIFRONS.—WHITE-FRONTED WEEPER.

Syn.   SIMIA ALBIFRONS (OUAVAPAVI).—Humb. Obs. Zool. p. 323 and 356,
          CEBUS ALBIFRONS.—Geoff. Ann. Mus. XIX.—Desm. Mam.—Fisch. Syn. Mam.
Icon.

SPECIFIC CHARACTERS.

THE HAIR dark grey above, lighter beneath; the top of the head greyish-black; limbs of a yellowish-brown.
THE FACE bluish-grey. THE FOREHEAD and ORBITS pure white.
INHABITS the Banks of the Orinoco.

This Weeper-Monkey is only known to us from the description of the Baron Humboldt. It inhabits the forests near the Cataracts of the Orinoco, is mild, active, and not so noisy as its congeners in that locality. Numbers of them are found together in troops. At Maypures, M. Humboldt found a domesticated individual, which every morning caught a Pig, and mounting on its back, rode during the whole day over the savannah which surrounded the cabins of its Indian masters. The same individual was likewise in the habit of riding on the back of a Cat which had been brought up along with it in the house of the missionary of Maypures, while puss suffered patiently the petulance of her more intelligent companion.

The White-fronted Weeper, or Ouavapavi of the Cataracts, may be distinguished at once by the contrast between the pure white of the forehead and orbits, with the bluish-grey of the remainder of the face. Its head is in the form of a greatly elongated oval. The hair of the body is dark grey, lighter towards the breast and belly, but darker towards the limbs, where it becomes of a yellowish-brown. The top of the head is of a grey approaching to black, forming a cowl; a greyish streak extends towards the nose along the centre of the forehead; the eyelashes are of a very dark grey. The eyes are large, brown, and very vivid; the ears have a margin, and are covered with hair; the tail is nearly as long as the body, ash-coloured above, whitish beneath, and of a dark brown at the point, which has no callosity. The nails are all rounded, and very slightly convex. A stripe of very dark grey extends along the dorsal line.

7. CEBUS FULVUS.—FULVOUS WEEPER.

Syn.   CEBUS FULVUS.—Desm. Mam.
          CEBUS FLAVUS.—Geoff. Ann. Mus. XIX.—C. ALBUS.—Ib.
Icon. SIMIA FLAVA.—Schreb. Säügth. pl. 31. B.
          CEBUS FULVUS (var.)—D'Orb.[1] Voy. pl. 3 (Mammifères).
          CEBUS UNICOLOR.—Spix, Sim. et Vespert. Bras. pl. 4.

SPECIFIC CHARACTERS.

THE HAIR entirely yellow or brownish-yellow, deeper on the top of the head.
THE FACE and HANDS dark violet-grey.
INHABITS Brazil.

The Yellow Weeper was first noticed by Marcgravius under the name of *Caitaia* (Bras. 227). Its body is robust, the head large and round, while the face is shortened. The colour of its hair is a uniform clear brown or brownish-yellow, becoming deeper along the centre of the back, and especially on the top of the head, but assuming a greyer tint towards the limbs and tail. The ears are short and naked; the face is of a dark violet-grey; the eyes light brown; the tail is thick, and nearly as long as the body; the nails are yellow. This description is taken from a male, found by M. Spix in the forests of Teffé, a branch of the Rio Solimöens.

M. D'Orbigny has lately published the figure of a Fulvous Weeper much brighter in its colour than the C. unicolor of Spix, but its description has not yet appeared. The Cebus albus of M. Geoffroy is an Albino variety of this species.

8. CEBUS ROBUSTUS.—GREATER WEEPER.

Syn.   DER BRAUNE MICO.—Pr. Max. Beitr.
          CEBUS ROBUSTUS.—Kuhl. Beitr.—Desm. Mam.
Icon. Cebus robustus.—Pr. Max. Abbild.
          CEBUS MACROCEPHALUS.—Spix, Sim. et Vespert. Bras. pl. 1.

SPECIFIC CHARACTERS.

THE HAIR of the head almost black; the limbs and tail brownish-black; elsewhere reddish-chestnut brown.
THE FACE grey flesh-colour. THE HANDS violet-grey.
INHABITS Brazil.

This animal, as its specific name denotes, is strong and muscular, especially in the limbs and tail, and very thickly covered with hair. Its body is thick and round, the face is broad, greyish flesh-coloured, and scantily covered with hair. On the top of the head, a small tuft appears slightly elevated. The head is black; the hands, the inner surface of the limbs, the fore-arms, the lower part of the legs, and the tail, are of a glossy brownish-black; the whole body is thickly covered with long, soft, and shining hairs of a reddish-chestnut brown, which, however, are grey at their roots, and the belly is but scantily covered. The face often becomes surrounded with grey hairs through age.

The Greater Weeper is common in Brazil, but has not been found to the South of the Rio Doce. The females often have a yellowish-red band across the shoulder. Their cry is very similar to that of the Common Horned Weeper.

9. CEBUS XANTHOSTERNUS.—YELLOW-BREASTED WEEPER.

Syn.   DER GELBBRUSTIGE AFFE.—Pr. Max. Beitr.
          CEBUS XANTHOSTERNUS.—Kuhl. Beitr.—Desm. Mam.
          CEBUS VARIEGATUS (young).—Geoff. Ann. Mus. XIX.
          SIMIA VARIEGATA.—Humb. Obs. Zool. p. 356.
Icon. CEBUS XANTHOSTERNUS.—Pr. Max. Abbild.
          CEBUS XANTHOCEPHALUS (fem.)—Spix, Sim. et Vespert. Bras. pl. 3. (young).
          LE SAI À GROSSE TÊTE (C. MONACHUS).—F. Cuv. et Geoff. Hist. Mam. (young).

SPECIFIC CHARACTERS,

THE HAIR on the top of the head, back of the neck, and tail, black; arms and legs brownish-black; breast, neck, belly, and upper-arms, reddish-yellow; the back brown.
THE FACE and HANDS dark violet-grey.
INHABITS Brazil.

The Yellow-breasted Weepers bear much similarity to the species just described. They are found in great numbers in the forests near Rio Janeiro, whence they make excursions upon the plantations of maize, oranges, and other fruits. In the Cebus xanthocephalus of Spix, the head and back of the neck are of the same colour as the breast, and in the Cebus

---

[1] D'ORB. VOY.—Voyage dans l'Amérique Méridionale, exécuté dans le cours des années 1826 à 1833, par M. Alcide D. D'Orbigny. Paris, 1834, et seq.

Monachus of Frederic Cuvier they are white,—variations which are in all probability owing to mere differences of age. Temminck and Prince Maximilian have verified this last observation. We find in the Cebus variegatus of Geoffroy another instance of the undue multiplication of species.

10. CEBUS BARBATUS.—BEARDED WEEPER.

Syn. CEBUS BARBATUS.—Geoff. Ann. Mus. XIX.—Desm. Mam.—Kuhl Beitr.

Icon. LE SAI (Var. B).—Audeb. Sing.

SPECIFIC CHARACTERS.

THE HAIR yellowish-brown; reddish beneath; a yellowish-white beard extending over the cheeks; the top of the head dark yellowish-brown.

THE FACE flesh-coloured. THE HANDS black.

INHABITS Guiana.

This animal differs from the Capuchin Weeper, in having its hair much more yellow; that which surrounds the head is of a yellowish-white; on the top of the head it is dark yellowish-brown; the hair covering the arms is yellow, and changes into red on approaching the fore-hands, which are black, as well as the hinder. The hair on the breast and belly is red, and that of the back and tail yellowish-brown, mixed with grey. On all parts of the body it is very long and silky. The above description was taken from a living specimen in the Paris Menagerie. This animal was perfectly familiar, and imitated all the gestures of its masters. When seated, it used to curl the tail round its body.

(B.) HORNED WEEPERS. (SAJOUS CORNUS.)

The remaining Weepers have the hair of the forehead differently disposed in tufts.

11. CEBUS FATUELLUS.—COMMON HORNED WEEPER.

Syn. LE SAJOU CORNU.—Briss. Reg. Anim. p. 195.—Cuv. Reg. Anim. I. 102.

CEBUS FATUELLUS.—Erxl.—Geoff. Ann. Mus. XIX.—Desm. Mam.

SIMIA FATUELLUS.—Lion. Gmel. I.

HORNED MONKEY.—Penn. Quadr. I. No. 138.

Icon. Sajou cornu.—Buff. Hist. Nat. Suppl. VII. pl. 29.—Audeb. Sing.

Cebus Fatuellus.—Pr. Max. Abbild.

Sajou cornu mâle.—F. Cuv. et Geoff. Hist. Mam. (var.)

GENERIC CHARACTERS.

THE HAIR brownish-black; paler on the shoulders and upper arms; a yellowish-white band on the margin of the cheeks; a crest of upright hair on the forehead, ending in a tuft on each side.

THE FACE and HANDS violet-coloured.

INHABITS Brazil.

The Horned Weepers, first described by Brisson, abound on the east coast of Brazil, especially in the neighbourhood of Rio de Janeiro, and in the great woods near Cabo Frio. They sometimes are found alone, or in pairs; more usually in small troops ascending the trees in search of fruit, but apparently in perpetual motion. Generally they are very lively, active, and quick in their movements; the young especially are exceedingly ludicrous, and easily become attached to their master. From the continual watchfulness of these animals, the hunters find it very difficult to surprise them, and this they usually effect by imitating their sonorous whistle with the mouth. On perceiving the enemy, the troop soon effects its escape by wide springs out of the reach of the guns. They become very fat during the cool season of the year, and are then considered excellent game.

These animals acquire a larger size than the other Weepers. They are of a sooty brown almost approaching to black upon the head, body, limbs, and tail, becoming a paler brown on the shoulders and upper-arms; a band of yellowish-white hairs runs along the margin of each cheek, and meets in a very narrow line beneath the chin. The entire skin, whether naked or covered with hair, is violet-coloured. Instead of lying back upon the head, the hairs of the front stand erect, and form a crest terminated at each extremity by a bunch of hairs much longer than the remainder, from which circumstance the animal has derived its specific name. All the lower parts of the body are much more scantily covered than the upper, and these again are much thicker in winter than in summer. Then the frontal bunches increase in size, the hairs on the cheeks grow longer, and the entire animal appears of a much larger size, and in some degree out of shape. The horns, egrets, or tufts, do not appear until the animal has acquired its canine teeth; that is to say, until it has become adult. The base of its nose is rather broader than in the rest of its congeners, and is folded longitudinally, so as to give the animal a morose appearance, although it is in reality very mild and affectionate.

12. CEBUS CIRRIFER.—CIRCLED WEEPER.

Syn. CEBUS CIRRIFER.—Geoff. Ann. Mus. XIX.—Desm. Mam.—Pr. Max. Beitr.—Fisch. Syn. Mam.

SIMIA CIRRIFERA.—Humb. Obs. Zool. p. 356.

CEBUS LUNULATUS (young).—Kuhl Beitr.—C. LUNATUS.—Desm. Mam.

Icon. CEBUS CIRRIFER.—Pr. Max. Abbild.

Variété du Sajou cornu.—F. Cuv. et Geoff. Hist. Mam.

SPECIFIC CHARACTERS.

THE HAIR dark brown above; lighter beneath; a margin of whitish hairs round the face: the hairs of the head erect, in the form of a Horse's shoe.

INHABITS Brazil.

This animal differs from the Common Horned Weeper, just described, in having the tints of its hair much darker; its back and sides of a dusky chestnut brown: the breast and belly of a paler brown; the fore-part of the arm, the neck, and the under part of the lower jaw, of a yellowish-brown: the arms, limbs, and tail, black, and the temples of a dirty white.

There is a variety of this species, figured by Fred. Cuvier, as a variety of the Horned Weeper; it differs in having the hair of the back black, of the shoulders brown, and of the summit of the head deep brown. White whiskers also ornament the cheeks, and ascend crescent-shaped to the centre of the forehead. These characteristic differences are not great; but yet, as remarked by M. Cuvier, it is only by comparing and contrasting all these varieties that the specific characters can finally be adopted, and this consideration imparts an importance to descriptions and correct drawings which otherwise they could not individually possess.

DOUBTFUL SPECIES.

1. THE FEARFUL MONKEY of Pennant (Quadr. I. No. 134), the Cebus trepidus of Erxleben and Geoffroy, is absolutely identical with C. frontatus of Kuhl and Desmarest. Its hair is chestnut brown; that on the head is still darker, rather long, and elevated in a crest. We find it figured by Audebert (Sing.) under the name of " Sajou, var. λ.," and in Edwards' Gleanings, pl. 312, under the title of the Bush-tailed Monkey. It is probably identical with some of the preceding, perhaps C. cirrifer.

2. THE CEBUS GRACILIS of Spix (Sim. et Vespert. Bras. pl. 5) is yellowish-brown above; whitish beneath; the top of the head and occiput of a deeper brown; the body slender. This is probably an individual of Cebus fulvus, altered in the stuffing.

3. THE CEBUS CUCULLATUS of Spix (Sim. et Vespert. Bras. pl. 6) has the hair of the head directed forwards; a white circle round the face; the head and back brownish; the limbs and tail dusky; the shoulder, throat, and breast, whitish; the remainder of the body of a rusty red. This is, perhaps, a female of one of the species already described.

4. THE CEBUS LIBIDINOSUS of Spix (lb. pl. 2) is of a reddish-yellow; a dark brown cowl on its head, and a white circle round the face. That peculiarity in its behaviour, denoted by the specific name, seems rather to have been accidental to the individual described by Spix.

5. THE ANTIGUA MONKEY (S. Antiguensis) of Shaw, Gen. Zool. I. p. 78, is blackish-fulvous, white beneath, with black limbs, the face black, with bearded cheeks; and the tail brown. It was exported from Antigua, but its real country was unknown.

IMAGINARY SPECIES.

1. CEBUS NIGER of Geoffroy, Desmarest, and Kuhl, derived from the Sajou nègre of Buffon, Suppl. VII. pl. 28, is a melanic variety of C. Apella, according to Humboldt.

2. C. VARIEGATUS of Kuhl, Geoffroy, and others, is the young of C. xanthosternus, according to Temminck. It is likewise the C. xanthocephalus of Spix.

3. C. ALBUS of Geoffroy and others is an albino variety of C. fulvus.

4. C. LUNULATUS of Kuhl (C. LUNATUS of Desmarest) is the young ot the C. cirrifer of Prince Maximilian.

5. SIMIA MORTA; and,

6. SIMIA SYRICHTA of Linnæus (Gmel. I. 38), are founded upon imperfect specimens.

7. C. FRONTATUS of Kuhl and others (the Bush-tailed Monkey of Edwards) is identical with C. trepidus, noticed above.

8. C. MACROCEPHALUS of Spix is the C. robustus of Prince Maximilian.

9. C. UNICOLOR of Spix is the same as the Fulvous Weeper (C. fulvus).

10. C. XANTHOCEPHALUS of Spix should be referred to the C. xanthosternus of Prince Maximilian.

11. C. FLAVUS or SIMIA FLAVA of systematic authors is the same as C. fulvus.

12. C. MONACHUS of F. Cuvier is the young of C. xanthosternus.

SUB-TRIBE II.—CHALANURA.[1]—SAGOINS.

SYNONYMS.

GEOPITHECI (Geopithèques).—Geoff. Ann. Mus. XIX. 112.
LES SAGOUINS[2] (in part).—Buff. Hist. Nat.
LES SAIMIRIS, LES SAKIS, et LES NOCTHORES.—Cuv. Reg. Anim. I. 193, 194.
CALLITHRIX (in part).—Erzl. p. 55.

CHARACTERS OF THE SUB-TRIBE.

THE TAIL not prehensile, or imperfectly so, generally bushy and long.
THE HABITS diurnal or nocturnal.   HANDS always pentadactylous.

The remainder of the American Apes either have the tail depressed, and imperfectly prehensile, or it is not at all prehensile, and hangs loosely downwards. The latter are known in general by the name of *Sakis.*

GENUS VI. CALLITHRIX—SQUIRREL-MONKEYS.

Syn.   CALLITHRIX.—Geoff. Ann. Mus. XIX.—Kuhl Beitr.—Desm. Mamm.
        CALLITHRIX (in part).—Erzl.
        SAGOUINUS.—Less. Mam.

GENERIC CHARACTERS.

THE HEAD round.   THE MUZZLE short.   THE FACIAL ANGLE about 60°.   THE EYES and EARS large.
THE BODY and LIMBS medium size.   HABITS diurnal.
THE TAIL longer than the body, and covered with short hairs.
THE NAILS short and flat on the thumbs; long and narrow on the fingers.
INHABIT South America.

The prettily coloured hair observed in some of the animals belonging to this Genus has occasioned the term *Callithrix, or beautiful hair,* to be applied to them, although it is not by any means applicable to all the species.   The manners of the greater part of them are but little known. It has been ascertained, however, that they are in general very intelligent; that they live on fruits, insects, and sometimes on small birds, or other animal food, and that they occur abundantly in troops or small families, in the equatorial parts of the New World. Some of the species seem to be delicately sensible to slight alterations of temperature and moisture, and they soon languish and die under the influence of the heats of the low-grounds near the coast.

Some authors, among whom is the Baron Cuvier, have separated the Varied Squirrel-Monkey (C. sciurea) from the remainder, on account of the superior development of its cranium; but this circumstance does not appear, in our opinion, to be sufficient to justify a generic distinction. The eyes in all of them are of considerable size, and the orbits are completely circular; the internal ear is supplied with large auditory chambers. But in all the species, excepting the first, the cerebral cavity is not very capacious; the occipital hole is placed more in arrear, and the interorbital partition is wholly osseous.

The dentition of the whole of this genus is very uniform. In the upper-jaw the incisor nearest the medial line is double, the size of the external one, their shape being alike, the lower margin is roundish, the outer side sloping, and the inner aspect much curved. The canine is of moderate size.   The first false molar is smaller than the other two, and they are all shaped alike. The last of the true grinders is smaller than the other two, and of a peculiar form, exhibiting two circular crests within its inner margin, and one on the outer.   The teeth of the lower jaw correspond with the description above supplied of the Howlers, with the exception of the last molar.

We are acquainted with eight species: some of them, however, are but imperfectly known. Many imaginary species have to be expunged from the lists of our predecessors.

In the first species (LES SAIMIRIS of the Baron Cuvier), the tail is depressed, and scarcely ceases to be prehensile; the head is very flat; and there is a membraneous expansion at the interorbital partition of the skeleton.

1. CALLITHRIX SCIUREA.—VARIED SQUIRREL-MONKEY.

Syn.   LE SAIMIRI.—Cuv. Reg. Anim. I. 103.
        CALLITHRIX SCIUREUS.—Geoff. Ann. Mus. XIX.—Desm. Mamm.
        SIMIA SCIUREA.—Linn. Gmel. I.
        LE TITI DE L'ORÉNOQUE (S. sciurea).—Humb. Obs. Zool. p. 332 and 357.

SAGOUINUS SCIUREUS.—Less. Mam.
ORANGE-MONKEY.—Penn. Quadr. I. No. 137.
SQUIRREL-MONKEY.—Shaw, Gen. Zool. I. 77.
Icon.   Saimiri.—F. Cuv. et Geoff. Hist. Mamm.
        Le Saimiri.—Audeb. Sing.—Buff. Hist. Nat. XV. pl. 10.
        Calitrix entomophagus.—D'Orb. Voy. pl. 4 (Mammifères).

SPECIFIC CHARACTERS.

THE HAIR greenish-yellow above; white beneath; the upper-arms and thighs grey; the lower-arms and legs bright orange.
THE FACE and HANDS flesh-coloured.   THE MUZZLE black.
INHABITS the Banks of the Orinoco.

The Varied Squirrel-Monkeys are very common in the regions to the south of the Cataracts of the Orinoco; but the smallest and handsomest are those of the Cassiquiare. Their bodies exhale a slight odour of musk. The expression of their countenance resembles that of a child; they have the same innocence of aspect, the same malicious smile, and they pass with equal rapidity from the extremity of delight to that of sadness. The Indians assert that they shed tears like a human being; and M. Humboldt remarked in one specimen that its eyes became moistened when it was uneasy or under the influence of fear. The Titi, as it is termed by the natives, is in continual motion, and all its movements are light and graceful. It seldom becomes irritated, like the Marmousets, but seems continually occupied in playing, leaping, and catching Insects or Spiders, the latter of which it prefers to all other food. It has the singular habit of looking attentively at the mouth of the person who speaks; and if it happen to sit on his shoulder, attempts to play with his teeth or tongue. It is a formidable enemy of all collectors of Insects, and however carefully they may conceal their specimens, the Varied Squirrel-Monkey is sure to devour them, without even wounding itself with the pins by which the spoils are fixed. As an instance of its sagacity, M. Humboldt informs us that one of these little animals could distinguish uncoloured plates of Insects from those which represented Quadrupeds or any other subject. Whenever it saw the engraving of a Grasshopper or a Wasp, its hand was instantly extended to grasp that object. Being accustomed to live in a moister and cooler climate than that of the coast, the Varied Squirrel-Monkey soon loses its liveliness when removed from the forests of the Orinoco to Cumana or Guayra, and it seldom survives this change of locality above a few months.

The upper parts of its body are of a greenish-yellow, which assumes a greyish tint upon the upper-arms and thighs, and changes into a bright orange on the fore-arms and legs. The tail has the same greenish tint at the back, but becomes black towards the point; the abdomen, breast, neck, cheeks, and a circle round the ears, are white; there is, however, a spot of greenish-yellow in the middle of the white hairs of the cheeks. The tip of the muzzle, the nostrils at the corners of the mouth, and the under part of the chin, are black; the remainder of the face and the other naked parts of the body are flesh-coloured; likewise the nails, which are black only at the points. The eyes are brown; the hands perfectly formed, but the thumbs are opposable only on the hinder; the nails of the thumbs are flat and broad, those of the fingers long and narrow. The scrotum is very voluminous, and the remainder of these parts very similar to the corresponding arrangements in the human species.

This animal never uses the tail for prehensile purposes, but is often rolls up the point, and, when any thing is conveniently placed, encircles the extremity of the tail round that object; but it is unable to do so with any considerable force. When seated, its hinder limbs are extended forwards, and the fore-hands rest upon them. It sleeps in this sitting posture,

---

[1] Chalanura, from χαλαίνω, to let loose, and ουρά, a tail.
[2] All the American Monkeys, having the tail not prehensile, are included by Buffon along with the *Ouistitis,* under the common name of Sagouins (CALLITHRIX, Erzl.) This name of Sagouin, or sagui, is applied in Brazil to all the smaller Quadrumanous animals having the tail not prehensile. M. Geoffroy assigns to his genus Callithrix (which forms only a subdivision of the Callithrix of Erxleben), and, to his genera Aotus and Pithecia, the common appellation of Geopitheci, or Ground-Monkeys.—*Note of the Baron Cuvier.*

merely resting the head between tne thighs, which then touch the ground. The thumb of the fore-hands is parallel to the fingers, that of the hinder is completely opposable. Its cry consists of a mild and acute whistle, which it repeats three or four times in rapid succession.

In the other species (LES CALLITHRIX of the Baron Cuvier), the tail is slender, and the teeth do not project outwards. They have for a long time been united with the Saimiris, but the head of these Sagoins is more elevated, and their canines much shorter.

2. CALLITHRIX PERSONATA.—MASKED SQUIRREL-MONKEY

Syn. LE SAGOUIN À MASQUE.—Cuv. Reg. Anim. I. 104.
CALLITHRIX PERSONATUS.—Geoff. Ann. Mus. XIX.—Kuhl, Beitr.—Desm. Mam.—Pr. Max. Beitr.
SIMIA PERSONATA.—Humb. Obs. Zool. p. 357.

Icon. SAUASSU (Callithrix personatus).—Pr. Max. Abbild.
Callithrix personata (mas.)—Spix, Sim. et Vespert. Bras. pl. 12.
CALLITHRIX NIGRIFRONS (mas.)—Ibid. pl. 15.

SPECIFIC CHARACTERS.

THE Hair greyish-brown or yellowish; the head black; the tail chestnut-brown; the back of the neck whitish in the male.

THE FACE and HANDS black.

INHABITS Brazil.

This is one of the most agreeable Monkeys in the Brazils. It was found by Spix near Rio de Janeiro, but abounds between the Rio Para-'hyba and Rio Doce. It is commonly seen in little communities consisting of two or three families. Its cry is sharp and loud, with a kind of rattle, occasioned by a peculiar form of the larynx. When domesticated, it becomes very gentle and sociable. The natives call it *Sauassu*, by which name it is also known to the Brazilians.

The Masked Squirrel-Monkey is at once known by the deep black hue of its head and hands, while the tail is chestnut-brown. Its hair is very bushy, deep brown near the roots, and chestnut-brown at the points. The black hair of the head hangs downwards to a considerable length around the face.

3. CALLITHRIX TORQUATA.—WHITE-THROATED SQUIRREL-MONKEY.

Syn. LE SAGOUIN EN DEUIL ou LA VEUVE.—Cuv. Reg. Anim. I. 104.
CALLITHRIX TORQUATA. — Hoffmannsegg in Mag.[1] Gesellsch. Naturf. Freund. X.—C. TORQUATUS, C. AMICTUS & C. LUGENS.—Geoff. Ann. Mus. XIX.—Kuhl, Beitr.—Desm. Mam.
SIMIA LUGENS (Viudus).—S. TORQUATA, & S. AMICTA.—Humb. Obs. Zool. p. 357.

Icon. Callithrix amicta (mas.)—Spix, Sim. et Vespert. Bras. pl. 13.

SPECIFIC CHARACTERS.

THE Hair chestnut-brown; the fore-hands yellowish-white; the hinder-hands and tail black; the neck and breast white.

THE FACE and HANDS brown.

INHABITS the banks of the Solimöens, and the mountains on the right bank of the Orinoco.

The White-throated Squirrel-Monkey is rare and much esteemed. It appears to be extremely mild, timid, and inoffensive; its eye denotes great vivacity, yet it remains for hours motionless, without sleep, and noticing attentively every thing passing around it. Often it refuses to eat, though very hungry; and seems to have a great aversion to be touched on the hands, hiding them under the belly whenever any one attempts to touch them. This mildness and timidity, however, are merely apparent. At the sight of a small bird it becomes furious, springs upon it like a Cat, and devours it in an instant. It is very fond of fresh meat, although it usually lives on fruits; and when it eats, raises both hands to the mouth at once, like the Sagoins. It does not associate with Monkeys of a different species, and the sight of a Varied Squirrel-Monkey puts it into a rage. It runs and leaps with prodigious agility, and, like the Monkey last mentioned, does not thrive on the coasts of South America.

The hair on the body of this Monkey is of a deep chestnut-brown; the hinder-hands, fore-arms, and tail, are of a shining black; the forehead, temples, and circle round the face, of a dusky grey. The throat and breast are white. The fore-hands yellowish-white. In the females and young, the tail is not so black, and the back, instead of being chestnut-brown, is more varied, and tends rather towards a light brown.

4. CALLITHRIX MELANOCHIR.—BLACK-HANDED SQUIRREL-MONKEY.

Syn. CALLITHRIX MELANOCHIR.—Pr. Max. Beitr.—Kuhl, Beitr.—Desm. Mam.

Icon. CALLITHRIX CINERASCENS, mas. (young).—Spix, Sim. et Vespert. Bras. pl. 14.
CALLITHRIX GIGOT.—Ibid. pl. 16.
Callithrix melanochir.—Pr. Max. Abbild.

SPECIFIC CHARACTERS.

THE Hair ash-grey; bright reddish-brown on the back; the tail spotted with white and yellowish-grey.

THE FACE and HANDS black.

INHABITS Brazil.

This is an agreeable little animal, common to the North of the Rio Doce on all the eastern coast of Brazil, and is even not rarely found in the interior. It is commonly known by the name of *Gigò*, and in its form and mode of life is not far different from the Masked Squirrel-Monkey already described. There is the same peculiarity in its voice. These animals reside on the highest trees of the primitive forests of Brazil in small troops of three to five individuals. When taken young they are easily tamed, and become very mild and confiding. They are sometimes hunted as game.

The bright reddish-brown of the back is peculiar to the adults; the young are entirely grey.

5. CALLITHRIX CUPREA.—COPPER-BREASTED SQUIRREL-MONKEY.

Syn. et Icon. CALLITHRIX CUPREA.—Spix, Sim. et Vespert. Bras. pl. 17 (fem.)

SPECIFIC CHARACTERS.

THE Hair brownish-grey above; the head, breast, limbs, and under part of the body, copper-coloured.

THE FACE and HANDS brown.

INHABITS the banks of the Rio Solimöens.

The Copper-breasted Squirrel-Monkey, called *Yapusa* by the natives, is of a brownish-grey above, and copper-colour beneath, as far as the feet. The head is furnished with short and thick hairs, the face and eye-lids are almost naked. The face is surrounded from the temples to the base of the cheeks with copper-coloured hairs directed forwards. The ears are naked behind, very prominent, and covered in front by some red hairs. In general, all the hairs are directed backwards, those of the back, shoulders, and thighs. are tolerably long, black, and intermixed with red. The tail is of a reddish-grey, becoming gradually lighter towards the extremity. The eyes are brown.

6. CALLITHRIX MOLOCH.—MOLOCH SQUIRREL-MONKEY.

Syn. CEBUS MOLOCH.—Hoffm. in Mag. Gesell. Naturf. Freund. X.
CALLITHRIX MOLOCH.—Geoff. Ann. Mus. XIX.—Kuhl, Beitr.—Desm. Mam.

Icon.

SPECIFIC CHARACTERS.

THE Hair ash-coloured, annulated with brown; the temples, cheeks, and belly, bright red; the tip of the tail and hands of a clear grey, almost white.

THE FACE and HANDS dusky and naked.

INHABITS Para.

This elegant little animal has all the upper parts of the body, the neck, and head, as well as the internal surface of all the limbs, covered with hairs, annulated with light grey and pale brown, occasioning these parts of the body to present an agreeable varied appearance. The hairs of the tail, which are very bushy at its base and short elsewhere, are annulated to a considerable extent with dark greyish-brown and dirty white. The inner surface of the limbs is of a clearer grey than the upper part of the body. The backs of the hands, especially the fore ones, are of a clear grey, almost white. The hairs on the top of the head are short and perpendicular. The cheeks, the upper part of the neck, the breast, belly, and the internal surface of the limbs, are of a reddish-yellow, tending rather to a bright red on the limits of the grey tint of the sides, where these colours are separated by a definite tint.

7. CALLITHRIX INFULATA.—MITRED SQUIRREL-MONKEY.

Syn. CALLITHRIX INFULATUS.—Kuhl, Beitr.—Desm. Mam.

Icon.

SPECIFIC CHARACTERS.

THE Hair grey above, reddish-yellow beneath, a white spot surrounded with black above the eyes; the base of the tail reddish-yellow, the point black.

INHABITS Brazil.

This animal, first named by Lichtenstein, and described by Kuhl, from a specimen in the Museum of Berlin, is known to us only by the above description.

---

[1] See Hoffmannsegg's Description of Four Quadrumanous Animals in the Mag. der Gesellsch. Naturforsch. Freund. zu Berlin, for 1809, vol. X. p. 8.

46

6. CALLITHRIX DONACOPHILA.—D'ORBIGNY'S SQUIRREL-MONKEY.

Syn. et Icon. CALLITHRIX DONACOPHILUS.—D'Orb. Voy. pl. 5 (Mammifères).

SPECIFIC CHARACTERS.

THE HAIR of a uniform greyish-brown, interspersed with white.
THE FACE and HANDS bluish-grey.
INHABITS South America.

A figure of this animal is given in D'Orbigny's Voyage, now in course of publication; but its description has not yet reached us. It appears to be fond of climbing the elevated reeds of those tropical regions.

IMAGINARY SPECIES.

1. C. NIGRIFRONS of Spix, is identical with C. personata.
2. C. CINERASCENS of Spix, is the young of C. melanochir.
3. C. AMICTA, and,
4. C. LUGENS, do not differ specifically from C. torquata.

GENUS VII. PITHECIA.—FOX-TAILS.

Syn. PITHECIA (Schweif Affe).—Illig. Prodr.—Kuhl, Beitr.—(Saki) Desm. Mam.—Geoff. Ann. Mus. XIX.
SIMIA (in part).—Linn. Gmel. L.—Humb. Obs. Zool.
CEBUS (in part).—Erxl.
BRACHYURUS (in part).—Spix, Sim. et Vespert. Bras.

GENERIC CHARACTERS.

THE HEAD round. THE MUZZLE short. THE FACIAL ANGLE about 60°.
THE EYES and EARS resembling those of Man.
THE CANINE TEETH powerful, and projecting forwards.
THE BODY and LIMBS medium size. HABITS diurnal or crepuscular.
THE TAIL usually long, and plentifully covered with long hairs.
THE NAILS short and curved.
INHABITS South America.

Several of the Sagoins have the tail long and bushy, for which reason they may be called *Fox-tailed Monkeys* (Singes à queue de renard), or briefly FOX-TAILS, and their teeth project forwards to a greater degree than in the other Monkeys. The above are the Pithecia of Desmarest and Illiger. One species, having the tail less than the body, is separated by Spix from the remainder to form his Genus *Brachyurus*.

The Fox-tails are also designated *Night Monkeys* in Cayenne, but inaccurately, as they are not really nocturnal but rather crepuscular, being busily astir principally in the evenings and at early dawn. They have a general resemblance to the Sapajous and the remaining Sagoins; but are distinguished from the former in not having the tail prehensile; and from the other genera of Sagoins, by the tail being supplied with long and bushy hair. Another, and very marked essential character, consists in the circumstance of their incisive teeth not preserving their parallelism with the canines, but being crowded together as if forced forwards by the great size of the canines. Their head is round, their muzzle short, and the facial angle ranges from about 60° to 52°. The size of the ear is moderate; the nasal bones are elevated and extended; the tail is somewhat shorter than the body; the hands are pentadactylous, with short and curved claws.

These animals usually reside in the depths of the forest, where they conceal themselves or sleep during the day, so that their manners are not well known. Their usual food is fruit and insects; they collect in small troops, and they are often pursued by some of the larger Monkeys, who seize their supplies, and beat them if they have the temerity to resist.

1. PITHECIA LEUCOCEPHALA.—WHITE-HEADED FOX-TAIL.

Syn. LE YARKÉ.—Cuv. Reg. Anim. I. 103.
PITHECIA LEUCOCEPHALA.—Geoff. Ann. Mus. XIX.—Desm. Mam. (No. 91).—Kuhl, Beitr.
SIMIA LEUCOCEPHALA (YARQUÉ).—Humb. Obs. Zool. p. 359.
PITHECIA OCHROCEPHALA (var).—Kuhl, Beitr.—Desm. Mam.

Icon. L'YARQUÉ.—Buff. Hist. Nat. XV. pl. 12.—Audeb. Sing.—Schreb. Saügth. pl. 32.

SPECIFIC CHARACTERS.

THE HAIR brownish-black; a broad yellowish-white circle round the face; no beard.
THE FACE and HANDS brown.
INHABITS Guiana.

This species has very generally been confounded with the Red-breasted Fox-tail (P. rufiventer), though it clearly differs in its markings, which, according to Audebert, are very uniform. Its face is brown; the sides of the head and neck are covered with a yellowish-white hair, short, and cut, as it were, with scissors; the body generally is brownish-black, covered with bushy hair, about four inches long. This remark applies also to the tail, which is nearly of the same length as the body; on the extremities the hair is short; the colour brown.

This is a rare species, frequenting the thickets of Guiana and Surinam. Its habits require further elucidation. De la Borde states that this animal goes in troops, consisting of from half-a-dozen to a dozen; whilst Stedman notices that it is the only species of Monkey which is not sociable, being always found solitary. He also remarks that it is generally persecuted by its congeners, who never fail to attack, and rob it of its stores. It feeds upon honey, rice, and the other grains used by Man. The female has usually only one at a birth, which for a time it carries on its back. Its cry is said to resemble that of the Weepers.

Sometimes these animals are found with the hair surrounding the face of a bright yellow-ochre tint (P. Ochrocephala of Kuhl and others), a variation in all probability belonging to the female or young.

2. PITHECIA HIRSUTA.—URSINE FOX-TAIL.

Syn. LE SAKI GRIS.—Cuv. Reg. Anim. I. 103.
Icon. PITHECIA HIRSUTA (mas.)—Spix, Sim. et Vespert. Bras. pl. 9.

SPECIFIC CHARACTERS.

THE HAIR greyish-black, very long and curly; yellowish on the hands; no beard.
THE FACE dusky. THE HANDS yellow.
INHABITS the banks of the Rio Solimöens and Rio Negro.

The Ursine Fox-tail was introduced to notice by Spix; and no additional information has hitherto been supplied. This, however, is the less to be regretted, as the statements of the above named Naturalist are both minute and interesting.

The native name of all these well-clad Monkeys is *Parauá*. Their coat in a remarkable degree resembles that of the Bear. At morning and evening they issue from the forests, collect in great numbers, and cause the air to resound with their piercing cries. They are ever watchful and alert, so that they cannot be caught without much trouble. On the slightest noise they retreat with the greatest rapidity, and plunge into the depths of the forests. At the same time, when once tamed and domesticated, they become exceedingly fond of their master; they retreat to him when alarmed, and rejoice to become his companion, especially at meal-time.

The fur on the body and tails of these Monkeys is very bushy and even frizzled. The species on which we are now dwelling, and to which the Indians give the name of the *Great Parauá*, to distinguish it from a smaller, is greyish-black. Its hair is about three inches long, crisp, here and there grizzly, occasionally of a light brown hue. On the back of the head it is peculiarly long, and has been compared to a hood. The hands and feet are nearly devoid of hair, and of a light flesh colour; the neck, too, is nearly naked. The tail is as long as the body. The young ones are of the same colour with the full-grown animals.

3. PITHECIA SATANAS.—BLACK FOX-TAIL.

Syn. LE SAKI NOIR.—Cuv. Reg. Anim. I. 103.
SIMIA SATANAS (COUXIO).—Humb. Obs. Zool. p. 315 and 358.
CEBUS SATANAS.—Hoffm. in Mag. Gesellsch. Naturf. Freund. X.
PITHECIA SATANAS.—Geoff. Ann. Mus. XIX.—Desm. Mam. No. 84.—Kuhl, Beitr.

Icon. Humb. Obs. Zool. pl. 27.
BRACHYURUS ISRAELITA, mas. (young).—Spix, Sim. et Vespert. Bras. pl. 7.

SPECIFIC CHARACTERS.
THE ADULT.

THE HAIR dusky black; scanty beneath; a very long beard.

THE YOUNG.

THE HAIR yellowish-brown; dusky black on the head and tail; the beard short.

THE FACE and HANDS reddish-grey.
INHABITS Para, Rio Negro, and the Orinoco.

For our acquaintance with this Monkey, we are indebted to the liberality of the Count of Hoffmannsegg, who dispatched the Naturalist Sieber to Brazil, and, among other newly discovered animals, received from him a specimen of this species at Berlin. The face and all the hands are of a reddish-grey colour, and naked; the mouth is large. The whole of the rest of the body is covered with long coarse hair of a deep brownish-black in the male, and of a yellowish-brown in the female.

With this account of the Black Fox-tail we have ventured to combine the description of the *Brachyurus Israelita* of Spix, under the conviction that it is the young of the same species. The external and other characters, as far as they have been noted, correspond; and the comparative shortness of the tail, upon which Spix chiefly insists, and readily explicable upon our supposition, is an insufficient ground for the formation of a new species. The point, however, requires further elucidation; and the more so, as the learned Naturalist assigns the forests of the Yapura, a tributary of the Solimõens, near Peru, as the habitat of his Brachyurus.

4. PITHECIA RUFIVENTER.—RED-BREASTED FOX-TAIL.

Syn. Le Saki à ventre roux.—Cuv. Reg. Anim. I. 103.
PITHECIA RUFIBARBATA.—Kuhl, Beitr.—Desm. Mam. No. 88.
SIMIA PITHECIA.—Linn. Gmel. I.
PITHECIA RUFIVENTER.—Geoff. Ann. Mus. XIX.—Desm. Mam. No. 86.—Kuhl, Beitr.
SIMIA RUFIVENTER.—Humb. Obs. Zool. p. 358.
Icon. Le Saki.—Buff. Suppl. VII. pl. 31.—Audeb. Sing.
PITHECIA CAPILLAMENTOSA (fem.)—Spix, Sim. et Vespert. Bras. pl. 11.

SPECIFIC CHARACTERS.

The Hair dark brown above; red beneath in the male, yellow beneath in the female; yellowish-brown on the head; black on the hands.
The Face and Hands dark flesh colour.
Inhabits Guiana.

This animal has been called the *Wigged* Monkey, a name to which it seems justly entitled; and Spix gives a particular description of no fewer than three wigs; the first flowing down the shoulders, the second forming a marked zone around the face, and the third ascending backward from the eyes. Under this profusion of hair, its ears, which are small, are entirely hid. Audebert informs us, that the tints of colouring vary.

The habits of this species are not accurately described. From its having received from Buffon the appellation of the Nocturnal Monkey, it has been inferred that it is taciturn, solitary, feeble, and timid. It lives in thickets; associates in small groups of eight or ten, and is not often met with.

5. PITHECIA BREVICAUDATA.—SHORT FOX-TAIL.

Syn. Le Couxte-queue Ouakary.—Spix, pl. 13.
Cebus Ouakary.—Fisch. Syn. Mam.
Icon. Brachyurus Ouakary (mas.)—Spix, Sim. et Vespert. Bras. pl. 8.

SPECIFIC CHARACTERS.

The Hair on the head, arms, and legs, black; on the back yellowish-brown; on the thighs and tail ferrugineous; no beard; the hair of the forehead distichous.
The Face and Hands dusky. The Tail short.
Inhabits the banks of the Rio Solimõens and Rio Iça.

To these characters, it is scarcely necessary to add any thing descriptive of the animal discovered by Spix. It is of moderate dimensions, and lives in considerable troops, which confine themselves to the woods which skirt the rivers. It is chiefly during the day that these animals make the forests resound with their piercing and savage cries.

6. PITHECIA MELANOCEPHALA.—BLACK-HEADED FOX-TAIL.

Syn. SIMIA MELANOCEPHALA (CACAJAO).—Humb. Obs. Zool. p. 316 and 359.
PITHECIA MELANOCEPHALA.—Geoff. Ann. Mus. XIX.—Desm. Mam. No. 92.—Kuhl, Beitr.
Icon. Humb. Obs. Zool. pl. 29.

SPECIFIC CHARACTERS.

The Hair of the head black; of the point of the tail, dark brown, elsewhere brown, varied with yellow.
The Face and Hands dusky.
Inhabits the banks of the Rio Negro and Cassaquiare.

Our acquaintance with this interesting animal we owe to the exertions of the indefatigable Humboldt. Its countenance has a resemblance to that of an infant, while its expression approximates that of an old Negro. The hair of the head is as if all combed forward; and bristles occupy the place of the eye-brows and beard. The ears are quite naked, very large, and more than any of the American Monkeys, like those of Man. The hair is long and shining, and generally copious, except round the neck, where it is nearly wanting.

This animal is very voracious, but dull and heavy, feeble, and extremely gentle. It eats all kinds of fruits, not excepting the sourest lemons. In

seizing an object, it extends its arms, and curves its back in a singular manner. As its fingers are exceedingly long and slender, its attempts at grasping are very awkward, and even its mode of feeding. It has a dread of the other Monkeys, and it trembles in every fibre at the sight of a Crocodile or Serpent. When irritated, a rare circumstance, it opens its mouth in an extraordinary manner, and utters convulsive cries. The little animal, which for a time was the companion of Humboldt, was of a delicate constitution, and died under the effects of a *Coup-de-Soleil*, notwithstanding all the means which were employed for its recovery.

7. PITHECIA MONACHUS.—HOODED FOX-TAIL.

Syn. PITHECIA MONACHUS (Moine).—Geoff. Ann. Mus. XIX.—Desm. Mam. No. 90.—Kuhl, Beitr.
SIMIA MONACHUS.—Humb. Obs. Zool. p. 359.
Icon. Buff. Hist. Nat. Suppl. VII. pl. 30.

SPECIFIC CHARACTERS.

The Hair variegated with large spots of brown and bright yellow; forming a cowl or hood on the top of the head.
Inhabits Brazil (probably).

The whole history of this species, noted by so many respectable authorities, is very obscure, and requires revision. It was introduced into the catalogue of Monkeys on the authority of M. Geoffroy-St-Hilaire, from a specimen in the Paris Museum, but of this specimen little was known or determined. Kuhl says it is the least of all the tribe; its habits and habitat are very doubtful, if not wholly unknown.

8. PITHECIA AZARÆ.—AZARA'S FOX-TAIL.

Syn. PITHECIA MIRIQOOUINA.—Geoff. Ann. Mus. XIX.—Desm. Mam.—Kuhl, Beitr.
Le Miriqoouina.—D'Azar. Quadr. Parag. II. p. 213.
SIMIA AZARÆ.—Humb. Obs. Zool. p. 359.
Icon.

SPECIFIC CHARACTERS.

The Hair brownish-grey, cinnamon colour beneath; two white spots beneath the eyes.
Inhabits Paraguay.

As the distinguished and indefatigable Azara is our only authority for this animal, as for many others of Southern America, and as no plate of it has been published, we give in detail the characters he supplies. The head is very small and almost round; the neck is uncommonly short, and seems even thicker than the head. The whole face to the very eyes is covered with hair, the eyelids and nose, which is prominent, being alone naked. The eye is large; the iris of a pale brown. The ear, too, is very large, round, hairy, somewhat elevated at the point. The fur is very soft, bushy, and erect, that of the tail alone lying close. The greater part of the body is of a grizzly colour; the lower parts are of cinnamon hue, and on the face, above the eye, upon the cheek, and under the chin, there is a white marking. The female is of the same colour as the male, and is a trifle less in size, and the young in no respect differs from the markings of its parents.

Azara drew his description from the examination of three females and one male. He had also seen it domesticated, and learned that it was very gentle and quiet.

9. PITHECIA CHIROPOTES.—CAPUCHIN FOX-TAIL.

Syn. SIMIA CHIROPOTES (Capuchin de l'Orénoque).—Humb. Obs. Zool. p. 311 and 358.
PITHECIA CHIROPOTES (CAPUCHIN).—Geoff. Ann. Mus. XIX.—Desm. Mam. No. 85.—Kuhl, Beitr.
Icon.

SPECIFIC CHARACTERS.

The Hair reddish-brown; distichous on the head; the beard very long.
The Face black and naked.
Inhabits the banks of the High Orinoco.

This Monkey, as stated by Humboldt, is one of the most remarkable of South America, though not mentioned by any preceding Naturalist. It is of a reddish-brown colour; its coat long and shining. Its head is oval shaped; the facial angle about 52°; the face and palms of the hand are black and naked. The forehead and top of the head are covered with thick and very long hair, lying forwards, and dividing itself over the eyes into two large tufts. The eyes are large and deep-set; the canine teeth very formidable, and the deep brown beard venerable, extended down the breast. The head, thighs, and tail, are of a deeper tint than the rest of the body.

Of all the animals of its kind, the Capuchin Fox-tail is that one whose features most resemble those of Man. His eyes have an expression of melancholy; not free from ferocity. As the chin is hid under his bushy beard, the facial line appears larger than it really is. He is a strong animal, agile, ferocious, and scarcely tameable. When irritated, he starts back on his hind feet, grinds his teeth, pulls the end of his beard, and leaps with threatening gestures around his assailant. In the fits of his ire I have often seen him, says Humboldt, fix his teeth deep in a wooden plank. He generally maintains a sullen sadness, which is interrupted only at the sight of some favourite food. He drinks but seldom, and in a way which differs remarkably from the other American Monkeys, who raise the cup presented to them to the lip. The Capuchin, on the contrary, drinks from the hollow of his hand, at the same time turning his head to a side. This is a tedious operation, which he performs with either hand, and only when he imagines he is unobserved. He becomes quite furious when any one wets his beard; and it would appear that it is to avoid this annoyance that he resorts to his peculiar mode of drinking.

These Monkeys do not live in troops, but in pairs only, in the forest. They are found in the vast deserts in the High Orinoco, to the south and east of the Cataracts, and appear to be unknown in most of the neighbouring provinces. The Priest Juan Gonzalis, who was intimately acquainted with the locality they frequent, informed Baron Humboldt that the native Indians devour these animals in great numbers at certain seasons of the year.

DOUBTFUL SPECIES.

1. PITHECIA INUSTA (Spix, Sim. et Vespert. Bras. pl. 10) is suspected by Temminck to be identical with P. hirsuta. It is about one-third smaller, and the head is wholly ferrugineous.

2. SIMIA SAGULATA, THE JACKETED MONKEY (Traill, in the Memoirs of the Wernerian Society, vol. iii.) is conjectured by Fischer to be a Pithecia Satanas.

To this Catalogue we have nothing to add, except that our examination inclines us to agree in the justness of the suspicions expressed by Temminck and Fischer.

IMAGINARY SPECIES.

1. P. OCHROCEPHALA (Kuhl, Beitr.) is the female or young of P. leucocephala.

2. P. RUFIBARBATA (Kuhl and Desm.) is absolutely identical with P. rufiventer. This is the Simia Pithecia of Linnæus.

3. P. CAPILLAMENTOSA of Spix is merely a duplicate of P. rufiventer.

4. BRACHYURUS ISRAELITA of Spix is the young of P. Satanas.

GENUS VIII. NYCTIPITHECUS.—NIGHT-MONKEYS.

Syn.   LES NOCTHORES.—F. Cuv. Hist. Mam.—Cuv. Reg. Anim. I. 104.
       AOTUS.—Illig. Prodr.
       NYCTIPITHECUS.—Spix, Sim. et Vespert. Bras.

GENERIC CHARACTERS.

THE HEAD round and broad. THE MUZZLE short. THE FACIAL ANGLE about 60°.
THE EYES very large and approximated. THE EARS very small.
THE NAILS short. HABITS nocturnal.
THE TAIL longer than the body.
INHABIT Guiana and Brazil.

To the Night-Monkeys Illiger has very improperly assigned the generic term Aotus (Earless). They differ from the other Sagoins, merely by their large nocturnal eyes, and their ears are partly concealed under the hair.

All these animals come from Guiana or Brazil.

It was the Baron Humboldt who proposed the establishment of this genus for the arrangement of the Douroucouli, which he discovered in the forests of the Orinoco. He designated it Aotes (ἄωτος), earless; but as this was a character which was inapplicable to the animal, Spix substituted the appellation Nyctipithecus (Night-Monkey), which, taken from one of its most striking characteristics, has been generally adopted.

The generic characters are distinctly marked. The head is round and very broad; the muzzle is short; the eyes nocturnal, very large, and near each other; the ears are very small; the tail is longer than the body, not prehensile, covered with hair; each foot has five toes, and the nails are flat. In all these particulars, the Nyctipitheci have a strong resemblance to the Loris of the ancient Continent. For a long time the animal introduced by Humboldt was the only species of the genus; lately, two more have been added by Spix, who thinks it highly probable there are others.

1. NYCTIPITHECUS TRIVIRGATUS.—HUMBOLDT'S NIGHT-MONKEY.

Syn.   LE DOUROUCOULI.—Cuv. Reg. Anim. I. 104.
       AOTUS HUMBOLDTII.—Illig. Prodr.—Schinz. Thierr.
       AOTUS TRIVIRGATA.—Geoff. Ann. Mus, XIX.
       SIMIA TRIVIRGATA (DOUROUCOULI).—Humb. Obs. Zool. p. 307 and 358.
Icon. Humb. Obs. Zool. pl. 28.
       Douroucouli (fem.)—F. Cuv. et Geoff. Hist. Mam.

SPECIFIC CHARACTERS.

THE HAIR grey, mixed with white, a brown band along the back.
THE FACE blackish. HANDS white.
INHABITS Guiana.

The Baron Humboldt, who discovered this animal, observes that it is the most remarkable he had met in Guiana. It differs from its congeners not only in the form of its teeth and ears, but still more in its habits, the size of its eyes, and in the whole of its physiognomy, which very much resembles one of the Loris (Stenops) of the Old World. It is strikingly characterized by its head being cat-like, by its large yellow eyes, which cannot support the light; by the smallness of the external ear, and by its unreprehensible tail being much longer than its body.

This Night-Monkey is generally of a grey colour mixed with white, with a brownish line running along the back; the lower parts of the body have an orange yellow tint. The head, and especially the forehead, is marked with three black streaks which descend to the eye. The face is covered with blackish hair; the beautifully yellow eyes are of an enormous size when compared with the magnitude of the animal. The mouth is surrounded with white and short bristles. The hands and soles of the feet are white. The tail, which exceeds the length of the body by about a half, is of the same colour as the back, and tipt with black. The whole fur is soft and pleasant to the touch, and is used by the natives for tobacco-pouches and such like purposes. M. Geoffroy-St.-Hilaire gives the vertebræ as follows; cervical 7; dorsal 14; lumbar 9; sacral 2; coccygeal 18. (Cours d'Hist. Nat. in loco.)

The Douroucouli sleeps throughout the whole day, and is much annoyed by the light. Hence it retires into some shady corner, or into the hollow of a tree. If roused during the day, it is not only sad, but lethargic. It often sits like a Dog, with its back bent, the four feet collected under it, and its head resting on the fore paws. It is gentle during the day-time, and may be handled with impunity. It is, however, as active during the night, as it is stupid during the day. Its vision now improves, and it preys upon Birds, and especially Insects. When in New Barcelona, I used, says Humboldt, to keep one in my bed-room, and it unceasingly vaulted about, and made a great noise. It also eats vegetables, especially sugar-cane, dates, and almonds, and flies, which it catches with great address. Upon the whole, however, it eats but little, and it has been observed not to drink for twenty or thirty days.

According to Humboldt, it is exceedingly difficult to tame this Monkey. At all events, says the Baron, my companion only snapt at all the caresses bestowed upon him. He puffed like a Cat, and violently struck with his claws. M. F. Cuvier's experience, however, on this point, was different; the individual which he possessed, a female, being very gentle. Its night-cry (mu, mu) resembles that of the Jaguar, and its strength is quite extraordinary for so small an animal. It has also other cries, which are very peculiar.

2. NYCTIPITHECUS FELINUS.—CAT-FACED NIGHT-MONKEY.

Syn.   NYCTIPITHECUS FELINUS (Le Singe-de-nuit à face-de-chat).—Spix, p. 25.
       —Less. in Dict. Class. XV.
       CEBUS FELINUS.—Fisch. Syn. Mam.
Icon. Spix, Sim. et Vespert. Bras. pl. 18 (fem.)

SPECIFIC CHARACTERS.

THE HAIR ash-coloured above, reddish beneath.
THE FACE and HANDS white. The Male has a beard. Tail longer than the body.
INHABITS Para.

Our acquaintance with this and the succeeding species, as already stated, we owe to Spix, who has given figures of both. This Night-Monkey is cat-faced; its visage is lean, its mouth large, and its eyes red and very large. The greater part of the face is white. The body is slender. The hair rising from the forehead and cheeks is black, and inclines backwards. The ears are conspicuous, oblong, naked, clad only at their margins. The male has a beard. The fur above is close, ash-coloured; beneath, reddish. The tail is somewhat larger than the body, and chiefly black.

Spix procured this animal in Para, and kept it long in domestication. It fed upon rice.

3. NYCTIPITHECUS VOCIFERANS.—NOISY NIGHT-MONKEY.

Syn. Nyctipithecus vociferans (Le Babillard Brun).—Spix, p. 25.—Less. in Dict. Class. XV.

Cebus vociferans.— Fisch. Syn. Mam.

Icon. Spix, Sim. et Vespert. Bras. pl. 19 (*fem.*)

SPECIFIC CHARACTERS.

The Hair wholly of a brown colour, paler underneath.

The Face and Hands brown. No beard. Tail scarcely longer than the body.

Inhabits the Forests of Solimoëns near Tabatinga.

As already stated, this species is introduced upon the authority of Spix, and adopted by Fischer and Lesson. We are led to understand it is smaller in size, is beardless, and that the tail is scarcely longer than the

body. It has the same slender body, but is almost entirely of a brown colour. The two species have much the same habits, sleeping throughout the day, and being active during the night, chattering loud in small companies. Though not easily caught, yet the Indians, on discovering their retreats, soon capture them, as the strong day-light almost blinds them.

This animal has a white marking both above and below the eye; there are some dark brown streaks on the forehead, and the fur is paler coloured under the belly. The ears are not so large as in the last species, nor is the tail so long. Both species are very timid, retiring, cleanly, and delicate.

The natives speak of another nearly allied species, which they call *Xupara*, and Spix considers it probable there may be several in the forests of Solimoëns.

FAMILY II. HAPALE.—MARMOUSETS.

Syn. Les Ouistitis.—Cuv. Reg. Anim. I. 104.

Simia (in part).—Linn. Gmel. I.

Hapale (from ἁπαλος, *soft*).—Illig. Prodr. p. 71.

Arctopitheci (Arctopithèques).—Geoff. Ann. Mus. XIX. 118.

GENUS HAPALE.—Illig.

CHARACTERS OF THE FAMILY AND GENUS.

General Form approaching to that of Man.

Claws on the thumbs of the fore-hands, and on all the fingers. Nails only on the hinder thumbs.

The Dental Formula $; \left| \frac{2+C+(3\ F+2)M}{2+C+(3\ F+2)M} \frac{16}{16} = 32. \right.$

Inhabit the tropical parts of America.

The Marmousets, forming a small group like the Sakis, have long been confounded with the great family of Apes or Monkeys. In fact, they resemble the Apes of America, in having the head round, the face flat, the nostrils lateral, and the buttocks hairy; they have no cheek-pouches, and their tail, like that of the Sakis, is not prehensile. They differ from them, however, in having only twenty molar teeth like the Apes of the Old Continent; all their nails or claws are compressed and pointed, excepting those of the hinder thumbs, while the thumbs of the fore-hands are so slightly separated from the fingers, that one would almost hesitate in applying the term *quadrumanous* to this family.

They are diminutive animals, of an agreeable form, and are easily tamed.

We have now in the foregoing pages taken a survey of the Apes of the Old World, and, latterly, of the majority of those of the other hemisphere. A small group still remains, which M. Geoffroy-St-Hilaire formed into a section of the Apes of the New World. This section, the *Arctopithèques* of the last named Naturalist—the *Hapale* of Illiger, has been subdivided by many Naturalists into the *Jacchus* and *Midas*, while Mikan, in his splendid work on the Fauna of Brazil (*Delectus Floræ et Faunæ Brasiliensis*), has arranged it into three minor divisions. Without doubting the existence of the minor distinctions pointed out by these celebrated writers, we think that the purposes of modern classification will all be satisfied by arranging them in one family; and, with Desmarest, Ranzani, and especially with M. Isidore Geoffroy-St-Hilaire, we shall consider them as forming a single genus.

Of the principal characteristics of the Ape family, namely, four vertical incisors in each jaw, flat nails on the fingers and toes, and a complete ossified case for the lodgment of the eyes, the family of the Marmousets possesses the last alone. Comparing this family, again, with the other Monkeys of the New World, we find, while the latter have 36 teeth, the former have only 32, agreeing herein with the Apes of the Old World. The form of their teeth, moreover, differs from those of both the foregoing groups. The incisors are oblique and prominent, more especially those of the upper jaw, which are also broad: those of the lower jaw are much longer and narrower; the lateral incisors are much shorter. The three false molars have a point at their external edge, and a heel on their internal; the two true molars of the upper jaw are tricuspidate, those of the lower have four tubercles. Not only are these Marmousets destitute of some leading characters of the Apes, but, literally, they do not deserve the name of Quadrumana. Their upper extremities are not true

hands; and this is not owing, as in the Sapajous and others, to the want or rudimentary state of the thumb, but because it is not sufficiently free, and hence cannot be opposed to the fingers; moreover, it is not armed with a nail but a claw. The tail is always longer than the body, and thickly clad. The fur is generally long, bushy, and very soft to the touch; its colours are usually different from the Sakis, and beautiful.

Like most other Monkeys they live among trees; and though destitute of the grasping hands of some, and the prehensile tails of others, the deficiency is made up by their claws, which enable them to climb like Birds, and to the very summits of the loftiest trees, where their more weighty and powerful associates and foes cannot follow them.

Little is known of the habits of these beautiful little creatures in their native haunts; but many of the species have been imported into Europe, and as here they thrive with due care, and even propagate, their manners are not wholly unknown. Interesting notices will appear under many of the species, and we shall here introduce only a few anecdotes illustrative of their mental powers as observed by M. Audouin in the Common Marmouset. Daily experience shows that a Dog placed before a mirror does not recognise his likeness, and is still less capable of receiving any peculiar impression from the most striking picture. M. Audouin, however, assures us from innumerable observations, that it was very different with his Monkeys; and that in a picture they could recognise not only their own likeness, but also that of other animals. Thus the picture of a Cat, and, which is even more remarkable, even that of a Wasp, would put them in terror, whilst if a Beetle or Lady-bird was represented on the canvas, they would dart upon it for their prey. This single fact seems to indicate very considerable intelligence, and it is supported by others. One day one of M. Audouin's pets, in eating a grape, squirted some of the juice into its eye; and never afterwards would it eat grapes but with its eyes shut. Alarmed at the picture of a Wasp, their panic, as will readily be supposed, is much greater for a real one. Thus we are told that one day a Wasp being attracted to their cage by a lump of sugar, the two Monkeys instantly retreated to the most remote corner. On this M. Audouin, having caught the Wasp, approximated it to them, when they violently shut their eyes, and hid their heads between their hands. They were exceeding fond of the smaller insects, which they seized with address: also of sugar, roasted apples, and eggs; they never would eat any kind of nuts, or acid fruits; they also declined meat; but they instantly seized and devoured small living Birds. Their sight was very acute; and their curiosity insatiable; they were very capricious, but became familiar with their keepers. Their cry was various, according to the different emotions which agitated them.

47

(A.) Proper Marmousets. (Jacchus.)

M. Geoffroy distinguishes the Ouistitis, properly so called (Jacchus), characterized by having their lower incisive teeth pointed, placed in a curved line, and equal to the canines. Their tail is bushy and annulated; their ears are usually ornamented with a bunch of hair. It is rather difficult to establish clear specific differences among these animals, differing from each other only in colour.

1. HAPALE JACCHUS.—COMMON MARMOUSET.

Syn.   L'Ouistiti commun.—Cuv. Reg. Anim. I. 105.
    Jacchus vulgaris.—Geoff. Ann. Mus. XIX.—Desm. Mam.
    Simia Jacchus.—Linn. Gmel. I. 39.—Humb. Obs. Zool. p. 360.
    Hapale Jacchus.—Kuhl. Beitr.—Pr. Max. Beitr.
    Striated Monkey.—Penn. Quadr. No. 142.

Icon.   Ouistiti mâle adulte—jeune femelle.—F. Cuv. et Geoff. Hist. Mam.
    Hapale Jacchus.—Pr. Max. Abbild.
    L'Ouistiti.—Buff. Hist. Nat. XV. pl. 14.—Audeb. Sing.—Schreb. Säügth. pl. 33.
    The Sanglin or Cagui Minor.—Edw. Gleam. pl. 218.

SPECIFIC CHARACTERS.

The Hair greyish-brown; the crupper and tail annulated with black and greyish-brown; a white spot on the forehead; very long whitish hairs on each side of the ears.

Inhabits the tropical parts of America.

The *Common Marmouset* has been long and familiarly known. The head is round, covered with black hair at the crown, and on the temples there are two remarkable tufts of long white hair; the ears resemble the human. The face is flesh-coloured and naked, as also the hands and feet; the eyes are reddish. The upper part of the body is covered with longish hair, in alternate stripes of black and greyish-brown. The ring-like markings are still more conspicuous on the tail, to the number of about twenty of each colour. The under parts of the body, and inside of the limbs, are brown. This beautiful species is about eight inches long, without including the tail, which is somewhat longer than the body. At birth their eyes are open; they are of a greyish colour; and immediately attach themselves to their mother, and hide themselves in her fur. M. F. Cuvier states concerning one in the Paris Ménagerie which had three at a birth, that she destroyed one before two of the others. Her maternal feelings were any thing but strong; and the male showed generally a greater affection for the young. Though very active and attentive to all that passes, they seem rather stupid, and are very distrustful. They never distinguish persons, even those most familiar with them; and they are very irritable, and apt to snap at all. They have a singular whistling sort of a cry, in which they particularly indulge.

This species is widely spread over both the American Continents; and as it bears the change of climate well, and readily propagates, it is very commonly met with in these countries.

Var. Rufus.—There is a variety of this species, with the tail annulated with red and ash colour.

2. HAPALE PENICILLATUS.—PENCILLED MARMOUSET.

Syn.   Jacchus Penicillatus.—Geoff. Ann. Mus. XIX.—Desm. Mam.
    Simia Penicillata.—Humb. Obs. Zool. p. 361.

Icon.   Ouistiti femelle à pinceau.—F. Cuv. et Geoff. Hist. Mam.
    Jacchus penicillatus.—Spix, Sim. et Vespert. Bras. pl. 26.
    Hapale penicillatus.—Pr. Max. Abbild.

SPECIFIC CHARACTERS.

The Hair grey; the crupper and tail annulated with dark and light grey; a white spot on the forehead; long dark brown or black pencils of hair in front of each ear; the head and upper parts of the neck black.

Inhabits Brazil.

This singular and graceful-looking animal is distinguished from the previous species principally by those remarkable pencil-formed tufts in front of its ears, from which it has received its specific name. Similar appendages are sometimes likewise found behind the ear, and on the back of the neck. The hair on the head is usually black, with a remarkable white marking on the forehead. The countenance generally is of a dark Ethiopic hue, and is nearly naked; it is surrounded with a tawny-coloured fur, which on the neck is nearly black. The fur of the body is beautifully striped light and dark grey and yellow; the tail is decidedly annulated, black and white; its tip is white. M. Cuvier's specimen was not six inches long. The habits of this species are but little known. Spix

states it is always found in small troops, and that the mother never carries the young either on her back or breast. It is one of the most common Monkeys in Brazil.

3. HAPALE LEUCOCEPHALUS.—WHITE-HEADED MARMOUSET.

Syn.   Jacchus Leucocephalus.—Geoff. Ann. Mus. XIX.—Desm. Mam.
    Simia Geoffroyi.—Humb. Obs. Zool. p. 360.

Icon.   Hapale Leucocephalus.—Pr. Max. Abbild.

SPECIFIC CHARACTERS.

The Hair black, spotted with dark grey; the tail annulated with black and dark grey; the head and breast white; long black pencils of hair on each side of the ears.

Inhabits Brazil.

This very beautiful little animal has a strong family likeness to the two foregoing species, and, like the last it has a broad pencil or rather tuft of black hair at the side of its head. Its other markings, however, completely distinguish it from its congeners. The whole head and front of the neck are white, while the rest of the body is black, spotted with dark grey, the tail annulated, but with very dark colours. This species is somewhat larger than the preceding. It is an inhabitant of Brazil; and Prince Maximilian states that he has witnessed the adult carrying one of its young on its back and another at its breast. Its favourite resorts are the lofty forests, and it is rather abundant. It is so much esteemed for its beauty that it is very often tamed, and made a household ornament.

4. HAPALE HUMERALIFER.—WHITE-ARMED MARMOUSET.

Syn.   Jacchus humeralifer (Le Camail).—Geoff. Ann. Mus. XIX.—Desm. Mam.
    Simia humeralifera.—Humb. Obs. Zool. p. 360.

Icon.

SPECIFIC CHARACTERS.

The Hair dusky brown; on the shoulders, breast, and arms, white; on the top of the head, dark brown; the tail indistinctly annulated with grey.

Inhabits Brazil?

This species differs somewhat more than the preceding from the Common Marmouset. The ring-like markings of the tail are not so distinct, and the back is of a dusky brown colour. The upper part of the head is also very dark, whilst the neck and the lower part of the body are dull white; the hind legs are of a speckled brown colour. It is somewhat smaller than the Common Marmouset, and the tail is proportionally larger. Its habitat has not been accurately ascertained, though it is suspected to be from Brazil.

5. HAPALE AURITUS.—BLACK MARMOUSET.

Syn.   Jacchus Auritus (Oreillard).—Geoff. Ann. Mus. XIX.—Desm. Mam.
    Simia Aurita.—Humb. Obs. Zool. p. 360.

Icon.

SPECIFIC CHARACTERS.

The Hair black, varied with red above; the tail annulated with dark brown; the upper part of the head and face marked in the middle with yellowish-white; the hands ash-coloured; the ears covered with long white pencils.

Inhabits Brazil.

This animal, of which little is known, is about the size of the Common Marmouset.

(B.) Tamarin Marmousets. (Midas.)

M. Geoffroy assigns the term *Midas* to those species having their lower incisors trenchant, placed nearly in a straight line, and equal to the canines. Their tail is more slender, and it is not annulated.

6. HAPALE ŒDIPUS.—RED-TAILED MARMOUSET.

Syn.   Le Pinche.—Cuv. Reg. Anim. I. 105.
    Midas Œdipus.—Geoff. Ann. Mus. XIV.—Jacchus Œdipus.—Desm. Mam.
    Simia Œdipus.—Linn. Gmel. I.—Humb. Obs. Zool. 361.
    Red-tailed Monkey.—Penn. Quadr. No. 144.

Icon.   Pinche mâle.—F. Cuv. et Geoff. Hist. Mam.
    Midas Œdipus. (var.)—Spix, Sim. et Vespert. Bras. pl. 23.
    Le Pinche.—Buff. Hist. Nat. XV. pl. 17.—Audeb. Sing.
    Little Lion Monkey.—Edw. Birds, pl. 195.

THE HAIR grey, mixed with brown; long white hairs on the head, hanging behind the ears; the tail red throughout its first half, black towards the end.

INHABITS the Banks of the River Amazon, Guiana, and Brazil.

This *Red-tailed* Marmouset has many well-marked characteristics, of which none is more striking than its long white crest which falls down about the neck. All the under parts of the body, and the inside and extreme parts of the extremities, are also white. The outer sides of the limbs are of a deep red colour, as is also the tail, which is tipt with black. The face is wholly black and naked, as is also the front of the neck; the eyes, too, are stated to be altogether black. Its size varies from eight to ten inches, and the tail is considerably longer than the body. A variety of the female has been figured by Spix (tab. 23), with the body striped black and dark yellowish-grey.

Though not very common, this species is found in the neighbourhood of Carthagena, at the mouth of the Rio-Sinu, and in Guiana. Humboldt states it is very savage in its temper, and is not tamed without much difficulty; but when once domesticated, lives a long time in its native country. One was brought to him, which he was anxious to preserve, but it obstinately refused all nourishment, and died in great wrath, squeaking like a Bat, and biting every one that approached it. This animal has by Edwards been designated the " *Little Lion Monkey,*" for which he assigns the following reason :—" When it prances about the room on its all-fours, and plays its tail over its back, it has very much the air of a little Lion."

7. HAPALE RUFIMANUS.—RED-HANDED MARMOUSET.

Syn. LE TAMARIN.—Cuv. Reg. Anim. I. 106.
　　MIDAS RUFIMANUS.—Geoff. Ann. Mus. XIX.—JACCHUS RUFIMANUS.—Desm. Mam.
　　SIMIA MIDAS.—Linn. Gmel. L.—Humb. Obs. Zool. p. 362.
　　GREAT-EARED MONKEY.—Penn. Quadr. No. 141.
Icon. TAMARIN À MAINS ROUSSES.—F. Cuv. et Geoff. Hist. Mam.
　　Le Tamarin.—Buff. Hist. Nat. XV. pl. 13.—Audeb. Sing.
　　LITTLE BLACK MONKEY.—Edw. Birds, pl. 196.

THE HAIR black; on the crupper, varied with ash colour; on the hands, red.

INHABITS Guiana.

This Red-handed Marmouset was first described by our countryman Edwards. It is one of the least species, usually not exceeding six or seven inches; its tail being twice as much; its bite, owing to its smallness, is not more offensive than the pinch from a Sparrow's bill; it is, however, very lively and full of action. The eyes are of a hazel colour, the face of a dark flesh, the nose scarcely rising at all; the upper lip slit like a hare's lip, the teeth very small, nearly approaching in shape to the human. The ears are large in proportion, of a blackish flesh colour, and thinly beset with short hairs. The hair on the head forms a peak on the forehead, and the face is nearly naked. The head, body, and tail, are covered with soft black hair, rather rough and shaggy; the hair on the lower part of the back stands erect, and is mixed with yellow coloured hairs; the hands are covered with short, sleek, deep orange-coloured hair; the fore-hands are not so human-like as in some other Monkeys, though it can still hold any thing in one hand.

Pennant and Buffon made the *Negro* Monkey a variety of this species; but M. F. Cuvier says they are evidently two distinct species. " I have had seven or eight individuals," he says, " of both, and the Negro Monkey has the fore arm invariably black, whilst in the other it is constantly orange-coloured."

The favorite resort of this species is the deep forests. They are bold, and do not flee at the approach of Man. The females have but one at a birth. They remain almost constantly upon the trees in large troops; their cry is a sharp whistle; though very choleric they are easily tamed, and delight in sitting upon their master's shoulders; they are full of pleasantry. Their flesh has a disagreeable taste; and, therefore, they are not used as food by the native tribes of South America.

8. HAPALE URSULUS.—NEGRO MARMOUSET.

Syn. LE TAMARIN NÈGRE.—Cuv. Reg. Anim. I. 106.
　　MIDAS URSULUS.—Geoff. Ann. Mus. XIX.—JACCHUS URSULUS.—Desm. Mam.
　　SIMIA URSULA.—Humb. Obs. Zool. p. 361.
Icon. Tamarin nègre femelle.—F. Cuv. et Geoff. Hist. Mam.
　　Buff. Hist. Nat. Suppl. VII. pl. 32.
　　Le Tamarin nègre.—Audeb. Sing.

THE HAIR black, slightly undulated on the back with red.
THE HANDS black.
INHABITS Para—South America.

The fur of this species is soft and thick, and is composed of only one kind of hair, which is wholly black upon the head, round the neck, on the extremities, and upon all the lower parts of the body, where it is more sparing than elsewhere. The back and flanks have a waved appearance—black and fawned colour. The face, ears, hands, and feet, are naked, and of an Ethiopic complexion ; the colour of the eyes brownish-yellow. The external ear is remarkably large and appears at its back part as if mutilated, in a way that is seldom witnessed in other animals.

M. F. Cuvier had one of these animals in his custody for some days, and satisfied himself as to the strong general resemblance it bore to the foregoing species. Its character was remarkable only for its extraordinary irritability. On the slightest movement being made, it showed its teeth, and bit with violence as soon as it was touched. Fortunately, however, its jaws were so weak, that it could not even penetrate the skin. Desmarest informs us that this species is found in Para, where it is very common.

9. HAPALE LABIATUS.—WHITE-LIPPED MARMOUSET.

Syn. LE TAMARIN À LÈVRES BLANCHES.—Cuv. Reg. Anim. I. 106.
　　MIDAS LABIATUS.—Geoff. Ann. Mus. XIX.—JACCHUS LABIATUS.—Desm. Mam.
　　SIMIA LABIATA.—Humb. Obs. Zool. p. 361.
Icon. MIDAS FUSCICOLLIS.—Spix, Sim. et Vespert. Bras. pl. 20.
　　MIDAS NIGRICOLLIS.—Ibid. pl. 21 (var.)
　　MIDAS MYSTAX (fem.)—Ibid. pl. 22 (var.)

THE HAIR dusky-brown ; beneath red ; on the head black ; on the nose and margins of the lips, white.
INHABITS Brazil.

This species, as noted above, was arranged by Spix in his genus *Midas;* and two others were added—the M. fuscicollis, and M. Mystax. Temminck, however, considers them only as varieties, a view which most Naturalists adopt. It is somewhat singular, however, that according to this view, the female (*M. Mystax,* Spix) should be furnished with great white moustaches, an ornament denied to the male.

The White-lipped Marmouset is very striking in its markings. The back, and outer parts of the arms and legs, are of a dusky-brown colour, speckled with rosy-white; the head, tail, hands, and feet, are black; and the inside of the extremities, and the under parts of the body and tail, are of a beautiful red colour. Finally, the neck is of a reddish-fawn colour, and the mouth is surrounded with a circle of white hair which forms a striking contrast with the neighbouring dark parts. It is of smaller dimensions than the Rufimanus.

This animal was found at Ollivenza, near a dark-coloured river, in the country of the Tocunos, between the Solimõens and the Iça.

10. HAPALE ROSALIA.—SILKY MARMOUSET.

Syn. LE MARIKINA.—Cuv. Reg. Anim. I. 106.
　　M'DAS ROSALIA.—Geoff. Ann. Mus. XIX.—JACCHUS ROSALIA.—Desm. Mam.
　　SIMIA ROSALIA.—Humb. Obs. Zool. p. 361.
　　SILKY MONKEY.—Penn. Quadr. No. 143.
Icon. Marikina mâle.—F. Cuv. et Geoff. Hist. Mam.
　　Le Marikina.—Buff. Hist. Nat. XV. pl. 16.—Audeb. Sing.

THE HAIR clear yellow, deeper about the neck ; a long mane.
INHABITS Guiana and Brazil.

The history of this beautiful little animal has scarcely been investigated in its native haunts, the forests of Brazil. The deficiency, however, has, to a certain extent, been supplied by the indefatigable F. Cuvier, who had frequent opportunities of becoming acquainted with it in Paris. Its elegant form, and easy and graceful movements, the intelligent expression which animates its look, its sweet voice, and especially its attachment to those about it, have always made it a favorite. Without the petulance, it has all the vivacity of its congeners. When imported into these cold regions, it must be protected with care from the inclemencies of weather; it must also be kept with a minute attention to cleanliness, for without this it speedily pines and dies. Accustomed to live in families, solitude appears intolerable; and, therefore, it is most desirable that two or more should be in company. The food they most affect is Insects and sweet fruits, but they may be habituated to live on biscuits and milk. The individual whose habits were studied by M. Cuvier, sought to hide itself on the least alarm, and expressed its fear by a continued whistle. It delighted in caresses, and testified affection, though not complete confidence; it came at the call of those it knew, and retreated from strangers, displaying its teeth, its only, though far from formidable weapons. Like many Birds it delighted to resort to the highest parts of its cage, descending but seldom, and eating but little.

This animal is generally of a beautiful clear yellow colour, somewhat more golden about its neck; its face is naked, and of a deep flesh colour, so are the paws, and is fact the whole skin over the body. The fur is all of one kind, composed of fine silky hair (hence Pennant's name, the *Silky Monkey*), much longer on the head and neck than in the other parts of the body. This supplies it with a great mane, and from this single point of resemblance it has received from many travellers the name of the *Little Lion Monkey*, and *Lêoncito*. Its tail is also covered on all sides with long hair. M. Isidore Geoffroy remarks, that soon after these animals arrive in the colder regions, their bright coat fades, and before death they are usually very pale, leading to the supposition that in extreme age they may become white.

VAR. GUYANENSIS.—There is a variety from Guiana, having the tail variegated with red and black.

VAR. BRASILIENSIS.—And another from Brazil, of a deeper red, but having the tail of a uniform colour.

11. HAPALE CHRYSOMELAS.—RED-HEADED MARMOSET.

Syn. LE MARIKINA NOIR,—Cuv. Reg. Anim. I. 106.
   JACCHUS CHRYSOMELAS.—Desm. Mam.
   MIDAS CHRYSOMELAS.—Kuhl Beitr.
Icon. HAPALE CHRYSOMELAS (Der Schwarze Lowen-Sahui), Pr. Max. Abbild.

SPECIFIC CHARACTERS.

THE HAIR black; on the arms and round the face, bright red; on the forehead and the upper part of the tail, light yellow.

INHABITS Brazil.

Although there are obscure notices of this very remarkable and beautiful animal in the writings of Desmarest and Kuhl, yet we believe it is chiefly to Prince Maximilian of Neuwied that we are indebted for accurate information regarding it. In his work on the Natural History of the Brazils, he remarks, " This exquisite *Sahui* (the Brazilian name) is one of the most beautiful ornaments of the great primordial forests of the Ilheos and the Rio Pardo. The traveller must proceed for four or five days' journey from the coast before he encounters them, but after this they will be often seen. These small animals, notwithstanding their insignificant dimensions, which amount to only six inches and a half for the body, and fifteen for the tail, have often contributed to our support when we were ranging these vast deserts."

The face, body, lower limbs, and greater part of the tail, is of a beautiful black colour. The hair which surrounds the face and that of the neck is extraordinarily long, and its general colour is of a fiery red, more or less mixed with yellow; such, too, is the colour of the fore-arm, of the tail, and, though darker, of the upper part of the foot. Near the ear the hue is chestnut colour, and a mixture of this shade pervades the chest. The Prince truly remarks, that could they be domesticated in this country, they would be regarded as beautiful pets.

12. HAPALE LEONINUS.—LEONINE MARMOSET.

Syn. MIDAS LEONINUS.—Geoff. Ann. Mus. XIX.—JACCHUS LEONINUS.—
   Desm. Mam.
   SIMIA LEONINA.—Humb. Obs. Zool. p. 36.
Icon. Humb. Obs. Zool. pl. 5.

SPECIFIC CHARACTERS.

THE HAIR olive-brown; on the back striped with yellowish-white; a thick mane of olive-brown.

THE FACE black; whitish on the nose and lips.

INHABITS the eastern plains of the Andes.

This species, which was discovered by Baron Humboldt, has been described by him as of the size of the Red-handed Marmouset; the upper part of the face is black, the lower, including a part of the nose, whitish. The fur, generally, is of an olive-brown, with a heavy mane of the same colour; the back is striped with yellowish-white. The tail, which is of the same length with the body, is black on its upper, and brown on its under side. The hands, feet, and nails, are deep black.

In its native district this little animal has received the appellation of *Leoncito de Mocoa*, and hence, probably, its specific name as given by Humboldt. It is a very rare species. It inhabits the plains of the eastern slope of the Cordilleras of the Andes, especially the fertile banks of the Putumayo and Coqueta; it never mounts even to the elevation of temperate regions. It is one of the smallest and most elegant of the Monkeys; it is gay and playful; but like many of its congeners, very irascible. When provoked, it bristles up its mane, so acquiring some kind of resemblance to the African Lion. Our traveller only saw two; they were kept in a cage, and their movements were so rapid and constant, that he could scarcely take a sketch of them. The Mocoa Indians breed them extensively in a domestic state. Their whistle is not unlike the singing of some small birds.

13. HAPALE CHRYSOPYGUS.—NATTERER'S MARMOSET.

Syn. MIDAS CHRYSOPYGUS.—Natterer.
Icon. JACCHUS CHRYSOPYGUS.—Mikan, Delect. Flor. et Faun. Bras.

SPECIFIC CHARACTERS.

THE HAIR black; on the buttocks, thighs, and inner surface of the legs, golden yellow.

INHABITS St Paolo, Brazil.

For our acquaintance with this very striking and elegant little animal, we are indebted to the active Naturalist Natterer, who has sent several specimens to Vienna. In that city it fell under the examination of M. Mikan, who has furnished a most beautiful drawing of it in his superb work on the Flora and Fauna of Brazil. Its size is between ten and eleven inches from the crown of the head to the origin of the tail, which is fourteen inches long. Its face, of a light olive hue, is nearly free from hair; its forehead is a bright orange. Its long flowing locks divide on the head, and descend gracefully over the back and shoulders. Its body and upper extremities are thickly clad with a shining black vestment; and its feet and tail are of the same colour: its trowsers, the only remaining part of its covering, are of a bright golden colour, and have conferred upon it the above specific name.

Of its peculiar habits M. Natterer has sent no accounts, and they are hence unknown.

14. HAPALE MELANURUS.—BLACK-TAILED MARMOSET.

Syn. JACCHUS MELANURUS.—Geoff. Ann. Mus. XIX.—Desm. Mam.
   SIMIA MELANURUS.—Humb. Obs. Zool. p. 360.
Icon.

SPECIFIC CHARACTERS.

THE HAIR yellowish-brown above, greyish-yellow beneath; on the tail, black.

INHABITS Brazil.

This animal, of whose habits nothing is yet known, is of a yellowish-brown above, becoming deeper on the lumbar region, and also on the head. The face is brown; beneath the neck, breast, and belly, the hair is of a greyish-yellow; the limbs are still browner than the head; and the anterior surface of the thighs is of a yellowish colour, which reaches to the haunches, and is divided off from the brown of the hinder parts by an oblique line; the tail is of a uniform brownish-black.

According to M. Kuhl, this species serves to connect the Proper Marmousets with the Tamarins.

DOUBTFUL SPECIES.

1. THE FAIR MONKEY (Penn. Quadr. No. 145), which is identical with LE MICO, illustrated by Buffon, Hist. Nat. XV. pl. 18, and by Audeb. Sing.,—the Jacchus argentatus of Geoffroy and Desmarest, and the SIMIA ARGENTATA of Linnæus and Humboldt—is very plausibly conjectured by Isidore Geoffroy St-Hilaire to be an albino variety of the Black-tailed Marmouset just described. It is of a uniform silvery white; the tail black; the face and hands reddish. Kuhl notices a specimen with the tail also white.

2. JACCHUS ALBICOLLIS (Spix, Sim. et Vespert. Bras. pl. 25) is conjectured by some writers to be a variety of the Common Marmouset. The auricular pencils of the hinder part of the head, neck, and throat, are white; the fore-part of the head is brown, mixed with white hairs.

3. MIDAS PYGMÆUS (Spix, Ibid. pl. 24, fig. 2), of a diminutive size, variegated with yellow and grey above, reddish beneath; the tail, which is longer than the body, is annulated with black and yellow. Found on the banks of the Solimoëns.

4. MIDAS BICOLOR (Spix, Ibid. pl. 24, fig 1), with the head, neck, breast, and fore limbs, white, the remainder brown; the tail ferrugineous. Found near the Rio Negro.

IMAGINARY SPECIES.

1. MIDAS FUSCICOLLIS (Spix, Ibid. pl. 20), a male.
2. MIDAS NIGRICOLLIS (Ibid. pl. 28), a male.
3. MIDAS MYSTAX (Ibid. pl. 29), a female.
These are varieties of Hapale labiatus already described.

FAMILY ·III. PROSIMJA.—MAKIS.

SYNONYMS.

Les Makis.—Cuv. Reg. Anim. I. 106.
Lemur.—Linn. Gmel. I. 41.
Prosimia.—Briss. Reg. Anim. p. 220.—Prosimii (Affer) et Macrotarsi (Längfüsser).—Illig. Prodr. p. 73.

CHARACTERS OF THE FAMILY.

General Form approaching to that of the Quadrupeds.

Claws always on the first finger of the hinder hands, and sometimes also on the second finger. Nails flat on all the other fingers, and on the thumbs.

The Makis, according to Linnæus, comprise all those Quadrumanous animals, which have their incisors either more or less than four in number, or at least otherwise directed than in the Apes or Monkeys. This negative character cannot fail to include animals differing rather considerably in their characters, while it does not even unite all that ought to be comprised in one division. M. Geoffroy has, however, established in this family several divisions, which are more distinctly characterized.

These animals have their four thumbs well developed and opposable, and the first finger of the hinder hands is always armed with a pointed and elevated claw; all the other fingers are [usually] covered with flat nails. The fur is woolly, and their teeth begin to exhibit sharp tubercles, locking into each other, as we find in the insectivorous quadrupeds.

GENUS I. LEMUR.—LEMURS.

Syn. Les Makis proprement dits.—Cuv. Reg. Anim. I. 107.
Lemur.—Geoff. Ann. Mus. XIX. 158.

GENERIC CHARACTERS.

The Head long and triangular. The Nostrils terminal. The Eyes medium size. The Ears short and hairy.

The Dental Formula $\frac{2+C+(3 \ F+3)M}{3+C+(3 \ F+3)M}\frac{18}{18}=36$.

The Tarsus shorter than the tibia. The Tail longer than the body, and covered with thick hair.

The Mammæ two. Habits diurnal.

The First Finger only of the hinder hand with a claw.

Inhabit Madagascar and adjacent Islands.

The Lemurs, or Proper Makis, have six incisors below, compressed, and sloping forwards, but only four above, placed straight, and the first incisors being separated from each other. The canines are trenchant; and there are six molars on each side above and six below.

These animals are very active, and have been termed Fox-Nosed Monkeys (Singes à museau de Renard), on account of their pointed faces. Their ears are not very large. The species of Lemurs are numerous; they live on fruits, and inhabit the Island of Madagascar, where they appear to occupy the place of the Monkeys, which, it is reported, are not to be found in that island. They differ from each other chiefly in their colours.

To these characters a few general remarks may be added. Their lower incisive teeth differ remarkably from those of the Monkeys both in form and position, being very long and slender, but directed horizontally, and not vertically, as usual. The first incisor of the lower jaw is of a different shape from those placed more internally, and is also larger; a fact the more important as, according to some authors, this last incisor should be regarded as the true canine, and the next would thus be the first molar. According to this view, many of the Lemurian animals would have precisely the same dental formula as the American Monkeys, and the anomaly which presents itself in these genera would be explained,—the superior canine being placed anterior to the lower one, an arrangement which is but seldom seen. Be this as it may, the so-called inferior canine is small, triangular, and very like a false molar. There are three true molars in each jaw.

The limbs of the Lemurs, especially the hinder ones, are long, and the thumbs are widely separated from the fingers, so that they are excellent instruments for grasping. No one has yet been discovered for the remark-

able claw on the index toe. The tail is longer than the body, and contributes to the gracefulness of the animal. The general form of the Lemurs is slender; and their head being long and their snout projecting, they have certainly some resemblance to the Fox. Their fur is generally woolly, very bushy, and abundant; their ears are short and clothed; their nostrils terminal and sinous; and their eyes are placed not anteriorly, as in Man, nor laterally, as in most animals, but in an intermediate position. The mammæ, two in number, are pectoral.

As their organization thus approximates them to the Apes, so also do their habits. They live upon trees, and vault with agility. M. F. Cuvier tells us of one which would spring to a branch ten feet from the ground, while its gait was at the same time constrained. They are not so petulant and impudent as the Monkeys, especially those of the Old World; and they never advance upon a stranger with threatening gestures and grimaces, or attempt to seize or bite hard. Though their manners have scarcely been at all studied in their native country,—Madagascar and the neighbouring isles,—yet, being often domesticated in Europe, we are by no means strangers to their dispositions. Many curious traits will be found in the accounts of the species; and we may here subjoin a few of the original remarks of M. F. Cuvier. He thinks, that however inoffensive and timid, they are not remarkable for their intelligence; and though frequently tamed, they but rarely form strong attachments. They are partly nocturnal, and spend much of the day rolled up in the form of a ball. They feed themselves with their hands, and, notwithstanding the length of their snout, drink by suction; when at ease, their cry is a feeble grunt, but when alarmed, it becomes deep and strong, and as they sing out in concert, the noise becomes insupportable. The different species fight furiously among themselves, biting savagely, and tearing off each other's hair with their hands. Two pair, M. Cuvier remarks, "which I possessed, could never regard each other with complacency. If I raised the partition which separated them, they were roused to fury, uttering acute, interrupted, and rapid cries. Unless the wires of the cage had separated them, they would certainly have injured each other; and the females were not more amicable than the males." These animals were fed with boiled roots and fruits, bread and milk, and they were preserved at a uniform and warm temperature. Provided they were kept clean, they enjoyed excellent health; they seemed less annoyed with their captivity, and the inclemency of the climate, than the Apes. The history of an individual, as afterwards detailed, was traced for nineteen years.

1. LEMUR CATTA.—RING-TAILED LEMUR.

Syn. Le Mococo.—Cuv. Reg. Anim. I. 107.
Lemur Catta.—Linn. Gmel. I. 43.—Geoff. Ann. Mus. XIX. 162.—Desm. Mam. 98.
Ring-tailed Lemur.—Shaw, Gen. Zool. I.—Ring-tailed Maucauco.—Penn. Quadr. I. No. 150.

Icon. Le Mococo.—Audeb. Sing.—Ménag. du Mus.—Buff. Hist. Nat. XIII. pl. 22.
Mococo mâle.—F. Cuv. et Geoff. Hist. Mam.
Maucauco.—Edw. Birds, pl. 197.—Copied in Schreb. pl. 41.

SPECIFIC CHARACTERS.

The Hair grey, reddish above, whitish beneath; on the tail annulated with black.

This Ring-tailed Lemur is the one of all others with which we are most familiar in Europe, and is remarkable for the beauty of its fur, the elegance of its form, and its familiarity. It is fourteen inches long from the snout to the origin of the tail. All the upper parts of its body are of a beautiful grey colour, which has a rosy hue on the back and shoulders; the summit of the head, and back of the neck, are black; as are the margin of the eyes and the snout; all the other parts of the body are white; and the tail is ringed throughout, alternately black and white,

48

to the number of about thirty. It almost invariably bears its tail elevated. The fur is always clear and shining.

It is of this animal that our countryman Edwards says, in his interesting description, " I kept one in my house for some time ; it was a very innocent, harmless creature, having none of the cunning or malice of the Monkey kind, though it has much of its shape and manner of sitting." M. Geoffroy informs us that he had traced the history of one of these animals for the period of nineteen years ; hence we may conclude that it can be brought to support the temperature of these northern climates. At the same time this individual was always much annoyed by the cold; hence he often rolled himself into the shape of a ball, and covered his back with his tail. In winter time his favourite resort was the fire, putting out his paws to warm them. So much did he enjoy the warmth, that he permitted his whiskers and face to be singed before he would retire; and often he did no more than turn his face aside. He also delighted to bask in the sun. He was allowed a certain degree of liberty, and made one of the workshops of the museum his home. Here he indulged in the liveliest curiosity; unceasingly in motion, he examined, pawed, overturned every thing. A shelf above the door of his chamber was his bed; before retiring to rest, he regularly amused himself with exercise, and for half an hour jumped and danced with heart and heel; this feat accomplished, he was asleep in a moment. He fed on bread, carrots, and fruit, of which he was exceedingly fond. He also ate eggs, and from his birth had a partiality for roast beef and brandy. He was gentleness itself, sensible of caresses, familiar with every one, though somewhat taciturn in his declining days; at the same time he had no partialities, and jumped on the knees or shoulders indifferently of every visitor.

M. F. Cuvier also studied the manners of this favourite animal; and he has recorded a few facts which we must not omit. The palm of the hand extends, so to speak, in a straight line, hid under the hair, to the middle of the fore-arm, where it reappears naked ; a somewhat singular occurrence. Again, when the arm of this Lemur is stretched out, its fingers are necessarily closed; accounting for the facility with which these animals hang from the branches. Many Naturalists have fallen into the mistake that the tongue is rough like the *Felinæ*, whereas it is smooth. It is to be added, that these animals, though they never use their teeth to bite or to cut, yet have the sociable instinct of using them to dress the vestments of their fellows ; in fact, they use them as a kind of comb; and, finally, says M. F. Cuvier, I have been able to verify the obse,va,ton of Linnæus, that, when at their ease and happy, they purr like the Cat.

Though thus well known in Europe, it would appear to be very different in their native haunts. At all events, no information of their native manners has been recorded. All we know on this point is the remark of Flaccourt, that they live upon trees, and congregate in troops to the number of thirty or forty.

2. LEMUR MACACO.—RUFFED LEMUR.

Syn. LE VARI.—Cuv. Reg. Anim. I. 107.
LEMUR MACACO.—Linn. Gmel.—Geoff. Ann. Mus.—Desm. Mam.
RUFFED LEMUR.—Shaw, Gen. Zool. I.—RUFFED MAUCAUCO.—Penn. Quadr. I. No. 151.
Icon. Maki Vari.—F. Cuv. et Geoff. Hist. Mam.
Le Vari.—Audeb. Sing.—Le Vari (Var. A.)—Ibid.—Buff. Hist. Nat. XIII. pl. 27.
BLACK MAUCAUCO.—Edw. Birds, pl. 217.

SPECIFIC CHARACTERS.

THE HAIR longest on the cheeks ; varied with large black and white spots ; on the tail entirely black.

Though this Ruffed Lemur has a specific name (*Macaco*), very much resembling a common appellation (*Mococo*) of the one preceding, yet there seems to be the widest difference in their natural disposition,—much greater, indeed, than in their external appearance. In its natural haunts this animal appears to be quite ferocious, and Flaccourt says they are furious like Tigers, and that two of them will make a noise which might pass for a hundred. They are also, he says, very difficult to tame, if not captured when quite young. This character is borne out by what is reported by M. P. Cuvier of one whose dispositions he had watched in a state of confinement. One of these Macacos was put into a cage with one of its congeners, where for a time they lived without hostility, if not with much cordiality. Ere long, however, they were removed into another cage, and in a different locality, upon which the Macaco murdered his companion during the night, and devoured him all but the skin.

The only specific characters which have been supplied of this animal

relate to the markings of the fur, which, after all, are by no means uniform. They differ somewhat in the sexes, though confined to black and white. The black prevails on the face, body, feet, and tail ; but it is strikingly contrasted with the white of the back of the head, of a band, ribbon-shaped, thrown across the body, and of the four limbs, mounting behind over the lower part of the crupper: the lower jaw, too, is white, and there is a white band on the snout. The males alone are white headed, the females superiorly being all black. The fur is remarkable for its beauty; it is very long and bushy, and remarkably soft to the touch. The Ruffed Lemur is about seventeen inches long, from the snout to the origin of the tail ; its tail has the same dimensions. At Malmaison, where Madame Bonaparte amused herself by collecting a number of objects of Natural History, this species bred occasionally. The eyes of the young were open at birth.

Audebert gives a variety, founded upon trifling differences of the markings ; sometimes the upper part of the body is all white.

3. LEMUR RUBER.—RED LEMUR.

Syn. LE MAKI ROUGE.—Cuv. Reg. Anim. I. 107.
LEMUR RUBER.—Geoff. Ann. Mus. XIX.—Desm. Mam.—Péron et Lesueur.
Icon. Maki roux femelle.—F. Cuv. et Geoff. Hist. Mam.

SPECIFIC CHARACTERS.

THE HAIR of a bright reddish marrone ; the face, hands, tail, belly, and the inner surfaces of the limbs, black ; a white spot on the back of the neck.

This, the most beautiful perhaps of all the Lemurs, was first noticed by the able and unfortunate Commerçon during his sojourn at Madagascar. He took a drawing of it, which lay long neglected among his papers. The interesting and indefatigable Péron, again, in his short visit to the same island, was struck with the appearance of the animal, and sent its fur to Paris, where it was preserved. France had the good fortune to receive the third specimen, which has been noticed in the *Annales des Sciences*, which animal was brought home alive in a merchant ship, and in the *Jardin des Plantes* fell under the observation of F. Cuvier, who gave a drawing and description of it in his *Mammifères*. This individual was a female, and probably the markings of the male are different; they are, however, unknown.

This individual is the most beautiful of the Lemurs hitherto described, both from its size and shape, and also from its brilliant colouring. In its general organization it resembles the other Lemurs. The upper parts of the body, including the back, the sides of the body and neck, the outer sides of the extremities, and the summit and sides of the head, are of a beautiful chesnut-red colour; whilst the face, hands, and feet, together with the inner sides of the limbs, and the under parts of the neck, chest, and belly, and the whole of the tail, are of the deepest black colour: there is besides a broad white marking on the back of the neck, and a band of the same colour over the instep and back of the head ; and the reddish tint is somewhat paler round the ears. The eyes are fawn-coloured. M. Cuvier remarks, that there are very few animals in which the colouring of the under parts of the body is of a deeper shade than that of the upper; and the Grison alone had previously been supposed to exhibit this anomaly.

This Red Lemur was very gentle and tame ; and though very agile, it was usually sad and somnolent ; it spent its days rolled up in the shape of a ball, and waked up only to eat. It never emitted any cry. It was seventeen inches long from the snout to the origin of the tail, which extended to eighteen inches.

4. LEMUR ALBIMANUS.—WHITE-HANDED LEMUR.

Syn. LEMUR ALBIMANUS.—Geoff. Ann. Mus. XIX.—Desm. Mam.
LEMUR COLLARIS.—Geoff. et Desm. ubi supra.
Icon. Le Mongouz.—Audeb. Sing.

SPECIFIC CHARACTERS.

THE HAIR greyish-brown above ; reddish marrone on the cheeks ; belly white ; all the hands white.

This animal is far from being well known; although it has been described by Brisson, from a specimen in the Museum of Réaumur, and also by Audebert, under what would now be regarded the inaccurate name of *Mongous* (Buffon). From snout to tail it measures about sixteen inches. Its snout is black; its ears round ; the hair of the face is short, and of a yellowish-grey colour, that of the temple and throat ferruginous. The top of the head, neck, shoulders, back, and outer sides of the limbs, are clothed with a deep brown-grey fur somewhat speckled; that of the chest, belly, and insides of the limbs, is of a lighter colour. The hands and feet are covered with whitish hair to the very nails, and hence its specific name. The tail, which is longer than the body, is covered with long hair, grey and grizzly.

5. LEMUR ALBIFRONS.—WHITE-FRONTED LEMUR.

Syn.   LE MONGOUS À FRONT BLANC.—Cuv. Reg. Anim. I. 107.
        LEMUR ALBIFRONS.—Geoff. Ann. Mus. XIX.—Desm. Mam.
Icon. LE MAKI À FRONT BLANC, mâle, femelle, et son pétit.—F. Cuv. et Geoff. Hist. Mam.—Audeb. Sing.

SPECIFIC CHARACTERS.

MALE.

THE HAIR chestnut-brown above ; olive-grey beneath ; the face, from the eyes to the muzzle, black ; a white band round the head ; the hands yellowish.

FEMALE.

THE HAIR paler than in the male ; no white band on the head.

The White-fronted Lemur was catalogued among the species by M. Geoffroy-St-Hilaire, and described and depicted by Audebert. Little, however, was known concerning it until the year 1816, when M. F. Cuvier obtained two pairs.

The male L. albifrons has the hair on the upper parts of the body, the outer sides of its limbs, and a third of its tail, of a golden chestnut-brown colour, when in a strong light ; the inferior parts of the body, and inner sides of the limbs, are olive-grey-brown. The tail towards the tip is black. The front part of the head, and as far back as the ears, the cheeks, and under part of the lower jaw, are white. The face, palms, and soles, are of the hue of an Ethiopian black ; and the iris is orange coloured. The only difference of colour in the female is, that those parts which are white in the male are of a dark grey colour, and that the rest of the fur is somewhat paler.

The animals under M. Cuvier's observation bred in Paris, and the period of gestation was about 3½ months. The young had the same markings, and was of the same colour as its dam ; its hair, at birth, was very short ; its eyes were open, and it was about the size of a small Rat. No sooner did the young one make its appearance, than it hid itself in its mother's bosom, and soon began to feed itself. For a long while it was scarcely possible to get a sight of it, so hid was it in its mother's fur, and she, on her part, always turned her back on all intruders, even those with whom she was most familiar. Previous to the birth of her young one, she had been extremely gentle and familiar ; she courted caresses, and licked the hand ; but the moment she had her little one, she became suspicious, retreated as far as possible from every one, and threatened those who approached. When her care of the young one became unnecessary, in the third month, her natural demeanour returned ; but throughout the nursing her care was most assiduous. When five and six weeks old, the young one began to eat the aliment presented to it, but it continued to suck for six months. Whenever in the slightest degree alarmed, it rapidly retreated to its mother's arms.

6. LEMUR MONGOZ.—MONGOOZ LEMUR.

Syn.   LE MONGOUS.—Cuv. Reg. Anim. I. 107.
        LEMUR MONGOZ.—Linn. Gmel. I.—Geoff. Ann. Mus. XIX.—Desm. Mam.
        WOOLLY MAUCAUCO.—Penn. Quadr. I. No. 149.
Icon. Mongous mâle, et tête de sa femelle.—F. Cuv. et Geoff. Hist. Mam.
        MONGOUZ.—Edw. Glean. pl. 216.
        Mongous.—Buff. Hist. Nat. XIII. pl. 26.

SPECIFIC CHARACTERS.

MALE.

THE HAIR brown fawn colour, with an olive or yellow shade ; the end of the tail black ; the face grey ; the top of the head black ; the cheeks bright brown.

FEMALE.

THE HAIR on the top of the head grey, otherwise resembling the male.

This name (Mongous), originally applied by Edwards and other early writers to nearly all the Lemurs, is now confined to a single species. The general colour of the fur, which is remarkably fine and thick, is of a brown fawn colour, with an olive or yellow shade, and this colour is nearly uniform both on the upper and lower parts of the body ; the tail is black at its extremity, and the summit of the head is entirely so in the male, while it is grey in the female. The lower parts of the cheeks are supplied with a ruff of a beautiful orange colour, and the face, ears, and

palms of the hands. are of a violet hue : the iris is orange. The form of the head of the male is not precisely similar to that of the female, and generally she is smaller and of a lighter hue than her mate.

When taken young. the Mongooz Lemur is easily tamed ; though it is not so gentle as its ring-tailed congener. Buffon, who of course speaks of an imprisoned specimen, describes it as a filthy animal, which gnawed its tail. The individual mentioned by him required to be chained ; it escaped into the neighbouring shops and houses, helping itself to all the fruits and sugar it could find : and was recaptured only with difficulty. It bit cruelly, making no exception, even of those who had the charge of it. It had a habitual insignificant grunt ; and when tired, it uttered a stronger cry, not unlike the croaking of Frogs. It was invariably chilly, and delighted in the warmth of a fire.

7. LEMUR NIGRIFRONS.—DARK-FRONTED LEMUR.

Syn.   SIMIA SCIURUS.—Petiv. Gazophyl. p. 26.
        LEMUR NIGRIFRONS.—Geoff. Ann. Mus. XIX.—Desm. Mam.
Icon. MAKI À FRONT NOIR (mâle).—F. Cuv. et Geoff. Hist. Mam.

SPECIFIC CHARACTERS.

THE HAIR greyish-brown above, ash-grey beneath. THE FOREHEAD and FACE blackish-brown.

This species was first described and represented by our countryman Petiver (*Gazophylacium*, p. 26, tab. 17, fig. 5), under the name of *Simia Sciurus*, and in this he was followed by Schreber. Some uncertainty. however, prevailing, and new opportunities of examination occurring in Paris, it was there re examined, and described by M. Geoffroy-St-Hilaire, under the specific name of *nigrifrons*, which it is now likely to retain. It is about the size of the Ring-tailed Lemur, and in external appearance differs but little from the *Mongoz*. Its ears are rather shorter than those of its congeners. Its forehead and cheeks are of a blackish-brown colour, gradually becoming lighter towards the snout, which is light grey. The upper part of the head and neck, the shoulders, and outer sides of the fore-legs, are of a greyish-brown colour, somewhat variegated with white and black. The back, flanks, and outer parts of the hind legs, are of a uniform brownish-grey ; the tail becomes darker as it approaches its tip ; the fur in front of the neck and chest is whitish ; the hands and feet are covered with short ash-coloured hair.

8. LEMUR FULVUS.—FULVOUS LEMUR.

Syn.   LEMUR FULVUS.—Geoff. Ann. Mus. XIX.—Desm. Mam.
Icon. LE MAKI BRUN.—Ménag. du Mus.
        GRAND MANGOUS.—Buff. Hist. Nat. Suppl. VII. pl. 33.

SPECIFIC CHARACTERS.

THE HAIR brown above, grey beneath. THE FOREHEAD elevated and prominent.

The Fulvous Lemur has not been long catalogued in any of our systems. and was first described by M. Geoffroy-St-Hilaire (Ménag. du Mus.) Care should be taken not to confound it with the Mongooz, than which it is about a third larger ; its head also is rounder, and its trunk more delicate ; its tail likewise is not so bushy or woolly, and becomes more slender towards its extremity. It is brown above, and ash-coloured below. The croup and hind-legs are of an olive tinge, and the hairs are here reddish at their points. The iris is of a faint orange hue ; the hair is entirely black, and the forehead is elevated and prominent. This animal has been exhibited in Paris as the Pig-lemur (*Cochon*).

9. LEMUR RUFUS.—RUFOUS LEMUR.

Syn.   LEMUR RUFUS.—Geoff. Ann. Mus. XIX.—Desm. Mam.
Icon. LE MAKI ROUX.—Audeb. Sing.—Copied in Schreb. pl. 39, C.

SPECIFIC CHARACTERS.

THE HAIR yellowish-red above, dull white beneath ; a white circle round the head ; a black line from the face to the hinder part of the head ; the tail black near the tip.

Care should be taken not to confound this *Reddish*, or rust-coloured Lemur, with the *ruber* or Red Lemur of M. F. Cuvier, which is of a far brighter colour. This species has been established only upon some stuffed specimens which exist in the Paris Museum, and therefore requires further elucidation. Audebert is not quite convinced that it differs from the Macaco, though he inclines to this belief. M. Geoffroy-St-Hilaire, on the other hand, is satisfied upon this point, though it may still be allied to some other species.

The Reddish Lemur is of the same dimensions as the Macaco ; its snout is black ; its ears short and round ; the summit of its head, temples, cheeks, and under part of the neck, are of a dull white ; a black line runs from the face, and extends to the crown of the head.　All the body is of a yellowish-red colour, and the tail, much more slender than that of the Macaco, is black at its extremity.

10. LEMUR CINEREUS.—GREY LEMUR.

Syn.　LEMUR CINEREUS.—Geoff. Mag. Encyc. I. p. 20.—Desm. Mam.
Icon.　LE GRISET.—Audeb. Sing.—Copied in Schreb. pl. 40, C.
　　　PETIT MAKI.—Buff. Hist. Nat. Suppl. VII. pl. 84.

SPECIFIC CHARACTERS.

THE HAIR grey, tipped with yellow above ; whitish beneath ; the point of the tail yellowish.

Some degree of doubt for some time hung over this last species, but it is now considered as unquestionably distinct. Stuffed specimens, we believe, are common in Paris ; and Buffon described it from an individual discovered by Sonnerat ; Audebert has also given us a description.

The Grey Lemur is a very pretty little animal, only ten inches long from the tip of the snout to the origin of the tail, which is somewhat longer. Its hair is mouse-grey towards its root, yellowish at its extremity, and frizzled like the wool of the Merino Sheep. Though its snout is not so prominent as that of the other Lemurs, its physiognomy is more delicate, and its movements lighter. The whole of the body is covered with this grey fur tipt with yellow ; the under parts are almost white ; the tail is yellow at its point.

DOUBTFUL SPECIES.

1. LEMUR ANJUANENSIS (Geoff. Ann. Mus. XIX.) was considered by M. Fred. Cuvier to be the female of the White-fronted Lemur. However, a pair of specimens of this latter animal, exhibited in London, resembled each other precisely (see Linn. Trans. XIII. p. 624), so that the question of their identity still remains doubtful.

2. LEMUR NIGER (Geoff. Ann. Mus. XIX.), entirely black, with long hairs hanging from the neck, is not very distinctly established. It is figured by Edwards (Gleanings, pl. 217) under the name of the BLACK MAUCAUCO.

3. LE MAKI À GORGE BLANCHE (F. Cuv. et Geoff. Hist. Mam.)

In the year 1834, M. F. Cuvier published a beautiful representation and a good description of a Lemur, of whose species he still remained doubtful. This animal was a female ; and though satisfied it was not a *Mococo*, a *Vari*, an *albifrons*, *Mongos*, nor *Red Lemur*, yet still it might be the mate of some other of the previously described species. This animal possessed the size, proportions, and general physiognomy of the Mongooz ; its snout was grey, with the exception of the muzzle, which was violet-coloured ; round the eyes it was black. The head, as far as the ears, the neck, shoulders, and upper extremity, were grey ; the lower part of the upper jaw, the sides of the head as far back as the ears, and the under part of the neck and chest, were white. The ears were of a dark flesh colour ; the back to the tail, the sides of the body, the belly, thighs, and legs, were fawn-coloured ; the hands and feet were greyish ; the first half of the tail was of a dull fawn grey, and the other half was blackish. All the naked parts of the body had a violet hue. We have been the more particular in tracing these external colourings, that others may assist in determining the species.

Like many of the female Lemurs, this individual was of an extremely sweet disposition. It was strongly attached to its owner—a lady, who was very fond of it, but obliged, however, to part with it, to their mutual regret, and so much did this affect the poor animal, that it sank under grief, but retaining its accustomed amiability to the last. The regret was manifested by its inactivity. It sat still with arms crossed, neglecting wanton amusement, and hanging the head on its breast. At first it ate a little, as in brighter days, but gradually its strength and appetite declined, cough supervened, and in a few days it died.

IMAGINARY SPECIES.

1. LEMUR COLLARIS (Le Maki à fraise) is a duplicate of Lemur albimanus described above.

It should be mentioned here, that many of the differences noted above as specific are, in all probability in some instances, only sexual. It is still more probable that many species still remain undescribed and unknown.

GENUS II. LICHANOTUS.—INDRIS.

Syn.　LES INDRIS.—Cuv. Reg. Anim. I. 108.
　　　LICHANOTUS.[1]—Illig. Prodr. 72.
　　　INDRI.—Geoff. Ann. Mus. XIX.—Desm. Mam.
　　　LEMUR (in part).—Linn. Gmel. I.

GENERIC CHARACTERS.

THE HEAD triangular.　THE MUZZLE pointed.
THE EARS short and rounded.　THE EYES directed forwards.
THE DENTAL FORMULA $\frac{2+C+(2\ F\ \ldots\ )M}{2+C+(.\ F\ \ldots\ )M}$ imperfectly known.
THE TARSUS shorter than the tibia.
THE FIRST FINGER only of the hinder hand with a claw.
THE MAMMÆ two.
INHABIT Madagascar.

The Indris, in respect to their dentition, coincide [as far as known] with the Lemurs, excepting that they have only four [incisors] in the lower jaw.

This genus is very readily distinguished from the neighbouring ones of the Lemurian family, by its having only four incisors in each jaw. Those of the upper jaw form pairs, the centre ones having their edge concave, whilst in the two lateral they are convex. The lower incisors are contiguous, and are especially remarkable as regards their direction, being almost quite horizontal ; the side ones are somewhat larger, and are rounded externally. The canines are slightly separated from the incisors.

1. LICHANOTUS BREVICAUDATUS.—SHORT-TAILED INDRI.

Syn.　LEMUR INDRI.—Linn. Gmel.
　　　INDRI BREVICAUDATUS.—Geoff. Ann. Mus. XIX.—Desm. Mam.
　　　INDRI NIGER.—Lacépède.
　　　INDRI MACAUCO.—Penn. Quadr. No. 147.
Icon.　L'INDRI.—Audeb. Sing.—Sonner.[2] Voy. II. pl. 88.

SPECIFIC CHARACTERS.

THE HAIR brownish-black ; a large spot on each side, reddish above, and yellowish below ; the crupper and tail white.
THE TAIL very short.

This species, the only one distinctly known, [almost] without a tail, is three feet high, black, with a grey face, and white buttocks. The inhabitants of Madagascar tame the short-tailed Indri, and even train it, like a Dog, for the chase.

The name *Indri*, in the language of Madagascar, denotes Man-of-the-Woods, and notwithstanding its inferior size, it possesses many claims to the appellation. M. Audebert, indeed, remarks, that the Indri, of all known animals, bears altogether the closest resemblance to Man ; and this not only in its general contour, but also in its several proportions. We must, however, add, that the differences are most conspicuous ; the head is shaped somewhat like that of a Fox ; there is a tail, though it is very short ; and the hind feet are truly hands, making it completely quadrumanous.

It has been stated above, that this Maki is tamed and reared to the chase. This circumstance is the more worthy of observation, because most of the animals which Man has domesticated and taught to assist him in hunting, are themselves of predatory habits, as, for example, the Dogs, the Weasels, the Cheetah, and the Falcons. The Indri, on the other hand, feeds wholly on vegetables, and is, moreover, a harmless creature, delighting in fruits, and having no thirst for blood.

This animal is three feet and a half high ; and the lower limbs are very nearly equal in length to the body ; its snout is long, and the ears are short and round ; and the tail is remarkably short. All the nails on the extremities are flat (with the exception of that of the first finger of the hinder hand, which is a strong claw), and terminate in a very acute point, in which respect they possibly differ from those of Man. In colour it is almost black ; its fur is silky and abundant. The muzzle, the arm-pits, and lower part of the abdomen, are grey, and the buttocks are white, where also the hair is woolly, and curled as in the Sheep. Its eye is white and very lively ; its cry like that of a weeping child.

---

[1] Lichanotus.—From λιχανός, the first finger.
[2] SONNER. VOY.—Voyage aux Indes Orientales et à la Chine fait par ordre du Roi, depuis 1774 à 1781, par M. Sonnerat. Paris, 1782.

1. L'INDRI À LONGUE QUEUE is described and figured by Sonnerat (Voy. II. pl. 89) under the name of Maquis à bourres (Flocky Lemur). This is the Lemur laniger of Linn. Gmel., the Maki fauve of Buffon, and Indris longicaudatus of most other authors. It is said to be yellow, with a very long tail; but the species itself requires revision, being probably identical with some of the Lemurs already described.

GENUS III. STENOPS.—LORISES.

Syn. LES LORIS.—Cuv. Reg. Anim. I. 108.
STENOPS.'—Illig. Prodr. I. 73.
LORIS and NYCTICEBUS.—Geoff. Ann. Mus. XIX. 162.—Desm. Mam.
LEMUR (in part).—Linn. Gmel. I.

GENERIC CHARACTERS.

THE HEAD round. THE MUZZLE short. THE EYES very large, approximated, and directed forwards. THE EARS short and hairy.
THE DENTAL FORMULA as in the genus Lemur (see page 189).
THE TARSUS and METATARSUS of equal length.
THE MAMMÆ four. HABITS nocturnal.
The first finger only of the hinder hand with a claw.
INHABIT the East Indies and Africa.

The Lorises, otherwise called Slow-paced Lemurs (Singes Paresseux), have the same dentition as the Lemurs, only the points of their molars are sharper. They have the abrupt muzzle of a Mastiff; the body slender; the tail wanting [or medium size]; large approximated eyes; and a rough tongue.

They feed on Insects, sometimes also on small Birds or Quadrupeds; they walk at an excessively slow pace; and their habits are nocturnal. Sir A. Carlisle has noticed that the arteries of their limbs are subdivided at the base into small branches, in the same manner as in the true Sloths.[2]

Two species are noticed from the East Indies [and one from Africa].

To this genus we assign, with Temminck, the Potto of Bosman. It thus comprises three species; but there are two others, the reality of whose existence requires further proof. The dentition appears to undergo some important changes during its progress to maturity.

1. STENOPS TARDIGRADUS.—SLOTH-LORIS.

Syn. LE LORIS PARESSEUX ou le PARESSEUX DE BENGALE.—Cuv. Reg. Anim. I. 108.
LEMUR TARDIGRADUS.—Linn. Gmel. I.—Raffles, in Linn. Trans. XIII. 247.
NYCTICEBUS BENGALENSIS.—Geoff. Ann. Mus. XIX. 164.—Desm. Mam.
SLOW-PACED LEMUR.—Shaw, Gen. Zool. I.
Icon. LE LORIS PARESSEUX.—Audeb. Sing.
LORIS DE BENGALE.—Buff. Hist. Nat. Suppl. VII. pl. 36.—Vosm.[3] Descr. Paress. (Amsterdam, 1770).
POUKAM.—F. Cuv. et Geoff. Hist. Mam.
TAIL-LESS MAUCAUCO.—Penn. Quadr. I. pl. 48.

SPECIFIC CHARACTERS.

THE HAIR reddish-brown, a dark brown line along the back; the tail apparently wanting; a white spot on the forehead.
INHABITS Bengal.

The Sloth-Loris has been long and pretty accurately known to Naturalists. Linnæus described it; as did Vosmaër, the celebrated Dutch Zoologist. D'Obsonville examined it in its native haunts, and Audebert furnished an account from the Paris Museum. Our distinguished countryman, Sir William Jones, supplied a truly classical description in the Asiatic Researches, Vol. IV., while Dr Shaw, and Sir A. Carlisle, the celebrated anatomist, have both examined it with care.

The following detailed account of the Sloth-Loris is from the pen of the learned and accomplished Sir William Jones :—" This male animal had four hands, each five-fingered; palms naked; nails round, except those in the indices behind, which were long, curved, pointed; hair very thick, especially on the haunches, extremely soft, mostly dark-grey, varied with brown, and a tinge of russet; darker on the back, paler about the face, and under the throat, reddish toward the rump; no tail; a dorsal stripe, broad, chestnut-coloured, narrower towards the neck; a head almost spherical; a countenance expressive and interesting; eyes round, large, approximated, weak in the day-time, glaring and animated at night; a white vertical stripe between them; eye-lashes black, short, ears dark, rounded, concave; great acuteness at night, both in seeing and hearing; a face hairy, flattish; a nose pointed, not much elongated, the upper lip cleft; canine teeth comparatively long, very sharp.

" In his manners he was for the most part gentle, except in the cold season, when his temper seemed wholly changed; and his Creator, who made him so sensible of cold, to which he must often have been exposed even in his native forests, gave him, probably for that reason, his thick fur, which we rarely see in animals in these tropical climates. To me, who not only constantly fed him, but bathed him twice a-week in water accommodated to the seasons, and whom he clearly distinguished from others, he was at all times grateful; but when I disturbed him in winter, he was usually indignant, and seemed to reproach me with the uneasiness which he felt, though no possible precautions had been omitted to keep him in a proper degree of warmth. At all times he was pleased with being stroked on the head and throat, and frequently suffered me to touch his extremely sharp teeth; but at all times his temper was quick, and, when he was unseasonably disturbed, he expressed a little resentment by an obscure murmur, like that of a Squirrel, or a greater degree of displeasure by a peevish cry, especially in winter, when he was often as fierce, on being much importuned, as any beast of the woods. From half an hour after sun-rise, to half an hour before sun-set, he slept without intermission, rolled up like a Hedgehog; and, as soon as he awoke, he began to prepare himself for the labours of his approaching day; licking and dressing himself like a Cat, an operation which the flexibility of his neck and limbs enabled him to perform very completely: he was then ready for a slight breakfast, after which he commonly took a short nap; but when the sun was quite set, he recovered all his vivacity. His ordinary food was the sweet fruit of the country; plantains always and mangos during the season; but he refused peaches, and was not fond of mulberries, or even of guaiavas: milk he lapped eagerly, but was contented with plain water. In general he was not voracious, but never appeared satisfied with Grasshoppers; and passed the whole night, while the hot season lasted, in prowling for them. When a Grasshopper, or any Insect, alighted within his reach, his eyes, which he fixed on his prey, glowed with uncommon fire; and, having drawn himself back to spring on it with greater force, he seized the prey with both his fore-paws, but held it in one of them, while he devoured it. For other purposes, and sometimes even for that of holding his food, he used all his paws indifferently as hands, and frequently grasped with one of them the higher part of his ample cage, while his three others were severally engaged at the bottom of it; but the posture of which he seemed fondest was to cling with all four of them to the upper wires, his body being inverted; and in the evening he usually stood erect for many minutes, playing on the wires with his fingers, and rapidly moving his body from side to side, as if he had found the utility of exercise in his unnatural state of confinement. A little before day-break, when very early hours gave me frequent opportunities of observing him, he seemed to solicit my attention; and if I presented my finger to him, he licked it with great gentleness, but eagerly took fruit when I offered it; though he seldom ate much at his morning repast: when the *day brought back his night*, his eyes lost their lustre and strength, and he composed himself for a slumber of ten or eleven hours.

" My little friend was, on the whole, very engaging; and, when he was found lifeless, in the same posture in which he would naturally have slept, I consoled myself with believing that he died without pain, and lived with as much pleasure as he could have enjoyed in a state of captivity.

" In India it is found in the *Garrow* mountains, in the woods on the Coast of *Coromandel*, and has likewise been transmitted from the Eastern Islands."

Little requires to be added to this truly graphic description. M. F. Cuvier remarks, that the length of the body of this Loris is about 14 or 16 inches, equal to the size of a small Cat; and when standing erect upon its paws, its shoulders are nearly six inches high. *Sprawling*, however, may be said to be the favourite gait of this animal; its extremities being wide asunder, and its chest and abdomen almost touching the ground; so that it has a very uncommon appearance. Regarding the dental system of this species, M. F. Cuvier remarks, that the crest, on the inner side of the true molars, projects more at the anterior than the posterior part; that the upper incisors are regularly placed at the side, not before each other, and that the inferior canine is round, and not flattened externally. D'Obsonville informs us he could readily distinguish the peculiar cries of this Loris, when it was happy and sad, when it was hungry or impatient; it is a kind of soft whistle. It appears susceptible of some education, ceasing to bite

[1] Stenops, from στενός, *narrow*, and ωψ, *visage.*
[2] The remarkably slow pace of these *Lorises* has led travellers to suppose them true Sloths, and hence some authors have asserted, contrary to Buffon and to the fact, that the genus of Sloths exists also in Asia.—*Note of the Baron Cuvier.*
[3] VOSM. DESCR.—Description de différens Animaux de la Ménagerie du Prince d'Orange, par P. Vosmaër. Amsterd. 1766—1787.
49

and snap, and becoming attached to its master. Its odour is far from being agreeable.

To this already somewhat extended account, we cannot omit a valuable contribution made to our knowledge, of the anatomical structure of this animal, by Sir Anthony Carlisle, and communicated by him to Dr Shaw. Becoming possessed of the body of a *tardigradus*, he injected the arterial system, and discovered an unusual appearance in the great arterial trunks proceeding to all the limbs. " Immediately," he remarks, " after the great artery from the body (subclavian) has penetrated the arm-pit, it is divided into twenty-three equal-sized cylinders, which closely surround the principal trunk of the artery, now diminished in size to an inconsiderable vessel. The cylindrical arteries accompany each other, and divide with the two principal branches of the fore-arm (the radial and ulnar), being distributed in their routes upon the muscles, each of which has one of these cylinders. The other branches, for example, the radial and ulnar, proceed like the arteries in general; disposing themselves upon the skin, membranes, bones, &c., in an arborescent form. The great artery of the inferior extremity, the iliac, in the same way divides itself on the margin of the pelvis into upwards of twenty equal-sized cylinders, also surrounding the main trunk; these vessels are also finally distributed as in the upper extremity; the cylinders wholly upon the muscles, and the arborescent branches on all the other parts. It would be of some importance," adds Sir Anthony, " to ascertain whether the other slow-moving quadrupeds have any peculiar arrangement of the arteries of their limbs. This solitary fact is hardly sufficient for the foundation of any theoretical explanation of the slow movement of these muscles; if, however, it should be corroborated by similar circumstances in other animals, a new light may be thrown upon muscular motion by tracing a connection between the kind of action produced in a muscle, and the condition of its vascularity or supply of blood."—(*Shaw's Gen. Zoology,* Vol. I. p. 91.)

These animals are sometimes found with two of the upper incisors wanting.

VAR. GRISEUS.—GREY SLOTH LORIS.

There is a larger variety, found in Bengal, called *Bru samundi* by the natives. It is grey, with the dorsal stripe entirely black.

2. STENOPS GRACILIS.—SLENDER LORIS.

Syn.   LE LORIS GRÊLE.—Cuv. Reg. Anim. I. 108.
       LORIS GRACILIS.—Geoff. Ann. Mus. XIX. 164.—Desm. Mam.
       LORIS CEYLONICUS.—Fisch.[1] Anat. Mak., pl. 7 (skeleton).
       LORIS.—Shaw, Gen. Zool. I.

Icon. LE LORIS GRÊLE.—Audeb. Sing.
       LORIS.—Buff. Hist. Nat. XIII., pl. 30.—Seb.[2] Thes. L, pl. 35, fig. 1 (male), fig. 2 (fem.)
       LEMUR TARDIGRADUS.—Schreb. Saügth, pl. 38.

SPECIFIC CHARACTERS.

THE HAIR reddish-brown above, whitish beneath; a white spot on the forehead; circle round the eyes red. THE TAIL wanting.

INHABITS Ceylon.

This animal is smaller than the preceding, and has its nose more elevated, owing chiefly to the projection of the intermaxillary bones. From this difference, M. Geoffroy was led to form his genus Nycticebus of the former species, and his genus Loris of the latter.

The information we possess concerning this Slender Loris is but scanty, more especially respecting its habits and mode of life. Seba remarks, that it has an acute sense of smell; lives upon the seeds of lofty trees, which the male always tastes before offering to his mate.

Audebert counted four mammæ upon the female, although there were two glands only. The dimensions of the animal were small, the head and trunk extending only to five inches. The head is flat, but when garnished with hair, appears capacious and round. The eyes are very large and prominent, and the eye-lashes conspicuous. The muzzle is about half an inch long; the snout prominent, projecting over the mouth, whose upper lip is somewhat cleft. The ears are large, round, very concave, and almost naked. The arms are very long and slender; the hands are only an inch long, and the fingers are armed with short and flat nails. The legs are as slender as the arms, and somewhat longer; the feet being twice as long as the hands. The great toe is very strong; and has a striking tubercle between it and the next, as may be also seen on the hand. This Loris has neither tail nor tubercle answering to it. The fur covering the head, neck, back, and external portions of the extremities, is

of a reddish-brown colour, and this colour surrounds the eyes; there is a grey spot in the middle of the forehead, extending to the temples and cheeks; the muzzle is naked and flesh-coloured. The fur upon the extremities is very thin; and the whole of the under part of the body of a light yellowish-grey colour.

3. STENOPS POTTO.—BOSMAN'S LORIS.

Syn. LEMUR POTTO.—Linn. Gmel. I. 42.
      NYCTICEBUS POTTO.—Geoff. Ann. Mus. XIX. 165.
      GALAGO GUINEENSIS.—Desm. Mam. No. 127.

Icon. POTTO.—Bosm.[3] Guin. II. pl. 4.

SPECIFIC CHARACTERS.

THE HAIR reddish in the adult; grey in the young. THE TAIL of medium length.

INHABITS Guinea.

To this genus we may refer the Potto of Bosman—an animal having the same remarkably slow movements as the Sloths and Lorises. [Cuvier considers it to be a Galago, and Temminck a Loris.]

This species seems very obscure, known only by Bosman's description and figure in his account of Guinea. He mentions that the animal is called Potto by the natives, and Sloth by Europeans, on account of the extreme slowness of its movements. He tells us it is scarcely able to walk ten paces a day; that it eats up all the fruit and leaves of a tree, thus becoming fat, after which it grows lean, and is in danger of starving, before it climbs a second tree. All this he narrates not from personal knowledge, but from the testimony of the Negroes. Its figure bears some resemblance to the S. tardigradus, but it is represented with a tail of some length. He adds, " This animal is so ugly and hideous, that I scarcely believe its match can be found in any part of the world. On the ground it crawls like a Reptile. Its hands bear a close resemblance to those of Man; its head is very large in proportion to its body. The robe of the young is of the same colour as that of the Rat, through which its smooth and glistening skin is seen; that of the old is red and tufted like wool."

DOUBTFUL SPECIES.

1. NYCTICEBUS JAVANICUS of Geoffroy (Ann. Mus. XIX,) and others, was found in Java by Leschenault de la Tour. It differs from the Sloth Loris merely in having the dorsal line deeper, and the muzzle more pointed. Probably it is only a variety of Stenops tardigradus.

2. NYCTICEBUS CEYLONICUS of Geoffroy (loc. cit.), figured in Seba's Thesaurus, I., pl. 47, under the name of *Cercopithecus Zeylonicus, seu tardigradus dictus major,*—is said to be dark brown approaching to black; the back entirely black; the tail very short. As its specific name denotes, it inhabits Ceylon.

GENUS IV. OTOLICNUS.—GALAGOES.

Syn. LES GALAGO (in part).—Cuv. Reg. Anim. I. 109.
      OTOLICNUS.—Illig. Prodr. 74.
      GALAGO (in part).—Geoff. Ann. Mus. XIX. 165.—Desm. Mam.

GENERIC CHARACTERS.

THE HEAD round. THE MUZZLE short. THE EYES very large, approximated, and directed forwards. THE EARS long, naked, and membraneous.

THE DENTAL FORMULA, as in the Genus Lemur (see page 189), sometimes, by abortion, $\frac{3|1+C+(3\ F+3)M}{3+C+(2\ F+3)M}=\frac{16}{18}=34.$

THE TARSUS three times the length of the metatarsus.

THE TAIL long and bushy.

THE MAMMÆ two. HABITS nocturnal.

THE FIRST FINGER only of the hinder hand with a claw.

INHABIT the African continent and Madagascar.

These animals have the teeth and the insectivorous diet of the Lorises; but their elongated tarsi give to their hinder limbs a disproportionate length. Their tail is long and bushy, their ears expanded and membraneous, while their large eyes indicate that their habits are strictly nocturnal. Several species are known, all from Africa.

In every part of their frame the Galagoes bear a close resemblance to

[1] FISCH. ANAT. MAK.—Gotthelf Fischer's Anatomie der Maki, und der ihnen verwandten Thiere. Frankfurt am Main, 1804.
[2] SEB. THES.—A. Sebæ locupletissimi rerum naturalium Thesauri accurata descriptio. Amsterdam, 1734; 1765.
[3] BOSM. GUIN.—Reise nach Guinea durch W. Bosman. Hamburg, 1708.

the Lemurs properly so called; a remark which requires particular application to the teeth, as it was long supposed there was a subgenus having only two incisors in the upper jaw. The truth is this: The incisors are very small, the upper canines again are particularly large, and the excessive development of these latter frequently displaces the neighbouring teeth. The most remarkable feature in the organization of this genus is the great length of the posterior extremities, approximating them to the Kangaroos. In the Quadrumana an essential character, as is well known, consists in the multiplication, the separation, and the distinct specification, so to speak, of all the parts of the foot. Now, the elongation of the hinder limb is, in the case before us, effected without in the slightest degree deranging the type of the order, and solely by a change in the volume of some of its parts. Of the seven bones which form the tarsus, two only, namely, the *Scaphoid* and *Calcaneum*, are lengthened; and notwithstanding the marked change thus produced, the common forms and use of the bones themselves are modified but to a trifling extent.

The great length of these limbs, and the size of the other eyes and external ears, all harmonize with the fact that the Galagoes are nocturnal and insectivorous. By means of the large auricle, whose folds it actually expands, it is advertised of the slightest noise, even to the flitting of an Insect through the air; and on perceiving one, darts upon it like a Hawk. This is does in two ways; seated in ambush, and hid beneath the foliage, it sometimes starts up only on its hind feet, without quitting the branch, in a moment it darts upon its victim, and clenches it: more frequently, however, like the Bat, it seizes its prey in the air, vaulting surprisingly, flying from branch to branch, and scarcely ever missing the object of pursuit.

Like most of the Bats, the Galagoes, during repose, escape from the annoyance which the extreme acuteness of their hearing might produce; for they have the remarkable power of closing their ears when asleep. These appendages contracting and folding at their base, retract to that extent that they even become invisible. When roused from sleep by any sudden noise, the animal unfolds, and we may almost say expands, every part of its ear, extending it in the direction whence the sound emanates. This appendage, then, it is interesting to observe, subserves a double purpose; expanded, it is an admirable acoustic instrument; and contracted, it completely plugs up the auditory foramen. The animal can thus at will make itself deaf, or nearly so; a most happy faculty during its hours of repose, when the animated and busy scenes around it are all active and noisy under the light of day. It has thus a kind of eye-lid to the ear, rendered the more necessary from the exquisite sensibility and great perfection of the sense.

The habits of the Galagoes resemble those of Monkeys and Squirrels. Generally quite gentle, they live perched upon trees, and cling to the branches almost like Birds. Their agility in pursuit of their living prey quite astonishes an observer; their motions are so rapid that the eye cannot follow them, and they are almost as quick in devouring their prey as in seizing it. They make a most comfortable bed for their young. The Negroes hunt them as an article of food.

1. OTOLICNUS CRASSICAÜDATUS.—GREAT GALAGO.

Syn. GALAGO CRASSICAUDATUS.—Geoff. Ann. Mus. XIX.—Desm. Mam.
GALAGO À QUEUE TOUFFUE.—Desm. Mam.

Icon. LE GRAND GALAGO.—Cuv. Reg. Anim. III. pl. 1, fig. 1.—Nouv. Dict. d'Hist. Nat. XIII. pl. E. 31.

SPECIFIC CHARACTERS.

THE HAIR reddish-grey. THE EARS two-thirds of the length of the head.

INHABITS Africa.

Of this Galago, distinguished as *Le Grand* by Cuvier, and catalogued by nearly all systematic writers, exceedingly little is known. It is of about the size of a Rabbit; the ears are oval, and equal two-thirds of the body in length; the fur is thick and silky, and of a reddish-grey colour; the tail is throughout bushy. Its habits are supposed to correspond with those of its congeners, and its precise locality has not been ascertained.

2. OTOLICNUS SENEGALENSIS.—SENEGAL GALAGO.

Syn. LE MOYEN (GALAGO).—Cuv. Reg. Anim. I. 109.
GALAGO SENEGALENSIS.—Geoff. Ann. Mus. XIX.—Desm. Mam.
GALAGO GEOFFROYI.—G. Fischer, in Act. Soc. Mosc. I. p. 25.
LEMUR GALAGO.—Shaw, Gen. Zool. I.

Icon. LE GALAGO.—Audeb. Sing.
GALAGO DU SENEGAL.—F. Cuv. et Geoff. Hist. Mam.—Schreb. Saügth. pl. 38, B.
GALAGO MOHOLI.—Smith,[1] Zool. S. Afr. pl. 8. (Msmm.)

SPECIFIC CHARACTERS.

THE HAIR yellowish-grey above; yellowish-white beneath; tending to reddish on the tail..

THE EARS as long as the head.

INHABITS Western and Southern Africa.

The most striking characters of this interesting-looking animal are its ears, equal in dimensions to its whole head; its posterior limbs greater in length than the body and head together, and the tail longer than both. The fur is rather long, bushy, and very soft; it is longest on the body; somewhat less so on the head, rather unequal on the lower part of the body, very short on the hands, and under the uesus. This Galago is yellowish-white beneath, and yellowish-grey above, tending to reddish on the tail; the points only of the hairs have the grey cast, the basal portion being of a blueish ash colour; the yellow commences on the extremities, whilst the head is wholly grey. A yellowish-white band pervades the whole chanfrin.

This, and probably the other species, are very common in Western Africa. The Moors, who frequently bring them to the coast, sell them to the Europeans under the name of *Gum animals*—a circumstance which has induced some to believe that they eat this article. It is owing merely, however, to the gum trees attracting Insects, and of course their devourers.

In captivity these creatures must be kept with all the care exercised towards Birds; for they are exceedingly apt to escape, and it is almost impossible again to catch them. Their vivacity, their extreme petulance, and the extent of their leaps, are truly surprising, and not less so the extensive motion of their ears.

Although the reasons assigned by Dr Smith do not appear to us sufficient for the establishment of the new species he proposes (Galago Moholi, which we anticipate further inquiry will assign to the present), yet we are happy to quote his excellent description of the animal so frequently seen in Southern Africa. "The first specimens we observed were upon two trees close to the Limpopo River, in about latitude 25° S., and from that parallel we continued to observe others as far as we travelled. During their movements they evince great activity; they spring from branch to branch, and even from tree to tree, with extraordinary facility, and always seize with one of their fore-feet the branch upon which they intend to rest. In their manners they manifest considerable resemblance to Monkeys, particularly in their propensity to the practice of ridiculous grimaces, gesticulations, &c. According to the natives it is a nocturnal animal, and is rarely to be seen during the day. The latter it spends in its nest; where the female rears her young, generally two at a birth. Its food consists principally of pulpy fruits, though there is reason to believe it also consumes Insects, as the remains of the latter were discovered in the stomachs of several individuals we examined."

3. OTOLICNUS MADAGASCARIENSIS.—MADAGASCAR GALAGO.

Syn. LE PETIT (Galago).—Cuv. Reg. Anim. I. 109.
GALAGO MADAGASCARIENSIS.—Geoff. Ann. Mus. XIX.—Desm. Mam.
OTOLICNUS MADAGASCARIENSIS.—Schinz Thierr. I. 147.
MURINE LEMUR.—Penn. Quadr. I. 232.
MICROCEBUS RUFUS.—Geoff. Cours. Leç. 11, p. 24.
LEMUR MURINUS.—Linn. Gmel. I. 44.

Icon. MAKI NAIN.—F. Cuv. et Geoff. Hist. Mam.
LEMUR PUSILLUS (Le Maki nain).—Audeb. Sing.
RAT DE MADAGASCAR.—Buff. Hist. Nat. Suppl. III. pl. 20.

SPECIFIC CHARACTERS.

THE HAIR dark-grey above; whitish beneath. THE EARS much less than the head.

INHABITS Madagascar.

This Galago, though of dimensions and general appearance such as very naturally procured for it the appellation of the Madagascar Rat, yet possesses a structure which removes it far from the order Rodentia, and places it in that one on which we are now dwelling. Its organs of motion, even to the crooked nail on the index fingers of the posterior extremity, agree precisely with those of its congeners. Its tail has been remarked to be somewhat less bushy; its ears are proportionally very decidedly smaller than the previous species; they are also rounder, but are membranaceous and naked; the eyes are of the same great size, and the pupil is round. The tongue is smooth. The whole of its body, except the muzzle and the extremity of its members, is covered with a thick fur, composed of wavy silk-looking hairs, soft and light. The forehead, back of the head, upper part of the neck, the shoulders, and superior portion of the arms, as well as the back and upper parts of the body, and the whole of the tail, are

---

[1] SMITH, ZOOL. S. AFR.—Illustrations of the Zoology of South Africa, consisting chiefly of Figures and Descriptions of the objects of Natural History, collected during an Expedition into the Interior of South Africa in the years 1834 to 1836, fitted out by the Cape of Good Hope Association for exploring Central Africa. By Andrew Smith, M.D. London, 1838, et seq.

of a uniform fawn-grey colour; while the under part of the lower jaw, the throat, lower part of the neck, the chest, the inner side of the arms, the belly, and external aspect of the hinder limbs, are white. The face and hands are flesh colour, with a white longitudinal spot between the eyes. There is no very manifest difference between the sexes.

The habits of this animal are very similar to those previously detailed. In confinement, he passes the entire day hid in a comfortable nest, rolled up into a ball, and sound asleep; but with the twilight he leaves his retreat, and is active throughout the night. It is now that he eats and amuses himself; exceedingly lively and active, he runs round his cage as if flying, and will leap six feet vertically.· He lives upon fruit, bread, and biscuits.

DOUBTFUL SPECIES.

1. THE LITTLE MAUCAUCO (Penn. Quadr. I. 233), figured in Brown,[1] Illustr. pl. 44, is either the young of the Senegal Galago, or the type of a new species. It differs from the Madagascar Galago in having the ears nearly as long as the head, and the tail reddish.

2. GALAGO DEMIDOFFII (G. Fisch. in Act. Soc. Mosc. I. p. 24, fig. 1) is said to have its fur reddish-brown; the muzzle blackish, and the ears half as long as the head; the tail is longer than the body, and ends in a tuft. This animal is thought to have come from Senegal. M. Geoffroy considers it to be the young of some other species.

GENUS V. TARSIUS.—TARSIERS.

Syn. LES TARSIERS.—Cuv. Reg. Anim. I. 109.
TARSIUS— Storr,[2] Prodr.—Geoff. Ann. Mus. XIX. 167.—Illig. Prodr. 74.—Desm. Mam.
DIDELPHIS (in part).—Linn. Gmel. I. 109.

GENERIC CHARACTERS.

THE HEAD round. THE MUZZLE short and pointed. THE EYES very large and approximated. THE EARS large, naked, and membraneous.

THE DENTAL FORMULA $\dfrac{2+C+(3\ F+3)M\ \ 18}{1+C+(3\ F+3)M\ \ 16}=34.$

THE TARSUS three times the length of the metatarsus.
THE TAIL very long, tufted at the end.
THE MAMMÆ two, ventral. HABITS nocturnal.
THE INDEX and MIDDLE FINGERS of the hinder hands armed each with a pointed nail.
INHABIT the East Indian Archipelago.

The Tarsiers have the elongated tarsi of the Galagoes, and resemble them in most of the details of their structure; but the interval between their true molars and incisors is occupied by several smaller teeth, and the first or middle incisors of the upper jaw are elongated and resemble the canines. Their muzzle is very short, and their eyes are still larger than any of the preceding.·

These animals are of nocturnal habits, and live on Insects. They come to us from the Moluccas [and other islands of the East Indian Archipelago].

1. TARSIUS SPECTRUM.—PODJE TARSIER.

Syn. LEMUR SPECTRUM.—Pall.[3] Glir. p. 275.
TARSIUS SPECTRUM.—Geoff. Ann. Mus. XIX.—Desm. Mam.
MACROTARSUS INDICUS.—Lacep.·
DIDELPHIS MACROTARSUS.—Linn. Gmel. I. 109.
TARSIER MAUCAUCO.—Buff. Quadr. I. 231.

Icon. LE TARSIER.—Buff. Hist. Nat. XIII. pl. 9.—Copied in Schreb. Säugth. pl. 155.
TARSIUS FUSCUS s. FUSCOMANUS.—G. Fisch. Anat. Mak. pl. 3, 4.
TARSIUS DAUBENTONII.—Audeb. Sing.
TARSIUS BANCANUS (young).—Horsf. Jav.

The Podje, the only well-ascertained species of this genus, is of the size of a Rat, measuring about six inches from the muzzle to the origin of the tail; this appendage is considerably longer. The head is round, the ears transparent and naked, and half the length of the head. The snout is short and pointed, the eyes are remarkably large, and the posterior extremities as long as the body and head taken together; the extremities almost naked, the nails short and flat, with the exception of those on the

index and middle fingers of the hinder hands, which are hooked. The tail is clothed with hair only at its base and tip, The fur, which lies close, is of a dark reddish-brown colour.

Of the habits of this animal but few particulars have been stated. It lives upon trees, and pursues Insects. The name *Podje* is applied to it by the natives of the island of Macassar, where it abounds.

IMAGINARY SPECIES.

. 1. T. BANCANUS of Dr Horsfield, figured in his work on Java, is the young of the Podje Tarsier according to Temminck.

2. T. DAUBENTONII; and,

· 3. T. FUSCOMANUS of G. Fischer (Anat. Mak.), are identical with T. Spectrum.

DOUBTFUL GENUS.—CHIROGALEUS.

Travellers should search for some animals, drawn by Commerçon [the originals are deposited in the Museum d'Histoire Naturelle], and engraved by Geoffroy (Ann. Mus. XIX. pl. 10) under the name of Cheirogaleus. These figures seem to indicate a new genus or sub-genus of Quadrumanous animals.

M. Geoffroy has the fullest conviction that these animals will turn out to belong to a distinct genus. The accurate Commerçon had carefully sketched them of their natural size; and this after having prepared a history of the Lemurs, and examined all their minute and distinguishing characteristics. These animals, like the Felinæ, have the head round, the nose and muzzle short, the lips armed with moustaches, the eyes large, prominent, and approximated, and the ears short and oval. Their tail is long, very bushy, regularly cylindrical, and generally curling forward, sometimes upon itself, and sometimes round the animal's body. All these traits correspond with those belonging to the Cat family. But to these we have to add, that the phalanges of the extremities, widely separated, and formed for grasping, like those of the Lemurs and the Chirogalei, have the four thumbs completely opposable, and apt for all their peculiar movements. They are, moreover, supplied with broad nails, which are short and flat. The nails, again, on all the other phalanges are straight, slender, acute, surpassing considerably the fleshy extremity. · These nails, however, are very different from the claws of Bears, Cats, &c., and in their form and position much resemble the awl-shaped nail which in the Lemurs is attached to the index finger of the hinder hand.

The respective dimensions of the three species which the celebrated traveller has sketched, supply the specific names which M. Geoffroy has provisionally supplied to them. They are—

1. CHIROGALEUS MAJOR.—Geoff., whose length is about twelve inches. It is of a dark brown colour, particularly about the chanfrin.·

2. CHIROGALEUS MEDIUS.—Geoff. Length nine inches. The colour is not so deep; a black circle surrounds the eyes, and the chanfrin is much lighter.

3. CHIROGALEUS MINOR.—Geoff. · Little more than seven inches long. The colour generally · is much lighter, especially about the eyes and chanfrin, which are both surrounded with a black circle.

Are not these different ages of the same animal?

Note.—Mr Waterhouse, in the Annals of Natural History, Vol. II. p. 468, has described the skins of several Quadrumanous animals, brought to the Zoological Society's Museum from Fernando Po. Not having had an opportunity of examining the skins themselves, or of procuring drawings, we have abstained from noticing them in the text.

1. COLOBUS PENNANTII, Waterh., seems not to differ specifically from the Bay Monkey of Pennant.

2. COLOBUS SATANAS, Waterh., greatly resembles the Colobus Guereza of Ruppell, if it be not absolutely identical therewith.

3. CERCOPITHECUS MARTINI, Waterh., founded upon two skins, of which *the face and hands were wanting*, resembles the Vaulting Guenon (C. nictitans).

4. CERCOPITHECUS ERYTHROTIS, Waterh., *wanting the face and hands*, seems to resemble the Moustache Guenon (C. Cephus).

5. 6. Two others, Colobus leucomeros and C. ursinus, are mentioned from the same locality.

---

[1] BROWN, ILLUSTR.—New Illustrations of Zoology, by Peter Brown.· London, 1776.
[2] STORR, PRODR.—Prodromus Methodi Mammalium. Auctore Theophilus C. C. Storr. Tub. 1780.
[3] PALL. GLIR.—Novæ Species Quadrupedum e Glirium ordine. Auctore Petro Sim. Pallas. Erlang. 1778.

END OF THE FIRST VOLUME.

LIST OF THE PLATES

OF THE

ANIMALS REPRESENTED IN VOLUME I. OF THIS WORK.

DIVISION I. VERTEBRATA.—CLASS I. MAMMALIA.

Plate.	Genus.		Figures.	Plate.	Genus.		Figures.
A.		Head, skeleton, and teeth, of Orang-outang, &c.	10	XII.	MUSTELA,	Weasels,	11
I.	PITHECUS,	Orang-outang,	2	XIV.*	CANIS,	Dogs,	8
I.	TROGLODYTES,	Chimpansee,	1	XIV.**	do.	do.	9
I.	HYLOBATES,	Gibbons,	2	XV.	do.	Wolves and Foxes,	7
II.	CERCOPITHECUS,	Guenons,	7	XV.*	do.	Dogs,	9
II. C.	NASALIS,	Proboscis Apes,	2	XVI.	VIVERRA,	Civets,	10
II. C.	SEMNOPITHECUS,	Solemn Apes,	5	XIX.	FELIS,	Cats,	9
II. D.	MACACUS,	Macacos,	6	XIX. B.	do.	do.	4
III.	CYNOCEPHALUS,	Baboons,	6	XIX. C.	do.	do.	6
III. B.	do.	do.	6	XXX.	SCIURUS,	Squirrels,	9
IV.	MYCETES,	Howlers,	6	XXXIX.	ELEPHAS,	Elephants,	3
IV. D.	CEBUS,	Weepers,	6	XLII.	SUS,	Hogs,	4
V.	HAPALE,	Marmousets,	6	XLII.	PHACOCHŒRUS,	Ethiopian do.	1
VI.	LEMUR,	Makis,	7	XLII.	DICOTYLES,	Peccaries,	2
VI. B.	LICHANOTUS,	Indris,	1	XLIII.	RHINOCEROS,	Rhinoceroses,	2
VI. B.	STENOPS,	Lorises,	2	XLVII.	CAMELUS,	Camels,	2
VI. B.	OTOLICNUS,	Galagoes,	1	L.	CERVUS,	Deer,	7
VI. B.	TARSIUS,	Tarsiers,	1	L. B.	do.	do.	5
VI. C.	GALEOPITHECUS,	Flying-Cats, male, female, and young,	3	LI.	do.	do.	6
VII. G.	VESPERTILIO,	True Bats,	5	LII.	CAMELOPARDALIS,	Giraffe,	3
VII. G.	NYCTICEIUS,	Roquet-Dog Bats,	3	LIII.	ANTELOPE,	Antelopes,	9
X.	URSUS,	Bears,	5	LVII.	BOS,	Oxen,	5
XI.	AILURUS,	Pandas,	1				
XI.	ICTIDES,	Benturongs,	2	35 Plates.		Total,	230
XI.	NASUA,	Coatis,	3				

DIVISION I. VERTEBRATA.—CLASS II. AVES.

Plate.	Genus.		Figures.	Plate.	Genus.		Figures.
I. B.	VULTUR,	Vultures,	4	L.	HIRUNDO,	Swallows,	6
III.	AQUILA,	Eagles,	4	LV.	PARUS,	Titmice,	9
IV.	HALIAETUS,	Fisher-Eagles,	3	LVII.	EMBERIZA,	Buntings,	6
VI.	CIRCUS,	Harriers,	4	LVIII.	do.	do.	3
VI.	SERPENTARIUS,	Snake-Eaters,	1	LVIII.	FRINGILLA,	Finches,	5
VII.	STRIX,	Owls,	4	LXII.	LOXIA,	Crossbills,	5
VIII.	do.	do.	4	LXXV.	GARRULUS,	Jays,	5
X.	LANIUS,	Shrikes,	7	LXXXV.*	CINNYRIS,	Soui-Mangas,	9
X. B.	do.	do.	5	LXXXVIII.	TROCHILUS,	Humming-Birds,	12
X. B.	LANIARIUS,	Thrush-Shrikes,	1	LXXXVIII. B.	do.	do.	7
X. C.	do.	do.	6	LXXXVIII. C.	do.	do.	7
XVIII.	TANAGRA,	Tanagers,	5	LXXXVIII. D.	do.	do.	6
XXIII.*	SETOPHAGA,	Gnat-Catchers,	9	LXXXIX.	do.	do.	12
XXIII.	VIREO,	Chats,	8	XCII.	UPUPA,	Hoopoes,	1
XXX.	MERULA,	Thrushes,	8	XCII.	EPIMACHUS,	Promerops,	1
XXXII.	PITTA,	Breves,	5	XCII. B.	do.	do.	3
XXXIV.	ORIOLUS,	Orioles,	6	XCIV.	MEROPS,	Bee-Eaters,	6
XLII.	SYLVIA,	Warblers,	9	XCIX.	BUCEROS,	Hornbills,	6
XLVI.	REGULUS,	Kinglets,	8	XCIX. B.	do.	do.	6
XLVII.	TROGLODYTES,	Wrens,	8	CI.	PICUS,	Woodpeckers,	9
XLIX.	EURYLAIMUS,	Eurylaimes,	5	CIX.	TROGON,	Couroucouis,	6

LIST OF THE PLATES.

Plate.	Genus.		Figures.	Plate.	Genus.		Figures.
CXII.	{ CALYPTORHYNCHUS,	Muffled-Cockatoos,	5	CXXVII.	ORTYX,	Quails,	6
	{ PSITTACUS,	Parrots,	4	CXXIX.	COLUMBA,	Pigeons,	9
CXIII.	{ LORIUS,	Lories,	1	CXXIX. B.	do.	do.	5
	{ PSITTACULA,	Parrakeets,	5	CXXIX. C.	do.	do.	6
CXXIII.	TETRAO,	Grouse,	5	CXXXIII.	STRUTHIO,	Ostriches,	2
CXXIII.*	do.	do.	5				
CXXIV.	LAGOPUS,	Ptarmigan,	5	48 Plates.		Total,	302

DIVISION II. INVERTEBRATA.—SECTION I. MOLLUSCA.

CLASS GASTEROPODA.—ORDER TRACHELIPODA.

Plates.	Genus.		Figures.
III.	CONUS,	Cones,	25
XLVI.	ACHATINA,	Agates,	10
		Total,	35

DIVISION II. INVERTEBRATA.—SECTION II. ARTICULATA.

CLASS INSECTA.

Plates.	Order.		Figures.
IV.	COLEOPTERA,	Beetles,	15
XII.	LEPIDOPTERA,	Butterflies,	17
		Total,	32

	Figures.	Plates.
Total of the Mammalia represented,	230	35
of the Birds represented,	302	48
of the Mollusca represented,	35	2
of the Insecta represented,	32	2
In all,	597	87

THE EDINBURGH PRINTING COMPANY, SOUTH ST DAVID STREET.

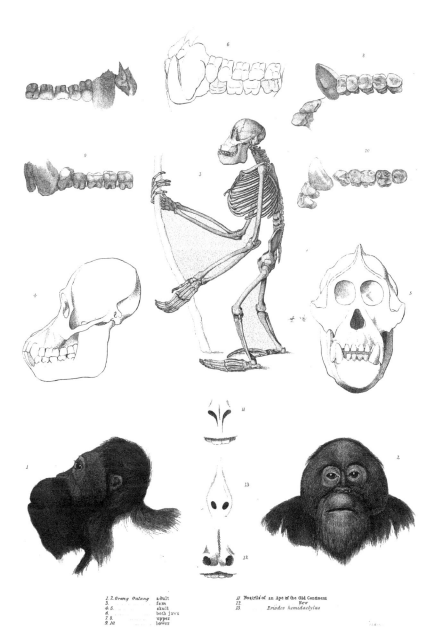

1.2 *Orang Outang*   adult
3.   fem.
4. 5.   skull
6.   both jaws
7. 8.   upper
9. 10   lower

11 Nostrils of an Ape of the Old Continent
12.   New
13.   *Eriodes hemidactylus*

PITHECI'Y, MEN OF THE WOODS
1.P.Satyrus.   Orang Outang.
2.   Gan.

CERCOPITHECUS. GYENONS.

1. C. Sabæus.   Green G.
                    Var.
3 _ Diana   Diana G.

APES. PROBOSCIS APES.
J. Chevalier.   Kahau.
SEMNOPITHECUS, SOLENT APES.

MACACUS BLACACUS.

CYNOCEPHALUS BABOONS.
Le Sphinx   Guinea
                          Young Female

CYNOCEPHALUS, BABOONS.
1. C.Babouin. var Anubis Little B.
PAPIO. MANDRILLS.
2. P.Maimon.   Great
3. —   young
4. — leucophœus.   Drill.
5. —   Female
6. —   young

Miles sc.—Turvey etc.

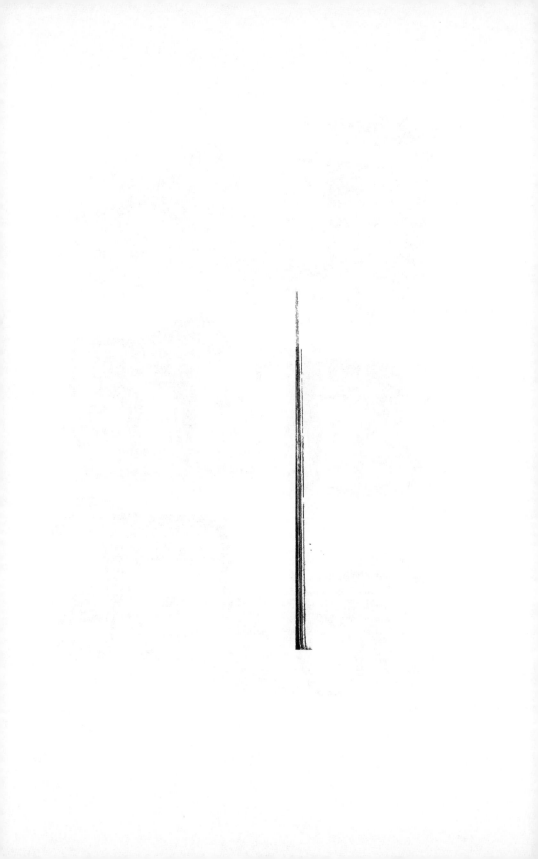

MYCETES, HOWLERS.
1. M.*seniculus* .. Royal H.
2. .. *ursinus* .. Ursine
3. .. *chrysurus* .. Golden-tailed
4. .. *niger* .. Black

CEBUS.   WEEPERS.
1. C. Apella   Common W.
2. ___ Vellerosus   Fearful
3. ___ Capucinus   Capuchin
4. ___ 10. Imputatus   White-throated.
5. barbatus   Bearded.
6. Fatuellus   Horned.

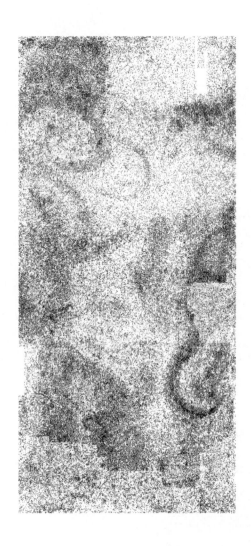

SAIMIRI. MARMOSETS.
1. S. sciureus   Common M.
2. Midas.   TAMARINS.
3. fuscicollis   Grey-necked. T
4. Œdipus   Red-crested
5. rosalia   Fair
6. Jacchus   Silky
7. leonina   Argus

LEMUR. MAXIS.
1. Laibimanus   White-handed M.
2. rufus   Reddish   —
3. albifrons   White-fronted —
4. Macaco   Buffed   —
5.   Var.
6. griseus   Grey   —

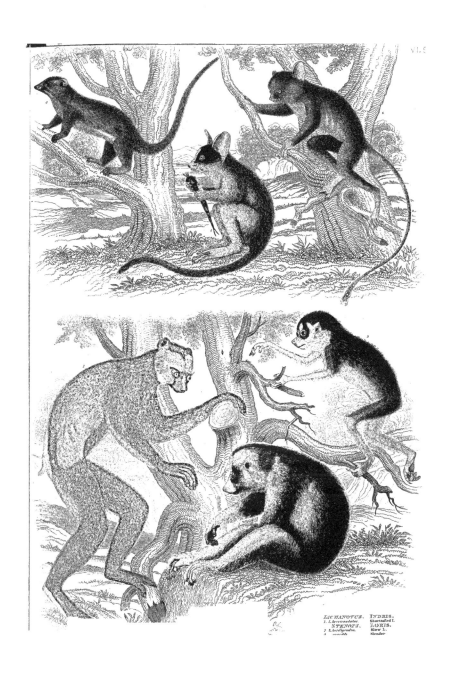

LICHANOTUS. INDRIS.
1. L. brevicaudatus. Shorttailed I.
STENOPS. LORIS.
2 L. tardigradus. Slow l.
3 — gracilis. Slender

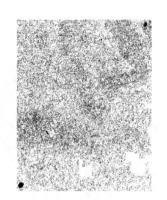

PLECOTUS. LONG-EARED BATS.
1.P. auritus.   Common Long-eared B.
23.   d°

AILURUS, PANDAS.
1.refulgens. . . Shining.
ICTIDES, BENTURONGS.
2.albifrons. White-fronted.
3.ater. . . Black.

XII.

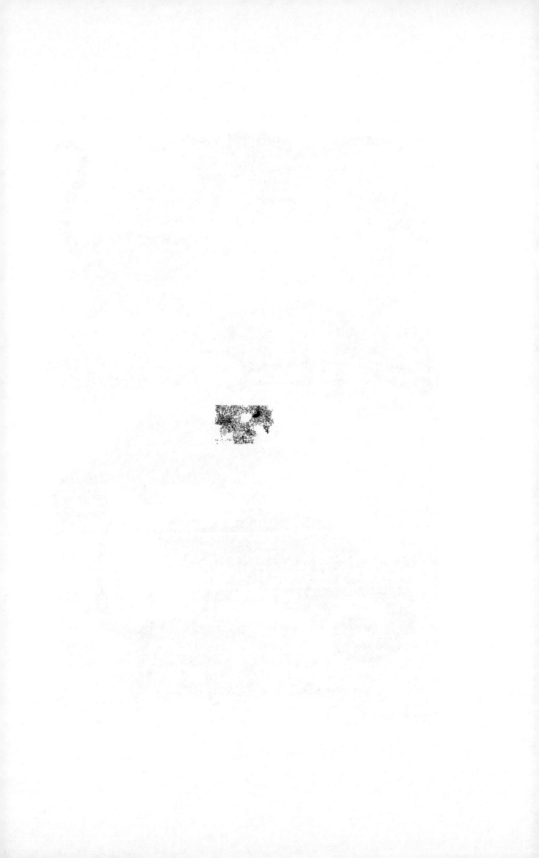

XII.

MUSTELA.WEASELS.
1.Polecat.   PUTORIUS Mærinus.
2.D?   ...... Iko
3.Hardwick's D? .... Hardwickii.
4.Weasel   .... Vulgaris.
5.Ermine   .... Erminea. Summer
6.D?   .... Erminea. Winter
7.Pine Marten MARTES Pinus.
8.White cheeked ...... Flavigula.
9.White eared .... leucotis.
10.Sable   .... Zibellina.
11.Vison   .... Vison.

CANIS. DOGS.
Canis Familiaris.

1 Blood Hound   5 Beagle
2 Stag Hound   6 English Terrier
3 Fox Hound   7 Scotch D.
4 Harrier   8 Isle of Skye D.

Painted by Gilpin, Edwards & Cap.n Brown --- Engraved by W.m Warwick.

CANIS. DOGS.
C. Familiaris Var.
1 Spanish Pointers
2 English   D.º
3 English Setter
4 Old English D.º
5 Springer
6 Cocker
7 King Charles D.º
8 Comforter

Painted by Gilpin; Reinagle, Chalon &c. — Engraved by W.m Warwick.

XV.

CANIS.WOLVES & FOXES.
1.Dusky Wolf.   C. nubilus. Var 3.
2.Prairie D°   — latrans — . 4.
3.Common D°   — lupus.
4.European Fox.   — vulpes.
5.Cross   D°   — fulvus. Var 1.
6.Silver   D°   — D° ——— 2.
7.Red.   D°   — D° ——— 3.

CANIS. DOGS.
1. Gray Wolf.   C. Occidentalis.
2. Arctic Fox.   — Lagopus.
3. Corsac.   — Corsac.
4. Jackal,   — aureus.
5. Cape D.º   — Kacumelas.
LYCAON.
6. Burchel's Lycaon. L. tricolor.
MEGALOTIS.
7. Bruce's Fennec. M. Brucii.
8. Smith's   D.º — — Smithii.
9. Lalande's D.º — — Lalandii.

Eng.d by W.m Warwick.

THE CIVETS.
1. Civet   VIVERRA Civetta.
2. Grey-headed   —— Poliocephala.
3. Zibet   —— Zibetha.
4. Luwak   —— Musanga.
5. Javanese   —— Rasse.
6. Genet   GENETTA Vulgaris.
7. Malacca D[o.]   —— ——Ver.
8. Fossane   —— Fossa.
9. Banded   —— Fasciata.
10. Slender Delundung   —— Gracilis.

FELIS CATS.

1 African Lion.
2 Lioness and Cubs.
3 Puma.
4 Tiger.
5 Clouded Tiger.
6 Leopard.

FELIS, CATS.
LE Uncia. Jaguar.

Turvey sc.—Mlno sc.

TRUE SQUIRRELS.
1 Malabar ..... Macrourus
2 Gray ..... Cinereus
3 American Black ..... Niger

Mordecai Del.

ELEPHAS. ELEPHANTS.
1st Indian E. Indicus.
2nd African. — Africanus. Fem.

Engraved by S.Milne

SUS.   HOGS.
1.S Scrofa. var Sinensis. Tem Chinese H.
2.......... Capensis.   Cape..... H.
3.__Indicus   M. Babyrussa.
4.   F
PHACOCHOERUS, ETHIOPIAN HOGS.
5.P. Æliani.   Ælian's H.
DICOTYLES,   PECCARIS.
6.D. torquatus.   Collared P.
7._ labiatus   White-lipped .

Turvey sc.__Milne etc.

RHINOCEROS.

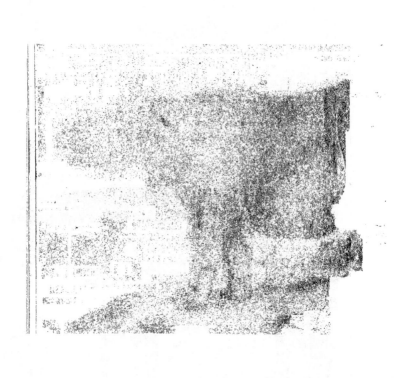

XLVII.

1. Bactrian Camel...C. Bactrianus.   2. Dromedary...Dromadarius.

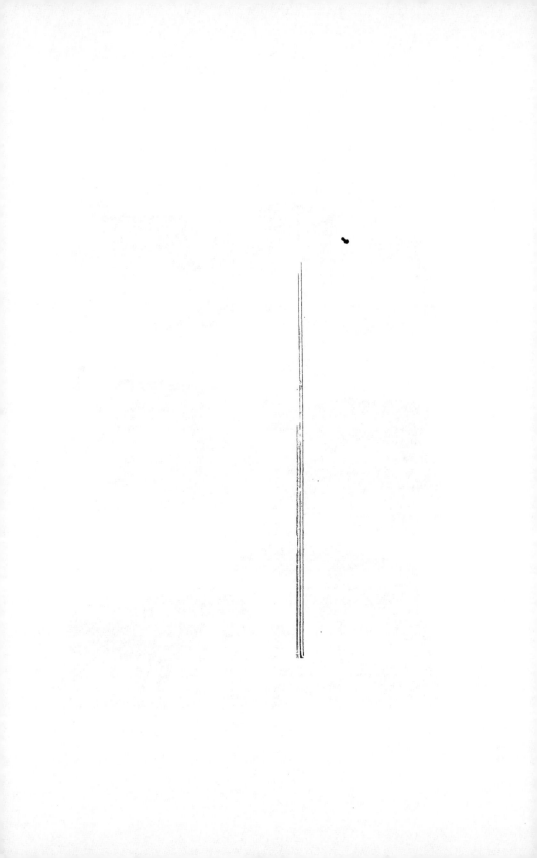

CERVUS. DEER.

1. *Canadensis.*   Wapiti D.
2. *Elaphus.*   Red.
3. *Bengalensis.*   Bengal.
4. *porcinus.*   Porcine.
5. *Timorensis.*   Timor.

Milne sc.

1. C.(Mulocerus) Dama . Fallow D.(Sumner)   4 & 5 (Axis)Molnocensis   Molnccan A.

Drawn by Captain Brown.

CAMELOPARDALIS GIRAFFA.
THE GIRAFFE.

Engraved by Wm Warwick.

ANTLOPE. ANTELOPES.
1 Chamois Antelope .. A.Rupicapra.
2 Duvaucel's .. .. Duvaucelli.
3 Four-tufted .. .. Quadriscopa.
4 Cambing Ootan .. Sumatrensis.
5 Prong-horned .. .. furcifer.
6 Vlackte Steenbock .. Bogesvres.
7 Four-horned .. .. Quadricornis.
8 Bontebock .. .. Pygmaea.
819

Drawn by Capt Brown.   Engraved by Wm Warwick & John Reid.

BOS. OXEN.

1. Indian Ox.   Bs. Taurus Var. Indicus.
2. Zebu   —— IIt.
3. Little Dº   —— IIt.
4. Hornless Dº   —— Iºt.
5. American Bison —— Americanus.

Eng.d by W.m Harvey

VULTURE VULTURES.

III.

AQUILA. EAGLES.
1. A. Chrysaëtos. Golden Eagle. Fem.
2. young
3. A. nævia. Plaintive E.
4. young

Rhind sc

HALIÆTUS. FISHER EAGLES.
Le Pygargue à tête blanche White-headed Eagle
Male   Young

VII.

STRIX OWLS.
1 Virginian Horned Owl. S. Virginiana.
2 Long-eared _____ ___ Otus.
3 Mottled _____ ___ Nævia. ♀.
4. D? _____ ___ D? Young Nest.

STRIX OWLS.
1. Tengmalm's Owl.   5. Tengmalmi
2. Great Cinereous   6. Cinerea
3. Hawkowl   7. Nisoria
4. Little   ...Passerina

LAVIN. SHRIKES.
1 L. ticrus.   Buff-breasted S.
2   " pendens.   Cruel ....

XVIII.

TANAGRA. TANAGERS.
1 Summer Redbird. T. Estiva. M
2 ———— Dᵒ ———— Fem.
3 Scarlet Tanager ———— Rubra. M.
4 ———— Dᵒ ———— F.
5 Louisiana ———— Columbiana

Drawn by A. Wilson. Engraved by B. Tanner.

XXIII.*

SETOPHAGA.
GNAT-CATCHERS.

1 Hooded Gnat-catcher ___ 5 Mitrata.
2 Green Blackcap ___ ___ Pusilla.
3 Yellow-tailed ___ ___ Ruticilla, Male.
d ___ Do ___ ___ ___ Fem.
d ___ Do ___ ___ ___ Young,
5 Selby's ___ ___ ___ Selbii. M.
7 Bonaparte's ___ ___ Bonapartii.
6 Small Blue-gray ___ ___ Cærulea. M.
8 ___ Do ___ ___ ___ Fem.

Eng.d by S. Milne.

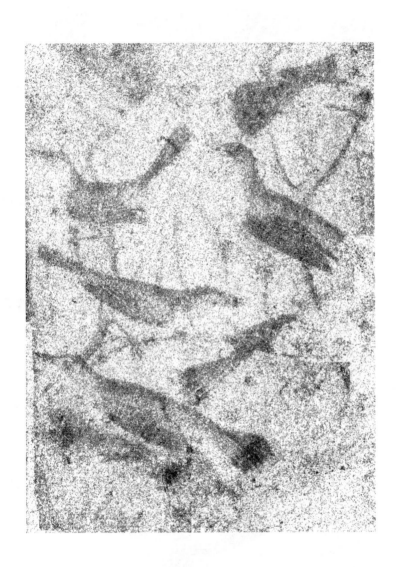

VIREO. CHATS.

1. Yellow-throated Chat   V. Flavifrons
2. Solitary   D.º   — Solitaria Male.
3. —   —   — Female.
4. Pine-swamp   D.º   — Sphagnosa.
5. Yellow-breasted D.º   — Polyglotta.
6. Red-eyed   D.º   — Olivacea.
7. White-eyed   D.º   — Noveboracensis.
8. Warbling   D.º   — Melodia.

A. Wilson.   George Crisson Sculp.

XXX.

MERULA. THRUSHES.
1.Little Tawny Thrush M.Minor.
2.Golden-crowned Do — Auricapilla.
3.Richardson's Do — Richardsonii.
4.Audubon's Do — Ludoviciana.
5.Tawny Do — Wilsonii.
6.Water Do — Aquatica.

PITTA. BREVES.
1. P. cyanoptera.   Giant.
2. ,, Macklotii.   Macklot's
3. ,, streptiana.   Noisy
4. ,, grenadina.   Grenadine

Miles sc.___Turvey sc.

ORIOLUS. ORIOLES.
1. Javanese   6. Leucogaster Male.
2. Golden   .... Galbula Male.
3. D.º   .... Fem.
4. Kink   .... Sinensis
5. Blackheaded   .... Melanocephalus
6. Two-coloured   .... Bicolor

Drawn by Capt.ⁿ Brown & Eng.ᵈ by J. Turvey.

SYLVIA. WARBLERS.

1.Palm Warbler.   3.Palmarum.
2.Blue Mountain.   — Tigrina.
3.Hemlock   — Parus.
4.Autumnal   — Autumnalis.
5.Blackthroated Green   — Arsaus.
6.Maryland Yellow-throat.   Marylandica. Male.
7.D⁰.   — Do.   Fem.
8.Kentucky   — Formosa.
9.Yellow throat   — Flavicollis.

A. Wilson Del⁹.   George Cranston Sculp⁹.

REGULUS, KINGLETS.

1 Ruby crowned	R. Calendulus Male
2 Do.	Do. Female
3 Cuvier's	Cuvierii
4 American	Americana M.
5 Do.	Do. F.
6 Byron's	Byronensis
7 All coloured	Omnicolor
8 European	Cristatus.

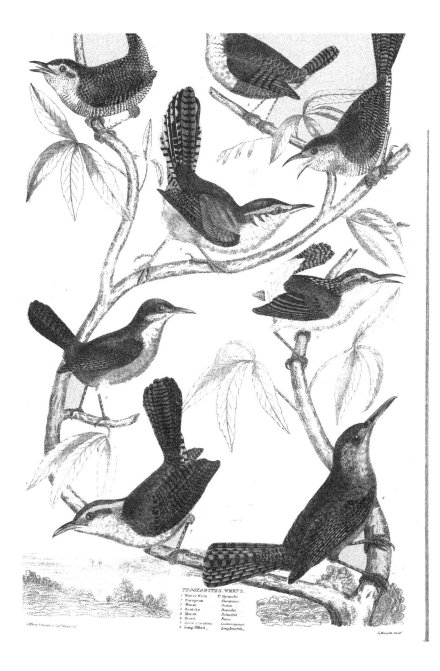

TROGLODYTES. WRENS.

1	Winter Wren	T. Hyemalis
2	European	Europœus
3	House	Œdon
4	Bewick's	Bewickii
5	Marsh	Palustris
6	Brush	Parve
7	Great Carolina	Ludovicianus
8	Long-billed	Longirostris

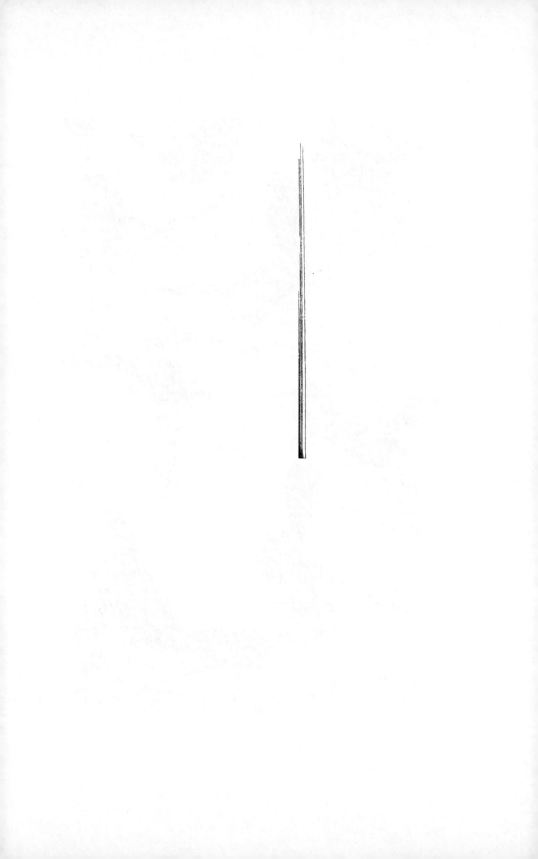

EURYLAIMES.
EURYLAIMES.
1.1.Corydon.   Corydon
2._nasutus   Great-billed
3._Horsfieldii   Horsfield's
4._   female
5._cucullatus   Hooded

Hilne sc._Darvey etc.

L.

HIRUNDO.
SWALLOWS.

1.Barn Swallow. H. rufa. Male.
2.   Do.   Fem.
3.Bank.   _ riparia.
4.White-bellied. _ bicolor.
5.CHT.   _ fulva.
6.White-collared. _ albicollis.

A.Wilson & T.B Peale Del.

S. Kilne Sc.

GENUS TITMICE.
1 Black-capt Titmouse.   2 Atricapilla.
3 Canadian   4 —

LVII.

EMBERIZA PARADISEA.
WHIDAH BUNTING, Male.
1. State of Plumage   10th Nov.
2. ――――――――   1st Oct.
3. ――――――――   14th Sept.
4. ――――――――   19th Jan.
5. ――――――――   20th June
6. Female

FRINGILLA, FINCHES.
1 Field-sparrow.   F.Pusilla.
2 Swamp   — Palustris.
3 Tree   — Arborla.
4 Song   — Melodia.
5 Chipping   — Socialis.
EMBERIZA, BUNTINGS.
6 Henslow's   E.Henslowi.
7 Lapland long-spurred — Lapponica.N
    D°   F.

Drawn by A.Wilson, A.Rider, & Cap.t Bean.   Engraved by S.Milne.

LOXIA
CROSSBILLS.
1. American Crossbill. *L. Americana.* Male.
2. ———— D.º ————————— Fem.
3. White-winged. ————————— *Leucoptera.* Male.
4. ———— D.º ————————— Young D.º
                        ——— Fem.

J. Rea Sc.

GARRULUS.   JAYS.
1. Glandarius   European J.
2. Floridanus   Florida
3. Cristatus   Blue
4. Stelleri   Steller's
5. Canadensis   Canada

CINNYRIS.
SOUI-MANGAS.

Cardinal.   C. Cardinalis Mâle.
D.°   D.°   Fem.
S. Orange   Orange.   H.

LXXXVIII.

TROCHILUS,
HUMMING-BIRDS.

1 Tufted-necked	*7* Ornatus Male
2 D.º	D.º Female
3 Azure-blue	*Lawdue*
4 Harlequin	*Multicolor*
5 Ruby-crested	*Moschitus*
6 Gould's	*Gouldii*
7 Great	*Gigas*
8 D.º	*Mostmus Male*
9 D.º	D.º Fem
10 White Striped	*Mesoleucus*
11 Evening	*Vesper*
12 Tri-coloured	*Tricolor*

Engraved by W.ᵐ Beale.

TROCHILUS.
HUMMING-BIRDS.

1 Modest _ _ _ _ T. modestus. Vie.
2 Mango _ _ _ _ _ _ Mango M.
3 D? _ _ _ _ _ _ _ _ _ Vayra M.
4 D? _ _ _ _ _ _ _ _ W. Fem. small
5 White eared. D? _ _ _ _ leucotis
6 Blue throated _ _ _ _ Lucelly.
7 Swainsons _ _ _ _ Swainsonii

Engraved by S.Milne.

TROCHILUS
HUMMING-BIRDS.
1. Azure-crowned. T. cyanocephalus. M.
2.   D°   —   Young.
3. Blue-fronted — clarisugia. F.
4. Tammninck's — Tamminckii.
5. Sapphire & Emerald — bicolor. M.
6.   D°   —   Young.
7. Clemencio's — Clemenciae. F.

Engraved by C.Milne.

TROCHILUS.
HUMMING-BIRDS.

TROCHILUS,
HUMMING-BIRDS.
1 Gigantic.   7 Sparkler
2 Violet-crowned — Stephanoides
3 Sickled   — Trochoii
4 Sephora   — Colubris Nob.
5 D°   fem
6   Nest of D°
7 Crested   Cristatus H.
8 D°   ♀
9 Purple   — Callgena.
10 Peacock   — Wayleri.

n Horned   T° cornutus ♂
12 Half-tailed   Palaetus

Engraved by WT Dash

UPUPA. HOOPOES.
1.V. Epops.   Common H.
2._ minor.   Lesser _
3._ Capensis. Madagascar _
EPIMACHUS. PROMEROPS.
1 M. fuscus. _ ..... Striped P.

Atlas 31

EPIMACHUS, PROMEROPS.

MEROPS. BEE EATERS.
1 M. Apiaster   Common B.
2 — coralocephalus Blue headed
3 — bicolor   Malimbic
4 — viridis   Indian
5 — erythropterus Red Winged

OZEOS. HORNBILLS. &c.
S. carrinatus   Grooved H.
   bicristatus   Trumpet
   hydrocorax   Bontini
   lunata   Double-beaked
   compala   Yarkunibel
   abediva   Black-crested

Fisk Hubert sc.

C1.

PICUS. WOODPECKERS.
1 Red-headed   P. Erythrocephalus
2 Yellow-bellied ___ Varius
3 Downy   ___ Pubescens
4 Bengal   ___ Bengalensis
5 Red-bellied   ___ Carolinus

TROGON COUROUCOU.

1 Golden Couroucou 7 Peronius
2 Resonent'k D Rosmarini
3 Blackarched D Gricellus
4 Mexican D Maccanus
5 Rosey D Narina Mela
6 D Female

CALYPTORHYNCHUS,
RUFFLED COCKATOOS.
1 C Leo.   Rosa
PSITTACUS PARROTS.
2 P. sclurius   Racket-tailed P.
3 Banksii   Buers
4 Petrii   Pecters
5 ustrictus   Mitred

LORIES, LORIES.
1. L. invittatus. Sparkling L.
PRITCHARD'S PARRAKEETS.
1. coronulata. Tirica P.
2. placens. Coquette.
3. Tea.

TETRAO. GROUSE.
1 Pinnated Grouse. T. Cupido M.
2   D°   F.
3 Spotted D°____ Canadensis M.
4   D°____   F.
5 Sharp-tailed D°. _ Phasianellus

CXXIII?

TETRAO. GROUSE.
Rocky Mountain Spotted Grouse. T.Frankliní. M.
     Do. .... Fe.
¹ Dusky. .... Sharpwew. M.
² Do. .... Fe.
³ Ruffed. .... Umbellus. C.

Drawn by A Wilson J Neder & Capt Lewis   Eng.d by J Allen

LAGOPUS PTARMIGAN.

1. Willow Ptarmigan.   *L.Saliceti. Male, Spring.*
2. ___ 1st ___   ___ ___ Summer.
3. ___ 1st ___   ___ Fem. Winter.
4. Ptarmigan ___   ___ Mutus ___ Summer.
5. 1st ___   ___ ___ Winter.

Drawn by Capt. Brown.   Eng.d by S. Milne.

ORTYX. QUAILS.

1.Virginian Quail.	O. Virginiana. M.
2. __ Dt. __	_ Virginiana. Fem.
3.Californian Dt. __	_ Californica. M.
4. __ Dt. __	_ __ F.
5.Long-tailed Dt. __	_ Macroura.
6.Montezuma's Dt. __	_ Montezumæ.

Drawn by A. Wilson & Capt.n Brown.   Engd by S. Milne.

COLUMBA. PIGEONS.

1. Blue-headed Ground P.   C. Cyanocephala.
2. Zenaida   Zenaida.
3. Purple-crowned   Purpurata.
4. Passenger   Migratoria.
5. Blue & Green   Cyano-virens.
6. African   Afra.
7. Ground   Passerina.
8. Black-capped   Melanocephala.
9. Great-crowned   Coronata.

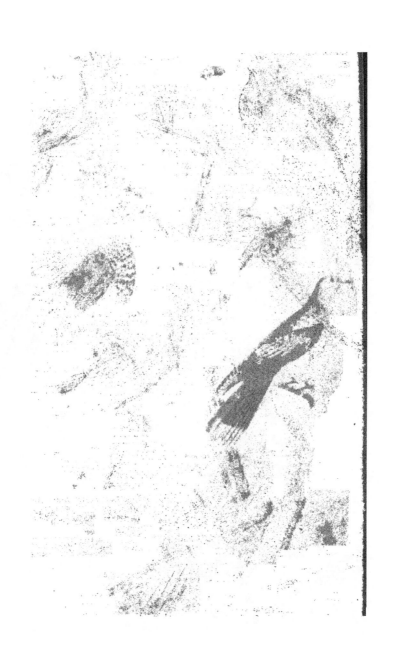

COLUMBA. PIGEONS.

1. *C. torqua* .... Speckled P.
2. *— lævensletta* .... Mottled
3. *— capistrata* ...... Grey-headed
4. *— dilopha* .... Double-crested
5. *— humeralis* ...... Red-collared

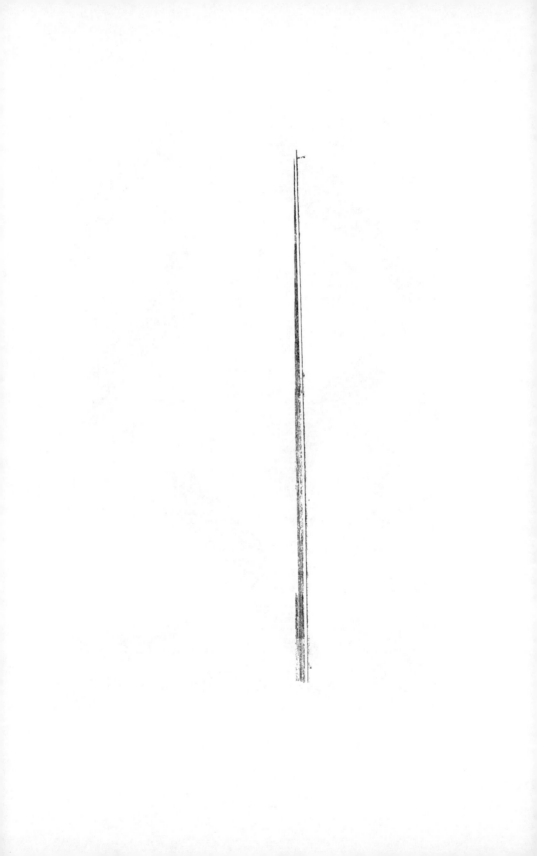

COLUMBA. PIGEONS.
1. nana . . . . . . Dwarf P .
2. perlata . . . . . Pearly .
3. rubicps . . . . . Red-headed .
4. leptogrammica . Banded .
5. cinerea . . . . Fem. Ash-coloured .
6. diademata . . . . Fem. Diadem . . . . . .

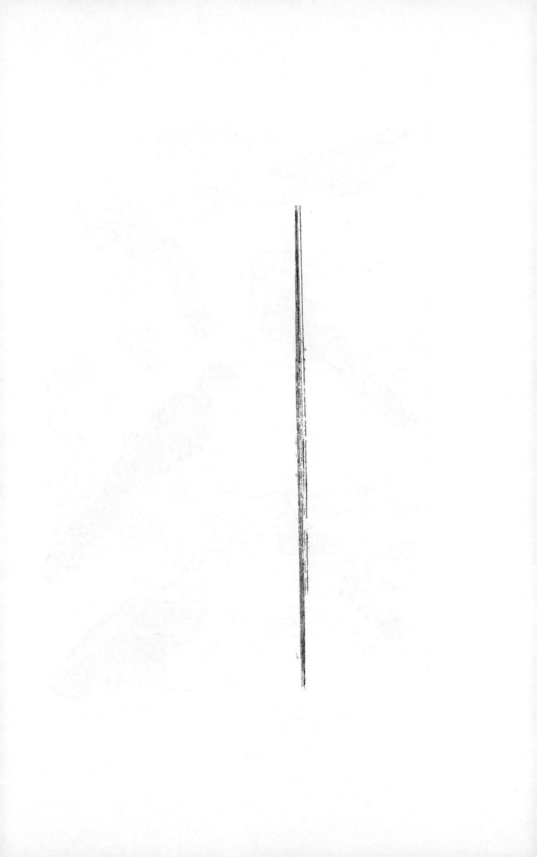

STRUTHIO. OSTRICHES.
1.S. Camelus.   The Ostrich.
2 .... Fem ....

Turvey sc. & etc.

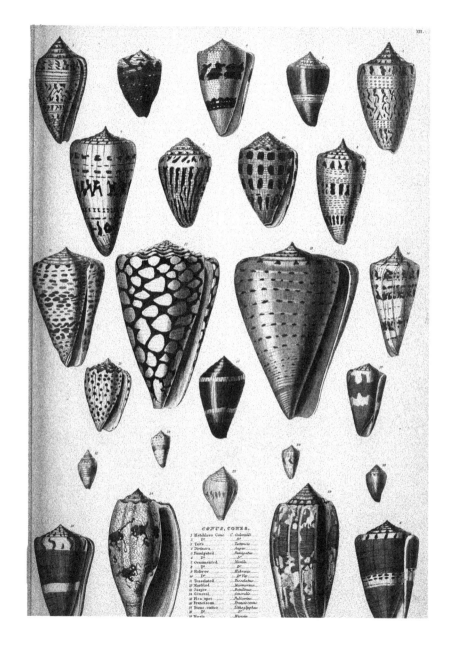

CONUS, CONES.

1 Matchless Cone ... C. Cedonulli
2 D.º .... D.º
3 Taits .... Taitensis
4 Diríndo's .... Augur
5 Fuscigated .... Fuscigatus
6 D.º .... D.º
7 Ornamented .... Monile
8 D.º .... D.º
9 Hebrew .... Hebræus
10 D.º .... D.º Top
11 Tessellated .... Tessellatus
12 Marbled .... Marmoreus
13 Jasper .... Bandinus
14 General .... Generalis
15 Flea-spot .... Pulicarius
16 Franciscan .... Francis canus
17 Stone-cutter ...... Litho glyphus
18 D.º .... D.º
19 Music .... Musicus

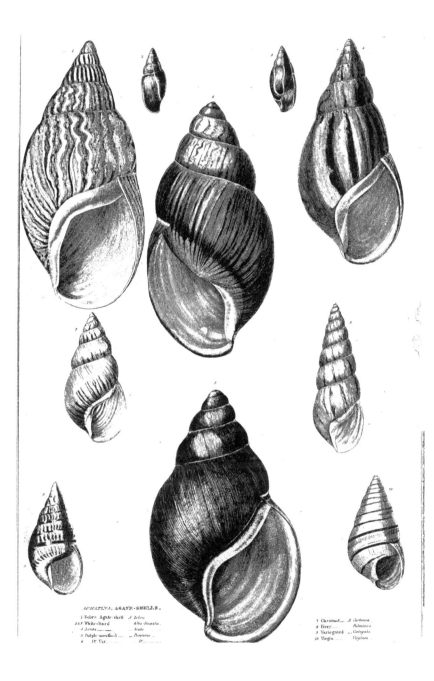

ACHATINA. AGATE-SHELLS.

1 Zebra Agate-shell	A Zebra
2&3 White-lined	Alba-lineata
4 Acute	Acuta
5 Purple-mouthed	Purpurea
6 Do. Var.	Do.

7 Chestnut	A Castanea
8 Fiery	Fulminea
9 Variegated	Variegata
10 Virgin	Virginea

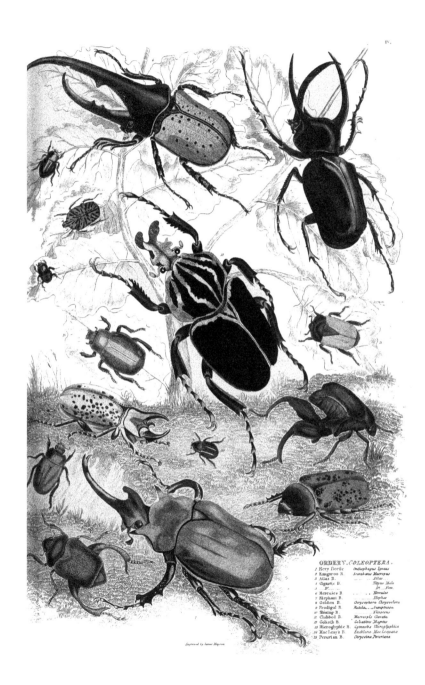

IV.

ORDER V. COLEOPTERA.

1 Kirry Beetle   Onthophagus Ignata
2 Kangaroo B.   Scarabæus Macropus
3 Atlas B.   Atlas
4 Gigantic B.   Thyus Male
5 Dº.   Do   Fem.
6 Hercules B.   Hercules
7 Elephant B.   Elephas
8 Golden B.   Chrysophora Chrysocelora
9 Prodigal B.   Rutelia... Prampinosa
10 Shining B.   Nitescens
11 Clubbed B.   Marcospls Clavata
12 Goliath B.   Gobiathus Magnus
13 Hieroglyphic B.   Cymnetis Hieroglyphica
14 MacLeay's B.   Exothora Macleayana
15 Peruvian B.   Chryseina Peruviana

Engraved by James Magsom

ORDER X. *LEPIDOPTERA* BUTTERFLIES.

1 Royal Butterfly   *Endymion Regalis*   8 Nictype   *Piers Nictype*
2 D°   9 D°

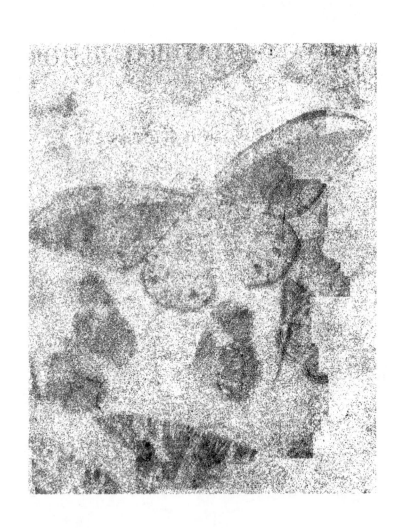

THE

EDINBURGH

ıL OF NATURAL HISTORY,

AND OF

THE PHYSICAL SCIENCES.

CONDUCTED BY

WILLIAM MACGILLIVRAY, A. M.,

CONSERVATOR OF THE ROYAL COLLEGE OF SURGEONS, EDINBURGH; FELLOW OF THE ROYAL SOCIETY
OF EDINBURGH; MEMBER OF THE WERNERIAN NATURAL HISTORY SOCIETY; OF THE
ACADEMY OF NATURAL SCIENCES OF PHILADELPHIA, &c. &c.

ASSISTED BY SEVERAL SCIENTIFIC AND LITERARY MEN.

VOLUME II.

A.D. 1839—1840.

EDINBURGH
PUBLISHED BY THE EDINBURGH PRINTING COMPANY, 12, SOUTH ST DAVID STREET.
GLASGOW AND THE WEST OF SCOTLAND: JOHN SMITH AND SON, 70, ST VINCENT STREET; AND
JOHN M'LEOD, 20, ARGYLE STREET. ABERDEEN: A. BROWN AND CO
NCHESTER: AINSWORTH AND SONS, 107, GREAT ANCOATS STREET. LEEDS: THOMAS AINSWORTH, JUNIOR,
50, QUEEN'S PLACE, COBURG STREET.
DUBLIN: GEORGE YOUNG, 9, SUFFOLK STREET.
PARIS: J. B. BALLIÈRE, RUE DE L'ECOLE-DE-MÉDECINE, 13, (BIS.)
LONDON: SMITH, ELDER, & CO., 65, CORNHILL.
M.DCCC.XL.

ND - #0007 - 090323 - C0 - 229/152/30 [32] - CB - 9780331819847 - Gloss Lamination